Anoxygenic Photosynthetic Bacteria

Advances in Photosynthesis

VOLUME 2

Advances in Photosynthesis provides an up-to-date account of research on all aspects of photosynthesis, the most fundamental life process on earth. *Photosynthesis* is an area that requires, for its understanding, a multidisciplinary (biochemical, biophysical, molecular biological, and physiological) approach. Its content spans from physics to agronomy, from femtosecond reactions to those that require an entire season, from photophysics of reaction centers to the physiology of the whole plant, and from X-ray crystallography to field measurements. The aim of this series of publications is to present to beginning researchers, advanced graduate students and even specialists a comprehensive current picture of the advances in the various aspects of photosynthesis research. Each volume focusses on a specific area in depth.

The titles to be published in this series are listed on the backcover of this volume.

Anoxygenic Photosynthetic Bacteria

Edited by

Robert E. Blankenship
Department of Chemistry and Biochemistry,
Arizona State University,
Tempe, Arizona, U.S.A.

Michael T. Madigan
Department of Microbiology,
Southern Illinois University,
Carbondale, Illinois, U.S.A.

and

Carl E. Bauer
Department of Biology,
Indiana University,
Bloomington, Indiana, U.S.A.

KLUWER ACADEMIC PUBLISHERS
DORDRECHT / BOSTON / LONDON

A C.I.P. Catalogue record for this book is available from the Library of Congress

ISBN 0-7923-3681-X

Published by Kluwer Academic Publishers,
P.O. Box 17, 3300 AA Dordrecht, The Netherlands.

Kluwer Academic Publishers incorporates
the publishing programmes of
D. Reidel, Martinus Nijhoff, Dr W. Junk and MTP Press.

Sold and distributed in the U.S.A. and Canada
by Kluwer Academic Publishers,
101 Philip Drive, Norwell, MA 02061, U.S.A.

In all other countries, sold and distributed
by Kluwer Academic Publishers Group,
P.O. Box 322, 3300 AH Dordrecht, The Netherlands.

The camera ready text was prepared
by Lawrence A. Orr, Center for the Study of Early Events in Photosynthesis
Arizona State University, Tempe, Arizona 85287-1604, U.S.A.

Printed on acid-free paper

Printed in the Netherlands

Contents

Part II: Molecular Structure and Biosynthesis of Pigments and Cofactors

Part III. Membrane and Cell Wall Architecture and Organization

Part V: Reaction Center Structure, Electron and Proton Transfer Pathways

Part VI: Cyclic Electron Transfer Components and Energy Coupling Reactions

Part VII: Metabolic Processes

Part VIII: Genetics and Genetic Manipulations

Part IX: Regulation of Gene Expression

Part X: Applications

Preface

The editors are proud to present *Anoxygenic Photosynthetic Bacteria* to students and researchers in the field of photosynthesis. We feel that this book will be the definitive volume on non-oxygen evolving photosynthetic bacteria for years to come, as it is literally loaded with the most recent information available in this field. Contributors were given the freedom to develop topics in depth, and although this has lead to a lengthy volume, readers can be assured that nothing of significance in the field of anoxygenic photosynthetic bacteria has been neglected.

We have organized the book along ten major themes: (1) Taxonomy, physiology and ecology, (2) Molecular structure of pigments and cofactors, (3)Membrane and cell wall structure, (4) Antenna structure and function, (5) Reaction center structure and electron/proton pathways, (6) Cyclic electron transfer, (7) Metabolic processes, (8) Genetics, (9), Regulation of gene expression, and (10) Applications. For each theme, several chapters, written by leading experts in the field, combine to paint a detailed picture of the current state of affairs of research in that area. The editors have also tried to provide ample cross-references between chapters to help guide the reader to all of the coverage on a particular topic.

The last time a comprehensive volume on anoxygenic photosynthetic prokaryotes was published was in 1978. This was the landmark volume *The Photosynthetic Bacteria*, edited by R. K. Clayton and W. R. Sistrom, and published by Plenum Press in New York. That book brought together in one place the state of current knowledge on the ecology, physiology, taxonomy, ultrastructure, biochemistry, biophysics and genetics of this diverse group of bacteria. As a testament to its value as a reference book, every copy of *The Photosynthetic Bacteria* that any of the three editors of the present volume have ever seen was tattered from years of almost constant use and stained with coffee and cell extracts of various colors. *The Photosynthetic Bacteria* had a tremendous impact on a generation of scientists in many disciplines and will remain a classic for many years to come.

Thus it was with some trepidation that we undertook what was in essence the same task that faced Clayton and Sistrom 17 years earlier. The field has changed dramatically in that time in nearly every one of the areas mentioned above, perhaps most noticeably in the area of genetics and related topics. A single chapter on this subject sufficed in 1978, whereas 14 of the 62 chapters in the present volume have 'gene' or 'genetic' in the title and nearly every chapter makes use of information obtained using mutants.

Another area of bacterial photosynthesis that has seen tremendous progress in 17 years is structural studies. There is only one protein structure determined using X-ray diffraction that appears as a figure in the 1978 volume, whereas dozens of structural figures are included in this volume, including reaction centers, antenna complexes, electron transfer proteins and porins. An even more dramatic difference is that there is only a single, partial amino acid sequence in the 1978 volume, while there are literally hundreds of sequences in this book. The emphasis in every area covered in this book, from taxonomy and ecology all the way to biophysics has continually shifted more and more toward analysis at the molecular level.

This is the second volume in a series titled 'Advances in Photosynthesis,' with Govindjee as the series editor. The series provides an up to date account of all aspects of the process of photosynthesis. A statement of the scope of the series and a list of previous and upcoming titles can be found on the back cover.

We would like to dedicate this volume to Howard Gest, who is unquestionably the father of modern research involving photosynthetic bacteria. He has also greatly influenced the careers of each of the editors as either a collaborator, mentor or colleague and to all of us as a friend.

Special thanks go to Larry Orr, without whose

extraordinary talents and hard work this effort could never have come to fruition. Larry prepared all the camera ready copy and is primarily responsible for the layout and 'look' of the book.

Finally, we thank our wives, Liz, Nancy and Chris and children, Larissa, Sam, Scott and Kevin for their patience and understanding of the time needed to complete this volume.

Robert E. Blankenship
Michael T. Madigan
Carl E. Bauer

Chapter 1

Taxonomy and Physiology of Phototrophic Purple Bacteria and Green Sulfur Bacteria

Johannes F. Imhoff
Institut für Meereskunde an der Universität Kiel, Düsternbrooker Weg 20, 24105 Kiel, Germany

Summary

Anoxygenic phototrophic bacteria have always attracted scientists because of their coloration and ability to perform photosynthesis in the absence of air and without producing oxygen. Despite this common feature of these bacteria, variation in morphological, physiological and molecular properties, including molecular structures of the photosynthetic pigments and the photosynthetic apparatus, is great. This chapter will give a short introduction into the diversity of green sulfur and phototrophic purple bacteria, list some important properties of the species, and indicate important physiological features.

I. Introduction

The most striking and common property of all purple and green phototrophic bacteria is their ability to carry out anoxygenic photosynthesis on the basis of bacteriochlorophyll mediated processes. The various anoxygenic phototrophic bacteria contain several types of bacteriochlorophylls and a variety of carotenoids as pigments, which function in the transformation of light into chemical energy and give cultures a distinct coloration varying with the pigment content from various shades of green, yellowish-green, brownish-green, brown, brownish-red, red, pink, purple, and purple-violet to even blue (carotenoidless mutants of some species containing bacteriochlorophyll *a*). Characteristic absorption maxima of bacteriochlorophylls and major carotenoid groups of anoxygenic phototrophic bacteria are shown in Tables 1 and 2. Photosynthesis in anoxygenic phototrophic bacteria depends on oxygen-deficient conditions, because synthesis of the photosynthetic pigments is repressed by oxygen (bacteria like *Erythrobacter longus* are exceptions to this rule). Unlike cyanobacteria (including *Prochloron* and related forms) and eukaryotic algae, anoxygenic phototrophic bacteria are unable to use water as an electron donor and oxygen is not produced. They use only one photosystem and require electron donors of lower redox potential than water. Most characteristically, sulfide and other reduced sulfur

R. E. Blankenship, M. T. Madigan and C. E. Bauer (eds): Anoxygenic Photosynthetic Bacteria, pp. 1–15.

Table 1. Characteristic absorption maxima of different bacteriochlorophylls in living cells[a]

Bacteriochlorophyll	Esterifying alcohol[b]	Characteristic absorption maxima (nm)
a	P,Gg	375, 590, 800-810, 830-890
b	P,Gg[c]	400, 605, 835-850, 1015-1035
c	F[d]	335, 460, 745-760
d	F	325, 450, 725-745
e	F	345, 450-460, 715-725
g	F	370, 419, 575, 670, 780-790

[a] Data are collected from Brockmann and Lipinski (1983), Gloe et al. (1975), Michalski et al. (1987) and Pfennig (1978).
[b] Esterifying alcohol: P, phytol; Gg, geranylgeraniol; F, farnesol.
[c] In *Ectothiorhodospira halochloris* and *Ectothiorhodospira abdelmalekii* phytadienol was found (Steiner et al., 1981; also R. Steiner, personal communication).
[d] In *Chloroflexus aurantiacus* a straight-chain aliphatic stearyl alcohol (C-18) is esterified to the propionic acid side chain (Gloe and Risch, 1978).

Table 2. Major carotenoid groups of anoxygenic phototrophic bacteria[a]

Biosynthetic group	*Major components*
Normal spirilloxanthin series	Lycopene, rhodopin, spirilloxanthin
Rhodopinal series	Lycopene, lycopenal, lycopenol, rhodopin, rhodopinal, rhodopinol
Alternative spirilloxanthin series	Hydroxyneurosporene, spheroidene, spheroidenone (spirilloxanthin)
Okenone series	Okenone
Isorenieratene series	*β*-Carotene, isorenieratene
Chlorobactene series	*γ*-Carotene, chlorobactene

[a] Data are taken from Schmidt (1978).

compounds, but also hydrogen and a number of small organic molecules, are used as photosynthetic electron donors. Quite recently even growth with reduced iron as electron donor has been demonstrated with some phototrophic purple bacteria (Widdel et al., 1993).

Besides this common theme of photosynthesis, anoxygenic phototrophic bacteria are an extremely heterogeneous eubacterial group, on the basis of morphological, physiological and molecular data. According to their phenotypic properties we distinguish between the green sulfur bacteria, the green nonsulfur bacteria, the purple sulfur bacteria, the purple nonsulfur bacteria and the heliobacteria. On the basis of 16S rRNA analyses major eubacterial groups have been defined (Woese et al., 1985a). Among these, one is represented by the cyanobacteria, one by the green sulfur bacteria, one by *Chloroflexus* and 'relatives', and one by the phototrophic purple bacteria and their 'relatives'. On the basis of their 16S rRNA, the recently discovered *Heliobacterium chlorum* (Gest and Favinger, 1983) and *Heliobacillus mobilis* (Beer-Romero et al., 1988) do not fit into the aforementioned groups, but are related to certain Gram-positive bacteria (Woese et al., 1985c; Beer-Romero and Gest, 1987). Heliobacteria are considered in Chapter 2 by Madigan and Ormerod.

A quite remarkable group of bacteria, containing bacteriochlorophyll, but unable to grow phototrophically under anaerobic conditions, is represented by a number of Gram-negative aerobic marine bacteria (Sato, 1978; Shiba et al., 1979; Nishimura et al., 1981; Trüper, 1989). The best studied of these bacteria is *Erythrobacter longus* (Shiba, 1991), which can synthesize bacteriochlorophyll *a*, form intracytoplasmic membranes, and has reaction center complexes similar to those of other purple bacteria (Harashima et al., 1980; Shimada et al., 1985; Iba et al., 1988). In contrast to all previously known phototrophic purple bacteria, synthesis of bacteriochlorophyll *a* and carotenoids in aerobic phototrophs is stimulated by oxygen (Harashima et al., 1980). Shiba (1984) demonstrated that *Roseobacter denitrificans* (formerly designated as *Erythrobacter* strain OCH 114) effectively uses light to increase the cellular ATP level and the rate of incorporation of

Abbreviations: Cm. – Chromatium; Ec. – Ectothiorhodospira; Rb. – Rhodobacter; Rp. – Rhodopseudomonas; Rs. – Rhodospirillum

CO_2. According to 16S rRNA analyses, *Erythrobacter longus* belongs to the alpha subgroup of the Proteobacteria, and appears distantly related to other bacteria of this group, such as *Rhodobacter* and *Rhodopseudomonas* species (Woese et al., 1984a). These bacteria and also a recently discovered bacteriochlorophyll containing *Rhizobium* sp. (Evans et al., 1990) are treated in separate chapters (see Chapters 6 and 7 by Shimada and Fleischmann et al., respectively). In this chapter we consider the green sulfur bacteria and the phototrophic purple bacteria.

II. General Aspects of Taxonomy

Traditionally, differentiation of species, genera and even families of phototrophic bacteria was based on a number of morphological properties, such as cell form and size, flagellation and intracytoplasmic membrane structures, on pigment composition, DNA base ratio, and also physiological properties such as carbon and nitrogen substrate utilization and ability to respire aerobically and anaerobically in the dark, among others (Pfennig and Trüper, 1974; Trüper and Pfennig, 1981; Imhoff and Trüper, 1989). Especially some significant features of dissimilatory sulfur metabolism, such as the ability to oxidize sulfide, the preference to grow photoautotrophically with sulfide as electron donor (compared to photoheterotrophic growth), the formation of sulfur globules inside or outside the cells, and the oxidation of extracellular elemental sulfur have been used as criteria to distinguish between major groups of the phototrophic purple and green bacteria (Molisch, 1907; Bavendamm, 1924; Pfennig and Trüper, 1971; Imhoff 1984a; Pfennig, 1989a, b; Gibson et al., 1984).

Also, chemotaxonomic methods have been applied to the taxonomy of phototrophic bacteria and have supported the heterogeneity of this group. The cellular composition of polar lipids, fatty acids, and quinones was found to be useful for their taxonomic characterization. Principal differences in the composition of these compounds were found among the major groups, and even the recognition of species was possible by comparing the composition of these membrane constituents (Imhoff et al., 1982; Imhoff, 1982, 1984b, 1988, 1991; Hansen and Imhoff, 1985). The lipids and quinones of phototrophic bacteria are considered in detail in Chapter 10 by Imhoff and Imhoff. Also, other molecular information, such as lipopolysaccharide structure, size and sequence of cytochromes *c*, and 16S rRNA oligonucleotide catalogues and sequence data can be used to determine similarity and taxonomic relations among species. The lipid A structure of the lipopolysaccharides shows significant differences among members of the purple nonsulfur bacteria (Weckesser et al., 1974, 1979; Mayer, 1984); it is quite similar among all the Chromatiaceae (Meißner et al., 1988b), though differences occur between *Ec. halophila* and less halophilic *Ectothiorhodospira* species (Meißner et al., 1988c; Zahr et al., 1992). Lipid A was absent from *Chloroflexus* (Meißner et al., 1988a).

With the aid of sequencing techniques of proteins and nucleic acids, molecular data became available which were used for the analysis of phylogenetic relatedness of bacterial strains. The best record of phylogeny is conserved in the primary sequences of nucleic acids and proteins, and attempts to compare phylogenetic relatedness have been made by quantitative comparison of sequence data of both kinds of macromolecules. Because of the available methods, proteins were the first macromolecules considered in the phylogenetic relationships of various bacteria. Among the phototrophic prokaryotes, *c*-type cytochromes and ferredoxins are two examples of such proteins. On the basis of primary sequence and tertiary structure of the *c*-type cytochromes a first 'phylogenetic tree' of the phototrophic purple bacteria was constructed (Dickerson, 1980).

More recently, nucleic acid sequences of the 16S rRNA have been widely used to trace bacterial phylogeny. This molecule is universally distributed among prokaryotes and is considered to be phylogenetically quite conservative. The method has developed from a mere comparison of oligonucleotide catalogues derived by digestion of the molecule with T1 RNase (Zablen and Woese, 1975). With the development of techniques to completely sequence the 16S rRNA molecule, total information of its nucleic acid sequence is now available for comparison. Although this method is superior to all previous attempts of tracing phylogenetic relations because of the universal distribution of ribosomal RNA and the amount of sequence information, it considers only a single molecule of the whole bacterium. To judge the systematic position of a bacterium, the combination of results obtained by different methods—with the support of 16S rRNA sequence data—appears to be the best approach.

III. General Aspects of Physiology

Besides the unifying property to perform light dependent and bacteriochlorophyll-mediated energy transduction, the metabolic capabilities of phototrophic eubacteria and, in particular, of phototrophic purple bacteria, show great diversity. High versatility is not only found among the different species of this group of bacteria, but also in single species which can adapt to a variety of different growth conditions.

Photoautotrophic growth is typical of purple and green sulfur bacteria, while photoheterotrophic growth is typical of purple nonsulfur bacteria. All green sulfur bacteria, almost all purple sulfur bacteria, and also many representatives of the purple nonsulfur bacteria grow under photoautotrophic conditions with either reduced sulfur compounds or hydrogen as electron donors. Many phototrophic purple bacteria are able to grow photoheterotrophically using organic substrates as electron donors and carbon sources. Also, chemolithotrophic growth (aerobically in the dark) has been demonstrated in several species (Madigan and Gest, 1979; Siefert and Pfennig, 1979; Kämpf and Pfennig, 1980). Chemoheterotrophic growth in the presence of oxygen is common among purple nonsulfur bacteria, but is also found in some purple sulfur bacteria.

While some species are very sensitive to oxygen, others grow equally well under aerobic conditions in the dark at the full oxygen tension of air. Under anaerobic dark conditions, growth of some species is also supported by respiratory electron transport in the presence of nitrate, nitrite, nitrous oxide, dimethylsulfoxide (DMSO), or trimethylamine-N-oxide (TMAO) as electron acceptors (for reviews see Ferguson et al., 1987, and Chapter 44 by Zannoni in this volume). In the absence of external electron acceptors a number of substrates allow for energy generation and slow growth by a fermentative metabolism.

In addition to these diverse methods for energy generation, there is considerable variation in the utilization of carbon, nitrogen and sulfur compounds for assimilatory and dissimilatory purposes, as well as in the enzymatic reactions and pathways involved in these processes.

Organic carbon sources have principally different functions under phototrophic, respiratory and fermentative conditions. Under phototrophic growth conditions they serve primarily as a source of cellular carbon, but in addition may function as an electron source for photosynthetic electron transport. In the presence of inorganic electron donors they may be exclusively photoassimilated. During phototrophic growth the citric acid cycle serves a biosynthetic purpose and a complete cycle is actually not required under these conditions. Indeed, a number of purple sulfur bacteria lack oxoglutarate dehydrogenase and a complete citric acid cycle (Kondratieva, 1979). During respiratory growth the major part of the carbon source is completely oxidized. Reactions of the citric acid cycle are involved in substrate oxidation and one would expect elevated enzymatic activities of the cycle under these conditions. Indeed, *Rb. capsulatus* activities of citrate synthase, isocitrate dehydrogenase and, in particular, oxoglutarate dehydrogenase are increased during respiratory growth (Beatty and Gest, 1981). During fermentative growth, substrates or storage products are oxidized incompletely on a large scale. Reduced organic compounds and/or hydrogen are excreted in order to achieve redox balance of the cell.

Various purple bacteria show great variation in their capability to photoassimilate organic carbon compounds and to grow photoheterotrophically. A group of metabolically specialized, obligately phototrophic and strictly anaerobic Chromatiaceae, including *Chromatium okenii, Cm. weissei, Cm. warmingii, Cm. buderi, Cm. tepidum, Thiospirillum jenense, Lamprocystis roseopersicina* and *Thiodictyon elegans,* photoassimilate only acetate and pyruvate and this only in the presence of CO_2 and sulfide. Endogenous fermentation, as it occurs in *Cm. vinosum* (Van Gemerden, 1968), should be possible, though this has not been demonstrated. Chemotrophic growth has not been found with *Cm. okenii* (Kämpf and Pfennig, 1980), but high respiration rates with intracellular sulfur as the electron source have been reported (Kristjansen, 1988). Also, *Cm. warmingii* and *Thiocapsa pfennigii*, both unable to grow chemotrophically, oxidize intracellular sulfur with oxygen as the terminal acceptor (Kämpf and Pfennig, 1986).

Others, such as the small-celled *Chromatium* species and the majority of the purple nonsulfur bacteria utilize a variety of different organic carbon sources. Intermediates of the tricarboxylic acid cycle in addition to acetate and pyruvate are generally used. Other organic acids, amino acids, fatty acids, alcohols, carbohydrates and even aromatic carbon compounds support growth of many purple nonsulfur bacteria. One of these metabolically highly versatile bacteria is *Rb. capsulatus*. This species can grow photoautotrophically with sulfide or hydrogen as the electron donor, chemoautotrophically with hydrogen, photoheterotrophically with a variety of organic compounds as electron and carbon sources, chemoheterotrophically at the expense of respiratory electron transport-driven energy generation with oxygen, nitrate, nitrous oxide, dimethylsulfoxide or TMAO as electron acceptors, and finally also by fermentation of organic substrates.

The diversity of the carbon metabolism of phototrophic bacteria is also demonstrated by the assimilation pathway of acetate, which is a substrate for green sulfur bacteria and almost all phototrophic purple bacteria. It is assimilated by almost all purple nonsulfur bacteria (*Rp. globiformis* does not grow with acetate) and even by the most specialized Chromatiaceae. The metabolism of acetate assimilation, however, is quite different among the species. In many phototrophic purple bacteria the primary reaction of acetate metabolism is the ATP-dependent formation of acetyl-CoA, which is the substrate for further reactions. In *Cm. vinosum*, as in most purple bacteria, the two key enzymes of the glyoxylate cycle, isocitrate lyase and malate synthase, are present and acetate is assimilated via this pathway. The cycle is, however, unusual in *Cm. vinosum* in so far as malate dehydrogenase is lacking, but the reaction catalyzed by malic enzyme followed by carboxylation of pyruvate (pyruvate carboxylase) results in the formation of oxalacetate. Unlike most of the acetate assimilating phototrophic purple bacteria *Rs. rubrum* does not have an active glyoxylate cycle. Isocitrate lyase is not present. Conversion of acetate to oxalacetate by two carboxylation reactions from acetate to pyruvate (with pyruvate synthase) and further to oxalacetate is possible (Buchanan et al., 1967). In *Rubrivivax gelatinosus*, however, photoassimilation of acetate is via the serine-hydroxypyruvate pathway (Albers and Gottschalk, 1976). In the green sulfur bacteria neither a Calvin Cycle nor an oxidative tricarboxylic acid cycle are operating (Kondratieva, 1979; Fuchs et al., 1980a, b). In *Chlorobium limicola* pyruvate is formed from Acetyl-CoA by a ferredoxin-dependent pyruvate synthase. Further reactions involve carboxylation to oxaloacetate (phosphoenolpyruvate synthetase) and reactions of the reductive tricarboxylic acid cycle and of gluconeogenesis.

Detailed discussions of the carbon, sulfur, and nitrogen metabolism of anoxygenic phototrophic bacteria is given in additional chapters of this volume.

IV. Description of the Groups

A. Green Sulfur Bacteria (Chlorobiaceae)

The phototrophic green sulfur bacteria form a tight phylogenetic group, separated from other phototrophic bacteria and also from known chemotrophic bacteria (Gibson et al., 1985). They are characterized by non-motile, spherical, ovoid or vibrio shaped cells. All species lack flagella; some have gas vesicles. Only *Chloroherpeton thalassium* is motile by gliding and in addition forms gas vesicles; a further difference from all other species of this group is the carotenoid composition, with γ-carotene as major component. Physiological properties and results of 16S rRNA analyses led to the inclusion of *Chloroherpeton* into this group (Gibson et al., 1985). Some properties of the green sulfur bacteria are shown in Table 3.

We distinguish green and brown species of green sulfur bacteria. The green species contain bacteriochlorophyll *c* or *d* and the carotenoid chlorobactene and OH-chlorobactene as light-harvesting pigments (Gloe et al., 1975; Schmidt, 1978). The brown species have bacteriochlorophyll *e* and the carotenoids isorenieratene and β-isorenieratene as light-harvesting pigments (Liaaen-Jensen, 1965). The different pigment content is responsible for differences in the light absorption properties. The brown species have a broader absorption range, between 480 and 550 nm, which apparently is of ecological significance (see Chapter 4 by Van Gemerden and Mas).

In contrast to the phototrophic purple bacteria, green sulfur bacteria and green nonsulfur bacteria do not have intracytoplasmic membrane systems. They contain special structures called chlorosomes (or chlorobium vesicles in the older literature), which are attached to the internal face of the cytoplasmic membrane, carry the light-harvesting pigments, and represent large and powerful antenna systems (Cohen-Bazire et al., 1964; Staehelin et al., 1978, 1980). Chlorosomes contain the major photosynthetic pigments of green sulfur bacteria (see Chapter 20 by Blankenship et al.). The reaction centers of phototrophic green bacteria are located in the cytoplasmic membrane at the attachment sites of the chlorosomes. The localization of the small amounts of bacteriochlorophyll *a* that are present in these bacteria is mainly restricted to the reaction centers and protein complexes of the so-called baseplate, which connects the chlorosome with the cytoplasmic

Table 3. Characteristic properties of green sulfur bacteria[a]

Species	Cell shape	Cell diameter (μm)	Color of culture	Major carotenoid	Major BChl	Gas vesicles	Salt response	G + C content (mol%)
Chlorobium								
limicola	rod	0.7-1.0	green	cl	c or d	-	F	51.0-58.1
vibrioforme	vibrio	0.5-0.7	green	cl	c or d	-	2%	52.0-57.1
phaeobacteroides	rod	0.6-0.8	brown	irt,β-irt	e	-	F	49.0-50.0
phaeovibrioides	vibrio	0.3-0.4	brown	irt,β-irt	e	-	2%	52.0-53.0
chlorovibrioides	vibrio	0.3-0.4	green	cl	c or d	-	2-3%	o
tepidum	rod	0.6-0.8	green	cl	c	-	F	56.5-58.2
Prosthecochloris								
aestuarii	sphere	0.5-0.7	green	cl	c	-	2-5%	50.0-56.0
phaeoasteroides	sphere	0.3-0.6	brown	irt	e	-	0.5-2%	52.2
Ancalochloris								
perfilievii	sphere	0.5-1.0	green	o	o	+	F	o
Pelodictyon								
luteolum	ovoid	0.6-0.9	green	cl	c or d	+	F	53.5-58.1
clathratiforme	rod	0.7-1.2	green	cl	c or d	+	F	48.5
phaeum	vibrio	0.6-0.9	brown	irt	e	+	3%	o
phaeoclathratiforme	rod	0.7-1.1	brown	irt,β-irt	e	+	F	47.9
Chloroherpeton								
thalassium	rod	1.0	green	γ-c	c	+	1-2%	45.0-48.2

Abbreviations: o – no data available; BChl – bacteriochlorophyll.
carotenoids: cl – chlorobactene; irt – isorenieratene; β-irt – β-isorenieratene; γ-c – γ-carotene.
salt response: F – fresh water isolates; numbers give optimum salinity of isolates from brackish or marine habitats.
[a] Data were collected from: Gibson et al. (1984); Gorlenko and Lebedeva (1971); Overmann and Pfennig (1989); Trüper and Pfennig (1981); Wahlund et al. (1991).

membrane (Staehelin et al., 1980; Amesz and Knaff, 1988). Small amounts (about 1% of the total) of bacteriochlorophyll *a* have also been found inside the chlorosomes (Gerola and Olson, 1986).

All green sulfur bacteria (Chlorobiaceae) have highly similar physiological capacities. They are metabolic specialists, strictly anaerobic and obligately phototrophic. All species grow photoautotrophically with CO_2 as sole carbon source. In the presence of sulfide and CO_2 only acetate and pyruvate are assimilated as organic carbon sources. CO_2 is assimilated via reactions of the reductive tricarboxylic acid cycle (Evans et al., 1966; Fuchs et al., 1980a, b; Ivanovsky et al., 1980; see also Chapter 40 by Sirevåg). Sulfide is used as electron donor and sulfur source and is oxidized to sulfate with the intermediate accumulation of elemental sulfur globules outside the cells. Some strains also use thiosulfate and molecular hydrogen as photosynthetic electron donors. Most species require vitamin B_{12} as growth factor.

B. Purple Bacteria

In all purple bacteria the photosynthetic pigments and the photosynthetic apparatus are located within a more or less extended system of intracytoplasmic membranes that is considered as originating from and being continuous with the cytoplasmic membrane. These intracytoplasmic membranes consist of small fingerlike intrusions, vesicles, tubules or lamellae. The major photosynthetic pigments are bacteriochlorophyll *a* or *b* and various carotenoids of the spirilloxanthin, rhodopinal, spheroidene (alternative spirilloxanthin), or okenone series (Schmidt, 1978).

Molisch (1907) distinguished two types of phototrophic purple bacteria as belonging to the order Rhodobacteria. In the Thiorhodaceae, all purple bacteria that are able to form globules of elemental sulfur inside the cells were grouped and in the Athiorhodaceae all purple bacteria which lack this ability were placed. These two groups have been maintained for a long time and were renamed later to Chromatiaceae (Bavendamm, 1924) and Rhodospirillaceae (Pfennig and Trüper, 1971), respectively.

Later, purple sulfur bacteria became known that accumulate elemental sulfur outside their cells (Pelsh, 1937; Trüper 1968). They were included in the Chromatiaceae as the genus *Ectothiorhodospira* in the eighth edition of Bergey's Manual (Pfennig and Trüper, 1974) but later considered as a separate family the Ectothiorhodospiraceae (Imhoff, 1984a), which comprise purple sulfur bacteria that form sulfur globules outside the cells. Now the Chromatiaceae exclusively comprise those phototrophic sulfur bacteria able to deposit elemental sulfur inside the cells (Imhoff, 1984a), which is in agreement with Molisch's (1907) definition of the 'Thiorhodaceae'.

The analysis of 16S rRNA sequences revealed deep branches among different groups of the phototrophic purple bacteria as well as the close relationship of some phototrophic purple bacteria with certain nonphototrophic chemoheterotrophic bacteria (Gibson et al., 1979; Stackebrandt and Woese, 1981). The name Proteobacteria, as a new class, has been proposed for all purple bacteria and their purely chemotrophic relatives (Stackebrandt et al., 1988). The Proteobacteria are subdivided into four subgroups called α, β, γ, δ (Woese, 1987; Woese et al., 1984a, b, 1985 b). The first three subgroups contain non-phototrophic as well as phototrophic representatives; for the δ subgroup phototrophic representatives are so far unknown. So far only one purely chemotrophic bacterium has been found which clusters within the closely related species of Chromatiaceae and Ectothiorhodospiraceae (Adkins et al., 1993). According to 16S rRNA analyses, both of these families form fairly well separated groups within the γ-subgroup of the Proteobacteria. In contrast, the purple nonsulfur bacteria are an extremely heterogeneic group of bacteria, which belong to the α- and β-subgroups of the Proteobacteria and which in many cases are more similar (on a 16S rRNA data basis) to purely chemotrophic bacteria than to themselves. Therefore, the use of a true family name has been abandoned and the popular name Purple Nonsulfur Bacteria proposed to be used for this group (Imhoff et al., 1984, Imhoff and Trüper, 1989).

The α-subgroup contains species of the genera *Rhodospirillum, Rhodopila, Rhodopseudomonas, Rhodomicrobium,* and *Rhodobacter* in different branches. Specific high similarities exist between *Paracoccus denitrificans* and *the Rhodobacter*-group (Gibson et al., 1979), between *Nitrobacter winogradskyi* and *Rp. palustris* (Seewaldt et al., 1982), and between *Aquaspirillum itersonii* and *Azospirillum brasilense* and the *Rhodospirillum*-group (Woese et al., 1984a). High similarities between *P. denitrificans and Rb. capsulatus* were also revealed by cytochrome c_2 amino acid sequences (Ambler et al., 1979). In addition, lipid A structures and internal membrane structures show great similarities between *Nb. winogradskyi* and *Rp. palustris* (Mayer et al., 1983).

The β-subgroup contains the species that have been rearranged into the genus *Rhodocyclus* (Imhoff et al., 1984). *Rhodocyclus gelatinosus* has been transferred recently to *Rubrivivax gelatinosus* (Willems et al., 1991). This species appears specifically related to *Sphaerotilus natans*, while *Rc. tenuis* is closely related to *Alcaligenes eutrophus* (Woese et al., 1984b). A bacterium that has been known as the '*Rhodocyclus gelatinosus*-like group' (Hiraishi and Kitamura, 1984) has been described as a new genus and species *Rhodoferax fermentans* (Hiraishi et al., 1991). Also this genus most likely belongs to the β-subgroup.

These data support the idea that ancestors of present day phototrophic prokaryotes are among the most ancient of eubacteria and they further point to several lines of development of non-phototrophic bacteria from phototrophic ancestors.

1. Chromatiaceae

The Chromatiaceae Bavendamm 1924 comprise those phototrophic bacteria that, under the proper growth conditions, deposit globules of elemental sulfur inside their cells (Imhoff 1984a). This family is a quite coherent group, as shown by the similarity of 16S rRNA molecules (Fowler et al., 1984) and lipopolysaccharides (Meißner et al., 1988b). All but one species (*Thiocapsa pfennigii)* have the vesicular type of intracytoplasmic membranes. Many species are able to grow under photoheterotrophic conditions, some species grow as chemolithotrophs, and a few species also grow chemoheterotrophically (Gorlenko, 1974; Kondratieva et al., 1976; Kämpf and Pfennig, 1980). Many species are motile by means of flagella, some have gas vesicles, and only *Thiocapsa* species completely lack motility. We distinguish two major physiological groups of Chromatiaceae, metabolically specialized and versatile species. Among the specialized species are *Chromatium okenii, Cm. weissei, Cm. warmingii, Cm. buderi, Cm. tepidum, Thiospirillum jenense, Thiocapsa pfennigii,* and the gas-vacuolated species of the genera *Lamprocystis* and *Thiodictyon*. These species depend on strictly anaerobic conditions and are obligately phototrophic. Sulfide is required; thiosulfate and hydrogen are not used as electron donors. Only acetate and pyruvate (or propionate) are photoassimilated in the presence of sulfide and CO_2. They do not grow with organic electron donors and chemotrophic growth is not possible. None of these species assimilates sulfate as a sulfur source. The versatile species photoassimilate a greater number of organic substrates and also grow in the absence of reduced sulfur sources with organic substrates as electron donors for photosynthesis. Most of these species are able to grow with sulfate as the sole sulfur source. They also can grow under chemolithotrophic conditions. Among these species are the small-celled *Chromatium* species, *Thiocystis violacea, Thiocapsa roseopersicina, Lamprobacter modestohalophilus, and Amoebobacter* species. Characteristic properties of Chromatiaceae are shown in Table 4.

2. Ectothiorhodospiraceae

The Ectothiorhodospiraceae Imhoff 1984 are phototrophic sulfur bacteria that, during oxidation of sulfide, deposit elemental sulfur outside the cells. They are distinguished from the Chromatiaceae by lamellar intracytoplasmic membrane structures, by significant differences in polar lipid composition (Imhoff et al., 1982), and by the dependence on saline and alkaline growth conditions (Imhoff, 1989). *Ectothiorhodospira halophila* is the most halophilic eubacterium known and even grows in saturated salt solutions. All species are motile by polar flagella and *Ectothiorhodospira vacuolata* in addition forms gas vesicles. The slightly halophilic species such as *Ec. mobilis, Ec. shaposhnikovii* and *Ec. vacuolata* and the moderately and extremely halophilic species such as *Ec. halophila, Ec. abdelmalekii* and *Ec. halochloris* form two distinct groups of species (Stackebrandt et al., 1984; Imhoff, 1989) that should be separated taxonomically. Some properties of *Ectothiorhodospira* species are shown in Table 5.

3. Purple Nonsulfur Bacteria

The group of the purple nonsulfur bacteria (Rhodospirillaceae Pfennig and Trüper 1971) is by far the most diverse group of the phototrophic purple bacteria (Imhoff and Trüper, 1989). This diversity is reflected in greatly varying morphology, internal membrane structure, carotenoid composition, utilization of carbon sources, and electron donors, among other features. The intracytoplasmic membranes are small, fingerlike intrusions, vesicles, or different types of lamellae. Most species are motile by flagella; gas vesicles are not formed by any of the known species. Some properties of purple nonsulfur bacteria are shown in Tables 6 A and B.

The preferred growth mode of all species is photoheterotrophic under anaerobic conditions in the light with various organic substrates. Many species also are able to grow photoautotrophically with either molecular hydrogen or sulfide as electron donor and CO_2 as the sole carbon source. Most of these species oxidize sulfide to elemental sulfur only (Hansen and Van Gemerden, 1972). In *Rhodobacter sulfidophilus* and *Rhodopseudomonas palustris* sulfate is the final oxidation product and is formed without accumulation of elemental sulfur as an intermediate (Hansen, 1974; Hansen and Veldkamp, 1973). Three species *Rhodobacter veldkampii, Rhodobacter adriaticus,* and *Rhodobacter euryhalinus*, are known to deposit elemental sulfur outside the cells, while oxidizing sulfide to sulfate (Hansen et al., 1975; Neutzling et al., 1984; Hansen and Imhoff, 1985; Kompantseva, 1985).

Most representatives can grow under microaerobic to aerobic conditions in the dark as chemoheterotrophs, a few also as chemolithotrophs. Some species are very sensitive to oxygen, but most species are quite tolerant of oxygen and grow well under aerobic conditions in the dark. Under these conditions synthesis of photosynthetic pigments is repressed and the cultures are faintly colored or colorless. Even under phototrophic growth conditions (anaerobic/light) many species exhibit considerable respiratory capacity. Respiration under these conditions is inhibited, however, by light. The fact that considerable respiratory activity is expressed

Table 4. Characteristic properties of Chromatiaceae[a]

Species	Cell shape	Cell diameter (μm)	Major carotenoids	Growth factor	Flagella	Gas vesicles	Chemolithotrophy	Utilization of Thiosulfate	Utilization of Hydrogen	Utilization of Sulfate	Utilization of Organic carbon	G + C content (mol%)
Thiospirillum												
jenense	spiral	2.5–4.5	rh,ly	B_{12}	+	–	–	–	–	–	–	45.5
Thiorhodovibrio												
winogradskyi	vibrio	0.7–2.1	rh,sp	–	+	–	o	+	o	o	(–)	61.0–62.4
Chromatium												
okenii	rod	4.5–6.0	ok	B_{12}	+	–	–	–	–	–	–	48.0–50.0
weissei	rod	3.5–4.0	ok	B_{12}	+	–	–	–	–	–	–	48.0–50.0
warmingii	rod	3.5–4.0	ra,ro	B_{12}	+	–	–	–	–	–	–	55.1–60.2
buderi	rod	3.5–4.5	ra	B_{12}	+	–	–	–	–	–	–	62.2–62.8
tepidum	rod	1.2	sp, rv	–	+	–	–	–	–	–	–	61.5
minus	rod	2.0	ok	–	+	–	+	+	–1	–	–	62.2
salexigens	rod	2.0–2.5	sp	B_{12}	+	–	+	+	–1	–	–	64.6
vinosum	rod	2.0	sp,ly,rh	–	+	–	+	+	+	+	+	61.3–66.3
violascens	rod	2.0	rh,ro,ra	–	+	–	+	+	+	+	+	61.8–64.3
gracile	rod	1.0–1.3	sp,ly,rh	–	+	–	+	+	+	+	+	68.9–70.4
minutissimum	rod	1.0–1.2	sp,ly,rh	–	+	–	+	+	+	+	+	63.7
purpuratum	rod	1.2–1.7	ok	o	+	–	o	+	o	o	+	68.9
Thiocystis												
violacea	sphere	2.5–3.5	ra,ro,rh	–	+	–	+	+	–1	+/–	+	62.8–67.9
gelatinosa	sphere	3.0	ok	–	+	–	+	–	–1	o	–	61.3
Lamprocystis												
roseopersicina	sphere	3.0–3.5	la,lo	–	+	+	–	o	o	o	–	63.8
Lamprobacter												
modesto-halophilus	rod	2.0–2.5	o	B_{12}	+	+	+	+	+	–	+	64.0
Thiodictyon												
elegans	rod	1.5–2.0	ra,rh	–	–	+		–	o	o	–	65.3
bacillosum	rod	1.5–2.0	ra	–	–	+	–	–	o	o	–	66.3
Amoebobacter												
roseus	sphere	2.0–3.0	sp	B_{12}	–	+	+	+	–	–	+	64.3
pendens	sphere	1.5–2.5	sp	B_{12}	–	+	–	+	–	–	+	65.3
pedioformis	sphere	2.0	sp	(B_{12})	–	+	+	+	–	–	+	65.5
purpureus	sphere	1.9–2.3	ok	o	–	+	+	+	–	–	+	63.4–64.1
Thiopedia												
rosea	ovoid	1.0–2.0	ok	B_{12}	–	+	–	–	–	–	+	o
Thiocapsa												
roseopersicina	sphere	1.2–3.0	sp	–	–	–	+	+	+	+	+	63.3–66.3
pfennigii	sphere	1.2–1.5	ts	–	–	–	–	–	–1	o	–	69.4–69.9
halophila	sphere	1.5–2.5	ok	B_{12}	–	–	+	+	+	–	+	65.9–66.6

Abbreviations: o – not determined; (B_{12}) – vitamin B_{12} strongly enhancing growth, but not absolutely required; –1 – hydrogenase present, but growth with H_2 not demonstrated. Utilization of organic carbon: minus sign (–) – only acetate and pyruvate (or propionate) are photoassimilated; plus sign (+) – other carbon sources are used as well. Carotenoids: la – lycopenal, lo – lycopenol, ly – lycopene, ok – okenone, ra – rhodopinal, rh – rhodopin, ro – rhodopinol, rv – rhodovibrin, sp – spirilloxanthin, ts – 3,4,3',4', tetrahydrospirilloxanthin.

[a] Data were collected from: Caumette et al. (1988, 1991); Eichler and Pfennig (1986, 1988); Gorlenko et al. (1979); Kämpf and Pfennig (1980); Madigan (1986); Overmann et al. (1992) Schmidt (1978); Trüper and Pfennig (1981).

Table 5. Characteristic properties of Ectothiorhodospiraceae[a]

Species	Cell shape	Cell diameter (μm)	Color of culture	Major carotenoids	Major BChl	Gas vesicle	Optimum salinity range	G + C content (mol%)
Ectothiorhodospira								
mobilis	rod-spiral	0.7–1.0	red	sp,rh	*a*	–	3–15%	62.0–69.9
shaposhnikovii	rod-spiral	0.8–0.9	red	sp,rh	*a*	–	1–3%	61.2–62.8
vacuolata	rod	1.5	red	sp	*a*	+	1–6%	61.4–63.6
halophila	spiral	0.8–0.9	red	sp	*a*	–	15–30%	64.3–69.7
halochloris	spiral	0.5–0.6	green	rhg*,rh	*b*	–	14–27%	50.5–52.9
abdelmalekii	spiral	0.9–1.2	green	**	*b*	–	12–20%	63.3–63.8
marismortui	pleomorphic rods	0.9–1.3	red	(sp)	*a*	–	3–8%	65

Abbreviations: carotenoids: rh – rhodopin, rhg* – rhodopin glucoside and derivatives, sp – spirilloxanthin, (sp) – most probably spirilloxanthin, ** – most probably similar as in *Ec. halochloris.*

[a] Data were collected from: Imhoff (1984b); Imhoff (1989); Imhoff and Trüper (1981); Imhoff et al. (1981); Oren et al. (1989); Schmidt (1978).

Table 6A. Characteristic properties of purple nonsulfur bacteria (Rhodospirillaceae): The genera *Rhodospirillum*, *Rhodopila*, *Rhodomicrobium*, and *Rhodobacter*

Species	Cell shape	Cell diameter (μm)	ICM	Major caro-tenoids	Sulfide oxidation to	Growth factors	Salt response	G + C content (mol%)
Rhodospirillum								
rubrum	spiral	0.8–1.0	V	sp, rv	S°	b	F	63.8–65.8
photometricum	spiral	1.2–1.5	S	rv, rh	–	YE	F	65.8
molischianum	spiral	0.7–1.0	S	ly, rh	–	AA	F	61.7–64.8
fulvum	spiral	0.5–0.7	S	ly, rh	–	paba	F	64.3–65.3
salexigens	spiral	0.6–0.7	L	sp	–	glutamate	H	64.0
salinarum	spiral	0.8–0.9	V	sp	–	YE	H	67.4–68.1
mediosalinum	spiral	0.8–1.0	V	sp	+	t, paba, n	H	66.6
*centenum**	spiral	1.0	L	o	o	b,B_{12}	F	68.3
sodomense	spiral	0.6–0.7	V	sp	–	complex	H	66.2
Rhodopila								
globiformis	sphere	1.6–1.8	V	kts	–	b, paba	F	66.3
Rhodomicrobium								
vannielii	ovoid-rod	1.0–1.2	L	rh,ly,sp	+	none	F	61.8–63.8
Rhodobacter								
capsulatus	rod	0.5–1.2	V	sn, se	S°	t (b,n)	F	65.5–66.8
veldkampii	rod	0.6–0.8	V	sn, se	S°/Sul	b, t, paba	F	64.4–67.5
sphaeroides	ovoid-rod	0.7	V	sn, se	S°	b, t, n	F	68.4–69.9
sulfidophilus	rod	0.6–0.9	V	sn, se	Sul	b, t, n, paba	M	67.0–71.0
euryhalinus	rod	0.7–1.0	V	se	S°/Sul	b, t, n, paba	M	62.1–68.6
adriaticus	rod	0.5–0.8	V	sn, se	S°/Sul	b, t	M	64.9–66.7

Abbreviations and references: see Table 6B.

* It has recently been proposed to transfer this species to a new genus, as *Rhodocista centenaria* (Kawaski et al., 1992).

also under phototrophic growth conditions enables these bacteria (e.g. *Rhodobacter capsulatus, Rhodobacter sphaeroides, Rubrivivax (Rhodocyclus) gelatinosus, Rhodospirillum rubrum*) to switch immediately from phototrophic to respiratory metabolism when environmental conditions change.

Some species also may perform a respiratory metabolism anaerobically in the dark with sugars and either nitrate, dimethyl-sulfoxide, or trimethyl-amine-N-oxide as electron sink, or though poorly – with metabolic intermediates as electron acceptors (fermentation).

Table 6B. Characteristic properties of purple nonsulfur bacteria (Rhodospirillaceae): The genera *Rhodopseudomonas*, *Rhodocyclus*, *Rhodoferax*, and *Rubrivivax*[a]

Species	Cell shape	Cell diameter (μm)	ICM	Major carotenoids	Sulfide oxidation to	Growth factors	Salt response	G + C content (mol%)
Rhodopseudomonas[1]								
palustris	rod	0.6–0.9	L	sp, rv, rh	Sul	paba (b)	F	64.8–66.3
viridis	rod	0.6–0.9	L	neu*, ly*	–	paba, b	F	66.3–71.4
sulfoviridis	rod	0.5–0.9	L	neu, sp	+	b, p, paba	F	67.8.–68.4
blastica	rod	0.6–0.8	L	sn, se	–	B_{12}, b, n, t	F	65.3
acidophila	rod	1.0–1.3	L	rh, rg, rag	–	none	F	62.2–66.8
marina	rod	0.7–0.9	L	sp	+	o	M	61.5–63.8
julia	rod	1.0–1.5	L	o	S°/Sul	none	F	63.5
cryptolactis	rod	1.0	L	o	o	n,B_{12},paba*	F	68.8
rosea	rod	1.0	L	o	–	n	F	66.0
Rhodocyclus								
purpureus	half circle	0.6–0.7	T	ra, rh	–	B_{12}	F	65.3
tenuis	spiral	0.3–0.5	T	ly, rh, ra	–	none	F	64.8
Rhodoferax								
fermentans	curved rod	0.6–0.9	N	sn,se,sp	–	b,t	F	59.8–60.3
Rubrivivax								
gelatinosus	rod	0.4–0.5	T	sn, se	–	b, t	F	70.5–72.4

Abbreviations: o – not determined; ICM – structure of intracytoplasmic membrane system: V – vesicles, L – lamellae, S – stacks, T – tubes, N – none detected. Salt response: F – fresh water species; M – marine species; H – halophilic species. Sulfide oxidation to: S° – elemental sulfur only; S°/Sul – sulfate with intermediate formation of extracellular elemental sulfur globules; Sul – sulfate without formation of S°. Vitamins: b – biotin, n – niacin, t – thiamine, paba – p-aminobenzoic acid, YE – yeast extract, AA – amino acids; vitamins in brackets are required only by some strains, paba* – p-aminobenzoic acid stimulatory. Carotenoids: kts – ketocarotenoids, ly – lycopene, ly* – 1,2 dihydrolycopene, neu – neurosporene, neu* – 1,2 dihydroneurosporene, ra – rhodopinal, rag – rhodopinal glucoside, rg – rhodopin glucoside, rh – rhodopin, rv – rhodovibrin, se – spheroidene, sn – spheroidenone, sp – spirilloxanthin.

[a] Data are collected from: Drews (1981); Eckersley and Dow (1980); Favinger et al. (1989); Hansen (1974); Hansen and Imhoff (1985); Hiraishi et al. (1991); Imhoff (1983); Janssen and Harfoot (1991); Kompantseva (1985); Kompantseva (1989); Kompantseva and Gorlenko (1984); Mack et al. (1993); Neutzling et al. (1984); Nissen and Dundas (1984); Schmidt (1978); Schmidt and Bowien (1983); Stadtwald-Demchick et al. (1990a), Stadtwald-Demchick et al. (1990b); Trüper and Pfennig (1981); Willems et al. (1991).

[1] *Rhodopseudomonas rutila* (Akiba et al., 1983) is now considered as a subjective synonym of *Rp. palustris* (Hiraishi et al., 1992).

One or more vitamins are generally required as growth factors, most commonly biotin, thiamine, niacin, and p-aminobenzoic acid; these compounds are rarely needed by species of the Chromatiaceae and Ectothiorhodospiraceae, which may require vitamin B_{12} as sole growth factor. Growth of most purple nonsulfur bacteria is enhanced by small amounts of yeast extract, and some species have complex nutrient requirements.

A number of changes have recently been made concerning the taxonomy of purple nonsulfur bacteria, and are noted here. *Rhodopseudomonas blastica* has been recognized as a species of the genus *Rhodobacter*, *Rhodobacter blastica* (Kawasaki et al., 1993). The correct name should be *Rhodobacter blasticus* because *Rhodobacter is a masculinum nomen*. *Rhodobacter marinus* has been described as a new species (Burgess et al., 1994) and in parallel, not considering this new species, the marine species of *Rhodobacter*, *Rb. adriaticus*, *Rb. euryhalinus* and *Rb. sulfidophilus* have been transferred to the new genus *Rhodovulum* with *Rhodovulum sulfidophilum* as type species (Hiraishi and Ueda, 1994a). *Rhodopseudomonas rosea* has been transferred to the new genus *Rhodoplanes*, as *Rhodoplanes roseus* and was designated as the type species of this genus (Hiraishi and Ueda, 1994b). In addition, a new species of this genus, *Rhodoplanes elegans* has been described (Hiraishi and Ueda, 1994b).

References

Adkins JP, Madigan MT, Mandelco L, Woese CR and Tanner RS (1993) *Arhodomonas aquaeolei* gen. nov., sp. nov., an aerobic halophilic bacterium isolated from a subterranean brine. Intl J Syst Bacteriol 43: 514–520

Albers H and Gottschalk G (1976) Acetate metabolism in *Rhodopseudomonas gelatinosa* and several other Rhodospirillaceae. Arch Microbiol 111: 45–49

Akiba T, Usami R and Horikoshi K (1983) *Rhodopseudomonas rutila*, a new species of nonsulfur purple photosynthetic bacteria. Intl J Syst Bacteriol 33: 551–556

Ambler RP, Daniel M, Hermoso J, Meyer TE, Bartsch RG and Kamen MD (1979) Cytochrome c_2 sequence variations among the recognised species of purple nonsulphur photosynthetic bacteria. Nature 278: 659–660

Amesz J and Knaff DB (1988) Molecular mechanism of bacterial photosynthesis. In: Zehnder AJB (ed) Biology of Anaerobic Microorganisms, pp 113–178. Wiley, Chichester

Bavendamm W (1924) Die farblosen und roten Schwefelbakterien des Süß- und Salzwassers. Fischer Verlag, Jena

Beatty JT and Gest H (1981) Biosynthetic and bioenergetic functions of citric acid cycle reactions in *Rhodopseudomonas capsulata*. J Bacteriol 148: 584–593

Beer-Romero P and Gest H (1987) *Heliobacillus mobilis*, a peritrichously flagellated anoxyphototroph containing bacteriochlorophyll *g*. FEMS Microbiol Lett 41: 109–114

Beer-Romero P, Favinger JL, and Gest H (1988) Distinctive properties of bacilliform photosynthetic heliobacteria. FEMS Microbiol Lett 49: 451–454

Brockmann H Jr and Lipinski A (1983) Bacteriochlorophyll *g*. A new bacteriochlorophyll from *Heliobacterium chlorum*. Arch Microbiol 136: 17–19

Buchanan BB, Evans MCW and Arnon DI (1967) Ferredoxin-dependent carbon assimilation in *Rhodospirillum rubrum*. Arch Microbiol 59: 32–40

Burgess JG, Kawaguchi R, Yamada A and Matsunaga T (1994) *Rhodobacter marinus* sp. nov.: A new marine hydrogen producing photosynthetic bacterium which is sensitive to oxygen and sulphide. Microbiology 140: 965–970

Caumette P, Baulaigue R and Matheron R (1988) Characterization of *Chromatium salexigens* sp. nov., a halophilic Chromatiaceae isolated from Mediterranean salinas. Syst Appl Microbiol 10: 284–292

Caumette P, Baulaigue R and Matheron R (1991) *Thiocapsa halophila* sp. nov., a new halophilic phototrophic purple sulfur bacterium. Arch Microbiol 155: 170–176

Cohen-Bazire G, Pfennig N, and Kunizawa R (1964) The fine structure of green bacteria. J Cell Biol 22: 207–225

Dickerson RE (1980) Evolution and gene transfer in purple photosynthetic bacteria. Nature 283: 210–212

Drews G (1981) *Rhodospirillum salexigens*, spec. nov., an obligatory halophilic phototrophic bacterium. Arch Microbiol 130: 325–327

Eckersley K and Dow CS (1980) *Rhodopseudomonas blastica* sp. nov.: A member of the Rhodospirillaceae. J Gen Microbiol 119: 465–473

Eichler B and Pfennig N (1986) Characterization of a new platelet-forming purple sulfur bacterium, *Amoebobacter pedioformis* sp. nov. Arch Microbiol 146: 295–300

Eichler B and Pfennig N (1988) A new purple sulfur bacterium from stratified fresh-water lakes, *Amoebobacter purpureus* sp. nov. Arch Microbiol 149: 395–400

Evans MCW, Buchanan BB and Arnon DI (1966) A new ferredoxin-dependent carbon reduction cycle in a photosynthetic bacterium. Proc Natl Acad Sci USA 55: 928–934

Evans WR, Fleischmann DE, Calvert HE, Pyati PV, Alter GM and Rao NSS (1990) Bacteriochlorophyll and photosynthetic reaction centers in *Rhizobium* strain BTAi 1. Appl Environ Microbiol 56: 3445–3449

Favinger J, Stadtwald R and Gest H (1989) *Rhodospirillum centenum*, sp. nov., a thermotolerant cyst-forming photosynthetic bacterium. Antonie van Leeuwenhoek 55: 291–296

Ferguson SJ, Jackson JB and McEwan AG (1987) Anaerobic respiration in the Rhodospirillaceae: characterisation of pathways and evaluation of roles in redox balancing during photosynthesis. FEMS Microbiol Rev 46: 117–143

Fowler VJ, Pfennig N, Schubert W and Stackebrandt E (1984) Towards a phylogeny of phototrophic purple sulfur bacteria — 16S rRNA oligonucleotide cataloguing of 11 species of Chromatiaceae. Arch Microbiol 139: 382–387

Fuchs G, Stupperich E and Jaenchen R (1980a) Autotrophic CO_2 fixation in *Chlorobium limicola*. Evidence against the operation of the Calvin cycle in growing cells. Arch Microbiol 128: 56–63

Fuchs G, Stupperich E and Eden G (1980b) Autotrophic CO_2 fixation in *Chlorobium limicola*. Evidence for the operation of a reductive tricarboxylic acid cycle in growing cells. Arch Microbiol 128: 64–71

Gerola PD and Olson JM (1986) A new bacteriochlorophyll *a*-protein complex associated with chlorosomes of green sulfur bacteria. Biochim Biophys Acta 848: 69–76

Gest H and Favinger JF (1983) *Heliobacterium chlorum*, an anoxygenic brownish-green bacterium containing a 'new' form of bacteriochlorophyll. Arch Microbiol 136: 11–16

Gibson J, Stackebrandt E, Zablen LB, Gupta R and Woese CR (1979) A phylogenetic analysis of the purple photosynthetic bacteria. Curr Microbiol 3: 59–64

Gibson J, Pfennig N and Waterbury JB (1984) *Chloroherpeton thalassium* gen. nov. et spec. nov., a nonfilamentous, flexing, and gliding green sulfur bacterium. Arch Microbiol 138: 96–101

Gibson J, Ludwig W, Stackebrandt E and Woese CR (1985) The phylogeny of the green photosynthetic bacteria: Absence of a close relationship between *Chlorobium* and *Chloroflexus*. Syst Appl Microbiol 6: 152–156

Gloe A and Risch N (1978) Bacteriochlorophyll c_s, a new bacteriochlorophyll from *Chloroflexus aurantiacus*. Arch Microbiol 118: 153–156

Gloe A, Pfennig N, Brockmann H Jr, and Trowitsch W (1975) A new bacteriochlorophyll from brown-colored Chlorobiaceae. Arch Microbiol 102: 103–109

Gorlenko VM (1974) Oxidation of thiosulphate by *Amoebobacter roseus* in darkness under microaerobic conditions. Microbiologiya 43: 729–731 (in Russian)

Gorlenko VM and Lebedeva EV (1971) New green sulphur bacteria with apophyses. Microbiologiya 40: 1035–1039 (in Russian)

Gorlenko VM, Krasilnikova EN, Kikina OG and Tatarinova N Ju (1979) The new motile purple sulphur bacteria *Lamprobacter modestohalophilus* nov. gen., nov. spec. with gas vacuoles. Biol Bull Acad Sci USSR 6: 631–642 (in Russian)

Hansen TA (1974) Sulfide als electronendonor voor Rhodospirillaceae. Doctoral thesis, University of Groningen, The Netherlands

Hansen TA and Imhoff JF (1985) *Rhodobacter veldkampii*, a new species of phototrophic purple nonsulfur bacteria. Intl J Syst Bacteriol 35: 115–116

Hansen TA and Van Gemerden H (1972) Sulfide utilization by purple nonsulfur bacteria. Arch Mikrobiol 86: 49–56

Hansen TA and Veldkamp H (1973) *Rhodopseudomonas sulfidophila* nov. spec., a new species of the purple nonsulfur bacteria. Arch Mikrobiol 92: 45–58

Hansen TA, Sepers ABJ and Van Gemerden H (1975) A new purple bacterium that oxidizes sulfide to extracellular sulfur and sulfate. Plant Soil 43: 17–27

Harashima K, Hayashi J-I, Ikari T and Shiba T (1980) O_2-stimulated synthesis of bacteriochlorophyll and carotenoids in marine bacteria. Plant Cell Physiol 21: 1283–1294

Hiraishi A and Kitamura H (1984) Distribution of phototrophic purple nonsulfur bacteria in activated sludge systems and other aquatic environments. Bull Jpn Soc Sci Fish 50: 1929–1937

Hiraishi A and Ueda Y (1994a) Intrageneric structure of the genus *Rhodobacter*. Transfer of *Rhodobacter sulfidophilus* and related marine species to the genus *Rhodovulum* gen. nov. Intl J Syst Bacteriol 44: 15–23

Hiraishi A and Ueda Y (1994b) *Rhodoplanes* gen. nov., a new genus of phototrophic bacteria including *Rhodopseudomonas rosea* as *Rhodoplanes roseus* comb. nov. and *Rhodoplanes elegans* sp. nov. Intl J Syst Bacteriol 44: 665–673

Hiraishi A, Hoshino Y and Satoh T (1991) *Rhodoferax fermentans* gen. nov., sp. nov., a phototrophic purple nonsulfur bacterium previously referred to as the 'Rhodocyclus gelatinosus- like' group. Arch Microbiol 155: 330–336

Hiraishi A, Santos TS, Sugiyama J and Komagata K (1992) *Rhodopseudomonas rutila* is a later subjective synonym of *Rhodopseudomonas palustris*. Intl J Syst Bacteriol 42: 186–188

Iba K, Takamiya K-I, Toh Y and Nishimura M (1988) Roles of bacteriochlorophyll and protein complexes in an aerobic photosynthetic bacterium, *Erythrobacter* sp. strain OCH 114. J Bacteriol 170: 1843–1847

Imhoff JF (1982) Taxonomic and phylogenetic implications of lipid and quinone compositions in phototrophic microorganisms. In: Wintermans JFGM and Kuiper PJC (eds) Biochemistry and Metabolism of Plant Lipids, pp 541–544. Elsevier Biomedical Press, Amsterdam

Imhoff JF (1983) *Rhodopseudomonas marina* sp. nov., a new marine phototrophic purple bacterium. Syst Appl Microbiol 4: 512–521

Imhoff JF (1984a) Reassignment of the genus *Ectothiorhodospira* Pelsh 1936 to a new family Ectothiorhodospiraceae fam. nov., and emended description of the Chromatiaceae Bavendamm 1924. Intl J Syst Bacteriol 34: 338–339

Imhoff JF (1984b) Quinones of phototrophic purple bacteria. FEMS Microbiol Lett 25: 85–89

Imhoff JF (1988) Lipids, fatty acids and quinones in taxonomy and phylogeny of anoxygenic phototrophic bacteria. In: Olson JM, Ormerod JG, Amesz J, Stackebrandt E and Trüper HG (eds) Green Photosynthetic Bacteria, pp 223–232. Plenum Press, New York

Imhoff JF (1989) Genus *Ectothiorhodospira*. In: Staley JT, Bryant MP, Pfennig N and Holt JC (eds) Bergey's Manual of Systematic Bacteriology, Volume 3, pp 1654–1658. Williams and Wilkins, Baltimore

Imhoff JF (1991) Polar lipids and fatty acids in the genus *Rhodobacter*. System Appl Microbiol 14: 228–234

Imhoff JF and Trüper HG (1981) *Ectothiorhodospira abdelmalekii* sp. nov., a new halophilic and alkaliphilic phototrophic bacterium. Zentralbl Bakteriol Hyg I Abt Orig C2: 228–234

Imhoff JF and Trüper HG (1989) The purple nonsulfur bacteria. In: Staley JT, Bryant MP, Pfennig N and Holt JC (eds) Bergey's Manual of Systematic Bacteriology, Vol 3, pp 1658–1661. Williams and Wilkins, Baltimore

Imhoff JF, Tindall B, Grant WD and Trüper HG (1981) *Ectothiorhodospira vacuolata* sp. nov., a new phototrophic bacterium from soda lakes. Arch Microbiol 130: 238–242

Imhoff JF, Kushner DJ, Kushwaha SC and Kates M (1982) Polar lipids in phototrophic bacteria of the Rhodospirillaceae and Chromatiaceae families. J Bacteriol 150: 1192–1201

Imhoff JF, Trüper HG and Pfennig N (1984) Rearrangement of the species and genera of the phototrophic 'purple nonsulfur bacteria'. Intl J Syst Bacteriol 34: 340–343

Ivanovsky RN, Sinton NV and Kondratieva EN (1980) ATP-linked citrate lyase activity in the green sulfur bacterium *Chlorobium limicola* forma *thiosulfatophilum*. Arch Microbiol 128: 239–241

Janssen PH and Harfoot CG (1991) *Rhodopseudomonas rosea* sp. nov., a new purple nonsulfur bacterium. Intl J Syst Bacteriol 41: 26–30

Kämpf C and Pfennig N (1980) Capacity of Chromatiaceae for chemotrophic growth. Specific respiration rates of *Thiocystis violacea* and *Chromatium vinosum*. Arch Microbiol 127: 125–135

Kämpf C and Pfennig N (1986) Isolation and characterization of some chemoautotrophic Chromatiaceae. J Basic Microbiol 9: 507–515

Kawasaki H, Hoshino Y, Kuraishi H and Yamasoto K (1992) *Rhodocista centenaria* gen. nov., sp. nov., a cyst-forming anoxygenic photosynthetic bacterium and its phylogenetic position in the Proteobacteria alpha group. J Gen Appl Microbiol 38: 541–551

Kawasaki H, Hoshino Y, Hirata A and Yamasato K (1993) Is intracytoplasmic membrane structure a generic criterion? It does not coincide with phylogenetic interrelationships among phototrophic purple nonsulfur bacteria. Arch Microbiol 160: 358–362

Kompantseva EJ (1985) *Rhodobacter euryhalinus* sp. nov., a new halophilic purple bacterial species. Mikrobiologiya 54: 974–982 (in Russian)

Kompantseva EJ (1989) A new species of budding purple bacterium: *Rhodopseudomonas julia* sp. nov. Microbiologiya 58: 254–259

Kompantseva EJ and Gorlenko VM (1984) A new species of moderately halophilic purple bacterium *Rhodospirillum mediosalinum* sp. nov. Microbiologiya 53: 775–781

Kondratieva EN (1979) Interrelation between modes of carbon assimilation and energy production in phototrophic purple and green bacteria. In: Quale JR (ed) Microbial Biochemistry, Vol 21, pp 117–175. University Park Press, Baltimore, MD

Kondratieva EN, Zhukov VG, Ivanovsky RN, Petushkova YP and Monosov EZ (1976) The capacity of phototrophic sulfur bacterium *Thiocapsa roseopersicina* for chemosynthesis. Arch Microbiol 108: 287–292

Kristjansen O (1988) Large Chromatiaceae: Part-time litho-autotrophs? Abstr VI Int Symp Photosynthetic Prokaryotes, Nordwijkerhout, Netherlands, p 162

Liaaen-Jensen S (1965) Bacterial carotenoids. XVIII. Aryl-carotenes from *Phaeobium*. Acta Chem Scand 19: 1025–1030

Mack EE, Mandelco L, Woese CR, and Madigan MT (1993)

Rhodospirillum sodomense, sp. nov, a Dead Sea *Rhodospirillum* species. Arch Microbiol 160: 363–371

Madigan MT (1986) *Chromatium tepidum* sp. nov, a thermophilic photosynthetic bacterium of the family Chromatiaceae. Intl J Syst Bacteriol 36: 222–227

Madigan, MT and Gest H (1979) Growth of the photosynthetic bacterium *Rhodopseudomonas capsulata* chemoautotrophically in darkness with H_2 as the energy source. J Bacteriol 137: 524530

Mayer H (1984) Significance of lipopolysaccharide structure for taxonomy and phylogenetical relatedness of Gram-negative bacteria. In: Haber E (ed) The Cell Membrane, pp 71–83. Plenum Press, New York

Mayer H, Bock E and Weckesser J (1983) 2,3-Diamino-2,3-dideoxy- glucose containing lipid A in the *Nitrobacter* strain X_{14}. FEMS Microbiol Lett 17: 93–96

Meißner J, Krauss JH, Jürgens UJ and Weckesser J (1988a) Absence of a characteristic cell wall lipopolysaccharide in the phototrophic bacterium *Chloroflexus aurantiacus*. J Bacteriol 170: 3213–3216

Meißner J, Pfennig N, Krauss JH, Mayer H and Weckesser J (1988b) Lipopolysaccharides of *Thiocystis violacea, Thiocapsa pfennigii* and *Chromatium tepidum*, species of the family Chromatiaceae. J Bacteriol 170: 3217–3222

Meißner J, Borowiak D, Fischer U, Weckesser J (1988c) The lipopolysaccharide of the phototrophic bacterium *Ectothiorhodospira vacuolata*. Arch Microbiol 149: 245 – 248

Michalski TJ, Hunt JE, Bowman MK, Smith U, Bardeen K, Gest H, Norris JR and Katz JJ (1987) Bacteriophytin *g*: Properties and some speculations on a possible primary role for bacteriochlorophylls *b* and *g* in the biosynthesis of chlorophylls. Proc Natl Acad Sci USA 84: 2590–2594

Molisch H (1907) Die Purpurbakterien nach neuen Untersuchungen. G. Fischer, Jena

Neutzling O, Imhoff JF and Trüper HG (1984) *Rhodopseudomonas adriatica* sp. nov., a new species of the Rhodospirillaceae, dependent on reduced sulfur compounds. Arch Microbiol 137: 256–261

Nishimura Y, Shimizu M and Iizuka H (1981) Bacteriochlorophyll formation in radiation-resistant *Pseudomonas radiora*. J Gen Appl Microbiol 27: 427 – 430

Nissen H and Dundas ID (1984) *Rhodospirillum salinarum* sp. nov., a halophilic photosynthetic bacterium from a Portuguese saltern. Arch Microbiol 138: 251–256

Oren A, Kessel M and Stackebrandt E (1989) *Ectothiorhodospira marismortui* sp. nov., an obligatory anaerobic, moderately halophilic purple sulfur bacterium from a hypersaline sulfur spring on the shore of the Dead Sea. Arch Microbiol 151: 524–529

Overmann J and Pfennig N (1989) *Pelodictyon phaeoclathratiforme* sp. nov., a new brown-colored member of the Chlorobiaceae forming net-like colonies. Arch Microbiol 152: 401–406

Overmann J, Fischer U and Pfennig N (1992) A new purple sulfur bacterium from saline littoral sediments, *Thiorhodovibrio winogradskyi* gen. nov. and sp. nov. Arch Microbiol 157: 329–335

Pelsh AD (1937) Photosynthetic sulfur bacteria of the eastern reservoir of Lake Sakskoe. Mikrobiologiya 6: 1090–1100

Pfennig N (1978) General physiology and ecology of photosynthetic bacteria. In: Clayton RE and Sistrom WR (eds) The Photosynthetic Bacteria, pp 3–18. Plenum Press, New York

Pfennig N (1989a) Green sulfur bacteria. In: Staley JT, Bryant MP, Pfennig N and Holt JC (eds) Bergey's Manual of Systematic Bacteriology, Volume 3, pp 1682–1683. Williams and Wilkins, Baltimore

Pfennig N (1989b) Multicellular filamentous green bacteria. In: Staley JT, Bryant MP, Pfennig N and Holt JC (eds) Bergey's Manual of Systematic Bacteriology, Vol 3, pp 1697. Williams and Wilkins, Baltimore

Pfennig N and Trüper HG (1971) Higher taxa of the phototrophic bacteria. Intl J Syst Bacteriol 21: 17–18

Pfennig N and Trüper HG (1974) The phototrophic bacteria. In: Buchanan RE and Gibbons NE (eds) Bergey's Manual of Determinative Bacteriology, pp 24–64. Williams and Wilkins, Baltimore

Sato K (1978) Bacteriochlorophyll formation by facultative methylotrophs, *Protaminobacter ruber* and *Pseudomonas* AM 1. FEBS Lett 85: 207–210

Schmidt K (1978) Biosynthesis of carotenoids. In: Clayton RK and Sistrom WR (eds) The Photosynthetic Bacteria, pp 729–750. Plenum Press, New York

Schmidt K and Bowien B (1983) Notes on the description of *Rhodopseudomonas blastica*. Arch Microbiol 136: 242

Seewaldt E, Schleifer K-H, Bock E and Stackebrandt E (1982) The close phylogenetic relationship of *Nitrobacter* and *Rhodopseudomonas palustris*. Arch Microbiol 131: 287–290

Shiba T (1984) Utilization of light energy by the strictly aerobic bacterium *Erythrobacter* sp. OCH 114. J Gen Appl Microbiol 30: 239–244

Shiba T (1991) *Roseobacter litoralis* gen. nov., sp. nov., and *Roseobacter denitrificans* sp. nov., aerobic pink-pigmented bacteria which contain bacteriochlorophyll *a*. Syst Appl Microbiol 14: 140–145

Shiba T, Simidu U and Taga N (1979) Distribution of aerobic bacteria which contain bacteriochlorophyll *a*. Appl Environ Microbiol 38: 43–45

Shimada K, Hayashi H and Tasumi M (1985) Bacteriochlorophyll-protein complexes of aerobic bacteria, *Erythrobacter longus* and *Erythrobacter* species OCH 114. Arch Microbiol 143: 244–247

Siefert E and Pfennig N (1979) Chemoautotrophic growth of *Rhodopseudomonas* species with hydrogen and chemotrophic utilization of methanol and formate. Arch Microbiol 122: 177–182

Stackebrandt E and Woese CR (1981) The evolution of procaryotes. In: Carlile MJ, Collins JR, and Moseley BEB (eds) Molecular and Cellular Aspects of Microbial Evolution, pp 1–31. Cambridge University Press, Cambridge

Stackebrandt E, Fowler VJ, Schubert W and Imhoff JF (1984) Towards a phylogeny of phototrophic purple bacteria – The genus *Ectothiorhodospira*. Arch Microbiol 137: 366–370

Stackebrandt E, Murray RGE and Trüper HG (1988) *Proteobacteria* classis nov., a name for the phylogenetic taxon that includes the 'purple bacteria and their relatives.' Intl J Syst Bacteriol 38: 321–325

Stadtwald-Demchick R, Turner FR and Gest H (1990a) Physiological properties of the thermotolerant photosynthetic bacterium, *Rhodospirillum centenum*. FEMS Microbiol Lett 67: 139–144

Stadtwald-Demchick, Turner FR and Gest H (1990b) *Rhodopseu-*

domonas cryptolactis, sp. nov., a new thermotolerant species of budding phototrophic purple bacteria. FEMS Microbiology Letters 71: 117–122

Staehelin LA, Fuller RC and Drews G (1978) Visualization of the supramolecular architecture of chlorosomes (chlorobium vesicles) in freeze-fractured cells of *Chloroflexus aurantiacus*. Arch Microbiol 119: 269–277

Staehelin LA, Golecki JR and Drews G (1980) Supramolecular organization of chlorosomes (chlorobium vesicles) and of their membrane attachment sites in *Chlorobium limicola*. Biochim Biophys Acta 589: 30–45

Steiner R, Schäfer W, Blos I, Wieschoff H and Scheer H (1981) 2, 10-Phytadienol as esterifying alcohol of bacteriochlorophyll *b* from *Ectothiorhodospira halochloris*. Z Naturforsch 36c: 417–420

Trüper HG (1968) *Ectothiorhodospira mobilis* Pelsh, a photosynthetic sulfur bacterium depositing sulfur outside the cells. J Bacteriol 95: 1910–1920

Trüper HG (1989) Genus *Erythrobacter*. In: Staley JT, Bryant MP, Pfennig N and Holt JG (eds) Bergey's Manual of Systematic Bacteriology, Vol 3, pp 1708–1709. Williams and Wilkins, Baltimore

Trüper HG and Pfennig N (1981) Characterization and identification of the anoxygenic phototrophic bacteria. In: Starr MP, Stolp H, Trüper HG, Balows A and Schlegel HG (eds) The Prokaryotes, pp 299–312. Springer Verlag, New York

Van Gemerden H (1968) On the ATP generation by *Chromatium* in darkness. Arch Microbiol 64: 118–124

Wahlund TM, Woese CR, Castenholz RW and Madigan MT (1991) A thermophilic green sulfur bacterium from New Zealand hot springs, *Chlorobium tepidum* sp. nov. Arch Microbiol 156: 8190

Weckesser J, Drews G, Mayer H and Fromme I (1974) Lipopolysaccharide aus Rhodospirillaceae, Zusammensetzung und taxonomische Relevanz. Zentralbl Bakteriol Hyg I Abt Orig A 228: 193–198

Weckesser J, Drews G and Mayer H (1979) Lipopolysaccharides of photosynthetic prokaryotes. Ann Rev Microbiol 33: 215–239

Widdel F, Schnell S, Heising S, Ehrenreich A, Assmus B and Schink B (1993) Ferrous iron oxidation by anoxygenic phototrophic bacteria. Nature 362: 834–836

Willems A, Gillis M and de Ley J (1991) Transfer of *Rhodocyclus gelatinosus* to *Rubrivivax gelatinosus* gen. nov., comb nov., and phylogenetic relationships with *Leptothrix, Sphaerotilus natans, Pseudomonas saccharophila,* and *Alcaligenes latus.* Intl J Syst Bacteriol 41: 65–73

Woese CR (1987) Bacterial evolution. Microbiol Rev 51: 221–271

Woese CR, Stackebrandt E, Weisburg WG, Paster BJ, Madigan MT, Fowler VJ, Hahn CM, Blanz P, Gupta R, Nealson KH and Fox GE (1984a) The phylogeny of purple bacteria: The alpha subdivision. Syst Appl Microbiol 5: 315 – 326

Woese CR, Weisburg WG, Paster BJ, Hahn CM, Tanner RS, Krieg NR, Koops H-P, Harms H and Stackebrandt E (1984b) The phylogeny of purple bacteria: The beta subdivision. Syst Appl Microbiol 5: 327–336

Woese CR, Stackebrandt E, Macke TJ and Fox GE (1985a) The phylogenetic definition of the major eubacterial taxa. Syst Appl Microbiol 6: 143–151

Woese CR, Weisburg WG, Hahn CM, Paster BJ, Zablen LB, Lewis BJ, Macke TJ, Ludwig W and Stackebrandt E (1985b) The phylogeny of purple bacteria: The gamma subdivision. Syst Appl Microbiol 6: 25–33

Woese CR, Debrunner-Vossbrink BA, Oyaizu H, Stackebrandt E and Ludwig W (1985c) Gram positive bacteria: Possible photosynthetic ancestry. Science 229: 762–765

Zablen L and Woese CR (1975) Procaryote phylogeny IV: Concerning the phylogenetic status of a photosynthetic bacterium. J Mol Evol 5: 25–34

Zahr M, Fobel B, Meyer H, Imhoff JF, Campos V P and Weckesser J (1992) Chemical composition of the lipopolysaccharides of *Ectothiorhodospira shaposhnikovii, Ectothiorhodospira mobilis,* and *Ectothiorhodospira halophila.* Arch Microbiol 157: 499–504

Chapter 2

Taxonomy, Physiology and Ecology of Heliobacteria[†]

Michael T. Madigan
Department of Microbiology, Southern Illinois University, Carbondale, IL 62901, USA

John G. Ormerod
Department of Biology, Division of Molecular Cell Biology, University of Oslo, Blindern, 0316 Oslo, Norway

Summary

Heliobacteria are anoxygenic phototrophs that contain bacteriochlorophyll *g* as their sole chlorophyll pigment. These organisms are primarily soil residents and are phylogenetically related to Gram-positive bacteria, in particular to the endospore-forming *Bacillus/Clostridium* line. Some species of heliobacteria produce heat resistant endospores containing dipicolinic acid and elevated Ca^{2+} levels. Heliobacteria can grow photoheterotrophically on a limited group of organic substrates and chemotrophically (anaerobically) in darkness by pyruvate or lactate fermentation; they are also active nitrogen-fixers. Their photosynthetic system resembles that of photosystem I of green plants but is simpler, containing a small antenna closely associated with the reaction center located in the cytoplasmic membrane; no chlorosomes typical of the green sulfur bacteria or differentiated internal membranes typical of purple bacteria are found in the heliobacteria. Heliobacteria are apparently widely distributed in rice soils and occasionally found in other soils. The ecology of heliobacteria may be tightly linked to that of rice plants, and the ability of heliobacteria to produce

[†] This chapter is dedicated to Professor Howard Gest within whose laboratory the heliobacteria were first discovered and whose infectious enthusiasm for the photosynthetic bacteria has stimulated many current workers in this field.

R. E. Blankenship, M. T. Madigan and C. E. Bauer (eds): Anoxygenic Photosynthetic Bacteria, pp. 17–30.

endospores probably has significant survival value in the highly variable habitat of rice soils. The unique assemblage of properties shown by the heliobacteria has necessitated creation of a new taxonomic family of anoxygenic phototrophic bacteria, the Heliobacteriaceae, to accommodate organisms of this type.

I. Introduction

For well over a hundred years microbiologists have used the enrichment culture technique successfully to search for new kinds of bacteria. On very rare occasions a bacterium may turn up in such a culture which has properties so unprecedented that a new taxonomic family has to be created to accommodate it. Such was the remarkable discovery of *Heliobacterium chlorum* (Gest and Favinger, 1983). The discovery was made through a combination of serendipity and insight, and for microbiologists and biochemists familiar with this field, the event was of great interest because the new organism was so unlike any of the known phototrophic bacteria. *Hb. chlorum* had a novel type of chlorophyll, BChl *g*, and it had neither the chlorosomes of green bacteria nor the membrane invaginations of most purple bacteria. And of profound importance was the revelation, on the basis of 16S rRNA sequences (Woese et al., 1985), that *Hb. chlorum* belonged to the Gram-positive bacteria and was thus phylogenetically related more to the Bacillaceae than to other phototrophic bacteria.

Heliobacterium chlorum was isolated from garden soil on the campus of Indiana University, Bloomington. Subsequently it was discovered that the soil of rice fields is a rich source of heliobacteria. *Heliobacillus mobilis* (Beer-Romero and Gest, 1987), *Heliobacterium gestii* (Ormerod et al., 1990), and several other heliobacteria strains were all isolated from the same sample of dry, rice field soil from Chainat province Thailand. Several isolates have also been obtained from hot springs. Today the heliobacteria may be viewed as being soil residents and primarily (but not exclusively) tropical organisms. They have simple nutritional requirements and are efficient nitrogen fixers (Kimble and Madigan, 1992a,b). A series of observations on various heliobacterial strains aroused suspicions that these organisms might have a resistant resting stage and eventually it was shown beyond any doubt that at least some heliobacteria can form heat resistant endospores (Ormerod et al., 1990).

One of the most striking characteristics of heliobacteria is the visible absorption spectrum of the cells (Fig. 1), which is due to the presence of BChl *g* (Brockmann and Lipinski, 1983; Michalski et al., 1987; see Fig. 2 for structure). This molecule, which was unknown prior to the discovery of heliobacteria, shows a number of interesting structural features. For example, BChl *g* resembles BChl *b* in having an ethylidene group on C8 (Fig. 2) which in the presence of oxygen and light isomerizes to give a vinyl group. Since BChl *g* has a second vinyl group on C3, the product of the isomerization is a molecule which is very similar to chlorophyll *a* (Brockmann and Lipinski, 1983; Michalski et al., 1987). This transition from BChl *g* to chlorophyll *a* presumably accounts for the dramatic change from brownish green to emerald green in cultures of heliobacteria exposed to oxygen (Gest and Favinger, 1983). Another interesting feature of BChl *g* is the long chain alcohol esterified to the propionic acid side chain at $C17^3$, which is farnesol (Michalski *et al.*, 1987). In all other known phototrophs, membrane bound chlorophylls (the cytoplasmic membrane is the location of BChl *g* in heliobacteria) contain a 20C alcohol, either phytol or the less saturated geranylgeraniol. Farnesol is only present in the chlorosome BChls of green sulfur bacteria. The significance of farnesol in the

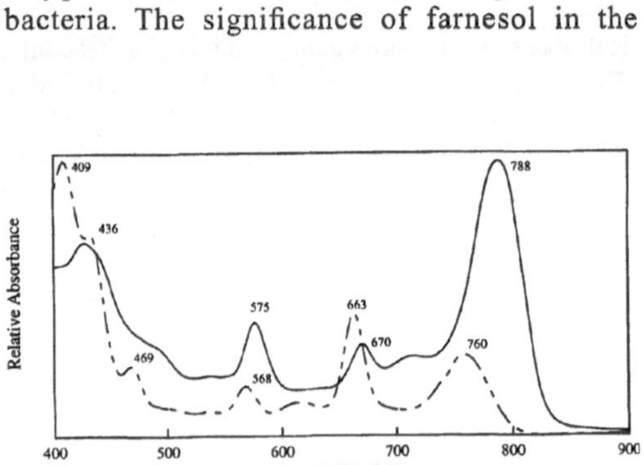

Fig. 1. Absorption spectra of intact cells (solid line) and acetone extracts (dashed line) of *Heliobacterium medesticaldum* strain Ice1. Spectra were performed anaerobically (from Kimble et al., 1995).

Abbreviations: BChl – bacteriochlorophyll; *Hb. – Heliobacterium; Hc. – Heliobacillus; Hp. – Heliophilum*; RC – reaction center membrane-bound BChl *g* of the heliobacteria is unknown.

The fact that the reaction center of heliobacteria is of the photosystem I type has important physiological consequences for these organisms. On the one hand, the highly reducing coenzyme(s) generated by the RC can drive biosynthetic reactions of a kind which require powerful reductants. On the other hand, these reduced coenzymes can react with oxygen to form toxic radicals. In the light, therefore, heliobacteria are very sensitive to oxygen and the effect is probably exacerbated by the presence in their membrane lipids of a very high proportion of unsaturated fatty acids (Imhoff, 1988; Beck et al., 1990; Aase et al., 1994). The structure of the heliobacterial reaction center is detailed in Chapter 31 by Amesz.

In this chapter we shall concentrate on aspects of the heliobacteria which are not dealt with in other chapters, such as taxonomy, physiology, and ecology. Consideration of the properties of the strains of heliobacteria that have already been isolated in pure culture indicates that many of their basic characteristics are very similar. This makes classification of heliobacteria difficult and there is now an acute need for a proper consideration of how to proceed with the taxonomy of this growing family of phototrophic bacteria. The next section describes the current state of affairs concerning the taxonomy of this group.

II. Taxonomy of Heliobacteria

The family Heliobacteriaceae (Beer-Romero and Gest, 1987) comprises all species of anoxygenic phototrophic bacteria that contain BChl *g*. All organisms that fit this definition lack BChls *a, b, c, d* or *e* (Madigan, 1992). At present two genera are recognized in the family Heliobacteriaceae, *Heliobacterium* and *Heliobacillus*. The genus *Heliobacterium* contains the species *Hb. chlorum*, the first of the heliobacteria discovered (Gest and Favinger, 1983), the endospore-forming species *Hb. gestii* and *Hb. fasciatum* ('*Hb. fasiculum*', Ormerod et al., 1990), and the thermophilic species, *Hb. modesticaldum* (Kimble et al., 1995). However, because of several unique properties including its phylogenetic position, the present authors will formally describe *Hb. fasciatum* as a new genus of heliobacteria, *Heliophilum* (as *Hp. fasciatum*), in a future publication, thus this organism will be referred to as such herein. In addition to containing BChl *g*, all heliobacteria are strict anaerobes, at least when grown phototrophically, are rod-shaped or spiral in morphology, and are motile by either gliding or flagellar means (Madigan, 1992, and Table 1).

A. General Description

The number of strains of heliobacteria isolated to date is probably in the region of fifty. Not all of these cultures have survived. Most heliobacteria are motile rods or spirilla with flagella, either peritrichous, as in *Hc. mobilis*, or polar/subpolar as in *Hb. gestii*. *Heliobacterium chlorum* is unique in being motile by gliding (Gest and Favinger, 1983). All heliobacteria that have been examined share a number of physiological properties. They have a characteristic absorption spectrum (Fig. 1) and *Hb. chlorum*, *Hb. gestii,* and *Hc. mobilis* are photosynthetically and spectroscopically indistinguishable (Table 1). Heliobacteria utilize a rather limited range of simple organic substrates as carbon sources photoheterotrophically. No heliobacteria have been shown to grow autotrophically but they can use hydrogen, and one strain can use H_2S (Starynin and Gorlenko, 1993) as reductant in the assimilation of organic substrates. All heliobacteria require biotin and some are dependent on a reduced source of sulfur (Beer-Romero, 1986; Stevenson, 1993).

Although the heliobacteria are phylogenetically Gram-positive, they do not stain as such. Their cell wall has no typical lipopolysaccharide components (Beck, et al., 1990; Aase et al., 1994). Many heliobacteria show a marked tendency to form spheroplasts and lyse in the late log or stationary phase of growth. The extent of this depends on the strain and in some cases the growth medium. The phenomenon of lysis may be related to the unusual nature of the cell wall; no outer membrane is present in these organisms and their peptidoglycan content is very low compared to that of other Gram-positive bacteria (Beer-Romero et al., 1988). That the peptidoglycan layer of heliobacteria is either not as thick or as strong as that of other Gram-positive bacteria is also apparent from the exquisite sensitivity shown to penicillin by these organisms: *Hb. chlorum* and *Hc. mobilis* are one thousand times more sensitive to penicillin G than is *Bacillus subtilis* (Beer-Romero et al., 1988). Additionally, it is possible that autolysins, induced around the time of sporulation to free the spore from the sporangium (as is known to occur in *Bacillus,* Joliffe et al., 1981), are also produced by

Table 1. Major properties of the *Heliobacteriaceae*

Property	*Heliobacterium chlorum*[a]	*Heliobacterium gestii*[b]	*Heliobacterium modesticaldum*[c]	*Heliophilum fasciatum*[b]	*Heliobacillus mobilis*[d]	Strain BR-4 Rod[e]
Cell morphology	Rod	Spirillum	Rod/curved rod	Rod	Rod	Rod
Cell size (μm)	1 × 7–9	1 × 7–10	1 × 2.5–6.5	0.8–1 × 8-20; associate in bundles	1 × 7–10	0.6–1 × 4–7
Pigments	Bacterio-chlorophyll *g*, neurosporene	Bacterio-chlorophyll *g*, neurosporene	Bacterio-chlorophyll *g*, neurosporene	Bacterio-chlorophyll *g*, carotenoids unknown	Bacterio-chlorophyll *g*, neurosporene	Bacterio-chlorophyll *g*, neurosporene
Absorption maxima in vivo (nm)	788, 718(S), 670, 575, 375	788, 718(S), 670, 575, 375	788, 718(S), 670, 575, 430	792, 723(S), 673,577, 481, 418	788, 718(S), 670, 575, 375	788, 670, 575, 375
Motility type	Gliding	Flagellar, subpolar flagella	Flagellar, polar or subpolar, or nonmotile	Flagellar, thick polar flagella flagella	Flagellar, peritrichous flagella	Flagellar, peritrichous
Carbon compounds photo-metabolized	Pyruvate, lactate, yeast extract	Pyruvate, lactate, butyrate (+CO_2), ethanol (+CO_2), acetate (+H_2), yeast extract	Pyruvate, lactate, acetate, yeast extract	Pyruvate, lactate, yeast extract	Pyruvate, lactate, butyrate (+CO_2), acetate, yeast extract	Pyruvate, lactate, acetate, butyrate (+CO_2), malate
Endospores observed	No	Yes	Yes	Yes	No	Yes
Alternative (non-Mo) nitrogenase[f]	No	Yes	Unknown	Unknown	No	Unknown
Required growth factors	Biotin	Biotin, reduced sulfur source	Biotin	Biotin	Biotin	Biotin
Optimum temperature (°C)	38-42	38-42	50–52	35-40	38-42	30-35
Optimum pH	6.2-7	6.2-7	6–7	7	6.2-7	7-8
GC content (mol%)	52	54.8	54.5–55	51.8	50.3	51.3

[a] Gest and Favinger (1983) and Madigan (1992)
[b] Ormerod et al. (1990) and Madigan (1992)
[c] Kimble et al., 1995
[d] Beer-Romero and Gest (1987); Beer-Romero et al. (1988), and Madigan (1992)
[e] Starynin and Gorlenko (1993)
[f] Kimble and Madigan (1992b)

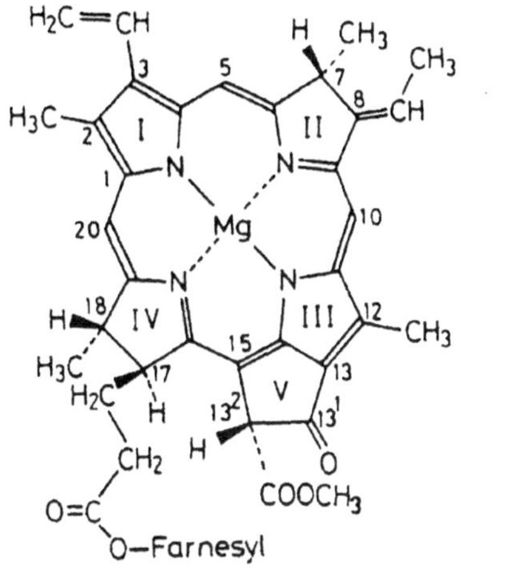

Fig. 2. Structure of bacteriochlorophyll *g* (modified after Kobayashi et al., 1991).

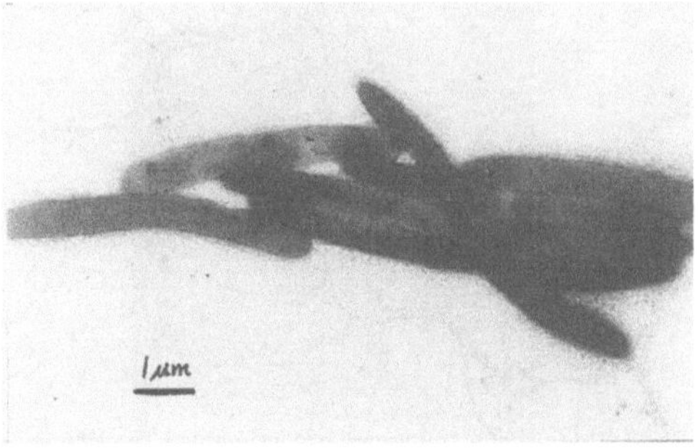

Fig. 3. Electron micrograph of a raft of *Heliophilum fasciatum* cells negatively stained with phosphotungstate.

certain heliobacteria and contribute to premature lysis. Cell lysis is a nuisance in experimental work with heliobacteria and some of the newer isolates are more stable in this regard (see below).

A large fraction of the heliobacteria that have been isolated to date have been observed to form endospores. However, it has been repeatedly observed that cultures of heliobacteria newly isolated from enrichment cultures may sporulate heavily on plates or in liquid medium, but that upon repeated transfer the occurrence of spores begins to diminish and eventually becomes very low. *Heliobacterium chlorum* and *Heliobacillus mobilis* have not been reported to produce endospores. However, because other species of heliobacteria do sporulate it is possible that conditions for sporulation of *Hb. chlorum* and *Hc. mobilis* have not yet been discovered. This possibility is supported by preliminary experiments in the laboratory of one of us (MTM) which have shown that a small fraction of cultures of *Hb. chlorum* and *Hc. mobilis* tested do survive pasteurization (80 °C for 15 min), suggesting that sporulation may occur in these species but only at low levels. Despite this, microscopic examination of hundreds of cultures of *Hb. chlorum* and *Hc. mobilis* has never revealed the presence of endospores.

Endospore-formation has been clearly demonstrated in *Hb. gestii* and *Hp. fasciatum* (Fig. 3, and Ormerod et al., 1990), in the *Hc. mobilis*-like isolate of Starynin and Gorlenko (1993), in strain HY3 of Kelly (Pickett et al., 1994), in the HD7 strain of *Hb. gestii* isolated by Pfennig (personal communication and own observations), and in the thermophilic species, *Hb. modesticaldum* (Kimble et al., 1995). In liquid cultures of *Hb. gestii*, disintegrating sporangia form clumps containing large numbers of spores (Fig. 4) and in some strains it has been observed that the sporangia are swollen and spindle-shaped like those of many clostridia (Ormerod, unpublished).

The process of sporulation has been studied in *Hb. gestii* (Torgersen, 1989; Ormerod et al., 1990) and in strain BR-4 (Starynin and Gorlenko, 1993). Endospores in *Hb. gestii* are subterminal and cylindrical, measuring about 1 × 2μm (Figs. 4 and 5), and are produced in various amounts in aged cultures. By contrast, spores of *Hp. fasciatum* are longer than those of *Hb. gestii*. Spores of strain BR-4 (Starynin and Gorlenko, 1993) are similar in morphology to those of *Hb. gestii* and *Hp. fasciatum* and also, as in these species, located subterminally. Heliobacterial endospores are heat resistant, withstanding 100 °C for a few minutes (Ormerod et al., 1990; Starynin and Gorlenko, 1993). In addition, spores of *Hb. gestii* have been shown to contain dipicolinic acid, the signature molecule of bacterial endospores, as well as six times the Ca^{2+} content of vegetative cells (Torgersen, 1989; Ormerod et al., 1990). These observations indicate that heliobacterial spores are similar to those of Bacillaceae. Starynin and Gorlenko (1993) showed that the number of spores produced in strain BR-4 was greatest at pH 9 and that ripening and release of the spores from the sporangium was stimulated by oxygen.

B. New Isolates

In the past few years one of us (MTM) has enriched

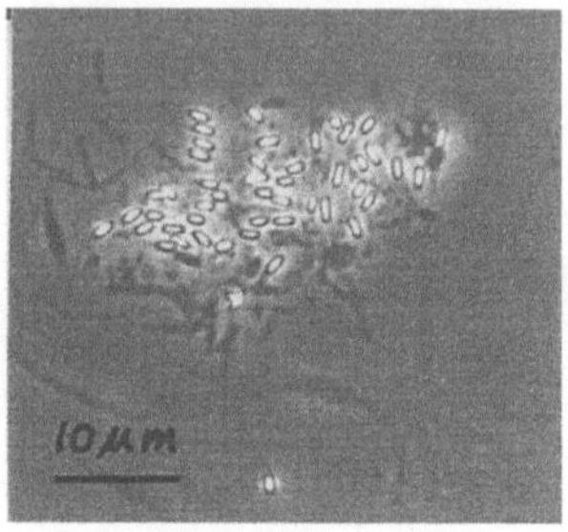

Fig. 4. Phase-contrast micrograph of accumulated endospores of *Heliobacterium gestii* with remains of sporangia.

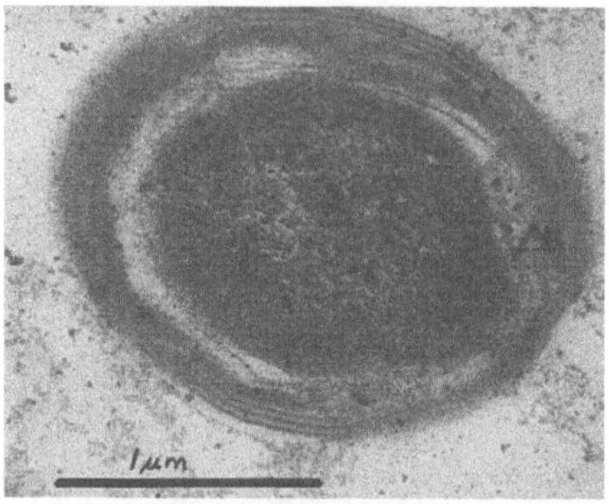

Fig. 5. Electron micrograph of a thin section of an endospore of *Heliobacterium gestii*. A multilayered wall similar to that found in Bacillaceae surrounds a typical cortex (courtesy of Y.A. Torgersen, 1989).

and isolated several new heliobacteria from soil habitats and hot springs (see Section IVB). Of 16 newly isolated strains, 14 were rod-shaped, resembling *Hc. mobilis*, and two were spirilla, resembling *Hb. gestii* (Stevenson, 1993). The new isolates were obtained from soils from the USA and from Thailand and Iceland, although positive enrichments (but not pure cultures) were also obtained with soils from Venezuela, Spain, Japan and Australia. Most of these isolates have not been thoroughly characterized except for their carbon and nitrogen nutrition. In brief, all of them grew photoheterotrophically on pyruvate or lactate and most also utilized acetate. Butyrate was used by about half of the isolates, and one strain of *Hb. gestii* (strain THAI 15-1 isolated from Thailand paddy soil), also grew photoheterotrophically on sugars. Strain THAI 15-1 grew on glucose, fructose, galactose and maltose, but grew to highest cell yield on fructose of all sugars tested. These results indicate that sugar-utilizing heliobacteria exist, although they are apparently not common. The two Icelandic isolates, one a rod and one a spirillum, were thermophilic, capable of growth above 50 °C; four similar strains were isolated from Yellowstone hot springs (Kimble and Madigan, 1994). The thermophilic isolates have been described as a new speices of the genus *Heliobacterium*, *Hb. modesticaldum* (Kimble et al., 1995). As has been found with recognized species of heliobacteria (Kimble and Madigan, 1992a), all new isolates grew on N_2 as sole nitrogen source as well as on ammonia. All of the isolates also used glutamine as a nitrogen source while the sugar-degrading strain THAI 15-1 also grew well on asparagine.

Most of the new isolates of heliobacteria described above were obtained from enrichment cultures that employed a pasteurization step (80 °C, 15 min) before incubation. Despite this, many of the pure cultures obtained were not observed to sporulate. Although most of the rod-shaped isolates resemble *Hc. mobilis*, assignment of any of them to genus and species rank will require further characterization.

C. The Family Heliobacteriaceae

With the description of the new genus and species *Heliobacillus mobilis* by Beer-Romero and Gest (1987), the family Heliobacteriaceae was established to group together anoxygenic phototrophs that contain BChl *g*. The establishment of family rank for these organisms was fully justified on the basis of the phenotypic and ecological properties of the heliobacteria, and was further confirmed by ribosomal RNA sequencing studies which showed that these phototrophs have Gram-positive evolutionary roots, unlike those of any other anoxygenic phototrophic bacteria (Woese et al., 1985; Woese, 1987). Comparisons of the 16S rRNA sequences of *Hb. chlorum*, *Hc. mobilis*, and *Hb. gestii*, show them all to be fairly closely related and to belong to the 'low GC' (*Bacillus*/*Clostridium*) subdivision of the Gram-positive bacteria (Woese et al., 1985; Beer-Romero, 1987; Beer-Romero and Gest, 1987; Woese, 1987), specifically to the *Desulfotomaculum* group (Redburn and Patel, 1993; Woese, personal communication to MTM). Phylogenetic analysis of *Hp. fasciatum* (Woese and Madigan, unpublished) showed it to be sufficiently distinct from all species of *Heliobacterium* and from *Hc. mobilis* to have its own genus, *Heliophilum*; formal publication of this name is forthcoming.

In summary (Table 1) heliobacteria can be defined as (1) anoxygenic phototrophic bacteria that show evolutionary relationships to Gram-positive bacteria, (2) that contain BChl *g* but no other BChls, (3) that lack the chlorosomes of green sulfur bacteria and *Chloroflexus* and the intracytoplasmic membranes of purple bacteria, (4) that are strictly anaerobic, and (5) that apparently reside chiefly in soil (see Section IVA). Other unifying properties of note in Table 1 are the fact that the G + C content of DNA from heliobacteria clusters tightly within the range of 50–55%, all species require biotin as a growth factor, and all have relatively high growth temperature optima as compared to most purple or green bacteria.

III. Physiology

A. Phototrophic Growth

Although the heliobacteria have been known for only a decade or so, a good deal of research has already been carried out on them. Much of this research has understandably been concerned with the structure and functioning of the unique reaction center of these organisms, which is of the photosystem I type. The antenna of heliobacteria is directly associated with the polypeptides of the RC in the cytoplasmic membrane and is limited to about 35 molecules of BChl *g* per RC (see Chapter 31 in this volume by Amesz). There are indications (Trost, 1990) that a light regulated peripheral antenna is absent but the results of Beer-Romero (1986) and of Torgersen (1989) suggest that total BChl content of the cells varies inversely with the light intensity, as in other anoxygenic phototrophs. As a result of the small size of the antenna, heliobacteria grow best at high light intensities (Beer-Romero, 1986; see also Section IVB).

The heliobacterial RC consists of a protein dimer. The primary electron donor is designated P798 and the first stable product of the RC is most likely a reduced ferredoxin-like coenzyme which acts as the main connecting link between the RC and the biosynthetic metabolism of the cell. P798 becomes reduced back to the ground state by a cytochrome, which is itself reduced by organic or inorganic electron donors. Examples of the latter are molecular hydrogen (Ormerod, unpublished) and sulfide (Starynin and Gorlenko, 1993). It is assumed that the RC can also create a proton motive force through cyclic electron transfer via the membrane cytochrome bc_1 complex (see Chapter 31 by Amesz). This provides the illuminated cell with a source of energy for motility, transport, and ATP synthesis.

Flagellated heliobacteria show the 'shock movement' typical of phototrophic purple bacteria—they reverse direction of swimming on encountering darkness. This phenomenon has been termed *scotophobotaxis* (H. Gest, personal communication). The large, rod shaped cells of *Hp. fasciatum* associate to form parallel bundles that move as a unit, by polar flagella (Fig. 3). The bundles, which consist of a few to hundreds of rods, show a characteristic rolling motion (Ormerod et al., 1990), and are scotophobotactic, indicating that some form of communication between individual cells in the bundle may exist. Microscopic observation of aggregating cells suggests that the flagella adhere to each other first, followed by the body of the cell in longitudinal fashion. An interesting observation concerning flagella is that in old cultures of flagellated heliobacteria objects are present which look like spirilla with pointed ends. On examination in the electron microscope, these

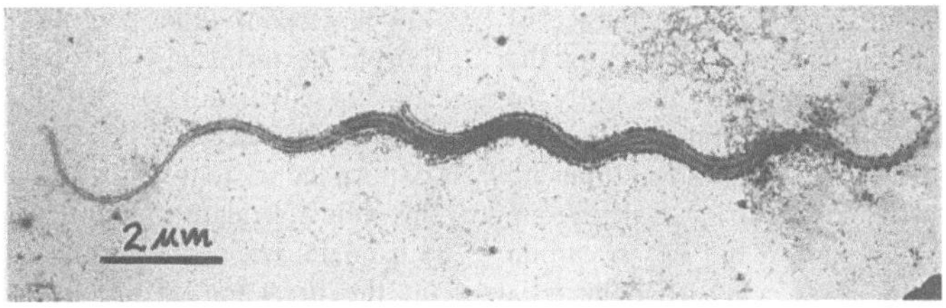

Fig. 6. Electron micrograph of an aggregate of cast off flagella from a lysed culture of *Hb. gestii* negatively stained with phosphotungstate.

can be seen to be accumulations of flagella (Fig. 6).

Phototrophic growth of heliobacteria depends on anaerobic conditions. Some strains of heliobacteria have been observed to swim away from oxygen, but in some strains, exposure to oxygen in bright light can lead to catastrophic damage to the cell protoplast in a matter of seconds. This has been observed microscopically in *Hb. gestii* and *Hp. fasciatum* as the sudden development of transverse striations in the cell, indicating destruction of the cytoplasmic membrane. Presumably strong reductants from the RC react with oxygen to form toxic oxygen species which destroy the unsaturated fatty acids in the membrane.

On the basis of 16S rRNA sequences (Woese et al., 1985) heliobacteria may be considered to be phototrophic clostridia. The main difference between the two is, of course, that the clostridia are chemotrophs and obtain chemical energy for growth by fermentation of organic substrates while heliobacteria, which also grow on organic substrates, are phototrophs, and get energy from light. In view of the distinct phylogenetic relationship between the two organisms, we should be on the lookout for clostridial traits in the carbon metabolism of heliobacteria. One such trait, fermentative growth (see Section IIIB), has already been discovered (Kimble et al., 1994).

With these considerations in mind, we can examine the possible assimilatory pathways for the various substrates heliobacteria are known to utilize and view these against the background of what is known of heliobacterial enzymes. One heliobacterial enzyme, malic dehydrogenase has been characterized (Charnock et al., 1992). Until recently nothing was known about other heliobacterial enzymes, but significant findings have been reported by Kelly and his coworkers (Pickett et al., 1994) and it is now possible to outline how carbon assimilation may proceed in heliobacteria.

Pickett et al. (1994) detected activities of pyruvate and 2-oxoglutarate synthases in cell-free extracts of heliobacteria. This finding is noteworthy because the products of these two reactions are quantitatively the most important of all the biosynthetic precursor metabolites in the cell. Since the two reactions are driven by reduced ferredoxin (or its equivalent), one can envisage direct coupling of the RC to carbon assimilation. Other enzymes detected in heliobacteria were PEP synthetase, PEP carboxykinase, malate dehydrogenase, fumarase, fumarate reductase, and isocitrate dehydrogenase; enzymes not detected included citrate synthase, ATP-citrate lyase, aconitase, pyruvate dehydrogenase, 2-oxoglutarate dehydrogenase, and ribulose bisphosphate carboxylase.

The range of carbon sources that heliobacteria can utilize is apparently very limited. The reason for this is not known, but the failure to observe autotrophic growth may be due to the apparent lack of key enzymes of the reductive pentose phosphate (Calvin) cycle or the reductive tricarboxylic acid cycle in extracts of cells of heliobacteria. Most heliobacteria that have been tested can grow on acetate. In the experiments of Pickett et al. (1994) the fate of the label from 2-^{13}C acetate was studied by mass spectrometry of amino acids obtained by hydrolysis of cell protein. Alanine and aspartate were labelled primarily in C_3, indicating that acetate assimilation proceeds initially as in *Chlorobium*, by reductive carboxylation to pyruvate and further carboxylation to oxaloacetate. However, the labeling in glutamate was in C_2 and C_4 with a smaller amount in C_3. If acetate had been assimilated reductively via succinate, as it is in *Chlorobium*, the labeling would have been equally distributed between C_3 and C_4. As discussed by Pickett et al. (1994), the labeling pattern suggests that acetate assimilation had occurred via citrate, even though citrate synthase and aconitase could not be detected in cell free extracts.

In *Hb. gestii* growth on acetate depends on the presence of molecular hydrogen (Ormerod, unpublished results). The conversion of acetate to glutamate requires reducing power, particularly if it proceeds via succinyl CoA, and hydrogen can supply the electrons for this. In heliobacteria that do not require an electron donor for acetate assimilation (e.g. *Hc. mobilis*), oxidation of some of the acetate beyond 2-oxoglutarate would have to occur to meet this requirement. How this is done is not clear. It is probably a sluggish process, as indicated by the fact that acetate accumulates during growth on pyruvate (Pickett et al., 1994). Such accumulations are not encountered in other phototrophic bacteria, such as purple bacteria, and may reflect the phylogenetic kinship of heliobacteria with the clostridia (Woese et al., 1985). Added 2-^{13}C pyruvate gave a labeling pattern consistent with a route of incorporation involving reductive assimilation to glutamate via succinate. The positions of ^{13}C in isoleucine formed in the presence of labelled acetate or pyruvate indicated synthesis via citramalate rather than the usual threonine (Pickett et al., 1994). This mechanism

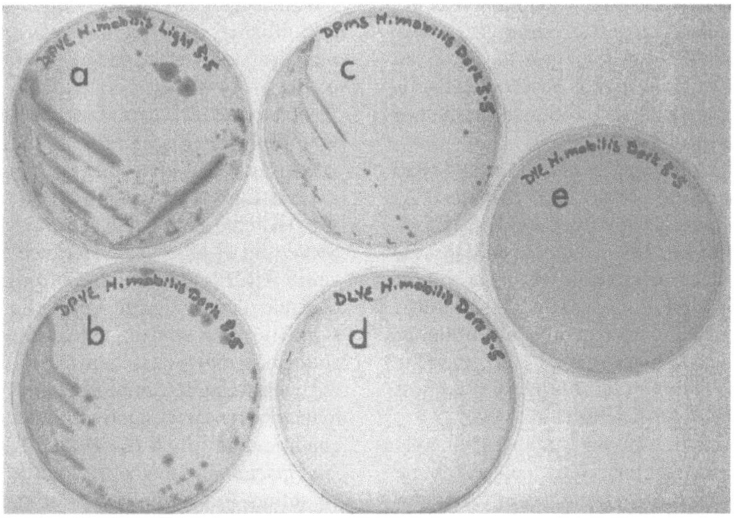

Fig. 7. Photograph of plate cultures of *Heliobacillus mobilis* (a) Colonies of photoheterotrophically grown cells (b-e) Colonies of chemotrophically grown cells (b) pyruvate/yeast extract medium; (c) pyruvate defined medium; (d) lactate/yeast extract medium; (e) yeast extract (only) medium. Plates for chemotrophic growth were incubated anaerobically in darkness for 4 days ($N_2/CO_2/H_2$ headspace). From Kimble et al. (1994).

has been found in a few other bacteria, including *Chlorobium* (Nesbakken et al., 1988).

Hb. gestii can grow photoheterotrophically on ethanol and CO_2 (Ormerod, unpublished results) and several heliobacteria can grow on butyrate plus CO_2, (see e.g. Madigan, 1992). Ethanol is presumably oxidized to acetyl CoA with the liberation of 4H which can then be used for assimilation via pyruvate synthase. Conversion of butyrate to two molecules of acetyl CoA would provide enough electrons for assimilation of only one of these to cell material (CH_2O) via pyruvate synthase. In this case, the rest of the reducing equivalents would have to come from the oxidation of some of the acetyl CoA. Lactate seems to be universally utilized by heliobacteria (Table 1) and its conversion into cell material is presumably via pyruvate; sugars are used by only a very few strains (Stevenson, 1993).

In summing up this section it may be said that heliobacteria grow photoheterotrophically at high light intensities on a limited range of organic substrates. Assimilation of these involves reductive carboxylations driven in large part by reduced ferredoxin from the RC. The electrons for this may come from oxidation of the substrates themselves or from H_2 or H_2S.

B. Chemotrophic Growth

In the original description of *Heliobacterium chlorum* Gest and Favinger (1983) concluded that this organism was an obligate phototroph. This was also the conclusion of Beer-Romero and Gest (1987) concerning *Heliobacillus mobilis* and of Ormerod et al. (1990) concerning *Heliobacterium gestii* and *Heliophilum fasciatum*. However, during the course of studies of N_2 fixation in heliobacteria (see Section IIIC and Kimble and Madigan, 1992a) it was discovered that heliobacteria were not obligate phototrophs, but can grow anaerobically in darkness at the expense of pyruvate (Fig. 7). Slow dark growth occurs in rich media containing 60–80 mM pyruvate, 0.2–0.5% yeast extract, and 40 mM of the organic buffer MOPS (3-N-morpholinopropane sulfonic acid) (Kimble et al., 1994).

From these experiments it was concluded that (1) dark growth of *Hc. mobilis* is pyruvate dependent (Fig. 7), (2) acidic and gaseous products including acetate, H_2 and CO_2 are produced, and (3) aerobic chemotrophic growth does not occur. Dark growth has also been studied by Kelly and his coworkers (Pickett et al., 1994) in *Heliobacterium* sp., strain HY3. They showed that in addition to pyruvate, lactate could also support dark growth of this strain (unlike in *Hc. mobilis*, see Fig. 7 and Kimble et al., 1994), and that dark growth of strain HY3 was almost as fast as phototrophic growth. From an ecological point of view, dark growth on lactate is significant because lactate is a common product of chemotrophic fermentation.

A survey of all known species of heliobacteria and several new unclassified strains (see Section IVA) showed them all capable of pyruvate-dependent dark growth (Kimble et al., 1994). Growth in darkness requires strictly anaerobic conditions, as is true of phototrophic growth, but it has been observed that heliobacteria exposed to microaerobic conditions in darkness retain viability for long periods; thus the strictly anaerobic nature of these organisms is most pronounced in the light (Kimble et al., 1994).

The energy-generating mechanism(s) of chemotrophic pyruvate-grown heliobacteria appears to be pyruvate fermentation yielding acetate. Acidic products produced by dark-grown cells lower the pH of even well-buffered media by 1 unit or more (Kimble et al., 1994). In addition, cultures of *Hc. mobilis* produce H_2 + CO_2 from the dark metabolism of pyruvate whereas *Hb. gestii* produces CO_2 but no H_2. Pickett et al. (1994) detected high levels of pyruvate: ferredoxin oxidoreductase activity in cells of chemotrophic pyruvate-grown *Hc. mobilis* and this is presumably the mechanism of energy conservation and acetate production. Dark diazotrophic growth of *Hc. mobilis* has also been achieved but growth rates and cell yields under these conditions are very low.

Heliobacteria thus have at least one metabolic strategy for growth in darkness and this makes good ecological sense considering that they are soil phototrophs and receive only intermittent illumination in nature. However, dark fermentation and growth of heliobacteria on pyruvate or lactate is also of benefit to basic researchers as it makes possible molecular genetic studies of photosynthesis in strains of these organisms that carry otherwise lethal mutations in photosynthesis genes.

C. Nitrogen Fixation

The original description of *Heliobacterium chlorum* indicated that the organism appeared in enrichment cultures designed to select for N_2-fixing anoxygenic phototrophs and that pure cultures grew on N_2 as sole source of nitrogen (Gest and Favinger, 1983). Since that time we have surveyed the nitrogen-fixing potential of heliobacteria (Kimble and Madigan, 1992a,b; Kimble et al., 1995; Kimble and Madigan unpublished results) and have shown that all representatives tested can fix N_2 but that the efficacy of the process varies dramatically among species. *Heliobacillus mobilis* and a new heliobacterial isolate that resembles *Hc. mobilis*, strain AR2P1 (from an Arkansas, USA, rice soil) grow rapidly on N_2 and express high levels of nitrogenase as measured in vivo (Kimble and Madigan, 1992a; Stevenson, 1993; Stevenson et al., 1993). Because *Hc. mobilis*, like strain AR2P1, was obtained from a rice soil, it is possible that a significant portion of the photosynthetic nitrogen fixation that occurs in paddy field soils (most of which is known to be due to free-living nitrogen-fixing bacteria; Buresh et al., 1980) is due to heliobacteria. By contrast, under the same growth conditions in which *Hc. mobilis* grew well on N_2, *Hp. fasciatum* grew only poorly, suggesting that conditions for diatrophy may differ among species of heliobacteria.

All the heliobacteria examined by Kimble and Madigan (1992a) were subject to regulation of nitrogenase activity by ammonia. This phenomenon, called the ammonia 'switch-off' effect, is a universal feature of anoxygenic phototrophic bacteria (Ludden and Roberts, 1989) and is presumably a means for rapid inactivation of nitrogenase activity in the presence of excess ammonia; such regulation serves to conserve reductant and ATP. Ammonia 'switch-off' of nitrogenase activity in heliobacteria was the first observation of this form of enzyme control in Gram-positive bacteria.

An alternative (non-molybdenum) nitrogenase was detected in cells of *Hb. gestii* but not in *Hb. chlorum* or *Hc. mobilis* (Kimble and Madigan, 1992b). In this study cells of *Hb. gestii* were transferred repeatedly in media containing only Fe^{3+} as a nitrogenase metal cofactor; vanadium did not stimulate iron-dependent growth (Kimble and Madigan, 1992b). Additional evidence for a Mo-independent nitrogenase system in *Hb. gestii* emerged from acetylene (C_2H_2) reduction experiments where ethane (C_2H_6) as well as ethylene (C_2H_4) was produced by cell suspensions grown on N_2 in the absence of Mo; C_2H_6 production was Mo repressible (Kimble and Madigan, 1992b). This pattern is typical of alternative nitrogenase systems of other diazotrophs (Bishop and Premakumar, 1992). Whether the alternative nitrogenase(s) of *Hb. gestii* are of the iron-only or vanadium-iron types could not be resolved in the studies of Kimble and Madigan (1992b), but a nitrogenase 3 (iron-only type) was suggested on the basis of the accumulated evidence.

Alternative nitrogenases apparently function as backup systems for diazotrophic growth of organisms whose habitats are periodically or permanently deficient in Mo. The alternative nitrogenase of *Hb. gestii* may thus have ecological significance for this organism in its natural habitat, paddy field soil (Ormerod et al., 1990), where Mo deficiency is often a problem due to metal leaching under acidic conditions (Bishop and Joerger, 1990).

IV. Ecology and Isolation of Heliobacteria

A. Habitats of Heliobacteria

The ecology and distribution of heliobacteria in nature is now becoming clear. On the basis of isolations made to date beginning with that of *Heliobacterium chlorum* (Gest and Favinger, 1983), virtually all reported isolates of heliobacteria have been obtained from soils (Beer-Romero and Gest, 1987; Ormerod et al, 1990; Madigan, 1992; Stevenson, 1993; Stevenson et al., 1993; Pickett et al., 1994). Indeed, enrichment studies (see below) suggest that with the exception of the hot springs isolate of Starynin and Gorlenko (1993), the Yellowstone hot springs isolates of Kimble and Madigan (Kimble and Madigan, 1994; Kimble et al., 1995) and a strain isolated from an Oregon hot spring by R.W. Castenholz (personal communication to JGO), soil may be the *only* habitat of heliobacteria.

In an extensive series of enrichment cultures carried out in the laboratory of MTM, positive heliobacterial enrichments were obtained only from soil. From over 100 soils tested, 22 were positive for heliobacteria. Of some 20 aquatic samples tested in the same way, including freshwater lakes, ponds, marsh waters, raw sewage, and several marine and salt marsh samples, none yielded heliobacteria (Stevenson et al., 1993). In addition, it should be mentioned that heliobacteria were never reported (before 1983) from the thousands of enrichment cultures for anoxygenic phototrophic bacteria that have been made by various workers over the years. However, it is possible that enrichments which had turned green may have contained heliobacteria that were mistaken for green algae or cyanobacteria and discarded. Thus, although all indications are that heliobacteria are terrestrial phototrophs, more enrichments using selective methods for heliobacteria but with aquatic samples are needed to further support this.

Rice field soils appear to be major habitats for heliobacteria. Of 16 new isolates obtained in enrichment studies, 9 were from such soils (Stevenson et al., 1993). In addition, *Heliobacillus mobilis*, *Heliobacterium gestii* and *Heliophilum fasciatum* all originated from rice soils (Beer-Romero and Gest, 1987; Ormerod et al., 1990). Dry paddy field soil in particular appears to be a good source of heliobacteria (Ormerod et al., 1990; Madigan, 1992; Stevenson et al., 1993). This suggests that these strictly anaerobic organisms have evolved strategies for surviving drying and any resultant exposure to oxygen; endospore production is clearly one such strategy. In recent enrichment studies (Stevenson, 1993) soil moisture, pH, and ammonia levels were measured in attempts to define the optimal physicochemical parameters in soils for heliobacteria. Heliobacteria-positive soil samples ranged from 0.1–66% in soil moisture content, 4.2–8.4 in pH, and 0.3–135 μg ammonia/g dry soil. Thus, dry soils are not required for recovery of heliobacteria (although most successful soils were dry), acidic soils can contain heliobacteria, and, although N_2 fixation may be of survival value to heliobacteria, they are also indigenous to soils containing sufficient ammonia to repress nitrogenase synthesis. In addition, although tropical soils (for example, Far Eastern and African paddy field soils) were the best sources of heliobacteria, the organisms were also present in temperate soils (for example Minnesota, Arkansas, Texas [USA] and Italy) and even in soils from near the Arctic Circle (Iceland; Stevenson et al., 1993; Kimble et al., 1995; Kimble and Madigan, unpublished).

Heliobacteria are thus clearly ecologically distinct from green and purple bacteria, which are generally found in aquatic environments (Madigan, 1988; Pfennig, 1989; see also Chapter 4 by Van Gemerden and Mas, this volume). Although nonsulfur purple bacteria are present in some soils (Gest et al., 1985), like purple and green sulfur bacteria, their primary habitats are aquatic (Madigan, 1988). The ability of at least some heliobacteria to sporulate is likely related to their soil habitat, as other endospore-forming bacteria are primarily soil organisms as well. Sporulation would allow heliobacteria to survive the alternate flooding and extreme drying of paddy field soils, and the changing moisture levels experienced in most other soils.

It may also be significant to the ecology of

heliobacteria that all species fix molecular nitrogen (Kimble and Madigan, 1992a,b). In rice soils, nitrogen fixation by heliobacteria may be an important factor in providing combined nitrogen for rice plants. Perhaps rice plants excrete organic substances that are used by heliobacteria in exchange for fixed nitrogen. If true, this would suggest interesting agricultural applications for heliobacteria in improving soil fertility.

B. Enrichment and Isolation of Heliobacteria

Enrichment of heliobacteria begins with soil as inoculum. Dry soil is suitable, and as discussed above, may even be desirable. From enrichment studies (Ormerod et al., 1990; Stevenson, 1993; Stevenson et al., 1993), it has become clear that dilute complex media at neutral pH or defined media containing lactate or acetate are most suitable for enrichment of heliobacteria. Yeast extract solution (0.2–0.3% w/v, pH 7) seems to be the best for primary isolation (Stevenson et al., 1993; Kimble et al., 1995).

In establishing enrichments for heliobacteria, one should take some precautions as to anaerobic conditions. When growing phototrophically, the heliobacteria are strict anaerobes (Madigan, 1992). Thus, media prereduced by storage for several days in contact with the $N_2 : H_2 : CO_2$ atmosphere of an anaerobic glove box, or alternatively, boiled for several minutes and sealed under an anaerobic gas before autoclaving (Hungate method), are helpful for obtaining positive enrichments. However, since heliobacterial spores are oxygen resistant, the job of removing the oxygen in enrichments can be left to aerobic heterotrophs in the inoculum; once these have used up the oxygen, any heliobacteria will begin to grow. If crude enrichments are obtained, however, strictly anaerobic conditions must be maintained thereafter in order to obtain pure cultures of heliobacteria.

Light and temperature can also be selective enrichment factors. Heliobacteria cannot grow at as low a light intensity as purple and green bacteria do, but they tolerate high light intensities quite well (Beer-Romero, 1987; Madigan, 1992; Kimble and Madigan, unpublished results; Ormerod, unpublished observations). Thus 5,000–10,000 lux (incandescent) illumination is recommended for primary isolations. Enrichment temperatures around 40 °C should be used as an additional selective factor for heliobacteria (all heliobacteria isolated thus far grow up to at least 42 °C) and to discourage growth of purple bacteria. A final selective factor which is frequently essential for isolating heliobacteria from many soils is to pasteurize the soil sample used in the enrichment. Pasteurization eliminates problems with competing organisms, in particular nonsulfur purple bacteria such as *Rhodopseudomonas palustris*. It has been observed that if an enrichment for heliobacteria gets overtaken by purple bacteria, heliobacteria never develop (Stevenson, 1993).

Positive enrichments for heliobacteria generally appear as slimy green clumps of cells typically located at the soil/liquid or glass/liquid interface; only weak growth is usually observed in the bulk medium (in fact the bulk medium may not become pigmented, thus one must look carefully for signs of heliobacteria before concluding that an enrichment is negative). Microscopically, pasteurized enrichments generally contain a variety of sporulating organisms including heliobacteria and clostridia. Endospore-forming heliobacteria are usually some of the largest cells in enrichments and spores, if present, are typically located either centrally or subterminally (Stevenson et al., 1993). If an enrichment is suspected of being positive for heliobacteria, it should immediately be streaked on plates of the same medium within an anaerobic chamber and then incubated photosynthetically in anaerobic jars (streaking plates in air followed by anaerobic incubation rarely yields heliobacteria). The presence of heliobacteria on plates is signaled within 2–3 days by the formation of greenish or brown-green colonies. These can be picked and restreaked (anaerobically) to obtain pure cultures. Most isolates can be purified by this method; however, some strains have required streaking on plates of medium solidified with washed agar or Gelrite®, a nonagar-based solidifying agent, in order to be obtained in axenic culture. Further details of enrichment methods for isolating heliobacteria and recipes for a variety of media for growing pure cultures of heliobacteria are given in Madigan (1992) and in Gest et al., (1985).

Finally, it should be mentioned that because enrichment cultures of heliobacteria generally contain chemotrophic spore-forming bacteria, the chances of some of these being present as contaminants in 'purified' cultures of heliobacteria, even after plating, must always be considered. Therefore, the presence of spores in some heliobacterial cultures does not necessarily indicate the identity of the spores. The surest way of determining this is to observe scotophobotaxis in motile sporangia. When the microscope light is shaded momentarily, a swimming heliobacterial sporangium cell usually reverses its direction. Having ascertained the identity of the spores, further steps in purification can be carried out as described above.

C. Contrasts of Heliobacteria with Purple and Green Bacteria

Any consideration of the ecology of heliobacteria must take into account other phototrophic bacteria which can compete with them in one way or another. Heliobacteria are primarily soil organisms preferring relatively high light intensities; this clearly separates them from the classical purple and green bacteria which are mainly aquatic and able to grow, particularly in the case of the green sulfur bacteria, at very low light intensities. Since they are most commonly found in the soil of rice fields, heliobacteria may have some sort of special relationship with rice plants. The considerable quantities of methane produced in rice field soils would in general preclude the presence of significant amounts of sulfide in them. The heliobacteria present in rice soils are therefore probably not exposed to competition from phototrophic sulfur bacteria in this environment. Instead, competition most likely exists with purple nonsulfur bacteria, which like heliobacteria live a photoheterotrophic lifestyle, and with cyanobacteria, which are common in rice soils (Buresh et al., 1980; Habte and Alexander, 1980).

Habte and Alexander (1980) showed that purple bacteria were present in rice field soils, and these may become dominant in enrichment cultures inoculated with such soils, especially if a fresh, moist sample is used. Rice fields are waterlogged and anaerobic during the growth season, so the ability to tolerate oxygen shown by some purple bacteria would scarcely be of advantage here, as it is in stratified lakes (De Wit and Van Gemerden, 1990). It may therefore be that the cardinal heliobacterial property in this regard is spore formation; endospores would allow for survival of heliobacteria during the dry season and upon return of flooded conditions would serve as inoculum for rapid development of a heliobacterial microflora.

The absorption properties of heliobacteria may also be of ecological advantage. Absorption spectra of heliobacteria have a major peak at about 790 nm. This absorption peak is unique for heliobacteria and would allow them to grow beneath a layer of cyanobacteria. It is also interesting to note that from a phylogenetic standpoint heliobacteria show closer relationships to the cyanobacteria than to any group of anoxygenic phototrophic bacteria (Woese, 1987). Perhaps the obligately anaerobic heliobacteria and the oxygen-producing cyanobacteria have more intimate ecological relationships in rice soil and other soils than one would at first suspect. The discovery of such relationships awaits more thorough understanding of the microbial ecology of photosynthesis in the rice soil environment.

Acknowledgments

The research of MTM on heliobacteria has been supported by the United States Department of Agriculture (USDA) Competitive Grants Program. JGO received support from the Research Council of Norway. We thank Linda Kimble for supplying unpublished results.

References

Aase B, Jantzen E, Bryn K and Ormerod JG (1994) Lipids of heliobacteria are characterized by a high proportion of monoenoic fatty acids with variable double bond positions. Photosynth Res 41: 67–74

Beer-Romero P (1986) Comparative studies on *Heliobacterium chlorum*, *Heliospirillum gestii* and *Heliobacillus mobilis*. MA. Thesis, Indiana University, Bloomington

Beer-Romero P, and Gest H (1987) *Heliobacillus mobilis*, a peritrichously flagellated anoxyphototroph containing bacteriochlorophyll *g*. FEMS Microbiol Lett 41: 109–114

Beer-Romero P, Favinger JL and Gest H (1988) Distinctive properties of bacilliform photosynthetic heliobacteria. FEMS Microbiol Lett 49: 451–454

Beck H, Hegeman GD and White D (1990) Fatty acid and lipopolysaccharide analyses of three *Heliobacterium* spp. FEMS Microbiol Lett 69: 229–232

Bishop PE and Joerger RD (1990) Bacterial alternative nitrogen fixation systems. Ann Rev Plant Physiol Plant Mol Bio. 41: 109–125

Bishop PE and Premakumar R (1992) Alternative nitrogen fixation systems. In: Stacey G, Burris RH and Evans HJ (eds) Biological Nitrogen Fixation, pp 736–762. Chapman and Hall, New York

Brockmann H and Lipinski A (1983) Bacteriochlorophyll *g*. A new bacteriochlorophyll from *Heliobacterium chlorum*. Arch Microbiol 136: 17–19

Buresh RJ, Casselman ME and Patrick WH, Jr (1980) Nitrogen fixation in flooded soil systems, a review. Adv Agron 33: 149–192

Charnock C, Refseth UH, Sirevåg R (1992) Malate dehydrogenase from *Chlorobium vibrioforme*, *Chlorobium tepidum* and *Heliobacterium gestii*: Purification, characterization and investigation of dinucleotide binding by dehydrogenases by use of empirical methods of protein sequence analysis. J Bacteriol 174: 1307–1313

De Wit R and Van Gemerden H (1990) Growth of the phototrophic purple sulfur bacterium *Thiocapsa roseopersicina* under oxic/anoxic regimes in the light. FEMS Microbiol Ecol 73: 69–76

Gest H and Favinger JL (1983) *Heliobacterium chlorum*, an anoxygenic brownish-green photosynthetic bacterium containing a 'new' form of bacteriochlorophyll. Arch Microbiol 136: 11–16

Gest H, Favinger JL and Madigan MT (1985) Exploitation of N_2 fixation capacity for enrichment of anoxygenic photosynthetic bacteria in ecological studies. FEMS Microbiol Ecol 31: 317–322.

Habte M and Alexander M (1980) Nitrogen fixation by photosynthetic bacteria in lowland rice culture. Appl Environ Microbiol 39: 342–347

Imhoff JF (1988) Lipids, fatty acids and quinones in taxonomy and phylogeny In: Olson JM, Ormerod JG, Amesz J, Stackebrandt E and Trüper HG (eds) Green Photosynthetic Bacteria, pp 223–232. Plenum Press, New York

Joliffe LK, Doyle RJ and Streips UN (1981). The energized membrane and cellular autolysis in *Bacillus subtilis*. Cell 25: 753–763

Kimble LK and Madigan MT (1992a) Nitrogen fixation and nitrogen metabolism in heliobacteria. Arch Microbiol 158: 155–161

Kimble LK and Madigan MT (1992b) Evidence for an alternative nitrogenase system in *Heliobacterium gestii*. FEMS Microbiol Lett 100: 251–256

Kimble LK and Madigan MT (1994) Isolation and characterization of thermophilic heliobacteria. Abstract I-9 of the American Society for Microbiology General Meeting, Las Vegas, NV

Kimble LK, Stevenson AK and Madigan MT (1994). Chemotrophic growth of heliobacteria in darkness. FEMS Microbiol Lett 115: 51–56

Kimble LK, Mandelco L, Woese CR and Madigan MT (1995) *Heliobacterium modesticaldum*, sp. nov., a thermophilic heliobacterium of hot springs and volcanic soils. Arch Microbiol 163: 259–267

Kobayashi M, Watanabe T, Ikegami I, van de Meent EJ and Amesz J. (1991) Enrichment of bacteriochlorophyll *g'* in membranes of *Heliobacterium chlorum* by ether extraction. Unequivocal evidence for its existence in vivo. FEBS Lett 284: 129–131

Ludden PW, and Roberts GP (1989) Regulation of nitrogenase activity by reversible ADP-ribosylation. In: Horecher B, Stadtman E, Chock PB and Levitzki A (eds) Current Topics in Cellular Regulation, Vol 30. Academic Press, Orlando, FL

Madigan MT (1988) Microbiology, physiology, and ecology of phototrophic bacteria. In: A.Z.B. Zehnder (ed) Biology of Anaerobic Microorganisms, pp 39–111. John Wiley and Sons, New York

Madigan MT (1992) The family Heliobacteriaceae. In: Balows A, Trüper HG, Dworkin M and Schleifer K-H (eds) The Prokaryotes, second edition pp 1981–1992, Springer-Verlag, New York

Michalski TJ, Hunt JE, Bowman MK, Smith U, Bardeen K, Gest H, Norris JR and Katz JJ (1987) Bacteriopheophytin *g*: Properties and some speculations on a possible primary role for bacteriochlorophylls *b* and *g* in the biosynthesis of chlorophylls. Proc Natl Acad Sci USA 84: 2570–2574

Nesbakken T, Kolsaker P and Ormerod J (1988) Mechanism of biosynthesis of 2-oxo-3-methylvalerate in *Chlorobium vibrioforme*. J Bacteriol 170: 3287–3290

Ormerod J, Nesbakken T and Torgersen Y (1990) Phototrophic bacteria that form heat resistant endospores. In: Baltscheffsky M(ed) Current Research in Photosynthesis, Vol. IV, pp 935–938. Kluwer Academic, Dordrecht

Pickett MW, Williamson MP and Kelly DJ (1994) An enzyme and ^{13}C-NMR study of carbon metabolism in heliobacteria. Photosynth Res 41: 75–88

Pfennig N (1989) Ecology of phototrophic purple and green sulfur bacteria. In: Schlegel HG and Bowien B (eds), Autotrophic Bacteria pp 97–116. Springer-Verlag, New York

Redburn AC and Patel BKC (1993) Phylogenetic analysis of *Desulfotomaculum thermobenzoicum* using polymerase chain reaction-amplified 16S rRNA-specific DNA. FEMS Microbiol Letts 113: 81–86

Starynin DA and Gorlenko VM (1993) Sulphide-oxidizing spore forming heliobacteria isolated from a thermal sulphide spring. Microbiology (English translation of Mikrobiologiya) 62: 343–347

Stevenson AK (1993) Isolation and characterization of heliobacteria from soil habitats worldwide. MA Thesis, Department of Microbiology, Southern Illinois University, Carbondale

Stevenson AK, Kimble LK and Madigan MT (1993) Isolation and characterization of heliobacteria from soil habitats worldwide. Abstract I-62 of the American Society for Microbiology General Meeting, Atlanta, GA

Torgersen YA (1989) Karakterisering av en obligat fototrof anaerob bakterie som inneholder bakterioklorfyll *g*. Cand Scient Thesis. Oslo University

Trost JR (1990) Characterization of the photosynthetic reaction center-core antenna complex from the heliobacteria. Ph.D. Dissertation, Arizona State University, Tempe

Woese CR (1987) Bacterial evolution. Microbiol Revs 51: 221–271

Woese CR, Debrunner-Vossbrinck BA, Oyaizu H, Stackebrandt E and Ludwig W (1985) Gram-positive bacteria: Possible photosynthetic ancestry. Science 229: 762–765

Chapter 3

Taxonomy and Physiology of Filamentous Anoxygenic Phototrophs[†]

Beverly K. Pierson
Biology Department, University of Puget Sound, Tacoma, WA 98416-0320, USA

Richard W. Castenholz
Biology Department, University of Oregon, Eugene, OR 97403-1210, USA

[†] We dedicate this chapter to William R. Sistron (deceased September, 1993) whose ideas and discussions contributed in many important ways to the discovery and characterization of the BChl-containing filamentous bacteria.

R. E. Blankenship, M. T. Madigan and C. E. Bauer (eds): Anoxygenic Photosynthetic Bacteria, pp. 31–47.

Summary

The filamentous anoxygenic phototrophs are a diverse group of photosynthetic bacteria that are of particular evolutionary significance. The best known species is the thermophilic *Chloroflexus aurantiacus*. This organism is a prominent member of hot spring microbial mat communities. Because it forms a deep division in the eubacterial line of descent and because it has an interesting combination of characteristics found in very different and diverse groups of phototrophic prokaryotes, it is of particular significance in addressing questions of evolutionary importance.

There are several other strains of filamentous photosynthetic bacteria from a wide range of environments that are substantially different from *Cf. aurantiacus* yet have enough similarity in fundamental photosynthetic features to be likely relatives. Sequence data (16S rRNA) are needed to define the phylogenetic range of the family Chloroflexaceae. Some of the interesting biology of the diverse filamentous phototrophs is discussed in this chapter along with the taxonomic and phylogenetic problems they present.

The physiology of *Cf. aurantiacus* is intriguing in several respects. The recently described autotrophic CO_2 fixation pathway involving 3-hydroxypropionate is unlike any other known autotrophic mechanism. *C. aurantiacus* is also unique among all groups of phototrophs in lacking the capacity for nitrogen fixation. The regulation of pigment synthesis in response to changing growth conditions is particularly interesting due to the presence of two different photosynthetic pigments located in different sub-cellular environments. The fact that *Cf. aurantiacus* is a thermophile provides another dimension of complexity to its physiology. It is also quite resistant to UV radiation. Some of its characteristics may be relicts from Precambrian ancestors.

I. Introduction

The anoxygenic filamentous phototrophic bacteria are a large and diverse group of microorganisms (Table 1). They are discussed together in this chapter primarily for convenience but also because they share some similarities in structure and metabolism. Although we are not prepared to argue strongly that the filamentous character is a particularly significant trait to the overall taxonomy of this diverse group of organisms, it does appear that many of these bacteria may in fact bear a relatively close relationship to each other. The organisms discussed in this review include the members of the family Chloroflexaceae and the other anoxygenic filamentous phototrophs of uncertain affiliation.

The family Chloroflexaceae includes the genera *Chloroflexus*, *Oscillochloris*, and *Chloronema*. Only one species, *Chloroflexus aurantiacus*, has been described in substantial detail from the extensive study of pure cultures. A new thermophilic species is currently under study and may be described soon (Satoshi Hanada, personal communication). We will review the taxonomy of this group and the physiology of all the genera, including the well-studied *Cf. aurantiacus*, the lesser known species of the 'green *Chloroflexus*' (Strain GCF), the mesophilic *Chloroflexus*, the marine and hypersaline *Chloroflexus*-like organisms (MCLOs), and the species of *Oscillochloris* and *Chloronema*.

We will also review the other anoxygenic filamentous phototrophs of uncertain affiliation. These organisms include the described species *Heliothrix oregonensis* (no relationship to the 'Heliobacteria'), the thermophilic red filamentous bacteria, and the hypersaline filamentous purple bacteria.

Our approach will be topical. For each topic considered, we will first discuss the well-studied *Cf. aurantiacus* followed by a review of what is known about the other filamentous phototrophs.

Much of the physiology of *Cf. aurantiacus* will be discussed in greater detail in other chapters which we will cross-reference where appropriate. We bring a particular perspective to this review. Our objective is to identify problems of interest in the ecophysiology and evolutionary biology of this group of organisms. We have long recognized the existence of many poorly understood and possibly related phototrophic bacteria that are important to our overall understanding of the phylogeny of phototrophs and the origin and evolution of photosynthesis. We hope to increase awareness of and interest in the study and successful isolation of a much greater diversity of phototrophs to increase the database for the development of future evolutionary theories.

Abbreviations: BChl – bacteriochlorophyll; BPh – bacteriopheophytin; *Cf.* – *Chloroflexus*; MCLO – Marine *Chloroflexus*-like organism

Table 1. Anoxygenic filamentous phototrophs[1]

Organism	Morphology	Pigments	Metabolism	Habitat	References
Cf. aurantiacus	gliding filaments 0.5-1.5 μm diam.	BChl *a* and *c*	photoheterotrophy some photoautotrophy (sulfide-dependent or hydrogen-dependent)	neutral to alkaline thermal springs undermat, surface mat, or unispecific mat	see Pierson and Castenholz, 1992 for review of references
var. *mesophilus*				mats in freshwater stratified lakes	Gorlenko, 1976; Pivovarova and Gorlenko, 1977
Marine *Chloroflexus*-like organisms	gliding filaments 1.0-2.0 μm diam.	BChl *a* and *c*	photoheterotrophy photoautotrophy?	marine intertidal salt marsh sediments	Mack and Pierson, 1988
Marine *Chloroflexus*-like organisms	gliding filaments variable diam.	BChl *c* BChl *a* ?	undetermined	hypersaline marine mats often in association with *Microcoleus chthonoplastes*	D'Amelio et al., 1989 Palmisano et al., 1989 Stolz, 1983, 1990
Marine *Chloroflexus*-like organisms	gliding filaments 1.2-3.0 μm diam	BChl *a* and *c* or *d*	photoautotrophy photoheterotrophy?	hypersaline marine mats (a) as distinct layer beneath surface layer of cyanobacteria (b) as surface layer mixed with *Spirulina* and *Beggiatoa*	Larsen et al., 1991 Pierson et al., 1994
Heliothrix oregonensis	gliding filaments 1.5 μm diam.	BChl *a*	photoheterotrophy	Alkaline thermal springs lacking sulfide. Surface mat above cyanobacteria	Pierson et al., 1984, 1985
Thermophilic red filaments	gliding filaments 1.5 μm diam.	BChl *a*	photoheterotrophy	alkaline thermal springs as undermat below cyanobacteria and *Chloroflexus*	Boomer et al., 1990; Castenholz, 1984
Marine purple filaments	gliding filaments 0.8-0.9 μm diam.	undetermined	undetermined	hypersaline mats in association with *Microcoleus chthonoplastes*	D'Amelio et al., 1987
Chloronema	gliding filaments 2 μm diam.	BChl *d*	undetermined	planktonic freshwater	Gorlenko, 1988
Oscillochloris	gliding filaments 1.0-5.5 μm diam.	BChl *c*	undetermined	surface of freshwater sulfide-containing muds marine?	Gorlenko, 1988 Keppen et al., 1993a

[1] All the organisms listed in this table are reviewed in more detail in Pierson and Castenholz, 1992.

II. Taxonomy and Phylogeny

A. Problems With This Group

Establishing a meaningful taxonomy and understanding the phylogeny of the filamentous phototrophs are two tasks beset with problems unique to this group of organisms. It is apparent from ecological studies and field observations of the natural history of several habitats that the anoxygenic filamentous phototrophs comprise a large and diverse group (Table 1). From this group only one species has been successfully isolated and sustained in pure culture and has been subjected to extensive analyses. Other 'species' have been studied as natural populations, in dual or mixed cultures, or in pure cultures which have been difficult to maintain and were subsequently lost, or in cultures that have not been readily available to many labs. Consequently, the broadly based phenotypic data needed to construct useful taxa and the arrays of nucleic acid sequences needed to construct meaningful phylogenies are lacking. Thus, while abundant data on natural history, microscopy, and ultrastructure tantalize us with the obvious recognition of numerous diverse species of interest, the very limited availability of pure cultures frustrates our understanding of this group.

B. Taxonomy of Chloroflexaceae

The thermophilic *Chloroflexus aurantiacus* (Pierson and Castenholz, 1974a) is still the only genus and species of the Chloroflexaceae that has been described on the basis of axenic cultures that are readily available. *Chloronema* (Dubinina and Gorlenko, 1975) and *Oscillochloris* (Gorlenko and Pivovarova, 1977) are the two other genera included in this family. The description of *Chloronema* was not based on cultured material. Recently a culture of *Oscillochloris trichoides* was isolated and made available (Keppen et al., 1993). Trüper (1976) proposed Chloroflexaceae as a family with affinities to the Chlorobiaceae (green sulfur bacteria) and included the two families in the suborder Chlorobiineae. Bacteria in the family Chloroflexaceae were described as filamentous, phototrophic bacteria with gliding motility, Gram-negative, flexible cell walls, and bacteriochlorophyll (BChl) *a* and BChl *c*, *d*, or *e*.

Problems with the description of this taxon exist. Although Gram staining reactions appear to be negative, neither the ultrastructural appearance nor the chemistry of the cell wall of *Cf. aurantiacus* is typical of other Gram-negative bacteria (Knudsen et al., 1982; Jürgens et al., 1987; see also Chapter by Weckesser and Golecki). '*Oscillochloris chrysea*' was Gram-positive and had a thick peptidoglycan layer with no outer membrane (Gorlenko, 1988, 1989a). Some of the marine *Chloroflexus*-like organisms we have been studying stain Gram-positive or Gram-variable and appear to lack an outer membrane (B. Pierson, unpublished).

In recognition of the probable significance of the true filamentous morphology of these phototrophs, we have referred to them as photosynthetic flexibacteria (Pierson and Castenholz, 1992). In *Bergey's Manual of Systematic Bacteriology*, Pfennig (1989) proposed the group 'multicellular filamentous green bacteria' to include the genera *Chloroflexus*, *Chloronema*, *Oscillochloris*, and *Heliothrix*. The genus *Heliothrix*, however, differs significantly from the other three genera in lacking chlorosomes and BChl *c*, *d*, or *e*. Consequently the appropriateness of this taxonomic grouping as a family is weakened. Further difficulties have arisen with the utility of this proposed grouping due to physiological constraints included in the description. The description emphasizes that all members of the group are facultatively aerobic and preferentially use organic substrates in their metabolism. It further states that reduced sulfur compounds are not important electron donors for laboratory cultures. These proposed taxa and less formal groupings may be unnecessarily restrictive at a time when very few laboratory cultures have actually been available for study. Field studies of apparently related organisms have revealed a tolerance for high levels of sulfide, as well as a capacity for sulfide dependent autotrophy. Studies on pure cultures have shown that one strain of *Chloroflexus*, the GCF strain (Giovannoni et al., 1987) is not facultatively aerobic.

Our discussion of the range of physiologies present in filamentous phototrophs presented in this chapter will support our suggestion that at this time the most useful grouping of these organisms is as photosynthetic flexibacteria or anoxygenic filamentous phototrophs. The term green non-sulfur bacteria should be dropped. It is potentially very misleading. The analogy to purple sulfur and non-sulfur bacteria breaks down on phylogenetic grounds. The two 'groups' of purple bacteria are closely related phylogenetically, at least as members of the vast array of Proteobacteria. The two groups of green bacteria are not. The term filamentous green bacteria is appropriate when referring to the three genera *Chloroflexus*, *Chloronema*, and *Oscillochloris* but not *Heliothrix*. The term 'gliding green bacteria' should be dropped because this would include *Choroherpeton* which is not filamentous, but does glide and is closely related to the green sulfur bacteria and not to *Chloroflexus*.

C. Phylogeny and Evolution

On the basis of 16S rRNA catalogs and complete sequence data, it was concluded that *Chloroflexus aurantiacus* forms a very deep division within the eubacterial line of descent (Oyaizu et al., 1987; Woese, 1987). Other organisms included in the same grouping with *Cf. aurantiacus* are *Herpetosiphon aurantiacus*, a filamentous gliding non-photosynthetic bacterium, and *Thermomicrobium roseum*, another non-photosynthetic bacterium. *Thermus thermophilus* may also be fairly closely related (Hartmann et al., 1989). The two most interesting aspects of the position of *Chloroflexus* on the eubacterial tree are its lack of phylogenetic closeness to any other groups of phototrophs and the depth of its branch on the tree. Its position as the deepest division of all the phototrophs suggests its ancestors diverged earlier than those of any of the other phototrophs.

Heliothrix oregonensis has been described on the basis of studies of natural populations and a co-culture with the non-phototrophic *Isosphaera pallida* (Pierson et al., 1984a). On the basis of 16S and 5S rRNA sequence data, *Heliothrix oregonensis* is phylogenetically close to *Cf. aurantiacus* and is not at all close to any of the other phototrophic bacteria (Pierson et al., 1985; Weller et al., 1992). *Heliothrix oregonensis*, however, is not a 'green' bacterium and is excluded from the taxon Chloroflexaceae as currently defined. Analysis of 5S rRNA from *Oscillochloris trichoides* also supports greater phylogenetic closeness to the *Chloroflexus* group than to any other phototrophic bacteria (Keppen et al., 1993b). The closeness of *Herpetosiphon aurantiacus* to *Chloroflexus* on the basis of 16S rRNA analysis reinforces the idea of the significance of filamentous morphology in defining this group and lends credence to our suggestion that for the time being all filamentous phototrophs should be grouped together. However, *Thermomicrobium roseum*, which is also in this group, is not filamentous, hence weakening this argument.

There are several serious problems plaguing these phylogenetic interpretations, however. Some could be solved quite readily. First, only two strains of *Cf. aurantiacus* have been analyzed for 16S rRNA sequences. It is desirable to do other strains since potentially significant phenotypic variations do exist, for example, between the two most studied strains, J-10-fl and OK-70-fl. The facultatively autotrophic and sulfide-dependent thermophilic strain, GCF (green *Chloroflexus*), has not been analyzed. Sequence data are also lacking for the mesophilic freshwater strain first described by Gorlenko (1976) and Pivovarova and Gorlenko (1977) and for *Chloronema* and *Oscillochloris*. Likewise, phylogenetic data are lacking for the uncultured marine strain (Mack and Pierson, 1988) and the hypersaline strains that are not yet axenic but will be briefly described in this chapter. With the availability of PCR technology, a comprehensive phylogenetic analysis of all of these filamentous anoxygenic phototrophs could be undertaken and would greatly enhance our understanding of the relationships of these organisms to each other and to other phototrophs.

The necessity of a more extensive analysis of 16S rRNA based phylogeny within this group is further enhanced by two other analyses which are not readily resolved with the current paucity of phylogenetic data. The relative position of some groups of phototrophs in the eubacterial tree has been challenged recently on the basis of 5S rRNA sequence data (Van den Eynde et al., 1990). The 5S RNA data corroborated the close relationship of *Cf. aurantiacus* to *Thermomicrobium roseum* but excluded *Herpetosiphon aurantiacus* from this group in contrast to the 16S RNA data. Furthermore, the 5S RNA data produced a tree in which the cyanobacterial-plastid cluster of phototrophs and the green sulfur bacterial cluster of phototrophs both formed earlier branches or deeper divisions than the *Chloroflexus* cluster. Analysis of 23S rRNA sequences (Woese et al., 1990), however, supports earlier 16S rRNA data suggesting a close relationship between the green sulfur bacteria and the flavobacterial group and reinforcing the early branch points suggested by the 16S rRNA data (Woese, 1987). These findings have particular significance for evolutionary interpretations. Such conflicts must be resolved in order to interpret phylogeny with confidence. One important step in solving these problems is to increase the resolution of the system by including a larger number of data sets. As it currently stands, the phylogenetic grouping with *Cf. aurantiacus* contains only two unequivocal members, both of which are thermophiles. Many of the other, perhaps related, bacteria that are not yet sequenced are not thermophiles, and it seems likely that thermophily will not be a unifying characteristic of the group. More analyses as suggested above could lead to a larger, more stable grouping, in which the branch point might be determined with greater confidence.

Further confusion in interpreting phylogenetic relationships, and certainly in interpreting the evolution of phototrophs and photosynthesis, has been introduced with recent analysis of evolutionary relationships among the reaction center genes (Blankenship, 1992). The two fundamental types of reaction centers are distributed among the various groups of phototrophs in ways that preclude any simple linear explanation of the evolution of the reaction centers within and among the different groups. Gene duplications followed by lateral gene transfer and divergence may account for the extant distribution of these pigment-protein complexes. *Cf. aurantiacus*, despite its apparent early divergence from the other eubacteria as suggested by the 16S rRNA data, does not appear to possess an ancestral reaction center. It would be most interesting to acquire data on the functional nature and molecular sequence of the reaction centers from the other filamentous anoxygenic phototrophs. The details of photosynthetic reaction centers from anoxygenic phototrophic bacteria are covered in several chapters of this book.

Thermophilic *Cf. aurantiacus* is but one species within a very large and probably quite diverse group of filamentous phototrophic bacteria that are likely to be significant to our understanding of the early evolution of photosynthesis. Its thermophily and its frequently cited chimeric nature in photosynthetic properties may prove to be the exceptions rather than the rule among these bacteria. We must bear in mind that these characteristics have been important in defining the taxa to which *Chloroflexus* belongs only because by chance it was the first, and remains so far the only, species that has been readily isolated and cultured. The very recent availability of pure cultures of *O. trichoides* will pose new challenges to the extant taxonomic systems (Keppen et al., 1993a).

III. Physiology

Most of our understanding of the physiology of the filamentous phototrophs is based on studies of *Cf. aurantiacus*. Studies of other organisms will also be mentioned where appropriate. The type species *Cf. aurantiacus* is a thermophile with an optimum temperature for growth between 52 and 60 °C (Pierson and Castenholz, 1974a). It is a common and abundant inhabitant of neutral to alkaline hot springs with or without sulfide over a wide temperature range up to about 70–72 °C. Occurrence above 66 °C has been observed only when associated with cyanobacteria that grow at higher temperatures (see Chapter 6 by Castenholz and Pierson). It is usually associated with cyanobacteria. Its distribution in natural habitats, general ecology, isolation, and laboratory culture, as well as some aspects of its physiology have recently been reviewed (Pierson and Castenholz, 1992; see Chapter 5 by Castenholz and Pierson). Green *Chloroflexus* (GCF) is found as nearly pure mats devoid of cyanobacteria in a limited number of hot springs containing sulfide over the temperature range of 50 to 66 °C (Giovannoni et al., 1987). *Heliothrix oregonensis* is also a thermophile found much less commonly than *Chloroflexus* in alkaline hot springs where it is occasionally very abundant. Its optimum temperature for metabolic activity is between 50 and 55 °C. All three of these thermophilic filamentous phototrophs are found naturally as major visible components of microbial mats that usually contain various unicellular or filamentous cyanobacteria. In springs devoid of sulfide, *Chloroflexus* commonly forms a distinct orange layer beneath a surface layer of oxygenic cyanobacteria. Since no sulfide-tolerant cyanobacteria are known above temperatures of 56 °C, mats with primary sulfide at this temperature and above are dominated by presumably autotrophic *Chloroflexus* which exists as a surface layer or a unispecific mat up to about 66 °C. *Heliothrix* often forms a flocculant surface layer above cyanobacteria in mats where sulfide is absent. For a short review of ecological conditions that support these phototrophs, see Castenholz (1988).

Mesophilic *Cf. aurantiacus* with a temperature optimum of 20–25 °C was found as a component of microbial mats growing on mud containing some sulfide at the bottom of stratified freshwater lakes (Gorlenko, 1976; Pivovarova and Gorlenko, 1977).

Marine *Chloroflexus*-like organisms (MCLOs) have been observed in a large number of marine and hypersaline environments containing biogenic sulfide (Pierson and Castenholz, 1992). They are prominent members of microbial mat communities containing other phototrophs such as purple and green sulfur bacteria and a top layer of cyanobacteria. Such habitats are characterized by steep and dynamic microgradients of oxygen, sulfide, and light. Species of *Beggiatoa* are frequently found in association with the MCLOs which may migrate on a diel basis in response to changing gradients. Although mesophilic, some of these populations are exposed to internal mat temperatures above 40 °C in shallow habitats exposed to full sun in the summer.

Oscillochloris species have been collected primarily from sulfide-containing freshwater mats over a temperature range of 10 to 20 °C, often in association with purple and green sulfur bacteria and species of *Beggiatoa* (Gorlenko, 1989a). Cultures of *Oscillochloris trichoides* were isolated from the sediment of a sulfide-containing spring in the Caucasus (Keppen et al., 1993a).

Unlike the other filamentous phototrophs, the known *Chloronema* species is mesophilic and planktonic, occurring in freshwater lakes with low sulfide and high ferrous iron content (Gorlenko, 1989b). It was found below the chemocline in an anoxic zone along with species of purple and green sulfur bacteria.

There are two other anoxygenic filamentous phototrophs that occur in microbial mat communities. BChl *a*-containing filaments form a prominent red layer deep in microbial mats in some alkaline hot springs (Castenholz, 1984). Like *Heliothrix*, these bacteria lack chlorosomes and BChls *c*, *d*, or *e* (Pierson et al., unpublished). The gliding filaments appear to be photoheterotrophic and contain abundant membranes similar in appearance to those of *Ectothiorhodospira* species (Boomer et al., 1990). Ultrastructurally similar filaments occur in hypersaline microbial mats in association with the cyanobacterium *Microcoleus chthonoplastes* (D'Amelio et al., 1987).

A. Carbon Metabolism

1. Heterotrophy vs. Autotrophy

Cf. aurantiacus grows most rapidly as a photoheterotroph. Autotrophic capability has been demonstrated in some strains (Madigan et al., 1974; Madigan and Brock, 1975), but is slower than photoheterotrophic growth. Both sulfide (Madigan and Brock, 1977a) and hydrogen (Holo and Sirevåg, 1986) can serve as electron donors for CO_2 fixation. Although most strains appear to grow slowly autotrophically in the lab, sulfide-dependent photoautotrophy has been demonstrated in natural hot spring populations of *Chloroflexus* (Giovannoni et al., 1987; Jørgensen and Nelson, 1988).

2. Autotrophic CO_2 Fixation Pathways

The pathway for CO_2 fixation has been difficult to identify (Madigan and Brock, 1977a; Sirevåg and Castenholz, 1979) until recently when it became clear that *Cf. aurantiacus* possesses a novel autotrophic pathway not found in any other organisms. Initial studies on the nature of this pathway suggested that acetyl-CoA was formed from CO_2 directly and that 3-hydroxypropionate was an intermediate in the pathway (Holo and Grace, 1987 and 1988; Holo, 1989). Carboxylation of acetyl-CoA was suggested as a means to produce pyruvate (Holo and Grace, 1988).

Further studies on the CO_2 fixation pathway in *Cf. aurantiacus* (DSM 636) which is the same strain (OK-70-fl) studied by Holo and associates, have led to a postulated pathway in which two carboxylation steps result in the net formation of glyoxylate which is assimilated into cellular carbon, and the regeneration of acetyl-CoA which is the primary CO_2 acceptor molecule (Strauss et al., 1992; Strauss and Fuchs, 1993; Eisenreich et al., 1993; see Fig. 3 in Chapter 40 by Sirevåg). A different pathway has been proposed from studies of *Chloroflexus* strain B-3 (Ivanovsky et al., 1993; see Fig. 2 in Chapter 40 by Sirevåg). The two proposed pathways are similar in their first and last steps, but the intermediary reactions, including the second carboxylation step, are different.

In both proposed pathways the first carboxylation reaction uses acetyl-CoA as the acceptor molecule. The product of this reaction in the 3-hydroxypropionate cycle (Strauss et al., 1992) is malonyl-CoA, which is converted via 3-OH-propionate to propionyl-CoA which is the acceptor molecule for the second carboxylation reaction. In the cycle proposed by Ivanosky (1993), the first carboxylation reaction produces pyruvate which is converted to phosphoenolpyruvate which serves as the acceptor molecule for the second carboxylation step. In the 3-hydroxypropionate cycle, propionyl-CoA is carboxylated to methylmalonyl-CoA which is converted to malyl-CoA, perhaps via succinate and malate.

Consistent with these proposed pathways is the presence in *Cf. aurantiacus* of a substantial corrinoid fraction, 85% of which is in the form of coenzyme B_{12} (Stupperich et al., 1990). This corrinoid could be the prosthetic group of a methylmalonyl-CoA mutase involved in the reaction sequence from propionyl-CoA to succinyl-CoA (Stupperich et al., 1990). In the last reaction of the proposed cycle, the malyl-CoA is cleaved to glyoxylate, regenerating the acetyl-CoA (Strauss et al., 1992).

In the cycle proposed by Ivanovsky et al. (1993), the second carboxylation reaction produces malate via oxaloacetate. Ivanovsky et al. (1993) have shown that the terminal reaction producing glyoxylate is dependent on ATP and CoA. The major differences in the two cycles are the pyruvate vs. 3-OH-propionate routes as the primary carboxylation products and the oxaloacetate vs. succinate paths to malate following the second carboxylation step. Further work on the enzymology of each of the specific reactions, including regulatory sites, is needed to clarify the pathway of autotrophy in *Cf. aurantiacus*. Both proposed CO_2 fixation pathways represent significant differences in autotrophic mechanisms from all other known organisms and point to the importance of *Cf. aurantiacus* in studies on the evolution of photosynthesis.

No evidence for a reversed TCA cycle, typical of green sulfur bacteria, or for the Calvin Cycle, typical of purple bacteria, has been found in *Cf. aurantiacus*. However, key enzymes of the Calvin Cycle (ribulosebisphosphate carboxylase and phosphoribulokinase) have recently been reported in autotrophically grown cells of *Oscillochloris trichoides*, suggesting that the filamentous green bacteria are metabolically diverse (Ivanovsky, 1993).

3. Other Aspects of Carbon Metabolism

Heterotrophy and other aspects of carbon metabolism have recently been reviewed (Pierson and Castenholz, 1992). Suggested pathways for the assimilation of autotrophically produced glyoxylate into cellular carbon and other aspects of carbon metabolism are reviewed in Chapter 40 by Sirevåg.

4. Carbon Metabolism in Other Filamentous Phototrophs

Little is known of the carbon metabolism in the other anoxygenic phototrophic filamentous bacteria. Field studies revealed photoheterotrophy in the thermophilic *Heliothrix oregonensis* (Pierson et al., 1984; Castenholz and Pierson, 1989). Photoheterotrophy appeared to be the dominant form of metabolism in *Cf. aurantiacus* var. *mesophilus* (Gorlenko, 1976). Both photoautotrophic and photoheterotrophic activity were demonstrated in a marine *Chloroflexus*-like organism in field studies (Mack and Pierson, 1988). Strong light-dependent uptake of ^{14}C-bicarbonate has been observed in hypersaline *Chloroflexus*-like organisms (Pierson et al., 1994), suggesting sulfide-dependent autotrophy as their major metabolism in situ.

B. Nitrogen Metabolism

1. Nitrogen Fixation

The process of nitrogen fixation in anoxygenic phototrophic bacteria is reviewed elsewhere in this volume (see Chapter 42 by Madigan and 43 by Ludden and Roberts). *Cf. aurantiacus* is most notable for its total lack of nitrogen fixing ability (Heda and Madigan, 1986), a characteristic that distinguishes it from all other major groups of phototrophs. While the antiquity of the *Chloroflexus* divergence remains in question, this absence of diazotrophy would be consistent with the idea that the ancestors of *Chloroflexus* diverged from the ancestral phototrophic bacteria before the evolution or acquisition of nitrogenase in the phototrophs.

Reports of nitrogen fixation in a recently isolated strain of *Oscillochloris trichoides* are of particular interest (Keppen et al., 1989; Keppen et al., 1993a, b). No nitrogenase activity occurred in the thermophilic strains of *Cf. aurantiacus*, but the strain Dg6 of *O. trichoides* had nitrogenase activity when grown in media with glycine, asparagine or in an atmosphere of N_2. Because the phylogenetic closeness of *O. trichoides* and *Cf. aurantiacus* remains to be determined, we are not certain what the significance of this finding is to the evolution of nitrogen metabolism among the filamentous phototrophs.

2. Other Aspects of Nitrogen Metabolism

Studies on nitrogen metabolism in *Cf. aurantiacus* (Heda and Madigan, 1986) revealed that nitrate did not support growth of the strains tested but that NH_4^+ did. Among organic nitrogen compounds several amino acids also functioned as nitrogen sources for some strains, but proline, adenine, and urea did not serve as nitrogen sources for any of the strains tested (Heda and Madigan, 1986).

Pathways of nitrogen metabolism have been relatively little studied in *Cf. aurantiacus*. The pathway for ammonia assimilation involves glutamine synthetase only (Kaulen and Klemme, 1983). The lack of glutamate synthase activity remains to be clarified (Klemme, 1989).

Amino acid metabolism and regulation have been the subject of some degree of research. Within the microbial mat environment a considerable array of organic substrates is available and *Chloroflexus* appears to be able to metabolize several amino acids for both carbon and nitrogen sources with good growth on casein hydrolysate or glutamate (Pierson and Castenholz, 1974) or aspartate (Madigan et al., 1974). Glutamate (glu), alanine, and isoleucine (ile) are the main constituents of the intracellular amino acid pool when cells are grown on minimal medium (Klemme et al., 1988). Although the concentration of glu is high, the ratio of ile to glu is unusually high in *Chloroflexus* (Klemme et al., 1990). Two different enzymes with L-threonine (L-serine) dehydratase activity were found (Laakmann-Ditges and Klemme, 1986 and 1988). The casein hydrolysate can be replaced by a mixture of glutamate and the amino acids of the 'aspartate' (met, lys, thr) and the 'ilv' (ile, leu, val) groups (Klemme et al., 1990). Aspartate kinase from *Chloroflexus* has a complex regulation system and is strongly inhibited by threonine (Klemme et al., 1990).

The utilization of amino acids in *Chloroflexus* grown in undefined media in batch culture or in chemostats at different dilution rates is quite complex. It is clear that physiological conditions of cultures will change when grown under these different conditions as preferential uptake of different amino acids occurs (Oelze et al., 1991). The relationship of growth phase in batch culture or the dilution rate in a chemostat to the uptake of glu is particularly significant since glu is the precursor to BChl synthesis. These authors also found that the commonly used buffer in medium for growing *Chloroflexus*, gly-gly, is preferentially consumed by batch cultures in the later stages of growth. Ser, ala, and glu each stimulated growth and pigment synthesis (Oelze and Söntgerath, 1992), and ser was the best substrate tested for both growth and bacteriochlorophyll synthesis.

Much remains to be learned about the enzymology and regulation of the amino acid and inorganic nitrogen metabolism in *Chloroflexus aurantiacus*. Very little is known about nitrogen metabolism in any of the other filamentous anoxygenic phototrophs.

C. Sulfur Metabolism

Cf. aurantiacus can use a variety of compounds including cysteine, glutathione, methionine, sulfide, and sulfate as a source of nutrient sulfur during photoheterotrophic growth (Krasil'nikova, 1987). Sulfate is the best source for biosynthesis although thiosulfate may also be used (Krasil'nikova, 1987; Kondrat'eva and Krasil'nikova, 1988). High levels of ATP sulfurylase activity were found with optimal activity at 60–70 °C (Krasil'nikova, 1987).

The use of sulfur compounds as electron donors in photosynthesis has been reviewed elsewhere (see Chapter 39 by Brune). Madigan and Brock (1975) and Giovannoni et al. (1987) showed that sulfide was oxidized to elemental sulfur that adhered to and accumulated outside the cells during both photoheterotrophic and photoautotrophic growth.

Sulfide can also be used as an electron donor for photoautotrophic growth of *Oscillochloris trichoides* (Keppen et al., 1993a) and appears to support photoautotrophy in the hypersaline marine *Chloroflexus*-like organisms (Pierson et al., 1994).

D. Aspects of Energy Metabolism

1. Phototrophy vs. Chemotrophy

Cf. aurantiacus appears to grow best in nature as a photoheterotroph. Its growth rates and yields in pure culture are highest when grown anaerobically in the light on complex organic media. Aerobic growth in the dark occurs but is slower and has lower yields (Pierson and Castenholz, 1974b). Krasil'nikova and Kondrat'eva (1987) reported that fermentation occurs anaerobically in the dark and products from the fermentation of glucose included acetate, pyruvate, lactate, malate, ethanol, and formate. Even those strains which will grow photoautotrophically (such as the GCF strain) grow best as photoheterotrophs (Giovannoni et al., 1987). The latter strain could tolerate exposure to oxygen but did not grow aerobically in the dark, appearing to be an obligate phototroph.

A thorough analysis of the capacity of *Cf. aurantiacus* for anaerobic respiration remains to be done. Thiosulfate may serve as an electron acceptor in some strains with the accompanying production of hydrogen sulfide (Kondrat'eva and Krasil'nikova, 1988).

The metabolic versatility of *Cf. aurantiacus* may be a definite advantage in its microbial mat habitat in thermal springs. The mat community usually includes oxygenic cyanobacteria growing above and within the *Chloroflexus* layer. Conditions in such a mat fluctuate on a diel basis (Revsbech and Ward, 1984). During the day, such mats are highly oxic. During darkness and low light periods, the mats become anoxic. The cyanobacteria may produce a large variety of organic compounds that can be assimilated by *Chloroflexus* both in the light and in the dark (Anderson et al., 1987; Bateson and Ward, 1988; Teiser and Castenholz, unpublished data; Ward et al., 1984). In hot springs containing primary sulfide, sulfide-tolerant cyanobacteria do not occur at temperatures above 56 °C and *Chloroflexus* may grow above this temperature (to 66 °C) alone as a photoautotroph (Castenholz, 1988; see also Chapter 5 by Castenholz and Pierson). Various physiologically significant environmental factors such as oxygen, light, and organic compounds are likely to have important roles directly or indirectly in regulating the metabolism of *Chloroflexus*.

2. ATPase

ATP synthase has been isolated from *Cf. aurantiacus* strain A (Microbiology Culture Collection, Moscow State University). The F_1 factor was isolated first and shown to have ATPase activity (Yanyushin, 1988). Further characterization revealed that it was very similar to the enzyme isolated from other eubacteria although the *Chloroflexus* enzyme had only four subunits rather than five in the catalytic F_1 portion (Yanyushin, 1991). The four subunits are of molecular masses 62, 53, 36, and 16 kD in a stoichiometry of 3: 3: 1: 1. The proton-channel F_0 factor was composed of three subunits. The enzyme was unusually stable (Yanyushin, 1988). While *Chloroflexus* appears to have a few differences in its ATP synthase from those of other organisms, the enzymes are fundamentally similar.

3. Hydrogenase

A hydrogenase that is loosely associated with the cell membranes occurs in *Cf. aurantiacus* strain A (Microbiology Culture Collection, Moscow State University) (Serebryakova et al., 1989). This hydrogenase is a constitutive enzyme that removes electrons from hydrogen during photoautotrophic growth and releases molecular hydrogen produced during fermentation of glucose and pyruvate by cultures grown anaerobically in the dark. Significant inactivation of cellular hydrogenase did not occur below 80 °C, reflecting an ability of the cell environment to stabilize the enzyme (Serebryakova et al., 1989). Maximum activity of the enzyme was at 65 °C (Serebryakova et al., 1989). The enzyme contains an iron-sulfur cluster and nickel and has ferredoxin as its natural electron donor (Serebryakova et al., 1990). It differs from the sulfide-inactivated membrane-bound enzyme described by Drutschmann and Klemme (1985) from strain OK-70-fl and from the hydrogenase of green sulfur bacteria (Serebryakova et al., 1990). The hydrogenase was inhibited by CO and NO but not by C_2H_2 (Serebryakova and Gogotov, 1991).

E. Pigmentation

One of the most interesting aspects of the physiology of photosynthesis in *Cf. aurantiacus* lies in the diversity, synthesis, and regulation of its pigments.

1. Diversity and Function of Pigments

Cf. aurantiacus contains BChl *c* in its chlorosomes as the primary light-harvesting pigment. It contains smaller amounts of BChl *a* (B800 and B865 complexes) in the membrane which also function in light-harvesting and in transfer of energy from the chlorosomes to the membrane-bound reaction centers which contain BChl *a* and BPh *a*. A specialized BChl *a* absorbing at 792 nm is associated with the baseplate of the chlorosome at the attachment site to the membrane. In addition to the bacteriochlorophylls, *Cf. aurantiacus* also contains carotenoids: β-carotene, γ-carotene, and hydroxy-γ-carotene-glucoside comprising 80–95% of the total carotenoids in anaerobically grown cells and echinenone and myxobactone constituting 75% of the carotenoids in aerobically grown cells (K. Schmidt, personal communication; Schmidt et al., 1980; Halfen et al., 1972).

Based on analyses of mat layers, the marine and hypersaline *Chloroflexus*-like organisms appear to contain BChl *c* absorbing between 747 and 755 nm or BChl *d* or *e* absorbing between 710 and 725 nm plus BChl *a* (Mack and Pierson, 1988; Pierson et al., 1994; Palmisano et al., 1989; Des Marais et al., 1992). Evidence for the same pigments has also been found in highly enriched cultures (Pierson et al., 1994). Evidence for γ-carotene as a biomarker for MCLOs in hypersaline microbial mats has been reported (Palmisano et al., 1989; Des Marais et al., 1992). *Chloronema* appears to contain BChl *d*, but also contains small amounts of BChl *c* or *e* and presumably BChl *a* (Gorlenko, 1989b). *Oscillochloris* contains BChl *c* and *a* (Keppen et al., 1993a).

Heliothrix oregonensis and the red thermophilic filaments of mats in Yellowstone National Park contain only BChl *a*. The latter contain a BChl *a* which has absorption maxima at 910 and 807 nm (Boomer et al., 1990).

The light-harvesting BChl *c* of *Cf. aurantiacus* is actually a diverse assemblage of several different isomers and is found as esters of five different alcohols: geranylgeraniol, phytol, and stearol as well as the non-isoprenoid alcohols cetol and oleol (Fages et al., 1990).

Knowledge of the identity, evolutionary origin, and adaptive variation in the light-harvesting chlorosome BChls of green sulfur bacteria and *Chloroflexus* has been progressing in recent years. It has become clearer that changes may occur in the structure of these chlorophylls in a species of *Chlorobium* when cultured for prolonged periods under low light conditions (Broch-Due and Ormerod, 1978; Bobe et al., 1990). Enzymatic methylation of BChl *d* may result in conversion to BChl *c* with a red-shift in absorption maxima from about 714 nm to around 740 nm (Bobe et al., 1990). This transition has been documented in some species of green sulfur bacteria while maintained in long-term culture at low light intensities. Because small amounts of BChl *d* have been detected in *Cf. aurantiacus* (Brune et al., 1987), it is suggested that a similar transition may have occurred in this species. It is not clear whether such changes represent adaptive responses or genetic evolutionary changes. The presence of such transitions and two major forms of BChl (*c* and *d*) in the same organism as well as the isomers and different alcohol esters provide a diversity of pigment molecules within *Cf. aurantiacus*. All of these pigments are found in the chlorosomes and the increased alkylation occurring at low light levels could increase the pigment levels attainable by providing more hydrophobic character and hence closer packing of the pigment molecules (Bobe et al, 1990).

2. Synthesis and Regulation of Pigments

Synthesis of bacteriochlorophylls in *Cf. aurantiacus* is via the C_5 pathway from glutamate as in the oxygenic cyanobacteria and green sulfur bacteria (Swanson and Smith, 1990; Oh-hama et al., 1991; Kern and Klemme, 1989; Avissar et al., 1989). This appears to be the more ancestral pathway for the synthesis of bacteriochlorophylls (Oh-hama et al., 1991).

Cf. aurantiacus grows under conditions of great environmental flux (see Section III. D.1 of this Chapter and Chapter 5 by Castenholz and Pierson). It is not surprising that these environmental parameters have a regulatory influence on the metabolism of *Cf. aurantiacus*. Pigment synthesis is affected by oxygen, light, growth rate, and organic substrates. Photosynthetic activity and growth are also affected by these parameters. The array of pigments present in *Cf. aurantiacus* and the alterations that occur to their relative specific contents due to changes in environmental parameters and stage of growth and rate of growth, result in highly varied absorption spectra. Thus, one must consider carefully the conditions of growth and stage of growth when reporting spectra for *Cf. aurantiacus* in culture (Pierson and Castenholz, 1974b; Kharchenko, 1992).

Natural populations of *Cf. aurantiacus* adapted to low light intensities show photoinhibition when incubated in higher light intensities (Madigan and Brock, 1977b). In culture, growth rate increased and bacteriochlorophyll synthesis decreased in response to higher light intensities (Pierson and Castenholz, 1974b). Other early studies on the effects of light and oxygen on cultures of *Cf. aurantiacus* (Sprague et al., 1981a, 1981b; Schmidt et al., 1980; Feick et al., 1982) showed that bacteriochlorophyll synthesis is reduced in the presence of oxygen and is induced under anoxic conditions, even in the dark. The structure of the photosynthetic apparatus is also affected by oxygen. Lowering the oxygen partial pressure increased the level of BChl *c* and the number and size of chlorosomes. However, the change in ratio of BChl *c* to BChl *a* indicated differential regulation of these two pigments. Increasing light intensity was accompanied by an increase in growth rate (Pierson and Castenholz, 1974b; Oelze and Fuller, 1987).

Using chemostat-grown cultures Oelze and Fuller (1987) showed that the specific contents of both BChl *a* and *c* increased with decreasing growth rate.

Golecki and Oelze (1987) showed that the number of chlorosomes and the percentage of cell membrane surface area covered by baseplates correlated directly with the specific content of BChl *a*. As the specific content of BChl *a* increased linearly, the specific content of BChl *c* increased exponentially. This increase in BChl *c* is structurally accommodated by increasing the volume of the chlorosomes, the density of the BChl *c* molecules within the chlorosomes, and the number of chlorosomes (Golecki and Oelze, 1987). While the effects of growth rate, light, and oxygen on pigment levels and ultrastructure are complex, results of more recent studies are beginning to clarify some of the physiological aspects of this system.

In an effort to elucidate the mechanisms of regulation by light and oxygen on synthesis of both bacteriochlorophylls *a* and *c*, Oelze (1992) has studied the impact of these factors on parts of the C_5 biosynthetic pathway. By using batch cultures and serine-limited growth in chemostat cultures, it was shown that gabaculine, which inhibits the first step in 5-aminolevulinic acid (ALA) synthesis, inhibits BChl *c* synthesis more than BChl *a* synthesis. Oxygen produced the same effect as gabaculine, thus reducing the BChl *c* to *a* ratio. Light also appeared to act as a control on ALA formation lowering the BChl *c* to *a* ratio.

Other parts of the pathway may also be affected by oxygen since addition of ALA to aerated cultures produced higher levels of coproporphyrin but not BChl (Oelze, 1992). Although the site of ALA synthesis is clearly one important regulatory site for oxygen and light, there may be other sites too, and little is known about the mechanism of action of light and oxygen or even whether they act directly or indirectly. It appears that BChl *c* synthesis competes with biomass synthesis for some common substrate such as glutamate (Oelze, 1992) so that when biomass levels are high, BChl *c* levels are low. It was further suggested that coproporphyrinogen excretion might be a valve to eliminate excess precursors of BChl *c* when a decrease in chlorosome production would lead to their accumulation (Oelze, 1992).

The regulation of the protein components of the chlorosomes may also be affected by oxygen. The BChl *c* is thought to be associated with a low molecular weight (5.7 kDa) protein. This protein is not synthesized as part of a large polypeptide precursor that is posttranslationally cleaved into functional fragments (Theroux et al., 1990). The authors isolated the gene encoding this protein, showing it to be a discrete gene (*csm*A) whose expression is regulated by oxygen either transcriptionally or posttranscriptionally. Transcriptional regulation would involve induction of message synthesis by anaerobiosis. Posttranscriptional regulation would require oxygen-dependent digestion of the *csm*A messages (Theroux et al., 1990).

A greater understanding of regulation of pigment synthesis requires isolation of key enzymes in the C_5 biosynthetic pathway and study of their regulation. Regulation of expression of the chlorosomal genes will also elucidate the possible multiple roles of oxygen in regulating the synthesis and assembly of the entire photosynthetic apparatus. The use of pigment mutants (Pierson et al., 1984b) could enhance the understanding of the role of pigments in the assembly of the photosynthetic apparatus.

F. Physical and Chemical Factors Affecting Growth and Distribution in the Environment

1. Temperature

Many aspects of thermophily have been studied in *C. aurantiacus* and only a few references are cited here. Optimal activity of several enzymes is between 45 °C and 70 °C. Isolated enzyme activities are sometimes lower than activities measured in cells. An amylase has been described with optimum activity at 71 °C (Ratanakhanokchai et al., 1992).

The photochemical reaction center of *Cf. aurantiacus* has thermal stability (Pierson et al., 1983; Nozawa and Madigan, 1991), a fact that has made it applicable to the study of energy-requiring transport phenomena in membranes from thermophilic anaerobic bacteria. The reconstituted membranes containing the thermostable reaction centers can develop a proton motive force in the presence of light under anaerobic conditions (Speelmans et al., 1993). The special pair band within the reaction center appears to be more sensitive to thermal denaturation than the accessory pigment band (Pierson et al., 1983; Nozawa and Madigan, 1991).

Membrane phenomena at high temperatures are dependent not only on thermal stability of proteins but on the stability of the inherent membrane structure which is influenced by lipid phase transitions. Oelze and Fuller (1983) determined the temperature characteristics of growth and the membrane-bound enzyme activities of NADH oxidase, succinate 2,6-dichlorophenolindophenol reductase, ATPase, and light-induced proton extrusion. Enzyme activities were maximal at 65–70 °C but membrane phenomena, such as proton extrusion, appeared to be significantly altered at temperatures above 60 °C. An apparent lipid phase transition occurred at suboptimal temperatures (near 40 °C in cells grown at 50 °C) which could contribute to a decrease in photosynthetic efficiency.

The presence of the polyamine, *sym*-homospermidine may contribute to thermal stability of some components in cells of *Cf. aurantiacus* (Norgaard et al., 1983), although this role remains to be confirmed.

2. Salinity

Marine and hypersaline *Chloroflexus*-like organisms occur at salinities ranging from 3.5% (Mack and

Pierson, 1988) to 15% (Pierson et al., 1994). The organisms of the latter habitat, when studied in mixed culture, had an optimum salinity for growth of 10% and did not grow at salinities as high as 15% even though they are exposed to such high salinities at least transiently in nature.

3. Radiation

a. Ultraviolet Radiation

Cf. aurantiacus is surprisingly resistant to ultraviolet radiation (Pierson et al., 1984b, 1993; Pierson, 1994). Cultures were grown under anoxic conditions in the light in the presence of continuous UV-C radiation with near normal growth rates but depressed yields (Pierson et al., 1993). *Cf. aurantiacus* grew very well in a continuous UV-C irradiance of 0.01 Wm^{-2}, a level that was lethal to *E. coli* (Pierson et al., 1993). Although nothing is yet known of the mechanisms of UV resistance in *Cf. aurantiacus*, such tolerance may represent a relict of specific adaptations to growth in shallow mat communities early in the Precambrian prior to the existence of oxygen and a protective ozone layer (Pierson, 1994; Pierson et al., 1993).

b. Visible and Near Infrared Radiation

As a phototroph *Cf. aurantiacus* is dependent on light to sustain its primary growth in the natural environment of hot spring microbial mats. With carotenoids, BChl *c* and *a* as important photosynthetic pigments, these cells seem well-adapted to using much of the visible and NIR parts of the spectrum. By using a fiber optic probe to measure spectral irradiance within the microbial mats (Pierson et al., 1990), it has been possible to demonstrate which wavelengths are available to sustain photosynthesis at a given depth in the mat community where *Chloroflexus* cells may be found and to show which wavelengths are attenuated by specific layers of bacteria including *Chloroflexus*. In most mat habitats wavelengths below 600 nm are attenuated just below the surface leaving primarily red and NIR wavelengths penetrating to the depths at which *Cf. aurantiacus* and other *Chloroflexus*-like organisms are found (see Figs. 7 and 8 in Chapter 5 by Castenholz and Pierson). Consequently in the hot spring environment it appears that *Chloroflexus* must rely on the absorption bands at 740, 805, and 865 nm from BChl *c* and *a* to sustain phototrophic growth.

The marine and hypersaline *Chloroflexus*-like organisms are also often found in microbial mats at depths which are exposed to irradiances containing only red and NIR wavelengths (Pierson et al., 1994).

When *Chloroflexus*-like organisms are in planktonic habitats, they are usually found fairly deep in anoxic zones where all NIR is eliminated. Consequently the photosynthetic activity of members of the Chloroflexaceae there will be dependent on the visible spectrum.

G. Motility and Tactic Responses

All of the filamentous bacteria discussed in this chapter appear to be capable of slow gliding motility, but few quantitative data are available. *Chloroflexus aurantiacus* and the broader filaments of *Heliothrix oregonensis* (F-2 and F-1, respectively, Pierson and Castenholz, 1971) moved at apparently maximum rates of 0.04 $\mu m\ s^{-1}$ and 0.4 $\mu m\ s^{-1}$, respectively, on agar in a uniform light field. *Heliothrix* displays a clumping phenomenon similar to that of many gliding filamentous cyanobacteria (see Pierson et al., 1984a). In the even broader filament type of *Oscillochloris chrysea*, gliding rates of 7 $\mu m\ s^{-1}$ were recorded (Gorlenko, 1989a). This is a very rapid rate even for fast gliding species of the cyanobacterial genus, *Oscillatoria*.

Tactic behavior has also been observed in many field populations of *Chloroflexus* and *Chloroflexus*-like organisms, but no rigorous tests of the nature of the stimulus have been made. *Chloroflexus*-like organisms are known to migrate to the surface of hot spring mats in darkness (Doemel and Brock, 1974) and this behavior has been attributed to positive aerotaxis. Similar behavioral responses may occur in *Heliothrix* and the red filamentous bacteria of many Yellowstone hot springs (Boomer et al., 1990; Pierson et al., unpublished). On the other hand, populations of *Chloroflexus*-like filaments in hypersaline mats appear to remain in anoxic underzones below both the sedentary and vertically migrating cyanobacterial populations (Garcia-Pichel et al., 1994).

H. Associations with Other Organisms

The little that is known about intimate associations with other organisms comes from studies of photo- or chemo-heterotrophic *Chloroflexus* and the photoautotrophic cyanobacteria of hot springs. These are discussed in Chapter 5 by Castenholz and Pierson.

Acknowledgment

We thank Kat Menear for preparation of the manuscript.

References

Anderson KL, Tayne TA and Ward DM (1987) Formation and fate of fermentation products in hot spring cyanobacterial mats. Appl Environ Microbiol 53: 2343–2352

Avissar YJ, Ormerod JG and Beale SI (1989) Distribution of δ-aminolevulinic acid biosynthetic pathways among phototrophic bacterial groups. Arch Microbiol 151: 513–519

Bateson MM and Ward DM (1988) Photoexcretion and fate of glycolate in a hot spring cyanobacterial mat. Appl Environ Microbiol 54: 1738–1743

Blankenship RE (1992) Origin and early evolution of photosynthesis. Photosynth Res 33: 91–111

Bobe FW, Pfennig N, Swanson KL and Smith KM (1990) Red shift of absorption maxima in chlorobiineae through enzymic methylation of their antenna bacteriochlorophylls. Biochemistry 29: 4340–4348

Boomer SM, Austinhirst R, Pierson BK (1990) New bacteriochlorophyll-containing filamentous phototrophs from hot spring microbial mats. Am Soc Microbiol, Annual Meeting, Anaheim, CA. Abstract

Broch-Due M and Ormerod JG (1978) Isolation of a BChl *c* mutant from *Chlorobium* with BChl *d* by cultivation at low light intensity. FEMS Microbiol Lett 3: 305–308

Brune DC, Nozawa T, and Blankenship RE (1987) Antenna organization in green photosynthetic bacteria. I. Oligomeric bacteriochlorophyll *c* as a model for the 740 nm absorbing bacteriochlorophyll *c* in *Chloroflexus aurantiacus* chlorosomes. Biochemistry 26: 8644–8652

Castenholz RW (1973) Ecology of blue-green algae in hot springs. In: Carr NG and Whitton BA (eds) The Biology of Blue-Green Algae, pp 379–414. Blackwell, Oxford

Castenholz RW (1984) Composition of hot spring microbial mats: a summary. Cohen Y, Castenholz RW and Halvorson HO (ed) Microbial Mats: Stromatolites, pp 101–119. Alan R Liss, Inc., New York

Castenholz RW (1988) The green sulfur and nonsulfur bacteria of hot springs. In: Olson JM, Ormerod JG, Amesz J, Stackebrandt E and Trüper HG (ed) Green Photosynthetic Bacteria, pp 243–255. Plenum Press, New York

Castenholz RW and Pierson BK (1989) Genus *Heliothrix*. In: Staley JT, Bryant MP, Pfennig N and Holt JG (ed) Bergey's Manual of Systematic Bacteriology, Vol 3, pp 1702–1703. Williams & Wilkins, Baltimore

D'Amelio ED, Cohen Y and Des Marais DJ (1987) Association of a new type of gliding, filamentous, purple phototrophic bacterium inside bundles of *Microcoleus chthonoplastes* in hypersaline cyanobacterial mats. Arch Microbiol 147: 213–220

D'Amelio ED, Cohen Y and Des Marais DJ (1989) Comparative functional ultrastructure of two hypersaline submerged cyanobacterial mats: Guerrero Negro, Baja California Sur, Mexico, and Solar Lake, Sinai, Egypt. In: Cohen Y and Rosenberg E (eds) Microbial Mats: Physiological Ecology of Benthic Microbial Communities, pp 97–113. Amer Soc Microbiol, Washington, DC

Des Marais DJ, D'Amelio ED, Farmer JD, Jørgensen BB, Palmisano AC and Pierson BK (1992) Case study of a modern microbial mat-building community: the submerged cyanobacterial mats of Guerrero, Negro, Baja California Sur, Mexico. In: Schopf JW and Klein C. The Proterozoic Biosphere, pp 325–333. Cambridge Univ Press, New York

Doemel WN and Brock TD (1974) Bacterial stromatolites: origin of laminations. Science 184: 1083–1085

Doemel WN and Brock TD (1977) Structure, growth, and decomposition of laminated algal-bacterial mats in alkaline hot springs. Appl Environ Microbiol 34: 433–452

Drutschmann M and Klemme J-H (1985) Sulfide-repressed, membrane-bound hydrogenase in the thermophilic facultative phototroph, *Chloroflexus aurantiacus*. FEMS Microbiol Lett 28: 231–235

Dubinina GA and Gorlenko VM (1975) New filamentous photosynthetic green bacteria containing gas vacuoles. Microbiology (Eng transl of Mikrobiologiya) 44: 452–458

Eisenreich W, Strauss G, Werz U, Fuchs G and Bacher A (1993) Retrobiosynthetic analysis of carbon fixation in the phototrophic eubacterium *Chloroflexus aurantiacus*. FEBS Eur J Biochem 215: 619–632

Fages F, Griebenow N, Griebenow K, Holzwarth AR, Schaffner K (1990) Characterization of light-harvesting pigments of *Chloroflexus aurantiacus* two new chlorophylls oleyl octadec-9-enyl and cetyl hexadecanyl bacteriochlorophyllides *c*. Journal of the Chemical Society Perkin Transactions I, O 10: 2791–2798

Feick RG, Fitzpatrick M and Fuller RC (1982) Isolation and characterization of cytoplasmic membranes and chlorosomes from the green bacterium *Chloroflexus aurantiacus*. J. Bacteriol 150: 905–915

Garcia-Pichel F, Mechling M and Castenholz RW (1994) Diel migrations of microorganisms within a benthic, hypersaline mat community. Appl Environ Microbiol 60: 1500–1511

Giovannoni SJ, Revsbech NP, Ward DM and Castenholz RW (1987) Obligately phototrophic *Chloroflexus:* primary production in anaerobic hot spring microbial mats. Arch Microbiol 147: 80–87

Golecki JR and Oelze J (1987) Quantitative relationship between bacteriochlorophyll content, cytoplasmic membrane structure and chlorosome size in *Chloroflexus aurantiacus*. Arch Microbiol 148: 236–241

Gorlenko VM (1976) Characteristics of filamentous phototrophic bacteria from freshwater lakes. Mikrobiologiya (Eng transl) 44: 682–684

Gorlenko VM (1988) Ecological niches of green sulfur and gliding bacteria. In: Olson JM, Ormerod JG, Amesz J, Stackebrandt E and Trüper HG (ed) Green Photosynthetic Bacteria, pp 257–267. Plenum Press, New York

Gorlenko VM (1989a) Genus 'Oscillochloris.' In: Staley JT, Bryant MP, Pfennig N and Holt JG (ed) Bergey's Manual of Systematic Bacteriology. Vol 3, pp 1703–1706. Williams and Wilkins, Baltimore

Gorlenko VM (1989b) Genus 'Chloronema.' In: Staley JT, Bryant MP, Pfennig N and Holt JG (ed) Bergey's Manual of Systematic Bacteriology. Vol 3, pp 1706–1707. Williams and Wilkins, Baltimore

Gorlenko VM and Pivovarova TA (1977) On the belonging of bluegreen alga *Oscillatoria coerulescens* Gicklhorn, 1921 to a new genus of Chlorobacteria *Oscillochloris* nov. gen. Izvestiya Akademii Nauk SSSR, Seriya Biologicheskaya 3: 396–409

Halfen LN, Pierson BK, Francis GW (1972) Carotenoids of a gliding organism containing bacteriochlorophylls. Arch. Mikrobiol. 82: 240–246

Hartmann RK, Wolters J, Kröger B, Schultze S, Specht T and Erdmann VA (1989) Does *Thermus* represent another deep eubacterial branching? System Appl Microbiol 11: 243–249

Heda GD and Madigan MT (1986) Utilization of amino acids and lack of diazotrophy in the thermophilic anoxygenic phototroph *Chloroflexus aurantiacus*. J. Gen. Microbiol. 132: 2469–2473

Holo H (1989) *Chloroflexus aurantiacus* secretes 3-hydroxy-propionate, a possible intermediate in the assimilation of CO_2 and acetate. Arch Microbiol 151: 252–256

Holo H and Grace D (1987) Polyglucose synthesis in *Chloroflexus aurantiacus* studied by ^{13}C-NMR. Arch Microbiol 148: 292–297

Holo H and Grace D (1988) A new CO_2 fixation mechanism in *Chloroflexus aurantiacus* studied by ^{13}C-NMR. In: Olson JM, Ormerod JG, Amesz J, Stackebrandt E and Trüper HG (ed) Green Photosynthetic Bacteria, pp 149–155. Plenum Press, New York

Holo H and Sirevåg R (1986) Autotrophic growth and CO_2 fixation of *Chloroflexus aurantiacus*. Arch Microbiol 145: 173–180

Ivanovsky RN (1993) Calvin cycle in a green mesophilic filamentous bacterium, *Oscillochloris trichoides* strain DG6. EMBO Workshop on Green and Heliobacteria, Nyborg, Denmark August 16–19, 1993, Abstract

Ivanovsky RN, Krasilinkova EN and Fal YI (1993) A pathway of the autotrophic CO_2 fixation in *Chloroflexus aurantiacus*. Arch Microbiol 159: 257–264

Jørgensen BB and Nelson DC (1988) Bacterial zonation, photosynthesis, and spectral light distribution in hot spring microbial mats of Iceland. Microb. Ecol. 16: 133–147

Jürgens UJ, Meißner J, Fischer U, König WA and Weckesser J (1987) Ornithine as a constituent of the peptidoglycan of *Chloroflexus aurantiacus*, diaminopimelic acid in that of *Chlorobium vibrioforme* f. *thiosulfatophilum*. Arch Microbiol 148: 72–76

Kaulen H and Klemme J-H (1983) No evidence of covalent modification of glutamine synthetase in the thermophilic phototrophic bacterium *Chloroflexus aurantiacus*. FEMS Microbiol Lett 20: 75–79

Keppen OI, Lebedeva NV, Troshina OYu, and Rodionov Yu V (1989) The nitrogenase activity of a filamentous phototrophic green bacterium. Mikrobiologiya 58: 520–521

Keppen OI, Baulina OI, Lysenko AM, and Kondratieva EN (1993a) New green bacterium belonging to family Chloroflexaceae . Mikrobiologiya (Eng transl) 62: 179–185

Keppen OI, Baulina OI, and Kondratieva EN (1993b) *Oscillochloris trichoides* neotype strain DG6. EMBO Workshop on Green and Heliobacteria, Nyborg, Denmark August 16–19, 1993. Abstract

Kern M and Klemme J-H (1989) Inhibition of bacteriochlorophyll biosynthesis by gabaculin (3-amino, 2,3-dihydrobenzoic acid) and presence of an enzyme of the C5-pathway of δ-aminolevulinate synthesis in *Chloroflexus aurantiacus*. Z Naturforsch 44c: 77–80

Kharchenko SG (1992) Characteristic features of changes in the absorption spectra of phototrophic green bacteria as an index of the stage of culture growth. Mikrobiologiya (Eng transl) 61: 305–309

Klemme J-H (1989) Organic nitrogen metabolism of phototrophic bacteria. Anton van Leeuwenhoek J Microbiol 55: 197–219

Klemme J-H, Laakmann-Ditges G and Mertschuweit J (1988) Ammonia assimilation and amino acid metabolism in *Chloroflexus aurantiacus*. In: Olson JM, Ormerod JG, Amesz J, Stackebrandt E and Trüper HG (ed) Green Photosynthetic Bacteria, pp 173–174. Plenum Press, New York

Klemme J-H, Laakmann-Ditges G and Mertschuweit J (1990) Cellular amino acid concentrations and regulation of aspartate kinase in the thermophilic phototrophic prokaryote *Chloroflexus aurantiacus*. Zeitschrift für Naturforschung C, Biosciences 45: 74–78

Knudsen E, Jantzen E, Bryn K, Ormerod JG and Sirevåg R (1982) Quantitative and structural characteristics of lipids in *Chlorobium* and *Chloroflexus*. Arch Microbiol 132: 149–154

Kondrat'eva EN and Krasil'nikova EN (1988) Utilization of thiosulfate by *Chloroflexus aurantiacus*. Mikrobiologiya (Eng transl) 57: 291–294

Krasil'nikova EN (1987) ATP sulfurylase activity in *Chloroflexus aurantiacus* and other photosynthesizing bacteria as a function of temperature. Mikrobiologiya (Eng transl) 55: 418–421

Krasil'nikova EN and Kondrat'eva EN (1987) Growth of *Chloroflexus aurantiacus* under anaerobic conditions in the dark and the metabolism of organic substrates. Mikrobiologiya (Eng transl) 56: 281–285

Laakmann-Ditges G and Klemme J-H (1986) Occurrence of two L-threonine (L-serine) dehydratases in the thermophile *Chloroflexus aurantiacus*. Arch Microbiol 144: 219–221

Laakmann-Ditges G and Klemme J-H (1988) Amino acid metabolism in the thermophilic phototroph, *Chloroflexus aurantiacus*: properties and metabolic role of two L-threonine (L-serine) dehydratases. Arch Microbiol 149: 249–254

Larsen M, Mack EE, and Pierson BK (1991) Mesophilic *Chloroflexus*-like organisms from marine and hypersaline environments. VII International Symposium on Photosynthetic Prokaryotes, Amherst, MA, USA. Abstract 169B

Mack EE and Pierson BK (1988) Preliminary characterization of a temperate marine member of the Chloroflexaceae. In: Olson JM, Ormerod JG, Amesz J, Stackebrandt E and Trüper HG (ed) Green Photosynthetic Bacteria, pp 237–241. Plenum Publ. Corp., New York

Madigan MT and Brock TD (1975) Photosynthetic sulfide oxidation by *Chloroflexus aurantiacus*, a filamentous, photosynthetic, gliding bacterium. J Bacteriol 122: 782–784

Madigan MT and Brock TD (1977a) CO_2 fixation in photo-synthetically-grown *Chloroflexus aurantiacus*. FEMS Microbiol Lett 1: 301–304

Madigan MT and Brock TD (1977b) Adaption by hot spring phototrophs to reduced light intensities. Arch Microbiol 113: 111–120

Madigan MT, Petersen SR and Brock TD (1974) Nutritional studies on *Chloroflexus*, a filamentous photosynthetic, gliding bacterium. Arch Microbiol 100: 97–103

Norgaard E, Sirevåg R and Eliassen KA (1983) Evidence for the occurrence of sym-homospermidine in green phototrophic bacteria. FEMS Microbiol Lett 20: 159–161

Nozawa T and Madigan MT (1991) Temperature and solvent effects on reaction centers from *Chloroflexus aurantiacus* and *Chromatium tepidum*. J. Biochem 110: 588–594

Oelze J (1992) Light and oxygen regulation of the synthesis of bacteriochlorophylls *a* and *c* in *Chloroflexus aurantiacus*. J Bacteriol 174: 5021–5026

Oelze J. and Fuller RC (1983) Temperature dependence of growth and membrane-bound activities of *Chloroflexus aurantiacus* energy metabolism. J Bacteriol 155: 90–96

Oelze J and Fuller RC (1987) Growth rate and control of development of photosynthetic apparatus in *Chloroflexus aurantiacus*. Arch Microbiol 148: 132–136

Oelze J and Söntgerath B (1992) Differentiation of the photosynthetic apparatus of *Chloroflexus aurantiacus* depending on growth with different amino acids. Arch Microbiol 157: 141–147

Oelze J, Jürgens UJ and Ventura S (1991) Amino acid consumption by *Chloroflexus aurantiacus* in batch and continuous cultures. Arch Microbiol 156: 266–269

Oh-hama T, Santander PJ, Stolowich NJ and Scott AI (1991) Bacteriochlorophyll *c* formation via the C_5 pathway of 5-aminolevulinic acid synthesis in *Chloroflexus aurantiacus*. FEBS Lett 281: 173–176

Oyaizu H, Debrunner-Vossbrinck B, Mandelco L, Studier JA and Woese CR (1987) The green non-sulfur bacteria: a deep branching in the eubacterial line of descent. System Appl Microbiol 9: 47–53

Palmisano AC, Cronin SE, D'Amelio ED, Munoz E and Des Marais DJ (1989) Distribution and survival of lipophilic pigments in a laminated microbial mat community near Guerrero Negro, Mexico. In: Cohen Y and Rosenberg E. (eds) Microbial Mats: Physiological Ecology of Benthic Microbial Communities. Amer Soc Microbiol, pp 138–152. Washington, DC

Pfennig N (1989) Multicellular filamentous green bacteria. In: Staley JT, Bryant MP, Pfennig N and Holt JG (ed) Bergey's Manual of Systematic Bacteriology. Vol 3 (p. 1697). Williams & Wilkins, Baltimore

Pierson BK (1994) The emergence, diversification and role of photosynthetic eubacteria. In: Bengston S (ed) Early Life on Earth, Nobel Symp. No. 84, pp 161–180. Columbia University Press, New York

Pierson BK and Castenholz RW (1971) Bacteriochlorophylls in gliding filamentous prokaryotes of hot springs. Nature 233: 25–27

Pierson BK and Castenholz RW (1974a) A phototrophic gliding filamentous bacterium of hot springs, *Chloroflexus aurantiacus*, gen. and sp. nov. Arch Microbiol 100: 5–24

Pierson BK and Castenholz RW (1974b) Studies of pigments and growth in *Chloroflexus aurantiacus*, a phototrophic filamentous bacterium. Arch Microbiol 100: 283–305

Pierson BK and Castenholz RW (1992) The family Chloroflexaceae. In: Balows A, Trüper HG, Dworkin M, Harder W and Schleifer KH (eds) The Prokaryotes. 2nd ed, pp 3754–3774. Springer-Verlag, New York

Pierson BK, Thornber JP and Seftor REB (1983) Partial purification, subunit structure and thermal stability of the photochemical reaction center of the thermophilic green bacterium *Chloroflexus aurantiacus*. Biochim Biophys Acta 723: 322–326

Pierson BK, Giovannoni SJ and Castenholz RW (1984a) Physiological ecology of a gliding bacterium containing bacteriochlorophyll a. Appl Environ Microbiol 47: 576–584

Pierson BK, Keith LM and Leovy JG (1984b) Isolation of pigmentation mutants of the green filamentous photosynthetic bacterium *Chloroflexus aurantiacus*. J Bacteriol 159: 222–227

Pierson BK, Giovannoni SJ, Stahl DA and Castenholz RW (1985) *Heliothrix oregonensis*, gen. nov., sp. nov., a phototrophic filamentous gliding bacterium containing bacteriochlorophyll a. Arch Microbiol 142: 164–167

Pierson BK, Sands VM and Frederick JL (1990) Spectral irradiance and distribution of pigments in a highly layered marine microbial mat. Appl Envir Microbiol 56: 2327–2340

Pierson BK, Mitchell HK and Ruff-Roberts AL (1993) *Chloroflexus aurantiacus* and ultraviolet radiation: implications for Archean shallow-water stromatolites. Origins of Life and Evolution of the Biosphere 23: 243–260

Pierson BK, Valdez D, Larsen M, Morgan E and Mack EE (1994) *Chloroflexus*-like organisms from non-thermal environments: distribution and diversity. Photosynth Res, 41: 35–52

Pivovarova TA and Gorlenko VM (1977) Fine structure of *Chloroflexus aurantiacus* var. *mesophilus* (Nom. prof.) grown in light under aerobic and anaerobic conditions. Mikrobiologiya (Eng transl) 46: 276–282

Ratanakhanokchai K, Kaneko J, Kamio Y and Izaki K (1992) Purification and properties of a maltotetraose- and maltotriose-producing amylase from *Chloroflexus aurantiacus*. Appl Envir Microbiol 58: 2490–2494

Revsbech NP and Ward DM (1984) Microelectrode studies of interstitial water chemistry and photosynthetic activity in a hot spring microbial mat. Appl Environ Microbiol 48: 270–275

Schmidt K, Maarzahl M. and Mayer F (1980) Development and pigmentation of chlorosomes in *Chloroflexus aurantiacus* strain Ok-70-fl. Arch Microbiol 127: 87–97

Serebryakova LT and Gogotov IN (1991) Inhibition of the hydrogenases of green bacteria by gaseous compounds. Biochemistry (Eng transl) 56: 171–175

Serebryakova LT, Zorin NA, Gogotov IN and Keppen OI (1989) Hydrogenase activity of the thermophilic green bacterium *Chloroflexus aurantiacus*. Mikrobiologiya (Eng transl) 58: 424–428

Serebryakova LT, Zorin NA and Gogotov IN (1990) Purification and properties of the hydrogenase of the green nonsulfur bacterium *Chloroflexus aurantiacus*. Biochemistry (Eng transl) 55: 277–283

Sirevåg R and Castenholz RW (1979) Aspects of carbon metabolism in *Chloroflexus*. Arch Microbiol 120: 151–153

Speelmans G, Hillenga D, Poolman B and Konings WN (1993) Application of thermostable reaction centers from *Chloroflexus aurantiacus* as a proton motive force generating system. Biochim Biophys Acta 1142: 269–276

Sprague SG, Staehelin LA, DiBartolomeis MJ and Fuller RC (1981a) Isolation and development of chlorosomes in the green bacterium *Chloroflexus aurantiacus*. J Bacteriol 147: 1021–1031

Sprague SG, Staehelin LA and Fuller RC (1981b) Semiaerobic induction of bacteriochlorophyll synthesis in the green bacterium *Chloroflexus aurantiacus*. J Bacteriol 147: 1032–1039

Stolz JF (1983) Fine structure of the stratified microbial community at Laguna Figueroa, Baja California, Mexico. I methods of in situ study of the laminated sediments. Precambrian Res 20: 479–492

Stolz JF (1990) Distribution of phototrophic microbes in the flat laminated microbial mat at Laguna Figueroa, Baja California, Mexico. BioSystems 23: 345–357

Strauss G and Fuchs G (1993) Enzymes of a novel autotrophic CO_2 fixation pathway in the phototrophic bacterium *Chloroflexus aurantiacus*, the 3-hydroxypropionate cycle. FEBS Eur J Biochem 215: 633–643

Strauss G, Eisenreich W, Bacher A and Fuchs G (1992) ^{13}C-NMR study of autotrophic CO_2 fixation pathways in the sulfur-reducing Archaebacterium *Thermoproteus neutrophilus* and in the phototrophic Eubacterium *Chloroflexus aurantiacus*. Eur J Biochem 205: 853–866

Stupperich E, Eisinger H-J and Schurr S (1990) Corrinoids in anaerobic bacteria. FEMS Microbiol Rev 87: 355–359

Swanson KL and Smith KM (1990) Biosynthesis of bacteriochlorophyll-*c* via the glutamate C-5 pathway in *Chloroflexus aurantiacus*, pp 1696–1697. J Chem Soc, Chem Commun

Theroux SJ, Redlinger TE, Fuller RC and Robinson SJ (1990) Gene encoding the 5.7-kilodalton chlorosome protein of *Chloroflexus aurantiacus*: regulated message levels and a predicted carboxy-terminal protein extension. J Bacteriol 172: 4497–4504

Trüper HG (1976) Higher taxa of the phototrophic bacteria: Chloroflexaceae fam. nov., a family for the gliding, filamentous, phototrophic 'green' bacteria. Intl J Syst Bacteriol 26: 74–75

Van Den Eynde H, Van De Peer Y, Perry J and De Wachter R (1990) 5SrRNA sequences of representatives of the genera *Chlorobium, Prosthecochloris, Thermomicrobium, Cytophaga, Flavobacterium, Flexibacter* and *Saprospira* and a discussion of the evolution of eubacteria in general. J Gen Microbiol 136: 11–18

Ward DM, Beck E, Revsbech NP, Sandbeck KA and Winfrey MR (1984) Decomposition of hot spring microbial mats. In: Cohen Y, Castenholz RW and Halvorson HO (ed) Microbial Mats: Stromatolites, pp 191–214. Alan R. Liss, Inc., New York

Weller R, Bateson MM, Heimbuch BK, Kopczynski, Ward DM (1992) Uncultivated cyanobacteria, *Chloroflexus*-like inhabitants, and spirochete-like inhabitants of a hot spring microbial mat. Appl Environ Microbiol 58: 3964–3969

Woese CR (1987) Bacterial evolution. Microbiol Rev 51: 221–271

Woese CR, Mandelco L, Yang D, Gherna R and Madigan MT (1990)The case for relationship of flavobacteria and their relatives to the green sulfur bacteria. System Appl Microbiol 13: 258–262

Yanyushin MF (1988) Isolation and characterization of F_1-ATPase from the green nonsulfur photosynthesizing bacterium *Chloroflexus aurantiacus*. Biochemistry (Eng transl) 53: 1120–1127

Yanyushin MF (1991) ATP synthase of the green nonsulfur photosynthetic bacterium *Chloroflexus aurantiacus*. Biochemistry (Eng transl) 56: 786–793

Chapter 4

Ecology of Phototrophic Sulfur Bacteria

Hans Van Gemerden
Department of Microbiology, University of Groningen, Kerklaan 30, 9751 NN Haren, The Netherlands

Jordi Mas
Department of Genetics and Microbiology, Autonomous University of Barcelona, 08193 Bellaterra, Spain

R. E. Blankenship, M. T. Madigan and C. E. Bauer (eds): Anoxygenic Photosynthetic Bacteria, pp. 49–85.

Summary

Phototrophic sulfur bacteria often form mass developments in aquatic environments, either planktonic or benthic, where anoxic layers containing reduced sulfur compounds are exposed to light. This chapter summarizes a number of reports from the literature, collecting the information on the abundance of these bacteria as well as on their contribution to primary production. From the point of view of population dynamics, the abundance of these organisms is the consequence of a certain balance between growth and losses. Both specific growth rates, and specific rates of loss through several processes are analyzed in several environments, in an attempt to generalize on the growth status of blooms of phototrophic sulfur bacteria. The information available indicates the existence of an upper limit for the production of these bacteria in nature, and seems to suggest the existence of an upper limit for biomass based in the balance between growth and losses.

The chapter also reviews the main variables affecting growth of phototrophic sulfur bacteria in nature, paying attention both to the in situ status of these variables and to the functional response of the organisms to each of them. All of this information is integrated in a section in which several case-studies are described, and which emphasizes the role fluctuations play on competition and coexistence between different phototrophic sulfur bacteria.

I. Introduction

Phototrophic sulfur bacteria are organisms commonly found in illuminated aquatic environments containing hydrogen sulfide. Since the first observations, made at the end of last century, their presence has been repeatedly reported in lakes and sediments all over the world, and a number of reviews has been published gathering information on several aspects of their biology (Pfennig, 1978, 1989; Van Gemerden, 1983; Van Gemerden and Beeftink, 1983; Madigan, 1988; Caumette, 1989; Drews and Imhoff, 1991).

Observations on phototrophic sulfur bacteria have been reported as early as 1888 (Pfennig and Trüper, 1992). Many of the reports constitute casual observations on their presence and abundance or short term studies. More recently, communities of phototrophic sulfur bacteria have been studied for an extended period of time, sometimes spanning several years. In these cases, a great deal of information regarding that particular environment can be found in a number of different publications. Such is the case of Lake Vechten in The Netherlands (Steenbergen, 1982; Steenbergen and Korthals, 1982; Steenbergen et al., 1987) or Lake Cisó in Spain (Guerrero et al., 1985; Van Gemerden et al., 1985; Mas et al., 1990; Gasol et al., 1991; Pedrós-Alió and Guerrero, 1993). Of interest are also the detailed observations made by different investigators in the former Soviet Union and compiled by Gorlenko et al. (1983), as well as the extensive work carried out in Japanese lakes (Takahashi and Ichimura, 1968, 1970; Matsuyama and Shirouzu, 1978; Matsuyama, 1987) or in several different lakes in the USA (Parkin and Brock, 1980a, 1980b, 1981).

The present chapter attempts to approach the ecology of these organisms by, first, providing background information on their abundance and activity in natural environments and trying to extract general patterns from the information available and, second, by reviewing some relevant data on their physiology, which allow the interpretation of a number of situations commonly found in nature.

II. Characteristic and Vertical Structure of the Habitats

Phototrophic sulfur bacteria thrive in two different types of environments, either in stratified lakes, where they can occupy the top of the anoxic layers, or on the top millimeters of shallow aquatic sediments, where they can sometimes be detected in the anoxic zone, underlying layers of oxygenic phototrophs. Their development requires the existence of the physical structure needed to avoid vertical mixing and to allow the establishment of an illuminated anoxic compartment. In planktonic environments, this structure is provided either by thermal stratification,

Abbreviations: 3-MPA – 3-mercapto propionate; A – affinity; BChl – bacteriochlorophyll; D – dilution rate; DMDS – dimethyl disulfide; DMS – dimethyl sulfide; DMSO – dimethyl sulfoxide; DMSP – dimetyl sulfonio propionate; Ks – saturation constant; MSH – methanethiol; RuBisCO – ribulose bisphosphate carboxylate; Y – yield factor; μ – specific growth rate; μ_{max} – maximum specific growth rate as in holomictic lakes, or by differences in density related to differences in the concentration of solutes, as in meromictic lakes. In holomictic lakes, thermal stratification develops during spring and summer when the sun warms up the surface layers, and disappears with the coming of cold weather. In meromictic lakes, stratification is permanent, lasting for more than one season and, thus, permits a more permanent community. Both thermal and chemical stratification provide a stable environment where, except for gentle variations in the vertical position of the interface which in most cases can be easily followed by the organisms, phototrophic sulfur bacteria enjoy a rather stable situation. An example of a planktonic community of phototrophic sulfur bacteria can be observed in Fig. 1 corresponding to Lake Vilar (Spain). The left panel of the figure shows vertical profiles of oxygen and hydrogen sulfide. The central panel displays the vertical distribution of purple and green bacteria. Purple sulfur bacteria are positioned forming a very narrow band (around 10 cm) at the interface, while green sulfur bacteria, with a lower biomass are found underneath. Measurements of the rates of carbon dioxide fixation and sulfide oxidation, shown in the third panel, indicate low activities at the bottom of the plate and comparably high activities at the top and at the middle of the plate. Because biomass at the top is considerably lower, the specific activity is higher, an observation which is in agreement with a higher irradiance at the top of the plate (Guerrero et al., 1985).

Benthic communities, more often referred to as microbial mats, occur at a different scale. Figure 2 shows a vertical profile of the distribution of phototrophic sulfur bacteria in a laminated sediment in Scapa Beach (Orkney Islands), as well as vertical profiles of oxygen and hydrogen sulfide at different times of the day. Although the overall structure of the system is quite similar to the structure observed in planktonic environments, bacteria in sediments face a rather different situation. As in the case of lakes, sediments also provide the physico-chemical conditions required for the development of phototrophic sulfur bacteria. However, due to the difference in scale involved, certain characteristics differ remarkably between both environments:

a) Sediments, especially in littoral areas, are much more subject to external factors such as tidal movements. The presence of the oxic layer often depends on the activity of oxygenic phototrophs occupying the upper part of the sediment. Since

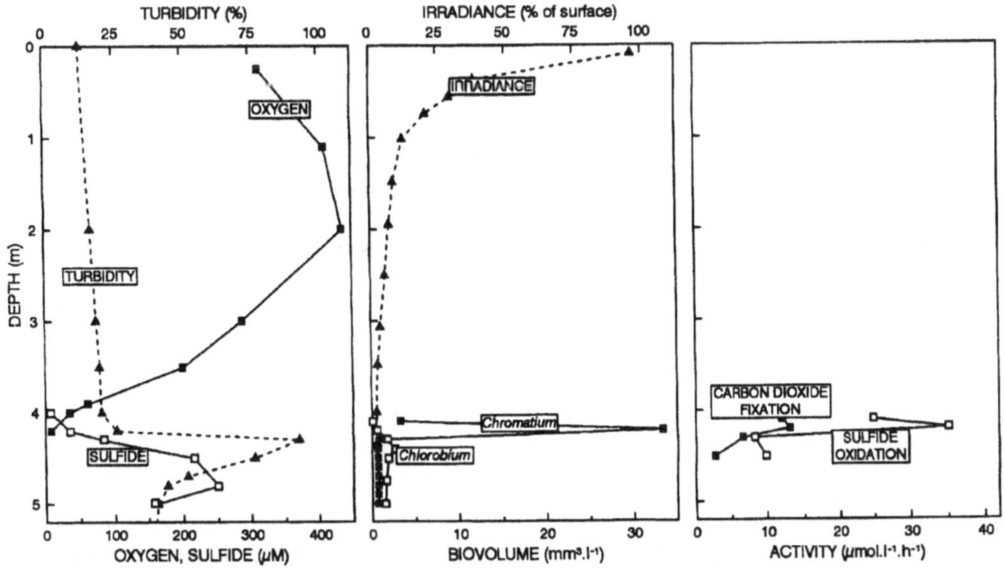

Fig. 1. Vertical profiles of oxygen, sulfide, and turbidity (left panel), irradiance and biovolumes of populations of purple and green sulfur bacteria (middle panel), and rates of CO_2 fixation and sulfide oxidation (right panel) in Lake Vilar, Spain. Measurements were performed at zenith in July, 1982. Redrawn from Guerrero et al. (1985).

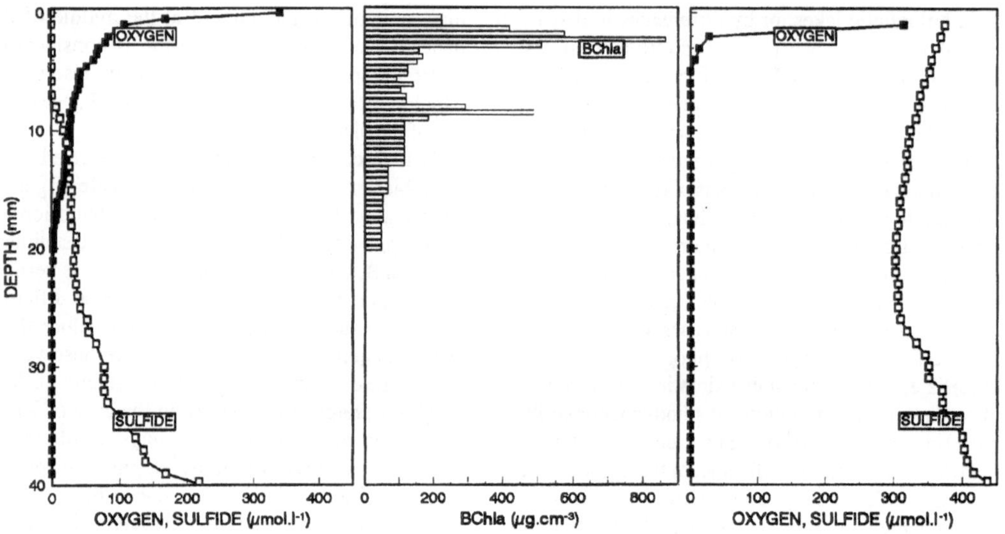

Fig. 2. Vertical profiles of sulfide and oxygen at 1 p.m. (left panel) and 7 p.m. (right panel) in the bloom of *Thiocapsa roseopersicina* developing on Scapa Beach, Orkney Islands, Scotland. The middle panel shows the vertical distribution of bacteriochlorophyll *a*, the upper maximum represents the present bloom, the maximum at 8–9 mm reflects an earlier, buried, bloom. Redrawn from Van Gemerden et al. (1989a).

production of oxygen by these organisms depends on light, these environments display dramatic variations in the position of the oxic/anoxic interface. These variations expose phototrophic sulfur bacteria living there to fluctuations in the levels of oxygen and hydrogen sulfide that, no doubt, influence the metabolic profile needed to successfully colonize these environments.

b) Benthic communities of microorganisms are characterized by an extremely high population density in a relatively thin layer. Because microbial activity is much more concentrated in these sediments than in planktonic layers, the biological contribution to the chemical processes occurring at the interface is also much more important. When comparing the contribution of biological processes to the oxidation of reduced sulfur compounds in oxic/anoxic interfaces, Jørgensen (1982) draws attention to the fact that in the Black Sea where the interface spans 35 meters, the oxidation of sulfide is mainly abiotic, while in a microbial mat where the interface occupies only 50 μm, the oxidation is mediated predominantly by microorganisms.

c) The fact that microbial mats have small vertical dimensions, are rich in organic matter, and occur in exposed surfaces, makes these communities potentially unstable, as they are exposed both to erosion, and grazing by a number of different predators. Thus, their occurrence is relatively precarious, being found in many cases only during reduced periods of time, or more permanently, in areas where the existence of extreme conditions reduces the presence and the impact of invertebrate predators (Awramik, 1984).

d) The spectrum of light available to phototrophic sulfur bacteria in planktonic environments is strongly modified due to absorption by water, becoming enriched with depth in the yellow-green part of the spectrum. Sediments, on the contrary, offer a light climate much more enriched in the near infrared. These differences in light quality are likely to influence the composition of the microbial community inhabiting each environment.

As a consequence of these factors, important differences in the composition of the community of phototrophic sulfur bacteria might be expected between planktonic and benthic environments. Many data have been gathered on their occurrence in lakes, unfortunately, the data available on microbial mats do not yet allow a generalization in that respect. However, as will be shown in the following sections, the physiological characteristics of some organisms make them good candidates to occupy the sediment niche.

III. Distribution and Abundance of Phototrophic Sulfur Bacteria in Nature

Somehow, most of the literature available describes the occurrence of phototrophic sulfur bacteria in lakes and, only recently, detailed descriptions of benthic systems have been published. Many old reports contain little information besides the mentioning of the organisms, and some estimates of their abundance, based on total counts under the microscope. More recent reports also give information on the type of Bchl as well as their concentrations. Information on the major carotenoids, although extremely interesting from an ecological and taxonomic point of view, is seldom provided although, when the organisms have been identified, it can easily be deduced. After going through a rather exhaustive examination of the literature, we made a selection of the cases in which, besides a report on the occurrence of a certain organism, estimates of its abundance were also provided. The information, for both planktonic and benthic environments, has been summarized in Table 1.

A. Planktonic Environments

The systems described are representative for all sorts of planktonic environments, and display a wide range of different physico-chemical variables. Habitats for phototrophic sulfur bacteria can be found ranging from freshwater, to marine or even hypersaline environments. Some of these have a karstic origin, while others are close to the sea shore and thus exposed to intrusions of sea water. From a preliminary examination of the cases listed in Table 1, no clear regularities can be readily deduced.

The depth at which phototrophic sulfur bacteria occur in lakes is a function of the depth of the oxic/anoxic interface. In most cases, the latter is a consequence of the morphometry of the water body, its exposure to wind, the degree of thermal stability and the surface to volume ratio (Hutchinson, 1957). The depth of the interface is likely to affect both the quantity and the quality of the light reaching the bacterial layer. Thus, certain organisms might be better adapted to the light climates predominant at higher depths, while others are better suited for growth closer to the surface. Since carotenoids are the main photosynthetic pigments responsible for light harvesting in planktonic environments, the different organisms listed in Table 1 were grouped according to their carotenoids. Their distribution with depth is represented in Fig. 3, in which the upper panel shows the occurrences reported of each carotenoid, and the lower panel displays their actual distribution with depth. In the case of purple sulfur bacteria, okenone seems to be the most frequent (36 cases), spirilloxanthin, lycopenal and rhodopinal occurring in a much lower number of cases (10, 7 and 4, respectively). In the case of purple bacteria, no clear pattern seems to emerge, except perhaps for the presence of lycopenal- and rhodopinal- containing organisms closer to the surface. However, the small number of cases involved does not allow to reach any firm conclusion. Within the green sulfur bacteria, a pattern which has already been reported and explained by Montesinos et al. (1983) emerges clearly. Green sulfur bacteria containing isorenieratene (brown-colored species) are usually found at deeper layers (average 14 m) than their chlorobactene containing counterparts (average 6 m).

The maximum abundance at which phototrophic sulfur bacteria are found in planktonic environments varies within several orders of magnitude. In Table 1 maximum abundance have been expressed either as the concentration of organisms or as the concentration of photopigments, in which BChl *a* and *b* correspond to purple sulfur bacteria and *c*, *d* and *e*, to green sulfur bacteria. In order to better visualize how often these bacteria are found at a certain abundance, the frequency distribution of the abundance has been plotted in Fig. 4. In the upper panel, apart from Mahoney Lake (British Columbia, Canada) in which population densities reach 4×10^8 cells·ml^{-1}, 50% of the cases lay in the interval between 2.4×10^4 and 4×10^6 cells·ml^{-1}. The lower panel shows the distribution of BChl which ranges from approximately 10 to 1000 μg·l^{-1}, with 50% of the cases being between 35 and 451 μg·l^{-1}. In some cases, abundance was available as the amount of BChl per square meter (Table 2). In some cases values between 1000 and 2000 mg BChl a·m^{-2} have been observed. These values, as will be discussed in section VII, are close

Table 1. Summary of the distribution and abundance of phototrophic sulfur bacteria in natural environments. Data have been sorted by lakes. Only reports giving an estimate of the abundance have been included

System	Organism	Depth (m)	BChl	μg BChl·l⁻¹	cells·ml⁻¹	Ref
PLANKTONIC						
Arcas–2	*Chromatium weissei*	9.25	a	100	1.0×10^4	(1)
Banyoles III	*Chromatium minus*	13	a	4	–	(2)
	Chlorobium phaeobacteroides	17	e	20	–	(2)
Banyoles IV	*Chromatium minus*	13	a	16	–	(2)
	Chlorobium phaeobacteroides	15	e	70	–	(2)
Banyoles VI	*Chromatium minus*	13	a	6	–	(2)
	Chlorobium phaeobacteroides	15	e	16	–	(2)
Belovod	*Chromatium okenii*	10–13	–	–	5.0×10^4	(3)
Biétri Bay	*Chromatium gracile*	3.5	a	65	–	(4,5)
	Chromatium violascens					
	Chromatium vinosum					
	Chlorobium vibrioforme	4	d	60	–	(4,5)
	Chlorobium phaeobacteroides	4	e	50	–	(4,5)
	Pelodictyon					
Big Soda Lake	*Ectothiorhodospira vacuolata*	21–22	a	150–200	–	(6,7)
Black Sea	*Chlorobium phaeobacteroides*	74	e	1	–	(8)
	Thiocapsa roseopersicina	160–200	–	–	2.3×10^5	(9)
Bol'shoi Kichier	*Ancalochloris perfilievii*	5.75	–	–	1.0×10^6	(28)
	Pelodictyon luteolum	6	–	–	4.0×10^6	(28)
Buchensee	*Chloronema giganteum*	8	d	15	–	(10)
Chernoe–Kucheer	*Chlorobium*	4	–	–	5.0×10^6	(29)
Chernyi Kichier	*Pelodictyon luteolum*	4	–	–	7.5×10^6	(28)
	Ancalochloris perfilievii	6	–	–	2.0×10^6	(28)
Cisó	*Chromatium*	2	a	450	7.6×10^5	(11,12)
	Amoebobacter	2.5	–	–	7.5×10^5	(11,12)
	Chlorobium	2–3	c, d, e	50	7.9×10^5	(11,12)
Cullera	*Chlorobium phaeovibrioides*	4	e	110	–	(13)
Deadmoose	–	–	a	1355	–	(14)
	Lamprocystis roseopersicina	9–9.2	a	120–250	–	(15,16)
Estanyol Nou	*Chromatium minus*	4	a	280	7.3×10^5	(17)
Faro	*Chlorobium phaeobacteroides*	12–15	–	–	4.0×10^6	(18)
Fayetteville Green	*Chlorobium phaeobacteroides*	18–20	–	–	1.2×10^6	(19)
Fidler	*Chlorobium limicola*	2.7–3	c	900	3.0×10^5	(30)
	Chlorobium	3	d	4000	4.0×10^6	(31)
Fish	*Thiopedia*	10	a	4–5	–	(20)
	Chlorobium	13	a, d	32	–	(21)
	Pelodictyon	11	d	13	–	(20)
Gek Gel	*Chlorobium phaeobacteroides*	30.5	–	–	3.5×10^6	(22)
Hamana	–	11	c	61	–	(23)
Haruna	–	13–14	a	24	–	(23)
Harutori	–	3.6	d, e	543	–	(23)
Holmsjön	*Chromatium* sp.	4	a	484	–	(24)
	–	4–4.5	d	4000	–	(24)
Kaiike	*Chromatium*	4.75–6	–	–	5.3×10^6	(32,33)
Kinneret	*Thiocapsa roseopersicina*	18	–	–	1.0×10^2	(25)
	Chlorobium phaeobacteroides	18	–	–	1.0×10^6	(25)
Kisaratsu	–	6.2	a	186	–	(23)
	–	6.2	d, e	828	–	(23)
	–	7	–	–	–	(23)
Knaack	*Clathrochloris*	1.75–2.5	c, d, e	550–575	–	(20,26)
	Chlorobium					
	Pelodictyon					
Konon'er	*Amoebobacter roseus*	10.75	–	–	4.0×10^6	(34)
Kuznechikha	*Chlorochromatium aggregatum*	6	–	–	2.6×10^5	(28)
	Chloronema giganteum					
Laguna de la Cruz	*Chromatium*	16	a	100	–	(27)
	Thiocapsa					
	Chlorobium	16	d, e	300	–	(27)
Lesnaya Lamba	*Pelodictyon clathratiforme*	4.3	–	–	2.0×10^5	(39)
	Pelodictyon luteolum					
	Chlorochromatium aggregatum					
	Chloronema giganteum					
Mahoney	*Amoebobacter purpureus*	6.68	a	20880	4.0×10^8	(40)
	–	7.5–8.3	a	7090	–	(41)
Mary	*Pelodictyon*	3.5	d	110	–	(21)
	Cathrochloris					
	Chlorobium					
Mirror	*Chlorobium*	10–10.5	a, d	145	–	(21,20)
	Lamprocystis	8.75	a	14	–	(20)
	Chromatium					
Myshin'er	*Amoebobacter roseus*	13.5	–	–	1.0×10^4	(28)
	Thiocapsa					
	Pelochromatium roseum					
Mittlerer Buchensee	*Amoebobacter purpureus*	9	a	50	–	(42)
	Pelodictyon phaeoclathratiforme	9	d, e	100	–	(42)
	Pelochromatium					
Mogilnoye	*Chlorobium phaeovibrioides*	9.5–10.5	–	–	2.4×10^7	(40)
	Pelodictyon phaeum					
	Prosthecochloris phaeoasteroidea	9.5–10.5	–	–	4.5×10^5	(40)

Table 1. Continued

System	Organism	Depth (m)	BChl	μg BChl·l^{-1}	cells·ml^{-1}	Ref
Mutek	*Chlorobium limicola*	4	c	184	–	(44)
Paul	*Prosthecochloris, Pelodictyon*	6.5	d	421	–	(21)
Peter	*Chlorobium*	10–14	a, d	11	–	(21)
Pluss–See	*Clathrochloris hypolimnica*			–	5.0×10^5	(45)
	Thiopedia rosea	5	–	–	5.0×10^4	(45)
	Pelochromatium roseum	6	–	–	5.0×10^4	(45)
Pomyaretskoe	*Chlorobium vibrioforme*	3.25	–	–	4.6×10^7	(46)
Popówka Mala	*Chlorobium limicola*	4	c	150	–	(47)
	Chlorobium limicola	4–6	c	538	–	(44)
Popówka Wielka	*Chlorobium limicola*	3.5	c	3410	–	(44)
Puddledock	*Chlorobium limicola*	4.5	c	410	–	(48)
	Thiopedia rosea	6	a	107	–	(48)
Repnoe	*Chlorobium phaeovibrioides*	5.75	–	–	3.5×10^7	(35)
Rose	*Pelodictyon,*	4	d	550	–	(21)
	Clathrochloris					
	Chlorobium					
Rotsee	*Pelochromatium roseum*	11.5	–	–	2.9×10^3	(49)
	Thiopedia rosea	7	–	–	3.1×10^5	(49)
	Lamprocystis roseopersicina	7.5–9.5	–	–	3.5×10^4	(49)
Schleinsee	*Chloronema giganteum*	6	d	34	–	(10)
Solar Lake	*Chromatium* sp.	2.5–2.8	a	20	–	(50)
	Chromatium violascens	1–3	–	–	1.0×10^6	(51)
	Prosthecochloris	3–4.5	–	–	2.0×10^6	(51)
Suga	–	8.6	a	120	–	(23)
	–	8.6	d, e	148	–	(23)
Suigetsu	*Chromatium*	7	–	–	900[a]	(36)
	–	8.5	a	81	–	(23)
	–	8.5	d, e	100	–	(23)
Vae San Juan	*Amoebobacter roseus*	10	–	–	2.2×10^4	(52)
	Pelodictyon phaeum	10	–	–	6.8×10^6	(52)
Vasikkalampi pond	*Chlorobium limicola*	4–5	c	300	–	(53)
	Pelodictyon luteolum					
	Chlorochromatium aggregatum					
Vechten	*Thiopedia*	6.2	a	3	50–150[b]	(37)
	Chloronema	6.2	d	15	7.0×10^3	(37)
	Chlorobium phaeobacteroides	7.6	e	20	–	(37)
	Chloronema giganteum	6	d	30	–	(10)
Veisovo	*Pelodictyon phaeum*	1.5	–	–	9.0×10^6	(38)
Vilar	*Chromatium*	4.2	a	900	3.4×10^6	(11)
	Chlorobium	4.4	c, d, e	200	3.9×10^6	(11)
Wadolek	*Chlorobium limicola*	6	c	400	–	(54)
Waku–Ike	–	2.5–3	c	79.5	–	(23)
Waldsea	*Chlorobium* sp.	7.8	d	2325	–	(14)
	Chlorobium sp.	7.9	d	2325	–	(16)
Zaca	*Thiopedia rosea*	6–8	a	203	–	(55)
System	**Organism**	**Depth (mm)**	**BChl**	**μg BChl·l^{-1}**	**cells·ml^{-1}**	**Ref**
BENTHIC						
Ebro Delta	*Chromatium gracile*	2.5–4	a	150–467	–	(56)
	Thiocapsa roseopersicina					
	Chromatium sp.					
Mellum	*Thiocapsa roseopersicina*	0–1.3	a	125–310	–	(57)
	Chromatium					
	Thiopedia					
	Ectothiorhodospira					
Scapa Bay	*Thiocapsa roseopersicina*	0–5	a	–	2.9×10^6	(58)
	Thiocapsa roseopersicina	2–2.5	a	900	–	(58)
Schiermonnikoog	*Thiocapsa roseopersicina*	3–3.5	a	100	–	(59)
Sippewissett (layer 3)	*Thiocapsa roseopersicina*	3–4	a	242	–	(60)
	Chromatium					
	Thiocystis					
Sippewissett (layer 4)	*Thiocapsa pfennigii*	5–6	b, c	52, 114	–	(60)
	Prosthecochloris aestuarii					
Sippewissett (layer 5)	*Prosthecochloris aestuarii*	6–7	b, c	40, 117	–	(60)
	Thiocapsa pfennigii					
Swanbister Bay	*Thiocapsa roseopersicina*	0–3	a	160–500	–	(61)
	Thiocapsa roseopersicina	0–5	a	–	2.4×10^7	(58)
Texel	*Thiocapsa roseopersicina*	0–5	a	–	1.0×10^7	(62)
	Thiocapsa roseopersicina	2.5–5	a	440	–	(63)

[a] mg C·m^{-3}·d^{-1}, [b] platelets·ml^{-1}, [c] most probable number counts, (1) Vicente et al.1991, (2) Garcia–Gil and Abellà1992, (3) Sorokin 1970, (4) Caumette et al. 1983, (5) Caumette1984, (6) Cloern et al. 1983a, (7) Cloern et al. 1983b, (8) Repeta et al. 1989, (9) Dickman and Artuz1978, (10) García–Gil and Abellà 1992b, (11) Guerrero et al.1985, (12) Pedrós–Alió and Guerrero, 1993, (13) Miracle and Vicente1985, (14) Lawrence et al. 1978, (15) Parker and Hammer 1983, (16) Parker et al.1983, (17) Abellà et al.1985, (18) Sorokin and Donato 1975, (19) Culver and Brunskill 1969, (20) Parkin and Brock 1980b, (21) Parkin and Brock 1980a, (22) Dubinina et al. 1973, (23) Takahashi and Ichimura 1968, (24) Lindholm et al. 1985, (25) Bergstein et al. 1979, (26) Parkin and Brock 1981, (27) Vicente and Miracle 1988, (28) Gorlenko et al. 1983, (29) Kuznetsov 1970, (30) Baker et al. 1985, (31) Croome 1986, (32) Matsuyama 1981, (33) Matsuyama and Shirouzu 1978, (34) Gorlenko and Kusnezow 1972, (35) Gorlenko et al. 1973, (36) Matsuyama 1978, (37) Steenbergen et al. 1989, (38) Gorlenko et al. 1974b, (39) Dubinina and Kuznetsov 1976, (40) Overmann et al. 1991, (41) Hall and Northcote 1990, (42) Overmann and Tilzer 1989, (43) Gorlenko et al. 1978, (44) Czeczuga 1965, (45) Anagnostidis and Overbeck 1966, (46) Gorlenko et al. 1974a, (47) Czeczuga 1966, (48) Banens 1990, (49) Kohler et al. 1984, (50) Jørgensen et al. 1979, (51) Cohen et al. 1977, (52) Romanenko et al. 1976, (53) Eloranta 1985, (54) Czeczuga 1968, (55) Folt et al. 1989, (56) Mir et al. 1991, (57) Stal et al. 1984, (58) Van Gemerden et al. 1989a, (59) De Wit et al. 1989, (60) Pierson et al. 1987, (61) Van Gemerden et al. 1989b, (62) Visscher and Van Gemerden 1991a, (63) Visscher et al. 1990

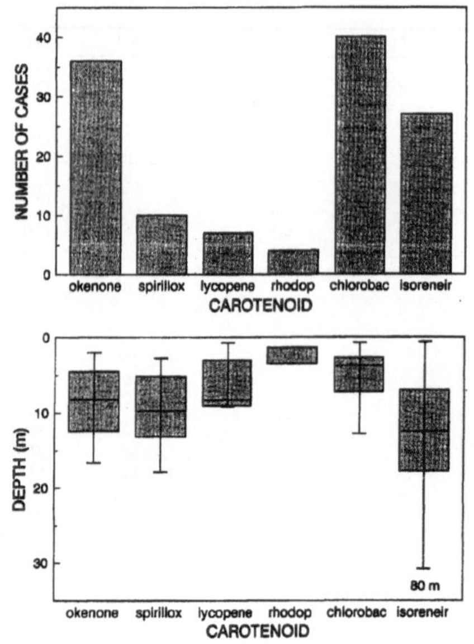

Fig. 3. Occurrence of blooms of phototrophic bacteria containing different carotenoids. The upper panel shows the number of cases in each category, the lower panel shows the depth of the bloom in the water column. The box encloses the depth interval at which 50% of the cases are found, the vertical bar indicates the upper and lower limits of the depth distributions, the horizontal line inside the box indicates the average depth of all cases. Based on literature data presented in Table 1.

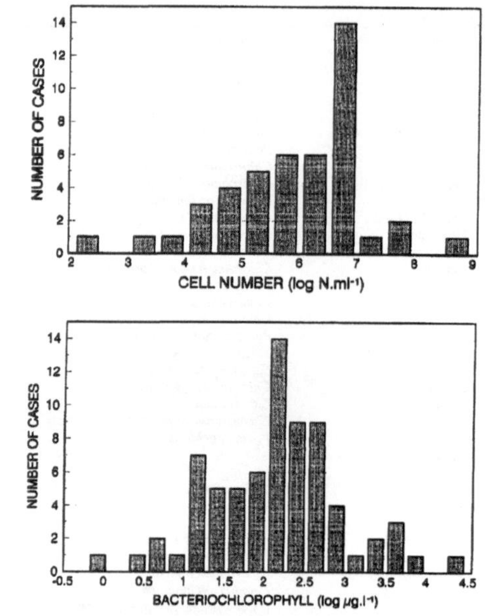

Fig. 4. Frequency distribution of cell number (upper panel) and bacteriochlorophyll concentration (lower panel) in blooms of phototrophic sulfur bacteria. Based on literature data presented in Table 1.

to the theoretical maximum for the biomass of phototrophic sulfur bacteria.

B. Benthic Environments

Most of the cases collected in Table 1 describing the presence of phototrophic sulfur bacteria in benthic environments correspond to sandy sediments close to the seashore, either beaches or tidal flats. The presence of these organisms has also been described in other benthic microbial communities such as those occurring in hypersaline lagoons like Laguna Figueroa (Stolz, 1983, 1990) or Laguna Guerrero Negro (Javor and Castenholz, 1981, 1984; Jørgensen and Des Marais, 1986b; D'Amelio et al, 1989), or in hot springs (Castenholz, 1984, 1988; Madigan, 1984).

Due to the rapid extinction of light in the sediments, development of phototrophic sulfur bacteria is usually limited to the upper 5 millimeters. The predominant organism in most systems seems to be *Thiocapsa roseopersicina* which can reach between 10^6 and 10^7 cells·cm^{-3}. Since these values correspond to most probable number viable counts, the actual numbers might be somewhat higher. Concentrations of bacteriochlorophyll lay in most cases between 100 and 500 μg·cm^{-3}. However, the maximum concentration found might depend to some extent on the resolution of the sampling, with finer resolution allowing for the detection of more narrow and concentrated layers.

IV. Contribution of Phototrophic Sulfur Bacteria to Primary Production

Besides the abundances mentioned in the previous section, some studies also provide data on CO_2 fixation. In some cases, the data have been integrated and are provided per square meter on a daily basis. In very few instances, the contribution to primary productivity is available on a yearly basis. The available rates (μg C·m^{-3}·d^{-1}) are shown in the upper panel of Fig. 5. Although the maximum value of 5750 μg C·m^{-3}·d^{-1} measured by Wetzel (1973) in Smith's Hole constitutes the highest measure reported

Table 2. Integrated amount of bacteriochlorophyll in diferent environments

System	mg BChl·m^{-2}	Ref
Big Soda Lake	300-480	(1)
Lake Cisó	1000-2000	(2)
Lake Hamana	72	(3)
Lake Haruna	36	(3)
Lake Harutori	587	(3)
Kisaratsu reservoir	25-352	(3)
Knaack Lake	300	(4)
Mahoney Lake	780-1050	(5,6)
Lake Suga	79-119	(3)
Lake Suigetsu	138-182	(3)
Lake Waku-Ike	235	(3)
Lake Waldsea	698	(7)
Scapa Beach	2000	(8)

(1) Cloern et al. 1983, (2) Mas et al. 1990, (3) Takahashi and Ichimura 1968, (4) Parkin and Brock 1981, (5) Overmann et al. 1991, (6) Hall and Northcote 1990, (7) Lawrence et al 1978, (8) Van Gemerden et al. 1989

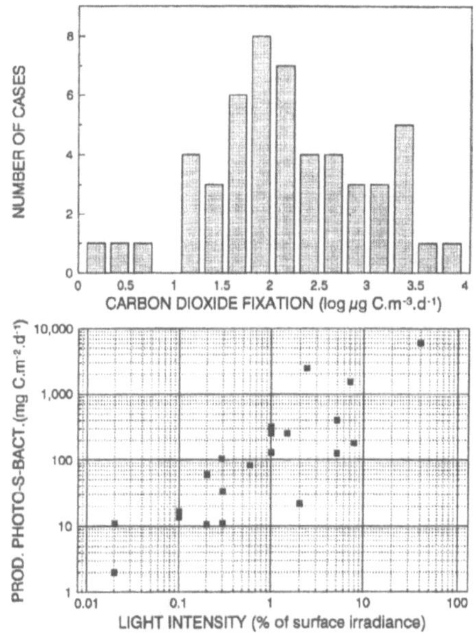

Fig. 5. Frequency distribution of the rate of CO_2 fixation in blooms of phototrophic sulfur bacteria (upper panel) and the productivity of these organisms in relation to the light intensity in % of the surface irradiance reaching the depth of the bloom (lower panel). Based on literature data presented in Table 1.

until now, the distribution in Fig. 5 seems to indicate that maximum rates usually are much lower, approximately between 50 and 500 μg C·m^{-3}·d^{-1}.

In order to have a better idea of the extent to which phototrophic sulfur bacteria contribute to the carbon cycle of their environments, the values available on their total annual production have been compiled in Table 3. The data indicate that phototrophic sulfur bacteria indeed can play a very active role as primary producers in some environments. However, since the occurrence of these organisms in anoxic layers containing hydrogen sulfide probably restricts the role of aerobic grazers, it is not clear whether the carbon they fix will have any impact on the overall trophic structure of their environments. Furthermore, measurements of the content of stable carbon and sulfur isotopes in consumers of several meromictic lakes containing abundant populations of phototrophic sulfur bacteria seem to rule out the hypothesis of a food web based mainly on bacterial primary production (Fry, 1986).

Because photosynthetic activity depends on light, and in situ irradiances decrease with depth and with the concentration of algae in the surface layers, production by phototrophic sulfur bacteria will be high both in shallow lakes (Lake Cisó, Smith Hole) and in lakes where, although deeper, contain little algae (Fayetteville Green Lake). A good correlation seems to exist between the fraction (in %) of the total primary production due to phototrophic sulfur bacteria, and the amount of light reaching the anoxic layers (Parkin and Brock, 1980a). A similar correlation can be established between the total amount of CO_2 fixed through anoxygenic photosynthesis by these organisms (per square meter and per day) and the amount of light (expressed as % of the surface irradiance) reaching the layer of phototrophic bacteria (Montesinos and Van Gemerden, 1986)(Fig. 5, lower panel). The good correlation observed suggests that light is indeed the main factor controlling primary production in blooms of phototrophic sulfur bacteria. Extrapolation of a regression line to 100% of surface irradiance actually indicates the existence of an upper limit for the production of phototrophic sulfur bacteria of approximately 10^4 mg C·m^{-2}·d^{-1}.

This relationship, however, does not necessarily indicate that phototrophic bacteria are limited by light in all the cases included in the graph, it only indicates the existence of an otherwise obvious stoichiometry between light absorption and carbon dioxide fixation.

Table 3. Annual productivity of purple and green sulfur bacteria in various lakes

System	Organism(s)	Productivity ($g\ C{\cdot}m^{-2}{\cdot}y^{-1}$)	Productivity (% of total)	Reference
Fayetteville Green Lake (USA)	*Chlorobium phaeobacteroides*	239	83	(1)
Smith Hole (USA)	*Chromatium* sp.	35	50	(2)
Waldsea (Canada)	*Chlorobium* sp.	32	46	(3, 4)
Deadmoose Lake (Canada)	*Lamprocystis roseopersicina*	14	17	(4, 5)
Big Soda Lake (USA)	*Ectothiorhodospira vacuolata*	50	10	(6)
Knaack Lake (USA)	*Pelodictyon* sp.	17	5	(7)
	Clathrochloris sp.			
Mittlerer Buchensee	*Amoebobacter purpureus*	–	4	(8)
(Germany)	*Pelodictyon phaeoclathratiforme*			
Lake Vechten (The Netherlands)	*Chloronema* sp.	6	4	(9)
	Thiopedia sp.			(10)
	Chromatium sp.			
Vae San Juan (Cuba)	*Amoebobacter roseus*	14	3	(11)
	Pelodictyon phaeum			
Lake Cisó (Spain)	*Amoebobacter* sp.	56	25	(12)
	Chromatium minus			

(1) Culver and Brunskill 1969, (2) Wetzel 1973, (3) Lawrence et al. 1978, (4) Parker et al. 1983, (5) Parker and Hammer 1983, (6) Cloern et al. 1983, (7) Parkin and Brock 1981, (8) Overmann and Tilzer 1989, (9) Steenbergen 1982, (10) Steenbergen and Korthals, (11) Romanenko et al. 1976, (12) Garcia-Cantizano 1992,

V. Growth Rates in Nature

In situ specific growth rates of phototrophic sulfur bacteria have been determined following two different approaches. In one of them, estimates are based on measurements of carbon dioxide fixation which are converted into specific growth rates using the following equation (Tilzer, 1984):

$$\mu = \ln\left(1 + \frac{P}{B}\right) \qquad (1)$$

in which P is the rate of carbon dioxide fixation and B is the biomass expressed as carbon. When measurements are carried out at different depths within the lake, estimates of μ obtained using this approach constitute a good indication of how the organism responds to the specific set of environmental conditions prevailing at each depth. Since this procedure is based on the assumption that all carbon fixed results in growth, it tends to overestimate the growth rate of organisms which synthesize storage compounds.

Production-based estimates of the specific growth rates described in the literature are rather close to the maximum specific growth rates measured in the laboratory (0.050–0.150 h^{-1}). Thus, Montesinos (1987) found a specific growth rate of 0.074 h^{-1} at the top of the bacterial layer in Lake Cisó where, although biomass was rather low, light irradiance was 850 $\mu E{\cdot}m^{-2}{\cdot}s^{-1}$. Similar measurements carried out also in Lake Cisó (Garcia-Cantizano, 1992) give specific growth rates between 0.094 and 0.144 h^{-1}. The much lower values reported for bacteria in the hypolimnion of Zaca Lake (0.0030–0.0005 h^{-1}) (Folt et al., 1989) can be attributed to the poor light conditions prevailing or, as the authors suggest, to the failure to locate the layer of maximum activity at the top of the bacterial plate.

A second approach for measuring growth rates in situ requires a detailed follow up of the biomass levels during an extended period of time, together with a careful analysis of the main loss processes such as sedimentation or predation which can effectively decrease population levels (Tilzer, 1984; Mas et al., 1990). This kind of studies assume that net biomass changes occur due to the combined effect of growth and a number of different loss processes. The overall growth balance is described by the following equation:

$$\frac{dB}{dt} = \left(\mu - \sum_{i=0}^{n} k_i\right) \cdot B \qquad (2)$$

in which k_i are the specific loss rates of the n different factors taken into account. The measurement of both growth and losses requires samples spaced by at least several days, and vertical movement of the interface and the microorganisms during this interval can occur. Therefore, it is extremely difficult to carry out this kind of study for different layers within the bacterial plate, and measurements usually refer to the population as a whole.

Specific growth rate estimates based on the analysis of population dynamics are much lower than their production based counterparts. The analysis of the dynamics of phototrophic sulfur bacteria in Lake Cisó reveals that during most of the time loss rates are unimportant (0.025 and 0.015 d^{-1}) and that population growth rates are also very low (0.063 and 0.037 d^{-1}), compensating for what little losses occur and maintaining a constantly high biomass (1–2 g BChl $a{\cdot}m^{-2}$) (Mas et al., 1990). High population growth rates are only observed when biomass levels decline after periods of intense washout which occurs approximately once a year, during late winter/early spring. Recovery of the initial biomass levels occurs before summer in a pattern which can be observed periodically during several years (Fig. 6, upper panel). In a similar study Montesinos and Esteve (1984) estimated population growth rates from the increase in total biomass during summer stratification in four Spanish lakes. The data obtained (0.008, 0.012, 0.019 and 0.024 d^{-1}) were somewhat lower, probably because the negative effect of loss processes had not been taken into account.

Considering the facts that the maximum values found (0.063 d^{-1}) are about 40 times lower than the maximum rates at which the organisms can grow ($\approx$ 2.4 d^{-1}) and that carbon dioxide fixation based estimates give values close to the maximum growth rates of the organisms, it appears that only a fraction of the population is growing at a certain moment. Measurement of the specific rates of CO_2 fixation at different depths within the bacterial plate certainly indicate that the bacteria in the top few centimeters of the plate are growing (Van Gemerden et al., 1985; Montesinos, 1987) while the cells in the remainder of the plate stay in a non-active yet viable state.

Determinations of photosynthetic activity and specific growth rates in microbial mats have been hampered by the technical problems involved. Some attempts have been made to measure photosynthetic rates in sediment slurries (Pierson et al., 1987), but besides showing that photosynthesis in the lower layers was saturated at lower irradiances that in the upper layers, interpretation of the results is difficult. Using a completely different approach Van Gemerden et al. (1989a) scrapped the upper layer of a *Thiocapsa* mat in the beach of Scapa Bay, and followed the recolonization of the exposed surface with time (Fig. 6, lower panel). The results indicate that growth went on at a specific rate of 0.306 d^{-1}, leveling up at a biomass of approximately 2 g of BChl $a{\cdot}m^{-2}$.

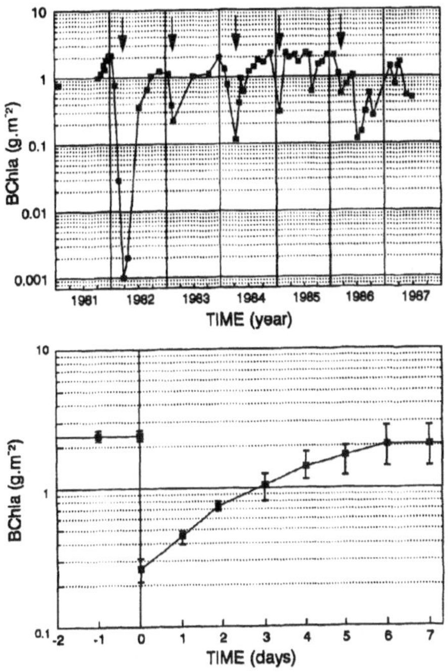

Fig. 6. Integrated values of BChl *a* in $g{\cdot}m^{-2}$ in Lake Cisó, Spain (upper panel). Growth, indicated for each year by an vertical arrow, is apparent only after the population density has been decreased during the winter months. During the summer, losses are balanced by growth resulting in rather constant population densities. A similar phenomenon can be observed when the population of purple sulfur bacteria on Scapa Beach, Orkney Islands, Scotland, has been removed artificially. Within a week the population density is restored. Redrawn from Mas et al (1990, upper panel) and Van Gemerden et al. (1989b, lower panel).

VI. Loss Processes

From a dynamic point of view, population levels are the consequence of an equilibrium between growth and losses. In planktonic environments losses by bacterial death, predation and sedimentation seem to play a prominent role, which in a few cases has been quantified. In benthic environments identical or similar processes are likely to occur although the studies carried out so far do not provide any quantitative data on their influence on the different populations of bacteria.

A. Death

Loss of activity and subsequent death can occur in populations of phototrophic sulfur bacteria exposed to adverse conditions during extended periods of time. One of the factors often summoned is exposure to oxygen (Dickman and Artuz, 1978; Folt et al., 1989), which is lethal for all green sulfur bacteria as well as for some of the large-celled Chromatiaceae. In stratified planktonic environments, exposure to oxygen occurs only during mixing. In contrast in benthic communities, in which the oxic/anoxic interface fluctuates, organisms experience a daily exposure to oxygen.

Leaving aside the deleterious effects of oxygen, the main cause of viability loss is insufficient supply of energy. When irradiance falls below a certain threshold level, phototrophic sulfur bacteria are unable to gather enough energy to fuel vital processes such as turnover of macromolecules, osmotic regulation or control of their internal pH. As a consequence, the organisms lose viability, and eventually die. Observations carried out in the hypolimnion of Lake Cisó show that, while cells in the top and peak of the bacterial plate are 100% viable, viability at the bottom of the plate decreases rapidly with depth (Van Gemerden et al., 1985). The decrease of viability is actually a function of the time the organisms have been in the dark. Given that the loss of viability occurs exponentially with depth at a rate of $-0.0142\ cm^{-1}$, and assuming that these bacteria were sinking at a rate of approximately $1\ cm \cdot d^{-1}$ (Pedrós-Alió et al., 1989), a death rate of $-0.014\ d^{-1}$ can be calculated. This value compares well with the maximum decomposition rates of $-0.016\ d^{-1}$ estimated for *Chromatium* (Mas et al., 1990).

B. Predation

Since phototrophic sulfur bacteria live in anoxic environments containing hydrogen sulfide, and this compound is toxic for most of the potential predators, it is generally assumed that predation does not play a very important role as a loss process. Yet, the presence of crustaceans or rotifers on top of the oxic/anoxic interface feeding on the phototrophic bacteria underneath, has been repeatedly reported (See Table 3). In several cases the existence of grazing was suggested by the red color of the zooplankton which was placed right on top of the bacterial layer. In some cases, the existence of grazing was confirmed either through experiments involving uptake of radioactively labeled bacteria (Takahashi and Ichimura, 1968; Sorokin, 1970; Matsuyama and Shirouzu, 1978) or analyzing the gut content of the predators (Caumette et al., 1983). None of these studies, however, provided any indication of the actual impact of this feeding on dynamics of the microbial community.

A quantitative study carried out in Crawford Lake (Mazumder and Dickman, 1989) showed feeding rates for the major predator, *Daphnia*, of 0.3 – 1.6% of the prey population per hour. Specific loss rates calculated per hour range between –0.003 and $-0.016\ h^{-1}$. Extrapolation of these rates to a 24 h period would yield values between –0.072 and $-0.384\ d^{-1}$. In a more recent study, Massana et al. (1994) determined the impact of predation on the microbial community of Lake Cisó during a bloom of *Daphnia pulex*. Their results clearly indicate that *Daphnia* feeds mainly on the microaerophilic protist *Cryptomonas phaseolus* and only marginally on phototrophic bacteria due to the presence of hydrogen sulfide which somehow limits the vertical distribution of *Daphnia* to the oxic layers. Even though phototrophic bacteria constituted only a small percentage of the diet of this organism, estimated loss rates for *Chromatium* and *Amoebobacter* were $-0.092\ d^{-1}$ and $-0.069\ d^{-1}$, respectively. These values are high compared to previously observed population growth rates in the same lake (Mas et al., 1990) but could explain the specific rates of biomass decrease ($-0.100\ d^{-1}$ and $-0.041\ d^{-1}$) observed during the period of the bloom (Massana et al., 1994). The study of the composition of stable sulfur and carbon isotopes in the different trophic components of several meromictic lakes containing phototrophic sulfur bacteria indicates that these organisms do not contribute substantially to the carbon and energy flow of those systems, again suggesting a marginal role as a food source for the zooplankton living there.

While the vertical distribution of macroscopic zooplankton is severely limited to the oxic layers, several groups of ciliates have been described which are well adapted to life in anoxic environments. These organisms have been found in coexistence with phototrophic sulfur bacteria (Dyer et al., 1986) and can be considered as potential predators (Table 4). Indications of grazing by ciliates on phototrophic sulfur bacteria come from the early work of Sorokin and Donato (1975), who observed the existence of red digestive vacuoles in several ciliates. More recently, Finlay et al. (1991) studied the anaerobic ciliates present in Lake Arcas-2 and estimated, on theoretical grounds, a consumption of approximately 6% of the prey biomass per day (equivalent to $-0.062\ d^{-1}$). In a similar study carried out in Lake Cisó (Massana and Pedrós-Alió, 1994) a much lower impact (0.1% of the prey per day) was found, as a consequence of the low numbers of ciliates present in that environment. The specific loss rate calculated from the feeding rates ($-0.001\ d^{-1}$) is too low to have any effect on the population dynamics of the prey.

The ability to prey on phototrophic sulfur bacteria is not limited to eukaryotic organisms and, in fact, a couple of prokaryotic organisms have been described (Guerrero et al., 1986) which seem to attack purple sulfur bacteria, forming lytic plaques (Esteve et al., 1992). One of them, an epibiont named *Vampirococcus*, attaches to the surface of the prey where it carries out its division. Electron micrographs suggest that during the process the cytoplasm of the infected cell undergoes major degradation (Esteve et al., 1983). The second type described, *Daptobacter*, enters the cytoplasm of the prey and divides inside. As in the previous case, the prey is killed. The presence of epibionts has been detected in several blooms of phototrophic sulfur bacteria in which they can infect a considerable fraction of the population (15–94%)(Table 4). In the upper parts of the bacterial plate where cells are active and viability is high, the percentage of infected cells is very low, virtually zero. In the layers below, where no light is available and viability actually decreases with depth, the percentage of infected cells increases (Guerrero et al., 1986). This seems to suggest that these bacteria are not operative at infecting healthy active individuals, but rather attack cells which are somehow impaired due to prolonged exposure to a limited energy supply. Thus, because they are preying on the non-reproductive part of the population, their impact on population dynamics is likely to be small. A different hypothesis has been put forward by Clarke et al. (1993) for the epibiont they found in Arcas-2. Since the organism does not form lytic plaques and attaches to healthy cells, they suggest that it actually grows heterotrophically on the organic carbon excreted by the phototrophs.

C. Sedimentation in Planktonic Environments

Sedimentation is one of the main mechanisms through which planktonic organisms are removed from the water column. The extent to which sedimentation affects population levels depends in fact on the sinking speed (v)of the organism which, according to Stoke's Law

$$v = \frac{2}{9} \cdot g \cdot r^2 \cdot (\rho_{Cell} - \rho_{H_2O}) \cdot \frac{1}{\eta} \qquad (3)$$

is a function of its size (*r*)(expressed as the radius of a sphere equivalent in volume to the cell), the difference between the density of the cell (ρ_{Cell}) and the density of the surrounding water (ρ_{H_2O}), and the viscosity of water (η).

Field measurements carried out during several months on the volume and density of the two main phototrophic sulfur bacteria present in Lake Cisó (Pedrós-Alió et al., 1989) allowed for calculations of the range of sinking speeds these organisms can experience. In the case of *Chromatium*, density remained rather constant between 1.13 and 1.16 $pg \cdot \mu m^{-3}$ while cell volumes ranged between 31 and 58 μm^3. *Amoebobacter* (referred to as *Lamprocystis* in the paper) changed its density between 1.07 and 1.13 $pg \cdot \mu m^{-3}$ and its volume (in this case the volume of multicellular aggregates) between 31 and 97 μm^3. Because sinking speeds, density, and viscosity of the water are affected by temperature, calculations were performed at the maximum and minimum temperatures (4 and 20 °C) measured in the lake. The calculations predicted maximum sinking speeds of 11.5 $cm \cdot d^{-1}$ for *Amoebobacter* and 18.4 $cm \cdot d^{-1}$ for *Chromatium*, while minimum sinking speeds were 1.6 and 6 $cm \cdot d^{-1}$ respectively. Actual measurements of the sinking speeds of the two organisms using sedimentation traps during a two year period (Pedrós-Alió et al., 1989) gave values of less than 1 $cm \cdot d^{-1}$, lower than the minimum predicted by Stokes' Law.

Table 4. Principal organisms reported to prey on populations of phototrophic sulfur bacteria

Lake	Predators	Abundance[a]	Observations	Reference
CRUSTACEANS AND ROTIFERS				
several lakes	Copepods	-	red color due to grazing on phototrophic bacteria	(1)
			uptake of ^{14}C labeled bacteria	
Fayetteville Green Lake	*Daphnia*	2-3	red color due to grazing on phototrophic bacteria	(2)
Lake Kaiike	*Acartia*	4	uptake of ^{14}C labeled *Chromatium*	(3)
	Oithona	727		
	Paracalanus	13		
	Brachionus	1126		
Biétri Bay	*Acartia*	8	Analysis of the gut content indicates feeding	(4)
	Brachionus	170	on phototrophic bacteria	
	Cyclopoids	6		
Lake Belovod	*Daphnia*	122-190	migration following the bacterial plate	(5)
	Copepods	30-56	red color due to grazing on phototrophic bacteria	
			uptake of ^{14}C labeled *Chromatium* cells	
Lake Cisó	*Daphnia*	80-130	red color due to the presence of hemoglobin	(6)
			8.8% of *Chromatium* consumed per day ($-0.092\ d^{-1}$)[b]	
			6.6% of *Amoebobacter* consumed per day ($-0.069\ d^{-1}$)[b]	
Crawford Lake	*Daphnia*	15	0.3-1.6% of prey consumed per hour	(7)
			equivalent to (-0.003 to $-0.016\ h^{-1}$)	
Devil's Hole	-	-	red color due to grazing on phototrophic bacteria	(8)
CILIATES				
Lake Faro	*Urostile* sp.	1 mg $ww \cdot l^{-1}$	red digestive vacuoles	(9)
	Trachelostyle			
Cisó	*Plagiopyla*	< 2000	0.1% prey consumed per day ($-0.001\ d^{-1}$)[b]	(10)
	Metopus			
Arcas-2	*Caenomorpha*	50000	6% prey consumed per day ($-0.062\ d^{-1}$)[b]	(11)
	Lacrymaria		theoretical estimate 0.4 $gC.m^{-2}.d^{-1}$	
	Lagynus			
Priest Pot	*Caenomorpha*	10000	*Caenomorpha* follows prey distribution	(12)
PREDATORY BACTERIA				
Cisó	epibiont	-	41% of prey infected[d]	(13,15)
Estanya	epibiont	-	60% of prey infected[c]	(14)
Estanya	epibiont	-	94% of prey infected[d]	(15)
Arcas-2	epibiont	-	15% of prey infected[d]	(16,17)

[a]Abundance expressed in number of individuals per liter, [b]Specific loss rates calculated from the feeding rate, [c]Estimates based on scanning electron microscope observations, [d]Estimates based on microscopy counts. (1) Takahashi and Ichimura, 1968, (2) Culver and Brunskill, 1969, (3) Matsuyama and Shirouzu, 1978, (4) Caumette et al., 1983, (5) Sorokin, 1970, (6) Massana et al., 1994, (7) Mazumder and Dickman, 1989, (8) Goehle and Storr, 1978, (9) Sorokin and Donato, 1975, (10) Massana and Pedrós-Alió, 1994, (11) Finlay et al., 1991, (12) Guhl and Finlay, 1993, (13) Esteve et al., 1983, (14) Esteve et al., 1992, (15) Guerrero et al., 1986, (16) Vicente et al., 1991, (17) Clarke et al., 1991

Sedimentation did occur at higher rates during certain periods, but never exceeded 10 $cm \cdot d^{-1}$. Sedimentation loss rates determined during the same period (Mas et al., 1990) gave maximum values of $-0.0149\ d^{-1}$ for *Chromatium* and $-0.0150\ d^{-1}$ for *Amoebobacter*. Similar measurements carried out in Lake Vechten (Steenbergen et al., 1987) gave higher sinking speeds (16–27 $cm \cdot d^{-1}$) corresponding to a rather high specific loss rate of $-0.22\ d^{-1}$.

Formation of large aggregates would actually increase theoretical sinking speeds by several orders of magnitude. It is a rather common observation that field samples containing dense populations of phototrophic sulfur bacteria form macroscopic aggregates when exposed to oxygen, which subsequently sediment to the bottom of the container in a few minutes. Thus, oxygenation of the layers where the populations of purple sulfur bacteria reside might bring up a similar phenomenon which would result in a rapid clearing of the water column. In the case of organisms containing gas vesicles, aggregate formation might constitute a mechanism to increase the speed at which they can position themselves at their optimal depth. For individual cells possessing flagella, the tendency to sink can be easily overcome through swimming. The swimming speeds reported by *Chromatium* range between 20 and 30 $\mu m \cdot s^{-1}$ (Matsuyama, 1987; Mitchell et al., 1991), meaning that individual cells could move between 170 and 260 $cm \cdot d^{-1}$. These values probably constitute an overestimation due to the random component of bacterial movement; however it indicates clearly that motile purple sulfur bacteria can easily overcome their tendency to sink.

VII. Growth Balance, Maximum Biomass and Production in Phototrophic Communities

In both the studies carried out in Lake Cisó (Mas et al., 1990) and in Scapa Beach (Van Gemerden et al., 1989a), apparent growth slows down to a virtual stop at between 1 and 2 g of BChl $a \cdot m^{-2} \cdot s^{-1}$. This observation, together with the systematic finding of lower integrated values of BChl in other systems (see Table 2), strongly suggests that these values actually constitute an upper limit for the biomass of phototrophic sulfur bacteria in nature.

The existence of an upper limit for the biomass of phototrophic sulfur bacteria can be predicted on theoretical grounds from the existence of an upper limit for phototrophic production discussed in section IV. Steady levels of biomass indicate that population growth rates match loss rates. An educated guess of the loss rates of phototrophic sulfur bacteria, based on the estimates reported in the previous section, is $\approx -0.1\ d^{-1}$ (corresponding to a half live time, in the absence of growth, of approximately one week). Maximum production based on Fig. 5 is 10 g $C \cdot m^{-2} \cdot d^{-1}$. Substituting the values for μ and production in equation 1, results in an apparent upper limit for biomass of 95 g $C \cdot m^{-2}$. Considering that the carbon content of the dry weight is approximately 50% (Van Gemerden, 1968; Göbel, 1978) and that the average content of BChl is approximately 2% of the dry weight (Takahashi and Ichimura, 1968), maximum levels of integrated BChl turn out to be approximately 4 g $BChl \cdot m^{-2}$. This value could be somewhat higher in environments, either planktonic or benthic, subject to lower loss rates. On the contrary, environments where loss rates are higher, or where bacteria are not close to the surface and thus, irradiance is not 100%, will only be able to support much lower biomass levels.

VIII. Environmental Factors Affecting Growth and Survival of Phototrophic Sulfur Bacteria

A. Light

1. In situ Light Climate

Development of phototrophic sulfur bacteria in nature requires the presence of light. At the same time, due to their anaerobic nature, these organisms are usually constrained to live in the anoxic parts of lakes or sediments, at depths where light penetration is severely hampered and irradiance is actually very low, on the order of a few $\mu Einstein \cdot m^{-2} \cdot s^{-1}$. On top of that, the spectral composition of this light is dramatically modified due to specific absorption of certain wavelengths by water itself, and very often by populations of algae or other phototrophic bacteria positioned above. In planktonic systems the infrared and UV part of the spectrum are rapidly attenuated due to absorption and scattering, while in sediments, infrared radiation penetrates deeper. This phenomenon is clearly illustrated in Fig. 7, using data from Lake Kinneret (upper panel) (Dubinsky and Berman, 1979) and from some sediments in San Francisco Bay (lower panel)(Jørgensen and Des Marais, 1986a). While in Lake Kinneret, as depth increases light is enriched in a narrow band around 550 nm, in the sediment, at a depth of 1 mm, most of the light consists of wavelengths above 850 nm. Similar data available for several other lakes and sediments and collected in Table 5 indicate a similar behavior. The higher percent of infrared light penetrating in sediments seems to be a consequence of both a high loss of blue wavelengths during the reflection in the particles of sediment and absorption by oxygenic phototrophs and little losses in the near infrared due to negligible absorption by water (Jørgensen and Des Marais, 1986a, 1988). The consequence of this differential absorption of light in lakes and in sediments is that, while organisms living in lakes have to rely mainly on their carotenoids for light harvesting, organisms in mats are probably forced to use the infrared light which is absorbed mainly by their BChl. Although it is not clear whether some organisms are better adapted than others to grow under either set of conditions, the composition of the light spectrum is a factor which will certainly have to be taken into account in order to explain the dominance of certain organisms in mats and in sediments. Thus, while okenone seems to be very

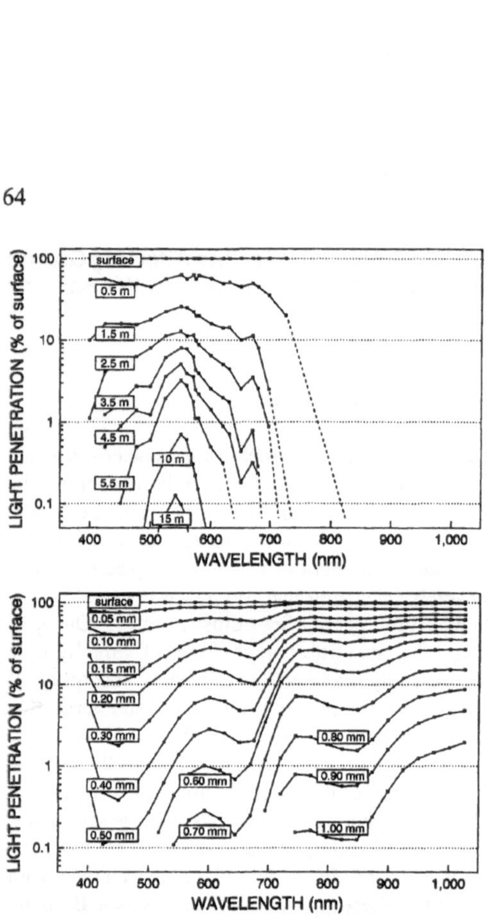

Fig. 7. Penetration of light of different wavelengths in a lake and in a sediment. The upper panel shows the spectral composition of light at increasing depths in Lake Kinneret, the dotted lines are based on patterns observed in various lakes. Depth horizons exceeding 10 m receive light of a very narrow wavelength range (500-600 nm). The lower panel shows the penetration of light in sediment of San Francisco Bay. Compared to lakes, light penetrates much less, and the spectral composition is shifted to the red part of the spectrum. Redrawn from Dubinsky and Berman, 1979 (upper panel) and Jørgensen and Des Marais, 1986 (lower panel).

abundant in aquatic environments (63% of the cases) benthic systems seem to be almost exclusively colonized by *Thiocapsa roseopersicina*, which contains spirilloxanthin. We have not been able to find any reference in the literature describing a benthic system populated by organisms containing okenone. It is, therefore, tempting to speculate on the existence of a relation between the type of carotenoid and the spectrum of the light to which the bacteria are exposed.

The adaptive role of carotenoids to grow at the light spectrum prevailing in certain planktonic environments has been well documented in Chlorobiaceae. Brown species of Chlorobiaceae contain isorenieratene and are predominant in deep bacterial plates overlaid by algae. On the contrary, green Chlorobiaceae are selected in systems in which the layer of Chlorobiaceae is overlaid by a layer of Chromatiaceae. This phenomenon is explained by the absorption spectra of both carotenoids. Isorenieratene-containing bacteria absorb in the 500–550 nm band, which are the predominant wavelengths at higher depths. Chlorobactene, on the contrary, absorbs around 450 nm, which are precisely the wavelengths available after filtering by absorption of a plate of purple sulfur bacteria close to the surface (Montesinos et al., 1983).

2. Physiological Responses to Light Limitation

The specific growth rate of phototrophic organisms changes as a function of light irradiance according to a saturation curve whose mathematical description is still a matter of discussion (Jassby and Platt, 1973; Iwakuma and Yasuno, 1983). Despite the model used, the curve is characterized by two parameters, μ_{max} which gives the maximum specific growth rate the organisms can achieve when light is saturating, and α, which is the initial slope of the μ vs. irradiance curve and corresponds to the maximum photosynthetic efficiency of the organisms under the experimental conditions used. Sometimes, instead of α the parameter I_k is used, which is in fact equivalent to μ_{max}/α, and which corresponds to the irradiance at which the organism would reach μ_{max} if the initial slope α where maintained all the way to saturation. I_k provides a rough indication of the irradiance below which growth is light limited. Because phototrophic organisms do have to invest energy into maintenance processes, even when growth does not occur, the growth vs. irradiance curve does not go through the origin, intersecting the irradiance axis at a certain irradiance. This irradiance, at which μ=0, gives an indication of the total amount of energy the organisms must spend in maintenance. Extrapolation of the curve towards the μ axis yields a negative value (usually referred to as μ_e), which is actually correlated to the maintenance threshold and gives an idea of the extent to which growth of the organism is slowed down due to allocation of energy to maintenance processes.

While the adaptation to limiting irradiances has been extensively studied in phytoplankton, little is known about phototrophic sulfur bacteria. In general, they seem able to adapt their light harvesting systems by increasing their specific content of pigments (Broch-Due et al., 1978). Recent observations indicate

Table 5. Penetration of light in lakes and sediments

System	Depth	Irradiance (% of surface)	Wavelength max	Wavelength low–high	Reference
Lake Fidler (Tasmania)	2.6 m	–	638	565–710	(1)
Fayetteville Green Lake (USA)	16.0 m	88.2	460–510	404–630	(2)
	18.0 m	36.5	460–520	404–630	
	18.5 m	2.4	583	545–630	
Mahoney Lake (Canada)	surface	100.0	675	400→700 [c]	(3)
	6.0 m	8.8	584	513–626	
Lake Kinneret (Israel)	10.0 m	0.75	550	450–650	(4)
	16.5 m	0.02	550	500–600	
Lake Vechten (The Netherlands)	2.0 m	–	584	433–720	(5)
	6.0 m	–	586	483–706	
Lake Banyoles (Spain)	5.0 m	23.0	540	440↔630 [d]	(6)
	plate	0.6	540	480–630	
Lake Vilar (Spain)	3.0 m	3.7	538	440↔630 [d]	(6)
	5.5 m	0.1	538	480–580	
Sediment San Francisco Bay (USA)	0.7 mm	4.7	>1000 [c]	725→1000 [c]	(7)
Sediment Guerrero Negro (USA)	1.2 mm	4.3	>1000 [c]	875→1000 [c]	(7)
Sediment Limfjorden [a] (Denmark)	0.8 mm	21.5	767	680–880	(8)
Sediment Limfjorden [b] (Denmark)	0.5 mm	42.8	761	700–860	(8)

[a] diatom mat, [b] *Oscillatoria* mat, [c] upper wavelength provided, [d] upper and lower wavelength provided.
(1) Croome 1986, (2) Culver and Brunskill 1969, (3) Overmann et al. 1991, (4) Dubinsky and Berman 1979, (5) Steenbergen et al. 1989, (6) Montesinos et al. 1983, (7) Jørgensen and Des Marais 1986, (8) Lassen et al. 1993.

that the increase in the specific content of pigments actually results in an increase in the photosynthetic efficiency (α) of the organism and do not have a significant effect on μ_{max} (O. Sánchez, personal communication). The extent to which this increase in the specific content of pigments occurs, depends in fact on the organisms. As can be seen in Table 6, green sulfur bacteria have high specific contents of BChl with maximum values of 915 and an average content of 214 μg BChl·mg^{-1} protein, while purple sulfur bacteria have much lower contents (maximum 85, average 41 μg BChl·mg^{-1} protein). Similar differences are found in μ_e, with green sulfur bacteria having values of μ_e around –0.001 h^{-1} and purple sulfur bacteria around –0.01 h^{-1} (see Table 7), ten times higher, indicating greater maintenance expenses. As a rule, green sulfur bacteria seem to be better adapted than purple sulfur bacteria to grow at low irradiances thanks to their higher light harvesting capacity together with their lower maintenance requirements. This is actually consistent with the fact that they are found at higher depths where light irradiance is likely to be very low, and also with the rather common mention of their presence right under the layers of purple sulfur bacteria when coexisting in the same habitat.

B. Electron Donors

By definition, anoxygenic phototrophic bacteria do not produce oxygen in photosynthesis because they lack Photosystem II and, consequently, are unable to use water as an electron donor. Instead these organisms use reduced forms of sulfur or hydrogen gas in the reduction of carbon dioxide. In addition, a limited number of small organic molecules may serve the same purpose or may be directly photoassimilated.

Chemolithotrophy, namely the oxidation of reduced sulfur compounds as sources of energy, has also been observed in several Chromatiaceae.

Table 6. Specific content of bacteriochlorophyll in several phototrophic sulfur bacteria.

Organism	μg BChl·mg^{-1}protein	Reference
CHLOROBIACEAE		
Chlorobium limicola (6 strains)	100–190	(1)
Chlorobium limicola 8327	110–220	(2)
Chlorobium limicola 8327	210–915	(3)
Chlorobium limicola 6230	323	(4)
Chlorobium vibrioforme CA4210	60–132	(5)
Chlorobium phaeobacteroides	80–625	(3)
Chlorobium phaeobacteroides MN1	51–221	(6)
Chlorobium phaeobacteroides 2430	29–124	(6)
CHROMATIACEAE		
Chromatium vinosum D	25–75	(7)
Chromatium vinosum CA1812	28–42	(5)
Chromatium vinosum DSM185	50–85	(8)
Chromatium vinosum DSM185	35	(9)
Chromatium vinosum DSM185	24–85	(10)
Thiocapsa roseopersicina	5–28	(5)
Thiocapsa roseopersicina	29–33	(11)

(1) Cohen–Bazire, 1963, (2) Broch–Due, 1978, (3) Montesinos, 1982, (4) Schmidt, 1980, (5) Matheron, 1976, (6) Overmann, 1992, (7) Takahashi et al., 1972, (8) Van Gemerden, 1980, (9) Mas and Van Gemerden, 1992, (10) O. Sánchez, personal communication. (11) De Wit, 1989

Table 7. Maintenance rate requirements of various purple and green bacteria. The value of μ_e is the ordinate intercept of the μ-light curve

Organism	μ_e	Reference
CHROMATIACEAE		
Chromatium vinosum D	–0.007	(1)
Chromatium vinosum DSM 185	–0.010	(2)
Chromatium vinosum DSM 185	–0.025	(3)
Thiocapsa roseopersicina K2	–0.010	(4)
CHLOROBIACEAE		
Chlorobium phaeovibrioides	<–0.001	(1)
Chlorobium phaeobacteroides	–0.001	(2)
Chlorobium phaeobacteroides K1	–0.001	(4)
Chlorobium phaeobacteroides 2430	–0.031	(5)
Chlorobium phaeobacteroides MN1	–0.001	(5)

(1) Biebl and Pfennig, 1978, (2)Van Gemerden, 1980, (3) O. Sánchez, personal communication, (4) Veldhuis and Van Gemerden, 1986, (5) Overmann et al., 1992.

1. Potential Photosynthetic Electron Donors for Purple and Green Sulfur Bacteria

Virtually all reduced inorganic forms of sulfur are used as electron donors by various purple and green sulfur bacteria. The one most commonly used in media to grow these organisms is sulfide (H_2S, HS^-, S^{2-}). As first demonstrated by Van Niel (1931), the oxidation of sulfide is stoichiometrically linked to the reduction of carbon dioxide. As a rule, sulfide oxidation results in the formation of zero-valent sulfur ('elemental sulfur', S^0), which is further oxidized to sulfate. In these equations, $<CH_2O>$ represents the simplified over-all composition of cell material at the level of carbohydrate.

$$CO_2 + 2\,H_2S \rightarrow\ <CH_2O> + 2\,S^0 + H_2O$$

$$3\,CO_2 + 2\,S^0 + 5\,H_2O \rightarrow 3\,<CH_2O> + 2\,H_2SO_4$$

$$4\,CO_2 + 2\,H_2S + 4\,H_2O \rightarrow 4\,<CH_2O> + 2\,H_2SO_4 \qquad (4)$$

The capacities of sulfide oxidation, intracellular or extracellular S^0 deposition, oxidation of S^0 to sulfate, and photopigmentation, are used to differentiate the major groups of anoxygenic phototrophic bacteria from each other. Members of the family Chromatiaceae deposit S^0 as refractile globules inside the cells, whereas those belonging to the family Ectothiorhodospiraceae form extracellular sulfur. Also green sulfur bacteria (family Chlorobiaceae) deposit S^0 outside the cells.

In the purple sulfur bacteria, S^0 appears not to be present in the form of S_8 rings, but rather in a more easily accessible form, presumably as long-chain polysulfides or polythionates. Because of these uncertainties, the term 'zero-valent sulfur' is to be preferred over 'elemental sulfur' (Steudel, 1989; Steudel et al., 1990). Chromatiaceae, like Chlorobiaceae, are able to oxidize extracellular S^0, but growth on flower of sulfur usually is slower than on soluble sulfur species.

Among the Chromatiaceae, the utilization of thiosulfate is not uncommon, in the green sulfur bacteria this ability is restricted to the forma *thiosulfatophilum* of *Chlorobium limicola* and *Cb. vibrioforme*. In the Chromatiaceae, as well as in *Cb. vibrioforme* f. *thiosulfatophilum*, thiosulfate oxidation results in the formation of S^0. It has been shown that in the Chromatiaceae S^0 is exclusively derived from the outer S atom (sulfane group) with the oxidation state -2 (Trüper and Pfennig, 1966).

The stoichiometry of thiosulfate oxidation, with CO_2 as carbon source, can be described as shown in the equations:

$$CO_2 + Na_2S_2O_3 + H_2O \rightarrow <CH_2O> + 2\,S^0 + 2\,Na_2SO_4$$

$$3\,CO_2 + 2\,S^0 + 5\,H_2O \rightarrow 3\,<CH_2O> + 2\,H_2SO_4 \quad +$$

$$4\,CO_2 + Na_2S_2O_3 + 6\,H_2O \rightarrow 4\,<CH_2O> + 2\,H_2SO_4 + 2\,Na_2SO_4 \qquad (5)$$

The simultaneous formation of thiosulfate and S^0 from sulfide was reported to occur in *Cb. limicola* f. *thiosulfatophilum* (Schedel, 1978; Trüper, 1984) and *Cb. vibrioforme* f. thiosulfatophilum (Trüper, 1984). Low concentrations of $S_2O_3^{2-}$ were also detected after sulfide incubation of *Cm. vinosum* (Steudel et al., 1990). In *Chromatium*, the final oxidation product of all inorganic sulfur compounds is sulfate, with SO_3^{2-} as intermediate.

Tetrathionate ($S_4O_6^{2-}$) appears not to be an important electron donor for phototrophic sulfur bacteria. However, $S_4O_6^{2-}$ is readily reduced abiotically to form $S_2O_3^{2-}$ according to

$$S_4O_6^{2-} + 2\,e^- \rightarrow 2\,S_2O_3^{2-} \qquad (6)$$

in which sulfide acts as the electron donor. The formation of tetrathionate in cultures of *Rhodomicrobium vannielii* was shown to follow this route (Hansen, 1974).

Polysulfides (S_x^{2-}) may be formed in abiotic reactions between S^0 and HS^-, and are stable at elevated pH values. The utilization of S_3^{2-} has been demonstrated for a few purple and green sulfur bacteria (Visscher and Van Gemerden, 1988; Steudel et al., 1990; Visscher et al., 1990).

In addition to reduced sulfur species, hydrogen is an electron donor for many phototrophic bacteria. This ability was first described in 1935 by Roelofsen. In the Chlorobiaceae, hydrogen utilization occurs in the majority of strains, however, due to the lack of assimilatory sulfate reduction, a reduced sulfur source is required during growth on H_2 (Lippert and Pfennig, 1969).

For a few purple sulfur bacteria, utilization of dimethyl sulfide (DMS, $(CH_3)_2S$) as electron donor has been demonstrated. *Thiocystis* sp. was reported to grow slowly in the simultaneous presence of H_2S and DMS; products formed were dimethylsulfoxide (DMSO, $(CH_3)_2S{=}O$) and methane (Zeyer et al., 1987). In contrast, *Thiocapsa roseopersicina* showed rapid growth on DMS as sole electron donor with DMSO as oxidation product. Also, DMS and sulfide were oxidized simultaneously, yielding DMSO and S^0, subsequently the latter was oxidized to sulfate (Visscher and Van Gemerden, 1991b).

Recently, biological CO_2 fixation in the light, in the absence of oxygen, occurring concomitantly with the oxidation of ferrous iron, was reported (Widdel et al., 1993). The organisms responsible were found to resemble *Rhodomicrobium vannielii*, *Rhodopseudomonas palustris*, and *Thiodictyon* sp., however, the latter isolate did not oxidize sulfide. The stoichiometry observed was found to be in agreement with

$$17\,FeCO_3 + 28\,H_2O + NH_4^+ \rightarrow (C_5H_8O_2N) + 17\,Fe(OH)_3 + H^+ \qquad (7)$$

Since the midpoint potential of the reaction center of purple non-sulfur bacteria and that of the purple sulfur bacteria are similar (Dutton and Prince, 1978), growth of the latter with ferrous iron should be possible as well. However, Fe^{II} and S^{2-} readily react to form FeS, but ferrous iron might act as electron donor in the absence of sulfide.

A detailed consideration of sulfur chemistry, sulfur compounds and their role as photosynthetic electron donors is given in Chapter 39.

2. Abundance of Electron Donors in Nature

The most important and selective environmental factors for the development of phototrophic sulfur bacteria are (1) the lack of oxygen, (2) the availability of light, and (3) the presence of reduced sulfur compounds. There is general agreement that in nature sulfide is an important electron donor for purple and green sulfur bacteria. For obvious reasons, the profiles of oxygen and sulfide show diel fluctuations. Sediment ecosystems often become completely anoxic during the night with sulfide reaching the surface layers (De Wit et al., 1989; Revsbech et al., 1989; Van Gemerden, 1990). In stratified lakes fluctuations occur at the depth horizon of blooms of anoxygenic phototrophic sulfur bacteria (Van Gemerden, 1985; Jørgensen et al., 1979).

As a rule, an overlap is observed in the profiles of sulfide and oxygen. In a comparison between three systems, Jørgensen (1982) mentioned the large differences in the vertical dimension of the oxygen-sulfide interface in different systems. In the Black Sea, sulfide and oxygen coexisted over 35 m depth or more (Jørgensen et al., 1991), in Solar Lake, over 10 cm, and in a *Beggiatoa*-dominated mat, over no more than 50 μm. The integrated sulfide oxidation rate over the whole column was of the same order of magnitude for all three systems (10–30 $mmol \cdot m^{-2} \cdot d^{-1}$). However, the peak rates varied over many orders of magnitude, and were 0.8 $\mu mol \cdot l^{-1} \cdot d^{-1}$ in the Black Sea, 250 $\mu mol \cdot l^{-1} \cdot d^{-1}$ in Solar Lake, and 250 000 $\mu mol \cdot l^{-1} \cdot d^{-1}$ in the microbial mat. The residence time of sulfide in the peak layer was calculated to be 5 d, 10–20 min, and 0.6 s, respectively. Consequently, the ratio between the biological and the chemical oxidation rate may vary considerably. It was concluded that the biological contribution to sulfide oxidation in the Black Sea was negligible, thus confirming earlier data of Sorokin (1964, 1972), amounted 30–50% in Solar Lake, and was 100% in the *Beggiatoa*-dominated mat (Jørgensen, 1982). The abiotic oxidation of S^{2-} by oxygen in seawater has been reported to result in the formation of $S_2O_3^{2-}$, with minor amounts of SO_3^{2-} (Chen and Gupta, 1972). In water devoid of O_2, S^{2-} oxidation may be attributed to a direct reaction with MnO_2 (Millero, 1991).

Except for the situation in hydrothermal vents, sulfide is generated principally by dissimilatory sulfate reduction: in marine sediments the contribution of sulfate-reducing bacteria to sulfide production has been estimated to be close to 100% (Jørgensen, 1977).

In addition to sulfide, a wide range of inorganic and organic sulfur species has been reported to occur in habitats colonized by sulfur bacteria. In microbial mats, relatively high concentrations of 'elemental' sulfur, polysulfides (S_x^{2-}), and the iron sulfides FeS and FeS_2 (pyrite) have been reported, while the concentrations of thiosulfate and polythionates ($S_3O_6^{2-}$, $S_4O_6^{2-}$, $S_5O_6^{2-}$) were found to be much lower (Van Gemerden et al., 1989b; Visscher et al., 1990; Visscher and Van Gemerden, 1993). High concentrations of polysulfides are often found in salt marsh sediments and microbial mats (Jørgensen et al., 1979; Aizenshtat et al., 1983; Lord and Church, 1983; Luther et al., 1986; Luther and Church, 1988).

For stratified freshwater lakes and sediments, sulfur speciation is less well documented, except for the occurrence of sulfide in the hypolimnion. The explanation probably lies in the much lower sulfate content of freshwater lakes.

Organic sulfur compounds may be formed abiotically in reactions between inorganic polysulfides and organic molecules. Relatively low molecular weight organic S-compounds result as well from the microbially mediated breakdown of protein and dimethyl-sulfoniopropionate (DMSP, $(CH_3)_2$-S-CH_2-CH_2-COOH) (Kiene and Taylor, 1988a; Kiene et al., 1990). When degraded, sulfur-containing amino acids (cysteine, methionine) yield methanethiol (MSH, CH_3-SH), dimethyl sulfide (DMS, CH_3-S-CH_3), and dimethyl disulfide (DMDS, CH_3-S-S-CH_3). DMSP, which is present in many strains of marine phytoplankton (Keller et al., 1989) and presumably acts as an osmolyte (Reed, 1983), yields DMS and acrylate (CH_2=CH_2COOH), or 3-mercaptopropionate (3-MPA, HS-CH_2-CH_2-COOH) and MSH upon bacterial degradation (Kiene and Visscher, 1987; Kiene and Taylor, 1988b; Kiene et al., 1990). Recently, evidence was obtained for DMSP-lyase activity in axenic cultures of the phytoplanktonic species *Phaeocystis* sp. (Stefels and Van Boekel, 1993) and *Emiliania huxleyi* (J. Stefels, pers. comm.), indicating that oceanic DMS formation not necessarily is mediated by bacteria. DMSP is also formed by the dominant cyanobacterium in mature microbial mats, *Microcoleus chthonoplastes*. Upon a down-shift in salinity, mimicking severe rainfall, 50% of the cellular DMSP was excreted (Visscher and Van Gemerden, 1991a), which likely resulted in the formation of DMS.

Potential DMS consumers are sulfate-reducing bacteria and methanogens (Kiene and Visscher, 1987), colorless sulfur bacteria (Smith and Kelly, 1988; Visscher et al., 1991), and phototrophs (Zeyer et al., 1987; Visscher and Van Gemerden, 1991b).

3. Substrate Affinities of Purple and Green Sulfur Bacteria

The scavenging capacity of organisms for a given substrate can be ranked from low to high by comparing their growth affinity for that substrate. The specific growth rate (μ) of microbes depends on the concentration (s) of the growth-rate limiting substrate. With non-inhibitory substrates the relation between μ and s is adequately described by the Monod equation, being :

$$\mu = \mu_{max} \cdot \frac{s}{K_s + s} \qquad (8)$$

in which μ_{max} is the maximum specific growth rate and K_s is the saturation constant equal to the substrate concentration at which $\mu = \frac{1}{2}\mu_{max}$. Plotting μ versus s results in a saturating type of curve, the steeper the initial slope the higher the affinity of the organism for the substrate involved. Since K_s is a function of μ_{max}, the magnitude of the affinity (usually) cannot directly be deduced from the saturation constant (Healey, 1980; Zevenboom, 1980; Van Gemerden, 1984). Under constant environmental conditions, the organism with the highest affinity can be expected to outcompete all others utilizing the same substrate. With the assumption that substrate concentrations in nature are low (in any case much lower than K_s), affinity can be described mathematically as $A_{s \to 0} = \mu_{max}/(K_s+s)$, being the initial slope of the Monod curve.

It appears that the affinity for sulfide of purple and green sulfur bacteria is related to the location of sulfur deposition. Chlorobiaceae and Ectothiorhodospiraceae, depositing S^0 extracellularly, invariably exhibit higher affinities for sulfide than Chromatiaceae, which store S^0 inside the cells (Table 8). *Rhodobacter capsulatus*, a purple nonsulfur bacterium depositing S^0 extracellularly (Hansen and Van Gemerden, 1972), also has a high sulfide affinity (Van Gemerden, 1984). Likewise, colorless sulfur bacteria have high sulfide and thiosulfate affinities (Beudeker et al., 1982: Visscher et al., 1992). It has been postulated that in organisms having a high affinity for sulfide, the primary acceptor for the electrons released in the oxidation of sulfide is situated at the outside of the cell membrane, whereas for organisms depositing S^0 intracellularly, the primary acceptor is located at the inside of the membrane. Active transport of sulfide across the membrane, only being necessary in organisms depositing S^0 intracellularly, might then be the limiting step (Van Gemerden, 1984). At present there are no exceptions to the rule, but conclusive evidence to support this hypothesis is still lacking.

In a few organisms, affinities have been estimated for different sulfur substrates (Table 9). *Cm. vinosum* and *T. roseopersicina* displayed similar affinities for S^{2-} and S_3^{2-}. If the undissociated forms would have to pass the membrane by passive diffusion, one would expect that the larger trisulfide molecule enters the cell at a lower rate than hydrogen sulfide, resulting in a lower affinity for the former. These results thus suggests that both sulfur species are translocated by active transport. In *Cb. limicola* f. *thiosulfatophilum* the affinities for S^{2-} and S_3^{2-} differed sixfold. For yet unknown reasons, all three organisms showed a lower μ_{max} on S_3^{2-}.

Few data are available on the utilization of other substrates than the ones the organisms were grown on. It was reported that in *Cm. vinosum* and *T. roseopersicina*, S_3^{2-} oxidation by S^{2-}-grown cells proceeds without lag (Steudel et al., 1990; Visscher et al., 1990). In contrast, *de novo* protein synthesis was required for the utilization of polysulfide by *Cb. limicola* (Visscher and Van Gemerden, 1988). Acetate-grown *Chromatium* cells oxidize sulfide without lag, which is explained by the presence of Calvin cycle enzymes in these cells (see section VIII.D).

Purple sulfur bacteria appear to be able to oxidize different substrates simultaneously, e.g. S^{2-} and S_3^{2-} (*Thiocapsa*), acetate and sulfide (*Chromatium*), or S^{2-} and DMS (*Thiocapsa*) (Visscher and Van Gemerden, 1991b). In the green sulfur bacterium

Table 8. Kinetic parameters for growth of purple and green sulfur bacteria on sulfide

Organism	Sulfur[1] iS^0/eS^0	μ_{max} [2] h^{-1}	μ_{max} [3] h^{-1}	K_s $\mu mol \cdot l^{-1}$	K_i $mmol \cdot l^{-1}$	Affinity[4] $h \cdot mmol^{-1} \cdot l$	Ref
CHROMATIACEAE							
Chromatium vinosum DSM 185	iS^0	0.130	0.117	7	2.5	18.6	(1)
Chromatium vinosum DSM 192	iS^0	0.115	0.097	7	0.85	16.4	(2)
Chromatium vinosum M9	iS^0	0.090	0.080	12	3	7.5	(3)
Chromatium weissei DSM 171	iS^0	0.050	0.040	10	0.7	5.0	(1)
Thiocapsa roseopersicina K2	iS^0	0.087	0.072	21	2.2	4.1	(1)
Thiocapsa roseopersicina M1	iS^0	0.091	0.086	8	8	11.4	(4)
ECTOTHIORHODOSPIRACEAE							
Ectothiorhodospira vacuolata DSM 2111	eS^0	0.138	0.129	3	2.8	46.0	(1)
Ectothiorhodospira sp. 80	eS^0	0.116	0.109	2	2	58.0	(1)
Ectothiorhodospira shaposhnikovii DSM 243	eS^0	0.089	0.086	1	4	89.0	(1)
Ectothiorhodospira shaposhnikovii M3	eS^0	0.062	0.059	2	4	31.0	(1)
CHLOROBIACEAE							
Chlorobium limicola C2	eS^0	0.130	0.124	2	4	65.0	(1)
Chlorobium limicola f.thiosulfatophilum DSM 249	eS^0	0.110	0.105	1.5	3	73.3	(5)
Chlorobium vibrioforme DSM 263	eS^0	0.070	0.068	1	3.8	70.0	(1)
Chlorobium phaeobacteroides C1	eS^0	0.105	0.100	2.2	3	47.7	(1)
Chlorobium phaeobacteroides K1	eS^0	0.097	0.092	0.9	2.5	107.8	(6)

[1] iS^0 = intracellular sulfur, eS^0 = extracellular sulfur, [2] theoretically maximum value if there were no substrate inhibition (data from Direct Linear Plot, Eisenthal and Cornish–Bowden 1974), [3] maximum attainable specific growth rate,[4] calculated with theoretically maximum specific growth (i.e. if K_i were infinite: no inhibition).

(1) Van Gemerden 1974, (2) Van Gemerden and Jannasch 1971, (3) Van Gemerden 1984, (4) Visscher et al. 1990, (5) Visscher and Van Gemerden 1988,(6) Veldhuis and Van Gemerden 1986.

Table 9. Kinetic parameters for phototrophic growth of purple and green sulfur bacteria on sulfide (S^{2-}) and polysulfide (S_3^{2-})

		Chromatium vinosum DSM 185		*Thiocapsa roseopersicina* M1		*Chlorobium limicola* f. *thiosulfatophilum* DSM 249	
Parameter	Dimension	Sulfide S^{2-}	Polysulfide S_3^{2-}	Sulfide S^{2-}	Polysulfide S_3^{2-}	Sulfide S^{2-}	Polysulfide S_3^{2-}
μ_{max} [a]	h^{-1}	0.130	0.052	0.091	0.065	0.110	0.071
μ_{max} [b]	h^{-1}	0.117	0.048	0.086	0.056	0.105	0.064
K_s	$\mu mol \cdot l^{-1}$	7	3.1	8	6.7	1.5	5.9
K_i	$mmol \cdot l^{-1}$	2.5	2 [c]	8	1.1	3	2 [c]
affinity	$h^{-1} \cdot mmol^{-1} \cdot l$	18.6	16.8	11.4	9.7	73.3	12.0

[a] theoretically maximum value if there were no substrate inhibition (infinite value for K_i), [b] maximum attainable specific growth rate, [c] tentative estimate

Cb. limicola, S_3^{2-} oxidation ceased upon the addition of S^{2-}, concomitantly S_3^{2-} was resynthesized. The amount of S^{2-} added was insufficient to explain the formation of S_3^{2-} from S^{2-} and S^0, and it was hypothesized that the initial oxidation of S_3^{2-} resulted in the formation of S_2^{2-} (Visscher and Van Gemerden, 1988). The inhibition of S_3^{2-} oxidation by S^{2-} suggests a common transport system.

For other reasons, in green sulfur bacteria acetate assimilation and sulfide oxidation proceed synchronously. These organisms fix CO_2 via the reversed TCA cycle (Ormerod and Sirevåg, 1983; and see Chapter 40). The first step in acetate assimilation is a reductive carboxylation which involves the participation of reduced ferredoxin. For the formation of the latter, sulfide is required.

4. Reduced Sulfur Compounds as Sources of Energy

Chemolithotrophy is the type of metabolism in which energy requirements are fulfilled by the respiration of an externally supplied inorganic compound. When occurring, these substrates usually are respired aerobically, i.e. with oxygen. Chemolithotrophy certainly is not a common characteristic of anoxygenic phototrophic sulfur bacteria. Green sulfur bacteria are obligate phototrophs, but among the Chromatiaceae several species have been shown to possess a chemolithotrophic metabolism. Experiments in batch cultures have revealed this capacity in several, but not all, strains of *Thiocapsa roseopersicina*, *Amoebobacter roseus*, *Thiocystis violacea*, *Chromatium vinosum*, *Cm. minus*, *Cm. violascens*, *Cm. gracile*, and *Thiorhodovibrio winogradskyi* (Pfennig, 1970; Bogorov, 1974; Gorlenko, 1974; Kondratieva et., 1976; Kämpf and Pfennig, 1980, 1986a,b; De Wit and Van Gemerden, 1987, 1990a,b; Overmann and Pfennig, 1992). Some species only grow at low pO_2, while others can do so at full atmospheric oxygen tension. All large-celled Chromatiaceae and a few small-celled species fail to grow in the presence of even low concentrations of oxygen (Kämpf and Pfennig, 1980). However, even species having the ability to grow chemotrophically appear to have a preference for phototrophy. Experimental evidence to support this hypothesis will be presented in section IX. Chlorobiaceae are unable to grow chemotrophically (Kämpf and Pfennig, 1980).

When incubated in the continuous presence of low to moderately high concentrations of oxygen, purple sulfur bacteria eventually become completely colorless (De Wit and Van Gemerden, 1987; Overmann and Pfennig, 1992). During chemotrophic growth (aerobic respiration of sulfide or thiosulfate), the yield is much lower that during phototrophic growth. The protein concentration in phototrophic cultures of *Thiocapsa roseopersicina* strain M1, when subjected to a shift from anoxic to oxic (52 $\mu mol \cdot l^{-1}$) conditions, decreased substantially; concomitantly, the cells became completely colorless. Approximately 61% of the thiosulfate was respired to provide energy and 39% resulted in the synthesis of cellular material (De Wit and Van Gemerden, 1987). The chemotrophic yield of phototrophic purple bacteria thus appeared to be somewhat higher than reported for colorless sulfur bacteria (Kelly, 1982).

Shift experiments as described above, were also carried out in media containing equimolar concentrations of $S_2O_3^{2-}$ and acetate (Overmann and Pfennig, 1992). Under these conditions, the protein concentration during chemotrophic growth of *T. roseopersicina* strain 6311 was 47% compared to full phototrophic growth, and the corresponding values for *Amoebobacter roseus* strain 6611 and *Thiocystis violacea* strain 2711 were 67% and 52%, respectively (Overmann and Pfennig, 1992). These organisms are unable to respire acetate. The higher yields of these organisms under chemotrophic conditions, compared to *T. roseopersicina* strain M1 are not to be explained by strain differences, but rather illustrate that much energy is required for the fixation of CO_2.

C. Oxygen

The impact of oxygen on growth and metabolism of anoxygenic phototrophic bacteria has been studied for a long time. The metabolism of the phototrophic purple and green sulfur bacteria is inhibited by oxygen in one way or another. These organisms thrive in the anoxic parts of sediments and freshwater ponds and lakes, provided light is available. For some species (green sulfur bacteria, large-celled Chromatiaceae) oxygen is lethal, while other species survive and may grow at relatively low concentrations of oxygen. Although the general statement that 'photosynthesis of anoxygenic phototrophic bacteria is dependent on oxygen-deficient conditions' (Drews and Imhoff, 1991) suggests otherwise, some species continue to photosynthesize in the presence of elevated concentrations of O_2, and a few are even capable of carrying out chemolithotrophy, in which a reduced sulfur species is oxidized aerobically to provide energy.

Oxygen, at relatively low partial pressure, specifically inhibits the synthesis of photopigments; this has not only been observed in purple non-sulfur bacteria (Cohen-Bazire et al., 1957), but in purple sulfur bacteria as well (Hurlbert, 1967) The lower limit for complete repression is not exactly known, and may well be different for different organisms. In *Chromatium vinosum*, reduced rates of BChl synthesis were observed at O_2 concentrations below 10 $\mu mol \cdot l^{-1}$ (Kämpf and Pfennig, 1986a). Also, *Thiocapsa roseopersicina* and *Thiocystis violacea* were reported to contain low levels of BChl and carotenoids at 'low' oxygen concentrations (Bogorov, 1974; Kämpf and Pfennig, 1980). The latter two species are ranked as the most tolerant members of the Chromatiaceae to perform anoxygenic photosynthesis in the presence of oxygen (Kondratieva et al., 1976; Kondratieva, 1979; Kämpf and Pfennig, 1980; Madigan, 1988).

D. Carbon Sources

Carbon dioxide is commonly used as the sole source of carbon in media for growing purple and green sulfur bacteria. The stoichiometry of the reduction of CO_2 and the oxidation of S^{2-} and $S_2O_3^{2-}$ was shown in Equations 4 and 5, respectively.

In the Chromatiaceae, the main route of CO_2 fixation is the Calvin cycle. Additional CO_2-fixing reactions have been reported (Fuller, 1978; Sahl and Trüper, 1977), but these do not seem to contribute substantially to the net conversion of CO_2 to cell material (Madigan, 1988). In the green sulfur bacteria, carbon dioxide is fixed through the reductive TCA cycle (Ormerod and Sirevåg, 1983). This metabolic route of CO_2 fixation is energetically less energy-demanding, which may, at least in part, explain why green sulfur bacteria are selectively enriched at low light intensities. Autotrophic growth of Chlorobiaceae depends on the presence of a reduced form of sulfur, such as sulfide, since these organisms lack the possibility of the reduction of SO_4^{2-} for assimilatory purposes.

Organic compounds are also assimilated by purple and green sulfur bacteria. In this respect, Chlorobiaceae show little versatility (Trüper, 1981). Except for acetate and pyruvate, for which the simultaneous presence of CO_2 and a reduced sulfur species (S^{2-}, $S_2O_3^{2-}$) is required for assimilation, no organic compounds stimulate growth. The Chromatiaceae show a large variation in the extent to what, and under what conditions, organic compounds are used as substrates for growth. The large-celled species (such as *Chromatium okenii, Cm. weissei, Thiospirillum jenense*, and a few others), are nutritionally rather restricted. Their photo-assimilation is limited to acetate and pyruvate. In addition, like the green sulfur bacteria, these organisms are unable to reduce sulfate for assimilatory purposes, and thus rely on the presence of sulfide. The small-celled Chromatiaceae, on the other hand, possess the assimilatory sulfate reduction route, and, in addition, have a much larger range of organic compounds that can be assimilated (Trüper, 1981).

Ribulose bisphosphate carboxylase (RuBisCO), being the key enzyme of the Calvin cycle, is most active during photoautotrophic growth. During photoheterotrophic growth the Calvin cycle enzymes are under metabolic control; in *Cm. vinosum* the synthesis of RuBisCO is repressed by 50–90% under photoheterotrophic conditions (Fuller, 1978). It was suggested that the considerable activity of the Calvin cycle enzymes still present under these conditions might play a role in the disposal of excess reducing power (Hurlbert and Lascelles, 1963). Under constant environmental conditions, lowering the cellular content of RuBisCO and other enzymes may be an effective adaptation, since the Calvin cycle is largely superfluous during growth on acetate. However, under fluctuating conditions, which are characteristic for most of the habitats of the phototrophic sulfur bacteria, such an 'adaptation' could be a serious drawback and might result in a reduced reactivity upon a shift from acetate assimilation to CO_2-fixation. However, acetate-grown *Chromatium* cells oxidize sulfide at similar specific rate as sulfide-grown cells. It thus appears that the organism has an overcapacity with respect to the Calvin-cycle enzymes. A similar overcapacity with respect to the content of BChl is discussed in section IX.C.

IX. Case Studies

The competitive position of (a group of) organisms cannot be judged from their own characteristics, but requires comparison with data of other organisms under ecologically relevant conditions. A few illustrative examples are discussed below.

A. Competition Among Purple Sulfur Bacteria

In aquatic environments, particularly in fresh-water lakes and ponds, blooms of large-celled *Chromatium* species (*Cm. okenii, Cm. weissei*) occur frequently. In such blooms, small-celled *Chromatium* species (*Cm. vinosum, Cm. minutissimum*) usually are present as well. If large-celled and small-celled species comprise different ecophysiological groups, is of interest to know what parameters are of decisive importance for bloom formation.

Large-celled *Chromatium* species grow slowly (have a relatively low μ_{max}), require vitamin B_{12}, and were reported to grow best in dim light and dark-light cycles (Pfennig, 1965; Pfennig and Trüper, 1992). Small-celled *Chromatium* species usually do not require B_{12} and, as a rule, are cultivated under continuous illumination (Table 10). Although the discovery of the B_{12} requirement was a breakthrough in the cultivation of the large *Chromatium* species (Pfennig, 1965; Pfennig and Lippert, 1966), it probably is not a factor of decisive importance in the competition between these species. Similarly, once pure cultures were available, most large-celled species were found to be culturable at higher light intensities.

In a study on the competition between the small-celled *Cm. vinosum* and the large-celled *Cm. weissei*, attempts to grow the organisms together in sulfide-limited chemostats in the light invariably resulted in *Cm. weissei* being outcompeted by *Cm. vinosum*, irrespective of the dilution rate employed. This is to be expected in view of their respective affinities for sulfide (see Table 8). However, when continuous illumination was replaced by a light-dark regimen, the two organisms coexisted, which could be explained by taking into consideration that, in addition to the growth parameters μ_{max} and K_s, the temporary storage of glycogen and S^0 are important factors (Van Gemerden, 1974).

The affinity for sulfide is calculated from μ_{max} and K_s data obtained in chemostats. Under these conditions, the concentration of sulfide permanently is in the μmolar range. In nature, substrate concentrations are not constant, but rather show diel fluctuations (Jørgensen et al., 1983; Revsbech and Ward, 1984; Van Gemerden et al., 1985; Nicholson et al., 1987; Pierson et al., 1987; De Wit et al., 1989; Lassen et al., 1992; Van Gemerden, 1993). Due to decreased oxidation rates and ongoing production rates, sulfide concentrations increase during nights, and decrease during the day. All purple sulfur bacteria analyzed so far exhibit a higher sulfide oxidation capacity than that required for growth. As a result, storage compounds accumulate intracellularly during periods in which excess substrate is available (See Chapter 45). It has been established for *Cm. vinosum* that the potential rate of electron-donor oxidation is virtually identical at all specific growth rates. Consequently, the storage capacity is maximal at low growth rates (Van Gemerden and Beeftink, 1978; Beeftink and Van Gemerden, 1979). Likewise, in *Cm. weissei* the maximum rates of electron-donor oxidation and CO_2-fixation exceed the maximum specific growth rate (Van Gemerden, 1974). The fact that slow or non-growing phototrophic bacteria have high reactivities would have little ecological meaning, unless substrate concentrations fluctuate.

Secondly, the maximum rate of sulfide oxidation of *Cm. vinosum* is approximately half that of *Cm. weissei*. This is not related to differences in the rate of CO_2 fixation, which is similar in the two organisms, but rather to the fact that the large-celled *Cm. weissei* preferentially oxidizes sulfide to S^0, whereas in the small-celled *Cm. vinosum* a larger fraction of the sulfide is directly oxidized to sulfate. Under fluctuating sulfide concentration this phenomenon could counteract a high sulfide affinity.

Sulfide is not oxidized in the dark. Applying a dark-light regimen, in combination with a constant sulfide supply, thus mimicking the environmental

Table 10. Comparison of general characteristics of large–celled and small–celled Chromatiaceae

	Large–celled Species (e.g. *Chromatium okenii*)	Small–celled Species (e.g. *Chromatium vinosum*)
Cell dimensions	4.5–6 × 8–15 μm	2.0 × 2.5–6 μm
Assimilatory SO_4^{2-} reduction	lacking	present
Organic substrates	acetate, pyruvate	many
Vitamin B_{12} requirement	yes	no
Light intensity	2 – 6 μEin·m^{-2}·sec^{-1}	4 – 40 μEin·m^{-2}·sec^{-1}
Light regimen	daylight, light – dark	continuous
Specific growth rate	≈ 0.04 h^{-1} (t_d ≈ 17 h)	≈ 0.12 h^{-1} (t_d ≈ 5 h)

conditions, results in a slow increase in the concentration of sulfide during the dark periods. Co-culturing *Cm. weissei* and *Cm. vinosum* under these conditions showed that in the early hours of the illumination period, a large fraction of the accumulated sulfide was oxidized by *Cm. weissei*, resulting in the intracellular storage of S^0 and glycogen. During the remainder of the light periods, in which the sulfide concentration was extremely low, most of the incoming sulfide was oxidized by *Cm. vinosum* due to its higher affinity. During these periods, *Cm. weissei* was growing predominantly at the expense of the previously stored S^0 and glycogen. Stable coexistence between *Cm. weissei* and *Cm. vinosum* was observed at different light-dark regimens, in which longer nights resulted in a more pronounced dominance of the large *Cm. weissei* (Van Gemerden, 1974).

The co-existence of the two *Chromatium* species under fluctuating conditions thus can be explained on the basis of (1) a higher growth response of the small-celled species at all sulfide concentrations, (2) a similar maximal rate of CO_2 fixation in both organisms, and (3) a faster storage of S^0 in *Cm. weissei* upon a temporary excess of sulfide.

A high sulfide affinity also appears to be of crucial importance for organisms lacking the ability to use sulfate as sulfur source for assimilatory purposes. However, the lack of assimilatory sulfate reduction, as observed in the large-celled species, is fully compensated by the extensive storage of sulfur, thus providing the organisms with a suitable source of sulfur for the synthesis of structural cell components.

Although μ_{max} and K_s data from many more strains are required before a general conclusion can be reached, this example illustrates that the environmental fluctuations in an organism's habitat should be taken into consideration before conclusions can be drawn on the (dis)advantage of a certain characteristic.

B. Competition Between Purple and Green Sulfur Bacteria

Purple sulfur bacteria and green sulfur bacteria have very similar nutritional requirements. It is, therefore, not surprising that these organisms can be grown in media of similar composition. Green sulfur bacteria, being obligate anaerobic phototrophs, have been isolated from marine environments, but seldom develop profusely. An exception is found in the multi-layered microbial mats of Great Sippewissett Salt Marsh, Cape Cod, USA (Nicholson et al., 1987; Pierson et al., 1987). In these systems, a layer of green sulfur bacteria developed underneath two distinct layers of purple sulfur bacteria which presumably prevented the downward diffusion of oxygen. No oxygen data were reported, but circumstantial evidence showed the layer of green sulfur bacteria to be permanently anoxic.

Green sulfur bacteria frequently bloom in freshwater lakes, often as mixed populations with purple sulfur bacteria. It is of interest to know what conditions are in favor of such a co-existence.

Numerous chemostat studies have shown that co-culturing of organisms with different substrate affinities, results in the competitive exclusion of the organism with the lower affinity. However, stable co-cultures of *Chlorobium limicola* f. *thiosulfatophilum* DSM 249 and *Chromatium vinosum* DSM 185 were obtained in sulfide-limited continuous cultures, despite the higher sulfide affinity of *Chlorobium* (see Table 8). With increasing dilution rates, the population density of *Chromatium* increased in proportion (Van Gemerden and Beeftink, 1981).

On theoretical grounds, stable coexistence of two competing organisms is feasible. One possibility is that there are at least two mutual substrates, and that the competing organisms have complementary μ-s relationships on these substrates (Taylor and Williams, 1975; Yoon et al., 1977; Gottschal, 1986). In the case of *Chlorobium / Chromatium*, sulfide obviously is one of the substrates for which *Chlorobium* has the better affinity. It has been suggested that S^0 acted as the second substrate (Van Gemerden and Beeftink, 1981). Extracellular S^0, produced by green sulfur bacteria, can be used equally well by purple and green sulfur bacteria. S^0 is a powerful electron donor because it represents six out of the eight electrons released in the oxidation of sulfide to sulfate. However, the hypothesis of S^0 as second substrate had to be abandoned because of the following pure culture studies.

In continuous cultures, the population density (x) depends on the concentration of the substrate in the inflowing medium ($S_{Reservoir}$), and can be mathematically described as $x = Y \cdot (S_{Reservoir} - s)$, in which Y is the yield factor. The concentration of the left-over substrate is mathematically described as $s = K_s \cdot D/(\mu_{max} - D)$, in which K_s and μ_{max} are the kinetic growth parameters (see section VIII.B.3) and D is the dilution rate. Under steady-state conditions, μ equals D. Thus, in contrast to x, the concentration of s is independent of $S_{Reservoir}$. However, a prerequisite is that the left-over substrate is dispersed in the medium. Consequently, in a culture of *Chromatium*, the mathematical description of s given above, is not expected to hold for the concentration of intracellular S^0. Indeed, raising the $S_{Reservoir}$ in *Chromatium* cultures, not only resulted in increased cell densities, but also in increased concentrations of S^0. As was to be expected, the specific content of S^0 (S^0/x) remained constant. Unanticipated, however, similar phenomena were observed in *Chlorobium* cultures. These data indicate that the S^0 produced by one individual is not detected by other individuals. For *Chromatium* this is not surprising, the observations in *Chlorobium* may be explained by the fact that S^0 somehow remains attached to the cell by which it is produced. The mechanistic explanation could be the presence of appendages as observed in other *Chlorobium* strains (Cohen-Bazire, 1963). The ecological implication of these findings is that the term 'extracellular sulfur' should be interpreted carefully, and should not be read as being fully available to other cells. Thus, it is unlikely that, in the coexistence of *Chromatium* and *Chlorobium*, extracellular S^0 acted as the second substrate (Van Gemerden, 1986).

This intriguing problem was solved by the finding that, in pure cultures of *Chlorobium*, polysulfides (S_3^{2-}) were formed abiotically (Visscher and Van Gemerden, 1988; Visscher et al., 1990). The polysulfide affinities of *Chlorobium* and *Chromatium* were very similar; however in *Chlorobium*, the utilization of polysulfide was inhibited by hydrogen sulfide, and, in addition, polysulfide oxidation required *de novo* protein synthesis (Visscher and Van Gemerden, 1988). In contrast, *Chromatium*, oxidized polysulfide and sulfide simultaneously, and polysulfide utilization occurred without lag (Steudel et al., 1990).

The balanced coexistence between *Chromatium* and *Chlorobium* observed in the laboratory thus can be explained on the basis of (1) a higher sulfide affinity of *Chlorobium*, (2) the formation of polysulfide from the abiotic reaction between H_2S, present in μmolar concentrations, and S^0 produced by *Chlorobium*, (3) the occurrence of sulfide and polysulfide as mutual substrates for *Chromatium* and *Chlorobium*, and (4) the fact that in *Chlorobium* the utilization of polysulfide required induction, in contrast to the oxidation of polysulfide by *Chromatium* (Van Gemerden, 1987). These interactions are visualized in Fig. 8. The relationships between the specific growth rate and the concentraiton of sulfide and polysulfide for each of the organisms are shown in Fig. 9.

The phenomena discussed above, in combination with the differences in absorption spectra of green and purple sulfur bacteria, could offer a clue to the co-existence of these phototrophic bacteria in natural environments. Unfortunately, data on polysulfide concentrations in lakes harboring both Chromatiaceae and Chlorobiaceae presently are lacking.

C. Competition Between Purple and Colorless Sulfur Bacteria

Under constant environmental conditions, chemotrophically growing purple sulfur bacteria appear to have little chance in the competition for sulfide and thiosulfate with genuine colorless sulfur bacteria. The kinetic parameters μ_{max} and K_s for *Thiocapsa roseopersicina* M1 and *Thiobacillus thioparus* T5 for growth on reduced sulfur compounds are shown in Table 11. Accordingly, the sulfide affinity of *Thiobacillus* is 47-fold higher than that of *Thiocapsa*, and the same is true for thiosulfate. However, purple sulfur bacteria compete best when growing phototrophically, and appear to have several other options

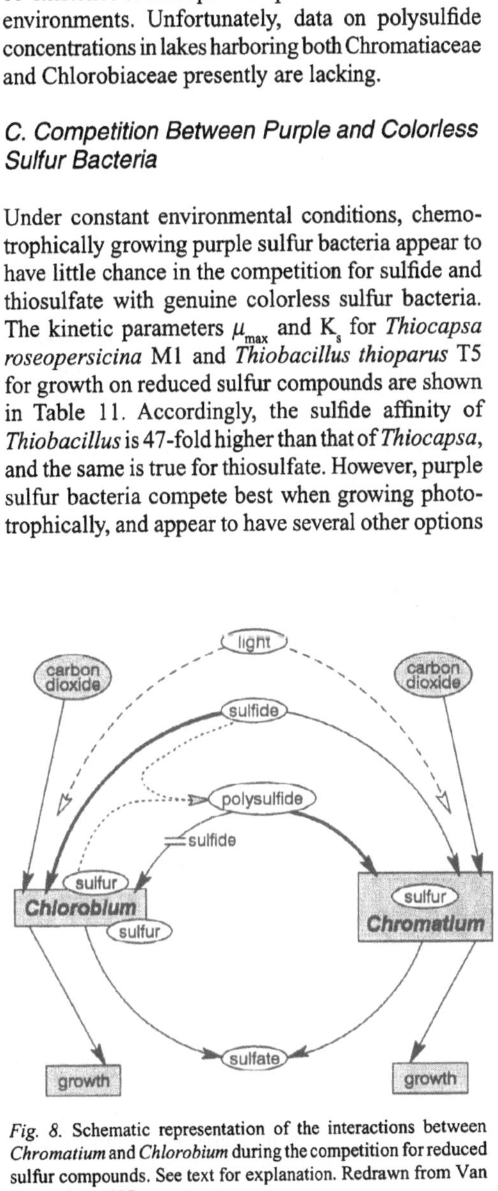

Fig. 8. Schematic representation of the interactions between *Chromatium* and *Chlorobium* during the competition for reduced sulfur compounds. See text for explanation. Redrawn from Van Gemerden, 1987.

Table 11. Comparison of kinetic growth parameters of *Thiocapsa roseopersicina* strain M1 and *Thiobacillus thioparus* strain T5 on reduced sulfur compounds and CO_2. Both organisms were isolated from a microbial mat

Organism	Type of metabolism	Electron donor	μ_{max} [a] (h^{-1})	μ_{max} [b] (h^{-1})	K_s ($\mu mol \cdot l^{-1}$)	K_i ($mmol \cdot l^{-1}$)	Affinity ($h^{-1} \cdot mmol^{-1} \cdot l$)	Yield (mg protein·mmol e^-)	Ref
Thiocapsa roseopersicina strain M1	phototrophic	sulfide	0.091	0.086	8	8	11.4	2.03	(1,7)
		thiosulfate	0.080	0.080	8	n.i. [c]	10.0	2.24	(2,4)
		polysulfide	0.065	0.056	6.7	1.1	9.7	2.1	(3)
		DMS	0.068	0.068	38	– [d]	1.8	2.62	(1)
	chemotrophic	thiosulfate	0.052	0.052	1.5	n.i. [c]	34.7	0.74	(4)
Thiobacillus thioparus strain T5	chemotrophic	sulfide	0.320	0.320	0.6	– [d]	533	0.42	(5)
		thiosulfate	0.336	0.336	0.8	n.i. [c]	420	0.51	(2)
		DMS	0.100	0.100	90	– [d]	1.1	1.01	(8)

[a] theoretical maximum if there were no inhibition, [b] maximum attainable specific growth rate, [c] not inhibitory, [d] not available, substrate assumed not to be inhibitory

(1) Visscher et al. 1990, (2) Visscher et al. 1992, (3) Visscher and Van Gemerden 1991, (4) De Wit and Van Gemerden 1987, (5) Van den Ende and Van Gemerden 1993, (6) Visscher et al. 1991, (7) Van Gemerden 1984, (8) De Wit and Van Gemerden 1990

then chemotrophic growth. Important features are (1) the possibility to migrate, (2) the ability to photosynthesize in the presence of oxygen, and (3) the ambient concentration of oxygen.

Migration away from oxic zones is a successful strategy to avoid inhibition of photopigment synthesis. Since in most ecosystems, sulfide and oxygen have complementary profiles with little overlap, photosynthesis can continue, provided sufficient light is available in the deeper layers. Particularly in sediment ecosystems, having small vertical dimensions compared to stratified lakes, migration can be considered an effective mechanisms.

It was observed in marine coastal sediments that in the early morning the purple sulfur bacteria rapidly moved away from the surface layers once the overlying cyanobacteria started to produce oxygen (Jørgensen, 1982). It was reported by Sorokin (1970) that in the stratified Lake Belovod (Russia), the maximum population density of motile purple sulfur bacteria (*Chromatium okenii*, Pfennig and Trüper, 1992) was at 6 m depth at 6 a.m., and at 13 m depth at 12 a.m., at each time coinciding with the oxygen/sulfide interface. Vertical migration over such long distances could not be observed in Wintergreen Lake (USA) and have been doubted to occur on the basis of maximum swimming rates (Caldwell and Tiedje, 1975). Vertical migration over a distance of 35 cm has been reported for *Chromatium minus* in Lake Cisó (Spain), but, in the same lake, no migration could be observed for the gas-vacuolated *Amoebobacter* M3 (Pedrós-Alió and Sala, 1990). In Rotsee (Switzerland), the gas-vacuolated *Thiopedia rosea* was reported to migrate over 1 m distances (Kohler et al., 1984), and in Lake Holmsjön (Finland), *Chromatium* sp. migrated downwards during the day and upwards during the night, the total distance being 30 cm (Lindholm et al., 1985). Thus, vacuolated or motile phototrophic sulfur bacteria appear to be able to position themselves in the chemocline, probably as a response to the prevailing sulfide and/or oxygen concentrations.

In the second place, cessation of photopigment synthesis does not necessarily imply cessation of photosynthesis. The Calvin-cycle enzymes of *Thiocapsa roseopersicina* are not repressed in the presence of oxygen (Kondratieva et al., 1976). It has been suggested that the organism would shift from phototrophy to chemotrophy when conditions become oxic (Madigan, 1988). However, although pigment synthesis does not occur in the presence of oxygen, pigments already present do not become inoperative, as illustrated below.

During phototrophy, electron donors are exclusively used for the synthesis of cell material. In contrast, during chemotrophy approximately two thirds of the pool of electron donors is oxidized to provide for the energy metabolism, and only one third results in the synthesis of cell material (see section VIII.B.4). Consequently, a judgment of the energy metabolism of *Thiocapsa* is facilitated by yield estimates. This purple sulfur bacterium

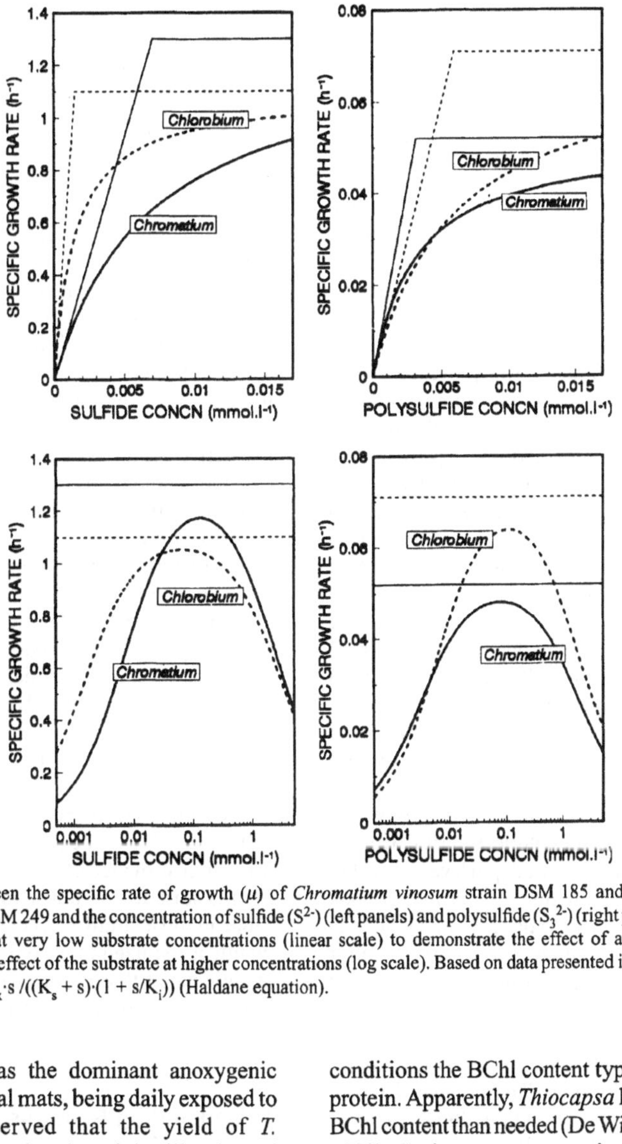

Fig. 9. Relationship between the specific rate of growth (μ) of *Chromatium vinosum* strain DSM 185 and *Chlorobium limicola* f. *thiosulfatophilum* strain DSM 249 and the concentration of sulfide (S^{2-}) (left panels) and polysulfide (S_3^{2-}) (right panels). Top panels show the specific growth rates at very low substrate concentrations (linear scale) to demonstrate the effect of affinity, the lower panels demonstrate the inhibitory effect of the substrate at higher concentrations (log scale). Based on data presented in Table 8, and calculated using the equation $\mu = \mu_{max} \cdot s / ((K_s + s) \cdot (1 + s/K_i))$ (Haldane equation).

frequently is found as the dominant anoxygenic phototroph in microbial mats, being daily exposed to oxygen. It was observed that the yield of *T. roseopersicina*, when incubated in illuminated continuous cultures exposed to a regime of 21 h oxic/ 3 h anoxic, was virtually identical to that observed during continuously anoxic conditions. In other words, the cells were still growing phototrophically during the 21 h oxic periods, despite the fact that synthesis of photopigments only occurred during the 3 h anoxic periods (De Wit and Van Gemerden, 1990). The content of BChl *a* varied from 4 to 7 $\mu g \cdot mg^{-1}$ protein, whereas during continuously anoxic conditions the BChl content typically is 30 $\mu g \cdot mg^{-1}$ protein. Apparently, *Thiocapsa* has a 4–7 fold higher BChl content than needed (De Wit and Van Gemerden, 1990). Such an excess capacity would not have any ecological relevance if the environmental conditions would allow the synthesis of photopigments at all times, but is of great advantage under fluctuating conditions. Despite the fact that no synthesis of photopigments occurs for prolonged periods of time, it enables *Thiocapsa* to grow phototrophically in the presence of oxygen. Conceivably, this capacity is functional as well during strongly fluctuating light intensities, due to the passing by of clouds.

Applying a regimen of 1 h anoxic/23 oxic, resulted in a BChl content varying between 0.8 and 2.5 $\mu g \cdot mg^{-1}$ protein, which is too low to allow for full phototrophic growth. However, even under these conditions *Thiocapsa* did not shift to full chemotrophy, since the yield was observed to be higher than during full chemotrophy, but lower than during full phototrophy, indicating that growth was the result of a mixed photo-chemotrophic metabolism (Schaub and Van Gemerden, 1993).

In natural environments, illumination usually coincides with the presence of oxygen, and darkness with anoxic conditions. Incubation of *Thiocapsa* at regimen in which illumination was provided during oxic periods only and not during anoxic periods, resulted in photopigment synthesis in the dark. It was calculated that ample energy was provided by the concomitant degradation of glycogen, which had been synthesized during the previous oxic/light period (De Wit and Van Gemerden, 1990). These phenomena are particularly relevant to immobile benthic organisms unable to migrate.

In the third place, it is of importance to evaluate the oxygen conditions at the depth horizon where sulfide oxidation takes place. The analogy between the three systems analyzed by Jørgensen (1982), is the prevailing low oxygen concentration at the site of maximal sulfide oxidation. As judged from the profiles of oxygen and sulfide in a microbial mat on the Frisian Island of Texel (The Netherlands), in which purple sulfur bacteria and colorless sulfur bacteria were most abundant in the same depth layer (Visscher et al., 1992), sulfide oxidation also primarily took place at low oxygen concentrations. At first sight, the much higher sulfide affinities of colorless sulfur bacteria compared to purple sulfur bacteria, seem to contradict the observed coexistence of *Thiocapsa* and *Thiobacillus*.

It has been demonstrated in chemostat experiments with pure cultures of *Thiobacillus thioparus* T5, that, in the presence of excess O_2, sulfide virtually completely is oxidized to sulfate. During oxygen shortage, the concentration of sulfide still remained below the detection limit; however various other reduced forms of sulfur, notably S^0 and $S_2O_3^{2-}$, were excreted into the medium. Under severe oxygen-limiting conditions, the concentration of sulfide in the culture was $\leq 1\ \mu mol \cdot l^{-1}$, but from the 7.1 $mmol \cdot l^{-1}$ sulfide supplied, only 1.87 $mmol \cdot l^{-1}$ was oxidized to sulfate, the remaining being present as reduced sulfur intermediates (Van den Ende and Van Gemerden, 1993). *Thiocapsa*, once introduced in such a *Thiobacillus* culture, thus is provided with ample substrates to grow. Although oxygen is supplied continuously, its concentration in the mixed culture is lowered to undetectable levels by *Thiobacillus*, and thus, pigment synthesis in *Thiocapsa* is not inhibited. The result is co-existence of the colorless sulfur bacterium and the phototrophic sulfur bacterium, each growing in its own preferred mode (F. P. van den Ende and A. Laverman, personal communication).

It may be argued that in the example described above, the two organisms did not actually compete for mutual substrates, since *Thiobacillus* oxidized all sulfide and *Thiocapsa* grew on the products formed thereof. However, the fraction of a mutual substrate utilized by each population not only depends on their respective affinities, but as well on the population densities. Potentially, *Thiocapsa* might have been able to claim part of the sulfide as a result of its growth on the other sulfur species. Since the formation of these were the result of the activities of *Thiobacillus*, *Thiocapsa* will not be able to outcompete the colorless sulfur bacterium (Van Gemerden, 1993). In a similar fashion, purple sulfur bacteria might profit from their chemotrophic potential in the competition with colorless sulfur bacteria (Kuenen, 1989). This mass effect also may play a role in the competition between autotrophic and heterotrophic nitrifying bacteria (Kuenen, 1989; Tiedje et al., 1982).

In conclusion, colorless sulfur bacteria and purple sulfur bacteria can coexist under conditions of oxygen limitation, provided of course, that the penetration of light is sufficient to allow for phototrophic growth. Since in most stratified ecosystems sulfide predominantly is oxidized in the oxygen-sulfide interface, it is anticipated that in many systems these different types of sulfur bacteria are able to thrive in the same depth layers.

X. Concluding Remarks

A proper understanding of the ecology of phototrophic sulfur bacteria requires a careful description of both bacterial populations and their environment. This understanding, however, is often hampered by the nature of the information available. As van Niel pointed out (1955):

> Unfortunately, the relationships between the characteristics of an environment and the flora and fauna found therein must often be deduced from

observations made at the time when the organisms are already present in large numbers. This is not always a satisfactory guide to an interpretation of ecological factors, because at such time the environment may have been considerably modified by the activities of the organisms themselves.

Underlying this statement is the idea that bacterial communities cannot be considered static systems and that a one time description of a fully developed environment actually says very little. What is actually relevant in order to understand how phototrophic sulfur bacteria bloom and decay is the analysis of the conditions, both biotic and abiotic, which regulate their growth, as well as the set of factors determining their removal from the system.

Since natural communities are complex and environmental conditions are often unpredictable, the study of their ecology can benefit from laboratory experiments using model systems. Although these models constitute extremely simplified versions of nature and the information they provide must be used cautiously, they constitute a unique tool for the testing of hypotheses both on the interaction between organisms, and on the interaction between the organisms and their environment. The successful development of a laboratory system, however, requires a detailed knowledge of the environmental conditions to which the organisms are exposed. Thus, it would be extremely helpful if field studies, in which so much time and effort is invested, provided such information.

From the analysis of the different case studies, it seems apparent that study of the ecology of phototrophic sulfur bacteria (actually the ecology of any organism) must contemplate the existence of fluctuations. There is a rather general agreement on the importance of fluctuations and a considerable body of supportive evidence has been gathered along the years. To assess which kind of physiological adaptations are better suited to deal with such fluctuations is the task of the experimental ecophysiologist.

Acknowledgments

The authors would like to thank Olga Sánchez and Pep Gasol for assistance with the references. The administration of the Biology Center in Haren provided extra space and facilities which made writing and discussion extremely comfortable. This work was supported by DGICYT grant PB91-0075-C02-02 to JM.

References

Abellà CA, Montesinos E and Turet J (1985) Colonization and dynamics of phototrophic bacteria in a recently formed lagoon in Banyoles karstic area (Girona, Spain). Scient Gerund 10: 33–49

Aizenshtat Z, Stoler A, Cohen Y and Nielsen H (1983) The geochemical sulphur enrichment of recent organic matter by polysulfides in the Solar Lake. In: Bjoroy M, Albrecht P, Cornford C, de Groot K, Eglington G, Galimov E Leythaeuser D, Pelet R, Rullkotter J and Speers G (eds) Advances in Organic Geochemistry, pp 279–288. Wiley, Chicester

Anagnostidis K and Overbeck J (1966) Methanoxydierer und hypolimnische Schwefelbakterien. Studien zur okölogischen Biocönotik der Gewässermikroorganismen. Ber Dtsch Bot Ges 79: 163–174

Awramik SM (1984) Ancient stromatolites and microbial mats. In: Cohen Y, Castenholz RW and Halvorson HO (eds) Microbial Mats: Stromatolites, pp 1–21. Alan R. Liss Inc., New York

Baker AL, Kromer Baker K and Tyler PA (1985) Fine-layer depth relationships of lakewater chemistry, planktonic algae and photosynthetic bacteria in meromictic Lake Fidler, Tasmania. Freshwater Biol 15: 735–747

Banens RJ (1990) Occurrence of hypolimnetic blooms of the purple sulfur bacterium, *Thiopedia rosea,* and the green sulfur bacterium, *Chlorobium limicola,* in an Australian Reservoir. Aust J Mar Freshwater Res 41: 223–235

Beeftink HH and Van Gemerden H (1979) Actual and potential rates of substrate oxidation and product formation in continuous cultures of *Chromatium vinosum* . Arch Microbiol 121: 161–167

Bergstein T, Henis Y and Cavari DZ (1979) Investigations on the photosynthetic sulfur bacterium *Chlorobium phaeobacteroides* causing seasonal blooms in Lake Kinneret. Ca. J Microbiol 25: 999–1007

Beudeker RF, Gottschal JC and Kuenen JG (1982) Reactivity versus flexibility in thiobacilli. Antonie van Leeuwenhoek 48: 39–51

Biebl H and Pfennig N (1978) Growth yields of green sulfur bacteria in mixed cultures with sulfur and sulfate reducing bacteria. Arch Microbiol 117: 9–16

Bogorov LV (1974) About the properties of *Thiocapsa roseopersicina* strain BBS, isolated from the estuary of the White Sea. Microbiology (Transl) 43: 275–280

Broch-Due M, Ormerod JG and Fjerdingen BS (1978) Effect of light intensity on vesicle formation in *Chlorobium* . Arch Microbiol 116: 269–274

Caldwell DE and Tiedje JM (1975) The structure of anaerobic bacterial communities in the hypolimnia of several Michigan lakes. Can J Microbiol 21: 377–385

Castenholz RW (1984) Composition of hot spring microbial mats: a summary. In: Cohen Y, Castenholz RW and Halvorson HO (eds) Microbial Mats: Stromatolites, pp101–119. Alan R. Liss Inc., New York

Castenholz RW (1988) The green sulfur and nonsulfur bacteria of hot springs. In: Olson JM, Ormerod JG, Amesz J, Stackebrandt E and Trüper HG (eds) Green Photosynthetic

Bacteria, pp 243–255. Plenum Press, New York

Caumette P (1984) Distribution and characterization of phototrophic bacteria isolated from the water of Bietri Bay (Ebrie Lagoon, Ivory Coast). Can J Microbiol 30: 273–284

Caumette P (1989) Ecology and general physiology of anoxygenic phototrophic bacteria in benthic environments. In: Cohen Y and Rosenberg E (eds) Microbial Mats: Physiological Ecology of Benthic Microbial Communities, pp 283–304. ASM, Washington DC

Caumette P, Pagano M and Saint-Jean L (1983) Répartition verticale du phytoplancton, des bactéries et du zooplancton dans un milieu stratifié en baie du Biétrie (lagune Ebrié, Cote d'Ivoire). Hydrobiologia 106: 135–148

Clarke KJ, Finlay BJ, Vicente E, Lloréns H and Miracle MR (1993) The complex life-cycle of a polymorphic prokaryote epibiont of the photosynthetic bacterium *Chromatium weissei* . Arch Microbiol 159: 498–505

Cloern JE, Cole BE and Oremland RS (1983a) Seasonal changes in the chemistry and biology of a meromictic lake (Big Soda Lake, Nevada, USA). Hydrobiologia 105:, 195–206

Cloern JE, Cole BE and Oremland RS (1983b) Autotrophic processes in meromictic Big Soda Lake, Nevada. Limnol Oceanogr 28: 1049–1061

Cohen Y, Krumbein WE and Shilo M (1977) Solar Lake (Sinai). 3. Bacterial distribution and production. Limnol Oceanogr 22: 621–634

Cohen-Bazire G (1963) Some observations on the organization of the photosynthetic apparatus in purple and green bacteria. In: Gest H, San Pietro A and Vernon LP (eds) Bacterial Photosynthesis, pp 89–110. The Antioch Press, Yellow Springs, Ohio

Cohen-Bazire G, Sistrom WR and Stanier RY (1957) Kinetic studies of pigment synthesis by non-sulfur purple bacteria. J Cellular Comp Physiol 49: 25–68

Croome RL (1986) Biological studies of meromictic lakes. In: De Deckker P and Williams WD (eds) Limnology in Australia, pp 113–130. CSIRO, Melbourne

Culver DA and Brunskill GJ (1969) Fayetteville Green Lake, New York. V. Studies of primary production and zooplankton in a meromictic marl lake. Limnol Oceanogr 14: 862–873

Czeczuga B (1965) *Chlorobium limicola* Nads. (Chlorobiaceae) and the distribution of chlorophyll in some lakes of the Mazur lake district. Hydrobiologia 25: 412–423

Czeczuga B (1966) An attempt to determine the primary production of the green sulphur bacteria, *Chlorobium limicola* Nads, (Chlorobacteriaceae). Hydrobiologia 31: 317–333

Czeczuga B (1968) Primary production of the green hydrosulfuric bacteria *Chlorobium limicola* Nads. (Chlorobiaceae). Photosynthetica (Prague) 2: 11–15

Chen KY and Gupta K (1972) Kinetics of oxidation of aqueous sulfide by O_2. Environ Sci Tech 6: 529–537

D'Amelio ED, Cohen Y and Des Marais DJ (1989) Comparative functional ultrastructure of two hypersaline submerged cyanobacterial mats: Guerrero Negro, Baja California Sur, Mexico, and Solar Lake, Sinai, Egypt. In: Cohen Y and Rosenberg E (eds) Microbial Mats: Physiological ecology of benthic microbial communities, pp 97–113. ASM, Washington DC

De Wit R (1989) Interactions between phototrophic bacteria in marine sediments. Ph.D. Thesis, University of Groningen, The Netherlands

De Wit R and Van Gemerden H (1987) Chemolithotrophic growth of the phototrophic sulfur bacterium *Thiocapsa roseopersicina* . FEMS Microbiol Ecol 45: 117–126

De Wit R and Van Gemerden H (1990) Growth and metabolism of the purple sulfur bacterium *Thiocapsa roseopersicina* under combined light/dark and oxic/anoxic regimens. Arch Microbiol 154: 459–464

De Wit R and Van Gemerden H (1990) Growth of the phototrophic purple sulfur bacterium *Thiocapsa roseopersicina* under oxic/anoxic regimens in the light. FEMS Microbiol Ecol 73: 69–76

De Wit R, Jonkers HM, Van den Ende FP and Van Gemerden H (1989) In situ fluctuations of oxygen and sulphide in marine microbial sediment ecosystems. Neth J Sea Res 23: 271–281

Dickman M and Artuz I (1978) Mass mortality of photosynthetic bacteria as a mechanism for dark lamina formation in sediments of the Black Sea Nature 275:, 191–195

Drews G and Imhoff JF (1991) Phototrophic sulfur bacteria. In: Shively JM and Barton LL (eds) Variations in Autotrophic Life, pp 51–97. Academic Press, London

Dubinina GA and Kuznetsov SI (1976) The ecological and morphological characteristics of microorganisms in Lesnaya Lamba (Karelia). Int Revue ges Hydrobiol 61: 1–19

Dubinina GA, Gorlenko VM and Suleimanov YI (1973) A study of microorganisms involved in the circulation of manganese, iron, and sulfur in meromictic Lake Gek-Gel. Mikrobiologiya 42: 918–924

Dubinsky Z and Berman T (1979) Seasonal changes in the spectral composition of downwelling irradiance in Lake Kinneret (Israel). Limnol Oceanogr 24: 652–663

Dutton PL and Prince RC (1978) Reaction-center-driven cytochrome interactions in electron and proton translocation and energy coupling. In: Clayton RK and Systrom WR (eds) The photosynthetic bacteria, pp 525–570. Plenum Press, New York

Dyer BD, Gaju N, Pedrós-Alió C, Esteve I and Guerrero R (1986) Ciliates from a freshwater sulfuretum. BioSystems, 19: 127–135

Eisenthal R and Cornish-Bowden A (1974) The direct linear plot. A new graphical procedure for estimating enzyme kinetic parameters. Biochem J 139: 715–720

Eloranta P (1985) Hypolimnetic chlorophyll maximum by algae and sulphur bacteria in one eutrophic pond. Arch Hydrobiol Suppl 71: 459–469

Esteve I, Guerrero R, Montesinos I and Abellà C (1983) Electron microscopy study of the interaction of epibiontic bacteria with *Chromatium minus* in natural habitats. Microbial Ecol 9: 57–64

Esteve I, Gaju N, Mir J and Guerrero R (1992) Comparison of techniques to determine the abundance of predatory bacteria attacking Chromatiaceae. FEMS Microbiol Ecol 86: 205–211

Finlay BJ, Clarke KJ, Vicente E and Miracle MR (1991) Anaerobic ciliates from a sulphide-rich solution lake in Spain. Europ J Protistol 27: 148–159

Folt CL, Wevers MJ, Yoder-Williams MP and Howmiller RP (1989) Field study comparing growth and viability of phototrophic bacteria. Appl Environ Microbiol 55: 78–85

Fry B (1986) Sources of carbon and sulfur nutrition for consumers in three meromictic lakes of New York state. Limnol Oceanogr 31: 79–88

Fuller RC (1978) Photosynthetic carbon metabolism in the green and purple bacteria. In: Clayton RK and Systrom WR (eds)

The photosynthetic bacteria, pp 691–705. Plenum Press, New York

García-Cantizano J (1992) Analisis funcional de la comunidad microbiana en ecosistemas planctónicos. PhD Thesis. Autonomous University of Barcelona. Spain

Garcia-Gil LJ and Abellà CA (1992a) Population dynamics of phototrophic bacteria in three basins of Lake Banyoles (Spain). Hydrobiologia 243/244: 87–94

García-Gil LJ and Abellà CA (1992b) Microbial ecology of planktonic filamentous phototrophic bacteria in holomictic freshwater lakes. Hydrobiologia 243/244: 79–86

Gasol JM, Guerrero R and Pedrós-Alió C (1991) Seasonal variations in size structure and procaryotic dominance in sulfurous Lake Cisó. Limnol Oceanogr 36: 860–872

Göbel F (1978) Quantum efficiencies of growth. In: Clayton RK and Systrom WR (eds) The Photosynthetic Bacteria, pp 907–925. Plenum Press, New York

Goehle KH and Storr JF (1978) Biological layering resulting from extreme meromictic stability, Devil's Hole, Abaco Island, Bahamas. Verh Internat Verein Limnol 20: 550–555

Gorlenko VM (1974) The oxidation of thiosulfate by *Amoebobacter roseus* in the dark under microaerophilic conditions. Microbiology (Transl) 43: 624–625

Gorlenko VM and Kusnezow SI (1972) Über die photosynthesierenden bakterien des Kononjer-Sees. Arch Hydrobiol 70: 1–13

Gorlenko VM, Chebotarev EN and Kachalkin VI (1973) Microbiological processes of oxidation of hydrogen sulfide in the Repnoe Lake (Slavonic lakes). Mikrobiologiya 42: 723–728

Gorlenko VM, Chebotarev EN and Kachalkin VI (1974a) Participation of microorganisms in the circulation of sulfur in Pomyaretskoe Lake. Mikrobiologiya : 908–914

Gorlenko VM, Chebotarev EN and Kachalkin VI (1974b) Microbial oxidation of hydrogen sulfide in Lake Veisovo (Slavyansk Lake). Mikrobiologiya 43: 530–534

Gorlenko VM, Vainstein MB and Kachalkin VI (1978) Microbiological characteristic of Lake Mogilnoye. Arch Hydrobiol 81: 475–492

Gorlenko VM, Dubinina GA and Kuznetsov SI (1983) The ecology of aquatic micro-organisms E. Schweizeizerbart'sche Verlagsbuchhandlung, Stuttgart

Gottschal JC (1986) Mixed substrate utilization by mixed cultures. In: Leadbetter ER and Poindexter JS (eds) Bacteria in Nature, Vol 2, pp 261–292. Plenum Press, New York

Guerrero R, Montesinos E, Pedrós-Alió C, Esteve I, Mas J, Van Gemerden H, Hofman PAG and Bakker JF (1985) Phototrophic sulfur bacteria in two Spanish lakes: Vertical distribution and limiting factors. Limnol Oceanogr 30: 919–931

Guerrero R, Pedrós-Alió C, Esteve I, Mas J, Chase D and Margulis L (1986) Predatory prokaryotes: Predation and primary consumption evolved in bacteria. Proc Natl Acad Sci USA 83: 2138–2142

Guhl BE and Finlay BG (1993) Anaerobic predatory ciliates track seasonal migrations of planktonic photosynthetic bacteria. FEMS Microbiol Lett 107: 313–316

Hall KJ and Northcote TG (1990) Production and decomposition processes in a saline meromictic lake. Hydrobiologia, 197: 115–128

Hansen TA (1974) Sulfide als electrondonor voor Rhodospirillaceae. PhD Thesis, University of Groningen, The Netherlands

Hansen TA and Van Gemerden H (1972) Sulfide utilization by purple nonsulfur bacteria. Arch Mikrobiol 86: 49–56

Healey FP (1980) Slope of the Monod equation as an indicator of advantage in nutrient competition. Microb Ecol 5: 287–293

Hurlbert RE (1967) Effect of oxygen on viability and substrate utilization in *Chromatium* . J Bacteriol 93: 1346–1352

Hurlbert RE and Lascelles J (1963) Ribulose diphosphate carboxylase in Thiorhodaceae. J Gen Microbiol 33: 445–458

Hutchinson GE (1957) A treatise on limnology. I. Geography, physics and chemistry of lakes. Wiley, New York

Iwakuma T and Yasuno M (1983) A comparison of several mathematical equations describing photosynthesis-light curve for natural phytoplankton populations. Arch Hydrobiol 97: 208–226

Jassby AD and Platt T (1976) Mathematical formulation of the relationship between photosynthesis and light for phytoplankton. Limnol Oceanogr 21: 540–547

Javor BJ and Castenholz RW (1981) Laminated microbial mats Laguna Guerrero Negro, Mexico. Geomicrobiol J 2: 237–274

Javor BJ and Castenholz RW (1984) Productivity studies of microbial mats Laguna Guerrero-Negro Mexico. In: Cohen Y, Castenholz RW and Halvorson HO (eds) Microbial Mats: Stromatolites, pp 149–170. Alan R. Liss Inc., New York

Jørgensen BB (1977) The sulphur cycle of a coastal marine sediment (Limfjorden, Denmark). Limnol Oceanogr 22: 814–832

Jørgensen BB (1982) Ecology of the bacteria of the sulphur cycle with special reference to the anoxic-oxic interface. Phil Trans R Soc London. 298: 543–561

Jørgensen BB and Des Marais DJ (1986a) A simple fiber-optic microprobe for high resolution light measurements: applications in marine sediments. Limnol Oceanogr 31: 1374–1381

Jørgensen BB and Des Marais DJ (1986b) Competition for sulfide among colorless and purple sulfur bacteria in cyanobacterial mats. FEMS Microbiol Ecol. 38: 179–186

Jørgensen BB and Des Marais DJ (1988) Optical properties of benthic photosynthetic communities: fiber-optic studies of cyanobacterial mats. Limnol Oceanogr 33: 99–113

Jørgensen BB, Kuenen J G and Cohen Y (1979) Microbial transformations of sulfur compounds in a stratified lake (Solar Lake, Sinai). Limnol Oceanogr 24: 799–822

Jørgensen BB, Revsbech NP and Cohen Y (1983) Photosynthesis and structure of benthic microbial mats: microelectrode and SEM studies of four cyanobacterial communities. Limnol Oceanogr 28: 1075–1093

Jørgensen BB, Fossing H, Wirsen CO and Jannasch HW (1991) Sulfide oxidation in the anoxic Black Sea chemocline. Deep-Sea Research 38: 1083–1103

Kämpf C and Pfennig N (1980) Capacity of Chromatiaceae for chemotrophic growth. Specific respiration rates of *Thiocystis violacea* and *Chromatium vinosum*. Arch Microbiol 127: 125–135

Kämpf C and Pfennig N (1986a) Chemoautotrophic growth of *Thiocystis violacea, Chromatium gracile* and *C. vinosum* in the dark at various O_2-concentrations. J Basic Microbiol 26: 517–531

Kämpf C and Pfennig N (1986b) Isolation and characterization of some chemoautotrophic Chromatiaceae. J Basic Microbiol 26: 507–515

Keller MD, Bellows WK and Guillard RRL (1989) Dimethyl sulfide production in marine phytoplankton. In: Saltzman ES and Cooper WJ (eds) Biogenic Sulfur in the Environment, pp 167–182. American Chemical Society Symp. Ser. 393, Washington DC

Kelly DP (1982) Biochemistry of the chemolithotrophic oxidation of inorganic sulphur. Phil Trans R Soc London 298: 499–528

Kiene RP and Taylor BF (1988a) Biotransformations of organosulfur compounds in sediments via 3-mercaptopropionate. Nature 332: 148–150

Kiene RP and Taylor BF (1988b) Demethylation of dimethylsulfoniopropionate and production of thiols in anoxic marine sediments. Appl Environ Microbiol 54: 2208–2212

Kiene RP and Visscher PT (1987) Production and fate of methylated sulfur compounds from methionine and dymethylsulfoniopropionate in anoxic salt marsh sediments. Appl Environ Microbiol 53: 2426–2434

Kiene RP, Malloy KD and Taylor BF (1990) Sulfur-containing amino acids as precursors of thiols in anoxic coastal sediments. Appl Environ Microbiol 56: 156–161

Kohler H-P, Ahring G, Abella C, Ingvorsen K, Keweloh H, Laczko E, Stupperich E and Tomei F (1984) Bacteriological studies on the sulfur cycle in the anaerobic part of the hypolimnion and in the surface sediments of Rotsee in Switzerland. FEMS Microbiol Lett 21: 279–289

Kondratieva EN (1979) Interrelation between modes of carbon assimilation and energy production in phototrophic purple and green bacteria. In: Quayle JR (ed) Microbial Biochemistry (Int. Rev. of Biochemistry, Vol. 21, pp 117–175). Univ. Park Press, Baltimore

Kondratieva EN, Zhukov, Ivanovsky RN, Petushkova YP and Monosov EZ (1976) The capacity of phototrophic sulfur bacterium *Thiocapsa roseopersicina* for chemosynthesis. Arch Microbiol 108: 287–292

Kuenen GJ (1989) Comparative ecophysiology of the nonphototrophic sulfide-oxidizing bacteria. In: Cohen Y and Rosenberg E (eds) Microbial Mats: Physiological Ecology of Benthic Microbial Communities, pp 349–365. ASM, Washington DC

Kuznetsov SI (1970) The microflora of lakes and its geochemical activities. University of Texas Press, Austin

Lassen C, Ploug H and Jørgensen BB (1992) A fibre-optic scalar irradiance microsensor: Application for spectral light measurements in sediments. FEMS Microbiol Ecol 86: 247–254

Lawrence JR, Haynes RC and Hammer UT (1978) Contribution of photosynthetic green sulfur bacteria to total primary production in a meromictic saline lake. Verh Internat Verein Limnol 20: 201–207

Lindholm T, Weppling K and Jensen HS (1985) Stratification and primary production in a small brackish lake studied by close-interval siphon sampling. Verh Internat Verein Limnol 22: 2190–2194

Lippert KD and Pfennig (1969) Die Verwertung von molekularem Wasserstoff durch *Chlorobium thiosulfatophilum* Wachstum and CO_2-Fixierung. Arch Mikrobiol 65: 29–47

Lord III CJ and Church TM (1983) The geochemistry of salt marshes: Sedimentary ion diffusion, sulfate reduction, and pyritization. Geochim Cosmochim Acta 47: 1381–1391

Luther III GW and Church TM (1988) Seasonal cycling of sulfur and iron in porewaters of a Delaware salt marsh. Mar Chem 23: 295–309

Luther III FW, Church TM, Scudlark JR and Cosman M (1986) Inorganic and organic sulfur cycling in salt-marxh pore waters. Science 232: 746–749

Madigan MT (1984) A novel photosynthetic purple bacterium isolated from a Yellowstone hot spring. Science 225: 313–315

Madigan MT (1988) Microbiology, physiology, and ecology of phototrophic bacteria. In: Zehnder AJB (ed) Biology of Anaerobic Microorganisms, pp 39–111. Wiley-Liss, New York

Mas J, Pedrós-Alió C and Guerrero R (1990) In situ specific loss and growth rates of purple sulfur bacteria in Lake Cisó. FEMS Microbiol Ecol 73: 271–281

Mas J and Van Gemerden H (1992) Phosphate-limited growth of *Chromatium vinosum* in continuous culture. Arch Microbiol 157: 135–140

Massana R, Gasol JM, Jürgens K and Pedrós-Alió C. submitted. Impact of *Daphnia pulex* on a metalimnetic microbial community. J Plankton Res 16: 1379–1399

Matheron R (1976) Contribution à l'étude écologique systématique et physiologique des Chromatiaceae et des Chlorobiaceae isolées des sédiments marins. Ph.D. Thesis. University Aix-Marseille III, France

Matsuyama M (1978) Limnological aspects of meromictic Lake Suigetsu: its environmental conditions and biological metabolism. Bull Fac Fish Nagasaki Univ. 44: 1–65

Matsuyama M (1981) Comparative aspects of a small coastal lake, Kaiike, on Kamikoshiki Island, Southern Kyushu, Japan. Verh Internat Verein Limnol 21: 979–986

Matsuyama M (1987) A large phototrophic bacterium densely populating the O_2-H_2S interface of Lake Kaiike on Kamikoshiki Island, Southwest Japan Acta Acad Aboensis 47: 29–43

Matsuyama M and Shirouzu E (1978) Importance of photosynthetic sulfur bacteria, *Chromatium* sp. as an organic matter producer in Lake Haiike. Jap J Limnol 39: 103–111

Mazumder A and Dickman MD (1989) Factors affecting the spatial and temporal distribution of phototrophic sulfur bacteria. Arch Hydrobiol 116: 209–226

Millero FJ (1991) The oxidation of sulfide in Black Sea waters. Deep Sea Res 38: 1139–1150

Mir J, Martínez-Alonso M, Esteve I and Guerrero R (1991) Vertical stratification and microbial assemblage of a microbial mat in the Ebro Delta (Spain). FEMS Microbiol Ecol 86: 59–68

Miracle MR and Vicente E (1985) Phytoplankton and photosynthetic sulphur bacteria production in the meromictic coastal lagoon of Cullera (Valencia, Spain). Verh Internat Verein Limnol 22: 2214–2220

Mitchell JG, Martínez-Alonso M, Lalucat J, Esteve I and Brown S (1991) Velocity changes, long runs and reversals in the *Chromatium minus* swimming response. J Bacteriol 173: 997–1003

Montesinos E (1982) Ecofisiología de la fotosíntesis bacteriana. Ph.D. thesis. Autonomous University of Barcelona, Spain

Montesinos E (1987) Change in size of *Chromatium minus* cells in relation to growth rate, sulfur content, and photosynthetic activity: A comparison of pure cultures and field populations. Appl Environ Microbiol 53: 864–871

Montesinos E and Esteve I (1984) Effect of algal shading on the net growth and production of phototrophic sulfur bacteria in lakes of the Banyoles karstic area. Verh Int Verein Limnol 22: 1102–1105

Montesinos E and Van Gemerden H (1986) The distribution and metabolism of planktonic phototrophic bacteria. In: Megusar

F and Gantar M (eds) Perspectives in Microbial Ecology, pp 344–359. Slovene Society for Microbiology, Ljubljana

Montesinos E, Guerrero R, Abellà C and Esteve I (1983) Ecology and physiology of the competition for light between *Chlorobium limicola* and *Chlorobium phaeobacteroides* in natural habitats. Appl Environ Microbiol 46: 1007–1016

Nicholson JAM, Stolz JF and Pierson BK (1987) Structure of a microbial mat at Great Sippewissett Marsh, Cape Cod, Massachusetts. FEMS Microbiol Ecol 45: 343–364

Ormerod JG and Sirevåg R (1983) Essential aspects of carbon metabolism. In: Ormerod JG (ed) The Phototrophic Bacteria: Anaerobic Life in the Light, pp 100–119. Blackwell Scientific Publications, Oxford

Overmann J and Pfennig N (1992) Continuous chemotrophic growth and respiration of Chromatiaceae species at low oxygen concentrations. Arch Microbiol 158: 59–67

Overmann J and Tilzer MM (1989) Control of primary productivity and the significance of photosynthetic bacteria in a meromictic kettle lake. Mittlerer Buchensee, West-Germany. Aquatic Sciences 51: 262–278

Overmann J, Beatty JT, Hall KJ, Pfennig N and Northcote TG (1991) Characterization of a dense, purple sulfur bacterial layer in a meromictic salt lake. Limnol Oceanogr 36: 846–859

Overmann J, Cypionka H and Pfennig N (1992) An extremely low-light adapted phototrophic sulfur bacterium from the Black Sea. Limnol Oceanogr 37: 150–155

Parker RD and Hammer UT (1983) A study of the Chromatiaceae in a saline meromictic lake in Saskatchewan, Canada. Int Revue ges Hydrobiol. 68: 839–851

Parker RD, Lawrence JR and Hammer UT (1983) A comparison of phototrophic bacteria in two adjacent saline meromictic lakes. Hydrobiologia 105: 53–61

Parkin TB and Brock TD (1980a) Photosynthetic bacterial production in lakes: the effects of light intensity. Limnol Oceanogr 25: 711–718

Parkin TB and Brock TD (1980b) The effects of light quality on the growth of phototrophic bacteria in lakes. Arch Microbiol 125:, 19–27

Parkin TB and Brock TBD (1981) Photosynthetic bacterial production and carbon mineralization in a meromictic lake. Arch Hydrobiol 91: 366–382

Pedrós-Alió C and Guerrero R (1993) Microbial ecology in Lake Cisó. Adv. Microb Ecol 13: 155–209

Pedrós-Alió C and Sala MM (1990) Microdistribution and diel vertical migration of flagellated vs. gas-vacuolate purple sulfur bacteria in a stratified water body. Limnol Oceanogr 35: 1637–1644

Pedrós-Alió C, Mas J, Gasol J M and Guerrero R (1989) Sinking speeds of free-living phototrophic bacteria determined with covered and uncovered traps. J Plankton Res 11: 887–905

Pfennig N (1965) Anreicherungskulturen für rote und grüne Schwefelbakterien. Zentr Bakteriol Parasitenk Abt I, Suppl I: 179–189

Pfennig N (1970) Dark growth of phototrophic bacteria under microaerophilic conditions. J Gen Microbiol 61: i

Pfennig N (1978) General physiology and ecology of photosynthetic bacteria. In: Clayton RK and Systrom WR (eds) The photosynthetic Bacteria, pp 3–18. Plenum Press, New York

Pfennig N (1989) Ecology of phototrophic purple and green sulfur bacteria. In: Schlegel HG and Bowien B. Autotrophic bacteria. Springer Verlag, New York pp 97–116

Pfennig N and Lippert KD (1966) Über das Vitamin B_{12}-Bedürfnis phototropher Schwefelbakterien. Arch Mikrobiol 55: 245–256

Pfennig N and Trüper HG (1992) The family Chromatiaceae. In: Balows A, Trüper HG, Dworkin M, Harder W and Schleifer KH (eds) The Prokaryotes (2nd ed), pp 3200–3231. Springer Verlag, New York

Pierson BK, Oesterle A and Murphy GL (1987) Pigments, light penetration, and photosynthetic activity in the multi-layered microbial mats of Great Sippewissett Salt Marsh, Massachusetts. FEMS Microbiol Ecol 45: 365–376

Reed RH (1983) Measurement and osmotic significance of ß-dimethylsulfoniopropionate in marine microalgae. Mar Biol Lett 4: 173–178

Repeta DJ, Simpson DJ, Jørgensen BB and Jannasch HW (1989) Evidence for the existence of anoxygenic photosynthesis from the distribution of bacteriochlorophylls in the Black Sea. Nature 342: 69–72

Revsbech NP and Ward DM (1984) Microelectrode studies of interstitial water chemistry and photosynthetic activity in a hot spring microbial mat. Appl Environ Microbiol 48: 270–275

Revsbech NP, Christensen PB and Nielsen LP (1989) Microelectrode analysis of photosynthetic and respiratory processes in microbial mats. In: Cohen Y and Rosenberg E (eds) Microbial Mats: Physiological Ecology of Benthic Microbial Communities, pp 153–162. ASM, Washington DC

Roelofsen PA (1935) On the metabolism of the purple sulphur bacteria. Kon Ned Acad Wet 37: 660–669

Romanenko VI, Peres Eiris M, Kudryavtsev VM and Pubienes MA (1976) Microbiological processes in meromictic Lake Vae de San Juan, Cuba. Mikrobiologiya 45: 539–546

Sahl HG and Trüper HG (1970) Enzymes of CO_2 fixation in Chromatiaceae. FEMS Microbiol Lett 2: 129–132

Schaub BEM and Van Gemerden H (1993) Simultaneous phototrophic and chemotrophic growth in the purple sulfur bacterium *Thiocapsa roseopersicina*. FEMS Microbiol Ecol 13: 185–196

Schedel M (1978) Untersuchungen zur anaeroben Oxidation reduzierter Schwefelverbindungen durch *Thiobacillus denitrificans*, *Chromatium vinosum* und *Chlorobium limicola*. PhD Thesis, University of Bonn, Germany

Schmidt K (1980) A comparative study on the composition of chlorosomes (*Chlorobium* vesicles) and cytoplasmic membranes from *Chloroflexus aurantiacus* strain Ok-70-fl and *Chlorobium limicola* f. *thiosulfatophilum* strain 6230. Arch Microbiol 124: 21–31

Smith NA and Kelly DP (1988) Isolation and physiological characterization of autotrophic sulphur bacteria oxidizing dimethyl disulphide as sole source of energy. J Gen Microbiol 134: 1407–1417

Sorokin YI (1964) On the primary production and bacterial activities in the Black Sea. Journal du Conseil 29: 41–60

Sorokin YI (1972) The bacterial population and the processes of hydrogen sulphide oxidation in the Black Sea. Journal du Conseil 34: 423–454

Sorokin YI (1970) Interrelations between sulfur and carbon turnover in leromictic lakes. Arch Hydrobiol 66: 391–446

Sorokin YI and Donato N (1975) On the carbon and sulfur metabolism in the meromictic Lake Faro (Sicily) Italy. Hydrobiologia 47: 241–252

Stal LJ, Van Gemerden H and Krumbein WE (1984) The simultaneous assay of chlorophyll and bacteriochlorophyll in natural microbial communities. J Microbiol Methods 2: 295–306

Steenbergen CLM (1982) Contribution of photosynthetic sulphur bacteria to primary production in Lake Vechten. Hydrobiologia 95: 59–64

Steenbergen CLM and Korthals HJ (1982) Distribution of phototrophic microorganisms in the anaerobic and micro-aerophilic strata of Lake Vechten (The Netherlands). Pigment analysis and role in primary production. Limnol Oceanogr 27: 883–895

Steenbergen CLM, Korthals HJ and Van Nes M (1987) Ecological observations on phototrophic sulfur bacteria and the role of these bacteria in the sulfur cycle of monomictic Lake Vechten (The Netherlands). Acta Acad Aboensis 47: 97–115

Steenbergen CLM, Korthals HJ, Baker AL and Watras CJ (1989) Microscale vertical distribution of algal and bacterial plankton in Lake Vechten (The Netherlands). FEMS Microbiol Ecol 62: 209–220

Stefels J and Van Boekel WHM (1993) Production of DMS from dissolved DMSP in axenic cultures of the marine phytoplankton species *Phaeocystis* sp. Mar Ecol Progr Ser. 97: 11–18

Steudel R (1989) On the nature of the 'elemental sulfur' (S^0) produced by sulfur-oxidizing bacteria — A model for S^0 globules. In: Schlegel HG and Bowien B (eds) Autotrophic Bacteria, pp 289–303. Springer Verlag, New York

Steudel R, Holdt G, Visscher PT and Van Gemerden H (1990) Search for polythionates in cultures of *Chromatium vinosum* after sulfide incubation. Arch Microbiol 153: 432–437

Stolz JF (1983) Fine structure of the stratified microbial community at Laguna Figueroa, Baja California, Mexico: I. Methods of in situ study of the laminated sediments. Precambrian Res. 20: 479–492

Stolz JF (1990) Distribution of phototrophic microbes in the flat microbial mat at Laguna Figueroa, Baja California, Mexico. Biosystems 23: 345–358

Takahashi M and Ichimura S (1968) Vertical distribution and organic matter production of photosynthetic sulfur bacteria in Japanese lakes. Limnol Oceanogr 13: 644–655

Takahashi M and Ichimura S (1970) Photosynthetic properties and growth of photosynthetic sulfur bacteria in lakes. Limnol Oceanogr 15: 929–944

Takahashi M, Shiokawa R and Ichimura S (1972) Photosynthetic characteristics of a purple sulfur bacterium grown under different light intensities. Can J Microbiol 18: 1825–1828

Taylor PA and Williams PJ LeB (1975) Theoretical studies on the coexistence of competing species under controlled flow conditions. Can J Microbiol 21: 90–98

Tiedje JM, Sexstone AG, Myrold DM and Robinson JA (1982) Denitrification: ecological niches, competition and survival. Antonie van Leeuwenhoek 48: 569–583

Tilzer MM (1984) Estimation of phytoplankton loss rates from daily photosynthetic rates and observed biomass changes in Lake Constance. J Plankton Res 6: 309–324

Trüper HG (1981) Versatility of carbon metabolism in the phototrophic bacteria. In: Dalton H (ed) Microbial Growth on C_1 Compounds, pp 116–121. Heyden and Son, London

Trüper HG (1984) Phototrophic bacteria and their sulfur metabolism. In: Müller A and Krebs B (eds) Studies in Inorganic Chemistry, Vol 5, Sulfur, Its Significance for the Geo-, Bio- and Cosmosphere and Technology, pp 367–382. Elsevier, Amsterdam

Trüper HG and Pfennig N (1966) Sulphur metabolism in Thiorhodaceae. III. Storage and turnover of thiosulfate sulphur in *Thiocapsa floridana* and *Chromatium* species. Antonie van Leeuwenhoek 32: 261–276

Van den Ende FP and Van Gemerden H (1993) Sulfide oxidation under oxygen limitation by a *Thiobacillus thioparus* isolated from a marine microbial mat. FEMS Microbiol Ecol 13: 69–78

Van Gemerden H (1968) Growth measurements of *Chromatium* cultures. Arch Mikrobiol 64: 103–110

Van Gemerden H (1974) Coexistence of organisms competing for the same substrate: An example among the purple sulfur bacteria. Microb Ecol 1: 104–119

Van Gemerden H (1980) Survival of *Chromatium vinosum* at low light intensities. Arch Microbiol 125: 115–121

Van Gemerden H (1983) Physiological ecology of purple and green bacteria. Ann Microbiol (Inst. Pasteur) 134: 73–92

Van Gemerden H (1984) The sulfide affinity of phototrophic bacteria in relation to the location of elemental sulfur. Arch Microbiol 139: 289–294

Van Gemerden H (1986) Production of elemental sulfur by green and purple sulfur bacteria. Arch Microbiol 146: 52–56

Van Gemerden H (1987) Competition between purple sulfur bacteria and green sulfur bacteria: Role of sulfide, sulfur and polysulfides. Acta Acad Aboensis 47: 13–27

Van Gemerden H (1990) Immobilized anoxygenic phototrophic bacteria in tidal areas. J de Bont, J Visser and J Tramper (eds) Physiology of immobilized cells. Elsevier, Amsterdam pp 37–48

Van Gemerden H (1993) Microbial mats: a joint venture. Mar Geol 113: 3–25

Van Gemerden H and Beeftink HH (1978) Specific rates of substrate oxidation and product formation in autotrophically growing *Chromatium vinosum* cultures. Arch Microbiol 119: 135–143

Van Gemerden H and Beeftink HH (1981) Coexistence of *Chlorobium* and *Chromatium* in a sulfide-limited continuous culture. Arch Microbiol 129: 32–34

Van Gemerden H and Beeftink HH (1983) Ecology of phototrophic bacteria. In: Ormerod JG (ed) The Phototrophic Bacteria: Anaerobic Life in the Light, pp 146–185. Blackwell Scientific Publications, Oxford

Van Gemerden H and Jannasch HW (1971) Continuous culture of Thiorhodaceae. Arch Mikrobiol 79: 345–353

Van Gemerden H, Montesinos E, Mas J and Guerrero R (1985) Diel cycle of metabolism of phototrophic purple sulfur bacteria in Lake Cisó (Spain). Limnol Oceanogr 30: 932–943

Van Gemerden H, De Wit R, Tughan CS and Herbert RA (1989a) Development of mass blooms of *Thiocapsa roseopersicina* on sheltered beaches on the Orkney Islands. FEMS Microbiol Ecol 62: 111–118

Van Gemerden H, Tughan CS, De Wit R and Herbert RA (1989b) Laminated microbial ecosystems on sheltered beaches in Scapa Flow, Orkney Islands. FEMS Microbiol Ecol 62: 87–102

Van Niel CB (1931) On the morphology and physiology of the purple and green sulfur bacteria. Arch Mikrobiol 3: 1–112

Van Niel CB (1955) Natural selection in the microbial world. J Gen Microbiol 13: 201–217

Veldhuis MJW and Van Gemerden H (1986) Competition between

purple and brown phototrophic bacteria in stratified lakes: sulfide, acetate, and light as limiting factors. FEMS Microbiol Ecol 38: 31–38

Vicente E and Miracle MR (1988) Physicochemical and microbial stratification in a meromictic karstic lake of Spain. Verh Internat Verein Limnol 23: 522–529

Vicente E, Rodrigo MA, Camacho A and Miracle MR (1991) Phototrophic prokaryotes in a karstic sulphate lake. Verh Internat Verein Limnol 24: 998–1004

Visscher PT and Van Gemerden H (1988) Growth of *Chlorobium limicola* f. *thiosulfatophilum* on polysulfides. In: Olson JM, Ormerod JG, Amesz J, Stackebrandt E and Trüper HG (eds) Green Photosynthetic Bacteria, pp 287–294. Plenum Press, New York

Visscher PT and Van Gemerden H (1991a) Production and consumption of dimethylsulfoniopropionate in marine microbial mats. Appl Environ Microbiol 57: 3237–3242

Visscher PT and Van Gemerden H (1991b) Photo-autotrophic growth of *Thiocapsa roseopersicina* on dimethyl sulfide. FEMS Microbiol Lett. 81: 247–250

Visscher PT and Van Gemerden H (1993) Sulfur cycling in laminated marine microbial ecosystems. In: Oremland RS (ed) Biogeochemistry of Global Change: Radiatively Active Trace Gases, pp 672–690. Chapman and Hall, New York

Visscher PT, Nijburg JW and Van Gemerden H (1990) Polysulfide utilization by *Thiocapsa roseopersicina*. Arch Microbiol 155: 75–81

Visscher PT, Quist P and Van Gemerden H (1991) Methylated sulfur compounds in microbial mats: In situ concentrations and metabolism by a colorless sulfur bacterium. Appl Environ Microbiol 57: 1758–1763

Visscher PT, Van den Ende FP, Schaub BEM and Van Gemerden H (1992) Competition between anoxygenic phototrophic bacteria and colorless sulfur bacteria in a microbial mat. FEMS Microbiol Ecol 101: 51–58

Wetzel RG (1973) Productivity investigations of interconnected marl lakes (I). The eight lakes of the Oliver and Walters chains, northeastern Indiana. Hydrobiol Stud 3: 91–143

Widdel F, Schnell S, Heising S, Ehrenreich A, Assmus B and Schink B (1993) Ferrous iron oxidation by anoxygenic phototrophic bacteria. Nature 362: 834–836

Yoon H, Klinzing G and Blanch HW (1977) Competition for mixed substrates by microbial populations. Biotechnol Bioeng, 19: 1193–1210

Zevenboom W (1980) Growth and nutrient uptake kinetics of *Oscillatoria agardhii*. PhD Thesis, University of Amsterdam, The Netherlands

Zeyer J, Eicher P, Wakeham SG and Schwarzenbach RP (1987) Oxidation of dimethyl sulfide to dimethyl sulfoxide by phototrophic sulfur bacteria. Appl Environ Microbiol 53: 2026–2032

Chapter 5

Ecology of Thermophilic Anoxygenic Phototrophs

Richard W. Castenholz
Biology Department, University of Oregon, Eugene, OR 97403-1210, USA

Beverly K. Pierson
Biology Department, University of Puget Sound, Tacoma, WA 98416-0320, USA

Summary

It is apparent that very few species of anoxygenic phototrophs occur or grow at high temperatures, particularly when compared to species numbers for thermophilic Archaea and non-photosynthetic Bacteria. *Chloroflexus* spp. are the most thermotolerant (up to ~70 °C), but none are in the hyperthermophilic category.

Recognizing that there may be some endemic populations of anoxygenic phototrophic bacteria that have not been dispersed among geographically disparate geothermal sites, the major factors affecting the distribution of these bacteria are temperature, pH, and concentration of sulfide. Oxygen may have an effect on the vertical distribution and the diel vertical migration of some species within mats. Facultative aerobic metabolism appears to be a property of many of the anoxygenic phototrophs (but not *Chlorobium* or *Heliobacillus*) in these dynamic habitats. Light quantity and quality are affected by the diversity of pigmentation within the vertically stratified communities and adaptation to low photon fluence rates is a necessity for many species.

R. E. Blankenship, M. T. Madigan and C. E. Bauer (eds): Anoxygenic Photosynthetic Bacteria, pp. 87–103.

I. Introduction

A. Definition of Thermophily

The terms thermophily and thermotolerance in phototrophic prokaryotes should encompass a different range of temperatures than those used to define thermophily in non-phototrophic organisms, both archaebacteria (Archaea) and eubacteria (Bacteria). Some contemporary biologists have proposed a high temperature environment for the origin of life and support this in part by pointing to the existence of living hyperthermophilic 'bacteria' that, according to parsimonious 16S rRNA sequence analyses, branch early in both eubacterial and archaebacterial phylogenetic 'trees.' Chlorophyll-based photosynthesis occurs only in branches of the eubacterial tree (which includes chloroplasts). Phototrophic bacteria do not exist above 73–74 °C, a relatively low temperature when compared to extreme hyperthermophilic non-phototrophic prokaryotes, such as *Thermotoga* and *Aquifex* which branch off early (Olsen et al., 1994). Thus, if chlorophyll-based photosynthesis has a monophyletic origin and appeared early (as indicated by the *Chloroflexus* branch) it did not evolve at extreme temperatures (i.e. >75 °C) unless subsequent loss of caldoactive photosynthesis has occurred through extinctions. Diversity in the upper temperature range for photosynthesis (63–73 °C) is limited, although it is possible that many genotypes (or species) of the highest temperature form of the cyanobacterium *Synechococcus* sp. exist. A relatively large number of species, genera, and higher categories of phototrophic bacteria inhabit waters between 53 and 63 °C. In addition, the highest temperature phototrophs known today (i.e. *Synechococcus* sp. and *Chloroflexus aurantiacus*) have not adapted fully to the upper several degrees of their temperature range (Meeks and Castenholz, 1971; Pierson and Castenholz, 1974a; Castenholz and Schneider, 1993). It is likely, therefore, that the origins of thermophilic cyanobacteria, at least, were at low or moderate temperatures and that evolutionary expansion occurred slowly and selectively up the temperature gradient.

Abbreviations: Cb. – Chlorobium; Cf. – Chloroflexus; Cm. – Chromatium; Hc. – Heliobacillus; HTF – high temperature form; LTF – low temperature form; MTF – moderate temperatrue form; *O. – Oscillatoria;* YNP – Yellowstone National Park

In contrast to at least 20 species of morphologically distinct cyanobacteria known in the 53–63 °C range (e.g. Castenholz, 1969) the anoxygenic phototrophic bacteria have few known thermophilic species. Only *Chloroflexus aurantiacus* (including some distinct genotypes) and at least one other species of *Chloroflexus* thrive at temperatures above 57 °C. In the range of 50–57 °C habitats, only one known species of the green sulfur bacteria, one purple bacterium, one *Heliothrix* species, one Heliobacterium, and an undescribed filamentous 'purple bacterium' are known.

In the context of phototrophs, then, we will use the term high temperature forms (HTFs) for those able to grow above 63 °C and moderate temperature forms (MTFs) for those that have upper limits below 63 °C but above 53 °C. Low temperature forms (LTFs) (upper growth limits between 43 and 53 °C) include many species with growth temperature optima well below this range. Thus, for the purposes of this discussion, those that are not able to exceed 52 °C will not be regarded as thermophiles.

B. Summary of Thermophilic Anoxygenic Phototrophs

1. Chloroflexus

Although it is now known that *Chloroflexus*-like organisms reach their greatest biomass in marine and hypersaline microbial mats (see Chapter 3 by Pierson and Castenholz), *Chloroflexus* was first discovered in hot springs, and the type species, *Cf. aurantiacus,* has been the only species studied extensively by physiologists, biochemists, and phylogenists. It grows best as a photoheterotroph, but some strains are capable of aerobic chemoheterotrophic or anaerobic photoautotrophic growth as well (see Chapter 3 by Pierson and Castenholz).

Chloroflexus aurantiacus, with at least a few genotypes (see Weller et al., 1992), occurs worldwide in hot springs of alkaline pH down to about pH 6.2 (Pierson and Castenholz, 1974a; Castenholz, 1984b, 1988; Table 1). Species determinations have not been made from populations in most hot spring areas, so with the limitation of only morphology, pigmentation, and temperature tolerance comparisons, all will be referred to simply as *Chloroflexus.* Recently Hanada et al. (1993) have isolated a new species of *Chloroflexus* from Japanese hot springs. Although quite similar to *Cf. aurantiacus* physiologically, a fairly large separation on the basis of 16S rRNA nucleotide sequences and DNA-DNA hybridizations (9–14%) is indicated.

In western North America (Castenholz, 1984b) in Yunnan, China (Yun, 1986), and presumably in other parts of Asia and in Africa, *Chloroflexus* populations are most prominent as distinct orange to salmon colored 'undermats' lying directly below a top layer of cyanobacteria (Fig. 1). Although less conspicuous, *Chloroflexus* is also invariably intermingled with the topmat species. These two-parted microbial mats may extend upwards in the thermal stream to the upper temperature limit of the cyanobacterial provider. This may be up to about 68–69 °C in North American and Asian springs where a high temperature form (HTF) of *Synechococcus* extends upward to a constant or mean temperature of ~ 73 °C (Fig. 2). *Chloroflexus* mixed with *Synechococcus* may extend in low numbers up to this temperature as well. In some areas (e.g. New Zealand, Iceland, Europe) *Chloroflexus* and cyanobacteria are not as conspicuously separated into undermat and topmat, and extend upward only to 63–64 °C, the highest growth temperature for the cyanobacteria of these areas. Since all of the above populations of *Chloroflexus* are growing as heterotrophs (photo- or chemo-) they are dependent on immediate or upstream reduced carbon, the sources of which are ultimately the photoautotrophic cyanobacteria. No diazotrophic strains of *Chloroflexus* are known, in contrast to all other known anoxygenic phototrophs (Heda and Madigan, 1986).

Chloroflexus strains are also capable of a slow sulfide-dependent photoautotrophy (Castenholz, 1973a; Madigan and Brock, 1975; Giovannoni et al., 1987), and in sulfide-rich springs, mats or biofilms of *Chloroflexus* extend up the temperature gradient beyond that of the cyanobacteria to about 66 °C (Table 1). Thus, mats lacking cyanobacteria exist and are either orange, salmon-colored, or green, depending primarily on the presence (orange) or absence (green) of O_2 (Fig. 3). In culture, *Cf. aurantiacus* (from 2 sources) also has the same upper temperature limit of ~ 66 °C, even as a photoheterotroph, and it is possible that field populations above ~ 68 °C intermingled with the cyanobacterium *Synechococcus* are incapable of phototrophic growth, and are thus relegated to chemoheterotrophy (Castenholz, unpublished observations).

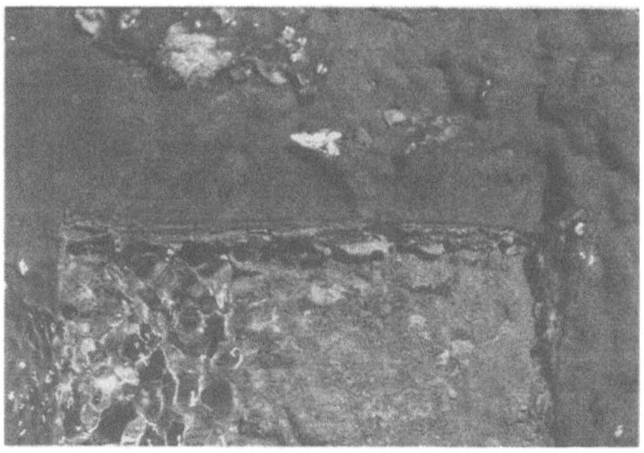

Fig. 1. (*See also Color Plate 1A*) Section through a microbial mat consisting of a top cover of the cyanobacterium *Synechococcus* underlaid by *Chloroflexus*. The cut is about 15 cm across and the water temperature ~60 °C. This is the source of *Chloroflexus aurantiacus*, strain OK-70-fl, Kahneeta Hot Springs, Oregon.

Fig. 2. (*See also Color Plate 1B*) Upper limits of *Chloroflexus* as a distinct undermat at ~68 °C. The outflow from the spring is at about 73 °C, and the siliceous sinter is dominated by *Synechococcus* until the edges cool to about 68 °C where the *Chloroflexus* undermat accretes (more orange on photograph). Buffalo Spring, White Creek drainage, YNP.

2. *Heliothrix and Possibly Related Filamentous Phototrophs*

On the basis of 5S and 16S rRNA sequence comparisons (Pierson et al., 1985; Weller at al., 1992), *Heliothrix oregonensis* falls within the same

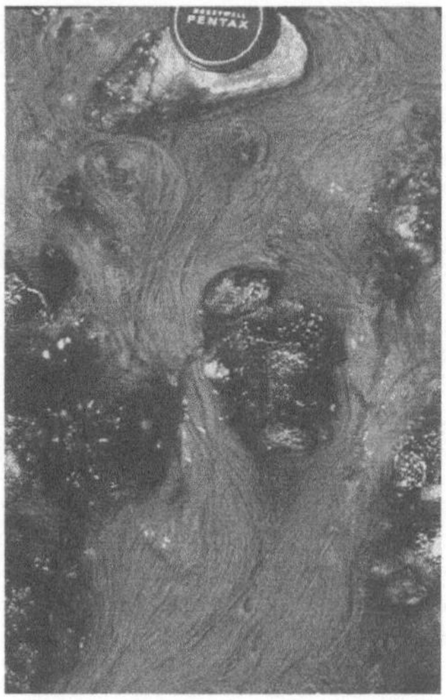

Fig. 3. (*See also Color Plate 1C*) *Chloroflexus* as a photoautotroph (orange-red streamers) growing at a temperature of ~64–62 °C on primary sulfide in Badstofuhver, Hveragerdi, Iceland. The sulfide is quite depleted by about 62 °C, at which point the green mat of the cyanobacterium, *Chlorogloeopsis* 'high temperature form' (formerly referred to as *Mastigocladus* 'HTF') appears (see Castenholz, 1973a).

cluster branch as *Chloroflexus* (see Chapter 3 by Pierson and Castenholz). The very apparent difference between the two is that *Heliothrix* does not possess chlorosomes, and therefore lacks the major light harvesting pigments of *Chloroflexus* (i.e. bacteriochlorophyll *c*, *d*, or *e*). BChl *a*, the sole chlorophyll, occurs in the cytoplasmic membrane only; the orange color is a result of carotenoids, primarily oxo-γ and OH-γ carotenes and an OH-γ carotene glycoside. The undescribed reddish filamentous phototroph of Yellowstone, distinguished from *Heliothrix* by having 'internal' lamellar thylakoids, also possesses only BChl *a* (see Chapter 3 by Pierson and Castenholz and Castenholz, 1984b).

Heliothrix oregonensis has been studied in Oregon hot springs (Pierson et al., 1984). It is in all known cases a photoheterotroph, tolerates and possibly requires O_2, and shows a field distribution and photoheterotrophic activity up to a temperature between 56 and 60 °C (Table 1, see also Fig. 4). *Heliothrix* is a wider diameter filament than *Chloroflexus* (~1.5 vs. 0.5–1.0 μm) and glides at a faster rate (~0.2 vs. 0.02 μm/s).

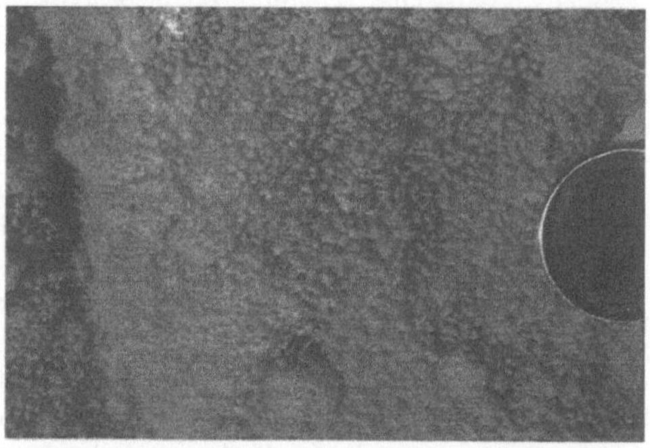

Fig. 4. (*See also Color Plate 1D*) *Heliothrix oregonensis* as an aerotolerant mat above a deep green cyanobacterial layer in a non-sulfidic Kahneeta hot spring (Oregon); temperature ~ 50 °C.

3. Chlorobium

The only thermophilic green sulfur bacterium known is *Chlorobium tepidum*, (Wahlund et al., 1991). It is barely a moderate temperature form (MTF), growing only up to about 55–56 °C in the few hot springs to which it is relegated and up to ~53 °C in the two culture isolates tested (Table 1). It is known from only a few springs in the Rotorua region of New Zealand (Castenholz et al., 1990, see Fig. 5). In culture, *Cb. tepidum* required at least 1 mM sulfide for growth and tolerated up to about 4 mM. Thiosulfate (2–8 mM) greatly stimulated growth and could be used as the sole photoreductant (Wahlund et al., 1991). A pH range for growth of 5.9–7.2 reflected the pH range of the source stream (strain TLS). *Cb. tepidum*, like most other anoxygenic bacteria, is capable of nitrogen fixation (Wahlund and Madigan, 1993).

The characteristic that most distinguishes *C. tepidum* from other species of *Chlorobium* is its thermophily. A few other features, such as large amounts of unique wax esters (Shiea et al., 1991) also help to distinguish this species. Its 'preference' for thiosulfate as an electron donor in photosynthesis, however, indicates a closeness to other thiosulfate-utilizing species of *Chlorobium*, such as *Cb. limicola*, and 16S rRNA sequence data confirm this (Wahlund et al., 1991).

4. Chromatium

Cm. tepidum (Madigan, 1986) is the only purple sulfur bacterium known to be thermophilic, at least

Table 1. Characteristics of Field Populations of Thermophilic Anoxygenic Phototrophs

Genus	Color of Field Populations	Principal Mode of Growth	Field Temperature Limits (°C)	Field pH Limits	Primary Sulfide in bathing water	O_2 in bathing water	Comments	Representative Springs
Chloroflexus	orange-salmon	photoheterotroph (& chemoheterotroph)	~35–69(73)	6.0– ~10	low or absent	+	With cyanobacteria &/or distinct bands underneath.	Worldwide
	orange-salmon	photoautotroph	< 66	6.2– >9.0	low-moderate > 0.02 mM	+	Upstream of cyanobacteria or above cyanobacteria (inverted).	esp. YNP, Iceland, New Zealand
	dull green	photoautotroph	< 66	6.2– ~7.0	moderate ~0.02–0.1 mM	absent	Upstream of cyanobacteria; independent, laminated mat.	Mammoth Springs, (YNP), & (formerly) Steamboat Springs, Nevada
Heliothrix	orange	photoheterotroph	~35– ~57	~8.0–8.5	low or absent	+	Distinct layer above cyanobacteria or mixed.	Western USA (incl. Oregon & YNP)
Chlorobium	dull green	photoautotroph	<56	~4.3– ~6.2	high (0.3–1.8mM)	absent	'Pure' unlaminated mat upstream of other phototrophs.	Rotorua region, New Zealand
Chromatium	red to pink	photoautotroph	<58	6.2– ~7.0	moderate > 0.02 mM	absent	Red-pink top cover with undermat of *Chloroflexus* upstream of cyanobacteria.	Mammoth Springs, (YNP), Thermopolis, Wyoming.
	red to pink	photoautotroph	<58	~7.5– >8.5	low or absent	+	Distinct pink to red band under cyanobacteria in sulfate-rich springs.	Western USA
'Purple non-sulfur bacteria'	unknown	photoheterotroph	< 47(?)	4.8– >8.5	?	+	Unknown as distinct populations.	Probably most springs < 45° C
Heliobacillus	unknown (dull brown-green)	photoheterotroph	≲53(?)	~7.5–8.2(?)	?	+	Unknown as distinct population (probably in strictly anoxic zone).	Hunter's Hot Springs, Oregon
Anoxygenic cyanobacteria *Oscillatoria 'amphigranulata'*	olive green	photoautotroph	< 57	5.9–9.2	low to high: 5 μM–3 mM	+ or absent	Distinct mat upstream of others, less sulfide-tolerant cyanobacteria.	North Island, New Zealand
Spirulina labyrinthiformis	deep blue-green	photoautotroph	< 52	6.2– ~7.0	moderate: ~0.02–0.1 mM	+ or absent	Distinct blue-green band near 45–51° C sources.	Mammoth Springs, (YNP)

YNP – Yellowstone National Park

Fig. 5. (*See also Color Plate 2A*) *Chlorobium tepidum* mat (green cover) in highly sulfidic 'Travelodge Stream', Rotorua, New Zealand. The temperature is 45 °C in the foreground, with a pH usually in the range of 5.3–6.2 (see Castenholz et al., 1990).

capable of growth above 50 °C (optimum 48–50 °C). Halophilic species of *Ectothiorhodospira* have mildly thermophilic tendencies, often with optima for growth up to 47 °C, but with upper limits of 50 °C or lower (e.g. Raymond and Sistrom, 1969; Imhoff and Trüper, 1981). Although several genotypes of *Chromatium tepidum* may exist, the general phenotypic characteristics of this species appear to be the same or very similar wherever it occurs. It is known from several hot spring complexes in western North America where it is often abundant up to temperatures of 57–58 °C (Table 1). In culture as well it shows no growth above 57 °C or below 34 °C, and thus would be classified as a moderate temperature form (MTF). It is known primarily as an 'undermat' form, presumably depending on biogenic sulfide, but in a few areas forms a surface mat when primary sulfide is abundant.

The diagnostic characteristics of *Cm. tepidum* are the higher temperature tolerance and optima for growth, rhodovibrin as the major carotenoid (in addition to spirilloxanthin), and the light-harvesting bacteriochlorophyll *a*/protein complex with a major absorption maximum at 918 nm in addition to 800 and 855 nm (Garcia et al., 1986).

5. Purple 'Non-Sulfur' Bacteria

At present, purple 'non-sulfur' bacteria have not been isolated in culture with growth temperature limits above 47 °C (see Resnick and Madigan, 1989; Stadtwald-Demchick et al., 1990 a,b). These few species will be regarded here as mildly thermophilic or thermotolerant species of *Rhodopseudomonas (Rp. cryptolactis)* and *Rhodospirillum (Rs. centenum)*. Little is known of their ecology except the nature of the hot spring pools from which they were isolated (Table 1). *Rhodospirillum centenum*, however produces high heat- and desiccation-resistant cysts which presumably explains the recovery of this species from thermal waters of 55 °C. Most earlier described species of the purple 'non-sulfur' bacteria in culture have temperature optima of 30–35 °C and only two species have optima as high as 40–42 °C (Imhoff and Trüper, 1989). N_2 fixation by several mildly thermophilic 'non-sulfur' purple bacteria has been demonstrated, indicating that this may be a means for enriching the mat environment with fixed nitrogen (Resnick and Madigan, 1989; Statwald-Demchick et al., 1990a,b).

Other studies of hot spring species composition and temperature limits have recovered purple 'non-sulfur' bacteria from temperatures exceeding 50 °C, perhaps up to 60 °C (Gorlenko et al., 1985, 1987). These studies have not demonstrated growth in the 50–60 °C range, since recovery was determined by plate dilutions and growth at lower temperatures (Yurkov et al., 1991). Using undiluted inoculum from hot springs of western N. America, enrichments for purple 'non-sulfur' bacteria, invariably resulted in success at 40 °C, sometimes at 45 °C, but never at 50 or 55 °C (Castenholz, unpublished data).

The purple 'non-sulfur' bacteria are infamous in being present almost everywhere (as revealed by selective/enrichment cultures). However, as in most other habitats, no identifiable, discrete populations of these organisms are apparent at any temperature in hot spring mats. Purple phototrophic bacteria of various types are represented in three phylogenetic branches of the Proteobacteria where extreme thermophily has not yet been found. The purple 'sulfur' bacterium, *Chromatium tepidum*, of the gamma branch may represent the highest temperature-tolerant thermophile of the Proteobacteria.

6. *Heliobacteria*

Heliobacteria, in general, are known as soil inhabitants and have been isolated mostly from dried soil samples, including those from rice paddies of southeastern Asia (Madigan, 1992). Essentially nothing is known of the ecology of this group of Gram positive-related bacteria. *Heliobacillus* sp., which appears quite similar morphologically to *Hc. mobilis* (Beer-Romero and Gest, 1987) has regularly become the dominant or sole phototroph in enrichments from wet (30–50 °C) inoculum from Hunter's Hot Springs (Lakeview, Oregon). Enrichments arose under incandescent lamps at 40–50 °C, although at 40–45 °C purple 'non-sulfur' bacteria sometimes arose instead (Castenholz, unpublished data). The original pyruvate-containing HAMP medium (Beer-Romero and Gest, 1987) usually 'selected' for this thermophilic *Heliobacillus*. Growth of the 45 °C enrichment culture continued when transferred to 50 or 53 °C but not at 55 or 60 °C. The abundance of *Heliobacillus* in Hunter's Hot Springs has not been estimated (Table 1). Several thermophilic heliobacteria have been isolated in pure culture from Yellowstone hot springs and Icelandic soils (Kimble and Madigan, 1994). These organisms grow up to 56 °C and have recently been described as a new species of the genus *Heliobacterium*, *Heliobacterium modesticaldum*, the species epithet meaning 'moderately hot' (Kimble et al., 1995). *Heliobacterium modesticaldum* appears similar to the isolates from Hunter's Hot Springs.

7. *Anoxygenic Cyanobacteria*

All known cyanobacteria retain the ability to perform oxygenic photosynthesis when environmental sulfide is low or absent. In addition to several non-thermophiles, anoxygenic photosynthesis with sulfide as the electron donor is known in some thermophilic or semi-thermophilic species of *Oscillatoria* (e.g. *O.* cf. *limnetica*: Cohen et al., 1986; *O.* cf. *amphigranulata*: Castenholz and Utkilen, 1984; *O.* cf. *boryana:* Castenholz et al., 1991) and also in *Spirulina* cf. *labyrinthiformis* (Castenholz, 1977). Not all cyanobacteria are capable of living in the presence of sulfide. Some that are incapable of anoxygenic photosynthesis are completely inhibited; some are merely tolerant and continue oxygenic photosynthesis; still others apparently perform both types simultaneously (see Cohen et al., 1986). The competent species are generally anoxygenic only when free sulfide is over a critical level; H_2S is more effective than HS^- (Howsley and Pearson, 1979), the latter proportionately greater at higher pH's. In a warm sulfide-rich pool in New Zealand, it was concluded that sulfide-dependent anoxygenic photosynthesis of benthic *Oscillatoria boryana* generally began each day; as light intensity increased the sulfide surrounding the cells was depleted to ~50 μM, at which point oxygenic photosynthesis resumed (Castenholz et al., 1991).

It is known that Photosystem II is inhibited by sulfide in many cyanobacteria capable of anoxygenic photosynthesis. It is merely conjecture, however, that the ability of PS I to accept electrons derived from sulfide represents a primitive ability inherited from the photosystem-flavocytochrome complex of green sulfur bacteria (see Blankenship, 1985). Nevertheless, PS I of cyanobacteria and the photosystem of green sulfur bacteria are similar in several respects, most specifically in having iron sulfur clusters as early acceptors in the reaction centers (see Blankenship, 1992).

The thermophilic, facultatively anoxygenic cyanobacteria are limited to temperatures below 57 °C. This temperature represents the upper limit for growth of *O.* cf. *amphigranulata* in sulfide-rich to low sulfide springs of New Zealand with pH ranges of 5.9–9.2, and also in cultures with no sulfide (Castenholz, 1976; Castenholz and Utkilen, 1984; see Fig. 6). *Spirulina labyrinthiformis* in hot springs of Yellowstone National Park is limited to temperatures below 51 °C under any conditions (Castenholz, 1977).

II. Chemical and Geographical Diversity of Thermal Habitats

Photosynthetic prokaryotes are limited to waters with pH levels above about 4.3. Very few species tolerate levels below 6, however. Thus, even slightly acidic springs have few photosynthetic components that are not eukaryotic.

Above the pH level of 6, however, springs with waters of great chemical diversity exist (e.g. Table 2; White et al., 1963; Castenholz, 1969; Brock, 1978). The largest number of springs or clusters of host

Fig. 6. (*See also Color Plate 2B*) Facultatively anoxygenic *Oscillatoria* cf. *amphigranulata* in small sulfidic spring WH-1 (Cirque-1), Whakarewarera, New Zealand. Temperature at the grey-green edge of the cyanobacterium was ~56 °C. *Thiobacillus*-like whitish-yellow streamers occurred upstream to about 68°C (see Castenholz, 1976).

springs are in areas of contemporary or geologically recent volcanism. Most of these have inorganic solute values well over 1000 and below 3000 mg/l, as compared to surface waters with mean values of ~150 mg/l (Table 2). Saline hot springs exist, but most of these are not associated with volcanism, but rather with regions of ancient salt deposits or oil field brines. Numerous hot springs also exist in areas of active tectonism and deep faulting. These, non-volcanic types usually have a lower solute content than those associated with volcanism (Table 2).

Neutral to alkaline springs associated with volcanism occur in many regions, most prominently in western North and South America, and much of the remainder of the Pacific rim, including Kamchatka, Japan, Indonesia, and New Zealand (Waring et al., 1965). Other large clusters exist in Mediterranean Europe, the Azores, Iceland, portions of central and eastern Africa, and many other smaller island groups worldwide. Although each volcanic area appears to have a characteristic chemistry as exemplified by differing concentrations and relative proportions of various ions, most of the extensive volcanic geothermal areas have chemical overlap with other geographically remote volcanic areas, so that endemism in species composition, if indeed it exists, may be a result of local speciation and poor dispersal capabilities of some types of micro-organisms.

The same overlap of chemistry occurs in the more scattered, and smaller spring clusters associated with tectonism (e.g. White et al., 1963; Waring et al., 1965). Some of these springs which drain limestone or dolomitic bedrocks are very similar to each other chemically in the relatively few regions where they occur (Table 2).

Although endemism appears to exist among thermophilic cyanobacteria, including the species known to be capable of anoxygenic photosynthesis, so little is known about the distribution of anoxygenic purple, green, and chloroflexean bacteria, that it would be premature to discuss this except in the context of the next section, the summaries of selected hot spring habitats.

III. Principal Habitats of Thermophilic Anoxygenic Phototrophs and the Environmental Requirements: Case Studies

A. Non-Sulfidic Springs of Yellowstone and Oregon: Photoheterotrophic Chloroflexus *and Other Filamentous Anoxygenic Phototrophs*

The ecology of *Chloroflexus* in two principal non-sulfidic hot springs will be summarized. These are Hunter's Hot Springs of south central Oregon and Octopus and nearby springs in the Lower Geyser Basin of Yellowstone National Park (YNP). Both of these spring systems are alkaline (pH ~ 8.0–8.5), but with obvious differences in ion content (Table 2). Nevertheless, the cyanobacterium *Synechococcus* cf. *lividus* occupies the substrate up to a relatively constant temperature of 73–74 °C in both systems. Free sulfide is very low or undetectable at this point in the outflow and *Chloroflexus aurantiacus* may be found associated with *Synechococcus* up to that temperature, and in crude culture up to about 70 °C. However, in these and other similar springs, a recognizable orange to yellowish undermat of *Chloroflexus* does not develop except below a mean temperature of 68–69° C (Table 1).

In these and most of the North American springs, the visually abrupt transition below ~68–69 °C, from the top cover of photoautotrophic *Synechococcus* to photoheterotrophic *Chloroflexus*, is obvious, i.e. from the 0.5–1.5 mm thick green to yellowish-green cyanobacterium cover to the distinctly orange to salmon-colored underlayer of *Chloroflexus* (Fig. 1). Only the uppermost few millimeters (or less) of most mats receive enough irradiance to drive phototrophic

Table 2. Partial chemical composition of selected hot springs (mg/l)

	pH	Na^+	K^+	Ca^{2+}	Mg^{2+}	Cl^-	SO_4^{2-}	HCO_3^-	SiO_2	HBO_2	H_2S
1. Octopus Spring, YNP	~8.0	310	<16	0.6	tr	256	23	340	254	10	<0.2
2. Hunter's Hot Springs Oregon	~8.0	210	8.5	13.0	<0.1	120	260	79	140	8	<0.2
3. Mammoth Springs, YNP	~6.6	130–150	40–70	210–430	60–85	160–185	475–860	525–990	45–70	~4	<3.2
4. Sulfide Springs, Rotorua, New Zealand	~5.5	170–245	15–21	Ca + Mg[a] 21 – 27		141–191	490–625	–	155–195	~7	52–58
5. Badstofuhver, Hveragerdi, Iceland	8.6	154	10	4.4	tr	155	68	–	270	–	2–3

Water collected at spring sources. 1. Brock (1978); Stauffer et al. (1980) 2. See Castenholz (1973b) 3. See White et al. (1963); Castenholz (1973a); 4. See Castenholz (1990) 5. See Castenholz (1973a). YNP – Yellowstone National Park.

[a] Combined analysis of both Ca^{2+} and Mg^{2+}.

growth, at least this is true in Hunter's Hot Springs, Oregon (Fig. 7). By a depth of 1 mm in this representative *Synechococcus* and *Chloroflexus* mat, all wavelengths below 550 nm were absorbed and only 0.01–0.001 of longer wavelength visible irradiance remained. The *Synechococcus* top layer (of unbound unicells) was only about 0.5 mm thick; at 1.0 mm depth, there was almost complete absorption by BChl *c* (740 nm) and approximately 0.05–0.1 of incident infrared irradiance remained (Fig. 7). In similar mats, topped by *Oscillatoria* cf. *terebriformis* in Hunter's Hot Springs (below 55 °C)(Richardson and Castenholz, 1987) and by *Synechococcus* sp. in Octopus Spring, microelectrode data of daytime O_2 profiles (Revsbech and Ward, 1984 a,b) showed that during most midday periods, O_2 diffuses throughout the active mixed and undermat *Chloroflexus* layer at levels often greatly exceeding saturation. Since characteristic BChl *c* and BChl *a* bands are still evident, and in the laboratory, *Chloroflexus* requires anoxic or at least microoxic conditions for the synthesis of these pigments, it is likely that the consistent anoxic conditions of nighttime are necessary for the synthesis of bacteriochlorophyll. With photopigments present in *Chloroflexus* during the oxic daytime period, photoheterotrophy probably predominates over aerobic chemoheterotrophy, since culture experiments have shown that light suppresses aerobic respiration when bacteriochlorophyll is present. Both modes apparently share the same electron transport system (see Pierson and Castenholz, 1974b, 1992).

Since anoxygenic photoautotrophy is generally excluded in these two spring systems because of the lack of sulfide or other suitable reductants, photo- or chemo-heterotrophy is a necessity for *Chloroflexus*. A direct supply of reduced carbon from *Synechococcus* or other cyanobacteria is possible but organic acids from 'intermediate' organisms, such as acetate or butyrate fermenters, may be the sources of carbon. Although Bateson and Ward (1988) gave evidence for glycolate as a main release compound of *Synechococcus* in Octopus Spring, presumably during conditions promoting photorespiration, Teiser and Castenholz have evidence that a great variety and quantity of low molecular weight compounds are released by strains of thermophilic *Synechococcus* (Teiser, 1993). These have been identified in axenic strains from Hunter's Hot Springs (Oregon), West Thumb (YNP), and Hunter Hot Springs (Montana).

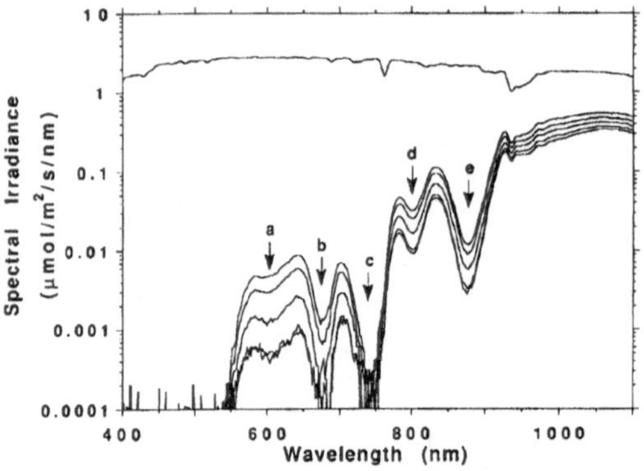

Fig. 7. Downwelling spectral irradiance profile through a hot spring microbial mat composed of cyanobacteria (*Synechococcus* cf. *lividus*) and *Chloroflexus aurantiacus*. The vertical core was taken at about 60 °C from Hunter's Hot Springs, Oregon, and methods were those of Pierson et al. (1990). The top spectrum is the incident solar irradiance at the surface of the mat. The other spectra were at 1.0 mm depth increments within the mat. Arrows indicate specific attenuation of irradiance due to pigments: a = phycocyanin; b = Chl *a*; c = BChl *c*; d & e = BChl *a*. Note that all wavelengths shorter than 550 nm are severely attenuated at a depth of 1.0 mm. The wavelengths that penetrate deeper are in yellow, red and near infrared portions of the spectrum.

They include succinate, beta-hydroxybutyrate, acetate, adonitol, and fructose as major release products; most of these can also be used by *Chloroflexus* for anaerobic phototrophic growth (Teiser, 1993).

The realization that *Chloroflexus aurantiacus* could grow entirely on products of *Synechococcus* came many years ago when dual cultures of the two organisms were first grown routinely on autotrophic agar medium (see Pierson and Castenholz, 1974a).

Hunter's Hot Springs, Oregon, differs dramatically from Octopus Spring, Yellowstone in the high concentration of sulfate (~2.7 mM – Oregon vs. 0.24 mM – Octopus; Table 2). This results in much biogenic sulfide production below about 58 °C in Hunter's Hot Springs which results in large populations of *Chromatium tepidum* also below this temperature, under a top cover of the cyanobacterium *Oscillatoria* cf. *terebriformis*. *Chloroflexus aurantiacus* is also present and in this situation may use photoautotrophy to some extent.

It is now known that more than one genotype of thermophilic *Chloroflexus* occurs, at least in Octopus Spring (Weller et al., 1992; Ruff-Roberts et al.,

1994). However, at this point phenotypic differences have not been determined. Other *Chloroflexus*-like organisms also occur in these two spring systems in Oregon and Yellowstone. *Heliothrix oregonensis* occurs below about 56° C in Hunter's Hot Springs and in Octopus Spring (unpublished observations). This aerotolerant, highly motile organism tends to form surface tufts, which may represent an aerotactic response. It can be fairly accurately recognized in the field and by microscopy (see section B.2). The greatest development of *Heliothrix*, however, is in summer, in unshaded springs of the Kahneeta group in eastern Oregon (Fig. 4; Pierson et al., 1984; Castenholz, 1984 a,b). In several of these small stable springs, a brilliant orange 'puffy' cover of *Heliothrix* forms on top of a dark green cyanobacterial mat (Fig. 4). It is assumed that its aerotolerance, low content of BChl *a*, and extremely high content of presumably photoprotective carotenoids, allow or promote its vertical position in these springs. Nevertheless, *Heliothrix* is known as a photoheterotroph only, and is presumably dependent on the cyanobacteria beneath it for reduced carbon.

Thick translucent mats of filamentous cyanobacteria (e.g. *Phormidium* spp.) occur in some hot springs in the Lower and Midway geyser regions of Yellowstone National Park. The *Chloroflexus* layer may occur as deep as 1.5 cm in the mat, followed even deeper by the undescribed red, filamentous bacterium containing thylakoids and BChl *a* (Castenholz, 1984b; see Chapter 3 by Pierson and Castenholz). It has been shown that primarily near infrared wavelengths are transmitted to the depths occupied by these organisms (Jørgensen et al., 1992)

B. Sulfidic Hot Springs with Photoautotrophic Chloroflexus

The most prominent mats of photoautotrophic *Chloroflexus* known are in Yellowstone National Park, Iceland, and New Zealand. The first two of these will be discussed.

In the Upper Terraces of Mammoth Hot Springs, hot springs emerge at 73–75 °C and at various temperatures below this. In all the source waters there is primary sulfide of about 30–130 μM and a pH of 6.2–6.8, depending on the spring. Prominent dull green mats of *Chloroflexus* occur in slow flowing springs that emerge in the range of 51–66 °C (Castenholz, 1977; Ward et al., 1989, 1992). Photoautotrophic strains of *Chloroflexus* from these springs are obligate phototrophs, and although autotrophic in the field, grow best as photoheterotrophs in organic medium in the laboratory (Giovannoni et al., 1987). The conditions which dictate the dominance of photoautotrophic *Chloroflexus,* rich in BChl *c*, are waters with significant concentrations of primary sulfide bathing the substrate or mat ($\gtrsim$30 μM), pH levels above 6.2, and temperatures between 52 and 66 °C. These sulfide levels exclude all known species of cyanobacteria in North America that would otherwise grow at temperatures above about 52 °C. The fact that sulfide in these springs is lost by gaseous diffusion, by abiological oxidation, and by biological usage soon after the waters emerge, limits the mats of *Chloroflexus* to sites near the sources of the few, small, low volume springs that discharge within this narrow temperature range (see Castenholz, 1977). These waters are anoxic at the sources and over most of the photoautotrophic mat (Giovannoni et al., 1987).

A remarkable aspect of some of the stable, green *Chloroflexus* mats is that they develop laminations as they accrete. However, the orange to brownish colored underlaminae also appear to be composed of *Chloroflexus* (Giovannoni et al., 1987; Ward et al., 1989, 1992). It is possible that the uppermost bands of this 'discolored' undermat are supported as photoheterotrophs by the top layer of photoautotrophic *Chloroflexus*. Although unknown, there is no requirement that these be different strains. A depth profile of downwelling spectral irradiance revealed only the absorption bands of pigments typical of *Chloroflexus* (Fig. 8).

Orange to salmon-colored photoautotrophic mats of *Chloroflexus* occur within the same temperature range as the dull green–colored mats in larger volume, higher temperature springs of the Mammoth terraces (Castenholz, 1973a,1977). These springs contain similar sulfide concentrations as the lower temperature springs, but by the point in the shallow outflow (< 1 cm) that temperature has dropped to 66 °C (the absolute upper limit for autotrophic growth of *Chloroflexus*), much of the sulfide has disappeared and some O_2 has been acquired. O_2 limits the synthesis of bacteriochlorophylls in *Cf. aurantiacus* even more than does high light intensity, (e.g. Sprague et al., 1981; Oelze, 1992). The orange to salmon–colored mats of *Chloroflexus* which occur upstream of all cyanobacteria appear to be reliable indicators of at

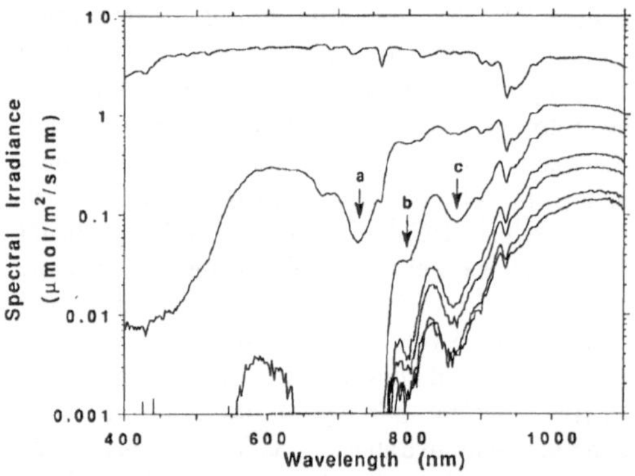

Fig. 8. Downwelling spectral irradiance profile through a hot spring microbial mat composed of 'green' *Chloroflexus* (see Giovannoni et al., 1987). The mat was collected from 'New Pit Spring', Mammoth Springs (YNP) at about 57 °C, and methods were those of Pierson et al., 1990. The top spectrum is the incident solar irradiance at the surface of the mat. The other spectra were at depths within the mat of 0.125, 0.5, 1.0, 2.0, 3.0, and 4.5 mm. Arrows indicate specific attenuation due to pigments: a = BChl *c*; b & c = BChl *a*. As in Fig. 2, only red and near infrared wavelengths penetrate deeply into the mat.

least micro-oxic conditions.

Chloroflexus mats of a similar salmon or orange color, upstream of the cyanobacteria, are prominent in a few other geothermal regions as well; particularly in sulfidic springs of Iceland and New Zealand (Castenholz, 1973a, 1976; Table 1). Since many of these springs have sources near 100 °C, the orange *Chloroflexus* mats do not develop near the sources, but considerable distances downstream. Nevertheless, in many of these volcanic springs of alkaline pH, substantial sulfide exists even at 66 °C (Fig. 3). Although photoautotrophic *Chloroflexus* usually forms mats well upstream of cyanobacteria in these springs, there are some notable examples in Iceland where *Chloroflexus* forms the top layer of mat with an undermat of cyanobacteria (Fig. 9). It was shown by Jørgensen and Nelson (1988) that oxygenic photosynthesis by sulfide-sensitive cyanobacteria occurred below ~61 °C in a 0.5 mm thick band 0.8 mm below the top of the *Chloroflexus*. The sulfide concentration at the surface of the *Chloroflexus* was ~60 μM, but by a depth of 0.8 mm within the mat the sulfide had decreased sufficiently for oxygenic

Fig. 9. (*See also Color Plate 2C*) Inverted *Chloroflexus* mat at about 61–62 °C in Ystihver Stream, Húsavik, North Iceland. The primary sulfide which supports the photoautotrophic *Chloroflexus* is lowered within the mat, so that an underlayer of oxygenic cyanobacteria (*Chlorogloeopsis* 'high temperature form') occurs at about 1 mm below the *Chloroflexus* (see Jørgensen and Nelson, 1988).

photosynthesis, presumably because of the photosynthetic use of sulfide by the top layer of *Chloroflexus*.

C. Chlorobium *in Acidic, Sulfide-rich Hot Springs of New Zealand*

High sulfide (0.3–1.8 mM) with moderate temperature (55–40 °C) and pH levels between ~4.3 and 6.2 is a combination that supports *Chlorobium tepidum* as the sole phototroph, at least in hot springs in New Zealand (Castenholz, 1988; Castenholz et al., 1990). The same conditions are unknown elsewhere (at least to these authors). This photoautotroph forms unlaminated slime covers on solid substrates in the few springs with the above combination of conditions (Fig. 5). There are, of course, no cyanobacteria at this temperature range in waters with a high sulfide content and this low a pH. *Chloroflexus* is also excluded, presumably by the low pH. Even in the Rotorua/Rotoiti region of the central volcanic region of the North Island, most sulfide-rich hot springs are either too acidic (< pH 4.3) or too alkaline (continuously over ~pH 6.2) to allow *Chlorobium tepidum* to predominate. Since the pH optimum of *Cb. tepidum* isolates is about 6.8–7.0 (Wahlund et al., 1991), the *Chlorobium* dominance in these springs is probably a result of merely a high tolerance of low pH rather than a reflection of its optimum. The restricted conditions for *Chlorobium* dominance relegate these mats to a few small natural springs in the Sulphur Point region of Rotorua, Whakarewarewa, and the south shore of western Lake Rotoiti (Castenholz et al., 1990). One spring with substantial discharge is Manupairua on Lake Rotoiti. The principal spring of study, however, was 'Travelodge Spring' which issued from a geothermal well of the Rotorua Travelodge at Sulphur Point in Rotorua (Fig. 5). In 1990, four years after the initial study, this spring had changed to the extent that little of the formerly extensive *Chlorobium* mat remained.

The *Chlorobium* mat, as originally described, was slightly yellow-green on the surface due to a thin (0.1–0.2 mm) dusting of elemental sulfur mixed with cells. Dislodging this layer with turbulence from a syringe revealed a deep green densely packed, slimy mass of *Chlorobium* which appeared microscopically to consist of one uniform cell type that corresponded to *Chlorobium*. This dense compaction of cells was about 3 mm thick over a 1–2 mm thick transition layer 'stained' black by metal sulfides (see Ward et al., 1989 for color photograph of vertical section). Photosynthetic usage of sulfide occurred to a depth of about 1.0 mm, at which point only about 0.001 % of the downward irradiance remained (Castenholz et al., 1990). Some species of *Chlorobium* are known for their ability to grow at lower photon fluence rates than any other phototrophic organism (see Overmann et al., 1991, 1992). Sulfide was also generated rapidly within the *Chlorobium* mat, at least in darkness, which may indicate anaerobic fermentation of elemental sulfur by *Chlorobium* (e.g. Paschinger et al., 1974), rather than activity by non-photosynthetic sulfur- or sulfate-reducing bacteria.

It appears that *Chlorobium tepidum* mats occur only in a restricted area of New Zealand. However, it is possible, or even probable, that a similar combination of chemical and physical conditions occurs in other volcanic regions. Therefore, an interesting question is: if such conditions exist will *Chlorobium tepidum* be there or is this non-spore forming organism an example of microbial endemism, where descent from *Chlorobium limicola* occurred over the several million year history of geothermal activity and geographic isolation in New Zealand?

D. Chromatium *in North American Hot Springs*

Chromatium tepidum, the only known thermophilic purple sulfur bacterium, appears to be distributed through many hot springs of North America, and possibly the world (see Madigan, 1986, 1988). It is known from various hot springs of eastern Oregon, northern Nevada and California, New Mexico, northwestern Wyoming (e.g. Thermopolis Springs, Thermopolis; and Mammoth Springs, Yellowstone National Park) (Fig. 10). Species or strains of *Chromatium* somewhat similar to *Cm. tepidum* occur in other hot springs, but there has not yet been a demonstration of growth in these above 51 °C. Such is the *Chromatium* isolated from hot springs of the Rotorua region of New Zealand (M. Madigan, personal comm.).

In a few small springs in the Mammoth area *Cm. tepidum* is the sole photoautotroph where sulfide occurs in emerging spring sources with a temperature in the mid 50s, and where pH is above 6.2 (Madigan, 1984, 1986, 1988; Madigan et al., 1989). Springs with primary sulfide at levels sufficient to support *Chromatium* over the appropriate pH and temperature range are not common, but the undermat habitat of numerous hot springs supports *Chromatium* with biogenic sulfide. High biological production of sulfide, however, depends on water and sediments with a high concentration of sulfate (see Table 2).

In some small springs in the Upper Terraces of Mammoth Hot Springs (YNP) appropriate conditions exist to support photoautotrophic *Chromatium* (see Ward et al., 1988 for color figure). The physical and chemical constraints are very restrictive and probably fairly constant (Tables 1 and 2). Primary sulfide bathing the substrate must exceed ~25 μM, the pH must exceed ~6.2, and the temperature must be between ~50 and 56 °C. Below pH 6.2, *Chlorobium tepidum* would not be expected, however, even if inoculum existed in the region. *Chlorobium tepidum*, in culture, requires at least 1 mM sulfide (Wahlund et al., 1991), which is rare in the Mammoth Springs. Below ~51 °C the facultatively anoxygenic cyanobacterium, *Spirulina* cf. *labyrinthiformis*, predom-

Fig. 10. (*See also Color Plate 3A*) Exposed *Chromatium tepidum* in a nighttime view of a 45–48 °C mat of Hunter's Hot Springs, Oregon. In daytime the *Chromatium* forms an underlayer beneath the brownish-red cyanobacterium *Oscillatoria terebriformis*, which descends below the *Chromatium* at night (see Richardson and Castenholz, 1987).

Fig. 11. (*See also Color Plate 3B*) Daytime view of the same area as Fig. 10., with *O. terebriformis* forming the mat surface, but with *Chromatium* swarming into the water in the can which is open at both ends, but which had been darkened by a top lid a few hours earlier.

inates in these springs, particularly when sulfide is in the range of 30– ~100 μM (Castenholz, 1977; see part E. below). Temperatures above 56 °C are out of the range of *Chromatium tepidum* and within the optimum range of photoautotrophic *Chloroflexus*. However, there is still an unidentifiable element that dictates the dominance of *Chromatium*. Why, for example, does not the photoautotrophic *Chloroflexus* predominate on the temperature range of ~ 45–56 °C? The upper half of this range, at least, is well within the optimal temperature range of the autotrophic *Chloroflexus* isolates. However, *Chloroflexus* grows very slowly as a sulfide supported photoautotroph, so perhaps the potentially more rapid growth rate of *Chromatium* allows its dominance. It is notable that even in these few springs with *Cm. tepidum* as the photoautotroph, there is still an undermat of *Chloroflexus*, presumably persisting as a photo- or chemo-heterotroph as it would under a top cover of photoautotrophic cyanobacteria.

Since undermat *Chromatium*, subsisting on biogenic sulfide, and perhaps organic acids from associated mat microorganisms, is common in several North American hot springs rich in sulfate, it is difficult to single out any one system. However, waters of Hunter's Hot Springs, Oregon, contains about 2.6 mM sulfate at the sources (Table 2) and a soft organic mat is produced with abundant *Chromatium* below about 55 °C under a top cover of the cyanobacterium *Oscillatoria* cf. *terebriformis*.

Cm. tepidum is very conspicuous at night in these springs since it swarms upward to the surface of the mat after dusk, at the same time that the *Oscillatoria terebriformis* makes its dusk decent into the soft mat material (see Jørgensen et al., 1992; Richardson and Castenholz, 1987; see Figs. 10 and 11). It is probable (but from still inconclusive data) that the *Chromatium* (which is laden with elemental sulfur from the incomplete, daytime, photosynthetic oxidation of sulfide) ascends to the surface at night in a positive aerotactic response. During darkness it oxidizes its 'internal' sulfur and/or environmental sulfide as a chemolithotroph and uses CO_2 as a chemoautotroph in the micro-oxic environment of the mat surface/water interface (Jue, 1990; Jue and Castenholz, unpublished data; see Figs. 10 and 11). The *Chromatium* descends into the mat again with light (using an undetermined tactic response) as the *O. terebriformis* ascends to the mat surface (see Jørgensen et al., 1992). The *Chromatium* then resumes anoxygenic photosynthesis, using the near IR radiation (which penetrates through the surface cover of cyanobacteria) and the free sulfide present beneath the cyanobacterial layer in daytime (see Richardson and Castenholz, 1987; Revsbech and Ward, 1984a). These mass nighttime movements are known also in non-thermophilic *Chromatium* (Jørgensen, 1982; Sorokin, 1970). Chemolithotrophy is known in several purple sulfur bacteria (e.g. de Wit and van Gemerden, 1987, 1990; Kämpf and Pfennig, 1986), but results were negative in the isolates of *Cm. tepidum* as described by Madigan (1986).

E. Anoxygenic Cyanobacteria in Hot Springs

The two best known examples of hot spring systems with anoxygenic cyanobacteria are sulfidic, alkaline springs of the Rotorua area of New Zealand and the neutral pH, sulfidic springs of the Mammoth Terraces of Yellowstone National Park.

The alkaline springs of New Zealand are unusual in that high sulfide levels are common (up to 3 mM), and at levels above ~5 μM *Oscillatoria* cf. *amphigranulata* and, rarely, *Synechococcus* sp. are the sole cyanobacteria present in the thermophilic range. Throughout the natural distribution of *O. amphigranulata* this species forms coherent mats up to its maximum growth temperature of 56 °C (Castenholz, 1976; Fig. 6). At high sulfide concentrations (1.0–1.5 mM) in culture, *O. amphigranulata* photosynthesizes anoxygenically, but at lower sulfide concentrations (0.3–0.6 mM) and high photon fluence rates, part of the photosynthetic activity could be ascribed to oxygenic photosynthesis (Castenholz and Utkilen, 1984). Several strains of this species were isolated from different hot springs, and capacity for anoxygenic photosynthesis varied in a predictable manner with the sulfide concentration characteristics of the springs of origin (Garcia-Pichel and Castenholz, 1990).

The *O. amphigranulata* springs of New Zealand are recognized by the dull olive-green mats which can be demonstrated, in all cases, to terminate upstream at the fairly constant 56–57 °C point. These mats are sometimes preceded upstream by delicate white streamers of *Thiobacillus*-like chemolithotrophs (Fig. 6). In New Zealand springs with little or no primary sulfide, oxygenic, sulfide-intolerant cyanobacteria form brighter green mats up to a constant temperature of 63–64 °C (see Castenholz, 1976).

In contrast, the Mammoth Springs of YNP have a much lower sulfide content, but the toxicity level may be as high because of the lower pH of 6.2– ~6.8 (e.g. Howsley and Pearson, 1979). *Spirulina* cf. *labyrinthiformis* forms nearly monotypic dark blue-green mats wherever springs issue at about 45–51 °C with sulfide levels of 30–100 μM (Castenholz, 1977). Although this species can hardly be regarded as a thermophile with an upper limit of only 51 °C, in the Yellowstone springs area there is apparently no cyanobacterium that grows at a higher temperature in the presence of sulfide above ~25 μM. As in the case of *O. amphigranulata, S. labyrinthiformis* is capable of strictly anoxygenic photosynthesis but only when sustained by sulfide of ~ 0.5–1.0 mM, levels higher than in the springs of origin (Castenholz, 1977).

References

Bateson MM and Ward DM (1988) Photoexcretion and fate of glycolate in a hot spring microbial mat. Appl Environ Microbiol 54: 1738–1743

Beer-Romero P and Gest H (1987) *Heliobacillus mobilis*, a peritrichously flagellated anoxyphototroph containing bacteriochlorophyll g. FEMS Microbiol Lett 41: 109–114

Blankenship RE (1985) Electron transport in green photosynthetic bacteria. Photosynth Res 6: 317–333

Blankenship RE (1992) Origin and early evolution of photosynthesis. Photosynth Res 33: 91–111

Brock TD (1978) Thermophilic microorganisms and life at high temperatures. Springer-Verlag, NY, Heidelberg

Castenholz RW (1969) Thermophilic blue-green algae and the thermal environment. Bacteriol Rev 33: 476–504

Castenholz RW (1973a) The possible photosynthetic use of sulfide by the filamentous phototrophic bacteria of hot springs. Limnol Oceanogr 18: 863–876

Castenholz RW (1973b) Ecology of blue-green algae in hot springs. In: Carr NG and Whitton BA (eds) The Biology of Blue-Green Algae, pp 379–414. Blackwell Publications, Oxford

Castenholz RW (1976) The effect of sulfide on the blue-green algae of hot springs I. New Zealand and Iceland. J Phycol 12: 54–68

Castenholz RW (1977) The effect of sulfide on the blue-green algae of hot springs II. Yellowstone Park. Microbial Ecol 3: 79–105

Castenholz RW (1984a) Habitats of *Chloroflexus* and related organisms. In: Klug MJ and Reddy CA (eds) Current Perspectives in Microbial Ecology, pp 196–200. American Society for Microbiology, Washington, DC

Castenholz RW (1984b) Composition of hot spring microbial mats: A summary. In: Cohen Y, Castenholz RW, Halvorson HO (eds), Microbial Mats: Stromatolites, pp 101–119. Alan R. Liss, New York

Castenholz RW (1988) The green sulfur and nonsulfur bacteria of hot springs. In: Olson JM, Ormerod JG, Amesz J, Stackebrandt E and Trüper HG (eds) Green Photosynthetic Bacteria, pp 243–255. Plenum Press, New York

Castenholz RW and Schneider AJ (1993) Cyanobacterial dominance at high and low temperatures: Optimal conditions or precarious existence? In: Guerrero R and Pedros-Alio C (eds), Trends in Microbial Ecology, pp 19–24. Spanish Society for Microbiology, Barcelona

Castenholz RW and Utkilen HC (1984) Physiology of sulfide tolerance in a thermophilic *Oscillatoria*. Arch Microbiol 138: 299–305

Castenholz RW, Bauld J and Jørgensen BB (1990) Anoxygenic microbial mats of hot springs: thermophilic *Chlorobium* sp. FEMS Microbiol Ecol 74: 325–336

Castenholz RW, Jørgensen BB, D'Amelio E and Bauld J (1991) Photosynthetic and behavioral versatility of the cyanobacterium

Oscillatoria boryana in a sulfide-rich microbial mat. FEMS Microbiol Ecol 86: 43–58

Cohen Y, Jørgensen BB, Revsbech NP and Poplowski R (1986) Adaptation to hydrogen sulfide of oxygenic and anoxygenic photosynthesis among cyanobacteria. Appl Environ Microbiol 51: 398–407

de Wit R and van Gemerden H (1987) Chemolithotrophic growth of the phototrophic sulfur bacterium *Thiocapsa roseopersicina*. FEMS Microbiol Ecol 45: 117–126

de Wit R and van Gemerden H (1990) Growth of the phototrophic purple sulfur bacterium *Thiocapsa roseopersicina* under oxic/anoxic regimens in the light. FEMS Microbiol Ecol 73: 69–76

Garcia D, Parot P, Vermeglio A and Madigan MT (1986) The light-harvesting complexes of a thermophilic purple sulfur photosynthetic bacterium *Chromatium tepidum*. Biochim. Biophys. Acta 850: 390–395

Garcia-Pichel F and Castenholz RW (1990) Comparative anoxygenic photosynthetic capacity in 7 strains of a thermophilic cyanobacterium. Arch Microbiol 153: 344–351

Giovannoni SJ, Revsbech NP, Ward DM and Castenholz RW (1987) Obligately phototrophic *Chloroflexus*: Primary production in anaerobic hot spring microbial mats. Arch Microbiol 147: 80–87

Gorlenko VM, Kompantseva EI and Puchkova NN (1985) Influence of temperature on the prevalence of phototrophic bacteria in hot springs. Mikrobiologiya 54: 848–853

Gorlenko VM, Bonch-Osmolovskaya EA, Kompantseva EI and Starynin DA (1987) Differentiation of microbial communities in connection with a change in the physicochemical conditions in thermophile spring. Mikrobiologiya 56: 314–322

Hanada S, Hiraishi A, Shimada K and Matsuura K (1993) New isolates of *Chloroflexus*-like bacteria from Japanese hot springs. Abstract, 1993 International Botanical Congress, Yokohama, Japan

Heda GD and Madigan MT (1986) Utilization of amino acids and lack of diazotrophy in the thermophilic anoxygenic phototroph *Chloroflexus aurantiacus*. J Gen Microbiol 132: 2469–2473

Howsley R and Pearson HW (1979) pH-dependent sulfide toxicity to oxygenic photosynthesis in cyanobacteria. FEMS Microbiol Lett 6: 288–292

Imhoff JF and Trüper HG (1981) *Ectothiorhodospira abdelmalekii* sp. nov., a new halophilic alkaliphilic phototrophic bacterium. Zentralbl Bakteriol Mikrobiol Hyg I Abt Orig. C2: 228–234

Imhoff JF and Trüper HG (1989) Purple nonsulfur bacteria. In: Staley JT, Bryant MP, Pfennig N and Holt JT (eds) Bergey's Manual of Systematic Bacteriology, Vol. 3: 1658–1682. Williams and Wilkins, Baltimore

Jørgensen BB (1982) Ecology of the bacteria of the sulfur cycle with special reference to anoxic-oxic interface environment. Phil Trans Royal Soc London B 298: 543–561

Jørgensen BB and Nelson DC (1988) Bacterial zonation, photosynthesis, and spectral light distribution in hot spring microbial mats of Iceland. Microbial Ecol 16: 133–147

Jørgensen BB, Castenholz RW and Pierson BK (1992) The microenvironment within modern microbial mats. In: Schopf JW and Klein C (eds) The Proterozoic Biosphere, a Multidisciplinary Study, pp 271–278. Cambridge University press, Cambridge, New York

Jue C-S (1990) The effect of aerobic environments on *Chromatium* cf. *tepidum*, a thermophilic purple sulfur bacterium. M.A. thesis, University of Oregon, Eugene

Kämpf C and Pfennig N (1980) Capacity of Chromatiaceae for chemotrophic growth. Specific respiration rate of *Thiocystis violacea* and *Chromatium vinosum*. Arch Microbiol 127: 125–135

Kimble LK and Madigan MT (1994) Isolation and characterization of thermophilic heliobacteria. Abstract I-7, General Meeting of the American Society for Microbiology

Kimble LK, Mandelco L, Woese CR and Madigan MT (1995) *Heliobacterium medesticaldum*, sp. nov., a thermophilic heliobacterium of hot springs and volcanic soils. Arch Microbiol 163: 259–267

Madigan MT (1984) A novel photosynthetic purple bacterium isolated from a Yellowstone hot spring. Science 225: 313–315

Madigan MT (1986) *Chromatium tepidum* sp. nov., a thermophilic photosynthetic bacterium of the family Chromatiaceae. Int J Syst Bacteriol 36: 222–227

Madigan MT (1988) Microbiology, physiology, and ecology of phototrophic bacteria. In: Zehnder AJB (ed) Biology of Anaerobic Microorganisms, pp 39–111. Wiley, Chichester

Madigan MT (1992) The family Heliobacteriaceae. In: Balows A, Trüper HG, Dworkin M, Harder W and Schleiffer K-H (eds) The Prokaryotes, second edition, Vol. 4: 1981–1992. Springer-Verlag, NY, Heidelberg

Madigan MT and Brock TD (1975) Photosynthetic sulfide oxidation by *Chloroflexus aurantiacus*, a filamentous, photosynthetic, gliding bacterium. J Bacteriol 122: 782–784

Madigan MT, Takigiku R, Lee RG, Gest H and Hayes JM (1989) Carbon isotope fractionation by thermophilic phototrophic sulfur bacteria: Evidence for autotrophic gowth in natural populations. Appl Environ Microbiol 55: 639–644

Meeks JC and Castenholz RW (1971) Growth and photosynthesis in an extreme thermophile, *Synechococcus lividus* (Cyanophyta). Arch Microbiol 78: 25–41

Oelze J (1992) Light and oxygen regulation of the synthesis of bacteriochlorophylls a and c in *Chloroflexus aurantiacus*. J Bacteriol 174: 5021–5026

Olsen GJ, Woese CR and Overbeek R (1994) The winds of (evolutionary) change: Breathing new life into microbiology. J Bacteriol 176: 1–6

Overmann J, Lehmann S and Pfennig N (1991) Gas vesicle formation and buoyancy regulation in *Pelodictyon phaeoclathratiforme* (Green sulfur bacteria). Arch Microbiol 157: 29–37

Overmann J, Cypionka H and Pfennig N (1992) An extremely low-light-adapted phototrophic sulfur bacterium from the Black Sea. Limnol Oceanogr 37: 150–155

Paschinger H, Paschinger J and Gaffron H (1974) Photochemical disproportionation of sulfur into sulfide and sulfate by *Chlorobium limicola* forma *thiosulfatophilum*. Arch Microbiol 96: 341–351

Pierson BK and Castenholz RW (1974a) A phototrophic gliding filamentous bacterium of hot springs, *Chloroflexus aurantiacus*, gen. and sp. nov. Arch Microbiol 100: 5–24

Pierson BK and Castenholz RW (1974b) Studies of pigments and growth in *Chloroflexus aurantiacus*, a phototrophic filamentous bacterium. Arch Microbiol 100: 283–305

Pierson BK and Castenholz RW (1992) The family Chloroflexaceae. In: Balows A, Trüper HG, Dworkin M, Harder W and Schleifer K-H (eds) The Prokaryotes, 2nd edition, Vol. 4, pp 3754–3774. Springer-Verlag, NY, Heidelberg

Pierson BK, Giovannoni SJ and Castenholz RW (1984)

Physiological ecology of a gliding bacterium containing bacteriochlorophyll *a*. Appl Environ Microbiol 47: 576–584

Pierson BK, Giovannoni SJ, Stahl DA and Castenholz RW (1985) *Heliothrix oregonensis*, gen. nov., sp. nov., a phototrophic filamentous gliding bacterium containing bacteriochlorophyll a. Arch Microbiol 142: 164–167

Pierson BK, Sands VM and Frederick JL (1990) Spectral irradiance and distribution of pigments in a highly layered marine microbial mat. Appl Environ Microbiol 56: 2327–2340

Raymond JC and Sistrom WR (1969) *Ectothiorhodospira halophila*: A new species of the genus *Ectothiorhodospira*. Arch Mikrobiol 69: 121–126

Resnick SM and Madigan MT (1989) Isolation and characterization of a mildly thermophilic nonsulfur purple bacterium containing bacteriochlorophyll *b*. FEMS Microbiol Lett 65: 165–170

Revsbech NP and Ward DM (1984a) Microelectrode studies of interstitial water chemistry and photosynthetic activity in a hot spring microbial mat. Appl Environ Microbiol 48: 270–275

Revsbech NP and Ward DM (1984b) Microprofiles of dissolved substances and photosynthesis in microbial mats measured with microelectrodes. In: Cohen Y, Castenholz RW, and Halvorson HO (eds) Microbial Mats: Stromatolites, pp 171–188. Alan R. Liss, New York

Richardson LL and Castenholz RW (1987) Diel vertical movements of the cyanobacterium *Oscillatoria terebriformis* in a sulfide-rich hot spring microbial mat. Appl Environ Microbiol 53: 2142–2150

Ruff-Roberts AL, Kuenen JG and Ward DM (1994) Distribution of cultivated and uncultivated cyanobacteria and *Chloroflexus*-like bacteria in hot spring microbial mats. Appl Environ Microbiol 60: 697–704

Shiea J, Brassell SC and Ward DM (1991) Comparative analysis of extractable lipids in hot spring microbial mats and their component photosynthetic bacteria. Org Geochem 17: 309–319

Sorokin YI (1970) Interrelations between sulphur and carbon turnover in meromictic lakes. Arch Hydrobiol 66: 391–446

Sprague SG, Staehelin LA and Fuller RC (1981) Semiaerobic induction of bacteriochlorophyll synthesis in the green bacterium *Chloroflexus aurantiacus*. J Bacteriol 147: 1032–1039

Stadtwald-Demchick R, Turner FR and Gest H (1990a) *Rhodopseudomonas cryptolactis*, sp. nov., a new thermotolerant species of budding phototrophic purple bacteria. FEMS Microbiol Lett 71: 117–122

Stadtwald-Demchick R, Turner FR and Gest H (1990b) Physiological properties of the thermotolerant photosynthetic bacterium, *Rhodospirillum centenum*. FEMS Microbiol Lett 67: 139–144

Stauffer RE, Jenny EA and Ball JW (1980) Chemical studies of selected trace elements in hot-spring drainages of Yellowstone National Park. Geohydrology of Geothermal Systems, Geological Survey Professional Paper 1044-F, U.S. Department of the Interior, Washington, DC

Teiser ML (1993) Extracellular low molecular weight organic compounds produced by *Synechococcus* sp. and their roles in the food web of alkaline hot spring microbial mat communities. Ph.D. Thesis, Dept. of Biology, University of Oregon, Eugene

Wahlund TM and Madigan MT (1993) Nitrogen fixation by the thermophilic green sulfur bacterium *Chlorobium tepidum*. J Bacteriol 175: 474–478

Wahlund TM, Woese CR, Castenholz RW and Madigan MT (1991) A thermophilic green sulfur bacterium from New Zealand hot springs, *Chlorobium tepidum* sp. nov. Arch Microbiol 156: 81–90

Ward DM, Weller R, Shiea J, Castenholz RW and Cohen Y (1989) Hot spring microbial mats: Anoxygenic and oxygenic mats of possible evolutionary significance. In: Cohen Y and Rosenberg E (eds) Microbial Mats, Physiological Ecology of Benthic Microbial Communities, pp 3–15. ASM, Washington, DC

Ward DM, Bauld J, Castenholz RW and Pierson BK (1992) Modern phototrophic microbial mats: anoxygenic, intermittently oxygenic/anoxygenic, thermal, eukaryotic, and terrestrial. In: Schopf JW and Klein C (eds) The Proterozoic Biosphere, a Multidisciplinary Study, pp 309–324. Cambridge University Press, Cambridge, New York

Waring GA (1965) Thermal springs of the United States and other countries of the world – a summary. Revised by Blankenship RR and Bentall R, Geological Survey Professional Paper 492, U.S. Government Printing Office, Washington, DC

Weller R, Bateson MM, Heimbuch BK, Kopczynski ED and Ward DM. (1992) Uncultivated cyanobacteria, *Chloroflexus*-like inhabitants, and spirochete-like inhabitants of a hot spring microbial mat. Appl Environ Microbiol 58: 3964–3969

White DE, Hem JD and Waring GA (1963) Chapter F. Chemical composition of sub-surface waters. In: Fleischer M (ed) Data of Geochemistry, Sixth edition. U.S. Government Printing Office, Washington, DC

Yun Z (1986) Thermophilic microorganisms in the hot springs of Tengchong Geothermal Area, West Yunnan, China. Geothermics 15: 347–358

Yurkov VV, Gorlenko VM, Mityushina LL and Starynin DA (1992) Effect of limiting factors on the structure of phototrophic associations in thermal springs. Mikrobiologiya 60: 129–138

Chapter 6

Aerobic Anoxygenic Phototrophs

Keizo Shimada

Department of Biology, Faculty of Science, Tokyo Metropolitan University, Minami-ohsawa 1-1, Hachioji, Tokyo 192-03, Japan

Summary

During the last 15 years, more than 20 strains of aerobic bacteria which possess bacteriochlorophyll (BChl) *a* have been found. They are distinguished from typical anaerobic (anoxygenic) phototrophs in that they synthesize BChl only under aerobic conditions and cannot grow without O_2 or other oxidants, even in the light. In some species, photosynthetic activities have been demonstrated. Reaction centers and light-harvesting complexes isolated from some species were shown to be similar to those of typical purple photosynthetic bacteria. The regulatory mechanism of synthesis of pigments and proteins of the photosynthetic apparatus are apparently opposite with respect to O_2 tension to that of typical anoxygenic phototrophs. The low content of BChl, unique composition of carotenoids and presence of non-photosynthetic carotenoids in most strains are other marked characteristics of these aerobic bacteria. Phylogenetically, they are not classified into single group. Species so far described are distributed rather widely within the α-subclass of Proteobacteria (purple

R. E. Blankenship, M. T. Madigan and C. E. Bauer (eds): Anoxygenic Photosynthetic Bacteria, pp. 105–122.

bacteria) in which most of the purple nonsulfur bacteria as well as many non-photosynthetic bacteria are included. Apparently, these aerobic BChl-containing bacteria represent an evolutionary transient phase from anaerobic phototrophs to aerobic non-phototrophs. However, some characteristic features distinct from anaerobic phototrophs suggest that most of them are in a evolutionary stable state.

I. Introduction

It has long been known that bacteriochlorophylls (BChls) are present in prokaryotes which perform anoxygenic photosynthesis under anaerobic-light conditions. Although some bacteria, such as species of *Rhodobacter*, are known to synthesize BChl *a* under semiaerobic conditions, their photosynthetic activities and BChl content are higher under anaerobic conditions. In 1978, however, bacteriochlorophyll *a* (BChl *a*) was detected in three strains of obligate aerobic bacteria (Sato, 1978; Harashima et al, 1978). Since then, more than 20 strains of aerobic bacteria which contain BChl *a* have been reported (Shiba et al, 1979a; Nishimura et al., 1981; Shiba and Simidu, 1982; Urakami and Komagata, 1984; Shiba and Abe, 1987; Evans et al., 1990; Shiba et al., 1991; Yurkov and Gorlenko, 1990,1991,1992; Yurkov et al., 1992; Fuerst et al., 1993; Yurkov and Van Gemerden, 1993b; Wakao et al., 1993). These bacteria were distinguished from 'typical' photosynthetic bacteria in the following points: (1) They are, in principle, obligate aerobes and cannot grow under anaerobic conditions, even in the light. (2) BChl synthesis occurs under aerobic conditions, not under anaerobic conditions.

The BChl content of these aerobic bacteria are, however, significantly lower than those of 'typical' photosynthetic bacteria (Table 1). Thus, the first question may be whether all of these aerobic BChl-containing strains have significant photosynthetic activity or not. In this article, all aerobic bacteria in which BChl has been found are described as aerobic anoxygenic phototrophs without evidence for their phototrophy or photosynthetic activity, since these properties have not been confirmed in most strains. Photosynthetic activities such as photo-stimulated CO_2 incorporation, photo-stimulated growth and photooxidation of cytochromes have been demonstrated only in a few species (Harashima et al., 1982; Shiba, 1984; Takamiya and Okamura, 1984; Harashima et al, 1987). Phototrophy and/or the advantage of photosynthesis to other non-phototrophic aerobic bacteria have not yet been assessed in their natural habitats. The second question concerns the in vivo state of BChl. All strains so far described seem to have a photosynthetic apparatus similar to that of typical purple photosynthetic bacteria. In some species, reaction centers and antenna complexes of a purple bacterial type have been isolated. In *Roseobacter (Ro.) denitrificans* (formerly *Erythrobacter* sp. strain OCh 114), which is the species investigated most extensively among aerobic anoxygenic phototrophs, the gene structure coding for the photosynthetic apparatus was also shown to be similar to those of typical purple bacteria. However, it is possible in some species that the genetic organization of the photosynthetic apparatus is incomplete compared with those of typical purple bacteria. Regulatory systems for synthesis of BChl and the photosynthetic apparatus, which must be different from those of anaerobic phototrophs, are still unclear. A third question concerns the phylogenetic position of aerobic anoxygenic phototrophs. Phylogenetic studies based on ribosomal RNA indicated that the strains of aerobic anoxygenic phototrophs so far studied are distributed among the α-subclass of Proteobacteria (purple bacteria) in which most of the 'purple nonsulfur bacteria' as well as many non-phototrophic aerobes are included. One may consider, therefore, that each aerobic anoxygenic phototroph is on a transient phase in evolution from an anaerobic purple bacterium to an aerobic non-phototrophic bacterium. In this chapter, available data are presented to discuss the above questions, although the data are still insufficient and are confined to limited species.

Abbreviations: ALA – 5 aminolevulinic acid; BChl – bacteriochlorophyll; *E.* – *Erythrobacter*; *Em.* – *Erythromicrobium;* LH – light-harvesting; *M.* – *Methylobacterium; Rb.* – *Rhodobacter*; RC – reaction center; *Ro.* – *Roseobacter*; TMAO – trimethylamine N-oxide;

II. Habitats and Culture

A. Habitats

Aerobic anoxygenic phototrophs so far described are all chemoheterotrophs and have been isolated mainly from aquatic environments, except for some

Table 1. Species of aerobic anoxygenic phototrophs and their cellular BChl content

Species	BChl-content (nmol/mg dry cell)	References
Erythrobacter longus	2.0	Harashima et al., 1980
Erythrobacter sp. OCh 175	1.5	Shiba and Abe, 1984
Roseobacter litoralis	NM*	Shiba, 1991
Roseobacter denitrificans	8.0	Harashima et al., 1980
Methylobacterium rhodesianum	0.06§	Sato et al., 1985
Methylobacterium radiotolerans	0.19	Nishimura et al., unpublished
Methylobacterium extorquens	NM	Urakami and Komagata, 1984
Methylobacterium zatmanii	NM	Urakami and Komagata, 1984
Methylobacterium fujisawaense	NM	Urakami and Komagata, 1984
Methylobacterium rhodinum	NM	Urakami and Komagata, 1984
Porphyrobacter neustonensis	NM	Fuerst et al., 1992
Rhizobium BTAi1	0.40	Evans et al., 1990
Acidiphilium rubrum	0.71	Wakao et al., 1993
Acidiphilium angustum	0.08	Wakao et al., 1993
Acidiphilium cryptum	0.04	Wakao et al., 1993
Erythromicrobium sibiricum	NM	Yurkov and Gorlenko, 1992
Erythromicrobium ezovicum	NM	Yurkov and Gorlenko, 1992
Erythromicrobium hydrolyticum	2.0†	Yurkov and Van Gemerden, 1993a
Erythromicrobium ursincola	NM	Yurkov et al., 1992
Erythromicrobium ramosum	NM	Yurkov and Gorlenko, 1992
Roseococcus thiosulfatophilum	NM	Yurkov and Gorlenko, 1991
Rhodobacter sphaeroides‡	20.0	Takemoto and Kao, 1977

* NM, Not measured.

§ Estimated from the concentration in the culture solution.

† Estimated from the content on protein basis.

‡ Typical anoxygenic photosyntheic bacterium for reference.

strains of facultative methylotrophs. Siba et al. (1979b,1991) have isolated *Erythrobacter, Roseobacter* and some other strains from the surface of sea weeds, coastal sands, cyanobacterial mat and water in high tidal zone. Occupation of 20 to 50% of the total aerobic heterotrophs by aerobic BChl-containing bacteria was observed in some tropical high-tidal zones (Shiba et al., 1991). However, isolation from sea water is, so far, unsuccessful (Shiba et al., 1979b). The marine bacterial group seem to contain many other species which have not yet been characterized (Shiba et al., 1991; Yurkov and Van Gemerden, 1993b). From fresh water, *Porphyrobacter, Acidiphilium, Erythromicrobium* and *Roseococcus* have been isolated (Fuerst et al., 1992; Yurkov and Gorlenko, 1990,1991,1992; Wakao et al., 1993). Species of the latter two genera were isolated from cyanobacterial mat formed downstream of alkaline hot springs. *Porphyrobacter,* which is phylogenetically close to *Erythrobacter,* was isolated from eutrophic fresh water. Facultative methylotrophic strains which were once classified as *Protomonas* and, later, included into *Methylobacterium,* are isolates from various sources such as foods, soils, leaf surfaces, etc. (Urakami and Komagata, 1984; Bousfield and Green, 1985). Although one may suppose that phototrophy would be advantageous in oligotrophic environments, most strains of aerobic BChl-containing bacteria have been found in rather eutrophic environments like many species of purple nonsulfur bacteria. Among them, however, species of *Acidiphilium* require relatively low concentration of nutrients (Wakao et al., 1993). Thus, it is possible that unknown oligotrophic strains are present in various environments. So far, all strains described are mesophilic as are typical purple nonsulfur bacteria.

B. Isolation and Culture

As stated above, most aerobic anoxygenic phototrophs grow under aerobic, mesophilic and eutrophic conditions. Therefore, isolation was also performed

Table 2. Composition of the medium for growth of *Roseobacter* (Shioi, 1986)

Ingredient	Concentration (g/liter)
I. Basal salts	
$NaCl^a$	20.0
$MgCl_2 \cdot 6H_2O^a$	5.0
$Na_2SO_4{}^a$	2.0
KCl^a	0.5
$CaCl_2 \cdot 2H_2O^a$	0.5
$NaHCO_3$	0.2
Ferric citrate[b]	0.1
II. Organic substrates	
Yeast extract (Difco)	2.0
Polypeptone	1.0
Casaminio acids	1.0
Glycerol	1.0 ml

All of the ingredients are mixed and autoclaved for 15 min at 1.0 $kg \cdot cm^{-2}$ after adjusting the pH to 7.5 with NaOH. The precipitate formed does not affect the bacterial growth.
[a] Can be stocked together as five strength solution.
[b] Added as solution (10 mg/ml); store in refrigerator.

aerobically at 20 to 30 °C on agar plates containing rather rich media. In the case of marine bacteria, media such as PPES-II, which are rich in peptones and yeast extract, have been used (Shiba et al., 1979b; Shioi, 1986, Table 2). Since BChl formation is inhibited by light as described below, the culture is kept in the dark or in some cases under intermittent light. Light is not necessary for growth in all strains so far investigated. However, optimum condition for growth and/or BChl-formation have not yet been established in most strains. Thus, it is possible that the very low content (or lack) of BChl in some aerobic phototrophs does not necessarily reflect the nature of the strain in its habitat. The isolated pigmented colonies were enriched and examined for the presence of BChl and the inability to grow under anaerobic conditions in the light. Batch cultures are usually grown on a shaker or in a bubbled bottle. The aeration rate should be controlled (depending on the strains) for the maximal formation of BChl, since saturated oxygen does not necessarily give a maximum yield of BChl (Shiba, 1987). Light conditions should also be surveyed for the yield of BChl, since the BChl-containing *Rhizobium* is known to synthesize BChl only under intermittent light (Evans et al., 1990; see Chapter by Fleishman et al.) and the BChl content of some species of *Methylobacterium* varies significantly by the interval of the intermittent light (Sato and Shimizu, 1979; see below).

III. Photosynthetic Apparatus

A. Reaction Center

The presence of reaction centers (RCs) in aerobic BChl-containing heterotrophs was first discovered in *Ro. denitrificans* and *Erythrobacter (E.) longus* by Harashima et al. (1982) using light induced absorption changes and an absorption spectrum of iridic chloride-oxidized membranes; the results were similar to those obtained with typical purple photosynthetic bacteria. Later, similar absorption changes were observed in *Methylobacterium(M.) rhodesianum* (formerly *Protaminobacter ruber*), *M. radiotolerans* (*Pseudomonas radiora*), and *Rhizobium* strain BTAi1(Takamiya and Okamura, 1984; Nishimura et al, 1989; Evans et al, 1990).

Purified RC complexes have been obtained only from *Ro. denitrificans* (Shimada et al., 1985; Takamiya et al., 1987). The absorption spectrum and the photo-induced absorption changes of purified RCs were almost identical to those of typical purple bacteria which have the cytochrome subunit (Fig. 1).

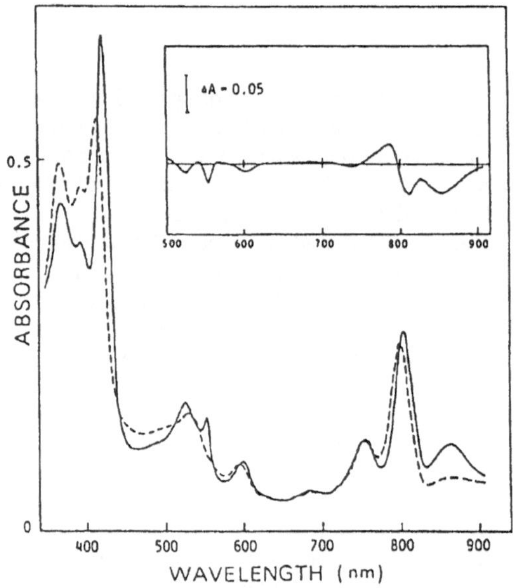

Fig. 1. Absorption and reduction-oxidation difference spectra (inset) of the RC of *Roseobacter denitrificans* (from Shimada et al., 1985).

The composition of cofactors in this preparation was also the same as that in purple bacterial RCs; 4 BChls, 2 bacteriopheophytins, 4 *c*-type hemes and 2 ubiquinones-10. The estimated midpoint potential (E_m) of the primary acceptor (Q_A), –44 mV, and its independence of pH, were also very similar to those of typical purple bacteria (Takamiya et al., 1987). The polypeptide composition was also similar to those of typical purple bacteria; L, M, and H polypeptides were present along with cytochrome subunits of apparent molecular mass of 26, 30, 32 and 42 kDa, respectively. Recently, primary structures of the L and M subunits were deduced by analysis of the nucleotide sequence of a gene cluster (*puf* operon) that contained the genes for the subunits of the RC, as well as those for the two subunits of the B870 light-harvesting complex (LH I) (Liebetanz et al., 1991). The gene structure was essentially the same as those of typical purple bacteria and the amino acid sequence homology of the subunits showed very high values, especially with those of *Rhodobacter* species (up to 70%). These results suggested close relationships between aerobic anoxygenic phototrophs and anaerobic purple photosynthetic bacteria. From *E. longus* only a trace amount of RC could be obtained, which contained three subunits; probably H, M and L (Shimada et al., 1985). The lack of cytochrome in the RC of this species was supported by the absorption difference spectra of RC-LH I complex and the light induced absorption change of the membranes (Matsuura and Shimada, 1990).

Unlike RC, a pigment-protein complex presumed as RC-LH I complex could be obtained rather easily from some species. A minor absorption band around 800 nm and a shoulder around 750 nm in this complex suggested the presence of a RC of the purple bacterial type in *M. radiotolerans* and *Erythromicrobium (Em.) ramosum* (Nishimura et al., 1989; Yurkov et al., 1992). From the RC-LH I complex of *M. radiotolerans*, the four presumed subunits of RC, i.e., L, M, H and the cytochrome subunit, were clearly resolved by SDS-polyacrylamide gel electrophoresis. The mobilities of these proteins were also very similar to those of the corresponding proteins in the RC of typical purple bacteria. The presence of a purple bacterial-type RC was also suggested in many other aerobic anoxygenic phototrophs from the absorption spectra of membranes or cells which showed the same profile as that of the RC-LH I complex, probably due to a lack of the peripheral antenna (LH II) in these species.

It may be noteworthy that the cytochrome subunit of RC has been found from at least two species of aerobic anoxygenic phototrophs, since these types of RCs have been found from relatively anaerobic species of purple bacteria and are rather rare in facultative species of purple bacteria such as *Rhodobacter* species (Matsuura and Shimada, 1990). Although the role of the cytochrome subunit of RC in purple bacteria is not necessarily understood, the presence of this subunit in aerobic anoxygenic phototrophs might give a clue for the elucidation of its significance.

B. Light-Harvesting System

The in vivo absorption spectra of BChl of all aerobic anoxygenic phototrophs so far described showed red-shifted spectra of BChl as in typical purple bacteria; this indicates that almost all BChl molecules are bound to pigment proteins, as in typical purple bacteria. As stated above, most strains were suggested to have a RC-LH I complex of the purple bacterial type. Thus, in terms of the photosynthetic apparatus, all aerobic anoxygenic phototrophs seem to have the ability to perform photosynthesis despite the low content of BChl. Antenna complexes have been obtained from *E. longus*, *M. radiotolerans* and *Ro. denitrificans* (Shimada et al, 1985; Nishimura et al., 1989; Figs 2 and 3). RC-LH I type complexes were obtained from all of these species as from all typical

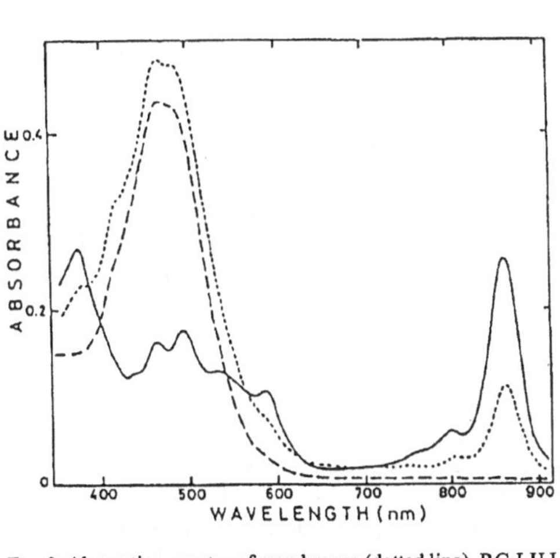

Fig. 2. Absorption spectra of membranes (dotted line), RC-LH I complex (solid line) and 'non-photosynthetic' carotenoids (broken line) of *Erythrobacter longus* (from Shimada et al., 1985).

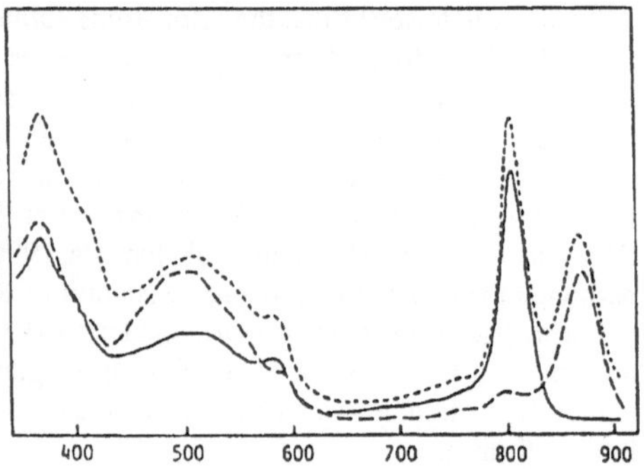

Fig. 3. Absorption spectra of membranes (dotted line), RC-LH I complex (broken line) and B806 (LH II) complex (solid line) of *Roseobacter denitrificans* (from Shimada et al., 1985).

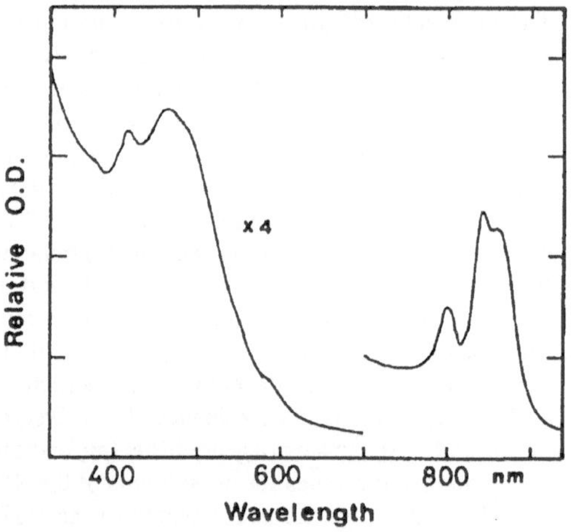

Fig. 4. Absorption spectra of intact cells of *Erythrobacter* sp. strain OCh 175 (from Shiba and Abe, 1987).

purple bacteria indicating a close similarity between the light-harvesting systems of the two groups of anoxygenic phototrophs. In *E. longus* and *M. radiotolerans*, there seem to be no pigment proteins other than the RC-LH I type complex (Fig. 2). The absorption spectra of cells or membranes of the *Methylobacterium* species, *Porphylobacter neustonensis*, *Em. ursincola*, *Roseococcus thiosulfatophilus* and the *Acidiphilium* species indicated that these organisms also lack an LH II-type complex as in *E. longus*. The lack of LH II in these species is apparently consistent with the hypothesis that aerobic anoxygenic phototrophs are on an evolutionary line to abandon photosynthesis (Woese et al., 1984), although some typical purple bacteria such as *Rhodospirillum rubrum* also lack LH II.

Thus, the presence of an LH II type complex seems to be somewhat rare in aerobic anoxygenic phototrophs. In vivo absorption spectra of some *Erythromicrobium* species and *Erythrobacter* sp. strain OCh 175 showed absorption bands other than those of the RC-LH I complex, indicating the presence of LH II type complexes in these species (Shiba and Abe, 1987; Yurkov and Gorlenko, 1990, 1992; Fig. 4). Species of *Roseobacter* also have another pigment protein in addition to the RC-LH I type complex. The pigment protein isolated from *Ro. denitrificans* showed only one peak at 806 nm in the near infrared region at room temperature (Fig. 3). Although an additional spectral component around 815 nm was resolved at 77 K, no similar antenna complexes have been found in typical purple bacteria (Shimada et al., 1992). In spite of its unique spectral nature, the *Ro. denitrificans* B806 complex is considered to correspond to the peripheral antenna (LH II) of typical purple bacteria, since it contained no RC and showed a Raman spectrum similar to LH II complexes of purple bacteria and transferred energy to the RC-LH I complex with high efficiency as described below (Hayashi et al., 1986; Shimada et al., 1990b). Whether the unique spectral characteristics of its B806 complex is correlated with the aerobic nature of *Roseobacter*, is not clear.

The apparent size of the antenna complexes and their constituent polypeptides of aerobic anoxygenic phototrophs so far examined were also similar to those of typical purple bacteria. Two polypeptides with apparent molecular masses of 5 to 7 kDa seemed to be the main protein components (Shimada et al., 1985; Nishimura et al., 1989). The deduced amino acid sequence of the two small polypeptides coded upstream of the RC components in the *puf* operon of *Ro. denitrificans* showed characteristic features of LH I polypeptides of typical purple bacteria (Liebetanz et al., 1991).

Efficiency of the light harvesting system of aerobic anoxygenic phototrophs has been investigated in only a few species. LH II (B806) complex of *Ro. denitrificans* showed no spectral components at longer than 820 nm which suggests low efficiency of excitation energy transfer to the LH I(B870) complex due to small spectral overlapping between the two antenna complexes. However, the transfer occurred within 6 ps, indicating a highly efficient light-harvesting system of this bacterium (Shimada et al.,

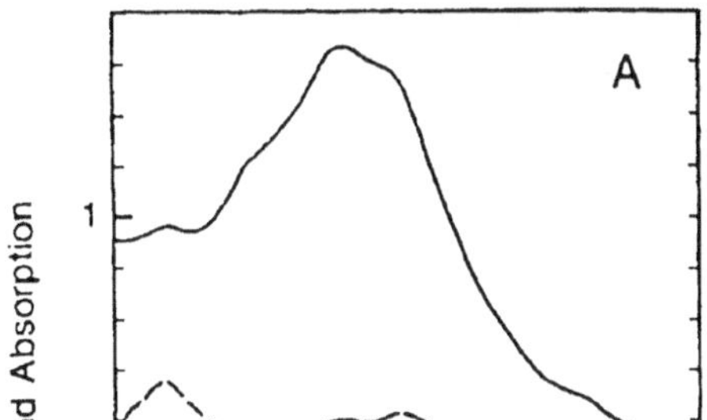

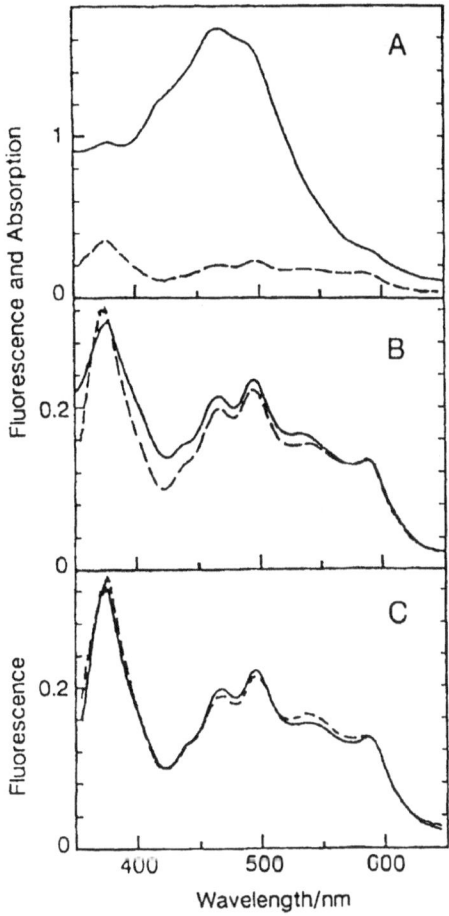

Fig. 5. Absorption and fluorescence excitation spectra of membranes and RC-LH I complex of *Erythrobacter longus*; (A) Absorption (solid line) and fluorescence excitation (broken line) spectra of membranes. (B) Absorption (solid line) and fluorescence excitation spectra (broken line) of RC-LH I complex. (C) Fluorescence excitation spectra of RC-LH I complex (solid line) and the membranes (broken line) depicted in the same scale (from Noguchi et al., 1990).

1990a). Energy transfer from carotenoids to BChl in this bacterium also occurred with high efficiency (Shimada et al., 1990b). The high efficiency of excitation energy transfer from carotenoids to BChl was also demonstrated in the RC-LH I complex of *E. longus* in spite of its unique composition of carotenoids (Noguchi et al., 1992; Fig. 5a). Thus, efficiency of the light-harvesting system of aerobic anoxygenic phototrophs is not necessarily low in spite of its low BChl content.

It may be remarkable that many species of aerobic anoxygenic phototrophs show higher ratios of carotenoid content to BChl than those of typical purple bacteria. It was suggested that only some carotenoids are bound to the light-harvesting complex or to RC in *E. longus* (Fig. 2). Excitation spectra for BChl fluorescence of membranes confirmed that the rest 'excess' carotenoids are not functioning as light-harvesting pigments (Fig. 5). In *M. radiotolerans*, a significant part of the carotenoid pool was not bound to the photosynthetic apparatus (Nishimura et al., 1989). Since such 'non-photosynthetic' carotenoids have not been found in typical purple bacteria, one of the roles of these carotenoids is likely to be concerned with protection from oxygen toxicity. As described below, these 'non-photosynthetic' carotenoids are highly polar and are structurally distinct from the carotenoids which are bound to the antenna complexes as light-harvesting pigments.

C. Membrane Structure

Aerobic anoxygenic phototrophs contain significantly lower amounts of BChl than typical purple bacteria which usually develop intracytoplasmic membranes coincidentally with BChl synthesis. Despite the low content of BChl, 'chromatophore-like' structures have been observed in *Ro. denitrificans*, *M. rhodesianum*, *Em. sibiricum* and *Rhizobium* strain BTAi1 (Harashima et al, 1982; Sato and Shimizu, 1979; Yurkov et al., 1991; Evans et al., 1990). In *Ro. denitrificans* and *M. rhodesianum*, these vesicular intracytoplasmic membranes, though poorly developed, can be observed in cells of relatively high BChl-content but not in cells of low BChl content (Harashima et al., 1982; Sato and Shimizu, 1979; Iba et al., 1988). These results suggest that the photosynthetic apparatus in these aerobes are mainly localized in the specific structure of membranes and the synthesis of such structures are regulated coincidentally with BChl synthesis as in typical purple bacteria.

IV. Pigments and Other Components

A. Bacteriochlorophyll

The only bacteriochlorophyll found in aerobic anoxygenic phototrophs thus far is BChl *a*. The esterified alcohol in all cases has been phytol

(Harashima and Takamiya, 1989; Shiba and Abe, 1987; Wakao et al, 1993). The reason that only BChl *a* has been found may be that BChl *a* is the most stable among the six chemical species of BChl in aerobic environments. However, there seems to be no reason for phytol as the exclusive esterified moiety, since some facultative purple bacteria such as *Rhodospirillum rubrum* have geranylgeraniol in place of phytol. It may thus be possible that aerobic bacteria containing BChl a_{gg} will be found in the future.

B. Carotenoids

The carotenoid composition of most aerobic anoxygenic phototrophs seems to be different from those of typical purple bacteria, as they probably have 'non-photosynthetic' carotenoids. In *E. longus*, more than 70% of total carotenoids do not function as light-harvesting pigments (Noguchi et al., 1992). These carotenoids were shown not to be bound to RC nor to the LH I complex, which is the sole antenna complex of this bacterium (Shimada, unpubl.; Harashima, 1989). The major components of these 'non-photosynthetic' carotenoids in *E. longus* have been shown to be novel sulfate-containing carotenoids, for example, erythroxanthin sulfate (Fig. 6) and caloxanthin sulfate (Takaichi et al., 1991a). Carotenoids bound to the RC-LH I complex of *E. longus* are less polar species: Among them, bacteriorubixanthinal is a novel carotenoid containing a cross-conjugated aldehyde group and a tertiary methoxy group which are unique to purple bacteria. This carotenoid also has an ionone ring which is rarely found in purple bacteria (Takaichi at al., 1988; Fig. 6). Zeaxanthin and other hydroxyl derivatives of *β*-carotene, which are common carotenoids of oxygenic phototrophs, are also major carotenoids in the pigment protein complex of *E. longus* (Takaichi et al., 1990; Fig. 6). It should be noted that spirilloxanthin, which is unique to purple bacteria, was also found in this RC-LH I complex, although the amount was small. Thus, in *E. longus*, three

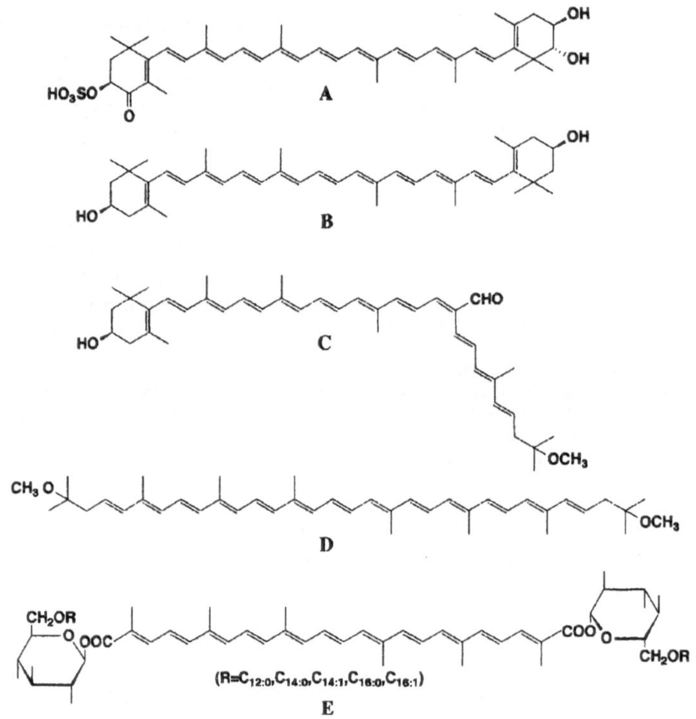

Fig. 6. Structure of characteristic carotenoids of aerobic anoxygenic phototrophs; (A) Erythroxanthin sulfate from *Erythrobacter longus* (B) Zeaxanthin from *E. longus* (C) Bacteriorubixanthinal from *E. longus* (D) Spirilloxanthin from *E. longus* (E) Diglucosyl C_{30}-carotenoid from *Methylobacterium rhodinum* .

classes of carotenoids were found; bicyclic, monocyclic and acyclic carotenoids (Fig. 6). Almost the same carotenoid composition has been postulated in *Em. ramosum* (Yurkov et al., 1993). The complex composition of carotenoids in *E. longus* seems to be consistent with its intermediary nature between typical purple bacteria and aerobic organisms. Elucidation of the gene organization for carotenoid synthesis in *E. longus* might give clues to the reason of such complex composition.

In species of *Methylobacterium*, two classes of carotenoids seem to be present. In *M. radiotolerans*, spirilloxanthin is the dominant species in the RC-LH I complex and a few polar carotenoids containing a carboxyl group constitute the 'non-photosynthetic' carotenoid group. Interestingly, the latter is suggested to be localized in the outer membrane (Saito et al., unpubl.). One of the major carotenoids of *M. rhodinum* (formerly *Pseudomonas rhodos*) was indicated to be a diglycosyl ester of C_{30} carotene-dioate (Kleinig et al., 1979; Fig. 6). The same carotenoid and its deglycosylated form also have been postulated to be the major carotenoids in *Roseococcus thiosulfatophilus* (Yurkov et al., 1994). In *M. rhodesianum*, a major carotenoid was also suggested to have a sugar moiety and a carbonyl group (Sato et al., 1982). The highly polar nature of these carotenoids suggests that they are 'non-photosynthetic' carotenoids which are not bound to the photosynthetic apparatus of these aerobes.

The carotenoid composition of *Roseobacter* is somewhat different from other aerobic anoxygenic phototrophs, since it lacks 'non-photosynthetic' carotenoids. The dominant carotenoid species is spheroidenone which is also the dominant species in semi-aerobically grown cells of some typical purple bacteria such as *Rhodobacter (Rb.) sphaeroides* (Shimada et al., 1985, 1990a; Harashima and Nakada, 1982). In this respect, *Roseobacter* seems to have closer relationships with typical purple bacteria than other aerobic anoxygenic phototrophs. It should be noted that reduction of a significant part of spheroidenone to 3,4-dihydrospheroidenone upon illumination in *Ro. denitrificans* was observed under anaerobic conditions (Takaichi et al., 1991b; see below).

C. Quinones and Lipids

Information on quinone composition is so far limited to *Roseobacter*, *Erythrobacter* and *Methylobacterium*. The dominant chemical species in all strains of these genera was ubiquinone-10 (Shiba and Simidu, 1982; Nishimura at al., 1983; Shiba, 1991; Urakami and Komagata, 1984). Except for small amounts of ubiquinone homologues, no other quinones, such as menaquinone and rhodoquinone, were found in these organisms. This seems to be consistent with the quinone composition of facultative purple bacteria in the alpha-subclass of Proteobacteria.

Cellular fatty acid composition has been studied in *Porphyrobacter* as well as in the above three genera. In all strains in the four genera, octadecenoic acid ($C_{18:1}$) was the dominant fatty acid (Urakami and Komagata, 1984; Urakami and Komagata, 1988; Shiba, 1991; Fuerst et al., 1993). The dominancy is especially high in BChl-containing strains of *Methylobacterium*. It may be noted that significant amounts of triterpenoids have been found in some strains of *Methylobacterium*, although this was not the case in *Erythrobacter* (Urakami and Komagata, 1986,1988).

V. Photosynthetic Activity and Electron Transfer System

A. Evidence for Photosynthetic Activity

The first question after the discovery of BChl-containing aerobic bacteria was whether the BChl molecules were functionally active or not. The first observation indicating photosynthetic activity was the reversible photo-oxidation of cytochromes and the reversible photobleaching of BChl in the presumed RC in *Ro. denitrificans* (Harashima et al., 1982). Photo-inhibition of respiration in *E. longus* and *Ro. denitrificans*, which suggested the operation of a photosynthetic electron transport system, was also observed in the same study. Similar activities were also demonstrated in *M. rhodesianum* (Takamiya and Okamura, 1984). Later, the photo-induced absorbance change of RC was also observed in *M. radiotolerans* and *Rhizobium* strain BTAi1 (Nishimura et al., 1989; Evans et al., 1990). Photo-inhibition of respiration has also been observed in *Acidiphilium rubrum* (Wakao et al., 1993).

Except for these observations, studies on photosynthetic activity of BChl-containing aerobic bacteria have been confined to *Ro. denitrificans* and *M. rhodesianum*. Shiba (1984) observed light-stimulated CO_2 uptake and increases in cellular ATP pools in *Ro. denitrificans* (Fig. 7). He also observed the positive effect of light on survival under conditions of starvation. Also, light-dependent ATP formation in membrane preparations was demonstrated in the above two species (Takamiya and Okamura, 1984; Okamura et al., 1986). It should be noted that the light-stimulated activities were observed only in the presence of O_2 or, in *Ro. denitrificans*, auxiliary oxidants as described below. Light-stimulated growth of *Ro. denitrificans* was demonstrated by Harashima et al. (1987); dark grown cells which had accumulated significant amounts of BChl subsequently showed a higher growth rate in the light than did control cells kept in the dark. Also, this activity was prominent in the presence of sufficient O_2 (Fig. 8). Recently, similar light-stimulated growth was observed in *Em. hydrolyticum* (Yurkov and Van Gemerden, 1993a). In spite of these photosynthetic activities, however, no autotrophic growth supported by H_2 and CO_2 has been observed in these bacteria in the light or in darkness.

B. Effect of O_2 and Auxiliary Oxidants

As stated above, aerobic conditions are necessary, in principle, for photosynthetic activities of these aerobes. Okamura et al. (1985) observed that light-induced oxidation of cytochromes as well as that of RC disappeared under anaerobic conditions both in membrane preparations and in intact cells of *Ro. denitrificans* (Fig. 9). Since these activities were recovered by aeration, the inactivation was considered to be caused by the over-reduction of the electron transfer system. In this respect, the higher midpoint potential (E_m) of the primary acceptor (Q_A) of the RC (+35 mV at pH 7) in aerobic phototrophs compared to that of typical purple bacteria, has been noted (Harashima and Takamiya, 1989). In *M. rhodesianum*, also, the high E_m was considered to be one of the reasons of the inactivation of photosynthetic activities under anaerobic conditions. In *Ro. denitrificans*, Harashima et al. (1982) observed the photo-reduction of a significant portion (up to 35%) of the predominant carotenoid, spheroidenone, under anaerobic conditions. This phenomenon was considered to be due to excess reducing power accumulated under anaerobic conditions. (Harashima and Takamiya, 1989; Takaichi et al., 1991b).

In 1988, Arata et al. (1988) and Shioi et al. (1988) reported that *Ro. denitrificans* could grow anaerobically in the presence of trimethylamine N-oxide (TMAO) or nitrate. Arata et al. (1988) and Takamiya et al. (1988) also observed reversible photo-oxidation of cytochrome *c* and light-driven proton translocation

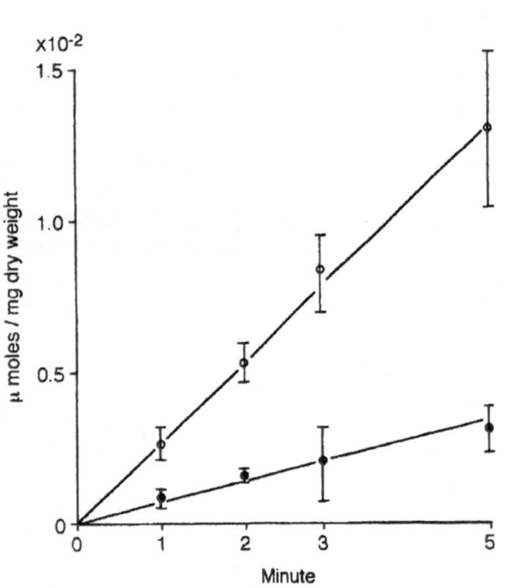

Fig. 7. Light-stimulated uptake of CO_2 by *Roseobacter denitrificans* in aerobic condition. Open circles, in the light; Closed circles, in the dark (from Shiba, 1984).

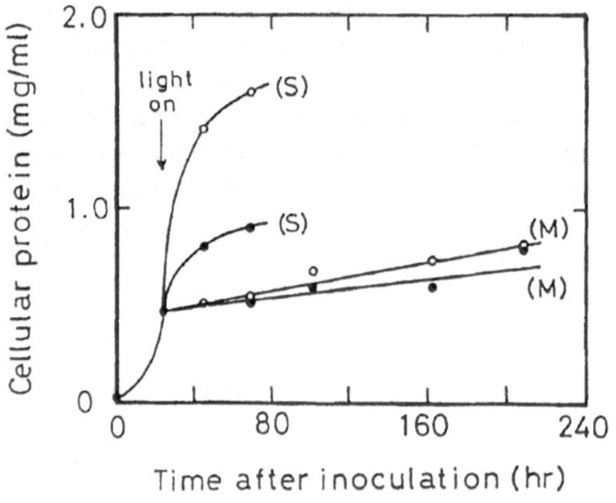

Fig. 8. Oxygen requirement for the light-stimulated growth of *Roseobacter denitrificans*. The bacterium was aerobically grown in the dark for 24 h and then either semi-aerobically (S) or microaerobically (M). Closed symbols, values in the dark; open symbols, values in the light (from Harashima et al., 1987)

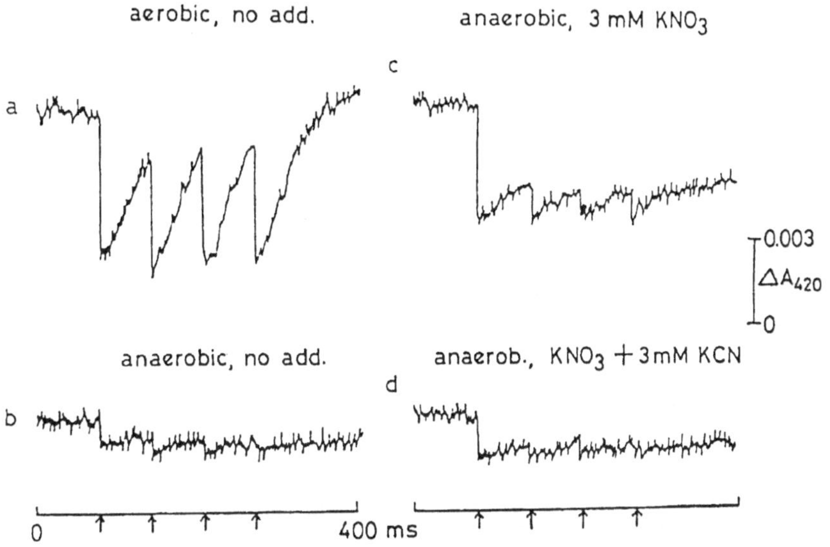

Fig. 9. Flash-induced oxidation of cytochrome *c* in *Roseobacter denitrificans*. (a) Under aerobic condition (b) under anaerobic condition (c) anaerobic condition in the presence of 3 mM KNO_3 (d) same as (c) + 3 mM KCN (from Arata et al., 1988).

under anaerobic condition in the presence of these oxidants (Fig. 9). These observations clearly indicated that these oxidants can be utilized as electron acceptors in *Ro. denitrificans*. Since only a small accumulation of BChl was observed in cells grown anaerobically with these auxiliary oxidants, light-stimulated growth was not observed under anaerobic conditions. However, when BChl had been accumulated previously by growth under aerobic-dark condition and, further, the corresponding reductase had been induced by either oxidants, light-stimulated growth could be observed in the presence of high concentration of the oxidant (Harashima and Takamiya, 1989).

C. Electron Transfer System

As described above, most strains of aerobic anoxygenic phototrophs seem to have purple bacterial-type RCs. The operation of this type of RC in the light has also been indicated in some of these aerobic bacteria. Therefore, one may consider that electron transfer systems similar to those of facultative purple bacteria are also functioning in aerobic anoxygenic phototrophs. However, not much is known about the electron transfer systems of aerobic anoxygenic phototrophs, especially about light-driven electron transfer. Although participation of *c*-type cytochromes in light-driven electron transport was indicated by light-induced absorption changes in *Ro. denitrificans*, *E. longus* and *M. rhodesianum*, involvement of the bc_1-type complex in the system is not clear. Involvement of the bc_1 complex in the respiratory electron transfer system of *Ro. denitrificans* has been suggested by the effect of antimycin on electron transport (Takamiya, 1989). Also, inhibition of electron transfer to the auxiliary oxidants by antimycin or myxothiazol suggested the participation of bc_1 complex in the anaerobic respiratory systems. In *Ro. denitrificans*, cytochrome *c*-551, which was shown to be homologous with cytochrome c_2 of purple bacteria (Okamura et al., 1987), was oxidized by illumination on the one hand, and was oxidized by O_2 or some auxiliary oxidants on the other hand. These results indicated that the bc_1 complex also participated in a light-driven electron transport system in this bacterium (Takamiya et al., 1988). The cytochrome subunit of the RC of *Ro. denitrificans*, *c*-554, was also shown to be a component of the photosynthetic electron transport system as the direct electron donor to the light-oxidized BChl (Matsuura and Shimada, 1990). As stated above, this bacterium has dissimilatory reductases for the auxiliary oxidants. One of them, nitrite reductase, was revealed to be a cytochrome cd_1-type reductase which could accept electrons from

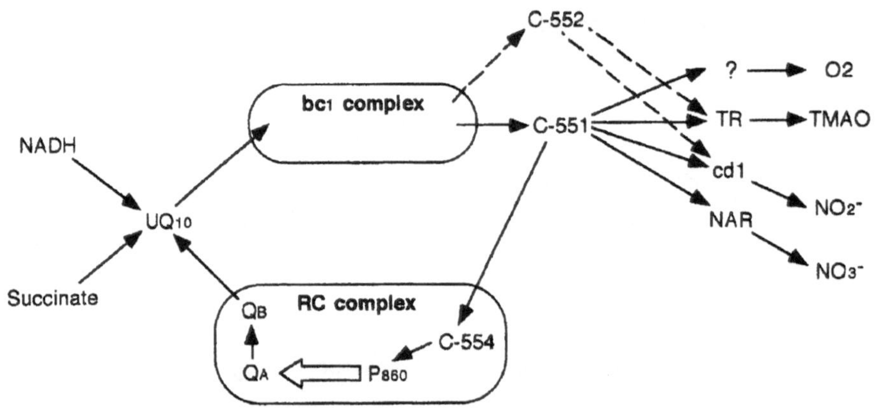

Fig. 10. A tentative scheme for the electron transfer system in *Ro. denitrificans* (from Takamiya, 1989).

cytochrome *c*-551 and also from another *c*-type cytochrome, *c*-552 (Doi and Shioi, 1989; Takamiya et al., 1993). The in vivo function of c-552 is, however, not clearly understood. Although the nature of cytochrome oxidase is still unclear, a tentative scheme for the electron transfer system in *Ro. denitrificans* is shown in Fig. 10.

From *E. longus*, an aa_3-type cytochrome oxidase as well as two cytochromes *c* (*c*-549 and *c*-550) have been isolated (Fukumori et al., 1987). However, presence of the bc_1 complex in this bacterium is unclear and, consequently, little is known about the components participating in the photosynthetic electron transport system in this bacterium. In *Methylobacterium*, an aa_3-type cytochrome oxidase has also been obtained from a strain of *M. extorquens* (formerly *Pseudomonas* AM1) (Fukumori et al., 1985). Although an outline of the respiratory chain in this bacterium has been elucidated (Fukumori and Yamanaka, 1987), the organization of the presumed photosynthetic electron transport system is totally unknown.

The anaerobic respiratory system for the dissimilatory reduction of the auxiliary oxidants in *Ro. denitrificans* is somewhat different from those in some strains of *Rhodobacter* which are also known to utilize some auxiliary oxidants. Dimethylsulfoxide, which is a common oxidant in *Rhodobacter* strains, was not reduced by *Ro. denitrificans*. Also, denitrification from nitrate or nitrite yield nitrous oxide (N_2O), indicating the absence of nitrous oxide reductase in *Ro. denitrificans* (Arata et al., 1988; Shioi et al., 1988).

VI. Regulation of Pigment Synthesis

A. Effect of Oxygen

Aerobic anoxygenic phototrophs can synthesize and accumulate BChl which is then integrated into the photosynthetic apparatus, though the amounts are significantly lower than those in the typical purple bacteria. The fundamental difference between aerobic anoxygenic phototrophs and typical purple photosynthetic bacteria is the requirement for O_2. Aerobic anoxygenic phototrophs can grow and synthesize BChl as well as the photosynthetic apparatus only under aerobic conditions. Under anaerobic conditions, no synthesis of BChl nor growth occurs, except in the case of *Ro. denitrificans* which can grow using auxiliary oxidants and can thus synthesize a small amount of photosynthetic apparatus anaerobically (Takamiya et al., 1992). In *E. longus*, BChl content was significantly reduced in semi-aerobic conditions compared with that of aerobic conditions, while the growth rate was only slightly decreased in the same growth conditions. This suggests that O_2 tension itself stimulated the synthesis of BChl in this bacterium (Harashima et al., 1980). In *Ro. denitrificans,* on the other hand, it was suggested that O_2 itself was not stimulative for BChl synthesis, since BChl content was almost constant down to an O_2 tension below which respiratory activity dropped significantly (Shiba, 1987; Sato et al., 1989). In typical purple bacteria, some facultative species are known to synthesize the photosynthetic apparatus under semi-aerobic conditions (Shioi and Doi, 1990).

However, these species synthesize the photosynthetic apparatus also under anaerobic conditions and show significant photosynthetic activity in the light as do the obligately anaerobic strains. Thus, the effect of O_2 on the synthesis of BChl and the photosynthetic apparatus in aerobic anoxygenic phototrophs is almost opposite to that of typical purple bacteria.

In aerobic anoxygenic phototrophs which have 'non-photosynthetic' carotenoids, it has been known that the content of carotenoids is rather insensitive to oxygen tension (Harashima et al., 1980). This can be explained by constitutive synthesis of the 'non-photosynthetic' species of carotenoids. On the other hand, the synthesis of the carotenoid species which are bound to the LH and RC complexes seems to be regulated, more or less, by O_2 tension. In *E. longus* the amount of 'photosynthetic' carotenoids was roughly equal to that of BChl in various growth conditions. However, the composition of the carotenoid species varied significantly depending on aeration rate (Shimada et al., unpubl.; Harashima, 1989). Whether the regulatory mechanism for the synthesis of these 'photosynthetic' carotenoids is the same with that of BChl is unclear.

Nothing is known so far about the regulatory mechanism for the synthesis of BChl and/or the photosynthetic apparatus by O_2 in aerobic anoxygenic phototrophs. The absence of the *puf*Q gene in the *puf* operon of *Ro. denitrificans* might be concerned with the lack of anaerobic synthesis of BChl (Liebetanz et al., 1991). Elucidation of the regulatory mechanism of the synthesis of pigments and proteins for the photosynthetic apparatus in typical purple bacteria (see Chapter 58 by Bauer) will likely lead to elucidation of the role of O_2 in the synthesis of pigments and photosynthetic apparatus in aerobic anoxygenic phototrophs.

B. Effect of Light

It has been known that high light depresses the pigment content of typical purple photosynthetic bacteria. This light effect was explained by the degradation of BChl and not the repression of BChl synthesis (Biel, 1986). In aerobic anoxygenic phototrophs, continuous light also clearly repressed accumulation of BChl, while many species could synthesize BChl in continuous dark in the presence of O_2 (Harashima et al., 1987; Shioi and Doi, 1988; Sato et al., 1989) In some species, however, maximal synthesis of BChl occurred under intermittent light. In *Rhizobium* BTAi1, BChl accumulation could be observed only under intermittent light (Evans et al., 1990; see Chapter 7 by Fleischman et al.). In *M. rhodesianum*, the synthesis of BChl depended on the light/dark condition in each growth phase (Sato and Shimizu, 1979; Sato et al., 1985b; Fig. 11). These results suggest that, in some species of aerobic phototrophs, light is not exclusively repressive but has a role of positive effector for BChl synthesis.

In *M. rhodesianum*, two isozymes for synthesis of 5-aminolevulinic acid (ALA) have been found (Sato et al., 1985a). These enzymes synthesize ALA from glycine and succinyl-CoA and, thus, are called ALA

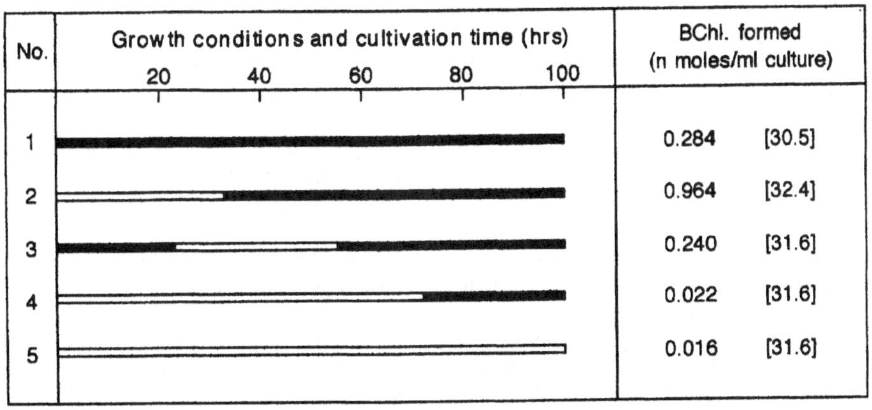
No.
Growth conditions and cultivation time (hrs)
20 40 60 80 100
BChl. formed
(n moles/ml culture)
1
2
3
4
5
0.284 [30.5]
0.964 [32.4]
0.240 [31.6]
0.022 [31.6]
0.016 [31.6]

Fig. 11. Effect of light on BChl-formation in *Methylobacterium rhodesianum*. Open bar, illuminated period; Closed bar, dark period (from Sato and Shimizu, 1979). The values in brackets represent the OD_{660} of the culture [see Sato and Shimizu (1979) for details of how growth was measured].

synthases. One of them was shown to be constitutive, while the other was shown to be inducible by the conditions under which BChl synthesis occurs. This result suggests that BChl synthesis in this bacterium is regulated by de novo synthesis of the ALA synthase isozyme. In *Ro. denitrificans*, an ALA synthase which is similar to that of *Rb. sphaeroides* has been found (Shioi and Doi, 1988). However, participation of this enzyme in the regulatory mechanism of BChl synthesis in this bacterium is not clear.

In *Ro. denitrificans*, Iba and Takamiya (1989) investigated the action spectrum of the inhibitory effect of light on BChl accumulation. They found that wavelengths between 400 to 500 nm were most effective and those in the near-infrared region were less effective. This blue light effect was also observed in the aerobic growth of a typical purple bacterium, *Rb. sphaeroides*. Since this effect of blue light on BChl accumulation was also observed in the anaerobic culture of *Ro. denitrificans* in the presence of TMAO, Takamiya et al. (1992) suggested the effect of light on BChl accumulation is not stimulation of oxidative degradation of BChl or its precursor, but is a regulatory effect on the biosynthesis of BChl. The stable nature of BChl integrated in the photosynthetic apparatus in the presence of light and O_2 also supported the absence of photochemical degradation of BChl.

C. Other Factors

In some species of aerobic anoxygenic phototrophs other factors are known to affect the synthesis of pigments. In the marine bacterium, *E. longus*, low salinity caused a significant reduction of BChl content without severely affecting growth (Sato et al., 1989). In *Porphyrobacter neustonensis*, reduction of BChl content in richer growth media has been observed (Fuerst et al., 1993). However, the mechanisms of these inhibitions remains to be investigated. An inhibitor of BChl synthesis, α,α'-dipyridyl, also inhibited the formation of intracytoplasmic membranes in *Ro. denitrificans*, suggesting that BChl synthesis accompanied the synthesis of other components of the photosynthetic apparatus as in typical purple bacteria (Iba et al., 1988).

The reasons why aerobic anoxygenic phototrophs show lower specific BChl contents than typical purple bacteria is not known (see Table I). Similarly, the reasons for the significant difference in BChl content among species of aerobic anoxygenic phototrophs is not known. It is possible, however, that the cultural conditions adopted for each species was not optimal for BChl formation. In addition, there may be unknown critical factors required for BChl formation in species which accumulate only a small amount of BChl in the growth condition currently adopted.

VII. Evolution and Taxonomy

A. Phylogeny Based on Ribosomal RNA Sequence

At present, up to 20 species and 8 genera are described as BChl-containing aerobic bacteria which are referred as aerobic anoxygenic phototrophs in this chapter (Table 1). More than half of the 20 species are newly described ones and others are newly recognized ones as the BChl-containing species. As seen in the fact that these descriptions have been made continually during these 15 years, further discovery of new species of aerobic anoxygenic phototrophs will continue in the future.

Phylogenetically, all species of aerobic anoxygenic phototrophs for which data on their ribosomal RNA sequence are available have been shown to belong to the α-subclass of Proteobacteria (Woese et al., 1984; Komagata, 1989; Fuerst et al., 1993). This seems to be consistent with the similarity of their photosynthetic apparatus to those of purple nonsulfur bacteria which belong mostly to the same subclass. In the α-subclass, aerobic anoxygenic phototrophs showed rather wide distribution as represented in Fig. 12 (see also Yurkov et al., 1994; Nishimura et al., 1994). This indicates that they are not necessarily closely related to each other. The phylogenetic diversity of them seems also to be consistent with their heterogeneity with respect to pigments and photosynthetic apparatus, such as carotenoid composition, the presence of 'non-photosynthetic' carotenoids and the possession of a cytochrome subunit in the RC.

B. Consideration on the Origin of Aerobic Anoxygenic Phototrophs

Based on phylogenetic studies and the heterogeneity of photosynthetic apparatus, it may be reasonable to assume that aerobic anoxygenic phototrophs were branched from some independent lines of purple photosynthetic bacteria. Furthermore, it seems that

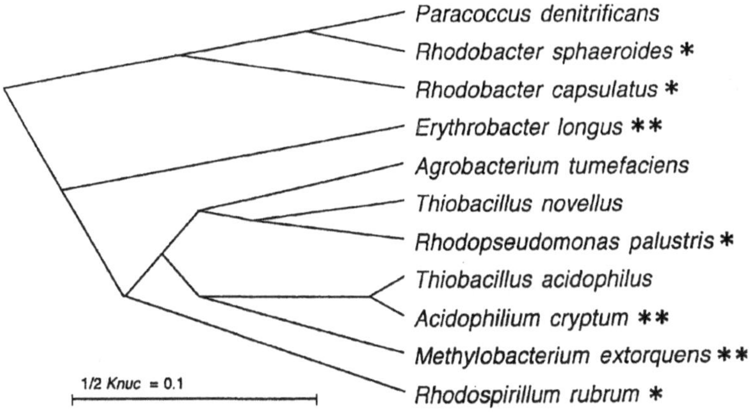

Fig. 12. A phylogenetic tree of α-subgroup of Proteobacteria based on 5S rRNAs from representative species (from Komagata, 1989). * phototrophic purple photosynthetic bacteria; ** aerobic phototrophic bacterium.

each branching is not necessarily a recent event, since each of the aerobic anoxygenic phototrophs shows significant differences from its presumed purple bacteria relative in carotenoid composition and antenna organization, for example. Although *Ro. denitrificans* seems to be closely related to *Rhodobacter* in terms of ribosomal RNA sequence, primary structure of a *c*-type cytochrome and the dominant carotenoid species, they still differ in the possession of a cytochrome subunit in the RC and in the spectral characteristics of LH II, in addition to the response to O_2. *E. longus* and *Porphyrobacter*, which are closely related to each other, have no extant purple bacteria relatives in their apparently isolated phylogenetic group (Fuerst et al., 1993). Thus, it is probable that at least some species of aerobic anoxygenic phototrophs are in a 'stable state' in the evolutionary process. The abundancy of aerobic anoxygenic phototrophs in some habitats supports their ecological and evolutionary stability.

Based on their reduced BChl content, one may consider that aerobic anoxygenic phototrophs are in a transient phase of evolving from purple photosynthetic bacteria to non-photosynthetic aerobes (Woese et al., 1984). However, the reason for the small content of BChl is not known. In spite of the small content of BChl, photosynthetic activities have been clearly shown in some species suggesting that they have an almost complete set of 'photosynthetic' genes. Therefore, the smaller amount of BChl compared with those of typical purple bacteria does not necessarily mean that the loss of the 'photosynthetic' genes in aerobic phototrophs has occurred. It is possible, however, that some 'non-phototrophic' aerobic bacteria, which are closely related to phototrophs, still have the photosynthetic genes or their relicts. Search for such strains by suitable nucleic acid probes may contribute to our understanding of evolutionary processes in prokaryotes.

Recently, occurrence of lateral transfer of photosynthetic genes has been postulated in some cases (Blankenship, 1992; Nagashima et al., 1993). It is also possible to consider that aerobic anoxygenic phototrophs originated by incorporation of a photosynthetic gene cluster into 'non-photosynthetic' aerobes. However, the heterogeneous nature of the photosynthetic apparatus of aerobic anoxygenic phototrophs described above suggests that the origin of such photosynthetic genes are also heterogeneous. It seems unlikely that lateral gene transfers occurred in many independent lines although such possibility in some cases cannot be ruled out.

C. Perspectives

As described above, accumulated data are still not enough to discuss extensively the phylogeny and evolution of aerobic anoxygenic phototrophs. Especially, data on the structure and sequence of 'photosynthetic' genes are lacking except for that of *Ro. denitrificans*. This may be due to a lack of suitable probes for the genes. However, recent progress of gene manipulation will make it possible to search and sequence 'photosynthetic' genes in other aerobic anoxygenic phototrophs and, possibly, in some 'non-photosynthetic' bacteria. Thus, further studies are needed to approach the origin of aerobic anoxygenic phototrophs as well as the origin of non-photosynthetic aerobes.

Thus far, species of aerobic anoxygenic phototrophs are confined to the alpha-subclass of Proteobacteria. However, it may not be unreasonable to expect the occurrence of aerobic anoxygenic phototrophs in other taxa, such as the γ-subclass. If they are present in other classes than Proteobacteria, the possibility of lateral gene transfer should be, then, taken into consideration, as in the case of *Chloroflexus* (Blankenship, 1992). Further investigation of new species may reveal a phylogenetically and ecologically wider distribution of aerobic anoxygenic phototrophs than is presently known.

And finally, it should be mentioned that nomenclature of the aerobic anoxygenic phototrophs is not consistent. Recently, Gest (1993) proposed a term for aerobic anoxygenic phototrophs, 'quasi-photosynthetic bacteria,' to distinguish them from typical anoxygenic photosynthetic bacteria. Another expression, 'paraphotosynthetic' by Madigan, was also referred to in the article. Currently, this group of bacteria is called by at least three names: 'aerobic photosynthetic bacteria,' a primary designation (Shiba, 1989); 'aerobic anoxygenic phototroph,' used in this book; and 'aerobic BChl-containing bacteria.' Thus, agreement in the nomenclature of these organisms may be an important future subject in this field.

Acknowledgments

The author is grateful to the following individuals who generously supplied reprints, illustrations and pre-publication material: G. Drews, K. Harashima, A. Hiraishi, K. Komagata, T. Noguchi, K. Sato, T. Shiba, Y. Shioi, K. Takamiya.

References

Arata H, Serikawa Y and Takamiya K (1988) Trimethylamine N-oxide respiration by aerobic photosynthetic bacterium, *Erythrobacter* sp. OCh 114. J Biochem 103: 1011–1015

Biel AJ (1986) Control of bacteriochlorophyll accumulation by light in *Rhodobacter capsulatus*. J Bacteriol 168: 655–659

Blankenship RE (1992) Origin and early evolution of photosynthesis. Photosynth Res 33: 91–111

Bousfield IJ and Green PN (1985) Reclassification of bacteria of the genus *Protomonas* Urakami and Komagata 1984 in the genus *Methylobacterium* (Patt, Cole and Hanson) emend. Green and Bousfield 1983. Int J Syst Bacteriol 35: 209

Doi M and Shioi Y (1989) Two types of cytochrome cd_1 in the aerobic photosynthetic bacterium, *Erythrobacter* sp. OCh 114. Eur J Biochem 184: 521–527

Evans WR, Fleischman DE, Calvert HE, Pyati PV, Alter GM and Rao NSS (1990) Bacteriochlorophyll and photosynthetic reaction centers in *Rhizobium* strain BTAi 1. Appl Environ Microbiol 56: 3445–3449

Fuerst JA, Hawkins JA, Holmes A, Sly LI, Moor CJ and Stackebrandt E (1993) *Porphyrobacter neustonensis* gen. nov., sp. nov., an aerobic bacteriochlorophyll-synthesizing budding bacterium from fresh water. Int J Syst Bacteriol 43: 125–134

Fukumori Y and Yamanaka T (1987) Cytochrome aa_3 from the aerobic photoheterotroph *Erythrobacter longus*: Purification, and enzymatic and molecular features. J Biochem 102: 777–784

Fukumori Y, Nakayama K and Yamanaka T (1985) Cytochrome *c* oxidase of *Pseudomonas* AM1: Purification and molecular and enzymatic properties. J Biochem 98: 493–499

Gest H (1993) Photosynthetic and quasi-photosynthetic bacteria. FEMS Microbiol Lett 112: 1–6

Harashima K (1989) Carotenoids, quinones and other lipids. In: Harashima, K, Shiba T and Murata N (eds) Aerobic Photosynthetic Bacteria, pp 125–148. Springer-Verlag, Berlin

Harashima K and Nakada H (1983) Carotenoids and ubiquinone in aerobically grown cells of an aerobic photosynthetic bacterium, *Erythrobacter* species OCh 114. Agric Biol Chem 47: 1627–1628

Harashima K and Takamiya K (1989) Photosynthesis and photosynthetic apparatus. In: Harashima K, Shiba T and Murata N (eds) Aerobic Photosynthetic Bacteria, pp 39–72. Springer-Verlag, Berlin

Harashima K, Shiba T, Totsuka T, Shimidu U and Taga N (1978) Occurrence of bacteriochlorophyll *a* in a strain of an aerobic heterotrophic bacterium. Agric Biol Chem 42: 1627–1628

Harashima K, Hayasaki J, Ikari T and Shiba T (1980) O_2-stimulated synthesis of bacteriochlorophyll and carotenoids in marine bacteria. Plant Cell Physiol 21: 1283–1294

Harashima K, Nakagawa M and Murata N (1982) Photochemical activity of bacteriochlorophyll in aerobically grown cells of heterotrophs, *Erythrobacter* species (OCh 114) and *Erythrobacter longus* (OCh 101). Plant Cell Physiol 23: 185–193

Harashima K, Kawazoe K, Yoshida I and Kamata H (1987) Light-stimulated aerobic growth of *Erythrobacter* species OCh 114. Plant Cell Physiol 28: 365–374

Hayashi H, Shimada K, Tasumi M, Nozawa T and Hatano M (1986) Circular dichroism and resonance Raman spectra of bacteriochlorophyll-protein complexes from aerobic bacteria, *Erythrobacter longus* and *Erythrobacter* species OCh 114. Photobiochem Photobiophys 10: 223–231

Iba K and Takamiya K (1989) Action spectra for inhibition by light of accumulation of bacteriochlorophyll and carotenoid during aerobic growth of photosynthetic bacteria. Plant Cell Physiol 30: 471–477

Iba K, Takamiya K, Toh Y and Nishimura M (1988) Roles of bacteriochlorophyll and carotenoid synthesis in formation of intracytoplasmic membrane systems and pigment-protein complexes in an aerobic photosynthetic bacterium, *Erythrobacter* sp. strain OCh 114. J Bacteriol 170: 1843–1847

Kleinig H, Schmitt R, Meister W, Englert G and Thommen H (1979) New C_{30}-carotenoic acid glucosyl esters from *Pseudomonas rhodos*. Z Naturforsch 34C: 181–185

Komagata K (1989) Taxonomy of facultative methylotrophs. In: Harashima K, Shiba T and Murata N (eds) Aerobic Photosynthetic Bacteria pp. 25–38 Springer-Verlag, Berlin

Liebetanz RL, Hornberger U and Drews G (1991) Organization of the genes encoding for the reaction-center L and M subunits and B870 antenna polypeptides alpha and beta from the aerobic photosynthetic bacterium *Erythrobacter* species OCH 114. Mol Microbiol 5: 1459–1468

Matsuura K and Shimada K (1990) Evolutionary relationships between reaction center complexes with and without cytochrome *c* subunits in purple bacteria. In: Baltscheffsky M (ed) Current Research in Photosynthesis, pp 193–196. Kluwer Academic Publishers, Dordrecht

Nagashima KVP, Shimada K and Matsuura K (1993) Phylogenetic analysis of photosynthetic genes of *Rhodocyclus gelatinosus*: possibility of horizontal gene transfer in purple bacteria. Photosynth Res 36: 185–191

Nishimura Y, Shimadzu M and Iizuka H (1981) Bacteriochlorophyll formation in radiation-resistant *Pseudomonas radiora*. J Gen Appl Microbiol 27: 427–430

Nishimura Y, Yamanaka S and Iizuka H (1983) Ubiquinone, fatty acid, polar lipid and DNA base composition in *Pseudomonas radiora*. J Gen Appl Microbiol 29: 421–427

Nishimura Y, Mukasa S Iizuka H and Shimada K (1989) Isolation and characterization of bacteriochlorophyll-protein complexes from an aerobic bacterium, *Pseudomonas radiora*. Arch Microbiol 152: 1–5

Nishimura Y, Muroga Y, Saito S, Shiba T, Takamiya K and Shioi Y (1994) DNA relatedness and chemotaxonomic feature of aerobic bacteriochlorophyll-containing bacteria isolated from coasts of Australia. J Gen Appl Microbiol 40: 287–296

Noguchi T, Hayashi H, Shimada K, Takaichi S and Tasumi M (1992) In vivo states and function of carotenoids in an aerobic photosynthetic bacterium, *Erythrobacter longus*. Photosynth Res 31: 21–30

Okamura K, Takamiya K and Nishimura M (1985) Photosynthetic electron transfer system is inoperative in anaerobic cells of *Erythrobacter* species strain OCh 114. Arch Microbiol 142: 12–17

Okamura K, Mitsumori F, Ito O, Takamiya K and Nishimura M (1986) Photophosphorylation and oxidative phosphorylation in intact cells and chromatophores of an aerobic photosynthetic bacterium, *Erythrobacter* sp. strain OCh 114. J Bacteriol 168: 1142–1146

Okamura K, Miyata T, Iwanaga S, Takamiya K and Nishimura M (1987) Complete amino acid sequence of cytochrome *c*-551 from *Erythrobacter* species strain OCh 114. J Biochem 101: 957–966

Sato K (1978) Bacteriochlorophyll formation by facultative methylotrophs, *Protaminobacter ruber* and *Pseudomonas* AM1. FEBS Lett 85: 207–210

Sato K and Shimizu S (1979) The conditions for bacteriochlorophyll formation and the ultrastructure of a methanol-utilizing bacterium, *Protaminobacter ruber*, classified as nonphotosynthetic bacteria. Agric Biol Chem 43: 1669–1675

Sato K, Mizutani T, Hiraoka M and Shimazu S (1982) Carotenoid containing sugar moiety from a facultative methylotroph, *Protaminobacter ruber*. J Ferment Technol 60: 111–115

Sato K, Ishida K, Shirai M and Shimizu S (1985a) Occurrence and some properties of two types of δ-aminolevulinic acid synthase in a facultative methylotroph, *Protaminobacter ruber*. Agric Biol Chem 49: 3423–3428

Sato K, Hagiwara K and Shimizu S (1985b) Effect of cultural conditions on tetrapyrrole formation, especially bacteriochlorophyll formation in a facultative methylotroph, *Protaminobacter ruber*. Agric Biol Chem 49: 1–5

Sato K, Shiba T and Shioi Y (1989) Regulation of the biosynthesis of bacteriochlorophyll. In: Harashima K, Shiba T and Murata N (eds) Aerobic Photosynthetic Bacteria, pp 95–124. Springer-Verlag, Berlin

Shiba T (1984) Utilization of light energy by the strictly aerobic bacterium *Erythrobacter* sp. OCh 114. J Gen Appl Microbiol 30: 239–244

Shiba T (1987) O_2 regulation of bacteriochlorophyll synthesis in the aerobic bacterium *Erythrobacter*. Plant Cell Physiol 28: 1313–1320

Shiba T (1989) Overview of the aerobic photosynthetic bacteria. In: Harashima K, Shiba T, Murata N (eds) Aerobic Photosynthetic Bacteria, pp 1–8. Springer-Verlag, Berlin

Shiba T (1991) *Roseobacter litoralis* gen. nov., and *Roseobacter denitrificans* sp. nov., aerobic pink-pigmented bacteria which contain bacteriochlorophyll *a*. System Appl Microbiol 14: 140–14

Shiba T and Abe K (1987) An aerobic bacterium containing bacteriochlorophyll-proteins showing absorption maxima of 802, 844 and 862 nm in the near infrared region. Agric Biol Chem 51: 945–946

Shiba T and Simidu U. (1982) *Erythrobacter longus* gen. nov., sp. nov., an aerobic bacterium which contains bacteriochlorophyll *a*. Int J Syst Bacteriol 32: 211–217

Shiba T, Simidu U and Taga N (1979a) Another aerobic bacterium which contains bacteriochlorophyll *a*. Bull Jpn Soc Sci Fish 45: 801

Shiba T, Simidu U, Taga N (1979b) Distribution of aerobic bacteria which contain bacteriochlorophyll *a*. Appl Environ Microbiol 38: 43–45

Shiba T, Shioi Y, Takamiya K, Sutton DC and Wilkinson CR (1991) Distribution and physiology of aerobic bacteria containing bacteriochlorophyll *a* on the east and west coast of Australia. Appl Environ Microbiol 57: 295–300

Shimada K, Hayashi H and Tasumi M (1985) Bacteriochlorophyll-protein complexes of aerobic bacteria, *Erythrobacter longus* and *Erythrobacter* sp. OCh 114. Arch Microbiol 143: 244–247

Shimada K, Hayashi H, Noguchi T and Tasumi M (1990a) Excitation and emission spectroscopy of membranes and pigment-protein complexes of an aerobic photosynthetic bacterium, *Erythrobacter* sp. OCh114. Plant Cell Physiol 31: 395–398

Shimada K, Yamazaki I, Tamai N and Mimuro M (1990b) Excitation energy flow in a photosynthetic bacterium lacking B850. Fast energy transfer from B806 to B870 in *Erythrobacter* sp. strain OCh 114. Biochim Biophys Acta 1016: 266–271

Shimada K, Hirota M, Nishimura Y, Yamazaki I and Mimuro M (1992) Excitation energy flow in *Roseobacter denitrificans* (*Erythrobacter* sp. OCh 114) at low temperature. In Murata N (ed) Research in Photosynthesis, pp 137–140. Kluwer Academic Publishers, Dordrecht

Shioi Y (1986) Growth characteristics and substrate specificity of aerobic photosynthetic bacterium, *Erythrobacter* sp. (OCh

114). Plant Cell Physiol 27: 567–572
Shioi Y and Doi M (1988) Control of bacteriochlorophyll accumulation by light in an aerobic photosynthetic bacterium, *Erythrobacter* sp. OCh 114. Arch Biochem Biophys 266: 470–477
Shioi Y and Doi M (1990) Aerobic and anaerobic photosynthesis and bacteriochlorophyll formation in *Rhodobacter sulfidophilus*. In: Baltscheffsky (ed) Current Research in Photosynthesis, pp 853–856. Kluwer Academic Publishers, Dordrecht
Shioi Y, Doi M, Arata H and Takamiya K (1988) A denitrifying activity in an aerobic photosynthetic bacterium, *Erythrobacter* sp. strain OCh 114. Plant Cell Physiol 29: 861–865
Takaichi S, Shimada K and Ishidsu J (1988) Monocyclic cross-conjugated carotenal from an aerobic photosynthetic bacterium, *Erythrobacter longus*. Phytochemistry 27: 3605–3609
Takaichi S, Shimada K and Ishidsu J (1990) Carotenoids from the aerobic photosynthetic bacterium, *Erythrobacter longus*: β-carotene and its hydroxyl derivatives. Arch Microbiol 153: 118–122
Takaichi S, Furihata K, Ishidsu J and Shimada K (1991a) Carotenoid sulphates from the aerobic photosynthetic bacterium, *Erythrobacter longus*. Phytochemistry 30: 3411–3415
Takaichi S, Furihata K and Harashima K (1991b) Light-induced changes of carotenoid pigments in anaerobic cells of the aerobic photosynthetic bacterium, *Roseobacter denitrificans* (*Erythrobacter* species OCh 114): reduction of spheroidenone to 3,4-dihydrospheroidenone. Arch Microbiol 155: 473–476
Takamiya K (1989) Cytochromes and respiratory systems. In: Harashima K, Shiba T, Murata N (eds) Aerobic Photosynthetic Bacteria, pp 73–90. Springer-Verlag, Berlin
Takamiya K and Okamura K (1984) Photochemical activities and photosynthetic ATP formation in membrane preparation from a facultative methylotroph, *Protaminobacter ruber* strain NR-1. Arch Microbiol 140: 21–26
Takamiya K, Iba K and Okamura K (1987) Reaction center complex from an aerobic photosynthetic bacterium, *Erythrobacter* species OCh 114. Biochim Biophys Acta 890: 127–133
Takamiya K, Arata H, Shioi Y and Doi M (1988) Restoration of the optimal redox state for the photosynthetic electron transfer system by auxiliary oxidants in an aerobic photosynthetic bacterium, *Erythrobacter* sp. OCh 114. Biochim Biophys Acta 935: 26–33
Takamiya K, Shioi Y, Shimada H and Arata H (1992) Inhibition of accumulation of bacteriochlorophyll and carotenoids by blue light in an aerobic photosynthetic bacterium, *Roseobacter denitrificans*, during anaerobic respiration. Plant Cell Physiol 33: 1171–1174
Takamiya K, Shioi Y, Morita M, Arata H, Shimizu M and Doi M (1993) Some properties and occurrence of cytochrome *c*-552 in the aerobic photosynthetic bacterium *Roseobacter denitrificans*. Arch Microbiol 159: 51–56
Takemoto J and Kao MYCH (1977) Effects of incident light levels on photosynthetic membrane polypeptide composition and assembly in *Rhodopseudomonas sphaeroides*. J Bacteriol 129: 1102–1109
Urakami T and Komagata K (1984) *Protomonas*, a new genus of facultative methylotrophic bacteria. Int J Syst Bacteriol 34: 188–201
Urakami T and Komagata K (1986) Occurrence of isoprenoid compounds in Gram-negative methanol-, methane- and methylamine-utilizing bacteria. J Gen Appl Microbiol 32: 317–341
Urakami T and Komagata K (1988) Cellular fatty acid composition with special reference to the existence of hydroxy fatty acids, and the occurrence of squalene and sterols in species of *Rhodospirillaceae* genera and *Erythrobacter longus*. J Gen Appl Microbiol 34: 67–84
Wakao N, Shiba T, Hiraishi A, Ito M and Sakurai Y (1993) Distribution of bacteriochlorophyll *a* in species of the genus *Acidiphilium*. Current Microbiol 27: 277–279
Wakao N, Nagasawa N, Matsuura T, Matsukura H, Matsumoto T, Hiraishi A, Sakurai Y and Shiota H (1994) *Acidiphilium multivorum* sp. nov., an acidophilic chemoorganotrophic bacterium from pyritic acid mine drainage. J Gen Appl Microbiol 40: 143–159
Woese CR, Stackebrandt E, Weisburg WG, Paster BJ, Madigan MT, Fowler VJ, Hahn CM, Blanz P, Gupta R, Nealson KH and Fox GE (1984) The phylogeny of purple bacteria: The alpha subdivision. System Appl Microbiol 5: 315–326
Yurkov VV and Gorlenko VM (1990) *Erythrobacter sibiricus* sp. nov., a new freshwater aerobic bacterial species containing bacteriochlorophyll *a*. Mikrobiologia 59: 120–126
Yurkov VV and Gorlenko VM (1991) A new genus of freshwater aerobic bacteriochlorophyll *a*-containing bacteria, *Roseococcus* gen. nov. Mikrobiologia 60: 902–90
Yurkov VV and Gorlenko VM (1992) New species of aerobic bacteria from the genus *Erythromicrobium* containing bacteriochlorophyll *a*. Mikrobiologia 61: 248–255
Yurkov VV and Van Gemerden H (1993a) Impact of light/dark regimen on growth rate, biomass formation and bacteriochlorophyll synthesis in *Erythromicrobium hydrolyticum*. Arch Microbiol 159: 84–89
Yurkov VV and Van Gemerden H (1993b) Abundance and salt tolerance of obligately aerobic, phototrophic bacteria in a marine microbial mat. Neth J Sea Res 31: 57–62
Yurkov VV, Mityushina LL and Gorlenko VM (1991) Ultrastructure of the aerobic bacterium *Erythrobacter sibiricus*, which contains bacteriochlorophyll *a*. Mikrobiologia 60: 339–344
Yurkov VV, Gorlenko VM and Kompantseva EI (1992) A new type of freshwater aerobic orange-colored bacterium *Erythromicrobium* gen. nov., containing bacteriochlorophyll *a*. Mikrobiologia 61: 256–260
Yurkov V, Gad'on N and Drews (1993) The major part of polar carotenoids of the aerobic bacteria *Roseococcus thiosulfatophilus* RB3 and *Erythromicrobium ramosum* E5 is not bound to the bacteriochlorophyll *a*-complexes of the photosynthetic apparatus. Arch Microbiol 160: 372–376
Yurkov V, Stackebrandt E, Holmes A, Fuerst JA, Hugenholtz P, Golecki J, Gad'on N, Gorlenko VM, Kompantseva EI and Drews G (1994) Phylogenetic positions of novel aerobic, bacteriochlorophyll *a*-containing bacteria and description of *Roseococcus thiosulfatophylus* gen. nov., sp. nov., *Erythromicrobium ramosum* gen. nov., sp. nov., and *Erythrobacter litoralis* sp. nov. Int J Syst Bacteriol 44: 427–434

Chapter 7

Bacteriochlorophyll-Containing *Rhizobium* Species

Darrell E. Fleischman
Department of Biochemistry and Molecular Biology, Wright State University, Dayton, OH 45435, USA

William R. Evans
Department of Biological Sciences, Wright State University, Dayton, OH 45435, USA

Iain M. Miller
Department of Biological Sciences, University of Cincinnati, Cincinnati, OH 45221, USA

R. E. Blankenship, M. T. Madigan and C. E. Bauer (eds): Anoxygenic Photosynthetic Bacteria, pp. 123–136.

Summary

The Green Revolution has allowed food production in many developing countries to keep pace with population growth. But the use of chemical fertilizers as a source of fixed nitrogen is expensive and creates environmental problems. The United States Agency for International Development sponsored a collaboration between U. S. and Indian scientists to develop sources of biologically-fixed nitrogen fertilizers. This effort led to the discovery that photosynthetic rhizobia are found in nitrogen-fixing nodules located on the stems of *Aeschynomene*, a legume used as a green manure in rice fields. Such rhizobia now have been isolated from *Aeschynomene* nodules and soils throughout the world. Photosynthetic rhizobia which nodulate a different genus, *Lotononis*, have been discovered recently.

The rhizobium has the photosynthetic properties of aerobic anoxygenic phototrophs. It will grow, produce the photosynthetic system and perform photosynthetic electron transport only under aerobic conditions. It can fix nitrogen in *ex planta* culture and grow in the absence of any other source of fixed nitrogen. Phylogenetic studies based on 16S rRNA sequences and numerical taxonomy suggest that all of the *Aeschynomene* isolates are closely related to each other and form a cluster with *Bradyrhizobium* and *Rhodopseudomonas palustris* in the alpha-2 subdivision of the class *Proteobacteria*.

Formation of the photosynthetic system is triggered by visible and far-red light, and is suppressed by visible light. The photosynthetic system of the stem nodule endophytes probably provides energy for nitrogen fixation, diminishing competition between carbon and nitrogen fixation and allowing for more efficient plant growth.

I. Introduction

A. Symbiosis Between Plants and Rhizobia

1. Rhizobiaceae

The Rhizobiaceae are a family of Gram-negative bacteria, most of whose members form symbiotic associations with plants. Members of four currently-recognized genera form nitrogen-fixing nodules on leguminous plants. Photosynthetic bacteria may have been ancestors of the Rhizobiaceae (Woese, 1987). Phylogenetic studies based on DNA-rRNA homology (Jarvis et al., 1986) and 16S rRNA sequences (Yanagi and Yamasato, 1993) indicate that species of the genus *Bradyrhizobium* (which nodulate soybeans) form a cluster on the *Rhodopseudomonas palustris* branch of the alpha-2 subdivision of the class *Proteobacteria*. *Bradyrhizobium* and *Rp. palustris* are much more distantly related to the other Rhizobiaceae and *Rhodopseudomonas* genera.

Abbreviations: *A.* – *Aeschynomene; B.* – *Bradyrhizobium; Rp.* – *Rhodopseudomonas; Rs.* – *Rhodospirillum;* rRNA – ribosomal RNA

2. The Symbiotic System

The legume nodule is a complex structure established through an elaborate exchange of chemical signals between the bacteria and the plant (Verma, 1992; Fisher and Long, 1992; Long and Staskawicz, 1993). Its function is to allow the resident rhizobia to fix nitrogen for export to the plant in return for fixed carbon. The rhizobia invade the roots, ultimately entering plant cells where they differentiate to form bacteroids, the endosymbiotic form of the rhizobia. Generally sucrose, formed photosynthetically in the leaves, is translocated to the root nodules, where it is converted to dicarboxylic acids such as malate. The bacteroids use the dicarboxylic acids as substrates for oxidative phosphorylation and as the source of the electrons required for the reduction of dinitrogen to ammonia. The ammonia nitrogen is incorporated into organic molecules such as amides and ureides which are fed to the plant by way of the xylem. Nitrogen fixation is energetically expensive, requiring at least 8 moles of ATP per mole of NH_3 formed, as well as 4 equivalents of electrons from low-potential donors. As much as half of the photosynthate produced by legumes may be allocated to nitrogen fixation.

3. Stem Nodulated Legumes and Their Agricultural Significance

Most members of the Rhizobiaceae genera *Rhizobium, Sinorhizobium* and *Bradyrhizobium* nodulate only the roots of plants. Members of a fourth genus, *Azorhizobium*, form nitrogen-fixing nodules on both the roots and stems of the tropical legume *Sesbania rostrata* (Dreyfus and Dommergues, 1981; Dreyfus et al., 1988). Nodules are found on above-ground parts of other plants as well, including plants of the genera *Aeschynomene* (Arora, 1954; Alazard, 1985) and *Neptunia* (Schaede, 1940). Occasionally other plants possessing aerial nodules are reported (Walter and Bien, 1989; Yatazawa et al., 1987), and under appropriate conditions nodules may form on the lower stems of such familiar legumes as field bean (Fyson and Sprent, 1980), soybean (Eaglesham and Ayanaba, 1984) and peanut (Nambiar et al., 1982). Such nodules are most frequently found on tropical legumes, where they form under flooded conditions. The nodules are usually formed on adventitious roots or their primordia. The circumstances under which it is appropriate to refer to them as stem rather than root nodules has been discussed by Ladha et al. (1992).

Sesbania and *Aeschynomene* have been of particular interest in rice-growing regions because they are flood-tolerant and so can serve as green manures—sources of biologically-fixed fertilizer nitrogen—for rice fields. When these plants are grown in flooded rice fields, the nodules on their stems can fix nitrogen even though their roots are submerged. The plants are then plowed into the fields before rice is planted, and release their fixed nitrogen into the soil. The use of green manures is of vital importance in developing countries. Most of the rice in these areas is grown by farmers who do not have convenient access to expensive chemical nitrogen fertilizers produced by the petroleum-consuming Haber-Bosch process. The expense of such fertilizers is certain to increase in the future. In addition, biological nitrogen fertilizers generate much less water and air pollution than do chemical nitrogen fertilizers.

Stem-nodulated legumes and the rhizobia that nodulate them have received special attention in India (Subba Rao et al., 1991), the Philippines (Ladha et al, 1990) and Senegal (Alazard, 1985). All aspects of stem nodulation and its applications have been reviewed recently by Ladha et al. (1992), and the symbiosis between *Azorhizobium caulinodans* and *Sesbania rostrata* has been discussed in detail by de Bruijn (1988).

B. Discovery of Bacteriochlorophyll-Containing Rhizobium *Species.*

1. Species That Nodulate Aeschynomene

In the 1980's the United States Agency for International Development sponsored a collaborative effort between scientists from India and the United States, the Indo-US Science and Technology Initiative. One of its objectives was the development of biological nitrogen fertilizers. N. S. Subba Rao, of the Indian Agricultural Research Institute, had noticed that stem nodules of *A. indica* plants growing in northern India appeared to contain unusual endophytes (Subba Rao, 1988). Under USAID STI sponsorship, he brought a collection of fixed nodules to the Charles F. Kettering Research Laboratory, where the late Harry Calvert examined them with the electron microscope. Minocher Reporter and others at the Kettering Laboratory pointed out that coccoid endophytes within some of the nodules contained chromatophore-like membranous vesicles (Evans et al., 1990), and suggested that they might be photosynthetic bacteria. Endophytes subsequently separated from stem nodules collected throughout southern India by Subba Rao and S. Shanmugasundaram, of Madurai Kamaraj University, were shown to contain bacteriochlorophyll and photosynthetic reaction centers (Fleischman et al., 1988, 1991).

Efforts to culture photosynthetic bacteria from the Indian nodules by anaerobic incubation in the light were unsuccessful. However, earlier, A. R. J. Eaglesham of the Boyce Thompson Institute had isolated a strain of rhizobium, designated BTAi 1, from a nodule which had formed on the stem of an *A. indica* plant growing in unsterilized sand obtained from West Virginia (Eaglesham and Szalay, 1983). The rhizobium had been examined in detail and found to have properties of both *Rhizobium* and *Bradyrhizobium* (Stowers and Eaglesham, 1983). Experiments with BTAi 1 revealed that it could form the photosynthetic system but only when grown aerobically under cyclic illumination (e. g., 16 h light/8 h dark) in the presence of specific carbon sources (Evans et al., 1990). The properties of BTAi 1 have been reviewed by Eaglesham et al. (1990).

Lorquin et al. (1993) found that of 126 strains of

rhizobia isolated in Senegal from stem nodules of *Aeschynomene* plants belonging to three cross-inoculation groups, 83 could synthesize bacteriochlorophyll under appropriate conditions. Bacteriochlorophyll-containing rhizobia could nodulate plants belonging to two of the cross-inoculation groups. The bacteria were described as light pink, dark pink or orange.

Several hundred strains of bacteriochlorophyll-containing rhizobia have now been isolated. J. K. Ladha of the International Rice Research Institute has isolated such strains from nine *Aeschynomene* species growing in the Philippines (Ladha et al., 1990; Ladha and So, 1994) and demonstrated that most form bacteriochlorophyll-containing stem nodule bacteroids. We have isolated strains in India from a tenth species (D. Fleischman, S. Shanmugasundaram, G. Gopalan and F. Schwelitz, unpublished). Peter van Berkum of the U. S. Department of Agriculture has obtained isolates that nodulate *A. indica* from soil samples collected on five continents, from sites as diverse as Maryland soybean fields and Zimbabwe *Aeschynomene* groves (Van Berkum et al., 1995). The isolates vary widely in color and colony morphology; some form pigment constitutively. Van Berkum et al. (1995) further report that bacteriochlorophyll was formed by 38 of the 79 strains of *A. indica*-nodulating rhizobia they isolated. Both pigmented and nonpigmented strains nodulated stems and formed nitrogen-fixing symbioses. DNA from all of the pigmented isolates tested, and from 3 of 5 nonpigmented isolates, hybridized with a DNA probe containing the genes for the photosynthetic apparatus of *Rhodopseudomonas capsulatus*.

2. Species That Nodulate Other Plants

Thus far, the only rhizobia reported to contain bacteriochlorophyll have been isolated from stem or root nodules of *Aeschynomene* species. Attempts to induce formation of bacteriochlorophyll in other rhizobia, including strains isolated from stem nodules of *Sesbania rostrata* and root nodules of *A. americana* (Evans et al., 1990; Ladha and So, 1994) and from stem nodules of *Sesbania procumbens* and *Neptunia* (S. Shanmugasundaram and D. Fleischman, unpublished) have been unsuccessful. However, W. Heumann, of Friedrich-Alexander Universität in Erlangen-Nürnberg, Germany, has found that rhizobia which form effective nodules on the hypocotyl and root, but not the stem, of *Lotononis bainesii* contain bacteriochlorophyll (personal communication). The absorption spectrum of the cells includes far-red absorption bands characteristic of the bacteriochlorophyll proteins of typical photosynthetic bacteria. Their chromosomal guanosine + cytosine content is 72.1% and their plasmid GC content is 65.3%, values much higher than those previously found in *Rhizobiaceae*. The *Lotononis* rhizobia appear to be rather distant taxonomically from the *Aeschynomene* stem nodule rhizobia. Their principal carotenoid is a C_{40} carotenoid acid.

Liesack and Stackebrandt (1992) have attempted to determine the diversity of the bacterial population in an Australian soil sample, not by isolating bacteria, but by using the polymerase chain reaction to amplify 16S ribosomal RNA genes collected directly from the soil. They found a surprising number of sequences that were highly homologous to the 16S rRNA of BTAi 1. Their observations, together with those of Heumann, suggest that photosynthetic rhizobia may not be an isolated occurrence, but may in fact be quite widespread.

II. Relationship of Bacteriochlorophyll-containing *Rhizobium* Species to Other Bacteria

A. Relationship Based on 16S Ribosomal RNA Sequences

The unusual nature of the presence of a photosynthetic system in a rhizobium led to the proposal that BTAi 1 be placed in a new genus, to be named *Photorhizobium* (Eaglesham et al., 1990). BTAi 1, the first isolated *Rhizobium* shown to possess a photosynthetic system, would be the type strain and named *Photorhizobium thompsonianum*.

Young et al. (1991) provided the first direct information about the relationship of BTAi 1 to other bacteria. Using the polymerase chain reaction, they determined partial 16S rRNA sequences of BTAi 1 and 12 bacteria in the alpha subdivision of the *Proteobacteria*. The BTAi 1 sequence examined differed by a single base from the corresponding sequence of *Bradyrhizobium japonicum* USDA 110, and by only 6 bases from that of *Rhodopseudomonas palustris*. The sequences obtained from members of the *Rhizobium* genus were quite different. Young et al. (1991) pointed out that the diversity within the whole *Bradyrhizobium*/*Rhodopseudomonas*/BTAi 1 cluster was no greater than that to be expected within a single genus.

Shortly afterward, the International Subcommittee for the Taxonomy of *Rhizobium* and *Agrobacterium* proposed minimal standards for the description of new genera and species of root-and stem-nodulating bacteria (Graham et al., 1991). They recommended the use of both phylogenetic and phenotypic traits, using a relatively large number of strains. Two recently completed studies performed with bacteriochlorophyll-containing rhizobia have made substantial progress toward meeting the Committee's criteria.

Wong et al. (1994) have determined nearly complete 16S rRNA sequences of several strains of bacteriochlorophyll-containing rhizobia isolated from stem nodules of *Aeschynomene* species growing in widely separated geographical areas and belonging to two cross-inoculation groups, along with those of two *Bradyrhizobium* strains. Rhizobia belonging to the same cross-inoculation group will nodulate the same population of host species. The 16S rRNA sequences of all of the bacteriochlorophyll-containing strains were highly similar. It was concluded that the photosynthetic rhizobia form a group 'within a distinctly branched alpha-2 cluster' which also includes several *B. japonicum* strains and, branching somewhat more deeply, *Rp. palustris*. But the cluster clearly contained bacteria belonging to genera other than *Bradyrhizobium* and *Rhodopseudomonas*. These authors concur that resolution of the taxonomic question will require both molecular sequence data and more detailed phenotypic data.

B. Relationship Based on Phenotype

Ladha and So (1994) have performed a multivariate statistical analysis of 150 phenotypic features of 52 strains of bacteriochlorophyll-containing rhizobia along with reference strains of *Rhizobium, Bradyrhizobium* and *Azorhizobium.* They conclude that the bacteriochlorophyll-containing rhizobia isolated from *Aeschynomene* nodules belong to a cluster that is separate from those of *Bradyrhizobium, Rhizobium* and *Azorhizobium*. Differences between cross-inoculation groups were apparent. Based on the phenotypic characteristics examined, the bacteriochlorophyll-containing rhizobia were distinguished from *Bradyrhizobium* at a similarity level which is borderline for intrageneric differences. They were also clearly distinguishable from rhizobia that could form effective stem and root nodules on *Aeschynomene* but that could not form bacteriochlorophyll. The latter had the properties of slow-growing rhizobia.

The bacteriochlorophyll-containing isolates formed bacteriochlorophyll only aerobically under intermittent light. They were diazotrophic, i. e. they could fix nitrogen *ex planta,* and grow in the absence of any other source of fixed nitrogen. Alazard (1990) had made a similar observation with other isolates.

It appears that further study will be necessary before it can be decided whether the bacteriochlorophyll-containing rhizobia belong in a separate genus or should be included within *Bradyrhizobium.* The issue is complicated by wide phylogenetic diversity within the *Bradyrhizobium* and *Rhodopseudomonas* genera and close phylogenetic similarity between members of these genera and other bacteria which are quite different phenotypically. Willems and Collins (1992), for example, report that there is a close genealogical relationship between *Bradyrhizobium, Blastobacter denitrificans* and *Afipia* (the agent which causes cat scratch disease).

The available evidence does suggest that the bacteriochlorophyll-containing rhizobia, *Bradyrhizobium* and *Rp. palustris* diverged from a common photosynthetic ancestor which is not shared by the other genera of Rhizobiaceae.

III. Characteristics of *Aeschynomene Rhizobium* BTAi 1 Grown *ex planta*

A. Utilization of Carbon Sources

Stowers and Eaglesham (1983) first described the capacity of BTAi 1 to utilize various carbon sources. Twenty different carbon sources were tested at a concentration of 1% (w/v). Growth was observed with sucrose and lactose, which is a characteristic associated with fast-growing rhizobia. Growth was not obtained with malate at the 1% level but growth is observed with this carbon source at lower concentrations (Evans et al., 1990). Both succinate and malate were shown to influence the oxygen-sparing effect of light (Wettlaufer and Hardy, 1992).

Growth was not obtained when CO_2 was employed as the sole carbon source (Eaglesham et al., 1990) in the light. BTAi 1 was one of 52 strains of bacteriochlorophyll-containing rhizobia surveyed for growth on 30 different carbon sources (Ladha and So, 1994) Galactose, gluconate, glucose, glycerol, malate, mannitol and mannose were utilized by all of the strains tested.

B. Regulation of Bacteriochlorophyll Accumulation

Evans et al. (1990) reported that BTAi 1 would form bacteriochlorophyll *ex planta* only if the medium contained an appropriate carbon source. Glutamate (A. R. J. Eaglesham, personal communication), malate, fructose and succinate allow bacteriochlorophyll formation. Little bacteriochlorophyll is formed with mannitol as the carbon source. The presence of arabinose, even added concomitantly with a dicarboxylic acid, suppresses bacteriochlorophyll formation.

When BTAi 1 cultures are grown in the presence of malate under cyclic illumination from tungsten lamps, bacteriochlorophyll accumulation begins only at the end of exponential growth (Fig. 1). Perhaps the photosynthetic system is formed when the carbon source becomes depleted in order to provide an alternate energy source to allow the bacteria to survive. Such behavior would explain the observation (Eaglesham et al., 1990) that illumination does not affect the growth rate of BTAi 1 cultures. Exponentially-growing cultures do not have the photosynthetic system. When glutamate is the carbon source, bacteriochlorophyll forms early in exponential phase (unpublished). Glutamate seems to be a poor carbon source. When cells are grown on glutamate they respire more slowly than in the presence of malate (Wettlaufer and Hardy, 1992) and grow more slowly (unpublished). Perhaps cells experience carbon limitation even in the presence of glutamate and this event triggers bacteriochlorophyll synthesis.

Illumination does slow the decline of viable cells after the exogenous carbon source had been utilized (Eaglesham et al., 1990). At this stage the photosynthetic system will have formed. Shiba (1984) has reported that illumination prolongs the viability of the aerobic photosynthetic bacterium *Roseobacter denitrificans* (formerly *Erythrobacter* OCH 114), and suggested that it forms the photosynthetic system to allow it to survive in the absence of energy-yielding substrates.

Evans et al. (1990) found that BTAi 1 would accumulate bacteriochlorophyll only if the cultures were grown under cyclic illumination, but not in continuous light or continuous darkness. Wettlaufer and Hardy (1993) report that bacteriochlorophyll will accumulate in the dark after cultures have been exposed to light for at least 30 minutes. They found that the action spectrum for initiation of bacteriochlorophyll accumulation included a blue component.

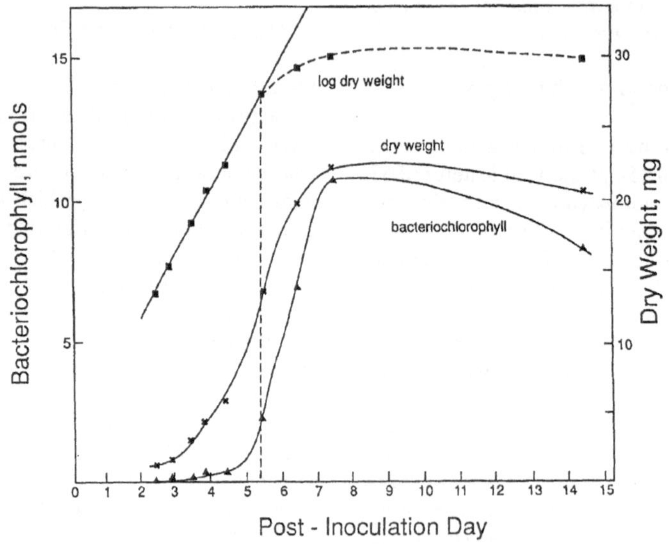

Fig. 1. Growth and bacteriochlorophyll accumulation of BTAi 1 cells. Cultures were grown under cyclic white illumination with malate ($1.5\ g\ l^{-1}$) as the carbon source.

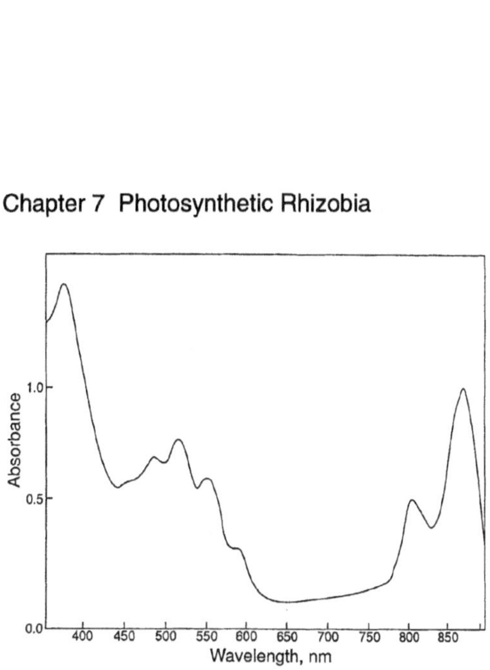

Fig. 2. Absorption spectrum of membranes from BTAi 1 cells grown in continuous red light with glutamate (1.5 g l^{-1}) as the carbon source.

We find that it also contains a far-red component (unpublished). As Wettlaufer and Hardy point out, the photoreceptor cannot be the photosynthetic light harvesting system. Light of wavelength beyond 830 nm is ineffective in inducing pigment accumulation, while the photosynthetic bacteriochlorophyll protein has a strong absorption band near 870 nm (Fig. 2). The best candidate may be a tetrapyrrole, perhaps bacteriochlorin.

Visible light inhibits bacteriochlorophyll accumulation in *Roseobacter denitrificans* (Iba and Takamiya, 1989). BTAi 1 behaves similarly (unpublished). Bacteriochlorophyll can accumulate in continuous light (Fig. 2), but only if wavelengths below 650 nm are excluded.

In summary, bacteriochlorophyll accumulation can be initiated by visible or far-red light at the end of exponential growth, but is inhibited by visible light. The conditions necessary for formation of the photosynthetic system *ex planta* are those expected within stem nodules: a dicarboxylic acid as carbon source and illumination with cyclic light (day/night) filtered through chloroplasts and leghemoglobin, which will transmit far-red light but absorb visible light.

C. The Photosynthetic System

The absorption spectrum of BTAi 1 (Evans et al., 1990; Fig. 2) resembles that of a typical purple nonsulfur bacterium or aerobic anoxygenic phototroph (Shimada,Chapter 6, this volume). The bacteriochlorophyll content of BTAi 1 cells is typically about 0.4 nmol·mg^{-1} (dry weight), a value similar to that found in the aerobic anoxygenic phototrophs. The major far-red absorption band of the light-harvesting bacteriochlorophyll protein is narrow, comparable to that of *Rhodospirillum rubrum*. Perhaps, like *Rs. rubrum*, BTAi 1 has only a single form of light-harvesting bacteriochlorophyll protein. Wettlaufer and Hardy (1992) found that light suppresses oxygen uptake by BTAi 1 cells growing on several carbon sources. Such behavior is expected of photosynthetic bacteria. The action spectrum of the effect was similar to the absorption spectrum of the photosynthetic light-harvesting system; light absorbed by the carotenoids was less effective than light absorbed by bacteriochlorophyll.

BTAi 1 membranes contain photochemically-active photosynthetic reaction centers (Evans et al., 1990). The difference spectrum and kinetics of the reversible light-induced absorbance changes resemble those of typical purple photosynthetic bacteria. The photosynthetic unit of BTAi 1 contains about 80 bacteriochlorophyll molecules, implying that these bacteria have a low bacteriochlorophyll content because they have few reaction centers, not because of a small photosynthetic unit size (Evans et al., 1990).

The inability of bacteriochlorophyll-containing rhizobia to grow photosynthetically in the absence of oxygen has been perplexing. In part this may be because their low bacteriochlorophyll content results in inefficient light capture. The growth of the aerobic anoxygenic phototroph *Roseobacter denitrificans*, for example, is stimulated by very intense illumination (Harashima et al., 1987). But in this bacterium electron transport does not occur under anaerobic conditions (Okamura et al., 1985). To determine whether this is true of BTAi 1, D. Kramer and A. Kanazawa (unpublished) used a sensitive kinetic spectrophotometer (Kramer et al., 1990) to measure flash-induced electron transfer in intact BTAi 1 cells both in the presence and absence of oxygen. Light-induced electron transfer could be detected only under aerobic conditions. A sensitive kinetic fluorimeter was then used to measure light-induced fluorescence yield changes under aerobic and anaerobic conditions. The fluorescence induction kinetics indicated that the quinone pool of BTAi 1 cells becomes completely reduced within seconds after oxygen has been depleted. Light-induced electron transport is no longer possible, since there are no (or at most no more than one) available electron acceptors. Kramer and Kanazawa were also able to observe the cytochrome bc_1 complex in membranes of BTAi 1.

The photosynthetic system of BTAi 1 thus seems to be quite similar to those of the aerobic anoxygenic phototrophs, both in composition and function.

D. Carbon Dioxide Uptake and Nitrogen Fixation

Experiments on the effect of light on CO_2 uptake with the bacteriochlorophyll-containing rhizobia have been carried out with BTAi 1. Uptake of CO_2 by BTAi 1 was stimulated severalfold under light for those cells grown under conditions which elicited bacteriochlorophyll formation versus cells grown under conditions which inhibited pigment development (Hungria et al., 1990). $^{14}CO_2$ uptake by BTAi 1 was sensitive to inhibitors of photosystem 2, but the sites of inhibition may have been the bacterial reaction center and cytochrome bc_1 complex. It seems unlikely that BTAi 1 contains an oxygen-evolving photosystem. The uncoupler NH_4Cl inhibited both $^{14}CO_2$ uptake and light-decreased O_2 uptake. However this strain could not be grown in medium lacking a carbon source (Hungria et al., 1990).

Increases of 10- to 30-fold by light have been observed in the rate of CO_2 fixation by *A. scabra* nodules (Hungria et al., 1992). Undoubtedly some of the increased rate of CO_2 fixation was a consequence of the presence of chloroplasts in the stem nodule, but the greater response of *A. scabra* stem nodules as compared to *S. rostrata* stem nodules correlated with the photosynthetic capacity of the *Aeschynomene* endophyte.

Eaglesham et al. (1990) found that cells grown for two days under cyclic illumination had 2–5 times as much N_2-fixing activity (assayed by acetylene reduction) as dark-grown cells. There was no immediate effect of light on the rate of N_2 fixation.

E. Ultrastructure

In general, cells of BTAi 1 grown in pure culture display fairly typical Gram-negative bacterial ultrastructure (Fig. 3a). They possess an outer membrane separated from the inner membrane, in the case of BTAi 1, by a rather pronounced periplasmic space. The cytoplasm is relatively featureless,

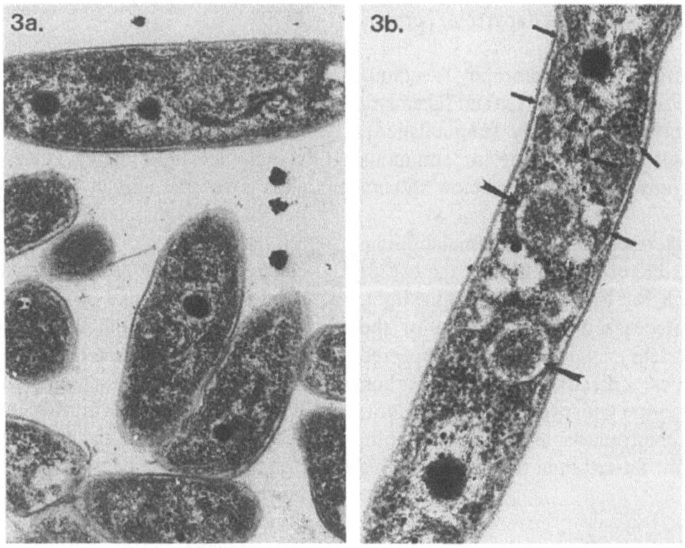

Fig. 3. (a). Transmission electron micrograph of a cell of BTAi 1 from a lag/early exponential phase culture showing typical Gram-negative ultrastructure. Apart from the fibrillar nucleoid with osmiophilic spheroids, the cytoplasm is relatively featureless. (b) TEM of a cell from a late exponential phase culture. Large intracytoplasmic photosynthetic vesicles are now found in the cytoplasm (large arrow). These vesicles are not free entities in the cytoplasm but rather represent a highly invaginated and convoluted inner membrane. Inner membrane invaginations are noticeable (small arrow).

containing only ribosomes and occasional poly-β-hydroxybutyrate granules. The nucleoid has a distinct fibrillar appearance. In some cells large osmiophilic spheroids are associated with the nucleoid. Conditions (described earlier) which trigger development of the photosynthetic system lead to changes in ultrastructure. In cells from late log phase cultures, one or more large photosynthetic vesicles can be found in the cytoplasm and undulations and invaginations of the inner membrane are common (Fig. 3b). The photosynthetic vesicles apparently arise from these invaginations of the inner membrane. During vesicle formation, a granular and relatively electron dense material accumulates in the periplasmic space and is eventually included in the vesicle. Although in cross-section the vesicles appear to be independent entities in the bacterial cytoplasm, observations from multiple and serial sectioning suggests that the vesicles remain attached at some point to the inner membrane. The vesicles therefore represent a complex, highly convoluted extension of the periplasmic space.

IV. Characteristics of Photosynthetic *Rhizobium* in the Symbiotic System

A. Gross Morphology of the Aeschynomene *Stem Nodule*

Depending on their location on the stem, *Aeschynomene* stem nodules are spherical or ellipsoidal in shape. The ellipsoidal forms are found more commonly on upper stems. The nodule consists of an inner infected bacterial region surrounded by a region of host cortex (Fig. 4). The bacterial region in effective nodules is colored pinkish-red due to the presence of leghemoglobin. The host cortical tissue contains chloroplasts and is therefore green.

B. Ultrastructure of the Symbiont

1. Comparison to Other Rhizobia

In mature nodules, the ultrastructure of photosynthetic rhizobial bacteroids is quite unlike that of bacteroids

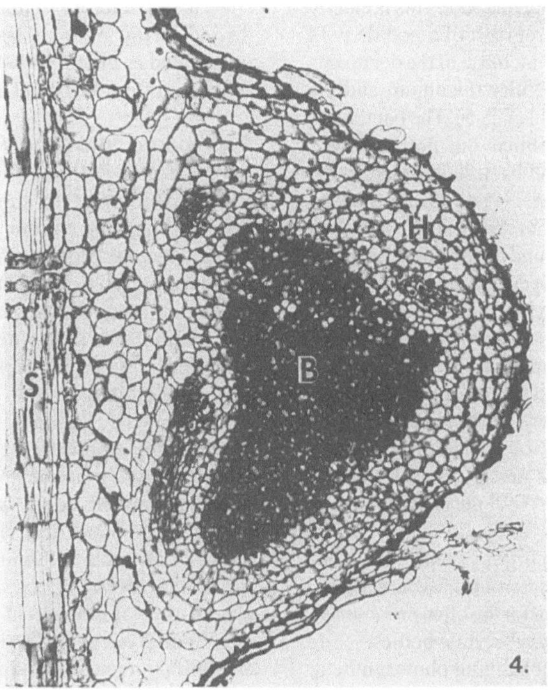

Fig. 4. Light micrograph of an upper stem nodule from *Aeschynomene indica.* Bacterial Region, B; host cortex, H; Stem Tissue, S.

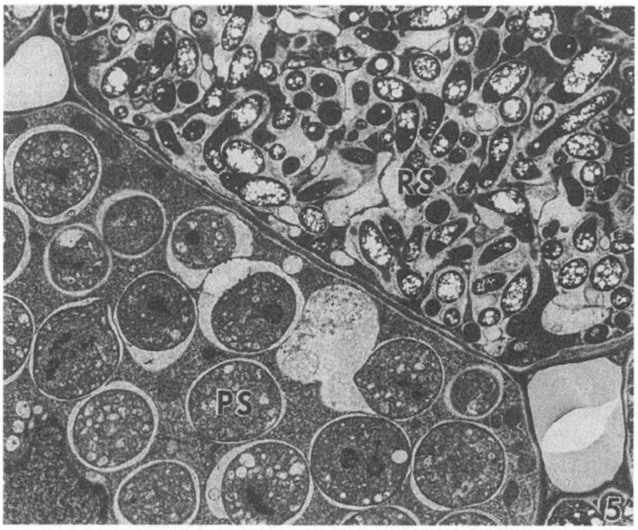

Fig. 5. Transmission electron micrograph of a stem nodule of *Aeschynomene indica* showing dual nodule occupancy (i.e., nodule is inhabited by two different rhizobial strains. In the natural habitat, 10–15% of leguminous nodules display dual nodule occupancy). The differences between regular rhizobial symbiosomes, RS, and photosynthetic rhizobial symbiosomes, PS, is striking. Note the extensive system of intracytoplasmic vesicular lamellae in the photosynthetic bacteroids, a feature absent in the regular bacteroids.

of regular nonphotosynthetic rhizobia. This is clearly visible in the electron micrograph of a nodule with dual nodule occupancy in which one of the occupants has the appearance of a regular rhizobium and the other a photosynthetic strain (Fig. 5). The bacteroids can be considered intracellular but not, however, intracytoplasmic. When the bacteria break through the wall during cell invasion, they do not continue on to penetrate the cell membrane. Rather, host cell membrane is laid down around the invading bacterial cells, isolating them from the cytoplasm. The resulting large cytoplasmic vesicles are known as symbiosomes. In mature nodule cells containing regular nonphotosynthetic rhizobia, each symbiosome consists of one or more irregular or pleiomorphic bacteroids (Sprent and Sprent, 1990). The bacteroids are embedded in a matrix of bacterially-produced polysaccharide. Bacteroids and polysaccharide are enclosed in the host-derived peribacteroid, or, symbiosome membrane.

Apart from their pleiomorphic shapes, the nonphotosynthetic rhizobial bacteroids have a relatively featureless cytoplasm except for the often pronounced deposition of poly-β-hydroxybutyrate bodies.

In mature nodule cells containing photosynthetic rhizobia, each symbiosome contains one large spherical bacteroid. Each bacteroid possesses a distinct array of cytoplasmic vesicles which are interpreted as photosynthetic apparatus. A lamellate septum is often also found in the bacteroid cytoplasm.

2. Comparison to Other Photosynthetic Bacteria

In sharp contrast to the above, photosynthetic rhizobia, both in culture and in symbiosis, show remarkable ultrastructural similarities to free-living photosynthetic bacteria. The pronounced periplasmic space, the intracytoplasmic membrane system and the distinct fibrillar nucleoid containing large osmiophilic spheroids seen in cultured BTAi 1 are almost identical to the ultrastructural features demonstrated by *Rhodopseudomonas palustris* when grown semiaerobically in the dark (Tauschel and Drews, 1968). The intracytoplasmic array of vesicular membranes seen in the symbiotic bacteroids are highly reminiscent of a similar membrane system found in anaerobic illuminated cultures of *Rhodospirillum rubrum* (Drews, 1960). From our own observations it is apparent that the intracytoplasmic membrane system in BTAi 1 develops as a complex and highly convoluted invagination of the inner membrane, in a manner identical to that described for *R. rubrum* (Oelze and Drews, 1973). In symbiosis, BTAi 1 forms large spherical bacteroids, a cellular shape very rarely found in other rhizobia but which occurs commonly in *Rhodopseudomonas* spp. We have found that BTAi 1 reproduces by asymmetrical division or 'budding' (unpublished), a feature it has again in common with *Rhodopseudomonas* spp. but which has never been reported for any other rhizobia.

C. Photosynthetic System of the Symbiont

1. Pigments and Electron Transport

Bacteroids isolated from *Aeschynomene* stem nodules usually contain bacteriochlorophyll (Fleischman et al., 1991; Ladha et al., 1990). The absorption spectrum of bacteroids isolated from *A. indica* stem nodules collected in India resembles that of free-living BTAi 1 (unpublished). They possess photochemically-active photosynthetic reaction centers whose spectrum and electron transfer kinetics resemble those of BTAi 1. A comparison of the absorption and fluorescence emission spectra of these bacteroids suggests that excitation energy transfer from the carotenoids to the bacteriochlorophyll is not very efficient (Fleischman et al., 1991). Wettlaufer and Hardy (1992) have concluded that light absorbed by the carotenoids of free-living BTAi 1 cells also drives photochemistry inefficiently.

2. Influence of Light on Nitrogen Fixation

Aeschynomene stem nodule bacteroids possess a photosynthetic system. Its most plausible function is to provide energy for nitrogen fixation rather than for carbon fixation. Conventional root nodule bacteroids consume photosynthate to generate the ATP and reductant needed for dinitrogen reduction. It would be futile for bacteroids to use photosynthetically-derived ATP and reductant to fix carbon, only to degrade it again to generate ATP and reductant for nitrogen fixation. The use of the bacteroid photosynthetic system to produce ATP and reductant directly within the cells where dinitrogen is reduced is more efficient than carbon fixation in leaf chloroplasts, translocation of the photosynthate to the nodules and its reoxidation in the bacteroids. More energy is required to form photosynthate than can be regained during its catabolism. In addition, the bacteroid photosynthetic system can use light in the 800–900 nm region which is not absorbed by chloroplasts. Although some chloroplast-generated photosynthate would still be needed as an electron source, photosynthetic nitrogen fixation in the bacteroids might decrease competition between carbon and nitrogen fixation substantially.

Plant cells in the stem nodule cortex contain chloroplasts. Eardley and Eaglesham (1985) have suggested that the stem nodule is an autonomous system, since it contains both nitrogen-fixing bacteroids and the chloroplasts needed to generate fixed carbon and O_2. They found that stem nodules indeed continued to fix nitrogen after defoliation of the plants. Illumination of stem nodules accelerated acetylene reduction (the standard assay for N_2 fixation), an effect they attributed to O_2 evolution by the chloroplasts.

In order to separate the effects of light on the chloroplasts and the bacteroids, Evans et al. (1990) examined the effect of far-red light on acetylene reduction by stem nodules.

Nodule-containing stem sections were illuminated through filters which transmitted light of wavelengths between 730 and 950 nm. Such light will drive electron transport in the bacteroids but not in the chloroplasts. The illumination consistently accelerated acetylene reduction. It appeared that acetylene reduction continued at a constant rate far longer in illuminated nodules than in nonilluminated ones. Perhaps illuminated bacteroids consume substrates more slowly, a corollary of the oxygen-sparing effect observed by Wettlaufer and Hardy (1992) in free-living cells. But the nodule behavior was quite variable, so measurements with a much larger number of stem sections will be necessary before conclusions about the effect of light can be made with confidence.

D. Mechanism of Infection Leading to the Symbiosis

1. Infection Methods

Legumes are infected by rhizobia in one of three ways. In 'root hair' infection, which is found in the most widely studied group of leguminous hosts (legume crops), the bacteria enter the plant by penetrating the wall of a root hair (reviewed by Bauer, 1981). Once inside the hair, a tubular structure made of host cell wall material is laid down around the advancing bacteria. This structure, the infection thread, is continuously laid down around the rhizobia as they make their way toward the host cortex.

In 'wound' or 'crack entry' infection, bacteria gain access to the cortical cells of the host plant *via* ruptures in the epidermis caused by the emergence of lateral or adventitious roots (Chandler, Date and Roughly, 1982). In 'epidermal' infection, rhizobia arrive in the cortex by penetrating the middle lamella between healthy epidermal cells (Faria, Hay and Sprent, 1988)

2. Nodule Development

Once inside the cortex, infection of cortical cells must occur to allow nodule development. This can happen in at least three ways (Sprent and Sprent, 1990). Firstly, each cell in the nodule is infected by bacteria being released from branches of the infection thread. Secondly, a certain number of nodule cells become infected by bacteria released from the infection thread. The infected cells then enlarge and undergo successive divisions to form a mature nodule. In the third scenario, no infection threads are formed. Some host cells simply become infected by some rhizobia entering through the cell wall. These infected cells then divide repeatedly to form the mature nodule. Nodules in *Aeschynomene* form by this third method.

3. What is the Infection Mechanism in Aeschynomene?

The method by which *Aeschynomene* becomes infected by the photosynthetic rhizobia is still unclear. Arora (1954) showed that root and stem nodules form in the vicinity of emerging lateral and adventitious roots, concluding that some form of wound entry was being used. In a later study, Yatazawa and Yoshida (1979) claimed that there are two types of stem nodules, which they designated lower and upper stem nodules. They agree with Arora that wound entry was the probable infection mechanism. However, they found no lateral roots in association with the upper stem nodules and surmised that some other form of infection may be taking place in the upper stem. In the absence of emerging roots in the vicinity of upper stem nodules, we concur with the conclusions of Yatazawa and Yoshida. We have noticed, however, that the upper stem of *Aeschynomeme* is sparsely populated with small glandular hairs of sub-epidermal origin (unpublished). It may be that some form of hair infection is occurring or else perhaps a special case of wound infection where the shaft of the glandular hair pierces the epidermis.

V. Conclusions

A. Role of the Rhizobial Photosynthetic System

The effect of far-red light on acetylene reduction by stem nodule bacteroids (Evans et al., 1990) is consistent with the hypothesis that the bacteroid photosynthetic system furnishes energy for nitrogen fixation, and so decreases the demand of the bacteroids for photosynthate. Since photosynthate availability often limits nitrogen fixation, this process may allow increased nitrogen fixation and plant growth.

The photosynthetic system of free-living cells may allow them to survive when the nutrient supply is limited. Adebayo et al. (1989) have found substantial populations of stem-nodulating rhizobia on the surfaces of leaves. Ladha et al. (1992) have suggested that the stems may be inoculated when rain washes these bacteria down the stems. An aerobic phototroph having a limited number of photosynthetic reaction centers should be ideally adapted for survival in such an aerobic, brightly-lighted, nutrient-limited environment.

B. Future Research Directions

Highest priority should be given to determining whether the bacteroid photosynthetic system really makes a substantial contribution to nitrogen fixation and plant growth. If it does, the symbiotic association could be optimized by selection of the most effective symbiotic partners for a given region. Such studies could lead to the development of exceptionally effective green manures. Symbiosis with photosynthetic rhizobia might be extended to other legumes, including *Sesbania*.

The mechanism by which the photosynthetic rhizobia regulate the development of the photosynthetic system is especially intriguing. A comparable process occurs in cyanobacteria (Chang et al., 1992). It seems likely that the bacteria contain a light sensor which can regulate gene expression. Many bacteria contain 'two component' regulatory systems in which an environmental sensor conveys a signal, perhaps by phosphorylation, to a second component which regulates gene expression (Gilles-Gonzalez et al., 1991). The photoreceptor that initiates bacteriochlorophyll accumulation may be one of these. A bacterium having the machinery necessary for the regulation of gene expression by light presents fascinating experimental possibilities.

Biological systems which can be manipulated by light have been extraordinarily useful research subjects. Photosynthesis research has yielded insights about electron transport and its coupling to phosphorylation that could scarcely have emerged from study of oxidative phosphorylation. Similarly, study of the visual system has provided the most detailed understanding of how G-protein-coupled hormone receptors work. Stem nodules, in which the development and function of the symbiotic partners can be manipulated by light, could provide a similar system for the study of symbiotic nitrogen fixation.

Acknowledgments

We thank Peter van Berkum, John Fuerst, W. Heumann, David Kramer and J. K. Ladha for permission to cite unpublished data. We also thank James Rogers for the light micrograph of an *Aeschynomene* upper stem nodule.

References

Adebayo A, Watanabe I and Ladha JK (1989) Epiphytic occurrence of *Azorhizobium caulinodans* and other rhizobia on host and nonhost legumes. Appl Environ Microbiol 55: 2407–2409

Alazard D (1985) Stem and root nodulation in *Aeschynomene* spp. Appl Environ Microbiol 50: 732–734

Alazard D (1990) Nitrogen fixation in pure culture by rhizobia isolated from stem nodules of tropical *Aeschynomene* species. FEMS Microbiol Lett 68: 177–182

Arora N (1954) Morphological development of the root and stem nodules of *Aeschynomene indica* L. Phytomorphology 4: 265–272

Bauer, WD (1981) Infection of legumes by rhizobia. Ann Rev Plant Physiol 32: 407–449

de Bruijn FJ (1988) The unusual symbiosis between the diazotrophic stem-nodulating bacterium *Azorhizobium caulinodans* ORS571 and its host, the tropical legume *Sesbania rostrata*. In: Kasuge T and Nester EW (eds) Plant Microbe Interactions Molecular and Genetic Perspectives, Vol. 3, pp 457–504. McGraw-Hill, New York

Chandler MR, Date RA and Roughley, RJ (1982) Infection and root nodule development in *Stylosanthes* species by *Rhizobium*. J Exp Bot 33: 47–57

Chang GG, Schaefer MR and Grossman AR (1992) Complementation of a red-light-indifferent cyanobacterial mutant. Proc Natl Acad Sci USA 89: 9415–9419

Drews G (1960) Untersuchungen zur Substruktur der 'Chromatophoren' von *Rhodospirillum rubrum* und *Rhodospirillum molischianum*. Arch Mikrobiol 36: 99–108

Dreyfus B and Dommergues YR (1981) Nitrogen-fixing nodules induced by *Rhizobium* on the stem of the tropical legume *Sesbania rostrata*. FEMS Microbiol Lett 10: 313–317

Dreyfus B, Garcia JL and Gillis M (1988) Characterization of *Azorhizobium caulinodans*, gen. nov., sp. nov., a stem-nodulating nitrogen-fixing bacterium isolated from *Sesbania rostrata*. Int J Syst Bacteriol 38: 89–98

Eaglesham ARJ and Ayanaba A (1984) Tropical stress ecology of *Rhizobium* root nodulation and legume N_2 fixation. In: Subba Rao NS (ed) Current Developments in Biological Nitrogen Fixation, pp 1–35. Oxford and IBH Publishing Co, New Delhi

Eaglesham ARJ and Szalay AA (1983) Aerial stem nodules on *Aeschynomene* spp. Plant Sci Lett 29: 265–272

Eaglesham ARJ, Ellis JM, Evans WR, Fleischman DE, Hungria M and Hardy RWF (1990) The first photosynthetic N_2-fixing *Rhizobium*. In: Gresshoff PM, Roth LE, Stacey G and Newton WE (eds) Nitrogen Fixation: Achievements and Objectives, pp 805–811. Chapman and Hall, New York, London

Eardly BD and Eaglesham ARJ (1985) Fixation of nitrogen and carbon by legume stem nodules. In: Evans HJ, Bottomley PJ and Newton WE (eds) Nitrogen Fixation Research Progress, p 324. Martinus Nijhoff, The Hague

Evans WR, Fleischman DE, Calvert HE, Pyati PV, Alter GM and Subba Rao NS (1990) Bacteriochlorophyll and photosynthetic reaction centers in *Rhizobium* strain BTAi 1. Appl Environ Microbiol 56: 3445–3449

Faria SM de, Hay GT and Sprent JI. (1988) Entry of rhizobia into roots of *Mimosa scabrella* Bentham occurs between epidermal cells. J Gen Microbiol 134: 2291–2296

Fisher RF and Long SR (1992) *Rhizobium*-plant signal exchange. Nature 357: 655–660

Fleischman D, Evans WR, Shanmugasundaram S and Shanmugasundaram S (1988) Induction of photosynthetic capability in a *Rhizobium*. Plant Physiology (Supplement) 86: 21

Fleischman DE, Evans WR, Eaglesham ARJ, Calvert HE, Dolan E Jr., Subba Rao NS and Shanmugasundaram S (1991) Photosynthetic properties of stem nodule rhizobia. In: Dutta SK and Sloger C (eds) Biological Nitrogen Fixation Associated with Rice Production, pp 39–46. Oxford and IBH Publishing, New Delhi

Fyson A and Sprent JI (1980) A light and scanning electron microscope study of stem nodules in *Vicia faba* L. J Exp Botany 31: 1101–1106

Gilles-Gonzalez MA, Ditta GS and Helinski DR (1991) A haemoprotein with kinase activity encoded by the oxygen sensor of *Rhizobium meliloti*. Nature 350: 170–172

Graham PH, Sadowsky MJ, Keyser HH, Barnet YM, Bradley RS, Cooper JE, De Ley DJ, Jarvis BDW, Roslycky EB Strijdom BW and Young JPW (1991) Proposed minimal standards for the description of new genera and species of root- and stem-nodulating bacteria. Int J Syst Bacteriol 41: 582–587

Harashima K, Kawazoe K, Yoshida I and Kamata H (1987) Light-stimulated aerobic growth of *Erythrobacter* species OCh 114. Plant Cell Physiol 28: 365–374

Hungria M, Ellis JM, Eaglesham, ARJ and Hardy RWF (1990) Light-driven $^{14}CO_2$ fixation, light-decreased O_2 uptake, and acetylene reduction activity by free-living *Rhizobium* strain BTAi 1. In: Greshoff PM, Roth LE, Stacey G and Newton WE (eds) Nitrogen Fixation: Achievements and Objectives, p 351. Chapman and Hall, New York, London

Hungria M, Eaglesham ARJ and Hardy RWF. (1992) Physiological comparisons of root and stem nodules of *Aeschynomene scabra* and *Sesbania rostrata*. Plant Soil 139: 7–13

Iba K and Takamiya K (1989) Action spectra for inhibition by light of accumulation of bacteriochlorophyll and carotenoid during aerobic growth of photosynthetic bacteria. Plant Cell Physiol 30: 471–477

Jarvis BDW, Gillis M and De Ley J (1986) Intra- and intergeneric similarities between the ribosomal ribonucleic acid cistrons of *Rhizobium* and *Bradyrhizobium* species and some related bacteria. Int J Syst Bacteriol 36: 129–138

Kramer DM, Robinson HR and Crofts AR (1990) A portable multi-flash kinetic fluorimeter for measurement of donor and acceptor reactions of Photosystem 2 in leaves of intact plants under field conditions. Photosynthesis Research 26: 181–193

Ladha JK and So RB (1994) Numerical taxonomy of phototrophic rhizobia nodulating *Aeschynomene* species. Intl J Syst Bacteriol 44: 62–73

Ladha JK, Pareek RP, So R and Becker M (1990) Stem nodule symbiosis and its unusual properties. In: Greshoff PM, Roth LE, Stacey G and Newton WE (eds) Nitrogen Fixation: Achievements and Objectives pp 633–640 Chapman and Hall, New York, London

Ladha JK, Pareek RP and Becker M (1992) Stem-nodulating legume-*Rhizobium* symbiosis and its agronomic use in lowland rice. Advances in Soil Science 20: 147–192

Liesack W and Stackebrandt E (1992) Occurrence of a novel group of the domain *Bacteria* as revealed by analysis of genetic material isolated from an Australian terrestrial environment. J Bacteriol 174: 5072–5078

Long SR and Staskawicz BJ (1993) Prokaryotic Plant Parasites. Cell 73: 921–935

Lorquin J, Molouba F, Dupuy N, Ndiaye S, Alazard D, Gillis M and Dreyfus B (1993) Diversity of photosynthetic *Bradyrhizobium* strains from stem nodules of *Aeschynomene* species. In: Palacios M, Mora J and Newton WE (eds) New Horizons in Nitrogen Fixation, pp 683–689. Kluwer Academic Publishers, Dordrecht

Nambiar PTC, Dart PJ, Rao BS and Rao VR (1982) Nodulation in the hypocotyl region of ground nut (*Arachis hypogeae*). Exptl Agric 18: 203–207

Oelze J and Drews G. (1973) Membranes of Photosynthetic Bacteria. In: Lascelles J. (ed) Microbial Photosynthesis, pp 137–144. Dowden, Hutchinson and Ross, Stroudsburg, PA

Okamura K, Takamiya K and Nishimura M (1985) Photosynthetic electron transfer system is inoperative in anaerobic cells of *Erythrobacter* species strain OCh 114. Arch Microbiol 142: 12–17

Schaede R (1940) Die knollchen der adventiven wasserwurzeln von *Neptunia oleracea* und ihre bakteriensymbiose. Planta 31: 1–21

Shiba T (1984) Utilization of light energy by the strictly aerobic bacterium *Erythrobacter* sp. OCH 114. J Appl Microbiol 30: 239–244

Sprent JI and Sprent P. (1990) Nitrogen Fixing Organisms: Pure and applied aspects. Chapman and Hall, London

Stowers MD and Eaglesham ARJ (1983) A stem-nodulating *Rhizobium* with physiological characteristics of both fast and slow growers. J Gen Microbiol 129: 3651–3655

Subba Rao NS (1988) Microbiological aspects of green manure in lowland rice soil. In: Green Manuring in Rice Farming, pp 131–149). International Rice Research Institute, Manila

Subba Rao NS, Gaur YD and Murthy A (1991) Biology of root and stem nodules of *Aeschynomene aspera* and *A. indica*, potential green manure plants. In: Dutta SK and Sloger C (eds) Biological Nitrogen Fixation Associated with Rice Production, pp 31–37. Oxford and IBH Publishing, New Delhi

Tauschel HD and Drews G. (1968) Thylakoid morphogenese bei *Rhodopseudomonas palustris*. Arch Mikrobiol 59: 381–404

van Berkum P, Tuly RE and Keister DL (1995) Nonpigmented and bacteriochlorophyll-containing bradyrhizobia isolated from *Aeschynomene indica*. Appl Environ Microbiol 61: 623–629

Verma DPS (1992) Signals in root nodule organogenesis and endocytosis of *Rhizobium*. Plant Cell 4: 373–382

Walter CA and Bien A (1989) Aerial root nodules in the tropical legume *Pentachlora macrolobata*. Oecologia 80: 27–31

Wettlaufer SH and Hardy RWF (1992) Effect of light and organic acids on oxygen uptake by BTAi 1, a photosynthetic rhizobium. Appl Environ Microbiol 58: 3830–3883

Wettlaufer SH and Hardy RWF (1993) The photosynthetic nitrogen-fixing microsymbiont *Bradyrhizobium*, strain BTAi 1: Light quality and intensity effects on bacteriochlorophyll production and accumulation. Plant Physiol (Suppl.) 102: 19

Willems A and Collins MD (1992) Evidence for a close genealogical relationship between *Afipia* (the causal organism of cat scratch disease), *Bradyrhizobium japonicum* and *Blastobacter denitrificans*. FEMS Microbiol Lett 96: 241–246

Woese CR (1987) Bacterial evolution. Microbiol. Rev. 51: 221-271.

Wong FYK, Stackebrandt E, Ladha JK, Fleischman DE, Date RA and Fuerst JA (1994) Phylogenetic analysis of *Bradyrhizobium japonicum* and photosynthetic stem-nodulatin bacteria from *Aeschynomene* species grown in separated geographical regions. Appl Environ Microbiol 60: 940–946

Yanagi M and Yamasato K (1993) Phylogenetic analysis of the family *Rhizobiaceae* and related bacteria by sequencing of 16S rRNA gene using PCR and DNA sequencer. FEMS Microbiol Lett 107: 115–120

Yatazawa M, Hambali GG and Wiriadinata H (1987) Nitrogen-fixing stem nodules and stem-warts of tropical plants. Biotropica Spec Publ 31: 191–205

Yatazawa M and Yoshida S. (1979) Stem nodules in *Aeschynomene indica* and their capacity of nitrogen fixation. Phys Plantarum 45: 293–295

Young JPW, Downer HL and Eardly BD (1991) Phylogeny of the phototrophic *Rhizobium* strain BTAi 1 by polymerase chain reaction-based sequencing of a 16S rRNA segment. J Bacteriol 173: 2271–2277

Chapter 8

Biosynthesis and Structures of the Bacteriochlorophylls

Mathias O. Senge and Kevin M. Smith
Department of Chemistry, University of California, Davis, CA 95616, USA

Summary

The current status of our knowledge on the structures and biosynthesis of the bacteriochlorophylls from photosynthetic bacteria is reviewed. While the structural chemistry of the bacteriochlorophylls is well established, not much is known about the biosynthesis of the chromophores. Except for the early biosynthetic steps the current insight rests mainly with mutant studies performed in the sixties. For the bacteriochlorophylls of the sulfur and non sulfur bacteria certain assumptions can be made about the biosynthesis on the basis of analogy to chemical reactions.

I. Introduction

Bacteriochlorophylls (BChls) are the central photopigments of the photosynthetic bacteria. As such they function as antenna pigments in the light harvesting complexes and as accessory and special pair chlorophylls in the reaction center. Chemically, all chlorophylls (Chls) are derivatives of the parent compound porphyrin (1). All Chls have in common the isocyclic pentanone ring V, a central magnesium ion, and a propionic ester group at C17. The nomenclature of these compounds has been standardized (IUPAC, 1987; Smith 1994) and the IUPAC system is used throughout this paper. Figure 1 shows as an example the numbering scheme for porphyrin (Fig. 1.1) and Chl *a* (Fig 1.2). We have however retained the use of the trivial names BChl *c*, *d*, *e* for the antenna Chls from green sulfur and non sulfur bacteria to prevent the use of confusing long systematic names. These names are actually misleading since the pigments are chlorins not bacteriochlorins. In recent years a number of reviews

R. E. Blankenship, M. T. Madigan and C. E. Bauer (eds): Anoxygenic Photosynthetic Bacteria, pp. 137–151.

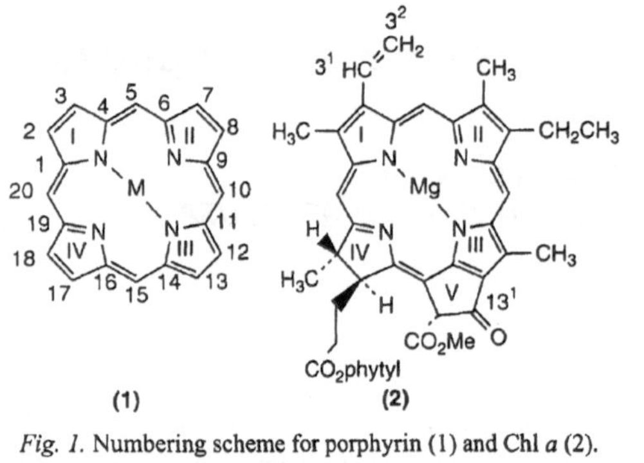

Fig. 1. Numbering scheme for porphyrin (1) and Chl *a* (2).

have been published on the structure and biosynthesis of the Chls (Rüdiger and Schoch, 1988, 1989, 1991; Beale and Weinstein, 1990; 1991; Leeper, 1991; Scheer, 1991; Smith, 1991; Senge, 1992a,b, 1993). Due to restricted space we have cited mostly leading references for the older literature and have concentrated more on giving complete literature references to papers with newer results. An entire special issue of *Photosynthesis Research* (Olson et al., 1994) was recently devoted to Green and Heliobacteria.

II. Structural Chemistry of the Bacteriochlorophylls

A. Bacteriochlorophylls a, b, and g

Bacteriochlorophyll *a* (Fig. 2.3) and *b* (Fig. 2.4) are the most widely distributed Chls found in photosynthetic bacteria, and their purple color results from the fact that the parent porphyrin macrocycle is 'reduced' not only in ring IV [as is the case in Chl *a* (2)], but also in ring II (Fig. 2). Therefore they are often termed 'tetrahydroporphyrins' or bacteriochlorins. Besides this difference in the degree of reduction another structural difference relative to Chl *a* is the presence of an acetyl group at C3 instead of a vinyl group. BChl *a* serves in most purple bacteria (*Rhodospirales*) as the sole antenna and reaction center pigment. In the green sulfur bacteria it functions in the same roles while its Mg-free derivative, bacteriopheophytin (BPheo) *a* (Fig. 2.5), serves as an accessory pigment in the photosynthetic reaction center. The structure elucidation for the Chls and BChls *a* and *b* has been reviewed by Brockmann (1978). Recently, a high precision crystal structure was published for a BChl *a* derivative, Bacteriopheophorbide (BPhide) *a* (Barkigia et al., 1989; Barkigia and Gottfried, 1994). BChl *b* (Fig. 2.4) is found in *Rhodopseudomonas viridis*, *Rhodopseudomonas sulfoviridis*, *Thiocapsa pfennigii*, *Ectothiorhodospira halochloris*, and *Ec. abdelmaleki* (Scheer, 1991) and is distinguished from BChl *a* by possessing an ethylidene group at C8 instead of an ethyl (Brockmann and Kleber, 1970; Scheer et al., 1974). Due to the ethylidene group the degree of reduction in BChl *b* can also be formulated as that of a dihydroporphyrin, since a simple migration of the ring II ethylidene double bond into ring II proper would actually yield a macrocycle at the same reduction level as Chl *a* (Fig. 1.2).

BChls generally show a higher variability in their esterifying alcohol than do the Chls. They often contain phytol, which is the most widely found esterifying alcohol (Fig. 3). However, BChl *a* from *Rhodospirillum rubrum* contains geranylgeraniol, while the BPheo *a* of the reaction center is esterified with phytol. BChl *b* from *Ec.halochloris* and *Ec. abdelmalekii* was found to contain Δ2,10-phytadienol and smaller amounts of phytol and a trienol (Rüdiger and Schoch, 1988).

BChl *g* (Fig. 4.6) is closely related to BChl *b* and possesses a vinyl at the 8-position instead of an acetyl group (Brockmann and Lipinsky, 1983) (Fig. 4). BChl *g* is found in *Heliobacterium chlorum* and contains farnesol as the esterifying alcohol (Michalski et al., 1987). It functions both as the antenna and reaction center pigment in the *Heliobacteria*. Two special derivatives have been proposed to function in the reaction center of photosynthetic bacteria besides BChl *a* and *b* and their pheophytins. These are BChl *g'* (Fig. 4.7), the 13^2-S-epimer of BChl *g* (Fig. 4.6) (Kobayashi et al., 1991) and 8^1-hydroxy Chl *a* (Fig. 4.8) (van de Meet et al., 1991). However, these compounds can be easily formed from the more abundant BChl *g*—the *a'* pigment by simple epimerization and the 8^1-hydroxy-Chl *a* by isomerization and oxidation in ring II of BChl *g*. Thus

Abbreviations: ALA – 5-Aminolevulinic acid; BChl – bacteriochlorophyll; BPheo – bacteriopheophytin; BPhide – bacteriopheophorbide; *Cb.* – *Chlorobium; Cf.* – *Chloroflexus*; Chl – chlorophyll; Chlide – chlorophyllide; *Cm.* – *Chromatium; Ec.* – *Ectothirhodospira*; Pheo – pheophytin; Phide – pheophorbide; Proto – protoporphyrin; *Rb.* – *Rhodobacter*; RNAase – Ribonuclease; SAM – S-adenosylmethionine

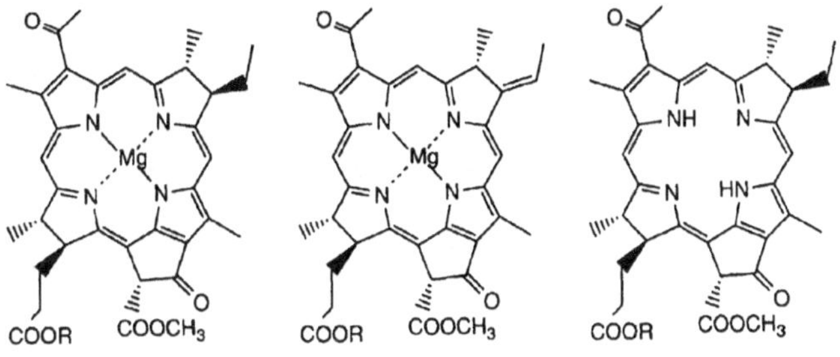

(3) (4) (5)

Fig. 2. Structural formulae for BChl *a* (3), BChl *b* (4), and BPheo *a* (5). R = mostly phytyl, see text for details.

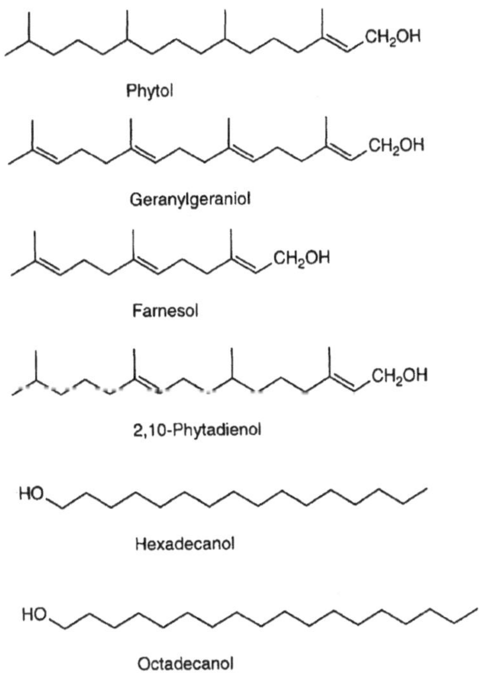

Fig. 3. Esterifying alcohols found in photosynthetic bacteria.

further studies are needed to clarify the situation.

B. Bacteriochlorophylls c, d, e, *and* f

The BChls *c* (Fig. 5.9), *d* (Fig. 5.10) and *e* (Fig. 5.11) present a complex mixture of photosynthetic pigments (Fig. 5). In contrast to the other Chls they can occur as more than one defined compound within a given organism (Holt et al., 1966). They function as antenna pigments in the green, brown, or red sulfur bacteria (*Chlorobiaceae*) and in Chloroflexaceae (Pfennig, 1977). Originally these pigments were named '*Chlorobium* chlorophylls' (Holt, 1966), a term still found sometimes in the Russian literature. In contrast

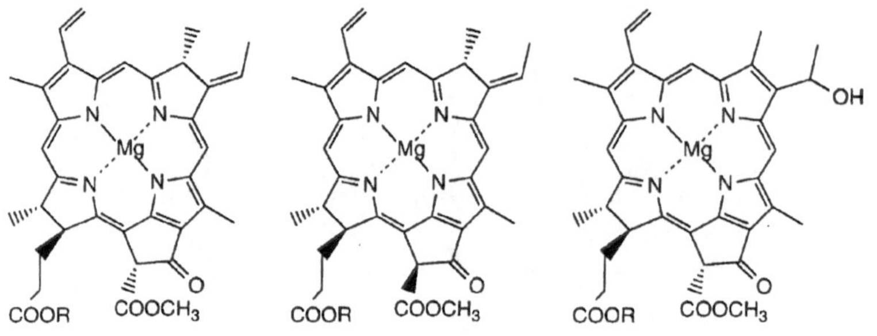

(6) (7) (8)

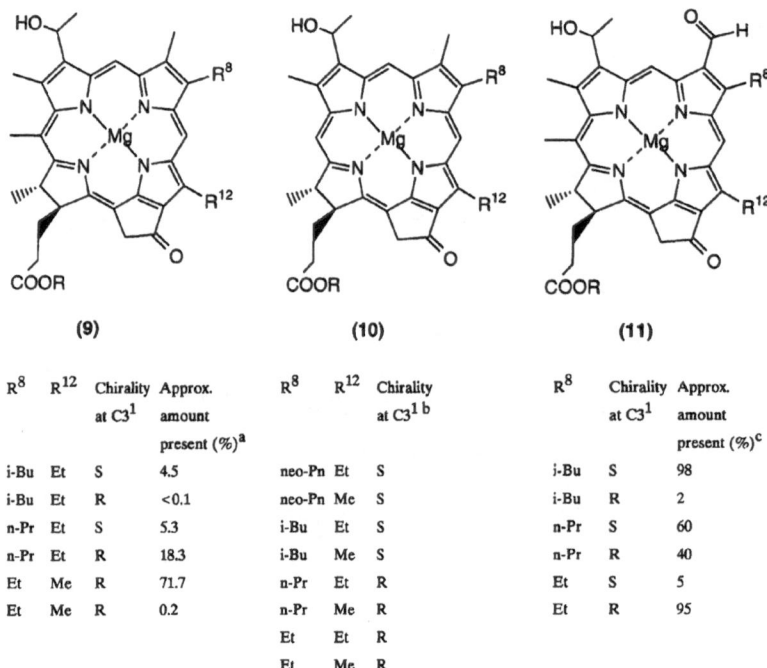

Fig. 4. Structural formulae of BChl *g* (6), BChl *g'* (7), and 8^1-hydroxy-Chl *a* (8). R = farnesyl.

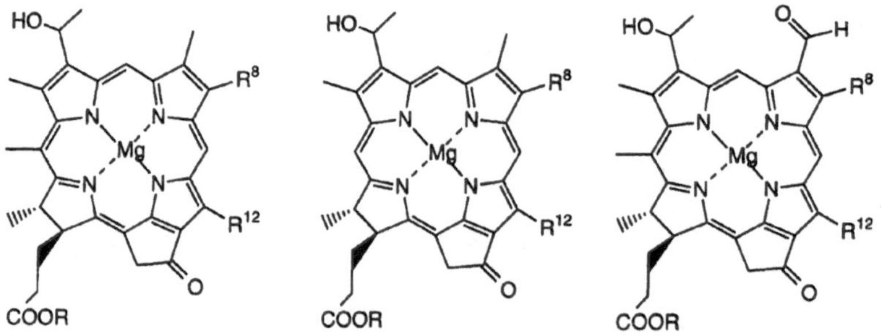

(9) (10) (11)

R^8	R^{12}	Chirality at $C3^1$	Approx. amount present (%)[a]
i-Bu	Et	S	4.5
i-Bu	Et	R	<0.1
n-Pr	Et	S	5.3
n-Pr	Et	R	18.3
Et	Me	R	71.7
Et	Me	R	0.2

R^8	R^{12}	Chirality at $C3^1$ [b]
neo-Pn	Et	S
neo-Pn	Me	S
i-Bu	Et	S
i-Bu	Me	S
n-Pr	Et	R
n-Pr	Me	R
Et	Et	R
Et	Me	R

R^8	Chirality at $C3^1$	Approx. amount present (%)[c]
i-Bu	S	98
i-Bu	R	2
n-Pr	S	60
n-Pr	R	40
Et	S	5
Et	R	95

[a]Smith et al. (1983b); [b]Smith & Goff, 1985; [c]Simpson & Smith, 1988.

Fig. 5. The structures of the bacteriochlorophylls *c* (9), *d* (10), and *e* (11). R = esterified alcohol, for details see text.

to BChl *a* and BChl *b*, these pigments have only one reduced pyrrole ring and are, therefore chlorins like Chl *a* or Chl *b*. The gross chemical structure of the BChls *c* (Fig. 5.9) and d (Fig. 5.10) and their similarity to the plant chlorophylls was established by Holt and coworkers (Holt et al., 1966) while subsequently their stereochemistry (Brockmann, 1976; Brockmann and Tacke-Karimdadian, 1979; Smith and Goff, 1985; Smith, 1991) and absolute structure (Smith et al., 1982, 1983) could be established. BChl *e* (Fig. 5.11) was first isolated during the seventies (Gloe et al., 1975) and subsequently its structure elucidated (Risch et al., 1978; Simpson and Smith, 1988). Recently the conformational flexibility of the macrocycles in Chls has come under active scrutiny (Barkigia et al., 1988; Senge, 1992b; Senge et al., 1994).

All these Chls have in common that they lack the 13^2-carbomethoxy group and bear at the 3-position a α-hydroxyethyl group. BChl *c* (Fig. 5.9) and *e* (Fig. 5.11) are distinguished from BChl *d* (Fig. 5.10) by an additional methyl group at C20. BChl *e* also has the meso (Fig. 5.12) methyl group as well as an aldehyde group at C7 and thus is similar to Chl *b*. The name BChl *f* has been assigned to a generic structure of a BChl *e* derivative without the C20 methyl group (analogous to BChl *c*– *d*), though this series of pigments has not yet been found in nature (Risch et al., 1988). All three Chls occur in nature as so called homologous mixtures. The single components of these mixtures are distinguished by the substituents at C8 and C12, the stereochemistry at $C3^1$ (R- or S-configuration) and the nature of the esterified alcohol at $C17^3$. On account of the various combinatory possibilities a large variety of different derivatives is possible. BChl *c* (Fig. 5.9) most often is found with an ethyl group at C8, as is the case in most natural tetrapyrroles. In *Chlorobium* strains, smaller amounts of pigments with a C8 n-propyl, iso-butyl or even neo-pentyl groups are found. The substituent at C12 is either ethyl or methyl in BChl *c* and d. So far BChl *e* has been isolated only with an ethyl group at C12. In most cases the esterified alcohol is farnesol (in *Chlorobium*) or stearol (in *Cf. aurantiacus*). Smaller amounts of other esterifying alcohols have been found in many cases (Caple et al., 1978). A recent example is a strain of *Cb. limicola* which produced BChl *c* esterified with six different alcohols (farnesol, geranylgeraniol, phytol, octadec-9-en-1-ol, hexadecanol, and octadecanol) (Fig. 3) (Fages et al., 1990).

III. Biosynthesis of the Bacteriochlorophylls

A. Formation of 5-Aminolevulinic Acid

All tetrapyrroles are built up in a complex sequence from small biosynthetic precursors. The first intermediate which is common to all tetrapyrroles is 5-aminolevulinic acid (ALA) (Fig. 6.18). Depending on the organism two biosynthetic pathways leading to the formation of ALA are known (Fig. 6). The ALA-synthase pathway was first found in *Rhodobacter sphaeroides* (formerly called *Rhodopseudomonas sphaeroides*) (Kikuchi et al., 1958) and serves in animals as the pathway for the formation of heme (see Chapter 9 by Beale). Because of the pioneering studies by Shemin on animals, this pathway is also called the Shemin-pathway. It involves the formation of ALA (Fig. 6.18) by condensation of succinyl-CoA (Fig. 6.16) and glycine (Fig. 6.17), catalyzed by the pyridoxal-dependent enzyme, ALA-synthase. Glycine is first bound as a Schiff-base to the pyridoxalphosphate cofactor, on which a succinate is transferred via succinyl-CoA. Subsequently CO_2 is split off and ALA is released from the enzyme. The other pathway, the so called C_5-pathway, was first found in plants and algae. This pathway involves the conversion of glutamate (Fig. 6.12) into ALA. First glutamate is bound by a glutamate-tRNA-synthase to a $tRNA^{Glu}$. This activated compound (Fig. 6.13) is utilized by a NADPH dependent hydrogenase for the formation of glutamate-1-semialdehyde (Fig. 6.14). The enzyme-bound aldehyde is then converted by a pyridoxalphosphate dependent aminotransferase into ALA (Jahn et al., 1992).

Which pathway is utilized in a given organism can be determined by feeding radioactive precursors. For example, feeding 1-[^{14}C]-glutamate as substrate and isolation of labeled ALA indicates the operation of the C_5-pathway. Similarly, addition of ribonuclease (RNAase), which degrades the glutamyl-$tRNA^{Glu}$, resulting in unlabeled ALA indicates the C_5-pathway. On the other hand the ALA-synthase pathway is operating if 2-[^{14}C]-glycine is incorporated in ALA in the presence of an RNAase. Such experiments have been performed e.g. for *Rhodopseudomonas palustris* indicating the ALA synthase route (Andersen et al., 1983).

Similar labeling experiments with ^{13}C-enriched precursors allow a direct analysis of the final tetrapyrrole of a given biosynthetic pathway. After isolation of (e.g.) the BChls they can then be investigated by ^{13}C-NMR for specifically enriched positions. Such experiments were performed for BChl *a* in *Rb. sphaeroides* (Oh-hama et al., 1985) again indicating the ALA synthase route. The methoxyl of the isocyclic ring methyl ester was derived from the C2 carbon of glycine, presumably via methionine. In *Cm. vinosum* the C_5-pathway was found (Oh-hama et al., 1986). Figure 7 shows the labeling pattern found for the macrocycle atoms of BChl *a* (Fig. 7.19) and *c* (Fig. 7.20). Comparative analyses of various organisms showed a large variability in the two pathways found in photosynthetic bacteria. The C_5-pathway is widely distributed and presumably is the

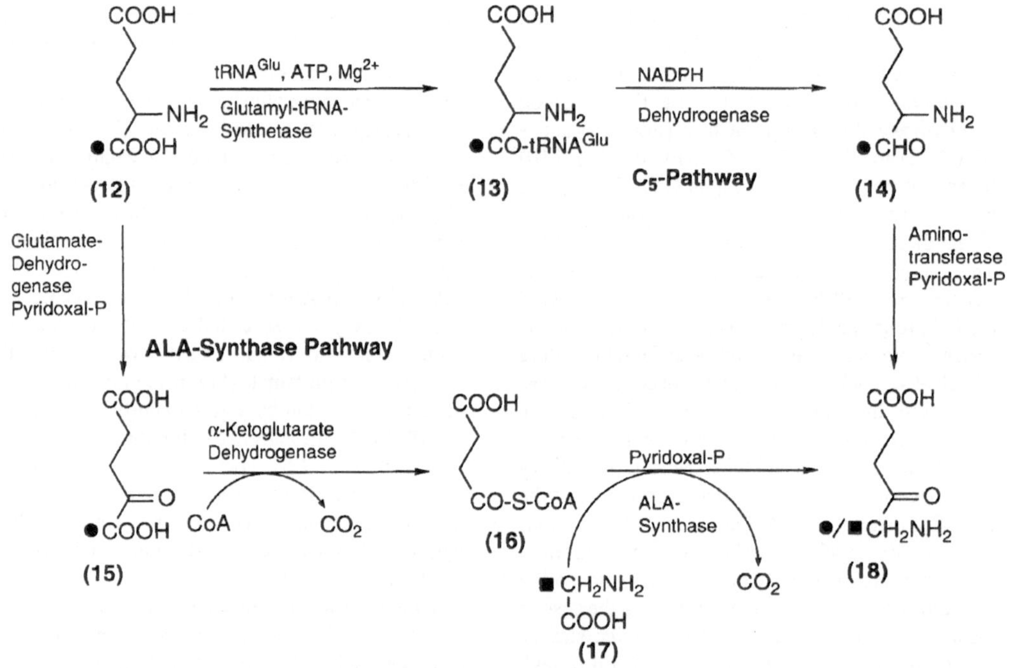

Fig. 6. Scheme of the two biosynthetic pathways for ALA found in photosynthetic bacteria. The distribution of specifically labeled C-atoms (from labeling experiments) is indicated by symbols (● C1 from glutamate; ■ C2 from glycine).

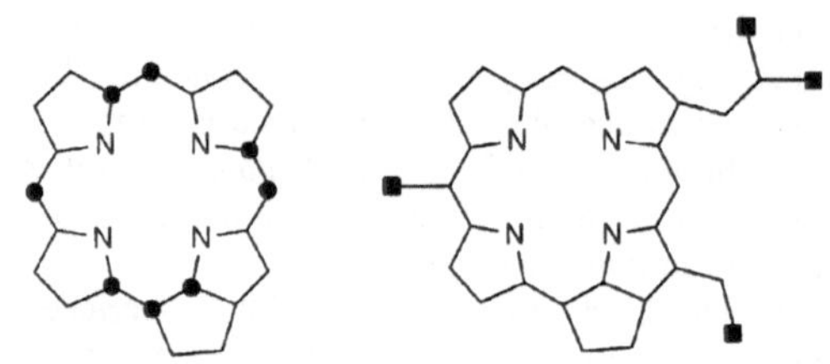

(19) **(20)**

Fig. 7. Labeling of the macrocycle atoms in BChl *a* from C5 of ALA (19) and in BChl *c* from methionine (20).

phylogenetic oldest biosynthetic pathway for Chls. It is found in the green sulfur bacteria, e.g. in *Chloroflexus*. The ALA-synthase pathway is found in the non-sulfur purple bacteria. A detailed analysis of the pathway distribution has been performed by Avissar et al. (1989) and a detailed description of the two pathways has been given by Beale and Weinstein (1991) and by Beale in Chapter 9.

B. Intermediate Steps to Protoporphyrin

The next biosynthetic step is catalyzed by the enzyme ALA-dehydratase. It involves the condensation of two ALA molecules, and by splitting off of two water molecules, leads to the formation of the first pyrrole, porphobilinogen. Besides detailed studies on the biosynthesis of heme in animals, this enzyme has been well studied in *Rb. sphaeroides* (van Heyningen and Shemin, 1971; Jordan and Seehra, 1980). A detailed description of the enzymes and mechanisms involved in the steps to Proto has been given by Leeper (1991) and by Beale in Chapter 9.

With the exception of the variability in ALA biosynthesis discussed above, there has been little progress in understanding BChl biosynthesis since early experiments using *Chlorella* and *Rb. sphaeroides* mutants were published. This work has been extensively reviewed (Jones, 1978). Not much is known about the intermediate steps in biosynthesis up to Proto. It can however be assumed that the biosynthesis of the BChl utilizes the same intermediates as were found in the biosynthesis of heme

(Fig. 8). The enzymes hydroxymethylbilane synthase and uroporphyrinogen III synthase catalyze the formation of uroporphyrinogen III from four molecules of porphobilinogen. This is the first biosynthetic intermediate containing a tetrapyrrole macrocycle. The uroporphyrinogen III decarboxylase then stepwise decarboxylates the four acetate residues, eventually leading to the formation of coproporphyrinogen III, which is converted by oxidative decarboxylations of the propionate residues at ring I and II to give the vinyls in protoporphyrinogen. The enzyme protoporphyrinogen oxidase oxidizes this intermediate by removal of 6 hydrogen atoms (2 NH, and the four H at C5, C10, C15, and C20) to Proto, the first tetrapyrrole with a conjugated ring system. While in higher plants dioxygen is required for the aromatization and is the source of the keto-oxygen in the isocyclic ring (Walker et al. 1989), this reaction occurs in photosynthetic bacteria anaerobically (Tait, 1972). The electron acceptor in photosynthetic bacteria has not been determined, yet. The biosynthesis of all known tetrapyrroles uses the same intermediates up to uroporphyrinogen III. After this, branching occurs, leading to the formation of the different tetrapyrroles (heme, chlorophylls, corrins, and factor F430).

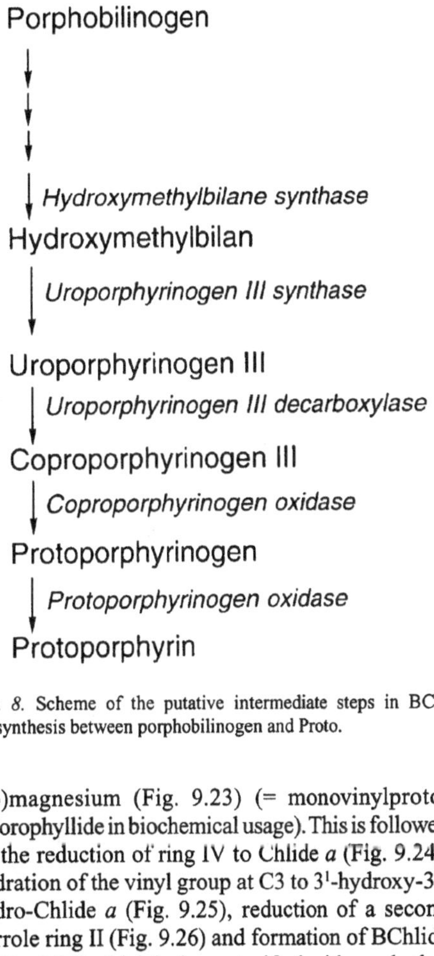

Fig. 8. Scheme of the putative intermediate steps in BChl biosynthesis between porphobilinogen and Proto.

C. Formation of Bacteriochlorophyll a, b *and* g

Since the early studies on *Rb. sphaeroides* in the sixties (reviewed in Jones, 1978) not much progress has been made with elucidation of the steps after Proto. The most important results were obtained by experiments with photosynthetic mutants with blocks in the BChl biosynthesis. (Protoporphyrinato monomethyl ester)magnesium (Fig. 9.21), several metal-free derivatives, Phide a, 3^1-hydroxy-3^2-hydro-Chlide *a* (Fig. 9.25) and -BChlide *a* (Fig. 9.26) and BChlide *a* (Fig. 9.27) have all been isolated after excretion from the mutants. Fig. 9 shows the currently accepted pathway from (protoporphyrinato monomethyl ester)magnesium to BChl *a* (Fig. 9.3) and this sequence is almost exclusively based on these mutant studies.

The first step is the incorporation of magnesium into Proto which is catalyzed by a magnesium chelatase. The product of this reaction is then converted by the enzyme S-adenosylmethionine: Mg-protoporphyrin-O-methyl transferase. After formation of the isocyclic ring a vinyl group is hydrated to (3^1,3^2-didehydro-13^2-carbomethoxy-phytoporphyrinato)magnesium (Fig. 9.23) (= monovinylprotochlorophyllide in biochemical usage). This is followed by the reduction of ring IV to Chlide *a* (Fig. 9.24), hydration of the vinyl group at C3 to 3^1-hydroxy-3^2-hydro-Chlide *a* (Fig. 9.25), reduction of a second pyrrole ring II (Fig. 9.26) and formation of BChlide *a* (Fig. 9.27) which is then esterified with an alcohol to give BChl *a* (Fig. 9.3).

A possible alternate pathway has also been proposed (Pudek and Richards, 1975) on account of the fact that an *Rb. sphaeroides* mutant has been shown to excrete a protein-pigment complex containing 3-desacetyl-3-vinyl-BPhide *a*. This suggests that reduction of chlorin to bacteriochlorin can take place before, or after hydration of the 3-vinyl in Chlide *a*, to 3^1-hydroxy-3^2-hydro-Chlide *a* (Fig. 9.25). Interestingly, BChl *g* (Fig. 9.6) possesses a 3-vinyl in place of the 3-acetyl found in BChl *a*, and thus would by necessity, require 3-vinyl precursors for its biosynthesis. These contradicting results indicate that the results from the mutant studies have only limited use in elucidating the exact biosynthetic pathway. The determination of the

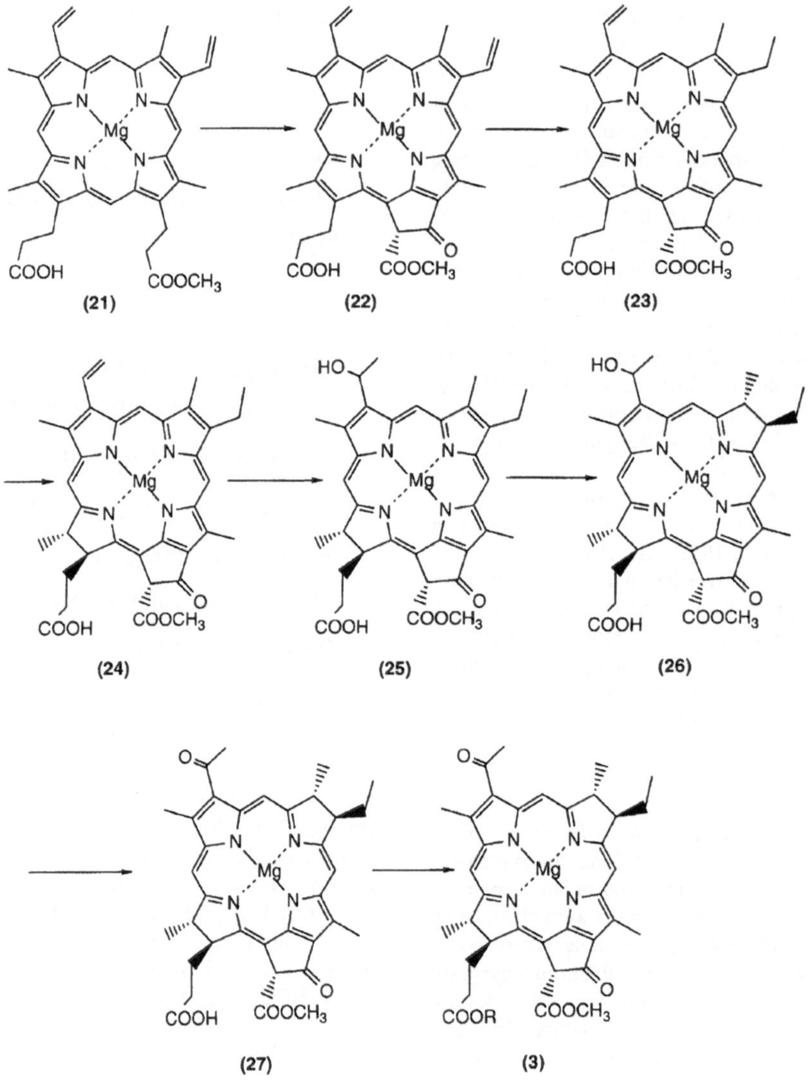

Fig. 9. Currently accepted biosynthetic pathway to BChl *a*.

individual biosynthetic steps has been hampered by the fact that it is extremely difficult to develop active, cell-free systems for enzymatic and labeling studies. Some indications for the individual steps can however be delineated from analogies with chemical syntheses in situ.

A recent analysis of mutant strains of *Rb. capsulatus* with blocks in its BChl *a* biosynthesis at different steps of the magnesium branch showed that (3^1,3^2,8^1,8^2-tetradehydro-13^2-carbomethoxy-phytoporphyrinato)magnesium (divinyl-protoChlide) and (3^1,3^2-didehydro-13^2-carbomethoxy-phytoporphyrinato)magnesium (monovinyl-protoChlide), 3-devinyl-3-hydroxyethyl-Chlide *a*, and BChlide *a* were all bound to a 32 kDa protein, which was identified as the outer membrane porin protein (Bollivar and

Bauer, 1992). For a wild-type *Rb. capsulatus* the excretion of coproporphyrin bound to a protein with the same N-terminal sequence was reported by Biel (1991). No information is yet available on formation, excretion, or specificity of these complexes. The protein accumulated also in cells grown in high-oxygen conditions, indicating that it is not involved in regulation of tetrapyrrole biosynthesis. Recent years have seen an increase in studies on the molecular biology of the BChl biosynthesis (Bauer et al., 1993; Bollivar et al., 1994a,b,c). This is the topic of the Chapter 50 by Alberti et al.

Using (2R,3R)-ALA-2,3-T_2-2,3-$^{14}C_2$, Emery and Akhtar (1985) have shown that the 8-ethyl group in BChl *a* in *Rb. sphaeroides* is formed by addition of hydrogen to the si-face of the corresponding double bond, resulting in an overall, formal, *trans*-addition. Chemical studies on model compounds have shown (Smith and Simpson, 1987; Simpson and Smith, 1988), that a formal reduction of the vinyl group to ethyl is possible by an initial reduction of the double bond in the corresponding pyrrole ring (Fig. 10.29) and subsequent migration of the double bond, resulting in the ethylidene derivative (Fig. 10.30) (Fig. 10). Another migration of the double bond leads, then, to the unsaturated ethyl compound (Fig. 10.31). A similar mechanism might be involved in the formation of the ethylidene group in BChl *b*. On the basis of observations on photoisomerism reactions of BPheo derived from BChl *g* of *Heliobacterium chlorum*, Michalski et al. (1987) have also suggested that all Chls are biosynthesized from a common intermediate bearing an ethylidene group as is present in BChl *b* (Fig. 2.4) and BChl *g* (Fig. 4.6).

The biosynthesis of phytol, which is esterified with the C17 propionic ester group, proceeds coupled to the biosynthesis of the chromophore (Brown and Lascelles, 1972). Starting with the specific C_5-precursor of the isoprenoid pathway, isopentenyl-diphosphate, geranylgeraniol diphosphate is formed, which is then esterified with the chromophore. The use of labeled ALA showed that the oxygen atoms of the ester groups at C17 and C13 both are derived from the carboxylate oxygen of ALA (Ajaz et al., 1985; Akhtar et al., 1984; Emery and Akhtar, 1987). This means that the ester groups in BChl *a* are formed by attack of the carboxylate groups of the chromophore on activated isoprenyl (and for C13, methyl) moieties. For Chl *a* it is known that first geranylgeraniol is bound by a chlorophyll synthetase to the pigment. Subsequently three double bonds are reduced in the side chain to give phytyl (Rüdiger and Schoch, 1988). The different esterified alcohols found in the BChls *c*, *d*, and *e* show, however, that the situation is much more complicated for the BChls.

The biosynthesis of BChl *b* (Fig. 2.4) presumably proceeds similarly to that of BChl *a* as far as the 3^1-hydroxy-3^2-hydro-Chlide *a* (Fig. 9.25) step. Reduction of a double bond in ring II, as shown in Fig. 10, leads to the formation of the respective 3-hydroxyethyl-8-vinylbacteriochlorin, which after migration of the double bond yields 3-hydroxyethyl-

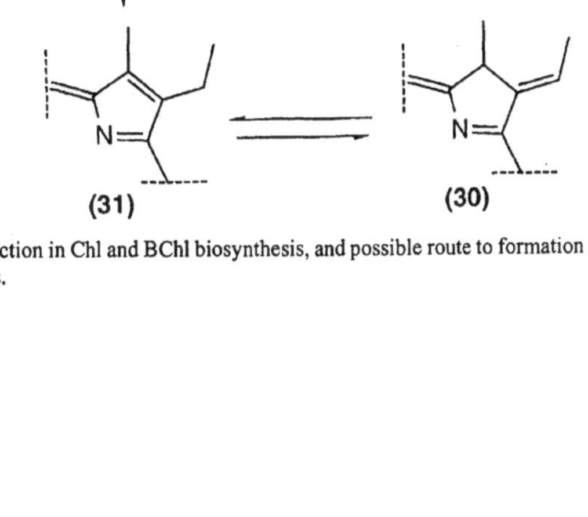

Fig. 10. Pathways for vinyl reduction in Chl and BChl biosynthesis, and possible route to formation of ring II ethylidene substituents in BChl *b* and BChl *g* biosynthesis.

8-ethylidenebacteriochlorin. Subsequent oxidation of the hydroxyethyl group to acetyl and esterification of the C17 acid with an isoprenoid alcohol would then yield BChl *b*. As of yet no studies on the biosynthesis of BChl *g* have been reported. It can be assumed that it follows in its biosynthesis that of BChl *a* up to Chlide *a* and that the branch point involves reduction of the Chlide *a* ring II double bond to give a vinyl bacteriochlorin (Fig. 10.29). Subsequent rearrangement of the double bond would yield an ethylidene compound analogous to (Fig. 10.30), followed by esterification with farnesol.

D. Biosynthesis of Bacteriochlorophyll c, d, and e

Similar to the situation found with BChl *a* and *b*, detailed information on the biosynthesis of the BChls *c* (Fig. 5.9), *d* (Fig. 5.10), and *e* (Fig. 5.11) is only available for the initial and ultimate steps. As described above, ALA is formed via the C_5-pathway in the green sulfur bacteria (*Chlorobium*) and the green nonsulfur bacteria (*Chloroflexus*). Detailed labeling experiments have been described for *Cb. limicola* (Andersen et al., 1983), *Cf. aurantiacus* (Swanson and Smith, 1990; Oh-hama et al., 1991), *Prosthecochloris aestuarii* (Oh-hama et al., 1986b) and *Cb. vibrioforme* (Smith and Huster, 1987; Smith and Bobe, 1987; Huster and Smith, 1990; Bobe et al., 1990). A discussion of the different results has been given by Smith (1991). No results are available on the biosynthesis of the reaction center pigments BChl *a* and *b*, which occur together with the BChls *c*, *d*, and *e* in these bacteria. A number of inhibition experiments have been performed recently. Earlier inhibition studies (see Jones, 1978) indicated a biosynthetic pathway closely resembling that of BChl *a* on the basis that intermediates similar to those of the BChl *a* biosynthesis were found. However, these results can also be reinterpreted in terms of total inhibition of BChl *c* biosynthesis very early in the pathway such that the metabolites being excreted into the medium were simply intermediates of normal BChl *a* biosynthesis. This problem could be partially overcome with the finding that anaesthetic gases, like N_2O, ethylene, and acetylene, inhibit BChl *d* (antenna pigment) formation but not that of BChl *a* (reaction center pigment) in *Cb. vibrioforme* (Ormerod et al., 1990). (Protoporphyrinato monomethyl ester)magnesium (Fig. 9.21) was found to accumulate under these conditions, suggesting that the biosynthetic pathway of BChl *a* and d branch prior to isocyclic ring formation. Another possibility might be different compartmentalization of the biosynthetic enzymes for both pigment types and/or interference of the inhibitors with chlorosome formation. Inhibition experiments with gabaculine also supported the operation of the C_5-pathway in *Cf. aurantiacus* (Kern and Klemme, 1989). For *Chlorobium* it was shown that cell extracts have RNA-dependent ALA formation from glutamate as describe above (Rieble et al., 1989).

Labeling experiments with ^{13}C-labeled precursors showed that all of the 'extra' carbon atoms in the side chains of the BChl *c* are derived from methionine (Fig. 7, formula 20) (Kenner et al., 1978; Smith and Huster, 1987; Huster and Smith, 1990). A lower level incorporation of porphobilinogen was also found.

Enzymatic studies are still lacking, due to the problems described above. However some insight into the biosynthesis can be obtained from analogy with chemical reactions. Not much is known with certainty about the latter stages in the biosynthesis of the BChls *c*, *d*, and *e*, so much of what follows should be regarded as conjecture rather than substantiated fact. For the biosynthesis of the different alkyl substituents at C8 two different pathways are feasible (Fig. 11) (Smith, 1991) which, by successive methylation and reduction of the side groups lead to the different homologues. The ethyl group at C8 can be formed analogous to that in BChl *a* (Fig. 10), i. e. by dehydrative decarboxylation of the propionate (Fig. 10.31) and subsequent reduction of the vinyl group (Fig. 11, steps 33 → 34). A methylation of the α-methyl group of the propionate by SAM and subsequent vinylation and reduction presents a possible scheme for the formation of the propyl homologue (Fig. 11, steps 35 → 37). Repetition of this sequence would lead to the isobutyl homologue (Fig. 11.40). Biosynthesis of the neopentyl group via this 'propionate alkylation pathway' would however involve a rather cumbersome concomitant decarboxylation and methylation of (Fig. 11.38) to (Fig. 11.44). The presumed more likely 'vinyl alkylation' pathway is shown as path B in Fig. 11. Starting with the 8-vinyl derivative (Fig. 11.33) a direct methylation with SAM would yield a carbocation which after deprotonation to the corresponding alkene (Fig. 11.36) and subsequent reduction would yield the propyl homologue (Fig. 11.37). Repetition of this process would give the corresponding isobutyl and neopentyl homologue (Fig. 11.41). A first indication

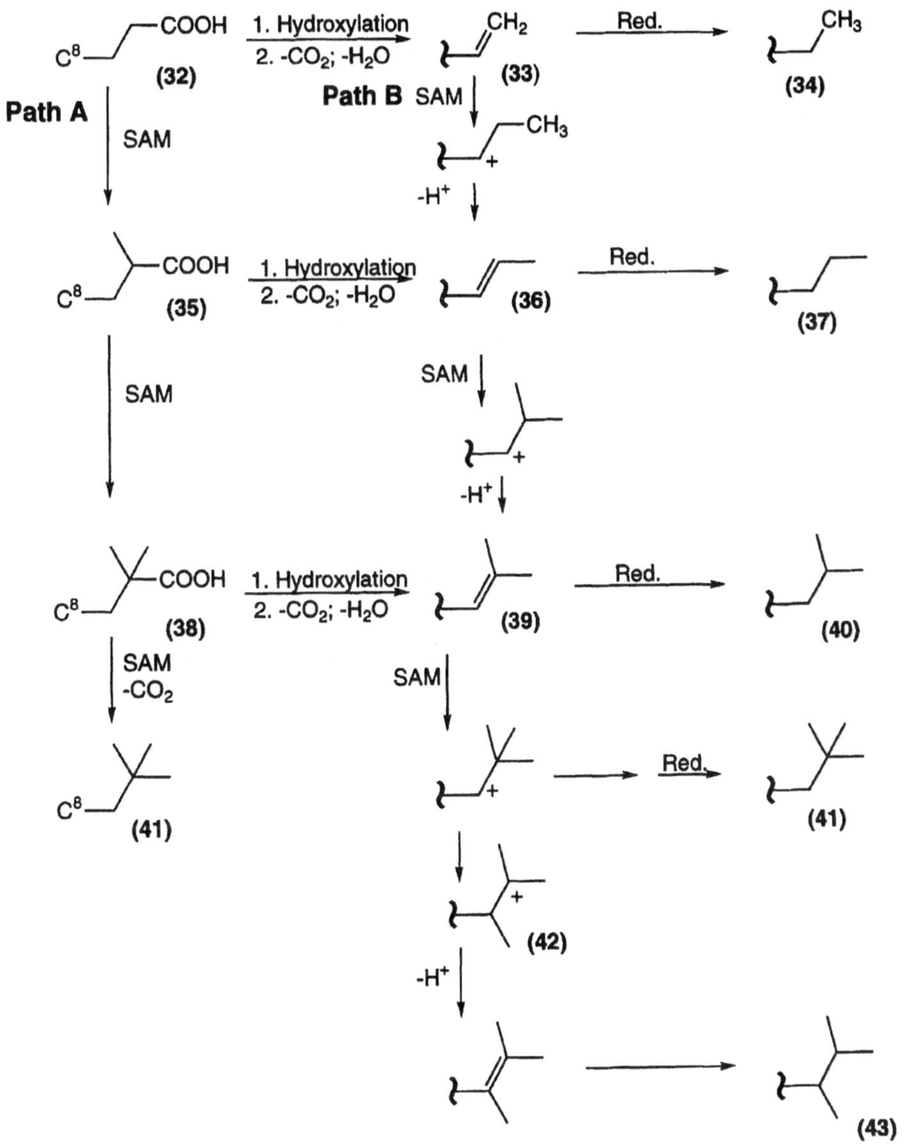

Fig. 11. Scheme of the propionate (A) and vinyl (B) alkylation pathway as a possible biosynthetic route for R^8 homologation in the BChl *c*, *d*, and *e*. Note that direct decarboxylation of intermediates (35) and (38) can lead directly to (37) and (40), respectively.

for the existence of this pathway was given by the identification of the rearranged homologue (Fig. 11.43) in small quantity from *Cb. vibrioforme* (Goff, 1984). The formation of this derivative can only be explained by a rearrangement of the carbocation which precedes the formation of the neopentyl homologue. Deprotonation of the tertiary carbocation (Fig. 11.42) and reduction yields then the isolated 1,2-dimethylpropyl homologue (Fig. 11.43).

The biosynthesis of the C12 homologues can be explained by a decarboxylation of the C12 acetic acid group (Fig. 12). This step occurs in the other Chls before the vinyl reduction of the C8 group and thus constitutes a possible early branch point in the biosynthesis of the chlorophylls. After decarboxylation the methylene group (Fig. 12.46) can be methylated by SAM to give an ethyl group (Fig. 12.47). This is corroborated by model studies on the oxidation of C12 ethyl to acetyl in BPheo *d* (Senge et al., 1991) and labeling experiments which show the direct incorporation of the $^{13}CD_3$-methyl group of SAM at this position (Kenner et al., 1978).

The BChls *c* and *d* are found with both possible stereochemical configurations at the C3-hydroxyethyl group (Fig. 5). The distribution of the different homologues shows that the predominant isomer depends on the size of the substituent at C8. If the residue at C8 is small (e.g. ethyl) the R-configuration is predominant. The S-configuration is preferred if the C8 substituent is large (e.g. neopentyl). This has been interpreted in terms or remote control of induced stereochemistry by the bulk of the C8 substituent, possibly involving rotamers of the presumed 2-vinyl precursor (Smith et al., 1983b). This requires that the formation of the different C8 homologues occurs prior to the introduction of the hydroxy group at C-3^1 (presumably by hydration of the respective vinylogous precursor).

The question occurs as to why the green sulfur bacteria produce only one BChl *a* or *b* but a multitude of different homologues of the antenna BChls *c*, *d*, and *e*. Investigation of the absorption spectra of BChl *c*, *d*, and *e* in nonpolar solvents revealed a strong dependence of the absorption maxima from the size of the substituent at C8 (Smith et al., 1983c). Similar results were observed in living cells. Since the light harvesting complex of the bacteria constitutes of oligomeric BChls and a protein backbone it is possible that modification of the homologue composition allows a variation of the wavelengths region in which light is absorbed. When the oligomer is small, the bathochromic shift (monomer to oligomer) in the optical spectrum of the antenna system is small, and *vice versa* for large oligomers. Thus, the size of the antenna aggregate is controlled indirectly by the size of the C8 substituent and directly by the chirality of the C3^1-center, because a major interaction holding the oligomer together involves the central magnesium ion of one BChl and the oxygen of the 3-(1-hydroxyethyl) group of another. Under low light intensities an activation of the corresponding methylating enzymes leads to modification of the absorption spectrum. Alkylation in newly developing cells causes a red shift in the antenna system which in better light utilization because of better penetration of low energy light through the medium, and because existing bacteria in the medium absorb at higher energy (Smith and

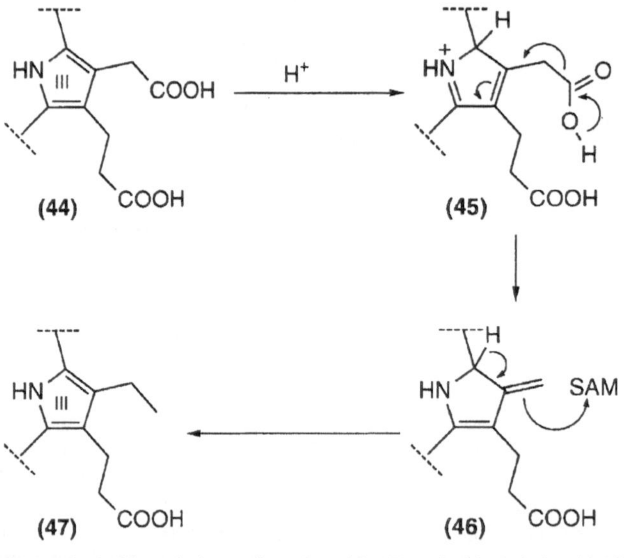

Fig. 12. Possible mechanism for the biosynthetic transformation of the 12-acetic side chain into 12-ethyl, mediated by SAM.

Bobe, 1987; Bobe et al., 1990). Dilution of the bacterial suspension results in a termination of the alkylation process and a decrease in the content of n-propyl and isobutyl containing pigments relative to 8-ethyl. Thus this adaptation process is reversible.

A similar adaptation to environmental conditions has been observed for the BChl *d* producing strain *Cb. vibrioforme* (Broch-Due and Ormerod, 1978). Under low light intensities this bacterium slowly converts BChl *d* to BChl *c* (Smith and Bobe, 1987; Bobe et al., 1990). This *meso*-alkylation shifts the absorption maxima another 20 nm into the red spectral region. The C20 methyl group is also derived from methionine via SAM.

IV. Degradation of the Bacteriochlorophylls

Besides the many unsolved questions concerning the biosynthesis of the BChls, the degradation of these pigments remains unsolved, too. So far it is known that the first steps in the catabolic pathway are connected to loss of magnesium, removal of the esterified alcohol at $C17^3$ and to oxidation reactions at ring V (Haidl et al., 1985). The enzymes of these steps remain unknown. For heme it is known that the ring system is broken up at a *meso*-bridge and then subsequently degraded (Brown et al., 1991). From model compounds it is known that C20-substituted BChls *c* undergo photochemical ring opening reactions at the C20 *meso*-bridge (Brockmann and Belter, 1979; Brown et al., 1980; Struck et al., 1990). The methyl group activates the molecule for such reactions and it was believed that introduction of a *meso*-substituent might be an essential part of Chl breakdown (Rüdiger and Schoch, 1989). Recent reports on Chl degradation show however that in plants the ring opening reaction occurs at the C5 *meso*-bridge (Kräutler et al., 1991; Engel et al., 1991). Thus the situation remains unclear for the BChls.

Both the catabolic and anabolic pathway remain largely unknown. Due to their easy cultivation and simple physiology, photosynthetic bacteria remain the premier objects for studies on photosynthesis. Thus they also serve as models for studying the photosynthesis in higher organisms. Therefore a more detailed analysis of their biochemistry is of central importance. A renewed effort with a combination of chemical, biological, and physiological methods might reveal more fascinating facets about these interesting compounds.

Acknowledgments

Work from UC Davis described herein was supported by grants from the National Science Foundation (CHE-93-05577) and the Deutsche Forschungsgemeinschaft (Se 543/2-1).

References

Ajaz AA, Corina DL and Akhtar M (1985) The mechanism of the $C\text{-}13^2$ esterification step in the biosynthesis of bacteriochlorophyll *a*. Eur J Biochem 150: 309–312

Akhtar M, Ajaz AA and Corina DL (1984) The mechanism of attachment of esterifying alcohol in bacteriochlorophyll *a* biosynthesis. Biochem J 224: 187–194

Andersen T, Briseid T, Nesbakken T, Ormerod J, Sirevåg R and Thorud M (1983) Mechanism of synthesis of 5-aminolevulinate in purple, green and blue-green bacteria. FEMS Microbiol Lett 19: 303–306

Avissar YJ, Ormerod JG and Beale SI (1989) Distribution of δ-aminolevulinic acid biosynthetic pathways among phototropic bacterial groups. Arch Microbiol 151: 513–519

Barkigia KM and Gottfried DS (1994) A new crystal form of methyl bacteriopheophorbide a. Acta Cryst C50: 2069–2072

Barkigia KM, Chantranupong L, Smith KM and Fajer J (1988) Structural and theoretical models of photosynthetic chromophores. Implications for redox, light absorption properties and vectorial electron flow. J Am Chem Soc 110: 7566–7567

Barkigia KM, Gottfried DS, Boxer SG and Fajer J (1989) A high precision structure of a bacteriochlorophyll derivative, methyl bacteriopheophorbide *a*. J Am Chem Soc 111: 6444–6446

Bauer CE, Bollivar DW and Suzuki JY (1993) Genetic analyses of photopigment biosynthesis in eubacteria: A guiding light for algae and plants. J Bacteriol 175: 3919–3925

Beale SI and Weinstein JD (1990) Tetrapyrrole Metabolism in Photosynthetic Organisms. In: Dailey HA (ed) Biosynthesis of Heme and Chlorophylls, pp 287–391. McGraw-Hill, New York

Beale SI and Weinstein JD (1991) Biosynthesis of 5-aminolevulinic acid in phototrophic organisms. In: Scheer H (ed) Chlorophylls, pp 385–406. CRC Press, Boca Raton

Biel N (1991) Characterization of a coproporphyrin-protein complex from *Rhodobacter capsulatus*. FEMS Microbiol Lett 81: 43–48

Bobe FW, Pfennig N, Swanson KL and Smith KM (1990) Red shift of absorption maxima in Chlorobiineae through enzymic methylation of their antenna bacteriochlorophyll. Biochemistry 29: 4340–4348

Bollivar DW and Bauer CE (1992) Association of tetrapyrrole intermediates in the bacteriochlorophyll *a* biosynthetic pathway with the major outer-membrane porin protein of *Rhodobacter capsulatus*. Biochem J 282: 471–476

Bollivar DW, Suzuki JY, Beatty JT, Dobrowlski J and Bauer CE (1994a) Directed mutational analysis of bacteriochlorophyll *a* biosynthesis in *Rhodobacter capsulatus*. J Mol Biol 237: 622–640

Bollivar DW, Wang S, Allen JP and Bauer CE (1994b) Molecular genetic analysis of terminal steps in bacteriochlorophyll *a* biosynthesis: Characterization of a *Rhodobacter capsulatus* strain that synthesizes geranylgeraniol esterified bacteriochlorophyll *a*. Biochemistry 33: 12763–12768

Bollivar DW, Jiang Z-Y, Bauer CE and Beale SI (1994c) Heterologous overexpression of the *bchM* gene product from *Rhodobacter capsulatus* and demonstration that it encodes for *S*-adenosyl-L-methionine:Mg-protoporphyrin methyltransferase. J Bacteriol 176: 5290–5296

Broch-Due M and Ormerod JG (1978) Isolation of a BChl-*c* mutant from *Chlorobium* with BChl-d by cultivation at low-light intensity. FEMS Microbiol Lett 3: 305–308

Brockmann H Jr (1976) Bacteriochlorophyll *e*: Structure and stereochemistry of a new type of chlorophyll from *Chlorobiaceae*. Philos Trans Roy Soc London, Ser B 273: 277–285

Brockmann H Jr (1978) Stereochemistry and Absolute Configuration of Chlorophylls and Linear Tetrapyrroles. In: Dolphin, D (ed) The Porphyrins, Vol II, pp 287–326. Academic Press, New York

Brockmann H Jr and Belter C (1979) Regioselektive Ringspaltung von Bakteriochlorophyll-Derivaten durch Photooxidation. Z Naturforsch 34b: 127–128

Brockmann H Jr and Kleber I (1970) Bacteriochlorophyll *b*. Tetrahedron Lett, 2195–2198

Brockmann H Jr and Lipinsky A (1983) Bacteriochlorophyll-G, a new bacteriochlorophyll from *Heliobacterium-chlorum*. Arch Microbiol 136: 17–19

Brockmann H Jr and Tacke-Karimdadian R (1979) Oxidativer Abbau von BChl *d*, Bestätigung der Konstitution und Bestimmung der absoluten Konfiguration. Liebigs Ann Chem, 419–430

Brown AE and Lascelles J (1972) Phytol and bacteriochlorophyll synthesis in *Rhodopseudomonas spheroides*. Plant Physiol 50: 747–749

Brown SB, Smith KM, Bisset GMF and Troxler RF (1980) Mechanism of photooxidation of bacteriochlorophyll *c* derivatives. A possible model for natural chlorophyll breakdown. J Biol Chem 255: 8063–8068

Brown SB, Houghton JD and Hendry GAF (1991) Chlorophyll breakdown. In: Scheer H (ed) Chlorophylls, pp 465–489. CRC Press, Boca Raton

Caple MB, Chow H and Strouse CE (1978) Photosynthetic Pigments of green sulfur bacteria. The esterifying alcohols of bacteriochlorophylls *c* from *Chlorobium limicola*. J Biol Chem 253: 6730–6736

Emery VC and Akhtar M (1985) Stereochemistry of the generation of the ethyl groups in bacteriochlorophyll *a* biosynthesis. J Chem Soc, Chem Commun, 1646–1647

Emery VC and Akhtar M (1987) Mechanistic studies on the phytylation step in bacteriochlorophyll *a* biosynthesis: An application of the ^{18}O induced isotope effect in ^{13}C NMR. Biochemistry 26: 1200–1208

Engel N, Jenny TA, Mooser V and Gossauer A (1991) Chlorophyll catabolism in *Chlorella protothecoides*. Isolation and structure elucidation of a red bilin derivative. FEBS Lett 293: 131-133

Fages FN, Griebenow N, Griebenow K, Holzwarth AR and Schaffner K (1990) Characterization of light-harvesting pigments of *Chloroflexus aurantiacus*. Two new chlorophylls: oleyl (octadec-9-enyl) and cetyl (hexadecanyl) bacteriochlorophyllides-c. J Chem Soc, Perkin Trans 1, 2791–2797

Gloe A, Pfennig N, Brockmann H Jr and Trowitzsch W (1975) A new bacteriochlorophyll from brown-colored Chlorobiaceae. Arch Microbiol 102: 103–109

Goff DA (1984) Synthetic and structural studies on the bacteriochlorophylls-*d*. Dissertation, University of California, Davis

Haidl H, Knödlmayer K, Rüdiger W, Scheer H, Schoch S and Ulrich J (1985) Degradation of bacteriochlorophyll *a* in *Rhodopseudomonas sphaeroides* R-26. Z Naturforsch 40c: 685–692

Holt AS, Purdie WS and Wasley JWF (1966) Structures of *Chlorobium* chlorophylls. Can J Chem 44: 88–93

Huster MS and Smith KM (1990) Biosynthetic studies of substituent homologation in bacteriochlorophylls *c* and *d*. Biochemistry 29: 4348–4355

IUPAC (1987) Nomenclature of Tetrapyrroles. Pure Appl Chem 59: 787–832

Jahn D, Verkamp E and Söll D (1992) Glutamyl-transfer RNA: A precursor of heme and chlorophyll biosynthesis. Trends Biochem Sci 17: 215–218

Jones OTG (1978) Biosynthesis of Porphyrins, Hemes, and Chlorophylls. In: Clayton RK and Sistrom WR (eds) The Photosynthetic Bacteria, pp 751–777. Plenum Press, New York

Jordan PM and Seehra JS (1980) Mechanism of action of 5-aminolevulinic acid dehydratase: Stepwise order of addition of the two molecules of 5-aminolevulinic acid in the enzymic synthesis of porphobilinogen. J Chem Soc Chem Commun, 240–242

Kenner GW, Rimmer J, Smith KM and Unsworth JF (1978) Structural and biosynthetic studies of the *Chlorobium* chlorophylls-660 (bacteriochlorophylls-*c*). Incorporation of methionine and porphobilinogen. J Chem Soc, Perkin Trans 1, 845–852

Kern M and Klemme JH (1989) Inhibition of bacteriochlorophyll biosynthesis by gabaculin (3-amino-2,3-dihydro-benzoic acid) and presence of an enzyme of the C_5-pathway of δ-aminolevulinate synthesis in *Chloroflexus aurantiacus*. Z Naturforsch 44c: 77–80

Kikuchi G, Kumar A, Talmage P and Shemin D (1958) The enzymatic synthesis of δ-aminolevulinic acid. J Biol Chem 233: 1214–1219

Kobayashi M, Watanabe T, Ikegami I, van de Meet EJ and Amesz J (1991) Enrichment of bacteriochlorophyll *g'* in membranes of *Heliobacterium chlorum* by ether extraction. Unequivocal evidence for its existence in vivo. FEBS Lett 284: 129–131

Kräutler B, Jaun B, Bortlik K, Schellenberg M and Matile P (1991) On the enigma of chlorophyll degradation. The constitution of a secoporphinoid catabolite. Angew Chem Int Ed Engl 30: 1315–1318

Leeper, FJ (1991) Intermediate steps in the biosynthesis of the chlorophylls. In: Scheer H (ed) Chlorophylls, pp 407–431. CRC Press, Boca Raton

Michalski TJ, Hunt JE, Bowman MK, Smith U, Bardeen K, Gest H, Norris JR and Katz JJ (1987) Bacteriopheophytin *g*–properties and some speculations on a possible primary role for Bacteriochlorophyll *b* and *g* in the biosynthesis of chlorophylls. Proc Natl Acad Sci USA 84: 2570–2574

Oh-hama T, Seto H and Miyachi S (1985) ^{13}C-Nuclear magnetic resonance studies on bacteriochlorophyll *a* biosynthesis in *Rhodopseudomonas spheroides*. Arch Biochem Biophys 237: 72–79

Oh-hama T, Seto H and Miyachi S (1986a) ^{13}C-NMR evidence of bacteriochlorophyll *a* formation by the C_5 pathway in Chromatium. Arch Biochem Biophys 246: 192–198

Oh-hama T, Seto H and Miyachi S (1986b) ^{13}C NMR evidence for bacteriochlorophyll *c* formation by the C_5-pathway in green sulfur bacterium, *Prosthecochloris*. Eur J Biochem 159: 189–194

Oh-hama T, Santander PJ, Stolowich NJ and Scott AI (1991) Bacteriochlorophyll *c* formation via the C_5 pathway of 5-aminolevulinic acid synthesis in *Chloroflexus aurantiacus*. FEBS Lett 281: 173–176

Ormerod JG, Nesbakken T and Beale SI (1990) Specific inhibition of antenna bacteriochlorophyll synthesis in *Chlorobium vibrioforme* by anaesthetic gases. J Bacteriol 172: 1352–1360

Olson JM, Amesz J, Ormerod JG and Blankenship RE (eds) (1994) Photosynth Res 41: 1–294 [Special issue on Green and Heliobacteria]

Pfennig N (1977) Phototrophic green and purple bacteria: A comparative study. Ann Rev Microbiol 31: 275–290

Pudek MR and Richards WR (1975) A possible alternate pathway of bacteriochlorophyll biosynthesis in a mutant of *Rhodopseudomonas spheroides*. Biochemistry 14: 3132–3137

Rieble S, Ormerod JG and Beale SI (1989) Transformation of glutamate to δ-aminolevulinic acid by soluble extracts of *Chlorobium vibrioforme*. J Bacteriol 171: 3782–3787

Risch N, Kemmer T and Brockmann H Jr (1978) Chromatographische Trennung und Charakterisierung der Bakteriomethylpheophorbide e. Liebigs Ann Chem, 585–594

Risch N, Küster B, Schormann A, Siemens T and Brockmann H Jr (1988) Bakteriochlorophyll *f*-Partialsynthese und Eigenschaften einiger Derivate. Liebigs Ann Chem, 343–347

Rüdiger W and Schoch S (1988) Chlorophylls. In: Goodwin GS (ed) Plant Pigments, pp 1–59. Academic Press, New York.

Rüdiger W and Schoch S (1989) Abbau von Chlorophyll. Naturwissenschaften 76: 453–457

Rüdiger W and Schoch S (1991) The last steps in chlorophyll biosynthesis. In: Scheer H (ed) Chlorophylls, pp 451–464, CRC Press, Boca Raton

Scheer H (1991) Structure and Occurrence of Chlorophylls. In: Scheer H (ed) Chlorophylls, pp 3–30. CRC Press, Boca Raton

Scheer H, Svec WA, Cope BT, Studier MH, Scott RG and Katz JJ (1974) Structure of Bacteriochlorophyll *b*. J Am Chem Soc 96: 3714-3716

Senge MO (1992a) Struktur und Biosynthese der Bakteriochlorophylle. Chem uns Zeit 26: 86–93

Senge MO (1992b) The conformational flexibility of tetrapyrroles –current model studies and photobiological relevance. J Photochem Photobiol B:Biol 16:3–36

Senge MO (1993) Recent advances in the biosynthesis and chemistry of the chlorophylls. Photochem Photobiol 57: 189–206

Senge MO and Smith KM (1994) Structure and conformation of photosynthetic pigments and related compounds. 7. On the conformation of the methyl ester of (20-methyl-phytochlorinato)nickel(II)-A bacteriochlorophyll *c* model compound. Phtochem Photobiol 60: 139–142

Senge MO, Bobe FW, Huster MS and Smith KM (1991) Preparation and Crystal Structure of Methyl[12-Acetyl-8-ethyl]-bacteriopheophorbide *d*. A New Bacteriochlorophyll Derivative. Liebigs Ann Chem, 871–874

Simpson DJ and Smith KM (1988) Structures and transformations of the bacteriochlorophylls *e* and their bacteriopheophorbides. J Am Chem Soc 110: 1753–1758

Smith KM (1991) The structure and biosynthesis of bacteriochlorophylls. In: Jordan PM (ed) Biosynthesis of Tetrapyrroles, pp 237–255. Elsevier Science Publ, Amsterdam

Smith KM (1994) Nomenclature of the bacteriochlorophylls *c*, *d* and *e*. Photosynth Res 41: 23–26

Smith KM and Bobe FW (1987) Light adaptation of bacteriochlorophyll-*d* producing bacteria by enzymic methylation of their antenna pigments. J Chem Soc, Chem Commun, 276–277

Smith KM and Goff DA (1985) Bacteriochlorophylls *d* from *Chlorobium vibrioforme*: Chromatographic separations and structural assignments of the methyl bacteriopheophorbides. J Chem Soc, Perkin Trans. 1, 1099–1113

Smith KM and Huster MS (1987) Bacteriochlorophyll-*c* formation via the glutamate C-5 pathway in *Chlorobium* bacteria. J Chem Soc, Chem Commun, 14–16

Smith KM and Simpson DJ (1987) Site-specific reduction of unsymmetrically substituted porphyrins to give isomerically pure chlorins. J Chem Soc, Chem Commun, 613–614

Smith KM, Goff DA, Fajer J and Barkigia KM (1982) Chirality and structures of bacteriochlorophylls *d*. J Am Chem Soc 104: 3747–3749

Smith KM, Goff DA, Fajer J and Barkigia KM (1983a) Isolation and characterization of two new bacteriochlorophyll *d* bearing neopentyl substituents. J Am Chem Soc 105: 1674–1676

Smith KM, Craig GW, Kehres LA and Pfennig N (1983b) Reversed phase HPLC and structural assignments of the bacteriochlorophylls *c*. J Chromatogr 281: 209–223

Smith KM, Kehres LA and Fajer J (1983c) Aggregation of the bacteriochlorophylls *c*, *d*, and *e*. Models for the antenna chlorophylls of green and brown photosynthetic bacteria. J Am Chem Soc 105: 1387–1389

Struck A, Cmiel E, Schneider S and Scheer H (1990) Photochemical ring-opening in *meso*-chlorinated chlorophylls. Photochem Photobiol 51, 217–222

Swanson, KL and Smith KM (1990) Biosynthesis of bacteriochlorophyll-*c* via the glutamate C-5 pathway in *Chloroflexus aurantiacus*. J Chem Soc, Chem Commun, 1696–1697

Tait GH (1972) Coproporphyrinogenase activities in extracts of *Rhodopseudomonas spheroides* and *Chromatium* strain D Biochem J 128: 1159–1169

Van de Meet EJ, Kobayashi M, Erkelens C, van Veelen PA, Amesz J and Watanabe T (1991) Identification of 8^1-hydroxychlorophyll *a* as a functional reaction center pigment in heliobacteria. Biochim Biophys Acta 1058: 356–362

van Heyningen S and Shemin D (1971) Quarternary structure of δ-aminolevulinate dehydratase from *Rhodopseudomonas spheroides*. Biochemistry 10: 4676–4682

Walker CJ, Mansfield KE, Smith KM and Castelfranco PA (1989) Incorporation of atmospheric oxygen into the carbonyl functionality of the protochlorophyllide isocyclic ring. Biochem J 257: 599–602

Chapter 9

Biosynthesis and Structures of Porphyrins and Hemes

Samuel I. Beale
Division of Biology and Medicine, Brown University, Providence, RI 02912, USA

Summary

Anoxygenic phototrophs are capable of synthesizing all three major types of tetrapyrroles: hemes, chlorophylls, and corrins. These molecules are derived from a common, branched biosynthetic pathway. Facultative aerobes in the α-proteobacterial group (purple nonsulfur bacteria), which includes the genera *Rhodobacter*, *Rhodopseudomonas*, and *Rhodospirillum*, have been used extensively as model systems for studying tetrapyrrole biosynthesis, and much is known about the biosynthetic pathway and the physical and regulatory properties of the enzymes in these organisms. Strictly anaerobic phototrophic species have been less thoroughly studied, but they are known to derive their tetrapyrroles from metabolic precursors that differ from those used by the facultative aerobes. This chapter describes the structures and biosynthesis of tetrapyrrole precursors, hemes, and related molecules.

R. E. Blankenship, M. T. Madigan and C. E. Bauer (eds): Anoxygenic Photosynthetic Bacteria, pp. 153–177.

I. Introduction

A. The Tetrapyrroles of Anoxygenic Phototrophic Bacteria

Tetrapyrroles are ubiquitous molecules that play essential roles in energy metabolism. The tetrapyrrole family includes hemes, chlorophylls, and structurally related compounds. In the anoxygenic phototrophs, tetrapyrroles are used in the fundamental roles of light absorption and electron transfer through which these organisms achieve phototrophy.

The anoxygenic phototrophs are versatile tetrapyrrole synthesizers: along with the ability to produce hemes and chlorophylls, a property which they share with plants, algae, and cyanobacteria, some of these organisms are capable of forming the corrin nucleus of coenzyme B_{12}. Moreover, these procaryotes contain several heme types that are not found in eucaryotic organisms. This chapter is concerned specifically with the structures and biosynthesis of hemes in anoxygenic phototrophs. However, inasmuch as all tetrapyrroles share a common biosynthetic pathway, much of the material covered here is applicable to the biosynthesis of other tetrapyrrole end products.

B. Outline of the General Tetrapyrrole Biosynthetic Pathway

All biological tetrapyrroles can be arranged as products of a single, branched biosynthetic pathway (Fig. 1). The biosynthetic steps from early precursors to protoporphyrin-based hemes constitute the major, common portion of the pathway, and other steps leading to specific groups of products can be considered to be branches off the common pathway. The pathway is a highly conserved one and, with few exceptions, the biosynthetic intermediates and enzyme-catalyzed reactions are very similar or identical in all organisms where they have been studied. The existence of a branched pathway with several categories of end products indicates the need for complex regulation to ensure that the products are synthesized in appropriate proportions in response to changing environmental and growth conditions.

Many major discoveries in the area of tetrapyrrole biosynthesis have been made using one group of anoxygenic phototrophs, the α-proteobacteria (nonsulfur purple bacteria), as model systems. For example, *Rhodobacter sphaeroides* and *Rhodospirillum rubrum* were among the first organisms of any kind to yield in vitro enzyme activities for early steps of the pathway. Compared to the α-proteobacteria, other groups of anoxygenic phototrophs are much less well characterized with respect to their tetrapyrrole biosynthetic capabilities. It is important to be aware of the different extent of knowledge about tetrapyrrole biosynthesis in the different groups of organisms and to avoid premature generalizations, since, as will be described below, there are some major differences in the actual biosynthetic steps used by the different organisms for producing the same end product.

Anoxygenic phototrophs, like other phototrophic organisms, are able to regulate the content and composition of their tetrapyrroles, especially those which are components of their photosynthetic apparatus, in response to environmental signals such as light intensity and spectral composition, nutritional status, and O_2 tension. Several of these topics are covered in Chapter 52 byBiel. In the present chapter, discussion of regulation will be primarily concerned with the regulatory properties of individual biosynthetic enzymes.

II. The Early Common Steps

A. Biosynthesis of ALA

δ-Aminolevulinic acid (ALA) can be considered to be the first universal, committed tetrapyrrole precursor (Fig. 1). Discussion of ALA formation is complicated by the fact that two different biosynthetic routes to ALA exist. The α-proteobacteria include the facultative phototrophic genera *Rhodopseudomonas*, *Rhodobacter* and *Rhodospirillum*, obligately aerobic bacteriochlorophyll-(BChl)-containing phototrophs that have been placed in the *Erythrobacter* and *Methylobacterium* groups, and non-chlorophyllic relatives such as *Agrobacterium*, *Rhizobium*, *Azorhizobium*, and *Bradyrhizobium*. These organisms, as well as eucaryotes that do not contain plastids (e.g., animals, yeasts, fungi), form ALA from glycine and succinate (Fig. 2). Plants, algae, and most groups

Abbreviations: ALA – δ-aminolevulinic acid; BChl – bacteriochlorophyll; *Cb.* – *Chlorobium; Cf.* – *Chloroflexus; Cm.* – *Chromatium; E. coli* – *Escherichia coli;* gabaculine – 3-amino-2,3-dihydrobenzoic acid; GSA – L-glutamate-1-semialdehyde; PBG – porphobilinogen; *Rb.* – *Rhodobacter; Rc.* – *Rhodocyclus;* RNase A – Ribonuclease A; *Rp.* – *Rhodopseudomonas; Rs.* – *Rhodospirillum*

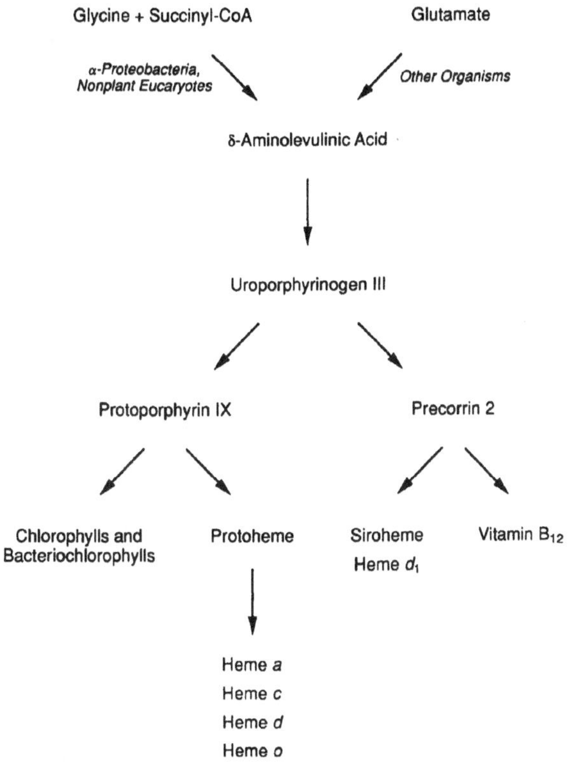

Fig. 1. Outline of tetrapyrrole biosynthesis.

of bacteria, including cyanobacteria, the strictly anaerobic phototrophs (green sulfur bacteria, green nonsulfur bacteria, and heliobacteria), and archaebacteria, form ALA by a different route (Fig. 3), wherein glutamate, rather than glycine and succinate, provides the C and N atoms.

1. The Glycine/Succinyl-CoA Route

Rb. sphaeroides and *Rs. rubrum* were among the first organisms of any kind to yield in vitro ALA-forming activity (Gibson, 1958; Kikuchi et al., 1958; Sawyer and Smith, 1958). ALA is formed in these cells by the condensation of succinyl-CoA and glycine, mediated by the pyridoxal phosphate-requiring enzyme, ALA synthase (succinyl-CoA:glycine C-succinyltransferase (decarboxylating) EC 2.3.1.37) (Fig. 2). In the reaction, the carboxyl carbon of glycine is lost as CO_2 and the remainder is incorporated into the ALA.

ALA synthase has been highly purified from *Rb. sphaeroides* (Warnick and Burnham, 1971; Fanica-Gaignier and Clement-Metral, 1973b; Nandi and Shemin, 1977). The native enzyme has an apparent molecular weight of 80,000 to 100,000 and is a homodimer consisting of two 41,000–45,000 molecular weight subunits. A native molecular weight of 61,000–65,000 was derived for ALA synthase from *Rhodopseudomonas palustris* (Viale et al., 1987a). One mole of pyridoxal-phosphate is bound per subunit of the *Rb. sphaeroides* ALA synthase (Davies and Neuberger, 1979b), but the relatively loose binding (the apparent K_m was calculated to be 3–5 μM), and the fact that pyridoxal-phosphate protects an essential thiol group, suggest that the cofactor is not bound covalently as a Schiff base, but instead may be complexed with a cysteine thiol. The existence of two distinct ALA synthases, called Fraction I and Fraction II, has been demonstrated in *Rb. sphaeroides* (Tuboi et al., 1970a,b; Fanica-Gaignier and Clement-Metral, 1973a,b). Fraction I and Fraction II are both repressed under high O_2

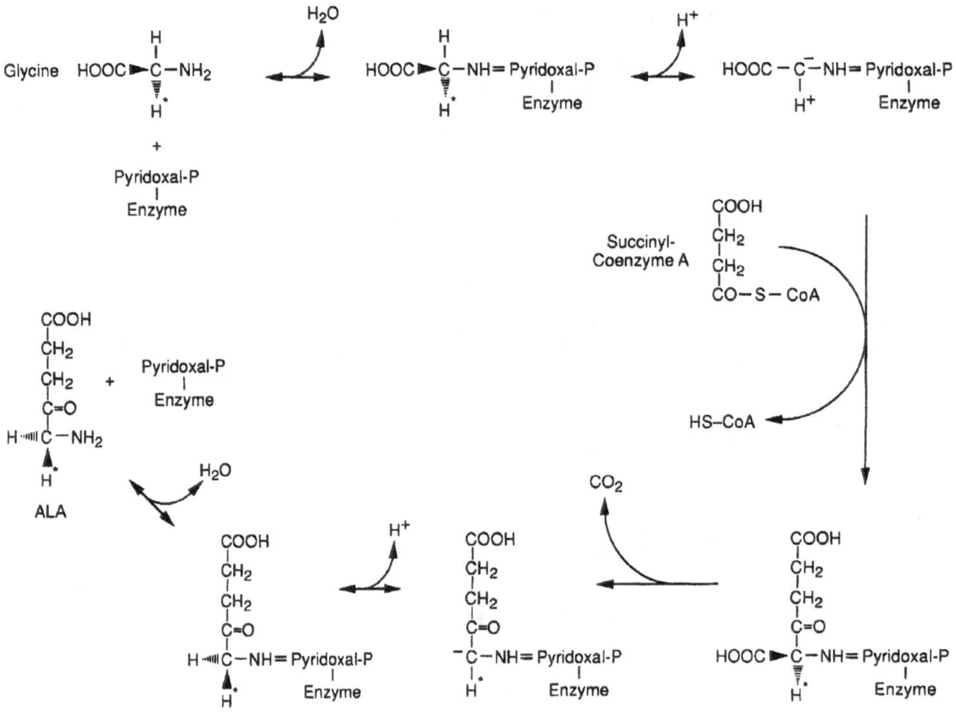

Fig. 2. ALA biosynthesis from glycine and succinyl CoA catalyzed by ALA synthase.

tension. Reduction of O_2 tension induces Fraction I, and induction of Fraction II requires both low O_2 tension and light (Tuboi et al., 1970). It has been reported that one form of ALA synthase is cytoplasmic and is functionally associated with dark metabolism, and the other is found in the chromatophores, is specifically induced in the light, and is functionally associated with photometabolism (Fanica-Gaignier and Clement-Metral, 1973a). The two enzyme forms may be encoded by different genes (see below).

The ALA synthase reaction mechanism has been extensively studied. Product inhibition studies with the *Rb. sphaeroides* enzyme indicate a bi-bi reaction sequence in which glycine binds first and ALA is released last (Fanica-Gaignier and Clement-Metral, 1973c). During the reaction, the glycine α-proton having the *pro-R* configuration is specifically removed and the *pro-S* proton occupies the pro-*S* position at C-5 of ALA (Zaman et al., 1973; Abboud et al., 1974). The retention of only one glycine α-proton in the product, and its stereospecific position in the product, indicate that the reaction of pyridoxal-bound glycine carbanion with succinyl-CoA precedes decarboxylation, i.e., that enzyme-bound α-amino-β-ketoadipic acid is an intermediate, and that the decarboxylation step occurs before the final product is released from the enzyme, which means that ALA synthase is a decarboxylase as well as a condensing enzyme. In the absence of succinyl-CoA, the enzyme catalyzes the exchange of the glycine *pro-R* α-proton with the solvent (Laghai and Jordan, 1976), and in the absence of either substrate, the enzyme catalyzes exchange of one of the C-5 protons of ALA with the solvent (Laghai and Jordan, 1977). Protoheme, a powerful allosteric inhibitor (see below), has little effect on the exchange reaction with glycine but does affect the exchange reaction with ALA.

Genes that encode ALA synthase have been cloned and sequenced from *Rhodobacter capsulatus* (Biel et al., 1988; Hornberger et al., 1990) and *Rb. sphaeroides* (Tai et al., 1988; Neidle and Kaplan,

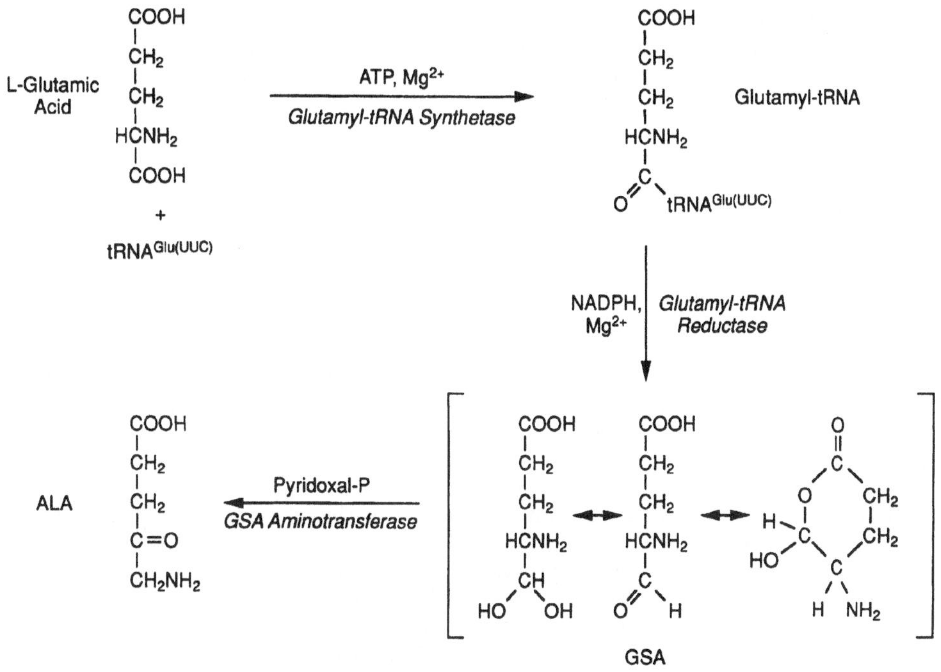

Fig. 3. ALA biosynthesis from glutamate via the tRNA-dependent five-carbon pathway. The intermediate, GSA, is shown in linear, hydrated, and cyclic forms, which probably coexist as an equilibrium mixture in solution.

1993a). Whereas *Rb. capsulatus* contains a single ALA synthase-encoding gene, *hemA*, *Rb. sphaeroides* contains two such genes, *hemA* and *hemT*, which respectively are located on the large and small chromosomes of these cells (Neidle and Kaplan, 1993b). The *Rb. capsulatus hemA* sequence predicts a 401 amino acid, 44,100 molecular weight peptide. The *Rb. sphaeroides hemA* and *hemT* genes encode peptides that are 53% identical to each other, and each contains 407 amino acids. The *Rb. capsulatus* and *Rb. sphaeroides* HemA peptides are 76% identical. These bacterial ALA synthase peptides are also significantly similar to ALA synthase peptides from other bacterial and eucaryotic species. Interestingly, there is detectable similarity to the peptide sequences of two other enzymes that catalyze similar condensation reactions, 7-keto-8-amino-pelargonic acid synthase (BioF) and 2-amino-3-ketobutyrate coenzyme A ligase (EC 2.3.1.29) (Neidle and Kaplan, 1993a). It has been proposed (Tai et al., 1988), but not directly proven, that the *hemA* and *hemT* gene products correspond to the two types of *Rb. sphaeroides* ALA synthase discussed above.

As might be expected for an enzyme that catalyzes the first committed step of tetrapyrrole biosynthesis, ALA synthase is highly regulated. Regulation is exerted at both the enzyme (allosteric activity modulation) and gene expression (transcriptional control) levels. *Rb. sphaeroides* ALA synthase is reversibly inhibited by protoheme at physiologically relevant concentrations (Burnham and Lascelles, 1963; Yubisui and Yoneyama, 1972). Greater than 50% inhibition was caused by 1 μM protoheme, and nearly complete inhibition was achieved at 10 μM protoheme. The inhibition appeared to be competitive with glycine. Unnatural protoheme analogs such as hematoheme, deuteroheme, and dimethylprotoheme also inhibited at concentrations comparable to that of protoheme (Yubisui and Yoneyama, 1972). Other physiological tetrapyrroles such as protoporphyrin IX and Mg-protoporphyrin IX inhibited at higher concentrations (Burnham and Lascelles, 1963; Yubisui and Yoneyama, 1972), but it is doubtful that these tetrapyrroles ever reach inhibitory concentrations in vivo.

Rb. sphaeroides ALA synthase is also inhibited by

ATP and inorganic pyrophosphate at physiologically relevant concentrations; purified enzyme was inhibited 60–88% by 1 mM ATP (Fanica-Gaignier and Clement-Metral, 1971; 1973c). The inhibition was competitive with glycine. It has been noted that exogenous ATP inhibits BChl formation in *Rb. sphaeroides*, and the suggestion was made that ATP may be a general repressor of the tetrapyrrole biosynthesis in this species (Fanica-Gaignier and Clement-Metral, 1971).

In addition to the inhibitory responses noted above, *Rb. sphaeroides* ALA synthase activity has been shown to be influenced by certain thiol-rich compounds. Initially, it was reported that fresh cell extracts from phototrophically-grown cells had low ALA synthase activity and that activity spontaneously increased upon preliminary purification (Kikuchi et al., 1958) or exposure of the extract to air (Marriott et al., 1970). It was suggested that activation requires the oxidation of some unidentified inhibitor. Disulfide-containing compounds, such as cystine, oxidized glutathione, and lipoic acid, were shown to activate ALA synthase in vitro (Tuboi and Hayasaka, 1972). Later, trisulfides such as cystine trisulfide and the mixed trisulfide of glutathione and cystine, but not disulfides, were reported to activate the enzyme (Sandy et al., 1975) and to supplant air activation (Davies and Neuberger, 1979a). In contrast to the activating effect of O_2 on ALA synthase in vitro, exposure of whole cells to O_2 reduces the level of extractable ALA synthase. Cellular levels of trisulfide compounds were found to decrease upon oxygenation, and the decrease was proposed to be responsible for the decrease in extractable active ALA synthase upon oxygenation of the cells. Finally, proteinaceous activators, such as cystathionase (EC 4.4.1.1) (Inoue et al., 1979; Oyama and Tuboi, 1979) and reduced thioredoxin (Clement-Metral, 1979) were shown to activate ALA synthase. It remains to be determined whether the influence of the sulfhydryl environment on ALA synthase activity is of importance in the control of BChl formation by O_2 tension.

Rb. sphaeroides and *Rs. rubrum* cells accumulate significantly more BChl when grown at low light intensity than at high intensity (Cohen-Bazire et al., 1957; Arnheim and Oelze, 1983; Oelze and Arnheim, 1983). It seems probable that the influence of light intensity on BChl content is an indirect one, perhaps mediated by effects of light on BChl stability (Biel, 1986) or on the rate of ALA synthesis via feedback inhibition, cellular ATP concentration, and/or thiol oxidation/reduction state. However, there has been one unconfirmed report that light directly inhibits *Rb. sphaeroides* ALA synthase in vitro (Oelze, 1986).

Relatively little is known about the regulation of the expression of ALA synthase-encoding genes. These genes are not located within the major photosynthetic gene clusters (Hornberger et al., 1990). ALA synthase is induced upon transfer of dark-grown, aerobic *Rb. sphaeroides* cells to anaerobic, light conditions (Kikuchi et al., 1958). Exogenous protoheme inhibits this induction (Goto et al., 1967). Oxygen completely blocks BChl formation in *Rb. sphaeroides* (Cohen-Bazier et al., 1957). Oxygenation of anaerobically-grown *Rb. sphaeroides* (Lascelles, 1960) and *Rp. palustris* (Viale et al., 1987b) cells diminished the level of extractable ALA synthase by approximately 70%. *Rb. capsulatus hemA-lacZ* fusion expression was inhibited by factors of 2–3 by high O_2 tension. The relatively small effect of O_2 on ALA synthase levels and the fusion expression suggests that transcriptional regulation of *hemA* is not a major factor in O_2-mediated control of tetrapyrrole biosynthesis in this species (Wright et al., 1991). In *Rb. sphaeroides*, disruption of both ALA synthase-encoding genes, *hemA* and *hemT*, was required to produce ALA auxotrophy (Neidle and Kaplan, 1993b). However, disruption of either *hemA* or *hemT* alone was sufficient to cause decreased cellular contents of BChl, carotenoids, and light-harvesting BChl-proteins, as well as lowered levels of *puc* and *puf* transcripts. Somewhat paradoxically, *hemT* transcripts could not be detected in wild-type cells that had an intact, expressible *hemA* gene, although *hemT* transcripts were present in *hemA* mutant cells (Neidle and Kaplan, 1993a). In wild-type and *hemT* mutants, the *hemA* transcript level was 3 times higher under phototrophic conditions than during aerobic conditions.

In contrast to facultative aerobes such as *Rb. sphaeroides* and *Rb. capsulatus* which form BChl only in the absence of O_2, several obligately aerobic BChl-containing species have been described (see Chapters 6 and 7). These include the marine species *Erythrobacter longus* (Harashima et al., 1978) and *Roseobacter denitrificans*. OCh114 (Harashima et al., 1980), the methylotrophs *Methylobacterium (Protomonas) ruber* (Sato, 1978) and *M. (Pseudomonas) radiora* (Nishimura et al., 1981), and a BChl-containing *Bradyrhizobium*-like organism, strain BTAi 1 (Evans et al., 1990). These organisms are all members of the α-proteobacterial group (Kawasaki et al., 1993) and all form ALA via the ALA synthase route (Sato et al., 1981; Shioi and Doi, 1988). Whereas O_2 prevents BChl synthesis in the facultatively aerobic species, the obligate aerobes require O_2 for BChl synthesis as well as growth (Chapters 6 and 7). In the methylotrophs, BChl is formed in the dark, but not under continuous illumination. After a shift from light to dark, the ALA synthase level increased in parallel with the rate of BChl synthesis (Sato et al., 1985). However, detectable ALA synthase activity was present in light-grown cells, even though BChl was not formed. The cells contain two ALA synthase isoenzymes, one of which is constitutive and the other is induced in the dark (Sato et al., 1985). Although the marine *Erythrobacter* species, like the methylotrophs, synthesize BChl only in the dark, the ALA synthase level does not correlate with BChl synthesis, and is actually higher in the light than in the dark (Shioi and Doi, 1988).

2. The Five-Carbon Pathway

Until the early 1970s, the only known route of ALA formation was *via* condensation of succinyl-CoA and glycine. The first evidence indicating an alternative biosynthetic route was the preferential incorporation of ^{14}C from exogenous five-carbon precursors, such as glutamate, and poorer incorporation of ^{14}C from glycine and succinate, into ALA by greening intact plant tissues that had been treated with the ALA analog, levulinic acid, which causes ALA to accumulate in vivo by inhibiting porphobilinogen (PBG) synthase (Beale and Castelfranco, 1974a, 1974b; Meller et al., 1975). Degradation studies indicated that the glutamate carbon skeleton was incorporated intact into ALA (Beale et al., 1975; Meller et al., 1975). Shortly after the initial reports on the plant studies appeared, similar results were reported for cyanobacteria (Meller and Harel, 1978; Avissar, 1980; Kipe-Nolt and Stevens, 1980). Later, incorporation of glutamate into ALA or BChl was shown to occur in the anoxygenic phototrophs *Chromatium vinosum* (Oh-hama et al., 1986a), *Prosthecochloris aestuarii* (Oh-hama et al., 1986b), *Chlorobium vibrioforme* (Andersen et al., 1983; Smith and Huster, 1987), and *Chloroflexus aurantiacus* (Swanson and Smith, 1990; Oh-hama et al., 1991).

Particulate-free cell extracts capable of converting glutamate to ALA were first obtained from plants (Gough and Kannangara, 1977) and algae (Wang et al., 1984; Weinstein and Beale, 1985a; Mayer et al., 1987; Breu and Dörnemann, 1988). Later, active cell-free preparations were obtained from cyanobacteria (Rieble and Beale, 1988; O'Neill et al., 1988) and the prochlorophyte *Prochlorothrix hollandica* (Rieble and Beale, 1988). Similar or identical reaction mechanisms appear to operate in all cases, and reaction components from some heterologous sources can be mixed to reconstitute activity in fractionated systems.

In a survey of ALA biosynthetic routes in anoxygenic phototrophs, in vitro RNA-dependent conversion of glutamate to ALA was measured in extracts of a green sulfur bacterium (*Cb. vibrioforme*), a green nonsulfur bacterium (*Cf. aurantiacus*), a Heliobacterium (*Heliospirillum* sp.), a phototrophic β-proteobacterium (*Rhodocyclus gelatinosus*), and a phototrophic γ-proteobacterium (*Cm. vinosum*) (Avissar et al., 1989). In contrast, ALA synthase-catalyzed condensation of glycine and succinyl-CoA was detected only in extracts from members of the α-proteobacterial group (*Rs. rubrum*, *Rb. sphaeroides*). Thus the five-carbon ALA biosynthetic route is the predominant one in anoxygenic phototrophs.

The currently accepted model for the transformation of glutamate to ALA requires three enzyme-catalyzed reactions (Fig. 3). In the first, glutamate is activated by ligation to tRNA in a reaction identical to the charging reaction in protein biosynthesis. Like aminoacyl-tRNA formation in general, this reaction requires ATP and Mg^{2+}. Next, the tRNA-bound glutamate is converted to a reduced form in a reaction that requires a reduced pyridine nucleotide. The product of this reduction has been characterized as glutamate-1-semialdehyde (GSA) (Hoober et al., 1988), the hydrated hemiacetal form of GSA (Hoober et al., 1988), or a cyclized form of GSA (Jordan et al., 1993). Finally, the positions of the nitrogen and oxo atoms of the reduced five-carbon intermediate are interchanged to form ALA. Consistent with this reaction model, the ALA-forming systems extracted from plants (Bruyant and Kannangara, 1987), algae (Weinstein et al., 1987), and cyanobacteria (Rieble and Beale, 1991) have been separated into four macromolecular components, all of which must be present to catalyze in vitro ALA formation from glutamate. Three of these components are enzymes and the fourth is a low molecular weight RNA.

Most of the available information on the five-carbon pathway has been derived from plants, algae, cyanobacteria, and three nonphotosynthetic bacteria that have well understood genetic systems, *Escherichia coli*, *Salmonella typhimurium*, and *Bacillus subtilis*. However, one anoxygenic phototroph, *Cb. vibrioforme*, has been examined in some detail: the overall properties of in vitro ALA formation have been described (Rieble et al., 1989) and a *Cb. vibrioforme* gene that encodes a key enzyme has been cloned (Avissar and Beale, 1990), expressed in *E. coli*, and sequenced (Majumdar et al., 1991). The results indicate that the *Cb. vibrioforme* system is similar to those that have been examined in more detail from other organisms that use the five-carbon pathway.

a. $tRNA^{Glu}$

ALA formation in extracts of organisms that form ALA from glutamate is blocked by preincubation with ribonuclease (RNase) A (Kannangara et al., 1984; Breu and Dörnemann, 1988; Huang et al., 1984; Weinstein and Beale, 1985b; Avissar et al., 1989; Rieble et al., 1989). Addition of the RNase inhibitor, RNasin, plus low molecular weight RNA from the same species restores activity. In all species that have been examined, the RNA required for ALA formation is $tRNA^{Glu(UUC)}$ (Schön et al., 1986; Schneegurt and Beale, 1988). The same tRNA is used for both ALA and protein synthesis. The first anticodon base of $tRNA^{Glu}$ from *E. coli* cells and barley chloroplasts is highly modified, to 5-methylaminomethyl-2-thiouridine (Schön et al., 1986). In *E. coli*, modification of this base is important for efficient charging with glutamate by glutamyl-tRNA synthetase (Sylvers et al., 1993). It has been proposed that the modified base may have regulatory significance, because oxidation of the 5-methyl-aminomethyl-2-thiouridine with I_2 inactivated the tRNA for both glutamate acceptance and ALA synthesis, and subsequent reduction with thiosulfate reactivated it (Kannangara et al., 1988). However, treatment of the tRNA derived from dark-grown barley leaves with thiosulfate did not increase its activity.

b. Glutamyl-tRNA Synthetase

The same glutamyl-tRNA synthetase (EC 6.1.1.17) is used to charge $tRNA^{Glu}$ for both protein and ALA synthesis (Bruyant and Kannangara, 1987; Rieble and Beale, 1991). Glutamyl-tRNA synthetase from *E. coli* (Lapointe and Söll, 1972) and *B. subtilis* (Proulx et al., 1983) is a monomeric enzyme of 56,000 and 63,000 molecular weight, respectively, and is encoded by the *gltX* gene. These enzymes co-purify with a 46,000 molecular weight regulatory peptide. Glutamyl-tRNA synthetase from *E. coli* and *B. subtilis* contains a tightly-bound Zn^{2+} atom that is required for native conformation and catalytic activity (Liu et al., 1993). The aminoacylation reaction requires ATP and Mg^{2+}. Although *Chlamydomonas reinhardtii* glutamyl-tRNA synthetase is inhibited by heme under some conditions (Chang et al., 1990), and the *Scenedesmus obliquus* enzyme is inhibited by protochlorophyllide (Dörnemann et al., 1989), the *Synechocystis* sp. PCC 6803 glutamyl-tRNA synthetase is insensitive to heme or protochlorophyllide at physiologically relevant concentrations (Rieble and Beale, 1991). Moreover, it was determined that *Synechocystis* sp. PCC 6803 $tRNA^{Glu}$ was always fully acylated in vivo, and that the cellular content of $tRNA^{Glu}$ was a constant fraction of the total cellular tRNA population under growth conditions in which chlorophyll content was modulated over a 10-fold range (O'Neill and Söll, 1990). It therefore seems unlikely that either glutamyl-tRNA synthetase or $tRNA^{Glu}$ has a regulatory role in tetrapyrrole biosynthesis.

c. Glutamyl-tRNA Reductase

Glutamyl-tRNA reductase catalyzes NADPH-linked reduction of tRNA-activated glutamate to GSA, which is the first committed step unique to ALA formation from glutamate. For technical reasons, glutamyl-tRNA reductase is usually measured in coupled enzyme assays where the substrate is generated in vitro from glutamate plus tRNA, and/or the product is converted *in situ* to ALA by GSA aminotransferase. However, glutamyl-tRNA reductase is physically separable from the other enzyme components by affinity chromatography (Weinstein et al., 1987; Rieble and Beale, 1991). Affinity-purified glutamyl-tRNA reductase is active in GSA formation from glutamyl-tRNA in the absence of glutamyl-tRNA synthetase and GSA aminotransferase (Rieble and Beale, 1991). However, there is some evidence that glutamyl-tRNA synthetase and glutamyl-tRNA reductase may form an association. In the presence of glutamyl-tRNA, *C. reinhardtii* glutamyl-tRNA synthetase and glutamyl-tRNA reductase form a complex that migrates as a single entity on glycerol gradient centrifugation (Jahn, 1992). A complex between the two enzymes may facilitate the channeling of glutamyl-tRNA toward ALA biosynthesis and competition with the protein synthesizing apparatus for glutamyl-tRNA.

Affinity-purified glutamyl-tRNA reductase from *Synechocystis* sp. PCC 6803 and *Chlorella vulgaris* requires a divalent metal such as Mg^{2+}, Mn^{2+}, or Ca^{2+} for activity, with optimum activity occurring at 15 mM Mg^{2+} (Mayer et al., 1993b).

It is of interest that glutamyl-tRNA reductase from a given source can use, as substrate, glutamyl-tRNA from some, but not other, sources. For example, the barley and *C. vulgaris* enzymes can use plant and algal, but not *E. coli*, yeast, or animal glutamyl-tRNA (Kannangara et al., 1984; Weinstein et al., 1986). On the other hand, the reductases from *C. reinhardtii* and *Cb. vibrioforme* can use *E. coli* glutamyl-tRNA (Huang and Wang, 1986; Rieble et al., 1989).

Native glutamyl-tRNA reductase isolated from different sources appears to have widely divergent physical properties. The *C. reinhardtii* enzyme was reported to be a monomer with a molecular weight of 130,000 (Chen et al., 1990). In contrast, *E. coli* appears to contain two different glutamyl-tRNA reductases with molecular weights of 45,000 and 85,000 (Jahn et al., 1991). *B. subtilis* glutamyl-tRNA reductase migrated as an oligomer with an apparent molecular weight of 230,000 (Schröder et al., 1992).

Glutamyl-tRNA reductase is encoded by the *hemA* gene in organisms that use the five-carbon pathway. Mutation of the *E. coli hemA* gene results in a deficiency of glutamyl-tRNA reductase (Avissar and Beale, 1989a). The tRNA substrate specificity of glutamyl-tRNA reductase in complemented *hemA* strains of *E. coli* resembled that of the species from which the complementing DNA was derived, which indicates that the *hemA* gene encodes a structural component of glutamyl-tRNA reductase (Avissar and Beale, 1990; Majumdar et al., 1991). The glutamyl-tRNA reductase-encoding *hemA* gene from most sources encodes a peptide with a molecular weight of approximately 45,000, although the *B. subtilis* gene product is somewhat larger at 50,800 molecular weight (Petricek et al., 1990). The *Cb. vibrioforme* predicted HemA peptide has a molecular weight of 46,174 and is 30% identical to that of *B. subtilis* (Majumdar et al., 1991).

Glutamyl-tRNA reductase from many sources is allosterically inhibited by physiologically relevant concentrations of protoheme. The *Cb. vibrioforme* enzyme was inhibited 50% by 0.2 μM protoheme (Rieble et al., 1989). Protoheme inhibition is likely to be significant in regulating the rate of ALA formation in response to the cellular demand for end-product tetrapyrroles. The sensitivity of *C. vulgaris* glutamyl-tRNA reductase to protoheme inhibition is increased several fold by physiologically relevant concentrations of GSH (Weinstein et al., 1993).

d. GSA

GSA has been chemically synthesized by several methods for use as a substrate for the enzyme-catalyzed conversion to ALA (Kannangara and Gough, 1978; Houen et al., 1983; Gough et al., 1989). Material identical to chemically synthesized GSA accumulates in greening leaves (Wang et al., 1981; Kannangara and Schouboe, 1985) and algal extracts (Breu and Dörnemann, 1988) that have been treated with 3-amino-2,3-dihydrobenzoic acid (gabaculine), a mechanism-based suicide inhibitor of ω-aminotransferases (see below) that blocks chlorophyll synthesis.

Jordan et al. (1993) have investigated the structure of GSA in aqueous solution by NMR and mass spectroscopy and concluded that it exists as the cyclic ester between the γ-carboxyl group and the hydrated aldehyde group, rather than the free aldehyde (Fig. 3). The cyclic structure does not contain free aldehyde or carboxylic acid functions, and is more compatible with previously reported properties of the chemically synthesized product (stability in aqueous solution, heat stability) than the free α-aminoaldehyde. The cyclic compound, and not GSA, was proposed to be the product of glutamyl-tRNA reductase and the substrate of GSA aminotransferase. It is likely that the cyclic and linear forms of GSA coexist in solution in an equilibrium ratio, analogously to the aldose sugars.

e. GSA Aminotransferase

GSA aminotransferase (EC 5.4.3.8) catalyzes the conversion of GSA to ALA. Enzymes capable of converting chemically synthesized GSA to ALA have been purified from several plant, algal, and bacterial sources. The transamination reaction requires no added substrate other than GSA. The enzyme contains bound pyridoxal-phosphate (Avissar and Beale, 1989b; Bull et al., 1990). GSA aminotransferase is inhibited by the mechanism-based suicide substrate analog gabaculine (Kannangara and Schouboe, 1985; Avissar and Beale, 1989b). Gabaculine reacts with enzyme-bound pyridoxal-phosphate to irreversibly form a secondary amine. Gabaculine-treated *C. vulgaris* GSA aminotransferase can be reactivated by gel filtration (to remove gabaculine-pyridoxal-phosphate adducts and excess gabaculine) followed by incubation with pyridoxal-phosphate (Avissar and Beale, 1989b).

A proposed reaction mechanism for GSA aminotransferase involves transfer of NH_2 from enzyme-pyridoxamine-phosphate to the terminal aldehyde carbon of GSA to form enzyme-pyridoxal-phosphate and 4,5-diaminovaleric acid, followed by transfer of the NH_2 at the 4-position of the intermediate back to the cofactor, thereby forming ALA and regenerating the pyridoxamine-phosphate form of the cofactor. Implicit in this mechanism is that the N atom of ALA is derived from a different precursor molecule than the C atoms. This prediction was tested by the use of a mixture of ^{13}C- and ^{15}N-labeled glutamate molecules as substrate for conversion to ALA by algal cell extracts. When the heavy isotope labels were present on separate substrate molecules, a significant proportion of the ALA product molecules contained two heavy atoms, indicating that the conversion occurs by intermolecular nitrogen transfer (Mau and Wang, 1988; Mayer et al., 1993a). This result supports the proposed reaction mechanism and indicates that the enzyme catalyzes two aminotransferase reactions, rather than an aminomutase reaction, even though the substrate and product have the same atomic composition.

Native GSA aminotransferase from *Synechocystis* sp. PCC 6803 has a molecular weight of 99,000 (Rieble and Beale, 1991b). Purified aminotransferase from *Synechococcus* sp. PCC 6301 has a molecular weight of 46,000 on denaturing SDS-PAGE (Grimm et al., 1989). Therefore, the native enzyme appears to be a homodimer, like other members of the aspartate aminotransferase enzyme family.

The GSA aminotransferase-encoding *hemL* gene (called *gsa* in plants and algae) has been cloned and sequenced from several plants, algae, and bacteria. These genes encode highly conserved peptides that have predicted molecular weights of approximately 46,000 and have recognizable similarity to other members of the aspartate aminotransferase enzyme family (Elliott et al., 1990; Matters and Beale, 1994). All *hemL/gsa*-encoded peptides have a conserved putative active site containing an essential lysine, which is at position 265 in the *Synechococcus* sp. PCC 6301 enzyme (Grimm et al., 1991a). It is believed that the pyridoxal-phosphate cofactor binds to this lysine. Mutagenesis of this lysine inactivates the enzyme (Grimm et al., 1992; Ilag and Jahn, 1992). A *Synechococcus* sp. PCC 6301 mutant that was selected for resistance to gabaculine has a GSA aminotransferase with a lower specific activity than the wild-type enzyme. The mutation that confers gabaculine resistance is $M_{248}F$ (Grimm et al., 1991b). Most, but not all GSA aminotransferases contain a methionine at this position (Matters and Beale, 1994).

B. Formation of the Tetrapyrrole Skeleton from ALA

1. Conversion of ALA to PBG

Formation of the first pyrrole in the pathway, PBG (Fig. 4), by asymmetric condensation of two ALA molecules is catalyzed by the enzyme PBG synthase (ALA dehydratase) (EC 4.2.1.24). PBG synthase from all sources examined, including plants and *Rb. sphaeroides*, has an octameric structure with native molecular weights ranging from 250,000 to 320,000 (Nandi and Shemin, 1968; Shibata and Ochiai, 1977; Huault et al., 1987; Jordan, 1991). An interesting difference among PBG synthase enzymes from different sources is the metal requirement: the animal, yeast, and *E. coli* enzymes require Zn^{2+} for activity (Jordan, 1991; Spencer and Jordan, 1993), the plant enzyme requires Mg^{2+} (Shibata and Ochiai, 1977; Liedgens et al., 1983), and the *Rb. sphaeroides* enzyme requires K^+ (Burnham and Lascelles, 1963; Nandi and Shemin, 1968). A comparative study concluded that PBG synthase from all sources has tightly-bound Zn^{2+}, but that a second, weaker metal-binding site binds Zn^{2+} in the animal, yeast, and *E. coli* enzymes and Mg^{2+} in the plant enzyme (Mitchell and Jaffe, 1993). The *E. coli* and plant enzymes were proposed to have a third metal-binding site which stimulates activity approximately two-fold when Mg^{2+} is bound (Mitchell and Jaffe, 1993). The *hemB* gene that encodes PBG synthase has been cloned from *Rb. sphaeroides* and expressed in *E. coli* (Delaunay et al., 1991). The expressed gene product has a molecular weight of 39,000.

Single turnover experiments with PBG synthase from *Rb. sphaeroides* have established that in the reaction, the first bound ALA molecule is the one

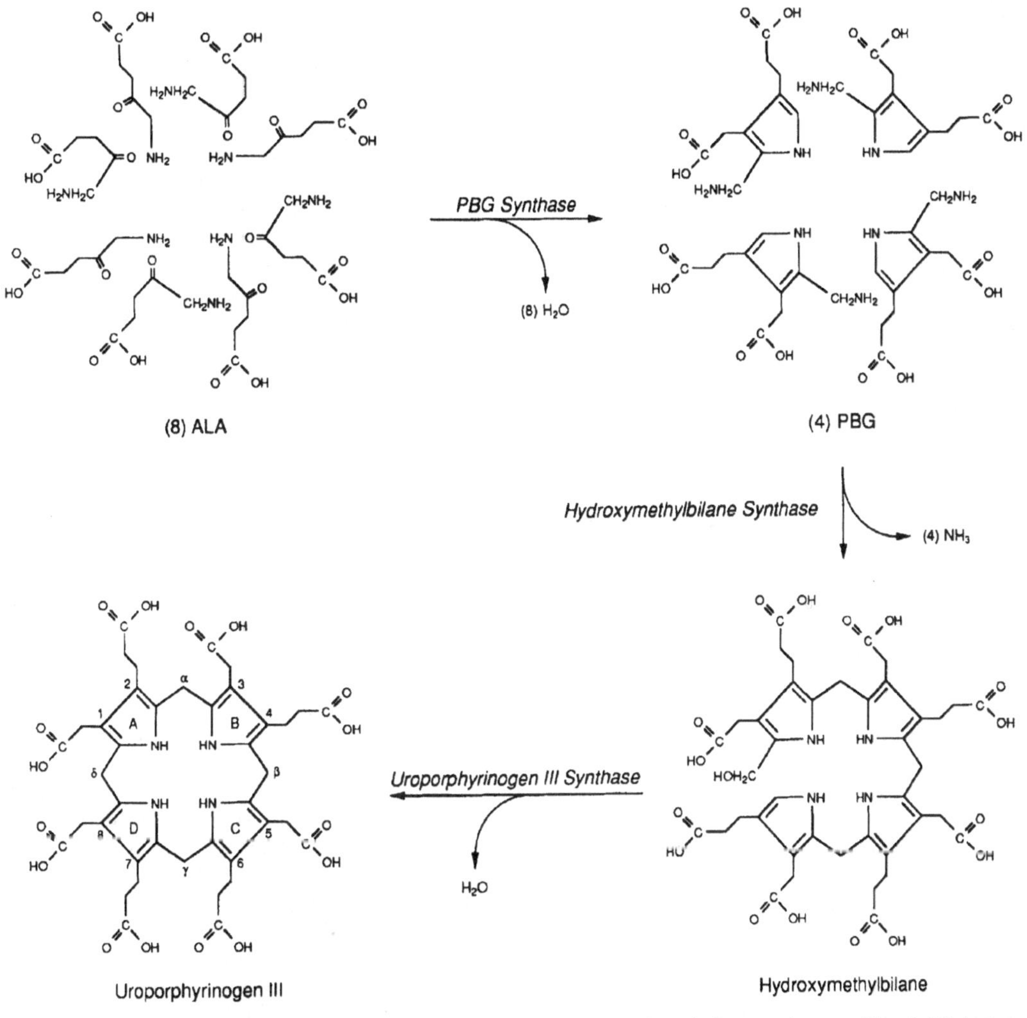

Fig. 4. Transformation of ALA to uroporphyrinogen III. Eight ALA molecules are shown in the approximate positions that their atoms occupy in the intermediates and final product. The conventional Fischer scheme for designating the pyrrole rings, peripheral pyrrole carbon atoms, and bridge carbons of tetrapyrroles is shown for uroporphyrinogen III.

that contributes the propionic acid side chain of the product (Jordan and Seehra, 1980). The fact that the *pro-S* hydrogen atom that is derived from the C_5 hydrogens of ALA is stereospecifically retained in the product indicates that during the in the formation of PBG, removal of hydrogen to form the aromatic pyrrole ring must occur on the enzyme (Abboud and Akhtar, 1976).

Some evidence exists for a regulatory role of PBG synthase in *Rb. capsulatus*. A *bchH* mutant strain of *Rb. capsulatus* cannot form BChl and accumulates protoporphyrin IX when grown anaerobically. Oxygen inhibited the protoporphyrin IX accumulation (Biel, 1992; Chapter 52). Exogenous PBG, but not ALA, overcame the O_2 inhibition. The results were interpreted to indicate that O_2 regulates the conversion of ALA to PBG.

2. Conversion of PBG to Uroporphyrinogen III

Uroporphyrinogen III, the last common precursor of all end-product tetrapyrroles, is synthesized from

PBG by the sequential action of two enzymes, hydroxymethylbilane synthase and uroporphyrinogen III synthase (Fig. 4).

a. Hydroxymethylbilane Synthase

Hydroxymethylbilane synthase (PBG deaminase) (EC 4.1.3.8) condenses four PBG molecules to form the first tetrapyrrole, uroporphyrinogen. The initial product of enzymic catalysis is the linear tetrapyrrole, hydroxymethylbilane (preuroporphyrinogen), which, in the absence of uroporphyrinogen III synthase, spontaneously cyclizes to form uroporphyrinogen I. Biosynthesis of the biologically relevant isomer, uroporphyrinogen III, requires the action of uroporphyrinogen III synthase during or immediately after release of hydroxymethylbilane from PBG deaminase.

Hydroxymethylbilane synthase from most sources is a monomer of approximately 40,000 molecular weight, and the enzyme does not require metal ions or other cofactors for activity (Jordan, 1991). The *Rp. palustris* enzyme was reported to have a native molecular weight of 74,000, and it may be a dimer (Kotler et al., 1987). The *hemC* gene, which encodes hydroxymethylbilane synthase, has been cloned from several bacteria including *E. coli* (Thomas and Jordan, 1986) and *Cb. vibrioforme* (Majumdar et al., 1991; Moberg and Avissar, 1994), and the *E. coli* enzyme has been crystallized and its structure determined by X-ray crystallography (Louie et al., 1992).

Hydroxymethylbilane synthase from *Rb. sphaeroides* and *Euglena gracilis* was used to establish that the order of assembly of the four PBG units is ABCD, as they appear in uroporphyrinogen (Fig. 4) (Jordan and Seehra, 1979; Battersby et al., 1979a). The nascent monopyrrole- through tetrapyrrole-enzyme complexes are bound *via* a dipyrrole cofactor (Battersby et al., 1979b; Scott et al., 1980; Jordan and Berry, 1981; Battersby et al., 1983; Hart et al., 1987; Jordan and Warren, 1987; Warren and Jordan, 1988). In hydroxymethylbilane synthase from *E. coli*, the dipyrrole cofactor is attached to a cysteine residue (C_{242} of the *E. coli* enzyme) after formation of the apoprotein, and it remains permanently attached to the enzyme, while the link between the cofactor and the nascent oligopyrrole chain is severed after the hexapyrrole stage is reached, which releases a tetrapyrrole and prepares the enzyme to accept new substrate molecules.

C. Uroporphyrinogen III Synthase

Hydroxymethylbilane is unstable in solution and it rapidly cyclizes to uroporphyrinogen I, a physiologically nonproductive end product. Uroporphyrinogen III synthase (EC 4.2.1.75) catalyzes closure of the tetrapyrrole macrocycle with inversion of ring D to form the type III product. The mechanism of ring inversion has been the subject of intensive investigation and is now generally believed to involve a *spiro* intermediate (Crockett et al., 1991, Stark et al., 1993). Uroporphyrinogen III synthase has been purified from several sources including *E. coli* (Jordan et al., 1988). The native enzyme from all sources is a monomer of approximately 29,000 molecular weight and contains no reversibly-bound cofactors or metal ions. The *hemD* gene, which encodes uroporphyrinogen III synthase, has been cloned from *E. coli*, sequenced, and overexpressed (Alwan et al., 1989; Sasarman et al., 1987; Jordan et al., 1988, Crockett et al., 1991). The *Cb. vibrioforme hemD* gene has been cloned and expressed in *E. coli* (Moberg and Avissar, 1994).

Hydroxymethylbilane synthase and uroporphyrinogen III synthase may form a complex that facilitates transfer of hydroxymethylbilane between the two enzymes. The presence of *E. gracilis* uroporphyrinogen III synthase influences the K_m of *E. gracilis* hydroxymethylbilane synthase for PBG (Battersby et al., 1979c). The sedimentation velocity of wheat germ hydroxymethylbilane synthase is also influenced by the presence of wheat germ uroporphyrinogen III synthase (Higuchi and Bogorad, 1975). The presence of *Rb. sphaeroides* uroporphyrinogen III synthase was reported to facilitate release of the tetrapyrrole product from hydroxymethylbilane synthase (Rosé et al., 1988).

D. The Branch from Uroporphyrinogen III to Reduced Hemes and Corrinoids

Siroheme (Fig. 5) is the prosthetic group of the assimilatory nitrite and sulfite reductases (ferredoxin-nitrite reductase, EC 1.7.7.1; NAD(P)H-nitrite reductase, EC 1.6.6.4; ferredoxin-sulfite reductase, EC 1.8.7.1; and NAD(P)H-sulfite reductase, EC 1.8.1.2) that are required for full reduction of highly oxidized forms of nitrogen and sulfur to NH_4^+ and S^{2-} (Siegel, 1978). Heme d_1 (Fig. 5) is another reduced heme that is a prosthetic group of the cytochrome cd_1

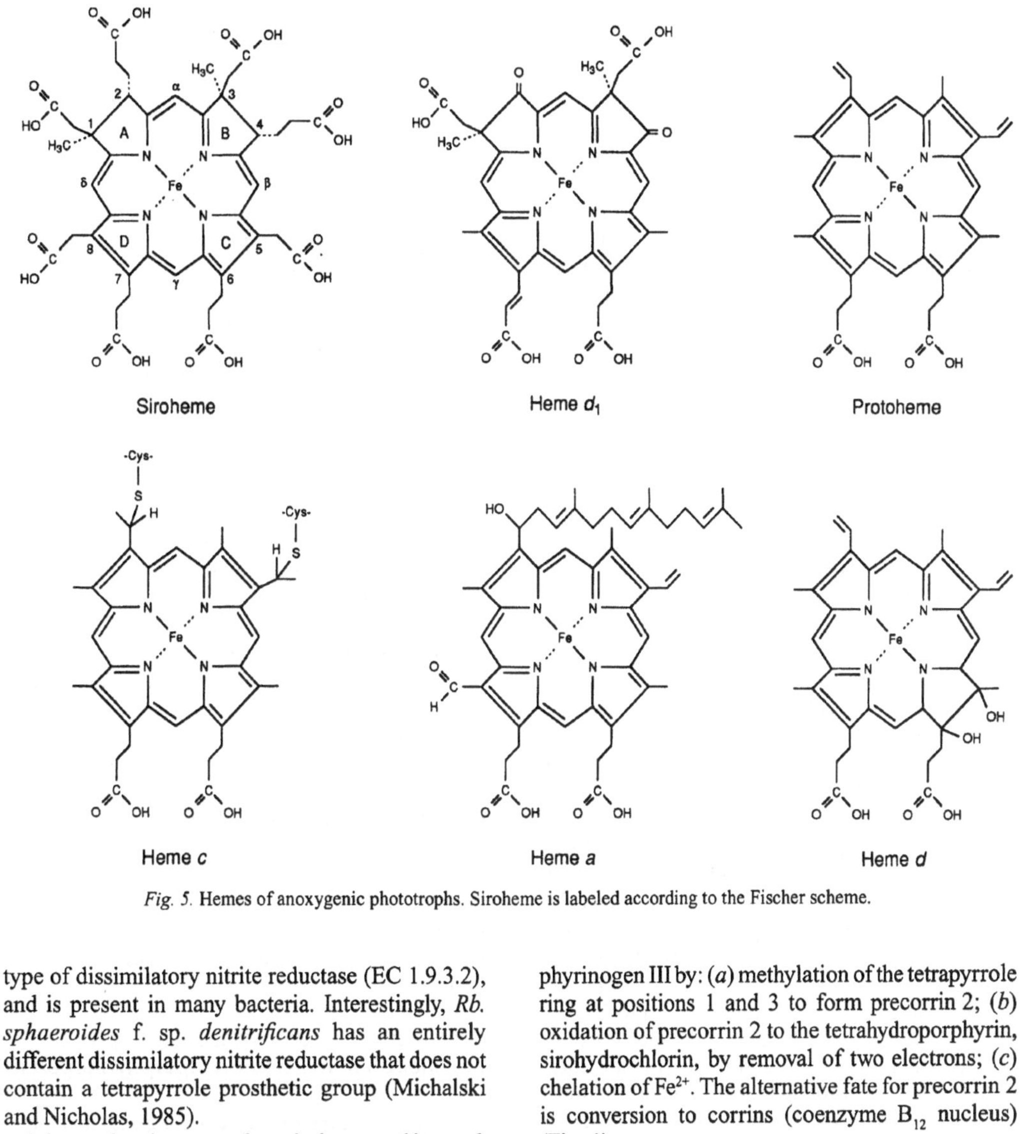

Fig. 5. Hemes of anoxygenic phototrophs. Siroheme is labeled according to the Fischer scheme.

type of dissimilatory nitrite reductase (EC 1.9.3.2), and is present in many bacteria. Interestingly, *Rb. sphaeroides* f. sp. *denitrificans* has an entirely different dissimilatory nitrite reductase that does not contain a tetrapyrrole prosthetic group (Michalski and Nicholas, 1985).

It is not clear how prevalent siroheme and heme d_1 prosthetic groups are in anoxygenic phototrophs, but siroheme-containing enzymes are likely to be present at least in those facultative and obligately aerobic species that must obtain their nitrogen and sulfur nutrients from oxygen-rich environments. In any case, many anoxygenic phototrophs are capable of coenzyme B_{12} biosynthesis, the initial step of which is shared with that of siroheme.

Formally, siroheme can be derived from uroporphyrinogen III by: (*a*) methylation of the tetrapyrrole ring at positions 1 and 3 to form precorrin 2; (*b*) oxidation of precorrin 2 to the tetrahydroporphyrin, sirohydrochlorin, by removal of two electrons; (*c*) chelation of Fe^{2+}. The alternative fate for precorrin 2 is conversion to corrins (coenzyme B_{12} nucleus) (Fig. 1).

Chemical arguments suggest that the order of the steps of siroheme formation from uroporphyrinogen III is probably that given above: first methylation, then oxidation, and finally Fe^{2+} chelation. Methylation of rings A and B of uroporphyrinogen III to produce precorrin 2 effectively limits subsequent oxidation beyond the tetrahydroporphyrin (dihydrochlorin) state. Oxidation of precorrin 2 to sirohydrochlorin produces a compound that has the aromaticity and metal-binding properties necessary for efficient chelation of Fe^{2+}. Heme d_1 could be formed from siroheme by the replacement of the propionate groups on rings A and B with oxygen atoms, decarboxylation of the acetate groups on rings C and D to methyl groups, and the oxidation of the 7-propionate to an acrylate group (Wu and Chang, 1987). The methyl groups at positions 1 and 3 of heme d_1 are in the same configuration as those of siroheme and, like the methyl groups of siroheme (see below), they are derived from *S*-adenosyl-L-methionine (Yap-Bondoc et al., 1990).

The methyl groups of siroheme were shown to be derived from methionine in *E. coli* cells (Siegel et al., 1977), and precorrin 2 is formed in vitro from *S*-adenosyl-L-methionine and uroporphyrinogen III in a reaction catalyzed by uroporphyrinogen methyltransferase (Warren et al., 1990). On the basis of the structure of a singly-methylated (1-methyl) product isolated from *Clostridium tetanomorphum* (Deeg et al., 1977), and the incorporation of 1-methyluroporphyrinogen III into cobyrinic acid by extracts of *Propionibacterium shermanii* (Brunt et al., 1989), the order of methylation was deduced to be first at the 1-position, and then at the 3-position. In *Pseudomonas denitrificans*, uroporphyrinogen methyltransferase is encoded by the *cobA* gene (Crouzet et al., 1990). A somewhat different methyltransferase is encoded by the *cysG* gene in *E. coli* (Warren et al., 1990) and *S. typhimurium* (Goldman and Roth, 1993). It is of interest that *cysG* is the only known genetic locus specifically associated with siroheme synthesis in enteric bacteria. *cysG* encodes a 52,000 molecular weight peptide. Its COOH-terminal region is similar to the smaller, 29,200 molecular weight peptide encoded by the *cobA* gene. Because *cysG* is the only known gene of *S. typhimurium* that is involved in siroheme synthesis, and because only a portion of its encoded peptide is similar to the smaller *cobA* product, the CysG protein may be a multifunctional enzyme that catalyzes all steps of the conversion of uroporphyrinogen III to siroheme in the enteric bacteria (Goldman and Roth, 1993). In any case, it appears that Fe^{2+} insertion into sirohydrochlorin is not catalyzed by the ferrochelatase that is responsible for protoheme formation (Powell et al., 1973; Hansson and Wachenfeldt, 1993). That enzyme is encoded by the *hemH* gene in *E. coli* (previously known as *visA*) (Frustaci and O'Brian, 1993), *B. subtilis* (Hansson and Hederstedt, 1992), and *Bradyrhizobium japonicum* (Frustaci and O'Brian, 1992) (see below).

E. The Branch from Uroporphyrinogen III to Protoporphyrin IX

1. Decarboxylation of Uroporphyrinogen III to Form Coproporphyrin III

Uroporphyrinogen decarboxylase (EC 4.1.1.37) catalyzes the decarboxylation of all four of the acetate residues on uroporphyrinogen to yield coproporphyrinogen, which contains methyls in their place. At physiological substrate concentrations, the decarboxylations occur in a specific sequence, beginning at ring D and proceeding clockwise around the macrocycle (Luo and Lim, 1993). At higher substrate concentrations, the decarboxylation sequence becomes random(Luo and Lim, 1993; Jones and Jordan, 1993). No metal requirements were detected. A stereochemical analysis of the reaction indicated that the decarboxylations proceed with retention of configuration about the α-carbon atoms, i.e., the lost carboxyl groups are replaced with solvent hydrogens in the same orientation (Barnard and Akhtar, 1975).

Uroporphyrinogen decarboxylase has been purified to homogeneity from *Rb. sphaeroides*, and the native enzyme was characterized as a 41,000 molecular weight monomer (Jones and Jordan, 1993). The enzyme from *Rp. palustris* was also reported to be a monomer, with a molecular weight of 46,000 (Koopman et al., 1986). The gene that encodes uroporphyrinogen decarboxylase has been cloned from several eucaryotes including yeast and mammals, as well as *B. subtilis*, in which it has been identified with the *hemE* locus (Hansson and Hederstedt, 1992). *B. subtilis hemE* encodes a 40,347 molecular weight peptide.

2. Oxidative Decarboxylation of Coproporphyrinogen III to Form Protoporphyrinogen IX

Coproporphyrinogen oxidase (EC 1.3.3.3) oxidatively decarboxylates the propionate groups at positions 2 and 4 of coproporphyrinogen III to vinyls, thereby producing protoporphyrinogen IX. The enzyme is specific for the III isomer of coproporphyrinogen over the nonphysiological I isomer, although chemically synthesized coproporphyrinogen IV is also decarboxylated by the enzyme (Mombelli et al., 1976). Evidence indicating that the 2-propionate is converted before the 4-propionate includes characterization of a 2-monovinyl intermediate in rat liver preparations (Elder et al., 1978) and the preferential action of the *E. gracilis* enzyme on the chemically synthesized 2-monovinyl porphyrin compared to the ring 4-monovinyl porphyrin (Cavaleiro et al., 1974).

In obligately aerobic organisms, coproporphyrinogen oxidase is an O_2-requiring reaction. Extracts of anaerobically-grown *Rb. sphaeroides* cells can carry out the reaction anaerobically in the presence of ATP, oxidized pyridine nucleotide, and methionine (Tait, 1972). Similar requirements were reported for anaerobic extracts of yeast (Poulson and Polglase, 1974) and *B. japonicum* (Keithly and Nadler, 1983). Involvement of *S*-adenosyl-L-methionine in the reaction is indicated by its inhibition by *S*-adenosyl-L-homocysteine (Tait, 1972).

Seehra et al. (1983) have studied the anaerobic as well as the aerobic coproporphyrinogen oxidase reactions in extracts of *Rb. sphaeroides*. The oxidative decarboxylations proceed with specific loss of the pro-*S* β-protons of the propionate groups and retention of the pro-*R* protons. A reaction mechanism was proposed involving pyrrolic N-assisted removal of single protons as hydride ions from the β-carbons of the propionate groups. The α-protons of the propionate groups do not appear to be involved: In the reaction catalyzed by an avian blood extract, both of the α-protons of both propionate groups were retained on the terminal carbon atoms of the protoporphyrinogen vinyl groups (Zaman and Akhtar, 1978).

Yeast coproporphyrinogen oxidase is a 70,000 molecular weight homodimer that contains two molecules of Fe (Camadro et al., 1986). A yeast gene (*HEM13*) encoding coproporphyrinogen oxidase has been cloned and sequenced (Zagorec et al., 1988), and a soybean nuclear gene that encodes a 43,000 molecular weight protein complements a yeast *HEM13* mutant (Madsen et al., 1993). The deduced soybean gene product is 50% similar to the yeast *HEM13* gene product. An *Rb. sphaeroides* gene has been cloned and sequenced that, when overexpressed in *E. coli*, causes increased coproporphyrinogen oxidase activity (Coomber et al., 1992). The *Rb. sphaeroides* gene product has a predicted molecular weight of 34,000 and is 44% identical to the yeast *HEM13* product. Because *Rb. sphaeroides* cells in which this gene was disrupted were unable to form BChl under anaerobic inducing conditions, but were able to grow aerobically, it was proposed that in *Rb. sphaeroides*, the gene product is an anaerobic coproporphyrinogen oxidase dedicated to BChl synthesis. However, it was not reported whether O_2 was required for activity of the recombinant coproporphyrinogen oxidase obtained from *E. coli* cells expressing the *Rb. sphaeroides* gene.

The *E. coli hemF* gene has been cloned by complementation of a yeast *HEM13* mutant (Troup et al., 1994). *E. coli hemF* encodes a 34,300 molecular weight peptide that is 43% identical to the yeast *HEM13* product. *S. typhimurium* contains two genes, *hemF* and *hemN*, either of which can support aerobic heme synthesis (Xu et al., 1992) The *hemF* gene encodes a peptide with a predicted molecular weight of 34,400 that is 44% identical to the yeast *HEM13* product and 90% identical to the *E. coli hemF* product (Xu and Elliott, 1993). Because *hemN* mutants accumulated coproporphyrinogen III and were auxotrophic for protoheme when grown anaerobically, the authors suggesed that the *hemN* product is an anaerobic coproporphyrinogen oxidase. The predicted *S. typhimurium hemN* product is a 52,800 molecular weight peptide that is 38% identical to the *Rb. sphaeroides hemF* product but has little if any similarity to the *S. typhimurium hemF* product (Xu and Elliott, 1994). Interestingly, the predicted amino acid sequence of the *Rb. sphaeroides hemF* and *S. typhimurium hemN* products are significantly similar (35% identity) to a portion of the *Rhizobium phaseoli* and *R. leguminosarum nifD* gene product, which is the α-subunit of nitrogenase.

3. Oxidation of Protoporphyrinogen IX to Form Protoporphyrin IX

Protoporphyrinogen oxidase (EC 1.3.3.4) catalyzes the removal of six electrons from the tetrapyrrole macrocycle to form protoporphyrin IX, the last biosynthetic step that is common to hemes and chlorophylls. In obligately aerobic organisms, O_2 is the electron acceptor and is required for enzyme activity. In contrast, the *Rb. sphaeroides* protoporphyrinogen oxidase cannot use O_2 directly as an oxidant, but instead, protoporphyrinogen IX oxidation is coupled to the respiratory electron transport chain (Jacobs and Jacobs, 1981). The same is true for the protoporphyrinogen oxidase reaction in anaerobic *E. coli* cells, which is coupled to the reduction of nitrate or fumarate (Jacobs and Jacobs, 1976).

There is an interesting stereochemistry in the removal of the four meso hydrogens from protoporphyrinogen IX. In the transformation of PBG to protoporphyrinogen IX, the meso-bridge methylene carbon atoms are derived from the aminomethyl carbon (C_{11}) of PBG (see Fig. 4). If the protoheme was formed from PBG that was labeled nonstereospecifically with ^3H at C_{11}, then all four meso positions of the porphyrin moiety were equally labeled with ^3H (Jones et al., 1984). However, if the PBG was labeled specifically in the *pro-S* hydrogen at C_{11}, then the product contained ^3H only at the β-meso position (the one between rings B and C). Given the fact that the incorporation of the four PBG groups into the original tetrapyrrole, hydroxymethylbilane, is an oligomerization that probably involves a single reaction mechanism, and given that cyclization of the hydroxymethylbilane to form the porphyrinogen macrocycle does not seem to cause loss of label from the γ and δ meso methylene hydrogens, it seems likely that the C_{11} *pro-S* hydrogen of PBG will be found on the same face of the porphyrinogen at all four meso positions. Therefore, the hydrogen that is removed from the β-meso position must be removed from the opposite face of the molecule than the ones that are removed from the α, γ, and δ meso positions. It was hypothesized that three of the meso hydrogens are lost by an oxidation process occurring on one face of the molecule, and that the fourth proton is lost from the other face by a tautomerization reaction (Jones et al., 1984). Confirmation of this hypothesis will require determination of whether the uroporphyrinogen III synthase reaction affects the stereochemistry of the γ and δ meso methylene hydrogens.

In eucaryotic organisms, protoporphyrinogen oxidase is a mitochondrial or chloroplast membrane-associated protein that has an apparent molecular weight of 210,000 and is composed of 36,000 molecular weight subunits (Jacobs and Jacobs, 1987). The *Rb. sphaeroides* enzyme is also associated with membranes (Jacobs and Jacobs, 1981). Detergent-solubilized protoporphyrinogen oxidase from the anaerobe *D. gigas* was reported to have a native molecular weight of 148,000 and to contain three dissimilar subunits of 12,000, 18,500, and 57,000 molecular weight (Klemm and Barton, 1987). In contrast to the above enzymes, the *B. subtilis* protoporphyrinogen oxidase is soluble (Dailey et al., 1994).

In *S. typhimurium*, the *hemG* gene was suggested to encode protoporphyrinogen oxidase, on the basis of the accumulation of protoporphyrin IX (as well as coproporphyrin III and uroporphyrin III) in *hemG* mutants (Xu et al., 1992). In *B. subtilis*, mutations at the *hemY* locus caused accumulation of either coproporphyrinogen III alone or both coproporphyrinogen III and protoporphyrinogen IX (Hansson and Hederstedt, 1992). It was originally proposed that the *B. subtilis hemY* product, which at 51,400 molecular weight is significantly larger than the *S. typhimurium* HemF and the 37,600 molecular weight yeast coproporphyrinogen oxidase that is encoded by the *HEM13* gene, may be a bifunctional enzyme that has both coproporphyrinogen oxidase and protoporphyrinogen oxidase activities (Hansson and Hederstedt, 1992). More recently, however, the *B. subtilis hemY* gene was expressed in *E. coli* and its product was determined to have protoporphyrinogen oxidase activity, but not coproporphyrinogen oxidase activity (Dailey et al., 1994). It was suggested that the *B. subtilis* gene designation be changed from *hemY* to *hemG*, to correspond with the protoporphyrinogen oxidase-encoding gene of *S. typhimurium*.

The *E. coli hemG* gene has been cloned, sequenced and expressed (Sasarman et al., 1993). The cloned gene complements a protoporphyinogen oxidase-deficient *E. coli* mutant and the expressed *hemG* product has protoporphyrinogen oxidase activity. *E. coli hemG* encodes a 21,202 molecular weight membrane-associated peptide that is consideralby smaller than the protoporphyrinogen oxidase peptides from barley, *B. subtilis*, yeast and mouse liver, which have molecular weights of 36,000, 51,400, 56,000 and 65,000, respectively.

F. Chelation of Iron to Form Protoheme

The last step of protoheme formation, insertion of Fe^{2+} into protoporphyrin IX, is catalyzed by ferrochelatase (protoheme ferrolyase, EC 4.99.1.1). In addition to its physiological substrates, ferrochelatase can use Zn^{2+} and Co^{2+} as the metal substrate and deuteroporphyrin IX, mesoporphyrin IX, and hematoporphyrin IX as the porphyrin substrate. Ferrochelatase from most sources, including *Rb. sphaeroides*, is an intrinsic membrane protein that requires detergents or chaotropic agents for solubilization (Dailey, 1982). An apparent exception is the *B. subtilis* ferrochelatase, which was reported to be a soluble enzyme (Hansson and Hederstedt, 1992). Detergent-solubilized ferrochelatase from *Rb. sphaeroides* is a 115,000 molecular weight monomer (Dailey, 1982). No chromophoric prosthetic groups were detected. Mg-protoporphyrin IX is a powerful inhibitor of *Rb. sphaeroides* ferrochelatase: It inhibits by 50% at 4 μM and nearly completely at 80 μM (Jones and Jones, 1970). However, it is uncertain whether the intracellular free Mg-protoporphyrin IX ever reaches inhibitory concentrations.

The *hemH* gene, which encodes ferrochelatase, has been cloned and sequenced from yeast and several animal sources, as well as from *E. coli* (Miyamoto et al., 1991; Frustaci and O'Brian, 1993), *S. typhimurium* (Xu et al., 1992), *B. japonicum* (Frustaci and O'Brian, 1992), and *B. subtilis* (Hansson and Hederstedt, 1992). The molecular weights of the encoded polypeptides from these bacterial genes range from 34,000 to 38,000 and are somewhat smaller than the polypeptides encoded by the animal and yeast genes.

G. Biosynthesis of Other Protoporphyrin IX-Based Hemes

1. Protoheme Ligation to Apoproteins to Form Heme c

All phototrophic organisms contain one or more types of cytochrome *c* which function as components of the photosynthetic and respiratory electron transport chains (Meyer and Cusanovich, 1989). Heme *c* is the covalently bound prosthetic group of *c*-type cytochromes (Fig. 5). Apocytochromes *c* incorporate protoheme by enzyme-catalyzed ligation of the vinyl groups to cysteine residues on the protein. A specific ligating enzyme, named cytochrome *c* heme lyase, has been described in yeast and *Neurospora crassa* (Taniuchi et al., 1983; Nicholson et al., 1987). In these eucaryotic cells, heme lyase is localized in the inner mitochondrial membrane (Nargang et al., 1988; Enosawa and Ohashi, 1986) and ligation requires reduced (ferro)protoheme as a substrate (Nicholson and Neupert, 1989). Genes for cytochrome *c* heme lyase have been cloned and sequenced from yeast (*CYC3*) and *N. crassa* (*cyt-2*) (Dumont et al., 1987; Drygas et al., 1989). Although the two encoded peptides are 32% identical, the yeast peptide is considerably smaller than the *N. crassa* peptide (molecular weights of 29,600 and 38,000, respectively). A second yeast heme lyase, cytochrome-c_1-heme lyase, is responsible for assembly of cytochrome c_1 (Zollner et al., 1992). This protein is encoded by the *CYT2* gene, which has been cloned and sequenced. The two yeast heme lyases are 35% identical. The yeast results suggest that each *c*-type cytochrome may require a different, specific heme lyase for its formation.

In gram-negative bacteria, *c*-type cytochromes are located in the periplasmic space. *Rb. sphaeroides* and *Rb. capsulatus* contain at least five distinct cytochromes *c* (Meyer and Cusanovich, 1989). In *Rb. capsulatus*, several genes have been identified that are necessary for the biogenesis of all cytochromes *c* but are not required for synthesis and secretion of apocytochrome *c* into the periplasmic space (Davidson et al., 1987; Biel and Biel, 1990; Beckman et al., 1992). One group of these genes, *helA*, *helB*, and *helC*, are similar to a class of transporter proteins and their products may be involved in exporting protoheme to the periplasmic space (Beckman et al., 1992). The products of two other genes, *ccl1* and *ccl2*, have no recognizable similarity to yeast heme lyases, but they do have similarity to mitochondrial and chloroplast open reading frames with unassigned function. The specific roles of *ccl1* and *ccl2* in cytochrome *c* formation have not been identified. The topic of heme attachment to cytochrome apoproteins is covered in more detail by Kranz and Beckman in Chapter 33.

2. Terminal Oxidase Hemes

Cytochrome oxidase (EC 1.9.3.1) of the aa_3 type is present in many obligate and facultative aerobic bacteria including *B. subtilis*, *Rb. sphaeroides* (Sasaki et al., 1970; Gennis et al., 1982; Garcia-Horsman et al., 1994; Chapter 44), and the BChl-containing obligate aerobe *Erythrobacter longus* (Fukumori et al., 1987). The heme *a* prosthetic group of cytochrome oxidase (Fig. 5) is formed from protoheme by ligation of a farnesyl group to the 2-vinyl group and oxidation or oxygenation of the 8-methyl group to a formyl group. The precursor status of protoheme has been shown in *B. subtilis*. Strains that are defective in enzymes that catalyze the later steps of protoheme formation (uroporphyrinogen decarboxylase, coproporphyrinogen oxidase, protoporphyrinogen oxidase, and ferrochelatase) could not form heme *a* unless protoheme was added to the medium, even though the same cells could form siroheme, presumably using a different ferrochelatase for its biosynthesis (Hansson and Wachenfeldt, 1993).

Cytochrome *o* is an O_2-reducing terminal oxidase present in many facultative aerobes including *E. coli* and the phototrophs *Rb. capsulatus* (Zannoni et al., 1976), *Rhodopseudomonas viridis* (Kämpf et al., 1987), *Rp. palustris*, and *Rb. sphaeroides* (which also contains cytochrome aa_3) (Sasaki et al., 1970;

Anraku and Gennis, 1987; Garcia-Horsman et al., 1994). Heme *o*, the prosthetic group of cytochrome *o*, has the structure of a possible heme *a* precursor in which the 2-farnesyl group has been added but the 8-methyl group has not been converted to a formyl group (Wu et al., 1992). The *cyoE* gene of *E. coli*, when overexpressed, caused conversion of protoheme to heme *o* in vivo (Saiki et al., 1992), and the overexpressed *cyoE* gene product catalyzed in vitro condensation of ferrous protoheme and farnesyl-pyrophosphate to form heme *o* (Saiki et al., 1993). Two *B. subtilis* genes, *ctaA* and *ctaB*, are required for heme *a* synthesis (Svensson et al., 1993). The *ctaB* gene complements *E. coli cyoE* mutants. *B. subtilis ctaA* mutants accumulate heme *o* instead of heme *a*. *B. subtilis ctaB* mutants accumulate neither heme *o* nor heme *a*. Interestingly, the *B. subtilis ctaA* gene, when expressed in *E. coli*, caused the accumulation of heme *a*, even though *E. coli* normally does not produce this heme. These results indicate that *ctaB* and *ctaA* encode a farnesyltransferase and a methyl oxidase or oxygenase, respectively. The results also suggest that farnesylation precedes formyl group formation in heme *a* biosynthesis.

A third type of O_2-reducing terminal oxidase, cytochrome *d*, is present in some facultative aerobes including *E. coli* (Gennis, 1987; Garcia-Horsman et al., 1994) and *Rp. viridis* (Kämpf et al., 1987). Heme *d*, the prosthetic group of cytochrome *d*, has a dihydroxychlorin macrocycle structure (Fig. 5) (Sotiriou and Chang, 1988) that differs greatly from the sirohydrochlorin structure of heme d_1. Although the active form of heme *d* is believed to be the free dihydroxyl form shown in Fig. 5, the 6-hydroxyl group forms a lactone with the 6-propionate under some isolation conditions. In addition to its presence in cytochrome *d*, heme *d* is also found as the prosthetic group of *E. coli* catalase (EC 1.11.1.6) type HPII. However, heme *d* from these two sources is subtly different: while the two ring C hydroxyl groups of the cytochrome *d* prosthetic group are in the relative *trans* configuration (Timkovich et al., 1985), the hydroxyls are *cis* in the catalase HPII heme *d* (Chiu et al., 1989). The absolute configurations of the hydroxyl groups have not been determined for either type of heme *d*. It can be deduced from the structure of heme *d* that it is derived from protoporphyrin IX or protoheme by hydroxylation. It was proposed (Timkovich and Bondoc, 1990) and later reported (Loewen et al., 1993) that *E. coli* catalase HPII catalyzes the formation of its own heme *d* prosthetic group from protoheme. It is not known whether a similar autologous conversion occurs for the prosthetic group of cytochrome *d*.

Acknowledgments

The author thanks Y. J. Avissar and D. W. Bollivar for critically reading the manuscript and making helpful suggestions. Research grant support from the National Science Foundation, the Department of Energy, and the U. S. Department of Agriculture is gratefully acknowledged.

References

Abboud MM and Akhtar M (1976) Stereochemistry of hydrogen elimination in the enzymatic formation of the C-2–C-3 double bond of porphobilinogen. J Chem Soc Chem Commun 1976: 1007–1008

Abboud MM, Jordan PM and Akhtar M (1974) Biosynthesis of 5-aminolevulinic acid: Involvement of a retention-reversion mechanism. J Chem Soc Chem Commun 1974: 643–644

Alwan AF, Mgbeje BI and Jordan PM (1989) Purification and properties of uroporphyrinogen III synthase (co-synthase) from an overproducing recombinant strain of *Escherichia coli* K-12. Biochem J 264: 397–402

Andersen T, Briseid T, Nesbakken T, Ormerod J, Sirevåg R and Thorud M (1983) Mechanisms of synthesis of 5-amino-levulinate in purple, green, and blue-green bacteria. FEMS Microbiol Lett 19: 303–306

Anraku Y and Gennis RB (1987) The aerobic respiratory chain of *Escherichia coli*. Trends Biochem Sci 12: 262–266

Arnheim K and Oelze J (1983) Differences in the control of bacteriochlorophyll formation by light and oxygen. Arch Microbiol 135: 299–304

Avissar YJ (1980) Biosynthesis of 5-aminolevulinate from glutamate in *Anabaena variabilis*. Biochim Biophys Acta 613: 220–228

Avissar YJ and Beale SI (1989a) Identification of the enzymatic basis for δ-aminolevulinic acid auxotrophy in a *hemA* mutant of *Escherichia coli*. J Bacteriol 171: 2919–2924

Avissar YJ and Beale SI (1989b) Biosynthesis of tetrapyrrole pigment precursors: Pyridoxal requirement of the amino-transferase step in the formation of δ-aminolevulinate from glutamate in extracts of *Chlorella vulgaris*. Plant Physiol 89: 852–859

Avissar YJ and Beale SI (1990) Cloning and expression of a structural gene from *Chlorobium vibrioforme* that complements the *hemA* mutation in *Escherichia coli*. J Bacteriol 172: 1656–1659

Avissar YJ, Ormerod JG and Beale SI (1989) Distribution of δ-aminolevulinic acid biosynthetic pathways among phototrophic bacterial groups. Arch Microbiol 151: 513–519

Barnard GF and Akhtar M (1975) Stereochemistry of porphyrinogen carboxy-lyase reaction in haem biosynthesis. J

Chem Soc Chem Commun 1975: 494–496

Battersby AR, Fookes CJR, Matcham GWJ and McDonald E (1979a) Order of assembly of the four pyrrole rings during biosynthesis of the natural porphyrins. J Chem Soc Chem Commun 1979: 539–541

Battersby AR, Fookes CJR, Gustafson-Potter KE, Matcham GWJ and McDonald E (1979b) Proof by synthesis that unrearranged hydroxymethylbilane is the product from deaminase and the substrate for cosynthetase in the biosynthesis of Uro'gen-III. J Chem Soc Chem Commun 1979: 1155–1158

Battersby AR, Fookes CJR, Matcham GWJ, McDonald E and Gustafson-Potter KE (1979c) Biosynthesis of the natural porphyrins: Experiments on the ring-closure steps with the hydroxy-analogue of porphobilinogen. J Chem Soc Chem Commun 1979: 316–319

Battersby AR, Fookes CJR, Matcham GWJ, McDonald E and Hollenstein R (1983) Biosynthesis of porphyrins and related molecules, part 20: Purification of deaminase and studies on its mode of action. J Chem Soc Perkin Trans I 1983: 3031–3040

Beale SI and Castelfranco PA (1974a) The biosynthesis of δ-aminolevulinic acid in higher plants, I: Accumulation of δ-aminolevulinic acid in greening plant tissues. Plant Physiol 53: 291–296

Beale SI and Castelfranco PA (1974b) The biosynthesis of δ-aminolevulinic acid in higher plants, II: Formation of ^{14}C-δ-aminolevulinic acid from labeled precursors in greening plant tissues. Plant Physiol 53: 297–303

Beale SI, Gough SP and Granick S (1975) The biosynthesis of δ-aminolevulinic acid from the intact carbon skeleton of glutamic acid in greening barley. Proc Natl Acad Sci USA 72: 2719–2723

Beckman DL, Trawick DR and Kranz RG (1992) Bacterial cytochromes c biogenesis. Genes Devel 6: 268–283

Biel AJ (1986) Control of bacteriochlorophyll accumulation by light in *Rhodobacter capsulatus*. J Bacteriol 68: 655–659

Biel AJ (1992) Oxygen-regulated steps in the *Rhodobacter capsulatus* tetrapyrrole biosynthetic pathway. J Bacteriol 174: 5272–5274

Biel SW and Biel AJ (1990) Isolation of a *Rhodobacter capsulatus* mutant that lacks *c*-type cytochromes and excretes porphyrins. J Bacteriol 172: 1321–1326

Biel SW, Wright MS and Biel AJ (1988) Cloning of the *Rhodobacter capsulatus hemA* gene. J Bacteriol 170: 4382–4384

Breu V and Dörnemann D (1988) Formation of 5-aminolevulinate *via* glutamate-1-semialdehyde and 4,5-dioxovalerate with participation of an RNA component in *Scenedesmus obliquus* mutant C-2A'. Biochim Biophys Acta 967: 135–140

Brunt RD, Leeper FJ, Grgurina I and Battersby AR (1989) Biosynthesis of vitamin B_{12}: Synthesis of (±)-[5-^{13}C]Faktor-1 ester: Determination of the oxidation state of precorrin-1. J Chem Soc Chem Commun 1989: 428–431

Bruyant P and Kannangara CG (1987) Biosynthesis of δ-aminolevulinate in greening barley leaves, VIII: Purification and characterization of the glutamate-tRNA ligase. Carlsberg Res Commun 52: 99–109

Bull AD, Breu V, Kannangara CG, Rogers LJ and Smith AJ (1990) Cyanobacterial glutamate 1-semialdehyde aminotransferase: Requirement for pyridoxamine phosphate. Arch Microbiol 154: 56–59

Burnham BF and Lascelles J (1963) Control of porphyrin biosynthesis through a negative-feedback mechanism: Studies with preparations of δ-aminolaevulate synthetase and δ-aminolaevulate dehydratase from *Rhodopseudomonas spheroides*. Biochem J 87: 462–472

Camadro J-M, Chambon H, Jolles J and Labbe P (1986) Purification and properties of coproporphyrinogen oxidase from the yeast *Saccharomyces cerevisiae*. Eur J Biochem 156: 579–587

Cavaleiro JAS, Kenner GW and Smith KM (1974) Pyrroles and related compounds, part XXXII: Biosynthesis of protoporphyrin-IX from coproporphyrinogen-III. J Chem Soc Perkin Trans I 1974: 1188–1194

Chang T-E, Wegmann B and Wang W-y (1990) Purification and characterization of glutamyl-tRNA synthetase: An enzyme involved in chlorophyll biosynthesis. Plant Physiol 93: 1641–1649

Chen M-W, Jahn D, Schön A, O'Neill GP and Söll D (1990) Purification of the glutamyl-tRNA reductase from *Chlamydomonas reinhardtii* involved in δ-aminolevulinic acid formation during chlorophyll biosynthesis. J Biol Chem 265: 4058–4063

Chiu JT, Loewen PC, Switala J, Gennis RB and Timkovich R (1989) Proposed structure for the prosthetic group of catalase HPII from *Escherichia coli*. J Am Chem Soc 111: 7046–7050

Clement-Metral JD (1979) Activation of ALA synthetase by reduced thioredoxin in *Rhodopseudomonas spheroides* Y. FEBS Lett 101: 116–120

Cohen-Bazire G, Sistrom WR and Stanier RY (1957) Kinetic studies of pigment synthesis by non-sulfur purple bacteria. J Cell Comp Physiol 49: 25–68

Coomber SA, Jones RM, Jordan PM and Hunter CN (1992) A putative anaerobic coproporphyrin III oxidase in *Rhodobacter sphaeroides*, I: Molecular cloning, transposon mutagenesis and sequence analysis of the gene. Mol Microbiol 6: 3159–3169

Crockett N, Alefounder PR, Battersby AR and Abell C (1991) Uroporphyrinogen III synthase: Studies on its mechanism of action, molecular biology and biochemistry. Tetrahedron 47: 6003–6014

Crouzet J, Cauchois L, Blanche F, Debussche L, Thibaut D, Rouyez M-C, Rigault S, Mayaux J-F and Cameron B (1990) Nucleotide sequence of a *Pseudomonas denitrificans* 5.4-kilobase DNA fragment containing five *cob* genes and identification of structural genes encoding *S*-adenosyl-methionine:uroporphyrinogen III methyltransferase and cobyrinic acid *a,c*-diamide synthase. J Bacteriol 172: 45968–5979

Dailey HA (1982) Purification and characterization of membrane-bound ferrochelatase from *Rhodopseudomonas sphaeroides*. J Biol Chem 257: 14714–14718

Dailey TA, Meissner P and Dailey HA (1994) Expression of a cloned protoporphyrinogen oxidase. J Biol Chem 269: 813–815

Davidson E, Prince RC, Daldal F, Hauska G and Marrs BL (1987) *Rhodobacter capsulatus* MT113: A single mutation results in the absence of *c*-type cytochromes and in the absence of the cytochrome bc_1 complex. Biochim Biophys Acta 890: 292–301

Davies RC and Neuberger A (1979a) Control of 5-aminolaevulinate synthetase activity in *Rhodopseudomonas spheroides*: Purification and properties of the high-activity form of the enzyme. Biochem J 177: 649–659

Davies RC and Neuberger A (1979b) Control of 5-aminolaevulinate synthetase activity in *Rhodopseudomonas spheroides*: Binding of pyridoxal phosphate to 5-aminolaevulinate synthetase. Biochem J 177: 661–671

Deeg R, Kriemler H-P, Bergmenn K-H and Müller G (1977) Zue Cobyrinsäure-Biosynthese: Neuartige, methylierte Hydroporphyrine und deren Bedeutung bei der Cobyrinsäure-Bildung. Hoppe-Seyler's Z Physiol Chem 358: 339–352

Delaunay A-M, Huault C and Balangé AP (1991) Molecular cloning of the 5-aminolevulinic acid dehydratase gene from *Rhodobacter sphaeroides*. J Bacteriol 173: 2712–2715

Dörnemann D, Kotzabasis K, Richter P, Breu V and Senger H (1989) The regulation of chlorophyll biosynthesis by the action of protochlorophyllide on ^{glu}t-RNA-ligase. Bot Acta 102: 112–115

Drygas ME, Lambowitz AM and Nargang FE (1989) Cloning and analysis of the *Neurospora crassa* gene for cytochrome *c* heme lyase. J Biol Chem 264: 17897–17906

Dumont ME, Ernst JF, Hampsey DM and Sherman F (1987) Identification and sequence of the gene encoding cytochrome *c* heme lyase in the yeast *Saccharomyces cerevisiae*. EMBO J 6: 235–241

Elder GH, Evans JO, Jackson JR and Jackson AH (1978) Factors determining the sequence of oxidative decarboxylation of the 2- and 4-propionate substituents of coproporphyrinogen III by coproporphyrinogen oxidase in rat liver. Biochem J 169: 215–223

Elliott T, Avissar YJ, Rhie G and Beale SI (1990) Cloning and sequence of the *Salmonella typhimurium hemL* gene and identification of the missing enzyme in *hemL* mutants as glutamate-1-semialdehyde aminotransferase. J Bacteriol 172: 7071–7084

Enosawa S and Ohashi A (1986) Localization of enzyme for heme attachment to apocytochrome *c* in yeast mitochondria. Biochem Biophys Res Commun 141: 1145–1150

Evans WR, Fleischman, DE, Calvert HE, Pyati PV, Alter GM and Subba Rao NS (1990) Bacteriochlorophyll and photosynthetic reaction centers in *Rhizobium* strain BTAi 1. Appl Environ Microbiol 56: 3445–3449

Fanica-Gaignier M and Clement-Metral JD (1971) ATP inhibition of aminolevulinate (ALA) synthetase activity in *Rhodopseudomonas spheroides* Y. Biochem Biophys Res Commun 44: 192–198

Fanica-Gaignier M and Clément-Métral J (1973a) Cellular compartmentation of two species of δ-aminolevulinic acid synthetase in a facultative photohetero-trophic bacterium (*Rps. spheroides Y.*). Biochem Biophys Res Commun 55: 610–615

Fanica-Gaignier M and Clement-Metral J (1973b) 5-Amino-levulinic-acid synthetase of *Rhodopseudomonas spheroides* Y: Purification and some properties. Eur J Biochem 40: 13–18

Fanica-Gaignier M and Clement-Metral J (1973c) 5-Amino-levulinic-acid synthetase of *Rhodopseudomonas spheroides* Y: Kinetic mechanism and inhibition by ATP. Eur J Biochem 40: 19–24

Frustaci JM and O'Brian MR (1992) Characterization of a *Bradyrhizobium japonicum* ferrochelatase mutant and isolation of the *hemH* gene. J Bacteriol 174: 4223–4229

Frustaci JM and O'Brian MR (1993) The *Escherichia coli visA* gene encodes ferrochelatase, the final enzyme of the heme biosynthetic pathway. J Bacteriol 175: 2154–2156

Fukumori Y, Watanabe K and Yamanaka T (1987) Cytochrome aa_3 from the aerobic photoheterotroph *Erythrobacter longus*: Purification, and enzymatic and molecular features. J Biochem 102: 777–784

Garcia-Horsman JA, Barquera B, Rumbley J, Ma J and Gennis RB (1994) The superfamily of heme-copper respiratory oxidases. J Bacteriol 176: 5587–5600

Gennis RB (1987) The cytochromes of *Escherichia coli*. FEMS Microbiol Lett 46: 387–399

Gennis RB, Casey RP, Azzi A and Ludwig B (1982) Purification and characterization of the cytochrome *c* oxidase from *Rhodopseudomonas sphaeroides*. Eur J Biochem 125: 189–195

Gibson KD (1958) Biosynthesis of δ-aminolaevulic acid by extracts of *Rhodopseudomonas spheroides*. Biochim Biophys Acta 28: 451

Goldman BS and Roth JR (1993) Genetic structure and regulation of the *cysG* gene in *Salmonella typhimurium*. J Bacteriol 175: 1457–1466

Goto K, Higuchi M, Sakai H and Kikuchi G (1967) Differential inhibition of induced syntheses of δ-aminolevulinate synthetase and bacteriochlorophyll in dark-aerobically grown *Rhodopseudomonas spheroides*. J Biochem 61: 186–192

Gough SP and Kannangara CG (1977) Synthesis of δ-aminolevulinate by a chloroplast stroma preparation from greening barley leaves. Carlsberg Res Commun 42: 459–464

Gough SP, Kannangara CG and Bock K (1989) A new method for the synthesis of glutamate 1-semialdehyde: Characterization of its structure in solution by NMR spectroscopy. Carlsberg Res Commun 54: 99–108

Grimm B, Bull A and Breu V (1991a) Structural genes of glutamate 1-semialdehyde aminotransferase for porphyrin synthesis in a cyanobacterium and *Escherichia coli*. Mol Gen Genet 225: 1–10

Grimm B, Smith AJ, Kannangara CG and Smith M (1991b) Gabaculine-resistant glutamate 1-semialdehyde aminotransferase of *Synechococcus*: Deletion of a tripeptide close to the NH_2 terminus and internal amino acid substitution. J Biol Chem 266: 12495–12501

Grimm B, Smith MA and von Wettstein D (1992) The role of Lys272 in the pyridoxal 5-phosphate active site of *Synechococcus* glutamate-1-semialdehyde aminotransferase. Eur J Biochem 206: 579–585

Hansson M and Hederstedt L (1992) Cloning and characterization of the *Bacillus subtilis hemEYH* gene cluster, which encodes protoheme IX biosynthetic enzymes. J Bacteriol 174: 8081–8093

Hansson M and von Wachenfeldt C (1993) Heme *b* (protoheme IX) is a precursor of heme *a* and heme *d* in *Bacillus subtilis*. FEMS Microbiol Lett 107: 121–126

Harashima K, Shiba T, Totsuka T, Simidu U and Taga N (1978) Occurrence of bacteriochlorophyll *a* in a strain of an aerobic heterotrophic bacterium. Agric Biol Chem 42: 1627–1628

Harashima K, Hayasaki J, Ikari T and Shiba T (1980) O_2-stimulated synthesis of bacteriochlorophyll and carotenoids in marine bacteria. Plant Cell Physiol 21: 1283–1294

Hart GJ, Miller AD, Leeper FJ and Battersby AR (1987) Biosynthesis of natural porphyrins: Proof that hydroxymethylbilane synthase (porphobilinogen deaminase) uses a novel binding group in its catalytic action. J Chem Soc Chem Commun 1987: 1762–1765

Higuchi M and Bogorad L (1975) The purification and properties of uroporphyrinogen I synthases and uroporphyrinogen III cosynthase: Interactions between the enzymes. Ann N Y Acad Sci 244: 401–418

Hoober JK, Kahn A, Ash D, Gough S and Kannangara CG (1988) Biosynthesis of δ-aminolevulinate in greening barley leaves, IX: Structure of the substrate, mode of gabaculine inhibition, and the catalytic mechanism of glutamate 1-semialdehyde aminotransferase. Carlsberg Res Commun 53: 11–25

Hornberger U, Liebetanz R, Tichy H-V and Drews G (1990) Cloning and sequencing of the *hemA* gene of *Rhodobacter capsulatus* and isolation of a δ-aminolevulinic acid-dependent mutant strain. Mol Gen Genet 221: 371–378

Houen G, Gough SP and Kannangara CG (1983) δ-Aminolevulinate synthesis in greening barley, V: The structure of glutamate 1-semialdehyde. Carlsberg Res Commun 48: 567–572

Huang D-D and Wang W-Y (1986) Chlorophyll synthesis in *Chlamydomonas* starts with the formation of glutamyl-tRNA. J Biol Chem 261: 13451–13455

Huang DD, Wang W-Y, Gough SP and Kannangara CG (1984) δ-Aminolevulinic acid-synthesizing enzymes need an RNA moiety for activity. Science 225: 1482–1484

Huault C, Aoues A and Colin P (1987) Reconsidération de la structure sous-unitaire de la δ-aminolévulinate déshydratase de feuilles d'épinard. C R Acad Sci Paris 305(III): 671–676

Ilag LL and Jahn D (1992) Activity and spectroscopic properties of the *Escherichia coli* glutamate 1-semialdehyde aminotransferase and the putative active site mutant K265R. Biochemistry 31: 7143–7151

Inoue I, Oyama H and Tuboi S (1979) On the nature of the activating enzyme of the inactive form of δ-aminolevulinate synthetase in *Rhodopseudomonas spheroides*. J Biochem 86: 477–482

Jacobs NJ and Jacobs, JM (1976) Nitrate, fumarate, and oxygen as electron acceptors for a late step in microbial heme synthesis. Biochim Biophys Acta 449: 1–9

Jacobs NJ and Jacobs, JM (1981) Protoporphyrinogen oxidation in *Rhodopseudomonas spheroides*, a step in heme and bacteriochlorophyll synthesis. Arch Biochem Biophys 211: 305–311

Jacobs JM and Jacobs, NJ (1987) Oxidation of protoporphyrinogen to protoporphyrin, a step in chlorophyll and haem biosynthesis: Purification and partial characterization of the enzyme from barley organelles. Biochem J 244: 219–224

Jahn D (1992) Complex formation between glutamyl-tRNA synthetase and glutamyl-tRNA reductase during tRNA-dependent synthesis of 5-aminolevulinic acid in *Chlamydomonas*. FEBS Lett 314: 77–80

Jahn D, Michelsen U and Söll D (1991) Two glutamyl-tRNA reductase activities in *Escherichia coli*. J Biol Chem 266 2542–2548

Jones MS and Jones OTG (1970) Ferrochelatase of *Rhodopseudomonas spheroides*. Biochem J 119: 453–462

Jones RM and Jordan PM (1993) Purification and properties of the uroporphyrinogen decarboxylase from *Rhodobacter sphaeroides*. Biochem J 293: 703–712

Jones C, Jordan PM and Akhtar M (1984) Mechanism and stereochemistry of the porphobilinogen deaminase and protoporphyrinogen IX oxidase reactions: Stereospecific manipulation of hydrogen atoms at the four methylene bridges during the biosynthesis of haem. J Chem Soc Perkin Trans I 1984: 2625–2633

Jordan PM (1991) The biosynthesis of 5-aminolaevulinic acid and its transformation into uroporphyrinogen III. In: Jordan PM (ed) Biosynthesis of Tetrapyrroles, pp 1–66. Elsevier, Amsterdam

Jordan PM and Berry A (1981) Mechanism of action of porphobilinogen deaminase: the participation of stable enzyme substrate covalent intermediates between porphobilinogen and the porphobilinogen deaminase from *Rhodopseudomonas spheroides*. Biochem J 195: 177–181

Jordan PM and Seehra JS (1979) The biosynthesis of uroporphyrinogen, III: Order of assembly of the four porphobilinogen molecules in the formation of the tetrapyrrole ring. FEBS Lett 104: 364–366

Jordan PM and Seehra JS (1980) Mechanism of action of 5-aminolevulinic acid dehydratase: Stepwise order of addition of the two molecules of 5-aminolevulinic acid in the enzymic synthesis of porphobilinogen. J Chem Soc Chem Commun 1980: 240–242

Jordan PM and Warren MJ (1987) Evidence for a dipyrromethane cofactor at the catalytic site of *E. coli* porphobilinogen deaminase. FEBS Lett 225: 87–92

Jordan PM, Mgbeje IAB, Thomas SD and Alwan AF (1988) Nucleotide sequence for the *hemD* gene of *Escherichia coli* encoding uroporphyrinogen III synthase and initial evidence for a *hem* operon. Biochem J 249: 613–616

Jordan PM, Cheung K-M, Sharma RP and Warren MJ (1993) 5-amino-6-hydroxy-3,4,5,6-tetrahydropyran-2-one (HAT): A stable, cyclic form of glutamate 1-semialdehyde, the natural precursor for tetrapyrroles. Tet Lett 34: 1177–1180

Kämpf C, Wynn RM, Shaw R and Knaff DB (1987) The electron transfer chain of aerobically grown *Rhodopseudomonas viridis*. Biochim Biophys Acta 894: 228–238

Kannangara CG and Gough SP (1978) Biosynthesis of δ-aminolevulinate in greening barley leaves: Glutamate 1-semialdehyde aminotransferase. Carlsberg Res Commun 43: 185–194

Kannangara CG and Schouboe A (1985) Biosynthesis of δ-aminolevulinate in greening barley leaves, VII: Glutamate 1-semialdehyde accumulation in gabaculine treated leaves. Carlsberg Res Commun 50: 179–191

Kannangara CG, Gough SP, Oliver RP and Rasmussen SK (1984) Biosynthesis of δ-aminolevulinate in greening barley leaves, VI: Activation of glutamate by ligation to RNA. Carlsberg Res Commun 49: 417–437

Kannangara CG, Gough SP, Bruyant P, Hoober JK, Kahn A and von Wettstein D (1988) $tRNA^{Glu}$ as a cofactor in δ-aminolevulinate biosynthesis: Steps that regulate chlorophyll synthesis. Trends Biochem Sci 13: 139–143

Kawasaki H, Hishino Y and Yamasato K (1993) Phylogenetic diversity of phototrophic purple bacteria in the *Proteobacteria* α group. FEMS Microbiol Lett 112: 61–66

Keithly JH and Nadler KD (1983) Protoporphyrin formation in *Rhizobium japonicum*. J Bacteriol 154: 838–845

Kikuchi G, Kumar A., Talmage P and Shemin D (1958) The enzymatic synthesis of δ-aminolevulinic acid. J Biol Chem 233: 1214–1219

Kipe-Nolt JA and Stevens SE Jr (1980) Biosynthesis of δ-aminolevulinic acid from glutamate in *Agmenellum quadruplicatum*. Plant Physiol 65: 126–128

Klemm DJ and Barton LL (1987) Purification and properties of protoporphyrinogen oxidase from an anaerobic bacterium, *Desulfovibrio gigas*. J Bacteriol 169: 5209–5215

Koopman GE, Juknat de Geralink AA and Batlle AM del C (1986) Porphyrin biosynthesis in *Rhodopseudomonas palustris*, V: Purification of porphyrinogen decarboxylase and some unusual properties. Int J Biochem 18: 935–944

Kotler ML, Fumagalli SA, Juknat AA and Batlle AM del C (1987) Porphyrin biosynthesis in *Rhodopseudomonas palustris*, VIII: Purification and properties of deaminase. Comp Biochem Physiol 87B: 601–606

Laghai A and Jordan PM (1976) A partial reaction of δ-aminolaevulinate synthetase from *Rhodopseudomonas spheroides*. Biochem Soc Trans 4: 52–53

Laghai A and Jordan PM (1977) An exchange reaction catalysed by δ-aminolaevulinate synthase from *Rhodopseudomonas spheroides*. Biochem Soc Trans 5: 299–300

Lapointe J and Söll D (1972) Glutamyl transfer ribonucleic acid synthetase of *Escherichia coli*, I: Purification and properties. J Biol Chem 247: 4966–4974

Lascelles J (1960) The synthesis of enzymes concerned in bacteriochlorophyll formation in growing cultures of *Rhodopseudomonas spheroides*. J Gen Microbiol 23: 487–498

Liedgens W, Lütz C and Schneider HAW (1983) Molecular properties of 5-aminolevulinic acid dehydratase from *Spinacia oleracea*. Eur J Biochem 135: 75–79

Liu J, Lin S-X, Blochet J-E, Pézolet M and Lapointe J (1993) The glutamyl-tRNA synthetase of *Escherichia coli* contains one atom of zinc essential for its native conformation and its catalytic activity. Biochemistry 32: 11390–11396

Loewen PC, Switala J, von Ossowski I, Hillar A, Christie A, Tattrie B and Nicholls P (1993) Catalase HPII of *Escherichia coli* catalyzes the conversion of protoheme to *cis*-heme *d*. Biochemistry 32: 10159–10164

Louie GV, Brownlie PD, Lambert R, Cooper JB, Blundell TL, Wood SP, Warren MJ, Woodcock SC and Jordan PM (1992) Structure of porphobilinogen deaminase reveals a flexible multidomain polymerase with a single catalytic site. Nature 359: 33–39

Luo J and Lim CK (1993) Order of uroporphyrinogen III decarboxylation on incubation of porphobilinogen and uroporphyrinogen III with erythrocyte uroporphyrinogen decarboxylase. Biochem J 289: 529–532

Madsen O, Sandal L, Sandal NN and Marcker KA (1993) A soybean coproporphyrinogen oxidase gene is highly expressed in root nodules. Plant Mol Biol 23: 35–43

Majumdar D, Avissar YJ, Wyche JH and Beale SI (1991) Structure and expression of the *Chlorobium vibrioforme hemA* gene. Arch Microbiol 156: 281–289

Marriott J, Neuberger A and Tait GH (1970) Activation of δ-aminolaevulate synthetase in extracts of *Rhodopseudomonas spheroides*. Biochem J 117: 609–613

Matters GL and Beale SI (1994) Structure and light-regulated expression of the *gsa* gene encoding the chlorophyll biosynthetic enzyme, glutamate-1-semialdehyde aminotransferase, in *Chlamydomonas reinhardtii*. Plant Mol Biol 24: 617–629

Mau Y-HL and Wang W-Y (1988) Biosynthesis of δ-aminolevulinic acid in *Chlamydomonas reinhardtii*: Study of the transamination mechanism using specifically labeled glutamate. Plant Physiol 86: 793–797

Mayer SM, Weinstein JD and Beale SI (1987) Enzymatic conversion of glutamate to δ-aminolevulinate in soluble extracts of *Euglena gracilis*. J Biol Chem 262: 12541–12549

Mayer SM, Gawlita E, Avissar YJ, Anderson VE and Beale SI (1993a) Intermolecular nitrogen transfer in the enzymatic conversion of glutamate to δ-aminolevulinic acid by extracts of *Chlorella vulgaris*. Plant Physiol 101: 1029–1038

Mayer SM, Rieble S and Beale SI (1993b) Magnesium is required for glutamyl-tRNA reductase activity in extracts of *Chlorella vulgaris* and *Synechocystis* sp. PCC 6803. Plant Physiol 101: S47

Meller E and Harel E (1978) The pathway of 5-aminolevulinic acid synthesis in *Chlorella vulgaris* and in *Fremyella diplosiphon*. In: Akoyunoglou G and Argyroudi-Akoyunoglou JH (eds) Chloroplast Development pp 51–57 Elsevier, Amsterdam

Meller E, Belkin S and Harel E (1975) The biosynthesis of δ-aminolevulinic acid in greening maize leaves. Phytochemistry 14: 2399–2402

Meyer TE and Cusanovich MA (1989) Structure, function and distribution of soluble bacterial redox proteins. Biochim Biophys Acta 975: 1–28

Michalski WP and Nicholas DJD (1985) Molecular characterization of a copper-containing nitrite reductase from *Rhodopseudomonas sphaeroides* forma sp. *denitrificans*. Biochim Biophys Acta 828: 130–137

Mitchell LW and Jaffe EJ (1993) Porphobilinogen synthase from *Escherichia coli* is a Zn(II) metalloenzyme stimulated by Mg(II). Arch Biochem Biophys 300: 169–177

Miyamoto K, Nakahigashi K, Nishimura K and Inokuchi H (1991) Isolation and characterization of visible light-sensitive mutants of *Escherichia coli* K12. J Mol Biol 219: 393–398

Moberg PA and Avissar YJ (1994) A gene cluster in *Chlorobium vibrioforme* encoding the first enzymes of chlorophyll biosynthesis. Photosynth Res 41: 253–260

Mombelli L, McDonald E and Battersby AR (1976) Enzymatic formation of a tricarboxylic porphyrin and protoporphyrin-XIII from coprogen-IV. Tet Lett 1976: 1037–1040

Nandi DL and Shemin D (1968) δ-Aminolevulinic acid dehydratase of *Rhodopseudomonas spheroides*, I: Isolation and properties. J Biol Chem 243: 1224–1230

Nandi DL and Shemin D (1977) Quaternary structure of δ-aminolevulinic acid synthase from *Rhodopseudomonas spheroides*. J Biol Chem 252: 2278–2280

Nargang FE, Drygas ME, Kwong PL, Nicholson DW and Neupert W (1988) A mutant of *Neurospora crassa* deficient in cytochrome *c* heme lyase activity cannot import cytochrome *c* into mitochondria. J Biol Chem 263: 9388–9394

Neidle EL and Kaplan S (1993a) Expression of the *Rhodobacter sphaeroides hemA* and *hemT* genes, encoding two 5-aminolevulinic acid synthase isoenzymes. J Bacteriol 175: 2292–2303

Neidle EL and Kaplan S (1993b) 5-Aminolevulinic acid availability and control of spectral complex formation in HemA and HemT mutants of *Rhodobacter sphaeroides*. J Bacteriol 175: 2304–2313

Nicholson DW, Köhler H and Neupert W (1987) Import of cytochrome *c* into mitochondria: Cytochrome *c* lyase. Eur J Biochem 164: 147–157

Nishimura Y, Shimadzu M and Iizuka H (1981) Bacteriochlorophyll formation in radiation-resistant *Pseudomonas radiora*. J Gen Appl Microbiol 27: 427–430

Oelze J (1986) Inhibition by light of 5-aminolevulinic acid synthase in extracts from *Rhodopseudomonas sphaeroides*. FEMS Microbiol Lett 37: 321–323

Oelze J and Arnheim K (1983) Control of bacteriochlorophyll formation by oxygen and light in *Rhodopseudomonas sphaeroides*. FEMS Microbiol Lett 19: 197–199

Oh-hama T, Seto H and Miyachi S (1986a) ^{13}C-NMR evidence of bacteriochlorophyll *a* formation by the C_5 pathway in *Chromatium*. Arch Biochem Biophys 246: 192–198

Oh-hama T, Seto H and Miyachi S (1986b) ^{13}C-NMR evidence for bacteriochlorophyll *c* formation by the C_5 pathway in green sulfur bacterium, *Prosthecochloris*. Eur J Biochem 159: 189–194

Oh-hama T, Santander PJ, Stolowich NJ and Scott AI (1991) Bacteriochlorophyll *c* formation via the C_5 pathway of 5-aminolevulinic acid synthesis in *Chloroflexus aurantiacus*. FEBS Lett 281: 173–176

O'Neill GP and Söll D (1990) Expression of the *Synechocystis* sp. PCC 6803 $tRNA^{Glu}$ gene provides tRNA for protein and chlorophyll biosynthesis. J Bacteriol 172: 6363–6371

O'Neill GP, Peterson DM, Schön A, Chen M-W and Söll D (1988) Formation of the chlorophyll precursor δ-aminolevulinic acid in cyanobacteria requires aminoacylation of a $tRNA^{Glu}$ species. J Bacteriol 170: 3810–3816

Oyama H and Tuboi S (1979) Occurrence of a novel high molecular weight activator of δ-aminolevulinate synthetase in *Rhodopseudomonas spheroides*. J Biochem 86: 483–489

Petricek M, Rutberg L, Schröder I and Hederstedt L (1990) Cloning and characterization of the *hemA* region of the *Bacillus subtilis* chromosome. J Bacteriol 172: 2250–2258

Poulson R and Polglase WJ (1974) Aerobic and anaerobic coproporphyrinogenase activities in extracts from *Saccharomyces cerevisiae*. J Biol Chem 249: 6367–6371

Powell KA, Cox R, McConville M and Charles HP (1973) Mutations affecting porphyrin biosynthesis in *Escherichia coli*. Enzyme 16: 65–73

Proulx M, Duplain L, Lacoste L, Yaguchi M and Lapointe J (1983) The monomeric glutamyl-tRNA synthetase from *Bacillus subtilis* 168 and its regulatory factor: Their purification, characterization, and the study of their interaction. J Biol Chem 258: 753–759

Rieble S and Beale SI (1988) Enzymatic transformation of glutamate to δ-aminolevulinic acid by soluble extracts of *Synechocystis* sp. 6803 and other oxygenic prokaryotes. J Biol Chem 263: 8864–8871

Rieble S and Beale SI (1991) Separation and partial characterization of enzymes catalyzing δ-aminolevulinic acid formation in *Synechocystis* sp. PCC 6803. Arch Biochem Biophys 289: 289–297

Rieble S, Ormerod JG and Beale SI (1989) Transformation of glutamate to δ-aminolevulinic acid by soluble extracts of *Chlorobium vibrioforme*. J Bacteriol 171: 3782–3787

Rosé S, Frydman RB, de los Santos C, Sburlati A, Valasinas A and Frydman B (1988) Spectroscopic evidence for a porphobilinogen deaminase-tetrapyrrole complex that is an intermediate in the biosynthesis of uroporphyrinogen III. Biochemistry 27: 4871–4879

Saiki K, Mogi T and Anraku Y (1992) Heme *o* biosynthesis in *Escherichia coli*: The *cyoE* gene in the cytochrome *BO* operon encodes a protoheme IX farnesyltransferase. Biochem Biophys Res Commun 189: 1491–1497

Saiki K, Mogi T, Ogura K and Anraku Y (1993) In vitro heme *o* synthesis by the *cyoE* gene product from *Escherichia coli*. J Biol Chem 268: 26041–26045

Sandy JD, Davies RC and Neuberger A (1975) Control of 5-aminolaevulinate synthetase activity in *Rhodopseudomonas spheroides*: A role for trisulfides. Biochem J 150: 245–257

Sasaki T, Motokawa Y and Kikuchi G (1970) Occurrence of both *a*-type and *o*-type cytochromes as the functional terminal oxidases in *Rhodopseudomonas spheroides*. Biochim Biophys Acta 197: 284–291

Sasarman A, Nepveu A, Echelard Y, Dymetryszyn J, Drolet M and Goyer C (1987) Molecular cloning and sequencing of the *hemD* gene of *Escherichia coli* K-12 and preliminary data on the Uro operon. J Bacteriol 169: 4257–4262

Sasarman A, Letowski J, Czaika G, Ramirez V, Nead MA, Jacobs JM and Morais R (1993) Nucleotide sequence of the *hemG* gene involved in the protoporphyrinogen oxidase activity of *Escherichia coli* K12. Can J Microbiol 39: 1155–1161

Sato K (1978) Bacteriochlorophyll formation by facultative methylotrophs, *Protaminobacter ruber* and *Pseudomonas* AM1. FEBS Lett 85: 207–210

Sato K, Ishida K, Kuno T, Mizuno A and Shimizu S (1981) Regulation of vitamin B_{12} and bacteriochlorophyll biosynthesis in a facultative methylotroph, *Protaminobacter ruber*. J Nutr Sci Vitaminol 27: 439–447

Sato K, Ishida K, Shirai M and Shimizu S (1985) Occurrence and some properties of two types of δ-aminolevulinic acid synthase in a facultative methylotroph, *Protaminobacter ruber*. Agric Biol Chem 49: 3423–3428

Sawyer E. and Smith RA (1958) δ-Aminolevulinate synthesis in *Rhodopseudomonas spheroides*. Bacteriol Proc 1958: 111

Schneegurt MA and Beale SI (1988) Characterization of the RNA required for biosynthesis of δ-aminolevulinic acid from glutamate: Purification by anticodon-based affinity chromatography and determination that the UUC glutamate anticodon is a general requirement for function in ALA biosynthesis. Plant Physiol 86: 497–504

Schön A, Krupp G, Gough S, Berry-Lowe S, Kannangara CG and Söll D (1986) The RNA required in the first step of chlorophyll biosynthesis is a chloroplast glutamate tRNA. Nature 322: 281–284

Schröder I, Hederstedt L, Kannangara CG and Gough SP (1992) Glutamyl-tRNA reductase activity in *Bacillus subtilis* is dependent on the *hemA* gene product. Biochem J 281: 843–850

Scott AI, Burton G, Jordan PM, Matsumoto H, Fagerness PE and Pryde LM (1980) N.M.R. spectroscopy as a probe for the study of enzyme-catalysed reactions: Further observations of preuroporphyrinogen, a substrate for uroporphyrinogen III cosynthetase. J Chem Soc Chem Commun 1980: 384–387

Seehra JS, Jordan PM and Akhtar M (1983) Anaerobic and aerobic coproporphyrinogen III oxidases of *Rhodopseudomonas spheroides*: Mechanism and stereochemistry of vinyl group formation. Biochem J 209: 709–718

Shibata H and Ochiai H (1977) Purification and properties of δ-aminolevulinic acid dehydratase from radish cotyledons. Plant Cell Physiol 18: 421–429

Shioi Y and Doi M (1988) Control of bacteriochlorophyll accumulation by light in an aerobic photosynthetic bacterium, *Erythrobacter* sp. OCh114. Arch Biochem Biophys 266: 470–477

Siegel LM (1978) Structure and function of siroheme and the

siroheme enzymes. In: Singer TP and Ondarza RN (eds) Developmental Biochemistry, Vol. 1, pp 201–214. Elsevier, Amsterdam

Siegel LM, Davis PS and Murphy MJ (1977) Incorporation of methionine-derived methyl groups into sirohaem by *Escherichia coli*. Biochem J 167: 669–674

Smith KM and Huster MS (1987) Bacteriochlorophyll-*c* formation via the glutamate C-5 pathway in *Chlorobium* bacteria. J Chem Soc Chem Commun 1987: 14–16

Sotiriou C and Chang CK (1988) Synthesis of the heme *d* prosthetic group of bacterial terminal oxidase. J Am Chem Soc 110: 2264–2270

Spencer P and Jordan PM (1993) Purification and characterization of 5-aminolevulinic acid dehydratase from *Escherichia coli* and a study of the reactive thiols at the metal-binding domain. Biochem J 290: 279–287

Stark WM, Hawker CJ, Hart GJ, Philippides A, Petersen PM, Lewis JD, Leeper FJ and Battersby AR (1993) Biosynthesis of porphyrins and related macrocycles, part 40: Synthesis of a spiro-lactam related to the proposed spiro-intermediate for porphyrin biosynthesis: Inhibition of cosynthetase. J Chem Soc Perkin Trans I 1993: 2875–2892

Svensson B, Lübben M and Hederstedt L (1993) *Bacillus subtilis* CtaA and CtaB function in haem A biosynthesis. Mol Microbiol 10: 193–201

Swanson KL and Smith KM (1990) Biosynthesis of bacteriochlorophyll-*c* via the glutamate C-5 pathway in *Chloroflexus aurantiacus*. J Chem Soc Chem Commun 1990: 1696–1697

Sylvers LA, Rogers KC, Shimizu M, Ohtsuka E and Söll D (1993) A 2-thiouridine derivative in $tRNA^{Glu}$ is a positive determinant for aminoacylation by *Escherichia coli* glutamyl-tRNA synthetase. Biochemistry 32: 3836–3841

Tai T-N, Moore MD and Kaplan S (1988) Cloning and characterization of the 5-aminolevulinate synthase gene(s) from *Rhodobacter sphaeroides*. Gene 70: 139–151

Tait GH (1972) Coproporphyrinogenase activities in extracts of *Rhodopseudomonas spheroides* and *Chromatium* strain D. Biochem J 128: 1159–1169

Taniuchi H, Basile G, Taniuchi M and Veloso D (1983) Evidence for formation of two thioether bonds to link heme to apocytochrome *c* by partially purified cytochrome *c* synthetase. J Biol Chem 258: 10963–10966

Thomas SD and Jordan PM (1986) Nucleotide sequence of the *hemC* locus encoding porphobilinogen deaminase of *Escherichia coli* K12. Nuc Acids Res 14: 6215–6226

Timkovich R and Bondoc LL (1990) Diversity in the structure of hemes. Adv Biophys Chem 1: 203–247

Timkovich R, Cork MS, Gennis RB and Johnson PY (1985) Proposed structure of heme *d*, a prosthetic group of bacterial terminal oxidases. J Am Chem Soc 107: 6069–6075

Troup B, Jahn M, Hungerer C and Jahn D (1994) Isolation of the *hemF* operon containing the gene for the *Escherichia coli* aerobic coproporphyrinogen III oxidase by in vivo complementation of a yeast *HEM13* mutant. J Bacteriol 176: 673–680

Tuboi S and Hayasaka S (1972) Control of δ-aminolevulinate synthetase activity in *Rhodopseudomonas spheroides*, II: Requirement of a disulfide compound for the conversion of the inactive form of Fraction I to the active form. Arch Biochem Biophys 150: 690–697

Tuboi S, Kim HJ and Kikuchi G (1970a) Occurrence and properties of two types of δ-aminolevulinate synthetase in *Rhodopseudomonas spheroides*. Arch Biochem Biophys 138: 147–154

Tuboi S, Kim HJ and Kikuchi G (1970b) Differential induction of Fraction I and Fraction II of δ-aminolevulinate synthetase in *Rhodopseudomonas spheroides* under various incubation conditions. Arch Biochem Biophys 138: 155–159

Viale AA, Wider EA and Batlle AM del C (1987a) Porphyrin biosynthesis in *Rhodopseudomonas palustris*, XI: Extraction and characterization of δ-aminolevulinate synthetase. Comp Biochem Physiol 87B: 607–613

Viale AA, Wider EA and Batlle AM del C (1987b) Porphyrin biosynthesis in *Rhodopseudomonas palustris*, XII: δ-aminolevulinate synthetase switch-off/on regulation. Int J Biochem 19: 379–383

Wang W-Y, Gough SP and Kannangara CG (1981) Biosynthesis of δ-aminolevulinate in greening barley leaves, IV: Isolation of three soluble enzymes required for the conversion of glutamate to δ-aminolevulinate. Carlsberg Res Commun 46: 243–257

Wang W-Y, Huang D-D, Stachon D, Gough SP and Kannangara CG (1984) Purification, characterization, and fractionation of the δ-aminolevulinic acid synthesizing enzymes from light-grown *Chlamydomonas reinhardtii* cells. Plant Physiol 74: 569–575

Warnick GR and Burnham BF (1971) Regulation of porphyrin biosynthesis: Purification and characterization of δ-aminolevulinic acid synthase. J Biol Chem 246: 6880–6885

Warren MJ and Jordan PM (1988) Further evidence for the involvement of a dipyrromethene cofactor at the active site of porphobilinogen deaminase. Biochem Soc Trans 16: 963–965

Warren MJ, Roessner CA, Santander PJ and Scott AI (1990) The *Escherichia coli cysG* gene encodes *S*-adenosylmethionine-dependent uroporphyrinogen III methylase. Biochem J 265: 725–729

Weinstein JD and Beale SI (1985a) Enzymatic conversion of glutamate to δ-aminolevulinate in soluble extracts of the unicellular green alga, *Chlorella vulgaris*. Arch Biochem Biophys 237: 454–464

Weinstein JD and Beale SI (1985b) RNA is required for enzymatic conversion of glutamate to δ-aminolevulinic acid by extracts of *Chlorella vulgaris*. Arch Biochem Biophys 239: 87–93

Weinstein JD, Mayer SM and Beale SI (1986) Stimulation of δ-aminolevulinic acid formation in algal extracts by heterologous RNA. Plant Physiol 82: 1096–1101

Weinstein JD, Mayer SM and Beale SI (1987) Formation of δ-aminolevulinic acid from glutamic acid in algal extracts: Separation into an RNA and three required enzyme components by serial affinity chromatography. Plant Physiol 84: 244–250

Weinstein JD, Howell RW, Leverette RD, Grooms SY, Brignola PS, Mayer SM and Beale SI (1993) Heme inhibition of δ-aminolevulinic acid synthesis is enhanced by glutathione in cell-free extracts of *Chlorella*. Plant Physiol 101: 657–665

Wright MS, Eckert JJ, Biel SW and Biel AJ (1991) Use of a *lacZ* fusion to study transcriptional regulation of the *Rhodobacter capsulatus hemA* gene. FEMS Microbiol Lett 78: 339–342

Wu W and Chang CK (1987) Structure of 'dioneheme': Total synthesis of the green heme prosthetic group in cytochrome cd_1 dissimilatory nitrite reductase. J Am Chem Soc 109: 3149–3150

Wu W, Chang CK, Varotsis C, Babcock GT, Puustinen A and Wikström M (1992) Structure of the heme *o* prosthetic group from the terminal quinol oxidase of *Escherichia coli*. J Am Chem Soc 114: 1182–1187

Xu K and Elliott T (1993) An oxygen-dependent coproporphyrinogen oxidase encoded by the *hemF* gene of *Salmonella typhimurium*. J Bacteriol 175: 4990–4999

Xu K and Elliott T (1994) Cloning, DNA sequence and complementation analysis of the *Salmonella typhimurium hemN* gene encoding a putative oxygen-dependent coproporphyrinogen III oxidase. J Bacteriol 176: 3196–3203

Xu K, Delling J and Elliott T (1992) The genes required for heme synthesis in *Salmonella typhimurium* include those encoding alternative functions for aerobic and anaerobic coproporphyrinogen oxidation. J Bacteriol 174: 3953–3963

Yap-Bondoc F, Bondoc LL, Timkovich R, Baker DC and Hebbler A (1990) C-methylation occurs during the biosynthesis of heme d_1. J Biol Chem 265: 13498–13500

Yubisui T and Yoneyama Y (1972) δ-Aminolevulinic acid synthetase of *Rhodopseudomonas spheroides*: Purification and properties of the enzyme. Arch Biochem Biophys 150: 77–85

Zagorec M, Buhler J-M, Treich I, Keng T, Gaurente L and Labbe-Bois R (1988) Isolation, sequence, and regulation by oxygen of the yeast *HEM13* gene coding for coproporphyrinogen oxidase. J Biol Chem 263: 9718–9724

Zaman Z and Akhtar M (1976) Mechanism and stereochemistry of vinyl-group formation in haem biosynthesis. Eur J Biochem 61: 215–223

Zaman Z, Jordan PM and Akhtar M (1973) Mechanism and stereochemistry of the 5-aminolaevulinate synthetase reaction. Biochem J 135: 257–263

Zannoni D, Melandri BA and Baccarini-Melandri A (1976) Energy transduction in photosynthetic bacteria, X: Composition and function of the branched oxidase system in wild type and respiration deficient mutants of *Rhodopseudomonas capsulata*. Biochim Biophys Acta 423: 413–430

Zollner A, Rödel G and Haid A (1992) Molecular cloning and characterization of the *Saccharomyces cerevisiae CYT2* gene encoding cytochrome-c_1-heme lyase. Eur J Biochem 207: 1093–1100

Chapter 10

Lipids, Quinones and Fatty Acids of Anoxygenic Phototrophic Bacteria

Johannes F. Imhoff and Ursula Bias-Imhoff
Institut für Meereskunde an der Universität Kiel, Düsternbrooker Weg 20, 24105 Kiel, Germany

R. E. Blankenship, M. T. Madigan and C. E. Bauer (eds): Anoxygenic Photosynthetic Bacteria, pp. 179–205.

Summary

Differences in profiles of polar lipids and quinones distinguish major groups of phototrophic prokaryotes. Also the fatty acid composition of representative species of purple sulfur bacteria, purple nonsulfur bacteria, green sulfur bacteria, *Chloroflexus aurantiacus* and *Heliobacterium* and *Heliobacillus* species are significantly different. Phospholipids, and specifically PG, are present in all phototrophic bacteria. The presence of ornithine lipids and the sulfolipid SQDG has been clearly established only in purple nonsulfur bacteria. Various glycolipids are major components in *Chloroflexus* and green sulfur bacteria and are also common in Chromatiaceae, but absent from Ectothiorhodospiraceae and *Heliobacterium*; occasionally, they occur in purple nonsulfur bacteria. The presence of MGDG has only been established for green sulfur bacteria. The specific distribution of major quinones in anoxygenic phototrophic bacteria and the variety of hopanoid triterpene structures found in these bacteria are considered in this chapter. A major part of the chapter is concerned with polar lipids of *Rhodobacter* species because they have been most intensively studied in this chapter. Also biosynthesis, lipid transfer activities, and incorporation of lipids into different membrane fractions are discussed. Special attention is paid to the influence of growth conditions on lipid and fatty acid composition and to possible differences in the composition of cellular membrane fractions (CM, ICM, OM).

I. Introduction

Lipids are important structural components of all biological membranes. They are essential for the formation of bilayer structures and for a number of membrane associated processes. The term 'lipids' is not clearly defined on the basis of a chemical structure, but lipids are characterized as a group of molecules which have only limited solubility in water and are more readily dissolved in organic solvents. Structurally and functionally completely different molecules are known as lipids (Ratledge and Wilkinson, 1988). In this chapter we will discuss *polar* lipids (including phospholipids, glycolipids, sulfolipids, aminolipids) and *apolar* lipids (including quinones and hopanoids). Some representative structures of lipids, quinones and hopanoids are shown in Fig. 1. Fatty acids will also be treated separately, although they are rarely present in the free form but constitute a molecular part of polar lipids and other membrane lipids. Carotenoids, bacteriochlorophylls and complex lipids such as lipid A and lipoproteins typical of Gram-negative bacteria are not considered in this chapter; the reader is referred to Chapter 8 by Senge and Smith.

Phototrophic purple bacteria have three distinct types of membranes, the cytoplasmic membrane (CM), the intracytoplasmic membranes (ICM, also called chromatophores) and the outer membrane (OM), which is characteristic of all Gram-negative bacteria. The green bacteria lack intracytoplasmic membranes, but have chlorosomes that are bound by a lipid membrane, which is proposed to be a lipid unilayer (Staehelin et al., 1980). Though bacterial membranes differ in composition and function, various kinds of lipids and fatty acids are principal constituents. Together with integral membrane proteins, lipids form the continuum of the cytoplasmic membrane that separates the internal space of the bacterial cells from the external one. The functional unit of CM, ICM and OM is a bilayer with the head groups of polar lipids facing the inner and outer surface of this bilayer and the fatty acid tails touching each other in the membrane interior. In this way a hydrophobic interior and a hydrophilic membrane surface can be formed. The integrity of the cytoplasmic membrane bilayer is crucial for energy generation of the cells, for maintenance of gradients of a variety of solutes and the success of transport processes and thus for the survival of each single cell.

A second property important to membrane function

Abbreviations: AL – aminolipid; APL – aminophospholipid; BChl – bacteriochlorophyll; *Cb.* – *Chlorobium; Cf.* – *Chloroflexus;* CK – chlorobiumquinone; CL – cardiolipin; CM – cytoplasmic membrane; *Cm.* – *Chromatium;* DGDG – digalactosyldiglyceride; *Ec.* – *Ectothiorhodospira;* GL – glycolipid; *Hb.* – *Heliobacterium;* ICM – intracytoplasmic membrane; MGDG – monogalactosyldiglyceride; MK – menaquinone; NMR – nuclear magnetic resonance; OA – ornithine amide; OM – outer membrane; PA – phosphatidic acid; PC – phosphatidylcholine; PE – phosphatidylethanolamine; PG – phosphatidylglycerol; PI – phosphatidylinositol; PL – phospholipid; PS – phosphatidylserine; Q – ubiquinone; *Rb.* – *Rhodobacter; Rc.* – *Rhodocyclus; Rm.* – *Rhodomicrobium; Rp.* – *Rhodopseudomonas;* RQ – rhodoquinone; *Rs.* – *Rhodospirillum; Rv.* – *Rubrivivax;* SL – sulfolipid; SQDG – sulfoquinovosyldiglyceride; TLC – thin layer chromatography

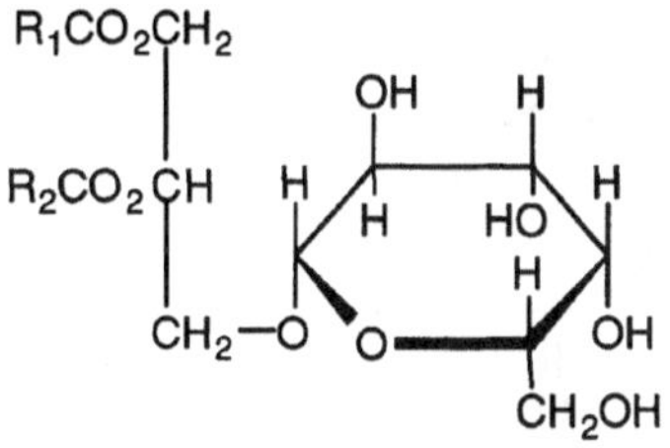

Monogalactosyldiglyceride (MGDG)

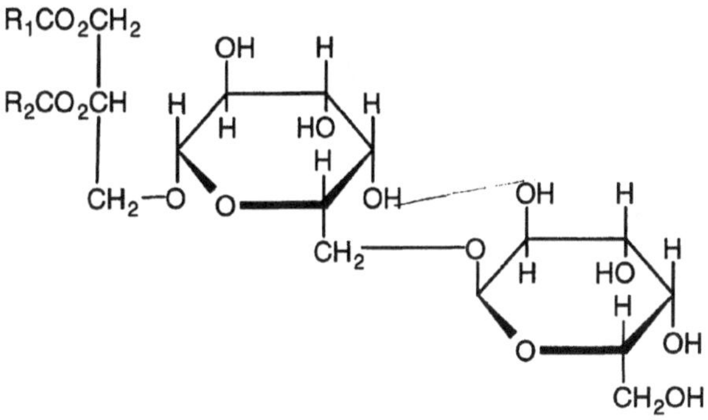

Digalactosyldiglyceride (DGDG)

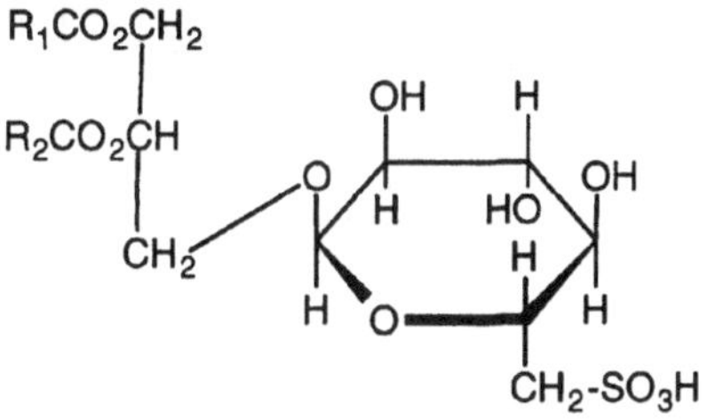

Sulfoquinovosyldiglyceride (SQDG)

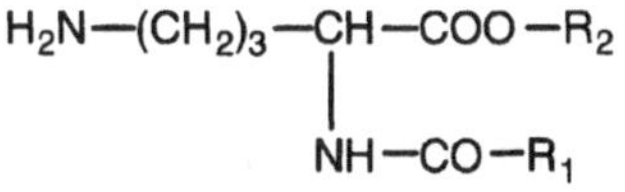

Ornithine amide of *Rb. sphaeroides* (Gorchein, 1968)

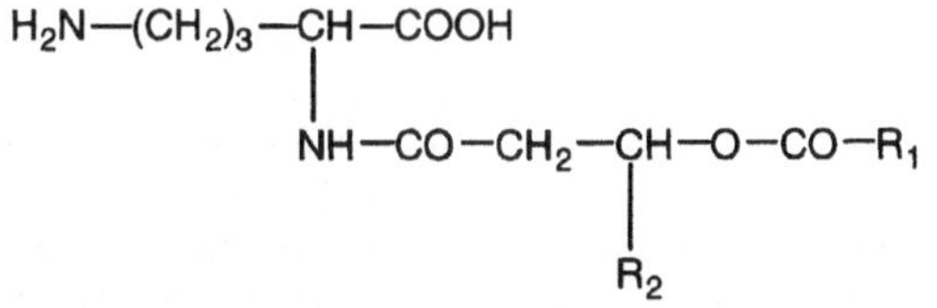

Ornithine amide of *R. rubrum* (Brooks and Benson, 1972)

Fig. 1. Representative structures of lipids, quinones and hopanoids.

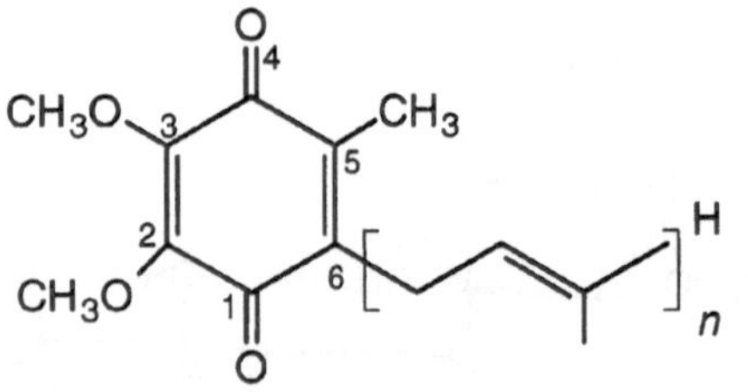

Ubiquinone

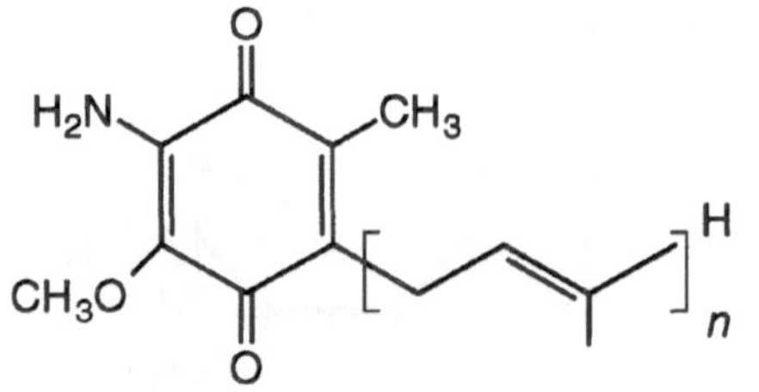

Rhodoquinone

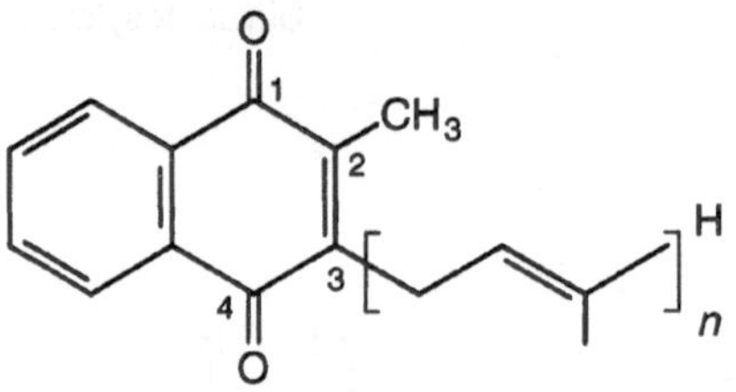

Menaquinone

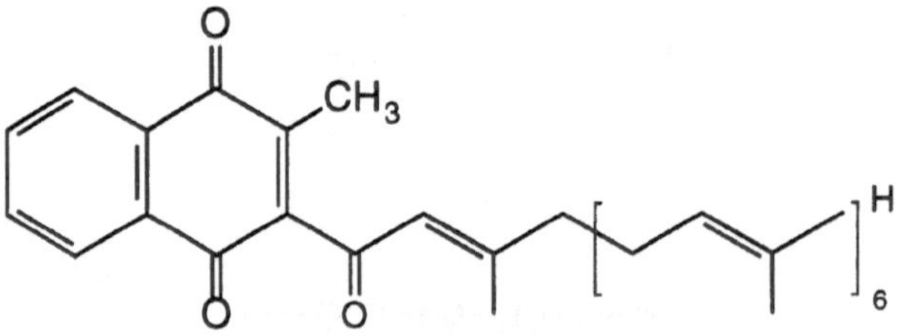

Chlorobiumquinone

Fig. 1. Continued. Quinones.

is the mobility of the membrane components, which significantly influences reaction kinetics within the membrane. Both bilayer stability and membrane fluidity are influenced by the composition of lipid headgroups, fatty acid structure, lipid-lipid and lipid-protein interactions, as well as external parameters such as temperature, pressure, salinity, pH and others. In order to maintain the functional integrity of their membranes, bacteria have developed mechanisms to adapt their membrane composition to varying environmental conditions. Therefore, the qualitative and quantitative lipid composition of the membranes of a specific bacterium is not constant but varies with growth conditions and also with the growth phase. Nevertheless, specific compositions of lipids, fatty acids and quinones are characteristic properties of numerous bacterial species and can be valuable tools in species identification and taxonomy. However,

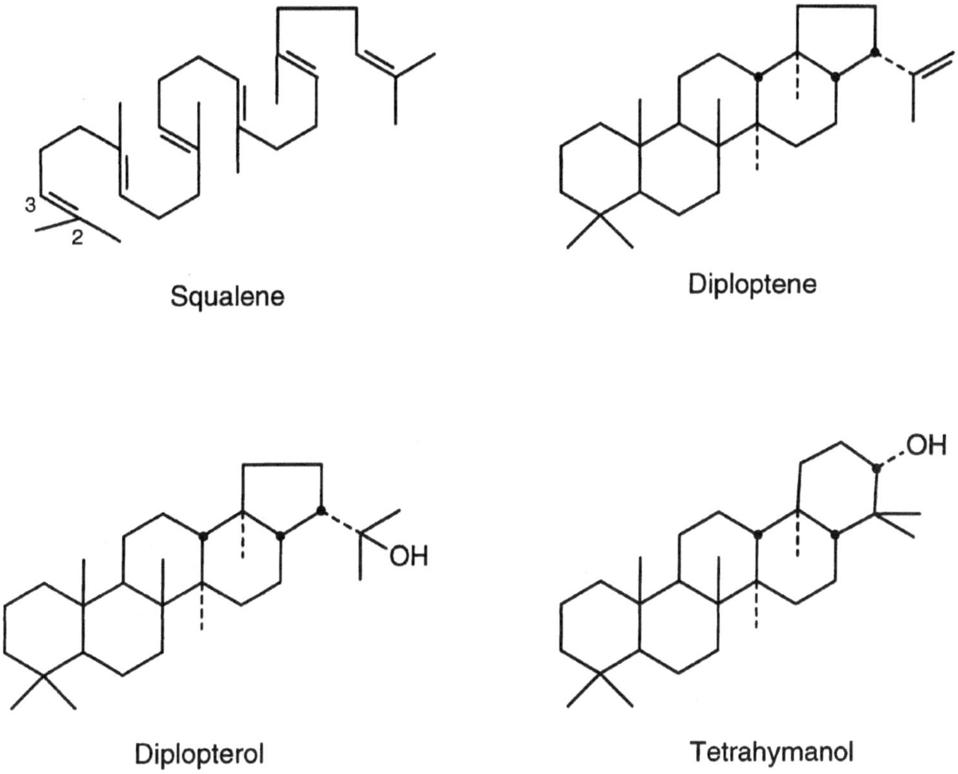

Fig. 1. Continued. Hopanoids.

whenever possible, standardized media and culture conditions have to be used for this purpose.

Indeed, lipids, fatty acids and quinones have been analyzed in recent years in a great number of bacteria. Their composition varies greatly among different bacteria and is used to differentiate bacterial groups and even closely related bacterial species. As would be expected from the great heterogeneity of anoxygenic phototrophic bacteria derived from physiological, structural as well as genetic information, the composition of lipids, fatty acids and quinones shows great variation among these bacteria.

This chapter will not give a comprehensive overview on the relevant literature of lipids, fatty acids and quinones of phototrophic bacteria. It rather presents a quite personal view of this topic and primarily will discuss the distribution of lipid components among the species and groups of anoxygenic phototrophic bacteria. A comprehensive review and a detailed compilation of early data on lipids and fatty acids of phototrophic bacteria can be found in Kenyon (1978).

II. Polar Lipids and Fatty Acids

A. Polar Lipids

Great variation occurs in the polar lipid structures of different phototrophic bacteria. However, analytical procedures employed often do not give precise information on their exact chemical structure, which therefore, is still unknown for a number of lipids of phototrophic bacteria. This applies in particular to some glycolipids and aminolipids.

Though the polar lipid composition under standard cultivation conditions is a stable property, the qualitative and quantitative composition may change with the growth conditions much more than that of quinones, but in general less than that of fatty acids. From the data available it is obvious that major groups of phototrophic bacteria can be differentiated on the basis of their polar lipid composition (Table 1). Qualitative and quantitative differences between *Heliobacterium chlorum, Chloroflexus aurantiacus*, green sulfur bacteria, Chromatiaceae, Ectothiorhodospiraceae and purple nonsulfur bacteria are depicted in respect to their content of different phospholipids, glycolipids, sulfolipids and aminolipids. Relative proportions of glycolipids are high in Chlorobiaceae and Chloroflexaceae, but considerable lower in Chromatiaceae, which have higher proportions of phospholipids. Phospholipids are clearly dominant in purple nonsulfur bacteria and glycolipids are absent from the majority of these bacteria as they are from Ectothiorhodospiraceae and *Hb. chlorum*.

1. Ectothiorhodospiraceae

The lipid composition of all Ectothiorhodospiraceae is characterized by the predominance of PG, PC and CL and minor but varying amounts of PE (Asselineau and Trüper, 1982; Imhoff et al., 1982). All species contain small to trace amounts of lyso-PE, and also PA, lyso-PC, and unidentified phospholipids have been found in some species. Glycolipids as found in Chromatiaceae, purple nonsulfur bacteria and green sulfur bacteria and ornithine lipids as known from purple nonsulfur bacteria, are absent in *Ectothiorhodospira* species (Asselineau and Trüper, 1982; Imhoff et al., 1982; Thiemann and Imhoff, 1991). In strains of *Ectothiorhodospira halophila*, both PG and CL separated into two distinct spots on two-dimensional TLC plates and were significantly different in their fatty acid composition (Thiemann and Imhoff, 1991). Nothing is known about

Table 1. Major polar lipids of representative phototrophic prokaryotes[a]

	PG	CL	PE	PC	OA	MGDG	DGDG	SQDG	Other lipids
Chloroflexaceae									
Cf. aurantiacus	+	(+)	–	–	–	–*	–*	o*	PI, GL(glu-gal)
Chlorobiaceae									
Cb. limicola f. thiosulf.	+	+	+	–	–	+	–*	–*	GLII, lipid 2, aminoglycosphingolipid
Pelodictyon luteolum	+	+	–	–	–	+	–*	–*	GLII, lipid 2, aminoglycosphingolipid
Heliobacteriaceae									
Heliobacterium chlorum	+	+	+	–	o	–	–	–	unknown aminolipids
Chromatiaceae									
Chromatium vinosum	+	+	+	–	–	–*	–	o*	lyso-PE, several glycolipids
Chromatium minus	+	+	+	–	–	o*	–	o*	lyso-PE, several glycolipids
Thiocystis gelatinosa	+	+	+	–	–	o*	o*	o*	lyso-PE, APL, several glycolipids
Ectothiorhodospiraceae									
Ec. halochloris	+	+	+	+	–	–	–	–	lyso-PE
Ec. halophila	+	+	+	+	–	–	–	–	lyso-PE
Ec. vacuolata	+	+	+	+	–	–	–	–	lyso-PE
Purple Nonsulfur Bacteria									
Rc. tenuis	+	+	+	–	+	–	–	–	lyso-PE, PL
Rs. rubrum	+	+	+	–	+	–	–	(+)	lyso-PE
Rp. marina	+	+	+	+	+	–	–	+	lyso-PE, PL
Rp. palustris	+	+	+	+	+	–	–	(+)	lyso-PE, PL, AL
Rp. viridis	+	+	+	+	+	–	–	–	SL
Rm. vannielii	+	–	+	+	+	–	–	tr	PA, ornithine-PG, bis-PA

* A lipid with similar chromatographic properties is present. + component present; – component absent; o not investigated or not definitely proven; (+) present in very small amounts or presence uncertain, has been detected in some but not in other studies; tr traces.
[a] Data from references cited in the text.

differences of localization and function of these two fatty acid variants in *Ec. halophila.*

2. Chromatiaceae

The phospholipids, PG, PE and CL are major components of Chromatiaceae species, none of which is clearly predominant. Also lyso-PE could be detected in the investigated species (Imhoff et al., 1982; Steiner et al., 1969; Takacs and Holt, 1971). Additional aminophospholipids were found in *Thiocapsa roseopersicina* and *Thiocystis gelatinosa*, but not in *Chromatium vinosum*, *Chromatium minus* and *Chromatium warmingii* (Imhoff et al., 1982). PC was absent from these species.

Several glycolipids were found in almost all Chromatiaceae (Imhoff et al., 1982; Imhoff, unpublished results). The number of glycolipid components and their chromatographic properties varied with the species. One of these glycolipids, which was characterized from *Chromatium vinosum*, had similar chromatographic and staining properties to MGDG but was shown to lack galactose and to be a monoglucosyldiglyceride (Steiner et al., 1969). Galactose was also absent from hydrolyzed total lipid extracts, proving the absence of galactolipids from this species. This glycolipid may be present in other species of the Chromatiaceae as well (Imhoff et al., 1982; Imhoff, 1988; Imhoff, unpublished results). Two other glycolipids from *Chromatium vinosum* have been described as mannosyl, glucosyl-diglyceride and as dimannosyl, glucosyl-diglyceride (Steiner et al., 1969). Whether these also occur in other Chromatiaceae is not known. Glycolipids from *Thiocapsa roseopersicina* were found to contain glucose and rhamnose (Takacs and Holt, 1971). SQDG has so far not been identified in Chromatiaceae species, although glycolipids with similar chromatographic properties were present (Imhoff et al., 1982).

3. Purple Nonsulfur Bacteria

The polar lipid composition of purple nonsulfur bacteria shows the greatest diversity of all groups of phototrophic bacteria. In many cases polar lipid compositions are characteristic of the species. Phospholipids are predominant in all species, ornithine lipids and other aminolipids are characteristic for these bacteria, whereas glycolipids are not common, but present in some species in smaller proportions (Kenyon, 1978; Imhoff et al., 1982; Imhoff, 1988). Among the lipids most commonly distributed in these bacteria are PG, PE, OA and CL, while PC and SQDG characteristically are present in some species but absent from others. Glycolipids with chromatographic properties similar to MGDG only occasionally have been found. The presence of such a glycolipid in *Rs. molischianum* reported by Constantopoulos and Bloch (1967) could not be verified later (Imhoff et al., 1982), but lipids with similar chromatographic properties have been found in *Rs. salexigens* and *Rp. sulfoviridis* (Imhoff, unpublished results). Their exact chemical structure is unknown.

a. Sulfolipids

Sulfolipids have obtained much interest because it was thought that they may play a functional role in photosynthesis. The first evidence of SQDG, the 'plant type sulfolipid', in an anoxygenic phototrophic bacterium was obtained for *Rs. rubrum* (Benson et al., 1959). Later this sulfolipid was found in *Rm. vannielii* at very low proportions of only 0.01% of cell dry weight (Park and Berger, 1967) and in *Rb. sphaeroides* containing 2–4% of this sulfolipid (Wood et al., 1965; Radunz, 1969; Russell and Harwood, 1979). Its presence in *Rb. capsulatus* at low proportions (Russell and Harwood, 1979) could not be reproduced by others (Wood et al., 1965; Imhoff et al., 1982; Imhoff, 1991). On the basis of incorporation of label from radioactive sulfur sources, however, sulfolipids were found in several other species including *Rb. capsulatus* at very low proportions (Imhoff, 1984b). Significant amounts of the plant type sulfolipid have been found in *Rp. marina*, *Rs. salexigens* and some *Rhodobacter* species (Tables 1 and 2) (Imhoff et al., 1982; Imhoff, 1984b). The identity of the head group of the sulfolipid from *Rb. sphaeroides* with that from spinach leafs was demonstrated by identical fragmentation pattern in fast atom bombardment mass spectrometry (Gage et al., 1992).

During studies with radioactively labelled sulfur sources, several sulfolipids were detected in purple nonsulfur bacteria (Imhoff, 1984b). Sulfate as well as reduced sulfur compounds (thiosulfate and cysteine) served as sulfur sources for the sulfolipids in these species. The relative proportions of the sulfolipid in *Rb. sulfidophilus* were not significantly influenced by the sulfur source, but reduced sulfur sources inhibited the incorporation of label from ^{35}S-sulfate (Imhoff, 1984b). This indicates the preference of reduced sulfur compounds over sulfate as a source of sulfur for this sulfolipid. Sulfolipids in high proportions were also found in *Rp. sulfoviridis* and *Rb. adriaticus*, which are unable to assimilate sulfate and depend on reduced sulfur sources. Both species incorporated sulfur from the reduced sulfur sources into their sulfolipids (Imhoff, 1984b). The major sulfolipids of *Rp. viridis* and *Rp. sulfoviridis* had R_f-values different from SQDG (Imhoff, 1984b).

b. Ornithine lipids

Many Gram-negative bacteria have ornithine lipids with amide-linked hydroxy fatty acids that show some structural similarity to the lipid A of lipopolysaccharides and may fulfill some of its functions (Wilkinson et al., 1982). Ornithine lipids have been found in *Rs. rubrum, Rp. palustris, Rubrivivax gelatinosa, Rp. viridis, Rm. vannielii* and others (Brooks and Benson, 1972; DePinto, 1967; Gorchein, 1968; Russell and Harwood, 1979; Wood et al., 1965). Apparently ornithine lipids are generally present in and characteristic for purple nonsulfur bacteria (Imhoff et al., 1982). The presence of ornithine-PG in several purple nonsulfur bacteria, claimed by Wood et al. (1965), was not confirmed later (Imhoff et al., 1982; Imhoff, 1991). A lipid that could be ornithine-PG according to chromatographic and staining properties, but was not clearly identified, was found in *Rm. vannielii* and *Rhodopila globiformis* (Imhoff et al., 1982). Ornithine-PG has been identified in *Rm. vannielii* (Park and Berger, 1967).

The ornithine lipid from *Rb. sphaeroides* has been shown to be an ester of N-acylornithine with a long-chain fatty acid residue forming an amide bond with the α-amino group of ornithine and an alcohol residue esterified to its carboxyl group (Gorchein, 1968). Another ornithine amide was identified in *Rs. rubrum* with a long chain fatty acid esterified to the hydroxyl group of a 3-hydroxy fatty acid, which in turn forms an amide bond with the α-amino group of ornithine (Brooks and Benson, 1972). Either of these compounds or a mixture of them may be present in other species of purple nonsulfur bacteria.

4. Polar Lipids of Rhodobacter *Species*

Most reports on composition and metabolism of phospholipids deal with one or the other species of the genus *Rhodobacter*, in particular with *Rb. sphaeroides*. Polar lipids of *Rb. sphaeroides* and *Rb. capsulatus* have been investigated repeatedly (Wood et al., 1965; Russell and Harwood, 1979; Marinetti and Cattieu, 1981; Gorchein, 1968; Imhoff et al., 1982) and also the polar lipid composition of *Rb. sulfidophilus* has been reported (Imhoff et al., 1982). A comparison of polar lipid and fatty acid compositions of the six described *Rhodobacter* species on the basis of 21 strains has been undertaken recently (Imhoff, 1991). All strains were grown under identical conditions anaerobically in the light at their respective optimum salinity. Each of the investigated species revealed a specific polar lipid composition (Table 2).

There is a general agreement upon the presence of PC, PG and PE and the absence of SQDG and CL in whole cell lipid extracts from *Rb. capsulatus* (see Kenyon, 1978), though the results of Russell and Harwood (1979) imply the presence of CL and SQDG in one strain (NCIB 8253) of this species. An aminolipid, which is most probably identical to the ornithine amide from *Rb. sphaeroides,* has not been reported in earlier literature, but has been regularly found later (Russell and Harwood, 1979; Imhoff et al., 1982; Imhoff, 1991). The presence of ornithine-PG postulated by Wood et al. (1965) could not be verified.

In *Rb. sphaeroides*, the presence of SQDG, CL and ornithine amide in addition to PC, PG and PE is generally agreed on. The ornithine lipid of this species has been characterized (see above) and may be common to all *Rhodobacter* species. It was named ornithine lipid 1 (OL1, Imhoff et al., 1982) and later aminolipid (AL, Imhoff, 1991). The analysis of the lipids of *Rb. sphaeroides* by Marinetti and Cattieu (1981) revealed the presence of three nitrogen bases, ethanolamine, ornithine and an unidentified amino-compound with chromatographic mobility as 1-amino-3-propanol. These authors concluded that an unknown aminolipid 'is a fatty acid amide of ornithine and of 1-aminopropanol or some other diamino compound'. Although the structure of this lipid needs to be established, it may well be that depending on the species and the growth conditions, much more structural variability among the aminolipids of purple nonsulfur bacteria exists than hitherto known.

During studies on cell-cycle specific accumulation of phospholipids into the intracytoplasmic membrane, an unknown phospholipid, later identified as N-acylphosphatidylserine (NAPS) was detected in *Rb. sphaeroides* strain M29-5 (Cain et al., 1981; Donohue

Table 2. Major polar lipids of *Rhodobacter* species[a]

	PG	CL	PE	PC	SQDG	OA	APL	PL1	PL2
Rb. sphaeroides	+	+	+	+	+	+	–	+	+
Rb. capsulatus	+	–	+	+	–	+	–	+	–
Rb. veldkampii	+	–	+	–	–	+	+	–	–
Rb. sulfidophilus	+	–	+	–	+	+	–	+	–
Rb. adriaticus	+	–	–	–	+	+	–	–	–
Rb. euryhalinus	+	+	+	–	+	+	–	–	–

[a]Data from Imhoff (1991)

et al., 1982a). An unknown phospholipid with similar properties to NAPS was also detected in *Rb. sphaeroides* strain NCIB 8253 (Onishi and Niederman, 1982). This lipid represented between 20–45% of the total cellular phospholipids in several strains of *Rb. sphaeroides* (except two wild type strains) and *Rb. capsulatus* grown either photoheterotrophically or chemoheterotrophically in Tris supplemented culture media (Donohue et al., 1982b). Its accumulation at the expense of PE ceased immediately upon the removal of Tris (Donohue et al., 1982b). Together with the other phospholipids, the accumulated NAPS was distributed to all membrane systems of the cell. Small levels (1.0 to 2.0%) of NAPS also were present in *Rs. rubrum* and *Rp. palustris* (Donohue et al., 1982b). From the labeling pattern observed in pulse-chase experiments it was concluded that NAPS is neither a product of the direct acylation of PS nor an obligate precursor in the PS biosynthesis, but is probably a metabolite of an undescribed branch of phospholipid biosynthesis (Cain et al., 1982; 1983).

a. Biosynthesis and Lipid Transfer Activities

Biosynthesis of PE and PG in *Rb. sphaeroides* apparently proceeds as in *E. coli* with CDP-diglyceride as intermediate reacting either with sn-glycerol-3-phosphate or L-serine to yield PG (via phosphatidylglycerophosphate) or PE (via PS), respectively (Cain et al., 1983). Enzymatic activities that transfer fatty acids from acyl-ACP to glycerol-3-phosphate forming lyso-PA and PA have been found in a particulate fraction from *Rb. sphaeroides* grown phototrophically (Lueking and Goldfine, 1975). In pulse-chase experiments with (2-^{3}H)glycerol, PA was the earliest identifiable glycerol-containing intermediate and was considered to be the product of the sn-glycerol-3-phosphate acyltransferase activity (Cain et al., 1983). PA labeling decreased with the increase in PS label, followed by accumulation of label in PE concomitant with a decrease in that of PS. After treatment of the cells with 50 μM hydroxylamine, a known inhibitor of PS decarboxylase of *E. coli* (Raetz and Kennedy, 1972), an accumulation of label in PS at the expense of that in PE was observed. Both the transient labeling of PS and the accumulation of label in PS in the presence of hydroxylamine emphasize the precursor-product relationship between these two phospholipids (Cain et al., 1983). Synthesis of PC by successive methylation of PE has been demonstrated by studies with methyl-^{14}C-methionine that yielded label in phosphatidyl-N-methylethanolamine, phosphatidyl-N,N-dimethylethanolamine and PC (Gorchein et al., 1968a).

Apparently lipids are synthesized in the CM or in distinct domains thereof and transferred from the location of biosynthesis to other membranes. Intracytoplasmic membranes of *Rb. sphaeroides* grown under photoheterotrophic conditions are considered to originate from the cytoplasmic membrane (Gorchein et al., 1968b; Drews and Oelze, 1981; Kaplan and Arntzen, 1982; Niederman and Gibson, 1978). When non-pigmented cells of *Rb. sphaeroides* double-labeled with ^{32}P and methyl-^{14}C-methionine were allowed to adapt to semi-anaerobic growth conditions to induce the development of intracytoplasmic membranes, lipids of the cytoplasmic membrane became incorporated into the ICM (Gorchein et al., 1968b). Thus, a complete de novo biosynthesis of ICM of *Rb. sphaeroides* was ruled out and it was concluded that the ICM and their lipids originate from the cytoplasmic membrane (Gorchein et al., 1968b). These findings were substantiated later by further detailed studies. Enzymes of phospholipid biosynthesis mainly were found located within the CM of *Rb. sphaeroides* (Cooper and Lueking, 1984; Cain et al., 1984; Radcliffe and Niederman, 1984; Radcliffe et al., 1985). A separation of the membranes from *Rb.*

sphaeroides strain NCIB 8253 by density gradient centrifugation verified the localization of enzyme activities catalyzing the first steps of phospholipid biosynthesis in membrane fractions of the cells (Radcliffe et al., 1985). Both in chemoheterotrophically and in photoheterotrophically grown cells, the majority of the activity of phosphatidylglycerophosphate synthase (EC 2.7.8.5.) was found in the CM and was distributed uniformly over this membrane (Radcliffe et al., 1985; 1989). Phosphatidylserine (PS) synthase activity (EC 2.7.8.8), however, was associated with a lower membrane fraction present in photo- and chemotrophically grown cells and located at the position of ICM of phototrophically grown cells. The authors suggested that PS synthase is localized in distinct domains of the cytoplasmic membrane, which after cell disruption and separation by density gradient centrifugation, form a band together with the ICM fraction (Radcliffe et al., 1985, 1989).

Phospholipid transfer activities have been detected in *Rb. sphaeroides* cells grown chemotrophically and phototrophically (Tai and Kaplan, 1984; 1985). The cellular distribution of these activities was similar under both conditions and the total cellular activity was directly related to the growth rate (Tai and Kaplan, 1985). Activities were found to be membrane-associated as well as in soluble cell fractions (15% membrane-associated, 32% periplasmic, 53% cytoplasmic at medium light intensity). Cells grown anaerobically at low light intensity, characterized by large amounts of ICM, had the highest membrane-associated transfer activity, whereas aerobically grown cells (lacking ICM) only showed minor activities of the membrane-associated enzyme, indicating that this activity could be involved in ICM biogenesis (Tai and Kaplan, 1985). Soluble transfer activity was low under low light conditions, but much higher under high light conditions (91.4% of total transfer activity) (Tai and Kaplan, 1985). Both cytoplasmic and periplasmic phospholipid transfer activities had different preferences for their substrates. The cytoplasmic transfer activities showed a clear preference for PG over PE and PC, while the periplasmic activity transferred phospholipids without a specific preference (Tai and Kaplan, 1985).

b. Incorporation of Lipids into ICM

During the normal cell cycle lipid incorporation into ICM apparently occurs only within a limited time interval in close relation to the division of the cells, which does, however, not reflect cyclic changes in lipid biosynthesis but rather transfer of lipids into the growing ICM from other 'places of the cell' (Lueking et al., 1978; Fraley et al., 1979; Cain et al., 1981). The discontinuous increase of net accumulation of phospholipids into intracytoplasmic membranes was in strong contrast to the continuous insertion of proteins and pigments into this membrane (Lueking et al., 1978). It was found in experiments with synchronously dividing cells of *Rb. sphaeroides* that the protein/phospholipid ratio of purified intracytoplasmic membrane preparations underwent cyclic changes of 35–40% during the normal cell division cycle (Lueking et al., 1978). This ratio increased to a maximum value in membranes shortly before cell division and decreased to a basic value with the cell division, indicating that phospholipids are only accumulated into the intracytoplasmic membrane at the time of cell division.

Similar cell-cycle specific fluctuations in the composition of intracytoplasmic membranes were also observed in synchronous cultures of phototrophically grown *Rs. rubrum* (Myers and Collins, 1986). Furthermore, in starvation-synchronized cells of this species that were grown under aerobic conditions and had no intracytoplasmic membranes, cell cycle specific changes in the protein/phospholipid ratio were found in the cytoplasmic membrane but not in the outer membrane (Myers and Collins, 1987). Asynchronous cultures, however, had a constant protein/phospholipid ratio.

In contrast to the discontinuous incorporation of phospholipids into ICM coupled to cell-division in synchronously dividing cells of *Rb. sphaeroides* (Lueking et al., 1978; Cain et al., 1981), a continuous incorporation of phospholipids into ICM was demonstrated under conditions which induce ICM formation and transiently repress cell division (Radcliffe et al., 1985).

5. Green Sulfur Bacteria

All Chlorobiaceae contain considerable amounts of glycolipids, the major part of which (more than 80%) is associated with the chlorosomes. Phospholipids dominate in the cytoplasmic membrane (Cruden and Stanier, 1970). On a molar basis chlorosomes have 1–3 moles of sugar per mol of phosphate, whereas membranes contain only 1 mol of sugar per 5–12 moles of phosphate.

First Constantopoulos and Bloch (1967) described glycosyl glycerides in the phototrophic green bacterium '*Chloropseudomonas ethylica*' (later found to be a mixed culture) as monogalactosyl diglyceride (MGDG) and glycolipid II (containing galactose, rhamnose and a third unidentified sugar). Cruden and Stanier (1970) substantiated these findings with pure cultures of *Chlorobium limicola* (strain 1230) and *Chlorobium limicola f. thiosulf.* (strains 6130 and 6230). They separated membranes and chlorosomes and found monogalactosyl diglyceride associated with the chlorosomes and the glycolipid II with the membranes. It has been suggested later that MGDG is specifically associated with the chlorosomes and forms a monolayer covering its cytoplasmic boundary (Staehelin et al., 1980). As a consequence of this model the cellular content of MGDG in green sulfur bacteria would be expected to vary with their content of chlorosomes and BChl. However, no such correlation could be found in cells of *Chlorobium vibrioforme f. thiosulf.* (strain NCIB 8327)* (Holo et al., 1985). These authors found the specific content of MGDG to be approximately constant and independent of light intensity and BChl content. Furthermore, the ratio of MGDG to BChl was higher in whole cells than in the chlorosomes and it was suggested that the chlorosomes are not the only site of MGDG in green sulfur bacteria (Holo et al., 1985).

Though only two glycolipids have been detected in green sulfur bacteria using the diphenylamine spray reagent (Constantopoulos and Bloch, 1967; Cruden and Stanier, 1970), later work has shown that this reagent gives no color reaction with several other glycolipids of these bacteria. At least five different glycolipids could be found in *Cb. vibrioforme f. thiosulf.* (strain 8327) and in *Cb. limicola f. thiosulf.* (strains 6230 and 6430) by using 1-naphtol, periodate-Schiff, and anisaldehyde spray reagents (Bias, 1985). Two of these were identical to MGDG and glycolipid II. Two others have been described as lipid 1 and lipid 2 from *Cb. vibrioforme f. thiosulf.* (strain 8327) and *Cb. limicola f. thiosulf.* (strains 6230 and 2230) by Knudsen et al. (1982). Lipid 1 demonstrates R_f-values similar to PI, but contains no phosphate, stains positive with ninhydrin and shows characteristic color reactions with 1-naphtol, anisaldehyde and periodate-Schiff reagents (Bias, 1985). This aminolipid was characterized from the membrane fraction of *Cb. limicola f. thiosulf.* strain 6230 and shown to be a tertiary or secondary amide containing one mole of myristic acid but no phosphorus or known amino acid (Olson et al., 1983; 1984). Jensen et al. (1991) purified this aminolipid and characterized it by TLC, IR-Spectroscopy, ^{1}H-NMR, ^{13}C-NMR, plasma desorption mass spectrometry and fast atom bombardment mass spectrometry. They revealed the lipid 1 to be an aminoglycosphingolipid which in contrast to previous reports (Olson et al., 1983), does not fluoresce in the pure state. The carbohydrate component was identified as neuraminic acid and the main fatty acid as myristic acid.

By proton NMR studies it was confirmed that only one mole of fatty acid is present per mole of aminolipid. As a possible structure the authors propose the neuraminic acid to be attached to the sphingosine backbone with the myristic acid linked to the sphingosine by an amide bond. The aminoglycosphingolipid of *Chlorobium* was exclusively found in the cytoplasmic membrane but not in the chlorosomes (Olson et al., 1983, 1984; Jensen et al., 1991). It was therefore speculated that this lipid might operate as a membrane anchor for the hydrophilic BChl *a* protein (Low and Slatid, 1988; Ferguson and Williams, 1988), that forms a two-dimensional crystal between the plasma membrane and the chlorosomes of *Chlorobium* (Olson, 1980; see also discussion of the Fenna–Matthews–Olson [FMO] protein in Chapter 20 by Blankenship et al.).

Lipid 2 moves somehow similar to DGDG on one-dimensional thin layer plates. The negative results of staining reactions with molybdate and choline reagents exclude this lipid to be PC. Lipid 2 also is diphenylamine and ninhydrin negative, but shows characteristic colorations with 1-naphtol, periodate-Schiff and anisaldehyde reagents. Its characteristic but atypical staining reactions demonstrate the unusual nature of this lipid (Bias, 1985). Knudsen et al. (1982) report on the presence of mannose and of two unidentified sugars in this lipid.

In addition, one or two minor glycolipids with chromatographic properties somehow similar to SQDG are present in *Chlorobium*. Because of its

* This strain was isolated by June Lascelles and was deposited in the NCIB as *Chlorobium limicola f. thiosulf.* However, its identity is uncertain and it is more likely to belong to *Chlorobium vibrioforme f. thiosulf.* (Norbert Pfennig, personal communication). This strain can be obtained from the DSM as a strain of *Chlorobium vibrioforme*. We are using the name *Cb. vibrioforme f. thiosulf.* in this chapter, though most of the literature refers to it as *Chlorobium limicola f. thiosulf.!*

similar R_f-values to the plant type SQDG, its negative staining reactions with ninhydrin, molybdate and diphenylamine reagents, and its positive reaction with the 1-naphtol, one of these lipids was suggested to be SQDG (Kenyon and Gray, 1974; Knudsen et al., 1982). However, comparison of the R_f-values of authentic SQDG from spinach with those of the glycolipids from *Cb. vibrioforme f. thiosulf.* 8327 under identical two-dimensional TLC conditions, showed slight differences (Bias, 1985). In addition, the color reactions with periodate Schiff and anisaldehyde-sulfuric acid reagents of the *Chlorobium* lipids were different from those of the spinach SQDG (Bias, 1985). The absence of SQDG in *Cb. vibrioforme f. thiosulf.* (strain 8327) and in *Cb. limicola f. thiosulf.* (strains 6230 and 6430) was further substantiated by culture experiments with radioactively labelled sulfur isotopes in cysteine and thiosulfate (labelled in sulfane or sulfone sulfur). All polar lipids inclusive of the SQDG-like glycolipid were present in the lipid extracts of cells from these experiments. Although growing cells did incorporate labelled sulfur from both substrates, no radioactivity could be found in the polar lipid fraction and in particular in the glycolipid chromatographically similar to SQDG (Bias, 1985).

CL and PG have been commonly identified as phospholipids of green sulfur bacteria. PE has been found in some cases, but not in others. Olson et al. (1983) working with *Cb. limicola f. thiosulf.* (strain 6230) mention that high proportions of PE in lipid extracts from whole cells were correlated with the presence of a colorless, motile contaminant in the cultures, though very low levels of PE were also found in the absence of the contaminant. PE has been identified in pure cultures of *Cb. vibrioforme f. thiosulf.* (strain 8327) and *Cb. limicola f. thiosulf.* (strains 6230 and 6430) (Knudsen et al., 1982; Bias, 1985; Imhoff, 1988). PE was absent from *Cb. vibrioforme* (6030, not thiosulfate-utilizing) and from *Pelodictyon luteolum* (Imhoff, 1988). Though data were not presented, the absence of PE from strains of *Pelodictyon luteolum, Cb. phaeobacteroides, Cb. phaeovibrioides, Cb. limicola, Cb. vibrioforme* and *Cb. vibrioforme f. thiosulf.* have been reported (Kenyon and Gray, 1974). In contrast to other reports these authors did not find PE in *Cb. limicola f. thiosulf.* strain 6230.

It may be concluded that the polar lipid composition of green sulfur bacteria is characterized by the presence of PG, CL, MGDG, glycolipid II, lipid 2 and the aminoglycosphingolipid (lipid 1). It also appears that PE, despite the discrepancies mentioned above, is present in *Cb. limicola f. thiosulf.* and *Cb. vibrioforme f. thiosulf.*, but may be absent in other green sulfur bacteria. PC, PI and sulfolipids are absent from Chlorobiaceae.

6. Chloroflexus aurantiacus

Chloroflexus aurantiacus also contains large proportions of glycolipids (Kenyon and Gray 1974; Knudsen et al., 1982; Imhoff 1988). Glycolipid II, lipid 1 (aminoglycosphingolipid) and lipid 2 as described for *Chlorobium* species were absent (Jensen et al., 1991; Imhoff, 1988). Instead two glycolipids have been found with R_f-values and staining behavior similar to MGDG and DGDG, respectively (Kenyon and Gray, 1974). Both of these lipids contain galactose and another sugar. The DGDG-like lipid contains galactose and glucose in equal amounts and is different from the lipid 2 of Chlorobiaceae by staining with 1-naphtol, although its R_f-value is similar to this lipid (Knudsen et al., 1982; Imhoff, 1988). Similar to the situation in Chlorobiaceae, the presence of SQDG has been postulated on the basis of negative results in staining with molybdate and ninhydrin and its R_f-value, but is doubtful. The lipid with a similar R_f-value to SQDG does not show the intense red color typical for SQDG by staining with 1-naphtol and 40% sulfuric acid (Imhoff, 1988).

Cf. aurantiacus appears to be the only phototrophic bacterium that contains PI as major phospholipid (Kenyon and Gray, 1974). Analysis of phospholipids in several strains of *Cf. aurantiacus* indicated the presence of PG and PI whereas PE and PS were absent (Kenyon and Gray, 1974; Knudsen et al., 1982; Imhoff, 1988). In contrast to other reports, CL has been tentatively identified in *Cf. aurantiacus* strain OK-70-fl, by R_f-value and positive staining reaction with molybdate reagent (Imhoff 1988).

Similar to the findings with green sulfur bacteria, also chlorosomes from *Chloroflexus* are enriched in glycolipids relative to the cell membranes. The chlorosomes of *Cf. aurantiacus* contain phosphate and sugar in nearly equal molarity, whereas its membranes have four times the amount of phosphate compared to sugars (Schmidt, 1980). In *Chloroflexus* cells grown aerobically in the dark, chlorosomes and BChl are absent, whereas substantial amounts of the lipid similar to MGDG are detectable (Sprague et al., 1981; Holo et al., 1985). The specific cellular content of this lipid was approximately constant and independent of light intensity and BChl content (Holo et al. 1985). This suggests that this lipid, similar to the MGDG of green sulfur bacteria, is enriched in but not exclusively bound to the chlorosomes.

7. Heliobacterium chlorum

The lipid composition of *Hb. chlorum* is completely different from that of the green sulfur and nonsulfur bacteria. Glycolipids are absent and among the phospholipids present, CL, PE and PG are dominant. Lyso-PE and three further unidentified aminolipids are present in smaller amounts (Imhoff, 1988).

B. Fatty Acids

Fatty acids of most phototrophic bacteria are mainly saturated and mono-unsaturated straight chain C-16 and C-18 fatty acids. Small amounts of C-14 and C-20 fatty acids are present in many phototrophic purple bacteria and also C-15, C-17 and C-19 fatty acids may occur in traces or small proportions. Branched-chain fatty acids represent a major fraction of the cellular lipids only in *Heliobacterium* species and related bacteria. 3-OH fatty acids are known as constituents of the lipopolysaccharides and are present in all phototrophic purple bacteria and in Chlorobiaceae. Also C-17 and/or C-19 cyclopropane fatty acids are present in many species. Altogether, there are significant differences in the fatty acid composition of the major groups of anoxygenic phototrophic bacteria represented by the purple sulfur bacteria, purple nonsulfur bacteria, green sulfur bacteria, *Chloroflexus* sp. and heliobacteria.

Fatty acids are most susceptible to dynamic changes during growth and to variations with growth conditions and media compositions. This is certainly the reason for greatly varying fatty acid compositions found by different research groups. It is therefore important to use standardized media composition and growth conditions whenever the fatty acid composition of different strains and species is to be compared. In this respect surveys of a large number of strains yield valuable data if attention has been paid to standardization of the methods. Such surveys have been published on purple nonsulfur bacteria (Kato et al., 1985; Urakami and Komogata, 1988). A congruent set of data was also obtained with more than 20 strains of all known *Rhodobacter* species (Imhoff, 1991).

1. Purple Sulfur Bacteria

All Chromatiaceae examined have well balanced proportions of 16:0, 16:1 and 18:1, with 18:1 as the major but not dominant component (Table 3).

Under all culture conditions major fatty acids of *Ectothiorhodospira* species were 18:1 and 16:0 (Asselineau and Trüper, 1982; Imhoff and Thiemann, 1991). These results are in accordance with previously described observations on *Ec. halophila* SL1 (see Kenyon, 1978). 18:0 and 16:1 were found as minor components in *Ec. mobilis* and the species requiring low salt concentrations. The content of 16:1 was even lower in *Ec. halophila* and other extremely halophilic *Ectothiorhodospira* species (Table 3). Various proportions of 19cyc were present in all *Ectothiorhodospira* species. The presence of branched-chain fatty acids in *Ec. mobilis* found by Oyewole and Holt (1976) could not be verified by others (Asselineau and Trüper, 1982; Imhoff and Thiemann, 1991, Imhoff, unpublished).

3-OH-14:0 is the only hydroxy fatty acid of Chromatiaceae analyzed so far (Meissner et al., 1988a). 3-hydroxy fatty acids of *Ectothiorhodospira* species are 3-OH-10:0 and 3-OH-12:0 (*Ec. vacuolata* and *Ec. shaposhnikovii),* 3-OH-10:0 (*Ec. mobilis*) and 3-OH-10:0, 3-OH-12:0, 3-OH-14:0 and 3-oxo-14:0 in *Ec. halophila* (Zahr et al., 1992; Meissner et al., 1988b).

2. Purple Nonsulfur Bacteria

The fatty acid composition of all purple nonsulfur bacteria is characterized by C-16 and C-18 straight-chain saturated and monounsaturated fatty acids, with minor amounts of C-14 fatty acids present in most species. Also C-15 and C-17 fatty acids may be present in trace amounts (Kato et al., 1985; Urakami and Komagata, 1988; Kenyon, 1978; Imhoff, 1988, 1991). Gradual differences in the proportions of 18:1, 16:1, 16:0 and 18:0 allow the distinction of several groups of species (Tables 4 and 5).

In contrast to all other purple nonsulfur bacteria, in species of the genera *Rhodocyclus, Rhodoferax* and *Rubrivivax* (β-subgroup of the Proteobacteria) C-16 fatty acids are present in higher proportions than C-18 fatty acids. 16:1 is the major component and also proportions of 16:0 are higher than those of 18:1 (Table 4b). Species of the genus *Rhodospirillum*, as shown in Table 4a, contain 18:1 as major but not dominant component (approx. 45–55%) and a total

Table 3. Major fatty acids of representative phototrophic purple sulfur bacteria

	14:0	16:0	16:1	18:0	18:1	19 cyc	Comments/References
Chromatiaceae							
Cm. purpuratum 5500	0.6	30.3	28.2	1.7	38.5		5% salts, Imhoff and Thiemann ,1991
Thiocapsa roseop. 9314	+	20	28	+	45		Takacs and Holt, 1971
Thiocapsa roseop. 6311	0.4	22.1	25.4	0.9	43.8		Imhoff, 1988
Cm. vinosum D	+	28.7	32.6	+	38.7		Haverkate et al., 1965
Cm. vinosum D	0.7	18.9	36.2	0.7	39.3		Imhoff, 1988
Cm. minus 1211	0.3	20.5	30.4	1.2	43.8		Imhoff, unpublished
Cm. warmingii 6512	1.3	25.7	36.5	0.8	32.3		Imhoff, unpublished
Thiocystis gelatinosa 2611	0.6	22.9	30.5	1.2	39.5		Imhoff, unpublished
Ectothiorhodospiraceae							
Ec. halophila 9628	0.8	25.4	1.2	8.6	63.8*		10% salts, Imhoff and Thiemann, 1991
Ec. halophila 9628	0.7	22.0	1.6	5.6	69.4*		20% salts, Imhoff and Thiemann, 1991
Ec. halophila 9628	0.9	24.9	2.2	6.9	63.8*		30% salts, Imhoff and Thiemann, 1991
Ec. halophila SL1	–	11.4	0.9	8.7	63.8	7.3	Imhoff, unpublished
Ec. halophila 9630	1	16	1	15	33	31	Asselineau and Trüper, 1982
Ec. halochloris 9850	0.4	18.9	0.4	7.5	69.7	1.0	Imhoff, 1988
Ec. halochloris 9850	tr	19	tr	7	51	18	Asselineau and Trüper, 1982
Ec. abdelmalekii 9840	0.3	24.7	1.8	5.4	58.9	6.3	Imhoff, unpublished
Ec. abdelmalekii 9840	0.5	16	3	4	57	10	Asselineau and Trüper, 1982
Ec. mobilis 9903	0.2	14.0	5.5	4.8	72.3	tr	5% salts, 27 °C, Imhoff and Thiemann, 1991
Ec. mobilis 9903	0.2	20.7	4.6	9.6	62.2	0.4	5% salts, 36 °C, Imhoff and Thiemann 1991
Ec. mobilis 9903	0.5	23.1	3.4	10.9	59.7	0.3	5% salts, 42 °C, Imhoff and Thiemann, 1991
Ec. shaposhnikovii N1	tr	17	8	tr	68	nd	Asselineau and Trüper, 1982
Ec. shaposhnikovii N1	0.1	13.1	6.8	2.0	74.7*		Imhoff, unpublished
Ec. vacuolata DSM 2111	–	15.3	4.8	2.4	73.5*		Imhoff, unpublished
Ec. vacuolata DSM 2111	tr	20	6	7	60	nd	Asselineau and Trüper, 1982

* 18:1 including 19 cyc, nd not detected

of approx. 50% C-16 fatty acids. The clear predominance of 18:1 is characteristic of species of *Rhodobacter, Rhodopseudomonas, Rhodopila* and *Rhodomicrobium* with 18:1 generally representing more than 70% and in some species (*Rm. vannielii, Rp. blastica*) reaching 90% or more. Species of the genus *Rhodobacter* can be distinguished on the basis of differences in their fatty acid pattern (Table 5). Cyclopropane fatty acids have been found in several species. They are usually present in quite small proportions that may increase considerably in stationary phase or under unfavorable growth conditions.

3-OH-10:0 is the dominant 3-hydroxy fatty acid in *Rhodocyclus* and *Rubrivivax* species, while 3-OH-8:0 predominates in *Rhodoferax fermentans* (Drews et al., 1978; Hiraishi et al., 1991). Major-3-hydroxy fatty acids of *Rhodobacter* species are 3-OH-10:0 and 3-OH-14:0, C-14 and C-16 chains are dominant in other purple nonsulfur bacteria either single or both together. 3-OH-14:0 and 3-OH-18:0 are present in *Rhodopila globiformis* (Drews et al., 1978; Urakami and Komagata, 1988; see also Imhoff and Trüper, 1989). 3-OH-18:0 and also 3-OH-12:0 occasionally have been found in other purple nonsulfur bacteria.

3. Green Sulfur Bacteria

Green sulfur bacteria contain straight-chain, saturated and monounsaturated fatty acids in the range of C-12 to C-18, with large amounts of 14:0, 16:0 and C 16:1 and small to trace amounts of 18:0, 14:1 and 18:1 (Kenyon and Gray, 1974; Knudsen et al., 1982). In comparison to purple bacteria, Chlorobiaceae have a relative higher ratio of saturated to unsaturated fatty

Table 4a. Major fatty acids of representative purple nonsulfur bacteria (α-subgroup of Proteobacteria)[a]

Species	Strains	14:0	16:0	16:1	18:0	18:1	References
Rp. palustris	ATCC 17001	–	7.6	2.7	7.1	78.9	Urakami and Komagata, 1988
	DSM 126	0.1	14.0	3.0	9.8	70.1	Urakami and Komagata, 1988
	ATCC 17009	0.1	5.7	1.6	7.3	76.5	Urakami and Komagata, 1988
	1a1	–	5.2	3.1	7.3	79.7	Imhoff, unpublished
Rp. acidophila	ATCC 25092	0.7	19.6	42.7	2.3	31.5	Kato et al., 1985
	ATCC 25092	0.8	14.8	37.2	0.8	46.0	Imhoff, unpublished
Rp. blastica	ATCC 33485	0.4	1.5	2.8	4.9	90.1	Imhoff, 1988
Rp. rutila	ATCC 33872	1.0	8.7	3.6	10.3	75.7	Imhoff, unpublished
Rp. viridis	DSM 133	0.3	12.7	10.0	0.7	74.0	Kato et al., 1985
	DSM 133	0.5	8.4	5.5	2.2	74.6	Imhoff, unpublished
	BN 170	0.5	6.7	5.6	1.1	76.7	Imhoff, unpublished
Rp. sulfoviridis	DSM 729	2.5	8.6	9.2	1.7	76.5	Imhoff, unpublished
Rp. marina	DSM 2698	0.4	1.9	0.5	14.1	69.0	Imhoff, 1988
	DSM 2780	0.1	3.2	0.5	21.3	67.8	5% salts, Imhoff and Thiemann, 1991
Rm. vannielii	ATCC 17100	2.4	3.7	0.6	3.6	85.6	Urakami and Komagata, 1988
	ATCC 17100	1.0	2.3	0.4	1.6	92.3	Imhoff, unpublished
	ns	1.8	5.4	1.0	0.1	88.2	Park and Berger, 1967
Rhodopila globiformis	DSM 161	5.8	9.3	4.7	1.0	74.4	Urakami and Komagata, 1988
Rs. rubrum	DSM 467	4.5	20.7	25.5	2.0	40.8	Urakami and Komagata, 1988
	IFO 3986	0.2	13.8	26.1	1.4	46.1	Urakami and Komagata, 1988
	DSM 467	2.1	14.0	27.1	1.3	54.8	Imhoff, 1988
	1761-1a	1.7	14.6	21.8	3.0	53.3	Imhoff, unpublished
Rs. photometricum	DSM 122	1.0	25.2	22.2	0.4	51.0	Imhoff, unpublished
Rs. molischianum	DSM 120	0.7	18.1	36.5	0.7	43.5	Imhoff, 1988
	ns	tr	21.9	29.8	tr	47.1	Constantopoulos and Bloch, 1967
Rs. fulvum	DSM 115	0.6	15.1	25.8	1.2	54.5	Imhoff, unpublished

[a] All values are from phototrophically grown cells.
ns – not specified

Table 4b. Fatty acids of purple nonsulfur bacteria (β-subgroup of Proteobacteria)

Species	Strains	14:0	16:0	16:1	18:0	18:1	References
Rv. gelatinosus	BN 156	5.0	34.6	41.3	2.3	16.2	Imhoff, 1988
	ATCC 17011	2.1	24.0	34.8	1.2	16.5	Urakami and Komagata, 1988
	BN 151	0.8	29.4	45.1	3.0	21.5	Imhoff, unpublished
	DSM 149	0.7	35.1	42.7	2.9	18.4	Imhoff, unpublished
	DSM 151	0.8	31.7	39.0	3.3	25.0	Imhoff, unpublished
Rc. tenuis	BN 230	1.4	33.5	49.5	0.5	14.7	Imhoff, 1988
	DSM 109	5	36	43	tr	15	Hiraishi et al., 1991
Rc. purpureus	DSM 168	1.4	33.5	44.5	1.3	18.2	Imhoff, unpublished
	DSM 168	5	35	40	tr	18	Hiraishi et al., 1991
Rhodoferax fermentans	FR 2	tr	35	54	tr	5	Hiraishi et al., 1991

tr – trace amounts

Table 5. Major fatty acids of *Rhodobacter* species

Species	strains	14:0	16:0	16:1	18:0	18:1	References
Rb. sphaeroides (5)*		0.2–0.3	3.9–8.3	1.2–2.0	1.9–16.2	72.0–77.9[b]	Imhoff, 1991
	2	–	1.7	1.0	4.8	90.8	Wood et al., 1965
	GA	tr	5.5	1.6	12	80	Marinetti and Cattieu, 1981
	17023	–	7.8	1.5	10.7	75.9	Urakami and Komagata, 1988
Rb. capsulatus (6)*		0.1–0.4	4.1–5.0	5.1–7.4	3.8–9.3	78.1–84.2[b]	Imhoff, 1991
	2,3,11	–	2.3	2.4	4.2	90.3	Wood et al., 1965
	3764	tr	1.9	2.6	1.3	93.6	Schröder et al., 1969
	NCIB 8254	–	13.8	15.0	4.0	63.5	Urakami and Komagata, 1988
Rb. veldkampii (1)*		0.1	4.3	17.5	6.5	69.4[b]	Imhoff, 1991
Rb. adriaticus(4)*		0.2	3.8–6.4	0.2–0.4	19.8–22.0	61.6–67.2[a]	Imhoff, 1991
Rb. sulfidophilus (4)*		0.1–0.3	11.6–14.6	0.7–1.5	7.6–9.8	72.8–75.0[a]	Imhoff, 1991
	DSM 1374		16.0	0.4	2.4	72.1	Urakami and Komagata, 1988
	DSM 2351		17.8	3.0	4.0	72.2	Urakami and Komagata, 1988
Rb. euryhalinus (1)*		0.1	4.9	2.5	21.8	67.5[b]	Imhoff, 1991

* number of strains investigated are in brackets, the range of proportion of all strains is given.
[a] 19 cyc is present, [b] 19 cyc not detected.

acids. The large amounts of 14:0 are characteristic for green sulfur bacteria (Table 6). This fatty acid was found as major component of the aminoglycosphingolipid from *Chlorobium* (Olson et al., 1983; Jensen et al., 1991). 17cyc was present in all strains examined though significant amounts were only found in strains of *Cb. limicola f. thiosulf.* and *Pelodictyon luteolum* (Kenyon and Gray, 1974; Knudsen et al., 1982). Hydroxy fatty acids (3-OH-14:0; 3-OH-16:0) were found in whole cells and phenol-water extracted phases (Knudsen et al., 1982). Whole cells and chlorosomes from *Cb. limicola f. thiosulf.* had a largely similar fatty acid composition apart from a lower content of 14:0 in chlorosomes (Schmitz, 1967). In the thermophilic *Chlorobium* species *Cb. tepidum* (Wahlund et al., 1991), C-16 and C-18 fatty acids were predominant (Shiea et al., 1991). Similar to *Chloroflexus aurantiacus, Cb. tepidum* also contained large amounts of wax esters, predominantly of C-30 and C-32 chain length (Shiea et al., 1991).

4. Chloroflexus aurantiacus

The thermophilic nature of *Cf. aurantiacus* certainly influences its fatty acid composition. Although mesophilic strains of *Chloroflexus* have been observed (Pirovarova and Gorlenko, 1977), nothing is known about their lipid and fatty acid composition. The fatty acid composition of *Cf. aurantiacus* is characterized by the presence of 16:0, 18:0 and 18:1 as major components. Significant amounts of 16:0, 17:0, 17:1, 19:0, 19:1, 20:0 and 20:1 were present in all strains investigated (Kenyon and Gray, 1974; Knudsen et al., 1982). In addition, fatty alcohols and wax esters (2.5–3.0% of cell dry weight) were present (Knudsen et al., 1982). The chain length of these saturated and monounsaturated wax esters was from C-28 to C-38, those of C-34 and C-36 being predominant (Knudsen et al., 1982). The results of the two research groups showed major differences in the proportions of various fatty acids (Table 7). These differences may be due to differences in growth temperature (45 °C by Kenyon and Gray, 1974 and 52 °C by Knudsen et al., 1982), but also due to different media composition. In particular the elevated proportion of 18:1 and the lower one of 16:0 at the higher temperature (Knudsen et al., 1982) resemble temperature-dependent changes observed in *Ectothiorhodospira* species (Imhoff and Thiemann, 1991) and could indicate similar temperature dependent adaptation phenomena in the two bacteria. Two different media have been used in the work of Knudsen et al. (1982), but it is not clear whether cells taken for fatty acid analysis were grown in the synthetic medium or not. Although these authors do not discuss discrepancies to the earlier work in regard

Table 6. Fatty acids of green sulfur bacteria

Species	Strain	12:0	14:0	14:1	16:0	16:1	17:cy	18:0	18:1	14:0 3-OH	16:0 3-OH
Cb. limicola f. thiosulf.	6330[1]		13	2	17	57	3	tr			
Cb. limicola f. thiosulf.	6230[1]		21	2	10	43	21	tr	1		
	6230[2]	1.1	27.1		20.3	37.3	7.3	tr	3.1	1.1	1.4
Cb. vibriodes f. thiosulf.	1930[1]		12	tr	23	52	3	1	2		
Cb.vibrioides f. thiosulf.	8327[2]	2.1	24.4		23	42.8	0.7	3.3	1.2	1.4	2.1
Cb. vibrioides f. thiosulf.	2230[2]	2.9	23.8		22.1	43.4	2.6	tr	3.1	1.4	1.6
Cb. phaeobacteroides	2430[1]		16	1	15	64	1	tr	1		
Cb. phaeovibrioides	2631[1]		10	tr	29	51	2	tr	2		
Pelodictyon luteolum	2530[1]		14	1	21	47	11	tr	tr		

Data from [1]Kenyon and Gray (1974) and [2]Knudsen et al. (1982).

Table 7. Fatty acids of *Chloroflexus aurantiacus*

Strain	16:0	16:1	17:0	17:1	18:0	18:1	18:2	19:0	19:1	20:0	20:1
OK-70-fl[1]	8	3	5	7	10	46	3	2	8	1	5
OK-70-fl[2]	23.7	2.5	12.3	3.1	28.6	9.0		3.6	1.3	0.8	2.5
J-10-fl[1]	17	2	3	1	27	34	3	1	2	1	5
J-10-fl[2]	20.3	2.2	9.1	2.3	29.5	9.1		3.6	1.6	1.1	2.9
OH-64-fl[1]	12	4	3	3	14	52	2	1	3	1	4

Data from [1]Kenyon and Gray (1974) and [2]Knudsen et al. (1982).

to the presence of 18:2, it could well be that the presence of complex nutrients led to the incorporation of 18:2 in one of the studies (Kenyon and Gray, 1974), but that *Chloroflexus* is unable to synthesize this fatty acid. This would explain the failure of Knudsen et al. (1982) to detect 18:2 in *Cf. aurantiacus* strains that presumably were grown in a synthetic medium.

5. Heliobacterium *and* Heliobacillus *Species*

Fatty acids from *Heliobacterium chlorum, Heliobacterium gestii* and *Heliobacillus mobilis* revealed significant differences to all other anoxygenic phototrophic bacteria (Beck et al., 1990). This group of bacteria is related to the Gram-positive bacteria according to 16S rRNA sequence similarities (Woese et al., 1985). Characteristic features of their fatty acid composition are the high content of branched-chain fatty acids (iso 13:0, anteiso 14:1, anteiso 14:0, iso 15:0, iso 16:1) and the dominance of 16:1 (Beck et al., 1990); 18:0 and 16:0 were minor components (Table 8).

C. Influence of Growth Conditions on Lipid and Fatty Acid Composition

Growth conditions may have considerable influence on the polar lipid and fatty acid composition of phototrophic bacteria. However, detailed studies on the influence of growth conditions have rarely been undertaken. Differences in lipid and fatty acid compositions obtained for the same species or even the same strain may be the result of different conditions applied to cultivate the bacteria much more than variations in analytical methods. In particular, variation of proportions of individual lipids and the detection of some minor constituents may depend upon the cultivation methods. However, little emphasis is given to this aspect in the literature and often media and growth conditions are reported insufficiently to evaluate the results in this respect.

The presence of Tris (20 mM) in media of *Rb. sphaeroides* and its effect on synthesis and proportions of N-acylphosphatidylserine (NAPS) well demonstrate the effect of minor changes in culture media on lipid proportions quite well. Very minor levels of 1–2% of NAPS increased to more than 40% of total

Table 8. Fatty acids of *Heliobacterium chlorum, Heliobacterium gestii* and *Heliobacillus mobilis*

Fatty acid	*Hb. chlorum*	*Hb. gestii*	*Hc. mobilis*
Iso 13:0	0.6	1.1	<5.0
anteiso 14:1	2.5	4.0	13.7
anteiso 14:0	4.8	3.5	10.2
16:1	35.9	48.6	23.5
Iso 15:0	10.7	2.9	<5.0
Iso 16:1	30.3	27.4	37.2
17:0	0.8	1.0	<5.0
18:1	12.1	9.7	<5.0
18:0	0.6	1.1	nd

[a] Data from Beck et al. (1990); nd – not detected

lipids in the presence of Tris (Cain et al., 1981; Donohue et al., 1982a; 1982b). In respect to the findings with *Rb. sphaeroides* it is of interest that *Ectothiorhodospira halophila* has been cultivated in media containing 20 mM Tris-HCl by Raymond and Sistrom (1969) and that these authors found a ninhydrin-positive phospholipid in addition to PE that could be NAPS (see Kenyon, 1978). This lipid was never found in our studies, in which media without Tris were used. It needs to be established whether the effect of Tris on the lipid composition of *Ec. halophila* is similar to that of *Rb. sphaeroides*.

Also phosphate concentrations of the media are important, because phospholipids may be replaced by other phosphate-free lipids under conditions of phosphate limitation. It has been shown recently that the proportions of the sulfolipid in *Rb. sphaeroides* increased under phosphate limiting growth conditions. Phosphate limitation caused reduced growth of mutants deficient in sulfolipid biosynthesis, but not of wild type cells (Benning et al., 1993), indicating a clear advantage of the sulfolipid under these conditions. Different concentrations of phosphate in growth media could be of relevance in explaining discrepancies in the literature in regard to sulfolipid presence within some purple nonsulfur bacteria.

Also complex nutrients that contain lipid components or precursor molecules must have an impact on the lipids and fatty acids incorporated into the bacterial membranes. For example, cells of *Rs. rubrum* contained PC when peptone was present in the medium but PC was absent when yeast extract was substituted for peptone as complex nutrient source (see Kenyon, 1978). The formation of PI may depend on the supplementation with inositol, which is present in yeast extract. Therefore, it was necessary to assure synthesis of PI in *Cf. aurantiacus* also in synthetic medium lacking yeast extract (Knudsen et al., 1982).

The growth phase also has significant influence on the fatty acid composition. Proportions of cyclopropane fatty acids are known to increase in stationary phase cells. In particular proportions of 19cyc increased at the expense of 18:1 in stationary phase cells of *Ec. mobilis* and other *Ectothiorhodospira* species (Oyewole and Holt, 1976; Imhoff and Thiemann, 1991; Imhoff, unpublished results). Proportions of 19cyc may be as high as 25–33% in stationary phase cells of *Ec. mobilis* 8113 (Oyewole and Holt, 1976) and as low as 2–3% in exponentially growing cells of *Ec. mobilis* 9903 (Ditandy and Imhoff, 1993).

Apart from this, an increase in temperature caused a decrease of C-18 and of unsaturated fatty acids in *Ec. mobilis* (strain 9903), while the contents of C-16 and saturated fatty acids increased (Imhoff and Thiemann, 1991; Table 3). These effects were principally due to changes of 16:0 and 18:1 and were regarded as mechanisms to regulate membrane fluidity and bilayer stability of the membranes in response to temperature changes (Imhoff and Thiemann, 1991).

1. Phototrophic Versus Chemotrophic Growth and Lipids of Cell Fractions

a. Lipids From Cell Fractions

Many investigations have attempted the analysis of differences in lipid and fatty acid composition of membrane fractions. The failure to detect such differences may be due to the fact that clear separation of CM, ICM and OM is difficult to achieve, and that significant differences in overall composition may not exist. Polar lipid compositions apparently were very similar in CM and ICM of *Rb. sphaeroides* (Gorchein, 1968b), and ICM and whole cell lipids from *Chromatium vinosum* (Haverkate et al., 1965) and *Thiocapsa roseopersicina* (Takacs and Holt, 1971). The analysis of pure fractions of CM, ICM and OM from *Ec. mobilis* (strain 9903) revealed no significant qualitative and quantitative differences in polar lipid and fatty acid composition between CM and ICM, though both were not identical and showed minor quantitative variations (e.g. content of C-14 and C-16 fatty acids) (Ditandy and Imhoff, 1993). Differences were, however, noted between OM and other membrane fractions. Lyso-PE was specifically and exclusively found in the OM fraction of *Ec. mobilis*, which also had higher proportions of CL compared to the other membrane fractions. Also in *Rs. rubrum*, relative proportions of individual phospholipids were similar in CM and OM fractions, though lyso-PE was specifically associated with the outer membrane fraction (Oelze et al., 1975). 18:1 was the major fatty acid of all membrane fractions from *Ec. mobilis*, although its proportion was significantly lower in OM, which in turn had higher proportions of all shorter chain and saturated fatty acids (14:0, 16:0, 18:0, 20:0) (Ditandy and Imhoff, 1993). The proportions of saturated fatty acids, in particular those of 14:0, were higher in the OM fraction of *Rhodospirillum rubrum* grown aerobically (chemoheterotrophically), whereas the CM fraction had higher amounts of 16:1 and 18:1 (Collins and Niederman, 1976; Oelze et al., 1975). Very similar trends were reported for OM fractions from *Rb. capsulatus* (Flamman and Weckesser, 1984), *Rp. palustris* (Weckesser et al., 1973) and *Cm. vinosum* (Hurlbert et al., 1974). It was concluded that this trend could be a general phenomenon in Gram-negative bacteria (Ditandy and Imhoff, 1993).

b. Phototrophic Versus Chemotrophic Growth

Because major differences between ICM and other cellular membranes were not observed, it was not surprising that no significant changes in qualitative and quantitative phospholipid compositions were observed in cells of *Rb. sphaeroides* grown aerobically or phototrophically at high light intensities (Onishi and Niederman, 1982). In phototrophically grown cells, the phospholipid composition was dependent upon the incident light intensity (Onishi and Niederman, 1982).

Nevertheless, relative proportions of PC, PG and PE of *Rb. capsulatus* were found to be different in cells grown phototrophically or chemotrophically (Steiner et al., 1970), although no significant differences in the polar lipid composition between CM and ICM were reported. Russell and Harwood (1979) found increasing proportions of PG in cells of *Rb. sphaeroides, Rb. capsulatus* and *Rs. rubrum* when cells were transferred from chemotrophic to phototrophic growth conditions. Proportions of PE (and in *Rb. sphaeroides* also PC) decreased concomitantly.

Kato et al. (1985) and Urakami and Komagata (1988) analyzed fatty acids from a number of strains and species of purple nonsulfur bacteria grown aerobically in the dark and anaerobically in the light. The cellular fatty acid composition was not much different in cells grown phototrophically or chemotrophically, but the results from Urakami and Komagata (1988) were different from those of Kato et al. (1985), although they came from the same lab. In the case of disagreement between the two papers results from Urakami and Komagata (1988) were included in Table 4. Similar findings have been made earlier with *Rs. rubrum* (Oelze et al., 1970; Brooks and Benson, 1972) and *Rb. sphaeroides* (Scheuer-brandt and Bloch, 1962). Differences in fatty acid composition were noted, however, by some other authors. The cellular content of saturated fatty acids was found to be lower in chemotrophically grown cells of *Rubrivivax gelatinosus, Rs. rubrum* and *Rp. palustris* (Hands and Bartley, 1962; Wood et al., 1965).

2. The Influence of Salt Concentration

Ectothiorhodospira species represent a group of halophilic phototrophic bacteria with a characteristic dependence on saline and alkaline growth conditions (Imhoff, 1989). Certain strains of the extremely halophilic *Ec. halophila* are the most halophilic eubacteria known and are able to grow in saturated salt solutions. Comparative analyses of salinity-dependent lipid composition of slightly, moderately, and extremely halophilic *Ectothiorhodospira* strains revealed strong effects on their lipid composition. Most significant salt-dependent changes were observed in the proportions of PG, CL and PE. These changes reflect phenomena of salinity-dependent PG/PE antagonism known from other Gram-negative eubacteria; increasing the salinity promotes the biosynthesis of PG relative to that of PE, CL and PC. Upon dilution stress responses were reversed and corresponded in a strong increase of PE synthesis (Thiemann and Imhoff, 1991). Significant increases in the proportions of PG and concomitant decreases in those of PE and CL resulted in a considerable increase of the excess negative charges with increasing salinity (Imhoff et al., 1991; Thiemann and Imhoff, 1991). This effect was most dramatic in the slightly halophilic *Ec. mobilis* (strain BN 9903) and quantitatively less significant in moderately and extremely halophilic strains of *Ec. halophila* (strains BN 9626 and BN 9630) which had, however, an overall higher percentage of negative charges of their lipids.

The most important aspect of structural adaptation of lipid composition upon increasing medium salinity was assumed to be the stabilization of the membrane bilayer. The portions of PC and PG, which are both regarded as lipids supporting the bilayer structure, steadily increased with salinity in all three *Ectothiorhodospira* strains and were expected to counterbalance the bilayer destabilizing effects of the increasing ionic strength (Thiemann and Imhoff, 1991).

Of crucial importance for integrity of the bilayer and its proper fluidity also are changes of the fatty acid composition. The salt concentration had strong influences on the fatty acid composition in *Ectothiorhodospira* species. Minimum values of C-16 and saturated fatty acids and maximum values of C-18 and unsaturated fatty acids were found at or close to the salt optimum. Similar changes and correlations with minima of C-16 and saturated fatty acids at the salt optima also were found in *Rb. adriaticus, Rp. marina* and *Chromatium purpuratum* (Imhoff and Thiemann, 1991). Despite the fact that the species-specific proportions of major fatty acids in these bacteria were significantly different, in principal similar changes occurred in respect to changing salt concentrations, when related to their growth optima. It was concluded that the underlying mechanisms were the same in all of these purple bacteria and that the data obtained were of relevance to the fatty acid composition of Gram-negative bacteria in general.

III. Apolar Lipids

A. Quinones

The characteristic feature of all quinones is their ability to function as redox carriers in electron transport processes within the membranes. While the important role of Q in photosynthetic electron transport is well recognized, the functions of RQ and MK are much less studied. Their crucial role in electron transport processes is discussed in other chapters of this book (see Chapters 14 and 44 by Vermeglio et al. and Zannoni, respectively). The chemical structure of quinones contains two elements, the quinoid ring system and an isoprenoid side chain. Variation occurs in both of these elements. According to the quinone ring structure we distinguish between menaquinones, ubiquinones and rhodoquinones. Variation of the side chain is obtained by the chain length (number of isoprenoid units) and in some cases by specific modifications of its chemical structure.

The presence of quinones with defined structures of ring system and side chain apparently is a characteristic feature of bacterial species and can be used for taxonomic classification. Also different species of the same genus are expected to have identical quinone composition. Minor variations of growth conditions have no or only small influence on the quinone composition. However, depending on the quinone composition and on their different functions in respiratory and photosynthetic electron transport, changes may be expected during adaptation of the more versatile species from aerobic dark to anaerobic light conditions.

1. Purple Nonsulfur Bacteria

The older literature on quinones in phototrophic purple bacteria has been reviewed by Collins and Jones (1981) and reveals scattered and sometimes conflicting results, which varied in particular regarding the presence of menaquinones and the determination of the predominant chain length of some species. More recently several investigations on the quinone composition of phototrophic purple bacteria have been published and give a quite comprehensive view of quinone occurrence in this group (Hiraishi and Hoshino, 1984; Hiraishi et al., 1984; Imhoff, 1984a; Kato et al., 1985). Available data are summarized in Tables 9 and 10.

All known purple nonsulfur bacteria contain ubiquinones. Some species have ubiquinones as sole quinone component and lack menaquinones and rhodoquinones. Others contain ubiquinone and rhodoquinone, ubiquinone and menaquinone or ubiquinone, menaquinone and rhodoquinone. Rhodoquinones, which were first discovered in *Rs.*

Table 9. Major quinones of purple nonsulfur bacteria[a]

Rhodospirillum			
Rs. rubrum	Q-10	–	RQ-10
Rs. photometricum	Q-8	–	RQ-8
Rs. molischianum	Q-9	MK-9	–
Rs. fulvum	Q-9	MK-9	–
Rs. salexigens	Q-10	MK-10	–
Rs. salinarum	Q-10	MK-10	–
Rhodopseudomonas			
Rp. palustris	Q-10	–	
Rp. acidophila	Q-10	MK-10	RQ-10
Rp. viridis	Q-9	MK-9	–
Rp. sulfoviridis	Q-8/10	MK-7/8	
Rp. marina	Q-10	MK-10	
Rp. rutila	Q-10	–	
Rp. blastica	Q-10	-	
Rhodobacter			
Rb. sphaeroides	Q-10	–	–
Rb. capsulatus	Q-10	–	–
Rb. veldkampii	Q-10	–	–
Rb. sulfidophila	Q-10	–	–
Rb. adriaticus	Q-10	–	–
Rhodopila			
Rhodopila globiformis	Q-9/10	MK-9/10	RQ-9/10
Rhodomicrobium			
Rm. vannielii	Q-10	–	RQ-10
Rhodocyclus			
Rc. tenuis	Q-8	MK-8	–
Rc. purpureus	Q-8	MK-8	–
Rubrivivax			
Rubrivivax gelatinosus[1]	Q-8	MK-8	–
Rhodoferax			
Rhodoferax fermentans	Q-8	–	RQ-8

[a] Data from Hiraishi and Hoshino, 1984; Hiraishi et al., 1984; Hiraishi et al., 1991; Imhoff, 1984a; Kato et al., 1985.

[1] *Rhodocyclus gelatinosus* (formerly *Rhodopseudomonas gelatinosa*) has been transferred to *Rubrivivax gelatinosus* (Willems et al., 1991)

rubrum (Glover and Threlfall, 1962) also were found in several other purple nonsulfur bacteria. RQ-8 dominates in *Rhodospirillum photometricum*, RQ-9 and RQ-10 in *Rhodopila globiformis*, RQ-10 in *Rs. rubrum, Rhodopseudomonas acidophila,* and *Rhodomicrobium vannielii* (Hiraishi and Hoshino, 1984; Imhoff, 1984a). In *Rs. rubrum* also epoxy derivatives of Q-10 have been found (Friis et al., 1967), although it was not clear whether these occur naturally or are formed during extraction and purification procedures.

Species grouped according to their quinone ring systems can be further differentiated on the basis of the isoprenoid chain length of their predominant quinone components. Regardless of the quinone ring structure, the side chains of the major components in an individual species normally are of the same length, though exceptions do exist. In the majority of the species quinones with one specific chain length predominate over all others with 90–99% of the total amount. In most, but not all cases the major component has the longest side chain. In general, the analogue with one isoprenoid unit less can be found in amounts of less than 10% and others in very low amounts or traces, with relative concentrations decreasing with the chain length. Only high-

Table 10. Major quinones of purple sulfur bacteria, *Chlorobiaceae*, *Chloroflexus* and *Heliobacterium*[a]

Ec. mobilis	Q-7	MK-7	
Ec. shaposhnikovii	Q-7	MK-7	
Ec. vacuolata	Q-7	MK-7	
Ec. halophila	Q-8	MK-8+4	
Ec. halochloris	Q-8	MK-8+4	
Ec. abdelmalekii	Q-8	MK-8+4	
Chromatium warmingii	Q-8	MK-8	
Chromatium vinosum	Q-8	MK-8	
Thiocapsa roseopersicina	Q-8	MK-8	
Thiocystis gelatinosa	Q-8	MK-8	
Chloroflexus aurantiacus	–	MK-10	–
Chlorobium limicola	–	MK-7	CK-7
Chlorobium vibrioforme	–	MK-7	CK-7
Pelodictyon luteolum	–	MK-7	CK-7
Heliobacterium chlorum	–	MK-9	–

[a]Data from Hale et al., 1983; Hiraishi, 1989; Imhoff, 1984a; Imhoff, 1988.

sensitivity mass spectrometry of ubiquinones from *Rs. rubrum* revealed the presence of Q-1 through Q-10 with extremely low levels even of Q-1 through Q-4 (Daves et al., 1967).

2. Purple Sulfur Bacteria

Much less information is available on quinones in other groups of phototrophic bacteria. From the data available it may be concluded that all Chromatiaceae species uniformly have Q-8 and MK-8 (Imhoff, 1984a), though more species have to be analyzed. Among Ectothiorhodospiraceae differences occur between species with low and high salt requirements (Imhoff, 1984a; Table 10).

3. Green Sulfur Bacteria

Quite a few reports deal with the quinones of green sulfur bacteria. The presence of MK-7 and 1'-oxo-MK-7 (also called chlorobiumquinone) was found to be highly specific for *Cb. limicola f. thiosulf.* (Frydman and Rappaport, 1963; Redfearn and Powls, 1968; Powls et al., 1968). MK-7 and chlorobiumquinone were later found in strains of *Cb. vibrioforme* (strain 6030), *Cb. vibrioforme f. thiosulf.* (strain 8327) and *Pelodictyon luteolum* (strain 2530) (Imhoff, 1988). These components may be characteristic for all species of the green sulfur bacteria but more species have to be analyzed to prove this.

4. Chloroflexus aurantiacus

Chloroflexus aurantiacus has menaquinone (MK-10) as the only quinone component but lacks Q, RQ and chlorobiumquinone (Hale et al., 1983; Imhoff, 1988).

5. Heliobacterium chlorum

The presence of menaquinones (MK-9 being the major component) but the absence of ubiquinones, rhodoquinones and chlorobiumquinone in *Heliobacterium chlorum* has been demonstrated (Imhoff, 1988; Hiraishi, 1989).

B. Hopanoids

Hopanoid triterpenes are present in many bacteria. They are structural analogues and putative phylogenetic precursors of the eukaryotic sterols, which are generally absent from prokaryotes. They posses a quasi-planar, rigid, amphiphilic structure similar to that of sterols and therefore also may be functional equivalents in bacteria (Rohmer et al., 1979; Ourisson et al., 1987). Major components always were C_{35} bacteriohopanepolyols with bacteriohopanetetrol as the most common one (Rohmer et al., 1984). In addition, the C_{30} hopanoids diploptene and diplopterol were found in almost all hopanoid-containing bacteria. Their biosynthesis is completely independent from molecular oxygen since no oxidation step is required as in sterol biosynthesis. Therefore hopanoid biosynthesis is possible under anoxic conditions and evolution of the biosynthetic pathways could have occurred in strictly anaerobic bacteria. Hopanoids are formed by direct cyclization of squalene by a squalene cyclase resulting in the formation of pentacyclic triterpenoids (Ourisson et al., 1987). Some squalene analogues that act as inhibitors of squalene cyclases were growth inhibitory to hopanoid producing bacteria (including *Rm. vannielii, Rp. acidophila and Rp. palustris*) at concentrations in the μM range, whereas hopanoid non-producers were not affected (Flesch and Rohmer, 1987). These results demonstrate the importance of hopanoids for life and survival of their producers.

Hopanoids were not detected (less than 10 μg/g dry weight) in the phototrophic green and purple bacteria *Chromatium vinosum* (strain D), *Thiocapsa roseopersicina* (strain 6311), *Amoebobacter roseus* (strain 6611), *Thiocystis violacea* (strain 2311), *Ectothiorhodospira mobilis* (strain 8115), *Ectothiorhodospira shaposhnikovii* (strain Moskau N1) and *Chlorobium limicola f. thiosulf.* (strain 6230) (Rohmer et al., 1984).

The first triterpene (paluol) in a phototrophic purple bacterium was found in *Rp. palustris* (Jensen, 1962). From inhibitor studies it had been concluded that sterols may occur in this bacterium (Aaronson, 1964). This author found that benzmalecene and triparanol inhibited growth of *Rp. palustris* and that ergosterol, squalene and oleic acid overcame this inhibition. In a screening of purple nonsulfur bacteria for squalene and hopanoid components, squalene was found only in *Rubrivivax gelatinosus* and hop-22(29)-ene and hopan-22-ol in *Rp. palustris* and *Rp. rutila* (Urakami and Komagata, 1988). Other authors found hopanoids in several purple nonsulfur bacteria, including *Rs. rubrum, Rb. sphaeroides, Rp. palustris, Rp. acidophila* and *Rm. vannielii* (Rohmer et al., 1984; Ourisson et al., 1987). The principal fraction was represented by bacteriohopanepolyols present in these species in amounts from 1.4–4.8 mg/g dry weight. Lower amounts (below 0.3 mg/g dry weight) of diploptene and diplopterol were also found. Diplopterol was absent from two strains of *Rp. acidophila* (Rohmer et al., 1984). The highest content of bacteriohopanepolyols were found in *Rm. vannielii* (Rohmer et al., 1984).

The structural variety of the triterpenoid components is great. For example, in cells from *Rhodomicrobium vannielii* grown phototrophically 16 different hopanoids and fernens have been isolated (Howard et al., 1984). In this report 3β-hydroxy-21-methylhopane, 29-hydroxy-3,21dimethylhopane and hop-22(29)-ene (diploptene) were identified as principal components (Howard et al., 1984). Further components were hop-21(22)-en-3-one, 22-hydroxy-hopan-3-one, and other minor components, some of which, however, may be isolation artifacts (Ourisson et al., 1987). The presence of 22-hydroxyhopan-3-one could not be confirmed by Neunlist et al. (1985), though reference material was available and sensitivity was at the microgram level. These authors recognized derivatives of bacteriohopanetetrol with an amino group in their side chain and linked to aminoacyl residues as the major hopanoids of *Rm. vannielii* (Neunlist et al., 1985). 35-aminobacteriohopane-32,33,34-triol and N-tryptophanyl and N-ornithyl derivatives thereof were identified and represented the first hopanoids known to contain an amino group in the side chain. Bacteriohopane-aminotriol and free aminopolyols also have been found in *Rp. palustris* (Ourisson et al., 1987). The first evidence of the triterpenoid tetrahymanol in a prokaryote has recently been obtained in *Rp. palustris* (Kleemann et al., 1990). Furthermore, nucleoside analogues, 30-(5'-adenosyl)-hopanes, were isolated and identified from *Rp. acidophila* (Neunlist et al., 1988). Another new hopanoid has been described from *Rs. rubrum* as a derivative of bacteriohopanetetrol having an glucuronopyranosyl residue linked alpha-glycosidically to the hydroxyl group of C_{35} (Lopitz et al., 1992). In addition to the principal structural variation in different bacteria, a strong dependence of the relative levels of different triterpenes on growth conditions was revealed in *Rs. rubrum* (Barrow and Chuck, 1990).

References

Aaronson S (1964) A role for a sterol and a sterol precursor in the bacterium *Rhodopseudomonas palustris*. J Gen Microbiol 37: 225–232

Asselineau J and Trüper HG (1982) Lipid composition of six species of the phototrophic bacterial genus *Ectothiorhodospira*. Biochim Biophys Acta 712: 111–116

Barrow KD and Chuck JA (1990) Determination of hopanoid levels in bacteria using high-performance liquid chromatography. Anal Biochem 184: 395–399

Beck H, Hegeman GD and White D (1990) Fatty acid and lipopolysaccharide analyses of three *Heliobacterium* ssp. FEMS

Microbiol Lett 69: 229–232

Benning C, Beatty JT, Prince RC and Sommerville CR (1993) The sulfolipid sulfoquinovosyldiacylglycerol is not required for photosynthetic electron transport in *Rhodobacter sphaeroides* but enhances growth under phosphate limitation. Proc Natl Acad Sci USA 90: 1561–1565

Benson AA, Daniel H and Wiser R (1959) A sulfolipid in plants. Proc Natl Acad Sci USA 45: 1582–1587

Bias U (1985) Zur Freisetzung von Sulfat, Verwertung von Cystein und Vorkommen von Sulfolipiden bei *Chlorobium*. Doctoral Thesis, University of Bonn

Brooks JL and Benson AA (1972) Studies on the structure of an ornithine-containing lipid from *Rhodospirillum rubrum*. Arch Biochem Biophys 152: 347–355

Cain BD, Deal CD, Fraley RT and Kaplan S (1981) In vivo intermembrane transfer of phospholipids in the photosynthetic bacterium *Rhodopseudomonas sphaeroides*. J Bacteriol 143: 1154–1166

Cain BD, Donohue TJ and Kaplan S (1982) Kinetic analysis of n-acylphosphatidylserine accumulation and implications for membrane assembly in *Rhodopseudomonas sphaeroides*. J Bacteriol 152: 607–615

Cain BD, Donohue TJ, Sheperd WD and Kaplan S (1984) Localization of the phospholipid biosynthetic enzyme activities in cell-free fractions derived from *Rhodopseudomonas sphaeroides*. J Biol Chem 259: 942–948

Cain BD, Singer M, Donohue TJ and Kaplan S (1983) In vivo metabolic intermediates of phospholipid biosynthesis in *Rhodopseudomonas sphaeroides*. J Bacteriol 156: 375–385

Collins MD and Jones D (1981) Distribution of isoprenoid quinone structural types in bacteria and their taxonomic implications. Microbiol Rev 45: 316–354

Collins MLP and Niederman RA (1976) Membranes of *Rhodospirillum rubrum*: Isolation and physicochemical properties of membranes from aerobically grown cells. J Bacteriol 126: 1316–1325

Constantopoulos G and Bloch K (1967) Isolation and characterization of glycolipids from some photosynthetic bacteria. J Bacteriol 93: 1788–1793

Cooper CL and Lueking DR (1984) Localization and characterization of the sn-glycerol-3-phosphate acyltransferase in *Rhodopseudomonas sphaeroides*. J Lipid Res 25: 1222–1232

Cruden DJ and Stanier RY (1970) The characterization of *Chlorobium* vesicles and membranes isolated from green bacteria. Arch Microbiol 72: 115–134

Daves GD Jr, Muraca RF, Whittick JS, Friis P and Folkers K (1967) Discovery of ubiquinones-1, -2, -3, and -4 and the nature of biosynthetic isoprenylation. Biochemistry 6: 2861–2866

DePinto JA (1967) Ornithine-containing lipid in *Rhodospirillum rubrum*. Biochim Biophys Acta 144: 113–117

Ditandy T and Imhoff JF (1993) Preparation and characterization of highly pure fractions of outer membrane, cytoplasmic and intracytoplasmic membranes from *Ectothiorhodospira mobilis*. J Gen Microbiol 139: 111–117

Donohue TJ, Cain BD and Kaplan S (1982a) Purification and characterization of an N-acylphosphatidylserine from *Rhodopseudomonas sphaeroides* . Biochemistry 21: 2765–2773

Donohue TJ, Cain BD and Kaplan S (1982b) Alterations in the phospholipid composition of *Rhodopseudomonas sphaeroides* and other bacteria induced by Tris. J Bacteriol 152: 595–606

Drews G and Oelze J (1981) Organization and differentiation of membranes of phototrophic bacteria. Adv Microbiol Physiol 22: 1–92

Drews G, Weckesser J and Mayer H (1978) Cell envelopes. In: Clayton RK and Sistrom WR (eds) The Photosynthetic Bacteria, pp 61–77. Plenum Press, New York

Ferguson MAJ, Williams AF (1988) Cell-free anchoring of proteins via glycosylphosphatidyl inositol structures. Annu Rev Biochem 57: 285–320

Flamman H and Weckesser J (1984) Characterization of cell wall and outer membrane of *Rhodopseudomonas capsulata*. J Bacteriol 159: 191–198

Flesch G and Rohmer M (1987) Growth inhibition of hopanoid synthesizing bacteria by squalene cyclase inhibitors. Arch Microbiol 147: 100–104

Fraley RT, Lueking DR and Kaplan S (1979) The relationship of intracytoplasmic membrane assembly to the cell division cycle in *Rhodopseudomonas sphaeroides*. J Biol Chem 254: 1980–1986

Friis P, Daves GD Jr and Folkers K (1967) New epoxyubiquinones. Biochemistry 6: 3618–3624

Frydman B and Rappaport H (1963) Non-chlorophyllous pigments of *Chlorobium thiosulfatophilum* in chlorobiumquinone. J Am Chem Soc 85: 823–825

Gage DA, Huang ZH and Benning C (1992) Comparison of sulfoquinovosyl diacylglycerol from spinach and the purple bacterium *Rhodobacter sphaeroides* by fast atom bombardment tandem mass spectrometry. Lipids 27: 632–636

Glover J and Threlfall DR (1962) A new quinone (rhodoquinone) related to ubiquinone in the photosynthetic bacterium *Rhodospirillum rubrum*. Biochem J 85: 14P–15P

Gorchein A (1968) Studies on the structure of an ornithine-containing lipid from non-sulphur purple bacteria. Biochim Biophys Acta 152: 358–367

Gorchein A, Neuberger A and Tait GH (1968a) Incorporation of radioactivity from (Me-^{14}C)methionine and (2-^{14}C)glycine into lipids of *Rhodopseudomonas sphaeroides*. Proc Roy Soc B 170: 299–310

Gorchein A, Neuberger A and Tait GH (1968b) Metabolic turnover of the lipids of *Rhodopseudomonas spheroides*. Proc Roy Soc B 170: 311–318

Hale MB, Blankenship RE and Fuller RC (1983) Menaquinone is the sole quinone in the facultatively aerobic green photosynthetic bacterium *Chloroflexus aurantiacus*. Biochim Biophys Acta 723: 376–382

Hands AR and Bartley W (1962) The fatty acids of *Rhodopseudomonas* particles. Biochem J 84: 238–246

Haverkate F, Teulings, FAG and van Deenen LLM (1965) Studies on the phospholipids of photosynthetic microorganisms. K Ned Acad Wet Proc Ser B 68: 154–159

Hiraishi A (1989) Occurrence of menaquinone as the sole isoprenoid quinone in the photosynthetic bacterium *Heliobacterium chlorum*. Arch Microbiol 151: 378–379

Hiraishi A and Hoshino Y (1984) Distribution of rhodoquinone in Rhodospirillaceae and its taxonomic implications. J Gen Appl Microbiol 30: 435–448

Hiraishi A, Hoshino Y and Kitamura H (1984) Isoprenoid quinone composition in the classification of Rhodospirillaceae. J Gen Appl Microbiol 30: 197–210

Hiraishi A., Hoshino, Y and Satoh, T (1991) *Rhodoferax fermentans* gen. nov., sp. nov., a phototrophic purple nonsulfur bacterium previously referred to as the '*Rhodocyclus gelatinosus* like' group. Arch Microbiol 155: 330–336

Holo H, Broch-Due M and Ormerod JG (1985) Glycolipids and the structure of chlorosomes in green bacteria. Arch Microbiol 143: 94–99

Howard DL, Simoneit BRT and Chapman DJ (1984) Triterpenoids from lipids of *Rhodomicrobium vannielii*. Arch Microbiol 137: 200–204

Hurlbert RE, Golecki JR and Drews G (1974) Isolation and characterization of *Chromatium vinosum* membranes. Arch Microbiol 101: 169–186

Imhoff JF (1984a) Quinones of phototrophic purple bacteria. FEMS Microbiol Lett 25: 85–89

Imhoff JF (1984b) Sulfolipids in phototrophic purple nonsulfur bacteria. In: Siegenthaler PA and Eichenberger W (eds) Structure, function and metabolism of plant lipids pp 175–178 Elsevier Science Publ., Amsterdam

Imhoff JF (1988) Lipids, fatty acids and quinones in taxonomy and phylogeny of anoxygenic phototrophic bacteria. In: Olson JM, Ormerod JG, Amesz J, Stackebrandt E and Trüper HG (eds) Green photosynthetic bacteria, pp 223–232. Plenum Press, New York

Imhoff JF (1989) The family Ectothiorhodospiraceae. In: Staley JT, Bryant MP, Pfennig N and Holt JG (eds) Bergey's Manual of Systematic Bacteriology, 1 st ed., Vol. 3, pp 1654–1658. Williams and Wilkens, Baltimore

Imhoff JF (1991) Polar lipids and fatty acids in the genus *Rhodobacter*. System. Appl Microbiol 14: 228–234

Imhoff JF and Thiemann B (1991) Influence of salt concentration and temperature on the fatty acid compositions of *Ectothiorhodospira* and other halophilic phototrophic purple bacteria. Arch Microbiol 156: 370–375

Imhoff JF and Trüper HG (1989) The purple nonsulfur bacteria. In: Staley JT, Bryant MP, Pfennig N and Holt JG (eds) Bergey's Manual of Systematic Bacteriology, Vol. 3, pp 1658–1661. Williams and Wilkens, Baltimore

Imhoff JF, Kushner DJ, Kushwaha SC and Kates M (1982) Polar lipids in phototrophic bacteria of the Rhodospirillaceae and Chromatiaceae families. J Bacteriol 150: 1192–1201

Imhoff JF, Ditandy T and Thiemann B (1991) Salt adaption of *Ectothiorhodospira*. In: Rodriguez-Valera F (ed), General and Applied Aspects of Halophilic Microorganisms, pp 115–120. Plenum Press, New York

Jensen MT, Knudsen J and Olson JM (1991) A novel aminoglycosphingolipid found in *Chlorobium limicola f. thiosulfatophilum* 6230. Arch Microbiol 156: 248–254

Jensen SL (1962) The constitution of some bacterial carotenoids and their bearing on biosynthetic problems. K Nor Vidensk Selsk Skr 8: 1–12

Kaplan S and Arntzen CJ (1982) Photosynthetic membrane structure and function. In: Govindjee (ed) Photosynthesis, Vol 1, pp 65–151. Academic Press, New York

Kato S-I, Urakami T and Komagata K (1985) Quinone systems and cellular fatty acid composition in species of Rhodospirillaceae genera. J Gen Appl Microbiol 31: 381–398

Kenyon CN (1978) Complex lipids and fatty acids of photosynthetic bacteria. In: Clayton RK and Sistrom WR (eds) The Photosynthetic Bacteria, pp. 281–313. Plenum Press, New York

Kenyon CN and Gray AM (1974) Preliminary analysis of lipids and fatty acids of green bacteria and *Chloroflexus*. J Bacteriol 120: 131–138

Kleemann G, Poralla K, Englert G, Kjosen H, Liaaen-Jensen S, Neunlist S and Rohmer M (1990) Tetrahymenol from the phototrophic bacterium *Rhodopseudomonas palustris*: First report of a gammacerane triterpene from a prokaryote. J Gen Microbiol 136: 2551–2553

Knudsen E, Jantzen E, Bryn K, Ormerod JG and Sirevåg R (1982) Quantitative and structural characteristics of lipids in *Chlorobium* and *Chloroflexus*. Arch Microbiol 132: 149–154

Lopitz P, Neunlist S and Rohmer M (1992) Prokaryotic triterpenoids alpha-D-glucuronopyranosylbacteriohopanetetrol a novel hopanoid from the bacterium *Rhodospirillum rubrum*. Biochem J 287: 159–161

Low MG and Slatid AE (1988) Structural and functional roles of glycosyl phosphatidyl inositol in membranes. Science 239: 268–275

Lueking DR and Goldfine H (1975) Sn-glycerol-3-phosphate acyltransferase activity in particulate preparations from anaerobic, light grown cells of *Rhodopseudomonas spheroides*. J Biol Chem 250: 8530–8535

Lueking DR, Fraley RT and Kaplan S (1978) Intracytoplasmic membrane synthesis in synchronous cell populations of *Rhodopseudomonas sphaeroides*. J Biol Chem 253: 451–457

Marinetti GV and Cattieu K (1981) Lipid analysis of cells and chromatophores of *Rhodopseudomonas sphaeroides*. Chem Phys Lipids 28: 241–251

Meissner J, Pfennig N, Krauss JH, Mayer H and Weckesser J (1988a) Lipopolysaccharides of *Thiocystis violacea, Thiocapsa pfennigii*, and *Chromatium tepidum*, species of the family Chromatiaceae. J Bacteriol 170: 3217–3222

Meissner J, Borowiak D, Fischer U and Weckesser J (1988b) The lipopolysaccharide of the phototrophic bacterium *Ectothiorhodospira vacuolata*. Arch Microbiol 149: 245–248

Myers CR and Collins MLP (1986) Cell-cycle-specific oscillation in the composition of chromatophore membrane in *Rhodospirillum rubrum*. J Bacteriol 166: 818–823

Myers CR and Collins MLP (1987) Cell-cycle-specific fluctuation in cytoplasmic membrane composition in aerobically grown *Rhodospirillum rubrum*. J Bacteriol 169: 5445–5451

Neunlist S, Holst O and Rohmer M (1985) Prokaryotic triterpenoids. The hopanoids of the purple non-sulphur bacterium *Rhodomicrobium vannielii*: An aminotriol and its aminoacyl derivatives, N-tryptophanyl and N-ornithinyl aminotriol. Eur J Biochem 147: 561–568

Neunlist S, Bisseret P and Rohmer M (1988) The hopanoids of the purple non-sulfur bacteria *Rhodopseudomonas palustris* and *Rhodopseudomonas acidophila* and the absolute configuration of bacteriohopanetetrol. Eur J Biochem 171: 245–252

Niederman RA and Gibson KD (1978) Isolation and physicochemical properties of membranes from purple photosynthetic bacteria. In. Clayton RK, Sistrom WR (ed) The Photosynthetic Bacteria, pp 79–118. Plenum Publishing Corp, New York

Oelze J, Schröder J and Drews G (1970) Bacteriochlorophyll, fatty acid, and protein synthesis in relation to thylakoid

formation in mutant strains of *Rhodospirillum rubrum*. J Bacteriol 101: 669–674
Oelze J, Golecki JR, Kleinig H and Weckesser J (1975) Characterization of two cell-envelope fractions from chemotrophically grown *Rhodospirillum rubrum*. Ant van Leeuwenhoek 41: 273–286
Olson JM (1980) Chlorophyll organization in green photosynthetic bacteria. Biochim Biophys Acta 594: 33–51
Olson JM, Shaw EK, Gaffney JS and Scandella CJ (1983) A fluorescent aminolipid from a green photosynthetic bacterium. Biochem. 22: 1819–1827
Olson JM, Shaw EK, Gaffney JS and Scandella CJ (1984) *Chlorobium* aminolipid. A new membrane lipid from green sulfur bacteria. In: Sybesma C (ed) Advances in photosynthesis research, Vol 3, pp 139–142. Nijhoff/Junk, The Hague
Onishi JC and Niederman RA (1982) *Rhodopseudomonas sphaeroides* membranes: Alterations in phospholipid composition in aerobically and phototrophically grown cells. J Bacteriol 149: 831–839
Ourisson G, Rohmer M and Poralla K (1987) Prokaryotic hopanoids and other polyterpenoid sterol surrogates. Ann Rev Microbiol 41: 301–333
Oyewole SH and Holt S (1976) Structure and composition of intracytoplasmic membranes of *Ectothiorhodospira mobilis*. Arch Microbiol 107: 167–182
Park CE and Berger LR (1967) Complex lipids of *Rhodomicrobium vannielii*. J Bacteriol 93: 221–229
Pirovarova TA and Gorlenko VM (1977) Fine structure of *Chloroflexus aurantiacus var. mesophilus* (nom. prof.) – Growth in light under aerobic and anaerobic conditions. Microbiologiya 46: 276–282 (Engl. transl.)
Powls R, Redfearn ER and Trippett S (1968) The structure of chlorobiumquinone. Biochem Biophys Res Comm 33: 408–411
Radcliffe CW and Niederman RA (1984) Intracellular localization of membrane-associated phosphatidylserine synthase in *Rhodopseudomonas sphaeroides*. In: Siegenthaler PA, Eichenberger W (eds) Structure, function and metabolism of plant lipids, pp 319–323. Elsevier Science Publishers, Amsterdam
Radcliffe CW, Broglie RM and Niederman RA (1985) Sites of phospholipid biosynthesis during induction of intracytoplasmic membrane formation in *Rhodopseudomonas sphaeroides*. Arch Microbiol 142: 136–140
Radcliffe CW, Steiner FX, Carman GM and Niederman RA (1989) Characterization and localization of phosphatidylglycerophosphate and phosphatidylserine synthase in *Rhodobacter sphaeroides*. Arch Microbiol 152: 132-137
Radunz A (1969) Über das Sulfochinovosyl-diacylglycerin aus höheren Pflanzen, Algen und Purpurbakterien. Hoppe-Seyler's Z Physiol Chem 350: 411–417
Raetz CRH and Kennedy EP (1972) The association of serine synthetase with ribosomes in extracts of *Escherichia coli*. J Biol Chem 247: 2008–2014
Ratledge C and Wilkinson SG (1988) An overview of microbial lipids. In: Ratledge C and Wilkinson SG (eds) Microbial lipids. Vol 1, pp 3–22. Academic Press, London
Raymond JC and Sistrom WR (1969) *Ectothiorhodospira halophila*: A new species of the genus *Ectothiorhodospira*. Arch Microbiol 69: 121–126
Redfearn ER and Powls R (1968) The quinones of green photosynthetic bacteria. Biochem J 106: 50P
Rohmer M, Bouvier P and Ourisson G (1979) Molecular evolution of biomembranes: Structural equivalents and phylogenetic precursors of sterols. Proc Natl Acad Sci USA 76: 847–851
Rohmer M, Bouvier-Nave P and Ourisson G (1984) Distribution of hopanoid triterpenes in prokaryotes. J. Gen. Microbiol. 130: 1137–1150
Russell NJ and Harwood JL (1979) Changes in the acyl lipid composition of photosynthetic bacteria grown under photosynthetic and non-photosynthetic conditions. Biochem J 181: 339–245
Scheuerbrandt G and Bloch K (1962) Unsaturated fatty acids in microorganisms. J Biol Chem 237: 2064–2073
Schmidt K (1980) A comparative study on the composition of chlorosomes (*Chlorobium* vesicles) and cytoplasmic membranes from *Chloroflexus aurantiacus* strain OK-70-fl and *Chlorobium limicola f. thiosulfatophilum* strain 6230. Arch Microbiol 124: 21–31
Schmitz R (1967) Über die Zusammensetzung der pigmenthaltigen Strukturen aus Prokaryonten. II. Untersuchungen an Chromatophoren von *Chlorobium thiosulfatophilum* Stamm Tassajara. Arch Microbiol 56: 238–247
Schröder J, Biederdermann M and Drews G (1969) Die Fettsäuren in ganzen Zellen, Thylakoiden und Lipopolysacchariden von *Rhodospirillum rubrum* und *Rhodopseudomonas capsulata*. Arch Microbiol 66: 273–280
Shiea J, Brassell SC, and Ward DM (1991) Comparative analysis of extractable lipids in hot spring microbial mats and their component photosynthetic bacteria. Org Geochem 17: 309-319
Sprague S, Staehelin A and Fuller RC (1981) Semiaerobic induction of bacteriochlorophyll synthesis in the green bacterium *Chloroflexus aurantiacus*. J Bacteriol 147: 1032–1039
Staehelin LA, Golecki JR and Drews G (1980) Supramolecular organization of chlorosomes and of their membrane attachment sites in *Chlorobium limicola*. Biochim Biophys Acta 589: 30–45
Steiner S, Conti SF and Lester RL (1969) Separation and identification of the polar lipids of *Chromatium* strain D. J Bacteriol 98: 10–15
Steiner S, Sojka GA, Conti SF, Gest H and Lester RL (1970) Modification of membrane composition in growing photosynthetic bacteria. Biochim Biophys Acta 203: 571–579
Tai S-P and Kaplan S (1984) Purification and properties of a phospholipid transfer protein from *Rhodopseudomonas sphaeroides*. J Biol Chem 259: 12178–12183
Tai S-P and Kaplan S (1985) Intracellular localization of phospholipid transfer activity in *Rhodopseudomonas sphaeroides* and a possible role in membrane biogenesis. J Bacteriol 164: 181–186
Takacs BJ and Holt SC (1971) *Thiocapsa floridana*: A cytological, physical and chemical characterization. II. Physical and chemical characteristics of isolated and reconstituted chromatophores. Biochim Biophys Acta 233: 278–295
Thiemann B and Imhoff JF (1991) The effect of salt on the lipid composition of *Ectothiorhodospira*. Arch Microbiol 156: 376–384
Urakami T and Komagata K (1988) Cellular fatty acid composition with special reference to the existence of hydroxy fatty acids, and the occurrence of squalene and sterols in species of

Rhodospirillaceae genera and *Erythrobacter longus*. J Gen Appl Microbiol 34: 67–84

Wahlund TM, Woese CR, Castenholz RW, and Madigan MT (1991) A thermophilic green sulfur bacterium from New Zealand hot springs, *Chlorobium tepidum* sp. nov. Arch Microbiol 156: 81-90

Weckesser J, Drews G., Fromme I and Meyer H (1973) Isolation and chemical composition of the lipopolysaccharides of *Rhodopseudomonas palustris* strains. Arch Microbiol 92: 123–138

Wilkinson BJ, Sment KA and Mayberry WR (1982) Occurrence, localization, and possible significance of an ornithine-containing lipid in *Paracoccus denitrificans*. Arch Microbiol 131: 338–343

Willems A, Gillis M and de Ley, J (1991) Transfer of *Rhodocyclus gelatinosus* to *Rubrivivax gelatinosus* gen. nov., comb. nov., and phylogenic relationships with *Leptothrix, Sphaerotilus natans, Pseudomonas saccharophila*, and *Alcaligenes latus*. Int J Syst Bacteriol 41: 65–73

Woese C, Debrunner-Vossbrink B, Oyaizu H, Stackebrandt E and Ludwig W (1985) Gram-positive bacteria: Possible photosynthetic ancestry. Science 229: 762–765

Wood JB, Nichols BW and James AT (1965) The lipids and fatty acid metabolism of photosynthetic bacteria. Biochim Biophys Acta 106: 261–273

Zahr M, Fobel B, Mayer H, Imhoff JF, Campos P and Weckesser J (1992) Chemical composition of the lipopolysaccharides of *Ectothiorhodospira shaposhnikovii, Ectothiorhodospira mobilis*, and *Ectothiorhodospira halophila*. Arch Microbiol 157: 499–504

Chapter 11

Anoxygenic Phototrophic Bacteria: Model Organisms for Studies on Cell Wall Macromolecules

Jürgen Weckesser
Institut für Biologie II, Mikrobiologie, Albert-Ludwigs-Universität, Schänzlestr. 1, D-79104 Freiburg im Breisgau, Germany

Hubert Mayer
Max-Planck-Institut für Immunbiologie, Stübeweg 51, D-79108 Freiburg im Breisgau, Germany

Georg Schulz
Institut für Organische Chemie und Biochemie der Albert-Ludwigs-Universität, Albertstr. 21-23, D-79104 Freiburg im Breisgau, Germany

R. E. Blankenship, M. T. Madigan and C. E. Bauer (eds): Anoxygenic Photosynthetic Bacteria, pp. 207–230.

Summary

The purple bacteria—being spread in the different subdivisions of the *Proteobacteria*—have a Gram-negative cell wall organization. Satisfactory matching of lipopolysaccharide composition, in particular of the conservative lipid A but also of the core-regions, with genetic studies confirms in many cases the phylogenetic relationship between distinct species of purple bacteria and their nonphototrophic relatives. The deep phylogenetical separation between the green sulfur bacteria (Chlorobiaceae) and the green non-sulfur bacteria (Chloroflexaceae) finds also a remarkable confirmation in the analyses of cell wall composition: The cell wall of Chlorobiaceae is typically Gram-negative with A1γ-type peptidoglycan and lipopolysaccharide, whereas that of *Chloroflexus aurantiacus* has characteristic properties in common with the Gram-positive cell wall type.

The different lipid A types (lipid A_{GlcN}, lipid A_{Hyb}, lipid A_{DAG}) found in purple bacteria have proven to be either highly toxic (*Rubrivivax gelatinosus*) or to exhibit little (*Rhodocyclus tenuis*) or no toxicity (*Rhodobacter capsulatus*, *Rhodobacter sphaeroides*, *Rhodopseudomonas viridis*). A comparison of their structure-endotoxic activity relationships may contribute to the knowledge of the structural requirements for exhibiting endotoxic activity as well as of the fate and site of action of endotoxin in the higher organism.

The complete three-dimensional structure of the porin of *Rhodobacter capsulatus* 37b4 (and recently also that of *Rhodopseudomonas blastica* DSM 2131) has been identified at atomic resolution by X-ray diffraction of crystals. The functional complex is a trimer built up by three identical subunits each forming a pore. Each subunit consists of an antiparallel 16-stranded β-barrel twisted in the usual right-handed manner. The strands are connected mainly by short loops. Among the longer loops, the largest one runs into the inside of the barrel and narrows the channel, defining the pore size. A strong charge-gradient exists within the pore. The porin of *Rhodobacter capsulatus* is assumingly a paradigm for the hydrophilic pores presently considered as nonspecific of the outer membrane of *Proteobacteria*.

I. Introduction

In this chapter the cell wall and the external layers of the anoxygenic phototrophic bacteria will be discussed. We will focus on more recent data; for respective earlier reviews see Weckesser et al. (1978), Weckesser et al. (1979), Mayer et al. (1985), and Mayer et al. (1990a). The cell walls of the oxygenic cyanobacteria (Weckesser and Jürgens, 1988; Jürgens and Weckesser, 1990), phylogenetically quite distant from the anoxygenic phototrophic bacteria, are not considered here.

A brief characterization of the cell wall and its constituents will be given first, then the various phylogenetic groups of anoxygenic phototrophic bacteria and of some of their nonphototrophic relatives will be discussed. Separate sections will be devoted to the contribution of their LPS's to endotoxin research, to the structure of their porins, and to the so far limited knowledge of their external surface layers.

II. Definition, Function, and Constituents of the Cell Wall

The term cell envelope comprises all layers surrounding the cytoplasm (Fig. 1): the cytoplasmic membrane, the cell wall (peptidoglycan plus outer membrane) and the external layers (capsule and slime materials, S-layers). The function of the cell wall may be summarized very briefly: The peptidoglycan (murein sacculus) contributes to the formation and maintenance of size and shape of the cell, it withstands the osmotic pressure of the protoplast. In the Gram-negative cell wall type, the outer membrane represents an effective permeability barrier in controlling the permeation of substrate and antibiotics as well as the export of substances. In addition, the outer membrane carries the molecules responsible for the serological O-specificity of the respective cells and their phage-receptors. With the exception of *Chloroflexus aurantiacus*, anoxygenic phototrophic bacteria are Gram-negative. The macro-

Abbreviations: Ara – D-arabinose; 4-AA – 4-amino-L-arabinose; *Cb.* – *Chlorobium*; *Cf.* – *Chloroflexus*; *Cm.* – *Chromatium*; DAG – 2,3-diamino-2,3-dideoxy-D-glucose; D-GalA – D-galacturonic acid; D-GlcN – D-glucosamine; D-GlcA – D-glucuronic acid; D,D-Hep – D-*glycero*-D-*manno*-heptose; DOC-PAGE – sodium deoxy-cholate-polyacrylamide gel-electrophoresis; L,D-Hep – L-*glycero*-D-*manno*-heptose; Kdo – 3-deoxy-D-manno-octulosonic acid; LPS – lipopolysaccharide; D-Man – D-mannose; m-A_2pm – *meso*-diaminopimelic acid; MurN – muramic acid; Neu5Ac – neuraminic (sialic) acid; Orn – L-ornithine; P – Phosphate; *Rb.* – *Rhodobacter*; *Rc.* – *Rhodocyclus*; *Rm.* – *Rhodomicrobium*; *Rp.* – *Rhodopseudomonas*; *Rs.* – *Rhodospirillum*; *Rv.* – *Rubrivivax*; Thr – threonine

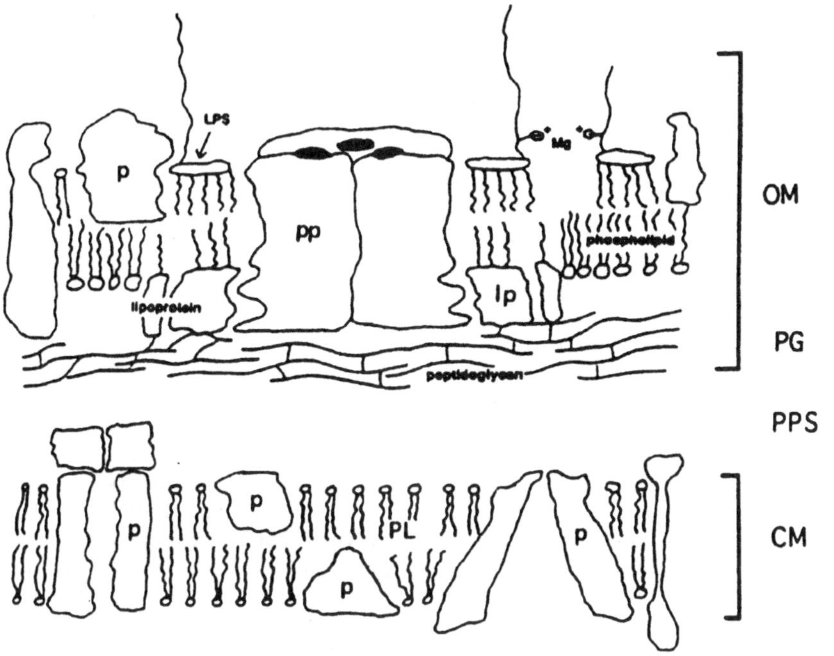

Fig. 1. The cell envelope of Gram-negative bacteria (Hancock, 1991). CM, cytoplasmic membrane; LPS, lipopolysaccharide; lp, lipoprotein; OM, outer membrane; p, proteins; PG, peptidoglycan; PL, phospholipids; pp, porin proteins; PPS, periplasmic space.

molecular constituents of the Gram-negative wall type are peptidoglycan of the A1γ-type, and the outer membrane constituents LPS (O-antigen, endotoxin), lipoprotein, porin and other proteins, and phospholipids.

The external layers, often being lost during growth under laboratory conditions, protect the cell from generally unfavorable conditions such as dryness or mechanical stress. Compounds often found as the most external layers are acidic polysaccharides of the more or less amorphous capsule or slime layers, or the regular arranged proteins or glycoproteins of the (surface) S-layers (Beveridge and Graham, 1991; Messner and Sleytr, 1992). The S-layers may serve as protective coats, molecular sieves and/or ion traps.

III. Phylogenetic Relevance of Cell Wall Constituents

A. Phylogenetic Considerations

The anoxygenic phototrophic bacteria are spread throughout the phylogenetic tree of Bacteria, phototrophic and nonphototrophic phenotypes are intermixed in many cases, in particular within the *Proteobacteria* (Woese, 1987; Stackebrandt et al., 1988a,b; Olsen and Woese, 1993). Restriction to only phototrophic bacteria in this article would bring about neglect of their intimate nonphototrophic relatives. Consequently, some nonphototrophic *Proteobacteria* will also be discussed briefly.

B. Major Groups of Anoxygenic Phototrophic Bacteria and Their Cell Wall Composition

1. Proteobacteria: Purple Bacteria and Relatives

As far as studied, the anoxygenic phototrophic *Proteobacteria* have a Gram-negative cell wall organization with lipoprotein likely bound to A1γ-type peptidoglycan, as revealed for *Rhodobacter sphaeroides* and *Rhodobacter capsulatus* (Drews et al., 1978; Bräutigam et al., 1988; Woitzik et al., 1989), *Rhodospirillum rubrum* (Newton, 1968), *Chromatium tepidum*, *Thiocystis violacea* (Meissner et al., 1988a) and *Ectothiorhodospira vacuolata* (Meissner et al., 1988b). Baumgardner et al. (1980)

described in *Rhodobacter sphaeroides* ATCC 17023 two different outer membrane proteins (app. M_r 8,000 and 10,000, respectively). They were suggested to be equivalents to the free and peptidoglycan-bound form, respectively, of the *Escherichia coli* lipoprotein. The rigid layer of both *Rb. sphaeroides* ATCC 17023 and *Rb. capsulatus* 37b4, however, did not react with antisera against lipoprotein of *E. coli* K12 (Drews et al., 1978; Woitzik et al., 1989). Antisera against heat-killed cells of *Rb. capsulatus*, *Rp. viridis* and *Rp. palustris* also showed little or no serological cross-reactivity with lipoprotein of *Escherichia coli* (Drews et al., 1978). In the peptidoglycan of *Rp. viridis*, *Rp. sulfoviridis*, and *Rp. palustris* there is a partial lack of N-acetyl substitution of the D-GlcN residues, rendering this peptidoglycan lysozyme-resistant (Schmelzer et al., 1982).

In general, *Proteobacteria* have LPS's in their outer membrane. From a phylogenetic point of view—aside from peptidoglycan—the composition of their lipid A regions is particularly of interest (Fig. 2). It is the most conserved domain of the heteropolymer; the core-structure is less conserved but in some cases also valuable as a taxonomical criterion. The O-chains are highly variable and confer the serological O-specificity to a given strain (Framberg et al., 1974; Drews et al., 1978; Weckesser et al., 1979), but they are absent in many species.

In the following, the purple bacteria and some of their nonphototrophic relatives will be discussed according to their grouping into the α, β, and γ subdivisions of *Proteobacteria* (Table 1; Woese et al., 1984 a,b; Woese et al., 1985; Woese, 1987; Stackebrandt et al., 1988a). The δ subdivision does not include phototrophic species. The analytical data are compiled in Table 1. For the definition of lipid A_{GlcN}, lipid A_{Hyb}, lipid A_{DAG}, and 'mixed' lipid A, see Section IV ('Lipid A') and Fig. 6.

a. α Subdivision

Subgroup α-1: Among the subgroup α-1, *Rhodospirillum rubrum*, *Rhodospirillum molischianum*, and *Rhodopila globiformis* have been investigated for the composition of their LPS's (Pietsch et al., 1990). *Rs. rubrum* and *Rs. molischianum* have lipid A_{GlcN}, while the lipid A of *Rhodopila globiformis*, more distantly related to *Rs. rubrum* and *Rs. molischianum*, contains in addition, DAG. It also contains the rare 3-OH-19:0, possibly in its iso- or anteiso-branched form. It is not yet clear whether we are dealing here with a lipid A_{Hyb} type or 'mixed' lipid A, or whether the GlcN found in the lipid A fraction is part of the polysaccharide. The latter would indicate the presence of lipid A_{DAG} in *Rhodopila globiformis*, otherwise found mainly in subgroup α-2 of *Proteobacteria*. The moderately halophilic *Rhodospirillum salexigens*, belonging also to the α subdivision (cited in Evers et al., 1986) lacks LPS (Evers et al., 1986), and therefore neither β-hydroxy fatty acids nor Kdo were found in hot phenol-water extracts and in whole cells. However, an outermost protein layer of hexagonally arranged subunits was observed in *Rs. salexigens* (Golecki and Drews, 1980). The protein (app. M_r = 68,000) forming the outermost layer had a high excess of acidic over basic amino

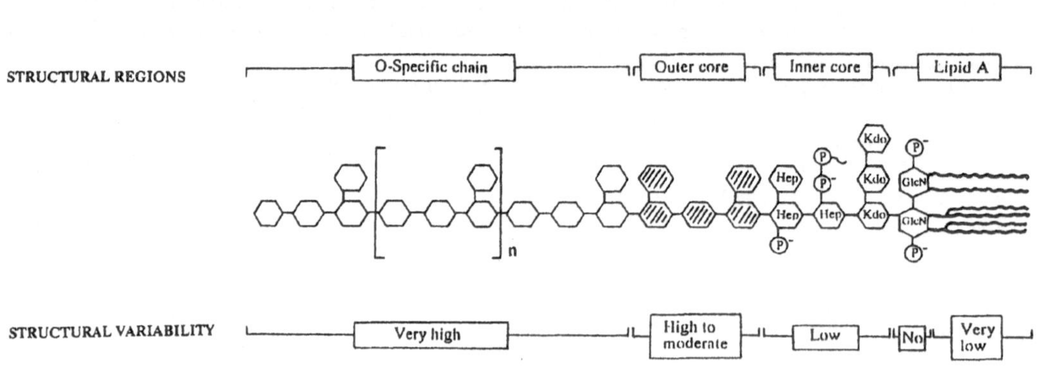

Fig. 2. Structure makeup of enterobacterial lipopolysaccharide (Rietschel et al., 1987, modified).

acids, which is in accord with its low isoelectric point of 4.4 (Evers et al., 1984). A thin, apparently outer membrane but no peptidoglycan layer was seen in ultrathin sections. Although peptidoglycan was described as a major constituent of a cell envelope fraction (Tadros et al., 1982), whole cells of *Rs. salexigens* were found to have a very low peptidoglycan content (0.17 μmol muramic acid/mg cell dry weight), which is apparently not sufficient to form a complete sacculus structure (Evers et al., 1986). A polysaccharide consisting of D-GlcN and a 2-amino-2-deoxy-pentose (likely ribosamine) was extracted into the water phase of phenol water extracts of whole cells. Removal of the cell surface layer exposed six proteins to labeling with radioactive iodine catalyzed by lactoperoxidase. They were suggested to be constituents of an outer membrane of *Rs. salexigens* (Evers, 1984; Evers et al., 1986). The cell wall of *Rs. mediosalinum* (Kompanseva and Gorlenko, 1985) shows a fine structure typical of Gram-negative bacteria, but chemical data are not yet available. An outermost layer of hexagonal subunits and a thick capsule were also observed.

Subgroup α-2: The finding of lipid A_{DAG} in *Rhodopseudomonas palustris*, *Rhodopseudomonas viridis*, *Rhodopseudomonas sulfoviridis*, *Nitrobacter winogradskyi*, *Pseudomonas diminuta*, and *Phenylobacterium immobile* (Fig. 3A) supports their common phylogenetic grouping (Mayer et al., 1983, 1984; Mayer, 1984; Mayer and Weckesser, 1984; Weckesser and Mayer, 1987). They all exhibit amide-bound 3-OH-14:0. The close relationship between *Rp. palustris* and *Nitrobacter winogradskyi* is additionally expressed by the common presence of L,D-Hep in the core-region, which is absent in other members of the subgroup α-2. It should be noted that lipid A_{DAG} is not uniformly distributed in all members of the subgroup α-2. It is found neither in *Rhodomicrobium vannielii* nor in *Rp. acidophila* and especially not in species of Rhizobiaceae (Weibgen et al., 1993). Lipid A_{GlcN} of *Rm. vannielii* and *Rp. acidophila* contain nonphosphorylated β-1,6-linked D-GlcN disaccharides, carrying D-Man as a non-acylated substituent at C'-4.

Recently, long-chained *n*-2-hydroxylated fatty acids, a new characteristic class of lipid A components, were detected in α-2 subgroup species (Hollingsworth and Carlson, 1989). In a comparative study, 27-OH-28:0 was found in lipid A of almost all species of the subgroup α-2 (Bhat et al., 1991). The distribution of 27-OH-28:0 and higher homologues was restricted to this subgroup; strains of the β and γ subdivisions, including also the enterobacterial LPS's, were devoid of this unusual fatty acid. The long-chained fatty acids are usually ester-bound to ester- or mostly amide-linked 3-OH-fatty acids. The diesters reach chain lengths which allow them to stretch through the entire outer membrane. Their functional OH-group at the penultimate position of the hydrocarbon chain may interact with components of the hydrophilic part of the inner leaflet of the outer membrane, thus, possibly contributing to membrane stabilization. Long-chained n-2-hydroxylated fatty acids are found in both, D-GlcN- or DAG-containing lipid A's. They have also been detected in the LPS of methane-utilizing (methanotrophic) bacteria (Bowman et al., 1991).

The polysaccharide moiety also allows to distinguish between the species of the subgroup α-2. The O-antigens of *Rp. palustris* strains have a core structure with Kdo and L,D-Hep, while those of *Rp. viridis* and *Rp. sulfoviridis* lack heptoses (Weckesser et al., 1979; Ahamed et al., 1982). The O-chains of *Rp. palustris* and *Rp. viridis* are complex and reveal a ladder-like pattern on DOC-PAGE, indicating possible S-layers in *Rp. palustris* (Krauss et al., 1988a).

Subgroup α-3: Most studies on the cell wall of anoxygenic phototrophic bacteria were performed with species of the subgroup α-3. Both *Rhodobacter capsulatus* 37b4 and *Rhodobacter sphaeroides* ATCC 17023 have lipid A_{GlcN} (see chapter A, 'Lipid A'). Both also share a rare variation in their fatty acid composition: The amide-bound 3-OH-14:0 is partly (*Rb. sphaeroides* ATCC 17023) or almost entirely (*Rb. capsulatus* 37b4; *Roseobacter denitrificans*) replaced by 3-oxo-14:0 (Salimath et al., 1983; Qureshi et al., 1988; 1983; Krauss et al. 1989). Both have ester-linked 3-OH-10:0, and this was recognized to be due to a highly specific acyltransferase (Williamson et al., 1991). Comparable fatty acid patterns have been found in nonphototrophic species of this cluster (Fig. 3A) such as *Paracoccus denitrificans* and *Thiobacillus versutus* (Wilkinson et al., 1986; Mayer et al., 1990a,b).

In all the species mentioned, the core region contains Kdo but no heptoses. It is worth mentioning that in a comparative study (Krauss et al., 1988b) Neu5Ac was found as a constituent of LPS only in the α-3 group, in particular only in *Rhodobacter*

Table 1. Characteristic lipid A and core constituents in lipopolysaccharides from anoxygenic phototrophic bacteria

Phylogenetic branch/ Species	Backbone amino sugar GlcN	DAG	P	Hydroxy (or 3-oxo) fatty acids Amide-linked	Ester-linked	Other compounds	KDO	Heptoses	Length of sugar chain[a]
α Subdivision of Proteobacteria:									
Subgroup α-1									
Rhodospirillum molischianum	+	–[b]	–	3-OH-16:0	3-OH-14:0	ND[c]	+	L,D-Hep	Middle
Rhodospirillum fulvum	+	–	–	3-OH-16:0	3-OH-14:0	GlcA; GalA	+	L,D-Hep; D,D-Hep	Middle
Rhodospirillum rubrum	+	–	+	3-OH-16:0	3-OH-14:0	ND	+	L,D-Hep	Long
Rhodospirillum salinarum	+	+	+	3-OH-14:0[e] 3-OH-18:0[e]		GlcA; GalA	+	L,D-Hep	Short
Rhodopila globiformis	Tr[d]	+	+	3-OH-14:0 3-OH-18:0	3-OH-19:0	ND	+	L,D-Hep	Long
Aquaspirillum itersonii subsp. *nipponicum*	+	–	ND	3-OH-14:0[e] 3-OH-16:0[e] 3-OH-18:0[e]		GlcA, GalA	+	L,D-Hep	Middle
Subgroup α-2									
Rhodopseudomonas viridis	–	+	–	3-OH-14:0	27-OH-28:0	–	+	–	Long
Rhodopseudomonas sulfoviridis	–	+	–	3-OH-14:0	ND	–	+	–	Long
Rhodopseudomonas palustris	–	+	ND	3-OH-14:0 3-OH-16:0	27-OH-28:0 28-OH-29:0	–	+	L,D-Hep	Long
Rhodopseudomonas acidophila	+	–	–	3-OH-16:0	3-OH-16:0	Man	+	–	Short
Rhodomicrobium vannielii	+	–	–	3-OH-16:0	3-OH-14:0	Man	+	–	Long
Nitrobacter winogradsky	–	+	–	3-OH-14:0	27-OH-28:0	–	+	L,D-Hep	Long
Subgroup α-3									
Rhodobacter capsulatus	+	–	+	3-oxo-14:0	3-OH-10:0	Neu5Ac	+	–	Middle
Rhodobacter sphaeroides	+	–	+	3-oxo-14:0 3-OH-14:0	3-OH-10:0	GlcA, Thr Neu5Ac	+	–	Middle
Rhodobacter sulphidophilus	+	–	+	3-OH-14:0	3-OH-10:0	GalA, Neu5Ac	+	–	ND
Rhodobacter veldkampii	+	–	+	3-oxo-14:0	3-OH-10:0	GalA, Neu5Ac	+	–	Middle
Rhodopseudomonas blastica	+	–	+	3-oxo-14:0	3-OH-10:0	–	+	–	Long
Roseobacter denitrificans	+	–	+	3-oxo-14:0	3-OH-10-0	–	+	–	Short
Paracoccus denitrificans	+	–	+	3-OH-14:0 3-oxo-14:0	3-OH-10:0	–	+	–	Short
β Subdivision of Proteobacteria:									
Subgroup β-1									
Rubrivivax gelatinosus	+	–	+	3-OH-10:0	3-OH-10:0	GalA	+	D,D-Hep	Short
Sphaerotilus natans	+	–	+	3-OH-10:0	3-OH-10:0	–	+	L,D-Hep	Short
Aquaspirillum aquaticum (syn. *Comamonas terrigena*)	+	–	+	3-OH-10:0[e] 3-OH-11:0[e]		–	+	–	Long

Table 1. Continued

Phylogenetic branch/ Species	Backbone amino sugar GlcN	DAG	P	Hydroxy (or 3-oxo) fatty acids Amide-linked	Ester-linked	Other compounds	KDO	Heptoses	Length of sugar chain[a]
Subgroup β-2									
Rhodocyclus tenuis	+	–	+	3-OH-10:0	3-OH-14:0	4-AA; Ara, GlcN	+	L,D-Hep; D,D-Hep[g]	Middle
Rhodocyclus purpureus	+	–	+	3-OH-10:0	3-OH-14:0	4-AA; Ara	+	L,D-Hep; D,D-Hep	Long
γ Subdivision of Proteobacteria:									
Subgroup γ-1									
Chromatium vinosum	+	(+)[h]	–	3-OH-14:0	3-OH-14:0	Man	+	D,D-Hep	Long
Chromatium tepidum	+	(+)	–	3-OH-14:0	3-OH-14:0	Man	+	D,D-Hep	Long
Thiocapsa roseopersicina	+	(+)	–	3-OH-14:0	3-OH-14:0	Man	+	D,D-Hep	Long
Thiocapsa pfennigii	+	(+)	–	3-OH-14:0	3-OH-14:0	Man	+	D,D-Hep	Short
Thiocystis violacea	+	(+)	–	3-OH-14:0	3-OH-14:0	Man	+	D,D-Hep	Long
Ectothiorhodospira vacuolata	(+)	+	+	3-OH-10:0 3-OH-12:0		GlcA, GalA	+	D,D-Hep	Short
Ectothiorhodospira shaposhnikovii	+	++[e]	+	3-OH-10:0 3-OH-12:0		GlcA, GalA	+	D,D-Hep	Short
Ectothiorhodospira mobilis	+	++	+	3-OH-10:0		GlcA, GalA	+	D,D-Hep	Short
Ectothiorhodospira halophila	Tr	++	+	3-OH-10:0 3-OH-14:0 3-oxo-14:0		GlcA, GalA Thr, Glc	+	D,D-Hep	Short
Green sulfur bacteria									
Chlorobium vibrioforme	+	Tr	–	3-OH-14:0 3-OH-16:0 iso-3-OH-18:0		ND	+	–	Short

[a] As determined by DOC-PAGE
[b] –, Absent
[c] ND, Not determined
[d] Tr, Trace amounts
[e] Discrimination between amide- and ester-linked fatty acids is not known
[f] ++, Dominating over GlcN
[g] In some strains only
[h] (+), Present, but less than 5% of total hexosamines

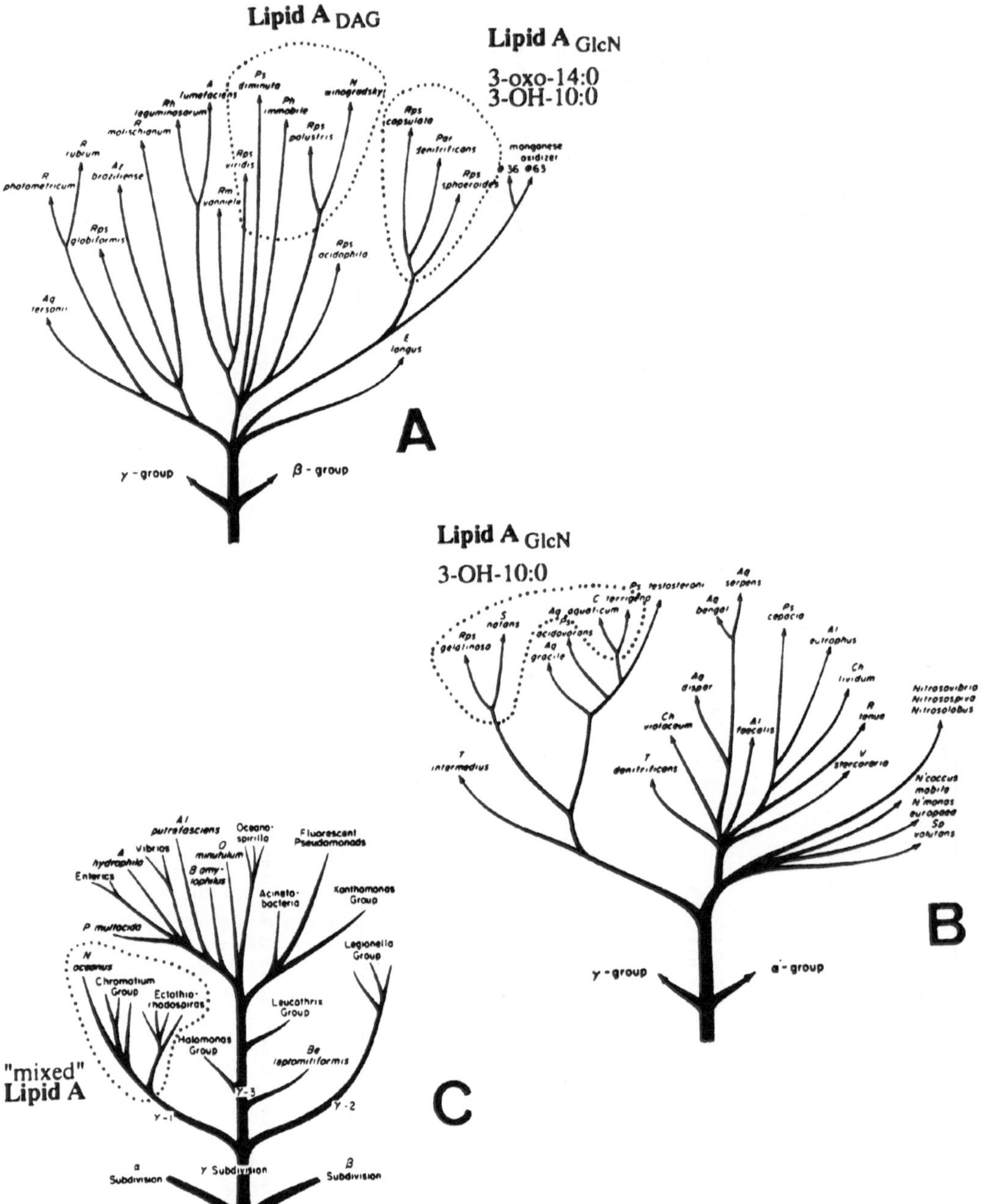

Fig. 3. Comparison of lipid A types and 16S rRNA catalogue data of the (A) α, (B) β, and (C) γ subdivisions of *Proteobacteria* (Mayer et al., 1989, modified). Note that 3-OH-10:0 is exclusively ester-linked in (A) and both amide- and ester-linked in (B).

species (*Rb. capsulatus*, *Rb. sphaeroides*, *Rb. sulfidophilus*, *Rb. veldkampii*).

Outer membrane lipoprotein with a fatty acid pattern of 18:1, 18:0, and 16:0 was isolated from *Rb. sphaeroides* ATCC 17023 (Woitzik et al., 1989). In accordance with lipoprotein of *Escherichia coli*, a major part of 18:1 is amide-bound and lysine possibly represents the C-terminal amino acid. In another study (Baumgardner et al., 1980), two different outer membrane proteins (app. M_r 8,000 and 10,000, respectively) in the same strain were suggested to be equivalents of the lipoprotein of *E. coli*, the app. M_r 8,000 protein to the free and the app. Mr 10,000 protein to the peptidoglycan-bound form of lipoprotein. The rigid layer of both *Rb. sphaeroides* ATCC 17023 and *Rb. capsulatus* 37b4 did not react with antisera against lipoprotein of *E. coli* K12 (Drews et al., 1978; Woitzik et al., 1989).

b. β Subdivision

Subgroup *β*-1: Lipid A_{GlcN} of *Rubrivivax (Rhodocyclus) gelatinosus* also resembles that of enterobacterial lipid A_{GlcN} (see Section IV 'Lipid A') in having a bisphosphorylated lipid A glcN. Characteristic are the amide- and ester-linked 3-OH-10:0 fatty acids, ester-bound are also 12:0 and 16:0 (Fig. 4). *Sphaerotilus natans*, closely related to *Rubrivivax gelatinosus* (Fig. 3B), has high structural similarities: also here 3-OH-10:0 is both amide- and ester-linked (Masoud et al., 1991a). All *Rubrivivax gelatinosus* strains studied so far have R-type LPS's (Krauss et al., 1988a) with D,D-Hep as the only heptose.

Subgroup *β*-2: Lipid A_{GlcN} of *Rhodocyclus tenuis* 2761 carries a third, unsubstituted D-GlcN unit in its backbone structure (see Section IV 'Lipid A'). In further contrast to *Rubrivivax gelatinosus*, the phosphate moieties at C-1 and C-4' are substituted with D-Ara*f* and 4-amino-L-arabinose (4-AA), respectively. Lipid A_{GlcN} from *Rhodocyclus purpureus* is similar to that of *Rc. tenuis* in also containing these two sugars (Weckesser et al., 1983). The ester-bound fatty acids of both *Rc. purpureus* and *Rc. tenuis* are 14:0 and 16:0.

In contrast to *Rubrivivax gelatinosus*, both *Rc. tenuis* and *Rc. purpureus* have L,D-Hep (in some cases together with D,D-Hep) and O-chains with typical repeating units as indicated by a ladder-like pattern on DOC-PAGE (Fig. 4; Krauss et al., 1988a).

c. γ Subdivision

Among the Chromatiaceae, five species (*Chromatium vinosum*, *Chromatium tepidum*, *Thiocapsa roseopersicina*, *Thiocapsa pfennigii*, *Thiocystis*

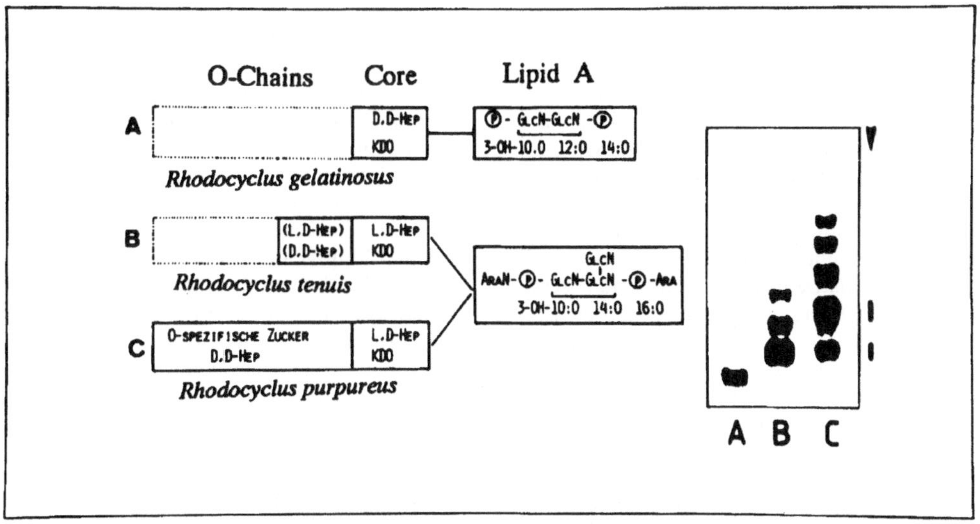

Fig. 4. Structural regions with their characteristic constituents, and DOC-PAGE of LPS's from *Rubrivivax* (*Rhodocyclus*) *gelatinosus* (short), *Rhodocyclus tenuis* (middle), and *Rhodocyclus purpureus* (long); in parenthesis are given the lengths of the sugar chains of the respective LPS's. Taken from Weckesser and Mayer (1987).

violacea) have been investigated for their LPS composition (Hurlbert et al., 1976; Hurlbert and Hurlbert, 1977; Meissner et al., 1988c). The so far incomplete data reflect the close phylogenetic relationship within the Chromatiaceae (Fig. 3C). Lipid A_{GlcN} from all of these species has uniformly a phosphate-free backbone with D-GlcN as the only amino sugar (only traces of DAG were found), and contain terminally-bound, nonacylated D-Man, probably attached to the C-4' position, but proven so far only for *Cm. vinosum* lipid A (Hurlbert et al., 1976). They all have amide-bound 3-OH-14:0 and 12:0 as the major ester-bound fatty acid. Other common characteristics are a core structure with D,D-Hep, as well as long O-chains, built up by repeating units as first demonstrated for a phototrophic bacterium by Hurlbert and Hurlbert (1977) using SDS-PAGE.

Several features in LPS composition have confirmed the separation of the Ectothiorhodospiraceae from the Chromatiaceae, as examined for the moderate halophilic species *Ectothiorhodospira vacuolata*, *Ectothiorhodospira mobilis*, *Ectothiorhodospira shaposhnikovii* and for the extreme halophilic *Ectothiorhodospira halophila* (Meissner et al., 1988d; Zahr et al., 1992). None of the species of Ectothiorhodospiraceae contain D-Man in their lipid A. The lipid A's from the Ectothiorhodospiraceae are phosphate-containing and have both D-GlcN and DAG with the diamino sugar being found in at least similar amounts as D-GlcN, revealing lipid A_{Hyb} or 'mixed' lipid A. In *Ec. halophila*, DAG is strongly dominating over D-GlcN, indicating the presence of lipid A_{DAG}. Lipid A_{DAG} has been found so far mostly within the subgroup α-2. Amide-bound 3-OH-14:0, as found with Chromatiaceae, is absent from the Ectothiorhodospiraceae. Within the Ectothiorhodospiraceae, the three moderately halophilic species can be differentiated from *Ec. halophila* by their LPS fatty acid spectra (Zahr et al., 1992). The similarities between the less moderately halophilic species confirm the data of 16S rRNA analyses (Stackebrandt et al. 1988b), revealing particular genome similarities between *Ec. shaposhnikovii, Ec. mobilis* and *Ec. vacuolata*. The finding of 3-oxo-14:0 in *Ec. halophila* is remarkable. This fatty acid has been found so far only in the species of the subgroup α-3 and in *Vibrio anguillarum* (cited in Mayer et al. 1990b). Also notable is the amide-linkage of 3-OH-10:0. It has been found elsewhere only in *Rhodocyclus* (Tharanathan et al. 1985) and in some *Aquaspirillum* species (Rau et al., 1993) belonging to the β subdivision of *Proteobacteria*.

Like the Chromatiaceae, *the* LPS's of all of the Ectothiorhodospiraceae species studied so far have D,D-Hep as the only heptose and Kdo in their core, but the Ectothiorhodospiraceae are lacking typical O-chains with repeating units.

*2. Green Sulfur Bacteria (*Chlorobiaceae*)*

Among the green sulfur bacteria (Chlorobiaceae), only *Chlorobium vibrioforme f. thiosulfatophilum* NCIB 8327 has been studied for its cell wall composition. Peptidoglycan with m-A_2pm of the A1γ-type was present (Jürgens et al., 1987), and LPS is found in their outer membrane (Meissner et al., 1987). The latter has a phosphate-containing lipid A_{GlcN} with D-GlcN (and only traces of DAG). It is free of D-Man, a fact which clearly differentiates lipid A_{GlcN} of Chlorobiaceae from that of Chromatiaceae. The fatty acid spectrum is very characteristic by having three different amide-bound β-hydroxy fatty acids (3-OH-14:0, 3-OH-16:0, iso-branched 3-OH-18) but no ester-bound fatty acids. The polysaccharide moiety has a core region with Kdo and heptoses (L,D-Hep and D,D-Hep). On DOC-PAGE no ladder-like pattern was seen, indicating an R-type LPS (Table 1).

*3. Green Non-Sulfur Bacteria (*Chloroflexaceae*)*

The deep phylogenetic separation of the thermophilic green-nonsulfur bacterium *Chloroflexus aurantiacus* from the Chlorobiaceae is remarkably confirmed by cell wall analyses, although both contain the characteristic chlorosome structures (Fig. 5): Firstly, *Cf. aurantiacus*, although staining Gram-negatively (the peptidoglycan layer is thin), has partly phosphorylated muramic acid in which m-A_2pm is completely replaced by L-Orn (Jürgens et al., 1987). Secondly, a large polysaccharide of complex composition is likely covalently linked to muramic acid-6-phosphate *via* phosphodiester bridges in this peptidoglycan. Thirdly, peptidoglycan-bound protein or lipoprotein was not found here. Fourthly, LPS is absent, no β-hydroxy fatty acids, nor Kdo or heptoses were found in hot phenol-water extracts of cells of *Cf. aurantiacus* (Meissner et al., 1988e). In addition, an outer membrane is not seen in ultrathin sections. These properties are characteristic of Gram-positive

CHLOROSOMES		
	CHLOROBIACEAE (*Chlorobium vibrioforme*)	CHLOROFLEXACEAE (*Chloroflexus aurantiacus*)
LPS:	KDO, Hep; D-GlcN, 3-OH-14:0, 3-OH-16:0, 3-OH-18:0	—
Peptidoglycan:	m-A_2pm; (Lipo)protein	L-Orn and MurN-6-P; Complex polysaccharide

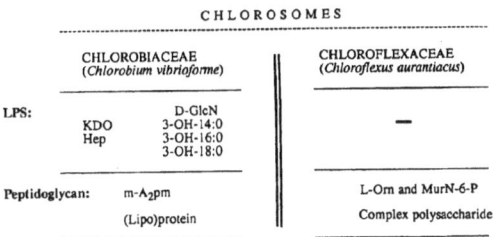

Fig. 5. Cell wall composition of Chlorobiaceae (Gram-negative type) and Chloroflexaceae (Gram-positive type). Taken from Stackebrandt et al. (1988b).

bacteria. It deserves mention that the cell wall of the nonphototrophic gliding bacterium *Herpetosiphon aurantiacus*, which together with *Cf. aurantiacus* belongs to the same phylum of eubacteria (Woese, 1987) and which shows clear, but not close, affinities with *Cf. aurantiacus* (Gibson et al., 1985), also was found to have characteristics of the Gram-positive cell wall organization comparable to those of *Cf. aurantiacus* (Jürgens et al., 1989).

4. Heliobacteriaceae

The cell wall of heliobacteria has properties reminiscent of the Gram-positive cell wall type (Madigan, 1992). *Heliobacterium chlorum* contains L,L- instead of m-A_2pm in its peptidoglycan, and glycine forms the interpeptide bridges according to preliminary studies (as stated in Beer-Romero et al., 1988). Both properties are found with Gram-positive but not with Gram-negative bacteria. Correspondingly, LPS was absent from both hot phenol-water or phenol-chloroform-petroleum ether (PCP) extracts of several species of heliobacteria (Beck et al., 1990). Instead, a polysaccharide was extractable by hot phenol-water. The walls of heliobacteria are unusually fragile and penicillin-sensitive, the reason for this is so far not known.

IV. Structure of Lipopolysaccharide (LPS)

A. Lipid A

The backbone amino sugar of lipid A in anoxygenic phototrophs can vary (Fig. 6), as first detected by Roppel et al. (1975) in *Rhodopseudomonas viridis*: 'Lipid A_{GlcN}' has D-GlcN as the only amino sugar,

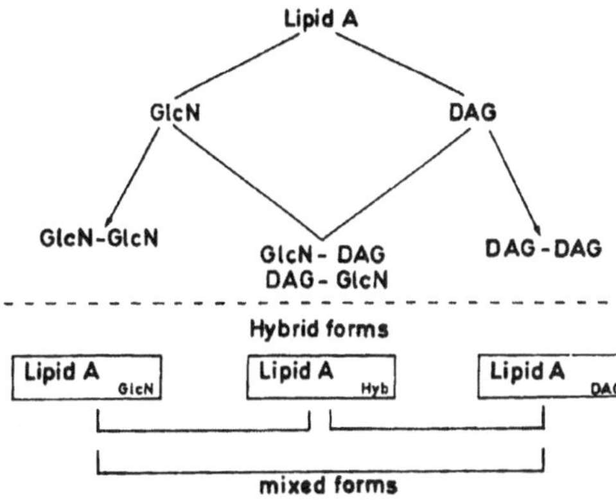

Fig. 6. Definition of lipid A_{GlcN}, lipid A_{Hyb}, lipid A_{DAG}, and 'mixed' lipid.

'lipid A_{Hyb}' (hybrid lipid A) has D-GlcN and DAG in the backbone (Moran et al., 1991), and 'lipid A_{DAG}' has exclusively DAG (Weckesser and Mayer, 1988). 'Mixed lipid A' indicates the concomitant occurrence of lipid A_{GlcN} and lipid A_{DAG} and sometimes also of hybrid forms in one and the same strain (Moran et al., 1991; Rietschel et al., 1992). Although lipid A_{Hyb} is likely present in distinct anoxygenic phototrophic bacteria, there is so far only a single proof for it, namely in *Rhodospirillum salinarum* (Heike Rau, unpublished data). The explanation for the structural variability in the backbone structure results probably from the reported lack of specificity of the lipid A synthetase (Raetz, 1992) as pointed out by Weckesser and Mayer (1988), and secondary processing of lipid A structure.

Lipid A_{GlcN} from *Rhodobacter sphaeroides* and *Rhodobacter capsulatus*: Lipid A_{GlcN} of both of these species (Fig. 7C and D) is structurally similar to enterobacterial lipid A_{GlcN} (Fig. 7A), which results also in a complete serological cross-reactivity (H. M. Kuhn, unpublished; Mayer et al., 1992). They all share a bisphosphorylated D-GlcN-β(1,6)D-GlcN backbone (Strittmatter et al., 1983; Krauss et al., 1989). In *Rb. sphaeroides* the phosphate groups are unsubstituted and 3-OH-10:0 is ester-linked at both the C-3 and C-3' positions. The acyloxyacyl group [(either 3-O-(*cis* Δ^7-14:1)-10:0 or 3-O-(14:0)14:0)] and the 3-oxo-14:0 or 3-OH-14:0 occupy the 2'- and 2-positions, respectively (Qureshi et al., 1988). In free lipid A, the hydroxyl groups at C-4 and C-6' are free (Salimath et al., 1983).

In contrast to *Rb. sphaeroides*, both amino groups

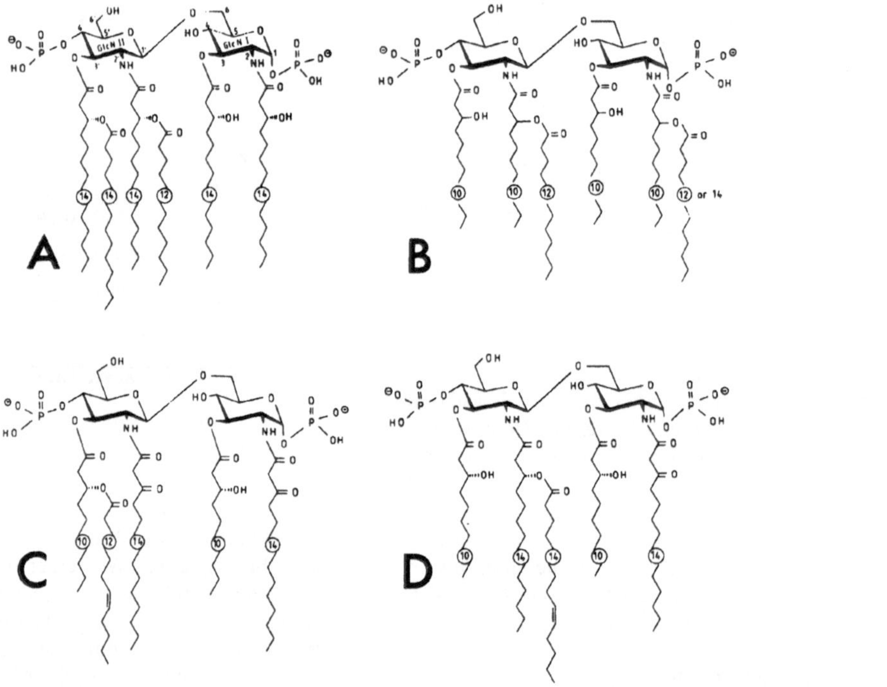

Fig. 7. Structures of toxic lipid A_{GlcN} from (A) *Escherichia coli* and (B) *Rubrivivax gelatinosus*, and of nontoxic lipid A_{GlcN} from (C) *Rhodobacter capsulatus* and (D) *Rhodobacter sphaeroides* (Weckesser and Mayer, 1988). Toxic forms with asymmetrical (A) and symmetrical (B) distribution of fatty acyl groups.

of the D-GlcN backbone are almost entirely substituted in *Rb. capsulatus* by 3-oxo-14:0, but like in *Rb. sphaeroides*, 3-OH-10:0 occupies both C-3 and C-3', the 3-OH-10:0 at C-3' being substituted with *cis*Δ^5-12:1 (T. Merkofer, unpublished diploma work), forming the diester 3-O(*cis*Δ^5-12:1)-10:0. In *Rb. capsulatus* 37b4 lipid A_{GlcN}, the C-4' and C-1 positions are partially substituted by ethanolamine and ethanolamine phosphate, respectively. Interestingly, lipid A_{GlcN} from *Rb. capsulatus* and *Rb. sphaeroides* are almost completely nontoxic (see next paragraph).

Lipid A_{GlcN} from *Rubrivivax (Rhodocyclus) gelatinosus*: The highly toxic lipid A_{GlcN} from *Ru. gelatinosus* (Galanos et al., 1977) is also structurally very similar to enterobacterial lipid A_{GlcN} but the six fatty acids are symmetrically distributed over both D-GlcN residues (Fig. 7B). The bisphosphorylated D-GlcN-β(1,6)D-GlcN backbone is in some strains substituted at the phosphate groups at C-1 and C-4' by ethanolamine (Masoud et al., 1990). The amino groups of D-GlcN are occupied by 3-OH-10:0, which in turn are substituted by 12:0 and 14:0 forming acyloxyacyl residues. Unsubstituted 3-OH-10:0 residues occupy the C-3 and C-3' positions.

Lipid A_{GlcN} from *Rhodocyclus tenuis*: The backbone of this lipid A_{GlcN} is the commonly found β-D-GlcN-(1,6)D-GlcN disaccharide, but at least in *Rc. tenuis* 2761, the C-4 position is substituted by a third, non O- or N-acylated D-GlcN. The positions C-1 and C-4' are substituted by the rare D-Ara*f* and by 4-AA*p*, respectively. The amino groups of both backbone D-GlcN units are substituted by 3-OH-10:0. In further contrast to *Ru. gelatinosus*, C-3 and C-3' are occupied by 14:0 and 16:0. In cultures grown at lower temperature (14 or 20 °C) the ratio of unsaturated/saturated fatty acids increased significantly (Masoud et al., 1990).

Lipid A_{GlcN} from *Rhodomicrobium vannielii*: This lipid A_{GlcN} has also a β-D-GlcN-(1,6)D-GlcN backbone which, however, is not phosphorylated (Holst et al., 1983). Instead, the OH-group at C-1 is free, that of

C-4' carries partially D-Man. A comparison of photo- and chemotrophically grown *Rm. vannielii* cells revealed no significant difference in fatty acid composition except for the presence of an additional palmitic acid in chemotrophically grown cells (Rózalski and Kotelko, 1985). The amino group of the reducing GlcN is substituted by 3-OH-16:0.

Lipid A_{DAG} from *Rhodopseudomonas viridis*: A first structural proposal of this lipid A_{DAG} has been published (Roppel et al.; 1975; Mayer et al., 1989). The backbone consists of nonphosphorylated DAG substituted by amide-linked 3-OH-14-OH. Although a monosaccharide backbone structure was proposed, it is very likely that we are dealing here also with a disaccharidic backbone of the diamino sugar similar to that of lipid A_{DAG} from *Pseudomonas diminuta* (Kasai et al., 1987). The only ester-linked fatty acids in lipid A_{DAG} of *Rp. viridis* (see Section III B, 1a) are the long-chained fatty acids such as 27-OH-28:0 and 28-OH-29:0. They are linked to the amide-linked 3-OH-14:0 (M. Busch and H. Mayer, unpublished data).

B. Polysaccharide Moiety

There are only a few structural studies available on the polysaccharide moieties of LPS's of phototrophic bacteria. One example is the core region of the LPS of *Rubrivivax gelatinosus* Dr2. It contains one residue of Kdo ketosidically linked to C-6' of the non-reducing D-GlcN of lipid A_{GlcN} (Masoud et al., 1991b). The Kdo unit is substituted at C-4 by an α-1,4-linked D-GalA disaccharide and at C-5 by a single unit of D,D-Hep partially substituted at C-7 by phosphate (Fig. 8). *Ru. gelatinosus* strains are lacking ADP-D,D-Hep-6-epimerase (Masoud et al., 1991b), which in Enterobacteriaceae converts ADP-D,D-Hep to ADP-L,D-Hep.

Rhodocylus tenuis strains have a common core composed of a branched trisaccharide of L,D-Hep and Kdo comparable to the inner core of enterobacterial LPS's, but there are strain-specific variations in the substituents (Fig. 8). They all share the D-*manno*-configuration (D-Man, L,D- and D,D-Hep) and are assumed to be transferred to the 4-position of the heptose unit of the L,D-heptosyl-Kdo-disaccharide by related or even identical transferases (Radziejewska-Lebrecht et al., 1981). Both, the core trisaccharide and the strain-specific regions are partially phosphorylated.

With the Neu5Ac-containing S- and R-type LPS's of *Rb. capsulatus* a structural investigation of the core or the core-adjacent region was performed (Krauss et al., 1992). With the deep R-type LPS of strain Sp 18 which lacks the outer core region, the nonreducing terminal position of Neu5Ac was determined. In the R- and S-type LPS's of strains Kb-1 and 37b4, however, Neu5Ac is chain-linked. A trisaccharide of D-Gal-α(1,6)D-Glc-β(1,7)Neu5Ac was proven in the core-oligosaccharide of strain 37b4. It represented the first demonstration of a 7-substituted Neu5Ac in nature (Carfield and Schauer, 1982).

The structure of the oligosaccharide moiety of the *Rb. sphaeroides* ATCC 17023 O-antigen has been elucidated by Salimath et al. (1984). It is a L-Thr-containing 1,4-linked trisaccharide of α-linked D-GlcA disaccharide units linked to the 4-position of Kdo. Phosphorylation of Kdo at C-8 contributes further to this high accumulation of negative charges. L-Threonine is linked to the carboxylic group of D-GlcA by its amino group (Fig. 8). The linkage between core and lipid A_{GlcN} is by a single 1,4-substituted unit of Kdo attached by a 2,6'-linkage to lipid A_{GlcN}. Together with the knowledge of the respective lipid A_{GlcN} (see Section IV 'Lipid A'), the complete structure of the *Rhodobacter sphaeroides* ATCC 17023 LPS is thus known.

V. Contribution of the Lipopolysaccharides (LPS) of Purple Bacteria to Endotoxin Research

The different structural variants of lipid A found in anoxygenic phototrophic bacteria (for some structures see Fig. 7) can be nontoxic, as first observed with *Rhodobacter capsulatus* 37b4 (Weckesser et al., 1972; Strittmatter et al., 1983), or highly toxic, as in case of *Rubrivivax gelatinosus* 29/1 (Galanos et al., 1977). A comparison of their structure-endotoxic activities relationship has been undertaken in order to elucidate the structural requirements of lipid A for exhibiting endotoxic activity. For recent publications on the mechanism of endotoxic action including the activation of monocytes by the LPS-binding protein (LBP) via the LBP-LPS-complex, and the CD-14-receptor of monocytes, cytokine release including the tumor necrosis factor (TNF) as the prime mediator of endotoxic activity, and other cytokines see Mathison et al. (1992), Rock and Lowry (1991), and Schumann (1990).

```
Rubrivivax gelatinosus Dr2          αD-GalA(1,4)αD-GalA(1,4)Kdo
                                                            5
                                                            |
                                                         D,D-Hep

Rhodocyclus tenuis                  L,D-Hep(1,3)L,D-Hep-Kdo        R: D,D-Hep(1,2)L,D-Hep          (strain 2761)
(common trisaccharide)                   7              4             L,D-Hep                      (strain EU1)
                                         |              |             D,D-Hep(1,2)D,D-Hep          (strain 3661)
                                      L,D-Hep           R             X(1,2)D-Man (X=unknown)      (strain GFUy)

Rhodobacter sphaeroides ATCC 17023  αD-GlcA(1,4)αD-GlcA(1,4)αD-GlcA(1,4)Kdo
                                         6
                                         |
                                        Thr
```

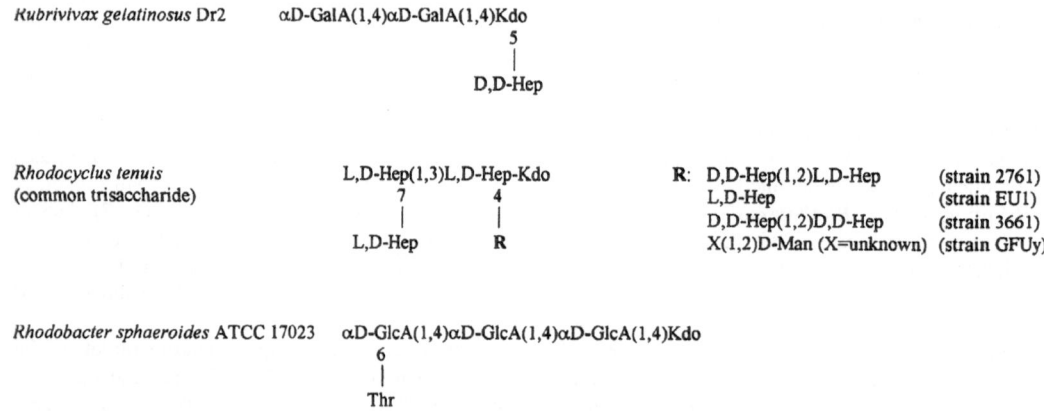

Fig. 8. Examples of core structures of LPS's from purple bacteria (according to Holst and Brade, 1992).

The nontoxic variants comprise two different groups, the first one (*Rb. capsulatus* 37b4, *Rb. sphaeroides* ATCC 17023) possessing a backbone structure comparable to enterobacterial lipid A_{GlcN}. The affinity of nontoxic lipid A_{GlcN} of *Rb. capsulatus* 37b4 and *Rb. sphaeroides* ATCC 17023 to LBP is only 1/100 and 1/40, respectively, of that of lipid A_{GlcN} of *Salmonella* sp. (Mayer et al., 1992). Lipopolysaccharide of *Rb. capsulatus* 37b4 induces mononuclear cell-derived interleukin-1 and interleukin-6 only at very high concentrations, and does not induce TNF activity. The *Rb. capsulatus* 37b4 LPS or the lipid A precursor Ia, or synthetic lipid A precursor ('compound 406') can act as endotoxin antagonists not only for lethality but virtually for all endotoxic activities (Loppnow et al., 1990; Loppnow et al., 1992). The mechanism of the antagonism is still unknown, but a strong inhibition of cytokine-induction by toxic LPS specimens from various bacteria was observed in co-incubation experiments in vitro (Loppnow et al., 1990).

Considering cytokine and especially TNF release as a measure of endotoxic activity, *Rb. sphaeroides* lipid A_{GlcN} has been recognized to specifically inhibit in vitro the biological action of toxic LPS's, including blocking the stimulatory effect of toxic LPS for $\beta 2$ integrin CD11b/CD18 expression on neutrophils in a variety of human and mouse cell types (Golenbock et al., 1991; Lynn et al., 1991; Lynn and Golenbock, 1992). The LPS of *Rb. sphaeroides* ATCC 17023 (Strittmatter et al., 1983) was used to serve as an antagonist in the activation of macrophages by toxic lipid A_{GlcN} by studies on competitive binding to the receptor sites on the macrophage. Its purified bisphosphoryl lipid A_{GlcN} was shown to block the induction of TNF by toxic Re-type LPS of *Escherichia coli* in macrophage cell lines; it does not activate B-lymphocytes and does not induce cytokines in macrophages (Takayama et al., 1989). Lipid A from *Rb. sphaeroides* ATCC 17023 also did not stimulate the murine pre-B cell line 70Z/3 to synthesize surface immunoglobulin or mRNA (Kirkland et al., 1991). It was, however, able to block specifically *Escherichia coli* LPS-induced activation of 70Z/3 cells. Kirkland et al. (1991) conclude that lipid A from *Rb. sphaeroides* ATCC 17023 may antagonistically compete with the toxic LPS for lipid A receptors on the 70Z/3 cell, but this is not a generally accepted hypothesis.

In vivo tolerance to endotoxin-shock in rats by lipid A_{GlcN} from *Rb. sphaeroides* was recently described by Carpati et al. (1993). Qureshi et al. (1993) reported lipid A of *Rb. sphaeroides* to induce corticosteroids. Chemical reduction of 3-oxo-14:0 and Δ^7-14:1 did not render the *Rb. sphaeroides* ATCC 17023 lipid A_{GlcN} toxic, which indicated that the lacking endotoxic activity is not due to the presence of 3-oxo and unsaturated fatty acids (Qureshi et al., 1991). The authors concluded that the lack of endotoxic activity may be either due to the presence of only five (instead of six in toxic lipid A_{GlcN}) fatty acyl groups or to the presence of unusually short ester-linked 3-OH-10:0 coexisting with much longer amide-linked fatty acids (3-oxo-14:0) at positions 3 and 3' and the higher variation in size of the fatty acids (Qureshi et al., 1993). Since lipid A_{GlcN} from both *Rb. capsulatus* 37b4 and *Rb. sphaeroides* ATCC 17023 express complete serological cross-reactivity with enterobacterial lipid A_{GlcN} using polyclonal (Galanos et al., 1977) and defined monoclonal antibodies against various lipid A_{GlcN} with known epitope specificities (Mayer et al., 1992; Kuhn et al., 1992), the lack of toxicity should be caused by structural differences residing in the hydrophobic (fatty acid) part. For discussion of a lamellar supramolecular structure in the nontoxic lipid A, *versus* an inverted structure in toxic lipid A's, see below.

The other group of nontoxic lipid A is represented by phosphate-free lipid A_{DAG} or lipid A_{Hyb} (see Section IV A), thus, having profound modifications in the hydrophilic part of lipid A. Lipid A_{DAG} from *Rp. viridis* does not cross-react with either of the other lipid A_{GlcN} variations mentioned.

Recent studies have indicated that the sole replacement of D-GlcN by DAG in a toxic lipid A, while keeping all other structural parameters identical does not lead to a decrease in lethality (Moran et al., 1991). Variation in the phosphate substitution pattern of the backbone disaccharide is accompanied by reduced lethality (Takada and Kotani, 1992), although it seems that the endotoxic activities are (partly) retained by an exchange of one of the backbone phosphate groups by D-GlcA as in *Pseudomonas vesicularis* (Arata et al., 1992), and as recently documented also by D-GalA in *Rs. fulvum* (Heike Rau, unpublished results). Using X-ray diffraction with synchrotron radiation as well as by Fourier Transform (FT) infrared spectroscopy with toxic (*Rv. gelatinosus*, *Salmonella* sp.) and nontoxic (*Rb. capsulatus*, *Rp. viridis*) LPS's, it was found that differences in the primary lipid A structure led to differences in their supramolecular aggregation. The toxic lipid A_{GlcN} variants have a strong tendency to form inverted, the nontoxic lipid A_{GlcN} variants as well as the nontoxic lipid A_{DAG} from *Rp. viridis*, however, lamellar structures even at low temperature (Brandenburg et al., 1993). As opposed to the toxic properties (lethality, pyrogenicity, cytokine induction), the anticomplementary activity is most pronounced for compounds with lamellar structure. These findings may indicate that receptor molecules such as CD-14 or distinct proteins with high affinity for LPS/lipid A complexes with LPS-binding protein (LBP) or bactericidal permeability increasing protein (BPI) (Gazzano-Santoro et al., 1992), may react preferentially with lipid A having inverted structures.

VI. Porin

A. Porins in Anoxygenic Phototrophic Bacteria

Porins are proteins of the outer membrane which form large water-filled channels which allow for diffusion of hydrophilic molecules into the periplasmic space (Nikaido, 1992). Among phototrophic bacteria, biochemical and functional characterizations of porins were performed with *Rhodobacter sphaeroides* ATCC 17023, *Rhodobacter capsulatus* strains St. Louis and 37b4 and *Rhodobacter blastica* DSM 2131 (Weckesser et al., 1984; Flammann and Weckesser, 1984; Benz et al., 1987; Woitzik et al., 1990; Butz et al., 1993). A major outer membrane protein (42 kDa) was described in *Chromatium vinosum* but not investigated for porin-function (Lane and Hurlbert, 1980a,b). The same is true for the major outer membrane proteins (app. Mr 43,000 and 14,000, respectively) of *Rhodospirillum rubrum* FR 1 (Oelze et al., 1975).

In contrast to porin from e.g. *Escherichia coli*, the oligomeric forms of the porins from both *Rb. sphaeroides* and *Rb. capsulatus* form monomeric porin upon treatment with 10 mM EDTA at 30 °C for 20 min (Weckesser et al., 1984; Flammann and Weckesser, 1984; Woitzik et al., 1990). The monomers chromatofocused revealing an isoelectrical point of about 4.0. With the oligomer, a single channel conductance of 3.15 nS in 1 M KCl was determined after reconstitution into artificial lipid bilayer membranes (Benz et al., 1987; Woitzik et al., 1990). The value was essentially in agreement with the pore diameter of the porin of *Rb. capsulatus* St. Louis, as determined by the liposome swelling assay (Flammann and Weckesser, 1984). The native porin of *Rb. sphaeroides* 2.4.1 (ATCC 17023) has been reported to be composed of heterologous subunits, an app. M_r 47,000 subunit and additional polypeptides of smaller molecular masses. The N-terminal alanine was found to be blocked due to bound fatty acids (Baumgardner et al., 1980; Deal and Kaplan, 1983a,b). Polyclonal antibodies against the major outer membrane protein of *Rb. sphaeroides* 2.4.1 (ATCC 17023), suggested to be a portion of the porin structure, reacted with outer membrane preparations of various *Rb. sphaeroides* and one *Rs. rubrum* strain, but not with those of *Rb. capsulatus* and *Paracoccus denitrificans* (Deal and Kaplan, 1983c). No serological cross-reaction was obtained with porins from *Rb. capsulatus* and *Rb. sphaeroides* (Woitzik et al., 1989).

The export of the porin of *Rb. capsulatus* 37b4 was studied with the use of an uncoupler of the electron transport chain, radioactive labeling and immunoprecipitation (Woitzik et al., 1989). A protease-sensitive preform porin was accumulated in the cells in the presence of the uncoupler; the exported form was resistant to the enzyme. A 32 kDa protein associated with tetrapyrrole intermediates in the bacteriochlorophyll *a* biosynthetic pathway was found to be identical to the porin of *Rb. capsulatus* in biochemical and serological features (Bollivar and Bauer, 1992). It was discussed by the authors that the intermediates excreted by the cell become non-specifically lodged within the porin complex.

B. Structure of Porins

The porin of *Rhodobacter capsulatus* 37b4 was first crystallized by Nestel et al. (1989), showing at that time an X-ray diffraction to 3.2 Å resolution and a trigonal R3 space group. The resolution was later on increased up to 1.8 Å (Kreusch et al., 1991; Weiss and Schulz, 1992). Recently, the structure of porin from *Rhodopseudomonas blastica* was also elucidated to atomic resolution (Fig. 9A). The primary structure of the porin from *Rb. capsulatus* 37b4 was determined after trypsin, CNBr and Asp-N protease cleavage (Schiltz et al., 1991). The peptides obtained were aligned to a total length of 301 residues with an Mr of 31,536. A high excess of 34 Asp and 17 Glu (16.9%) over 10 Lys, 7 Arg, and 2 His (6.3%) confirmed the low isoelectric point of 3.9. The sequence was used as the basis for the interpretation of the electron density map derived from X-ray diffraction of porin crystals (Weiss et al., 1991a,b). Using the above mentioned techniques, the complete three-dimensional structure of the *Rb. capsulatus* 37b4 porin was elucidated (Weiss et al., 1991a,b). The functional complex is a trimer of three identical subunits each forming a separate pore. The three pores merge on the cell external surface side. Each monomer is a 16-stranded β-barrel (Fig. 9B). The β-strands are all antiparallel, the barrel is twisted in a right-handed manner, all strands are connected to their direct neighbors. Three of the 15 connecting loops contain short α helices. The barrel rim, orientated to the external surface side, contains quite irregularly folded chain segments (no loop is shorter than 3 residues) and is called the 'rough side'. The largest connecting loop contains 43 residues and is between the fifth and the sixth β-strand at the rough side of the barrel. It runs into the inside of the barrel where it narrows the channel at the approximately half height and defines the pore size. In contrast, the lower rim of the barrel at the periplasmic space side of the pore is formed by 7 β-strand loops, 5 of them with short two-residue loops and 2 with three or more residue loops in a rather regular way. It is therefore called the 'smooth side'. The smooth side contains the N- and C-termini of the porin protein. Within the pore, 12 negatively charged amino acids (Asp and Glu) are localized at the large 43 residue loop. Opposite to them in this area, at the other inside of the pore, are located 6 positively charged amino acids (Arg, His, Lys), producing a strong charge-gradient across the pore (Fig. 9 C and D).

Thus, the trimer contains 3 distinct functional channels separated from each other by walls with an axial length of about 20 Å. If one divides one subunit at the barrel equator one observes only 7 negative and 7 positive charges in the lower half whereas there are as many as 44 negative and 12 positive charges in the upper half. The large excess of negative charges in the upper half is obvious and therefore the channel is electrostatically asymmetric along the axis. Around the equatorial circumference of the trimer there are non-polar side chains contacting the non-polar inner part of the membrane, whereas the side-chains lining the channel are polar.

VII. External Envelope Layers

A. Capsules and Slimes

In *Rhodobacter capsulatus* St. Louis, capsule and slime polysaccharides with different chemical compositions and which differed in sugar composition from the respective O-chains of LPS were found (Omar et al., 1983a,b,c). The capsule was found to be strongly attached to the lipoprotein-peptidoglycan forming an unusually firm capsule polysaccharide-protein-peptidoglycan complex, insoluble in boiling sodium dodecyl sulfate and hot phenol-water (Bräutigam et al., 1988). The assumed capsule polysaccharide-protein bond was sensitive to hydrofluoric or hydrochloric acid, and to alkaline hydrolysis. Digestion of the complex with pronase solubilized the capsule polysaccharide moiety with a characteristic amino acid pattern (Gly, Glu, Ser and Ala) remaining associated with it. On treatment of the protein-peptidoglycan moiety with lysozyme, the protein with peptidoglycan residues bound to it was solubilized. In the phage-resistant mutant *Rb. capsulatus* St. Louis $RC1^-$, in which the capsule polysaccharide is present in a free form, the same protein was obtained by lysozyme digestion without preceding acid treatment. The capsule-free *Rb. capsulatus* 37b4 contained the protein-peptidoglycan complex but not the capsule polysaccharide moiety.

B. S-Layers

Surface (S)-layers are planar arrays of proteinaceous or glycoproteinaceous subunits which can be aligned as regular (crystalline) layers on the cell surface. They are found in many taxonomic groups of both Bacteria and Archaea species, but can be lost on cultivation under laboratory conditions. Amongst anoxygenic phototrophic bacteria, they were found in strains of the Chromatiaceae (*Chromatium okenii, Cm. weissei, Cm. warmingii, Cm. buderi, Cm. gracile, Thiocapsa floridana, Amoebobacter* sp.), and in the Ectothiorhodospiraceae (the moderate halophilic *Ectothiorhodospira mobilis* and the extremely halophilic *Ectothiorhodospira halochloris*). Layers, especially those of hexagonal symmetry, are widely distributed in nonsulfur purple bacteria of the subgroups α-1 (*Rhodospirillum rubrum, Rhodospirillum molischianum*, likely also in *Rhodospirillum salexigens*) and in the subgroup α-2 (*Rhodopseudomonas palustris, Rhodopseudomonas acidophila*). They have been at least tentatively identified in the green bacteria (*Pelodictyon* sp., *Chlorochromatium aggregatum*), as summarized by Messner and Sleytr (1992). The DOC-PAGE pattern of LPS preparations can provide valuable hints for the presence of S-layers as exemplified with a number of anoxygenic phototrophic bacteria (Krauss et al., 1988a). Detailed structural studies on S-layers from phototrophic bacteria, however, are to our knowledge not yet available.

VIII. Concluding Remarks

It has been urgently requested to use chemotaxonomic or other phenotypic characteristics for proving or disproving the validity of relatedness of species based on nucleic acid sequence analyses (Murray et al., 1990). The analyses of cell wall macromolecules of anoxygenic phototrophic bacteria generally confirm ribosomal RNA sequencing investigations. One impressive example is the confirmation of the disparate phylogeny of the green sulfur and the green non-sulfur bacteria by their completely different cell wall organization. The cell wall of the thermophilic *Chloroflexus aurantiacus* may represent a character phylogenetically more ancient than the Gram-negative cell wall organization of the purple and the green sulfur bacteria, or the cyanobacteria.

A particularly attractive feature of phototrophic *Proteobacteria* and green-sulfur bacteria for comparative cell wall studies is their polyphyletic origin. Thus, they can be used as model organisms for study of the cell wall chemistry of nonphototrophic bacteria as well. The relatedness is especially expressed and confirmed within the α subdivision of Proteobacteria. Matching of lipid A composition and phylogenetic position of species is recognizable for most of the phototrophic bacteria studied so far, confirming in particular the phylogenetic relationship existing between phototrophic bacteria and their nonphototrophic relatives. Since bacterial photosynthesis is considered to have developed relatively early in evolution, it is not surprising that some of the LPS's of recent phototrophic bacteria may represent analogues of biosynthetic precursors or variants of the R-core and lipid A of more developed nonphototrophic species as discussed with examples by Mayer (1984) and Weckesser and Mayer (1988).

The spectrum of lipid A structural variants found with phototrophic bacteria (including the cyanobacteria), with a high variability of their biological properties, forms a reservoir of naturally occurring lipid A variants. This is of pharmacological and even actual clinical interest. Structural analyses of nontoxic and toxic lipid A's can contribute to answers concerning structure and endotoxic activity, in particular to the differentiation between the wanted biological properties of LPS, such as increase in unspecific resistance, and undesired properties, such as toxicity or endotoxic shock. The interest is focused today on those lipid A's which show highly reduced toxicity, for their possible use as endotoxin antagonists; in this regard lipopolysaccharide of *Rhodobacter sphaeroides* has recently become commercially available (List Biological Laboratories,

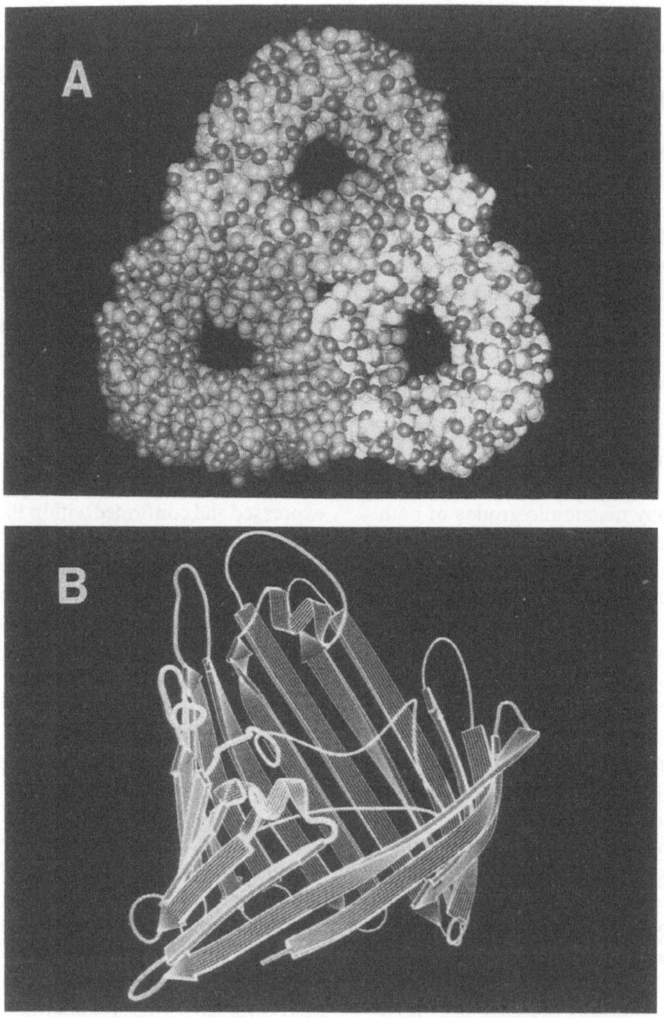

Fig. 9. (See also Color Plate 4.) Structure representations of porins from *Rhodopseudomonas blastica* and *Rhodobacter capsulatus*: (A) Space-filling model of the porin from *Rp. blastica* DSM 2131 as established at 2.0 Å resolution (Kreusch et al., 1994). The view is perpendicular to the membrane plane, the three pore eyelets are clearly visible. Hydrogens are not considered in this model. (B) Secondary structure of one subunit of the major porin from *Rb. capsulatus* 37b4 as viewed from the threefold symmetry axis. The external medium is at the top, the periplasm at the bottom. The 16 β-strands are given by blue arrows and the 3 short α-helices by orange helices. The long 43-residue loop inserting into the barrel is obvious. The interface part of the barrel is rather low while the membrane-facing part is high. The 'rough' barrel end at the top and the 'smooth' barrel end at the bottom can be visualized clearly.

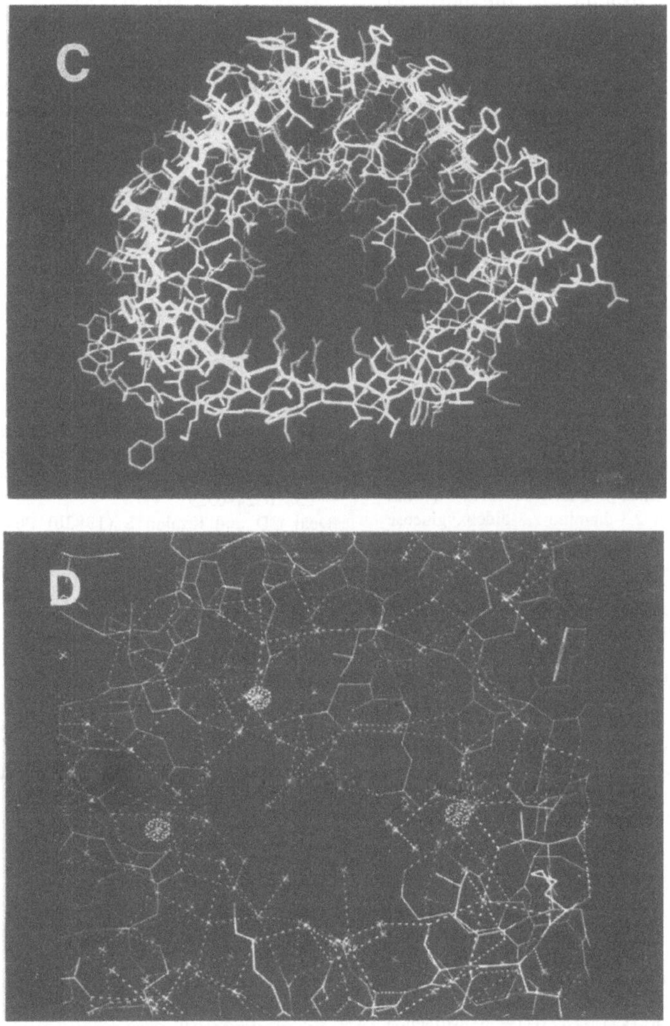

Fig. 9. Continued (See also Color Plate 5.) (C) Color-coded projection of one subunit from the *Rb. capsulatus* 37b4 porin onto the membrane plane. Aromatic side chains are yellow; positively and negatively charged side chains are red and dark blue, respectively. The interface and the threefold symmetry axis are at the bottom. Obviously, the rim of the pore eyelet is lined by positively charged side chains protruding from the interface area of the barrel juxtaposed to negatively charged side chains from the 43-residue loop running into the barrel interior. Most likely these charges form an electric field that facilitates permeation of polar solutes and hinders nonpolar ones. (D) Detail of the construction of the pore eyelet of the *Rb. capsulatus* porin. The amino acid residues are given in the usual color code, the fixed water molecules by crosses and the calcium ions by balls. Hydrogen bonds are indicated by dashed lines. 'Fixation' of the water molecules means that they are clearly visible in the X-ray structure analysis at the high resolution of 1.8 Å. This implies that water molecules are accommodated at the water positions most of the time. Still, these water molecules can move away. It should be noted that there is a defined network of hydrogen bonded water molecules at the eyelet leaving a rather small nonstructured cross-section in the center.

Inc., Campbell, CA, USA).

Knowledge of the structure of the porin of *Rhodobacter capsulatus* 37b4 at atomic resolution was achieved by X-ray diffraction studies. The main properties of the *Rb. capsulatus* 37b4 porin have been used to predict those of the OmpF and PhoE porins of *Escherichia coli* (Welte et al., 1991), which were later essentially confirmed (Cowan et al., 1992), indicating that the porin from *Rb. capsulatus* 37b4 can be used as a model for the non-specific hydrophilic pores of the outer membrane of species of *Proteobacteria*. Thus, anoxygenic phototrophic bacteria have been proven to be useful as model organisms for cell wall studies contributing to a number of questions of general biological interest.

References

Ahamed NM, Mayer H, Biebl H and Weckesser J (1982) Lipopolysaccharide with 2,3-diamino-2,3-dideoxyglucose containing lipid A in *Rhodopseudomonas sulfoviridis*. FEMS Microbiol Lett 14: 27–30

Arata S, Shiga T, Mizutami T, Mashimo J, Kasai N and Zabo L (1992) Structure of a new type of lipid A from *Pseudomonas vesicularis*. Second Conference of the International Endotoxin Society, Aug. 17–20, 1992, Vienna, Austria

Baumgardner D, Deal C and Kaplan S (1980) Protein composition of *Rhodopseudomonas sphaeroides* outer membrane. J Bacteriol 143: 265–273

Beck H, Hegemann GD and White D (1990) Fatty acid and lipopolysaccharide analyses of three *Heliobacterium* spp. FEMS Microbiol Lett 69: 229–232

Beer-Romero P, Favinger JL and Gest H (1988) Distinctive properties of bacilliform photosynthetic heliobacteria. FEMS Microbiol Lett 49: 451–454

Benz R, Woitzik D, Flammann HT and Weckesser J (1987) Pore forming activity of the major outer membrane protein of *Rhodobacter capsulatus* in lipid bilayer membranes. Arch Microbiol 148: 226–230

Beveridge TJ and Graham LL (1991) Surface layers of bacteria. Microbiol Rev 55: 684–705

Bhat UR, Carlson RW, Busch M and Mayer H (1991) Distribution and phylogenetic significance of 27-hydroxy-octacosanoic acid in lipopolysaccharides from bacteria belonging to the alpha-2 subgroup of *Proteobacteria*. Int J Syst Bacteriol 41: 213–217

Bollivar DW and Bauer CE (1992) Association of tetrapyrrole intermediates in the bacteriochlorophyll *a* biosynthetic pathway with the major outer-membrane porin protein of *Rhodobacter capsulatus*. Biochem J 282: 471–476

Bowman JP, Skerratt JH, Nichols PD and Sly LI (1991) Phospholipid fatty acid and lipopolysaccharide fatty acid signature lipids in methane-utilizing bacteria. FEMS Microbiol Ecol 85: 15–22

Bräutigam E, Fiedler F, Woitzik D, Flammann HT and Weckesser J (1988) Capsule polysaccharide-protein-peptidoglycan complex in the cell envelope of *Rhodobacter capsulatus*. Arch Microbiol 150: 567–573

Brandenburg K, Mayer H, Koch MHJ, Weckesser J, Rietschel ET and Seydel U (1993) Influence of the supramolecular structure of free lipid A on its biological activity. Eur J Biochem 218: 555–563

Butz S, Benz R, Wacker T, Welte W, Lustig A, Plapp R and Weckesser J (1993) Biochemical characterization and crystallization of porin from *Rhodopseudomonas blastica*. Arch Microbiol 159: 301–307

Carfield AP and Schauer R (1982) Occurrence of sialic acids. In: Schauer R (ed) Sialic Acids, pp 5–39. Cell Biol Monogr 10, Springer, Wien

Carpati CM, Astiz ME, Saha DC and Rackow EC (1993) Diphosphoryl lipid A from *Rhodopseudomonas sphaeroides* induces tolerance to endotoxin-shock in the rat. Critical Care Medicine 21: 753–758

Cowan SW, Schirmer T, Rummel G, Steiert M, Gosh R, Pauptit RA, Jansonius JN and Rosenbusch JP (1992) Crystal structures explain functional properties of two *E. coli* porins. Nature 358: 727–733

Deal CD and Kaplan S (1983a) Solubilization, isolation, and immunochemical characterization of the major outer membrane protein from *Rhodopseudomonas sphaeroides*. J Biol Chem 258: 6524–6529

Deal CD and Kaplan S (1983b) Physical and chemical characterization of the major outer membrane protein of *Rhodopseudomonas sphaeroides*. J Biol Chem 258: 6530–6536

Deal CD and Kaplan S (1983c) Immunochemical relationship of the major outer membrane protein of *Rhodopseudomonas sphaeroides* 2.4.1 to proteins of other photosynthetic bacteria. J Bacteriol 154, 1015–1020

Drews G, Weckesser J and Mayer H (1978) Cell envelopes. In: Clayton RK and Sistrom WR (eds) The Photosynthetic Bacteria, pp 61–77. Plenum Publishing Corporation, New York

Evers D, Weckesser J and Drews G (1984) Protein on the cell surface of the moderately halophilic phototrophic bacterium *Rhodospirillum salexigens*. J Bacteriol 160: 107–111

Evers D, Weckesser J and Jürgens UJ (1986) Chemical analyses on cell envelope polymers of the halophilic, phototrophic *Rhodospirillum salexigens*. Arch Microbiol 145: 245–258

Flammann HT and Weckesser J (1984) Porin isolated from the cell envelope of *Rhodopseudomonas capsulata*. J Bacteriol 159: 410–412

Framberg K, Mayer H, Weckesser J and Drews G (1974) Serologische Untersuchungen an isolierten Lipopolysacchariden aus *Rhodopseudomonas palustris* Stämmen. Arch Microbiol 98: 239–250

Galanos C, Roppel J, Weckesser J, Rietschel ET and Mayer H (1977) Biological activities of lipopolysaccharides and lipid A from Rhodospirillaceae. Infect Immun 16: 407–412

Gazzano-Santoro H, Parent JB, Grinna L, Horwitz A, Parsons T, Theofan G, Elsbach P, Weiss J and Conlon PJ (1992) High-affinity binding of the bactericidal/permeability-increasing protein and a recombinant amino-terminal fragment to the lipid A region of lipopolysaccharide. Infect Immun 60: 4754–4761

Gibson J, Ludwig W, Stackebrandt E and Woese CR (1985) The phylogeny of the green photosynthetic bacteria: Absence of a close relationship between *Chlorobium* and *Chloroflexus*. Syst Appl Microbiol 6: 152–156

Golecki JR and Drews G (1980) Cellular organization of the halophilic phototrophic bacterium strain WS 68. Eur J Cell Biol 22: 654–660

Golenbock DT, Hampton RY, Qureshi N, Takayama K and Raetz CRI (1991) Lipid A-like molecules that antagonize the effects of endotoxins on human monocytes. J Biol Chem 266: 19490–19498

Hancock RFW (1991) Bacterial outer membranes: Evolving concepts. ASM-News 57: 175–182

Hollingsworth RI and Carlson RW (1989) 27-Hydroxy-octacosanoic acid is a major structural fatty acyl component of the lipopolysaccharide of *Rhizobium trifolii* ANU 843. J Biol Chem 264: 9300–9303

Holst O, Borowiak D, Weckesser J and Mayer H (1983) Structural studies on the phosphate-free lipid A of *Rhodomicrobium vannielii* ATCC 17100. Eur J Biochem 137: 325–332

Holst O and Brade M (1992) Chemical structure of the core region of lipopolysaccharides. In: Morrison DC and Ryan JL (eds) Bacterial Endotoxic Lipopolysaccharides, Vol I, Molecular Biochemistry and Cellular Biology, pp 132–204. CRC Press, Boca Raton

Hurlbcrt RE, Hurlbert IM (1977) Biological and physicochemical properties of the lipopolysaccharide of *Chromatium vinosum*. Infect Immun 16: 983–994

Hurlbert RE, Weckesser J, Mayer H and Fromme I (1976) Isolation and characterization of the lipopolysaccharide of *Chromatium vinosum*. Eur J Biochem 68: 365–371

Jürgens UJ and Weckesser J (1990) The cell envelope of cyanobacteria. In: Kumar HD (ed) Phycotalk, Vol 2, pp 37–57. Rastogi and Co, Meerut, India

Jürgens UJ, Meissner J, Fischer U, König WA and Weckesser J (1987) Ornithine as a constituent of the peptidoglycan of *Chloroflexus aurantiacus*, diaminopimelic acid in that of *Chlorobium vibrioforme f. thiosulfatophilum*. Arch Microbiol 148: 72–76

Jürgens UJ, Meissner J, Reichenbach H and Weckesser J (1989) L-Ornithine containing peptidoglycan-polysaccharide complex from the cell wall of the gliding bacterium *Herpetosiphon aurantiacus*. FEMS Microbiol Lett 60: 247–250

Kasai N, Arata S, Mashima JI, Akiyama Y, Tanaka C, Egawa K and Tanaka S (1987) *Pseudomonas diminuta* LPS with a new endotoxic lipid A structure. Biochem Biophys Res Commun 142: 972–978

Kirkland TN, Qureshi N and Takayama K (1991) Diphosphoryl lipid A derived from lipopolysaccharide (LPS) of *Rhodo-pseudomonas sphaeroides* inhibits activation of 70Z/3 cells by LPS. Infect Immun 59: 131–136

Kompanseva EI and Gorlenko VM (1985) A new species of the moderately halophilic purple bacterium *Rhodospirillum mediosalinum* sp. nov. Microbiology 53: 775–781

Krauss JH, Weckesser J and Mayer H (1988a) Electrophoretic analysis of lipopolysaccharides of purple nonsulfur bacteria. Int J Syst Bacteriol 38: 157–163

Krauss JH, Reuter G, Schauer R, Weckesser J and Mayer H (1988b) Sialic acid-containing lipopolysaccharides in purple nonsulfur bacteria. Arch Microbiol 50: 584–589

Krauss J, Seydel U, Weckesser J and Mayer H (1989) Structural analysis of the nontoxic lipid A of *Rhodobacter capsulatus* 37b4. Eur J Biochem 180: 519–526

Krauss JH, Himmelspach K, Reuter G, Schauer R and Mayer H (1992) Structural analysis of a novel sialic-acid-containing trisaccharide from *Rhodobacter capsulatus* 37b4 lipopolysaccharide. Eur J Biochem 204: 217–223

Kreusch A, Weiss MS, Welte W, Weckesser J and Schulz GE (1991) Crystals of an integral membrane protein diffracting to 1.8 A resolution. J Mol Biol 217: 9–10

Kreusch A, Neubüser A, Schiltz E, Weckesser J and Schulz ER (1994) The structure of the membrane channel porin from *Rhodopseudomonas blastica* at 2.0 Å. J Protein Science 3: 58–63

Kuhn HM, Brade I, Appelmelk BJ, Kusomoto S, Rietschel TTh and Brade H (1992) Characterization of the epitope specificity of murine monoclonal antibodies directed against lipid A. Infect Immun 60: 2201–2210

Lane BC, Hurlbert RE (1980a) Characterization of the cell wall and the cell wall proteins of *Chromatium vinosum*. J Bacteriol 141: 1386–1398

Lane BC, Hurlbert RE (1980b) Isolation and partial characterization of the major outer membrane protein of *Chromatium vinosum*. J Bacteriol 143: 349–354

Loppnow H, Libby P, Freudenberg M, Krauss JH, Weckesser J and and Mayer H (1990) Cytokine induction by lipopolysaccharide (LPS) corresponds to lethal toxicity and is inhibited by nontoxic *Rhodobacter capsulatus* LPS. Infect Immun 58: 3743–3750

Loppnow H, Rietschel ET, Brade H, Schönbeck U, Libby P, Wang MH, Heine H, Feist W, Dürrbaum-Landmann I, Ernst M, Brandt E, Grage-Griebenow E, Ulmer AJ, Campos-Portuguez S, Schade U, Kirikae T, Kusumoto S, Krauss J, Mayer H, Flad HD (1993) Lipid A precursor Ia (compound 406) and *Rhodobacter capsulatus* lipopolysaccharide: Potent endotoxin antagonists in the human system in vitro. In: Levin J, Alving CR, Munford RS, Stütz PL (eds) Bacterial Endotoxin: Recognition and Effector Mechanisms, pp 337–348. Excerpta Medica, Amsterdam

Lynn WA and Golenbock DT (1992) Lipopolysaccharide antagonists, Immunology Today 13: 271–276

Lynn WA, Raetz CRH, Qureshi N and Golenbock (1991) Lipopolysaccharide-induced stimulation of CD11b/CD18 expression on neutrophils. Evidence of specific receptor-based response and inhibition by lipid A-based antagonists. J Immunol 147: 3072–3079

Madigan MT (1992) The family Heliobacteriaceae. In: Balows A, Trüper H, Dworkin M, Harder W and Schleifer KH (eds) The Prokaryotes. A Handbook on the Biology of Bacteria: Ecophysiology, Isolation, Identification, Applications, pp 1981–1992. Springer, Heidelberg

Masoud H, Lindner B, Weckesser J and Mayer H. (1990) The structure of the lipid A component of *Rhodocyclus gelatinosus* Dr2 lipopolysaccharide. Syst Appl Microbiol 13: 227–233

Masoud H, Urbanik-Sypniewska T, Lindner B, Weckesser J and Mayer H. (1991a) The structure of the lipid A component of *Sphaerotilus natans*. Arch Microbiol 156: 167–175

Masoud H, Mayer H, Kontrohr T, Holst O and Weckesser J (1991b) The structure of the core region of the lipopolysaccharide from *Rhodocyclus gelatinosus* Dr2. Syst Appl Microbiol 14: 222–227

Mathison JC, Tobias PS, Wolfson E and Ulevitch RJ (1992) Plasma lipopolysaccharide (LPS)-binding protein. A key component in macrophage recognition of Gram-negative LPS. J Immunol 149: 200–206

Mayer H (1984) Significance of lipopolysaccharide structure for questions of taxonomy and phylogenetical relatedness of Gram-negative bacteria. In: Haber E (ed) The Cell Membrane. Its Role in Interaction with the Outside World, pp 71–83. Plenum Publishing Corporation, New York

Mayer H and Weckesser J (1984) 'Unusual' lipid A': Structures, taxonomical relevance and potential value for endotoxin research. In: Rietschel ET (ed) Handbook of Endotoxin, Vol 1, Chemistry of Endotoxin, pp 221–247. Elsevier Publishers, Amsterdam

Mayer H, Bock E and Weckesser J (1983) 2.3-Diamino-2.3-dideoxglucose containing lipid A in the *Nitrobacter* strain X14. FEMS Microbiol Lett 17: 93–96

Mayer H, Salimath PV, Holst O and Weckesser J (1984) Unusual lipid A types in phototrophic bacteria and related species. Rev Infect Diseases 6: 542–545

Mayer H, Tharanathan RN and Weckesser J (1985) Analysis of lipopolysaccharides of Gram-negative bacteria. Meth Microbiol 18: 157–207

Mayer H, Masoud H, Urbanik-Sypniewska T and Weckesser J (1989) Lipid A composition and phylogeny of Gram-negative bacteria. Bull Japan Federation Culture Collections 5: 19–25

Mayer H, Campos-Portuguez SA, Busch M, Urbanik-Sypniewska T and Bhat UR (1990a) Lipid A variants—or, how constant are the constant regions in lipopolysaccharides? In: Nowotny A, Spitzer JJ and Ziegler EJ (eds) Cellular and Molecular Aspects of Endotoxin Reactions, pp 111–120. Elsevier Science Publications, Amsterdam

Mayer H, Krauss JH, Yokota A and Weckesser J (1990b) Natural variants of lipid A. In: Friedman H, Klein W, Nakano M and Nowotny A (eds) Endotoxin, pp 45–70. Plenum Publishing, New York

Mayer H, Rau H, Campos-Portuguez S, Kuhn HM, Tobias PS and Freudenberg MA (1992) Differences and correlations between lipid A and lipid A variants in serological specificity, LBP reactivity and lethality. Second Conference of the International Endotoxin Society. Aug. 17-20, 1992, Vienna, Austria

Meissner J, Fischer U and Weckesser (1987) The lipopolysaccharide of the green sulfur bacterium *Chlorobium vibrioforme f. thiosulfatophilum*. Arch Microbiol 149: 125–129

Meissner J, Jürgens UJ, Weckesser J (1988a) Chemical analysis of peptidoglycans from species of *Chromatiaceae* and *Ectothiorhodospiraceae*. Z Naturforsch 43c: 823–826

Meissner J, Jürgens UJ and Weckesser J (1988b) Chemical analysis of peptidoglycans from species of *Chromatiaceae* and *Ectothiorhodospiraceae*. Z Naturforsch 43c: 823–826

Meissner J, Pfennig N, Krauss JH, Mayer H and Weckesser J (1988c) Lipopolysaccharides of *Thiocystis violacea*, *Thiocapsa pfennigii*, and *Chromatium tepidum*, species of the family *Chromatiaceae*. J Bacteriol 170: 3217–3222

Meissner J, Borowiak D, Fischer U, Weckesser J (1988d) The lipopolysaccharide of the phototrophic bacterium *Ectothiorhodospira vacuolata*. Arch Microbiol 149: 245–248

Meissner J, Krauss JH, Jürgens UJ, and Weckesser J (1988e) Absence of a characteristic cell wall lipopolysaccharide in the phototrophic bacterium *Chloroflexus aurantiacus*. J Bacteriol 170: 3213–3216

Messner P and Sleytr UB (1992) Crystalline bacterial cell-surface layers. Adv Microb Physiol 33: 213–275

Moran AP, Zähringer U, Seydel U, Scholz D, Stütz P and Rietschel ET (1991) Structural analysis of the lipid A component of *Campylobacter jejuni* CCUG 10936 (serotype O:2) lipopolysaccharide. Description of a lipid A containing a hybrid backbone of 2-amino-2-deoxy-D-glucose and 2,3-diamino-2,3-dideoxy-D-glucose. Eur J Biochem 198: 459–469

Murray RGE, Brenner DJ, Colwell RR, De Vos P, Goodfellow M, Grimont PAD, Pfennig N, Stackebrandt E and Zavarzin GA (1990) Report of the ad hoc committee on approaches to taxonomy within the Proteobacteria. Int J Syst Bacteriol 40: 213–215

Nestel U, Wacker T, Woitzik D, Weckesser J, Kreutz W and Welte W (1989) Crystallization and preliminary X-ray analysis of porin from *Rhodobacter capsulatus*. FEBS Lett 242: 405–408

Newton JW (1968) Linkages in the walls of *Rhodospirillum rubrum* and its bacilliform mutants. Biochim Biophys Acta 165: 534–537

Nikaido H (1992) Porins and specific channels of bacterial outer membranes. Molec Biol 6: 435–442

Oelze J, Golecki JR, Kleinig H and Weckesser J (1975). Characterization of two cell-envelope fractions from chemotrophically grown *Rhodospirillum rubrum*. Antonie van Leeuwenhoek 41: 273–286

Olsen GJ and Woese CR (1993) Ribosomal RNA: A key to phylogeny. FASEB J 7: 113–123

Omar AS, Flammann HT, Golecki JR and Weckesser J (1983a). Detection of capsule and slime polysaccharide layers in two strains of *Rhodopseudomonas capsulata*. Arch Microbiol 134: 114–117

Omar AS, Flammann HT, Borowiak D and Weckesser J (1983b) Lipopolysaccharides of two strains of the phototrophic bacterium *Rhodopseudomonas capsulata*. Arch Microbiol 134: 212–216

Omar AS, Weckesser J and Mayer H (1983c) Different polysaccharides in the external layers (capsule and slime of the cell envelope of *Rhodopseudomonas capsulata* Sp 11. Arch Microbiol 136: 291–296

Pietsch K, Weckesser J, Fischer U and Mayer H (1990) The lipopolysacccharides of *Rhodospirillum rubrum*, *Rhodospirillum molischianum*, and *Rhodopila globiformis*. Arch Microbiol 154: 433–437

Qureshi N, Honovich JP, Hara H, Cotter RJ and Takayama K (1988) Location of fatty acids in lipid A obtained from lipopolysaccharide of *Rhodopseudomonas sphaeroides* ATCC 17023. J Biol Chem 263: 5502–5504

Qureshi N, Takayama K, Meyer KC, Kirkland TN, Bush A, Chen L, Wang R and Cotter RJ (1991) Chemical reduction of 3-oxo and unsaturated groups in fatty acids of diphosphoryl lipid A from the lipopolysaccharide of *Rhodopseudomonas sphaeroides*. J Biol Chem 266: 6532–6538

Qureshi N, Takayama K, Hofman J, Zuckerman SH (1993) Diphosphoryl lipid A obtained from the nontoxic lipopolysaccharide of *Rhodopseudomonas sphaeroides* is an LPS antagonist and an inducer of corticosteroids. In: Levin J, Alving CR, Munford RS, Stütz PL (eds) Bacterial Endotoxin: Recognition and Effector Mechanisms, pp 361–371. Excerpta Medica, Amsterdam

Radziejewska-Lebrecht J, Feige U, Mayer H and Weckesser J (1981) Structural studies on the heptose-region of lipopolysaccharides from *Rhodospirillum tenue*. J Bacteriol 145: 138–144

Raetz CRH (1992) Biosynthesis of lipid A. In: Morrison DC and

Ryan JL (eds) Bacterial Endotoxic Lipopolysaccharides, Vol I, Molecular Biochemistry and Cellular Biology, pp 67–80. CRC Press, Boca Raton

Rau H, Yokota A, Sakane T and Mayer H (1993) Isolation and chemical characterization of lipopolysaccharides from four *Aquaspirillum* species (*Aquaspirillum itersonii* subsp. *nipponicum* IFO 13615, *Aquaspirillum polymorphum* IFO 13961, *Aquaspirillum aquaticum* IFO 14918, *Aquaspirillum metamorphum* IFO 13960 and *Aquaspirillum metamorphum* IFO 13960 mutant strain 12-3). J Gen Appl Microbiol 39: 547–557

Rietschel ET, Brade L, Schade U, Seydel U, Zähringer U, Kusumoto S and Brade H (1987) Bacterial endotoxin: Properties and structure of biologically active domains. In: Schrinner E, Richmond MH, Seibert G and Schwarz U (eds) Surface Structures of Microorganisms and their Interactions with the Mammalian Host, pp 1–41. Proc. Workshop Conf., Hoechst. Verlag Chemie, Weinheim

Rietschel ET, Brade L, Lindner B and Zähringer U (1992) Biochemistry of lipopolysaccharides. In: Morrison DC and Ryan JL (eds) Bacterial Endotoxic Lipopolysaccharides, Vol I, Molecular Biochemistry and Cellular Biology, pp 3–41. CRC Press, Boca Raton

Rock CS and Lowry SF (1991) Tumor necrosis factor-α. J Surgical Res 51: 434–445

Roppel J, Mayer H and Weckesser J (1975) Identification of a 2,3-diamino-2,3-dideoxyhexose in the lipid A component of lipopolysaccharides of *Rhodopseudomonas viridis* and *Rhodopseudomonas palustris*. Carbohyd Res 40: 31–40

Rózalski A and Kotelko K (1985) Comparison of lipopolysaccharides of *Rhodomicrobium vannielii* grown in photo- and chemotrophic conditions. Archivum Immunologiae et Therapiae Experimentalis 33: 387–395

Salimath PV, Weckesser J, Strittmatter W and Mayer H (1983) Structural studies on the non-toxic lipid A from *Rhodopseudomonas sphaeroides* ATCC 17023. Eur J Biochem 136: 195–200

Salimath PV, Tharanathan RN, Weckesser J and Mayer H (1984) The structure of the polysaccharide moiety of *Rhodopseudomonas sphaeroides* ATCC 17023 lipopolysaccharide. Eur J Biochem 144: 227–232

Schiltz E, Kreusch A, Nestel U, Schulz GE (1991) Primary structure of porin from *Rhodobacter capsulatus*. Eur J Biochem 199: 587–594

Schmelzer E, Weckesser J, Warth R and Mayer H (1982) Peptidoglycan of *Rhodopseudomonas viridis*: Partial lack of N-acetyl substitution of glucosamine. J Bacteriol 149: 151–155

Schumann RR (1990) Function of the lipopolysaccharide (LPS)-binding protein (LBP) and CD14, the receptor for LPS/LBP complexes: A short review. Res. Immunol 143: 11–15

Stackebrandt E, Murray RGE and Trüper HG (1988a). *Proteobacteria* classis nov., a name for the phylogenetic taxon that includes the 'purple bacteria and their relatives'. Int J Syst Bacteriol 38: 312–325

Stackebrandt E, Embley M and Weckesser J (1988b) Phylogenetic, evolutionary, and taxonomic aspects of phototrophic eubacteria. In: Olson JM, Ormerod JG, Amesz J, Stackebrandt E and Trüper HG (eds) Green Photosynthetic Bacteria, pp 201–215. Plenum Publishing Corporation, New York

Strittmatter W, Weckesser J, Salimath PV and Galanos C (1983) Nontoxic lipopolysaccharide from *Rhodopseudomonas sphaeroides* ATCC 17023. J Bacteriol 155: 153–158

Takada H and Kotani S (1992) Structure-function relationships of lipid A. In: Morrison DC and Ryan JL (eds) Bacterial Endotoxic Lipopolysaccharides, Vol I, Molecular Biochemistry and Cellular Biology, pp 107–134. CRC Press, Boca Raton

Takayama K, Qureshi N, Beutler B and Kirkland TN (1989) Diphosphoryl lipid A from *Rhodopseudomonas sphaeroides* ATCC 17023 blocks induction of cachectin in macrophages by lipopolysaccharide. Infect Immun 57: 1336–1338

Tadros MH, Drews G and Evers D (1982) Peptidoglycan and protein, the major cell wall constituents of the obligate halophilic bacterium *Rhodospirillum salexigens*. Z Naturforsch 37c: 210–212

Tharanathan RN, Salimath PV, Weckesser J and Mayer H (1985) The structure of lipid A from the lipopolysaccharide of *Rhodopseudomonas gelatinosa* 29/1. Arch Microbiol 141: 279–283

Weckesser J and Jürgens UJ (1988) Cell walls and external layers. Meth Enzymol 167: 173–186

Weckesser J and Mayer H (1987) Lipopolysaccharide aus phototrophen Bakterien: Ein Beitrag zur Phylogenie und zur Endotoxin-Forschung. Forum Mikrobiologie 10: 242–248

Weckesser J and Mayer H (1988) Different lipid A types in lipopolysaccharides of phototrophic and related nonphototrophic bacteria. FEMS Microbiol Rev 54: 143–154

Weckesser J, Drews G and Ladwig R (1972) Localisation and biological and physicochemical properties of cell wall lipopolysaccharide of *Rhodopseudomonas capsulata*. J Bacteriol 110: 346–353

Weckesser J, Drews G and Mayer H (1979) Lipopolysaccharides of photosynthetic prokaryotes. Annu. Rev. Microbiol. 33: 215–239

Weckesser J, Mayer H, Metz E and Biebl H (1983) Lipopolysaccharide of *Rhodocyclus purpureus*: Taxonomic implication. Int J Syst Bacteriol 33: 53–56

Weckesser J, Zalman LS and Nikaido H (1984) Porin from *Rhodopseudomonas sphaeroides*. J Bacteriol 159: 199–205

Weibgen U, Russa R, Yokota A and Mayer H (1993) Taxonomic significance of the lipopolysaccharide composition of the three biovars of *Agrobacterium tumefaciens*. Syst Appl Microbiol 15: 177–182

Weiss MS. and Schulz GE (1992) Structure of porin refined at 1.8 Å resolution. J Mol Biol 227: 493–509

Weiss MS, Kreusch A, Schiltz E, Nestel U, Welte W, Weckesser J and Schulz GE (1991a) The structure of porin from *Rhodobacter capsulatus* at 1.8 Å resolution. FEBS Lett 280: 379–382

Weiss MS, Abele U, Weckesser J, Welte W, Schiltz E and Schulz GE (1991b) Molecular architecture and electrostatic properties of a bacterial porin. Science 254: 1627–1630

Welte W, Weiss MS, Nestel U, Weckesser J, Schiltz E and Schulz GE (1991) Prediction of the general structure of OmpF and PhoE from the sequence and structure of porin from *Rhodobacter capsulatus*. Orientation of porin in the membrane. Biochim Biophys Acta 1080: 271–274

Wilkinson BJ, Hindahl MS, Galbraith L and Wilkinson SG (1986) Lipopolysaccharide of *Paracoccus denitrificans* ATCC 13543. FEMS Microbiol Lett 37: 63–67

Williamson JM, Anderson MS and Raetz CRH (1991) Acyl-ACP specificity of UDP-GlcNAc acyltransferases from Gram-negative bacteria: Relationship to lipid A structure. J Bacteriol 173: 3591–3596

Woese CR, Stackebrandt E, Weisburg WG, Paster BJ, Madigan MT, Fowler VJ, Hahn CM, Blanz P, Gupta R, Nealson KH and Fox GE (1984a) The phylogeny of the purple bacteria: The alpha-subdivision. Syst Appl Microbiol 5: 315–326

Woese CR, Weisburg WG, Paster BJ, Hahn CM, Tanner RS, Krieg NR, Koops HP, Harms H and Stackebrandt E (1984b) The phylogeny of the purple bacteria: The beta-subdivision. Syst Appl Microbiol 5: 327–336

Woese CR, Stackebrandt E, Weisburg WG, Paster BJ, Madigan MT, Fowler VJ, Hahn CM, Blanz P, Gupta R, Nealson KH and Fox GE (1985) The phylogeny of purple bacteria: The gamma-subdivision. Syst Appl Microbiol 6: 25–33

Woese CR (1987) Bacterial evolution. Microbiol Rev 51: 221–271

Woitzik D, Dierstein R and Weckesser J (1989) The lipoprotein and peptidoglycan of *Rhodobacter sphaeroides*. Z Naturforsch 44c: 749–753

Woitzik D, Weckesser J and Dierstein R (1989) Export of porin to the outer membrane of the phototrophic bacterium *Rhodobacter capsulatus* 37b4. Arch Microbiol 152: 196–200

Woitzik D, Weckesser J, Benz R, Stevanovic S, Jung G and Rosenbusch JP (1990) Porin of *Rhodobacter capsulatus*: Biochemical and functional characterization. Z Naturforsch 45c: 576–582

Zahr M, Mayer H, Imhoff JF, Campos-Pardo V and Weckesser J (1992) Chemical composition of the lipopolysaccharides of *Ectothiorhodospira shaposhnikovii*, *Ectothiorhodospira mobilis*, and *Ectothiorhodospira halophila*. Arch Microbiol 157: 499–504

Chapter 12

Structure, Molecular Organization, and Biosynthesis of Membranes of Purple Bacteria

Gerhart Drews and Jochen R. Golecki
Institute of Biology 2, Microbiology, Albert-Ludwigs-University, Schaenzlestrasse 1, 79104 Freiburg, Germany

Summary

Photosynthetic Proteobacteria are used as model systems for many fields of modern biological research. The exciting results increased the progress considerably: Transduction of light energy into an electrochemical gradient of protons and electron- and proton-translocation in reaction centers and other membrane bound complexes has been studied on the molecular and atomic level with purple bacteria. Theories on the phylogeny of the energy metabolism have been stimulated and supported by 16S rRNA evolutionary analyses and biochemical investigations. Facultative purple bacteria have also been used to study cell and membrane differentiation. In this chapter it will be shown that most members of the anaerobic, anoxygenic photosynthetic bacteria develop a continuous system of cytoplasmic (CM)-intracytoplasmic membranes (ICM). The ICM is organized in characteristic, species-specific patterns of vesicles, tubules, thylakoid-like membrane sacs or highly organized membrane stacks. This development is strictly regulated by external growth factors like oxygen tension and light intensity. The ICM is formed under low oxygen tension by invaginations of the CM. At high oxygen tension and high light intensities the formation of ICM is suppressed. Although CM and ICM

R. E. Blankenship, M. T. Madigan and C. E. Bauer (eds): Anoxygenic Photosynthetic Bacteria, pp. 231–257.

form a membrane continuum, they are not homogenous in their composition. Membrane fractions differ in buoyancy density (lipid/protein ratio), pigment content, and biochemical activities. The photochemical apparatus is concentrated in the ICM and the enzymes of the respiratory chain and transport systems in the CM, respectively. Functional complexes of the photosynthetic apparatus have been visualized by freeze-fracture electron microscopy. Photochemical reaction centers are surrounded by oligomers of the core antenna complexes (B870) in stoichiometric ratios. The working hypothesis that core particle and ubiquinone-cytochrome bc_1 complexes form supramolecular structures in the membrane and that the dense population of membranes with proteins limits the free lateral mobility of proteinous complexes in the fluid-crystalline phase of the CM and ICM, is a challenge for future research on protein-protein interactions in membranes. Another interesting observation is the characteristic size and structure of ICM in the form of vesicles, tubules or stacks of flat membrane sacs. It has been proposed but not proved that distinct proteins are morphogenetic factors.

The morphogenesis of the photosynthetic apparatus is mainly regulated on the transcriptional level, but the signal chains, from sensing the external oxygen-partial pressure or light intensity to activation of transcription, is known only in parts, and an interesting field for future research. Targeting and translocation of pigment-binding proteins to and across the membrane is proposed to be supported by chaperonene and membrane-bound assembling proteins. Exchange or deletion of highly conserved amino acid positions in the pigment-binding proteins resulted in an inhibition of stabile insertion and assembly. The cofactors bacteriochlorophyll, carotenoids and cytochromes are important factors in the assembling process. A blockade of their biosynthesis inhibits formation of stable complexes. Biosynthesis of the photosynthetic apparatus is presumably regulated via a signal chain and many activators which initiate transcription of supraoperonal DNA structures. Since no large precursor pools for phospholipids, membrane proteins, pigments, hemes and quinones have been observed, the process of membrane synthesis and differentiation must be highly coordinated. It is a challenge for future research to analyze and discover this process on transcriptional and posttranscriptional levels.

I. Introduction

Most of the anoxygenic photosynthetic purple bacteria have in addition to the cytoplasmic membrane (CM), which encloses the protoplast, intracytoplasmic membranes (ICM), which have a characteristic organization and function (Oelze and Drews, 1972; Collins and Remsen, 1991). The third type of membranes, the outer membrane, is a part of the cell wall of Gram-negative bacteria and has been described in Chapter 11 by Weckesser et al. ICMs are characteristic of but not restricted to phototrophic proteobacteria. Obligate methylotrophic bacteria and nitrifying bacteria also contain ICMs (Drews, 1992). Obviously, ICMs have developed during evolution as a functional adaptation to environmental conditions and as a process of functional differentiation of membrane-bound energy-transducing systems. Structure, function and biogenesis of the CM-ICM system has been studied extensively in *Rhodospirillum rubrum, Rhodobacter capsulatus*, and other facultative photosynthetic bacteria, which grow under chemotrophic and phototrophic conditions as well (Fig. 1). These bacteria are excellent models to study membrane biosynthesis and differentiation under well defined external growth conditions. The major components of the ICM are the membrane-bound complexes of the photosynthetic apparatus.

The reader of this chapter will be introduced by the classical static view of electron microscopy, which, along with biochemical studies, opened the field of modern research on photosynthetic bacteria about forty years ago. Electron microscopical studies of thin sections from cells cultivated under defined conditions showed very early that formation of ICM and bacteriochlorophyll (BChl) and increase of photophosphorylating activity run parallel as a dynamic process. Quantitative freeze-fracture electron microscopy opens the interior of membranes and exposes the integral membrane particles, showing the asymmetry of the fracture halves of the lipid layer and distinct size classes of particles which have

Abbreviations: BChl – bacteriochlorophyll; CM – cytoplasmic membrane; EF – exoplasmic fracture face; ICM – intracytoplasmic membrane; LH – light-harvesting; PF – plasmic fracture face; *Rb. – Rhodobacter*; RC – reaction center; *Rc. – Rhodocyclus; Rp. – Rhodopseudomonas; Rs. – Rhodospirillum*

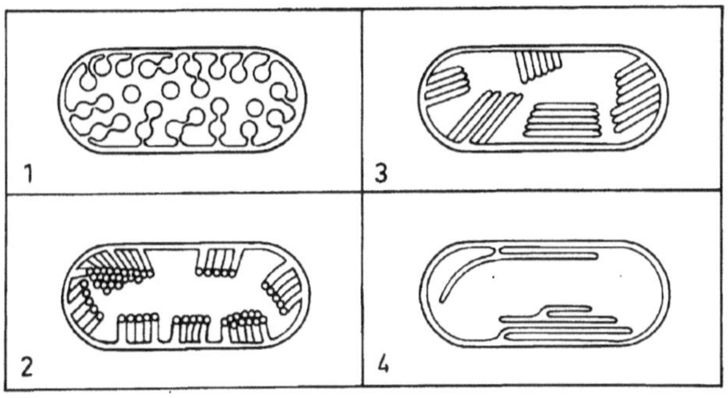

Fig. 1. Scheme of structure and arrangement of intracytoplasmic membranes (ICM) bearing the photosynthetic apparatus. 1) Vesicle type, realized in *Rhodospirillum rubrum* (see also Fig. 12), *Rhodobacter capsulatus* (Figs. 7, 8, 15), *Chromatium vinosum* (Figs. 5, 6), *Thiocapsa roseopersicina* and other species. 2) ICM consisting of tubuli in *Thiocapsa pfennigii* and some other bacteria. 3) ICM are flat thylakoid-like membranes organized in regular stacks, present in *Rhodospirillum molischianum*, *Ectothiorhodospira shaposhnikovii* (Fig. 4) and other species. 4) Large thylakoids, partially stacked and irregularly arranged, are presented in *Rhodopseudomonas palustris* (Fig. 13) and *Rhodopseudomonas viridis* (Reproduced from G. Drews and J. F. Imhoff, 1991).

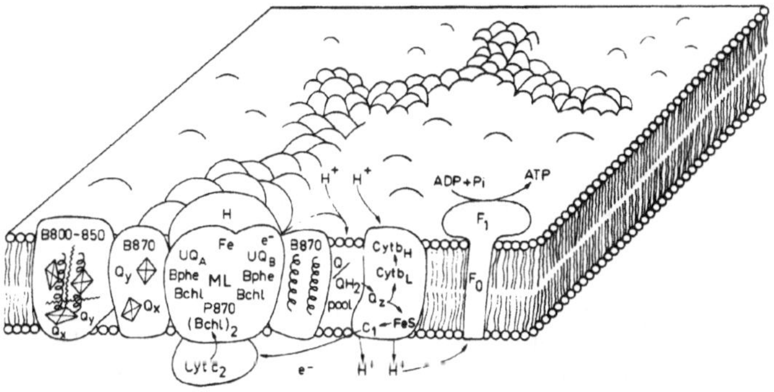

Fig. 2. Scheme of the photosynthetic apparatus in the intracytoplasmic membrane of *Rhodobacter capsulatus* and related bacteria. The reaction center (RC), consisting of subunits M, L and H, is surrounded by the light-harvesting complex LH I (B870) forming the core complex. The core complexes are connected by LH II (B800–850) antenna complexes. B870 has 2 mol BChl per monomer and B800–850 3 mol BChl per monomer (1α-, 1β-polypeptide). All LH complexes form oligomers (≈ hexamers, octamers). *Rb. capsulatus*, *Rb. sphaeroides* and *Rhodospirillum rubrum* have no tetra-heme cytochrome unit bound to RC. The electrons are shuttled by cytochrome c_2 from the ubiquinone-cytochrome bc_1 complex to RC. The functional subunits of the respiratory chain, such as NADH dehydrogenase and the terminal oxidases, which are present in CM and ICM, have been omitted for clarity.

been identified as pigment protein complexes. High-resolution electron microscopy of isolated defined complexes confirmed and extended results from freeze-fracture electron microscopy. Today many of the functional complexes of CM and ICM have been isolated, the primary structure of their proteins determined by protein or DNA sequencing, and the structure, topography and organization (Fig. 2) analyzed to different levels. The most exciting work on biosynthesis and assembly of membranous functional complexes and their regulation is in an early state. The first results will be reported at the end of this chapter.

II. Structure and Supramolecular Organization of Cytoplasmic and Intracytoplasmic Membranes

A. Ultrastructure

The outer membrane of the cell wall is characterized in ultrathin sections as a typical double track layer

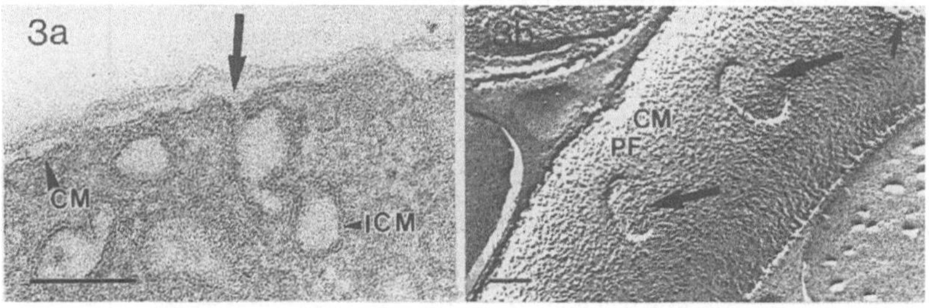

Fig. 3. Invaginations (↗) of the cytoplasmic membrane (CM) to form intracytoplasmic membranes (ICM) demonstrating the continuity of both membranes. a) Ultrathin section of *Rhodobacter capsulatus*. b) Freeze-fracture micrograph of the plasmic fracture face (PF) of the cytoplasmic membrane of *Rhodospirillum salexigens* (reproduced from Golecki and Drews, 1980). Arrow in the upper right corner of the freeze-fracture micrograph indicates, as in all following freeze-fracture pictures, the direction of shadowing. Bars represent 100 nm.

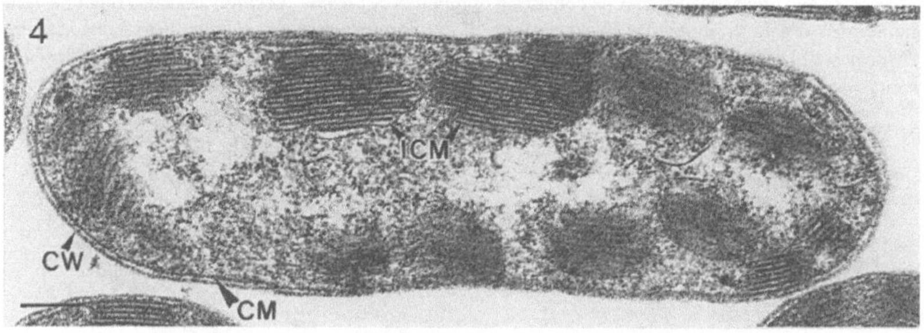

Fig. 4. Ultrathin section of *Ectothiorhodospira shaposhnikovii* exposing several lamellar stacks of ICM. Cell wall (CW). Bar represents 200 nm.

with two electron-dense lines separated by an electron transparent middle zone (Figs. 3, 4; see also Figs. 8, 15). The thickness of the outer membrane is approximately 8 nm.

In this report we only want to mention a connection of the outer membrane with the cytoplasmic membrane by adhesion sites which could be detected in ultrathin sections of some purple bacteria (Golecki, unpublished results). The adhesion sites described for *Escherichia coli* and *Salmonella typhimurium* (Bayer, 1968; Mühlradt et al., 1974) have been interpreted to be export sites of newly synthesized lipopolysaccharides into the outer membrane. In respect to the presence of lipopolysaccharides in the outer membrane of purple bacteria (Weckesser et al., 1979), these adhesion sites may have the same function as in enteric bacteria.

The CM underlying the cell wall surrounds the protoplast and represents a selective barrier to the environment. In ultrathin sections the CM presents the same ultrastructure of a double track layer (7–8 nm of thickness) as the outer membrane (Figs. 3, 4; see also Figs. 8, 15). In some micrographs the CM appears to be separated from the cell wall by an electron transparent periplasmic space. This seems to be an artifact because the periplasmic space is a densely populated space and not empty. In the CM the components of the respiratory chain, other energy transducing complexes and transport systems are localized, whereas the photosynthetic activities generally are housed in the ICM (Fig. 2)(Bowyer et al., 1985). Most phototrophically grown purple bacteria form ICM that carry out photosynthetic activities. The continuity of the CM and ICM has been demonstrated in several facultative phototrophic bacteria (Drews and Giesbrecht, 1963; Cohen-Bazire and Kunisawa, 1963; Boatman 1964; Holt and Marr 1965a; Tauschel and Drews, 1967).

The ICM is formed by invagination of the CM (Fig. 3)(Drews and Giesbrecht, 1963; Hickman and Frenkel, 1965; Tauschel and Drews, 1967). In spite of the morphological continuity and possible morphogenetic relationships, both membrane types present clear differences in chemical and structural composition, physical properties, function and kinetics of biosynthesis (Section II.B–D).

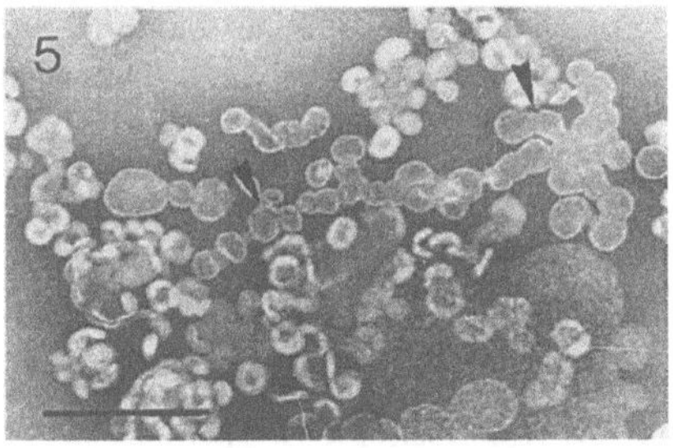

Fig. 5. Negative stained preparation of intracytoplasmic membranes of *Chromatium vinosum* isolated by a mild procedure. Large parts of the ICM vesicles are still connected to each other (➤). Bar represents 200 nm (Reproduced from Hurlbert et al., 1974; courtesy of J.R. Golecki).

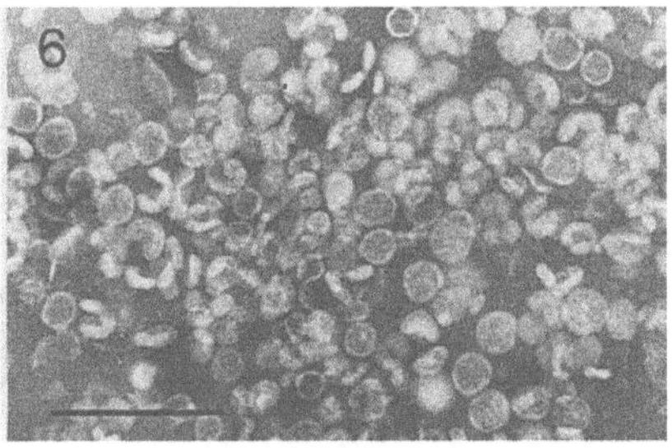

Fig. 6. Negatively stained ICM vesicles (= chromatophores) of *Chromatium vinosum* after French press isolation. The connections between the vesicles are interrupted. Bar represents 200 nm (Reproduced from Hurlbert et al., 1974; courtesy of J.R. Golecki).

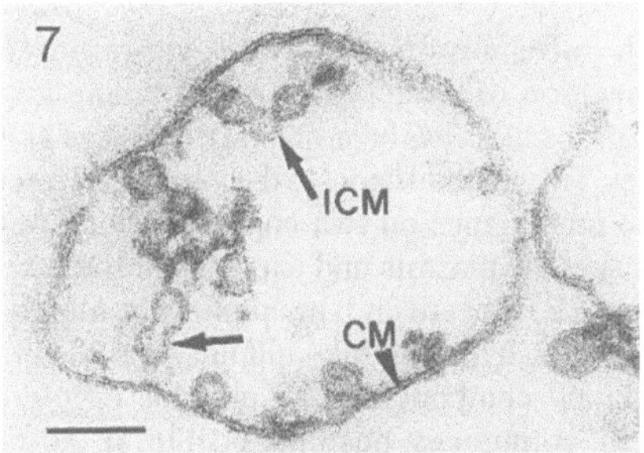

Fig. 7. Ultrathin section of a spheroplast preparation of *Rhodobacter capsulatus* exposing the transition of tubular into vesicular ICM types (↖). The sample was taken early after induction of membrane differentiation. Bar represents 100 nm (Reproduced from Dierstein et al., 1981; courtesy of J.R. Golecki).

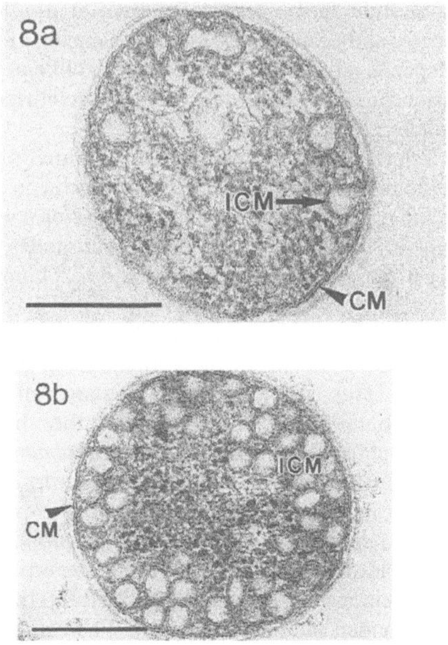

Fig. 8. Ultrathin sections of *Rhodobacter capsulatus*. (a) grown at high light conditions exhibiting only a small number of ICM vesicles. (b) grown at low light conditions with numerous ICM of vesicular type. Bars represent 200 nm.

Among the numerous members of the purple bacteria, several types of ICM have been discovered (Fig. 1; Bachofen and Wiemken, 1986; Sprague and Varga, 1986). The ICMs consist of a network of vesicles (Figs. 5, 6, 7, 8; see also Figs. 12, 15), tubules (single or arranged in bundles), thylakoid-like flat lamellae sometimes forming grana-like stacks (Fig. 4; see also Fig. 13), or irregular non-stacked membranes (Oelze and Drews, 1972). The ICM are connected to each other (Fig. 5), at least during their morphogenesis, and form a large membrane continuum together with the cytoplasmic membrane (Holt and Marr, 1965 a; Hurlbert et al., 1974; Dierstein et al., 1981).

During isolation procedures, the small tubular connections between the ICM vesicles or lamellae are interrupted and single ICM vesicles are obtained, which have been described as 'chromatophores' (Fig. 6). Due to the process of invagination during morphogenesis, the interior of the ICM structures is an extra-cytoplasmic space. Therefore the intra-membrane components of the ICMs are oriented inversely (see Figs. 12 and 15) in comparison to cytoplasmic membranes (Hochman et al., 1975; Michels and Konings, 1978). The extension of ICM is dependent on several external factors like oxygen tension, light intensity, and some other growth factors (Holt and Marr, 1965 b).

Under strictly aerobic, chemotrophic growth conditions, purple bacteria generally exhibit no ICM. One exception is *Rhodobacter capsulatus* which forms a few polar tubular ICMs even under high oxygen tensions (Lampe and Drews, 1972; Lampe et al., 1972; Dierstein et al., 1981; Fig. 7).

Turning to phototrophic growth conditions (i.e. anaerobic in the light) the formation of ICM is induced (Figs. 3, 7). In the early stages of ICM differentiation (i.e. 1 to 4 h after the shift to phototrophic growth conditions) in *Rb. sphaeroides*, the newly formed ICM often has the form of flat sheets or broad tubules with a high lipid-to-protein ratio as indicated by the low particle number per membrane area (Chory et al., 1984). Later on during the adaptation the broad tubules or flat sheets became more rounded and reduced in size. When at the end of the adaptation period the cells were growing photosynthetically, the ICM is composed of small vesicles (40 to 70 nm in diameter) seen in photosynthetically grown control cells. Similar observations were obtained with *Rb. capsulatus* after the shift from chemotrophic to phototrophic growth conditions (Dierstein et al., 1981; Kaufmann et al., 1982; Fig. 7). The amount of ICM produced under anaerobic conditions depends on light intensity (Cohen-Bazire and Kunisawa, 1963; Holt and Marr, 1965 b; Golecki et al., 1980; Reidl et al., 1983, 1985): Under high-light conditions only a few vesicles are detectable at the periphery of the protoplast in cells of*Rs. rubrum* and *Rb. capsulatus* (Fig. 8a). Under low light conditions, however, the bacteria contain a much higher number of vesicles filling most of the cytoplasmic space (Fig. 8b). Cells of *Rb. capsulatus* grown at low light conditions (7 W m^{-2}) contained 1150 ICM vesicles per cell, which is 6.3 fold more than in cells grown at high light intensities (2000 W m^{-2})(Golecki et al., 1980).

Due to an increase of the number of vesicles under low light conditions the total surface area of ICM per cell is enlarged to approximately 6 μm^2; this represents an area 2.7-fold larger than the area of the whole CM. Concomitant with the enlargement of the ICM area in low light grown cells, the BChl content as well as the size and concentration of photosynthetic units increases (a photosynthetic unit is one photochemical reaction center (RC) plus surrounding antenna pigment-protein complexes and is calculated as mol total BChl per mol RC).

In contrast to most nonsulfur purple bacteria, which adapt from chemotrophic to phototrophic growth conditions by forming photosynthetically active ICMs, cells of *Rubrivivax gelatinosus*, *Rhodocyclus purpureus*, and, in particular, *Rhodocyclus tenuis*, exhibit only occasional small single tubular intrusions of the CM (De Boer, 1969). This lack of ICM is compensated by a larger area of CM (Wakim et al., 1978) and higher density of photosynthetic units.

Occasionally ICMs with non-photosynthetic activities have been described for purple bacteria. These mesosomes, also described for other non-photosynthetic bacteria, were defined as sac-like invaginations of the cytoplasmic membrane (Reavely and Burge, 1972). They were thought to be involved in cell processes like replication and segregation of the chromosomes and in cell division. However, modern electron microscopic preparation techniques (Ebersold et al., 1981; Dubochet et al., 1983) gave evidence that classical mesosomes are artifacts of the preparation technique. Another type of ICM with a myelin-like configuration was observed in cells of*Rs. rubrum* after long-term anaerobic dark growth (Schön and Ladwig, 1970; Uffen et al., 1971). It was suggested that these myelin-like deposits resulted from the degradation of photosynthetically active membranes (chromatophores).

B. Membrane Architecture Shown by Freeze-Fracture Electron Microscopy

Much information about the supramolecular organization of photosynthetic membrane systems in purple bacteria has been obtained by freeze-fracture studies. This method exposes the internal architecture of the membranes on two complementary fracture faces (i.e. the plasmic and exoplasmic fracture face [PF and EF, respectively]) by splitting the membranes along their hydrophobic middle zone. Therefore membrane components like proteins or pigment-protein complexes become visible as so-called intramembrane particles of different size. The number of the particles per membrane area is an indication of the numerous biochemical activities of the membrane (Figs. 9, 10, 11).

Participation of the CM in the ICM-forming process has been demonstrated by freeze-fracture for several purple bacteria. In chemotrophically (dark)

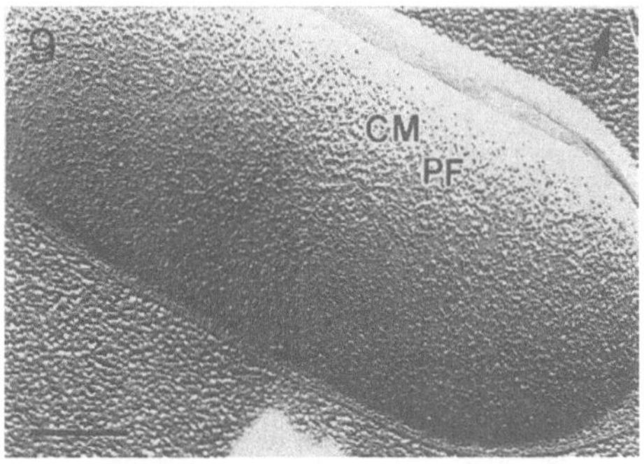

Fig. 9. Plasmic fracture face (PF) of the cytoplasmic membrane of *Rhodobacter sphaeroides* grown under chemotrophic conditions aerobically in the dark. A high population of homogeneously distributed intramembrane particles is visible on the PF of the cytoplasmic membrane. Bar represents 200 nm.

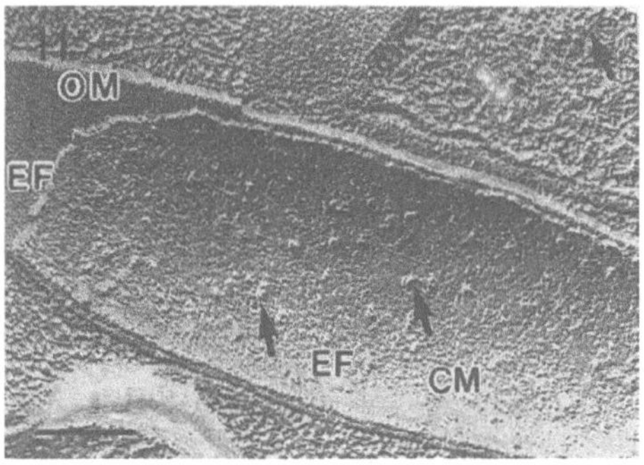

Fig. 11. Freeze-fracture micrograph of *Rhodospirillum rubrum* exposing numerous invaginations (➘) on the EF of the CM after induction of ICM formation by shifting to photosynthetic growth. Bar represent 200 nm (Reproduced from Golecki and Oelze, 1975).

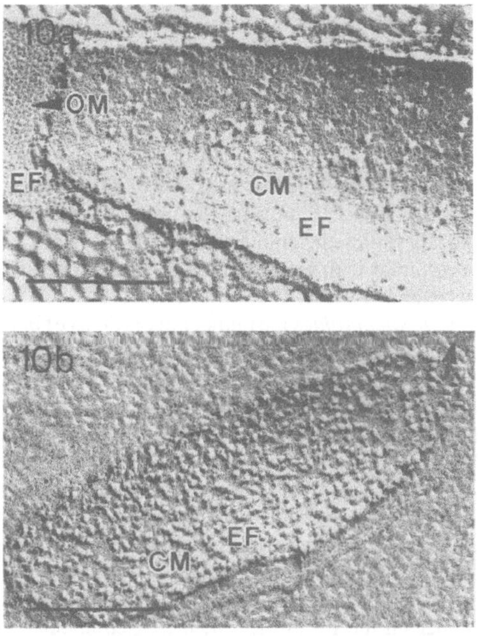

Fig. 10. Freeze-fracture micrographs of *Rhodocyclus tenuis* exposing the exoplasmic fracture face (EF) of the cytoplasmic membrane. (a) Under chemotrophic growth conditions the cytoplasmic membrane is covered only by a small number of intramembrane particles. The exoplasmic fracture face (EF) of the outer membrane (OM) is also exposed. (b) Under phototrophic growth conditions photosynthetically active complexes are incorporated in a high number into the exoplasmic leaflet of the CM. Bars represent 200 nm (Reproduced from Wakim et al., 1978; courtesy J.R. Golecki).

grown cells of *Rs. rubrum*, *Rb. sphaeroides*, *Rc. tenuis* and *Rb. capsulatus* (Golecki and Oelze, 1975, 1980; Wakim et al., 1978; Golecki et al., 1979) the plasmic fracture face (PF) is usually densely covered with intramembrane particles of varying sizes (Fig. 9; see also Fig. 13), while the exoplasmic fracture face (EF)(Fig. 10a) of the CM is sparsely covered with large particles. Apart from some minor differences in the number and size of the intramembrane particles among the different strains, the fracture faces of chemotrophically grown purple bacteria are not distinguishable from those of other bacteria like *E. coli* (Fig. 9). Clear differences become visible when chemotrophically grown cells are transferred to phototrophic growth conditions (Fig. 10, 11). Among the different members of photosynthetic bacteria a great variation was demonstrated concerning the participation of the CM in the formation of photosynthetically active structural components (Golecki and Oelze, 1980).

Cells of the facultative purple bacteria *Rhodocyclus tenuis*, *Rubrivivax gelatinosus* and *Rc. purpureus* are distinguished from other phototrophic bacteria by their poorly developed ICM. They contain only small single tubular intrusions of the CM. This lack of ICM system is compensated by a larger area of CM and a higher density and diameter of intramembrane particles in the CM of phototrophic compared with chemotrophic cells. The extension of the CM area results from incorporation of photosynthetic units in the existing electron transport system of chemotrophic cells (Wakim et al., 1978; Wakim and Oelze, 1980). Cells of *Rhodocyclus* species have a lower diameter than most other purple bacteria. The incorporation of the photosynthetic apparatus is not accompanied by invaginations but by extension of CM. In a comparative study by freeze-fracture electron microscopy, Golecki and Oelze (1980) demonstrated the striking difference in intramembrane particle patterns between the ICM-forming species *Rhodospirillum rubrum* and *Rhodobacter sphaeroides* and the CM extension-type represented by *Rhodocyclus tenuis*.

The number and size of intramembrane particles in the CM of*Rs. rubrum* and *Rb. sphaeroides* remain relatively constant under various culture conditions, but in the CM of *Rhodocyclus tenuis* the particle size and number change strongly upon induction of the photosynthetic apparatus. The newly formed photosynthetic units are placed in the outer leaflet of the CM. On the EF of the CM the number of intramembrane particles increases by a factor 6.9 after transfer to phototrophic conditions, whereas the particle number on the PF is nearly constant under both conditions. The increased number of particles on the EF was mainly the result of an increased occurrence of particles with diameters of ≥ 10 nm (Fig. 10b).

In*Rs. rubrum*, however, the number of particles on the EF of the CM stayed nearly constant, but on the PF the particle number increased from 4,700 to 6,200 particles per μm^2 after transfer to phototrophic conditions. The increase in number was largely caused by an enhanced occurrence of 10 nm particles.

Invaginations of the CM to form the ICM could be demonstrated as indentations and protuberances (Fig. 11) on both complementary fracture faces of the CM (Golecki and Oelze, 1975). Before the first invaginations could be observed, BChl was detected in the cells. The synthesis of BChl before forming ICM indicates that BChl and BChl-binding proteins become incorporated into the CM during the initial phase of adaptation. In a second phase, the BChl content increased and invaginations of the CM occurred and increased in number. In a later stage the number of invaginations and the cellular BChl content do not increase in the same rate. In*Rs. rubrum* the particular involvement of the CM in the process of BChl incorporation and ICM formation could be demonstrated by particle size histograms of freeze-fractured cells (Golecki and Oelze, 1980). After transfer to phototrophic conditions, the number of 10 nm particles increased on the PF of the CM by 33%. The 10 nm class is the predominant class of particles

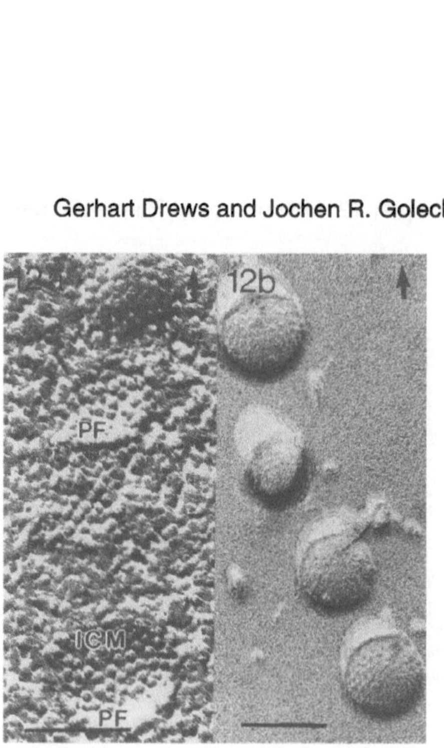

Fig. 12. ICM vesicles of ***Rhodospirillum rubrum*** exposing regularly arranged intramembrane particles of approx. 10 nm in diameter. (a) on the plasmic fracture face of the ICM and (b) on the outer surface of frozen-etched ICM vesicles. Bar represent 100 nm.

visible on the plasmic fracture faces of ICM vesicles (Fig. 12). These 10 nm particles could be extracted with lauryl dimethylamine oxide from ICM vesicles concomitant with the extraction of reaction centers (Oelze and Golecki, 1975).

In *Rb. sphaeroides* the number of particles of both fracture faces of the CM stayed nearly constant after transfer to phototrophic conditions. The mean diameter appeared to be slightly increased under phototrophic conditions. Particles in CM and ICM of phototrophically grown cells were of similar diameter (7.0 – 8.0 nm). In cells of *Rb. sphaeroides* invaginations were demonstrated on fracture faces of the CM under phototrophic and under chemotrophic conditions at low oxygen tension.

In *Rb. capsulatus* (Golecki et al., 1979) the particle number on the PF of the CM increased about 20–55%; after the induction of BChl synthesis the 9.5 nm particle class dominates under all conditions. The distribution of particles on the fracture faces of the CM is homogenous. The same was observed in other purple bacteria.

The CM of photosynthetically grown cells of *Rhodopseudomonas palustris*, however, is spatially differentiated into regions of extremely high particle density and areas of lower particle densities (Varga and Staehelin, 1983). Areas with a higher particle

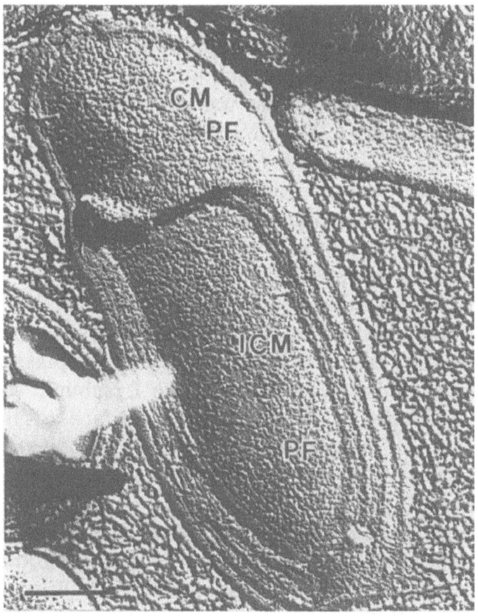

Fig. 13. Rhodopseudomonas palustris exposing plasmic fracture faces of the cytoplasmic membrane and of the lamellar arranged ICM membranes. Bar represents 200 nm.

density correspond to the adhesion sites between CM and ICM. The stacked ICM of *Rp. palustris* occupied the same areas of high particle density as detected on the CM (Fig. 13). This implies that the CM too is differentiated for photosynthesis in these regions. On the respective fracture faces of the ICM stacks, changes in particle size distribution could be demonstrated in response to changes in the light intensity supporting growth. When light intensity was decreased from 8500 to 100 lx, the average particle diameter on the PF of stacked ICM and CM increased in size. Four prominent particle classes of 7.5, 10.0, 12.5, and 15.0 nm were measured. When light intensity decreased, the portion of larger particles increased, suggesting that after lowering the light intensity subunits of discrete size are being added to the core particle. The smaller particles (7.5–10.0 nm), which are the most prevalent at high light conditions, were interpreted to represent the core particles formed by reaction center and light-harvesting I (LH I) complexes (Figs. 13, 14). The larger particles (10.0–15.0 nm), only seen in appreciable amounts in low-light cells, were proposed to be formed by adding light-harvesting II (LH II) complexes in defined portions to the core particles

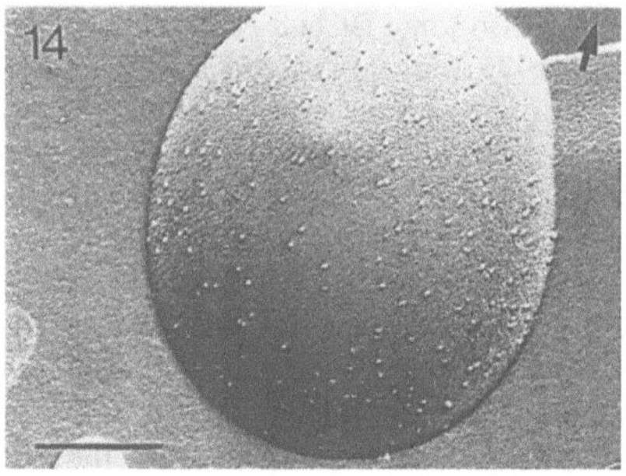

Fig. 14. Freeze-fracture micrograph of a liposome with incorporated reaction center-LH I core complexes of *Rhodobacter capsulatus*. Bar represents 200 nm.

(Fig. 2). The arrangement of the LH I and LH II complexes around the reaction centers results in a six-fold axis of symmetry.

The influence of different light intensities on the size and number of intramembrane particles on the fracture faces of ICM was demonstrated also for *Rb. sphaeroides* (Yen et al., 1984) and *Rb. capsulatus* (Golecki et al., 1979, 1980). An increasing size of intramembrane particles on the PF of the ICM concomitant with decreasing light intensities was shown also for *Rb. sphaeroides* (Yen et al., 1984). This is similar to *Rp. palustris*, where the RC and LH I complexes are in a fixed stoichiometry while the number of LH II complexes vary inversely as a function of light intensity.

Different particle sizes of the LH I-core complexes in purple bacteria containing either BChl *a* or BChl *b*, were explained by a different polypeptide composition (Meckenstock et al., 1992). The number of LH polypeptides per reaction center was calculated to be 36 in BChl *b*-containing bacteria compared to 24 in BChl *a*-containing bacteria. The size of the BChl *a*-containing LH I-core complex was believed to be smaller. This is in good agreement with the RC-to-RC distance of 10.2 and 10.0 nm for the BChl *a*-containing core complexes of *Rp. marina* (Meckenstock et al., 1992) and *Rs. rubrum* (Oelze and Golecki, 1974), respectively, compared with RC-to-RC distances of 13.0 nm for the BChl *b*-containing bacterium *Rhodopseudomonas viridis* (Stark et al., 1984).

The intramembrane particles of the photosynthetic apparatus showed a hexagonal lattice in several BChl *b*-containing (Giesbrecht and Drews, 1966; Miller, 1979; Wehrli and Kübler, 1980; Engelhardt et al.,

1983) and also some BChl *a*-containing (Oelze and Golecki, 1975, Meyer et al., 1981; Merckenstock et al., 1992) purple bacteria. The ordered arrangement was shown by freeze-fracturing (Fig. 12), negative staining, and shadow casting. From the concomitant disappearance of 10 nm particles and RC it was concluded that the 10 nm particles of*Rs. rubrum* are formed by the reaction center and LH I complexes.

Generally the particle density (= particle number per defined membrane area) on fracture faces of photosynthetic ICMs is higher than the particle density on the respective fracture faces of the CM (Lommen and Takemoto, 1978 a; Golecki et al., 1979; Golecki and Oelze, 1980; Varga and Staehelin, 1983). In *Rb. capsulatus*, a decrease in light intensity leads to an increase in the number of particles in the ICM (Golecki et al., 1980), whereas in *Rp. palustris*, the particle number decreases with reduction of light intensity while particle size increases (Varga and Staehelin, 1983).

The analysis of freeze-fractured or negative stained preparations of ICMs with image processing methods like Fourier filtration techniques increased the resolution and gave new information about the supramolecular organization of the RC-LH I core complexes. This was demonstrated with *Rp. viridis* (Miller, 1979, 1982; Wehrli and Kübler, 1980; Stark et al., 1984) and some other BChl *b*-containing species (Engelhardt et al., 1983) and the BChl *a*-containing *Rhodopseudomonas marina* (Meckenstock et al., 1992).

A combination of image analysis and immunoelectron microscopy with monoclonal antibodies against the H-polypeptide of the reaction center (Stark et al., 1986) leads to the following model of the photosynthetic unit of *Rp. viridis*: the reaction center is surrounded by 12 LH complexes. The H-polypeptide is located in the central core with an asymmetrical lateral distribution on the CM surface. The LH units are arranged in a ring of 24 transmembrane α-helices. It is speculated that each of these α-helices belong to a B1015-α or B1015-β polypeptide and their N-termini are located on the plasmic surface.

C. Identification and Organization of Membrane Particles

Immunoelectron microscopical methods using specific antibodies against defined components of the photosynthetic apparatus were applied to get information on the topography of defined components of the photosynthetic apparatus in the ICM (Valkirs and Feher, 1982; Collins and Remsen, 1984; Crook et al., 1986). The ATPase was demonstrated on the surface of the CM of *Rb. sphaeroides* (Reed and Raveed, 1972) and*Rs. rubrum* (Löw and Afzelius, 1964) as knob-like particles of 9 nm in diameter. Labeling with ferritin-conjugated antibodies against H , L and M proteins of the RC demonstrated the asymmetrical distribution of the reaction center across the *Rb. sphaeroides* membrane(Reed et al., 1975; Valkirs and Feher, 1982). The H and M subunits were labeled at both the cytoplasmic and periplasmic surface of the ICM, whereas the L subunit was labeled only at the periplasmic surface of the membrane. These data in combination with iodination experiments indicated very early that RC complexes span the membrane. Using gold-labeled antibodies against the LH complex of *Rb. sphaeroides,* the presence of LH antigens on the PF of the chromatophore membrane was demonstrated (Collins and Remsen, 1984).

The asymmetrical distribution of RC and light-harvesting proteins on cytoplasmic and periplasmic surfaces has been demonstrated for *Rp. viridis* (Jay et al., 1983). A cytoplasmic location of the reaction center cytochrome, H, M and L-chain and of the light-harvesting polypeptides α, β and gamma was observed on isolated ICM vesicles by the use of ferritin-conjugated antibodies in combination with surface-specific iodination and protease treatment. In addition, an exoplasmic location of polypeptide H was noted.

Subchromatophore particles with photosynthetic activity were detected in Triton X-100 treated preparations of *Rhodopseudomonas* sp. (Garcia et al., 1968). The particle had an outer diameter of 13.5 nm and showed an inner core which was 6.0 nm in diameter. A pure fraction of native photoreceptor units isolated from ICMs of *Rp. viridis* could be shown in negative stained preparations (Jay et al., 1984, Stark et al., 1984). The isolated photoreceptor unit presented the same symmetry and dimensions as the particles in the intact ICM.

Defined membrane particles, for example RC + LH I core complexes, have been incorporated into liposomes to study size and structure (Fig. 14). Pigment-protein complexes from *Rp. palustris* were inserted into phospholipid liposomes (Varga and Staehelin, 1985). In freeze-fracture preparations of the liposomes the different particles could be classified in distinct size classes which could be related to specific components of the photosynthetic apparatus. The following correlations were observed: 5 nm particles represent free RC or LH I tetramers; 7.5 nm particles LH I or LH II octamers (or both); 10 nm particles RC-LH I core complexes (1 RC plus 12 LH I)(Fig. 14) or large LH II oligomers (or both), and large particles of 12.5 and 15 nm LH II associated with the RC-LH I core complex (Varga and Staehelin, 1985). The purified reaction centers of*Rs. rubrum* presented after incorporation into liposome membranes in freeze-fracture preparations, slightly smaller (9–11 nm) particles than in native chromatophores (Meyer et al., 1981). The functional interaction and the lateral topography of photosynthetically active components were examined by combined electron microscopic and functional test methods with reconstituted complexes of the photosynthetic apparatus (Baciou et al., 1991; Takemoto et al., 1985; Crielaard et al., 1989). Reaction centers from *Rp. viridis* reconstituted into liposomes were frozen for freeze-fracturing at temperatures between, above, and below the phase transition temperatures of the lipids forming the liposomes (Baciou et al., 1991). It was shown that the phase transition affects the thermodynamic parameters associated with the electron-transfer process in the reaction center.

Isolated chromatophores of *Rb. capsulatus* and *Rb. sphaeroides* were fused with liposomes. The fused chromatophores were examined electron microscopically after freeze-fracturing and the kinetics of excitonic energy transfer were studied. It was shown that lipid dilution lowered efficiencies of energy transfer between the LH I complex and the reaction center in *Rb. capsulatus* (Takemoto et al., 1985), and in a decrease of cytochrome *b*-561 reduction (Snozzi and Crofts, 1984) and a slightly reduced absorption due to the LH I complex (Theiler and Niederman, 1991) in *Rb. sphaeroides*.

D. The Influence of Proteins on the Structure of Intracytoplasmic Membranes

The different morphological types of ICM (Fig. 1) were investigated with respect to a possible morphogenetic effect of defined components of the photosynthetic apparatus on the ICM formation. In the carotenoid- and LH II-less mutant strain R 26 of *Rb. sphaeroides* (Lommen and Takemoto, 1978 b) both vesicular and large lamellar ICM has been observed instead of small regular vesicles normally seen in the wild type. According to experiments with several mutants of *Rb. sphaeroides* having defects in the LH II-complex it was supposed that incorporation of the α and β polypeptides of the LH II-complex is essential for molding and maturation of the ICM (Hunter et al., 1988; Kiley et al., 1988). In these mutants the ICM consists mainly of long tubes spanning the cell in the longitudinal axis. It was suggested that in absence of the LH II-complex the ICM was arrested at a tubular stage.

However, the LH II-deficient mutant 19 of *Rb. sphaeroides* exhibited in ultrathin sections not only regularly shaped ICM vesicles (average diameter of 43 nm), but also long tubes of ICM with diameters of 60–95 nm (Golecki et al., 1989, 1991). Freeze-fracture preparations revealed that these tubes were densely covered with intramembrane particles regularly arranged in rows. Therefore these tubes do not represent lipid-enriched ICM-invaginations as discussed for other mutants of *Rb. sphaeroides* with

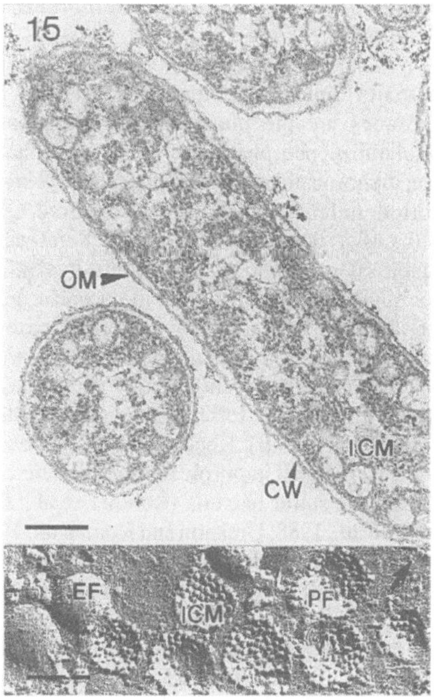

Fig. 15. Ultrathin section of *Rhodobacter capsulatus* mutant A1 a⁺, lacking the LH II complex. The cells contain ICM vesicles of larger diameter (mean diameter = 72 ± 9 nm) than wild type. Bar represents 200 nm. Inset: ICM vesicles of *Rb. capsulatus* A1 a⁺ exposing intramembrane particles on the PF. OM, outer membrane. Freeze-fracture preparation. Bar represents 100 nm.

similar tubular ICM (Kiley et al., 1988). Phototrophic and RC positive, but LH I and/or LH II negative mutants of *Rb. capsulatus* (Dörge et al., 1990; Stiehle et al., 1990) still retain vesicular ICM (Golecki et al., 1989, 1991; Golecki unpublished results; Fig. 15). *Rs. rubrum*, which does not contain the LH II complex, forms regularly shaped ICM vesicles with diameters ranging from 60 to 100 nm. It is remarkable that all ICM vesicles from LH II-lacking purple bacteria are larger than vesicles from bacteria with LH II besides the core complex (Figs. 8, 15). A defined LH II deletion mutant (Δ*pucBACDE*) of *Rb. capsulatus* contains vesicles of irregular size but no tubules (Golecki and Drews, unpublished).

III. Biosynthesis of Membranes and the Assembly of the Functional Complexes of the Photosynthetic Apparatus

A. Coordinated Biosynthesis of Phospholipids, Proteins and Pigments during Membrane Formation

The major lipid compounds of CM and ICM membranes are the phospholipids phosphatidyl ethanolamine, phosphatidyl choline, phosphatidyl serine, diphosphatidyl glycerol, phosphatidyl inositol and ornithine lipids, which are esters of C16:0, C16:1, C18:0, C18:1; fatty acids (Kenyon, 1978 and see also Chapter 10 on quinones and lipids by Imhoff). Glycolipids appear to be absent in nonsulfur purple bacteria; they were found in Chromatiaceae and green sulfur bacteria. Sulfolipids (sulfoquinovosyl-diacylglycerol) are present in many photosynthetic bacteria (Imhoff et al., 1982; and see also Chapter 10 by Imhoff and Imhoff). Hopanoid pentacyclic triterpenoids were found in purple nonsulfur bacteria but not in purple sulfur bacteria (Rohmer et al., 1984; Neunlist et al., 1988; Urakami and Komagata, 1988). Ubiquinone-10 is the major mobile electron carrier in the membrane. Ubiquinone-9, menaquinone-8, 9 or 10, and rhodoquinone-10 have been found occasionally (Urakami and Komagata, 1988; and see also Chapter 10 by Imhoff and Imhoff).

The phospholipid biosynthetic enzymes, i.e. CDP-diglyceride synthetase, phosphatidylglycerophosphate synthetase, phosphatidylglycerophosphate phosphatase, phosphatidylserine (PS) synthase, PS decarboxylase, and S-adenosyl-L-methionine: phosphatidylethanolamine N-methyltransferase, were detected in the soluble fraction of cells of *Rb. sphaeroides* while CDP-diglyceride synthetase, phosphatidylglycerophosphate synthase, and PS decarboxylase was localized in the CM (Cain et al., 1984). The pathway for synthesis of phosphatidyl-ethanolamine and phosphatidylglycerol in purple bacteria seems to be the same as in other eubacteria (Cain et al., 1984). A phospholipid transfer protein and an acyl carrier protein have been isolated and purified (Tai and Kaplan, 1984; Cooper et al., 1987).

Kaplan's group has shown that phospholipids are synthesized at the CM and from there transferred to the ICM just prior to cell division (Lueking et al., 1978; Cain et al., 1984). Proteins and pigments, however, are inserted into the CM and ICM continuously throughout the cell cycle (Fraley et al., 1977, 1979; Kaplan et al., 1983; Myers and Collins, 1987). As a consequence, the protein/phospholipid ratio and the specific density and fluidity of the membrane fluctuate with the cell cycle. The discontinuous accumulation of cellular phospholipids during synchronous growth is directly attributable to observed changes in the rate of phospholipid synthesis (Fraley et al., 1979). The number of photosynthetic units per membrane area can vary by nearly a factor of two over the course of the cell cycle (Kaplan et al., 1983). Under constant growth conditions the average size and density of intramembrane particles remains unchanged during the cell cycle (Yen et al., 1984). The cell-cycle-dependent fluctuation in membrane composition, shown for *Rb. sphaeroides*, has been likewise confirmed for CM preparations of *Rs. rubrum* (Myers and Collins, 1987).

Shifts of light intensity or oxygen tension do not change significantly the patterns of phospholipids and fatty acids in the membrane fractions of non-synchronized cells (Kaufmann et al., 1982; Onishi and Niederman, 1982; Campbell and Lueking, 1983).

B. Characteristics of Membrane Fractions

Several membrane fractions have been isolated by sucrose density gradient centrifugation from various purple bacteria induced to form the photosynthetic apparatus. These membrane fractions differ in specific density and composition. Components of the respiratory chain, photosynthetic apparatus, transport systems, penicillin-binding proteins, and other functional units and phospholipids are present in variable but different amounts in the different membrane fractions (Hurlbert et al., 1974; Barrett et

Table 1. Composition of membrane fractions isolated from chemotrophically and phototrophically grown cells of *Rhodobacter capsulatus*

Membrane fraction	Growth conditions	BChl	RC nmol·mg protein	PSU mol BChl·mol RC^{-1}	NADH DH	Cyt *c* oxidase	Photophos-phorylation nmol ATP·mg prot·min^{-1}	Fluorescense emission peaks[c]	ICM vesicles per cell	Cyt c_2[d]	UQ[d]	*Cyt* $b_{561} + b_{566}$[d]
					nmol substrate·mg prot·min^{-1}					moles per mol RC		
light[a]	chemotrophic	0.048	0.005	9	11.4	54	0.00005	907 nm				
medium	chemotrophic	0.25	0.034	7	246	500						
heavy	chemotrophic	0.37	0.030	12	n.d.	n.d.	39.3					
light[a]	phototrophic	0.013	0.001	13	5.2	5	0.05	880, 910 nm				
heavy	phototrophic	15.5	0.21	74	39.6	320	500	915 nm				
total[b] membrane fraction	high light phototrophic	4.7	0.118	40	70		220		32	2.5	64	2.28 + 1.28
	low light phototrophic turbidostat	25.9	0.205	126	46		260		624	0.69	19	0.62 + 0.33

[a] data from Kaufmann et al., 1982;
[b] data from Reidl et al., 1983, 1985;
[c] data from Garcia et al., 1981
[d] data from Garcia et al., 1987
Abbreviations: Cyt – cytochrome; DH – dehydrogenase; RC – reaction center; UQ – ubiquinone

al., 1978; Dierstein et al., 1981; Garcia at al., 1981; Inamine and Niederman, 1982; Kaufmann et al., 1982; Inamine et al., 1984). Penicillin-binding proteins have been found exclusively in the CM (Shepherd et al., 1981). Table 1 illustrates the compositional heterogeneity of membrane fractions isolated from cells of *Rb. capsulatus* grown chemotrophically and phototrophically, respectively. A generalization is not possible, because data are available from few species only.

The isolated 'light' membrane fraction consists of small membrane pieces (Lampe et al., 1972; Kaufmann et al., 1982). The concentration of bacteriochlorophyll-protein complexes is low and the ratio of reaction center plus light-harvesting complex I (LHI) : LHII is much higher than in the 'heavy' fraction (Reilly and Niederman, 1986). The transfer of excitation energy from LHII to LHI is not coupled efficiently. Rates of photophosphorylation and photoreduction of *b*-type cytochromes are low (Hunter et al., 1979; Garcia et al., 1981; Kaufmann et al., 1982). Barrett et al. (1978) found cytochrome *c* and NADH-cytochrome *c* reductase activity enriched in the light fraction. The light fraction seems to be identical with the CM or those parts of the CM where the insertion of photosynthetic units is initiated (Reilly and Niederman, 1986).

The low activities of oxidative phosphorylation, NADH oxidation, cytochrome oxidase and other enzymes of electron transport chains in the light fraction seems to be caused by harsh treatment of cells during disruption and isolation. The separation of CM from ICM and outer membrane is a difficult task, especially when phototrophically grown cells are used (Oelze and Drews, 1972; Hurlbert et al., 1974; Kaplan and Arntzen, 1982). Small pieces of CM, formed during cell disruption, do not form vesicles. Washing of membranes reduces the level of soluble cytochrome c_2 and of enzyme constituents loosely bound to the membrane. Even when CM is isolated under mild and controlled conditions it remains difficult to quantify light-driven and respiratory coupled functions because CM vesicles are at best 70% right-side out, while ICM vesicles can be 100% inside out with a higher content of cytochrome c_2 than CM vesicles (Lommen and Takemoto, 1978; Michels and Konings, 1978; Garcia et al., 1987). Moreover, apparent activity of cytochrome oxidase and amount of oxidase protein, detected by immunoelectrophoresis, do not follow the same kinetics when cells are shifted from phototrophic to chemotrophic growth conditions (Hüdig and Drews, 1985; Hüdig et al., 1987). Enzymatic activity and synthesis of membrane bound components seem to be regulated differently.

The light membrane fractions of *Chromatium vinosum* and *Ectothiorhodospira mobilis* contained the highest BChl content and were not contaminated with cell wall constituents (Hurlbert et al., 1974; Ditandi and Imhoff, 1993). Obviously the photosynthetic apparatus of both species is extended to the CM, which may be caused by the obligately or primarily phototrophic way of life in these species. The fractions of highest density are enriched in cell wall constituents.

The heavy membrane fraction of facultative phototrophic bacteria consists of ICM vesicles (i.e. chromatophores) which are enriched in the photosynthetic apparatus (Fig. 6). Chromatophores catalyze light-driven proton translocation into the vesicles and photophosphorylation by the H^+-ATPase (Frenkel, 1954; Lampe and Drews, 1972; Lampe et al., 1972).

The morphological continuity of the CM-ICM system has been described in Section II.A of this chapter (Figs. 3, 5). The continuity of the membranes implies that ICM vesicles have an inverted orientation after isolation; they have an inside-out orientation, while protoplasts and CM vesicles should have right-side out orientation. This can be shown by localization of the F_1 part of ATPase on the outer surface of ICM vesicles and on the inner surface of CM vesicles. Light-induced proton translocation is directed into the ICM vesicles but from the cytoplasm to the periplasmic space in whole cells and spheroplasts. Likewise, light-driven amino acid transport is directed oppositely in chromatophores and CM vesicles, (Michels and Konings, 1978). Opposite membrane asymmetry of the CM and ICM was also detected by freeze-fracture electron microscopy (see Section II.B).

Since membranes are fluid-crystalline structures it cannot be excluded that the connections between ICM vesicles or lamellae and CM and ICM are transient. From the heterogeneity of membrane fractions it can be inferred that a barrier must exist between membranous domains which inhibits or slows down free lateral diffusion of phospholipids and proteins in the membrane continuum (Lavergne and Joliot, 1991).

C. The Formation of Intracytoplasmic Membranes

Rhodospirillum rubrum and *Rhodobacter sphaeroides* are facultative photosynthetic bacteria which do not contain either BChl or ICM vesicles when cultivated under strictly aerobic conditions in the dark. Upon induction of the photosynthetic apparatus by lowering the oxygen tension or shifting to anaerobic light conditions, invaginations of the CM and BChl synthesis appear simultaneously (Cohen-Bazire and Kunisawa, 1963; Drews and Giesbrecht, 1963; Peters and Cellarius, 1972; Golecki and Oelze, 1975). The formation of ICM by invaginations of the CM was also visualized by immunochemical means (Crook et al., 1986). A shift of newly incorporated radioactivity from CM into the ICM fraction was demonstrated by pulse-chase experiments (Oelze and Drews, 1969).

It was suggested that an 'upper pigmented band' (light membrane fraction) represents peripheral regions of immature ICM invaginations (Niederman et al., 1979; Inamine and Niederman, 1982; Inamine et al., 1984). This fraction was found to be enriched in *b*- and *c*-type cytochromes, in succinate-cytochrome *c* oxidoreductase and in reaction center (Bowyer et al., 1985; Reilly and Niederman, 1986). The incorporation of newly synthesized proteins into this membrane fraction was demonstrated in pulse-chase experiments (Niederman et al., 1979, 1981).

The H-subunit of RC, present in the membrane fraction of aerobically grown cells of *Rb. sphaeroides* (Chory et al., 1984) and *Rb. capsulatus* (N. Cortez, unpublished), was proposed to function as a structure to initiate assembly of the core complex (RC + LHI). A rapid shift from aerobic, chemotrophic growth to anaerobiosis in the presence of light in *Rb. sphaeroides* results in an immediate cessation of cell growth, and of protein, DNA and phospholipid accumulation. But the synthesis of phosphatidylcholine begins midway through the lag period and well before the resumption of total phospholipid synthesis. Small indentations within the CM appear early in the lag phase and seem to be converted into discrete ICM invaginations and mature ICM vesicles. It was speculated that specific regulatory sites of phospholipid synthesis at the level of phosphatidylethanolamine methyltransferase and acyl carrier protein, besides H-subunit and other proteins, are locally enriched where indentations in the membrane are formed as initiation sites for assembly of pigment-proteins and invaginations of the CM or ICM (Chory et al., 1984). Throughout the cell cycle the level of LH complex II remains low in the light membrane fraction. It was shown by pulse-labeling with [^{35}S] methionine that the incorporation of proteins into the light membrane fraction was much higher than in the heavy membrane fraction. Pulse-chase experiments showed a shift of radioactivity from the light into the heavy membrane fraction. These results were interpreted as support of the idea that core complexes of the photosynthetic apparatus were incorporated preferentially into sites of membrane growth initiation which were visualized as small indentations of the CM (Chory et al., 1984; Reilly and Niederman, 1986). Later, when the indentation became CM invaginations and ICM vesicles, light-harvesting complexes II were incorporated and due to a higher density, these vesicles were isolated with the heavy membrane fraction.

The molecular mechanism of this local membrane growth and differentiation resembles the budding process at the plasma membrane of eukaryotic cells during assembly of enveloped viruses. In this process host membrane proteins are displaced by virus proteins which serve as anchor proteins for virus assembly.

The insertion of pigment- or redox carrier-binding proteins of the photosynthetic apparatus into the membrane lowers the fluidity of the membrane. This might be a signal to transfer phospholipids to ICM. New ICM vesicles are formed not only by invagination of CM but also by incorporation of new photosynthetic units into buds of preexistent ICM vesicles or thylakoids (Drews and Giesbrecht, 1963; Oelze and Drews, 1969, 1972; Hurlbert al., 1974; Dierstein et al., 1981) or flat membranes (thylakoids) which appears as branching of membranes in *Rp. palustris* (Tauschel and Drews, 1967).

D. The Influence of Oxygen Tension and Light Intensity on Membrane Differentiation

The external signals of oxygen partial pressure and light intensity are the major factors which determine membrane differentiation and formation of the photosynthetic apparatus (Lascelles, 1959; Cohen-Bazire and Kunisawa, 1963; Biedermann et al., 1967; Marrs and Gest, 1973; Dierstein and Drews, 1975; Drews 1978, 1988, 1991; Schumacher and Drews, 1979; Arnheim and Oelze, 1983a,b; Donohue and

Kaplan, 1986; Kiley and Kaplan, 1987; Oelze, 1988; Reidl et al., 1983, 1985; Chory et al., 1984; Klug et al., 1985, 1991; Lee and Kaplan, 1992 a,b). In addition, temperature and substrate composition and concentration influence the morphogenesis of the photosynthetic apparatus (Kaiser and Oelze, 1980a,b; Grether-Beck and Oelze, 1987; Gardiner et al., 1992).

Lowering the oxygen tension in a dark culture of facultative phototrophic bacteria induces the formation of the photosynthetic apparatus. The threshold value of oxygen partial pressure which is critical for triggering the morphogenetic process is different in the species which have been examined. Cells of*Rs. rubrum* and *Rb. sphaeroides* are free of ICM and BChl when cultivated in air saturated culture medium. The threshold value for these species is relatively sharp at about 5 mm Hg = 666 Pa = 6.66 mbar oxygen partial pressure (Biedermann et al., 1967).

Cultures of *Rb. capsulatus* always contain small amounts of BChl and some polar tubular ICM, even when grown at 500 kPa pO_2 (Lampe et al., 1972; Fig. 7). The BChl concentration increases strongly when in chemotrophically grown cells of *Rb. capsulatus* the oxygen tension is lowered below 1 kPa (Dierstein and Drews, 1974, 1975). In other species, like *Rb. sulfidophilus* (Doi et al., 1991) or *Rs. centenum* (Yildiz et al., 1991) changes in oxygen tension are much less effective on the formation of the photosynthetic apparatus.

Lowering of the oxygen tension below the threshold value stimulates, in species responsive to oxygen, not only synthesis of the photosynthetic apparatus but also formation of ICM. Both processes are light-independent and have been observed even in anaerobic dark cultures of *Rb. capsulatus* (Madigan et al., 1982). Activities of respiratory enzymes are higher in cells grown chemotrophically than phototrophically (King and Drews, 1975; Drews and Oelze, 1981; Cox et al., 1983; Hüdig et al., 1987). Thus, variation of oxygen tension has an opposite effect on synthesis of components of the photosynthetic versus respiratory apparatus. Oxygen stimulates synthesis and/or activity of NADH-dehydrogenase and cytochrome *c* oxidases, but the redox components of electron transport, which are active for respiratory and light-driven electron transport, like the bc_1-complex, cytochrome c_2 and the ubiquinone pool, are regulated differently (Kaufmann et al., 1982; Garcia et al., 1987).

BChl synthesis is regulated on the level of key enzymes, i.e. the aminolevulinic acid synthase and Mg-chelatase, the branching point of tetrapyrrole synthesis to heme and BChl synthesis, respectively (Lascelles, 1959, 1978; Gorchein, 1973). Transcriptional and posttranscriptional regulatory processes are involved (see Chapters 9 by Beale and 52 by Biel).

When cells of facultative phototrophic bacteria are shifted from anaerobic light to aerobic dark conditions, the synthesis of pigments, pigment-binding proteins of the photosynthetic apparatus and formation of ICM are immediately suppressed (Marrs and Gest, 1973; Drews et al., 1987). The components of the photosynthetic apparatus are not degraded but remain in the ICM. Instead of these components, structures of the respiratory apparatus and other housekeeping components are incorporated into the CM. Hence the relative concentration of photosynthetic structures in the membrane gradually dilutes out. Since phototrophically grown cells of*Rs. rubrum* and *Rb. capsulatus* and other facultative phototrophic bacteria have under all growth conditions a potentially active respiratory system, they continue to grow after shift from phototrophic to chemotrophic conditions. Growth and membrane differentiation are separately regulated (Marrs and Gest, 1973; Baccarini-Melandri and Zannoni, 1978; Zannoni et al., 1978; Arnheim and Oelze, 1983a,b; Oelze and Arnheim, 1983; Drews et al., 1987; Kiley and Kaplan, 1988).

In all phototrophic organisms light is the energy source for growth and also an external factor which determines morphogenesis. The quantum efficiency of light on growth has been studied in chemostat cultures (Göbel, 1978) and the influence of light on growth and morphogenesis by Oelze's group (Arnheim and Oelze, 1983a; Grether-Beck and Oelze, 1987). Changes of light intensity modify the composition and activity of the photosynthetic apparatus. Lowering of light intensity under anaerobiosis increases the BChl content of cells and membranes and the ratio of the pigment-binding proteins of LH II/RC + LH I (Cohen-Bazire et al., 1957; Cohen-Bazire and Kunisawa, 1963; Drews and Giesbrecht, 1963; Lien and Gest, 1973; Schumacher and Drews, 1979; Golecki et al., 1980; Reidl et al., 1985; Garcia et al., 1987; Oelze, 1988 and Fig. 2). Cells exposed to a downshift in light intensity adapt to a new energy level of light irradiation. In cells of *Rb. capsulatus* and *Rb. sphaeroides* this process takes about 2–3 generations.

During adaptation, the growth rate is lowered due to energy limitation. The amount of ICM vesicles and the number of photosynthetic units per cell increase (Fig. 8, Table 1). The size of the photosynthetic unit (mol BChl/mol RC) increases due to the relative increase of the peripheral LH complex II (Reidl et al., 1985). In species like*Rs. rubrum* and *Rp. viridis*, which have one light-harvesting complex only, the number of photosynthetic units and the ICM increase (Oelze et al., 1969). In species like *Rp. palustris, Rp. acidophila* and *Cm. vinosum*, which contain different types of peripheral LH complexes, the development of single species of LH complexes is differently regulated by light intensity, oxygen tension and temperature (Firsow and Drews, 1977; Mechler and Oelze, 1978 a,b,c; Hayashi et al., 1982; Cogdell et al., 1983; Evans et al., 1990; Deinum et al., 1991; Gardiner et al., 1992; Takaichi et al., 1992). The primary acceptor for light energy transduction is the pigment system of the photosynthetic apparatus. The sensor for light effecting morphogenesis is unknown. Blue light seems to inhibit BChl synthesis (Drews and Jaeger, 1963; Oelze, 1986; Takamiya et al., 1992).

Variation of light intensity influences not only the synthesis of pigment-proteins of the photosynthetic apparatus, but also the concentration of redox carriers. The molar ratios of ubiquinone and cytochromes c_2, c_1, *b*-561 and *b*-566 per reaction center were three to five fold higher in membranes from high-light than from low-light grown cells (Table 1; Garcia et al., 1987) and the photochemical and respiratory activities change accordingly (Table 1; Reidl et al., 1983, 1985; Garcia et al., 1987; Hüdig et al., 1987).

Membrane differentiation is regulated on different levels, transcriptionally and posttranscriptionally, which has been reviewed by Kiley and Kaplan (1988) and Sganga and Bauer (1992) and detailed in several chapters of this volume.

E. Assembly of the Photosynthetic Apparatus

The complexes of the photosynthetic and of the respiratory apparatuses are multicomponent, supramolecular integral membrane-bound structures. From several complexes we have detailed information on the primary structure of proteins, localization and orientation of the bound cofactors, and organization and topography of the complex in the membrane (see Chapters 18, 23, 37 and 38, by Cogdell and Zuber, Roy et al., Gromet-Elhanon, and Jackson, respectively). The organization of proteins and cofactors has been optimized during evolution to an efficient communication between the cofactors within the complex and redox carriers in the lipid layer of the membrane (cofactors e.g. quinones). Furthermore, the complexes have to interact with ions, substrates, or redox carriers on the cytoplasmic and periplasmic side of the membrane, respectively. Excitation energy or electrons have to be efficiently translocated. All these functions and the formation of a membrane potential and an electrochemical gradient of protons across the membrane are dependent on the correct assembly of subunits resulting in a functional organization and a correct orientation of the complexes and cofactors in the membrane. All immunochemical, ultrastructural, and biochemical studies have confirmed that complexes are uniformly oriented when synthesized in native membranes, but they may be more or less randomly oriented when incorporated in liposomes (Tadros et al., 1987; Crielaard et al., 1989; Güner et al., 1991; Brunisholz and Zuber, 1992).

F. Targeting and Insertion of Proteins into the Membrane

Targeting of proteins to the membrane surface is the first important step of the assembly process which is initiated by distinct signals, the signal sequences on the so-called targeting protein (Saier et al., 1989; Müller and MacFarlane, 1994). Signal sequences of proteins which are translocated across the membrane are cleaved off after translocation. Proteins which are inserted into the CM generally have non-cleavable signal sequences. These non-cleavable signal sequences are difficult to identify because they are not split off and the sequence is not well defined (Saier et al., 1989). The N-terminal region of the LH I polypeptide α of *Rb. capsulatus* has some features in common with typical signal sequences, i.e. two positively charged amino acids at the end of the N-terminus followed by a short sequence of hydrophobic amino acids and two other positively charged amino acids immediately before the central hydrophobic membrane spanning domain of the polypeptide. Exchange of four positively charged with four negatively charged amino acids on the N-terminal domain of the α subunit inhibits a stable insertion of α and β proteins into the membrane and the assembly of the LHI complex (Dörge et al., 1990; Stiehle et al., 1990). Interestingly, the β polypeptide of the LHI complex, which has a net negative charge in its N-terminal region and no similarity to a signal sequence, is inserted earlier into the membrane than the LHI α polypeptide. Moreover, the β polypeptide inserts in the absence of α but does not remain stably integrated as it does in presence of α (Richter and Drews, 1991). LHI α in absence of β is also integrated into the membrane (A. Meryandini, personal communication). Site-directed mutagenesis of other highly conserved amino acids in the N-terminal region of the α and β LHI polypeptides interfered with stable formation of a light-harvesting complex I, indicating that the N-termini of the LHI polypeptides contains important topogenic information (Dörge et al., 1990; Richter et al., 1991, 1992).

Intensive studies on insertion and translocation of proteins into and across the CM of *Escherichia coli* have shown that the process is generally *sec*-dependent (Müller and MacFarlane, 1994). Sec A, Sec B and Sec Y are proteins which support targeting (Sec B+A) or translocation (Sec Y) of proteins across the CM of *E. coli* (Müller and MacFarlane, 1994). Sec A- and Sec B-activities for protein export have not been found in cells of *Rb. capsulatus*, but proteins similar in function are present (Müller et al., unpublished). In vitro experiments using a cell-free homologous system of *Rb. capsulatus* have shown that the incorporation of LHI α and β and of the H-protein of the RC into the ICM fraction is more efficient in the presence than in absence or posttranslationally addition of membranes (Troschel and Müller, 1990; Troschel et al., 1992). A low salt extract from ICM-vesicles contains one or more proteins which support targeting and translocation of precytochrome c_2 and LHI $\alpha\beta$ (Wieseler and Müller, 1993; A. Meryandini, unpublished). These experiments suggest but do not prove that targeting and insertion of LH and RC proteins are cotranslational processes. They indicate that the competence of these proteins for insertion is diminished when membranes are added posttranslationally (Troschel and Müller, 1990). LHI α and β polypeptides seem to interact before insertion. In absence of DnaK, LHI α and β were not synthesized in a cell-free translation system. When GroEL was removed from the system, the LHI proteins were not stably inserted into the membrane (A. Meryandini and G. Drews unpublished).

LH-proteins of purple bacteria are very small, (about 50 residues) and they span the membrane only once (Drews, 1985). The translocation of the M13 coat protein and some other small proteins is Sec A or Sec Y independent (Kuhn and Troschel, 1992). The translocation of leader peptidase became Sec A independent when the protein was shortened (Andersson and von Heijne, 1993). This, and the failure to detect Sec functions in *Rb. capsulatus*, suggest that all topogenic information necessary for targeting and insertion already exist in the primary structure of LH proteins. However, several observations support the idea that protein targeting and membrane insertion in *Rb. capsulatus* is assisted by a system of proteins. Trypsin-treatment of ICM interfered with integration of LHI α and β and the H-subunit of the RC into ICM (Troschel et al., 1992; Troschel and Müller, unpublished) indicating that proteins exposed on the membrane surface assist targeting and insertion. LHI α polypeptide is not stably inserted into membranes in the absence of LHI β, although its N-terminus has a positive net charge, suggesting that α and β bind to each other before or during targeting. This could be shown recently in a cell-free translation system of *Rb. capsulatus* (A. Meryandini and G. Drews, unpublished). The N-terminal region of LHI β has a negative net charge which interferes with the negative charge of the phospholipid head groups on the membrane surface during targeting. A surface exposed membrane protein could overcome this problem and act as a receptor for targeting and insertion. Gene products which are not identical with the pigment-binding proteins have been identified which are essential for correct and stable assembly (Tichy et al., 1989, 1991; Drews, 1992). Others which may have a more general function for insertion and assembly of proteins in the membranes of phototrophic bacteria are yet to be detected. LHI α becomes phosphorylated during or after incorporation into the membrane of *Rb. capsulatus* (Cortez et al., 1992; Garcia et al., 1994).

G. Translocation and Assembly of Antenna-Pigment-Proteins

The N-terminal regions of LH α and β and RC proteins M and L of several species become exposed on the cytoplasmic side of the membrane after insertion and translocation (Tadros et al., 1987; Brunisholz and Zuber, 1992). Translocation of the hydrophobic α-helical region of these proteins across the membrane depends on the proton motive force. Addition of uncouplers completely inhibits translocation (Dierstein and Drews, 1986). The sequence of steps in the assembly of the LH-complexes is not known. The binding of BChl and carotenoids to the proteins of the LH-complexes seems to be an early step that influences the conformation of proteins and therefore assembly. In mutant strains blocked at different stages of BChl synthesis, LHI α, LHI β, and LHII β proteins are incorporated into the membrane, as shown by pulse experiments, but disappear during the chase period (Klug et al., 1986; Brand and Drews, unpublished). The *pufQ* gene product seems to be of importance for assembly of BChl and proteins (see Chapters 57, 58 and 59 by Beatty, Bauer, and Klug, respectively). LHII α polypeptide, however, remains stable in the membrane when BChl is absent. The steps of BChl synthesis, catalyzed by Mg-chelatase and methyl-transferase, seem to be membrane bound (Gorchein, 1973). The pool of unbound BChl in the membrane and whole cells is very small (Beck and Drews, 1982). Precursors of BChl can be accumulated in mutant strains. They are excreted as free precursors or nonspecifically bound to proteins (Drews et al., 1971; Bolivar and Bauer, 1992). They do not form LH complexes. However, pseudo complexes may be formed by interaction with other LH proteins in the absence of the specific ones (Tadros et al., 1989). The LH-complexes in the membrane are oligomers of the basic α-β-pigment-subunit. This has been shown by molecular mass determination of isolated complexes in lipid micelles or in crystals using different methods (Shiozawa et al., 1982; Guthrie et al., 1992; Kleinekofort et al., 1992) and by in vitro resolution and reconstitution of isolated complexes (see Chapter 16 by Zuber and Cogdell). The monomeric subunits differ from oligomers in their spectroscopic properties (van Grondelle et al., 1992).

Besides BChl, the carotenoids are a morphogenetic factor for LHII complexes. Carotenoid-less mutants of *Rb. capsulatus* are unable to form the stable, wild-type LHII complex (Kaufmann et al., 1984; Zsebo and Hearst, 1984; Dörge et al., 1987; Lang and Hunter, 1994).

Within the transmembrane α-helical segment of both α and β antenna polypeptides a strong conserved histidine residue, serving as a fifth ligand for BChl, is positioned. Other conserved amino acids in a distance of approximately one helix turn [Ala-X-X-X-His-X-X-X-Leu (Trp, Tyr, Phe)] (His + 4 and His – 4) are candidates to interact and stabilize the position of BChl in the periplasmic bilayer half of the membrane (Brunisholz and Zuber, 1992). Exchange of histidine or one of the conserved amino acids in position ± 4 for another amino acid impair or inhibit not only BChl binding but also formation of the LHI complex and stable insertion of α and β polypeptides (Bylina et al., 1988; M. Brand, G. Drews, unpublished). Interestingly, an exchange of Ala-34 and His-38 of LHI β does not impair the assembly of LHI or alter the wild type in vivo absorption spectrum. Deletion of the highly conserved His-20 in LHI β did not inhibit LHI formation but resulted in a broadening of the absorption peak in one membrane fraction (M. Brand, G. Drews, unpublished). Pulse chase experiments have shown that inhibition of BChl binding to α or β polypeptides by mutation of amino acids very often impairs insertion and translocation of the mutated and of the partner protein (Klug et al., 1986; M. Brand, G. Drews, unpublished). These multiple effects of mutations in amino acid positions are possibly caused by changes in conformation and topogenic signals of proteins and, therefore, influence targeting, translocation and assembly.

The interpretation of results of the influence of pigments on the synthesis and assembly of pigment-binding proteins is conflicting in the literature. The light-harvesting chlorophyll-binding protein of chloroplast thylakoid membranes was stably incorporated into the CM of *E. coli* after addition of a bacterial signal peptide, although chlorophyll was not present (Kohorn and Auchincloss, 1991). In eukaryotic systems, the chlorophyll apoprotein expression and accumulation was found to be influenced by the level of chlorophyll (Eichacker et al., 1992; Herrin et al., 1992).

The LHII complex of *Rb. capsulatus* contains, besides the α and β polypeptides, a third protein, called γ which does not bind BChl. It can be removed from the membrane by an alkaline wash (Tadros et al., 1990) or by trypsin treatment (Feick and Drews, 1978) and is not necessary for the function of LHII (Tadros et al., 1990). Deletion of *pucE*, the gene coding for the gamma protein, and of *pucD*, a gene of unknown function, affects the formation and stability of LHII (Tichy et al., 1991). Cells which are suppressed in transcription of *pucD*, and *E*, form less LHII than in the wild type and the absorption maximum at 800 nm decreased (Tichy et al., 1991). It has been shown that the gamma protein stabilizes the LHII complex stronly (C. Kortlüke, F. Weber, G. Drews, unpublished). The gene product of *pucC* is essential for expression of the *puc* operon and for synthesis and assembly of LHII proteins (Tichy et al., 1991; LeBlanc and Beatty, 1993). The function of PucC is unknown but it is neither a structural component of LHII nor a protein which assists targeting and assembly. Presumably it contributes to regulation of gene expression (Tichy et al., 1991); under control of the T7 promoter and a strong Shine-Dalgarno sequence, *pucC* was overexpressed in *E. coli* (C. Kortlüke, unpublished).

Phosphorylation of BChl-binding proteins and other membrane proteins have been observed in several eukaryotic and prokaryotic photosynthetic organisms (Bennet, 1977; Owens and Ohad, 1982; Holuigue et al., 1985; Holmes and Allen, 1988; Pairoba and Vallejos, 1989; Harrison et al., 1991; Cortez et al., 1992). The intensity of phosphorylation of LHIα of *Rb. capsulatus* was found to be redox-controlled which indicates that phosphorylation and dephosphorylation regulate the organization and function of the antenna system (Cortez et al., 1992). The LHIα protein is phosphorylated presumably by a Ser/Thr kinase after insertion into the membrane (Garcia et al., 1994; M. Brand, unpublished).

I. Assembly of Other Membrane-Bound Complexes

RC, like LHII of *Rb. capsulatus*, contains a third subunit, called the H-subunit, which does not bind pigments (see Chapter 23 by Lancaster et al.). The gene for the H subunit is located some 35 kb upstream from the *puf*-operon (coding for RC-M and RC-L) in the *puhA* locus (Youvan et al., 1984; Socket et al., 1989). Deletion of the *puhA* gene results in a photosynthetically incompetent mutant strain of *Rb. sphaeroides* which synthesizes only trace amounts of RC and no LHI complex, although *puf* mRNA is synthesized. It was speculated that the H subunit is necessary for either insertion of RC M and RC L polypeptides into the ICM or for stabilization of RC complexes (Sockett et al., 1989). Since the H subunit is present in membranes of chemotrophically grown cells, it was proposed that the H subunit directs M and L subunits to the proper site and facilitates assembly (Sockett et al., 1989). The effect of a deletion in *puhA* on the expression of *pufB* may be due to an inhibition of the expression of the open reading frame F1696. An insertion mutation in F1696 affects a 67% reduction in LHI in *Rb. capsulatus* (Bauer et al., 1991). The primary structure of F1696 is very similar to that of *pucC* (Bauer et al., 1991; Tichy et al., 1991) These proteins may have a similar function in activation of gene expression. Until now there is no direct evidence for proteins (chaperones), besides GroEL, which keep the pigment-binding proteins of the LH- and RC-complexes in an unfolded state after translation and before insertion, or which assist targeting, translocation, and assembly in the membrane.

The ubiquinol:cytochrome *c* oxidoreductase, an ubiquitous proton pump, comprises three catalytic subunits, a two-heme Cyt *b*, a cytochrome *c*, and a Rieske 2Fe-2S protein (see Chapter 35 by Gray and Daldal). The Rieske subunit from *Rb. sphaeroides* has been expressed in *Escherichia coli* and also in a strain of *Rb. sphaeroides* lacking the other subunits of the bc_1 complex. The Rieske subunit forms a Rieske-like iron-sulfur cluster and assembles in the CM showing the characteristic EPR signal (van Doren et al., 1993). The data are consistent with models in which the Rieske subunit is bound to the membrane and assembled via a single membrane-spanning α helix and independently of the other subunits of the bc_1 complex (van Doren et al., 1993). Otherwise the apoproteins for cytochromes *b* and c_1 are stably assembled in the ICM of *Rb. capsulatus* mutants containing no Rieske subunit. The two-protein subcomplex was functional with intact Q_i and Q_o sites (Davidson et al., 1992). These data indicate that synthesis, cofactor binding and assembly of subunits of the bc_1 complex seem to be relatively independent from each other. Specific factors for targeting and incorporation of the subunits into the membrane are not known.

Cytochrome c_2 is a soluble periplasmic redox carrier involved in both aerobic and photosynthetic electron transport in *Rhodobacter* (see Chapter 34 by Meyer and Donohue). The N-terminal sequence of 21 amino acids is absent in the mature polypeptide and contains all of the features of a typical signal peptide (Brandner et al., 1991). The signal peptidase and cleaving of the signal sequence has been demonstrated in an in vitro system in *Rb. capsulatus* (Wieseler et al., 1992). Translocation of pre-cytochrome c_2 across the ICM requires the proton-motive force and proceeds at a higher efficiency when cell-free membranes are present cotranslationally. Peripheral membrane proteins, which can be removed from the membrane by low-salt treatment, support translocation. Deletion of the signal peptide does not prevent export, heme attachment, or function of cytochrome c_2 (Brandner and Donohue, 1994). The extract of peripheral proteins does not contain activities that in vitro could replace the *E. coli* transport factors Sec A and Sec B (Wieseler and Müller, 1993). From *cycA-phoA* translational fusion experiments it was concluded that cytochrome c_2 export does not require heme attachment as was shown for Cyt *c* import into mitochondria (Brandner et al., 1991). Heme is covalently attached to two cysteines in the Cyt c_2 polypeptide via thioether linkages. His-19 and Met-100 are noncovalent axial ligands to the iron in the porphyrin ring. Heme can be attached to the Cyt *a* apoprotein which lacks as many as 94 C-terminal amino acids, including Met-100 (Brandner et al., 1991).

Acknowledgments

The work of the authors described in this article was supported by Deutsche Forschungsgemeinschaft, Land Baden-Württemberg, and Fonds der Chemischen Industrie. We thank Ms Kerstin Jakobs and Brunhilde Schmidt for typing the manuscript.

References

Andersson H and von Heijne G (1993) *Sec* dependent and *sec* independent assembly of *E. coli* inner membrane proteins: The topological rules depend on chain length. EMBO J 12: 683–691

Arnheim K and Oelze J (1983a) Differences in the control of bacteriochlorophyll formation by light and oxygen. Arch Microbiol 135: 299–304

Arnheim K and Oelze J (1983b) Control by light and oxygen of B875 and B850 pigment-protein complexes in *Rhodobacter sphaeroides*. FEBS Lett 162: 57–60

Baccarini-Melandri A and Zannoni D (1978) Photosynthetic and respiratory electron flow in the dual functional membrane of facultative photosynthetic bacteria. J Bioenerg Biomembr 10: 109–138

Bachofen R and Wiemken V (1986) Topology of the chromatophore membranes of purple bacteria. In: Staehelin LA and Arntzen CJ (eds) Photosynthesis III, pp 620–631. Springer Publ Berlin, Heidelberg

Baciou L, Gulik-Krzywicki T and Sebban P (1991) Involvement of the protein-protein interactions in the thermodynamics of the electron-transfer process in the reaction centers from *Rhodopseudomonas viridis*. Biochemistry 30: 1298–1302

Barrett J, Hunter CN and Jones OTG (1978) Properties of a cytochrome c-enriched particulate fraction isolated from the photosynthetic bacterium *Rhodopseudomonas capsulata*. Biochem J 174: 267–275

Bauer CE, Buggy JJ, Yang Z and Marrs B (1991) The superoperonal organization of genes for pigment biosynthesis and reaction center proteins. Mol Gen Genet 228: 433–447

Bayer ME (1968) Areas of adhesion between cell wall and membrane of *Escherichia coli*. J Gen Microbiol 53: 395–401

Beck J and Drews G (1982) Tetrapyrrole derivatives shown by fluorescence emission and excitation spectroscopy in cells of *Rhodopseudomonas capsulata* adapting to phototrophic conditions. Z Naturforsch 37c: 199–204

Bennet J (1977) Phosphorylation of chloroplast membrane polypeptides. Nature 269: 4344–4346

Biedermann M, Drews G, Marx R and Schröder J (1967) Der Einfluß des Sauerstoffpartialdruckes und der Antibiotica Actinomycin und Puromycin auf das Wachstum, die Synthese von Bacteriochlorophyll und die Thylakoidmorphogenese in Dunkelkulturen von *Rhodospirillum rubrum*. Arch Mikrobiol 56: 133–147

Boatman ES (1964) Observations on the fine structure of spheroplasts of *Rhodospirillum rubrum*. J Cell Biol 20: 297–311

Bollivar DW and Bauer CE (1992) Association of tetrapyrrole intermediates in the bacteriochlorophyll a biosynthesis pathway with the major outer-membrane porin protein of *Rb. capsulatus*. Biochem J 282: 471–476

Bowyer JR, Hunter CN, Ohnishi T and Niederman RA (1985) Photosynthetic membrane development in *Rhodopseudomonas sphaeroides*. J Biol Chem 260: 3295–3304

Brandner JP and Donohue TJ (1994) The *Rhodobacter sphaeroides* cytochrome c_2 signal peptide is not necessary for export and heme attachment. J Bacteriol 176: 602–609

Brandner JP, Stabb EV, Temme R and Donohue TJ (1991) Regions of *Rhodobacter sphaeroides* cytochrome c_2 required for export, heme attachment and function. J Bacteriol 173: 3958–3965

Brunisholz RA and Zuber H (1992) Structure, function and organization of antenna polypeptides and antenna complexes from the three families of *Rhodospirillaneae*. J Photochem Photobiol B: Biol 15: 113–140

Bylina EJ, Robles SJ and Youvan DC (1988) Directed mutations affecting the putative bacteriochlorophyll-binding sites in the light-harvesting I antenna of *Rhodobacter capsulatus*. Israel J Chem 128: 73–78

Cain BD, Donohue TJ, Shepherd WD and Kaplan S (1984) Localization of phospholipid biosynthetic enzyme activities in cell-free fractions derived from *Rhodopseudomonas sphaeroides* J Biol Chem 259: 942–948

Campbell TB and Lueking DR (1983) Light-mediated regulation of phospholipid synthesis in *Rhodopseudomonas sphaeroides*. J Bacteriol 155: 806–816

Chory J, Donohue TJ, Varga AR, Staehelin LA and Kaplan S (1984) Induction of the photosynthetic membranes of *Rhodopseudomonas sphaeroides*: Biochemical and morphological studies. J Bacteriol 159: 540–554

Cogdell RJ, Durrant I, Valentine J, Lindsy JG and Schmidt K (1983) The isolation and partial characterisation of the light-protein complement of *Rhodopseudomonas acidophila*. Biochim Biophys Acta 722: 427–435

Cohen-Bazire G and Kunisawa R (1963) The fine structure of *Rhodospirillum rubrum*. J Cell Biol 16: 401–419

Cohen-Bazire G, Sistrom WR and Stanier RY (1957) Kinetic studies of pigment synthesis by non-sulfur purple bacteria. J Cell Comp Physiol 49: 25–68

Collins MLP and Remson ST (1984) Immunogold detection of chromatophore antigens on the surface of *Rhodopseudomonas sphaeroides* spheroplasts. Current Microbiol 11: 269–274

Collins MLP and Remsen CC (1991) The purple phototrophic bacteria. In: Stolz JF (ed) Structure of Phototrophic Prokaryotes, pp 49–77. CRC Press, Boca Raton

Cooper CL, Boyce SG and Lueking DR (1987) Purification and characterization of *Rhodobacter sphaeroides* acyl carrier protein. Biochemistry 26: 2740–2746

Cortez N, Garcia AF, Tadros MH, Gad'on N, Schiltz E and Drews G (1992) Redox-controlled, in vivo and in vitro phosphorylation of the α subunit of the light-harvesting complex I in *Rhodobacter capsulatus*. Arch Microbiol 158: 315–319

Cox JC, Beatty T and Favinger JL (1983) Increased activity of respiratory enzymes from photosynthetically grown *Rhodopseudomonas capsulata* in response to small amounts of oxygen. Arch Microbiol 134: 324–328

Crielaard W, Hellingwerf KJ and Konings WN (1989) Reconstitution of electrochemically active pigment-protein complexes from *Rb. sphaeroides* into liposomes. Biochim Biophys Acta 973: 205–211

Crook SM, Treml SB and Collins MLP (1986) Immunochemical ultrastructural analysis of chromatophore membrane formation in *Rhodospirillum rubrum*. J Bacteriol 167: 89–95

Davidson E, Ohnishi T, Tokito M and Daldal F (1992) *Rhodobacter capsulatus* mutants lacking Rieske FeS form a stable cytochrome bc_1 subcomplex with an intact quinone reduction site. Biochemistry 31: 33351–33357

DeBoer WE (1969) On ultrastructure of *Rhodopsudomonas gelatinosa* and *Rhodospirillum tenue*. Antonie van Leeuwenhoek J Microbiol Serol 35: 241–242

Deinum G, Otte SCM, Gardiner AT, Aartsma TJ, Cogdell R and Amesz J (1991) Antenna organization of *Rhodopseudomonas acidophila*: A study of the excitation migration. Biochim Biophys Acta 1060: 125–131

Dierstein R and Drews G (1974) Nitrogen-limited continuous culture of *Rhodopseudomonas capsulata* growing photosynthetically or heterotrophically under low oxygen tensions. Arch Microbiol 99: 117–128

Dierstein R and Drews G (1975) Control of composition and activity of the photosynthetic apparatus of *Rhodopseudomonas capsulata* grown in ammonium-limited continuous culture. Arch Microbiol 106: 227–235

Dierstein R and Drews G (1986) Effect of uncoupler on assembly pathway for pigment-binding protein of bacterial photosynthetic membranes. J Bacteriol 168: 167–172

Dierstein R, Schumacher A and Drews G (1981) On insertion of pigment-associated polypeptides during membrane biogenesis in *Rhodopseudomonas capsulata*. Arch Microbiol 128: 376–383

Ditandi T and Imhoff JF (1993) Preparation and characterization of highly pure fractions of outer membrane, cytoplasmic and intracytoplasmic membranes from *Ectothiorhodospira mobilis*. J Gen Microbiol 139: 111–117

Doi M, Shioi Y, Gad'on N, Golecki JR and Drews G (1991) Spectroscopical studies on the light-harvesting pigment protein complex II from dark aerobic and light anaerobic grown cells of *Rhodobacter sulfidophilus*. Biochim Biophys Acta 1058: 235–241

Donohue TJ and Kaplan S (1986) Synthesis and assembly of bacterial photosynthetic membranes. In: Staehelin LA and Arntzen CJ (eds) Photosynthesis III, pp 632–639. Springer Publ, Berlin

Dörge B, Klug G and Drews G (1987) Formation of the B800–850 antenna pigment-protein complex in the strain GK2 of *Rhodobacter capsulatus* defective in carotenoid synthesis. Biochim Biophys Acta 892: 68–74

Dörge B, Klug G, Gad'on N, Cohen SN and Drews G (1990) Effects on the formation of antenna complex B870 of *Rb. capsulatus* by exchange of charged amino acids in the N-terminal domain of the α and β pigment-binding proteins. Biochemistry 29: 7754–7758

Drews G (1978) Structure and development of the membrane system of photosynthetic bacteria. In: Sanadi DR, Vernon LP (eds) Current Topics Bioenerg Photosynthesis, Vol 8 B, pp 161–207. Academic Press, New York

Drews G (1985) Structure and functional organization of light-harvesting complexes and photochemical reaction centers in membranes of phototrophic bacteria. Microbiol Rev 49: 59–70

Drews G (1988) Effect of oxygen partial pressure on formation of the bacterial photosynthetic apparatus. In: Acker H (ed) Oxygen Sensing in Tissues, pp 3–11. Springer Verlag, Berlin

Drews G (1991) Regulated development of the photosynthetic apparatus in anoxygenic bacteria. In: Bogorad L and Vasil IK (eds) The Photosynthetic Apparatus: Molecular Biology and Operation, pp 113–148. Academic Press, New York

Drews G (1992) Intracytoplasmic membranes in bacterial cells: Organisation, function and biosynthesis. In: Mohan S, Dow C and Cole JA (eds) Prokaryotic Structure and Function: A New Perspective. Soc Gen Microbiol Symp, Vol 47, pp 249–274. Cambridge Univ Press

Drews G and Giesbrecht P (1963) Zur Morphogenese der Bakterien Chromatophoren und zur Synthese des Bakteriochlorophylls bei *Rhodopseudomonas spheroides* und *Rhodospirillum rubrum*. Zbl Bakt Parasitenkd Infekt.Krankh. und Hygiene I Orig 190: 508–536

Drews G, and Imhoff JF (1991) Phototrophic purple bacteria. In: Shively JM and Barton LL (eds) Variations in Autotrophic Life, pp 51–97. Academic Press, London

Drews G and Jaeger K (1963) Influence of light on the biosynthesis of bacteriochlorophyll by *Rhodopseudomonas spheroides*. Nature 199: 1112–1113

Drews G and Oelze J (1981) Organization and differentiation of membranes of phototrophic bacteria. Adv Microb Physiol 22: 1–92

Drews G, Klug G, Liebetanz R and Dierstein R (1987) Regulation of gene expression and assembly of the photosynthetic pigment-protein complexes. In: Biggins J (ed) Progress in Photosynthesis Research, Vol IV, pp 691–697. Marinus Nijhoff Publ, Dordrecht

Drews G, Leutiger I and Ladwig R (1971) Production of protochlorophyll, protopheophytin and bacteriochlorophyll by the mutant A1a of *R. capsulata*. Arch Mikrobiol 76: 349–363

Dubochet J, McDowall AW, Menge B, Schmid EN and Lickfeld KG (1983) Electron microscopy of frozen-hydrated bacteria. J Bacteriol 155: 381–390

Ebersold HR, Cordier JL and Lüthy P (1981) Bacterial mesosomes: Method dependent artifacts. Arch Microbiol 130: 19–22

Eichacker L, Paulsen H and Rüdiger W (1992) Synthesis of chlorophyll a regulates translation of chlorophyll a apoprotein P 700, CP47, CP43 and D2 in barley etioplasts. Eur J Biochem 205: 17–24

Engelhardt H, Baumeister W and Saxton WO (1983) Electron microscopy of photosynthetic membranes containing

bacteriochlorophyll *b*. Arch Microbiol 135: 169–175
Evans MB, Hawthornthwaite AM and Cogdell RJ (1990) Isolation and characterization of the different B800–850 light-harvesting complexes from low- and high-light grown cells of *Rhodopseudomonas palustris*. Biochim Biophys Acta 1016: 71–76
Feick R and Drews G (1978) Protein subunits of bacteriochlorophyll B802 and B855 of the light-harvesting complex II of *Rhodopseudomonas capsulata*. Z Naturforsch 34c: 196–199
Fraley RT, Lueking DR and Kaplan S (1977) Intracytoplasmic membrane synthesis in synchronous cell populations of *Rhodopseudomonas sphaeroides*. J Biol Chem 253: 458–464
Fraley RT, Lueking DR and Kaplan S (1979) The relationship of intracytoplasmic membrane assembly to the cell division cycle in *Rhodopseudomonas sphaeroides*. J Biol Chem 254: 1980–1986
Frenkel AW (1954) Light induced phosphorylation by cell-free extracts of photosynthetic bacteria. J Amer Chem Soc 76: 5568–5569
Garcia A, Vernon LP, Ke B and Mollenhauer H (1968) Some structural and photochemical properties of *Rhodopseudomonas* species NHTC 133 subchromatophore particles obtained by treatment with Triton X-100. Biochemistry 7: 326–332
Garcia AF, Drews G and Reidl HH (1981) Comparative studies of two membrane fractions isolated from chemotrophically and phototrophically grown cells of *Rhodopseudomonas capsulata*. J Bacteriol 145: 1121–1128
Garcia AF, Venturoli G, Gad'on N, Fernández-Velasco JG, Melandri BA and Drews G (1987) The adaptation of the electron transfer chain of *Rhodopseudomonas capsulata* to different light intensities. Biochim Biophys Acta 890: 335–345
Garcia AF, Meryandini A, Brand M, Tadros MH and Drews G (1994) Phosphorylation of the α and β polypeptides of the light-harvesting complex I (B870) of *Rhodobacter capsulatus* in an in vitro translation system. FEMS Microbiol Lett 124: 87–92
Gardiner AT, MacKenzie RC, Barrett SJ, Kaiser K and Cogdell R (1992) The genes for the peripheral antenna complex apoproteins from *Rhodopseudomonas acidophila* 7050 form a multigene family. In: Murata N (ed) Research in Photosynthesis, Vol I, pp 77–80. Kluwer Academic Publishers, Dordrecht
Giesbrecht P and Drews G (1966) Über die Organisation und die makromolekulare Architektur der Thylakoide lebender Bakterien. Arch Microbiol 54: 297–330
Göbel F (1978) Quantum efficiency of growth. In: Clayton RK, Sistrom WR (eds) The Photosynthetic Bacteria, pp 907–925. Plenum Press, New York
Golecki JR and Oelze J (1975) Quantitative determination of cytoplasmic membrane invaginations in phototrophically growing *Rhodospirillum rubrum*. J Gen Microbiol 88: 253–258
Golecki JR and Oelze J (1980) Differences in the architecture of cytoplasmic and intracytoplasmic membranes of three chemotrophically and phototrophically grown species of the Rhodospirillaceae. J Bacteriol 144: 781–788
Golecki JR, Drews G and Bühler R (1972) The size and number of intramembrane particles in cells of the photosynthetic bacterium *Rhodopseudomonas capsulata* studied by freeze-fracture electron microscopy. Cytobiology 18: 381–389
Golecki JR, Schumacher A and Drews G (1980) The differentiation of the photosynthetic apparatus and the intracytoplasmic membrane in cells of *Rhodopseudomonas capsulata* upon variation of light intensity. Eur J Cell Biol 23: 1–5
Golecki JR, Tadros MH, Ventura S and Oelze J (1989) Intracytoplasmic membrane vesiculation in light-harvesting mutants of *Rhodobacter sphaeroides* and *Rhodobacter capsulatus*. FEMS Microbiol Lett 65: 315–318
Golecki JR, Ventura S and Oelze J (1991) The architecture of unusual membrane tubes in the B800–850 light-harvesting bacteriochlorophyll-deficient mutant 19 of *Rhodobacter sphaeroides*. FEMS Microbiol Lett 77: 335–340
Gorchein, A (1973) Control of magnesium-protoporphyrin chelatase activity in *Rhodopseudomonas spheroides*. Biochem J 134: 833–845
Grether-Beck S and Oelze J(1987) The development of the photosynthetic apparatus and energy transduction in malate-limited phototrophic cultures of *Rhodobacter capsulatus*. Arch Microbiol 149: 70–75
Güner S, Robertson DE, Yu L, Quin ZH, Yu CA, Knaff DB (1991) The *Rhodospirillum rubrum* cytochrome bc_1 complex: Redox properties, inhibitor sensitivity and proton pumping. Biochim Biophys Acta 1058: 269–279
Guthrie N, MacDermott G, Cogdell RJ, Freer AA, Isaacs NW, Hawthornthwaite AM, Halloren E and Lindsay JG (1992) Crystallization of the B800–820 light-harvesting complex from *Rhodopseudomonas acidophila* strain 7750. J Mol Biol 224: 527–528
Harrison MA, Tsinoremas NF and Allen JF (1991) Cyanobacterial thylakoid membrane proteins are reversibly phosphorylated under plastoquinone-reducing conditions in vitro. FEBS Lett 282: 144–148
Hayashi H, Nakano M and Morita S (1982) Comparative studies of protein properties and bacteriochlorophyll contents of bacteriochlorophyll-protein complexes from spectrally different types of *Rhodopseudomonas palustris*. J Biochem 92: 1805–1811
Herrin DL, Battey JF, Greer K and Schmidt GW (1992) Regulation of chlorophyll apoprotein expression and accumulation. J Biol Chem 167: 8260–8269
Hickman DD and Frenkel AW (1965) Observations on the structure of *Rhodospirillum rubrum*. J Cell Biol 25: 279–291
Hochman A, Friedberg I and Carmeli C (1975) The location and function of cytochrome c_2 in *Rhodopseudomonas capsulata* membranes. Eur J Biochem 58: 65–72
Holmes NG and Allen JF (1988) Protein phosphorylation in chromatophores from *Rhodospirillum rubrum*. Biochim Biophys Acta 935: 72–78
Holt SC and Marr AG (1965a) Location of chlorophyll in *Rhodospirillum rubrum*. J Bacteriol 89: 1402–1412
Holt SC and Marr AG (1965b) Effect of light intensity on the formation of intracytoplasmic membrane in *Rhodospirillum rubrum*. J Bacteriol 89: 1421–1429
Holuigue L, Lucero HA and Vallejos RH (1985) Protein phosphorylation in the photosynthetic bacterium *Rhodospirillum rubrum*. FEBS Lett 181: 103–107
Hüdig H and Drews G (1985) Kinetic studies on formation of cytochrome oxidase of *Rhodopseudomonas capsulata* after a shift from phototrophic to chemotrophic growth. J Bacteriol 162: 897–901

Hüdig H, Stark G and Drews G (1987) The regulation of cytochrome c oxidase of *Rhodobacter capsulatus* by light and oxygen. Arch Microbiol 149: 12–18

Hunter CN, Holmes NG, Jones OTG and Niederman RA (1979) Photochemical properties of a fraction enriched in newly synthesized bacteriochlorophyll a-protein complexes. Biochim Biophys Acta 548: 253–266

Hunter CN, Pennoyer JD, Sturgis JN, Farrelly D and Niederman RA (1988) Oligomerization states and associations of light-harvesting pigment-protein complexes of *Rhodobacter sphaeroides* as analyzed by lithiumdodecyl sulfate-polyacrylamide gel electrophoresis. Biochemistry 27: 3459–3467

Hurlbert RE, Golecki JR and Drews G (1974) Isolation and characterization of *Chromatium vinosum* membranes. Arch Microbiol 101: 169–186

Imhoff JF, Kushner DJ, Kushwaha SC and Kates M (1982) Polar lipids in phototrophic bacteria of the *Rhodospirillaceae* and *Chromatiaceae* families. J Bacteriol 150: 1192–1201

Inamine GS and Niederman RA (1982) Development and growth of photosynthetic membranes of *Rhodospirillum rubrum*. J Bacteriol 150: 1145–1153

Inamine GS, van Houton J and Niederman RA (1984) Intracellular localization of photosynthetic membrane growth initiation sites in *Rhodopseudomonas capsulata*. J Bacteriol 158: 425–429

Jay F, Lambillotte M, Mühlethaler K (1983) Localization of *Rhodopseudomonas viridis* reaction centre and light-harvesting proteins using ferritin antibody labelling. Eur J Cell Biol 30: 1–8

Jay F, Lambillotte M, Stark W and Mühlethaler K (1984) The preparation and characterization of the native photoreceptor units from thylakoids from *Rhodopseudomonas viridis* EMBO J 3: 773–776

Kaiser I and Oelze J (1980a) Growth and adaptation to phototrophic conditions of *Rhodospirillum rubrum* and *Rhodopseudomonas sphaeroides* at different temperatures. Arch Microbiol 126: 187–194

Kaiser I and Oelze J (1980b) Temperature dependence in *Rhodospirillum rubrum* and *Rhodobacter sphaeroides*. Arch Microbiol 126: 195–200

Kaplan S and Arntzen CJ (1982) Photosynthetic membrane structure and function. In: Govindjee (ed) Photosynthesis: Energy Conversion By Plants and Bacteria, Vol II, pp 65–157. Academic Press, New York

Kaplan S, Cain BD, Donohue TJ, Shepherd WD and Yen GSL (1983) Biosynthesis of the photosynthetic membranes of *Rhodopseudomonas sphaeroides*. J Cell Biochem 22: 15–29

Kaufmann N, Reidl HH, Golecki JR, Garcia AF and Drews G (1982) Differentiation of the membrane system in cells of *Rhodopseudomonas capsulata* after transition from chemotrophic to phototrophic growth conditions. Arch Microbiol 131: 313–322

Kaufmann N, Hüdig H and Drews G (1984) Transposon Tn5 mutagenesis of genes for the photosynthetic apparatus in *R. capsulata*. Mol Gen Genet 198: 153–158

Kenyon CN (1978) Complex lipids and fatty acids of photosynthetic bacteria. In: Clayton RK and Sistrom WR (eds) The Photosynthetic Bacteria, pp 281–313. Plenum Press, New York

Kiley PJ and Kaplan S (1988) Molecular genetics of photosynthetic membrane biosynthesis in *Rhodobacter sphaeroides*. Microbiol Rev 52: 50–69

Kiley PJ, Varga A and Kaplan S (1988) Physiological and structural analysis of light-harvesting mutants of *Rhodobacter sphaeroides*. J Bacteriol 170: 1103–1115

King MT and Drews G (1975) The respiratory electron transport system of heterotrophically-grown *Rhodopseudomonas palustris*. Arch Microbiol 102: 219–231

Kleinekofort W, Germeroth L, Van den Brock JA, Schubert D and Michel H (1992) The light-harvesting complex II from *Rhodospirillum molischianum* is an octamer. Biochim Biophys Acta 1140: 102–104

Klug G, Kaufmann N and Drews G (1985) Gene expression of pigment-binding proteins of the bacterial photosynthetic apparatus:Transcription and assembly in the membrane of *Rhodopseudomonas capsulata*. Proc Natl Acad Sci USA 82: 6485–6489

Klug G, Liebetanz R and Drews G (1986) The influence of bacteriochlorophyll biosynthesis on formation of pigment-binding proteins and assembly of pigment protein complexes in *Rhodopseudomonas capsulata*. Arch Microbiol 146: 284–291

Klug G, Gad'on N, Jock S and Narro ML (1991) Light and oxygen effects share a common regulatory DNA sequence in *Rhodobacter capsulatus*. Mol Microbiol 5: 1235–1239

Kohorn BD and Auchincloss AH (1991) Integration of a chlorophyll-binding protein into *E. coli* membranes in the absence of chlorophyll. J Biol Chem 266: 12048–12052

Kuhn A, Troschel D (1992) Distinct steps in the insertion pathway of bacteriophage coat proteins. In: Neupert W and Lill R (eds) Membrane Biogenesis and Protein Targeting, pp 33–47. Elsevier, Amsterdam

Lampe HH and Drews G (1972) Die Differenzierung des Membransystems von *Rhodopseudomonas capsulata* hinsichtlich seiner photosynthetischen und respiratorischen Funktionen. Arch Mikrobiol 84: 1–19

Lampe HH, Oelze J and Drews G (1972) Die Fraktionierung des Membransystems von *Rhodopseudomonas capsulata* und seine Morphogenese. Arch Mikrobiol 83: 78–94

Lang HP and Hunter CN (1994) The relationship between carotenoid biosynthesis and the assembly of the light-harvesting LH2 complex in *Rhodobacter sphaeroides*. Biochemistry, 298: 197–205

Lascelles J (1959) Adaptation to form bacteriochlorophyll in *Rhodobacter sphaeroides*, changes in activity of enzymes concerned pyrrole synthesis. Biochem J 72: 508–518

Lascelles J (1978) Regulation of pyrrole synthesis. In: Clayton RK and Sistrom WR (eds) The Photosynthetic Bacteria, pp 795–808. Plenum Press, New York

Lavergne J and Joliot PC (1991) Restricted diffusion in photosynthetic membranes. TIBS 16: 129–134

Le Blanc HN and Beatty T (1993) *Rhodobacter capsulatus* puc operon; promoter location, transcript sizes and effects of deletions on photosynthetic growth. J Gen Microbiol 139: 101–109

Lee JK and Kaplan S (1992a) *cis*-acting regulatory elements involved in oxygen and light control of *puc* operon transcription in *Rhodobacter sphaeroides*. J Bacteriol 174: 1146–1157

Lee JK and Kaplan S (1992b) Isolation and characterization of trans-acting mutations involved in oxygen regulation of *puc* operon transcription in *Rhodobacter sphaeroides*. J Bacteriol 174: 1158–1171

Lien S and Gest H (1973) Regulation of chlorophyll synthesis in photosynthetic bacteria. Bioenergetics 4: 423–434

Lommen MAJ and Takemoto J (1978a). Comparison, by freeze fracture electron microscopy, of chromatophores, spheroplast-derived membrane vesicles, and whole cells of *Rhodopseudomonas sphaeroides*. J Bacteriol 136: 730–741

Lommen MAJ and Takemoto J (1978b) Ultrastructure of carotenoid mutant strain R-26 of *Rhodopseudomonas sphaeroides*. Arch Microbiol 118: 305–308

Löw H and Afzelius AB (1964) Subunits of the chromatophore membrane in*Rs. rubrum*. Exp Cell Res 85: 431–434

Lueking DR, Fraley RT and Kaplan S (1978) Intracytoplasmic membrane synthesis in synchronous cell populations of *Rhodopseudomonas sphaeroides*. J Biol Chem 253: 451–457

Madigan MT, Cox JC and Gest H (1982) Photopigments in *Rhodopseudomonas capsulata* cells grown anaerobically in darkness. J Bacteriol 150: 1422–1429

Marrs B and Gest H (1973) Regulation of bacteriochlorophyll synthesis by oxygen in respiratory mutants of *Rhodopseudomonas capsulata*. J Bacteriol 114: 1052–1057

Mechler B and Oelze J (1978a) Differentiation of the photosynthetic apparatus of *Chromatium vinosum*, strain D. Arch Microbiol 187: 91–97

Mechler B and Oelze J (1978b) Differentiation of the photosynthetic apparatus of *Chromatium vinosum*, strain D. II, Structural and functional differences. Arch Microbiol 187: 99–108

Mechler B and Oelze J (1978c) Differentiation of the photosynthetic apparatus of *Chromatium vinosum*, strain D. III, Analyses of spectral alterations. Arch Microbiol 187: 109–114

Meckenstock RU, Krusche K, Brunisholz RA and Zuber H (1992) The light-harvesting core-complex and the B820 subunit from *Rhodopseudomonas marina*. FEBS Lett 311: 135–138

Meyer R, Snozzi M, and Bachofen R (1981) Freeze fracture studies of reaction centers from *Rhodospirillum rubrum* in chromatophores and liposomes. Arch Microbiol 130: 125–128

Michels PAM and Konings WN (1978) Structural and functional properties of chromatophores and membrane vesicles from *Rhodopseudomonas sphaeroides*. Biochim Biophys Acta 507: 353–368

Miller KR (1979) Structure of a bacterial photosynthetic membrane. Proc Natl Acad Sci USA 76: 6415–6419

Miller KR (1982) Three-dimensional structure of a photosynthetic membrane. Nature 300: 53–55

Mühlradt PF, Menzel J, Golecki JR and Speth V (1974) Lateral mobility and surface density of lipopolysaccharide in the outer membrane of *Salmonella typhimurium*. Eur J Biochem 43: 533–539

Müller M and MacFarlane J (1994) Membrane assembly in bacteria. In: Maddy AH and Harris JP (eds) Subcellular Biochemistry: Membrane Biogenesis, Vol 22, pp 327–359. Plenum Press, New York

Myers CR and Collins MLP (1987) Cell cycle-specific fluctuation in cytoplasmic membrane composition in aerobically grown *Rhodospirillum rubrum*. J Bacteriol 169: 5445–5451

Neunlist S, Bisseret P and Rohmer M (1988) The hopanoids of the purple non-sulfur bacteria *Rhodopseudomonas palustris* and *Rhodopseudomonas acidophila* and the absolute configuration of bacteriohopanetetrol. Eur J Biochem 171: 245–252

Niederman RA, Mallon DE and Parks LC (1979) Isolation of a fraction enriched in newly synthesized bacteriochlorophyll a protein complexes. Biochim Biophys Acta 555: 210–220

Oelze J (1978) Proteins exposed at the surface of chromatophores of *Rhodospirillum rubrum*. The orientation of isolated chromatophores. Biochim Biophys Acta 509 : 450–461

Oelze J (1986) Inhibition by light of 5-aminolevulinic acid synthase in extracts from *Rhodopseudomonas sphaeroides*. FEMS Microbiol Lett 37: 321–323

Oelze J (1988) Regulation of tetrapyrrole synthesis by light in chemostat cultures of *Rhodobacter sphaeroides*. J Bacteriol 170: 4652–4657

Oelze J and Arnheim K (1983) Control of bacteriochlorophyll formation by oxygen and light in *Rhodopseudomonas sphaeroides*. FEMS Microbiol Lett 19: 197–199

Oelze J and Drews G (1969) Die Kinetik der Thylakoidsynthese nach Markierung der Membranen mit $[2^{14}C]$ Azetat. Biochim Biophys Acta 173: 448–455

Oelze J and Drews G (1972) Membranes of photosynthetic bacteria. Biochim Biophys Acta 265: 209–239

Oelze J and Golecki JR (1975) Properties of reaction center depleted membranes of *Rhodospirillum rubrum*. Arch Microbiol 102: 59–64

Oelze J, Biedermann M, Freund-Mölbert E and Drews G (1969) Bakteriochlorophyllgehalt und Proteinmuster der Thylakoide von *Rhodospirillum rubrum*. Arch Mikrobiol 66: 154–165

Onishi JC and Niederman RA (1982) *Rhodopseudomonas sphaeroides* membranes: Alterations in phospholipid composition in aerobically and phototrophically grown cells. J Bacteriol 149: 831–839

Owens GC and Ohad I (1982) Phosphorylation of *Chlamydomonas reinhardii* chlorophyll membrane protein in vivo and in vitro. J Cell Biol 93: 712–718

Pairoba C and Vallejos RH (1989) Protein phosphorylation in purple photosynthetic bacteria. Biochemie 71: 1039–1041

Peters GA and Cellarius RA (1972) Photosynthetic membrane development in *Rhodopseudomonas sphaeroides*. J Bioenerg 3: 345–359

Reaveley DA and Burge RE (1972) Walls and membranes in bacteria. In: Rose AH and Tempest DW (eds) Adv Microb Physiol Vol 7 pp 1–81. Academic Press, New York

Reed DW and Raveed D (1972) Some properties of the ATPase from chromatophores of *Rhodopseudomonas sphaeroides* and its structural relationship to the bacteriochlorophyll proteins. Biochim Biophys Acta 283: 79–91

Reed DW, Raveed D and Reporter M (1975) Localization of photosynthetic reaction centers by antibody binding to chromatophore membranes from *Rhodopseudomonas spheroides* strain 26. Biochim Biophys Acta 387: 368–378

Reidl H, Golecki JR and Drews G (1983) Energetic aspects of photophosphorylation capacity and reaction center content of *Rhodopseudomonas capsulata* grown in a turbidostat under different irradiances. Biochim Biophys Acta 725: 455–463

Reidl H, Golecki JR and Drews G (1985) Composition and activity of the photosynthetic system of *Rhodopseudomonas capsulata*. The physiological role of the B800–850 light-harvesting complex. Biochim Biophys Acta 808: 328–333

Reilly PA and Niederman RA (1986) Role of apparent membrane growth initiation sites during photosynthetic membrane development in synchronously dividing *Rhodopseudomonas sphaeroides*. J Bacteriol 167: 153–159

Richter P and Drews G (1991) Incorporation of light-harvesting complex I α and β polypeptides into the intracytoplasmic membrane of *Rhodobacter capsulatus*. J Bacteriol 173: 5336–5345

Richter P, Cortez N and Drews G (1991) Possible role of the highly conserved amino acids Trp-8 and Pro-13 in the N-terminal segment of the pigment-binding polypeptide LHI α of *Rhodobacter capsulatus*. FEBS Lett 285: 80–84

Richter P, Brand M and Drews G (1992) Characterization of LHI$^-$ and LHI$^+$ *Rhodobacter capsulatus pufA* mutants. J Bacteriol 174: 3030–3041

Rohmer M, Bouvier-Nave P and Ourisson G (1984) Distribution of hopanoids triterpenes in prokaryotes. J Gen Microbiol 130: 1137–1150

Saier Jr MH, Werner PK and Müller M (1989) Insertion of proteins into bacterial membranes: Mechanism, characteristics and comparisons with the eukaryotic process. Microbiol Rev 53: 333–366

Schön G and Ladwig R (1970) Bacteriochlorophyllsynthese und Thylakoidmorphogenese in anaerober Dunkelkultur von *Rhodospirillum rubrum*. Arch Microbiol 74: 356–371

Schumacher A and Drews G (1979) Effects of light intensity on membrane differentiation in *Rhodopseudomonas capsulata*. Biochim Biophys Acta 547: 417–428

Sganga MW and Bauer CE (1992) Regulatory factors controlling photosynthetic reaction center and light-harvesting gene expression in *Rhodobacter capsulatus*. Cell 68: 945–954

Shepherd WD, Kaplan S and Park JT (1981) Penicillin-binding proteins of *Rhodopseudomonas sphaeroides* and their membrane localization. J Bacteriol 147: 354–362

Shiozawa JA, Welte W, Hodapp N and Drews G (1982) Studies on the size and composition of the isolated light-harvesting B800–850 pigment-protein complex of *Rhodopseudomonas capsulata*. Arch Biochem Biophys 213: 473–485

Snozzi M and Crofts AR (1984) Electron transport in chromatophores from *Rhodopseudomonas sphaeroides* GA fused with liposomes. Biochim Biophys Acta 766: 451–463

Sockett RE, Donohue TJ, Varga AR and Kaplan S (1989) Control of photosynthetic membrane assembly in *Rhodobacter sphaeroides* mediated by *puhA* and flanking sequences. J Bacteriol 171:436–446

Sprague SG and Varga AR (1986) Membrane architecture of anoxygenic photosynthetic bacteria. In: Staehelin LA and Arntzen CJ (eds) Photosynthesis III, pp 603–619. Springer, Berlin

Stark W, Kühlbrandt W, Wildhaber I, Wehrli E and Mühlethaler K (1984) The structure of the photoreceptor unit of *Rhodopseudomonas viridis*. EMBO J3: 777–783

Stark W, Jay F and Mühlethaler K (1986) Localization of reaction centre and light-harvesting complexes in the photosynthetic unit of *Rhodopseudomonas viridis*. Arch Microbiol 146: 130–133

Stiehle H, Cortez N, Klug G and Drews G (1990) A negatively charged N-terminus in the α polypeptide inhibited formation of the light-harvesting complex I in *Rhodobacter capsulatus*. J Bacteriol 172: 7131–7137

Tadros MH, Frank R, Dörge B, Gad'on N, Takemoto JY and Drews G (1987) Orientation of the B800–850, B870 and reaction center polypeptides on the cytoplasmic and periplasmic surfaces of *Rhodobacter capsulatus* membranes. Biochemistry 26: 7680–7687

Tadros MH, Garcia AF, Gad'on N and Drews G (1989) Characterization of a pseudo-B870 light-harvesting complex isolated from the mutant strain Az1$^+$ pho$^-$ of *Rhodobacter capsulatus* which contains B800–850 type polypeptides. Biochim Biophys Acta 976: 161–167

Tadros MH, Garcia AF, Drews G, Gad'on N and Skatchkov MP (1990) Isolation and characterization of a light-harvesting complex II lacking the gamma-polypeptide from *Rhodobacter capsulatus*. Biochim Biophys Acta 1019: 245–249

Tai, S-P and Kaplan S (1984) Purification and properties of a phospholipid transfer protein from *Rhodopseudomonas sphaeroides*. J Biol Chem 259: 12178–12183

Takaichi S, Gardiner AT and Cogdell R (1992) Pigment composition of light-harvesting pigment-protein complexes from *Rhodopseudomonas acidophila*. Effect of light intensity. In: Murata N (ed) Research in Photosynthesis, Vol I, pp 149–152. Kluwer Academic Publishers, Dordrecht

Takamiya K, Shioi Y, Shimada H and Arata H (1992) Blue-light inhibition of accumulation of photosynthetic pigments in *Roseobacter denitrificans* under anaerobic conditions. In: Murata N (ed) Research in Photosynthesis, Vol III, pp 91–94. Kluwer Academic Publishers, Dordrecht

Takemoto JY, Schonhardt T, Golecki JR and Drews G (1985) Fusion of liposomes and chromatophores of *Rhodopseudomonas capsulata*: Effect on photosynthetic energy transfer between B875 and reaction center complexes. J Bacteriol 162: 1126–1134

Tauschel HD and Drews G (1967) Thylakoidmorphogenese bei *Rhodopseudomonas palustris*. Arch Mikrobiol 59: 381–404

Theiler R and Niederman RA (1991) Localization of chromatophore proteins of *Rhodobacter sphaeroides*. I. Rapid Ca - induced fusion of chromatophores with phosphatidyl-glycerol liposomes for proteinase delivery to the luminal membrane surface. J Biol Chem 266: 23157–23162

Tichy HV, Oberlé B, Stiehle H, Schiltz E and Drews G (1989) Genes downstream from *pucA* and *pucA* are essential for formation of the B800–850 complex of *Rhodobacter capsulatus*. J Bacteriol 171: 4914–4922

Tichy HV, Albien KV, Gad'on N and Drews G (1991) Analysis of the *Rhodobacter capsulatus puc* operon: The *pucC* gene plays a central role in the regulation of LHII (B800–850) complex expression. EMBO J 10: 2949–2955

Troschel D and Müller M (1990) Development of a cell-free system to study the membrane assembly of photosynthetic proteins of *Rhodobacter capsulatus*. J Cell Biol. 111: 87–94

Troschel D, Eckhardt S, Hoffschulte HK and Müller M (1992) Cell-free synthesis and membrane-integration of the reaction center subunit H from *Rhodobacter capsulatus*. FEMS Microbiol Lett 91: 129–134

Uffen RL, Sybesma C and Wolfe RS (1971) Mutants of *Rhodospirillum rubrum* obtained after long-term anaerobic, dark growth. J Bacteriol 108: 1348–1356

Urakami T and Komagata K (1988) Cellular fatty acid composition with special reference to the existence of hydroxy fatty acids, and the occurrence of squalene and sterols in species of *Rhodospirillaceae genera* and *Erythrobacter longus*. J Gen Appl Microbiol 34: 67–84

van Doren SR, Yun CH, Crofts AR and Gennis RB (1993) Assembly of the Rieske iron-sulfur subunit of the cytochrome bc_1 complex in the *E. coli* and *Rb. sphaeroides* membranes independent of the cytochrome b and c_1 subunits. Biochemistry 32: 628–636

van Grondelle R, van Mourik F, Visschers RW, Somson OJG and

Valkunas L (1992) The bacterial photosynthetic light-harvesting antenna: Aggregation state, spectroscopy and excitation energy transfer. In: Murata N (ed) Research in photosynthesis, Vol I, pp 9–16. Kluwer Academic Publishers, Dordrecht

Valkirs GE and Feher G (1982) Topography of reaction center subunits in the membrane of the photosynthetic bacterium, *Rhodopseudomonas sphaeroides*. J Cell Biol 95: 179–188

Varga AR and Staehelin LA (1983) Spatial distribution in photosynthetic and non-photosynthetic membranes of *Rhodopseudomonas palustris*. J Bacteriol 154: 1414–1430

Varga AR and Staehelin LA (1985) Pigment-protein complexes from *Rhodopseudomonas palustris*: Isolation, characterization, and reconstitution into liposomes. J Bacteriol 161: 921–927

Wakim B and Oelze J (1980) The unique mode of adjusting the composition of the photosynthetic apparatus to different environmental conditions by *Rhodospirillum tenue*. FEMS Microbiol Lett 7: 221–223

Wakim B, Golecki JR and Oelze J (1978) The unusual mode of altering the cellular membrane content by *Rhodospirillum tenue*. FEMS Microbiol Lett 4:199–201

Weckesser J, Drews G and Mayer H (1979) Lipopolysaccharides of photosynthetic prokaryotes. Ann Rev Microbiol 33: 215–239

Wehrli E and Kübler O (1980) The two-dimensional lattice of the photosynthetic membrane of *Rhodopseudomonas viridis*. In: Baumeister W and Vogell W (eds) Electron Microscopy of Molecular Dimensions, pp 56–88. Springer Publ, Berlin

Wieseler B and Müller M (1993) Translocation of precytochrome c_2 into intracytoplasmic membrane vesicles of *Rhodobacter capsulatus* requires a peripheral membrane protein. Mol Microbiol 7: 167–176

Wieseler B, Schiltz E and Müller M (1992) Identification and solubilization of a signal peptidase from the phototrophic bacterium *Rhodobacter capsulatus*. FEBS Lett 298: 273–276

Yen GSL, Cain BD and Kaplan S (1984) Cell-cycle specific biosynthesis of the photosynthetic membrane of *Rhodopseudomonas sphaeroides*. Biochim Biophys Acta 777: 41–55

Yildiz FH, Gest H and Bauer CE (1991) Attenuated effect of oxygen on photopigment synthesis in *Rhodospirillum centenum*. J Bacteriol 173: 5502–5506

Youvan DC, Bylina EJ, Alberti M, Begusch H and Hearst JE (1984) Nucleotide and deduced polypeptide sequences of the photosynthetic reaction center, B870 and antenna, and flanking polypeptides from *Rhodobacter capsulata*. Cell 37: 949–957

Zannoni D, Jasper P and Marrs B (1978) Light-induced oxygen reduction as a probe of electron transport between respiratory and photosynthetic components in membranes of *Rhodopseudomonas capsulata*. Arch Biochem Biophys 191: 625–631

Zsebo K and Hearst JE (1984) Genetic physical mapping of a photosynthetic gene cluster from *Rhodopseudomonas capsulata*. Cell 37: 937–947

Chapter 13

Membranes and Chlorosomes of Green Bacteria: Structure, Composition and Development

Jürgen Oelze and Jochen R. Golecki
Institut für Biologie II (Mikrobiologie), Universität Freiburg, Schänzlestrasse 1, D-79104 Freiburg

Summary

Although belonging to evolutionary distantly related groups, cells of members of the Chlorobiaceae and the Chloroflexaceae exhibit structurally and functionally comparable substructures. While the photochemical reaction center complex plus a light-harvesting unit are housed in the peripheral cytoplasmic membrane (CM) system, an accessory light-harvesting unit is localized in specialized structures, the chlorosomes, underlying the CM. In this chapter, the present knowledge on the fine structure of chlorosomes and the CM is reviewed. After a description of methods commonly employed to isolate chlorosomes and CM, data of chemical analyses of both subcellular fractions are detailed. In spite of considerable similarities in the overall ultrastructural and

R. E. Blankenship, M. T. Madigan and C. E. Bauer (eds): Anoxygenic Photosynthetic Bacteria, pp. 259–278.

functional properties, the chemical composition reveals significant differences between chlorosomes and CM, when isolated from *Chlorobium* and *Chloroflexus*, respectively. The same holds true with respect to the supramolecular organization, particularly of chlorosomes and elements involved in their connection to the CM. Since the photosynthetic apparatus of green bacteria is composed of two different moieties, i.e. the chlorosomes and the CM-bound unit, the important questions arise how these different units are synthesized, how the synthesis of individual constituents is controlled and how the syntheses of chlorosomes and the CM-bound units are coordinated. In the present contribution these problems are approached on the basis of the pigments characteristic of both units. In addition, the knowledge on polypeptide formation is presented. These data are combined with the respective changes in number and size of chlorosomes. In spite of a considerable amount of detailed information available as yet, it is finally concluded that considerable research efforts are still required in order to understand the development of the biologically unique type of photosynthetic apparatus characteristic of the green bacteria.

I. Introduction

Early studies on the ultrastructure of several members of the purple phototrophic bacteria revealed the presence of an intracytoplasmic membrane continuum, which upon cell breakage formed closed vesicles. These vesicles contained the entire photosynthetic apparatus and were named chromatophores according to their pigmentation (Oelze and Drews, 1972). Any attempts to identify similar structures within cells of the green bacteria, however, remained unsuccessful. Instead, Cohen-Bazire (1963) was able to show that thin sections of cells of *Chlorobium limicola* and *Cb. thiosulfatophilum* exhibit the presence of oblong inclusion bodies adjacent to the CM. These structures surrounded by a thin layer were named *Chlorobium* vesicles. Subsequently, similar structures containing high amounts of bacteriochlorophyll (BChl) could be isolated from homogenates of these and related organisms. On this basis, it was proposed that, in members of the green bacteria, *Chlorobium* vesicles were the major sites of the photosynthetic apparatus (Cohen-Bazire, 1963) including the photochemical reaction centers (Fowler et al., 1971). The latter suggestion, however, was not supported by subsequent investigations. Instead, it could be shown that the photochemical reaction center, containing BChl *a*, was localized in the CM, while *Chlorobium* vesicles house accessory light-harvesting BChl units (Boyce et al., 1976; Olson et al., 1976; Olson 1980).

Abbreviations: ALA – 5-aminolevulinic acid; BChl – bacteriochlorophyll; *Cb.* – *Chlorobium*; *Cf.* – *Chloroflexus*; CM – cytoplasmic membrane; CW – cell wall; D – dilution rate; EF – exoplasmic fracture face; MGDG – monogalactosyl diglyceride; Pa – Pascal; PF – plasmic fracture face; SDS-PAGE – sodium dodecyl sulphate polyacrylamide gel electrophoresis

In 1974 it became apparent that *Chlorobium* vesicles were formed not only by members of the Chlorobiaceae but also by the distantly related and newly discovered *Chloroflexus aurantiacus* (Pierson and Castenholz, 1974a,b). Therefore, it was proposed to more generally name such vesicles chlorosomes (= green bodies) (Staehelin et al., 1978). In this contribution we will compile the present knowledge on the structure and composition as well as on the development of the photosynthetic apparatus of green bacteria comprising the chlorosomal and the CM-bound units.

II. Ultrastructure

A. Membranes

The CM of green bacteria is visualized by electron microscopy of ultrathin sections of cells as a double track layer underlying the cell wall. This membrane with a thickness of approximately 8 nm represents the outer layer of the protoplast (Fig. 1). The periplasmic space becomes visible in thin sections as an electron transparent space localized between the CM and the cell wall. In freeze-fracture electron micrographs of chemoorganotrophically grown *Cf. aurantiacus* (Sprague et al., 1981b) the CM exposes the well known architecture of other bacteria (Golecki and Oelze, 1980) that is to say, a plasmic fracture face (PF) densely covered with intramembrane particles of different size and an exoplasmic fracture face (EF) sparsely covered with larger particles. However, in phototrophically grown cells of both *Cb. limicola* and *Cb. tepidum*, distinct areas enriched in larger intramembrane particles (> 9 nm) were demonstrated on both fracture faces of the CM (Fig. 2

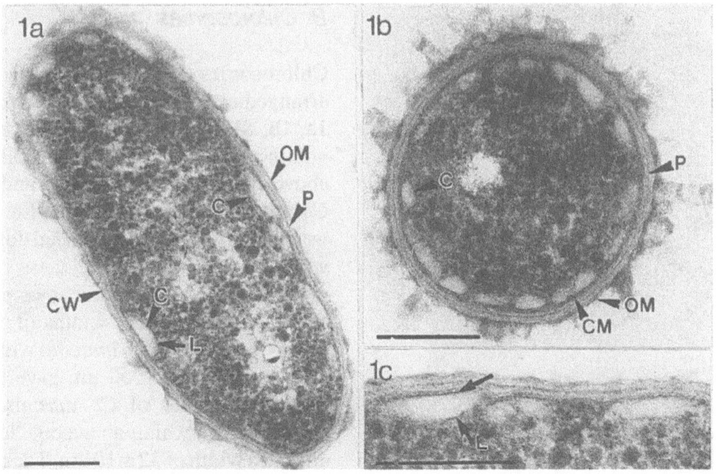

Fig. 1. Ultrathin sections of *Chlorobium vibrioforme* strain 6230 exhibiting the cytoplasmic membrane (CM) with underlying chlorosomes (C) as well as the cell wall (CW) including the outer membrane (OM) and the peptidoglycan (P) layer. Bar represents 200 nm. Sections in (a) longitudinal (reproduced from Staehelin et al., 1980, with permission) and (b) horizontal axis of the cell; (c) chlorosomes at higher magnification presenting the baseplate (←) between the chlorosome and the CM as an electron dense layer, L = chlorosome envelope.

and Staehelin et al, 1980). These areas of the CM are localized at the presumptive sites of chlorosome attachment. The larger particles were interpreted as structural equivalents of the photosynthetic units.

Both fracture faces of the CM of phototrophically grown cells of *Chloroflexus* and *Chlorobium* exhibit, in addition, characteristic chlorosomes (Fig. 3 and Staehelin et al., 1978). Because of the close association of chlorosomes and the CM, the fracture plane can be deflected very often from the CM into the underlying chlorosome.

In *Chloronema* (Dubinina and Gorlenko, 1975), *Oscillochloris* (Sprague and Fuller, 1991) and in some mesophilic *Chloroflexus*-like organisms from marine environments, the CM forms invaginations studded with chlorosomes (Larsen et al., 1991). Internal membranes originally described as mesosomes (Cohen-Bazire et al., 1964) may be the results of artificial preparation conditions as demonstrated with several other bacterial species (Ebersold et al., 1981). Additional artifacts may be the lipid bilayer structures, which can be identified on freeze-fracture preparations of the thermophilic *Cf. aurantiacus* strain J-10-fl as large flattened vesicles without detectable substructure (Staehelin et al., 1978). It appears likely that these vesicles arise in the CM by phase separation induced by a shift-down in temperature.

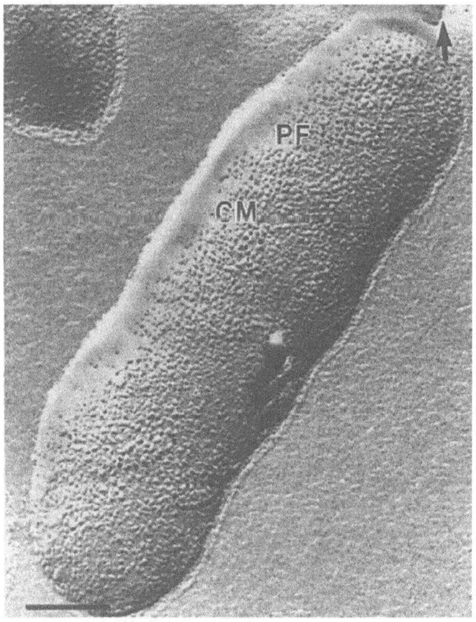

Fig. 2. Freeze-fracture micrograph of *Chlorobium tepidum* exposing the plasmic fracture face (PF) of the cytoplasmic membrane (CM). The membrane fracture face reveals the presence of elongated patches of mostly larger (> 9 nm) intramembrane particles corresponding to the attachment sites of the underlying chlorosomes. Some of these patches have been outlined (black dots). Arrow in the upper right corner indicates direction of shadowing. Bar represents 200 nm.

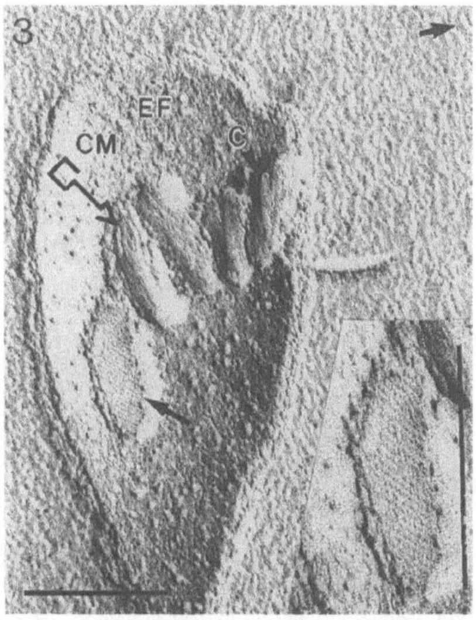

Fig. 3. Freeze-fracture micrograph of *Chlorobium tepidum* showing the exoplasmic fracture face (EF) of the cytoplasmic membrane with associated chlorosomes. The crystalline baseplate (➔) exhibits striations with a periodicity of 6 nm. The ridges of the baseplate make an angle of 40–60° with the longitudinal axis of the chlorosome. Each ridge possesses a regular, globular substructure with a periodicity of approximately 3.5 nm. In addition some rod elements (⊏➔) of 10 nm in diameter are visible in the core of the chlorosome. Inset: baseplate at higher magnification. Arrow in the upper right corner indicates direction of shadowing. Bar represents 200 nm.

In the filamentous cells of *Chloroflexus*, the CMs of two neighbouring cells are connected by 10–30 microplasmadesmata, which can be identified within the central area of the septum (Staehelin et al., 1978).

Cell envelopes of Gram-negative bacteria are composed of the CM and the peptidoglycan layer as well as the outer membrane of the cell wall. All of these structural elements can be identified on ultrathin sections of *Chlorobium* strains (Fig. 1). However, in *Cf. aurantiacus* it is difficult to identify an outer membrane or a defined peptidoglycan layer. These morphological findings, together with biochemical data such as the the lack of both lipoprotein and lipopolysaccharides and the presence of L-ornithine in the peptidoglycan, revealed that *Chloroflexus* forms a cell wall resembling that of Gram-positive rather than that of Gram-negative bacteria (Jürgens et al., 1987.)

B. Chlorosomes

Chlorosomes are elliptical structures, which are arranged in a single layer underlying the CM (Figs. 1a, 1b, 8). Usually chlorosomes are not detectable within the cytoplasm of the cells. As mentioned above, however, *Oscillochloris* and some mesophilic *Chloroflexus*-like organisms (Larsen et al., 1991) were reported to form internal foldings of the CM with aligned chlorosomes.

On the average, chlorosomes exhibit lengths of about 70–260 nm and widths of 30–100 nm. With chlorosomes from *Cb. limicola* widths of 40–100 nm and lengths of 70–260 nm have been determined. The chlorosomes of *Cf. aurantiacus* seem to be smaller. They exhibit an average length of 106 ± 24 nm and a width of 32 ± 10 nm. The height of the latter chlorosomes is 10–20 nm (Golecki and Oelze, 1987). The lengths and widths of chlorosomes of other members of the green bacteria are generally within the same range: for '*Chloropseudomonas ethylicum*': 130–150 nm and 30–50 nm (Holt et al., 1966), *Cb. thiosulfatophilum*: 100–150 nm and 30–40 nm (Cohen-Bazire et al., 1964), *Chloroherpeton thalassium*: 140 nm and 33 nm (Gibson et al., 1984), *Cb. tepidum*: 100–180 nm and 40–60 nm (Wahlund et al., 1991), *Cb. vibrioforme*, *Pelodictyon luteolum* and *Pelodictyon phaeum*: 70–150 nm and 30–80 nm (Puchkova et al., 1975). However, external conditions may have a significant influence on the size and, thus, the volume of chlorosomes (Broch-Due et al., 1978; Golecki and Oelze, 1987). A more detailed description of the effects of external factors on the size and development of chlorosomes will be given in Section V-B of this chapter.

As yet, detailed studies on the ultrastructure of chlorosomes have been performed only with cells of either *Cb. limicola* or *Cf. aurantiacus*. In these studies, freeze-fracture electron microscopy has been employed to obtain insights into the internal architecture and organization of chlorosomes and the underlying membrane (Staehelin et al., 1978 and 1980; Sprague et al., 1981a,b; Golecki and Oelze, 1987). The results revealed that the chlorosomes of *Chloroflexus* and *Chlorobium* exhibit essentially the same ultrastructure with the following comparable elements; i.) a distinct area within the CM referred to as the membrane attachment site, ii.) a highly ordered baseplate between the chlorosome core and the CM, iii.) internal rod elements in the core of chlorosomes, iv.) an envelope forming the cytoplasmic boundary of the chlorosome.

As mentioned above, in *Chlorobium*, the attachment sites are visible as distinct oblong areas on the plasmic (PF) and exoplasmic (EF) fracture faces of the CM enriched in larger intramembrane particles (> 9 nm) in a non-crystalline order (Fig. 2). The size of these non-crystalline ordered particles is larger than the average size of the other particles on the respective fracture face of the CM. In *Cf. aurantiacus*, on the other hand, the attachment site presents a highly ordered array of particles. The periodicity of the lattice is 6 nm with an orientation perpendicular to the long axis of the chlorosome (Staehelin et al., 1980).

According to the original definition, the baseplate connects the chlorosome to the CM (Staehelin et al., 1980). In ultrathin sections, the baseplate appears as an electron dense layer of 5–6 nm thickness (Figs. 1 and 8). The supramolecular architecture of the baseplate, however, is detectable only with freeze-fracture preparations. The baseplates of *Cb. limicola* and *Cb. tepidum* are detectable on the exoplasmic fracture-face of the CM as elliptical, striated plate-like structures exhibiting a crystalline substructure (Fig. 3 and Staehelin et al., 1980). The ridges of the baseplate lattice (6 nm distance) form an angle of 40° to 60° with the longitudinal axis of the plate. Each ridge exhibits a fine granular substructure with a periodicity of 3.3–3.5 nm. Baseplates are nearly as large as complete chlorosomes. In *Chloroflexus*, baseplates exhibit a crystalline arrangement with fine striations of 6 nm periodicity (Fig. 4d, Staehelin et al., 1978; Sprague et al., 1981a,b). In contrast to *Chlorobium*, these striations are oriented at right angles to the longitudinal axis of the baseplate. With *Chloroflexus* an average baseplate size of 3,200 ± 800 nm^2 was determined (Golecki and Oelze, 1987).

The cores of *Chlorobium* and *Chloroflexus* chlorosomes are filled with a variable number of closely packed rod-shaped elements with diameters of 10 and 5.2 nm, respectively (Fig. 4 and Staehelin et al., 1978 and 1980). The rod elements are embedded in an unetchable matrix, oriented longitudinally to

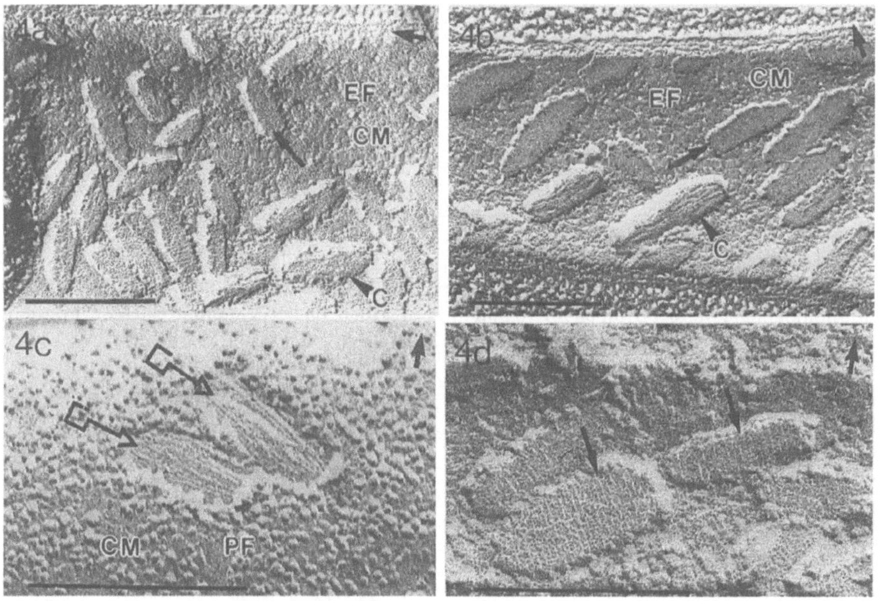

Fig. 4. Freeze-fracture micrographs of cells of *Chloroflexus aurantiacus*. The fracture faces of the CM show baseplates (➜) and chlorosomes (➤) with fractures along the cytoplasmic face of the envelope layer. (a) Exoplasmic fracture face (EF) exhibiting small chlorosomes (Table 1, culture 1C); (b) EF exhibiting long chlorosomes (Table 1, culture 2C); (c) plasmic fracture face (PF) exhibiting the rod elements (⇨)of chlorosomes; (d) baseplate with a lattice order with a periodicity of 6 nm. Arrow in the upper right corner indicates direction of shadowing. Bar represents 200 nm.

the long axis and spanning the full length of the chlorosomes (Staehelin et al., 1980). In *Chlorobium* the presence of 10 to 30 rod elements per chlorosome was described. In *Chloroflexus*, however, the number of rod elements seems to vary in a wider range as a response to changes in external growth factors and, thus, to changes in BChl *c* contents (Sprague et al., 1981 a; Golecki and Oelze, 1987).

On their plasmic sides, chlorosomes are bounded by a layer which exhibits a thickness of 2–3 nm (Fig. 1). This boundary appears as a single electron dense layer in ultrathin sections.

III. Chemical Composition

A. Isolation of Chlorosomes and Membranes

1. Chlorosomes

Pierson and Castenholz (1978) compiled the methods employed to isolate subcellular fractions from green bacteria. Since then a number of improvements have been developed. However, differences in isolation procedures have yielded chlorosome preparations of different composition. This in turn, may be considered as one of the primary reasons for the formulation of controversial hypotheses on the functional architecture of chlorosomes. Basic principles of the most commonly used methods are detailed below.

Feick et al. (1982) suspended washed cells of *Cf. aurantiacus* in Tris-hydrochloride buffer (1 g wet weight per 5 ml, 10 mM, pH 8.0) and passed the suspension three times through a precooled French pressure cell (138 MPa). After a low spin centrifugation, the homogenate was centrifuged for 90 min at 180,000 × g. The pellet, containing more than 95% of the total cellular BChl, was treated with sodium iodide and the dipolar detergent Deriphat (N-lauryl-β-iminodipropionate) in order to detach chlorosomes from CM fragments. Subsequent centrifugation yielded a supernatant (referred to as supernatant I) and a floating pellet, which was enriched in chlorosomes. The floating pellet was withdrawn and dialysed against buffer. After addition of the dipolar detergent Miranol, the material was separated by sucrose gradient centrifugation. Chlorosomes banded at 27% (w/w) of sucrose. Gradient centrifugation was repeated if contaminations of CM-bound BChl *a* absorbing at 866 nm were observed. Alternatively, the whole membrane pellet of the first high spin centrifugation was directly treated with Miranol and separated by sucrose centrifugation.

Chlorosomes isolated by the method of Feick et al. (1982) contained BChl *c* and small, yet constant, amounts of BChl *a* in a ratio of 25: 1.

Gerola and Olson (1986) published a widely accepted method to isolate chlorosomes from *Cb. limicola* f. *thiosulphatophilum* by sucrose density gradient centrifugation. This method has subsequently been used to successfully isolate chlorosomes from *Cf. aurantiacus* as well. The procedure includes the chaotropic agent NaSCN in the presence of ascorbate rather than of detergents. NaSCN was shown to stabilize *Chlorobium* chlorosomes, which are much more fragile than those of *Cf. aurantiacus*. Stable chlorosomes from *Chlorobium* isolated by the Gerola/Olson-method contain BChls *c* and *a* at a ratio of approximately 90.

Chlorosomes completely free of any detectable amounts of BChl *a* were isolated from *Cf. aurantiacus* by Griebenow and Holzwarth (1989) (Fig. 5). After sonication of cell suspensions, chlorosomes were isolated from homogenates by a combination of native gel electrophoresis and gel filtration (i.e. gel-electrophoretic filtration) in the presence of the Li salt of the detergent dodecyl sulfate. In addition, it turned out that *Chloroflexus* chlorosomes obtained by the Gerola/Olson-method can be purified from residual amounts of BChl *a* by treatment with sodium dodecyl sulfate (Olson et al., 1990).

2. Membranes

All of the relevant methods employed as yet to isolate membrane fractions from cell homogenates of green bacteria include sucrose gradient centri-

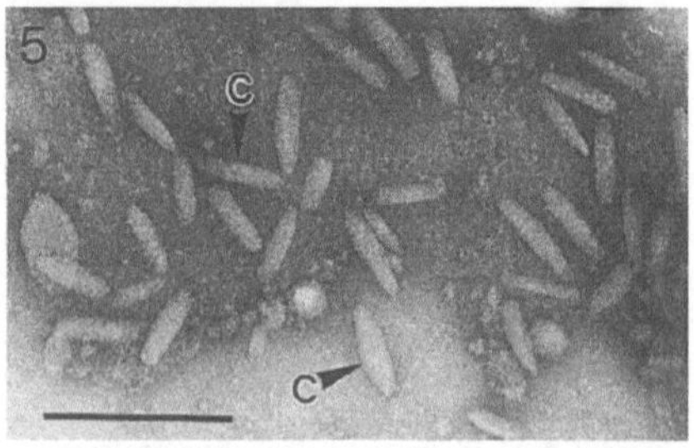

Fig. 5. Negative staining preparation of isolated chlorosomes from *Chloroflexus aurantiacus*. The intact chlorosomes present the typical oblong shape. Bar represents 200 nm.

fugation. According to Schmidt (1980), CM from *Cf. aurantiacus* band at a density corresponding to 40–45% (w/w) of sucrose. CM from *Cb. limicola* f. *thiosulfatophilum*, on the other hand, bands at 35%. Interestingly, CM fractions from *Cf. aurantiacus* were reported to exhibit flat, irregular pieces, while fractions from *Chlorobium* appear as vesicular and regular structures.

Probably the purest CM preparations depleted of any detectable amounts of the chlorosomal BChl *c* were isolated by Feick et al. (1982). These membranes were purified from supernatant I (mentioned above in context of chlorosome isolation) by sucrose density gradient centrifugation in the presence of sodium iodide and the detergent Deriphat.

B. Composition of Chlorosomes and Membranes

Cell breakage as well as isolation procedures may affect severely the structure and composition of subcellular structures. Thus, when evaluating results of biochemical analyses, it should be kept in mind not only that essential constituents might have been lost or decomposed, but also that other cellular constituents may have become attached during the isolation of subcellular fractions. Moreover, these artifacts may affect comparable fractions from different organisms to different extents. An example for the latter is given by chlorosomes from *Chlorobium* species, which appear much more fragile than chlorosomes from *Chloroflexus*.

1. Chlorosomes

a. Pigments

Chlorophylls. BChls are the most prominent constituents of chlorosomes. Depending on the bacterial species, the chlorosomes house the majority of BChls *c, d* or *e*. These different types of magnesium tetrapyrroles can be further subdivided into different homologues. In the cases of BChls *e* and *d*, respectively, at least eight and six homologues are known. These homologues are characterized on the basis of different substituents at the chlorine nucleus. In addition, there are indications that the most common farnesyl ester side chains of BChls *c* and *d* may be substituted to a minor extent by several other alcohols (Smith et al., 1983; Bobe et al., 1990; Otte et al., 1993).

It should also be noted that different basic types of BChl may be formed by one organism as a response to changes in culture conditions. This latter effect has been observed with *Cb. vibrioforme* f. *thiosulfatophilum*, which synthesizes BChl *c* instead of BChl *d* when adapting from high to low light conditions (Broch-Due and Ormerod, 1978; Huster and Smith, 1990). It has been suggested that methylation of BChl *d* to BChl *c* increases the hydrophobicity, giving rise to a higher degree of packing of antenna BChl molecules in the chlorosome (Bobe et al., 1990). On this basis it is not surprising that, recently, Otte et al. (1993) reported on the presence of significant amounts of BChl *c* together with BChl *d* in chlorosomes of *Cb. vibrioforme*.

BChl *c* of chlorosomes of *Cf. aurantiacus* (strain OK-70-fl) was identified as a mixture of at least ten derivatives. In contrast to BChl *c* homologues of *Chlorobium*, the BChls of *Cf. aurantiacus* differ on the basis of five different esterifying alcohol side chains, each of which has two epimeric forms (either 2aR, 7S, 8S or 2aS, 7S, 8S, Fages et al., 1990). In addition, *Cf. aurantiacus* strain J-10-fl synthesizes small amounts of BChl *d* (Bobe et al., 1990).

As mentioned above, minor amounts of BChl *a* have been reported for chlorosomes from various Chlorobiaceae as well as from *Cf. aurantiacus*. However, while a BChl *a* polypeptide complex with an absorption peak at 794 nm apparently forms an integral part of chlorosomes of *Chlorobium*, a comparable BChl *a* complex absorbing at 790 nm can be removed from chlorosome preparations from *Cf. aurantiacus* (Griebenow and Holzwarth, 1989; Gerola and Olson, 1986).

Carotenoids. Schmidt (1980) analyzed the carotenoid composition of *Cb. limicola* f. *thiosulfatophilum* and *Cf. aurantiacus*. Most of the carotenoids synthesized by both species were localized in the chlorosome fraction. Specific carotenoid contents of chlorosomes from both species were 19 and 33 μg per mg of protein, respectively. About 80% of the total carotenoids of *Cf. aurantiacus* are composed of almost equal amounts of β- and γ-carotene. In addition, minor amounts of lycopene, 4-oxo-γ or β- carotene as well as hydroxy-γ-carotene and its glucoside were identified. The predominant carotenoids of *Chlorobium* chlorosomes are chlorobactene (69%) and neurosporene (21%).

b. Lipids

Glycolipids are the major lipids of chlorosomes of Chlorobiaceae and Chloroflexaceae (Cruden and Stanier, 1970; Schmidt, 1980). Chlorosomes from

Chlorobium preferentially contain monogalactoside diglyceride (MGDG), while those of *Cf. aurantiacus* contain digalactosyl diglyceride as well (Holo et al., 1985).

c. Polypeptides

Chlorosomes isolated from different members of the green bacteria contain significant amounts of protein (Schmidt, 1980; Feick and Fuller, 1984; Holo et al., 1985). *Chloroflexus* chlorosomes were shown by SDS-PAGE to comprise four major polypeptides of 18, 11, 5.8 and 3.7 kDa (Feick and Fuller, 1984). The 5.8 kDa polypeptide belongs to the baseplate, which, depending on the isolation method employed, may co-purify with chlorosomes. Amino acid analysis of the 3.7 kDa polypeptide revealed a mass of 5,592 Da. Therefore, in the following and in the literature, this polypeptide is referred to as 5.6 kDa polypeptide (Zuber and Brunisholz, 1991). In spite of the large body of results obtained on the structure and molecular genetics of the 5.6 kDa polypeptide, it was claimed that this polypeptide was a degradation product (Griebenow et al., 1990). This claim, however, was not supported by subsequent investigations (Eckhardt et al., 1990; Niedermeier et al., 1992; Wullink et al., 1991). Nevertheless, it gave rise to the development of, until then, unknown concepts on the organization of photosynthetic antenna systems (see Chapter 20 by Blankenship et al.).

A 6.3 kDa polypeptide with 30% sequence homology to the 5.6 kDa polypeptide of *Cf. aurantiacus* was isolated from different species of green sulfur bacteria (Wagner-Huber et al., 1988). In addition, a second polypeptide of 7.5 kDa molecular mass and an amino acid sequence significantly different from the 6.3 kDa polypeptide, was isolated from chlorosomes of *Cb. limicola* (Gerola et al., 1988). The sequence homology of this 7.5 kDa polypeptide with the 5.6 kDa *Chloroflexus* polypeptide was only 17% (Gerola et al., 1988). Lack of cross reactivity of antisera to chlorosome peptides from *Chloroflexus* with chlorosome polypeptides from various green sulfur bacteria support the low degree of homology described on the basis of the amino acid sequence (Stolz et al., 1990).

2. Membranes

The chemical composition of CM isolated from *Cf. aurantiacus* and *Cb. limicola* f. *thiosulfatophilum* was compared by Schmidt (1980). Interestingly, CM from *Cf. aurantiacus* contained significantly lower amounts of gluco- and phospholipids than those from *Chlorobium*. With respect to the cellular composition and distribution of carotenoids it is interesting to note that typical carotenoids of both species are present as glucosides, while chlorosomes contain the corresponding carotenoids in a non-glucosylated form. Moreover, while membranes from *Cf. aurantiacus* exhibited both NADH-oxidase and succinate-dehydrogenase activities, the latter activity could not be detected in preparations from *Chlorobium*.

All species of green bacteria synthesize a CM-bound moiety of the photosynthetic apparatus characterized by the presence of BChl *a* as the major tetrapyrrole pigment. In *Cf. aurantiacus*, this BChl is associated with both light-harvesting antenna units and photochemical reaction center complexes with absorption properties very similar to the corresponding complexes of purple bacteria. However, reaction centers from *Cf. aurantiacus* are composed of two polypeptides, which do not show any amino acid sequence homology to known reaction center polypeptides from purple bacteria (Shiozawa et al., 1987). This difference is supported on the basis of the pigment composition of reaction centers, which, in *Cf. aurantiacus*, contains the primary electron donor BChl *a* dimer (P 865), an additional BChl *a*, three bacteriopheophytins *a* and two menaquinones (Pierson and Thornber, 1983; Blankenship and Fuller, 1986).

The CM-bound antenna complex of *Cf. aurantiacus*, designated B808-866 (Feick and Fuller, 1984), contains BChl, carotene and two different polypeptides (Zuber and Brunisholz, 1991; Wechsler et al., 1991). This complex resembles that of purple bacteria not only on the basis of spectral properties (Vasmel et al., 1986) but also on the homology of the two polypeptides. The proportion of light-harvesting BChl *a* per reaction center was shown to stay fairly constant at (20–25): 1 (Feick et al., 1982; Feick and Fuller, 1984).

Since *Cf. aurantiacus* is a facultatively phototrophic organism, it is able to perform a respiratory energy metabolism. Constituents of the respiratory chain are assembled together with the photochemical electron transport chain in the CM. However, when growing strictly chemotrophically, i.e. aerobically in the dark, the photosynthetic apparatus is missing in membranes from *Cf. aurantiacus* (Foster et al., 1986).

The latter applies not only to BChl complexes but also to cytochromes like the high potential cytochrome *c*-554, which is assumed to function as electron donor to the reaction center. Nevertheless, membranes isolated from chemotrophically grown cells contain at least three *c*-type and one *b*-type cytochromes in addition to menaquinone (Wynn et al., 1987). According to Zannoni and Fuller (1988) aerobic cells of *Cf. aurantiacus* exhibit a branched electron transport chain from NADH to two different terminal oxidases as based on their different sensitivities toward KCN and CO (see also Chapter 44 by Zannoni).

The CM-bound moiety of the photosynthetic apparatus of green sulfur bacteria exhibits a high degree of relatedness to photosystem I of plants. The presence of a quinone pool, of photo-reducible cytochrome *b* and of photo-oxidizable cytochrome *c* suggests participation of a cytochrome *bc* complex in photochemical electron transport (Hurt and Hauska, 1984). The analysis of a recently isolated reaction center complex from *Chlorobium* revealed the presence of about 40 molecules of BChl *a* and about four cytochromes *c*-553 per primary donor, P840 (Feiler et al., 1992). It has been suggested that the primary electron donor BChl P840 is associated with a chlorophyll-like primary acceptor as well as with a set of acceptors comprising a ubiquinone-type compound and three FeS centers (Nitschke et al., 1990). Recently, the primary acceptor, originally designated $BChl_{663}$, has been identified as an isomeric form of chlorophyll *a* (van de Meent et al., 1992).

IV. Supramolecular Organization of Chlorosomes

A. Chlorosome Envelope

As mentioned above, results of freeze-fracture studies suggest that a single layered envelope with a thickness of 2–5 nm forms the cytoplasmic boundary of chlorosomes. This together with the localization of MDGD in chlorosomes has led to the hypothesis that the chlorosome envelope is composed of a monolayer of galactolipids (Staehelin et al., 1980). Indeed, the presence of galactose moieties pointing to the cytoplasmic outside of the envelope was demonstrated by the agglutination of isolated chlorosomes from *Cb. limicola* and *Cf. aurantiacus* with Ricinus lectin (Holo et al., 1985). Agglutination was enhanced after trypsin treatment of chlorosomes, suggesting in situ masking of galactose moieties by protein. The presence of a lipid monolayer with polar groups on the outside infers that the core of chlorosomes is largely hydrophobic. A detailed study on the localization of polypeptides in chlorosomes from *Cf. aurantiacus* gave rise to a hypothetical model structure, in which the 18 and the 11 kDa polypeptides are constituents of the envelope (Feick and Fuller, 1984). Immunogold-labeling electron microscopy performed subsequently showed, however, that each of the three antisera raised against the 18, the 11, or the 5.6 kDa polypeptides of *Chloroflexus* chlorosomes reacted with isolated chlorosomes (Wullink et al., 1991). Consequently, the 5.6 kDa polypeptide is now considered a constituent of the envelope as well.

B. Chlorosome Core

Originally, it was proposed that the rod elements of the chlorosome core were composed of regularly arranged antenna BChl *c*-polypeptide complexes (Feick and Fuller, 1984). However, through numerous studies, performed more recently, the hypothesis has been developed that the proper arrangement of BChl *c* molecules within the chlorosome does not require the presence of polypeptides (Holzwarth et al., 1990, 1992; Matsuura and Olson, 1990; Matsuura et al., 1993; see Chapter 20 by Blankenship et al.).

C. Baseplate

It is assumed that the baseplates of *Chloroflexus* chlorosomes contain the chlorosome-specific BChl *a*-polypeptide complex of 5.8 kDa (Sprague and Fuller, 1991). (This complex should not be confused with the 5.6 kDa polypeptide believed to be localized in the chlorosome envelope). As mentioned above, the baseplate BChl *a*-polypeptide usually co-purifies with chlorosomes. It can, however, be detached from chlorosome preparations without impairing their structure and function.

Chlorosomes from different members of the green sulfur bacteria, like *Cb. limicola* f. *thiosulfatophilum*, *Cb. vibrioforme*, *Cb. phaeobacteroides* and *Prosthecochloris aestuarii*, contain small but significant amounts of BChl *a* with an absorption peak near 800 nm (Gerola and Olson, 1986; Otte et al., 1991). This chlorosomal BChl *a* seems to be involved in energy transfer from the pool of major antenna pigments in the chlorosome to the CM-bound moiety of the photosynthetic apparatus. In contrast to chlorosomes of *Cf. aurantiacus*, however, the chlorosomal BChl *a* of Chlorobiaceae absorbing near 800 nm appears to be functionally connected to a second type of BChl *a*-protein complex. This complex, with an absorption peak at 809 nm, is commonly referred to as the 'water soluble BChl a-protein'. Originally, this water soluble complex was postulated to form the baseplate of green sulfur bacteria (Staehelin et al., 1980). However, in more recent models, BChl *a* moieties of chlorosomes rather than the water soluble BChl *a*-complexes of Chlorobiaceae are referred to as baseplates because of their functional similarities to BChl *a* (790 nm) complexes of baseplates of *Cf. aurantiacus* (Otte et al., 1991).

V. Development of the Photosynthetic Apparatus

A. Control of Bacteriochlorophyll Synthesis

Most of the BChl *a* of green bacteria is localized in the CM and most of the BChls *c, d* or e in chlorosomes. Therefore, formation of these different BChl species can be used, in a first approximation, to follow the formation of the two structurally as well as functionally defined units of the photosynthetic apparatus. Moreover, the ratio of BChl *a/c* is a basis for estimating the ratio of the CM-bound to the chlorosomal moieties of the photosynthetic apparatus of *Cf. aurantiacus*.

For obvious reasons, light may be considered the physiologically most relevant external factor involved in the control of the development of the photosynthetic apparatus. However, factors like oxygen or the proper supply of substrates and cofactors may become important, too. In the following, the effects of these physiological factors are detailed as well as those of physiologically probably less important inhibitors.

1. Light

The first studies on the regulation of BChl synthesis in green bacteria were performed with '*Chloropseudomonas ethylicum*' strain 2-K, which was later identified as *Prosthecochloris aestuarii* (Holt et al., 1966; Trüper and Pfennig, 1978). These cultures increased their specific BChl *c* contents by a factor of 5–8 when illumination was increased from about 0.1 to 110 klx. (Interestingly, the specific growth rate of the culture was hardly affected by this variation in light energy supply). A comparable, however less pronounced effect, was observed by Broch-Due et al. (1978) upon analyzing light dependent changes in BChl *d* contents of *Cb. limicola*. Apparently, BChl *a* contents stayed largely constant under the various conditions tested by Broch-Due et al. (1978). As described above, *Cb. limicola* and *Cb. vibrioforme* were shown to eventually shift from BChl *d* to BChl *c* formation when adapting to conditions of low illumination (Broch-Due and Ormerod, 1978; Bobe et al., 1990). These shifts are believed to increase the packing of BChl within the chlorosome. Obviously, synthesis and the proper packing of an enormous amount of light-harvesting antenna BChl enables members of the Chlorobiaceae to grow at extremely low illumination. This conclusion is supported by the fact that a strain of *Cb. phaeobacteroides* (containing BChl *e*) could be isolated from a level of the Black Sea as deep as 80 m (Overmann et al., 1992).

The first more systematic analysis of the control of BChl *a* and *c* synthesis in *Cf. aurantiacus* was performed by Pierson and Castenholz (1974b). When illumination was increased from 0.25 to 54 klx, specific BChl *c* contents decreased in a negatively exponential fashion from 23.42 to 0.63 (μg per mg dry weight of cells), while specific BChl *a* contents decreased from 2.27 to 0.78. Thus the molar ratio of BChl *c/a* decreased from 10.3 to 0.8. These results show that specific BChl contents vary in inverse proportion with variations in illumination. However, BChl *c* contents are subjected to a significantly higher degree of variation than are BChl *a* contents.

Schmidt et al. (1980), analyzing effects of different illuminations at the level of isolated chlorosomes, showed that a decrease in illumination from 2 to 0.4 klx caused an increase in the specific BChl *c* content of chlorosomes from 69 to 348 (μg per mg of protein), while the carotenoid content stayed almost constant at about 30 μg per mg of protein. Consequently, chlorosomes from highly illuminated cells exhibited a three-times higher ratio of carotenoid to BChl *c* than chlorosomes from cells grown at low illumination. The functional implications of these observations are not as yet understood.

All of the results mentioned above were obtained with batch cultures of *Cf. aurantiacus* grown in a complex medium. Therefore, to avoid the possibility of uncontrolled culture conditions in studying phenomena of control, the effects of light were studied with serine-limited chemostat cultures of *Cf. aurantiacus* (Oelze, 1992). Increasing illumination from 5 to 25 klx caused decreases in specific steady state contents of both BChls *a* and *c*. Since, however, BChl *c* formation was more affected by the shift-up in illumination, the molar ratio of BChl *c/a* decreased from 9.5 to 6.

Studies performed with members of the non-sulfur purple bacteria on the control of BChl synthesis gave rise to the hypothesis that external factors, like light and oxygen, exert their controlling effects at the level of the initial reaction of tetrapyrrole synthesis, i.e. the formation of 5-aminolevulinic acid (ALA). To confirm this hypothesis in *Cf. aurantiacus*, serine-limited chemostat cultures adapted to 50 klx were supplied with 10 mM ALA (Oelze, 1992). This resulted in an increase of the initially low specific BChl *c* content by a factor of four (Fig. 6). The effect was less pronounced at the level of BChl *a*, so that the steady state molar ratio of BChl *c/a* increased from 5 to 12. This means that, in spite of high illumination (50 klx), addition of ALA to the culture resulted in a BChl *c/a* ratio which is typical of low-light cells (see also, Pierson and Castenholz, 1974b). As expected, addition of 10 mM of the growth limiting substrate, serine, increased the steady state biomass level. However, addition of serine stimulated the level of BChl *c* to a much lesser extent than did the addition of ALA.

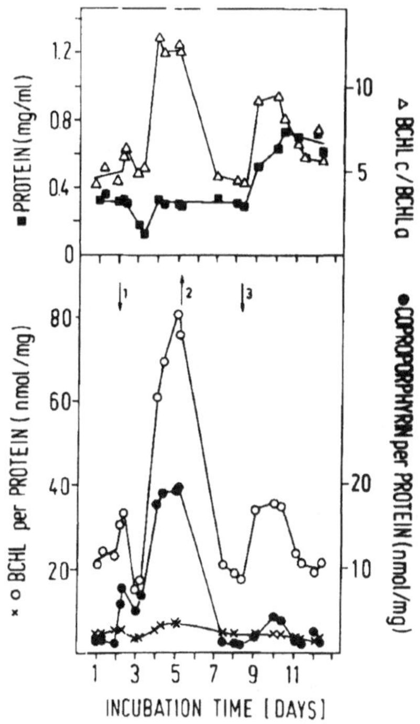

Fig. 6. Dependence of cell protein levels and tetrapyrrole formation by *Chloroflexus aurantiacus* on ALA (10 mM) or on increasing the serine supply from 5 to 15 mM. Serine-limited chemostat cultures were grown at 50 klx and a constant dilution rate of D = 0.068 [h^{-1}]. At the times indicated, ALA was added to the culture (arrow 1), ALA supply was stopped (arrow 2), and serine supply was increased (arrow 3). BChl *a* (×), BChl *c* (○). (Reproduced from Oelze, 1992 with permission).

2. Oxygen

In contrast to members of the green sulfur bacteria, *Cf. aurantiacus* can grow not only anaerobically in the light (i. e. phototrophically) but also with aeration in the dark (i.e., chemoheterotrophically) (Madigan and Brock, 1977; Foster et al., 1986). Under these latter conditions, the photosynthetic apparatus is not required. Nevertheless, the organisms still synthesized BChl when growing with low aeration. Vigorous aeration, adjusting the dissolved oxygen concentration to about 180 μM, is required to completely repress BChl synthesis. On the other hand, when the dissolved oxygen was adjusted to concentrations in the range of 1 to 50 μM, BChl was still synthesized in the dark (Foster et al., 1986). Importantly, during the first phase of adaptation from high to low aeration, BChl *a* was synthesized at a higher rate than BChl *c*. This result was in contrast to a previous observation (Fuller and Redlinger, 1985). It appears possible, however, that this discrepancy may result from the fact that the two cited investigations were started with cultures containing different initial levels of BChl.

In the phototrophic purple bacteria, formation of accessory light-harvesting units is generally more sensitive toward oxygen than formation of the photochemical reaction center complex (Oelze, 1981). The same principle can be observed with respect to the accessory antenna BChl *c* units and the reaction center BChl *a* complex of *Cf. aurantiacus*. This was shown recently, when phototrophic chemostat cultures adapted to anaerobiosis were suddenly exposed to air (Oelze, 1992). Upon aeration the specific rates of BChl synthesis decreased from 0.086 h^{-1} to 0.052 h^{-1} for BChl *a* yet to 0.039 h^{-1} for BChl *c*. In contrast to the inhibition of BChl *c* synthesis by high illumination, the effect of oxygen could not be compensated for by addition of ALA. Obviously, this means that oxygen affects BChl *c* synthesis not only at the level of ALA formation.

3. Supply of Nutrients

Phototrophic chemostat cultures of *Cf. aurantiacus* supplied with the complex medium of Pierson and Castenholz (1974a) showed a negatively exponential decrease in specific BChl *c* contents with increasing the growth rate, while specific BChl *a* contents decreased linearly (Oelze and Fuller, 1987). Accordingly, the molar ratio of BChl *c/a* decreased with increasing growth rate, although the external conditions of illumination were not changed. Actually, variations in illumination did not influence specific BChl contents adjusted to a given growth rate. Thus, it was concluded that the growth rate or factors constituting the growth rate, like the rate of substrate supply, were involved in the control of BChl synthesis. To further analyze the basis of these observations, the consumption of the most prominent substrates of the complex medium, i. e. amino acids, was determined (Oelze et al., 1991). This revealed a characteristic pattern of amino acid utilization by batch and chemostat cultures of *Cf. aurantiacus*. In particular, it seems important to note that consumption of Glx (representing glutamate and glutamine) was high at low growth rates and decreased with increasing growth rates. The opposite was true with respect to consumption of Asx (i.e. aspartate and asparagine). In evaluating these results, it should be remembered that *Cf. aurantiacus* employs the C5 pathway of tetrapyrrole synthesis, which is initiated by the conversion of glutamic acid into ALA (Avissar et al., 1989; Kern and Klemme, 1989). This suggests that the BChl *c/a* ratio decreased because consumption of glutamate (i. e. the precursor of BChl) decreased with increasing growth rate on complex medium as well. Again, the shortage in BChl precursors affected formation of BChl *c* to a significantly higher extent than formation of BChl *a*. Biomass production, however, was warranted under these conditions by the increased consumption of amino acids like Asx.

An explanation for the decreasing rate of Glx utilization became apparent when *Cf. aurantiacus* was grown phototrophically in glutamate-limited chemostat culture. Upon increasing the growth rate, increasing steady state amounts of glutamate became detectable in the culture fluid (Oelze and Söntgerath, 1992) suggesting the presence of a cellular uptake system with relatively low affinity for glutamate.

Serine, on the other hand, is an amino acid, which, at any growth rate tested, was almost completely taken up by *Cf. aurantiacus*. Consequently, steady state specific BChl *c* contents as well as the molar ratio of BChl *c/a* of serine-limited chemostat cultures increased with increasing growth rate.

In addition to growth substrates, vitamins have been reported to be essential for the proper formation of BChl complexes. For instance, *Cf. aurantiacus* produced significantly higher amounts of BChls *c* and *a* when grown in the presence of the vitamin solution introduced by Madigan et al. (1974) than in its absence (Oelze and Söntgerath, 1992). Very recently, BChl formation has been studied with the vitamin B_{12}-dependent strain 1230 of *Cb. limicola* (Fuhrmann et al., 1993). Cells grown in the presence of 20 μg of vitamin B_{12} per liter exhibited, respectively, 5 and 2.4 times higher specific BChl *c* and BChl *a* contents than control cells grown in the absence of the vitamin (see Fig. 10). Slight effects of vitamin B_{12} on BChl *d* and *a* synthesis were observed, as well, with the vitamin B_{12}-independent strain 8327 of *Cb. vibrioforme*.

4. Non-Physiological Inhibitors

Gabaculine is an inhibitor of transaminase reactions and, thus, of the initial step of tetrapyrrole formation via the C5 pathway (Kern and Klemme, 1989). When *Cf. aurantiacus* was grown in the presence of gabaculine, formation of BChl *c* was inhibited with considerably higher sensitivity than was BChl *a*, while biomass production was hardly affected by the inhibitor (Oelze, 1992). According to the site of inhibition, addition of ALA completely abolished the inhibitory effect of gabaculine (Fig. 7), which indicates that BChl *c* synthesis was stimulated to a much higher extent than BChl a synthesis.

Several anesthetic gases have been described to comprise a group of inhibitors of BChl synthesis in *Cb. vibrioforme* (Ormerod et al., 1990). When added to phototrophic cultures of *Cb. vibrioforme*, gases such as nitrous oxide (N_2O), ethylene, and acetylene inhibited the synthesis of the accessory BChl *d*, but not of BChl *a*. Interestingly, mutants could be isolated which, except for N_2O, were resistant to the gases mentioned above.

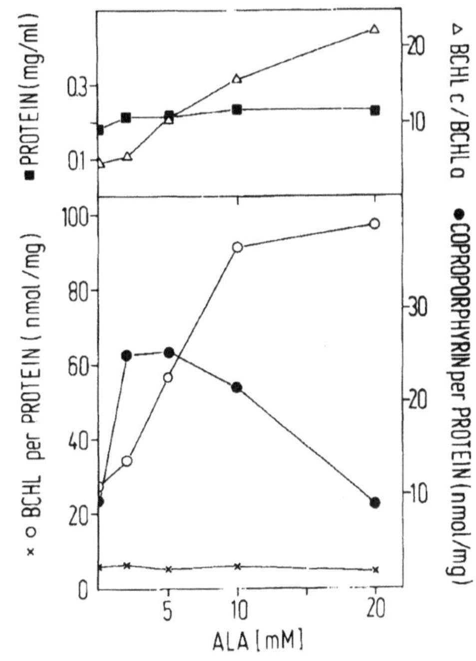

Fig. 7. Effects of different concentrations of ALA on phototrophic batch cultures of *Chloroflexus aurantiacus*. The organisms were grown at 25 klx in batch culture in the presence of 2 μM of gabaculine for 72 h. Cell protein concentrations, the molar ratio of BChl *c/a*, and specific contents of BChl *a* (×), BChl *c* (O), and coproporphyrin are shown. (Reproduced from Oelze, 1992, with permission).

5. Conclusions

All of the results reviewed above on the control of BChl synthesis in green bacteria show that formation of the accessory BChls *c, d* or *e* is much more affected by controlling factors than is formation of BChl *a*. The physiological significance of this effect is evident: Formation of the central unit of the photosynthetic apparatus, i.e. the photochemical reaction center BChl *a* complex, occurs as far as possible under conditions of restricted BChl synthesis. However, the mechanisms channeling tetrapyrrole precursors into the formation of either BChl *a* or the accessory types of BChl are not as clear. In an ALA synthetase-deficient mutant of *Rhodobacter sphaeroides* it has been shown that formation of the reaction center complex becomes saturated at a much lower rate of ALA supply than formation of the accessory light-harvesting unit (Oelze, 1983). Operation of a comparable mechanism may explain the differential stimulation of BChl *a* and BChl *c* synthesis in *Cf. aurantiacus* upon addition of ALA to cultures inhibited in BChl synthesis either by high illumination or by the presence of gabaculine. This might mean that formation of the CM-bound reaction center BChl *a* complex requires only low levels of tetrapyrrole precursors, while the resulting surplus of precursors is converted into accessory BChl *c* units. In this context it is interesting to note that, at very high supplies of ALA, cells excrete tetrapyrrole precursors into the growth medium (Oelze, 1992). This suggests that packing of BChl *c* within the cells may become limited because of limited availability of chlorosomes or of space within the chlorosomes. This in turn, leads to questions as to the coordination of BChl *c* synthesis and chlorosome formation.

B. Development of Chlorosomes

Variations in the cellular contents of BChls *c, d,* or *e* mean that the light-harvesting capacity of the cells is subject to variation. Since by far most, if not all, of the three types of BChl is located in chlorosomes, the question arises whether variation of the cellular levels of the chlorosomal pigments is accompanied by i.) changes in the size of the already existing chlorosomes, ii.) changes in the number of chlorosomes per unit of cell, or iii.) both changes in number and size of chlorosomes. Evidence available as yet suggest that all of the three possibilities may take place if the cells are grown with different illuminations or other factors controlling the synthesis of chlorosomal pigments (Foidl et al., 1994).

The first report on the effect of light on the photosynthetic apparatus of green bacteria suggested that regulation of the cellular BChl content of *Prosthecochloris aestuarii* (formerly *Chloropseudomonas ethylica*) was achieved by changes in numbers as well as by changes in BChl content of chlorosomes (Holt et al., 1966). As far as known, the latter change is accompanied by changes in the size of chlorosomes. A light dependent change in the size of chlorosomes was observed with *Cb. limicola*, as well (Broch-Due et al., 1978). In the latter case, however, it became apparent that chlorosome preparations isolated from cells grown at a given light intensity exhibited already considerable variations in size. Similar results were obtained with *Cf. aurantiacus* as well. More detailed studies on this aspect performed with *Cf. aurantiacus* revealed that changes in size are largely due to changes in length as well as in height rather than to changes in the width of chlorosomes (Figs. 8, 9) (Sprague et al., 1981a; Golecki and Oelze, 1987). Yet, like other members of the green phototrophic bacteria, *Cf. aurantiacus* changes not only the size but also the cellular number of chlorosomes (Schmidt et al., 1980; Sprague et al., 1981b; Golecki and Oelze, 1987).

In contrast, Pierson and Castenholz (1974b) suggested in their early paper that the cellular BChl *c* content of *Cf. aurantiacus* was regulated by the number of chlorosomes. This suggestion was taken up by Feick et al. (1982) who concluded that cells of *Cf. aurantiacus* increased only the number of chlorosomes upon adaptation from high to low light conditions. This conclusion was based, first, on the isolation of chlorosomes exhibiting identical ratios of BChl *c/a* irrespective of the light regimen employed to grow the cultures and, secondly, on the assumption that the BChl *c/a* ratio represents the size of chlorosomes.

On the other hand, an increase in only the size of chlorosomes has been observed very recently in *Cb. phaeobacteroides* strain MN1, which showed increased BChl *e* contents after adaptation to low light (Fuhrmann et al., 1993). In addition, resting as well as chemostat cultures of *Cf. aurantiacus* supplied with ALA increased the cellular content of BChl *c* as well as the volumes of chlorosomes rather than their numbers (Figs. 4a,b, 8 and Table 1). As demonstrated in Fig. 9, these changes in volume were largely due to changes in the length of chlorosomes, while the height varied only in the range of 15 to 17 nm.

Moreover, the data compiled in Table 1 demonstrate that both types of chlorosome development mentioned above, i.e. increases in number and/or size, may take place under appropriate conditions (Foidl et al., 1994).

In chemostat cultures growing under steady state conditions the steady state number of chlorosomes per area of membrane stayed constant (Table 1, culture 1C). This means that new chlorosomes of constant volume were steadily formed at a constant rate. In contrast, resting cells did not form any additional chlorosomes. Upon stimulation of BChl *c* synthesis by addition of ALA, however, chemostat cells and resting cultures increased the volume of chlorosomes. The possibility of an increase in the number of chlorosomes without significant changes in volume is demonstrated with cultures of comparable BChl *c/a* ratios as well as chlorosome sizes but different BChl *c* contents (cultures 1C and 2B of Table 1).

On the basis of all of the results reported above, we propose that growing cells steadily form new chlorosomes and fill them up with pigments. Under these conditions the average volume of chlorosomes stays constant. If, however, this equilibrium becomes impaired, e. g. by an increased supply of ALA, the average volume of chlorosomes increases because of enhanced BChl *c* incorporation.

C. Formation of the Photosynthetic Apparatus

The ability of *Cf. aurantiacus* to grow both phototrophically in the light and chemotrophically in the dark makes them unique organisms to study de novo formation of the photosynthetic apparatus of green bacteria. However, cultures of *Cf. aurantiacus* need not be transferred from chemotrophic to phototrophic conditions in order to synthesize the

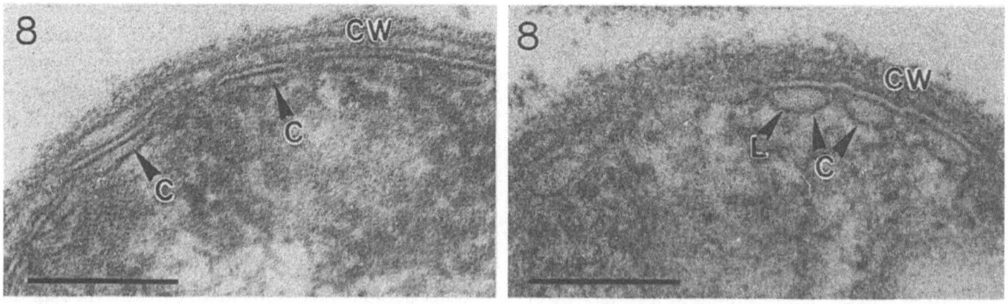

Fig. 8. Ultrathin sections of *Chloroflexus aurantiacus* with low BChl *c* contents (7.7 nmole·mg of protein^{-1}) and flat chlorosomes (left) and with high BChl *c* contents (48.5 nmole·mg of protein^{-1}) and thick chlorosomes (right). Cell wall (CW). Bar represents 100 nm.

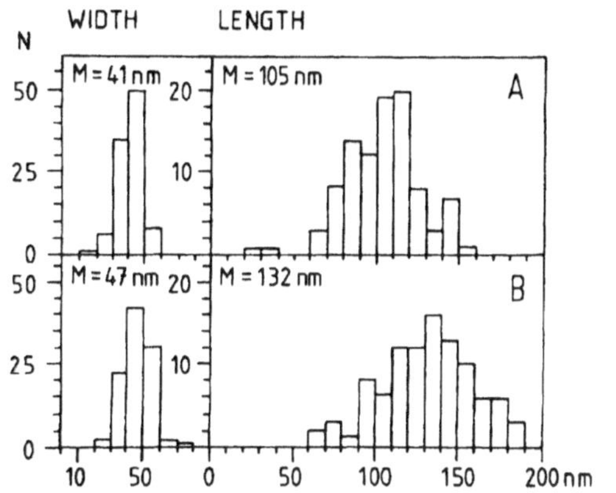

Fig. 9. Size histograms of baseplates of cultures of *Chloroflexus aurantiacus* grown in the presence (B) and absence (A) of ALA. Cultures 1C and 2C of Table 1 were analyzed.

photosynthetic apparatus. As is known for the photoheterotrophic purple bacteria, *Cf. aurantiacus* is also able to induce the photosynthetic apparatus when transferred from conditions of high to low aeration in the dark (Sprague et al., 1981a; Foster et al., 1986). Importantly, a comparative study of the induction of the synthesis of BChls *a* and *c* revealed that it is not possible to distinguish the effect of low oxygen from that of light (Fuller and Redlinger, 1985; Redlinger and Fuller, 1985).

Electron micrographs of cell samples from highly aerated dark cultures were reported to be free of any detectable chlorosomes (Sprague and Fuller, 1991). A more precise definition of oxygen concentrations showed that cultures adapted to darkness and dissolved oxygen concentrations near 180 μM O_2 were completely depleted of both BChls *a* and *c* as well as of cytochrome *c*-554, the primary electron donor to the photochemical reaction center (Foster et al., 1986). BChls *a* and *c* as well as cytochrome *c*-554 were induced in parallel and became detectable as early as 1 h after the oxygen concentration was reduced to a value near 50 μM. In an earlier study it was reported that chlorosome polypeptides became detectable 8–12 h after induction of the photosynthetic apparatus (Fuller and Redlinger, 1985; Redlinger and Fuller, 1985). On the basis of immunological results, it was suggested that the three chlorosomal polypeptides of 18, 11, and 3.7 kDa (determined by SDS-PAGE) were transcribed and translated as precursor polyproteins of 60 and 47 kDa. While the former protein reacted with antisera against all of the three chlorosomal polypeptides, the latter reacted only with antisera against the 18 and the 11 kDa polypeptides. Both polypeptide precursors were proposed to be processed into chlorosomal polypeptides in the presence of BChl, which in turn is controlled by light or oxygen (Redlinger and Fuller, 1985). Yet, a more recent study on the molecular biology of the chlorosomal polypeptides could not identify a genetic basis for the occurrence of a precursor polyprotein (Theroux et al., 1990).

A promising system to study the development of the photosynthetic apparatus in green sulfur bacteria has recently been introduced by Fuhrmann et al. (1993). These authors reported that strain 1230 of *Cb. limicola* forms chlorosomes (Fig. 10) as well as significant amounts of BChls *a* and *c* only if supplied with vitamin B_{12}. Cells grown in the absence of the vitamin are depleted of BChls and of any detectable chlorosomes. However, when subsequently supplied with B_{12}, it takes no more than 10 h to fully develop BChls and chlorosomes.

Little is known about the coordination of the synthesis of chlorosomes and the synthesis of the CM-bound moiety of the photosynthetic apparatus. According to Redlinger and Fuller (1985) as well as to Foster et al.(1986), the antenna BChl *c*-chlorosome complex is assembled by posttranslational protein processing, while the synthesis of BChls *a* and *c*, of the reaction center BChl *a*- and the antenna BChl *a*-

Table 1. Chlorosome development upon addition of 10 mM of 5-aminolevulinic acid (ALA) to serine-limited chemostat (C) and resting batch (B) cultures of *Chloroflexus aurantiacus*. In addition, specific BChl contents of cells are given. Cultures 1 and 2 were grown in the absence and presence of ALA, respectively (Foidl et al., 1994).

Culture	No. of chlorosomes per μm^2 of CM	Area of baseplates per area of CM	Volume of chlorosomes [nm^3]	BChl *a*	BChl *c*	BChl *c/a*
				nmol per mg protein		
1 C	48	0.18	45,900	3.2	16	5
2 C	48	0.26	75,300	5.0	60	12
1 B	76	0.20	23,800	5.3	11.5	2.2
2 B	60	0.26	48,800	4.8	22	4.6

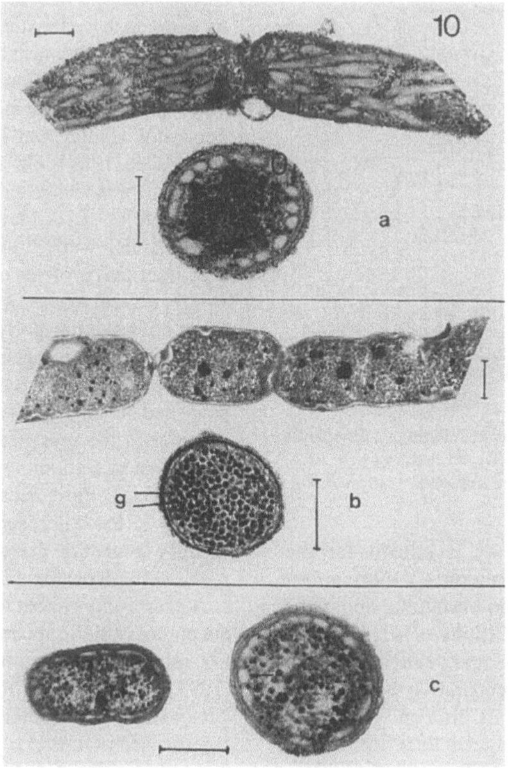

Fig. 10. Ultrathin section of *Chlorobium limicola* strain 1230 grown under different culture conditions: (a) cells grown with 20 μg vitamin B_{12}/L (growth time : 3 d), (b) vitamin B_{12}-deficient cells (growth time: 21 d), (c) addition of 20 μg/L vitamin B_{12} to the culture lacking vitamin B_{12} (growth time : 10 d). Each bar represents 0.2 μm; arrows point to chlorosomes; g: 'granular structures'. (Reproduced from Fuhrmann et al. 1993 with permission).

complexes as well as of cytochrome *c*-554, are under coordinated control of oxygen. Coordination of the synthesis of both BChls *a* and *c*, on the one hand, and, on the other, of complexes containing BChl *a* and chlorosomes was suggested by results on the relationship between cellular BChl *a* levels and chlorosome contents of *Cf. aurantiacus*. This showed a direct proportionality of specific BChl *a* contents of cells and the percentage of CM covered with chlorosomes as well as of numbers of chlorosomes per unit of CM (Golecki and Oelze, 1987).

VI. Open Questions

All of the light-harvesting pigment units of the photosynthetic apparatus of different organisms investigated so far are made up of BChl-polypeptide holochromes. With respect to chlorosomes of green bacteria, however, the picture emerges that these units may be active even if depleted of their typical polypeptides (see Chapter 20 by Blankenship et al.). In addition, it is possible to re-assemble structures from lipophilic fractions extracted from native chlorosomes which exhibit structural as well as functional properties of the original chlorosomes which lack polypeptides. This questions the significance of polypeptides in native chlorosomes. It has been stated, for example, that the 5.6 kDa polypeptide is required for a proper arrangement of pigment molecules within the chlorosome. Reconstitution of chlorosomal structures without participation of polypeptides, however, does not favor this interpretation.

An alternative function of chlorosomal polypeptides may be based on their participation in the process of chlorosome biogenesis. As detailed above, chlorosomes are defined structures underlying the CM membrane. Thus the question arises of how pigments and lipids become incorporated into the interior of chlorosomes. As far as is known on the basis of results obtained with other bacterial species, pigments like BChl and carotenoids, as well as lipids, are formed in close association with the CM. If this is also true with the green bacteria, then the chlorosome precursors have to be transferred from the CM into the chlorosomes. These processes might require participation of polypeptides. Alternatively, chlorosomal polypeptides might function in the final steps of the synthesis of chlorosomal constituents. At present, however, practically nothing is known about the function, composition, and intracellular localization of biosynthetic pathways leading to individual constituents of the chlorosomes.

Another closely related problem concerns the site of chlorosome biogenesis. Chlorosomes are always found attached to the CM. Currently it is assumed that the attachment site represents the site of the CM-bound unit of the photosynthetic apparatus. However, evidence available indicates that in cells adapting from high to low aeration, the CM-bound unit is formed before the chlorosomes. This could mean that components of the CM-bound unit function as targets in the early assembly of chlorosomes. If this is the case, the question arises how cells control the assembly of the CM-bound unit within defined patches. Electron micrographs suggest the presence of such patches. It is not known, however, whether all of the CM-bound units of the photosynthetic apparatus are confined to patches or whether all of the patches represent sites of chlorosome attachment.

All of the questions posed above require more detailed insights into various aspects of chlorosome biogenesis. We believe that the time is ready for these studies, which should yield new and important information on the assembly of biological structures.

Acknowledgments

The original investigations by J.O. were financially supported by the Deutsche Forschungsgemeinschaft. The authors would like to thank M.D. Kamen and M. Foidl for their critical comments on the manuscript.

References

Avissar YJ, Ormerod JG and Beale SI (1989) Distribution of 5-aminolevulinic acid biosynthetic pathways among phototrophic bacterial groups. Arch Microbiol 151: 513–519

Blankenship RE and Fuller RC (1986) Membrane topology and photochemistry of the green photosynthetic bacterium *Chloroflexus aurantiacus*. In: Staehelin LA and Arntzen CJ (eds) Encyclopedia of Plant Physiology, Photosynthesis III: Photosynthetic Membranes and Light-Harvesting Systems, Vol. 19, pp 390–399

Bobe FW, Pfennig N, Swanson KL and Smith KM (1990) Red shift of absorption maxima in Chlorobiineae through enzymic methylation of their antenna bacteriochlorophylls. Biochemistry 29: 4340–4348

Boyce CO, Oyewole SH and Fuller RC (1976) Localization of photosynthetic reaction center in *Chlorobium limicola*. Brookhaven Symp Biol 28: 365

Broch-Due M and Ormerod JG (1978) Isolation of a BChl *c* mutant from *Chlorobium* with BChl *d* by cultivation at low light. FEMS Microbiol Lett 3: 305–308

Broch-Due M, Ormerod JG and Fjerdingen B (1978) Effect of light intensity on vesicle formation in *Chlorobium*. Arch Microbiol 116: 269–274

Cohen-Bazire G (1963) Some observations on the organization of the photosynthetic apparatus in purple and green bacteria. In: Gest H, San Pietro A and Vernon LP (eds) Bacterial Photosynthesis, pp 89–119. Antioch Press, Yellow Springs, Ohio

Cohen-Bazire G, Pfennig N and Kunisawa R (1964) The fine structure of green bacteria. J Cell Biol 22: 207–225

Cruden DL and Stanier RY (1970) The characterization of Chlorobium vesicles and the membranes isolated from green bacteria. Arch Microbiol 72: 115–134

Dubinina GA and Gorlenko VM (1975) New filamentous photosynthetic green bacteria containing gas vacuoles. Microbiology USSR (English translation) 44: 511–517

Ebersold HR, Cordier JL and Lüthy P (1981) Bacterial mesosomes: method dependent artifacts. Arch Microbiol 130: 19–22

Eckhardt A, Brunisholz R, Frank G and Zuber H (1990) Selective solubilization of chlorosome proteins in *Chloroflexus aurantiacus*. FEBS Lett 267: 199–202

Fages F, Griebenow N, Griebenow K, Holzwarth AR and Schaffner K (1990) Characterization of light-harvesting pigments of *Chloroflexus aurantiacus*. Two new chlorophylls: oleyl (octadec-9-enyl) and cetyl (hexadecanyl) bacteriochlorophyllides-c. J Chem Soc Perkin Trans 1: 2791–2797

Feick RG and Fuller RC (1984) Topography of the photosynthetic apparatus of *Chloroflexus aurantiacus*. Biochemistry 23: 3693–3700

Feick RG, Fitzpatrick M and Fuller RC (1982) Isolation and characterization of cytoplasmic membrane and chlorosomes from the green bacterium *Chloroflexus aurantiacus*. J Bacteriol 150: 905–915

Feiler U, Nitschke W and Michel H (1992) Characterization of an improved reaction center preparation from the photosynthetic green sulfur bacterium *Chlorobium* containing the FeS centers F_A and F_B and a bound cytochrome subunit. Biochemistry 31: 2608–2614

Foidl M, Golecki JR and Oelze J (1994) Bacteriochlorophyll *c* formation and chlorosome development in *Chloroflexus aurantiacus*. Photosynth Res 41: 145–150

Foster JM, Redlinger TE, Blankenship RE and Fuller RC (1986) Oxygen regulation of development of the photosynthetic membrane system in *Chloroflexus aurantiacus*. J Bacteriol 167: 655–659

Fowler CF, Nugent NA and Fuller RC (1971) The isolation and characterization of a photochemically active complex from *Chloropseudomonas ethylica*. Proc Natl Acad Sci USA 68: 2278–2282

Fuhrmann S, Overmann J, Pfennig N and Fischer U (1993) Influence of vitamin B_{12} and light on the formation of chlorosomes in green- and brown-colored *Chlorobium* species. Arch Microbiol 160: 193–198

Fuller RC and Redlinger TE (1985) Light and oxygen regulation of the development of the photosynthetic apparatus in *Chloroflexus aurantiacus*. In: Steinbeck KE, Bonitz S, Arntzen CL and Bogorad L (eds) Molecular Biology of the Photosynthetic Apparatus, pp 155–162. Cold Spring Harbor Laboratory, Cold Spring Harbor, New York

Gerola PD and Olson JM (1986) A new bacteriochlorophyll *a*-protein complex associated with chlorosomes of green sulfur bacteria. Biochim Biophys Acta 848: 69–76

Gerola PD, Hojrup P and Olson JM (1988) A comparison of the bacteriochlorophyll *c*-binding proteins of *Chlorobium* and *Chloroflexus*. In: Scheer H and Schneider S (eds) Photosynthetic Light-Harvesting Systems, Organization and Function, pp 129–139. de Gruyter, Berlin, New York

Gibson J, Pfennig N and Waterbury JB (1984) *Chloroherpeton thalassium* gen. nov. et spec. nov., a non filamentous, flexing and gliding bacterium. Arch Microbiol 156: 96–101

Golecki JR and Oelze J (1980) Differences in the architecture of cytoplasmic and intracytoplasmic membranes of three chemotrophically and phototrophically grown species of the Rhodospirillaceae. J Bacteriol 144: 781–788

Golecki JR and Oelze J (1987) Quantitative relationship between bacteriochlorophyll content, cytoplasmic membrane structure and chlorosome size in *Chloroflexus aurantiacus*. Arch Microbiol 148: 236–241

Griebenow K and Holzwarth AR (1989) Pigment organization and energy transfer in green bacteria. 1. Isolation of native chlorosomes free of bound bacteriochlorophyll *a* from *Chloroflexus aurantiacus* by gel-electrophoretic filtration. Biochim Biophys Acta 973: 235–240

Griebenow K, Holzwarth AR and Schaffner K (1990) The 5.6-kilodalton protein in isolated chlorosomes from *Chloroflexus aurantiacus* strain Ok-70-fl is a degradation product. Z Naturforsch 45c: 823–828

Holo H, Broch-Due M and Ormerod JG (1985) Glycolipids and the structure of chlorosomes in green bacteria. Arch Microbiol 143: 94–99

Holt SC, Conti SF and Fuller RC (1966) Photosynthetic apparatus in the green bacterium *Chloropseudomonas ethylicum*. J Bacteriol 91: 311–323

Holzwarth AR, Griebenow K and Schaffner K (1990) A photosynthetic antenna system which contains a protein-free chromophore aggregate. Z Naturforsch 45c: 203–206

Holzwarth AR, Griebenow K and Schaffner K (1992) Chlorosomes, photosynthetic antennae with novel self-organized pigment structures. J Photochem Photobiol A: Chem 65: 61–71

Hurt CH and Hauska G (1984) Purification of membrane-bound cytochromes and a photoreactive P840 protein complex of the green sulfur bacterium *Chlorobium limicola* f. *thiosulfatophilum*. FEBS Lett 168: 149–153

Huster MS and Smith KM (1990) Biosynthetic studies of substituent homologation in bacteriochlorophylls *c* and *d*. Biochemistry 29: 4348–4355

Jürgens UJ, Meißner J, Fischer U, König WA and Weckesser J (1987) Ornithine as a constituent of the peptidoglycan of *Chloroflexus aurantiacus*, diaminopimelic acid in that of *Chlorobium vibrioforme* f. *thiosulfatophilum*. Arch Microbiol 148: 72–76

Kern M and Klemme J-H (1989) Inhibition of bacteriochlorophyll biosynthesis by gabaculin (3-amino, 2,3-dihydrobenzoic acid) and presence of an enzyme of the C_5-pathway of 5-aminolevulinate synthesis in *Chloroflexus aurantiacus*. Z Naturforsch 44: 77–80

Larsen M, Mack EE and Pierson BK (1991) Mesophilic *Chloroflexus*-like organisms from marine and hypersaline environments. VII International Symp Photosyn Prokaryotes, Amherst, Mass USA, Abstracts p 169

Madigan MT and Brock TD (1977) Chlorobium-type vesicles of photosynthetically-grown *Chloroflexus aurantiacus* observed using negative staining techniques. J Gen Microbiol 102: 279–285

Madigan TM, Petersen SR and Brock TD (1974) Nutritional studies on *Chloroflexus*, a filamentous photosynthetic, gliding bacterium. Arch Microbiol 100: 97–103

Matsuura K and Olson JM (1990) Reversible conversion of aggregated bacteriochlorophyll *c* to the monomeric form by 1-hexanol in chlorosomes from *Chlorobium* and *Chloroflexus*. Biochim Biophys Acta 1019: 233–238

Matsuura K, Hirota M, Shimada K and Mimuro M (1993) Spectral forms and the orientation of bacteriochlorophylls *c* and *a* in chlorosomes of the green photosynthetic bacterium *Chloroflexus aurantiacus*. Photochem Photobiol 57: 92–97

van de Meent EJ, Kobayashi M, Erkelens C, van Veelen PA, Otte SCM, Inoue K, Watanabe T and Amesz J (1992) The nature of the primary electron acceptor in green sulfur bacteria. Biochim Biophys Acta 1102: 371–378

Niedermeier G, Scheer H and Feick RG (1992) The functional role of protein in the organization of bacteriochlorophyll *c* in chlorosomes of *Chloroflexus aurantiacus*. Eur J Biochem 204: 685–692

Nitschke W, Feiler U and Rutherford AW (1990) Photosynthetic reaction center of the green sulfur bacteria studied by EPR. Biochemistry 29: 3834–3842

Oelze J (1981) Composition and development of the bacterial photosynthetic apparatus. In: Roodyn DB (ed) Subcellular Biochemistry, Vol. 8, pp 1–73. Plenum Press, New York and London

Oelze J (1983) Control of the formation of bacteriochlorophyll, and B 875- and B 850-bacteriochlorophyll complexes in *Rhodopseudomonas sphaeroides* mutant strain H5. Arch Microbiol 136: 312–316

Oelze J (1992) Light and oxygen regulation of the synthesis of bacteriochlorophylls *a* and *c* in *Chloroflexus aurantiacus*. J Bacteriol 174: 5021–5026

Oelze J and Drews G (1972) Membranes of photosynthetic bacteria. Biochim Biophys Acta 265: 209–239

Oelze J and Fuller RC (1987) Growth rate and control of the development of the photosynthetic apparatus in *Chloroflexus aurantiacus*. Arch Microbiol 148: 132–136

Oelze J and Söntgerath B (1992) Differentiation of the photosynthetic apparatus of *Chloroflexus aurantiacus* depending on growth with different amino acids. Arch Microbiol 157: 141–147

Oelze J, Jürgens UW and Ventura S (1991) Amino acid consumption by *Chloroflexus aurantiacus* in batch and continuous culture. Arch Microbiol 156: 266–269

Olson, JM (1980) Chlorophyll organization in green photosynthetic bacteria. Biochim Biophys Acta 594: 33–51

Olson JM, Giddings TH and Shaw EK (1976) An enriched reaction center preparation from green photosynthetic bacteria. Biochim Biophys Acta 449: 197–208

Olson JM, Brune DC and Gerola PD (1990) Organization of chlorophyll in chlorosomes. In: Drews G and Dawes EA (eds) Molecular Biology of Membrane-Bound Complexes in Phototrophic Bacteria, pp 227–234. Plenum Press, New York

Ormerod JG, Nesbakken T and Beale SI (1990) Specific inhibition of antenna bacteriochlorophyll synthesis in *Chlorobium vibrioforme* by anesthetic gases. J Bacteriol 172: 1352–1360

Otte SCM, van der Heiden JC, Pfennig N and Amesz J (1991) A comparative study of the optical characteristics of intact cells of photosynthetic green sulfur bacteria containing bacteriochlorophyll *c*, *d* or *e*. Photosyn Res 28: 77–87

Otte SCM, van de Meent EJ, van Veelen PA, Pundsnes AS and Amesz J (1993) Identification of the major chlorosomal bacteriochlorophylls of the green sulfur bacteria *Chlorobium vibrioforme* and *Chlorobium phaeovibrioides*; their function in lateral energy transfer. Photosynth Res 35: 159–169

Overmann J, Cypionka H and Pfennig N (1992) An extremely low light adapted phototrophic sulfur bacterium from the Black Sea. Limnol Oceanogr 37: 150–155

Pierson BK and Castenholz RW (1974a) A phototrophic gliding filamentous bacterium of hot springs, *Chloroflexus aurantiacus*. Arch Microbiol 100: 5–24

Pierson BK and Castenholz RW (1974b) Studies of pigments and growth in *Chloroflexus aurantiacus*, a phototrophic filamentous bacterium, Arch Microbiol 100: 283–305

Pierson BK and Castenholz RW (1978) Photosynthetic apparatus and cell membranes of the green bacteria. In: Clayton RK and Sistrom WR (eds) The Photosynthetic Bacteria, pp 179–197. Plenum Press, New York, London

Pierson BK and Thornber JP (1983) Isolation and spectral characterization of photochemical reaction centers from the thermophilic green bacterium *Chloroflexus aurantiacus* strain J-10-fl. Proc Natl Acad Sci USA 80: 80–84

Puchovka NN, Gorlenko VM and Pivovarova TA (1975) A comparative structural study of vibroid green sulfur bacteria. Microbiology USSR (English translation) 44: 89–94

Redlinger TE and Fuller RC (1985) Protein processing as a regulatory mechanism in the synthesis of the photosynthetic antenna in *Chloroflexus aurantiacus*. Arch Microbiol 141: 344–347

Schmidt K, Maarzahl M and Mayer F (1980) Development and pigmentation of chlorosomes in *Chloroflexus aurantiacus* OK-70-fl. Arch Microbiol 127: 87–97

Schmidt K (1980) A comparative study on the composition of chlorosomes (Chlorobium vesicles) and the cytoplasmic membrane from *Chloroflexus aurantiacus* strain OK-70-fl and *Chlorobium limicola* f. *thiosulfatophilum* strain 6230. Arch Microbiol 124: 21–31

Shiozawa JA, Lottspeich F and Feick R (1987) The photochemical reaction center of *Chloroflexus aurantiacus* is composed of two structurally similar polypeptides. Eur J Biochem 167: 595–600

Smith KM, Craig GW, Kehres LA and Pfennig N (1983) Reversed-phase high performance liquid chromatography and structural assignments of the bacteriochlorophylls *c*. J Chromatogr 281: 209–223

Sprague SG and Fuller RC (1991) The green phototrophic bacteria and heliobacteria. In: Stolz JF (ed) Structure of Phototrophic Prokaryotes, pp 79–103. CRC Press, Boca Raton

Sprague SG, Staehelin LA and Fuller RC (1981a) Semiaerobic induction of bacteriochlorophyll synthesis in the green bacterium *Chloroflexus aurantiacus*. J Bacteriol 147: 1032–1039

Sprague SG, Staehelin LA, DiBartolomeis MJ and Fuller RC (1981b) Isolation and development of chlorosomes in the green bacterium *Chloroflexus aurantiacus*. J Bacteriol 147: 1021–1031

Staehelin LA, Golecki JR and Drews G (1980) Supramolecular organization of chlorosomes (Chlorobium vesicles) and of their membrane attachment sites in *Chlorobium limicola*. Biochim Biophys Acta 589: 30–45

Staehelin LA, Golecki JR, Fuller RC and Drews G (1978) Visualization of the supramolecular architecture of chlorosomes (Chlorobium type vesicles) in freeze-fractured cells of *Chloroflexus aurantiacus*. Arch Microbiol 119: 269–277

Stolz JF, Fuller RC and Redlinger TE (1990) Pigment-protein diversity in chlorosomes of green phototrophic bacteria. Arch Microbiol 154: 422–427

Theroux SJ, Redlinger TE, Fuller RC and Robinson SJ (1990) Gene encoding the 5.7-kilodalton chlorosome protein of *Chloroflexus aurantiacus*; regulated message level and a predicted carboxy-terminal protein extension. J Bacteriol 172: 4497–4504

Trüper HG and Pfennig N (1978) Taxonomy of the Rhodospirillales. In: Clayton RK and Sistrom WR (eds) The Photosynthetic Bacteria, pp 19–27. Plenum Press, New York

Vasmel H, Dorssen RJ, de Vos GJ and Amesz J (1986) Pigment organization and energy transfer in the green photosynthetic bacterium *Chloroflexus aurantiacus;* I. The cytoplasmic membrane. Photosynth Res 7: 281–294

Wahlund TM, Woese CR, Castenholz RW and Madigan M (1991) A thermophilic green sulfur bacterium from New Zealand hot springs, *Chlorobium tepidum* sp. nov. Arch Microbiol 156: 81–90

Wagner-Huber R, Brunisholz R, Frank G and Zuber H (1988) The BChl *c/e*-binding polypeptides from chlorosomes of green photosynthetic bacteria. FEBS Lett 239: 8–12

Wechsler TD, Brunisholz RA, Frank G and Zuber H (1991) Isolation and protein chemical characterization of the B806-866 antenna complex of the green thermophilic bacterium *Chloroflexus aurantiacus*. J Photochem Photobiol B: Biol 8: 189–197

Wullink W, Knudsen J, Olson JM, Redlinger TE and van Bruggen

EFJ (1991) Localization of polypeptides in isolated chlorosomes from green phototrophic bacteria by immuno-gold labeling electron microscopy. Biochim Biophys Acta 1060: 97–105

Wynn RM, Redlinger TE, Foster JM, Blankenship RE, Fuller RC, Shaw RW and Knaff DB (1987) Electron-transport chains of phototrophically and chemotrophically grown *Chloroflexus aurantiacus*. Biochim Biophys Acta 981: 216–226

Zannoni D and Fuller RC (1988) Functional and spectral characterization of the respiratory chain of *Chloroflexus aurantiacus* grown in the dark under oxygen-saturated conditions. Arch Microbiol 150: 368–373

Zuber H and Brunisholz R (1991) Structure and function of antenna polypeptides and chlorophyll-protein complexes: principles and variability. In: Scheer H (ed) Chlorophylls, pp 627–703. CRC Press, Boca Raton

Chapter 14

Organization of Electron Transfer Components and Supercomplexes

André Verméglio
CEA, DPVE/SBC CE de Cadarache 13108 Saint Paul-lez-Durance Cedex, France

Pierre Joliot and Anne Joliot
IBPC, 13 Rue Pierre-et-Marie Curie 75005 Paris, France

Summary

Purple nonsulfur photosynthetic bacteria are probably the most versatile of all microorganisms. Besides growing photoautotrophically or photoheterotrophically, they can also develop chemotrophically in darkness under aerobic conditions. Moreover, some of these bacteria are capable of dark anaerobic respiration. The photosynthetic and respiratory chains are localized in two different regions of the membrane, the intracytoplasmic and the cytoplasmic parts, respectively. This variety in bioenergetic pathways allows the bacteria to accommodate changes in the available sources of energy and in environmental factors. A first level of regulation concerns the biosynthesis of electron transfer components. A second regulation concerns the interactions between these different processes. In this chapter, emphasis is made on the interactions and the organization of these different electron transport chains. The bacteria utilize preferentially the light as energy source. In darkness, the use of the electron acceptor with the highest redox potential allows the bacteria to recover the maximum free energy. Two different mechanisms are responsible for these interactions. First the proton motive force, delocalized on the internal membrane, exerts a thermodynamic back pressure on the first complexes of respiratory chains. Second, modulation is mediated by changes in the redox state of electron carriers involved in the different bioenergetic processes. Two distinct pools of cytochrome c_2, a periplasmic electron carrier, have been found. A first pool, localized in the periplasmic space, is connected to the respiratory chains but can be photooxidized by the small number of reaction centers present in the cytoplasmic part of the membrane. This photooxidation inhibits the respiratory activities. The second pool is associated with the intracytoplasmic membrane. One cytochrome c_2, two reaction centers and one cytochrome bc_1 complex are organized in a supercomplex where the electron transfer is confined. This supermolecular organization allows for a very efficient photoinduced cyclic electron transfer not limited by the diffusion of the reactants. The stability of the supercomplex depends upon different factors like the redox state of cytochrome c_2, the pH and the presence of divalent cations. Different mechanisms for their formation are discussed.

R. E. Blankenship, M. T. Madigan and C. E. Bauer (eds): Anoxygenic Photosynthetic Bacteria, pp. 279–295.

I. Introduction

Anoxygenic photosynthetic bacteria, as with all living cells, derive energy from their environment. The universal cellular energy vector, ATP, is synthesized by three metabolic pathways: fermentation, respiration (aerobic and anaerobic) and photosynthesis. Contrary to fermentation, respiration and photosynthesis involve electron carrier chains formed by transmembrane complexes connected by lipid (quinone) and water (cytochrome) soluble diffusible components. These photosynthetic and respiratory chains catalyze electron and proton transport across the membrane and induce the formation of a transmembrane potential, the main component of the proton motive force (Junge and Jackson, 1982). This proton motive force is recognized as the central energetic intermediate that induces ATP synthesis, and indirectly participates in the translocation of ions and metabolites, motility, and other energy consuming processes. In photosynthetic bacteria, the membrane potential perturbs the absorption spectra of the light-harvesting carotenoids (Wraight et al., 1978). The identity of the carotenoid band shift induced by photosynthesis or respiration (Chance and Smith, 1955) is a strong indication that the membrane potential is delocalized on the same energy-transducing membrane where the electron transfer components coupled to these chains are located.

The large number of bacterial species with different energetic pathways is the evolutionary response to variability of the environment. Anoxygenic photosynthetic bacteria have been previously classified into three families of purple bacteria (Rhodospirillaceae, Chromatiaceae, Ectothiorhodospiraceae) and two families of green bacteria (Chlorobiaceae, Chloroflexaceae) (Trüper and Pfennig, 1981). With the advance of studies on ribosomal ribonucleic acid (rRNA), a new classification on a phylogenetic basis has been proposed (Woese, 1987). The purple bacteria and their relatives belong to a new class, the Proteobacteria (Stackebrandt et al., 1988). The α subclass of this phylum contains most of the species of the Rhodospirillaceae (nonsulfur purple bacteria), *Rhodobacter, Rhodospirillum, Rhodopseudomonas* and the strictly aerobic *Erythrobacter* species (Harashima et al., 1989). Other members of the Rhodospirillaceae, *Rhodocyclus* and *Rubrivivax*, are found in the β subclass. Members of these two subclasses are widely distributed in aquatic environments and soils exposed to light and many of them can also develop in the presence of air. They include various aerobic and anaerobic respiratory pathways in addition to the photosynthetic chain (Ferguson et al., 1987). The purple sulfur bacteria (Chromatiaceae and Ectothiorhodospiraceae) are included in the γ subclass.

Green sulfur (Chlorobiaceae) and nonsulfur bacteria (Chloroflexaceae) are not closely related in this new classification and form two distinct phylogenetic classes (Woese, 1987). The green nonsulfur bacteria can develop photoautotrophically, photoheterotrophically or, in the dark, chemoorganotrophically while the green sulfur (Chlorobiaceae) and purple sulfur (Chromatiaceae and Ectothiorhodospiraceae) bacteria are rather strict anaerobes (Madigan, 1988). The photosynthetic electron transfer chain operates according to cyclic or linear electron flow, depending on the presence of an exogenous reductant such as reduced sulfur compounds. The growth of green and purple sulfur bacteria is primarily phototrophic. Chemotrophic growth has never been observed under any nutritional conditions for Chlorobiaceae. Some Chromatiaceae species are capable of chemoorganotrophic or chemolithotrophic growth at very low oxygen tension but the growth rates are extremely slow. This suggests that these metabolic pathways do not play an essential role in their energetic metabolism. Little is known about the organization of the electron transfer chains in green and purple sulfur bacteria.

Most of our present knowledge concerns the members of the α subclass of Proteobacteria. We will discuss in this chapter the mechanism of interactions between the different electron transfer chains and the structural and functional information on their organization. It is generally assumed that the soluble carriers, namely the quinones and cytochrome (Cyt), or protons and ions are able to diffuse over long distances, mediating a redox equilibration over large membrane areas. In this case, kinetics of electron transfer between components of the respiratory and photosynthetic chains will be essentially independent of their relative localization and of the supramolecular

Abbreviations: BChl – bacteriochlorophyll; Cyt – cytochrome; Cyt c_t – cytochrome c_2 + cytochrome c_1; DCPIP – 2,6-dichlorophenolindophenol; DMSO – dimethylsulfoxide; P – primary electron donor; RC – reaction center; *Rc.* – *Rhodocyclus*; *Rb.* – *Rhodobacter*; *Rp.* – *Rhodopseudomonas*; *Rs.* – *Rhodospirillum*; *Rv.* – *Rubrivivax*; TMAO – trimethylamine-*N*-oxide; TMPD – *N,N,N',N'*-tetramethyl-*p*-phenylenediamine organization of proteins within the membrane. However several lines of evidence suggest that, in some instances, diffusion of mobile carriers is restricted to membrane domains of various sizes. These domains may correspond to the association of a few protein complexes with reproducible structure and stoichiometry, in which soluble carriers are trapped (i.e. supercomplexes). Such restricted diffusion is a way to favor one electron transfer pathway relative to another.

II. The Versatile Metabolism of the Rhodospirillaceae

Contrary to mitochondria and chloroplasts which function in a highly controlled environment, photosynthetic bacteria can be found in a large number of quite different habitats. They are therefore subjected to large changes in available source of energy (light or electron donors and acceptors). These organisms have developed multiple pathways to accommodate these versatile growth conditions. Besides photosynthesis and aerobic respiration, some species of photosynthetic bacteria are able to use a variety of acceptors in anaerobic electron transfer reactions (Ferguson et al., 1987). As a consequence, the purple nonsulfur bacteria are probably the most flexible group of microorganisms in biology. This flexibility will be discussed from examples taken essentially from *Rb. sphaeroides*, *Rb. capsulatus* and *Rp. viridis*.

Figure 1 represents the different electron transport systems which can occur in these species (Vignais et al., 1985; Ferguson et al., 1987; Jackson, 1988). Three main chains can be present: the photosynthetic chain, the aerobic respiratory electron transfer system, and the anaerobic respiratory chains with trimethylamine-*N*-oxide (TMAO), dimethyl sulfoxide (DMSO) or nitrous oxide (N_2O) as electron acceptors. In addition, some strains of a few species (two strains of *Rp. palustris*, *Roseobacter denitrificans* and *Rb. sphaeroides* forma sp. *denitrificans*) are able to perform complete denitrification, i.e. to reduce nitrate (NO_3^-) into dinitrogen (N_2) via nitrite (NO_2^-), and also probably nitric oxide (NO) and nitrous oxide (N_2O) (Satoh et al., 1976; Klemme et al., 1980; Shio et al., 1988). Each of these electron transfer chains induces a proton motive force across the same membrane. This proton motive force, created by a given chain, may influence the functioning of the electrogenic complexes of the other chains. A pool of quinone molecules, a small, lipid soluble electron carrier, serves as a turntable for these different electron transfer chains. The periplasmic soluble Cyt c_2 is also involved in many of these different bioenergetic pathways. Diffusion of these two electron carriers within the hydrophobic domain (ubiquinone) and

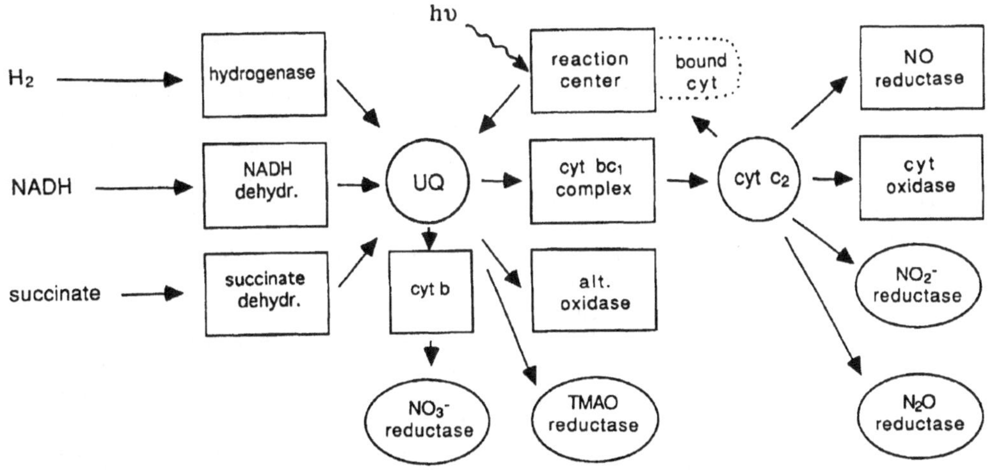

Fig. 1. The different electron transfer pathways of *Rhodobacter sphaeroides* and *Rhodobacter capsulatus*. The components in boxes are membrane proteins while the components in ovals are soluble periplasmic enzymes. The components in circles are the membranous (UQ) and periplasmic (Cyt c_2) turntables. See details in the text.

the periplasmic space (Cyt c_2) may provide interconnections between the different electron pathways. These points will be discussed in Section IV.

In general two types of antenna complexes, LH I and LH II, serve as light-harvesting complexes in Rhodospirillaceae. The LH I complex absorbs at around 870–880 nm and at around 1010 nm for bacteriochlorophyll (BChl) *a*- and BChl *b*-containing species, respectively. The LH II is only found in BChl *a* containing species and presents two absorption bands centered around 800 and 850 nm. Two types of reaction center (RC) can be found in Rhodospirillaceae. Most of these bacterial species contain a RC which possesses a tightly bound tetrahemic Cyt as in *Rp. viridis* (Deisenhofer et al., 1985; Deisenhofer and Michel, 1989; Bartsch, 1991). This Cyt serves as immediate electron donor to the primary donor P (Dracheva et al., 1986; Shopes et al., 1987). In this case, the cyclic electron transfer between RC and the Cyt bc_1 complex may be completed by Cyt c_2 (Garcia et al., 1993) or other soluble donors such as High Potential Iron sulfur Protein (HiPIP), Cyt *c*-553 or copper proteins depending on the species (Bartsch, 1991; Meyer, 1991). Other species do not have a RC bound Cyt and the cyclic electron transfer is completed by the Cyt bc_1 complex and the soluble Cyt c_2 acting as secondary donors (Dutton, 1986; Jackson, 1988; Prince, 1990).

Under dark conditions, the proton motive force is generated by the reduction of different acceptors at the expense of substrate oxidation. Input of electrons into the transfer chains is realized by membrane polypeptide complexes, namely NADH dehydrogenase, succinate dehydrogenase or hydrogenase. Two membrane proteins, both including a Cyt *b*, the Cyt oxidase and the ubiquinol oxidase, catalyze electron transfer from Cyt c_2 or from the ubiquinone pool, respectively, to oxygen during aerobic respiration (Zannoni and Baccarini-Melandri, 1980; Fenoll and Ramirez, 1984). In addition to oxygen, several compounds can be used as alternative electron acceptors such as DMSO, TMAO, nitrate (NO_3^-) or N_2O (Ferguson et al., 1987). These reactions are realized by soluble periplasmic enzymes (Sawada and Satoh, 1980; Urata et al., 1982; McEwan et al., 1984a).

The different electron transfer chains described in Fig. 1 are regulated at two different levels. A first level concerns control of the biosynthesis of the enzymes and components involved in these different processes. The second level concerns short term regulation, with no change in the relative concentration of the electron transfer carriers and allows for the bacteria to utilize efficiently the available energy by selecting the most favorable pathway.

III. Regulation of the Biosynthesis of the Electron Transfer Components

Some of the electron transfer components described above are genetically controlled to adapt the bacteria to environmental changes (Kiley and Kaplan, 1988; Drews, 1991). Nevertheless, under most growth conditions the different electron transfer chains described in Fig. 1 are present. The respiratory chain is expressed even under strict anaerobic conditions and the development of the photosynthetic apparatus does not require light. The terminal reductase of DMSO/TMAO respiration is constitutive although present in low concentration. This is also true for the various enzymes involved in denitrification in the few species able to reduce nitrate (Michalski and Nicholas, 1984; Sabaty et al., 1994a).

The relative ratio between the different bioenergetic chains is regulated by light intensity, oxygen concentration and the presence of acceptors like nitrate. The highest concentration of RC and light-harvesting complexes per cell is observed at low light intensity under anaerobic conditions. When light intensity is increased, the synthesis of antennae and RCs is partly repressed. The presence of low oxygen concentration in the growth medium at a given light intensity also partially represses the synthesis of the photosynthetic chain. Therefore, high light intensity and the presence of oxygen increase the ratio between Cyt c_t (c_2 plus c_1) and RC (Garcia et al., 1987; Verméglio et al., 1993). The synthesis of antenna and RCs is totally repressed only under dark and highly aerated cultures (Cohen-Bazire et al., 1957; Zhu and Hearst, 1986). In the case of *Rb. sphaeroides* but not for *Rb. capsulatus* or *Rs. rubrum*, these dark aerobic conditions induce the synthesis of a terminal oxidase similar to the aa_3 Cyt oxidase complex of mitochondria (Connelly et al., 1973).

In *Rb. sphaeroides denitrificans* and in *Rb. capsulatus*, a partial repression of the photosynthetic apparatus is also observed when cells are grown in the presence of nitrate or nitrous oxide (Michalski and Nicholas, 1984; Matsuura et al., 1988; Richardson et al., 1989). Consequently the ratio between Cyt c_t and RC is more than three times higher for cells grown photosynthetically in the presence than in the absence of nitrate (Matsuura et al., 1988). Under these conditions, the enzymes of the denitrification pathway are strongly expressed even at oxygen concentration as high as 37% air saturation (Sabaty et al., 1994a).

The structural genes coding for the α and β subunits of the LH I complex and the two of the RCs polypeptides, L and M, are adjacent in the genome (Williams et al., 1986; Youvan et al., 1984) and transcribed as a polycistronic message (Zhu and Kaplan, 1985; Zhu et al., 1986). All these photosynthetic genes are therefore members of a single operon, denoted *puf*, and the stoichiometry between the LH I complexes and the RC is constant whatever the growth conditions are (Aagaard and Sistrom, 1972). In agreement with this constant stoichiometry between the LH I complexes and RC, a supramolecular organization between the components has been characterized for a few species (Miller, 1979; Engelhardt et al., 1983; Meckenstock et al., 1992). These results will be discussed in section V. On the other hand, the structural genes of the α and β subunits of the LH II complex are linked to the *puc* operon. The *puf* operon is regulated by oxygen at the transcriptional level while the *puc* operon is regulated by both light and oxygen. Therefore the relative amounts of LH I and LH II complexes are highly dependent upon growth conditions. The large decrease in the concentration of the photosynthetic chain components per cell induced by dark semiaerobic growth and high light intensity or the presence of nitrate, is also accompanied by a decrease in the LH II/LH I ratio (Aagaard and Sistrom, 1972; Michalski and Nicholas, 1984; Reidl et al., 1985).

Such regulation, since it affects the content of the complexes of the internal membrane, also induce large changes in the ultrastructure of the cells (Sprague and Varga, 1986; Drews, 1991). A relationship between the amount of photosynthetic chains and the amount of intracytoplasmic membranes (chromatophores) is observed for *Rb. sphaeroides*, *Rb. capsulatus* and *Rs. rubrum*. The amount of intracytoplasmic membrane is minimal under aerobic conditions. The cellular content in chromatophores decreases when the incident light intensity increases (Kiley and Kaplan, 1988). The addition of nitrate or oxygen in the growth medium induces a large decrease in the number of chromatophores (Sawada and Satoh, 1980). This is shown in Fig. 2 where the ultrastructure of *Rb. sphaeroides* forma sp. *denitrificans* cells, grown in the absence or in the presence of nitrate, are compared. The induction of long tubular structures, which may impair cell division, is also observed when this species is grown in the presence of nitrate (Sabaty et al., 1994b) (Fig. 2). The formation of these tubules is correlated with the decrease in the LH II/LH I ratio observed in these growth conditions, since different mutants lacking the LH II complex have a similar ultrastructure (Hunter et al., 1988; Kiley et al., 1988).

In the case of *Rc. tenuis* or *Rubrivivax gelatinosus*, internal membranes are present as only a few irregular invaginations. When adapting from chemotrophic to phototrophic growth conditions, changes in the cellular content is achieved by modifications in the length and the diameter of the cell and in the density of particles in the cytoplasmic membrane (Drews and Oelze, 1981; see also Chapter by Drews and Golecki).

The finding that intracytoplasmic membranes are formed when the photosynthetic apparatus is induced

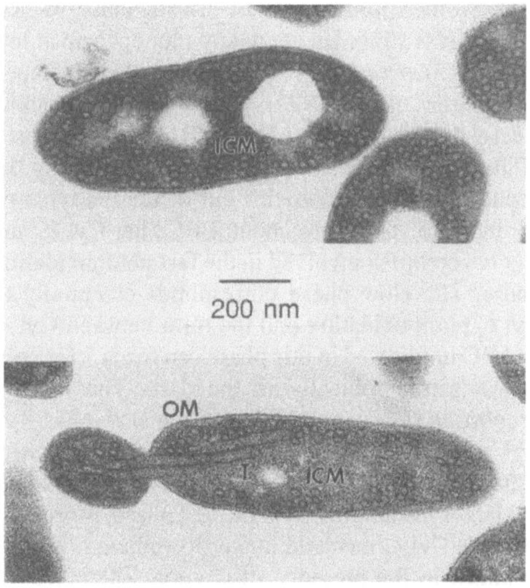

Fig. 2. Thin-section electron micrographs of *Rhodobacter sphaeroides* forma sp. *denitrificans*. The top and bottom panels show cells grown under photosynthetic conditions in the absence or in the presence of nitrate, respectively. Outer membrane (OM), intracytoplasmic membrane (ICM) and tubules (T) are indicated by arrows. Note that the tubules are running between two parts of a dividing cell. The bar corresponds to 200 nm.

has been taken as evidence that respiratory and photosynthetic chains are separated and localized in the cytoplasmic and in the intracytoplasmic parts of the membrane, respectively. A large number of biochemical investigations have confirmed this model. The main evidence is the following: (1) cytoplasmic membranes isolated from phototrophically grown cells of *Rb. sphaeroides* are largely depleted in BChl complexes but contained the enzymes of the respiratory activity; (2) the photosynthetic apparatus is localized in the purified intracytoplasmic membranes; (3) the segregation of these different bioenergetic activities between cytoplasmic and intracytoplasmic membranes is correlated with differences in the lipid composition of the two types of membranes (Kaufman et al., 1982).

Other arguments based on the analysis of photooxidation of Cyt c_t are also in agreement with the intracytoplasmic and cytoplasmic parts of the membrane having different functions. In an early study, Chance and Smith (1955) noticed that oxygenation of *Rs. rubrum* cells only partly oxidized the Cyt c_2 but that its complete oxidation required continuous illumination. We have more recently shown that Cyt c_t and RC photooxidation by continuous illumination of whole cells of *Rb. sphaeroides* strain Ga are nearly monophasic at low pH (< 7.5) over a large photon flux density range (Verméglio et al., 1993). These kinetics become clearly biphasic at high pH (> 8.5) or in the presence of divalent cations such as Mg^{++}, demonstrating the occurrence of two domains out of thermodynamic equilibrium. There are about 2 RCs per Cyt c_2 and Cyt bc_1 complex involved in the fast photooxidation phase. The slow phase corresponds essentially to Cyt c_2 photooxidation and the ratio between Cyt c_2 and RC implicated in this phase can reach 7 for cells grown semiaerobically in the dark. The relative proportion between the fast and the slow phases of Cyt c_t photooxidation depends upon the growth conditions: the larger the number of chromatophores, the larger the amplitude of the fast phase. Moreover, only the Cyt c_2 involved in the slow phase is readily oxidized in the presence of oxygen. Therefore the slowly photooxidized Cyt c_2 is coupled to the cytoplasmic part of the membrane while the fast one is connected to the intracytoplasmic part (Verméglio et al., 1993). Similarly, two distinct pools of Cyt c_2 have been detected in *Rb. sphaeroides* forma sp. *denitrificans* whole cells (Matsuura et al., 1988). The first pool is photooxidized by the first two flashes of a series. The relative stoichiometry of the electron transfer components involved in this reaction is similar to the one determined for the fast photooxidation of Cyt c_t of *Rb. sphaeroides* Ga, i.e. 2 RCs, 1 Cyt c_2 and 1 Cyt bc_1 complex (Sabaty et al., 1994b). This reaction is present for cells grown both in the presence or in the absence of nitrate and is not affected by an increase of the medium viscosity (Matsuura et al., 1988). The second pool of Cyt c_2 is observed only when the denitrificaton pathway is induced. Its full photooxidation requires a large number of actinic flashes and their efficiency is strongly decreased in the presence of glycerol (Matsuura et al., 1988). Moreover, this second pool of Cyt c_2 is specifically oxidized by respiratory or denitrification activities (Sabaty et al., 1994b).

This series of results suggests that one pool of Cyt c_2 is tightly connected to RC and the Cyt bc_1 complex. These electron transfer components are localized in the intracytoplasmic membrane and are rapidly photooxidized whatever the conditions are. Clear cut evidence of the organization of the components involved in photosynthesis into supercomplexes will be presented in more detail in Section V. The other pool of Cyt c_2 is free to diffuse between the photosynthetic, respiratory or the denitrification chains, only at low pH or in a low viscosity medium. At high pH or when the diffusion is restricted, this Cyt c_2 is photooxidized by a small fraction of RC present in the cytoplasmic part of the membrane. A schematic representation of the organization and localization of the electron carriers is depicted in Fig. 3.

IV. Short Term Interactions Between Photosynthetic and Respiratory Activities

Short term interactions regulate electron transfer between the different chains. Photosynthesis inhibits all the other bioenergetic processes, namely respiration (Nakamura, 1937; Cotton et al., 1983), denitrification (Satoh, 1977; McEwan et al., 1982; Sabaty et al., 1993) and reduction of DMSO and TMAO (King et al., 1987). Aerobic or anaerobic respiration can only occur at a very low rate under continuous illumination.

In the dark, photosynthetic bacteria use the available acceptor which possesses the highest redox potential, i.e. the most energetically favorable. Thus

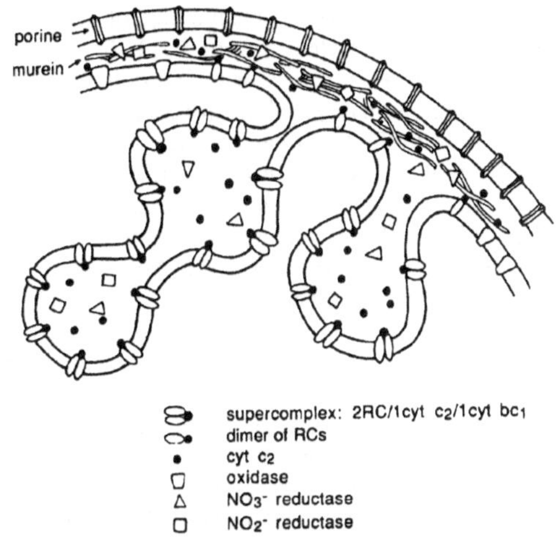

Fig. 3. Schematic representation of the organization and location of electron transfer components in *Rhodobacter sphaeroides* cells. See text for details.

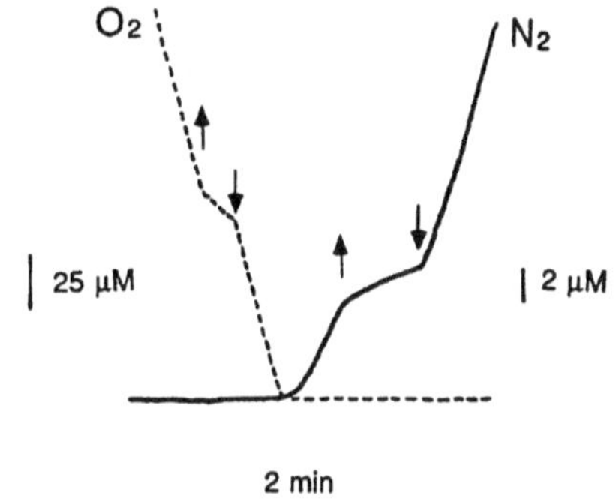

Fig. 4. Inhibition by light of the respiratory and denitrification activities and inhibition by oxygen of the denitrification pathway. The evolution of gases (O_2) and ($^{15}N_2$) was followed by a mass spectrometer for a suspension of *Rhodobacter sphaeroides* forma sp. *denitrificans* in the presence of 2 mM $K^{15}NO_3$. Reduction of nitrate is effective only when all of the oxygen has been consumed by the bacteria. Light (arrow up: light on, arrow down: light off) strongly inhibits both respiration and denitrification (from Sabaty et al., 1993).

O_2/H_2O (Em_7 = +820 mV) is reduced in preference to NO_3^-/NO_2^- (Em_7 = +420 mV) or DMSO/DMS (Em_7 = +160 mV). This allows the bacteria to recover a free energy of 226, 146 or 92 kJ/mol when NADH is oxidized at the expense of oxygen, nitrate or DMSO reduction, respectively. These interactions between photosynthesis, aerobic respiration and denitrification are illustrated in Fig. 4. We will now discuss the mechanisms of the regulation of these bioenergetic activities.

A. Light-Inhibition of Aerobic and Anaerobic Respirations

The light inhibition of respiration has been extensively studied. In intact cells of *Rb. capsulatus*, inhibition by continuous illumination is observed whatever the oxidase, alternative oxidase, or Cyt *c* oxidase, implicated in the respiratory activity (Richaud et al., 1986). Two types of mechanism have been proposed to explain the inhibition of respiration by light. The first type is the result of thermodynamic control by the light induced proton motive force on the proton translocating complexes, i.e. the dehydrogenases and the Cyt bc_1 complex, involved in the electron flow (Cotton et al., 1983; Richaud et al., 1986). In the second type, changes in the redox state of electron transfer components common to the different bioenergetic chains, i.e. ubiquinone and Cyt c_2, play an important role in the modulation of respiration by light. Both types of mechanisms are an efficient way for the bacteria to respond rapidly to variations in environmental factors and therefore, to use the most favorable bioenergetic process.

Although non exclusive, these two types of interpretations have often been opposed. Both interpretations have important structural implications. The inhibition of respiration by the light induced membrane potential implies that both respiratory and photosynthetic electron components are localized on the same continuous internal membrane and experience the same proton motive force, but do not necessarily share redox components. Such connection between electron transfer chains is of course, mandatory in the second mechanism. Our present view is that both types of mechanism exist. Their relative contribution depends upon the type of illumination (continuous light or flash excitation) and the oxidases involved.

The involvement of the proton motive force in the control of electron transfer processes can be demonstrated by measuring the effect of uncouplers. The inhibition of respiration by continuous light is prevented by the addition of uncouplers like CCCP (Ramirez and Smith, 1968; Cotton et al., 1983; Richaud et al., 1986). The modulation of the respiratory activity by the light induced membrane potential is therefore predominant under continuous illumination. The light induced proton motive force plays an important role in inducing a reverse electron flow between the quinone pool and NAD^+ (Knaff and Kämpf, 1987). This explanation is an extension of Mitchell's description of respiratory control in mitochondria, where the proton motive force exerts a back pressure on the proton translocating electron components and consequently, decreases the respiration rate. The proton motive force may also exert a control on the Cyt bc_1 complex involved in the respiratory activity. On the other hand, this mechanism is not operative for the terminal Cyt *c* oxidase. Light inhibition is prevented in the presence of exogenous electron donors like TMPD, although the light-induced membrane potential is maximal in these conditions.

Interaction between respiratory and photosynthetic chains involving common electron carriers has been observed under specific conditions. The role of the ubiquinone pool as a branching point between photosynthetic and respiratory chains has been clearly demonstrated by the light induced respiratory activity of the ubiquinol oxidase (photooxidase activity). This stimulation of the respiratory activity is observed under continuous illumination when the ubiquinone pool is reduced at the expense of artificial electron donors like TMPD or DCPIP (Zannoni et al., 1978). Interaction between photosynthetic chains and anaerobic respiratory chains at the level of ubiquinone has also been established by the effect of electron acceptors such as TMAO or NO_3^-. These acceptors provide a sink for reducing power during photosynthetic growth and maintain an optimal redox poise for the photosynthetic electron transfer components (Takamiya et al., 1988; McEwan et al., 1985). Indeed photoinduced cyclic electron transfer is maximal in a narrow range of redox potential around +50 mV (Culbert-Runquist et al., 1973; Van der Berg et al., 1983). Over-reduction of the primary electron acceptor or over-oxidation of the quinone pool will induce a decrease of the rate of electron flow (Dutton and Jackson, 1972) and consequently of the rate of ATP synthesis (Van der Berg et al., 1983).

Direct interaction at the level of electron transfer components has been demonstrated during a series of short excitation flashes (Verméglio and Carrier, 1984; Richaud et al., 1986). The respiratory activity is inhibited on each odd flash and reactivated after even flashes. This oscillatory pattern is not affected by the addition of uncouplers and has therefore been interpreted as due to the channeling of electrons from the respiratory chain toward the photosynthetic chain at the level of Cyt c_2 or the ubiquinone pool. The oscillations in respiration have been shown to be linked only to the Cyt oxidase activity, respiration coupled to the activity of the alternative oxidase not being affected by low frequency flash excitation (Richaud et al., 1986). A single flash reduces the rate of respiration by less than 10% (Verméglio and Carrier, 1984; Richaud et al., 1986). This result is in agreement with the observation that only a small fraction of RCs, compared to Cyt c_2, is localized in the cytoplasmic part of the membrane where the respiratory chains reside (see Section III).

Since most of the enzymes of the denitrification chain are periplasmic and coupled to Cyt c_2 (see Fig. 1), interaction at the level of this electron carrier plays an important role in the modulation of denitrification by the photosynthetic chain. This has been demonstrated in whole cells of *Rb. sphaeroides* forma sp. *denitrificans* where the inhibition by light of each step of the reduction of nitrate into dinitrogen is prevented by the addition of an exogenous electron donor, although the light induced membrane potential is maximal under these conditions (Sabaty et al., 1993).

B. Interactions Between Different Respiratory Activities

Complete inhibition of electron transport to either NO_3^-, N_2O or TMAO by oxygen is observed for whole cells of *Rb. capsulatus* and *Rb. sphaeroides* forma sp. *denitrificans* (McEwan et al., 1984b; King GF et al., 1987; Sabaty et al., 1993). As already stated, this allows the bacteria to use the available electron acceptor with the highest redox potential and therefore, to recover the maximum free energy. Contrary to what is observed for the light inhibition, the inhibition of both TMAO reduction or denitrification by oxygen is not reversed by the addition of uncouplers (McEwan et al., 1984b; King et al., 1987). This implies that inhibition by oxygen is not a consequence of a thermodynamic control exerted by the respiration induced proton motive force. The more likely hypothesis is that the different electron chains interact via some common electron carriers. Sabaty et al. (1993) have shown that the oxidation of Cyt c_1 induced by respiratory activity is responsible for the inhibition of NO_2^- and N_2O reduction. This mechanism cannot be applied to the inhibition by oxygen of NO_3^- or TMAO reduction since Cyt c_2 and the Cyt bc_1 complex are not involved in these processes (Fig. 1). In this case the redox state of the quinone pool must be the monitoring factor. The reduction of NO_3^- or TMAO may require a large reduction of the ubiquinone pool. By oxidizing this pool, respiratory activity may inhibit the reduction of NO_3^- or TMAO. Similar modulation of the activity of the alternative oxidase by the redox state of the quinone pool has been demonstrated (Zannoni and Moore, 1990). This enzyme is not engaged until this pool is at least 25% reduced. On the other hand, a linear relationship is observed between the respiratory rate of the Cyt oxidase and the reduction level of the quinone pool (Zannoni and Moore, 1990).

Two main conclusions concerning the interactions between the different bioenergetic chains can be drawn:

(1) The light induced membrane potential exerts a thermodynamic control on the NADH dehydrogenase. During continuous illumination the ubiquinone pool is oxidized at the expense of the reduction of NAD^+. This indirect modulation of the redox state of the ubiquinone pool explains how the respiratory activity of the alternative oxidase, the first step of denitrification and TMAO reduction, are inhibited by light, since these reactions necessitate a large reduction of this pool.

(2) Direct interactions between the different bioenergetic chains have also been clearly demonstrated at the level of Cyt c_2 and the ubiquinone pool. Oxidation of Cyt c_2 is involved in oxygen inhibition of denitrification and light inhibition of Cyt oxidase activity. The light stimulation of the alternative oxidase is mediated by the photoreduction of the ubiquinone pool. This implies diffusion of reduced quinone along the internal membrane and the photooxidation of Cyt c_2 linked to the respiratory activities by the small number of RCs present in the cytoplasmic part of the membrane.

V. Organization of the Electron Carrier Proteins into Supercomplexes

Information on the lateral organization of the photosynthetic complexes within the membrane have been obtained for BChl *b*-containing species, such as *Rp. viridis* and *Ectothiorhodospira halochloris* (Miller, 1979; Engelhardt et al., 1983) and one BChl *a*-containing species, *Rp. marina* (Meckenstock et al., 1992). This has been possible because of three favorable circumstances: the above species possess large flat membranes, they do not possess the LH II antenna and the amount of Cyt bc_1 complex is very low. This allows electron microscopy and image processing of negatively stained membranes. A large central structure corresponding to the RC, surrounded by six smaller LH I units, has been identified. Similar hexagonal arrays of RC surrounded by LH I have also been observed by freeze fracture of chromatophores of *Rs. rubrum* (Meyer et al., 1981). In the case of mutants of *Rb. sphaeroides* lacking the LH II complexes, freeze-etching micrographs reveal a very ordered organization of the membrane complexes. Particles, which have dimensions of about 100 × 200 Å, are arrayed helically along the longitudinal axis of the tubules (Robert A Niederman, personal communication) (Fig. 5). Similar structures were observed on tubules present in the cells of *Rb. sphaeroides* forma sp. *denitrificans* grown in the presence of nitrate (Fig. 2) (Sabaty et al., 1994b). It is hoped that the biochemical characterization and image processing of negative staining micrographs will reveal the organization of the electron transfer components of this region of the membrane.

Functional analysis is the main source of information on the supramolecular organization of the electron carrier proteins within the membrane. These approaches directly address the structural properties of functional relevance, but rely heavily on models with several assumptions. In some instances, the diffusion of mobile carriers is restricted to domains of limited size. These domains could correspond to supercomplexes defined as the association of a few protein complexes in which mobile carriers are possibly trapped. Within these domains, local thermodynamic equilibration rapidly occurs (ms range) while equilibration between domains occurs over a much longer time range (several seconds). A thermodynamic approach based on the idea that a restricted diffusion of mobile electron carriers confined within isolated domains will prevent a global equilibration of the photochemically generated redox equivalents has been developed. This method has been applied to the photosynthetic electron transfer chain, including primary and secondary donors in *Rb. sphaeroides*

Fig. 5. Freeze-etching electron micrograph of the tubules present in cells of *Rhodobacter sphaeroides* mutant M_{21} lacking the LH II complex. Specific associations (possibly dimeric) between membrane particles induce a regular array. The bar represents 100 nm (Courtesy of R. Niederman).

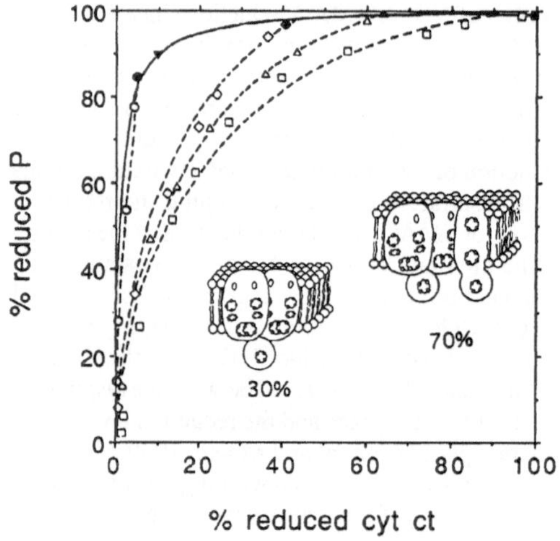

Fig. 6. Percentage of reduced Cyt c_t as a function of reduced P for a suspension of *Rhodobacter sphaeroides* R_{26} for different redox states of the cells. Open symbols: during the time course of continuous illumination. Filled symbols: after 2 min of dark adaptation. The solid line corresponds to the theoretical curve for an equilibrium constant of 70 between the primary donor P and the secondary acceptor, Cyt c_t. The dashed lines correspond to theoretical curves computed supposing that 70% of super-complexes involved 2RC, 1 Cyt c_2 and 1 Cyt bc_1 complex and 30% of supercomplexes involved 2RC and 1 Cyt c_2 (see insert for a schematic representation) (adapted from Joliot et al., 1989).

R_{26} cells, lacking the peripheral antenna LH II (Joliot et al., 1989). The kinetics of oxidation of the primary donor (P) and Cyt c_t have been measured under continuous illumination in the presence of myxo-thiazol, which blocks the reduction of the donors. In Fig. 6 is plotted the redox state of P versus that of Cyt c_t during the course of illumination. Whatever the initial redox conditions, the [P] versus [Cyt c_t] functions measured during the course of continuous illumination are far from the equilibrium curve computed from the redox titration data (solid line). These functions are unchanged for light intensities varying from 10 to 100 photons per RC per second. These results show that the primary and secondary donors are not in thermodynamic equilibrium despite the rate of electron transfer not being limited by dark processes.

A theoretical analysis of these thermodynamic anomalies has been made assuming that components of the photosynthetic electron transfer chain are included in small isolated domains (Lavergne et al., 1989). Under continuous illumination, photons are randomly distributed among domains according to a Poisson distribution, which implies that the rates of donor oxidation differ among domains. Therefore, at a given time, the redox state of the electron transfer components is not identical in the different domains. This explains why these carriers are not in thermodynamic equilibrium during the course of illumination. By comparing the oxidation of electron donors induced by either continuous or saturating flash excitation, a structural model in which RC, Cyt c_2 and the Cyt bc_1 complex form a supercomplex has been proposed. The key features of this model are that Cyt c_2 is trapped in the supercomplex which then behaves as an isolated domain and that the contact between two RCs allows the binding of only one Cyt c_2 per dimeric RCs. This model provides a simple interpretation for the biphasic reduction of P^+ observed after flashes of short duration (Dutton et al., 1975; Overfield et al., 1979). Following a saturating flash, the two P's of the RC dimer are oxidized. About half of the RCs have a Cyt c_2 bound at the time the flash occurs. For this fraction of centers, P^+ is reduced in a few microseconds (μs). The reduction of the second P^+ follows a first order process ($t_{1/2}$=110 μs) which is essentially limited by the electron transfer from Cyt c_1 to Cyt c_2. From the relative extents of the three different phases of P^+ reduction (few μs, 110 μs and 1 s phases), we deduced that about 70 % of the supercomplexes include 2 RC, 1 Cyt c_2 and 1 Cyt bc_1 complex (see insert of Fig. 6) while the remaining ones included only 2 RC and 1 Cyt c_2. Dimeric association of RCs has also been suggested from the analysis of the energy transfer between antenna and RC in membranes of *Rb. sphaeroides* (Vos et al., 1988).

The model of supercomplexes has been challenged by Fernández-Velasco and Crofts (1991), who addressed the same problem in isolated chromatophores of *Rb. sphaeroides* Ga, using a different functional approach based on inhibitor titrations. These authors studied the flash induced redox changes of Cyt c_t in the presence of subsaturating concentrations of inhibitors (myxothiazol or stigmatellin) of the quinol oxidizing site of the Cyt bc_1 complex. When 80% of the complexes are inhibited, most of Cyt c_t can still be rapidly reduced by quinol (Fig. 7). Thus, the fraction of uninhibited complexes (20%) can catalyze the reduction of most of the oxidized Cyt c_t. Under these conditions, uninhibited centers undergo several turnovers and thus, the half-time for Cyt c_t reduction increases with the concentration of inhibited Cyt bc_1 complexes. The authors conclude that the oxidation of quinol is a second order process and that Cyt c_2 freely diffuses in large domains including several RCs and Cyt bc_1 complexes. These domains probably correspond to chromatophores (Fernández-Velasco and Crofts, 1991). Although these experiments do not exclude the possibility that the membrane components (RCs and Cyt bc_1 complexes) are associated in supercomplexes, they imply that Cyt c_2 is not trapped in the supercomplex.

We have used the same experimental approach to test the validity of the supercomplex model in the case of whole cells of *Rb. sphaeroides* R_{26} (Joliot et al., 1993). Following a flash excitation, we observed that subsaturating concentrations of myxothiazol decreased in the same proportion as the fraction of active Cyt bc_1 complexes and the amplitude of the fast phase of Cyt c_t reduction, with no change in the half-time of this reaction. This behavior is opposite

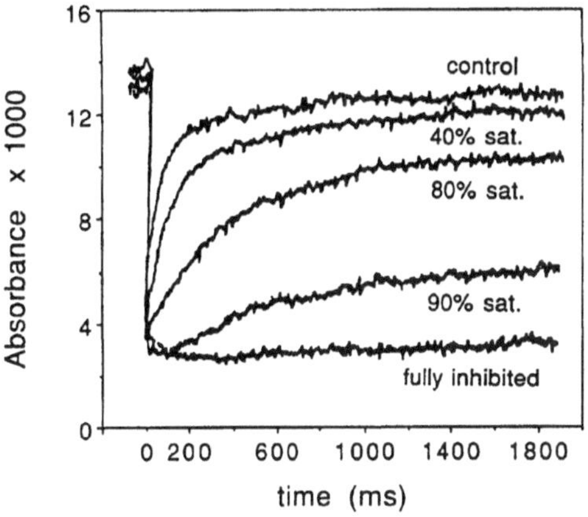

Fig. 7. Kinetics of reduction of Cyt c_t at different concentrations of stigmatellin following a single saturating excitation flash for chromatophores isolated from *Rhodobacter sphaeroides* Ga. The redox potential of the suspension was poised at 200 mV (from Fernández-Velasco and Crofts, 1991).

to that observed in isolated chromatophores and clearly supports the supercomplex model. We recently addressed the same problem in the case of *Rb. sphaeroides* Ga cells, which have a normal LH II content. In this strain, the kinetics of Cyt c_t reduction are multiphasic, which makes their interpretation difficult. We therefore analyzed the flash induced kinetics of the carotenoid bandshift. In the absence of inhibitor, a slow increasing phase occurs in the ms range. This slow phase is associated with ubiquinol oxidation and Cyt c_t reduction catalyzed by the Cyt bc_1 complex. Addition of subsaturating concentrations of myxothiazol decreases both the amplitude and the initial rate of the slow electrogenic phase to the same extent, with no significant change in the half-time (Fig. 8). Moreover, the same decrease in the rate and amplitude of the slow electrogenic phase is observed for saturating or subsaturating flashes. We therefore conclude that the diffusion of Cyt c_2 is strictly restricted to domains including a single Cyt bc_1 complex, in agreement with the supercomplex model proposed for the organization occurring in *Rb. sphaeroides* R_{26} cells. It must be appreciated, however, that chromatophore extraction perturbs the supercomplex structure (Fernández-Velasco and Crofts, 1991). As discussed in section III, the photosynthetic electron transfer chains organized in supercomplexes are located within the intracytoplasmic membranes (Fig. 3). Since most of the slow phase of the carotenoid

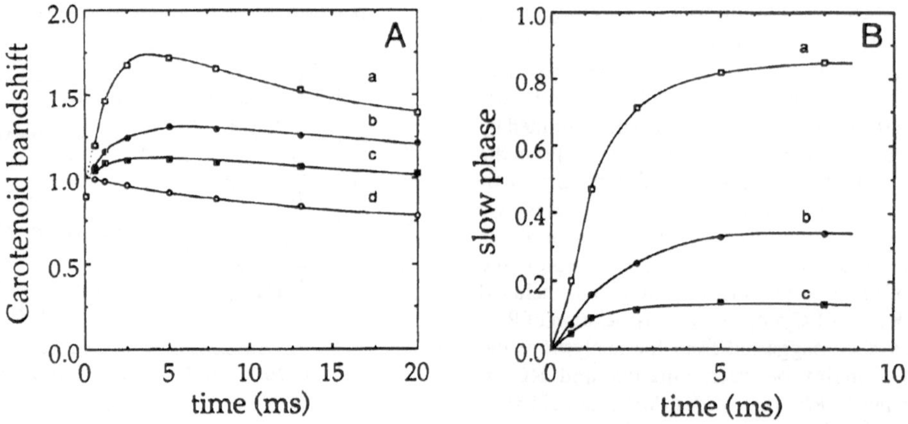

Fig. 8. Kinetics of the carotenoid bandshift in the presence of different concentrations of myxothiazol: (a) no addition (b) 0.08 μM (c) 0.12 μM and (d) 20 μM. Graph A: total absorption changes. Graph B: amplitude of the slow electrogenic phases after subtraction of the myxothiazol insensitive part and correction for the decay of the membrane potential, assuming an exponential decrease (from Verméglio et al., 1993).

bandshift corresponds to the activity involved in the supercomplexes, the small fraction of RCs and Cyt bc_1 complexes included in the cytoplasmic membrane do not give rise to a measurable electrogenic phase in the ms range.

In the case of *Rb. capsulatus*, Snozzi and Crofts (1985) have studied the dependence of the binary oscillations in the two electron gate of the acceptor side of the RC. They found that these oscillations titrated out with the midpoint potential of Cyt c_2 within chromatophores with more than 0.5 of this Cyt per RC. The binary oscillations could be observed at much lower redox potential in spheroplasts or in chromatophores with a low Cyt c_2 content. They deduced that the association of two RCs in dimers is stabilized by one molecule of reduced Cyt c_2. The study of different mutants of *Rb. capsulatus* has also been an important source of information. Although *Rb. sphaeroides* and *Rb. capsulatus* are closely related, differences in the phenotype have appeared when deletion mutants in the Cyt c_2 have been obtained for both species. In the case of *Rb. sphaeroides*, the loss of Cyt c_2 induces a photosynthetically incompetent phenotype (Donohue et al., 1988), demonstrating that this Cyt is the sole and obligatory electron carrier between the Cyt bc_1 complex and RCs. Analysis of spontaneous revertants has shown that they accumulate a soluble Cyt *c* (isocyt c_2), 20 to 40% of the level of Cyt c_2 found in the wild type cells (Fitch et al., 1989; Rott et al., 1992). On the other hand, the deletion of Cyt c_2 in *Rb. capsulatus* does not impair the photosynthetic activity (Daldal et al., 1986). Subsequent work by Jones et al. (1990) and Daldal and coworkers (Zannoni et al., 1992; Jenney and Daldal, 1993) demonstrate that connection between RC and the Cyt bc_1 complex is performed by a new Cyt denoted Cyt c_x or c_y. Mutants deleted in both Cyt c_y and Cyt c_2 are photosynthetically incompetent but regain this competence by complementation with either Cyt c_2 or Cyt c_y (Jenney and Daldal, 1993). The primary structure of Cyt c_y has been determined by sequencing of a complementary DNA fragment. Hydropathy analyses of Cyt c_y indicate that the amino-terminal portion is hydrophobic while the carboxy-terminal is hydrophilic and equivalent to Cyt *c* (Jenney and Daldal, 1993). These data together with biochemical analyses strongly support the interpretation that Cyt c_y is a membrane bound *c*-type Cyt able to mediate electron transfer between two integral membrane proteins, the RC and the Cyt bc_1 complex. A fast electron transfer (half-time faster than 100 μs) has been observed between these different electron transfer components and the RC (Prince et al., 1986; Jones et al., 1990). This suggests that a strong association between these components occurs in the membrane, in agreement with the supercomplex model we have proposed in the case of *Rb. sphaeroides*. Strong associations between the Cyt bc_1 complex, the Cyt oxidase and a membrane bound Cyt *c* have already been demonstrated in the case of *Paracoccus denitrificans* (Berry and Trumpower, 1984).

The organization into supercomplexes, in which Cyt c_2 is trapped, does not appear as a general characteristic of species of Rhodospirillaceae. As expected, the Cyt c_2 is a freely diffusing species when the stoichiometry between RCs and the Cyt bc_1 complex largely exceeds 2. This is the case for *Rs. rubrum* (Van Grondelle et al., 1976; Joliot et al., 1990) and *Rp. viridis* (Garcia et al., 1993) and in the cytoplasmic membrane of *Rb. sphaeroides* (Verméglio et al., 1993). In the case of *Rs. rubrum*, electron transfer from Cyt c_2 to the reaction center is a pure diffusional process occurring in the 100 μs time range, which implies that no tightly bound Cyt c_2 is present in dark adapted cells. Under saturating flash excitation, about half of the photooxidized primary donors are reduced in less than 1 ms, while the remainder is reduced in tens of ms. The complex kinetics of Cyt c_2 oxidation and P^+ reduction are correctly described by a model in which RCs are associated in dimers. The affinity of Cyt c_2 for the RCs is low when it is reduced and high when oxidized. Electrostatic or steric interactions prevent the binding of two Cyt c_2 molecules on the same dimer of RC (Joliot et al., 1990). In *Rp. viridis* whole cells, the viscosity of the medium does not affect the rate of electron transfer between Cyt c_2 and high potential heme *c*-556 of the tetrahemic Cyt, while it greatly slows down the reduction of the photooxidized Cyt c_2. Moreover, a full reduction of Cyt c_2 is observed in the presence of subsaturating concentrations of myxothiazol while the half time of the reaction is increased (Garcia et al., 1993). It was concluded that reduced Cyt c_2 forms a stable complex with the tetrahemic Cyt and is rereduced after photooxidation by the Cyt bc_1 complex according to a diffusion limited process.

In the future it is important to search for similar dimeric association of the RC in other species which do not possess tetrahemic Cyt and contain a low concentration of Cyt bc_1 complex. The organization between the dimeric RC, soluble (or membranous) Cyt *c* and the Cyt bc_1 complex must be studied for species such as *Rb. capsulatus* or *Rp. palustris* in which the ratio between Cyt bc_1 complex and RC is close to 0.5.

As discussed above, the organization of the photosynthetic chain into a supercomplex is not mandatory for its functioning. Such organization is not present in *Rp. viridis* and *Rs. rubrum* cells or in the cytoplasmic membrane of *Rb. sphaeroides* Ga or *Rb. sphaeroides* forma sp. *denitrificans* cells. However, in conditions where the diffusion of mobile carriers is the limiting step, i.e. under rapid electron flow, the organization into a supercomplex of RC, Cyt c_2 and Cyt bc_1 complexes would allow a more efficient photoinduced cyclic electron transfer, not limited by the diffusion of the reactants. In the cytoplasmic part of the membrane, the same Cyt c_2 and Cyt bc_1 complex is involved in photosynthesis, respiratory, or denitrification activities and therefore the organization into supercomplexes is not possible. The small proportion of RC located in the cytoplasmic membrane is able to photooxidize all the Cyt c_2 involved in respiration or denitrification. This allows a rapid modulation of these bioenergetic processes by light in optimizing the rate of electron transfer of the photosynthetic pathway relative to the others.

Several questions have not yet been addressed: How are the supercomplexes formed and what are the forces involved in these associations? The PufX polypeptide is encoded by a gene located downstream of the L and M genes of RC on the *puf* operon (Zhu et al., 1986). Although the *pufX* gene is part of the polycistronic photosynthesis gene messages, no specific function has yet been described for the PufX polypeptide. This gene is not present, at least at the same position on the genome, in *Rp. viridis* and *Rs. rubrum* cells, species which do not possess supercomplex association (Laußermair and Oesterhelt, 1992). Deletion of this polypeptide leads to photosynthetically incompetent mutants of *Rb. sphaeroides* (Farchaus et al., 1990) and *Rb. capsulatus* (Lilburn et al., 1992) species. The electron transfer from the RC to the Cyt bc_1 complex is impaired beyond the second flash of a series. The PufX polypeptide may play an important role in the formation of supercomplexes and therefore in the photosynthetic competence of these species (Farchaus et al., 1992; Lilburn et al., 1992). This hypothesis deserves further examination.

In *Rb. sphaeroides*, the fraction of RCs involved in supercomplexes is regulated by the relative amount of intracytoplasmic and cytoplasmic membranes. This is a long term modulation. It may also be necessary for the bacteria to adapt rapidly to environmental changes. This can be achieved by at least two different ways. We have already seen that stabilization of the association of the Cyt c_2 within the supercomplex is modulated by the redox state of Cyt c_2 and the pH or the presence of divalent cations

(Section III). This modulation occurs probably without dissociation of the interactions between the membranous complexes. Association or dissociation of these complexes involve small energy differences. Changes in phosphorylation states of membrane proteins, such as antenna complexes or RC, may play an important role in such regulation. Several membrane proteins have been found to be phosphorylated (Holuigue et al., 1985; Turner and Mann, 1986; Holmes et al., 1986). Phosphorylation and dephosphorylation of antenna has been observed in photosynthetic bacteria depending upon the redox state of the quinone pool (Holmes and Allen, 1988). Contrary to the situation in green plants and microalgae, this phenomenon is not well understood and the location and nature of the different complexes involved is still unresolved (Allen, 1992). In *Chlamydomonas* cells, it has been clearly shown that reduction of the plastoquinone pool, i.e. phosphorylation of light-harvesting complexes, induces the movement of Cyt b_6f complexes from the grana to the lamellae. This reduction also favors a strong association between the Cyt b_6f complexes and Photosystem I complexes (Wollman and Bulté, 1989). This regulates the relative contribution between linear and cyclic electron transfer. Regulation of the amount of supercomplexes, i.e. modulation between the different bioenergetic pathways in purple nonsulfur bacteria, may involve similar mechanisms of phosphorylation and migration of membrane proteins between cytoplasmic and intracytoplasmic membranes.

References

Aagaard J and Sistrom WR (1972) Control of synthesis of reaction center bacteriochlorophyll in photosynthetic bacteria. Photochem Photobiol 15: 209–225

Allen JF (1992) Protein phosphorylation in regulation of photosynthesis. Biochim Biophys Acta 1098: 275–335

Bartsch RG (1991) The distribution of soluble metallo-redox proteins in purple phototrophic bacteria. Biochim Biophys Acta 1058: 28–30

Berry EA and Trumpower BL (1985) Isolation of ubiquinol oxidase from *Paracoccus denitrificans* and resolution into cytochrome bc_1 and cytochrome aa_3 complexes. J Biol Chem 260: 2458–2467

Chance B and Smith L (1955) Respiratory systems of *R. rubrum*. Nature 175: 803–806

Cohen-Bazire G, Sistrom WR and Stanier RW (1957) Kinetic studies of pigment synthesis by non-sulphur purple bacteria. J Cell Comp Physiol 49: 25–68

Connelly JL, Jones OTG, Saunders VA and Yates DW (1973) Kinetic and thermodynamic properties of membrane-bound cytochromes of aerobically and photosynthetically grown *R. spheroides*. Biochim Biophys Acta 292: 644–653

Cotton NPJ, Clark AJ and Jackson JB (1983) Interaction between the respiratory and photosynthetic electron transport chains of intact cells of *Rhodopseudomonas capsulata* mediated by membrane potential. Eur J Biochem 130: 581–587

Culbert-Runquist JA, Hadsell RM and Loach PA (1973) Dependency on environmental redox potential of photophosphorylation in *Rhodopseudomonas spheroides*. Biochemistry. 12: 3508–3514

Daldal F, Cheng S, Applebaum J, Davidson E and Prince RC (1986) Cytochrome c_2 is not essential for phototrophic growth of *Rhodopseudomonas capsulata*. Proc Natl Acad Sci USA 83: 2012–2016

Deisenhofer J and Michel H (1989) The photosynthetic reaction centre from the purple bacterium *Rhodopseudomonas viridis*. EMBO J 8: 2149–2170

Deisenhofer J, Epp O, Miki K, Huber R and Michel H (1985) Structure of the protein subunits in the photosynthetic reaction center of *Rhodopseudomonas viridis* at 3 Å resolution. Nature 318: 19–26

Donohue TJ, McEwan AG, Van Doren S, Crofts AR and Kaplan S (1988) Phenotypic and genetic characterization of cytochrome c_2 deficient mutants of *Rhodobacter sphaeroides*. Biochemistry 27: 1918–1925

Dracheva SM, Drachev LA, Konstantinov AA, Semenov AY, Skulachev VP, Arutjunjan AM, Shuvalov VA and Zaberezhnaya SM (1988) Electrogenic steps in the redox reactions catalysed by photosynthetic reaction-centre complex from *Rhodopseudomonas viridis*. Eur J Biochem 1871: 253–264

Drews G (1991) Regulated Development of the Photosynthetic Apparatus in Anoxygenic Bacteria. In: Cell Culture and Somatic Cell Genetics of Plants, Vol 7B, pp 113–148. Academic Press, New York

Drews G and Oelze J (1981) Organization and differentiation of membranes of phototrophic bacteria. Adv Microbiol Physiol 22: 1–97

Dutton PL (1986) Energy Transduction in Anoxygenic Photosynthesis. In: Staehelin LA and Arntzen CJ (eds) Encyclopedia of Plant Physiology, New series, Vol 19, pp 197–237. Springer Verlag, Berlin

Dutton PL and Jackson JB (1972) Thermodynamic and kinetic characterization of electron transfer components in situ in *Rhodopseudomonas sphaeroides* and *Rhodospirillum rubrum*. Eur J Biochem 30: 495–510

Dutton PL, Petty KM, Bonner HS and Morse SD (1975) Cytochrome c_2 and reaction center of *Rps. sphaeroides* Ga membranes. Extinction coefficients, content, half-reduction potentials, kinetics and electric field alterations. Biochim Biophys Acta 387: 536–556

Engelhardt H, Baumeister W and Saxton O (1983) Electron microscopy of photosynthetic membranes containing bacteriochlorophyll *b*. Arch Microbiol 35: 169–175

Farchaus JW, Barz WP, Grünberg H and Oesterhelt D (1992) Studies on the expression of the *puf*X polypeptide and its requirement for photoheterotrophic growth in *Rhodobacter sphaeroides*. EMBO J 11: 2779–2788

Fenoll C and Ramirez JM (1984) Simultaneous presence of two

terminal oxidases in the respiratory system of dark aerobically grown *Rhodospirillum rubrum*. Arch Microbiol 137: 42–46

Ferguson ST, Jackson JB and McEwan AG (1987) Anaerobic respiration in the Rhodospirillaceae: Characterisation of pathways and evaluation of roles in redox balancing during photosynthesis. FEMS Microbiol Rev 46: 117–143

Fernández-Valasco J and Crofts AR (1991) Complex or supercomplexes: Inhibitor titration show that electron transfer in chromatophores from *Rhodobacter sphaeroides* involves a dimeric UQH_2: cytochrome c_2 oxidase, and is delocalized. Biochem Soc Trans, 19: 588–593

Fitch J, Cannac V, Meyer TE, Cusanovich MA, Tollin G, Van Beeumen J, Rott MA and Donohue TJ (1989) Expression of a cytochrome c_2 isozyme restores photosynthetic growth of *Rhodobacter sphaeroides* mutants lacking the wild-type cytochrome c_2 gene. Arch Biochem Biophys 271: 502–507

Garcia AF, Venturoli G, Gad'on N, Fernández-Velasco JG, Melandri BA and Drews G (1987) The adaptation of the electron transfer chain of *Rhodopseudomonas capsulata* to different light intensities. Biochim Biophys Acta 890: 335–345

Garcia D, Richaud P and Verméglio A (1993) The photoinduced cyclic electron transfer in whole cells of *Rhodopseudomonas viridis*. Biochim Biophys Acta 1144: 195–301

Harashima K, Shiba T and Murata N (1989) Aerobic Photosynthetic Bacteria. Springer Verlag, Tokyo

Holuigue L, Lucero HA and Vallejos RH (1985) Proteins phosphorylation in the photosynthetic bacterium *Rhodospirillum rubrum*. FEBS Lett 181: 103–108

Holmes NG and Allen JF (1988) Proteins phosphorylation in chromatophores from *Rhodospirillum rubrum*. Biochim Biophys Acta 935: 72–78

Holmes NG, Sanders CE and Allen JF (1986) Membrane protein phosphorylation in the purple photosynthetic bacterium *Rhodopseudomonas sphaeroides*. Biochem Soc Trans 14: 67–68

Hunter CN, Pennoyer JD, Sturgis JN, Farrely D and Niederman RA (1988) Oligomerization states and associations of light-harvesting pigment-protein complexes of *Rhodobacter sphaeroides* as analysed by lithium dodecyl sulfate-polyacrylamide gel electrophoresis. Biochemistry 27: 3459–3467

Jackson JB (1988) Bacterial photosynthesis. In: Anthony C (ed) Bacterial Energy Transduction, pp 317–375. Academic Press, London

Jenney FE and Daldal F (1993) A novel membrane-associated *c*-type cytochrome, Cyt c_y, can mediate the photosynthetic growth of *Rhodobacter capsulatus* and *Rhodobacter sphaeroides*. EMBO J 12: 1283–1292

Joliot P, Verméglio A and Joliot A (1989) Evidence for supercomplexes between reaction centers, cytochrome c_2 and cytochrome bc_1 complex in *Rhodobacter sphaeroides* whole cells. Biochim Biophys Acta 975: 336–345

Joliot P, Verméglio A and Joliot A (1990) Electron transfer between primary and secondary donors in *Rhodospirillum rubrum*: Evidence for a dimeric association of reaction centers. Biochemistry 29: 4355–4361

Joliot P, Verméglio A and Joliot A (1993) Supramolecular membrane protein assemblies in photosynthesis and respiration. Biochim Biophys Acta 1141: 151–174

Jones MR, McEwan AG and Jackson JB (1990) The role of c-type cytochromes in the photosynthetic electron pathway of *Rhodobacter capsulatus*. Biochim Biophys Acta 109: 59–66

Junge W and Jackson JB (1982) The development of electrochemical potential gradient across photosynthetic membranes. In: Govindjee (ed) Photosynthesis, Vol 1, pp 589–646. Academic Press, New York

Kaufmann N, Reidl H, Golecki JR, Garcia AF and Drews G (1982) Differentiation of the membrane system in cells of *Rhodopseudomonas capsulata* after transition from the chemotrophic to phototrophic growth conditions. Arch Microbiol 131: 313–322

Kiley PJ and Kaplan S (1988) Molecular genetics of photosynthetic membrane biosynthesis in *Rhodobacter sphaeroides*. Microbiol Rev 52: 50–69

Kiley PJ, Varga A and Kaplan S (1988) Physiological and structural analysis of light-harvesting mutants of *Rhodobacter sphaeroides*. J Bacteriol 170: 1103–1115

King GF, Richardson DJ, Jackson JB and Ferguson SJ (1987) Dimethylsulfoxide and trimethylamine-N-oxide as bacterial electron transport acceptors: use of nuclear magnetic resonance to assay and characterise the reductase system in *Rhodobacter capsulatus*. Arch Microbiol 149: 47–51

Klemme JH, Chyla, I and Preuss M (1980) Dissimilatory nitrate reduction by strains of the facultative phototrophic bacterium *Rhodopseudomonas palustris*. FEMS Microbiol Lett 9: 137–140

Knaff DB and Kämpf C (1987) Substrate Oxidation and NAD^+ Reduction by Phototrophic Bacteria. In: Amesz J (ed) Photosynthesis, New Comprehensive Biochemistry, Vol 15, pp 199–211. Elsevier, Amsterdam, New York, Oxford

Lavergne J, Joliot P and Verméglio A (1989) Partial equilibration of photosynthetic electron carriers under weak illumination: a theoretical and experimental study. Biochim Biophys Acta 975: 346–354

Laußermair E and Oesterhelt D (1992) A system for site-specific mutagenesis of the photosynthetic reaction center in *Rhodopseudomonas viridis*. EMBO J 11: 777–783

Lilburn TG, Haith CE, Prince RC and Beatty JT (1992) Pleiotropic effects of *pufX* gene deletion on the structure and function of the photosynthetic apparatus of *Rhodobacter capsulatus*. Biochim Biophys Acta 1100: 160–170

Madigan MT (1988) Microbiology, Physiology, and Ecology of phototrophic Bacteria. In: Zehnder AJB (ed) pp 39–111. John Wiley and Sons, New York

Matsuura K, Mori M and Satoh T (1988) Heterogeneous pools of cytochrome c_2 in photodenitrifying cells of *Rhodobacter sphaeroides* forma sp. *denitrificans*. J Biochem 104: 1016–1020

McEwan AG, George CL, Ferguson SJ and Jackson JB (1982) A nitrate reductase activity in *Rhodopseudomonas capsulata* linked to electron transfer and generation of a membrane potential. FEBS Lett 150: 277–280

McEwan AG, Jackson JB and Ferguson SJ (1984a) Rationalization of properties of nitrate reductases in *Rhodopseudomonas capsulata*. Arch Microbiol 137: 344–349

McEwan AG, Cotton NPJ, Ferguson SJ and Jackson JB (1984b) The inhibition of nitrate reduction by light in *Rhodopseudomonas capsulata* is mediated by the membrane potential but the inhibition by oxygen is not. Adv Photosynth Res 2, 449–452

McEwan AG, Cotton NPJ, Ferguson SJ and Jackson JB (1985)

The role of auxiliary oxidants in the maintenance of balanced redox poise for photosynthesis in bacteria. Biochim Biophys Acta 810: 140–147

Meckenstock RU, Krusche K, Brunisholz RA and Zuber H (1992) The light-harvesting core-complex and the B820-subunit from *Rhodopseudomonas marina*. Part II. Electron microscopic characterisation. FEBS Lett 311: 135–138

Meyer R, Snozzi M and Bachofen R (1981) Freeze fracture studies of reaction centers from *Rhodospirillum rubrum* in chromatophores and liposomes. Arch Microbiol 130: 125–128

Meyer TE (1991) Evolution of cytochromes and photosynthesis. Biochim Biophys Acta 1058: 31–34

Michalski WP and Nicholas DJD (1984) The adaptation of *Rhodopseudomonas sphaeroides* forma sp. *denitrificans* for growth under denitrifying conditions. J Gen Microbiol 130: 155–165

Miller KR (1982) Three-dimensional structure of a photosynthetic membrane. Nature 300: 53–55

Nakamura H (1937) Über die photosynthese bei der schelfreien purpurbakterie *Rhodobazillus palustris*. Beiträge zur stoffwechselphysiologie der purpurbakterie. Acta Phytochim. 9: 189–234

Overfield RE, Wraight CA and DeVault D (1979) Microsecond photooxidation kinetics of cytochrome c_2 from *Rhodopseudomonas sphaeroides*: in vivo and solution studies. FEBS Lett 105: 137–142

Prince RC (1990) Bacterial photosynthesis: from photons to the Δp. In: Krulwich TA (ed) Bacterial Energetics, The Bacteria, Vol 12, pp 111–150. Academic Press, New York

Prince RC, Davidson E, Haith CE and Daldal F (1986) Photosynthetic electron transfer in the absence of cytochrome c_2 in *Rhodopseudomonas sphaeroides*: Cytochrome c_2 is not essential for electron flow from the cytochrome bc_1 complex to the photochemical reaction center. Biochemistry 25: 5208–5214

Ramirez J and Smith L (1968) Synthesis of adenosine triphosphate in intact cells of *Rhodospirillum rubrum* and *Rhodopseudomonas sphaeroides* on oxygenation or illumination. Biochim Biophys Acta 153: 466–475

Reidl H, Golecki JR, and Drews G (1985) Composition and activity of the photosynthetic system of *Rhodobacter capsulatus*. Biochim Biophys Acta 808: 328–333

Richardson DJ, McEwan AG, Jackson JB and Ferguson SJ (1989) Electron transport pathways to nitrous oxide in *Rhodobacter* species. Eur J Biochem 185: 659–669

Richaud P, Marrs BL and Verméglio A (1986) Two modes of interaction between photosynthetic and respiratory electron chains in whole cells of *Rhodopseudomonas capsulata*. Biochim Biophys Acta 850: 256–263

Rott MA, Fitch J, Meyer TE and Donohue TJ (1992) Regulation of a cytochrome c_2 isoform in wild-type and cytochrome c_2 mutant strains of *Rhodobacter sphaeroides*. Arch Biochem Biophys 292: 576–582

Sabaty M, Gans P and Verméglio A (1993) Inhibition of nitrate reduction by light and oxygen in *Rhodobacter sphaeroides* forma sp. *denitrificans*. Arch Microbiol 159: 153–159

Sabaty M, Gagnon J and Verméglio A (1994a) Induction by nitrate of cytoplasmic and periplasmic proteins in the photodenitrifier *Rhodobacter sphaeroides* forma sp. *denitrificans* in anaerobic and aerobic conditions. Arch Microbiol 162: 335–343

Sabaty M, Jaffé J, Olive J and Verméglio A (1994b) Organization of electron transfer components in *Rhodobacter sphaeroides* forma sp. *denitrificans*. Biochim Biophys Acta 1187: 313–323

Satoh T (1977) Light-activated, -inhibited and -independent denitrification by a denitrifying phototrophic bacterium. Arch Microbiol 115: 293–298

Satoh T, Hoshino Y and Kitamura H (1976) *Rhodopseudomonas sphaeroides* forma sp. *denitrificans*, a denitrifying strain as a subspecies of *Rhodopseudomonas sphaeroides*. Arch Microbiol 108: 265–269

Sawada E and Satoh T (1980) Periplasmic location of dissimilatory nitrate and nitrite reductases in a denitrifying phototrophic bacterium, *Rhodopseudomonas sphaeroides* forma sp. *denitrificans*. Plant Cell Physiol 21: 205–210

Shioi Y, Doi M, Arata H and Takamiya K (1988) A denitrifying activity in an aerobic photosynthetic bacterium *Erythrobacter* sp. strain OCh 114. Plant Cell Physiol 29: 861–865

Shopes RJ, Levine LMA, Holten D and Wraight CA (1987) Kinetics of the bound cytochromes in reaction centers from *Rhodopseudomonas viridis*. Photosynth Res 12: 165–180

Snozzi M and Crofts AR (1985) Kinetics of the *c*-cytochromes in chromatophores from *Rhodopseudomonas sphaeroides* as a function of the concentration of cytochrome c_2. Influence of this concentration on the oscillation of the secondary acceptor of the reaction center Q_B. Biochim Biophys Acta 809: 260–270

Sprague SC and Varga AR (1986) Membrane architecture of anoxygenic photosynthetic bacteria. In: Staehelin LA and Arntzen (eds) Encycl. Plant Physiol, Vol 19, pp 303–319. Springer Verlag, Berlin, Heidelberg and New York

Stackebrandt E, Murray RGE and Trüper HG (1988) Proteobacteria classis nov., a name for the phylogenetic taxon that includes the 'Purple bacteria and their relatives'. Int J Syst Bacteriol 38: 321–325

Takamiya K, Arata H, Shioi Y and Doi M (1988) Restoration of the optimal redox state for the photosynthetic electron transfer system by auxiliary oxidants in an aerobic photosynthetic bacterium *Erythrobacter* sp. OCh 114. Biochim Biophys Acta 935: 26–33

Trüper HG and Pfennig N (1981) Characterization and identification of the anoxygenic phototrophic bacteria. In: Starr MP, Stolp H, Trüper HG, Balows A and Schlegel HG (eds) The Prokaryotes, a Handbook on Habitats, Isolation and Identification of Bacteria, pp 299–312. Springer Verlag, Berlin, Heidelberg and New York

Turner AM and Mann NH (1986) Protein phosphorylation in *Rhodomicrobium vannielli*. J Gen Microbiol 132: 3433–3440

Urata K, Shimada K and Satoh T (1982) Periplasmic location of nitrous oxide reductase in a photodenitifier, *Rhodopseudomonas sphaeroides* forma sp. denitrificans. Plant Cell Physiol 23: 1121–1124

Van der Berg WH, Bonner WD and Dutton PL (1983) Redox potential of photophosphorylation and electron transfer in continuous illumination of *Rhodopseudomonas sphaeroides* chromatophores. Arch Biochem Biophys 222: 299–309

Van Grondelle R, Duysens LNM and Van der Wal HN (1976) Function of three cytochromes in photosynthesis of whole cells of *Rhodospirillum rubrum* as studied by flash spectroscopy. Biochim Biophys Acta 449: 169–187

Verméglio A and Carrier JM (1984) Photoinhibition by flash and continuous light of oxygen uptake by intact photosynthetic bacteria. Biochim Biophys Acta 764: 233–238

Verméglio A, Joliot P and Joliot A (1993) The rate of cytochrome c_2 photooxidation reflects the subcellular distribution of reaction centers in *Rhodobacter sphaeroides* Ga cells. Biochim Biophys Acta 1183: 352–360

Vos M, Van Dorssen RJ, Amesz J, Van Grondelle R and Hunter CN (1988) The organization of the photosynthetic apparatus of *Rhodobacter sphaeroides*: studies of antenna mutants using singlet-singlet quenching. Biochim Biophys Acta 933: 132–140

Vignais PM, Colbeau A, Willison JC and Jouanneau Y (1985) Hydrogenase, nitrogenase and hydrogen metabolism in the photosynthetic bacteria. Adv Microb Physiol 26: 155–234

Williams JC, Steiner LA and Feher G (1986) Primary structure of the reaction center from *Rhodopseudomonas sphaeroides*. Proteins 1: 312–325

Woese CR (1987) Bacterial Evolution. Microbiol Rev 51: 221–271

Wollman FA and Bulté L (1989) Towards an understanding of the physiological role of states transitions. In: Hall DO and Grassi G (eds) Photosynthetic Processes for Energy and Chemicals, pp 198–207. Elsevier, Amsterdam

Wraight CA, Cogdell RJ and Chance B (1978) Ion transport and electrochemical gradients in photosynthetic bacteria. In: Clayton RK and Sistrom RW (eds) The Photosynthetic Bacteria, pp 471–502. Plenum Press, New York and London

Youvan DC, Bylina EJ, Alberti M, Begusch H and Hearst JE (1984) Nucleotide and deduced polypeptide sequences of the photosynthetic reaction center, B870 antenna and flanking sequences from *Rhodopseudomonas capsulata*. Cell 37: 949–957

Zannoni D and Baccarini-Melandri A (1980) Respiratory electron flow in facultative photosynthetic bacteria. In: Knowles CJ (ed) The Diversity of Bacterial Respiratory Systems, Vol 2, pp 183–202. CRC Press, Cleveland

Zannoni D and Moore AL (1990) Measurement of the redox state of the ubiquinone pool in *Rhodobacter capsulatus* membrane fragments. FEBS Lett 271: 123–127

Zannoni D, Jasper P and Marrs BL (1978) Light-induced absorption changes in intact cells of *Rhodopseudomonas sphaeroides*. Evidence for interaction between photosynthetic and respiratory electron transfer chains. Biochim Biophys Acta, 191: 625–631

Zannoni D, Venturoli G and Daldal F (1992) The role of membrane bound cytochromes of *b*- and *c*-type in the electron transport chain of *Rhodobacter capsulatus*. Arch Microbiol 157: 367–374

Zhu YS and Hearst J (1986) Regulation of expression of genes for light-harvesting antenna proteins LHI and LHII; reaction centers polypeptides RC-L, RC-M, and RC-H; and enzymes of bacteriochlorophyll and carotenoid biosynthesis in *Rhodobacter capsulatus* by light and oxygen. Proc Natl Acad Sci USA 83: 7613–7616

Zhu YS and Kaplan S (1985) Effects of light, oxygen, and substrates on steady-state levels of mRNA coding for ribulose 1,5-bisphosphate carboxylase and light-harvesting and reaction center polypeptides in *Rhodopseudomonas sphaeroides*. J Bacteriol 162: 925–932

Zhu YS, Kiley PJ, Donohue TJ and Kaplan S (1986) Origin of the mRNA stoichiometry of the *puf* operon in *Rhodobacter sphaeroides*. J Bacteriol 261: 10366–10374

Chapter 15

Theory of Electronic Energy Transfer

Walter S. Struve
Ames Laboratory-USDOE and Department of Chemistry, Iowa State University, Ames, IA 50010, USA

Summary

This chapter deals with theoretical mechanisms for electronic excitation transfer among photosynthetic pigments. General multipole expansions of intermolecular electrostatic potential energies are described in detail. We review the Golden Rule derivation of the Förster-Dexter theory for dipole-dipole and exchange transfer in the weak-coupling limit; emphasis is placed on its underlying assumptions and limitations. Extension of the Golden Rule to higher-order perturbation theory permits the description of spin-forbidden (e.g. singlet-triplet) energy transfer, as well as remote heavy-atom effects on intersystem crossing. The concepts of the density operator and density matrix are introduced. We demonstrate the need for a unified energy transfer theory that treats coherence as well as excited state population decay. The stochastic Liouville equations (SLE) are solved for the prototype case of a strongly interacting pair of chromophores, and expressions are obtained for the time-dependent anisotropy *r(t)* in optical fluorescence and pump-probe experiments in strongly coupled systems. While time-dependent SLE solutions are required to predict anisotropy decays, evaluation of the *initial* anisotropy *r(0)* requires only knowledge of the excitation conditions and the exciton component transition moments. Current techniques are described for the simulation of temperature dependence in Förster energy transfer rates, via the influence of electron-phonon and electron-vibration coupling on pigment absorption and fluorescence spectra.

R. E. Blankenship, M. T. Madigan and C. E. Bauer (eds): Anoxygenic Photosynthetic Bacteria, pp. 297–313.

I. Introduction

During the last ten years, many excellent reviews have addressed the theory of electronic excitation transfer and its applications to photosynthetic systems (Knox, 1975; van Grondelle, 1984; van Grondelle and Amesz, 1986; Geacintov and Breton, 1987). In the present short review, we begin by examining some of the physical assumptions underlying the well-established Golden Rule theories for dipole-dipole and exchange energy transfer in the weak-coupling limit. The latter limit holds when the intermolecular interactions are small compared to the bandwidths of the molecular electronic transitions. We then briefly consider extensions of the Golden Rule to higher-order perturbations, which can provide mechanisms for spin-forbidden (e.g. singlet-triplet) energy transfer.

The recent acquisition of femtosecond laser technology by a growing segment of the photosynthesis community has prompted intense interest in coherence effects on energy transfer, and especially about their manifestation through optical anisotropies in time-resolved fluorescence and pump-probe experiments. These topics are the subject of Section III. One of our goals is to make these calculations widely accessible to the photosynthesis community, so that others may be encouraged to tailor similar calculations to protein-pigment systems of interest. In Section IV, we review some techniques for simulation of temperature-dependent rate constants in Förster energy transfer.

While most of the chapters in this book have adopted SI units as a standard, this chapter has been written in cgs (centimeters-grams-seconds) units, because the vast majority of previous accounts of energy transfer theory have been based on these units. This is intended to facilitate comparisons of this chapter with the other sources in the literature.

II. The Weak Coupling Limit

A. Multipole Expansion of Electrostatic Interactions

The exact classical electrostatic energy of two molecules D and A is given by the sum of pairwise Coulomb energies,

$$V_e = \sum_i \sum_j \frac{q_i q_j}{r_{ij}} \tag{1}$$

where r_{ij} is the separation between charges q_i and q_j, and the indices i and j run over all pairs of charges (electrons and nuclei) on molecules D and A collectively. This energy may be broken up into the partial sums

$$V_e = V_{mol} + V_i \tag{2}$$

Here the molecular *self-energy*

$$V_{mol} = \sum_{i,j \in D} \frac{q_i q_j}{r_{ij}} + \sum_{i,j \in A} \frac{q_i q_j}{r_{ij}} \tag{3}$$

is summed over all pairs of charges (i,j) on molecules D and A, respectively. In the evaluation of the intermolecular *interaction energy*

$$V_i = \sum_{\substack{i \in D \\ j \in A}} \frac{q_i q_j}{r_{ij}} \tag{4}$$

the indices i, j indicate charges on molecules D and A, respectively. If the intermolecular separation R is large compared to the molecular size, so that the overlap between the respective electronic charge distributions is small, it is useful to expand the interaction energy V_i in a power series in $1/R$ (for example, see Hirschfelder and Meath, (1967)),

$$V_i = \sum_{n=1}^{\infty} V_n / R^n \tag{5}$$

In this *multipole expansion*, the coefficients V_n are given by

$$V_n = \sum_{l_D=0}^{n-1} \sum_{m=-l_<}^{l_<} \frac{(-1)^{l_D}(n-1)!\,Q_{l_D}^{m}(D)\,Q_{l_A}^{-m}(A)}{[(l_D-m)!(l_D+m)!(l_A-m)!(l_A+m)!]^{1/2}} \tag{6}$$

where $l_<$ is the lesser of l_D and $l_A = (n-1-l_D)$. The Q_l^m are the irreducible tensor components of the electrostatic multipole operators for the interacting molecules D, A (see, for example, Gottfried (1966)). For example, Q_0^0(D) and Q_0^0(A) are the total molecular charges q_D and q_A, respectively. The tensor components Q_l^m for $l = 1$ are linear combinations of

Abbreviations: FMO – Fenna-Matthews-Olson bacteriochlorophyll protein; RHAE – remote heavy atom effect; SLE – stochastic Liouville equation

Cartesian components of the electric dipole moment operators μ,

$$Q_1^{-1}(D) = -e\sum_i (x_i - iy_i)/\sqrt{2} = (\mu_{Dx} - i\mu_{Dy})/\sqrt{2}$$
$$Q_1^{0}(D) = -e\sum_i z_i = \mu_{Dz}$$
$$Q_1^{1}(D) = e\sum_i (x_i + iy_i)/\sqrt{2} = -(\mu_{Dx} + i\mu_{Dy})/\sqrt{2} \qquad (7)$$

where e is the magnitude of the electronic charge, and the z-axis points from the center of molecule D to the center of molecule A. In Eqs. (7), the sums run over all of the charges in molecule D; analogous expressions may be written for molecule A. For general $l > 0$, the irreducible tensor components for the charge distribution on molecule D may be expressed in terms of the spherical harmonic functions via

$$Q_{l_D}^{m}(D) = -\sum_i \left(\frac{4\pi}{2l_D + 1}\right)^{1/2} (er_i)^{l_D} Y_{l_D}^{m}(\theta_i, \phi_i) \qquad (8)$$

and similarly for molecule A. The angles θ_i, ϕ_i are defined with respect to a right-handed Cartesian coordinate system with its origin at the center of molecule D, and with its z-axis parallel to the intermolecular axis. After substitution of explicit expressions for the spherical harmonic functions into Eq. (8), the multipole expansion of the electrostatic interaction energy (Eq. (5)) assumes the form

$$V_i = \frac{q_D q_A}{R} + \frac{q_D(\mu_A \cdot \hat{R}) - q_A(\mu_D \cdot \hat{R})}{R^2} + \frac{\mu_D \cdot \mu_A - 3(\mu_D \cdot \hat{R})(\mu_A \cdot \hat{R})}{R^3} + \frac{q_D[Q_{xx}(A) + Q_{yy}(A) - 2Q_{zz}(A)]}{R^3} + \frac{q_A[Q_{xx}(D) + Q_{yy}(D) - 2Q_{zz}(D)]}{R^3} + \cdots \qquad (9)$$

Here the vector **R** points from the center of molecule D to the center of molecule A, and the quantities Q_{kl} represent Cartesian components of the molecular electric quadrupole tensor **Q** (see, for example, Jackson (1966)). The successive terms in Eq. (9) represent the charge-charge, charge-dipole, dipole-dipole, and charge-quadrupole contributions to the total electrostatic energy. An advantage of this multipole expansion stems from the fact that in the long-range (large-R) limit, the power series in $1/R$ converges rapidly, and can be well approximated by its leading nonzero term. In the special case where both molecules are uncharged, the terms involving q_D and q_A drop out. The lowest-order nonvanishing terms in the multipole expansion of V_i then correspond to the dipole-dipole and dipole-quadrupole interactions,

$$V_i = \frac{\mu_D \cdot \mu_A - 3(\mu_D \cdot \hat{R})(\mu_A \cdot \hat{R})}{R^3} + \frac{3[Q_{xz}(D)\mu_x(A) + Q_{yz}(D)\mu_y(A) - Q_{xz}(A)\mu_x(D) - Q_{yz}(A)\mu_y(D)]}{R^4} + \frac{3[Q_{xx}(D) + Q_{yy}(D) - 2Q_{zz}(D)]\mu_z(A)}{2R^4} \frac{-3[Q_{xx}(A) + Q_{yy}(A) - 2Q_{zz}(A)]\mu_z(D)}{2R^4} + \cdots \qquad (10)$$

In cases where the donor and acceptor electronic charge distributions overlap appreciably, the multipole expansion of Eq. (5) remains valid. However, its utility then becomes limited by the fact that many higher-order terms may need to be included in order to accurately describe medium- to short-range interactions. In such cases, it may be simpler to proceed from the exact expression for V_e, given by Eq. (1).

B. Dipole-Dipole Energy Transfer

In the limit of weak intermolecular coupling, the probability for energy transfer from donor molecule D to acceptor molecule A is given by Dirac's Golden Rule (Gottfried, 1966),

$$P_{D \to A} = \frac{2\pi}{\hbar}\left|\langle D^* A | V_i | D A^* \rangle\right|^2 \rho(E) \qquad (11)$$

where the donor and acceptor molecules are electronically excited in the initial and final states $|i\rangle = |D^*A\rangle$ and $|f\rangle = |DA^*\rangle$, respectively. The total (electronic plus nuclear) energies of the donor and acceptor in their electronic ground states are denoted ε_D and ε_A; the corresponding quantities for the electronically excited species are $\varepsilon_D{}^*$ and $\varepsilon_A{}^*$. E is the initial donor excitation energy $\varepsilon_D{}^* - \varepsilon_D$; $\rho(E)$ is the density of states isoenergetic with the initial level in the final manifold of states $|f\rangle$. Integration over all allowed combinations of initial and final donor and acceptor energies then yields

$$P_{D\to A} = \frac{2\pi}{\hbar}\frac{1}{g_D^{\bullet} g_A}\int dE \int d\varepsilon_A p_A(\varepsilon_A) \bullet \int d\varepsilon_D^{\bullet} p_D^{\bullet}(\varepsilon_D^{\bullet}) \left|\langle D^{\bullet} A | V_e | D A^{\bullet}\rangle\right|^2 \quad (12)$$

Here p_D^* and p_A denote the initial energy distributions in the donor and acceptor molecules; g_D^* and g_A are the electronic state degeneracies of the excited donor and ground-state acceptor; and $E = (\varepsilon_D^* - \varepsilon_D) = (\varepsilon_A^* - \varepsilon_A)$. The electronic ground state of the isolated n-electron donor molecule may be expanded in Slater determinants of spin-orbitals u_i (e. g. Levine, 1991),

$$|D\rangle = \frac{1}{\sqrt{n!}}\begin{vmatrix} u_1(1) & u_2(1) & \ldots & u_n(1) \\ u_1(2) & u_2(2) & \ldots & u_n(2) \\ \ldots & \ldots & \ldots & \ldots \\ u_1(n) & u_2(n) & \ldots & u_n(n) \end{vmatrix} \equiv |u_1 u_2 u_3 \ldots u_n| \quad (13)$$

while the excited state $|D^*\rangle$ may be similarly expanded in Slater determinants of spin-orbitals u_i'. Analogous expressions may be written for the electronic wavefunctions of acceptor molecule A. According to Eq. (10), the leading term in the matrix element required to evaluate the transition probability for energy transfer between uncharged molecules becomes

$$\left|\langle D^* A | V_i | D A^*\rangle\right|^2 = \frac{\kappa^2}{R^6}\left|\langle D^* | \mu_D | D\rangle\right|^2 \left|\langle A | \mu_A | A^*\rangle\right|^2 \quad (14)$$

where κ^2, the dipole-dipole orientational factor, is given by

$$\kappa^2 = [\hat{\mu}_D \cdot \hat{\mu}_A - 3(\hat{\mu}_D \cdot \hat{R})(\hat{\mu}_A \cdot \hat{R})]^2 \quad (15)$$

It may be shown (Birks, 1970) that the probability for spontaneous electric dipole emission of a photon of energy E by excited donor D^* is

$$A_D(E) = \frac{4E^3 n^3}{3\hbar^4 c^3 g_D^*}\int p_D^*(\varepsilon_D^*) d\varepsilon_D^* \left|\langle D^* | \mu_D | D\rangle\right|^2 \quad (16)$$

where c is the speed of light and n is the medium refractive index at the emission wavelength. Since the donor molecule's radiative lifetime is

$$\tau_0 = \frac{1}{\int A_D(E) dE} \quad (17)$$

it follows that

$$\int p_D^*(\varepsilon_D^*) d\varepsilon_D^* \left|\langle D^* | \mu_D | D\rangle\right|^2 = \frac{3\hbar^4 c^3 g_D^*}{4E^3 n^3 \tau_0} f_D(E) \quad (18)$$

where $f_D(E)$ is the donor emission spectrum normalized to unity. The Einstein coefficient for electric dipole absorption of a photon with energy E by the acceptor molecule A is

$$B_A(E) = \frac{2\pi}{3\hbar^2 g_A}\int p_A(\varepsilon_A) d\varepsilon_A \left|\langle A | \mu_A | A^*\rangle\right|^2 \quad (19)$$

This differs from the molecular absorption cross section $\sigma_A(E)$ by the factor nhE/c (Birks, 1970). Substitution of Eqs. (18) and (19) into Eq. (12), with the help of Eq. (14), then leads to

$$P_{D\to A} = \frac{9c^4 \kappa^2 \phi_D}{8\pi n^4 R^6 \tau}\int d\omega f_D(\omega) \sigma_A(\omega) / \omega^4 \equiv \frac{3}{2}\frac{\kappa^2}{\tau}\left(\frac{\overline{R}_0}{R}\right)^6 \quad (20)$$

The factor of 3/2 is conventionally introduced so that when $P_{D\to A}$ is averaged over randomly oriented molecular pairs, it assumes the form $P_{D\to A} = (\overline{R}_0 / R)^6 / \tau$. Here we have used the fact that the empirical fluorescence lifetime of a noninteracting donor chromophore differs from its radiative lifetime τ_0 by the factor ϕ_D, its fluorescence quantum yield; ω denotes the circular frequency $E/\hbar = 2\pi\upsilon$. The use of Eqs. 18 and 19 in deriving the transition probability in terms of donor fluorescence and acceptor absorption spectra implicitly assumes that donor vibrational relaxation occurs on a faster timescale than energy transfer. This is probably valid for the original situations visualized by Förster (energy transfers in dye solutions, in which the chromophores were typically separated by tens of Å). Many single-step antenna energy transfers occur within hundreds of fs (Du et al., 1993; Savikhin et al., 1993), and so this assumption may be questionable in photosynthetic systems. Donor and acceptor pigment spectra in photosynthetic antennae frequently exhibit 100–200 cm^{-1} inhomogeneous broadening. In cases where a narrow laser excitation bandwidth (e.g. ~ 50 cm^{-1} for ~ 1 ps laser pulses) excites only a subpopulation within an inhomogeneously broadened donor spectrum, the use of the full donor pigment fluorescence spectrum in Eq. (20) appears to be inappropriate. The quantity $\bar{R}_0$ (which has units of length) is the *Förster parameter* that characterizes the strength of the dipole-dipole contribution to the energy transfer probability. An alternative Förster parameter, R_0, is defined by using the donor radiative lifetime τ_0 instead of its empirical fluorescence lifetime τ in Eq. (20). Since the two Förster parameters $\bar{R}_0$ and R_0 then differ by the factor $\phi_D^{1/6}$, it is important to ascertain which definition of 'R_0' is being used when accessing values from the literature (Knox, 1975). Experimental chemical physicists have tended to use $\bar{R}_0$ because τ is directly measurable. In the photosynthesis community, R_0 is widely used instead. According to Eqs. (16), (19), and (20), dipole-dipole energy transfer occurs with large probability only if the D*→D and A→A* electronic transitions are strongly electric dipole-allowed. Since the electric dipole operators μ_D and μ_A in Eq. (14) do not operate on the electron spin functions, dipole-dipole energy transfer necessarily conserves spin on both donor and acceptor molecules. The electronic ground states of photosynthetic antenna pigments are closed-shell singlet states, so this restricts Förster energy transfer to singlet-singlet processes. The orientational factor κ^2, which can assume values from 0 to 4, strongly influences the transition probability; its importance for energy transfer between molecules in ordered pigment systems has been discussed by Knox (1975). Since the molar decadic absorption coefficient $\varepsilon(E)$ is obtained from the molecular absorption cross section $\sigma(E)$ by multiplying the former by $10^3\ ln\ 10\ /N_{Av}$ (where N_{Av} is Avogadro's number), a working expression for the Förster parameter $\bar{R}_0$ is (Förster, 1959; Knox, 1975)

$$\bar{R}_0 = \left[\frac{9000c^4\phi_D \ln 10}{8\pi n^4 N_{Av}} \int f_D(\omega)\varepsilon_A(\omega)d\omega / \omega^4\right]^{1/6} \qquad (21)$$

Van Grondelle and Amesz (1986) have tabulated representative values of $\bar{R}_0$, τ, and ϕ_D for energy transfers among commonly encountered pigments, including Chl *a* and *b*, BChl *a* spectral forms in purple bacteria, several carotenoids, and open tetrapyrrole chromophores found in phycocyanin and allophycocyanin antennae.

C. Exchange Mechanism for Energy Transfer

At close separations, for which the molecular electronic charge distributions overlap appreciably, many terms may contribute significantly to the power series for the interaction energy V_I, and the advantages of using the multipole expansion are lost. At the same time, the magnitudes of the inter- and intramolecular interactions V_i and V_{mol} may become comparable. We therefore revert to the original expression for V_e (Eq. (1)) in the transition probability,

$$P_{D\to A} = \frac{2\pi}{\hbar}\left|\sum_{ij}\left\langle D^* A\left|\frac{q_i q_j}{r_{ij}}\right|DA^*\right\rangle\right|^2 \rho(E) \qquad (22)$$

Here the summation is carried out over all pairs of charges on D and A. The matrix elements now contain the *two-electron* operators $O(ij) = 1/r_{ij}$, evaluated between initial and final electronic states. For isolated donor and acceptor molecules containing *n* and *m* electrons respectively, each of the antisymmetrized electronic states may be expanded as (linear combinations of) Slater determinants with *n*! and *m*! terms respectively (cf. Eq. (13)). However, in the presence of substantial overlap between donor and acceptor charge distributions, the electronic states of the molecule pair must be completely antisymmetrized by expanding them in terms of Slater determinants with *(n+m)*! terms, i.e.

$$\begin{aligned} \left|D^* A\right\rangle &= \left|u_1 u_2 u_3 \ldots u_n ; v_1 v_2 v_3 \ldots v_m\right| \\ \left|DA^*\right\rangle &= \left|u_1 u_2 u_3 \ldots u_n' ; v_1 v_2 v_3 \ldots v_m'\right| \end{aligned} \qquad (23)$$

The evaluation of matrix elements like those in Eq. (22) is then greatly simplified by the *Condon-Slater rules* for matrix elements of two-electron operators between determinant wavefunctions (for example, see Levine (1991)). Each spin-orbital u_k represents a product of a space function ϕ_k and a spin function σ_k, i.e. $u_k \equiv \phi_k \sigma_k$. The spin functions σ_k are either α (spin up) or β (spin down). We assume in Eqs. (23) that the ground and excited states on the donor molecule D differ by only *one* spin-orbital ($u_n \neq u_n'$), and similarly for acceptor molecule A ($v_m \neq v_m'$). In this case, application of the Condon-Slater rules for matrix elements of a two-electron operator leads immediately to

$$\begin{aligned}\sum_{ij}\left\langle D^{*}A\left|\frac{1}{r_{ij}}\right|DA^{*}\right\rangle &= \left\langle u_n'(1)v_m'(2)\left|\frac{1}{r_{12}}\right|u_n(1)v_m(2)\right\rangle \\ &-\left\langle u_n'(2)v_m'(1)\left|\frac{1}{r_{12}}\right|u_n(1)v_m(2)\right\rangle \\ &=\left\langle \phi_n'(1)\phi_m'(2)\left|\frac{1}{r_{12}}\right|\phi_n(1)\phi_m(2)\right\rangle\left\langle\sigma_n'(1)|\sigma_n(1)\right\rangle\left\langle\sigma_m'(2)|\sigma_m(2)\right\rangle \\ &-\left\langle \phi_n'(2)\phi_m'(1)\left|\frac{1}{r_{12}}\right|\phi_n(1)\phi_m(2)\right\rangle\left\langle\sigma_m'(1)|\sigma_n(1)\right\rangle\left\langle\sigma_n'(2)|\sigma_m(2)\right\rangle\end{aligned} \tag{24}$$

Here the first and second terms on the right-hand side are termed the *direct* and *exchange* contributions to the matrix element, respectively. The numbers 1,2 are dummy indices that denote the coordinates of electrons 1 and 2. The *direct* term is responsible for the Förster dipole-dipole matrix element discussed in the previous section. Its contribution to the total transition probability reduces to the Förster rate for energy transfer between uncharged molecules in the limit of large separations R when the dipole-dipole transition probability is appreciable. At smaller separations, it will also contain substantial contributions from higher-order multipole interactions (dipole-quadrupole, etc.). For the direct term, the spin-function overlaps clearly require that $\sigma_n = \sigma_n'$ and that $\sigma_m' = \sigma_m$. This implies that spin must be conserved on both donor and acceptor molecules during energy transfer, and hence that only singlet-singlet energy transfers arise from this term in antennae (since the electronic ground states of antenna pigments are singlets). The direct term can also account for excitation trapping at oxidized reaction centers ($D^* + P^+ \rightarrow D + (P^+)^*$), which frequently occurs with high efficiency. Like singlet-singlet energy transfers in antennae, this direct singlet-doublet energy transfer process is spin-allowed, because the ground and excited acceptor states P^+ and $(P^+)^*$ are both doublet states. In contrast, the *exchange* term requires that $\sigma_m' = \sigma_n$ and that $\sigma_n' = \sigma_m$. This term is therefore nonzero in antennae when (a) the electronically excited molecules D* and A* are both singlet states, or (b) D* and A* are both triplet states; it vanishes otherwise. This exchange term thus appears to provide a major mechanism for triplet-triplet energy transfer (Dexter, 1953). It may also provide a route for singlet-singlet energy transfer when Förster transfer is symmetry-forbidden, as in energy transfers from 2^1A_g carotenoids to BChl *a* pigments (Frank et al., 1993). Inspection of the spatial factor in the exchange term shows that its R-dependence is dominated not so much by the factor $1/r_{12}$, but by the fact that appreciable contributions to this matrix element arise only from regions where the two-center products $\phi_n'\phi_m$, $\phi_m'\phi_n$ are large. Substantial spatial overlap between donor and acceptor orbitals is therefore a prerequisite for rapid exchange energy transfer. Two-center exchange integrals like the one in Eq. (24) exhibit exceedingly complicated R-dependence, even for the prototypical simple case where the electronic wavefunctions ϕ_m, ϕ_m', ϕ_n, ϕ_n' are hydrogenlike atomic orbitals - not to mention porphyrin molecular orbitals (Slater, 1963). Following Dexter (1953), R-dependence in the exchange integral is often coarsely parameterized in the form exp (- R/L), where L is a constant (with dimensions of length) that is on the order of the molecular size. The spatial overlap requirement thus tends to limit the exchange mechanism to short-range energy transfers between closely situated chromophores. This situation contrasts markedly with the case of Förster transfer, where (according to Eq. (20)) an excited donor exhibits equal probabilities of intramolecular decay and nonradiative excitation transfer when $\kappa^2 = 2/3$ and $R = \bar{R}_0$. For dipole-dipole transfers between identical Chl *a* chromophores, $\bar{R}_0$ can be as large as 60–70 Å, which considerably exceeds the molecular size (Knox, 1975).

Knox and Davidovich (1979) considered the form of the spectral overlap integral appropriate for triplet-triplet exchange energy transfer. The zeroth-order chromophore electronic states $|D\rangle$, $|D^*\rangle$, $|A\rangle$, and $|A^*\rangle$ may be chosen to be eigenstates of the Born-Oppenheimer electronic Hamiltonian H_{el}, which includes all intramolecular electrostatic interactions and the electronic (but not nuclear) kinetic energies. Since H_{el} does not contain electron spin operators, these Born-Oppenheimer states will be either pure singlet or pure triplet states. Perturbation of H_{el} by the intramolecular spin-orbit coupling H_{so} mixes these zeroth-order states, so that the perturbed electronic states $|D'\rangle$, $|D'^*\rangle$, $|A'\rangle$, and $|A'^*\rangle$ are either singlet states containing small admixtures of triplet states, or vice versa. This provides one mechanism (mechanism 1) through which the otherwise spin-forbidden radiative processes of singlet-triplet absorption and phosphorescence can occur. A second mechanism (mechanism 2) is first- and/or second-order spin-orbit-vibronic coupling. If the absorption and phosphorescence spectra *are dominated by mechanism 1*, the triplet-triplet exchange energy transfer rate is given by

$$P_{D\to A} = \frac{2\pi e^4}{\hbar g_D^* g_A} \left| \left\langle \phi_n'(2)\phi_m'(1) \left| \frac{1}{r_{12}} \right| \phi_n(1)\phi_m(2) \right\rangle \right|^2 \int f_D(E) F_A(E) dE \tag{25}$$

Here $f_D(E)$ and $F_A(E)$ are the normalized donor phosphorescence and the acceptor singlet-triplet absorption spectra,

$$f_D(E) = \frac{E^{-3} A_D(E)}{\int E^{-3} A_D(E) dE}$$
$$F_A(E) = \frac{E^{-1} \sigma_A(E)}{\int E^{-1} \sigma_A(E) dE} \tag{26}$$

If, however, the spectra contain overlapping components arising from mechanism 2, these components do not contribute to the energy transfer rate; using the empirical spectra in Eq. (25) would then overestimate the rate constant.

D. Higher-Order Processes: Spin-Forbidden Energy Transfer

The Golden Rule rate in Eq. (11) only expresses the leading term in a perturbation expansion of the radiationless transition rate. Extension of this treatment to higher orders under a perturbing Hamiltonian V yields the more general result (Gottfried, 1966)

$$P_{D\to A} = \frac{2\pi}{\hbar} |T_{D\to A}|^2 \rho(E) \tag{27}$$

where the T-matrix is given to third order by

$$T_{D\to A} = \langle D^* A | V | DA^* \rangle + \sum_{D'A'} \frac{\langle D^* A | V | D'A' \rangle \langle D'A' | V | DA^* \rangle}{E_{D^*A} - E_{D'A'}} + \sum_{\substack{D'A' \\ D''A''}} \frac{\langle D^* A | V | D'A' \rangle \langle D'A' | V | D''A'' \rangle \langle D''A'' | V | DA^* \rangle}{(E_{D^*A} - E_{D'A'})(E_{D^*A} - E_{D''A''})} + \cdots \tag{28}$$

These summations are carried out over all donor and acceptor states (D'A', D"A", ...) other than the initial state D^*A. The perturbation V is not limited to the electrostatic interaction V_e, but may include additional perturbations such as the intra- and/or intermolecular spin-orbit Hamiltonian H_{so} (Avouris et al., 1977; Struve, 1989). For example, two second-order terms in $T_{D\to A}$ that may contribute to the spin-forbidden singlet-triplet energy transfer process ${}^1D^* A \to D\,{}^3A^*$ are

$$T^{(2)}_{D\to A} = \frac{\langle {}^1D^* A | H_{so} | {}^3D^* A \rangle \langle {}^3D^* A | V_e | D^3A^* \rangle}{E_{{}^1D^*A} - E_{{}^3D^*A}} + \frac{\langle {}^1D^* A | V_e | D^1A^* \rangle \langle D^1A^* | H_{so} | D^3A^* \rangle}{E_{{}^1D^*A} - E_{D^1A^*}} \tag{29}$$

In the first term, the donor excited singlet state ${}^1D^*$ is viewed as being coupled (through H_{so}) to a virtual donor triplet state ${}^3D^*$; the latter interacts in turn with the acceptor triplet state ${}^3A^*$ through the exchange mechanism. (As explained in Section II.C, the exchange mechanism alone cannot account for spin-forbidden energy transfer between a pure singlet state and a pure triplet state.) Such a mechanism may be relevant in cases where the acceptor molecule has no energetically accessible singlet states, but has triplet levels lying below the donor state ${}^1D^*$. In the second term, the donor excited singlet state interacts with an acceptor singlet state (e.g. by the Förster mechanism); the latter is coupled through H_{so} to an acceptor triplet state. In the presence of spin-orbit coupling, the acceptor triplet states are not pure triplets, but contain small admixtures of acceptor singlet states. Hence, these admixtures may alternatively be viewed as being coupled to the donor singlet state through the dipole-dipole interaction.

Third-order terms in the T matrix appear to be responsible for the remote heavy-atom effect (RHAE), in which intramolecular singlet → triplet intersystem crossing on a 'donor' molecule is prompted by intermolecular spin-orbit coupling, due to the presence of a heavy atom on a nearby chromophore. Such processes have recently been characterized in dimeric peptides that contain a fluorescing β-(1'-naphthyl)-alanine chromophore perturbed by p-bromo-L-phenylalanine (Basu et al., 1993).

III. Unified Theory for Energy Transfer

A. The Density Operator and Density Matrix

We consider a quantum mechanical system whose time-dependent wavefunction $|\psi(t)\rangle$ can be expressed as a linear superimposition of orthonormal basis states $|\chi_n\rangle$,

$$|\psi_n(t)\rangle = \sum_n c_n(t)|\chi_n\rangle \tag{30}$$

The time-dependent expectation value of a quantum mechanical observable in this state is

$$\langle A(t)\rangle = \langle\psi(t)|\hat{A}|\psi(t)\rangle = \sum_{n,m} c_n^*(t)c_m(t)\langle\chi_n|\hat{A}|\chi_m\rangle$$
$$\equiv \sum_{n,m} c_n^*(t)c_m(t)A_{nm} \tag{31}$$

In this representation, all of the information concerning the state's time evolution is given by the set of time-dependent products of expansion coefficients $c_n^*(t)c_m(t)$. If we define the *density operator*

$$\rho(t) \equiv |\psi(t)\rangle\langle\psi(t)| \tag{32}$$

the matrix elements of this new operator in the basis states $|\chi_n\rangle$ will be given by

$$\rho_{nm}(t) = \langle\chi_n|\rho(t)|\chi_m\rangle$$
$$\equiv \langle\chi_n|\psi(t)\rangle\langle\psi(t)|\chi_m\rangle = c_m^*(t)c_n(t) \tag{33}$$

and will thus contain the dynamic information of interest. The set of these matrix elements $\rho_{nm}(t)$ make up the *density matrix*.

Oftentimes a system is not prepared in a pure quantum mechanical state $|\psi(t)\rangle$, but in a mixture of states $|\psi_k(t)\rangle$ in such a way that the probability that the system in state k is $p_k(t)$. In this case, we may define the partial density operator

$$\rho_k = |\psi_k\rangle\langle\psi_k| \tag{34}$$

for state $|\psi_k\rangle$. The corresponding expression for the full density operator that yields the correct expressions for the expectation values and the density matrix is then given by (Cohen-Tannoudji, Diu, and Laloë, 1977)

$$\rho = \sum_k p_k\rho_k \tag{35}$$

The physical significance of the *diagonal* density matrix elements may be seen by evaluating

$$\rho_{nn} = \sum_k p_k[\rho_k]_{nn} = \sum_k p_k\langle\chi_n|\psi_k(t)\rangle\langle\psi_k(t)|\chi_n\rangle$$
$$= \sum_k p_k\left\langle\chi_n\left|\sum_m c_m^{(k)}(t)\chi_m\right\rangle\left\langle\sum_{m'} c_{m'}^{(k)}(t)\chi_{m'}\right|\chi_n\right\rangle$$
$$= \sum_k p_k\left|c_n^{(k)}(t)\right|^2 \tag{36}$$

It is clear from this that the diagonal density matrix element $\rho_{nn}(t)$—which is nonnegative definite — is the probability that the system is in the basis state $|\chi_n\rangle$, statistically averaged over the probabilities p_k that the system is in the various superimposition states $|\psi_k(t)\rangle$. It is frequently referred to as the *population* in state $|\chi_n\rangle$. A similar calculation shows that the *off-diagonal* matrix elements are

$$\rho_{nm} = \sum_k p_k\, c_m^{(k)*}(t)\, c_n^{(k)}(t) \tag{37}$$

The complex-valued cross terms appearing in each term of this sum reflect interferences between the basis states $|\chi_n\rangle$ and $|\chi_m\rangle$. When statistically weighted by the p_k over the superimposition states $|\psi_k(t)\rangle$, these interferences may average to a nonzero value —in which case the quantum mechanical phase coherence among the basis states is at least partially maintained in the ensemble averaging. The off-diagonal matrix elements in Eq. (37) are sometimes termed *coherences*.

B. Master equations in the site and delocalized representations

The conventional master equations for the probabilities $\rho_{11}(t)$, $\rho_{22}(t)$ that electronic excitation will be found on one of two identical molecules *1* and *2* after laser pumping are given by (Förster, 1948)

$$\frac{d\rho_{11}}{dt} = -\frac{\rho_{11}}{\tau} - F(\rho_{11}-\rho_{22})$$
$$\frac{d\rho_{22}}{dt} = -\frac{\rho_{22}}{\tau} - F(\rho_{22}-\rho_{11}) \tag{38}$$

Here $1/\tau$ is the total (radiative plus nonradiative) decay rate for either of the isolated molecules, and F is the rate of excitation transfer between molecules. (In the weak-coupling limit, F is given by expressions developed in Section II.) Equations 38 predict that at long times $t >> 1/F$, the excited state populations will converge to the same equilibrium value, i.e. $(\rho_{11} - \rho_{22}) \rightarrow 0$.

Alternatively, the interaction energy J may be so large that it is more useful to describe the excitations in terms of the (delocalized) exciton states

$$|+\rangle = 2^{-1/2}(|1\rangle + |2\rangle)$$
$$|-\rangle = 2^{-1/2}(|1\rangle - |2\rangle) \tag{39}$$

This interaction lifts the original degeneracy between the localized states $|1\rangle$ and $|2\rangle$, yielding exciton component states $|+\rangle, |-\rangle$ whose energies are separated by $2J$. In the strong coupling limit, the time-dependent probabilities ρ_{++}, ρ_{--} that the respective delocalized states are populated following laser excitation are governed by the master equations (Knox and Gülen, 1993)

$$\frac{d\rho_{++}}{dt} = -\frac{\rho_{++}}{\tau_+} - \gamma\rho_{++} + \gamma'\rho_{--}$$
$$\frac{d\rho_{--}}{dt} = -\frac{\rho_{--}}{\tau_-} - \gamma'\rho_{--} + \gamma\rho_{++} \tag{40}$$

Here γ and γ' are rates for upward and downward transitions between the exciton components. These must be unequal at finite temperatures, in view of the requirement that $\gamma'/\gamma = \exp(-2J/kT)$ according to detailed balancing (Rahman et al., 1979). Furthermore, the fluorescence lifetimes τ_+, τ_- of the two exciton components $|+\rangle, |-\rangle$ will be different unless the transition moments of the two chromophores *1, 2* happen to be perpendicular (Chambers et. al., 1974). The master equations 40 in the delocalized representation reasonably predict that at long times, the ratio ρ_{--}/ρ_{++} of equilibrium populations in the exciton states will be given by the ratio γ/γ' of upward to downward transition rates.

C. The Need for a Unified Theory

A paradox arises when one attempts to transform the master equations for the site representation in the weak coupling limit (Eqs. (38)) into those for the delocalized representation in the strong coupling limit (Eqs. (40)). In view of the transformation (Eqs. (39)) between localized and delocalized states, it follows that

$$|+\rangle\langle+| = \frac{1}{2}(|1\rangle + |2\rangle)(\langle 1| + \langle 2|)$$
$$= \frac{1}{2}(|1\rangle\langle 1| + |2\rangle\langle 1| + |1\rangle\langle 2| + |2\rangle\langle 2|) \tag{41}$$

The corresponding density matrix elements ρ_{kl} analogously obey

$$\rho_{++} = \frac{1}{2}(\rho_{11} + \rho_{21} + \rho_{12} + \rho_{22}) \tag{42}$$

The diagonal matrix elements ρ_{kk} give the probability that excitation resides in state $|k\rangle$. Similar equations can be written for the other density matrix elements ρ_{--}, ρ_{+-}, and ρ_{-+} in the delocalized representation in terms of the matrix elements ρ_{11}, ρ_{21}, ρ_{12}, and ρ_{22} in the site basis (Rahman et al., 1979),

$$\rho_{--} = \frac{1}{2}(\rho_{11} - \rho_{21} - \rho_{12} + \rho_{22})$$
$$\rho_{+-} = \frac{1}{2}(\rho_{11} + \rho_{21} - \rho_{12} - \rho_{22}) \tag{43}$$
$$\rho_{-+} = \frac{1}{2}(\rho_{11} - \rho_{21} + \rho_{12} - \rho_{22})$$

It is clear that substitution of these expressions into the strong-coupling master equations (40) *cannot* yield the weak-coupling master equations (38) in the site representation, because this would introduce cross terms (involving the off-diagonal matrix elements ρ_{12} and ρ_{21}) that are not present in the weak-coupling master equations. In a similar way, it may be shown (by using the inverse of the transformation in Eqs. 42–43) that the weak-coupling master equations (small J, large γ) do not go over into the strong-coupling master equations (large J). Hence, both theories are fundamentally incomplete, through their omission of these cross terms. According to Eq. (43), the probability that excitation resides in the delocalized state $|+\rangle$ is equal to

$$|c_+|^2 = |c_1|^2 + c_2^*c_1 + c_1^*c_2 + |c_2|^2 \tag{44}$$

While the quantum-mechanical phases in the localized basis states $|1\rangle, |2\rangle$ automatically cancel in the computation of the absolute values-squared in the diagonal terms in Eq. (44), they do not cancel in the off-diagonal terms. This phase information will persist in the statistical averaging that yields the off-diagonal elements of the full density matrix, unless it is destroyed by dephasing processes (e.g. arising from interactions of the electronic states with random protein motions). Hence, the cross terms are connected with coherence decay in the laser-prepared excited states.

D. The Stochastic Liouville Equations

Rahman et al. (1979) showed that the correct master equations in the delocalized representation for the excited state dynamics in a homodimer (i.e. a pair of identical chromophores) have the generic form

$$\frac{d\rho_{++}}{dt} = -\gamma\rho_{++} + \gamma'\rho_{--} - \frac{(1+\cos\theta)\rho_{++}}{\tau_0} - \left(\frac{1}{\tau} - \frac{1}{\tau_0}\right)\rho_{++}$$
$$\frac{d\rho_{--}}{dt} = -\gamma'\rho_{--} + \gamma\rho_{++} - \frac{(1-\cos\theta)\rho_{--}}{\tau_0} - \left(\frac{1}{\tau} - \frac{1}{\tau_0}\right)\rho_{--}$$
$$\frac{d\rho_{+-}}{dt} = -2iJ\rho_{+-} - \left(2A + \frac{1}{\tau}\right)\rho_{+-} - (\Gamma + B - A)(\rho_{+-} - \rho_{-+})$$
$$\frac{d\rho_{-+}}{dt} = 2iJ\rho_{-+} - \left(2A + \frac{1}{\tau}\right)\rho_{-+} - (\Gamma + B - A)(\rho_{-+} - \rho_{+-}) \tag{45}$$

Here the rate constants γ and γ' for upward and downward transitions between the exciton components depend on local fluctuations in the excited state energies $E_{1(2)}$; their mean is Γ. The monomer radiative and total decay rates are $1/\tau_0$ and $1/\tau$, respectively. The angle between the transition dipole moments of chromophores *1,2* is θ. The constants A and B are related to local fluctuations in the interaction J, and therefore to J^2. In the weak coupling limit, the Förster rate for excitation transfer between the two chromophores is given by $F = A + 2J^2/(2\Gamma + 2B + 1/\tau)$. Using the relationships (Eqs. (42–43)) between the density matrices in the delocalized and site representations, these master equations can be transformed into (more complicated) master equations in the site representation. A major simplifying feature of the master equations in the delocalized representation is that the diagonal matrix elements ρ_{++}, ρ_{--} are not coupled to the off-diagonal matrix elements ρ_{+-}, ρ_{-+}.

The solutions to these master equations are readily obtained through standard Laplace transform techniques. In particular, the diagonal density matrix elements behave as

$$\rho_{++}(t) = a_+(t)\rho_{++}(0) + b_+(t)\rho_{--}(0)$$
$$\rho_{--}(t) = a_-(t)\rho_{--}(0) + b_-(t)\rho_{++}(0) \tag{46}$$

with

$$a_\pm(t) = \frac{r_1 e^{-r_1 t} - r_2 e^{-r_2 t} + \left(e^{-r_2 t} - e^{-r_1 t}\right)\left(\gamma_\pm + (1 \mp \cos\theta)/\tau_0 + \left(\frac{1}{\tau} - \frac{1}{\tau_0}\right)\right)}{r_1 - r_2} \tag{47}$$

and

$$b_\pm(t) = \frac{\gamma_\pm\left(e^{-r_2 t} - e^{-r_1 t}\right)}{r_1 - r_2} \tag{48}$$

Here we have set $\gamma_+ = \gamma'$ and $\gamma_- = \gamma$. The quantities r_1, r_2 (which have units of s^{-1}) are given by

$$r_1, r_2 = \Gamma + \frac{1}{\tau} \pm \frac{1}{2}\sqrt{\left(\gamma - \gamma' + \frac{2\cos\theta}{\tau_0}\right)^2 + 4\gamma\gamma'} \tag{49}$$

The off-diagonal density matrix elements may be expressed as

$$\rho_{+-}(t) = c_+(t)\rho_{+-}(0) + d_+(t)\rho_{-+}(0)$$
$$\rho_{-+}(t) = c_-(t)\rho_{-+}(0) + d_-(t)\rho_{+-}(0) \tag{50}$$

where

$$c_\pm(t) = \frac{s_1 e^{-s_1 t} - s_2 e^{-s_2 t} + \left(e^{-s_2 t} - e^{-s_1 t}\right)\left(A + B + \Gamma + \frac{1}{\tau} \mp 2iJ\right)}{s_1 - s_2}$$
$$d_\pm(t) = \frac{(\Gamma + B - A)\left(e^{-s_2 t} - e^{-s_1 t}\right)}{s_1 - s_2} \tag{51}$$

Here the roots s_1 and s_2 are given by

$$s_1, s_2 = A + B + \Gamma + \frac{1}{\tau} \pm i\sqrt{4J^2 - (\Gamma + B - A)^2} \tag{52}$$

It is clear from Eqs. (46–48) and (50–51) that the time-dependent functions $a_\pm$(t) and $c_\pm$(t) initialize to unity at zero time. Similarly, the functions $b_\pm$(t) and $d_\pm$(t) initialize to zero. In general, $a_\pm$(t) and $b_\pm$(t) — which control the exciton level populations, cf. Eqs. (46) —are biexponential functions of time, because the numbers r_1, r_2 are always real. However, the functions $c_\pm$(t) and $d_\pm$(t) – which control the coherence decays according to Eqs. (50) – may display oscillatory behavior ('quantum beats'), since the numbers s_1, s_2 are complex when $4J^2 > (\Gamma + B - A)^2$.

E. Optical Anisotropy in Fluorescence and Pump-Probe Experiments

The result of an optical anisotropy experiment in many-chromophore systems such as photosynthetic antennae is directly sensitive to the presence of exciton couplings and coherence decay. Since the delocalized exciton states $|+\rangle$ and $|-\rangle$ in a homodimer are linear combinations of the localized excited states $|1\rangle$ and $|2\rangle$ (Eqs. 40), the transition moment orientations μ_+, μ_- of the exciton transitions differ in both direction and magnitude from those in the site representation, μ_1 and μ_2. The initial values of the density matrix elements in the delocalized representation, promptly following electric-dipole absorption of a photon with polarization $\mathbf{e}_i$, are given by (Rahman et al., 1979)

$$\rho_{++}(0) = K_+B_+(\hat{e}_i \cdot \hat{\mu}_+)^2$$
$$\rho_{--}(0) = K_-B_-(\hat{e}_i \cdot \hat{\mu}_-)^2$$
$$\rho_{+-}(0) = \rho_{-+}(0) = (K_+K_-B_+B_-)^{1/2}(\hat{e}_i \cdot \hat{\mu}_+)(\hat{e}_i \cdot \hat{\mu}_-) \tag{53}$$

Here K_+, K_- are the incident light intensities at the wavelengths for absorptive transitions to the two exciton states, and B_+, B_- are the corresponding Einstein coefficients. The fluorescence intensity subsequently emitted by the homodimer at time t with polarization $\mathbf{e}_f$ will then be

$$I(t) = A_+\rho_{++}(t)(\hat{e}_f \cdot \hat{\mu}_+)^2 + A_-\rho_{--}(t)(\hat{e}_f \cdot \hat{\mu}_-)^2 + (A_+A_-)^{1/2}(\rho_{+-}(t) + \rho_{-+}(t))(\hat{e}_f \cdot \hat{\mu}_+)(\hat{e}_f \cdot \hat{\mu}_-) \tag{54}$$

where A_+, A_- are the corresponding Einstein coefficients for spontaneous emission from the exciton components. According to Eq. 53, selective excitation of *one* of the exciton components (e.g. $K_+ = 0$) yields vanishing initial values $\rho_{+-}(0)$, $\rho_{-+}(0)$ of the off-diagonal density matrix elements. Since the diagonal and off-diagonal matrix elements are uncoupled in the stochastic Liouville equations in this representation, the off-diagonal elements (the coherences) then remain equal to zero at all later times. Equations (54) then imply that the total fluorescence spectrum and polarization will simply evolve as incoherent, weighted sums of the respective quantities for the two exciton components. For *spectrally broad* excitation ($K_+ = K_-$), the off-diagonal elements will initialize to large values. The resulting interference terms in Eq. (54) can then markedly affect the fluorescence polarization, with time dependence that reflects processes responsible for the decay in ρ_{+-} and ρ_{-+}. These coherence decays may be observed in ultrafast fluorescence or pump-probe spectroscopy, since femtosecond excitation pulses are inherently spectrally broad.

In a typical fluorescence anisotropy measurement, the excitation laser is polarized along the laboratory x-axis, while the fluorescence is detected with analyzers polarized along either the x- or y-axis. The polarized fluorescence intensities emitted by a sample of randomly oriented homodimers (in which the transition moments of chromophores 1,2 are separated by the well-defined angle θ) are then obtained by combining Eqs. (46), (50), (53), and (54) :

$$\begin{aligned} I_\parallel &= K_+A_+B_+a_+(t)\left\langle(\hat{x}\cdot\hat{\mu}_+)^4\right\rangle + K_-A_+B_-b_+(t)\left\langle(\hat{x}\cdot\hat{\mu}_-)^2(\hat{x}\cdot\hat{\mu}_+)^2\right\rangle \\ &+K_-A_-B_-a_-(t)\left\langle(x\cdot\mu_-)^4\right\rangle + K_+A_-B_+b_-(t)\left\langle(\hat{x}\cdot\hat{\mu}_+)^2(\hat{x}\cdot\hat{\mu}_-)^2\right\rangle \\ &+(K_+K_-A_+A_-B_+B_-)^{1/2}(c_+(t)+d_+(t)+c_-(t)+d_-(t))\left\langle(\hat{x}\cdot\hat{\mu}_+)^2(\hat{x}\cdot\hat{\mu}_-)^2\right\rangle \end{aligned}$$

and

$$\begin{aligned} I_\perp &= K_+A_+B_+a_+(t)\left\langle(x\cdot\mu_+)^2(\hat{y}\cdot\hat{\mu}_+)^2\right\rangle + K_-A_+B_-b_+(t)\left\langle(\hat{x}\cdot\hat{\mu}_-)^2(\hat{y}\cdot\hat{\mu}_+)^2\right\rangle \\ &+K_-A_-B_-a_-(t)\left\langle(\hat{x}\cdot\hat{\mu}_-)^2(\hat{y}\cdot\hat{\mu}_-)^2\right\rangle + K_+A_-B_+b_-(t)\left\langle(\hat{x}\cdot\hat{\mu}_+)^2(\hat{y}\cdot\hat{\mu}_-)^2\right\rangle \\ &+(K_+K_-A_+A_-B_+B_-)^{1/2}(c_+(t)+d_+(t)+c_-(t)+d_-(t))\cdot \\ &\left\langle(\hat{x}\cdot\hat{\mu}_+)(\hat{x}\cdot\hat{\mu}_-)(\hat{y}\cdot\hat{\mu}_+)(\hat{y}\cdot\hat{\mu}_-)\right\rangle \end{aligned}$$

Here the brackets $\langle\,\rangle$ indicate averaging over the (random) orientational distribution of homodimers. The transition moments μ_+, μ_- in a symmetric homodimer are mutually orthogonal, regardless of the angle θ between the chromophore transition moments μ_1, μ_2 (Chambers et al., 1974). This fact may be used in performing the required orientational averages (van Amerongen and Struve, 1993), yielding

$$\begin{aligned} I_\parallel(t) &= \frac{3}{15}K_+A_+B_+a_+(t) + \frac{3}{15}K_-A_-B_-a_-(t) + \\ &\frac{1}{15}K_-A_+B_-b_+(t) + \frac{1}{15}K_+A_-B_+b_-(t) \\ &+\frac{1}{15}(K_+K_-A_+A_-B_+B_-)^{1/2}(c_+(t)+d_+(t)+c_-(t)+d_-(t)) \end{aligned} \tag{55a}$$

and

$$I_{\perp}(t)=\frac{1}{15}K_+A_+B_+a_+(t)+\frac{1}{15}K_-A_-B_-a_-(t)$$
$$+\frac{2}{15}K_-A_+B_-b_+(t)+\frac{2}{15}K_+A_-B_+b_-(t)$$
$$-\frac{1}{30}(K_+K_-A_+A_-B_+B_-)^{1/2}(c_+(t)+d_+(t)+c_-(t)+d_-(t)) \tag{55b}$$

In view of the fact that $a_{\pm}(0) = c_{\pm}(0) = 1$ and $b_{\pm}(0) = d_{\pm}(0) = 0$, the fluorescence anisotropy at time $t = 0$ becomes

$$r(0)=\frac{I_{\parallel}(0)-I_{\perp}(0)}{I_{\parallel}(0)+2I_{\perp}(0)}$$
$$=\frac{2K_+A_+B_+ + 2K_-A_-B_- + 3(K_+K_-A_+A_-B_+B_-)^{1/2}}{5(K_+A_+B_+ + K_-A_-B_-)} \tag{56}$$

In a homodimer for which θ is the angle between transition moments μ_1 and μ_2, the Einstein coefficients $A_{\pm}$ and $B_{\pm}$ for spontaneous emission and absorption are proportional to $(1 \pm \cos\theta)$. In the limit of *spectrally broad* excitation ($K_+ = K_-$), in which all of the density matrix elements are 'created equal' by broadband light — the initial anisotropy therefore simplifies into

$$r(0)=\frac{7+\cos^2\theta}{10(1+\cos^2\theta)} \tag{57}$$

Hence, the initial anisotropies are bounded between 0.4 and 0.7 in the case of broadband excitation of a homodimer (Knox and Gülen, 1993). The minimum $r(0) = 0.4$ is attained when the chromophore transition moments μ_1 and μ_2 are parallel; in this case, only one of the exciton components is electric-dipole accessible from the homodimer ground state. The maximum $r(0) = 0.7$ occurs when $\theta = 90°$, in which case the two perpendicular transition moments μ_+ and μ_- have equal magnitude. The initial anisotropy thus depends on the symmetry of the transition moment distribution among the exciton components that are excited. Equation 57 agrees with one obtained by Wynne and Hochstrasser (1993) using Redfield theory (1965). These results are of course strongly influenced by the excitation and detection conditions; empirical anisotropies will be 'filtered' by the excitation bandwidth and by the detector's spectral sensitivity.

It is straightforward to show from Eqs. 46–52 that in the special case where $\theta = 90°$, the anisotropy function $r(t)$ evolves as

$$r_{90}(t)=\frac{1}{10}\left(1+3e^{-2\Gamma t}+3\xi(t)\right) \tag{58}$$

It may be similarly shown that for general θ,

$$r(t)=\frac{r_{90}(t)+\frac{1}{10}\left(3+e^{-2\Gamma t}-3\xi(t)\right)\cos^2\theta+\frac{2}{5}\left(1-e^{-2\Gamma t}\right)\frac{(\gamma'-\gamma)}{2\Gamma}\cos\theta}{1+e^{-2\Gamma t}\cos^2\theta+\left(1-e^{-2\Gamma t}\right)\frac{(\gamma'-\gamma)}{2\Gamma}\cos\theta} \tag{59}$$

when the excited state decay rate $1/\tau$ is small compared to A, B, Γ, and J. The function $\xi(t)$, which equals unity at zero time, is

$$\xi(t)=e^{-(A+B+\Gamma)t}\cdot\left[(B+\Gamma-A)\frac{\sin\sqrt{4J^2-(B+\Gamma-A)^2}\,t}{\sqrt{4J^2-(B+\Gamma-A)^2}}+\cos\sqrt{4J^2-(B+\Gamma-A)^2}\,t\right]$$

In accordance with detailed balancing, the factor $(\gamma' - \gamma)/2\Gamma$ must equal $\tanh(J/kT)$. Exciton splittings $2J$ on the order of several hundred cm^{-1} are common in pigment-protein complexes, and hence this factor is significant even at room temperature. The function $\xi(t)$ exhibits sinusoidal character in the strong-coupling (large J) limit, causing the occurrence of 'quantum beats' (Knox and Gülen, 1993). In the weak-coupling limit (where $J \sim 0$ and $(B+\Gamma-A) \sim \Gamma$, the argument of the square root in Eqs. (51–52) becomes negative, and $\xi(t)$ no longer contains sinusoidal terms. The third term on the right-hand side of Eq. 58 then approaches $3\exp(-2Ft)$, and so this decay component in $r_{90}(t)$ corresponds to ordinary Förster energy transfer between the two chromophores. Equations 57 and 59 predict that $r(t)$ should decay rapidly from $r(0) \geq 0.4$ with lifetime $(2\Gamma)^{-1}$, and then decay more slowly from $r \sim 0.4$ with dynamics that resemble Förster kinetics in the weak-coupling limit.

It appears that the stochastic Liouville equations for N-chromophore systems (N > 2) are not simple extensions of the homodimer master equations (Kenkre, 1982). However, the prediction of *initial* anisotropies $r(0)$ does not require knowledge of the time-dependent density matrix, but may be obtained directly from initial conditions similar to those used in Eqs. (53–56). For example, each protein subunit in an FMO trimer from the BChl *a* baseplate antennae in green sulfur bacteria contains 7 BChl *a* chromophores. Each of the 7 monomer exciton components becomes split into three sublevels in the trimer: a nondegenerate level polarized parallel to the threefold axis, and a doubly degenerate, mutually perpendicular pair polarized in the plane of the trimer. The Einstein absorption coefficients $B_\perp$ of the two degenerate transitions are equal, while the relative magnitude $B_\parallel$ of the third Einstein coefficient depends on the projection of the unsplit monomer exciton component's transition moment on the threefold axis. Uniform excitation of these three FMO trimer exciton components alone would then lead to the initial anisotropy function

$$r(0) = \frac{7+2x^2+6x}{10+5x^2} \tag{60}$$

where $x = B_\parallel/B_\perp$ is the ratio of parallel to perpendicular absorption coefficients. In the limit where $x \rightarrow 0$ (uniform excitation of two equally intense, mutually perpendicular transitions), $r(0) \rightarrow 0.7$. In the limit where $x \rightarrow$ large (absorption dominated by a single linearly polarized transition), $r(0) \rightarrow 0.4$. In the special case where $x = 1$ (uniform excitation of three equally intense, mutually perpendicular transitions), $r(0) = 1.0$. In the latter case, the total weighted transition moment for exciton components excited by the laser is exactly parallel to the laser polarization — with the result that the perpendicular fluorescence component $I_\perp(0)$ vanishes and $r(0)$ becomes unity. Hence, these three limiting cases illustrate examples of linear, planar, and spherical oscillators, respectively. In a recent pump-probe study of FMO trimers from the green sulfur bacterium *Chlorobium tepidum* (Savikhin et al., 1993), initial anisotropies $r(0)$ in excess of 0.4 were found at 796 and 821 nm. These decayed to ≤ 0.4 within ~ 100 fs. A subsequent, slower anisotropy decay occurred at both wavelengths, with lifetime 1.7–2.0 ps. These fast component decays are consistent with spectral hole-burning experiments on FMO trimers from *Prosthecochloris aestuarii* (Johnson and Small, 1991), in which the level widths of the higher-lying Q_y exciton components corresponded to a timescale of 100 fs for relaxation between exciton components. The slower (~ 2 ps) components are similar to Q_y pump-probe anisotropy decays observed under lower time resolution by Lyle and Struve (1991). Anisotropies r(0) ~ 0.7 have recently been observed in magnesium tetraphenylporphyrin (Galli et al., 1993). In this case, the prototype homodimer model of Rahman et al. (1979) is germane, because the two 'chromophores' are provided by the components of a doubly degenerate excited state (in effect, an intramolecular version of the intermolecular excited state dynamics treated by Rahman et al.).

IV. Temperature Dependence of Energy Transfer

In the context of the first-order Golden Rule treatment of dipole-dipole and exchange energy transfer in the weak coupling limit, their thermal effects originate from temperature changes in the spectral overlap integral. Hence, temperature modeling of Förster rate constants has been focused on simulating the temperature dependence in pigment absorption and fluorescence spectra. Such temperature dependence is caused by electron-phonon and electron-vibrational interactions, arising from differences between nuclear mode equilibrium geometries and frequencies in the pigment's electronic ground and excited states. For definiteness, we consider a particular harmonic vibrational mode of frequency ω_s. Its coordinate equilibrium position is shifted by Δ, and its vibrational frequency changes by $\Delta\omega_s$ from ω_s to ω_s', when a ground-state chromophore is promoted to the electronic excited state. In the special case where Δ and $\Delta\omega_s$ both vanish (zero electron-vibration coupling), the vibrational quantum number obeys the strict selection rule $\Delta v = v' - v'' = 0$ in dipole-allowed electronic transitions (Birks, 1970; Struve, 1989). All vibrational bands in the electronic spectrum then occur at the same wavelength irrespective of temperature T, and this mode contributes no T dependence to the electronic absorption spectrum. A nonzero shift Δ in equilibrium geometry exerts two major effects on the absorption spectrum. It leads to a series of nonzero Franck-Condon factors (which are proportional to vibrational band intensities) over a *progression* of wavelengths, corresponding to different vibrational state changes Δv in the electronic band spectrum. It also causes *hot bands* to appear at wavelengths longer than that of the vibrationless electronic transition. The hot band intensities are weighted by the initial thermal vibrational level populations in the electronic ground state. Similar effects arise when the ground and excited state vibrational frequencies are unequal.

A commonly used measure of electron-vibration (or electron-phonon) coupling strength is the dimensionless Huang-Rhys parameter S. If the ground and excited state vibrational Hamiltonians are given by

$$H_g = \frac{-\hbar^2}{2\mu}\frac{\partial^2}{\partial Q^2} + \frac{1}{2}\mu\omega_s^2 Q^2$$
$$H_e = \frac{-\hbar^2}{2\mu}\frac{\partial^2}{\partial Q^2} + \frac{1}{2}\mu(\omega_s')^2(Q-\Delta)^2 \quad (61)$$

the Huang-Rhys parameter is defined by

$$S = \frac{\frac{1}{2}\mu\,\omega_s^2\Delta^2}{\hbar\omega_s} = \frac{1}{2}\frac{\mu\,\omega_s}{\hbar}\Delta^2 \quad (62)$$

The temperature-dependent single-site absorption profile for an electronic transition in which the electron-vibration (or electron-phonon) coupling is strongly dominated by a *single* mode of frequency ω_s is (Harris, Mathies, and Pollard, 1986)

$$\alpha(\omega) = 2e^{-S\coth(\omega_s/2kT)}\,.$$
$$\int_0^\infty e^{-\gamma t/2} e^{S\coth(\omega_s/kT)\cos\omega_s t}\cos(\omega t - S\sin\omega_s t)\,dt \quad (63)$$

Here γ is the Lorentzian broadening parameter, equal to the sum of the inverse excited state radiative and nonradiative lifetimes. This model does not consider vibrational frequency shifts $\Delta\omega_s$. The behavior of this function is illustrated at several temperatures in Fig. 1 for $S = 0.2$ and $S = 0.8$. The profiles for $S = 0.8$ exhibit the longer Franck-Condon progressions of vibrational bands, in consequence of the larger equilibrium geometry shift Δ. The relative intensities of the hot bands (which occur at lower frequencies than the 0-0 band) increase with temperature, to an extent that correlates with the magnitude of the Huang-Rhys parameter. The linewidths of the individual vibrational bands are proportional to γ. Hence, Eq. (63) concisely expresses the Franck-Condon sequence of vibrational band intensities for electronic transitions between displaced harmonic oscillators, with hot band intensities weighted by appropriate Boltzmann factors.

Realistic simulations of pigment absorption profiles require simultaneous treatment of all Franck-Condon active phonons and vibrational modes. Spectral broadening due to site inhomogeneity must also be taken into account. In a spectral hole-burning study of the antenna Chl *a* Q_y spectrum in PSI-200 particles from spinach, Gillie et al. (1989) measured the frequencies, Huang-Rhys factors, and frequency shifts for some 41 Chl *a* vibrations. The ground state frequencies of these intramolecular modes are distributed from 262 to 1524 cm^{-1}, and exhibit small S (≤ 0.044) along with fractional frequency shifts of $\leq 3\%$. A highly Franck-Condon active phonon mode ($\omega_s = 22$ cm^{-1}) was characterized with $S = 0.8$, arising from a large pigment-protein interaction. Hence, this phonon mode is expected to dominate the temperature dependence in the Chl *a* Q_y absorption spectrum. Jia et al. (1992) incorporated these data in a simulation of the Chl *a* temperature-dependent absorption spectrum in a full 42-mode quantum-mechanical calculation, using

$$\alpha(\omega) = \int_{-\infty}^{+\infty} e^{-\gamma t} e^{-\Gamma t^2}\left\langle e^{iH_e t} e^{-iH_g t}\right\rangle e^{-i\omega t}\,dt \quad (64)$$

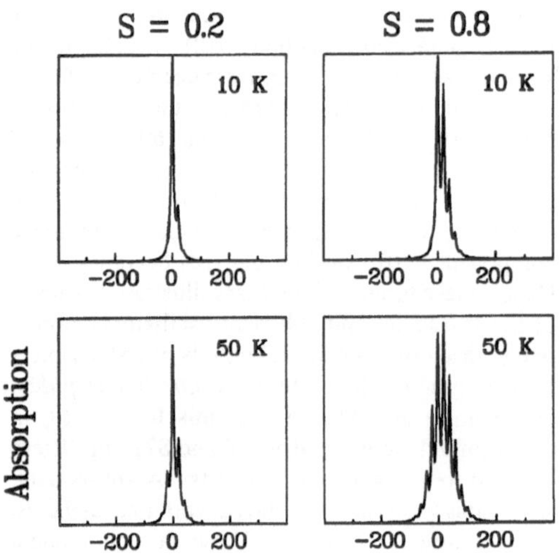

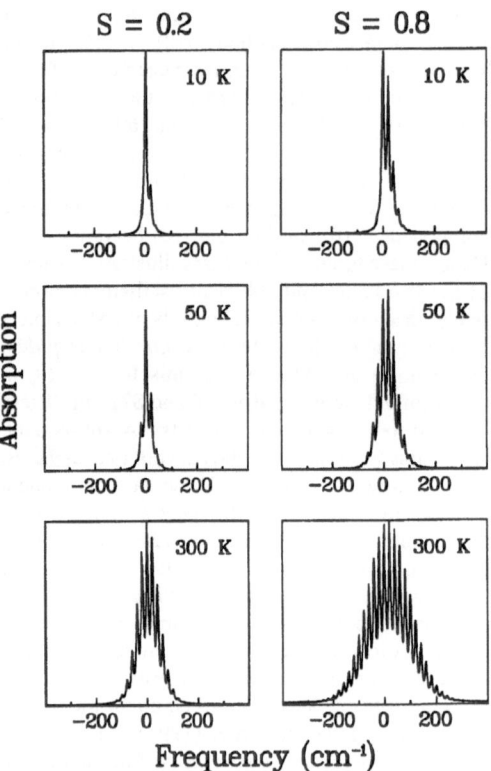

Fig. 1. Single-site absorption spectra calculated using Eq. (63) at several temperatures for a molecule with one Franck-Condon active mode of frequency 20 cm^{-1}: (a) $S = 0.2$; (b) $S = 0.8$.

Here the vibrational Hamiltonians H_e and H_g are 42-mode generalizations of Eqs. (61), and the brackets < > represent a thermal average over the phonon and vibrational levels in the ground electronic state. This integrand includes a Gaussian factor that simulates 200 cm^{-1} inhomogeneous broadening (Hayes et al., 1988). Aside from the presence of Gaussian broadening, this expression coincides with Eq. (63) in the limit where only one mode is Franck-Condon active and exhibits no frequency shift. The fluorescence spectrum required in the calculation of the Förster transition rate (cf. Eqs. 20, 21) was derived from the mirror image of the absorption spectrum (Birks, 1970), a procedure that is valid when the Franck-Condon active modes do not exhibit large frequency shifts. In view of the 200 cm^{-1} inhomogeneous broadening, the total Chl *a* absorption band fwhm in this simulation increases only by a factor of ~ 2 (from ~ 200 to ~ 450 cm^{-1}) when the temperature is raised from 0 to 300 K.

An alternative formulation of the single-site absorption profile for a Franck-Condon active phonon mode at low temperature (Personov, 1983; Hayes et al., 1988) is

$$\alpha_p(\omega) = e^{-S} l_0(\omega) + \sum_{r=1}^{\infty} \frac{S^r e^{-S}}{r!} l_r(\omega - r\omega_s) \qquad (65)$$

where the values of r correspond to zero-, one-, etc., phonon transitions. The r-phonon absorption profile $l_r(\omega)$ is obtained by convoluting the one-phonon profile with itself (r – 1) times. The coefficient $S^r\exp(-S)/r!$ is the Franck-Condon factor for the r-phonon band, which is centered at frequency $r\omega_s$. If N vibrational modes are incorporated with Huang-Rhys factors S_v and frequencies ω_v, the low-temperature single-site absorption profile becomes

$$\alpha(\omega) = \sum_{v=1}^{N} \sum_{t=0}^{\infty} \frac{S_v^t e^{-S_v}}{t!} \alpha_p(\omega - t\omega_v) \qquad (66)$$

which is a superimposition of phonon profiles centered at the t-vibration transition frequencies $t\omega_v$ and weighted by vibrational Franck-Condon factors. At finite temperatures, hot band terms must be included with appropriate thermal and Franck-Condon weights; the Huang-Rhys factors S are superseded everywhere by a temperature-dependent parameter $S(T)$. Using a suitable choice (Hayes et al., 1988) for the zero-phonon profile $l_0(\omega)$, T-dependent single-site absorption spectra are readily calculated. Examples of single-site Chl *a* absorption spectra evaluated using Eq. (66) are shown in Fig. 2. These must be convoluted with a site distribution function to simulate inhomogeneous broadening.

Jia et al. raised the issue of whether the site inhomogeneities of the donor and acceptor pigments (modeled using a 200 cm^{-1} Gaussian distribution) are truly uncorrelated as assumed in their model. In the absence of such correlation, the energy transfer rates are comparatively insensitive to temperature between 0 and 300 K, because the total Q_y absorption linewidth only increases by a factor of ~ 2. However,

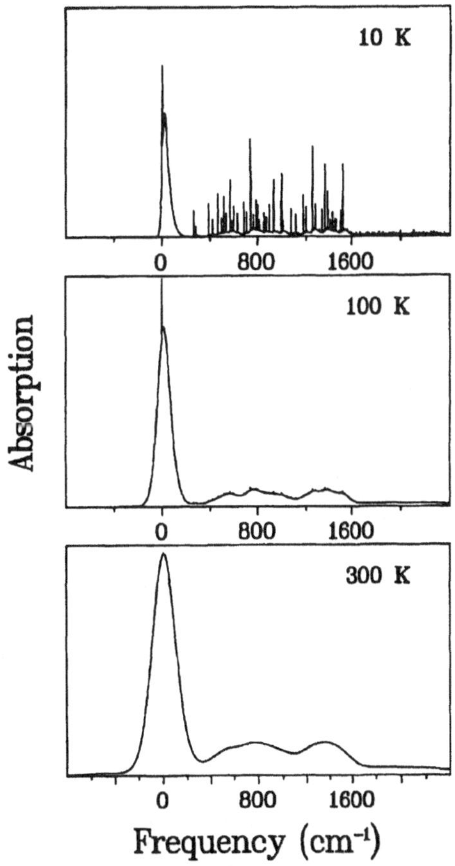

Fig. 2. Simulations of Chl *a* single-site absorption spectra using Eq. (66). Frequencies and Huang-Rhys factors for phonon and vibrational modes are from spectral hole-burning work of Gillie et al. (1989). Temperature are (in order of increasing fwhm) 10, 100, and 300 K.

it is possible *a priori* that some degree of correlation exists between the site inhomogeneities of donor and acceptor pigments, in such a way that the donor-acceptor energy gap ε exhibits a significantly narrower site distribution function than that of the empirical donor fluorescence or acceptor absorption spectrum. Such a situation would produce much stronger variation of energy transfer rates in this temperature range. Experimental temperature dependence studies of single-step energy transfer kinetics are expected to shed light on this question in the near future.

Acknowledgments

The author has benefited greatly from extensive conversations and correspondence with Drs. Robert Knox and Herbert van Amerongen. I am indebted to Dr. Tonu Pullerits for the computer algorithm and program used to calculate the spectra in Fig. 2. The Ames Laboratory is operated by Iowa State University for the U.S. Department of Energy under Contract No. W-7405-Eng-82. This work was supported by the Division of Chemical Sciences, Office of Basic Energy Sciences.

References

Avouris P, Gelbart WM and El-Sayed M (1977) Nonradiative electronic relaxation under collision-free conditions. Chem Rev 77: 793–834

Basu G, Kubasik M, Anglos D and Kuki A (1993) Spin-forbidden excitation transfer and heavy-atom induced intersystem crossing in linear and cyclic peptides. J Phys Chem 97: 3956–3967

Birks JB (1970) Photophysics of Aromatic Molecules. Wiley-Interscience, New York

Chambers RW, Kajiwara T and Kearns DR (1974) Effect of dimer formation on the electronic absorption and emission spectra of ionic dyes. Rhodamines and other common dyes. J Phys Chem 78:380–387

Cohen-Tannoudji C, Diu B and Laloë F (1977) Quantum Mechanics. Wiley-Interscience, New York.

Davidovich MA and Knox RS (1979) On the rate of triplet excitation transfer in the diffusive limit. Chem Phys Letts 68: 391–394

Dexter DL (1953) A theory of sensitized luminescence in solids. J Chem Phys 21:836–850

Du M, Xie X, Jia Y, Mets L and Fleming GR (1993) Direct observation of ultrafast energy transfer in PSI core antenna. Chem Phys Letts 201: 535–542

Förster T (1948) Intermolecular energy transfer and fluorescence. Ann Phys (Leipzig) 2: 55–75

Förster T (1959) Transfer mechanisms of electronic excitation. Discuss Faraday Soc 27: 7–17

Galli C, Wynne K, LeCours SM, Therien MJ and Hochstrasser RM (1993) Direct measurement of electronic dephasing using anisotropy. Chem Phys Letts 206:493–499

Geacintov NE and Breton J (1987) Energy transfer and fluorescence mechanisms in photosynthetic membranes. CRC Critical Rev Plant Sci 5: 1–44

Gottfried K (1966) Quantum Mechanics, Vol I: Fundamentals. Benjamin, New York

Frank HA, Farhoosh R, Aldema ML, DeCoster B, Christensen RL, Gebhard R and Lugtenburg J (1993) Carotenoid-to-bacteriochlorophyll singlet energy transfer in carotenoid-incorporated B850 light-harvesting complexes of *Rhodobacter sphaeroides* R-26.1. Photochem Photobiol 57: 49–55

Gillie JK, Small GJ and Golbeck JH (1989) Nonphotochemical hole burning of the native antenna complex of photosystem I (PSI-200). J Phys Chem 93: 1620–1627

Harris RA, Mathies RA and Pollard WT (1986) Simple interpretation of dephasing in absorption and resonance Raman theory. J Chem Phys 85:3744–3748

Hayes JM, Gillie JK, Tang D and Small GJ (1988) Theory for spectral hole burning of the primary electron donor state of photosynthetic reaction centers. Biochim Biophys Acta 851: 75–85

Hirschfelder JO and Meath WJ (1967) The nature of intermolecular forces. In: Hirschfelder JO (ed) Intermolecular Forces, Advances in Chemical Physics, Vol 12, pp 3–106. Wiley-Interscience, New York

Jackson JD (1962) Classical Electrodynamics. Wiley-Interscience, New York

Jia Y, Jean JM, Werst MM, Chan C-K and Fleming GR (1992) Simulations of the temperature dependence of energy transfer in the PSI core antenna. Biophys J 63: 259–273

Johnson SG and Small GJ (1991) Excited state structure and energy transfer dynamics of the bacteriochlorophyll *a* antenna complex from *Prosthecochloris aestuarii*. J Phys Chem 95: 471–479

Kenkre VM (1982) Exciton dynamics in molecular crystals and aggregates: The master equation approach. In: Kenkre VM and Reineker P (eds) Exciton Dynamics in Molecular Crystals and Aggregates, pp 1–109. Springer-Verlag Tracts in Modern Physics, Vol. 94, Hohler G (ed) Springer-Verlag, Berlin, Heidelberg, New York

Knox RS (1975) Excitation energy transfer and migration: Theoretical considerations. In: Govindjee (ed) Bioenergetics of Photosynthesis, pp 183–221. Academic Press, New York

Knox RS and Gülen (1993) Theory of polarized fluorescence from molecular pairs: Förster transfer at large electronic coupling. Photochem Photobiol 57: 40–43

Levine IN (1991) Quantum Chemistry, Fourth Edition. Prentice-Hall, Englewood Cliffs, NJ

Lyle PA and Struve WS (1990) Evidence for ultrafast exciton localization in the Q_y band of the bacteriochlorophyll *a*-protein from *Prosthecochloris aestuarii*. J Phys Chem 94: 7338–7339

Personov RI (1983) Site selection spectroscopy of complex molecules in solutions and its applications. In: Agranovich VM and Hochstrasser RM (eds) Spectroscopy and Excitation Dynamics of Condensed Molecular Systems, Vol. 4, pp 555–580. North-Holland, Amsterdam

Rahman TS, Knox RS and Kenkre VM (1979) Theory of depolarization of fluorescence in molecular pairs. Chem Phys 44: 197–211

Savikhin S, Zhou W, Blankenship RE and Struve WS (1994) Femtosecond energy transfer and spectral equilibration in bacteriochlorophyll *a*-protein antenna trimers from the green bacterium *Chlorobium tepidum*. Biophys J 66: 110–114

Slater JC (1963) Quantum Theory of Molecules and Solids, Vol I. Electronic Structure of Molecules, pp 263–276. McGraw-Hill, New York

Struve WS (1989) Fundamentals of Molecular Spectroscopy. Wiley-Interscience, New York

Turro NJ (1978) Modern Molecular Photochemistry, pp 298–361. Benjamin/Cummings, Menlo Park, CA

Van Amerongen H and Struve WS (1995) Polarized optical spectroscopy of chromoproteins. Meth Enzymol 246, in press

Van Grondelle R (1984) Excitation energy transfer, trapping, and annihilation in photosynthetic systems. Biochim Biophys Acta 811: 147–195

Van Grondelle R and Amesz J (1986) Excitation energy transfer in photosynthetic systems. In: Govindjee, Amesz J and Fork DC (eds) Light Emission by Plants and Bacteria, pp 191–223. Academic Press, New York

Wynne K and Hochstrasser RM (1993) Coherence effects in the anisotropy of optical experiments. Chem Phys 171: 179–188

Chapter 16

Structure and Organization of Purple Bacterial Antenna Complexes

Herbert Zuber
Institut für Molekularbiologie und Biophysik, ETH-Hönggerberg, HPM, CH-8093 Zürich, Switzerland

Richard J. Cogdell
Department of Botany, University of Glasgow, Glasgow G 12 8QQ, Scotland, U.K.

R. E. Blankenship, M. T. Madigan and C. E. Bauer (eds): Anoxygenic Photosynthetic Bacteria, pp. 315–348.

Summary

This chapter presents a comprehensive overview of what is currently known about the structure of purple bacterial antenna complexes, with special emphasis upon modeling their three-dimensional structure and intramembrane organization.

Purple bacterial antenna complexes can be grouped into two major types, the 'core' B875-type which forms a close association with the reaction centers and the 'variable' B800-850- or B800-820-type. Both these types of complex are constructed on a similar modular principle and are oligomers of α and β-apoproteins. These antenna apoproteins have been sequenced form a range of species and form a rather homologous group of proteins. The light-absorbing pigments, BChl and carotenoids, are non-covalently bound to these apoproteins.

As is described below, in a few cases these antenna complexes have been crystallized and structure determinations are under way. However, in the absence of three-dimensional structural information a great deal can be deduced by a comparison of their primary structures. The main part of this chapter illustrates the type of models for antenna structure that can be constructed by a detailed analysis of the conserved structural elements which can be seen by such an examination of the database of antenna apoprotein primary structures.

I. Structure and Function of Antenna Complexes from Purple Bacteria

A. The Purple Bacterial Photosynthetic Unit

The major light-absorbing pigments in purple bacteria are the bacteriochlorophylls (*a* and *b*) and the carotenoids. These pigments are non-covalently attached to two types of integral membrane proteins forming on the one hand the photochemical reaction centers and on the other the light-harvesting or antenna complexes (Hawthornthwaite and Cogdell, 1991; Zuber and Brunisholz, 1991). The majority of these pigments perform a light-harvesting role and serve to funnel absorbed solar radiation to a specialized few which form the reaction center. The combination of antenna complexes with a reaction center constitutes the photosynthetic unit (PSU). For most commonly studied purple bacteria the size of the PSU is variable. Usually, however, it is in the range of between 20 and 200 BChl *a* molecules per reaction center (Aagaard and Sistrom, 1972). This arrangement of reaction centers surrounded by an antenna system ensures that each reaction center is kept well supplied with incoming solar energy and effectively increases their cross-sectional area for photon capture. The detailed structure and function of the reaction center is described in section V of this volume. The rest of this contribution will concentrate upon the antenna complexes.

B. Intracytoplasmic Membranes

In the purple bacteria the PSU is located in the intracytoplasmic membranes. Most species can survive either by respiration or photosynthesis. In the presence of oxygen, the cells tend to be unpigmented and have a smooth cell membrane. When they become anaerobic, they switch photosynthetic mode and the photosynthetic apparatus develops de novo. During this changeover, the cells

Abbreviations: Cm. – Chromatium; Ec. – Ectothiorhodospira; LDAO – lauryl dimethylamine-N-oxide; PSU – Photosynthetic unit; *Rb. – Rhodobacter; Rc – Rhodocyclus; Rp. – Rhodopseudomonas; Rs. – Rhodospirillum*

become highly pigmented and the cell membrane invaginates to form an extensive system of intracytoplasmic membranes. All of the newly synthesized components of the light-reaction of photosynthesis are 'housed' in and on these membranes.

The structure of these membranes varies. In some species they are present as lamellar structures, others as tubular structures and still others as vesicular invaginations (Trüper and Pfennig, 1981). What regulates which membrane shape develops is entirely unclear. Interestingly, some mutants of *Rhodobacter (Rb.) sphaeroides* which lack carotenoids and the peripheral antenna complex (B800–850) show intracytoplasmic membranes with an altered morphology. The reason for this pleotropic effect is a mystery but does show that an analysis of mutants may provide a way of unraveling this problem of the control of membrane shape. The internal architecture of the intracytoplasmic membranes is one of the major taxonomic characteristics that has been used in the recent reclassification of the purple bacteria (Imhoff et al., 1984; Trüper and Pfennig, 1981).

C. Isolation and Characterization of Antenna complexes

When BChl *a* is dissolved in an organic solvent such as 7:2 v/v acetone:methanol its near IR absorption band is at 772 nm. This is typical of monomeric BChl *a*. However, when it is non-covalently bound into a light-harvesting complex its NIR absorption band is red shifted to between 800 and 900 nm depending upon the type of complex. For BChl *b* this red shift is even larger with the in vivo NIR absorption peak in *Rhodopseudomonas (Rp.) viridis,* for example, located at 1012 nm (Hawthornthwaite and Cogdell, 1991). In most species this red shift is also associated with an increase in spectral complexity, with several peaks/shoulders clearly visible in the absorption spectrum. This red-shift arises from pigment-protein and pigment-pigment interactions within the native antenna complexes and is routinely used as a way of identifying them and judging their integrity.

Some species of purple bacteria, such as *Rhodospirillum (Rs.) rubrum* and *Rp. viridis*, only have a single strong NIR absorption band, corresponding to a single antenna type (B880 or B1012 respectively) (Hawthornthwaite and Cogdell, 1991; Zuber and Brunisholz, 1991). Others, such as *Rb. sphaeroides* or *Rb. capsulatus*, contain two types of antenna complexes (B875 which is equivalent to the B880 from *Rs. rubrum*, and B800–850). Still others, such as *Rp. acidophila* and *Rp. cryptolactis*, contain additional antenna types, for example, the B800–820 complexes. At first glance this may seem a very confusing situation, however it is now possible to draw a simplified picture which clarifies this situation.

In purple bacteria there are two major types of antenna complex (Hawthornthwaite and Cogdell, 1991; Zuber and Brunisholz, 1991). The first type is the B880 from *Rs. rubrum* or the B875 from *Rb. sphaeroides*. This forms the so-called 'core' complex and all wild-type purple bacteria contain it. The 'core' complex is closely associated with the reaction center and forms a stoichiometric complex with it (e.g. there are 24 molecules of BChl *a* per reaction center, in an intact RC-'core' antenna supramolecular complex, see Section V.C.2). In some species such as *Rp. viridis* or *Rs. marina* this supramolecular complex has been visualized by electron microscopy (Stark et al., 1984; Meckenstock et al., 1992). The antenna component forms a doughnut-like structure around the reaction center (Fig. 16). All other antenna types, such as B800–850, B800–880, B830, etc., form the so-called variable or peripheral antenna complexes. This heterogeneous class of complexes, as their names suggest, are located further away from the reaction centers than the 'core' complexes and are present in variable amounts with respect to the number of reaction centers present. The amount present depends upon a variety of environmental factors such as light-intensity, temperature, carbon source, etc.

Isolation of a purple bacterial antenna complex, since they are integral membrane proteins, begins with the solubilisation of the photosynthetic membrane with a suitable detergent (Cogdell and Hawthornthwaite, 1993). Then the individual complexes are purified from the solubilized mixture by standard methods of protein purification. The choice of the detergent is very important. It must be 'strong' enough to solubilize the antenna complexes yet 'mild' enough not to denature them. In most cases we have found that the detergent lauryl dimethylamine-N-oxide (LDAO) is the most suitable and it is also very cheap! LDAO also has the advantage that if you need the complex in a different detergent for subsequent studies then it is easily exchanged for other detergents both by dialysis and on small ion exchange columns.

We have found that for most purple bacteria the easiest procedure to purify the antenna complexes is to initially fractionate them by sucrose gradient centrifugation. This conveniently separates them into RC-B875 and B800–850 or B800–820 complexes. They can then be further purified by DE cellulose anion exchange chromatography and molecular sieve chromatography. The integrity of the isolated antenna complexes can be easily assessed from their absorption spectra (as described above) and indeed by eye.

All of the purple antenna complexes are built upon the same modular principle (Papiz et al., 1989). The pigments are non-covalently bound to two low molecular weight (typically 5–7Kd) very hydrophobic apoproteins. These are called the α- and β-apoproteins (or polypeptides) and they aggregate to produce the native pigmented antenna complexes (see Sections V.A–D).

D. Approaches to Determine the Structure of Antenna Complexes

For most proteins with molecular masses of 20 kD and above, determination of their 3D structure requires the production of highly ordered crystals, either 2D or 3D, which can then be subjected to analysis by electron or X-ray crystallography. The antenna complexes are no exception to this, they are just difficult to crystallize since they are membrane proteins.

Several RC-B880 complexes have been crystallized as 2D arrays (Meckenstock et al., 1994; Ghosh et al., 1993). Indeed, in some cases, such as *Rp. viridis,* they naturally occur in 2D-arrays within the photosynthetic membrane, and these have been subjected to high-resolution EM and image analysis to generate low resolution (~20Å) maps of their structure. In recent years bacteriorhodopsin (Henderson and Unwin, 1975) and now the major Chl *a/b* protein from peas (Kühlbrandt, 1984) have been very successfully studied by an electron diffraction of 2D arrays. The methodology for this type of crystallographic analysis is now well developed, and indeed may be more generally applicable for membrane proteins than X-ray diffraction of 3D-crystals.

Table 1 lists the purple bacterial antenna complexes which have been crystallized 2D arrays. These are now actively being studied and improved, and it is hoped that they will go on to yield high-resolution structural information.

Several B800–850 complexes have been crystallized as large, single 3-D crystals (Table 2). Two in particular, the B800–850 complexes from *Rp. acidophila* strain 10050 (Papiz et al., 1989) and from *Rs. molischianum* (Michel, 1990) have been shown to diffract X-rays to high resolution. In both cases heavy atom searches have begun and it is hoped that these will soon yield the phase information needed to solve their structures. However, initial analyses of the native data sets, together with molecular weight analysis and electron microscopy, has revealed that these two complexes have different oligomeric states. The *Rp. acidophila* complex is an $\alpha_6\beta_6$ oligomer (Papiz et al., 1989) while the *Rs. molischianum* complex appears to be $\alpha_8\beta_8$ (Kleinekofort et al., 1993; Germeroth et al., 1993). It is most likely that the fundamental building block is an $\alpha_2\beta_2$ tetramer which repeats three times in *Rp. acidophila* and four times in *Rs. molischianum*.

E. Modeling the Structure of the B800–850 Complex from Rp. acidophila Strain 10050 Based upon Native Crystallographic Diffraction Data.

In the absence of suitable heavy atom derivatives that would allow the phases to be solved and therefore the three-dimensional structure of this antenna complex to be determined, some structural information can be obtained from the native crystallographic data. It must be emphasized, though, that this is just modelling and must be treated with caution. Figure 1 shows a slice through Patterson space calculated from the full native data set to a resolution of 3.2 Å. The section depicted is through the a, b plane. Very simply, this type of calculation takes each atom in turn, puts it at the origin of the projection, and calculates the vector between it and every other atom in the molecule. Clearly, regular structures will build up density in this type of calculation. As is described below (see section 3b) it is thought that 40–50% of the antenna apoproteins are folded up as an α-helix. The pattern seen in Fig. 1 can be interpreted by considering what type of α helical packing might be present in the complex. The density at the origin goes up and down and 'c' axis (that is at right-angles to the projection) for 22–23Å. If we assume that this density comes from the α-helices, then they must be lined up in the crystal almost parallel to the 'c' axis and be 44–46Å in length. Clearly, this is an oversimplification but it

Table 1. 2D crystals of B880 antenna complexes

Species	Type of complex	Reported resolution	Reference
Rs. rubrum	B880	~40Å	Ghosh et al., 1993
Rs. marina	B880	~20Å	Meckenstock et al., 1994

Table 2. Three-dimensional crystals of B800–850 and B800–820 complexes

Species	Complex	Resolution	Reference
Rp. acidophila			
10050	B800–850	2.7Å	Papiz et al (1989)
7750	B800–820	4.3Å	Guthrie et al (1989)
Rs. molischianum	B800–850	2.4Å	Michel (1990)
Rb. capsulatus	B800–850	8Å	Welte et al (1985)
Rs. salexigens	B800–850	8Å	Wacker et al (1988)
Rp. palustris	B800–850	7Å	Welte et al (1985)

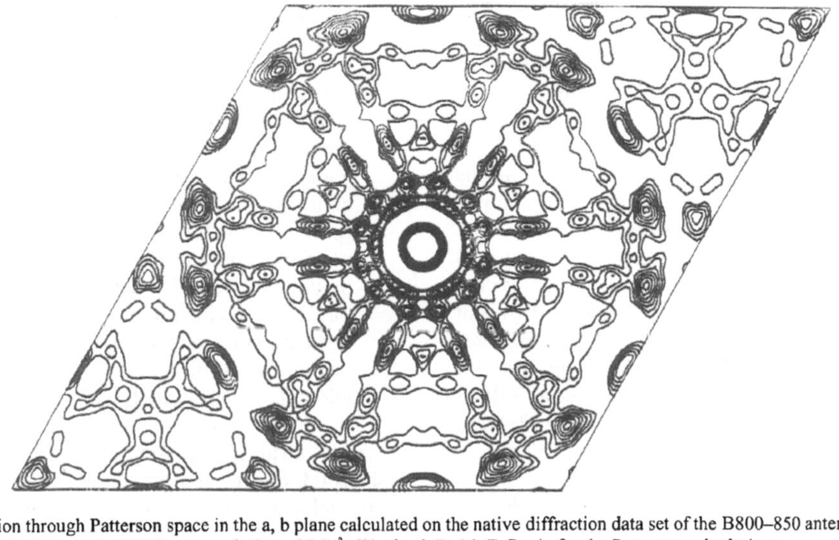

Fig. 1. Section through Patterson space in the a, b plane calculated on the native diffraction data set of the B800–850 antenna complex from *Rp. acidophila* strain 10050 to a resolution of 3.2 Å. We thank Dr M. Z. Papiz for the Patterson calculations.

does allow an interesting, testable model to be developed. The rings of density progressing outwards from the origin at 10.5Å, 14.2Å and 24Å can be explained by reference to Fig. 2. If two a-helical rods lie next to and parallel to each other, then the distance between the center of gravity of one and the center of gravity of the other will be 10.5Å. If four parallel a-helical rods are arranged in a square, then the distance across the diagonal, from the center of gravity of one helix to the center of gravity of the next is 14.2Å. At 24Å in the space group R32 six peaks would be expected from the crystallographic symmetry. However, 18 peaks are seen. Hence there must be an additional non-crystallographic three-fold axis of symmetry. All these features can be accounted for by assuming that the intact antenna complex contains three bundles of four a-helices arranged as in Fig. 2. It will be interesting to see when the structure of this antenna complex is finally determined how accurate this model is.

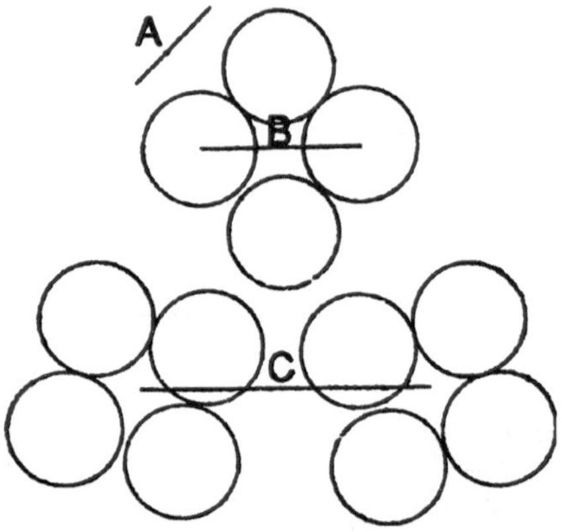

Fig. 2. A model for the arrangements of the α-helices based on the Patterson calculation shown in Fig. 1. (A) 10.5 Å, distance between helices, (B) 14.2 Å, diagonal packing distances, (C) 24 Å, distance between the bundles of four helices.

II. Structure and Function of Antenna Polypeptides from Purple Bacteria. General Aspects

The antenna systems of all photosynthetic organisms surround the reaction center and form a multimolecular unit which functions as a regulated, energy-uptake and energy-transfer system. The physical requirements of the energy transfer processes within the antenna system and to the reaction center result in the structure and organization of antenna complexes and their various pigment molecules displaying a set of general features, which may be outlined as follows (Zuber, 1987, 1993; Zuber et al., 1987; Zuber and Brunisholz, 1991):

- In the antenna system pigment molecules are highly ordered and form pigment clusters of a given size and symmetry.
- Pigment clusters in the antenna complexes are the basic units of the directed, heterogeneous transfer of energy within and between the antenna complexes (hierarchy of pigment organization).
- Directed energy transfer to the reaction center is influenced and regulated by the spatial separation of the pigment clusters (minimizing random walk).

Important to the formation of the energy transfer system of organized pigment molecules in pigment-protein complexes are specific, structural and functional interactions, that is pigment-pigment, pigment-polypeptide and polypeptide-polypeptide. Here the lipid environment of antenna complexes in the photosynthetic membrane also plays a decisive role. The structure and function of antenna polypeptides and their corresponding hierarchical organization is of fundamental importance for the organization of pigment molecules. The polypeptides form the main basis of the spectral and energy-transfer characteristics of the pigments. A large body of structural data shows that:

- Pigment molecules are bound specifically to polypeptides and form defined antenna complexes. They determine the position, distance, orientation and environment of the pigment molecules. That is their essential function.
- Polypeptides have specific association characteristics in forming oligomeric or multimeric units (pigment-protein complexes), i.e. antenna complexes (including subcomplexes) and the entire antenna system. Antenna polypeptides associate in an hierarchical order: a) formation of microdomains (subcomplexes of 2–4 polypeptides), b) formation of antenna complexes with larger pigment clusters through association of microdomains, c) formation of the entire antenna system through the association of antenna complexes.

These structural principles of pigment and polypeptide organization are particularly recognizable in the relatively simple antenna systems of purple bacteria. (See also the reviews: Thornber et al., 1983; Cogdell and Valentine, 1983; Drews, 1985; Zuber 1985a,b; Cogdell, 1986; Thornber, 1986; Zuber, 1987, 1993; Zuber et al., 1987; Zuber and Brunisholz, 1991). Characteristic of these and indicating the structural principles are: 1.) the existence of two main types of pigment-protein complexes, the core complex B870/B890 (BChl *a*), or B1020 (BChl *b*) in proximity to the reaction center, and the peripheral complex B800–850 (800–820) (see Section I this chapter) and, 2.) the existence of two main populations of BChl molecules in antenna complexes, one group having absorption bands between 820 and 860 nm (peripheral complex), or 870–1020 nm (core complex), and a second group having absorption bands at 800 nm (peripheral complex only) (Thornber et al., 1983). Both structural/functional features come to fruition in the three-dimensional structure of the entire antenna system, that is, in the efficient transfer of energy to the reaction center. As the three-dimensional structures of polypeptides are based upon their primary structure (amino acid sequence), the primary structure of these polypeptides already contains the information which determines the structure and organization of the antenna complexes and the arrangement of the pigments in the entire antenna system.

III. Amino Acid Sequences (Primary Structure) of Antenna Polypeptides. Sequence Homologies, Structural and Functional Features

A. Isolation and Sequence Analysis of Antenna Polypeptides

In order to solubilize and isolate membrane polypeptides from the photosynthetic membrane (chromatophores) or from isolated antenna complexes, organic solvents such as chloroform (or methylene chloride)-methanol (ratio 1:1) have been used (Tonn et al., 1977). This solvent mixture plus 0.1 M ammonium acetate (with (5–10%) and without acetic acid) proved especially useful in solubilization and fractionation (gel filtration and ion exchange chromatography by stepwise elution with acetic acid) of the antenna polypeptides in purple bacteria (Brunisholz et al., 1981; Brunisholz et al., 1984a; Brunisholz et al., 1985; Theiler et al., 1984). More recently the use of reverse phase chromatography (matrix: octyl-silica or butyl-silica, solvent 50–85% acetonitrile) has proved to be very quick and effective and is now the method of choice for separating large numbers of various types of antenna polypeptides (6 and more) which are formed in some species of purple bacteria (Brunisholz and Zuber, 1987).

Although today DNA sequencing is frequently the method of choice for determining the primary structure of polypeptides (for example for the of antenna polypeptides (see Chapter 22 by Hunter), protein sequencing is also necessary in order to obtain a full understanding of the structure and function of antenna polypeptides for the following reasons:

- Identification of the sites of processing of the antenna polypeptides, e.g. sites of posttranslational cleavage of the polypeptide or modification of amino acid side chains
- Assignment of individual antenna polypeptides to particular antenna complexes
- Determination of the stoichiometry of antenna polypeptides in the antenna complex
- Determination of the position and orientation of antenna polypeptides in the photosynthetic membrane by means of enzymatic cleavage and chemical labeling
- Analysis of the relation of structure to spectral characteristics by use of chemical labels and limited proteolytic hydrolysis
- Crosslinking of antenna polypeptides

B. Comparison of Primary Structures (Amino Acid Sequences) of Antenna Polypeptides in Purple Bacteria: Derivation of Important, General Structural Elements

During the course of the past ten years, the primary structures of a great number of antenna polypeptides, both of the B870/B890 core complexes and of the peripheral B800–850 (820) complexes, were determined (Zuber and Brunisholz, 1991) (Figs. 3, 4). It was found that all antenna complexes are composed of α and β polypeptides which are of approximately the same size (40–70 amino acid residues) and which have homologous amino acid sequences (Zuber, 1985, a, b; 1987). They do differ, however, in their characteristic structural features. The polarity (Capaldi and Vaanderkooi, 1972) of these membrane polypeptides lies at 23–37%. The low degree of sequence homology, or distant phylogenetic relation, between the α and β polypeptides of about 5% is an indication of an early evolutionary separation of these polypeptides Fig. 5A). On the other hand, a sequence homology of 13–28% between the α or β polypeptides of the core and peripheral complexes shows them to be definitely related phylogenetically (evolution of the entire antenna). Species differences of antenna polypeptides of the core and peripheral antenna complexes in

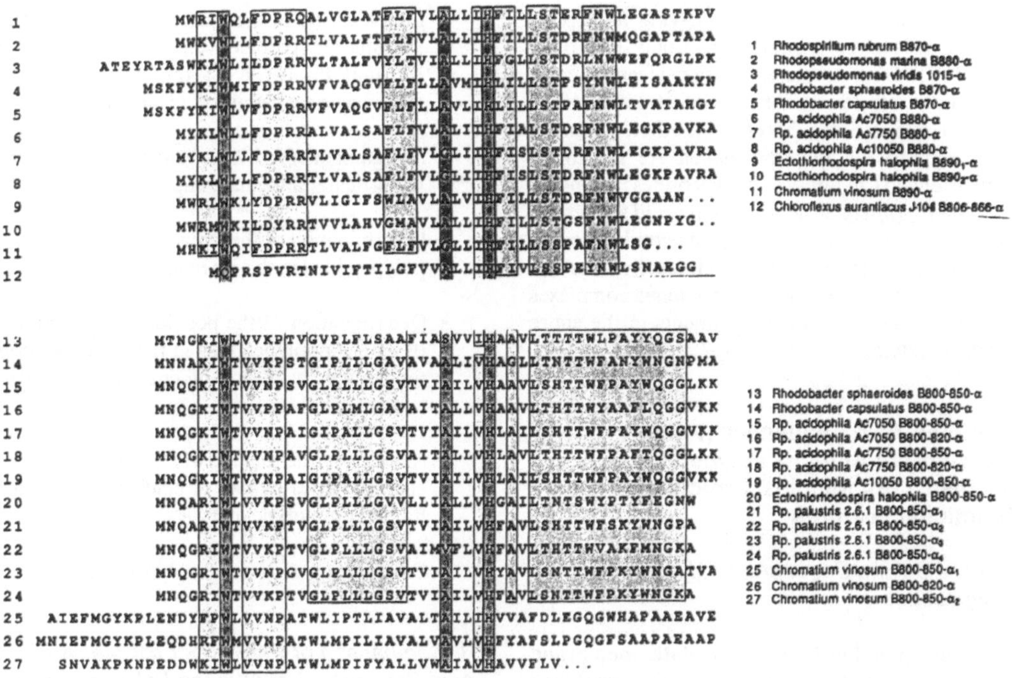

Fig. 3. Amino acid sequences of the α-antenna polypeptides of the core (1–12) and peripheral (13–27) antenna complexes of purple bacteria. Sequence homologous amino acid residues or complex specific clusters of amino acid residues with pronounced homology are shaded.

various purple bacteria mean that the degree of homologies lie between 28 and 78%. The characteristic structural features of the amino acid sequences in the antenna polypeptides are both conserved single amino acid residues and clusters of conserved amino acid residues. These are characteristic both of the α and β polypeptides and of the B870/B890 and B800–850 (820) complexes, which indicates that they play a special structural and functional role (Figs. 3, 4 and 5B). Especially striking is the large number of basic amino acid residues (Arg, Lys) in the N-terminal region of the α polypeptide chain and acidic amino acid residues (Glu, Asp) in the N-terminal region of the β polypeptide chain. A specific, electrostatic interaction of the α and β polypeptide chains in this region (ion pairs) therefore seems likely. Perhaps, however, the most important structural characteristic common to all antenna polypeptides in purple bacteria is their three-domain structure, which is found in both the linear amino acid sequence and in their three-dimensional structure (Brunisholz et al., 1984a; Zuber, 1985, a, b). We can distinguish between the N-terminal, polar domain (α polypeptide chain: core and peripheral antennae 12–14 residues; β polypeptide chain: core antenna complex: 17–25 residues, peripheral antenna complex: 13–23 residues), the polar C-terminal domain (α polypeptide chain, core and peripheral antenna complexes: 16–20 residues, β polypeptide chain, core and peripheral antenna complexes: 5 residues, with the exception of e.g. *Rs. rubrum, Rp. viridis, Rp. acidophila*: 11/12 residues) and a central, hydrophobic domain (21–23 residues). The two polar domains lie on the surface or in the polar head region of the membrane. The primary structure in the N-terminal domain suggests that there is an amphipatic helix parallel to the membrane plane (Theiler and Zuber, 1984). The long C-terminal domain of the α polypeptide chain, especially in the case of *Rhodocyclus (Rc.) gelatinosus* (36 residues, oligo-Ala segments), has a particular sequence which indicates that this may represent a second domain spanning the membrane (Zuber and Brunisholz, 1991). In the center of the sequence of the antenna polypeptides lies the hydrophobic domain which

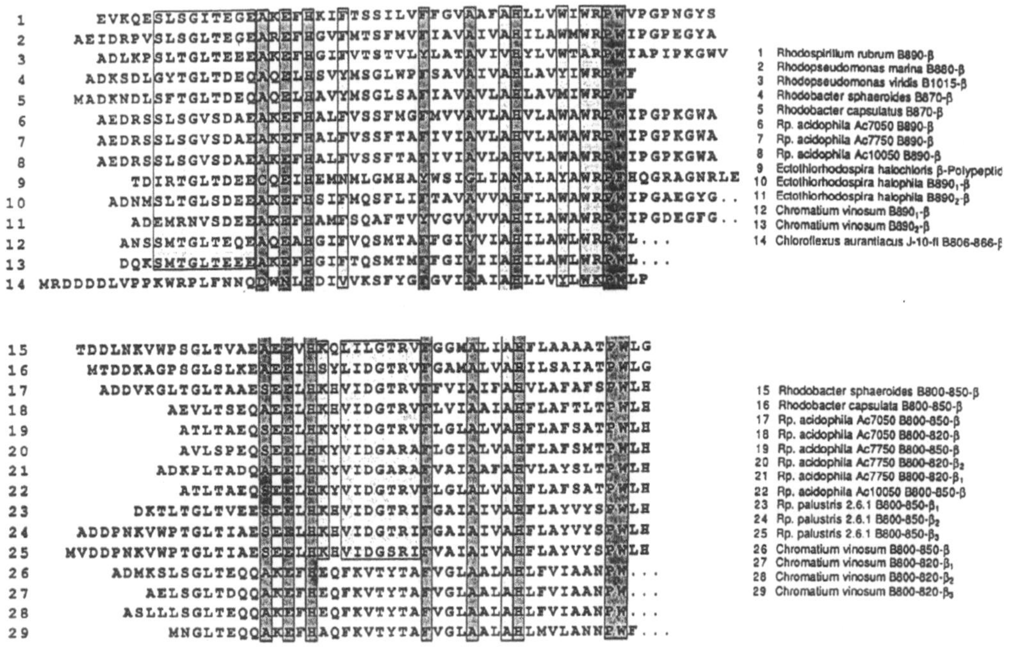

Fig. 4. Amino acid sequences of the β-antenna polypeptides of the core (1–14) and peripheral (15–29) antenna complexes of purple bacteria. Sequence homologous amino acid residues or complex specific clusters of amino acid residues with pronounced homology are shaded.

spans the membrane as an α helix (tilted α helix: average 30–35°) (IR (UV), CD spectra, labeling experiments) (Jay et al., 1983; Breton and Nabedryk, 1984). Proteolytic digestion and labeling experiments applied to the B870 core complex of *Rs. rubrum* showed the orientation of the antenna polypeptides (α helix) in the membrane, i.e. the location of the N-terminal domain of the α and β polypeptides at the cytoplasmic side (proteinase K, trypsin, chymotrypsin) (Wiemken et al., 1983; Brunisholz et al. 1984a, 1986). Similarly, the N-terminal regions of the L and M polypeptides of the reaction center were also localized at the cytoplasmic side. The structural role of hydrophobic or hydrophilic amino acid residues, particularly with respect to the assembly of the antenna polypeptides and the formation of the antenna complexes in the membrane, was analyzed by mutagenesis of the α and β-polypeptides (Richter et al., 1992). Redox-controlled in-vivo and in-vitro phosphorylation of the α-polypeptides of the core complex (*Rb. sphaeroides*) was found, which indicates a membrane-bound protein kinase (Cortez et al., 1992).

C. Relationship Between the Data on the Primary Structure of Antenna Polypeptides and the Function (Spectral Characteristics) of the Antenna Complexes

The three-domain structure of the antenna polypeptides also provides the framework for considering possible structural-spectral (functional) relationships, i.e. the binding, organization and environment of the BChl molecules in the three-dimensional structure of the transmembrane helix. The most conserved amino acid residue of the α and β polypeptides from the core, as well as from peripheral antenna complexes, is a His residue. This central His residue is 4 (α) or 6 (β) amino acid residues (1–2, α-helix turns) from the boundary between the hydrophobic domain and the polar C-terminal domain which extends towards the periplasmic side. This conserved His most probably forms the central binding site for BChl (central BChl), and is especially important in the energy transfer within and between the antenna complexes and to the reaction center (absorption maximum 850, 820 or 1020 nm) (Brunisholz et al.,

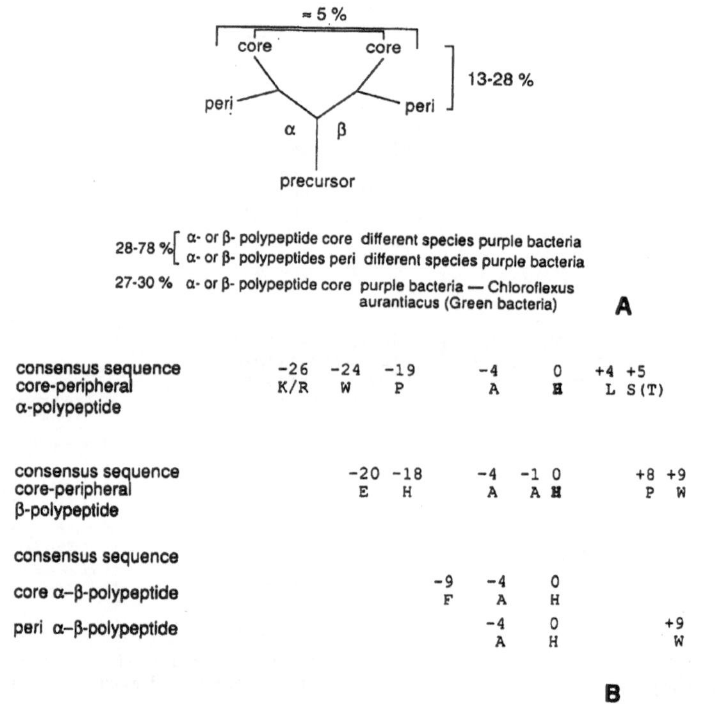

Fig. 5. A: possible phylogenetic relationship (% sequence homology) of α- and β-antenna polypeptides of purple bacteria. B: consensus sequences of the α- or β-polypeptides between the core-and peripheral antenna complexes and between the α- and β-polypeptides of the core or peripheral antenna complexes.

1981, 1984a; Theiler et al., 1984). This was confirmed by resonance-Raman spectroscopy (Robert and Lutz, 1985) and mutagenesis experiments (Bylina et al., 1986). A second, usually conserved His residue in the β chain near the N-terminal domain has been suggested to be the binding site for monomeric BChl (absorption maximum: 800 nm) (Zuber, 1985b, 1987). On the other hand, this conserved His residue is also found in the β chain of the B870/B890 antenna complexes, even though this additional BChl is absent in this core complex. Interestingly, as well resonance-Raman studies also did not find that the BChl (800 nm)-absorbing in the peripheral antenna complexes (Robert and Lutz, 1985) had an imidazole ligand. In *Rc. gelatinosus,* (B800–850, β-polypeptide), His is replaced by Gln, which could also, in principle, bind BChl (Zuber and Brunisholz, 1991). In the β chain of core antenna complex B806–866 in the green photosynthetic bacterium *Chloroflexus aurantiacus* (related phylogenetically to purple bacteria), the second conserved His residue was also found, and the BChl bound by this His could be responsible for the absorption maximum of 806 nm (coplanar orientation of the BChl with respect to the membrane) (Wechsler et al., 1985, 1987).

The folding of the α helices, the relative organization of α and β polypeptides (heterodimers) and the order of the diverse side chains of amino acid residues in the linear polypeptide sequence create a particular micro-environment for the His residues, or BChl molecules respectively, which by means of pigment-polypeptide interactions determines the function of the BChl molecules (spectral characteristics, energy transfer). Of special interest are the conserved amino acid residues found at a certain distance from the His residues in the amino acid sequence, that is, in relation to the folded polypeptide chain of the α helix (Figs. 3, 4). These may form secondary binding sites for the of BChl molecules and could also control the specific orientation of the tetrapyrrole ring. Amino acid residues around the central His, His-4 (Ala, α and β polypeptide chains, small hydrophobic residue) and His +4 (Leu, α polypeptide chain) seem to be of special importance here (Theiler and Zuber, 1984). The spectral characteristics of the BChl molecules bound to the central His residue are also influenced by: 1.) the totality of hydrophobic amino acid residues (Ala, Val, Ile, Leu, Phe) in the region of His ±4 and 2.) the conserved aromatic amino acid residues (Tyr, Trp) (Brunisholz et al., 1984a): at a distance from the central His residue of His +4, +6 and +9 in the β chain of the core complex B870/B890, or His +9 in the β chain of the peripheral complex B800–850 (820) and in the α polypeptide chain of His +11 of the core complex B870/B890 and of His +9 and +14 of the peripheral complex B800–850. In the case of Trp (His +14) of the peripheral antenna complex in *Rp. acidophila* (strain 7050, 7750), it was postulated that the substitution of Trp by Leu or Thr was the origin of the spectral shift of the 850 nm band of B800–850 to the 820 nm band of B800–820 (Brunisholz et al., 1984, a; Brunisholz and Zuber, 1992). This postulated functional role of aromatic amino acid residues was then demonstrated by means of site-directed mutagenesis of the B870 complex in *Rb. capsulatus* (Babst et al., 1991), through substitution of Trp (His +11, position 43) by Ala, Leu, Tyr (blue shift 8–11 nm) and the peripheral B800–850 complex in *Rb. sphaeroides* (Fowler et al., 1992) through substitution of Trp (His +13) and (His +14) by Phe, or Phe, Leu (double mutant, blue shift 11 and 24 nm, respectively). Interestingly, there is a Trp residue in the α polypeptide of the extremely long-wave length absorbing core complex B1015 of *Rp. viridis* at position His +12 (next to Trp His +11) (Brunisholz et al., 1985). This aromatic amino acid residue, and possibly the aromatic amino acid residue (Tyr) in this region in the γ polypeptide (see 3. e.), cause this large red shift. Strong (NIR)-CD signals were reported on the core antenna complexes from *Rs. rubrum, Rp. viridis* and *Rp. acidophila.* (biphasic, with zero crosspoint at 880 nm) (Fig. 6). In contrast, CD signals of the core complex from *Rb. sphaeroides, Rb. capsulatus* and *Rc. gelatinosus* are much weaker. It was suggested that these spectral differences are determined by the greater length of C-terminal domains of the β polypeptide chains and the existence of Trp, or Tyr residues, in this lengthened polypeptide chain area in *Rs. rubrum, Rp. viridis* and *Rp. acidophila.* (Zuber and Brunisholz, 1991). In addition, there is in the α polypeptide chain of the core complex from the first group of organisms (above) a Phe residue in place of Leu at position His +1.

A surprising finding was made with regard to the functional role of aromatic amino acid residues when comparing the aromatic clusters of the antenna polypeptides (α, β, core antenna) with those in the L and M polypeptides of the photosynthetic reaction center (*Rp. viridis*) (that is close to the central His residues of the α, β-antenna polypeptides, or the His residues of the special pair of the L and M subunits (D helix)) (Brunisholz and Zuber, 1988, a, b). The BChl molecules, bound by His residues of both 'special pairs', show similar absorption characteristics. Proteolytic cleavage experiments (*Rs. rubrum)* showed that the N terminus of the antenna polypeptides and the L and M polypeptides of the reaction center lies on the cytoplasmic side. However, due to the identical position of the two 'special pairs' towards the periplasm and because of the differing number of transmembrane helices (the antenna polypeptides (α, β) have one helix each, the reaction center polypeptides 4 helices up to helix D), the amino acid sequences of the α and β antenna polypeptides and the D helices in the membrane of the reaction center must run antiparallel (N $\rightarrow$ C and C $\rightarrow$ N, respectively, Fig. 7B). A comparison of these sequences, taking account of the fact that they are antiparallel, showed an interesting structural similarity with regard to the arrangement of clusters of aromatic amino acid residues in proximity to both 'special pairs'. 8 of 11 aromatic amino acid residues in the special pair binding pocket of the reaction center are in the same position in the core antenna complex (Fig. 7A). This finding may represent a phylogenetic relationship of a primordial reaction center and the antenna system (Fig. 7B). The third polypeptide component of the B1015 complex, the γ polypeptide from *Rp. viridis*, also has these aromatic amino acid residues, which may also be close to the BChl pair and indicate that this polypeptide plays a role in terms of the large red shift of this complex. These findings indicate analogous functional roles of the aromatic amino acid residues in both of the complexes of the cooperative antenna-reaction center conjugate. An ever increasing amount of data is now being accumulated which demonstrates that the aromatic amino acid residues interact functionally with BChl molecules. A future task will be to visualize this structural situation in the three-dimensional structure of antenna complexes as well, and to understand the theoretical basis of the spectroscopic consequences of these interactions.

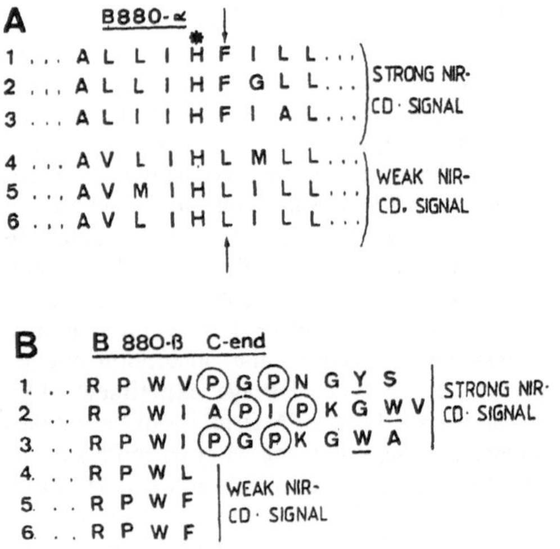

Fig. 6. Correlation of structural elements of the α- and β-antenna polypeptides with the intensity of their NIR-CD-signals (S = strong CD signal, W = weak CD signal). CD characterization taken from Brunisholz and Zuber 1988, 1988a. A: Amino acid sequence in the vicinity of the central His residue of the core α-polypeptides. B: Amino acid sequence of the C-terminal domains of the core β-polypeptides. 1.) *Rs. rubrum*, 2.) *Rp. viridis*, 3.) *Rp. acidophila*, 4.) *Rc gelatinosus*, 5.) *Rb. sphaeroides*, 6.) *Rb. capsulatus*. Figure from Brunisholz and Zuber, 1988a:

D. Studies on the Influence of the Environment of the BChl Molecules on their Spectral Properties: Limited Proteolysis of Isolated Peripheral Antenna Complexes

In addition to mutagenesis experiments modifying the α and β polypeptide chains of antenna complexes (see Chapter 22 by Hunter), limited proteolysis of these polypeptides in the antenna complex is a method well suited to studying the influence of the polypeptide environment on the spectral properties. Especially promising is hydrolytic cleavage of the polypeptide chain with enzymes at the C-terminus near the central BChl molecules (Brunisholz and Zuber, 1993). When the B802–858 (B800–850) complex from *Rp. acidophila* (10050) is detergent solubilized (LDAO) it is relatively stable. Such a preparation was treated with elastase and carboxypeptidase A and B, and spectral changes were analyzed in relation to the changes in amino acid sequence of the antenna polypeptides. Use of elastase resulted in a strongly decreased 858 nm absorption and a small blue shift (856 nm). This enzyme cleaved the α polypeptide chain at the C-terminus, removing the amino acid residues Lys-Ala-Ala, and the β polypeptide chain, removing Leu-His or Trp-Leu-His. Carboxypeptidase A treatment shifted the absorption maximum from 858 nm to 853 nm, and this was connected with a strong decrease in 858 nm absorption and in the (NIR)-CD signal. In this proteolytic cleavage in the α polypeptide chain Lys-Ala-ALa and in the β polypeptide chain Trp-Leu-His are split off at the C-terminus. Limited proteolysis using carboxypeptidase A plus B resulted in a further blue shift to 837 nm of the absorption maximum and CD signal and to a further hypochromic shift (extinction coefficient of about 50 to 60%) of the 800 nm absorption. The increased spectral changes result from a further cleavage of the α polypeptide chain up to the region of Trp His +14. These findings also show the important contribution of the aromatic Trp residue to spectral characteristics of BChl (absorption, extinction and CD signal).

E. Antenna Polypeptides of Core Antenna Complexes Binding BChl b

The core antenna complexes of photosynthetic bacteria containing BChl *b*, e.g. *Rp. viridis, Ectothiorhodospira (Ec.) halochloris*, show unusual functional and structural features: 1.) an extreme red shift to 1015 – 1020 nm of the absorption maximum and 2.) a marked formation of regular 2D arrays of antenna complexes in the membrane which are clearly visible in the electron microscope (Engelhardt et al., 1983; Stark et al. 1984). These features must be due to the special structure of the antenna polypeptides. The primary structures of the α and β antenna polypeptides, which are very similar to analogous antenna polypeptides of the other purple bacteria, have additional aromatic amino acid residues at two important positions of the α and β polypeptide chains: in the α polypeptide chain Trp (His +12) and Tyr (His-9) and in the β polypeptide chain Tyr (His +1) and Tyr (His-8) (Figs. 3, 4). These may be the origin of the large red shift. But of very special significance, both structurally and functionally, is the presence of the additional γ polypeptides (Brunisholz et al., 1985; Brunisholz and Zuber, 1992). These were isolated, characterized and in part sequenced from *Rp. viridis, Ec. halochloris* and *Ec. abdelmalekii* (Fig. 8A.) The γ polypeptides and the α and β polypeptides are present in the B1015 complex of *Rp. viridis* in a ratio of 1:1:1. Two

reaction center polypeptides:	C-terminus ← His$_{SP}$ → N-terminus
light-harvesting polypeptides:	N-terminus ← His$_{TM}$ → C-terminus

		A-1 A-2 A-3 A-4 A-5
Rhodopseudomonas viridis	RC-L	..F.F.....H...Y.W..Y.Y.F...W
	RC-M	..Y.F...F.HW..YYF..Y...F...W
Rhodobacter sphaeroides	RC-L	..FFF.....H...Y.F..Y.Y.....W
	RC-M	..Y.F.....HF..YF...H...F...W
Rhodobacter capsulatus	RC-L	..FF......HF..Y.F..Y.Y.....W
	RC-M	..Y.......HF..YF...H...F...W
Rhodopseudomonas viridis	α	.Y........HF.........WW.F...
	β	..Y.......HY..W....W.......W
	γ	W..W..........W....Y........
Rhodobacter sphaeroides	α	.F.F......H........Y.W......
	β	..F.......H...Y.W..WF.......
Rhodobacter capsulatus	α	.F.F......H........F.W......
	β	..F.......H.....W..WF.......
Rhodocyclus gelatinosus	α	.F........H........F.W......
	β	..F.......HF..W.W..W........
Rhodospirillum rubrum	α	.F.F......HF.......F.W......
	β	..FF......H...W.W..W......Y.
Rhodopseudomonas acidophila	α	.F.F......HF.......F.W......
	β	..F.......H...W.W..W......W.
Rhodopseudomonas marina	α	.F.F......HF.......F.W......
	β	..F.......H...W.W..W......Y.

A

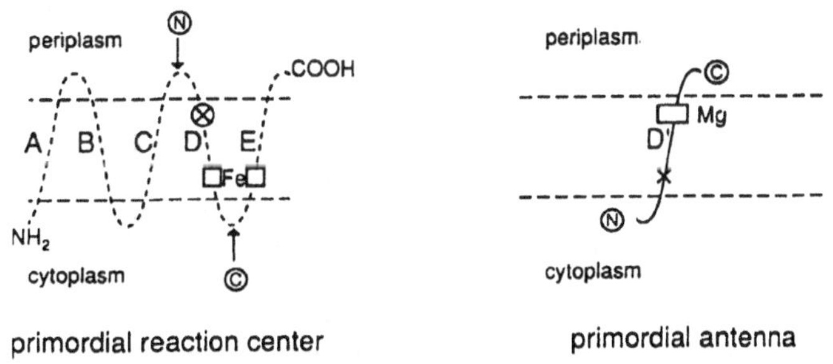

D = α reaction center
D' = α' antenna

B

Fig. 7. A: structural relatedness of the cluster of aromatic residues between the antenna polypeptides (α, β, γ) and the reaction center polypeptides L and M (RC-LRC-M, *Rp. viridis*) in the neighborhood of the central His residue binding the central BChl, or the His residue binding the special pair BChl, respectively. Fig. from Zuber and Brunisholz, 1991. B: Arrangement and orientation of the α-helices in the reaction center (also primordial reaction center) and of the antenna polypeptides in the antenna complexes (or the primordial antenna) in the membrane. Inverse orientation of the D helix (N → C) and of the α-helix of the antenna polypeptides (N → C) in the membrane. D and D' primordial polypeptides.

hydrophobic γ polypeptide variants in a ratio of 2:1 were isolated from this core antenna complex and sequenced (36 amino acid residues). They differ only in position 34 (Thr, Val). Also isolated and partially sequenced (29 amino acid residues) were a number of γ polypeptides (three types—one main type and two minor types) from *Ec. halochloris* and *Ec. abdelmalekii* (unpublished). All the γ polypeptides

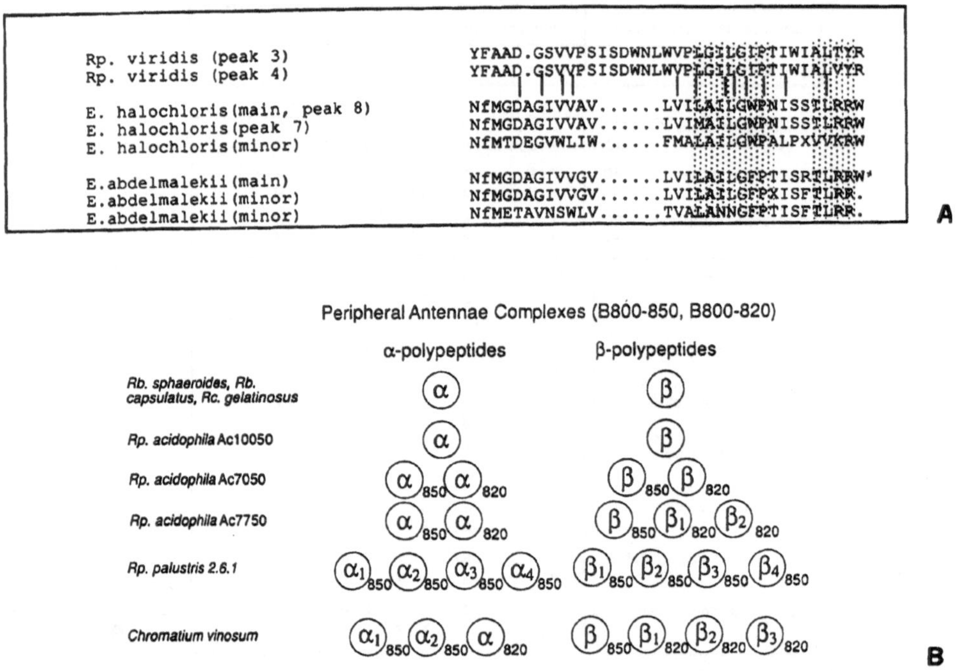

Fig. 8. A: Amino acid sequences of the γ-polypeptides of *Rp. viridis* , *Ec. halochloris* and *Ec. abdelmalikii*. Hatched parts indicate conserved portions. Nf: presumable formylated amino terminus. Reference: Brunisholz R, Habilitationschrift, ETH Zürich, 1993). B: Multiplicity of α- and β-antenna polypeptides of the peripheral antenna complexes in various purple bacteria (also see Figs. 3 and 4).

are clearly closely related and show common structural features. Of great importance here is a special sequence-homologous region in the C-terminus with its structurally important Gly, Pro and the aliphatic amino acid residues (Ile, Leu), or Arg, Trp and Tyr residues. These C-terminal, aromatic amino acid residues appear to be of particular functional significance. Crucial to understanding the structural/functional role of these conserved amino acid residues is the position and orientation of γ polypeptides in the membrane and their interaction with the α and β polypeptides. It has been found that the orientation of the N-terminal region of the γ polypeptide from *Rp. viridis* is towards the cytoplasm (unpublished results). As a consequence, hydrophobic γpolypeptides most probably form a transmembrane helix reaching to the periplasmic side of the membrane. The role of the γ polypeptide could be, on the one hand, that of a linker polypeptide, which together with the BChl *b*-carrying α and β polypeptides forms the core complex. The particular structure of the BChl *b*-containing core complexes on the membrane surface could reflect this possibility. On the other hand, the specific position of the Trp and Tyr residues at the C-terminus could, when we consider the length of the γ polypeptides (36 or 29 amino acid residues), allow interactions of these aromatic amino acid residues with Trp residues in the C-terminal domains of the α and β polypeptide chains (α: Trp His +11 and +12, and β: Trp His +4 and +9). This type of interaction could increase the functional coupling of these residues with BChl *b*- and therefore be responsible for the large red shift.

IV. Structural and Functional Variability of the Antenna Polypeptides and Pigments within the Antenna Complex

A. Variability of the BChl-Binding Sites

It was observed in some cases that BChl-binding sites of the α and β polypeptide chains vary, both in regard to type (nucleophilic amino acid side chain) and to position. In the β polypeptide chain of the core antenna complex from *Ec. halochloris*, the central His residues is replaced by Asn (BChl *b*-binding site) (Wagner-Huber et al., 1988). In the β polypeptide chain of the peripheral antenna complex from *Rc. gelatinosus* (DSM 149, 151), the second His (near the cytoplasm) is substituted by Gln (Zuber and Brunisholz, 1991). In peripheral antenna complexes B800–850 of Type II from *Cm. vinosum, Rp. acidophila* and *Rp. palustris*, the absorption band at 800 nm is spectrally dominant (extinction, specific (IR)-CD signal) compared to absorption at 850 nm or 820 nm. The structural basis of this spectral characteristic is as yet unknown—possible alternative BChl interactions (dimeric BChl molecule) are currently being considered (Zuber and Brunisholz, 1991).

B. Variability and Multiplicity of Antenna Polypeptides. Role of the Binding Site B800

1. Core Antenna Complexes

Members of the family of Ectothiorhodospiraceae, moderately to extremely halophilic purple photosynthetic bacteria, show in their antenna systems characteristic, and quite different, spectral features compared to those seen in the Rhodospirillaceae. These must also be due to the variation in the types and organization of their antenna complexes. *Ec. halophila*, for example, is characterized by NIR absorption bands of 800, 850 and 890 nm, with the 890 nm as the dominant band (Leguijt and Hellingwerf, 1991). On the other hand, the BChl *b*-containing antenna system from *Ec. halochloris* and *Ec. abdelmalekii* has, in addition to a typical absorption maximum of 1020 nm, further absorption maxima at 800 nm and 830 nm (and a very weak one at 880 nm) (Steiner and Scheer, 1985).

How can these spectral characteristics be explained by the organization and structure of these antenna complexes, and in particular, the structure of the antenna polypeptides. In the case of *Ec. halochloris*, it was possible to isolate a fraction of antenna complexes having absorption maxima at 800 nm, 850 nm and 1020 nm (Steiner and Scheer, 1985). This fraction contains, according to its (NIR)-CD signals, at least 5 strongly interacting BChl *b*-molecules. Especially characteristic of this antenna system is the strong biphasic CD-signal with a zero crosspoint at around 800 nm (which interestingly is also typical of the peripheral antenna complex in *Chromatium (Cm.) vinosum* (Hayashi et al., 1981)). Five polypeptides having the apparent molecular weight of 4.5, 6.0, 6.5, 15.5 and 35 kDa were identified by means of SDS-PAGE (Steiner and Scheer, 1985). As yet it has not been possible to dissociate these antenna complexes into subfractions, that is, into discrete antenna complexes with absorption maxima at 800 nm plus 830 nm and 1020 nm. In spite of this, attempts have been made to analyze the structure of diverse antenna polypeptides present, because such analysis might be expected to yield insight into the nature of these antenna complexes (Wagner-Huber et al., 1988, 1992). Using organic solvent extraction and subsequent molecular sieve and reversed phase chromatography, 2α, 2β antenna polypeptide types and a fifth, very small polypeptide were isolated from the membrane of *Ec. halochloris*. From *Ec. halophila* it was possible to isolate 3 α and 2 β antenna polypeptide types. Here, for the first time in a core antenna complex, a multiplicity of α and β antenna polypeptides was found. A marked degree of multiplicity of antenna polypeptides was also found in peripheral antenna complexes of *Cm. vinosum* (Bissig, 1989). Primary structure analysis of these *Ectothiorhodospira* polypeptides again revealed the characteristic three-domain structure which is so diagnostic of antenna polypeptides as found in the Rhodospirillaceae. The α and β polypeptides could not, unfortunately, be ascribed directly to core or peripheral antenna complexes, as these polypeptides could only be obtained from the photosynthetic membrane and not from isolated antenna complexes. However, comparison of the primary structures of diverse antenna polypeptides from *Ec. halochloris* and *Ec. halophila* did allow, upon the basis of their structural characteristics, their possible assignment to core or peripheral antenna complexes. The two types of α- polypeptides (N-terminal sequence MWR, MWK (*Ec. halochloris*) or MWRL, MWRM (*Ec. halophila*), and β-polypeptides (N-terminal sequence AND, TDI (*Ec. halochloris*) or ADEM, ADNM (*Ec. halophila*) exhibited a pronounced sequence homology to diverse antenna polypeptides of the core complexes from Rhodospirillaceae (Wagner-Huber et al., 1992). Only the α polypeptide with the N-terminal sequence MNQA (*Ec. halophila*) was clearly related to the antenna polypeptides of 'classical', peripheral complexes. But why then are there two types of α and β antenna polypeptides (α, β) types in the core complex? The proportion of α to

β antenna polypeptides, isolated from the membrane, is 1:1:1:1. On the basis of the possible structure of the core antenna complex, analogous to Rhodospirillaceae, for which a cyclical arrangement of 6 tetrameric units ($\alpha_2\beta_2$) was postulated (see V.C.), there are three possible arrangements of the α_1 and α_2, or β_1 and β_2 antenna polypeptides, respectively:

- There are two different core antenna complexes, each having 6 $\alpha_1\beta_1$ tetramers, or $\alpha_2\beta_2$ tetramers, respectively.
- There is only one core antenna complex, having mixed α_1/α_2, or β_1/β_2 polypeptide units ($\alpha_1\alpha_2/\beta_1\beta_2$).
- There is only one core antenna complex, for example $(\alpha_1\alpha_1)_2$, whereby the α_2 and β_2 polypeptides surrounding the core complex form the peripheral antenna complexes. It is also conceivable here that the α_1 or β_1 and α_2 or β_2 antenna polypeptides could form both the core and also the peripheral antenna complex. This would lead, in principle, to antenna complexes with mixed polypeptide units, within the entire antenna system.

The existence of an α polypeptide with the characteristic structural elements of the peripheral antenna complex in *Ec. halophila* indicates the existence of peripheral antenna complexes, possibly having absorption maxima from 800–850 nm. This does not, however, rule out the possibility that a part of the α_1-(α_2-), or β_1-(β_2)-antenna polypeptides also form peripheral antenna complexes (see above) and are located between the actual core complex B890 and the B800–850 antenna complex (heterogeneous energy transfer system).

2. Peripheral Antenna Complexes

In contrast to the classic Rhodospirillaceae (for example *Rs. rubrum, Rp. viridis, Rb. sphaeroides* and *Rb. capsulatus*) which have only one type of α and β antenna polypeptides (in a ratio of 1:1), some species of Rhodospirillaceae have peripheral antenna complexes with several types of α and β polypeptides which differ in their amino acid sequence (Zuber and Brunisholz, 1991) (Fig.3, 4, Fig. 8B). The appearance of differing types of antenna polypeptides is often accompanied by variability in spectral characteristics, whereby the proportion of the different polypeptide types or the spectral characteristics of the antenna complexes can be changed (or regulated) through the environmental factors of light and temperature. In this sense, the diversity of antenna polypeptides is possibly of functional significance with regard to the regulatory adaptation of antenna complexes to the variable conditions of the environment. The various types of α and β polypeptides still, however, possess the characteristic features of α and β polypeptides (which allow the α or β type classification). This allows the suggestion that even given the diversity of these polypeptides the α and β polypeptides form the basis of α/β heterodimers, just as is the case in the antenna complexes of classic Rhodospirillaceae with only one type of α or β polypeptide. However, the question remains open here (as in the core complexes as well) as to whether the individual types of α or β polypeptides build specific antenna complexes or hybrid antenna complexes with differing types of α and β polypeptides. The diversity of antenna polypeptides is based on the diversity of genes (gene families) (Tadros and Waterkamp, 1989). This corresponds to the situation with other peripheral antenna complexes, for example LHC II in algae and in higher plants, e.g. Dunsmuir et al., 1983. In all these cases it is difficult, however, to clearly recognize or experimentally demonstrate the functional-adaptive reason for the diversity of polypeptides, due to the fact that the diversity of genes in these organisms today could have other origins (e.g. evolutionary reasons.)

A definite relationship between the variable pattern of antenna polypeptides and variable spectral characteristics has been demonstrated, for example with the peripheral antenna complexes in *Rp. acidophila*. *Rp. acidophila* strain 7050 changes its NIR absorption maxima and its carotenoid composition depending upon the light intensity (Heinemeyer and Schmid, 1983; Cogdell et al., 1983). Under high light (2000 lux), the core antenna complex B880 and the peripheral antenna complex B800–850 (type 1) are synthesized. Under low light conditions (200 lux), 800 nm absorption is greatly decreased, and a blue shift of 850 to 825 takes place. Under low light the B800–850 complex is replaced by a B800–820 complex. Both antenna complexes have only 1 α and 1 β polypeptide. Primary structure analysis of the α and β polypeptides of the peripheral B800–850 complex (α: 53 residues, β: 48 residues) and of the B800–820 (825) complex (α: 53 residues, β: 42 residues) revealed in the C-terminal domain a substitution of Tyr^{44} Trp^{45} by Phe^{44}-Leu^{45}, which most probably causes the blue shift from 850 to 820 nm (Brunisholz et al., 1987; Bissig, 1989; Brunisholz and Zuber, 1988a; Zuber and Brunisholz, 1991). An analogous amino acid substitution (Tyr^{44} Trp^{45} to Phe^{44} Thr^{45}) resulting in a blue shift of 850 to 820 nm (that is, the transition from B800–850 complex to B800–820 complex) was unexpectedly observed in *Rp. acidophila* strain 7750 when the growth temperature was changed from 30 °C to temperatures below 25 °C (optimum 22 °C). (Schmidt and Brunisholz, 1986). This temperature effect (towards lower temperatures) produces the B800–820 complex (arising from the B800–850 complex having 1 type each of α and β polypeptide chain), with two different β polypeptide types. A multiplicity of antenna polypeptides, e.g. in B800–850 3 α- and 3 β-polypeptides, was found in *Rp. acidophila* (strain 10050). *Rp. palustris* also synthesizes multiple forms of antenna complexes and antenna polypeptides (Brunisholz et al., 1989, 1990). Under high light conditions, a B800–850 complex (type 1) is formed in *Rp. palustris* (strain 'French') in accordance with the absorption maxima, as in classical Rhodospirillaceae. On the other hand, the 850 nm band decreases greatly under low light conditions. Both in the high and low light forms of the peripheral antenna complex, several types of α and β polypeptides are found (a total of 7 polypeptides: 4 α and 3 β polypeptides). The α polypeptide chains differ especially in the C-terminal domain, and the β polypeptide chains in the N-terminal domain, in a number of amino acid substitutions. At the genetic level as well, a diversity of genes were isolated and sequenced from *Rp. palustris* (four sets of α and β polypeptides) (Tadros and Waterkamp, 1989). Through comparison of the DNA sequences with the protein sequences derived by protein sequencing, it was shown that in the C-terminal domain, cleavage of the polypeptide chain occurs post-translationally. The situation in *Cm. vinosum* appears to be even more complex in terms of spectral features and patterns of antenna polypeptides. Five different absorption maxima of BChl were found in *Chromatium* chromatophores: 795 nm, 805 nm, 820 nm, 850 nm, (845 nm) and 890 nm (Thornber, 1970; Hayashi et al., 1981). Here also the spectral features are dependent upon light intensity and temperature (30–40 °C) (Mechler and Oelze, 1978). In *Cm. vinosum*, at least 10 different antenna polypeptides are synthesized independently of culture conditions (Bissig, 1989; Bissig et al., 1989). The core complex B890 contains 3 polypeptides $\alpha/\beta_1/\beta_2$ in a ratio of 2:1:1. From the peripheral antenna complex 3 α polypeptides and 4 β polypeptides were isolated and sequenced. They show the typical three-domain structure of α and β polypeptides of the peripheral antenna complex. In addition, however, they also have characteristic structural elements that differentiate them from antenna polypeptides in Rhodospirillaceae; e.g., which are located in the elongated N-terminal domain of the α-polypeptides.

V. Possible Structural Organization of Polypeptides and BChl Molecules in the Antenna Complex Derived from Primary Structure Data (Primary Structure Models): Considerations on the Three-Dimensional Structure of Antenna Complexes

A. General Structural Features of the Antenna Polypeptides Important for the 3-D Structure of Antenna Complexes

The specific structural-functional features of antenna polypeptides derived from primary structure data must ultimately be the basis of the three-dimensional structure of the antenna complexes and, finally, of the entire antenna system of purple bacteria. The following characteristic structural features of antenna polypeptides indicate that there is a specific organization of the antenna polypeptides and BChl molecules in the 3-D structure (Zuber, 1987, 1993; Zuber et al., 1987; Zuber and Brunisholz, 1991):

- Two different types of polypeptides (α, β) bind functionally different α and β BChl molecules. These polypeptides or BChl pairs form, as heterodimers, the basic structural and functional unit of the antenna complexes and the entire antenna system.

- The α and β antenna polypeptides form a homologous family and are structurally closely related. However, they do show specific amino acid clusters which clearly separate them into core and peripheral antenna complexes. We can therefore expect to find in the three-dimensional structure a similar arrangement of these polypeptides but having a complex-specific form.

- The three-dimensional structure of the α and β polypeptides makes it likely that the central, transmembrane α helix (length 31 or 34 Å, diameter approximately 10 Å) plays an important role in the organization of antenna polypeptides in the antenna system (Fig. 9). The α-helical structure of antenna polypeptides in antenna complexes was postulated on the basis of IR, UV and CD spectra (Breton and Nabedryk, 1984) and labeling experiments (Jay et al., 1983). IR dichroism studies also showed that the α helices are tilted (on the average, at 30 to 35° away from the membrane plane) (Breton, 1986). In the tilted α helices of the α-β polypeptide heterodimers, the central BChl molecules can lie quite close together (up to 10 –15 Å) and thus be exciton-coupled (Sauer and Austin, 1978). The arrangement of the tilted helices is the basis for the tilted helices model.
- The present status of crystallographic studies indicates that the α-helices of the α- and β-polypeptides are not tilted and form a four helix bundle of parallel helices. (Basis of the four helix bundle model, Cogdell and Hawthornthwaite, 1993; Donnelly and Cogdell, 1993).
- Proteolytic cleavage experiments (Brunisholz et al., 1984b) and labeling experiments (Wiemken et al., 1983; Peters and Drews, 1983a; Jay et al., 1983, 1984) yielded a vertical asymmetry of the α and β polypeptide chain, with the N-terminal domain on the cytoplasmic side and the C-terminal domain on the periplasmic side. Thus the central α- β-BChl pair also lies on the periplasmic side and forms a two-dimensional energy transfer system in

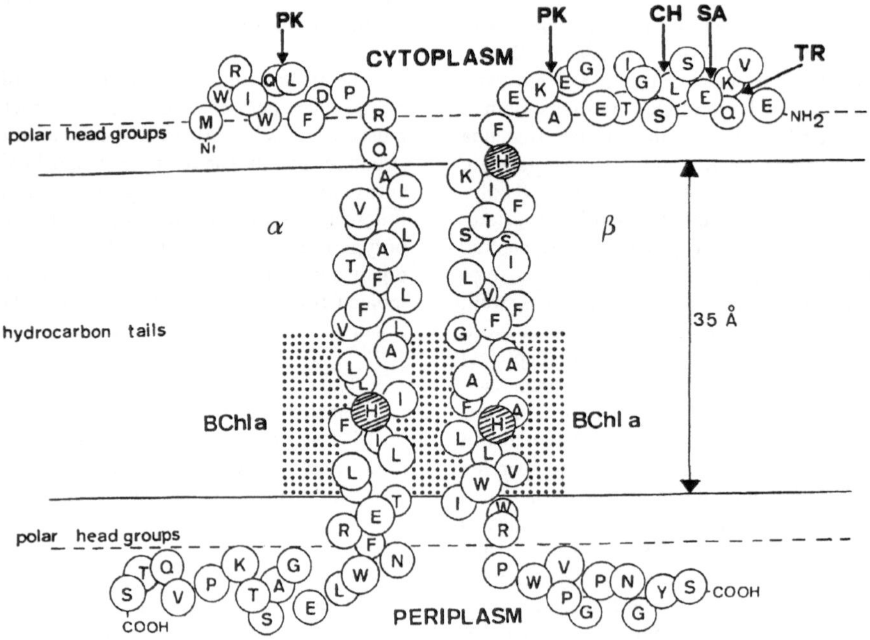

Fig. 9. Transmembrane orientation of the α- and β-antenna polypeptides (example *Rs. rubrum*). 3-domain structure with polar N- and C-terminal domain and a hydrophobic central domain (α-helix). PK, TR, CH and SA: splitting points with proteinase K, Trypsin, Chymotrypsin, staph. proteinase (see further text). Fig. from Zuber, 1987.

the environment of the special pair of the reaction center, which is also located on the periplasmic side.

B. Transmembrane Helices of the α-β Heterodimer Form the Minimal Structural Unit of the Antenna Complexes and the Entire Antenna System

Data from various sources (see below V.E) indicate that the relatively small antenna polypeptides form large polypeptide aggregates and are thus the fundamental building blocks from which the functional network of the BChl molecules is assembled. It is assumed that the specific interactions of the α helices of the α and β polypeptide chains, the packing of the helices in the lipid environment, and the specific BChl polypeptide and BChl-BChl interactions are the decisive factors in the formation of the polypeptide aggregates in the antenna (Zuber, 1987, 1993; Zuber and Brunisholz, 1991;). Complementary to this, the globular structures of the N and C terminal domains on the membrane surfaces form interaction sites within and between the antenna complexes.

1. Four Tilted Helices Model

With regard to the specific aggregation of the α helices, this model predicts that a crucial role is played by the specific interactions among the amino acid side chains on the surface of the helices, helically distributed in four rows along the axis of the transmembrane helices (Fig. 10A) (Zuber, 1987; Zuber and Brunisholz, 1991). These interactions make possible an optimal packing of the tilted helices, in the formation of both α-β heterodimers and larger polypeptide assemblies in the lipid environment. Here it is important that both (α- and β-) helices possess a transverse asymmetry with more polar amino acid residues on the N-terminal (cytoplasmic) side (Fig. 11C) and more hydrophobic amino acid residues including the central BChl binding site on the C-terminal (periplasmic) side (Fig. 11P) of the helices. This leads to structural (packing), as well as functional (energy transfer) asymmetry. Polar (charged) amino acid residues on the N-terminal side can form four possible regions of interaction (hydrogen bonds) between the α helices, which results in a defined orientation of the α helices both in the α-

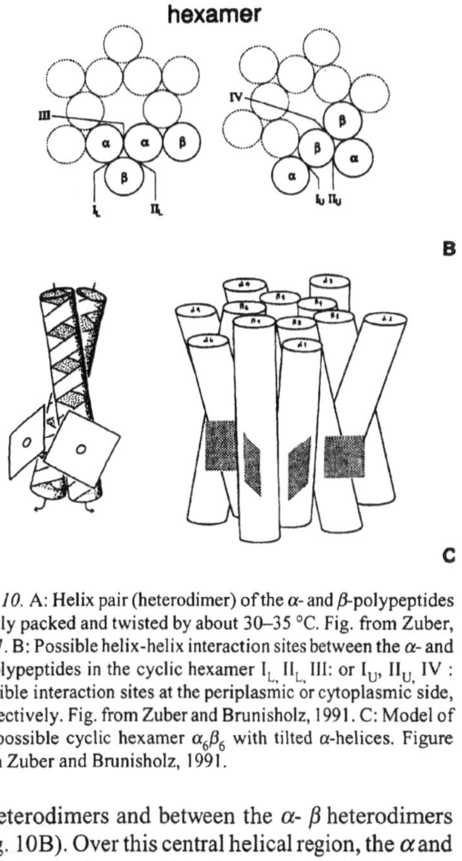

Fig. 10. A: Helix pair (heterodimer) of the α- and β-polypeptides tightly packed and twisted by about 30–35 °C. Fig. from Zuber, 1987. B: Possible helix-helix interaction sites between the α- and β-polypeptides in the cyclic hexamer I_L, II_L, III: or I_U, II_U, IV : possible interaction sites at the periplasmic or cytoplasmic side, respectively. Fig. from Zuber and Brunisholz, 1991. C: Model of the possible cyclic hexamer $\alpha_6\beta_6$ with tilted α-helices. Figure from Zuber and Brunisholz, 1991.

β heterodimers and between the α- β heterodimers (Fig. 10B). Over this central helical region, the α and β polypeptides associate in an hierarchical order, leading to the formation of α-β heterodimers (interaction in region I_L, I_U) and the association of the heterodimers via regions II_L, II_U (β-α), III_L, III_U (α-α) and IV_L, IV_U (β-β)) to cyclic arrays, e.g. cyclic hexamers (of heterodimers = 12 polypeptides) (Fig. 10B, C). Thus, through the tilt and the sideness of the helices, a system of specific polypeptide interaction results, with polar (charged) and hydrophobic residues on the cytoplasmic side (upper ($_U$) level) and hydrophobic residues and BChl-BChl interactions on the periplasmic side (lower ($_L$) level).

2. The Use of Fourier Transform Methods to Predict the α-Helical Structure of Antenna Complexes. Four Helix Bundle Model

Recently Donnelly and Cogdell (1993) used a Fourier transform method to try and predict the conformation and orientation of the conserved transmembrane

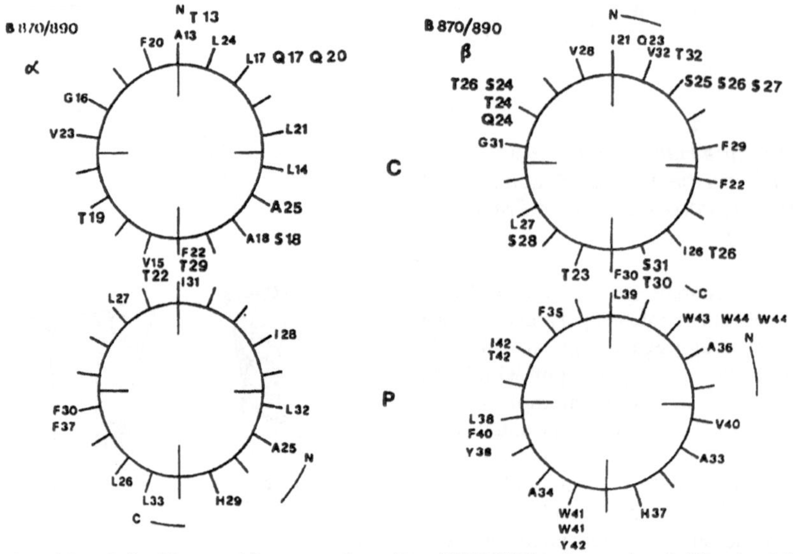

Fig. 11. Cross section of the α-helix of the α- and β-antenna polypeptides of B870/B890 at the cytoplasmic (C) and periplasmic (P) side. C: positions of the polar residues S, T, Q that form up to four specific helix-helix interaction sites. Compilation of all polar residues of the antenna polypeptides of *Rs. rubrum, Rp. viridis, Rb. sphaeroides, Rb. capsulatus, Rc. gelatinosus* and *Rp. acidophila*. P : H 29 (α) and H 37 (β) possible BChl binding site. Fig. from Zuber and Brunisholz, 1991.

segments of the purple bacterial antenna complexes. Similar methods had been successful with bacterial reaction centers and bacteriorhodopsin (Donnelly and Cogdell, 1993) where the 'real' structures are known. The modelling was carried out on 60 antenna sequences, using primary structural information from both 'core' and 'peripheral' antenna complexes. As part of the modelling procedure the direction of the hydrophobic face of the helix was calculated with reference to the helical wheel. The data could be best reconciled by assuming a 4 helix bundle (see Fig. 12). Interestingly, the Fourier analysis suggested that the length of the helical rods was at least 30 residues, and that part way along the helix the hydrophobic face, which is to the outside, 'flips' and moves to the inside. This is just the sort of behavior that would be expected for a helix which extends from the hydrophobic regions of the membrane out into the polar head group region, and indeed into the aqueous phase.

C. Dimers of Heterodimers $\alpha_2\beta_2$: Basic Structural and Functional Units for Energy Transfer Within and Between the Cyclic Antenna Complexes

1. Cyclic Peripheral Antenna Complexes (Hexamers)

With regard to the four tilted heliced model with optimally packed and interacting peripheral antenna complexes (B800–850, B800–820) surrounding the core complexes with the reaction center, the smallest functional, active cyclical aggregate of α-β heterodimers appear to be the cyclical hexamers (Fig. 13). In the case of the peripheral complex B850–880 from *Rs. molischianum*, however, an octameric structure was postulated (Kleinekofort et al., 1992). In the case of cyclical hexamers not only are the α and β polypeptides packed into a cyclical system, but these cyclical hexamers are packed optimally in 'lakes' as well (Fig. 14). As a result of this, two types of BChl-BChl interactions are possible in the two-dimensional energy transfer system on the periplasmic side: (1) BChlα-BChl β- interactions for cyclical energy transfer within the cyclical hexamers and (2) BChl β—BChl β- interactions for the transfer of energy between the cyclical hexamers. Here BChlα-BChl β- as well as BChl β—BChl β- can be exciton-coupled. The important structural and

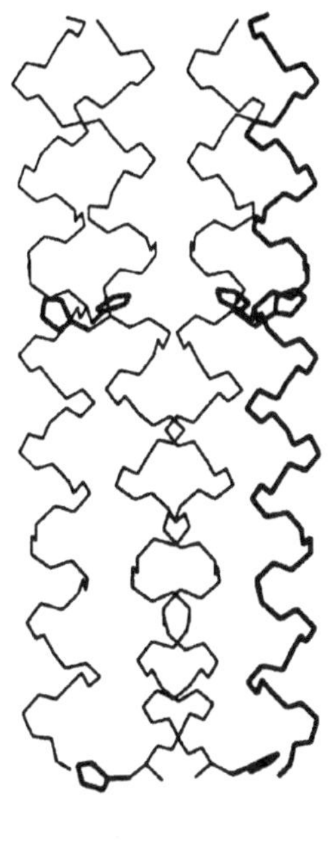

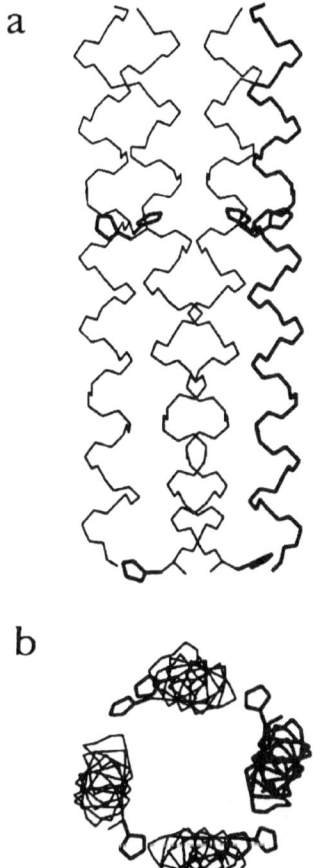

Fig. 12. A 3-D model of the $\alpha_2\beta_2$ tetrameric building block of a purple bacterial antenna complex viewed (a) from within and (b) perpendicular to the plane of the membrane. One of the α subunits and the conserved histidines is shown in bold.

functional units in the optimal packing of the hexamers are $\alpha_2\beta_2$ dimers of the heterodimer ($\alpha_2\beta_2$ units) (polypeptides, or BChl molecules) (Fig. 15). Through these $\alpha_2\beta_2$ units, the cyclical transfer of energy within hexamers is transformed to energy transfer between hexamers. With regard to the $\alpha_2\beta_2$ units, the cyclical hexamers show a three-fold symmetry. Two or three hexamers can be coupled through $\alpha_2\beta_2$ units (Fig. 15A, B).

With regard to the four helix bundle model see V.B.2 and Fig. 12.

2. Cyclic Core Antenna Complexes Close to the Reaction Center: Four Tilted Helices Model

The principle of a cyclical arrangement of the α-β heterodimers, or $\alpha_2\beta_2$ units, seems to apply not only to the peripheral complexes but also to the core complexes B870 or B890. Certain indications of this are also found in the primary structures of antenna polypeptides of peripheral and core complexes, which are sequence-homologous (structural basis for the formation of α-β-heterodimers or $\alpha_2\beta_2$-units). On the other hand, complex-specific sequence differences may lead to the somewhat different arrangement of the $\alpha_2\beta_2$ units in the core complex surrounding the reaction center. For the various core complexes with BChl *a*, ratios of BChl to the reaction center were found of between 21:1 and 41:1 (Dawkins et al., 1988), while for the core complexes containing BChl *b*-, a ratio of 24:1 was postulated. Hypothetically, there are three possible reasons for this variability of the ratio of BChl to reaction center: (1) In the core reaction center complex the reaction center is unstable. Therefore, the ratio of BChl to the reaction center varies. (2) 24 polypeptides with 24 central BChls, bound to the central His, and a variable amount of BChls, bound to the β polypeptides, surround the reaction center (maximum 36 BChl). In the corresponding structure model (Fig. 16B), 2 hexamers (with 12 polypeptides each) would always be coupled over an $\alpha_2\beta_2$ unit (Fig. 15A). (3) 36 polypeptides with 36 central BChls, bound to the central His, surround the reaction center. The variability of BChls is caused by loss of BChl occurring during isolation or a lower BChl content of the core complex. In this structure model (Fig. 16A) 3 hexamers (with 12 polypeptides each) are coupled over 3 $\alpha_2\beta_2$ units, as in the peripheral antenna complex (Fig. 15B). This structure model (Fig. 16A) seems to be rather unlikely for BChl *a* containing core complexes. Recent electron microscopic studies indicate that in all core complexes investigated, only 24 polypeptides seem to exist (Cogdell, unpublished data). In the BChl *b*- containing complexes, however, with the additional γ polypeptide, 36 polypeptides have to be organized in the core complex (see below).

In the case of the core complex in *Rp. viridis* containing BChl *b*- with 24 BChl, circular structures having a diameter of 130 Å were found in the

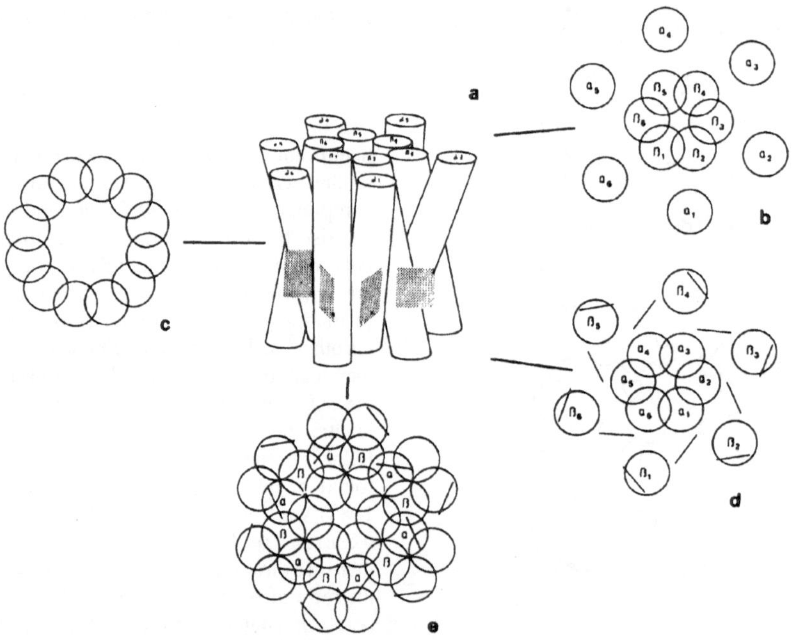

Fig. 13. Model of the cyclical hexamer of the peripheral antenna complex B800–850 (B800–820). Fig. from Zuber, 1987.

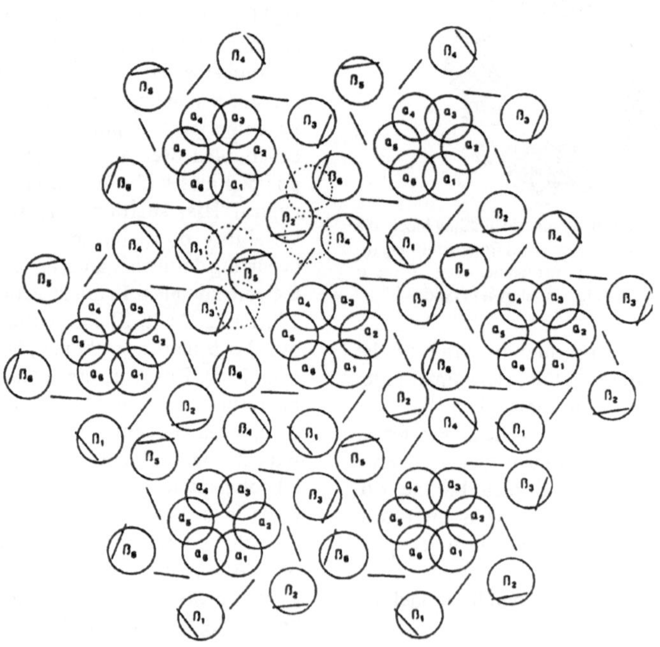

Fig. 14. 'Lake' of cyclic hexamers (see Figs. 10 and 13) with six hexamers surrounding a central hexamer and densely packed by specific helix-helix interactions between the hexamers. Basis for the directed energy transfer within and between the hexamers. Circles: α-helices of the α- and β-polypeptides at the periplasmic level. Bars: BChl molecules. Fig. from Zuber and Brunisholz, 1991.

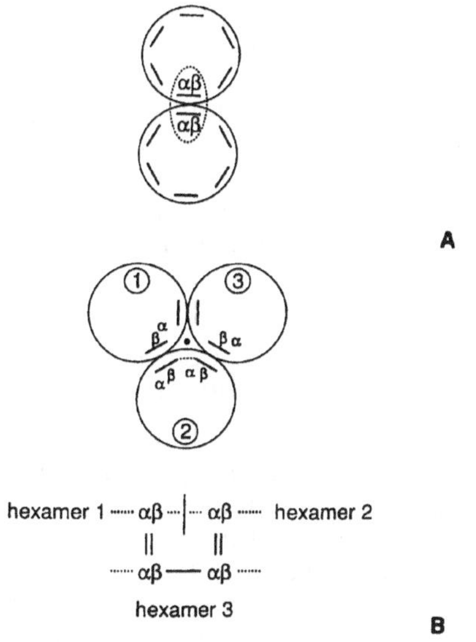

Fig. 15. A: Basic structural and functional units of $\alpha_2\beta_2$ (dimers of heterodimers) for energy transfer within and between the hexamers (two hexamers: structure model 1 (Fig. 16B) and three hexamers: structure model 2 (Fig. 16A)). B: Interaction unit between 3 hexamers via the $\alpha_2\beta_2$ unit.

photosynthetic membrane (chromatophores) with the electron microscope (Engelhardt et al., 1983; Stark et al., 1984; Jay et al., 1984; Stark et al., 1986) (Fig. 17). They appear to consist of a core region with a reaction center and a ring region (width 20 Å) of antenna complexes with 12 (Fig. 17A) or 6 (Fig. 17B) peripheral peaks. They may represent the 12 α-β heterodimers or the 6 $\alpha_2\beta_2$ units. The γ polypeptides without BChl, which exist in a ratio to the α(1) and β(1) polypeptides of 1:1:1, are not visible. There are two possible ways that the 6 $\alpha_2\beta_2$ units and the γ polypeptides (12 α, 12 β and 12 γ polypeptides) can be organized relative to the reaction center (Fig. 17):

- The $\alpha_2\beta_2$ units are organized as in the hexamer of the peripheral complex (Fig. 18A, C; structure model Fig. 16A). Three hexamer units, including the γ polypeptides, surround the reaction center. Through this, six $\alpha_2\beta_2$ units surround the reaction center, and six γ polypeptides form a linker polypeptide complex. Interaction of the BChl of the antenna system with the special pair of the reaction center is possible.

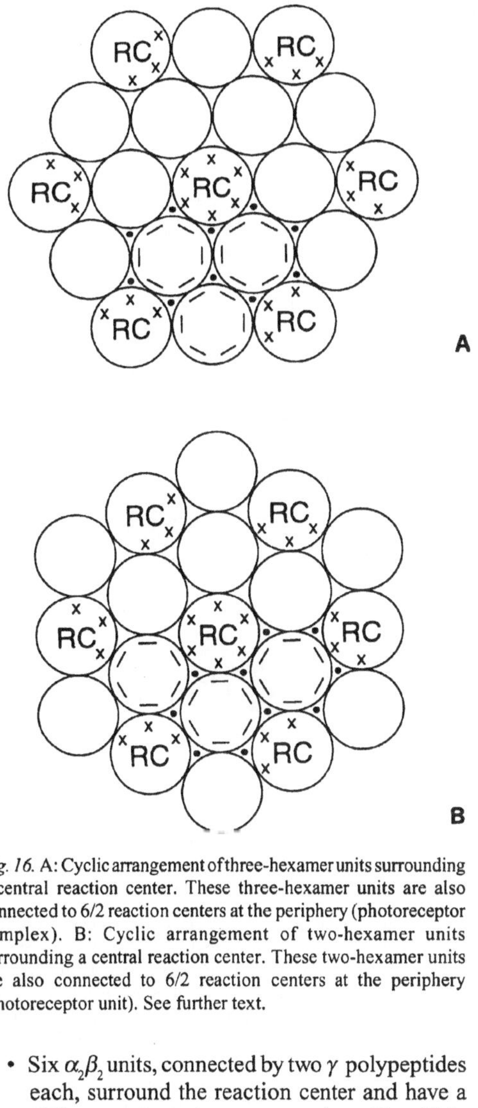

Fig. 16. A: Cyclic arrangement of three-hexamer units surrounding a central reaction center. These three-hexamer units are also connected to 6/2 reaction centers at the periphery (photoreceptor complex). B: Cyclic arrangement of two-hexamer units surrounding a central reaction center. These two-hexamer units are also connected to 6/2 reaction centers at the periphery (photoreceptor unit). See further text.

- Six $\alpha_2\beta_2$ units, connected by two γ polypeptides each, surround the reaction center and have a different orientation compared to Fig. 18A. (Fig. 18B; structure model Fig. 16B). Through this, the β BChls of two $\alpha_2\beta_2$ units can interact between the core complexes. Interaction of BChl of the antenna complexes with the reaction center is more difficult due to the larger distance.

A B880 core complex containing BChl *a* was isolated from *Rp. marina* and characterized (molecular weight, α, β polypeptide ratio, BChl *a* : α or β polypeptide ratio) (Meckenstock et al., 1992a;

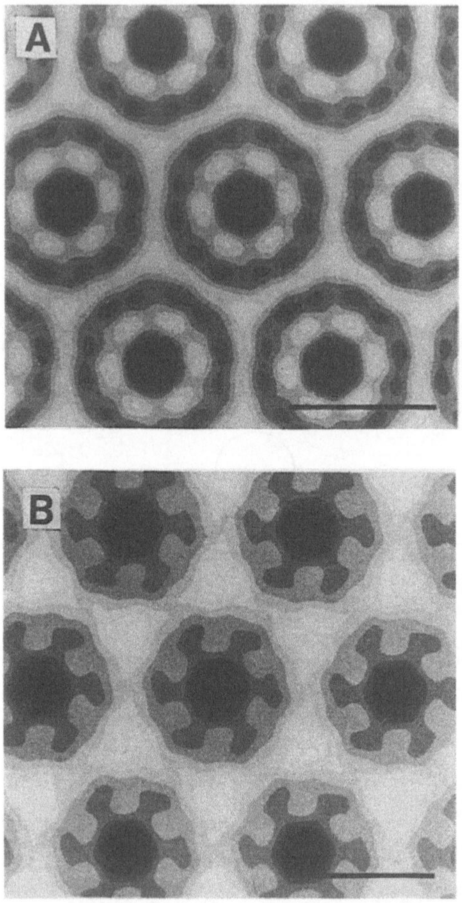

Fig. 17. Structure of the core complex surrounding the reaction center of *Rp. viridis*. Electron micrographs and Fourier processed images of tantalum/tungsten rotary shadowed, Triton X-100-treated membranes (Stark et al. 1984). A: Single membrane sheet with the plasmic surface upwards. B: Double membrane with the exoplasmic surface uppermost. Hexagonal lattice with a repeat distance of ~ 130 Å. The reaction center (diameter ~ 45 Å) is surrounded by 12 or 6 subunits on the plasmic or exoplasmic side, respectively. Scale of bar = 100 Å. Figure from Zuber, 1987.

see also Ghosh et al, 1993) (*Rs. rubrum*). The core complex consists of 24 polypeptides with 24 BChl *a* molecules, or of 6 B820 subcomplexes having 4 polypeptides each, or 4 BChl. Two-dimensional crystals reveal in the electron microscope a six-fold symmetry of the ringlike B880 complex (image processing, 20 Å resolution) (Meckenstock et al., 1992b, 1994) (Fig. 19A, B). They exhibit a hexagonal lattice with a lattice constant of 102 ± 3 Å. The $\alpha_2\beta_2$ units could be organized correspondingly to the structural model in Fig. 16B (6/3 hexamers = 24 polypeptides or 24 BChl) analogously, as in the core complex of *Rp. viridis*, which contains BChl *b* (Fig. 18B), but without connecting γ polypeptides, which would explain the smaller lattice constant of only 102 Å.

3. Cyclic Core Antenna Complexes Close to the Reaction Center: Four Helix Bundle Model

In this model the tetramer $\alpha_2\beta_2$-unit should be organized similarly to the structure model in Fig. 16B for the BChl *a* containing core complexes. The bars in the cyclic hexamers (Fig. 16A, B, and Fig. 15A) represent the external α- or β-BChl interacting with other antenna hexamers or the special pairs of the reaction center.

D. Possible Structural and Functional Organization of the Core and Peripheral Antenna Complexes in the Entire Antenna

Energy transfer to the reaction center depends on 1.) the state of the reaction centers (open and closed reaction centers) and, 2.) since the photochemical reaction is a rate limiting step, relative to the fast energy transfer processes within the antennae, on the ratio of the number of BChl *a* as final energy traps of the antenna to the special pair energy trap. Of particular importance in this respect are the structural (steric) conditions for the energy transfer from the core antenna to the special pair, which is particularly determined by the twofold symmetry of the special pair or the L and M subunits of the reaction center in relation to the six-fold symmetry of the surrounding core complex. Due to the non-central location of the special pair in the reaction center, six orientations of the special pair are possible within the cycle of the core complex (see Fig. 16).

In the case of the four tilted helices model the special pair could interact with an α-β heterodimer of the core complex in the form of an $\alpha_2\beta_2$ unit (hybrid between the α-β antenna special pair (heterodimer) and the special pair). A statistical distribution of the orientation of the special pair

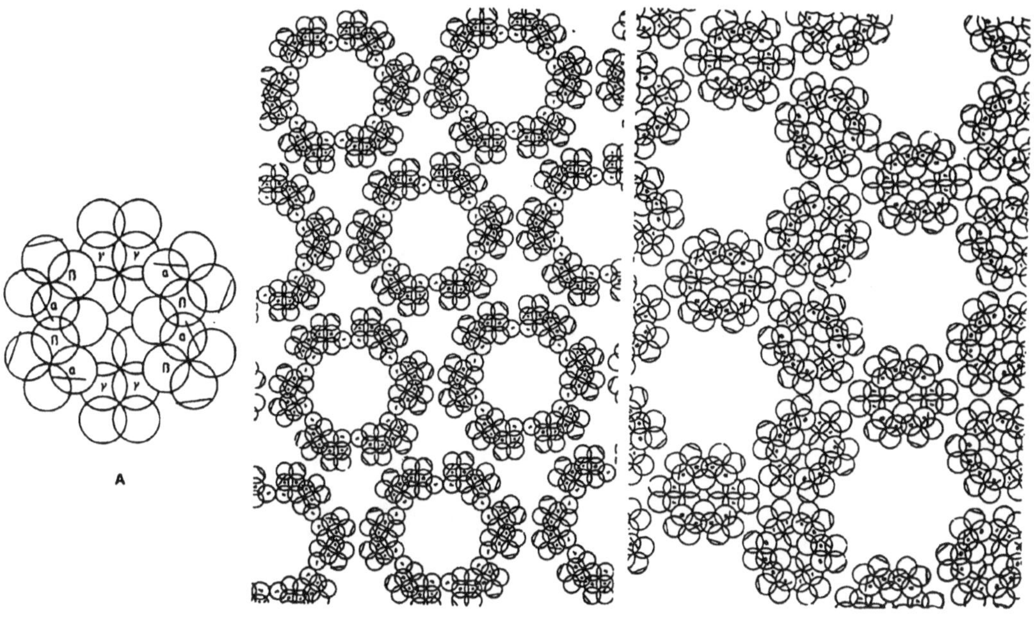

Fig. 18. Possible arrangements of the antenna polypeptides of the core complex B1015 (eg. *Rp. viridis*) surrounding the reaction center. The helices represent the α-helices (three cross-sections per tilted helix) of the α-, β- and γ polypeptide. A: Possible basis for the formation of the $\alpha_2\beta_2$-units that surround the reaction center: substitution of two $\alpha\beta$-heterodimers by 2 x 2 γpolypeptides. B: Cyclic dodecamer ($\alpha_{12}\beta_{12}\gamma_{12}$) or ($\alpha_2\beta_2\gamma_2/6$) surrounds the reaction center. The six $\alpha_2\beta_2\gamma_2$ units associate primarily via the γpolypeptides and the β-BChl pair, or C: the γ polypeptides forming hexamers of linker polypeptides. Fig. from Zuber and Brunisholz, 1991.

would, in the cycle of the six α-β heterodimers of the core complex pointing to the reaction center, always result in such a single $\alpha_2\beta_2$ unit transferring energy from the antenna system to the reaction center (one nearest-neighbouring antenna lattice site; (Pearlstein, 1982, 1992; Zhang et al., 1992). In the instance of the four helix bundle model this functional interaction would also take place via a BChl pair of the core complex and of the special pair.

The core-antenna complex forms in the entire antenna system large aggregates (photoreceptor units), in which, according to the structural model in Fig. 16B, 108 BChl (12 × 12 = 144 − 36 BChl) interact and are in contact with the next photoreceptor unit and with 4 reaction centers (1 + 6/2). This situation is perhaps especially the case in the photosynthetic membrane of purple bacteria, which contain only the B880(890) or B1015 complex.

In the entire antenna system of purple bacteria with core and peripheral complexes, two different arrangements of the peripheral antenna complex relative to the core complex and the reaction center are possible:

- The core antenna complexes and the peripheral antenna complexes are distinct entities. The peripheral antenna complexes surround the core complexes (Fig. 20). From the point of view of heterogeneous energy transfer, this arrangement should be favored above the mixed type below.

- An antenna system surrounds the reaction centers in which the core and peripheral antenna complexes are mixed.

- The entire antenna system may be made up of 10 – 20 photoreceptor units of the core complex with 40 – 80 reaction centers and > 1000 – 2000 BChl (see below, 0.5–1.0 μ diameter). Light intensity, or spectral conditions, determine the

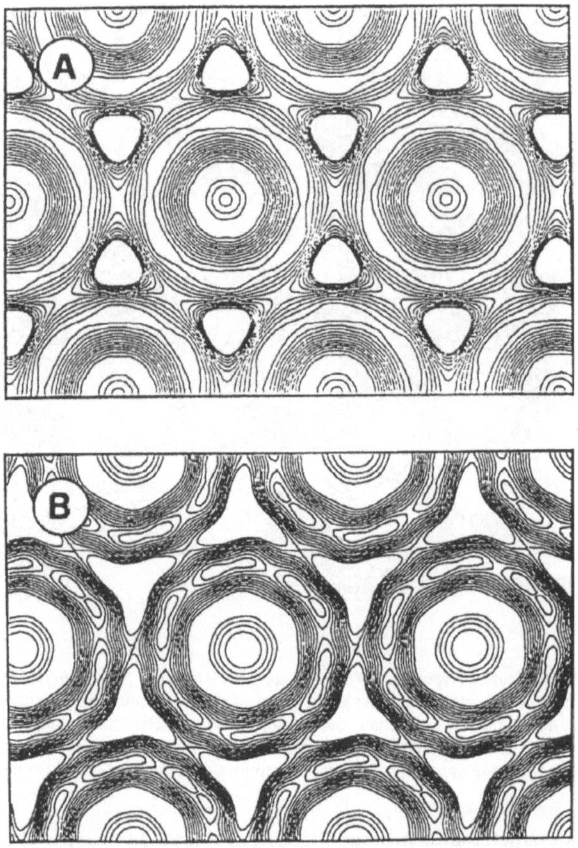

Fig. 19. A: Electron micrograph of a two dimensional crystal of the B880 core complex of *Rp. marina* negatively stained with Uranyl acetate. Image processing and six fold symmetrization of image. Each B880 core complex is shown as a hexagonal ring consisting of 6 subunits. B: Electron micrograph of a two dimensional crystal. Image processing and six fold symmetrization of image of a frozen hydrated preparation. The six subunits are further divided into two identical, minor subunits, possibly $\alpha_2\beta_2$-units. Fig. from Meckenstock et al., 1994.

size of the entire antenna and the ratio of core : peripheral antennae (up to 1:3) (Aagaard and Sistrom, 1972).

E. Comparison of the Structural Models of the Core and Peripheral Antenna Complexes with Experimental Data

The antenna models above should satisfy the following biochemical data:

- Composition and size of the isolated pigment-protein complexes, BChl content, αpolypeptide: βpolypeptide ratio, BChl : antenna polypeptides ratio, BChl (antenna polypeptide): reaction center ratio (Cogdell and Valentine, 1983; Drews, 1985; Kaplan and Arntzen, 1982; Zuber, 1985a, 1987; Zuber et al., 1987; Zuber and Brunisholz, 1991; Barber, 1987; Miller et al. 1987; Ghosh et al., 1988). There is, however, disagreement with respect to the mol. weight of the B820 subunit ($\alpha\beta$ heterodimer or $\alpha_2\beta_2$ unit) (Ghosh et al., 1988; Meckenstock et al., 1992a; Jirsakova et al., 1992).
- Crosslinking experiments establishing lateral topography. In *Rb. capsulatus* (B800–850) and

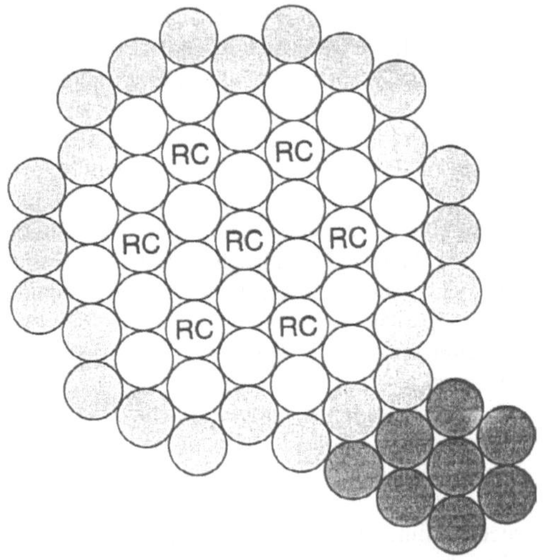

Fig. 20. Possible arrangement of the peripheral antenna complexes surrounding the reaction centers (RC). Core and peripheral antenna complexes are separated. Peripheral antenna complexes (cyclic hexamers): shaded circles.

Rp. viridis (B1015), heterooligomers (α-β) and homooligomers (α-α, β-β) as crosslinking products have been identified (Peters et al., 1983, 1984; Peters and Drews, 1983b; Ludwig and Jay, 1985). In *Rb. capsulatus*, in addition to crosslinking between the α-α and β-β polypeptides of the B870 core complex, a crosslinking product between the α polypeptide of the B800–850 complex and the H subunit of the reaction center has been found (Peters et al., 1983).

- Dissociation and reassociation experiments with antenna complexes and antenna polypeptides, including the isolation and characterization of antenna subcomplexes (Miller et al., 1987; Ghosh et al., 1988; Parkes-Loach et al., 1988, 1990; Heller and Loach, 1990; Berger et al., 1992; Visschers et al., 1992).

- The structural model must also be in agreement with the following spectroscopic data (Bolt and Sauer, 1978; Sauer and Austin, 1978; Breton et al., 1981; Kramer et al., 1984; Breton and Nabedryk, 1984):

- UV, CD and IR data which show the trans-membrane orientation. The four tilted helices model agrees with experimental data which shows a tilt of the α-helices, but the four helix bundle model does not.

- Strong CD signals suggest exciton interaction between pairs of BChl molecules.

- Linear dichroism data demonstrate that BChl Q_y transitions (850 nm) are approximately parallel to the membrane plane.

- Q_x transitions (590 nm) of BChl *a* are approximately perpendicular to the membrane plane.

- The orientations of the Q_y and Q_x transitions suggest that the porphyrin rings of BChl *a* 850 are approximately perpendicular to the membrane plane.

- On the other hand, the Q_y and Q_x transitions of BChl *a* 800 lie approximately in the plane of the membrane.

The structural models of the B800–850 complex, therefore, fit in principle (and particularly in the instance of the four helix bundle model) with a model of a minimal functional unit of the B800–850 complex of four antenna polypeptides ($\alpha_2\beta_2$) with four BChl 850, two BChl 800, established on the basis of spectroscopic data and the size of antenna polypeptides and the ratio of BChl: carotenoid: polypeptide (Kramer et al., 1984).

Fluorescence kinetic measurements (picosecond time-domain in coherent singlet excitation and trapping kinetics transfer) in the core complex of *Rs. rubrum* also suggest aggregates of four BChls, which could correspond to the $\alpha_2\beta_2$ minimal functional unit (Pullerits and Freiberg, 1992).

In contrast, on the basis of fluorescence polarization measurements and low temperature absorption spectroscopy of a subunit (B820) of the core complex and of the entire core complex of various purple bacteria, it was postulated that the B820 subunit is a dimer of interacting BChl (exciton coupling) (Visschers et al. 1991, 1993; Van Mourik et al., 1991; see also chapter by Loach and Parkes-Loach).

The structural models are also consistent with data on the size of functional domains, i.e. the cluster of connected BChl among which the excitations can be freely transferred. With the singlet-singlet annihilation method (Van Grondelle et al., 1983); Kramer et al., 1984; Van Grondelle, 1985; Vos et al., 1986; Deinum et al., 1991) using chromatophores or isolated antenna complexes, it has been found that, e.g. at 4K, B870 (*Rb. sphaeroides, Rs. rubrum*) consists of clusters of 100 – 150 functionally connected BChl, and the peripheral B800–850 of clusters of 30 BChl 850. At 298K, energy transfer takes place within these complexes in larger lakes of >1000 BChl 870 or >365 BChl 850, respectively. The low-temperature form of the functional domain having approximately 100 BChl would correspond to the photoreceptor unit of the core complex of the structural models above, and the 298K form would correspond to the size of the entire antenna of the structure model with at least 10 photoreceptor units (10 × 100 BChl = 1000 BChl). In this connection there is a finding of great importance: in two-dimensional crystals of the core antenna complex in *Rp. marina*, the giant CD spectrum, which shows a large aggregate of BChl, appears in two-dimensional crystals of a size of 0.5–1 μ. (Meckenstock et al., 1994). This would be in agreement with the structural model with an aggregate of approximately 10 photoreceptor units. Interestingly, the size of these aggregates lies within the region of the wavelength of the absorption maximum of these complexes (880 nm)—possibly an important fact with regard to the theoretical basis of the functional state of the entire antenna during energy transfer, or of the giant CD spectra.

Finally, the data from X-ray structural analyses of antenna complexes (see Fig. 1.) and from electron microscopy (see above) should be congruous with the antenna model described above.

The structure model of the core complex B820 with six α-β heterodimers derived from spectroscopic measurements does not fit the data of electron microscopy since for sterical reasons six α-β heterodimers are to small to form a core complex with the diameter of 103 Å or 140 Å (+ γ) (see also core-complex, *Rb. sphaeroides*, mutant) (Nunn et al., 1992).

F. Other Antenna Models and Related Theoretical Studies on the Energy Transfer System of Light-Harvesting Antennae of Purple Bacteria

Scherz and Parson (1986) developed a model for the structure of the antenna complexes based upon spectroscopic data. They assumed that two heterodimers coupled to a tetramer represents the minimal unit of the B850 antenna complex. In principle, this concept agrees with the structural models described above. However, the arrangement of the tetramers ($\alpha_2\beta_2$) in the B870/B890 and B800–850 complexes is different in the four tilted helices model. There is no agreement between these structural models discussed above and the structural model of Hunter et al. (1988), which assumes there are α-β polypeptide trimers ($(\alpha\beta)_3$) in the core antenna complex B875 and the peripheral complex B800–850 (see Chapter 22 by Hunter).

On the basis of the various antenna models of purple bacteria, efforts have been made to try and analyze theoretically exciton effects and electronic state energy transfer (Pearlstein, 1991). The diverse structural-functional models of antenna complexes and of the entire antenna system do not, however, allow a more detailed theoretical analysis to be performed, as they only describe structural-functional marginal conditions. In order to surmount this problem, in the future, three-dimensional structural analyses of antenna complexes will be necessary which show the exact position, orientation and spacing of BChl and carotenoid molecules.

Note Added in Proof

Since this chapter was written, the structure of the B800-850 complex from *Rp. acidophilia* strain 10050 has been determined to a resolution of 2.5 Å (McDermott et al., 1995). It is an $\alpha_9\beta_9$ oligomer as shown in Fig. 21. The nine α-apoproteins are arranged in the middle with the nine β-apoproteins around the edge. The view seen in Fig. 21 looks down on this complex as it would lie in the plane of the membrane, from the cytoplasmic side, while Fig. 22 shows it from a view perpendicular to the symmetry axis. The α-helices are clearly seen. All the pigments are located within the space between the two rings of helices. The top and bottom of the structure is formed by the N- and C-termini folding over. At no point do the α- or β-apoprotein α-helices interact. The BChl *a* molecules are organized into two groups. A ring of nine monomeric 800 nm-absorbing BChl *a*'s lying flat in the plane of the membrane between the β-apoprotein α-helices. A second ring of 18 850 nm-absorbing BChl *a* molecules form a tightly coupled ensemble towards the periplasmic side of the complex.

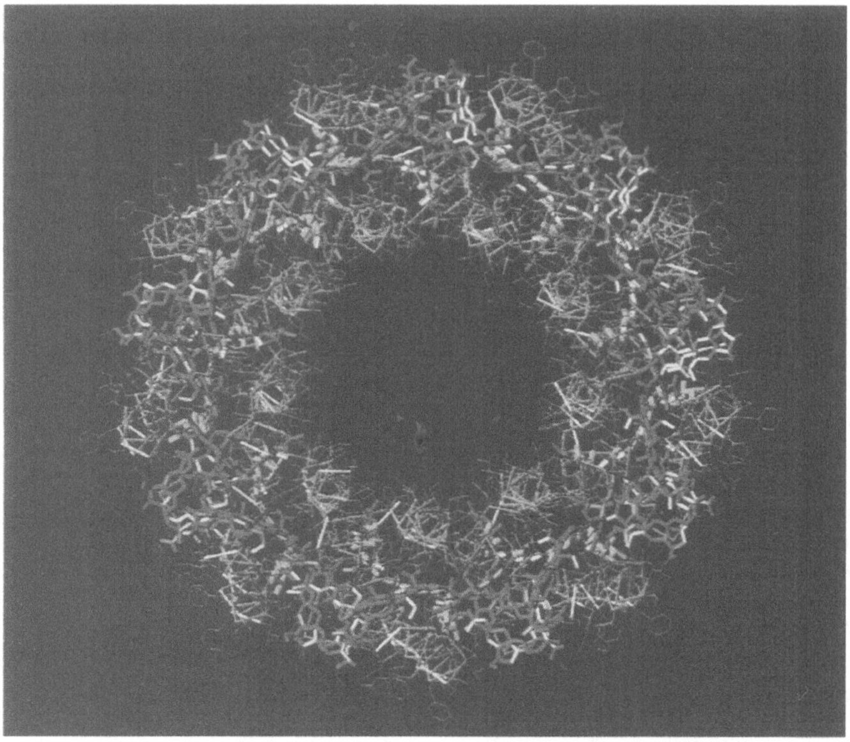

Fig. 21. (See Color Plate 6A). Cytoplasmic view of the LH2 antenna complex from *Rhodopseudomonas acidophila*. The protein is shown in purple, with BChl molecules in green and carotenoid in yellow. For further details see McDermott et al. (1995).

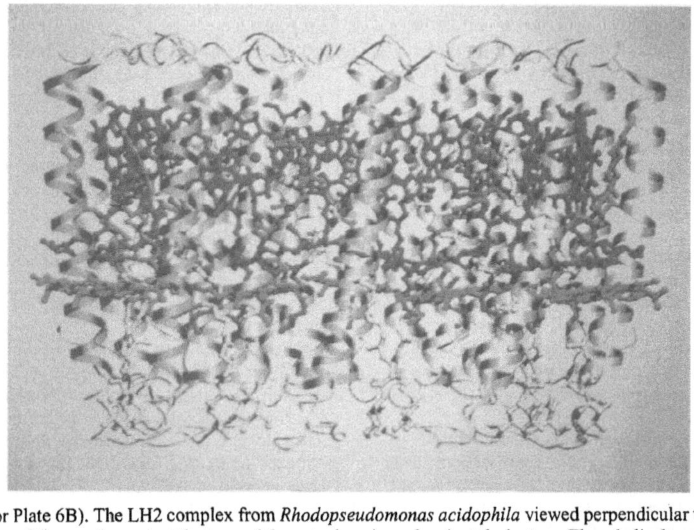

Fig. 22. (See Color Plate 6B). The LH2 complex from *Rhodopseudomonas acidophila* viewed perpendicular to the symmetry axis. The periplasmic surface of the membrane is at the top and the cytoplasmic surface is at the bottom. The α helical parts of the protein are shown as purple ribbons. BChl B850 molecules are shown in red, with B800 molecules in green and carotenoids in yellow. For further details see McDermott et al. (1995).

The carotenoids connect these two rings.

It is clear that this structure was not well modeled by any of the models described in this chapter. However, it is interesting, especially since modeling the structure of membrane proteins is a very popular pastime, to describe why this is. All of the models presented in this chapter were based upon $\alpha\beta$ helix-helix interation. This does not occur in the real structure. The real structrue is dominated by pigment-pigment and pigment-protein interactions—40% by weight of the complex is pigments. It is clear that the modeling process did not and could not allow for this and, hence, the models failed.

Acknowledgments

HZ wishes to thank Mrs. E. Zollinger for the careful preparation of the manuscript. His work reported in this chapter was supported by the Swiss National Foundation (Project 3.286-0.82, 3207-0.85, 31-25179.88, 31-32071.91) and by the Eidgenössische Technische Hochschule, Zürich.

References

Aagaard J and Sistrom WR (1972) Control of synthesis of reaction center bacteriochlorophyll in photosynthetic bacteria. Photochem Photobiol 15: 209–225

Babst M, Albrecht H, Wegmann, I, Brunisholz RA and Zuber H (1991) Single amino acid substitutions in the α- and β-870 light-harvesting polypeptides of *Rhodobacter capsulatus*: Structural and spectral effects. Eur J Biochem: 282, 277–284

Barber J (ed) (1987) The Light Reactions, Topics in Photosynthesis. Vol. 8, Elsevier Science, Amsterdam

Berger G, Andrianambinintsoa S, Kleo J, Grison S, Dejonghe D, and Breton, J (1992) Dissociation and reconstitution studies by high performance liquid chromatography of the light harvesting complex of *Rhodospirillum rubrum*. J Liq Chromatogr 15: 585–602

Bissig I (1989) Die Strukturanalyse der Antennenpolypeptide von *Rhodopseudomonas acidophila*, *Chromatium vinosum* und *Ectothiorhodospira halophila*. Ph.D. thesis No. 8945, Eidg Technische Hochschule, Zürich

Bissig I, Brunisholz RA, Suter F, Cogdell RJ and Zuber H (1988) The complete amino acid sequence of the B800–850 antenna polypeptides from *Rhodopseudomonas acidophila* strain 7750. Z Naturforsch 43c: 77–83

Bissig I, Wagner-Huber R, Brunisholz RA, Frank G and Zuber H (1989) Multiple forms of light-harvesting polypeptides in *Chromatium vinosum*. In: Drews G and Daves EA (eds) Molecular Biology of Membrane-Bound Complexes of Phototrophic Bacteria, pp 199–210. Plenum Press, New York

Bolt J and Sauer K (1978) Linear dichroism of light-harvesting bacteriochlorophyll proteins from *Rhodopseudomonas sphaeroides* in stretched polyvinyl alcohol films. Biochim Biophys Acta: 546, 54–63

Breton J (1986) Molecular orientation of pigments and the problem of energy trapping in photosynthesis. In: Staehelin LA and Arntzen CJ (eds) Encyclopedia of Plant Physiology, Vol. 19, pp 319–336. Springer, Berlin

Breton J, and Nabedryk E (1984) Transmembrane orientation of α-helices and the organization of chlorophyll in photosynthetic pigment protein complexes. FEBS Lett 176: 355–359

Breton J. Vermeglio A, Garrigos M and Paillotin G (1981) Orientation of the chromophores in the antenna system of *Rhodopseudomonas sphaeroides*. In: Akoyunoglou G (ed) Photosynthesis, Vol. III, pp 445–459. Balaban Int. Science Services, Philadelphia

Brunisholz RA and Zuber H (1987) Application of FPLC-reversed-phase chromatography for the purification of small hydrophobic light-harvesting polypeptides. Experientia 43: 672

Brunisholz RA and Zuber H (1988a) Primary structure analyses of bacterial antenna polypeptides: Correlation of aromatic amino acids with spectral properties. Structural similarities with reaction center polypeptides. In: Scheer H and Schneider S (eds) Photosynthetic Light-Harvesting Systems, pp 103–114. Walter de Gruyter, Berlin

Brunisholz RA and Zuber H (1988b) Bacterial intramembrane-bound antenna complexes: Structural and spectral correlation. In: Stevens SE Jr. and Bryant DA (eds) Light-Energy Transduction in Photosynthesis: Higher Plant and Bacterial Models, pp 47–61. American Society of Plant Physiologists, Rockville, MD

Brunisholz R and Zuber H (1992) Structure, function and organization of antenna polypeptides and antenna complexes from the three families of Rhodospirillaneae. Eur J Photochem Photobiol B Biol 15: 113–140

Brunisholz R and Zuber H (1993) Spectral modifications of bacterial antenna complexes by limited proteolysis. Photochem Photobiol 57: 6–12

Brunisholz RA, Cuendet PA, Theiler R and Zuber H (1981) The complete amino acid sequence of the single light-harvesting protein from chromatophores of *Rs. rubrum* G-9^{+}. FEBS Lett 129: 150–154

Brunisholz RA, Wiemken V, Suter F, Bachofen R and Zuber H (1984a) The light harvesting polypeptides of *Rhodospirillum rubrum*: I. The amino acid sequence of the second light-harvesting polypeptide B880-β of *Rhodospirillum rubrum* S1 and the carotenoidless mutant G9^{+}. Aspects of the molecular structure of the two light-harvesting polypeptides B880α (B870α) and B880β (B870β) and of the antenna complex B880 (B870) from *Rhodospirillum rubrum*. Hoppe-Seyler's Z Physiol Chem 365: 675–688

Brunisholz RA, Wiemken V, Suter F, Bachofen R and Zuber H (1984b) The light-harvesting polypeptides of *Rhodospirillum rubrum*: II. Localization of amino terminal regions of the light-harvesting polypeptides B870-α and B870-β and the reaction center subunit L at the cytoplasmic side of the photosynthetic membrane of *Rs. Rubrum* G-9^{+}. Hoppe-Seyler's Z Physiol Chem 365: 689–701

Brunisholz RA, Jay F, Suter F and Zuber H (1985) The light harvesting polypeptides of *Rhodopseudomonas viridis*: The complete amino acid sequences of B 1015-α, B 1015-β and B

1015-γ. Biol Chem Hoppe-Seyler 366: 87–89

Brunisholz RA, Zuber H, Valentine J, Lindsay JG, Woolley KJ and Cogdell RJ (1986) The membrane location of the B890-complex from *Rs. rubrum* and the effect of carotenoid on the conformation of its two apoproteins at the cytoplasmic surface. Biochim Biophys Acta 849: 295–303

Brunisholz RA, Bissig I, Niederer E, Suter F and Zuber H (1987) Structural studies on the light-harvesting polypeptides of *Rp. acidophila*. In: Biggins J (ed) Photosynthesis, II.1, p13. Martinus Nijhoff, The Hague

Brunisholz R, Evans MB, Cogdell RJ, Frank G and Zuber H (1989) The peripheral antenna polypeptides of *Rp. palustris*: Synthesis of multiple forms and structural variability as a consequence of light intensity. Physiologia Plantarum A 68

Brunisholz R, Evans MB, Cogdell RJ and Zuber H (1990) The peripheral antenna polypeptides of *Rhodopseudomonas palustris*. Synthesis of multiple forms and structural variability as a consequence of light intensity. In: Current Research in Photosynthesis, Vol. II, 4, pp 61–64. Kluwer Academic Publishers, Dordrecht

Bylina EJ, Ismail S and Youvan DC (1986) Site specific mutagenesis of bacteriochlorophyll-binding sites affects biogenesis of the photosynthetic apparatus. In: Youvan DC and Daldal F (eds) Current Communications in Molecular Biology. Microbial Energy Transduction. Genetics, Structure and Function of Membrane, pp 63–70. Cold Spring Harbor Laboratory, New York

Capaldi RA and Vanderkoii G (1972) The low polarity of many membrane proteins. Proc Natl Acad Sci USA 69: 930–932

Cogdell RJ (1986) Light-harvesting complexes in purple photosynthetic bacteria. In: Staehelin IA and Arntzen CJ (eds) Encyclopedia of Plant Physiology, Vol. 19, pp 252–259. Springer, Berlin

Cogdell RJ and Hawthornthwaite AM (1993) The preparation, purification and crystallisation of purple bacterial antenna complexes. In: Norris JR and Deisenhofer J (eds) The Photosynthetic Reaction Center, Vol. 1, pp 23–42. Academic Press, New York

Cogdell RJ and Valentine J (1983) Yearly Review. Bacterial Photosynthesis. Photochem Photobiol: pp. 769–772

Cogdell RJ, Durant I, Valentine J, Lindsay JG and Schmidt K (1983) The isolation and partial characterization of the light-harvesting pigment protein complexes of *Rp. acidophila*. Biochem Biophys Acta 722: 427–435

Cortez N, Garcia AF, Tadros MH, Gad-on N, Schiltz E and Drews G (1992) Redox-controlled in-vivo and in-vitro phosphorylation of the alpha subunit of the light-harvesting complex I in *Rhodobacter capsulatus*. Arch Microbiol 158: 315–319

Dawkins DJ, Ferguson LA and Cogdell R (1988) The structure of the 'core' of the purple bacterial photosynthetic unit. In: Scheer H and Schneider S (eds) Photosynthetic Light-Harvesting Systems, pp 115–127. Walter de Gruyter, Berlin

Deinum G, Otte SCM, Gardiner AT, Aartsma TJ, Cogdell RJ, Amesz J (1991) Antenna organization of *Rhodopseudomonas acidophila* a study of the excitation migration. Biochim Biophys Acta 1060: 125–131

Drews G (1985) Structure and functional organization of light-harvesting complexes and photochemical reaction centers in membranes of phototrophic bacteria. Microbiol Rev 49: 59–70

Donnelly D and Cogdell RJ (1993) Predicting the point at which transmembrane helices protrude from the bilayer: a model of the antenna complexes from photosynthetic bacteria. Protein Engineering 6: 629–635

Dunsmuir P, Smith SM and Bedbrook J (1983) The major chlorophyll *a/b* binding protein of petunia is composed of several polypeptides encoded by a number of distinct nuclear genes. J Mol Appl Genet 2: 285–300

Engelhardt H, Baumeister W and Saxton WO (1983) Electron microscopy of photosynthetic membranes containing bacteriochlorophyll *b*. Arch Microbiol 135: 169–175

Fowler GJS, Visschers RW, Grief GG, Van Grondelle R and Hunter CN (1992) Genetically modified photosynthetic antenna complexes with blue shifted absorbance bands. Nature 355: 848–850

Germeroth L, Lottspreich F, Robert B and Michel H (1993) Unexpected similarities of the B800–850 light-harvesting complex from *Rs. molischianum* to the B870 light-harvesting complexes from other purple photosynthetic bacteria. Biochemistry 32: 5615–5621

Ghosh R, Hauser H and Bachofen R (1988) Reversible dissociation of the B873 light harvesting complex from *Rs. rubrum* G-9$^+$. Biochemistry 27: 1004–1014

Ghosh R, Hoenger A, Hardmeyer A, Mihailescu D, Bachofen R, Engel A and Rosenbusch JP (1993) Two-dimensional crystallization of the light-harvesting complex from *Rhodospirillum rubrum*. J Mol Biol 231: 501–504

Guthrie N, McDermott G, Cogdell RJ, Freer AA, Isaacs NW, Hawthornthwaite AM, Halloren E and Lindsay JG (1989) Crystallization of the B800–850 light-harvesting complex from *Rp. acidophila* strain 7750. J Mol Biol 224: 527–528

Hayashi H, Nozawa T, Hatano M and Morita S (1981) Circular dichroism of bacteriochlorophyll a in light-harvesting bacteriochlorophyll-protein complexes from *Chromatium vinosum*. J Biochem 89: 1853–1861

Hawthornthwaite AM and Cogdell RJ (1991) Bacteriochlorophyll-binding proteins. In: Scheer H (ed) The Chlorophylls, pp 493–528. CRC Press Inc., Boca Raton, Florida

Heinemeyer EA and Schmidt K (1983) Changes in carotenoid biosynthesis caused by variations of growth conditions in cultures of *Rp. acidophila* strain 7050. Arch Microbiol 134: 217–221

Heller BA and Loach PA (1990) Isolation and characterization of a subunit form of the B 875 light-harvesting complex from *Rhodobacter capsulatus*. Photochem Photobiol 51: 621–638

Henderson R and Unwin PNT (1975) Three dimensional model of purple membrane obtained by electron microscopy. Nature 257: 28–32

Hunter CN, Pennoyer JD, Sturgis JN, Farrelly D and Niederman RA (1988) Oligomerization states and associations of light-harvesting pigment protein complexes of *Rhodobacter sphaeroides* as analyzed by lithium dodecyl sulfate polyacryl amide gel electrophoresis. Biochemistry 27: 3459–3467

Imhoff JF, Trüper HG and Pfennig N (1984) Rearrangement of the species and genera for the phototrophic 'purple non-sulphur bacteria'. Int J Syst Bacteriol 34: 340

Jay F, Lambillotte M and Mühlethaler K (1983) Localization of *Rhodopseudomonas viridis* reaction center and light-harvesting proteins using ferritin-antibody labeling. Eur J Cell Biol 30: 1–8

Jay F, Lambillotte M, Stark E and Mühlethaler K (1984) The preparation and characterization of native photoreceptor units from the thylakoids of *Rhodopseudomonas viridis*. EMBO J 3: 773–776

Jirsakova V, Agalidis I, Reiss-Husson F (1992) Purification of the LH1 core antenna from *Rc. gelatinosus*. Photosynth Research 34: 109

Kaplan S and Arntzen CJ (1982) Photosynthetic membrane structure and function. Photosynthesis 1: 65–74

Kleinekofort W, Germeroth L, Van den Broek JA, Schubert D, Michel H (1992) The light-harvesting complex II B800–850 from *Rhodospirillum molischianum* is an octamer. Biochim Biophys Acta 1140: 102–104

Kramer HJM, van Grondelle R, Hunter CN, Westerhuis WHJ and Amesz J (1984) Pigment organization of the B800–850 antenna complex of *Rp. sphaeroides*. Biochim Biophys Acta 765: 156–165

Kühlbrandt W (1984) Three dimensional structure of the light-harvesting chlorophyll *a/b* protein complex. Nature 307: 478–480

Leguijt T and Hellingwerf KJ (1991) Characterization of reaction center/antenna complexes from bacteriochlorophyll *a* containing Ectothiorhodospiraceae. Biochim Biophys Acta 1057: 353–360

Ludwig FR and Jay FA (1985) Reversible cross-linking experiments of the light-harvesting polypeptides of *Rp.. viridis*. Eur J Biochem 151: 83–87

McDermott G, Prince SM, Freer AA, Hawthornthwaite-Lawless AM, Papiz MZ, Cogdell RJ and Isaacs NW (1995) Crystal structure of an integral membrane light-harvesting complex from photosynthetic bacteria. Nature 374: 517–521

Meckenstock R, Brunisholz R and Zuber H (1992a) The light-harvesting core-complex and the B820-subunit from *Rhodopseudomonas marina*. Part I. Purification and characterization. FEBS Lett 311: 128–134

Meckenstock R, Krusche K, Brunisholz R and Zuber H (1992b) The light-harvesting core-complex and the B820-subunit from *Rhodopseudomonas marina*. Part II. Electron microscopic characterization. FEBS Lett 311: 135–138

Meckenstock R, Krusche K, Staehelin LA, Cyrklaff M, Brunisholz R and Zuber H (1994) The six-fold symmetry of the B880 light-harvesting membranes of *Rhodopseudomonas marina*. Biol Chem Hoppe-Seyler 375: 429–438

Mechler B and Oelze J (1978) Differentiation of the photosynthetic apparatus of *Chromatium vinosum*, strain D III. Analyses of spectral alterations. Arch Microbiol 118: 109–114

Michel H (1990) General and practical aspects of membrane protein crystallization. In: Michel H (ed) Crystallization of Membrane Proteins, pp 74–88. CRC Press, Boca Raton, Florida

Miller JF, Hinchingeri PS, Parkes-Loach PM, Callahan JR, Springle JR, Riccoboni JR and Loach PA (1987) Isolation and characterization of a subunit form of the light-harvesting complex of *Rs. rubrum*. Biochemistry 26: 5055–5062

Nunn RS, Artymiuk PJ, Baker PJ, Rice DW, Hunter CN (1992) Purification and crystallization of the light harvesting LH1 complex from *Rhodobacter sphaeroides*. J Mol Biol 228: 1259–1262

Papiz MZ, Hawthornthwaite AM, Cogdell RJ, Woolley KJ, Wightman PA, Ferguson LA and Lindsay JG (1989) Crystallization of the B800–850 light-harvesting complex from *Rp. acidophila* strain 10050 and the determination of its oligomeric state. J Mol Biol 209: 833–835

Parkes-Loach PS, Springle, JR and Loach PA (1988) Reconstitution of the B 873 light-harvesting complex of *Rhodospirillum rubrum* from separately isolated α- and β-polypeptides and bacteriochlorophyll a. Biochemistry 27, 2718–2727

Parkes-Loach PS, Michalski TJ, Bass WJ, Smith V and Loach PA (1990) Probing the bacteriochlorophyll binding site by reconstitution of the light-harvesting complex of *Rhodospirillum rubrum* with BChl *a* analogues. Biochemistry 29: 2951–2960

Pearlstein RM (1982). Exciton migration and trapping in photosynthesis. Photochem Photobiol 35: 835–844

Pearlstein RM (1991) Theoretical interpretation of antenna spectra. In: Scheer H (ed) Chlorophylls, pp 1047–1078. CRC Press, Boca Raton

Pearlstein RM (1992) Kinetics of exciton trapping by monocoordinate reaction centers. J Lumin 51: 139–147

Peters JD and Drews G (1983a) The transverse membrane orientation of the light-harvesting and reaction center polypeptides of Rhodopseudomonas capsulata. FEBS Lett 162: 57–60

Peters J and Drews G (1983b) Chemical crosslinking studies of the light- harvesting pigment-protein complex B800–850 of *Rhodopseudomonas capsulata*. Eur J Cell Biol 29, 115–120

Peters J, Takemoto J and Drews G (1983) Spatial relationships between the photochemical reaction center and the light-harvesting complexes in the membrane of *Rhodopseudomonas capsulata*. Biochemistry 22: 5660–5667

Peters J, Welte W and Drews G (1984) Topographical relationship of polypeptides in the photosynthetic membrane of *Rhodopseudomonas viridis* investigated by reversible chemical cross-linking. FEBS Lett 171: 267–270

Pullerits T and Freiberg A (1992) Kinetic model of primary energy transfer and trapping in photosynthetic membranes. Biophys J 63: 879–896

Richter P, Brand M and Drews G (1992) Characterization of LHI-minus and LHI-plus *Rhodobacter capsulatus* PufA mutants. J Bacteriol 174: 3030–3041

Robert B and Lutz M (1985) Structures of antenna complexes of several *Rhodospirallales* from their resonance *Raman spectra*. Biochim Biophys Acta 807: 10–23

Sauer K and Austin LA (1978) Bacteriochlorophyll-protein complexes from the light-harvesting antenna of the photosynthetic bacteria. Biochemistry 17: 2011–2019

Scherz A and Parson WW (1986) Interactions of bacteriochlorophylls in antenna chlorophyll-protein complexes of photosynthetic bacteria. Photosynth Res 9: 21–32

Schmidt K and Brunisholz R (1986) unpublished observation

Stark W, Kühlbrandt K, Wildhaber I, Wehrli E and Mühlethaler K (1984) The structure of the photoreceptor unit of *Rp. viridis*. EMBO J 3: 777–783

Stark W, Jay F and Mühlethaler K (1986) Localization of reaction center and light-harvesting complexes in the photosynthetic unit of *Rp. viridis*. Arch Microbiol 146: 130–133

Steiner R and Scheer H (1985) Characterization of a B800/1020 antenna from the photosynthetic bacteria *Ectothiorhodospira halochloris* and *Ec. abdelmalekii*. Biochim Biophys Acta 807: 278–284

Tadros MH and Waterkamp K (1989) Multiple copies of the coding regions for the light-harvesting B800–850 a- and b-polypeptides are present in the *Rp.. palustris* genome. EMBO J 8: 1303–1308

Theiler R and Zuber H (1984) The light harvesting polypeptides of *Rhodopseudomonas sphaeroides* R-26-1. II. Conformational analyses by attenuated total reflection infrared spectroscopy and possible molecular structure of the hydrophobic domain of the B 850 complex. Hoppe-Seyler's Z Physiol Chem 365: 721–729

Theiler R, Suter F, Wiemken V and Zuber H (1984) The light-harvesting polypeptides of *Rp. sphaeroides* R-26.1. I. Isolation purification and sequence analyses. Hoppe-Seyler's Z Physiol Chem 365: 703–719

Thornber JP (1986) Biochemical characterization and structure of pigment proteins of photosynthetic organisms. Encyl Plant Physiol New Series 19: 98–142

Thornber JP (1970) Photochemical reactions of purple bacteria as revealed by studies of three spectrally different caroteno-bacteriochlorophyll-protein complexes isolated from chromatium, strain D Biochemistry 9: 2688–2698

Thornber JP, Cogdell RJ, Pierson BK and Seftor REB (1983) Pigment-protein complexes of purple photosynthetic bacteria: an overview. J Cell Biochem 23: 159–169

Tonn SJ, Gogel GE and Loach PA (1977) Isolation and characterization of an organic solvent soluble polypeptide component from photoreceptor complexes of *Rhodospirillum rubrum*. Biochemistry 16: 877–885

Trüper HG and Pfennig N (19981) Characterization and identification of the anoxygenic phototrophic bacteria. In: Scheer H, Stolp MP, Trüper HG, Balows A and Schlegel HG (eds) The Prokaryotes: a Handbook on Habitats, Isolation and Identification of Bacteria, Vol. I, Springer Verlag, Berlin

Van Grondelle R (1985) Excitation energy transfer, trapping and annihilation in photosynthetic systems. Biochim Biophys Acta 811: 147–195

Van Grondelle R, Hunter CN, Bakker JGC and Kramer HJM (1983) Size and structure of antenna complexes of photosynthetic bacteria as studied by singlet-singlet quenching of the bacteriochlorophyll fluorescence yield. Biochim. Biophys Acta 723: 30–36

Van Mourik F, Van der Oord CJR, Visschers KJ, Parkes-Loach PS and Van Grondelle R (1991) Exiton interactions in the light-harvesting antenna of photosynthetic bacteria studied with triplet-singlet spectroscopy and singlet-triplet annihilation on the B820 subunit form of *Rhodospirillum rubrum*. Biochim Biophys Acta 1059: 111–119

Visschers RW, Chang MC, Van Mourik F, Parkes-Loach PS, Heller BA, Loach PA and Van Grondelle R (1991) Fluorescence polarization and low-temperature absorption spectroscopy of a subunit form of light-harvesting complex I from purple photosynthetic bacteria. Biochemistry 30: 5734–5742

Visschers RW, Nunn R, Calkoen F, Van Mourik F, Hunter CN, Rice DW and Van Grondelle R (1992) Spectroscopic characterization of B820 subunits from light-harvesting complex I of *Rhodospirillum rubrum* and *Rhodobacter sphaeroides* prepared with the detergent N octyl-rac-2 3-dipropylsulfoxide. Biochim Biophys Acta 1100: 259–266

Visschers RW, Van Mourik F, Monshouwer R and Van Grondelle R (1993) Inhomogeneous spectral broadening of the B820 subunit form of LH1. Biochim Biophys Acta 1141: 238–244

Vos M, Van Grondelle R, Van der Kooij FW, Van de Poll D, Amesz J and Duysens LNM (1986) Singlet-singlet annihilation at low temperatures in antenna of purple bacteria. Biochim Biophys Acta 850, 501–512

Wacker T, Gad'on N, Steck K, Welte W and Drews G (1988) Isolation of reaction center and antenna complexes from the halophilic purple bacterium *Rhodospirillum salexigens*. Crystallization and spectroscopic investigation of the B800–850-complex. Biochim Biophys Acta 933: 299–305

Wagner-Huber R, Brunisholz RA, Bissig I, Frank G and Zuber H (1988) A new possible binding site for bacteriochlorophyll *b* in a light-harvesting polypeptide of the bacterium *Ectothiorhodospira halochloris*. FEBS Lett 233: 7–11

Wagner-Huber R, Brunisholz R, Bissig I, Frank G and Zuber H (1992) The primary structure of the antenna polypeptides of *Ectothiorhodospira halochloris* and *Ectothiorhodospira halophila*. Four core-type antenna polypeptides in *Ec. halochloris* and *Ec. halophila*. Eur J Biochem 205: 917–925

Wechsler TD, Brunisholz RA, Suter F, Fuller RC and Zuber H (1985) The complete amino acid sequence of a bacteriochlorophyll *a* binding polypeptide isolated from the cytoplasmic membrane of the green photosynthetic bacterium *Chloroflexus aurantiacus*. FEBS Lett 191: 34–38

Wechsler TD, Brunisholz RA, Frank G, Suter F and Zuber H (1987) The complete amino acid sequence of the antenna polypeptide B806–866 β from the cytoplasmic membrane of the green bacterium *Chloroflexus aurantiacus*. FEBS Lett 210: 189–194

Welte W and Wacker T (1990) Protein-detergent unicellular solutions for the crystallization of membrane proteins: some general approaches and experiences with the crystallization of pigment-protein complexes from purple bacteria. In: Michel H (ed) Crystallization of Membrane Proteins, pp 118–123. CRC Press, Boca Raton, Florida

Welte W, Wacker T, Leis M, Kreutz W, Shiozawa J, Gad'on N and Drews G (1985) Crystallization of the photosynthetic light-harvesting pigment protein complex B800–850 of *Rp. capsulata*. FEBS Lett 182: 260–264

Wiemken V, Brunisholz RA, Zuber H and Bachofen R (1983). Topology of chromatophore membrane proteins studies with a chemical marker and with proteinases in *Rhodospirillum rubrum*. FEBS Microbiol Lett 16: 297–301

Zhang FG, Gillbro T, Van Grondelle R and Sundström V (1992) Dynamics of energy transfer and trapping in the light-harvesting antenna of *Rhodopseudomonas viridis*. Biophys J 61: 694–703

Zuber H (1985a) Structure and function of light-harvesting complexes and their polypeptides. Photochem Photobiol 42: 821–844

Zuber H (1985b) Structural organization of tetrapyrrole pigments in light-harvesting pigment-protein complexes. In: Blauer G and Sund H (eds) Optical Properties and Structure of Tetrapyrroles, pp 425–441. Walter de Gruyter, Berlin

Zuber H (1986) Primary structure and function of the light-harvesting polypeptides from cyanobacteria, red algae and purple photosynthetic bacteria. In: Staehelin LA and Arntzen CJ (eds) Photosynthesis III. Photosynthetic Membranes and Light-Harvesting Systems, Encyclopedia of Plant Physiol. New Series. Vol. 19, pp 238–251. Springer, Berlin

Zuber H (1987) The structure of light-harvesting pigment-protein complexes. In: Barber J (ed) The Light Reactions, Topics in Photosynthesis, Vol. 8, pp 197–259. Elsevier Science, Amsterdam

Zuber H (1993) Structural features of photosynthetic light-harvesting systems. In: Deisenhofer J and Norris JR (eds) The Photosynthetic Reaction Center. Vol, 1 pp 43–100. Academic Press

Zuber H and Brunisholz RA (1991) Structure and function of antenna polypeptides and chlorophyll-protein complexes: Principles and variability. In: Scheer H (ed) Chlorophylls, pp 627–703. CRC Press, Boca Raton

Zuber H, Brunisholz R and Sidler W (1987) Structure and function of light-harvesting pigment-protein complexes. In: Amesz J (ed) Photosynthesis, pp 233–271. Elsevier Science, Amsterdam

Chapter 17

Kinetics of Excitation Transfer and Trapping in Purple Bacteria

Villy Sundström
Department of Chemical Physics, Lund University, S-22100 Lund, Sweden

Rienk van Grondelle
Department of Biophysics, Free University of Amsterdam, De Boelelaan 1081, The Netherlands

Summary

The transfer and trapping of excitation energy in photosynthetic purple bacteria, as it appears from time-resolved and steady-state spectroscopy, are discussed. As a background to the dynamics we first briefly describe the structure and organization and spectroscopy of purple bacterial antenna pigments. The concept of spectral inhomogeneous broadening and its consequences for understanding the spectroscopy and dynamics is thoroughly discussed. We start the discussion of the dynamics by describing the overall energy equilibration and trapping processes in antenna-reaction center systems of varying complexity. These processes occur on the time scale of several picoseconds to tens of picoseconds. The most conspicuous results are that excitation energy is equilibrated very rapidly, ≤10 ps, over the entire antenna, but transfer of the energy from the antenna to the reaction center (the special pair) is a relatively slow process (30–40 ps), both at room temperature and 77 K. The reaction center is not a perfect trap for the excitation energy; at room temperature ~25% of the energy returns

R. E. Blankenship, M. T. Madigan and C. E. Bauer (eds): Anoxygenic Photosynthetic Bacteria, pp. 349–372.

to the antenna in *Rhodospirillum rubrum*, upon selective excitation of the reaction center pigments. At low temperatures (≤77 K) the back transfer to the antenna is negligible. With recently available subpicosecond and femtosecond laser pulses the most fundamental steps of energy transfer has been resolved. It is shown that energy transfer between a pair of bacteriochlorophyll molecules typically occur on the timescale 0.2–0.5 ps. On this extremely short timescale it becomes necessary to also take into account various intramolecular processes, like vibrational relaxation and vibrational energy redistribution. For such very fast energy transfer processes it may be necessary to question the applicability of conventional Förster theory of electronic energy transfer. In order to test particular structural and dynamical aspects of the energy transfer and trapping, molecular biology methods are invaluable tools to introduce specific changes into the pigment-protein systems. Several examples of advanced spectroscopy on genetically modified systems are given.

I. Introduction

Photosynthetic purple bacteria are particularly well suited for studying structure-dynamics-function relationships of the primary photosynthetic processes, since these organisms have relatively simple pigment systems. A variety of well characterized light-harvesting and reaction center preparations exist, and specific mutations can be performed to introduce well defined site-selective alterations of individual amino acids. During the last approximately five years there has been a very rapid development in time resolved studies of excited state dynamics in bacterial pigment systems. Before 1985 very few studies existed, but with the advent of low pulse energy high repetition rate dye lasers there has been a virtual explosion in the number of works using picosecond transient absorption and fluorescence to study energy transfer and trapping dynamics. At present, researchers are engaged in the second step of this revolution where it is now becoming possible to study dynamics in photosynthetic systems on a sub-100 femtosecond timescale, using femtosecond lasers to generate highly stable, very short (10–100 fs) and broadly tunable pulses. Simultaneously, there has been a strong development of high resolution non-photochemical spectral hole burning techniques as an alternative to study very fast dynamics in photosynthetic pigment systems at very low temperatures. This development has now brought us to the point where we very accurately can study energy and electron transfer dynamics over a wide timescale, representing the characteristic times ranging from that of energy or electron transfer processes between individual molecules to those of energy equilibration and overall trapping (representing a large number of individual steps).

Abbreviations: BChl – bacteriochlorophyll; *Cm.* – *Chromatium;* LH1, LH2 – light-harvesting complex 1 or 2; *Rb.* – *Rhodobacter;* RC – reaction center; *Rp.* – *Rhodopseudomonas; Rs.* – *Rhodospirillum*

II. Pigment-Protein Organization in LH1 and LH2

Purple photosynthetic bacteria contain bacteriochlorophyll *a* (BChl *a*) or bacteriochlorophyll *b* (BChl *b*) as the dominant pigment in the light-harvesting antenna and the reaction center (RC). Of the BChl *a*-containing bacteria many different species have been characterized, whereas for BChl *b*-containing bacteria only *Rhodopseudomonas (Rp.) viridis* has been reasonably well studied. The high-resolution crystallographic structures of *Rp. viridis* (Deisenhofer et al. 1984; Deisenhofer et al., 1986) and *Rb. sphaeroides* (Allen et al., 1987a,b; Yeates et al., 1987) reaction centers suggest that reaction centers of different purple bacteria are quite similar. The reaction center is surrounded by the LH1 core antenna, which consists of pairs of small transmembrane polypeptides (α and β) to which BChl is non-covalently attached in a 1:1 stoichiometry (Zuber and Brunisholz, 1991; Brunisholz and Zuber, 1992). Most likely, the highly conserved histidines in the α and β polypeptides are used as 5th ligand (Robert and Lutz, 1985; Zuber, 1985; Zuber and Brunisholz, 1991; Brunisholz and Zuber, 1992). The BChl *b* containing bacteria have an additional γ-polypeptide, which is not involved in pigment binding (Brunisholz et al., 1985). In vivo, the LH1-RC core has a constant number of 24 BChl per RC (12α and 12β), although some uncertainty and variation in this number is possible (Aagard and Sistrom, 1972; Dawkins et al. 1988; Beekman et al., 1993). Electron micrographs of membranes of *Rp. viridis* and other BChl *b*-containing species (Miller, 1982; Engelhardt et al., 1983; Stark et al., 1984; Stark et al., 1986) and, recently, the BChl *a*-containing species *Rp. marina* (Meckenstock et al., 1992b) show that LH1 surrounds the RC in a circular way. The diameter of the RC-LH1 complex was estimated to be about 13 nm for the BChl *b*-containing and 10 nm for the BChl *a*-containing species; the difference is possibly due to the absence of the γ-polypeptide in the latter. A recent electron microscopy and image analysis study of purified RC-LH1 core particles of *Rhodospirillum (Rs.) molischianum* resulted in very similar structure, in which the 12 $\alpha\beta$-pairs can be distinguished (Boonstra et al., unpublished). Ring-like structures with approximately the same dimensions have also been observed for concentrated solutions of purified LH1 (Meckenstock et al., 1992b). For the purified LH1 complex isolated from *Rb. sphaeroides* M2192 in detergent solution (octylglucoside) the ring may have collapsed and the particle appears to consist of two closely linked $(\alpha\beta)_6$ units (Boonstra et al., 1993). All these results suggest models for the organization of the α and β polypeptides in the membrane in which the minimal structural unit is a $(\alpha\beta BChl_2)$ complex (Zuber, 1985; Hunter et al., 1989; see also Chapter 16 by Zuber and Cogdell and Chapter 22 by Hunter in this volume). These minimum units form a circular structure with the centers of all BChl pigments and their Q_y-transition moments in a single plane parallel to the membrane (Breton and Vermeglio, 1982; van Grondelle, 1985). Spectroscopic measurements have shown that the RC-LH1 cores aggregate further into extended networks containing more than 10 RCs over which, at least at room temperature, the excitation can freely diffuse (Vredenberg and Duysens, 1963; Clayton, 1966; Bakker et al., 1983; Kingma et al., 1983; Vos et al., 1986; Vos et al., 1988; Deinum et al., 1989; Trissl et al., 1990; Deinum et al., 1991, 1992a). At low temperature the excitation diffusion is restricted to, possibly, a single RC-LH1 core (Vos et al., 1988; Deinum et al., 1991; van Mourik et al., 1993).

Various models have been suggested for the detailed structural organization of the α and β polypeptides of the LH1 antenna (Theiler et al., 1984a,b, Theiler and Zuber, 1984c; Hunter et al., 1989; Scherz and Rosenbach-Belkin, 1989, Sherz et al., 1990; Braun and Sherz, 1991; Zuber and Brunisholz, 1991; Brunisholz and Zuber, 1992). It is believed that pairs of α and β-polypeptides form a minimum unit which aggregates further to form the fully functional in vivo antenna. Such a minimal unit was first isolated by Loach and coworkers (Loach et al., 1985; Miller et al., 1987; Parkes-Loach et al., 1988) and spectroscopically analyzed in detail (see section III.D). A molecular weight determination of this so-called B820 subunit by gel filtration showed that B820 had a considerably lower molecular weight than the reassociated B873 form. Similar subunits could be purified from various other photosynthetic purple bacteria (Ghosh et al., 1988; Chang et al., 1990; Heller and Loach, 1990; Meckenstock et al., 1992a; Jirsakova et al., 1992). Recently, a B820 subunit was prepared from *Rp. marina* (Meckenstock et al., 1992a) with a reported molecular mass of 33 kDa, about 1/6 of the molecular mass of purified LH1 from the same species. From these biochemical results the $(\alpha\beta)_2$-form was considered to be the most likely aggregation form for B820 (Meckenstock et al., 1992a,b). However, a careful evaluation of all the molecular weight measurements strongly suggests that B820 arises from a single $\alpha\beta$-pair, in agreement with the spectral properties (see Chapter 21 by Loach and Parkes-Loach).

Several species of purple bacteria contain, in addition to the LH1 core antenna, a peripheral LH2 antenna at higher excited state energy with major BChl absorption between 820 and 860 nm depending on the species, and a second major BChl *a* Q_y-transition around 800 nm (see Hawthornthwaite and Cogdell, 1991 for an overview). The ratio of LH2 to LH1/RC depends strongly on the growth conditions. For instance, in *Rb. sphaeroides* grown under low-light conditions LH2 is by far the dominant spectral species. It is now well established that some species (*Rp. palustris*, *Rp. acidophila*, *Cm. vinosum*) have the ability to change the antenna polypeptide composition (and consequently their absorption) in response to light and/or temperature (Zuber and Brunisholz, 1991; Hawthornthwaite and Cogdell, 1991; Brunisholz and Zuber, 1992). Simultaneously, they can adapt the relative amount of pigment absorbing around 800 nm. LH2 may have a complex carotenoid composition and the BChl to carotenoid ratio may vary between 1:1 and 3:1 (Thornber et al., 1983; Hawthornthwaite and Cogdell, 1991); for instance in the case of LH2 of *Rb. sphaeroides* this ratio is 2:1 (Evans et al., 1988). The LH2 protein backbone also consists of transmembrane α and β-polypeptides, which bind three BChl *a* per $\alpha\beta$-pair.

The organization of the pigments in the LH2 of *Rb. sphaeroides* (B800–850) was extensively studied (Bolt and Sauer, 1979; Breton et al., 1981; van Grondelle et al., 1982; Kramer et al., 1984a) and detailed arrangements of the pigments have been proposed (see below), although up to now no agreement exists about the aggregation state and minimum functional unit ($\alpha\beta$, $\alpha_2\beta_2$, etc.). A unit composed of $(\alpha\beta)_2$ containing 6 BChl *a* molecules and three carotenoids explains most of the steady-state and time-resolved spectroscopic properties of LH2 (Kramer et al., 1984a; van Grondelle, 1985). Zuber and coworkers (Zuber, 1985; Zuber and Brunisholz, 1991; Brunisholz and Zuber, 1992) have proposed the cyclic $(\alpha\beta \text{BChl } a_3)_6$ structure based on structural considerations. An electron microscopic and image analysis study of LH2 from *Rb. sphaeroides* isolated in the detergent lauryl dimethyl amine oxide (LDAO) gave similar result (Boonstra et al., 1993). Crystals of the B800–850 LH2 complex of *Rp. acidophila* suggest a trimeric $(\alpha\beta)_2$ structure as the basic unit (Hawthornthwaite et al., 1989), which are in their turn arranged in a circle, in agreement with a suggestion originally proposed by Scherz and co-workers (Scherz and Rosenbach-Belkin, 1989, Sherz et al., 1990; Braun and Sherz, 1991).

In vivo, light absorbed by LH2 is efficiently transported to the RC probably via LH1, although a mutant of *Rb. sphaeroides* lacking LH1 but containing LH2 and RCs shows trapping kinetics closely similar to those observed for the RC-LH1 species (Hess et al., 1993a). For *Rb. sphaeroides* LH2 connects cores of LH1/RC units, containing at most a few RCs (Monger and Parson, 1977; Vos et al., 1986). Recently, it was shown for *Rp. acidophila* that the LH2 antenna interconnects individual LH1/RC units. In case of a B800–820 connecting antenna, the LH1/RC units were effectively isolated (even at room temperature), while in case of a B800–850 antenna excitations were able to migrate between different LH1/RC units (Deinum et al., 1991).

III. Spectroscopy of the LH1 and LH2 Light-Harvesting Antennae

A. *In vivo* Spectral Properties of LH1 and LH2

The major BChl (*a* or *b*) absorption in LH1 and LH2 is strongly red-shifted relative to the absorption of BChl in organic solvents. In case of BChl *a*-containing LH1, room temperature absorption maxima are observed at about 870–880 nm, which shift even further to the red upon lowering the temperature. For some species, e.g. *Ectothiorhodospira mobilis* (Trüper, 1968; Leguijt et al., 1992) and *Cm. tepidum* (Garcia et al., 1986), the long wavelength maximum may be above 900 nm. For each species the precise maximum depends somewhat on the particular conditions; for instance the absence (or type) of carotenoid in LH1 of *Rb. sphaeroides* may shift the absorption maximum almost 10 nm to the blue (Parkes-Loach et al., 1988; Olsen unpublished). For the BChl *b* containing species the major absorption at room temperature is around 980 nm, which sharpens and shifts to about 1035 nm upon cooling to 77 K (or 4 K).

Typically, the major LH2 absorption in purple bacteria is between 820 and 860 nm with a second peak around 800 nm. For instance, LH2 of *Rb. sphaeroides* absorbs at about 800 and 850 nm; LH2 of *Rp. acidophila* 7050, grown under dim light conditions absorbs maximally around 800 and 820 nm (Zuber and Brunisholz, 1991; Brunisholz and Zuber, 1992; Hawthornthwaite and Cogdell, 1991). The precise location of the peaks is also here sensitive to the presence and type of carotenoid. Recently, Fowler et al. (1992) succeeded in blue shifting the major absorption transition of LH2 of *Rb. sphaeroides* by site-directed mutation of the double tyrosine motif (αTyr44, Tyr45) in the α-polypeptide. The single (αTyr44, Tyr45 $\rightarrow$ Phe, Tyr) and double (αTyr44, Tyr45 $\rightarrow$ Phe, Leu) mutants absorbed around 838 and 826 nm, respectively. This mutation was inspired by the PheLeu motive in the 800–820 complex of *Rp. acidophila* and showed that this pair of residues is indeed responsible for the spectral tuning, possibly by the formation of an H-bridge between αTyr44 residues and the c-2-acetyl group of one of the BChl molecules (Fowler, unpublished).

Both LH1 and LH2 are believed to have a quite similar organization of their pigments. Consequently, their spectroscopic properties are also closely related. In particular, the organization of the BChl *a* of LH2 absorbing at 820–860 nm is believed to be rather similar to that of BChl *a* of LH1, with a dimer of BChl *a* as basic structure. All BChl molecules in LH1 and LH2 have their major Q_y transitions in the plane of the membrane (Bolt and Sauer 1979, 1981; Bolt et al., 1981; Breton et al., 1981; Kramer et al., 1984a,b). Within this plane there is, however, no preferential orientation. The Q_x-transitions belonging to B870 in LH1 and B820–850 in LH2 absorb around 590 nm and are oriented perpendicular to the membrane plane (or make at most a small angle with the normal). In LH2 of *Rb. sphaeroides* the

B800:B850 ratio is 1:2, but the B800 peak is markedly sharper than the B850 peak and asymmetric in low temperature absorption spectra (Clayton and Clayton, 1981; van Grondelle et al., 1982). The B800 pigment is assumed to be 'monomeric' and is possibly not liganded to histidine (Robert and Lutz, 1985), but rather weakly attached to the LH2 structure (Clayton and Clayton, 1982; Kramer at al, 1984a; Chadwick et al., 1987). The B800 absorption shows a predominant orientation in the plane (Breton et al., 1981; Kramer et al., 1984a). Species with an increased absorption around 800 nm (e.g., *Rp. palustris*) usually exhibit complex CD, LD and polarized fluorescence excitation spectra reflecting stronger interactions between the B800 pigments (Hayashi et al., 1982a,b,c; Leguijt et al., 1992; van Mourik et al., 1992c).

B. Site Inhomogeneous Broadening in LH1 and LH2

Previously, the low temperature (<100K) spectroscopic and excitation transfer properties of the purple bacterial antenna have been described in terms of the 'two-pool' hypothesis, in which the LH1 spectrum of *Rb. sphaeroides* was proposed to be composed of a major form (B880) and a minor long-wavelength form (B896) (Borisov et al., 1982; Kramer et al., 1984b; Bergström et al., 1988a; Deinum et al., 1989; Shimada et al., 1989; Hunter et al., 1990; Reddy et al., 1992a). Fast excitation energy transfer was proposed to take place from B880 to B896, leading to strong depolarization of the fluorescence. Similarly, for LH2 of *Rb. sphaeroides* the existence of a major B850 and a minor B870 form has been forwarded (van Dorssen et al., 1988; Reddy et al., 1992a).

Recently, however, low-temperature fluorescence (van Mourik et al., 1992a; Visschers et al., 1993), time-resolved fluorescence (Timpmann et al., 1991; Pullerits et al., 1994) and absorption (van Mourik et al., 1993) and spectral hole-burning experiments (van der Laan et al., 1990; Reddy et al., 1991, 1992a,b) have demonstrated that the LH1 and LH2 absorption spectra are in fact inhomogeneously broadened. Van Mourik et al. (1992a) applied narrow-banded laser excitation on purified LH1 complexes at 4 K and measured as a function of the excitation wavelength (λ_{exc}) the position of the LH1 emission maximum (λ_{max}) and the fluorescence polarization. It was observed that λ_{max} was unchanged at 904 nm for $\lambda_{exc} \leq 886$ nm. For $\lambda_{exc} > 886$ nm, λ_{max} shifted linearly with λ_{exc} to the red and the polarization increased from about 0.1 to close to maximal (0.5). Similar results were also obtained with purified LH2 (Visschers et al., 1993). For LH1-RC complexes and intact membranes of *Rp. viridis* the increase in fluorescence polarization in the long-wavelength part of the absorption spectrum (1050 nm) was also detectable using narrow-banded laser excitation and occurred only upon excitation in the very red edge (Monshouwer et al., unpublished).

These results are not consistent with the two-pool hypothesis but can be fully explained by assuming inhomogeneous broadening as the dominant factor in determining the shape of the LH1 and LH2 absorption spectra. Within the framework of this concept, LH1 and LH2 may be regarded as small clusters of weakly coupled BChl *a* dimers among which fast excitation energy transfer takes place. Each individual cluster is supposed to be a random sample from the total inhomogeneously broadened pigment pool. For small clusters, with efficient energy transfer leading to rapid thermalization of the excited state, the effects of site-selection depend on the cluster size only and, consequently, site-selective excitation yields information about the number of coupled BChl *a* molecules in a cluster. It was demonstrated that this model explains the observed emission polarization excitation spectrum unequivocally (van Mourik et al., 1992a). The cluster-size dependence of the inhomogeneously broadened LH1 antenna was confirmed by the correlation of the value of the excitation wavelength at which the fluorescence polarization started to increase with the LH1 cluster size (van Dorssen et al., 1988; Visschers et al., 1993).

Spectral holeburning experiments at 1.2 K have recently been reported for LH1 and LH2 of *Rb. sphaeroides* (Reddy et al., 1991, 1992a) and *Rp. acidophila* (Reddy et al. 1992b). Excitation in the middle or blue wing of the main LH1 (~875 nm) or LH2 (~850 nm) absorption bands only resulted in broad, featureless spectra, whereas excitation of the red wings of these absorption bands resulted in narrow holes (3.2 cm^{-1}), implying a 5–10 ps lifetime. Plotting burning efficiency as a function of laser frequency mapped out a profile with a width of 70 cm^{-1} peaking around 895 nm in LH1 and around 870 nm in LH2. The narrow (5–10 ps) hole was always accompanied by a rather broad feature, with a width of about 200–210 cm^{-1}, close to half the bandwidth of the original LH1/2 absorption band, and a strong absorption increase (anti-hole) in the blue wing. In contrast to the 'weak coupling of dimers' model mentioned above (van Mourik et al., 1992a), the B850 results were interpreted in terms of a 'strong exciton coupling' model in which all ($\alpha\beta BChl_2$) pairs are organized in a cyclic unit cell and strongly exciton coupled. Further (weaker) coupling was proposed to occur between the interacting cells (note that such a view has also been proposed for the spectroscopic properties of the BChl *a*-containing Fenna-Matthew-Olson (FMO) complex of green sulphur bacteria (Johnson and Small, 1989, 1991)). Within the context of this model, B896 and B870 are the lowest exciton states of the LH1 and LH2 unit cells and the 70 cm^{-1} profile reflects the inhomogeneous linewidth. In contrast to the FMO complex, no further exciton level structure was observed on the ground state and hole-burning spectrum. The broad profile observed in the holeburning spectrum was ascribed to a special form of homogeneous broadening caused by the close spacing of the many exciton levels, the lifetime broadening of each exciton level due to fast inter-exciton level relaxation (~200 fs) and disorder (Johnson and Small, 1991; Reddy et al., 1992a,c). Alternatively, the excitation profile of the narrow hole could be related to the probability of exciting a lowest energy state in a cluster. It is of interest to note that the limiting wavelength for observing the narrow hole in LH1 and LH2 coincides with the onset wavelength of the fluorescence redshift and polarization increase observed in fluorescence site-selection experiments described above (van Mourik et al., 1992a; Monshouwer et al., unpublished). In that case the broad hole arises from a broad phonon wing accompanying the zero-phonon hole obtained at the burning wavelength (Pullerits et al., 1994; Pullerits et al., unpublished).

Holeburning experiments on LH2 complexes from various organisms have shown that the B800 absorption band is inhomogeneously broadened. The holewidth in the red edge of the band yields a linewidth that corresponds with the estimated time-constant for B800 $\rightarrow$ B850 excitation energy transfer (see below), which is 2.4 ps (Γ_{hom} = 70 GHz (van der Laan et al., 1990; Reddy et al., 1991, 1992a,b; van der Laan et al., 1993) in LH2 of *Rb. sphaeroides* between 1.2 and 30K. If the burning wavelength is tuned to the blue edge of the band the holes become progressively broader, reaching a value of 250–300 GHz upon burning at 795 nm or lower (De Caro et al., unpublished). In the case of blue excitation in LH2 of *Rp. acidophila* a broad hole is obtained in the red edge of the B800 band. All these features are interpreted as resulting from subpicosecond downhill energy transfer within the B800 band, consistent with steady-state fluorescence polarization measurements (Kramer et al., 1984a) and with more recent time-resolved data (Visscher et al., 1993a, 1993b; Hess et al., 1993b).

C. Stark Spectroscopy of LH1 and LH2

Stark spectroscopy of BChl *a* in films has yielded a value for the difference (between excited and ground state) dipole moment of $|\Delta\mu_A|$ = 1.3–2.0 D/f (Gottfried et al., 1991), in which f is the local field correction factor. For BChl *a* in the FMO complex rather similar values were obtained, suggesting that the BChl *a* molecules in FMO behave as a set of non-interacting pigments with respect to the electric field.

For LH2 of *Rb. sphaeroides* the B800 band has $|\Delta\mu_A|$ = 0.8–0.9 D/f, while the B850 band shows an unusually large $|\Delta\mu_A|$ = 3.4 D/f (Gottfried et al., 1991). Stark spectra of both components have anomalous lineshapes and the spectra do not match the second derivative of the absorption spectrum (as expected for a single transition with a single value for $\Delta\mu_A$). The complex spectrum may (in part) be explained by assuming that red-shifted species are hidden under the LH2 absorption profile with enhanced charge transfer character. The amplitude of the B850 Stark spectrum shows that the BChl *a* chromophores are strongly coupled, giving rise to enhancement of $|\Delta\mu_A|$. $|\Delta\mu_A|$ for B850 is not affected by the disappearance of B800 upon lithium dodecyl sulfate (LDS) treatment. The B800 Stark effect is consistent with the idea that B800 represents weakly coupled BChl *a*, susceptible to perturbations in the external environment. Surprisingly, LH2 shows an unprecedented large decrease in fluorescence yield in an applied electric field. It was suggested that the proposed charge transfer character of (part of) the B850 absorption is responsible.

For B875, the Stark absorption and emission spectra are dominated by a strong first derivative contribution (Gottfried et al., 1991). Fitting the Stark absorption spectrum to a sum of derivative components yielded $|\Delta\mu_A|$ = 3.3 ± 0.2 D/f and in addition a very large polarizability difference. Although the Stark absorption spectrum does not hint at special red states in the B875 profile, it is still possible that the observed effects are due to overlap of transitions within the LH1 absorption band, with different responses to the applied electric field. In comparison with the special pair of the RC ($|\Delta\mu_A| = 7$ D/f) (Lockhart and Boxer, 1988) the Stark effect is 'weak', which may indicate that for LH1 not all of the redshift is due to an increase in electronic coupling, but rather to 'local' effects (see next section).

D. Spectral Properties of the B820 LH1 Subunit and the Nature of the Redshift

The isolation of a subunit form, B820, of LH1 of *Rs. rubrum* and *Rb. sphaeroides* (Loach et al., 1985; Miller et al., 1987; Parkes-Loach et al., 1988; Gosh et al., 1988; Chang et al., 1990; Heller and Loach, 1990; Meckenstock et al., 1992a; Jirsakova et al., 1992; Visschers et al., 1992) has added significantly to our understanding of the spectroscopy of LH1 in relation to the aggregation state. From absorption, CD and fluorescence polarization spectra it was concluded that two excitonically interacting BChl *a* molecules are responsible for the spectral properties of B820. Recent triplet-minus-singlet spectroscopy, singlet-triplet annihilation and fluorescence site-selection studies (van Mourik et al., 1991; Visschers et al., 1992, 1993a) have demonstrated that the B820 spectral properties are those of a BChl *a* dimer, with little or no interaction between different dimers. The dimer was proposed to have a head-to-tail arrangement, putting most of the oscillator strength in the red-shifted transition (Visschers et al., 1991). The high-energy component could only be observed by a small dip in the polarized excitation spectrum around 790 nm. The 'monomer' transition is located at 808 nm, which indicates that in B820 more than 2/3 of the redshift (from 777 nm) is non-excitonic. Within the B820 dimer the monomer transition dipoles are quasi-equivalent and they make a small angle of about 20° to account for the observed CD (Visschers et al., 1991; Koolhaas et al., unpublished). Low-temperature site-selected fluorescence experiments (Visschers et al., 1993) have shown that the B820 absorption band is inhomogeneously broadened and in fact behaves like a 'single state' system, as expected for a strongly coupled dimer. In addition, the major B820 absorption band is narrower with a factor of about $\sqrt{2}$ than the estimated monomer transition, which is consistent with the idea that each B820 dimer is composed of two BChl *a* monomers that are randomly selected from the total inhomogeneous distribution.

Recently, the time-resolved association of B820 into B875 using stopped flow has strongly suggested that only two B820 units are required to form B875 (van Mourik et al., 1992b). A spectroscopic intermediate could not be detected and further aggregation only marginally affected the spectrum. The CD spectrum of the reassociated B873 is, apart from the red-shift, not very different in shape and intensity from that of B820, which suggests that the interactions within the dimer have not been altered much during the aggregation. These experiments have been taken to imply that the nature of the red-shift of BChl in LH1 (and probably also in LH2) is largely non-excitonic. This is supported by low-temperature triplet-minus-singlet spectra (van Mourik et al., 1993), which appear to be red-shifted copies of the B820 triplet-minus-singlet spectrum and do not show the appearance of a strong monomer absorption band far to the blue of the original LH1 absorption.

An earlier model that was proposed to explain the in vivo LH1 and LH2 spectra was based on model studies, which showed that artificially prepared dimers of BPheo *a* have absorption and CD spectra that are strongly reminiscent to those of LH1 and LH2 (Scherz and Parson, 1984, 1986; Scherz and Rosenbach-Belkin, 1989; Scherz et al., 1990). On the basis of these results it has been suggested that the major species of LH1 and LH2 consists of dimers of BChl *a* in a configuration similar to that of the special pair of the bacterial RC. The red-shift would be purely excitonic. The strong CD of LH1 and in particular LH2 was ascribed to much weaker interaction between various dimers. However, also in the case of the special pair of the RC, the origin of the redshift is certainly not purely excitonic, since even the proposed high energy exciton component (Breton, 1985) is still very red-shifted relative to the absorption of BChl *a* in organic solvents. Experiments by Picorel and co-workers (Picorel et al., 1986; Gingras and Picorel, 1990) provide further evidence that this model is not correct. It was demonstrated that the cross-section for γ-radiation inactivation of the spectroscopic unit of LH1 corresponds with the size of the ($\alpha\beta$-BChl$_2$) pigment-protein complex. The product of the inactivation has the size of a single polypeptide plus one molecule of BChl *a*. Despite its small size, the product formed by the radiation inactivation of LH1 still has a highly red-shifted absorption spectrum.

Then what is the cause for the redshift observed in LH1 and LH2 and how is it related to the redshift of B820 upon formation of B873? In the B820 complex a molecule of the detergent octylglucoside may occupy the position of the carotenoid (Visschers et al., 1993 b). Consequently, the interaction between the detergent molecule and BChl *a* may have some effect on the BChl *a* absorption band. As discussed earlier, Fowler et al. (Fowler et al., 1992) have shown that in LH2 of *Rb. sphaeroides* the B850 absorption band can be shifted more than 20 nm to the blue upon mutation of a Tyr pair in the α-polypeptide. Resonance Raman spectra have shown that the spectral shift upon formation of B820 and in LH2 upon mutagenesis of the Tyr pair in the α-polypeptide are accompanied by dramatic changes in the H-bonding pattern of the BChl molecules (Visschers et al., 1993b; Fowler et al., unpublished). Similar effects have been observed in LH1 (Olsen, unpublished). These observations are in line with the view that non-excitonic factors, such as charge effects (Davis, et al., 1981; Eccles and Honig, 1983), deviations in planarity of the porphyrin ring system (Gudowska-Nowak et al., 1991) or the change in dielectric constant in close environment of the pigments (Altmann et al., 1992) may largely determine the site-absorption of the pigments. Similar ideas have been put forward for the interpretation of the spectrum of the FMO-complex (Gudowska-Nowak, et al. 1991; Pearlstein, 1992; Lu and Pearlstein, 1993).

IV. Energy Transfer in LH2-Less Purple Bacteria

A. Rhodospirillum rubrum *and* Rhodobacter sphaeroides *(LH1-RC Only Mutants)*

Rs. rubrum and RC-LH1-only mutants of *Rb. sphaeroides* may be regarded as photosynthetic systems in which excitation energy transfer and trapping occur in a 'homogenous' core antenna. The original results of Vredenberg and Duysens (1963) have shown that at room temperature such systems should be viewed as 'lakes' in which many reaction centers are connected through very efficient excitation energy transfer. These ideas were further substantiated by singlet-singlet and singlet-triplet annihilation experiments (Monger and Parson, 1977; Paillotin et al., 1979; Bakker et al., 1983; van Grondelle et al., 1983; Vos et al. 1986).

The actual time required to trap an excitation by photochemically active ('open') reaction centers at room temperature was measured to be 50–70 ps with both picosecond absorption (Nuijs et al., 1985, 1986; Sundström et al., 1986) and fluorescence (Freiberg et al., 1984, 1987; Borisov et al., 1985) techniques. In 'closed' centers (with oxidized primary electron donor), the excited state lifetime lengthens to about 200 ps, showing that even in the state P^+ the reaction center efficiently quenches the excitations in the antenna. Picosecond absorption kinetics (Sundström et al., 1986) also displayed an additional 20–50 ps decay component, which at that time was ascribed to the equilibration of excitation density among different spectral forms in the LH1 antenna, but also may have included singlet-triplet annihilation (Freiberg et al., 1990; Valkunas et al. 1991; Timpmann et al., 1993).

1. Trapping and Detrapping of the Excitation Energy

From the functional dependence on distance, orientation and spectral parameters of energy transfer it can be expected that the rate and efficiency of trapping and detrapping (escape) of excitation energy by the reaction center is a sensitive probe of pigment organization and relative excited state energies of antenna and special pair. Earlier models based on homogeneous lattices have suggested a fast transfer rate from neighbouring LH1 to P and a high escape probability for an excitation arriving on P, as a consequence of the near degeneracy of P and LH1 and of the assumed fast transfer rate from neighboring LH1 to P (Duysens, 1979; Bakker et al., 1983; Borisov et al., 1985). On the other hand, early measurements of fluorescence excitation spectra of purple bacteria in which the reaction center absorption was well separated from the major LH1 band indicated a relatively low probability of observing the LH1 fluorescence through excitation of the reaction center, also at room temperature (Olson and Clayton, 1986; Wang and Clayton, 1971). These results were recently confirmed and extended at low temperature (Otte et al., 1993) where the escape probability was found to be <5% in all cases.

Selective picosecond excitation and probing in the red wing of the LH1 absorption band has shown that in the temperature range 100–200 K energy is transferred to the RC with a time-constant of 35–40 ps (Bergström et al., 1989; Visscher et al., 1989), which was considered to reflect excitation transfer from (one of) the nearest neighbor(s) of the RC to P. At room temperature, a slightly shorter time constant was estimated. The weak temperature dependence suggests that LH1 is nearly resonant with P; the localization of the excitation energy on P, however, may be slightly more favorable at lower temperatures due to the spectral inhomogeneity of the LH1 antenna (see below). Thus, the excitations may be spread more homogeneously over all pigments at higher temperatures. From the measured rate constant a distance of about 3 nm between the nearest neighbor and P was calculated, in reasonable agreement with the distance estimated on the basis of the dimensions of the RC. Similar results were obtained by Timpmann et al. (1991) from a picosecond fluorescence measurement at 77 K. These results on trapping/detrapping by the reaction center were further substantiated by measuring the amplitude of antenna bleaching or antenna fluorescence decay components in a time resolved experiment with selective reaction center excitation, and comparing it with the corresponding amplitudes with direct antenna excitation (Timpmann et al., 1993). It was concluded that with photochemically active reaction centers in the state PIQ, excitation energy is transferred back to the antenna with a yield of 25 ± 5% at room temperature. The corresponding figure for reaction centers in the state PIQ^- was 40 ± 5%. With this measured yield, the antenna excited state lifetime of 60 ps, and the charge separation time of 3 ps, a simple kinetic model simulation showed that trapping (i.e. energy transfer from antenna to the reaction center) occurs with a time constant of 35–40 ps, and that detrapping occurs with a characteristic time constant of 7.5 ± 1.5 ps. Assuming a six-fold antenna coordination of the reaction center, these overall time constants correspond to pairwise trapping and detrapping time constants of approx. 35 and 45 ps, respectively (Timpmann et al., 1993). For a monocoordinate reaction center, as has been suggested on the basis of shape and symmetry of the reaction center (Pearlstein, 1992a), the pairwise rates would be a factor of six faster than this.

That energy transfer from the antenna to the reaction center is a slow rate limiting step in the exciton dynamics was recently also shown in experiments by Beekman et al. (1994) who studied the rate of excitation trapping in RC-LH1-only mutants of *Rb. sphaeroides* in which the Tyr M210 residue in the RC was mutated into Phe, Leu or His. The first two of these mutations induced a reduction of the rate of charge separation in the isolated reaction center by a factor of 3–5, probably by raising the energy level of P^+I^- relative to that of P^*. However, the trapping time increased only by about 15% in membranes of these mutants, while on the other hand a phase in the decay of the LH1 excited state reflecting the electron transfer from I^- to Q increased in amplitude. Consequently, we suggest that the excitation transfer from a neighbouring LH1 to P and not the rate of charge separation dominates the trapping time. This would make the trapping in LH1-RC cores of photosynthetic purple bacteria 'diffusion limited', although the rate of excitation diffusion in LH1 itself is exceptionally high.

A comparison of the trapping time obtained from the measurements of Timpmann et al. (1993) with that predicted from a homogeneous random-walk model (Pearlstein, 1982) suggested that the LH1 antenna in the energy migration process behaves as if it were organized in clusters of BChl *a* molecules. The number of BChl molecules in each cluster was concluded to be at least two and could be as large as four, which would correspond to $\alpha\beta BChl_2$ to $(\alpha\beta BChl_2)_2$ as a minimum unit in this respect. Considering the difficulties associated with calculations of spectral overlap integrals and combining all available results on this matter (see above), it seems likely that $\alpha\beta BChla_2$ constitutes the minimum pigment-protein unit of the LH1 antenna.

The detrapping yield in *Rp. viridis* was also obtained, by measuring the time resolved antenna fluorescence with selective reaction center excitation and comparing the amplitude of fluorescence decay to the corresponding decay with direct antenna excitation (Timpmann et al., 1995). From the observed detrapping yield of 20 ± 5% at room temperature with photochemically active reaction centers (PI), it was concluded that trapping and detrapping dynamics in *Rp. viridis* is quite similar to that in the BChl *a*-containing purple bacteria.

Spectrally redshifted minor LH pigments have often been suggested to play a role in the trapping process, as specialized pigments concentrating the energy to the vicinity of the reaction center and making an optimized contact/entry between the reaction center and the antenna (Deinum et al., 1989; Bergström et al., 1988; Hunter at al, 1990). The need for such specialized contact pigments was examined by studying the trapping dynamics in a mutant of *Rb. sphaeroides* containing reaction centers and LH2 as the sole LH pigment, i.e. lacking the LH1 core antenna (Hess et al. 1993a). The fact that the overall trapping time at room temperature was identical within experimental error to the trapping time in *Rb. sphaeroides* WT, and only approx. two times longer at 77 K, suggested that the energy transfer from the antenna to the special pair primarily is controlled by the spectral overlap between the participating pigments, and that specialized antenna-RC contact pigments only play a minor role in the trapping process.

It has already been mentioned that the oxidized primary donor P^+ is an efficient quencher of excitation energy from the antenna; in most purple bacterial species an antenna exciton lifetime approximately four times the lifetime with active reaction centers, i.e. 200–250 ps, is observed at both room temperature and 77 K (Sundström et al. 1986; Borisov et al. 1985; van Grondelle et al. 1987), but the mechanism of this efficient quenching is poorly understood. By measuring the P^+ quenching rate in a mutant of *Rb. sphaeroides* containing LH2 as the only antenna pigment, it was shown that this quenching is a factor of 1.5–2 less efficient for excitation energy delivered by LH2 as compared to LH1 (Hess et al. 1993a). It was suggested that the spectral overlap between antenna fluorescence and the broad P^+ absorption in the near-infrared region (~900–1300 nm) constitutes the basis for a relative efficient Förster energy transfer from the antenna to P^+. The fact that the primary donor is preoxidized and cannot transfer electrons implies that the excitation energy is dissipated as heat in a radiationless process, or emitted as infrared fluorescence (>1300 nm) which will be difficult to detect with the detectors conventionally used.

2. Spectral Inhomogeneity of LH1

Early low temperature fluorescence measurements have indicated that below 100K the fluorescence yield of (open or closed) *Rs. rubrum* starts to rise dramatically (Rijgersberg et al. 1980) and that the fluorescence polarization of LH1 and LH1-RC cores of all purple bacteria shows a marked dependence on the excitation wavelength (Bolt et al., 1981; Kramer et al., 1984b), suggesting that below 100K the LH1 antenna behaves like a spectrally inhomogeneous system. Low temperature (77 K) picosecond absorption (van Grondelle et al., 1987), fluorescence (Freiberg et al., 1987; Freiberg et al., 1988a, 1988b; Timpmann et al., 1991) and absorption anisotropy (van Grondelle et al., 1987; Bergström et al., 1988a; Hunter et al., 1990) kinetics can also be given the same interpretation. Early fluorescence and absorption decays indicated the presence of a 10–20 ps component on the high energy side of the absorption/emission spectrum that precedes trapping. Hence, LH1 is spectrally heterogeneous and consists of at least two, but possibly many spectral forms, as was discussed in section III.B. In all models fast (≤10 ps) energy transfer occurs from the major BChl *a* pool to the low-energy fraction.

At ultralow temperatures (4 K), the trapping efficiency for excitations by P drops to about 60% of its value at >100 K (Rijgersberg et al., 1980). Picosecond absorption measurements indicated an increase in the trapping time upon far-red excitation (Visscher et al., 1989). 4 K annihilation experiments showed that the major part of the fluorescence quenching occurred in the red wing of the emission spectrum (Deinum et al., 1989). Time-resolved fluorescence measurements at 4 K (Timpmann et al., 1991) on *Rs. rubrum* membranes, in which the primary donor was accumulated in the triplet state, revealed at least 10–20 ps components in the blue part of the emission spectrum, 100–200 ps components in the center and very slow components (1 ns) in the red-edge. All these results can well be explained by the 'inhomogeneous broadening' model described in III.B (Pullerits and Freiberg, 1991, 1992; Valkunas et al., 1992; van Mourik et al., 1993; Pullerits et al., 1994).

In contrast to the relatively slow transfer between P and its nearest neighbors (~30–40 ps at room temperature) and the slow equilibration phenomena observed at 77 K and 4 K, there are strong indications that the transfer among identical LH1 dimers (or tetramers) is much faster. From singlet-singlet annihilation measurements, the residence time of an excitation on a B880 molecule was estimated to be less than 1 ps (Bakker et al., 1983; Vos et al., 1986) and a recent detailed analysis of the annihilation experiment suggested a τ_{hop} of 0.65 ps (Valkunas et al., 1989). Time-resolved absorption anisotropy measurements of *Rs. rubrum* and *Rb. sphaeroides* showed that at room temperature during the first few picoseconds after excitation the polarization of the excited state is largely lost, independent of the excitation and probing wavelengths (Sundström et al., 1986; Bergström et al., 1988a; Hunter et al., 1990). It is very probable that this initial depolarization occurs as a result of ultrafast (sub-ps) energy transfer among spectrally similar BChl *a* dimers (assuming that the dimer is the basic spectroscopic structure of LH1—see above). The 'initial' value of the anisotropy of 0.1 is in agreement with earlier polarized spectroscopy results and consistent with models proposed for LH1 (Breton et al., 1981; Kramer et al., 1984b; Hunter et al., 1989). Some further depolarization took place on a slower time-scale (10–30 ps), possibly reflecting the diffusion of the excitation through the LH1-domain. Also at 77 K the initial absorption anisotropy was about 0.1 in the main LH1 absorption band and its blue wing. Only upon excitation and probing in the low-energy wing ($\lambda > 900$ nm) highly polarized decays were observed ($r(0) = r(\infty) = 0.25$) (van Grondelle et al., 1987; Bergström et al., 1988a). This agrees with earlier fluorescence polarization measurements (Kramer et al., 1984b) and supports the idea that a minor fraction of the pigments in the low-energy wing of the LH1 absorption band acts as the final energy acceptor in LH1. Finally, the transient absorption measurements after picosecond excitation exhibit a strong bleaching at the red side of the LH1 absorption band and strong excited state absorption in the blue wing. Around the isobestic point fast additional absorption changes have been observed (van Grondelle et al., 1987; Bergström et al., 1988a, 1988b; Pullerits et al., 1994) which, due to the duration of the excitation pulse could not be time-resolved, but indicated an equilibration process within LH1 on a timescale of at most a few picoseconds.

Recently, Pullerits and Freiberg (1992) modeled energy transfer and trapping by a master equation approach, on the basis of inhomogeneously broadened LH1 and RC spectra. A hexameric arrangement of six $(\alpha\beta BChl_2)_2$ units around the RC analogous to the electron micrographs of *Rp. viridis* was taken as the structural basis of this model. The spectral properties of each unit were chosen randomly from the total inhomogeneously broadened pool and only three fitting parameters remained: the interunit hopping time, the hopping time to the RC and the homogeneous linewidth. At 77 K, a satisfactory fit was obtained with a single hopping time of 60 ps (implying a mean residence time of 12 ps) and a hopping time to the RC of 70 ps; extrapolating these results to room temperature yielded a perfect fit of the fluorescence data. At 4 K the fit was only qualitatively correct, in part due to the uncertainty about the homogeneous absorption and emission spectra at 4 K. The width of the inhomogeneous distribution function (20 nm) was estimated to be about half of the bandwidth (40 nm).

The extrapolated hopping time of 60 ps between tetrameric clusters of BChl *a*, however, seemed too slow to be consistent with the annihilation and absorption anisotropy experiments. It was argued (van Mourik et al., 1993) that due to the use of periodic boundary conditions the simulation of Pullerits and Freiberg missed a phase in the relaxation due to energy transfer between different LH1-RC cores. Consequently, Pullerits and Freiberg had to attribute the observed 10–20 ps decay to an intra-LH1-RC core energy transfer process, which resulted in the slow hopping time. More recent accurate picosecond transient absorption and fluorescence measurements at 77 K (Pullerits et al., 1994), exciting LH1 in the blue wing and detecting the antenna transient absorption or fluorescence signals at various wavelengths over the absorption/fluorescence spectrum, showed that the decay of antenna excited states in the blue wing of the spectrum actually occurs with a 3–5 ps time constant and is coupled to an equally fast rise time of antenna bleaching/fluorescence in the red part of the spectrum. A master equation kinetic simulation of these experimental data, applying periodic boundary conditions to a unit containing several LH1-RC cores, resulted in a satisfactory fit of the 77 K picosecond absorption and fluorescence data using a hopping time of 2–3 ps (Pullerits et al., 1994). Although still larger than the hopping time extrapolated from annihilation experiments, the relative values are sufficiently close to make us confident about the correctness of the basic concepts. In this respect, a reevaluation of the annihilation experiment using a spectrally and spatially inhomogeneous antenna would be appropriate.

Very recent measurements using sub-100 fs pulses revealed that this picture is essentially correct, but that the fastest energy transfer processes occur on the few hundred femtosecond timescale (Chachisvilis et al., 1994). Even faster processes (<100 fs) may also be present, representing relaxation between exciton states in a strongly coupled pigment (Pullerits et al., 1994b). In experiments with very short pulses (40 fs) an interesting longlived oscillatory decay of the antenna excited state population was observed (Chachisvilis et al., 1994) and related to vibrational coherence in the BChl *a* pigment molecules. In the future it will be interesting to investigate the nature of this coherence and its role in the energy transfer processes.

B. Rhodopseudomonas viridis

Until quite recently there was an almost complete lack of knowledge about the energy transfer dynamics in the LH1 antenna of *Rp. viridis*. The main reason for this was the absence of a suitable picosecond laser source in the wavelength range of BChl *b* absorption (960–1030 nm). At present, however, modelocked dye lasers and solid state lasers have become available that operate in the desired part of the spectrum and consequently, many new results on *Rp. viridis* may be expected in the near future. Using the picosecond photo-induced electric gradient technique, Trissl and coworkers (Deprez et al., 1986; Trissl et al., 1990) reported a trapping time of 40 ps in *Rp. viridis* whole cells with active reaction centers. Bittersmann et al. (1990) employed low-intensity excitation pulses and time-resolved fluorescence detection and reported a 80 ps fluorescence decay time. This difference in lifetimes may at least partially be ascribed to the fact that the electric measurements by Trissl et al. were performed with high intensity pulses due to which the excited state lifetime could have been shortened by singlet-singlet annihilation. Picosecond absorption measurements using low-intensity, tunable infrared (960–1020 nm) pulses provided more detailed results about the *Rp. viridis* excitation transfer dynamics (Zhang et al., 1992a). It was shown that the room temperature excited state lifetime is 60 ps in open centers, 90 ps in centers with reduced secondary acceptor (PIA_A^-) and 150 ps in centers with oxidized primary electron donor. Measurement of the time-resolved absorption anisotropy and isotropic absorption kinetics at several wavelengths at 77 K did not reveal any multi-exponential decay at the blue edge of the LH1 absorption spectrum (as was observed for BChl *a*-containing purple bacteria). The observation of a low, on a picosecond timescale constant polarization, ($r(0) = r(\infty) \leq 0.1$), is consistent with rapid (sub)picosecond energy transfer between differently oriented BChl *b* antenna molecules, and agrees with earlier steady-state measurements (Breton et al., 1985). Motivated by the observed single exponential antenna excited state decay and fast depolarization, the energy transfer and trapping in *Rp. viridis* was modeled by a homogeneous random-walk model (Pearlstein, 1982). Good agreement between experiment and prediction was obtained provided the P^+IQ^--PIQ bleaching spectrum measured in *Rp. viridis* membranes (Kleinherenbrink et al., 1992) was used to obtain the absorption spectrum of P in calculating the Förster spectral overlap integrals for energy transfer to/from P (Zhang et al., 1992a). It is known that the absorption spectrum of isolated reaction centers is blueshifted by about 20 nm at room temperature (Kleinherenbrink et al., 1992), which severely influences the overlap ratio for trapping and detrapping. However, the use of the bleaching spectrum is not completely free from problems since it may contain some contributions from a BChl *b* bandshift (see below). From this analysis (Zhang et al., 1992a) it was also concluded that energy transfer from antenna to the special pair (and escape from the reaction center) in *Rp. viridis* is fast, approx. 6 ps, based on the assumption that the antenna is organized as monomeric BChl *b* molecules and the reaction center has a six-fold coordination; with a mono-coordinate reaction center the trapping time would have to be approx. 1 ps in order to fit the experimental results. For the BChl *a*-containing purple bacteria it was mentioned above that the antenna most likely is organized in small clusters of $\alpha\beta$-polypeptides holding 2–4 BChl molecules. If this also is the case for *Rp. viridis*, which is reasonable to believe, trapping and detrapping dynamics would in fact be very similar to that of the BChl *a*-containing purple bacteria, with relatively slow (25–30 ps) energy transfer from the antenna to the reaction center and very fast energy equilibration within the antenna.

A study of the efficiency of charge separation in *Rp. viridis* LH1-RC complexes as a function of temperature showed, (Kleinherenbrink et al., 1992) surprisingly, that the quantum yield of charge separation was independent of the excitation wavelength at 300 K and at 6 K, but the absolute efficiency, as measured by the amount of P^+ formed upon flash excitation, dropped to 55% upon cooling to 6 K. The fluorescence excitation spectrum showed no contribution from direct RC excitation in the wavelength range 780–860 nm. The effective rate of transfer from neighbouring pigments to the special pair was estimated to be $(1.3\ ns)^{-1}$ at 6 K. Kleinherenbrink et al. (1992) suggested that this rate can only be compatible with the Förster theory for electronic energy transfer if the long-wavelength absorption bands are homogeneously broadened. However, low-temperature polarized fluorescence experiments on isolated LH1-RC complexes and membranes (Monshouwer et al., unpublished) revealed highly polarized emission only upon excitation (with a narrow-banded laser) in the very red wing of the absorption spectrum, suggesting a similar behavior as the LH1 absorption bands of BChl *a*-containing purple bacteria (Kramer et al., 1984b; van Mourik et al., 1992a). Inhomogeneous broadening of the main absorption bands remained so far undetected and will further complicate the problem of excitation trapping at low temperatures. In membranes, the polarization effect is limited to the very red tail of the absorption band and therefore is probably difficult to detect with broad-band excitation (Deinum, 1991). Since time-resolved absorption (Zhang et al., 1992a) and fluorescence experiments at 23 K (Kleinherenbrink et al., 1993) only yielded a single decay time, the time for excitation energy transfer to these lowest states must be faster than a few picoseconds.

In general, we believe that the observed trapping kinetics and temperature dependence of the quantum yield for charge separation are not compatible with the proposed large energy difference between P* and LH1* (Trissl et al., 1990; Zhang et al., 1992a; Kleinherenbrink et al., 1993). The calculation of this energy difference and of the relevant overlap integrals using the P^+-P difference spectrum is possibly not correct. The difference spectrum measured in *Rp. viridis* membranes contains an unknown contribution of a BChl *b* bandshift, which may have caused an apparent blue shift to 1010 nm relative to the main absorption band (max at 1040 nm at 6 K). Also, the use of the spectrum of the isolated RC is not correct because that certainly is blue-shifted in position. The fact that trapping still has an efficiency of 55% at 6 K suggests that on the average P* and LH1* are about equal in energy (apart from the entropic contribution). If both P and LH1 are inhomogeneously broadened, the number 55% may simply reflect the fraction of LH1-RC cores in which P* is the lowest energy state.

Excitation annihilation experiments (Deprez et al., 1986; Deinum et al., 1992a) have indicated significant competition between trapping by reaction centers and annihilation. The general theory for trapping, loss and annihilation (Den Hollander et al., 1983; van Grondelle, 1985) describes these results correctly if the finite duration of the laser pulse is accounted for (Deinum et al., 1991). At room temperature, the domains for annihilation are large (>15 RC-LH1 cores). At ultralow temperature, the annihilation becomes less efficient (which could be due to inhomogeneity), but still sizable annihilation domains ($N_D > 100$) are estimated. This is consistent with the observation that the increase in fluorescence polarization occurs only in the very red edge of the absorption spectrum. Three effects may contribute to the relatively efficient energy transfer and annihilation at 4 K in *Rp. viridis*: (i) the LH1-RC structure is highly organized and probably well connected, (ii) the width of the inhomogeneous distribution function is probably relatively small in the case of *Rp. viridis* (Monshouwer et al., unpublished) mitigating the effects due to lowering the temperature, and *(iii)*, the Stokes shift of *Rp. viridis* is probably quite small, resulting in a high average transfer rate. Since in the excited state decay no 10–20 ps component is observed reflecting energy transfer within the inhomogeneous band (Zhang et al., 1992a; Kleinherenbrink et al., 1993) we have to conclude that this energy transfer process occurs at a rate of less than a few ps, even at 4 K.

V. Energy Transfer in LH2-Containing Purple Bacteria

A. Energy Transfer Pathways and Trapping in Rhodobacter sphaeroides

Rhodobacter (Rb.) sphaeroides is the best studied LH2 containing purple bacterial species. At room temperature and with open centers, picosecond absorption (Sundström et al., 1986) and fluorescence (Freiberg et al., 1988; Freiberg et al., 1989; Shimada et al., 1989) techniques monitoring the kinetics within the B850 or B875 absorption/emission bands, generally revealed a non-exponential decay with lifetime components of 20–40 and 70–100 ps. In fluorescence studies, a faster (< 10 ps) energy transfer component was dominant while a minor 40 ps phase corresponded to the decay observed in the single-color transient absorption measurements. The bi- (or multi-) exponentiality of the decay is a consequence of the complex antenna structure of this bacterium, and reflects the equilibration of energy between LH1 and LH2. Using a simple two-step equilibrium model involving LH2, LH1 and the RC it was shown (Sundström et al., 1986) that the measured trapping time of about 100 ps corresponds to an actual trapping time of 60 ps for the LH1-RC core, i.e., the same value as observed for *Rs. rubrum* (Sundström et al., 1986) and LH1-RC only mutants of *Rb. sphaeroides*

(Hunter et al., 1990). The very fast energy transfer step from B800 to B850 in LH2 is discussed in section VI.B.

In centers with oxidized primary electron donor (P^+), charge separation cannot occur and the excitation lifetime increases to 200–250 ps (Sundström et al., 1986; Freiberg et al., 1988; Freiberg et al., 1989; Shimada et al., 1989) similar to what was observed for *Rs. rubrum*. This shows again that the closed RC quenches antenna excitations rather efficiently. The LH1 ↔ LH2 equilibration appeared not to be affected by the RC redox state. Time-resolved absorption anisotropy experiments revealed a low 'initial' anisotropy ($r(0) \approx 0.1$) of the major LH2 (850 nm) absorption band. Such a behavior was also observed for the main LH1 absorption band of *Rs. rubrum* and is consistent with (sub-)picosecond energy transfer between neighbouring BChl molecules in LH1 and LH2 (Sundström et al., 1986; van Grondelle et al., 1987).

Measurements of energy transfer kinetics at low temperature (77 K) with improved spectral resolution permitted a much better resolution of the individual energy transfer steps (narrower spectral bands) and effectively unidirectional energy transfer. Recently, two-color pump-probe experiments were reported for *Rb. sphaeroides* at 77 K, in which the probing wavelength was varied over the 790–910 nm wavelength range (Zhang et al., 1992b). Excitation and detection in the B800 band showed that the approximate lifetime of the B800* excited state was between 1–2 ps at 77 K. The absorption changes in the B800 band were moderately polarized ($r(0) \approx 0.25$). Exciting the high-energy pigments (B800) and probing the arrival of the excitations on the low-energy pigments through the rise-time of the 910 nm ground state bleaching/stimulated emission revealed a major time-constant of about 10–12 ps, while the remaining part had a considerably slower rise of about 30–40 ps. Selective excitation of B850 in the wavelength range 830–860 nm and probing the arrival of excitations at 885 nm (in the middle of the LH1 band) or at 910 nm (at the edge of LH1 spectrum) showed that this biphasic risetime is the result of the spectral heterogeneity of B850, because excitation in the blue wing of the B850 spectrum resulted in a slower energy transfer to LH1. This wavelength dependent energy transfer can be represented by two spectral components with two different transfer times of 10 and 30–40 ps, respectively (Zhang et al., 1992b). However, it does not necessarily imply that there exist two different forms of B850. The more likely explanation is that excitation in the blue wing of B850 results in more and possibly slower transfer steps than excitation of the main peak. Time-resolved fluorescence measurements have been interpreted in a similar way (Freiberg et al., 1988; Freiberg et al., 1989; Shimada et al., 1989) and Freiberg et al. (1989) proposed a model with two parallel pools of LH2, i.e. an inhomogeneity of both B800 and B850.

The 40 ps time-constant mentioned above, characterizing the B850 heterogeneity, was also observed in earlier single-wavelength pump-probe experiments (Sundström et al., 1986; van Grondelle et al., 1987) and on the basis of these experiments a sequential model was proposed, involving a single 30–40 ps step from LH2 to LH1. However, with the more detailed information from the two-color experiments (Zhang et al., 1992b) it is now clear that this relatively slow step involves only a minor part of the excitations and that the major part of the LH2 → LH1 transfer is fast (< 5 ps). An accurate analysis of the absorption kinetics of *Rb. sphaeroides* (van Grondelle et al., 1987) and studies of the polarized spectroscopic (van Dorssen et al., 1988) and time-resolved absorption properties (Hunter et al., 1990) of an LH2-only mutant of *Rb. sphaeroides* suggested that at 77 K there is a pool of red pigments in LH2 (called B870) involved in the coupling of LH2 and LH1. It is highly likely that this 'B870' fraction originates from the inhomogeneous distribution of B850. Most time-resolved measurements available for *Rb. sphaeroides* thus show that energy equilibration over the entire antenna is fast ≤10 ps at both room temperature and 77 K. In combination with results from measurements of the trapping rate in the temperature interval 100–200 K, and results from measurements of detrapping yield and rate (Timpmann et al., 1993) in *Rs. rubrum* (see IV.A.1) this shows that energy transfer from the antenna to the reaction center is the rate limiting step in the overall energy migration and trapping process.

B. Energy Equilibration and Trapping in Other LH2-Containing Species

We will briefly summarize the results obtained with other LH2 containing purple bacteria. *Cm. minutissimum*, which has a pigment system very similar to that of *Rb. sphaeroides*, was studied with time-resolved fluorescence spectroscopy (Freiberg et al., 1988; Freiberg et al., 1989) and the results were very similar to those obtained with *Rb. sphaeroides* on all essential points. Membranes of low-light *Rp. palustris*, in which the B800 band shows a pronounced exciton structure (possibly due to a tight packing of several B800 pigment molecules - see below) show ultrafast relaxation within the 800 nm exciton band, followed by excitation transfer via B850 to LH1 in at most a few ps (van Mourik et al. 1992c). The bacterium *Erythrobacter* sp. strain Och 114 having the unusual pigmentation B806–870, where B870 serves as LH1 and B806 as LH2, was investigated with time-resolved fluorescence spectroscopy (Shimada et al., 1990; Shimada et al. 1992). Despite the large energy difference between the two antenna species, energy was found to be transferred from B806 to B870 within the time resolution of the experiment (about 6 ps) at room temperature (Shimada et al., 1990), but significantly slower at 77 K (Shimada et al., 1992). A low-energy pigment component of LH1 (B888), contributing approximately 3 BChl/RC was also observed, with properties closely related to those reported for the low-energy components of *Rs. rubrum* and *Rb. sphaeroides*. The observed risetime of the B888 fluorescence of 9 ps agrees well with the observed rise-time of red-edge excitation in *Rb. sphaeroides* (Zhang et al., 1992b). A slower decay component in the B806 fluorescence, which was not observed as a corresponding risetime in the B888 fluorescence, suggested a heterogeneous decay pattern of B806, somewhat reminiscent of the situation for LH2 in *Rb. sphaeroides*. Low temperature experiments suggested that also in the B806 band energy transfer to low energy pigments precedes transfer to B870 (Shimada et al., 1992).

From a comparison of a five-component global analysis fit and a so-called global target analysis kinetic simulation to time-resolved fluorescence kinetic data of *Rb. capsulatus* membranes at room temperature, Holzwarth et al. (1993) arrived at conclusions that on several important points are in conflict with those described above for the very similar bacterium *Rb. sphaeroides*. Thus, it was concluded that, *i.* the overall antenna exciton dynamics is limited by the rate of charge separation (i.e. trap limited dynamics), and that the excited states of the LH1 core antenna and the special pair electron donor of the reaction center are equilibrated within less than 3 ps; and *ii.* LH2 → LH1 energy transfer at room temperature occurs with a time constant of 9 ps. Both of these conclusions are difficult to reconcile with most of the available data for *Rb. sphaeroides* and similar bacteria. Thus, from direct measurements of trapping (Bergström et al., 1989; Visscher et al., 1989) and detrapping (in *Rs. rubrum* (Timpmann et al., 1993)) the energy transfer from the antenna to the special pair was shown to be slow, 35–40 ps, both at room temperature and lower temperature (100–200 K). Further, the suggestion of a ≤ 3 ps equilibration time of the excitation energy over the LH1-special pair pigments is expected to lead to an approximately 1/N population probability of the special pair (N = number of LH1 antenna molecules per reaction center). Following selective excitation of P this would consequently lead to a much higher detrapping yield than the 25% experimentally found for *Rs. rubrum* at room temperature (Timpmann et al., 1993) (see also IV.A.1). This comparison relies on the reasonable assumption that *Rb. sphaeroides* and *Rs. rubrum* are similar in this respect. The LH2 → LH1 energy transfer time constant (9 ps) deduced from the analysis of the fluorescence data is considerably longer than what has been concluded for *Rb. sphaeroides* from transient absorption (van Grondelle et al., 1987; Zhang et al., 1992b) and time resolved fluorescence measurements at 77 K (Freiberg et al., 1988, 1989; Shimada et al., 1989), pointing to a LH2 → LH1 transfer time of ≤5 ps; at room temperature the transfer is most likely faster than this.

VI. Energy Transfer in Isolated Light-Harvesting Complexes and Mutants

A. LH1

If detergent-purified LH1 is analyzed on an LDS-polyacrylamide gel, a discrete set of complexes $(\alpha\beta)_n$ is observed with n ≥ 3 (Westerhuis et al., 1987; Hunter et al., 1988). All complexes absorb around 870 nm and the absorption maxima shift slightly further to the red with increasing n (Westerhuis et al., 1987, 1992). Excitation annihilation (van Grondelle et al., 1983) and polarized fluorescence (Westerhuis et al., 1987, 1992) experiments have shown that the various BChl molecules in these LH1 complexes are well coupled. For the LH1-only mutant of *Rb. sphaeroides* M2192, annihilation experiments indicated large domains at room temperature (N_D > 100) (Vos et al., 1988).

The excited state decay of LH1 is slow, even at room temperature, and takes about 650 ps for the detergent-isolated form (Sebban et al., 1985; Bergström et al., 1988a; Freiberg and Timpmann, 1992) and 450 ps for the LH1-only mutant M2192 (Hunter et al., 1990), which is consistent with what is expected for a spectrally 'homogeneous' pigment decoupled from any energy acceptors. At 77 K, excitation of LH1 of *Rb. sphaeroides* in the major absorption peak resulted in a biphasic decay with a fast ~20 ps phase dominant in the blue edge of the spectrum followed by a slow decay of the excited state with maximum amplitude in the red wing of the spectrum (900 ps in the detergent isolated LH1 complex, 400 ps in M2192 mutant of *Rb. sphaeroides*). Time-resolved absorption anisotropy of detergent isolated LH1 (Bergström et al., 1988a) and the LH1-only mutant M2192 (Hunter et al., 1990) showed that (sub)picosecond depolarization occurs, probably as a result of very fast ($\geq 10^{12}$ s^{-1}) energy transfer within a minimum unit of LH1. More recent and more accurate transient absorption measurements of *Rs. rubrum* chromatophores at 77 K with blue wing excitation at 870 nm and probing in the range 885–905 nm showed that the decay of bleaching in the blue part of the spectrum occurs with a lifetime of approx. 3–5 ps and is correlated with an equally fast risetime of bleaching/stimulated emission on the red side of the spectrum (Pullerits et al., 1994), see section IV.A.2. Time- and spectrally resolved fluorescence experiments of detergent-isolated LH1 of *Rs. rubrum* were performed at low temperatures down to 4 K (Freiberg and Timpmann, 1992). The observed kinetics depended strongly on the excitation wavelength; the shortest decay times of 10–20 ps were observed on the blue side of the emission spectrum, whereas above 940 nm the emission was essentially monoexponential with a decay time of 940 ps at 4 K. In all cases the kinetics were taken to reflect energy transfer in a spectrally heterogeneous system. Measurement of the polarized LH1 fluorescence spectrum as a function of narrow-banded excitation at 4 K showed a characteristic shift of the emission maximum which was interpreted to reflect energy transfer within an inhomogeneously broadened cluster of coupled BChl *a* molecules (van Mourik et al., 1992a).

B. LH2

Energy transfer properties of a variety of LH2 complexes have been studied extensively over the past few years, and they all demonstrate rather similar characteristics. Both stationary and time-resolved absorption and fluorescence experiments have indicated that energy transfer among the B850 pigments is fast (< 1 ps) and leads to ultrafast depolarization of the excited state. At 77 K spectral equilibration phenomena were observed to take place on a 10–20 ps time-scale (Bergström et al., 1988b), very similar to LH1.

The most conspicuous feature of LH2 is the energy transfer process from B800 to B850, which may be one of the best studied photosynthetic energy transfer processes. At room temperature, the emitting B800 excited state is fully equilibrated with the B850 excited state (van Grondelle et al., 1982). From early fluorescence yield measurements it was estimated that at 4 K this energy transfer process occurs within 2–3 ps (van Grondelle et al., 1982). Later experiments with picosecond time-resolution indicated a time-constant of 1–2 ps at 77 K and somewhat faster rate at room temperature (Bergström et al., 1986, 1988b). In the so-called B800–820 complex of *Rp. acidophila* strain 7050 the energy transfer from B800 to B820 was observed to take place within 0.5 ps at 77 K, consistent with the improved spectral overlap for B800 → B850 transfer (Bergström et al., 1988b). In a transient absorption experiment with 0.2 ps time-resolution using the B800–850 complex of *Rb. sphaeroides*, Shreve et al. (1991) concluded that the room temperature energy transfer time is 0.6 ± 0.1 ps upon direct B800 excitation. Spheroidene excitation yielded a transfer time of about 2–3 ps (Trautman et al., 1990), which was later interpreted as sphaeroidene → B800 excitation transfer (Shreve et al., 1991). Direct sphaeroidene → B850 excitation transfer would still result in sub-ps kinetics (Shreve et al., 1991). Visscher et al. (1993a, 1993b) extended these measurements to LH2 complexes of several other species and to low temperature (77 K), by using low-intensity one-color pump-probe spectroscopy with 0.5 ps pulses. These measurements unequivocally showed that the B800 → B850 energy transfer time at room temperature is 0.7 ± 0.1 ps in *Rb. sphaeroides*. At 77 K the transfer rate slows down by a factor of 3–4, and was observed to be 2.4 ps in a chromatophore preparation of *Rb. sphaeroides* (Visscher et al., 1993a, 1993b). The same LH2 complex isolated and solubilized with LDAO displayed a slightly faster B800 → B850 transfer, 1.7 ps at 77 K, probably as a result of more extensive aggregation in the isolated complex than in vivo. Visscher et al. (1993a, 1993b) also measured the B800 excited state lifetime in membranes of *Rp. palustris* HL and LL and in the detergent isolated B800–820 complex of *Rp. acidophila* strain 7050. In both species of *Rp. palustris* the excited state lifetime and thus the B800 → B850 energy transfer time constant, was found to be slightly shorter than in LH2 of *Rb. sphaeroides*, i.e. 1.5 ± 0.1 ps. For B800–820 the energy transfer at 77 K was even faster, ~0.8 ps. The temperature dependence of the B800 → B850 transfer time and variation with bacterial species suggested that the B800 → B850 transfer is principally controlled by the spectral overlap between the B800 donor and B850 acceptor pigments, at least at room temperature and 77 K. This was verified by subpicosecond absorption measurements on LH2 mutants of *Rb. sphaeroides* (Hess et al., 1994) in which the TyrTyr motif of the α-polypeptide at positions 44 and 45 had been replaced successively by PheTyr and PheLeu, resulting in spectral blueshifts to 839 and 826 nm, respectively (Fowler et al., 1992). The latter configuration is the same as that found in B800–820 of *Rp. acidophila*, and the energy transfer time from B800 measured in this mutant agreed very well with that observed in the native complex. The single mutations (Phe44, Tyr45 or Tyr44,Phe45) having an absorption band at 839 nm, intermediate between the 850 and 826 nm bands, displayed an intermediate excited state lifetime between that of B800–850 and B800–820, i.e. 1.5 ps. Spectral hole burning experiments were consistent with a B800 → B850 transfer time of 2.3–2.5 ps between 1.2 and 30 K in membranes of *Rb. sphaeroides* (van der Laan et al., 1990; Reddy et al. 1991, 1992a) and of 1.8 ps in membranes of *Rp. acidophila* (Reddy et al., 1992b). Hole burning experiments on the spectrally blueshifted LH2 mutants mentioned above (B839 and B826) at 1.6 K (van der Laan et al., 1993), exhibited a smaller variation of energy transfer time than what was observed in time resolved measurements with 500 fs pulses (Visscher et al., 1993a,b) and a much smaller variation than could be expected on the basis of the spectral overlap between the B800 homogeneous fluorescence spectral bandwidth and the B850 absorption. Consequently, it was suggested that energy transfer occurs from B800 to a broad vibronic structure of B850. A similar mechanism for B800 → B850 energy transfer at very low temperatures was earlier suggested by Reddy et al. (1992a) to account for hole burning results on B800 → B850 energy transfer of *Rb. sphaeroides*.

In the original model of Kramer et al. (1984a), energy transfer among (at least two) B800 molecules was assumed to take place on a timescale (~1 ps) that competes with the B800 → B850 transfer to explain the observed low polarization of the B800 emission upon selective Q_x-excitation. Subsequent room temperature and low-temperature absorption anisotropy measurements using single wavelength pump-probe spectroscopy resulted in different r(0) values, somewhat dependent on the wavelength of excitation/probing, on the species and on temperature (Bergström et al., 1986, 1988b). However, a high (0.3–0.35) and time independent anisotropy was generally observed in the B800 band, suggesting only limited B800 ↔ B800 energy transfer, in apparent disagreement with the model of Kramer et al. (1984a). Anisotropy measurements with 0.5 ps pulses within the B800 absorption band of several species (*Rb. sphaeroides*, *Rp. palustris* HL and LL, and *Rp. acidophila* B800–820) at 77 K identified and discussed problems associated with obtaining dynamical information from the observed anisotropies (Visscher et al., 1993b). It was concluded that the initial B800 anisotropy in all studied species was high (≥0.35), but that in most cases the rate and extent of anisotropy decay could not be unambiguously ascertained due to interference from strong and depolarized B850 excited state absorption. Nevertheless, the results suggested at the most partial depolarization on the time scale of the B800 excited state lifetime, i.e. within 5–10 ps. This implies that at 77 K B800 ↔ B800 energy transfer within the inhomogeneously broadened B800 band is slower than B800 → B850 transfer, or that transfer occurs between B800 molecules with close to parallel transition dipole moments. The dynamics within the 800 nm band were examined further in very recent time resolved absorption measurements with better time resolution (<100 fs). An accurate description of the contribution from excited state absorption to the measured anisotropy kinetics (Hess et al., 1993b), confirmed and substantiated the earlier results obtained with lower time resolution (Visscher et al., 1993a, 1993b). Thus, at room temperature a 0.9–1.5 ps decay of anisotropy was observed in both *Rb. sphaeroides* and *Rp. palustris*, which is considerably slower than the isotropic B800 excited state decay (0.6–0.8 ps). At 77 K the anisotropy decay was slowed down to the extent that no decay was clearly discernible within the available observation window of approx. 5 ps, set by the B800 excited state lifetime. Simultaneously, the magic-angle decay of the B800 excited state, measured in the blue wing of the 800 nm band, was observed to be strongly nonexponential (Hess et al., 1993b), with a 200–300 fs decay component in addition to the previously detected 0.6–0.8 ps or ~2.4 ps lifetime at room temperature and 77 K, respectively. This very fast relaxation was found to be almost independent of temperature and, interestingly, it was not correlated to a similar decay of anisotropy, showing that if this process reflects energy transfer between spectrally dissimilar B800 molecules ($B800^{blue} \rightarrow B800^{red}$) the transition dipoles of the participating molecules have to be close to parallel. Early hole burning experiments on LH2 of *Rb. sphaeroides* did not resolve any wavelength dependence of the holewidth across the 800 nm band or broad holes on the red side of the spectrum (following blue excitation) indicative of downhill $B800^{blue} \rightarrow B800^{red}$ energy transfer (van der Laan et al., 1990; Reddy et al., 1991, 1992a). However, studies of *Rp. acidophila* 7750 (B800–850) showed that blue excitation produced such broad holes in the red wing of the B800 band (Reddy et al., 1992b). An analysis of the efficiency of spectral hole burning vs. the burning wavelength and vs. laser power in LH2 of *Rb. sphaeroides* showed that a narrow 2.4 ps hole was only present in the red wing of the B800 band (van der Laan et al., 1993). Upon burning at shorter wavelengths the holewidth gradually increased, indicating a shorter excited state lifetime in the blue wing of the 800 nm band. The shortest lifetime observed in that experiment was about 600 fs. Thus, independent results from femtosecond time resolved, hole burning and steady-state fluorescence polarization spectroscopies show that there exist fast relaxation processes within the B800 band, in addition to the decay associated to B800 → B850 energy transfer. Energy transfer from 'blue' to 'red' B800 pigments ($B800^{blue} \rightarrow B800^{red}$) is a very likely origin of this relaxation process, but there still remains some ambiguity as to the relative orientation of the transition dipoles participating in the transfer process.

Excited state relaxation times on the several hundred femtosecond time scale have frequently been observed in various dye molecules in solution and attributed to vibrational relaxation or cooling and solvation processes (Elsaesser and Kaiser, 1992). Such processes are also to be expected in the pigment molecules of a biological pigment-protein complex, although little quantitative information so far is available. However, time resolved measurements on various porphyrin molecules (Bilsel et al., 1993) as well as on bacteriochlorophyll *a* in solution (Becker et al., 1990; Åkesson and Sundström, unpublished) showed the presence of approx. 0.5–1 ps relaxation processes, that most likely reflect vibrational relaxation and cooling of the excited molecules. The necessity to consider such relaxation processes, in order to account in full detail for the dynamics observed in a biologically active system of dipole coupled chromophores, exhibiting energy transfer on the picosecond to subpicosecond time scale was therefore discussed (Visscher et al., 1993b).

References

Aagard J and Sistrom WR (1972) Control of synthesis of reaction center and antenna bacteriochlorophyll in photosynthetic bacteria. Photochem Photobiol 15: 209–225

Allen JP, Feher G, Yeates TO, Komiya H and Rees DC (1987a). Structure of the reaction center from *Rhodobacter sphaeroides R-26*: The cofactors. Proc Natl Acad Sci USA 84: 5730–5734

Allen JP, Geher G, Yeates TO, Komiya H and Rees DC (1987b). Structure of the reaction center from *Rhodobacter sphaeroides R-26*: The protein subunits. Proc Natl Acad Sci USA 84: 6162–6166

Altmann RB, Renge I, Kador L and Haarer D (1992) Dipole moment differences of nonpolar dyes in polymeric matrices: Stark effect and photochemical hole burning. I., J Chem Phys 97: 5316–5322

Becker M, Nagarajan V, Middendorf D, Shield MA and Parson WW (1990) Excited state properties of bacteriochlorophyll *a* and of bacterial photosynthetic reaction centers as revealed by picosecond absorption studies. Baltscheffsky M (ed) Current Research in Photosynthesis, Vol 1, 101–104. Kluwer Academic Publishers, Dordrecht

Bakker JGC, Van Grondelle R and Den Hollander WTF (1983). Trapping, loss and annihilation of excitations in a photosynthetic system. II. Experiments with the purple bacteria *Rhodospirillum rubrum* and *Rhodopseudomonas capsulata*. Biochim Biophys Acta 725: 508–518

Beekman LMP, Visschers RW, Visschers KJ, Althuis B, Barz W, Oesterhelt D, Sundström V and van Grondelle R (1993) Excitation energy transfer in mutants of *Rb. sphaeroides*: The effects of changes in the core antenna size. In Martin JL, Migus A, Mourou G and Zewail AH (eds) Ultrafast Phenomena VIII, Vol 55, pp 552–554. Springer Series in Chemical Physics Springer-Verlag, Berlin

Beekman LMP, van Mourik F, Jones MR, Visser HM, Hunter CN and van Grondelle R (1994) Trapping kinetics in mutants of the photosynthetic purple bacterium *Rhodobacter sphaeroides:* Influence of charge separation rate and consequences for the rate-limiting stop in the light-harvesting process. Biochmeistry 33: 3143–3147

Bergström H, Sundström V, van Grondelle R, Åkesson E and Gillbro T (1986) Energy transfer within the isolated light-harvesting B800-850 pigment-protein complex of *Rhodobacter sphaeroides*. Biochim Biophys Acta 852: 279–287

Bergström H, Westerhuis WHJ, Sundström V, van Grondelle R, Niederman RA and Gillbro T (1988a) Energy transfer within the isolated B875 light-harvesting pigment-protein complex of *Rhodobacter sphaeroides* at 77 K studied by picosecond absorption spectroscopy. FEBS Lett 233: 12–16

Bergström H, Sundström V, van Grondelle R, Gillbro T and Cogdell R (1988b) Energy transfer dynamics of isolated B800-850 and B800-820 pigment-protein complexes of *Rhodobacter sphaeroides* and *Rhodopseudomonas acidophila*. Biochim Biophys Acta 936: 90–98

Bergström H, van Grondelle R and Sundström V (1989) Characterization of excitation energy trapping in photosynthetic purple bacteria at 77 K. FEBS Lett 250: 503–508

Bilsel O, Milam SL, Girolami GS, Suslick KS and Holten D (1993) Ultrafast electronic deactivation and vibrational dynamics of photoexcited uranium (IV) porphyrin sandwich complexes. J Phys Chem 97: 7216–7221

Bittersmann E, Blankenship RE and Woodbury N (1990) Picosecond fluorescence studies of *Rhodopseudomonas viridis*. In: Baltscheffsky M (ed) Current Research in Photosynthesis, Vol II, pp 169–172. Kluwer Academic Publishers, Dordrecht

Bolt J and Sauer K (1979) Linear dichroism of light harvesting bacteriochlorophyll proteins from *Rhodopseudomonas sphaeroides* in stretched polyvinyl alcohol films. Biochim Biophys Acta 546: 54–63

Bolt JD and Sauer K (1981) Fluorescence properties of the light-harvesting bacteriochlorophyll protein from *Rhodopseudomonas sphaeroides R-26*. Biochim Biophys Acta 637: 342–347

Bolt JD, Hunter CN, Niederman RA and Sauer K (1981) Linear and circular dichroism and fluorescence polarization of the B875 light-harvesting bacteriochlorophyll-protein complex from *Rhodopseudomonas sphaeroides*. Photochem Photobiol 34: 653–656

Boonstra AF, Visschers RW, Calkoen F, van Grondelle R, van Bruggen EFJ and Boekema EJ (1993) Structural characterization of the B800-850 and B875 light-harvesting antenna complexes from *Rhodobacter sphaeroides* by electron microscopy. Biochim Biophys Acta 1142: 181–188

Borisov AY, Gadonas R, Danielius R, Piskarskas A and Razhivin AP (1982) Minor component B-905 of light-harvesting antenna in *Rhodospirillum rubrum* chromatophores and the mechanism of singlet-singlet annihilation as studied by difference selective picosecond spectroscopy. FEBS Lett 138: 25–28

Borisov AY, Freiberg AM, Godik VI, Rebane K and Timpmann K (1985) Kinetics of picosecond bacteriochlorophyll luminescence in vivo as a function of the reaction center state. Biochim Biophys Acta 807: 221–229

Braun P and Scherz A (1991) Polypeptides and bacteriochlorophyll organization in the light-harvesting complex B850 of *Rhodobacter sphaeroides R-26.1*. Biochemistry 30: 5177–5184

Breton J (1985) Orientation of the chromophores in the reaction center of *Rhodopseudomonas viridis*. Comparison of low-temperature linear dichroism spectra with a model derived from X-ray crystallography. Biochim Biophys Acta 810: 235–245

Breton J and Vermeglio A (1982) Orientation of photosynthetic pigments in vivo. In: Govindjee (ed) Photosynthesis: Energy conversion by Plants and Bacteria, Vol 1, 153–194. Academic Press, New York

Breton J, Vermeglio A, Garrigos M and Paillotin G (1981) Orientation of the chromophores in the antenna system of *Rhodopseudomonas sphaeroides*. In: Akoyunoglou (ed) Photosynthesis III. Structure and Molecular Organization of the Photosynthetic Apparatus, pp 445–459. Balaban International Science Services, Philadelphia, PA

Breton J, Farkas DL and Parson WW (1985) Organization of the antenna bacteriochlorophylls around the reaction center of *Rhodopseudomonas viridis* investigated by photoselection techniques. Biochim Biophys Acta 808: 421–427

Brunisholz RA and Zuber H (1992) Structure, function and organisation of antenna polypeptides and antenna complexes from the three families of *Rhodospirillanae*. J Photochem Photobiol 15: 113–140

Brunisholz RA, Jay F, Suter F and Zuber H (1985) The light-harvesting polypeptides of *Rhodopseudomonas viridis*. The complete amino-acid sequences of B1015-.alpha., B1015-.beta. and B1015-.gamma. Biol Chem Hoppe-Seyler 366: 87–98

Chachisvilis M, Pullerits T, Jones MR, Hunter CN and Sundström V (1994) Vibrational dynamics in the light-harvesting complexes of the photosynthetic bacterium *Rhodobacter sphaeroides*. Chem Phys Lett 224: 345–351

Chadwick BW, Zhang C, Cogdell RJ and Frank HA (1987) The effects of lithium dodecyl sulfate and sodium borohydride on the absorption spectrum of the B800-850 light-harvesting complex from *Rhodopseudomonas acidophila 7750*. Biochim Biophys Acta 893: 444–451

Chang MC, Meyer L and Loach PA (1990) Isolation and characterization of a structural subunit from the core light-harvesting complex of *Rhodobacter sphaeroides 2.4.1* and *puc 705-BA*. Photochem Photobiol 52: 873–881

Clayton RK (1966) Relations between photochemistry and fluorescence in cells and extracts of photosynthetic bacteria. Photochem Photobiol 5: 807–827

Clayton RK and Clayton BJ (1981) B850 pigment-protein complex of *Rhodopseudomonas sphaeroides*: Extinction coefficients, circular dichroism, and the reversible binding of bacteriochlorophyll. Proc Natl Acad Sci USA 78: 5583–5587

Davis RC, Ditson SL, Fentiman AF and Pearlstein RM (1981) Reversible wavelength shifts of chlorophyll induced by a point charge. J Am Chem Soc 103: 6823–6826

Dawkins DJ Ferguson LA and Cogdell RJ (1988) The structure of the 'core' of the purple bacterial photosynthetic unit. In: Scheer H and Schneider S (eds) Photosynthetic Light-Harvesting Systems, pp 115–127. Walter de Gruyter, Berlin

Deinum G (1991) Excitation migration in photosynthetic antenna systems. Ph.D. Thesis, State University of Leiden, The Netherlands

Deinum G, Aartsma TJ, Van Grondelle R and Amesz J (1989). Singlet-singlet excitation annihilation measurements on the antenna of *Rhodospirillum rubrum* between 300 and 4 K. Biochim Biophys Acta 976: 63–69

Deinum G, Otte CM, Gardiner AT, Aartsma TJ, Cogdell RJ and Amesz J (1991) Antenna organization of *Rhodopseudomonas acidophila*: A study of the excitation migration. Biochim Biophys Acta 1060: 125–131

Deinum G, Aartsma TJ and Amesz J (1992a) Fluorescence yield and singlet-singlet annihilation measurements in *Rhodopseudomonas viridis*. In: Murata (ed) Research in Photosynthesis, Vol 1, pp 161–164. Kluwer Academic Publishers, Dordrecht

Deinum G, Kleinherenbrink FAM, Aartsma TJ and Amesz J (1992b) The fluorescence yield of *Rhodopseudomonas viridis* in relation to the redox state of the primary electron donor., Biochim Biophys Acta 1099: 81–84

Deisenhofer J, Epp O, Miki K, Huber R and Michel H (1984) X-ray structure analysis of a membrane protein complex. Electron density map at 3 Å resolution and a model of the chromophores of the photosynthetic reaction center from *Rhodopseudomonas viridis*. J Mol Biol 180: 385–398

Deisenhofer J, Epp O, Miki K, Huber R and Michel H (1986) Structure of the protein subunits in the photosynthetic reaction center of *Rhodopseudomonas viridis* at 3.Å. resolution. Nature (London) 318: 618–624

Den Hollander WTF, Bakker JGC and van Grondelle R (1983) Trapping, loss and annihilation of excitations in photosynthetic systems. I. Theoretical aspects. Biochim Biophys Acta 725: 492–507

Deprez J, Trissl HW and Breton J (1986) Excitation trapping and primary charge stabilization in *Rhodopseudomonas viridis* cells, measured electrically with picosecond resolution. Proc Natl Acad Sci USA 83: 1699–1703

Duysens LNM (1979) Transfer and trapping of excitation energy in photosystem II. In: Chlorophyll Organization and Energy Transfer in Photosynthesis, CIBA Foundation Symposium 61 (new series), pp 323–340. Elsevier, Amsterdam

Eccles J and Honig B (1983) Charged amino acids as spectroscopic determinants for chlorophyll in vivo. Proc Natl Acad Sci USA 80: 4959–4962

Elsaesser T and Kaiser W (1991) Vibrational and vibronic relaxation of large polyatomic molecules in liquids. Ann Rev Phys Chem 42: 83–107

Engelhardt H, Baumeister W and Saxton WO (1983) Electron microscopy of photosynthetic membranes containing bacteriochlorophyll *b*. Arch Microbiol 135: 169–175

Evans MB, Cogdell RJ and Britton G (1988) Determination of the bacteriochlorophyll:carotenoid ratios of the B890 antenna complex of *Rhodospirillum rubrum* and the B800-850 complex of *Rhodobacter sphaeroides*. Biochim Biophys Acta 935: 292–298

Fowler GJS, Visschers RW, Grief GG, van Grondelle R and Hunter CN (1992) Genetically modified photosynthetic antenna complexes with blue-shifted absorbance bands. Nature 355: 848–850

Freiberg A and Timpmann K (1992) Picosecond fluorescence spectroscopy of light-harvesting antenna complexes from *Rhodospirillum rubrum* in the 300-4 K temperature range. Comparison with the data on chromatophores. J Photochem Photobiol B 15: 151–158

Freiberg A, Godik VI and Timpmann K (1984) Excitation energy transfer in bacterial photosynthesis studied by picosecond laser spectrochronography. In Sybesma C (ed) Progress in Photosynthesis, Vol 1, pp 45–48. Martinus Nijhoff/ Dr W Junk Publishers, Dordrecht

Freiberg A, Godik VI and Timpmann K (1987) Spectral dependence of the fluorescence lifetime of *Rhodospirillum rubrum*. Evidence for inhomogeneity of B880 absorption band. In: Biggins J (ed) Progress in Photosynthesis Research, Vol 1, pp 45–48. Marinus Nijhoff Publisher, Dordrecht

Freiberg A, Godik VI, Pullerits T and Timpmann KE (1988a) Directed picosecond excitation transport in purple photosynthetic bacteria. Chem Phys 128: 227–235

Freiberg A, Pullerits T and Timpmann K (1988b) Picosecond excitation transport in photosynthesis: factors for optimization of light harvesting. In Yayima T, Yoshihara K, Harris CB and Shionoya S (eds) Ultrafast Phenomena, Vol VI, pp 593–595. Springer-Verlag, Berlin, Heidelberg

Freiberg A, Godik VI, Pullerits T and Timpman K (1989) Picosecond dynamics of directed excitation transfer in spectrally heterogeneous light-harvesting antenna of purple bacteria. Biochim Biophys Acta 973: 93–104

Freiberg A, Godik VI, Pullerits T and Timpmann K (1990) Excitation transport and quenching in photosynthetic bacteria at normal and cryogenic temperatures. In: Baltscheffsky M (ed) Current Research in Photosynthesis, Vol II, pp 157–160. Kluwer Academic Publishers

Garcia D, Parot P, Verméglio A and Madigan MT (1986) The light-harvesting complexes of a thermophilic purple sulfur photosynthetic bacterium *Chromatium tepidum*. Biochim Biophys Acta 850: 390–395

Ghosh R, Hauser H and Bachofen R (1988) Reversible dissociation of the B873 light-harvesting complex from *Rhodospirillum rubrum G9+*. Biochemistry 27: 1004–1014

Gingras G and Picorel R (1990) Supramolecular arrangement of *Rhodospirillum rubrum* B880 holochrome as studied by radiation inactivation and electron paramagnetic resonance. Proc Natl Acad Sci USA 87: 3405–3409

Gottfried DS, Stocker JW and Boxer SG (1991) Stark effect spectroscopy of bacteriochlorophyll in light-harvesting complexes from photosynthetic bacteria. Biochim Biophys Acta 1059: 63–75

Gudowska-Nowak E, Newton MD and Fajer J (1990) Conformational and environmental effects on bacteriochlorophyll optical spectra: Correlations of calculated spectra with structural results. J Phys Chem 94: 5795–5801

Hawthornthwaite AM and Cogdell RJ (1991) Bacteriochlorophyll-binding proteins. In: Scheer H (ed), Chlorophylls, pp 493–528. CRC Press, Boca Raton, Florida

Hayashi H, Miyao M and Morita S (1982a) Absorption and fluorescence spectra of light-harvesting bacteriochlorophyll-protein complexes from *Rhodopseudomonas palustris* in the near-infrared region. J Biochem (Tokyo) 91: 1017–1027

Hayashi H, Nakano M and Morita S (1982b) Comparative studies of protein properties and bacteriochlorophyll contents of bacteriochlorophyll-protein complexes from spectrally different types of *Rhodopseudomonas palustris*. J Biochem (Tokyo) 92: 1805–1811

Hayashi H, Nozawa T, Hatano M and Morita S (1982c) Circular dichroism of bacteriochlorophyll a in light-harvesting bacteriochlorophyll-protein complexes from *Rhodopseudomonas palustris*. J Biochem (Tokyo) 91: 1029–1038

Heller BA and Loach PA (1990) Isolation and characterization of a subunit form of the B875 light-harvesting complex from *Rhodobacter capsulatus*. Photochem Photobiol 51: 621–627

Hess S, Visscher K, Ulander J, Pullerits T, Jones MR, Hunter CN and Sundström V (1993a) Direct energy transfer from the peripheral LH2 antenna to the reaction center in a mutant of *Rhodobacter sphaeroides* that lacks the core LH1 antenna. Biochemistry 32: 10314–10322

Hess S, Feldshtein F, Babin A, Nurgaleev I, Pullerits T, Sergeev A and Sundström V (1993b) Femtosecond energy transfer within the LH2 peripheral antenna of the photosynthetic purple bacteria *Rhodobacter sphaeroides* and *Rhodopseudomonas palustris* LL. Chem Phys Lett 216: 247–257

Hess S, Visscher KJ, Pullerits T and Sundström V (1994) Enhanced rates of subpicosecond energy transfer in blue-shifted light harvesting LH2 mutants of *Rhodobacter sphaeroides*. Biochemistry 33: 8300–8305

Hunter CN, Pennoyer JD, Sturgis JN, Farrelly D and Niederman

RA (1988) Oligomerization states and associations of light-harvesting pigment-protein complexes of *Rhodobacter sphaeroides* as analyzed by lithium dodecyl sulfate-polyacrylamide gel electrophoresis. Biochemistry 27: 3459–3467

Hunter CN, van Grondelle R and Olsen JD (1989) Photosynthetic antenna proteins: 100 ps before photochemistry starts. Trends Biochem Sci (Pers Ed) 14: 72–76

Hunter CN, Bergström H, van Grondelle R and Sundström V (1990) Energy-transfer dynamics in three light-harvesting mutants of *Rhodobacter sphaeroides*: a picosecond spectroscopy study. Biochemistry 29: 3203–3207

Jiraskova V, Agalidis I and Reiss-Husson F (1992) Characterization of the core light harvesting complex B875 of *Rhodocyclus gelatinosus* and its B820 derivative. In: Murata M (ed) Research in Photosynthesis, Vol 1, pp 33–36. Kluwer Academic Publishers, Dordrecht

Johnson SG and Small GJ (1989) Spectral hole burning of a strongly exciton-coupled bacteriochlorophyll *a* antenna complex. Chem Phys Lett 155: 371–375

Johnson SG and Small GJ (1991) Excited-state structure and energy-transfer dynamics of the bacteriochlorophyll a antenna complex from *Prosthecochloris aestuarii*. J Phys Chem 95: 471–479

Kingma H, Duysens LNM and Van Grondelle R (1983) Magnetic field-stimulated luminescence and a matrix model for energy transfer. A new method for determining the redox state of the first quinone acceptor in the reaction center of whole cells of *Rhodospirillum rubrum*. Biochim Biophys Acta 725: 434–443

Kleinherenbrink FAM, Deinum G, Otte SCM, Hoff AJ and Amesz J (1992) Energy transfer from long-wavelength absorbing antenna bacteriochlorophylls to the reaction center. Biochim Biophys Acta 1099: 175–181

Kleinherenbrink FAM, Cheng P, Amesz J and Blankenship RE (1993) Lifetimes of bacteriochlorophyll fluorescence in *Rhodopseudomonas viridis* and *Heliobacterium chlorum* at low temperatures. Photochem Photobiol 57: 13–16

Kramer HJM, Van Grondelle R, Hunter CN, Westerhuis WHJ and Amesz J (1984a) Pigment organization of the B800-850 antenna complex of *Rhodopseudomonas sphaeroides*. Biochim Biophys Acta 765: 156–165

Kramer HJM, Pennoyer JD, Van Grondelle R, Westerhuis WHJ, Niederman RA and Amesz J (1984b) Low-temperature optical properties and pigment organization of the B875 light-harvesting bacteriochlorophyll-protein complex of purple photosynthetic bacteria. Biochim Biophys Acta 767: 335–344

Leguijt T, Visschers RW, Crielaard W, van Grondelle R and Hellingwerf KJ (1992) Low-temperature fluorescence and absorption spectroscopy of reaction center/antenna complexes from *Ectothiorhodospira mobilis*, *Rhodopseudomonas palustris* and *Rhodobacter sphaeroides*. Biochim Biophys Acta 1102: 177–185

Loach PA, Parkes PS, Miller JF, Hinchigeri S and Callahan PM (1985) Structure-function relationships of the bacteriochlorophyll-protein light-harvesting complex of *Rhodospirillum rubrum*. In: Arntzen C, Bogorad L, Bonitz S and Steinback K (eds), Molecular Biology of the Photosynthetic Apparatus, pp 197–209. Cold Spring Harbor Laboratory, Cold Spring Harbor, NY

Lockhart DJ and Boxer SG (1988) Stark effect spectroscopy of *Rhodobacter sphaeroides* and *Rhodopseudomonas viridis* reaction centers. Proc Natl Acad Sci USA 85: 107–111

Lu X and Pearlstein RM (1993) Simulations of *Prostechochloris* bacteriochlorophyll a-protein. Optical spectra improved by parametric computer search. Photochem Photobiol 57: 86–91

Meckenstock RU, Brunisholz RA and Zuber H (1992a) The light-harvesting core-complex and the B820-subunit from *Rhodopseudomonas marina*. Part I. Purification and characterisation. FEBS Lett 311: 128–134

Meckenstock RU, Krusche K, Brunisholz RA and Zuber H (1992b) The light-harvesting core-complex and the B820-subunit from *Rhodopseudomonas marina*. Part II. Electron microscopic characterisation. FEBS Lett 311: 135–138

Miller KR (1982) Three-dimensional structure of a photosynthetic membrane. Nature 300: 53–55

Miller JF, Hinchigeri SB, Parkes-Loach PS, Callahan PM, Sprinkle JR, Riccobono JR and Loach PA (1987) Isolation and characterization of a subunit form of the light-harvesting complex of *Rhodospirillum rubrum*. Biochemistry 26: 5055–5062

Monger TG and Parson WW (1977) Singlet-triplet fusion in *Rhodopseudomonas sphaeroides* chromatophores. Biochim Biophys Acta 460: 393–407

Müller MG, Drews G and Holzwarth AR (1993) Excitation transfer and charge separation kinetics in purple bacteria. (1) Picosecond fluorescence of chromatophores from *Rhodobacter capsulatus* wild type. Biochim Biophys Acta 1142: 49–58

Nuijs AM, Van Grondelle R, Joppe HLP, Van Bochove AC and Duysens LNM (1985) Singlet and triplet excited carotenoid and antenna bacteriochlorophyll of the photosynthetic purple bacterium *Rhodospirillum rubrum* as studied by picosecond absorbance difference spectroscopy. Biochim Biophys Acta 810: 94–105

Nuijs AM, van Grondelle R, Joppe HLP, van Bochove AC and Duysens LNM (1986) A picosecond-absorption study on bacteriochlorophyll excitation, trapping and primary-charge separation in chromatophores of *Rhodospirillum rubrum*. Biochim Biophys Acta 850: 286–293

Olson JM and Clayton RK (1966) Sensitization of photoreactions in Eimhjellen's *Rhodopseudomonas* by a pigment absorbing at 830 nm. Photochem Photobiol 5: 655–660

Otte SCM, Kleinherenbrink FAM and Amesz J (1993) Energy transfer between the reaction center and the antenna in purple bacteria., Biochim Biophys Acta 1143: 84–90

Paillotin G, Swenberg CE, Breton J and Geacintov E (1979) Analysis of picosecond laser-induced fluorescence phenomena in photosynthetic membranes utilizing a master equation approach. Biophys J 25: 513–533

Papiz MZ, Hawthornthwaite AM, Cogdell RJ, Woolley KJ, Wightman PA, Ferguson LA and Lindsay JG (1989) Crystallization and characterization of two crystal forms of the B800-850 light-harvesting complex from *Rhodopseudomonas acidophila* strain 10050. J Mol Biol 209: 833–835

Parkes-Loach PS, Sprinkle JR and Loach PA (1988) Reconstitution of the B873 light-harvesting complex of *Rhodospirillum rubrum* from the separately isolated α- and β-polypeptides and bacteriochlorophyll *a*. Biochemistry 27: 2718–2727

Pearlstein RM (1982) Chlorophyll singlet excitations. In Govindjee (ed) Photosynthesis, pp 293–330. Academic Press, New York

Pearlstein RM (1992a) Kinetics of exciton trapping by monocoordinate reaction centers. J Lum 51: 139–147

Pearlstein RM (1992b) Theory of the optical spectra of the

bacteriochlorophyll *a* antenna protein trimer from *Prosthecochloris aestuarii*. Photosynth Res 31: 213–226

Picorel R, L'Ecuyer A, Potier M and Gingras G (1986) Structure of the B880 holochrome of *Rhodospirillum rubrum* as studied by the radiation inactivation method. J Biol Chem 261: 3020–3024

Pullerits T and Freiberg A (1992) Kinetic model of primary energy transfer and trapping in photosynthetic membranes. Biophys J 63: 879–896

Pullerits T, Visscher KJ, Hess S, Sundström V, Freiberg A and van Grondelle R (1994a) Energy transfer in the inhomogeneously broadened core antenna of purple bacteria: A simultaneous fit of low intensity picosecond absorption and fluorescence kinetics. Biophys J 66: 236–248

Pullerits T, Chachisvilis M, Jones MR, Hunter CN and Sundström V (1994b) Exciton dynamics in the light-harvesting complexes of *Rhodobacter sphaeroides*. Chem Phys Lett 224: 355–365

Reddy NRS, Small GJ, Seibert M and Picorel R (1991) Energy transfer dynamics of the B800-B850 antenna complex of *Rhodobacter sphaeroides*: a hole burning study. Chem Phys Lett 181: 391–399

Reddy NRS, Picorel R and Small GJ (1992a) B896 and B870 components of the *Rhodobacter sphaeroides* antenna: A hole burning study. J Phys Chem 96: 6458–6464

Reddy NRS, Cogdell RJ, Zhao L and Small GJ (1992b) Nonphotochemical hole-burning of the B800–B850 antenna complex of *Rhodopseudomonas acidophila*. Photochem Photobiol 57: 35–39

Reddy NRS, Lyle PA and Small GJ (1992c) Applications of spectral hole burning spectroscopies to antenna and reaction center complex. Photosynth Res 31: 167–194

Rijgersberg CP, Van Grondelle R and Amesz J (1980) Energy transfer and bacteriochlorophyll fluorescence in purple bacteria at low temperature. Biochim Biophys Acta 592: 53–64

Robert B and Lutz M (1985) Structures of antenna complexes of several *Rhodospirillales* from their resonance Raman spectra. Biochim Biophys Acta 807: 10–23

Scherz A and Parson WW (1984) Excitation interactions in dimers of bacteriochlorophyll and related molecules. Biochim Biophys Acta 766: 666–678

Scherz A and Parson WW (1986) Interactions of Bacteriochlorophylls in antenna Bacteriochlorophyll-protein complexes of photosynthetic bacteria. Photosynth Res 9: 21–32

Scherz A and Rosenbach-Belkin V (1989) Comparative study of optical absorption and circular dichroism of bacteriochlorophyll oligomers in Triton X-100, the antenna pigment B 850, and primary donor P-860 of photosynthetic bacteria indicates that all are similar dimers of bacteriochlorophyll *a*. Proc Natl Acad Sci USA 86: 1505–1509

Scherz A, Rosenbach-Belkin V and Fisher JRE (1990) Distribution and self-organization of photosynthetic pigments in micelles: Implication for the assembly of light-harvesting complexes and reaction centers in the photosynthetic membrane. Proc Natl Acad Sci USA 87: 5430–5434

Sebban P, Robert B and Jolchine G (1985) Isolation and spectroscopic characterization of the B875 antenna complex of a mutant of *Rhodopseudomonas sphaeroides*. Photochem Photobiol 42: 573–578

Shimada K, Mimuro M, Tamai N and Yamazaki I (1989) Excitation energy transfer in *Rhodobacter sphaeroides* analyzed by the time-resolved fluorescence spectroscopy. Biochim Biophys Acta 975: 72–79

Shimada K, Yamazaki I, Tamai N and Mimuro M (1990) Excitation energy flow in a photosynthetic bacterium lacking B850. Fast energy transfer from B806 to B870 in *Erythrobacter sp. strain OCh 114*. Biochim Biophys Acta 1016: 266–271

Shimada K, Hirota M, Nishimura Y, Yamazaki I and Mimuro M (1992) Excitation energy flow in *Roseobacter denitrificans* (*Erythrobacter* sp. Och 114) at low temperature. In: Murata M (ed) Research in Photosynthesis, Vol 1, pp 137–140. Kluwer Academic Publishers, Dordrecht

Shreve AP, Trautman JK, Frank HA, Owens TG and Albrecht AC (1991) Femtosecond energy-transfer processes in the B800-850 light-harvesting complex of *Rhodobacter sphaeroides 2.4.1*. Biochim Biophys Acta 1058: 280–288

Stark W, Kühlbrandt W, Wildhaber I, Wehrli E and Mühlethaler K (1984) The structure of the photoreceptor unit of *Rhodopseudomonas viridis*. EMBO J 3: 777–783

Stark W, Jay F and Muehlethaler K (1986) Localization of reaction center and light harvesting complexes in the photosynthetic unit of *Rhodopseudomonas viridis*. Arch Microbiol 146: 130–133

Sundström V, van Grondelle R, Bergström H, Åkesson E and Gillbro T (1986) Excitation-energy transport in the bacteriochlorophyll antenna systems of *Rhodospirillum rubrum* and *Rhodobacter sphaeroides*, studied by low-intensity picosecond absorption spectroscopy. Biochim Biophys Acta 851: 431–446

Theiler R and Zuber H (1984) The light-harvesting polypeptides of *Rhodopseudomonas sphaeroides R-26.1*. II. Conformational analyses by attenuated total reflection infrared spectroscopy and the possible molecular structure of the hydrophobic domain of the B 850 complex. Hoppe-Seyler's Z Physiol Chem 365: 721–729

Theiler R, Suter F, Wiemken V and Zuber H (1984a) The light-harvesting polypeptides of *Rhodopseudomonas sphaeroides R-26.1*. I. Isolation, purification and sequence analyses. Hoppe-Seyler's Z Physiol Chem 365: 703–719

Theiler R, Suter F, Zuber H and Cogdell RJ (1984b) A comparison of the primary structures of the two B800-850-apoproteins from wild-type *Rhodopseudomonas sphaeroides strain 2.4.1.* and a carotenoidless mutant *strain R26.1*. FEBS Lett 175: 231–237

Thornber JP, Cogdell RJ, Pierson BK and Seftor REB (1983) Pigment-protein complexes of purple photosynthetic bacteria: an overview. J Cell Biochem 23: 159–169

Timpmann K, Freiberg A and Godik VI (1991) Picosecond kinetics of light excitations in photosynthetic purple bacteria in the temperature range of 300-4 K. Chem Phys Lett 182: 617–622

Timpmann K, Zhang FG, Freiberg A and Sundström V (1993) Detrapping of excitation energy from the reaction center in the photosynthetic purple bacterium *Rhodospirillum rubrum*. Biochim Biophys Acta 1183: 185–193

Timpmann K, Freiberg A and Sundström V (1995) Energy trapping and detrapping in the photosynthetic bacterium *Rhodopseudomonas viridis*: Transfer-to-trap-limited dynamics. Chem Phys, in press

Trautman JK, Shreve AP, Violette CA, Frank HA, Owens TG and Albrecht AC (1990) Femtosecond dynamics of energy transfer in B800-850 light-harvesting complexes of *Rhodobacter sphaeroides*. Proc Natl Acad Sci USA 87: 215–219

Trissl HW, Breton J, Deprez J, Dobek A and Leibl W (1990) Trapping kinetics, annihilation, and quantum yield in the photosynthetic purple bacterium *Rp. viridis* as revealed by electric measurement of the primary charge separation. Biochim Biophys Acta 1015: 322–333

Trüper HG (1968) *Ectothiorhodospira mobilis* Pelsh, a photosynthetic sulphur bacterium depositing sulphur outside cells. J Bacteriol 95: 1910–1920

Valkunas L, Liuolia V and Freiberg A (1991) Picosecond processes in chromatophores at various excitation intensities. Photosynth Res 27: 83–95

Valkunas L, van Mourik F and van Grondelle R (1992) On the role of spectral and spatial antenna inhomogeneity in the process of excitation energy trapping in photosynthesis. J Photochem Photobiol B 15: 159–170

van der Laan H, Schmidt T, Visschers RW, Visscher KJ, Van Grondelle R and Volker S (1990) Energy transfer in the B800-850 antenna complex of purple bacteria *Rhodobacter sphaeroides*: a study by spectral hole-burning. Chem Phys Lett 170: 231–238

van der Laan H, deCaro C, Schmidt TH, Visschers RW, van Grondelle R, Fowler GJS, Hunter CN and Vöker S (1993) Excited-state dynamics of mutated antenna complexes of purple bacteria studied by hole-burning. Chem Phys Lett 212: 569–581

van Dorssen RJ, Hunter CN, Van Grondelle R, Korenhof AH and Amesz J (1988) Spectroscopic properties of antenna complexes of *Rhodobacter sphaeroides* in vivo. Biochim Biophys Acta 932: 179–188

van Grondelle R, Bergström H, Sundström V and Gillbro T (1987) Energy transfer within the bacteriochlorophyll antenna of purple bacteria at 77 K studied by picosecond absorption recovery. Biochim Biophys Acta 894: 313–326

van Grondelle R, Kramer HJM and Rijgersberg CP (1982) Energy transfer in the B800–850 - carotenoid light-harvesting complex of various mutants of *Rhodopseudomonas sphaeroides* and of *Rhodopseudomonas capsulata*., Biochim Biophys Acta 682: 208–215

van Grondelle R, Hunter CN, Bakker JGC and Kramer HJM (1983) Size and structure of antenna complexes of photosynthetic bacteria as studied by singlet-singlet quenching of the bacteriochlorophyll fluorescence yield. Biochim Biophys Acta 723: 30–36

van Mourik F (1993) Spectral inhomogeneity of the bacterial light-harvesting antennae: Causes and consequences. Doctoral thesis, Free University, Amsterdam

van Mourik F, van der Oord CJR, Visscher KJ, Parkes-Loach PS, Loach PA, Visschers RW and van Grondelle R (1991) Exciton interactions in the light-harvesting antenna of photosynthetic bacteria studied with triplet-singlet spectroscopy and singlet-triplet annihilation on the B820 subunit form of *Rhodospirillum rubrum*. Biochim Biophys Acta 1059: 111–119

van Mourik F, Visschers RW and van Grondelle R (1992a) Energy transfer and aggregate size effects in the inhomogeneously broadened core light-harvesting complex of *Rhodobacter sphaeroides*. Chem Phys Lett 193: 1–7

van Mourik F, Corten EPM, van Stokkum IHM, Visschers RW, Loach PA, Kraayenhof R and van Grondelle R (1992b) Self-assembly of the LH-1 light-harvesting antenna of *Rhodospirillum rubrum*: A time-resolved study of the aggregation of the B820 subunit. In: Murata M (ed) Research in Photosynthesis, Vol 1, pp 101–104. Kluwer Academic Publishers, Dordrecht

van Mourik F, Hawthornthwaite AM, Vonk C, Evans MB, Cogdell RJ, Sundström V and van Grondelle R (1992c) Spectroscopic characterization of the low-light B800–850 light-harvesting complex of *Rhodopseudomonas palustris strain 2.1.6*. Biochim Biophys Acta 1140: 85–93

van Mourik F, Visschers RW, Mulder JM and van Grondelle R (1993) Spectral inhomogeneity of the light-harvesting antenna of *Rhodospirillum rubrum* probed by T-S spectroscopy and singlet triplet annihilation at low temperatures., Photochem Photobiol 57: 19–23

Visscher KJ, Bergström H, Sundström V, Hunter CN and van Grondelle R (1989) Temperature dependence of energy transfer from the long wavelength antenna BChl-869 to the reaction center in *Rhodospirillum rubrum, Rhodobacter sphaeroides* (*w.t.* and *M21* mutant) from 77 to 177 K, studied by picosecond absorption spectroscopy. Photosynth Res 22: 211–217

Visscher KJ, Gulbinas V, Cogdell RJ, van Grondelle R and Sundström V (1993a) Ultrafast energy transfer within the light-harvesting antenna of photosynthetic purple bacteria. In: Martin J-L, Migus A, Mourou, GA and Zewail AH (eds) Ultrafast Phenomena VIII, Springer Series in Chemical Physics, Vol. 55, pp 559–561 Springer-Verlag. Berlin, Heidelberg

Visscher KJ, Hess S, Pullerits T, Feldshtein F, Babin A, Gulbinas V, van Grondelle R and Sundström V (1994) Ultrafast energy transfer in LH2 antenna complexes of the photosynthetic purple bacteria *Rhodobacter sphaeroides, Rhodopseudomonas palustris* and *Rhodopseudomonas acidophila*. Lith J Phys 34: 79–88

Visschers RW, Chang MC, van Mourik F, Parkes-Loach PS, Heller BA, Loach PA and van Grondelle R (1991) Reversible dissociation of the B873 light-harvesting complex from *Rhodospirillum rubrum G9+*. Biochemistry 27: 1004–1014

Visschers RW, Nunn R, Calkoen F, van Mourik F, Hunter CN, Rice DW and van Grondelle R (1992) Spectroscopic characterization of B820 subunits from light-harvesting complex I of *Rhodospirillum rubrum* and *Rhodobacter sphaeroides* prepared with the detergent n-octyl-rac-2,3-dipropylsulfoxide. Biochim Biophys Acta 1100: 259–266

Visschers RW, van Mourik F, Monshouwer R and van Grondelle R (1993a) Inhomogeneous spectral broadening of the B820 subunit form of LH1. Biochim Biophys Acta 1141: 238–244

Visschers RW, van Grondelle R and Robert B (1993b) Resonance Raman spectroscopy of the B820 subunit of the core antenna from *Rhodospirillum rubrum* G9. Biochim Biophys Acta 1183: 369–373

Vos M, van Grondelle R, van der Kooij FW, van de Poll D, Amesz J and Duysens LNM (1986) Singlet-singlet annihilation at low temperatures in the antenna of purple bacteria. Biochim Biophys Acta 850: 501–512

Vos M, van Dorssen RJ, Amesz J, van Grondelle R and Hunter CN (1988) The organization of the photosynthetic apparatus of *Rhodobacter sphaeroides*: studies of antenna mutants using singlet-singlet quenching. Biochim Biophys Acta 933: 132–140

Vredenberg W and Duysens LNM (1963) Transfer of energy from bacteriochlorophyll to a reaction center during bacterial photosynthesis. Nature 197: 335–357

Wang RT and Clayton RK (1971) The absolute yield of bacteriochlorophyll fluorescence in vivo. Photochem Photobiol 13: 215–224

Westerhuis WHJ, Vos M, van Dorssen RJ, van Grondelle R, Amesz J and Niederman RA (1987) Supramolecular organization of light-harvesting pigment-protein complexes of *Rhodobacter sphaeroides* studied by excitation energy transfer and singlet-singlet annihilation at low temperature in phospholipid-enriched membranes. In: Biggins J (ed) Progress in Photosynthesis Research, Vol 1, pp 29–32. Marinus Nijhoff Publisher, Dordrecht

Westerhuis WHJ, Xiao Z and Niederman RA (1992) Oligomerization-state dependent spectroscopic properties of the B850 light-harvesting complex of *Rhodobacter sphaeroides R-26.1*. In: Murata M (ed) Research in Photosynthesis, Vol 1, pp 37–40. Kluwer Academic Publishers, Dordrecht

Yeates TO, Komiya H, Rees DC, Allen JP and Feher G (1987) Structure of the reaction center from *Rhodobacter sphaeroides R-26*: Part 3. Membrane-protein interactions. Proc Natl Acad Sci USA 84: 6438–6442

Zhang FG, Gillbro T, van Grondelle R and Sundström V (1992a) Dynamics of energy transfer and trapping in the light-harvesting antenna of *Rhodopseudomonas viridis*. Biophys J 61: 694–703

Zhang FG, van Grondelle R and Sundström V (1992b) Pathways of energy flow through the light-harvesting antenna of the photosynthetic purple bacterium *Rhodobacter sphaeroides*. Biophys J 61: 911–920

Zuber H (1985) Structure and function of light-harvesting complexes and their polypeptides. Photochem Photobiol 42: 821–844

Zuber H and Brunisholz RA (1991) Structure and function of antenna polypeptides and chlorophyll-protein complexes: principles and variability. In: Scheer H (ed) The Chlorophylls, pp 627–703. CRC Press, Boca Raton, FL

Chapter 18

Singlet Energy Transfer from Carotenoids to Bacteriochlorophylls

Harry A. Frank
Department of Chemistry, University of Connecticut, Storrs, CT 06269-3060, USA

Ronald L. Christensen
Department of Chemistry, Bowdoin College, Brunswick, ME 04011, USA

Summary

The close association of carotenoids and bacteriochlorophylls in anoxygenic photosynthetic bacteria facilitates energy transfer processes between these molecules. A better understanding of the association between these two groups of pigments has been derived from significant, recent advances in: (1) The biochemical methodology for the separation and purification of pigment-protein complexes from photosynthetic bacteria; (2) The chemical procedures for the extraction, identification, and reconstitution of carotenoids into these complexes; (3) The analytical techniques for the chromatographic separation and purification of the carotenoid pigments; (4) The steady-state and time-resolved absorption and fluorescence spectroscopic techniques for exploring the nature of the excited states of carotenoids; and (5) Our understanding of the theoretical framework in which the photochemical properties of carotenoids may be cast. The present chapter will discuss these advances and attempt to provide a connection between the unique electronic structure of carotenoids and the manner in which they function as antenna pigments in photosynthesis.

R. E. Blankenship, M. T. Madigan and C. E. Bauer (eds): Anoxygenic Photosynthetic Bacteria, pp. 373–384.

I. Introduction

The abundance of carotenoids in photosynthetic preparations bears witness to the essential roles these pigments play in photosynthesis. Carotenoids are capable of at least three types of photochemical reactions in the photosynthetic apparatus: (1) Quenching of chlorophyll triplet states; (2) Quenching of chlorophyll singlet states; and (3) Singlet energy transfer from carotenoid to chlorophyll. The first two of these processes provide mechanisms by which carotenoids protect photosynthetic systems from singlet oxygen (generated by chlorophyll triplet states) and the harmful effects of excess light energy. We shall focus on singlet energy transfer from carotenoids, paying particular attention to prokaryotic photosynthetic bacteria. Recent advances in this area have come from systematic studies of the energies and decay rates of carotenoid excited states and measurements of the efficiencies of carotenoid-to-chlorophyll singlet energy transfer in vivo. In this chapter we shall explore the relationships between carotenoid electronic structure and the efficiency of energy transfer.

II. The Electronic Structure of Carotenoid Excited States

The photochemistry of carotenoids can be traced to the unique electronic properties of their conjugated π-electron frameworks (Zechmeister, 1962; Kohler 1993). Most carotenoids contain C_{40} carbon skeletons, corresponding to eight isoprene units. The number of conjugated C=C bonds in naturally occurring carotenoids ranges from 3 (e. g. phytoene) to 13 (e. g. spirilloxanthin) with 9–11 conjugated double bonds being typical for many of the carotenoids involved in photosynthesis (neurosporene, spheroidene, β-carotene, etc.). In addition, carotenoids with up to 19 double bonds have been obtained synthetically (Karrer and Eugster, 1951). The energies and dynamics of polyene excited states (the lowest energy excited singlet (S_1) and triplet (T_1) states are of particular importance in understanding energy transfer) are sensitive to the length of conjugation. The connection between biological function and the carotenoid structure is a topic of considerable interest.

Abbreviations: Rb. – Rhodobacter; Rp. – Rhodopseudomonas

Our current understanding of the low energy electronic states of carotenoids, in large part, is based on experimental and theoretical studies of shorter model systems (Hudson et al., 1982 and 1984; Kohler, 1991). Experimental studies of simple polyenes have exploited their relatively high fluorescence yields and their ability to be incorporated into low temperature n-alkane mixed crystals (Granville et al., 1979; D'Amico et al., 1980; Snyder et al., 1985; Simpson et al., 1987; Kohler et al., 1988). These crystals provide single, well-defined polyene conformations and homogeneous distributions of solvent/solute interactions resulting in vibronically-resolved absorption, fluorescence, and fluorescence excitation spectra. Such spectra lead to precise assignments of electronic and vibrational states and accurate determinations of electronic energies. Since energy transfer from carotenoids originates from vibrationally relaxed, electronically excited states, the 'zero-point' levels of these states are of particular relevance in understanding carotenoid/chlorophyll interactions. The high resolution experiments recently have been extended to model polyenes in molecular beams (Leopold et al., 1984; Heimbrook et al., 1984; Buma et al., 1990, 1991, 1992; Bouwman et al., 1990; Petek et al., 1991, 1992, 1993), resulting in an unprecedented view of the electronic and vibrational states of cold, isolated molecules. Although similar experiments are not yet feasible for carotenoids, these high-resolution results are critical for interpreting the typically broad optical spectra obtained from carotenoid solutions.

It is important to emphasize the profound importance of fluorescence spectroscopy in detecting and understanding the nature of the lowest energy S_1 states of polyenes/carotenoids. The high fluorescence quantum yields of intermediate length polyenes (e.g., octatetraene has $\phi_f \sim 0.6$ in low temperature glasses and mixed crystals (Gavin et al., 1978) has been a major factor in establishing these molecules as prototypical, linearly conjugated molecules. In contrast, shorter polyenes (e.g., unsubstituted dienes and trienes) and longer polyenes (e.g., β-carotene, spheroidene, and many other carotenoids of biological relevance) have fluorescence yields of 10^{-4} or less (Petek et al., 1992; Bondarev et al., 1988 and 1989; Gillbro et al., 1989; Cosgrove et al., 1990). The extremely low fluorescence yields of these compounds have presented significant barriers to the detection of their lowest excited singlet states (S_1).

This turns out to be a critical issue, since the lowest energy, $S_0 \rightarrow S_1$ absorptions in polyenes are extremely weak and difficult to detect on the tails of the $S_0 \rightarrow S_2$ absorptions responsible for the intense colors of the carotenoid family.

The $S_1 \rightarrow S_0$ and $S_0 \rightarrow S_1$ transitions have been directly detected in polyene/n-alkane mixed crystals using fluorescence and fluorescence excitation techniques, and it is to these experiments that we owe much of our current understanding of the nature of the S_1 states in linearly conjugated systems. The most characteristic signature of these spectra is the large (~2000–7000 cm^{-1}) shift in energy ('Stokes-shift') between spectral origins (0-0) of the strongly allowed, $S_0 \rightarrow S_2$ absorptions and the $S_1 \rightarrow S_0$ emissions. In addition, the S_1 states exhibit anomalously long fluorescence lifetimes (Hudson et al., 1972, 1973, 1982 and 1984). These features can be explained by the following ordering of singlet electronic energies: $S_0\,(1^1A_g) < S_1\,(2^1A_g) < S_2\,(1^1B_u)$. See Fig. 1. The fluorescence thus corresponds to a symmetry-forbidden, $2^1A_g \rightarrow 1^1A_g$ transition which lies below the strong, symmetry-allowed $1^1A_g \rightarrow 1^1B_u$ absorption. The Stokes-shifts thus provide accurate measures of the $1^1B_u \rightarrow 2^1A_g$ energy differences. The above energy scheme has been verified by high-level calculations on model systems (Schulten et al., 1972 and 1976; Ohmine et al., 1978; Tavan et al., 1979 and 1986; Orlandi et al., 1991). Theory shows that simple molecular orbital models (e.g., Hückel theory which incorrectly predicts a $1^1A_g < 1^1B_u < 2^1A_g$ ordering for the lowest electronic states) do not adequately treat the effects of electron-electron repulsion of the π electrons in the polyene chains. Though the u and g symmetry labels (and the concepts of strictly 'allowed' or 'forbidden' transitions) apply only to polyenes with centers of symmetry (e.g., β-carotene), all polyenes seem to exhibit the same electronic energy orderings and similar relative transition strengths. We shall retain the idealized symmetry labels in discussing the low lying energy levels of carotenoids.

Extending the optical studies of model systems to carotenoids has been hampered by their broad, almost featureless spectra and extremely low fluorescence yields. Early reports of fluorescence from β-carotene indicated an emission origin coinciding with the origin of the strongly allowed, $1^1A_g \rightarrow 1^1B_u$ absorption (Cherry et al., 1968; van Riel et al., 1983; Haley and Koningstein, 1983; Watanabe et al., 1986; Bondarev et al., 1988, 1989). It is important to note that van

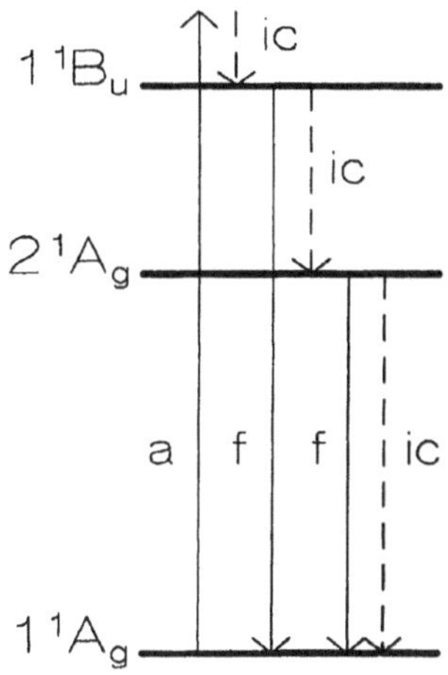

Fig. 1. Schematic representation of the ordering of excited state energies and the intramolecular photochemical processes (a, absorption; f, fluorescence; and ic, internal conversion) carotenoids and polyenes can undergo. The states are labeled by the idealized group theoretical representations in the point group C_{2h}.

Riel et al. (1983) and Bondarev et al. (1988) obtained fluorescence excitation spectra of the weak emission, thus correcting a fundamental shortcoming of the reports of β-carotene fluorescence by Cherry et al. (1968), Haley and Koningstein (1983) and Watanabe et al. (1986). The good agreement between the $1^1A_g \rightarrow 1^1B_u$ absorption and the fluorescence excitation spectra proved the existence of β-carotene fluorescence. Later work of Gillbro and Cogdell (1989) and Cosgrove et al. (1990) confirmed the presence of $1^1B_u \rightarrow 1^1A_g$ ($S_2 \rightarrow S_0$) emissions in β-carotene and related carotenols. The detailed studies of shorter polyenes allows us to identify the emission spectra of β-carotene and other long carotenoids as being due to $S_2 \rightarrow S_0$ fluorescence. $S_1 \rightarrow S_0$ ($2^1A_g \rightarrow 1^1A_g$) emission has not yet been detected in β-carotene, spheroidene, or other carotenoids with more than nine conjugated double bonds, leaving the location of their 2^1A_g states subject to indirect detection.

An early attempt to locate the low-lying 2^1A_g state in β-carotene was made by Thrash et al. (1977, 1979) using resonance Raman excitation profiles. This work suggested that the 2^1A_g state was ~3500 cm^{-1} below the 1^1B_u state placing the S_1 (2^1A_g) states of β-carotene and other carotenoids well above the S_1 (Q_y) states of most chlorophylls. These energetics have been widely cited in discussions of the antenna function of carotenoids in photosynthesis (Siefermann-Harms, 1985). However, a relatively small S_2-S_1 energy difference for β-carotene was not consistent with trends noted in shorter polyenes (Snyder et al., 1985; Cosgrove et al., 1990). Furthermore, a later investigation of the Raman excitation profiles of β-carotene (Watanabe, et al., 1987) could not reproduce the Thrash et al. (1977 and 1979) results. This showed that Raman excitation profiles shared the same limitations as standard absorption measurements in detecting weak $1^1A_g \rightarrow 2^1A_g$ transitions on top of strong $1^1A_g \rightarrow 1^1B_u$ absorption tails. Location of the 2^1A_g state in β-carotene and other carotenoids thus has relied on the extrapolation of trends observed in the $2^1A_g \rightarrow 1^1A_g$ transition energies of shorter, more fluorescent polyenes.

Cosgrove et al. (1990) carried out a systematic study of the connection between the fluorescence properties of shorter polyenes (relatively strong emissions from S_1 (2^1A_g)) and longer carotenoids (weak emissions from S_2 (1^1B_u)). A series of carotenols with 7–11 conjugated double bonds was purified using HPLC techniques, and absorption, fluorescence, and fluorescence excitation spectra were obtained in 77 K glasses. The shorter, less conjugated members of this series exhibited the Stokes-shifted $S_1 \rightarrow S_0$ emissions of short model polyenes. Optical spectra of the carotenols are significantly broader due to well-understood complications brought about by twisting of the cyclohexenylidene end group (Christensen and Kohler 1973; Hemley and Kohler 1977). Nevertheless, as illustrated in the comparison of the fluorescence excitation and emission spectra of β-apo-12'-carotenol and hexadecaheptaene (Fig. 2), there is sufficient resolution in the carotenol spectra to allow the accurate identification of electronic origins. An important feature of Fig. 2 is the almost identical $2^1A_g \rightarrow 1^1A_g$ transition energies of the two heptaenes. This illustrates that in β-apo-12'-carotenol the loss in conjugation due to the nonplanarity between the ring and side-chain is offset by the stabilizing effects of the isoprenoid structure, making hexadecaheptaene an unexpectedly appropriate model for the more complicated carotenol. (By implication, we might predict that the electronic states of β-carotene will have energies quite similar to those of corresponding states of model polyenes with eleven co-planar double bonds.)

Cosgrove et al. (1990) found that the fluorescence of carotenols with more than eight conjugated double bonds was dominated by 'anti-Kasha' $S_2 \rightarrow S_0$ fluorescence as previously had been observed for β-carotene (van Riel et al., 1983; Watanabe et al., 1986; Bondarev et al., 1988; Gillbro et al., 1989). The cross-over from $S_1 \rightarrow S_0$ to $S_2 \rightarrow S_0$ emissions was attributed to a larger $S_2 \rightarrow S_1$ energy difference and the resultant decrease in $S_2 \rightarrow S_1$ radiationless decay (following the 'energy-gap law' of Englman and Jortner, 1970). This allows S_2 emission to compete with internal conversion. The lack of S_1 fluorescence prohibits the direct observation of the S_1 state in longer carotenoids. Nevertheless, trends noted in the model polyenes suggested 2^1A_g energies considerably lower than those originally suggested by the work of Thrash et al. (1977 and 1979).

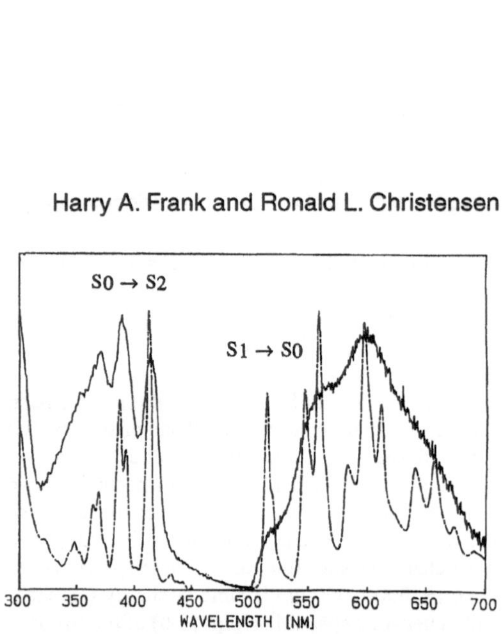

Fig. 2. Comparison of fluorescence and fluorescence excitation spectra of all-trans-2,4,6,8,10,12,14-hexadecaheptaene (dashed lines) and all-trans-β-apo-12'-carotenol (solid lines). The fluorescence spectrum of hexadecaheptaene in *n*-pentadecane at 77 K was obtained by exciting at 414 nm, and the fluorescence excitation spectrum monitored the emission intensity at 558 nm. The fluorescence spectrum of β-apo-12'-carotenol in 77 K EPA was obtained by exciting at 413 nm. The excitation spectrum was obtained by monitoring the fluorescence intensity at 650 nm. This figure from Cosgrove et al. (1990) was reproduced with permission.

DeCoster et al. (1992) extended the Cosgrove (1990) study to provide a systematic comparison of the energy levels of a series of α,ω-dimethylpolyenes (for which the vibronic resolution allows accurate measurement of electronic energies) with a series of iso-structural spheroidene analogs varying only in their extent of conjugation (7, 8, 9 and 10 double bonds). See Fig. 3. The energies of the S_2 and S_1 electronic origins are summarized in Fig. 4. The Cosgrove (1990) study points to the following conclusions: (1) The S_2-S_1 energy difference increases with increasing conjugation, (2) The S_2 and S_1 energies

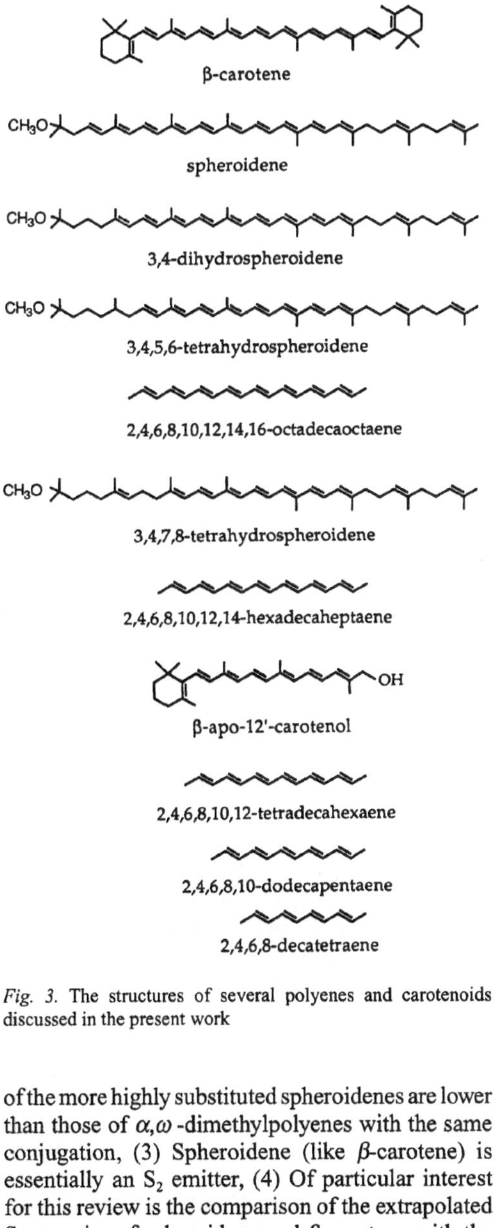

Fig. 3. The structures of several polyenes and carotenoids discussed in the present work

of the more highly substituted spheroidenes are lower than those of α,ω-dimethylpolyenes with the same conjugation, (3) Spheroidene (like β-carotene) is essentially an S_2 emitter, (4) Of particular interest for this review is the comparison of the extrapolated S_1 energies of spheroidene and β-carotene with the S_1 energies of bacteriochlorophylls and chlorophylls. These extrapolations place the S_1 (2^1A_g) energy of spheroidene at 14,000–14,500 cm^{-1}, above the S_1 levels of bacteriochlorophylls, allowing singlet energy transfer to occur via the spheroidene 1^1A_g state. However, the data presented in Fig. 4 strongly suggest the possibility that the S_1 state of β-carotene (11 double bonds) lies below the S_1 state (Q_y) of its chlorophyll acceptor.

Recent estimates of the 2^1A_g energies of spheroidene, β-carotene and other long carotenoids appear to span a rather wide range. For example, Watanabe et al. (1993) suggest a 2^1A_g energy for spheroidene of 14,900 cm^{-1} (hexane)/14,600 cm^{-1} (CS_2). Mimuro et al. (1993) have reported the observation of the $2^1A_g \rightarrow 1^1A_g$(0-0) transition for neurosporene at ~16,000 cm^{-1}, an energy considerably above the 15,300 cm^{-1} obtained by DeCoster et al. (1992) from fluorescence spectra (Fig. 4). The fluorescence assigned to $2^1A_g \rightarrow 1^1A_g$ emission by Watanabe et al. (1993) has almost all of its intensity in a single vibronic band (673 nm) analogous to a porphyrin derivative. This contrasts with the extended vibronic development of typical $2^1A_g \rightarrow 1^1A_g$ polyene emissions (e. g. Fig. 2). Analysis of the weak, long wavelength absorption spectrum of neurosporene (Mimuro et al., 1993) probably overestimates the 2^1A_g zero-point energy. Franck-Condon envelopes of polyene $2^1A_g \leftarrow 1^1A_g$ absorptions show a steep monotonic rise in vibronic intensities with increasing energies due to the $(\Delta E)^{-2}$ energy dependence of vibronic mixing between the 2^1A_g and 1^1B_u states (Petek et al., 1991). This is clearly evident in high resolution spectra of model polyenes (Simpson et al., 1987; Kohler et al., 1988). The $2^1A_g \leftarrow 1^1A_g$(0-0) bands are thus extremely weak compared to higher energy vibronic bands, making them difficult to detect in low resolution experiments.

Similar uncertainties in estimates of the 2^1A_g energy of β-carotene are probably of more relevance, given the possibility that this state may not have enough energy to participate in energy transfer to chlorophyll a (S_1 (Q_y) energy of ~15,000 cm^{-1} (Sauer (1975)). For example, Andersson et al. (1992a) have obtained $S_1 \rightarrow S_0$ fluorescence spectra from the shorter members of a series of β-carotene analogs. Based on estimates of (0-0) energies of shorter members of this series (unlike the spectra of spheroidenes and model polyenes, the fluorescence spectra of the carotene analogs are not sufficiently resolved to observe electronic origins), these authors estimate a β-carotene 2^1A_g energy of ~ 14,500 cm^{-1}. This is significantly higher than the value predicted from extrapolation of the data presented in Fig. 4 and might permit energy transfer from 2^1Ag. Andersson et al. (1992a) also suggest that the $S_2 \rightarrow S_1$ energy gap does not increase with increasing conjugation, a

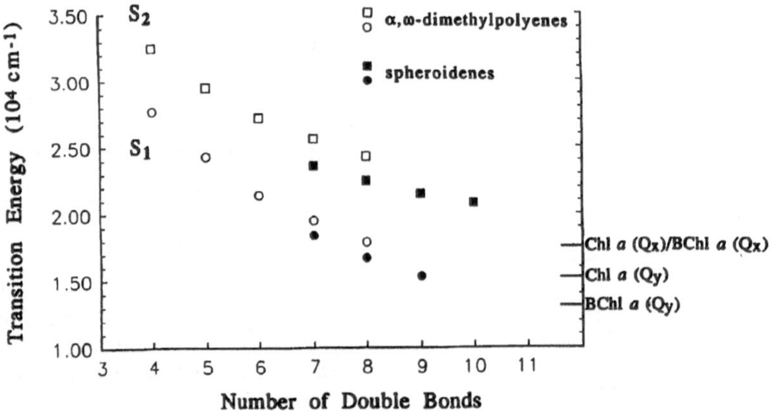

Fig. 4. $S_0 \rightarrow S_1$ ($1^1A_g \rightarrow 2^1A_g$) and $S_0 \rightarrow S_2$ ($1^1A_g \rightarrow 1^1B_u$) transition energies for α,ω-dimethylpolyenes and spheroidenes as a function of conjugated length. Energies of the S_1 states are the electronic origins (0-0 bands) observed for $S_1 \rightarrow S_0$ fluorescence in room temperature methanol. Energies of the S_2 states are the electronic origins (0-0 bands) observed for $S_0 \rightarrow S_2$ absorption in room temperature methanol. Dimethyl polyene data are from Morey and Christensen (unpublished data) and Kohler et al., (1988). Chlorophyll a (Q_x = 575 nm, Q_y = 660 nm) and bacteriochlorophyll a (Q_x = 573 nm, Q_y = 769 nm) transition energies in diethyl ether at room temperature are taken from Sauer (1975). This figure from DeCoster et al. (1992) was reproduced with permission.

result inconsistent with trends noted both in model polyenes and the spheroidenes. Haley et al. (1992) recently suggested a β-carotene 2^1A_g energy of ~14,200 ± 400 cm^{-1}, based on the *maximum intensity* of the reflectance spectrum in zeolite. However, as discussed above, a broad spectrum does not facilitate the identification of a relatively weak electronic origin which very likely lies on the long wavelength tail of the 700 nm peak. Another intriguing result is the recent observation of a weak absorption background in the inverse Raman (Raman loss) spectrum of canthaxanthin (Jones et al., 1992). This shows evidence for a low lying absorption at ~600–700 nm, though the identification of the (0-0) band of the $2^1A_g \leftarrow 1^1A_g$ transition again is subject to considerable uncertainty. Though the earlier (Thrash et al., 1977, 1979) estimates of the 2^1A_g energy of β-carotene clearly should be abandoned, higher resolution spectroscopic work will be needed to accurately locate the $2^1A_g \leftarrow 1^1A_g$ electronic origins in β-carotene and other long carotenoids.

III. The Dynamics of Carotenoid Excited States

A wide variety of methods have been employed for measuring the excited state dynamics of carotenoids. Indirect methods include ground state depletion (Dallinger et al., 1981b; Wylie and Koningstein 1984), lifetime broadening of resonance-Raman spectral lines (Haley and Koningstein 1983), time resolved resonance-Raman spectroscopy (Noguchi et al., 1989; Hashimoto and Koyama 1990; Kuki et al., 1990) and emission yield determinations (Shreve et al., 1991a; Gillbro and Cogdell 1989; Cosgrove et al., 1990, Katoh et al., 1991; Mimuro et al., 1991; and Andersson et al., 1991). The direct methods are picosecond (Wasielewski and Kispert 1986; Gillbro and Cogdell 1989) and femtosecond (Shreve et al., 1991b and 1991c, Frank et al., 1993) time resolved transient absorption measurements.

A. Dynamics of S_1 States

The lifetimes of the 2^1A_g states of a range of carotenoids have now been investigated. Using picosecond transient absorption spectroscopy, S_1 lifetimes of 8.4 ± 0.6 ps for β-carotene, 5.2 ± 0.6 ps for canthaxanthin and 25.4 ± 0.2 ps β-8'-apocarotenal in toluene were found (Wasielewski and Kispert, 1986). Wasielewski et al. (1989) also measured the temperature dependence of the lowest excited singlet state lifetime of all-trans-β-carotene and fully deuterated all-trans-β-carotene in 3-methylpentane. In this work the authors reported an 8.1 ± 0.5 ps 2^1A_g lifetime for all-trans-β-carotene and a 10.5 ± 0.6 ps 2^1A_g lifetime for fully deuterated all-trans-β-carotene.

The weak dependence of the 2^1A_g lifetime on isotopic substitution suggests that conformational changes in the C-H bending and stretching modes are not primary factors in the decay of the 2^1A_g state of carotenoids to the ground state. It was postulated that changes in the frequencies of the C-C stretching modes along the carbon backbone of the carotenoid provide the appropriate accepting modes for nonradiative decay of the 2^1A_g state.

Shreve et al. (1991a,b) used femtosecond ground state depletion/recovery techniques to record an 11 ps lifetime of the 2^1A_g state of β-carotene in CS_2. These authors extended their studies to carotenoids extracted from photosynthetic bacteria and found a 9.1 ps lifetime of the S_1 state of spheroidene in cyclohexane. Noguchi et al. (1989), Hashimoto and Koyama (1990) and Kuki et al. (1990) used picosecond time-resolved resonance Raman spectroscopy to examine the 2^1A_g states of β-carotene and spheroidene and concluded that the lifetime of the 2^1A_g states of these molecules must be at least 10 ps. Thus, several workers have converged on the range 10 ± 2 ps for the lifetimes of the 2^1A_g states of both β-carotene and spheroidene. (See also below.) In addition, Gillbro and Cogdell (1989) used picosecond ground state depletion/recovery experiments and measured the lifetime of the 2^1A_g state of spheroidenone, a carbonyl-containing derivative of spheroidene, to be 15 ps in CS_2.

Monitoring time-resolved fluorescence from the 2^1A_g state is another manner in which one may probe the S_1 dynamics of shorter carotenoids. The time-resolved fluorescence decay profile of fucoxanthin (n = 8) in CS_2 displayed a major kinetic component of 41 ps, while that of β-8'-apocarotenal in the same solvent was 16 ps (Katoh et al., 1991). Andersson et al. (1992) measured a 2.0 ns lifetime for emission from the 2^1A_g state of a β-carotene analog having 5 double bonds.

In an attempt to understand the controlling features of nonradiative decay of the 2^1A_g state of carotenoids, Frank et al. (1993) analyzed a series of four *all-trans*-carotenoids: 3,4,7,8-tetrahydrospheroidene, 3,4,5,6-tetrahydrospheroidene, 3,4-dihydrospheroidene, and spheroidene (Fig. 3). These molecules have π-electron conjugations that systematically increase from 7 to 10 carbon-carbon double bonds. Otherwise, they are structurally identical. The $S_1 \rightarrow S_n$ absorptions decayed with single-exponential kinetics revealing S_1 lifetimes of $\tau = 407 \pm 23$ ps for 3,4,7,8-tetrahydrospheroidene, 85 ± 5 ps for 3,4,5,6-tetrahydrospheroidene, 25.4 ± 0.9 ps for 3,4-dihydrospheroidene, and 8.7 ± 0.1 ps for spheroidene in petroleum ether. The data were analyzed in terms of the energy gap law for radiationless transitions (Englman and Jortner, 1970).

$$k_{ic} = A \cdot \exp(-\gamma \Delta E / \hbar\omega_M) \tag{1}$$

where k_{ic} is the internal conversion rate constant, in this case very well approximated by $1/\tau$ because k_{ic} for polyenes is much larger than the S_1 radiative rate constant, $k_f \sim 10^6$ to $10^7\,s^{-1}$. A is a pre-exponential factor that depends on the S_1-S_0 energy difference, ΔE,

$$A = \frac{C^2\sqrt{2\pi}}{\hbar\sqrt{\hbar\omega_M \Delta E}} \tag{2}$$

For the four spheroidenes A can be safely approximated as a constant. $\hbar\omega_M$ is the energy of the high frequency 'acceptor' modes (presumably C=C stretches of ~ 1600 cm^{-1}), and γ can be related to the relative displacement of the potential surfaces in the two electronic states. The $S_1 \rightarrow S_0$ transition energies deduced for 3,4,5,6-tetrahydrospheroidene, 3,4,7,8-tetrahydrospheroidene and 3,4-dihydrospheroidene compounds from fluorescence experiments (DeCoster et al., 1992) were 18,400 cm^{-1}, 16,700 cm^{-1} and 15,300 cm^{-1}, respectively. The $S_1 \rightarrow S_0$ transition energy for spheroidene has not previously been assigned. A fit of the energy gap law expression to these data yielded a 2^1A_g (S_1) state energy value of 14,100 cm^{-1} for spheroidene, in good agreement with the value determined by extrapolation from fluorescence studies (~14,000 cm^{-1}).

In a similar fashion, Andersson and Gillbro (1992b) studied a systematic series of β-carotene analogs. In this study, analogs of β-carotene possessing 5, 7, 9, 11 (β-carotene), 15 and 19 conjugated double bonds were analyzed by femtosecond transient absorption experiments. The S_1 decay rates were found to increase with the number of carbon-carbon double bonds in qualitative agreement with the energy gap law. The variation in the 2^1A_g energies for this series of carotenoids was so extensive, however, that the full energy dependence including the explicit $1/\sqrt{\Delta E}$ functionality in the pre-exponential term (Eq. (2)) must be considered in treating the data.

B. Dynamics of S_2 States

The lifetimes of the 1^1B_u states of a range of carotenoids have been directly measured. Shreve et al. (1991a,b) used femtosecond transient absorption measurements and determined the room temperature lifetimes of the 1^1B_u states of β-carotene and spheroidene to be 200–250 fs and 340 fs, respectively. The lifetimes varied slightly depending upon the solvent. This group also found that the 1^1B_u state decayed into a second excited state which had a lifetime of 11 ps for β -carotene and 9.1 ps for spheroidene. This second state is in all probability the 2^1A_g state.

C. Connection Between Direct and Indirect Measurements of S_1 and S_2 Dynamics

The fluorescence yield of any state can be expressed as:

$$\Phi_f = \frac{\tau}{\tau_r} \tag{3}$$

where τ is the lifetime and τ_r is the natural radiative lifetime of the state (Strickler and Berg, 1962). The natural radiative lifetime of the S_2 (1^1B_u) state (but not the S_1 (2^1A_g) state) of a carotenoid can be estimated by integrating its absorption profile. This procedure yields a value of about 1 ns for the S_2 radiative lifetime of β-carotene (Gillbro and Cogdell 1989; Shreve et al., 1991a). Fluorescence quantum yields lead to lifetimes of ~200 fs for the 1^1B_u states of β-carotene and spheroidene. This number agrees very well with the direct kinetic measurements of Shreve et al. (1991a,b).

The yield of S_1 emission from the previously described β-carotene analog having 5 carbon-carbon double bonds was $7 \pm 3 \times 10^{-3}$ (Andersson et al., 1992). This value, combined with the 2.0 ns lifetime of this state, yields a natural radiative lifetime for the 2^1A_g state of 300 ns If similar natural radiative lifetimes are assumed for the β-carotene analog with 7 carbon-carbon double bonds and for β-carotene (11 carbon-carbon double bonds), then fluorescence yield measurements indicate lifetimes of their 2^1A_g states of ~240 ps and 3-10 ps, respectively. These values are in good agreement with the lifetimes determined by direct time-resolved absorption measurements and ground state recovery experiments (Andersson and Gillbro, 1992).

IV. Implications for Energy Transfer

The efficiency, ε, of energy transfer between two molecules is defined by

$$\varepsilon = \frac{k_{ET}}{k_{ET} + \sum_i k_i} \tag{4}$$

where k_{ET} is the rate constant for energy transfer and k_i represents any one of several alternative excited state decay pathways. Energy transfer from carotenoids to chlorophylls can easily be demonstrated by comparing the absorption and chlorophyll fluorescence excitation profiles from antenna pigment-protein complexes isolated from photosynthetic bacteria or algae. Normalization of the two spectral profiles at a specific wavelength (e. g. the maximum of the bacteriochlorophyll Q_x transition near 600 nm in the B800–850 complex from photosynthetic bacteria) allows the calculation of ε. Efficiencies measured in this manner range from greater than 90% for spheroidene in the B800–850-complex from *Rhodobacter* (*Rb.) sphaeroides* wild type strain 2.4.1 (Cogdell et al., 1981; Noguchi et al., 1990; van Grondelle et al., 1982) to ~ 25% for the carotenoids, spirilloxanthin, rhodopin, and lycopene in the B880 complex from *Rhodopseudomonas* (*Rp.) acidophila* (Angerhofer et al., 1986). In general, the longer chain carotenoids show lower efficiencies of energy transfer to chlorophylls (Frank and Cogdell, 1993).

These steady state experiments are particularly interesting when comparing a systematic series of complexes. For example, Frank et al. (1993) analyzed the B850 complex from the carotenoidless mutant *Rb. sphaeroides* R-26 after it had been reconstituted with the four spheroidene analogs discussed above (Fig. 3). This study attempted to provide a systematic approach to exploring the effect of excited state energies, spectral overlap and excited state lifetimes on the efficiency of carotenoid-to-bacteriochlorophyll singlet energy transfer. The efficiencies of energy transfer for the four reconstituted complexes could be rationalized from changes in spectral overlaps and an increase in the rate of $2^1A_g \rightarrow 1^1A_g$ internal conversion with increasing conjugation. The work suggested that the general trend of lower efficiencies of energy transfer from carotenoids to bacteriochlorophyll for longer chain carotenoids (e. g. spirilloxanthin) may be due to the 2^1A_g states of these molecules lying below the S_1 state of bacteriochlorophyll. Augmenting these studies with transient dynamics data carried out directly on the isolated pigment-protein complexes would test this hypothesis.

Upon excitation of the B800–850-complex from *Rp. acidophila* strain 7750, Wasielewski et al. (1986b) observed a bleaching of the carotenoid absorption, which was restored with a time constant of 5.6 ± 0.9 ps. The rise-time of a bleaching of a bacteriochlorophyll band at 860 nm occurred in 6.1 ± 0.9 ps indicating the arrival of the carotenoid excitation at the bacteriochlorophyll site. Similar experiments were carried out by Gillbro et al. (1988) on the B800–850-complex from *Rps. acidophila* strain 7050. Energy transfer from the carotenoid to the bacteriochlorophyll took place in 3 ± 1 ps. Femtosecond time-resolved experiments were carried out (Trautman et al., 1990b) on the thylakoid membrane preparations from the diatom *Phaeodactylum tricornutum* and from *Nannochlorpsis sp.* The time for the carotenoid-to-chlorophyll energy transfer in the latter species was found to be < 240 ± 40 fs (*i. e.* instrument limited). Energy transfer in the diatom was bi-exponential with transfer times 500 ± 100 fs and 2.0 ± 0.5 ps and relative amplitudes of 1.7 ± 0.7:1.

Using femtosecond time-resolved techniques, Trautman et al. (1990a) and Shreve et al. (1991b) studied the B800–850 complex isolated from *Rb. sphaeroides* strain 2.4.1. The data were interpreted within the framework of a model whereby the carotenoids transfer energy to both the 800 nm and 850 nm-absorbing bacteriochlorophylls with the majority of the carotenoids present in the complex transferring their energy directly to the 850 nm bacteriochlorophylls in ~300–400 fs. The remainder of the carotenoids transfer their energy via the 800 nm bacteriochlorophylls to the 850 nm-absorbing molecules. The best fit of the dynamics data requires that energy transfer can originate from both the 1^1B_u and the 2^1A_g states of the carotenoid. The suggestion that energy transfer from carotenoids may originate from their 1^1B_u (S_2) states opens up the possibility of S_2 transfer in higher plant systems where S_1 might be too low to participate in energy transfer to chlorophyll *a*.

A fundamental goal of research in this area is to relate the efficiencies and kinetics of carotenoid/chlorophyll energy transfer to the electronic structures and excited state dynamics of the donors and acceptors. Energy transfer between unlike molecules usually is explained using one of two established models. These are the dipole-dipole or coulomb (Förster) mechanism or the electron exchange (Dexter) mechanism. (Förster 1948, 1965; Dexter, 1953; see also Chapter 15 by Struve in this volume). Which of these applies to energy transfer between antenna carotenoids and chlorophylls will depend critically on the answers to the following questions: (1) How do specific molecular features (e. g. distance, geometry, structure, extent of conjugation, stereochemistry, functional groups, etc.) control energy transfer? (2) Which carotenoid excited electronic states (1^1B_u or 2^1A_g) participate in the transfer and how is the overall energy transfer efficiency partitioned between them? (3) Which chlorophyll electronic states (those associated with the Q_x or Q_y transitions or other states) are involved in the transfer?

In principle, which mechanism (coulomb or exchange) operates between carotenoids and chlorophylls can be determined by analyzing how various molecular features affect the efficiency of energy transfer. Singlet energy transfer measurements using steady state and direct dynamics techniques need to be carried out on several different types of complexes including: (1) Well-defined pigment-protein complexes where a range of different carotenoids can be incorporated into a single controlled binding site. As alluded to above, this has recently been provided by Frank et al. (1993), but more work in this area is needed; (2) A variety of antenna complexes whose structures are known and have one carotenoid molecule bound in a single site. In experiments of this sort, the distances between donor and acceptor molecules can be correlated with the measured rates of the energy transfer. 3-dimensional structures of antenna complexes that contain chlorophylls and carotenoids are forthcoming (Cogdell et al., 1985; Papiz et al., 1989; Guthrie et al., 1992). With data on these complexes, a thorough understanding of the factors that control energy transfer between carotenoids and chlorophylls will ultimately be possible.

Acknowledgments

The authors wish to thank Veeradej Chynwat and Susan Lund for their assistance in the preparation of this review. This work was supported in the laboratory of H. A. F. by grants from the National Institutes of Health (GM-30353) and the University of Connecticut Research Foundation. R. L. C. acknowledges the donors of the Petroleum Research Fund, administered by the American Chemical Society, and a DuPont Fund grant to Bowdoin College for support of this research.

References

Andersson PO, Gillbro T, Ferguson L and Cogdell RJ (1991) Absorption spectral shifts of carotenoids related to medium polarizability. Photochem Photobiol 54:353–360

Andersson PO, Gillbro T, Asato AE and Liu RSH (1992a) Dual singlet state emission in a series of mini-carotenes. J Lumin 51: 11–20

Andersson PO and Gillbro T (1992b) Ultrafast radiationless relaxation in macro-carotenes and application of the energy gap law. Laser Spectroscopy of Biomolecules. 1921: 48–56

Angerhofer A, Cogdell RJ and Hipkins MF (1986) A spectral characterization of the light-harvesting pigment-protein complexes from *Rhodopseudomonas acidophila*. Biochim Biophys Acta 848: 333–341

Bondarev SL, Dvornikov SS and Bachilo SM (1988) Fluorescence of *β*-carotene at 77 and 4.2 K. Opt Spectrosc (USSR) 64: 268–270

Bondarev SL, Bachilo SM, Dvornikov SS and Tikhomirov SA (1989) $S_2 \rightarrow S_0$ fluorescence and transient $S_n \leftarrow S_1$ absorption of all-*trans*-*β*-carotene in solid and liquid solutions. J Photochem Photobiol A: Chemistry 46:315–322

Bouwman W, Jones A, Phillips D, Thibodeau P, Friel C and Christensen R (1990) Fluorescence of gaseous tetraenes and pentaenes. J Phys Chem 94: 7429

Buma WJ, Kohler BE and Song K (1990) Location of the 2^1A_g state in hexatriene. J Chem Phys 92: 4622–4623

Buma WJ, Kohler BE and Song K (1991a) Lowest energy excitation singlet state of isolated *cis*-hexatriene. J Chem Phys 94: 6367–6376

Buma WJ, Kohler BE and Song K (1991b) Lowest energy excitation singlet state of isomers of alkyl substituted hexatrienes. J Chem Phys 94: 4691–4698

Cherry RJ, Chapman D and Langelaar J (1968) Fluorescence and phosphorescence of *β*-carotene. Trans Far Soc 64: 2304–2307

Christensen RL and Kohler BE (1973) Low resolution optical spectroscopy of retinyl polyenes: Low-lying electronic levels and spectral broadness. Photochem Photobiol 18: 293

Cogdell RJ, Hipkins MF, MacDonald W and Truscott TG (1981) Energy transfer between the carotenoid and bacteriochlorophyll within the B800–850 light-harvesting pigment-protein complex of *Rps. sphaeroides*. Biochim Biophys Acta 634: 191–202

Cogdell RJ, Woolley K, Mackenzie RC, Lindsay JG, Michel H, Dobler J and Zinth W (1985) Crystallization of the B800850 complex from *Rhodopseudomonas acidophila* strain 7750 In: M E Michel-Beyerle (ed) Antennas and Reaction Centers of Photosynthetic Bacteria, Vol 42, pp 85–91. Springer-Verlag, Berlin

Cosgrove SA, Guite MA, Burnell TB and Christensen RL (1990) Electronic relaxation in long polyenes. J Phys Chem 94: 8118–8124

D'Amico KL, Manos C and Christensen RL (1980) Electronic energy levels in a homologous series of unsubstituted linear polyenes. J Am Chem Soc 102: 4671–4675

Dallinger RF, Woodruff WH and Rodgers MA (1981) The lifetime of the excited singlet state of *β*-carotene: Consequences to photosynthetic light harvesting. Photochem Photobiol 33: 275–277

DeCoster B, Christensen RL, Gebhard R, Lugtenburg J, Farhoosh R and Frank HA (1992) Low-lying electronic states of carotenoids. Biochim Biophys Acta 1102: 107–114

Dexter DL (1953) A theory of sensitised luminescence in solids. J Chem Phys 21: 836–860

Englman R and Jortner J (1970) The energy gap law for radiationless transitions in large molecules. Mol Phys 18: 145–164

Förster TH (1948) Intermolecular energy transfer and fluorescence. Ann Phys 2: 55–75

Förster TH (1965) Action of light and organic crystals In: Sinanoglu O (ed) Modern Quantum Chemistry, Part III, pp 93–137. Academic Press, New York

Frank HA and Cogdell RJ (1993) Photochemistry and function of carotenoids in photosynthesis. In: Young A and Britton G (eds) Carotenoids in Photosynthesis, pp 253–326. Chapman and Hall, London

Frank HA, Farhoosh R, Gebhard R, Lugtenburg J, Gosztola D and Wasielewski MR (1993) The dynamics of the S_1 excited states of carotenoids. Chem Phys Lett 207: 88–92

Gavin RM, Weisman C, McVey JK and Rice SA (1978) Spectroscopic properties of polyenes. III. 1,3,5,7-Octatetraene. J Chem Phys 68: 522–529

Gillbro T and Cogdell RJ (1989) Carotenoid fluorescence. Chem Phys Lett 158: 312–316

Gillbro T, Cogdell R and Sundström V (1988) Energy transfer from carotenoid to bacteriochlorophyll a in the B800-820 antenna complexes from *Rhodopseudomonas acidophila* strain 7050. FEBS Lett 235: 169–172

Granville MF, Holtom GR, Kohler BE, Christensen RL and D'Amico KL (1979) Experimental confirmation of the dipole forbidden character of the lowest excited singlet state in 1,3,5,7-octatetraene. J Chem Phys 70: 593–597

Guthrie N, MacDermott G, Cogdell RJ, Freer AA, Isaacs NW, Hawthornthwaite AM, Halloren E and Lindsay JG (1992) Crystallization of the B800–820 light-harvesting complex from *Rhodopseudomonas acidophila* strain 7750. J Mol Biol 224: 527–528

Haley LV and Koningstein JA (1983) Space and time-resolved resonance-enhanced vibrational Raman spectroscopy from femtosecond-lived singlet excited state of *β*-carotene. Chem Phys 77: 1–9

Hashimoto H and Koyama Y (1990) The $2^1A_g^-$ state of a carotenoid bound to spinach chloroplasts as revealed by picosecond transient Raman spectroscopy. Biochem Biophys Acta 1017: 181–186

Heimbrook LA, Kohler BE and Levy IJ (1984) Fluorescence from the 1^1B_u state of *trans, trans*-1,3,5,7-octatetraene in a free jet. J Chem Phys 81: 1592–1597

Hemley R and Kohler B (1977) Electronic structure of polyenes related to the visual chromophore: A simple model for the observed band shapes. Biophys J 20: 377

Hudson BS and Kohler BE (1972) A low lying weak transition in the polyene *α,ω*-diphenyloctatetraene. Chem Phys Lett 14: 299

Hudson BS and Kohler BE (1973) Polyene spectroscopy: The lowest energy excited single state of diphenyloctatetraene and other linear polyenes. J Chem Phys 59: 4984–5002
Hudson BS and Kohler (1984) Electronic structure and spectra of finite linear polyenes. Synthetic Metals 9: 241–253
Hudson BS, Kohler BE and Schulten K (1982) Linear polyene electronic structure and potential surfaces. In: Lim EC (ed) Excited States, Vol 6, pp 1–95. Academic Press, New York
Jones PF, Jones WJ and Davies B (1992) Direct observation of the 2^1A_g-electronic state of carotenoid molecules by consecutive two-photon absorption spectroscopy. J Photochem Photobiol A: Chem 68: 59–75
Karrer P and Eugster CH (1951) Carotinoidsynthesen. VIII. Synthese des Dodecapreno-β-carotins. Helv Chim Acta 34: 1805
Katoh T, Nagashima U and Mimuro M (1991) Fluorescence properties of the allenic carotenoid fucoxanthin: Implication for energy transfer in photosynthetic systems. Photosyn Res 27: 221–226
Kohler BE (1991) Electronic properties of linear polyenes. In: Brédas JL and Silbey R (eds) Conjugated Polymers: The Novel Science and Technology of Conducting and Nonlinear Optically Active Materials, pp 405–434. Kluwer Academic Publishers, Dordrecht
Kohler BE (1993) Carotenoid electronic structure. In: Pfander H, Liaaen-Jensen S and Britton G (eds) Carotenoids, Vol 1B. Birkhäuser Verlag AG, Basel (in press)
Kohler B, Spangler C and Westerfield C (1988) The 2^1A_g state in the linear polyene 2,4,6,8,10,12,14,16-octadecatetraene. J Chem Phys 89: 5422–5428
Kuki M, Hashimoto H and Koyama Y (1990) The $2^1A_g^-$-state of a carotenoid bound to the chromatophore membrane of *Rhodobacter sphaeroides* 2.4.1 as revealed by transient resonance Raman spectroscopy. Chem Phys Lett 165: 417–422
Leopold DG, Vaida V and Granville MF (1984a) Direct absorption spectroscopy of jet cooled polyenes, I. The $1^1B_u \leftarrow 1^1A_g$ transition of *trans, trans*-1,3,5,7-octatetraene. J Chem Phys 81: 4210–4217
Leopold DG, Pendley RD, Roebber JL, Hemley RJ and Vaida V (1984b) Direct absorption spectroscopy of jet cooled polyenes, II. The $1^1B_u \leftarrow 1^1A_g$ transition of butadienes and hexatrienes. J Chem Phys 81: 4218–4229
Michel H (1982) Three-dimensional crystals of a membrane protein complex. The photosynthetic reaction centre from Rhodopseudomonas viridis. J Mol Biol 158: 567–572
Mimuro M, Nishimura Y, Yamazaki I, Katoh T and Nagashima U (1991) Fluorescence properties of the allenic carotenoid fucoxanthin: Analysis of the effect of keto carbonyl group by using a model compound, all-*trans*-β—apo-8'-carotenal, J Luminescence 50: 1–10
Mimuro M, Nagashima U, Nagaoka S, Takaichi S, Yamazaki I, Nishimura Y and Katoh T (1993) Direct measurement of the low-lying singlet excited (2^1A_g) state of a linear carotenoid, neurosporene, in solution. Chem Phys Lett 204: 101–105
Noguchi T, Kolaczkowski S, Arbour C, Armaki S, Atkinson GH, Hayashi H and Tasumi M (1989) Resonance Raman spectrum of the excited 2^1A_g state of β-carotene. Photochem Photobiol 50: 603–609
Noguchi T, Hayashi H and Tasumi T (1990) Factors controlling the efficiency of energy transfer from carotenoids to bacteriochlorophyll in purple photosynthetic bacteria. Biochim Biophys Acta 1017: 280–290
Ohmine I, Karplus M and Schulten K (1978) Renormalized configuration interaction method for electron correlation in the excited states of polyenes. J Chem Phys 68: 2298–2318
Orlandi G, Zerbetto F and Zgierski MZ (1991) Theoretical analysis of spectra of short polyenes. Chem Rev 91: 867–891
Papiz MZ, Hawthornthwaite AM, Cogdell RJ, Woolley KJ, Wightman PA, Ferguson LA and Lindsay JG (1989) Crystallization and characterization of two crystal forms of the B800–850 light-harvesting complex from *Rhodopseudomonas acidophila* strain 10050. J Mol Biol 209: 833–835
Petek H, Bell AJ, Yoshihara K and Christensen RL (1991) Spectroscopic and dynamical studies of the S_1 and S_2 states of decatetraene in supersonic molecular beams. J Chem Phys 95: 4739–4750
Petek H, Bell AJ, Christensen RL and Yoshihara K (1992a) Fluorescence excitation spectra of the S_1 states of isolated trienes. J Chem Phys 96: 2412–2415
Petek H, Bell AJ, Kandori H, Yoshihara K and Christensen RL (1992b) Spectroscopy and dynamics of a model polyene decatetraene: A study of non-radiative pathways in S_1 and S_2 states under isolated conditions. In: Takahashi H (ed) Time resolved vibrational spectroscopy V, Vol 5, pp 198–199. Springer, Berlin
Petek H, Bell AJ, Choi YS, Yoshihara K, Tounge BA and Christensen RL (1993) The 2^1A_g state of *trans, trans*-1,3,5,7-octatetraene in free jet expansions. J Chem Phys 98: 3777–3794
Sauer K (1975) Primary events and the trapping of energy. In: Govindjee (ed), Bioenergetics of Photosynthesis, pp 115–181. Academic Press, New York
Schulten K and Karplus M (1972) On the origin of a low-lying forbidden transition in polyenes and related molecules. Chem Phys Lett 14: 305–309
Schulten K, Ohmine I and Karplus M (1976) Correlation effects in the spectra of polyenes. J Chem Phys 64: 4422–4441
Shreve AP, Trautman JK, Owens TG and Albrecht AC (1991a) Determination of the S_2 lifetime of β-carotene. Chem Phys Lett 178: 89–96
Shreve AP, Trautman JK, Frank HA, Owens TG and Albrecht AC (1991b) Femtosecond energy-transfer processes in the B800–850 light- harvesting complex of Rhodobacter sphaeroides 2.4.1 Biochim Biophys Acta 1058: 280–288
Shreve AP, Trautman JK, Owens TG and Albrecht AC (1991c) A femtosecond study of in vivo and in vitro electronic state dynamics of fucoxanthin and implications for photosynthetic carotenoid-to-chlorophyll energy transfer mechanisms. Chem Phys 154: 171–178
Siefermann-Harms D (1985) Carotenoids in photosynthesis. I. Location in photosynthetic membranes and light-harvesting function. Biochim Biophys Acta 811: 325–355
Snyder R, Arvidson E, Foote C, Harrigan L and Christensen RL (1985) Electronic energy levels in long polyenes: $S_2 \rightarrow S_0$ emission in all-*trans*-1,3,5,7,9,11,13-tetradecaheptaene. J Am Chem Soc 107: 4117–4122
Simpson JH, McLaughlin L, Smith DS and Christensen RL (1987) Vibronic coupling in polyenes: High resolution optical spectroscopy of all-*trans*-2,4,6,8,10,12,14-hexadecaheptaene. J Chem Phys 87: 3360–3365
Strickler SJ and Berg RA (1962) Relationship between absorption

intensity and fluorescence lifetime of molecules. J Chem Phys 37: 814–822

Tavan P and Schulten K (1979) The 2^1A_g-1^1B_u energy gap in the polyenes: An extended configuration interaction study. J Chem Phys 70: 5407–5413

Tavan P and Schulten K (1986) The low-lying electronic excitations in long polyenes: A PPP-MRD-CI study. J Chem Phys 85: 6602–6609

Thrash RJ, Fang H and Leroi GE (1977) The Raman excitation profile spectrum of *β*-carotene in the preresonance region: Evidence for a low-lying singlet state. J Chem Phys 67: 5930–5933

Thrash RJ, Fang H and Leroi GE (1979) On the role of forbidden low-lying excited state of light-harvesting carotenoids in energy transfer in photosynthesis. Photochem Photobiol 29: 1049–1050

Trautman JK, Shreve AP, Violette CA, Frank HA, Owens TG and Albrecht AC (1990a) Femtosecond dynamics of energy transfer in B800–850 light-harvesting complexes of *Rhodobacter sphaeroides*. Proc Natl Acad Sci USA 87: 215–219

Trautman JK, Shreve AP, Owens TG and Albrecht AC (1990b) Femtosecond dynamics of carotenoid-to-chlorophyll energy transfer in thylakoid membrane preparations from *Phaeodactylum tricornutum* and *Nannochlorisepsis sp.* Chem Phys Lett 166: 369–376

van Grondelle, R., Kramer, H. J. M. and Rijgersberg, C. P. (1982) Energy transfer in the B800–850 carotenoid light-harvesting complex of various mutants of *Rhodopseudomonas sphaeroides* and of *Rhodopseudomonas capsulata.* Biochim Biophys Acta 682: 208–215

van Riel M, Kleinen-Hammans J, van de Ven M, Verwer W, and Levine Y (1983) Fluorescence excitation profiles of *β*-carotene in solution and in lipid/water mixtures. Biochem Biophys Res Comm 113: 102–107

Wasielewski MR and Kispert LD (1986) Direct measurement of the lowest excited singlet state lifetime of all-trans-*β*-carotene and related carotenoids. Chem Phys Lett 128: 238–243

Wasielewski MR, Tiede DM and Frank HA (1986) Ultrafast electron and energy transfer in reaction center and antenna proteins from photosynthetic bacteria. In: Fleming GR and Siegman AE (eds) Ultrafast phenomena, pp 388–392. Springer-Verlag, Berlin

Wasielewski MR, Johnson DG, Bradford EG and Kispert LD (1989) Temperature dependence of the lowest excited singlet-state lifetime of all-*trans-β*-carotene and Fully Deuterated all-*trans-β*-carotene. J Chem Phys 91: 6691–6697

Watanabe J, Kinoshita S, and Kushida T (1986) Non-motional narrowing effects in excitation profiles of second-order optical processes: Comparison between Stochastic Theory and experiments in *β*-carotene. Chem Phys Lett 126: 197–200

Watanabe J, Kinoshita S and Kushida T (1987) Effects of nonzero correlation time of system-reservoir interaction on the excitation profiles of second-order optical processes in *β*-carotene. J Chem Phys 87: 4471–4477

Watanabe Y, Kameyama T, Miki Y, Kuri M and Koyama Y (1993) The $2^1A_g^-$-state and two additional low-lying electronic states of spheroidene newly identified by fluorescence and fluorescence-excitation spectroscopy at 170 K. Chem Phys Lett 206: 62–68

Wylie IW and Koningstein JA (1984) Photoisomerization and time-resolved Raman studies of 15,15'-*cis-β*-carotene and 15,15'-*trans-β*-carotene. J Phys Chem 88: 2950–2953

Zechmeister L, ed (1962) *Cis-trans* isomeric carotenoids, vitamin A, and arylpolyenes. Academic Press, New York

Chapter 19

Coupling of Antennas to Reaction Centers

Arvi Freiberg
Institute of Physics, Estonian Academy of Sciences, Riia 142, EE-2400 Tartu, Estonia

Summary

Recent years have nullified an old illusion that all that is needed for a complete picture of photosynthetic excitation transfer and trapping dynamics is detailed structural information about the system. This still seems to be only a prerequisite. The present chapter highlights the problems concerning the coupling of light-harvesting antennas to reaction centers. Structural, kinetic as well as functional aspects of the coupling are considered. As the antenna-reaction center coupling is largely governed by the specific protein medium, interactions of pigments with their protein surroundings and the elastic properties of these proteins are studied by employing high hydrostatic pressure technique. Excitation transfer from the antenna to the reaction center takes place in tens of picoseconds. The reaction center is a rather perfect trap. In conditions of photosynthesis, only about a quarter of the energy entering the reaction center is transferred back to the antenna. A conclusion is made that future progress in the field is expected by a more detailed study of the dynamical properties of pigment-protein complexes.

R. E. Blankenship, M. T. Madigan and C. E. Bauer (eds): Anoxygenic Photosynthetic Bacteria, pp. 385–398.

I. Introduction

In the previous encyclopedia about the photosynthetic bacteria, issued 15 years ago, the contemporary understanding of the antenna-reaction center coupling problem was reviewed by Zankel (1978). His conclusions were briefly as follows: Energy transfer from antenna pigments to the reaction center (RC) is channeled through the longest-wavelength antenna component. Transfer from the antenna to the RC takes place in tenths of nanoseconds. The same time is characteristic of the transfer between long-wavelength antenna components. Energy in an active RC becomes trapped in about 7 ps. Within this time a fraction of the energy entering the RC is transferred back to the antenna. A spectrally homogeneous model was considered sufficient to interpret most of the available experimental data, though Zankel realized that this was an oversimplification. The results of emerging picosecond time-resolved kinetic studies remained contradictory and confusing at that time.

The aim of this chapter is to review recent developments in the study of the coupling between the antenna and the RC in photosynthetic bacteria. In this respect, purple photosynthetic bacteria represent without doubt the most extensively studied and best characterized systems. We shall see, however, that even in these relatively simple objects ('hydrogen atoms of photosynthesis') our knowledge is far from being complete and further studies are necessary. The presentation given here is selective rather than comprehensive. The author has focused on what are, from his point of view, some of the most interesting and relevant items of the problem, trying to examine what is understood and what is not.

The contents of this chapter are closely related to (and sometimes overlapping) that of several others in the book, which are referenced where appropriate.

II. The Concept of a Photosynthetic Unit

The concept of a photosynthetic unit (PSU) forms the basis of the modern perceptions of the primary processes of photosynthesis. It originates from the experiments by Emerson and Arnold (1932). They have determined that, on the average, one oxygen molecule per 2500 chlorophyll molecules is formed when *Chlorella* cells are illuminated with a single light flash of saturating intensity. Dividing this number by 8, the minimum number of light quanta per oxygen molecule evolved, one reaches a conclusion that about 300 chlorophyll molecules cooperate in producing oxygen. In order to explain this cooperation it was postulated that the absorbed light quantum can transfer between chlorophyll molecules until it becomes trapped at a specific site where the energy is used to drive the photochemical reaction. The trap was given a special name: reaction center.

The statistical ensemble consisting of the RC and all the light-absorbing pigment molecules contributing excitation energy to this RC, has been defined as a PSU. The organization of the photosynthesis apparatus as PSUs increases the effective optical cross section of the RC (for review, see Mauzerall and Greenbaum, 1989).

The photochemistry leading to the evolution of oxygen and its coupled dark reactions occur within milliseconds, after which the original state is restored. In case of a moderate solar energy fluence every chlorophyll molecule absorbs a photon on the average once per second. This means that in the antenna of an approximate size of a hundred chlorophyll molecules, virtually each photon absorbed can effectively be utilized by a single RC.

The major experimental proof of the PSU concept was given in the early 1950s by Duysens (1952). Energy transfer between different pigments was observed in fluorescence studies in which the energy absorbed by one type of pigments sensitized emission from a spectrally distinct pigment. A small decrease of the near infrared absorption band was observed in the bacteria under illumination, which was correctly attributed to excitation energy trapping by the mechanism of photochemical bleaching of a small fraction of bacteriochlorophyll (BChl) molecules.

According to the PSU concept, the primary steps of photosynthesis comprise: (i) absorption of a photon by antenna pigments; (ii) transfer of excitation energy through the antenna to the RC pigments; (iii) trapping of energy at the RC by light-driven oxidation reaction of a special RC complex; (iv) a sequence of electron transfer steps within the RC.

Abbreviations: BChl – bacteriochlorophyll; *Cm. – Chromatium*; PSU – photosynthetic unit; *Rb. – Rhodobacter*; RC – reaction center; *Rp. – Rhodopseudomonas*; *Rs. – Rhodospirillum*; *Tc. – Thiocapsa*

III. Structure and Organization of Real Photosynthetic Systems

A. Antennas

An insight into the molecular organization of the photosynthetic apparatus has been given by a number of interdisciplinary studies involving spectroscopy, crystallography, biochemical and genetic analyses. Although much is known about the structure of RCs (see Chapter 23 by Lancaster et al.), similar information about the antenna systems is still very incomplete.

Antenna systems, like RCs, represent biopolymers containing pigment molecules. BChl-containing antennas are located either within a cytoplasmic membrane, like in purple bacteria (see Chapter 16 by Zuber and Cogdell), or within extra-membrane vesicles (chlorosomes) attached to the membrane, like in green bacteria (see Chapter 20 by Blankenship et al.). While the membrane part of the antenna of green bacteria is similar to that in purple bacteria, the composition and organization of chlorosomes is completely different. The recently discovered heliobacteria are unique in the sense that their photosynthetic apparatus is fully contained in the cell membrane (see Chapter 31 by Amesz).

Mauzerall and Greenbaum (1989) motivate the great variability of antenna structures with the limitations to the size of the PSU which is set by the finite radiative lifetime. Nature takes advantage of several macromolecular organizational principles to minimize the loss in efficiency of energy utilization. Reduction of the dimensionality of energy transfer, as it takes place in the chlorosomes, is one of the possibilities. The specific organization of membrane-bound pigment-protein complexes, ensuring the energy funneling towards RCs, as can be found in some purple bacteria where the RC is surrounded by core antenna complexes which in turn are surrounded by peripheral antenna complexes absorbing at higher energy, is an another one. By these means the size (number of antenna pigments per RC) of the PSU can be increased to 1000 molecules and more.

Difficulties with the crystallization of membrane antenna proteins do not allow an application of powerful X-ray structural analyses in full as yet. Electron microscopy (Miller, 1982; Welte and Kreutz, 1982) of freeze-fractured thylakoid membranes of *Rhodopseudomonas* (*Rp.*) *viridis* indicates an ordered pattern (with a lattice constant about 10–12 nm) of hexagonal particles. A large central RC structure (5–7 nm in diameter) protrudes from both sides of the membrane and is surrounded by six smaller antenna structures. In another purple bacterium, ***Rp. marina***, cyclic structures with an approximate diameter of 10 nm for both membranes with RCs and RC-less core antenna complexes have been revealed (Meckenstock et al., 1993).

The photosynthetic membranes of bacteria are not actually crystalline, despite a good periodic arrangement of morphological complexes seen in averaged electron-microscopic pictures (Engelhardt et al., 1985). The real structural nonuniformities may be blurred by instrumental reasons. Specifically, the RC complex is an ellipsoid with an approximate cross-section of 4 times 7 nm rather than a symmetric structure (see Chapter 23 by Lancaster et al.).

Many of the various antenna pigment-protein complexes of bacteria can now be isolated in a spectrally and functionally intact form. It appears that the RC is always associated with core antenna complexes of an almost fixed stoichiometric ratio: 20–40 antenna BChl molecules per RC (Varga and Staehelin, 1985). The peripheral antenna forms the variable component of the photosynthetic apparatus depending on growth conditions. In an early stage of the membrane development the mean distance between the complexes is larger than in a mature organism. In such membranes the energy transfer between the peripheral antenna and the core is inhibited (Pradel et al., 1978).

It is possible to tackle the spatial relationship problem from the other side by trying to restore the excitation transfer function by incorporation of the isolated complexes into artificial liposome membranes. Such attempts have been proved successful in the case of antennas from *Chromatium* (*Cm.*) *minutissimum* (Moskalenko et al., 1992). However, the energy transfer conditions created were less optimal than those in the native state.

Fundamental for the formation of functionally active antenna pigment clusters are protein polypeptides that are responsible for the hierarchical organization of the entire antenna based on a specific polypeptide aggregation. In purple bacteria the antenna consists of pairs of chemically different hydrophobic polypeptides, α- and β-polypeptides (see Chapter 16 by Zuber and Cogdell). Each of these membrane-spanning polypeptides hosts 50–60 amino acid residues. Protein polypeptides not only provide the environment, but also determine the position and orientation of pigment molecules as well as the distance between them (1–2 nm in an $\alpha\beta$-pair). Two to three BChl and one to two carotenoid molecules are non-covalently bound to an $\alpha\beta$-heterodimer. The actual size of the in vivo $\alpha\beta$-aggregates is still under debate. The reconstitution of the antenna from separately isolated α- and β-polypeptides and pigments has been demonstrated (see Chapter 21 by Loach and Parkes-Loach). It has been proposed that the basic building element is $\alpha\beta(BChl)_2$ and two basic elements are sufficient to form a complex which has spectral characteristics similar to the native core antenna complex. It is likely a weakly coupled aggregate of two dimers where two BChls are strongly coupled (see Chapter 17 by Sundström and van Grondelle). The BChl dimer may be similar to the RC primary electron donor.

B. Structural Aspects of Antenna-Reaction Center Coupling

Little is known about the spatial relationships and interactions of RC polypeptides with light-harvesting complexes. Three different arrangements of peripheral antenna complexes and core complexes with respect to RCs in the photosynthetic membrane plane have been proposed: (i) peripheral antenna complexes surround and structural connect core complexes, which in their turn host the pool of RCs (Monger and Parson, 1977); (ii) an isolated RC, surrounded by core complexes, in a pool of peripheral antenna complexes (Deinum et. al., 1991b); (iii) an arrangement where both peripheral and core complexes have close spatial relationships with the RC (Drews, 1985).

The ultimate purpose of these arrangements is to form an effective heterogeneous excitation energy transfer system towards the RC. Some of the peripheral complexes, however, seem to be more remote and, correspondingly, less effective energy mediators (Moskalenko and Toropygina, 1988; Freiberg et al., 1989). Reversible chemical cross-linking studies (Drews, 1985) give strong evidence that close spatial contacts (about 1 nm) exist between both the α and β polypeptides of the core antenna complex and the H protein subunit of the RC. The H subunit does not bind pigments. Its protein backbone spans the membrane once (five times in the case of L and M subunits) and forms a globular structure outside the membrane that makes extensive contact with the cytoplasmic side of the L and M subunits (see Chapter 23 by Lancaster et al.).

A close spatial contact of core antenna complexes with the RC is a prerequisite of an effective energy utilization. The dilution of the membrane by lipids results in the uncoupling of the antenna and RC with respect to energy transfer (Takemoto et al., 1985).

In membranes the core antenna BChl molecules bound to $\alpha\beta$-polypeptides lie asymmetrically on the periplasmic side. The special pair of the RC is also localized on the periplasmic side. This specific location is the basis for two-dimensional energy transfer parallel to the membrane plane within and between antenna complexes and the RC. The pigments of the core antenna, like the primary donor pigments, have their molecular plane oriented perpendicular to the membrane plane. The long-wavelength Q_y transition dipoles of antenna and RC pigments are believed to lie approximately parallel to the membrane plane (see Chapter 16 by Zuber and Cogdell).

IV. Trapping of Energy by the Reaction Center

The migration of singlet excited states from the antenna to the RC has intrigued theorists since the pioneering work of Franck and Teller in 1938 (for discussion see Robinson, 1967). The first clear experimental demonstration of an effective trapping of antenna excitations by RCs was given by Vredenberg and Duysens (1963). They also observed an increase of the fluorescence yield from photosynthetic bacteria when the photochemical machinery became saturated by the incoming light energy.

As to the kinetics of the excitation transfer through the antenna and from the antenna to the RC, the early experimental evidence was scanty and inconsistent. Thus, Zankel and Clayton (1969) and Zankel (1978) concluded that the average trapping time constant is rather slow, about a nanosecond. This was supported by the data of Govindjee et. al. (1972) who found that the fluorescence lifetime of bacteria increases from 0.3–1 ns as the RC traps are closed. Borisov and Godik (1970), however, have proposed much shorter lifetimes, around 50 ps for the photoactive RCs and 0.5 ns for the inactive, closed ones. This conclusion was based on a single-frequency phase fluorimetric measurements. The possibility to resolve picosecond decays by this technique was obscure and the data fell under criticism (see, e.g., Zankel, 1978). Another source of difficulties is the dependence of kinetics on the excitation intensity that is crucial when single picosecond pulses are used for excitation (for review, see van Grondelle, 1985; Valkunas et al., 1991).

These inconsistencies start to find clarification with the use of a new-generation of picosecond lasers and recording techniques (for review, see Freiberg, 1986). An important feature of such instruments is a high pulse repetition rate. In experiments employing single picosecond pulses, it is almost always the case that excitation intensities are considerably higher than in natural photosynthesis. Singlet-singlet type excitation annihilation processes may then prevail over photosynthetically relevant excitation quenching processes (Campillo and Shapiro, 1978). New spectrometers allow a low-fluence excitation more characteristic of natural photosynthesis, together with picosecond and even subpicosecond time resolution.

A. The Effect of the Reaction Center State

It was first shown by Sebban and Moya (1983) using the bacterium *Rhodobacter (Rb.) sphaeroides* and Freiberg et al. (1984) using *Rhodospirillum (Rs.) rubrum* that the variable fluorescence lifetime of the core antenna, associated with the redox state of the RC, changes from 50–60 ps in the case of active RCs to about 250 ps on closing the RCs. Later, a similar behavior was observed in a number of other bacteria.

Borisov et al. (1985) have summarized how the picosecond fluorescence from the antenna of *Rb. sphaeroides* and *Rs. rubrum* reacts to the closing of RCs. Almost a single fluorescence emission decay lifetime at different incident light levels was observed which simply increased as the RC traps were oxidized due to the increased light level. In contrast to these studies, Woodbury and Bittersmann (1990) found that the decay of fluorescence emission was nonexponential at all excitation intensities. The increase of the overall emission decay time was not only due to an increase in the initial fast excitation trapping time, but also to a substantial contribution from a second, longer-lived emission component which grew in as the traps were inactivated by high light level. Only recently, it was established (A. Freiberg and N. Woodbury, unpublished) that most of the described discrepancies can be explained by different excitation conditions used by different groups. In the case of a very small excitation volume as used in the measurements of Woodbury and Bittersmann (1990), diffusion of excited PSUs out of the illuminated volume and unexcited PSUs into this volume should be taken into account in order to obtain the correct kinetics. All the relevant data, including these corrections, seem to verify that the fluorescence yield as well as the fluorescence lifetime of an antenna are directly governed by the rate of energy trapping on active RCs.

Vredenberg and Duysens (1963) assumed that closed RCs neither trap nor quench antenna excitations. This assumption was based on the observed bleaching of the long-wavelength absorption band of photooxidized RCs, resonant with the antenna fluorescence (Duysens, 1952). Energy transfer from the antenna to the RC presumably takes place via the Förster dipole-dipole mechanism (see Chapter 15 by Struve). Because the absorption of the RC is almost lost, they assumed that the Förster overlap integral will become close to zero and so will the trapping/quenching rate.

Later studies proved that the reality is more complex. There are a number of different inactive (closed) RC states, not just a single one. All of them are able to quench excitations in the antenna (Godik and Borisov, 1977; Heathcote and Clayton, 1977). The coupling mechanism in specific cases, however, needs further clarification. Depending on the redox potential of the medium, the excitation light intensity, the availability of electron donors, temperature, and other factors, the following RC states have been quantified (Godik and Borisov, 1977):

active state	C P I Q,
inactive states	$C^{+}P^{+}I\,Q$,
	$C^{+}P^{+}I\,Q^{-}$,
	$C\,P\,I\,Q^{-}$,
	$C^{+}P\,I^{-}Q^{-}$.

Letters denote the important molecular constituents of the RC: C is the cytochrome electron donor complex, P is the BChl dimer primary donor, I is the bacteriopheophytin intermediate electron acceptor, Q is the primary quinone electron acceptor. When P is photoexcited, it becomes oxidized and the electron transfer proceeds via I to Q, and from there to a secondary quinone acceptor until the electron

Table 1. Antenna fluorescence lifetimes in the picosecond range in *Rs. rubrum* and *Tc. roseopersicina* for different reaction center states at room temperature

	Rs. rubrum	*Tc. roseopersicina*
RC state	lifetime (ps)	
C P I Q	60 ± 10	75 ± 10
C^+P^+ I Q	250 ± 20	210 ± 20
C^+P^+ I Q^-	120 ± 10	
C P I Q^-	80 ± 10	90 ± 10
C^+P I^-Q^-		280 ± 20

eventually returns to P. This electron transfer sequence results in trapping and stabilizing the energy of light excitations harvested by the antenna (for review, see Bixon et al., 1992).

Different RC states quench antenna excitations with different efficiencies as shown in Table 1, where the data relevant to the bacteria *Rs. rubrum* and *Thiocapsa (Tc.) roseopersicina* from various publications are gathered together (Freiberg et al., 1984; Godik et al., 1985; Borisov et al., 1985; Godik, 1989). When compared with the BChl lifetime in isolated core antenna complexes or in a RC-less mutant (about 650 ps at room temperature, according to Sebban et al. (1985) and Freiberg and Timpmann (1992)), it appears that when present RCs are always efficient antenna fluorescence quenchers.

Recently, it was found that in two bacteria, the BChl *g* containing *Heliobacterium chlorum* and the BChl *b* containing *Rp. viridis*, fluorescence is quenched more strongly when the primary donor is oxidized than when it is reduced (Deinum et al., 1991a, 1992). Accordingly, the fluorescence lifetime decreases after the closing of RCs (Kleinherenbrink et al., 1993), instead of increasing, which was observed in the 'normal' BChl *a* containing bacteria. This was interpreted as due to the fact that antenna levels are lower in energy than those of a reduced primary donor. In these conditions the Förster overlap integral may well increase and the antenna fluorescence decrease upon photooxidation of the primary donor when compared to the reduced state of the donor. Although this is in principle compatible with the Förster mechanism of energy transfer, the quantitative aspects of this interpretation remain to be proved. The same concerns the quantitative aspects of the coupling mechanism in 'normal' cases. Some of the problems encountered by the Förster mechanism of quenching have been underlined by Sebban (1985). The primary electron donor in higher plant Photosystem II, P680, also appears to be an equally effective quencher of chlorophyll fluorescence in the oxidized and in the reduced state (see, e.g. Freiberg et al., 1992, 1994).

B. 'Up-Hill' Energy Trapping

Because of the Stokes shift of the donor emission spectrum the Förster overlap integral has a larger value if the acceptor molecule absorption spectrum is red-shifted with respect to that of the donor. As a result, in the inhomogeneous ensemble excitations tend to accumulate on the molecules with lowest energies. When the pigments with redmost absorption spectra belong to the primary donor or to antenna pigments adjacent to it, the spectral uni-directionality results in a spatially directed energy transfer towards the RC. This is how the photosynthetic systems should be and generally are arranged, so far as the peripheral antenna is concerned. Otherwise, excitations may end up in a local minimum apart from the RC and get lost.

In many cases, the primary donor is not an energetic trap compared to all the core antenna pigments present. *Rp. viridis* is probably the most studied representative of this class of systems. In natural membranes the primary donor absorbs at 985 nm and the antenna pigments, at 1015 nm at room temperature (Kleinherenbrink et al., 1992). Note that *Rp. viridis* like *Rs. rubrum* possesses only the core antenna proteins, arranged concentrically in a single circle around the RC. In a structure optimized in such way excitations always stay adjacent to the RC until they are trapped. The trapping rate stays relatively high until the thermal energy, kT, and the gap between the 0-0 transition energies (not just between the absorption maxima) of the antenna and the RC pigments are comparable. In *Rp. viridis* the primary donor is at 150 cm^{-1} or about 217 K higher energy with respect to the antenna pigments. Thus, trapping via the long-wavelength antenna does not kinetically impede an efficient utilization of light energy at room temperature. The quantum yield of charge separation is even more conservative. This follows from the well-known relation between the yield of the primary photochemistry, Y, and the rate constants of trapping, k_t, and the competing loss processes, k_l: $Y = k_t/(k_t+k_l)$. The quantum yield stays high as long as $k_t > k_l$. This may explain the almost 50% quantum yield of charge separation in *Rp. viridis* at 6 K (Kleinherenbrink et al., 1992).

Picosecond time-resolved measurements on purple bacteria have shown that the energy equilibration between the proximal and the core antenna as well as over the core antenna is very fast and it is essentially complete within 10 ps (Freiberg et al., 1987, 1989; van Grondelle et al., 1987; Timpmann et al., 1991; Müller et al., 1993; Pullerits et al., 1994). The transfer rate between the antenna and the RC (the pairwise trapping rate) is probably slower, considering the several times longer lifetime of antenna fluorescence, a short charge separation time (3 ps) and the generally-accepted spatial organization of the antenna and the RC in the membrane. The latter aspect is very important, though it does not necessarily allow a complete picture of excitation transfer dynamics (Valkunas et al., 1992).

At room temperature the trapping rate constant cannot be measured directly, because of the interference with thermal equilibration processes within the antenna. Measurements at 77 K have given the following estimates for trapping time constants from the long-wavelength antenna: 35 ps in *Rb. sphaeroides* and 75 ps in *Rs. rubrum*. The trapping time is constant over the temperature range 177–77 K and it is believed to be not much different at room temperature (Visscher et al., 1989). Crude estimates for trapping times in *Rp. viridis* are about 10 ps (Zhang et al., 1992) and 1.3 ns (Kleinherenbrink et al., 1992), respectively at room temperature and at 6 K. Analyses of the room-temperature femtosecond pump-probe data in membranes of *Rb. capsulatus* lead Xiao et al. (1994) to the conclusion that the excitation energy transfer and trapping is limited by the overall rate of energy transfer between the antenna and the RC. The viewpoint that there exist some rate-limiting energy transfer steps in the antenna or from antenna to RC is shared by several others (van Grondelle et al., 1987; Freiberg et al., 1989). At the same time, Müller et al. (1993) are confident that their picosecond time-resolved fluorescence emission data of *Rb. capsulatus* could only be interpreted if the equilibration time for the excited states of antenna and RC is very short, <3 ps. These opposite views show that at this point our knowledge is, perhaps, still incomplete.

It has been proposed by Trissl (1993) that the possible reason for the development of long-wavelength antenna and spectrally broad heterogeneous absorption bands is an advantage which is evolutionarily obtained via the adaptation of photosynthetic organisms to the spectrally filtered light caused by self-absorption. Trissl (1993) has also achieved a good agreement between the measured long-wavelength fluorescence lifetime and the theoretically predicted one, assuming a rate-limited trapping from thermally equilibrated antenna states. Small (<2 nm) and uniform pigment-pigment distances have been utilized which is inconsistent with the known PSU organization described above. The original Förster formula (see, e.g. Förster, 1965) contains a number of uncertain parameters when photosynthetic systems are considered which limits its direct use. These have been discussed by Pullerits and Freiberg (1992). Here it suffices to note that in the formula the refractive index appears in the fourth power, but even the applicability of this average macroscopic constant on the needed microscopic level is not obvious. We will return to this specific problem later. Usually, these 'details' are overlooked and the distances between pigments are adjusted in order to get the needed transfer rates. Naturally, such a 'determination' of the structure is arbitrary and should be taken with caution.

V. Detrapping of Excitation Energy from the Reaction Center

The process opposite to trapping–detrapping, is even less well characterized. Some early measurements of the antenna fluorescence yield upon excitation at the wavelengths corresponding to the absorption from RC pigments, appear to suggest that not very extensive back-transfer of excitation energy from the RC to the antenna occurs in purple bacteria at room temperature (Clayton and Siström, 1966; Olson and Clayton, 1966; Ebrey, 1971; Wang and Clayton, 1971). However, the relatively limited knowledge of the pigment origin and not so well defined experimental conditions at the time of these measurements make these results at the best qualitative in nature. Much more recent steady-state fluorescence yield measurements on *Rp. viridis* and *Rs. rubrum* (Kleinherenbrink et al., 1992) suggest that at least at 6 K RCs behave as absolute traps. The extent of detrapping is very low, only some few per cent. On the basis of time-resolved absorption measurements with antenna-RC complexes from *Cm. minutissimum* (Abdourakhmanov et al., 1989) the energy detrapping efficiency at room temperature was estimated to be 10–20%.

It is important to have more detailed and quantitative information on the detrapping process,

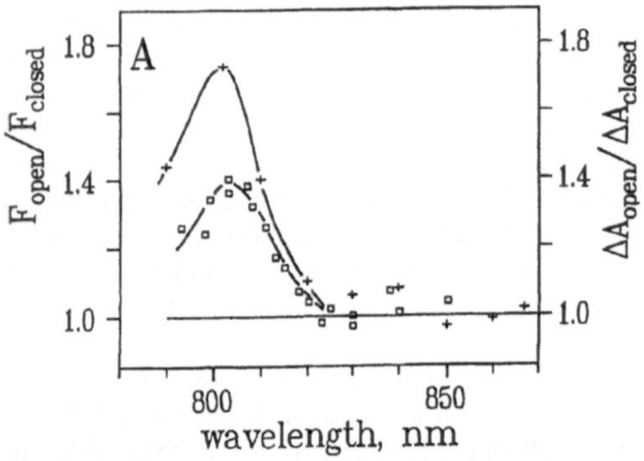

Fig. 1. The antenna-signal amplitude ratio measured at the peak of the decay curves of *Rs. rubrum* chromatophores with open and closed RC as a function of the excitation wavelength: absorption data (squares, right-hand scale); fluorescence data (crosses, left-hand scale).

because of at least two reasons. First, it tells us about the efficiency of the RC as an energy trap, and secondly, is of additional help in assessing the trapping rate, since the two processes—trapping and detrapping, are related to each other. The possibility of learning something about trapping via detrapping is particularly interesting because the detrapping can be tested at room temperature, and thus it yields the trapping rate at room temperature, which cannot be directly measured, as noted before.

Timpmann et al. (1993) have measured the energy back-transfer efficiency from the RC to the antenna in *Rs. rubrum* chromatophores at room temperature by using two types of picosecond time-resolved measurements, absorption and fluorescence methods. Due to the additional possibility to discriminate against unwanted signals, the time-resolved measurements are superior over the steady-state measurements. The RC pigments in *Rs. rubrum* can be almost selectively excited through the 800 nm absorption band of the accessory monomeric BChl. Figure 1 offers compelling evidence that there is a substantial back-transfer of excitation energy from the RC to the antenna. The excitation of the chromatophores within the 800 nm absorption band of the RC results in picosecond absorption and fluorescence signals, demonstrating the presence of antenna excitations. Different amplitudes of the signals result from different dark-adapted RC states (PBIQ in absorption and $PBIQ^-$ in fluorescence experiments). After taking into account the antenna absorption in the 800 nm region, the part of energy detrapped from the RC back to the antenna is determined to be 25 ± 5% for active RCs and 40 ± 5% when the primary quinone acceptor was prereduced. The yield obtained for active RCs is in remarkably good agreement with the room-temperature data, derived in other bacteria: 10–20% in *Cm. minutissimum* (Abdourakhmanov et al., 1989), 15 ± 10% in *Rb. capsulatus* (Xiao et al., 1994) and 20 ± 5% in *Rp. viridis* (Timpmann et al., unpublshed).

From the back-transfer data for active RCs in the state PBIQ the following rate constants corresponding to the trapping/detrapping kinetic model of Fig. 2 have been found: $k_1 = (40 \pm 5\ \mathrm{ps})^{-1}$ and $k_{-1} = (7.5 \pm 1.5\ \mathrm{ps})^{-1}$. For RCs in the state $PBIQ^-$, respectively $k_1 = (35 \pm 5\ \mathrm{ps})^{-1}$ and $k_{-1} = (6.5 \pm 1.5\ \mathrm{ps})^{-1}$. The pairwise detrapping rate, k_D, is related to the total detrapping rate, k_{-1}, from the RC to all the nearest q antenna molecules coordinating the RC by

$$k_D = k_{-1} / q \qquad (1)$$

The pairwise trapping rate, k_T, could be derived, in good approximation at room temperature, from the following equation (N is the number of spectrally equivalent pigments or aggregates of pigments):

$$k_1 = k_T\, q/N \qquad (2)$$

The obtained results, when combined with the

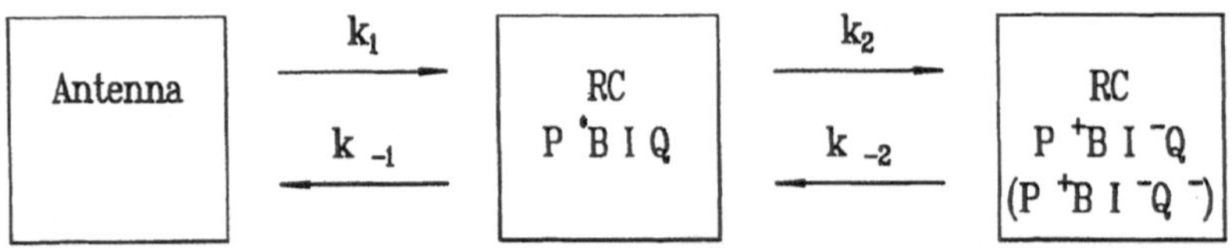

Fig. 2. A simplified kinetic model of energy transfer between the antenna and the RC. k_2 represents the charge separation rate in the RC and k_{-2}, the charge recombination rate. Slow loss rate constants have been omitted.

previous knowledge of the energy transfer rates in purple bacteria, suggest that the most crucial step of the overall energy transfer process in the photosynthetic pigment system of the purple bacteria is the energy transfer from the core antenna molecules to the RC ones. It is rather arbitrary whether the energy transfer in this situation is considered diffusion-limited or trapping-limited.

VI. Pigment-Protein Couplings According to High Hydrostatic Pressure Studies

The dynamic behavior and functioning of pigment-protein complexes is among other factors governed by the properties of protein due to important couplings between protein polypeptides and pigments with the protein surrounding (Pancoska et al., 1990). The protein matrix provides a large number of local conformations which lead to inhomogeneous broadening effects on spectra and support various dynamic processes, e.g. electronic excitation and electron transfer processes. Also, different low-frequency modes in proteins and their postulated glass-like structure (Iben et al., 1989) may have a profound effect on the dynamics of photosynthetic systems. Until now, relatively little attention has been paid to these problems, especially, as far as photosynthetic pigment-protein complexes are considered.

Therefore, and also because of the connection with the current actively disputed subject about the origin of the large spectral red shift of the near-infrared absorption bands of BChls in photosynthetic pigment-protein complexes when compared to those in organic solvents, we will consider the elastic properties of photosynthetic pigment-proteins in some detail here (Freiberg et al., 1993; Tars et al., 1994).

A. Solvent- Versus Pressure-Induced Spectral Shifts

It is well known that when a molecule is dissolved in a solvent, its spectrum shifts, usually towards lower energy. The shift is evidently induced by the interaction between the solute molecule and the surrounding medium. In the case of a nonpolar molecule in a nonpolar isotropic medium the major contribution to the shift is due to the dispersion interaction, i.e. an interaction via the fields of the induced dipole moments. This contribution to the spectral solvent shift (expressed in cm^{-1}) is given by the following expression (see, e.g. Andersson et al., 1991):

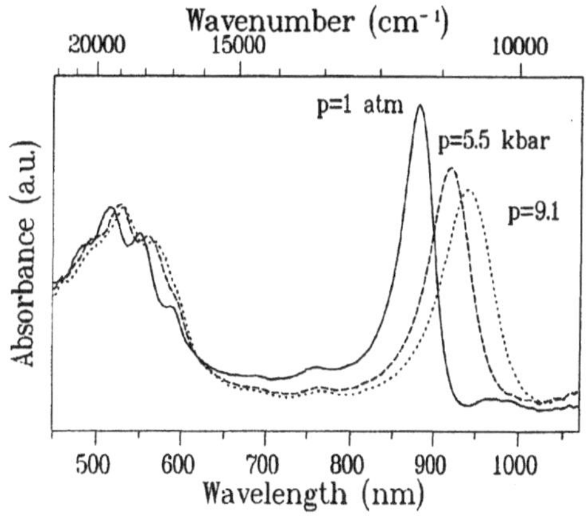

Fig. 3. Absorption spectra of light-harvesting pigment-protein complexes from membranes of the purple bacteria ***Rs. rubrum***, measured at room temperature under pressures of 1 atm, 5.5 kbar and 9.1 kbar.

$$\nu - \nu_0 = \frac{3(\alpha_g - \alpha_e)}{2hca^3} \frac{II'}{I + I'} \frac{n^2 - 1}{n^2 + 2} \quad (3)$$

Here ν and ν_0 are the transition frequencies of a molecule in the medium and in the gas phase; α_g and α_e are electronic polarizabilities of the molecule in the ground and in the excited electronic state, h is Planck's constant, c is the velocity of light; a is the Onsager cavity radius of the solute, I and I' are the ionization potentials of the solute and the medium molecules, and n is the refractive index of the medium at a given transition wavelength. Since molecules in the excited state are generally more polarizable than in the ground state ($\alpha_e > \alpha_g$ in Eq.(3)), a red solvent shift commonly occurs. A linear relationship between the absorption band maximum ν in solution and the Lorentz-Lorenz function $(n^2 - 1) / (n^2 + 2) = (4/3)\pi N\alpha_s$ characterizing the electronic polarizability of the medium (α_s is the electronic polarizability of the solvent molecules and N is their number in the molar volume) has been established for many solute-solvent systems. In particular, this relationship has proved to hold in the case of both chlorophylls (Renge, 1992) and carotenoids (Andersson et al., 1991) in nonpolar media.

Pressure leads to changes in the interaction between the molecule and the surrounding medium and,

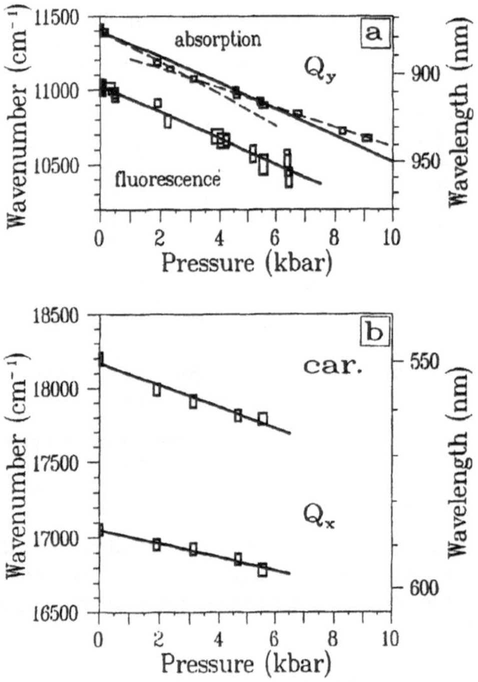

Fig. 4. Shifts of the spectral band peaks as a function of pressure at room temperature. The width and height of the rectangles denoting the experimental points characterize experimental inaccuracies. Straight lines serve as guides for the eye. (a) $S_1 \leftarrow S_0$ transition in BChl absorption and fluorescence emission spectra. (b) $S_2 \leftarrow S_0$ transitions in the BChl and carotenoid absorption spectra (the O-O band shift is shown for carotenoid only).

consequently, to the shift of spectral bands. However, it is not easy to see from Eq.(3) the functional dependence of this shift.

Figure 3 shows the room-temperature absorption spectra of light-harvesting pigment-protein complexes between 450 and 1050 nm at three different pressures: 1 atm, 5 kbar and 9.1 kbar (1 bar = 10^5 N/m^2 = 0.99 atm). The most intensive band at 880 nm belongs to the Q_y transition in aggregated BChl molecules in the protein surrounding. A much weaker band around 590 nm corresponds to the Q_x transition in the same molecule. Three blue bands (at 485, 513 and 549 nm) represent vibronic transitions between the ground and the second electronic excited state in the spirilloxanthin molecule. One can see in Fig. 3 a substantial red shift of the Q_y absorption band and of spirilloxanthin bands with the increase of pressure. The shift of the Q_y band is accompanied by a band broadening, the integral absorption remaining practically unchanged. The position of the Q_x band seems to be relatively less sensitive to pressure than the spirilloxanthin bands.

Although the pressure dependence of the spectral band shifts has both linear and quadratic terms, the latter contribution is relatively small (Fig. 4; Tars et al., 1994). The pressure coefficients, $\Delta\nu$ / ΔP, determined as slopes of a single straight line fitted to the data points in the plot of the band peak position versus the pressure were found to have the following values: -86 ± 5 cm^{-1}/kbar for the BChl Q_y band, -45 ± 5 cm^{-1}/kbar for the Q_x band and -74 ± 5 cm^{-1}/kbar for all the spirilloxanthin vibronic bands.

From the pressure-induced shifts of the spirilloxanthin spectral bands the local refractive index in the surrounding of the spirilloxanthin molecule can be calculated following Eq.(3) and by using the solvatochromic data. At atmospheric pressure n = 1.51 and it increases to n = 1.58 at 5.5 kbar. It has been suggested that the surrounding of the spirilloxanthin molecule may to a great extent consist of the aromatic amino acid residues of the protein (Andersson et al., 1991).

Some years ago a simple expression was deduced (Laird and Skinner, 1989) for the pressure-induced shifts of spectral bands of impurity molecules dissolved in solid media:

$$\Delta\nu(P) = 2\kappa(\nu_m - \nu_0)\,\Delta P \qquad (4)$$

Here $\kappa = -(1/V)(dV/dP)_T$ is the isothermal compressibility of the medium in the neighborhood of the guest molecule and $\nu_0 - \nu_m$ is the solvent (matrix) shift at normal pressure (see Eq.(3)) with respect to the inhomogeneous band maximum ν_m. In deriving Eq. (4), the elastic properties of the matrix were considered as isotropic and homogeneous. This is certainly not the case for proteins, as they are both heterogeneous and anisotropic. S. H. Lin (unpublished) has considered the case where the heat bath molecules are not isotropically distributed around the solute molecule. The result is similar to the Eq. (4), but without the numerical coefficient 2. Evidently, the linear pressure dependence given by these equations holds for sufficiently low pressures only. If either the compressibility or the solvent shift is known, the other can be determined from the pressure-induced spectral shift experiment.

B. Compressibility of the Protein Matrix

The gas phase transition frequencies for different carotenoids are known from solvatochromic data. For the 0-0 vibronic transition of the spirilloxanthin $\nu_0 = 21700 \pm 700$ cm^{-1} (P. O. Andersson, unpublished), which gives for the solvent shift $\nu_m - \nu_0 = -3550 \pm 700$ cm^{-1}. With the given pressure coefficient and the equation derived by Lin (unpublished) we calculate κ(car) = 20 ± 4 Mbar^{-1}. The gas phase frequencies of native BChl aggregates are not available for understandable reasons and neither are the corresponding solvent shifts. Here, in a limiting case, which presumably gives a larger solvent shift and thus a diminished compressibility, we will use the frequencies determined from solvatochromic measurements of monomeric BChl molecules (Renge, 1992): $\nu_0 = 13341 \pm 24$ cm^{-1} for Q_y and 17660± 79 cm^{-1} for Q_x transitions. With these data we have κ(BChl Q_y) = 44 ± 4 Mbar^{-1}. From the pressure shift of Q_x band we obtain κ(BChl Q_x) = 50 ± 10 Mbar^{-1}.

As can be seen, within the limits of experimental accuracy rather close compressibility values have been obtained from different transitions in the same BChl molecule. On the contrary, different molecules, BChl and spirilloxanthin, give clearly different compressibility values.

From the present data and the relatively short range of dispersion forces it seems evident that the chromophores probe the local structure of the protein matrix surrounding them and that the protein elastic properties are locally specific. A somewhat higher flexibility of BChl location in the light-harvesting protein complexes may stem from the specific binding of pigment molecules to protein polypeptides and the package of $\alpha\beta$ heterodimers into larger oligomers, leaving easily compressible voids. Both α and β polypeptides carry one BChl molecule, but the binding site of the spirilloxanthin molecule is shared by the two polypeptides (see Chapter 16 Zuber and Cogdell). Also, pressure-induced changes in excitonic coupling between chromophores in dimers and larger aggregates, resulting in the additional red shift of the lower excitonic component, could contribute to an overestimated compressibility in this case.

The perturbation induced by the external factor in the described experiments is substantial. At the highest applied pressure it constitutes 20–40% of the 'internal pressure' which could be defined as the pressure one would have to apply in order to achieve a band shift comparable to the solvent shift. In other terms, it corresponds to the decrease of the average distance between protein molecules by 3–9%.

The estimated local compressibilities at room temperature fall into the typical range of bulk compressibilities of various proteins, organic solvents, polymers and molecular crystals. These organics are approximately by an order of magnitude more compressible than observed in inorganic crystals. In crystals, a single set of elastic moduli characterizes all the elastic properties of the crystal. In protein macromolecules, various local elastic constants are needed in order to understand their properties and functioning. A quantitative correlation between the observed pressure effects and structural changes is still not available.

VII. Concluding Remarks

Retracing the Introduction to this chapter, one can conclude that the 15-years-ago understanding of the antenna-RC coupling problem was qualitatively correct. Although modern experimental techniques have allowed the specification of many important details and made the picture much richer, the basic concepts have remained essentially unchanged. However, the field is still highly descriptive and progress in quantifying the underlying mechanisms is too slow. Some of the pending important problems were pointed out in the main text.

Recent years have destroyed an old illusion that all we need for a complete picture of excitation transfer and trapping dynamics is detailed structural information about the system. This structural information still seems to be only a prerequisite for a deeper understanding of excitation transfer, which will require the application of many approaches.

One of the achievements of the reviewed period, unfortunately not paid due attention in this chapter, is the realization that different inhomogeneous effects constitute a primary characteristic of the protein medium. At the same time, both the kinetics and the dynamics of the antenna-RC coupling are largely governed by the specific protein medium. Future progress is expected through exploiting this knowledge.

Acknowledgments

It is my pleasant duty to thank all my collaborators, first of all V. I. Godik, T. Pullerits and K. Timpmann together with whom we have been searching the puzzles of photosynthesis and whose excellent contribution to the subject I have partly reviewed here. I am grateful to L. Juhansoo and E. Vaik for their help in preparation of the manuscript.

References

Abdourakhmanov IA, Danielius RV and Razjivin AP (1989) Efficiency of excitation trapping by reaction centers of complex B890 from *Chromatium minutissimum*. FEBS Lett 245: 47–50

Andersson PO, Gillbro T, Ferguson L and Cogdell RJ (1991) Absorption spectral shifts of carotenoids related to medium polarizability. Photochem Photobiol 54: 353–360

Bixon M, Fajer J, Feher G, Freed JH, Gamiel D, Hoff AJ, Levanon H, Möbius K, Nechushtai R, Norris JR, Scherz A, Sessler JL and Stehlik D (1992) Primary events in photosynthesis: Problems, speculations, controversies, and future trends. Israel J Chem 32: 369–518

Borisov AYu and Godik VI (1970) Fluorescence lifetime of bacteriochlorophyll and reaction center photooxidation in a photosynthetic bacterium. Biochim Biophys Acta 223: 441–443

Borisov AYu, Freiberg AM, Godik VI, Rebane KK and Timpmann KE (1985) Kinetics of picosecond bacteriochlorophyll luminescence in vivo as a function of the reaction center state. Biochim Biophys Acta 807: 221–229

Campillo AJ and Shapiro SL (1978) Picosecond fluorescence studies of exciton migration and annihilation in photosynthetic systems. Photochem Photobiol 28: 975–989

Clayton RK and Sistrom WR (1966) An absorption band near 800 nm associated with P870 in photosynthetic bacteria. Photochem Photobiol 5: 661–668

Deinum G, Kramer H, Aartsma TJ, Kleinherenbrink FAM and Amesz J (1991a) Fluorescence quenching in *Heliobacterium chlorum* by reaction centers in the charge separated state. Biochim Biophys Acta 1058: 339–344

Deinum G, Otte SCM, Gardiner AT, Aartsma TJ, Cogdell RJ and Amesz J (1991b) Antenna organization of *Rhodopseudomonas acidophila*: A study of the excitation migration. Biochim Biophys Acta 1060: 125–131

Deinum G, Kleinherenbrink FAM, Aartsma TJ, and Amesz J (1992) The fluorescence yield of *Rhodopseudomonas viridis* in relation to the redox state of the primary electron donor. Biochim Biophys Acta 1099: 81–84

Drews G (1985) Structure and functional organization of light-harvesting complexes and photochemical reaction centers in membranes of phototrophic bacteria. Microbiol Rev 49: 59–70

Duysens LNM (1952) Transfer of excitation energy in photosynthesis. Thesis. State University of Utrecht

Ebrey TG (1971) Anomalous energy transfer behaviour of light absorbed by bacteriochlorophyll in several photosynthetic bacteria. Biochim Biophys Acta 253: 385–395

Emerson R and Arnold WA (1932) The photochemical reaction in photosynthesis. J Gen Physiol 16: 191–205

Freiberg A (1986) Primary processes of photosynthesis studied by fluorescence spectroscopy methods. Laser Chem 6: 233–252

Freiberg A, Godik VI and Timpmann K (1984) Excitation energy transfer in bacterial photosynthesis studied by picosecond laser spectrochronography. In: Sybesma C (ed) Advances in Photosynthesis Research, Vol 1, pp 45–48. Martinus Nijhoff/Dr W Junk Publishers, Dordrecht

Freiberg A, Godik VI and Timpmann K (1987) Spectral dependence of the fluorescence lifetime of Rhodospirillum rubrum. Evidence for inhomogeneity of B880 absorption. In: Biggins J (ed) Progress in Photosynthesis Research, Vol 1, pp 45–48. Martinus Nijhoff/Dr W Junk Publishers, Dordrecht

Freiberg A, Godik VI, Pullerits T and Timpmann K (1989) Picosecond dynamics of directed excitation transfer in spectrally heterogeneous light-harvesting antenna of purple bacteria. Biochim Biophys Acta 973: 93–104

Freiberg A, Timpmann K, Moskalenko AA and Kuznetsova NYu (1992) Pico-nanosecond fluorescence kinetics of Photosystem II reaction centre and its complex with CP47 antenna. In: Murata N (ed) Research in Photosynthesis, Vol 2, pp 65–68. Kluwer Academic Publishers, Dordrecht

Förster TH (1965) Delocalized excitation and excitation transfer. In: Sinanoglu O (ed) Modern Quantum Chemistry, Vol III, pp 93–137. Academic Press, New York

Freiberg A, Ellervee A, Kukk P, Laisaar A, Tars M and Timpmann K (1993) Pressure effects on spectra of photosynthetic light-harvesting pigment-protein complexes. Chem Phys Lett 214: 10–16

Freiberg A, Timpmann K, Moskalenko AA and Kuznetsova N Yu (1994) Pico- and nanosecond fluorescence kinetics of Photosystem II reaction centre and its complex with CP47 antenna. Biochim Biophys Acta 1184: 45–53

Godik VI (1989) Kinetics of the primary processes of bacterial photosynthesis. Thesis. Moscow State University

Godik VI and Borisov AYu (1977) Excitation quenching by different states of photosynthetic reaction centers. FEBS Lett 82: 355–358

Godik VI, Timpmann KE, Freiberg AM, Borisov AYu and Rebane KK (1985) Picosecond fluorescence decay kinetics of *Thiocapsa roseopersicina* at different reaction center states. Dokl Akad Nauk SSSR 284: 491–494

Govindjee, Hammond JH and Merkelo H (1972) Lifetime of the excited state in vivo. II. Bacteriochlorophyll in photosynthetic bacteria at room temperature. Biophys J 12: 809–814

Heathcote P and Clayton RK (1977) Reconstituted energy transfer from antenna pigment-protein to reaction centres isolated from *Rhodopseudomonas sphaeroides*. Biochim Biophys Acta 459: 506–515

Iben IET, Braunstein D, Doster W, Frauenfelder H, Hong MK, Johnson JB, Luck S, Ormos P, Schulte A, Steinbach PJ, Xie AH and Young RD (1989) Glassy behaviour of a protein. Phys Rev Lett 62: 1916–1919

Kleinherenbrink FAM, Deinum D, Otte SCM, Hoff AJ and Amesz J (1992) Energy transfer from long-wavelength absorbing antenna bacteriochlorophylls to the reaction center. Biochim Biophys Acta 1099: 175–181

Kleinherenbrink FAM, Cheng P, Amesz J and Blankenship RE (1993) Lifetimes of bacteriochlorophyll fluorescence in

Rhodopseudomonas viridis and *Heliobacterium chlorum*. Photochem Photobiol 57: 13–18

Laird BB and Skinner JL (1989) Microscopic theory of reversible pressure broadening in hole-burning spectra of impurities in glasses. J Chem Phys 90: 3274–3281

Mauzerall D and Greenbaum NL (1989) The absolute size of a photosynthetic unit. Biochim Biophys Acta 974: 119–140

Meckenstock RU, Brunisholz RA and Zuber H (1993) The light-harvesting core-complex and the B820-subunit from *Rhodopseudomonas marina*. Part II. Electron microscopic characterization. FEBS Lett 311: 135–138

Miller KR (1982) Three-dimensional structure of a photosynthetic membrane. Nature 300: 53–55

Monger TG and Parson WW (1977) Singlet-triplet fusion in *Rhodopseudomonas sphaeroides* chromatophores. A probe of the organization of the photosynthetic apparatus. Biochim Biophys Acta 460: 393–407

Moskalenko AA and Toropygina OA (1988) Lateral position of pigment-protein complexes in the membrane of sulphur photosynthetic bacterium *Chromatium minutissimum*. Dokl Akad Nauk SSSR 296: 483–486

Moskalenko AA, Toropygina O, Godik VI, Timpmann K and Freiberg A (1992) Investigation of spatial relationships and energy transfer between complexes B800–850 and B890–RC from *Chromatium minutissimum* reconstituted into liposomes. FEBS Lett 308: 133–136

Müller MG, Drews G and Holzwarth AR (1993) Excitation transfer and charge separation kinetics in purple bacteria. (1) Picosecond fluorescence of chromatophores from *Rhodobacter capsulatus* wild type. Biochim Biophys Acta 1142: 49–58

Olson JM and Clayton RK (1966) Sensitization of photoreactions in Eimhjellen's *Rhodopseudomonas* by a pigment absorbing at 830 nm. Photochem Photobiol 5: 655–660

Pancoska P, Urbanova M, Bednarova L, Vacek K, Paschenko VZ, Vasiliev S, Malon P and Kral M (1990) Models of pigment-protein interactions in photosynthetic systems: tetraphenylporphyrin complexes with polycationic sequential polypeptides. Absorption, circular dichroism and fluorescence properties. Chem Phys 147: 401–413

Pradel J, Lavergne J and Moya I (1978) Formation and development of photosynthetic units in repigmenting *Rhodopseudomonas sphaeroides* wild type and 'profil' mutant strain. Biochim Biophys Acta 502: 169–182

Pullerits T and Freiberg A (1992) Kinetic model of primary energy transfer and trapping in photosynthetic membranes. Biophys J 63: 879–896

Pullerits T, Visscher KJ, Hess S, Sundström V, Freiberg A, Timpmann K and van Grondelle R (1994) Energy transfer in the inhomogeneously broadened core antenna of purple bacteria: A simultaneous fit of low-intensity picosecond absorption and fluorescence kinetics. Biophys J 66: 236–248

Renge I (1992) On the determination of molecular polarizability changes upon electronic excitation from the solvent shifts of absorption band maxima. Chem Phys 167: 173–183

Robinson GW (1967) Excitation transfer and trapping in photosynthesis. Brookhaven Symp Biol 19: 16–48

Sebban P (1985) Transfer and trapping of the light excitation energy in purple bacteria. Physiol Veg 23: 449–462

Sebban P and Moya I (1983) Fluorescence lifetime spectra of in vivo bacteriochlorophyll at room temperature. Biochim Biophys Acta 722: 436–442

Sebban P, Rober B and Jolchine G (1985) Isolation and spectroscopic characterization of the B875 antenna complex of a mutant of *Rhodopseudomonas sphaeroides*. Photochem Photobiol 42: 573–578

Takemoto JY, Schonhardt T, Golecki and Drews G (1985) Fusion of liposomes and chromatophores of *Rhodopseudomonas capsulata*: Effect of photosynthetic energy transfer between B875 and reaction center complexes. J Bacteriol 1126–1134

Tars M, Ellervee A, Kukk P, Laisaar A, Saarnak A and Freiberg A (1994) Photosynthetic proteins under high pressure. Lithuanian J Phys 34: 320–328

Timpmann K, Freiberg A and Godik VI (1991) Picosecond kinetics of light excitations in photosynthetic purple bacteria in the temperature range of 300–4 K. Chem Phys Lett 182: 617–622

Timpmann K, Zhang FG, Freiberg A and Sundström V (1993) Detrapping of excitation energy from the reaction centre in the photosynthetic purple bacterium *Rhodospirillum rubrum*. Biochim Biophys Acta 1183: 185–193

Trissl H-W (1993) Long-wavelength absorbing antenna pigments and heterogeneous absorption bands concentrate excitons and increase absorption cross section. Photosynth Res 35: 247–263

Valkunas L, Liuolia V and Freiberg A (1991) Picosecond processes in chromatophores at various excitation intensities. Photosyn Res 27: 83–95

Valkunas L, van Mourik F and van Grondelle R (1992) On the role of spectral and spatial antenna inhomogeneity in the process of excitation energy trapping in photosynthesis. J Photochem Photobiol B: Biol 15: 159–170

van Grondelle R (1985) Excitation energy transfer, trapping and annihilation in photosynthetic systems. Biochim Biophys Acta 811: 147–195

van Grondelle R, Bergström H, Sundström V and Gillbro T (1987) Energy transfer within the bacteriochlorophyll of purple bacteria at 77 K, studied by picosecond absorption recovery. Biochim Biophys Acta 894: 313–326

Varga AR and Staehelin LA (1985) Pigment-protein complexes from *Rhodopseudomonas palustris*: Isolation, characterization, and reconstitution into liposomes. J Bacteriol 161: 921–927

Visscher KJ, Bergström H, Sundström V, Hunter CN and van Grondelle R (1989) Temperature dependence of energy transfer from the long wavelength antenna BChl-896 to the reaction center in *Rhodospirillum rubrum*, *Rhodobacter sphaeroides* (w.t. and M21 mutant) from 77 to 177 K, studied by picosecond absorption spectroscopy. Photosynth Res 22: 211–217

Vredenberg WJ and Duysens LNM (1963) Transfer of energy from bacteriochlorophyll to a reaction centre during bacterial photosynthesis. Nature 4865: 355–357

Wang RT and Clayton RK (1971) The absolute yield of bacteriochlorophyll fluorescence in vivo. Photochem Photobiol 13: 215–224

Welte W and Kreutz W (1982) Formation, structure and composition of a planar hexagonal lattice composed of specific protein-lipid complexes in the thylakoid membranes of *Rhodopseudomonas viridis*. Biochim Biophys Acta 692: 479–488

Woodbury N and Bittersmann E (1990) Time-resolved measurements of fluorescence from the photosynthetic membranes of *Rhodobacter capsulatus* and *Rhodospirillum rubrum*. In: Baltscheffsky M (ed) Current Research in Photosynthesis, Vol 2, pp 165–168. Kluwer Academic Publishers, Dordrecht

Xiao W, Lin S, Taguchi KW and Woodbury NW (1994) Femtosecond pump-probe analyses of energy and electron transfer in photosynthetic membranes of *Rhodobacter capsulatus*. Biochemistry 33: 8313–8322

Zankel KL (1978) Energy transfer between antenna components and reaction centers. In: Clayton RK and Sistrom WR (ed) The Photosynthetic Bacteria, pp 341–347. Plenum Press, New York

Zankel KL and Clayton RK (1969) 'Uphill' energy transfer in a photosynthetic bacterium. Photochem Photobiol 9: 7–15

Zhang FG, Gillbro T, van Grondelle R and Sundström V (1992) Dynamics of energy transfer and trapping in the light-harvesting antenna of *Rhodopseudomonas viridis*. Biophys J 61: 694–703

Chapter 20

Antenna Complexes from Green Photosynthetic Bacteria

Robert E. Blankenship
Department of Chemistry and Biochemistry, Center for the Study of Early Events in Photosynthesis, Arizona State University, Tempe, AZ 85287-1604, USA

John M. Olson* and Mette Miller
Institute of Biochemistry, Odense University, DK-5230 Odense M, Denmark

* Current address: Department of Biochemistry and Molecular Biology, University of Massachusetts, Amherst, MA 01003, USA

R. E. Blankenship, M. T. Madigan and C. E. Bauer (eds): Anoxygenic Photosynthetic Bacteria, pp. 399–435.

Summary

Green photosynthetic bacteria contain antenna complexes known as chlorosomes. These complexes are appressed to the cytoplasmic side of the inner cell membrane and function to absorb light and transfer the energy to the photochemical reaction center, where photochemical energy storage takes place. Chlorosomes differ from all other known photosynthetic antenna complexes in that the geometrical arrangement of pigments is determined primarily by pigment-pigment interactions instead of pigment-protein interactions. The functional role of the proteins found in chlorosomes is not well understood. The bacteriochlorophyll *c, d* or *e* pigments found in chlorosomes form large oligomers with characteristic spectral properties significantly perturbed from those exhibited by monomeric pigments. Because of their close spatial interaction, the pigments are thought to be strongly coupled electronically, and many of the optical properties result from exciton interactions.

In addition to chlorosomes, the green sulfur bacteria contain another unique antenna complex known as the bacteriochlorophyll *a* protein, or the Fenna-Matthews-Olson (FMO) protein. This complex was the first pigment-containing photosynthetic complex to have its high resolution structure determined. It has been intensely studied by spectroscopic and theoretical methods.

This review summarizes existing knowledge on the chemical composition and properties of chlorosomes, the evidence for the oligomeric nature of chlorosome pigment organization and proposed structures for the oligomers, the properties of and possible functions of chlorosome proteins, the kinetics and mechanisms of energy transfer in chlorosomes, and the structure and spectroscopic properties of the FMO protein.

I. Introduction

Green photosynthetic bacteria contain two unique antenna complexes. Chlorosomes are found in both the green sulfur bacteria and the green filamentous bacteria. In addition to chlorosomes, a bacteriochlorophyll *a*-containing protein known as the Fenna-Matthews-Olson (FMO) protein is found in the green sulfur bacteria. Together, they constitute the bulk of the antenna pigments in these organisms. Interestingly, the two types of complexes represent the extremes of pigment organization, with the FMO protein being a classic pigment-protein, where the protein serves as a scaffold to which pigments are attached, while the chlorosome is organized on a completely different principle, with self-assembled pigment oligomers and relatively little direct pigment binding by the few proteins that are present.

Chlorosomes are membrane-associated antenna complexes found only in green photosynthetic bacteria. Cohen-Bazire et al. (1964) first reported chlorosomes and called them *Chlorobium* vesicles. A comprehensive account of the early work on chlorosomes is given by Pierson and Castenholz

Abbreviations: BChl – bacteriochlorophyll; *Cb.* – *Chlorobium;* CD – circular dichroism; *Cf* – *Chloroflexus*; FMO protein – Fenna-Matthews-Olson BChl *a* protein; LD – linear dichroism; *Pc.* – *Prosthecochloris; Pd.* – *Pelodictyon*

(1978), which should be consulted for references to the older literature. Staehelin et al. (1978, 1980) proposed that the name of this complex should be changed to chlorosome (green body) in recognition of the fact that these structures are more widely distributed than is implied by the name *Chlorobium* vesicle, and this term has become universally adopted. Previous reviews of chlorosome antennas can be found in Olson (1980a), Blankenship et al. (1988b), Zuber and Brunisholz (1991), Holzwarth et al. (1992) and van Grondelle et al. (1994).

There are two families of green photosynthetic bacteria, the green sulfur bacteria (Chlorobiaceae) and the green filamentous bacteria (Chloroflexaceae) (Trüper and Pfennig, 1992; Pierson and Castenholz, 1992). The two groups are only distantly related to each other by the criterion of 16S rRNA comparison (Gibson et al., 1985; Woese, 1987) and have very different reaction centers and integral membrane antenna complexes, as well as entirely distinct metabolic and ecological characteristics (see Chapters 1, by Imhoff, 3 by Pierson and Castenholz, 4 by Van Gemerden and Mas, 30 by Feiler and Hauska, 32 by Feick et al. and 40 by Sirevåg for further discussion of these characteristics). Yet these widely separated bacterial lines contain chlorosomes, remarkably similar antenna complexes, the presence of which can be considered to be their primary common characteristic. The possible evolutionary significance of this unusual situation has been discussed by Pierson and Olson (1989) and Blankenship (1992), and is considered below in section IX of this chapter.

Chlorosomes contain bacteriochlorophyll (BChl) *c*, *d* or *e*, usually as a complex mixture of structurally similar yet distinct structures (see below and Chapter 8 by Senge and Smith). BChl *c* is the main chlorosome pigment in many green sulfur bacteria, and is also found in all known green filamentous bacteria. BChls *d* and *e*, which are structurally related to BChl *c*, are found only in various species of green sulfur bacteria. All these pigments were formerly known as *Chlorobium* chlorophylls, and collectively are now referred to as chlorosome chlorophylls (Smith, 1994). BChl *c* (*d* or *e*) is found inside the chlorosome and is thought to be associated with rod-elements ca. 5–10 nm in diameter that have been visualized by electron microscopy (see Chapter 13 by Oelze and Golecki).

It is now quite clear that the mechanism of pigment organization in chlorosomes must be substantially different than is found in more familiar pigment-proteins. Chlorosomes contain surprisingly little protein, relative to the large amount of pigment present. Although there is considerable variation in the values reported in the older literature, much of this difference is almost certainly due to varying degrees of purity of the chlorosome preparations used (Pierson and Castenholz, 1978; Schmidt, 1980). Feick and Fuller (1984) reported a protein to BChl *c* ratio of approximately 2.2:1 (w/w) in *Chloroflexus (Cf.) aurantiacus*. A recent determination gave ratios of protein to BChl of about 0.5:1 (w/w) in *Chlorobium (Cb.) tepidum* (Chung et al., 1994). These values are in contrast to ratios much greater than one invariably found in other pigment-proteins. For example, the FMO protein from *Cb. tepidum* has seven BChl *a* molecules (911.5 Da each) associated with a monomer of protein of molecular mass 40,165 Da, (Dracheva et al., 1992) giving a protein to pigment ratio of 6.3:1.

The role of the small amount of protein in the chlorosome remains controversial. Some workers maintain that protein is necessary in some way to stabilize the chlorophyll structures in the chlorosome while others maintain that proteins serve only to give the chlorosome the correct shape. Possible structural and mechanistic consequences of this relatively small protein content are discussed in detail below in sections III, IV and V.

II. Chlorosome Composition

Purified chlorosome preparations can readily be obtained from cell homogenates either simply after cell breakage (Schmidt, 1980), or after further treatment with detergents (Feick and Fuller, 1984) or chaotropes (Gerola and Olson, 1986). Chlorosomes from both types of green bacteria contain lipids, proteins and carotenoids as well as chlorophylls.

A. *Bacteriochlorophylls* c, d *and* e

1. Structure and Distribution

All green bacteria contain either BChl *c*, *d* or *e*. The structures of these pigments are shown in Fig. 1 (also see Chapter 8 by Senge and Smith). The numbering used in Fig. 1 is the IUPAC system. Much of the older literature uses the Fischer system, in which position 3 is numbered 2, 3^1 is 2a, 8 is 4, 12 is 5, 13^1 is 9, 13^2 is 10 and 20 is δ. The chlorosome chlorophylls have

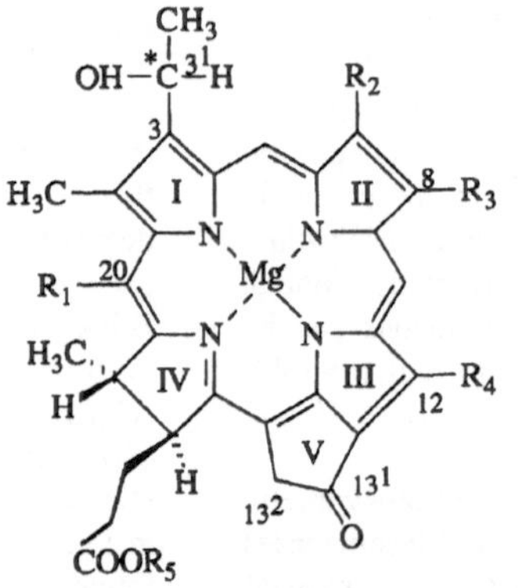

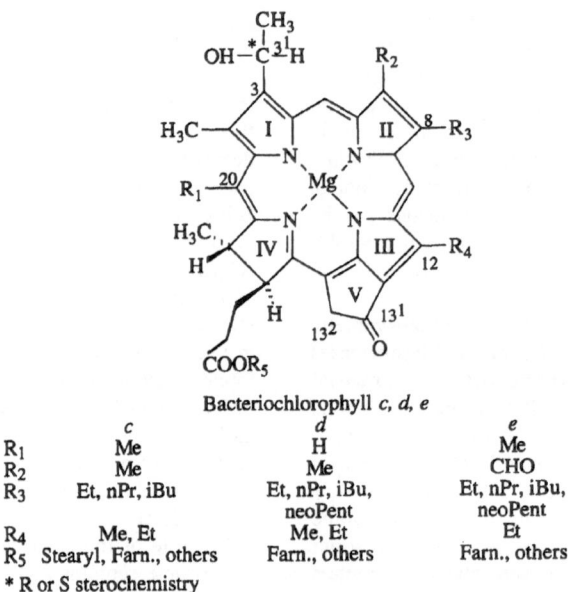

Bacteriochlorophyll *c, d, e*

	c	*d*	*e*
R_1	Me	H	Me
R_2	Me	Me	CHO
R_3	Et, nPr, iBu	Et, nPr, iBu, neoPent	Et, nPr, iBu, neoPent
R_4	Me, Et	Me, Et	Et
R_5	Stearyl, Farn., others	Farn., others	Farn., others

* R or S sterochemistry

Fig. 1. Structures of the chlorosome chlorophylls, bacteriochlorophylls *c*, *d* and *e*.

several distinctive structural features in common. These include the OH group at the 3^1 position and the lack of the carboxymethyl substituent found in most chlorophylls at the 13^2 position. The stereochemistry at the 3^1 position can be either R or S, and is often found as a mixture of diastereomers. There is a trend toward S stereochemistry at the 3^1 position as the substituents at the 8 and 12 positions become more bulky (Bobe et al., 1990; Smith, 1994; Chapter 8 by Senge and Smith). In contrast to the Chloroflexaceae which contain almost exclusively BChl *c*, the major light-harvesting pigments in different types of Chlorobiaceae are BChl *c*, *d* or *e*. A given organism usually contains a mixture of different homologs of only one of the three pigments, although in some strains of *Cb. vibrioforme* and *Cb. limicola*, both BChl *c* and *d* are often found (see below).

Bacteriochlorophylls *c*, *d* and *e* can be distinguished by the absorption maxima of their Q_y band both in vivo and in vitro (Oelze, 1985; Scheer, 1991; Hawthornthwaite and Cogdell, 1991; Hoff and Amesz, 1991). A characteristic feature of the bacteriochlorophyll in the chlorosomes is that its red absorption maximum is red shifted 70–80 nm with respect to that of the monomeric form observed in organic solvents. This spectral shift is largely due to aggregation, and is discussed in detail below. Spectroscopic properties of chlorosome chlorophylls in a variety of solvents have been reported by Fetisova and Borisov (1973) and Brune et al. (1988b).

In cell suspensions of different strains of green sulfur bacteria the long wavelength Q_y absorption for the BChl *c* containing species lies between 745 nm and 760 nm. For the BChl *d* and BChl *e* containing species the corresponding values are 725–745 nm and 715–725 nm, respectively (Gloe et al., 1975; Hawthornthwaite and Cogdell, 1991; Hoff and Amesz, 1991; Chapter 1 by Imhoff). At 5 K Otte et al. (1991) found that intact cells of *Prosthecochloris (Pc.) aestuarii*, *Cb. vibrioforme* and *Cb. phaeovibrioides* which contain BChl *c*, *d*, and *e* respectively, have absorption maxima at 758, 746 and 728 nm.

The biosynthetic pathways of BChls *c*, *d* and *e* have not been studied in the same detail as those of other chlorophyll-type pigments. Suggestions that there may be some unusual steps in these pathways are provided by the observations that anesthetic gases specifically inhibit biosynthesis of these pigments (Ormerod et al., 1990) and that there is a specific effect of vitamin B_{12} on chlorosome formation (Fuhrmann et al., 1993). Smith and co-workers have investigated the unique methylation that results in much of the pigment diversity found in green bacteria (Bobe et al. , 1990). None of the proteins involved in pigment biosynthesis in green bacteria have been isolated nor the genes that code for them identified.

2. Variations of Bacteriochlorophylls in Chlorobiaceae

In some green sulfur bacteria, the degree of methylation of the chlorin ring is dependent on the growth condition of the culture. Broch-Due and Ormerod (1978) observed that when *Cb. vibrioforme* was grown under low light intensities for four to five months there was a change in the in vivo absorption maximum in the red from 742 nm to 756 nm. Based on spectra of methanol extracts of the cells they concluded that the cells had mutated from a BChl *d*-containing form to a BChl *c*-containing form (Fig 1). Later, Smith and coworkers (Smith and Bobe, 1987; Bobe et al., 1990) found that cultures of *Cb. vibrioforme* strain B1-20 and *Cb. vibrioforme* strain NCIB 8327 responded to reduced light conditions in the culture by methylating substituents at the 8- and 12-positions of the bacteriochlorophylls. Eventually, BChl *d* was converted to BChl *c* in the last strain mentioned.

Concomitant with the increased methylation of the bacteriochlorophylls in the 8-, 12-, and 20-positions a red shift of the in vivo antenna pigment system is observed. In nature this could possibly be of advantage for the bacteria in competing for available light. Huster and Smith (1990) suggested that the homologation process increases the hydrophobic interactions between the individual bacteriochlorophyll molecules, giving rise to larger aggregates in the chlorosome antenna and thereby increasing the chance for light capture. Nozawa et al. (1992b) reported that the aggregation properties of different homologues of purified BChl *d* are strongly dependent on the nature of the peripheral substituents. Recently, similar results were reported for pigments from *Cb. limicola* (Uehara et al., 1994).

Otte et al. (1991) spectrally resolved at least two different pools of bacteriochlorophyll within the chlorosomes of green sulfur bacteria. The results were interpreted in terms of a lateral gradient promoting directional energy transfer to the antenna components in the membrane. Causgrove et al. (1992) came to a similar conclusion for BChl *e*-containing chlorosomes through analysis of energy transfer rates in isolated chlorosomes. Other data related to this question in both families of green bacteria are discussed in Section VII.

In all three types of bacteriochlorophyll isolated from green sulfur bacteria the major esterifying alcohol is the C-15 isoprenoid, farnesol. Thus, in batch cultures of *Cb. limicola* grown for 21 days, Caple et al.(1978) found that 91% of the esterifying alcohols of the BChl *c* molecules was *trans, trans*-farnesol with minor amounts of five additional alcohols, where *cis*-9-hexadecen-1-ol and *trans*-phytol each comprising 3% of the total. Otte et al. (1993) identified the esterifying alcohols in *Cb. vibrioforme* and *Cb. phaeovibrioides* by mass spectrometry. In addition to farnesol, various minor esterifying alcohols were found, including phytol, oleol, cetol, 4-undecyl-2-furanmethanol and decenol.

3. Variations of Bacteriochlorophylls in Chloroflexaceae

The BChl *c* composition of Chloroflexaceae consists only of the 8-Et, 12-Me homologue (Gloe and Risch, 1978; Brune et al., 1987b; Bobe et al., 1990). The main esterifying alcohol is stearol with phytol and all-*trans* geranylgeraniol as minor components (Risch et al., 1979). Fages et al. (1990) and Larsen et al. (1994) reported the presence of two additional alcohols in *Cf. aurantiacus*; hexadecanol and octadec-9-enol. It has been observed that *Cf. aurantiacus* can take up exogenous long chain fatty alcohols and incorporate them into the BChl *c* as the esterifying alcohol, thus changing the homolog distribution. By addition of short chain (C10, C12) or long chain (C20) alcohols novel BChl *c* homologs could also be formed (Larsen et al., 1995). Chlorosomes from *Cf. aurantiacus* contain trace quantities of BChl *d*, as well as substantial amounts of both the R and S diastereomers at the 3^1 position of BChl *c*, with a ratio that is somewhat dependent on culture conditions (Brune et al., 1987b; Bobe et al., 1990; Fages et al., 1990; Larsen et al., 1994).

Why are the bacteriochlorophyll molecules within the chlorosomes of one species of green bacteria present in so many different homologous forms? Is this a functional adaptation or is it simply a result of 'slippage' in the biosynthesis of these pigments? Are the chemically distinct pigments uniformly distributed within the chlorosome or is there a spatial and energetic gradient with the shorter wavelength-absorbing pigments further from the membrane, thus 'funneling' excitations toward the membrane and increasing the efficiency of trapping by the reaction center? If this is so, how is such a nonuniform distribution of pigments achieved during assembly of the chlorosome? Unfortunately, none of these questions can be answered definitively, and for many of them almost no information is available.

In conclusion, it is clear that chlorosomes are not homogeneous with respect to pigment composition, as indicated by the wide variety of bacteriochlorophylls present with different esterifying alcohols and peripheral substituents in the chlorin ring, as well as different stereochemistries at the 3^1 position, all of which are sensitive to culture conditions.

B. *Bacteriochlorophyll* a

In highly purified chlorosomes from *Cf. aurantiacus* Schmidt (1980) found a specific BChl *a*-complex with a characteristic absorption maximum at about 790–795 nm in amounts corresponding to about 5% of the BChl *c* content of the chlorosomes (Fig. 2). Similar results were subsequently reported by Sprague et al. (1981a), and Feick et al. (1982). There is now a considerable body of evidence that shows that the molar ratio of BChl *c* to BChl *a*, both in whole cells and in isolated chlorosomes depends strongly on growth conditions (Pierson and Castenholz, 1978; Sprague et al., 1981b; Feick and Fuller, 1984; Oelze, 1992; Lehmann et al., 1994a; Larsen et al., 1994; Foidl et al., 1994; see also Chapter 13 by Oelze and Golecki). The molar ratio of BChl *c* to Bchl *a* in isolated chlorosomes ranges from approximately 25:1 in cells of *Cf. aurantiacus* grown under low light intensity (Feick et al., 1982) to 2:1 in log-phase cells grown at high light intensity (Lehmann et al., 1994a). This special form of BChl *a*, known as B790 or B795 after the wavelength of the Q_y absorption band, was proposed to be associated with a M_r 5.8-kDa protein (Feick and Fuller, 1984). The evidence concerning this assignment is discussed in Section II E.

Chlorosomal BChl *a* has also been observed in the green sulfur bacteria, where most of the BChl *a* is found in the water-soluble FMO-protein (section VI). Fig. 3 shows absorption and fluorescence spectra of chlorosomes isolated from three species of green sulfur bacteria. In highly purified chlorosomes isolated from *Cb. limicola*, the level of BChl *c* is much higher than that of BChl *a*, with a molar ratio of approximately 100:1 (Gerola and Olson, 1986). This is contrast to the substantially higher amounts

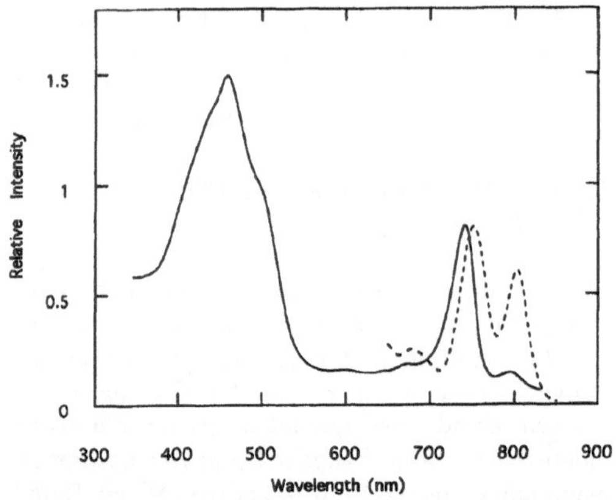

Fig. 2. Absorption (solid line) and fluorescence emission spectra (dashed line) of isolated chlorosomes from *Cf. aurantiacus*. Fluorescence excitation was at 460 nm. Figure courtesy of Judy Zhu.

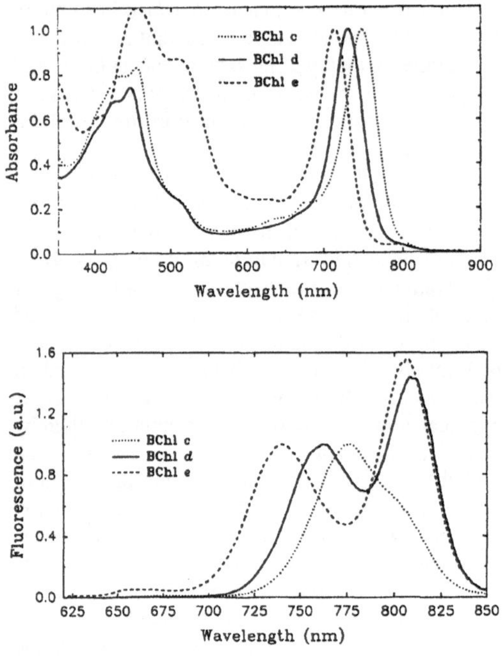

Fig. 3. Absorption (A) and fluorescence (B) of chlorosomes isolated from *Chlorobium limicola* (BChl *c*), *Chlorobium vibrioforme* (BChl *d*) and *Pelodictyon phaeoclathratiforme* (BChl *e*). Figure reproduced with permission from Causgrove et al. (1992).

of BChl *a* found in chlorosomes of *Cf. aurantiacus*, as indicated above. The presence of chlorosomal BChl *a* was detected in intact cells of three species of Chlorobiaceae by absorbance and fluorescence spectroscopy at 5K (Otte et al., 1991) and in isolated chlorosomes and membranes by time-resolved fluorescence spectroscopy at room temperature (Causgrove et al., 1992). The chlorosomal BChl *a* in these species not only had similar spectroscopic properties but also was present in approximately the same ratio to the total BChl. Similar results were reported recently by Van Noort et al. (1994).

B795 is generally believed to function as an intermediate in the energy transfer between the chlorosome and the membrane-embedded antennas and reaction centers. This conclusion is based primarily on spectroscopic data in *Cf. aurantiacus* (Betti et al., 1982; Van Dorssen and Amesz, 1988; Mimuro et al., 1989; Holzwarth et al., 1990b; Müller et al., 1990, 1993; Miller et al., 1991) and various species of green sulfur bacteria (Otte et al., 1991; Causgrove et al., 1990b, 1992; Van Noort et al., 1994). While this position in the energy transfer sequence very strongly suggests that the BChl *a* is indeed physically located in the region of the chlorosome that is proximal to the membrane, this has not been demonstrated directly. More work is needed to clarify the environment of BChl *a* in chlorosomes, especially in light of reports that suggest multiple spectral forms of BChl *a* in chlorosomes (Mimuro et al., 1994b; Savikhin et al., 1994b).

C. Carotenoids

Isolated chlorosomes from both types of green bacteria contain substantial amounts of carotenoids (Schmidt, 1980). In the Chlorobiaceae the major carotenoid is the aromatic chlorobactene (Liaaen Jensen et al., 1964), which is also detected in the thermophilic *Cb. tepidum* (Wahlund et al., 1991). In the chlorosome fraction of *Cb. limicola* chlorobactene comprises 69%, neurosporene 21% and lycopene 8% of the total carotenoid content (Schmidt, 1980). In the brown-colored *Cb. phaeobacteroides* the carotenoids are solely isorenierate/beta-isorenierate absorbing at 505 nm in the intact cell (Overmann et al., 1992).

In chlorosomes isolated from *Cf. aurantiacus* β- and γ-carotene and a hydroxy derivative of the latter comprise 36%, 43%, and 12% respectively of the total (Schmidt, 1980). These are the principal carotenoids in whole cells (Halfen et al., 1972). Schmidt (1980) reported that two-thirds of the hydroxy-β-carotene in the chlorosomes was glucosidated. The glucoside of hydroxy-β-carotene increased relative to the other carotenoids in the chlorosomes isolated from cells grown at either high light-intensities or at suboptimal temperatures and may therefore possibly have a role in protecting the photochemical systems from photodamage (Schmidt et al., 1980). Interestingly, some of the glucosides of the hydroxy-β-carotene has been found to be acylated with either C16:0 or C16:1 fatty acids. Some of these compounds are probably located in the envelope of the chlorosome, because they are lost from the chlorosome fraction after SDS treatment (K. Matsuura and S. Takaichi, personal communication).

Much of the carotenoid in *Cf. aurantiacus* is not associated with either the chlorosome or the cytoplasmic membrane. This carotenoid is localized in structures known as wax oleosomes (Beyer et al, 1983), and is recovered as a copious bright orange floating pellet upon cell breakage followed by ultracentrifugation. The amount of this fraction is greatly increased in cells grown at higher light intensity, imparting a distinct gold color to the cultures, hence the name *aurantiacus* (Pierson and Castenholz, 1974a,b). The function of this carotenoid is not known, although it probably does not transfer energy to bacteriochlorophyll due to the very short distance dependence of carotenoid to chlorophyll energy transfer (see Chapter 18 by Frank and Christensen). It is therefore probably ineffective at driving photosynthesis. The efficiency of carotenoid to BChl energy transfer in whole cells of any species of green bacteria apparently has not been measured quantitatively, although Van Dorssen et al. (1986b) found an apparently high efficiency of energy transfer from carotenoid to BChl *c* in isolated chlorosomes.

D. Lipids

The primary distinguishing characteristic of the lipid content of chlorosomes from both families of green bacteria is the presence of large amounts of glycolipids, which are rare or absent from most purple bacteria. The glycolipids are significantly enriched in the chlorosomes compared to the cytoplasmic membranes (see Chapter 10 by Imhoff and Bias-Imhoff).

In total lipid extracts of *Cf. aurantiacus* cells, wax esters comprised 45 mol %, polar lipids 36 mol %

and chlorophylls and carotenoids 11 and 8 mol %, respectively (Beyer et al., 1983). Wax esters are rarely found in bacteria, but in *Cf. aurantiacus* they are dominating both in oleosomes and spindle-shaped bodies found in a floating fraction after cell breakage and also in the chlorosomes and plasma membrane fractions (Beyer et al., 1983). The dominating wax ester in *Cf. aurantiacus* is the fully saturated C_{36} (Knudsen et al., 1982). Because the wax-esters are very hydrophobic molecules, the authors speculated that they might be a component of the presumed hydrophobic interior of the chlorosomes, although this has not been examined in isolated chlorosomes. Interestingly, the mesophilic *Cb. limicola* does not possess any wax esters, in contrast to the thermophilic *Cb. tepidum* (Knudsen et al., 1982; Wahlund et al., 1991). This suggests a possible role for the wax esters in maintaining a structural integrity of membranes or hydrophobic domains during growth at elevated temperature.

The polar lipid fraction from chlorosomes contains phospholipids and glycolipids in addition to the bacteriochlorophylls (Schmidt, 1980). Monogalactosyldiglycerol (MGDG) is found in chlorosomes from both types of green bacteria and *Cf. aurantiacus* also contains a diglycosyldiglycerol with glucose and galactose as carbon moieties (Knudsen et al., 1982). It has been suggested that the MGDG has a structural role in the chlorosomes by forming a monolayer covering the cytoplasmic boundary of the chlorosome (Cruden and Stanier, 1970). However, Holo et al. (1985) were not able to correlate the content of bacteriochlorophyll and galactolipids in cells grown under different conditions where different amounts of bacteriochlorophyll are produced, even though it has been reported that an enlargement of the average chlorosome volume up to a factor of 1.9 can take place in cells of *Cf. aurantiacus* (Golecki and Oelze, 1987). The work of Holo et al. (1985) also indicated that the chlorosome proteins are masking the galactose moieties on the galactolipids by comparing the agglutination pattern with *Ricinus* lectins before and after tryptic digestion of the chlorosome proteins.

The preferred location of glycolipids in chlorosomes in both families of green bacteria suggests that there may be an important role for these lipids in chlorosome structure and function. Recently, Uehara et al. (1994) have demonstrated that aggregated BChl *c* with spectral properties very close to those of the pigments in vivo can be assembled in aqueous suspensions if MGDG is included, lending strong support to this idea. Similar results were reported by Hirota et al. (1992) and Miller et al. (1993a,b) using protein-free fractions from cell extracts. See also Section IIIB. Thus, there is the suggestion that these lipids have a specific role in the pigment organization in chlorosomes. More work is needed on the localization and composition of lipids in chlorosomes and the properties of pigment aggregates as a function of lipids.

E. Proteins

The protein composition of chlorosomes is now reasonably well established for *Cf. aurantiacus* and is beginning to be so for the green sulfur bacteria. The possible functional role(s) of these proteins continues to be a controversial subject, as is discussed in Section VI.

Purified chlorosomes from *Cf. aurantiacus* contain three major proteins with molecular weights of 18, 11 and 3.7 kDa based on electrophoretic mobility on SDS-PAGE, and smaller amounts of a 5.8 kDa protein (Feick and Fuller, 1984). The 3.7 kDa M_r polypeptide was chemically sequenced by Wechsler et al. (1985b), who reported a molecular mass of 5.6 kDa and proposed a structural model for how this protein might bind BChl *c* to seven Gln and Asn residues. Later, the protein was determined to have an actual molecular weight of 5.7 kDa and to be processed at the carboxyl terminus from a longer preprotein (Theroux et al., 1990). The gene that encodes this protein is known as *csmA* (Theroux et al., 1990; Bryant, 1994). Assertions that this protein is a degradation product of much larger cellular proteins (Griebenow et al., 1990) cannot be correct, given the gene sequence.

Feick and Fuller (1984) proposed that the minor 5.8 kDa peptide observed in chlorosomes from *Cf. aurantiacus* is associated with the B795 BChl a found in all chlorosomes (see Section IIB). This assignment was made on the basis of protease digestion experiments that showed a correlation between the 790 nm absorption and the amount of this protein. In addition, the amount of this protein correlated with chlorosomal BChl *a* content in samples derived from cells grown under differing light intensity (Feick and Fuller, 1984). This putative 5.8 kDa pigment-protein has never been purified and biochemically characterized. Whether it is actually the binding site for chlorosomal BChl *a* has recently been questioned by Lehmann et al. (1994a). These authors reported a partial amino acid sequence of a peptide in this molecular mass range. A lack of correlation between the amount of the 5.8 kDa peptide and BChl *a* content was observed by Foidl et al. (1994) in cultures of *Cf. aurantiacus* grown under a variety of culture conditions. No homolog to the 5.8 kDa protein has yet been identified in green sulfur bacteria, although the much lower content of BChl *a* in these chlorosomes might well make its identification difficult.

Much of the protein in chlorosomes from six strains of green sulfur bacteria is accounted for by a 6.2–6.3 kDa polypeptide (Wagner-Huber et al., 1988, 1990; Chung et al., 1994). This protein is almost certainly the homolog to the 5.7 kDa CsmA protein found in *Cf. aurantiacus*. The sequence identity among the 6.2 kDa polypeptides isolated from *Cb. tepidum, Cb. vibrioforme, Pelodictyon luteolum, Prosthecochloris aestuarii, Cb. limicola* and *Cb. phaeovibrioides* is more than 90% and the sequence identity of these polypeptides to the 5.7 kDa polypeptide of *Cf. aurantiacus* is 30% (Fig. 4) (Wagner-Huber et al., 1988; Zuber and Brunisholz, 1991; Chung et al, 1994).

The sequence similarity between the CsmA proteins from the two families of green bacteria, although modest, clearly suggests a functional similarity and evolutionary relationship between these proteins in the chlorosomes of both green filamentous bacteria and green sulfur bacteria. The proteins are found in cellular structures that are very similar in terms of both overall morphology and function. In addition, all CsmA proteins from all green bacteria that have been examined so far are synthesized as precursor proteins that are post-translationally processed to cut 20–27 amino acids from the carboxyl terminus of the preprotein (Theroux et al., 1990; Chung et al., 1994). This type of processing is very unusual, so the similarity of this feature in the two groups of green bacteria is probably significant. Finally, the gene organization surrounding the *csmA* gene is similar in two green sulfur bacteria, *Cb. vibrioforme* and *Cb. tepidum*, as compared to *Cf. aurantiacus* (Theroux et al., 1990; Chung et al., 1994). All three organisms contain similar upstream and downstream open reading frames, although in the green sulfur bacteria an additional gene is found, known as *csmC*. This gene is not present in this location in *Cf. aurantiacus* and there is so far no evidence in this organism for any chlorosome protein with a similar sequence.

The sequences of all known CsmA proteins are compared in Fig. 4. None of the seven Asn and Gln residues proposed by Wechsler et al. (1985b) as possible BChl *c* binding sites in *Cf. aurantiacus* are

```
                                   10    ↓  ↓ 20     ↓   ↓ 30 ↓ ↓    40 ↓      50
Cf. aurantiacus          M†ATRGWFSES SAQVAQIGDI MFQGHWQWVS NALQATAAAV DNINRNAYPG
Cb. vibrioforme          M†SGGGVFTDI LAAAGRIFEV MVEGHWETVG MLFDSLGKGT MRINRNAYGS
Cb. tepidum              M SGGGVFTDI LAAAGRIFEV MVEGHWETVG MLFDSLGKGT MRINRNAYGS
Cb. limicola             M SGGGVFTDI LAAAGRIFEV MVEGHWETVG MLFDSLGKGT MRINRNAY/.
Cb. phaeovibrioides      M NGGGVFTDI LAASGRIFEV MVEGHWATVG YLFDSLGKGV SRINQNAYGN
Pd. luteolum             M SGGGVFTDI LAAAGRIFEV MVEGHWETVG MLFDSLGKGT MRINRNAYGS
Pc. aestuarii            M NGGGVFTDI LAAAGRIFEV MVEGHWETVG MLFDSLGKGT MRINRNAY/.

                 60         70         80
          VSR†SGSGEGA FSSSPSNGFR PKRIRSRFNR
          MG. GGSLRGS†SPEVSGYAVP TKEVESKFAK
          MG. GGSLRGS†SPEVSGYAVP TKEVESKFAK
          ... ....... .......... ..........
          M/. ....... .......... ..........
          MG. GGTSLRG S......... ..........
          ... ....... .......... ..........
```

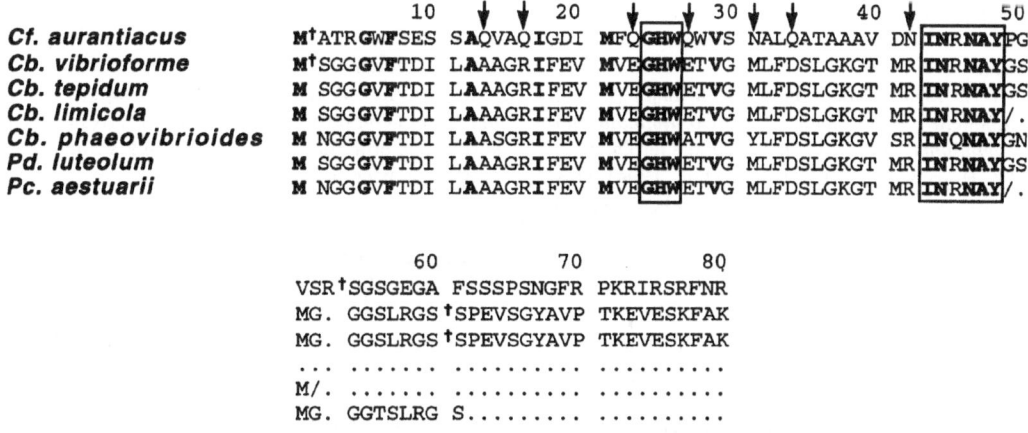

Fig. 4. Amino acid sequences of the CsmA proteins. Conserved residues are indicated by boldface and conserved stretches are boxed. Protein processing is indicated by †. Incomplete sequences are indicated by /. Residues proposed to be BChl *c* binding sites by Wechsler et al. (1985b) are indicated with arrows. The sequence for *Cf. aurantiacus* is from Wechsler et al. (1985b) and Theroux et al (1990), *Cb. vibrioforme* and *Cb. tepidum* from Chung et al. (1994), *Cb. limicola, Cb. phaeovibrioides, Pd. luteolum* and *Pc. aestuarii* from Wagner-Huber et al. (1988).

conserved in the CsmA proteins from green sulfur bacteria (arrows in Fig. 4), strongly suggesting that this assignment is incorrect. However, there are two conserved regions in the aligned sequences, which are boxed in Fig. 4. Lehmann et al. (1994a) have proposed that the His in the first of these regions could bind to the Mg of a BChl *c* in pigment oligomers.

A comparison of the hydropathy plot of the 6.3 kDa *Pelodictyon* protein with that of the 5.7 kDa *Cf. aurantiacus* protein supports this idea (Fig. 5). Both proteins have a hydrophilic region between residues 21 – 29 surrounded by hydrophobic segments on each side. The hydrophilic segments have a conserved amino acid sequence: Gly His Trp. In both proteins the hydrophobic region closest to the N terminus has a high potential for forming an alpha helix (Garnier et al., 1978), and in the case of the 5.7 kDa protein from *Cf. aurantiacus* the estimated length of this alpha-helix is about half of what is found for membrane-spanning proteins.

The sequences of the genes encoding the 11 and 18 kDa proteins have recently been determined (Niedermeier et al., 1994). The two genes, *csmM* and *csmN*, are cotranscribed as part of an operon (see Chapter 54 by Shiozawa). The sequences of the 11 and 18 kDa CsmM and CsmN proteins from *Cf. aurantiacus* are shown in Fig. 6 (Niedermeier et al., 1994). The two proteins exhibit a substantial degree of identity in the central region, and have similar hydropathy plots, suggesting that they are the descendants of an ancestral protein for which the gene underwent duplication and subsequent divergence.

A 7.5 kDa polypeptide is also a major protein component in chlorosomes from *Cb. limicola* and *Cb. tepidum* (Gerola and Olson, 1986; Gerola et al., 1988; Højrup et al., 1991; Chung et al., 1994). This protein has been sequenced (Højrup et al., 1991) and the gene that codes for it has been given the designation *csmB* (Bryant, 1994). Analysis of the amino acid sequence of the CsmB and CsmA proteins shows that there is less than 18% identity between the sequence of the 7.5 kDa CsmB protein of *Cb. limicola* and of either the 6.2 kDa CsmA polypeptide from the Chlorobiaceae or the 5.7 kDa CsmA protein from *Cf. aurantiacus*. However, Gerola et al. (1988) reported that the hydropathy plots and secondary structure profiles for the 7.5 kDa and 5.7 kDa proteins are similar. The 7.5 kDa polypeptide could not be found in detergent prepared chlorosomes from Chlorobiaceae lacking the baseplate (Wagner-Huber et al., 1988) and it has been proposed that this protein is part of the baseplate binding the chlorosomes to the FMO protein (Højrup et al., 1991).

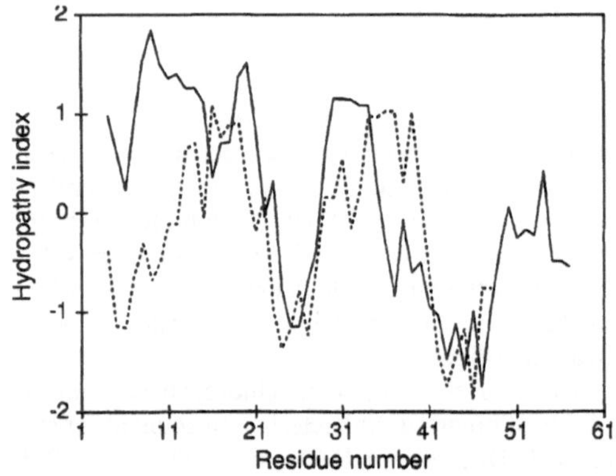

Fig. 5. Hydropathy plots of the CsmA proteins from *Cf. aurantiacus* and *Pd. luteolum*. Sequences taken from Wechsler et al. (1985b), Theroux et al (1990) and Wagner-Huber et al. (1988).

Some additional proteins with molecular weight between 9 and 38 kDa have been detected in chlorosomes from *Cb. limicola* (Gerola and Olson, 1986) whereas in *Cb. phaeobacterioides* and in *Cb. tepidum* only a 34.5 kDa protein was detected in addition to the low molecular weight polypeptide (Stolz et al., 1990). In contrast, chlorosomes isolated from *Cb. tepidum* by Chung et al. (1994) were found to contain a significantly larger number of proteins. These authors also noted that the sensitivity of different proteins to silver staining versus Coomassie blue staining was very different. This suggests that the protein composition of chlorosomes from green sulfur bacteria may be more complex than has previously been thought.

III. In vitro Aggregates of Chlorosome Chlorophylls

A large body of evidence has been accumulated that, taken as a whole, very strongly indicates that some sort of aggregated or oligomeric pigment complex is present in chlorosomes in vivo. An important component of this evidence is studies of the spectral properties of BChl *c* and related pigments in a variety of solvent systems in which pigment aggregates form spontaneously. These aggregates exhibit a substantial red shift of their absorbance spectra,

```
                       10                   20                   30                   40
CsmM  M M T E S E G E V R  V R S V P V R R N D  S F V E S A M E F G  G G I V R L G F S I
CsmN                       M S N E T T N E R D  G L F E M A A G F V  G G A T R I G L T V
                                             *        *   *     *  * *       *

                       50                   60                   70                   80
CsmM  F T L P L A L L P P  E S R Q H M H N A T  K E L M Y A F A S L  P R D F A E I A G K
CsmN  A S V P L V L L P R  N S R R R V R R A M  A E V A M A V V A F  P K E L A N V S E R
            * *   * * *      * *             *    *       *        *       *

                       90                  100                  110                  120
CsmM  S I E K W A E E G E  E P K G E A K
CsmN  V V D D I F A A D P  P Q I N L P S P Q R  V G E Q V R S F T E  R L A R A A E E L G

                      130                  140                  150
CsmN  T S F S R A A G R A  A D A V E Q G A A K  V D E W V E T P P K  T P P A P
```

Fig. 6. Protein sequences of the CsmM and CsmN proteins from *Cf. aurantiacus*. Sequences taken from Niedermeier et al. (1994).

which arises from strong exciton-type interactions among the pigments. Depending on the details of the solvent system, the pigment concentration and the precise pigment involved, a range of aggregated species can be formed, ranging from dimers, through intermediate species to very large oligomers that readily precipitate from solution.

A. A. Krasnovsky was a pioneer in the study of chlorophyll aggregation in solid films and in colloidal suspension, and first proposed the concept of aggregated pigments as models for pigment organization in chlorosomes of green bacteria. Krasnovsky and coworkers (Krasnovsky and Pakshina, 1959; Krasnovsky et al., 1961, 1962, 1966; Bystrova et al., 1979; Krasnovsky and Bystrova, 1980) showed that purified BChl *c* from green sulfur bacteria exists in solid films in a 740-nm form very similar to the spectral form seen in native chlorosomes and whole cells. Formation of the aggregated pigment required the presence of Mg in the macrocycle (Krasnovsky et al., 1966), but did not require water or other polar groups (Bystrova et al., 1979), which are required to form long-wavelength absorbing aggregated forms of other photosynthetic pigments (Katz et al., 1991). They further showed by IR spectroscopy that the 740-nm form was an aggregate involving the 13^1-keto groups and also that the 3^1 OH group was involved in the aggregation. While the six-coordinated Mg structural model that they proposed (Bystrova et al., 1979) has since been superseded by other models discussed below, this seminal work was instrumental in shaping the thinking of all subsequent workers.

A. Aggregates in Nonpolar Solvents

Shortly after the proposals by Krasnovsky discussed above, Smith et al. (1983) demonstrated that the same type of spectra could be achieved with BChls *c*, *d* and *e* and Zn methyl bacteriopheophorbides *c* dissolved in n-hexane: methylene chloride (200:1). The addition of methanol converted the long-wavelength aggregate to monomers. Metal-free bacteriopheophorbides did not form aggregates under any conditions. An oligomeric aggregate was proposed with coordination of the 3^1-hydroxyl on one BChl *c* to the Mg of another BChl *c*, but this model did not propose any involvement of the 13^1 carbonyl group (see Section VA). Olson et al. (1985) reported absorption and CD spectra of BChl *c* in the same solvent system and carried out exciton calculations assuming the Krasnovsky structural model.

Brune et al. (1987b, 1988a) carried out similar experimental studies with BChl *c* from *Cf. aurantiacus*. (This pigment is a complex mixture of homologs, the major component of which is 8-ethyl-12-methyl BChl *c* stearyl ester in a racemic mixture of diastereoisomers with different configurations at the 3^1 chiral center). BChl *c* in hexane formed an oligomer with absorption and fluorescence emission spectra closely resembling those of *Cf. aurantiacus* chlorosomes. FTIR studies showed that the 13^1-keto group of BChl *c* is involved in forming the oligomer. Pyrochlorophyll *a*, which is structurally very similar to BChl *d*, but lacks the polar OH group at the 3^1 position, did not form long wavelength aggregates under the same conditions in which the chlorosome chlorophylls readily form aggregates.

Olson and Pedersen (1988, 1990) showed that 4-isobutyl homologs of BChl *c*, which exist mainly in the 3^1-S configuration, form aggregates in CCl_4 whose absorption (747 nm maximum) and circular dichroism spectra resemble those of chlorosomes from green sulfur bacteria. These aggregates in water-saturated CCl_4 are oligomers consisting of 20–40 chlorophyll molecules (Uehara and Olson, 1992). FT-IR studies (Uehara et al., 1991) showed a keto-carbonyl stretching peak at 1641–3 cm^{-1}, tentatively assigned to a carbonyl group hydrogen bonded to a 3^1-hydroxyethyl group in the oligomer.

The fluorescence emission spectrum of the oligomer in CCl_4 peaks at ca. 760 nm (Olson and Pedersen, 1990), and two fluorescence decay times of, 19 and 67 ps have been measured (Causgrove et al., 1990a). These two decay times suggest that the oligomer/polymer is probably a mixture of more than one species, both of which absorb at ca. 747 nm.

Tamiaka et al. (1992, 1994) also showed that synthetic Zn chlorins in n-hexane-methylene chloride form aggregates with absorption maxima at 740 nm. Cheng et al. (1993) have demonstrated that Zn and Mg methyl bacteriopheophorbides *d* form aggregates with maxima at ca. 730 nm. These results show that the long hydrocarbon tail of the chlorophyll molecule is not required for oligomer formation.

Chiefari et al. (1995) recently reported an extensive study of aggregation of BChl *c* isolated from *Cf. aurantiacus*. They found that in dichloromethane the S diastereomer at the 3^1 position, but not the R, formed the 750 nm aggregate as monitored by UV-Vis, FTIR and NMR spectroscopy.

B. Aggregates in Aqueous Systems

BChl *c* will also form aggregates spontaneously in an aqueous medium in the presence of detergents or lipids (Hirota et al., 1992; Miller et al., 1993a,b; Uehara et al., 1994). In the two former studies, chlorosomes from *Cb. limicola* and *Cf. aurantiacus* were extracted with chloroform/methanol. The crude extracts, containing BChl *a* and *c* and lipids but no protein, were dissolved in methanol and then dispersed in buffer. The latter study used purified pigments and lipids, in particular MGDG. The extracts spontaneously formed aggregates with the spectral properties similar to those of BChl *c* in the chlorosome. Electron micrographs of aggregates from *Cb. limicola* showed a rather heterogeneous distribution of stain-excluding bodies with sizes ranging from 40 to 400 nm. The aggregates from *Cf. aurantiacus* were somewhat more uniform in size with a range of diameters between 70 and 170 nm. (The dimensions of the aggregates are comparable to those of chlorosomes).

The aggregates show fluorescence, CD and LD spectra similar to those for intact chlorosomes. In addition, treatment of the aggregates with 1-hexanol causes a reversible shift from 740 nm to 670 nm just as in intact chlorosomes (see Section IV F).

IV. Spectroscopic Probes of Chlorosome Structure

A. Linear Dichroism and Fluorescence Polarization

A number of studies have all clearly demonstrated that the Q_y transition dipole moment of the BChl *c* in chlorosomes is oriented generally parallel to the long axis of the chlorosome. Early work on chlorosomes from *Cf. aurantiacus* using stretched polyvinyl alcohol films (Betti et al., 1982) or compressed gels (van Dorssen et al., 1986b) came to this general conclusion, but did not give as precise a value for the angle between the transition dipole and the chlorosome axis as did later measurements (Van Amerongen et al., 1988; Griebenow et al., 1991a; Van Amerongen et al., 1991). The early measurements almost certainly suffered from incomplete orientation of the chlorosomes and therefore provided only lower limits of the degree of BChl *c* organization in chlorosomes. Van Amerongen et al. (1988) measured the linear dichroism (LD) of chlorosomes from *Cf. aurantiacus* between 250 and 800 nm and found that the transition moments corresponding to the 741-nm and 461-nm BChl *c* absorption bands make angles of 20° and 30° respectively with the long axis of the chlorosome. Subsequently, Griebenow et al. (1991a) measured an average angle of 15 ± 10° for the 740-nm BChl *c* absorption band. The results of Van Amerongen et al. (1991) using fluorescence polarization gave very similar results.

Matsuura et al. (1993) analyzed chlorosomes from *Cf. aurantiacus* by both three-dimensional LD and circular dichroism (CD). Three-dimensional orientation of chlorosomes in polyacrylamide gels was carried out by uniaxial compression along the x-axis and uniaxial stretching along the z-axis without changing the y-axis. It was assumed that the 10-nm axis of the chlorosomes oriented along the x-axis of the gel, and that the 30-nm and 100-nm axes of the chlorosomes oriented along the y- and z-axes respectively. The results confirmed the conclusions from earlier work that the Q_y transition moment is oriented along the long axis of the chlorosome, and that the BChl *a* component in the baseplate has its Q_y transition moment oriented nearly perpendicular to the long axis of the chlorosome.

Two slightly different spectral components of BChl *c* were indicated by the difference in the shapes of the main Q_y-bands in $A_{\parallel}$ and $A_{\perp}$ in the experiments of Matsuura et al. (1993). The absorption spectra were deconvoluted and two major components were found at 727 and 744 nm. The polarization values calculated for these two components were 0.59 and 0.76 respectively measured from the y-axis. These two components could also rationalize the various CD spectra observed in terms of two simple exciton-type spectra with the opposite order of positive and negative components (see Section IV B). BChl *c*744 was partially converted to BChl *c*727 by the addition of 0.5–0.6% 1-hexanol.

Considerably less work has been reported on the LD of chlorosomes from green sulfur bacteria. Fetisova et al. (1986) measured the LD of chlorosomes from *Cb. limicola* and found that the transition moment corresponding to the 730-nm band was parallel to the long axis of the chlorosome. The angle reported was $0 \pm 7°$ for the ca. 740-nm band. Whether this value is significantly different from the $15 \pm 10°$ values reported above for *Cf. aurantiacus* is not yet clear.

Kinetic studies of fluorescence polarization and transient photodichroism are discussed in section VI E.

B. Circular Dichroism (CD)

Whereas LD gives information about long-range order in the chlorosome, CD gives information about the local environment of the pigment molecules. The CD spectra reported for chlorosomes and oligomeric pigments arise primarily from excitonic interactions among the strongly coupled pigments. Surprisingly, the CD spectra that have been reported are extremely variable in intensity, number of components and even sign. These dramatically different CD spectra often arise from samples with essentially identical absorption spectra. There is as yet no quantitative understanding of these effects.

Early CD spectra of chlorosomes were reported by Betti et al. (1982), Olson et al. (1985) and Van Dorssen et al. (1986b). Brune et al. (1990) investigated the CD spectra of chlorosomes from both *Cb. limicola* f. *thiosulfatophilum* and *Cf. aurantiacus* and found that both had similar spectra with positive peaks at 735 nm (*Cf.*) or 747 nm (*Cb.*) and negative peaks at 752 nm (*Cf.*) or 770 nm (*Cb.*).

Treatment of chlorosomes with 0.3% sodium dodecylsulfate removes absorbance attributable to BChl *a* but does not change the CD spectrum of the remaining BChl *c* (Brune et al., 1990). This suggests that in the presence of this concentration of detergent the BChl *c* sees the same local environment as in the native chlorosome.

Griebenow et al. (1991a) obtained both CD and LD spectra of *Cf. aurantiacus* chlorosomes. Intact chlorosomes were compared with so-called 'GEF-chlorosomes' (GEF = Gel-Electrophoretic-Filtration) that are essentially free of BChl *a* and proteins. The LD spectra were very similar, but the CD spectra of various preparations showed pronounced differences in the magnitude as well as the shape of the spectra and the number of maxima. The set of different CD spectra obtained could be simulated by linear combinations of two basic spectra. This suggests a variation in the relative amounts of two different species in the chlorosome samples.

Niedermeier et al. (1992) reported that proteolytically modified chlorosomes exhibit a 10–20-fold larger CD signal than do native chlorosomes, but the published CD spectra were sign-reversed in comparison with those of Blankenship et al. (1988a) and Brune et al. (1990). Lehmann et al. (1994b) reported an extremely large CD in chlorosomes from *Cf. aurantiacus* that had been treated with hexanol and proteinase K and the hexanol subsequently removed by dialysis. They correlated this large CD with the presence of large structures apparent in electron micrographs, that appeared to be aggregates of chlorosomes. The giant CD effect is well known in systems with large macromolecular assemblies whose dimensions are comparable to the wavelength of light (Keller and Bustamante, 1986). This may possibly be the explanation for at least part of the extreme variability in reported CD spectra of chlorosomes and pigment oligomers.

The factors that influence the extremely variable CD spectra of chlorosomes and aggregates are not well understood, and this represents an area where considerable progress is needed, both in biochemically defining the conditions needed to obtain reproducible spectra and theoretically interpreting those spectra in the context of structural models of pigment organization.

C. Hole Burning Spectroscopy

Fetisova and coworkers (Fetisova and Mauring, 1992, 1993, Fetisova et al., 1994) examined whole cells of *Cb. limicola, Cf. aurantiacus* and *Cb. phaeovibrioides* at 1.8 K using holeburning spectroscopy in fluorescence excitation and emission spectra and found persistent hole spectra that were consistent with a strongly exciton-coupled BChl *c* (*d* or *e*) chromophore system. The 0-0 band of the lowest exciton state was directly detected in each organism, with wavelengths of 774 nm, 752 nm and 739 nm, respectively for the three organisms. The linewidths of the 0-0 bands are relatively narrow, 90–100 cm^{-1}, indicating that the amount of inhomogeneous broadening in the absorption spectra is relatively small. This result seems to argue against suggestions from other groups that much of the spectral width of the Q_y absorption band arises from unresolved distinct spectral components in the chlorosome chlorophylls, although it is not yet clear if there is really a discrepancy that is quantitatively important, as the different techniques measure the pigment distribution rather differently. This point is discussed further in Section VII.

D. Resonance Raman and Infrared Spectroscopy

Vibrational spectroscopy, especially resonance Raman, has been extremely valuable in narrowing down the possibilities for the structural arrangements of pigments in chlorosomes. A reasonably consistent picture of the environment of various functional groups is now available, in particular, the Mg, the 13^1 keto group and the 3^1 hydroxyl group.

Lutz and van Brakel (1988) studied chlorosomes from both *Cb. limicola* and *Cf. aurantiacus* and found a single, sharp band at 1639–1640 cm^{-1} (*Cb.*) or 1641–1642 cm^{-1} (*Cf.*). They assigned this band to most of the C-13^1 carbonyls (BChl *c*) sharing identical interaction sites. A single, sharp band at 1607 cm^{-1} (*Cb.*) or 1609 cm^{-1}(*Cf.*) was assigned to methine bridge stretching which is sensitive to the coordination number of the Mg atom. From the position of the band it was concluded that the Mg coordination number was five. This assignment was supported by the finding of a second band at 1552 cm^{-1}, also sensitive to coordination number. Lutz and van Brakel concluded that in chlorosomes of *Cb. limicola*, in particular, BChl *c* occurs as oligomers involving a single C-13^1 carbonyl ligand on each Mg atom. They did not assign a role to the 3^1 OH, as this group is not directly observable using resonance Raman.

Using FTIR, Blankenship et al. (1988a) and Brune et al. (1988a) found that a 1650 cm^{-1} band assigned to the 13^1 keto group in *Cf. aurantiacus* chlorosomes shifted to 1680 cm^{-1} upon exposure to pyridine vapors. This technique had been introduced earlier by Bystrova et al. (1979) in studies on oligomeric pigments isolated from green bacteria. Both groups proposed models for pigment organization, invoking six-coordinate Mg, with ligands from the 3^1 OH group and the 13^1 keto group. Brune et al. (1988a) also proposed structural models that maintained five coordinate Mg, but involved both the 3^1 OH group and the 13^1 keto group in intermolecular interactions. These models are discussed in more detail in Section V A.

Nozawa et al. (1990a) essentially confirmed the resonance Raman results of Lutz and van Brakel (1988), concluding that the Mg is five coordinate, but they interpreted their 1642 cm^{-1} (*Cf. aurantiacus*) band in terms of a 13^1 carbonyl either H-bonded to an OH group or coordinated to an Mg atom.

Hildebrandt at al. (1991) also used chlorosomes from *Cf. aurantiacus* and observed peaks at 1595, 1605, 1645 and 1715 cm^{-1}. The 1595 cm^{-1} peak was suggested as a possible indication of six coordination for a portion of the BChl *c* molecules, and the 1715 cm^{-1} peak was taken as evidence for a second conformational state of the 13^1-carbonyl group. The 1645 cm^{-1} peak was assigned to the 13^1-carbonyl group coordinated to a Mg atom in agreement with Lutz and van Brakel (1988). More recently, the same group, taking advantage of a wider range of model compounds reinterpreted the data of Hildebrandt at al. (1991) in terms of only five coordinate Mg (Hildebrandt et al., 1994) . There now appears to be consensus that essentially all the chlorosome chlorophylls in chlorosomes are five coordinate.

Feiler et al. (1994) found that the 13^1 keto frequencies from three green sulfur bacteria were essentially identical to each other, and slightly different from the spectra in *Cf. aurantiacus*, suggesting a somewhat different arrangement of pigments in the chlorosomes from the two classes of organisms. They also found that the C-7 formyl group of BChl *e* in *Cb. phaeobacterioides* chlorosomes does not enter into any intermolecular interactions.

E. Nuclear Magnetic Resonance (NMR)

Nozawa and coworkers (1990b, 1991, 1992a, 1994) used the technique of Cross Polarization/Magic Angle Spinning (CP/MAS) ^{13}C NMR to investigate the 13^1-carbonyl group of BChl *c* in *Cf. aurantiacus* and BChl *c* ring current shifts in *Cb. tepidum*. They found evidence for an H-bond between the 13^1-carbonyl group and the 3^1-hydroxyl group. On the basis of CP/MAS NMR experiments, Nozawa et al. (1991) proposed a stepped, parallel chain model in which the 3^1-hydroxyl group ligates the Mg atom of one BChl *c* while simultaneously H-bonding to the 13^1-carbonyl group of another. More recent results from the same group (Nozawa et al., 1992a,b, 1993, 1994) have been interpreted in terms of a 'ring overlap' model, which is discussed in Section V B.

F. Effects of Polar Molecules

Bystrova et al. (1979) showed that the addition of polar molecules such as methanol or pyridine broke up the long wavelength absorbing in vitro aggregates of BChl *c* in nonpolar solvents. The action of these polar species is almost certainly to coordinate to the Mg in the center of the BChl *c* macrocycle, thus displacing the 3^1 OH (or 13^1 keto) groups from adjacent molecules and converting the oligomer into monomeric pigments.

This same concept was applied to dried films of chlorosomes by Blankenship et al. (1988a) and Brune et al. (1988a), and the disaggregation followed by absorption spectroscopy and FTIR. Brune et al. (1987b) introduced the amphipathic molecule hexanol, which has similar effects on BChl *c* organization in chlorosomes in aqueous buffer solutions. They found that the effect could be partially reversed by dilution of the hexanol into buffer. More complete reversibility of the hexanol effect was demonstrated by Matsuura and Olson (1990).

Matsuura et al. (1993) showed that the characteristic linear dichroism of BChl *c* in chlorosomes (see Section IV A) was completely abolished upon the hexanol treatment. Remarkably, the dichroism returned to the control level upon removal of the hexanol. This suggests that an oligomeric structure that was very similar to that in untreated chlorosomes self-assembled from the disrupted pigments when the hexanol was removed.

The environment of pigment in hexanol-treated chlorosomes is not yet well understood. The absorption spectrum appears to be very similar to that of monomeric pigment. However, it is clearly still contained within the chlorosome and is still able to transfer energy to the BChl *a* in the chlorosome baseplate, yet exhibits no long range ordering or evidence for exciton-type interactions characteristic of closely packed pigments (Matsuura and Olson, 1990; Matsuura et al., 1993). The pigment is converted upon hexanol treatment from the highly packed form present in the untreated chlorosome into a state that exhibits no spectral evidence for order or pigment-pigment interactions, yet is still contained within the envelope of the chlorosome and carries out efficient energy transfer to BChl *a* in the baseplate.

V. Structural Models for Pigment and Protein Organization in Chlorosomes

A wide range of structural models for BChl *c* organization in oligomers and chlorosomes have been proposed. The detailed structure of the oligomeric complexes formed by BChl *c* and related pigments is not yet known, and there may be several related but distinct structures that are found in both in vitro systems and in chlorosomes.

It is interesting to compare the BChl *c* oligomers with the 740 nm absorbing chlorophyll oligomer that forms in wet hydrocarbon solvents (Katz et al., 1991). The structure of the 740 nm Chl species is known in detail (Chow et al., 1975; Kratky and Dunitz, 1975, 1977). The spectral properties of this species are in many ways similar to those of the BChl *c* oligomers, in particular the very large absorption red shift. However, the structure of the Chl oligomer is by necessity different from that of the BChl *c* oligomer, because the former structure involves bridging by water between the C-13^2 carboxymethyl substituent and the Mg of an adjacent molecule. BChl *c* lacks the carboxymethyl substituent at the 13^2 position, (as does pyrochlorophyll) so this arrangement is not possible. Furthermore, water is not required for formation of BChl *c* oligomers (see below). It is likely that the 3^1 OH group of BChl *c* has the same function of coordinating to the Mg as does the bridging water in the 740 nm Chl species. However, in case of BChl *c* the group is internal to the molecule rather than part of the solvent, so the details of the geometry of the pigment organization must be different.

A. Models Invoking Local Interactions Among Functional Groups

Figure 7 compares some of the local structural models that have been proposed by various workers. The original Krasnovsky model (Fig. 7A) includes coordination of both the 3^1 OH and the 13^1 keto group, and proposes that the Mg is hexacoordinate. A slight modification of this was proposed by Blankenship et al. (1988a) in which the Mg was hexacoordinate but out of the plane of the macrocycle (Fig. 7C). We now know from resonance Raman that essentially all the BChl molecules have pentacoordinate Mg, so these models have largely been abandoned (see Section IV D).

The Smith et al. (1983) model (Fig. 7B) and the Lutz and van Brakel (1988) model (Fig. 7D) have pentacoordinate Mg, but in the former case the 13^1 keto group is left free, while in the latter the 3^1 OH is free. A variety of evidence, summarized in Sections III and IV, indicates that both these groups are directly involved in oligomer formation.

Two models that are consistent with the absorbance, FTIR, and resonance Raman data were proposed by Brune et al. (1988a) and are shown in Fig. 7E and F. In both models the Mg atom of BChl *c* is 5-coordinated as suggested by the resonance Raman data discussed in Section IV D. In the model shown in Fig. 7E the BChl *c* molecules are arranged in antiparallel chains, with the molecules in each chain linked by H-bonds between the 3^1-hydroxyl and the 13^1 keto groups. BChl *c* molecules in opposite chains are attached by hydroxyl ligation of the central Mg atoms. In the model shown in Fig. 7F the interacting BChl *c* molecules are arranged in parallel, stepped chains. Each BChl *c* molecule is linked to the next by 3^1-hydroxyl-to-Mg ligation and to the one after the next by 3^1-hydroxyl-to-13^1-keto H-bonding. Most of the recent large scale models of pigment organization discussed in the next section are similar to one or the other of these two models. Alden et al. (1992) favored the parallel chain model of Brune et al. (1988a) because of exciton calculations that indicate that it and not the antiparallel chain model could more easily explain the magnitude of the observed red shift in oligomers. However, Nozawa et al. (1993) favored an extension of the antiparallel chain model that included more than a single row of pigments in the oligomer, also on the basis of exciton calculations. This has been termed the 'ring overlap model'. Recently, Chiefari et al. (1995) proposed an extension of the parallel chain model that also includes more than two chains of pigments.

Nozawa et al. (1991, 1992a, 1993, 1994) have proposed a number of variations on these models in which the pigments are arranged either face-to-face, back-to-back or piggyback, and have calculated the geometries of the pigments. Some of these models are also similar to a model proposed on the basis of NMR by Smith et al. (1986) for the solution structure of a BChl *d* dimer.

In organic solvent systems of somewhat higher polarity than those that favor the formation of long wavelength absorbing oligomers, a highly fuorescent intermediate-sized aggregate of BChl *d* with absorption maximum at 693 nm (710 nm for BChl *c*) is observed (Brune et al., 1987b, 1988a,b; Olson and Peterson, 1990; Olson and Cox, 1991; Uehara and Olson, 1992; Cheng et al., 1993). Olson and Cox (1991) proposed that this species is a tetramer, on the basis of log concentration vs. log concentration plots. In contrast, Causgrove et al. (1993) proposed a hexameric trimer of dimers model for this species. They argued that this species is probably not a building block of the large oligomers that mimic the in vivo environment in chlorosomes.

B. Models of Larger Scale Pigment Organization

Several attempts have been made to synthesize the local structure models discussed above with the larger scale information that comes primarily from electron microscopic investigations. While it is not yet clear which if any of these models is correct for either BChl *c* oligomers in solution or for BChl *c* in chlorosomes, they represent a new and important phase in this subject in which the molecular level interactions probed primarily by spectroscopy are correlated and reconciled with the larger scale structures visualized by microscopic techniques. Several of these models are reproduced in Fig. 8.

An interesting model for BChl *c* in chlorosomes has been developed by Katz and coworkers (Worcester et al., 1986, 1993; Katz et al., 1991). In their model system, aggregates of various chlorophylls are formed by addition of stoichiometric amounts of water to

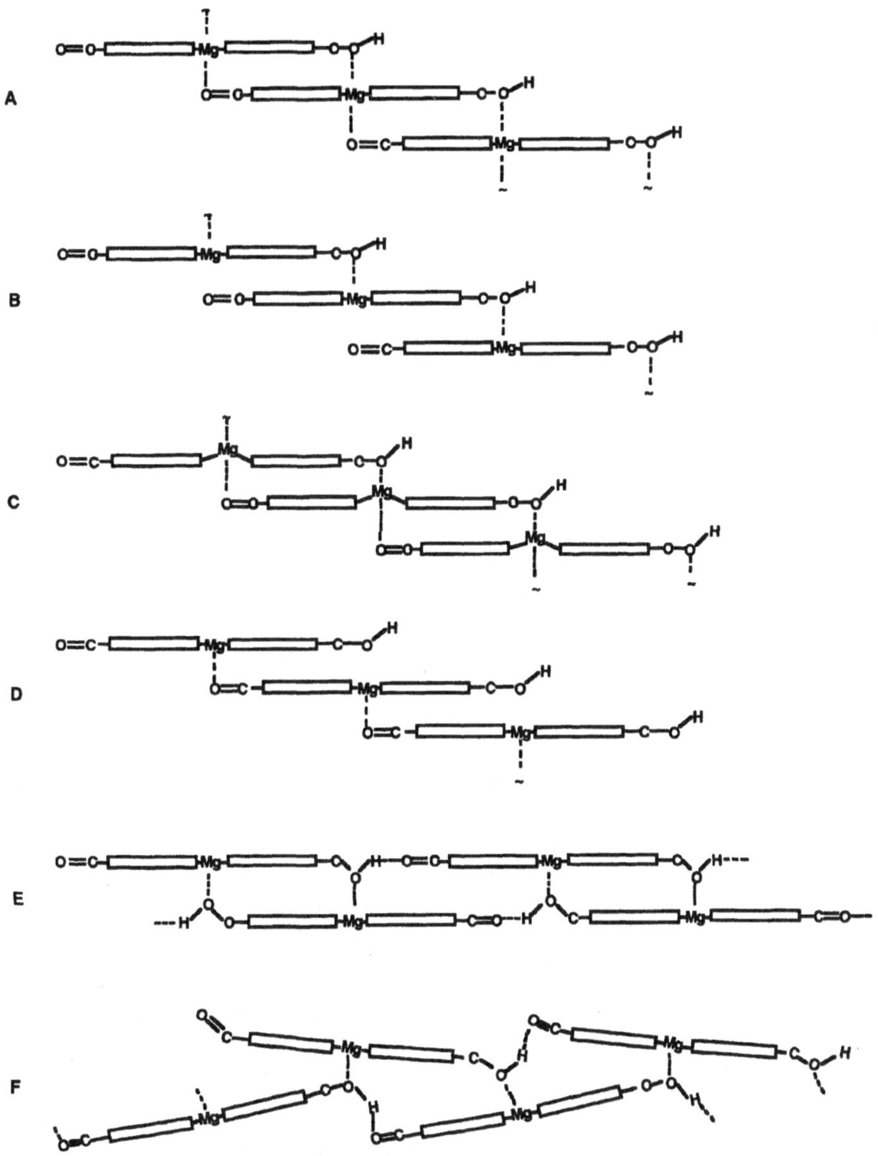

Fig. 7. Structural models for the 740 nm BChl *c* oligomer. A. Bystrova et al., 1979; B. Smith et al., 1983; C. Blankenship et al., 1988a; D. Lutz and van Brakel, 1988; E,F. Brune et al., 1988a. Reproduced with permission from Alden et al. (1992).

rigorously dry solutions of a particular chlorophyll in nonpolar organic solvents such as D-toluene and D-octane. However, with BChl *c* no water is required for aggregate formation (see Section III A). The aggregates formed by BChl *c* from *Chlorobium* in deuterated octane/toluene (1:1) are hollow cylinders 11–12 nm in diameter (determined by neutron scattering), and the absorption peak is at 750 nm. In the model of Worcester et al. (1986), the BChl *c* cylinders resemble the rod elements in chlorosomes

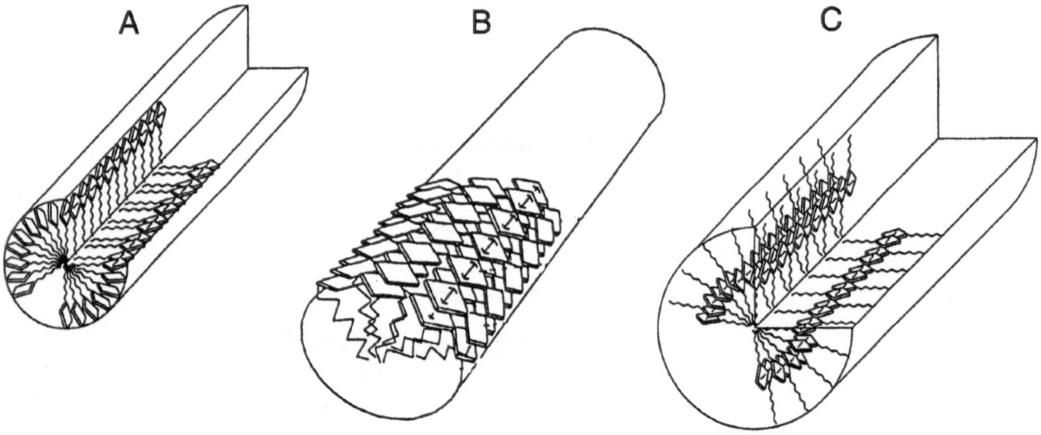

Fig. 8. Structural models for pigment organization into rodlike structures. Reproduced with permission from: Model A, Nozawa et al., 1993; Model B, Matsuura et al, 1993; Model C, Nozawa et al., 1994.

of green sulfur bacteria in shape and dimensions. However, the Q_y transition dipoles of the pigments in the Worcester et al. (1986) model are not parallel to the long axis of the tube. This appears to be inconsistent with the linear dichroism results in chlorosomes discussed in Section IV A, so this model is probably not a valid representation of the pigment structure in chlorosomes.

Several rod element models that are essentially a three dimensional extension into a tube of the antiparallel chain model first presented by Brune et al. (1988a) have been presented by Nozawa et al. (1992a,b, 1994) and Matsuura et al. (1993). The macrocycles are oriented in such a way that the tails are either all arranged towards the inside of the tube (Fig. 8A) (Nozawa et al., 1992a, 1993), or with alternate tails toward the inside and outside (Fig. 8C) (Nozawa et al., 1994). In most cases, the planes of the pigment macrocycles are essentially perpendicular to the surface of the tube, although in others, the pigments are covering the surface of the tube like scales on a snake (Fig. 8B) (Mimuro et al., 1992; Matsuura et al., 1993).

Perhaps the most ambitious and interesting of this new generation of larger scale structural models has been proposed by Holzwarth and Schaffner (1994). These authors used computer molecular modeling and energy minimization techniques to build up structures involving as many as 40 molecules of BChl-ide *d*. The basic local geometry is essentially the parallel chain model shown in Fig. 7e. In the Holzwarth and Schaffner model, the pigment macrocycles are perpendicular to the surface of the tube, and the hydrocarbon tails are oriented toward the outside of the tube, leaving a large hole of about 50Å diameter down the center. It is not clear what might fill this hole in the chlorosome. The C-8 and C-12 aliphatic substituents are oriented towards the hole, so this region is predicted to be very hydrophobic.

C. Conclusions from in vitro Models and Spectral Studies on Chlorosomes

While it is not yet possible to give a definitive structure for the pigment oligomers present in the in vitro aggregates or in chlorosomes, certain conclusions can be safely drawn. 1). Chlorosomes contain pigment oligomers in which the basic geometry of the oligomer is determined by pigment-pigment interactions (The oligomers are possibly associated as a group with proteins, see Section V). 2). The Mg atoms in the BChl *c* molecules in chlorosomes are essentially all pentacoordinate. 3). The 13^1-carbonyl of each BChl *c* interacts strongly with a functional group on another BChl *c*. 4). The 3^1-hydroxyethyl group is involved in oligomer formation and is most likely to be directly coordinated to a Mg atom and hydrogen bonded to the 13^1-carbonyl group. 5). The BChl *c* molecules are close together and oriented in such a way that they exhibit strong excitonic interactions. 6). BChl *c* molecules show long range ordering with the Q_y transition moments aligned parallel to the long axis of the chlorosome.

D. Pigment-Protein Organization in Chlorosomes

While essentially all workers now agree that some sort of pigment oligomer is present in chlorosomes, there is still no consensus on the role(s) of proteins in chlorosome structure and function. The role of proteins in the organization of the bacteriochlorophyll molecules within the chlorosome has been a controversial subject for many years. Feick and Fuller (1984) presented a model for *Cf. aurantiacus* chlorosomes with the 5.7 kDa protein placed inside the rod elements forming a framework for the BChl *c* molecules and with the 11 and 18 kDa polypeptides located in the chlorosome envelope. Wechsler et al. (1985b) determined the primary structure of the 5.7 kDa CsmA protein and proposed that each protein binds seven BChl *c* molecules. They further suggested that each of the globular subunits in the rod elements observed in freeze-fracture EM are composed of 12 of these 5.7 kDa polypeptides. However, this view now seems unlikely, based on sequence comparisons of CsmA proteins from a number of organisms (see Section II E).

In contrast, Griebenow and Holzwarth (1989) and Holzwarth et al. (1990a,b) by means of detergent treatment and gel-electrophoresis obtained a protein-free preparation (so-called 'GEF chlorosomes') from *Cf. aurantiacus* cells in which the spectral properties of the BChl *c* molecules were similar to those of native protein-containing chlorosomes (Holzwarth et al., 1992). Furthermore, Wullink et al. (1991) localized antigenic sites of the 5.7 kDa protein to the envelope of the *Cf. aurantiacus* chlorosomes. These results were interpreted as indicating that the 5.7 kDa polypeptide is not embedded in the rod-structures participating in the organization of BChl *c* molecules as predicted in the models of Feick and Fuller (1984) and Wechsler et al. (1985b).

Niedermeier et al. (1992) investigated the effect of thermolysin treatment on the protein composition of *Cf. aurantiacus* chlorosomes. The 11 and 18 kDa proteins were degraded within 20 seconds after addition of the protease. Longer treatment also lead to the digestion of the 5.7 kDa protein and simultaneously a change in the absorption intensity and rotational strength of the BChl *c* molecules. However, the authors were not able to correlate the decrease in the 740 nm absorption and the disappearance of the 5.7 kDa chlorosome protein and suggested that at least some of the BChl *c* molecules within the native chlorosomes could be arranged as aggregates of BChl *c* molecules. Lehmann et al. (1994a) found that the ratio of pigment to protein in *Cf. aurantiacus* chlorosomes changed substantially when cells are grown under different light intensities and concluded on the basis of protease treatment experiments and CD spectra that the 5.7 kDa protein binds to pigment oligomers, possibly by way of the conserved His residue (Fig. 4).

Treatment of chlorosomes from *Cf. aurantiacus* with low concentrations of LDS resulted in a fraction with a full complement of proteins and antenna BChl *c* absorbing at 740 nm but at the same time the characteristic ellipsoid shape of the chlorosome was changed to a more spherical form (Miller et al., 1993b). At higher concentrations of the detergent a protein-free fraction possessing the same absorption spectrum was obtained. This fraction appeared in electron micrographs as stain excluding spheres with the same overall dimension as intact chlorosomes. Miller et al. (1993b) concluded that the protein-free spheres probably are micelles of BChl *c* molecules since the same structures could be produced when dispersing purified BChl *c* molecules into buffer.

Currently we are left with the question of the relationship between these protein-free structures or 'GEF chlorosomes' (Griebenow and Holzwarth, 1989) produced after detergent treatment and native chlorosomes. It may be that the BChl *c* molecules in the intact chlorosome are organized in oligomeric structures similar to those observed in artificial aggregates with no protein involvement at all. On the other hand, it is possible that removal of proteins either by detergent treatment or proteolysis disrupts the normal pigment organization and the BChl *c* molecules then spontaneously form aggregates which coincidentally have the same spectral properties as those of intact chlorosomes, and possibly a similar structure. We think, however, that the name 'chlorosome' should be restricted to subcellular preparations that have the characteristic ellipsoidal shape and which contain proteins, lipids, carotenoids, etc.

E. Structural Model for Overall Chlorosome Organization

Figure 9 shows a schematic model presenting our current picture of the organization of proteins and bacteriochlorophyll in the chlorosome antenna of *Cf. aurantiacus*. The model is based on data obtained

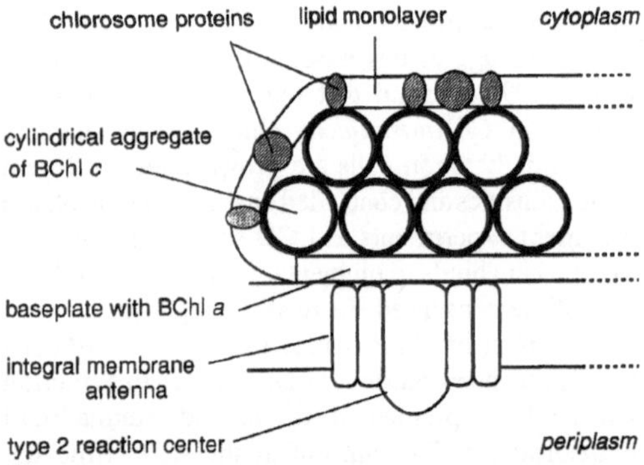

Fig. 9. Proposed model of the organization of proteins and pigments in chlorosomes of *Cf. aurantiacus*. For details see text.

from freeze-fracture electron microscopy, cross-linking experiments, surface-exposed localization of antigenic sites of the chlorosome proteins as well as the basic conclusions about the pigment organization and involvement of proteins from many lines of evidence, as discussed above.

The model depicts the BChl *c* molecules as being organized in rod-forming polymeric structures extending the full length of the chlorosome. No proteins are involved in the structural arrangement of the chlorophyll molecules: the polymeric structures are formed solely by pigment-pigment interactions as in artificial aggregates of BChl *c* molecules. However, in the model it is envisaged that the 5.7 kDa chlorosome protein penetrates deeper into the chlorosome core than do the 11 and 18 kDa proteins which are more exposed at the surface of the envelope of the chlorosome. In the 5.7 kDa protein the α helix region near the N terminus is long enough to penetrate the proposed monolayer of lipids forming the envelope of the chlorosome so that the conserved hydrophilic section with the His residue contacts the Mg-atom of a BChl *c* molecule in the rod element nearest the envelope. The 5.7 kDa 'aggregate-binding' CsmA protein might also be responsible for breaking up the aggregates into small domains of about 30 BChl *c* molecules, the unit of functionally connected BChl *c* molecules as determined by annihilation studies in *Cf. aurantiacus* using picosecond spectroscopy (Miller et al., 1991). The striation of the rod elements observed in the freeze-fracture EM may appear where the 5.7 kDa protein separates two pigment domains.

VI. Kinetics and Pathways of Energy Transfer in Chlorosomes

A. Overall Pathway of Energy Transfer

The overall energy transfer pathway of green bacteria is described by the scheme:

$$\text{carotenoid} \rightarrow \text{BChl } c \rightarrow \text{BChl } a_{\text{baseplate}}$$

$$\rightarrow \text{BChl } a_{\text{cytoplasmic membrane}} \rightarrow \text{BChl } a_{\text{reaction center}}$$

The latter steps of the energy transfer process are different in the green sulfur bacteria compared to the filamentous green bacteria, due to the substantial differences in the reaction centers and membrane-bound antenna complexes. Also, the filamentous green bacteria do not contain the FMO complex that is clearly an intermediate in the energy transfer pathway in all green sulfur bacteria.

The chlorosome antenna system is a classic example of the 'funnel' concept of photosynthetic antennas. Almost every successive pigment species in the scheme above has progressively red-shifted absorption and fluorescence spectra. The descending energy levels thus provide an energetic gradient of excitations into the reaction center. Absorption and fluorescence spectra and lifetimes, especially when coupled with careful biochemical resolution and separation procedures, can give insight into the pathway and mechanism of excitation transfer and trapping. Phycobilisomes found in cyanobacteria and red algae are another example of a peripheral antenna complex that has a substantial energy gradient, although there seems to be no structural relationship between the two classes of antennas, either in terms of pigments or proteins, and they are almost certainly independent evolutionary innovations.

Early fluorescence spectra that demonstrated energy transfer from BChl *c* to BChl *a* in green sulfur bacteria were measured by Krasnovsky et al. (1962, 1963) and the first quantitative measurements of this process were made by Sybesma and Olson (1963). Betti et al. (1982) and Van Dorssen et al (1986b) reported fluorescence spectra of isolated chlorosomes from *Cf. aurantiacus* and Van Dorssen et al. (1986a) measured spectra from *Cb. limicola* chlorosomes. The role of B795 BChl *a* in the chlorosome baseplate in the energy transfer pathway was proposed on the basis of these and other studies.

B. Anomalies in the Fluorescence Yields and Lifetimes in Chlorosomes

A number of studies have pointed out a curious phenomenon in chlorosomes (in particular those from *Cf. aurantiacus*), that the intensity of BChl *c* and even in some cases the B795 BChl *a* fluorescence emission is remarkably insensitive to the integrity of the later steps in the energy transfer pathway. Normally, one expects that if the acceptor that receives excitations is removed, then the fluorescence intensity of the donor is greatly increased. This is because the major deactivation pathway for the excited state of the energy donor is energy transfer to the acceptor, so that when this pathway is lost, other pathways, such as fluorescence, must assume a larger role in the excited state decay. A dramatic example of this effect is given by Brune et al. (1990), in which the B795 baseplate was eliminated using SDS treatment, a method introduced by Griebenow and Holzwarth (1989). The fluorescence due to the B795 pigments is completely lost, but the intensity of the 750 nm fluorescence arising from the BChl *c* is virtually unaffected. Steady-state fluorescence measurements can be deceptive, in that low amplitude but long-lived components that are quantitatively not important in excited state decay pathways can dominate the steady state fluorescence spectra. However, time-resolved measurements also clearly show that the principal 10-15 ps lifetime of the BChl *c* excited states is virtually unaffected by the presence or absence of the B795 pigments (Holzwarth et al., 1990b; Miller et al., 1991).

There are at least two possible explanations for this unexpected result. First, the treatments that disrupt the energy transfer pathway may introduce new decay processes that were not present in the intact sample. These new quenching processes would then take the place of the energy transfer process that normally deactivate the excited state, with the result that the observed intensities and kinetics of BChl *c* emission are not greatly affected. While this effect can explain the observed results, it does so in an ad hoc way and requires the apparently fortuitous generation of just enough quenchers to almost precisely compensate for the loss of the energy transfer acceptors. However, it is not clear what happens to the B795 pigments upon SDS treatment, so that the existence of a chemically damaged pigment whose absorption is buried under the large BChl *c* absorption envelope yet was still able to quench excitations, is certainly possible. It is also clear that chlorosomes are highly susceptible to excited state quenching, so this explanation seems plausible.

The other major way of explaining the near independence of the BChl *c* emission on the integrity of the energy transfer pathway is to propose that the observed BChl *c* emission arises from a pool of pigments that are not directly involved in the energy transfer to B795, and that the mainstream energy transfer is so rapid that the majority of BChl *c* antenna pigments contribute almost nothing to the steady-state emission spectrum. This explanation is also somewhat ad hoc, in that it proposes a functional heterogeneity in the BChl *c* and a putative ultrafast energy transfer process that is not yet well documented. However, some recent measurements do suggest that there is such an ultrafast component and that some excitations absorbed in BChl *c* are transferred to pigments absorbing in the 800 nm region on the femtosecond time scale (Savikhin et al., 1994b). Whether or not this can quantitatively account for the observations discussed above is not yet clear.

C. Kinetics of Energy Transfer within Chlorosomes

Borisov et al. (1977) and Fetisova and Borisov (1980a,b) made early time-resolved fluorescence measurements on subcellular complexes of *Cb. limicola*, and concluded that about 90% of the BChl *c* molecules transferred energy to BChl *a* in 20–50 ps. Brune et al. (1987a) made the first measurements of fluorescence lifetimes in isolated chlorosomes, and reported lifetimes of less than 30 ps for BChl *c* emission in *Cf. aurantiacus*.

A number of recent studies of the kinetics of energy trapping in green bacterial systems have been carried out by several groups using time-resolved fluorescence and absorption methods, with samples ranging from freshly isolated whole cells under physiological conditions to isolated complexes of various sorts at both room and low temperatures to purified oligomeric pigments. The data generally support a sequential energy transfer pathway from the BChl *c* (*d* or *e*) pigments in the body of the chlorosome to the baseplate to the membrane-bound antenna complexes and finally to the reaction centers. However, there are some inconsistencies in the available data that have not been resolved, so that a complete picture has not yet emerged. In particular, the question of whether or not a graded series of different spectral forms of BChl *c* creates an energetic funneling within the chlorosome, is not at all clear. While there is substantial evidence that there are different spectral forms of BChl *c* within the chlorosome, some data argues that they are not important in the main energy transfer pathway. Also, recent studies suggest that there are ultrafast excited state decay and rise components in these systems that were not observed in the majority of reported studies due to insufficient time resolution.

Excitation annihilation in the chlorosome system is a consideration that must be addressed, in particular for the baseplate pigments, where energy tends to accumulate. This effect is a particular concern in this system because so many pigments are coupled together in a single chlorosome, and the use of high powered lasers will inevitably introduce multiple excitations into a single chlorosome. Depending on the details of the experimental arrangement, either singlet-singlet or singlet-triplet annihilation may dominate. While in many experiments annihilation effects may introduce undesirable complications in the observed fluorescence yields and excited state kinetics, this type of experiment can also yield useful information about pigment organization. Unfortunately, these effects have not yet been very thoroughly studied in the chlorosome system and much remains to be learned from this sort of experiment.

Vos et al. (1987), observed clear evidence for annihilation in chlorosomes from both *Cf. aurantiacus* and *Cb. limicola*, using flash-induced fluorescence yield measurements, and determined domain sizes for excitation diffusion at room temperature and 4K. In *Cb. limicola*, the domain size at room temperature was essentially the entire chlorosome, when dithionite was included, while it was much smaller if dithionite was not added or at low temperatures. For *Cf. aurantiacus,* the domain size was much smaller and could not be accurately determined at room temperature. Gillbro et al. (1988) and Miller et al. (1991) reported no intensity effects on BChl *c* excited state lifetimes in isolated chlorosomes from *Cb. limicola* and *Cf. aurantiacus*. In *Cf. aurantiacus,* the amplitude of the absorbance changes were affected, however, and a domain size of 30 BChl *c* was determined (Miller et al., 1991). Van Noort et al (1994) observed evidence for annihilation effects in the baseplate region of chlorosomes from *Cb. vibrioforme*.

1. Energy Transfer Kinetics in Chloroflexus aurantiacus

In *Cf. aurantiacus* a major component of the BChl *c* excited state decay exhibits a lifetime of 10–15 ps, along with small amplitudes of longer components. This pattern has been observed by a number of groups using both time-resolved fluorescence and absorption techniques (Mimuro et al., 1989; Blankenship et al., 1990; Holzwarth et al., 1990b; Miller et al., 1991; Lin et al., 1991; Müller et al., 1993; Savikhin et al., 1994b). In addition to this component, Savikhin et al. (1994b) found substantial femtosecond timescale decay components in the BChl *c* decay.

Several, but not all, studies have found that the 10–15 ps BChl *c* decay is matched by a rise component in the baseplate BChl *a* 795. Mimuro et al. (1989) failed to observe such a correspondence in the decay of the BChl *c* emission and the rise of the BChl *a* emission at 805 nm, either by single photon timing or streak camera techniques. However, Müller et al. (1990) Holzwarth et al. (1990b) and Causgrove et al. (1990b) did observe corresponding rise and decay components, although the amplitudes of the decay and rise components were significantly different. Savikhin et al. (1994b) observed complex multi-exponential kinetics in the 800 nm region, including an ultrafast bleaching, interpreted as a femtosecond timescale excitation transfer from BChl *c* to B795 (or to a small pool of extremely red-shifted BChl *c* pigments). A slower rise was observed on the red edge of the B795 band, which was interpreted as energy transfer between two distinct pools of B795. Several earlier studies have also suggested that the B795 pigments have more than one spectral component. Mimuro et al. (1989, 1992, 1994a,b).

2. Energy Transfer Kinetics in Green Sulfur Bacteria

In the green sulfur bacteria the lifetimes of BChl *c, d* or *e* excited states are generally somewhat longer than in *Cf. aurantiacus*, as long as the redox-activated quenchers are reduced (see Section IX). Lifetimes of BChl *c, d* or *e* for all the green sulfur bacteria are in the 50–100 ps range (Fetisova et al., 1987, 1988; Gillbro et al., 1988; Blankenship et al., 1990; Causgrove et al., 1990b, 1992). Some of the early literature on energy transfer in these organisms may have mistakenly assigned the very short lifetimes observed under aerobic conditions to energy transfer processes, while it now seems more likely that quenching processes are responsible for the short lifetimes observed under such conditions (see Section IX).

D. Minor Spectral Species as Energy Transfer Intermediates Within Chlorosomes

In phycobilisome antenna complexes, clear evidence for sequential energy transfers from more blue-shifted to more red-shifted pigments has been obtained (see Van Grondelle et al., (1994) for recent review). Whether such a 'funneling' takes place within chlorosomes is as yet unclear. While a number of steady-state studies using various types of spectroscopy have clearly indicated the presence of multiple pigment species in chlorosomes (Otte et al., 1991; Griebenow et al., 1991a; Matsuura et al., 1993; Mimuro et al., 1994a,b), it is much less clear whether or not these intermediates are important in mainstream energy transfer. Holzwarth et al. (1990b) observed a 5 ps fluorescence component in *Cf. aurantiacus* chlorosomes in which the BChl *a* of the baseplate had been removed, which was interpreted in terms of an intermediate BChl *c* species in the energy transfer in intact chlorosomes (Müller et al., 1993). However, transient absorption studies with femtosecond time resolution on intact chlorosomes do not reveal any evidence for spectral development of the sort that would be expected if this were indeed the case (Savikhin et al., 1994b), and it may be that this component is only present in samples that have had the baseplate removed. The important question of whether or not energetic funneling processes take place within chlorosomes remains unresolved.

E. Kinetics of Energy Transfer Measured with Polarized Light

A number of time-resolved measurements have been made in chlorosome systems using polarized light, and the results from different laboratories and using a range of organisms are generally consistent both with each other and with steady-state measurements. These kinetic studies generally support the ideas of pigment organization discussed in Sections IV and V. Fetisova et al. (1987, 1988) found that the polarization of fluorescence from BChl *c* in both *Cf. aurantiacus* and *Cb. limicola* was high ($p = 0.41$–0.42) and nearly constant during the entire BChl *c* excited state lifetime. Similar results were reported by Freiberg et al. (1988) and Mimuro et al. (1994a). Gillbro et al. (1988) reported lower polarizations ($p = 0.2$) in *Cb. limicola*, as did a later report from the same group (Miller et al., 1991). Lin et al., (1991) using absorption difference techniques, observed an initial anisotropy $r(0)$ of 0.4 and also observed a 4–7 ps component of the anisotropy decay that was somewhat wavelength dependent, with a final $r(\infty)$ of 0.24 to 0.32. Savikhin et al., (1994b), using the highest time resolution absorption difference instrumentation employed in this system to date, found for *Cf. aurantiacus* $r(0)$ of 0.39–0.40 and $r(\infty) = 0.30$–0.37 throughout the BChl *c* absorption band. They observed a faster anisotropy decay (0.3–1.5 ps) than observed by Lin et al. (1991). In addition, Savikhin et al. (1994b) observed coherent oscillations in the BChl *c* region of the spectrum that damped within 1 ps, which were interpreted in terms of vibrational coherences in the pigment aggregates.

No evidence for $r(0) > 0.4$ has been obtained in any studies reported to date, despite the use of instrumentation with femtosecond time resolution. Such high values have been predicted in certain situations by Knox and Gülen (1993) (discussed in Chapter 15 by Struve) and are observed in FMO trimers (Section VIIE) (Savikhin et al., 1994a; Savikhin and Struve, 1994). While the significance of these observations is not yet entirely clear, the differences may reflect the much stronger electron coupling of pigments in chlorosomes compared to FMO trimers.

In contrast to the observations discussed above that there is little depolarization of excited states within the lifetime of the BChl c (*d* or *e*) pigments, transfer to the baseplate BChl a leads to substantial depolarization. This is in agreement with numerous steady-state fluorescence and linear dichroism studies that concluded that these pigments have substantially different orientations (Section IVA). Fetisova and coworkers found in *Cb. limicola* (Fetisova et al., 1987, 1988; Freiberg et al., 1988) and *Cf. aurantiacus* (Fetisova et al., 1988) that this depolarization was complete ($r(\infty) = 0$) in 120 ps. Miller et al. (1991) observed an anisotropy decay in *Cf. aurantiacus* at 800 nm of 7.6 ps using transient absorption, but found a larger residual anisotropy $r(\infty) = 0.13$. Mimuro et al. (1994a) reported fluorescence anisotropy decay times of 70 ps in *Cf. aurantiacus* and $r(\infty) = -0.09$. Savikhin et al. (1994b) observed 8–9 ps anisotropy decay in transient absorption in the 800 nm region in *Cf. aurantiacus*, with r(∞) = 0.04–0.12, depending on wavelength of excitation and detection. Van Noort et al. (1994) observed 20 ps depolarization with r(∞) = 0 in the 800 nm region of *Cb. vibrioforme.*

F. Kinetics of Energy Transfer Within Membranes and into the Reaction Center

The major component of decay of the BChl *a* 795 in *Cf. aurantiacus* is about 40 ps at room temperature, (Mimuro et al., 1989; Causgrove et al., 1990b; Müller et al., 1990, 1993) which is matched by the rise of the emission of the B808–866 complex in the membrane (Causgrove et al., 1990b; Müller et al., 1990, 1993). This process is substantially interrupted at low temperatures (Mimuro et al., 1992, 1994b).

The energy transfer time from B808 to B866 is extremely rapid and has not been resolved in intact systems, although Vasmel et al. (1986) estimated it to be 6 ps, on the basis of steady-state fluorescence spectroscopy. This transfer time has been measured as 5 ps in isolated B808–866 complexes (Griebenow et al., 1991b). This ultrafast energy transfer is similar to the results in the analogous B800–850 complexes found in purple bacteria (see Chapter 17 by Sundström and Van Grondelle). The chemical sequences of the B808–866 peptides have been determined by Wechsler et al. (1985a, 1987).

The energy transfer time from B866 to P870, the reaction center special pair has been measured both in samples with oxidized and reduced P870. Brune et al. (1987a) observed a time of 218 ps for this process at 77 K. In room temperature measurements, Nuijs et al. (1986) found a time of 200 ps, while Mimuro et al. (1989) reported 250 ps, and Müller et al. (1993) observed 200 ps. In all these experiments the reaction centers were almost certainly in the oxidized state, so this time represents trapping by $P870^+$. Müller et al (1990) reported 106 ps for this process, although the redox state of the trap was not known with certainty.

In samples where P870 is reduced, the trapping times of $B866^*$ by P870 are substantially faster. Causgrove et al. (1990b) observed 43 ps for this process, while Müller et al. (1993) reported it to be 70–90 ps. While this apparent discrepancy has not yet been resolved, the former time is more in agreement with what is found for the same process in purple bacteria with open traps (see Chapter 17 by Sundström and Van Grondelle).

VII. The Fenna-Matthews-Olson (FMO) Bacteriochlorophyll *a* Protein

A water-soluble BChl *a*-containing protein was isolated and characterized from two green sulfur bacteria by Olson and coworkers (Olson and Romano, 1962; Olson, 1966, 1971, 1978, 1980a,b, 1994; Thornber and Olson, 1968; JM Olson et al., 1969, 1973, 1976a,b; RA Olson et al., 1969a,b). The FMO protein was the first chlorophyll-containing protein to have its structure determined to atomic resolution. For this reason, much of our current understanding of the principles of pigment-proteins derives from studies of this system. This includes the fact that the pigments are held precisely in place in the protein by specific interactions with protein groups, that His is usually the fifth ligand to Mg, that the complex exhibits threefold rotational symmetry and that exciton coupling is important in determining the absorption and CD spectra of pigment-proteins. These features are now known to be generally true for most pigment-proteins. However, in other respects, the FMO protein is an atypical antenna complex, in that no other complex has been isolated that is similar in overall protein fold.

The FMO protein is found in all green sulfur bacteria that have been surveyed, but is clearly not present in the filamentous green bacteria. In the green sulfur bacteria, some FMO protein remains associated with the reaction center even after extensive purification (Feiler et al., 1992; Okkels et al., 1992; Oh-Oka et al., 1993; Kusumoto et al., 1994). The chlorosome, the FMO protein and the reaction center core itself apparently constitute the complete set of antenna complexes in green sulfur bacteria. Earlier work (Swarthoff and Amesz, 1979; Swarthoff et al. 1981) had suggested that there might be an additional membrane-bound antenna complex, although this now appears not to be the case. It is, however, unclear why a fraction of the complexes remain tightly associated with the reaction centers, while the majority of them are readily dissociated. The tightly associated FMO complexes have the same apparent molecular mass and identical N-terminal amino acid sequence as does the easily dissociated pool of FMO protein. Also, DNA hybridization experiments in *Cb. tepidum* indicate that only one gene with sequence similar to the FMO protein is present, arguing against multiple forms of the protein with slightly different primary structure and therefore slightly different properties (Dracheva et al, 1992).

A. X-ray Structure

The structure of the FMO-protein from *Pc. aestuarii* 2K has been determined to a resolution of 1.9 Å by Matthews and coworkers (Fenna and Matthews, 1975; Matthews et al., 1979; Matthews and Fenna, 1980; Tronrud et al., 1986). The amino acid sequence was determined by Daurat-Larroque et al. (1986), and the combined structure has recently been described by Tronrud and Matthews (1993). The crystallization of the FMO protein from *Cb. tepidum* has been reported, and an X-ray structure of the protein from this organism is in progress (J. Allen, personal communication).

A

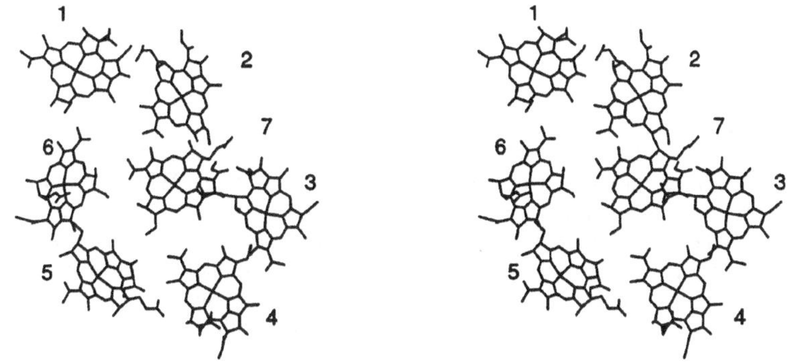

B

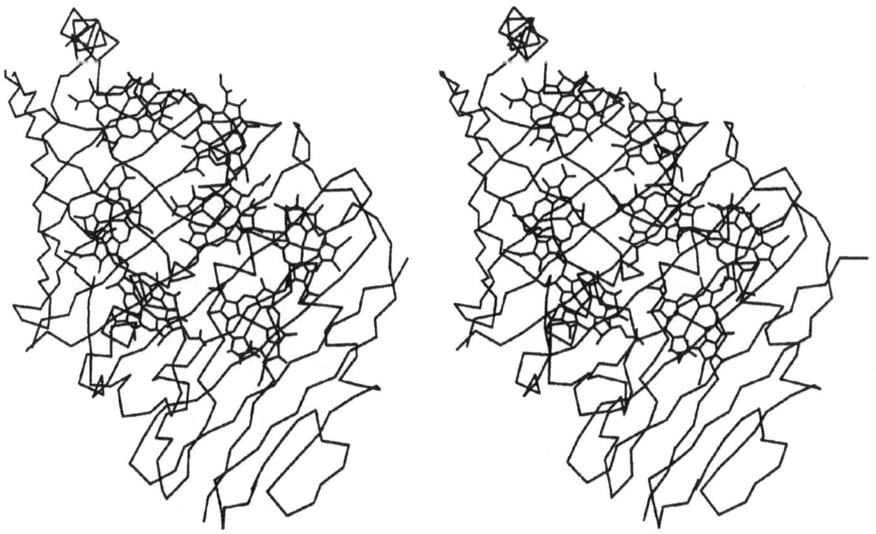

Fig. 10. Subunit of the FMO-protein from *Pc. aestuarii*. A. BChl *a* pigments (phytol tails removed for clarity). B. BChl *a* pigments and α carbon backbone. Drawn from coordinates supplied by Tronrud and Matthews (1993). Figure courtesy of Y.F. Li and J.P. Allen.

The FMO-protein is a trimer, and each subunit consists of a protein fold variously described as a string bag, taco shell or a distorted hollow cylinder. The protein surrounds seven BChl *a* molecules (See Fig. 10). One end of the cylinder is open, but this opening is covered in the trimer. The wall of the cylinder that faces the outside of the trimer is composed almost entirely of 15 strands of β-sheet. The side of the cylinder in contact with adjacent subunits consists of short lengths of α-helix together with regions of irregular conformation.

The seven BChl *a* molecules occupy the space within an ellipsoid of axial dimensions, 45 × 35 × 15 Å. The ligands to five of the BChls are His (110, 145, 290, and 298). BChl 2 is liganded to a water molecule, and BChl 5 is liganded to the backbone carbonyl of residue 242. The average center-to-center distance between porphine rings is 12 Å for nearest neighbors. The phytyl chains lie close together. The tails of BChls 4, 5 and 6 are parallel and form a planar structure between porphine rings 5 and 6 and the β-sheet of the outer wall. The tails of BChls 2, 3 and 7 lie in extended conformation in the inner space between porphine rings. The tail of BChl 1 is bent into a U-shaped loop.

Each BChl molecule is anchored to the protein through extensive hydrogen bonding and liganding to the Mg atom in addition to hydrophobic interactions through the phytyl tail. The position and orientation of each BChl molecule are controlled mainly by specific interaction with protein, rather than with other BChl molecules.

The subunits are tightly packed in the trimer, which is approximated by an oblate ellipsoid of revolution 57 Å × 83 Å (Fenna et al., 1974). The closest distance between BChl molecules in adjacent subunits is 24 Å. In trigonal crystals, the trimers are close packed, and the closest distance between BChl molecules in adjacent trimers is about 30 Å. In vivo, the trimers appear to form a two-dimensional trigonal crystal plate between the chlorosome and the plasma membrane (Olson, 1980a).

B. Amino Acid Sequence

The polypeptide chain of the FMO-protein from *Pc. aestuarii* consists of 366 amino acids (Daurat-Larroque et al., 1986). In *Cb. tepidum* the sequence of the FMO-gene has been completed, and the deduced amino acid sequence shows 78% identity with the FMO-protein from *Pc. aestuarii* (Dracheva et al., 1992). The six residues that ligand BChl molecules in *Pc. aestuarii* are conserved in *Cb. tepidum*. The sequences of the FMO proteins from these two organisms are compared in Fig. 11.

C. Absorption, Fluorescence and Circular Dichroism Spectra

The FMO-proteins from *Pc. aestuarii* and *Cb. limicola* show quite similar absorption and CD spectra at 20–25 °C, but distinctly different spectra at temperatures of 77 K and below (Olson et al., 1976a). These low temperature differences show that the excitonic interactions between the BChl molecules in the two FMO-proteins are not exactly the same. Combining the absorption and CD data at 5K for the two FMO-proteins, Whitten et al. (1980) deduced that all the spectra were consistent with seven excitonic components at 793, 801–2, 806, 810, 814–6, 819–20, and 823–25 nm. Extinction coefficients ($mM^{-1} \cdot cm^{-1}$) at the absorption maxima at 20–25° C are listed as follows: 37 (267 nm), 49 (343 nm), 67 (370.5 nm), 28.4 (603 nm), 13.4 (745 nm) and 154 (809 nm) (Olson, 1978).

The fluorescence emission peak at 20–25 °C is at 818 nm (Sybesma and Olson, 1963; Olson, 1966). At 77 K it sharpens and shifts to 831 nm, and the fluorescence yield increases from 0.19 at 293 K to 0.29 at 77 K (Olson, 1971).

D. Theoretical Analysis, Transient and Hole-Burning Spectroscopies

Because of the wealth of structural and spectral data available, the FMO protein has been very intensely studied using theoretical methods. In a series of calculations of increasing levels of sophistication (Philipson and Sauer, 1972; Olson et al., 1976a; Pearlstein and Hemenger, 1978; Gudowska-Nowak et al., 1990; Pearlstein, 1988, 1991, 1992; Lu and Pearlstein, 1993), a reasonable theoretical fit to the absorption and CD spectral data for *Pc. aestuarii* has finally been obtained. The seven pigments within a subunit of the trimer are strongly coupled by exciton interactions, with coupling strengths of from 100–375 cm^{-1}. Coupling between trimers is weaker, with coupling strengths of 20 cm^{-1} or less. Each of the seven pigments in a subunit has a slightly different site energy as a result of pigment-protein interactions. The 21 coupled pigments give rise to two sets of states, a set of 7 polarized parallel to the symmetry

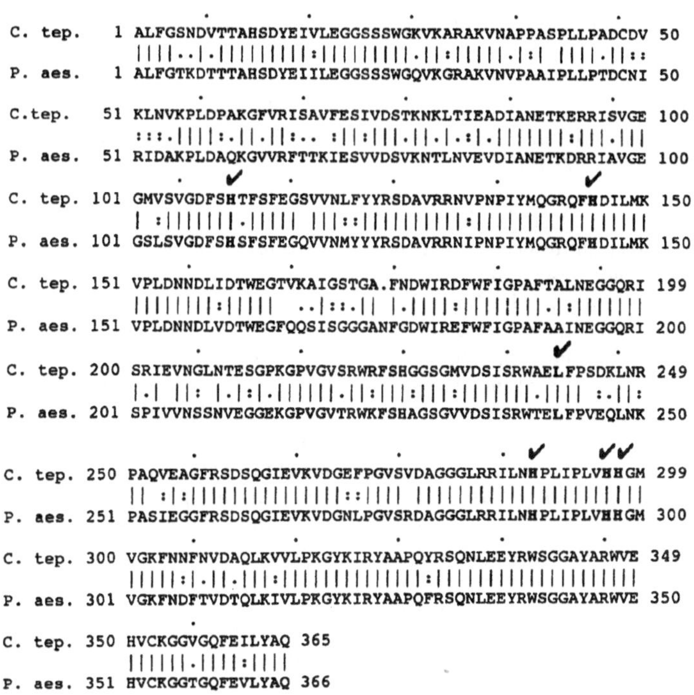

C. tep. 1 ALFGSNDVTTAHSDYEIVLEGGSSSWGKVKARAKVNAPPASPLLPADCDV 50
P. aes. 1 ALFGTKDTTTAHSDYEIILEGGSSSWGQVKGRAKVNVPAAIPLLPTDCNI 50
C.tep. 51 KLNVKPLDPAKGFVRISAVFESIVDSTKNKLTIEADIANETKERRISVGE 100
P. aes. 51 RIDAKPLDAQKGVVRFTTKIESVVDSVKNTLNVEVDIANETKDRRIAVGE 100
C. tep. 101 GMVSVGDFSHTFSFEGSVVNLFYYRSDAVRRNVPNPIYMQGRQFHDILMK 150
P. aes. 101 GSLSVGDFSHSFSFEGQVVNMYYYRSDAVRRNIPNPIYMQGRQFHDILMK 150
C. tep. 151 VPLDNNDLIDTWEGTVKAIGSTGA.FNDWIRDFWFIGPAFTALNEGGQRI 199
P. aes. 151 VPLDNNDLVDTWEGFQQSISGGGANFGDWIREFWFIGPAFAAINEGGQRI 200
C. tep. 200 SRIEVNGLNTESGPKGPVGVSRWRFSHGGSGMVDSISRWAELFPSDKLNR 249
P. aes. 201 SPIVVNSSNVEGGEKGPVGVTRWKFSHAGSGVVDSISRWTELFPVEQLNK 250
C. tep. 250 PAQVEAGFRSDSQGIEVKVDGEFPGVSVDAGGGLRRILNHPLIPLVHHGM 299
P. aes. 251 PASIEGGFRSDSQGIEVKVDGNLPGVSRDAGGGLRRILNHPLIPLVHHGM 300
C. tep. 300 VGKFNNFNVDAQLKVVLPKGYKIRYAAPQYRSQNLEEYRWSGGAYARWVE 349
P. aes. 301 VGKFNDFTVDTQLKIVLPKGYKIRYAAPQFRSQNLEEYRWSGGAYARWVE 350
C. tep. 350 HVCKGGVGQFEILYAQ 365
P. aes. 351 HVCKGGTGQFEVLYAQ 366

Fig. 11. Sequence comparison of FMO proteins from *Pc. aestuarii* (Daurat-Larroque et al., 1986) and *Cb. tepidum* (Dracheva et al., 1992). The residues that are involved in coordination of BChl *a* are indicated with check marks. Figure reproduced with permission from Dracheva et al. (1992).

axis and a set of seven pairs of doubly degenerate states polarized perpendicular to the symmetry axis. The lowest energy exciton states are primarily localized on BChl 7, and give rise to the low temperature absorption band centered at 825 nm. Hole-burning and triplet state spectroscopy have indicated that the 825 nm band is actually composed of two distinct transitions, which are assigned to the parallel and perpendicular transitions of the lowest energy exciton states (Johnson and Small, 1991; Van Mourik et al., 1994).

Several time-resolved spectroscopic studies on isolated FMO trimers have been reported, (Causgrove et al., 1988; Gillbro et al., 1988; Lyle and Struve, 1990; Van Amerongen and Struve, 1991; Van Mourik et al., 1992; Louwe and Aartsma, 1994; Savikhin et al., 1994a; Savikhin and Struve, 1994; Zhou et al., 1994). After excitation of the FMO protein from *Cb. tepidum* into either the Q_x or Q_y absorption bands, at room temperature the system relaxes to an equilibrated excited state distribution with a time constant of about 500 fs (Savikhin et al., 1994a). Relaxation at low temperature is complex (A Freiberg, personal communication). The relaxed excited state lifetime is multiphasic at room temperature, with components of about 60–80 ps and 2–2.5 ns (Gillbro et al., 1988; Zhou et al., 1994). The relative amount of the two phases was found to depend on redox potential, with reducing conditions increasing the longer lifetime at the expense of the shorter (Zhou et al., 1994). This effect is discussed further in Section VIII.

The FMO protein from *Cb. tepidum* exhibits initial excited state anisotropy values $r(0) > 0.4$ (see section VIA and Chapter 15 by Struve), but rapidly decays to < 0.4 in less than 100 fs (Savikhin et al., 1994a; Savikhin and Struve, 1994). Anisotropy decays of 70–100 fs and 1–2 ps are observed at most wavelengths, with $r(\infty)$ values less than 0.1. Interestingly, the anisotropy decays are not reflected in the isotropic decay processes. The slower anisotropy decay processes probably result from equilibration among equivalent lowest energy pigments on different subunits in the trimer, while the physical explanation for the faster processes is not yet clear.

VIII. Redox Modulation of Energy Transfer in Green Photosynthetic Bacteria

A growing body of evidence suggests that the energy transfer system in the green sulfur bacteria may be regulated in vivo by redox potential. The first indications for this effect were reported by Karapetyan et al. (1980), who observed an increase in the fluorescence of the FMO protein upon addition of dithionite. Van Dorssen et al., (1986a) and Vos et al., (1987) observed dramatic effects of dithionite on chlorosomes.

A. Effects on Efficiency of Energy Transfer

In 1990, a redox-activated regulation of energy transfer in the green sulfur bacteria was proposed (Wang et al., 1990; Blankenship et al, 1990). This effect appears to involve a direct chemical titration of redox-active groups in the chlorosome antennas, and was observed in whole cells, isolated membranes and purified chlorosomes. There is no evidence for nor reason to suspect phosphorylation in this system, as the effect is easily observed in purified systems with no phosphorylation substrates present. In the oxidized form, the redox-active groups efficiently quench excited states in the antenna system, reducing the overall energy transfer efficiency from nearly 100% to 10% or less. Redox titrations of fluorescence in isolated chlorosomes gave a midpoint potential of −146 mV vs NHE at pH 7, with a pH dependence of −59 mV per pH unit (Blankenship et al., 1990, 1993).

B. Effects on Excited State Lifetimes in Chlorosomes

The excited state lifetimes of BChl *c* (*d* or *e*) are dramatically affected by the redox potential. At low potential (which has generally been achieved by addition of dithionite) there is clear evidence for energy transfer from BChl *c* (*d* or *e*) to BChl *a* in the baseplate, and an excited state lifetime of 50–100 ps depending on the species (Blankenship et al., 1990, 1993; Causgrove et al., 1990b). Under oxidizing conditions these lifetimes are reduced to 10–15 ps and no characteristic rise is observed in the baseplate BChl *a*. The simplest explanation for the effects on both the energy transfer efficiency and the excited state lifetime is that quenchers are formed in the chlorosomes at high redox potentials. These quenchers provide a rapid nonradiative decay pathway that effectively competes with energy transfer. Possible mechanisms of this effect are considered in Section VIII D.

C. Effects in the FMO Protein

A similar redox modulation of the excited state lifetime is present in the Fenna-Matthews-Olson (FMO) protein (Karapetyan et al., 1980; Blankenship et al., 1993; Zhou et al., 1994). In this case the excited state lifetime is dominated by a 60 ps decay component at high redox potential and a 2 ns component at low potential (Zhou et al., 1994). The latter time is typical for an isolated antenna complex (Van Grondelle et al., 1994), while the shorter time is thought to reflect quenching processes within the protein.

D. Possible Mechanisms and Functional Significance

The redox modulation effect appears to operate on at least two levels, within the chlorosome itself and in the FMO protein. Whether these two effects have similar molecular mechanisms is not yet clear. Do these redox effects on the antenna system reflect a real cellular control mechanism? It is too soon to reach a final decision on this question, but a plausible case can be made that such an effect would serve to protect the cell from transient exposure to oxidizing conditions, in particular to oxygen. The green sulfur bacteria contain a reaction center that has very low potential iron sulfur centers as early acceptors and reduces ferredoxin directly in a manner similar to Photosystem I (see Chapter 30 by Feiler and Hauska). The reduced ferredoxin is freely diffusible in the cell cytoplasm and it is utilized in addition to NAD(P)H as the reducing substrate for its carbon reduction cycle, which is a reverse TCA cycle rather than the Calvin cycle (for reviews see Buchanan, 1992 and Chapter 40 by Sirevåg). The reduced ferredoxins will readily react with oxygen to form superoxide (Orme-Johnson and Beinert, 1969; Asada, 1994), which leads to a variety of damaging photooxidative products, including hydroxyl radicals. The system is thus very vulnerable to autooxidation, probably more so than Photosystem I, which contains a complex system of protective enzymes including superoxide dismutase and ascorbate peroxidase (Asada, 1994).

Green sulfur bacteria are obligate anaerobes and do not possess any respiratory activity. In nature they are found in a variety of environments, often just below the chemocline in stratified lakes (Trüper and Pfennig, 1992; Ormerod, 1992). Under these conditions they are likely to be exposed occasionally to moderate oxygen levels and a mechanism that provided even partial protection from oxidative damage would be of enormous adaptive advantage. Of course, there is an important difference between respiratory activity and protection from oxidative damage, and the two do not have to go together. *Cb. limicola* has been reported to contain superoxide dismutase activity (Asada et al., 1977), although it is not clear from the original paper that a pure culture was used rather than the co-culture with sulfate-reducing bacteria that was utilized extensively in the older literature.

By preventing charge separation under oxygenic conditions by quenching excitations in the antenna system, green bacterial cells avoid producing these toxic substances entirely. Thus, while these organisms apparently do not have an active mechanism for removing oxygen from their environment, they may have a system that permits them to survive transient exposures to oxygen without incurring cellular damage. No information is available as to whether green sulfur bacteria possess any of the other components of the system that protects oxygenic organisms from photooxidative stress.

The redox effects are largely missing in both whole cells and isolated chlorosomes of the green nonsulfur bacterium *Cf. aurantiacus*. (Vos et al., 1987; Wang et al., 1990), although some preparations of isolated chlorosomes do exhibit some redox-induced fluorescence increase (J. Zhu and R. E. Blankenship, unpublished observations). *Cf. aurantiacus* has an entirely different physiology and ecology with respect to oxygen. Most strains respire and usually lives in close proximity to cyanobacteria, and therefore is often under aerobic, even hyperoxic conditions (Pierson and Castenholz, 1992). However, *Cf. aurantiacus* is not vulnerable to oxidative damage in the same way that the green sulfur bacteria are. It has a reaction center that is similar to that found in the purple bacteria, and therefore does not directly reduce ferredoxin (see Chapter 32 by Feick). Its carbon reduction cycle, while not fully understood, clearly does not involve free ferredoxin as does the reverse TCA cycle (see Chapter 40 by Sirevåg). Therefore, a redox modulation effect such as has been found in green sulfur bacteria would provide no selective advantage for *Cf. aurantiacus*, and actually would prevent them from carrying out photosynthesis under most natural conditions.

The chemical mechanism for the redox effects on energy transfer in green bacteria is not yet understood and the mechanism active in chlorosomes may be entirely different from that in FMO proteins. Aggregated pigments formed from pigment extracts exhibit dithionite effects on fluorescence similar to those observed in chlorosomes (Matsuura et al., 1992). Similar results have been observed with HPLC purified pigments, where the magnitude of the redox effect correlates with the time that has elapsed since the purification (P. I. Van Noort and R. E. Blankenship, unpublished observations). This suggests that a trace impurity can cause such a result, although it does not prove that this is the mechanism for the effect in vivo. In FMO proteins, it has been suggested that modified amino acids may be involved in the redox effect, although no direct evidence has been found for such an effect (Zhou et al., 1994).

IX. Evolutionary Considerations

The very similar morphology, pigment content and functional role of the chlorosomes in the two families of green bacteria strongly suggests an evolutionary relationship between them, despite the fact that there is considerable evolutionary distance between the two families by almost any other criterion (see above). This disjunction between the chlorosome properties and other characteristics has prompted suggestions that the information coding for chlorosome components may have been transferred between otherwise unrelated organisms (Olson and Pierson, 1987; Pierson and Olson, 1989; Blankenship, 1992).

The recent availability of sequences coding for some chlorosomal proteins gives new information with which to make this comparison. As discussed above, the 30% sequence identity (as well as processing and gene organization) between the 5.7 kDa CsmA protein from *Cf. aurantiacus* and the 6.3 kDa CsmA protein from the green sulfur bacteria is highly suggestive of an evolutionary connection between these two proteins. However, this is the only case in which such a relationship has been found so far. To some extent, this may reflect the fact that the information needed to make such comparisons is only just now beginning to become available. The picture will undoubtedly become much clearer when more sequences of chlorosome proteins have been determined, as well as information concerning proteins involved in biosynthesis of other chlorosome components such as BChl *c*, *d* and *e*, as well as glycolipids.

Finally, gene organization patterns may be revealing. If, for example, the genes needed for synthesis of critical chlorosome components are clustered and in a similar overall order in the two families, it is highly likely that at least that part of the genome derive from a common ancestor. This sort of gene clustering is found in purple bacteria, where almost all the information needed to synthesize the photosynthetic apparatus is contained in a conserved order in a 43 kB cluster of genes (see Chapters 50 and 58 by Alberti et al. and Bauer, respectively). Some information of this sort may be forthcoming in the green bacteria, as genome mapping in several species is in progress (R. Sirevåg, personal communication).

Acknowledgments

REB gratefully acknowledges continuing support from the US Department of Energy, Division of Energy Biosciences. JMO and MM gratefully acknowledge support from the Danish Natural Science Research Council. This is publication #226 from the ASU Center for the Study of Early Events in Photosynthesis.

References

Alden RG, Lin SH and Blankenship RE (1992) Theory of spectroscopy and energy transfer of oligomeric pigments in chlorosome antennas of green photosynthetic bacteria. J Lumin 51: 51–66

Asada K (1994) Production and action of active oxygen species in photosynthetic tissues. In: Foyer CH and Mullineaux PM (eds) Causes of Photooxidative Stress and Amelioration of Defense Systems in Plants, pp 77–104. CRC Press, Boca Raton

Asada K, Kanematsu S and Uchida K (1977) Superoxide dismutases in photosynthetic organisms: Absence of the cuprozinc enzyme in eukaryotic algae. Arch Biochem Biophys 179: 243–256

Betti JA, Blankenship RE, Natarajan LV, Dickinson LC and Fuller RC (1982) Antenna organization and evidence for the function of a new antenna pigment species in the green photosynthetic bacterium *Chloroflexus aurantiacus*. Biochim Biophys Acta 680: , 194–201

Beyer P, Falk H and Kleining H (1983) Particulate fractions from *Chloroflexus aurantiacus* and distribution of lipids and polyprenoid forming activities. Arch Microbiol 134: 60–63

Blankenship RE (1992) Origin and early evolution of photosynthesis. Photosynth Res 33: 91–111

Blankenship RE, Brune DC, Freeman JM, King GH, McManus JD, Nozawa T and Wittmershaus BP (1988a) Energy trapping and electron transfer in *Chloroflexus aurantiacus* In: Olson JM, Ormerod JG, Amesz J, Stackebrandt E and Trüper HG (eds) Green Photosynthetic Bacteria, pp 57–69. Plenum Press, New York

Blankenship RE, Brune DC and Wittmershaus BP (1988b) Chlorosome antennas in green photosynthetic bacteria. In: Stevens SE Jr and Bryant DA (eds) Light Energy Transduction in Photosynthesis: Higher Plant and Bacterial Models, pp 32–64. American Society of Plant Physiologists, Rockville, MD

Blankenship RE, Wang J, Causgrove TP and Brune DC (1990) Efficiency and kinetics of energy transfer in chlorosome antennas from green photosynthetic bacteria. In: Baltscheffsky M (ed) Current Research In Photosynthesis, Vol II, pp 17–24. Kluwer, Dordrecht

Blankenship RE, Cheng P, Causgrove TP, Brune DC, Wang SHH, Choh JU and Wang J (1993) Redox regulation of energy transfer efficiency in antennas of green photosynthetic bacteria. Photochem Photobiol 57: 103–107

Bobe FW, Pfennig N, Swanson KL and Smith KM (1990) Red shift of absorption maxima in Chlorobiaceae through enzymic methylation of their antenna bacteriochlorophylls. Biochemistry 29: 4340–4348

Borisov AYu, Fetisova ZG and Godik VI (1977) Energy transfer in photoactive complexes obtained from green bacterium *Chlorobium limicola*. Biochim Biophys Acta 461: 500–509

Broch-Due M and Ormerod JG (1978) Isolation of a BChl *c* mutant from *Chlorobium* with BChl *d* by cultivation at low light intensities. FEMS Microbiol Lett 3: 305–308

Brune DC, King GH, Infosino A, Steiner T, Thewalt MLW and Blankenship RE (1987a) Antenna organization in green photosynthetic bacteria. 2. Excitation transfer in detached and membrane-bound chlorosomes from *Chloroflexus aurantiacus*. Biochemistry 26: 8652–8658

Brune DC, Nozawa T and Blankenship RE (1987b) Antenna organization in green photosynthetic bacteria. 1. Oligomeric bacteriochlorophyll *c* as a model for the 740 nm absorbing bacteriochlorophyll *c* in *Chloroflexus aurantiacus* chlorosomes. Biochemistry 26: 8644–8652

Brune DC, King GH and Blankenship RE (1988a) Interactions between Bacteriochlorophyll *c* molecules in oligomers and in chlorosomes of green photosynthetic bacteria. In: Scheer H and Schneider S (eds) Photosynthetic Light-Harvesting Systems, pp 141–151. Walter de Gruyter, Berlin

Brune DC, Blankenship RE and Seely GR (1988b) Fluorescence quantum yields and lifetimes for bacteriochlorophyll *c*. Photochem Photobiol 47: 759–763

Brune DC, Gerola PD and Olson JM (1990) Circular dichroism of green bacterial chlorosomes. Photosynth. Res. 24: 253–263.

Bryant DA (1994) Gene nomenclature recommendations for green photosynthetic bacteria and heliobacteria. Photosynth Res 41: 27–28

Buchanan BB (1992) Carbon dioxide assimilation in oxygenic

and anoxygenic photosynthesis. Photosynth Res 33: 147–162

Bystrova MI, Mal'gosheva IN and Krasnovskii AA (1979) Study of molecular mechanism of self-assembly of aggregated forms of BChl *c*. Mol Biol (English Trans) 13: 582–594

Caple MB, Chow H and Strouse CE (1978) Photosynthetic pigments of green sulfur bacteria. J. Biol. Chem. 253: 6730–6737

Causgrove TP, Yang S and Struve WS (1988) Polarized pump-probe spectroscopy of exciton transport in bacteriochlorophyll *a*-protein from *Prosthecochloris aestuarii*. J Phys Chem 92: 6790–6795

Causgrove TP, Brune DC, Blankenship RE and Olson JM (1990a) Fluorescence lifetimes of dimers and higher oligomers of bacteriochlorophyll *c* from *Chlorobium limicola*. Photosynth Res 25: 1–10

Causgrove TP, Brune DC, Wang J, Wittmershaus BP and Blankenship RE (1990b) Energy transfer kinetics in whole cells and isolated chlorosomes of green photosynthetic bacteria. Photosynth Res 26: 39–48

Causgrove TP, Brune DC and Blankenship RE (1992) Förster energy transfer in chlorosomes of green photosynthetic bacteria. J Photochem Photobiol B 15: 171–179

Causgrove TP, Cheng P, Brune DC and Blankenship RE (1993) Optical spectroscopy of a highly fluorescent aggregate of bacteriochlorophyll *c*. J Phys Chem 97: 5519–5524

Cheng P, Liddell PA, Ma SXC and Blankenship RE (1993) Properties of Zn and Mg methyl bacteriopheophorbide *d* and their aggregates. Photochem Photobiol 58: 290–295

Chiefari J, Griebenow K, Griebenow N, Balaban TS, Holzwarth AR and Schaffner K (1995) Models for the pigment organization in the chlorosomes of photosynthetic bacteria: Diastereo-selective control of in-vitro bacteriochlorophyll c_s aggregation. J Phys Chem 99: 1357–1365

Chow H-C, Serlin R and Strouse CE (1975) The crystal and molecular structure and absolute configuration of ethyl chlorophyllide dihydrate. J Am Chem Soc 97: 7230–7237

Chung S, Frank G, Zuber H and Bryant DA (1994) Genes encoding two chlorosome components from the green sulfur bacteria *Chlorobium vibrioforme* strain 8327D and *Chlorobium tepidum*. Photosynth Res 41: 261–275

Cohen-Bazire G Pfennig N and Kunisawa R (1964) The fine structure of green bacteria. J Cell Biol 22: 207–225

Cruden DL and Stanier RY (1970) Characterization of *Chlorobium* vesicles and membranes isolated from green bacteria. Arch Mikrobiol 72: 115–34

Daurat-Larroque ST, Brew K and Fenna RE (1986) The complete amino acid sequence of a bacteriochlorophyll *a*-protein from *Prosthecochloris aestuarii*. J Biol Chem 261: 3607–3615

Dracheva S, Williams JC and Blankenship (1992) Sequencing of the FMO-protein from *Chlorobium tepidum*. In: Murata N (ed), Research in Photosynthesis, Vol I, pp 53–56. Kluwer, Dordrecht

Fages F, Griebenow N, Griebenow K, Holzwarth A and Schaffner K (1990) Characterization of light-harvesting pigments of *Chloroflexus aurantiacus*. Two new chlorophylls: Oleyl (octadec-9-enyl) and cetyl (hexadecanyl) bacteriochloro-phyllides-*c*. J Chem Soc Perkin Trans I, 1990: 2791–2797

Feick RG and Fuller RC (1984) Topography of the photosynthetic apparatus of *Chloroflexus aurantiacus*. Biochemistry 23: 3693–3700

Feick RG, Fitzpatrick M and Fuller RC (1982) Isolation and characterization of cytoplasmic membranes and chlorosomes from the green bacterium *Chloroflexus aurantiacus*. J Bacteriol 150: 905–15

Feiler U, Nitschke W and Michel H (1992) Characterization of an improved reaction center preparation from the photosynthetic green sulfur bacterium *Chlorobium* containing the FeS centers F(A) and F(B) and a bound cytochrome subunit. Biochemistry 31: 2608–2614

Feiler U, Albouy D, Lutz M and Robert B (1994) Pigment interactions in chlorosomes of various green bacteria. Photosynth Res 41: 175–180

Fenna RE and Matthews (1975) Chlorophyll arrangement in a bacteriochlorophyll protein from *Chlorobium limicola*. Nature 258: 573–577

Fenna RE, Matthews BW, Olson JM and Shaw EK (1974) Structure of a bacteriochlorophyll-protein from the green photosynthetic bacterium *Chlorobium limicola*: Crystallo-graphic evidence for a trimer. J Mol Biol 84: 231–240

Fetisova ZG and Borisov AYu (1973) Intrinsic lifetimes of bacterioviridin-660 and chlorophyll *a* in different solvents. J Photochem 2: 151–159

Fetisova ZG and Borisov AYu (1980a) Fluorescence quantum yield and lifetime of antenna pigments of green bacterium *Chlorobium limicola*. Stud Biophys 80: 93–96

Fetisova ZG and Borisov AYu (1980b) Picosecond time scale of heterogeneous excitation energy transfer from accessory light-harvesting bacterioviridin antenna to main bacteriochlorophyll *a* antenna in photoactive pigment-protein complexes obtained from *Chlorobium limicola*, a green bacterium. FEBS Lett 114: 323–326

Fetisova ZG and Mauring K (1992) Experimental evidence of oligomeric organization of antenna bacteriochlorophyll *c* in green bacterium *Chloroflexus aurantiacus* by spectral hole burning. FEBS Lett 307: 371–374

Fetisova ZG and Mauring K (1993) Spectral hole burning study of intact cells of green bacterium *Chlorobium limicola*. FEBS Lett 323: 159–162

Fetisova ZG, Kharchenko and Abdourakhmanov IA (1986) Strong orientational ordering of the near-infrared transition moment vectors of light-harvesting antenna bacterioviridin in chromatophores of the green photosynthetic bacterium *Chlorobium limicola*. FEBS Lett, 199: 234–236

Fetisova ZG, Freiberg AM and Timpmann KE (1987) Investigations by picosecond polarized fluorescence spectro-chronography of structural aspects of energy transfer in living cells of the green bacterium *Chlorobium limicola*. FEBS Lett 223: 161–164

Fetisova ZG, Freiberg AM and Timpmann KE (1988) Long-range molecular order as an efficient strategy for light harvesting in photosynthesis. Nature 334: 633–634

Fetisova ZG, Mauring K and Taisova AS (1994) Strongly exciton-coupled BChl *e* chromophore system in the chlorosomal antenna of intact cells of the green bacterium *Chlorobium phaeo-vibrioides*: A spectral hole burning study. Photosynth Res 41: 205–210

Foidl M, Golecki JR and Oelze J (1994) Bacteriochlorophyll *c* formation and chlorosome development in *Chloroflexus aurantiacus*. Photosynth Res 41: 145–150

Freiberg AM, Timpmann KE and Fetisova ZG (1988) Excitation energy transfer in living cells of the green bacterium *Chlorobium limicola* studied by picosecond fluorescence spectroscopy. In:

Olson JM, Ormerod JG, Amesz J, Stackebrandt E and Trüper HG (eds) Green Photosynthetic Bacteria, pp 81–90. Plenum Press, New York

Fuhrmann S, Overmann J, Pfennig N and Fischer U (1993) Influence of vitamin-B_{12} and light on the formation of chlorosomes in green-colored and brown-colored *Chlorobium* species. Arch Microbiol 160: 193–198

Garnier J, Osguthorpe DJ and Robson B (1978) Analysis of the accuracy and implications of simple methods for predicting the secondary structure of globular proteins. J Mol Biol 120: 97–120

Gerola PD and Olson JM (1986) A new bacteriochlorophyll *a*-protein complex associated with chlorosomes of green sulfur bacteria. Biochim Biophys Acta 848: 69–76

Gerola PD, Højrup P, Knudsen J, Roepstorff P and Olson JM (1988) The bacteriochlorophyll *c*-binding protein from chlorosomes of *Chlorobium limicola* f. *thiosulfatophilum*. In: Olson JM, Ormerod JG, Amesz J, Stackebrandt E and Trüper HG (eds) Green Photosynthetic Bacteria, pp 43–52. Plenum Press, New York

Gibson J, Ludwig W, Stackebrandt E and Woese, CR (1985) The phylogeny of the green photosynthetic bacteria: Absence of a close relationship between *Chlorobium* and *Chloroflexus*. System Appl Microbiol 6: 152–156

Gillbro T, Sandström Å, Sundström V and Olson JM (1988) Picosecond energy transfer kinetics in chlorosomes and bacteriochlorophyll *a*-proteins of *Chlorobium limicola*. In: Olson JM, Ormerod JG, Amesz J, Stackebrandt E and Trüper HG (eds) Green Photosynthetic Bacteria, pp 91–96. Plenum Press, New York

Gloe A and Risch N (1978) Bacteriochlorophyll c_s, a new bacteriochlorophyll from *Chloroflexus aurantiacus*. Arch Microbiol 118: 153–156

Gloe A, Pfennig N, Brockmann H and Trowitzsch W (1975) A new bacteriochlorophyll from brown-colored Chlorobiaceae. Arch Microbiol 102: 103–109

Golecki JR and Oelze J (1987) Quantitative relationship between bacteriochlorophyll content, cytoplasmic membrane structure and chlorosome size in *Chloroflexus aurantiacus*. Arch Microbiol 148: 236–241

Griebenow K and Holzwarth AR (1989) Pigment organization and energy transfer in green bacteria. I. Isolation of native chlorosomes free of bound bacteriochlorophyll *a* from *Chloroflexus aurantiacus* by gel-electrophoretic filtration. Biochim Biophys Acta 973: 235–240

Griebenow K, Holzwarth AR and Schaffner K (1990) The 5.6 kilodalton protein in isolated chlorosomes of *Chloroflexus aurantiacus* strain Ok-70-fl is a degradation product. Z Naturforsch 45C: 823–828

Griebenow K, Holzwarth AR, van Mourik F and van Grondelle (1991a) Pigment organization and energy transfer in green bacteria. 2. Circular and linear dichroism spectrum of protein-containing and protein-free chlorosomes isolated from *Chloroflexus aurantiacus* strain Ok-70-fl. Biochim Biophys Acta 1058: , 194–202

Griebenow K, Müller MG and Holzwarth AR (1991b) Pigment organization and energy transfer in green bacteria .3. Picosecond energy transfer kinetics within the B806-866 bacteriochlorophyll *a* antenna complex isolated from *Chloroflexus aurantiacus*. Biochim Biophys Acta 1059: 226–232

Gudowska-Nowak E, Newton MD and Fajer J (1990) Conformational and environmental effects on bacteriochlorophyll optical spectra: Correlations of calculated spectra with structural results. J Phys Chem 94: 5795–5801

Halfen LN, Pierson BK and Francis GW (1972) Carotenoids of a gliding organism containing bacteriochlorophylls. Arch Microbiol 82: 240–246

Hawthornthwaite AM and Cogdell RJ (1991) Bacteriochlorophyll-binding proteins. In: Scheer H (ed) Chlorophylls, pp 493–528. CRC Press, Boca Raton

Hildebrandt P, Griebenow K, Holzwarth AR and Schaffner K (1991) Resonance Raman spectroscopic evidence for the identity of the bacteriochlorophyll *c* organization in protein-free and protein-containing chlorosomes from *Chloroflexus aurantiacus*. Z Naturforsch 46c: 228–232

Hildebrandt P, Tamiaki H, Holzwarth AR and Schaffner K (1994) Resonance Raman spectroscopic study of metallochlorin aggregates. Implications for the supramolecular structure in chlorosomal BChl *c* antennae of green bacteria. J Phys Chem 98: 2192–2197

Hirota M, Moriyama T, Shimada K, Miller M, Olson JM and Matsuura K (1992) High degree of organization of bacteriochlorophyll *c* in chlorosome-like aggregates spontaneously assembled in aqueous solution. Biochim Biophys Acta 1099: 271–274

Hoff AJ and Amesz J (1991) Visible absorption spectroscopy of chlorophylls. In: Scheer H (ed) Chlorophylls, pp 723–738. CRC Press, Boca Raton

Højrup P, Gerola P, Hansen HF, Mikkelsen JM, Shahed AE, Knudsen J, Roepstorff P and Olson JM (1991) The amino acid sequence of a major protein component in the light harvesting complex of the green photosynthetic bacterium *Chlorobium limicola* f. *thiosulfatophilum*. Biochim Biophys Acta 1077: 220–224

Holo H, Broch-Due M and Ormerod JG (1985) Glycolipids and the structure of chlorosomes in green bacteria. Arch Microbiol 143: 94–99

Holzwarth AR and Schaffner K (1994) On the structure of bacteriochlorophyll molecular aggregates in the chlorosomes of green bacteria. A molecular modeling study. Photosynth Res 41: 225–233

Holzwarth AR, Griebenow K and Schaffner K (1990a) A photosynthetic antenna system which contains a protein-free chromophore aggregate. Z Naturforsch 45c: 203–206

Holzwarth AR, Müller, MG and Griebenow K (1990b) Picosecond energy transfer kinetics between pigment pools in different preparations of chlorosomes from the green bacterium *Chloroflexus aurantiacus*. J Photochem Photobiol B 5: 457–465

Holzwarth AR, Griebenow K and Schaffner K (1992) Chlorosomes, photosynthetic antennae with novel self-organized pigment structures. J Photochem Photobiol A 65: 61–71

Huster MS and Smith KM (1990) Biosynthetic studies of substituent homologation in bacteriochlorophylls *c* and *d*. Biochemistry 29: 4348–4355

Johnson SG and Small GJ (1991) Excited state structure and energy transfer dynamics of the bacteriochlorophyll *a* protein from *Prosthecochloris aestuarii*. J Phys Chem 95: 471–479

Karapetyan NV, Swarthoff T, Rijgersberg CP and Amesz J (1980) Fluorescence emission spectra of cells and subcellular preparations of a green photosynthetic bacterium. Effects of dithionite on the intensity of the emission bands. Biochim Biophys Acta, 593: 254–60

Katz JJ, Bowman MK, Michalski TJ and Worcester DL (1991) Chlorophyll aggregation: Chlorophyll/water micelles as models for in vivo long-wavelength chlorophyll. In: Scheer H (ed) Chlorophylls, pp 211–235. CRC Press, Boca Raton

Keller D and Bustamante C (1986) Theory of the interaction of light with large inhomogeneous molecular aggregates. II. Psi-type circular dichroism. J Chem Phys 84: 2972–2980

Knox RS and Gülen D (1993) Theory of polarized fluorescence from molecular pairs: Förster transfer at large electronic coupling. Photochem Photobiol 57: 40–43

Knudsen E, Jantzen E, Bryn K, Ormerod JG and Sirevåg R (1982) Quantitative and structural characteristics of lipids in *Chlorobium* and *Chloroflexus*. Arch Microbiol 135: 149–154

Krasnovsky AA and Bystrova MI (1980) Self-assembly of chlorophyll aggregated structures. BioSystems 12: 181–194

Krasnovsky AA and Pakshina EV (1959) The photochemical and spectral properties of bacterioviridin of green sulfur bacteria. Doklady Acad Nauk SSSR (English Trans) 127: 215–218

Krasnovsky AA, Erokhin YuE and Federovich IB (1961) The fluorescence of green photosynthesizing bacteria and the state of the bacterioviridin in them. Doklady Acad Nauk SSSR (English Trans) 134: 225–227

Krasnovsky AA, Erokhin YuE and Yü-ch'ün H (1962) Fluorescence of aggregated forms of bacterioviridin and chlorophyll in relation to the state of the pigments in photosynthesizing organisms. Doklady Acad Nauk SSSR (English Trans) 143: 250–252

Krasnovsky AA, Erokhin YuE and Gulyaev BA (1963) Temperature dependence of luminescence of bacterioviridin and state of this pigment in photosynthetic bacteria. Doklady Acad Nauk SSR (English Trans) 152: 1191–1194

Krasnovsky AA, Bystrova MI and Pakshina EV (1966) Effect of the magnesium atom of the pigment molecule on the spectral properties of aggregated forms of chlorophyll analogs. Doklady Acad Nauk SSSR (English Trans) 167: 109–112

Kratky C and Dunitz JD (1975) Comparison of the results of two independent analyses of the ethyl chlorophyllide *a* dihydrate. Acta Cryst B31 1586–1589

Kratky C and Dunitz JD (1977) Ordered aggregation states of chlorophyll *a* and some derivatives. J Mol Biol 113: 431–442

Kusumoto N, Inoue K, Nasu H and Sakurai H (1994) Preparation of a photoactive reaction center complex containing photoreducible Fe-S centers and photooxidizable cytochrome *c* from the green sulfur bacterium *Chlorobium tepidum*. Plant Cell Physiol 35: 17–25

Larsen KL, Cox RP and Miller M (1994) Effects of illumination intensity on bacteriochlorophyll c homolog distribution in *Chloroflexus aurantiacus* grown under controlled conditions. Photosynth Res 41: 151–156

Larsen KL, Miller M and Cox RP (1995) Incorporation of exogenous long-chain alcohols into bacteriochlorophyll *c* homologs by *Chloroflexus aurantiacus*. Arch Microbiol 163: 119–123

Lehmann RP, Brunisholz RA and Zuber H (1994a) Structural differences in chlorosomes from *Chloroflexus aurantiacus* grown under different conditions support the BChl *c*-binding function of the 5.7 kDa polypeptide. FEBS Lett 342: 319–324

Lehmann RP, Brunisholz RA and Zuber H (1994b) Giant circular dichroism of chlorosomes from *Chloroflexus aurantiacus* treated with 1-hexanol and proteolytic enzymes. Photosynth Res 41: 165–173

Liaaen Jensen S, Hegge E and Jackman LM (1964) Bacterial carotenoids. XVII. The carotenoids of photosynthetic green bacteria. Acta Chem Scand 18: 1703–1718

Lin S, Van Amerongen H and Struve WS (1991) Ultrafast pump-probe spectroscopy of bacteriochlorophyll *c* antennae in bacteriochlorophyll *a*-containing chlorosomes from the green photosynthetic bacterium *Chloroflexus aurantiacus*. Biochim Biophys Acta 1060: 13–24

Louwe RJW and Aartsma TJ (1994) Optical dephasing and excited state dynamics in photosynthetic pigment-protein complexes. J Luminesc 58: 154–157

Lu XY and Pearlstein RM (1993) Simulations of *Prosthecochloris* Bacteriochlorophyll *a*-Protein Optical Spectra Improved by Parametric Computer Search. Photochem Photobiol 57: 86–91

Lutz M and van Brakel G (1988) Ground-state molecular interactions of bacteriochlorophyll *c* in chlorosomes of green bacteria and in model systems: A resonance Raman study. In: Olson JM, Ormerod JG, Amesz J, Stackebrandt E and Trüper HG (eds) Green Photosynthetic Bacteria, pp 23–34. Plenum Press, New York

Lyle PA and Struve WS (1990) Evidence for ultrafast exciton localization in the Q_y band of bacteriochlorophyll *a*-protein from *Prosthecochloris aestuarii*. J Phys Chem 94: 7338–7339

Matsuura K and Olson JM (1990) Reversible conversion of aggregated bacteriochlorophyll *c* to monomeric form by 1-hexanol in chlorosomes from *Chlorobium* and *Chloroflexus*. Biochim Biophys Acta 1019: 233–238

Matsuura K, Hirota M, Moriyama T, Shimada K, Nishimura Y, Yamazaki I and Mimuro M (1992) Pigment orientation and energy transfer kinetics in chlorosomes of green photosynthetic bacteria. In: Murata N (ed) Research in Photosynthesis, Vol I, pp 113–116) Kluwer, Dordrecht.

Matsuura K, Hirota M, Shimada K and Mimuro M (1993) Spectral forms and orientation of bacteriochlorophylls *c* and *a* in chlorosomes of the green photosynthetic bacterium *Chloroflexus aurantiacus*. Photochem Photobiol 57: 92–97

Matthews BW and Fenna RE (1980) Structure of a green bacteriochlorophyll protein. Accts Chem Res 13: 309–317

Matthews BW, Fenna RE, Bolognesi MC, Schmid MF and Olson JM (1979) Structure of a bacteriochlorophyll *a*-protein from the green photosynthetic bacterium *Prosthecochloris aestuarii*. J Mol Biol 131: 259–285

Miller M, Cox RP and Gillbro T (1991) Energy transfer kinetics in chlorosomes from *Chloroflexus aurantiacus*; studies using picosecond absorbance spectroscopy. Biochim Biophys Acta 1057: 187–194

Miller M, Gillbro T and Olson JM (1993a) Aqueous aggregates of bacteriochlorophyll *c* as a model for pigment organization in chlorosomes. Photochem Photobiol 57: 98–102

Miller M, Simpson D and Redlinger TE (1993b) The effect of detergent on the structure and composition of chlorosomes isolated from *Chloroflexus aurantiacus*. Photosynth Res 35: 275–283

Mimuro M, Nozawa T, Tamai N, Shimada K, Yamazaki I, Lin S, Knox RS. Wittmershaus BP, Brune DC and Blankenship RE (1989) Excitation energy flow in chlorosome antennas of green photosynthetic bacteria. J Phys Chem 93: 7503–7509

Mimuro M, Hirota H, Shimada K, Nishimura Y and Yamazaki I

(1992) Excitation energy transfer processes in green photosynthetic bacteria: Analysis in a three dimensionally oriented system in the picosecond time range. In: Murata N (ed) Research in Photosynthesis, Vol I, pp 17–24. Kluwer, Dordrecht

Mimuro M, Hirota H, Nishimura Y, Moriyama T, Yamazaki I, Shimada K, and Matsuura K (1994a) Molecular organization of bacteriochlorophyll in chlorosomes of the green photosynthetic bacterium *Chloroflexus aurantiacus*: Studies of fluorescence depolarization accompanied by energy transfer processes. Photosynth Res 41: 181–191

Mimuro M, Nozawa T, Tamai N, Nishimura Y and Yamazaki I (1994b) Presence and significance of minor antenna components in the energy transfer sequence of the green photosynthetic bacterium *Chloroflexus aurantiacus*. FEBS Letters 340: 167–172

Müller MG, Griebenow K and Holzwarth AR (1990) Fluorescence lifetime measurements of energy transfer in chlorosomes and living cells of *Chloroflexus aurantiacus* OK 70-fl. In: Baltscheffsky M (ed) Current Research In Photosynthesis, Vol II, pp 177–180. Kluwer, Dordrecht

Müller MG, Griebenow K and Holzwarth AR (1993) Picosecond energy transfer and trapping kinetics in living cells of the green bacterium *Chloroflexus aurantiacus*. Biochim Biophys Acta 1144: 161–169

Niedermeier G, Scheer H and Feick RG (1992) The functional role of protein in the organization of bacteriochlorophyll *c* in chlorosomes of *Chloroflexus aurantiacus*. Eur J Biochem 204: 685–692

Niedermeier G, Shiozawa JA, Lottspeich F and Feick RG (1994) Primary structure of two chlorosome proteins from *Chloroflexus aurantiacus*. FEBS Lett 342: 61–65

Nozawa T, Noguchi T and Tasumi M (1990a) Resonance Raman studies on the structure of bacteriochlorophyll *c* in chlorosomes of *Chloroflexus aurantiacus*. J Biochem 108: 737–740

Nozawa T, Suzuki M, Kanno S and Shirai S (1990b) CP/MAS ^{13}C-NMR studies on the structure of bacteriochlorophyll *c* in chlorosomes from *Chloroflexus aurantiacus*. Chem Lett: 1805–1808

Nozawa T, Suzuki M, Ohtomo K, Morishita Y, Konami H and Madigan MT (1991) Aggregation structure of bacteriochlorophyll *c* in chlorosomes from *Chlorobium tepidum*. Chem Lett: 1641–1644

Nozawa T, Ohtomo K, Suzuki M, Morishita Y and Konami H (1992a) CP/MAS ^{13}C NMR studies on antenna structure in green bacteria. In: Murata N (ed) Research in Photosynthesis, Vol I, pp 97–100. Kluwer, Dordrecht

Nozawa T, Ohtomo K, Takeshita N, Morishita Y and Madigan M (1992b) Substituent effects on the aggregation of bacteriochlorophyll *d* homologues purified from *Chlorobium limicola*. Bull Chem Soc Jpn 65: 3493–3494

Nozawa T, Ohtomo K, Suzuki M, Morishita Y and Madigan MT (1993) Structures and organization of bacteriochlorophyll *c*'s in chlorosomes from a new thermophilic bacterium *Chlorobium tepidum*. Bull Chem Soc Jpn 66: 231–237

Nozawa T, Ohtomo K, Suzuki M, Nakagawa H, Shikama Y, Konami H and Wang Z-Y (1994) Structures of chlorosomes and aggregated BChl *c* in *Chlorobium tepidum* from solid state high resolution CP/MAS ^{13}C NMR. Photosynth Res 41: 211–223

Nuijs AM, Vasmel H, Duysens LNM and Amesz J (1986) Antenna and reaction-center processes upon picosecond-flash excitation of membranes of the green photosynthetic bacterium *Chloroflexus aurantiacus*. Biochim Biophys Acta 849: 316–324

Oelze J (1985) Analysis of bacteriochlorophylls. Meth Microbiol 18: 257–284

Oelze J (1992) Light and oxygen regulation of the synthesis of bacteriochlorophyll-*a* and bacteriochlorophyll *c* in *Chloroflexus aurantiacus*. J Bacteriol 174: 5021–5026

Oh-Oka H, Kakutani S, Matsubara H, Malkin R and Itoh S (1993) Isolation of the photoactive reaction center complex that contains three types of Fe-S centers and a cytochrome *c* subunit from the green sulfur bacterium *Chlorobium limicola* f. *thiosulfatophilum*, strain Larsen. Plant Cell Physiol 34: 93–100

Okkels JS, Kjaer B, Hansson O, Svendsen I, Møller BL and Scheller HV (1992) A membrane-bound monoheme cytochrome-*c*551 of a novel type is the immediate electron donor to P840 of the *Chlorobium vibrioforme* photosynthetic reaction center complex. J Biol Chem 267: 21139–21145

Olson JM (1966) Chlorophyll-protein complexes derived from green photosynthetic bacteria. In: Vernon LP and Seely GR (eds.) The Chlorophylls (pp 413–425) Academic Press, New York.

Olson JM (1971) Bacteriochlorophyll-protein of green photosynthetic bacteria. In: San Pietro A (ed) Methods in Enzymology Vol 23 Part A, pp 636–639. Academic Press, New York

Olson JM (1978) Bacteriochlorophyll *a*-proteins from green bacteria. In: Clayton RK and Sistrom RS (eds) The Photosynthetic Bacteria, pp 161–178. Plenum Press, New York

Olson JM (1980a) Chlorophyll organization in green photosynthetic bacteria. Biochim Biophys Acta 594: 33–51

Olson JM (1980b) Bacteriochlorophyll *a*-proteins of two green photosynthetic bacteria. Meth Enzymol 69: 336–344

Olson JM (1994) Reminiscence about *'Chloropseudomonas ethylicum'* and the FMO protein. Photosynth Res 41: 3–5

Olson JM and Cox RP (1991) Monomers, dimers, and tetramers of 4-normal-propyl-5-ethyl farnesyl bacteriochlorophyll *c* in dichloromethane and carbon tetrachloride. Photosynth Res 30: 35–43

Olson JM and Pedersen JP (1988) Bacteriochlorophyll *c* aggregates in carbon tetrachloride as models for chlorophyll organization in green photosynthetic bacteria. In: Scheer H and Schneider S (eds) Photosynthetic Light-Harvesting Systems, pp 365–373 Walter de Gruyter, Berlin

Olson JM and Pedersen JP (1990) Bacteriochlorophyll *c* momomers, dimers, and higher aggregates in dichloromethane, chloroform, and carbon tetrachloride. Photosynth Res 25: 25–37

Olson JM and Pierson BK (1987) Evolution of reaction centers in photosynthetic prokaryotes. Int Rev Cytol 108: 209–248

Olson JM and Romano CA (1962) A new chlorophyll from green bacteria. Biochim Biophys Acta 59: 726–728

Olson JM, Koenig DF and Ledbetter MC (1969) Model of the bacteriochlorophyll-protein from green photosynthetic bacteria. Arch Biochem Biophys 129: 42–48

Olson JM, Philipson KD and Sauer K (1973) Circular dichroism and absorption spectra of bacteriochlorophyll-protein and reaction center complexes from *Chlorobium thiosulfatophilum*. Biochim Biophys Acta 292: 206–217

Olson JM, Ke B and Thompson KH (1976a) Exciton interaction among chlorophyll molecules in bacteriochlorophyll *a* proteins and bacteriochlorophyll *a* reaction center complexes from green bacteria. Biochim Biophys Acta 430: 524–537 Errata 763

Olson JM, Shaw EK and Englberger FM (1976b) Comparison of bacteriochlorophyll *a*-proteins from two green bacteria. Biochem J 159: 769–774

Olson JM, Gerola PD, van Brakel GH, Meiburg RF and Vasmel H (1985) Bacteriochlorophyll *a*- and *c*-protein complexes from chlorosomes of green sulfur bacteria compared with Bacteriochlorophyll *c* aggregates in CH_2Cl_2-hexane. In: Michel-Beyerle ME (ed) Antennas and Reaction Centers of Photosynthetic Bacteria, pp 67–73 Springer-Verlag, Berlin

Olson JM, Ormerod JG, Amesz J, Stackebrandt E and Trüper HG (eds) (1988) Green Photosynthetic Bacteria, Plenum, New York

Olson RA, Jennings WH and Olson JM (1969a) Chlorophyll orientation in crystals of bacteriochlorophyll-protein from green photosynthetic bacteria. Arch Biochem Biophys 129: 30–41

Olson RA, Jennings WH and Hanna CH (1969b) Paracrystalline aggregates of bacteriochlorophyll protein from green photosynthetic bacteria. Arch Biochem Biophys 130: 140–147

Orme-Johnson WH and Beinert H (1969) On the formation of the superoxide anion radical during the reaction of reduced iron-sulfur proteins with oxygen. Biochem Biophys Res Comm 36: 905–911

Ormerod JG (1992) Physiology of the photosynthetic prokaryotes. In: Photosynthetic Prokaryotes, NH Mann and NG Carr (eds), pp 93–120. Plenum, New York.

Ormerod JG, Nesbakken T and Beale SI (1990) Specific inhibition of antenna bacteriochlorophyll synthesis in *Chlorobium vibrioforme* by anesthetic gases. J Bacteriol 172: 1352–1360

Otte SCM, van der Heiden JC, Pfennig N and Amesz J (1991) A comparative study of the optical characteristics of intact cells of photosynthetic green sulfur bacteria containing bacteriochlorophyll *c*, *d* or *e*. Photosynth Res 28: 77–87

Otte SCM, van de Meent EJ, Van Veelen PA, Pundsnes AS and Amesz J (1993) Identification of the major chlorosomal bacteriochlorophylls of the green sulfur bacteria *Chlorobium vibrioforme* and *Chlorobium phaeovibrioides*; their function in lateral energy transfer. Photosynth Res 35: 159–169

Overmann J, Cypionka H and Pfennig N (1992) An extremely low-light-adapted phototrophic sulfur bacterium from the Black Sea. Limnol Oceanogr 37: 150–155

Pearlstein RM (1988) Interpretation of optical spectra of bacteriochlorophyll antenna complexes. In: Scheer H and Schneider S (eds) Photosynthetic Light-Harvesting Systems, pp 555–566. Walter de Gruyter, Berlin

Pearlstein (1991) Theoretical interpretation of antenna spectra. In: Scheer H (ed) Chlorophylls, pp 1047–1078. CRC Press, Boca Raton

Pearlstein RM (1992) Theory of the optical spectra of the bacteriochlorophyll *a* antenna protein trimer from *Prosthecochloris aestuarii*. Photosynth Res 31: 213–226

Pearlstein RM and Hemenger RP (1978) Bacteriochlorophyll electronic transition moment directions in bacteriochlorophyll *a*-protein. Proc Natl Acad Sci USA 75: 4920–4924

Philipson KD and Sauer K (1972) Exciton interaction in a bacteriochlorophyll-protein from *Chloropseudomonas ethylica*. Absorption and circular dichroism at 77 K. Biochemistry 11: 1880–1885

Pierson BK and Castenholz RW (1974a) Phototrophic gliding filamentous bacterium of hot springs, *Chloroflexus aurantiacus*. Arch Microbiol 100: 5–24

Pierson BK and Castenholz RW (1974b) Pigments and growth in *Chloroflexus aurantiacus*, a phototrophic filamentous bacterium. Arch Microbiol 100: 283–305

Pierson BK and Castenholz RW (1978). Photosynthetic apparatus and cell membranes of green bacteria. In: Clayton RK and Sistrom WR (eds) The Photosynthetic Bacteria, pp 161–178. Plenum, New York

Pierson BK and Castenholz RW (1992) The family Chloroflexaceae. In: Balows A, Trüper HG, Dworkin M, Schliefer KH and Harder W (eds) The Prokaryotes, pp 3754–3774. Springer-Verlag, Berlin

Pierson BK and Olson JM (1989) Evolution of photosynthesis in anoxygenic photosynthetic procaryotes. In: Cohen Y and Rosenberg E (eds) Microbial Mats: Physiological Ecology of Benthic Microbial Communities, pp 402–427. Am Soc Microbiol, Washington

Risch N, Brockmann H and Gloe A (1979) Structuraufklarung von neuartigen bacteriochlorophyllen aus *Chloroflexus aurantiacus*. Liebigs Ann Chem: 408–418

Savikhin S and Struve WS (1994) Ultrafast energy transfer in FMO trimers from the green bacterium *Chlorobium tepidum*. Biochemistry 33: 11200–11208

Savikhin S, Zhou W, Blankenship RE and Struve WS (1994a) Femtosecond energy transfer and spectral equilibration in bacteriochlorophyll *a*-Protein antenna trimers from the green bacterium *Chlorobium tepidum*. Biophys J 66: 110–113

Savikhin S, Zhu Y, Lin S, Blankenship RE and Struve WS (1994b) Femtosecond spectroscopy of chlorosome antennas from the green photosynthetic bacterium *Chloroflexus aurantiacus*. J Phys Chem 98: 10322–10334

Scheer H (1991) Structure and occurrence of chlorophylls. In: Scheer H (ed) Chlorophylls, pp 3–30. CRC Press, Boca Raton

Schmidt K (1980) A comparative study on the composition of chlorosomes (Chlorobium vesicles) and cytoplasmic membranes from *Chloroflexus aurantiacus* Strain Ok-70-fl and *Chlorobium limicola* f. *thiosulfatophilum* Strain 6230. Arch Microbiol 124: 21–31

Schmidt K, Maarzahl M and Mayer F (1980) Development and pigmentation of chlorosomes in *Chloroflexus aurantiacus* strain Ok-70-fl. Arch Microbiol 127: 87–97

Smith KM (1994) Nomenclature of the bacteriochlorophylls *c*, *d* and *e*. Photosynth Res 41: 23–26

Smith KM and Bobe FW (1987) Light adaptation of bacteriochlorophyll *d* producing bacteria by enzymic methylation of their antenna pigments. J Chem Soc Chem Commun 276–277

Smith KM, Kehres LA and Fajer J (1983) Aggregation of the bacteriochlorophylls *c*, *d*, and *e*. Models for the antenna chlorophylls of green and brown photosynthetic bacteria. J Am Chem Soc 105: 1387–1389

Smith KM Bobe FW, Goff DA and Abraham RJ (1986) NMR spectra of porphyrins 28. Detailed solution structure of a bacteriochlorophyllide *d* dimer. J Am Chem Soc 108: 1111–1120

Sprague SG, Staehelin LA, DiBartolomeis MJ and Fuller RC (1981a) Isolation and development of chlorosomes in the green bacterium *Chloroflexus aurantiacus.* J Bacteriol 147: 1021–1031

Sprague SG, Staehelin LA and Fuller RC (1981b) Semiaerobic induction of bacteriochlorophyll synthesis in the green bacterium *Chloroflexus aurantiacus.* J Bacteriol 147: 1032–1039

Staehelin LA, Golecki JR, Fuller RC and Drews G (1978) Visualization of the supramolecular architecture of chlorosomes (*Chlorobium* type vesicles) in freeze-fractured cells of *Chloroflexus aurantiacus*. Arch Microbiol 119: 269–277

Staehelin LA, Golecki JR and Drews G (1980) Supramolecular organization of chlorosomes (*Chlorobium* vesicles) and of their membrane attachment sites in *Chlorobium limicola.* Biochim Biophys Acta 589: 30–45

Stolz JF, Fuller RC and Redlinger TE (1990) Pigment-protein diversity in chlorosomes of green phototrophic bacteria. Arch Microbiol 154: 422–427

Swarthoff T and Amesz J (1979) Photochemically active pigment-protein complexes from the green photosynthetic bacterium *Prosthecochloris aestuarii.* Biochim Biophys Acta 548: 427–432

Swarthoff T, Amesz J, Kramer HJM and Rijgersberg CP (1981) The reaction center and antenna pigments of green photosynthetic bacteria. Isr J Chem 21: 332–337

Sybesma C and Olson JM (1963) Transfer of chlorophyll excitation energy in green photosynthetic bacteria. Proc Natl Acad Sci USA 49: 248–253

Tamiaki H, Holzwarth AR and Schaffner K (1992) A synthetic zinc chlorin aggregate as a model for the supramolecular antenna complexes in the chlorosomes of green bacteria. J Photochem Photobiol 15: 355–360

Tamiaki H, Holzwarth AR and Schaffner K (1994) Dimerization of synthetic zinc aminochlorins in non-polar organic solvents. Photosynth Res 41: 245–251

Theroux SJ, Redlinger TE, Fuller RC and Robinson SJ (1990) Gene encoding the 5.7 kilodalton chlorosome protein of *Chloroflexus aurantiacus*: Regulated message levels and a predicted carboxy-terminal protein extension. J Bacteriol 172: 4497–4504

Thornber JP and Olson JM (1968) The chemical composition of a crystalline bacteriochlorophyll-protein complex isolated from the green bacterium *Chloropseudomonas ethylicum.* Biochemistry 7: 2242–2249

Tronrud DE and Matthews BW (1993) Refinement of the structure of a water-soluble antenna complex from green photosynthetic bacteria with incorporation of the chemically determined amino acid sequence. In: Norris J and Deisenhofer H (eds) The Photosynthetic Reaction Center, Vol. 1, pp 13–21. Academic Press, San Diego

Tronrud DE, Schmid MF and Matthews BW (1986) Structure and x-ray amino acid sequence of a bacteriochlorophyll *a* protein from *Prosthecochloris aestuarii* refined at 1.9 Å resolution. J Mol Biol 188: 443–454

Trüper HG and Pfennig N (1992) The family Chlorobiaceae. In: Balows A, Trüper HG, Dworkin M, Schliefer KH and Harder W (eds) The Prokaryotes, pp 3583–3592. Springer-Verlag, Berlin,

Uehara K and Olson JM (1992) Aggregation of bacteriochlorophyll *c* homologs to dimers, tetramers, and polymers in water-saturated carbon tetrachloride. Photosynth Res 33: 251–257

Uehara K, Ozaki Y, Okada K and Olson JM (1991) FT-IR studies on the aggregation of bacteriochlorophyll *c* from *Chlorobium limicola.* Chem Lett: 909–912

Uehara K, Mimuro M, Ozaki Y, and Olson JM (1994) The formation and characterization of the in vitro polymeric aggregates of bacteriochlorophyll c homologs from *Chlorobium limicola* in aqueous suspension in the presence of monogalactosyl diglyceride. Photosynth Res 41: 235–243

Van Amerongen H and Struve WS (1991) Excited-state absorption in bacteriochlorophyll *a*-protein from the green bacterium *Prosthecochloris aestuarii*: Reinterpretation of the absorption difference spectrum. J Phys Chem 95: 9020–9023

Van Amerongen H, Vasmel H and van Grondelle R (1988) Linear dichroism of chlorosomes from *Chloroflexus aurantiacus* in compressed gels and electric fields. Biophys J 54: 65–76

Van Amerongen H, Van Haeringen B, Van Gurp M and Van Grondelle R (1991) Polarized fluorescence measurements on ordered photosynthetic antenna complexes - chlorosomes of *Chloroflexus aurantiacus* and B800-B850 antenna complexes of *Rhodobacter sphaeroides*. Biophys J 59: 992–1001

Van Dorssen RJ and Amesz J (1988) Pigment organization and energy transfer in the green photosynthetic bacterium *Chloroflexus aurantiacus*. III. Energy transfer in whole cells. Photosynth Res 15: 177–189

Van Dorssen RJ Gerola PD Olson JM and Amesz J (1986a) Optical and structural properties of chlorosomes of the photosynthetic green sulfur bacterium *Chlorobium limicola.* Biochim Biophys Acta 848: 77–82

Van Dorssen RJ, Vasmel H and Amesz J (1986b) Pigment organization and energy transfer in the green photosynthetic bacterium *Chloroflexus aurantiacus*. II. The chlorosome. Photosynth Res 9: 33–45

Van Grondelle R, Dekker JP, Gillbro T and Sundstrom V (1994) Energy transfer and trapping in photosynthesis. Biochim Biophys Acta 1187: 1–65

Van Mourik F, Verwijst RR, Mulder JM and Van Grondelle R (1992) Excitation transfer dynamics and spectroscopic properties of the light-harvesting BChl *a* complex of *Prosthecochloris aestuarii*. J. Luminesc 53: 499–502

Van Mourik F, Verwijst RR, Mulder JM and Van Grondelle R (1994) Singlet—triplet spectroscopy of the light-harvesting BChl *a* complex of *Prosthecochloris aestuarii*. The nature of the low-energy 825 nm transition. J Phys Chem 98: 10307–10312

Van Noort PI, Francke C, Schoumans N, Otte SCM, Aartsma TJ and Amesz J (1994) Chlorosomes of green bacteria: Pigment composition and energy transfer. Photosynth Res 41: , 193–203

Vasmel H, Van Dorssen RJ, De Vos GJ and Amesz J (1986) Pigment organization and energy transfer in the green photosynthetic bacterium *Chloroflexus aurantiacus*. I. The cytoplasmic membrane. Photosynth Res 7: 281–294

Vos M, Nuijs AM, van Grondelle R, van Dorssen RJ, Gerola PD and Amesz J (1987) Excitation transfer in chlorosomes of green photosynthetic bacteria. Biochim Biophys Acta 891: 275–285

Wagner-Huber R, Brunisholz R, Frank G and Zuber H (1988) The bacteriochlorophyll *c/e*- binding polypeptides from chlorosomes of green photosynthetic bacteria. FEBS Lett. 239: 8–12

Wagner-Huber R, Brunisholz R, Frank G and Zuber H (1990) The primary structure of the presumable BChl *d*-binding polypeptide of *Chlorobium vibrioforme* f. *thiosulfatophilum*. Z Naturforsch 45c: 818–822

Wahlund TM, Woese CR, Castenholz RW and Madigan MT (1991) A thermophilic green sulfur bacterium from New Zealand hot springs, *Chlorobium tepidum* sp. nov. Arch Microbiol 156: 81–90

Wang J, Brune DC and Blankenship RE (1990) Effects of oxidants and reductants on the efficiency of excitation transfer in green photosynthetic bacteria. Biochim Biophys Acta 1015: 457–463

Wechsler T, Brunisholz R, Suter F, Fuller RC and Zuber H (1985a) The complete amino acid sequence of a bacteriochlorophyll *a* binding polypeptide isolated from the cytoplasmic membrane of the green photosynthetic bacterium *Chloroflexus aurantiacus*. FEBS Lett, 191: 34–38

Wechsler T, Suter F, Fuller RC and Zuber H (1985b) The complete amino acid sequence of the bacteriochlorophyll *c* binding polypeptide of the green photosynthetic bacterium *Chloroflexus aurantiacus*. FEBS Lett 181: 173–178

Wechsler TD, Brunisholz RA, Frank G, Suter F and Zuber H (1987) The complete amino acid sequence of the antenna polypeptide B806-866-β from the cytoplasmic membrane of the green bacterium *Chloroflexus aurantiacus*. FEBS Lett 210: 189–194

Whitten WB, Olson JM and Pearlstein RM (1980) Seven-fold exciton splitting of the 810-nm band in bacteriochlorophyll *a*-proteins from green photosynthetic bacteria. Biochim Biophys Acta 591: 203–207

Woese CR (1987) Bacterial evolution. Microbiol Rev 51: 221–271

Worcester DL, Michalski TJ and Katz JJ (1986) Small-angle neutron scattering studies of chlorophyll micelles: Models for bacterial antenna chlorophyll. Proc Natl Acad Sci USA 83: 3791–3795

Worcester DL, Michalski TJ and Katz JJ (1993) Cylindrical aggregates of protochlorophyll-*a* compared with aggregates of bacteriochlorophyll-*c*: Models for antenna chlorophyll in *Chloroflexus* & *Chlorobium*. Biophys J 64: A215

Wullink W, Knudsen J, Olson JM, Redlinger TE and Van Bruggen EFJ (1991) Localization of polypeptides in isolated chlorosomes from green phototrophic bacteria by immunogold labeling electron microscopy. Biochim Biophys Acta 1060: 97–105

Zhou W, LoBrutto R, Lin S and Blankenship RE (1994) Redox effects on the bacteriochlorophyll *a*-containing Fenna-Matthews-Olson protein from *Chlorobium tepidum*. Photosynth Res 41: 89–96

Zuber H and Brunisholz RA (1991) Structure and function of antenna polypeptides and chlorophyll-protein complexes: principles and variability. In: Scheer H (ed) Chlorophylls, pp 628–692. CRC Press, Boca Raton

Chapter 21

Structure-Function Relationships in Core Light-Harvesting Complexes (LHI) As Determined by Characterization of the Structural Subunit and by Reconstitution Experiments

Paul A. Loach and Pamela S. Parkes-Loach

Department of Biochemistry, Molecular Biology and Cell Biology, Hogan 2-100, Northwestern University, Evanston, IL 60208–3500, USA

R. E. Blankenship, M. T. Madigan and C. E. Bauer (eds): Anoxygenic Photosynthetic Bacteria, pp. 437–471.

Summary

The preparation and characterization of a dissociated intermediate of LH1 has been achieved for each of five different photosynthetic bacteria and several of their mutants. On the basis of nearly identical spectroscopic and biochemical properties, together with the demonstration that in each case native LHI can be formed by association, it is concluded that the isolated complex represents a fundamental structural subunit of the core light-harvesting complex and probably occurs universally in bacterial LHI complexes. The composition of the subunit is assigned to $\alpha_1\beta_1\cdot$2BChl on the basis of biochemical and spectroscopic characterization. Upon association to form LHI, this subunit complex initially dimerizes and then further associates.

The subunit complexes prepared from these bacteria can be fully dissociated to their fundamental components and the subunit complex and LHI reformed by subsequent reassociation. Beginning with separately isolated α- and β- polypeptides and BChl *a*, reconstitution to form the native subunit complex was demonstrated for each of three BChl *a*-containing bacteria (*Rhodospirillum rubrum*, *Rhodobacter sphaeroides*, and *Rhodobacter capsulatus*). A similar reconstitution of the subunit complex of a BChl *b*-containing organism (*Rhodopseudomonas viridis*) has also been accomplished. In addition, LHI has been reconstituted as well for the three BChl *a*-containing systems. Using properties of the subunit- and LHI-type complexes as assays, hybrid reconstitution experiments were conducted, the results of which indicate the importance of a conserved core region of amino acids for subunit and LHI formation, as well as the likelihood of ion-pairing as stabilizing interactions in the N-terminal region of the polypeptides. Chemical and enzymatic modification of the polypeptides have provided further support for these tentative conclusions and minimal structural features required for formation of the subunit complex and LHI were also better defined by these modifications.

The BChl binding site has been probed by use of structural analogs of BChl in the reconstitution experiments with polypeptides of *Rhodospirillum rubrum* and *Rhodobacter sphaeroides*. Structural requirements including the Mg, carbonyl group at C3^1 and the carbomethoxy group at C13^2 were indicated. Comparative association constants for many reconstituted systems were measured and used to initially evaluate the contribution of various structural elements to the highly specific formation of the subunit and LHI complexes. Based on these experimental results and analyses of the stabilizing interactions, certain structural features of the subunit complex are proposed that involve cross-hydrogen bonding and a pseudo-plane of symmetry.

Reconstitution of wild-type LHI has been successfully achieved with the α-and β-polypeptides from *Rhodospirillum rubrum*, BChl *a* and spirilloxanthin as well as with the α-and β-polypeptides from *Rhodobacter sphaeroides*, BChl *a* and spheroidene. Appropriate structure and function was demonstrated by measuring absorption, CD and fluorescence excitation spectra and by examining protection of BChl from degradation by the combination of light and air.

Beginning with the isolated or reconstituted subunit complex and isolated reaction centers (RC) of *Rhodospirillum rubrum*, the photoreceptor complex was formed by association of these complexes. This LHI-RC complex exhibited native absorbance and CD spectra as well as a quantum yield of 1.0 for energy utilization from BChl of LHI to cause charge separation in the RC.

Abbreviations: BChl – bacteriochlorophyll; BChl *a* is implied unless BChl *b* is indicated; Car – carotenoid; CD – circular dichroism; Chl – chlorophyll; *Cm. – Chromatium;* cmc – critical micelle concentration; Da – daltons; K_{Assoc}, K_A – association constant for binding; kcal – kilocalories; LHI – core light-harvesting complex, also called B875, B890, etc. depending on the far-red absorption maximum; LHII – peripheral light-harvesting complex, also called B800–850, for example; OG – n-octyl β-D-glucopyranoside; PRC – photoreceptor complex which contains LHI and RC; *Rb. – Rhodobacter;* RC – reaction center; *Rc. – Rhodocyclus; Rp. – Rhodopseudomonas; Rs. – Rhodospirillum*

I. Introduction

In his thesis research, Duysens (1952) determined that the energy absorbed by the bacteriochlorophyll (BChl) pigments was efficiently transferred from shorter wavelength species to BChl species absorbing at longer wavelengths, downhill in energy flow. This energy was shown to be ultimately transferred to the reaction center (RC) where photochemistry was initiated (Clayton, 1962). The longest wavelength light-harvesting component was found to occur in a fixed stoichiometry to the RC whereas the shorter wavelength components varied in their stoichiometry to the RC in response to environmental conditions (Aagaard and Sistrom, 1972). Monger and Parson (1977) proposed that the long wavelength components, called the core light-harvesting complexes (LHI) were the most intimately associated with and interconnected the RC. They further suggested that the shorter wavelength complexes (peripheral LH, also called LHII) were more distant from the RC and surrounded the LHI-RC system. This arrangement would best facilitate efficient energy flow. The modern era of photophysics has seen the confirmation of the sequential path of energy migration (Hunter et al. 1989) and the era of membrane biochemistry has confirmed the existence of the expected peripheral and core BChl-protein complexes (Thornber et al., 1983; Drews, 1985; Cogdell, 1986).

Because LHI complexes can now be isolated and are well-characterized (Feick and Drews, 1978; Sauer and Austin, 1978; Cogdell and Thornber, 1979; Broglie et al., 1980; Cogdell et al., 1982; Picorel et al., 1983; Drews, 1985; Sundström and van Grondelle, 1991; Zuber and Brunisholz, 1991; Brunisholz and Zuber, 1992), there is great interest in determining their structure and how their structure relates to function. The goal of the research described in this review is to determine the interactions that stabilize this BChl-protein complex and to understand how the structure enables its function. The question of whether all LHI of photosynthetic bacteria have fundamentally the same structure is timely to consider. Each wild-type, BChl *a*-containing LHI has a long-wavelength absorption maximum which is red-shifted to about the same extent (the range is between 99 and 119 nm relative to BChl in acetone except for thermophiles such as *Chromatium (Cm.) tepidum* (Jensen et al., 1964; Zuber and Brunisholz, 1991; Hawthornthwaite and Cogdell, 1991)). Even though the magnitude of the shifts are similar, one can not assume that all these wavelength shifts have the same structural origin (Gottstein and Scheer, 1983; Scherz and Parson, 1986; Brunisholz and Zuber, 1988; Pearlstein, 1988; Braun and Scherz, 1991; Thompson et al., 1991; Thompson and Fajer, 1992). Moreover, the circular dichroism (CD) spectra of LHI complexes show considerable variation among the photosynthetic bacteria (Bolt et al., 1981a,b; Kramer et al., 1984; Cogdell and Scheer, 1985), indicating some differences in structure do exist. Whatever these differences may be, however, they seem to have no effect on function since all LHI complexes for which there are appropriate measurements seem to show very high efficiency of excited singlet state transfer to the RC (Loach and Sekura, 1968; Sebban, 1985).

Important information relating to the question of a common structure has been obtained from biochemical analyses of isolated LHI. LHI typically consist of associated complexes of two small polypeptides, α and β, which occur in a 1:1 ratio, and in the ratio of 2 BChl per $\alpha\beta$-pair (Broglie et al., 1980; Cogdell et al., 1982; Picorel et al., 1983; Theiler et al., 1985; Drews, 1985; Picorel and Gingras, 1988). The uniform existence of these small polypeptides, their fixed stoichiometry to BChl, and the fact that many such units, approximately twelve in number, must associate to form the LHI which associates with one RC, suggests that the same fundamental structure may exist in all LHI.

The determination of amino acid sequences for the α- and β-polypeptides of LHI from many photosynthetic bacteria (Zuber and Brunisholz, 1991; Brunisholz and Zuber, 1992) and the advent of molecular genetics methodology (Bérard et al., 1986; Scolnik and Marrs, 1987; Bélanger and Gingras, 1988; Kiley and Kaplan, 1988; Coleman and Youvan, 1990; Wiessner et al., 1990; Wellington et al., 1992) have made possible the first real insights into structure-function relationships. Recently, structural subunits formed by dissociation of the LHI have been isolated from several different bacteria (Miller et al., 1987; Chang et al., 1990b; Heller and Loach, 1990, Meckenstock et al., 1992). Comparison of these subunits has shown that they all exhibit extremely similar absorption and CD properties as well as biochemical behavior. Thus, it appears that there may be a fundamental building block that is the same in all LHI of photosynthetic bacteria. In addition, both the subunit complex (also referred to as B820) and LHI can be reversibly dissociated and reconstituted from individual components (Parkes-Loach et al., 1988; Loach et al., 1994).

The ability to reconstitute the complex from its individual components is a uniquely powerful tool for evaluating structural relationships in LHI (Parkes-Loach et al., 1990; Loach et al., 1994). Use of this methodology allows assessment of structural requirements of the BChl and carotenoid (Car), as well as the protein. In combination with spectroscopic measurements, it should be possible to obtain detailed structure-function information for LHI complexes as well as for their interaction with RC, peripheral LH complexes and each other. This review will summarize results obtained using the methodology for subunit preparation and reconstitution. Considering these results along with other knowledge of LHI, some comments about possible structural elements for the subunit complex and LHI will be offered.

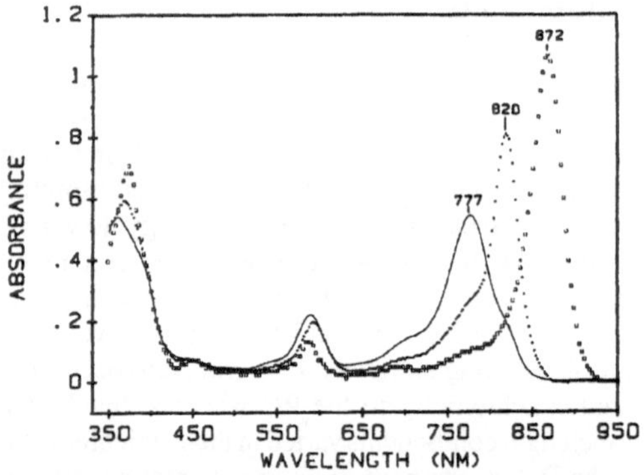

Fig. 1. Absorption spectra of the dissociated subunit complex of *Rs. rubrum* in 4.5% OG (λ_{max} = 777 nm), the subunit complex formed by lowering the OG concentration to 0.72% (λ_{max} = 820 nm), and reassociated LHI formed by further decrease in OG concentration to 0.36% (λ_{max} = 872 nm). The subunit complex and LHI spectra were corrected for sample dilution. Figure reproduced from Parkes-Loach et al. (1988) with permission.

II. The Structural Subunit of LHI

Structural subunits of the LHI of several bacteria have been isolated by using detergent methodology (Loach et al., 1985; Miller et al., 1987; Heller and Loach, 1990; Chang et al., 1990b; Meckenstock et al., 1992; Visschers et al., 1992). These subunit complexes are smaller in size and have blue-shifted Q_y absorption maxima relative to LHI, but are still substantially red-shifted relative to free BChl in detergent (Fig. 1). Upon further addition of detergent, the subunit complexes can be dissociated into their individual polypeptides and free BChl. By decreasing the detergent concentration, the subunit and the LHI complexes can be quantitatively reassociated (Parkes-Loach et al., 1988; Heller and Loach, 1990; Chang et al., 1990b).

A. Preparation

1. General Procedures

The first bacteria from which structural subunits of LHI were isolated were wild-type *Rhodospirillum (Rs.) rubrum* and its G-9 carotenoidless mutant (Loach et al., 1985; Miller et al., 1987). Membrane fractions from these bacteria required a wash step in Triton X-100 and EDTA before they were susceptible to forming the subunit complex. On the other hand, membrane preparations from *Rhodobacter (Rb.) sphaeroides* (wild-type and a LHII(–) mutant) did not require extensive treatment, but simply Car removal by extraction with a hydrocarbon solvent. In general, as Cogdell and Thornber once concluded about isolation of RC and LH complexes from different photosynthetic organisms, each system is unique and procedures are not necessarily interchangeable (Cogdell and Thornber, 1979; Thornber, 1986). This has been our experience for the preparation of the subunit complex of LHI from different bacterial species, and might also be extended to include various mutants of a single species. Even so, some general observations seem to apply. In the bacteria we have examined, it has been necessary to remove Car, or work with a carotenoidless mutant to successfully prepare high yields of the subunit complex. After the membranes are appropriately prepared, they are susceptible to formation of the subunit complexes upon addition of increasing amounts of n-octyl β-D-glucopyranoside (OG). For BChl *a*-containing organisms, the subunit complex of LHI is separated from RC and LHII (if present) by gel filtration chromatography on a Sephadex G100 column.

An interesting question to address is what role Car plays in LHI structure. Use of carotenoidless mutants, or Car removal prior to exposure to detergent, was required for quantitative formation of a stable subunit complex in the four bacteria we have studied (*Rs. rubrum*, Miller et al., 1987; *Rb. sphaeroides*, Chang et al., 1990b; *Rb. capsulatus*, Heller and Loach,

1990; *Rhodopseudomonas (Rp.) viridis*, Parkes-Loach et al., 1994). Prior Car removal was also necessary for the preparation reported by Meckenstock et al. (1992) of the subunit isolated from *Rp. marina*. In a recent report of a subunit preparation from *Rs. rubrum*, the Car was not removed prior to subunit formation, but none was found in the isolated subunit complex (Visschers et al., 1992). On the other hand, two preliminary reports indicate the presence of Car in subunit-containing fractions *Cm. purpuratum* (Kerfeld et al., 1994) and *Rhodocyclus (Rc.) gelatinosus* (now called *Rubrivivax gelatinosus*) (Jirsakova and Reiss-Husson, 1993), although the amount of Car was greatly reduced in the latter. Further characterization of such systems will be of interest.

All experimental results are consistent with the following equilibria where the lower case 'd' refers to detergent that solvates the component.

$$\alpha_d + \beta_d + BChl_d \rightleftharpoons B820_d + d \quad (1)$$

$$Car_d + B820_d \rightleftharpoons \text{wild-type } LHI_d + d \quad (2)$$

Thus, addition of Car shifts equilibrium (2) toward LHI at all detergent concentrations and decreases the concentration of $B820_d$. Depending on the value of the equilibrium constants for reactions (1) and (2), B820 may or may not be observed as the detergent concentration is varied (Ghosh et al., 1988a; Heller and Loach, 1990). The early work of Ogawa and Vernon (1969) suggested to us that removal of Car might be a useful step in our goal to reversibly dissociate LHI.

Some caution should be noted with regard to carotenoid extraction. The most effective solvent (petroleum ether or benzene) and the behavior of the membrane fractions are different for preparations from different bacteria. Variables such as the age of the lyophilized chromatophores or photoreceptor complex (PRC)-enriched membrane fractions, the temperature, and the ratio of membrane material to organic solvent are important. Each of these parameters must be evaluated for each membrane system to which extraction is applied. Excellent reproducibility was found when the exact same set of extraction parameters were applied to membrane preparations from one bacterium.

Another key factor for preparation of the subunit complex has been the choice of detergent. Some of the earliest biochemical experiments designed to isolate membrane complexes were conducted with membranes from photosynthetic organisms (Gingras, 1978; Thornber et al., 1978 and references therein). One reason for this is that these membranes have built-in monitors of structural integrity (the pigments of LH complexes and the redox centers of RC). Development of early methodology focused on isolation of intact LH and RC complexes. From these studies it was found that various detergents, organic solvents and chaotropic agents seemed to fall into the following categories: too gentle for separation of complexes (cholate, most chaotropic agents), somewhat effective and gentle (detergents such as deoxycholate and Deriphat), quite effective and not too disruptive (Triton X-100 and OG), effective but moderately disruptive (lauryl dimethylamine-N oxide, lithium dodecyl sulfate, cetyl dimethylammonium bromide, low concentrations of butanol) and too harsh (sodium dodecyl sulfate, high concentrations of organic solvents). For subunit preparation of LHI, OG is exceptionally effective. The reason for this is unclear, although size, hydrophilic/hydrophobic balance and ability of the polar moiety to participate in multiple hydrogen bonding are likely to be important. In our experience, few detergents can be successfully used to reversibly dissociate LHI. We found the following order of effectiveness: OG > n-octyl β-D-thioglucopyranoside > n-nonyl β-D-glucopyranoside > octanoyl, nonanoyl and decanoyl N-methylgluconamide, lauryl maltoside (Miller et al., 1987). In work from other laboratories, subunit complexes were observed in systems using LDAO and sodium cholate (Ghosh et al., 1988a,b) and subunit complexes were isolated using *n*-octyl-*rac*-2,3-dipropylsulfoxide (Visschers et al., 1992), sucrose mono-N-decanoate in Tris buffer containing EDTA and 60% ammonium sulfate (Jirsakova and Reiss-Husson, 1993) and n-dodecyl maltoside and OG (Kerfeld et al., 1994).

2. *Rhodospirillum rubrum*

The first organism with which we worked to isolate LH and RC complexes was *Rs. rubrum*. It was chosen because it has only one LH complex, the LHI, and its photosynthetic unit size (number of BChl/RC) was one of the smallest known. From our earlier experiments in which PRC were isolated (Loach et al., 1970a) we found that LHI and RC withstood a number of procedures which were used to remove other components from the membrane (Hall et al.,

1973). Our strategy was to first remove everything from the membrane that was not required to maintain native LH and RC physical properties and function. Among the treatments that were effective in removing other polypeptides (assayed by SDS-PAGE) was a wash step with high Triton X-100 concentration (3%) and EDTA. This treatment seemed to effectively solubilize many of the components of the bc_1 complex, F_o and F_1 components of ATP synthase and perhaps metabolite transport components.

The enriched-PRC preparations were used for development of procedures to quantitatively and reversibly separate LHI from RC. Many procedures were tried that gave poor yields (less than 50%) and/or were largely irreversible. The motivation for requiring quantitative separation and reversibility was twofold, both relating to eventual extrapolation back to the in vivo system. First, by obtaining quantitative yields, it would be possible to account for all LH and RC complexes in the membrane. It is prudent to note that nearly all commonly used RC preparations give low yields of between 20% to 50% (Feher and Okamura, 1978; Gingras, 1978; Vadeboncoeur et al., 1979) so that the question can be raised as to whether there might be other important components or structural differences in the intact membrane that were missed because of the low yield. Secondly, as Racker (1985) pointed out in his pioneering work with bioenergetic membranes, the only way to understand how the in vivo complexes really work is to reconstitute them from their individual components. We felt that a less than a quantitative accounting of all components would compromise eventual conclusions.

Upon addition of OG to the PRC-enriched membranes of the carotenoidless G-9 mutant, the long wavelength absorption band of BChl *a* in LHI underwent a systematic transition to an 820-nm-absorbing species (Loach et al. 1985, Miller et al. 1987). After extraction of Car from the PRC-enriched membranes prepared from wild-type cells, the same transition could be demonstrated. The 820-nm-absorbing material of both systems could be quantitatively reassociated to reform the 873-nm-absorbing complex by lowering the OG concentration. Thus, quantitatively-reversible conversion was achieved.

In membranes in which LHI was converted to the 820-nm form, light energy absorbed at this wavelength was poorly transferred to the RC. Therefore, it was likely that physical separation of LHI subunits from the RC had occurred. The two complexes were separated by gel-filtration chromatography under conditions where the 820 nm form was stable. This purification step could be refined to achieve > 95% yield of both the subunit of LHI (based on reassociation of B820 to form LHI) and the RC (Miller et al., 1987). Only the α- and β-polypeptides of LHI were present in the B820 material and a stoichiometry of $\alpha_1\beta_1 \cdot 2$BChl was determined. Preparations of the subunit complex in detergent solution were stable for several days if kept anaerobic and dark, but they degraded rapidly in light and air. The preparation can be lyophilized and stored at −15 °C for months to years without degradation. The long-wavelength maximum for the subunit preparation at 20 °C is 820 ± 1 nm which shifts to 825 nm on cooling to 77 K (Visschers et al., 1991).

The subunit complex of wild-type *Rs. rubrum* has also been prepared using *n*-octyl-*rac*-2,3-dipropylsulfoxide (Visschers et al., 1992). In this case, the membranes were not treated with Triton X-100/EDTA nor was the Car first removed, but Car was not observed spectrally in the subunit. The yield of the subunit complex was not as high as for that described above using OG, but the spectral properties were similar to those of the OG preparation.

3. Rhodobacter sphaeroides

After developing the procedures that were successful for preparation of the LHI subunit from *Rs. rubrum*, we extended the methodology to another bacterium, *Rb. sphaeroides* (Chang et al., 1990b). This organism contains the peripheral LH complex, B800–850, as well as LHI. The Triton X-100-EDTA wash step used with membranes from *Rs. rubrum* could not be used with *Rb. sphaeroides* membranes because it resulted in solubilization of large quantities of BChl-containing material. It was found that washing wild-type membrane vesicles with 0.01 M Tris buffer (pH 7.3) containing 0.05 M EDTA was effective in yielding a membrane preparation in which Car could then be extracted and that formed a subunit complex upon OG addition. For a LHII$^-$ mutant, PUC705-BA, this Tris-EDTA wash was unnecessary.

The above membrane preparations could then be extracted with petroleum ether (wild-type cells) or benzene (PUC705-BA cells) to remove Car without removing BChl. The LHI of these extracted systems could then be titrated with OG to produce a subunit complex and RC (and also LHII with wild-type cells). Separation of products was achieved in a manner similar to the procedure used for the subunit preparation with *Rs. rubrum* material.

The overall yield of the subunit complex from wild-type cells could be optimized to obtain over 80% of both LHI and RC while somewhat lower yields (about 50%) were obtained from PUC705-BA (LHII$^-$). HPLC analyses, together with determination of the N-terminal amino acid sequences of HPLC-purified polypeptide components, demonstrated that only the LHI α-and β-polypeptides were present in a stoichiometry of 1:1. The stabilities of the subunit preparations were similar to those of *Rs. rubrum*. The long-wavelength maximum for the subunit preparation at 20 °C, as well as at 77 K, was 825 ± 1 nm (Chang et al., 1990b; Visschers et al., 1991).

The subunit complex of *Rb. sphaeroides* was also prepared from an LHII$^-$ mutant, M2192, by dialyzing isolated LHI against the detergent *n*-octyl-*rac*-2,3-dipropylsulfoxide, but the subunit was not separated from remaining LHI and solubilized BChl (Visschers et al., 1992). Absorption and CD spectra of this subunit were similar to those formed with OG. Fluorescence polarization and transient triplet-singlet spectra indicated a similar BChl *a* dimer interaction for both detergent preparations.

4. Rhodobacter capsulatus

As was the case with membrane preparations from *Rb. sphaeroides*, the use of Triton X-100 and EDTA severely modified the membranes of *Rb. capsulatus*, resulting in irreversible degradation of LHI upon OG addition (Heller and Loach, 1990). In the case of the membrane preparation from *Rb. capsulatus*, no wash step was necessary prior to Car removal. Benzene was used to remove Car from membrane fractions of both wild-type cells of *Rb. capsulatus* and a LHII$^-$ mutant, MW442. The preparations could then be titrated with OG and the subunits isolated using the same procedure as with the *Rs. rubrum* and *Rb. sphaeroides* material.

Curiously, the membrane fraction of a LHII$^-$ carotenoidless mutant of *Rb. capsulatus*, SB203E, did not behave like the comparable carotenoidless mutant, G–9, of *Rs. rubrum*. Addition of OG to this sample caused formation of monomer BChl only (absorbing at 777 nm), although subsequent dilution did result in a small amount of subunit formation. Successful reversal from the 777-nm material to LHI was demonstrated (Heller and Loach, 1990).

The overall yields of subunit complex were typically about 45% with material isolated from the mutant MW442 (LHII$^-$) and much lower (about 20%) with that from wild-type cells (Heller and Loach, 1990). The isolated subunit complex of *Rb. capsulatus* was not as stable as those from *Rs. rubrum* and *Rb. sphaeroides*, which probably resulted in increased loss of BChl during isolation by gel-filtration chromatography. Only the LHI α-and β-polypeptides were present in the subunit complex and were found to be in a stoichiometry of 1:1 as determined by HPLC analyses, identification of the N-terminal amino acid sequences of HPLC-purified polypeptide components and UV absorption of the protein. The long-wavelength maximum for the subunit preparation at 20 °C was consistently 815 ± 1 nm. This shorter wavelength probably reflects the smaller association constant than those of subunit complexes from *Rs. rubrum* and *Rb. sphaeroides* because, for solutions of the subunit complex of *Rb. capsulatus*, more free BChl absorbing at 777 nm was present. If the absorption contribution due to free BChl was subtracted, the long-wavelength absorption maximum of the subunit complex was near 819 nm. The latter value was also observed in reconstitution experiments where formation of the subunit complex was optimized by using high protein concentration and low BChl concentration.

5. Rhodopseudomonas viridis

Rp. viridis was selected as an interesting organism to study and compare to *Rs. rubrum*, not because of the similarities (a single LHI antenna and 50% homology in the α-and β-polypeptides) but because of the differences. These differences include: (1) the chromophore, which is BChl *b* in *Rp. viridis* rather than BChl *a* as in *Rs. rubrum*, (2) the additional LHI polypeptide, γ, in *Rp. viridis*, and (3) lamellar stacking of membranes in *Rp. viridis* rather than the vesicular membrane structure of *Rs. rubrum*.

Because of the instability of BChl *b*, it has not been possible to obtain high reversibility for reassociation of LHI from the subunit complex. However, the conversion of LHI to a structural subunit has been achieved in high yield as well as the partial reversal (25%) to reform LHI (Parkes-Loach et al., 1994). As with the first three bacteria, Car removal was necessary before OG was effective in causing an absorption change reflecting formation of the subunit. As in the case of membrane material from *Rb. sphaeroides* and *Rb. capsulatus*, no treatment prior to benzene extraction of Car was required. OG titrations of the Car-depleted membranes resulted in a similar set of absorption changes as was observed in the other bacteria, except that the sample had to be kept anaerobic, with reductant present and at low temperature (0 °C). Because of the difficulty in maintaining these conditions during the steps required for isolation of the subunit (such as column chromatography), such preparation has not yet been attempted.

The most effective method for converting LHI into the subunit complex was by addition of OG to Car-extracted membrane preparations at room temperature (22 °C). This treatment resulted in the dissociation of LHI into free BChl *b*, absorbing at 809 nm, and the polypeptides. Upon chilling to 0 °C, high yields of the subunit complex (greater than 80% using expected extinction coefficients based on those from BChl-*a*-containing subunit complexes (Chang et al., 1990a)) were obtained (Parkes-Loach et al., 1994). The long-wavelength maximum for this BChl-*b*-containing subunit complex was found to be at 895 ± 5 nm, representing an identical shift in energy relative to free BChl *b* and LHI of *Rp. viridis* as was determined for subunit complexes of BChl-*a*-containing bacteria relative to free BChl *a* and their LHI.

6. Rhodopseudomonas marina

Rp. marina has a LHI, but no peripheral LH and its intracytoplasmic membrane structure is lamellar like *Rp. viridis* instead of vesicular like *Rs. rubrum*, *Rb. sphaeroides* and *Rb. capsulatus* (Zuber and Brunisholz, 1991). LHI from *Rp. marina* was found to be very stable upon isolation (Meckenstock et al., 1992). In comparison to *Rs. rubrum* LHI, the α-polypeptide of *Rp. marina* is 75% homologous and the β-polypeptide is 53% homologous (Brunisholz et al., 1989). In an initial step to form the subunit complex of *Rp. marina*, Meckenstock et al. (1992) prepared membranes containing LHI, RC, and cytochromes in 0.5% LDAO which were then dialyzed against 50 mM potassium phosphate buffer, pH 8.0 and subsequently lyophilized. The dried material was extracted several times with benzene to remove Car and then resuspended in 1% OG to convert LHI to the subunit form. The subunit complex was purified in the same manner as for the other subunit preparations summarized above. The biochemical composition of this subunit preparation was identical to that reported for *Rs. rubrum*; the α- and β-polypeptides were present in a stoichiometry of 1:1 with two BChl *a* per $\alpha\beta$ pair. Absorption, CD, and fluorescence emission spectra were also very similar to those of the subunit complex of *Rs. rubrum*. The % yield of subunit was not reported.

7. Rhodocyclus gelatinosus

Jirsakova and Reiss-Husson (1993) have described the formation of a subunit complex from a LHI isolated from *Rc. gelatinosus*. The subunit was formed in 0.2% sucrose mono-N-decanoate in 0.1 M Tris-HCl, 1 mM EDTA, pH 8.0 and 65% ammonium sulfate and separated from remaining LHI by chromatography on Sepharose 6B-CL gel. Absorption and CD properties were similar to those of subunit preparations from other bacteria but without addition of carotenoid, the subunit could not be reassociated to reform LHI. A unique feature of this subunit complex is that it was formed from LHI that still contained Car. Most of the Car was lost as the subunit was formed but a variable amount of about 20% still remained with the subunit-containing fraction. Upon addition of the native carotenoid (hydroxyspheroidene) of this bacterium to the subunit complex, a native-like LHI complex could be formed upon dialysis (Jirsakova and Reiss-Husson, 1994). Addition of different carotenoids that were structurally similar to the hydroxyspheroidene gave little or no LHI formation, indicating that the native carotenoid seems to be very important in the LHI structure in *Rc. gelatinosus*.

8. Chromatium purpuratum

Kerfeld et al. (1994) have described the preparation of LHI from *Cm. purpuratum*, a purple sulfur bacterium, in which 1.5% n-dodecyl maltoside and 1.5% OG were used to solubilize the membrane. Upon purification, a membrane fraction separated into an RC-LHI complex and a fraction with a typical subunit complex absorption spectrum. Car was present in the LHI before treatment and was still present in the subunit-containing fraction. The subunit preparation could not be induced to reform the native LHI complex by dilution of the detergent concentration. These are very interesting results because *Chromatium* are purple sulfur organisms whereas all other bacteria from which subunit complexes have so far been prepared are purple nonsulfur bacteria.

B. Characterization of Subunit Structure

1. Composition

Each subunit preparation that has been well-characterized (*Rs. rubrum*, *Rb. sphaeroides*, *Rb. capsulatus*, *Rp. marina*, *Rc. gelatinosus*) contains only the LHI α-and β-polypeptides, which occur in a 1:1 stoichiometry. The ratio of 2 BChl per $\alpha\beta$-pair has been well-established for the preparations from *Rs. rubrum* and *Rp. marina*. Lipids and other factors have not been reported in these preparations, although such analyses have not been exhaustive.

2. Subunit Size

The first indication of the molecular weight of the subunit complex was from gel-filtration chromatography of the subunit complex prepared from *Rs. rubrum* (Miller et al., 1987). In these experiments, blue dextran, which eluted at the void volume, separated from and preceded the RC which in turn preceded and separated from the subunit complex. The subunit could also be separated from free BChl and the uncomplexed α-or β-polypeptides which followed it. From the average of many experiments, the peak of the subunit fraction of *Rs. rubrum* eluted at 37,500 Daltons (with values varying within ± 3,500 Da) when compared to hydrophilic proteins used as molecular weight markers. Integral membrane proteins bind large amounts of detergent relative to water soluble proteins (Guidotti, 1978). If it is assumed that the subunit complex binds an amount of detergent equal to its molecular weight, which is a reasonable estimate based on its relative hydrophobic surface and based on detergent binding for known integral membrane proteins (for example, rhodopsin is reported to bind 1.1 mg Triton X-100/ mg protein; Guidotti, 1978), then the subunit complex molecularity could be either $\alpha_1\beta_1$·2BChl (M. Wt. = 14,000 Da) or $a_2\beta_2$·4BChl (M.Wt. = 28,000 Da), although the data suggest the former.

Several other studies of the subunit complex have resulted in suggestions that it has an $a_2\beta_2$·4BChl structure. In an ultracentrifugation study of LHI from *Rs. rubrum*, Ghosh et al. (1988a) observed several different molecular weight species in the octyl-pentaoxyethylene detergent system used for these measurements. These species were presumed to be different aggregate states of the subunit for which they concluded that the $a_2\beta_2$·4BChl (24–27 kDa) species best described the subunit complex (Ghosh et al. 1988b). A similar conclusion was reached by Meckenstock et al. (1992) for the subunit complex of *Rp. marina* based on gel filtration experiments. The molecular weight they determined was 32,000 Da not corrected for detergent binding, which is similar to the value of 37,500 Da obtained with *Rs. rubrum* subunit complexes.

Sturgis and Robert (1994) have investigated the equilibrium between the subunit and its dissociated products absorbing at 777 nm. On the basis of low temperature resonance Raman spectra and the lack of effect of added BChl on the equilibrium, it was concluded that BChl was still bound to the α- and/or β-polypeptide in the dissociated state that absorbs at 777 nm, and that the subunit complex was an $\alpha_2\beta_2$·4BChl structure. However, the lack of effect on the equilibrium by added BChl is counter to the experience of our laboratory in conducting reconstitution experiments and in measuring association

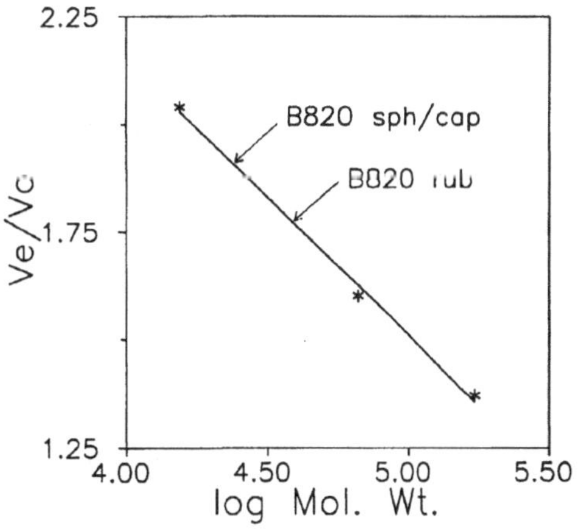

Fig. 2. Molecular weight of the subunit complex determined by gel filtration chromatography on Sephadex G-100. V_e = elution volume at peak of sample; V_o = elution volume at peak of blue dextran. Elution buffer was 0.05 M potassium phosphate, pH 7.5 which contained 5 mM $MgSO_4$ and 0.75% OG. The three protein standards (asterisks) were horse heart cytochrome *c* (mol. wt. = 12 kDa), bovine serum albumin (mol. wt. = 66 kDa), and γ-globulin (mol. wt. = 156 kDa). B820 rub indicates the value for the subunit complex of *Rs. rubrum* (data from Miller et al., 1987; Parkes-Loach et al., 1988) and B820 *sph/cap* indicates the values for the subunit complexes of *Rb. sphaeroides* (data from Chang et al., 1990b) and *Rb. capsulatus*, data from Heller (1992).

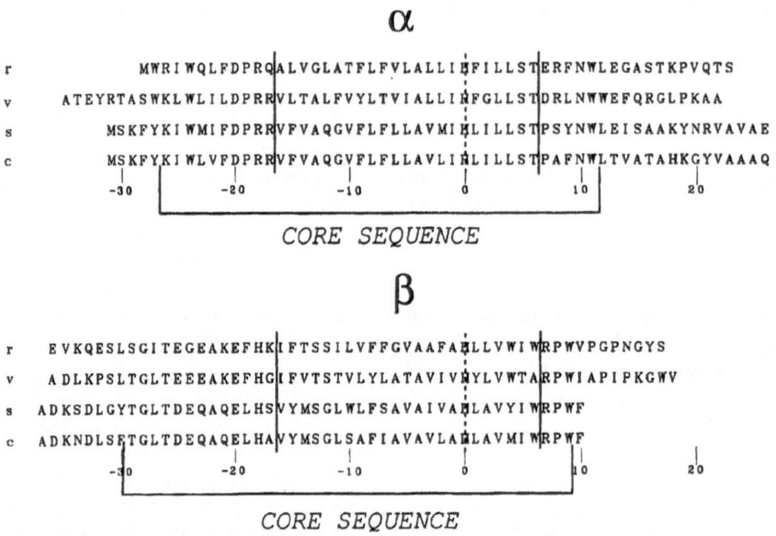

Fig. 3. Amino acid sequences of the α-and β-polypeptides of four bacteria. *r, Rs. rubrum* (Brunisholz et al. 1981, Gogel et al. 1983, Brunisholz et al. 1984, Bérard et al. 1986); *v, Rp. viridis* (Brunisholz et al. 1985, Wiessner et al. 1990); *s, Rb. sphaeroides* (Theiler et al. 1985, Kiley et al. 1987); *c, Rb. capsulatus* (Tadros et al. 1984,1985; Youvan et al. 1984). For ease of comparison, the histidine residue nearest the C-terminus was used for alignment and defined as the zero position. The core sequence defines the area of highly-conserved residues.

constants (see Section IV.C). The explanation of the differences in these observations remain to be determined.

In experiments conducted using our preparations of subunit complexes from *Rb. sphaeroides* and *Rb. capsulatus* (Chang et al., 1990b; Heller and Loach, 1990; Heller, 1992; Chang and Heller, unpublished results), we have observed that the subunit complexes consistently eluted later than the comparable complex from *Rs. rubrum*. A comparison of these molecular weight determinations is shown in Fig. 2, where the value for the subunit complex from *Rs. rubrum* is shown to be 37,500 Da and that for *Rb. sphaeroides* and *Rb. capsulatus* is 24,500 Da. The variation in values obtained with different preparations from the latter two organisms was only about ± 1,500 Da. The difference in the apparent molecular weight values for the *Rb. sphaeroides* and the *Rb. capsulatus* subunit complexes compared to the *Rs. rubrum* subunit complex may be a reflection of the difference in amount of detergent bound. From a comparison of the amino acid sequences of the α-and β-polypeptides (Fig. 3) it is apparent that the membrane-spanning segments of both polypeptides from *Rs. rubrum* are less polar than those of *Rb. sphaeroides* and *Rb. capsulatus*. Because this is the region that is expected to bind detergent (Deisenhofer and Michel, 1991), a somewhat higher percentage of detergent binding to the subunit complex of *Rs. rubrum* would be expected. It might also be noted that in OG titrations of LHI, higher detergent concentrations were required to stabilize the subunit complex of *Rs. rubrum* compared with those necessary to stabilize the other two (typically, 0.70% vs. 0.64%).

In their experiments with *Rc. gelatinosus*, Jirsakova and Reiss-Husson (1993) also reported an apparent molecular weight in the 18–25 kDa range as determined by gel filtration chromatography. If the assumption is made that the subunit complexes isolated from *Rb. sphaeroides*, *Rb. capsulatus* and *Rc. gelatinosus* bind an amount of detergent equal to their weight, then the corrected value of about 12,500 Da would agree well with the molecular weight expected for an $\alpha_1\beta_1$·2BChl structure. Due to the uncertainty of detergent binding and the extent to which the subunit complexes deviate from the assumed spherical shape, the conclusion that the composition of the subunit complex is an $\alpha_1\beta_1$·2BChl is only tentatively made. A further discussion of the size of the subunit complex based on spectroscopic evidence is found in the next section.

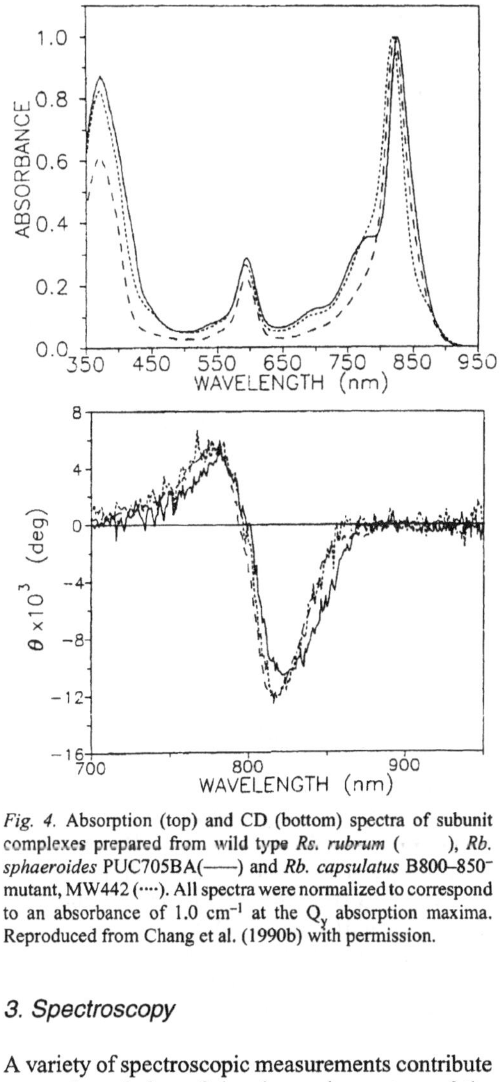

Fig. 4. Absorption (top) and CD (bottom) spectra of subunit complexes prepared from wild type *Rs. rubrum* ( ), *Rb. sphaeroides* PUC705BA(——) and *Rb. capsulatus* B800–850⁻ mutant, MW442 (····). All spectra were normalized to correspond to an absorbance of 1.0 cm^{-1} at the Q_y absorption maxima. Reproduced from Chang et al. (1990b) with permission.

3. Spectroscopy

A variety of spectroscopic measurements contribute to our knowledge of the size and structure of the subunit complex. Based on CD measurements, at least 2 BChl interact strongly, thus implying a minimal $\alpha_1\beta_1$·2BChl unit (Fig. 4; Chang et al., 1990a; Visschers et al., 1991). From fluorescence polarization (Visschers et al., 1991), triplet-singlet spectroscopy and singlet-triplet annihilation (van Mourik et al., 1991; van Mourik et al., 1993) and the relationship between the emission maximum and the excitation wavelength as a function of temperature (Visschers et al., 1993a), the subunit spectral properties clearly arise from strongly-coupled BChl dimers (exciton pairs) which cannot transfer energy to other such pairs. This makes the $a_2\beta_2$·4BChl structure very unlikely because two dimeric pairs would exist within energy transfer distance. Because the fluorescence is highly polarized, any such pair of dimers must either not transfer excited state energy between them or both dimers must exist in an essentially parallel configuration. Furthermore, since singlet-triplet quenching is not observed, further restrictions are placed on the possible distance between such pairs. The only subunit structure consistent with these restrictions is that containing one set of polypeptides and one pair of BChl (i.e., an $\alpha_1\beta_1$·2BChl).

From resonance Raman spectroscopy, the Mg^{2+} in BChl is 5-coordinate and is unchanged in the subunit complex of *Rs. rubrum* relative to LHI (Chang et al., 1990a). Presumably the 5th ligand is provided by a histidine in each case (Robert and Lutz, 1985). There are significant differences in the resonance Raman and FTIR spectra of LHI of *Rs. rubrum* compared to its subunit complex (Chang et al., 1990a; Visschers et al., 1993b; Sturgis and Robert, 1994). In each case, these changes in resonance Raman spectra were completely reversible upon reforming LHI.

C. Association of Subunits to Form LHI

1. Procedures

The subunit complexes of *Rs. rubrum*, *Rb. sphaeroides* and *Rb. capsulatus* are formed only in OG concentrations near the critical micelle concentration (cmc) and lowering that OG concentration results in association to form a LHI-type complex.

$$nB820_d \rightleftharpoons B873_d + d \qquad (3)$$

In our work, reassociation has most often been accomplished by chilling the subunit to 0 °C (at this temperature the cmc of OG has increased to above the actual OG concentration), or by dilution of the OG concentration in the sample. As was noted earlier (Eq. (2)), the equilibrium is also displaced toward LHI by addition of Car. Increasing the concentration of the subunit also displaces the equilibrium towards LHI and more detergent is required to prevent LHI formation. The value of n in Eq. (3) is at least 2 and probably 3 or larger, which explains the high sensitivity to the subunit concentration.

2. Properties of the Reassociated LHI

The NIR absorption spectra of reassociated LHI of *Rs. rubrum*, *Rb. sphaeroides* and *Rb. capsulatus* match very closely the absorption spectra of the corresponding in vivo Car-free LHI complexes. The long-wavelength absorption bands of these LHI complexes are very similar to each other in λ_{max} (typically between 868 and 875 nm), extinction coefficients, and half-widths. The major function of excited singlet state energy transfer is also restored as demonstrated by the highly efficient energy transfer from BChl of reconstituted LHI of *Rs. rubrum* to the RC of *Rs. rubrum* (see section XII).

Of special interest are the CD spectra of the reassociated LHI. Although the CD spectra of the subunit complexes of *Rs. rubrum*, *Rb. sphaeroides* and *Rb. capsulatus* with BChl *a* are extremely similar (Fig. 4), the CD spectra of the in vivo LHI complexes from different species are quite different (Bolt et al., 1981a,b; Kramer et al., 1984; Cogdell and Scheer, 1985). Thus, even though the absorption spectra and energy transfer function are nearly identical, structural subtleties exist that greatly affect the CD spectra. It is therefore quite significant that each LHI, prepared from reassociation of its respective subunit complex, displayed its own particular native CD spectrum (Parkes-Loach et al., 1988; Chang et al., 1990a,b; Heller and Loach, 1990). Since these CD spectra are unique to each LHI, they are useful as criteria for probing the quality of reassociated and reconstituted systems and indicate that the structure of the reassociated complexes are identical to the native forms. Other spectroscopic data that indicate reassociated LHI has a native structure are those of resonance Raman spectra in which changes that occurred upon forming the subunit complex from LHI of *Rs. rubrum*, were reversed upon subsequent reassociation to LHI (Chang et al., 1990a).

The relationship between the subunit and LHI provides insight into the structure of LHI. When, after appropriate treatment of the membrane, LHI is converted into the subunit complex by the addition of OG, a set of isosbestic points can be observed through about half of the titration and then a second set seems to exist through the rest of the titration (Miller et al., 1987; Heller and Loach, 1990; Chang et al., 1990b; Parkes-Loach et al. 1994). This observation suggests that only two species are in equilibrium initially and at the end of the titration, but an intermediate may be present between the subunit complex and LHI. Experimental results that bear directly on the relationship between the subunit and LHI were obtained by a stopped-flow dilution experiment in which the kinetics of subunit association to form LHI were found to be bimolecular in the subunit concentration with no other apparent intermediates (van Mourik et al., 1992). It should be noted, however, that the maximum wavelength reached in these experiments was approximately 860 nm. In our experience, the shift in the wavelength maximum to about 860 nm is fast (within a few seconds), but the final 12 to 15 nm shift is much slower (requiring hours to days). Thus, although the first product may result from a dimerization of the subunit, further association and/or rearrangement is probably involved in forming LHI. Fluorescence polarization measurements of *Rs. rubrum* LHI (Deinum et al. 1989) and *Rs. rubrum* LHI reassociated from its subunit complex (Visschers et al., 1991) were in good agreement, giving low polarization values, and indicating that in LHI, excitation energy is shared by a number of BChl molecules with non-parallel configuration. These results further suggest that after initial dimerization, a larger cluster of subunits forms.

If the long-wavelength band shift to approximately 860 nm is primarily associated with dimerization of the subunit, then the next question to address is whether there is a change in structure of the BChl binding sites upon dimerization and LHI formation. In this regard, it is of interest that both the subunit complex and LHI of *Rs. rubrum* exhibit inhomogeneous spectral broadening (Visschers et al., 1993a; van Mourik et al., 1993). This would be expected for a simple association with little perturbation in the fundamental subunit structure, but of course, it does not constitute proof.

As with most structural elements that become well-defined, it is possible to look back at earlier experimental data and possibly explain observations that were not understood at the time. This is certainly true with the structural subunit of LHI, since many shorter wavelength forms have been observed as a result of detergent or other treatment of photosynthetic bacterial membranes (Vernon and Garcia, 1967; Thornber, 1971; Cuendet and Zuber, 1977; Thornber et al., 1978, for example). These shorter wavelength species have, in general, been irreversibly formed. Among such early observations, a spectral intermediate absorbing near 820 nm was observed upon oxidative degradation of chromatophores. In these experiments, the membrane fraction of *Rs. rubrum* was oxidized with K_2IrCl_6 to selectively degrade BChl (Loach et al., 1963). As degradation progressed, the long wavelength absorption band shifted from 881 to shorter wavelengths, forming a band near 820 nm as an intermediate. In these experiments, if degradation of BChl was random, and if interacting subunits give rise to the LHI spectra, it might be expected that partial degradation of BChl might allow direct observation of some surviving subunit complexes.

The oxidation of BChl in LHI was further studied in isolated complexes by Picorel et al. (1984). In these studies, it was possible to reversibly oxidize part of the BChl in the LHI preparation and demonstrate the existence of the cation radical by EPR. The EPR signal was narrowed to 3.5 G (relative to 13.0 G for monomeric BChl cation radical in organic solvent (McElroy et al., 1969)). The narrowed signal could be explained by the sharing of the unpaired electron by a group of approximately 12 BChl molecules (Picorel et al., 1984). These results suggest six $\alpha\beta$·2BChl subunit complexes are in close association in LHI. The suggestion of a close association of six $\alpha\beta$·2BChl subunits (three $\alpha_2\beta_2$·4BChl) is also consistent with the preliminary results of crystallographic studies of LHII (Papiz et al. 1989), B800–820 (Guthrie et al. 1992) and LHI (Nunn et al., 1992). The following equilibria would be consistent with stopped-flow dilution data (van Mourik et al., 1992) and observations suggesting a $(\alpha_2\beta_2\cdot 4\text{BChl})_3$ structure:

$$2\alpha_1\beta_1\cdot 2\text{BChl} \rightleftharpoons \alpha_2\beta_2\cdot 4\text{BChl} \rightleftharpoons (\alpha_2\beta_2\cdot 4\text{BChl})_3 \quad (4)$$

(820 nm) (860 nm) (873 nm)

D. Conclusions

LHI is constructed from a fundamental unit, $\alpha_1\beta_1$·2BChl, which can be isolated as a stable detergent complex from many photosynthetic bacteria. The two BChl molecules of the subunit form a strongly-interacting exciton pair which are proposed to be arranged as shown in Fig. 5 and are about 11 Å apart (center-center) (Visschers et al., 1991). The structural suggestions in Fig. 5 will be further discussed in section X. The removal of detergent to form LHI, results in an initial dimerization (van Mourik et al., 1992). Further subunit association most likely occurs before the properties of LHI are fully attained. Thus, the red-shift from 820 nm to 873 nm would be ascribed to additional protein interaction due to close packing of additional α-helices of both α- and β-polypeptides and the existence of other subunit BChl pairs at distances between 8 to 15 Å (BChl exciton dimer center to BChl exciton dimer center). When present, Car binds to further stabilize the association of the subunits to form LHI, interacting between associated $\alpha_1\beta_1$·2BChl units. For further extrapolation to the structure of native LHI, it is desirable to have more information on the associated intermediate state(s) as well as fully formed LHI.

III. Reconstitution of Subunit Complexes and LHI from Separately-Isolated Components

A. Reversible Dissociation of Subunit Complexes

One of the original goals in dissociating the LHI was to develop methodology to dissociate the complex fully to its fundamental components and then reassociate them to reform LHI in high yield (> 90%). It was therefore with considerable satisfaction that we found the subunit complex of *Rs. rubrum* could be further dissociated to its fundamental components by increasing the OG concentration and then the subunit subsequently reformed by reducing the effective OG concentration (Parkes-Loach et al., 1988). Such reversible dissociation was also demonstrated for the subunit complexes of *Rb. sphaeroides* (Chang et al., 1990b), *Rb. capsulatus* (Heller and Loach, 1990) and *Rp. viridis* (Parkes-Loach et al., 1994), the last of which is a BChl-*b*-containing organism.

B. Reconstitution of the Subunit Complex and LHI from Separately-Isolated Polypeptides and BChl a

Since the subunit complex could be dissociated into its component polypeptides and BChl, it was clear that it should be possible to reconstitute the subunit and LHI complexes with separately-isolated polypeptides and BChl, unless some additional undetected component present in dissociated

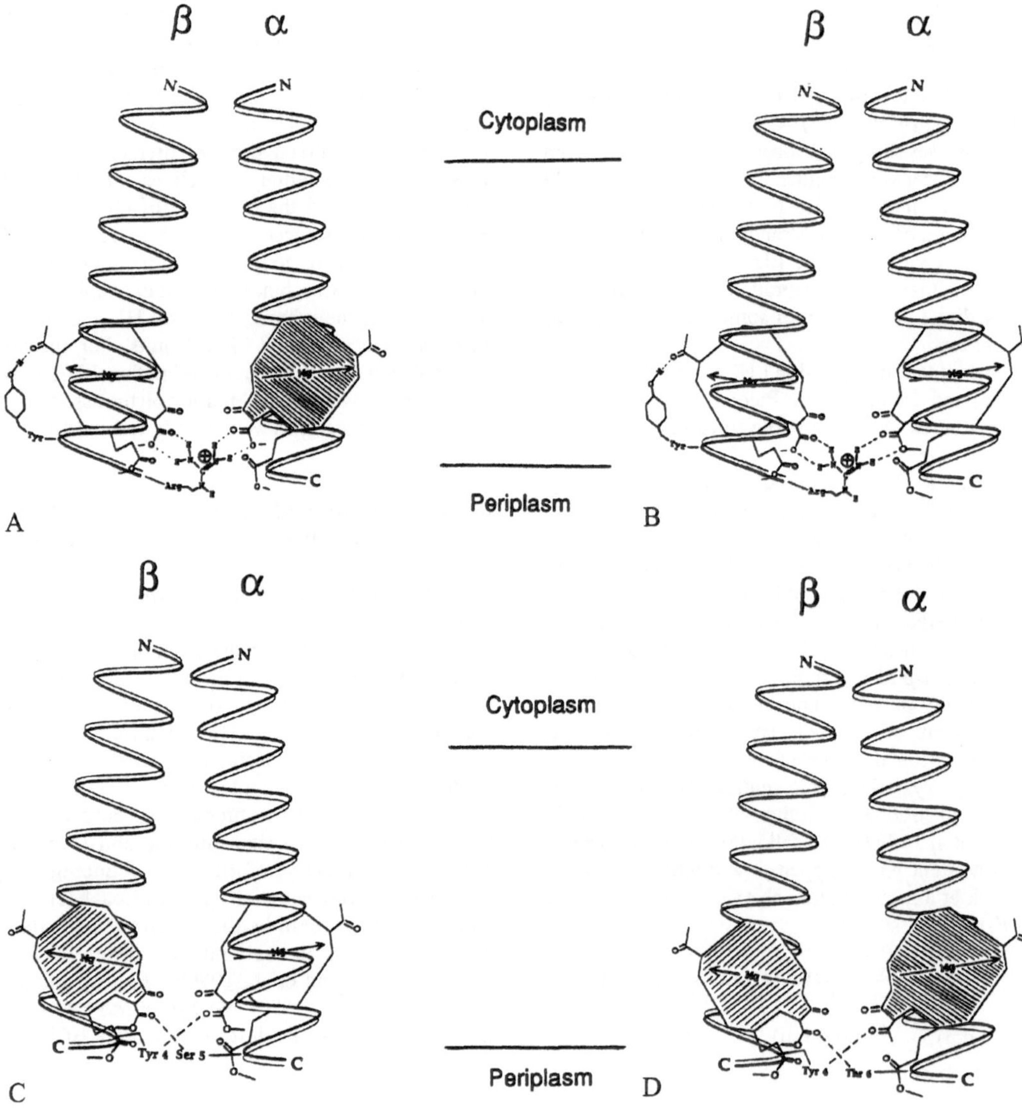

Fig. 5. Four variations of a model of the structural subunit of LHI. The sinusoidal lines represent the membrane-spanning α-helical portion of the α- and β-polypeptides without sidechains. The two BChl are represented by their outlines with Mg at their centers. The Q_y transition of each BChl molecule is depicted by the arrows running from pyrrole ring III to pyrrole ring I. The centers of the two BChl molecules are 11 Å apart. Each of the two BChl molecules are drawn with their Q_y axes rotated (one clockwise 8.5° and one counterclockwise 8.5°) along an axis passing through their Mg atoms, parallel to the membrane plane and perpendicular to the direction of the Q_y transition. Although not shown, one of the BChl *a* molecules is also to be rotated 17° along an axis perpendicular to the membrane plane and passing through its Mg atom. Two further small displacements of one of the BChl by about 1 Å results in a final angle of 24° between the two Q_y axes. These exciton BChl dimer representations are modeled after that presented in Visschers et al. (1991). The α-helical membrane-spanning regions are drawn with an angle of 17° between them. In (A) and (C), one of the polypeptides is intended to be closer to the viewer than is its bound BChl so that the BChl is on the back surface of one polypeptide and on the front surface of the other. In (B) the BChl are intended to be on the back surface of both polypeptides, while in (D) they are intended to be on the front surface. In each of the four dimers, different edges of the α- and β-polypetides α-helices are adjacent. In (A) and (B), the two amino acid side chain structures shown represent Tyr 4 and Arg 7 of the β-polypeptide of *Rb. sphaeroides*.

membrane material was required. The polypeptides used for reconstitution were extracted from the membranes with organic solvents, separated from lipid and other components by gel filtration chromatography, dialyzed against water, lyophilized and often further purified by reverse-phase HPLC. The pure polypeptides were resuspended in OG-containing buffer, BChl *a* was added and the OG concentration was adjusted to optimally stabilize the subunit complex. After the subunit complex was formed, LHI-type complexes were then formed by using the same procedures as was previously described for reassociating the subunit complex. Reconstitution of subunit and LHI-type complexes with BChl *a* has been demonstrated using the α-and β-polypeptides of LHI of *Rs. rubrum* (Parkes-Loach et al., 1988; Loach et al., 1994), *Rb. sphaeroides* (Loach et al., 1994), *Rb. capsulatus* (Loach et al., 1994), and *Rp. viridis* (Parkes-Loach et al., 1994).

From studies of these reconstituted systems, two important conclusions can be made: (1) only mature protein coded for by the *puf* A and *puf* B genes is required for *in vitro* LHI assembly and (2) the amino acid sequence of each polypeptide contains sufficient structural information to direct specific complex formation with BChl. These conclusions should not be interpreted to mean that no other protein factors are involved in the in vivo assembly of LHI in the membrane, but it does establish that such factors may not be required, and if an additional factor is involved, it likely plays no role in the eventual structure of LHI.

C. Formation of a Subunit-type Complex with the β-Polypeptide alone and BChl a

Because the subunit and LHI complexes could be reconstituted from individual components, it was possible to evaluate the role of each component and begin to probe structure-function relationships. For the α-polypeptides isolated from *Rs. rubrum*, *Rb. sphaeroides*, *Rb. capsulatus*, and *Rp. viridis*, no direct spectroscopic evidence for BChl binding was obtained when the α-polypeptide alone was reconstituted with BChl (Parkes-Loach et al., 1988; Loach et al., 1994; Parkes-Loach et al., 1994); that is, under conditions where the subunit complex readily forms when both polypeptides are present, no red-shift in the Q_y band (or any absorption change of BChl) was observed in the presence of the α-polypeptides alone. In addition, no change in the very weak molar ellipticity of monomeric BChl (Sauer, 1978; Parkes-Loach et al., 1988) in detergent solutions was noted in the presence of the α-polypeptides, even at high concentrations of the peptide (unpublished results). On the other hand, BChl-BChl oligomerization was significantly retarded in the presence of α-polypeptides under LHI-forming conditions, indicating that the α-polypeptide has an effect on BChl-BChl aggregation. As the detergent concentration is lowered to LHI-forming conditions, the BChl apparently binds to His of the α-polypeptide with a higher affinity than that exhibited for BChl oligomer formation.

In the case of the β-polypeptide alone of *Rs. rubrum*, *Rb. sphaeroides* and *Rb. capsulatus*, a very specific complex is formed at typical subunit-forming conditions. This complex has absorption, CD, and fluorescence polarization properties identical to those of the subunit complexes formed when the respective α-polypeptides are also present (Parkes-Loach et al., 1988; Visschers et al., 1991; Loach et al., 1994). However, the ratio of BChl:β-polypeptide is 1:1 in such complexes prepared from the β-polypeptide of *Rs. rubrum* (Loach et al., 1989) or from *Rb. sphaeroides* (unpublished data) rather than 2:1 as is observed in each normal heterodimer subunit complexes. Because the β-polypeptides all have two conserved His (whereas all α-polypeptides have only one), and since a His is thought to provide the ligand coordinated to each BChl (Roberts and Lutz, 1985), one might suggest that, in the subunit complex, only the β-polypeptides are involved in coordinating BChl, one molecule on each of its two His. In this case, the structure of the β-only complex would be ascribed to a β_2·2BChl complex where one β-polypeptide plays a role usually fulfilled by the α-polypeptide and the other one binds a pair of BChl. However, the spectral properties of the exciton pair for such a BChl dimer would be expected to be significantly different from those observed. In addition, recent reconstitution experiments with a truncated β-polypeptide from *Rs. rubrum* containing only one histidine, His 0 (see Fig. 3 for amino acid numbering), demonstrated that the subunit complex can still be formed (see section VII.C). Therefore, each β-polypeptide in the β-only subunit-type complex most probably has a BChl coordinated to the imidazole side chain of its His at position 0. When the α-polypeptide is present, the native subunit complex, αβ·2BChl, would have one

BChl coordinated to His 0 of the α-polypeptide and the other BChl coordinated to His 0 of the β-polypeptide.

IV. Probing the BChl Binding Site by Reconstitution with BChl Analogs

A. Common Binding Requirements

Using reconstitution of the subunit complex and LHI as assays for specific BChl binding, several BChl *a* analogs have been evaluated primarily with α- and β-polypeptides from *Rs. rubrum* and from *Rb. sphaeroides*, but also to some extent with the α- and β-polypeptides of *Rb. capsulatus* and *Rp. viridis* (Parkes-Loach et al., 1990; unpublished results). Most of the analogs tested gave similar results with each set of polypeptides and these will be discussed first. For reference, Fig. 6 shows the structure of BChl *a*. Neither a subunit complex or a LHI-type complex was formed with either the *Rs. rubrum* or *Rb. sphaeroides* polypeptides when analogs lacked Mg^{2+}, the carbomethoxy group at $C13^2$ or the bacteriochlorin oxidation state of the macrocycle. In addition, when the carbonyl group at $C3^1$ was absent, no LHI-type complex was formed. In this latter case, a subunit-type complex could be formed with polypeptides from *Rs. rubrum* but not with those from *Rb. sphaeroides*. This difference will be discussed in section IV.B.

Changes in structure that were compatible with forming both subunit-type and LHI-type complexes with polypeptides from either bacteria included a change in the side chain at C8 to a double bond (BChl *b*) and variation of the esterifying alcohol of the BChl at $C17^3$. In the latter case, even with an ethyl group as the esterifying alcohol, both the subunit complex and LHI were formed.

From these analog studies, it can be concluded that the Mg^{2+}, $C13^2$ carbomethoxy group and $C3^1$ carbonyl group are important structural features for BChl binding with α-and β-polypeptides from both *Rs. rubrum* and *Rb. sphaeroides*. The Mg atom is clearly involved in coordinating a group from the protein (most likely a histidine imidazole group (Robert and Lutz, 1985)). It is likely that the oxygen atoms of the $C3^1$ carbonyl and the $C13^2$ carbomethoxy groups are involved in hydrogen bonds with donor groups from the protein. We have so far not been able to test the carbonyl group at $C13^1$ specifically because of the lack of appropriate analogs.

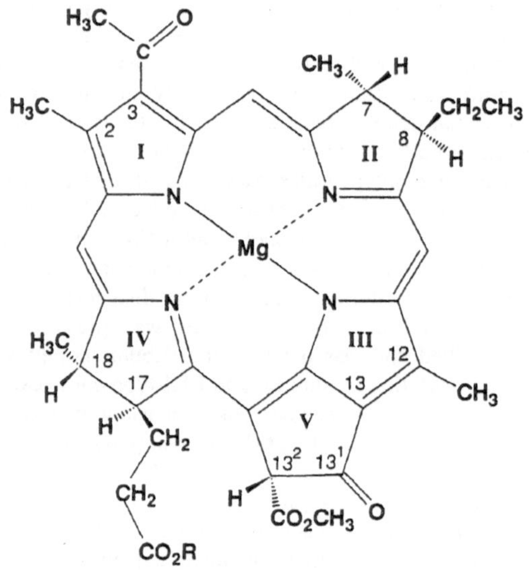

Fig. 6. Structure of BChl *a*. R is geranylgeranyl for BChl *a* isolated from *Rs. rubrum* and phytyl for BChl *a* isolated from *Rb. sphaeroides*.

B. Differences in Binding Requirements between Rs. rubrum and Rb. sphaeroides

Two of the BChl analogs behaved differently in reconstitution depending on the polypeptides used. One of these, $C13^2$ OH-BChl *a*, under the assay conditions chosen, forms both a subunit-type complex and a LHI-type complex with the polypeptides from *Rb. sphaeroides*, but the complexes were not as stable as when native BChl *a* was used (Davis et al., unpublished). However, with polypeptides from *Rs. rubrum*, only a small red-shift in the wavelength maximum (from 778 nm to 781 nm) was observed under subunit-forming conditions and no LHI-type complex formed (Parkes-Loach et al., 1990). It must be concluded that different binding constraints exist at the $C13^2$ carbomethoxy site with the two sets of polypeptides. Thus, under the conditions selected for the comparison, the addition of the hydroxyl group at $C13^2$ destabilizes complex formation with polypeptides from *Rb. sphaeroides* and prevents formation of a subunit or LHI complex with polypeptides from *Rs. rubrum*.

The second analog that behaves differently with the two sets of polypeptides is 3-vinyl BChl *a*. In this case, no evidence for interaction was found for the polypeptides of *Rb. sphaeroides*, but a subunit-type complex formed with *Rs. rubrum* polypeptides (Davis et al., unpublished). This subunit-type complex with *Rs. rubrum* polypeptides could not be converted to a LHI-type complex. From these data, it may be concluded that the carbonyl group at C3[1] is required for formation of LHI for both bacteria but not for formation of the subunit with *Rs. rubrum* polypeptides. These results showing differences between binding will be especially useful in identifying binding elements between the polypeptides and BChl and will be discussed further in section VIII.

C. Measurement of Association Constants for Formation of the Subunit Complex

Because the subunit complex can be reversibly formed from LHI and reversibly dissociated to its individual components, association constants for each of these equilibria can presumably be measured. Although these constants are highly dependent on the detergent concentration, relative values can be determined at fixed detergent concentration. The availability of such information is helpful in understanding the contributions made by some of the interactions that stabilize the subunit complex and that play a role in LHI formation.

Some association constants for forming the subunit complex are listed in Table 1. The molecularity of the reaction measured is assumed to be:

$$\alpha_d + \beta_d + 2BChl_d \rightleftharpoons \alpha\beta{\cdot}2BChl_d + d$$

or

$$2\beta_d + 2BChl_d \rightleftharpoons \beta_2{\cdot}2BChl_d + d$$

From the values of these constants, the high affinity of binding is apparent. Less than μM concentrations of the polypeptides and BChl are sufficient to quantitatively form many of the subunit complexes, even in the presence of 0.75 % OG. Thus, the high affinity of binding in forming the subunit complex and LHI allow the subtleties of binding interactions to be probed.

It is interesting that the relative stability of these subunit complexes formed with only the β-polypeptides of *Rs. rubrum*, *Rb. sphaeroides* or *Rb. capsulatus* vary more than 200-fold. In fact, even comparing the K_{Assoc} values of the β-polypeptides of *Rb. sphaeroides* and *Rb. capsulatus* which have 80% identity in their amino acid sequences, there is a 200-fold difference (Table 1). Moreover, the wavelength maximum of the Q_y absorption band of the subunit seems to reflect the strength of binding with the greater red shift correlating with stronger binding (*Rb. sphaeroides*, 825 nm > *Rs. rubrum*, 819 nm > *Rb. capsulatus*, 816 nm). The role of the α-polypeptide of *Rs. rubrum* in stabilizing the subunit complex of *Rs. rubrum* is also apparent, increasing the K_{Assoc} 6- to 10-fold. Contributions to the binding energy by various structural elements will be discussed in more detail in section VIII.

D. Competitive Inhibitors to BChl Binding

In testing BChl *a* analogs, the criteria for complex formation were a red shift in the long-wavelength absorption band and characteristic CD properties at specific concentrations of protein, BChl and OG. These properties are very sensitive to the way in which the two molecules of BChl in the dimeric pair interact with each other. Many of the analogs tested may have a high affinity for the polypeptides, but fail to show a red shift in the Q_y absorption band, presumably either because of changes in interaction

Table 1. Association constants for formation of the subunit complex[1]

Bacteria	Polypeptide(μM)	K_A (M^3) × 10^{-16}	% Variation
Rs. rubrum	β(3)	3.09	13
	β(6)	2.13	20
	α(3) + β(3)	19.9	12
Rb. sphaeroides	β(3)	22.1	22
	β(6)	73.0	12
	α(3) + β(3)	12.5	14
Rb. capsulatus[2]	β(6)	0.28	

[1] Unpublished results. Equilibria were measured in 0.05 M potassium phosphate buffer, pH 7.5, containing 0.75% OG, 22 °C. The molecularity assumed for the reaction is: $\alpha + \beta + 2BChl \rightleftharpoons \alpha_1\beta_1{\cdot}2BChl$.

[2] These data are less reliable because of the weak complex formed. The value given represents the largest value of those obtained.

in the dimer due to a different geometry or due to differing properties of the molecular orbitals in the ground and excited states of the analog molecule compared to BChl *a*. One of the analogs we expected to have some affinity for the binding site was 3-acetyl-Chl *a*, even though its chlorin oxidation state has a slightly different geometry than the bacteriochlorin oxidation state of BChl; the two molecules are otherwise structurally the same. However, experimentally, no red-shifted species were observed. If a complex were forming but exhibited no red-shift in the Q_y band, such binding could be determined by evaluating the extent to which the analog might compete with BChl *a* for its binding site(s). A variety of analogs were examined in this way and their relative effectiveness determined. Examples of analogs tested for competitive inhibition were vinyl BChl *a*, chlorophyll *a* (Chl *a*), chlorophyll *b* (Chl *b*), hemin dimethyl ester, manganese protoporphyrin IX dimethyl ester, and manganese hematoporphyrin IX dimethyl ester. Each of these molecules is able to coordinate an imidazole group of His and each contains several oxygen atoms in locations similar to those that occur in BChl *a*. All were effective in preventing formation of the normal subunit complex with BChl *a* at concentrations of analog comparable to that of BChl *a*. For comparison to the association constants of Table 1, we determined the competitive binding constant of Chl *a* with the *Rb. sphaeroides* β-polypeptide (3 μM), to be 1.47×10^{17}. Thus, by the use of competitive binding methodology, a wide variety of analogs can be used to probe the nature of BChl binding.

E. Reconstitution Studies Using DMSO

A different approach to reconstitution and competitive binding was used by Berger et al. (1992). They started with LHI isolated from *Rs. rubrum* and dissociated the complex in 0.02 M Tris buffer, pH 8.0 containing 0.3% LDAO and 20 to 30% DMSO. Reconstitution was carried out by adding concentrated BChl *a* or analog in DMSO, at about 5 times the original BChl concentration. After one hour at 4 °C, the sample was diluted 10-fold with 1 M NaCl, 0.02 M Tris buffer, pH 8.0. The reconstituted LHI-type complexes were separated from excess BChl, or analog, by gel filtration HPLC using Tris buffer (0.02 M, pH 8.0) containing 1 M NaCl and 1 to 5% cholate as eluent.

In these experiments the subunit complex was not observed as an intermediate either upon dissociation or reassociation. It is unclear whether the α-and β-polypeptides dissociated during the treatment, but with excess BChl *a* addition under reassociation conditions, good yields of the 873-nm-absorbing complex were obtained. Berger et al. (1992) found that BChl a_p, BChl a_{gg}, and BChl *b* formed red-shifted LH-type complexes while pyro BChl a_{gg}, 13^2-OH-BChl a_{gg}, Chl *a* and Chl *b* did not. These results were similar to those reported by Parkes-Loach et al (1990) for reconstitution assays with these analogs in OG. In addition, Berger et al. (1992) found no evidence for binding of 3 (α-hydroxy ethyl) BChl a_{gg}. Because the amount of analog and remaining BChl a_{gg} were determined analytically after each reconstitution, the relative effectiveness of the analogs in displacing BChl a_{gg}, and therefore presumably binding to the protein, could be compared. They found the order of effectiveness to be pyroBChl a_{gg} > BChl a_p > 13^2OH BChl a_{gg} > BChl *b* > Chl *a*, Chl *b* >> 3 (α-Hydroxy ethyl) BChl a_{gg}.

V. Reconstitution of Car to Form Wild-Type LHI with BChl *a* and Polypeptides of *Rb. sphaeroides* and *Rs. rubrum*

Car is found in all wild-type bacteria; only mutants grown under laboratory conditions can exist without carotenoids (Ke, 1971; Cogdell and Frank, 1987). One or two Car molecules are bound per αβ polypeptide pair in LHI, depending on the bacterial species (Cogdell, 1985a; Cogdell, 1986). Since the presence of Car displaces the subunit complex/LHI equilibrium towards LHI, all reconstitution experiments where it was desirable to form the subunit complex were conducted without Car. When the goal was to reconstitute the wild-type LHI, Car was added to the BChl and the α-and β-polypeptides under conditions where the subunit complex was in equilibrium with free BChl *a* and α-and β-polypeptides (Davis et al., 1995). The amount of subunit complex decreased and LHI was formed under these initial mixing conditions which are normally unfavorable for LHI formation. Upon decreasing the detergent concentration as is usually done to form LHI, the reconstituted LHI-type complex had an additionally red-shifted, long-wavelength BChl absorption maximum (usually to between 879 and 883 nm compared to 871 to 875 nm without Car). As further evidence for the near native

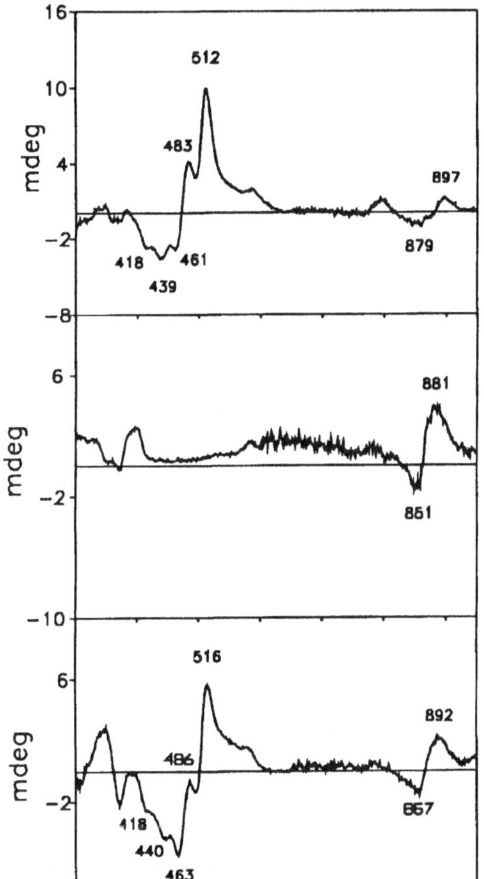

Fig. 7. CD spectra of *Rb. sphaeroides* PUC705BA (B800–850⁻) chromatophores (top) and reconstituted LHI without carotenoid (middle) and with carotenoid (bottom). Data from Davis et al. (1995). Note that the material measured for the top spectrum still contains RC while the other two systems do not.

structure of Car-containing, reconstituted LHI complexes, a red shift in the CD spectra in the Q_y region of BChl occurred (Fig. 7). Both of these properties were consistent with those displayed by in vivo wild-type LHI. In addition, the reconstituted Car had a characteristic red shift in its absorption spectrum and displayed CD properties in the 400 to 550 nm region similar to those of the in vivo state (Fig. 7). This is an especially sensitive probe of the binding environment since Car in organic solvent is not optically active (Cogdell et al., 1976).

Successful Car reconstitution has been achieved with BChl *a* and the α-and β-polypeptides of two photosynthetic bacteria, *Rs. rubrum* and *Rb. sphaeroides* (Davis et al., 1995). In addition to the absorption and CD spectra, measurement of Car function was also conducted. The functional role of Car seems to be two-fold (Cogdell and Frank, 1987): to act as an accessory pigment and to protect BChl from degradation. As an accessory pigment, it absorbs light in a wavelength region where BChl has little absorption, and then transfers this excitation energy to BChl of LHI. The efficiency of transfer of such Car excited singlet state energy to BChl can be estimated by measuring fluorescence excitation spectra (Goedheer, 1959). We chose two bacteria that differ greatly in their efficiency: for *Rs. rubrum* LHI the efficiency of transfer is 30–32% (Goedheer, 1959; Noguchi et al., 1990; Cogdell et al., 1992) while for *Rb. sphaeroides* LHI it is near 70% (Kramer et al., 1984; Cogdell et al., 1992). Spirilloxanthin is the native Car of *Rs. rubrum* S1 (van Neil et al., 1956; Schmidt, 1978). When this Car was reconstituted with the α-and β-polypeptides of *Rs. rubrum* and BChl *a*, a 30% efficiency for transfer of excited singlet state energy was determined from fluorescence excitation spectra (Davis et al., 1995). This compares very favorably to that of the intact membranes, indicating excellent reconstitution of function as well as structure. Interestingly, if spheroidene (the major native Car of *Rb. sphaeroides* 2.4.1 (Schneour, 1962; Schmidt, 1978) was used as the reconstituted Car with the α-and β-polypeptides of *Rs. rubrum*, absorption and CD properties of both BChl and Car were similar to those of wild-type LHI of *Rb. sphaeroides* and the relative efficiency of transfer of excited singlet state energy was increased to 75% (Davis et al., 1995), essentially the same as the value found in LHI of *Rb. sphaeroides*.

Spheroidene affects the BChl absorption and CD spectra the same way in both the *Rs. rubrum* and *Rb. sphaeroides* systems. Also, spheroidene's absorption maxima are red-shifted the same amount in both systems, the CD spectra have the same properties, and spheroidene's efficiency of energy transfer is identical in both systems, even though spirilloxanthin in the native *Rs. rubrum* system has a much lower yield of energy transfer. Therefore, the way in which spheroidene is bound in *Rs. rubrum* and in *Rb. sphaeroides* must be very similar. In this instance, the efficiency of energy transfer reflects the fundamental properties of the specific Car molecules (the energy level of the $2^1Ag(S_1)$ states relative to that of the BChl Q_y absorption band and the lifetime of the Car excited state (Frank et al., 1993)) and does not seem to be perturbed significantly by the protein. This is in contrast to the results reported by Noguchi et al. (1990) for *Chromatium* light-harvesting complexes where the protein did perturb the properties of the Car.

The second in vivo role of the Car is to quench the excited triplet state of BChl when it is produced (Cogdell, 1985b; Cogdell and Frank, 1987). This quenching occurs by transfer of excited triplet state energy from BChl *a* to the Car, which in turn, rapidly dissipates the energy and returns to the ground state. If the excited state occurs on BChl in the absence of Car, it can interact with oxygen and the BChl is subsequently degraded. A simple assay for evaluating this protective role by Car can be used in which the photodegradation of BChl is measured. Because triplet energy transfer is highly sensitive to distance and orientation between the donor and acceptor molecules, this assay is an effective probe for in vivo-like structure. If membrane vesicles (chromatophores) are illuminated with high-intensity radiation in air, BChl is degraded over time (Fig. 8; Davis et al., 1995). The data for chromatophores of wild-type cells shows the protective role of Car as compared to chromatophores of G9 cells, a carotenoidless mutant. Also shown in Fig. 8 is the relative stability of BChl in reconstituted LHI with and without Car. It can be seen that LHI reconstituted with Car resulted in marked stabilization of BChl toward photodegradation. This effective photoprotection indicates that the Car-BChl distance and orientation are very similar in reconstituted and in vivo complexes. Thus, beginning with the isolated α-and β-polypeptides, BChl *a* and Car, wild-type LHI has been reconstituted and displays in-vivo-like structure and function.

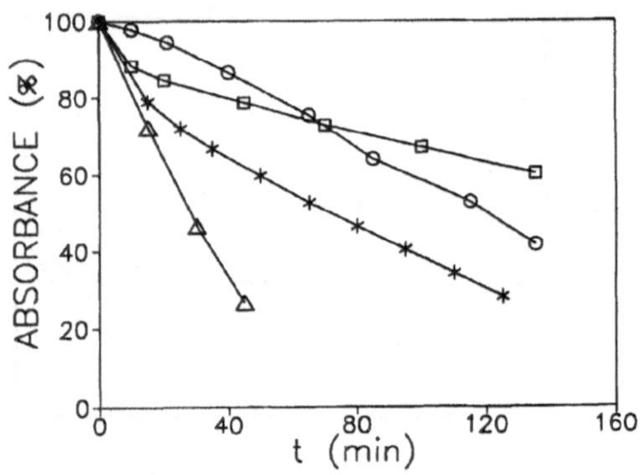

Fig. 8. Time dependence of LHI BChl degradation as measured at the λ_{max} of the Q_y band for wild type *Rs. rubrum* chromatophores (□), *Rs. rubrum* G–9 chromatophores which lack carotenoid (∗), and reconstituted *Rs. rubrum* LHI with (○) and without (△) carotenoid.

VI. Reconstitution of LHI-type Complexes Using Mixed Polypeptides

The LHI polypeptides have been isolated from many photosynthetic bacteria and their amino acid sequences have been determined (Zuber and Brunisholz, 1991; Brunisholz and Zuber, 1992). Certain structural elements are highly conserved, such as His 0 in the α-polypeptides, the two His residues in the β-polypeptides and the stretch of apolar amino acids in each polypeptide that have been assigned the role of spanning the membrane as an α-helix (Zuber, 1986). By comparison of these sequences, a core region can be identified in which there is substantial sequence homology as well as highly conservative changes (see core regions marked in Fig. 3). Because of these conserved core regions, it might be expected that the α-polypeptide of one bacterium could be paired with the β-polypeptide of another bacterium to reconstitute hybrid subunit and LHI complexes. Furthermore, such mixed polypeptide experiments might be useful in highlighting important structural elements necessary for complex formation.

Since reconstitution of subunit complexes from pure components has been demonstrated for three different bacteria (*Rs. rubrum*, *Rb. sphaeroides*, and *Rb. capsulatus*), these polypeptides were used for hybrid reconstitution experiments. The polypeptides of *Rp. viridis* were also employed since they have also been well-characterized in homologous reconstitution experiments with BChl *a* and BChl *b* (Parkes-Loach et al., 1994). Several conclusions may be drawn from the hybrid reconstitutions with these four sets of polypeptides (Loach et al., 1994). (1) Because half the hybrid combinations of α-and β-polypeptides with BChl *a* successfully formed subunit and LHI-type complexes, the experimental results support the supposition that a core region of the amino acid sequence may be sufficient for formation of functional complexes. The expectation that regions of the α-and β-polypeptides that show little or no homology are not required for subunit and LHI formation was confirmed. The amino acid sequences outside of the core region may be important for the assembly of LHI in vivo where membrane insertion and assembly with the RC are important. (2) A region sensitive to charge exists between amino acids at positions −17 to −26 of both the α-and β-polypeptides (see Fig. 3 for amino acid sequences and numbering). In the hybrid reconstitution experiments there were two general cases where reconstitutions of a LHI were unsuccessful. In one of these, the β-polypeptide of *Rs. rubrum* did not form LHI with any α-polypeptide other than its own. A possible explanation of this result is that the existence of an additional cationic residue in the core region of the *Rs. rubrum* β-polypeptide at position −17 could result in charge repulsion of the *Rs. rubrum* β-polypeptide with all α-polypeptides other than its own. In the other general case, the α-polypeptide of *Rp. viridis* failed to support LHI formation with any β-polypeptide other than that of *Rp. viridis*. The reasons for this are not obvious.

VII. Probing Structural Requirements for Formation of the Subunit Complex and LHI by Reconstitution with Modified Polypeptides

On the basis of the reconstitution results from the experiments with heterologous combinations of polypeptides, and from other reported experiments in which the importance of certain amino acids were evaluated by site-directed-mutagenesis (Bylina et al., 1988; Dörge et al., 1990; Stiehle et al., 1990; Babst et al., 1991; Fowler et al., 1992), it appeared useful to selectively modify polypeptides by chemical and enzymatic treatment. In the results to be summarized in this section, the amino and carboxyl groups of α-and β-polypeptides were modified chemically. Also, the α-polypeptide was shortened in the N-terminal region by chemical means and both the α-and β-polypeptides were shortened from the N- and C-terminal ends by enzymatic modification.

A. Chemical Modification of Charged Groups

A traditional biochemical approach in evaluating the importance of charged amino acid residues is to chemically modify them and determine the effect on biological activity. By selecting specific polypeptides whose amino acid sequences contained a single lysine (e.g., the α-polypeptide of *Rs. rubrum*), the group could be modified, for example, by acetylation, and the effect on reconstitution of the subunit complex and LHI evaluated. Conditions for quantitative modification of the ε-amino group of the lysine residue were developed for this simple system and then applied to polypeptides containing multiple lysine residues (Heller, 1992). Using this approach, the lysine side chains of the α-and β-polypeptides of *Rs. rubrum* and *Rb. capsulatus* were modified by acetylation and the carboxyl groups of the same polypeptides were modified by esterification (converting the carboxyl groups to carboxymethyl groups); both of these procedures removed charged groups. Conditions were selected to achieve as complete a modification as possible of all susceptible groups. Each modified polypeptide was isolated by HPLC and tested in reconstitution assays. The results can be summarized in the following way. In cases where the modified residues were located outside the core region, derivatization of these charged residues had no effect on reconstitution. Examples include the single lysine in the α-polypeptide of *Rs. rubrum* and the single lysine on the β-polypeptide of *Rb. capsulatus*. These results are consistent with those reported by Gogel et al. (1986) in which the single lysine of the α-polypeptide was chemically modified with dansyl chloride and found to have no effect on LHI properties. However, in cases where conserved lysine residues were modified within the core region (the α-polypeptide of *Rb. capsulatus* and the β-polypeptide of *Rs. rubrum*), the ability of these polypeptides to support formation of LHI with their homologous partner was eliminated or significantly impaired.

For reconstitution systems containing only the β-polypeptide and BChl *a*, the subunit complex formed well in the case of the acetylated β-polypeptide of *Rb. capsulatus* but it was destabilized in the case of the acetylated β-polypeptide of *Rs. rubrum*. These results are parallel to those obtained for LHI formation in that only in the case where the modification was of lysine residues outside the core sequence was there no effect on formation of the subunit complex.

In the case of the carboxyl groups, esterification of each polypeptide was conducted but it was not easy to determine whether the esterification reaction had gone to completion. In addition, there are more potential carboxyl groups to derivatize, making isolation of modified polypeptides more difficult. Some general trends in the results were observed and followed expectations according to whether the modified groups were in the core region of amino acid sequence (Heller, 1992)

The conclusions from these experiments clearly indicate the likely involvement of cationic amino acid side chains of the core region in stabilization of both the subunit complex and LH1. The results of esterification are less definite, but are consistent with a presumed role of several Glu and/or Asp in the core region providing anions to ion-pair with the conserved cationic groups.

B. Chemical Modification by Limited-Acid Hydrolysis

All of the α-polypeptides from LHI of bacteria where the amino acid sequences are known (Zuber and Brunisholz, 1991) contain the sequence ...Asp-Pro... in their hydrophilic N-terminal region (Fig. 3). The peptide bond connecting these two amino acids is known to be much more labile to acid than other peptide bonds (Landon, 1977). This bond can be specifically cleaved by limited acid hydrolysis, and the truncated α-polypeptide (e.g., of *Rs. rubrum*, Pro-Arg-Gln-Ala....) isolated by HPLC (Meadows et al., 1995). It should be noted that three highly conserved amino acids are removed in such a split (Arg or Lys at position –26; Trp at position –24; and Asp at position –20). When this truncated α-polypeptide of *Rs. rubrum* was tested for its ability to support formation of the subunit complex and LHI with the normal β-polypeptide of *Rs. rubrum*, each complex formed but the stability of the LHI-type complex was markedly reduced.

C. Truncation of Polypeptides by Hydrolysis with Proteolytic Enzymes

Among the family of specific proteases are endoproteinase Lys-C and endoproteinase Glu-C which specifically hydrolyze the bond on the carbonyl side of Lys and Glu, respectively. In the experiments summarized below, we found these enzymes to maintain their activity and specificity in 2.0% OG (Meadows et al., 1995). By selecting hydrolysis conditions (amount of enzyme, length of hydrolysis time, etc.), it appeared to be possible to isolate by HPLC all products for each polypeptide treated. Some of the polypeptides prepared in this way are illustrated below (see Fig. 3 to compare to the native polypeptide sequences). The sequence shown is the polypeptide remaining after protease treatment and isolation. The dots indicate that the amino acid sequence toward the amino- or carboxy-terminus is the same as that of the native polypeptide.

C-terminal truncated *Rs. rubrum* α-polypeptides

....RFNWLEGASTK
+18

....RFNWLE
+13

N-terminal truncated *Rs. rubrum* β-polypeptides

EFHKIFTSSI..
–20

FHKIFTSSI..

IFTSSI..

N-terminal truncated *Rb. sphaeroides* β-polypeptide

LHSVYM..
–19

N-terminal truncated *Rp. viridis* β-polypeptide

FHGIF..
–19

The two truncated C-terminal α-polypeptides were completely active in reconstitution experiments with native *Rs. rubrum* β-polypeptide. Thus, the C-terminus of the α-polypeptide can be shortened by 10 amino acids without affecting either subunit or LHI formation when tested with the native β-polypeptide. This result is consistent with expectations, since the amino acids removed are not highly conserved. Two of the N-terminal truncated β-polypeptides of *Rs. rubrum*, Glu-Phe-His-Lys.... and Phe-His-Lys-Ile...., when reconstituted with the native α-polypeptide of *Rs. rubrum*, were found to form normal subunit complexes but only partially formed LHI-type complexes, which were unstable. Of considerable interest is the fact that the N-terminal truncated Ile-Phe-Thr-Ser... from the *Rs. rubrum* β-polypeptide alone or with the normal *Rs. rubrum* α-polypeptide still formed the subunit complex, although the ΔG for formation was reduced by about 2 kcal mol^{-1} (Meadows et al., 1995). Even though a highly-conserved N-terminal region including the His at position −18 was removed, the subunit complex could still be formed. Clearly, although His −18 may be involved in stabilizing the subunit complex, it is not an absolute requirement. In contrast, no LHI-type complex could be demonstrated with this truncated β-polypeptide together with the native *Rs. rubrum* α-polypeptide. Presumably, too many stabilizing interactions had been lost, including those involving His −18.

The most stable subunit complex that we have examined is that prepared with the β–polypeptide only of *Rb. sphaeroides*. We were therefore especially interested in whether the N-terminal truncated β-polypeptide, Leu-His-Ser-Val... from *Rb. sphaeroides*, would still form a stable β-only subunit complex. This was indeed found to be the case, indicating the importance of groups within and immediately flanking the nonpolar, membrane-spanning α-helical region in forming the β-only subunit complex. The LHI-type complex with the native *Rs. rubrum* α-polypeptide (used in this experiment because of its strong interaction with all β-polypeptides) and the N-terminal truncated β-polypeptide of *Rb. sphaeroides*, Leu-His-Ser-Val... was quantitatively formed but absorbed at 847 nm and was quite unstable.

VIII. Analysis of Stabilizing Interactions in Subunit Complexes

In order to fully understand structure-function relationships in the subunit complex and LHI, it is important to evaluate the contributions made by specific binding elements. Using the experimental data from the research reviewed here, it is possible to tentatively identify and estimate the importance of some specific interactions. The ability to isolate and reconstitute subunit complexes is of immense help in this regard.

Stabilization of the subunit complex will be addressed first. The stabilization energy of the subunit complex in 0.75% OG is very large, 24 kcal mol^{-1}, calculated from $K_{Assoc} = 19.9 \times 10^{16}$ for formation of the subunit complex using the α-and β-polypeptides of *Rs. rubrum* and BChl *a* (Table 1). Possible interactions contributing to the stabilization of the subunit complex might include the following: (1) coordination of the ligand to the Mg atom of BChl, (2) hydrogen bonding between the protein and BChl, (3) hydrogen bonding between amino acid side chains of the polypeptides, (4) ion-pair interaction between the polypeptides, and (5) knob-hole (packing) interaction between polypeptide α-helices (Dunker and Jones, 1978).

The first of these interactions clearly contributes to BChl binding since, on the basis of resonance Raman data, it has been concluded that the imidazole group of His is coordinated to BChl (Robert and Lutz, 1985) and reconstitution with BPh does not form either subunit complexes or LHI (Parkes-Loach et al., 1990). From model studies (Cotton, 1976), the strength of Mg coordination to imidazole can be estimated to be between 4 and 6 kcal mol^{-1} in nonpolar solvents such as carbon tetrachloride. Since there are two BChl per subunit complex, each coordinated to an imidazole group of His, this interaction is responsible for approximately half the binding energy in the subunit complex.

The experimental data from reconstitutions using BChl *a* analogs indicated that at least two hydrogen bonds between the protein and BChl were important with polypeptides from *Rb. sphaeroides* (involving the $C3^1$ carbonyl and the $C13^2$ carbomethoxy groups of the BChl), while at least one seemed important with the polypeptides from *Rs. rubrum* (involving the $C13^2$ carbomethoxy group). Each such bond may contribute between 2 and 4 kcal mol^{-1} to complex formation in the nonpolar environment of these integral membrane proteins (Tinoco et al., 1985). Since there are two BChl molecules in the subunit complex of *Rs. rubrum*, two hydrogen bonds (one in each molecule) to a carbomethoxy group at $C13^2$ might be expected. From the comparison of reconstitution results with β-polypeptides from different bacteria, the special stability displayed by the subunit-type complex prepared with the β-polypeptide only of *Rb. sphaeroides* can only be attributed to hydrogen bond interaction in the region of BChl binding. This is because a much less stable subunit complex was formed with the highly homologous β-polypeptide only of *Rb. capsulatus* whose amino acid sequence only differs significantly from that of *Rb. sphaeroides* by the loss of two hydrogen bonding side chains (Loach et al., 1994 and section X). In addition to hydrogen bonding involving Arg 7, cross hydrogen bonding from groups such as Tyr (Trp) 4 and/or Trp 6 of the β-polypeptide to BChl coordinated to the α-polypeptide and/or from Ser 5 and/or Thr 6 of the α-polypeptide to BChl coordinated to the β-polypeptide would further stabilize the dimeric structure.

From hybrid reconstitution experiments and reconstitution with truncated or chemically modified polypeptides, at least one ion-pair (with hydrogen bonding) appears to contribute to stabilization (Loach et al., 1994; Meadows et al., 1995). This contribution is approximately 3 kcal mol^{-1} when calculated from the weakening of association observed in the absence of this pair of ions. Finally, comparison of the subunit complex of *Rs. rubrum* formed with the BChl analog containing an ethyl ester at position $C17^3$ instead of geranylgeranyl indicated that the native esterifying alcohol contributed about 1.5 to 2.0 kcal mol^{-1} to stabilize complex formation (unpublished results).

If the energies of stabilization for the four known interactions discussed above are added together for the *Rs. rubrum* system (2 × 5 kcal mol^{-1} for coordinated ligands + 4 × 2 kcal mol^{-1} for hydrogen bonding to BChl + 3 kcal mol^{-1} for an ion pair + 2 kcal mol^{-1} for the esterifying alcohol contributions = 23 kcal mol^{-1}), the elements primarily responsible for stabilization of the subunit complex in 0.75% OG are largely defined. These estimates indicate that there is little hydrogen bond interaction between polar amino acid side chains of the two polypeptides except for the above mentioned ion-paired groups, nor is there extensive stabilization by knob-hole packing between the polypeptides in the subunit complex in 0.75% OG. Thus, there seem to be four kinds of interactions that stabilize the subunit complex; in descending order of the magnitude of their contribution to the stabilization energy, they are ligand coordination to Mg, cross hydrogen bonding to BChl, ion pair formation, and packing assisted by the esterifying alcohol.

IX. Analysis of Stabilizing Interactions in LHI

Estimating the relative contributions of structural elements in the stabilization of LHI is more difficult due to the uncertain molecularity of the association reaction. As a first step, the assumption might be made that all interactions that stabilize the subunit complex are also important in formation of LHI. In the absence of the small amphipathic detergent molecules that are important in stabilizing the subunit complex, some of the subunit interactions could be stronger, perhaps by a few kcal mol^{-1}.

Equilibria between the subunit complex and LHI could be estimated from data obtained at 0.73% OG for reconstitution of LHI at high protein and BChl *a* concentration (Loach et al., 1989). If, in consideration of the results of the stopped flow dilution experiments of van Mourik et al. (1992), the initial association reaction is assumed to be a dimerization of the subunit complex

$$2\ \alpha_1\beta_1\cdot 2\mathrm{BChl} \rightleftharpoons \alpha_2\beta_2\cdot 4\mathrm{BChl} \qquad (5)$$

then,

$$K_{Assoc} = [\alpha_2\beta_2\cdot 4\mathrm{BChl}]/[\alpha_1\beta_1\cdot 2\mathrm{BChl}]^2$$

Using the data from reconstituting LHI at high protein and BChl *a* concentration (Loach et al., 1989), $K_{Assoc} = 1.7 \times 10^5$ corresponding to a $\Delta G° = -7.1$ kcal mol^{-1}. If further association then occurs to form a species with spectral properties equivalent to LHI, which we feel is likely to be the case, one might consider

$$3\ \alpha_2\beta_2\cdot 4\mathrm{BChl} \rightleftharpoons (\alpha_2\beta_2\cdot 4\mathrm{BChl})_3 \qquad (6)$$

for which,

$$K_{Assoc} = (\mathrm{LHI})/((\alpha_2\beta_2\cdot 4\mathrm{BChl})_3)^3$$

A conservative estimate of this latter K_{Assoc} would be $\approx 10^{10}$, corresponding to an additional $\Delta G° \approx -14$ kcal mol^{-1}.

In reconstitutions where Car is present, the wild-type LHI complex is formed at 0.73% OG at approximately ten-fold lower concentrations of subunit complex than in the absence of Car (Davis et al., 1995). Thus, for the reconstitutions with polypeptides from *Rs. rubrum* with BChl and Car

$$2\ \alpha_1\beta_1\cdot 2\mathrm{BChl} + 2\ \mathrm{Car} \rightleftharpoons (\alpha_2\beta_2\cdot 4\mathrm{BChl})\cdot 2\mathrm{Car} \qquad (7)$$

for which

$$K_{Assoc} \approx 1/[6 \times 10^{-7}]^3$$

corresponding to a $\Delta G° \approx -25$ kcal mol^{-1}. Further association presumably occurs to form a wild-type LHI complex. Clearly, even in the presence of 0.73

% OG, when Car is present, the subunit complexes strongly associate to form wild-type LHI.

If the reconstitution is conducted in the presence of phospholipids (phospholipid:BChl = 20:1), the LHI forms quantitatively even at the lowest concentrations of protein, BChl and Car that we can measure (approximately 1×10^{-7} M). One could conservatively estimate that the $\Delta G°$ for this initial step in association of the subunit complex in the presence of lipid is ≈ -10 kcal mol^{-1} in the absence of Car (Eq. (5)), and ≈ -30 kcal mol^{-1} in its presence (Eq. 7).

From the results obtained with the hybrid reconstitution experiments and with the truncated polypeptide experiments in the absence of Car, the following insights may be offered. Because most of the N-terminal truncated α-or β-polypeptides of *Rs. rubrum* supported formation of LHI-type complexes, but gave much lower yields, and because the complexes were less stable, specific interaction between these N-terminal regions most likely makes a major contribution to the stability of LHI. There are three pairs of strictly-conserved amino acids that may form specific ion pairs and/or hydrogen bonds (see further discussion in the next section). If one of these is involved in stabilizing the subunit complex, then the other two may be involved in stabilizing the initial associated product and each contribute about 3 kcal mol^{-1}. These interactions would account for a significant fraction of the $\Delta G°$ estimated for the association indicated by Eq. (5). It is interesting to note that unlike truncation of the N-terminus, truncation of the C-terminal region of the α-polypeptide of *Rs. rubrum* by ten amino acids did not appear to have a significant effect on stability of the reconstituted LHI (Meadows et al., 1995).

In addition to the possible ion pair interaction in the hydrophilic N-terminal region, hydrogen bonding between polypeptides (possibly including the BChl) in the otherwise nonpolar, membrane-spanning regions may be important. The special stability displayed by the subunit-type complex prepared with the β-polypeptide only of *Rb. sphaeroides* was attributed to this kind of interaction (section VIII)

Also, knob-hole packing of the α-helices is probably quite important to the stabilization to LHI. Such packing does not seem to be a very significant factor in the detergent-stabilized subunit complex, probably because the mixed micelles formed, composed of protein and detergent, satisfy the major hydrophobic driving force to minimize the exposure of hydrophobic surface to water. Decreasing the detergent concentration to below that necessary to support the mixed protein-detergent micelle structure would result in a strong hydrophobic force for subunit association. When the detergent concentration is decreased in the presence of phospholipid, the subunit complex presumably incorporates into the lipid bilayer of the proteoliposome with the lipid replacing detergent on the protein's hydrophobic surface. However, more restrictions to motion occur in the hydrocarbon tails of the phospholipid in comparison to detergent binding, especially in the region near the lipid head group. Thus, phospholipid may not effectively compete with polypeptide-polypeptide interaction in the bilayer of the proteoliposome and, as a result, extensive association of the subunit complexes occurs.

Of special interest is the stabilization due to the presence of Car. Although Car causes only a small perturbation in the structure of the BChl in LHI, as indicated by a small red-shift in the absorption and CD spectra in the Q_y-band region, its presence results in a significant increase in stability of the complex. Such stability presumably stems from specific packing between the rigid conjugated double bond chain of the Car molecule with its periodically protruding methyl groups and the $\alpha_1\beta_1 \cdot 2$BChl units.

X. Implications for the Structure of the Subunit Complexes

The models for the subunit complex suggested in Fig. 5 are based on a consideration of the results of biochemical and spectroscopic characterization of the subunit complexes from different bacteria, and from reconstitutions with systematically modified components. Essential features are the following:

(1) The composition of the subunit complex is $\alpha_1\beta_1 \cdot 2$BChl.

(2) One BChl is coordinated to the imidazole group of the His 0 of the α-polypeptide and one is coordinated to His 0 of the β-polypeptide.

(3) The two polypeptides are closely packed with the two BChl 11 Å apart (center to center), the planes of their macrocyclic rings are parallel to the axes of the α-helical polypeptides and there is an angle of about 24° between their Q_y axes (axis

through rings I and III). The Q_y axes are approximately perpendicular to the α-helical axes.

(4) Two or more hydrogen bonds/BChl are proposed to further fix the molecules in space and provide their specific orientations. The tail-to-tail orientation of the Q_y axes was chosen for all models because this maximizes the possibilities for cross-hydrogen bonding between BChl bound to one polypeptide and amino acid side chains of an adjacent polypeptide in the subunit complex $\alpha_1\beta_1 \cdot 2$BChl. Also significant in our thinking is that this orientation would provide a reason why the subunit is such a stable entity rather than a mixture of oligomers as one might expect in head-to-tail orientations. In addition, these subunit complexes are constructed with an element of symmetry. The location of possible hydrogen bond donors in the amino acid sequence are indicated in Fig. 9.

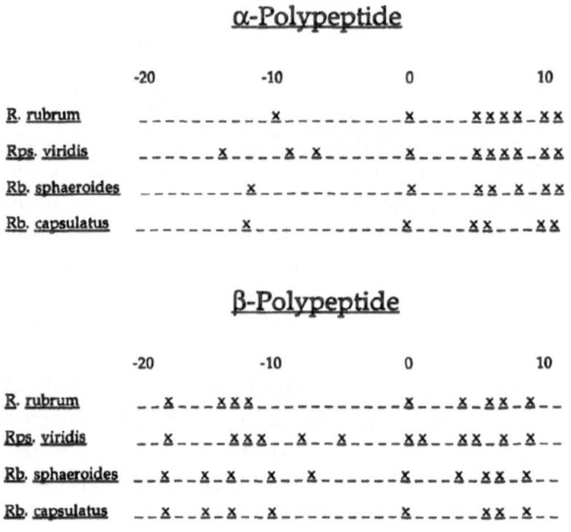

Fig. 9. Potential hydrogen bond participating side chains in the region −18 to +11.

A key feature of models (A) and (B) in fig. 5 is that Arg 7 of the β-polypeptide is suggested to provide a bridge between the two BChl molecules by hydrogen bonding, for example, to the carbomethoxy oxygens at C13^2 of both BChl. The charge on the cationic guanidino group of Arg may also contribute to the red-shift of the long-wavelength band of BChl (Davis et al., 1981; Eccles and Honig, 1983; Fowler et al., 1993). It has been found in experiments with site-directed-mutations of Arg 7 (in *Rb. capsulatus*) that normal-appearing LHI was formed (Babst et al., 1991). Although this result may be viewed as evidence that this highly conserved amino acid does not play a key role in formation of LHI, as noted below, if this group is merely one of many stabilizing features in LHI, its absence may not be sufficient to prevent LHI formation. In this regard, it will be important to quantitatively evaluate its role in formation of the subunit complex.

Two different groups may also contribute to cross-hydrogen bonding; one hydrogen bond donor on the α-polypeptide and one on the β-polypeptide, both at position 6 (for the sake of clarity, these are not shown in Figs. 5A and 5B). One of these could be contributed by Thr 6 of the α-polypeptide interacting with an oxygen of BChl which is coordinated to His 0 of the β-polypeptide. The second could be Trp 6 of the β-polypeptide (except for *Rp. viridis*; recall that the weakest subunit complex is formed with *Rp. viridis b*-polypeptide) contributing a hydrogen to bond with an oxygen atom of the BChl coordinated to His 0 of the *a*-polypeptide. A further possible hydrogen bond could involve Tyr (or Trp) 4 of the β-polypeptide and the carbonyl at position C3^1 of the BChl molecule coordinated to His 0 of the same β-polypeptide (see Figs. 5A and 5B).

Alternative models featuring multiple cross-hydrogen bonding are shown in Figs. 5C and 5D. These two models are not meant to exclude a role for Arg 7 as a bridging ligand, but the focus here is on Tyr (Trp) 4. The involvement of Tyr (or Trp) 4 in such hydrogen bonding could explain why the β-polypeptide of *Rb. sphaeroides* forms a highly stable subunit complex while that of *Rb. capsulatus* (whose Met at this position is incapable of donating a hydrogen to hydrogen bond formation) does not (Chang et al., 1990b; Heller and Loach, 1990; Loach et al., 1994); there are only two significant differences in the amino acid sequences in the apolar core region of these two bacteria (the other is at position −7). Furthermore, the fact that the β-only subunit complex formed with the β-polypeptide of *Rb. sphaeroides* is about ten-fold more stable than that formed with the β-polypeptide of *Rs. rubrum* could reflect the greater hydrogen bond strength of Tyr 4 compared to Trp 4, respectively. Also in the models of Figs. 5C and 5D, the conserved Ser 5 and Thr 6 of the α-polypeptides, respectively, are proposed to be involved in a hydrogen bond to the carbomethoxy group at position C13[2] or the keto group at group at position C13[1] of the BChl. Other possible hydrogen bond donor groups of the protein further toward the C-terminus that are highly conserved include Trp 9 of the β-polypeptide and Asn 10 and Trp 11 of the α-polypeptide. Interestingly, a change of this latter residue to Phe in *Rb. sphaeroides* was reported to result in a normal expression of LHI, but caused a blue-shift of the Q_y λ_{max} to 853 nm (Olsen et al., 1994).

Because BChl is very likely bound within the membrane but near the aqueous interface (Loach et al., 1985), hydrogen bonding between water molecules and the oxygen atoms of BChl are also important to consider. From this perspective, it should be noted that BChl is protected from deleterious interactions with oxygen, hydrogen ions and hydroxide ions in both the subunit complex and LH1 form relative to BChl in OG (unpublished results). Such marked stability to extremes of pH and oxygen presumably means that the C-terminal portion of the polypeptides cover the otherwise exposed edge of the BChl that would be at the aqueous interface and, as a consequence, protect it from interacting with water. Even so, a sequestered water molecule may be an important structural element required to stabilizing the binding site and could also act as a bridge between the two BChl.

(5) The two esterifying alcohol groups are suggested to immediately turn after the ester linkage resulting in the isoprenoid chain running towards the N-terminus next to the apolar α-helical regions of the polypeptides, one in front and one behind the approximate plane of the BChl dimer.

(6) Three sets of ion pairs may occur between groups in the N-terminal region of the α-and β-polypeptides. Arg-18, Asp-20 and Lys(Arg)-26 of the α-polypeptide may be paired with Glu-20, His-18 and Glu(Asp)-25, respectively, of the β-polypeptide. Polypeptide conformations in this region are probably not extensively α-helical because of the Pro in the α-polypeptides and several Gly in the β-polypeptides in their N-terminal amino acid regions. Reconstitution experiments utilizing α-or β-polypeptides with truncated N-terminal regions indicate that only the postulated ion pair involving Asp-20 of the α-polypeptide and His-18 of the β-polypeptides may be important for stabilizing the subunit complex. On the other hand, evidence from these experiments indicates that all three interactions are of importance in stabilizing LHI.

It is also necessary to account for the subunit-type complex formed with only the β-polypeptide. In fact, the ability to explain the formation of a highly similar environment for an exciton pair of BChl with only the β-polypeptide is a key point of the models presented in Fig. 5. The structure of this complex is suggested to be basically the same as that for the native subunit complex in which either Arg 7 is used to bridge between the two BChl molecules ((A) and (B)), or Tyr 4 is cross-hydrogen bonding as in (C) and (D). Then, in a manner similar to the native subunit complex, Trp 6 (except for *Rp. viridis*) of a second β-polypeptide could provide hydrogens for hydrogen bonds to BChl coordinated to His 0 of the other polypeptide; the lack of Trp 6 in *Rp. viridis* is consistent with this suggestion since the latter fails to form a stable subunit complex without an α-polypeptide.

Further stabilization of the β-only subunit complexes of *Rs. rubrum*, *Rb. sphaeroides*, and *Rb. capsulatus* could come from an ion pair formed between His 18 of one polypeptide and Glu 20 of the other. Both are highly conserved amino acids, which may stabilize the subunit-type complex in the same way as was suggested for His 18 of the β-polypeptide and Asp 20 of the α-polypeptide in the native subunit complex.

Very recently, the structure of LHII from *Rp. acidophila* has been determined in crystals prepared from LHII isolated in OG (McDermott et al., 1995). Many of the structural features of LHII undoubtedly have relevance to the structure of LHI and its subunit complex reviewed here. Because this crystallographic information became available very late in the final proof stage of this review, an in-depth analysis cannot be attempted. It is interesting to point out, however, that the BChl molecules apparently responsible for the 850 nm absorption band of LHII are coordinated to His 0 of the α- and β-polypeptides as predicted, and two different fundamental repeating $\alpha_1\beta_1\cdot$2BClhl units can be considered. In one of these repeating units, the distance and orientation of the two BChl are strikingly similar to those suggested in Fig. 5 for the subunit comples of LHI. In addition, in both LHII and LHI, Tyr and Trp side chains in the C-terminal region of the α-polypeptides are involved in hydrogen bonds to the $C3^1$ keto group of BChl. While no hydrogen bonds to the protein were found between the oxygen-containing groups at positions $C13^1$, $C13^2$, or $C17^3$ of the BChl of LHII, such bonding may be unique to LHI and responsible for the stability of the subunit complex of this species. (Also see Chapter 16 by Zuber and Cogdell.)

XI. Minimal Structural Requirements to Form LHI

In examining structural features involved in the formation of the subunit complex, a variety of binding elements could be probed because the detergent concentration was sufficiently high to satisfy all hydrophobic surfaces without recourse to unspecific self-aggregation. However, under conditions for LHI formation, the detergent concentration is purposely maintained below that necessary to stabilize the subunit-detergent, mixed micelle. Thus, the driving force for LHI formation is fundamentally the hydrophobic force which is very large and would result in non-specific aggregation if specific association did not occur first. Recognizing that self-aggregation will occur under LHI-forming conditions, the question we wish to address is what interactions are responsible for forming the very specific, associated state, LHI.

It is reasonable to speculate that, for LHI formation, all BChl molecules may be required to be coordinated to a His at position 0 because of the large stabilization energy involved in this interaction and the restrictive geometry of the coordination sites. On the other hand, the much smaller energy of a hydrogen bond, the six oxygen atoms available on the periphery of BChl and the possibly more flexible environment in the region of these oxygens probably means that some differences exist in the groups participating in hydrogen bonding in different bacteria. It seems unlikely that all of the oxygen atoms on BChl would be involved in hydrogen bonding or that so many bonds would be necessary for stabilizing the subunit complex. On the other hand, evidence summarized in this review indicates that in LHI at least three of the four oxygen-containing groups are likely to be involved in hydrogen bonding interactions with the protein. Thus, caution should be exercised in evaluating the results of site-directed-mutation experiments; a native-like LHI complex could form in a mutant even with the removal of a normal hydrogen bond participant. In this regard, measuring K_{Assoc} for appropriate reactions may be essential for proper evaluation.

A special role in forming LHI from the subunit complex is suggested for the carbonyl group at the $C3^1$ position, a group unique to BChl. From resonance Raman measurements (Robert and Lutz, 1985; Chang et al., 1990a; Visschers et al., 1993c) and BChl analog studies (Parkes-Loach et al., 1990; unpublished results) this keto group is required and is strongly interacting with the protein in LHI but not (or much less so) in the subunit complex. This group contains the only potential hydrogen bonding oxygen atom on this side of the molecule; the other five oxygen atoms are clustered together on the other side of the molecule. In the suggestions for the structure of the subunit complex shown in Fig. 5, the carbonyl group at $C3^1$ is ideally suited to direct association of the subunit complex to form LHI. Of particular interest in this regard are potential hydrogen-bond-donating amino acids such as Trp 9 of the β-polypeptide and Trp 11 of the α-polypeptide. Consistent with this suggestion, a site-directed mutant reported by Olsen et al. (1994), in which this latter residue was changed to Phe, resulted in a 23 nm blue-shift of the Q_y λ_{max} indicating a significant effect on LHI structure. It may also be purposeful to use an amino acid side chain as far away from the center of BChl binding as Trp 9 of the β-polypeptide and Trp 11 of the α-polypeptide in order to use the flexibility in the C-terminal portion of the polypeptide to allow it to fold back over the BChl edge, thus fulfilling an important role in protecting BChl from the aqueous environment.

Because the β-polypeptides alone of *Rs. rubrum*, *Rb. sphaeroides* and *Rb. capsulatus* can form the subunit complex but not LHI, and if the structure of LHI is basically due to the subunit in an associated state, it might be expected that a modified β-polypeptide or a native β-polypeptide of some other species might be able by itself to form LHI. The properties of the β-polypeptide of *Rp. viridis* fulfill this expectation as does the truncated β-polypeptide of *Rb. sphaeroides* (LHSVYM.....).

In characterizing the subunit complex and the

reconstitution behavior of the α-and β-polypeptides of *Rp. viridis*, it was found that with only the β-polypeptide and BChl *a* (not BChl *b*), a LHI-type complex was readily formed (Parkes-Loach et al., 1994). Therefore, the *Rp. viridis* β-polypeptide contains all the structural requirements necessary to form a LHI-like complex with BChl *a*. On the other hand, the subunit complex was not appreciably formed with only the β-polypeptide of *Rp. viridis*, although it was stabilized by the presence of any α-polypeptide (Loach et al., 1994). Interestingly, the truncated β-polypeptide of *Rp. viridis*, FHGIF....., still showed behavior nearly identical to that of the native β-polypeptide even though most of the N-terminal region was removed (Meadows et al., 1995). This result again suggests that there are strong interactions in the membrane spanning segment of the polypeptide that primarily direct the nature of association when the detergent concentration is decreased so that the hydrophobic force driving association causes a LHI-type complex to form.

The other β-polypeptide that we found to form a stable LHI-type complex was the N-terminal truncated β-polypeptide of *Rb. sphaeroides*, Leu-His-Ser-Val... . As indicated before, this truncated polypeptide does form a stable subunit-type complex as well. This example is especially interesting because the extra hydrogen bonding capability in the apolar region (relative to the β-polypeptide of *Rb. capsulatus*) seems to be responsible for formation of a very stable subunit complex. Perhaps these hydrogen bonding groups are also helpful in LHI formation when the hydrophilic N-terminus is removed.

To our surprise, two modified α-polypeptides have been observed to form LHI-type complexes by themselves with BChl *a*, the N- and C-terminal-truncated α-polypeptides of *Rs. rubrum*. Whereas the native α-polypeptide of *Rs. rubrum* will not form either a subunit-type complex or LHI-type complex without the β-polypeptide, the N-terminal truncated α-polypeptide, Pro-Arg-Gln-Ala... and the C-terminal truncated α-polypeptide .....Leu-Ser-Thr-Glu still did not form subunit complexes, but both formed stable LHI-type complexes (Meadows et al., 1995). These results can be explained by, and lend support to the structural suggestions shown in Fig. 5. For example, in Figs. 5(A) and 5(B) Arg 7 is used as a bridging group providing hydrogen bonds to both BChl molecules of the exciton dimer. An Arg is located at position +8 in the α-polypeptide of *Rs. rubrum* compared to the strictly conserved position +7 in all β-polypeptides. On the other hand, consistent with the suggestions indicated in Figs. 5C and 5D, Ser 5 and Thr 6 of the truncated α-polypeptides could perhaps play the role of Trp 4 and Trp 6 of the *R. rubrum* β-polypeptide.

Consideration of the results showing formation of a LHI-type complex with a single polypeptide, along with information discussed in earlier parts of this review, lead to the following suggestions for minimal structural requirements for formation of LHI:

(1) a single membrane spanning α-helical polypeptide that can associate with itself (reasonable knob-hole fit and without charge repulsion).

(2) a histidine residue approximately one-third of the way into the apolar region that serves to fix one dimension of BChl binding, holding this amphipathic molecule in its most stable location relative to the aqueous interface.

(3) three or more cross-hydrogen bonding elements that serve to link two BChl molecules in a dimeric structure.

If the above are valid conclusions, then the question arises as to why living organisms seem to have opted to build the subunit structure and LHI with heterodimer polypeptides instead of homodimer polypeptides. Use of the latter would have been simpler. Perhaps the answer to this question is to be found in the need for specificity and control, both of which are hallmarks of living systems. The diversification in using two different polypeptides, especially such small ones, considerably enhances the uniqueness of the subunit complex. Highly specific recognition is presumably required for proper binding of LHI to RC as well as LHI to LHI and in some species, LHI to peripheral LH complexes (LHII). There is also the possibility of a built-in control feature for energy distribution and utilization in the form of sites for phosphorylation (Allen, 1992a,b). Another possible need for specific recognition is to provide binding for Car molecules, an important component of wild-type cells. This latter consideration can actually be tested in reconstitution experiments using appropriately-modified polypeptides capable of forming LHI-type complexes either as homodimers or heterodimers.

From the point of view of designing macromolecules (proteins) to perform specific functions, it is instructive that polypeptides as small as those of the LHI contain so much information. Whereas a polypeptide containing as few as 30 amino acids seems to be sufficient to specifically bind BChl and form a unique associated system, requirements for other forms of recognition and binding (e.g., to the RC) may require the additional polypeptide structure.

XII. Interaction of LHI with RC

The preparation of the subunit complex and the ability to reconstitute this complex and the LHI from their fundamental components offers an exceptional opportunity to examine the interaction of LHI with RC. By addition of RC of *Rs. rubrum* and phospholipid to the subunit complex of *Rs. rubrum* at 0.80% OG at room temperature, proteoliposomes containing photoreceptor complexes (PRC) were then formed by chilling the sample to 4° C (Bustamante and Loach, 1994). These proteoliposomes were stabilized at room temperature by first decreasing the OG concentration at low temperature and then warming the sample. The absorption and CD properties of this reassociated system appeared to reproduce those of membrane vesicles prepared from whole cells. In addition, energy absorbed by BChl of the LHI was transferred to the RC with ideal efficiency; that is with a quantum yield = 1.0.

In these reassociation experiments, at the concentration employed, there was no evidence for interaction between the subunit complex and the RC in 0.70% OG at room temperature (Bustamante and Loach, 1994). These are the conditions, as described earlier, for the dissociation of LHI from the RC in the subunit preparation procedure. Energy transfer from BChl of the subunit complex to the RC was very low, indicating that large distances separate the two complexes. The interaction energy between these complexes is in a delicate balance in the presence of 0.70% OG as indicated by the fact that the equilibrium can be shifted toward formation of PRC by increasing the concentration of either the RC or the subunit complex of LHI. In addition, binding of the LHI to the RC did not result in major changes in the absorption and CD spectra of either complex.

Evidence that there is specific binding in vivo between LHI and RC comes from several sources. According to liposome fusion experiments, the peripheral LH complex, but not the LHI, can be easily separated from the RC (Westerhuis et al., 1989). Secondly, by the judicious choice of detergents and separation methods, the PRC complex can be isolated intact from several organisms (Loach et al., 1970a,b; Hall et al., 1973; Loach et al., 1980; Jay et al., 1984; Dawkins et al., 1988; Hara et al., 1990; Agalidis et al., 1991, Vasil'ev, 1991). Finally, in the subunit-RC reassociation experiments, the efficiency of transfer of excitation energy from BChl of the LHI to the RC was only 20% (quantum yield of 0.2) when a hybrid reassociation was attempted with the subunit complex of *Rb. sphaeroides* and RC from *Rs. rubrum* (Bustamante and Loach, 1994). In further studies it should be possible to establish interactions responsible for specificity in binding by using selectively-modified, reconstituted LHI.

In considering the types of interaction that may occur between LHI and RC to stabilize the PRC, the following are possible: (1) ion pairing, (2) hydrogen bonding especially in the nonpolar region, (3) knob-hole packing of nonpolar regions, and (4) other specific binding of amino acid side chains or polypeptide structural motifs that would most likely occur external to the membrane-spanning α-helical segments. It seems pertinent to consider that amino acid residues of the RC with their side chain groups exposed to the membrane (where they would interact with lipid, LHI and possibly other protein complexes, but do not contact the cofactors or adjacent polypeptides of the RC) are very poorly conserved (16% homology among known reaction center sequences). On the other hand, 42% of residues are conserved for those amino acid residues contacting the cofactors or adjacent RC polypeptides (Yeates et al., 1987)). This result indicates a lower level of specific interaction likely exists in knob-hole packing with LHI in the α-helical, membrane-spanning regions.

In some of our reassociation experiments, the RC was subjected to proteinase K treatment before addition to the subunit complex. This treatment resulted in total hydrolysis of the H-polypeptide (perhaps leaving the membrane-spanning segment intact) and removal of much of the hydrophilic portions of the L and M polypeptides. Even so, no effect on subsequent association with LHI was observed. These results, together with the fact that there are few hydrophilic regions of LHI, indicates that any ion pair stabilization must occur near the membrane interface, not further out in the aqueous environment. Also, there would not seem to be enough protein left to form a highly specific LHI receptor site. These considerations suggest that initial studies should focus on the membrane interface region where there is a higher probability of specific structural elements important for LHI–RC interaction. Preparation of certain site-directed mutations of the RC, together with reconstitution of selectively modified LHI, should allow identification of the LHI-RC binding region(s).

Acknowledgments

We gratefully acknowledge Peggy L. Bustamante, Christine M. Davis, Barbara A. Heller, Kouji Iida, Kelley A. Meadows, Mamoru Nango and Paul A. Recchia for their contributions to this manuscript in the forms of research data and insightful discussions. This work was supported by grant GM 11741 to PAL from the US Public Health Service.

References

Aagaard J and Sistrom WR (1972) Control of synthesis of reaction center bacteriochlorophyll in photosynthetic bacteria. Photochem Photobiol 15: 209–225

Agalidis I and Reiss-Husson F (1991) Resolution of *Rhodocyclus gelatinosus* photoreceptor unit components by temperature-induced phase separation in the presence of decyltetra-oxyethylene. Biochem Biophys Res Commun 177: 1107–1112

Allen JF (1992a) How does protein phosphorylation regulate photosynthesis? Trends Biochem Sci 17: 12–17

Allen JF (1992b) Protein phosphorylation in regulation of photosynthesis. Biochim Biophys Acta 1098: 275–335

Babst M, Albrecht H, Wegmann I, Brunisholz R and Zuber H (1991) Single amino acid substitutions in the B870 α and β light-harvesting polypeptides of *Rhodobacter capsulatus*. Eur J Biochem 202: 277–284

Bélanger G and Gingras G (1988) Structure and expression of the puf operon messenger RNA in *Rhodospirillum rubrum*. J Biol Chem 263: 7639–7645

Bérard J, Bélanger G, Corriveau P and Gingras G (1986) Molecular cloning and sequence of the B880 holochrome gene from *Rs. rubrum*. J Biol Chem 261: 82–87

Berger G, Andrianambinintsoa S, Kleo J, Grison S, Dejonghe D and Breton J (1992) Dissociation and reconstitution studies by high performance liquid chromatography of the light-harvesting complex of *Rhodospirillum rubrum*. J Liq Chrom 15: 585–602

Bolt JD, Hunter CN, Niederman RA and Sauer K (1981a) Linear and circular dichroism and fluorescence polarization of the B875 light-harvesting bacteriochlorophyll-protein complex from *Rhodopseudomonas sphaeroides*. Photochem Photobiol 34: 653–656

Bolt JD, Sauer K, Shiozawa JA and Drews G (1981b) Linear and circular dichroism of membranes from *Rhodopseudomonas capsulatus*. Biochim Biophys Acta 635: 535–541

Braun P and Scherz A (1991) Polypeptides and bacteriochlorophyll organization in the light-harvesting complex B850 of *Rhodobacter sphaeroides* R-26.1. Biochemistry 30:5177–5184

Broglie RM, Hunter CN, Delepelaire P, Niederman RA, Chua NH and Clayton RK (1980) Isolation and characterization of the pigment-protein complexes of *Rhodopseudomonas sphaeroides* by lithium dodecyl sulfate/polyacrylamide gel electrophoresis. Proc Natl Acad Sci USA 77: 87–91

Brunisholz RA and Zuber H (1988) Primary structure analyses of bacterial antenna polypeptides: Correlation of aromatic amino acids with spectral properties. Structural similarities with reaction center polypeptides. In: Scheer H and Schneider S (eds) Photosynthetic Light-Harvesting Systems: Organization and Function, pp 103–114. Walter de Gruyter and Co., New York

Brunisholz RA and Zuber H (1992) Structure, function and organization of antenna polypeptides and antenna complexes from the three families of Rhodospirillaneae. J Photochem Photobiol B: Biol 15: 113–140

Brunisholz RA, Cuendet PA, Theiler R and Zuber H (1981) The complete amino acid sequence of the single light harvesting protein from chromatophores of *Rhodospirillum rubrum* G–9+. FEBS Lett 129: 150–154

Brunisholz RA, Suter F and Zuber H (1984) The light-harvesting polypeptides of *Rhodospirillum rubrum*. I. The amino-acid sequence of the second light-harvesting polypeptide B 880-β (B 870-β) of *Rhodospirillum rubrum* S 1 and the carotenoidless mutant G–9+. Aspects of the molecular structure of the two light-harvesting polypeptides B880-α (B 870-α) and B 880-β (B 870-β) and of the antenna complex B 880 (B 870) from *Rhodospirillum rubrum*. Hoppe-Seyler's Z Physiol Chem 365: 675–688

Brunisholz R, Jay F, Suter F and Zuber H (1985) The light-harvesting polypeptides of *Rhodopseudomonas viridis*: The complete amino acid sequences of B1015-α, B1015-β and B1015-γ. Biol Chem Hoppe-Seyler 366: 87–98

Brunisholz RA, Bissig I, Wagner-Huber R, Frank G, Suter F, Niederer E and Zuber H (1989) The primary structures of the core antenna polypeptides from *Rhodopseudomonas marina*. Z Naturforsch C 44: 407–414

Bustamante PL and Loach PA (1994) Reconstitution of a functional photosynthetic receptor complex with isolated subunits of core light-harvesting complex and reaction centers. Biochemistry 33: 13329–13339

Bylina EJ, Robles SJ and Youvan DC (1988) Directed mutations affecting the putative bacteriochlorophyll-binding sites in light-harvesting I antenna of *Rhodobacter capsulatus*. Israel J Chem 28: 73–78

Chang MC, Callahan PM, Parkes-Loach PS, Cotton T and Loach PA (1990a) Spectroscopic characterization of the light-harvesting complex of *Rhodospirillum rubrum* and its structural subunit. Biochemistry 29: 421–429

Chang MC, Meyer L and Loach PA (1990b) Isolation and characterization of a structural subunit from the core light-harvesting complex of *Rhodobacter sphaeroides* 2.4.1 and *puc*705-BA. Photochem Photobiol 52: 873–881

Clayton RK (1962) Primary reactions in bacterial photosynthesis–I. The nature of light-induced absorbency changes in chromatophores; evidence for a special bacteriochlorophyll component. Photochem Photobiol 1: 2-1–210

Cogdell RJ (1985a) Carotenoid-bacteriochlorophyll interactions. Springer Ser Chem Phys 42: 62–66

Cogdell RJ (1985b) Carotenoids in photosynthesis. Pure and Appl Chem 57: 723–728

Cogdell RJ (1986) Light-harvesting complexes in the purple photosynthetic bacteria. Encycl Plant Physiol, New Ser 19: 252–259

Cogdell RJ and Frank HA (1987) How carotenoids function in photosynthetic bacteria. Biochim Biophys Acta 895: 63–79

Cogdell RJ and Scheer H (1985) Circular dichroism of light-harvesting complexes from purple photosynthetic bacteria. Photochem Photobiol 42: 669–678

Cogdell RJ and Thornber JP (1979) The preparation and characterization of different types of light-harvesting complexes from some purple bacteria. Ciba Found Symp 61 (new series): 61–79

Cogdell RJ and Thornber JP (1980) Light-harvesting pigment-protein complexes of purple photosynthetic bacteria. FEBS Lett 122: 1–8

Cogdell RJ, Parson WW and Kerr MA (1976) The type, amount, location and energy transfer properties of the carotenoid in reaction centres from *Rhodopseudomonas sphaeroides*. Biochim Biophys Acta 460: 83–93

Cogdell RJ, Lindsay G. Valentine J and Durant I (1982) A further characterization of the B890 light-harvesting pigment-protein complex from *Rhodospirillum rubrum* strain S1. FEBS Lett 150: 151–154

Cogdell RJ, Andersson PO and Gillbro T (1992) Carotenoid singlet states and their involvement in photosynthetic light-harvesting pigments. J Photochem Photobiol B: Biol 15: 105–112

Coleman WJ and Youvan DC (1990) Spectroscopic analysis of genetically modified photosynthetic reaction centers. Annu Rev Biophys Biophys Chem 19: 333–367

Cotton TM (1976) Spectroscopic investigations of chlorophyll *a* as donor and acceptor: A basis for chlorophyll *a* interactions in vivo. Ph.D. Thesis, Northwestern University, Evanston, Illinois, USA

Cuendet PA and Zuber H (1977) Isolation and characterization of a bacteriochlorophyll-associated chromatophore protein from *Rhodospirillum rubrum* G-9. FEBS Lett 79: 96–100

Davis CM, Bustamante PL and Loach PA (1995) Reconstitution of the bacterial core light-harvesting complexes of *Rhodobacter sphaeroides* and *Rhodospirillum rubrum* with isolated α- and β-polypeptides, bacteriochlorophyll and carotenoid. J Biol Chem 270: 5793–5804

Davis RC, Ditson SL, Fentiman AI and Pearlstein RM (1981) Reversible wavelength shifts of chlorophyll induced by a point charge. J Am Chem Soc 102: 6823–6826

Dawkins DJ, Ferguson LA and Cogdell RJ (1988) The structure of the 'core' of the purple bacterial photosynthetic unit. In: Scheer H and Schneider S (eds) Photosynthetic Light-Harvesting Systems, pp 115–127. Walter de Gruyter and Co., New York

Deinum G, Aartsma TJ, van Grondelle R and Amesz J (1989) Singlet-singlet excitation annihilation measurements on the antenna of *Rhodospirillum rubrum* between 300 and 4 K. Biochim Biophys Acta 976:63–69

Deisenhofer J and Michel H (1991) High resolution structures of photosynthetic reaction centers. Annu Rev Biophys Biophys Chem 20: 247–266

Dörge B, Klug G, Gad'on N, Cohen SN and Drews G (1990) Effects on the formation of antenna complex B870 of *Rhodobacter capsulatus* by exchange of charged amino acids in the N-terminal domain of the α-and β-pigment-binding proteins. Biochemistry 29: 7754–7758

Drews G (1985) Structure and functional organization of light-harvesting complexes and photochemical reaction centers in membranes of phototrophic bacteria. Microbiol Rev 49: 59–70

Dunker AK and Jones TC (1978) Proposed knobs-into-holes packing for several membrane proteins. Membr Biochem 2: 1–16

Duysens LNM (1952) Transfer of excitation energy in photosynthesis. Doctoral thesis, State University of Utrecht, The Netherlands

Eccles J. and Honig B (1983) Charged amino acids as spectroscopic determinants for chlorophyll in-vivo. Proc Natl Acad Sci USA 80: 4959–4962

Feher G and Okamura MY (1978) Chemical composition and properties of reaction centers. In: Clayton RK and Sistrom WR (eds) The Photosynthetic Bacteria, pp 349–386. Plenum Press, New York

Feick R and Drews G (1978) Isolation and characterization of light-harvesting bacteriochlorophyll-protein complexes from *Rhodopseudomonas capsulatus*. Biochim Biophys Acta 501: 499–513

Fowler GJS, Visschers RW, Grief GG, van Grondelle R and Hunter CN (1992) Genetically modified photosynthetic antenna complexes with blueshifted absorbance bands. Nature (London) 355: 848–850

Fowler GJS, Crielaard W, Visschers RW, van Grondelle R and Hunter CN (1993) Site-directed mutagenesis of the LHII light-harvesting complex of *Rhodobacter sphaeroides*: Changing βLys23 to Gln results in a shift in the 850 nm absorption peak. Photochem Photobiol 57: 2–5

Frank HA, Farhoosh R, Gebhard R, Lugtenburg J, Gosztola D and Wasielewski MR (1993) The dynamics of the S_1 excited states of carotenoids. Chem Phys Lett 207: 88–92

Ghosh R, Hauser H and Bachofen R (1988a) Reversible dissociation of the B873 light-harvesting complex from *Rhodospirillum rubrum* G9+. Biochemistry 27: 1004–1014

Ghosh R, Rosatzin T and Bachofen R (1988b) Subunit structure and reassembly of the light-harvesting complex from *Rhodospirillum rubrum* G9+. In: Scheer H and Schneider S (eds) Photosynthetic Light-Harvesting Systems, pp 93–102. Walter de Gruyter and Co., New York

Gingras G (1978) A comparative review of photochemical reaction center preparations from photosynthetic bacteria. In: Clayton RK and Sistrom WR (eds) The Photosynthetic Bacteria, Plenum Press, New York

Goedheer JC (1959) Energy transfer between carotenoids and bacteriochlorophylls in chromatophores of purple bacteria. Biochim Biophys Acta 35: 1–8

Gogel GE, Parkes PS, Loach PA, Brunisholz RA and Zuber H (1983) The primary structure of a light-harvesting bacteriochlorophyll-binding protein of wild-type *Rhodospirillum rubrum*. Biochim Biophys Acta 746: 32–39

Gogel GE, Michalski M, March H, Coyle S and Gentile L (1986) Covalent modification of lysines of the B880 light-harvesting protein of *Rs. rubrum*. Biochemistry 25: 7105–7109

Gottstein J and Scheer H (1983) Long-wavelength-absorbing forms of bacteriochlorophyll *a* in solutions of Triton X-100.

Proc Natl Acad Sci USA 80: 2231–2234

Guidotti G (1978) Membrane proteins: Structure and arrangement in the membrane. In: Andreoli TE, Hoffman JE and Fanestil JE (eds) Physiology of Membrane Disorders, pp 49–60. Plenum Medical, New York

Guthrie N, MacDermott G, Cogdell RJ, Freer AA, Isaacs NW, Hawthornthwaite AM, Halloren E and Lindsay JG (1992) Crystallization of the B800–850 Light-harvesting complex from *Rhodopseudomonas acidophila* strain 7750. J Mol Biol 224: 527–528

Hall RM, Kung MC, Fu M, Hales BJ and Loach PA (1973) Comparison of phototrap complexes from chromatophores of *Rhodospirillum rubrum*, *Rhodopseudomonas sphaeroides*, and the R-26 mutant of *Rhodopseudomonas sphaeroides*. Photochem Photobiol 18: 505–520

Hara M, Namba K, Hirata Y, Majima T, Kawamura S, Asada Y and Miyake J (1990) The photoreaction unit in *Rhodopseudomonas viridis*. Plant Cell Physiol. 31: 951–960

Hawthornthwaite AM and Cogdell RJ (1991) Bacteriochlorophyll binding proteins. In: Scheer H (ed) Chlorophylls, pp 493–528. CRC Press, Boca Raton, Florida

Heller BA (1992) Isolation and characterization of the core light-harvesting complex structural subunit of *Rb. capsulatus* and reconstitution of the subunit and complex using native and chemically modified polypeptides from *Rb. capsulatus* and *Rs. rubrum*. Ph.D Thesis, Northwestern University, Evanston, Illinois, USA

Heller BA and Loach PA (1990) Isolation and characterization of a subunit form of the B875 light-harvesting complex from *Rhodobacter capsulatus*. Photochem Photobiol 51: 621–627

Hunter CN, van Grondelle R and Olsen JD (1989) Photosynthetic antenna proteins: 100 ps before photochemistry starts. Trends Biol Sci 14: 72–76

Jay F, Lambillotte M, Stark W and Mühlethaler K (1984) The preparation and characterization of native photoreceptor units from the thylakoids of *Rhodopseudomonas viridis*. EMBO J 3: 773–776

Jensen A, Aasmundrud O and Eimhjellen KE (1964) Chlorophylls of photosynthetic bacteria. Biochim Biophys Acta 88: 466–479

Jirsakova V, and Reiss-Husson F (1993) Isolation and characterization of the core light-harvesting complex B875 and its subunit form, B820, from *Rhodocyclus gelatinosus*. Biochim Biophys Acta 1138: 301–308

Jirsakova V and Reiss-Husson (1994) A specific carotenoid is required for reconstitution of the *Rubrivivax gelatinosus* B875 light harvesting complex from its subunit form B820. FEBS Lett 353: 151–154

Ke B (1971) Carotenoproteins. Meth Enzymol 23: 624–636

Kerfeld CA, Yeates TO and Thornber JP (1994) Biochemical and spectroscopic characterization of the reaction center-LHI complex and the carotenoid-containing B820 subunit of *Chromatium purpuratum*. Biochim Biophys Acta 1185: 193–202

Kiley PJ and Kaplan S (1988) Molecular genetics of photosynthetic membrane biosynthesis in *Rhodobacter sphaeroides*. Microbiol Rev 52: 50–69

Kiley PJ, Donohue TJ, Havelka WA and Kaplan S (1987) DNA sequence and in vitro expression of the B875 light-harvesting polypeptides of *Rb. sphaeroides*. J Bacteriol 169: 742–750

Kramer HJM, Pennoyer JD, van Grondelle R, Westerhuis WHJ, Niederman RA and Amesz J (1984) Low-temperature optical properties and pigment organization of the B875 light-harvesting bacteriochlorophyll-protein complex of purple photosynthetic bacteria. Biochim Biophys Acta 767: 335–344

Landon M (1977) Cleavage at aspartyl-prolyl bonds. Meth Enzymol 47: 145–149

Loach PA (1980) Bacterial reaction center (RC) and photoreceptor complex (PRC) preparations. Meth Enzymol 69: 155–172

Loach PA and Sekura DL (1968) Primary photochemistry and electron transport in *Rhodospirillum rubrum*. Biochemistry 7: 2642–2649

Loach PA, Androes GM, Maksim AF and Calvin M (1963) Variation in electron paramagnetic resonance signals of photosynthetic systems with the redox level of their environment. Photochem Photobiol 2: 443–454

Loach PA, Hadsell RM, Sekura DL and Stemer A (1970a) Quantitative dissolution of the membrane and preparation of photoreceptor subunits from *Rhodospirillum rubrum*. Biochemistry 9: 3127–3135

Loach PA, Sekura DL, Hadsell RM and Stemer A (1970b) Quantitative dissolution of the membrane and preparation of photoreceptor subunits from *Rhodopseudomonas sphaeroides*. Biochemistry 9: 724–733

Loach PA, Parkes PS, Miller JF, Hinchigeri S and Callahan PM (1985) Structure-function relationships of the bacteriochlorophyll-protein light-harvesting complex of *Rhodospirillum rubrum*. In: Arntzen C, Bogorad L, Bonitz S and Steinbeck K (eds) Molecular Biology of the Photosynthetic Apparatus, pp 197–209. Cold Spring Harbor Laboratory, Cold Spring Harbor, New York

Loach PA, Parkes-Loach PS, Chang MC, Heller BA, Bustamante PL and Michalski T (1989) Comparison of structural subunits of the core light-harvesting complexes of photosynthetic bacteria. In: Drews G and Dawes EA (eds) Molecular Biology of Membrane-Bound Complexes in Phototrophic Bacteria, pp 235–244. Plenum Press, New York

Loach PA, Parkes-Loach PS, Davis CM and Heller BA (1994) Probing protein structural requirements for formation of the core light-harvesting complex of photosynthetic bacteria using hybrid reconstitution methodology. Photosynth Res 40: 231–245

McDermott G, Prince SM, Freer AA, Hawthornthwaite-Lawless AM, Papiz MZ, Cogdell RG and Isaacs NW (1995) Crystal structure of an integral membrane light-harvesting complex from photosynthetic bacteria. Nature 374: 517–521

McElroy JD, Feher G and Mauzerall DC (1969) On the nature of the free radical formed during the primary process of bacterial photosynthesis. Biochim Biophys Acta 172: 180–183

Meadows KA, Iida K, Recchia PA, Heller BA, Antonio B, Nango M and Loach PA (1995) Enzymatic and chemical cleavage of the core light-harvesting polypeptides of photosynthetic bacteria: toward the determination of the minimal polypeptide size and structure required for subunit and light-harvesting complex formation, Biochemistry34: 1559–1574

Meckenstock RU, Brunisholz RA and Zuber H (1992) The light-harvesting core-complex and the B820-subunit from *Rhodopseudomonas marina*. Part 1. Purification and characterization. FEBS Lett 311: 128–134

Miller JF, Hinchigeri SB, Parkes-Loach PS, Callahan PM, Sprinkle JR, Riccobono JR and Loach PA (1987) Isolation and characterization of a subunit form of the light-harvesting complex of *Rhodospirillum rubrum*. Biochemistry 26: 5055–5062

Monger TG and Parson WW (1977) Singlet-triplet fusion in *Rhodopseudomonas sphaeroides* chromatophores. A probe of the organization of the photosynthetic apparatus. Biochim Biophys Acta 460: 393–407

Noguchi T, Hayashi H and Tasumi M (1990) Factors controlling the efficiency of energy transfer from carotenoids to bacteriochlorophyll in purple photosynthetic bacteria. Biochim Biophys Acta 1017: 280–290

Nunn RS, Artymiuk PJ, Baker PJ, Rice DW and CN Hunter (1992) Purification and crystallization of the light harvesting LHI complex from *Rhodobacter sphaeroides*. J Mol Biol 228: 1259–1262

Ogawa T and Vernon LP (1969) A fraction from *Anabaena variabilis* enriched in the reaction center chlorophyll P_{700}. Biochim Biophys Acta 180: 334–346

Olsen JD, Sockalingum GD, Robert B and Hunter CN (1994) Modification of a hydrogen bond to a bacteriochlorophyll molecule in the light-harvesting 1 antenna of *Rhodobacter sphaeroides*. Proc Natl Acad Sci USA 91: 7124-7128

Papiz MZ, Hawthornthwaite AM, Cogdell RJ, Woolley KJ, Wightman PA, Ferguson LA and Lindsay JG (1989) Crystallization and characterization of two crystal forms of the B800–850 light-harvesting complex from *Rhodopseudomonas acidophila* strain 10050. J Mol Biol 209: 833–835

Parkes-Loach PS, Sprinkle JR and Loach PA (1988) Reconstitution of the B873 light-harvesting complex of *Rhodospirillum rubrum* from the separately-isolated α-and β-polypeptides and bacteriochlorophyll *a*. Biochemistry 27: 2718–2727

Parkes-Loach PS, Michalski TJ, Bass W, Smith U and Loach PA (1990) Probing the bacteriochlorophyll binding site by reconstitution of the light-harvesting complex of *Rhodospirillum rubrum* with bacteriochlorophyll *a* analogues. Biochemistry 29: 2951–2960

Parkes-Loach PS, Jones SM and Loach PA (1994) Probing the structure of the core light-harvesting complex (LHI) of *Rhodopseudomonas viridis* by dissociation and reconstitution methodology. Photosynth Res 40: 247–261

Pearlstein RM (1988) Interpretation of optical spectra of bacteriochlorophyll antenna complexes. In: Scheer H and Schneider S (eds) Photosynthetic Light-Harvesting Systems, pp 555–566. Walter de Gruyter and Co., New York

Picorel R and Gingras G (1988) Preparative isolation and characterization of the B875 complex from *Rhodobacter sphaeroides* 2.4.1. Biochem Cell Biol 66: 442–448

Picorel R, Bélanger G and Gingras G (1983) Antenna holochrome B880 of *Rhodospirillum rubrum* S1. Pigment, phospholipid, and polypeptide composition. Biochemistry 22: 2491–2497

Picorel R, Lefèbvre S and Gingras G (1984) Oxido-reduction of B800–850 and B880 holochromes isolated from three species of photosynthetic bacteria as studied by electron-paramagnetic resonance and optical spectroscopy. Eur J Biochem 142: 305–311

Racker E (1985) Reconstitutions of Transporters, Receptors, and Pathological States. Academic Press, New York

Robert B and Lutz M (1985) Structures of antenna complexes of several Rhodospirillales from their resonance Raman spectra. Biochim Biophys Acta 807: 10–23

Sauer K (1978) Photosynthetic membranes. Acc Chem Res 11: 257–264

Sauer K and Austin LA (1978) Bacteriochlorophyll-protein complexes from the light-harvesting antenna of photosynthetic bacteria. Biochemistry 17: 2011–2019

Scherz A and Parson WW (1986) Interactions of the bacteriochlorophylls in antenna bacteriochlorophyll-protein complexes of photosynthetic bacteria. Photosynth Res 9: 21–32

Schneour EA (1962) Carotenoid pigment conversion in *Rhodopseudomonas sphaeroides*. Biochim Biophys Acta 62: 534–540

Schmidt K (1978) Biosynthesis of carotenoids. In: Clayton RK and Sistrom WR (eds) The Photosynthetic Bacteria, pp 729–750. Plenum Press, New York

Scolnik PA and Marrs BL (1987) Genetic research with photosynthetic bacteria. Annu Rev Microbiol 41: 703–726

Sebban P (1985) Transfer and trapping of the light excitation energy in purple bacteria. Physiol Vég 23: 449–462

Stiehle H, Cortez N, Klug G and Drews G (1990) A negatively charged N-terminus in the α-polypeptide inhibits formation of light-harvesting complex I in *Rhodobacter capsulatus*. J Bacteriol 172: 7131–7137

Sturgis JN and Robert B (1994) Thermodynamics of membrane polypeptide oligomerization in light-harvesting complexes and associated structural changes. J Mol Biol 238: 445–454

Sundström V and van Grondelle R (1991) Dynamics of excitation energy transfer in photosynthetic bacteria. In: Scheer H (ed) Chlorophylls, pp 1098–1124. CRC Press, Boca Raton, Florida

Tadros MH, Suter F, Seydewitz H, Witt I, Zuber H and Drews G (1984) Isolation and complete amino-acid sequence of the small polypeptide from light-harvesting pigment-protein complex I (B870) of *Rhodopseudomonas capsulata*. Eur J Biochem 138: 209–212

Tadros MH, Frank G, Zuber H and Drews G (1985) The complete amino acid sequence of the large bacteriochlorophyll-binding polypeptide B870-a from the light-harvesting complex B870 of *Rb. capsulata*. FEBS Lett 190: 41–44

Theiler R, Suter F, Pennoyer JD, Zuber H and Niederman RA (1985) Complete amino acid sequence of the B875 light-harvesting protein of *Rhodopseudomonas sphaeroides* strain 2.4.1. Comparison with R26.1 carotenoidless-mutant strain. FEBS Lett 184: 231–236

Thompson MA and Fajer J (1992) Calculations of bacteriochlorophyll *g* primary donor in photosynthetic heliobacteria. How to shift the energy of the phototrap by 2000 cm^{-1}. J Phys Chem 96:2933–2935

Thompson MA, Zerner MC and Fajer J (1991) A theoretical examination of the electronic structure and excited states of the bacteriochlorophyll *b* dimer from *Rhodopseudomonas viridis*. J Phys Chem 95:5693–5700

Thornber JP (1971) The photochemical reaction centre of *Rhodopseudomonas viridis*. Meth Enzymol 23: 688–691

Thornber JP (1986) Biochemical characterization and structure of pigment-proteins of photosynthetic organisms. Encycl Plant Physiol, New Ser 19: 98–142

Thornber JP, Trosper TL and Strouse CE (1978) Bacteriochlorophyll in vivo: Relationship of spectral forms to specific membrane components. In: Clayton RK and Sistrom WR (eds) The Photosynthetic Bacteria, pp 133–160. Plenum Press, New York

Thornber JP, Cogdell RJ, Pierson BK, and Seftor REB (1983) Pigment-protein complexes of purple photosynthetic bacteria: an overview. J Cell Biochem 23: 159–169

Tinoco Jr. I, Sauer K and Wang JC (1985) Physical Chemistry: Principles and Applications in Biological Sciences. Prentice-Hall, Inc., Englewood Cliffs, NJ

Vadeboncoeur C, Noël H, Poirier L, Cloutier Y and Gingras G (1979) Photoreaction center of photosynthetic bacteria. 1. Further chemical characterization of the photoreaction center from *Rhodospirillum rubrum*. Biochemistry 18: 4301–4308

van Mourik F, van der Oord CJR, Visscher KJ, Parkes-Loach PS, Loach PA, Visschers RW and van Grondelle R (1991) Exciton interactions in the light-harvesting antenna of photosynthetic bacteria studied with triplet-singlet spectroscopy and singlet-triplet annihilation of the B820 subunit form of *Rhodospirillum rubrum*. Biochim Biophys Acta 1059: 111–119

van Mourik F, Corten EPM, van Stokkum IHM, Visschers RW, Loach PA, Kraayenhof R and van Grondelle R (1992) Self assembly of the LH-1 antenna of *Rhodospirillum rubrum*: A time-resolved study of the aggregation state of the B820 su unit form. In: Murata N (ed) Research in Photosynthesis, Vol I, pp 101–104. Kluwer Academic Publishers, Dordrecht

van Mourik F, Visscher KJ, Mulder JM and van Grondelle R (1993) Spectral inhomogeneity of the light-harvesting antenna of *Rhodospirillum rubrum* probed by triplet-minus-singlet spectroscopy and singlet-triplet annihilation at low temperatures. Photochem Photobiol 57: 19–23

van Niel CB, Goodwin TW and Sissins ME (1956) Studies in carotenogenesis 21. The nature of the changes in carotenoid synthesis in *Rhodospirillum rubrum* during growth. Biochem J 63: 408–412

Vasil'ev BG (1991) Heterogeneity of the photoreceptor complex RC-B890 in the sulfur purple bacterium *Chromatium minutissimum*. Biol Membr 8: 240–248

Vernon LP and Garcia AF (1967) Pigment-protein complexes derived from *Rhodospirillum rubrum* chromatophores by enzymatic digestion. Biochim Biophys Acta 143: 144–153

Visschers RW, Chang MC, van Mourik F, Parkes-Loach PS, Heller BA, Loach PA and van Grondelle R (1991) Fluorescence polarization and low-temperature absorption spectroscopy of a subunit form of light-harvesting complex I from purple photosynthetic bacteria. Biochemistry 30: 5734–5742

Visschers RW, Nunn R, Calkoen F, van Mourik F, Hunter CN, Rice DW and van Grondelle R (1992) Spectroscopic characterization of B820 subunits from light-harvesting complex I of *Rhodospirillum rubrum* and *Rhodobacter sphaeroides* prepared with the detergent *n*-octyl-*rac*-2,3-dipropylsulfoxide. Biochim Biophys Acta 1100: 259–266

Visschers RW, van Mourik F, Monshouwer R and van Grondelle R (1993a) Inhomogeneous spectral broadening of the B820 subunit form of LHI. Biochim Biophys Acta 1141: 238–244

Visschers RW, van Grondelle R and Robert B (1993b) Resonance Raman spectroscopy of the B820 subunit of the core antenna from Rhodospirillum rubrum G9. Biochim Biophys Acta (Visschers RW, PhD. Thesis, pp 77–86. Free University of Amsterdam)

Visschers RW, van Grondelle R and Robert B (1993c) Resonance Raman spectroscopy of the B820 subunit of the core antenna from *Rhodospirillum rubrum* G9. Biochim Biophys Acta 1183: 369-373

Wellington CL, Bauer CE and Beatty JT (1992) Photosynthesis gene superoperons in purple nonsulfur bacteria: The tip of the iceberg? Can J Microbiol 38: 20–27

Westerhuis WHJ, Vos M, van Dorssen RJ, van Grondelle R, Amesz J and Niederman RA (1989) Associations of pigment-protein complexes in phospholipid enriched bacterial photosynthetic membranes. In: Biacs PA, Gruiz K and Kremmer T (eds) Biological Role of Plant Lipids, pp 227–231. Plenum Publishing Corp., New York

Wiessner C, Dunger I and Michel H (1990) Structure and transcription of the genes encoding the B1015 light-harvesting complex β and a subunits and the photosynthetic reaction center L, M, and cytochrome c subunits from *Rhodopseudomonas viridis*. J Bacteriol 172: 2877–2887

Yeates TO, Komiya H, Rees DC, Allen JP and Feher G (1987) Structure of the reaction center from *Rhodobacter sphaeroides* R-26: Membrane-protein interactions. Proc Natl Acad Sci USA 84: 6438–6442

Youvan DC, Alberti M, Begusch H, Bylina EJ and Hearst JE (1984) Reaction center and light-harvesting I genes from *Rhodopseudomonas capsulatus*. Proc Natl Acad Sci USA 81: 189–192

Zuber H (1986) Structure of light-harvesting antenna complexes of photosynthetic bacteria, cyanobacteria and red algae. Trends Biol Sci 11: 414–419

Zuber H and Brunisholz RA (1991) Structure and function of antenna polypeptides and chlorophyll-protein complexes: Principles and variability. In: Scheer H (ed) Chlorophylls, pp 627–703. CRC Press, Boca Raton, Florida

Chapter 22

Genetic Manipulation of the Antenna Complexes of Purple Bacteria

C. Neil Hunter
Krebs Institute and Robert Hill Institute for Photosynthesis, Department of Molecular Biology and Biotechnology, University of Sheffield, Sheffield, S10 2UH, U.K.

Summary

Photosynthetic bacteria have relatively simple photosystems, and they are therefore attractive models for the study of the absorption of light, and the transfer and trapping of excitation energy. This simplicity extends to the organization and expression of genetic information, and it would be a missed opportunity if molecular genetics were not used to examine the assembly, organization and function of light harvesting complexes. Furthermore, directed mutagenesis of genes encoding structural subunits of the antenna, in combination with various forms of spectroscopy, provides a means of testing structural models for these complexes, and of enhancing our understanding of existing crystallographic data.

The potential of using genetic techniques to provide partial photosystems was realized over thirty years ago

R. E. Blankenship, M. T. Madigan and C. E. Bauer (eds): Anoxygenic Photosynthetic Bacteria, pp. 473–501.

by, among others, Clayton and Sistrom, and indeed the carotenoidless, LH2- deficient mutant R26 isolated by Clayton in 1961 is still used today. The availability of mutants lacking either the peripheral LH2 complex or the core LH1 complex provides a means of simplifying spectroscopic studies that would otherwise be complicated by overlapping signals, as well as a way of rapidly and efficiently obtaining a given complex using detergent-based isolation procedures. This chapter provides an account of the way in which various antenna mutants have been used for such purposes, concentrating on the bacteria for which an array of molecular genetic tools have been developed, *Rhodobacter (Rb.) capsulatus* and *Rb. sphaeroides*. The latter part of the chapter concentrates on more recent developments in which protein engineering has been used to modify the assembly and the spectroscopic properties of LH1 and LH2, by altering various aminoacids singly, or in groups. Further progress in this rapidly moving field is aided by new developments in spectroscopy, as well as by the application of combinatorial mutagenesis and the heterologous expression of LH genes. It is expected that molecular genetics will play an increasingly important role in the study of light harvesting complexes.

I. Introduction

A. Early Examples of Valuable Mutants

The use of genetic techniques to study bacterial light harvesting complexes has a long history, and the early work of Clayton and Sistrom contains several examples of using a genetic approach to advance the study of photosynthesis. This chapter will briefly survey this early work before moving on to the possibilities recently opened up by the application of molecular genetic and cloning techniques. Such a survey is particularly appropriate, since the forerunner of this volume was *The Photosynthetic Bacteria*, edited by Clayton and Sistrom in 1978.

The use of bacterial mutations has had a profound impact on biochemistry in general, and mutational analysis of *Escherichia coli*, for example, has established the order of many biochemical pathways and has facilitated the eventual identification of hundreds of genes. The small genome size of *E. coli* and the availability of gene transfer techniques lend themselves to this kind of genetic analysis, and so it would be surprising, and a missed opportunity, if similar approaches could not be used to analyze the most important and fundamental process on Earth, photosynthesis. The existence of bacteria that can photosynthesize would seem to offer almost limitless possibilities to explore the molecular biology of this process. This is certainly the case now for the study of light harvesting complexes. However, thirty years ago, there were restrictions imposed by the lack of molecular cloning techniques. Despite this, several elegant experiments were performed that turned out to have lasting significance.

The earliest example of the production of mutant antennas was in 1956, when Griffiths and Stanier isolated around 50 mutants of *Rhodobacter* (*Rb.*) (then *Rhodopseudomonas* (*Rp.*)) (Griffiths and Stanier, 1956; Sistrom et al., 1956). They noticed that some of these, such as UV-33, had alterations in carotenoid content, and that this was reflected in the absorbance spectrum in the near infrared arising from the light harvesting bacteriochlorophylls. In particular, the major absorbance peaks at 800 and 850 nm, attributable to the peripheral LH2 complex, were absent; this arose from a blue-green strain with the major peak at approximately 870 nm, which we now know to be the light harvesting LH1 complex. Interestingly, a minor peak of 800 nm could be seen which arises from the monomeric BChl800 molecules within the reaction center, so this was an early sighting of reaction center absorbance.

At the time, the concept of a reaction center, first advanced by Duysens in 1952, was a relatively recent one. His studies involved *Rhodospirillum* (*Rs.*) *rubrum*, which can be thought of as a naturally occurring $LH2^-$ strain. Indeed, the contribution of $LH2^-$ strains was to have a crucial bearing on subsequent progress toward the isolation of the reaction center (Reed 1969; Clayton and Wang, 1971; Feher, 1971) , and the eventual elucidation of the structure by X-ray crystallography (Deisenhofer et al., 1985; Allen et al., 1987, Chang et al., 1991). In each case the purification of the reaction center was facilitated either by the use of another naturally-

Abbreviations: B800, 850, 875 – bacteriochlorophyll absorbance in vivo at 800, 850 and 875 nm; *E. – Escherichia;* FP – fluorescence polarisation; I – reaction center pheophytin acceptor; LDAO – lauryl dimethyl amine oxide; LDS, SDS – lithium, sodium dodecyl sulfate; LH – light harvesting; P – reaction center primary donor; PCR – polymerase chain reaction; Q – the reaction center quinone, Q_A; *Rb. – Rhodobacter;* RC – reaction center; *Rc. – Rhodocyclus; Rp. – Rhodopseudomonas; Rs. – Rhodospirillum;* WT – wild-type occurring LH2⁻ strain, *Rp. viridis* or by the LH2⁻ mutant R26, because the absence of LH2 eliminates the need to remove this complex in a purification protocol. R26 has occupied an important role in bacterial photosynthesis for these reasons, although it was originally isolated as strain CC1R26 in a study of photooxidative killing in *Rb. sphaeroides* (Clayton and Smith, 1961).

The work of Sistrom and Clayton in 1963 produced a number of other interesting mutations in the bacterial photosynthetic apparatus; the mutant PM-8 was shown to lack the photochemical activity of the reaction center and contained diminished levels of the LH1 light harvesting complex. This study of PM-8 provided important evidence for the role of the reaction center 'P870' in photosynthesis (Sistrom and Clayton, 1963). As part of this study a derivative of PM-8, PM-8:bg58, was obtained which also lacked reaction center activity as well as the major absorbance peaks of 800 and 850 nm attributable to the major light harvesting complex, LH2. Thus PM8:bg58 was the first LH1-only mutant. Subsequently, Clayton and Haselkorn (1972) demonstrated that PM8 lacked the H, M and L subunits of the reaction center. PM8, or at least a more pigmented variant called PM8dp, was to find further use as the source of membranes for reconstitution in vitro with purified reaction centers. It was already known that addition of reaction centers to membranes from the bacteriochlorophyll-deficient mutant 01 resulted in restoration of cyclic light-driven electron flow (Hunter and Jones, 1979); the performance of the system could be improved by adding either the purified light harvesting LH2 complex to the reconstitution, or by using membranes from PM8dp; this demonstrated that the reaction centers can insert into the PM8dp membrane and receive excitation energy from the endogenous antenna (Hunter et al., 1979b).

At this stage, the genetics of *Rb. sphaeroides* were much less well understood than for *Rb. capsulatus* and although various kinds of gene transfer had been successfully carried out with *Rb. sphaeroides* (for example, Sistrom, 1977; Miller and Kaplan, 1978; Pemberton and Bowen, 1981) the availability of the gene transfer agent which is specific for *Rb. capsulatus* (Yen et al., 1979) facilitated genetic mapping of a region of the chromosome known to contain most of the information for photosynthesis (Yen and Marrs., 1976). Thus, light harvesting mutants, lacking either LH2 or LH1 complexes were reported for *Rb. capsulatus* quite early on (for example, Feick and Drews, 1978), well before further advances in genetics enabled gene-specific deletions and site directed mutagenesis to be undertaken.

B. A Brief Summary of the Molecular Genetics Underlying the Production of Mutant Antennas

A detailed consideration of the molecular genetics of the purple bacteria is outside the scope of this chapter, and for this information the reader should refer to Chapters 57, 58 and 59 by Beatty, Bauer and Klug in this volume. It is intended to be a quick summary for those whose main interest is in antenna function, rather than gene regulation. Thus, it will suffice to say that the genes for the major objects of study, LH1 and LH2, have been cloned, mapped and sequenced. This includes the LH2 genes of *Rb. sphaeroides* (Ashby et al., 1987; Kiley and Kaplan, 1987) and of *Rb. capsulatus* (Youvan and Ismail, 1985). LH1 genes were sequenced by Kiley et al., (1987) for *Rb. sphaeroides*, by Youvan et al. (1984) for *Rb. capsulatus*, and by Bérard et al. (1986) for *Rs. rubrum*. In the case of LH1, the genes (*pufBA*) lie upstream of the *pufLM* genes encoding two of the bacterial reaction center subunits and these *puf* genes form part of a single transcriptional unit (Zhu et al., 1986; Belasco et al., 1985; Bélanger and Gingras, 1988). In both *Rb. capsulatus* and *Rb. sphaeroides* the LH1-RC genes lie at one end of the photosynthesis gene cluster, the other end of which is conveniently defined by *puhA* encoding the reaction center H subunit (Zsebo and Hearst, 1984, Coomber and Hunter, 1989). The LH2 genes (*pucBA*) are situated approximately 18 kb from *puhA* in *Rb. sphaeroides* (Suwanto and Kaplan, 1989) but lie over 1000 kb away on the other side of the chromosome in *Rb. capsulatus* (Fonstein and Haselkorn, 1993).

Many spectroscopic studies on antenna mutants in the last ten years have been conducted on point mutants where the mutation was known to reside within a given gene, although the precise location of the alteration was unknown. This situation has now improved somewhat, so that genetically defined deletions or site directed alterations can be studied. Moreover, protein engineering studies on light harvesting (or reaction center) complexes require a well-defined genetic background: the altered genes are expressed on plasmids, that is independently of the chromosome, and the native copies of the gene have been excised from the chromosome and have been replaced by a selectable marker which confers resistance to a particular antibiotic. This is an important point, since the spectral effects arising from alterations in the plasmid-borne copy of the gene should not be superimposed on those from the normal version of the gene. A point mutation which merely inactivates the chromosomal copy of the gene but leaves it in place will not suffice, because it still leaves potential sites for recombination with the plasmid-borne altered copy since the sequences are still likely to be more than 99% homologous. Such events in the purple bacteria have been documented and cannot be ruled out since recombination-deficient strains are not widely available. Recombination between an inactive chromosomal copy containing a point mutation, and a plasmid-borne copy of a gene with a site-directed alteration could in theory produce a normal copy.

Youvan and co-workers were the first to construct deletion strains lacking either LH2 genes, or LH1-RC genes (Youvan et al., 1985) and they went on to use the LH1-RC deletion-complementation system to great effect, by producing a series of fascinating and important site-directed mutants in the reaction center complex, reviewed by Coleman and Youvan (1990). The LH2 genes (*pucBA*) of *Rb. sphaeroides* were deleted by Burgess et al., (1989) who replaced *pucBA* with a kanamycin resistance cassette, to produce strain DBC1. Subsequently, DBCΩ was constructed by using a streptomycin resistance cassette instead (Jones et al., 1992a). In each case the strain is LH2$^-$ LH1$^+$ RC$^+$. Conversely, the LH1-RC genes of *Rb. sphaeroides* (*pufBALMX*) have been deleted and replaced with a kanamycin resistance gene to produce strain DPF2, which is therefore LH2$^+$ LH1$^-$ RC$^-$. Finally, a combination of the DBCΩ and DPF2 deletions produced strain DD13 which is LH2$^-$ LH1$^-$ RC$^-$ (Jones et al., 1992a). A summary of these deletions, which in principle also applies to *Rb. capsulatus*, appears in Fig. 1. Despite its outwardly featureless and uninteresting phenotype, DD13 is valuable since it acts as a null background for the expression of LH2, LH1 or RC genes either singly or in any combination. In addition, three different carotenoid variants of DD13 are possible in which phytoene, neurosporene or spheroidenone are the major pigments. Thus many different phenotypes are available for the examination of site directed alterations in photosynthetic complexes (Jones et al., 1992a). Some of these possibilities are represented in Fig. 2. A comparable degree of flexibility exists for *Rb. capsulatus*, particularly since a double deletion strain has recently been constructed (unpublished; referred to in Goldman and Youvan, 1992). The main reason for constructing these various deletion strains is to provide a 'clean' background for the expression of mutated genes, although it is worth noting that they are interesting in their own right because they provide 'partial' photosystems for study, as will be mentioned later on.

The plasmids that are used to express mutated genes are generally based on derivatives of the broad host range plasmid RK2; the significance of the host range is that the plasmid will replicate equally well in either *E. coli* or, for example, *Rb. sphaeroides*. RK2 is a large (~54 kb) plasmid and so relatively small derivatives have been constructed, such as RK415 which is 10.5 kb in size (Keen et al., 1988). This plasmid confers tetracycline resistance on the recipient and contains several useful cloning sites. Once the gene(s) of interest is cloned into such a vector, it would normally be desirable to perform several rounds of directed mutagenesis in order to introduce a series of unique restriction sites flanking each individual gene so that each one can be removed, mutated and sequenced independently of the others (Jones et al., 1992a). The gene can be mutated by standard procedures such as that described by Kunkel (1985), reintroduced into the expression plasmid, and then transferred, usually by conjugation, into the recipient deletion strain.

It should be mentioned that as a result of expressing a given gene chromosomally or extrachromosomally there are now two different ways to obtain, for example, an LH2-only strain: either the LH1-RC genes could be deleted from the chromosome, leaving the LH2 genes, or an LH2-only expression plasmid could be introduced into a double deletion (LH2$^-$ LH1$^-$ RC$^-$) background. The only difference would be that the latter strain would probably express the LH2 complex to higher levels since the number of plasmids per cell will typically be 4–6 for pRK415 derivatives (Davis et al., 1988), compared to the copy number of one for the single, chromosomal copy of the LH2 genes. In a direct comparison of the levels of antenna in two such strains, LH1-only in this case, the cellular level of LH1 was 2–3-fold higher for the plasmid-expressed genes than for strain M2192 containing a single genomic copy (Hunter et al., 1991). It is not unlikely that at this level of abundance the membrane could run into 'capacity constraints' when faced with such high levels of antenna. However, plasmid-based expression is still a useful way of

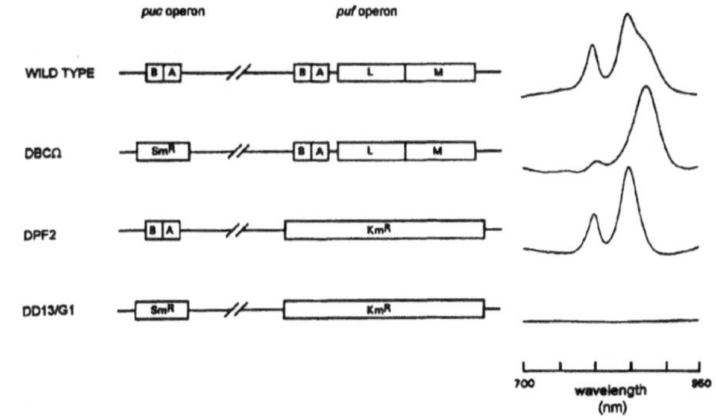

Fig. 1. A scheme showing deletions of the *puc* (LH2) and *puf* (LH1-RC) operons and the attendant affects on absorbance spectra.

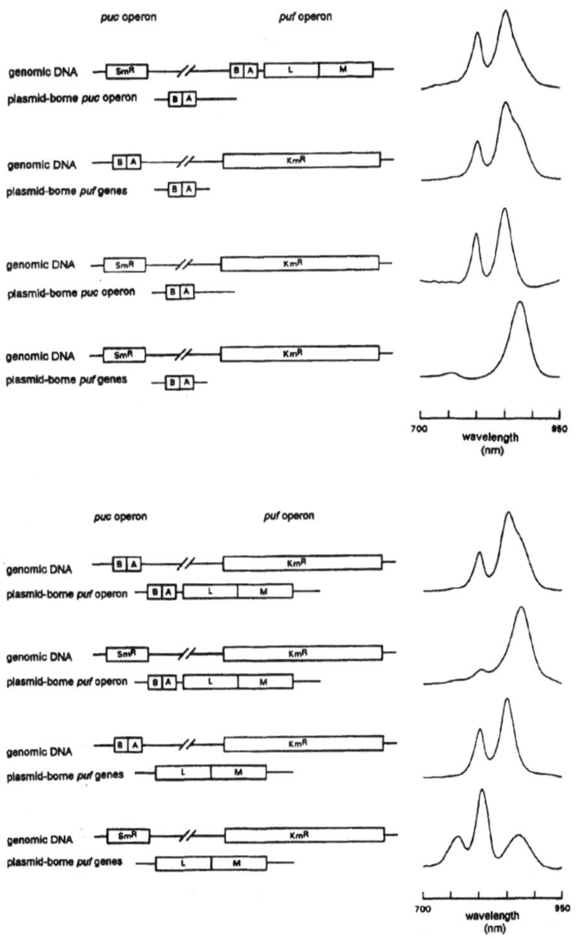

Fig. 2. A scheme showing the complementation of deletion strains deficient in LH2 and LH1-RC complexes, by plasmid-borne genes. The restoration of the appropriate complex can be seen by inspection of the absorbance spectra.

increasing the levels of complex, for example, crystallization trials (see section IIIA).

II. The Effect of Genetic Alteration of Antennas on the Morphology of Intracytoplasmic Membranes

Alteration of the normal composition of the photosynthetic unit, notably by abolishing the synthesis of the LH2 complex, has a radical effect on the morphology of the intracytoplasmic membranes. A more detailed discussion of the organization of cellular membranes appears in Chapter 12 by Drews and Golecki in this volume; for the purposes of discussing mutated antennas it is sufficient to note that the normal membrane morphology can vary widely between different genera. In *Rb. sphaeroides* and *Rb. capsulatus* a fully pigmented cell contains hundreds of spherical or near-spherical invaginations. The loss of the LH1-RC core appears to have little effect on the presence of these invaginations, as judged by electron microscopy of strain NF57, for example (Hunter et al., 1988). However, the loss of the LH2 complex can produce gross distortions resulting in flattened lamellae or tubular membranes that can sometimes spiral round the cytoplasm (Kiley et al., 1988.; Hunter et al., 1988). These observations were made earlier by Lommen and Takemoto (1978), who showed that the carotenoidless $LH2^-$ strain R26 also displays lamellar membrane morphology. In apparent contradiction of the observations on carotenoid containing $LH2^-$strains such as the point mutant M21, Oelze and co-workers found a normal spherical morphology (Golecki et al., 1991) in the $LH2^-$ strain 19. More recent work by Gibson et al. (1992) employed deletion and insertion mutagenesis of the LH2 structural genes *pucBA* (encoding the β and α subunits of LH2, respectively), as well as *pucC* which lies downstream of *pucBA*, and is essential for LH2 synthesis. The morphology of intracytoplasmic membranes was disrupted under conditions where the *pucBAC* operon is disrupted, but no clear correlation was observed between morphology and the position of the mutation. Whereas it is true to say that loss of LH2 can markedly affect membrane morphology, this is not always the case and it will be necessary to construct further mutations in order to define the relationship between LH2 biosynthesis and membrane morphology.

Finally, it should be mentioned that although LH2 could exert some effect on membrane morphology in some species of photosynthetic bacteria, the naturally LH2-deficient species *Rs. rubrum* still contains spherical invaginations (Drews and Oelze., 1981), so it might not be possible to draw any far-reaching conclusions on membrane morphology from a study of *Rhodobacter*.

III. The Use of LH1-Only and LH1-RC Mutants to Facilitate Biochemical and Structural Studies

This is an area of research that is to some extent under-exploited. In theory, the availability of mutants capable of synthesizing only one of the three possible pigment-protein complexes should be an invaluable aid for the purification of antenna and reaction center complexes. In practice, however, biochemical techniques are able to overcome these problems by physically separating the complexes by, for example, column chromatography or by electrophoretic techniques. In addition, some detergents can selectively destabilize a particular unwanted complex, which removes one purification problem from the protocol. This is illustrated by the observation that some of the most commonly used detergents such as lauryldimethylamine oxide (LDAO) destabilize the LH1 complex, leaving only LH2 and the RC, which can be easily separated and purified. Thus the RC and LH2 complexes of *Rb. sphaeroides* were purified relatively early on (Clayton and Wang, 1971; and Clayton and Clayton, 1972) whereas another eight years elapsed before a relatively mild procedure was found for the purification of LH1 from the wild-type, using lithium dodecyl sulphate-polyacrylamide gel electrophoresis in the dark at 4 °C (Broglie et al., 1980). It has already been mentioned in section I how the $LH2^-$ $LH1^+$ RC^+ carotenoidless mutant R26 was able to provide pure reaction centers relatively easily, because once LDAO had destroyed LH1, RCs are the only complex present. In subsequent work, Sauer and Austin (1978) were able to turn R26 to further advantage by purifying the carotenoidless LH1 complex. Other $LH2^-$ $LH1^+$ RC^+ mutants have been reported subsequently. For example, Kaufmann et al. (1984) used transposon Tn5 mutagenesis to disrupt LH2 biosynthesis and produce an $LH1^+$ RC^+ strain, NK3. Bachmann et al. (1983) used *Rb.*

sphaeroides strain 19, which contains RCs and LH1 as well as carotenoids, as a means of purifying the LH1 complex, and presented an amino acid analysis of the LH1 subunits. The LH1 complex was characterised further by Sebban et al. (1985) who used a mutant strain 3P17 of *Rb. sphaeroides* as a source of material. This mutant, although non-photosynthetic, contained reaction centers and very small amounts of LH2 in addition to the major complex, LH1. Treatment of membranes from this mutant by lauryldimethylamine oxide (LDAO) followed by column chromatography yielded an LH1 complex, and the lack of other absorbance peaks demonstrated the value of using a mutant as the source of LH1.

A. Structural Studies

Conventional biochemical techniques have been able to yield pure LH2 (and RCs) for many years, and therefore it would be surprising if X-ray crystallographic studies, which are notorious for consuming vast quantities of pure protein, did not progress first with these two complexes. Thus, Cogdell and co-workers have reported the crystallization of the B800–850 and B800–820 LH2 complexes of *Rp. acidophila* (Papiz et al., 1989;Guthrie et al., 1992) and Michel has reported diffracting crystals of the LH2 complex from *Rs. molischianum* (Michel, 1991). The *Rb. capsulatus* LH2 has also been crystallized (Welte et al., 1985). Clearly, the construction of an LH1-only strain could be of real benefit to structural studies of the LH1 complex, as long as suitably mild detergents are employed, and the LH1-only *Rb. sphaeroides* mutant M2192 has been used to provide large quantities of LH1 for X-ray crystallography, using β-octyl glucoside as the detergent (Nunn et al., 1992). In addition, *Rs. rubrum* has furnished LH1 for 2-dimensional crystallisation work on the LH1 complex (Ghosh et al., 1993; Karrasch et al., 1995).

At this point, strain M2192 will be briefly described, since it has been extremely useful for biochemical and spectroscopic studies (see also section V). The parental strain for M2192 is the mutant M21 which contains a point mutation at some undefined position in the *puc* operon; transconjugants of M21 containing this operon exhibit a normal phenotype (Ashby et al., 1987). Transposon Tn5 mutagenesis of the *Rb. sphaeroides* wild type had already identified several constructs which would selectively disrupt the *pufLM* gene pair encoding reaction center subunits (Hunter, 1988). One of these was introduced into strain M21, a manipulation which therefore produced an LH1-only strain, named M2192 (Hunter et al., 1989).

Mutant M2192 has been used as a source of LH1 for several purposes. For example, a recent electron microscopy study of the structure and aggregation behavior of LH1 (and LH2) used M2192 (Boonstra et al., 1993). In this work the LH2 complexes have a molecular mass of 150 kDa, and the particles corresponded to a ring-shaped structure consistent with 4–6 $\alpha\beta$ heterodimers. The LH1 complexes were isolated as 360 kDa dimeric particles and each particle was 5.2 nm in diameter and 6.2 nm in height, similar to the LH2 particle. Image analysis suggested a three or six-fold symmetry and an $\alpha_6\beta_6$ configuration. The crystallographic studies on LH complexes mentioned above are often held back by the lack of isomorphous heavy atom derivatives necessary to solve the phase problem. Again, the application of molecular genetics is able to make a contribution, this time by using the plasmid-encoded expression system mentioned earlier in section IA, in which LH genes are expressed in an $LH2^-$ $LH1^-$ RC^- genetic background (Jones et al., 1992a). Creation of potential heavy atom binding sites, through the introduction of cysteine residues for example, can provide a possible way forward. Moreover, this approach can be coupled with the heterologous expression of the other antenna complexes in *Rb. sphaeroides* for example, to facilitate the introduction of heavy atom binding sites in LH2 complexes from *Rp. acidophila* and *Rc. gelatinosus* (G.J.S. Fowler, S.J. Barrett, R.C. Mackenzie, R.J. Cogdell and C.N. Hunter, unpublished; see also section VI). Recently, the structure of the LH2 complex from *Rp. acidophila* has been determined by X-ray crystallography (McDermott et al., 1995) using a conventional approach for obtaining heavy atom derivatives. However, molecular genetics may still prove to be invaluable for other structural work.

B. The Production of α and β Subunits to Study the in vitro Assembly of the LH1 Complex

The study of bacterial antenna complexes in general and LH1 complexes in particular was advanced considerably by the isolation of the B820 particle (see Chapter 21 by Loach and Parkes-Loach). In this respect, LH1 appears to be more stable to such a fractionation process and little equivalent progress has been made with the LH2 complex. In most cases so far, the starting material has been an LH2-deficient background; in fact it is possible to generate the B820 particle using membranes as the starting material even if reaction centers are also present. Thus, the $LH2^- LH1^+ RC^+$ species *Rs. rubrum* proved to be the ideal starting point for this work and Miller et al. (1987) published a characterization of the subunits of the core antenna of *Rs. rubrum*, following on from the first report of this approach by Loach et al., (1985). They showed that dissociation of the complex from its normal oligomeric $(\alpha\beta)_{12}$ state (for example) resulted in a reversible shift of the absorption maximum from 873 nm to 820 nm after addition of 0.8% N-octyl-β-D-glucopyranoside (βOG). At even higher detergent concentrations, a second reversible blue shift to 777 nm is observed (Parkes-Loach et al., 1988). It appears to be important to remove carotenoids prior to detergent treatment, although the use of N-octyl-rac 2, 3, dipropylsulfoxide can circumvent this problem (Visschers et al., 1992). This approach has been extended to the $LH2^-$ mutants MW442 of *Rb. capsulatus*, and puc705BA of *Rb. sphaeroides*. (Heller and Loach, 1991; Chang et al., 1990a). Ghosh et al., (1988) have suggested that this B820 subunit form is an $(\alpha\beta)_2$ particle, and its relatively small size provides a good model system for various kinds of spectroscopy. Thus it has been shown that in spectroscopic terms at least, B820 behaves as an excitonically coupled $\alpha\beta$ dimer; measurements of low-temperature polarisation, CD and triplet-singlet spectroscopy are in good agreement with a dimer model in which a pair of excitonically coupled BChl *a* molecules are bound to the $\alpha\beta$ subunits (Chang et al., 1990b; van Mourik et al., 1991; Visschers et al., 1991, 1993a). The recent application of resonance Raman spectroscopy to the B820 species demonstrates that dissociation of the LH1 complex correlates with the replacement of internal hydrogen bonds by bonding to the detergent (Visschers et al., 1993b).

These results all show that the LH1 complexes studied so far are possibly built up as oligomers of an $\alpha\beta$ dimer, and apart from using molecular genetics to provide LH1-only strains for this kind of approach, it might be possible in the future to couple dissociation experiments with the production of site-specific alterations in LH complexes, in order to improve the spectroscopic characterisation of LH1 mutants.

IV. The Use of Antenna Mutants to Simplify Spectroscopic Studies of Antenna Complexes in a Membrane Environment

Despite the kind of advance made possible by the use of detergents to break down the oligomeric state of antenna complexes in order to simplify spectroscopic studies, summarised above, it is clear that there may be attendant effects of detergents which complicate spectroscopic work. This would arise from the creation of unwanted aggregation or dissociation states, or by altering or shifting absorbance peaks. At the same time, 'mixed' antenna systems containing both LH1 and LH2 are difficult to investigate because of overlapping signals. The availability of mutants deficient in one or two pigment proteins have made a valuable contribution to both problems, since the antenna is present in a spectroscopically pure state in its native lipid environment. This section will summarize the results of work undertaken on such mutants.

A. LH2-Only Mutants

This section will include some early work on both LH2-only and LH1-only mutants of *Rb. capsulatus* before moving on to summarize more recent work on a *Rb. sphaeroides* LH2- only mutant.

Several mutants of *Rb. capsulatus* provided valuable experimental material for spectroscopic studies, in particular the LH2-only mutants Y5 and the LH1-RC strain Ala^+ both of which were used as the starting point for the purification of the respective antenna complexes (Feick and Drews, 1978; Tadros et al., 1982). They demonstrated that the LH2 complex in this species has three subunits of 8, 10 and 14 kDa, the latter having no involvement in pigment binding. The amino acid composition was also determined for the subunits. In the case of LH1 the carotenoidless nature of Ala^+ may have contributed to the lability of the detergent-solubilized complex. The fluorescence emission properties of the same mutants were reported by Feick et al., (1980). Subsequently, Y5 and Ala^+ were studied using circular and linear dichroism, and Bolt et al. (1981a) were able to use these mutants to resolve the spectroscopic properties of the separate antennae into two classes. In particular, a strong CD signal reflecting exciton interactions was found in Y5 but was attenuated more than ten-fold in Ala^+; they concluded that these two LH complexes have significantly different chromophore arrangements or local environments. The same two mutants were used to provide a pure source of antenna for determination of the primary sequences of LH2 and LH1 subunits (Tadros et al., 1983;1984).

The LH2-only mutant Y5 of *Rb. capsulatus* was used as a source of B800–850 complex in a study of the temperature-dependence of energy transfer in the range 4–300K (van Grondelle et al., 1982). This work, which also included LH2 complexes purified from *Rb. sphaeroides* strains with different carotenoid contents, concluded that the energy transfer efficiencies from carotenoid to B800 and B850 are greater than 90% at all temperatures, and at 4K carotenoids transfer energy preferentially to B850. The yield of emission from B800 at 4K was used to calculate the B800–850 dipole-dipole distance as ≤ 21Å, a figure which was subsequently used in the model of Kramer et al. (1984a).

Thus, these *Rb. capsulatus* mutants were useful in ascribing basic properties to the antenna chlorophylls in LH complexes. Scolnik et al., (1980) provided an elegant example of how LH1 and LH2-only mutants could be used to examine the spectral properties of the carotenoids in these complexes. It was already known that the 'natural' carotenoid spheroidenone was unsuitable for conducting such a study due to its broad and relatively featureless absorption in the visible region, so the gene transfer agent was used to introduce a *crtD* mutation, producing neurosporene plus its methoxy and hydroxy derivatives. Thus the original LH2-only strain Y142 became BY1424 and the LH1+RC strain MW442 became MW4422. The study of these mutants demonstrated that carotenoids bound to LH1 are relatively blue-shifted, that those in LH2 are red-shifted and that only the LH2 carotenoids undergo a bandshift. However, the energy transfer efficiency from carotenoid to BChl was similar in each complex.

Another way to disrupt the synthesis of the LH1-RC core is to mutate the gene for the reaction center H subunit, *puhA* (Youvan et al., 1983; Hunter and van Grondelle, 1988; Sockett et al., 1989), although the reasons for the effect on the LH1 complex at least are not understood. Mutant NF57 of *Rb. sphaeroides* is a point mutant, which is complemented by a 1.4 kb *Bam*HI fragment encoding *puhA*, although the precise location of the mutation has never been determined. The preferable approach of constructing a defined deletion was carried out by Sockett et al., (1989) and mutant PUHA1 also lacked both the LH1 and RC complexes. The interesting mechanisms of gene expression and complex assembly that accompany these observations are discussed more fully in Sockett et al., (1989). Recently, a chromosomal deletion has been described in which only sequences internal to *puhA* are removed, which also produces an LH1$^-$ phenotype. Thus the loss of LH1 can be assigned to the deletion of *puhA* (J.T. Beatty, personal communication). The LH2 complex of mutant NF57 was analyzed by a variety of spectroscopic and biochemical techniques since it provided a membrane-bound form of this antenna. Low-temperature absorbance, fluorescence emission and polarised light spectroscopy (van Dorssen et al., 1988) showed that in many respects the LH2 of NF57 was as expected from earlier results on the LDAO-solubilised complex (Kramer et al., 1984a), except that the 'B850' absorbance band was broader due to the presence of a minor pool of red-shifted bacteriochlorophylls. There was no indication of a residual LH1 complex to explain this broadening, especially at the levels of red-shifted absorbance seen of approximately 15% of the B850 band. It was significant that a linear combination of the NF57 (LH2) and M21 (LH1+RC) spectra could be fitted to the wild type profile in a satisfactory manner, but that a combination of the respective detergent-solubilized complexes could not (van Dorssen et al., 1988). The analysis of singlet-singlet annihilation in mutant NF57 demonstrated that at room temperature it contains domains of more than 370 B850 BChls connected for energy transfer, a figure which is reduced to 30 B850 BChls at 4K when thermal barriers to energy transfer help to 'resolve' the domains into smaller structures (Vos et al., 1988); see Fig. 3. Since NF57 consists of domains of approximately 30 B850 BChls at 4K, this corresponds to ~ 45 (B800 + and B850) pigments, and an oligomeric state of $(\alpha\beta)_{15}$. This is in good agreement with the size of one of the detergent-solubilised species of LH2 measured as $(\alpha\beta)_{14}$ by electrophoretic mobility in LDS-PAGE (Hunter et al., 1988); a direct comparison of the aggregation state of LDS- solubilized and membrane-bound LH2 is discussed in Hunter and van Grondelle (1988). Thus it appeared that approximately 30 BChl 850 + 15 BChl 800 is a fundamental unit of LH2 structure, although this is 18 BChl 850 + 9 BChl 800, at least for the detergent-solubilized LH2 complex from *Rp. acidophila* (McDermott et al., 1995). It is worthwhile

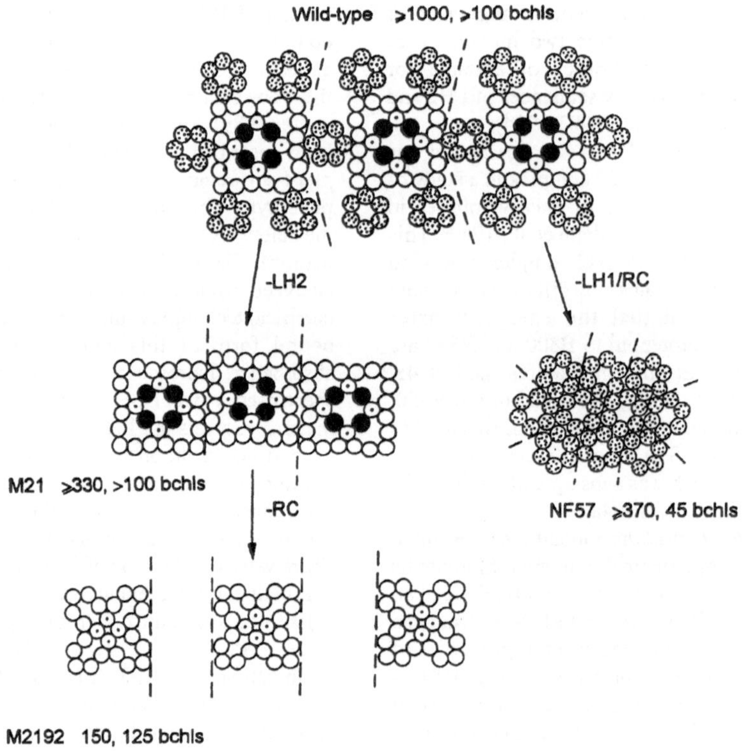

Fig. 3. Models for the in vivo functional sizes of light harvesting complexes (domains) from wild-type and mutant strains of *Rb. sphaeroides*, using the data of Vos et al. (1988), Hunter and van Grondelle (1988) and Hunter et al. (1989). These models depict the domains estimated from singlet-singlet annihilation data, but do not attempt to predict the two-dimensional arrangements of the complexes. Mutant M21 is $LH2^-$, NF57 is $LH1^-RC^-$ and M2192, $LH2^-RC^-$. The figures after each mutant represent the domain sizes at room temperature and 4K, respectively; the 4K domains are indicated by dashed lines. The individual circles represent: ● RC; ◎ LH2 - 6BChl850, 3BChl800; ○ LH1 - 6BChl875; ⊙ LH1 - 6BChl896, using a two-pool approximation of spectroscopic behavior at temperatures ≥ 77K.

to mention that although the singlet-singlet annihilation technique does not necessarily produce an accurate number at all times, the domain size for the water-soluble BChl *a* complex of *Prosthecochloris aestuarii* was determined to be 24 ± 4 BChl *a* (van Grondelle et al., 1983); the size determined by X-ray crystallography is 21 BChl *a* (Fenna and Matthews, 1975).

NF57 was put to further use, to obtain in situ extinction coefficients for LH2 BChls, which were determined to be 226 ± 10 and 170 ± 5 mM^{-1} cm^{-1} respectively for BChl 800 and BChl 850 (Sturgis et al., 1988). Also, picosecond absorption spectroscopy was used to investigate energy transfer dynamics within the membrane-bound LH2. The inhomogeneity within the B850 absorbance band was again noted and the transfer of energy to what at 77K appears as a red-shifted pool was determined to be 39 ± 9 ps (Hunter et al., 1990). It should be pointed out that this two-pool model has been re-evaluated in terms of an inhomogeneously broadened band, at least for the LH1 complex at 4K (van Mourik et al., 1992a). The use of NF57 has been discontinued since the availability of a defined LH2- only deletion strain, DPF2, (Jones et al., 1992a). However, Reddy et al. (1992) used NF57 as part of a wider study of the application of spectral hole burning to the study of several antenna complexes.

More recently, the alternative method of generating LH2-only mutants, by introducing plasmid-borne LH2 genes into an $LH2^-$ $LH1^-$ RC^- background, has produced a great deal of new information when coupled with Fourier transform resonance Raman spectroscopy and ultrafast time-resolved absorption spectroscopy. However, since this work was carried out as part of a site directed mutagenesis program, it will be summarized in section V C1.

B. LH1-Only and LH1-RC Mutants

The availability of LH1-only mutants is a great advantage, because the purification of this complex from wild-type membranes is hampered by the presence of other complexes. As noted earlier in section IIIA, biochemically-based studies such as X-ray crystallographic and reconstitution projects benefit greatly from the availability of large amounts of LH1. Spectroscopic projects are less demanding to an extent, since there is no particular necessity to remove LH1 from the membrane and so one can examine the complex in situ unless the manipulation of the aggregation state is of particular interest.

The LH1-only mutant M2192 of *Rb. sphaeroides* has been mentioned already in section IIIA. It has been used for a variety of spectroscopic studies for investigating the aggregation state of LH1, and the energy transfer dynamics within this antenna. A comparison of the M2192 absorbance spectrum with the LH1 complex purified by LDS-PAGE reveals some broadening and a slight blue shift of the long wavelength absorbance maximum for the mutant (Hunter et al., 1989). A comparison with the LH1-RC parental strain M21 showed a striking difference in fluorescence polarization (FP) in the 880–920 nm region: the FP of M2192 hardly increases over the wavelength region where the FP of M21 reaches a maximum value of 0.45. This work suggested a model in which the absence of reaction centers causes the LH1 complexes to adopt an altered conformation so that the depolarization resulting from energy transfer between red-shifted pools of pigment, which is normally prevented by the presence of reaction centers, can now take place. This would suggest that the red-shifted pools were located near to the reaction center. It is interesting to note that the rise in FP seen in both the LH1-RC mutant M21 and in the wild type is qualitatively similar to that seen for the LDS-solubilized LH1 complex; in this latter case the complex is sufficiently small that red-shifted pools of pigment cannot transfer energy among similar units. Vos et al. (1988) measured the annihilation behavior of M21 and M2192 to investigate the aggregation state of the LH1 antenna, with and without reaction centers. It was concluded that M21 contained rather larger domains than did M2192 (≥330 and 150 BChl *a* respectively), and that although the domains of M21 were resolved into units of around 100 BChl *a* at 4K the number for M2192 barely changed, to 125 (Fig. 3) This demonstrates two things: firstly that the LH1-RC 'core' in vivo probably contains a functional unit of up to 150 LH1 BChl *a* associated with 2, 3 or even 4 reaction centers; second, that the absence of reaction centers exerts an effect on the ability of the core antenna to aggregate, a phenomenon probably connected with the FP results summarised above. This large LH1 domain that constitutes the core antenna is made up of smaller fundamental units, which have been investigated by spectroscopic and biochemical techniques. The results of van Grondelle et al., (1983) suggest 6–8 BChl *a* as the likely size of the LDS-solubilised complex, which would correspond to $(\alpha\beta)_3$ or $(\alpha\beta)_4$. This figure is also predicted from the electrophoretic mobility studies of Hunter et al. (1988), which indicated two species of 54.6 ± 4.4 and 48.0 ± 3.0 kilodaltons corresponding also to $(\alpha\beta)_4$ and $(\alpha\beta)_3$ respectively.

As was the case for mutant NF57, strain M21 was useful for determining the extinction coefficient for the relevant antenna which was found to be 118 ± 5 $mM^{-}cm^{-1}$ for the B875 absorbance band (Sturgis et al., 1988).

The energy transfer dynamics within LH1 have been studied intensively and a complete survey of this topic appears in Chapter 17 by Sundström and van Grondelle. For the reasons mentioned already, the naturally occurring LH1-RC species *Rs. rubrum* is extensively used for such work, since it provides a relatively simple system for study. Thus, Borisov et al. (1982) had proposed early on that there may be an energetically low-lying component in the LH1 antenna, which would correspond to the B896 component identified in FP measurements (Bolt et al., 1981b; Kramer et al., 1984b), and by the fluorescence lifetime measurements performed on the reaction center-less mutant C71 (Sebban et al., 1984). The presence of what at 77K can be described as a small pool of energetically low-lying pigments does not appear to depend on the presence of the reaction center, since it can be seen in both the purified LH1 complex, as well as in mutant M2192 (Kramer et al., 1984b; Hunter et al., 1989).

The single wavelength picosecond absorption recovery measurements of Hunter et al., (1990) showed that at 77K mutant M21 behaves similarly to *Rs. rubrum*, and that excitation energy is transferred to the red-shifted pool with a time constant of about 22 ps. The lifetime of the excited state measured from the terminally emitting pigments in the red wing of the LH1 absorbance band was approximately 168, 335 and 670 ps for M21, M2192 and the LDS-B875 complex respectively, which could reflect differences in the aggregation state of the red-shifted pool caused by either the lack of reaction centers (in M2192) or the lack of reaction centers in combination with separation into smaller units in the LDS-B875 complex, as noted in Hunter et al., (1989). It is interesting to note the differences in time-resolved anisotropy, which reflect depolarization of excited states arising from relatively slow energy transfer in the red-shifted pool, and which provide a complementary approach to the FP measurements in Hunter et al, (1989). In M2192, the anisotropy is time dependent which could reflect energy transfer among similarly red-shifted pigments, whereas in the LDS-solubilised small units this parameter is high, and time-independent (Bergström et al., 1988).

A study of the temperature dependence of energy transfer from the red-shifted pool to the reaction center was carried out between 77 and 177K (Visscher et al., 1989). In this case, the reaction center traps were kept open. It was shown that for both the *Rb. sphaeroides* wild type and the M21 mutant the absorption changes were virtually independent of temperature in the range studied. A calculation was made of the activation energy for the final transfer step to the reaction center, which was 30 cm^{-1} or less; the transfer time would correspond to a distance of 26–31Å between the donor LH1 BChl and the special pair within the RC.

Although the 'two-pool' model of LH1 explains many experimental observations at 77K quite satisfactorily (van Grondelle et al., 1988), it was insufficient to explain the results of Timpmann et al. (1991) who measured time-resolved fluorescence at 4K (discussed in Valkunas et al., 1992). The decay of fluorescence contained several components, the relative contributions of which depended on the wavelength of detection. The two-pool (B875/B896) model was analyzed further by van Mourik et al., (1992) who used site-selected fluorescence emission spectroscopy at 4K on the LH1 complex purified from mutant M2192 to demonstrate that the wavelength of maximum emission depends on the wavelength of excitation. This would not be the case if all the pigments present were strongly coupled so that all emission took place from a single red-shifted pool; thus the absorption spectrum is now proposed to arise from the inhomogeneous broadening of the 'B875' absorption band. At 4K, each part of the total absorbance band contains a contribution from a series of weakly coupled clusters of pigments; the most significant parameter determining the wavelength of the maximum emission is the number of interacting pigments within a cluster. Thus, this measurement contains information on the number of connected pigments within such a cluster, which was calculated to be composed of 8 closely interacting BChl *a* dimers (van Mourik et al., 1992). This would correspond to two of the 6–8 BChl entities reported from biochemical and annihilation experiments (Hunter et al., 1988., van Grondelle et al., 1983). More recent data shows that, as expected, the emission maximum is only weakly dependent on excitation wavelength in intact membranes of M2192 at 4.2K, which suggests strongly-coupled clusters of larger size. At temperatures higher than 4K, such as 77K, energy transfer between clusters will become more feasible, so that the concept of a 'terminal emitter' from the absorbance band as a whole can be introduced; thus the two-pool model will still suffice at 77K. It appears likely that recent progress in crystallography, spectroscopy and molecular genetics will provide further information on this topic in the next few years. As an example of the two latter techniques, the combination of ultrafast (~40 fs) spectroscopy and molecular genetics has resulted in the first demonstration of coherent nuclear motion and exciton state relaxation in light harvesting complexes from *Rb. sphaeroides* (Chachisvilis et al., 1994; Pullerits et al., 1994).

C. R26 and R26.1

The carotenoidless mutant R26 of *Rb. sphaeroides* has already been mentioned in section 1A, as has its value as a source of pure reaction centers. Davidson and Cogdell (1981) noticed that an alteration to the absorbance spectrum had occurred at some time since the isolation of the mutant in 1961, in which the 870 nm maximum shifted down to 855–860 nm. They demonstrated that this blue-shifted strain, designated R26.1, contained not only LH1 but also an altered LH2 complex, in which the B800 peak is absent. This explained earlier spectroscopic observations that R26.1 contains two types of antenna complex (for example, Rijgersberg et al., 1980) and this was confirmed by subsequent work (Robert et al., 1984). Theiler et al. (1985) determined the amino acid sequence of the LH1 polypeptides from R26.1 and a comparison with the wild type LH1 revealed a βLeu29 → Pro change, which was proposed to weaken the $\alpha\beta$ heterodimer. In addition an αVal24 → Phe alteration in LH2 was observed (Theiler et al., 1984; Theiler and Zuber 1984). The accumulation of these mutations might be considered to be of no importance and indeed a reason to discard R26.1 as a source of antenna complexes, but R26.1 has been used several times as a model system, and the αVal24 → Phe mutation is interesting in its own right.

Braun and Scherz (1991) fractionated the 'B850' complex of R26.1 using SDS gel electrophoresis, and obtained several bands which were examined by absorption spectroscopy and by circular dichroism. Models of the organization of LH polypeptides were presented based on these data in which the basic unit is a $(\alpha\beta)_2$ tetramer. They proposed that three tetramers form a dodecamer having C_3 symmetry, and that larger hexagonal assemblies could assemble from these dodecamers.

Recently, the αVal24 → Phe change has been introduced into LH2 by site-directed mutagenesis, as part of an attempt to explain the reversion of R26 to R26.1 (G.J.S. Fowler and C.N. Hunter, unpublished). When expressed in a phytoene background the level of LH2 complex in the membrane is consistently higher when the αVal24 → Phe mutation is present and ultrastructural analysis of the transconjugant strains shows an enhanced ability to form intracytoplasmic membranes. A speculative sequence of events could be firstly that the lack of carotenoid destabilizes the BChl800 binding site and the LH2 complex as a whole and that the LH2 polypeptides are unfavorable in some way for the viability of the cell. Pressure for suppression mutations arises from the need to provide additional light harvesting capacity; the αVal24 → Phe mutation might improve matters by fostering additional interactions to counterbalance the loss of stability resulting from the lack of carotenoid and B800.

D. Mutants Containing LH2 and RCs, with No LH1

It is possible to generate LH1$^-$ mutants, either by random chemical mutagenesis or by a more deliberate process of deleting LH1 genes. Examples of both methods exist, and these will be summarized below. The absence of LH1 raises some fascinating questions about the 'uniqueness' of LH1 as a donor of energy to the reaction center, and the structural similarity between LH1 and LH2.

The first report of such a mutant was in 1985, when Meinhardt et al. published a spectral characterization of the *Rb. sphaeroides* mutant RS103. They measured the relative efficiency of energy transfer from LH2 to the reaction center and concluded that RS103 exhibited only 24% of the efficiency of the wild type, even though the photosynthetic unit was larger than normal. In a subsequent paper Kiley et al. (1988) demonstrated that RS103 contains approximately 25% of the normal complement of the LH1 α polypeptide; the amount of the β polypeptide was not determined. It was clear that in this mutant the lack of intact LH1 complex did not inhibit the assembly or function of the reaction centers, in contrast to the situation found in the LH1 mutant PAS108 of *Rb. capsulatus* (Jackson et al., 1986, 1987) where reaction centers are apparently assembled but do not receive excitation energy from the LH2 complex. In this mutant LH1 polypeptides could form but they were turned over relatively rapidly; indeed it is possible that low levels of the LH1 complex are present at all times in PAS108 (Jackson et al., 1986). It was concluded that in the absence of LH1, reaction centers of *Rb. capsulatus* assemble but are not capable of useful work. A direct comparison of *Rb. sphaeroides* RS103 and *Rb. capsulatus* PAS108 concluded that the two species must differ in their requirements for the correct insertion of reaction centers, and that in this respect LH1 is essential in *Rb. capsulatus* but not in *Rb. sphaeroides* (Jackson et al., 1987). However, later work on site-directed LH1 mutants of *Rb. capsulatus* (see also section IVB) produced some LH1$^-$ RC$^+$ phenotypes in a U43 background, which is LH2$^-$ by virtue of a point mutation in *pucC* downstream of structural genes for the LH2 complex. (Tichy et al., 1989; 1991). Thus Bylina et al. (1988), Dorge et al. (1990), Richter et al. (1991, 1992) and Garcia et al. (1991) have all reported *Rb. capsulatus* strains which were LH2$^-$ LH1$^-$ RC$^+$, and which would appear to contradict the results of Jackson et al. (1987). Finally, though, it should be mentioned that Babst et al. (1991) report some site-directed alterations to LH1 which do affect RC assembly.

The availability of defined deletion strains in *Rb. sphaeroides* enabled Jones et al. (1992a) to construct a mutant similar to RS103 of Meinhardt et al. (1985), but with the LH1 genes deleted. This was accomplished by introducing a plasmid containing only the *pufQLM* (and *X*) genes encoding reaction center subunits and from which the LH1 genes had been deleted, into strain DPF2 which contains a genomic copy of the LH2 genes and a deletion of *pufBALMX*. The resulting strain, DPF2 (pRKEH2) grows relatively slowly at low light intensity with a doubling time of 8 hours, compared with 4 hours for the wild type. However, it grows more rapidly than a strain which is completely devoid of antenna complexes, DD13/G1 (pRKEH2), which contains only reaction centers and which has a doubling time of 12 hours. This simple experiment shows firstly that reaction centers do assemble in the absence of the LH1 complex as first demonstrated by Meinhardt et al. (1985), and that there must be at least some productive energy transfer from LH2 to the RC. Recently the kinetics of energy transfer and trapping in DPF2 (pRKEH2) were examined by time-resolved absorption and fluorescence methods (Hess et al., 1993). Following excitation at 840 nm, the recovery of ground-state bleaching was monitored at 880 nm, which had a time constant of 55 ± 5ps. A long-lived bleaching was observed which is characteristic of the charge separated states P^+I^-Q/P^+IQ^-. This demonstrates that trapping and charge separation in DPF2 in (pRKEH2) occur with a time constant of 55 ± 5ps at room temperature; in the same experiment the results for the wild type were 60 ± 5ps. Time-resolved fluorescence measurements demonstrated that the decay of antenna fluorescence with open reaction centers in DPF2 (pRKEH2) is largely (>80%) dominated by a 52 ± 5ps decay component, in good agreement with the absorption measurements. Similar results were obtained for the wild type. These results demonstrate directly that LH2 can transfer energy efficiently and rapidly to reaction centers in this LH1-deficient mutant.

It is remarkable that the trapping of excitation energy by the reaction center appears to be unaffected by the removal of the LH1 core antenna. It would be reasonable to assume an optimized design of LH1-reaction center contacts and interchromophore distances, which could not be easily reproduced for the peripheral LH2 antenna. The fact that the rate of trapping is virtually the same in DPF2 (pRKEH2) and the wild type therefore suggests that the part of the core antenna surrounding the reaction center is not highly specialized, with finely tuned antenna-reaction center distances and a number of antenna pigments dedicated to energy transfer to each reaction center. Rather, it appears that when antenna polypeptides pack around the reaction center a suitable organization is formed more or less spontaneously; the exact organization of the antenna polypeptides around the reaction center is not known. It could also be the case that the architecture and organization of LH1 and LH2 may be similar, a point which is being addressed by progress on the crystallography of these complexes (McDermott et al., 1995; Karrasch et al., 1995).

The LH2-RC mutant DPF2 (pRKEH2) provides other opportunities, such as the study of fast energy transfer within LH2, without the complications of LH1 being present (Hess et al., 1993). This follows on from an earlier characterisation of membrane-bound LH2 only systems such as NF57 (Hunter et al., 1990). The measurements on DPF2 (pRKEH2) in this instance were conducted with reaction centers closed in the state $P870^+$, which are normally relatively efficient quenchers of antenna excitations, although this mechanism is poorly understood. Substitution of the LH1 core antenna by the LH2 complex provided an opportunity to gain new insight into the mechanism of antenna quenching by $P870^+$. The 330±30 ps time constants of LH2 absorption decay at 860 nm contrasts with the 200 ps lifetime observed for the wild type under similar conditions. The ΔA spectrum for reaction center charge separation (P^+IQ^- minus PIQ) shows that $P870^+$ develops an absorption band at wavelengths greater than 900 nm, where P870 has very little absorption initially. This absorption has very good spectral overlap with the fluorescence from LH1, but less overlap with LH2 fluorescence. This directly demonstrates that there are possibilities for energy transfer from the LH1 antenna to $P870^+$, which probably dissipates the energy as heat through a radiationless process. The smaller spectral overlap for LH2 $\rightarrow$ $P870^+$ energy transfer also explains why the LH2 antenna within DPF2 (pRKEH2) with closed reaction centers has a considerably longer excited state lifetime than the wild type. The broad $P870^+$ absorption band is consistent with the observation that the quenching time with closed reaction centers is virtually temperature independent in the species of purple bacteria studied so far.

The degree of spectral inhomogeneity within the B850 absorption band of DPF2 (pRKEH2) was assessed by exciting the antenna with a picosecond pulse at 835–845 nm at 77K with closed reaction centers and by monitoring the absorption changes in the 835–880 nm range. The kinetics at the different probe wavelengths were well described by two lifetime distributions centered at ~10 ps and 350 ps, the first of which corresponds to fast (1–5 ps) hopping of the excitation energy between small clusters of BChl molecules from the blue-absorbing to red-absorbing pigments within LH2. Anisotropy measurements show that very fast depolarizing energy transfer occurs on a subpicosecond timescale. Very recently these processes have been resolved through the application of 80 fs pulses (Hess et al., 1994).

E. Reaction Center-Only Mutants

Reaction center structure and function is an extremely large subject, and is outside the scope of this chapter. It is relevant to mention, though, that the correct series of deletions in light harvesting genes will result in a strain containing only reaction centers. This has been found many times for *Rb. capsulatus*, on occasions where mutations inactivating LH1 assembly are expressed in a U43 (LH2$^-$) background (Bylina et al., 1988; Garcia et al., 1991). Expression of reaction centers is then occurring in a background where small but unstable quantities of both LH1 and LH2 polypeptides can be incorporated into the membrane (Tichy et al., 1989), but a more recent modification of this approach by Jones et al., (1992a) relied on the expression of a plasmid bearing RC (*puf LMX*$^+$) genes in an LH2 LH1 RC background. The resulting strain, RCO1, has been characterised (Jones et al., 1992b), and is an ideal vehicle for the study of various mutations on reaction center structure and function in its native membrane environment.

F. PufX

The *pufX* gene is located at the 3' end of the *puf* operon, and is mentioned briefly here, since its function, at least in part, appears to concern the association of the LH1 complex with the reaction center in some way. Farchaus and Oesterhelt (1989) constructed a strain of *Rb. sphaeroides* lacking *pufL*, *M*, and *X* and complemented it with a plasmid containing only *pufB*, *A*, *L* and *M*, a manipulation which effectively produces a *pufX*$^-$ strain. Although all of the photosynthetic complexes are still present, the transconjugant strain is non-photosynthetic (Farchaus et al., 1990). Lilburn et al., (1992) have demonstrated that in a *Rb. capsulatus pufX*$^-$ strain, electron transfer from the RC to the cytochrome bc_1 complex is impaired. In addition, Westerhuis et al. (1993) suggest that the *pufX* gene product may limit the aggregation state of the LH1 complex. However, it appears that the PufX protein does not directly facilitate electron transfer between the RC and the cytochrome bc_1 complex, because there is no requirement for PufX for photosynthetic growth in an RC-only strain (McGlynn et al., 1994). Recently, the fascinating observation has been made that intergenic suppressors can restore photosynthetic growth to strains lacking *PufX* and that many of the revertants sequenced so far contained a modification of αTrp43 in the LH1 α subunit where a stop codon was introduced instead (Barz and Oesterhelt, 1994). Thus, truncation of this LH1 subunit compensates in some unknown way for the loss of *PufX*, which adds another level of complexity to the study of LH1-reaction center interactions.

V. Site-Directed Mutagenesis of Bacterial Light Harvesting Complexes

A. The Rationale for the Design of Site-Directed Mutations in Bacterial Light Harvesting Complexes

It is important to introduce some element of choice into the construction of site-directed mutants, unless this element is abolished and a highly discriminatory screening step is applied, as in random mutagenesis methods (see section V C4). The availability of a number of polypeptide sequences has provided important clues as to the conformation of the antenna polypeptides, and for a detailed account of the conclusions that can be drawn from these alignments, the reader is referred to Brunisholz and Zuber (1992) and Chapter 16 by Zuber and Cogdell in this volume. The generally accepted model, where a central membrane-spanning α helical region is flanked by polar N- and C-terminal domains, has been verified by X-ray crystallographic analysis (McDermott et al., 1995). For convenience, the $\alpha\beta$ polypeptide pairs of *Rb. sphaeroides* LH2, LH1 and *Rb. capsulatus* LH1 have been represented in this form to facilitate the discussion of the site directed mutants in sections V B and C.

An inspection of aligned sequences in Fig. 4 reveals several interesting similarities and differences; it is immediately apparent that the α polypeptides of

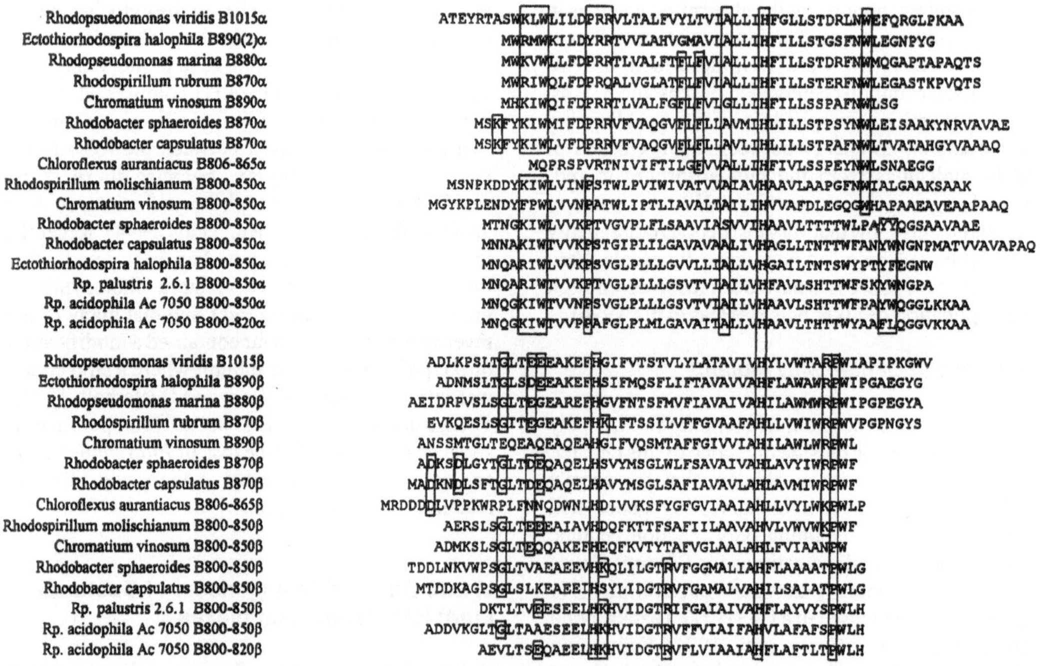

Fig. 4. Alignments of sequences of bacterial light harvesting complexes, using the data of Brunisholz and Zuber (1992). The boxes indicate those features which have been explored by site-directed mutagenesis techniques, and which are reviewed in sections V B and C.

LH1 and LH2 are rather similar, as are the β polypeptides. In the α polypeptides, the KIW motif is a possible membrane anchor, and the highly conserved proline is a likely location for a turn (see McDermott et al., 1995). In the membrane spanning region, the histidine is a ligand to a bacteriochlorophyll molecule and in the C-terminal region there is a cluster of conserved aromatic residues. Mutagenesis of these regions of the LH1α subunit will be summarized in section V B.

In the β polypeptide alignment in Fig. 4 there is a conserved histidine, which participates in the binding site of BChl 800 within LH2; however it is also present in LH1 which has no such pigment. In the LH2 sequences an arginine at position 29 in *Rb. sphaeroides*, *Rb. capsulatus*, *Rp. palustris* and *Rp. acidophila* is also involved in B800 binding and it is absent in all LH1 sequences. Again, there is a histidine in the membrane-spanning domain which is a likely ligand to BChl *a* and on the C-terminal (periplasmic) side there is a totally conserved proline-tryptophan pair. Mutagenesis of the first mentioned histidine, as well as the arginine in the transmembrane region will be summarized in the next section.

B. The LH1 Antenna Complex

A simple model of the LH1 complex is presented in Fig. 5; since most of this section deals with *Rb. capsulatus* it is this complex which is depicted in Fig. 5. Much of the information contained therein shows the locations of the site-directed mutations summarized below. The first report of the site-directed mutagenesis of any bacterial light harvesting complex was published by Youvan's group (Bylina et al., 1988), working on *Rb. capsulatus*. The object of study was the putative bacteriochlorophyll binding sequence Ala-X-X-X-His (see above and Chapter 16 by Zuber and Cogdell); αHis32 was changed to Asn, Asp, Gln, Thr, Arg, and Pro, and αAla28 was changed to Gly, Ser, Cys, Val, Asp, Glu, Phe and His. They concluded that no other residue in this list would

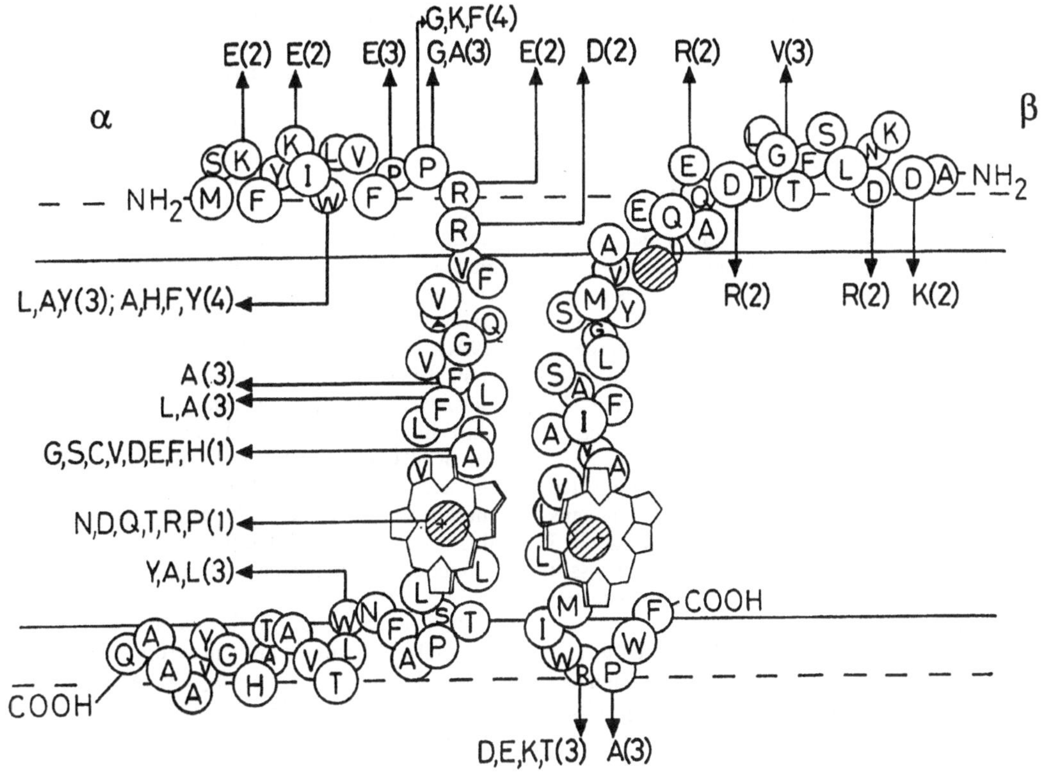

Fig 5. Representation of the $\alpha\beta$ dimer of the *Rb. capsulatus* LH1 complex, showing the positions of the site-directed mutations reviewed in section VB. The original references can be found as follows: (1) Bylina et al., 1988; (2) Dorge et al., 1990; (3) Babst et al., 1991; (4) Richter et al., 1991. The His residues are represented by hatched circles; some residues are hidden by the BChl molecules and the sequences flanking these regions are: α- AVLI**H**LILLST; β- VAVLA**H**LAVMIW.

substitute for histidine and that residues with side chains smaller than valine can substitute for alanine, but larger ones cannot.

In the last few years, there have been a series of studies which have mainly concentrated on those residues which have some bearing on the process of the integration and assembly of the LH1 polypeptides. It has been known for several years that intrinsic membrane proteins conform to the 'positive-inside' rule (von Heijne, 1986) in which the balance of charge on periplasmically and cytoplasmically exposed domains is with the positive residues in greater number on the cytoplasmic face. Indeed, it has been possible to reverse the orientation of leader peptidase by reversing the charge balance (von Heijne, 1989; Nilsson and von Heijne., 1990). Another determinant of membrane protein stability appears to be the location of aromatic groups at the membrane interface (Schiffer, 1992). A series of publications from Drews' group have addressed these and other aspects of LH1 assembly (Fig. 5). Dorge et al. (1990) examined the role of charged amino acids in the N-terminal regions of the α and β subunits. The changes Lys → Glu at positions α3 and α6 or βAsp2, βAsp5 → Lys, Arg had little effect on LH1 assembly, but the additional exchange of αArg14, αArg15 → Glu, Asp and/or βAsp13, βGlu14 → Arg, Arg had a strongly inhibitory effect. Stiehle et al. (1990) examined these mutations constructed in the α or β subunits separately and found that the positively charged residues of the N-terminus of α and the negatively charged residues of the N-terminus of β influence LH1 assembly differently. For example the α mutations (Lys3, Lys6, αArg14, αArg15 → Glu, Glu, Glu, Asp) resulted in a transient insertion of LH1 polypeptides whereas the β mutations (βAsp2, βAsp5, βAsp13, βGlu14 → Lys, Arg, Arg, Arg) resulted in a much slower degradation of LH1 polypeptides. This second series therefore permits the accumulation of small amounts of the LH1 complex.

Richter and Drews (1991) examined the stability of LH1α in the absence of β, and vice-versa, by deleting either *pufB* or *A*, followed by expression in a U43 background. In neither case did an LH1 complex assemble, so pulse chase experiments were used to find transient levels of LH1 subunits. In the 'wild type' control LH1β was found to appear in membranes before LH1α; moreover the β subunit could insert transiently in the absence of α but the reverse was not observed. These experiments were complicated by the presence of transiently expressed LH2 subunits in the U43 strain, as well as another labelled polypeptide in the background, but nevertheless it appears that LH1 α and β are not inserted simultaneously.

The studies of Babst et al., (1991) targeted several residues, and the effects seen among the eighteen mutations varied widely, even at a single position. The three tryptophan exchanges at position α8, for example, resulted in either the absence of a core complex, that is, the abolition of LH1 and RC assembly (αTrp8 → Leu), the absence of the core antenna (αTrp8 → Ala) or a reduction in the carotenoid content (αTrp8 → Tyr). Richter et al. (1991) also noted that αTrp8 is important for the insertion of LH1α into the membrane: αTrp8 → Ala abolished LH1 assembly, in agreement with Babst et al. (1991). However, substitution of αTrp8 by Phe or Tyr appeared to have little effect on assembly or on carotenoid content, in contrast with Babst et al. (1991); the introduction of αTrp8 → His yielded a less stable LH1 complex. Several other alterations produced an elevated amount of bacteriochlorophyll in LH1, such as αPhe23 → Ala and βGly10 → Val which produced a 1.3-fold higher level: this was attributed to the presence of additional BChl molecules not attached at the normal co-ordination sites. The change βArg45 → Asp destabilized the LH1 antenna considerably, but surprisingly βArg45 → Glu did not. Several of the changes studied produced a small blue shift in the absorbance maximum which is normally at 878 nm, notably those at αTrp43 → Tyr (867 nm), Ala (868 nm) and Leu (870 nm). The authors suggested that this residue interacts with BChl *a* and that it modulates the absorbance behavior of this pigment. It is interesting to note that αTrp43 is conserved in all LH1 antennae, and that in *Rb. sphaeroides* it is the site of the intergenic suppressor mutation provoked by the reversion of *pufX*$^-$ mutants to photosynthetic growth (Barz and Oesterhelt, 1994). Moreover, Fourier transform resonance Raman spectroscopy of αTrp43 mutants of *Rb. sphaeroides* demonstrates that this residue is hydrogen bonded to the 2 acetyl carbonyl of one of the BChl *a* molecules in LH1 (Olsen et al., 1994). The bond is abolished in an αTrp43 → Phe mutant, but strengthened in an αTrp43 → Tyr mutant; thus, at least one mechanism underlying these spectral shifts in vivo is likely to be by hydrogen bonding to the BChl. One interesting consequence arising from the demonstration of this hydrogen bond is that there must be a turn towards the N-terminal end of LH1α; this is consistent with the presence of a conserved proline residue in this region in both *Rb. capsulatus* and *Rb. sphaeroides*.

Richter et al. (1992) further examined the role of the LH1α N-terminus in the insertion and stability of the LH1 complex by deleting amino acids 6–8, 9–11, 12 and 13, or 14 and 15. In addition the hydrophobic stretch of amino acids 7–11 was lengthened by the addition of hydrophobic or hydrophilic amino acids. All of these mutations abolished the formation of LH1; however, a photosynthetic revertant of Δ Leu9-Val10-Phe11 contained an intragenic suppression mutation which arose from an in-frame duplication of an existing sequence, Lys6, Ile7, Trp8. In summary, the wild-type sequence is: Met-Ser-Lys-Phe-Tyr-Lys-Ile-Trp-Leu-Val-Phe-Asp. The revertant was Met-Ser-Lys-Phe-Tyr-(Lys-Ile-Trp)$_2$-Asp. The revertant strain had double the normal carotenoid content of an LH1 complex with respect to BChl, and an increase in the size of the photosynthetic unit. This interesting experiment supports the notion that αTrp8 in particular could be part of a carotenoid binding motif, as well as a determinant of LH1 stability. Preliminary work on the mutagenesis of αTrp8 in *Rb. sphaeroides* supports this: mutation of αTrp8 → His and Tyr had no effect on LH1 assembly, but αTrp8 → Leu, Arg and Ala did (McMaster et al., 1992). In contrast with the observation of Babst et al. (1991) that αTrp8 → Ala abolished reaction center assembly, no effect was observed, and it seems likely that different factors govern the stability of LH1-RC cores in *Rb. capsulatus* and *Rb. sphaeroides*, as noted earlier by Jackson et al. (1987). The flexibility of the genetic system of Jones et al. (1992a) facilitates the examination of a given antenna mutation in three different carotenoid backgrounds and therefore the possibility that αTrp8 and carotenoids interact in some way could be examined. For example, it was observed that the insertion and assembly of LH1 took place in the αTrp8 → Leu mutant in a neurosporene background, but not when phytoene was the major carotenoid.

C. The LH2 Antenna Complex

At the time of writing rather fewer residues of the LH2 complex have been altered by mutagenesis, compared to LH1. This is partly because until recently there was no LH2⁻ LH1⁻ RC⁻ (Δ*puc*, Δ*puf*) double deletion strain of *Rb. capsulatus*, although a single deletion LH2⁻ (Δ*puc*) strain has been available since 1985 (Youvan et al., 1985). Another reason is that the work on *Rb. sphaeroides* has been conducted with the aim of presenting a thorough spectral characterisation for each mutant which lengthens the turnover time of the whole process. Fig. 6 shows a simple model of the *Rb. sphaeroides* complex in the style of Fig. 5 which indicates the locations of the site-directed changes made.

1. Blue-Shifted Complexes Produced by Mutagenesis of αTyr44 and Tyr45

The first example of the mutagenesis of an LH2 complex was in 1992 when Fowler et al. reported the effects of altering αTyr44, Tyr45 firstly to Phe, Tyr then Phe, Leu. The rationale for this was that the mechanisms underlying the red-shifts of bacteriochlorophylls in vivo are poorly understood, although

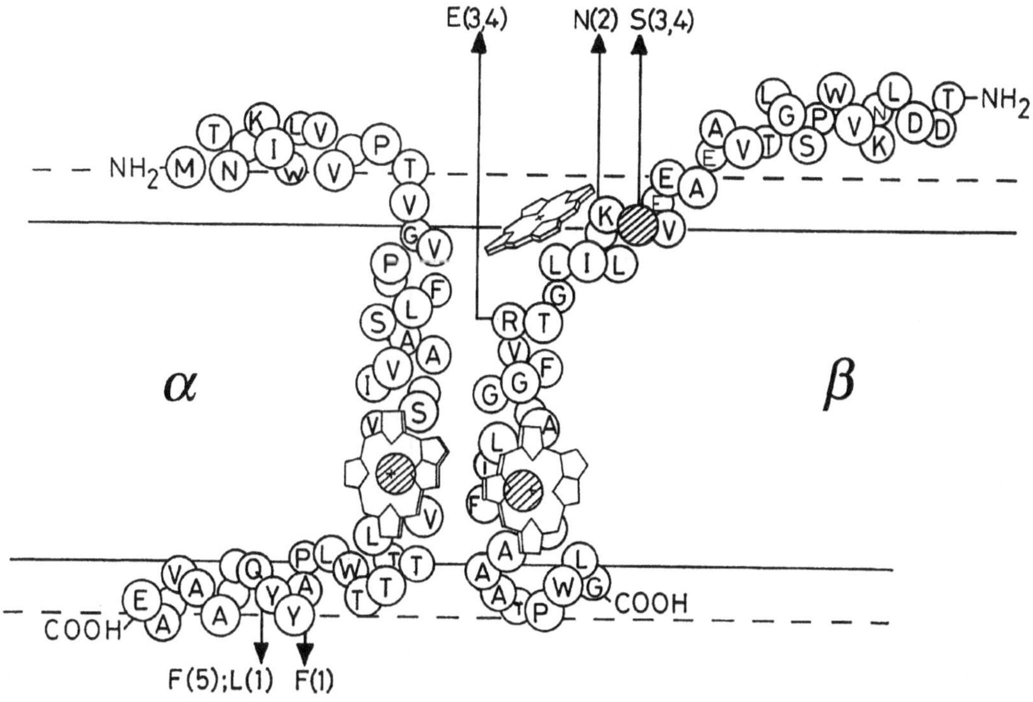

Fig. 6. Representation of the αβ dimer of the *Rb. sphaeroides* LH2 complex showing the positions of the site-directed mutations reviewed in section VC. The original references can be found as follows: (1) Fowler et al., 1992; (2) Fowler et al., 1993; (3) Crielaard et al., 1994; (4) Visschers et al., 1994; (5) Hunter et al., 1993. The His residues are represented by hatched circles; some residues are hidden by the BChl molecules and the sequences flanking these regions are: α- ASVVIHAAVLT; β- MALIAHFLAAA.

it is clear that aggregation effects such as those examined in vitro are important (Visschers et al., 1991). The B800–820 LH2α subunit from *Rp. acidophila* strain 7050 has a Phe, Leu pair where the *Rb. sphaeroides* B800–850 LH2 pair is Tyr, Tyr. This suggests a simple mutagenesis experiment, in which Tyr, Tyr is changed first to Phe, Tyr then to Phe, Leu, to see if this modulates the absorbance behavior of the complex. The altered genes were expressed in both single deletion (LH2$^-$, Δ*puc*) and double deletion (LH2$^-$, LH1$^-$, RC$^-$ (Δ*puc*, Δ*puf*) backgrounds. In the latter case, an unequivocal account of the effects on LH2 could be obtained: at 77K the absorbance maximum of the 850 nm band is shifted by 11 nm to 839 nm (Phe, Tyr) whereas in the Phe, Leu mutant there was a further shift to 826 nm. The intensity of this band decreased to the point where the absorbance spectrum of the Phe, Leu mutant did indeed resemble that of the *Rp. acidophila* B800–820 complex (Fig. 7). Excitation spectra demonstrated that in these mutants BChl800 was fully able to transfer energy to, and elicit fluorescence from the 850, 839 or 826 nm component. Expression of the altered genes in an LH1$^+$ RC$^+$ background resulted in a 'pseudo' wild type, in which the altered LH2 was completely efficient in energy transfer to the native core. Circular dichroism in the near infrared, which is a sensitive indicator of perturbations to the relative geometry of the pigments, showed the same, strong conservative signal in each core albeit a progressively blue-shifted one. Thus the absorbance behavior of these BChls is modulated by the protein framework of the antenna, by the use of the Tyr, Tyr motif in this case. The mechanism underlying this effect was examined by further mutations to Tyr, Phe, and Tyr, Leu (Hunter et al., 1993) and in summary the main shift arises from a Tyr → Phe change. Therefore, interactions between π-π^* orbitals of aromatic residues and the bchl macrocycles probably do not cause blue shifts, but rather it is hydrogen bonding between Tyr and the BChl (Hunter et al., 1993). A resonance Raman study of these mutants is in support of this, and demonstrates the disruption of at least one hydrogen bond to the C2 acetyl carbonyl of one of the B850 BChls in the mutants (Fowler et al., 1994). The X-ray crystallographic structure for the *Rp. acidophila* LH2 complex is consistent with the proposed hydrogen bonding arrangement (McDermott et al., 1995). This work runs parallel with the Raman assignment of LH1 αTrp43 mutants (Olsen et al., 1994). As in the case of LH1, this bond in LH2 necessitates a turn

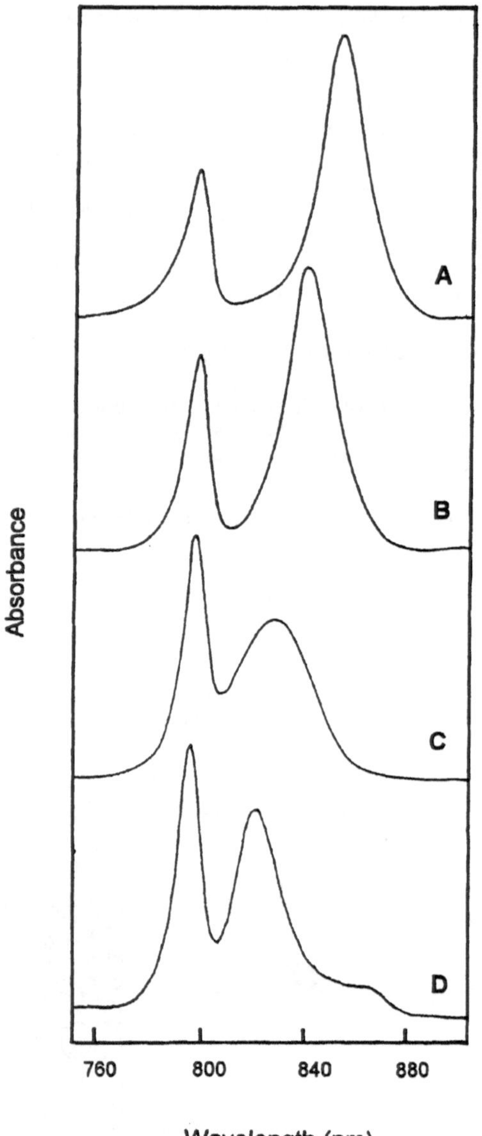

Fig. 7. Absorption spectra of blue-shifted LH2 complexes of *Rb. sphaeroides* and the *Rp. acidophila* B800–820 complex, recorded at 4K. The *Rb. sphaeroides* complexes are in the native membrane in a LH1$^-$RC$^-$ background; the *Rp. acidophila* complex is solubilized and purified in detergent.
A - *Rb. sphaeroides* wild type LH2
B - αTyr44 → Phe
C - αTyr44, Tyr45 → Phe, Leu
D - *Rp. acidophila* B800–820 complex

near the predicted membrane interface, at the sequence Thr-Trp-Leu-Pro-Ala (Fig. 6). More recent work shows that there are several ways to produce the same blue-shifted phenotype: for example, alteration of βLys22 → Gln (wrongly numbered as βLys22 in Fowler et al., 1993) produces a B800–837 complex, in this case possibly by affecting the relative disposition of the α and β subunits, and therefore the pigments as well. The mechanism underlying this change has not yet been determined.

The availability of LH2 complexes with progressively blue-shifted absorbance bands prompted an investigation of energy transfer dynamics within LH2 by spectral hole-burning (van der Laan et al., 1993). These measurements were performed on the BChl800 band at very low temperature (1.6–4K) and the measured hole-widths were related to the rate of B800 → B850 energy transfer. For wild-type LH2 complexes, good agreement was found with the B800* lifetimes obtained by time-resolved absorption techniques at 77K (Hess et al., 1994), with both techniques giving approximately 2.4 ps. The rates determined by hole-burning experiments on the mutants did not follow a clearly increasing trend, unlike the time-resolved study which produced rates of approximately 1.6 and 0.8 ps for the B800–839, 800–826 mutants, respectively. Thus, the time-resolved measurements at 77K correlated with the Förster description of the energy transfer rate whereas the hole-burning measurements at 4K did not. A discussion of the causes of this discrepancy is beyond the scope of this chapter, but these results emphasise the value of spectrally modified LH complexes to the study of energy transfer dynamics in antenna systems.

2. Alteration of the Absorbance Properties of the B800 Bacteriochlorophyll

LH2 contains an extra absorbance band (B800), when compared to LH1 and it is already known that this band is relatively labile in the presence of the detergent lithium dodecyl sulfate, for example, (Clayton and Clayton, 1981; Kramer et al., 1984). This band presents an attractive target for mutagenesis, and this then offers the prospect of being able to re-examine the conclusions of Kramer et al. (1984), who proposed a model of LH2 based on experiments conducted on a modified LH2 complex of *Rb. sphaeroides* lacking B800.

The residues chosen for this study were βHis21 and βArg29 (Visschers et al., 1994). In the former case, this histidine was an attractive site for the study of BChl binding, notwithstanding the fact that BChl800 is absent from LH1 which possesses a histidine residue at the analogous position. In addition, resonance Raman spectroscopy found no evidence for histidine coordination of BChl800 (Robert and Lutz, 1985). βArg29 was chosen because it is conserved in most LH2 complexes, and is not found at the analogous position in LH1 (see fig 4). Both mutations (βHis21 → Ser; βArg29 → Glu) lead to severe changes in the spectroscopic characteristics of the BChl800 pigment. The βHis21 → Ser mutant has a greatly attenuated BChl800 absorbance band, and membranes of the mutated strain have only a B850 absorbance peak which is shifted to the red by 2–3 nm. The βArg29 → Glu mutant results in a blue shift to 787 nm and a broadening of the absorbance band, suggesting that this residue also plays a role in determining the absorbance properties of BChl800. Circular dichroism reveals no perturbations of the BChl850 binding sites in either mutant. The residual BChl800 pigment in these mutants is still fully efficient for energy transfer to BChl850, and when expressed in an LH1-RC background energy transfer from the modified LH2 complexes to LH1 is still efficient. Linear dichroism of the Q_x region shows that the Q_x transitions of the B800 pigments are in the plane of the membrane, in good agreement with Kramer et al. (1984).

The conclusion from this work is that both βHis21 and βArg29 have some influence on the properties of BChl800, although results from studying whole cells are instructive, in that rather more BChl800 absorbance is preserved. Preparation of membranes results in some loss of BChl800, which demonstrates that site-directed mutants can be labile even at this stage; one would expect that even more spectral losses and modifications would occur if a detergent purification protocol was added. Nonetheless, it is interesting that a modification such as the loss of BChl800 is specific and does not prevent the assembly and function of the LH2 complex. The conclusions drawn from the study of these mutants is consistent with the crystallographic data, which show that these residues are in close proximity to the BChl 800 pigment (McDermott et al., 1995).

3. Alteration of the Binding, Function and Electrochromic Behavior of the Carotenoids

The same changes specified above, βHis21 → Ser

and βArg29 → Glu, had pronounced effects on the carotenoids in the LH2 complex (Crielaard et al., 1994). The altered LH2 genes were expressed in a double deletion (LH2⁻LH1⁻RC⁻) background which carries a mutation in the *crtD* locus leading to the synthesis of neurosporene and its methoxy-and-hydroxy derivatives (Coomber et al., 1990). The distinct spectral features of carotenoids in the green LH2 antenna are preferable to the rather featureless spectrum of spheroidenone in the wild-type. Difference spectra in the visible region (LH2-WT vs. βHis21 → Ser and LH2-WT vs. βArg29 → Glu) clearly showed that the loss of the histidine residue results in a broadening and a blue shift of 3 nm in all carotenoid peaks. A similar blue-shift of 1 nm was seen in the βArg29 mutant although no broadening was observed. In neither case was there any indication of a lower carotenoid binding capacity, despite the attenuation or loss of BChl800. Fluorescence excitation spectra at 77K showed that no disturbance of energy transfer to BChl had occurred, so that energy was being transferred directly to B850. Linear dichroism in the visible region indicated that no rearrangement of carotenoids had occurred in the mutants. With regard the model of Kramer et al., (1984a) these results show that an energy transfer pathway from carotenoid → BChl800 is not obligatory, at least in the mutants, which is not consistent with earlier data (Trautmann et al., 1990; Crielaard et al., 1992). More data is required on site-directed mutants to examine this point further.

Diffusion potentials were applied to membranes from the LH2-WT, βArg29 → Glu and βHis21 → Ser strains; all 3 contained electrochromically active carotenoids, but in the βHis21 mutant the difference spectrum was shifted 6 nm to the blue; a similar, smaller shift of 3.5 nm was observed for the βArg29 mutant. Finally, the relationship between the carotenoid absorption change and the membrane potential was examined. Each time a linear relationship was observed, demonstrating that a local field is still present, but in both mutants, particularly βArg29 → Glu, there was a severely reduced response. In all cases, an inverse potential reversed the sign of the absorption changes. The hypothesis that there is a large permanent field in the vicinity of the reactive carotenoids is strengthened by the observation that changing the positively charged βArg29 to glutamate has a major effect on the response to an applied membrane potential, indicating a change in the permanent field near the carotenoid. From the structural data of McDermott et al. (1995), it appears that βArg29 exerts an effect on the carotenoid despite the intervening presence of BChl 800; the creation of new mutations at this position should clarify the nature of the side chain-BChl-carotenoid interactions involved.

4. A Combinatorial Library of Localized Point Mutations

A careful inspection of the sequence alignments in Fig. 4 reveals a large number of potential point mutations which would yield interesting structural information, provided that the spectroscopy brought to bear on these mutants is sufficiently thorough. Nonetheless, even this 'rational' approach might represent potentially 20^{20} mutations, a number which would deter any spectroscopist. The second point to make is that there will be interesting mutations that are not readily suggested from sequence comparisons, and there is no way of predicting in advance what these will be. Douglas Youvan and co-workers have developed a different approach to mutagenesis, which involves a means of applying random, or semi-random mutagenesis to a target sequence which generates a large number of potential mutants (Goldman and Youvan, 1992; also see Chapter 61 by Goldman and Youvan). A typical library of mutants could consist of several thousand colonies, and then the pressure falls upon the speed and discrimination of screening, which in this case is facilitated by digital imaging spectroscopy. The spectral resolution of this technique is sufficient to allow the classification of 3 major classes of mutants in the case of LH2 : pseudo wild-type absorbing at 800 and 860 nm; pseudo LH1 showing reduced absorbance at 800 nm and a maximum near 875 nm, and finally mutants with no appreciable absorbance in the near infra-red. There are several ways to generate the mutations, for example by using error-based PCR of a whole subunit (McMaster and Hunter, unpublished), by combinatorial cassette mutagenesis which uses all 20 amino acids at each position, or by the aforementioned target set mutagenesis (Goldman and Youvan, 1992). In the latter case various parameters such as sequence alignments or structural considerations are used to limit the possible number of options; this has demonstrable benefits for the screening stage later on. The other limitation is the width of the target 'window' which was 15 residues in Goldman and Youvan (1992), sufficient to encompass almost half the length of a LH subunit. In this case, the window was located with the BChl-binding βHis-39 at the center. The DNA sequence is determined following the screening procedure, instead of beforehand as in conventional site-directed mutagenesis.

More recently Delagrave et al., (1993) described recursive ensemble mutagenesis, which is the application of successive rounds of combinatorial cassette mutagenesis (Goldman and Youvan, 1992), so that a list is compiled of 'acceptable' amino acids at each mutated position. This list allows another cassette to be designed which can increase the frequency of 'positive' mutants 30-fold, when compared to combinatorial cassette mutagenesis and 30,000-fold compared to random mutagenesis.

It is clear that the application of random and semi-random techniques will produce mutants which could not be designed or predicted merely from a study of aligned sequences, and therefore it is likely that this approach will make a valuable contribution to the study of antenna complexes in the future.

VI. Heterologous Expression of LH Genes in *Rhodobacter sphaeroides*

It is clear from the sequence alignments in Fig. 4 that nature has provided a large number of natural variants in light harvesting sequences among different species. Not many of these interesting bacteria can be examined by a full array of molecular genetic techniques, however. This difficulty can be overcome by expressing foreign LH genes in *Rb. sphaeroides*, for example. In Fig. 8 the absorbance spectrum of the *Rp. acidophila* LH2 antenna complex is shown, in an $LH2^-LH1^-RC^-$ *Rb. sphaeroides* background. Transcription is driven by the *Rb. sphaeroides puc* promoter (McGlynn and Hunter 1992) on a broad host range plasmid designed for expression of *Rb. sphaeroides* genes. There are several benefits from this approach: it is possible to observe and measure the absorbance of a *Rp. acidophila* LH2 complex for the first time in isolation and in a membrane environment, Second, *Rp. acidophila* has a family of many LH2 genes and this technique has allowed us to examine the complexes that result from previously 'silent' gene pairs. Third, the fact that the expressed complex appears to accommodate the host's carotenoid, neurosporene, in place of the normal pigment rhodopin poses interesting questions about the specificity of carotenoid binding in these antenna.

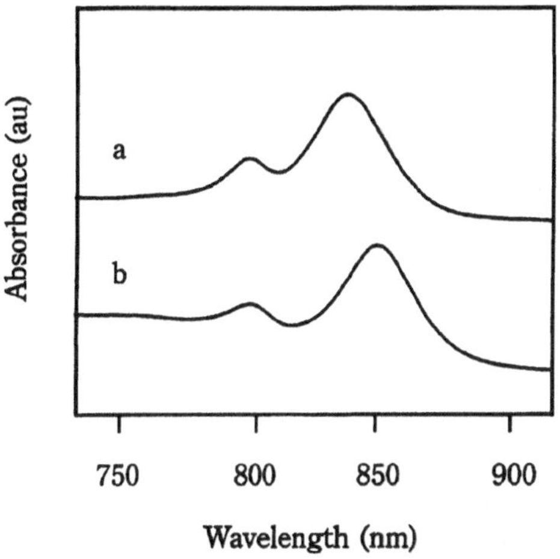

Fig. 8. Absorbance spectrum of *pucBA* gene pairs from (a) *Rp. acidophila* and (b) *Rc. gelatinosus* expressed in a $LH2^-LH1^-RC^-$ *Rb. sphaeroides* background (G. J. S. Fowler, A. T. Gardiner, R. C. Mackenzie, S. J. Barratt, A. E. Simmons, R. J. Cogdell and C. N. Hunter, unpublished).

Finally, when expressed in an $LH1^+RC^+$ *Rb. sphaeroides* background a mixed photosynthetic unit is obtained which raises new possibilities for productive contacts between membrane proteins.

All of the above can just as easily be applied to complexes from *Rc. gelatinosus* (see Fig. 8) and this approach of heterologous expression could in principle be extended further to examine other spectroscopically interesting antennae.

VII. Concluding Remarks

The objective of this review was to establish the value of genetic alterations to the study of bacterial antennas, firstly as a means of generating 'partial' systems, secondly for site-directed mutagenesis and finally for the study of antenna complexes for genetically poorly characterised bacteria. It is undoubtedly true that altered antenna systems muddy the waters a little. To echo the sentiments expressed in Marrs (1978): 'Why do you study crippled strains? It's hard enough to figure out how the normal ones work'. However, a thorough and patient spectroscopic examination of genetically altered complexes will, in combination with structural studies, reveal many of the molecular details of these complexes within the next few years. By the time this chapter is published, the structure of the *Rp. acidophila* LH2 complex will have been determined by X-ray crystallography (McDermott et al., 1995). With this new information, we look forward to using mutagenesis techniques to examine the role of various amino acid residues in determining the properties of this antenna.

References

Allen JP, Feher G, Yeates TO, Komiya H and Rees DC (1987) Structure of the reaction center from *Rhodobacter sphaeroides* R-26: The protein subunits. Proc Natl Acad Sci USA 84: 6162–6166

Ashby MK, Coomber SA and Hunter CN (1987) Cloning, nucleotide sequence and transfer of genes for the B800–850 light harvesting complex of *Rhodobacter sphaeroides*. FEBS Lett 209: 83–86

Babst M, Albrecht H, Wegmann I, Brunisholz R and Zuber H (1991) Single amino acid substitutions in the B870 α and β light-harvesting polypeptides of *Rhodobacter capsulatus*. Structural and spectral effects. Eur J Biochem 202: 277–284

Bachmann RC, Tadros MH, Oelze J and Takemoto JY (1983) Purification and characterization of the B875 light-harvesting complex of *Rhodopseudomonas sphaeroides*. Biochem Int 7: 629–634

Barz WP and Oesterhelt D (1994) Photosynthetic deficiency of a *pufX* deletion mutant of *Rhodobacter sphaeroides* is suppressed by point mutations in the light harvesting complex genes *pufB* or *pufA*. Biochemistry 33: 9741–9752

Bélanger G and Gingras G (1988) Structure and expression of *puf* operon messenger RNA in *Rhodospirillum rubrum*. J Biol Chem 263: 7639–7645

Belasco JG, Beatty JT, Adams CW, von Gabain A and Cohen SN (1985) Differential expression of photosynthesis genes in *Rhodopseudomonas capsulata* results from segmental differences in stability within the polycistronic *rxcA* transcript. Cell 40: 171–181

Bérard J, Bélanger G, Corriveau P and Gingras G (1986) Molecular cloning and sequence of the B880 holochrome gene from *Rhodospirillum rubrum*. J Biol Chem 261: 82–87

Bergström H, Westerhuis WHJ, Sundström V, van Grondelle R, Niederman RA and Gillbro T (1988) Energy transfer within the isolated B875 light-harvesting pigment-protein complex of *Rhodobacter sphaeroides* at 77K studied by picosecond absorption recovery. FEBS Lett 233: 12–16

Bolt JD, Sauer K, Shiozawa JA and Drews G (1981a) Linear and circular dichroism of membranes from *Rhodopseudomonas capsulata*. Biochim Biophys Acta 635: 535–541

Bolt JD, Hunter CN, Niederman RA and Sauer K (1981b) Linear and circular dichroism and fluorescence polarisation of the B875 light harvesting bacteriochlorophyll-protein complex from *Rhodopseudomonas sphaeroides*. Photochem Photobiol 34: 653–656

Boonstra AF, Visschers RW, Calkoen F, van Grondelle R, van Bruggen EFJ and Boekema EJ (1993) Structural characterization of the B800–850 and B875 light-harvesting antenna complexes from *Rhodobacter sphaeroides* by electron microscopy. Biochim Biophys Acta 1142: 181–188

Borisov AY, Gadonas RA, Danielius RV, Piskarkas AS and Razjivan AP (1982) Minor component B-905 of light harvesting antenna in *Rhodospirillum rubrum* chromatophores and the mechanisms of singlet-singlet annihilation as studied by difference selective picosecond spectroscopy. FEBS Lett 138: 25–28

Braun P and Scherz A (1991) Polypeptides and bacteriochlorophyll organization in the light-harvesting complex B850 of *Rhodobacter sphaeroides* R26.1 Biochemistry 30: 5177–5184

Broglie RM, Hunter CN, Delepelaire P, Niederman RA, Chua N-H and Clayton RK (1980) Isolation and characterization of the pigment-protein complexes of *Rhodopseudomonas sphaeroides* by lithium dodecyl sulfate polyacrylamide gel electrophoresis. Proc Natl Acad Sci USA 77: 87–91

Brunisholz RA and Zuber H (1992) Structure, function and organization of antenna polypeptides and antenna complexes from the three families of Rhodospirillaceae. J Photochem Photobiol B Biol 15: 113–140

Burgess JG, Ashby MK and Hunter CN (1989) Chromosomal deletion of genes encoding B800–850 (LH2) polypeptides of *Rhodobacter sphaeroides*. J Gen Microbiol 135: 1809–1816

Bylina EJ, Robles SJ and Youvan DC (1988) Directed mutations affecting the putative bacteriochlorophyll-binding sites in the light-harvesting I antenna of *Rhodobacter capsulatus*. Israel J Chem 28: 73–78

Chachisvilis M, Pullerits T, Jones MR, Hunter CN and Sundström V (1994) Vibrational dynamics in the light harvesting complexes of the photosynthetic bacterium *Rhodobacter sphaeroides*. Chem Phys Lett 224: 345–351

Chang C-H, El-Kabbani O, Tiede D, Norris J and Schiffer J (1991) Structure of the membrane-bound protein photosynthetic reaction center from *Rhodobacter sphaeroides*. Biochemistry 30: 5352–5360

Chang MC, Mayer L and Loach PA (1990a) Isolation and characterisation of a structural subunit from the core light harvesting complex of *Rhodobacter sphaeroides* 2.4.1 and *puc* 705-BA. Photochem Photobiol 52: 873–881

Chang MC, Callahan PM, Parkes-Loach PS, Cotton TM and Loach PA (1990b) Spectroscopic characterisation of the light-harvesting complex of *Rhodospirillum rubrum* and its structural subunit. Biochemistry 29: 421–429

Clayton RK and Clayton BJ (1972) Relations between pigments and proteins in the photosynthetic membranes of *Rhodopseudomonas sphaeroides*. Biochim Biophys Acta 283: 492–504

Clayton RK and Clayton BJ (1981) B850 pigment-protein complex of *Rhodopseudomonas sphaeroides*: extinction coefficients, CD and the reversible binding of bacteriochlorophyll. Proc Natl Acad Sci USA 78: 5583–5387

Clayton RK and Haselkorn R (1972) Protein components of bacterial photosynthetic membranes. J Mol Biol 68: 97–105

Clayton RK and Smith C (1960) *Rhodopseudomonas sphaeroides*: high catalase and blue-green double mutants. Biochem Biophys Res Commun 3: 143–145

Clayton RK and Wang RT (1971) Photochemical reaction centres from *Rhodopseudomonas sphaeroides*. Methods Enzymol 23: 696–704

Coleman WJ and Youvan DC (1990) Spectroscopic analysis of genetically modified photosynthetic reaction centers. Ann Rev Biophys Biophys Chem 19: 333–367

Coomber SA and Hunter CN (1989) Construction of a physical

map of the 45-kb photosynthetic gene cluster of *Rhodobacter sphaeroides*. Arch Microbiol 151: 454–458

Coomber SA, Chaudhri M, Connor A, Britton G and Hunter CN (1990) Localised transposon Tn5 mutagenesis of the photosynthetic gene cluster of *Rhodobacter sphaeroides*. Mol Microbiol 4: 977–989

Crielaard W, van Mourik F, van Grondelle R, Konings WN, Hellingwerf KJ (1992) Spectral identification of the electronically active carotenoids of *Rhodobacter sphaeroides* in chromatophores and reconstituted liposomes. Biochim Biophys Acta 1100: 9–14

Crielaard W, Visschers RW, Fowler GJS, van Grondelle R, Hellingwerf KJ and Hunter CN (1994) Probing the B800 bacteriochlorophyll binding site of the accessory light harvesting complex from *Rhodobacter sphaeroides* using site directed mutants: I Mutagenesis, effects on carotenoid binding, functional and electrochromic behavior. Biochim Biophys Acta 1183: 473–482

Davis J, Donohue TJ and Kaplan (1988) Construction, characterization and complementation of a Puf$^-$ mutant of *Rhodobacter sphaeroides*. J Bacteriol 170: 320–329

Davidson E and Cogdell R (1981) The polypeptide composition of the B850 light-harvesting pigment and protein complex from *Rhodopseudomonas sphaeroides*, R26.1. FEBS Lett 132: 81–84

Deisenhofer J, Epp O, Miki K, Huber R and Michel H (1985) Structure of protein subunits in the photosynthetic reaction centre of *Rhodopseudomonas viridis* of 3Å resolution. Nature 318: 618–624

Delagrave SD, Goldman ER and Youvan DC (1993) Recursive ensemble mutagenesis. Protein Engineering 6: 327–331

Dorge B, Klug G, Gad'on N, Cohen SN and Drews G (1990) Effects on the formation of antenna complex B870 of *Rhodobacter capsulatus* by exchange of charged amino acids in the N-terminal domain of the α and β pigment binding proteins. Biochemistry 29: 7754–7758

Drews G and Oelze J (1981) Organization and differentiation of membranes of phototrophic bacteria. Adv Microb Physiol 22: 1–92

Duysens LNM (1952) Transfer of excitation energy in photosynthesis. Thesis, University of Utrecht pp 1–96

Farchaus JW and Oesterhelt (1989) A *Rhodobacter sphaeroides pufL*, *M* and *X* deletion mutant and its complementation in *trans* with a 5.3 kb *puf* operon shuttle fragment. EMBO J 8: 47–54

Farchaus JW, Gruenberg H and Oesterhelt D (1990) Complementation of a reaction centre-deficient *Rhodobacter sphaeroides pufLMX* deletion strain in *trans* with *puf BALM* does not restore the photosynthesis-positive phenotype. J Bacteriol 172: 977–985

Feher G (1971) Some chemical and physical properties of a bacterial reaction center particle and its primary photochemical reactants. Photochem Photobiol 14: 373–387

Feick R and Drews G (1978) Isolation and characterisation of light harvesting bacteriochlorophyll protein complexes from *Rhodopseudomonas capsulata*. Biochim Biophys Acta 501: 499–513

Feick R, van Grondelle R, Rijgersberg CP and Drews G (1980) Fluorescence emission by wild-type and mutant strains of *Rhodopseudomonas capsulata*. Biochim Biophys Acta 593: 241–253

Fenna BR and Matthews BW (1975) Chlorophyll arrangement in a bacteriochlorophyll protein from *Chlorobium limicola*. Nature 258: 573–577

Fonstein M and Haselkorn R (1993) Chromosomal structure of *Rhodobacter capsulatus* strain SB1003-cosmid encyclopaedia and high resolution physical and genetic map. Proc Natl Acad Sci USA 90: 2522–

Fowler GJS, Visschers RW, Grief GG, van Grondelle R and Hunter CN (1992) Genetically modified photosynthetic antenna complexes with blue-shifted absorbance bands. Nature 355: 848–850

Fowler GJS, Crielaard W, Visschers RW, Grondelle R and Hunter CN (1993) Site-directed mutagenesis of the LH2 light-harvesting complex of *Rhodobacter sphaeroides*; changing βLys23 to Gln results in a shift in the 850 nm absorption peak. Photochem Photobiol 57: 2–6

Fowler GJS, Sockalingum GD, Robert B and Hunter CN (1994) Blue shifts in bacteriochlorophyll absorbance correlate with changed hydrogen bonding patterns in light-harvesting 2 mutants of *Rhodobacter sphaeroides* with alterations at α-Tyr-44 and α-Tyr-45. Biochem J 299: 695–700

Garcia AF, Mantele W, Gadon N, Tadros MH and Drews G. (1991) Growth and photosynthetic activities of wild-type and antenna deficient mutant strains of *Rhodobacter capsulatus*. Arch Microbiol 155: 205–209

Ghosh R, Hauser H and Bachofen R (1988) Reversible dissociation of the B873 light harvesting complex from, *Rhodospirillum rubrum* G9+. Biochemistry 27: 1004–1014

Ghosh R, Hoenger A, Hardmeyer A, Mihailescu D, Bachofen R, Engel A and Rosenbusch JP (1993) Two-dimensional crystallisation of the light-harvesting complex from *Rhodospirillum rubrum* J Mol Biol 231: 501–504

Gibson LCD, McGlynn P, Chaudhri M and Hunter CN (1992) A putative anaerobic coproporphyrinogen III oxidase in *Rhodobacter sphaeroides*. II. Analysis of a region of the genome encoding *hemF* and the *puc* operon. Molecular Microbiology 6: 3171–3186

Goldman ER and Youvan DC (1992) An algorithmically optimized combinatorial library screened by digital imaging spectroscopy. Biotechnology 10: 1557–1561

Golecki JR, Ventura S, Oelze J (1991) The architecture of unusual membrane tubes in the B800–850 light-harvesting bacteriochlorophyll-deficient mutant 19 of *Rhodobacter sphaeroides*. FEMS Microbiol Lett 77: 335–340

Griffiths M and Stanier RY (1956) Some mutational changes in the photosynthetic pigment system of *Rhodopseudomonas sphaeroides*. J Gen Microbiol 14: 698–715

Guthrie N, MacDermott G, Cogdell RJ, Freer AA, Isaacs NW, Hawthornthwaite AM, Halloren E and Lindsay JG (1992) Crystallisation of the B800–820 light-harvesting complex from *Rhodopseudomonas acidophila* strain 7750. J Mol Biol 224: 527–528

Heller BA and Loach PA (1991) Isolation and characterization of a subunit form of the B875 light harvesting complex from *Rhodobacter capsulatus*. Photochem Photobiol 51: 621–628

Hess S, Visscher K, Ulander J, Pullerits T, Jones MR, Hunter CN and Sundström V (1993) Direct energy transfer from the peripheral LH2 antenna to the reaction centre in a mutant of *Rhodobacter sphaeroides* which lacks the core LH1 antenna. Biochemistry 32: 10314–1032

Hess S, Visscher KJ, Sundström V, Fowler GJS and Hunter CN

(1994) Enhanced rates of subpicosecond energy transfer in blue-shifted light harvesting LH2 mutants of *Rhodobacter sphaeroides*. Biochemistry 33: 8300–8305

Hunter CN (1988) Transposon Tn5 mutagenesis of genes encoding reaction center and light harvesting LH1 polypeptides of *Rhodobacter sphaeroides*. J Gen Microbiol 134: 1481–1489

Hunter CN and Jones OTG (1979) The kinetics of flash induced electron flow in bacteriochlorophyll-less mutants of *Rhodopseudomonas sphaeroides* reconstituted with reaction centres. Biochim Biophys Acta 545: 339–351

Hunter CN and van Grondelle R (1988) The use of mutants to investigate the organization of the photosynthetic apparatus of *Rhodobacter sphaeroides*. In: Scheer H and Schneider S (eds) Structure and Function in Photosynthetic Light Harvesting Systems, pp 247-260 De Gruyter, Berlin

Hunter CN, van Grondelle R, Holmes NG and Jones OTG (1979) The reconstitution of energy transfer in membranes from a bacteriochlorophyll-less mutant of *Rhodopseudomonas sphaeroides* by addition of light harvesting and reaction centre pigment-protein complexes. Biochim Biophys Acta 548: 458–470

Hunter CN, Pennoyer JD, Sturgis JN, Farrelly D and Neiderman RA (1988) Oligomerisation states and associations of light-harvesting pigment protein complexes of *Rhodobacter sphaeroides* as analyzed by lithium dodecyl sulphate-polyacrylamide gel electrophoresis. Biochemistry 27: 3459–3467

Hunter CN, van Grondelle R and van Dorssen RJ (1989) Construction and properties of M2192, a mutant of *Rhodobacter sphaeroides* with the B875 antenna as the sole pigment protein. Biochim Biophys Acta 973: 383–389

Hunter CN, Bergström H, van Grondelle R and Sundström V (1990) Energy transfer dynamics in three photosynthetic mutants of *Rhodobacter sphaeroides* : A picosecond spectroscopy study. Biochemistry 29: 3203–3207

Hunter CN, McGlynn P, Ashby MK, Burgess JG and Olsen JD (1991) DNA sequencing and complementation/deletion analysis of the *bchA-puf* operon region of *Rhodobacter sphaeroides*: in vivo mapping of the oxygen-regulated *puf* promoter. Molecular Microbiology 5: 153–179

Hunter CN, Fowler GJS, Grief GG and Olsen JD (1993) Protein engineering of bacterial light harvesting complexes. Biochem Soc Trans 21: 41–43

Jackson WJ, Prince RC, Stewart GJ and Marrs BC (1986) Energetic and topographic properties of a *Rhodopseudomonas capsulata* mutant deficient in the B870 complex. Biochemistry 25: 8440–8446

Jackson WJ, Kiley PJ, Haith CE, Kaplan S and Prince RC (1987) On the role of the light-harvesting B880 in the correct insertion of the reaction center of *Rhodobacter capsulatus* and *Rhodobacter sphaeroides*. FEBS Lett 215: 171–174

Jones MR, Fowler GJS, Gibson LCD, Grief GG, Olsen J, Crielaard W and Hunter CN (1992a) Construction of mutants of *Rhodobacter sphaeroides* lacking one or more pigment protein complexes and complementation with reaction centre, LH1, and LH2 genes. Molecular Microbiology 6: 1173–1184

Jones MR, Visschers RW, van Grondelle R and Hunter CN (1992b) Construction and characterisation of a mutant of *Rhodobacter sphaeroides* with the reaction centre as the sole pigment-protein complex. Biochemistry 31: 4458–4465

Karrasch S, Bullough PA and Ghosh R (1995) The 8.5 Å projection map of the light harvesting complex I from *Rhodospirillum rubrum* reveals a ring composed of 16 subunits. EMBO J 14: 631–638

Kaufmann N, Hudig H and Drews G (1984) Transposon Tn5 mutagenesis of genes for the photosynthetic apparatus in *Rhodopseudomonas capsulata*. Mol Gen Genet 198: 153–158

Keen NT, Tamaki S, Kobayashi D and Trollinger D (1988) Improved broad-host-range plasmids for DNA cloning in Gram-negative bacteria. Gene 70: 191–197

Kiley PJ and Kaplan S (1987) Cloning, sequence and expression of the *Rhodobacter sphaeroides* light-harvesting B800–850 α and B800–850 β genes. J Bacteriol 169: 3268–3275

Kiley PJ, Donohue TJ, Havelka WA and Kaplan S (1987) DNA sequence and in vitro expression of the B875 light-harvesting polypeptides of *Rhodobacter sphaeroides*. J Bacteriol 169: 742–750

Kiley PJ, Varga A and Kaplan S (1988) Physiological and structural analysis of light-harvesting mutants of *Rhodobacter sphaeroides*. J Bacteriol 170: 1103–1115

Kramer HJM, van Grondelle R, Hunter CN, Westerhuis WHJ and Amesz J (1984a) Pigment organization of the B800–850 antenna complex of *Rhodopseudomonas sphaeroides*. Biochim Biophys Acta 765: 156–165

Kramer HJM, Pennoyer JD, van Grondelle R, Westerhuis WHJ, Niederman RA and Amesz J (1984b) Low temperature optical properties and pigment organization of the B875 light-harvesting bacteriochlorophyll protein complex of the purple photosynthetic bacteria. Biochim Biophys Acta 767: 335–344

Kunkel TA (1985) Rapid and efficient site-specific mutagenesis without phenotypic selection. Proc Natl Acad Sci USA 82: 488–492

Lilburn TG, Haith CE, Prince RC and Beatty JT (1992) Pleiotropic effects of *pufX* gene deletion on the structure and function of the photosynthetic apparatus of *Rhodobacter capsulatus* Biochim Biophys Acta 1100:160–170

Loach PA, Parkes PS, Miller JF, Hinchigeri S and Callahan PM (1985) Structure-function relationships of the bacterio-chlorophyll-protein light-harvesting complex of *Rhodo-spirillum rubrum*. In: Steinback KE, Bonitz S, Arntzen CJ and Bogorad L (eds) Molecular Biology of the Photosynthetic Apparatus, pp 197–207. Cold Spring Harbor Laboratory, New York

Lommen MAJ and Takemoto J (1978) Ultrastructure of carotenoid mutant strain R-26 of *Rhodopseudomonas sphaeroides*. Arch Microbiol 118: 305–308

Marrs, BL (1978) Mutations and genetic manipulations as probes of bacterial photosynthesis. Curr Top Bioenerg 8: 261–294

McDermott G, Prince SM, Freer AA, Hawthornthwaite-Lawless AM, Papiz MZ, Cogdell RJ and Isaacs NW (1995) Crystal structure of an integral membrane light-harvesting complex from photosynthetic bacteria. Nature 374: 517–521

McGlynn P and Hunter CN (1992) Isolation and characterisation of a putative transcription factor involved in the regulation of the *Rhodobacter sphaeroides puc BA* operon. J Biol Chem 267: 11098–11103

McGlynn P, Hunter CN and Jones MP (1994) The *Rhodobacter sphaeroides* PufX protein is not required for photosynthetic competence in the absence of a light harvesting system. FEBS lett 349: 349–353

McMaster L, Olsen JD, Jones MR and Hunter CN (1992) Site directed mutagenesis studies of the highly conserved Trp-8 residue of the α polypeptide of LH1 in *Rhodobacter sphaeroides*: effects on assembly of LHI and RC complexes. Photosynth Res 34: 186

Meinhardt SW, Kiley PJ, Kaplan S, Crofts AR and Harayama S (1985) Characterization of light harvesting mutants of *Rhodopseudomonas sphaeroides*. I. Measurement of the efficiency of energy transfer from light harvesting complexes to the reaction center. Arch Biochem Biophys 236: 130–139

Michel H (1991) General and practical aspects of membrane protein crystallisation. In: Michel H (ed) Crystallisation of Membrane Proteins, pp 82–83. CRC Press, Boca Raton, FL

Miller JF, Hinchingeri PS, Parkes-Loach PM, Callahan JR, Sprinkle JR, Riccobono JR, and Loach PA (1987) Isolation and characterisation of a subunit form of light harvesting complex of *Rhodospirillum rubrum*. Biochemistry 26: 5055–5062

Miller L and Kaplan S (1978) Plasmid transfer and expression in *Rhodopseudomonas sphaeroides*. Arch Biochem Biophys 187: 229–234

Nilsson I and von Heijne G (1990) Fine-tuning the topology of a polytopic membrane protein role of positively and negatively charged amino acids. Cell 62: 1135–1141

Nunn RS, Artymiuk PJ, Baker PJ, Rice DW and Hunter CN (1992) Purification and crystallization of the light harvesting LH1 complex from *Rhodobacter sphaeroides*. J Mol Biol 228: 1259–1262

Olsen JD, Sockalingum GD, Robert B and Hunter CN (1994) Modification of a hydrogen bond between a bound chromophore and the α subunit of the light harvesting LHI antenna of *Rhodobacter sphaeroides*. Proc Natl Acad Sci USA 91: 7124–7128

Papiz MZ, Hawthornthwaite AM, Cogdell RJ, Woolley KJ, Wightman PA, Ferguson LA and Lindsay JG (1989) Crystallisation and characterisation of two crystal forms of the B800–850 light-harvesting complex from *Rhodobacter acidophila* strain 10050. J Mol Biol 209: 833–835

Parkes-Loach PS, Sprinkle JR and Loach PA (1988) Reconstitution of the B873 light harvesting complex of *Rhodospirillum rubrum* with the separately isolated α and β polypeptides and BChla. Biochemistry 27: 2718–2727

Pemberton JM and Bowen ARSG (1981) High-frequency chromosome transfer in *Rhodopseudomonas sphaeroides* promoted by the broad host range plasmid RP1 carrying the mercury transposon Tn501. J Bacteriol 147: 110–117

Pullerits T, Chachisvilis M, Jones MR, Hunter CN and Sundström V (1994) Exciton dynamics in the light-harvesting complexes of *Rhodobacter sphaeroides*. Chem Phys Lett 224: 355–365

Reddy NRS, Lyle PA and Small GJ (1992) Application of spectral hole burning spectroscopies to antenna and reaction centre complexes. Photosynth Res 31: 167–194

Reed DW (1969) Isolation and composition of a photosynthetic reaction center complex from *Rhodopseudomonas sphaeroides*. J Biol Chem 244: 4396–4941

Richter P and Drews G (1991) Incorporation of light-harvesting complex I α and β polypeptides into the intracytoplasmic membrane of *Rhodobacter capsulatus* J Bacteriol 173: 5336–5345

Richter P, Cortez N and Drews G (1991) Possible role of the highly conserved amino acids Trp-8 and Pro-13 in the N-terminal segment of the pigment-binding polypeptide LHIα of *Rhodobacter capsulatus* FEBS Lett 285: 80–84

Richter P, Brand M and Drews G (1992) Characterisation of LHI^- and LHI^+ *Rhodobacter capsulatus pufA* mutants. J Bacteriol 174: 3030–3041

Rijgersberg CP, van Grondelle R and Amesz J (1980) Energy transfer and bacteriochlorophyll fluorescence in purple bacteria at low temperature. Biochim Biophys Acta (1980) 592: 53–64

Robert B and Lutz M. (1985) Structure of antenna complexes of several *Rhodospirillaceae* from their resonance Raman spectra. Biochim Biophys Acta 807: 10–23

Robert B, Vermeglio A and Lutz M (1984) Structural characterisation and comparison of antenna complexes of R26 and R26.1 mutants of *Rhodopseudomonas sphaeroides*. Biochim Biophys Acta 766: 259–262

Sauer K and Austin LA (1978) Bacteriochlorophyll-protein complexes from the light harvesting antenna of photosynthetic bacteria. Biochemistry 17: 2011–2019

Schiffer M, Chang C-H and Stevens FJ (1992) The functions of tryptophan residues in membrane proteins. Protein Engineering 5: 213–214

Scolnik PA, Zannoni D and Marrs BL (1980) Spectral and functional comparisons between the carotenoids of the two antenna complexes of *Rhodopseudomonas capsulata*. Biochim Biophys Acta 593: 230–240

Sebban P, Jolchine G and Moya I (1984) Spectra fluorescence lifetime and intensity of *Rhodopseudomonas sphaeroides* at room and low temperature. Comparison between wild-type the C71 reaction center-less mutant and the B800–850 pigment protein complex. Photochem Photobiol 39: 247–253

Sebban P, Robert B and Jolchine G (1985) Isolation and spectroscopic characterisation of the B875 antenna complex of a mutant of *Rhodopseudomonas sphaeroides*. Photochem Photobiol 42: 573–577

Sistrom WR (1977) Transfer of chromosomal genes mediated by plasmid R68.45 in *Rhodopseudomonas sphaeroides*. J Bacteriol 131: 526–532

Sistrom WR and Clayton RK (1963) Studies on a mutant of *Rhodopseudomonas sphaeroides* unable to grow photosynthetically. Biochim Biophys Acta 88: 61–73

Sistrom WR, Griffiths M and Stanier RY (1956) The biology of a photosynthetic bacterium which lacks coloured carotenoids. J Cell Comp Physiol 48: 473–575

Sockett RE, Donohue TJ, Varga AR and Kaplan S (1989) Control of photosynthetic membrane assembly in *Rhodobacter sphaeroides* medicated by *puhA* and flanking sequences. J Bacteriol 171: 436–446

Stiehle H, Cortez N, Klug G and Drews G (1990) A negatively charged N terminus in the α polypeptide inhibits formation of light-harvesting complex I in *Rhodobacter capsulatus*. J Bacteriol 172: 7131–7137

Sturgis JN, Hunter CN and Niederman RA (1988) Spectra and extinction coefficients of near-infrared absorption bonds in membranes of *Rhodobacter sphaeroides* mutants lacking light-harvesting and reaction center complexes. Photochem Photobiol 48: 243–247

Suwanto A and Kaplan S (1989) Physical and genetic mapping of the *Rhodobacter sphaeroides* 2.4.1. genome: genome size, fragment identification and gene localisation. J Bacteriol 171: 5850–5859

Tadros MH, Zuber H and Drews G (1982) The polypeptide components from light-harvesting pigment-protein complex II (B800–850) of *Rhodopseudomonas capsulata*. Solubilisation,

purification and sequence studies. Eur J Biochem 127: 315–318
Tadros MH, Suter F, Drews G and Zuber H (1983) The complete amino-acid sequence of the large bacteriochlorophyll-binding polypeptide from light-harvesting complex II (B800–850) of *Rhodopseudomonas capsulata*. Eur J Biochem 129: 533–536
Tadros MH, Suter F, Seydewitz HH, Witt I, Zuber H and Drews G (1984) Isolation and complete amino-acid sequence of the small polypeptide from light-harvesting pigment-protein complex I (B870) of *Rhodopseudomonas capsulata*. Eur J Biochem 138:209–212
Theiler R and Zuber H (1984) The light-harvesting polypeptides of *Rhodopseudomonas sphaeroides* R26.1 II. Conformational analyses of attenuated total reflection infrared spectroscopy and the possible molecular structure of the hydrophobic domain of the B850 complex. Hoppe-Seyler's Z Physiol Chem 365: 721–729
Theiler R, Suter F, Wiemken V and Zuber H (1984) The light-harvesting polypeptides of *Rhodopseudomonas sphaeroides* R26.1. I. Isolation, purification and sequence analyses. Hoppe-Seyler's Z Physiol Chem 365: 703–719
Theiler R, Suter F, Pennoyer JD, Zuber H and Neiderman RA (1985) Complete amino acid sequence of the B875 light-harvesting protein of *Rhodopseudomonas sphaeroides* strain 2.4.1. Comparison with R26.1 carotenoidless-mutant strain. FEBS Lett 184: 231–236
Tichy HV, Oberle B, Stiehle H, Schiltz E and Drews G (1989) Genes downstream from *pucB* and *pucA* are essential for formation of the B800–850 complex of *Rhodobacter capsulatus*. J Bacteriol 171: 4914–4922
Tichy HV, Albien KU, Gadion N and Drews G (1991) Analysis of the *Rhodobacter capsulatus puc operon*: the *pucC* gene plays a central role in the regulation of LHII (B800–850 complex) expression. EMBO J 10: 2949–2955
Timpmann K, Freiberg A and Godik VF (1991) Picosecond kinetics of light excitations in photosynthetic purple bacteria in the temperature range 300–4K. Chem Phys Lett 182: 617–622
Trautmann JK, Shreve AP, Violette CA, Frank HA, Owens TG and Albrecht AC. Femtosecond dynamics of energy transfer in B800–850 light-harvesting complexes of *Rhodobacter sphaeroides*. Proc Natl Acad Sci USA 87: 215–219
Valkunas L, van Mourik F and van Grondelle R (1992) On the role of spectral and spatial antenna inhomogeneity in the process of excitation energy transfer in photosynthesis. J. Photochem. Photobiol B Biol 15: 159–170
Van der Laan H, de Caro C, Schmidt T, Visschers RW, van Grondelle R, Fowler GJS, Hunter CN and Völker S (1993) Excited-state dynamics of mutated antenna complexes of purple bacteria studied by hole burning. Chem Phys Lett 212: 569–580
Van Dorssen RJ, Hunter CN, van Grondelle R, Korenhof AH and Amesz J (1988) Spectroscopic properties of antenna complexes of *Rhodobacter sphaeroides* in vivo. Biochim Biophys Acta 932: 179–188
Van Grondelle R, Kramer HJM and Rijgersberg CP (1982) Energy transfer in the B800–850 carotenoid light-harvesting complex of various mutants of *Rhodopseudomonas sphaeroides* and of *Rhodopseudomonas capsulata*. Biochim Biophys Acta 682: 208–215
Van Grondelle R, Hunter CN, Bakker JGC and Kramer HJM (1983) Size and structure of antenna complexes of photosynthetic bacteria as studied by singlet-singlet quenching of the bacteriochlorophyll fluorescence yield. Biochim Biophys Acta 723: 30–36
Van Grondelle R, Bergström H, Sundström V, van Dorssen RJ, Vos M, and Hunter CN (1988) Excitation energy transfer in the light harvesting antenna of purple photosynthetic bacteria: the role of the long wavelength absorbing pigment B896. In: Scheer H and Schneider S (eds) Structure and Function in Photosynthetic Light Harvesting Systems, pp 519–530 De Gruyter, Berlin
Van Mourik F, van der Oord CJR, Visscher KJ, Parkes-Loach PS, Loach PA, Visschers RW and van Grondelle R (1991) Excitation interactions in the light-harvesting antenna of photosynthetic bacteria studies with triplet-singlet annihilation on the B820 subunit from *Rhodospirillum rubrum*. Biochim Biophys Acta 1059: 111–119
Van Mourik F, Visschers RW and Van Grondelle R (1992) Energy transfer and aggregate size effects in the inhomogeneously broadened core light harvesting complex of *Rhodobacter sphaeroides* Chem Phys Lett 193: 1–7
Visschers KJ, Bergström H, Sundström V, Hunter CN and van Grondelle R (1989) Temperature dependence of energy transfer from the long wavelength antenna BChl-896 to the reaction center in *Rhodospirillum rubrum* and *Rhodobacter sphaeroides* (wt and M21 mutant) from 77 to 177K studied by picosecond absorption spectroscopy. Photosynth Res 22: 211–217
Visschers RW, Chang MC, van Mourik F, Parkes-Loach PS, Heller BA, Loach PA and van Grondelle R (1991) Fluorescence polarisation and low temperature absorption spectroscopy of a subunit form of light harvesting complex I from purple photosynthetic bacteria. Biochemistry 30: 2951–2960
Visschers RW, Nunn R, Calkoen F, van Mourik F, Hunter CN, Rice DW and van Grondelle R (1992) Spectroscopic characterisation of B820 subunits from light harvesting complex I of *Rhodospirillum rubrum* and *Rhodobacter sphaeroides* prepared with the detergent N-octyl-rac-2, 3 dipropylsulfoxide. Biochim Biophys Acta 1100: 259–266
Visschers RW, van Mourik F, Monshouwer R and van Grondelle R (1993a) Inhomogeneous spectral broadening of B820 subunit form of LH1. Biochim Biophys Acta 1141: 238–244
Visschers RW, van Grondelle R and Robert B (1993b) Resonance Raman spectroscopy of the B820 subunit of the core antenna from *Rhodospirillum rubrum* G9. Biochim Biophys Acta 1183: 369–373
Visschers RW, Crielaard W, Fowler GJS, Hunter CN and van Grondelle R (1994) Probing the B800 bacteriochlorophyll binding site of the peripheral light-harvesting complex from *Rhodobacter sphaeroides* using site directed mutants: II. A low temperature spectroscopy study of structural aspects of pigment-protein conformation. Biochim Biophys Acta 1183: 483–490
von Heijne G (1989) Control of topology and mode of assembly of a polytopic membrane protein by positively charged residues. Nature 341: 456–458
Vos M, van Dorssen RJ, Amesz J, van Grondelle R and Hunter CN (1988) The organization of the photosynthetic apparatus of *Rhodobacter sphaeroides*: studies of antenna mutants using singlet-singlet quenching. Biochim Biophys Acta 933: 132–140
Welte W, Wacker T, Leis M, Krentz W, Shiozawa J, Gad'on N

and Drews G (1985) Crystallisation of the photosynthetic light-harvesting pigment-protein complex B800–850 of *Rhodobacter capsulatus*. FEBS Lett 182: 260–264

Westerhuis WHJ, Farchaus JW and Niederman RA (1993) Altered spectral properties of the B875 light harvesting pigment-protein complex in a *Rhodobacter sphaeroides* mutant lacking *pufX*. Photochem Photobiol 53: 460–463

Yen H-C and Marrs B (1976) Map of genes for carotenoid and bacteriochlorophyll biosynthesis in *Rhodopseudomonas capsulata*. J Bacteriol 126: 619–629

Yen H-C, Hu NT and Marrs BL (1979) Characterization of the gene transfer agent made by an overproducer mutant of *Rhodopseudomonas capsulata*. J Mol Biol 131: 157–168

Youvan DC and Ismail S (1985) Light-harvesting (B800–850 complex) structural genes from *Rhodopseudomonas capsulata*. Proc Natl Acad Sci USA 82: 58–62

Youvan DC, Hearst JE and Marrs BL (1983) Isolation and characterization of enhanced fluorescence mutants of *Rhodopseudomonas capsulata*. J Bacteriol 154: 748–755

Youvan DC, Bylina EJ, Alberti MH, Begusch H and Hearst JE (1984) Nucleotide and deduced polypeptide sequences of the photosynthetic reaction-center B870 antenna and flanking polypeptides from *Rhodopseudomonas capsulata*. Cell 37: 937–947

Youvan DC, Ismail S and Bylina EJ (1985) Chromosomal deletion and plasmid complementation of the photosynthetic reaction center and light harvesting genes from *Rhodopseudomonas capsulata*. Gene 38: 19–30

Zhu YS, Kiley PJ, Donohue TJ and Kaplan S (1986) Origins of the mRNA stoichiometry of the *puf* operon in *Rhodobacter sphaeroides*. J Biol Chem 261: 10366–10374

Zsebo KM and Hearst JE (1984) Genetic-physical mapping of a photosynthetic gene cluster from *Rhodopseudomonas capsulata*. Cell 37: 937–947

Chapter 23

The Structures of Photosynthetic Reaction Centers from Purple Bacteria as Revealed by X-Ray Crystallography

C. Roy D. Lancaster, Ulrich Ermler and Hartmut Michel
Max-Planck-Institut für Biophysik, Abteilung Molekulare Membranbiologie, Heinrich-Hoffmann-Str. 7, D-60528 Frankfurt am Main, Germany

Summary

Photosynthetic reaction centers from purple bacteria are the best known membrane protein complexes. In this review, X-ray crystal structures of the reaction centers from *Rhodopseudomonas* (*Rp.*) *viridis* and *Rhodobacter* (*Rb.*) *sphaeroides* are compared on the basis of data quality and quantity, maximum resolution limits, and structural features. In contrast to earlier comparisons and on the basis of the most recent, best defined *Rb. sphaeroides* structure, a number of the reported differences between the two species cannot be confirmed. Not only the overall architecture of the reaction centers and the relative positions and orientations of the cofactors, but also specific structural features are well conserved. For example, the hydrogen-bonding pattern between the protein and the monomeric bacteriochlorophylls, the bacteriopheophytins, and the primary quinone, Q_A, are identical. Therefore, the asymmetry between the A- and B- branch is maintained. However, there are small conformational differences which might provide a basis for the explanation of observed spectral and functional discrepancies between the two species.

A particular focus in this review is on the binding site of the secondary quinone (Q_B), where electron transfer is coupled to the uptake of protons from the cytoplasm. For the description of the Q_B-binding site, an improved coordinate set for the *Rp. viridis* reaction center has been included. In particular, a new understanding of the role of Ser L223 in the Q_B-binding site has become apparent. In addition, chains of ordered water molecules are found leading from the cytoplasm to the Q_B site in the best defined structures of both *Rp. viridis* and *Rb.*

R. E. Blankenship, M. T. Madigan and C. E. Bauer (eds): Anoxygenic Photosynthetic Bacteria, pp. 503–526.

sphaeroides reaction centers. New insights concerning the binding of triazine-type electron transfer inhibitors to the *Rp. viridis* Q_B site include the identification of additional hydrogen bonds and of two tightly bound water molecules. Finally, recent developments regarding the structures of mutant reaction centers are discussed.

I. Introduction

The first steps of photosynthetic light-energy conversion take place in the photosynthetic membranes from various bacteria or organelles. Light is absorbed first by light-harvesting antenna complexes (see Chapters 15–22 of this volume for reviews). Upon transfer of the energy from the light-harvesting antenna to the photosynthetic reaction center, or direct excitation by light of the reaction center itself, an excited singlet state of the 'primary electron donor' is created. This event leads to the primary charge separation and a subsequent vectorial electron transfer in the photosynthetic reaction center across the photosynthetic membrane. The photosynthetic reaction centers are complexes containing several integral membrane proteins and a number of pigment molecules. Best characterized are those from the purple bacteria (for earlier reviews, see Okamura and Feher, 1992; Breton and Vermeglio, 1992; Deisenhofer and Norris, 1993). In general, they consist of three protein subunits, which are called H (heavy), M (medium) and L (light) according to their apparent molecular weights as determined by sodium dodecylsulfate polyacrylamide gel electrophoresis. The H subunit can frequently be removed without impairing the primary photochemistry (see e.g. Debus et al., 1985). The L and M subunits bind the photosynthetic pigments (cofactors) which are one carotenoid molecule, 2 bacteriochlorophylls (a or b), one ubiquinone (or menaquinone) as a primary electron acceptor Q_A, one non-heme iron, and another ubiquinone as the secondary electron acceptor Q_B. The L and M subunits possess a sequence homology of about 25–35% (Youvan et al., 1984; Williams et al., 1984; Michel et al., 1986a; Williams et al., 1986). The reaction centers from most purple bacteria contain a fourth protein subunit, a cytochrome *c* with four covalently bound heme groups. Among these is the reaction center from *Rhodopseudomonas (Rp.) viridis*, which possesses bacteriochlorophylls and bacteriopheophytins of the *b* type, and menaquinone-9 as Q_A. The carotenoid is 1,2-dihydroneurosporene (Sinning, 1988). The reaction center from *Rhodobacter (Rb.) sphaeroides* contains bacteriochlorophylls and bacteriopheophytins of the a type, ubiquinone-10 as Q_A, and spheroidene as the carotenoid. It does not possess a tightly bound cytochrome subunit.

The isolated photosynthetic reaction center from the filamentous green aerobic bacterium *Chloroflexus (Cf.) aurantiacus*, consists of an L and M subunit which show sequence similarities to the reaction centers from purple bacteria (Ovchinikov et al., 1988a, 1988b, Shiozawa et al., 1989). There is no evidence for the presence of an H subunit. However, *in vivo* a four-heme containing cytochrome subunit, similar to that from *Rp. viridis*, is present (Dracheva et al., 1991). Much of what will be said about the three-dimensional structure of the photosynthetic reaction center core from purple bacteria will also be valid for the reaction center from *Cf. aurantiacus*. The known amino acid sequences of reaction center subunits from purple bacteria have been compiled and compared with those from *Cf. aurantiacus* recently (Nagashima et al., 1993, 1994). The photosynthetic reaction centers from the anaerobic green sulphur bacteria are completely different (see Chapter 30 by Feiler and Hauska).

The photosynthetic reaction centers from purple bacteria are the best known membrane protein complexes: The crystal structures of the reaction centers from the purple bacteria *Rp. viridis* and *Rb. sphaeroides* have been determined as the first membrane protein complexes. Initially, the structure of the reaction center from *Rp. viridis* was solved by the multiple isomorphous replacement method (Deisenhofer et al., 1984, 1985) and refined to a crystallographic R-factor of 19.3% up to a resolution of 2.3 Å (Deisenhofer and Michel, 1989, Deisenhofer et al., 1995). The structure of the reaction center from *Rb. sphaeroides* was then determined by three different groups using a partially refined coordinate set of the *Rp. viridis* reaction center by the molecular replacement method. All three groups used an orthorhombic crystal form with slightly different cell dimensions for structure determination (Allen et al., 1986, Chang et al., 1986, Arnoux et al., 1989). The structures were refined at a resolution around 3

Abbreviations: Cf. – Chloroflexus; Rb. – Rhodobacter; Rc. – Rhodocyclus; Rp. – Rhodopseudomonas; Rs. – Rhodospirillum

Å with crystallographic R-factors between 20 and 22%.

Although the overall structure of the *Rb. sphaeroides* reaction center was published to be very similar to that from *Rp. viridis*, a number of substantial differences in cofactor binding have been reported (Yeates et al., 1988, Allen et al., 1988) However, there are also differences between the published structures of the *Rb. sphaeroides* reaction centers (El-Kabbani et al., 1991). Recently, a trigonal crystal form has been obtained which diffracts X-rays to a resolution of at least 2.65 Å (Buchanan et al., 1993). Structure determination using this crystal form and the refinement to a crystallographic R-factor of 18.6% has now been completed (Ermler et al., 1994). The Protein Data Bank ID code for this structure is 1PCR. Due to the improved resolution, more and better data and the lower R-factor, the new structural model of the *Rb. sphaeroides* reaction center is considerably better defined. The average coordinate error is now reduced to 0.3 Å compared to 0.4 Å or 0.5 Å for the previously published structures whereas it is 0.26 Å for the *Rp. viridis* reaction center (cf. Table 1). In addition, as for the *Rp. viridis* reaction center, more than 160 firmly bound water molecules could be identified and added to the atomic model. For the description of the Q_B-binding site we will also use an improved coordinate set for the *Rp. viridis* reaction center which was obtained after reconstitution of Q_B with ubiquinone-2, data collection, and refinement (Lancaster and Michel, to be published). The Q_B site in the *Rp. viridis* reaction center crystals is normally only 30% occupied, resulting in a 'mixed' structure around Q_B.

Table 1 lists the coordinate sets, which are available either from the Brookhaven Protein Data Bank (Bernstein et al., 1977; Peitsch et al., 1995) or from the authors upon request. The publications from one group originally describing the *Rb. sphaeroides* reaction center structure (Allen et al., 1987; Yeates et al., 1988, Allen et al., 1988; Komiya et al., 1988) are based on a coordinate set (1RCR) which has now been replaced by 4RCR without an updated publication. Also, very recently, another set of coordinates named 1PSS for the *Rb. sphaeroides* reaction center from wild type strain WS 231, has been deposited (Chirino et al., 1994). Due to the higher accuracy we will mainly use the coordinate set obtained with the new trigonal crystal form of *Rb. sphaeroides* reaction center in the following comparison of the reaction center from both species.

II. Structural and Functional Overview

Figure 1 shows an overview of the photosynthetic reaction center from *Rp. viridis* similar to that published by Deisenhofer et al. (1985). In such a representation the reaction center from *Rb. sphaeroides* would look nearly identical with the exception that the cytochrome subunit shown at the top (green) would be missing. The protein subunits are shown in the form of a smoothed ribbon diagram, the cofactors as atomic models with yellow carbon atoms, blue nitrogen and red oxygen atoms. The reaction center from *Rp. viridis* has an overall length of 130 Å. This is in the direction perpendicular to the membrane. Parallel to the membrane, the maximum width is about 70 Å. The central core of the reaction center is formed by the L subunit (shown in brown) and the M subunit (shown in blue) which possess five membrane spanning helices each. Both subunits are closely associated and bind the bacteriochlorophylls, the bacteriopheophytins, the quinones, the non-heme iron, and the carotenoid. With the exception of the carotenoid, the cofactors and the L and M subunits show a high degree of local twofold symmetry with the symmetry axis perpendicular to the membrane. On the external (periplasmic) and internal (cytoplasmic) side this central core forms flat surfaces. The cytochrome subunit with its four covalently bound heme groups is attached to this core on the periplasmic side, the H subunit on the cytoplasmic side. In addition the H subunit is anchored to the membrane by a single membrane spanning helix. The cytochrome subunit also possesses a membrane anchor. This membrane anchor is made up by two C_{18}-fatty acids which are covalently bound to the N-terminal Cys-residue via a glycerol moiety (Weyer et al., 1987a). It is not visible in the electron density map. Most likely it is disordered in the detergent micelle which surrounds the hydrophobic trans-membrane helices, as indicated by neutron crystallography (Roth et al., 1989).

The pigments form two branches (see also Fig. 2) each consisting of two bacteriochlorophylls, one bacteriopheophytin and one quinone, which both cross the membrane starting from a 'special pair' of two closely associated bacteriochlorophylls (one of which belongs to each branch) near the periplasmic side, followed by the 'accessory' bacteriochlorophyll, one bacteriopheophytin and a quinone. Only the branch more closely associated with L subunit is used in the light-driven electron transfer. According

Table 1. Reaction Center Structures[a]

	Resolution limit [Å] (max.)	No. of unique reflections	Completeness[b] [%]	No. of non-zero occupancy atoms	n_{obs}/n_{par} [c]	R-factor [%]	Mean coordinate error (max.)[d]	Reference
Rp. viridis								
1PRC	2.3	95,762	75.4	10,045	2.38	19.3	0.26	Deisenhofer & Michel, 1989
Q282	2.45	79,397	76.5	10,102	1.96	18.5	0.3	Lancaster & Michel, unpubl.
DGM	2.3	102,117	81.4	10,307	2.47	18.6	0.25	triazine inhibitor complex Lancaster & Michel, unpubl.
Rb. sphaeroides								
2RCR	3.1	13,493	50.8	6921	0.63	22.0	0.5	Chang et al., 1991
4RCR	2.8	21,992	60.0	6764	0.81	22.7	0.4	Yeates et al., 1988
1PSS	3.0	21,518	68.9	6789	0.79	22.3	0.5	Chirino et al., 1994
1PCR	2.65	56,141	90.4	7340	1.91	18.6	0.3	Ermler et al., 1994

[a] Structures 1PRC, 2RCR, 4RCR, 1PSS and 1PCR are named according to their four-character entry code in the Brookhaven Protein Data Bank (PDB). Names Q282 and DGM are unofficial and refer to the quinone-reconstituted complex and the triazine inhibitor complex, respectively.
[b] These values were calculated for the number of reflections actually used in refinement and may differ from those given in the appropriate references.
[c] n_{obs} = number of observed unique reflections (only the reflections used for refinement are considered)
n_{par} = number of parameters necessary to define the model. This includes 3 parameters (x,y,z coordinates) per non-zero-occupancy atom plus one (isotropic atomic B-factor) where applicable (the 2RCR structure is reported without atomic B-factors), plus additional parameters (occupancy) for alternate conformations.
[d] estimate of the upper limit of the mean coordinate error from a Luzzati (1952) plot.

to the nomenclature proposed by Hoff (1988) it is called the A (active)-branch, the inactive one the B-branch. The active branch ends with Q_A, the inactive one with Q_B. Halfway between both quinones a non-heme iron atom is located. The carotenoid is associated with the second bacteriochlorophyll of the inactive branch. Which branch is used in electron transfer, could be decided by optical spectroscopy with polarized light on single crystals (Zinth et al., 1983) in combination with the structural data (Michel and Deisenhofer, 1986) keeping in mind that the bacteriopheophytin involved in electron transfer absorbs light at longer wavelengths than the other (Vermeglio and Paillotin, 1982). The final electron transfer from Q_A to Q_B occurs parallel to the membrane. After or during a second electron transfer Q_B becomes doubly protonated. This reaction is discussed in Chapter 26 by Okamura and Feher.

III. Arrangement of the Cofactors

Figure 2 shows the arrangement of the cofactors for the *Rp. viridis* (green) and the *Rb. sphaeroides* reaction centers (red). The two pigment branches A and B can clearly be seen, starting with the bacteriochlorophylls D_A and D_B constituting the primary electron donor ('special pair'), then the accessory bacteriochlorophylls B_A and B_B, the bacteriopheophytins ϕ_A and ϕ_B, the primary electron acceptor Q_A (menaquinone is *Rp. viridis*, ubiquinone in *Rb. sphaeroides)*, the secondary electron acceptor Q_B (ubiquinone in both species) and the non-heme iron halfway between Q_A and Q_B. The carotenoid is in van der Waals contact with B_B and disrupts the twofold symmetry. The linear row of the four heme groups in the *Rp. viridis* reaction center is seen at the top.

The arrangement of the cofactors is very similar in both reaction centers. The head groups of the cofactors are in nearly identical positions except for Q_B which is discussed below. The small differences of the cofactor arrangement are difficult to discuss because they depend on the coordinate errors and the way of superposition. Both reaction centers were superimposed with the program O (Jones et al., 1991) using only Cα atoms of the L, M, and H subunits. The relative centers of mass of the ring systems of the cofactors relative to the polypeptide chain are displaced by less than 1 Å.

The arrangement of the bacteriochlorophyll monomers D_A and D_B may be slightly different in both reaction centers (Ermler et al., 1994). In projection, the overlapping area of the two pyrrole

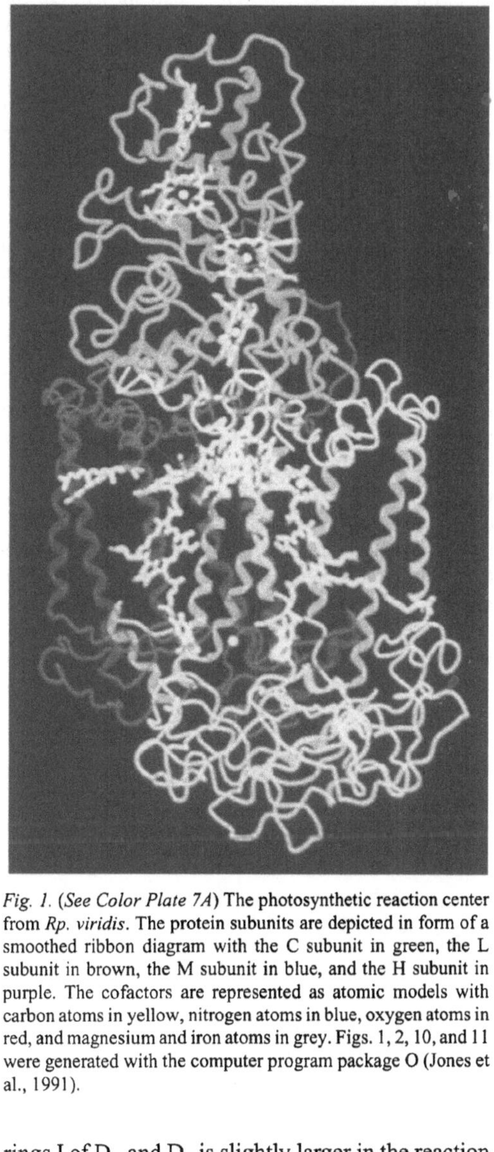

Fig. 1. (*See Color Plate 7A*) The photosynthetic reaction center from *Rp. viridis*. The protein subunits are depicted in form of a smoothed ribbon diagram with the C subunit in green, the L subunit in brown, the M subunit in blue, and the H subunit in purple. The cofactors are represented as atomic models with carbon atoms in yellow, nitrogen atoms in blue, oxygen atoms in red, and magnesium and iron atoms in grey. Figs. 1, 2, 10, and 11 were generated with the computer program package O (Jones et al., 1991).

rings I of D_A and D_B is slightly larger in the reaction center from *Rp. viridis* than in that from *Rb. sphaeroides*. Furthermore, the distance between the pyrrole rings I of D_A and D_B is 3.4 Å in *Rb. sphaeroides* and 3.2 Å in *Rp. viridis*.

The accessory bacteriochlorophyll molecule B_B of the B-branch in the *Rb. sphaeroides* reaction center seems to be located closer to the periplasmic side than in the *Rp. viridis* reaction center. Consequently,

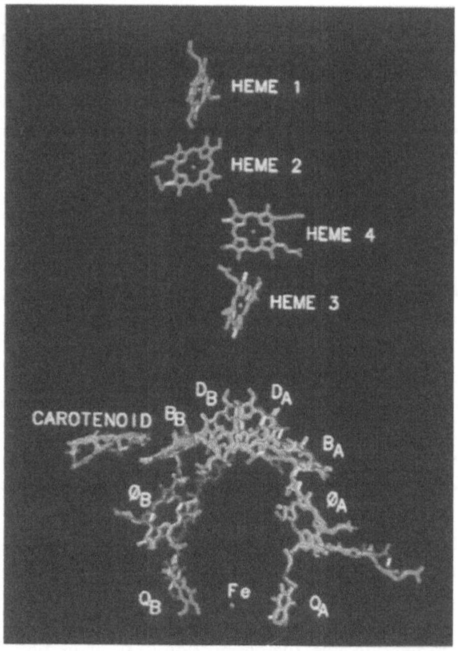

Fig. 2. (*See Color Plate 7B*) Superposition of the reaction center cofactors from *Rp. viridis* (green) and *Rb. sphaeroides* (red). The phytyl side chains of the bacteriochlorophyll and bacteriopheophytin molecules have been omitted for the sake of clarity.

in the *Rb. sphaeroides* reaction center the distance between the centers of D_B and B_B is 0.6 Å shorter, between the centers of B_B and ϕ_B 0.5 Å larger. The Q_B position is shifted substantially and is located about 3.5 Å further away from Q_A when compared to the reaction center from *Rp. viridis* and the older *Rb. sphaeroides* reaction center structures.

A nearly perfect agreement between both reaction centers is obtained when the deviation angles from the twofold axis for the cofactors of the A- and B-branch are compared (Michel and Deisenhofer, 1989; Ermler et al., 1994). An optimal superposition of D_A and D_B is obtained by a rotation of 179.7° in the *Rp. viridis* reaction center and 177.9 in the *Rb. sphaeroides* reaction center, of B_A and B_B by a rotation of −175.8° and −176.8°, respectively, and of ϕ_A and ϕ_B by a rotation of −173.2° and −171.9°, respectively.

Large conformational differences between both reaction centers are found for the phytyl side chains of the cofactors, particularly along the B branch. The arrangements of the phytyl side chains in the A and B branches deviate from the local twofold symmetry. This might be partially due to the carotenoid molecule attached to the B branch. In both species the crystallographic temperature factors, which are a measure for the rigidity of the structure, are considerably higher along the B branch than along the A branch.

IV. The Structure of the Protein Subunits

A. The L Subunit

The structures of the L subunits of the reaction centers from *Rp. viridis* and *Rb. sphaeroides* are shown in Fig. 3 in the form of $C\alpha$-traces. The dominant features are the five long membrane-spanning helices (A-E). They possess a length of 21 (helix A), 24 (helices C and E) or 28 (helices B and D) residues. The segments connecting the transmembrane helices and the C-terminus are also partly helical. On the periplasmic side, the connection of transmembrane helices C and D contains a helix with eleven residues, and the C-terminal stretch one of 9 residues. On the cytoplasmic side, the connection of transmembrane helices D and E contains a helix of 12 residues which partly intrudes into the membrane. This region of the structure forms the binding site of the secondary electron acceptor Q_B, which is also shown. On the cytoplasmic side, the L and M subunits are tightly interwoven. In projection, viewed from the top of the membrane, the transmembrane helices A-E form a half circle, in the order A,B,C,E and D. Polar interactions between transmembrane helices seem to be weak, there are only up to two hydrogen bonds per pair of helices. The cytoplasmic ends of the transmembrane helices frequently contain charged residues. Transmembrane helix A is an exception which starts with a Phe-Phe pair already in the hydrophobic part of the membrane. Transmembrane helices A, B and D are straight, helix E is smoothly curved, and helix C possesses a kink of more than 30° which is due to a Pro residue. However, helix C contains another Pro which does not lead to such a kink, but to only minor distortions from the regular helix geometry. On the periplasmic side there is an excess of negatively charged residues, on the cytoplasmic side of positively charged residues, leading to a permanent electric dipole moment across the membrane. When the L subunits from *Rp. viridis* and *Rb. sphaeroides* are compared, an additional 8 amino acids are found at the C-terminus in *Rb. sphaeroides*.

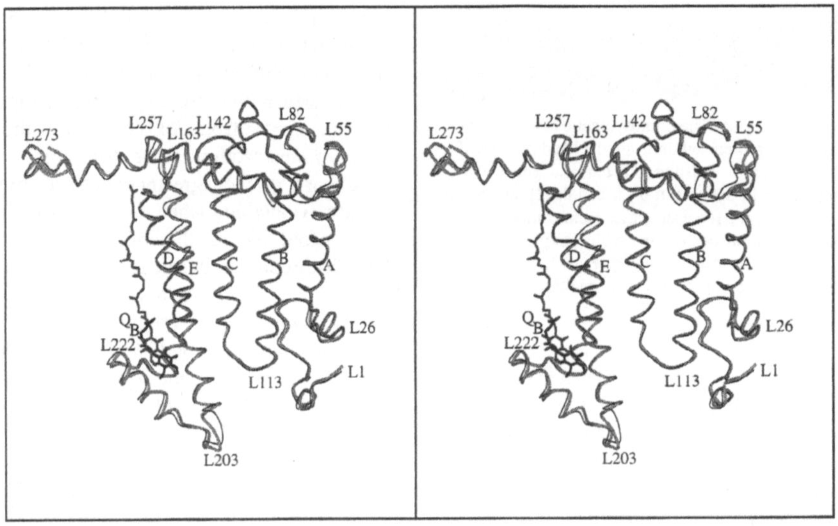

Fig. 3. Stereo pair of the $C\alpha$-traces of the *Rp. viridis* (thin black line) and the *Rb. sphaeroides* (thick grey line) L subunits and the Q_B molecules of *Rp. viridis* (black) and *Rb. sphaeroides* (grey). The letters 'A' to 'E' designate the five transmembrane helices. Figures 3 to 9 and Fig. 12 were made with the computer program MOLSCRIPT (Kraulis, 1991).

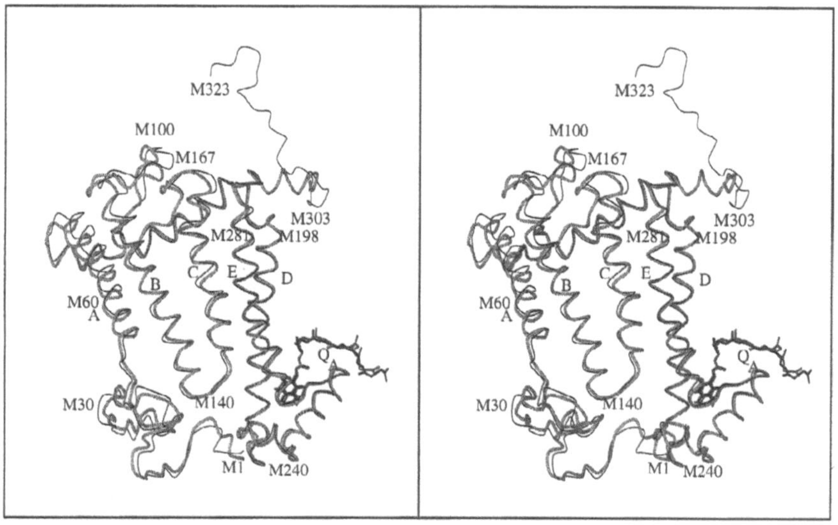

Fig. 4. Stereo pair of the Cα-traces of the *Rp. viridis* (thin black line) and the *Rb. sphaeroides* (thick grey line) M subunits and the Q_A molecules of *Rp. viridis* (black) and *Rb. sphaeroides* (grey). The labels refer to *Rp. viridis* residues.

B. The M Subunit

In a similar way the peptide chains of the M subunits of the reaction centers from *Rp. viridis* and *Rb. sphaeroides* are displayed in Fig. 4. As indicated already by the sequence identity of around 30% between the L and M subunits the overall protein fold is very similar. In *Rp. viridis* the five transmembrane helices of the M subunit possess a length of 24 (C), 25 (A, E), 26 (D) or 27 (B) residues. The short helices in the connections of transmembrane helices C and D (12 residues) near the C-terminus (7 residues) on the periplasmic side, and in the connection of the helices D and E (14 residues), forming part of the Q_A–binding site, are also present. Accompanied by an insertion of seven amino acids (compared to the L subunits) each, short additional helices are formed in the connections of trans-membrane helices A and B (7 residues) on the periplasmic side, and of transmembrane helices D and E (6 residues) on the cytoplasmic side. Much of what has been said for the L subunit is also valid for the M subunit: Transmembrane helices A, B and D are straight, E is curved and C contains a kink due to a Pro-residue. Negatively charged side chains dominate on the periplasmic side, positively charged ones on the cytoplasmic side. When the L and M subunits are compared the N-termini of the M subunits are 26 (*Rp. viridis*) or 25 (*Rb. sphaeroides*) residues longer than the N-termini of the L subunits. At the C-terminus the M subunit from *Rb. sphaeroides* is 9 amino acids shorter than the L subunit. The C-terminus of the M subunit from *Rp. viridis* possesses an additional 18 amino acids, which interact with the cytochrome subunit (see also Fig. 1). The M subunit from *Rb. sphaeroides* compared to that from *Rp. viridis* contains an insertion of one residue after amino acid M38, and another one after residue M104 (*Rp. viridis* numbering), so that the equivalent residue between residues M38 and M104 (*Rp. viridis* numbering) differ by one number and beyond M104 by two numbers (see Michel et al., 1986a).

C. The H Subunit

The H subunits from *Rp. viridis* and *Rb. sphaeroides* possess a sequence identity of 39% only with several insertions and deletions (Williams et al., 1986). Nevertheless, the protein fold is very similar with the exception of several loop regions as can be seen in Fig. 5. The N-terminus is located on the periplasmic side of the membrane. Residues H12 to H35 form a membrane-spanning helix, which is an α-helix at its beginning but a π-helix at its very end. Around the C-terminus of the helix a cluster of seven charged residues is found in *Rp. viridis* (residues H33-H39),

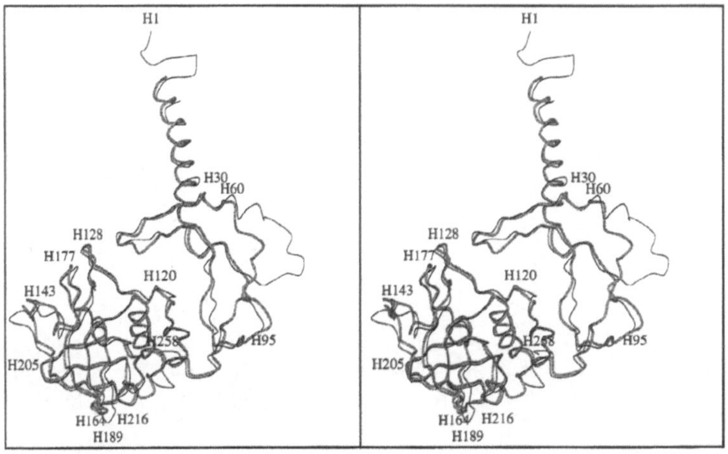

Fig. 5. Stereo pair of the Ca-traces of the *Rp. viridis* (thin black line) and the *Rb. sphaeroides* (thick grey line) H subunits. The labels refer to *Rp. viridis* residues. Residues H47 to H53 of the *Rp. viridis* H subunit are not observed in the electron density and no statement regarding their position is made. This region on the right of the figure is nevertheless included in order to facilitate chain tracing.

but of only three in *Rb. sphaeroides* (residues H34, H37, H38). The next 70 residues are preferentially in contact with the LM complex. A globular region follows that contains an extended system of antiparallel and parallel β-sheets, and an α-helix close to the C-terminus. The β-sheets form a pocket which is lined with hydrophobic amino acid residues. The significance of this pocket is unclear.

The function of the H subunit is not well defined at present. It is not involved in pigment binding. Its cytoplasmic domain helps to shield the Q_A-binding site from contact with the cytoplasm. Several glutamic acid residues of the H subunit line up the channel of bound water molecules that leads to the Q_B-binding site (see below). There is also some evidence that the H subunit is involved in the assembly of the reaction center (Chory et al., 1984). It may form a matrix for the sequential assembly of the reaction center components. However, there is no evidence for the presence of an H subunit in the green aerobic bacterium *Cf. aurantiacus*.

D. The Cytochrome Subunit

As mentioned in the introduction the reaction centers from most purple bacteria do possess a bound cytochrome subunit containing four covalently bound heme groups (see Chapter 36 by Nitschke and Dracheva). This is not the case for the reaction centers from the most frequently studied species *Rb. sphaeroides*, *Rb. capsulatus,* and *Rhodospirillum (Rs.) rubrum*, which also can grow very well under aerobic conditions. The amino acid sequences are known for the cytochrome subunits from *Rp. viridis* (Weyer et al., 1987b), from *Rhodocyclus (Rc.) gelatinosus*, now called *Rubrivivax gelatinosus* (Nagashima et al., 1993, 1994), and from the green aerobic bacterium *Cf. aurantiacus* (Dracheva et al., 1991). Only in *Rp. viridis* the cytochrome subunit is so firmly bound to the reaction center, that it is also present in the isolated reaction center. A recognition site for the covalent attachment of a diglyceride and removal of the signal peptide by signal peptidase II is present in *Rp. viridis* and *Rc. gelatinosus*, but not in *Cf. aurantiacus.*

The structure of the cytochrome subunit from the *Rp. viridis* reaction center has been described in detail (Deisenhofer et al., 1995). It consists of five segments. The N-terminal segment (residues C1-C66) from the Cys-residue with the attached diglyceride, to the beginning of the first heme binding segment (at the top of the reaction center) with the binding sites for heme 1 and heme 2 (see Figs. 1,2,6). This first segment contains two strands of antiparallel β-sheet parallel to the membrane surface followed by an α-helix (residues C25-C34) and irregular secondary structure. The first heme binding segment (residues C67-C142) with the binding sites for heme

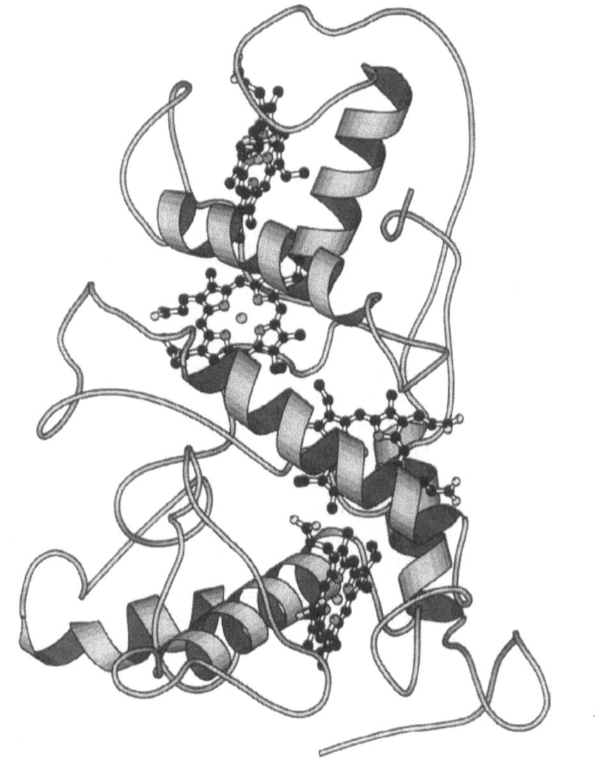

Fig. 6. The C subunit of the *Rp. viridis* reaction center. The protein backbone is shown in a smoothed ribbon diagram, the cofactor heme groups are displayed as atomic models.

1 and heme 2 is followed by a connecting segment (residues C143-C225) with little regular secondary structure. This connecting segment comprises around 60% of the contact surface with the M subunit. Then the second heme binding segment with the binding sites for heme 3 and heme 4 (residues C226-C315), and the C-terminal segment (residues C316-C336) follow. The heme binding segments and their attached heme groups are related by local twofold symmetry.

The hemes and the heme-binding segments make up the core of the cytochrome subunit. Each heme-binding site consists of an α-helix that runs parallel to the heme plane, a loop and the heme attachment site with the sequence Cys-X-Y-Cys-His. The Cys-residues form thioether bonds with the heme groups, the His is the fifth ligand to the heme iron. The Met residues C74, C110 and C233 in the parallel helices are the sixth ligands to hemes 1, 2 and 3, the sixth ligand to heme 4 is His C124, which is located in the loop region of the heme 2 binding site.

The redox midpoint potentials of the four heme groups follow the order low, high, high, low in the sequence, or high, low, high, low if the hemes are ordered with increasing distance (see Fig. 2) from the primary electron donor (Shopes et al., 1987; Alegria and Dutton, 1987; Dracheva et al., 1988; Fritzsch et al., 1989, Vermeglio et al., 1989; Shinkarev et al., 1990). Heme 3, closest to the primary electron donor has the highest redox potential (370 mV), heme 4 has 10 mV, heme 2 has 300 mV, and heme 1, the most distant one, has −60 mV. A recent theoretical investigation (Gunner and Honig, 1991) taking into account the different solvent exposures of the heme groups, the electrostatic effects of the protein environment and the heme arrangement reproduced the experimentally determined redox potentials with reasonable accuracy. The function of the alternate spatial order of high and low potential heme groups is unknown. Under physiological conditions of cyclic electron flow, only the high potential hemes are reduced. It is most likely heme 2, the second from the top (see Fig. 2), which is reduced by a water soluble cytochrome c_2. It transfers its electron to heme 3, from where the photooxidized primary electron donor is re-reduced.

V. Cofactor Conformation and Protein-Cofactor Interactions

A. The Primary Electron Donor

The primary electron donor ('special pair') is located at the interface of the L and M subunits near the periplasmic side (see Figs. 1 and 7). It interacts with residues of the C, D and E transmembrane helices and the connections of the C and D helices. The special pair bacteriochlorophylls D_A and D_B in both species are held in their position by specific interactions with the protein matrix (Fig. 7). The Nε atoms of His L173 and His M200 (M202 in *Rb. sphaeroides*) bind strongly to the central Mg atoms with distances of 1.9–2.3 Å in both species. The ring I acetyl group of D_A is hydrogen bonded to His L168 in the *Rb. sphaeroides* (3.2 Å) as well as in the *Rp. viridis* reaction center (2.8 Å), in contrast to an earlier publication (Yeates et al., 1988). The symmetry-related amino acid residue near D_B is Phe M197 in *Rb. sphaeroides*. Thus, no hydrogen bond can be formed. In *Rp. viridis* it is Tyr M195 forming a hydrogen bond (3.0 Å) with the acetyl carbonyl oxygen of ring I. Thr L248 forms a hydrogen bond (2.6 Å) with the ring V keto carbonyl of D_A in *Rp. viridis*. It is absent in *Rb. sphaeroides*. Neither the ester nor the keto carbonyl oxygens of ring V from D_B undergo polar interactions with the polypeptide chain in the *Rb. sphaeroides* reaction center, in contrast to published results (El Kabbani et al., 1991).

Substantial conformational differences between both reaction centers are found around rings V of D_A and D_B. In *Rp. viridis* ring V is bent towards Thr L248 (Fig. 7), whereas it is oriented into the opposite direction in *Rb. sphaeroides*. This difference is presumably due to the presence of bulky side chains on opposite sides of D_A in *Rp. viridis* and in *Rb. sphaeroides* (Ermler et al., 1994). Met L127 in *Rp. viridis* corresponds to a smaller Ala in *Rb. sphaeroides* on one side. *Vice versa* Gly L247 and Thr L248 in *Rp. viridis* are replaced by the larger Cys L247 and Met L248 in *Rb. sphaeroides* on the other side. Similar but less pronounced conformational differences are also seen at ring V of D_B.

B. The Accessory Bacteriochlorophylls

The accessory bacteriochlorophylls B_A and B_B are located between D_A and D_B and the bacteriopheophytins ϕ_A and ϕ_B respectively and are in van der Waals contacts with both. Although the intermediate position and the favorable redox properties of B_A suggest a direct role as very first electron acceptor (Zinth and Kaiser, 1993) its role is still under discussion (Kirmaier and Holten, 1993; also see Chapter 24 by Woodbury and Allen in this volume).

Significant conformational differences between the reaction centers from *Rp. viridis* and *Rb. sphaeroides* are only found at the ethyl groups of ring II (see Fig. 7) caused by the structural difference between the ethyl group of bacteriochlorophyll *a* and the ethylidene group of bacteriochlorophyll *b*. The polar interactions with the protein matrix are symmetric and identical in both reaction centers. The Nε atoms of His L153 (distance 2.2 Å in *Rp. viridis*, 2.3 Å in *Rb. sphaeroides*) and His M180 (182) (2.1 Å in *Rp. viridis*, 2.2 Å in *Rb. sphaeroides*) are the ligands to the central Mg atoms. The Nδ atom of His L153 is hydrogen bonded to the peptide carbonyl oxygens of Gly L149 and Ile L150. This kind of interaction between the Mg^{2+} and His L153 was not observed in one of the published *Rb. sphaeroides* reaction center structures (Yeates et al., 1988). In both reaction centers the ring V carbonyl oxygen atoms are hydrogen bonded via a water molecule to His L173 and His M200(M202) respectively. In contrast to earlier studies on the orthorhombic crystal forms of the *Rb. sphaeroides* reaction center (Yeates et al., 1988; El-Kabbani et al., 1991) the hydroxyl group of Ser L178 is at 4.3 Å too far away from the ester carbonyl oxygen to form a hydrogen bond.

The carotenoid molecule makes van der Waals contacts with B_B and with hydrophobic residues of the helices A, B, C and the connection of helices C and D. The carotenoid molecule possesses a kinked conformation with its conjugated π-system parallel to the membrane. The kink can be caused by a *cis* bond in position 13 or 15. A definite assignment has to await refinement at higher resolution. A 13 *cis* bond would contradict the results obtained by Raman spectroscopy (Arnoux et al., 1989; deGroot et al., 1992) which indicates a *cis* bond at position 15.

The carotenoid protects the reaction center against photooxidation by quenching triplet states of the primary electron donor. Its proximity to B_B suggests that the energy transfer from the triplet state of the primary electron donor occurs via B_B as proposed

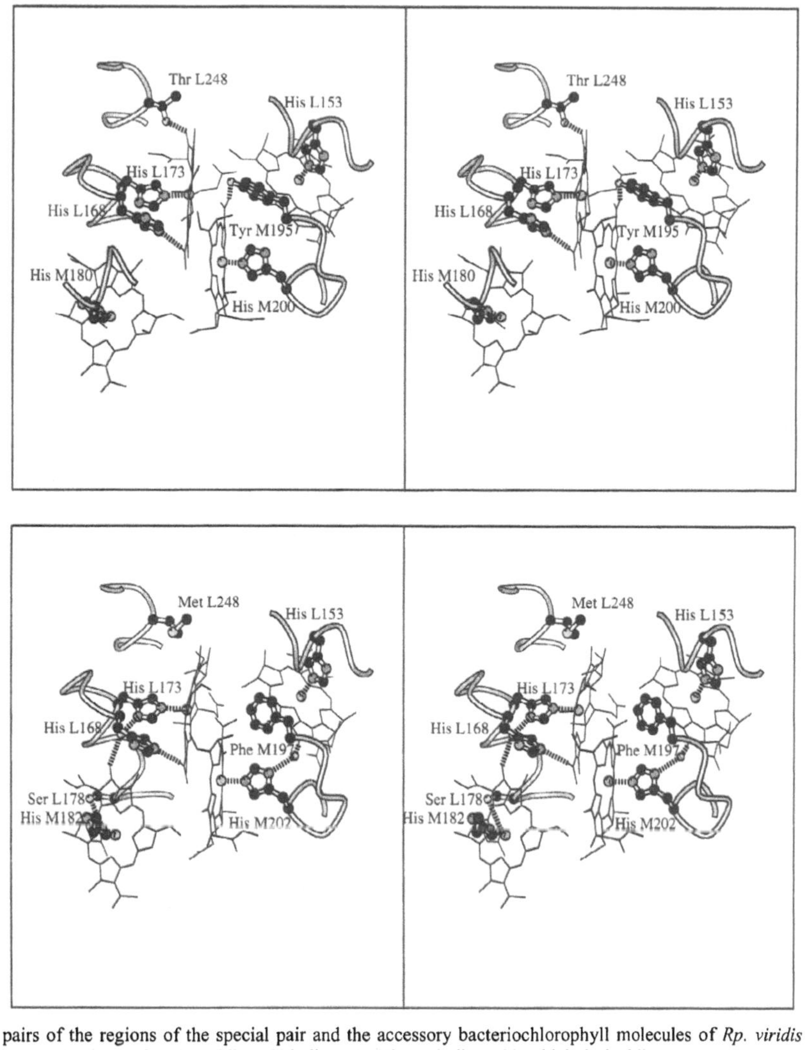

Fig. 7. Stereo pairs of the regions of the special pair and the accessory bacteriochlorophyll molecules of *Rp. viridis* (top) and *Rb. sphaeroides* (bottom). Hydrogen bonds and ligand binding Mg-His are indicated by thick dashed lines.

from spectroscopic measurements (Schenck et al., 1984).

C. The Bacteriopheophytins

Figure 8 shows for both reaction centers the location of the bacteriopheophytin ϕ_A between B_A and Q_A. At the top Tyr M208(M210) seems to be of importance since it is in van der Waals contact with D_B, D_A and B_A. The conformations of ϕ_A and ϕ_B are identical in both reaction centers except at the side chains of pyrrole rings II due to the chemical difference of the ethyl and ethylidene side chains. The pattern of hydrogen bonds formed by ϕ_a and ϕ_B with the protein matrix is identical in both species. Trp L100 and TrpM127 (129) respectively form a hydrogen bond with the ester carbonyls of ring V of ϕ_A (3.0 Å in *Rp. viridis*, 2.8 Å in *Rb. sphaeroides*) and ϕ_B (2.7 Å in *Rp. viridis*, 2.8 Å in *Rb. sphaeroides*) respectively (Michel et al., 1986b, Ermler et al., 1994). The carboxyl group of Glu L104 is probably in the protonated state and forms a short hydrogen bond (2.7 Å) to the ring

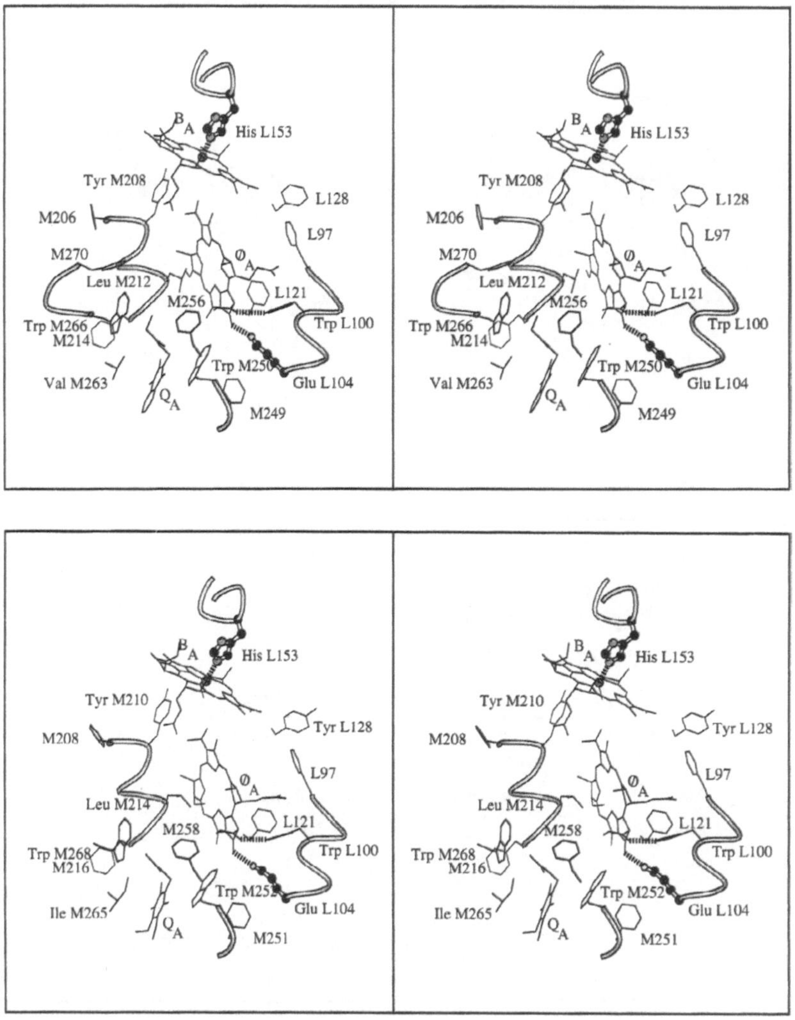

Fig. 8. Stereo pairs of the regions of the bacteriopheophytin molecules ϕ_A of *Rp. viridis* (top) and *Rb. sphaeroides* (bottom), respectively. Hydrogen bonds and ligand binding Mg-His are indicated by thick dashed lines. Not all of the phenylalanine residues in these regions are displayed for reasons of clarity. The phenylalanine residues are labeled by chain designator and residue number only.

V keto carbonyl in both reaction centers. No such hydrogen bond is found for ϕ_B. Replacement of Glu L104 by neutral amino acids in *Rb. capsulatus*, which is closely related to *Rb. sphaeroides*, influences the spectral properties of the reaction center, but only scarcely the dynamics and directionality of the electron transfer (Bylina et al., 1988). In the M subunit of the *Rb. sphaeroides* reaction center the residue corresponding to Glu L104 is Thr M131. In contrast to the results from the orthorhombic crystal forms (Yeates et al., 1988; El-Kabbani et al., 1991) the hydroxyl group of Thr M131 is at 3.9 Å too far away from the ring V carbonyl group to form a hydrogen bond. The bacteriopheophytin ϕ_A is surrounded by an astonishing number of phenylalanine residues (see Fig. 8). These bulky residues may contribute to the rigidity of the structure in the environment of ϕ_A. Around ϕ_B they are replaced to a

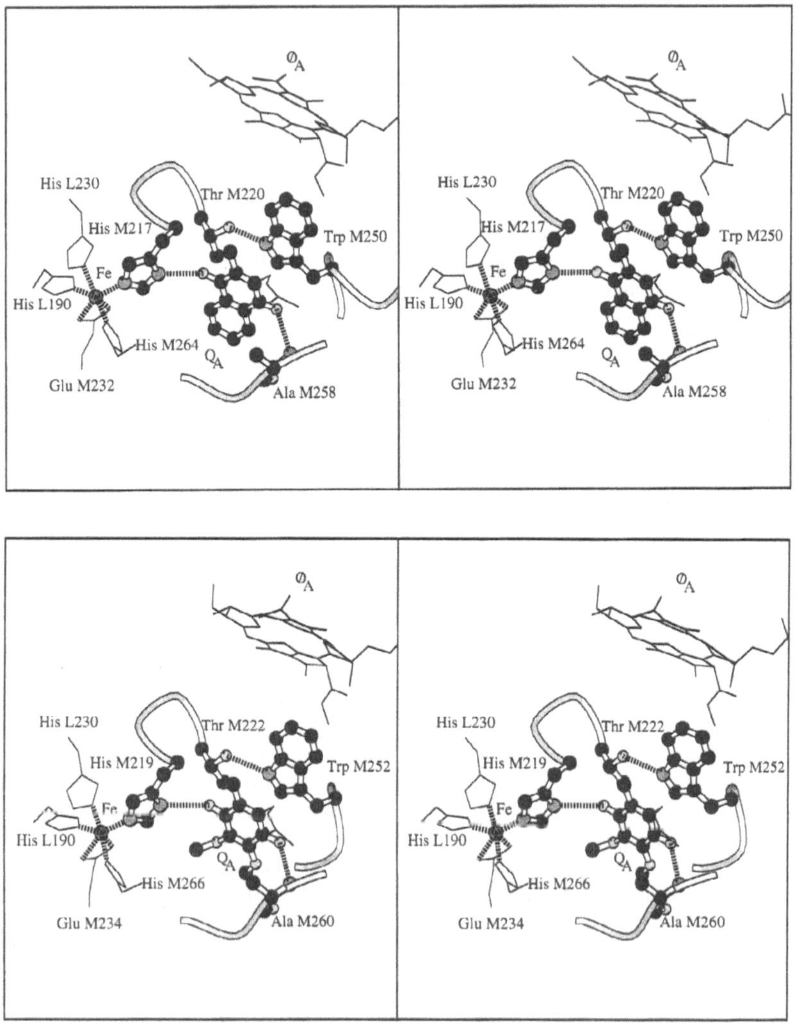

Fig. 9. Stereo pairs of the regions of the Q_A molecules and non-heme iron atoms of *Rp. viridis* (top) and *Rb. sphaeroides* (bottom), respectively.

large extend by smaller amino acid residues. As seen in Figs. 8 and 9 Trp M250(M252) with its large aromatic side chain bridges the gap between ϕ_A and Q_A in both reaction centers. In the symmetry-related position a phenylalanine is found. This observation implies a role for Trp M250(M252) in the electron transfer from ϕ_A to Q_A. A 'superexchange mechanism' has been postulated for the electron transfer from ϕ_A to Q_B with Trp M250(M252) as superexchange mediator (Plato et al., 1989).

D. The Primary Electron Acceptor Q_A and the Non-Heme Iron

Figure 9 shows the binding site of the first quinone Q_A for both reaction centers. Q_A is located on the A side where the L subunit dominates, but the quinone ring system exclusively interacts with residues of the M subunit. The Q_A-binding site clearly is more hydrophobic than the Q_B-binding site. A major part of the Q_A-binding site is made up by Trp M250-

(M252). The ring systems of the tryptophan and Q_A are parallel. Trp M250(M252) forms a hydrogen bond with Thr M220(M222). The structural difference between the menaquinone as Q_A in *Rp. viridis* and the ubiquinone as Q_A in *Rb. sphaeroides* causes only minor rearrangements of the Q_A-binding site, which are not shown in Fig. 9 for clarity.

In both species Q_A interacts with the protein surrounding in the same way. The first quinone oxygen atom is hydrogen bonded to the side chain of His M217(M219) (distance 3.1 Å in *Rp. viridis and 3.2 Å in* Rb. sphaeroides), the second one to the peptide nitrogen of the M258(M260) (distance 3.1 Å in *Rp. viridis* and 2.8 Å in *Rb. sphaeroides*). A shorter bond length for one hydrogen bond would be consistent with the finding that one carbonyl group is dominant for Q_A binding (Warncke and Dutton, 1993). Earlier studies on the orthorhombic crystal form (Allen et al., 1988, El-Kabbani et al., 1991) of the *Rb. sphaeroides* reaction center indicated a hydrogen bond between the first quinone oxygen and Thr M222. This result is excluded on the basis of the analysis of the trigonal crystal form.

His M217(M219) is one of four histidine ligands to the non-heme iron (Fig. 9), His M264(M266), His L190 and His L230 are the other three. Glu M232(M234) is a bidentate ligand to the non-heme iron. Both oxygen atoms are in contact with the iron. As a result the ligands to the non-heme iron form a distorted octahedron in both reaction centers. In the region around the iron both reaction centers can be superimposed with an rms deviation of better than 0.5 Å. Despite this hint about the importance of the non-heme iron, its role remains an enigma.

E. The Secondary Electron Acceptor Q_B

The binding site of the secondary acceptor quinone Q_B is formed exclusively by residues of the L subunit. In particular, it involves residues of the transmembrane helices D (L189, L190, L193) and E (L226, L229, L232), of the connecting helix DE (L212, L213, L216) and of the loop region connecting the latter two helices (L222 -L225). The profile of the pocket is conserved in all characterized reaction center structures. Q_B in the *Rb. sphaeroides* model based on the trigonal crystal form (1PCR) is displaced by approximately 5 Å from the other Q_B models (cf. Fig. 10) and displays high isotropic temperature factors (Ermler et al., 1994). It is mostly omitted from the following comparative discussion, because it seems likely that this Q_B site contains the quinol and not the quinone. Since it combines the requirements of high quinone occupancy and a sufficiently high n_{obs}/n_{par} ratio (cf. Table 1), the structure of the ubiquinone-2-reconstituted *Rp. viridis* reaction center complex (Q282) will be used as a guideline for the following discussion of the Q_B site.

The proximal carbonyl oxygen of Q_B, i.e. the one closer to the non-heme iron, accepts a hydrogen bond from the Nδ atom of His L190, whereas the distal carbonyl oxygen accepts two hydrogen bonds from the peptide nitrogens of Ile L224 and Gly L225. The side chain Oγ of Ser L223 is also within hydrogen-bonding distance of the distal carbonyl oxygen and was originally inferred to donate a hydrogen bond to the quinone carbonyl (Deisenhofer and Michel, 1989). However, in the refined quinone-reconstituted structure, the O$\gamma$of Ser L223 donates a hydrogen bond to the side chain Oδ of Asn L213 (cf. Fig. 10). In the absence of a hydrogen bond, the presence of the Ser L223 Oγso close to the distal carbonyl oxygen is electrostatically unfavorable for quinone binding. This finding explains the observed 9-fold increase in quinone affinity for the *Rp. viridis* mutant T1, where one of the mutations is Ser L223 → Ala (Sinning et al., 1989). While not favorable for quinone binding, the close association of the two oxygen atoms is compatible with the importance of Ser L223 in stabilizing the reduced quinone and for efficient proton donation to Q_B as demonstrated for the *Rp. viridis* reaction center by Leibl et al. (1993) and for the *Rb. sphaeroides* reaction center by Paddock et al. (1990). Apart from this new insight regarding the role of Ser L223, the hydrogen-bonding pattern is similar in the structures 1PRC, 4RCR, and 1PSS (cf. Fig. 10 and Table 2), although in the latter structure the hydrogen bond between Gly L225 N and the distal carbonyl oxygen is lost. The displacement of the Q_B model in the 1PCR structure has already been discussed. In the 2RCR structure, the quinone ring plane is oriented nearly perpendicular (65°–80°) to the other Q_B rings.

As Trp M250 (M252 in the *Rb. sphaeroides* reaction center) for Q_A, Phe L216 forms a significant part of the Q_B-binding pocket. In the ubiquinone-2-reconstituted structure Q282, the ring planes of Phe L216 and Q_B are approximately parallel, with an angle between the two planes of only 5.6° (cf. Table 3). The distance between the two planes is approximately 3.5 Å. However, as the rings are displaced in the planes (cf. Fig. 10), the centroid-to-centroid distance

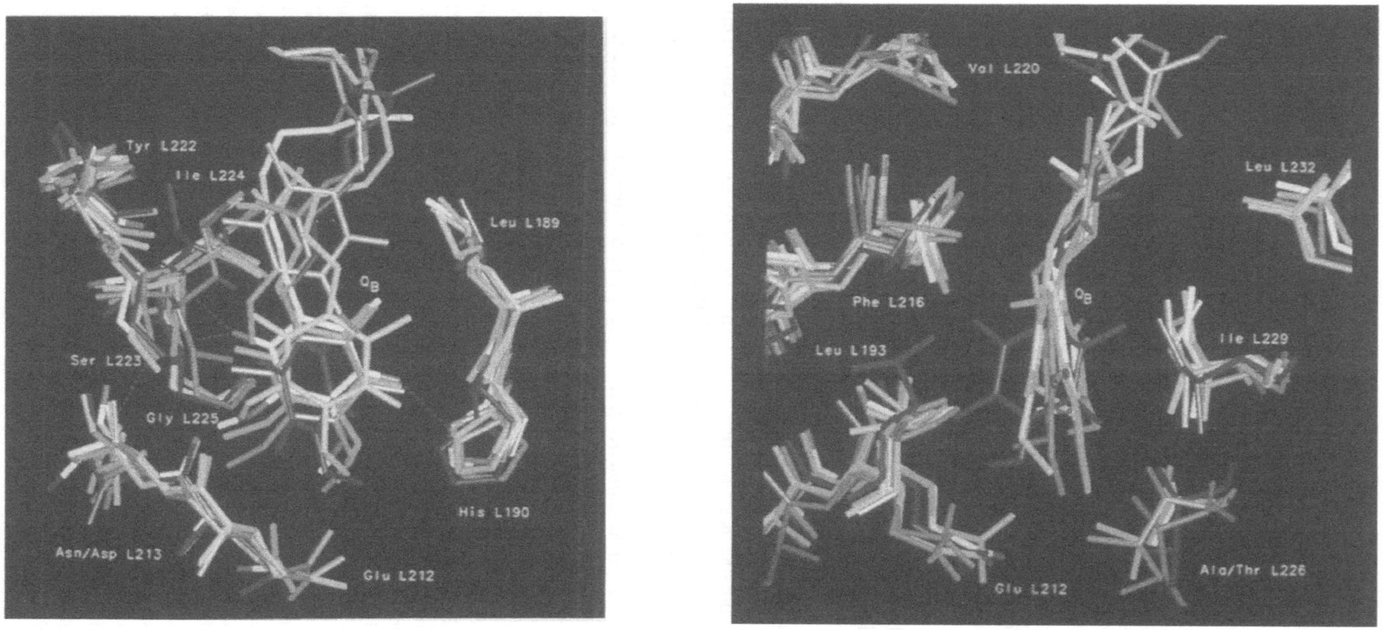

Fig. 10. (See Color Plate 8) Orthogonal views of the Q_B site: Six different structures are superimposed. The color coding is as follows: Q282 (multi-colored: carbon atoms are yellow, nitrogen atoms are blue, oxygen atoms are red), 1PRC (green), 2RCR (dark blue), 4RCR (brown), 1PSS (light blue) and 1PCR (purple). The dashed red lines indicate hydrogen bonds in the Q282 structure.

Table 2. Distances Between Prospective H-Bonding Donors and Acceptors in the Q_B-site

		Structure				
		Rp. viridis		*Rb. sphaeroides*		
Donor	Acceptor	Q282	1PRC	2RCR	4RCR	1PSS
His L190 δN	Q_B O2*	2.7	2.7	2.9	2.6	2.8
His L190 δN	Q_B O3*	3.3	3.0	(4.0)	3.4	(4.3)
Ser L223 γO	Q_B O5*	2.9	2.7	3.0	2.7	3.4
Ile L224 N	Q_B O5*	3.1	3.2	(5.2)	3.3	3.0
Gly L225 N	Q_B O5*	3.0	2.7	(5.5)	3.3	(5.5)

* O2 = proximal carbonyl oxygen; O3 = proximal methoxy oxygen; O5 = distal carbonyl oxygen
Distances in parentheses are too long for hydrogen-bonding interactions

Table 3. Angles Between Phe L216 and Q_B ring planes

Structure	Q282	1PRC	2RCR	4RCR	1PSS
Angle between planes	5.6°	12.9°	68.4°	78.7°	22.1°

of 5.3 Å is significantly larger. The importance of Phe L216 for quinone binding is illustrated by a reduced affinity for quinone in a PheL216 →Ser *Rp. viridis* mutant. (Sinning et al., 1989). The most striking discrepancy concerning the Phe L216 residues of the other structures (cf. Fig 10) is the 60°–70° plane deviation in the 4RCR structure.

A major difference between the binding sites of Q_A and Q_B is the presence of protonable residues at the Q_B site. Some evidence has been provided for the residues Glu L212, Asp L213 (in *Rb. sphaeroides*, Asp M43 in *Rp. viridis*), and Ser L223 to be involved in the transfer of protons to Q_B (Paddock et al., 1990; Takahashi and Wraight, 1992; Leibl et al., 1993; Rongey et al., 1993). Glu L212 forms a large part of the bottom of the Q_B site. A major difference in the orientation of the Glu L212 side chain is observed between the fully occupied quinone-reconstituted structure (Q282) and the original structure with 30% Q_B occupancy (1PRC). In this region, the latter structure is similar to a quinone-depleted structure (Lancaster and Michel, unpublished), in which one of the water molecules replacing the quinone is hydrogen-bonded to Glu L212. This seems to indicate that the 70% Q_B-depleted reaction centers in the 1PRC crystals dominate this part of the structure.

The side chain of Tyr L222 has an indirect effect on Q_B binding, as is illustrated by the reduced Q_B affinity of a Tyr L222 →Phe *Rp. viridis* mutant (Sinning et al., 1989). The nature of this indirect effect and the structural consequences of the mutation are discussed below (cf. Section VI).

The residues Leu L189, Leu L193, Val L220, Ala/Thr L226, Ile L229, and Leu L232 form a large part of the Q_B-binding pocket (cf. Fig. 10). The importance of some of these residues for Q_B binding has been demonstrated by the characteristics of the following mutants. A Val L220 →Leu mutation in *Rp. viridis* (with a second-site mutation Arg L217 →His) reduces the affinity for Q_B by a factor of eight (Ewald, 1992). A mutation of Thr L226 →Ala in *Rb. capsulatus* (i.e. from the residue also present in *Rb. sphaeroides* to the one present in the *Rp. viridis* reaction center) leads to a threefold increase in the affinity for Q_B (Baciou et al., 1993). An Ile L229 →Ser mutant of *Rb. capsulatus* (Baciou et al., 1993) and an Ile L229 →Met mutant of *Rb. sphaeroides* (Paddock et al., 1988) both display a reduced affinity for Q_B. The only major discrepancy among the different structures regarding these residues is observed for the side chain of Leu L193 (cf. Fig. 10) in the 2RCR structure. This is a consequence of the large deviation of the Q_B molecule in this structure compared to the others.

In summary, the improved model of the *Rp. viridis* Q_B site displays only small but important differences to the original (1PRC) model. In particular, this applies to the hydrogen bonding of the quinone and the role of Ser L223. The discrepancies between the four *Rb. sphaeroides* Q_B sites are substantially greater than those between the two *Rp. viridis* Q_B site models. The 1PCR Q_B is most likely reduced, but the discrepancies among the other *Rb. sphaeroides* structures underline the requirement for more and higher resolution data in order to resolve ambiguities. These could be resolved by additional measurements on the new trigonal crystal form. Considering the conservation of the overall profile of the Q_B site between *Rp. viridis* and *Rb. sphaeroides*, it may well turn out that the Q_B sites are more similar than can presently be discussed.

F. The 'Water Chain' to the Q_B Site

The Q_B site is deeply buried within the reaction center complex. For the protonation of Q_B associated with its reduction to quinol, protons must therefore be transferred from the cytoplasm through the protein to the Q_B site. This is likely to involve protonatable residues together with water molecules. The protons may move along a chain of proton donors and acceptors connected by hydrogen bonds. In both *Rp. viridis* reaction center structures considered (1PRC and Q282) and the *Rb. sphaeroides* reaction center structure based on the trigonal crystal form (1PCR), chains of water molecules from the cytoplasmic surface to the Q_B site are found. Most of them are in hydrogen bonding distance to one another (Fig. 11). Surprisingly, more water molecules can be observed in the case of the *Rb. sphaeroides* structure than for the *Rp. viridis* structures, even though the latter are based on higher resolution data. The water molecules are in extensive contact with mainly protonatable amino acid residues. Only in the *Rb. sphaeroides* structure is the row of water molecules uninterrupted over a length of 23 Å from the cytoplasmic surface to the Q_D molecule. In the *Rp. viridis* structures, 'detours' via protonatable amino acid residues must be taken in order to establish an uninterrupted chain.

At least some of the protonatable residues lining the 'water chain' have been shown to be functionally relevant for the proton transfer process. Based on the spectroscopic analysis of site-specific mutants (reviewed by Okamura and Feher, 1992; chapter in this volume) it was suggested that the first proton to Q_B is transferred in the reaction center of *Rb. sphaeroides* via Asp L213 and Ser L223, the second via Asp L213 and Glu L212. In *Rp. viridis*, residue L213 is an asparagine. However, its role can apparently be complemented (Rongey et al., 1993) by Asp M43 (which is Asn M44 in *Rb. sphaeroides*). In the *Rp. viridis* structures, Asp M43 and Glu L212 are within hydrogen-bonding distance to members of the 'water chain', whereas in the *Rb. sphaeroides* this is the case for Glu L212, but not for Asp L213. Another residue shown to be functionally relevant to the proton transfer process is Arg M231 (M233 in *Rb. sphaeroides*), as observed in the case of photocompetent revertants of mutants incapable of proton transfer. Exchanging Arg M233 to cysteine in *Rb. sphaeroides* (Okamura et al., 1992) and Arg M231 to leucine in *Rb. capsulatus* (Hanson et al., 1993) restores the photosynthetic activity to a level similar to that of the wild type. In *Rp. viridis*, Arg M231 forms salt bridges to Asp H125 and Glu H235, in *Rb. sphaeroides*, Arg M233 interacts with the corresponding residues Glu H122 and GluH230. In the *Rb. sphaeroides* reaction center, Glu H122 directly interacts with water molecules of the water chain. In the *Rp. viridis* reaction center, this applies to both Asp H125 and Arg M231 itself. On the other hand other residues, such as Glu H173 (Takahashi unpubl., cited by Shinkarev and Wraight, 1993) and Asp L210 (Paddock et al., 1992) in *Rb. sphaeroides*, corresponding to Glu H177 and Glu L210 in *Rp. viridis*, are possibly nonobligatory participants in the proton transfer pathways.

The many fixed water molecules observed in the 'water chain' may, however, solely be a consequence of the increased local polarity caused by a large number of charged residues. At present, there is no evidence for the use of this 'water chain' in proton transfer. It will be interesting to see whether an interruption of the 'water chain' is possible by site-directed mutagenesis and what effect this might have on the protonation of Q_B.

G. The Binding of Triazine Inhibitors to the Reaction Center

The Q_B site is also a well-established site of action of electron transfer inhibitors. Of particular interest in this respect are herbicides. Over 50% of commercially available herbicides function by inhibition of higher plants at the Q_B site on the D1 polypeptide of the photosystem II (PS II) reaction center (Percival and Baker, 1991). Many of them are triazines. The probably best available model to date in this reaction center is based on the X-ray structure of the reaction center from *Rp. viridis* (Michel and Deisenhofer, 1988). Although the triazine binding site has been localized by X-ray crystallographic analysis of the reaction center-terbutryn complex at 2.9 Å resolution (Michel et al., 1986b), description of the exact nature of the mode of binding and the effect of binding on the structure of the protein has to await the refinement of high resolution structures. The refined structure at 2.3 Å resolution of a *Rp. viridis* reaction center complex with the chiral atrazine derivative DG-

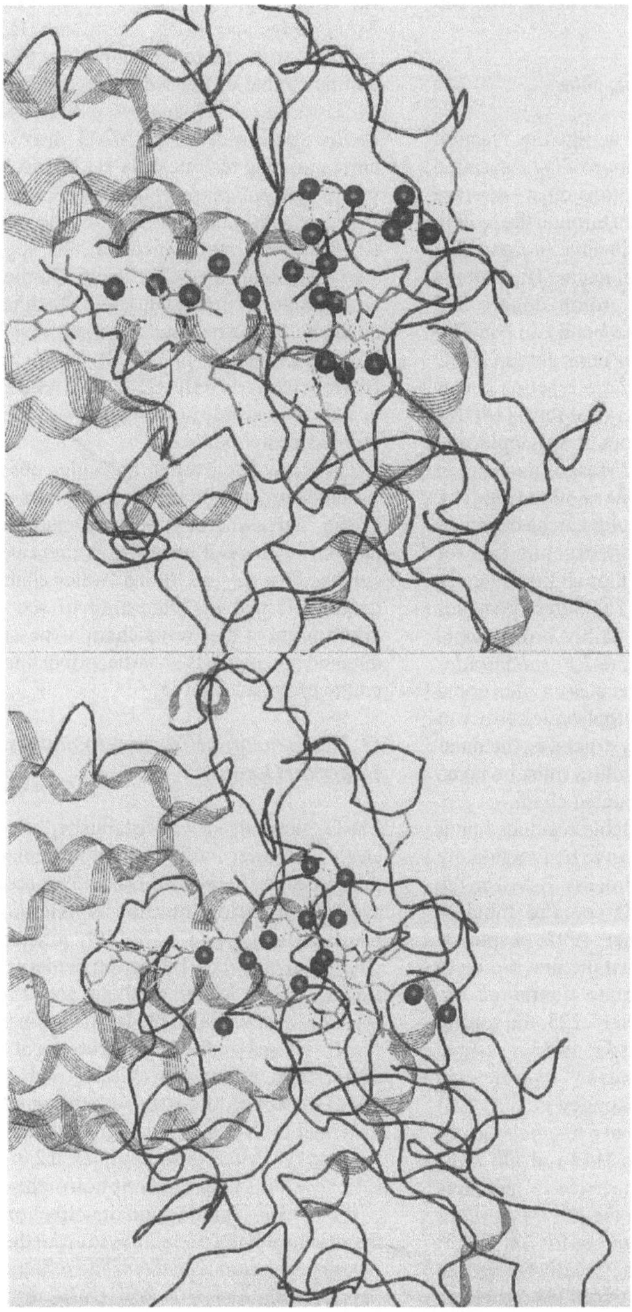

Fig. 11. (See also Color Plate 9) The 'water chain' from the cytoplasm to the Q_B site in the reaction centers of *Rp. viridis* (left) and *Rb. sphaeroides* (right), respectively. The protein backbones of the subunits are displayed in green, the water molecules in red. Also shown are the Q_B molecules and the protonatable amino acid residues lining the 'water chain'.

420314 (Lancaster and Michel, unpublished), has yielded a highly reliable model of triazine inhibitor binding based on over 100,000 unique reflections (cf. Table 1). Novel general principles of triazine class inhibitor binding include a tightly-bound water molecule involved in accepting a hydrogen bond from the Nδ atom of His L190 and donating a second hydrogen bond to a nitrogen of the triazine ring (cf. Fig. 12). A second tightly-bound water molecule is involved in completing the hydrogen-bonding web with hydrogen bonds to the first water molecule and to the Oε of Glu L212. In addition, a subtle change is observed in the conformation of the loop region (L220–L226) connecting helices DE and E, bringing the backbone carbonyl oxygen of tyrosine L222 within hydrogen bonding distance of the triazine. Also shown in Fig. 12 are the two hydrogen bonds observed previously (Michel et al., 1986b) between the backbone N of Ile L224 and the N5 of the triazine ring and between the aminoethyl N and Ser L223 Oγ, respectively. This means that the triazine molecule forms three hydrogen bonds with the protein on the distal side of the molecule and is only indirectly bound via water molecules on the proximal side. Finally, compared to the quinone-reconstituted complex of the reaction center, the aromatic ring of Phe L216 is reoriented by 18.2° to be aligned approximately parallel (with an angle between planes of 2.4°) to the plane of triazine ring, which is tilted by 14.5° compared to the quinone ring plane (Fig. 12).

VI. The Structure of Mutant Reaction Centers

Considering the importance of the photosynthetic reaction center it is not amazing that several groups have tried to establish systems for site-directed mutagenesis of photosynthetic reaction centers. Reviews of this recently developed field are already available (Coleman and Youvan, 1990; Diner et al., 1991). Work started with the photosynthetic reaction center from *Rb. capsulatus*, which is genetically very well characterized, and able to grow non-photosynthetically under aerobic conditions, as well as under anaerobic conditions using e.g. dimethylsulfoxide as electron acceptors. Most importantly, under these latter conditions the photosynthetic apparatus is fully induced. Unfortunately, the reaction center from *Rb. capsulatus* could not be crystallized, so that proper inspection for structural changes cannot be performed on *Rb. capsulatus* reaction centers. The closely related *Rb. sphaeroides* can be grown under similar non-photosynthetic conditions, so that site-directed mutagenesis is also straightforward. Also, structural changes in mutant reaction centers can be investigated by X-ray crystallography (Chirino et al., 1994) even more now with the availability of the new, better ordered trigonal crystal form. Site-directed mutagenesis of the structurally best characterized reaction center from *Rp. viridis* is now also possible (Laußermair and Oesterhelt, 1992). However, *Rp. viridis* can grow only under photosynthetic and, very slowly, under microaerophilic conditions. Under microaerophilic conditions the photosynthetic apparatus is not induced. Photosynthetic growth conditions exert a selection pressure for revertants and suppressor mutants if the reaction centers are functionally impaired. However, very interesting herbicide-resistant mutants were obtained by classical selection procedures, with mutations some of which would not have been made by site-directed mutagenesis (Sinning et al., 1989; Ewald et al., 1990; Sinning, 1992).

Many amino acids which were considered to be of importance for pigment binding or electron transfer (Michel et al., 1986b) were changed in *Rb. sphaeroides* reaction centers, and characterized by spectroscopic methods. Most outstanding is the mutation Tyr M210 →Phe. In the mutant reaction center the rate of the initial electron transfer is slowed down by a factor of 4–6 (Finkele et al., 1990; Nagarajan et al. 1990). The X-ray crystallographic analysis (Chirino et al., 1994) did not reveal any significant structural changes except for the absence of the O atom, so that the decrease of the electron transfer rate can be ascribed to removal of the oxygen atom. Other mutations such as GluL212 →Gln, AspL213 →Asn, with the protonatable residues presumably involved in protonation of Q_B (see Chapter 26 by Okamura and Feher) and HisM219 → Cys, removing a ligand to the non-heme iron, also do not lead to detectable structural changes. (Chirino et al., 1994). However, the resolution of the latter data set was limited to 4 Å. When the residues His L173 and His M202 liganded to the special pair bacteriochlorophylls D_A and D_B are replaced by Leu residues, bacteriopheophytins are incorporated as D_A and D_B respectively. *Vice versa*, in the LeuM214 →His mutant, a bacteriochlorophyll is incorporated as ϕ_A instead of a bacteriopheophytin. Using the new trigonal crystal form, the reaction center mutants Trp

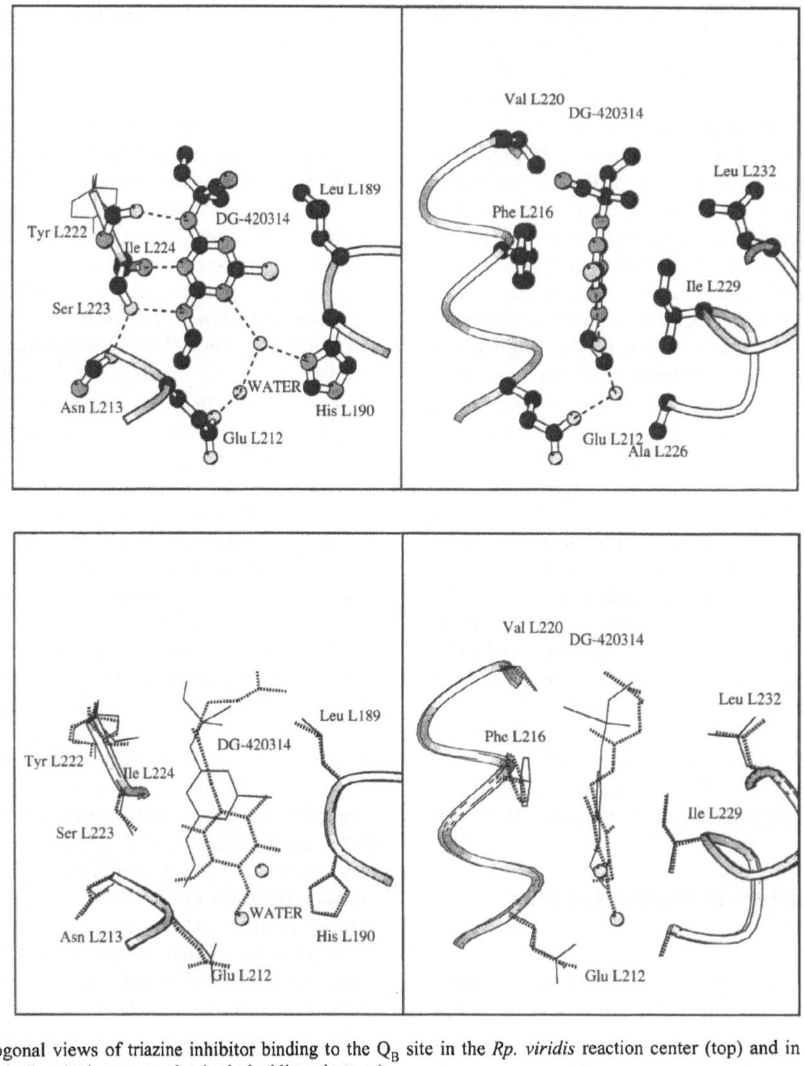

Fig. 12. Orthogonal views of triazine inhibitor binding to the Q_B site in the *Rp. viridis* reaction center (top) and in comparison to ubiquinone-2 binding (quinone complex in dashed lines, bottom).

M252 (see Figs. 8 and 9) →Phe, Trp M252 →Tyr, and Thr M222 →Val (Stilz et al., 1990) have been analyzed. In these mutants the electron transfer from ϕ_A to Q_A has been slowed down by a factor of three. The structural analysis (Merckel M, Ermler U, Fritzsch G, Michel H, Stilz U and Oesterhelt D, manuscript in preparation) indicates some structural changes in the Trp M252 →Tyr mutant due to the fact that tyrosine is longer than the original tryptophan which forms a hydrogen bond to Thr M222. This hydrogen bond cannot be formed by tyrosine for sterical reasons. The tyrosine evades towards Q_A, pushing back also Q_A. On the other side, Phe M151 is pulled in the same direction by the newly created tyrosine, so that the van der Waals and aromatic interactions are maintained. The mutations Trp M252 →Phe and Thr 222 →Val do not lead to detectable structural changes.

Some of the herbicide resistant mutants of the reaction center from *Rp. viridis* have also been analyzed by X-ray crystallography (Sinning et al., 1990; Sinning, 1992; Lancaster et al., unpublished). The double mutant (ArgL217 →His, SerL223 → Ala) shows some structural changes: The side chain of AsnL213 which is hydrogen-bonded to Ser L223 in the wild type is rotated towards the cavity which is created by the replacement of Arg L217 by the smaller His. At the same time, Q_B becomes more firmly bound (Sinning et al., 1989), as discussed above. The mutation TyrL222 →Phe unexpectedly leads to resistance against the herbicide terbutryn. In the wild type, TyrL222 forms a hydrogen bond with the peptide carbonyl oxygen of AspM43. Since this hydrogen bond is now missing, a stretch of the M subunit (M25–50) moves into a new position. The side chain of PheL222 rotates by 90° into the herbicide binding site (see above) thereby preventing the binding of terbutryn by steric hindrance. The Q_B binding pocket becomes slightly wider. Work on other mutants is in progress.

VII. Conclusions

As the result of the structure determination of the reaction centers from purple bacteria and sequence comparisons it could be concluded that the core of the reaction center from photosystem II is made up of the D1 and D2 subunits (Michel and Deisenhofer, 1986; Trebst, 1986; Michel and Deisenhofer, 1988). This proposal could be verified (Nanba and Satoh, 1987). In addition, the availability of a structural model of the photosynthetic reaction centers from two purple bacteria has led to a much better understanding of the primary charge separation and the electron transfer across the membrane. Physicists and theoreticians now have a solid basis for their experimental and theoretical investigations. Geneticists and biochemists start to understand the role of individual amino acids. As a result of the combined efforts one may have a satisfactory understanding of the first steps in photosynthetic electron transfer in the near future.

References

Alegria G and Dutton PL (1987) Construction and characterization of monolayer films of the reaction center cytochrome-*c* protein from *Rhodopseudomonas viridis*. In: Papa D, Chance B and Ernster L (eds) Cytochrome Systems, pp 601–608. Plenum Press, New York

Allen JP, Feher G, Yeates TO, Rees DC, Deisenhofer J, Michel H and Huber R (1986) Structural homology of reaction centers from *Rhodopseudomonas sphaeroides* and *Rhodopseudomonas viridis* as determined by X-ray diffraction. Proc Natl Acad Sci USA 83: 8589–8593

Allen JP, Feher G, Yeates TO, Komiya H and Rees, DC (1987) Structure of the reaction center from *Rhodobacter sphaeroides* R-26: the cofactors. Proc Natl Acad Sci USA. 84: 5730–5734

Allen JP, Feher G, Yeates TO, Komiya H and Rees DC (1988) Structure of the reaction center from *Rhodobacter sphaeroides* R-26 and 2.4.1: Protein-cofactor (bacteriochlorophyll, bacteriopheophytin, and carotenoid) interactions. Proc Natl Acad Sci USA 85: 8847–8491

Arnoux B, Decruix A, Reiss-Husson F, Lutz M, Norris J, Schiffer M and Chang CH (1989) Structure of spheroidene in the photosynthetic reaction center from Y *Rhodobacter sphaeroides*. FEBS Lett 258: 47–50

Baciou L, Bylina EJ and Sebban P (1993) Study of wild type and genetically modified reaction centers from *Rhodobacter capsulatus*: Structural Comparison with *Rhodopseudomonas viridis* and *Rhodobacter sphaeroides*. Biophys J 65: 652–660.

Bernstein FC, Koetzle TF, Williams GJB, Meyer EF, Brice MD, Rodgers JR, Kennard O, Shimanouchi T and Tasumi M (1977) The Protein Data Bank: A computer-based archival file for macromolecular structures. J Mol Biol 112: 535–542

Breton J and Verméglio A, eds. (1992) The Photosynthetic Bacterial Reaction Center II, NATO-ASI Series A, Life Sciences 237, Plenum Press, New York

Buchanan SK, Fritzsch G, Ermler U and Michel H (1993) A new crystal form of the photosynthetic reaction centre from *Rhodobacter sphaeroides* of improved diffraction quality. J Mol Biol 230: 1311–1314

Bylina EJ, Kirmaier C, McDowell L, Holten D and Youvan DC (1988) Influence of an amino-acid residue on the optical properties and electron transfer dynamics of a photosynthetic reaction centre complex. Nature 336: 182–184

Chang CH, Tiede D, Tang J, Smith U, Norris J and Schiffer M (1986) Structure of *Rhodopseudomonas sphaeroides* R-26 reaction center. FEBS Lett 205: 82–86

Chang CH, El-Kabbani O, Tiede D, Norris J and Schiffer M (1991) Structure of the membrane-bound protein photosynthetic reaction center from *Rhodobacter sphaeroides*. Biochemistry 30: 5353–5360

Chirino AJ, Lous EJ, Huber M, Allen JP, Schenck CC, Paddock ML, Feher G and Rees D (1994) Crystallographic Analyses of Site-Directed Mutants of the Photosynthetic Reaction Center from *Rhodobacter sphaeroides*. Biochemistry 33: 4584–4593

Chory J, Donohue TJ, Warga AR, Staehelin LA and Kaplan S (1984) Induction of the photosynthetic membranes of *Rhodopseudomonas sphaeroides*: biochemical and morphological studies. J Bacteriol 159: 540–554

Coleman WJ and Youvan DC (1990) Spectroscopic Analysis of Genetically Modified Reaction Centers. Annu Rev Biophys Biophys Chem 19: 333–367

Debus RJ, Feher G and Okamura MY (1985) LM Complex of Reaction centers from *Rhodopseudomonas viridis* R-26: Characterization and Reconstitution with H subunit. Biochemistry 24: 2488–2500

De Groot HJM, Gebhard R, van der Hoef I, Hoff AJ, Lugtenburg J, Violette CA and Frank HA (1992) ^{13}C Magic angle spinning NMR evidence for a 15,15'-cis configuration of the spheroidene in the *Rhodobacter sphaeroides* photosynthetic reaction center. Biochemistry 31: 12446–12450

Deisenhofer J and Michel H (1989) The photosynthetic reaction centre from the purple bacterium *Rhodopseudomonas viridis*. EMBO J 8: 2149–2169

Deisenhofer J and Norris JR, eds (1993) The Photosynthetic Reaction Center, Academic Press, San Diego

Deisenhofer J, Epp O, Miki F, Huber R and Michel H (1984) X-ray structure analysis of a membrane protein complex: electron density map at 3 Å resolution and a model of the chromophores of the photosynthetic reaction center from *Rhodopseudomonas viridis*. J Mol Biol 180: 385–398

Deisenhofer J, Epp O, Miki R, Huber R and Michel H (1985) Structure of the protein subunits in the photosynthetic reaction centre of *Rhodopseudomonas viridis* at 3 Å resolution. Nature 318: 618–624

Deisenhofer J, Epp O, Sinning I and Michel H (1995) Crystallographic refinement at 2.3 Å resolution and refined model of the photosynthetic reaction center from *Rhodopseudomonas viridis*. J Mol Biol 246: 429–457

Diner BA, Nixon PJ and Farchaus JW (1991) Site directed mutagenesis of photosynthetic reaction centers. Curr Opinion Struct Biol 1: 546–554

Dracheva SM, Drachev LA, Konstantinov AA, Semenov AY, Skulachev VP, Arutjunjan AM, Shuvalov VA and Zaberezhnaya SM (1988) Electrogenic steps in the redox reactions catalyzed by photosynthetic reaction-centre complex from *Rhodopseudomonas viridis*. Eur J Biochem 171: 253–264

Dracheva S, Williams JC, Van Driessche G, Van Beeumen JJ and Blankenship RE (1991) The primary structure of cytochrome *c*-554 from the green photosynthetic bacterium *Chloroflexus aurantiacus*. Biochemistry 30: 11451–11458

Ducruix A and Reiss-Husson F (1987) Preliminary characterization by X-ray diffraction of crystals of photochemical reaction centres from wild-type *Rhodopseudomonas sphaeroides*. J Mol Biol 193: 419–421

El-Kabbani O, Chang CH, Tiede D, Norris J and Schiffer M (1991) Comparison of reaction centers from *Rhodobacter sphaeroides* and *Rhodopseudomonas viridis*: Overall architecture and protein-pigment interactions. Biochemistry 30: 5361–5369

Ermler U, Fritzsch G, Buchanan SK and Michel H (1994) Structure of the photosynthetic reaction centre from *Rhodobacter sphaeroides* at 2.65 Å resolution: Cofactors and protein-cofactor interactions. Structure 2: 925–936

Ewald G (1992) Selektion, funktionelle und strukturelle Charakterisierung herbizidresistenter Mutanten von *Rhodopseudomonas viridis*. Thesis, Johann Wolfgang Goethe-University, Frankfurt am Main

Ewald G, Wiessner C and Michel H (1990) Sequence analysis of four atrazine-resistant mutants from *Rhodopseudomonas viridis*. Z Naturforsch 45c: 459–462

Finkele U, Lauterwasser C, Zinth W, Gray KA and Oesterhelt D (1990) Role of tyrosine M210 in the initial charge separation of reaction centers of *Rhodobacter sphaeroides*. Biochemistry 29: 8517–8521

Fritzsch G, Buchanan SK and Michel H (1989) Assignment of cytochrome hemes in crystallized reaction centers from *Rhodopseudomonas viridis*. Biochim Biophys Acta 977: 157–162

Gunner MR and Honig B (1991) Electrostatic control of midpoint potentials in the cytochrome subunit of the *Rhodopseudomonas viridis* reaction center. Proc Natl Acad Sci USA 88: 9151–9155

Hanson DK, Tiede DM, Nance SL, Chang CC and Schiffer M (1993) Site-specific and compensatory mutations imply unexpected pathways for proton delivery to the Q_B binding site of the photosynthetic reaction center. Proc Natl Acad Sci USA 90: 8929–8933

Hoff AJ (1988) Nomen est omen. A note on nomenclature. In: Breton J and Verméglio A (eds) The photosynthetic Bacterial Reaction Center: Structure and Dynamics, pp 98–99. Plenum Press, New York

Jones TA, Zou JY, Cowan SW and Kjeldgaard M (1991) Improved methods for building protein models in electron density maps and the location of errors in these models. Acta Cryst A47: 110–119

Kirmaier C and Holten D (1993) Electron transfer and charge recombination reactions in wild-type and mutant bacterial reaction centers. In: Deisenhofer J and Norris J (eds) The Photosynthetic Reaction Center, Vol II, pp 49–70. Academic Press, San Diego

Komiya H, Yeates TO, Rees DC, Allen JP and Feher G (1988) Structure of the reaction center from *Rhodobacter sphaeroides* R-26 and 2.4.1: Symmetry relations and sequence comparisons between different species. Proc Natl Acad Sci USA 85: 9012–9016

Kraulis PJ (1991) MolScript: a program to produce both detailed and schematic plots of protein structures. J Appl Cryst 24: 946–950

Laußermair E and Oesterhelt D (1992) A system for site-specific mutagenesis of the photosynthetic reaction center in *Rhodopseudomonas viridis*. EMBO J 11: 777–783

Leibl W, Sinning I, Ewald G, Michel H and Breton J (1993) Evidence that serine L223 is involved in the proton transfer pathway to Q_B in the photosynthetic reaction center from *Rhodopseudomonas viridis*. Biochemistry 32: 1958–1964

Luzzati V (1952) Traitement statistique des erreurs dans la détermination des structures cristallines. Acta Cryst 5: 802–810.

Michel H and Deisenhofer J (1986) X-ray diffraction studies on a crystalline bacterial photosynthetic reaction center: A progress report and conclusions on the structure of Photosystem II reaction centers. In: Staehelin LA and Arntzen CJ (eds) Encyclopedia of Plant Physiology, New Series, Vol 19, Photosynthesis III, pp 371–381. Springer, Berlin, Heidelberg

Michel H and Deisenhofer J (1988) Relevance of the photosynthetic reaction center from purple bacteria to the structure of photosystem II. Biochemistry 27: 1–7

Michel H, Weyer KA, Gruenberg H, Dunger I, Oesterhelt D and Lottspeich F (1986 a) The 'light' and 'medium' subunits of the photosynthetic reaction centre from *Rhodopseudomonas viridis*. isolation of the genes, nucleotide and amino acid sequence. EMBO J 5: 1149–1158

Michel H, Epp O and Deisenhofer J (1986b) Pigment-protein interactions in the photosynthetic reaction centre from *Rhodopseudomonas viridis*. EMBO J 5: 2445–2451

Nagarajan V, Parson WW, Gaul D and Schenk CC (1990) Effect of specific mutations of tyrosine-(M)210 on the primary photosynthetic electron-transfer process in *Rhodobacter sphaeroides*. Proc Natl Acad Sci USA 87: 7888–7892

Nagashima KVP, Shimada K and Matsuura K (1993) Photosynthetic analysis of photosynthetic genes of *Rhodocyclus gelatinosus*: Possibility of horizontal gene transfer in purple bacteria. Photosynth Res 36: 185–191

Nagashima KVP, Matsuura K, Ohyama S and Shimada K (1994) Primary Structure and Transcription of Genes Encoding B870 and Photosynthetic Apoproteins from *Rubrivivaxe gelatinosus*. J Biol Chem 269: 2477–2484

Nanba O and Satoh K (1987) Isolation of photosystem II reaction center consisting of D-1 and D-2 polypeptides and cytochrome b-559. Proc Natl Acad Sci USA 84: 109–112

Okamura MY and Feher G (1992) Proton transfer in reaction centers from photosynthetic bacteria. Annu Rev Biochem 61: 861–869

Okamura MY, Paddock ML, McPherson PH, Rongey S and Feher G (1992) Proton transfer in bacterial reaction centers: second site mutations Asn M44 →Asp or Arg M233 →Cys restore photosynthetic competence to Asp L213 →Asn mutants in RCs from *Rb. sphaeroides*. In: Murata N (ed) Research in Photosynthesis , Vol. I, pp 349–356. Kluwer Academic Publishers, Dordrecht

Ovchinnikov YA, Abdulaev NG, Zolotarev AS, Shmukler B, Zargarov AA, Kutuzov M, Telezhinskaya I and Levina NB (1988a) Photosynthetic reaction centre of *Chloroflexus aurantiacus* I. Primary structure of L subunit. FEBS Lett 231: 237–242

Ovchinnikov YA, Abdulaev NG, Shmuckler BE, Zargarov AA, Kutuzov MA, Telezhinskaya IN, Levina NB and Zolotarev A (1988) Photosynthetic Reaction Centre of *Chloroflexus auranticans*. Primary Structure of M subunit. FEBS Lett 232:364–368

Paddock ML, Rongey SH, Abresch EC, Feher G and Okamura MY (1988) Reaction centers from three herbicide-resistant mutants of *Rhodobacter sphaeroides* 2.4.1: sequence analysis and preliminary characterization. Photosynth Res 17: 75–96

Paddock ML, McPherson PH, Feher G and Okamura MY (1990) Pathway of proton transfer in bacterial reaction centers: replacement of serine-L223 by alanine inhibits electron and proton transfers associated with reduction of quinone to dihydroquinone. Proc Natl Acad Sci USA 87: 6803–6807

Paddock ML, Juth A, Feher G and Okamura MY (1992) Electrostatic effects of replacing Asp-L210 with Asn in bacterial RCs from *Rb. sphaeroides*. Biophys J 61: 153a

Peitsch MC, Wells TNC, Stampf DR and Sussman JL (1995) The Swiss-3D image collection and PDB-Browser on the World-Wide Web. Trends Biochem Sci 20:82–84

Percival MP and Baker NR (1991) Herbicides and photosynthesis. In: Baker NR and Percival MP (eds) Herbicides, pp 1–26. Elsevier Science Publishers, Amsterdam

Plato M, Michel-Beyerle ME, Bixon M and Jortner J (1989) On the role of tryptophan as a superexchange mediator for quinone reduction in photosynthetic reaction centers. FEBS Lett 249: 70–74

Rongey SH, Paddock ML, Feher G and Okamura MY (1993) Pathway of proton transfer in bacterial reaction centers: Second-site mutation Asn M44 →Asp restores electron and proton transfer in reaction centers from the photosynthetically deficient Asp L213 →Asn mutant of *Rhodobacter sphaeroides*. Proc Natl Acad Sci USA 90: 1325–1329

Roth M, Lewit-Bentley A, Michel H, Deisenhofer J, Huber R and Oesterhelt D (1989) Detergent structure in crystals of a bacterial photosynthetic reaction centre. Nature 340: 659–662

Schenk CC, Mathis P and Lutz M (1984) Triplet formation and triplet decay in reaction centers from the photosynthetic bacterium *Rhodopseudomonas sphaeroides*. Photochem Photobiol 39: 407–417

Shinkarev VP and Wraight CA (1993) Electron and proton transfer in the acceptor quinone complex of reaction centers of Phototrophic Bacteria. In: Deisenhofer J and Norris J (eds) The Photosynthetic Reaction Center, Vol. I., pp 193–255. Academic Press, San Diego

Shinkarev VP, Drachev AL and Dracheva SM (1990) The thermodynamic characteristic of four-heme cytochrome *c* in *Rhodopseudomonas viridis* reaction centers, as derived from a quantitative analysis of the differential absorption spectra in α-domain. FEBS Lett 261: 11–13

Shiozawa JA, Lottspeich F, Oesterhelt D and Feick R (1989) The primary structure of the *Chloroflexus auranticans* reaction-center polypeptides. Eur J Biochem 180: 75–84

Shopes RJ, Levine LMA, Holten D and Wraight CA (1987) Kinetics of oxidation of the bound cytochromes in reaction centers from *Rhodopseudomonas viridis*. Photosynth Res 12: 165–180

Sinning I (1988) Der Elektronenakzeptorkomplex im photosynthetischen Reaktionszentrum des Purpurbakteriums *Rhodopseudomonas viridis*. Thesis, University of Munich

Sinning I (1992) Herbicide binding in the bacterial photosynthetic reaction center. Trends Biochem Sci 17: 150–154

Sinning I, Michel H, Mathis P and Rutherford AW (1989) Characterization of four herbicide-resistant mutants of *Rhodopseudomonas viridis* by genetic analysis, electron paramagnetic resonance and optical spectroscopy. Biochemistry 28: 5544–5553

Sinning I, Koepke J, Schiller B and Michel H (1990) First glance on the three-dimensional structure of the photosynthetic reaction center from a herbicide-resistant *Rhodopseudomonas viridis* Mutant. Z Naturforsch 45c: 455–458

Stilz HU, Finkele U, Holzapfel W, Lauterwasser C, Zinth W and Oesterhelt D (1990) Site-directed mutagenesis of threonine M222 and tryptophan M252 in the photosynthetic reaction center of *Rhodobacter sphaeroides*. In: Michel-Beyerle ME (ed) Reaction centers of Photosynthetic Bacteria 6, pp 265–271. Springer Verlag, Berlin

Takahashi E and Wraight CA (1992) Proton and electron transfer in the acceptor quinone complex of *Rhodobacter sphaeroides* reaction centers: characterization of site-directed mutants of the two ionizable residues, Glu^{L212} and Asp^{L213}, in the Q_B binding site. Biochemistry 31: 855–866

Trebst A (1986) The topology of the plastoquinone and herbicide binding peptides of photosystem II in the thylakoid membrane. Z Naturforsch 41c: 240–245

Verméglio A and Paillotin G (1982) Structure of *Rhodopseudomonas viridis* reaction centers, absorption and photoselection at low temperature. Biochim Biophys Acta 681: 32–40

Verméglio A, Richaud P and Breton J (1989) Orientation and assignment of the four cytochrome hemes in *Rhodopseudomonas viridis* reaction centers. FEBS Lett 243: 259–263

Warncke K and Dutton PL (1993) Experimental resolution of the free energies of aqueous solvation contributions to ligand-protein binding: quinone-Q_A site interactions in the photosynthetic reaction center protein. Proc Natl Acad Sci USA 90: 2920–2924

Weyer KA, Schafer W, Lottspeich F and Michel H (1987a) The cytochrome subunit of the photosynthetic reaction center from *Rhodopseudomonas viridis* is a lipoprotein. Biochemistry 26: 2909–2914

Weyer KA, Lottspeich F, Gruenberg H, Lang FS, Oesterhelt D and Michel H (1987b) Amino acid sequence of the cytochrome subunit of the photosynthetic reaction centre from the purple bacterium *Rhodopseudomonas viridis*. EMBO J 6: 2197–2202

Williams JC, Steiner LA, Feher G and Sinnen MI (1984) Primary Structure of the L subunit of the reaction center from *Rhodopseudomonas sphaeroides*. Proc Natl Acad Sci USA 81: 7303–7307

Williams JC, Steiner L.A. and Feher G (1986) Primary structure of the reaction center from *Rhodopseudomonas sphaeroides*. Proteins 1: 312–325

Yeates TO, Komiya H, Chirino A, Rees DC, Allen JP and Feher G (1988) Structure of the reaction center from *Rhodobacter sphaeroides* R-26 and 2.4.1: Protein-cofactor (bacteriochlorophyll, bacteriopheophytin, and carotenoid) interactions. Proc Natl Acad Sci USA 85: 7993–7997

Youvan DC, Bylina EJ, Alberti M, Begusch H and Hearst JE (1984) Nucleotide and deduced polypeptide sequences of the photosynthetic reaction center, B870 antenna, and flanking polypeptides from *Rb. capsulatus*. Cell 37: 949–957

Zinth W and Kaiser W (1993) Time-resolved spectroscopy of primary electron transfer in reaction center of *Rhodobacter sphaeroides* and *Rhodopseudomonas viridis*. In: Deisenhofer J and Norris J (eds) The Photosynthetic Reaction Center, Vol. II, 71–88. Academic Press, San Diego

Zinth W, Kaiser W and Michel H (1983) Efficient photochemical activity and strong dichroism of single crystals of reaction centers from *Rhodopseudomonas viridis*. Biochim Biophys Acta 723: 128–131

Chapter 24

The Pathway, Kinetics and Thermodynamics of Electron Transfer in Wild Type and Mutant Reaction Centers of Purple Nonsulfur Bacteria

Neal W. Woodbury and James P. Allen
Department of Chemistry and Biochemistry and the Center for the Study of Early Events in Photosynthesis, Arizona State University, Tempe, AZ 85287–1604, USA

R. E. Blankenship, M. T. Madigan and C. E. Bauer (eds): Anoxygenic Photosynthetic Bacteria, pp. 527–557.

Summary

The light driven electron transfer reactions of photosynthesis are best characterized in the reaction centers of purple nonsulfur bacteria such as *Rhodobacter sphaeroides*. These reactions are extraordinarily fast: the first of them taking place within a few picoseconds. Because photosynthetic electron transfer is initiated by light, and the redox active cofactors have distinct spectral signatures, both the kinetics and thermodynamics of the reaction can be studied on the timescale of the actual reactions. This, in combination with the availability of detailed structural information, has made the electron transfer reactions of purple nonsulfur bacteria some of the most completely characterized electron transfer reactions in biological systems. In addition to the spectroscopic tools which have been used to characterize the electron transfer reactions, the reaction center has also been the subject of mutagenesis studies. The ease with which genetic engineering can be performed in these bacteria, coupled with the structurally robust nature of the reaction center, has resulted in many mutations which affect the early electron transfer reactions. Critical protein-cofactor interactions have now been identified and analyzed which affect the metal content, redox potentials, electronic structure and intermolecular coupling of the bacteriochlorophylls and bacteriopheophytins of the reaction center. In combination, these studies have suggested various possible mechanisms for early electron transfer which involve both the electronic states of the cofactors as well as nuclear conformational changes in the surrounding protein.

I. Introduction

Light absorption by the photosynthetic apparatus of purple nonsulfur bacteria results in the formation of the lowest excited singlet state of bacteriochlorophyll (BChl). Subsequently, this energy is transferred from one BChl-complex to another in the photosynthetic membrane until it arrives at the reaction center, a special BChl containing protein-cofactor complex which is the site of the initial photosynthetic electron transfer reactions. One of the most important constraints on these early electron transfer reactions is the brief lifetime of the BChl excited singlet state. In organic solvents, the lifetime of the excited singlet state of BChl is about 3 nanoseconds (Seely and Connolly, 1986). Thus, the initial electron transfer reactions in the bacterial photosynthetic reaction center must take place on a timescale much shorter than this in order to achieve a high quantum yield. The requirement for ultrafast electron transfer in the reaction center is made even more imperative by the competition between forward electron transfer and backward energy transfer to the antenna, a process which is usually thermodynamically favorable due to the large number of antenna pigments per reaction center (Dawkins et al., 1988; Woodbury and Parson, 1986).

The initial electron transfer reactions of photosynthesis have been best studied in the reaction centers of the purple nonsulfur bacteria (Kirmaier and Holten, 1987, 1993; Feher et al., 1989; Parson, 1991; Martin and Vos, 1992). Though many of the mechanistic details of this process remain unclear, the overall kinetics are well known. The initial charge separation takes place within a few picoseconds and is followed by two subsequent electron transfer reactions that take place on the hundred picosecond and hundred microsecond timescales, respectively. The state formed at each step is progressively more stable, starting with the excited singlet state, which has an intrinsic lifetime that is probably less than a nanosecond (see below) and resulting in a final charge-separated state with a lifetime of roughly a second and a quantum yield of nearly 100%. Of course, there is a price to be paid for this stability and reaction specificity; the final charge-separated state of the purple bacterial reaction center conserves only about one third of the standard free energy available in the initial excited singlet state, relative to the ground state.

The success of studying photosynthetic electron transfer in the purple nonsulfur bacteria comes from the many technical advantages of working with the reaction centers from these organisms. The reaction centers of several species are readily isolated in high

Abbreviations: B_A – A-side BChl monomer; B_B – B-side BChl monomer; BChl – bacteriochlorophyll; BPhe – bacteriopheophytin; H_A – A-side Bphe; H_B – B-side Bphe; P – BChl dimer; P* – excited singlet state of P; P^+ – cation of P; P_A – A-side BChl of P; P_B – B-side BChl of P; Q_A – A-side quinone; Q_B – B-side quinone; *Rb.* – *Rhodobacter; Rp.* – *Rhodopseudomonas; Rs.* – *Rhodospirillum* yield and in most cases are stable for days in the dark at room temperature or indefinitely when frozen (Feher and Okamura, 1978; Prince and Youvan, 1987; Snozzi and Bachofen, 1979). They have a small, well defined number of absorbing cofactors and, importantly, the electron transfer function of the reaction center can be cleanly separated from the energy gathering function of the antenna, making the performance and interpretation of optical spectroscopy much simpler than it is in systems where the antenna and the reaction center have not or cannot be separated. Finally, genetic manipulation of the reaction center genes has become routine in recent years, and the reaction center of purple nonsulfur bacteria has proven to be a surprisingly robust system for directed mutagenesis studies (Coleman and Youvan, 1990; Bylina and Youvan, 1991).

This chapter is devoted to a brief, and necessarily incomplete, summary of what has been learned, primarily during the past 10 years, about the pathway, kinetics and thermodynamics of early electron transfer in the reaction centers of purple nonsulfur bacteria and their mutants. The chapter starts with a discussion of the structural aspects of the reaction center that are known to affect the early electron transfer reactions and a brief description of some of the mutations that have been constructed and characterized. This leads to a description of reaction center spectroscopy and then to a discussion of possible mechanisms of electron transfer, particularly for the initial charge separation reaction. While there is at present no consensus on the mechanism of initial electron transfer, an attempt is made to distinguish between possible mechanisms and to point out the complex aspects of the experimental observations that have proven difficult to unambiguously explain using simple kinetic models.

II. The Reaction Center of *Rhodobacter sphaeroides*

Spectroscopically, the best characterized of the purple bacterial reaction centers is the complex isolated from *Rhodobacter (Rb.) sphaeroides*. The three dimensional structures of both this reaction center and the reaction center from *Rhodopseudomonas (Rp.) viridis* have been determined by X-ray diffraction studies as described in detail in Chapter 23 by Lancaster et al. Briefly, the structure of the reaction center of *Rb. sphaeroides* has two symmetry related subunits, L and M, that each have five transmembrane helices, and a third subunit, H, that has one transmembrane helix and a large globular domain (Fig. 1A; Allen et al., 1987; Chang et al., 1991; Chirino et al., 1994; Ermeler et al., 1994). The reaction center BChls, bacteriopheophytins (BPhes) and quinones are divided into two branches, labeled A and B, that are related by the same C_2 symmetry axis as the L and M subunits (Fig. 1A). Apparent in the structure is a closely interacting dimer of BChls that serves as the primary electron donor (P). Symmetrically placed with respect to P are two BChl monomers (B_A and B_B), two BPhes (H_A and H_B), and two quinones (Q_A and Q_B). Between the two quinones is a non-heme iron atom that appears to serve a structural role in stabilizing the complex. In carotenoid containing strains of *Rb. sphaeroides*, there is also a symmetry breaking carotenoid molecule situated near B_B.

The two BChls which make up P overlap at the ring I position with a separation of approximately 3 Å (Fig. 1A,B). The twofold symmetry of the dimer is broken by molecular interactions involving the protein. Some distances involving symmetry related residues are not equivalent. For example, the distances between each of the magnesiums of the initial electron donor BChls and the ligating Nε of histidines L173 and M202 in *Rb. sphaeroides* are different, respectively 2.1 Å and 2.6 Å (Chirino et al., 1994). Also, different side chains are found at some symmetry related positions. For example, the symmetry related residue of His L168, that forms a hydrogen bond with the acetyl group of ring I of the A-side BChl of the dimer, is Phe M197 which cannot form such a bond. The asymmetry in the protein interactions has led to the suggestion that these structural differences should establish the asymmetry in electron transfer (Plato et al., 1988; Scherer and Fischer, 1989; Parson et al., 1990; Thompson et al., 1991).

Electron transfer proceeds from the excited state of the primary donor, P*, to H_A in approximately 3.5 ps. The closest distance between P and H_A is approximately 7 Å. To explain the fast rate of electron transfer over this distance, bridging intermediates have been suggested, most commonly B_A located between P and H_A (see Fig. 1A). Near H_A is the primary quinone (Q_A) that receives the electron in approximately 200 ps. The electron is subsequently transferred to the secondary quinone (Q_B) in about 200 μs.

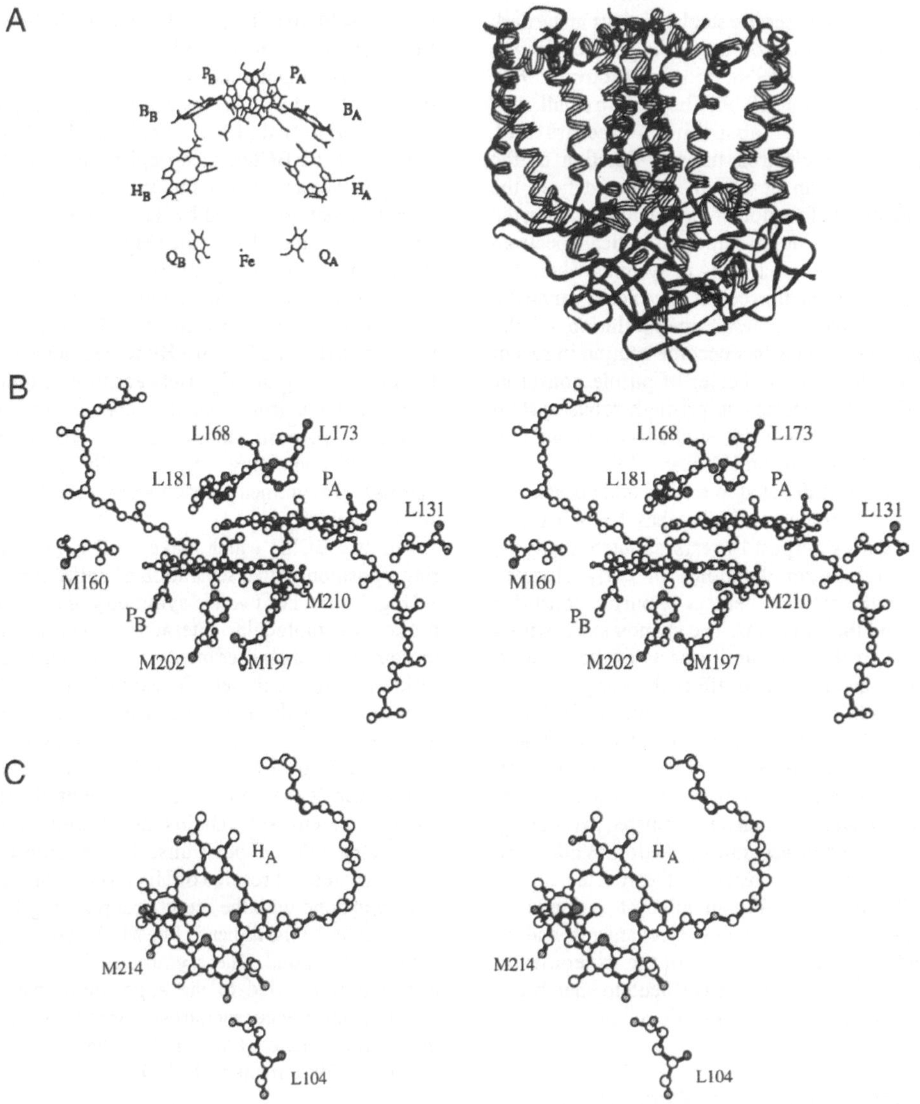

Fig. 1. (A) The structure of the cofactors (left side of panel) and a ribbon diagram of the L, M and H protein subunits (right side of panel) in reaction centers from *Rb. sphaeroides* R-26 reaction centers (Brookhaven Data Base, file 4RCR). The two BChls that make up the closely interacting pair of BChls (P) at the top of the cofactor structure are labeled P_A and P_B. The two monomer BChls are labeled B_A and B_B. The two Bphes are labeled H_A and H_B. The two quinones are labeled Q_A and Q_B. An approximate C_2 rotational axis runs from the center of P to the nonheme iron atom at the bottom of the figure. The subunit ribbon diagram is shown in the same orientation as the cofactors, and the C_2 symmetry extends to the positions of the helices between the L (right side of the ribbon diagram) and M (left side of the ribbon diagram) subunits. In vivo, the reaction center is embedded in the photosynthetic membrane such that the C_2 axis is perpendicular to the plane of the membrane. The two possible transmembrane electron transfer pathways, comprised of the A-side and B-side cofactors, respectively, utilize exactly the same cofactors with very similar spatial positions and orientations, yet electron transfer appears to occur nearly exclusively along the A-side branch. (B) A stereo view of the BChl dimer, P, and some of the amino acids which interact closely with it. This is the view obtained by looking from the top of panel A along the C_2 axis. Mutations have been made at all of these sites as discussed in the text. (C) A stereo view of the two Bphes, H_A and H_B, and two closely associated amino acids which have been the target of mutagenesis studies (see text).

III. Reaction Center Mutants

A number of mutations have been constructed involving residues near the dimer and monomer BChls in *Rb. sphaeroides* and a closely related bacterium, *Rb. capsulatus* (Fig 1B). These mutations fall into 4 general classes: 1) mutations of the histidines that provide ligands for the Mg of the BChls, 2) mutations of the aromatic residues, 3) alterations of hydrogen bonds to the dimer, and 4) large scale symmetry changes of the M subunit to corresponding regions of the L subunit and vice versa. In addition, two mutations have been constructed near one of the BPhes, H_A, (Fig. 1C) that: 1) introduce a potential Mg ligand and 2) break an existing hydrogen bond donor to the 9-keto group of ring V. The general characteristics of these mutations are described briefly below.

A. Heterodimer Mutants

Histidines L173 and M202 (or M200 in *Rb. capsulatus*) that provide ligands for P have been altered in both *Rb. sphaeroides* and *Rb. capsulatus* (Bylina and Youvan, 1988; Kirmaier et al., 1988; McDowell et al., 1991a,b). Changing either of these histidines to leucine results in the formation of a BChl-BPhe heterodimer in place of the normal BChl-BChl homodimer. Based upon the higher midpoint potential of the BPhe, the unpaired electron would be expected to be localized primarily on the BChl side of the pair forming $BChl^+$ $BPhe^-$. In both of the heterodimer mutants the yield of electron transfer is decreased by roughly a factor of two. The measured decay rates of the excited state are 42 ps, 18 ps and 14 ps for the L173 and M202 mutants in *Rb. sphaeroides*, and the M200 mutant in *Rb. capsulatus*, respectively. The rate and yield of transfer from H_A^- to the primary quinone is essentially unchanged. The M202 and L173 heterodimer mutants of *Rb. sphaeroides* have been characterized by X-ray diffraction to resolutions of 3.0 Å and 4.0 Å, respectively (Chirino et al., 1994). The structure of the M202 mutant has been refined and shows that the BPhe is substituted into a position similar to that normally occupied by the BChl. Displacements and rotations of amino acid residues, including Phe L181 and Phe M197, are observed as well as an apparent loss of a nearby water molecule. The limited resolution of the L173 data prevented refinement of the structure although features similar to that of the M202 mutant were observed in difference electron density maps.

B. Mutation of Aromatic Residues

Two symmetry related aromatic residues Phe L181 and Tyr M210 (M208 in *Rb. capsulatus*) are located near the dimer, BChl monomer, and BPhe on the A and B sides, respectively. These residues have been altered to a variety of other residues in various combinations by several groups in both *Rb. capsulatus* and *Rb. sphaeroides* (Nagarajan et al., 1990, 1993; Finkele et al., 1990; Chan et al., 1991a; Jia et al., 1993). The replacement of Tyr M210 by a Phe resulted in an electron transfer rate to H_A approximately five-fold slower than wild type. Replacement of the symmetrically related residue, Phe L181, with a Tyr resulted in a roughly two-fold increase in the P* decay rate. The mutations result in only small changes in the reaction center structure as measured by resonance Raman, linear dichroism, and X-ray diffraction (Mattioli et al., 1991a; Gray et al., 1990; Chirino et al., 1994). Theoretical calculations have suggested that the observed changes in rates for the mutants principally arise from altered energy levels of the charge-separated states due to changes in the electrostatic field (Parson et al., 1990; Nagarajan et al., 1993).

C. Hydrogen Bond Mutants

The BChl dimer has carbonyl groups in the conjugated system of rings I and V of both BChls. In wild type *Rb. sphaeroides* reaction centers, only one of these carbonyl groups forms a hydrogen bond to a nearby amino acid (His L168). Residues near the other three carbonyl groups, Leu L131, Leu M160, and Phe M197 have all been mutated to His in *Rb. sphaeroides* (Williams et al., 1992a,b,c) and Phe M195 in *Rb. capsulatus*, which corresponds to M197 in *Rb. sphaeroides*, has also been changed to His (Stocker et al., 1992). In addition, the one hydrogen bond present in the wild type has been removed by the mutation of the His at L168 to a Phe (Murchison et al., 1993). In each case that a histidine was inserted near a ring I or ring V carbonyl group, a new hydrogen bond resulted (Mattioli et al., 1994, 1995; Nabedryk et al., 1993). The addition of each hydrogen bond to the BChl dimer resulted in an increase of 50–120 mV in the midpoint potential of P in *Rb. sphaeroides* (Williams et al., 1992b; Murchison et al., 1993; Lin et al., 1994) and at least 100 mV in the $Phe^{M195} \rightarrow His$

mutant of *Rb. capsulatus* (Stocker et al., 1992). Associated with the decrease in the P^*/P^+H^- free energy difference for each mutant is a decrease in the rate of P* decay ranging from 1.5 to 4-fold at 295 K (Williams et al., 1992b; Murchison et al., 1993). The loss of the existing hydrogen bond in the $His^{L168} \rightarrow$ Phe mutant resulted in a corresponding decrease in the P/P^+ midpoint potential of 80 mV but the rate of electron transfer was essentially unchanged at 295 K (Murchison et al., 1993).

D. Large-Scale Symmetry Mutants

Large scale symmetry changes have been made by exchanging regions of the L subunit with the corresponding region of the M subunit, and vice versa, in *Rb. capsulatus* (Robles et al., 1990; Taguchi et al., 1992). Replacing residues M192 to M217 that form the D-helix of the M subunit with the corresponding residues of the L subunit, L165 to L190, results in a reaction center that is not capable of electron transfer due to the loss of H_A (the 'D_{LL} mutant', Robles et al., 1990). The corresponding replacement of the D-helix of the L subunit with M subunit sequences has also been constructed as has the complete D-helix swap between the L and M subunits (Robles et al., 1990). In the D_{LL} mutant, the excited state of P, P*, has a lifetime of 190 ps, allowing a detailed spectroscopic characterization of this state (Breton et al., 1990). Another symmetry mutant, *sym1*, was constructed by replacing residues M187 to M203 of the M subunit with the corresponding residues of the L subunit, L160 to L176. This mutant has a decreased quantum yield of electron transfer (65–70%), a three-fold decrease in the rate of P* decay, and an increase in midpoint potential of at least 100 mV (Stocker et al., 1992; Taguchi et al., 1992). Many of these functional changes were attributed to the specific change of Phe to His at M197 that results in the introduction of a hydrogen bond to the dimer (see above). None of the large scale symmetry mutants, nor any of the single site mutants that have been characterized to date, result in clearly demonstrable electron transfer involving the inactive (B-side) cofactors of the reaction center.

E. Mutations Near H_A

The introduction of a potential Mg ligand at M214 in *Rb. sphaeroides* by the mutation Leu to His results in a mutant with a BChl replacing H_A (the 'β' mutant; Kirmaier et al., 1991). The decay rate of the stimulated emission from P* in this mutant is 6.4 ps and the electron transfer rate from the β cofactor to the primary quinone occurs in approximately 350 ps with a 60% quantum yield of $P^+Q_A^-$ (Kirmaier et al., 1991). The standard free energy difference between P* and $P^+H_A^-$ (see below) was 85 mV less negative in the mutant than the value determined for wild type reaction centers.

A second mutation near H_A involves the loss of a hydrogen bond between Glu L104 and the ring V keto group of the BPhe. This mutation, $Glu^{L104} \rightarrow$ Leu, resulted in a 12 nm shift of the Q_X band of $BPhe_A$ and a slightly slower rate of 4.8 ps for the decay of P* with no change in yield compared to wild type (Bylina et al., 1988).

IV. The Ground State Optical Spectrum of Reaction Centers

The ground state optical spectrum of reaction centers isolated from *Rb. sphaeroides* strain 2.4.1 is shown in Fig. 2 at 295 K and at 20 K. The small number of bacteriochlorins (4 BChls and 2 BPhes) makes the isolated reaction center much simpler to study by optical spectroscopy than the intact photosynthetic membrane, and essentially all of the direct measurements of electron transfer discussed in this chapter have been performed on isolated complexes.

A great deal of experimental and theoretical investigation has been directed towards the assignments of the different reaction center spectral bands. Though the spectral features are complex (Fig. 2), and it is very likely that there is substantial mixing between the electronic transitions of the different cofactors affecting both the energy and oscillator strengths of the individual transitions, the transitions seen at 865, 802, 760, 590, and 540 nm in the 295 K spectrum of *Rb. sphaeroides* (960, 830, 790, 600, and 540 nm in *Rp. viridis*) can be largely assigned to specific cofactors or types of cofactors (Zinth et al., 1985; Parson, 1987, 1991; Breton, 1988; Breton et al., 1989; Feher et al., 1989). It is clear that the 865 nm band seen in the 295 K spectrum of *Rb. sphaeroides* is predominantly the Q_Y transition of the pair of BChls, P. The band at 802 nm in the 295 K spectrum of Fig. 2 is thought to be largely due to the Q_Y transitions of the monomer BChls. At 20 K, this band resolves into at least two transitions (the 802 nm transition becomes asymmetric, Fig. 2) at 802 and

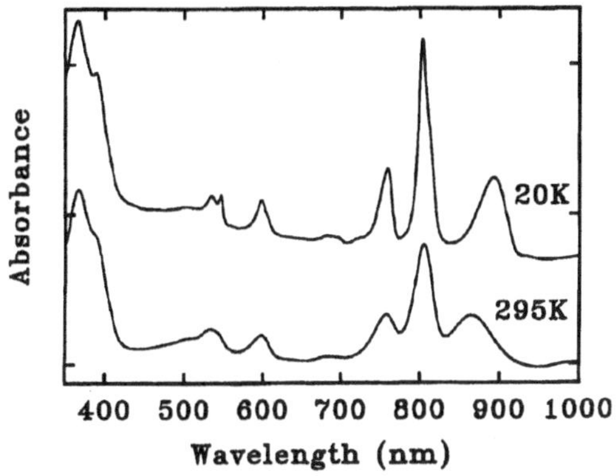

Fig. 2. The ground state absorption spectrum of reaction centers isolated from a strain of *Rb. sphaeroides* containing wild type reaction center subunit genes. The spectrum is shown both at 295 K and 20 K. The low temperature spectrum is identical to that published by Williams et al., 1992b.

810 nm, possibly representing the two monomer BChls in distinct environments (Beese et al., 1988; Williams et al., 1992b), though contributions from other transitions in this region are possible as well (Breton, 1988; Breton et al., 1989). The band at 760 nm in the 295 K spectrum of Fig. 2 is predominantly due to the Q_Y transition of the BPhes. At low temperature, this band becomes more asymmetric in *Rb. sphaeroides*, due, at least in part, to the two different environments of the BPhes as shown by specific reduction of the individual BPhes by photopumping methods (Robert et al., 1985). This is more easily seen in reaction centers from *Rp. viridis* where the two BPhe Q_Y bands are well resolved at low temperature (Tiede et al., 1987; Kellogg et al., 1989). The band near 600 nm in Fig. 2 is due to the overlapping Q_X transitions of all of the reaction center BChls. The band near 540 nm at room temperature in Fig. 2 is the combined Q_X transitions of the BPhes. The Q_X bands of the BPhes resolve quite cleanly into two transitions at low temperature. The BPhe which is predominantly involved in the electron transfer gives rise to the narrower 546 nm peak at 20 K while the broader 533 nm transition shows much less photobleaching activity (Fig. 3 and Bylina et al., 1988). Linear dichroism studies of single crystals (Zinth et al., 1985) and specific mutagenesis near H_A affecting its spectral properties (Bylina et al., 1988) have shown that the red absorbing BPhe is H_A. Mutagenesis studies have indicated that the spectral shift between the BPhes is largely due to

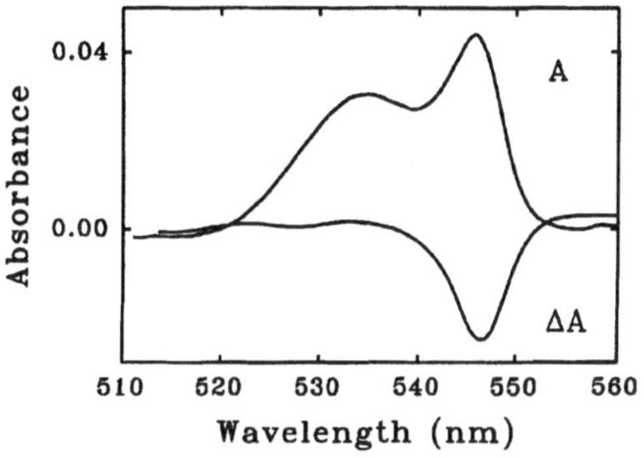

Fig. 3. The upper curve represents the 510 to 560 nm region of the ground-state absorption spectrum of reaction centers from the carotenoidless strain of *Rb. sphaeroides*, R-26, at 20 K. The lower curve shows the difference absorption spectrum of the charge-separated species formed during a 15 ps time range transient absorption measurement at 20 K. This difference spectrum was determined as the constant term in a global exponential fit of the transient absorption changes occurring over that time period at 40 wavelengths. In this case the fit was to an exponential term and a constant: $A_1(\lambda)\exp(-kt) + A_2(\lambda)$ where λ is the wavelength, $A_i(\lambda)$ are amplitudes allowed to vary freely at each wavelength during the fit and k is a rate constant which is wavelength independent. The lower curve in the figure is $A_2(\lambda)$. (This type of analysis and the apparatus used is described by Woodbury et al., 1994). The reaction centers were excited with a roughly 200 fs duration pulse with a 10 nm spectral band width centered at 880 nm. Note that the transient absorption spectrum was offset such that there was a zero absorbance change on average between 510 and 520 nm. There was a small (roughly 0.01) positive absorption in this region before the adjustment. Barring accidental absorption change cancellations, the amount of bleaching seen in the transient absorption spectrum due to the 533 nm H_B spectral transition in the ground state spectrum is less than 5% for reaction centers isolated from this strain. These results are similar to those published previously for R-26 reaction centers (Kirmaier et al., 1985a; Lockhart et al., 1990).

a glutamic acid in the L-subunit which hydrogen bonds to H_A (Bylina et al., 1988), though further mutagenesis in this region and at the symmetrically located site in the M subunit has suggested that other factors in the protein environment of the BPhes may significantly affect these transition energies as well (Williams and Allen, unpublished results).

V. The Spectral and Kinetic Properties of P*

The only known photochemically active excited singlet state of the reaction center is the lowest excited singlet state of the initial electron donor, P*.

This is also the state that is produced upon energy transfer from the antenna, and direct excitation into any of the other BChls or BPhes in the reaction center rapidly results in energy transfer to P again forming P* (Breton et al., 1986; Breton et al., 1988; Du et al., 1992; Xiao et al., 1994).

*A. Spontaneous Emission from P**

Though in wild type reaction centers the quenching of the excited singlet state of P is predominantly via charge separation, P* can also decay by other nonradiative pathways (see below) or by emitting a photon. The emission spectrum for P* is shown in Fig. 4B. The natural radiative rate constant of P* is estimated to be about 8×10^7 s^{-1} (Woodbury and Parson, 1984). Characterizing the decay of the spontaneous emission from P* is a useful method for determining its population as a function of time. Several laboratories have used picosecond and subpicosecond resolution transient spectroscopy to study the kinetics of the spontaneous emission from P*. The kinetic behavior of P* decay determined by these measurements is complex. It does not show single exponential kinetic behavior as one would expect for a simple irreversible, first order reaction. Fig. 5a shows the multiexponential decay of the spontaneous emission on the picosecond timescale in *Rb. sphaeroides* reaction centers (Du et al., 1992; Muller et al., 1992; Hamm et al., 1993). Fig. 5b shows that emission is also observed on timescales as long as nanoseconds, when forward electron transfer to the quinone is blocked (Schenck et al., 1982; Sebban and Moya, 1983; Woodbury and Parson, 1984; Peloquin et al., 1994; Ogrodnik et al., 1994). When fit as a series of exponential decay terms, at least three time constants are required to accurately describe the P* decay in reaction centers with unreduced quinones. At least five rate constants are probably required to describe the decay of P* in reaction centers with reduced or removed quinones.

*B. Stimulated Emission and Absorption Changes Associated with P**

It is also possible to follow the decay of P* by observing the transient absorption spectrum of reaction centers as a function of time. One of the features of the transient difference spectrum of P* often used to follow the decay of this state is the stimulated emission in the 915 nm region (Woodbury

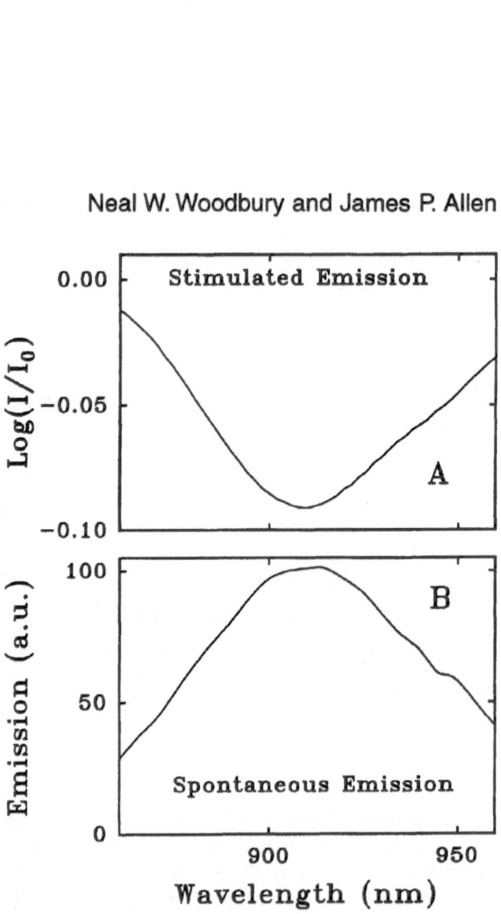

Fig. 4. (A) The 295 K spectrum of the transient absorption changes believed to be due to stimulated emission from the excited singlet state of P, P*. The data shown was taken using reaction centers of *Rb. sphaeroides*, strain R-26. The curve in panel A is the amplitude spectrum, $A_1(\lambda)$, determined from global exponential decay analysis of the absorption changes at 50 wavelengths between 860 and 960 nm on a 15 ps timescale. The data was fit was to one exponential decay term and a constant (using the fitting equation in the legend to Fig. 3) following excitation with an 840 nm, 200 fs pulse. The amplitude spectrum shown is that of the absorption change signal which is decaying on the 3.5 ps timescale in this spectral region. Most of the rapidly changing signal in this wavelength region is believed to be due to stimulated emission, as described in the text. Similar spectra have been published previously (Nagarajan et al., 1993; Taguchi et al., 1992; Williams et al., 1992a). (B) The 295 K spectrum of the spontaneous emission from quinone-reduced R-26 reaction centers. This was measured as the integrated emission during the first 2 ns after excitation using a time-correlated single photon counting apparatus. Data taken from Woodbury and Parson (1984).

et al., 1985; Martin et al., 1986; Kirmaier and Holten, 1987, 1988b; Fleming et al., 1988; Holzapfel et al., 1989). Fig. 4 compares the spectrum of the steady-state spontaneous emission (Fig. 4B) with the spectrum of the stimulated emission (Fig. 4A) from *Rb. sphaeroides* reaction centers as determined by exponential decay analysis of multiwavelength ultrafast transient absorption measurements in the

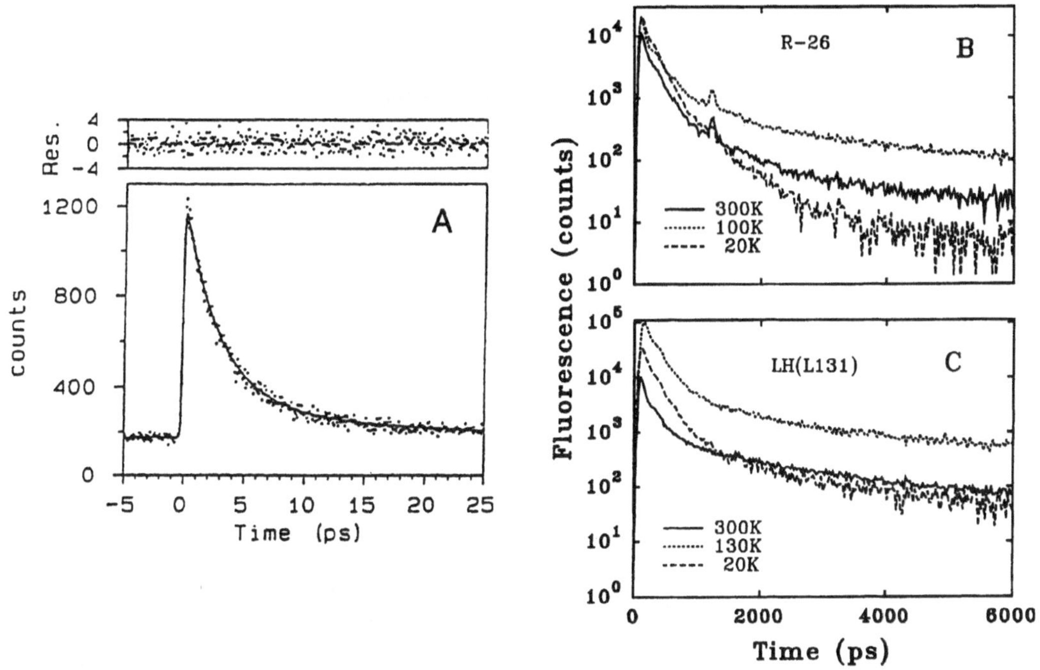

Fig. 5. (A) Spontaneous emission decay of P* on the picosecond timescale in reaction centers from *Rb. sphaeroides* strain R-26 at 295 K. Excitation was at 850 nm with a 160 fs duration pulse and emission was detected at 940 nm by gating the emission with a second pulse at a particular time relative to excitation using nonlinear two-photon mixing of the emission photon and the gating photon in a $LiIO_3$ crystal (so called 'up-conversion'). The points represent experimental results for different relative delays of the gating pulse. The solid line represents the best fit to two exponentials, $A_1\exp(-k_1t) + A_2\exp(-k_2t)$, where t is time, A_i are the preexponential amplitudes used as free parameters in the fit and k_i are the rate constants which are also free parameters in the fitting routine. For this fit, $A_1 = 0.808$, $A_2 = 0.192$ ($A_1 + A_2$ has been normalized to 1.0), $k_1 = 1/(2.7$ ps), and $k_2 = 1/(12.1$ ps). This figure was reproduced from Du et al. (1992). (B) Spontaneous emission decay of P* on the tens of picosecond to nanosecond timescale in R-26 reaction centers measured by time-correlated single photon counting. Decay components are present on the 100 ps, 800 ps, and several nanosecond timescales in the curves at each temperature. Note that the abscissa has a logarithmic gauge, while the ordinate is linear. (C) Same as panel B except for the Leu[L131]→His mutant of *Rb. sphaeroides*. The data of panels B and C were taken from Peloquin et al. (1994) and the experimental details are described there.

900 nm region. One disadvantage in measuring the decay kinetics of P* by following the stimulated emission instead of the spontaneous emission is that sensitivity to small stimulated emission signals is poor in comparison to that obtainable by following the spontaneous emission. This is because a low level of stimulated emission is indistinguishable from small absorption changes which underlie the stimulated emission band in the difference absorption spectrum. Despite these difficulties, multiexponential decay of the stimulated emission has been observed which closely matches that of the spontaneous emission from P* over the first 20 ps (Vos et al., 1992; Woodbury et al., 1994).

The state P* also has other characteristic features in the time-dependent difference absorption spectrum recorded on subpicosecond timescales by fast transient absorption spectroscopy (Kirmaier and Holten, 1987; Kirmaier and Holten, 1993). The time evolution of the absorption change spectrum in the 700–860 nm region is shown in Fig. 6A at room temperature. Fig. 6B shows a single difference spectrum at 0.4 ps that represents primarily the state P*. There is a broad absorption throughout the 700–820 nm region with small, but distinct, features in the 800 nm region (Fig. 6B). There is also a large bleaching of the Q_Y band of P as would be expected for the excited state of this cofactor. This can be seen most clearly in Fig. 6C which shows the difference absorption spectra as a function of time over a broader

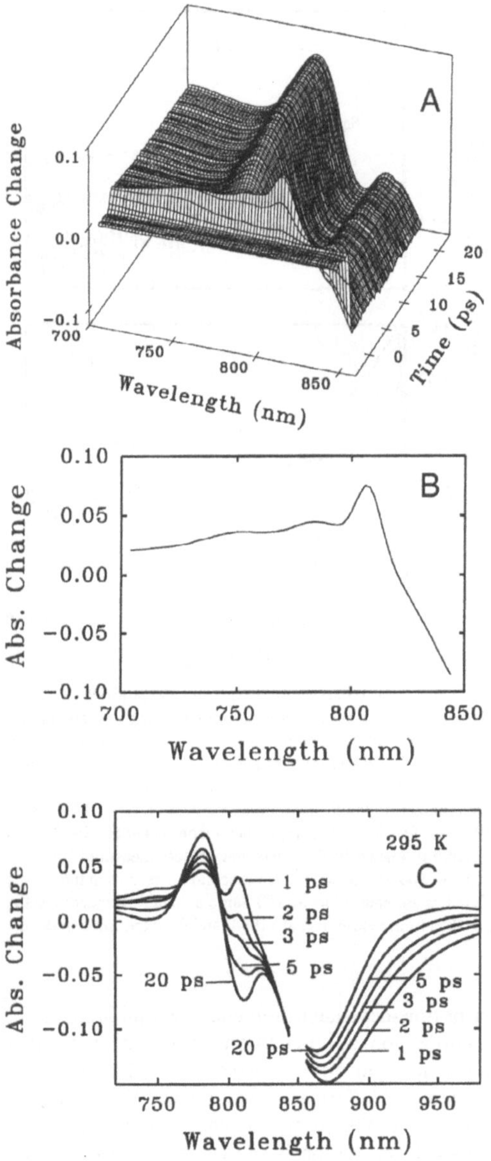

Fig. 6. (A) An absorption change *vs.* time and wavelength surface taken using reaction centers from *Rb. sphaeroides* strain R-26 at 295 K. Data points were taken every 200 fs over a 20 ps time range and every 2 nm between 705 and 845 nm. Excitation was with a roughly 200 fs pulse at 860 nm. (B) A single spectral slice through the surface of part A at 0.4 ps after excitation. This is thought to represent the absorption changes associated with the initial excited singlet state, P*. (C) 295 K transient absorption difference spectra of R-26 reaction centers as a function of time taken over a larger wavelength range. Excitation was with a roughly 200 fs pulse at 850 nm.

spectral range. The bleaching at early times in the 870 to 940 nm region is the superposition of the stimulated emission discussed above on the bleaching of the ground state Q_Y band of P.

C. The Intrinsic Lifetime and Electronic Characteristics of P*

Much of what is known about the electronic properties of P* comes from studies of the transition between the ground state and P*. For example, electric field effects on the energy of the P to P* transition provide information about the magnitude and direction of the difference dipole moment between P and P*. Low temperature hole-burning experiments provide insights into the mechanism of broadening of the Q_Y absorption band of P and about the degree of inhomogeneity and the vibronic structure of the excited singlet state. These issues have been considered in detail in recent reviews (Boxer, 1993; Reddy et al., 1992).

There is also complimentary information available about the inherent properties of P* from fast transient absorption measurements of both wild type and mutant reaction centers. One very important property of P* from a kinetic standpoint is its intrinsic lifetime in the absence of electron transfer. One can easily assign a minimum value to this number simply by considering the yield of initial electron transfer. The quantum yield of stable photooxidation of P has been measured to be at least 98% (Wraight and Clayton, 1973). If the observed lifetime of P* is about 3 ps at room temperature (just using the single exponential decay lifetime), this means that the intrinsic lifetime of P* must be at least 150 ps. Without a more precise measurement of the quantum yield of the first electron transfer step, it is difficult to do more than put a lower limit on the intrinsic lifetime in the wild type reaction centers. However, this value is much better known in several mutants with quantum yields of the initial electron transfer reaction that are low enough to accurately measure. The simplest example of such a mutant is the D_{LL} mutant (Robles et al., 1990) which does not undergo electron transfer at all. P* decays in about 190 ps in the D_{LL} mutant, consistent with the minimum value estimated above (Breton et al., 1990). This in itself is an interesting result, because monomeric BChl in solution typically decays on a timescale more than ten times this long (Seely and Connolly, 1986). Further information about the mechanism behind the rapid decay of P* comes from measurements of other mutants that perturb the symmetry of this state. The best studied of these are the heterodimer mutants (Bylina and Youvan, 1988; Kirmaier et al., 1988; McDowell et al., 1991a,b). In both heterodimer mutants the yield of electron transfer drops dramatically due to both a decrease in the rate of initial electron transfer and to a decrease in the intrinsic lifetime of P* (McDowell et al., 1991a; Kirmaier et al., 1988). For example, when the histidine at M200 is replaced by leucine in the *Rb. capsulatus* reaction center, the intrinsic lifetime of P*, as calculated from the yield and the observed lifetime of P*, is about 30 ps (Kirmaier et al., 1988). This has been rationalized in terms of the stabilization of the charge transfer state of the heterodimer, $BChl^+ BPhe^-$, relative to the homodimer charge transfer state, $BChl^+ BChl^-$, in the wild type. Model studies have indicated that when the excited state of a dimer has a large charge transfer character, nonradiative decay of the excited singlet state is more rapid (e.g. Bilsel et al., 1990). The results from these mutants have led to the suggestion that P* in wild type reaction centers has an intrinsic lifetime much shorter than monomeric BChl in solution, possibly due to contributions to P* from the charge transfer state. However, it is noteworthy that in one mutation ($Tyr^{M210} \rightarrow$ Phe in *Rb. sphaeroides*) the intrinsic lifetime of P* is thought to be more than 300 ps at room temperature and in the nanosecond range at low temperature (Nagarajan et al., 1990). The variability of the intrinsic lifetime of P* among different mutants makes an accurate estimate of this important number in wild type reaction centers difficult.

*D. Coherence Dephasing and Vibrational Relaxation of P**

It has recently been possible to directly observe an oscillatory time dependence of both absorption changes (Vos et al., 1991, 1993, 1994a, 1994b, 1994c) and emission decay profiles (Stanley and Boxer, 1994) shortly after P* formation. These are apparently temporal manifestations of the vibrational modes that have been observed by resonance Raman and photochemical hole-burning techniques (Shreve et al., 1991; Reddy et al., 1992; Cherepy et al., 1994). The Fourier transform of the oscillatory parts of these decays reveals two broad frequency regions. In wild type reaction centers, one of the modes strongly coupled to the P to P* transition is centered near 120 cm^{-1}at 10 K or 140 cm^{-1} at 295 K, the other is in the 15 to 30 cm^{-1} range (Vos et al., 1994a,b,c; Stanley and Boxer, 1994). The frequency of the lower energy mode is rather uncertain due to the fact that the oscillation is on the same timescale as the electron transfer in wild type (Stanley and Boxer, 1994). Fig. 7 shows the oscillatory behavior of the stimulated emission in the D_{LL} reaction center mutant of *Rb. capsulatus*. Here, the P* decay time is much longer than the period of the lower frequency mode, making the determination of the mode frequency by Fourier transform analysis more certain. In all studies performed on this phenomenon in reaction centers, coherence dephasing occurs on the 1–2 ps timescale (Vos et al., 1991, 1993, 1994a, 1994b, 1994c; Stanley and Boxer, 1994).

There is presently no direct measurement of the timescale of the vibrational relaxation of P*. The observation of vibrational coherence indicates that at least some of the modes coupled to P* are not well coupled to the bath on timescales comparable to the overall rate of initial electron transfer. In addition, results from hole-burning measurements of reaction centers suggest that thermal equilibration with the bath is not instantaneous (Middendorf et al., 1991), and in vitro measurements of monomer and dimer porphyrins have indicated that vibrational relaxation of porphyrin excited states can occur on a timescale of picoseconds to tens of picoseconds (Rodriguez and Holten, 1989; Rodriguez et al., 1991). There are also changes in the transient absorption difference spectrum on the subpicosecond timescale which could be interpreted as a change in the state P* (Vos et al., 1992). Recently, Vos et al. (1994c) have shown that the spectral features of the stimulated emission evolve dramatically during the first few hundred femtoseconds after light absorption. One would expect that dipolar relaxation of the protein around the excited state would give rise to a dynamic Stokes shift (Pierce and Boxer, 1992). The spectral changes seen in the stimulated emission from P* do imply an overall displacement of the excited state potential surface relative to the ground state potential surface (Vos et al., 1994c), though the excited state evolution observed appears to be a coherent movement and may not represent the complete vibrational relaxation process (Vos et al., 1994b). There is also recent evidence that in a reaction center mutant (Tyr^{M210} Æ Phe) with a somewhat longer P* lifetime than that observed in wild type there is a shift of the stimulated emission spectrum to higher energies as a function of time (Nagarajan et al., 1993). This has been

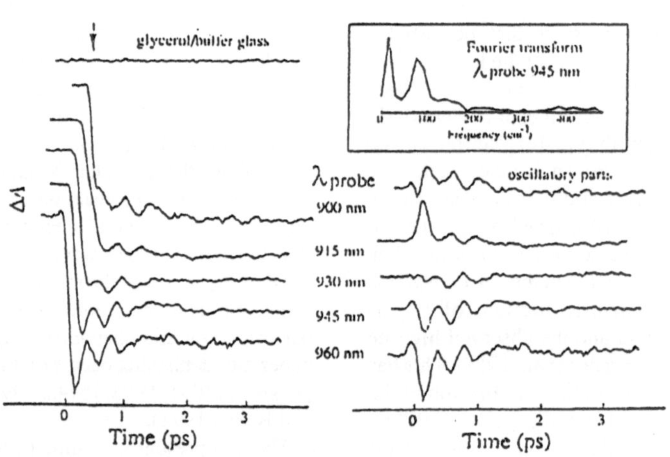

Fig. 7. Coherent oscillations in the stimulated emission from D_{LL} mutant reaction centers in antennaless membrane preparations from *Rb. capsulatus*. The results of observation at several different probe wavelengths are shown. The traces are normalized to the average amplitude of the overall signal. The oscillatory part at each probe wavelength is obtained by taking the residuals of a fit to a step function convoluted with the instrument response function. Discrete Fourier transformation describes the input signal I(t) by its amplitudes A and phases ϕ as a function of the frequency ν as

$$\sum_n \mathrm{Re}\left(A_n e^{i(2\pi\nu_n t-\phi_n)}\right)$$

(Re: real part) and was taken, for all probe wavelengths, over a 2.13 ps data interval (64 points), yielding a spectral resolution of 15 cm^{-1}. The Fourier transform (FT) window was started at t=0.10 ps, excluding the pump-probe overlap time interval. The 80 fs pump pulse duration determines the vibrational bandwidth of the experiment at about 70 cm^{-1}. The spectral region above about 200 cm^{-1} represents the noise level of the data. The FT amplitudes of the oscillatory parts all peak at 15 cm^{-1} and 77 cm^{-1}; the inset shows the FT amplitude for probe wavelength = 945 nm. Upper left: control experiment at 10 K (probe wavelength = 930 nm) using the same cuvette containing only the buffer/glycerol mixture. The arrow indicates t = 0. The trace is plotted on the same amplitude scale as the protein sample (930 nm trace). This figure is reproduced from Vos et al. (1993), and this reference should be consulted for further details.

interpreted as a relaxation of P* possibly involving changes in underlying absorption transitions of the excited state or changes in the oscillator strength of the stimulated emission.

VI. The State P^+

A. The Spectral Properties of P^+

The oxidized state P^+ can be formed as a stable state either by freezing reaction centers in the presence of light (forming $P^+Q_B^-$) or by addition of an oxidant such as ferricyanide (forming P^+). One can also cause nearly complete conversion to the state $P^+Q_B^-$ by steady-state illumination of reaction centers at room temperature. The P/P^+ difference absorption spectrum in the near infrared is dominated by a bleaching of the Q_Y band of P (865 nm in *Rb. sphaeroides*, Fig. 2) and a shift to higher energy of the 802 nm transition, presumably due to an electrochromic shift of one or both of the monomer BChls in this region. These features also dominate the $P^+Q_A^-$ and $P^+H_A^-$ spectra (see below). In addition, at 1250 nm in *Rb. sphaeroides* there is a small absorption band which appears upon P^+ formation, and at 2600 cm^{-1} an additional P^+ transition has been detected by Fourier transform infrared spectroscopy (Breton et al., 1992).

The stability of P^+ produced either chemically or photochemically has allowed it to be characterized by electron paramagnetic resonance (EPR) and electron nuclear double resonance (ENDOR) (reviewed in Lubitz, 1991). The initial modeling of the electron donor as a BChl dimer was based on the EPR and ENDOR spectra of P^+ in *Rb. sphaeroides* (Feher et al., 1975; Norris et al., 1975). In these models, the delocalization of the unpaired electron over a symmetrical dimer reduces the width of the EPR spectrum by a factor of $\sqrt{2}$ and reduces the hyperfine coupling constants of the ENDOR spectra by a factor of 2 compared to the spectra of $BChl^+$ in organic solvents. More refined models for the distribution of the electron spin density of P^+ have been developed using molecular orbital calculations based upon the three dimensional structure of the reaction center (reviewed in Plato et al., 1991). Analyses of these data indicate an unequal distribution of the electron spin density over the dimer halves, favoring P_A by approximately 2:1 in both *Rb. sphaeroides* and *Rp. viridis* (Lendzian et al., 1988, 1993). Essentially identical ENDOR spectra have been reported for P^+ in chromatophores of *Rb. sphaeroides, Rb. capsulatus, Rhodosprillum (Rs.) centenum* and *Rs. rubrum* (Rautter et al., 1994). While reaction centers isolated from *Rb. sphaeroides* and *Rs. rubrum* showed ENDOR spectra similar to those observed in chromatophres, reaction cnters isolated from *Rb. capsulatus* and *Rs. centenum* had significantly altered ENDOR spectra that indicate a 5:1 ratio of spin density favoring P_A rather than the 2:1 ratio observed in chromatophores (Rautter et al., 1994). In addition, the availability of reaction center crystals has allowed determination of the anisotropic components of the hyperfine coupling constants and consequently a significant improvement in accuracy of the calculated spin density of P^+ for reaction centers from *Rb. sphaeroides* (Lendzian et al., 1993).

Changes in the spin density due to altered interactions with the protein can be measured by EPR, ENDOR and FT Raman. The heterodimer mutants show large changes in the EPR and ENDOR signals due to the predominant localization of the unpaired electron on the BChl side of the dimer (Huber et al., 1990; Bylina et al., 1990). The *sym1* mutant also shows a broadened EPR signal indicating a more localized cation state than wild type (Stocker et al., 1992). Analyses of ENDOR experiments on the hydrogen bond mutants of *Rb. sphaeroides* have enabled the quantitative determination of the relative spin densities on the two BChls of the dimer. The addition of each hydorgen bond causes a significant redistribution of spin density yielding a series of mutants with P_A/P_B spin density ratios ranging from 0.3 to 5.0 (Rautter et al., 1992, 1995). These spin density changes have been modeled in terms of preferential stabilization of P_A^+ or P_B^+ due to the introduction of each hydrogen bond. The magnitude of the oxidation induced upshift of one of the dimer carbonyl Raman bands is also an indicator of the localization of the electron spin density (Mattioli et al., 1991a). This upshift provides estimates that are consistent with the ENDOR values (Mattioli et al., 1994, 1995)

B. The P/P+ Midpoint Potential

The P/P^+ midpoint potential can be measured in reaction centers by monitoring the amplitude of the Q_Y dimer band while titrating the sample either by chemical oxidants and reductants or by electrochemical titration. In wild type reaction centers from *Rb. sphaeroides*, the measured midpoint potential is approximately 500 mV (Williams et al., 1992b; Nagarajan et al., 1993). A number of mutations near the dimer have been characterized that alter the P/P^+ midpoint potential by 10 to 50 mV (Du et al., 1992; Nagarajan et al., 1993). Two types of reaction center mutations have been found that result in significantly larger midpoint potential changes. First, due to the altered cofactor composition, the *Rb. sphaeroides* heterodimer, $His^{M202} \rightarrow$ Leu, has a P/P^+ midpoint potential that is 160 mV above that of wild type (Davis et al., 1992). Second, addition or removal of hydrogen bonds between P and the surrounding protein results in large midpoint potential changes. The loss of the hydrogen bond between the ring I acetyl group and His L168 in *Rb. sphaeroides* reaction centers results in an 80 mV decrease in potential (Murchison et al., 1993). The gain of a hydrogen bond between a histidine and the other acetyl group or one of the keto groups of the dimer results in an increase of 55 mV to 125 mV (Williams et al., 1992a,b,c; Stocker et al., 1992). The combination of hydrogen bonds is additive (Lin et al., 1994) and correlated with the total change in hydrogen bonding energy (Mattioli et al., 1995). The combination of mutations yielding hydrogen bonds to histidine at all four keto and acetyl groups of P results in a net increase in the P/P^+ midpoint potential of 260 mV compared to wild type (Lin et al., 1994). The formation of a hydrogen bond between a Tyr and the acetyl group of ring I results in a smaller increase in the potential (30 mV, Wachtveitl et al., 1994). This suggests that the large increase in potential due to the formation of hydrogen bonds with histidines is due either to the formation of a strong hydrogen bond or to another specific property of the histidine, such as electrostatic interaction (Williams et al., 1992b; Mattioli et al., 1994, 1995).

VII. The Product of Initial Charge Separation

A. The Formation and Spectral Properties of the State $P^+H_A^-$

As one might expect from the complexity of the decay of the initial electron donor excited singlet state, the kinetics of formation of the first fully populated charge-separated state are complex as well (Holzapfel et al., 1989; Dressler et al., 1991; Kirmaier and Holten, 1990; Kirmaier and Holten, 1991; Chan et al., 1991b). Time constants between 0.9 and 4 ps have been reported at room temperature, depending on the wavelength measured and the method of analysis (single exponential *vs.* multiexponential). Complex kinetics are also obtained at low temperature, with lifetimes determined by exponential decay analysis on the 0.6 to 2.5 ps timescale as well as on the roughly 15 ps timescale (Kirmaier and Holten, 1990; Lauterwasser et al., 1991; Woodbury et al., 1994). Possible explanations for the complex spectral and kinetic behavior during the first 10–15 ps are discussed later in this chapter along with a consideration of several mechanistic models involving intermediate states. Here, we will consider the spectral properties of the charge-separated state present on the 15–20 ps timescale.

Though there is considerable disagreement about the mechanism of formation of the state present 15 ps after light absorption, there is general agreement about the spectral properties of the state that is formed. Examples of difference absorption spectra as a function of time for *Rb. sphaeroides* reaction centers are shown in Fig. 6A and 6C at room temperature and in Fig. 8A at 20 K. The difference spectrum on subpicosecond timescales is mostly that of P* (Fig. 6B). This then evolves with time to form the state shown at 20 ps in Fig. 6C. The absorption decreases in the 760 nm (Fig. 6C) and 546 nm (Fig. 3) bands of the reaction center at this time have been used to argue that the state present is predominantly a charge-separated state made up of the cation of P and the anion of the A-side BPhe (H_A; see below for a discussion of the relative yields of charge separation along the A and B cofactors), though the absorption changes near 800 nm complicate this assignment (Kirmaier and Holten, 1987). The extent of the 760 and 800 nm absorption changes associated with formation of the anion can be seen most easily when the contribution of the absorption changes due to P^+ have been subtracted.

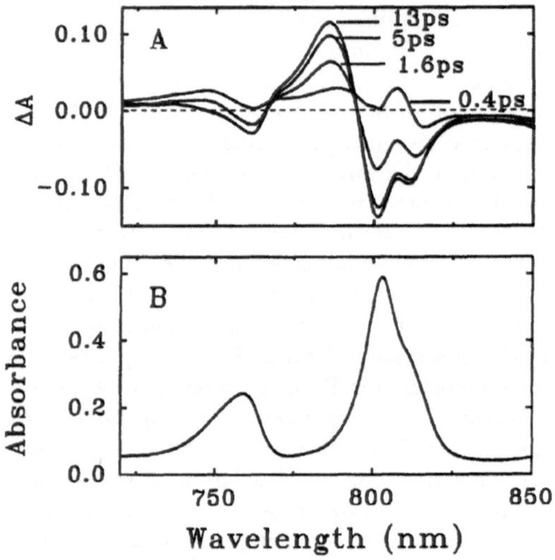

Fig. 8. (A) Transient absorption change spectra at 20 K from R-26 reaction centers (*Rb. sphaeroides*). Excitation was with a roughly 200 fs pulse at 880 nm. The difference spectra shown are representative spectral slices through a 20 K absorption change vs. time and wavelength surface similar to that shown in Fig. 6A at 295 K. (B) The 20 K ground state spectrum of R-26 reaction centers over the same wavelength region as in panel A.

Since the quinone does not absorb significantly in the near infrared, the $P^+Q_A^-$ difference spectrum is usually taken to represent absorption changes due to P^+. Fig. 9A compares the $P^+H_A^-$ absorbance change spectrum (15 ps) and the $P^+Q_A^-$ absorbance change spectrum (660 ps). Fig. 9B shows the result of subtracting the $P^+Q_A^-$ spectrum from the $P^+H_A^-$ spectrum. In the resulting difference spectrum there is a large bleaching not only at 760 nm presumably corresponding to H_A^-, but also a large bleaching near 800 nm. The concept that this state might be an equilibrium mixture of two states, $P^+H_A^-$ and $P^+B_A^-$, was proposed (Shuvalov and Parson, 1981) but later thought improbable due to the fact that the relative absorption changes between the 760 and 800 nm regions were not very temperature dependent (Kirmaier and Holten, 1987). Since BPhe has a less negative midpoint potential, it seemed reasonable that, at least at low temperature where thermal population of two states with different enthalpies is unlikely, $P^+H_A^-$ would predominantly remain. As will be discussed below, however, the energetics of initial charge separation is still a matter of debate and its temperature dependence is not entirely understood.

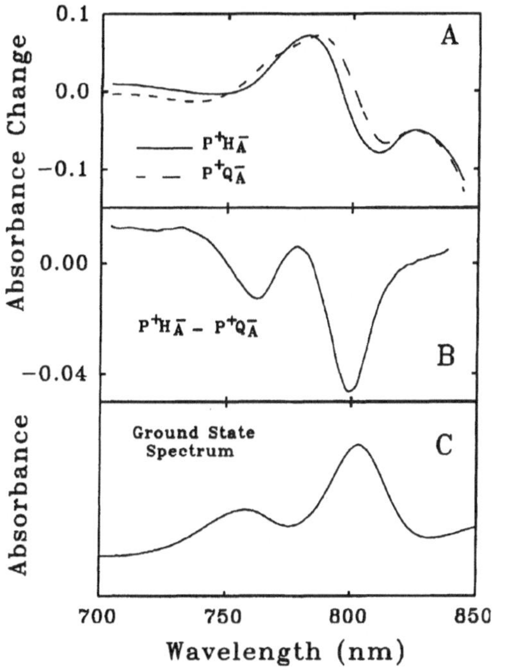

Fig. 9. (A) 295 K transient absorption change difference spectra of reaction centers from *Rb. sphaeroides* strain R-26 at 20 ps (solid line) and 660 ps (dashed line) thought to correspond to the states $P^+H_A^-$ and $P^+Q_A^-$, respectively. Spectra taken as described in Fig. 6. (B) The difference between the two spectra in panel A thought to represent the contribution from H_A^-. C) The ground state spectrum of R-26 reaction centers in the same wavelength region as the difference spectra in panels A and B.

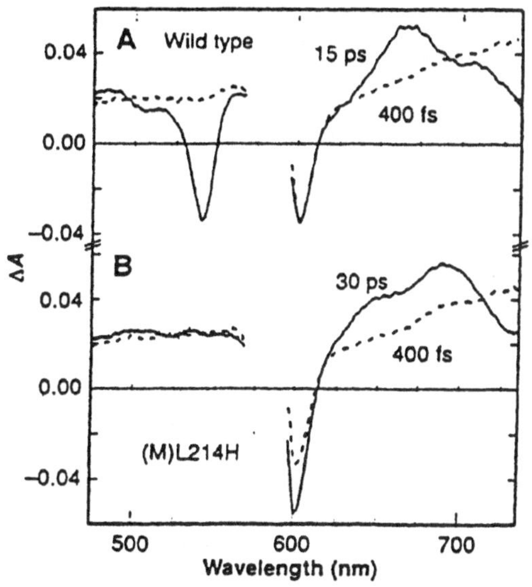

Fig. 10. Absorption difference spectra for (A) wild type reaction centers from *Rb. sphaeroides* at 0.4 ps and 15 ps and (B) Leu^{L214}→ His mutant reaction centers (the β mutant) at 0.4 ps and 30 ps. The absorption change spectra at 0.4 ps are thought to represent the initial excited state, P*. The longer time spectra shown are thought to represent the charge separated state $P^+H_A^-$ in the case of wild type and $P^+\beta_A^-$ in the case of the β mutant. Note that there are no absorbance changes observed in the 530 to 550 nm region in the β mutant, in sharp contrast to the wild type. This figure was reproduced from Kirmaier et al. (1991), and further details can be found in that publication.

Mutagenesis has also helped in understanding the charge-separated state. In the β mutant, the spectral characteristics of the state present after initial electron transfer are dramatically altered (Fig. 10), again implying that H_A is probably playing a predominant role in the formation of the 15 ps charge-separated state in the wild type (Kirmaier et al., 1991). This evidence coupled with detailed kinetic and photodichroism studies of a transient absorption increase near 650 nm thought to represent the anion of the charge-separated state (Fig. 10; Kirmaier et al., 1985a; Kirmaier and Holten, 1991) has led to the general acceptance of the assignment of the state present 15 ps after light absorption as being predominantly $P^+H_A^-$ (Kirmaier and Holten, 1987).

B. Charge Recombination in the State $P^+H_A^-$

It is possible to measure the intrinsic decay time of the state $P^+H_A^-$ in reaction centers where the subsequent transfer of an electron to Q_A has been blocked by reduction or removal of the quinones. At room temperature, the lifetimes reported for this state vary from 10 to 20 ns, depending on the sample conditions and the state of the quinone (reduced or removed, Schenck et al., 1982; Ogrodnik et al., 1982; Chidsey et al., 1984). There are three pathways for decay of $P^+H_A^-$ (Scheme (1)). First, it is possible to have recombination reforming the excited singlet state which then decays by mechanisms discussed above to the ground state. This probably has very little role in the decay of $P^+H_A^-$ in wild type reaction centers, since the free energy of the state $P^+H_A^-$ is substantially below that of P* on the nanosecond timescale (see below). It does appear to play a role in the decay of $P^+H_A^-$ in some mutants in which the midpoint potential of the special pair has been greatly increased, resulting in a much smaller free energy gap between P* and $P^+H_A^-$ (Taguchi et al., 1992; Peloquin et al., 1994).

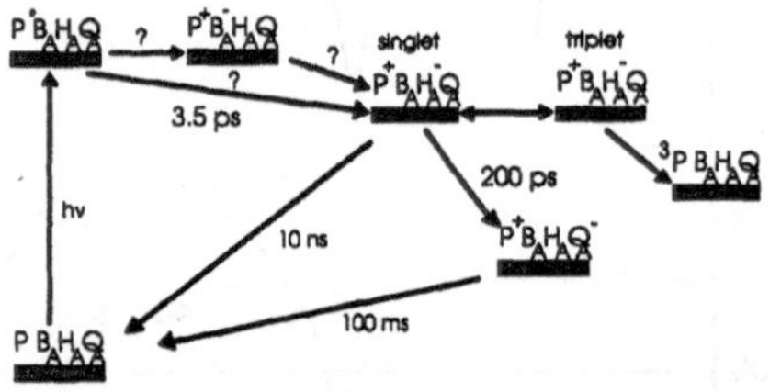

(1)

The second possible decay route for $P^+H_A^-$ is direct recombination to the ground state. This occurs with a time constant of roughly 10 ns in *Rb. sphaeroides* reaction centers where electron transfer to the quinone is blocked by reduction (Schenck et al., 1982). This is the dominant decay path in wild type reaction centers at room temperature; nearly 90% of the decay follows this route (Parson et al., 1975). This reaction is interesting in its own right, since it represents electron transfer between H_A and P, much as does the forward reaction, except that different molecular orbitals of P and H_A are involved.

The final decay path for $P^+H_A^-$ involves triplet formation. After charge separation, the coupling between the electron spins of the radical pair is small enough so that over a period of nanoseconds the spins can dephase resulting in the formation of radical pair states with triplet character (Parson et al., 1975). Charge recombination then results in the formation of triplet P, ^3P, with a yield of 5–10% at 295 K. At 77 K, the recombination reaction in the singlet state slows down and ^3P formation becomes the dominant decay product of $P^+H_A^-$ when electron transfer to the quinone is blocked.

VIII. Electron Transfer to the Primary Quinone

A. The Absorption Changes Associated with Electron Transfer to Q_A

As described above, most investigators believe that the state present 15 ps after excitation of the reaction center is predominantly $P^+H_A^-$. During the subsequent several hundred picoseconds, this state decays to form the state $P^+Q_A^-$. This process can be followed by observing the absorption changes at 546 or 760 nm due to recovery of the ground state absorption of the BPhe as the electron is transferred to the quinone or by observing the change in the absorption near 650 nm where the anion of BPhe absorbs (Kirmaier and Holten, 1987). As was the case with the initial electron transfer reaction, the rate at which these absorption changes occur depends on the wavelength of measurement (Kirmaier et al., 1985b; Kirmaier and Holten, 1990). Single exponential fits of the absorption changes yield time constants ranging from 80 ps to 320 ps in *Rb. sphaeroides* reaction centers. There are various possible ways of explaining the complex kinetics, but because there are no obvious intermediate charge-separated states between the BPhe and the primary quinone, it has been suggested that multiple conformational states of the reaction center, either as a static distribution or a dynamic progression, give rise to the kinetic and spectral complexity observed. Like the kinetics of the initial electron transfer reaction, the rate of electron transfer to Q_A is only weakly temperature dependent, increasing by about a factor of two as the temperature is lowered from 295 K to 4 K (Kirmaier et al., 1985b).

B. $P^+Q_A^-$ Charge Recombination

If electron transfer to the secondary quinone is blocked, $P^+Q_A^-$ decays to the ground state in about 100 ms at room temperature in *Rb. sphaeroides* reaction centers. Charge recombination in this state is faster at low temperature, occurring in about 30 ms at 77 K (Parson, 1987). There is a substantial body of literature that considers the mechanism of this recombination reaction. Much of this work has utilized conditions, mutations or quinone substitutions that change the midpoint potential of Q_A or P, followed by measurement of the effects of those changes on the charge recombination kinetics (Gunner et al., 1982; Gopher et al., 1985; Popovic et al., 1986a,b; Woodbury et al., 1986; Lösche et al., 1987; Franzen et al., 1990, 1993; Williams et al., 1992c; Franzen and Boxer, 1993; Lin et al., 1994). The thermodynamics of this state and the effect of changing the redox chemistry of the quinone on charge recombination will be considered in a later section of this chapter. However, measurements in *Rb. sphaeroides* reaction centers point to the concept that with ubiquinone in the Q_A pocket at room temperature, most of the recombination occurs utilizing $P^+H_A^-$ as a virtual intermediate in a superexchange mechanism (Franzen et al., 1993). When quinones with more negative redox potentials are substituted for ubiquinone, an activated route, probably also involving $P^+H_A^-$ as a thermally accessible intermediate, becomes dominant (Woodbury et al., 1986; Gunner et al., 1982). The thermally activated path is also observed in native reaction centers of *Rp. viridis* (Shopes and Wraight, 1987).

IX. Electron Transfer from the Primary to Secondary Quinone

In many respects, the most complex reaction that occurs in the reaction center is electron transfer between the two quinones. Unlike the other electron transfer reactions that occur between one-electron carriers in a nearly solid-state configuration, Q_B is a two electron carrier and electron transfer is linked both to proton uptake and binding of quinone to the Q_B site. The details of this reaction are the subject of Chapter 26 by Okamura and Feher.

X. The Thermodynamics of Reaction Center Electron Transfer

A. The Standard Free Energy Difference Between P and $P^+H_A^-$*

Two primary techniques have been used to measure the standard free energy difference between P* and early charge-separated states. One of these involves monitoring the yield and decay of ^3P. The predominant path to ^3P formation in the reaction center is intersystem crossing in the state $P^+H_A^-$ followed by charge recombination (Scheme (1)); Parson et al., 1975). The yield of formation and the rate of decay of the triplet state are sensitive to the free energy difference between $P^+H_A^-$ and ^3P (Chidsey et al., 1985). Combined with the known energies of ^3P and P* relative to the ground state, the free energy gap between P* and $P^+H_A^-$ has been estimated as about 0.26 eV in *Rb. sphaeroides* reaction centers (Takiff and Boxer, 1988; Goldstein et al., 1988; Ogrodnik et al., 1988). The limitation of this technique is that it can only be used to determine the free energy gap tens of nanoseconds or microseconds after electron transfer has occurred.

The other common method for estimating the free energy gap between P* and the early charge-separated states is to monitor the fluorescence of P* as a function of time after excitation and to compare the initial fluorescence amplitude of P* (the initial concentration of P*) to the amplitude of the fluorescence due to back electron transfer from early charge-separated states (Schenck et al., 1982). This method has the advantage of giving information about electron transfer energetics on the timescale of electron transfer. However, the analysis is model dependent, requiring the assumption that the small amplitude fluorescence components are due to thermal repopulation of P* from $P^+H_A^-$. Though this interpretation is largely accepted on the nanosecond timescale, there are alternative, kinetically consistent, interpretations of the very early time complexities of P* decay that involve an intrinsic static heterogeneity in the reaction center population giving rise to several different rates of P* decay (discussed below; see also Ogrodnik et al., 1994).

If one accepts the model that the complex kinetic decay of the excited singlet state is due to thermal repopulation of P* from a series of charge-separated states with different energies (this model for the decay kinetics of P* is discussed in more detail below), then this leads to the conclusion that the standard free energy gap between P* and the early charge-separated state(s) depends strongly on the timescale of the measurement. On the nanosecond timescale, in reaction centers in which electron transfer past $P^+H_A^-$ is blocked by reduction of the quinones, the standard free energy gap between P* and $P^+H_A^-$ calculated from the long-lived fluorescence amplitude is nearly 0.20 eV (Woodbury and Parson, 1984; Schenck et al., 1982; Woodbury et al., 1986), approaching that determined by the longer timescale triplet P technique. However, on the timescale of a few tens of picoseconds, the free energy gap estimated from the P* fluorescence decay is only about 0.12 eV (Williams et al., 1992b). If one interprets the multiexponential decay of P* on the picosecond timescale in terms of equilibration with a charge-separated state, then the free energy gap calculated is a few tens of meV or less, depending on the temperature (Woodbury et al., 1994). Obviously, the issue of the driving force for electron transfer at early times is not completely understood, and the estimates one obtains for this number are dependent on the model one uses to describe the kinetics of electron transfer.

One of the most confusing and controversial aspects

of the thermodynamics of electron transfer is its temperature dependence. If one estimates the standard free energy change between P* and $P^+H_A^-$ on microsecond timescales by monitoring the ^3P state, one finds that the free energy determined is nearly temperature independent (Takiff and Boxer, 1988; Ogrodnik et al., 1988). This implies that the driving force for electron transfer is largely enthalpic. However, if one estimates the free energy change between P* and $P^+H_A^-$ from the fluorescence decay as described above, then one finds that the standard free energy change calculated is very temperature dependent, decreasing from approximately 120 meV on the ten picosecond timescale at 295 K to less than 10 meV on the same timescale at 20 K (Peloquin et al., 1994). A plot of the time evolution of the standard free energy difference between P* and the early product(s) of charge separation as a function of temperature, calculated from the fluorescence decay, is shown in Fig. 11 for one of the high potential hydrogen bond mutants (Leu^{L131} → His). Similar results are obtained for R-26 reaction centers, though the level of the long-lived fluorescence is lower (see below). According to this analysis, the free energy becomes very small as one approaches zero time and zero temperature. A naive analysis of the temperature effect on the free energy gap might imply that the driving force for electron transfer was largely entropic, contrary to the experiments performed on longer timescales. However, a more careful consideration of the temperature dependence of the P* decay shows that the relationship between free energy and temperature is not linear over a very large temperature region as would be expected for a purely entropic process (Woodbury and Parson, 1984; Peloquin et al., 1994). Instead, the slope of the temperature dependence varies dramatically, particularly near 200 K and 100 K. This, coupled with the fact that the free energy appears to be smaller at early times, has led to the suggestion that the free energy of the system is strongly affected by movements of the cofactors and their protein environment, and that these movements are frozen out at low enough temperature or simply have not had time to occur at early times (Peloquin et al., 1994; Woodbury et al., 1994). Another possibility is that the time and temperature dependence of the fluorescence decay results from a broad distribution of charge-separated state free energies which is static on the time scale of

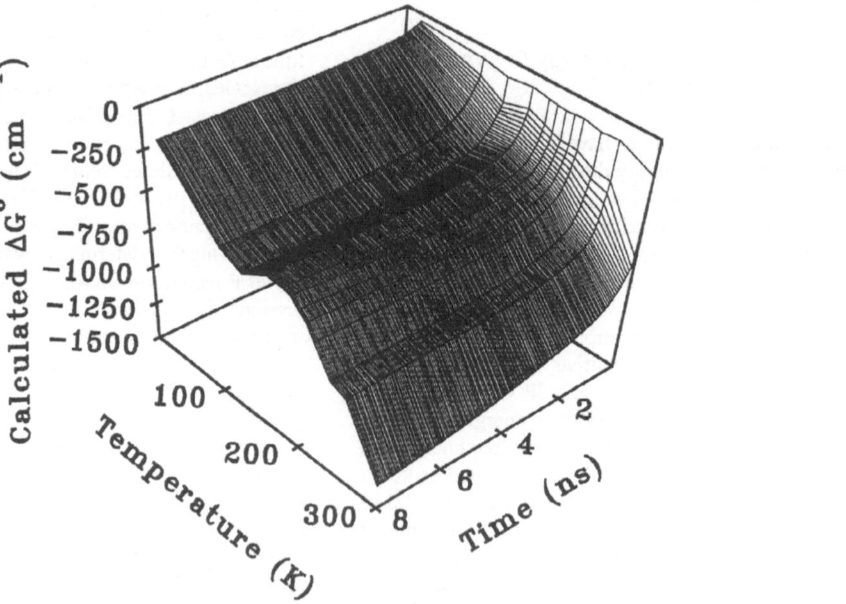

Fig. 11. The time and temperature dependence of the standard free energy difference between P* and $P^+H_A^-$ for Leu^{L131}→ His mutant reaction centers. The free energy gap was calculated as described in Peloquin et al. (1994). The figure is reproduced from Peloquin et al. (1994).

these experiments (Ogrodnik et al., 1994).

One further way of exploring the thermodynamics of the early electron transfer reactions is to consider the effects of mutations known to change the steady-state P/P^+ midpoint potential. For the hydrogen bond mutants that change the P/P^+ reduction potential over a nearly 400 mV range (Williams et al., 1992a,b; Murchison et al., 1993; Lin et al., 1994), one finds that as the P/P^+ midpoint potential is increased, more long-lived fluorescence due to back electron transfer is observed. This is consistent with the idea that much of the long-lived fluorescence is due to back electron transfer reforming P*, since one would expect an increase in the P/P^+ midpoint potential to be correlated with a decrease in the free energy gap between $P^+H_A^-$ and P* making thermal repopulation of P* more favorable.

B. The Relative Standard Free Energy of $P^+B_A^-$

Little experimental evidence exists on the free energy of $P^+B_A^-$. However, there is one recent paper which addresses this question within the model of an initial charge separation reaction forming a true $P^+B_A^-$ state followed by a faster reaction forming $P^+H_A^-$ (see below). In this work, Schmidt et al. (1994) used a transient absorbance increase near 1050 nm as a indication of the population of B_A^-. By performing a replacement of H_A with pheophytin (instead of bacteriopheophytin) they were able to increase the relative free energy of $P^+H_A^-$. This resulted in thermal repopulation of $P^+B_A^-$ and P^* from $P^+H_A^-$ which allowed them to estimate energies for these states. From the level of the apparent anion absorbance near 1050 nm under these conditions they calculated a standard free energy difference of about 50 meV between $P^+B_A^-$ and P^*. It is important to note that, like the fluorescence methods discussed above, the use of thermal repopulation to calculate the free energy of $P^+B_A^-$ in this system depends explicitly on the model employed.

C. The Relative Standard Free Energy of $P^+Q_A^-$

The decay of the much weaker long-lived (millisecond) fluorescence from P* in thermal equilibrium with $P^+Q_A^-$ has given rise to estimates for the standard free energy of this state as well. $P^+Q_A^-$ is estimated to be about 0.86 eV below P* in free energy or about 0.52 eV above the ground state in *Rb. sphaeroides* (Arata and Parson, 1981). The latter value very nearly matches what one obtains by using the measured midpoint potentials of P and Q_A. Similar measurements have placed $P^+Q_B^-$ about 60 meV below $P^+Q_A^-$ at neutral pH (Arata and Parson, 1981). A similar value is also obtained by a consideration of the decay kinetics of these two states (Kleinfeld et al., 1984; Maróti and Wraight, 1988). A much larger free energy difference between $P^+Q_A^-$ and $P^+Q_B^-$ is observed in reaction centers of *Rp. viridis* (Gao et al., 1991).

Similar estimates of the standard free energy difference between P* and $P^+Q_A^-$ have been made on a series of reaction centers in which Q_A has been removed and replaced by a quinone other than ubiquinone (Woodbury et al., 1986). As with the series of mutations described above that alter the P/P^+ midpoint potential, one would expect that substitution with quinones of higher reduction potential than ubiquinone would result in a $P^+Q_A^-$ state that is closer to P* in energy and thus would give rise to more thermal reformation of P* during the lifetime of $P^+Q_A^-$. A clear correlation is observed between the amplitude of the long-lived (millisecond) fluorescence decay and the reduction potential of the quinone used, supporting the validity of the estimates of the relative free energy of $P^+Q_A^-$ using the fluorescence measurements.

A summary of the energetics of the excited and charge-separated states of *Rb. sphaeroides* reaction centers is given in Fig. 12.

XI. Reaction Center Symmetry and the Direction of Electron Transfer

One of the most striking features of the reaction center crystal structure is the approximate 2-fold rotational symmetry which encompasses both the cofactors and the two core protein subunits, L and M (Fig. 1A; Deisenhofer et al., 1984; Allen et al., 1987; Chang et al., 1991). As alluded to above, however, the predominant cofactors involved in the initial electron transfer events appear to be those on the A-side of the reaction center. The exact ratio of A *vs.* B side electron transfer is not agreed upon. Values have been published ranging from about 3.6:1 favoring the A-side to 200:1 (Horber et al., 1986a,b; Tiede et al., 1987; Kellogg et al., 1989; Michel-Beyerle et al., 1988; Kirmaier et al., 1985a; Breton et al., 1986; Bylina et al., 1988; Lockhart et al., 1990). The evidence for electron transfer directionality comes primarily from monitoring the absorption

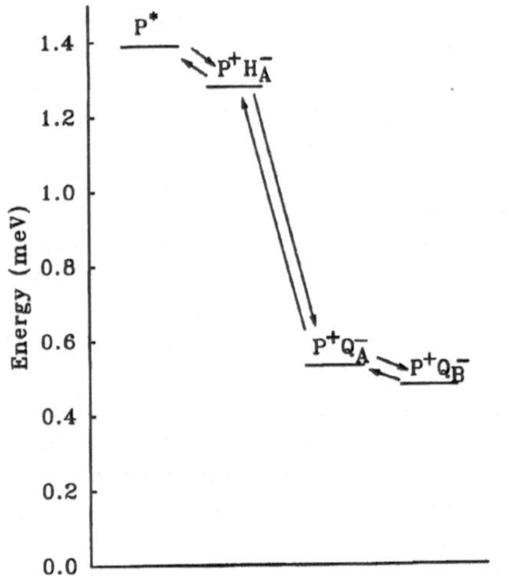

Fig. 12. Free energy diagram of the states P*, $P^+H_A^-$, $P^+Q_A^-$ and $P^+Q_B^-$ relative to the ground state (set at zero on the diagram) at 295 K. The energy of P* relative to the ground state was determined for the estimated zero-zero transition energy (Woodbury and Parson, 1984). The standard free energy difference between P* and $P^+H_A^-$ was determined from the long-lived fluorescence amplitude on the roughly 20 ps timescale (Peloquin et al., 1994; Williams et al., 1992a; Woodbury and Parson, 1984; Schenck et al., 1982). The standard free energy differences between P* and $P^+Q_A^-$ and between P* and $P^+Q_B^-$ were determined from the amplitude of the millisecond fluorescence (Arata and Parson, 1981).

changes due to reduction of the two BPhes which can be distinguished spectrally at low temperature (Fig. 3). Steady state photopumping experiments have been performed in which the rate or yield of H_B reduction is measured and compared to that of H_A reduction (Tiede et al., 1987; Kellogg et al., 1989). Also, direct measurements of absorption changes in the H_B spectral bands have been performed on the picosecond timescale (Fig. 3; Kirmaier et al., 1985a,b; Breton et al., 1986; Bylina et al., 1988; Lockhart et al., 1990). All of these have shown electron transfer strongly favoring formation of H_A^- though the magnitude and interpretation of the small spectral changes in the H_B absorption bands varies between different reports and using different samples and conditions.

It is clear from the photopumping measurements that it is possible to form the state $P^+H_B^-$, implying that this state does not lie very far above P* in energy. However, it is not clear at this point whether the major determining factor in the directionality of electron transfer is the kinetics or the thermodynamics of the process. Because the mechanistic reasons underlying the predominant A-side electron transfer are not known and the formation of $P^+H_B^-$ is clearly at least possible under certain conditions, it is difficult to extrapolate accurately between measurements at cryogenic temperatures (where the BPhe spectral bands are well resolved) in isolated reaction centers and the actual working of the photosynthetic apparatus at physiological temperatures in an energized biological membrane. However, the indication from the available spectroscopic data is that electron transfer in vivo predominantly involves A-side cofactors.

Several attempts have been made to alter the symmetry of the reaction center by mutagenesis. The first, and perhaps most relevant of these to our present understanding of the importance of symmetry in electron transfer, are the two heterodimer mutants. One might think that the symmetry-breaking aspect of the reaction center that is important for the observed asymmetry in electron transfer would be the environment around the special pair itself. Since there is no unambiguous evidence for B-side electron transfer on any timescale, it makes sense to suggest that the asymmetry of importance exists in the very initial state of the system, P*. It also seems reasonable to suppose that changing one of the BChl molecules of the special pair to a BPhe would skew the electron density in this state to one side (as it does in the state P^+; Huber et al., 1990; Bylina et al., 1990), therefore predisposing the system to undergo electron transfer along one of the branches. However, neither heterodimer mutant significantly affects the directionality of electron transfer (McDowell et al., 1991b).

Further evidence that the environment of the special pair does not play a predominant role in determining the direction of initial electron transfer comes from the *Rb. capsulatus* mutant, *sym1*, which symmetrizes a large fraction of the amino acids in the vicinity of P (Taguchi et al., 1992; Stocker et al., 1992). Though a small fraction of the B-side BPhe Q_X band appears to bleach in this mutant at 20 K, the effect is no larger than that observed in wild type grown and isolated under the same conditions (Woodbury et al., unpublished results). Apparently, the symmetry of the electron distribution of the special pair is not the only determining factor in the direction of electron flow.

This points to the environment of cofactors further along the chain as potentially important in determining the electron transfer asymmetry. Several different mutations have been made that alter the environment of one or the other monomer BChl in the reaction center. One of the amino acids predicted to significantly affect the energy of B_A is the tyrosine at M210 in *Rb. sphaeroides* (Parson et al., 1990). Changing this to a phenylalanine, the identity of the symmetrically related amino acid at L181, increases the lifetime of the excited singlet state of P, but does not appear to affect the direction of electron flow (Nagarajan et al., 1990; Finkele et al., 1990). Similarly, changing the phenylalanine at L181 to a tyrosine or preparing the double mutant in which the tyrosine and phenylalanine are exchanged between the two subunits affects the lifetime of P*, but does not appear to dramatically alter the direction of electron transfer (Chan et al., 1991b).

One other mutation that speaks to this issue is the β mutant that changes the A-side BPhe to a BChl (Kirmaier et al., 1991). This mutation cleanly removes any bleaching due to H_A in the BPhe Q_X region of the spectrum, and by perturbing the energy of the A-side electron acceptor, it could potentially force the electron transfer down the other branch, particularly if the determining factor in the electron transfer asymmetry was the thermodynamics of the charge-separated state (Kirmaier et al., 1991). Despite these changes, however, there is no significant bleaching in the Q_X absorption band of H_B in this mutant (Fig. 10; Kirmaier et al., 1991).

Finally, large scale mutations were prepared in which a number of amino acids in the D-helix near the monomer BChls and the BPhes have been either made the same between the L and M subunits or have been exchanged (Robles et al., 1990). These mutations do have large effects on electron transfer, but no evidence has been obtained for B-side electron transfer. Of particular note in this regard is the D_{LL} mutant that appears to completely lack the A-side BPhe. There is no evidence, even with the A-side completely nonfunctional, for significant electron transfer along the B-side (Vos et al., 1991).

XII. The Pathway and Mechanism of Initial Electron Transfer

The pathway and mechanism of the initial electron transfer reaction has been the source of considerable controversy (Holzapfel et al., 1989, 1990; reviewed by Kirmaier and Holten, 1993; Chan et al., 1991a; Martin and Vos, 1992) and as yet no consensus has been reached. As described above, the kinetics of both P* decay and of the formation of subsequent charge-separated states is complex. P* decay occurs on many timescales with multiple kinetic components as determined by exponential decay analysis of the spontaneous emission. The absorption changes associated with charge separation are also complex, involving multiple exponential decay components and wavelength dependent kinetics, at least at physiological temperatures. Since P* can be monitored by following the emission from this state, its population can be detected even at very low levels, and therefore, its decay kinetics have been characterized over longer timescales than charge-separated state formation. Thus, models to describe the decay of the excited state will be considered first.

A. Kinetic Models for P* Decay

There are two general types of models that have been suggested to explain the multiexponential decay of the excited singlet state. One is that the different kinetic components of the P* decay represent different subpopulations of reaction centers with different rate constants for electron transfer (Kirmaier and Holten, 1990; Muller et al., 1992). This could explain the complexity of the P* decay on the picosecond to hundred picosecond timescales if one assumed a very broad distribution of electron transfer rate constants within the reaction center population. However, it is unlikely that this could be used to account for P* decay times longer than a few hundred picoseconds because the intrinsic lifetime of P* in the absence of electron transfer is thought to be only 100–300 ps in the wild type (see above) and even shorter in some mutants that still show substantial levels of long-time fluorescence (Williams et al., 1992a; Peloquin et al., 1994). At least on the longer timescales, one needs to invoke the participation of substantial back electron transfer reforming P* from the charge-separated state(s).

The concept of reversible electron transfer brings up the second class of possible explanations. The complex kinetics of P* could be generated by reversible reactions between P* and several different charge-separated states or multiple nuclear configurations of the same charge-separated state. This could manifest itself as several possible parallel products of P* decay (Muller et al., 1992; Hamm et al., 1993) or as a linear sequence of product states

(Kirmaier et al., 1985b; Du et al., 1992; Woodbury et al., 1994). The latter model has been used for the estimation of free energy gaps from fluorescence decay data (discussed above in the section on thermodynamics; Schenck et al., 1982; Woodbury and Parson, 1984; Peloquin et al., 1994). The one requirement of any model involving substantial backwards electron transfer is that the states involved must be close enough in free energy so that significant thermal repopulation of the initial state is possible. This last point has important implications. A linear series of charge-separated states in a homogeneous reaction center population can only give rise to the complex P* decay observed if significant back electron transfer occurs, reforming P* from the charge-separated state(s). A spectral analysis of the reaction center absorption changes that are concurrent with the various decay components of P* is consistent with the notion that a series of protein conformational changes altering the free energy of the state $P^+H_A^-$ occurs as a function of time (Woodbury et al., 1994). A similar model also explains the temperature dependence of the long-lived fluorescence from reaction centers (Peloquin et al., 1994). However, since the decay of P* is complex even at low temperature, this model requires that the free energy difference between P* and the form of $P^+H_A^-$ present shortly after electron transfer is very small, a few meV at 20 K (discussed above; see also Peloquin et al., 1994; Woodbury et al., 1994).

None of the simple models described above to explain the complex P* decay are wholly satisfactory. It is not clear either how electron transfer at low temperature could occur with driving forces of a few meV or why there would be a distribution of productive electron transfer reaction rates within the reaction center population spanning two to three orders of magnitude. Before considering mechanistic issues further, the experimental evidence from transient absorption studies for the involvement of intermediate states in this reaction will be considered.

B. The Role of B_A in Initial Electron Transfer

The absorption changes associated with the early charge separation reaction involve both multiple exponential decay components and wavelength-dependent kinetics, at least at room temperature. Perhaps the simplest way to explain a wavelength-dependent reaction timecourse is to introduce an intermediate state with a significantly different spectrum from the initial or final states. Due to the position of the A-side monomer BChl in the reaction center structure (Fig. 1A), the state $P^+B_A^-$ is an obvious possibility for such a state as illustrated in the following kinetic scheme:

$$P^* \xrightarrow{3.1ps} P^+B_A^- \xrightarrow{0.9ps} P^+H_A^- \qquad (2)$$

where the time constants shown are for fits performed at 295 K of the data in Fig. 6A. Absorption change kinetics at some key wavelengths that have been used as evidence for a $P^+B_A^-$ intermediate are shown in Fig. 13. As shown in Fig. 14A, multiexponential fits of this transient absorption data over a broad wavelength range result in a fast (0.9 ps) change in these regions followed by a slower (3.1 ps) kinetic component. It is particularly striking that in some spectral regions these two changes are actually in opposite directions (see 788 nm in Fig. 13). This has been taken as strong evidence that there are more

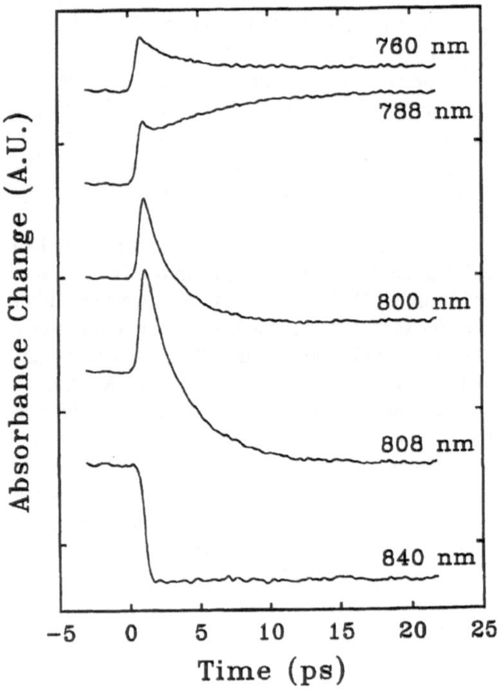

Fig. 13. The kinetics of absorption changes at several representative wavelengths between 760 and 840 nm using reaction centers from *Rb. sphaeroides* strain R-26 at 295 K. These curves represent temporal slices through the absorption change surface of Fig. 6A.

than two spectrally independent components involved in the reaction.

The involvement of anion states of bacteriochlorophyll has also been inferred from transient absorbance measurements in the 1000 to 1100 nm region where the anion of bacteriochlorophyll is expected to absorb. Zinth and coworkers have observed a short-lived absorbance increase in this region which they have assigned to the low steady-state population of $P^+B_A^-$ using a model similar to kinetic Scheme (2) above (Arlt et al., 1993; Schmidt et al., 1994).

Note that in Scheme (2) the initial reaction is slower than the second reaction (i.e., $k_1<k_2$). This is because the spectral evidence for the intermediate state involves relatively small absorption changes compared to those expected from the state $P^+B_A^-$. Fig. 14A shows the amplitude spectra of the absorption changes associated with the 0.9 ps, 3.1 ps and constant terms which result from an exponential fit of the absorption change surface in Fig. 6A.

Using the global exponential decay analysis of Fig. 14A, one can take a specific kinetic scheme, such as Scheme (2), and calculate the difference spectra for the proposed intermediate states. The results of such an analysis for Scheme (2) at 295 K are shown in Fig. 14B. Similar analyses have been performed by Holzapfel et al. (1990) on a series of kinetic traces with less spectral resolution. The calculated spectrum of the $P^+B_A^-$ intermediate state in this model does show substantial bleaching near 800 nm, as one might expect for a BChl anion. However, the spectra derived from this model are complex, and both $P^+B_A^-$ and $P^+H_A^-$ have significant absorption changes in all of the Q_Y spectral transitions of the reaction center. Thus, identification of the two spectra predicted by this model as $P^+B_A^-$ and $P^+H_A^-$ is not unambiguous. In any case, an irreversible scheme, such as Scheme (2), could not be entirely correct because it predicts a single exponential decay of P*, contrary to what is observed.

At low temperature (20 K), the complex kinetics of electron transfer still remain (Lauterwasser et al., 1991; Woodbury et al., 1994), but the wavelength dependence of the kinetics is much less pronounced (Kirmaier and Holten, 1990) and isobestic points are maintained throughout the initial electron transfer reaction (Kirmaier and Holten, 1988a). Singular value decomposition analysis (which is model independent) demonstrates that at 20 K there are only two spectrally distinct states required to describe the early electron transfer reactions of R-26 reaction centers (Fig. 15 and Woodbury et al., 1994). This argues against the concept that there is a distinct $P^+B_A^-$ intermediate at 20 K since one would expect this state to have a spectrum different from either P* or a pure $P^+H_A^-$ state. It is possible that the electron transfer pathway is different at low and high temperatures (Chan et al., 1991b). However, the fact that the complex kinetics are present at both temperatures suggests that there may be an explanation other than a significantly populated $P^+B_A^-$ intermediate for the multiexponential kinetics of the absorption changes associated with early charge separation.

C. Other Possible Reaction Intermediates

Because the kinetic model in Scheme (2) is not unequivocally supported at 295 K and appears not to hold at 20 K, several other possible explanations for the complex absorbance change kinetics at early times have been considered. First, one could assume that the 0.9 ps component (using the 295 K kinetics) represents the first reaction and the 3.1 ps component the second reaction as shown in Scheme (3).

$$P_1^* \xrightarrow{0.9\,ps} P_2^* \xrightarrow{3.1\,ps} P^+H_A^- \qquad (3)$$

Here, due to the small size of the absorption changes associated with the fast kinetic component, one possible intermediate state in this model could be an altered form of P* (Holzapfel et al., 1989). A spectral analysis of the same kinetic decays used for Scheme (2) is also shown in Fig. 14C for Scheme (3). Here one can see that the proposed intermediate state indeed looks similar to the initial P* in spectral features, though there are changes in the magnitude of the relative extinction coefficient as a function of wavelength. Evidence for changes in the state P* during the first few hundred femtoseconds after light absorption has been presented by Vos et al. (1991, 1994a,c). This model could explain some of the early time complexity in the decay kinetics of the fluorescence from P* assuming that P_2* was also able to fluoresce. Of course, it does not explain the components of the fluorescence decay on the several picosecond to hundreds of picosecond timescale (see above).

D. Reaction Center Heterogeneity

The fact that at low temperature, the wavelength

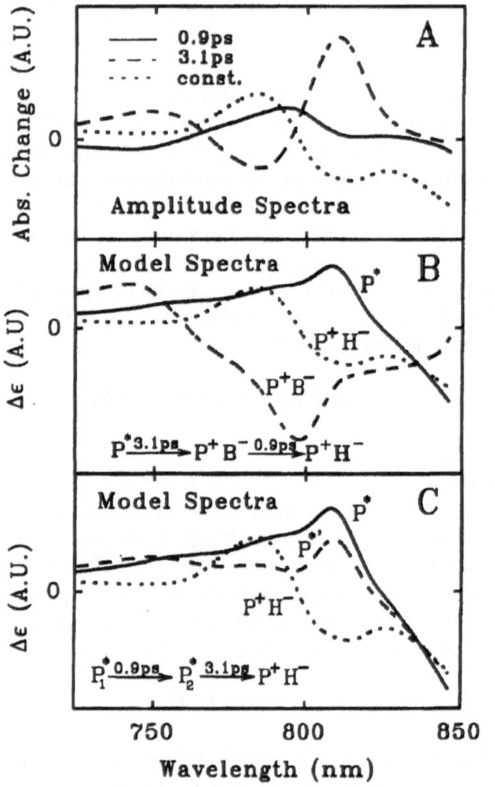

Fig. 14. (A) Preexponential amplitude spectra determined by fitting a time vs. wavelength absorbance change surface of Fig. 6A to a sum of two exponential terms and a constant, $A_1(\lambda)\exp(-k_1t) + A_2(\lambda)\exp(-k_2t) + A_3(\lambda)$, where λ is the wavelength, t is time, $A_i(\lambda)$ are the wavelength dependent preexponential amplitudes resulting from the fit and k_i are first order exponential decay rate constants. In this fit, k_1 was 1/(0.9 ps) and its associated preexponential amplitude spectrum, $A_1(\lambda)$, is represented by the solid line, k_2 was 1/(3.1 ps) and its preexponential amplitude spectrum, $A_2(\lambda)$, is given by the line with long dashes, and the amplitude spectrum of the constant term, $A_3(\lambda)$, is given by the line with short dashes. Seventy wavelengths were included in each fit over the wavelength region indicated. One hundred time points separated by 150 fs were collected at each wavelength. (B) The wavelength dependence of the model dependent relative difference extinction coefficient ($\Delta\varepsilon$) spectra were calculated (see Woodbury et al., 1994) from the amplitude spectra of panel A for each of the states (A, B and C) in a three state irreversible electron transfer model with the first step being slower than the second as described by Holzapfel et al. (1990). Holzapfel et al. (1990) have assigned the state A to P*, the state B to $P^+B_A^-$ and the state C to $P^+H_A^-$. (C) As in panel B except that the model used involves a fast (0.9 ps) irreversible reaction followed by a slower (3.1 ps) one.

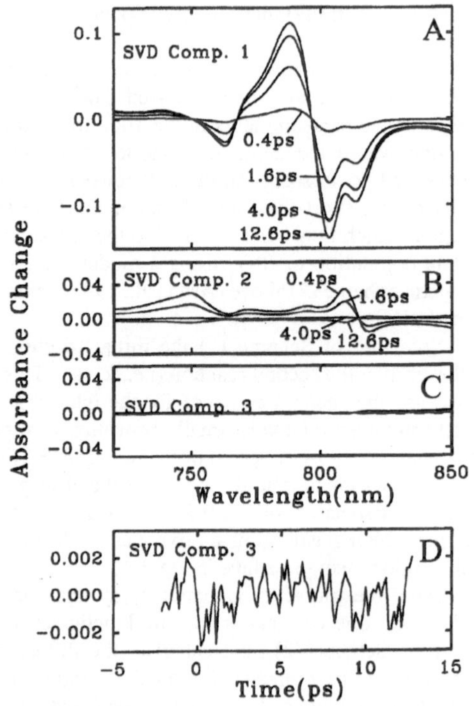

Fig. 15. Singular value decomposition analysis of the time vs. wavelength absorption change surface associated with electron transfer in reaction centers from *Rb. sphaeroides* strain R-26 at 20 K. The singular value decomposition was performed using the algorithms of Press et al. (1992). Panel A represents spectral slices through the portion of the time vs. wavelength absorption change surface which is described by matrix multiplication of the column vector representing the first spectral component of the singular value decomposition and the row vector representing the first temporal component weighted by the first diagonal element of the weighting matrix (see Press et al., 1992). Similarly, panels B and C are spectral slices through the surfaces defined by matrix multiplication of spectral and temporal vectors of the second (B) and third (C) components of the decomposition, respectively. It is important to note that the absorption change scale shown in these panels corresponds to the true absorption change associated with each singular value decomposition component. The sum of the surfaces represented by panels A, B and C is exactly equal to the original time vs. wavelength surface represented by the data which was used for this analysis, minus a small contribution from the fourth and higher singular value decomposition components. Panel D is the row vector associated with the third temporal singular value decomposition component. The abscissa of this panel has been scaled so that the timecourse is representative of absorption changes occurring in the 820 nm region. The size of absorption changes in the 790–810 nm region that are associated with this singular value decomposition component are even smaller. The data for this were taken from Woodbury et al. (1994) and this publication should be consulted for further details.

dependence of the kinetics essentially disappears and one finds only two spectrally distinct states involved in the initial electron transfer reaction, presumably P* and $P^+H_A^-$ (Kirmaier and Holten, 1988a; Kirmaier and Holten, 1990; Woodbury et al., 1994), has suggested that the complex spectral changes in the first few picoseconds may be due to a reaction center population heterogeneity that is either static or dynamic (Kirmaier et al., 1985b; Kirmaier and Holten, 1990; Muller et al., 1992; Small et al., 1992; Jia et al., 1993; Wang et al., 1993; Ogronik et al., 1994; Woodbury et al., 1994; Bixon et al., 1995). There could be either a static heterogeneity in the rate of initial electron transfer or a dynamic heterogeneity due to a series of $P^+H_A^-$ states formed sequentially each with a different free energy relative to P*, as described above to explain the complex P* decay. For example, the data of Figs. 6 and 13 could be explained in terms of a model in which there are two distinct reaction center subpopulations each undergoing electron transfer with a different rate. In this model, one needs to assume that the two subpopulations undergo electron transfer at different rates and that the states formed are significantly different in their spectral properties in the 800 nm region at 295 K but not at 20 K (Kirmaier and Holten, 1990; Woodbury et al., 1994). Alternatively, a model in which the free energy difference between P* and $P^+H_A^-$ is small and time dependent can be used, accounting for the complex absorption change kinetics on the picosecond timescale and for the complex emission decay kinetics on timescales from picoseconds to nanoseconds. However, this model requires a very small free energy difference between P* and $P^+H_A^-$, particularly at low temperature (Woodbury and Parson, 1984; Peloquin et al., 1994; Woodbury et al., 1994; see also sections above on P* decay and the thermodynamics of charge separation).

E. Pathway and Mechanism

As described above, there are several possible ways to describe the complex kinetic features associated with the initial electron transfer reaction, and no single mechanism for this reaction has yet been widely accepted. From the present understanding of electron transfer theory in these systems, it is clear that all three bacteriochlorins that comprise the A-side electron transfer chain must play some role in the early electron transfer pathway. Those who support the notion that the kinetic complexity during the first few picoseconds can be explained by the formation of $P^+B_A^-$ think of this state as a true populated intermediate between P* and $P^+H_A^-$ (Holzapfel et al., 1989; Arlt et al., 1993; Schmidt et al., 1994). On the other hand, those who use other mechanisms to explain the complexity of the transient absorption data usually assign the role of $P^+B_A^-$ as that of a virtual state, modifying the coupling between P* and $P^+H_A^-$ but never itself being populated (Kirmaier and Holten, 1991). Alternatively, it is possible that the excited and charge-separated states involved in electron transfer are close in energy and strongly coupled and that none of these states exist in pure form at any time during the course of the reaction (Woodbury et al., 1994; Small et al., 1995), though present theoretical estimates of the electronic coupling between the states involved in electron transfer argue against this concept (Plato et al., 1988; Scherer and Fischer, 1989).

XIII. The Role of the Reaction Center Protein in Electron Transfer

The discussion above has concentrated on the kinetics, thermodynamics and pathway of electron transfer in terms of the reaction center BChls, BPhes and quinones. Because the visible and near infrared spectroscopy of the reaction center is determined by the electronic structure of these cofactors, the cofactors have occupied center stage in electron transfer models. However, there is growing evidence that the protein may play an extremely important role in driving the reaction forward and preventing side reactions which might otherwise occur.

There are a number of studies which have supported the concept that there are changes in nuclear structure during the charge separation (as suggested above), and that the result of these movements either stabilizes the charge-separated state or prevents recombination to form the ground state. For example, the $P^+Q_A^-$ to $P^+Q_B^-$ electron transfer reaction will not occur at cryogenic temperatures if the sample is cooled in the dark. However, cooling the sample in the light results in a significant population of reaction centers apparently frozen in a conformation that allows this reaction to proceed (Kleinfeld et al., 1984). As described above, the multiexponential decay kinetics of P* can be explained in terms of protein motions progressively stabilizing the charge-separated state (Woodbury and Parson, 1984; Peloquin et al., 1994;

Woodbury et al., 1994). Similarly, some of the complex absorption changes observed during electron transfer to Q_A have been suggested to be due to nuclear motion (Kirmaier and Holten, 1990).

Recently, concerted protein motion has been used to explain the oscillatory behavior of the picosecond and subpicosecond absorption signals in the stimulated emission region of the spectrum (Vos et al., 1993, 1994a,c), opening up the possibility for a direct role of protein motion (or at least direct involvement of protein vibrational modes) in the electron transfer event. Finally, there have been indications from FTIR and resonance Raman measurements of changes in protein-protein or protein-cofactor interactions upon formation of P^+ or Q^- (Breton et al., 1991; Nabedryk et al., 1990a,b; Mattioli et al., 1991b). Ultrafast absorption changes in the infrared spectral region hold promise for further direct observation of protein motion during the early electron transfer reactions (Maiti et al., 1993).

Acknowledgments

The authors would like to acknowledge Drs. A. Taguchi and J. Williams for critical reading of the manuscript and Drs. S. Lin, J. Peloquin, R. Alden, W. Parson, R. Blankenship, C. Kirmaier, D. Holten and S. Boxer for helpful discussions. Some of the data presented here has not been published elsewhere and funding for this work came from grants DMB89-17729 and DMB91-58251 from the National Science Foundation and grants GM41300 and GM45902 from the National Institutes of Health. Instrumentation was purchased with funds from NSF grant DIR-8804992 and Department of Energy grants DE-FG-05-88-ER75443 and DE-FG-05-87-ER75361. This is publication No. 215 from the Arizona State University Center for the Study of Early Events in Photosynthesis.

References

Allen JP, Feher G, Yeates TO, Komiya H and Rees DC (1987) Structure of the reaction center from *Rhodobacter sphaeroides* R-26: the cofactors. Proc Natl Acad Sci USA 84: 5730–5734

Arata H and Parson WW (1981) Delayed fluorescence from *Rhodopseudomonas sphaeroides* reaction centers. enthalpy and free energy changes accompanying electron transfer from *P*-870 to quinones. Biochim Biophys Acta 638: 201–209

Arlt T, Schmidt S, Kaiser W, Lauterwasser C, Meyer M, Scheer H and Zinth W (1993) The accessory bacteriochlorophyll: A real electron carrier in primary photosynthesis. Proc Natl Acad Sci USA 90: 11757–11761

Beese D, Steiner R, Scheer H, Angerhofer A, Robert B and Lutz M (1988) Chemically modified photosynthetic bacterial reaction centers: circular dichroism, Raman resononance, low temperature absorption, fluorescence and ODMR spectra and polypeptide composition of borohydride treated reaction centers from *Rhodobacter sphaeroides* R26. Photochem Photobiol 47: 293–304

Bilsel O, Rodriguez J and Holten D (1990) Picosecond relaxation of strongly coupled porphyrin dimers. J Phys Chem 94: 3508–3512

Bixon M, Jortner J and Michel-Beyerle ME (1995) A kinetic analysis of the primary charge separation in bacterial photosynthesis. energy gaps and static heterogeneity. Chem Phys (in press)

Boxer SG (1993) Photosynthetic reaction center spectroscopy and electron transfer dynamics in applied electric fields. In: Deisenhofer J and Norris JR (eds) The Photosynthetic Reaction Center, Vol. II, pp 179–220. Academic Press, San Diego

Breton J (1988) Low temperature linear dichroism study of the orientation of the pigments in reduced and oxidized reaction centers of *Rps. viridis* and *Rb. sphaeroides*. In: Breton J and Vermeglio A (eds) The Photosynthetic Bacterial Reaction Center: Structure and Dynamics, pp 59–69. Plenum Press, New York

Breton J, Martin J-L, Migus A, Antonetti A and Orszag A (1986) Femtosecond spectroscopy of excitation energy transfer and initial charge separation in the reaction center of the photosynthetic bacterium *Rhodopseudomonas viridis*. Proc Natl Acad Sci USA 83: 5121–5125

Breton J, Martin J-L, Fleming GR and Lambry J-C (1988) Low-temperature femtosecond spectroscopy of the initial step of electron transfer in reaction centers from photosynthetic purple bacteria. Biochemistry 27: 8276–8284

Breton J, Bylina EJ and Youvan DC (1989) Pigment organization in genetically modified reaction centers of *Rhodobacter capsulatus*. Biochemistry 28: 6423–6430

Breton J, Martin J-L, Lambry J-C, Robles SJ and Youvan DC (1990) Ground state and femtosecond transient absorption spectroscopy of a mutant of *Rhodobacter capsulatus* which lacks the initial electron acceptor bacteriopheophytin. In: Michel-Beyerle M-E (ed) Reaction Centers of Photosynthetic Bacteria, Feldafing-II-Meeting, pp 293–302. Springer-Verlag, Berlin and Heidelberg

Breton J, Thibodeau DL, Berthomieu C, Mantele W, Vermeglio A and Nabedryk E (1991) Probing the primary quinone environment in photosynthetic bacterial reaction centers by light-induced FTIR difference spectroscopy. FEBS Lett 278: 257–260

Breton J, Nabedryk E and Parson WW (1992) A new infrared electronic transition of the oxidized primary electron donor in bacterial reaction centers: a way to assess resonance interactions between the bacteriochlorophylls. Biochemistry 31: 7503–7510

Bylina EJ and Youvan DC (1988) Directed mutations affecting spectroscopic and electron transfer properties of the primary donor in the photosynthetic reaction center. Proc Natl Acad Sci USA 85: 7226–7230

Bylina EJ and Youvan DC (1991) 3.7: Protein-chromophore

interactions in the reaction center of *Rhodobacter capsulatus*. In: Scheer H (ed) Chlorophylls, pp 705–719. CRC Press, Boca Raton

Bylina EJ, Kirmaier C, McDowell L, Holten D and Youvan DC (1988) Influence of an amino-acid residue on the optical properties and electron transfer dynamics of a photosynthetic reaction centre complex. Nature 336: 182–184

Bylina EJ, Kolaczkowski SV, Norris JR and Youvan DC (1990) EPR characterization of genetically modified reaction centers of *Rhodobacter capsulatus*. Biochemistry 29: 6203–6210

Chan C-K, Chen L X-Q, Dimagno TJ, Hanson DK, Nance SL, Schiffer M, Norris JR and Fleming GR (1991a) Initial electron transfer in photosynthetic reaction centers of *Rhodobacter capsulatus* mutants. Chem Phys Lett 176: 366–372

Chan C-K, DiMagno TJ, Chen L X-Q, Norris JR and Fleming GR (1991b) Mechanism of the initial charge separation in bacterial photosynthetic reaction centers. Proc Natl Acad Sci USA 88: 11202–11206

Chang C-H, El-Kabbani O, Tiede D, Norris J and Schiffer M (1991) Structure of the membrane-bound protein photosynthetic reaction center from *Rhodobacter sphaeroides*. Biochemistry 30: 5352–5360

Cherepy NJ, Shreve AP, Moore LJ, Franzen S, Boxer SG and Mathies RA (1994) Near-infrared resonance Raman spectroscopy of the special pair and the accessory bacteriochlorophylls in the photosynthetic reaction centers. J Phys Chem 98: 6023–6029

Chidsey CED, Kirmaier C, Holten D and Boxer SG (1984) Magnetic field dependence of radical-pair decay kinetics and molecular triplet quantum yield in quinone-depleted reaction centers. Biochem Biophys Acta 776:424–437

Chidsey CED, Takiff L, Goldstein RA and Boxer SG (1985) Effect of magnetic fields on the triplet state lifetime in photosynthetic reaction centers: evidence for thermal repopulation of the initial radical pair. Proc Natl Acad Sci USA 82: 6850–6854

Chirino AJ, Lous EJ, Huber M, Allen JP, Schenck CC, Paddock ML, Feher G and Rees DC (1994) Crystallographic analyses of site-directed mutants of the photosynthetic reaction center from *Rhodobacter sphaeroides*. Biochemistry 33: 4584–4593

Coleman W J and Youvan DC (1990) Spectroscopic analysis of genetically modified photosynthetic reaction centers. Annu Rev Biophys Biophys Chem 19: 333–367

Davis D, Wong A, Caughey WS and Schenck CC (1992) Energetics of the oxidized primary donor in wild type and heterodimer mutant reaction centers. Biophys J 61: A153

Dawkins DJ, Ferguson LA and Cogdell R (1988) The structure of the 'core' of the purple bacterial photosynthetic unit. In: Scheer H and Schneider S (eds) Photosynthetic Light-Harvesting Systems: Organization and Function, pp 115–127. Walter de Gruyter and Co., Berlin and New York

Deisenhofer J, Epp O, Miki K, Huber R and Michel H (1984) X-ray structure analysis of a membrane protein complex. electron density map at 3Å resolution and a model of the chromophores of the photosynthetic reaction center from *Rhodopseudomonas viridis*. J Mol Biol 180: 385–398

Dressler K, Umlauf E, Schmidt S, Hamm P, Zinth W, Buchanan S and Michel H (1991) Detailed studies of the subpicosecond kinetics in the primary electron transfer of reaction centers of *Rhodopseudomonas viridis*. Chem Phys Lett 183: 270–276

Du M, Rosenthal SJ, Xie X, DiMagno TJ, Schmidt M, Hanson DK, Schiffer M, Norris JR and Fleming GR (1992) Femtosecond spontaneous-emission studies of reaction centers from photosynthetic bacteria. Proc Natl Acad Sci USA 89: 8517–8521

Ermler U, Fritzsch G, Buchanan SK and Michel H (1994) Structure of the photosynthetic reaction centre from *Rhodobacter sphaeroides* at 2.65 A resolution: cofactors and protein-cofactor interactions. Structure 2: 925–936

Feher G and Okamura MY (1978) Ch. 19: Chemical composition and properties of reaction centers. In: Clayton RK and Sistrom WR (eds) The Photosynthetic Bacteria, pp 349–386. Plenum Press, New York

Feher G, Hoff AJ, Isaacson RA and Ackerson LC (1975) ENDOR experiments on chlorophyll and bacteriochlorophyll in vitro and in the photosynthetic unit. Ann NY Acad Sci USA 244: 239–259

Feher G, Allen JP, Okamura MY and Rees DC (1989) Structure and function of bacterial photosynthetic reaction centres. Nature 339: 111–116

Finkele U, Lauterwasser C, Zinth W, Gray KA and Oesterhelt D (1990) Role of tyrosine M210 in the initial charge separation of reaction centers of *Rhodobacter sphaeroides*. Biochemistry 29: 8517–8521

Fleming GR, Martin JL and Breton J (1988) Rates of primary electron transfer in photosynthetic reaction centres and their mechanistic implications. Nature 333: 190–192

Franzen S and Boxer SG (1993) Temperature dependences of the electric field modulation of Recombination in photosynthetic reaction centers. J Phys Chem 97: 6304–6318

Franzen S, Goldstein RF and Boxer SG (1990) Electric field modulation of electron transfer reaction rates in isotropic systems: long-distance charge recombination in photosynthetic reaction centers. J Phys Chem 94: 5135–5149

Franzen S, Goldstein RF and Boxer SG (1993) Distance dependence of electron-transfer reactions in organized systems: the role of superexchange and non-Condon effects in photosynthetic reaction centers. J Phys Chem 97: 3040–3053

Gao J-L, Shopes RJ and Wraight CA (1991) Heterogeneity of kinetics and electron transfer equilibria in the bacteriopheophytin and quinone electron acceptors of reaction centers from *Rhodopseudomonas viridis*. Biochim Biophys Acta 1056: 259–272

Goldstein R A, Takiff L and Boxer S G (1988) Energetics of initial charge separation in bacterial photosynthesis: the triplet decay rate in very high magnetic fields. Biochim Biophys Acta 934: 253–263

Gopher A, Blatt Y, Schonfeld M, Okamura MY, Feher G and Montal M (1985) The effect of an applied electric field on the charge recombination kinetics in reaction centers reconstituted in planar lipid bilayers. Biophys J 48: 311–320

Gray KA, Farchaus JW, Wachtveitl J, Breton J and Oesterhelt D (1990) Initial characterization of site-directed mutants of tyrosine M-210 in the reaction centre of *Rhodobacter sphaeroides*. EMBO J 9: 2061–2070

Gunner MR, Liang Y, Nagus DK, Hochstrasser RM and Dutton PL (1982) Variation in rates of electron transfer in photosynthetic reaction centers with the primary ubiquinone substituted with other quinones. Biophys J 37: 226a

Hamm P, Gray KA, Oesterhelt D, Feick R, Scheer H and Zinth W (1993) Subpicosecond emission studies of bacterial reaction centers. Biochim Biophys Acta 1142: 99–105

Holzapfel W, Finkele U, Kaiser W, Oesterhelt D, Scheer H, Stilz HU and Zinth W (1989) Observation of a bacteriochlorophyll anion radical during the primary charge separation in a reaction center. Chem Phys Lett 160: 1–7

Holzapfel W, Finkele U, Kaiser W, Oesterhelt D, Scheer H, Stilz HU and Zinth W (1990) Initial electron-transfer in the reaction center from *Rhodobacter sphaeroides*. Proc Natl Acad Sci USA 87: 5168–5172

Horber JKH, Gobel W, Ogrodnik A, Michel-Beyerle ME and Cogdell RJ (1986a) Time-resolved measurements of fluorescence from reaction centres of *Rhodopseudomonas viridis* and the effect of menaquinone reduction. FEBS Lett 198: 268–272

Horber JKH, Gobel W, Ogrodnik A, Michel-Beyerle ME and Cogdell RJ (1986b) Time-resolved measurements of fluorescence from reaction centres of *Rhodopseudomonas sphaeroides* R26.1. FEBS Lett 198: 273–278

Huber M, Lous EJ, Isaacson RA, Feher G, Gaul D and Schenck CC (1990) EPR and ENDOR studies of the oxidized donor in reaction centers of *Rhodobacter sphaeroides* strain R-26 and two heterodimer mutants in which histidine M202 or L173 was replaced by leucine. In: Michel-Beyerle M-E (ed) Reaction Centers of Photosynthetic Bacteria, Feldafing-II-Meeting, pp 219–228. Springer-Verlag, Berlin and Heidelberg

Jia Y, DiMagno TJ, Chan C-K, Wang Z, Du M, Hanson DK, Schiffer M,. Norris JR, Fleming GR and Popov MS (1993) Primary charge separation in mutant reaction centers of *Rhodobacter capsulatus*. J Phys Chem 97: 13180–13191

Kellogg EC, Kolaczkowski S, Wasielewski MR and Tiede DM (1989) Measurement of the extent of electron transfer to the bacteriopheophytin in the M-subunit in reaction centers of *Rhodopseudomonas viridis*. Photosynth Res 22: 47–59

Kirmaier C and Holten D (1987) Primary photochemistry of reaction centers from the photosynthetic purple bacteria. Photosynth Res 13: 225–260

Kirmaier C and Holten D (1988a) Subpicosecond characterization of the optical properties of the primary electron donor and the mechanism of the initial electron transfer in *Rhodobacter capsulatus* reaction centers. FEBS Lett 239: 211–218

Kirmaier C and Holten D (1988b) Subpicosecond spectroscopy of charge separation in *Rhodobacter capsulatus* reaction centers. Israel J Chem 28: 79–85

Kirmaier C and Holten D (1990) Evidence that a distribution of bacterial reaction centers underlies the temperature and detection-wavelength dependence of the rates of the primary electron-transfer reactions. Proc Natl Acad Sci USA 87: 3552–3556

Kirmaier C and Holten D (1991) An assessment of the mechanism of initial electron transfer in bacterial reaction centers. Biochemistry 30: 609–613

Kirmaier C and Holten D (1993) Electron transfer and charge recombination reactions in wild-type and mutant bacterial reaction centers. In: Deisenhofer J and Norris JR (eds) The Photosynthetic Reaction Center, Vol. II, pp 49–70. Academic Press, San Diego

Kirmaier C, Holten D and Parson WW (1985a) Picosecond-photodichroism studies of the transient states in *Rhodopseudomonas sphaeroides* reaction centers at 5 K: effects of electron transfer on the six bacteriochlorin pigments. Biochim Biophys Acta 810: 49–61

Kirmaier C, Holten D and Parson WW (1985b) Temperature and detection-wavelength dependence of the picosecond electron-transfer kinetics measured in *Rhodopseudomonas sphaeroides* reaction centers. resolution of new spectral and kinetic components in the primary charge-separation process. Biochim Biophys Acta 810: 33–48

Kirmaier C, Holten D, Bylina EJ and Youvan DC (1988) Electron transfer in a genetically modified bacterial reaction center containing a heterodimer. Proc Natl Acad Sci USA 85: 7562–7566

Kirmaier C, Gaul D, DeBey R, Holten D and Schenck CC (1991) Charge separation in a reaction center incorporating bacteriochlorophyll for photoactive bacteriopheophytin. Science 251: 922–927

Kleinfeld D, Okamura MY and Feher G (1984) Electron transfer in reaction centers of *Rhodopseudomonas sphaeroides*. I. Determination of the charge recombination pathway of $D^+Q_AQ_B^-$ and free energy and kinetic relations between $Q_A^-Q_B$ and $Q_AQ_B^-$. Biochim Biophys Acta 766: 126–140

Lauterwasser C, Finkele U, Scheer H and Zinth W (1991) Temperature dependence of the primary electron transfer in photosynthetic reaction centers from *Rhodobacter sphaeroides*. Chem Phys Lett 183: 471–477

Lendzian F, Lubitz W, Scheer H, Hoff AJ, Plato M, Trankle E and Möbius K (1988) ESR, ENDOR, and TRIPLE resonance studies of the primary electron donor cation radical P^++960 in the photosynthetic bacterium *Rps. viridis*. Chem Phys Lett 148: 377–385

Lendzian F, Huber M, Isaacson RA, Endeward B, Plato M, Bonigk B, Möbius K, Lubitz W and Feher G (1993) The electronic structure of the primary donor cation radical in *Rhodobacter sphaeroides* R-26: ENDOR and TRIPLE resonance studies in single crystals of reaction centers. Biochim Biophys Acta 1183: 139–160

Lin X, Murchison HA, Nagarajan V, Parson WW, Allen JP and Williams JC (1994) Specific alteration of the oxidation potential of the electron donor in reaction centers from *Rhodobacter sphaeroides*. Proc Natl Acad Sci USA 91: 10265–10269

Lockhart DJ, Kirmaier C, Holten D and Boxer SG (1990) Electric field effects on the initial electron-transfer kinetics in bacterial photosynthetic reaction centers. J Phys Chem 94: 6987–6995

Lösche M, Feher G and Okamura MY (1987) The Stark effect in reaction centers from *Rhodobacter sphaeroides* R-26 and *Rhodopseudomonas viridis*. Proc Natl Acad Sci USA 84: 7537–7541

Lubitz W (1991) EPR and ENDOR studies of chlorophyll cation and anion radicals. In: Scheer H (ed) Chlorophylls, pp 903–944. CRC Press, Boca Raton

Maiti S, Cowen BR, Diller R, Iannone M, Moser CC, Dutton PL and Hochstrasser RM (1993) Picosecond infrared studies of the dynamics of the photosynthetic reaction center. Proc Natl Acad Sci USA 90: 5247–5251

Maróti P and Wraight CA (1988) Flash-induced H^+ binding by bacterial photosynthetic reaction centers: influences of the redox states of the acceptor quinones and primary donor. Biochim Biophys Acta 934: 329–347

Martin J-L and Vos MH (1992) Femtosecond biology. Ann Rev Biophys Biomol Struct 21: 199–222

Martin J-L, Breton J, Hoff AJ, Migus A and Antonetti A (1986) Femtosecond spectroscopy of electron transfer in the reaction center of the photosynthetic bacterium *Rhodopseudomonas sphaeroides* R-26: Direct electron transfer from the dimeric bacteriochlorophyll primary donor to the bacteriopheophytin acceptor with a time constant of 2.8 +/– 0.2 psec. Proc Natl Acad Sci USA 83: 957–961

Mattioli TA, Gray KA, Lutz M, Oesterhelt D and Robert B (1991a) Resonance Raman characterization of *Rhodobacter sphaeroides* reaction centers bearing site-directed mutations at tyrosine M210. Biochemistry 30: 1715–1722

Mattioli TA, Hoffman A, Robert B, Schrader B and Lutz M (1991b) Primary donor structure and interactions in bacterial reaction centers from near-infrared Fourier transform resonance Raman spectroscopy. Biochemistry 30: 4648–4654

Mattioli TA, Williams JC, Allen JP and Robert B (1994) Changes in primary donor hydrogen bonding interactions in mutant reaction centers from *Rb. sphaeroides*: Identification of the vibrational frequencies of all the conjugated carbonyl groups Biochemistry 33: 1636–1643

Mattioli TA, Lin X, Allen JP and Williams JC (1995) Correlation between multiple hydrogen bonding and alteration of the oxidation potential of the bacteriochlorophyll dimer of reaction centers from *Rhodobacter sphaeroides*. Biochemistry (in press)

McDowell LM, Gaul D, Kirmaier C, Holten D and Schenck CC (1991a) Investigation into the source of electron transfer asymmetry in bacterial reaction centers. Biochemistry 30: 8315–8322

McDowell LM, Kirmaier C and Holten D (1991b) Temperature-independent electron transfer in *Rhodobacter capsulatus* wild-type and His ^{M200}→Leu photosynthetic reaction centers. J Phys Chem 95: 3379–3383

Michel-Beyerle ME, Plato M, Deisenhofer J, Michel H, Bixon M and Jortner J (1988) Unidirectionality of charge separation in reaction centers of photosynthetic bacteria. Biochim Biophys Acta 932: 52–70

Middendorf TR, Mazzola LT, Gaul DF, Schenck CC and Boxer SG (1991) Photochemical hole-burning spectroscopy of a photosynthetic reaction center mutant with altered charge separation kinetics: properties and decay of the initially excited state. J Phys Chem 95: 10142–10151

Muller MG, Griebenow K and Holzwarth AR (1992) Primary processes in isolated bacterial reaction centers from *Rhodobacter sphaeroides* studied by picosecond fluorescence kinetics. Chem Phys Lett 199: 465–469

Murchison HA, Alden RG, Allen JP, Peloquin JM, Taguchi AKW, Woodbury NW and Williams JC (1993) Mutations designed to modify the environment of the primary electron donor of the reaction center from *Rhodobacter sphaeroides*: phenylalanine to leucine at L167 and histidine to phenylalanine at L168. Biochemistry 32: 3498–3505

Nabedryk E, Andrianambinintsoa S, Berger G, Leonhard M, Mantele W and Breton J (1990a) Characterization of bonding interactions of the intermediary electron acceptor in the reaction center of Photosystem II by FTIR spectroscopy. Biochim. Biophys Acta 1016: 49–54

Nabedryk E, Bagley KA, Thibodeau DL, Bauscher M, Mantele W and Breton J (1990b) A protein conformational change associated with the photoreduction of the primary and secondary quinones in the bacterial reaction center. FEBS Lett 266: 59–62

Nabedryk E, Allen JP, Taguchi AKW, Williams JC, Woodbury NW and Breton J (1993) Fourier transform infrared study of the primary donor in chromatophores of *Rhodobacter sphaeroides* with reaction centers genetically modified at residues M160 and L131. Biochemistry 32: 13879–13885

Nagarajan V, Parson WW, Gaul D and Schenck CC (1990) Effect of specific mutations of tyrosine-(M)210 on the primary photosynthetic electron-transfer process in *Rhodobacter sphaeroides*. Proc Natl Acad Sci USA 87: 7888–7892

Nagarajan V, Parson WW, Davis D and Schenck CC (1993) Kinetics and free energy gaps of electron-transfer reactions in *Rhodobacter sphaeroides* reaction centers. Biochemistry 32: 12324–12336

Norris JR, Scheer H and Katz JJ (1975) Models for antenna and reaction center chlorophylls. Ann NY Acad Sci USA 244: 260–280

Ogrodnik A, Kruger HW, Orthuber H, Haberkorn R, Michel-Beyerle ME, and Scheer H (1982) Recombination dynamics in bacterial photosynthetic reaction centers. Biophys J 39: 91–99

Ogrodnik A, Volk M, Letterer R, Feick R and Michel-Beyerle ME (1988) Determination of free energies in reaction centers of *Rb. sphaeroides*. Biochim Biophys Acta 936: 361–371

Ogrodnik A, Keupp W, Volk M, Aumeier G and Michel-Beyerle M.E (1994) Inhomogeneity of radical pair energetics in photosynthetic reaction centers revealed by differences in recombination dynamics of $P^+H_A^-$ when detected in delayed emission and in absorption. J. Phys. Chem. 98: 3432–3439.

Parson WW (1987) The bacterial reaction center. In: Amesz J (ed) Photosynthesis, pp 43–61. Elsevier, Amsterdam

Parson WW (1991) Reaction centers. In: Scheer H (ed) Chlorophylls, pp 1153–1180. CRC Press, Boca Raton

Parson WW, Clayton RK and Cogdell RJ (1975) Excited states of photosynthetic reaction centers at low redox potentials. Biochim Biophys Acta 387: 265–278

Parson WW, Chu Z-T and Warshel A (1990) Electrostatic control of charge separation in bacterial photosynthesis. Biochim Biophys Acta 1017: 251–272

Peloquin JM, Williams JC, Lin X, Alden RG, Murchison HA, Taguchi AKW, Allen JP and Woodbury NW (1994) Time-dependent thermodynamics during early electron transfer in reaction centers from *Rhodobacter sphaeroides*. Biochemistry 33: 8089–8100

Pierce DW and Boxer SG (1992) Dielectric relaxation in a protein matrix. J Phys Chem 96: 5560–5566

Plato M, Möbius K, Michel-Beyerle ME, Bixon M and Jortner J (1988) Intermolecular electronic interactions in the primary charge separation in bacterial photosynthesis. J Am Chem Soc 110: 7279–7285

Plato M, Möbius K and Lubitz W (1991) 4.10: Molecular orbital calculations on chlorophyll radical ions. In: Scheer H (ed) Chlorophylls, pp 1015–1046. CRC Press, Boca Raton

Popovic ZD, Kovacs GJ, Vincett PS, Alegria G and Dutton PL (1986a) Electric field dependence of recombination kinetics in reaction centers of photosynthetic bacteria. Chem Phys 110: 227–237

Popovic ZD, Kovacs GJ, Vincett PS, Alegria G and Dutton PL (1986b) Electric-field dependence of the quantum yield in reaction centers of photosynthetic bacteria. Biochim Biophys Acta 851: 38–48

Press WH, Teukolsky SA, Vetterling WT and Flannery BP (1992) Numerical Recipes in C, the Art of Scientific Computing, Cambridge University Press, Cambridge

Prince RC and Youvan DC (1987) Isolation and spectroscopic properties of photochemical reaction centers from *Rhodobacter capsulatus*. Biochim Biophys Acta 890: 286–291

Rautter J, Gessner C, Lendzian F, Lubitz W, Williams JC, Murchison HA, Wang S, Woodbury NW and Allen JP (1992) EPR and ENDOR studies of the primary donor cation radical in native and genetically modified bacterial reaction centers. In: Breton J and Vermeglio A (eds) The Photosynthetic Bacterial Reaction Center II, Structure, Spectroscopy and Dynamics, pp 99–108. Plenum, New York

Rautter J, Lendzian F, Lubitz W, Wang S and Allen JP (1994) Comparative study of reaction centers from photosynthetic purple bacteria: electron paramagnetic resonance and electron nuclear double resonance spectroscopy. Biochemistry 33: 12077–12084

Rautter J, Lendzian F, Schulz C, Fersch A, Kohn M, Lin X, Williams JC, Allen JP and Lubitz W (1995) ENDOR-studies of the primary cation radical in mutant reaction centers of *Rhodobacter sphaeroides* with altered hydrogen-bond interactions. Biochemistry, in press

Reddy NRS, Lyle PA and Small GJ (1992) Applications of spectral hole burning spectroscopies to antenna and reaction center complexes. Photosynth Res 31: 167–194

Robert B, Lutz M and Tiede DM (1985) Selective photochemical reduction of either of the two bacteriopheophytins in reaction centers of *Rp. sphaeroides* R-26. FEBS Lett 183: 326–330

Robles SJ, Breton J and Youvan DC (1990) Partial symmetrization of the photosynthetic reaction center. Science 248: 1402–1405

Rodriguez J and Holten D (1989) Ultrafast vibrational dynamics of a photoexcited metalloporphyrin. J Chem Phys 91: 3525–3531

Rodriguez J, Kirmaier C, Johnson MR, Friesner RA, Holten D and Sessler JL (1991) Picosecond studies of quinone-substituted monometalated porphyrin dimers: evidence for superexchange-mediated electron transfer in a photosynthetic model system. J Am Chem Soc 113: 1652–1659

Schenck CC, Blankenship RE and Parson WW (1982) Radical-pair decay kinetics, triplet yields and delayed fluorescence from bacterial reaction centers. Biochim Biophys Acta 680: 44–59

Scherer POJ and Fischer SF (1989) Quantum treatment of the optical spectra and the initial electron transfer process within the reaction center of *Rhodopseudomonas viridis*. Chem Phys 131: 115–127

Schmidt S, Arlt T, Hamm P, Huber H, Nagele T, Wachtveitl J, Meyer M, Scheer H and Zinth W (1994) Energetics of the primary electron transfer reaction revealed by ultrafast spectroscopy on modified bacterial reaction centers. Chem Phys Lett 223: 116–120

Sebban P and Moya I (1983) Fluorescence lifetime spectra of in vivo bacteriochlorophyll at room temperature. Biochim Biophys Acta 722: 436–442

Seely GR and Connolly JS (1986) Fluorescence of photosynthetic pigments in vitro. In: Govindjee, Amesz J and Fork DC (eds) Light Emission by Plants and Bacteria, pp 99–133. Academic Press, Orlando

Shopes RJ and Wraight CA (1987) Charge recombination from the $P^+Q_A^-$ state in reaction centers from *Rhodopseudomonas viridis*. Biochim Biophys Acta 893: 409–425

Shreve AP, Cherepy NJ, Franzen S, Boxer SG and Mathies RA (1991) Rapid-flow resonance Raman spectroscopy of bacterial photosynthetic reaction centers. Proc Natl Acad Sci USA 88: 11207–11211

Shuvalov VA and Parson WW (1981) Energies and kinetics of radical pairs involving bacteriochlorophyll and bacteriopheophytin in bacterial reaction centers. Proc Natl Acad Sci USA 78: 957–961

Small GJ (1995) On the validity of the standard model for primary charge separation in the bacterial reaction center. Chem Phys (in press)

Small GJ, Hayes JM and Silbey RJ (1992) The question of dispersive kinetics for the initial phase of charge separation in bacterial reaction centers. J Phys Chem 96: 7499–7501Snozzi M and Bachofen R (1979) Characterisation of reaction centers and their phospholipids from *Rhodospirillum rubrum*. Biochim Biophys Acta 546: 236–247

Stanley RJ and Boxer SG (1994) Oscillations in the spontaneous fluorescence from photosynthetic reaction centers. J. Phys. Chem 99: 859–863

Stocker JW, Taguchi AKW, Murchison HA, Woodbury NW and Boxer SG (1992) Spectroscopic and redox properties of sym1 and (M)F195H: *Rhodobacter capsulatus* reaction center symmetry mutants which affect the initial electron donor. Biochemistry 31: 10356–10362

Taguchi AKW, Stocker JW, Alden RG, Causgrove TP, Peloquin JM, Boxer SG and Woodbury NW (1992) Biochemical characterization and electron-transfer reactions of *sym1*, a *Rhodobacter capsulatus* reaction center symmetry mutant which affects the initial electron donor. Biochemistry 31: 10345–10355

Takiff L and Boxer SG (1988) Phosphorescence from the primary electron donor in *Rhodobacter sphaeroides* and *Rhodopseudomonas viridis* reaction centers. Biochim Biophys Acta 932: 325–334

Thompson MA, Zerner MC and Fajer J (1991) A theoretical examination of the electronic structure and excited states of the bacteriochlorophyll *b* dimer from *Rhodopseudomonas viridis*. J Phys Chem 95: 5693–5700

Tiede DM, Kellogg E and Breton J. (1987) Conformational changes following reduction of the bacteriopheophytin electron acceptor in reaction centers of *Rhodopseudomonas viridis*. Biochim Biophys Acta 892: 294–302

Vos MH, Lambry J-C, Robles SJ, Youvan DC, Breton J and Martin J-L (1991) Direct observation of vibrational coherence in bacterial reaction centers using femtosecond absorption spectroscopy. Proc Natl Acad Sci USA 88: 8885–8889

Vos MH, Lambry J-C, Robles SJ, Youvan DC, Breton J and Martin J-L (1992) Femtosecond spectral evolution of the excited state of bacterial reaction centers at 10 K. Proc Natl Acad Sci USA 89: 613–617

Vos MH, Rappaport F, Lambry J-C, Breton J and Martin J-L (1993) Visualization of coherent nuclear motion in a membrane protein by femtosecond spectroscopy. Nature 363: 320–325

Vos MH, Jones MR, Hunter CN, Breton J, Lambry J-C and artin J-L (1994a) Coherent dynamics during the primar electron-transfer reactions in membrane-bound reaction center of *Rhodobacter sphaeroides*. Biochemistry 33: 6750–6757.

Vos MH, Jones MR, Hunter CN, Breton J and Martin J-L (1994b) Coherent nuclear dynamics at room temperature in bacterial reaction centers. Proc Natl Acad Sci USA 91: 12701–12705.

Vos MH, Jones MR, McGlynn P, Hunter CN, Breton J and Martin J-L (1994c) Influence of the membrane environment on vibrational motions in reaction centres of Rhodobacter sphaeroides. Biochim Biophys Acta 1186: 117–122

Wachtveitl J, Farchaus JW, Dos R, Lutz M, Robert B and

Mattioli TA (1994) Structure, spectroscopic and redox properties of *Rhodobacter sphaeroides* reaction centers bearing point mutations near the primary electron donor. Biochemistry 32: 12875–12886

Wang Z, Pearlstein RM, Jia Y, Fleming GR and Norris JR (1993) Chem Phys 176:421–425

Williams JC, Alden RG, Coryell VH, Lin X, Murchison HA, Peloquin JM, Woodbury NW and Allen JP (1992a) Changes in the oxidation potential of the bacteriochlorophyll dimer due to hydrogen bonds in reaction centers from *Rhodobacter sphaeroides*. In: Murata N (ed) Research in Photosynthesis, Vol. 1, pp 377–380. Kluwer Academic Publishers, Dordrecht

Williams JC, Alden RG, Murchison HA, Peloquin JM, Woodbury NW and Allen JP (1992b) Effects of mutations near the bacteriochlorophylls in reaction centers from *Rhodobacter sphaeroides*. Biochemistry 31: 11029–11037

Williams JC, Woodbury NW, Taguchi AKW, Peloquin JM, Murchison HA, Alden RG and Allen JP (1992c) Mutations that affect the donor midpoint potential in reaction centers from *Rhodobacter sphaeroides*. In: Breton J and Vermeglio A (eds) The Photosynthetic Bacterial Reaction Center II, Structure, Spectroscopy and Dynamics, pp 25–31. Plenum Press, New York

Woodbury NWT and Parson WW (1984) Nanosecond fluorescence from isolated photosynthetic reaction centers of *Rhodopseudomonas sphaeroides*. Biochim Biophys Acta 767: 345–361

Woodbury NW and Parson WW (1986) Nanosecond fluorescence from chromatophores of *Rhodopseudomonas sphaeroides* and *Rhodospirillum rubrum*. Biochim Biophys Acta 850: 197–210

Woodbury NW, Becker M, Middendorf D and Parson WW (1985) Picosecond kinetics of the initial photochemical electron-transfer reaction in bacterial photosynthetic reaction centers. Biochemistry 24: 7516–7521

Woodbury NW, Parson WW, Gunner MR, Prince RC and Dutton PL (1986) Radical-pair energetics and decay mechanisms in reaction centers containing anthraquinones, naphthoquinones or benzoquinones in place of ubiquinone. Biochim Biophys Acta 851: 6–22

Woodbury NW, Peloquin JM, Alden RG, Lin X, Lin S, Taguchi AKW, Williams JC and Allen JP (1994) Relationship between thermodynamics and mechanism during photoinduced charge separation in reaction centers from *Rhodobacter sphaeroides*. Biochemistry 33: 8101–8112

Wraight CA and Clayton RK (1973) The absolute quantum efficiency of bacteriochlorophyll photooxidation in reaction centres of *Rhodopseudomonas sphaeroides*. Biochim Biophys Acta 333: 246–260

Xiao W, Lin S, Taguchi AKW and Woodbury NW (1994) Femtosecond pump-probe analysis of energy and electron transfer in photosynthetic membranes of *Rhodobacter capsulatus*. Biochemistry 33: 8313–8322

Zinth W, Sander M, Dobler J, Kaiser W and Michel H (1985) In: Michel-Beyerle M-E (ed) Antennas and Reaction Centers in Photosynthetic Bacteria, pp 97–102. Springer-Verlag, Berlin and New York

Chapter 25

Theoretical Analyses of Electron-Transfer Reactions

William W. Parson
Department of Biochemistry, University of Washington, Seattle, WA 98195, USA

Arieh Warshel
Department of Chemistry, University of Southern California, Los Angeles, CA 90007, USA

Summary

Photosynthetic bacterial reaction centers provide a rich territory for exploring the nature of biological electron-transfer reactions. The crystal structures of reaction centers from two species of bacteria are known, and the structures can be modified by site-directed mutagenesis or by substituting other pigments for the natural bacteriochlorophylls, bacteriopheophytins or quinones. Femtosecond laser excitation pulses of various wavelengths can be used to prepare the initial excited state in a variety of conditions, and the kinetics of the subsequent charge separation and recombination reactions can be measured with extremely high time resolution. Measurements can be made under an extraordinarily wide range of conditions, including low temperatures and the presence of external electrical or magnetic fields. Further, there are many intriguing observations to explain. Several of the reactions occur extremely rapidly and become even faster with decreasing temperature; electron transfer along the active 'L' branch of the pigments is much faster than that along the 'M' branch; and seemingly drastic changes in the structure sometimes have little effect on the directionality or rates of the reactions. What is it, then, that determines the speed and temperature dependence of an intermolecular electron-transfer reaction in a protein?

This chapter discusses electron-transfer reactions from a theoretical perspective. Beginning with the picture provided by simple time-dependent perturbation theory, we first set down general expressions that relate electron-transfer rate constants to the energies of the vibrational states of the reactants and products. We then

R. E. Blankenship, M. T. Madigan and C. E. Bauer (eds): Anoxygenic Photosynthetic Bacteria, pp. 559–575.

consider some of the phenomenological expressions that can be used to describe electron transfer under various simplifying assumptions. These expressions range from the classical Marcus equation to the quantum mechanical equation of Kubo and Toyozawa. Many of the expressions have been applied in discussions of reaction centers and have been used to fit experimental data on rate constants as functions of temperature or of the overall free energy change in the reaction. We then turn to microscopic calculations of electron-transfer parameters on the basis of the actual structure of the protein. We show how molecular-dynamics simulations using potential energy functions allow one to calculate free-energy changes and reorganization energies and to evaluate the vibrational modes that are coupled to a reaction. These parameters can be used in several types of dynamic simulations of electron transfer, including fully quantum mechanical density-matrix treatments that incorporate vibrational relaxations of intermediate or product electronic states. The chapter ends with a brief survey of microscopic calculations of electronic coupling factors.

I. Introduction

Imagine a system consisting of an electron donor (D) and acceptor (A) that are very far apart. Let's designate by α the state of the system in which D is reduced and A oxidized, and by β the state in which D is oxidized and A reduced. We'll neglect electronic spins, which give rise to singlet and triplet substates of β, and for the moment we'll also neglect vibrational substates of α and β as well as the interactions of the molecules with their surroundings. We can represent states α and β by wavefunctions ψ_α and ψ_β, with $\psi_\alpha = \psi_D \psi_A$ and $\psi_\beta = \psi_{D^+} \psi_{A^-}$. As long as A and D are too far apart to interact, transitions between states α and β do not occur. In this case, ψ_α and ψ_β are eigenfunctions of the Hamiltonian operator, H_0, and either wavefunction satisfies the Schrödinger equation, $H_0 \psi = i\hbar \partial\psi / \partial t$.

Now suppose we bring D and A closer together so that the electrical fields from electrons on each molecule begin to contribute to the potential energies of the electrons on the other. These interactions introduce a new term to the Hamiltonian, $H_{\alpha\beta}$, which makes our original solutions to the Schrödinger equation no longer satisfactory. ψ_α and ψ_β now are *diabatic* states that do not diagonalize the complete Hamiltonian. However, if the interactions between D and A are still relatively weak, reasonably satisfactory wavefunctions can be written in the form of linear combinations of ψ_α and ψ_β, with coefficients (C_α and C_β) that reflect the extent to which the state of the system resembles α or β:

$$\psi = C_\alpha \psi_\alpha + C_\beta \psi_\beta \tag{1}$$

When this expression for ψ is inserted in Schrödinger equation along with the full Hamiltonian $H_0 + H_{\alpha\beta}$ (see, e.g., Merzbacher, 1970; Atkins, 1983), the coefficients C_α and C_β are found to oscillate with time. The frequency of the oscillation depends on the *interaction matrix element* or *electronic coupling factor*, $H_{\beta\alpha}$, which is the integral

$$H_{\beta\alpha} = \langle \psi_\beta | H_{\alpha\beta} | \psi_\alpha \rangle \tag{2}$$

The nomenclature $\langle b|O|a \rangle$ here means that operator O operates on function a, the result is multiplied by the complex conjugate of function b, and the product is integrated over all space. The amplitude of the oscillations of C_α and C_β depends on both $H_{\beta\alpha}$ and the magnitude of the energy difference $\Delta E_{\beta\alpha}$ between states α and β, and it peaks sharply when $\Delta E_{\beta\alpha}$ goes to zero. If we start out with molecule D reduced ($C_\alpha = 1$ and $C_\beta = 0$), the probability of finding that an electron has moved to A ($C_\beta{}^* C_\beta$, or $|C_\beta|^2$) will begin to grow with time according to the expression

$$\left|C_\beta(t)\right|^2 = 2\left|H_{\beta\alpha}\right|^2 \frac{1-\cos\left[\Delta E_{\beta\alpha} t / \hbar\right]}{\left(\Delta E_{\beta\alpha}\right)^2} \quad \text{(for } t \approx 0\text{)} \tag{3}$$

The initial rate of the transition is obtained by differentiating Eq. (3) with respect to time:

$$\frac{\partial}{\partial t}\left|C_\beta(t)\right|^2 = (2/\hbar)\left|H_{\beta\alpha}\right|^2 \frac{\sin\left[\Delta E_{\beta} t / \hbar\right]}{\Delta E_{\beta\alpha}} \quad \text{(for } t \approx 0\text{)} \tag{4}$$

Abbreviations: B_L and B_M – accessory bacteriochlorophylls on the L (photochemically active) and M (inactive) sides of the purple bacterial reaction center; H_L and Q_A – bacteriopheophytin and quinone that act as the initial electron acceptors; P – bacteriochlorophyll dimer that serves as the primary electron donor; *Rb.* – *Rhodobacter; Rp.* – *Rhodopseudomonas*

In general, an electron-transfer reaction involves a variety of nuclear vibrational and rotational substates of the product electronic state β, so $\Delta E_{\beta\alpha}$ can take on different values. Interactions with the surroundings broaden the distribution of energies by causing the initial product to decay into more stable forms. To obtain an expression for the electron-transfer rate constant (k_{et}), we therefore must integrate Eq. (4) over a range of energies. However, the term $\sin[\Delta E_{\beta\alpha}t/\hbar]/\Delta E_{\beta\alpha}$ drops off very rapidly as $\Delta E_{\beta\alpha}$ departs from zero.

The contributions of transitions to various nuclear substates can be assessed by writing the time-dependent wavefunction of the system as a sum of products of the electronic wavefunctions (ψ_α and ψ_β) and nuclear wavefunctions (χ_i and χ_f):

$$\Psi = \sum_i C_{i(\alpha)}\psi_\alpha\chi_i + \sum_f C_{f(\beta)}\psi_\beta\chi_f \quad (5)$$

The matrix elements for the individual vibronic transitions then are given by

$$H_{f(\beta)i(\alpha)} = \langle\psi_\beta\chi_f \,|\, H_{\alpha\beta} \,|\, \psi_\alpha\chi_i\rangle \approx \langle\chi_f \,|\, \chi_i\rangle H_{\beta\alpha} \quad (6)$$

The square of the overlap integral of a pair of nuclear wavefunctions, $|\langle\chi_f|\chi_i\rangle|^2$, is termed a *Franck-Condon factor*. If the reacting system is in thermal equilibrium with its surroundings, summing up the contributions of all the possible vibrational transitions and integrating Eq. (4) over the widths of the vibrational energy levels gives

$$k_e = (2\pi/\hbar)|H_{\beta\alpha}|^2 \sum_i \frac{e^{-E_i/k_BT}}{Z} \sum_f |\langle\chi_f|\chi_i\rangle|^2 \delta_{E_f,E_i} \quad (7)$$

where $Z = \sum_i e^{-E_i/k_BT}$

and k_B is the Boltzmann constant. The factor $e^{-E_i/k_BT}/Z$ weights the contribution from vibrational level i of α according to the population of this level at temperature T. The delta function δ_{E_f,E_i} is 1 if the energy of level f of β (E_f) is the same as that of α's level i (E_i), and zero otherwise.

Electron transfer thus can occur at an appreciable rate only if the reactants have a thermally accessible nuclear state whose energy is the same as that of a nuclear state of the products, and if the Franck-Condon factor for this combination of states is not zero. The semiclassical statement of these conditions is that electron transfer occurs at points where the potential-energy surfaces of the reactants and products intersect (Fig. 1). It is only at these intersections that the system can move from the reactant to the product surface while conserving both energy and momentum. The electron-transfer rate depends on the frequency at which random structural fluctuations of the reactants and their surroundings bring the system to such crossing points, and on the strength of the coupling factor that then can induce transitions to the product state. If $H_{\beta\alpha}$ is relatively small, so that the

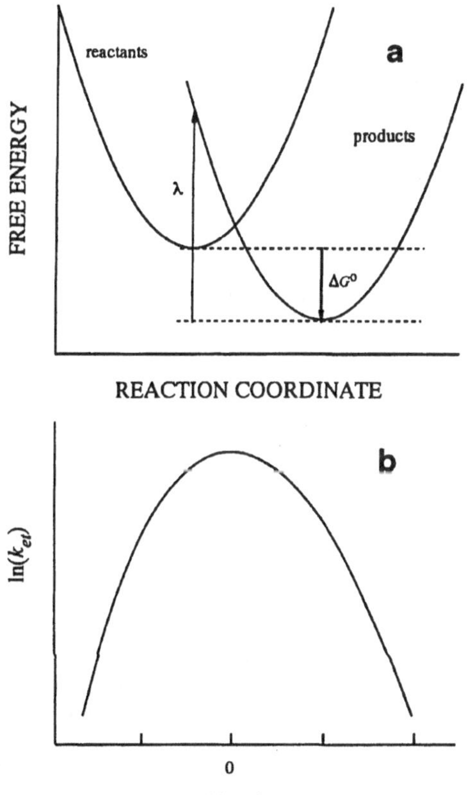

Fig. 1. (*a*) A schematic depiction of the free-energy surfaces for the reactant and product states in an electron-transfer reaction. The abscissa is the reaction coordinate (x). Electron transfer can occur when structural fluctuations bring the system to a region where the reactant and product surfaces intersect. The free energy change (ΔG^o) and the reorganization energy of the reaction (λ) are indicated. (*b*) According to the Marcus Eq. (9), the rate of an electron-transfer reaction is maximal when $\Delta G^o = -\lambda$. The two free-energy surfaces then intersect at the bottom of the reactant surface.

system usually passes through the crossing point without evolving into the products, the reaction is said to be *nonadiabatic*. Most of the electron-transfer steps in photosynthetic reaction centers probably are nonadiabatic, with the possible exception of the initial reduction of bacteriopheophytin H_L. If $H_{\beta\alpha}$ is large enough so that a transition to the product state occurs virtually every time the system reaches a crossing point, the reaction is termed *adiabatic*. In this case, the approximations inherent in the analysis outlined above become severe, and the problem usually is better treated by starting with wavefunctions that diagonalize the full Hamiltonian (see Warshel and Hwang, 1986).

The electronic coupling factor $H_{\beta\alpha}$ depends on the overlap of the molecular orbitals of the electron donor and acceptor, and so falls off approximately exponentially with the distance between the two molecules (Moser et al., 1992). It can be increased by *superexchange* coupling through virtual intermediate states. In superexchange, the virtual intermediate state (γ) connects the initial and final states (α and β) by mixing quantum mechanically with both of them. The strength of this mixing depends, in part, on the energy difference $\Delta E_{\gamma\alpha}$ between γ and α (or β) at the point where $\Delta E_{\beta\alpha} = 0$. If $\Delta E_{\gamma\alpha}$ is large relative to the interaction matrix elements $H_{\gamma\alpha}$ and $H_{\gamma\beta}$, then

$$H_{\beta\alpha} \approx 2H_{\gamma\alpha} H_{\gamma\beta} / \Delta E_{\gamma\alpha} \tag{8}$$

II. Phenomenological Descriptions of Electron-Transfer Reactions

Attempts to apply Eq. (7) to an actual system such as the reaction center encounter both conceptual and practical problems. For one thing, we have assumed that the reactants are in thermal equilibrium with their surroundings, and that the various vibrational modes of the system are excited stochastically, i.e., with random phases. Studies with femtosecond excitation flashes indicate that such an equilibrium may not be attained during the short lifetime of P* in the reaction center (Vos et al., 1993). Second, we have neglected the matter of what happens to the product state β after it forms. This amounts to assuming that the vibrational substates of β that are populated in the reaction relax by transferring energy to the surroundings rapidly enough so that back-reactions to α are insignificant, but not so rapidly as to give a large uncertainty-broadening of their energies. Perhaps most importantly, Eq. (7) leaves us with the practical problem of finding the energies and Franck-Condon factors for all the vibrational modes of the system. This is not yet feasible for a system as large as the reaction center, or indeed for any protein. Let us therefore consider some possible simplifications of the problem.

In a classic series of papers, Marcus (1956, 1965; reviewed by Marcus and Sutin, 1985) showed that the electron-transfer rate constant often can be related to the overall free energy change in the reaction, ΔG^o, and a phenomenological parameter termed the *reorganization energy*. The reorganization energy (λ) is the change in free energy associated with distorting the system from the mean nuclear configuration of the reactants to that of the products, or (equivalently, if the curvatures of the reactant and product free-energy surfaces are the same) from the mean nuclear configuration of the products back to that of the reactants. Marcus assumed that the free energies of the reactants and products were harmonic (quadratic) functions of the displacement of the system along the reaction coordinate (see Fig. 1a). The relationship between the reaction coordinate and the actual nuclear coordinates of the system was not clearly defined initially, but such a definition was not necessary at this level of the theory. (More precise definitions of the reaction coordinate and the free energy surfaces are given in Section III.) The assumption of classical, harmonic free-energy surfaces led to the expression

$$k_{et} = \frac{2\pi}{\hbar} \frac{|H_{\beta\alpha}|^2}{\sqrt{4\pi\lambda k_B T}} \exp\left\{-\frac{(\Delta G^o + \lambda)^2}{4\lambda k_B T}\right\} \tag{9}$$

Here as in Fig. 1, λ is defined to be a positive quantity, and $\Delta G^o = G^o_{products} - G^o_{reactants}$; many authors use the reverse convention with regard to ΔG^o.

Equation (9) has several important implications. First, in order for k_{et} to be greater than zero λ must be nonzero, except possibly in the event that ΔG^o also is zero. This means that there must be a difference in nuclear geometry between the equilibrium structures of the reactants and products. Second, the activation free energy for the reaction, $\Delta G^{\ddagger}$, is given by $(\Delta G^o+\lambda)^2/4\lambda$, and thus depends quadratically on $(\Delta G^o+\lambda)$. If $\Delta G^o=-\lambda$, $\Delta G^{\ddagger}$ should be zero and the rate should be essentially independent of temperature. Further, a plot of ln(k) vs ΔG^o should go through a maximum when $\Delta G^o = -\lambda$ (Fig. 1b). Making the reaction more exothermic should speed up the reaction in the region where $\Delta G^o > -\lambda$, but (perhaps more surprisingly) should slow down the reaction in the 'inverted region' where $\Delta G^o < -\lambda$. Marcus' prediction concerning the inverted region lacked experimental support for many years, but the qualitative trend eventually was observed in studies of covalently linked donor-acceptor pairs (Miller et al., 1984; Wasielewski et al., 1985; Closs and Miller, 1988).

In the reaction center, ΔG^o for the reactions $P^*H_L \rightarrow P^+H_L^-$ and $P^+Q_A^- \rightarrow PQ_A$ has been varied by substituting other quinones for ubiquinone as Q_A (Gunner et al., 1986; Gunner and Dutton, 1989), by external electrical fields (Popovic et al., 1986; Feher et al., 1988; Boxer et al., 1989; Franzen et al., 1990; Franzen and Boxer, 1993), and by genetic modifications (Finkele et al., 1990; Nagarajan et al., 1990, 1993; Chan et al., 1991; Williams et al., 1992a, 1992b). Phenomenological fits of the rate constant to the Marcus equation have been used to extract values of λ and $|H_{\beta\alpha}|$. The notion that $\Delta G^o \approx -\lambda$ has been invoked to account for the observation that the activation energies for these reactions are zero. (Both reactions actually speed up somewhat with decreasing temperature.) However, this analysis fails to explain why the $P^+Q_A^- \rightarrow PQ_A$ reaction still speeds up with decreasing temperature even if ΔG^o is altered substantially (Gunner et al., 1986; Gunner and Dutton, 1989; Nagarajan et al., 1993). Also, although the $P^+Q_A^- \rightarrow PQ_A$ reaction does slow down when ΔG^o is made more negative in the inverted region, the decrease is not as pronounced as Eq. (9) predicts. As we will discuss below, this discrepancy probably reflects the fact that the classical Marcus equation neglects quantum mechanical nuclear tunnelling (Warshel et al., 1989a,b).

Several authors have developed expressions that incorporate quantum mechanical effects and thus should be better suited than Eq. (9) for treating reactions at low temperatures. Probably the most general expression is due originally to Lax (1952) and Kubo and Toyozawa (1955), who showed that the thermally-weighted Franck-Condon factors in Eq. (7) can be expressed analytically if the vibrational modes are assumed to be harmonic and to have the same frequencies in the reactants and products. With these assumptions (and with the same assumptions made above concerning thermal equilibrium and irreversibility), the rate constant takes the form of a Fourier transform:

$$k_{et} = |(H_{\beta\alpha}/\hbar)|^2 \int_{-\infty}^{\infty} \exp\{(i/\hbar)\langle\Delta E_{\beta\alpha}\rangle_\alpha t + \gamma_Q(t)\}dt \tag{10}$$

with

$$\gamma_Q(t) = \frac{1}{2}\sum_j \Delta_j^2[(\bar{n}_j+1)e^{i\omega_j t} + \bar{n}_j e^{-i\omega_j t} - (2\bar{n}_j+1)]$$

Here $\langle\Delta E_{\beta\alpha}\rangle_\alpha$ is the mean value of $\Delta E_{\beta\alpha}$ when the system is in the reactant state α, ω_j is the frequency of vibrational mode j; and $\bar{n}_j = 1/[\exp(\hbar\omega_j/k_BT)-1]$. Δ_j is the difference between the mean nuclear coordinates for mode j in states α and β, which is defined to be dimensionless by setting $\Delta_j^2 = 2\lambda_j/\hbar\omega_j$. The problem of choosing appropriate values of the ω_j and Δ_j for an actual system such as the reaction center may appear to be intractable because the total number of vibrational modes available to a protein of this size can exceed 10^5. However, the assumption of harmonic vibrational modes simplifies this task significantly. We will return to this point below.

In the multimode model of Kubo and Toyozawa, the overall reorganization energy is

$$\lambda = \frac{1}{2}\sum_j \Delta_j^2\hbar\omega_j \tag{11}$$

A particular vibrational mode contributes to λ, and to the rate constant, only if Δ for that mode is nonzero. Vibrational modes that meet this test are *coupled* to the reaction. The dimensionless factor $\Delta_j^2/2$ is sometimes referred to as the Huang-Rhys factor. Motions along coordinates for which Δ is zero affect the potential energies of both the reactant and the product states equally, without helping to bring the system to a crossing point.

The quantum mechanical Franck-Condon factors in Eq. (10) allow nuclear transitions that are forbidden classically. Such nuclear tunnelling to excited vibrational levels of the product electronic state becomes particularly important when $\langle\Delta E_{\beta\alpha}\rangle_\alpha$ is more negative than λ, and this makes electron transfer in the inverted region faster than predicted by the classical Eq. (9) (Warshel et al., 1989a,b).

Alden et al. (1992) have derived an expression that gives a good approximation to Eq. (10) but avoids the need to evaluate the Fourier transform. A possible further simplification is to treat the high-frequency vibrational modes (e.g., those with $\hbar\omega_j \gtrsim 50\ cm^{-1}$) quantum mechanically, and the lower-frequency modes classically. In principle, this should be satisfactory as long as $k_BT >> \hbar\omega$ for the low-frequency modes. Another simplification is to treat all of the vibrational modes (or all the high-frequency modes) as a single mode with effective values of ω, Δ and $\bar{n}$. Following this approach, Levich (1966), Jortner (1976, 1980), Kestner et al. (1974) and Efrima and Bixon (1976) obtained expressions of the form

$$k_{et} = (2\pi \mid H_{\beta\alpha} \mid^2 / \omega\hbar^2) \exp[-\Delta^2(\bar{n} + 1/2)] \times$$

$$I_p \{\Delta^2[\bar{n}(\bar{n}+1)]^{1/2}\}[(\bar{n}+1)/\bar{n}]^{p/2} \quad (12)$$

where

$$p = \mid \langle \Delta E_{\beta\alpha} \rangle_\alpha \mid / \hbar\omega$$

and

$$I_p\{y\} = (y/2)^p \sum_{q=0}^{\infty} (y^2/4)^q / q!(p+q)!$$

Equation (12) reduces to the classical Eq. (9) in the high-temperature limit, as does Eq. (10). Kakitani and Kakitani (1981) have considered the case that the effective frequency ω differs in the reactants and products. The result in this case is formally the same as Eq. (12) except that a temperature-dependent free energy change (ΔG^o) replaces the energy gap $\langle \Delta E_{\beta\alpha} \rangle_\alpha$. Sarai (1979, 1980) presented a treatment that retained a small number of discrete modes instead of just one effective mode.

Small et al. (1992) have developed expressions that collect the vibrational modes into one or two effective modes but consider a heterogeneous sample with a Gaussian distribution of energy differences between the reactant and product states. If the mean energy gap is $\langle \Delta E_{\beta\alpha} \rangle_\alpha$ as before, and the width of distribution of $\Delta E_{\beta\alpha}$ is $2D$,

$$k_{et} = \frac{2\pi \mid H_{\beta\alpha} \mid^2}{\hbar[2\pi(D^2 + \lambda W)]^{1/2}} \exp\left\{-\frac{(\langle \Delta E_{\beta\alpha} \rangle_\alpha + \lambda)^2}{2(D^2 + \lambda W)}\right\} \quad (13)$$

where $W = \hbar\omega\coth(\hbar\omega/2k_BT)$. Besides including an energy distribution, this expression differs from the Marcus Eq. (9) primarily in the substitution of $\hbar\omega\coth(\hbar\omega/2k_BT)$ for k_BT, which incorporates the main quantum mechanical effect of the zero-point vibrational energy (Hopfield, 1974). For the initial electron-transfer step in reaction centers, Small et al. (1992) argue that D^2 is not large enough relative to λW to have much effect on the kinetics.

As mentioned above, Eqs. (7) through (13) all assume that the reacting species are in thermal equilibrium with their surroundings. Marcus (1988), Almeida and Marcus (1990) and Marcus and Almeida (1990) have considered the case of a reaction that proceeds through a very short-lived intermediate state, which reacts in a second step before it has much opportunity to relax. They refer to such a process as a 'nonsuperexchange coherent' mechanism, and derive phenomenological expressions that take into account vibrational coherence between the intermediate and final states. Kitzing and Kuhn (1990) have derived expressions for an electron-transfer reaction that is adiabatic in nature but is limited by the rate of thermal relaxation.

III. Microscopic Calculations of Electron-Transfer Parameters

Although the phenomenological expressions discussed in the previous section provide a great deal of insight into the nature of electron-transfer reactions, the significance of fitting experimental kinetic results to any of these equations is questionable because such fits rarely determine the parameters uniquely. Equivalent fits usually can be obtained by choosing various combinations of ΔG^o and λ or various vibrational frequencies, or by making some of the parameters temperature dependent. Further, these expressions do not tell us how to *predict* the rate of a reaction for a given molecular system, because they do not specify how to calculate the values for any of the parameters on the basis of the system's microscopic structure. However, new computational approaches and advances in computer power are making such calculations possible for increasingly large and complex systems. In the following sections, we will describe microscopic calculations of free energy changes, reorganization energies, and the frequencies and displacements of vibrational modes that are coupled to electron-transfer processes in photosynthetic reaction centers. We also will indicate how the same computational approaches allow one to simulate electron-transfer dynamics using either semiclassical or quantum mechanical formalisms. For more detailed reviews that also discuss electron-transfer reactions in solution, see Warshel and Parson (1991) and Parson and Warshel (1993).

A. Electrostatic Energies and the Importance of a Complete Description of the Micro-environment

As an illustration of a microscopic calculation of ΔG^0 for an electron-transfer reaction, consider the energy of $P^+B_L^-$ in the reaction center of *Rhodopseudomonas* (*Rp.*) *viridis*. Attempts to detect $P^+B_L^-$ experimentally have given controversial results, and it presently is unclear whether this state is a real intermediate in electron transfer from P* to H_L, or whether it simply mediates the electronic coupling of P* and H_L by superexchange. Accurate calculations of the energy of $P^+B_L^-$ should help to resolve this issue.

The free energy change associated with moving an electron from P to B_L is:

$$\Delta G^o = \Delta E^{\pm} + \Delta\Delta G_{sol}(P^+B^-)^{pro} \quad (14)$$

where $\Delta E^{\pm}$ is the energy of forming the $P^+B_L^-$ ion pair in the gas phase and $\Delta\Delta G_{sol}(P^+B^-)^{pro}$ represents the difference between the solvation free energies of P^+B^- and PB by their surroundings in the protein $[\Delta\Delta G_{sol}(P^+B^-)^{pro} = \Delta G_{sol}(P^+B^-)^{pro} - \Delta G_{sol}(PB)^{pro}]$.

The evaluation of the gas-phase energy $\Delta E^{\pm}$ is immediately problematic. In principle, $\Delta E^{\pm}$ could be obtained by gas-phase quantum mechanical calculations, and this approach has been taken by Scherer and Fischer (1989a,b), Thompson and Zerner (1991) and Marchi et al. (1993). However, such calculations are not yet able to give accurate results for the ionization energies and electron affinities of such large molecular systems as P and B. At present, it seems more reliable to use a thermodynamic cycle that incorporates experimental information on the midpoint redox potentials of P and B_L in the protein, or where necessary, of bacteriochlorophyll in solution (Creighton et al., 1988; Parson et al., 1990). This can be done by writing:

$$\Delta E^{\pm} = \Delta E^{vac} + \Delta V_{QQ} \quad (15a)$$

$$\Delta E^{vac} = -\mathcal{F}[E_m^B - E_m^P] - \Delta\Delta G_{sol}(P^+ + B^-)^{ref} \quad (15b)$$

Here ΔE^{vac} is the energy of forming $P^+B_L^-$ with P and B_L separated to infinite distance in a vacuum, ΔV_{QQ} is the calculated change in electrostatic interaction between P and B_L at their actual positions in the reaction center (still without any dielectric shielding by the protein or the surroundings), $\mathcal{F}$ is the Faraday constant, E^B_m and E^P_m are the redox potentials of the reactants, and $\Delta\Delta G_{sol}(P^+ + B^-)^{ref}$ is the calculated change in solvation energy for forming the individual ions under the reference conditions of the redox measurements $[\Delta\Delta G_{sol}(P^+ + B^-)^{ref} = \Delta G_{sol}(P^+ + B^-)^{ref} - \Delta G_{sol}(P + B)^{ref}]$. The thermodynamic cycle composed of these steps is shown in Fig. 2.

Combining (14) and (15) gives

$$\Delta G^o = -\mathcal{F}[E_m^B - E_m^P] + \Delta V_{QQ} + \Delta\Delta G_{sol}(P^+B^-)^{pro} - \Delta\Delta G_{sol}(P^+ + B^-)^{ref} \quad (16)$$

This expression circumvents the problem of calculating quantum mechanical gas-phase energies

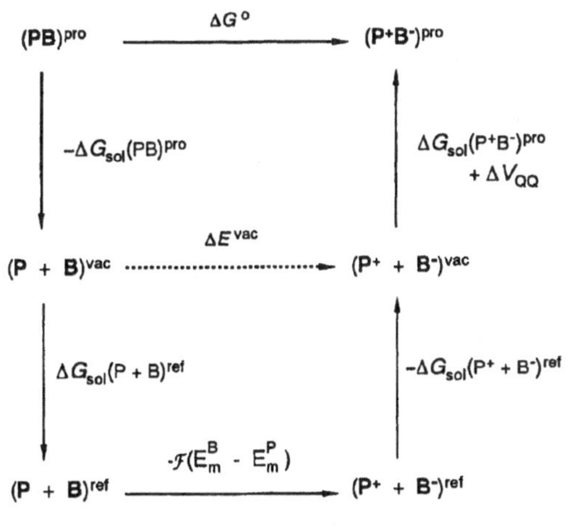

Fig. 2. The free energy change for electron transfer from P to B_L in the reaction center (ΔG^o) can be obtained from a thermodynamic cycle that includes a combination of experimentally measured midpoint redox potentials of P and B_L (E_m values), calculated solvation energies (ΔG_{sol}), and the calculated direct Coulombic interaction between P^+ and B_L^- (ΔV_{QQ}). Superscript *pro* refers to the environment in the reaction center protein; *vac*, to a vacuum; *ref*, to the reference conditions of the redox titrations. Using the redox potentials allows one to bypass quantum calculations of gas-phase energy differences ($\Delta E^{\pm}$ or ΔE^{vac}); in addition, the final result depends on differences between the solvation energies for the protein and the reference conditions, rather than the absolute energies. In practice, because the E_m of B_L has not been measured experimentally, it is necessary to use the E_m and ΔG_{sol} for monomeric bacteriochlorophyll in solution (see Parson et al., 1990).

directly. Calculations of solvation energies, though subject to difficulties we will discuss below, are presently more reliable and are necessary anyway for evaluation of the term $\Delta\Delta G_{sol}(P^+B^-)^{pro}$ in Eq. (14). In addition, Eq. (16) reduces the problem largely to a calculation of the *difference* between the solvation energy of $P^+B_L^-$ and the sum of the solvation energies for P^+ and B_L^-. This is helpful because some types of errors tend to cancel out in the result.

Given the molecular structure and rudimentary descriptions of the molecular orbitals of P and B_L, ΔV_{QQ} can be obtained straightforwardly by summing the interactions of the partial charges on the atoms of the two molecules, using a dielectric constant of 1. Evaluating the solvation energies requires more discussion. One approach here would be to use a macroscopic reaction-field treatment. The solvation energy of $P^+B_L^-$, for example, could be expressed in the form:

$$\Delta\Delta G_{sol}(P^+B^-)^{pro} = -C\frac{|\mu|^2}{b^3}(\varepsilon-1)/(2\varepsilon+1) \qquad (17)$$

where $|\mu|$ is the dipole moment of the $P^+B_L^-$ dipole, b is the effective radius of the protein cavity around the dipole, ε is the effective dielectric constant of the protein, and C = 332 kcal mol^{-1} Å (electronic charge)$^{-2}$. However, it is not clear what value of ε to use inside the protein (Warshel and Åqvist, 1991), nor how to choose the appropriate value of b (Luzhkov and Warshel, 1991). Since $\Delta\Delta G_{sol}(P^+B^-)^{pro}$ is proportional to b^{-3}, Eq. (17) cannot be used to obtain a reliable value for ΔG^o. In addition, resort to such a macroscopic treatment would dash any hopes of exploring how microscopic structural features give rise to differences in electron-transfer rates on the L and M sides of the reaction center.

The three main options are molecular-dynamics simulations with an all-atom model (Warshel et al., 1986; King and Warshel, 1989; Beveridge and DiCapua, 1989), the protein-dipole-Langevin-dipole (PDLD) model (Warshel and Russell, 1984; Lee et al., 1993) and the discretized-continuum model (Sharp and Honig, 1990). In the PDLD model, the solvation energy is broken into a sum of terms for the changes in (*i*) direct electrostatic interactions of P and B with the protein ($\Delta V_{Q\mu}$), (*ii*) interactions with induced dipoles in the protein (ΔV_{ind}), (*iii*) interactions with mobile water (ΔV_{H_2O}) and/or phospholipid side chains (ΔV_{memb}), (*iv*) interactions with solvent ions (ΔV_{ions}), and (*v*) interactions with solvent outside the sphere of protein that is treated microscopically (ΔV_{Born}):

$$\Delta\Delta G_{sol}(P^+B^-)^{pro} = \Delta V_{Q\mu} + \Delta V_{ind} + \Delta V_{H_2O} + \Delta V_{memb} + \Delta V_{ions} + \Delta V_{Born} \qquad (18)$$

The direct interactions ($\Delta V_{Q\mu}$) are evaluated using a dielectric constant of 1; ΔV_{ind}, ΔV_{H_2O}, ΔV_{memb} and ΔV_{ions} are evaluated by iterative procedures that seek self-consistent solutions for the potentials, charges and induced dipoles at all atoms of the pigments and protein and on a grid of points surrounding the protein. ΔV_{Born} is treated with macroscopic expressions due to Born and Onsager. The term $\Delta\Delta G_{sol}(P^+ + B^-)^{ref}$ in Eqs. (15b) and (16), which pertains to the redox measurements, includes additional reorganization energies that are obtained as described in the following section.

Calculations of electrostatic energies in proteins can easily go astray as a result of the use of improper boundary conditions or the omission of the solvent or other key components of the system (Warshel and Åqvist, 1991). A particularly important concern is the assignment of charges to ionizable amino acid residues. Calculations that take all the ionizable residues to be in their charged forms, but neglect the effects of mobile solvent molecules in and around the protein (Marchi et al., 1993; Treutlein et al., 1992), lead to a much higher energy for $P^+B_L^-$ than calculations that treat the ionizable groups as neutral or well solvated (Creighton et al., 1988; Parson et al., 1990). In the former treatment, the results also are very sensitive to the size of the system that is included in the analysis. Since most of the ionizable residues are near the surface of the protein including larger fractions of the reaction center brings more of these residues into the system. If the charged groups are not completely solvated, the calculated electrostatic energies will continue to fluctuate until the system contains essentially the entire protein. In our experience, this is an artifact of using an insufficiently complete model. Experimentally, observed interactions between charged groups in proteins usually are relatively weak because the charged atoms are well shielded by solvent or by induced dipoles of the protein (Warshel et al., 1986; Warshel and Åqvist, 1991). Conversely, ionizable groups that are not well solvated by the solution or by other groups in the protein are unlikely to be charged.

The reaction center ordinarily is embedded in a phospholipid bilayer or a belt of detergent molecules (Roth et al., 1989). Yet there may also be many water molecules around the protein. Although some water molecules are seen in the crystal structure, others probably are too mobile to be resolved. In the regions of B_L and B_M, the surface of the protein has grooves that could contain water because they fall well inside the cavity defined by placing straight phospholipid chains around the protein. Including a polarizable solvent in this region lowers the calculated free energy of $P^+B_L^-$ substantially with respect to $P^+H_L^-$. It therefore seems essential to consider the polarizabiltiy of the solvent in any discussion of electrostatic energies in the reaction center, and to use methods with proper boundary conditions (e.g., Lee et al., 1993) that give accurate calculations of absolute pK_a values and related properties of other proteins.

B. Free Energy Surfaces and Reorganization Energies

Microscopic calculations of reorganization energies require evaluating the free energy surfaces of the reactant and product states in terms of the actual structure of the system. This can be done by using semiempirical potential-energy functions of the atomic coordinates ($V(\mathbf{r})$) to treat the vibrations, rotations, and van der Waals and electrostatic interactions of all the atoms within a sphere surrounding the electron carriers. As the microscopic reaction coordinate (X), it is advantageous to choose the energy difference $\Delta V_{\beta\alpha}$ between states α and β ($\Delta V_{\beta\alpha} = V_\beta - V_\alpha$) (Warshel, 1982; Warshel and Parson, 1991). Free energy functions then can be defined as

$$\Delta g_\alpha(X) = -k_B T \ln[P(X)_\alpha] \qquad (19)$$

where $P(X)_\alpha$ is the probability that a system in state α has a value of $\Delta V_{\beta\alpha}$ equal to X. $P(X)_\alpha$ can be obtained by keeping track of $\Delta V_{\beta\alpha}$ as a function of time during a molecular-dynamics simulation (Fig. 3). In such simulations, the atoms in the computer model are initially assigned random kinetic energies corresponding to a selected temperature. The structure then is allowed to fluctuate a series of small time steps according to the classical Newtonian equations of motion (Jorgensen, 1989; Beveridge and DiCapua, 1989; Kollman and Merz, 1990; Warshel, 1991; van Gunsteren and Berendsen, 1990; Straatsma and McCammon, 1992).

A molecular-dynamics trajectory on the potential-energy surface of the initial state (V_α) is unlikely to yield reliable information about the activation energy of a reaction if $\Delta G^\ddagger$ is much larger than $k_B T$, because the probability that the trajectory will reach the crossing point between V_α and V_β ($X = 0$) then is very small. However, this problem can be overcome by free-energy-perturbation and umbrella-sampling procedures (Warshel, 1982; Hwang and Warshel, 1987; Kuharski *et al.*, 1988; King and Warshel, 1990). The idea is to force the system to move through regions of configuration space near $X = 0$. This is accomplished by running trajectories on hybrid potential surfaces of the form

$$V_m = (1 - m)V_\alpha + mV_\beta \qquad (0 \leq m \leq 1) \qquad (20)$$

From the values of $\Delta V_{\beta\alpha}$ obtained from simulations with different values of m, one can determine Δg_α and Δg_β for each value of X (Warshel, 1982; Hwang and Warshel, 1987; King and Warshel, 1990; Kuharski *et al.*, 1988). The free-energy functions $\Delta g_\alpha(X)$ and $\Delta g_\beta(X)$ are the microscopic equivalents of the Marcus parabolas shown in Fig. 1a, and the reorganization energy (λ) can be obtained directly from the Δg functions (Warshel, 1982; Hwang and Warshel, 1987). This approach has been used to obtain $\Delta g_\alpha(X)$, $\Delta g_\beta(X)$ and λ for the formation of $P^+B_L^-$ and $P^+H_L^-$ in mutant and wild-type reaction centers (Creighton et al., 1988; Parson et al., 1991).

Although they are expressed as functions of a single reaction coordinate whose choice may seem odd ($X = \Delta V_{\beta\alpha}$), Δg_α and Δg_β reflect the multi-dimensional, anharmonic fluctuations of the actual structure, at least to the extent that the starting crystal structure and the semiempirical potential energy functions V_α and V_β are reliable. It is important to note that, as in the calculations of ΔG° discussed above, the potential energy functions are not chosen specifically to fit data for a particular electron-transfer reaction of interest; they are parametrized on the basis of a large body of spectroscopic and thermodynamic data for many different systems including both macromolecules and small molecules. Of course, calculations on reaction centers are subject

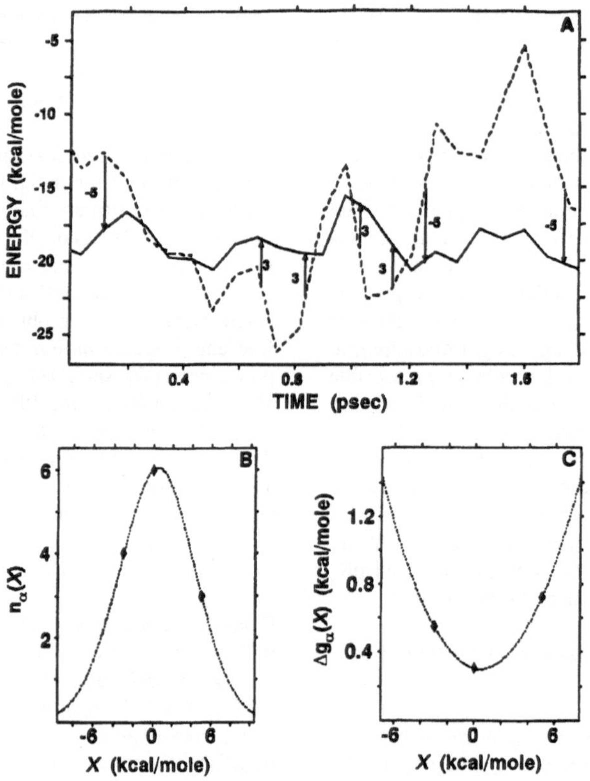

Fig. 3. The relationship between the time-dependent energy gap and the free energy functions for an electron-transfer reaction. (*A*) A typical history of the potential energies V_α (*solid line*) and V_β (*dashed line*) during a molecular-dynamics trajectory on V_α. V_α and V_β fluctuate as a result of the motions of the pigments, protein and solvent. Vertical arrows indicate values of $\Delta V_{\beta\alpha}$ at various points of the trajectory. The number function $n_\alpha(X)$ is obtained by counting the number of times that $\Delta V_{\beta\alpha} = X$ for each value of X. In this example, $n_\alpha(-5) = 3$, $n_\alpha(0) = 6$ and $n_\alpha(3) = 4$. (*B*) A plot of $n_\alpha(X)$. (*C*) The free energy function $\Delta g_\alpha(X)$ is obtained from $n_\alpha(X)$ by Eq. (19). $\Delta g_\beta(X)$ can be obtained in the same manner from a trajectory on V_β. (Reprinted from Warshel and Parson (1991).)

to the same uncertainties discussed in the previous section concerning the charges on ionizable residues and the effects of the surrounding solvent.

C. Semiclassical-Trajectory Simulations

Molecular-dynamics trajectories can provide the information needed for several types of dynamic simulations of electron-transfer reactions. In a *semiclassical-trajectory* or *surface-hopping* simulation (Tully and Preston, 1971; Miller and George, 1972; Warshel, 1982; Warshel and Hwang, 1986), a transition from the reactant to the product state can occur whenever the trajectory reaches an intersection of V_α and V_β. The probability that a system starting in α will be in β at time t is given by $|C_\beta(t)|^2$, with

$$C_\beta(t) = \int_0^t (-i/\hbar) C_\alpha(\tau) H_{\beta\alpha} \times \exp\{((-i/\hbar)\int_0^\tau \Delta V_{\beta\alpha}(\tau')d\tau'\}d\tau \qquad (21)$$

(Warshel, 1982). As in Eqs. (3)–(13) above, $C_\alpha(t)$ is assumed to be ≈ 1, and conversions to the product state are considered to be effectively irreversible. If

$H_{\beta\alpha}$ is small, the average rate constant is

$$k_{et} \approx \lim_{t \to \infty} ((1/t)\langle |C_\beta(t)|^2 \rangle_\alpha) \tag{22}$$

where $\langle\ \rangle_\alpha$ means an average over many trajectories on V_α. The exponent in Eq. (21) is a rapidly oscillating function that contributes to the rate constant only when $\Delta V_{\beta\alpha} \approx 0$. Thus k_{et} can be obtained by running a single long trajectory and collecting the transition probability every time V_α and V_β intersect (Warshel, 1982; Warshel and Hwang, 1986). For harmonic potential surfaces, the high-temperature limit of Eqs. (21) and (22) is identical to the high-temperature limit of the exact quantum mechanical expression (Eq. 10) (Warshel and Hwang, 1986):

$$k_{et} \approx |H_{\beta\alpha}/\hbar|^2 (\pi\hbar^2/k_B T\lambda)^{1/2} \exp[-\Delta V^\ddagger/k_B T]$$

$$\Delta V^\ddagger = (\Delta V^0_{\beta\alpha} + \lambda)^2/4\lambda \tag{23}$$

Here $\Delta V^0_{\beta\alpha}$ is the difference between the average potential energies of states β and α ($\Delta V^0_{\beta\alpha} = \langle V_\alpha \rangle_\beta - \langle V_\alpha \rangle_\alpha$; and the reorganization energy λ is $\langle \Delta V_{\beta\alpha} \rangle_\alpha - \Delta V^0_{\beta\alpha}$. Equation (23) can be viewed as the product of the probability that the system has the energy required in the transition state ($\exp[-\Delta V^\ddagger/k_B T]$), and the average rate of crossing from the reactant to the product surface in this state. If potential energies are replaced by free energies, Eq. (23) is identical to the Marcus Eq. (9).

Equations (21) and (22) provide more than an alternative derivation of Eq. (9) because they also can be used in the high-temperature limit for *anharmonic* systems. For an anharmonic system, the activation energy $\Delta V^\ddagger$ in Eq. (23) is replaced by the activation free energy $\Delta g^\ddagger$ obtained from the Δg_α and Δg_β functions, so that the factor $\exp[-\Delta g^\ddagger/k_B T]$ reflects the actual probability of reaching the transition state. This application is important because we presently have no analytical recipe for quantization of a multidimensional, anharmonic system such as a protein.

Creighton et al. (1988) have described semiclassical-trajectory simulations of *Rhodobacter (Rb.) sphaeroides* reaction centers. Transitions from P* to $P^+B_L^-$ and from $P^+B_L^-$ to $P^+H_L^-$ were evaluated in separate trajectories by means of Eq. (21), and were incorporated into coupled differential equations for the concentrations of P*, $P^+B_L^-$ and $P^+H_L^-$. The results agreed reasonably well with the experimental measurements. $P^+H_L^-$ rose with a time constant of about 3.5 ps, with the build-up of a small amount of $P^+B_L^-$ during the first 1 ps. Figures 4 a and b show the results of a similar simulation (Warshel and Parson, 1991). The autocorrelation function of the energy gap between P* and $P^+B_L^-$ is shown in Fig. 4c.

Molecular-dynamics simulations of the *Rp. viridis* reaction center have been described by Treutlein et al. (1992) and Marchi et al. (1993). Treutlein et al. focused on the structural reorganization that follows the simulated formation of $P^+H_L^-$ from P*. They found that most of the relaxation occurred within 0.2 ps, which is consistent with the rapid decay of the autocorrelation function shown in Fig. 4c. Similar rapid relaxations have been obtained in other simulations of charge-transfer processes in solutions (Warshel and Hwang, 1986; Hwang et al., 1988) and proteins (Churg and Warshel, 1985; Warshel et al., 1988b). Marchi et al. (1993) note that the relaxations of $P^+H_L^-$ also appear to include a much slower component.

D. Quantum Mechanical Simulations and Vibrational Relaxations

Semiclassical-trajectory simulations neglect the contributions to electron-transfer processes from the nuclear tunnelling that can occur even when the energy of the system is not sufficient to reach the transition state. A practical way to examine these classically forbidden transitions is provided by the 'dispersed polaron' model (Warshel and Hwang, 1986; Warshel *et al.*, 1989a,b). This model is based formally on the exact quantum mechanical expression for electron transfer in a multi-dimensional harmonic system (Eq. 10). The fluctuations of the time-dependent energy gap $\Delta V_{\alpha\beta}(t)$ during a molecular-dynamics simulation can be related to the fluctuations that would occur in a multidimensional harmonic system by equating the Fourier transforms of $\Delta V_{\alpha\beta}(t)$. In the Fourier transform of the harmonic model, each vibrational mode contributes a peak at its characteristic frequency (ω_j), with an amplitude proportional to the corresponding displacement (Δ_j). The time-dependent energy gap for the actual system thus provides all the necessary input for Eq. (10) except for $H_{\beta\alpha}$. If the dependence of $H_{\beta\alpha}$ on the fluctuations of the system is known, or is assumed to be negligible, a molecular-dynamics simulation for a single temperature yields all the information needed to

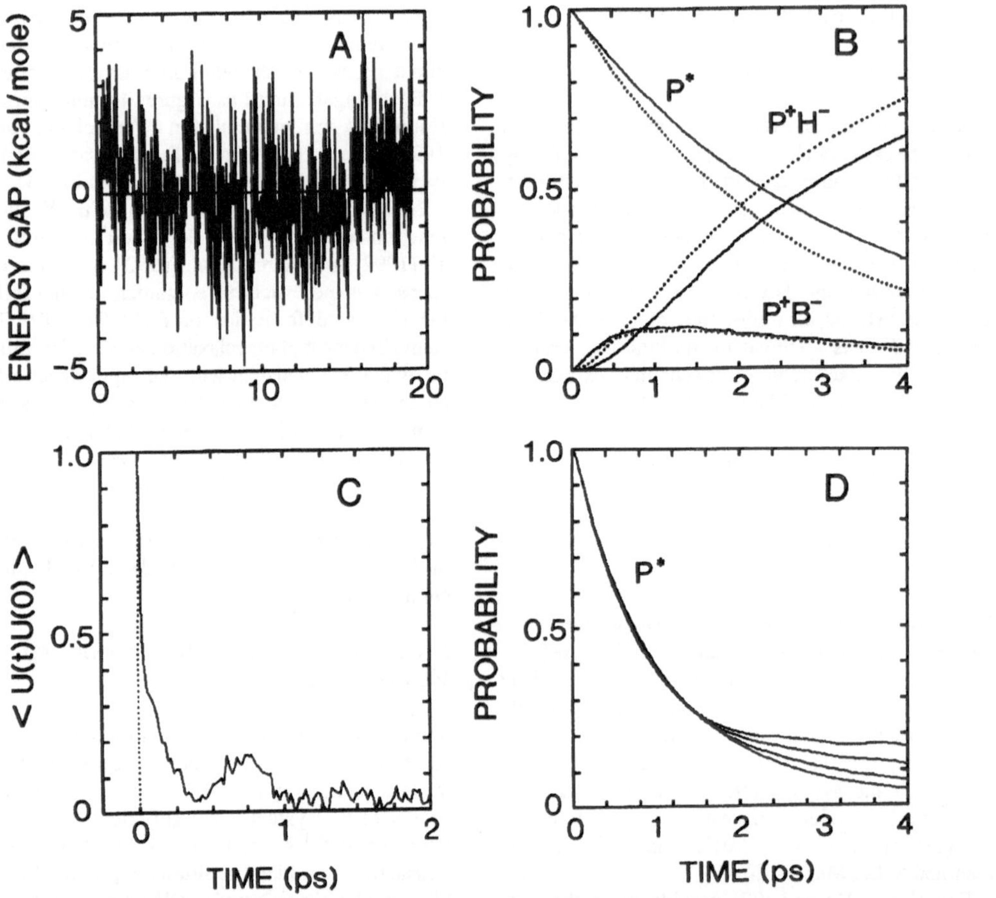

Fig. 4. (*A*) The time-dependent energy gap between P* and $P^+B_L^-$ in *Rb. sphaeroides* reaction centers, calculated during a molecular-dynamics simulation similar to that described by Creighton *et al.* (1988). (*B*) A semiclassical-trajectory simulation of the reaction P* → $P^+H_L^-$, using the energy gap of (*A*) for P* → $P^+B_L^-$ and a similar energy gap from a trajectory on $P^+B_L^-$ for $P^+B_L^- \rightarrow P^+H_L^-$. The electronic couplings were 25 cm^{-1} between P* and $P^+B_L^-$ and 80 cm^{-1} between $P^+B_L^-$ and $P^+H_L^-$. The smooth curves are obtained by concatenating the 20-ps record for each energy gap to make a repeating waveform, and averaging 200 simulations with the fluctuations of each of the energy gaps starting at random times in the two waveforms. The simulation shown with solid curves considered only the pathway through $P^+B_L^-$ as a real intermediate; that shown with dotted curves also included superexchange. Shifting the average energy of $P^+B_L^-$ or $P^+H_L^-$ by ±2.5 kcal/mole can change the calculated rate considerably. (*C*) The autocorrelation function of the energy gap shown in (*A*). The autocorrelation function

$$[\langle U(t)U(0)\rangle = \int_0^\infty U(\tau+t)U(\tau)d\tau \Big/ \int_0^\infty U(\tau)U(\tau)\,d\tau, \quad \text{where} \quad U(\tau) = \Delta V_{\beta\alpha}(\tau) - \langle \Delta V_{\beta\alpha}(\tau)\rangle_\alpha]$$

reflects the dynamics of nuclear relaxations on the classical potential energy surface of state α. (*A*, *B* and *C*: reprinted from Warshel and Parson, 1991.) (*D*) A density-matrix simulation of the reaction P* → $P^+B_L^- \rightarrow P^+H_L^-$ at 50 K. The Franck-Condon factors were obtained from a dispersed-polaron analysis. (The low-frequency region of the Fourier transform of the energy gap shown in (*A*) was divided among vibrational modes of 20, 40, 60, 80 and 100 cm^{-1} with $\Delta = 0.97, 0.57, 0.73, 0.75$ and 0.86, respectively. The same modes were used for both electron-transfer steps.) The electronic couplings were 20 cm^{-1} between P* and $P^+B_L^-$ and 80 cm^{-1} between $P^+B_L^-$ and $P^+H_L^-$. ΔG^o was set $\approx \lambda$ for each step. A total of 288 vibronic levels were included. The four curves show the disappearance of P* (initially populated with thermally equilibrated vibrational levels for this simulation), as calculated with a time constant (T_1) of (from top to bottom) 2.0, 1.0, 0.5 or 0.2 ps for vibrational equilibration within each of the three electronic states. The time constant for pure dephasing (T_2*) was 20 ps. Superexchange is included automatically in the density-matrix formulation. (Reprinted from Parson and Warshel, 1993.)

describe the kinetics of electron transfer at any other temperature, including the effects of nuclear tunneling. The temperature dependence of the rate constant enters Eq. (10) naturally through the $\bar{n}_i$. Schulten and Tesch (1991), Treutlein et al. (1992) and Marchi et al. (1993) recently have adopted a simulation strategy they term a 'spin boson' treatment, which is essentially identical to the dispersed polaron treatment.

The dispersed polaron model has been used to examine the reaction in which an electron moves from H_L^- to the first quinone (Q_A) in the reaction center, and also the much slower back-reaction between Q_A^- and P^+ (Warshel *et al.*, 1989a,b). The calculated rate of electron transfer from H_L^- to Q_A was essentially independent of temperature, in agreement with the experimental observations on *Rp. viridis*. By reducing the value of $\langle \Delta V_{\alpha\beta} \rangle_\alpha$, the reaction could be made to speed up with decreasing temperature, as is seen experimentally in *Rb. sphaeroides* (Kirmaier et al., 1985). The back-reaction was found to be relatively insensitive to the free energy difference between $P^+Q_A^-$ and the ground state, in agreement with studies in which the free energy of $P^+Q_A^-$ was changed experimentally (Gunner *et al.*, 1986; Gunner and Dutton, 1989). As was discussed above, the classical Eq. (9) predicts that the rate constant would decrease more abruptly when $\Delta G^o << -\lambda$.

The information on vibrational frequencies and displacements provided by the dispersed-polaron model also can be incorporated directly into fully quantum mechanical simulations. The idea here is to write the time-dependent wavefunction of the system as a sum of products of electronic and vibrational wavefunctions as in Eq. (5). This expression can be expanded to include an arbitrary number of electronic states. The nuclear overlap integrals that enter into the transition matrix elements between the various vibronic levels (Eq. 6) can be calculated straightforwardly if the vibrational energies and displacements are known.

The best way to treat the relaxation processes that result from interactions of the system with its surroundings probably is to construct the density matrix from the time-dependent coefficients. The elements of the reduced density matrix of the system, ρ, are defined by $\rho_{mn} = C_m^* C_n$. Diagonal terms of ρ represent the time-dependent probabilities of finding the system in particular vibronic states, while off-diagonal terms can convey information about coherent electronic or vibrational oscillations of an ensemble of many molecules (Redfield, 1965; Silbey, 1989; Hu and Mukamel, 1989; Jean et al., 1992). The initial values of the coefficients can be chosen to model a variety of situations, such as a thermally equilibrated excited state or a nonequilibrated state created by a femtosecond excitation flash with a given spectrum of wavelengths. The subsequent evolution of the system then is described by the set of simultaneous differential equations

$$\frac{\partial}{\partial t}\rho_{mn} = (-i/\hbar)\sum_k (H_{mk}\rho_{kn} - \rho_{mk}H_{kn}) + \sum_j \sum_k \Gamma_{mn;jk}\rho_{jk} \quad (24)$$

Here Γ, the relaxation matrix, includes both the stochastic thermal equilibration of vibrational levels within each electronic state and the 'pure dephasing' processes due to fluctuations of the energies of the vibronic states. The density-matrix treatment is particularly appropriate for the two-step reaction pathway from P* to $P^+H_L^-$ through P^+B^- because it provides a way of dealing with the relaxation of the intermediate state.

The simultaneous equations for the ρ_{mn} can be solved either by Runge-Kutta integration or by unfolding the $N \times N$ matrix of derivatives to a vector of N^2 elements and diagonalizing the corresponding $N^2 \times N^2$ matrix (Silbey, 1989; Sugawara et al., 1990). However, procedures for calculating the terms of Γ on the basis of microscopic simulations have not yet been developed for systems as complicated as proteins. The best that can be done at present probably is to treat the relaxation terms as phenomenological time constants (T_1 and T_2*) that are related to an autocorrelation function such as that shown in Fig. 4c.

An early density-matrix treatment that considered intramolecular vibrational modes of the pigments was described by Friesner and Wertheimer (1982) before the Franck-Condon factors for the protein were available. Recently, Sugawara et al. (1990) and Alden et al. (1992) also have used a density-matrix treatment based on the vibrational modes of the pigments. In their formulation of the problem, transitions between different electronic states are included in the relaxation matrix Γ, rather than in H. The density-matrix approach can be extended to include the vibrational modes of the protein by using the dispersed-polaron treatment to obtain the vibrational frequencies and displacements (Warshel and Parson, 1991; Parson and Warshel, 1993).

Figure 4d shows a density-matrix simulation of the initial electron-transfer step in the *Rp. viridis* reaction center. The simulated behavior is similar to that obtained with the surface-hopping model (Fig. 4b). The small oscillatory components reflect coherent electronic oscillations that are incompletely damped by the relaxation terms. Their magnitude decreases if one shortens the phenomenological time constants for thermal equilibration (T_I) or pure dephasing (T_2*) in Γ. The oscillations also become less distinct with increases in the number of vibrational levels that are included, and they would be attenuated if the vibronic energies were broadened by a distribution function. Oscillations have been seen experimentally in the absorption and stimulated-emission signals from excited reaction centers, particularly in a mutant strain that is missing the initial electron acceptor, H_L (Vos et al., 1993). However, these appear to represent coherent vibrational motions, rather than the electronic oscillations depicted in Fig. 4d.

E. Electronic Coupling Factors

Microscopic calculations of electronic coupling factors for intermolecular electron transfer reactions are still in early stages of development. To a reasonable approximation, $H_{\alpha\beta}$ decreases exponentially with the edge-to-edge distance between the donor and acceptor (Moser et al., 1992). Yet, within a series of closely related donor-acceptor pairs, the observed coupling factor can depart from this simple relationship by several orders of magnitude, depending apparently on the details of the individual molecular structures (Beratan et al., 1991, 1992). For widely separated donors and acceptors, the relative magnitude of $H_{\alpha\beta}$ can be calculated remarkably well by considering the network of possible connecting pathways through covalently bonded atoms and placing a penalty on jumps through space (Beratan et al., 1991, 1992). In this picture, the overall coupling factor for a particular pathway is a product of superexchange matrix elements for a series of microscopic steps. Attempts to include the molecular orbitals of the intervening amino acids explicitly in a supermolecular quantum mechanical description of the donor-acceptor complex also have met with some success (Broo and Larsson, 1989).

Several efforts have been made to calculate the coupling factors for electron transfer from P* to B_L or from B_L^- to H_L in bacterial reaction centers (Parson et al., 1987; Warshel et al., 1988; Plato et al., 1988; Scherer and Fischer, 1989; Thompson and Zerner, 1991). Here, the electron donors and acceptors seem close enough together for electrons to jump directly from one carrier to the other without extensively involving the surrounding protein. The central difficulty is that $H_{\alpha\beta}$ depends on the overlap of the molecular orbitals of the donor and acceptor in the intermolecular region where the orbital amplitudes are small and not reliably characterized. In addition, the calculated coupling factors could be sensitive to small errors in the crystal structures. Whereas calculations using the *Rp. viridis* crystal structure have given results of the right order of magnitude to account for the observed rate of electron transfer, those calculated with the *Rb. sphaeroides* structures are considerably smaller. However, it seems clear for both species that B_L must play a role in mediating electron transfer from P* to H_L because the coupling factor calculated for a direct reaction without the aid of B_L is too small by many orders of magnitude.

Acknowledgments

We thank R. Alden, Z.-T. Chu and V. Nagarajan for helpful discussions and NIH grant GM-40283 and NSF grant PCM-86188563 for support.

References

Alden RG, Cheng WD and Lin SH (1992) Vibrational relaxation and coherence and primary electron transfer in photosynthetic reaction centers. Chem Phys Lett 194: 318–326

Almeida R and Marcus RA (1990) Dynamics of electron transfer for a nonsuperexchange coherent mechanism. J Phys Chem 94: 2978–2985

Atkins, P. W. (1983) Molecular Quantum Mechanics (pp. 187–202) Oxford Univ. Press, Oxford

Beratan DN, Betts JN and Onuchic JN (1991) Protein electron transfer rates set by the bridging secondary and tertiary structure. Science 252: 1285–1288

Beratan DN, Onuchic JN, Winkler JR and Gray HB (1992) Electron-tunneling pathways in proteins. Science 258: 1740–1741

Beveridge DL and DiCapua FM (1989) Free energy via molecular simulation: Applications to chemical and biomolecular systems. Ann Rev Biophys Biophys Chem 18: 431–492

Boxer SG, Goldstein RA, Lockhart DJ, Middendorf TR, and Takiff L. (1989) Excited states, electron-transfer reactions, and intermediates in bacterial photosynthetic reaction centers. J Phys Chem 93: 8280–8294

Broo A and Larsson S (1989) Calculation of electron transfer rates in proteins. Int J Quant Chem: Quant Biol Symp 16: 185–198

Chan C-K, Chen LX-Q, DiMagno TJ, Hanson DK, Nance SL, Schiffer M, Norris JR and Fleming GR (1991) Initial electron transfer in photosynthetic reaction centers of *Rhodobacter capsulatus* mutants. Chem Phys Lett 176: 366–372

Churg AK and Warshel A (1985) Modeling the activation energy and dynamics of electron transfer reactions in proteins. In: Clementi E, Corongiu G, Sarma MH and Sharma RH (eds) Structure and Motion: Membranes, Nucleic Acids and Proteins pp 361–374 Adenine Press, Guilderland, NY

Closs GL and Miller JR (1988) Intramolecular long-distance electron transfer in organic molecules. Science 240: 440–447

Creighton S, Hwang J-K, Warshel A, Parson WW and Norris J (1988) Simulating the dynamics of the primary charge separation process in bacterial photosynthesis. Biochem. 27: 774–781

Deisenhofer J and Michel H (1989) The photosynthetic reaction center from the purple bacterium *Rhodopseudomonas viridis*. Science 245: 1463–1473

Efrima S and Bixon M (1976) Vibrational effects in outer-sphere electron-transfer reactions. Chem Phys 13: 447–460

Feher G, Arno TR and Okamura MY (1988) The effect of an electric field on the charge recombination rate of $D^+Q_A^- \rightarrow DQ_A$ in reaction centers from *Rhodobacter sphaeroides* R-26. In: Breton J and Verméglio A (eds) The photosynthetic bacterial reaction center pp 271–287 Plenum Press, New York

Finkele U, Lauterwasser C, Zinth W, Gray KA and Oesterhelt D (1990) Role of tyrosine M210 in the initial charge separation of reaction centers of *Rhodobacter sphaeroides*. Biochem 29: 8517–8521

Franzen S and Boxer SG (1993) Temperature dependence of the electric field modulation of electron transfer rates: Charge recombination in photosynthetic reaction centers. J Phys Chem 97: 6304–6318

Franzen S, Goldstein RF and Boxer SG (1990) Electric field modulation of electron transfer rates in isotropic systems: Long-distance charge recombination in photosynthetic reaction centers. J Phys Chem 94: 5135–5149

Friesner R and Wertheimer R (1982) Model for primary charge separation in reaction centers of photosynthetic bacteria. Proc Natl Acad Sci USA 79: 2138–2142

Gunner MR and Dutton PL (1989) Temperature and $-\Delta G^o$ dependence of the electron transfer from $BPh^{\cdot-}$ to Q_A in reaction center protein from *Rhodobacter sphaeroides* with different quinones as Q_A. J Am Chem Soc 111: 3400–3412

Gunner MR, Robertson DE and Dutton PL (1986) Kinetic studies on the reaction center protein from *Rhodopseudomonas sphaeroides*: The temperature and free energy dependence of electron transfer between various quinones in the Q_A site and the oxidized bacteriochlorophyll dimer. J Phys Chem 90: 3783–3795

Hopfield JJ (1974) Electron transfer between biological molecules by thermally activated tunneling. Proc Natl Acad Sci. USA 71: 3640-3644

Hu Y and Mukamel S (1989) Chem Phys Lett 18: 410–416

Hwang J-K and Warshel A (1987) Microscopic examination of free-energy relationships for electron transfer in polar solvents. J Am Chem Soc 109: 715–720

Hwang J-K, Creighton S, King G, Whitney D and Warshel A (1988) Effects of solute-solvent coupling and solvent saturation on solvation dynamics of charge transfer reactions. J Chem Phys 89: 859–865

Jean JM, Friesner RA and Fleming GR (1992) Application of a multilevel Redfield theory to electron transfer in condensed phases. J Chem Phys 110: 5827–5842

Jorgensen WL (1989) Free-energy calculations - a breakthrough for modeling organic chemistry in solution. Acc Chem Res 22: 184–189

Jortner J (1976) Temperature dependent activation energy for electron transfer between biological molecules. J Chem Phys 64: 4860–4866

Jortner J (1980) Dynamics of the primary events in bacterial photosynthesis. J Am Chem Soc 102: 6676–686

Kakitani T and Kakitani H (1981) A possible new mechanism of temperature dependence of electron transfer in photosynthetic systems. Biochim Biophys Acta 635: 498–514

Kestner NR, Logan J and Jortner J (1974) Thermal electron transfer reactions in polar solvents. J Phys Chem 78: 2148–2166

King G and Warshel A (1986) A surface constrained all-atom solvent model for effective simulations of polar solutions. J Chem Phys 91: 3647–3661

King G and Warshel A (1990) Investigation of the free energy functions for electron transfer reactions. J Chem Phys 93: 8682–8692

Kirmaier C, Holten D and Parson WW (1985) Temperature and detection-wavelength dependence of the picosecond electron-transfer kinetics measured in *Rhodopseudomonas sphaeroides* reaction centers. Resolution of new spectral and kinetic components in the primary charge-separation process. Biochim Biophys Acta 810: 33–48

Kitzing EV and Kuhn H (1990) Primary electron transfer in photosynthetic reaction centers. J Phys Chem 94: 1699–1702

Kollman PA and Merz KM (1990) Computer modeling of the interactions of complex molecules. Acc Chem Res 23: 246–252

Kubo R and Toyozawa Y (1955) Application of the method of generating function to radiative and non-radiative transitions of a trapped electron in a crystal. Prog Theor Phys 13: 160–182

Kuharski RA, Bader JS, Chandler D, Sprik M, Klein ML and Impey RW (1988) Molecular model for aqueous ferrous-ferric electron transfer. J Chem Phys 89: 3248–3256

Lax M (1952) The Franck-Condon principle and its application to crystals. J Chem Phys 20: 1753–1760

Lee FS, Chu ZT and Warshel A (1993) Microscopic and semimicroscopic calculations of electrostatic energies in proteins by the POLARIS and ENZYMIX programs. J Comp Chem 14: 161–185

Levich VG (1966) Present state of the theory of oxidation-reduction in solution (bulk and electrode reactions). Adv Electrochem Electrochem Eng 4: 249–371

Luzhkov V and Warshel A (1991) Microscopic models for quantum mechanical calculations of chemical processes in solutions: LD/AMPAC and SCAAS/AMPAC calculations of solvation energies. J Comp Chem 13: 199–213

Marchi M, Gehlen JN, Chandler D and Newton M (1993) Diabatic surfaces and the pathway for primary electron transfer in a photosynthetic reaction center. J Am Chem Soc 115: 4178–4190

Marcus RA (1956) On the theory of oxidation-reduction reactions involving electron transfer. I. J Chem Phys 24: 966–978

Marcus RA (1965) On the theory of oxidation-reduction reactions involving electron transfer. VI. Unified treatment for homogeneous and electrode reactions. J Chem Phys 43: 679–701

Marcus R and Almeida R (1990) Dynamics of electron transfer for a nonsuperexchange coherent mechanism. 1. J Phys Chem 94: 2973–2977

Marcus RA and Sutin N (1985) Electron transfers in chemistry and biology. Biochim Biophys Acta 811: 265–322

Merzbacher E (1970) Quantum Mechanics pp 450–486 Wiley, New York

Miller WH and George TF (1972) Semiclassical theory of electronic transitions in low energy atomic and molecular collisions involving several nuclear degrees of freedom. J Chem Phys 56: 5637–5652

Miller JR, Calcaterra LT and Closs GL (1984) Intramolecular long-distance electron transfer in radical anions. The effects of free energy and solvent on the reaction rate. J Am Chem Soc 106: 3047–3049

Moser CC, Keske JM, Warnicke K, Farid RS and Dutton PL (1992) Nature of biological electron transfer. Nature 355: 796–802

Nagarajan V, Parson WW, Gaul D and Schenck CC (1990) Effect of specific mutations of tyrosine-(M)210 on the primary photosynthetic electron-transfer process in *Rhodobacter sphaeroides*. Proc Natl Acad Sci USA 87: 7888–7892

Nagarajan V, Parson WW, Davis D and Schenck CC (1993) Kinetics and free energy gaps of electron-transfer reactions in *Rhodobacter sphaeroides* reaction centers. Biochem 32: 12324–12336

Parson WW and Warshel A (1993) Simulations of electron transfer in bacterial reaction centers. In: Deisenhofer J and Norris JR (eds) The Photosynthetic Reaction Center, Vol II, pp 23–47. Academic Press, New York

Parson WW, Creighton S and Warshel A (1987) Calculations of spectroscopic properties and electron transfer kinetics of photosynthetic bacterial reaction centers. In: Kobayashi T (ed) Primary Reactions of Photobiology, pp 43–51. Springer-Verlag, New York

Parson WW, Chu Z-T and Warshel A (1990) Electrostatic control of charge separation in bacterial photosynthesis. Biochim Biophys Acta 1017: 251–272

Parson WW, Nagarajan V, Gaul D, Schenck CC, Chu Z-T and Warshel A (1991) Electrostatic effects on the speed and directionality of electron transfer in bacterial reaction centers: The special role of tyrosine M-208. In: Michel-Beyerle ME (ed) Reaction Centers of Photosynthetic Bacteria, pp 239–249. Springer-Verlag, Berlin

Plato M, Möbius K, Michel-Beyerle ME, Bixon M and Jortner J (1988) Intermolecular electronic interactions in bacterial photosynthesis. J Am Chem Soc 110: 7279–7285

Popovic ZD, Kovacs GJ, Vincett PS, Alegria G and Dutton PL (1986) Electric field dependence of recombination kinetics in reaction centers of photosynthetic bacteria. Chem Phys 110: 227–237

Redfield AG (1965) The theory of relaxation processes. Adv Mag Reson 1: 1–32

Roth M, Lewit-Bentley A, Michel H, Deisenhofer J, Huber R and Oesterhelt D (1989) Detergent structure in crystals of a bacterial photosynthetic reaction centre. Nature 340: 680–662

Sarai A (1979) Energy and temperature dependence of electron and excitation transfer in biological systems. Chem Phys Lett 63: 360–366

Sarai A (1980) Possible role of protein in photosynthetic electron transfer. Biochim Biophys Acta 589: 71–82

Scherer POJ and Fischer SF (1989a) Quantum treatment of the optical spectra and the initial electron transfer process within the reaction center of *Rhodopseudomonas viridis*. Chem Phys 131: 115–127

Scherer POJ and Fischer SF (1989b) Long range electron transfer via energy tuning of molecular orbitals within the hexamer of the photosynthetic reaction center of *Rhodopseudomonas viridis*. J Phys Chem 93: 1633–1637

Schulten K and Tesch M (1991) Coupling of protein motion to electron transfer: Molecular dynamics and stochastic quantum mechanics study of photosynthetic reaction centers. Chem Phys 158: 421–446

Sharp KA and Honig B (1990) Electrostatic interactions in macromolecules. Annu Rev Biophys Biophys Chem 19: 301–332

Silbey R (1989) Relaxation theory applied to scattering of excitations and optical transitions in crystals and solids. In: Fünfschilling J (ed) Relaxation Processes in Molecular Excited States, pp 243–276. Kluwer Academic Publishers, Dordrecht

Small GJ, Hayes JM and Silbey RJ (1992) The question of dispersive kinetics for the initial phase of charge separation in bacterial reaction centers. J Phys Chem 96: 7499–7501

Straatsma TP and McCammon JA (1992) Computational alchemy. Ann Rev Phys Chem 43: 407–435

Sugawara M, Fujimura Y, Yeh CY and Lin SH (1990) Application of the density matrix method to the primary electron transfer in photosynthetic reaction centers. J Photochem Photobiol, A Chem 54: 321–331

Thompson MA and Zerner MC (1991) A theoretical examination of the electronic structure and spectroscopy of the photosynthetic reaction center from *Rhodopseudomonas viridis*. J Am Chem Soc 113: 8210–8215

Treutlein H, Schulten K, Brunger AT, Karplus M, Deisenhofer J and Michel H (1992) Chromophore-protein interactions and the function of the photosynthetic reaction center: A molecular dynamics study. Proc Natl Acad Sci USA 89: 75–79

Tully JC and Preston RM (1971) Trajectory surface hopping approach to nonadiabatic molecular collisions: The reaction of H^+ with D_2. J Chem Phys 55: 562–572

van Gunsteren WF and Berendsen HJC (1990) Computer simulation of molecular dynamics: Methodology, applications, and perspectives in chemistry. Angew Chem, Int Ed Engl 29: 992–1023

Vos MH, Rappaport F, Lambry J-H, Breton J and Martin J-L (1993) Visualization of coherent nuclear motion in a membrane protein by femtosecond spectroscopy. Nature 363: 320–325

Warshel A (1982) Dynamics of reactions in polar solvents. Semiclassical trajectory studies of electron-transfer and proton-transfer reactions. J Phys Chem 86: 2218–2224

Warshel A (1991) Computer modeling of chemical reactions in enzymes and solutions. John Wiley, New York

Warshel A and Åqvist J (1991) Electrostatic energy and macromolecular function. Ann Rev Biophys Biophys Chem 20: 267–298

Warshel A and Hwang J-K (1985) Quantized semiclassical trajectory approach for evaluation of vibronic transitions in harmonic molecules. J Chem Phys 82: 1756–1771

Warshel A and Hwang J-K (1986) Simulation of the dynamics of electron transfer reactions in polar solvents: Semiclassical trajectories and dispersed polaron approaches. J Chem Phys 84: 4938–4957

Warshel A and Parson WW (1991) Computer simulations of electron-transfer reactions in solution and in photosynthetic reaction centers. Ann Rev Phys Chem. 42: 279–309

Warshel A and Russell (1984) Electrostatic interactions in biological systems and solutions. Quart Rev Biophys 17, 283–422

Warshel A, Sussman F and King G (1986) Free energy of charges in solvated proteins: Microscopic calculations using a reversible charging process. Biochem 25: 8368–8372

Warshel A, Creighton S and Parson WW (1988a) Electron-transfer pathways in the primary event of bacterial photosynthesis. J Phys Chem 92: 2696–2701

Warshel A, Sussman F and Hwang J-K (1988b) Evaluation of catalytic free energies in genetically modified proteins. J Mol Biol 201: 139–159

Warshel A, Chu Z-T and Parson WW (1989a) Microscopic Simulation of Quantum Dynamics and Nuclear Tunneling in Bacterial Reaction Centers. Photosynth Res 22: 39–46

Warshel A, Chu Z-T and Parson WW (1989b) Dispersed polaron simulations of electron transfer in photosynthetic reaction centers. Science 246: 112–116

Wasielewski MR, Niemczyk MP and Svec WA and Pewitt EB (1985) Dependence of rate constants for photoinduced charge separation and dark charge recombination on the free energy of reaction in restricted porphyrin-quinone molecules. J Am Chem Soc 107: 1080–1082

Williams JC, Woodbury NW, Taguchi AKW, Peloquin JM, Murchison HA, Alden RG (1992a) Mutations that affect the donor midpoint potential in reaction centers from *Rhodobacter sphaeroides*. In: Breton J and Verméglio A (eds) The Photosynthetic Bacterial Reaction Center II, pp 25–31. Plenum Press, New York

Williams JC, Alden RG, Murchison HA, Peloquin JM, Woodbury NW and Allen JP (1992b) Effects of mutations near the bacteriochlorophylls in reaction centers from *Rhodobacter sphaeroides*. Biochem 31: 11029–11037

Chapter 26

Proton-Coupled Electron Transfer Reactions of Q_B in Reaction Centers from Photosynthetic Bacteria

M. Y. Okamura and G. Feher
Department of Physics, University of California, San Diego, La Jolla, California 92093-0319, USA

Summary

The reaction center from purple bacteria mediates the initial steps of a light driven proton pump, coupling light induced electron transfer to proton uptake. The key steps in this reaction involve the two electron reduction of the secondary quinone Q_B with the concomitant uptake of 2 protons:

$$2H^+ + 2e^- + Q_B \rightarrow Q_BH_2$$

Recent advances in studies of reaction centers from purple bacteria have provided insight into the pathway of proton transfer into the RC and the mechanism of coupling proton and electron transfer reactions. Studies using site directed mutagenesis have identified 3 residues near Q_B that are important for proton transfer in RCs from *Rb. sphaeroides*. The mutation of Ser L223 and Asp L213 to Ala and Asn, respectively, block proton transfer and the second electron transfer involved in Q_B reduction. Mutation of Glu L212 to Gln blocks proton transfer associated with Q_B reduction but 2 electron reduction of Q_B still occurs. The model explaining these results is that Ser L223 and Asp L213 are involved in the proton transfer pathway for uptake of the first proton which is bound prior to the second electron transfer, whereas Glu L212 is involved in the pathway for the second proton which is bound after the second electron transfer. These results clearly show that specific residues play important roles in proton transfer. However other functionally active mutant RCs have been obtained in which either Asp L213 , Ser L223 or Glu L212 were absent indicating that the pathways for proton transfer are not

R. E. Blankenship, M. T. Madigan and C. E. Bauer (eds): Anoxygenic Photosynthetic Bacteria, pp. 577–594.

unique. The structure of the RC near the Q_B site suggests the involvement of water molecules in addition to protein residues in the proton transfer chain to Q_B. In addition, negatively charged residues near Q_B increase the proton coupled electron transfer rate by stabilizing a proton in the interior of the RC.

I. Introduction

In photosynthetic bacteria light energy is transformed into chemical energy by the action of a light driven proton pump coupled to electron transfer (see Fig. 1). The bacterial reaction center (RC) plays a key role in this process by performing the initial light-activated electron transfer reactions and the initial proton uptake reactions of the proton pump (Feher et al., 1989; Breton and Vermeglio, 1992). The key reaction required in purple bacteria for coupling proton transfer with electron transfer involves the two electron reduction and concomitant binding of two protons by the secondary quinone Q_B on the RC:

$$2e^- + 2H^+ + Q_B \rightarrow Q_BH_2 \quad (1)$$

The two electrons in this reaction arise from successive photoionizations of the primary electron donor, a bacteriochlorophyll dimer, D which transfers the electron to Q_B through a series of electron acceptors (bacteriopheophytin and Q_A) (see Chapter 24 by Woodbury and Allen). The two protons in this reaction originate from the aqueous solution of the cytoplasm and are transferred to Q_B through a chain

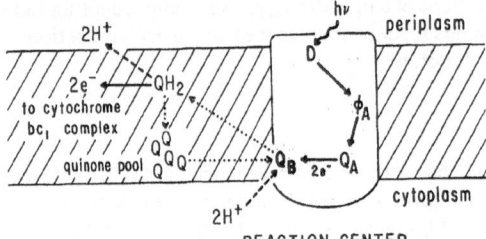

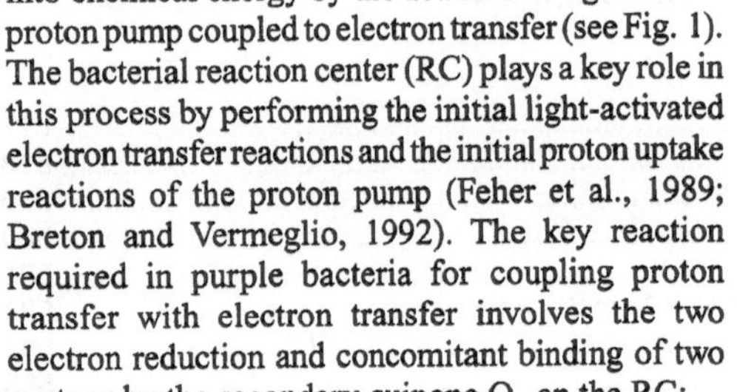

Fig. 1. Schematic representation of electron and proton transfer in bacterial photosynthesis. Electron transfer steps are indicated by solid lines: proton transfer steps by dashed lines; and diffusion steps by dotted lines. Proton transfer is coupled to electron transfer via the proton coupled electron transfer reaction of Q_B (From Paddock et al., 1989).

Abbreviations: Cyt c – cytochrome c; D – primary donor; IR – infrared; Q_A – primary quinone; Q_B – secondary quinone; Rb. – Rhodobacter; RC – reaction center; Rp. – Rhodopseudomonas

of proton acceptor/donor groups which include several protonatable residues. The exact sequence of the proton and electron transfer steps remains to be determined. After the two electron reduction, the doubly reduced quinone subsequently dissociates from the RC and is reoxidized by the cytochrome bc_1 complex releasing protons on the periplasmic side of the membrane (Fig. 1). The net result of these reactions is the vectorial transport of protons across the membrane driven by electron transfer, as proposed by Mitchell, 1961. This proton transport is used to drive ATP synthesis (reviewed in Crofts and Wraight, 1983; Cramer and Knaff, 1990).

Recently, the molecular basis for proton and electron transfer have begun to be elucidated as the result of two major developments in RC research. The first was the determination of the X-ray crystal structure of the RCs from *Rhodopseudomonas (Rp.) viridis* by Deisenhofer et al. (1984) and *Rhodobacter (Rb.) sphaeroides* by Allen et al.(1986) and Chang et al.(1986)(see Chapter 23 by Lancaster et al.). The second was the use of molecular biology in studying RCs by site directed mutagenesis (Bylina et al., 1989; Paddock et al., 1989; Takahashi and Wraight, 1990). The X-ray crystal structure revealed that Q_B is located in the interior of the protein, out of contact with the aqueous solution and suggested the possibility that protonatable amino acid side chains from the protein were responsible for proton transport to Q_B. Site-directed mutagenesis of several of these residues to non-protonatable groups resulted in loss of proton transport to Q_B and conclusively demonstrated that the protein plays an important role in proton transport and coupling between electron and proton transfer. Proton transfer to Q_B in bacterial RCs has been recently reviewed by several authors (Takahashi et al., 1992; Okamura and Feher, 1992; Feher et al., 1992; Takahashi and Wraight, 1994). In this review, we will focus on the coupling between electron and proton transfer reactions of Q_B in bacterial RCs, emphasizing the results from site-directed mutagenesis. Earlier reviews of quinones in bacterial RCs have been published (Parson, 1978; Okamura et al., 1982; Wraight, 1982; Morisson et al., 1982).

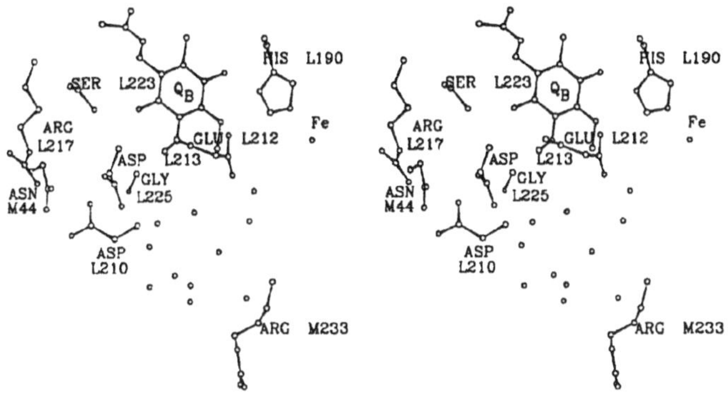

Fig. 2. Stereo view of the structure of the RC from *Rb. sphaeroides* near Q_B showing the residues affecting proton coupled electron transfer to Q_B (Allen et al., 1988). The free circles represent the predicted positions of internal water molecules (Beroza et al., 1992).

II. Structure of the Bacterial Reaction Center Near the Q_B Site

The structure of the RC from *Rb. sphaeroides* in the region near Q_B is shown in Fig. 2 (Allen et al., 1988). The Q_B binding site consists of a loop of amino acid residues spanning the region between the D and E transmembrane helices of the L subunit. One carbonyl oxygen of Q_B is hydrogen bonded to His L190 which is liganded to the Fe^{2+}. The other carbonyl oxygen is hydrogen bonded to Ser L223. The quinone ring is near the hydrophobic amino acid residues Ile L229, Val L220, and Phe L216. Also nearby are the acidic residues Glu L212, Asp L213, Asp L210 and the basic residue Arg L217. These acidic and basic residues could serve to modify the electrostatic potential at Q_B and/or serve as proton donor groups in a proton transfer pathway to Q_B.

Several residues which have been implicated in Q_B function are shown in Fig. 2. The residues Asp L213 (Takahashi and Wraight, 1990; Paddock et al., 1993), Ser L223 (Paddock et al., 1990a) and Glu L212 (Paddock et al., 1989; Takahashi and Wraight, 1992) have been shown by site-directed mutagenesis to be involved in proton transfer to Q_B. Near Q_B is the hydrophilic residue Asn M44. This residue is homologous to the Asp M43 in RCs from *Rp. viridis* (Deisenhofer and Michel, 1989) and several other photosynthetic bacteria that contain an Asn (instead of Asp) at position L213. The Asp at L213 could serve the same function as the Asp at M43 as suggested by recent experiments (Hanson et al., 1992b; Rongey et al., 1993). In the region below Q_B that defines an interface between the H subunit and the LM core of the RC are found numerous acidic and basic residues from the M and H subunits. Among these residues is Arg M233. The mutation of Arg M233[1] to Cys has been shown to partially restore the proton and electron transfer rates in deficient mutants containing the Asp L213 → Asn mutation (Okamura et al., 1992). A similar mutation of the homologous Arg M231 → Leu was shown to confer photosynthetic competence to deficient RCs from *Rb. capsulatus* containing the mutations Asp L213 → Ala and Glu L212 → Ala (Hanson et al., 1992a).

The X-ray crystal structure of *Rp. viridis* shows many internal water molecules including several in the region near Q_B that may be involved in proton transfer (Deisenhofer and Michel, 1989). The lower resolution structure of the RC from *Rb. sphaeroides* does not show the position of water molecules. However, the void spaces in the structure suggest the presence of water molecules. A computational approach to predict water in the *Rb. sphaeroides* structure was used by Beroza et al. (1992). They used the algorithm of Rashin et al. (1986) based on finding sites at positions capable of providing hydrogen bond interactions to stabilize water molecules in the interior of the RC. Interestingly, a considerable number of water molecules are predicted to be near Q_B in the *Rb. sphaeroides* RC. A high density of water is located near the methoxy groups below the quinone in a region called the 'methoxy pocket.'

[1] This mutation was inadvertently mislabeled Arg M223 by Okamura et al. (1992).

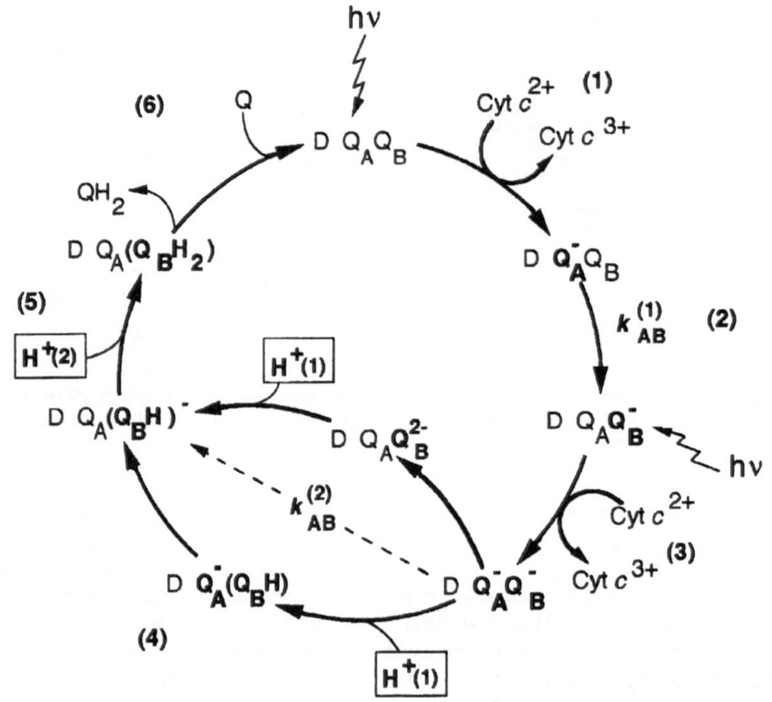

Fig. 3. The quinone reduction cycle. Light induced electron transfer to Q_A (steps 1 and 3) are followed by electron transfer to Q_B (steps 3 and 4) coupled to proton uptake (see squares) (steps 4 and 5). Two possible mechanisms are indicated for step 4. In the upper, electron transfer precedes protonation, in the lower proton transfer precedes electron transfer. The final step is quinone exchange (step 6). All steps are reversible. The cycling time in native RCs is ~1 ms.

These water molecules may play a role in proton transfer to Q_B as well as in stabilizing the charge on the reduced quinone. Recently, a chain of 14 water molecules extending 23 Å from Q_B to the cytoplasmic surface of the RC was reported in the structure of RCs from *Rb. sphaeroides* determined from X-ray diffraction studies at 2.65 Å resolution (Ermler et al., 1994).

III. The Quinone Reduction Cycle

The light-induced electron and proton transfer reactions that couple electron transfer to the uptake of protons in the RC are represented by the quinone reduction cycle shown in Fig. 3. This cycle summarizes the results of electron transfer (Vermeglio, 1977; Wraight, 1977; Wraight, 1979), and proton transfer (Wraight, 1979) as well as site-directed mutagenesis (Paddock et al., 1990a; Takahashi and Wraight, 1992). Two electrons are transferred and two protons are bound to Q_B as a result of two separate photochemical events resulting in the oxidation of two cytochrome molecules.

The cycle starts with the RC in the initial state DQ_AQ_B. The sequence of steps shown in Fig. 2 are as follows:

1) *First Light-Induced Q_A Reduction.* The first step in the cycle involves the transfer of an electron to Q_A and concomitant cytochrome oxidation. This involves 3 rapid successive electron transfer steps not explicitly shown in Fig. 3. These are: (a) Electron transfer from the primary donor to the bacteriopheophytin molecule. (b) Electron transfer from bacteriopheophytin to Q_A. (c) Electron transfer from a cytochrome c_2 molecule to D^+.

2) *First Electron Transfer to Q_B*, $k^{(1)}_{AB}$. The second reaction in the cycle involves the transfer of the first electron from Q_A^- to Q_B. This reaction is reversible and gives rise to an equilibrium between the states $Q_A^-Q_B$ and $Q_AQ_B^-$. In *Rb. sphaeroides* RCs $k^{(1)}_{AB} \approx 10^4$ s^{-1} (Vermeglio, 1977; Wraight, 1979; Kleinfeld et al., 1984).

3) *Second Light-Induced Q_A Reduction.* The third step is a repeat of step 1), a light induced electron transfer that gives rise to the diradical state $DQ_A^-Q_B^-$.

4) *Second Electron Transfer to Q_B*, $k^{(2)}_{AB}$. The fourth reaction is an electron transfer coupled to proton uptake that gives rise to the singly protonated doubly reduced state $DQ_A(Q_BH)^-$. This reaction could proceed through two possible paths: Proton uptake followed by electron transfer or electron transfer followed by proton uptake. These two mechanisms are shown as parallel paths in step 4 in Fig. 4. In *Rb. sphaeroides* RCs $k^{(2)}_{AB} \approx 10^3$ s^{-1} (Wraight, 1979; Kleinfeld et al., 1985).

5) *Second Proton Uptake.* Following the formation of the $DQ_A(Q_BH)$-state, Q_B takes up the second proton in reaction 5 to form the dihydroquinone $DQ_AQ_BH_2$. The rates for the individual steps in reactions 4 and 5 have not been resolved, but the overall time for reaction 4 and reaction 5 is about 1 ms (pH 7.5) (Wraight, 1979, Kleinfeld et al., 1985). Results from site-directed mutagenesis suggest that the two protons $H^+(1)$ and $H^+(2)$ are taken up via 2 separate pathways (Paddock et al., 1990b; Takahashi and Wraight, 1992).

6) *Quinone Exchange.* After reduction, the dihydroquinone dissociates from the RC and is replaced by an oxidized quinone (McPherson et al., 1990), thereby completing the cycle. The exchange times for the quinone depend on the nature of the quinone and detergent composition, but are generally in the ms range.

The electron transfer rate of the cycle, which is the turnover rate for cytochrome photo-oxidation by RCs under saturating light intensity in the presence of excess cytochrome and quinone, is about 10^3 s^{-1}. The values of the rates for electron and proton transfer reactions for RCs from different bacteria and mutants are shown in Table 1.

IV. Properties of the Quinones

Despite the roughly 2-fold symmetry of the RC about an axis bisecting the two quinones (Allen et al., 1988) the electron transfer functions of the quinones Q_A and Q_B are quite different. The primary quinone Q_A undergoes a rapid one electron reduction, taking an electron rapidly from the reduced bacteriopheophytin acceptor with a rate of 10^{10} s^{-1} (for a discussion see Chapter 24 by Woodbury and Allen). On the other hand, Q_B undergoes a slower two electron reduction by electron transfer from Q_A in two one-electron steps with rates of 10^3–10^4 s^{-1} with a concomitant uptake of 2 protons. Some properties of the quinones in RCs are summarized below.

The one-electron redox potential for Q_A/Q_A^- was found to be pH dependent with a value of –65 to –80 mV for *Rb. sphaeroides* at pH 8 (Prince and Dutton, 1978). The redox potentials for the Q_B/Q_B^- and Q_B^-/Q_B^{2-} couples (40 and –40 mV respectively at pH 8) (Rutherford and Evans, 1980) for *Rb. sphaeroides* are somewhat higher, favoring the electron transfer from Q_A^- to Q_B. A convenient assay for the free energy difference between the $Q_A^-Q_B \rightleftharpoons Q_AQ_B^-$ states is the recombination rate k_{BD} for electron transfer from Q_B^- to D^+ which has been shown to proceed in *Rb. sphaeroides* RCs through the reaction $D^+Q_AQ_B^- \rightleftharpoons D^+Q_A^-Q_B \rightarrow DQ_AQ_B$ (Kleinfeld et al., 1984). Thus, a small k_{BD} indicates a stabilization of Q_B^-. Values of k_{DD} for RCs from different mutants are shown in Table 1. The structural basis for the difference in redox potentials between the two quinones must lie in the protein structure, since both quinones are ubiquinone. Interestingly, the electron transfer equilibrium favors Q_B^- over Q_A^- despite the preponderance of negatively charged residues near Q_B. Possible reasons for the stabilization of Q_B^- may be the effect of helix dipoles (Gunner, 1993) or internal water molecules (Beroza et al., 1992).

The semiquinone states Q_A^- and Q_B^- have been shown to be unprotonated by a number of spectroscopic techniques, i.e. optical (Vermeglio, 1977; Wraight, 1977), EPR (Hales and Case, 1981), ENDOR (Lubitz et al., 1985; Feher et al., 1985) and IR spectroscopy (Breton et al., 1991b, 1991a). The transient protonation of the semiquinone may play a role in the mechanism of proton coupled electron transfer.

Proton uptake accompanying the reactions $Q_A \rightarrow Q_A^-$ and $Q_B \rightarrow Q_B^-$ exhibits a proton uptake stoichi-

Table 1. Rates (s^{-1}) of electron and proton tranfer reactions in RCs from photosynthetic bacterial strains containing mutations affecting proton transfer (pH 7–8)

Species	RC		$k_{AB}^{(1)}$	$k_{AB}^{(2)}$	k_H	Turnover	k_{BD}	Photosynthetic Growth	References
Rb. sphaeroides	NATIVE		6000	1300	1200	>500	0.7	yes	1,2,3
	Glu L212	→Gln	3500	750	6	12	1.0	no	4,5,6
		→Asp	1750	>500	>500	≈ 200	8	yes	4,7
	Ser L223	→Ala	15000	4	4	8	0.62	no	8
		→Asn	>300	≈ 10	–	≈10	6	no	9
		→Thr	~6000	8000	–	>150	6	yes	8
		→Asp	~5000	9000	–	>150	8	yes	9
	Asp L213	→Asn	350	0.3	0.3	0.5	0.04	no	8,10,11
		→Leu	300	–	–	0.6	0.04	no	11
		→Ser	500	0.5	–	1.0	0.04	no	11
		→Thr	500	0.5	–	2.0	0.04	no	11
		→Glu	5000	15	15	35	0.02	no	11
		→His	500	0.5	–	1.0	0.03	no	11
		→Lys	5000	0.5	–	5	0.02	no	9
	Asp L210	→Asn	800	600	–	290	0.6	yes	12
	Arg L217	→Gln	4000	–	>200	>200	–	yes	9
		→Leu	3200	–	–	>150	–	yes	9
	Asn M44	→Asp	1800	8000	–	400	3	yes	13
	{ Asp L213 Glu L212	→Asn/ →Gln	4000	0.4	–	0.6	0.05	no	5
	{ Asp L213 Asn M44	→Asn/ →Asp	2000	1200	1000	500	0.5	yes	13
	{ Asp L213 Arg M233	→Asn/ →Cys	3000	50	–	100	2	yes	14
	{ Ser L223 Tyr L222	→Ala/ →Cys	–	–	–	≥20	0.9	yes	15
Rp. viridis	NATIVE		4 x 10^4	2 x 10^4	–	–	0.95	yes	16,17
	{ Ser L223 Arg L217	→Ala/ →His	4 x 10^5	25	–	–	1.3	yes	16,17
	{ Val L220 Arg L217	→Leu/ →His	1 x 10^5	3 x 10^3	–	–	–	yes	16
Rb. capsulatus	NATIVE		5 x 10^4	2000	1000		0.8	yes	18,19,23
	Ser L223	→Ala	–	–	–	–	–	no	20
		→Thr	–	–	–	–	–	yes	21
		→Glu	–	–	–	–	–	yes	21
		→Gly	–	–	–	–	–	yes	21
	Glu L212	→Ala	–	–	–	–	–	yes	22
	{ Asp L213 Glu L212	→Ala/ →Ala	–	0.4	1.0	–	0.08	no	19,23
	{ Asp L213 Glu L212 Asn M43	→Ala/ →Ala/ →Asp	–	10	10	–	–	yes	20,23
	{ Asp L213 Glu L212 Arg M231	→Ala/ →Ala/ →Leu	–	20	20	–	0.2	yes	19,23
	{ Asp L213 Glu L212 Gly L225	→Ala/ →Ala/ →Asp	–	–	–	–	–	yes	22

References: 1 (Vermeglio and Clayton, 1977), 2 (Wraight, 1979), 3 (Kleinfeld et al., 1984) 4 (Paddock et al., 1989), 5 (Takahashi and Wraight, 1992), 6 (McPherson et al., 1993), 7 (Paddock et al., 1990), 8 (Paddock et al., 1990), 9 (M. L. Paddock, unpublished), 10 (Takahashi and Wraight, 1990), 11 (Paddock et al., 1993), 12 (Paddock et al., 1992), 13 (Rongey et al., 1993), 14 (Okamura et al., 1992), 15 (S. H. Rongey, unpublished), 16 (Leibl et al., 1993), 17 (Baciou et al., 1991), 18 (Tiede and Hanson, 1992), 19 (Hanson et al., 1992a), 20 (Bylina et al., 1989), 21 (Bylina and Wong, 1992), 22 (Schiffer et al., 1992), 23 (Maróti et al., 1994).

ometry of < $1H^+/e^-$. (Wraight, 1979; Maróti and Wraight, 1988; McPherson et al., 1988). This proton uptake has been attributed to proton binding by acidic residues whose pK_as have been shifted due to interactions with the negative charge on the quinone. Recent IR studies have shown that Glu L212 near Q_B is involved in proton uptake upon Q_B^- formation (Hienerwadel et al., 1992). Proton uptake measurements on the $Q_B \rightarrow Q_BH_2$ reaction have shown the binding of 2 $H^+/2e^-$ (Wraight, 1979; Maróti and Wraight, 1988; McPherson et al., 1993). However, a decrease in proton binding at pH > 8.5 has been observed. This decrease has been attributed to the ionization of bound Q_BH_2 or nearby protein residues (McPherson et al., 1993).

V. Effects of Site-Directed Mutagenesis on Electron and Proton Transfer Rates

The study of mutants with changes of amino acid residues near Q_B has served to elucidate the mechanisms of proton and electron transfer to Q_B. The results of replacing proton donor residues with non-proton donor residues were used to identify amino acid residues involved in the pathway for proton transfer to Q_B. The effect of modifying the environment of Q_B helped establish the role of the electrostatic potential in the mechanism of electron and proton transfer to Q_B. Mutations to residues affecting proton coupled electron transfer were obtained by 1) site-directed mutagenesis in *Rb. sphaeroides* (Paddock et al., 1989, 1990b; Takahashi and Wraight, 1992) 2) selection of second site mutations that compensated for detrimental effects of initial mutations (Hanson et al., 1992b; Okamura et al., 1992; Bylina and Wong, 1992) 3) Herbicide resistance mutants (Leibl et al., 1993). RCs containing the altered residues were isolated and characterized by measuring the cytochrome turnover rate (i.e. the rate of cytochrome oxidation during continuous illumination in the presence of excess cytochrome *c* and quinone), as well as by measuring electron transfer rates ($k_{AB}^{(1)}$, $k_{AB}^{(2)}$ and k_{BD}) and proton uptake (k_H). The characteristics of several RC mutants are described below.

A. Mutations Affecting Proton Transfer

1. Glu L212

Mutation of Glu L212, located near the Q_B binding site (see Fig. 2) to Gln in RCs from *Rb. sphaeroides* reduced the cytochrome turnover rate by a factor of ≈70 following the fast oxidation of 3 cytochromes (Paddock et al., 1989) (see Fig. 4).

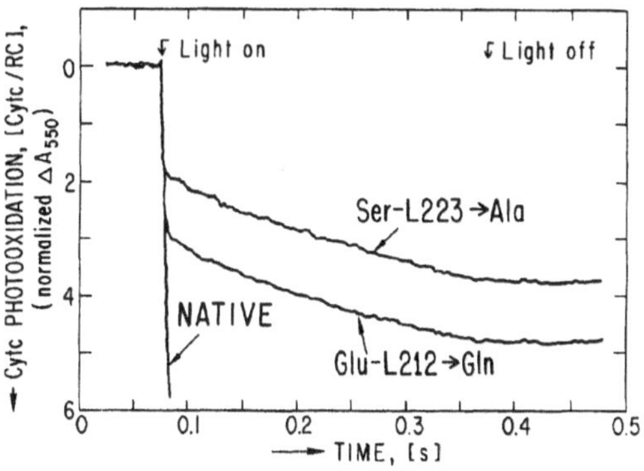

Fig. 4. Cytochrome photo-oxidation rates in native and mutant RCs. Native RCs showed a rapid turnover. Mutant RCs showed slow rates of turnover indicating a block in the quinone reduction cycle after oxidation of 2 or 3 cytochromes for the Ser L223 → Ala and Glu L212 → Gln mutants, respectively (from Okamura and Feher, 1992).

This result was explained by a reduction in the rate of transfer of the second proton $H^+(2)$ (see step 5 in Fig. 3) which slows down the turnover of the quinone reduction cycle. Since this block occurs after electron transfer of the second electron to Q_B, the RC can absorb a third photon to form the fully reduced quinone complex $Q_A(Q_BH)$ resulting in the oxidation of 3 cytochrome molecules. This hypothesis was verified by direct measurements of the electron and proton transfer rates, which showed that both the electron transfer rates $k_{AB}^{(1)}$, $k_{AB}^{(2)}$ were not appreciably changed, whereas the proton transfer rate was. When RCs were illuminated with a multi-turnover flash, biphasic proton uptake was observed with fast uptake of about one proton and slow uptake of a second proton (see Fig. 5) (Feher et al., 1990; McPherson et al., 1994).

The proton uptake rate k_H, measured using a pH sensitive dye, was biphasic with the slow phase reduced by a factor of about 200 to a rate of 6 s^{-1} (pH 7.5). This slow proton uptake rate accounted for the slow cytochrome turnover 12 s^{-1} (2 cytochromes/turnover). Control experiments in which Glu L212 was changed to Asp gave RCs with high turnover rates (Paddock et al., 1989, 1990a). The Glu L212 → Gln mutant was also studied by Takahashi and Wraight, (1992) who found a biphasic rate of electron transfer $k_{AB}^{(2)}$ after two flashes at pH above 7. The slow

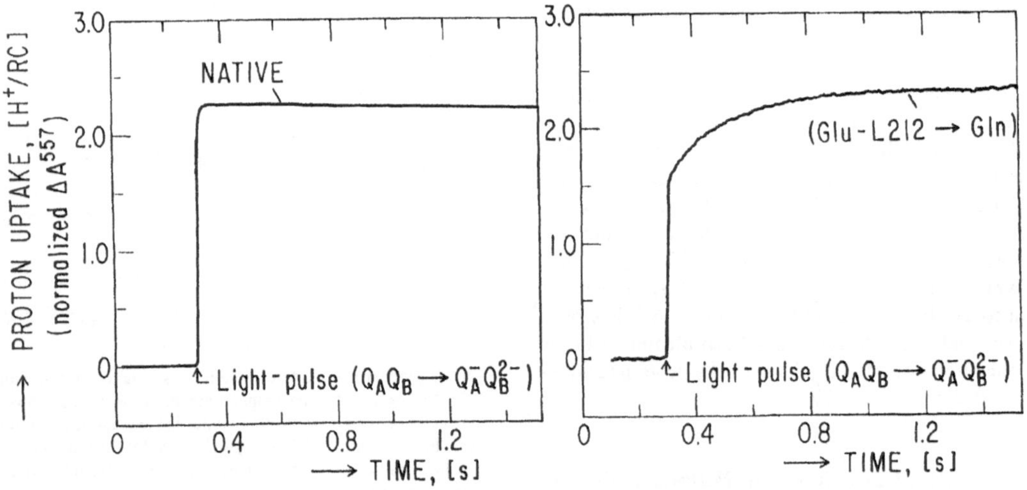

Fig. 5. Proton uptake measurements in Native (left) and mutant (right). RCs in the presence of excess cytochrome *c* after a strong saturating light pulse ($t \approx 1$ ms). The slow proton uptake step in the Glu L212 → Gln mutant indicates a bottleneck in a proton transfer step (from Okamura and Feher, 1992).

phase $k \approx 0.3\ s^{-1}$ (pH 7.5) was attributed to a slow proton uptake by the Glu L212 → Gln mutant. The difference between the rate observed by Takahashi and Wraight, (1992) and that measured by Feher et al., (1990) is at least partly due to the reduction of Q_A to Q_A^- under multi turnover conditions. The RCs containing Q_A^- exhibit a faster rate of protonation than those containing Q_A (McPherson et al., 1994). Shinkarev et al., (1993) used flash induced electrogenicity measurements on chromatophores from the Glu L212 → Gln mutant to show that the doubly reduced Q_B remains bound to the RC and inaccessible to the cytochrome bc_1 complex on the ms time scale. This shows that proton uptake is required before the release of Q_BH_2.

The change of Glu L212 → Gln is structurally conservative since the size and shape of the two residues are approximately the same. Preliminary X-ray crystallography measurements have indicated that no large changes (<0.5 Å) in the structure of the Glu L212 mutant are observed (Chirino et al.,1994). The pH dependence of the rates $k_{AB}^{(1)}$ and k_{BD} was eliminated in the mutant. Native RCs show a pH dependence with inflection points near pH 9.5 (Paddock et al., 1989; Takahashi and Wraight, 1992). This was explained by an apparent pK_a of 9.5 for Glu L212. However, recent results from IR spectroscopy indicate that Glu L212 is at least partially unprotonated at pH 7, indicating that a more complicated model for protonation must be invoked (Hienerwadel et al., 1992). Two models based on electrostatic calculations have been proposed to explain the results. In one model presented by Gunner and Honig, (1992), Glu L212 is fully ionized over the whole pH range when Q_B is oxidized. Formation of Q_B^- results in full proton uptake by Glu L212. Compensating proton release by other interacting residues account for the small value for the net proton uptake. In another model, (Beroza et al., 1995) Glu L212 is partially ionized from pH 7 to 9 and becomes fully ionized at higher pH. Formation of Q_B^- results in the fractional proton uptake by Glu L212, which increases with increasing pH. Further work is required to resolve the protonation state of Glu L212 and other interacting residues.

2. Ser L223

Ser L223, which provides a hydrogen bond to Q_B, was changed to Ala in RCs from *Rb. sphaeroides* (Paddock et al., 1990b). The cytochrome turnover rate slowed by a factor of 60 after an initial fast oxidation of 2 cytochromes (see Fig. 4). This was explained by a reduced rate for the transfer of the first proton, $H^+(1)$, in the quinone reduction cycle (step 4 – Fig. 3). A block in the uptake of this proton would reduce the rate of the second electron transfer and lead to the accumulation of the state $Q_A^- Q_B^-$ after oxidation of only 2 cytochromes. This interpretation was supported by the result that both proton uptake and electron transfer rates, $k^{(2)}_{AB}$, were found to be decreased by a factor of 400 (pH 7.5) (Table I). The change of Ser L223 → Ala does not lead to large structural changes as indicated by the lack of change in the electron transfer rates for the first electron $k^{(1)}_{AB}$ and the recombination k_{BD} and by an unchanged EPR spectrum of $Q_B^-Fe^{2+}$ (Paddock et al., 1990b). Control experiments in which Ser L223 was changed to the protonatable residues Thr and Asp gave RCs with normal turnover rates while substitution with Asn resulted in slow turnover (see Table 1). A second-site mutation of Tyr L222 → Cys was found to compensate for the Ser L223 → Ala change, restoring photosynthetic competence in *Rb. sphaeroides* (S. H. Rongey, unpublished).

The importance of Ser L223 was also shown by mutant RCs from *Rp. viridis* selected for herbicide resistance. Leibl et al., (1993) found that a mutation of Ser L223 → Ala was accompanied by a large reduction in the second electron transfer rate similar to that associated with this mutation in RCs from *Rb. sphaeroides*. The *Rp. viridis* RCs contained both the Ser L223 → Ala change and a change of Arg L217 → His. Evidence that the slow second electron transfer rate was due to the Ser L223 change and not the Arg L217 change came from a second double mutant containing the Arg L217 → His mutation along with a Val L220 → Leu change in which the second electron transfer rate was close to the native rate. In RCs from *Rb. capsulatus*, Bylina et al., (1992) found that the Ser L223 → Ala mutation resulted in a loss of photosynthetic competence, consistent with the slow rates of electron and proton transfer measured in RCs from *Rb. sphaeroides* and *Rp. viridis*. However, mutations of Ser L223 to Thr, Glu, Gly in *Rb. capsulatus* resulted in photosynthetically competent strains (Bylina et al., 1992). The Ser → Gly mutant is interesting since it lacks a proton donor group to replace the OH group in Ser. This mutation in RCs from *Rb. sphaeroides* did not decrease the proton coupled electron transfer rate $k^{(2)}_{AB}$ (Paddock et al., 1995). The X-ray crystal structure of the Ser L223 → Gly mutant RC showed that there were no changes in conformation (Axelrod et al., 1995). These results were interpreted in terms of a water molecule replacing the Ser side chain as the proton donor to Q_B.

Several Photosystem II RC herbicide resistant mutants have been obtained with alterations at Ser L264 on the D1 protein, homologous to Ser L223 in bacterial RCs (Oettmeier, 1992). These include Ser → Gly (Hirschberg and McIntosh, 1983), Ser → Ala (Erickson et al., 1984), Ser → Thr (Sato et al., 1988) and Ser → Asn (Pay et al., 1988). Interestingly, a Photosystem II mutant from *Amaranthus hybridus* containing the Ser → Gly at position 264 on the D1 subunit homologous to the Ser L223 site in purple bacteria, was also found to be photosynthetically competent and displayed normal electron transfer kinetics (Taoka and Crofts, 1990).

3. Asp L213

The Asp L213 residue located near Q_B was changed to Asn in RCs of *Rb. sphaeroides* (Takahashi and Wraight, 1990, 1992; Paddock et al., 1990b, 1994). The cytochrome turnover rate was decreased by a factor of 250 after a fast oxidation of 2 cytochromes similar to the effect of the Ser L223 → Ala mutation indicating a block in proton and electron transfer associated with the second electron. The electron transfer rate $k^{(2)}_{AB}$ coupled to proton transfer was found to be reduced by a factor of 1500 to ≈ 0.3 s^{-1} at pH 7.5. The change of Asp L213 → Asn produced significant changes in other rates besides $k^{(2)}_{AB}$. Both the first electron transfer rate $k^{(1)}_{AB}$ and recombination rate k_{BD} were decreased by a factor of ≈10 at pH 7.5 (see Table 1). These changes in rate are likely due to electrostatic effects associated with the loss of a negative charge on Asp L213 whose pK_a is estimated to be below 7. In addition, a double mutant with the changes Asp L213 → Asn and Glu L212 → Gln was constructed (Takahashi and Wraight, 1992; Takahashi et al., 1992). RCs from this mutant exhibited slow rates of cytochrome turnover and electron and proton transfer associated with the second electron $k^{(2)}_{AB}$, similar to the Asp L213 → Asn mutant. However, the rate of the first electron transfer $k^{(1)}_{AB}$ was fast and pH independent, similar to the Glu L212 → Gln mutant.

The reductions in proton and electron transfer rates described above can be explained if Asp 213 is involved in proton transfer of the first proton $H^+(1)$ that is necessary for the second electron transfer (see Fig. 3 step 4). Evidence for a proton donor role for Asp L213 comes from the finding that its replacement by Ser, Thr, Leu as well a Asn gives slow rates for $k^{(2)}_{AB}$ ≈ k_H ≈ 0.5 s^{-1}. However, replacement with the proton donor Glu gives a faster rate $k^{(2)}_{AB}$ ≈ 15 s^{-1} (Paddock et al., 1994). This rate is slower than the rate observed in native RCs $k^{(2)}_{AB}$ ≈ 10^3 s^{-1}, which is attributed to an electrostatic effect, as indicated by a slow k_{BD} (Table 1). Evidence that the reduction in proton uptake and electron transfer rates in the Asp L213 → Asn mutant are due to a proton transfer bottleneck comes from the observation by Takahashi and Wraight (1991) that these rates are increased by addition of azide and other weak acids. Paddock et al. (1994) found that azide also increases the recombination rate k_{BD} in Asp L213 → Asn RCs, indicating a destabilization of Q_B^-. This finding suggests a model in which binding of the azide anion near the Q_B site provides a negative charge which facilitates proton transfer. A similar increase in proton transfer rate was observed by addition of azide to proton transfer mutants of bacteriorhodopsin (Tittor et al., 1989; Otto et al., 1989).

The Asp L213 residue is the closest proton donor residue to Glu L212. It has, therefore, been suggested that it is involved in proton transfer to Glu L212 (McPherson et al., 1991; Takahashi and Wraight, 1992; Shinkarev et al., 1993). Supporting evidence of the involvement of Asp L213 in the proton transfer pathway to Glu L212 comes from experiments on the Asp L213 → Asn mutant in which the rate $k_{AB}^{(1)}$ for the reaction $Q_A^-Q_B \rightarrow Q_AQ_B^-$ is slower than in native RCs (see Table 1) (McPherson et al., 1991; Takahashi and Wraight, 1992). One explanation for the slower rate postulates a rate limiting proton transfer to ionized Glu L212 as a process that is required for electron transfer to Q_B. The slower rate in the mutant RCs lacking Asp L213 suggests that Asp L213 is involved in the proton transfer to Glu L212 (McPherson et al., 1991). Shinkarev et al. (1993) have presented a model for the pH dependence of $k_{AB}^{(1)}$ which accounts for the results from native RCs and the Glu L212 → Gln mutant but not the Asp L213 → Asn mutant. The failure of this model to account for the electron transfer rate in the Asp L213 → Asn mutant was attributed to a slow protonation rate in the Asp L213 → Asn mutant. This result was interpreted to indicate the central role of Asp L213 in proton transfer to residues near Q_B.

4. Glu H173

Recent work has implicated a residue on the H subunit in proton coupled electron transfer. The mutation of Glu H173 → Gln in RCs from *Rb. sphaeroides* has been found to decrease $k_{AB}^{(2)}$ by 20–50 (pH 7.5) indicating a possible role in proton coupled electron transfer (Takahasi and Wraight, 1995; Rongey et al., 1995). However, in contrast to the Asp L213 → Asn mutant, the back reaction rate k_{BD} was increased, indicating a destabilization of Q_B^-. These results could, in principle, be attributed to a slower rate limiting proton transfer in mutant RCs. However, measurements of the effects of substituting low potential quinones in the Q_A site on $k_{AB}^{(2)}$ showed that electron transfer, not proton transfer, was the rate limiting step. Thus, the mutation of Glu H173 → Gln slows down the rate of electron transfer or decreases the stability of the protonated semiquinone (M. S. Graige et al., 1995; and unpublished results).

B. Second-Site Mutants Compensating for Asp L213 (Asn M44, Arg M233, Gly L225)

Several second-site mutations which compensate for the loss of Asp L213 in *Rb. sphaeroides* and Asp L213 and Glu L212 in *Rb. capsulatus* have been obtained. These mutations give important information about the factors responsible for fast proton and electron transfer rates.

An interesting second-site mutation involves the Asn M44 → Asp change in *Rb. sphaeroides* (Rongey et al., 1993) and the homologous Asn M43 → Asp change in *Rb. capsulatus* (Hanson et al., 1992b, 1993; Maróti et al., 1994). These changes restore photosynthetic competence to the non-photosynthetic parent strains lacking Asp L213. The rates of the electron transfer reactions $k_{AB}^{(1)}$, $k_{AB}^{(2)}$ and k_{BD} for the double mutant in *Rb. sphaeroides* were found by Rongey et al., (1993) to be very similar to those in native RCs (see Table 1). The compensatory Asn M43 → Asp change in RCs from *Rb. capsulatus* containing the double mutation Asp L213→ Asn and Glu L212 → Ala increased $k_{AB}^{(2)}$ and proton transfer by about 10-fold compared to the low rates in the double mutant; however, these rates were considerably less (by $\approx 10^2$) than in native RCs (see Table 1; Maróti et al., 1994). The observation that either Asp L213 or Asp M44 can serve the same function in proton transfer explains the puzzling fact that Asp L213 is not conserved in different strains of photosynthetic bacteria. An Asp at either L213 or M44[2] is found in all photosynthetic bacteria in which RCs have been sequenced. RCs from *Rb. sphaeroides* (Williams et al., 1983, 1984), *Rb. capsulatus* (Youvan et al., 1984), and *Erythrobacter* OCH114 (Liebetanz et al., 1991) contain Asp at position L213 and an Asn at position

[2] Or the homologous residues, eg. M43 or M231 in *Rb. capsulatus*

M44, while RCs from *Rp. viridis* (Michel et al., 1986), *Rhodospirillum rubrum* (Belanger et al., 1988) and *Chloroflexus aurantiacus* (Ovchinnikov et al., 1988a,b) contain Asn at position L213 and Asp at position M44[2].

Another interesting second-site mutation involves the change of Arg M233 → Cys in *Rb. sphaeroides*[1] (Okamura et al., 1992) and the homologous change of Arg M231 → Leu in *Rb. capsulatus* (Hanson et al., 1992a). These changes confer photosynthetic competence to the parent strain lacking Asp L213. The mutant showed an increase in the recombination rate k_{BD} which is indicative of a negative charge near Q_B. The value of $k^{(2)}_{AB}$ in the second site mutant RCs from *Rb. sphaeroides* is larger than in RCs lacking Asp L213 but still slower than in native RCs (see Table 1). The unusual feature of the M233 site is its large distance to Q_B (≈ 15 Å) (see Fig. 2). The question, therefore, arises how a change at the M233[2] position can restore the rapid proton coupled electron transfer rate to Q_B. Possible explanations proposed involve the effect of altered electrostatics on a cluster of charged residues spanning the region between M233 and Q_B and the role of water molecules as proton donors (Hanson et al., 1992a; Okamura et al., 1992).

Other second-site mutations that restore photosynthetic competence to RCs from *Rb. capsulatus* lacking Asp L213 and Glu L212 are the changes of Gly L225 → Asp (Schiffer et al., 1993), Arg L217 → Cys and Asn M5 → Aso (Hanson et al., 1995). An important feature of this mutation as in all other second-site mutations that restore activity to RCs lacking Asp L213 is the increase in negative charge. These mutations show the importance of electrostatic effects on proton-coupled electron transfer.

C. Mutations Not Affecting Proton Transfer (Arg L217, Asp L210, His L190)

Mutations that did not produce large changes in the proton transfer rate include several conservative replacements that preserved the proton donating ability (see Table 1) eg. Glu L212 → Asp (Paddock et al., 1990), and Ser L223 → Thr (Paddock et al., 1990a). Several other mutations that did not result in a significant loss of activity were Arg L217 → Gln (M. L. Paddock unpublished) and Asp L210 → Asn (M. L. Paddock unpublished). Preliminary results on a mutant in which the residue His L190, which provides the second H bond to Q_B, was changed to Gln showed a near normal cytochrome turnover rate (J. Williams, unpublished). The role of H-bonded protons Q_B^- was studied by ENDOR measurement of their rates of exchange with deuterium (Paddock et al., 1995b). One of the protons, assigned to the H-bond from His L190, exchanged slowly ($\tau \approx 20$ s) with a rate that was independent of quinone turnover. The result was interpreted as indicating that this proton is not part of the proton transfer chain.

VI. Mechanism of Proton Coupled Electron Transfer

The coupling between electron transfer and proton uptake occurs in the second electron transfer to Q_B. The proposed mechanism for this step involves the uptake of two protons $H^+(1)$ and $H^+(2)$. The first proton uptake could occur either before or after electron transfer as shown in the upper and lower paths of Eq. (1) (Maróti and Wraight, 1990; Takahashi and Wraight, 1992; Paddock et al., 1994).

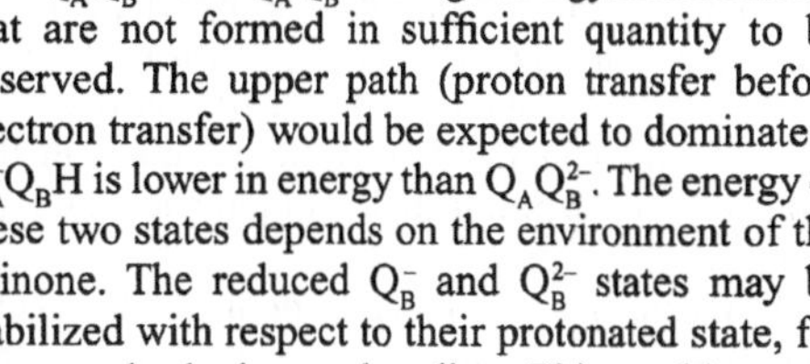

(2)

A difficulty in deciding between the two paths is that both Q_AQ_BH and $Q_AQ_B^{2-}$ are high energy intermediates that are not formed in sufficient quantity to be observed. The upper path (proton transfer before electron transfer) would be expected to dominate if $Q_A^-Q_BH$ is lower in energy than $Q_AQ_B^{2-}$. The energy of these two states depends on the environment of the quinone. The reduced Q_B^- and Q_B^{2-} states may be stabilized with respect to their protonated state, for instance, by hydrogen bonding. This could permit the electron transfer to proceed before protonation. It is also possible that proton transfer and electron transfer are highly cooperative and tightly coupled. In this case, electron transfer and proton transfer may proceed simultaneously. Experimentally, the individual steps have not yet been resolved kinetically, except for the case of the mutant (eg. Glu L212 → Gln) in which the second proton transfer is slow.

The mechanism involving the protonated semiquinone (upper path) can explain the observed values for $k^{(2)}_{AB}$ as follows (Paddock et al., 1994; Takahashi and Wraight, 1992). If the proton transfer rate $k^{(1)}_{H}$ is fast, $k^{(1)}_{H} >> k^{(2)}_{AB} \approx 10^3\,s^{-1}$, the observed rate would be given by $k^{(2)}_{AB} = k^{(2)}_{et}\, F(Q_BH)$ where $k^{(2)}_{et}$ is the rate of electron transfer to the protonated semiquinone shown in Eq. (2), and $F(Q_BH)$ is the fraction of protonated semiquinone.[3] We estimate $k^{(2)}_{et}$ to be $\approx 10^4\,s^{-1}$, approximately the same as $k^{(1)}_{AB}$. This assumes that similar free energy changes are associated with $k^{(2)}_{et}$ and $k^{(1)}_{AB}$ (see Eq. (3) below). For this value of $k^{(2)}_{et}$ and a $pK_a \approx 4$ for the protonated semiquinone the pH dependence $k^{(2)}_{AB}$ would approach an asymptotic limit of $\approx 10^4\,s^{-1}$ near pH 4, in agreement with the observed value. Various estimates for the pK_a of the protonated semiquinone Q_BH have been made. The pK_a for protonated ubisemiquinone in aqueous solution has been found to be near 5–6 (Swallow, 1982). The pK_a of the protonated semiquinone state $Q_A^-Q_BH$ in *Rb. sphaeroides* RCs was estimated to be < 7.4 (Kleinfeld et al., 1984, Takahashi and Wraight, 1992). Although the expected values for the pK_a of the protonated semiquinone is in reasonable agreement with those required by the mechanism of the upper path in Eq. (1), further evidence for the protonated semiquinone is needed to support the mechanism. The protonated semiquinone state Q_AQ_BH has not been observed even at relatively low pH ($\approx$ 5) (P. McPherson, unpublished results). Alternatively, a mechanism through the upper path could involve a rate limiting proton transfer $k^{(1)}_{H} < k^{(2)}_{et}$ at pH > 5. This would account for the pH dependence of $k^{(2)}_{AB}$ approaching the maximum value of $k^{(2)}_{et}$ at low pH. The mechanisms through the upper path can also explain the slow rate for $k^{(2)}_{AB}$ seen in the Asp L213 → Glu mutant by either a pK_a shift of the proton donor Glu L213 or by a reduction in the population of the protonated semiquinone (Paddock et al., 1994).

The mechanism involving the doubly reduced Q_B^{2-} state (lower path) cannot be excluded at present. However, estimates of the magnitude of the rate discussed below suggest that it may be less likely. The electron transfer rate $k'^{(2)}_{et}$ for the reaction $Q_A^- Q_B^- \rightarrow Q_AQ_B^{2-}$ involving the unprotonated semiquinone, depends on the free energy change of the reaction, $\Delta G^{o\prime}_{et}$. The dependence of $k'^{(2)}_{et}$ on $\Delta G^{o\prime}_{et}$ can be obtained from the Marcus equation (Marcus and Sutin, 1985) which gives the rate of electron transfer as a function of an intrinsic rate k_o and a reorganization energy λ:

$$k_{et} = k_o\, e^{-(-\Delta G^o - \lambda)^2/4\lambda k_B T} \qquad (3)$$

where k_B is the Boltzmann constant and T the absolute temperature. The value for k_o depends on the electron transfer pathway between the donor and acceptor. The value of λ depends on the change in energy due to structural changes associated with electron transfer. We assume that the values of k_o and λ are the same for the first and second electron transfer between Q_A and Q_B. Furthermore, under the conditions that $\lambda >> |\Delta G|$ we can obtain the electron transfer rate $k'^{(2)}_{et}$ involving the unprotonated semiquinone by comparing it with $k^{(1)}_{AB}$ and correcting for the difference in free energy using Eq. (4) (Paddock et al., 1991).

$$k'^{(2)}_{et} = k^{(1)}_{AB}\, e^{-\delta G/2k_B T} \qquad (3)$$

where $\delta G = \Delta G^{o\prime(2)}_{et} - \Delta G^{o(1)}_{AB}$ is equal to the difference in free energies of the two reactions involving the second and first electrons, respectively. ($\Delta G^{o(1)}_{AB} \approx -100$ mev, $k^{(1)}_{AB} \approx 10^4\,s^{-1}$).

One estimate of the limiting value for $\Delta G^{o\prime(2)}_{et}$ comes from the equilibrium constant of the reaction $Q_A^- Q_B^- \rightleftharpoons Q_AQ_B^{2-}$ in RCs (where the protonation state of Q_B^{2-} is not known). At high pH (> 10.5) the equilibrium constant for this reaction approaches a value of $\approx$ 0.02. Assuming that the quinol is unprotonated above pH 10.5, the value of $K'^{(2)}_{et} \approx 0.02$ and $\Delta G^{o\prime(2)}_{et} \approx 100$ meV (Kleinfeld et al., 1984; Takahashi and Wraight, 1992; McPherson et al., 1993, 1994). However, the $pK_a \approx 10.5$ for the ionization of the second proton from Q_BH_2 is much lower than observed for fully reduced ubiquinone in solution which has a pK_a for the first proton near 13 (Swallow, 1982; Morrison et al., 1982). The second proton would have an even higher pK_a. A smaller value ($< 2\times10^{-4}$) for the equilibrium constant $K'^{(2)}_{et}$ involving the unprotonated Q_B^{2-} was estimated by Paddock et al., (1994). This gave a lower limit of $\Delta G^{o\prime}_{et} > 220$ meV. The $k'^{(2)}_{et}$ values calculated from Eq. (4) using these estimated free energies were respectively ~ 1 and 2 orders of magnitude smaller than the observed values of $k^{(2)}_{AB} \cong 10^3\,s^{-1}$ at pH 7.5. This suggests that the mechanism does not involve the unprotonated Q_B^{2-} state. Another observation (Paddock et al., unpublished) which argues against the Q_B^{2-} mechanism is that use of quinone analogues with substitution of Br

[3] For simple titrations $F(Q_BH) = 0.5$ at pH = pK_a. However, the functional form of $F(Q_BH)$ in the RC may be more complicated due to interactions between interacting residues (see Paddock et al., 1994).

for the ring proton in Q_0 decreases the value of $k^{(2)}_{AB}$. Bromine-substituted quinones are expected to stabilize Q_B^{2-}, which would result in an increase of $k^{(2)}_{AB}$ of the lower path. This is in disagreement with the observation and therefore argues in favor of the upper path. (See note added in proof.)

VII. Pathways for Proton Transfer

A model for the proton transfer pathways in RCs from *Rb. sphaeroides* based on site directed mutagenesis studies is shown in Fig. 6 (Okamura and Feher, 1992; Paddock et al., 1990b; Takahashi and Wraight, 1994). In this model two protons $H^+(1)$ and $H^+(2)$ are transferred from solution to Q_B along two pathways. The first proton $H^+(1)$ is transferred along a pathway involving Ser L223 and Asp L 213. The second proton is transferred along a pathway involving Glu L212 and Asp L213. Inspection of the X-ray crystal structure shows that protons from solution have access to the region near Asp L213 with Asp L210 or Arg L217 possibly acting as intermediates, although evidence for the direct involvement of these residues is lacking. In addition, a void in the protein structure large enough to hold 5–6 water molecules is seen in the X-ray crystal structure in a region bordered by the methoxy groups of Q_B, Asp L213 and Glu L212 (Beroza et al., 1992). This pocket is likely to contain water molecules which could play an important role in the proton transfer to Q_B.

The pathway shown in Fig. 6 is based on the assumption that the large changes in proton transfer rate due to mutations of Glu L212, Ser L223 and Asp L213 result from the properties of the mutated residue and not from conformational changes. In addition, the model makes the assumption that the loss of activity due to mutation of a protonatable residue indicates its role as a proton donor in a proton transfer chain. Other interpretations of the results from site-directed mutagenesis are possible. These include: a) short range steric effects or long range conformational effects not related to the role of the mutated residue as a proton donor; b) electrostatic effects on either the rate or the equilibrium of the proton transfer reaction.

Although the mutagenesis results indicate that specific amino acid residues play important roles in facilitating proton transfer, the results from second-site mutations indicate that proton transfer pathways are not unique and that alternate proton transfer pathways are possible. The general requirement for fast proton transfer is the availability of suitably oriented proton donor/acceptor groups with appropriate pK_as. A likely possibility is that in some cases internal water molecules may be involved in proton transport. In addition, negatively charged groups

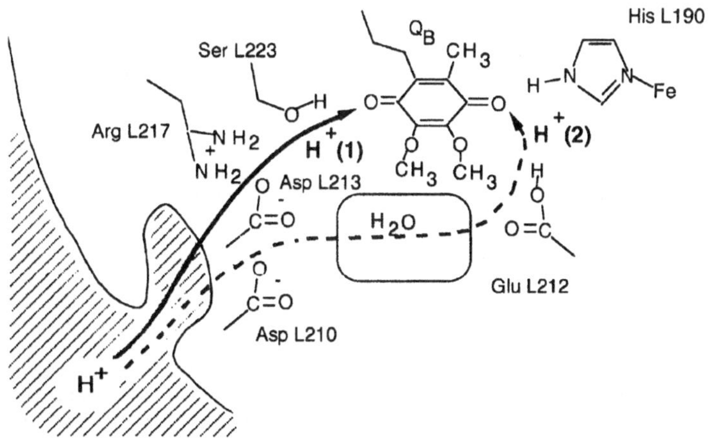

Fig. 6. Pathways for Proton Transfer in RCs from *Rb. sphaeroides*. The pathway for the first proton $H^+(1)$, to the quinone carbonyl H-bonded to Ser L223, involves Asp L213 and Ser L223. The pathway for the second proton $H^+(2)$, to the carbonyl H-bonded to His L190 involves Glu L212 and possibly Asp L213. A cavity near the methoxy groups of Q_B presumably containing internal water is likely to play a role in the proton transfer (from Okamura and Feher, 1992).

appear to be necessary to facilitate proton transfer by stabilizing protons in the interior of the protein.

The results obtained in bacterial RCs may also be applicable to RCs in oxygen-evolving species. The RC of Photosystem II in these organisms contains a bound Q_B site with properties similar to those observed in RCs from purple bacteria. In Photosystem II there are no obvious homologies to the Glu L212 and Asp L213 residues. However, a bound bicarbonate (Govindjee and van Rensen, 1993) has been shown to increase the rate of the second electron transfer reaction to Q_B. Diner and Petrouleas (1990) have suggested that the bicarbonate binds to the non-heme Fe in the PS II RC and plays a role in proton transfer similar to the acid groups in the RCs from purple bacteria. Takahashi and Wraight (1991) have suggested that the bicarbonate effect in Photosystem II RCs is similar to the azide effect in purple bacterial mutants deficient in proton transfer. It is interesting that Arg residues 233 and 251 in the D2 protein have been implicated in bicarbonate binding (Cao et al., 1991). These residues may have some homology with the Arg M233 found to affect proton transfer in bacterial RCs (Hanson et al., 1992a; Okamura et al., 1992). In view of the diversity in proton transfer pathways in bacterial RCs and mutants it is not surprising that specific residues responsible for fast proton transfer to Q_B in bacterial RCs are not conserved in the Photosystem II RC. Nevertheless, the Photosystem II RC may utilize the same principles for increasing proton transfer, e.g. the presence of protonatable groups and negative charges near Q_B to facilitate proton transfer.

Note Added in Proof

The mechanism of proton coupled electron transfer for the second electron $k^{(2)}_{AB}$ was studied in native RCs by substituting low potential quinones for Q_{10} in the Q_A site while retaining Q_{10} in the Q_B site (Graige et al., 1995). This should change the rate of electron transfer but not the rate of proton transfer. The values for $k^{(2)}_{AB}$ were found to increase with increasing driving force with a dependence consistent with the Marcus theory, indicating that electron transfer was the rate limiting step. This result, along with those previously discussed, indicates a mechanism in which rate limiting electron transfer follows rapid proton transfer ($Q_A^-Q_B^- + H^+ \rightleftharpoons Q_A^-Q_BH \rightarrow Q_AQ_BH^-$).

Acknowledgements

We would like to acknowledge M. L. Paddock, P. H. McPherson, S. H. Rongey, P. Beroza for helpful discussions and contributions to our understanding of proton transfer in bacterial reaction centers. The work from our lab was supported by grants from NIH and NSF.

References

Allen JP, Feher G, Yeates TO, Rees DC, Deisenhofer J, Michel H and Huber R (1986) Structural Homology of reaction centers from *Rb. sphaeroides* and *Rp. viridis* as determined by X-ray diffraction. Proc Natl Acad Sci USA 83: 8589–8593

Allen JP, Feher G, Yeates TO, Komiya H and Rees DC (1988) Structure of the reaction center from *Rb. sphaeroides* R-26: Protein-cofactor (quinones and Fe^{2+}) interactions. Proc Natl Acad Sci USA 85: 8487–8491

Axelrod HL, Abresch E, Paddock ML, Okamura MY, Feher G and Rees DC (1995) X-ray crystallographic analysis of a site-directed proton transfer mutant Ser-L223 → Gly of the bacterial RC from *Rb. sphaeroides*. Biophys J 68: A247

Baciou L, Sinning I and Sebban P (1991) Study of Q_B^- stabilization in herbicide-resistant mutants from the purple bacterium *Rhodopseudomonas viridis*. Biochemistry 30: 9110–9116

Belanger G, Berard J, Corriveau P and Gingras G (1988) The structural genes coding for the L and M subunits of *Rhodospirillum rubrum* photoreaction center. J Biol Chem 263: 7632–7638

Beroza P, Fredkin DR, Okamura MY and Feher G (1992) Proton transfer pathways in the reaction center of *Rhodobacter sphaeroides*: A computational study. In: Breton J and Vermeglio A (eds) The Photosynthetic Bacterial Reaction Center II, 363–374. Plenum Press, New York

Breton J and Vermeglio A (1988) The Photosynthetic Bacterial Reaction Center: Structure and Dynamics Plenum, New York

Breton J and Vermeglio A, Eds (1992) The Photosynthetic Bacterial Reaction Center II; Structure Spectroscopy and Dynamics, Vol 237. Plenum Press, New York

Breton J, Thibodeau DL, Berthomieu C, Mantele W, Vermeglio A and Nabedryk E (1991) Probing the primary quinone environment in photosynthetic bacterial reaction centers by light induced FTIR difference spectroscopy. FEBS Lett 278: 257–260

Bylina EJ and Wong R (1992) Analysis of spontaneous herbicide resistant revertants derived from *Rhodobacter capsulatus* in which serine L223 of the reaction center is replaced with alanine. In: Murata N (ed) Research in Photosynthesis, Vol I, pp 369–372. Dordrecht: Kluwer Academic Publ.

Bylina EJ, Jovine RVM and Youvan DC (1989) A genetic system for rapidly assessing herbicides that compete for the quinone binding site of photosynthetic reaction centers. Biotechnology 7: 69–74

Cao JC, Vermaas WFJ and Govindjee (1991) Arginine residues in the D2 polypeptide may stabilize bicarbonate binding in photosystem II of *Synechocystis* sp. PCC6803 Biochim Biophys Acta 1059: 171–180

Chang C-H, Tiede D, Tang J, Smith U, Norris J and Schiffer M (1986) Structure of *Rb. sphaeroides* R-26 reaction center. FEBS Lett 205: 82–86

Chirino AJ, Lous EJ, Huber M, Allen JP, Schenck CC, Paddock ML, Feher G and Rees DC (1994) Biochemistry 33: 4584–4593

Cramer WA and Knaff DB (1990) Energy Transduction in Biological Membranes Springer-Verlag, New York

Crofts AR and Wraight CA (1983) The electrochemical domain of photosynthesis. Biochim Biophys Acta 726: 149–185

Deisenhofer J and Michel H (1989) The photosynthetic reaction center from the purple bacterium *Rhodopseudomonas viridis*. EMBO J 8: 2149–2170

Deisenhofer J, Epp O, Miki K, Huber R and Michel H (1984) X-ray structure analysis of a membrane protein complex: Electron density map at 3 Å resolution and a model of the chromophores of the photosynthetic reaction center from *Rhodopseudomonas viridis*. J Mol Biol 180: 385–398

Diner BA, Petrouleas V and Wendoloski JJ (1991) The iron-quinone electron acceptor complex of photosystem II. Physiol Plant 81: 423–36

Ermler U, Fritzsch G, Buchanan SK and Michel H (1994) Structure of the photosynthetic reaction centre from *Rhodobacter sphaeroides* at 2.65 Å resolution: Cofactors and protein-cofactor interactions. Structure 2: 925–936

Erickson JM, Rahire M, Bennoun P, Delepelaire P, Diner B and Rochaix J-D (1984) Herbicide resistance in *Chlamydomonas reinhardtii* results from a mutation in the chloroplast gene for the 32 kilodalton protein of photosystem II. Proc Natl Acad Sci USA 81: 3617–3621

Feher G, Isaacson RA, Okamura MY and Lubitz W (1985) ENDOR of semiquinones in reaction centers from *Rhodopseudomonas sphaeroides*. In: Michel-Beyerle ME (eds) Antennas and Reaction Centers of Photosynthetic Bacteria, pp 174–189. Springer-Verlag, Berlin

Feher G, Allen JP, Okamura MY and Rees DC (1989) Structure and function of bacterial photosynthetic reaction centers. Nature 339: 111–116

Feher G, McPherson PH, Paddock M, Rongey S, Schoenfeld M and Okamura MY (1990) Protonation of quinones in reaction centers from *Rb. sphaeroides*, In: Baltscheffsky M (ed.) Current Research in Photosynthesis I, pp 39–46. Kluwer Academic Publishers, Dordrecht

Feher G, Paddock ML, Rongey SH and Okamura MY (1992) Proton transfer pathways in photosynthetic reaction centers studied by site-directed mutagenesis. In: Pullman A, Jortner J and Pullman B (eds) Membrane Proteins : Structures, Interactions and Models, pp 481–495. Kluwer Academic Publishers, Dordrecht

Govindjee and van Rensen JJS (1993) Photosystem II reaction centers and bicarbonate In: Deisenhofer J and Norris JK (eds) The Photosynthetic Reaction Center, pp 357–389. Academic Press, New York

Graige MS, Paddock ML, Labahn A, Bruce JM, Feher G and Okamura MY (1995) The mechanism of proton-coupled electron transfer to Q_B in reaction centers from *Rb. sphaeroides*. Biophys J 68: A246

Gunner MR (1993) Calculations of proton uptake coupled to electron transfer in bacterial photosynthetic reaction centers. Biophys J 64: A375

Gunner MR and Honig B (1992) Calculations of proton uptake in *Rhodobacter sphaeroides* RCs. In: Breton J and Vermeglio A (eds) The Photosynthetic Bacterial Reaction Center II, 403–410. Plenum Press, New York

Hales BJ and Case EE (1981) Immobilized radicals IV. Biological semiquinone anions and neutral semiquinones. Biochim Biophys Acta 637: 291–302

Hanson DK and Schiffer M (1995) Electrostatic effects and proton conduction in bacterial reaction center protein. Biophys J 68: A246

Hanson DK, Baciou L, Tiede DM, Nance SL, Schiffer M and Sebban P (1992a) In bacterial reaction centers, protons can diffuse to the secondary quinone by alternative pathways. Biochim Biophys Acta 1102: 260–265

Hanson DK, Nance SL and Schiffer M (1992b) Second site mutation at M43 (Asn → Asp) compensates for the loss of two acidic residues in the Q_B site of the reaction center. Photosynth Res 32: 147–153

Hanson DK, Tiede DM, Nance SL, Chang CH and Schiffer M (1993) Site-specific and compensatory mutations imply unexpected pathways for proton delivery to the Q_B binding site of the photosynthetic reaction center. Proc Natl Acad Sci USA 90: 8929–8933

Hienerwadel R, Nabedryk E, Paddock ML, Rongey SH, Okamura MY, Mäntele W and Breton J (1992) Proton transfer mutants of *Rb. sphaeroides*: Characterization of reaction centers by infrared spectroscopy. In: Murata N (ed.) Research in Photosynthesis, Vol I, pp 437–440. Dordrecht: Kluwer Academic Publishers

Hirschberg J and McIntosh L (1983) Molecular basis of herbicide resistance in *Aramanthus hybridus*. Science 222: 1346–1349

Kleinfeld D, Okamura MY and Feher G (1984) Electron transfer in reaction centers of *Rb. sphaeroides*. I. Determination of the charge recombination pathway of $D^+Q_AQ_B^-$ and free energy relations between $Q_A^-Q_B$ and $Q_AQ_B^-$. Biochim Biophys Acta 766: 126–140

Kleinfeld D, Okamura MY and Feher G (1985) Electron transfer in reaction centers from *Rb. sphaeroides*. II Free energy and kinetic relations between the acceptor states $Q_A^-Q_B^-$ and $Q_AQ_B^{2-}$. Biochim Biophys Acta 809: 291–310

Leibl W, Sinning I, Ewald G, Michel H and Breton J (1993) Evidence that serine L223 is involved in the proton transfer pathway to Q_B in the photosynthetic reaction center of *Rhodopseudomonas viridis*. Biochemistry 32: , 19581964

Liebetanz R, Hornberger U and Drews G (1991) Organization of the genes coding for the reaction center L-subunit and M-subunit and B870 antenna polypeptides alpha and polypeptide beta from the aerobic photosynthetic bacterium *Erythrobacter* species OCH114. Molec Microbiol 5: 1459–1468

Lubitz W, Abresch EC, Debus RJ, Isaacson RA, Okamura MY and Feher G (1985) Electron nuclear double resonance of semiquinone in reaction centers of *Rhodopseudomonas sphaeroides*. Biochim. Biophys. Acta 808: 464–469

Marcus RA and Sutin N (1985) Electron transfer in chemistry and biology. Biochim. Biophys. Acta 811: 265–322

Maróti P and Wraight CA (1988) Flash-induced H^+ binding by bacterial photosynthetic reaction centers: Influences of the redox states of the acceptor quinones and primary donor. Biochim Biophys Acta 934: 329–347

Maróti P and Wraight CA (1990) Kinetic correlation between H^+ binding, semiquinone disappearance and quinol formation in reaction centers of *Rb. sphaeroides*. In: Baltscheffsky M (ed)

Current Research in Photosynthesis, Vol 1, pp 165–168. Kluwer Academic Publishers, Dordrecht

Maróti P, Hanson DK, Baciou L, Schiffer M and Sebban P (1994) Proton conduction within the reaction centers of *Rhodobacter capsulatus*: The electrostatic role of the protein. Proc Natl Acad Sci 91: 5617–5621

McPherson PH, Okamura MY and Feher G (1988) Light induced proton uptake by photosynthetic reaction centers from *Rhodobacter sphaeroides* R-26.1 Protonation of the one-electron states $D^+Q_A^-$, DQ_A^-, $D^+Q_AQ_B^-$, and $DQ_AQ_B^-$. Biochim Biophys Acta. 934: 348–368

McPherson PH, Okamura MY and Feher G (1990) Electron transfer from the reaction center of *Rb. sphaeroides* to the quinone pool: Doubly reduced Q_B leaves the reaction center. Biochim Biophys Acta 1016: 289–292

McPherson PH, Rongey SH, Paddock ML, Feher G and Okamura MY (1991) The rate of electron transfer $Q_A^-Q_B \rightarrow Q_AQ_B^-$ in RCs from *Rb. sphaeroides* in which Asp-L213 is replaced with Asn. Biophys J 59: 142a

McPherson PH, Okamura MY and Feher G (1993) Light-induced proton uptake by photosynthetic reaction centers from *Rhodobacter sphaeroides* R-26.1 II. Protonation of the state $DQ_AQ_B^{2-}$. Biochim Biophys Acta 1144: 309–324

McPherson PH, Schönfeld M, Paddock ML, Okamura MY and Feher G (1994) Protonation and free energy changes associated with formation of Q_BH_2 in native and Glu-L212 → Gln mutant reaction centers from *Rhodobacter sphaeroides*. Biochemistry 33: 1181–1193

Michel H, Weyer KA, Gruenberg H, Dunger I, Oesterhelt D and Lottspeich F (1986) The 'light' and 'medium' subunits of the photosynthetic reaction center from *Rhodopseudomonas viridis*: isolation of the genes, nucleotide and amino acid sequence. EMBO J 5: 1149–1158

Mitchell P (1961) Coupling of phosphorylation to electron and proton transfer by a chemi-osmotic type of mechanism. Nature, 191: 144–148

Morrison LE, Schelhorn JE, Cotton TM, Bering CL and Loach PA (1982) Electrochemical and spectral properties of ubiquinone and synthetic analogs: relevance to bacterial photosynthesis. In: Trumpower BL (ed) Function of Quinones in Energy Conserving Systems, 35–58. Academic Press, New York

Oettmeier W (1992) Herbicides of photosystem II. In: Barber J (ed) The Photosystems: Structure, Function and Molecular Biology, pp 349–408. Elsevier Science Pub., Amsterdam

Okamura MY and Feher G (1992) Proton transfer in reaction centers from photosynthetic bacteria. Ann Rev Biochem 61: 861–96

Okamura MY, Debus RJ, Kleinfeld D and Feher G (1982) Quinone binding sites in reaction centers from photosynthetic bacteria. In: Trumpower B (ed) Function of Quinones in Energy Conserving Systems, 299–317. Academic Press, New York

Okamura MY, Paddock ML, McPherson PH, Rongey S and Feher G (1992) Proton transfer in bacterial reaction center: Second site mutations Asn M44 → Asp or Arg M233 → Cys restore photosynthetic competence to Asp L213 → Asn Mutants in RCs from *Rb. sphaeroides*. In: Murata N (ed) Research in Photosynthesis, vol. I, pp 349–356. Kluwer Academic Publishers, Dordrecht

Otto H, Marti T, Holz M, Mogi T, Lindau M, Khorana HG and Heyn MP (1989) Aspartic acid–96 is the internal proton donor in the reprotonation of the Schiff base of bacteriorhodopsin. Proc Natl Acad Sci USA 86: 9228–9232

Ovchinnikov YA, Abdulaev NG, Shmukler BE, Zargarov AA, Kutuzov MA, Telezhinskaya IN, Levina NB and Zolotarev AS (1988a) Photosynthetic reaction center of *Chloroflexus aurantiacus*: primary structure of the M subunit. FEBS Lett 232: 364–368

Ovchinnikov YA, Abdulaev NG, Zolotarev AS, Shmukler BE, Zargarov AA, Kutuzov MA, Telezhinskaya IN and Levina NB (1988b) Photosynthetic reaction center of *Chloroflexus aurantiacus*: Primary structure of the L subunit. FEBS Lett 231: 237–42

Paddock ML, Rongey SH, Feher G and Okamura MY (1989) Pathway of proton transfer in bacterial reaction centers: Replacement of Glu 212 in the L subunit inhibits quinone(Q_B) turnover. Proc Natl Acad Sci (USA) 86: 6602–6606

Paddock ML, Feher G and Okamura MY (1990a) pH dependence of charge recombination in RCs from *Rb. sphaeroides* in which Glu-L212 is replaced with Asp Biophys J 57: 569a

Paddock ML, McPherson PH, Feher G and Okamura MY (1990b) Pathway of proton transfer in bacterial reaction centers: Replacement of serine-L223 by alanine inhibits electron and proton transfers associated with the reduction of quinone to dihydroquinone. Proc Natl Acad Sci (USA) 87: 6803–6807

Paddock ML, Feher G and Okamura MY (1991) Reaction centers from three herbicide resistant mutants of *Rhodobacter sphaeroides* 2.4.1: Kinetics of electron transfer reactions Photosynth Res 27: 109–119

Paddock ML, Juth A, Feher G and Okamura MY (1992) Electrostatic effects of replacing Asp-L210 with Asn in bacterial RCs from *Rb. sphaeroides*. Biophys J 61: 153a

Paddock ML, Rongey SH, McPherson PH, Juth A, Feher G and Okamura MY (1994) Pathway of proton transfer in bacterial reaction centers: The Role of Asp-L213 in proton transfers associated with the reduction of quinone to dihydroquinone. Biochemistry 33: 734–745

Paddock ML, Feher G and Okamura MY (1995a) Suggested role of an internal water molecule in proton transfer in a Ser-L223 → Gly site-directed mutant reaction center from *Rb. sphaeroides*. Biophys J 68: A246

Paddock ML, Abresch E, Isaacson RA, Feher G and Okamura MY (1995b) The role of hydrogen bonded protons in the photochemical reduction of Q_B to Q_BH_2 in reaction centers of *Rb. sphaeroides*. Biophys J 68: A246

Parson WW (1978) Quinones as secondary acceptors. In: Clayton RK and Sistrom WR (ed) The Photosynthetic Bacteria, 455–469. Plenum Press, New York

Pay A, Smith MA, Nagy F and Marton L (1988) Sequence of the psbA gene from wild type and triazin resistant *Nicotiana plumbaginfolea*. Nucleic Acids Res 16: 8176

Prince RC and Dutton PL (1978) Protonation and the reducing potential of the primary electron acceptor. In: Clayton RK and Sistrom WR (ed) The Photosynthetic Bacteria, 439–453. Plenum Press, New York

Rashin AA, Iofin M and Honig B (1986) Internal cavities and buried waters in globular proteins. Biochemistry 25: 3619–3629

Rongey S, Paddock ML, Juth AL, McPherson PH, Feher G and Okamura MY (1991) Pathway of proton transfer in bacterial RCs from *Rb. sphaeroides*: Replacement of AspL213 with

Asn inhibits electron and proton transfer to the secondary quinone. Biophys J 59: 142a

Rongey SH, Paddock ML, Feher G and Okamura MY (1993) Pathway of proton transfer in bacterial reaction centers: second site mutation Asn-M44 → Asp restores electron and proton transfer in reaction centers from the photosynthetically deficient Asp-L213 → Asn mutant of *Rb. sphaeroides*. Proc Natl Acad Sci USA 90: 1325–1329

Rongey SH, Juth AL, Paddock ML, Feher G and Okamura MY (1995) Importance of a carboxylic acid at H173 for proton-coupled electron transfer in RCs of *Rb. sphaeroides*. Biophys J 68: A247

Rutherford AW and Evans MCW (1980) Direct measurement of the redox potential of the primary and secondary quinone electron acceptors in *Rb. sphaeroides* (wild type) by EPR spectrometry. FEBS Lett 110: 257–261

Sato F, Shigematsu Y and Yamada Y (1988) Selection of an atrazine-resistant tobacco cell line having a mutant *psba* gene. Mol Gen Genet 214: 358–360

Schiffer M, Chan C-K, Chang C-H, DiMagno TJ, Fleming GR, Nance S, Norris J, Snyder S, Thurnauer M, Tiede DM and Hanson DK (1992) Study of reaction center function by analysis of the effects of site-specific and compensatory mutations. In: Breton J and Vermeglio A (eds) The Photosynthetic Bacterial Reaction Center II, pp 351–361. Plenum Press, New York

Shinkarev VP, Takahashi E and Wraight CA (1993) Flash-induced electric potential generation in wild type and L212EQ mutant chromatophores of *Rhodobacter sphaeroides:* QBH2 is not released from L212EQ mutant reaction centers. Biochim Biophys Acta 1142: 214–216

Swallow AJ (1982) Physical chemistry of semiquinones. In: Trumpower BL (eds) Function of Quinones in Energy Conserving Systems, pp 59–72. Academic Press, New York

Takahashi E and Wraight CA (1990) A crucial role for Asp L213 in the proton transfer pathway to the secondary quinone of reaction centers from *Rhodobacter sphaeroides*. Biochim Biophys Acta 1020: 107–111

Takahashi E and Wraight CA (1991) Small weak acids stimulate proton transfer events in site-directed mutants of the two ionizable residues, Glu L212 and Asp L213, in the Q_B-binding site of *Rhodobacter sphaeroides* reaction center. FEBS Letters 283: 140–144

Takahashi E, Maróti P and Wraight CA, (1992) Coupled proton and electron transfer pathways in the acceptor quyinone complex of reaction centers from Rb. sphaeroides. In: Muller A, Ratajczak H, Junge W and Diemann E (eds) Electron and Proton Transfer in Chemistry and Biology, pp 219–236. Elsevier Publ., Amsterdam

Takahashi E and Wraight CA (1992) Proton and electron transfer in the acceptor quinone complex of *Rhodobacter sphaeroides* reaction centers: characterization of site-directed mutants of the two ionizable residues Glu L212 and Asp L213 in the Q_B binding site. Biochemistry 31: 855–866

Takahashi E and Wraight CA (1994) Molecular genetic manipulation and characterization of mutant photosynthetic reaction centers from purple non-sulfur bacteria. In: Barber J (ed) Advances in Molecular and Cell Biology: Molecular Processes in Photosynthesis, pp 197–251. JAI Press, Greenwich

Taoka S and Crofts AR (1990) Two electron gate in triazine resistant and susceptible *Aramanthus hybridus*. In: Baltscheffsky M (ed.) Current Research in Photosynthesis, Vol I, pp 547–550 Kluwer Academic Publishers, Dordrecht

Tiede DM and Hanson DK (1992) Protein relaxation following quinone reduction in *Rb. capsulatus*: Detection of likely protonation linked optical absorbance changes of the chromophores. In: Breton J and Vermeglio A (eds) The Photosynthetic Bacterial Reaction Center, Vol II, pp 341–350. Plenum Press, New York

Tittor J, Soell C, Oesterhelt D, Butt HJ and Steiner M (1989) A defective proton pump, point-mutated bacteriorhodopsin Asp 96 → Asn is fully reactivated by azide. EMBO J 8: 3477–3482

Vermeglio A (1977) Secondary electron transfer in reaction centers of *Rhodopseudomonas sphaeroides*: Out of phase periodicity of 2 for the formation of ubisemiquinone and fully reduced ubiquinone. Biochim Biophys Acta 459: 516–524

Vermeglio A and Clayton RK (1977) Kinetics of electron transfer between the primary and the secondary electron acceptor in reaction centers from *Rhodopseudomonas sphaeroides*. Biochim Biophys Acta 461: 159–165

Williams JC, Steiner LA, Ogden RC, Simon MI and Feher G (1983) Primary structure of the M subunit of the reaction center from *Rhodopseudomonas sphaeroides*. Proc Natl Acad Sci USA 80: 6505–6509

Williams JC, Steiner LA, Feher G and Simon MI (1984) Primary structure of the L subunit of the reaction center from *Rhodopseudomonas sphaeroides*. Proc Natl Acad Sci USA 81: 7303–7307

Wraight CA (1977) Electron acceptors of photosynthetic bacterial reaction centers: Direct observation of oscillatory behavior suggesting two closely equivalent ubiquinones. Biochim Biophys Acta 459: 525–531

Wraight CA (1979) Electron acceptors of bacterial photosynthetic reaction centers II H+ binding coupled to secondary electron transfer in the quinone acceptor complex. Biochim Biophys Acta 548: 309–327

Wraight CA (1982) The involvement of stable semiquinones in the two electron gates of plant and bacterial photosystems. In: Trumpower BL (ed) Function of Quinones in Energy Conserving Systems, pp 181–197. Academic Press, New York

Youvan DC, Bylina EJ, Alberti M, Begusch H and Hearst JE (1984) Nucleotide and deduced polypeptide sequences of the photosynthetic reaction center, B870 antenna, and flanking polypeptides from *Rb. capsulatus*. Cell 37: 949–957

Color Plates

A

B

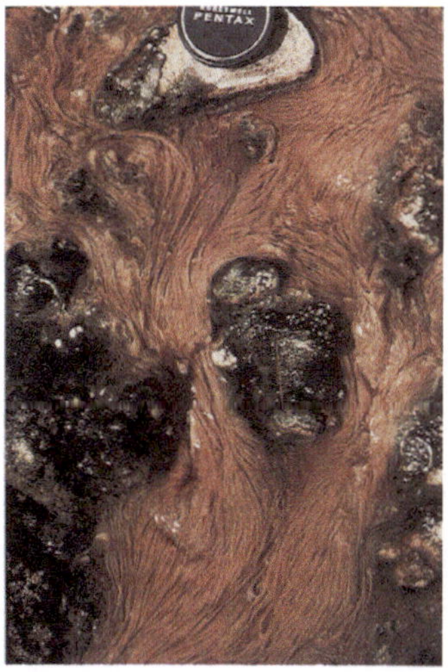

C

D

Color Plate 1. (A) Section through a microbial mat consisting of a top cover of the cyanobacterium *Synechococcus* underlaid by *Chloroflexus*. The cut is about 15 cm across and the water temperature ~60 °C. This is the source of *Chloroflexus aurantiacus*, strain OK-70-fl, Kahneeta Hot Springs, Oregon. (B) Upper limits of *Chloroflexus* as a distinct undermat at ~68 °C. The outflow from the spring is at about 73 °C, and the siliceous sinter is dominated by *Synechococcus* until the edges cool to about 68 °C where the *Chloroflexus* undermat accretes (more orange on photograph). Buffalo Spring, White Creek drainage, YNP. (C) *Chloroflexus* as a photoautotroph (orange-red streamers) growing at a temperature of ~ 64–62 °C on primary sulfide in Badstofuhver, Hveragerdi, Iceland. The sulfide is quite depleted by about 62 °C, at which point the green mat of the cyanobacterium, *Chlorogloeopsis* 'high temperature form' (formerly referred to as *Mastigocladus* 'HTF') appears (see Castenholz, 1973a). (D) *Heliothrix oregonensis* as an aerotolerant mat above a deep green cyanobacterial layer in a non-sulfidic Kahneeta hot spring (Oregon); temperature ~ 50 °C. (See Chapter 5, pp 89–90, Figs. 1–4.)

R. E. Blankenship, M. T. Madigan and C. E. Bauer (eds): Anoxygenic Photosynthetic Bacteria, pp. CP1–CP10.

A

B

C

Color Plate 2. (A) ***Chlorobium tepidum*** mat (green cover) in highly sulfidic 'Travelodge Stream', Rotorua, New Zealand. The temperature is 45 °C in the foreground, with a pH usually in the range of 5.3–6.2 (see Castenholz et al., 1990). (B) Facultatively anoxygenic *Oscillatoria* cf. *amphigranulata* in small sulfidic spring WH-1 (Cirque-1), Whakarewarera, New Zealand. Temperature at the grey-green edge of the cyanobacterium was ~56 °C. *Thiobacillus*-like whitish-yellow streamers occurred upstream to about 68°C (see Castenholz, 1976). (C) Inverted *Chloroflexus* mat at about 61–62 °C in Ystihver Stream, Húsavik, North Iceland. The primary sulfide which supports the photoautotrophic *Chloroflexus* is lowered within the mat, so that an underlayer of oxygenic cyanobacteria (*Chlorogloeopsis* 'high temperature form') occurs at about 1 mm below the *Chloroflexus* (see Jørgensen and Nelson, 1988). (See Chapter 5, pp. 92–98, Figs. 5, 6, 8.)

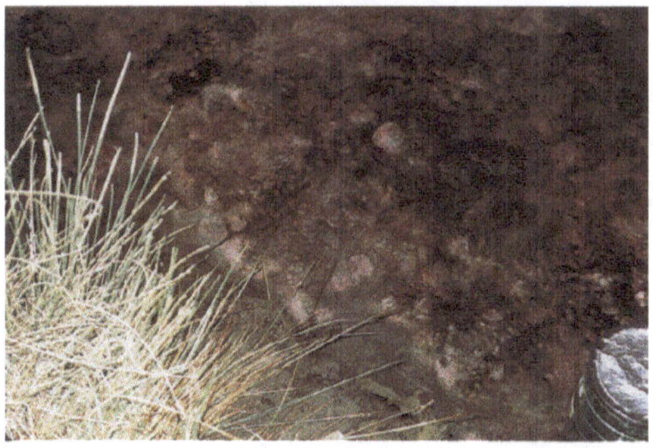

A

B

Color Plate 3-Upper. (A) Exposed *Chromatium tepidum* in a nighttime view of a 45–48 °C mat of Hunter's Hot Springs, Oregon. In daytime the *Chromatium* forms an underlayer beneath the brownish-red cyanobacterium *Oscillatoria terebriformis,* which descends below the *Chromatium* at night (see Richardson and Castenholz, 1987). (B) Daytime view of the same area as (A), with *O. terebriformis* forming the mat surface, but with *Chromatium* swarming into the water in the can which is open at both ends, but which had been darkened by a top lid a few hours earlier. (See Chapter 5, p. 100, Figs. 10–11.)

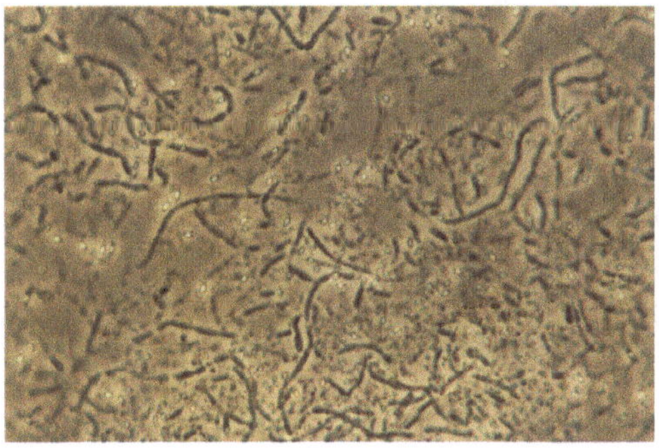

A

B

Color Plate 3-Lower. (A) Phase contrast optical microscope image of Nile Blue stained cells of *Bacillus cereus.* (B) Same as A, but activated by 450 nm light showing specific fluorescence of PHA. Spores do not fluoresce and note that not all cells contain PHA. This is a very specific reaction and can give a quantitative measure of PHA from fluorescence intensity. (See Chapter 60, p. 1247, Fig. 1.)

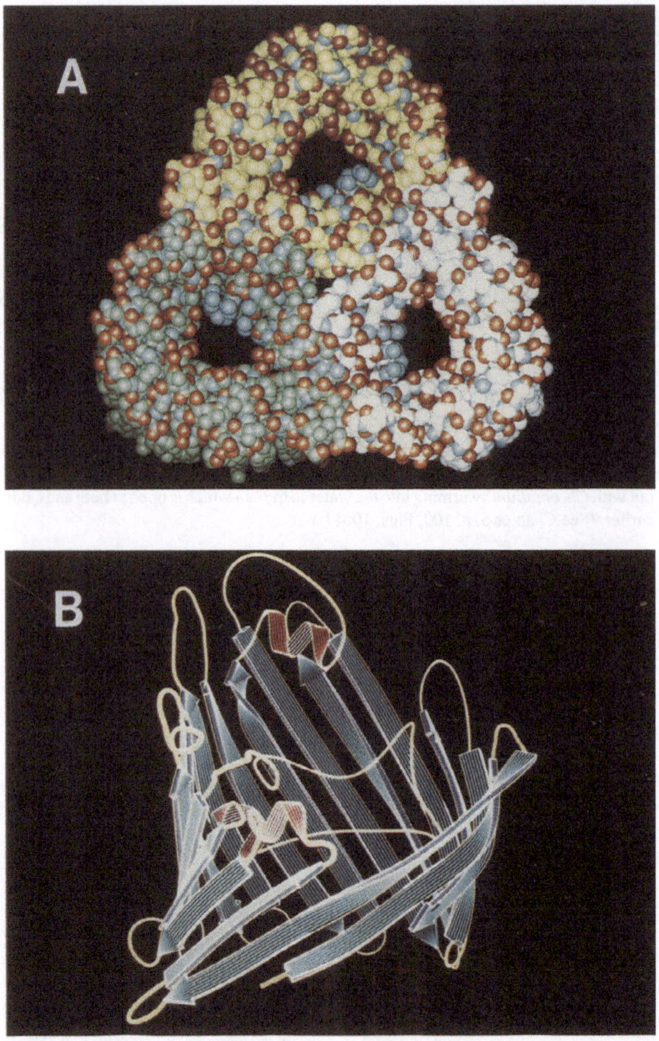

Color Plate 4. Structure representations of porins from *Rhodopseudomonas blastica* and *Rhodobacter capsulatus*: (A) Space-filling model of the porin from *Rp. blastica* DSM 2131 as established at 2.0 Å resolution (Kreusch et al., 1994). The view is perpendicular to the membrane plane, the three pore eyelets are clearly visible. Hydrogens are not considered in this model. (B) Secondary structure of one subunit of the major porin from *Rb. capsulatus* 37b4 as viewed from the threefold symmetry axis. The external medium is at the top, the periplasm at the bottom. The 16 β-strands are given by blue arrows and the 3 short α-helices by orange helices. The long 43-residue loop inserting into the barrel is obvious. The interface part of the barrel is rather low while the membrane-facing part is high. The 'rough' barrel end at the top and the 'smooth' barrel end at the bottom can be visualized clearly. (See Chapter 11, p. 224, Fig. 9.)

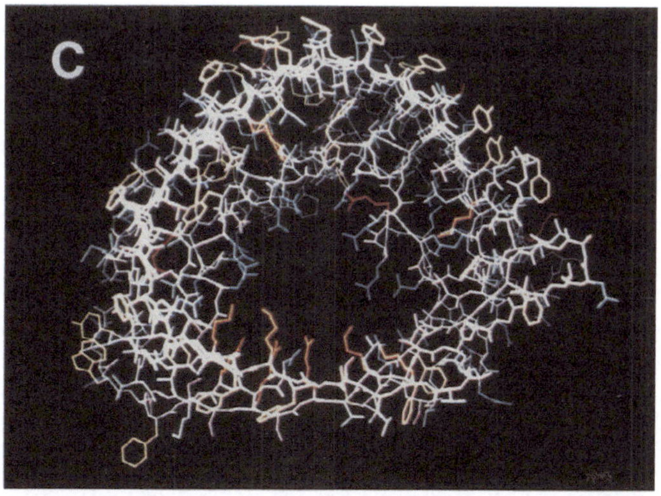

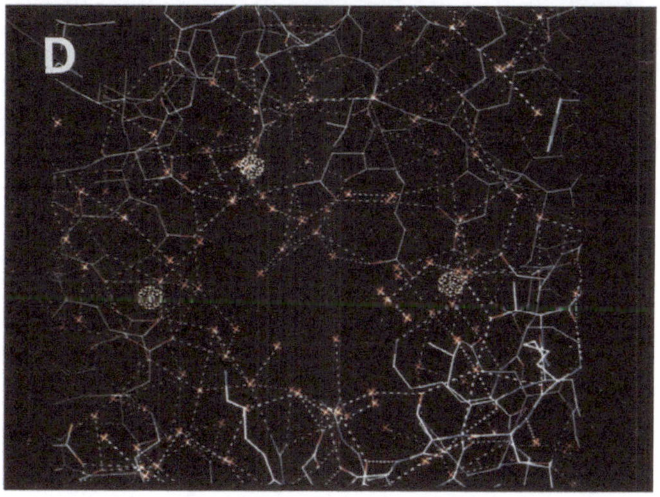

Color Plate 5. (Continued from Color Plate 4.) (C) Color-coded projection of one subunit from the *Rb. capsulatus* 37b4 porin onto the membrane plane. Aromatic side chains are yellow; positively and negatively charged side chains are red and dark blue, respectively. The interface and the threefold symmetry axis are at the bottom. Obviously, the rim of the pore eyelet is lined by positively charged side chains protruding from the interface area of the barrel juxtaposed to negatively charged side chains from the 43-residue loop running into the barrel interior. Most likely these charges form an electric field that facilitates permeation of polar solutes and hinders nonpolar ones. (D) Detail of the construction of the pore eyelet of the *Rb. capsulatus* porin. The amino acid residues are given in the usual color code, the fixed water molecules by crosses and the calcium ions by balls. Hydrogen bonds are indicated by dashed lines. 'Fixation' of the water molecules means that they are clearly visible in the X-ray structure analysis at the high resolution of 1.8 Å. This implies that water molecules are accommodated at the water positions most of the time. Still, these water molecules can move away. It should be noted that there is a defined network of hydrogen bonded water molecules at the eyelet leaving a rather small nonstructured cross-section in the center. (See Chapter 11, p. 224, Fig. 9.)

A

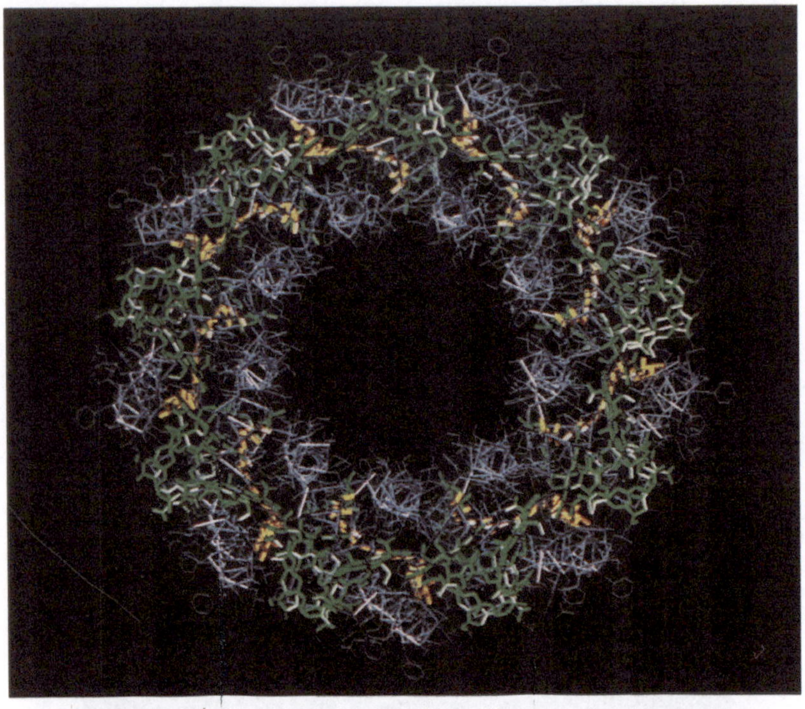

B

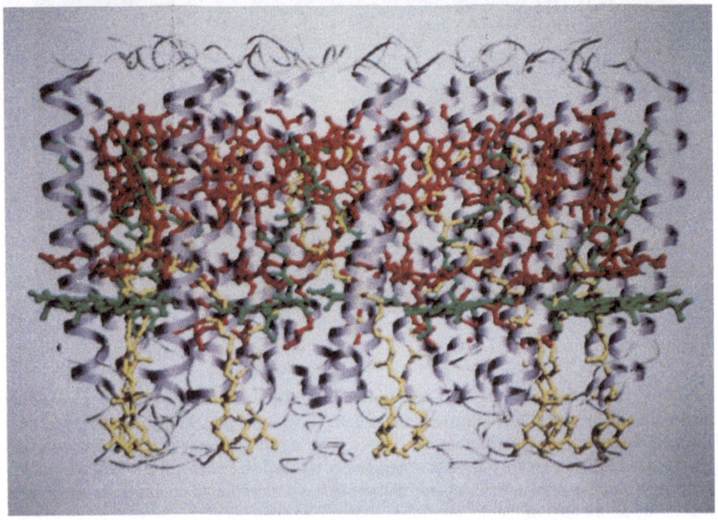

Color Plate 6. (A) Cytoplasmic view of the LH2 antenna complex from *Rhodopseudomonas acidophila*. The protein is shown in purple, with BChl molecules in green and carotenoid in yellow. For further details see McDermott et al. (1995). (B) The LH2 complex from *Rhodopseudomonas acidophila* viewed perpendicular to the symmetry axis. The periplasmic surface of the membrane is at the top and the cytoplasmic surface is at the bottom. The α helical parts of the protein are shown as purple ribbons. BChl B850 molecules are shown in red, with B800 molecules in green and carotenoids in yellow. For further details see McDermott et al. (1995). (See Chapter 16, p. 343, Figs. 21–22.)

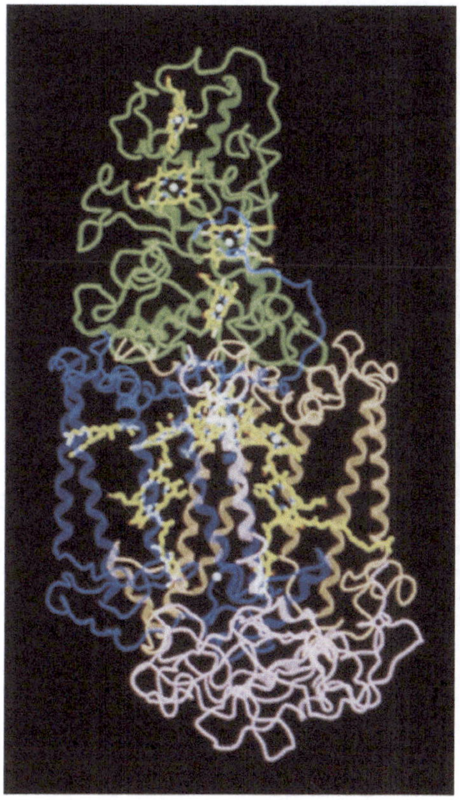

A

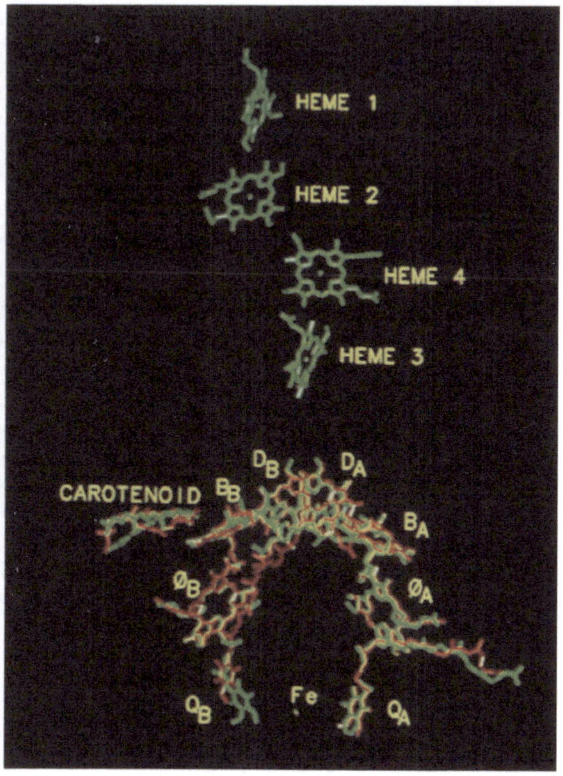

B

Color Plate 7. (A) The photosynthetic reaction center from *Rp. viridis*. The protein subunits are depicted in form of a smoothed ribbon diagram with the C subunit in green, the L subunit in brown, the M subunit in blue, and the H subunit in purple. The cofactors are represented as atomic models with carbon atoms in yellow, nitrogen atoms in blue, oxygen atoms in red, and magnesium and iron atoms in grey. Figs. 1, 2, 10, and 11 were generated with the computer program package O (Jones et al., 1991). (B) Superposition of the reaction center cofactors from *Rp. viridis* (green) and *Rb. sphaeroides* (red). The phytyl side chains of the bacteriochlorophyll and bacteriopheophytin molecules have been omitted for the sake of clarity. (See Chapter 23, p. 507, Figs. 1–2.)

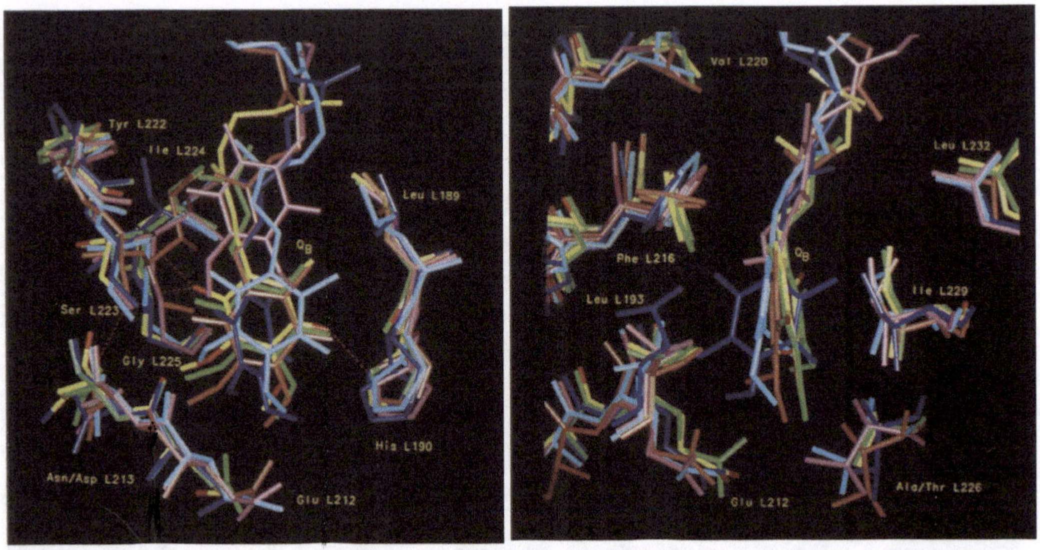

Color Plate 8-upper. Orthogonal views of the Q_B site: Six different structures are superimposed. The color coding is as follows: Q282 (multi-colored: carbon atoms are yellow, nitrogen atoms are blue, oxygen atoms are red), 1PRC (green), 2RCR (dark blue), 4RCR (brown), 1PSS (light blue) and 1PCR (purple). The dashed red lines indicate hydrogen bonds in the Q282 structure. (See Chapter 23, p. 517, Fig. 10.)

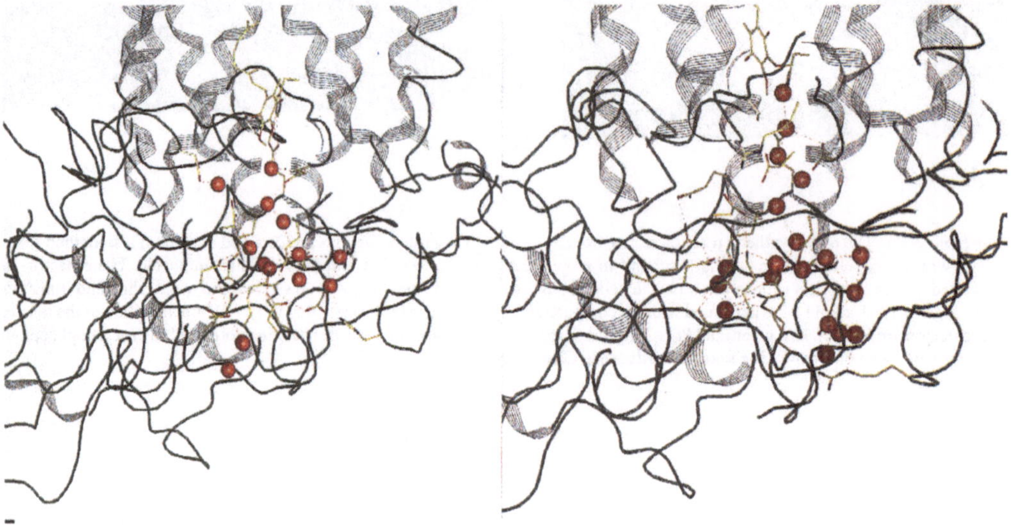

Color Plate 8-lower. The 'water chain' from the cytoplasm to the Q_B site in the reaction centers of *Rp. viridis* (left) and *Rb. sphaeroides* (right), respectively. The protein backbones of the subunits are displayed in green, the water molecules in red. Also shown are the Q_B molecules and the protonatable amino acid residues lining the 'water chain'. (See Chapter 23, p. 520, Fig. 11.)

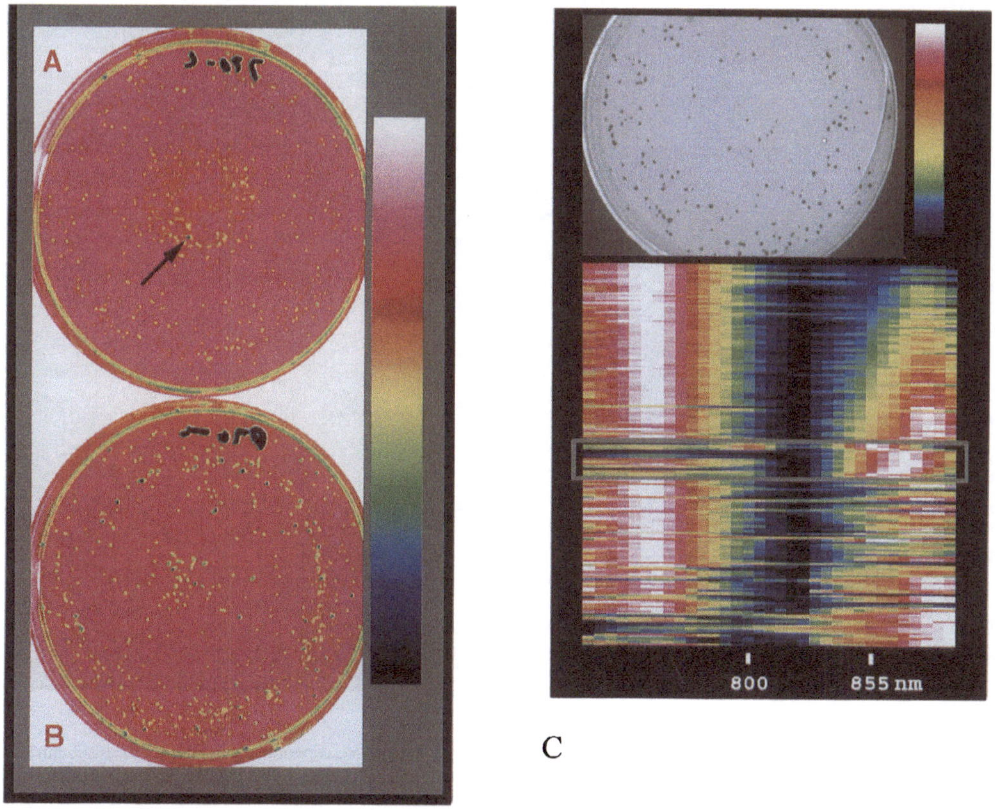

Color Plate 9. Pseudocolored image of bacterial colonies expressing mutagenized LHII complexes, taken at 860 nm, wherein TSM and conventional CCM are compared. (A) shows a spread of colonies from the conventional seven site CCM library with one positive colony (indicated by the arrow). This was the only positive colony observed in a set of similar spreads totaling approximately 4000 colonies. (B) shows a typical spread from the analogous TSM library. Approximately 6% of the colonies are classified as positive. Images were recorded using a radiometrically calibrated CCD camera. Gray scale values were rescaled and linearly mapped to pseudocolors to enhance the differences in optical density, i.e., higher ODs are mapped to darker pseudocolors. (C) Color contour map showing the spectral diversity of a seventeen site combinatorial library affecting amino acid residues near the β-subunit BChl (dimer) binding site. The upper left panel shows a monochrome image taken at 400 nm of a typical spread of colonies. The upper right hand panel shows a color bar; black corresponds to the lowest absorbance and white corresponds to the highest absorbance. The lower panel is a color contour map generated by DIS where the horizontal axis corresponds to wavelength (730–890 nm) and the vertical axis is colony number. Each horizontal row represents a spectrum encoded by pseudocolor. Spectra of BChl binding mutants are enclosed in the gray box. Mutants are displayed in 'full deflection' mode so each row scales from black to white. Nine percent of the colonies (15 out of 168) were judged to assemble LH antennae. Several categories of spectra are observed: 1) pseudo WT (i.e., dimer peak at 855 nm, monomer peak at 800 nm), 2) pseudo-WT mutants with a 5 nm red shift of the dimer band, 3) pseudo-LHI (i.e., 10–15 nm shift of the dimer band and reduced or missing monomer peak), 4) a mutant showing a single peak at 855 nm, and 5) mutants absorbing mainly at 760 nm (due to free BChl in the membrane) which are classified as 'nulls'. (See Chapter 61.)

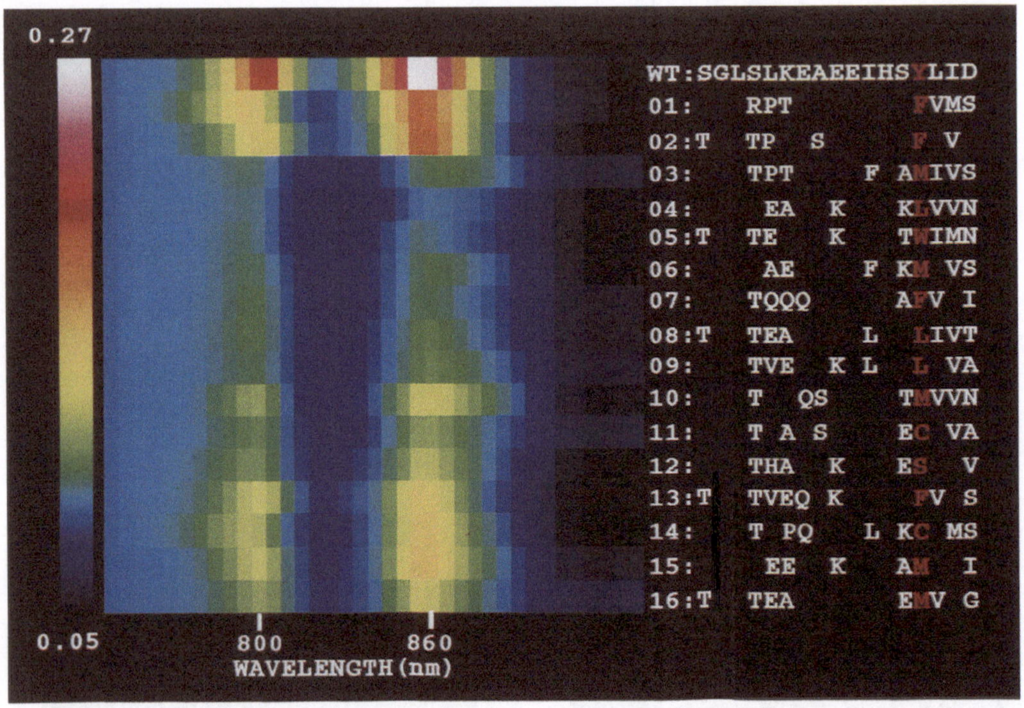

Color Plate 10. Correlation of sequence data with spectra from a combinatorial mutagenesis experiment on LHII near the β-subunit BChl (monomer) binding site. This color contour map is displayed in 'absolute' mode with the highest absorbance on the image set to white (OD = 0.72) and the lowest absorbance on the image set to black (OD = 0.05). The top spectra is WT which shows stronger absorption than these mutants. The amino acid sequence for each mutant is displayed to the right of its spectrum. Only amino acids different from the WT sequence are listed. The position indicated in red did not include the WT Tyr in the target set. (See Chapter 61.)

Chapter 27

The Recombination Dynamics of the Radical Pair P^+H^- in External Magnetic and Electric Fields

Martin Volk, Alexander Ogrodnik and Maria-Elisabeth Michel-Beyerle
Institut für Physikalische und Theoretische Chemie,
Technische Universität München, Lichtenbergstr. 4,
85747 Garching, Germany

R. E. Blankenship, M. T. Madigan and C. E. Bauer (eds): Anoxygenic Photosynthetic Bacteria, pp. 595–626.

Summary

When electron transfer to the primary quinone is blocked, the radical pair P^+H^-(P: primary donor, H: bacteriopheophytin at the A-branch) recombines on the 10 ns time scale either to the ground state P or, after hyperfine-induced singlet-triplet-mixing, to the triplet state $^3P^*$. An external magnetic field hinders singlet-triplet-mixing, thus reducing the yield of $^3P^*$ and slowing the recombination of P^+H^-. Magnetic field dependent measurements of the recombination dynamics allow the determination of the recombination rates k_S and k_T and the exchange interaction J. From these parameters free energies, electronic matrix elements and reorganization energies relevant for the fast charge separation and slow charge recombination processes in the reaction center can be determined. In many cases, recombination data constitute the sole experimental access to such parameters, which constitute the basis for the theoretical treatment of electron transfer processes.

In this review, results obtained on quinone-depleted reaction centers of *Rhodobacter sphaeroides*, *Rhodobacter capsulatus* and *Chloroflexus aurantiacus* are discussed in the context of (i) the similarity of reaction centers from different organisms, (ii) the mechanism of primary charge separation, (iii) the distinction between structural and energetic effects of genetic alterations of the reaction center, and (iv) the effects of an external electric field, which shifts the energy of the charge separated states. Furthermore, different recombination dynamics observed in transient absorption and delayed fluorescence reveal an inhomogeneous broadening of radical pair energies in the reaction center. This energetic broadening allows us to understand a variety of phenomena: (a) the observed multiphasic electron transfer kinetics, (b) the unexpectedly weak electric field effects on the fluorescence and (c) the discrepancies of energetics determined by delayed fluorescence and transient absorbance measurements. As a consequence, absorption measurements are better suited to determine the average of the energetic distribution of the radical pair P^+H^-.

I. The Magnetic Field Effect

Excitation of reaction centers (RCs) from *Rhodobacter (Rb.) sphaeroides* in which the primary quinone Q at the active A-branch was reduced showed the formation of a state P^F (Parson et al., 1975; Cogdell et al., 1975). This state was identified as the radical pair P^+H^-, P being the primary donor and H the bacteriopheophytin (BPheo) at the active branch. At room temperature, P^F was found to decay on the 10 ns time scale, partly to the ground state and partly to the state P^R, identified as the lowest triplet state $^3P^*$. From its unusual spin polarization (Dutton et al., 1973; Uphaus et al., 1974; Thurnauer et al., 1975) this triplet state was recognized (Blankenship et al., 1977; Haberkorn and Michel-Beyerle, 1977) to be formed via the magnetic field dependent radical pair mechanism (Michel-Beyerle et al., 1976; Schulten and Schulten, 1977; Werner et al., 1978; Haberkorn and Michel-Beyerle, 1979). In accordance with this mechanism, the yield of $^3P^*$ (Blankenship et al., 1977; Hoff et al., 1977) and the lifetime of P^+H^- (Michel-Beyerle et al., 1979, 1980) were found to be affected by an external magnetic field. Since the early eighties, RCs depleted of the primary quinone were predominantly investigated (Ogrodnik et al., 1982), because the interpretation of the results for Q-reduced RCs suffers from several difficulties: (i) the complex spin dynamics due to the coupling of the radical electron spin on H to the electron spin on Q (Lang, 1991), which in turn is coupled to the fast relaxing spin on the iron ion (Okamura et al., 1975), (ii) the distortion of the energies of the charge-separated states by the negative charge on Q which affects the electron transfer kinetics (Breton et al., 1986) and (iii) the destabilization of the Q binding site by reduction of Q (Scheidel, 1989). For earlier reviews see (Hoff, 1981, 1986; Boxer et al., 1983).

A. Hyperfine Induced Singlet-Triplet-Mixing

After formation of P^+H^- from the singlet state $^1P^*$ on the ps timescale (Kaufmann et al., 1975; Rockley et al., 1975; Woodbury et al., 1985; Martin et al., 1986; Fleming et al., 1988; Holzapfel et al., 1989), the radical electron spins on P and H are in a singlet configuration. In RCs in which the electron transfer from H to Q is blocked the radical pair P^+H^- recombines with a lifetime on the 10 ns timescale (Parson and Cogdell, 1975; Ogrodnik et al., 1982; Schenck et al., 1982; Chidsey et al., 1984; Ogrodnik et al., 1988). During this time the spins of the radical electrons on P and H undergo singlet-triplet-mixing (S-T-mixing). S-T-mixing is due to the hyperfine interaction (HFI) of the radical electron spins with their respective nuclear spin environments. In a simplified picture it can be viewed as a precession of the electron spins around the effective magnetic fields originating from the nuclear spins. Since the two electrons have different nuclear environments, in general the precession frequencies of the two electron spins will be different, resulting in a coherent oscillation between the original singlet and the triplet spin configuration.

B. Reaction Scheme of Radical Pair Recombination

The complete scheme for the recombination of P^+H^- is shown in Fig. 1. In addition to recombination to the

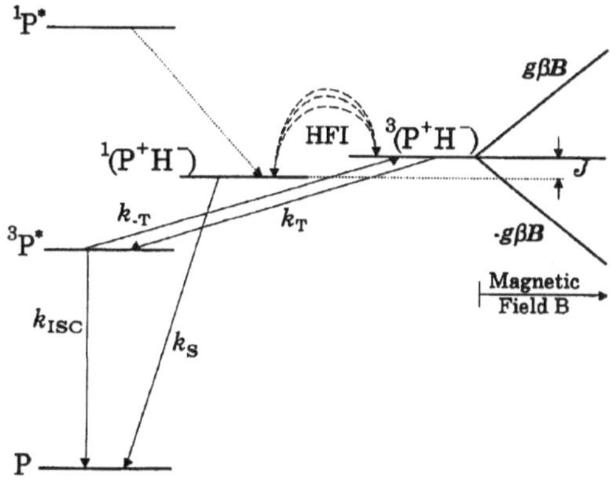

Fig. 1. Kinetic scheme for the recombination of P^+H^- in Q-depleted or Q-reduced RCs. The energetic splitting of $^1(P^+H^-)$ and $^3(P^+H^-)$ due to exchange interaction (J) and Zeeman interaction ($\pm g\beta B$) are shown considerably enlarged.

singlet ground state P, HFI induced S-T-mixing opens a second recombination channel from the triplet radical pair state to $^3P^*$. Also shown are the exchange interaction J, i.e. the energy splitting of $^1(P^+H^-)$ and $^3(P^+H^-)$ ($\approx 10G \approx 10^{-7}eV$),[1] and the Zeeman splitting of the triplet states in an external magnetic field B. The dipole interaction between the two radical electron spins, causing additional splitting of the triplet radical pair energies, is small ($D = 5.5G$) due to their large distance (17Å, as deduced from the X-ray structure (Deisenhofer et al., 1984; Chang et al., 1986; Allen et al., 1987)) and can be neglected for the discussion of radical pair recombination (Ogrodnik et al., 1985).[2]

The isoenergetic S-T-mixing is hindered by an energetic splitting of the singlet and triplet radical pair levels. For a magnetic field $B_{max} = J$ one of the triplet states is isoenergetic to the singlet state, thus one expects the maximum triplet yield (ϕ_T) for this field. The width ΔB of the magnetic field dependence of ϕ_T around resonance is determined by inhomogeneous broadening due to hyperfine interaction and homogeneous broadening reflecting the lifetime of singlet and triplet levels, with the lifetime broadening constituting the dominant contribution in RCs (Haberkorn and Michel-Beyerle, 1977, 1979; Werner et al., 1978).

Abbreviations: B – BChl monomer at the active branch; BChl – bacteriochlorophyll; BPheo – bacteriopheophytin; Car – carotenoid; *Cf.* – *Chloroflexus;* H – BPheo at the active branch; HFI – hyperfine interaction; MARY – MAgnetic field dependence of the Reaction Yield; P – primary donor (BChl dimer); PVA – polyvinyl alcohol; Q – quinone at the active branch; *Rb.* – *Rhodobacter;* RC – reaction center; *Rp.* – *Rhodopseudomonas;* RYDMR – Reaction Yield Detected Magnetic Resonance; S-T-mixing – singlet-triplet-mixing

[1] In this article, J is given in units of G, corresponding to the value of $J/g\beta$, with g the g-factor and β Bohr's magneton.

[2] In this reference a factor of 0.5 was erroneously dropped, thus the correct value of D is 5.5G.

In general, the rates for recombination via the two channels are different. For Q-reduced and Q-depleted RCs of *Rb. sphaeroides* it was found that the rate k_S for recombination via the singlet channel is smaller than the rate k_T for recombination via the triplet channel (Michel-Beyerle et al., 1980; Norris et al., 1982; Ogrodnik et al., 1982; Wasielewski et al., 1983a,b, 1984; Chidsey et al., 1984; Ogrodnik et al., 1988). This conclusion immediately follows from the observed increase of the lifetime τ_{RP} of P^+H^- upon application of a magnetic field which increases the yield of recombination via the slow singlet channel. Also, from the lifetime broadening of the radical pair levels, as measured by the width ΔB, the sum of the rates k_S and k_T can be estimated to be on the order of 1 ns^{-1}. Since τ_{RP} is on the order of 10 ns, the larger rate has to be k_T, with S-T-mixing constituting the bottleneck for the recombination via the fast triplet channel. The value of ΔB, therefore, is a convenient measure for the triplet recombination rate k_T.

C. Methods: MARY and RYDMR

The measurement of the magnetic field dependence of the yield ϕ_T of $^3P^*$ has been given the name MARY-(MAgnetic field dependence of the Reaction Yield)-spectroscopy (Lersch et al., 1982). As will be shown in the next section, from MARY spectra the parameters k_T and J can be determined, while the measurement of the radical pair lifetime allows the determination of k_S.

In addition to the static magnetic field a microwave field can be applied. If the frequency ω_{MW} of this field is in resonance with the Zeeman splitting of the triplet states of P^+H^-, it couples these states, thus modulating the triplet recombination yield (RYDMR–Reaction Yield Detected Magnetic Resonance) (Frankevich et al., 1977). The method has been applied to RCs by Bowman et al. in 1981. The modulation of ϕ_T depends on the external magnetic field B_0 as well as on the strength B_1 of the microwave field. As in MARY spectroscopy, a quantitative analysis of the experimental results allows the determination of k_T and J. The determination of J even is possible in a range of values which is not accessible to MARY spectroscopy (Lersch, 1987). However, this extended range for the determination of J is relatively small and, in contrast to MARY spectroscopy, RYDMR spectroscopy shows a significant selectivity in the case of sample inhomogeneity (see section II.D.2.c). For a recent review see (Lersch and Michel-Beyerle, 1989).

II. Temperature Dependent Recombination Dynamics of P^+H^- in Native RCs of Various Photosynthetic Bacteria

The parameters directly accessible to transient absorbance measurements are the radical pair lifetime τ_{RP}, the triplet and singlet recombination yields ϕ_T and ϕ_S and their magnetic field dependence. ϕ_S is the portion of RCs recombining to the ground state P and can best be determined by monitoring the nanosecond-recovery of the ground state absorbance band at 865 nm, normalized to the initial bleaching directly after the formation of P^+H^-. The remaining bleaching of this band after recombination is completely finished is due to $^3P^*$, thus being a measure for ϕ_T.

A. Previous Measurements

Table 1 summarizes previously reported results for the radical pair lifetime τ_{RP} and the triplet yield ϕ_T in Q-reduced and Q-depleted RCs. Most of these results refer to RCs of *Rb. sphaeroides R26*. As mentioned before, the application of an external magnetic field leads to a decrease of ϕ_T and an increase of τ_{RP}. Upon lowering the temperature, both ϕ_T and τ_{RP} increase. Also given are results for the magnetic field $B_{1/2}$ at which half of the maximum magnetic field effect on ϕ_T is found in the MARY spectrum, which is a measure for k_T, and values of k_S, k_T and J obtained from MARY- and RYDMR-measurements.

Additional information has been obtained for Q-depleted RCs of *Rb. sphaeroides* at room temperature from measurements in high magnetic fields, which cause an increase of the triplet yield (accompanied by a decrease of τ_{RP}) because of the enhanced S-T-mixing due to the different g-factors of the radical electrons (Δg-effect) (Chidsey et al., 1980; Boxer et al., 1982a,b,c). At 50 kG values of $\phi_T = 0.4$ and $\tau_{RP} =$ 9 ns were measured (Chidsey et al., 1984). Above 120 kG saturation at a triplet yield of 0.5 was observed (Goldstein et al., 1988). At such high magnetic fields a radical pair lifetime of 5 ns was extracted from fluorescence measurements (Goldstein and Boxer, 1989a).

The Δg-induced S-T-mixing in high magnetic fields was found to be pronouncedly anisotropic. In principle, conclusions on the RC structure can be

Table 1. Previously reported results

	0 G		500–1000 G	
Q-reduced RCs				
RT	τ_{RP} = 12–13 ns		τ_{RP} = 18 ns	(Shuvalov and Parson, 1981; Schenck et al., 1982; Budil et al., 1987)
	$\tau_{RP}\approx$15 ns			(*Rp. viridis*, Holten et al., 1978)
	ϕ_T = 0.15		ϕ_T = 0.08	(Parson et al., 1975; Michel-Beyerle et al., 1980; Schenck et al., 1982)
		$B_{1/2}$ = 40–60 G		(Michel-Beyerle et al., 1979)
		$k_T = 4 \times 10^8\ s^{-1}$		(Moehl et al., 1985)
90K	τ_{RP} = 20 ns		τ_{RP} = 30 ns	(Schenck et al., 1982)
	ϕ_T = 0.4		ϕ_T = 0.35	(Schenck et al., 1982)
		J = 10 G		(*R. rubrum*, Hore et al., 1993)
5K	τ_{RP} = 20 ns		τ_{RP} = 30 ns	(Schenck et al., 1982)
	ϕ_T = 0.7		ϕ_T = 0.6	(Parson and Cogdell, 1975; Schenck et al., 1982)
Q-depleted RCs				
RT	τ_{RP} = 13–14 ns		τ_{RP} = 16–18 ns	(Schenck et al., 1982; Chidsey et al., 1984; Norris et al., 1987a; Ogrodnik et al., 1988)
	ϕ_T = 0.3–0.32		ϕ_T = 0.15–0.19	(Schenck et al., 1982; Chidsey et al., 1984; Ogrodnik et al., 1988)
		$\phi_T(B)/\phi_T(0)$ = 0.5		(Ogrodnik et al., 1982; Boxer et al., 1982a,b; Roelofs et al., 1982)
		$B_{1/2}$ = 42–47 G		(Ogrodnik et al., 1982; Roelofs et al., 1982; Boxer et al., 1983; Wasielewski et al., 1984; Norris et al., 1987 a,b; Ogrodnik et al., 1987; Goldstein et al., 1988)
		$k_S = 5\text{–}7 \times 10^7\ s^{-1}$		(Norris et al., 1982; Chidsey et al., 1984; Budil et al., 1987)
		$k_T = 4\text{–}5 \times 10^8\ s^{-1}$		(Norris et al., 1982; Moehl et al., 1985; Goldstein and Boxer, 1989a)
		J = 10–15 G		(Norris et al., 1982, 1987a,b)
90–100K	τ_{RP} = 21 ns		τ_{RP} = 34 ns	(Ogrodnik et al., 1988)
	ϕ_T = 0.71		ϕ_T = 0.52	(Ogrodnik et al., 1988)
		$\phi_T(B)/\phi_T(0)$ = 0.6		(Boxer et al., 1982b)
		$B_{1/2}$ = 55 G		(Ogrodnik et al., 1987; Norris et al., 1987a)
		$k_S = 1 \times 10^7 s^{-1}$		(Budil, et al., 1987)
		J = 10–15 G		(Norris et al., 1987a)

τ_{RP} radical pair lifetime; ϕ_T triplet yield; $B_{1/2}$ magnetic field at which half of the magnetic field modulation of ϕ_T is observed in the MARY spectrum; k_S, k_T singlet and triplet recombination rate; J exchange interaction of P^+H^-. If not stated otherwise, results refer to RCs of *Rb. sphaeroides R26* (RT: room temperature).

drawn from this anisotropy of the Δg tensor (Boxer et al., 1983). In contrast, the HFI-induced S-T-mixing in low fields shows only a small anisotropy due to the small anisotropy of HFI and dipole interaction (Boxer et al., 1982a,c; Ogrodnik et al., 1985).

High pressure was found to have only small effects on the recombination dynamics of P^+H^- in Q-reduced RCs at room temperature, reducing its lifetime from 12 ns to approx. 8 ns at 345 MPa (Windsor and Menzel, 1989).

B. Measurements with High Accuracy on Quinone-Depleted RCs of Rhodobacter sphaeroides R26, Rhodobacter capsulatus *and* Chloroflexus aurantiacus

In the following we summarize the results of highly precise measurements on Q-depleted RCs of the purple bacteria *Rb. sphaeroides R26* (Table 2) and *Rb. capsulatus* (Table 4) and of the green bacterium *Chloroflexus (Cf.) aurantiacus* (Table 3), all recently obtained in our lab. The results for the RC of *Rb.*

Table 2. Results of (a) MARY- and (b) RYDMR-measurements on Q-depleted native RCs of *Rb. sphaeroides R26* and Q-depleted mutagenetically altered RC of *Rb. sphaeroides GA* ($Tyr^{M210} \rightarrow$ Phe).

(a) MARY results

		R26		$Tyr^{M210} \rightarrow$ Phe		R26[2]							
delay		electronic		electronic		optical							
T [K]		90	290	90	290	90	120	150	185	230	250	270	290
τ_{RP} [ns]	0 G	22.7	13.5	–	–	23.7	21.3	19.5	19.3	17.5	15.6	13.8	13.3
870 nm	700 G	38.4	19.1	–	–	39.8	32.4	27.6	25.5	23.0	19.2	17.4	16.4
τ_{RP} [ns]	0 G	18.5	13.6	18.5	11.2	–	–	–	–	–	–	–	–
750 nm	700 G	35.6	20.0	39.2	19.7	–	–	–	–	–	–	–	–
ϕ_T	0 G	0.63	0.32	(0.7)	(0.6)	0.71	0.61	0.52	0.45	0.39	0.34	0.31	0.30
	700 G	0.39	0.16	–	–	0.50	0.45	0.38	0.34	0.26	0.23	0.20	0.18
$B_{1/2}$ [G]		55	38	63	33	55	51	52	51	46	45	43	42
ΔB [G]		45	28	–	–	45	41	42	41	36	35	33	32
k_S [10^7 s^{-1}]		1.6	4.0	1.1	3.0	1.3	1.6	1.9	2.2	3.0	3.8	4.2	4.7
τ_3 [μs]	0 G	135	54.4	–	–	–	–	–	116[1]	94[1]	78[1]	65[1]	51[1]
	700 G	133	71.8	–	–	–	–	–	119[1]	101[1]	89[1]	76[1]	68[1]
ΔG_T [meV]		–	155	–	–	–	–	–	131	148	146	152	155

[1] Data from (Chidsey et al., 1985) (τ_3 at 0 G and 500 G)

[2] The results for R26 with optical delay have been published in Ogrodnik et al. (1988). The slightly different values given here are due to the consideration of data after 5 ns only and the determination of k_S from Eq. (2), different from the methods used in Ogrodnik et al. (1988).

(b) RYDMR results

T[K]	95	120	138	166	219	232	243	255	273	285
k_T [10^9 s^{-1}]	0.84	0.60	0.48	0.35	0.55	0.58	0.48	0.50	0.44	0.46
J[G]	8.5	11.2	10.9	9.3	9.6	10.9	9.7	10.7	10.1	9.8

τ_{RP} lifetime of P^+H^- (measured in the Q_y-absorbance band of P around 870 nm or in the Q_y-absorbance band of H around 750 nm) (±1 ns); ϕ_T recombination yield of ${}^3P^*$ (±0.01); $B_{1/2}$ magnetic field at which half of the maximum magnetic field modulation of ϕ_T is observed (±1G); ΔB halfwidth of the MARY spectrum (hwhm) (±1 G); τ_3 lifetime of ${}^3P^*$ (±2 μs); k_S singlet recombination rate (±0.2 × 10^7 s^{-1}); k_T triplet recombination rate (±0.1 × 10^9 s^{-1}); ΔG_T free energy gap between P^+H^- and ${}^3P^*$ (±8 meV); J exchange interaction of P^+H^- (±2 G). MARY-results were determined with two set-ups using an optical and an electronic delay between exciting and probing laser pulse, see text.

sphaeroides given here are the first set of all parameters of interest obtained on the same sample for a wide range of temperatures. They will be compared to the results obtained for the very similar RC of *Rb. capsulatus* (Prince and Youvan, 1987) and for the RC of *Cf. aurantiacus*, which has a different pigment content (BPheo at the position of the bacteriochlorophyll (BChl) at the inactive B-branch) (Blankenship et al., 1983; Pierson and Thornber, 1983) and, in spite of a high overall homology of the primary structure, contains different amino acid residues at various significant positions (Ovchinnikov et al., 1988a,b; Shiozawa et al., 1989).

The lifetime and MARY data in Tables 2a, 3a and 4a were obtained in spectrometers for time-resolved measurement of absorbance changes using either an optical delay line with a maximum delay of 92 ns (*Rb. sphaeroides*, *Rb. capsulatus*) or an electronic delay (*Rb. sphaeroides*, *Cf. aurantiacus*). The spectrometers have been described in Ogrodnik et al. (1988) and Volk et al. (1992), respectively. The electronic delay makes delay times in the nano- to millisecond range accessible, thus allowing the simultaneous measurement of radical pair recombination, the lifetime τ_3 of ${}^3P^*$ and of the residual fraction of RCs containing Q (Volk et al., 1992), see Fig. 2. The RYDMR data in Tables 2b, 3b and 4b were measured as described in Lersch et al. (1989).

For *Rb. sphaeroides* and *Cf. aurantiacus* the radical pair lifetime τ_{RP} was determined from a mono-

Table 3. Results of (a) MARY- and (b) RYDMR-measurements on Q-depleted RCs of *Cf. aurantiacus*

(a) MARY results

T [K]		75	90	120	150	180	200	230	250	270	290	310
τ_{RP} [ns]	0 G	24.2	23.8	24.7	23.1	22.2	21.0	17.2	14.6	12.9	11.4	11.5
	23 G							16.2	13.5	12.1	10.7	10.6
	700 G	32.6	32.5	33.8	32.7	32.1	29.7	22.4	19.1	16.2	14.1	14.5
ϕ_T	0 G	0.49	0.50	0.53	0.54	0.53	0.50	0.42	0.36	0.31	0.28	0.32
	23 G							0.44	0.39	0.34	0.30	0.34
	700 G	0.26	0.26	0.27	0.28	0.28	0.26	0.21	0.17	0.14	0.14	0.15
B_{res} [G]		–	–	–	–	–	–	22	22	21	22	20
$B_{1/2}$ [G]		124	115	104	95	90	83	67	62	62	55	53
ΔB [G]		103	94	83	74	69	62	46	41	41	34	32
k_S [10^7 s^{-1}]		2.4	2.4	2.3	2.3	2.3	2.5	3.6	4.3	5.2	6.0	5.8
τ_3 [μs]	0 G	103	103	101	101	100	102	96	93	87	80	74
	700 G	102	102	101	102	101	103	98	99	92	87	83
ΔG_T [meV]		–	–	–	–	–	–	186	182	198	196	206

(b) RYDMR results

T[K]		170	192	202	223	243	263	283
k_T [10^9 s^{-1}]	low m.w.p.	–	–	1.13	0.89	0.79	0.70	0.62
	high m.w.p.	2.16	2.08	1.62	1.50	1.17	1.05	0.97
J [G]	low m.w.p.	–	–	24.3	25.1	25.6	24.3	22.5
	high m.w.p.	20.3	17.8	21.8	23.9	25.4	24.0	22.8

τ_{RP} lifetime of P^+H^- (measured in the Q_y-absorbance band of P around 870 nm) (±1 ns); ϕ_T recombination yield of $^3P^*$ (±0.01); B_{res} resonance magnetic field of MARY spectrum (±1 G); $B_{1/2}$ magnetic field at which half of the maximum magnetic field modulation of ϕ_T is observed (±1 G); ΔB halfwidth of the MARY spectrum (hwhm) (±1 G); τ_3 lifetime of $^3P^*$ (±2 μs); k_S singlet recombination rate (±0.2 × 10^7 s^{-1}); k_T triplet recombination rate (±0.1 × 10^9 s^{-1}); ΔG_T free energy gap between P^+H^- and $^3P^*$ (±8 meV); J exchange interaction of P^+H^- (±2 G); m.w.p. = microwave power.

Table 4. Results of (a) MARY- and (b) RYDMR-measurements of Q-depleted native and mutagenetically altered RCs of *Rb. capsulatus* (Trp^{M250} → Leu, Glu, Phe).

(a) MARY results

		native RC		Trp^{M250}→ Leu		Trp^{M250}→ Glu		Trp^{M250}→ Phe	
T [K]		80	280	80	280	80	280	80	280
τ_{RP} [ns]	0 G	17.7	14.6	18.5	20.0	18.5	12.3	18.2	16.2
	700 G	29.3	18.7	27.4	22.0	29.0	16.5	29.8	19.7
ϕ_T	0 G	0.76	0.41	0.64	0.19	0.72	0.49	0.71	0.44
$B_{1/2}$ [G]		43	33	35	24	43	43	41	21
ΔB [G]		33	23	25	14	33	33	31	11
k_S [10^7 s^{-1}]		1.5	4.3	2.1	4.2	1.6	4.9	1.7	4.4

(b) RYDMR results

	native RC	Trp^{M250}→ Leu	Trp^{M250}→ Glu	Trp^{M250}→ Phe
T[K]	277	277	277	277
k_T [10^9 s^{-1}]	0.29	0.23	0.30	0.22
J [G]	10.8	11.0	11.3	10.4

τ_{RP} lifetime of P^+H^- (measured in the Q_y-absorbance band of H around 750 nm) (±2 ns); ϕ_T recombination yield of $^3P^*$ (±0.03); $B_{1/2}$ magnetic field at which half of the maximum magnetic field modulation of ϕ_T is observed (±2 G); ΔB halfwidth of the MARY spectrum (hwhm) (±2 G); k_S singlet recombination rate (±0.3 × 10^7 s^{-1}); k_T triplet recombination rate (±0.05 × 10^9 s^{-1}); J exchange interaction of P^+H^- (±2 G).

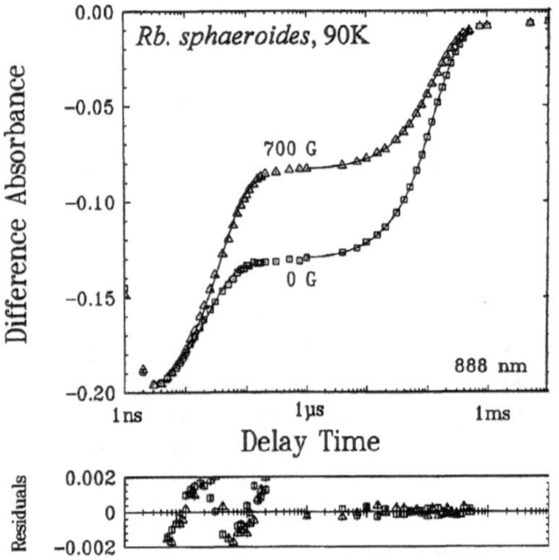

Fig. 2. Time-dependent P absorbance bleaching at 888 nm in Q-depleted RCs of *Rb. sphaeroides* at 90 K, 0 G and 700 G. (—) monoexponential fits in the range 5–200 ns and 1–500 μs, respectively.

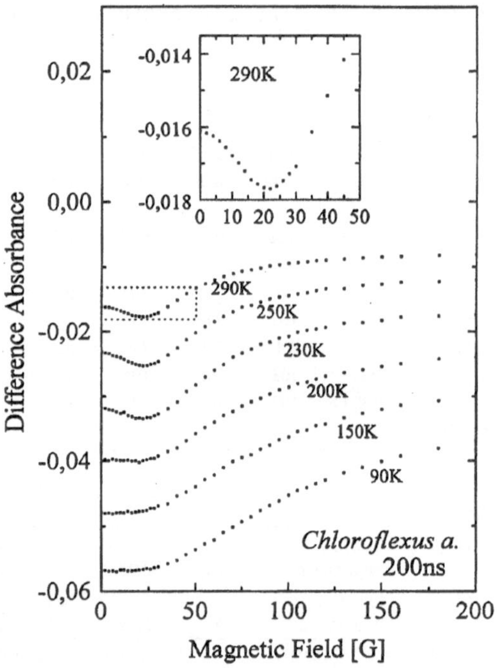

Fig. 3. Magnetic field dependence of the Q_y-absorbance bleaching of P in Q-depleted RCs of *Chloroflexus aurantiacus* at different temperatures, 200 ns after excitation (MARY spectrum). Inset: enlargement of MARY spectrum at 290 K.

exponential fit of the time dependent recovery of the P absorbance in the maximum of the Q_y-band, as shown in Fig. 2 for the case of *Rb. sphaeroides* at 90K. For *Rb. sphaeroides* also the recovery of the Q_y-band of H around 750 nm was measured, yielding slightly different time constants, see Table 2a. The observed small deviations from monoexponentiality (compare the residuals for the monoexponential fits of radical pair recombination and decay of $^3P^*$ in Fig. 2) and the wavelength dependence of τ_{RP} can be interpreted in a quantitative way, as will be discussed in Section II.C. The triplet recombination yield ϕ_T was determined from the bleaching of P after radical pair recombination is complete, normalized to the initial amplitude at 0 ns. Both the radical pair lifetime and the triplet yield have similar values in *Rb. sphaeroides* and *Cf. aurantiacus* (Tables 2a and 3a).

The MARY spectra were obtained by measuring the bleaching of the Q_y-absorbance of P at 92 ns and 200 ns in the spectrometer with optical and electronic delay, respectively. Figure 3 shows some of the spectra obtained for *Cf. aurantiacus* at different temperatures. Above 230K a pronounced resonance structure was observed, allowing the direct determination of the exchange interaction *J*. At lower temperatures no such resonance can be resolved due to the increasing width of the MARY spectrum. For RCs of *Rb. sphaeroides* we observed a weakly resolved resonance structure (with a maximum triplet yield near 10G) only at 230K. For the cases without resonance structure the halfwidth ΔB of the MARY spectra was determined by subtracting the value of J (10G for *Rb. sphaeroides*, 21G for *Cf. aurantiacus*, Section II.D.3, both essentially temperature independent), where resonance is expected to occur, from the value $B_{1/2}$ of the magnetic field at which half of the maximum magnetic field modulation of ϕ_T is observed. ΔB is almost temperature independent with values between 28G and 45G in the case of *Rb. sphaeroides*, while it increases from 34G at 290K to 94G at 90K in the case of *Cf. aurantiacus*.

The measurements on RCs of *Rb. capsulatus* are handicapped by the presence of carotenoids (Car). As in carotenoid containing RCs of *Rb. sphaeroides* (Schenck et al., 1984), fast energy transfer (within approx. 100 ns) from $^3P^*$ to $^3Car^*$ could be observed here. Since the decay of $^3P^*$ proceeds on a similar time scale as radical pair recombination, τ_{RP} was determined from the recovery of the bleaching of the H absorbance band at 750 nm, where neither $^3P^* \rightarrow P$ nor Car $\rightarrow$ $^3Car^*$ cause substantial absorbance changes. The magnetic field dependence of ϕ_T was measured indirectly by monitoring $^3Car^*$ in its strong absorbance band at 560 nm. No resonance was found in the MARY spectra at 80K and 280K, ΔB again was determined by subtracting the value of J (10G, from RYDMR measurements) from $B_{1/2}$. Due to the uncertainties of the $^3Car^*$ extinction coefficients, the absolute value of ϕ_T could not be determined directly, but was concluded upon from the magnetic field dependence of τ_{RP} and the magnetic field modulation of ϕ_T with the use of Eq. (1). All observed parameters have very similar values as the ones for *Rb. sphaeroides* (because of the inherent wavelength dependence of the observed time constants, (see Section II.C), only the values of τ_{RP} determined at corresponding wavelengths should be compared).

The values for k_T and J determined from RYDMR measurements on reaction centers of *Rb. sphaeroides*, *Cf. aurantiacus* and *Rb. capsulatus* (Lersch et al., 1990; Lang et al., 1990) are given in Tables 2b, 3b and 4b (see section II.D.2.b. for a more detailed description of the theoretical treatment of RYDMR data).

C. Inherent Inhomogeneity of the Singlet-Triplet-Mixing

The recombination of P^+H^- shows small, but significant deviations from monoexponentiality (Fig. 2), which were observed for all RCs investigated. Also, the observed time constants were found to depend on the probing wavelength (Table 2a). Such effects can be ascribed to structural or energetic heterogeneities. Here, however, they mostly arise from an inherent 'physical' inhomogeneity of the effective hyperfine coupling due to the random orientation of nuclear spins (Feick et al., 1990). This inhomogeneity of the HFI leads to a distribution of the frequency of S-T-mixing (Schulten and Wolynes, 1978; Goldstein and Boxer, 1987). Since RCs with strong S-T-mixing preferentially recombine via the fast triplet channel, while RCs with weak S-T-mixing recombine via the slower singlet channel, the inhomogeneous HFI leads to a distribution of radical pair lifetimes. Indeed, the observed deviations from monoexponentiality are more pronounced for RCs of bacteria with high triplet recombination yield, that is with considerable recombination via the triplet channel which introduces the inhomogeneity.

The distribution of τ_{RP} ranges from $\tau_{RP} \approx 2/(k_S+k_T)$ for the case of fast S-T-mixing (leading to a quasi-equilibration of singlet and triplet levels) to $\tau_{RP} = 1/k_S$ for the case of zero S-T-mixing. This is a rather broad distribution, ranging from 3 ns to 60 ns in RCs of *Rb. sphaeroides* at 90K. It can be approximated in an analytical form (Volk, 1991) using the one-proton model (Haberkorn and Michel-Beyerle, 1979) and assuming a Gaussian distribution of the effective HFI constant (Schulten and Wolynes, 1978; Goldstein and Boxer, 1987), Fig. 4.

The inhomogeneity of radical pair recombination observed at a particular wavelength, and consequently the average time constant measured at this wavelength, depend on the spin state of the recombination products which are monitored at that wavelength. For example, simulations of the radical pair recombination using the stochastic Liouville-Equation, Eq. (3), and the Hamiltonian, Eq. (4), for $k_S = 2 \times 10^7\ s^{-1}$ and $k_T = 1 \times 10^9\ s^{-1}$ (similar to the values for RCs of *Rb. sphaeroides* at 90K) yield average time constants of 22.0 ns, 14.4 ns and 16.8 ns for the recovery of the singlet ground state, the formation of $^3P^*$ and the overall radical pair decay, respectively. In measurements of the recovery of the

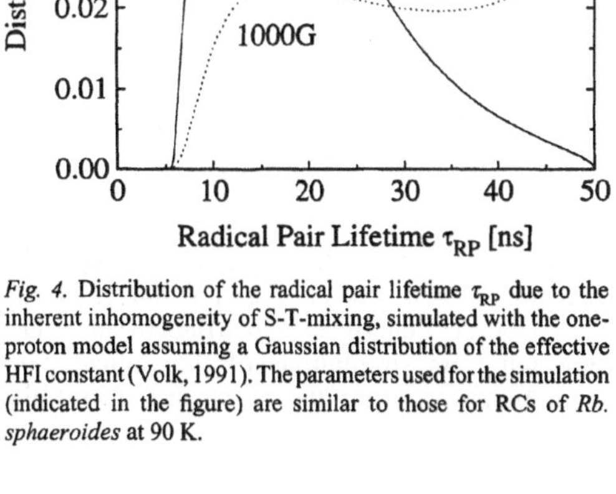

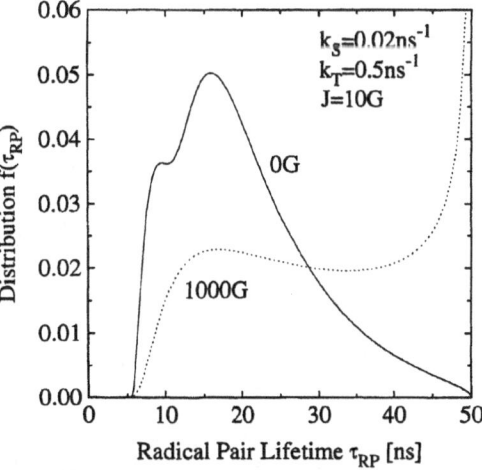

Fig. 4. Distribution of the radical pair lifetime τ_{RP} due to the inherent inhomogeneity of S-T-mixing, simulated with the one-proton model assuming a Gaussian distribution of the effective HFI constant (Volk, 1991). The parameters used for the simulation (indicated in the figure) are similar to those for RCs of *Rb. sphaeroides* at 90 K.

P ground state absorbance at 887 nm, the formation of $^3P^*$ absorbance at the isosbestic point of the P^+H^- difference spectrum at 552 nm and the recovery of the H absorbance at 762 nm we could determine these time constants experimentally, yielding values of 22.7 ns, 18.2 ns and 18.5 ns for RCs of *Rb. sphaeroides* at 90K (M. Volk, unpublished). The difference between the values at different wavelengths is significant, but not as pronounced as expected from the simulations. As shown recently (Ogrodnik et al., 1994), the energy of P^+H^- is inhomogenously broadened, probably due to different conformational substates of the protein (section V.C.). This leads to an inhomogeneity of the recombination rates k_T and k_S which also contributes to the observed deviations from monoexponentiality and was shown to partially cancel the effect of the inherent inhomogeneity of the S-T-mixing on the wavelength dependence of τ_{RP} (Aumeier, 1993).

Due to the deviations from monoexponentiality and the wavelength dependence of the observed time constants, an unambiguous determination of the average radical pair lifetime is not straightforward. Simulated radical pair recombination data, obtained with the stochastic Liouville equation, were employed to find the best procedure (Volk, 1991). They show that monoexponential fits of the recovery of the P ground state absorbance indeed yield time constants close to the average radical pair lifetime for the typical parameters observed in bacterial RCs. For the recombination parameters found in D1D2Cyt*b*-559-RCs of Photosystem II, however, the distribution of the radical pair lifetimes is much broader because of the unusually small value of k_S and complicates the determination of the average radical pair lifetime from the experimental data (Volk et al., 1993).

A more comprehensive discussion of the inherent inhomogeneity of S-T-mixing and its effects on the recombination dynamics of P^+H^- will be given elsewhere (M. Volk et al., unpublihsed).

D. Determination of the Recombination Rates and the Exchange Interaction

For RCs of *Rb. sphaeroides* at room temperature ϕ_T was found to be smaller than (i) any value compatible with the reaction scheme, Fig. 1 (Chidsey et al., 1984; Goldstein and Boxer, 1989b) and (ii) the value expected from simulations of the radical pair recombination using the stochastic Liouville equation, Eq. (3), (Volk, 1991). These discrepancies can be explained self-consistently by equilibration between $^3P^*$ and the triplet state of the BChl monomer in the inactive branch on the time-scale of P^+H^- recombination or faster. The triplet state of this BChl is expected to have an energy higher than $^3P^*$ by about 0.02–0.03 eV (Shuvalov and Parson, 1981; Schenck et al., 1984) and was shown to be an intermediate for the triplet energy transfer to the carotenoid in Car-containing RCs (Frank and Violette, 1989). With this extension of the reaction scheme, at higher temperatures the observable concentration of $^3P^*$ after completion of P^+H^- recombination will be smaller than the overall portion of RCs recombining via the triplet channel. In agreement with this scheme, no discrepancies are observed for RCs of *Rb. sphaeroides* at low temperatures and for RCs of *Cf. aurantiacus* even at high temperatures, since here the BChl at the inactive branch is replaced by a BPh (Ovchinnikov et al., 1988a,b; Shiozawa et al., 1989). The conclusions described below are not affected by neglecting this extension to the reaction scheme, since only relative triplet yields will be considered.

This extension contrasts to the previously proposed reaction scheme (Goldstein and Boxer, 1989b), which assumed fast recombination of P^+H^- preceding nuclear relaxations on the time scale of a few nanoseconds. Such an explanation does not hold, however, since in measurements of the recombination dynamics of P^+H^- with ps-time resolution (Chidsey et al., 1984) no corresponding recovery of the Q_y-absorbance of P in the first nanoseconds was observed.

1. The Singlet Recombination Rate k_S

The average radical pair lifetime is given (Haberkorn and Michel-Beyerle, 1979) by

$$\tau_{RP} = \frac{1-\phi_T}{k_S} + \frac{\phi_T}{k_T} \qquad (1)$$

The determination of τ_{RP} and ϕ_T at zero magnetic field and at a magnetic field B of several hundred G allows the determination of k_S from Eq. (1):

$$k_S = \frac{\phi_T(0)/\phi_T(B) - 1}{\tau_{RP}(B)\,\phi_T(0)/\phi_T(B) - \tau_{RP}(0)} \qquad (2)$$

The values of k_S obtained with Eq. (2) are included in Tables 2a, 3a and 4a. Within the error of measurement the singlet recombination rate is identical in RCs of *Rb. sphaeroides* and *Rb. capsulatus*, decreasing from $4 \times 10^7 s^{-1}$ to $1.5 \times 10^7 s^{-1}$ upon lowering the temperature from 290 K to 90 K. This is in agreement with previously published data (Norris et al., 1982; Ogrodnik et al., 1982, 1988; Chidsey et al., 1984; Budil et al., 1987). The values for k_S in RCs of *Cf. aurantiacus* are only slightly larger and are independent of the temperature below 200K.

In principle, the singlet recombination rate has a contribution from recombination via $^1P^*$ which is thermally activated. The absence of a significant activation of k_S for bacterial RCs shows that even at room temperature this recombination path is negligible (Ogrodnik et al., 1988).

2. The Triplet Recombination Rate k_T

a. Approximate Determination from the Width of the MARY Spectrum

A direct determination of k_T from τ_{RP} and ϕ_T, comparable to the determination of k_S, Eq. (2), as has been attempted in (Chidsey et al., 1984), is not possible, since the second term in Eq. (1) is much smaller than the first and, therefore, k_T has no significant direct contribution to the value of τ_{RP}. As described above, the width ΔB of the MARY spectrum is a measure for the triplet recombination rate k_T. However, there are other factors contributing to the width of the MARY spectrum like the HFI interaction. For a quantitative determination of k_T the measured MARY and RYDMR spectra have to be fitted to theoretical ones as described below. These fits indeed yield results for k_T which are very similar to the ones obtained directly from ΔB. Therefore, ΔB indeed is a good measure for k_T and will be used for the discussion of the temperature dependence of k_T, see Fig. 5.

b. Determination from Numerical Simulations

The dependence of the triplet yield on static magnetic fields and/or microwaves can theoretically be calculated by the solution of the stochastic Liouville equation for the density matrix ρ (Werner et al., 1978; Haberkorn and Michel-Beyerle, 1979; Lersch and Michel-Beyerle, 1983)

$$\frac{d}{dt}\rho = -\frac{i}{\hbar}[\mathbf{H},\rho] - \frac{k_S}{2}\left(P^S\rho + \rho P^S\right) - \frac{k_T}{2}\left(P^T\rho + \rho P^T\right) \tag{3}$$

for the appropriate Hamilton operator

$$\mathbf{H} = \sum_n^P \mathbf{I}_n^P \mathbf{A}_n^P \mathbf{S}^P + \sum_n^H \mathbf{I}_n^H \mathbf{A}_n^H \mathbf{S}^H + g_P\beta\left(B_0 + B_1(t)\right)\mathbf{S}^P + g_H\beta\left(B_0 + B_1(t)\right)\mathbf{S}^H - JP^S \tag{4}$$

Here P^S and P^T represent the projection operators on the singlet and the triplet subspaces, respectively, $\mathbf{I}_n^P$ and $\mathbf{I}_n^H$ the nth nuclear spin of P or H with the respective HFI tensor $\mathbf{A}_n^P$ or $\mathbf{A}_n^H$, $\mathbf{S}^P$ and $\mathbf{S}^H$ the spin of the radical electron on P or H with g-factors g_P and g_H, β Bohr's magneton, B_0 the static magnetic field, $B_1(t)$ the magnetic field of the applied microwaves and J the exchange interaction. Due to their large distance, the dipole interaction between the two radical electron spins can be neglected for the calculation of ϕ_T (Ogrodnik et al., 1985). Similarly, the anisotropy of the HFI tensors is neglected because of the small observed anisotropy of the HFI induced triplet yield (Boxer et al., 1982a,c).

Because of the large number of nuclear spins contributing to the HFI an exact solution of the Liouville equation is not feasible. In the one-proton model the manifold of nuclear spins is replaced by only one representative spin with a suitably chosen HFI constant, giving analytical expressions for ϕ_T (Haberkorn and Michel-Beyerle, 1979). However, this model does not account for the effects from the random orientation of nuclear spins. In a semiclassical model the sum of the nuclear spins, weighted with their hyperfine constants, is replaced by a classical magnetic field. The strength of this field can be modeled by a Gaussian distribution due to the large number of nuclear spins with random orientation (Schulten and Wolynes, 1978).

Using the semiclassical model, the measured MARY and RYDMR spectra[3] can be fitted with k_T and J as free parameters, inserting the values for k_S given above. The width of the distribution of the magnetic field modelling the HFI corresponds to the HFI-induced width of the EPR spectrum of the

[3] In view of the discrepancies with respect to the absolute triplet yield ϕ_T (Section II.D), only relative MARY and RYDMR spectra (i.e. the shape of these spectra) were considered here.

respective radical. These were determined to be 9.5G for P^+ in *Rb. sphaeroides* (McElroy et al., 1972), 9.8G for P^+ in *Cf. aurantiacus* (Bruce et al., 1982) and 13.0G for H^- (Okamura et al., 1979). For the magnetic fields employed here, the difference of the g-factors ($g_H - g_P \approx 0.001$ (McElroy et al., 1972; Okamura et al., 1979)) as well as their anisotropy (van der Est et al., 1993) can be neglected.

Fits of the MARY spectra yield values of 0.8–0.95 ns^{-1} for k_T in the case of *Rb. sphaeroides*, independent of the temperature. In the case of *Cf. aurantiacus* a value of 0.9–1 ns^{-1} was obtained at 290 K, increasing to 2–2.5 ns^{-1} upon lowering the temperature to 90 K. No fits of the MARY data for *Rb. capsulatus* were performed so far; however, due to the close similarity of the data obtained for RCs of *Rb. capsulatus* and *Rb. sphaeroides*, almost identical results are expected for k_T in the RCs of these bacteria.

These values are almost the same as those calculated from the MARY-width ΔB under the assumption of ΔB being the purely homogeneous width due to k_T (Fig. 5). The latter values are slightly larger due to the neglect of other contributions to ΔB, the largest of which is expected to be HFI with values of 9.5 G for P^+ and 13 G for H^-.

Values for k_T resulting from fits of the RYDMR spectra are given in Tables 2b, 3b and 4b. For *Rb. sphaeroides* the value is approx. 0.5 ns^{-1}, independent of the temperature. A somewhat smaller value is obtained for *Rb. capsulatus* at 277K. For RCs of *Cf. aurantiacus* a pronounced increase of k_T by a factor of 2 upon lowering the temperature from 280K to 170K was deduced from RYDMR data; the value of k_T obtained from the fits, however, depended on the microwave power. The cause of this discrepancy and of the different values obtained by the different methods will be discussed in the following section.

c. Inconsistency of the Values Obtained by Different Methods

The values of k_T determined in MARY and RYDMR measurements differ by a factor of up to 2. Similarly, RYDMR measurements on RCs of *Cf. aurantiacus* at different microwave powers could not be simulated with the same value of k_T (Table 3b).

As discussed in section V.C., from the measurement of the magnetic field dependent recombination dynamics of P^+H^- in delayed fluorescence one can conclude that there is an inhomogeneous distribution of the energy of P^+H^-. This yields a distribution of the recombination rates k_S and k_T, explaining the inconsistency of the values for k_T determined with the different techniques. It can be shown that the amplitude of the RYDMR spectra decreases rapidly with increasing k_T, particularly at low microwave powers (Lersch, 1987). Therefore, the overall signal from a sample inhomogeneous with respect to k_T will not be sensitive to contributions from RCs with large k_T.

Another method for the determination of the value of k_T is the measurement of τ_{RP} in very high magnetic fields (>100 kG). At these magnetic fields S-T-mixing is predominantly caused by the different g factors of P and H and ultimately leads to an equilibration of singlet and triplet levels, implying

$$\tau_{RP} = \frac{2}{k_S + k_T} \quad (5)$$

A value of $k_T = 0.4$ ns^{-1} was determined with this method for RCs of *Rb. sphaeroides* at room temperature from time resolved fluorescence measurements (Goldstein and Boxer, 1989a). This corresponds well with the value determined from RYDMR spectra (0.46 ns^{-1}), but is in disagreement with the MARY spectra, yielding $k_T \approx 0.8$–0.95 ns^{-1}. However, again the method is not sensitive to RCs with large rates of k_T, because large values of k_T yield lifetimes that could not have been resolved, particularly in view of the fast fluorescence components that are present even at low magnetic fields.

Therefore, the observed discrepancies can be resolved and even confirm the inhomogeneity of the RCs with respect to k_T. In fact, simulations of the MARY spectrum for *Rb. sphaeroides* at 290 K under the assumption of two populations of RCs with values of k_T of 0.4 ns^{-1} (70%) and 2.5 ns^{-1} (30%) reproduce the measured spectrum significantly better than under the assumption of only one value of k_T (Volk, 1991).

3. The Exchange Interaction J

As can be seen in Fig. 1, one of the triplet levels of P^+H^- is isoenergetic with the singlet level for an external magnetic field of $B_{res} = J$. S-T-mixing is most efficient for this magnetic field. The observation of a pronounced resonance structure in the MARY spectrum, therefore, allows the determination of the exchange interaction J. If the width of the MARY-spectrum is much larger than J, however, no resonance is observed. If only a weak resonance is found, the maximum triplet yield is not observed at $B = J$, since S-T-mixing to the complementary triplet level distorts the shape of the resonance. In this case, the value of J has to be obtained from a numerical fit. For *Cf. aurantiacus* the MARY spectra yield a value for J of 21 G at 310 K increasing to 25 G at 230 K. For the other RCs no resonance is found and the MARY spectra only allow the determination of an upper limit for the value of J.

For these cases a determination of J is possible from the fit of the RYDMR spectra. An almost temperature independent value of J = 10 G was found for *Rb. sphaeroides* (Table 2b) and *Rb. capsulatus* (Table 4b). For *Cf. aurantiacus* the best fits were obtained for values of J between 20 G and 25 G (Table 3b), in agreement with the MARY results.

In the literature, a value of $J = (14 \pm 1)$ G has been reported for *Rb. sphaeroides* at temperatures between 296 K and 222 K from the magnetic field dependent measurement of τ_{RP} with less resolution than the one employed here (Norris et al., 1987a). From the anisotropic electron spin polarization of Q^- in Q-reduced RCs of *Rhodospirillum rubrum*, which are very similar to RCs of *Rb. sphaeroides*, a value of J of 10 G was determined (Hore et al., 1993).

E. Determination of the Free Energies of P^+H^- and P^+B^-

From the parameters of the P^+H^- recombination dynamics and the magnetic field dependent lifetime of $^3P^*$, the free energy difference between $^3P^*$ and P^+H^- can be deduced (Chidsey et al., 1985; Goldstein et al., 1988; Ogrodnik et al., 1988). At low temperatures the triplet state $^3P^*$ decays on the 100 μs time scale to the singlet ground state P via intersystem crossing with the rate k_{ISC}. At higher temperatures the lifetime decreases and becomes magnetic field dependent. This magnetic field dependence has been shown to be parallel to that of the triplet recombination yield, implying a decay mechanism that involves thermal repopulation of $^3(P^+H^-)$ from $^3P^*$ with the rate k_{-T} followed by triplet-singlet-mixing and subsequent recombination to the ground state. A thorough treatment yields the following equation for the effective decay rate of $^3P^*$ (Chidsey et al., 1985):

$$1/\tau_3(B) = k_{OBS}(B) = k_{ISC} + \frac{1}{3}\frac{k_S}{k_T}\phi_T(B)\,k_{-T} \qquad (6)$$

The magnetic field dependence of k_{OBS} allows the separation of the two contributions and the determination of k_{-T}. From the ratio of k_T and k_{-T} the free energy difference ΔG_T between $^3P^*$ and P^+H^- can be obtained. Measurements on Q-depleted RCs of *Rb. sphaeroides* at room temperature yielded values of (0.173 ± 0.004) eV (Goldstein et al., 1988) and (0.165 ± 0.008) eV (Ogrodnik et al., 1988).

The apparatus with electronic delay allows the simultaneous determination of the lifetime τ_3 of $^3P^*$ and of the recombination parameters. The values of τ_3 and ΔG_T obtained for RCs of *Rb. sphaeroides* (Ogrodnik et al., 1988) and *Cf. aurantiacus* (Volk, 1991) are included in Table 2a and 3a, see also Fig. 8. The value of ΔG_T for RCs of *Cf. aurantiacus* is significantly larger (0.20 eV compared to 0.16 eV in *Rb. sphaeroides* at room temperature). Such a difference has been predicted from electrostatic calculations (Michel-Beyerle et al., 1988b; Parson et al., 1990a) due to the exchange of the polar (protonated) glutamic acid at position L104 in *Rb. sphaeroides* to the unpolar glutamine in *Cf. aurantiacus*.

The value of ΔG_T shows only a weak temperature dependence. A linear fit of the data yields values for the enthalpy difference $\Delta H_T = (0.10 \pm 0.03)$ eV and the entropy difference $\Delta S_T = -(2 \pm 1) \times 10^{-4}$ eV/K for *Rb. sphaeroides* and $\Delta H_T = (0.13 \pm 0.05)$ eV and $\Delta S_T = -(2.5 \pm 1) \times 10^{-4}$ eV/K for *Cf. aurantiacus*. The value of ΔH_T for *Rb. sphaeroides* determined here is more precise than the value determined in (Goldstein et al., 1988) from the temperature dependence of τ_3, where the temperature dependence of ϕ_T and k_S was not available.

For *Rb. sphaeroides* a value of $\Delta H(^1P^* – ^3P^*)$ = 0.41 eV was obtained from fluorescence and phosphorescence spectra (Takiff and Boxer, 1988) and from the temperature dependence of microsecond delayed fluorescence (Shuvalov and Parson, 1981). Because the entropy difference between $^1P^*$ and $^3P^*$ should be negligible, $\Delta G(^1P^* – ^3P^*) = \Delta H(^1P^* – ^3P^*)$, thus with $\Delta G_T = \Delta G(P^+H^- – ^3P^*)$ = 0.16 eV a free energy gap for charge separation to P^+H^- of $\Delta G(^1P^* – P^+H^-)$ = 0.25 eV can be calculated. At room temperature this value agrees with the one obtained from delayed fluorescence data (Hörber et al., 1986).[4] Fluorescence measurements with better time

[4] This result also can be derived from the data in (Woodbury et al., 1986), if refering to the slowest fluorescence component (Ogrodnik, 1990).

resolution yield a somewhat lower value of 0.21 eV (Ogrodnik et al., 1994). However, due to the energetic inhomogeneity of P^+H^-, the apparent value of $\Delta G(^1P^*-P^+H^-)$, as determined from fluorescence, strongly decreases upon lowering the temperature, see Section V.C.

In the carotenoid-containing RCs of *Rb. capsulatus*, energy is transferred from $^3P^*$ to the triplet state of the carotenoid on the 100 ns timescale; therefore, ΔG_T cannot be determined directly. From the close similarity of the temperature dependence of k_T in RCs of *Rb. capsulatus* to that in RCs of *Rb. sphaeroides*, however, one can conclude that the energy of P^+H^- is similar in both RCs, see Section II.F.5.

The recombination dynamics of P^+H^- also gives access to the free energy difference ΔG_{23} between P^+H^- and P^+B^- (B: BChl monomer at the active branch). Fast hopping between P^+H^- and P^+B^- affects various observables of the recombination dynamics (Haberkorn et al., 1979; Bixon et al., 1987; Ogrodnik et al., 1982, 1988; Michel-Beyerle and Ogrodnik, 1990; Ogrodnik, 1990). In this context the additional triplet recombination channel via $^3(P^+B^-)$, leading to an effective recombination rate

$$k_T = k_T^H + k_T^B \, e^{-\frac{\Delta G_{23}}{k_B T}} \tag{7}$$

gives the best access to ΔG_{23}. Here, k_T^H and k_T^B denote the rate of recombination to $^3P^*$ from $^3(P^+H^-)$ and $^3(P^+B^-)$, respectively. From the experimentally found inverted temperature dependence of k_T, Fig. 5, the activated second term in Eq. (7) can be estimated to contribute less than 20% even at room temperature in RCs of *Cf. aurantiacus* and of *Rb. sphaeroides*, corresponding to less than 0.2 ns^{-1}.

The electronic coupling between $^3(P^+B^-)$ and $^3P^*$ is similar to the coupling between $^1(P^+B^-)$ and $^1P^*$, which was estimated from the activationless rate of primary charge separation to be ≃ 20 cm^{-1} in RCs of *Rb. sphaeroides*(Bixon et al., 1988, 1989). This is larger than the electronic coupling between $^3(P^+H^-)$ and $^3P^*$ by a factor of 20, see section II.F.4. The free energy of P^+B^- is above that of P^+H^- and the triplet recombination from P^+H^- is near the activationless case, thus k_T^B is in the inverted region, but close to the activationless case as long as P^+B^- does not have a very large free energy. Therefore, the Franck-Condon-factor for k_T^B is expected to be similar to that of k_T^H, a decrease by at most a factor of 2 will be assumed here. It can be concluded that $k_T^B \geq (20)^2/2k_T^H \simeq$ 200 ns^{-1}.

With these values, a lower limit for ΔG_{23} of 0.17 eV can be estimated for *Rb. sphaeroides*. Due to the slightly slower primary charge separation in RCs of *Cf. aurantiacus*, the corresponding calculation yields $\Delta G_{23} > 0.16$ eV in this case.

As a consequence, the free energy gap $\Delta G_1 = \Delta G(^1P^* - P^+B^-)$ for charge separation to P^+B^- can be estimated to be smaller than 0.05 eV in *Cf. aurantiacus* and smaller than 0.08 eV in *Rb. sphaeroides*.

In a recent investigation (Schmidt et al., 1994) effort has been made to determine ΔG_1 in RCs of *Rb. sphaeroides* with the bacteriopheophytin exchanged with pheophytin. In this case, ΔG_{23} becomes so small that the electron transfer step $P^+B^- \rightarrow P^+H^-$ slows down to $k_2 \simeq$ 1/300 ps. This should give rise to delayed fluorescence from P^+B^-. From the amplitude of the experimental fluorescence component with a lifetime corresponding to that of P^+B^-, $\Delta G_1 \simeq 450$ cm^{-1} = 0.06 eV has been derived. However, (1) contributions to this fluorescence component not originating from recombination of P^+B^- and (2) an inhomogeneity of ΔG_1 (see Eq. (18)) would lead to an underestimation of the true average value of ΔG_1. Together with the limit derived above, ΔG_1 can thus be confined to 0.06 eV < ΔG_1 < 0.08 eV. We include this result in Fig. 8.

F. Conclusions

The recombination of P^+H^- cannot compete with forward electron transfer to Q, which in native RCs occurs with a quantum yield close to unity (Wraight and Clayton, 1973; Volk et al., 1991). This high quantum yield only is possible because the recombination to the ground state is slow due to the large energy gap and the fast recombination via the triplet channel is hindered by the 'bottleneck' of HFI-induced S-T-mixing. On the other hand, the parameters obtained from the measurement of the recombination dynamics, i.e. k_S, k_T and J, can be discussed in relation to the charge separation steps from $^1P^*$ to P^+H^- and allow the determination of free energies, as was shown in the previous section, as well as electronic matrix elements and reorganization energies.

1. Nonadiabatic Electron Transfer Theory

In the framework of nonadiabatic electron transfer theory, the simplest expression for the electron transfer rate (considering only protein vibronic modes with $\hbar\omega << kT$) is given by the electronic coupling matrix element V, the free energy ΔG of the reaction and the reorganization energy λ (Jortner, 1980a,b; Marcus and Sutin, 1985; see Chapter 25 by Parson and Warshel):

$$k_{ET} = \frac{2\pi}{\hbar} \frac{V^2}{\sqrt{4\pi\lambda kT}} e^{-\frac{(\Delta G+\lambda)^2}{4\lambda kT}} \tag{8}$$

The matrix element relevant for the recombination of P^+H^- is expected to be dominated by superexchange coupling via P^+B^- (Michel-Beyerle et al., 1988a; Bixon et al., 1991):

$$V = \frac{V_{PB} V_{BH}}{\delta E(P^+B^- - P^+H^-)} \tag{9}$$

with V_{PB} denoting the coupling between P and P^+B^-, V_{BH} the coupling between P^+B^- and P^+H^- and $\delta E(P^+B^- - P^+H^-)$ the vertical energy separation between P^+B^- and P^+H^- at the nuclear coordinates where the transition takes place. In Section V.C. experimental evidence for the predominance of the superexchange coupling mechanism for the recombination of P^+H^- is described.

2. k_S—A Rate in the Marcus Inverted Region

From the energies given in Section II.E., together with the known energy of $^1P^*$, the free energy ΔG_S for the singlet recombination process can be estimated to be $\approx$8000 cm^{-1}. The reorganization energy λ_S is expected to be similar to the one for the triplet recombination process, which will be shown below to be 1000–1500 cm^{-1}. Therefore, k_S is a rate deep in the Marcus inverted region ($-\Delta G >> \lambda$). For this case, a reduction of the rate by many orders of magnitude is expected upon lowering the temperature from 290 K to 90 K according to Eq. (8). In all three bacterial RCs discussed here, however, the singlet recombination rate k_S shows only a weak temperature dependence, decreasing by a factor of less than 3 upon decreasing the temperature from room temperature to 90 K. This can be reconciled with electron transfer theory by including the coupling of high energy vibronic modes to the recombination process, as has been discussed in (Bixon and Jortner, 1991). Such couplings strongly influence electron transfer rates in the inverted region and can diminish their temperature dependence to a large extent.

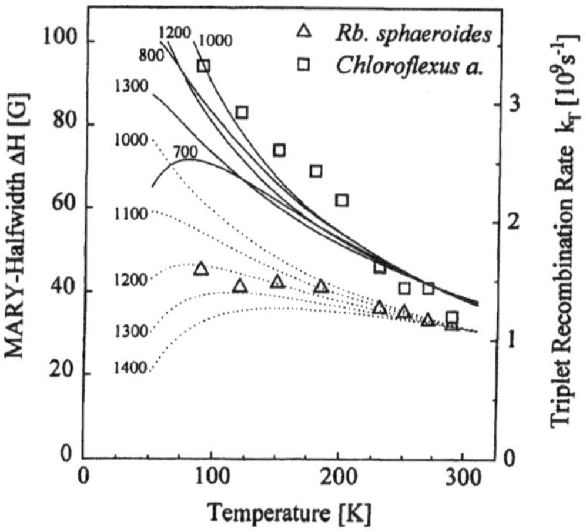

Fig. 5. Temperature dependence of the MARY halfwidth ΔB (left ordinate) for Q-depleted RCs of *Rb. sphaeroides* and *Chloroflexus aurantiacus*. On the right ordinate ΔB is converted to the triplet recombination rate k_T under the assumption of the lifetime broadening δE due to k_T being the only contribution to ΔB ($k_T = 2\pi/h\ \delta E = 2\pi/h\ 2g\beta\Delta B$). (·········) and (—): simulations of k_T with Eq. (8), using the values of ΔH_T and ΔS_T for *Rb. sphaeroides* and *Chloroflexus aurantiacus* given in Section II.E, values of λ_T given in the figure (in [cm^{-1}]), values of V_T chosen to scale the curves to the same value at high temperatures.

3. Determination of the Reorganization Energy λ_T

Figure 5 shows the temperature dependence of the MARY halfwidth ΔB for *Rb. sphaeroides* and *Cf. aurantiacus*. On the right hand ordinate of Fig. 5, ΔB is converted into the triplet recombination rate k_T. It shows an inverted temperature dependence, which according to Eq. (8) only is possible for an (pseudo)activationless (Bixon and Jortner, 1989) process, i.e. $-\Delta G_T \approx \lambda_T$. This discussion neglects possible thermal contraction effects altering the value of V_T upon changing the temperature. However, in the light of the small changes of the radical pair lifetime observed under extremely high pressure (Windsor and Menzel, 1989), such effects are expected to be small.

The curves in Fig. 5 are simulations of the temperature dependence of k_T with Eq. (8) (scaled to k_T at room temperature) for various values of λ_T with the values of $\Delta G_T(T) = \Delta H_T - T\Delta S_T$ as determined in Section II.E. It can be seen that for both species the measured temperature dependence of k_T is well reproduced by the simulations for a value of λ_T of 1000–1200 cm^{-1}. For *Cf. aurantiacus* these simulations indicate an activationless rate k_T ($-\Delta G_T = \lambda_T$) around 100 K and a small activation at higher temperatures, while in *Rb. sphaeroides* the activationless case is expected at the higher temperatures due to the different value of ΔH_T. Taking into account the uncertainty of the temperature dependent values of ΔG_T, the value of λ_T can be concluded to be in the range 1000–1500 cm^{-1}.

4. Determination of the Electronic Coupling V_T

With the values of k_T, ΔG_T and λ_T as determined above, the matrix element V_T for the triplet recombination process can be determined from Eq. (8). Taking into account the uncertainties of ΔG_T, λ_T and k_T, values of $V_T = (0.85 \pm 0.2)$ cm^{-1} and $V_T = (1.0 \pm 0.2)$ cm^{-1} are obtained for RCs of *Rb. sphaeroides* and *Cf. aurantiacus*, respectively.

5. Similarity of the Recombination Dynamics of P^+H^- in the Different RCs

The recombination dynamics of P^+H^- in the RC of *Rb. sphaeroides* and *Rb. capsulatus* is almost identical, while the RC of *Cf. aurantiacus* shows a somewhat different behavior. The triplet recombination rate k_T is similar in all RCs at room temperature, but, in contrast to *Rb. sphaeroides*, shows a pronounced increase to lower temperatures in *Cf. aurantiacus*. In addition, the value of the exchange interaction J in RCs of *Cf. aurantiacus* is more than twice as large as compared to *Rb. sphaeroides*. This is directly seen from the resonance structure of the MARY spectra, Fig. 3.

As discussed in Section II.F.3, the different temperature dependence of k_T in RCs of *Rb. sphaeroides* and *Cf. aurantiacus* is due to the different free energy gap ΔG_T (Fig. 5). In *Cf. aurantiacus* $-\Delta G_T$ is close to λ_T at 100K, resulting in an activationless electron transfer rate k_T, while in *Rb. sphaeroides* $-\Delta G_T$ is smaller than λ_T by several 100 cm^{-1}, resulting in a small activation energy. Upon increasing the temperature, k_T becomes slightly activated in *Cf. aurantiacus*, while it approaches the activationless case in *Rb. sphaeroides*, which explains the different temperature dependence of k_T in the two RCs. No further assumptions, such as different values of λ_T or effects of thermal contractions have to be made.

In RCs of *Rb. capsulatus*, on the other hand, the temperature dependence of k_T is almost identical to that in RCs of *Rb. sphaeroides*, as can be concluded from the temperature dependence of ΔB, Table 2a and Table 4a. Therefore, in reversal of the foregoing discussion, one expects ΔG_T to be very similar in *Rb. sphaeroides* and *Rb. capsulatus*.

From the absolute values of k_T, the electronic coupling V_T was estimated to be slightly larger in *Cf. aurantiacus* than in *Rb. sphaeroides*.[5] Indeed, if the electronic coupling for recombination is dominated by superexchange coupling via P^+B^-, one expects it to be larger in *Cf. aurantiacus* than in *Rb. sphaeroides*, due to the higher free energy of P^+H^-. The vertical energy separation $\delta E(P^+B^-{-}P^+H^-)$ can be estimated from the sum of $\Delta G(P^+B^-{-}P^+H^-)$ and the reorganization energy λ_2 for $P^+B^- \rightarrow P^+H^-$ to have a value of approx. 3000 cm^{-1} (here a value of approx. 1000 cm^{-1} was assumed for λ_2, close to the value determined for λ_T). The different free energy of P^+H^-, as determined in Section II.E, leads to a correspondingly different value of $\delta E(P^+B^-{-}P^+H^-)$. An electronic coupling that is 10% larger in *Cf. aurantiacus* than in *Rb. sphaeroides*, indeed, is expected from Eq. (9).

The different free energy of P^+H^- also can explain the larger value of J in RCs of *Cf. aurantiacus* compared to the value found in RCs of *Rb. sphaeroides*. This notion is based on several arguments. (i) The stronger electronic coupling in the case of *Cf. aurantiacus* results in a larger value of J. (ii) The shift J_S of $^1(P^+H^-)$ due to its interaction with $^1P^*$ (Eq. 10b) has an energy denominator which is smaller in the case of *Cf. aurantiacus* than in *Rb. sphaeroides*. (iii) The absolute value of ΔG_T is slightly smaller than λ_T in *Rb. sphaeroides*, therefore the energy of $^3P^*$ at the equilibrium nuclear coordinate of P^+H^- is higher than that of P^+H^-. The interaction of the two states results in a shift of $^3(P^+H^-)$ to lower energies, parallel to the energetic shift of $^1(P^+H^-)$,

[5] Within the uncertainty given in Section II.F.4, V_T is the same in both RCs. However, this uncertainty of the absolute value of V_T mostly arises from the uncertainty of λ_T. If one assumes the same value for λ_T for both RCs, V_T can be shown to be different in the two RCs by approx. 10%.

leading to a partial compensation of the two shifts. In *Cf. aurantiacus*, on the other hand, $|\Delta G_T|$ is larger than λ_T, so that $^3(P^+H^-)$ is shifted to higher energies and the two shifts add.

From this argument, the sign of J can be concluded to be positive in RCs of *Cf. aurantiacus*, which means that $^3(P^+H^-)$ is higher in energy than $^1(P^+H^-)$. A similar conclusion for *Rb. sphaeroides* is not possible, because it is not clear whether the singlet or the triplet contribution, both shifting P^+H^- to lower energies, predominates.

Summarizing, the differences in the recombination dynamics of *Rb. sphaeroides* and *Cf. aurantiacus* result from the different energy of P^+H^- presumably caused mostly by the exchange of the polar (protonated) glutamic acid at position L104 in *Rb. sphaeroides* to glutamine in *Cf. aurantiacus* (Michel-Beyerle et al., 1988b; Parson et al., 1990a). All other parameters seem to be very similar in the two RCs.

III. The Effect of Genetic Mutations on the Recombination Dynamics of P^+H^-

A. Mutagenesis of Tryptophan^{M250}

The amino acid residue tryptophan^{M250} (Trp^{M250}, in *Rhodopseudomonas (Rp.) viridis* Trp^{M252}), located between H and Q in RCs of *Rp. viridis* (Deisenhofer et al., 1984) and *Rb. sphaeroides* (Chang et al., 1986; Allen et al., 1987), is conserved in all bacterial RCs sequenced so far as well as in Photosystem II (Komiya et al., 1988). It is essential for the binding of Q (Coleman et al., 1990a,b) and has been suspected to enhance the electronic coupling for the electron transfer $P^+H^- \rightarrow P^+Q^-$ with the rate k_q via the superexchange mechanism (Plato et al., 1989). Indeed, the mutagenic replacement of Trp^{M250} by amino acid residues with high lying energy levels, as leucine or phenylalanine in RCs of *Rb. capsulatus* (Coleman et al., 1990a,b) or tyrosine or phenylalanine in RCs of *Rb. sphaeroides* (Stilz et al., 1990), leads to a significant decrease of k_q due to a reduction of the electronic coupling. However, this reduction could be due to structural changes affecting H and/or Q and/or changed energetics of the amino acid in its role as superexchange mediator.

To investigate possible structural changes of the H site, the recombination dynamics of P^+H^- were measured in Q-depleted RCs of *Rb. capsulatus* in which Trp^{M250} had been replaced by leucine (Leu), glutamic acid (Glu) or phenylalanine (Phe) (Coleman et al., 1990a; Lang et al., 1990; Lersch et al., 1990; Volk, 1991). Lifetime and MARY data as well as RYDMR data are almost unaffected by the mutation, Table 4. In all mutant RCs k_T increases slightly upon decreasing the temperature, similar to the native RCs of *Rb. capsulatus* and *Rb. sphaeroides*. This observation shows that ΔG_T is essentially unaffected by the mutation. The similarity of the values of k_S, k_T and J in the mutant and native RCs leads to the conclusion that mutation induced structural effects at the sites of P and H are small.

Conclusions on the specific role of Trp^{M250} as a superexchange mediator could not be drawn, since structural changes at the Q binding site might be predominant. This view is supported by (i) a decreased Q binding affinity after the mutation (Coleman et al., 1990a,b; Stilz et al., 1990) and (ii) the fact, that the aromatic Phe has a similar effect on k_q as the aliphatic Leu.

B. Mutagenesis of Tyrosine^{M210}

The amino acid residue tyrosine^{M210} (Tyr^{M210}, in *Rp. viridis* Tyr^{M208}) is in van-der-Waals-distance to P, B and H on the active branch in RCs of *Rp. viridis* (Deisenhofer et al., 1984) and *Rb. sphaeroides* (Chang et al., 1986; Allen et al., 1987). It is conserved in all purple bacterial RCs sequenced so far. Electrostatic calculations (Parson et al., 1990a) predict an increase of the free energy of P^+B^- by 1000–1300 cm^{-1} and a smaller increase of the free energy of P^+H^- upon replacement of Tyr^{M210} by phenylalanine (Phe). Such energetic shifts are expected to cause significant effects on the rate of primary charge separation. Indeed, the mutagenic replacement of Tyr^{M210} by Phe, tryptophan, isoleucine or leucine in RCs of *Rb. sphaeroides* slows primary charge separation considerably (Finkele et al., 1990; Parson et al., 1990b; Nagarajan et al., 1993). This could be due to effects of the mutation on the energies of the states involved in primary charge separation and/or mutation-induced structural changes affecting the electronic couplings. Measurements of the recombination dynamics of P^+H^- on Q-depleted RCs of *Rb. sphaeroides GA* (Tyr$^{M210} \rightarrow$ Phe) were performed to distinguish between these two possibilities (Volk et al., 1993a).

Neither the radical pair lifetime τ_{RP} nor the

magnetic field dependence of the triplet yield were found to be strongly affected by the mutation, Table 2a (because of the inherent wavelength dependence of the observed time constants (Section II.C.), only the values of τ_{RP} determined at the same wavelengths should be compared)[6]. The magnetic field $B_{1/2}$, at which half of the maximum modulation of ϕ_T is observed, was found to be larger in the mutant at 90K, but smaller at 290K. Since the genetic modification had been constructed in strain GA of *Rb. sphaeroides*, energy transfer from $^3P^*$ to $^3Car^*$ was found to occur on the 100 ns timescale between 90K and 290K. Consequently, the determination of the free energy gap between $^3P^*$ and P^+H^- from the lifetime of $^3P^*$ (Section II.E)was not possible.

However, the temperature dependence of the triplet recombination rate k_T (estimated from $B_{1/2}$) in the Tyr^{M210} → Phe mutant, increasing by a factor of almost 2 upon lowering the temperature from 290K to 90K mirrors the temperature dependence of k_T found in RCs of *Cf. aurantiacus*, see Fig. 5. In analogy to the discussion in Section II.F.5, this allows the conclusion that the mutation causes an increase of the free energy of P^+H^- to a value similar to that in *Cf. aurantiacus*, which was found to be shifted by 0.04 eV compared to *Rb. sphaeroides*, see Section II.E. A similar increase of the free energy of P^+H^- by 0.046 eV due to Tyr^{M210} → Phe was concluded upon from the amplitude of delayed fluorescence (Nagarajan et al., 1993). However, the conclusions on the energetics of P^+H^- from fluorescence measurements suffer from uncertainties with respect to the energetic inhomogeneity of P^+H^- (Section V.C.). The major contribution to the altered energetics of P^+H^- seems to arise from the altered redox potential of P/P^+, which is increased by 30 mV (Nagarajan et al., 1993).

From the difference of the absolute values of k_T between the native and the mutated RCs, it can be estimated that the replacement of Tyr^{M210} by Phe results in a decrease of the electronic coupling V_T by 10%. Such a reduction of the electronic couplings also is supported by results from RYDMR measurements (Lang, 1991) revealing reduced values of k_T and J.

[6] In (Nagarajan et al., 1993) an increase of τ_{RP} at 295K is reported due to the replacement of Tyr^{M210} by Phe. However, these results do not allow conclusions on the majority of RCs, since they originate from delayed fluorescence measurements which might be dominated by a minority of RCs (Section V). Furthermore, they refer to Q-reduced RCs.

The singlet recombination rate k_S was determined from Eq. (2) to be $1.1 \times 10^7 s^{-1}$ at 90 K and $3.0 \times 10^7 s^{-1}$ at 290 K, which is 30% smaller than in native RCs of *Rb. sphaeroides R26*. The temperature dependence of k_S is unaffected by the mutation. This is to be expected for a rate in the Marcus inverted region strongly coupled to high energy modes, which is insensitive to small variations of the free energy gap. The smaller value of k_S in the modified RCs, consequently, indicates that there has been a decrease of the electronic coupling V_S by a factor of 0.85.

The major contribution to the electronic couplings V_S and V_T for the recombination of P^+H^- is expected to arise from superexchange via P^+B^-, Eq. (9). The vertical energy separation $\delta E(P^+B^- - P^+H^-)$ was estimated above to have a value of approx. 3000 cm^{-1}. From this, it can be calculated that the predicted increase of the energy of P^+B^- by 1000 cm^{-1}, together with the predicted and indeed observed increase of the energy of P^+H^- by 400 cm^{-1}, should lead to a reduction of the superexchange couplings by a factor of approx. 0.85. This is in full agreement with the effect actually observed.

Summarizing, the effects of the replacement of Tyr^{M210} by Phe on the recombination dynamics of P^+H^- can be understood with the predicted shifts of the free energies of P^+B^- and P^+H^- (Section VI.B), without assuming any mutation-induced structural changes.

IV. Recombination Dynamics of P^+H^- in an External Electric Field

External electric fields are a powerful tool for experimentally altering the energies of charge separated states due to their large dipole moments (Popovic et al., 1986; Feher et al., 1988; Lockhart et al., 1990; Ogrodnik and Michel-Beyerle, 1992; Ogrodnik, 1993). The ion radical pair P^+H^- has a considerable electric dipole moment which can be calculated from the X-ray structure of the RC (Deisenhofer et al., 1984; Chang et al., 1986; Allen et al., 1987) to be 82 D. The interaction of this dipole moment with the external electric field leads to a shift of the energy of P^+H^-. Depending on the orientation of the dipole moment in the electric field, this shift can adopt a value of up to 800 cm^{-1} in a field of 6×10^5 V/cm. Recently, the recombination dynamics of P^+H^- in an external electric field was measured (Volk et al., 1993b). The results necessitated the development of an extended theoretical treatment for the exchange interaction J.

A. Results

Nanosecond time-resolved absorbance measurements in an external electric field (6×10^5 V/cm) were performed on quinone-depleted RCs of *Rb. sphaeroides* in PVA at 90 K and 210 K (Volk et al., 1993b). Within the experimental uncertainty, τ_{RP} was essentially unaffected by the electric field. Similarly, the shape of the MARY-spectra were hardly altered ($B_{1/2}$ was reduced from 52 G to 51 G at 90 K and from 55 G to 48 G at 210 K in the electric field).

However, significant effects of the electric field were observed on the P absorbance bleaching. At a delay time of 0 ns (± 1 ns) a reduction of this bleaching by approximately 8% was found both at 90 K and at 210 K. This effect, evolving on the picosecond time scale (Ogrodnik et al., 1992), reflects the reduction of the quantum yield of P^+H^- formation in the electric field. On the 10 ns time scale the relative electric field effect on the bleaching of P changes, mirroring an electric field effect on the triplet yield ϕ_T. At 90 K an electric-field-induced reduction of ϕ_T by 6% is observed at 0 G and at 700 G, while at 210 K there is a slight increase of ϕ_T in the electric field by 3% at 0 G and almost no effect at 700 G.

The electric-field-induced reduction of the overall yield of $^3P^*$ at 90 K, described in Volk et al. (1993b), agrees with that reported by Boxer and coworkers (Franzen et al., 1992; Boxer et al., 1992) at 80 K. In contrast to the above findings, Boxer and coworkers, however, did not obtain a difference between the electric field effects on P^+H^- and $^3P^*$, implying no effect on the recombination process yielding ϕ_T. Taking into account the increase of the relative field effect with a time constant of approximately 10 ns, due to the limited time resolution (excitation pulse width 8 ns) of the setup used by Boxer and coworkers caused them to fail to observe this difference.

B. The Effect of the Electric Field on the Recombination Rates

An energetic shift of P^+H^- alters the free energy differences ΔG_S and ΔG_T for the charge recombination rates k_S and k_T. The triplet recombination rate k_T, being almost activationless, is expected to decrease upon any changes of ΔG_T. Indeed, the slight reduction of $B_{1/2}$ at 210 K might reflect this decrease of k_T. The lack of such effects at 90 K suggests that the decrease of k_T in some RCs is compensated by the increase in other RCs, which is expected if $-\Delta G_T$ is not exactly equal to λ_T, as it is the case in *Rb. sphaeroides* at 90 K. The singlet recombination rate k_S, on the other hand, is a rate in the Marcus inverted region strongly coupled to high energy vibronic modes, which is expected to depend only weakly on the free energy (Bixon and Jortner, 1991). In summary, the observed small electric field effects on $B_{1/2}$ and τ_{RP} can be understood in terms of the energetic shift of P^+H^- by the electric field.

C. The Effect of the Electric Field on the Exchange Interaction—Extended Theoretical Treatment of J

1. Effects Expected from a Perturbation Theoretical Treatment of J

In a perturbation theoretical approach, the energy separation between the singlet and the triplet phased radical pair $^1(P^+H^-)$ and $^3(P^+H^-)$ arises from the different energetic shifts J_S of $^1(P^+H^-)$ and J_T of $^3(P^+H^-)$ due to their interaction with the manifolds of other singlet and triplet states:

$$J = J_T - J_S \tag{10}$$

The dominant contributions to J_S and J_T result from the interaction with $^1P^*$ and $^3P^*$:

$$J_T = V_T^2 / \delta E_T^0 \tag{11a}$$

$$J_S = V_S^2 / \delta E_S^0 \tag{11b}$$

with $V_{T,S}$ denoting the electronic coupling and $\delta E_{T,S}^0$ the vertical energy separation (at the equilibrium nuclear coordinates of P^+H^-) between $^3P^*$ and $^3(P^+H^-)$, and $^1P^*$ and $^1(P^+H^-)$, respectively. The activationless nature of k_T implies that $|\delta E_T^0|$ is smaller than ≈400 cm^{-1} (Michel-Beyerle et al., 1988a; Bixon et al., 1988, 1989). The singlet vertical energy separation δE_S^0, on the other hand, can be estimated to be approx. 3000 cm^{-1} from the sum of the free energy difference $\Delta G(^1P^*-P^+H^-) \approx 2000$ cm^{-1} and the reorganization energy $\lambda_S = \lambda(^1P^*- P^+H^-)$ which, in turn, is expected to be similar to $\lambda_T \approx 1000$ cm^{-1}. Due to these differences in the vertical energy separation, energetic shifts of P^+H^- by the electric field on the order of up to 800 cm^{-1} should strongly affect J_T, while they do not alter J_S considerably.

For the sake of clarity, in the following it will be assumed that $J_S << J_T$. For similar singlet and triplet electronic couplings, this is reasonable, because the triplet vertical energy separation is much smaller. However, the conclusions described here are valid also for $J_S \approx J_T$. With $|\delta E_T^0| \leq 400$ cm^{-1} in zero electric field, an electric field of 6×10^5 V/cm causes a shift of $|\delta E_T^0|$ to values under 200 cm^{-1} in 25% of the isotropically oriented RCs (Volk et al., 1993b). With $V_T \approx 1$ cm^{-1} the absolute value of J_T in these RCs is larger than 5×10^{-3} cm^{-1} according to Eq. (11a). This is more than the width of the MARY spectrum ($\Delta B \approx$ 45G $\approx 4 \times 10^{-3}$cm^{-1}) and, therefore, S-T-mixing to all three triplet levels should be effectively inhibited. On the other hand, in RCs with orientations leading to a reduction of J in the electric field, no significant enhancement of S-T-mixing occurs, because the value of J at zero electric field (1×10^{-3} cm^{-1}) already is small. Therefore, after averaging over all orientations, the electric field is expected to lead to a significant reduction (of at least 25%) of the overall triplet yield.

These expected large electric field effects on the triplet yield were not observed. In the electric field ϕ_T was found to be reduced by less than 10% at 90 K and even to increase at 210K. In addition to the effects on the triplet yield, the wide distribution of J in an isotropic sample of RCs in an external electric field as expected from Eq. (10) should have resulted in pronounced effects on the shape of the MARY spectrum which were not observed. In principle, the inhibition of S-T-mixing in 25% of the RCs could be compensated by an increase of ϕ_T due to electric field effects on the recombination rates k_S and k_T in the remaining RCs. This can be ruled out on the basis of the small electric field effects on MARY spectrum, radical pair lifetime and monoexponentiality of radical pair recombination (Volk et al., 1993b).

In conclusion, the observed electric field effects on the recombination dynamics of P^+H^- are much smaller then expected from the simple perturbation theoretical treatment of J, Eq. (11). In the next section we will describe an extension to this theory that reconciles theory and experiment.

2. *Extended Theoretical Treatment of* J

The shift J_T of $^3(P^+H^-)$ arising from the interaction with $^3P^*$ is very sensitive to the vertical energy separation δE_T between P^+H^- and $^3P^*$, which for Eq. (11a) is taken at the equilibrium position of P^+H^-. However, the nuclear coordinates are not fixed at this position, but oscillate around it. These oscillations lead to pronounced fluctuations of the vertical energy separation, especially in the domain of small δE_T^0.

Assuming displaced harmonic potentials for P^+H^- and $^3P^*$ with the same vibrational frequency ω, the vertical energy separation δE_T at the (generalized) nuclear coordinate q can be calculated:

$$\delta E_T(q) = \delta E_T^0 - \sqrt{2\lambda_T \hbar\omega}\ q \qquad (12)$$

Here $q = 0$ denotes the equilibrium position of P^+H^-. In the Born-Oppenheimer approximation, the energetic shift of $^3(P^+H^-)$ due to its interaction with $^3P^*$ can be calculated at each nuclear coordinate q from Eq. (11a). The effective triplet shift then is given by averaging over all values of q. With the wavefunction $\psi(q)$ for the harmonic oscillation (Davydov, 1965) in the nuclear vibronic ground state of P^+H^- this averaging yields

$$J_T = \int \frac{V_T^2}{\delta E_T(q)} |\psi(q)|^2\, dq \qquad (13)$$

For $T > 0$, the same average over q can be calculated for all excited nuclear vibronic states of P^+H^- and the value of J_T then is obtained by thermal averaging over these values.

For the sake of simplicity, the Born-Oppenheimer approximation also is used near the intersection of the two potential energy surfaces, although this may not well be justified in general. However, the results obtained here are in agreement with those of a thorough quantum mechanical treatment (Bixon et al., 1993).

The typical oscillation of the nuclear coordinates ($q \approx 1$) in an harmonic oscillator lead to fluctuations of the vertical energy separation δE_T between P^+H^- and $^3P^*$ around δE_T^0 of the order of $\sqrt{2\lambda_T\hbar\omega}$, which for typical parameters ($\lambda_T = 1000$ cm^{-1}, $\hbar\omega = 100$ cm^{-1}) has a value of ≈ 500 cm^{-1}. If the energy separation δE_T^0 at the equilibrium position of P^+H^- is smaller than those fluctuations, a partial cancellation of negative and positive contributions to J_T is expected. For the case $\delta E_T^0 = 0$, the negative and positive contributions to J_T should cancel completely in this approximation.

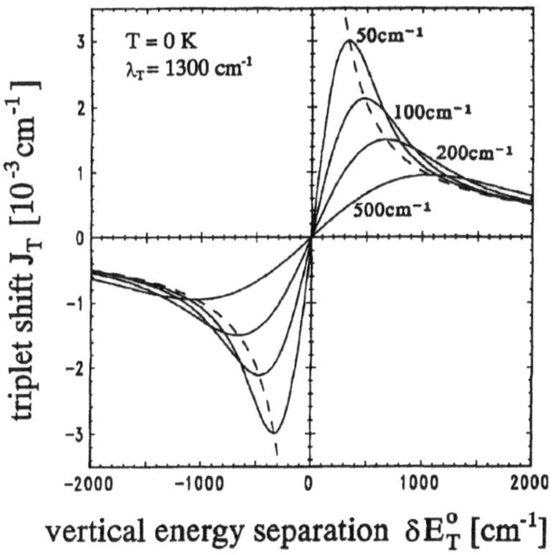

Fig. 6. Triplet shift J_T: (——) calculated from Eq. (13) for T = 0 K, λ_T = 1300 cm^{-1} and $\hbar\omega$ = 50 cm^{-1}, 100 cm^{-1}, 200 cm^{-1}, 500 cm^{-1}; (----) calculated from Eq. (11a); V_T = 1 cm^{-1} in all cases. From (Volk et al., 1993b).

Figure 6 shows the dependence of J_T on δE^0_T calculated with Eq. (13) for typical values of $\hbar\omega$ and λ_T. For comparison, the result obtained with Eq. (11a) also is given. For values of $|\delta E^0_T|$ smaller than 500–1000 cm^{-1}, the results of the extended theory indeed deviate considerably from Eq. (11a). At higher values, both approaches yield very similar results. As expected, for $\delta E^0_T = 0$ a value of J_T of 0 is obtained with Eq. (12). Thus the inclusion of nuclear oscillations removes the unrealistic discontinuity of $J_T(\delta E^0_T)$ found with Eq. (11a) at this point.

With the extended treatment the discrepancies between predicted and observed electric field effects on the triplet yield and the shape of the MARY spectrum are removed. No matter to which values δE^0_T is shifted by the electric field, $|J_T|$ calculated with Eq. (13) does not take values larger than 3–5 × 10^{-3} cm^{-1} (≈ 30–50 G) (for V_T = 1 cm^{-1}), comparable to the width of the MARY-spectrum. In addional, the small values of $|J_T|$ calculated with the extended treatment also allow to rationalize the small value of the experimentally determined exchange interaction, being smaller than 25 G (≈ 2.3 × 10^{-3} cm^{-1}) in all RCs investigated so far, in spite of considerable differences in the energy of P^+H^-.

V. Observation of the Recombination Dynamics of P+H− in Delayed Fluorescence: Evidence for an Energetic Inhomogeneity of P+H−

A nonuniform RC population with an energetic inhomogeneity with respect to P^+H^- can be revealed by comparing the recombination kinetics monitored by transient absorbance and by delayed fluorescence (due to repopulation of $^1P^*$ from P^+H^-). While transient absorbance measurements are sensitive to all RCs, the fluorescence signal will be dominated by RCs with an energetically higher-lying radical pair. From the observed different magnetic field dependence of delayed fluorescence and transient absorbance in RCs of *Rb. sphaeroides* it was concluded that the free energy of P^+H^- is inhomogeneously broadened (Ogrodnik et al., 1994).

A. Expected Magnetic Field Dependence of the Delayed Fluorescence

The delayed fluorescence due to thermal repopulation of $^1P^*$ mirrors the concentration of the singlet radical pair $^1(P^+H^-)$. Therefore, it monitors the lifetime $\tau_{S\text{-}RP}$ of $^1(P^+H^-)$ defined as

$$\tau_{S\text{-}RP}(B) = \int_0^\infty {}^1\left[P^+H^-\right](t,B)\,dt \tag{14}$$

with $[^1(P^+H^-)]$ denoting the concentration of the singlet radical pair (normalized to 1 at t = 0) and B the magnetic field. Note, that $\tau_{S\text{-}RP}$ is not the same as the overall radical pair lifetime τ_{RP} (see Section II.C).

Equation (14) allows us to relate the delayed fluorescence to transient absorbance measurements, because the yield of recombination to the singlet ground state of P ($\phi_S = 1-\phi_T$) is given by the product of the singlet recombination rate k_S with the same integral (Haberkorn and Michel-Beyerle, 1979):

$$\phi_S(B) = k_S\int_0^\infty {}^1\left[P^+H^-\right](t,B)\,dt = k_S\,\tau_{S\text{-}RP}(B) \tag{15}$$

Therefore, the magnetic field dependence of $\tau_{S\text{-}RP}$, as monitored by delayed fluorescence, and of the singlet or triplet recombination yield, as monitored by the transient absorbance at delay times sufficiently longer than τ_{RP}, are expected to have the same lineshape.

B. Results

A multiexponential decay of the time resolved fluorescence is observed at all temperatures for Q-depleted RCs of *Rb. sphaeroides*. Being in agreement with previous observations (Woodbury et al., 1986; Goldstein and Boxer, 1989a), further components could be revealed with higher time resolution (Ogrodnik et al., 1994). For example, at 275 K the fit yields a fast initial decay (within the temporal resolution of the set-up) followed by components with 135 ps, 550 ps, 2.6 ns and 11 ns lifetime with continuously decreasing amplitudes. The time constant of 11 ns compares fairly well with the lifetime τ_{RP} of P^+H^- (see Table 2a), therefore this component certainly relates to delayed fluorescence arising from thermal repopulation of $^1P^*$ from $^1(P^+H^-)$. Only this component is expected to be affected by the application of an external magnetic field, while all other lifetimes and all amplitudes are necessarily independent of the magnetic field. This is indeed confirmed by measurements of the fluorescence in an external magnetic field of 600 G.

Due to the multiple decay components it is difficult to isolate the lifetime $\tau_{S\text{-}RP}(B)$ in a fitting procedure with sufficient resolution to obtain its magnetic field dependence. A more precise determination of $\tau_{S\text{-}RP}$, Eq. (14), can be obtained by integrating over the decay of the delayed fluorescence signal $F_d(t,B)$ (normalized to 1 at $t = 0$):

$$\tau_{S\text{-}RP}(B) = \int_0^\infty F_d(t,B)\,dt \qquad (16)$$

Because contributions from the other fluorescence components (summarized as $F_0(t)$) in the experimentally accessible total fluorescence $F(t,B) = F_0(t) + F_d(t,B)$ are independent of the magnetic field, the magnetic field dependence of $\tau_{S\text{-}RP}$ can be obtained from:

$$\frac{\int_0^\infty F(t,B)\,dt - \int_0^\infty F(t,B=0)\,dt}{\int_0^\infty F(t,B=0)\,dt} = \alpha\frac{\tau_{S\text{-}RP}(B) - \tau_{S\text{-}RP}(B=0)}{\tau_{S\text{-}RP}(B=0)}$$
$$= \alpha\frac{\phi_S(B) - \phi_S(B=0)}{\phi_S(B=0)}$$

$$\text{with } \alpha = \int_0^\infty F_d(t,B=0)\,dt \Big/ \int_0^\infty F(t,B=0)\,dt \qquad (17)$$

The magnetic field dependence of the integrated fluorescence at different temperatures is shown in Fig. 7 and compared with the MARY spectra of the singlet yield ϕ_S (= $1-\phi_T$) obtained from transient absorption measurements. There are significant differences in the linewidths of the curves, particularly at low temperatures. While in transient absorbance measurements $B_{1/2}$ is almost temperature independent, increasing from 38 G to 55 G upon lowering the temperature from 290 K to 90 K (Table 2a), $B_{1/2}$ as measured in fluorescence increases from 65 G to 130 G in the same temperature range. Apparently, RCs with different recombination dynamics are monitored in fluorescence and absorbance measurements. Because $B_{1/2}$ is a measure for k_T, one can conclude that those RCs predominantly detected in fluorescence have larger rates k_T than the RCs that are detected in absorption.

Simulations of the recombination dynamics and its magnetic field dependence using the stochastic Liouville equation, Eq. (3), were carried out to search for the best match to the magnetic field dependence of ϕ_S and $\tau_{S\text{-}RP}$, respectively. At low temperatures a perfect match to the fluorescence data could be obtained, at high temperatures slight deviations had to be accepted. The values for k_T obtained from the simulations of the fluorescence data were significantly larger than the values obtained from the best fits of the absorption data (3.8 ns^{-1} compared to 1.0 ns^{-1} at 90 K, 1.8 ns^{-1} compared to 0.8 ns^{-1} at 290 K). Simulations of the time dependence of the overall concentration of P^+H^- were matched to the absorption recovery at 760 nm and simulations of the concentration of $^1(P^+H^-)$ were matched to the delayed fluorescence decay to obtain values for k_S. Again the k_S values obtained from the fluorescence data are significantly larger than those from the absorption data ($5.5 \times 10^7\,s^{-1}$ compared to $1.5 \times 10^7\,s^{-1}$ at 90 K, $6.5 \times 10^7\,s^{-1}$ and $3.5 \times 10^7\,s^{-1}$ at 290 K).

C. Energetic Inhomogeneity of P^+H^-

The delayed fluorescence signal is proportional to the singlet radical pair concentration weighted by the Boltzmann factor for repopulation of $^1P^*$ and thus has a preferential selectivity to radical pairs energetically close to $^1P^*$, while the absorption signal is related directly to the radical pair concentration.

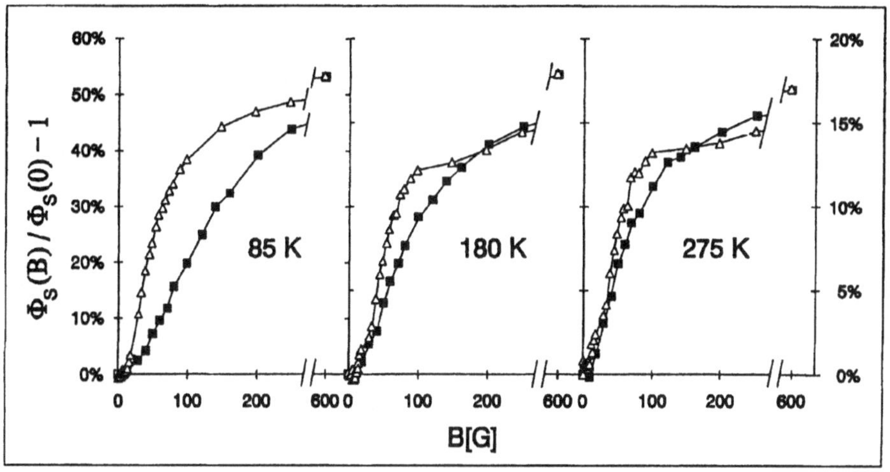

Fig. 7. Comparison of MARY spectra detected in absorption (△) and emission (■) in Q-depleted RCs of *Rb. sphaeroides*. (△) relative increase of the singlet recombination yield $\phi_S = 1-\phi_T$ by a magnetic field, determined from the transient absorbance measurements in the Q_y-band of P described in Section II.B. (■) relative increase of $\tau_{S\text{-}RP}$ by a magnetic field (Eq. 17), scaled to the same modulation as the increase of ϕ_S. The ordinate on the left refers to 85 K, the ordinate on the right to 180 K and 275 K. From (Ogrodnik et al., 1994).

Therefore, the experimental finding that fluorescence and absorbance measurements probe different RCs implies an energetic inhomogeneity of the free energy gap between $^1P^*$ and P^+H^-. In contrast to transient absorbance that monitors the majority of RCs, most of the delayed fluorescence can result from a small minority of RCs. At lower temperatures, the delayed fluorescence selects radical pairs even closer to $^1P^*$ than those selected at higher temperatures.

A significant energetic inhomogeneity of the state $^1P^*$ should lead to excitation wavelength dependent rates of primary charge separation. In photochemical hole-burning spectra in the Q_y-absorbance band of P, however, no dependence of the lifetime of $^1P^*$ on the burn frequency was found (Lyle et al., 1993). Furthermore, an energetic inhomogeneity of $^1P^*$ is not expected to have significant effects on the recombination of P^+H^- to the ground state. Therefore, the inhomogeneity of $\Delta G(^1P^*\text{–}P^+H^-)$ observed here has to arise from an energetic inhomogeneity of P^+H^-.

Differences in fluorescence and absorbance measurements will result from an energetic inhomogeneity of P^+H^- only, if at least one of the relevant recombination parameters changes with the free energy of P^+H^-. The k_T value of the majority of RCs was concluded to be near the activationless limit. Shifting the radical pair energy to higher values should result only in a small increase of the Franck-Condon factor when reaching this limit and a subsequent decrease upon a further shift. The difference in the apparent values of k_T measured at low temperatures in fluorescence and absorption, however, is distinctly larger, suggesting that not only the Franck-Condon factor, but also the electronic coupling V_T is changed with energy. The notion of a significant dependence of the recombination matrix element on the energy of P^+H^- is confirmed by the inhomogeneity of k_S which is found to parallel that of k_T. The Franck-Condon factor of this rate is expected to depend only weakly on the free energy, because the reaction is deep in the Marcus inverted region. Thus, a dependence of V_S on the free energy of P^+H^- corresponding to that of V_T can be deduced.

Indeed, the energetic inhomogeneity of P^+H^- is expected to affect V_S and V_T, if the superexchange mechanism prevails for these couplings, see Eq. (9). The expected dominance of superexchange coupling for the recombination rates k_S and k_T (Michel-Beyerle et al., 1988a; Bixon et al., 1991), thus, can be supported from experimental results.

From the temperature dependence of the delayed fluorescence, the width of the energetic distribution of P^+H^- can be estimated (Ogrodnik et al., 1994). When neglecting the energetic inhomogeneity, the activation barrier for the recombination process from P^+H^- to $^1P^*$ can be calculated from the ratio of the amplitudes of the prompt and the 10 ns-delayed fluorescence components to be 0.2 eV at 290 K and 0.08 eV at 90 K. However, due to the energetic inhomogeneity of P^+H^-, the delayed fluorescence amplitude is proportional to the average of the Boltzmann factor for recombination of P^+H^- to $^1P^*$ over the distribution of the free energy of P^+H^-. The apparent value ΔG_{app} as determined from the observed fluorescence amplitude will be smaller than the average value of $\Delta G(P^+H^- - {^1P^*})$, because energetically higher radical pairs are weighted more strongly by the Boltzmann factor. Assuming the energy of P^+H^- to have a Gaussian distribution of width 2σ around a maximum value ΔG_0, a simple approximate expression was derived for ΔG_{app}, which holds as long as σ is small compared to ΔG_0:

$$\Delta G_{app} = \Delta G_0 - \frac{\sigma^2}{2kT} \quad (18)$$

With the values of 0.2 eV and 0.08 eV for ΔG_{app} at 290 K and 90 K, respectively, the width σ of the energetic distribution of P^+H^- is calculated to be approx. 0.05 eV. The maximum of the distribution is at ΔG_0 = 0.25 eV, in full agreement with the value obtained from transient absorbance measurements.

VI. Recombination Dynamics as a Unique Diagnostic Tool: Conclusions

The unique merits of the elucidation of the recombination dynamics in absorption and emission experiments refer to two issues: (i) the demonstration of inhomogeneous broadening of radical pair energies and (ii) the determination of radical pair energetics for the majority of RCs, Fig. 8. Due to the specificity of delayed fluorescence experiments to energetically high lying radical pair states, as discussed in Section V, they may not yield information on the bulk of the RCs, unless suitable extrapolation from high temperature data is made. Thus, information derived from MARY absorption data constitutes in many cases the only direct access to parameters basic for the treatment of rates in the framework of electron transfer theories.

A. Implications of the Energetic Inhomogeneity of P^+H^-

Until recently, electron transfer dynamics in photosynthetic reaction centers were assumed to be single-exponential. Considering the complexity of the system with the numberless degrees of freedom of the protein, it was not surprising when growing evidence for deviations from homogeneous kinetics appeared with increasing experimental precision. Bi- and multiphasic kinetics were observed for the recombination of $P^+Q_A^-$ (Kleinfeld et al., 1984; Parot et al., 1987; Sebban and Wraight, 1989; Franzen et al., 1990) and $P^+Q_B^-$ (Gao et al., 1990), for the primary charge separation (Feick et al., 1990; Becker et al., 1991; Vos et al., 1991, 1992; Du et al., 1992; Hamm et al., 1993) and for delayed fluorescence (Sebban and Barbet, 1984; Woodbury and Parson, 1984; Woodbury et al., 1986; Hörber et al., 1986). Indications for inhomogeneity were also derived from kinetics that were found to depend on the probing wavelength (Kirmaier and Holten, 1990). The described inhomogeneities referred to phenomenological nonuniformities of the measured dynamics or to spectral information, but could not directly address the origin of these inhomogeneous features, so that various assumptions have been made on this issue (Woodbury and Parson, 1984; Goldstein and Boxer, 1989b; Feick et al., 1990; Hamm et al., 1993).

As discussed in Section V.C, the different magnetic field dependence of the recombination dynamics observed in delayed fluorescence and in transient absorbance measurements proves an energetic inhomogeneity of the radical pair P^+H^-. It is reasonable to suspect that this energetic inhomogeneity originates from differences of the interaction of the charged cofactors with their protein surroundings in different configurational states (Frauenfelder et al., 1988). If the protein matrix is thus responsible for the proven energetic inhomogeneity of P^+H^-, the assumption of a similar inhomogeneity for P^+B^- or any other radical pair seems justified.

Primary charge separation is discussed to form either P^+H^- or P^+B^- in the first electron transfer step. The energetic inhomogeneity of these radical pairs yields the physical basis for interpreting the multiphasic kinetics detected in the charge separation process (Wang et al., 1993; Ogrodnik et al., 1994). Due to the influence of the energy on the electron transfer rate via the Franck-Condon factor, a continuous distribution of the lifetime of the primary donor state $^1P^*$ is expected, which should become broader at lower temperatures. We propose that the multiphasic decay of $^1P^*$ detected experimentally in stimulated emission or fluorescence upconversion results from this inhomogeneity. This multiphasic decay extends with continuously decreasing amplitudes to considerably longer times, which have been detected in photon counting experiments. These prompt fluorescence lifetimes are limited by the internal conversion rate of $^1P^*$, which was determined to be faster than 1 ns (Eberl et al., 1992). Thus, we are inclined to assign the fluorescence components up to about 1 ns to very slow charge separation processes.

The influence of an external electric field on the fluorescence has been studied with the intention to clarify the mechanism of primary charge separation, in particular to assign the identity of the primary radical pair state to either P^+H^- or P^+B^- (Lockhart et al., 1988; Ogrodnik et al., 1991; Ogrodnik and Michel-Beyerle, 1992; Ogrodnik, 1993). Compared to theoretical expectations (Bixon et al., 1989), the observations yielded a weak dependence of the fluorescence on the electric field, both in steady-state and time resolved measurements. These weak effects now can be understood by taking into account the energetic inhomogeneity of the state P^+B^- and/or P^+H^-. Due to the interaction of the dipole moments of the radical pairs P^+H^- and P^+B^- with the electric field, these states will be shifted by a value in the order of 0.1 eV in the electric fields employed. Because the shift depends on the orientation of the dipole in the electric field, and isotropic RC samples were used, a continuous distribution of the radical pair energies around the levels at zero field will be induced by the field. If the radical pairs have a well-defined energy at zero field, this will result in a significant spread of the charge separation rate, leading to a significant increase of the average $^1P^*$ lifetime and thus of the prompt fluorescence quantum yield. However, if the energy levels of P^+B^- and/or P^+H^- already are broadened in absence of the field, the electric field mainly causes a redistribution of the energy levels and only results in an additional broadening. In this case, it is not surprising to find the field induced changes of the prompt or delayed fluorescence partially masked by the initial inhomogeneity.

Another important consequence of the energetic inhomogeneity of P^+H^- is the impossibility of measuring the energetics in the majority of RCs by delayed fluorescence directly. Since those RCs with a higher energy of P^+H^- will contribute predominantly, delayed fluorescence will yield an apparent energy gap between $^1P^*$ and P^+H^- which is smaller than the mean value, see Eq. (18). This effect is stronger at lower temperatures, therefore it is not surprising that a significant decrease of $\Delta G(^1P^*–P^+H^-)$ with decreasing temperature was derived from delayed fluorescence measurements. In contrast to delayed fluorescence, transient absorbance measurements are not expected to be affected by the energetic inhomogeneity of P^+H^- and indeed do not show the strong temperature dependence of ΔG, Table 2a. On the other hand, the temperature dependence of the apparent free energy difference determined from the delayed fluorescence allows one to estimate the width of the energetic distribution of P^+H^-, as described in Section V.C, and to extrapolate to the mean energy of the majority of RCs.

Summarizing, the energetic inhomogeneity of radical pair states, which was proven by the comparison of P^+H^- recombination dynamics measurements in delayed fluorescence and transient absorbance, allows us to understand the multiphasic kinetics observed for electron transfer processes in the RC. It also explains the unexpectedly weak dependence of prompt and delayed fluorescence on an external electric field and discrepancies in the energetics of P^+H^- determined by delayed fluorescence and transient absorbance measurements.

B. Energetics and Mechanism of Primary Charge Separation

After excitation of a RC of *Rb. sphaeroides*, an electron is transferred from the primary donor P to H within 3–4 ps (Woodbury et al., 1985; Martin et al., 1986; Fleming et al., 1988; Holzapfel et al., 1989). The detailed mechanism of this primary charge separation, especially the role of the accessory BChl monomer B, is still discussed controversially (Holzapfel et al., 1989; Kirmaier and Holten, 1990; Lauterwasser et al., 1991; Chan et al., 1991). Under debate are models, in which electron transfer either occurs in two steps via B or directly to H with B enhancing the electronic interaction between P and H via the superexchange mechanism (See Chapter 24 by Woodbury and Allen and Chapter 25 by Parson and Warshel).

In principle, both mechanisms will contribute in parallel to the primary charge separation. The relative contribution of each mechanism is determined mostly by the free energy gap ΔG_1 between $^1P^*$ and P^+B^-. At physiological temperatures the two-step-mechanism is expected to dominate as long as the free energy of P^+B^- is not above that of $^1P^*$ (Bixon et al., 1991), while at 100 K the two-step-mechanism still should prevail, if P^+B^- is more than 300 cm^{-1} under $^1P^*$. This is just the upper limit of ΔG_1 derived from the recombination dynamics of P^+H^- in RCs of *Rb. sphaeroides*, Fig. 8A. Therefore, the energetic results from the recombination dynamics are fully compatible with both superexchange and two-step charge separation even at 100 K. In the case of *Cf.*

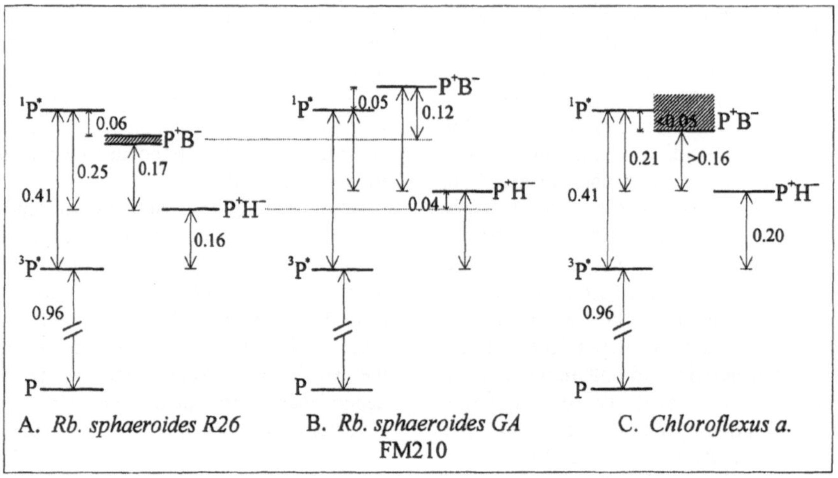

Fig. 8. Comparative energetics for Q-depleted RCs of (A) *Rb. sphaeroides R26*, (B) the $Tyr^{M210} \rightarrow$ Phe mutant of *Rb. sphaeroides GA*, and (C) *Chloroflexus aurantiacus*, as determined in Section II.E. and Section III.B.

aurantiacus, the upper limit of ΔG_1 is 150 cm^{-1} (Fig. 8C), again compatible with both mechanisms of charge separation, at least at physiological temperatures (however, the analysis by Bixon et al. (1991) strictly only refers to *Rb. sphaeroides*, because it is based on various assumptions and parameters taken from *Rb. sphaeroides*).

Besides picosecond time resolved spectroscopy and hole burning experiments (Reddy et al., 1992), both yielding information about the rate of charge separation, also the exchange interaction J of P^+H^- has been used to draw conclusions on the mechanism of primary charge separation (Marcus 1987, 1988; Michel-Beyerle et al., 1988a; Bixon et al., 1988, 1989, 1991). Even before the X-ray crystal structure of the RC revealed the location of B between P and H (Deisenhofer et al., 1984; Chang et al., 1986; Allen et al., 1987), the observed small value of J led to the proposal of two-step charge separation (Haberkorn et al., 1979). This was necessary to reconcile the small value of J with the strong electronic interaction concluded from fast electron transfer.

From the activationless rate of primary charge separation the relevant matrix element could be estimated to be approx. 20 cm^{-1}. Under the assumption of one-step (superexchange) charge separation, this is the electronic coupling between $^1P^*$ and P^+H^-. This coupling is also relevant for the calculation of the singlet shift J_S of P^+H^- due to its interaction with $^1P^*$. The value of J_S estimated this way is on the order of several 100 G, much larger than the experimentally observed value of J. Still, two possibilities resolving this discrepancy were discussed so that one step charge separation could not be ruled out solely on the basis of recombination data (Bixon et al., 1988, 1989; Michel-Beyerle et al., 1988a). (i) A fortuitous compensation of J_S by an equivalently large shift J_T of $^3(P^+H^-)$ in the same direction. However, the maximum values of J_T calculated with the extended treatment including nuclear motion, Eq. (13), are not large enough to compensate the expected large values of J_S. (ii) A fast decrease of the coupling between P and H due to some configurational relaxation after charge separation.

In a two-step model, the strong electronic interaction derived from the primary charge separation rate refers to the coupling between $^1P^*$ and P^+B^-. Theoretical calculations based on the known structure predict that the coupling between P^+B^- and P^+H^- is larger than that between $^1P^*$ and P^+B^- by a factor of 6 (Plato et al., 1988). Together with an estimate of the vertical energy gap between P^+H^- and P^+B^- ($\delta E(P^+B^- - P^+H^-) \approx 3000$ cm^{-1}, see Section II.F.5), the superexchange coupling between $^1P^*$ and P^+H^- can be calculated[7] to be on the order of 1 cm^{-1}, in accordance with the value determined for the coupling V_T between $^3P^*$ and P^+H^- (Section II.F.4). This coupling

[7] Even in the case of two-step charge separation, charge recombination is a one-step process dominated by superexchange coupling, Eq. (9), see Section II.F.1.

yields a value of J_S of several G, which is very compatible with the observed value of J.

Summarizing, the results obtained from the recombination dynamics of $P^{+}H^{-}$ in native RCs of *Rb. sphaeroides* and *Cf. aurantiacus* can only be reconciled with the superexchange mechanism of charge separation, if configurational relaxation leads to a substantial decrease of the electronic coupling between P and H after charge separation. The energetics of $P^{+}B^{-}$ as determined from the recombination dynamics of $P^{+}H^{-}$ are compatible with both mechanisms. It has to be stressed that the conclusions described are based on energetics derived from transient absorbance measurements. Therefore, they refer to the average values of the majority of RCs and are not distorted by the energetic inhomogeneity of the radical pair states discussed in Section V.C.

In the context of the investigation of mutagenic or chemical modifications of the RC, the measurement of the magnetic field dependent recombination dynamics of $P^{+}H^{-}$ yields important information. In particular, conclusions about the effects on the energies of the charge-separated states can be derived, see Section III. In contrast to other investigations referring only to mutationaly induced changes of the oxidation potential of P, information derived from recombination data includes energetic changes of H^{-} as well. The altered energies have to be taken into account for the discussion of the effects of any modification of the RC on the rate and mechanism of primary charge separation.

Acknowledgments

The biochemical basis of our work with respect to native bacterial reaction centers and their manipulation has developed in a long-standing cooperation with R. Feick (Max-Planck-Institut für Biochemie, Martinsried), to whom we are especially grateful. Mutagenetically altered reaction centers were designed by and investigated with two different biochemical laboratories: mutants of *Rb. capsulatus* with D.C. Youvan and W.W. Coleman (MIT, Cambridge), mutants of *Rb. sphaeroides* with D. Oesterhelt and K.A. Gray (Max-Planck-Institut für Biochemie, Martinsried).

The theoretical understanding of spin dynamics in radical pairs, esp. its manifestations in RYDMR spectroscopy, has been developed by W. Lersch. The interpretation of exchange interaction and electron transfer phenomena has been worked out in close collaboration and stimulating discussions with M. Bixon and J. Jortner (Tel Aviv University).

Finally, we want to thank all the people in the lab involved in recombination studies referred to in this review: R. Letterer, G. Aumeier, T. Langenbacher, T. Häberle, W. Neumann, E. Lang, G. Rousseau.

Since we were summing up over a central part of our studies within the last decade, it is a special pleasure to thank at this occasion E. W. Schlag (Institut für Physikalische Chemie, TU München) for his continuous interest and scientific support.

The totality of this work has been funded by the Deutsche Forschungsgemeinschaft in the framework of the Sonderforschungsbereich 143 ('Elementarprozesse der Photosynthese').

References

Allen JP, Feher G, Yeates TO, Komiya H and Rees DC (1987) Structure of the reaction center from *Rhodobacter sphaeroides* R-26: The cofactors. Proc Natl Acad Sci USA 84: 5730–5734

Aumeier G (1994) Rekombinationsdynamik des spinkorrelierten Radikalpaars $P^{+}H^{-}$ in photosynthetischen Reaktionszentren. Eine Gegenüberstellung von Simulationen und Messungen an *Rb. sphaeroides* und Photosystem II. PhD-Thesis. Technische Universität München

Becker M, Nagarajan V, Middendorf D, Parson WW, Martin JE and Blankenship RE (1991) Temperature dependence of the initial electron-transfer kinetics in photosynthetic reaction centers of *Chloroflexus aurantiacus*. Biochim Biophys Acta 1057: 299–312

Bixon M and Jortner J (1989) Activationless and pseudo-activationless primary electron transfer in photosynthetic reaction centers. Chem Phys Lett 159: 17–20

Bixon M and Jortner J (1991) Non-Arrhenius temperature dependence of electron-transfer rates. J Phys Chem 95: 1941–1944

Bixon M, Jortner J, Michel-Beyerle ME, Ogrodnik A and Lersch W (1987) The role of the accessory bacteriochlorophyll in reaction centers of photosynthetic bacteria: Intermediate acceptor in the primary electron transfer? Chem Phys Lett 140: 626–630

Bixon M, Michel-Beyerle ME and Jortner J (1988) Formation dynamics, decay kinetics and singlet-triplet splitting of the (bacteriochlorophyll dimer)$^{+}$(bacteriopheophytin)$^{-}$ radical pair in bacterial photosynthesis. Isr J Chem 28: 155–168

Bixon M, Jortner J, Michel-Beyerle ME and Ogrodnik A (1989) A superexchange mechanism for the primary charge separation in photosynthetic reaction centers. Biochim Biophys Acta 977: 273–286

Bixon M, Jortner J and Michel-Beyerle ME (1991) On the mechanism of the primary charge separation in bacterial photosynthesis. Biochim Biophys Acta 1056: 301–315

Bixon M, Jortner J and Michel-Beyerle ME (1993) The singlet-triplet splitting of the primary radical pair in the bacterial photosynthetic reaction center. Zeitschr Phys Chem 180: 193–208

Blankenship RE, Schaafsma TJ and Parson WW (1977) Magnetic field effects on radical pair intermediates in bacterial photosynthesis. Biochim Biophys Acta 461: 297–305

Blankenship RE, Feick R, Bruce BD, Kirmaier C, Holten D and Fuller RC (1983) Primary photochemistry in the facultative green photosynthetic bacterium *Chloroflexus aurantiacus*. J Cell Biochem 22: 251–261

Bowman MK, Budil DE, Closs GL, Kostka AG, Wraight CA and Norris JR (1981) Magnetic resonance spectroscopy of the primary state, P^f, of bacterial photosynthesis. Proc Natl Acad Sci USA 78: 3305–3307

Boxer SG, Chidsey CED and Roelofs MG (1982a) Anisotropic magnetic interactions in the primary radical ion-pair of photosynthetic reaction centers. Proc Natl Acad Sci USA 79: 4632–4636

Boxer SG, Chidsey CED and Roelofs MG (1982b) Use of large magnetic fields to probe photoinduced electron-transfer reactions: An example from photosynthetic reaction centers. J Am Chem Soc 104: 1452–1454

Boxer SG, Chidsey CED and Roelofs MG (1982c) Dependence of the yield of a radical-pair reaction in the solid state on orientation in a magnetic field. J Am Chem Soc 104: 2674–2675

Boxer SG, Chidsey CED and Roelofs MG (1983) Magnetic field effects on reaction yields in the solid state: An example from photosynthetic reaction centers. Ann Rev Phys Chem 34: 389–417

Boxer SG, Franzen S, Lao K, Lockhart DJ, Stanley R, Steffen M and Stocker JW (1992) Electric field effects on the quantum yields and kinetics of fluorescence and transient intermediates in bacterial reaction centers. In: Breton J and Vermeglio A (eds) The Photosynthetic Bacterial Reaction Center II, pp 271–282. Plenum Press, New York

Breton J, Martin JL, Migus A, Antonetti A and Orszag A (1986) Femtosecond spectroscopy of excitation energy transfer and initial charge separation in the reaction center of the photosynthetic bacterium *Rhodopseudomonas sphaeroides*. In: Fleming GR and Siegmann AE (eds) Ultrafast Phenomena V, pp 393–397. Springer-Verlag, Berlin

Bruce BD, Fuller RC and Blankenship RE (1982) Primary photochemistry in the facultatively aerobic green photosynthetic bacterium *Chloroflexus aurantiacus*. Proc Natl Acad Sci USA 79: 6532–6536

Budil DE, Kolaczkowski SV and Norris JR (1987) The temperature dependence of electron back-transfer from the primary radical pair of bacterial photosynthesis. In: Biggins J (ed) Progress in Photosynthesis Research, Vol I, pp 25–28. Martinus Nijhoff Publishers, Dordrecht

Chan C-K, DiMagno TJ, Chen LX-Q, Norris JR and Fleming GR (1991) Mechanism of the initial charge separation in bacterial photosynthetic reaction centers. Proc Natl Acad Sci USA 88: 11202–11206

Chang C-H, Tiede D, Tang J, Smith U, Norris J and Schiffer M (1986) Structure of *Rhodobacter sphaeroides* R-26 reaction center. FEBS Lett 205: 82–86

Chidsey CED, Roelofs MG and Boxer SG (1980) The effect of large magnetic fields and the g-factor difference on the triplet population in photosynthetic reaction centers. Chem Phys Lett 74: 113–118

Chidsey CED, Kirmaier C, Holten D and Boxer SG (1984) Magnetic field dependence of radical-pair decay kinetics and molecular triplet quantum yield in quinone-depleted reaction centers. Biochim Biophys Acta 766: 424–437

Chidsey CED, Takiff L, Goldstein RA and Boxer SG (1985) Effect of magnetic fields on triplet state lifetime in photosynthetic reaction centers: Evidence for thermal repopulation of the initial radical pair. Proc Natl Acad Sci USA 82: 6850–6854

Cogdell RJ, Monger TG and Parson WW (1975) Carotenoid triplet states in reaction centers from *Rhodopseudomonas sphaeroides* and *Rhodospirillum rubrum*. Biochim Biophys Acta 408: 189–199

Coleman WJ, Youvan DC, Aumeier W, Eberl U, Volk M, Lang E, Siegl J, Heckmann R, Lersch W, Ogrodnik A and Michel-Beyerle ME (1990a) How conclusive is mutagenic replacement of Trp M250 in photosynthetic reaction centers? In: Baltscheffsky M (ed) Current Research in Photosynthesis, Vol I, pp 153–156. Kluwer Academic Publishers, Netherlands

Coleman WJ, Bylina EJ, Aumeier W, Siegl J, Eberl U, Heckmann R, Ogrodnik A, Michel-Beyerle ME and Youvan DC (1990b) Influence of mutagenic replacement of tryptophan M250 on electron transfer rates involving primary quinone in reaction centers of *Rb. capsulatus*. In: Michel-Beyerle ME (ed) Reaction Centers of Photosynthetic Bacteria, pp 273–282. Springer-Verlag, Berlin

Davydov AS (1965) Quantum Mechanics. Pergamon Press, Oxford

Deisenhofer J, Epp O, Miki K, Huber R and Michel H (1984) X-ray structure analysis of a membrane protein complex. Electron density map at 3 Å resolution and a model of the chromophores of the photosynthetic reaction center from *Rhodopseudomonas viridis*. J Mol Biol 180: 385–398

Du M, Rosenthal SJ, Xie X, DiMagno TJ, Schmidt M, Hanson DK, Schiffer M, Norris JR and Fleming GR (1992) Femtosecond spontaneous emission studies of reaction centers from photosynthetic bacteria. Proc Natl Acad Sci USA 89: 8517–8521

Dutton PL, Leigh JS and Reed DW (1973) Primary events in the photosynthetic reaction centre from *Rhodopseudomonas spheroides* strain R26: Triplet and oxidized states of bacteriochlorophyll and the identification of the primary electron acceptor. Biochim Biophys Acta 292: 654–664

Eberl U, Gilbert M, Keupp W, Langenbacher T, Siegl J, Sinning I, Ogrodnik A, Robles SJ, Breton J, Youvan DC and Michel-Beyerle ME (1992) Fast internal conversion of the primary donor in modified reaction centers. In: Breton J and Vermeglio A (eds) The Photosynthetic Bacterial Reaction Center II. Structure, Spectroscopy and Dynamics, pp 253–260. Plenum Press, New York

Feher G, Arno TR, Okamura MY (1988) The effect of an electric field on the charge recombination rate of $D^+Q_A^- \rightarrow DQ_A$ in reaction centers of *Rhodobacter sphaeroides* R-26. In: Breton J and Vermeglio A (eds) The Photosynthetic Bacterial Reaction Center, Structure and Dynamics, pp 271–287. Plenum Press, New York

Feick R, Martin JL, Breton J, Volk M, Scheidel G, Langenbacher T, Urbano C, Ogrodnik A and Michel-Beyerle ME (1990) Biexponential charge separation and monoexponential decay of P^+H^- in reaction centers of *Chloroflexus aurantiacus*. In: Michel-Beyerle ME (ed) Reaction Centers of Photosynthetic

Bacteria, pp 181–188. Springer-Verlag, Berlin
Finkele U, Lauterwasser C, Zinth W, Gray KA and Oesterhelt D (1990) Role of tyrosine M210 in the initial charge separation of reaction centers of *Rb. sphaeroides*. Biochemistry 29: 8517–8521
Fleming GR, Martin JL and Breton J (1988) Rates of primary electron transfer in photosynthetic reaction centres and their mechanistic implications. Nature 333: 190–192
Frank HA and Violette CA (1989) Monomeric bacteriochlorophyll is required for the triplet energy transfer between the primary donor and the carotenoid in photosynthetic bacterial reaction centers. Biochim Biophys Acta 976: 222–232
Frankevich EL, Pristupa AI and Lesin VI (1977) Magnetic resonance of short-lived triplet exciton pairs detected by fluorescence modulation at room temperature. Chem Phys Lett 47: 304–308
Franzen S, Goldstein RF and Boxer SG (1990) Electric field modulation of electron transfer reaction rates in isotropic systems: long-distance charge recombination in photosynthetic reaction centers. J Phys Chem 94: 5135–5149
Franzen S, Lao K-Q, Stanley B and Boxer SG (1992) Electric-field-induced quantum yield failure of the primary charge separation step of photosynthetic reaction centers. Biophys J 61: A153
Frauenfelder H, Parak F and Young RD (1988) Conformational substates in proteins. Ann Rev Biophys Biophys Chem 17: 451–479
Gao J-L, Shopes RJ and Wraight CA (1990) Heterogeneity of kinetics and electron transfer equilibria in the bacteriopheophytin and quinone electron acceptors of reaction centers from *Rhodopseudomonas viridis*. Biochim Biophys Acta 1056: 259–272
Goldstein RA and Boxer SG (1987) Effects of nuclear spin polarization on reaction dynamics in photosynthetic bacterial reaction centers. Biophys J 51: 937–946
Goldstein RA and Boxer SG (1989a) The effect of very high magnetic fields on the delayed fluorescence from oriented bacterial reaction centers. Biochim Biophys Acta 977: 70–77
Goldstein RA and Boxer SG (1989b) The effect of very high magnetic fields on the reaction dynamics in bacterial reaction centers: Implications for the reaction mechanism. Biochim Biophys Acta 977: 78–86
Goldstein RA, Takiff L and Boxer SG (1988) Energetics of initial charge separation in bacterial photosynthesis: The triplet decay rate in very high magnetic fields. Biochim Biophys Acta 934: 253–263
Haberkorn R and Michel-Beyerle ME (1977) Mechanism of triplet formation in photosynthesis via hyperfine interaction. FEBS Lett 75: 5–7
Haberkorn R and Michel-Beyerle ME (1979) On the mechanism of magnetic field effects in bacterial photosynthesis. Biophys J 26: 489–498
Haberkorn R, Michel-Beyerle ME and Marcus R (1979) On spin-exchange and electron-transfer rates in bacterial photosynthesis. Proc Natl Acad Sci USA 76: 4185–4188
Hamm P, Gray KA, Oesterhelt D, Feick R, Scheer H and Zinth W (1993) Subpicosecond emission studies of bacterial reaction centers. Biochim Biophys Acta 1142: 99–105
Hoff AJ (1981) Magnetic field effects on photosynthetic reactions. Quart Rev Biophys 14: 599–665
Hoff AJ (1986) Magnetic interactions between photosynthetic reactants. Photochem Photobiol 43: 727–745
Hoff AJ, Rademaker H, van Grondelle R and Duysens LNM (1977) On the magnetic field dependence of the yield of the triplet state in reaction centers of photosynthetic bacteria. Biochim Biophys Acta 460: 547–554
Holten D, Windsor MW, Parson WW and Thornber JP (1978) Primary photochemical processes in isolated reaction centers of *Rhodopseudomonas viridis*. Biochim Biophys Acta 501: 112–126
Holzapfel W, Finkele U, Kaiser W, Oesterhelt D, Scheer H, Stilz HU and Zinth W (1989) Observation of a bacteriochlorophyll anion radical during the primary charge separation in a reaction center. Chem Phys Lett 160: 1–7
Hörber JKH, Göbel W, Ogrodnik A, Michel-Beyerle ME and Cogdell RJ (1986) Time-resolved measurements of fluorescence from reaction centers of *Rhodobacter sphaeroides* R26. FEBS Lett 198: 273–278
Hore PJ, Riley DJ, Semlyen JJ, Zwanenburg G and Hoff AJ (1993) Analysis of anisotropic electron spin polarization in the photosynthetic bacterium *Rhodospirillum rubrum*. Evidence that the sign of the exchange interaction in the primary radical pair is positive. Biochim Biophys Acta 1141: 221–230
Jortner J (1980a) Dynamics of the primary events in bacterial photosynthesis. J Am Chem Soc 102: 6676–6686
Jortner J (1980b) Dynamics of electron transfer in bacterial photosynthesis. Biochim Biophys Acta 594: 193–230
Kaufmann KJ, Dutton PL, Netzel TL, Leigh JS and Rentzepis PM (1975) Picosecond kinetics of events leading to reaction center bacteriochlorophyll oxidation. Science 188: 1301–1304
Kirmaier C and Holten D (1990) An inhomogeneous distribution of bacterial reaction centers underlies the observed temperature and detection wavelength dependence of the rates of the primary electron transfer reactions. Proc Natl Acad Sci USA 87: 3552–3556
Kleinfeld D, Okamura MY and Feher G (1984) Electron-transfer kinetics in photosynthetic reaction centers cooled to cryogenic temperatures in the charge-separated state: Evidence for light-induced structural changes. Biochemistry 23: 5780–5786
Komiya H, Yeates TO, Rees DC, Allen JP and Feher G (1988) Structure of the reaction center from *Rhodobacter sphaeroides* R-26 and 2.4.1: Symmetry relations and sequence comparisons between different species. Proc Natl Acad Sci USA 85: 9012–9016
Lang E (1991) Zeitaufgelöste magnetische Resonanzspektroskopie am Radikalpaar P^+H^- der bakteriellen Photosynthese mit optischem Nachweis über die Rekombinationsausbeuten. PhD-Thesis. Technische Universität München
Lang E, Lersch W, Tappermann P, Coleman WJ, Youvan DC, Feick R and Michel-Beyerle ME (1990) High power RYDMR spectra of P^+H^- in reaction centers of photosynthetic bacteria. In: Baltscheffsky M (ed) Current Research in Photosynthesis, Vol I, pp 137–140. Kluwer Academic Publishers, Dordrecht
Lauterwasser C, Finkele U, Scheer H and Zinth W (1991) Temperature dependence of the primary electron transfer in photosynthetic reaction centers from *Rhodobacter sphaeroides*. Chem Phys Lett 183: 471–477
Lersch W (1987) Magnetische Resonanz an kurzlebigen Radikalpaaren mit optischem Nachweis über die Rekombinationsausbeuten. PhD-Thesis. Technische Universität München
Lersch W and Michel-Beyerle ME (1983) Magnetic field effects

on the recombination of radical ions in reaction centers of photosynthetic bacteria. Chem Phys 78: 115–126

Lersch W and Michel-Beyerle ME (1989) RYDMR-Theory and applications. In: Hoff AJ (ed) Advanced EPR, Applications in Biology and Biochemistry, pp 685–705. Elsevier, Amsterdam

Lersch W, Ogrodnik A and Michel-Beyerle ME (1982) On the influence of microwaves and static magnetic fields on the recombination of radical ions in reaction centers of photosynthetic bacteria. Z Naturforsch 37a: 1454–1456

Lersch W, Lendzian F, Lang E, Feick R, Möbius K and Michel-Beyerle ME (1989) High-Power RYDMR with a loop-gap resonator. J Magn Res 82: 143–149

Lersch W, Lang E, Feick R, Coleman WJ, Youvan DC and Michel-Beyerle ME (1990) Determination of the exchange interaction in the primary radical ion pair in reaction centers. In: Jortner J and Pullman B (eds) Perspectives in Photosynthesis, pp 81–90. Kluwer Academic Publishers, Netherlands

Lockhart DJ, Goldstein RF and Boxer SG (1988) Structure-based analysis of the initial electron transfer step in bacterial photosynthesis: Electric field induced fluorescence anisotropy. J Chem Phys 89: 1408–1415

Lockhart DJ, Kirmaier C, Holten D and Boxer SG (1990) Electric field effects on the initial electron-transfer kinetics in bacterial photosynthetic reaction centers. J Phys Chem 94: 6987–6995

Lyle PA, Kolaczkowski SV and Small GJ (1993) Photochemical hole-burned spectra of protonated and deuterated reaction centers of *Rhodobacter sphaeroides*. J Phys Chem 97: 6924–6933

Marcus RA (1987) Superexchange versus an intermediate $BChl^-$ mechanism in reaction centers of photosynthetic bacteria. Chem Phys Lett 133: 471–477

Marcus RA (1988) An internal consistency test and its implications for the initial steps in bacterial photosynthesis. Chem Phys Lett 146: 13–22

Marcus RA and Sutin N (1985) Electron transfers in chemistry and biology. Biochim Biophys Acta 811: 265–322

Martin JL, Breton J, Hoff AJ, Migus A and Antonetti A (1986) Femtosecond spectroscopy of electron transfer in the reaction center of the photosynthetic bacterium *Rhodobacter sphaeroides* R-26: Direct electron transfer from the dimeric bacteriochlorophyll primary donor to the bacteriopheophytin acceptor with a time constant of 2.8 ± .2 ps. Proc Natl Acad Sci USA 83: 957–961

McElroy JD, Feher G and Mauzerall DC (1972) Characterization of primary reactants in bacterial photosynthesis. I. Comparison of the light-induced EPR signal (g = 2.0026) with that of a bacteriochlorophyll radical. Biochim Biophys Acta 267: 363–374

Michel-Beyerle ME and Ogrodnik A (1990) Views on primary charge separation in reaction centers of photosynthetic bacteria. In: Baltscheffsky M. (ed.) Progress in Photosynthesis Research, Vol I, pp 19-26. Kluwer Academic Publishers, Dordrecht.

Michel-Beyerle ME, Haberkorn R, Bube W, Steffens E, Schröder H, Neusser HJ, Schlag EW and Seidlitz H (1976) Magnetic field modulation of geminate recombination of radical ions in a polar solvent. Chem Phys 17: 139–145

Michel-Beyerle ME, Scheer H, Seidlitz H, Tempus D and Haberkorn R (1979) Time-resolved magnetic field effect on triplet formation in photosynthetic reaction centers of *Rhodobacter sphaeroides* R-26. FEBS Lett 100: 9–12

Michel-Beyerle ME, Scheer H, Seidlitz H and Tempus D (1980) Magnetic field effect on triplets and radical ions in reaction centers of photosynthetic bacteria. FEBS Lett 110: 129–132

Michel-Beyerle ME, Bixon M and Jortner J (1988a) Interrelationship between primary electron transfer dynamics and magnetic interactions in photosynthetic reaction centers. Chem Phys Lett 151: 188–194

Michel-Beyerle ME, Plato M, Deisenhofer J, Michel H, Bixon M and Jortner J (1988b) Unidirectionality of charge separation in reaction centers of photosynthetic bacteria. Biochim Biophys Acta 932: 52–70

Moehl KW, Lous EJ and Hoff AJ (1985) Low-power, low-field RYDMAR of the primary radical pair in photosynthesis. Chem Phys Lett 121: 22–27

Nagarajan V, Parson WW, Davis D and Schenck CC (1993) Kinetics and free energy gaps of electron-transfer reactions in *Rhodobacter sphaeroides* reaction centers. Biochemistry 32: 12324–12336

Norris JR, Bowman MK, Budil DE, Tang J, Wraight CA and Closs GL (1982) Magnetic characterization of the primary state of bacterial photosynthesis. Proc Natl Acad Sci USA 79: 5532–5536

Norris JR, Budil DE, Tiede DM, Tang J, Kolaczkowski SV, Chang CH and Schiffer M (1987a) Relating structure to function in bacterial photoreaction centers. In: Biggins J (ed) Progress in Photosynthesis Research, Vol I, pp 363–369. Martinus Nijhoff Publishers, Dordrecht

Norris JR, Lin CP and Budil DE (1987b) Magnetic resonance of ultrafast chemical reactions. J Chem Soc, Faraday Trans 1, 83: 13–27

Ogrodnik A (1990) The free energy difference between the excited primary donor $^1P^*$ and the radical pair state P^+H^- in reaction centers of *Rhodobacter sphaeroides*. Biochim Biophys Acta 1020: 65–71

Ogrodnik A (1993) Electric field effects on steady state and time resolved fluorescence from photosynthetic reaction centers. Mol Cryst Liq Cryst 230: 35–56

Ogrodnik A and Michel-Beyerle ME (1992) Testing primary charge separation in photosynthetic reaction centers with external electric fields. In: Kochanski E (ed) Photoprocesses in Transition Metal Complexes, Biosystems and Other Molecules. Experiment and Theory, pp 349–373. Kluwer Academic Publishers, Dordrecht

Ogrodnik A, Krüger HW, Orthuber H, Haberkorn R and Michel-Beyerle ME (1982) Recombination dynamics in bacterial photosynthetic reaction centers. Biophys J 39: 91–99

Ogrodnik A, Lersch W, Michel-Beyerle ME, Deisenhofer J and Michel H (1985) Spin dipolar interactions of radical pairs in photosynthetic reaction centers. In: Michel-Beyerle ME (ed) Antennas and Reaction Centers of Photosynthetic Bacteria: Structure, Interaction and Dynamics, pp 198–206. Springer-Verlag, Berlin

Ogrodnik A, Remy-Richter N and Michel-Beyerle ME (1987) Observation of activationless recombination in reaction centers of *Rhodobacter sphaeroides*. A new key to the primary electron-transfer mechanism. Chem Phys Lett 135: 576–581

Ogrodnik A, Volk M, Letterer R, Feick R and Michel-Beyerle ME (1988) Determination of free energies in reaction centers of *Rb. sphaeroides*. Biochim Biophys Acta 936: 361–371

Ogrodnik A, Eberl U, Heckmann R, Kappl M, Feick R and Michel-Beyerle ME (1991) Excitation dichroism of electric field modulated fluorescence yield for the identification of primary electron acceptor in photosynthetic reaction center. J Phys Chem 95: 2036–2041

Ogrodnik A, Langenbacher T, Bieser G, Siegl J, Eberl U, Volk M and Michel-Beyerle ME (1992) Electric field-induced decrease of quantum yield of charge separation in photosynthetic reaction centers. Chem Phys Lett 198: 653–658

Ogrodnik A, Keupp W, Volk M, Aumeier G and Michel-Beyerle ME (1994) Inhomogeneity of radical pair energies in photosynthetic reaction centers revealed by differences in recombination dynamics of P^+H^- when detected in delayed emission and in absorption. J Phys Chem 98: 3432–3439

Okamura MY, Isaacson RA and Feher G (1975) Primary acceptor in bacterial photosynthesis: Obligatory role of ubiquinone in photoactive reaction centers of *Rhodopseudomonas spheroides*. Proc Natl Acad Sci USA 72: 3491–3495

Okamura MY, Isaacson RA and Feher G (1979) Spectroscopic and kinetic properties of the transient intermediate acceptor in reaction centers of *Rhodobacter sphaeroides*. Biochim Biophys Acta 546: 394–417

Ovchinnikov YA, Abdulaev NG, Zolotarev AS, Shmukler BE, Zargarov AA, Kutuzov MA, Telezhinskaya IN and Levina NB (1988a) Photosynthetic reaction centre of *Chloroflexus aurantiacus*. I. Primary structure of L-subunit. FEBS Lett 231: 237–242

Ovchinnikov YA, Abdulaev NG, Shmukler BE, Zargarov AA, Kutuzov MA, Telezhinskaya IN, Levina NB and Zolotarev AS (1988b) Photosynthetic reaction centre of *Chloroflexus aurantiacus*. Primary structure of M-subunit. FEBS Lett 232: 364–368

Parot P, Thiery J and Vermeglio A (1987) Charge recombination at low temperature in photosynthetic bacteria reaction centers: evidence for two conformational states. Biochim Biophys Acta 893: 534–543

Parson WW and Cogdell RJ (1975) The primary photochemical reaction of bacterial photosynthesis. Biochim Biophys Acta 416: 105–149

Parson WW, Clayton RK and Cogdell RJ (1975) Excited states of photosynthetic reaction centers at low redox potentials. Biochim Biophys Acta 387: 265–278

Parson WW, Chu Z-T and Warshel A (1990a) Electrostatic control of charge separation in bacterial photosynthesis. Biochim Biophys Acta 1017: 251–272

Parson WW, Nagarajan V, Gaul D, Schenck CC, Chu Z-T and Warshel A (1990b) Electrostatic effects on the speed and directionality of electron transfer in bacterial reaction centers: The special role of tyrosine M-208. In: Michel-Beyerle ME (ed) Reaction Centers of Photosynthetic Bacteria, pp 239–249. Springer-Verlag, Berlin

Pierson BK and Thornber JP (1983) Isolation and spectral characterization of photochemical reaction centers from the thermophilic green bacterium *Chloroflexus aurantiacus* strain J-10-fl. Proc Natl Acad Sci USA 80: 80–84

Plato M, Möbius K, Michel-Beyerle ME, Bixon M and Jortner J (1988) Intermolecular electronic interactions in the primary charge separation in bacterial photosynthesis. J Am Chem Soc 110: 7279–7285

Plato M, Michel-Beyerle ME, Bixon M and Jortner J (1989) On the role of tryptophan as a superexchange mediator for quinone reduction in photosynthetic reaction centers. FEBS Lett 249: 70–74

Popovic ZD, Kovacs GJ, Vincett PS, Alegria G and Dutton PL (1986) Electric field dependence of the quantum yield in reaction centers of photosynthetic bacteria. Biochim Biophys Acta 851: 38–48

Prince RC and Youvan DC (1987) Isolation and spectroscopic properties of photochemical reaction centers from *Rhodobacter capsulatus*. Biochim Biophys Acta 890: 286–291

Reddy NRS, Lyle PA and Small GJ (1992) Applications of spectral hole burning spectroscopies to antenna and reaction center complexes. Photosynth Res 31, 167–194

Rockley MG, Windsor MW, Cogdell RJ and Parson WW (1975) Picosecond detection of an intermediate in the photochemical reaction of bacterial photosynthesis. Proc Natl Acad Sci USA 72: 2251–2255

Roelofs MG, Chidsey CED and Boxer SG (1982) Contributions of spin-spin interactions to the magnetic field dependence of the triplet quantum yield in photosynthetic reaction centers. Chem Phys Lett 87: 582–588

Scheidel G (1989) Rekombinationsdynamik des spinkorrelierten Radikalpaars P^+BPh^- in nativen und mutagenetisch veränderten Reaktionszentren photosynthetischer Bakterien. Diploma Thesis. Technische Universität München

Schenck CC, Blankenship RE and Parson WW (1982) Radical-pair decay kinetics, triplet yields and delayed fluorescence from bacterial reaction centers. Biochim Biophys Acta 680: 44–59

Schenck CC, Mathis P and Lutz M (1984) Triplet formation and triplet decay in reaction centers from the photosynthetic bacterium *Rhodopseudomonas sphaeroides*. Photochem Photobiol 39: 407–417

Schmidt S, Arlt T, Hamm P, Huber H, Nägele T, Wachtveitl J, Meyer M, Scheer H and Zinth W (1994) Energetics of the primary electron transfer reaction revealed by ultrafast spectroscopy on modified bacterial reaction centers. Chem Phys Lett 223: 116–120

Schulten Z and Schulten K (1977) The generation, diffusion, spin motion, and recombination of radical pairs in solution in the nanosecond time domain. J Chem Phys 66: 4616–4634

Schulten K and Wolynes PG (1978) Semiclassical description of electron spin motion in radicals including the effect of electron hopping. J Chem Phys 68: 3292–3297

Sebban P and Barbet JC (1984) Intermediate states between P^* and Pf in bacterial reaction centers, as detected by the fluorescence kinetics. FEBS Lett 165: 107–110

Sebban P and Wraight CA (1989) Heterogeneity of the $P^+Q_A^-$ recombination kinetics in reaction centers from *Rhodopseudomonas viridis*: the effects of pH and temperature. Biochim Biophys Acta 974: 54–65

Shiozawa JA, Lottspeich F, Oesterhelt D and Feick R (1989) The primary structure of the *Chloroflexus aurantiacus* reaction-center polypeptides. Eur J Biochem 180: 75–84

Shuvalov VA and Parson WW (1981) Energies and kinetics of radical pairs involving bacteriochlorophyll and bacteriopheophytin in bacterial reaction centers. Proc Natl Acad Sci USA 78: 957–961

Stilz HU, Finkele U, Holzapfel W, Lauterwasser C, Zinth W and Oesterhelt D (1990) Site-directed mutagenesis of threonine M222 and tryptophan M252 in the photosynthetic reaction center of *Rb. sphaeroides*. In: Michel-Beyerle ME (ed) Reaction Centers of Photosynthetic Bacteria, pp 265–271. Springer-Verlag, Berlin

Takiff L and Boxer SG (1988) Phosphorescence from the primary electron donor in *Rhodobacter sphaeroides* and *Rhodopseudomonas viridis* reaction centers. Biochim Biophys Acta 932: 325–334

Thurnauer MC, Katz JJ and Norris JR (1975) The triplet state in bacterial photosynthesis: Possible mechanisms of the primary photo-act. Proc Natl Acad Sci USA 72: 3270–3274

Uphaus RA, Norris JR and Katz JJ (1974) Triplet states in photosynthesis. Biochem Biophys Res Commun 61: 1057–1063

van der Est A, Bittl R, Abresch EC, Lubitz W and Stehlik D (1993) Transient EPR spectroscopy of perdeuterated Zn-substituted reaction centres of *Rhodobacter sphaeroides* R-26. Chem Phys Lett 212: 561–568

Volk M (1991) Die Rekombinationsdynamik des intermediären Radikalpaares P^+H^- in photosynthetischen Reaktionszentren. PhD-Thesis. Technische Universität München

Volk M, Scheidel G, Ogrodnik A, Feick R and Michel-Beyerle ME (1991) High quantum yield of charge separation in reaction centers of *Chloroflexus aurantiacus*. Biochim Biophys Acta 1058: 217–224

Volk M, Aumeier G, Häberle T, Ogrodnik A, Feick R and Michel-Beyerle ME (1992) Sensitive analysis of the occupancy of the quinone binding site at the active branch of photosynthetic reaction centers. Biochim Biophys Acta 1102: 253–259

Volk M, Neumann W, Ogrodnik A, Gray KA, Oesterhelt D and Michel-Beyerle ME (1993a) The effect of site-directed mutagenesis at position M210 on the energies of the charge separated states P^+B^- and P^+H^- in reaction centers of *Rb. sphaeroides*. Biophys J 64: A18

Volk M, Häberle T, Feick R, Ogrodnik A and Michel-Beyerle ME (1993b) What can be learned from the singlet-triplet splitting of the radical pair P^+H^- in the photosynthetic reaction center? Conclusions from electric field effects on the P^+H^- recombination dynamics. J Phys Chem 97: 9831–9836

Volk M, Gilbert M, Rousseau G, Richter M, Ogrodnik A and Michel-Beyerle ME (1993c) Similarity of primary radical pair recombination in Photosystem II and bacterial reaction centers. FEBS Lett 336: 357–362

Vos MH, Lambry J-C, Robles SJ, Youvan DC, Breton J and Martin J-L (1991) Direct observation of vibrational coherence in bacterial reaction centers using femtosecond absorption spectroscopy. Proc Natl Acad Sci USA 88: 8885–8889

Vos MH, Lambry J-C, Robles SJ, Youvan DC, Breton J and Martin J-L (1992) Femtosecond spectral evolution of the excited state of bacterial reaction centers at 10 K. Proc Natl Acad Sci USA 89: 613–617

Wang Z, Pearlstein RM, Jia Y, Fleming GR and Norris JR (1993) Inhomogeneous electron transfer kinetics in reaction centers of bacterial photosynthesis. Chem Phys 176: 421–425

Wasielewski MR, Bock CH, Bowman MK and Norris JR (1983a) Nanosecond time-resolved magnetic resonance of the primary radical pair state P^F of bacterial photosynthesis. J Am Chem Soc 105: 2903–2904

Wasielewski MR, Bock CH, Bowman MK and Norris JR (1983b) Controlling the duration of photosynthetic charge separation with microwave radiation. Nature 303: 520–522

Wasielewski MR, Norris JR and Bowman MK (1984) Time-domain magnetic resonance studies of short-lived radical pairs in liquid solution. Faraday Discuss Chem Soc 78: 279–288

Werner HJ, Schulten K and Weller A (1978) Electron transfer and spin exchange contributing to the magnetic field dependence of the primary photochemical reaction of bacterial photosynthesis. Biochim Biophys Acta 502: 255–268

Windsor MW and Menzel R (1989) Effect of pressure on the 12 ns charge recombination step in reduced bacterial reaction centers of *Rhodobacter sphaeroides* R-26. Chem Phys Lett 164: 143–150

Woodbury NW and Parson WW (1984) Nanosecond fluorescence from isolated photosynthetic reaction centers of *Rhodopseudomonas sphaeroides*. Biochim Biophys Acta 767: 345–361

Woodbury NW, Becker M, Middendorf D and Parson WW (1985) Picosecond kinetics of the initial photochemical electron-transfer reaction in bacterial photosynthetic reaction centers. Biochemistry 24: 7516–7521

Woodbury NW, Parson WW, Gunner MR, Prince RC and Dutton PL (1986) Radical-pair energetics and decay mechanisms in reaction centers containing anthraquinones, naphtoquinones or benzoquinones in place of ubiquinone. Biochim Biophys Acta 851: 6–22

Wraight CA and Clayton RK (1973) The absolute quantum efficiency of bacteriochlorophyll photooxidation in reaction centers of *Rhodobacter sphaeroides*. Biochim Biophys Acta 333: 246–260

Chapter 28

Infrared Vibrational Spectroscopy of Reaction Centers

Werner Mäntele*
Institut für Biophysik und Strahlenbiologie, Universität Freiburg, Albertstrasse 23, 79104 Freiburg, Germany

Summary

In the past 10 years infrared spectroscopy has become a valuable tool for the analysis of the primary processes in reaction centers. Because it analyzes vibrational properties of the entire molecule, it can yield information on the protein which is otherwise difficult to obtain. In this chapter, the basic concepts of IR spectroscopy for macromolecules and the reaction-induced difference techniques using Fourier transform and single-wavelength techniques are briefly reviewed. The strategies used for the assignment of vibrational modes are described. Light-induced Fourier transform difference spectra monitoring charge-separated states and electrochemically-induced spectra monitoring the redox transitions of individual cofactors are reported. A major section deals with the vibrational spectra of the quinones in the RC, and information on the modes of the quinones in situ and of their protein sites is discussed. Finally, time-resolved IR experiments at individual wavelengths showing quinone reduction and concomitant protonation processes are reviewed.

* Current address: Universität Erlangen-Nürnberg, Institut für Physikalische und Theoretische Chemie, Egerlandstraße 3, 91058 Erlangen, Germany

R. E. Blankenship, M. T. Madigan and C. E. Bauer (eds): Anoxygenic Photosynthetic Bacteria, pp. 627–647.

I. Introduction: What Can We Learn About Reaction Centers From Infrared Spectroscopy?

Infrared spectroscopy analyzes vibrational properties of a molecule and can thus address other properties of reaction centers (RC) than those monitored by optical spectroscopy in the UV, visible, and near-infrared region. The vibrational spectrum of the RC is composed of individual and coupled modes of the numerous bonds of the protein, the pigments, the quinones, lipids, and water constituting the entire molecule, and thus contains abundant information on structural details and functional properties. In this chapter we shall deal with the information that can be drawn from this 'messy' overlap of multiple IR bands. This subject is also covered by recent reviews by Hoff (1992) and Mäntele (1993b).

In infrared spectroscopy, photons with energies that match those of vibrational sublevels of the molecule (typically 0.5–0.05 eV) are directly absorbed. These low-energy photons do not cause photochemistry and thus can be used at high intensities. A large variety of techniques in infrared spectroscopy have emerged in recent years, and most of them can be applied to obtain useful information on structural and functional properties of the RC.

Infrared spectroscopy probes vibrational modes of the pigments, the quinones, the polypeptide backbone and amino acid side chains without any selectivity and may thus be used to address questions on pigment and quinone in situ properties and interactions, on the role of the protein mediating electron transfer and coupling proton transfer to electron transfer. This is complementary to Raman spectroscopy, where resonance enhancement can yield spectra of individual pigments even in a complex photosynthetic membrane. A problem of the non-selectivity of IR spectroscopy is the background absorption of the bulk part of the RC, of lipids, water, and buffer, which necessitates difference techniques. Since the size of the RC protein does not allow 'classical' difference spectroscopies where two different samples are compared, 'reaction-induced' IR difference techniques on a single sample have to be used. The result are highly structured difference spectra which represent the sum of all molecular changes associated with the induced reaction.

IR spectroscopy relies on physical processes with short intrinsic lifetimes and thus allows very high time resolution for the study of intermediate states in electron transfer as well as of the dynamic processes in the protein. The past few years have seen developments especially in time-resolved IR spectroscopy which have given access to all steps of electron transfer, from picoseconds to several seconds. To date, quite a few processes in RCs have been detected in real-time in this entire time range which are functionally related to the electron transfer and proton transfer phenomena, but kinetically uncoupled from these. Vibrational spectroscopy can thus provide us with a detailed molecular view of the role of the protein in mediating electron transfer.

Abbreviations: Cm. – Chromatium; FTIR – Fourier transform infrared; IR – infrared; *Rb. – Rhodobacter;* RC – reaction center; *Rc. – Rhodocyclus; Rp. – Rhodopseudomonas; Rs. – Rhodospirillum*

II. Technical Aspects of Infrared Spectroscopy of Reaction Centers

The above-mentioned non-selectivity of IR spectroscopy necessitates difference techniques on a single sample, rather than subtracting the IR absorbance spectra obtained from two different samples. The basic concept of this reaction-induced IR difference spectroscopy for the study of protein function and reaction mechanisms has been recently discussed (Mäntele, 1993). The technical developments of the past 20 years have increased the sensitivity of IR spectroscopy to a level where the contributions of individual bonds in the RC to its total IR spectrum are detectable. These include the practical applicability, availability, and 'affordability' of Fourier transform infrared techniques which have been predominantly applied to RCs. Further developments include sensitive semiconductor quantum detectors and tunable IR laser sources for CW and pulsed applications. Although the main purpose of this chapter is to review the molecular information obtained from vibrational spectroscopy, we shall briefly 'dive' into the technical aspects.

A. Steady-State FTIR Spectroscopy

FTIR spectroscopy of stationary states in the RC has become the easiest and most widespread technique and relies on commercially available FTIR instruments. An FTIR spectrophotometer makes use of an interferometer, rather than scanning the individual wavelength elements dispersed by a prism or a grating.

The detector intensity as a function of the interferometer position is called the interferogram; its Fourier transform yields the IR spectrum. The theory, basic principles, and applications of FTIR spectroscopy to the study of proteins and biological membranes has been described in some recent reviews (Griffiths and DeHaseth, 1986; Susi and Byler, 1986, Braiman and Rothschild, 1988; Arrondo et al., 1993; Rothschild, 1993; Mäntele, 1993). FTIR spectroscopy results in a number of advantages over dispersive spectroscopy, like *Jacquinot's* advantage (the high energy throughput because of the lack of resolution-determining slits), and *Conne's* advantage (the high wavelength accuracy because of the laser reference interferometer). The major improvement, however, is the simultaneous recording of all spectral elements (*Felgett's* advantage), which takes less than a second even for a simple FTIR spectrophotometer and only a few milliseconds for sophisticated 'rapid scan' interferometers. *Felgett's* advantage can be either used toward a gain in sensitivity by accumulating many interferometer scans before performing the Fourier transform, or toward time-resolved studies using a series of single interferometer scans.

A steady-state FTIR experiment on RCs consists of recording a single beam IR spectrum of the RC in an initial state, inducing a triggering reaction, and recording a single beam IR spectrum of the final state. The resulting difference spectra should arise from the particular structural changes induced by the triggering reaction, with the absorption from the bulk of the protein being removed, and is selective for the functionally important part rather than for the global structure. The 'trigger' should provide a minimum disturbance of the sample, but specifically and quantitatively start the desired reaction.

In RC, mostly the intrinsic photoreactions have been applied to create a reaction-induced ('light-induced') IR difference spectrum. The photo-stationary state of an RC sample in the FTIR spectrophotometer beam can be easily modified by additional bleaching light, which does not perturb the measurement, since the spectral regions (VIS/NIR for excitation, mid-IR for detection) are well-separated. Light-induced FTIR difference spectra reflecting charge-separation between P and Q_A or Q_B are easily obtained at moderate bleaching intensity. These spectra reflect the differences for the $PQ_{A,B} \rightarrow P^+Q_{A,B}^-$ transition and are termed $P^+Q_A^-/PQ_A$ or $P^+Q_B^-/PQ_B$ difference spectra, with the understanding that the bands of the disappearing state, $PQ_{A,B}$, are negative, and those of the appearing state, $P^+Q_{A,B}^-$ are positive. However, care needs to be taken with higher intensities of the bleaching light, and correct filtering is recommended, since even minor heating of the sample results in changes of hydration, leading to unspecific signals which can be much larger than those caused by light-induced charge separation. In general, it is advantageous to add a number of light-dark difference cycles rather than to accumulate a high number of scans for the dark-adapted or the light-adapted state. Efficient thermostating of the samples is a prerequisite for high-quality spectra.

Light-induced difference spectroscopy combined with the use of exogenous donors can be used to obtain difference spectra between further states of the RC. Rapid rereduction of P^+ at low redox potential conditions leads to formation of H^-, a procedure which has been used for *Rhodopseudomonas (Rp.) viridis* RC and *Chromatium (Cm.) vinosum* membrane particles, and, recently, for *Rhodobacter (Rb.) sphaeroides* RC. At moderately reducing conditions, but in the presence of a redox mediator to shuttle electrons rapidly to the photooxidized P^+, reduction of Q_A or Q_B has been achieved without oxidizing a detectable fraction of P^+. In RCs with bound cytochromes, adjusting the redox potential for the IR samples has led to light-induced FTIR difference spectra where charge separation between the hemes and the quinone acceptors can be observed.

All steady-state FTIR difference spectra that have been obtained by use of the intrinsic photochemical reactions, however, present signals from a mixture of states such as of P, P^+, Q, and Q^- in spectra reflecting charge separation. Difference spectra in which an exogenous donor has been used seem to change only one cofactor's redox state. However, the difference spectra due to the redox reaction of the exogenous donor adds to that of the respective RC cofactor, although convincing evidence comes from control experiments that the difference bands of the exogenous donor are broad and featureless or simply in a different spectral region.

This limitation has been successfully overcome with the development of ultra-thin-layer electro-chemical cells that allow redox reactions in the RC to be triggered at a transparent electrode in combination with UV-VIS-NIR and FTIR spectroscopy. Electron transfer at an electrode is more specific than light as a trigger, and different cofactors of the RC could be selected by the choice of the appropriate potential ('dial-a-cofactor'). These 'redox-induced' difference spectra are as detailed and sensitive as are light-induced difference spectra; moreover, individual bands can be titrated and thus assigned to a specific redox transition. It should be mentioned that both, redox-induced and light-induced difference techniques, can be combined on one RC sample to start photochemical reactions from a well-defined redox clamp.

B. Time-Resolved FTIR Spectroscopy

Time-resolved FTIR studies can be carried out by three different techniques. In *rapid-scan* FTIR spectroscopy (Braiman et al., 1987), the interferometer mirror is moved as fast as possible (and as fast as digitization of the detector signal allows) to successively record and store interferograms (which may be transformed afterwards) in the course of a reaction. For RCs, IR spectra spaced some tens of milliseconds can be recorded after flash excitation. With the *stroboscope technique* (Braiman et al., 1991), a series of interferograms is collected, each with a different delay time with respect to the triggering event. This set of interferograms is then used to construct a set of time-resolved spectra in the microsecond-to millisecond time domain, with (partial) loss of the multiplex advantage. The *'step-scan'* technique (Uhmann et al., 1991) uses a stepwise (instead of continuously) moving interferometer to record the time dependence at fixed mirror positions and reconstructs time-dependent IR spectra from them, without a loss of the multiplex advantage.

C. Single Wavelength Techniques, Diode Lasers, PS Techniques

As an alternative to FTIR, time-resolved IR studies are possible by single-wavelength techniques, either using thermal IR sources with a monochromator (Siebert et al., 1980; Iwata and Hamaguchi, 1990) or tunable lead salt IR diode lasers (Mäntele et al., 1990a). The latter can be continuously tuned over 100–400 cm^{-1} in the 2000–1000 cm^{-1} range and provide sufficient power (up to a few mW) to allow sensitive detection of transient changes of band intensities. Investigations of dynamic processes in the RC in the ns to ms time domain by this technique will be discussed below. CO lasers, a principal alternative IR source of abundant power, have not yet been used for time-resolved studies at longer (ms to ns) time scales because of their limited amplitude stability and tunability. Recently, several laboratories have applied picosecond techniques for ultrashort IR studies on RC. The first approach (mainly from the Hochstrasser group) used a CO laser for the generation of IR probing light, with 'upconversion' techniques to detect the transient ps IR signal in the visible spectral range (Diller et al, 1992). The second approach (from the Zinth group) used the generation of an ultrashort IR pulse from visible light by mixing techniques with detection in the IR. The latter authors elegantly used the spectral width of their IR pulse (~ 60 cm^{-1}) for multiwavelength detection with an IR multichannel analyzer (Hamm et al., 1995)

D. Reaction Center Samples for Infrared Spectroscopy

The IR range accessible with most FTIR spectrophotometers spans from > 4000 to < 600 cm^{-1}. However, most difference spectroscopic studies are confined to the 2000–1000 cm^{-1} range, where the majority of bands with diagnostic relevance are located. The strong water H-O-H bending mode centered around 1650 cm^{-1} with an extinction coefficient of ca. 20 l mol^{-1} cm^{-1} necessitates thin layers for aqueous samples, because the absorbance from water at this frequency is around 1 for a pathlength of 10 μm. As an additional problem, the extinction coefficients for most of the protein modes (at most 1000 l mol^{-1} cm^{-1}) are one or two orders of magnitude smaller than the electronic transition moments of pigments, quinones, or hemes. RC IR samples thus need to be of much higher concentration than samples used for optical transmission spectroscopies.

IR windows used for studies on RC are almost always from CaF_2, a compromise between transmission (from 190 nm to 10 μm), stability against aqueous solutions, and price. IR cells with CaF_2 windows allow spectroscopic investigations of the same samples from the UV to the mid-IR, or excitation in the VIS/NIR range and detection in the mid-IR. The considerably differing extinction coefficients for the pigments in the NIR (for example for the monomeric BChls) and for the RC protein (the amide I band) lead to comparable absorbance at 800 nm and at approx. 1650 cm^{-1}, a fact which facilitates excitation coaxially with the IR beam, or simultaneous VIS/NIR and mid-IR spectroscopic investigations such as described in a recent review (Mäntele, 1993b).

Two major types of samples have been used for FTIR and kinetic IR investigations. The first type are thin, semihydrated films dried on CaF_2 windows from solutions or suspensions of RCs in detergent, RCs reconstituted in lipid vesicles, or from membranous particles containing RC. Salt, buffer, and detergent concentrations in the suspensions need to be reduced considerably to avoid excessively high salt concentrations or detergent enrichment in the films upon drying, and controlled hydration of the films is essential to maintain full activity. Correct buffering and ionic strength are almost impossible to maintain, and there has been a continuous concern that these films at reduced water content might not correspond to RCs in dilute solutions. In addition, the RCs become oriented upon drying, leading to polarization effects even for samples perpendicular to the IR beam. Careful control of the structural and functional integrity by measurement of the electron transfer rates is recommended, and has provided evidence that the electron transfer activity of the RC at high hydration (for example at equilibrium with 98% relative humidity) is indistinguishable from that in dilute solution.

The second type of IR samples used are concentrated solutions or suspensions. Detergent-solubilized RCs can be concentrated to 1–2 mM in ultrafiltration cells and transferred to demountable IR microcells with pathlengths as low as 5–10 μm. Ionic strength can be easily controlled this way, and orientation does not occur, however, the pH may be offset toward the isoelectric point of the RC if the concentration of the buffer is too low with respect to that of the RC. This type of sample has been routinely used for electrochemical investigations, where diffusibility of the RC is a prerequisite.

Finally, the amounts of protein needed for either sample type have frequently been overestimated, a misjudgment from the 'old days' of IR spectroscopy. Quantities of ca. 2–5 μL of a 0.5–1 mM solution are sufficient for a film or concentrated suspension IR sample, and correspond to what is conveniently used at higher dilution for UV and VIS spectroscopy.

III. From Bands to Bonds: Strategies for Band Assignments

Ideally, the bands in an IR spectrum are assigned to the vibrational modes of the molecule by performing a normal mode calculation and matching the force field parameters by replacing all atoms with stable isotopes, a procedure far out of reach for the RC. Even without the limitation of computer capacity, a force field yielding consistent modes would be difficult to obtain. In the case of the RC, the concept of group frequencies, i.e. of vibrational modes uncoupled from others, has proven to be very useful. This concept assumes the vibrational spectrum of the RC to be composed of modes from the backbone, from amino acid side chain groups, from cofactors, lipids, detergent and buffer molecules, and water. Coupling of such modes is possible, but will only result in small shifts depending on the interactions. For each of these group frequencies, an empirical assignment then becomes possible by isotopic labeling and comparison of the observed with the calculated frequency shifts.

The most simple isotope labeling experiment is substitution of ^1H_2O by ^2H_2O, easily done by resuspending concentrated RC or membrane suspensions in a ^2H_2O buffer and equilibrating. The resulting exchange of accessible N-, O-, or S- bound protons leads to distinct shifts in the spectrum, some of which will be discussed below. The probability for a given group to exchange depends on the global accessibility for water to the protein domain, on the local environment, and on the pK of the group. The extent of exchange can be monitored at the relative intensities of the amide II and the amide II' (the corresponding mode of the amide I group in ^2H_2O) modes.

Other mass labels, like ^{13}C, ^{15}N, or ^{18}O, can be introduced biosynthetically, but will not be incorporated specifically. Isotope-labeled amino acids, introduced into the RC via the nutrient, in combination with site-directed mutations, present the tedious yet clear tool to identify and to assign specific vibrational modes of the protein. In general, the remarkable potential of site-directed mutagenesis of RCs is a tempting tool for IR assignments. However, any mutant resulting in severe blocking of the function renders reaction-induced IR difference spectroscopy useless. A further problem of site-directed mutagenesis, in the case of charged residues, is the perturbation of entire charge patterns, for which IR difference spectroscopy (which monitors the changes of dipole moments) is a very sensitive technique. The balance seems to require generation of mutants with sufficiently small perturbations, but with retained general functional properties. In this line, natural variants can be considered to be of a similar relevance.

In RCs, the exchange of cofactors has long been limited to the secondary and the primary quinone acceptor, and only recently exchange of the monomeric bacteriochlorophylls and the bacteriopheophytins by analogs has been reported (see Chapter by Scheer and Hartwich). Nevertheless, the comparison of light- or redox-induced FTIR difference spectra with the spectra of the isolated cofactors, i.e. bacteriochlorophylls, bacteriopheophytins, and quinones and numerous analogs, all in their neutral and relevant ion radical, dianion, or protonated forms, has created a basic data set of vibrational modes. Table 1 summarizes the IR frequencies that can be expected in reaction induced difference spectra of the RC in the 2000 cm^{-1} to 1000 cm^{-1} range.

IV. Light-Induced Difference Spectroscopy of Charge Separation in Reaction Centers

Steady-state light-induced FTIR difference spectra reflecting the difference between the neutral PQ and the charge-separated P^+Q^- state have been the first and most widespread data used to analyze the molecular processes occurring upon charge separation. Mostly RCs from *Rb. sphaeroides*, but also RC, membrane particles and chromatophores from *Rp. viridis, Rb. capsulatus, Cm. vinosum, Rhodospirillum (Rs.) rubrum, Rhodocyclus (Rc.) gelatinosus*, and *Rp. palustris* were investigated (Mäntele et al., 1985a; Hayashi et al., 1986, 1990; Nabedryk et al., 1986, 1987, 1990; Gerwert et al., 1988; Buchanan et al., 1990a,b, 1992).

The light-induced difference spectra consist of sharp difference bands, indicating that predominantly intramolecular processes are observed. The maximum size of the difference bands is small, only about 0.5 – 1 % of the total absorbance at the strongest band of the amide I mode at around 1650 cm^{-1}. A first and general conclusion from all authors was that no major protein conformational change occurred upon charge separation and stabilization. Even though the signal amplitude in these difference spectra is small, the sensitivity of FTIR difference spectroscopy shows that even minor and subtle band structures are highly reproducible in amplitude and in frequency. All signals arise from primary charge separation and present a key to the molecular processes around the pigments and the quinone(s).

The highest-frequency band in these difference spectra is a positive band (we use the notation that positive bands correspond to the light-induced P^+Q^- state, while negative bands correspond to the quiescent PQ state) at approx. 1750 cm^{-1}. In the absence of model compounds, and, later, using IR spectra of isolated pigments, this band was assigned to the 13^2 ester C=O group of the primary donor BChls (Mäntele et al., 1985a, 1988b). A differential signal at 1712(+)/1683–) cm^{-1} was assigned to the 13^1-keto C=O group. In the region from 1660 cm^{-1} to 1600 cm^{-1}, several small bands were candidates for the 3 acetyl C=O group.

The clear assignment of the rich band structures in these spectra awaited numerous models studies. The cation-minus-neutral difference spectra of isolated BChls generated electrochemically in a variety of solvents clearly reproduce the band structures of the light-induced FTIR difference spectra of RCs in the carbonyl region (Leonhard et al., 1989a,b). However, the exact band positions depend on the solvation of the pigments and may thus be used for the mapping of the in vivo interaction of the pigments. The electrochemically-generated BChl a^+ /BChl *a* FTIR difference spectra insufficiently describe the light-induced FTIR difference spectra in the 1600–1400 cm^{-1} region. Nevertheless, the assignment of a band doublet at 1566 cm^{-1} and at 1546 cm^{-1} to C=C/C-C modes appears justified, although these bands are shifted with respect to model compound spectra (Mäntele et al., 1990c).

Clearly, the comparison of light-induced FTIR difference spectra with model spectra of monomeric pigments is not very satisfying, and it would be highly desirable to either analyze the vibrational modes of the light-induced spectra in comparison with a model dimer or to extract pigments and reconstitute with isotopically labeled or structurally different analogs. A dimeric model suitable as a maquette for IR spectroscopy would have to be precisely defined, not only for its neutral state, but also for the cation radical form. Besides, it should be very difficult to mimic the heterogeneous environment of the pigments in the RC. As for the second point, exchange of pigments has actually only been possible for the monomers and the pheophytin (Struck and Scheer, 1990; Struck et al., 1990). FTIR spectra of RC with modified monomers have been obtained (Leonhard, 1991), and revealed only minor effects on the primary donor cation vibrational spectrum.

The quinone contribution in light-induced P^+Q^-/PQ difference spectra could not be separated easily, mainly because the strong modes of the pigments mask the weaker quinone modes. We shall discuss the approaches to generate 'pure' Q^-/Q difference

Table 1. Vibrational modes from the RC constituents in the 2000–1000 cm^{-1} range

Component	Assigned bond(s) and vibrational modes	Frequency [cm^{-1}]
Polypeptide backbone		
amide I	mainly C=O stretching	1620–1690
amide II	C=O stretching, N-H bending	ca. 1550
amide III	C-N stretching, N-H bending	ca. 1300
Amino acid side chains		
ASP, GLU	COOH carboxyl C=O stretching	1750–1710
	COO^- antisymmetric stretching	1620–1570
	COO^- symmetric stretching	1450–1350
ASN, GLN	C=O stretching	ca. 1680
	NH_2 deformation	ca. 1620
	NH_3^+ deformation	ca. 1470
ARG	$CN_3H_5^+$ antisymmetric stretching	1680–1670
	$CN_3H_5^+$ symmetric stretching	1630–1640
LYS	NH_3^+ antisymmetric deformation	1640–1610
	NH_3^+ symmetric deformation	ca. 1525
TYR	TYR-O-H ring stretching	ca. 1600
	TYR-O-H ring stretching	1520–1510
	C-O-H stretching	ca. 1270
	TYR-O^- ring stretching	ca. 1600
	TYR-O^- ring stretching	ca. 1500
	TYR-O^- C-O^- stretching	ca. 1270
HIS	NH_2^+ deformation	ca. 1660
	ring stretching	ca. 1600
PHE		ca. 1600
	ring stretching	ca. 1495
TRP	ring stretching	1550–1530
Bacteriochlorophyll and Bacteriopheophytin	10a-, 7c- ester C=O stretching	1750–1720
	9a-keto C=O stretching	1710–1670
	2a-acetyl C=O stretching	1650–1620
	skeletal C-C, C-N	ca. 1600
	skeletal C-C, C-N	ca. 1560
	skeletal	ca. 1520
	CH_3 C-H bending	ca. 1375, 1340
Quinone	quinone C=O stretching	1660–1650
	quinone C=C stretching	1610–1600
	semiquinone anion C-O stretching	ca. 1490–1470
	semiquinone anion C-C stretching	ca. 1490–1460
Water	H-O-H bending	ca. 1660–1640

spectra in a separate section.

The availability of site-directed mutants in the primary donor environment was recently used to probe binding and interaction of the primary donor BChls (Nabedryk et al., 1992a-c). For the 'hetero-dimer' mutants His M200 → Leu and His L173 → Leu, the most prominent differences were found in the 1200–1600 cm^{-1} region, with strong bands missing, thus supporting the previous view based on comparison of light-induced difference spectra with monomeric BChl that bands in this range are characteristic for a dimeric BChl. The comparison of RCs with mutations near the primary donor like Leu M160 → His, Leu L131 → His, and His L168 → Phe, which add one or more hydrogen bonds to the primary donor, had a large effect in the keto C=O region. For a detailed discussion of the effect of mutations around the primary electron donor, the reader is referred to a previous review by Hoff (1992).

While the spectra discussed above were obtained in steady-state upon continuous illumination, Thibodeau et al. (1990a,b) have recorded rapid-scan FTIR spectra to follow the decay after a pulse of light to create a charge-separated state. The bands of P^+ (and those of Q^-) decayed in a concerted manner, and no long-term molecular changes could be observed.

Recently, Burie et al. (1993) have used the *step-scan* technique to record light-induced spectra in the microsecond time domain. The spectral features characteristic for P^+ formation could already be detected 10 μs after flash excitation and were quite comparable to those found in steady-state spectra. The same authors used RCs with a chemically prereduced primary quinone acceptor to generate the triplet state of the primary electron donor. This triplet state, which gives rise to characteristic IR absorbance changes decaying with half-times of 130 μs, compares well with the triplet difference spectrum ($^3P/P$) obtained from the same group by steady-state FTIR spectroscopy at low temperature (Breton and Nabedryk, 1993a,b). In the steady triplet spectrum, bands at 1648 cm^{-1} and 1619 cm^{-1} were assigned to the free and H-bonded 3 acetyl C=O groups of the primary electron donor, respectively. Peaks at 1749 and at 1740 cm^{-1} were assigned to the ester (13^2) C=O groups, and peaks at 1695 and at 1682 cm^{-1} to the 13^1-keto C=O group. It was noted that all these bands were downshifted by ^3P formation. The major conclusion given by the authors was that the triplet state is mostly delocalized over the two bacteriochlorophyll molecules.

Ultrafast time-resolved measurements of charge separation in RC have recently been performed by the Hochstrasser group (Hochstrasser et al., 1992; Cowen et al., 1993; Maiti et al., 1993) and by the Zinth group (Hamm et al., 1995). IR difference spectra at 70 ps and at 600 ps (Hochstrasser et al., 1992; Cowen et al., 1993) showed mainly band features which can be constructed from the static P^+, H^-, and Q_A^- spectra, with some additional bands not observed in static spectra. At present, it is unclear whether the spectral features observed at shorter time intervals can be interpreted in terms of a transient reduction of the monomeric BChl. Hamm et al. (1995) have recorded the full spectral range from 1800 to 1000 cm^{-1} in the time domain from 200 fs to 1000 ps, and constructed spectra at t=1 ps (mainly P^*/P), 10 ps (mainly P^+H^-/PH) and 1000 ps ($P^+Q_A^-/PQ_A$). The 1 ps spectrum exhibits a shift of a band at 1684 cm^{-1} to 1666 cm^{-1}(attributed to a shift of the keto C=O group to lower frequencies) and several strong positive bands at 1532 cm^{-1}, 1458 cm^{-1}, and between 1300 and 100 cm^{-1}. However, no effect of P^* formation on the ester C=O modes was noted.

V. Electrochemically-Induced FTIR Difference Spectra

Unlike light-induced FTIR difference spectra, which always involve a donor/acceptor couple, electrochemically-induced FTIR difference spectra offer the opportunity to investigate single cofactors and their redox transitions, in equilibrium with a potential applied via an electrode (Moss et al., 1990). The electrochemically-induced FTIR difference spectrum of the primary donor of *Rb. sphaeroides* and *Rp. viridis* RC was analyzed by Leonhard and Mäntele (1993). A difference spectrum (P^+/P) of *Rb. sphaeroides* RCs is reproduced in Fig. 1 (courtesy of Dr. M. Leonhard). It should be emphasized that this vibrational difference spectrum exclusively represents the primary donor reactions. This can be shown by titration of the IR difference spectrum, which yields the correct midpoint potential for all bands.

The difference spectrum obtained by electrochemical titration was extremely similar to the light-induced spectrum, a result which gives confidence to the previous assignments of the bands in the 1750 cm^{-1} to 1660 cm^{-1} region to the C=O modes of the primary electron donor. The sharp band at 1748 cm^{-1} can be assigned to the 13^2 ester C=O mode in the P^+ state. The apparent lack of a negative band for the P state was proposed to be explained by assuming strong hydrogen bonding in this state, resulting in extreme broadening. The bands at 1682 cm^{-1} and at 1702 cm^{-1} have been assigned to the 13^1 keto C=O modes of the primary electron donor in the P and P^+ states, respectively. Both the negative and the positive

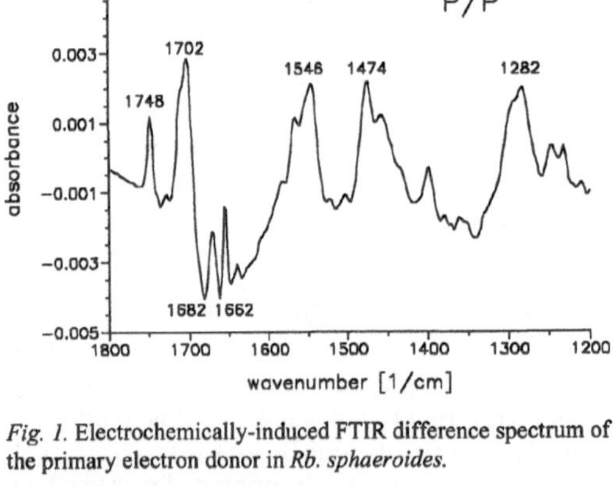

Fig. 1. Electrochemically-induced FTIR difference spectrum of the primary electron donor in *Rb. sphaeroides*.

bands have some high-frequency shoulder.

Equilibration of the RC in ^2H_2O did not change the signals above approx. 1680 cm^{-1}. This allowed to exclude contributions from protonation/deprotonation of carboxylic groups at neutral pH, where these experiments were performed. However, effects of ^2H_2O were observed below 1680 cm^{-1}: a broad signal between 1680 cm^{-1} and 1600 cm^{-1} is removed in ^2H_2O, hand in hand with appearing signals centered around 1450 and 1200 cm^{-1}. The authors explained this by the disappearance of an ^{1}H-O-^{1}H mode and the appearance of corresponding ^{1}H-O-^{2}H and ^{2}H-O-^{2}H. This effect, however, was only observed in *Rb. sphaeroides* RCs.

The direct accessibility of the vibrational difference spectra for the BChls in situ by these electrochemical techniques probably presents the most direct access to the impact of the protein on the primary donor pigments, since it yields spectra free from contributions of other cofactors. For further work, this approach should be of use in combination with RCs from site-directed mutants, with RC containing isotopically labeled amino acids, and with RCs with pigments removed and exchanged by chemically-modified pigments.

VI. Vibrational Modes of the Quinones and Their Host Sites

Infrared difference spectra of the quinones in RCs (Q_A^-/Q_A or Q_B^-/Q_B) without contributions from other cofactors have been obtained by various techniques:

(1) Electrochemically, by direct electrochemical reduction of the primary electron acceptor quinone Q_A by protein electrochemistry in a thin-layer cell (Mäntele et al., 1990b,d; Bauscher, 1991; Bauscher et al., 1993a,b). Selective reduction of Q_B in reaction centers containing a full Q_B complement has not been possible, mainly because Q_B in part became doubly reduced, but also because of equilibration of Q_B^- with Q_A.

(2) Photochemically/electrochemically, by subtraction of an electrochemically generated P^+/P difference spectrum from a light-induced $P^+Q_A^-/PQ_A$ or $P^+Q_B^-/PQ_B$ difference spectrum, both obtained from the same sample under identical conditions (Mäntele et al., 1990b; Bauscher, 1991; Bauscher et al., 1993a,b).

(3) Photochemically, by illumination of an RC sample in the presence of a strong electron donor and a mediator to rapidly rereduce P^+, resulting in Q_A^-/Q_A difference spectra for RC with the Q_B site blocked by a herbicide or in Q_B^-/Q_B difference spectra with RC containing a full Q_B complement at much weaker illumination. Principally, these spectra contain contributions from the electron donor which becomes partly oxidized and undergoes chemical reactions through this procedure, however, convincing evidence could be presented that these signals were either broad (arising from a reaction in aqueous solution) or negligible in the spectral region under consideration (Breton et al., 1991a,b).

(4) Photochemically/ electrochemically, by flash-induced formation of $P^+Q_A^-/PQ_A$ or $P^+Q_B^-/PQ_B$ (whichever functioned as the terminal electron acceptor in the RC), rapid rereduction of P^+ by cytochrome (Cyt) c_2 present in the sample, and rereduction of the oxidized Cyt c_2 by protein electrochemistry, resulting in net Q_A^-/Q_A or Q_B^-/Q_B difference spectra (Bauscher, 1991; Bauscher et al., 1993a,b, 1994).

In addition to these 'single' Q_A^-/Q_A or Q_B^-/Q_B difference spectra, double difference spectra $Q_A^-Q_B/Q_AQ_B^-$ were obtained by various techniques:

(5) By illumination of an RC sample at different cryogenic temperatures to obtain $P^+Q_A^-/PQ_A$ difference spectra (at approx. 100 K) or $P^+Q_B^-/PQ_B$ difference spectra (at ~ 260 K) (Nabedryk et al., 1990; Bagley et al., 1990).

(6) By time-resolved FTIR spectroscopy making use of the rapid-scan technique and the different reaction half times of approx. 60 ms for charge recombination from Q_A and approx. one second for recombination from Q_B (Thibodeau et al., 1990a,b). In this case, FTIR spectra recorded immediately (within 100 ms) after charge separation by a light pulse yielded difference spectra from all charge-separated states ($P^+Q_A^-/PQ_A$ and $P^+Q_B^-/PQ_B$) in RCs containing an intermediate (ideally 50%) Q_B complement, and spectra recorded at extended time intervals (> 500 ms) yielded spectra reflecting predominantly the $P^+Q_B^-/PQ_B$ state. Double difference spectra (late minus early) were obtained using a normalization

procedure to make P and P^+ bands disappear.

(7) By recording light-induced $P^+Q_A^-/PQ_A$ and $P^+Q_B^-/PQ_B$ difference spectra from two different samples (one with a full Q_B complement, the other with the Q_B site empty or blocked with a herbicide) and using a similar normalization to remove the P and P^+ contributions (Buchanan et al., 1990, 1992).

(8) A further technique using time-resolved IR spectroscopy in the microsecond time domain (Mäntele et al., 1990) also yields double difference spectra. Here, the transient IR signals after laser flash illumination measured at different IR probing wavelengths provided by tunable IR diode lasers were recorded straightforwardly and analyzed for reaction time constants in the μs to ms time domain, which can be ascribed to $Q_A \rightarrow Q_B$ electron transfer.

The approaches to generate IR difference spectra which give access to the vibrational modes of Q_A, Q_B, and their respective protein host sites differ considerably, and we should keep in mind that deviations in the band structures can arise from the experimental conditions rather than being due to experimental inaccuracies. For example, the methods (1) to (7) assess the relaxed state of the protein after Q_A or Q_B formation, whereas the approach using microsecond kinetics (8) monitors 'real-time' kinetic processes upon electron transfer to Q_A and from Q_A to Q_B. A second difference is the involvement of charge-separated states in the methods (2), (5), (6), (7) and (8), which may result in contributions from more distant residues because of the formation of transprotein dipoles. Yet, the common feature of all different methods is that the strong vibrational modes of the primary electron donor, which dominate in light-induced difference spectra of charge separation and make the detection of quinone or quinone site modes in the latter difficult (Nabedryk et al., 1990) or impossible, could be suppressed.

As an example for the IR signals obtained upon Q_A or Q_B reduction, figure 2 shows the Q_A^-/Q_A and Q_B^-/Q_B FTIR difference spectra obtained according to method (4) described above, i.e. with cytochrome c as secondary electron donor in a reaction probably closest to native RC electron transfer. These seemingly complicated FTIR difference spectra, at first glimpse, allow a number of conclusions:

(1) The changes in the vibrational spectrum upon Q_A or Q_B reduction are minute, at most ≈ 0.5 mOD or ca. 5×10^{-4} of the total absorbance, and in general much smaller. This size of absorbance changes is in agreement with the small IR extinction coefficients of, for example, C=O or C=C bonds, and supports the view that the signals in fact correspond to alterations of individual bonds which, however, can be followed reproducibly by the FTIR techniques described.

(2) The spectra obtained for Q_A or Q_B reduction are significantly different, although the same type of quinone (UQ-10) occupies both sites in *Rb. sphaeroides* RCs. It is thus evident that the changes in the vibrational spectrum are largely dominated by the site rather than arising from the quinone molecule itself.

(3) This point (2) is illustrated by comparison of these spectra with the electrogenerated Q^-/Q difference spectrum of UQ-10 in the non-protic solvent tetrahydrofuran (Bauscher and Mäntele, 1992), which is relatively simple and shows the disappearance of the C=O and C=C modes of the neutral quinone at approx. 1650 cm^{-1} and 1610 cm^{-1}, respectively, as well as the appearance of C-O and C-C modes from the semiquinone anion in a broad band around 1488 cm^{-1}. The difference spectra shown in figure 2 contain a substantial number of bands in excess of those that have been found in these model spectra.

Thus, the message is that Q_A or Q_B reduction has a considerable impact on their respective protein sites, and possibly even on more distant residues. This impact seems to be much higher at the quinone side than at the donor side (see above), a fact which may arise from the spatial density of the excess negative charge on the acceptor side, which is considerably higher than that of the 'hole' on the donor side. We may hypothesize that the changes in the vibrational spectrum reflect the solvation of the negative charge.

The signal amplitudes in Q_A^-/Q_A or Q_B^-/Q_B difference spectra are small, at least a factor of five smaller for the strongest bands as compared to the strongest bands of the difference spectra obtained from the donor side. This may explain why the quinone signals have proven elusive in earlier light-induced difference spectra.

Model spectra of quinones have been an essential tool for the 'fishing' of assignments in these quinone spectra, especially since FTIR difference spectra between neutral quinones and their electrochemically generated anions (as well as dianions and quinols) became available for a number of relevant quinones and their analogs which can be incorporated in the Q_A or Q_B sites (Bauscher and Mäntele, 1988, 1992; Bauscher et al., 1989, 1990a,b, 1991; Bauscher, 1991). Most of these model difference spectra are remarkably simple in comparison with the highly structured and complex Q_A^-/Q_A or Q_B^-/Q_B difference spectra. For the further discussion, the complicated spectra with numerous protein modes necessitate to distinguish between Q_A, Q_A^-, Q_B, and Q_B^- as the respective quinone 'species' or as quinone-protein 'states'.

A. Genuine and Non-Genuine Quinone Vibrational Modes

Most Q_A^-/Q_A and Q_B^-/Q_B difference spectra have been analyzed in the 1800–1000 cm^{-1} region, though band structures also appear above and below this range. We can discern four major frequency regions where sharp and structured bands appear. The strongest signals are found in the 1700–1600 cm^{-1} region, where the quinone C=O and C=C modes, but also peptide C=O modes and modes from some amino acid side chain residues are expected. This region is rather complex and contains at least 6–8 signals, the strongest at 1684 (–), 1666 (–), 1647 (–), 1627 (–) cm^{-1} in Q_A^-/Q_A difference spectra and at 1684 (–), 1661 (–), 1650 (+), 1638 (–) and 1616 cm^{-1} (–) in Q_B^-/Q_B difference spectra. (we follow here the convention that negative bands are counted to the Q_A (Q_B) state, and positive bands to the state of the corresponding semiquinone anion radicals). According to model quinone spectra (Bauscher and Mäntele, 1988, 1992; Bauscher, 1991, Bauscher et al. 1989, 1990a,b, 1991), the C=O and the C=C modes of the neutral species should appear as negative bands in this region between approx. 1670 and 1630 cm^{-1} (C=O) and at approx. 1610–1600 cm^{-1} (C=C). Although one is tempted to ascribe one of the strong negative bands in the Q_A^-/Q_A or Q_B^-/Q_B difference spectra to the quinone C=O or C=C mode, detailed analysis using RCs with isotopically-labeled quinones or structurally different quinones has shown that the majority of the bands in this range arises from the quinone site rather than from the quinone itself.

In the 1600 cm^{-1}–1500 cm^{-1} region, mainly signals from aromatic amino acid side chain residues (ring stretching modes, see Table 1) are expected. A second possible origin of signals is the amide II mode (coupled C=O stretching and N-H bending). Difference bands arising from the amide II mode should show some correlation with signals from the amide I mode, i.e. whenever an amide I mode is perturbed resulting in a signal in the 1600 cm^{-1} to 1700 cm^{-1} range, a corresponding amide II signal should appear.

The spectral region below 1500 cm^{-1} is governed by the C-O and C-C modes of Q^- between approx. 1500 cm^{-1} and 1460 cm^{-1} and thus can be used to probe the in situ binding and interaction of the semiquinone anion.

The 1700 cm^{-1} – 1750 cm^{-1} range is potentially the most interesting one, since it contains the COOH modes of protonated Asp and Glu residues and should provide a means of following protonation/deprotonation of amino acid side chain residues in response to electron transfer. A positive signal in this range would correspond to protonation ($COO^- + H^+ \rightarrow COOH$), a negative signal to deprotonation ($COOH \rightarrow COO^- + H^+$). The location of the C=O group in this range depends on solvation: a noninteracting group would be located at the upper end of the scale; H-bonding and polar environment would lower the C=O frequency to as low as approx. 1700 cm^{-1}. In addition, the location and width of the band contain information on the degree of conformational flexibility and accessibility. Apart from protonation and deprotonation, difference signals in this range could also be generated by changes of environment without changing the degree of protonation. In addition to signals from Asp and Glu side chains, there is a chance that quinone reduction might affect neighboring pigments (the bacteriopheophytin on the L or M branch) by electrostatic effects and cause signals from their ester or keto C=O groups.

B. Assignment of Genuine and Non-Genuine Quinone Modes

Assignments for bands arising from the modes of the quinone molecule itself have been proposed on the basis of quinone extraction and reconstitution with either isotopically labeled (Bagley et al., 1990; Thibodeau et al., 1990; Breton et al., 1994) or structurally different quinones (Breton et al, 1991a,b, 1992b; Bauscher et al., 1993a,b).

In a first paper by Bagley et al., $P^+Q_A^-/PQ_A$ and $P^+Q_B^-/PQ_B$ difference spectra were compared for

Rb. sphaeroides RC containing native and ^{13}C or ^{18}O labeled UQ-10. It was noted that a strong band in Q_A^-/Q_A difference spectra at 1649 cm^{-1} did not shift upon isotope substitution, thus excluding a vibration from Q_A itself. Small effects were seen at around 1625 cm^{-1} and 1618 cm^{-1} for ^{13}C and ^{18}O labeling. Thus, a C=O stretching mode could not be detected unambiguously. However, a band at 1604 cm^{-1} in native RCs still present for ^{18}O-UQ-10 RCs, but absent for ^{13}C-UQ-10 RCs was tentatively assigned to the quinone C=C mode for Q_A. In a second study, Thibodeau et al. (1990) used time-resolved FTIR spectroscopy to construct $Q_A^-Q_B/Q_AQ_B^-$ double difference spectra according to method (6) described above. Bands at 1670 cm^{-1} and at 1652 cm^{-1} in these spectra remained at the same position when Q_A and Q_B were replaced by ^{13}C UQ–10 and thus could be excluded as candidates for *genuine* quinone modes. However, a complex negative band at 1493–1480 cm^{-1} for the native UQ-10 shifted to 1434 cm^{-1} and was thus attributed to the C-C modes of the semiquinone anions, without excluding the possibility of C-O contributions.

A more thorough analysis using isotopes was recently carried out by Breton et al. (1994), giving final hints for the identification of the C=O and C=C mode(s) on the basis of constructed double-difference spectra from pairs of Q_A^-/Q_A spectra obtained from RC reconstituted with $^{18}O/^{16}O$ UQ-6 and UQ-1, $^{13}C/^{12}C$ UQ-8, and $^{13}C^{18}O/^{12}C^{16}O$ UQ-8. The double difference spectra revealed two C=O bands at 1660 cm^{-1} and at 1628 cm^{-1} and one C=C band at 1601 cm^{-1} for the neutral quinone. The inequivalence of the two C=O bands was explained by the asymmetrical bonding properties of the site as already proposed from X-ray crystallography and quinone exchange experiments. In particular, the isotopic shifts observed for the various quinone C=O and C=C bands indicated that the mixing of C=O and C=C modes is affected by the site, and that the 1628 cm^{-1} and the 1601 cm^{-1} bands of UQ-6 in vivo are highly mixed. Since no splitting of the C-O mode after photoreduction was observed, it was concluded that more symmetrical bonding was prevalent after photoreduction.

Replacement of the native UQ-10 at the Q_A site of *Rb. sphaeroides* RC by naphthoquinone analogs (Breton et al., 1992b) and by duroquinone, anthraquinone and 2-Cl-anthraquinone (Bauscher et al., 1993b) has also been used for tentative Q_A C=O and C=C assignments, which were in agreement with the isotope replacement studies discussed above. In general, these replacements result in minor perturbations of the spectra in the 1700 cm^{-1} to 1600 cm^{-1} range, but show substantial differences with respect to the native quinone in the range between 1500 and 1400 cm^{-1}, where the C-C and C-O bands of the semiquinone anion appear. Interestingly, the Q_A^-/Q_A spectra of *Rb. sphaeroides* RC reconstituted with vitamin K_1 are very similar in this range to Q_A^-/Q_A spectra of *Rp. viridis* RCs (Breton et al., 1992b). It is important to note that the alterations of the quinone site by bulkier quinones or quinones with Cl substituents, whenever they are functioning as electron acceptors, do not affect strongly the conformationally-sensitive 1700–1600 cm^{-1} range. Moreover, the spectral range above 1700 cm^{-1}, which reflects signals from protonated Asp and Glu side chain groups, is not altered at all. This range will be described in more detail below.

An analysis of the *non-genuine* quinone modes, i.e. of those arising from the amino acids lining the binding site(s) and from more distant groups is still in its beginning. Clearly, this task needs isotopically-modified amino acids entered biosynthetically and a multitude of site-directed mutations for a complete analysis, and appears to be a Sisyphean task, even if shared among many research groups[1].

The first assignment of a protein signal associated with quinone reduction was made by Nabedryk et al. (1990). A signal at 1650 cm^{-1} present in $P^+Q_A^-/PQ_A$ difference spectra and not in the $P^+Q_B^-/PQ_B$ difference spectra from *Rb sphaeroides* RCs, but in both spectra for *Rp. viridis* RCs, was interpreted in terms of a conformational change of the protein backbone near Q_A and tentatively assigned to the peptide C=O of a conserved alanine residue in the Q_A pocket.

Further possible tentative assignments of bands to protein modes were proposed by Bauscher et al (1993a) on the basis of the proximity of amino acids to the primary and secondary quinone, and confirmed or rejected by analyzing the effect of $^1H_2O \rightarrow {}^2H_2O$ substitution (Bauscher et al., 1994). A small shift of a differential signal at 1552(–)/1560(+) to 1548(–)/1556(–) cm^{-1} upon $^1H \rightarrow {}^2H$ exchange was used for assignment to the aromatic stretching mode of a Trp residue, and Trp M252 in van-der-Waals contact

[1] This reminds me of an FTIR paper which we had submitted some time ago, and which contained some highly structured difference spectra. It returned with the following reviewers comments: '... I have counted approx. 50 bands in your spectra, of which you have assigned only five. Please resubmit the paper when you have asigned the rest of the peaks ...'

with Q_A was proposed as a candidate. Several bands between 1600 cm^{-1} and 1700 cm^{-1} could tentatively be assigned to amide I modes, since they experience only small shifts or none for $^1H \rightarrow {}^2H$ exchange. The positive band at 1538 cm^{-1} in Q_B^-/Q_B difference spectra only present in ^1H_2O, was assigned to an amide II mode. Overall, these tentative assignments present reasonable hints and may help to select and to reduce the number of site-directed mutations needed for a conclusive picture of the microstructural changes at the Q_A and Q_B site.

C. Protonation/Deprotonation of Amino Acid Side Chains upon Quinone Reduction in the Reaction Centers

The bands in FTIR difference spectra between 1700 cm^{-1} and 1780 cm^{-1} are not affected by the substitution of quinones by (even exotic) analogs, provided that these function as electron acceptors. However, band positions and intensities are changed by $^1H \rightarrow {}^2H$ exchange (Bauscher et al., 1994). The fact that model quinones do not show any mode in this range has led to the assumption that the appearance and disappearance of C=O modes of Asp and Glu side chains is predominantly responsible for spectral changes in this range, thus making it an ideal spectral region for probing protonation/deprotonation reactions in the RC. The major difference signals for Q_A^-/Q_A FTIR difference spectra are a negative peak at 1732 cm^{-1} and positive peaks at 1724 and 1712 cm^{-1}, respectively. In Q_B^-/Q_B difference spectra, a negative signal at 1738 cm^{-1} and a positive signal at 1725 cm^{-1} are observed (the numbers refer to the spectra shown in Fig. 2; peak intensities and positions may vary slightly for other spectra reported). As a simple working hypothesis, one could assume that the reduction of Q_A results in partial deprotonation of a group absorbing at 1732 cm^{-1} and partial protonation of groups absorbing at 1724 and 1712 cm^{-1}. Consequently, reduction of Q_B results in deprotonation of a group absorbing at 1738 cm^{-1} and protonation of a group absorbing at 1725 cm^{-1}. In the case of Q_B reduction, the residues involved could be the ones lining the binding sites; for Q_A, however, one would have to assume rather distant residues (the closest Asp or Glu residues are at least 8–10 Å away).

In addition to these steady-state FTIR spectra which clearly reflect the relaxed state of the protein upon reduction of Q_A and Q_B, time-resolved IR spectroscopy using tunable diode lasers has provided

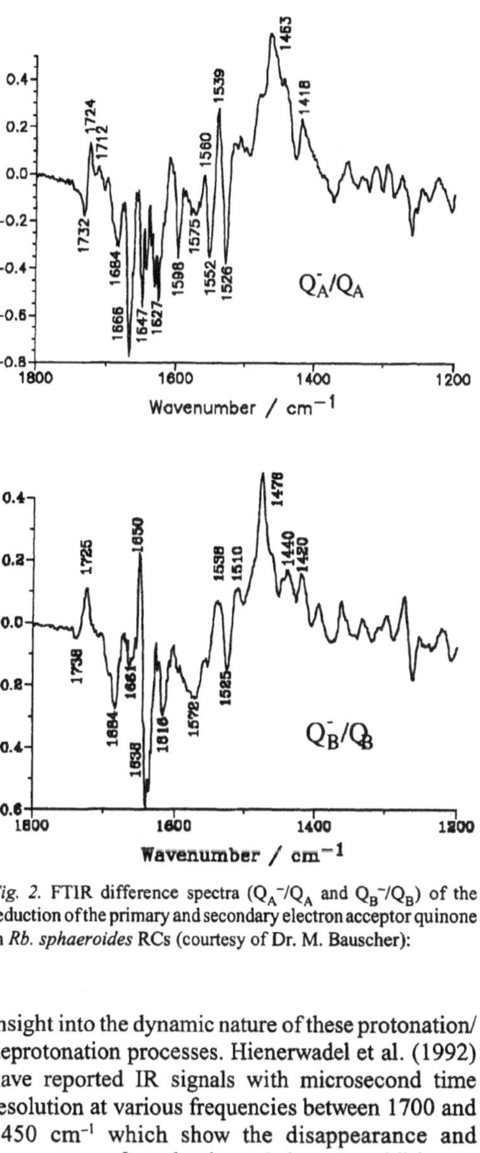

Fig. 2. FTIR difference spectra (Q_A^-/Q_A and Q_B^-/Q_B) of the reduction of the primary and secondary electron acceptor quinone in *Rb. sphaeroides* RCs (courtesy of Dr. M. Bauscher):

insight into the dynamic nature of these protonation/deprotonation processes. Hienerwadel et al. (1992) have reported IR signals with microsecond time resolution at various frequencies between 1700 and 1450 cm^{-1} which show the disappearance and appearance of modes in real-time. In addition to signals rising/decaying immediately, which were attributed to primary charge separation $PQ_A \rightarrow P^+Q_A^-$, signals with half-times of 60–100 ms and approximately 1 sec, which were attributed to charge recombination from Q_A and Q_B, respectively, and signals with half-times of approx. 100 μs which were attributed to modes changing upon $Q_A \rightarrow Q_B$ electron transfer, it was noted that transient IR signals in the 1700 to 1750 cm^{-1} range exhibit half times of approx.

1 ms. These signals were proposed to arise from Asp or Glu side chain protonation/deprotonation reactions.

Recently, Hienerwadel et al. (1995) have reported a detailed analysis of the temporal and spectral distribution of these signals in the 1700 cm^{-1} to 1760 cm^{-1} range. Figure 3a shows the major signal at 1725 cm^{-1}. The fast rising component originates from primary charge separation $PQ_A \rightarrow P^+Q_A^-$ and was suppressed for the display of the three-dimensional representation of the transient signals shown in Fig. 3b. Adopting the above working hypothesis, namely that the absorbance changes in this range predominantly represent protonation/deprotonation reactions at Asp or Glu residues, this representation corresponds to a *real-time video* of proton movements in the RC upon $Q_A^-Q_B \rightarrow Q_AQ_B^-$ electron transfer.

A total of two negative signals (1698 cm^{-1} and 1716 cm^{-1}, respectively) and three positive signals (1706 cm^{-1}, 1725 cm^{-1}, and 1732 cm^{-1}, respectively) is seen. Global fit analysis of these signals showed that they can be described by two kinetic processes with approximately 1 ms and approx. 180 μs half time at wavelength-dependent amplitude proportions. Both kinetic components depend strictly on electron transfer between Q_A and Q_B and are thus absent in terbutryn-treated RC (see Fig. 3a). The only evidence for transient signals in the absence of Q_B is around 1730–1740 cm^{-1}, where weak components in the μs domain were detected (Hienerwadel et al., 1995). The attribution of the major signal at 1725 cm^{-1} to an Asp or Glu COOH group is further confirmed by the absence of the signal at this frequency for the COO^2H group of RCs equilibrated in ^2H_2O. Analysis of the signal for RCs from site-directed mutants, where a Glu L212 Gln mutant showed a complete absence of the signal, while it was little affected in an Asp L213 Asn mutant (Hienerwadel et al., 1995). This provided the final clue to identify the residue involved as Glu L212. It was further concluded from the sign and amplitude of the signal that Glu L212 takes up approx. 0.3 to 0.6 H^+ upon formation of Q_B^-. This calculation of the stoichiometry was based upon an the (integrated) extinction coefficient of Glu side chain residues in model compounds, which may not be a good approximation for a Glu residue embedded in the protein. The proton uptake was found to show little variation with pH, in disagreement with a Henderson-Hasselbalch titration curve expected for a single titrating residue, and thus taken as evidence for the titration of Glu L212 in a cluster of strongly interacting ionizable groups (Hienerwadel et al., 1995).

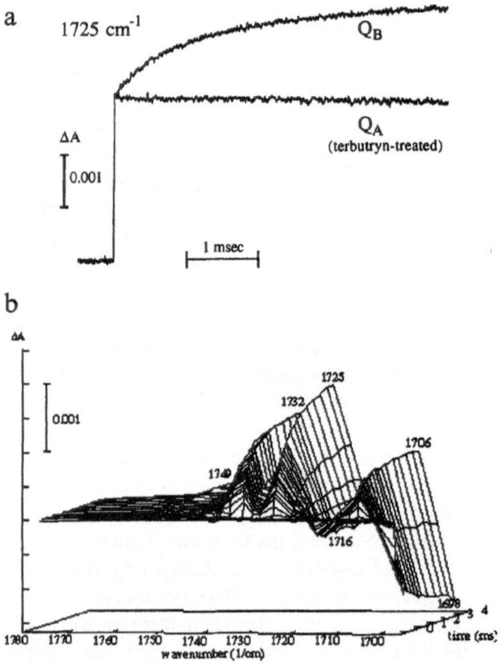

Fig. 3. a) kinetic IR signal at 1725 cm^{-1}. b) 3D representation of the kinetic IR signals between 1690 cm^{-1} and 1760 cm^{-1}. (Courtesy of Dr. R. Hienerwadel.)

The appearance of a COOH mode from Glu L212 should be strictly correlated with the disappearance of the antisymmetric and symmetric COO^- modes. Hienerwadel et al. (1992b) have proposed that a negative transient signal at 1617 cm^{-1} which closely matches the kinetics of the 1725 cm^{-1} signal and is also absent in RC lacking Q_B, could arise from the disappearing COO^- mode.

One might be tempted to attribute the entire set of transient IR signals between 1700 cm^{-1} and 1760 cm^{-1} to deprotonation and protonation processes at specific Asp/Glu residues or to shifts of the C=O modes of protonated Asp/Glu residues upon Q_B^- formation. This may be oversimplified, since the signals can be partly explained assuming IR absorbance changes upon reduction of Q_A which then decay upon $Q_A \rightarrow Q_B$ electron transfer. In fact, the transient IR difference spectrum between 1760 cm^{-1} and 1690 cm^{-1} can be modeled by the sum of a Q_B^-/Q_B and of an inverted Q_A^-/Q_A FTIR difference spectrum. As already discussed above, an alternative but less likely interpretation of signals in this range is that shifts of ester and keto C=O modes from the pigments lead to additional signals.

Further kinetic IR and steady-state FTIR studies on RC from site-directed mutants have indicated that a another group involved in the cluster of ionizable residues is Asp L210, presumably associated with the smaller positive IR signal at 1706 cm^{-1} (see Fig. 3b.). In ongoing experiments, a complete pH dependency of the transient IR signals for several mutant RCs is recorded for each mutant RC, since the removal of the ionizable group may affect the pK values of other groups in the cluster.

The question is whether a conclusive interpretation can be given for the signals in the 1700 cm^{-1} to 1600 cm^{-1} range. The comparison with quinone model compounds has indicated that only a small fraction is due to modes from the quinone itself, and the majority presumably arises from peptide carbonyl groups. One possible explanation is that reduction of Q_A or Q_B strongly polarizes or depolarizes the peptide C=O bonds of neighboring groups, depending on their orientation relative to the charge density on the respective quinone. Gunner (personal communication) has pointed out the role of the backbone carbonyl groups in stabilizing buried charges in the protein, and it may well be that the signals discussed here are a manifestation of this process.

D. Can IR Spectroscopy Provide a Model for H^+ Transfer to Q_B?

The data presently available from steady-state FTIR difference spectroscopy and time-resolved IR spectroscopy on the molecular processes at the Q_B site upon Q_B^- formation suggest a role of the site in maintaining a favorable electrostatic situation. The movements of protons in the site and the net proton uptake of the RC upon Q_B^- formation may represent a rearrangement of charges to form local countercharges for the negative charge on the quinone (Q_B^-), which is predominantly localized on the quinone C=O oxygens. On the basis of the present data, which have identified Glu L212 (and tentatively also Asp L210) as groups that increase protonation, the following scenario summarized in Fig. 4 is proposed.

In this scenario, we assume Q_B to be surrounded by a cluster of several ionizable residues, which interact strongly and form a pattern of charges around Q_B. Note that the arrangement and the distances of the three residues with respect to Q_B shown in Fig. 4 are purely schematic. This cluster may include Glu L212, Asp L213, and Asp L 210, as well as other residues identified by X-ray structural analysis to be

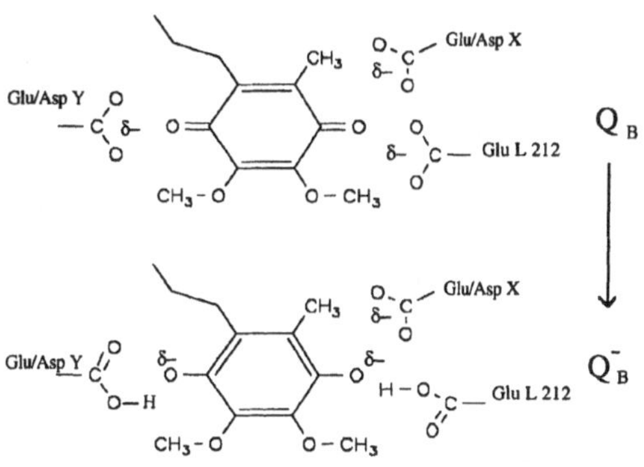

Fig. 4. Schematic representation of Q_B surrounded by a cluster of ionizable residues forming a delocalized countercharge.

close to Q_B. Positively charged amino acid residues may also be part of the cluster; however, they are omitted in this simplified scheme.

In this scenario, the residues located closest to the quinone carbonyls are proposed to be partly or fully ionized for neutral Q_B. Upon formation of Q_B^-, excess negative charge resides on either quinone carbonyl oxygen. It is now conceivable that this excess charge induces pK shifts on the neighboring carboxyl group residues. These shifts of pK may lead to internal reequilibration of protons in the cluster as well as to a net proton uptake. Hienerwadel et al. (1995) have hypothesized that the fast kinetic components that correspond to those of the electron transfer (approx. 100 μs) could be due to movements of protons directly correlated to electron transfer, and the slow kinetic components (approx. 1 ms) could be due to net proton uptake.

In order to provide details of this scenario, the rates of electron transfer, proton transfer, and proton uptake need to be determined separately. Hienerwadel (1993) has analyzed the rate of reduction of Q_B by following the rate of appearance of its semiquinone C-O mode at 1478 cm^{-1}. This rate was found to be much higher than the shift of the pheophytin signal at 750–760 nm, which is the 'classical' monitor for $Q_A \rightarrow Q_B$ electron transfer. The half-time found (approx. 150 μs at 6 °C) was much shorter than that of the protonation of Glu L212 (approx. 1 ms at 6 °C) and was almost temperature-independent. The protonation of Glu L212 and the shift of the pheophytin signal, however, match perfectly in their kinetic parameters and their temperature dependence. Consequently, Hienerwadel et al. assumed that the pheophytin absorbance responds to changes of protonation rather than to the charge of the quinone.

It may be difficult to fully separate protonation and electron transfer events, since any movement of charges in the RC will probably result in a concerted electrostatic reaction. In this line, we could hypothesize that the transient IR signals seen in the absence of Q_B represent alterations in the empty Q_B site caused by formation of Q_A^-. One could imagine these alterations as a means of 'preparing' a favorable electrostatic situation for $Q_A \rightarrow Q_B$ electron transfer.

For the first turnover of the RC and Q_B^- formation, the protonation and deprotonation processes seen by kinetic IR and FTIR spectroscopy seem to reflect the response of an *electrostatic reaction field* which account for local electroneutrality. The groups forming this cluster may also be involved in supplying the protons for protonation of Q_B upon transfer of the second electron to form the quinol. Kinetic IR experiments are in progress to follow the fate of the Glu L212 for the second turnover; however, they are complicated by the fact that a rapid secondary donor to the RC is involved which can contribute to the difference spectra, and that quinone/quinol exchange at the Q_B site has to proceed before the system can be repetitively excited by double flashes. Nevertheless, we may be confident that within the next years kinetic IR and FTIR spectroscopy will bring some light into the pathway of protons to Q_B.

VII. RC-Associated Cytochromes

In the cytochrome subunits of RC from *Rp. viridis*, *Chloroflexus (Cf.) aurantiacus*, *Rc. gelatinosus* and others, c-type hemes are bound and given a wide range (almost 0.5 V) of redox potentials (see Chapter 36 by Nitschke and Dracheva). This presents an example of how a protein can modulate the spectral and redox properties of a cofactor by varying charged residues near the hemes, placing water molecules into the heme pocket, and playing with the ionization state of the heme propionates. Gunner and Honig (1990, 1991) have calculated the electrostatic potential in the cytochrome subunit of the *Rp. viridis* RC and pointed out the weight of the factors determining the midpoint potential.

IR spectroscopy has been used to seek for experimental evidence for the nature of the heme-protein interactions and for the role of the protein in stabilizing a heme redox state. Nabedryk et al. (1991) have obtained light-induced $Cyt_{ox}Q^-/Cyt_{red}/Q$ FTIR difference spectra and subtracted the Q^-/Q contributions by normalizing to bands previously assigned to quinone reduction. They obtained reduced-minus-oxidized difference spectra of the two high-potential hemes (c-559 and c-556) as well as of one low-potential heme (c-552). Fritz(1995) has used electrochemically-induced FTIR spectroscopy to directly generate the reduced-minus-oxidized difference spectra of all four hemes of the *Rp. viridis* reaction center and of the isolated tetraheme subunit from *Cf. aurantiacus* reaction centers. The difference spectra obtained from the *Rp. viridis* cytochrome by either method agree in their main band features. The differences observed for smaller peaks may be the result of the different states of the RC used: the first approach involves charge-separated states, while in the second approach the hemes are in thermodynamic equilibrium with the applied potential.

In general, the band amplitudes in the reduced-minus-oxidized IR difference spectra are about 5 times smaller than those observed for primary donor oxidation on comparable samples, and on the order of the signals observed for the quinones. Again, we may conclude that no major rearrangement occurs in the protein structure as a result of the redox transitions. The difference spectra have a common signature in the conformationally-sensitive 1700 cm^{-1} to 1600 cm^{-1} spectral region: two negative bands at 1658 cm^{-1} and at 1686 cm^{-1} and two positive bands at 1634 cm^{-1} and at 1676 cm^{-1} are observed. Further deviations among the different hemes were observed in the amide II region, where the c-559 high-potential heme exhibits a negative band doublet at 1558 and 1540 cm^{-1}.

In the difference spectra, fairly strong signals appear between 1680 and 1710 cm^{-1}, a region characteristic for the heme propionates. Signals in this region were also observed in reduced-minus-oxidized FTIR difference spectra of different cytochromes *c* (Moss et al., 1990) and of cytochromes c_2 from *Rp. viridis, Rb. sphaeroides*, and *Rs rubrum* (Baymann, 1991). and tentatively assigned to the heme propionate C=O. This immediately implies that one or both propionate groups are at least partially protonated at neutral pH, where these experiments were performed. Furthermore, a differential signal would imply that the protonation state changes upon the redox transition. At present, the problem seems unresolved for the *Rp. viridis* RC, where electrostatic calculations (Gunner and Honig, 1990, 1991 and

Gunner, personal communication) rather favor ionized heme propionates, while the FTIR spectra provide some evidence for propionate C=O signals.

In contrast to *Rp. viridis* RC, the cytochrome subunit of *Cf. aurantiacus* can be removed from the RC and investigated separately. Fritz (1995) has analyzed the FTIR difference spectra of the isolated subunit obtained by electrochemical oxidation/reduction. In this case, the IR difference bands could be titrated and thus assigned to the individual hemes. As in the case of the *Rp. viridis* cytochrome, each heme exhibits a characteristic set of IR signals, including difference bands which can be tentatively attributed to the propionate C=O group.

In the case of the isolated tetraheme cytochrome subunit of *Cf. aurantiacus*, the titration of the IR difference signals offered an interesting additional application. In this tetraheme subunit, all hemes exhibit α-bands at the same wavelength, which makes an optical titration of all four hemes almost impossible. Fritz (1995) has thus used the titration of IR difference signals to determine the redox midpoint potentials of the four hemes. In the IR difference spectra, bands common to all four hemes, to three or two hemes only, but also characteristic for individual hemes were detected, thus allowing very precise titrations.

We currently see the studies of the cytochrome subunits at an early stage, and particularly weak on the assignments of vibrational modes. Nevertheless, it appears that the heme protein interactions and the stabilization of redox states translate to the vibrational difference spectra and make this a worthwhile approach.

VIII. Conclusions

The past decade has seen vibrational IR spectroscopy of RCs developed as a new technique which had earlier gained interest only for the probing of CO binding in heme proteins and for retinal proteins. The major outcome seems to be that IR spectroscopy can be used to probe the relaxation of the protein in electron transfer processes (not only in photosynthetic RCs). Even within these ten years, the progress from easy-to-obtain light-induced FTIR difference spectra to the more sophisticated IR difference spectra of single cofactors or to transient spectra showing slow or even ultrafast electron transfer steps in real-time was breathtaking (at least for me).

Among the lessons learned from IR spectroscopy, the one that no large conformational changes occur with electron transfer was the first one. It taught us to look closer and search for microconformational changes in the vicinity of cofactors, which are clearly present and most pronounced (and best characterized) for the quinones. Although the scenario shown in Fig. 4 may not fully correspond to reality, it may illustrate why the protein is sometimes called an 'optimized solvent,' and I find it very helpful for the understanding of the role of the protein pocket of a cofactor.

The deficit of IR spectroscopy of RCs is certainly the question of the band assignment (here I agree with a reviewer's opinion mentioned in the footnote in section VI.B). It is clear that further progress will have to include more experiments using site-directed mutagenesis and isotope labeling, but I doubt that we shall ever be able to fully explore the wealth of information contained in an IR difference spectrum. Yet, focusing on 'hot' spectral regions like the carboxyl group ranged between 1700 cm^{-1} and 1750 cm^{-1} is most promising, and I am optimistic that a full mapping of this range is within reach.

Ultrafast IR spectroscopy of RC is just at its beginnings: The present data range from several hundred femtoseconds to nanoseconds and open the possibility to follow the role of the protein in the primary steps. Soon these experiments will have evolved from mostly being technical highlights, and the search for protein signals in primary charge separation will be a rewarding task. Altogether, the use of IR spectroscopy on the RC over a wide wavelength range and, even more important, over more than twelve orders of magnitude in the time domain, will help us to understand the role of the protein matrix in charge separation, stabilization and photosynthetic energy conversion.

Acknowledgments

I would like to thank my coworkers involved in the work summarized here and the many colleagues in the field for valuable discussion and for providing information on unpublished work and on work in press. Our own work in this field has only been possible because of continuous financial support from the Deutsche Forschungsgemeinschaft through several grants and a Heisenberg fellowship.

References

Armstrong FA, Hill HAO and Walton NJ (1986) Reactions of electron-transfer proteins at electrodes. Quarterly Review of Biophysics 18: 261–322

Arrondo JLr, Muga A, Castresana J and Goñi FM (1993) Quantitative studies of the structure of proteins in solution by Fourier-transform infrared spectroscopy. Prog Biophys Molec Biol 59: 23–56

Bagley K, Abresch E, Okamura MY, Feher G, Bauscher M, Mäntele W, Nabedryk E, and Breton J (1990) FTIR studies of the $D^+Q_A^-$ and $D^+Q_B^-$ states in reaction centers from *Rb. sphaeroides*. In: Baltscheffsky M (ed) Current Research in Photosynthesis, Vol I, pp 77–80. Kluwer Academic Publishers, Dordrecht

Bartel K, Mäntele W, Siebert F, and Kreutz W (1985) Time-resolved infrared studies of light-induced processes in thylakoids and bacterial chromatophores. Evidence for the function of water molecules and the polypeptides in energy dissipation. Biochim Biophys Acta 808: 300–315

Bauscher M (1991) Elecktrochemische und infrarotspektroskopische Untersuchungen der Redoxreaktionen von Chinonen in vitro und in photosynthetischen Reaktionszentren. PhD Thesis, Faculty of Chemistry, University of Freiburg, Germany

Bauscher M and Mäntele W (1988) Fourier transform infrared spectroelectrochemistry in a novel thin-layer cell: Quinone models for photosynthetic electron acceptors. Proceedings of the Fifth Bioenergetics Conference, Aberystwyth, p 165

Bauscher M. and Mäntele W. (1992) Electrochemical and infrared spectroscopic characterization of redox reactions of para-quinones. J Phys Chem 96:11101–11108

Bauscher M, Nabedryk E, Breton J, and Mäntele W (1989) IR thin-layer spectroelectro-chemistry of models for ubiquinones involved in biological electron transfer. In: Bertoluzza A, Fagnano C and Monti P (eds) Spectroscopy of Biological Molecules - State of the Art, pp 397–398. Societa Editrice Esculapio, Bologna

Bauscher M, Bagley K, Nabedryk E, Breton J, and Mäntele W (1990a) Models for ubiquinones and their anions, involved in photosynthetic electron transfer, characterized by thin-layer electrochemistry and FTIR/UV/VIS spectroscopy. In: Baltscheffsky M (ed) Current Research in Photosynthesis, Vol I, pp 81–84. Kluwer Academic Publishers, Dordrecht

Bauscher M, Nabedryk E, Bagley K, Breton J, and Mäntele W (1990b) Investigation of models for photosynthetic electron acceptors: Infrared spectroelectrochemistry of ubiquinone and its anions. FEBS Lett 261: 191–195

Bauscher M, Leonhard M, Moss DA, and Mäntele W (1993a) Binding and interaction of the primary and the secondary electron acceptor quinones in bacterial photosynthesis. An infrared spectroelectrochemical study of *Rhodobacter sphaeroides* reaction centers. Biochim Biophys Acta 1183: 59–71

Bauscher M, Mäntele W, and Dutton PL (1993b) Infrared spectroscopic investigation of the Q_A site of photosynthetic reaction centers by replacing the native quinone with structurally different analogs. In: Theophanides T, Anastassopoulou J and Fotopoulos N (eds) Fifth International Conference on the Spectroscopy of Biological Molecules, pp 319–320. Kluwer Academic Publishers, Dordrecht

Bauscher M, Dutton PL, and Mäntele W (1994) Exchangeable protons in photosynthetic reaction center revealed by the effects of $^1H_2O \rightarrow {}^2H_2O$ exchange on the vibrational spectra of the Q_A and Q_B sites. Biochim Biophys Acta, submitted

Baymann F. (1991) Mechanismen des Elektronentransports bakterieller Cytochrome c. Diploma Thesis, Faculty of Biology, Universität Freiburg, Germany

Bellamy LJ (1968) Advances in infrared group frequencies, Vol. 2. Chapman and Hall, London

Braiman MS and Rothschild KJ (1988) Fourier transform infrared techniques for probing membrane protein structure. Ann Rev Biophys Biophys Chem 17: 541–570

Braiman M, Ahl P, and Rothschild KJ (1987) Millisecond Fourier-transform infrared spectra of bacteriorhodopsin's M_{412} photoproduct. Proc Natl Acad Sci USA 84: 5221–5225

Braiman MS, Bousché O and Rothschild KJ (1991) Protein dynamics in the bacteriorhodopsin photocycle: Submillisecond Fourier transform infrared spectra of the L, M and N photointermediates. Proc Natl Acad Sci USA 88: 2388–2392

Breton J, and Nabedryk E (1993a) FTIR difference spectrum of the triplet state of the primary electron donor in photosynthetic bacterial reaction centers. In: Theophanides T, Anastassopoulo J and Fotopoulos N (eds) Fifth International Conference on the Spectroscopy of Biological Molecules, pp 309–310. Kluwer Academic Publishers, Dordrecht

Breton J, and Nabedryk E (1993b) $S_0 \rightarrow T_1$ infrared difference spectrum of the triplet state of the primary electron donor in *Rb. sphaeroides* photosynthetic bacterial reaction centers. Chem Phys Lett 213: 571–575

Breton J, Thibodeau DL, Berthomieu C, Mäntele W, Vermeglio A and Nabedryk E (1991a) Probing the primary quinone environment in photosynthetic bacterial reaction centers by light-induced FTIR difference spectroscopy. FEBS Lett 278: 257–260

Breton J, Berthomieu C, Thibodeau DL, and Nabedryk E (1991b) Probing the secondary quinone (Q_B) environment in photosynthetic bacterial reaction centers by light-induced FTIR difference spectroscopy. FEBS Lett 288: 109–113

Breton J, Nabedryk E, and Parson WW (1992a) A new infrared transition of the oxidized primary electron donor in bacterial reaction centers: a way to assess resonance interactions between the bacteriochlorophylls. Biochemistry 31:7503–7510

Breton J, Burie JR, Berthomieu C, Thibodeau DL, Andrianambinintsoa S, Dejonghe D, and Nabedryk E (1992b) Light-induced charge separation in photosynthetic bacterial reaction centers monitored by FTIR difference spectroscopy: the Q_A vibrations. In: Breton J and Verméglio A (eds) The Photosynthetic Bacterial Reaction Center: Structure, Spectroscopy and Dynamics, NATO ASI series, Vol. 237, pp 155–162. Plenum, New York

Breton J, Burie JR, Berthomieu C, Berger G and Nabedryk E (1994) The binding sites of quinones in photosynthetic bacterial reaction centers investigated by light-induced FTIR difference spectroscopy: Assignment of the Q_A vibrations in *Rb. sphaeroides* using ^{18}O- or ^{13}C-labeled ubiquinones and vitamin K_1. Biochemistry 33: 4953–4965

Buchanan S, Michel H, Hess B, and Gerwert K (1990a) FTIR studies of light-induced intramolecular processes on crystals and reconstituted reaction centers from *Rhodopseudomonas viridis*. In: Baltscheffsky M (ed) Current Research in Photosynthesis, Vol. I, pp 69–72. Kluwer Academic Publishers, Dordrecht

Buchanan S, Michel H, and Gerwert K (1990b) Investigation of quinone reduction in *Rhodopseudomonas viridis* by FTIR difference spectroscopy and X-ray diffraction analysis. In: Michel-Beyerle ME (ed) Reaction Centers of Photosynthetic Bacteria, Springer Series in Biophysics, Vol. 6, pp 75–85. Springer Verlag, Berlin

Buchanan S, Michel H, and Gerwert K (1992) Light-induced charge separation in *Rhodopseudomonas viridis* reaction centers monitored by Fourier transform infrared difference spectroscopy: The quinone vibrations. Biochemistry 31: 1314–1322

Burie JR, Leibl W, Nabedryk E and Breton J (1993) Step-scan FT-IR spectroscopy of electron transfer in the photosynthetic bacterial reaction center. Applied Spectroscopy 47: 1401–1404

Chirgadze YN, Fedorov OV, and Trushina NP (1975) Estimation of amino acid residue side chain absorption in the infrared spectra of protein solutions in heavy water. Biopolymers 14: 679–694

Chang CH, El-Kabbani O, Tiede D, Norris J, and Schiffer M (1991) Structure of the membrane-bound protein photosynthetic reaction center from *Rhodobacter sphaeroides*. Biochemistry 30: 5352–5360

Cowen BR, Walker GC, Maiti S, Moser CC, Dutton PL, and Hochstrasser RM (1993) Protein motions and electron transfer in photosynthetic reaction centers studied by ultrafast infrared spectroscopy. Biophys J 64:A213

Deisenhofer J, and Michel H (1989) The photosynthetic reaction centre from the purple bacterium *Rhodopseudomonas viridis*. EMBO J 8: 2149–2169

Diller R, Iannone M, Bogomolni R , and Hochstrasser M (1991) Ultrafast infrared spectroscopy of bacteriorhodopsin. Biophys J 60: 286–289

Diller R, Iannone M, Cowen BR, Maiti S, Bogomolni R, and Hochstrasser RM (1992) Picosecond dynamics of bacteriorhodopsin, probed by time-resolved infrared spectroscopy. Biochemistry 31:5567–5572

Fritz F (1995) Elektrochemische und spektroskopische untersuchungen der multihämcytochrom-untereinheiten von *Chloroflexus aurantiacus* und *Rhodopseudomonas viridis*. PhD Thesis, Faculty of Biolgoy, University of Freiburg, Germany

Gerwert K, Hess B, Michel H, and Buchanan S (1988) FTIR studies on crystals of photosynthetic reaction centers. FEBS Lett 232: 303–307

Giangiacomo KM, and Dutton PL (1989) In photosynthetic reaction centers, the free energy for electron transfer between quinones bound at the primary and secondary sites governs the observed secondary site specificity. Proc Natl Acad Sci USA 86: 2658–2662

Griffiths PR and de Haseth JA (1986) Fourier transform infrared spectrometry. In: Chemical Analysis, Vol 83, pp 386–425. Wiley/Interscience, New York

Gunner M, and Honig B (1990) Electrostatic analysis of the midpoints of the four hemes in the bound cytochrome of the reaction center of *Rhodopseudomonas viridis*. In: Jortner J and Pullman B (eds) Perspectives in Photosynthesis, pp 53–90. Kluwer Academic Publishers, Dordrecht

Gunner M, and Honig B (1991) Electrostatic control of midpoint potentials in the cytochrome subunit of the *Rhodopseudomonas viridis* reaction center. Proc Natl Acad Sci USA 88: 9151–9155

Hamm P, Zureck M, Mäntele W, Meyer M, Scheer H and Zinth W (1995) Femtosecond infrared spectroscopy of reaction centers from *Rhodobacter sphaeroides* between 1000 and 1800 cm^{-1}. Proc Natl Acad Sci 92: 1826–1830

Hayashi H, Go M, and Tasumi M (1986) Light-induced changes in the infrared spectra of reaction centers from *Rhodopseudomonas sphaeroides* in H_2O and D_2O solutions. Chem Lett 1986: 1511–1514

Hayashi H, Morita EH, Tasumi M (1990) Structural changes of the bacteriochlorophyll dimer in reaction centers of photosynthetic bacteria as studied by infrared spectroscopy. In: Baltscheffsky M (ed) Current Research in Photosynthesis, Vol I, pp 73–76. Kluwer Academic Publishers, Dordrecht

Hienerwadel R (1993) Ladungstransportvorgänge in photosyntetischen Reaktionszenten. PhD Thesis, Faculty of Chemistry, University of Freiburg, Germany

Hienerwadel R, Kreutz W, and Mäntele W (1989) Time-resolved infrared spectroscopy using tunable diode lasers: Characterization of intermediates in light-induced electron transfer of photosynthesis. In: Bertoluzza A, Fagnano C and Monti P (eds) Spectroscopy of Biological Molecules - State of the Art, pp 315–316. Societa Editrice Esculapio, Bologna

Hienerwadel R, Thibodeau DL, Lenz F, Nabedryk E, Breton J, Kreutz W, and Mäntele W (1992a) Time-resolved vibrational analysis of quinone binding and interaction in the electron transfer of bacterial photosynthetic reaction centers. In: Takahashi H (ed) Time-Resolved Vibrational Spectroscopy V, Springer Proceedings in Physics, Vol 68, pp 83–86. Springer, Berlin

Hienerwadel R, Thibodeau DL, Lenz F, Nabedryk E, Breton J, Kreutz W, and Mäntele W (1992b) Time-Resolved Infrared Spectroscopy of Electron Transfer in Bacterial Photosynthetic Reaction Centers: Dynamics of Binding and Interaction upon Q_A and Q_B Reduction. Biochemistry 31: 5799–5808

Hienerwadel R, Nabedryk E, Breton J, Kreutz W, and Mäntele W (1992c) Time-resolved infrared and static FTIR studies of $Q_A \rightarrow Q_B$ electron transfer in *Rhodopseudomonas viridis* reaction centers. In: Breton J and Verméglio A (eds) The Photosynthetic Bacterial Reaction Center: Structure, Spectroscopy and Dynamics. NATO ASI Series A237, pp 163–172. Plenum, New York

Hienerwadel R, Nabedryk E, Paddock ML, Rongey S, Okamura MY, Mäntele W, and Breton J (1992d) Proton transfer mutants of *Rhodobacter sphaeroides*: Characterization of reaction centers by infrared spectroscopy. In: Murata M (ed) Research in Photosynthesis, Vol. I, pp 437–440. Kluwer Academic Publishers, Dordrecht

Hienerwadel R, Paddock M, Okamura MY, Nabedryk E, Breton J, and Mäntele W (1993) Coupling of proton transfer to electron transfer in photosynthetic reaction centers: Time-resolved IR signals from carboxyl group protonation upon reduction of Q_B. In: Theophanides T, Anastassopoulo, J, and Fotopoulos N (eds) Fifth International Conference on the Spectroscopy of Biological Molecules, pp 313–314. Kluwer Academic Publishers, Dordrecht

Hienerwadel R, Grzybek S, Fogel C, Kreutz W, Okamura MY, Paddock ML, Breton J, Nabedryk E and Mäntele W (1995) Protonation of Glu L212 following Q_B-formation in the photosynthetic reaction center of *Rhodobacter sphaeroides*: Evidence from time-resolved infrared spectroscopy. Biochemistry 34: 2832–2843

Hochstrasser RM, Diller R, Maiti S, Lian T, Locke B, Moser CC, Dutton PL, Cowen BR, and Walker GC (1992) Ultrafast infrared spectroscopy of protein dynamics. In: Martin L, Migus G, Mouru M and Zewail H (eds) Ultrafast Phenomena VIII, Springer Verlag

Hoff AJ (1992) Infrared spectroscopy of reaction centers. Isr J Chem 32: 405–412

Iwata K and Hamaguchi H (1990) Construction of a versatile microsecond time-resolved infrared spectrometer. Appl Spectrosc 44: 1431–1437

Leonhard M (1991) Redoxreaktionen von Chlorophyllen und Pheophytinen in photosynthetichen Reaktionszentren und in vitro: Untersuchungen mit spektroskopischen Methoden. PhD Thesis, Faculty of Chemistry, University of Freiburg, Germany

Leonhard M, Nabedryk E, Berger G, Breton J, and Mäntele W (1989a) Infrared spectroscopy and electrochemistry of pyrochlorophylls and pyropheophytins: Model compound studies for the interaction of photosynthetic pigments in their native environment. In: Bertoluzza A, Fagnano C and Monti P (eds) Spectroscopy of Biological Molecules - State of the Art, pp 121–122. Societa Editrice Esculapio, Bologna

Leonhard M, Wollenweber AM, Berger G, Kleo J, Nabedryk E, Breton J, and Mäntele W (1989b) Infrared spectroscopy and electrochemistry of chlorophylls: Model compound studies on the interaction in their native environment. In: Barber J and Malkin R (eds) Techniques and New Developments in Photosynthesis Research, pp 115–118. Plenum Publishing Corporation, New York

Leonhard M, Nabedryk E, Berger G, Breton J, and Mäntele W (1990) Model compound studies of pigments involved in photosynthetic energy conversion: Infrared (IR)-spectro-electrochemistry of chlorophylls and pheophytins. In: Baltscheffsky M (ed) Current Research in Photosynthesis, Vol I, pp 89–92. Kluwer Academic Publishers, Dordrecht

Leonhard M and Mäntele W (1993) Fourier transform infrared spectroscopy and electrochemistry of the primary electron donor in *Rhodobacter sphaeroides* and *Rhodopseudomonas viridis* reaction centers: Vibrational modes of the pigments in situ and evidence for protein and water modes affected by P^+ formation. Biochemistry 32: 4532–4538

Lutz M, and Mäntele W (1991) Vibrational Spectroscopy of Chlorophylls. In: Scheer H (ed) Chlorophylls, pp 855–902. CRC Press, Boca Raton, Florida

Maiti S, Cowen BR, Diller , Iannone M, Moser CC, Dutton PL, and Hochstrasser RM (1993) Picosecond infrared studies on the dynamics of the photosynthetic reaction center. Proc Natl Acad Sci USA 90:5247–5251

Mäntele W (1993a) Reaction-induced infrared difference spectroscopy for the study of protein function and reaction mechanisms. Trends in Biochem Sci 18: 197–202

Mäntele W (1993b) Infrared vibrational spectroscopy of the photosynthetic reaction center. In: Deisenhofer J and Norris J (eds) The Photosynthetic Bacterial Reaction Center, Vol. II, pp 239–283. Academic Press

Mäntele W, Nabedryk E, Tavitian BA, Kreutz W, and Breton J (1985a) Light-induced Fourier transform infrared (FTIR) spectroscopic investigations of the primary donor oxidation in bacterial photosynthesis. FEBS Lett 187: 227–232

Mäntele W, Nabedryk E, Tavitian BA, and Breton J (1985b). Fourier-transform infrared (FTIR) spectroscopic investigations in bacterial photosynthesis: intermediary acceptor reduction. In: Alix A, Bernard L, and Manfait M (eds) Spectroscopy of Biological Molecules, pp 370–373. John Wiley and Sons, Chichester

Mäntele W, Wollenweber AM, Nabedryk E, Breton J, Rashwan F, Heinze J, and Kreutz W (1987) Fourier-transform infrared (FTIR) spectroelectrochemistry of bacteriochlorophylls. In: Biggins J (ed) Progress in Photosynthesis Research, pp 329–332. Martinus Nijhoff Publishers, Dordrecht

Mäntele W, Wollenweber AM, Rashwan F, Heinze J, Nabedryk E, Berger G, and Breton J (1988a) Fourier transform infrared spectroelectrochemistry of the bacteriochlorophyll a anion radical. Photochem Photobiol 47: 451–456

Mäntele W, Wollenweber AM, Nabedryk E, and Breton J (1988b) Infrared spectroelectro-chemistry of chlorophylls: Models for their interaction *in vivo*. In: Schmid E, Schneider FW, and Siebert F (eds) Spectroscopy of Biological Molecules-New Advances, pp 317–321. Wiley Publishers, Chichester

Mäntele W, Wollenweber AM, Nabedryk E, and Breton J (1988c) Infrared spectroelectro-chemistry of bacteriochlorophylls and bacteriopheophytins: Implications for the binding of the pigments in the reaction center from photosynthetic bacteria. Proc Natl Acad Sci USA 85: 8468–8472

Mäntele W, Leonhard M, Bauscher M, Nabedryk E, and Breton J (1989) Infrared spectroscopic characterization of chlorophyll radicals in the primary processes of photosynthesis and *in vitro*. In: Bertoluzza A, Fagnano C and Monti P (eds) Spectroscopy of Biological Molecules - State of the Art, pp 93–96. Societa Editrice Esculapio, Bologna

Mäntele W, Hienerwadel R, Lenz F, Riedel WJ, Grisar R, and Tacke M (1990a) Application of tunable infrared diode lasers for the study of biochemical reactions: Time-resolved spectroscopy of intermediates in the primary process of photosynthesis. Spectroscopy International 2: 29–35

Mäntele W, Leonhard M, Bauscher M, Nabedryk E, Breton, J and Moss DA (1990b) Infrared difference spectroscopy of electrochemically generated redox states in bacterial reaction centers. In: Michel-Beyerle ME (ed) Reaction Centers of Photosynthetic Bacteria, Structure and Dynamics, Springer Series in Biophysics, Vol. 6, pp 31–44. Springer Verlag, Berlin

Mäntele W, Hienerwadel R, Bauscher M, Leonhard M, Moss DA, Nabedryk E, Thibodeau D, and Breton J (1990c) Infrared difference spectroscopy of pigments and redox components in the bacterial reaction center: Comparison with model compounds. In: Baltscheffsky M (ed) Current Research in Photosynthesis, Vol I, pp 85–88. Kluwer Academic Publishers, Dordrecht

Mäntele W, Leonhard M, Bauscher M, Nabedryk E, Berger G, and Breton J (1990d) The binding and interaction of pigments and quinones in bacterial reaction centers studied by infrared spectroscopy. In: Drews G and Dawes EA (eds) Molecular Biology of Membrane-Bound Complexes in Phototrophic Bacteria, pp 313–321. Plenum Publishing Corporation, New York

Michel H, Epp O, and Deisenhofer J (1986) Pigment-protein interactions in the photosynthetic reaction centre from *Rhodopseudomonas viridis*. EMBO J 5: 2445–2451

Moss DA, Nabedryk E, Breton J, and Mäntele W (1990) Redox-linked conformational changes in proteins detected by a combination of infrared spectroscopy and electrochemistry: Evaluation of the technique with cytochrome *c*. Eur J Biochem 187: 565–572

Moss DA, Leonhard M, Bauscher M, and Mäntele W (1991) Electrochemical redox titration of cofactors in the reaction center from *Rhodobacter sphaeroides*. FEBS Lett 283: 33–36

Nabedryk E and Breton J (1994) FTIR difference spectroscopy of the intermediary electron acceptor in *Rb. sphaeroides* bacterial reaction center. Biophys J 66: A272

Nabedryk E, Tavitian BA, Mäntele W, and Breton J (1985) Light-induced Fourier-transform infrared spectroscopy of the primary donor oxidation in bacterial photosynthesis. In: Alix A, Bernard L, and Manfait M (eds) Spectroscopy of Biological Molecules, pp 357–370. John Wiley and Sons, Chichester

Nabedryk E, Mäntele W, Tavitian BA, and Breton J (1986) Light-induced Fourier transform infrared spectroscopic investigation of the intermediary electron acceptor reduction in bacterial photosynthesis. Photochem Photobiol 43: 461–465

Nabedryk E, Tavitian BA, Mäntele W, Kreutz W, and Breton J (1987) Fourier-Transform Infrared (FTIR) Spectroscopic Investigations of the Primary Reactions in Purple Photosynthetic Bacteria. In: Biggins J (ed) Progress in Photosynthesis Research, pp 177–180. Martinus Nijhoff Publishers, Dordrecht

Nabedryk E, Andrianambinintsoa S, Mäntele W, and Breton J (1988) FTIR spectroscopic investigations of the intermediary electron acceptor reduction in purple photosynthetic bacteria and green plants. In: Breton J and Verméglio A (eds) The Photosynthetic Bacterial Reaction Center: Structure and Dynamics, pp 237–248. Plenum Publishing Corporation, New York

Nabedryk E, Mäntele W, and Breton J (1989) FTIR investigations on orientation of protein secondary structures and primary reactions in photosynthesis. In: Barber J and Malkin R (eds) Techniques and New Developments in Photosynthesis Research, pp 17–34. Plenum Publishing Corporation, New York

Nabedryk E, Bagley KA, Thibodeau DL, Bauscher M, Mäntele W, and Breton J (1990) A protein conformational change associated with the photoreduction of the primary and secondary quinones in the bacterial reaction center. FEBS Lett 266: 59–62

Nabedryk E, Berthomieu C, Verméglio A, and Breton J (1991) Photooxidation of the high-potential (c559, c556) and the low potential (c552) hemes in the cytochrome subunit of *Rhodopseudomonas viridis*. Characterization by FTIR spectroscopy. FEBS Lett 293: 53–58

Nabedryk E, Robles SJ, Goldman E, Youvan DC, and Breton J (1992a) Probing the primary donor environment in the histidine M200 → leucine and histidine L173 → leucine heterodimer mutants of *Rhodobacter capsulatus* by light-induced Fourier transform infrared difference spectroscopy. Biochemistry 31:10852–10858

Nabedryk E, Breton J, Allen J, Murchison H, Taguchi A, and Woodbury N (1992b) FTIR characterization of Leu M160 → His, Leu L131 → His and His L168 → Phe mutations near the primary electron donor in *Rb. sphaeroides* reaction centres. In: Breton J and Verméglio A (eds) The Photosynthetic Bacterial Reaction Center II: Structure, Spectroscopy and Dynamics, NATO ASI Series, Vol. 237, pp 141–145. Plenum, New York

Nabedryk E, Breton J, Wachtveitl J, Gray KA, Oesterhelt D (1992c) FTIR spectroscopy of the $P^+Q_A^-/PQ_A$ state in Met L248 → Thr, Ser L244 → Gly, Phe M197 → Tyr, Tyr M210 → Phe, Tyr M210 → Leu, Phe L181 → Tyr and Phe L181-Tyr M210 → Tyr L181-Phe M210 mutants of *Rb. sphaeroides*. In: Breton J and Verméglio A (eds) The Photosynthetic Bacterial Reaction Center II: Structure, Spectroscopy and Dynamics, NATO ASI Series, Vol. 237, pp 147–153. Plenum, New York

Nabedryk E, Goldman E, Robles SJ, Youvan DC, and Breton J (1993) FTIR analysis of genetically modified photosynthetic reaction centers of *Rb. capsulatus*. In: Theophanides T, Anastassopoulo, J, and Fotopoulos N (eds) Fifth International Conference on the Spectroscopy of Biological Molecules, pp 311–312. Kluwer Academic Publishers, Dordrecht

Rothschild KJ (1992) FTIR difference spectroscopy of bacteriorhodopsin: Toward a molecular model. J Bioenerg Biomemb 24: 147–167

Siebert F, Mäntele W, and Kreutz W (1980) Flash-induced kinetic infrared spectroscopy applied to biochemical systems. Biophys Struct Mech 6:139–146

Slifkin MA, and Walmsley RH (1970) Infra-red studies of quinhydrone-type complexes. Spectrochimica Acta 26A: 1237–1242

Struck A, and Scheer H (1990) Modified reaction centers from *Rhodobacter sphaeroides* R26: Exchange of monomeric bacteriochlorophyll with 13^2-hydroxy-bacteriochlorophyll. FEBS Lett 261: 385–388

Struck A, Cmiel E, Katheder I, and Scheer H (1990) Modified reaction centers from *Rhodobacter sphaeroides* R26: 2. Bacteriochlorophylls with modified C-3 substituents at sites B_A and B_B. FEBS Lett 268: 180–184

Susi H and Byler DM (1986) Resolution-enhanced Fourier transform infrared spectroscopy of enzymes. Meth Enzymol 130: 290–311

Thibodeau DL, Breton J, Berthomieu C, Mäntele W, and Nabedryk E (1990a) Steady-state and time-resolved FTIR spectroscopy of quinones in bacterial reaction centers. In: Michel-Beyerle ME (ed) Reaction Centers of Photosynthetic Bacteria, Structure and Dynamics, Springer Series in Biophysics, Vol. 6, pp 87–98. Springer Verlag Berlin

Thibodeau DL, Nabedryk E, Hienerwadel R, Lenz F, Mäntele W, and Breton J (1990b) Time-resolved FTIR spectroscopy of quinones in the *Rb. sphaeroides* reaction centers. Biochim Biophys Acta 1020: 253–259

Thibodeau DL, Nabedryk E, Hienerwadel R, Lenz F, Mäntele W, and Breton J (1992) Time-resolved FTIR difference spectroscopy of photosynthetic bacterial reaction centers. In: Takahashi H (ed) Time-Resolved Vibrational Spectroscopy, pp 79–82. Springer Verlag Berlin

Uhmann W, Becker A, Taran C, and Siebert F (1991) Time-resolved FT-IR absorption spectroscopy using a step-scan interferometer. Applied Spectroscopy 45: 390–397

Venyaminov SY, and Kalnin NN (1990) Quantitative IR spectrophotometry of peptide compounds in water (H_2O) Solutions. I. Spectral parameters of amino acid residue absorption bands. Biopolymers 30: 1243–1257

Chapter 29

Bacterial Reaction Centers with Modified Tetrapyrrole Chromophores

Hugo Scheer and Gerhard Hartwich
Botanisches Institut der Universität, Menzinger Str. 67, D-80638 München, Germany

Summary

Bacteriochlorophylls at sites $B_{A,B}$ and bacteriopheophytins at sites $H_{A,B}$ can be exchanged against modified pigments in reaction centers of purple bacteria. The exchange can be carried out at both branches simultaneously (exchange of B_A and B_B, of H_A and H_B), or selectively at either the A- or B-branch. Pigment modifications include both the peripheral substituents of the terapyrroles and the central metal. The structural requirements for all four binding sites are discussed, as well as the stabilization of the reaction centers by the cofactors. The modified reaction centers were used to assign overlapping absorption bands, as well as to study pigment-protein- and pigment-pigment-interactions, singlet and triplet energy transfer, and light-induced electron transfer.

R. E. Blankenship, M. T. Madigan and C. E. Bauer (eds): Anoxygenic Photosynthetic Bacteria, pp. 649–663.

I. Introduction

Cofactor modifications complement site-directed mutagenesis in evaluating the structural and functional determinants of reaction centers (RC); the two techniques can be used separately or in combination. For the preparation of RCs with modified tetrapyrrolic pigments, there exist currently three different routes: 1) Exchange of chromophores into native or partly denatured RCs by chemical or enzymatic means. 2) Modification of binding sites to incorporate different chromophores, e.g. by mutagenesis. 3) Mutagenic modifications of the pigments' biosynthetic pathways. Alternative routes which have been applied to other photosynthetic pigment-protein complexes, involving the chemical or enzymatic modification of chromophores within the RC[1] or the complete reassembly from the components (apoproteins and cofactors) have as yet been unsuccessful. This chapter deals with recent results obtained by a method of type (1) which was developed several years ago in our laboratory (Struck, 1990; Scheer and Struck, 1993). It allows us to introduce, singly or in combination, modified pigments into the sites B_A and B_B of the monomeric bacteriochlorophylls (BChl) or the sites H_A or H_B of the bacteriopheophytins (BPhe). A first example of method (3) has been reported recently[2], and method (2) is discussed in the Chapter 24 by Woodbury and Allen.

II. Selectivity of the Exchange

The structural modifications to BChl and BPhe are summarized in Fig. 1. Most of the pigments used for exchange experiments have been modified only at a single position, however, metalation and hydroxylation at C–13^2 have frequently been combined with other structural changes. Preparative procedures for most of the pigments have been summarized by Scheer and Struck (1993).

A. Influence of Central Metal

Chlorophylls (Chl) and BChls contain Mg as the central metal. Although formally four coordinate, it is coordinatively unsaturated and is normally five or six coordinate by binding of extraneous ligands above or below the tetrapyrrole plane. This affinity is so high, that in dry, nonpolar solvents, it is a major driving force for (B)Chl aggregation via ligation to peripheral C=O or OH groups of neighboring molecules (Katz et al., 1978). A fifth ligand has also been identified in all (B)Chl protein complexes whose structures are known with sufficient resolution. A variety of such ligands have been characterized: Histidine[3] is the common one in RC, but it can be replaced by site-directed mutagenesis with glutamine, asparagine, serine and threonine without loss of the BChl in the assembled RC (Coleman and Youvan, 1990; Bylina et al., 1991; Scheer and Struck, 1993). In the water-soluble Fenna-Matthews-Olson (FMO)-protein from *Prosthecochloris aestuarii*, which is involved in energy transfer from the chlorosome to the RC (Olson, 1978), five of the seven BChls are also ligated to histidines, one to a backbone peptide carbonyl, and one to water (Fenna et al., 1977) (see Chapter 20 by Blankenship et al.).

Yet another motif has recently been characterized in the LHCII antenna from green plants (Kühlbrandt et al., 1993). Here, at least two of the chlorophylls

[1] See Struck et al., 1991 for the discussion of an attempt along this line. It should also be mentioned that under the exchange conditions used by us (Scheer and Struck, 1993), small amounts of BChl *a*' (the 13^2-epimer of BChl *a*) and of 13^2-hydroxy-BChl *a* are extracted from treated reaction centers. Since these pigments are not extracted from untreated samples, they should be formed within the reaction centers during the treatment. However, they are present in non-stoichiometric amounts and their binding sites are not specified, so this method is presently not useful for preparative work.

[2] First examples of pigment modification by manipulating the biosynthesis have recently been achieved (C. Bauer, private communication, 1993). By systematically inactivating the genes of the BChl biosynthesis cluster, they obtained mutants containing BChl a_{gg} (through inhibition of the geranyl-geraniol dehydrogenase) and [8-vinyl]-BChl a_p (through inhibition of the 8-vinyl-hydrogenase). Pigments of the latter type have also been found naturally in prochlorophytes (Goericke, 1990; Goericke and Repeta, 1992; Chisholm et al., 1992). Aside from interesting biogenetic implications, the mutants had no functional phenotype, and reactions centers isolated from the second mutant had the same kinetics as the wild type ones (J. Allen, private communication, 1993).

[3] The early notion of too long histidine-N-to-Mg distances at the P-sites in *Rb. sphaeroides* (Allen et al., 1987) has been corrected by newer structures with higher resolution (Chang et al., 1991; Ermler et al., 1994).

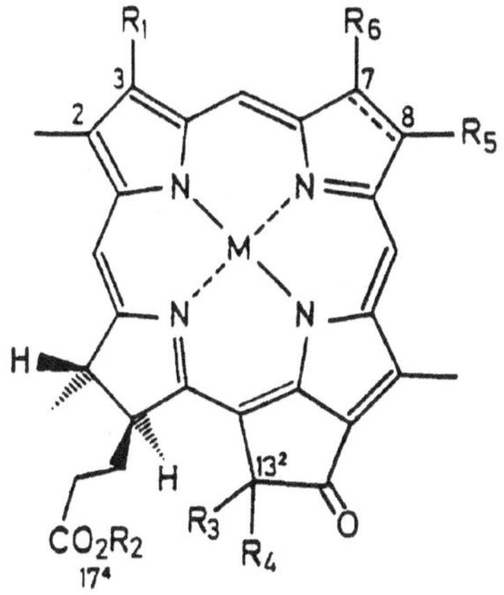

M = Mg; Zn; Ni	M = 2H
R_1 = $\underline{-COCH_3}$; $-CH(OH)CH_3$; $-CHCH_2$	R_1 = $\underline{-COCH_3}$; $-CH(OH)CH_3$; $CHCH_2$
R_2 = $\underline{-C_{20}H_{39}}$(phytyl); $-C_{20}H_{33}$ (geranylgeranyl); $-CH_3$; $-CH_2CH_2F$; $-(CH_2)_2CF_3$	R_2 = $\underline{-C_{20}H_{39}}$(phytyl); $-C_{20}H_{33}$ (geranylgeranyl); $-CH_2CF_3$
R_3 = $\underline{-H}$; $-OH$	R_3 = $\underline{-H}$; *-OH*
R_4 = $\underline{-CO_2CH_3}$	R_4 = $\underline{-CO_2CH_3}$; $-H$
R_5 = $\underline{-CH_2CH_3}$	R_5 = $\underline{-CH_2CH_3}$; $=CH-CH_3$
R_6 = $\underline{-CH_3}$	R_6 = $\underline{-CH_3}$; *-CHO*; $-CH_2OH$
$\Delta_{7,8}$ = $\underline{2\ H}$	$\Delta_{7,8}$ = $\underline{2H}$; double bond

Fig. 1. Modifications in (bacterio) chlorins that are accepted in the binding sites of *Rb. sphaeroides*. Left: BChl-derivatives that exchange into B_A and/or B_B. Right: (B)Phe-derivates that exchange into $H_{A,B}$ (normal print) or selectively into H_A (*italic* print). The substituents R present in native BChl *a* (M = Mg) and BPhe *a* (M = 2H) are underscored. Stereochemistry at C-13^2 is generally undefined, unless mentioned in the text.

(Chl *a* or *b*) have an aspartate ligand, which in turn is bound to an arginine. Interestingly, this motif has a precedence in the biliproteins, where it does not bind to the central metal of a cyclic tetrapyrrole, but rather to two nitrogens located in the 'inner' rings (B,C) of a linear tetrapyrrole (Schirmer et al., 1987; Dürring et al., 1990,1991. The two rings (pyrrole and pyrrolenine), are in a Z,s-syn-geometry, and the latter is believed to be protonated (Scharnagl and Schneider, 1989,1991), this partial structure thus constitutes one half of the dication of a cyclic tetrapyrrole. The similar binding then would support the similarity between metal complexes and dications of tetrapyrroles.

Bacterial RCs of type II[4] possess binding sites for BChl (= Mg-complexes) and BPhe (= free bases). A major distinction between these two types of sites is the presence of a histidine in a favorable position to bind to the central Mg as a fifth ligand in the BChl binding sites $P_{A,B}$ and $B_{A,B}$, and an isoleucine in the respective position of the BPhe binding sites $H_{A,B}$. This distinction holds also for the RC from *Chloroflexus (Cf.) aurantiacus*, in which histidine M180 in the B_B-pocket is replaced by isoleucine, and the normally bound BChl *a* is replaced by BPhe, and for a variety of site-directed mutants of several bacteria. Polar residues capable of ligation result in the binding of BChl, and large, nonpolar residues in the binding of BPhe. A complementary distinction has been found for the pigments during exchange experiments: Bacteriochlorophylls are always exchanged into the BChl pockets $B_{A,B}$, irrespective of the substituents at the periphery, and (bacterio)-pheophytins lacking the central Mg are always exchanged into the BPhe pockets $H_{A,B}$.

The peripheries of the binding pockets seem to play a lesser role in the distinction between BChl and BPhe, if judged from the ready change of pigment upon change of the 'central' amino acid; the exchange experiments with modified pigments indicate that this is also true for at least part of the periphery of the pigments (see below). The aforementioned interaction between the central metal and the ligand provides a ready explanation for the selection of BChls: the N-Mg binding energy is large enough to select for the Mg-complexes only. The positive selection of BPhe

Abbreviations: B_{AB} – binding sites for the monomeric BChls; BChl – bacteriochlorophyll; BPhe – bacteriopheophytin; *Cf.* – *Chloroflexus*; Chl – chlorophyll; H_{AB} – binding sites for the BPhes, in each case the subscripts A and B refer to the 'active' and 'inactive branch', respectively; P_{AB} – binding sites for the primary donor BChls; Phe – pheophytin; RC – reaction center; *Rb.* – *Rhodobacter*; *Rp.* – *Rhodopseudomonas*; *Rs.* – *Rhodospirillum*

[4] Type I and II refers to reaction centers which are structurally related to RC from Photosystem (PS) I (like those from heliobacteria and Chlorobiaceae) and to RC from PS II (like those from purple bacteria and Chloroflexaceae), respectively.

and rejection of BChls in the $H_{A,B}$ pockets is more difficult to understand. Probably, it does not simply require the absence of an appropriately placed ligating amino-acid, but also the absence of sufficient space to introduce an extraneous ligand, like e.g. water. There are two lines of evidence for this: Firstly, the FMO-protein provides already an example that water can act as ligand in a BChl protein; if this could be introduced into $H_{A,B}$ in RCs, binding of BChl may be expected. Secondly, there is recent direct evidence for this concept from site-directed mutagenesis of the D1 protein of the PS II-RC (B. Diner and P. Nixon, private communication, 1993). Replacement of His^{D1-168} (which is a presumed ligand to Chl *a* in the P-site) with alanine yields functioning RCs in which neither the content of Phe *a* is increased, nor the content of Chl *a* is decreased. On the other hand, the mutation $His^{D1-168} \rightarrow$ Ile (the equivalent of which produces a heterodimer in bacterial RC (Coleman and Youvan, 1990)) has a PS^- phenotype. The small size of the alanine in the former mutant may allow the introduction of a molecule of water as a fifth ligand to Mg, while this is not possible in the latter. Histidine-to-alanine mutations have been done with bacterial RCs and support, at least in part, this selection mechanism (S. Boxer, personal communication, C. Schenck, personal communication).

B. Replacement of Central Magnesium

Replacement of Mg with other metals allows in porphyrins a systematic variation of the redox potentials (Fuhrhop et al., 1974; Watanabe and Kobayashi, 1991). A series of metallo-bacteriochlorophylls (which more precisely should be termed metallo-bacteriopheophytins) has been recently prepared by us (Hartwich et al., 1995b; see Tables 1 and 2 for their spectroscopic properties and redox potentials). Several pigments in which the central Mg has been replaced by other metals are exchangeable to a high degree of selectivity into the $B_{A,B}$ pockets. These include [Zn]- and [Ni]-BChl *a* (Hartwich, 1994). Zn complexes of porphyrins have been classified by Buchler (1975) as square-pyramidal (=5 coordinate), Ni-porphyrins as square planar (=4 coordinate). In view of the above discussion, an exchange of [Ni]-BChl into both the $B_{A,B}$ and the $H_{A,B}$ sites might be expected. The selective exchange into $B_{A,B}$ and the failure of exchange into $H_{A,B}$ is then surprising and may indicate the operation of additional steric, electronic or other factors besides ligation. It should also be mentioned, that the coordination of Ni is quite variable, and EXAFS studies have indicated that [Ni]-BChl *a* is five coordinated both in solvents of intermediate polarity and in the RC (Chen et al., 1995). Paulsen et al. (1990, 1991) studied the reconstitution of the LHCII antenna complex from plants with several [metallo]-Chls. Their results also indicate that pigments with central metals other than Mg can replace at least part of the 13–14 molecules of Chl *a* and *b* present. No data on the site of replacement, the type of pigment replaced, or metal coordination states in these reconstituted LHCII complexes are yet available.

The biosynthetic origin of (B)Phe in type-II RCs is unknown. All (B)Phes derive from Mg-porphyrins, which must then be demetalated. Mg-complexes of bacteriochlorins and, even more so of chlorins, are labile, they are readily demetalated by acid. There are also several reports on dechelatase activities (Ziegler et al., 1988; Owens and Falkowski, 1982), but the postulated enzymes have not been purified, no information on their active sites is available, and it is not clear if their function is to provide a pool of (B)Phe which is utilized for supplying the RC, or if they are involved in (B)Chl breakdown[5]. Principally, an alternative mechanism is possible for (B)Phe insertion: BChl (e.g. the Mg-complex) is first bound to all six tetrapyrrole binding pockets, and the pigments at $H_{A,B}$ are subsequently demetalated. None of the exchange experiments support such a mechanism, but it could also mean that once a BPhe is bound, it is no longer exchangeable with BChl. However, other available evidence also points against an in situ demetalation. Walter et al. (1979) identified different esterifying alcohols in BPhe *a* and BChl *a* from *Rhodospirillum (Rs.) rubrum* RCs. Since exchange experiments with these RCs did not strongly discriminate between pigments bearing the two alcohols, viz. phytol or geranylgeraniol (Beese, 1989; Struck et al., 1990; Struck, 1990), this indicates a dedicated synthesis (and possibly even delivery) of the BPhe. The mutagenesis experiments (see e.g. Table 1 in Scheer and Struck, 1993) point in the same direction, unless the exchange of the amino acid next

[5] Several degradation products of Chl *a* (Engel et al., 1991; Iturraspe et al., 1991; Kräutler et al., 1992; Mühlecker and Kräutler, 1993) and Chl *b* (Iturraspe et al., 1994) have been recently reported, but it is not yet clear if the ring-opening reaction involves the Mg-complexes as substrates, or if a previous demetalation is required.

Table 1. Midpoint-potentials of metallobacteriochlorophylls in THF (in [V] versus Ag/AgCl-electrode)[1]

	$E^{3-}_{1/2}$	$E^{2-}_{1/2}$	$E^{-}_{1/2}$	$E^{+}_{1/2}$	$E^{2+}_{1/2}$
BChl*a*		−1.61	−1.29	0.29	0.65
3-α-hydroxyethyl-BChl *a*		−1.79	−1.46	0.27	0.66
13²-OH-BChl *a*		−1.57	−1.25	0.34	0.74
Ni-BChl *a*	−2.8*	−1.31	−0.95	0.36	0.76
13²-OH-Ni-BChl *a*	−2.76*	−1.33	−0.95	0.36	0.78
Zn-BChl *a*		−1.52	−1.15	0.38	0.77
13²-OH-Zn-BChl *a*		−1.48	−1.13	0.42	0.85
Cu-BChl *a*	−2.62	−1.42	−1.09	0.42	0.86#
13²-OH-Cu-BChl *a*		−1.31	−0.98	0.59	1.06#
Pd-BChl *a*		−1.37	−0.94	0.66	1.06#

[1] C. Geskes, G. Hartwich, I. Katheder, H. Scheer, A. Scherz, W. Mäntele and J. Heinze, unpublished
* Quasi-reversible peak
# EEC-reaction, peak almost irreversible

Table 2. Spectral properties of metallo-bacteriochlorphylls. All wavelengths are in [nm], modified from Hartwich et al., 1995b.

Compound	$\lambda_{max, abs}$ (298K)[a]				$\lambda_{max,emi.}$[b](77 K)	¹H-NMR[c] (298 K)	FAB-MS[d]	
	B_y	B_x	Q_x	Q_y				
BPhe *a*	357	384	525	749	759[1]	+	888	
13²-OH-BHhe *a*	358	383	521	750	755	+	904	
Cd-BChl *a*	359	389	576	760	775	+	1000	(¹¹⁴Cd)
13²-OH-Cd-BChl *a*	360	390	570	760	773	+	1016	(¹¹⁴Cd)
Ni-BChl *a*	336	391	531	780	–[2]	–[3]	944	(⁵⁸Ni)
13²-OH-Ni-BChl *a*	336	390	526	779	–[2]	–[3]	960	(⁵⁸Ni)
Co-BChl *a*	336	387	530	764	–[2]	–	945	(⁵⁹Co)
13²-OH-Co-BChl *a*	335	388	528	763	–[2]	–	961	(⁵⁹Co)
Mn-BChl *a*	363	389	586	769	–[2]	–	941	(⁵⁵Mn)
13²-OH-Mn-BChl *a*	364	389	582	769	–[2]	–	957	(⁵⁵Mn)
Cu-BChl *a*	342	390	537	772	–[2]	–	949	(⁶³Cu)
13²-OH-Cu-BChl *a*	341	390	534	772	–[2]	–	965	(⁶³Cu)
Zn-BChl *a*	353	389	559	762	772	+	950	(⁶⁴Zn)
13²-OH-Zn-BChl *a*	354	389	555	762	770	+	965	(⁶⁴Zn)
Pd-BChl *a*	329	385	527	755	–[2]	+	992	(¹⁰⁶Pd)
13²-OH-Pd-BChl *a*	331	385	524	755	–[2]	+	1008	(¹⁰⁶Pd)
BChl *a*	357	391	573	771	778[1]	+	910	(²⁴Mg)
13²-OH-BChl *a*	357	391	568	771	778	+	926	(²⁴Mg)

[a] In diethylether
[b] In EPIP = diethylether/petroleumether/isopropanol (5:5:2, vol/vol/vol)
[c] In pyridine: '+' indicates sharp signals, '–' extensive line broadening due to the paramagnetic central metal
[d] Ionization in m-hydroxy-bencylic alcohol matrix
[1] Data taken from A. P. Losev et al., 1990
[2] No fluorescence detectable (SPEX Fluorolog 221)
[3] Sharp signals in CH_3CN

to the central N4-cavity at any one of the position confers indiscriminately dechelatase activities to such sites.

C. Peripheral Substituents

The influence of modified peripheral substituents on the exchange of (B)Chl or (B)Phe capacity can give new insights to the interaction between amino-acids of the protein with the pigments. However, being a purely operational criterion, only positive results are meaningful. A previous overview of all modifications tested until the end of 1992, has been given (Scheer and Stuck, 1993). Briefly, pigments modified at C-3, or C-17^4, or hydroxylated at C–13^2, alone or in combination, exchange into both the B_A and the B_B pockets of carotenoid-less *Rhodobacter (Rb.) sphaeroides* RCs, and into B_A only in wild-type RCs. Also, the exchange into B_B is faster than into B_A (see below).

Tinkering with the isocyclic ring other than hydroxylation at C–13^2, inhibited an exchange. This includes removal of the 13^2-$COOCH_3$-substituent, as well as enlargement or an opening of the ring. Even the hydroxylation is highly stereospecific: 13^2(S)-hydroxy-BChl *a*, which has the same orientation of the $COOCH_3$-substituent as BChl *a*[6], exchanges and is functionally competent, whereas the 13^2(R)-epimer does not exchange (F. Storch, H. Scheer, unpublished, 1993). This discrimination is so strong, that only the traces of the S-in preparations of the R-epimer are taken up during exchange experiments, rather than the main product with the 'wrong' orientation of the $COOCH_3$ group. A similar selectivity also applies to 13^2-hydroxy-BPhe *a* (see below). Interestingly, 13^2(S)-BChl *a*, the epimer of the native 13^2(R)-BChl *a* can be found in all 'blank' samples of RCs subjected to the exchange conditions in the absence of extraneous pigments. It is not clear at which site it is bound, but the tight binding and its absence in untreated RCs using the same extraction conditions indicates that it is indeed incorporated into the RC and not merely adsorbed. If the stereoselectivity of the $B_{A,B}$ sites found for 13^2-hydroxy-BChl *a* holds also for BChl *a*, the epimer might then be bound at the $P_{A,B}$ site. It should be mentioned in this context, that both the chlorophyll synthetase (Helfrich et al., 1994) and chlorophyllase (Fiedor et al. 1992; Y. Nishiyama, M. Kitamura, S. Tamura, T. Watanabe, unpublished, 1993) are also highly stereospecific with regard to the 13^2-position.

The bacteriochlorin conjugated system appears to be essential for an exchange into the B_{AB}-sites, since neither Chl *a* nor [3-acetyl]-Chl *a* have been found exchangeable. The latter differs from BChl *a* 'only' by the removal of the two 'extra' hydrogens at C-7 and C-8. Inspection of the crystal structure (Chang et al., 1991) does not indicate an obvious reason for the failure to incorporate chlorins instead of bacteriochlorins, but it may be due to a changed flexibility of the macrocycle (Scheer and Katz, 1975; Senge et al., 1993). The $H_{A,B}$-sites are much more tolerant in this respect. Phe *a*, Phe *b* and [7-hydroxymethyl]-Phe *a* are exchangeable, they differ from BPhe *a* not only in ring B being unsaturated, but also in the nature of the C-7 substituent. It is hydrophobic in Phe *a*, an H-bond acceptor in Phe *b*, and both acceptor or donor in [7-hydroxymethyl]-Phe *a* (Shkurupatov and Shuvalov, 1993; Meyer, 1992; Scheer et al., 1992; Meyer and Scheer, 1995; A. Y. Shkurupatov, M. Meyer, H. Scheer and V. A. Shuvalov, unpublished).

D. Asymmetric Exchanges

B_A and the symmetry-related site B_B do not strongly discriminate between the pigments tested up to now. There is, however, a kinetic discrimination: The first molecule is exchanged considerably faster than the second one, and there are several lines of evidence for the assignment of the faster exchanging site as B_B. One comes from kinetic analysis of the electron transfer in a partially exchanged sample: After exchange of ≈80% of BChl at sites $B_{A,B}$ with [3-vinyl]-BChl *a*, there were about 40% RC left in which the kinetics of the primary charge separation were unchanged, viz. which contained BChl *a* at site B_A. Even considering the error limits of the analysis, which are in the range of ±10%, this would indicate a larger exchange in B_B than in B_A.

Selective exchange into B_A is possible in wild-type reaction centers. Here, the carotenoid located close to B_B inhibits an exchange into the latter (Hartwich, 1994). This assignment of the exchange site was originally tentative, but it has recently been

[6] Due to the Cahn-Ingold-Prelog nomenclature rules, the configuration of C-13^2 changes by replacing the H-substituent present in BChl *a* with the OH-substituent present in 13^2-hydroxy-BChl *a*, even though the arrangement of all other substituents remains unchanged.

confirmed by optical-electron-double resonance experiments (see below). The selective exchange into the B_B-site has been accomplished in a more roundabout sequence exemplified here for [3-Vinyl]-13^2-hydroxy-BChl *a* (G. Hartwich, 1994). First, both B-sites have been exchanged with [3-Vinyl]-13^2-hydroxy-BChl *a* in RC from *Rb. sphaeroides* R26. This is followed by a reconstitution with spheroiden(on)e to protect the B_B-site, and in the subsequent step the B_A site only is brought back 'to normal' by a second exchange with the native pigment, BChl *a*.

The discrimination between the A-and B-sites is much stronger for $H_{A,B}$. 13^2-OH-and 13^2-demethoxycarbonyl-bacteriopheophytins do only exchange for one BPhe. There is even a strong discrimination for the alcohol: BPhe a_{gg} (the subscript denoting geranylgeraniol rather than phytol as the esterifying alcohol), exchanges only into one site. There is presently only circumstantial evidence as to which of the sites is affected by the exchange, and whether it is the same in all cases, because the spectra of the aforementioned pigments are very similar and allow no distinction in the Q_X-region. Our working hypothesis is that it is H_B, because there is also a distinctly more rapid exchange at this position (as compared to H_A) with pigments having sufficiently different spectra that the exchange can be followed spectroscopically. There is a selective interaction of glutamine L104 with the 13^1 CO of the pigment in H_A (Allen et al., 1987; Chang et al., 1991; El-Kabbani et al., 1991; Michel and Deisenhofer, 1988; Lutz and Mäntele, 1991) leading to a red-shift of the Q_X-band of this pigment. This interaction may also account for the failure to exchange pigments bearing a 13^2-OH-or lacking the 13^2-$COOCH_3$-substituent. In these pigments, enolization of the β-ketoester system at ring V is inhibited, and enolization (Bocian et al., 1987) or at least strong H-bonding between the 13^1-keto-group of BPhe *a* and the neighboring glutamate (see Lubitz, 1991) in H_A has been suggested as the reason for the spectral shift. Because the enolizable β-ketoester system is found in all tetrapyrroles known to participate in electron transfer in RCs, there may be a functional and structural significance at least in this site (as contrasted from $B_{A,B}$, see above). However, it should be noted that $Glu^{L104} \rightarrow$ Gln mutagenesis abolished the shift, but does neither inhibit the binding of BPhe *a* nor the function of these RCs (Coleman and Youvan, 1990).

III. Static Spectroscopy of Reaction Centers with Modified Pigments

All pigments introduced into reaction centers are subject to spectral modifications (absorption shifts, hyperchromism, induced circular dichroism). The origin of these can formally be separated in two components: those induced by the protein moiety, and those induced by pigment-pigment interaction. Both have been discussed by Struck and Scheer (1993) and are summarized here only briefly:

After self exchange experiments, i.e. incubation under exchange conditions with native pigments added exogenously, RCs show that the protein subunit is unaltered as judged by UV/VIS absorption, CD, linear dichroism, magnetic resonance (ENDOR) and vibrational spectroscopy, and by light-induced bleaching of the P-absorption at 870 nm. Accordingly, the treatment leaves the RC largely unchanged.

The absorption spectra (e.g. in diethylether) of many of the pigments used for exchange differ from those of the native pigments (see Scheer, 1988 and Table 2). If the spectra of a series of such pigments in monomeric solution are compared with the spectra of the same pigments incorporated in the RC, a surprisingly constant environment-induced red shift (EIRS) was found: it amounts to $\approx$500 cm^{-1} (Q_y) and $\approx$700 cm^{-1} (Q_x) for the sites $B_{A,B}$ and to $\approx$150 cm^{-1} (Q_y) and $\approx$450 cm^{-1} (Q_x) for the sites $H_{A,B}$ (Scheer and Struck, 1993; Scheer et al., 1993). In view of the short distances among all pigments in the RC the problem of EIRS has received considerable attention and it is not clear whether this is a result of protein-chromophore or chromophore-chromophore interaction, or both (Scherz et al. 1991; Scherer and Fischer, 1991). Unchanged EIRS in chromophore-exchanged RCs argue against specific protein-chromophore interactions, at least as far as the modified groups are concerned. However, they are compatible with more global changes, e.g. distortions of the macrocyclic system from planarity.

Introduction of spectrally different pigments also leads to negligible changes of the absorption spectral features assigned to P[7]. This indicates that none of the chromophore substituents modified hitherto is strongly involved in pigment-pigment interactions with this site. This is further supported by CD-spectroscopy which is very sensitive to pigment

[7] The position of this band is considerably affected by the other environmental conditions like detergent, temperature, etc.

interaction and geometry. Besides shifts of the spectral position due to band shifts of the modified pigments, only minor variations of the non-exchanged pigments are observed. ENDOR shows that there is also little to no interaction between the primary donor in its doublet state ($P^{+\bullet}$) and any of the other pigments (Struck, 1990). The situation is more complex and not yet well explored for interactions among the $H_{A,B}$ and $B_{A,B}$ sites. Here, small but distinct shifts are always observed if a neighboring pigment is exchanged, which is most obvious if the newly introduced pigment does not absorb in the same spectral region. Based on kinetic studies with RC in which BPhe *a* has been replaced by Phe *a*, Shkurupatov and Shuvalov (1993) have recently proposed a considerable contribution of the pigments in $B_{A,B}$ to the ≈540 nm band assigned generally to BPhe at the $H_{A,B}$ sites.

The differential absorptions in the Q_X-region of BPhe *a* t H_A and H_B was essential in defining and exploring the asymmetry of electron transfer. The former absorbs at longer wavelengths (≈545 nm) than the latter (≈535 nm) (Clayton and Yamamoto, 1976; Vermeglio et al., 1978; Breton 1985; Kirmaier et al., 1985; Kellog et al., 1989). The two bands overlap at room temperature but can be deconvoluted; at low temperatures they are well resolved. This differentiation has been observed, too, for all replacement pigments in the H-sites, and can be used to establish asymmetric echanges. No similar differentiation existed for the B-sites. Static spectroscopy after site-selective, asymmetric exchanges at B_A and/or B_B then allowed the spectral assignment and differentiation of the B-sites. Taking advantage of the partial and asymmetric triplet energy transfer from the carotenoid to B_B, BChl at this site could be tagged (see also below). This was explored by low-temperature absorption, ADMR and MIA spectroscopy (Hartwich, 1994; Hartwich et al., 1995c) in RCs, in which B_A and/or B_B were exchanged with [3-vinyl]-13²-OH-BChl *a*, a BChl *a*-derivative where the Q_X and the Q_Y transitions in solution as well as in the RCs shifted to the blue. This allowed the following assignments: BChl *a* at B_A absorbs at 803(Q_Y) and 595 nm (Q_X), whereas BChl *a*-B_B absorbs at 812 and 600 nm, respectively. There is furthermore additional evidence from these experiments for locating the upper exciton Q_Y-band of the primary donor (P_+) at 808 nm (see also section E), and the Q_X absorption of P at 603 nm.

A. Triplet Energy Transfer

The main previous evidence for specific interaction of the primary donor in its triplet state ($^3P^*$ or P^T) with pigments in $B_{A,B}$ was derived from microwave-induced absorption difference spectra (MIA, often also referred to as singlet-triplet difference, Lous and Hoff, 1987) of RCs, in which the BChl at both sites $B_{A,B}$ was replaced by pigments with progressively blue-shifted absorptions: there is one band in the MIA spectrum which shifts like the absorption band of the exchanged pigments in $B_{A,B}$ from ~800 to ~770 nm.

Selective exchange experiments of the type mentioned above also allowed a detailed study of the triplet energy transfer within the RC (Hartwich et al., 1995c). The B_B-exchanged sample showed almost the same $^3P^*$-MIA spectra as the $B_{A,B}$-exchanged sample, indicating that there is no strong interaction between $^3P^*$ and B_A. The B_A-exchanged sample, however, shows not only the expected features around 815 nm identical to native RCs, but also the 776 nm band indicative of the modified substituent at B_A. This contrasts with the suggestion that $^3P^*$ interacts selectively with B_B, and rather indicates that there is an interaction of the primary donor triplet with both B-sites, at least at liquid He temperatures.

In wild-type RCs, B_B is close to the carotenoid (Car) site, which breaks the high symmetry of the RC. It has a protective function, and the otherwise long-lived $^3P^*$ (≈30 μs), is quenched by fast energy transfer (≈30 ns) to Car (Frank, 1993). The latter triplet lies energetically below $^1O_2$ and is hence photochemically safe, and it furthermore relaxes quickly to the singlet ground state (Parson and Monger, 1977). It is then suggestive, that the partial triplet energy transfer from $^3P^*$ monitored by MIA, is related to the fast and complete transfer to Car, and hence to the protection of RC under conditions where the acceptor Q_A is reduced. This has been supported by two lines of experiments:

In the first one (Hartwich et al., 1995c), MIA spectra of the carotenoid triplet (^{3}Car) from wild-type RC from *Rb. sphaeroides* have been investigated. Besides the signal at 550 nm characteristic for ^{3}Car, they show an additional signal in the long-wavelength region (812 nm) characteristic for BChl at B_B. This has been confirmed by using RCs where B_B has been selectively exchanged with [3-vinyl]-13²-OH-BChl, where this signal is shifted to 775 nm. Modification of BChl at the B_A-site has no effect on the MIA of

^{3}Car. Thus only the pigment at B_B experiences interaction with ^{3}Car, indicating this pigment as a mediator or even an intermediate in energy transfer from $^3P^*$ to ^{3}Car. These experiments allowed at the same time an unequivocal assignment of the two exchange sites (see above).

A second experiment (Frank et al., 1993) was based on the suggestion that the triplet transfer via B_B is an activated process, and that specifically the activation barrier is determined by the energy difference of ^{3}P and ^3B_B. The origin of this suggestion is the temperature dependence of the triplet transfer, which stops in wild-type RC from *Rb. sphaeroides* at ≈35K (Schenck et al., 1984). The onset of triplet transfer has now been measured in RCs derived from the R26 mutant, in which first BChl *a* in both B_A and B_B had been exchanged against modified pigments, and which have then been (re)constituted with Car. The onset of triplet transfer to Car in these RCs, followed the estimated increase in the triplet energy of the modified BChls, thus confirming the activation hypothesis, and furthermore suggesting B_B as a real intermediate.

It can be concluded that unidirectionality of triplet energy transfer is conditioned by interaction of $^3P^*$ with $B_{A,B}$ and selective interaction of ^{3}Car with B_B, and that this transfer is an activated process. This asymmetry of triplet energy transfer to the B-branch may be compared to the well known and still not understood asymmetry of electron transfer to the A-branch.

B. Singlet Excitation Energy Transfer

Reaction centers where BChl *a* in (B_A or) $B_{A,B}$ is replaced by [Ni]-BChl *a* seem to be a powerful tool in explaining the role of the accessory BChls. They are capable of charge separation, but with a lower quantum yield, which is furthermore dependent on the excitation wavelength. The preliminary analysis indicates that both energy and electron transfer in these RCs are strongly impaired (Hartwich, 1994).

Energy transfer was studied by fluorescence spectroscopy. The excitation spectrum for the emission of $^1P^*$ (Fig. 2) shows a much decreased quantum yield (<30%) for all spectral regions where the Ni-BChl or the BPhe are excited, as compared to direct excitation of P. It is concluded, that i) [Ni]-BChl in $B_{A,B}$ is a trap for excitation energy and that ii) after excitation of BPhe in $H_{A,B}$ energy is transported, at least partly, via the accessory BChls to the primary

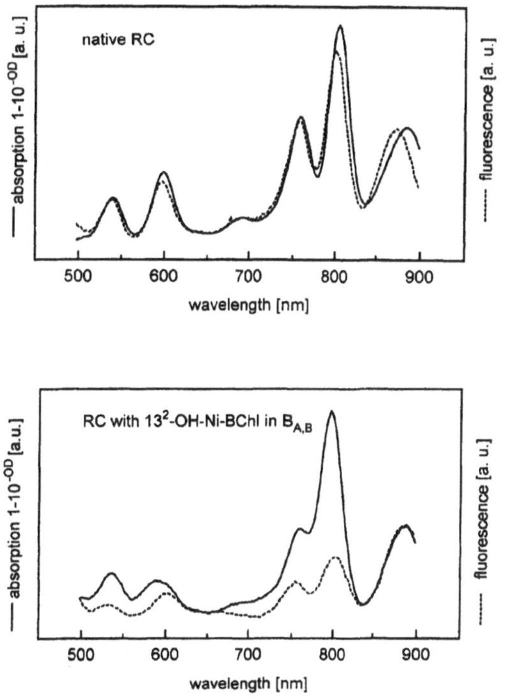

Fig. 2. Simultaneously recorded absorption (——) and excitation (- - - -) spectra of *Rb. sphaeroides* R26 RC (upper panel) and of RC with 13^2-OH-Ni-BChl in $B_{A,B}$ (bottom panel) at T = 85 K. Fluorescence is detected at 930 nm and normalized to the primary donor absorption.

donor (Hartwich et al., 1995a). Further measurements (e. g. transient absorption etc.) will be necessary to elucidate the effect of [Ni]-BChl in B_A and/or B_B on the dynamics and unidirectionality of charge separation. The energy trapping of [Ni]-BChl can be explained by the electronic structure of the pigment. In solution, excitation energy in the S_1-state of the π-system is transferred within <1 ps to a d-orbital of the Ni, and then to the S_0 ground-state within less than 30 ps (Kobayashi et al., 1979). By using an upper limit for the excitation transfer in RC of ~100 fs (Breton et al., 1986), the S_1-to-S_d relaxation must be faster than 50 fs to explain the decreased fluorescence yield. Fluorescence is the competition to charge separation, which is believed to take place only along the active branch (A-branch)[8]. This can be taken as for additional support that exchange of

[8] Strictly speaking, this is true only of charge separation to Q, whereas transient charge separation to B_B as a 'parking process' has been discussed (see Zinth and Kaiser, 1993).

bacteriochlorophylls in wild-type RCs from *Rb. sphaeroides* is restricted to the B_A-site. These RCs show the same reduction of fluorescence quantum yield as the $B_{A,B}$-exchanged sample. Even if an additional charge separation in the $B_{A,B}$ -exchanged sample via the B-branch is assumed, one would expect a higher fluorescence quantum yield in the B_A-exchanged sample or for a nonselective exchange in wild-type RCs, detectable within a limit of error of 10%.

IV. Stabilization

Self-organization of (B)Chl is a widely studied process (Katz et al., 1978, 1991, Scherz et al., 1991), and it has been suggested that the pigments also play a role in formation and stabilization of the native pigment-protein complexes. Experimental evidence to this comes also from reconstitution studies with antenna complexes from green plants (Plumley and Schmidt, 1987, Paulsen et al., 1990, 1991) and purple bacteria (Parkes-Loach et al., 1988, 1990; Ghosh et al., 1988; see chapter by Loach and Parkes-Loach). Several lines of evidence indicate a structural role of the bound tetrapyrroles for the integral reaction center complex. The first comes from temperatrure studies carried out in order to optimize the exchange reaction. In the presence of an excess of exogenous pigment(s), the RC can be heated 3–5 degrees higher than in their absence. This can be rationalized in terms of an equilibrium between RCs with open and occupied binding sites, which is shifted towards the former, more stable ones by the mass action of exogenous pigment(s). Further evidence comes from recent attempts to remove the quinones from modified RCs: the procedure of Okamura et al. (1975) gives much lower yields with such RCs. Interestingly, there is also a stabilizing effect of the quinone(s). If they are removed prior to the exchange of BPhe or BChl, then the yield of the subsequent exchange procedure is much reduced (M. Meyer, G. Hartwich and H. Scheer unpublished results). There is, furthermore, a distinct difference among wild-type reaction centers from *Rb. sphaeroides* 2.4.1 and the carotenoidless mutant R26: the latter tolerates only a 3 degree lower temperature for exchange, viz. it is considerably less stable due to the loss of the carotenoid, 15,15'-*cis*-spheroiden(on)e. The stabilization of reaction centers by their various cofactors, therefore, is clearly a cumulative effect. Only a single exchange has hitherto been done with RCs modified by site-directed mutagenesis ($Tyr^{M208} \rightarrow$ Phe, Göbel, 1993). The stability of wild-type RCs (*Rb. sphaeroides* 2.4.1) under the reaction conditions was comparable to that of non-mutated RCs from the carotenoidless mutant R26.1.

V. Time-Resolved Spectroscopy

RC are energy transducers, they convert excitation into electrochemical energy. Modifications of pigments may help to clarify the assignment of kinetic processes to the different molecular events which are currently discussed controversially (Kirmaier and Holton, 1993; Zinth and Kaiser, 1993; Shuvalov, 1993; Martin and Vos, 1992; DiMagno and Norris, 1993; see Chapter 24 by Woodbury and Allen). If judged from the light-induced bleaching, all reaction centers mentioned in the previous sections are competent for light-induced electron transfer. The quantum yield and recovery of P870 in these samples shows some variations, but none of the recombination times are strongly affected.

The first kinetic experiments of the forward reaction have been done with 'blank' samples which had been subjected to the exchange conditions in the absence of extraneous pigment (Finkele et al., 1992). Spectra and kinetics of these RCs were identical to those of native ones, indicating no significant damage to those RCs which are isolated from the exchange mixture (see above). Only few pigment modifications have hitherto been characterized kinetically. RCs containing 13^2(S)-hydroxy-BChl *a* at sites B_{AB} have within the limits of error identical kinetics to native ones (Finkele et al., 1992). Hence, this modification is 'neutral' as far as the charge separation is concerned. At the same time, it supports again that structural changes due to the exchange procedure are negligible. RCs containing [3-vinyl]-13^2-hydroxy-BChl *a* at sites B_{AB} gave heterogeneous kinetics (Finkele et al., 1992). At all wavelengths studied, the kinetics could be fit by four components, viz. 0.9, 3.5, 30 and 200 ps. Native reaction centers lack the 30 ps component. The amplitude of this component varied considerably in the 700–800 nm spectral range and showed an inverse relationship to the 3.5 ps one: The former was highest upon excitation at 776 nm, which is the absorption maximum of [3-vinyl]-13^2-hydroxy-BChl *a*, while the latter is maximum in the absorption region of BChl *a* at sites $B_{A,B}$, around 800 nm. The most straight-forward explanation is a heterogeneity of the preparation due to only partial exchange at the B_A site: One fraction consisted of (kinetically) native RCs (0.9, 3.5 and 200 ps), the other of modified RCs, in which the 3.5 ps component was replaced by a 30 ps one[9]. If one assumes that the 13^2-OH substituent alone has no influence on the kinetics (see above), then the tenfold decrease of the intermediate rate is the result of a replacement of the 3-acetyl- by a vinyl-substituent. In solution, the redox potential of the modified pigment is estimated to be 150 mV more negative (Watanabe and Kobayashi, 1991), one would then expect that the step leading to B_A-reduction becomes slower. However, such an effect would also be expected for a superexchange mechanism. A more direct evidence for a stepwise mechanism is that in exchanged RCs there is no fast component with a high amplitude: if this fast component (0.9 ps in native RC) were to precede the slow component, its amplitude should increase, whereas it is expected to decrease if this step is subsequent to the intermediate one. Since it is no longer observable in regions where the exchanged RC dominate the spectrum, this result is consistent with a stepwise electron transfer (Holzapfel et al., 1989; Finkele et al., 1990). There is some evidence for a kinetic component with $\tau \approx 4$ ps in the modified reaction centers, but this needs further work (Finkele, 1992).

The third detailed characterization has been done with RC in which BPhe *a* has been replaced by Phe *a* (Shkurupatov and Shuvalov, 1993; Arlt et al., 1993). Although its redox potential in solution is ≈160 mV more positive (Watanabe and Kobayashi, 1991), the kinetics are not much changed: The most significant difference is the increase of the 200 ps component to ≈270 ps, this time relates to the transfer of the electron from H_A to Q_A, and might reflect the changed redox potential. However, the changed redox potential has yet another, more profound, effect of raising the energy of the state $P^{+\cdot}H^{-\cdot}$ to a value close to that of $P^{+\cdot}B^{-\cdot}$. Because the kinetics of interconversion of these states are much faster than the decay of the former, this allows for thermal equilibration and a cumulative decay of the states. This equilibration does even extend to P*: it stays populated to ≈5% during the lifetime of $P^{+\cdot}H^{-\cdot}$, thus allowing an estimate of the energetic separation between P* and $P^{+\cdot}H^{-\cdot}$ as ≈400 cm^{-1}. Taken together, it appears that modification of pigments at the B-site(s) affect the kinetics of electron transfer much more profoundly than those at the H-site(s).

[9] This heterogeneity was supported by pigment analysis and the relatively high intensity at the 800 nm band, indicating the presence of >20% of BChl *a* at site B_A.

VI. Open Problems and Outlook

Although the method of pigment exchange has so far been tested only with a limited number of pigment modifications and RC species, it seems to be a valuable complement to site-directed mutagenesis, because it allows a direct manipulation of the pigments. Realizing the limited knowledge of bacteriochlorophyll chemistry, a major effort of future work will have to go into the synthesis and the collection of basic data of the modified pigments, not only in situ, but also in solution. (Hartwich, 1994; Hartwich et al., 1994; Feiler et al., 1994; Käss et al., 1994; Teuchner et al., 1994; and unpublished results referred to in the text). Only then can the effects in the native environment be fully evaluated. A potential which is hitherto hardly explored, is selective (isotope) labeling for magnetic resonance or vibrational spectroscopic work.

The exchange method is restricted so far to the monomeric BChl and BPhe, and it appears currently unclear if the principle can be extended to the primary donor as well. There is hope, however, that the recent advances in the folding of previously unwieldy proteins, will render this site accessible, too, in the foreseeable future. A further challenge will be the extension to other RC, in particular those from PS II, but also to the variety of other (B)Chl protein complexes.

Appendix: Updated Procedure for Pigment Exchange

The exchange procedure has been summarized by Struck and Scheer (1993). This is an update for asymmetric exchanges and exchanges with metallo-bacteriochlorophylls.

For replacement of bacteriochlorophylls in the $B_{A,B}$ sites, RC from *Rb. sphaeroides* R26 in TL-buffer (10 mM tris, 0. 1% LDAO, pH = 8) are incubated under Ar with a 10-fold molar excess of exogenous pigment in MeOH at 43 °C for 90 min. Recovery of RCs with exchanged chromophores is

30–50%. The concentration of MeOH should not exceed 10% of the volume of the reaction mixture. Under these reaction conditions, which are close to the melting point of the RCs, internal motions become larger and allow for equilibration between the external, modified pigments and the original ones bound to the protein.

A selective substitution at the B_A-site is possible when RC from the carotenoid-containing wild-type strain 2.4.1 of *Rb. sphaeroides* are used, and the temperature is raised to 47 °C in order to account for the increased stability of these RCs. The selectivity is due to the spheroiden(on)e, which is located asymmetrically and adjacent to the B_B-pocket. This pigment probably is responsible for the increased thermal stability of the RC, and seems to block the B_B site from exchange of the chromophore.

For an exchange at the B_B-site only, the starting material are RCs from the carotenoid-less R26 mutant. First BChl *a* is replaced symmetrically at the sites B_A and B_B with exogenous pigment under the standard exchange conditions. After purification, carotenoid (re)constitution was done with spheroidene dissolved to maximal concentration in diethylether and diluted with 3 parts (v/v) methanol. This solution was added to the RC (OD = 1 cm^{-1} at 865 nm) to a concentration of the organic solvent of 10% (v/v), and the mixture incubated for 50 min at 43 °C under Ar. By this (re)constitution, the B_B-pocket becomes blocked, and native BChl *a* can finally be re-exchanged into the B_A-pocket under the conditions described above for the wild-type strain 2.4.1.

For the exchange with some metallo-bacterio-chlorophylls (e. g. [Ni]-BChla) as well as for the replacement of (bacterio-)pheophytins into R26 RC, the reaction conditions were similar, but acetone was used instead of methanol, because of poorer solubility and/or pigment aggregation effects in the latter, and the reaction time was 1 h at 43.5 °C.

Protein and pigment analyses of modified RCs were done as detailed in Struck and Scheer (1993). Extraction of pigments is done on DEAE-columns with methanol as denaturant of the protein moiety and $CHCl_3$ as eluant for the pigments. Exhaustive extraction has to be ensured for quantitative pigment analysis.

Acknowledgments

Referenced work of the authors was supported by the Deutsche Forschungsgemeinschaft, Bonn (SFB 143 'Elementarprozesse der Photosynthese', TP A9). G. H. acknowledges a Ph. D. stipend Freistaat Bayern.

References

Allen JP, Feher G, Yeates TO, Komiya H and Rees DC (1987) Structure from the reaction center from *Rhodobacter sphaeroides* R26 The cofactors. Proc Natl Acad Sci USA 84: 5730–5734

Arlt T, Schmidt S, Kaiser W, Lauterwasser C, Meyer M, Scheer H and Zinth W (1993) Bacteriochlorophyll – a real electron carrier in primary photosynthesis. Proc Natl Acad Sci USA 90: 11757–11761

Beese D (1989) Untersuchungen zur Modifikation der Bacteriochlorophyll-Chromophore in bakteriellen photosynthetischen Reaktionszentren. Dissertation, Universität, München

Bocian DF, Boldt NJ, Chadwick BW and Frank HA (1987) Near-infrared-excitation resonance Raman-spectra of bacterial photosynthetic reaction centers—implications for path-specific electron-transfer. FEBS Lett 214: 92–96

Breton J (1985) Orientation of the chromophores in the reaction center of *Rhodopseudomonas viridis*. Comparison of low-temperature linear dichroism spectra with a model derived from X-ray crystallography. Biochim Biophys Acta 810: 235–245

Breton J, Martin JL, Petrich J, Migus A and Antonetti A (1986) The absence of a spectroscopically resolved intermediate state P^+B^- in bacterial photosynthesis. FEBS Lett 209: 37–43

Buchler JW (1975) Static coordination chemistry of metalloporphyrins. In: Smith KM (ed) Porphyrins and Metalloporphyrins, pp 157–232. Elsevier, Amsterdam

Bylina EJ and Youvan DC (1991) Protein-chromophore interactions in the reaction center of *Rhodobacter capsulatus*. In: Scheer H (ed) Chlorophylls, pp 705–722. CRC-Press, Boca Raton

Chang CH, El-Kabbani O, Tiede D, Norris J and Schiffer M (1991) Structure of the membrane-bound protein photosynthetic reaction center from *Rhodobacter sphaeroides*. Biochemistry 30: 5352–5360

Chen LX, Wang Z, Hartwich G, Katheder I, Scheer H, Tiede DM, Montano DA, Scherz A and Norris JR (1995) An X-ray absorption study of chemically modified bacterial photosynthetic reaction center. Chem Phys Lett 234: 437–444

Chisholm SW, Frankel SL, Goericke R, Olson RJ, Palenik B, Waterbury J B, West-Johnsrud L and Zettler ER (1992) *Prochlorococcus marinus* nov. gen. nov. spec.; an oxyphototrophic marine prokaryote containing divinyl chlorophyll *a* and *b*. Arch Microbiol 157: 297–300

Clayton RK and Yamamoto T (1976) Photochemical quantum efficiency and absorption spectra of reaction centers from *Rhodopseudomonas sphaeroides* at low temperature. Photochem Photobiol 24: 67–70

Coleman WJ and Youvan DC (1990) Spectroscopic analysis of genetically modified photosynthetic reaction centers. Ann Rev Biophys Biophys Chem 19: 333–367

DiMagno TJ and Norris JR (1993) Initial electron transfer events in photosynthetic bacteria. In: Norris JR and Deisenhofer J (eds) Photosynthetic Reaction Centers, pp 105–132. Academic Press, San Diego

Dürring M, Huber R, Bode W, Ruembeli R and Zuber H (1990) Refined three-dimensional structure of phycoerythrocyanin from the cyanobacterium *Mastigocladus laminosus* at 2.7 Å. J Mol Biol 211: 633–644

Dürring M, Schmidt GB and Huber R (1991) Isolation, crystallization, crystal structure analysis and refinement of constitutive C-Phycocyanin from the chromatically adapting cyanobacterium *Fremyella diplosiphon* at 1.66 Å resolution. J Mol Biol 217: 577–592

El-Kabbani O, Chang CH, Tiede D, Norris J and Schiffer M (1991) Comparison of reaction centers from *Rhodobacter sphaeroides* and *Rhodopseudomonas viridis:* Overall architecture and protein-pigment interactions. Biochemistry 30: 5361–5369

Engel N, Jenny TA, Mooser V and Gossauer A (1991) Chlorophyll catabolism in *Chlorella protothecoides*. Isolation and structure elucidation of a red bilin derivative. FEBS Lett 293: 131–133

Ermler U, Fritzsch G, Buchanan SK and Michel H (1994) Structure of the photosynthetic reaction center from *Rhodobacter sphaeroides* at 2.65 Å resolution: Cofactors and protein-cofactor interactions. Structure 2: 925–936

Feiler U, Mattioli TA, Katheder I, Scheer H, Lutz M and Robert B (1994) Effects of vinyl substituions on resonance Raman spectra of (bacterio)chlorophylls. J Raman Spectrosc 25: 365–370

Fenna RE, Ten Eyck LF and Matthews BW (1977) Atomic coordinates for the chlorophyll core of a bacteriochlorophyll *a*-protein from green photosynthetic bacteria. Biochem Biophys Res Comm 75: 751–756

Fiedor L, Rosenbach-Belkin V and Scherz A (1992) The stereospecific interaction between chlorophylls and chlorophyllase; possible implication to chlorophyll biosynthesis and degradation. J Biol Chem 267: 22043–22047

Finkele U, Dressler K, Lauterwasser C and Zinth W (1990) Analysis of transient absorption data from reaction centers of purple bacteria. In: Michel-Beyerle ME (ed) Reaction Centers of Photosynthetic Bacteria, pp 127–134. Springer Verlag, Berlin

Finkele U, Lauterwasser C, Struck A, Scheer H and Zinth W (1992) Primary electron transfer kinetics in bacterial reaction centers with modified bacteriochlorophylls at the monomeric sites B (AB). Proc Natl Acad Sci USA 89: 9514–9518

Frank HA (1993) Carotenoids in photosynthetic bacterial reaction centers: Structure, spectroscopy and photochemistry. In: Norris JR and Deisenhofer J (eds) Photosynthetic Reaction Centers, pp 221–239. Academic Press, San Diego

Frank HA, Chynwat V, Hartwich G, Meyer M, Katheder I and Scheer H (1993) Carotenoid triplet state formation in *Rhodobacter sphaeroides* R26 reaction centers exchanged with modified-bacteriochlorophyll pigments and reconstituted with sphaeroidene. Photosynth Res 37: 193–203

Fuhrhop JH, Kadish K and Davis DG (1974) Redox behavior of metallo octaethylporphyrins. J Amer Chem Soc 95: 5140–5145

Ghosh R, Hauser H and Bachofen R (1988) Reversible dissociation of the B873 light-harvesting complex from *Rhodospirillum rubrum* G9+. Biochemistry 27: 1004–1014

Göbel A (1993) Vergleichende Pigmentaustausche an Reaktionszentren aus Wildtyp 2.4.1 und der Mutante FM210 von *Rhodobacter sphaeroides*. Diplomarbeit, Universität München

Goericke R (1990) Pigments as ecological tracers for the study of the abundance and growth of marine phytoplankton. Ph.D. thesis, Harvard University, Cambridge

Goericke R and Repeta DJ (1992) The pigments of *Prochlorococcus marinus*: the presence of divinyl chlorophyll *a* and *b* in a marine prokaryote. Limnol Oceanogr 37: 425–433

Hartwich G (1994) Dynamik des Electrontransfers in Reaktionszentren photosynthatischer bakterien: Beeinflussung durch gezielten Pigmentaustansch. Dissertation, Technische Universität München

Hartwich G, Griese M, Ogrodnik A, Scheer H and Michel-Beyerle ME (1995a) Ultrafast loss channel competing with excitation energy transfer from the accessory bacteriochlorophyll in photosynthetic RCs. Biophys J 68: A367

Hartwich G, Cmiel E, Katheder I, Schäfer W, Scherz A and Scheer H (1995b) Transmetalated bacteriochlorophylls: Preparation and influence of metal and coordination on spectra. J Am Chem Soc, in press

Hartwich G, Scheer H, Aust V and Angerhofer A (1995c) Absorption and ADMR studies of bacterial photosynthetic reaction centers with modified pigments. Biochim Biophys Acta, in press

Helfrich M, Schoch S, Lempert U, Cmiel E and Rüdiger W (1994) Chlorophyll synthetase cannot synthesize chlorophyll *a*'. Eur J biochem 219: 267–273

Holzapfel W, Finkele U, Kaiser W, Oesterhelt D, Scheer H, Stilz HU and Zinth W (1989) Observation of a bacteriochlorophyll anion radical during the primary charge separation in a reaction center. Chem Phys Lett 160: 1–7

Iturraspe J and Gossauer A (1991) Dependence of the regioselectivity of photo-oxidative ring opening of the chlorophyll macrocycle on the complexed metal ion. Helv Chim Acta 74: 1713–1717

Iturraspe J, Engel N and Gossauer A (1994) Chlorophyll catabolism. 5. Isolation and structure elucidation of chlorophyll *b* catabolites in *Chlorella protothecoides*. Phytochem 35: 1387–1390

Käss H, Rautter J, Zweygart W, Struck A, Scheer H and Lubitz W (1994) EPR, ENDOR, and TRIPLE resonance studies of modified bacteriochlorophyll cation radical. J Phys Chem 98: 354–363

Katz JJ, Bowman MK, Michalski TJ and Worcester DL (1991) Chlorophyll aggregation: Chlorophyll-water micelles as models for in vivo long-wavelength chlorophyll. In: Scheer H (ed) Chlorophylls, pp 211–236. CRC Press, Boca Raton

Katz JJ, Shipman LL, Cotton TM and Janson TR (1978) Chlorophyll aggregation: Coordination interactions in Chlorophyll monomers, dimers and oligomers. In: Dolphin D (ed) The Porphyrins, Vol V, pp 402–458. Academic Press, New York

Kellog EC, Kolaczkowski S, Wasielewski NR and Tiede DM (1989) Measurement of the extent of electron transfer to the bacteriopheophytin in the M-subunit in reaction centers of *Rhodopseudomonas viridis*. Photosynth Res 22: 47–59

Kirmaier C, Holten D and Parson WW (1985) Picosecond photodichroism studies of the transient states in *Rhodopseudomonas sphaeroides* reaction centers at 5K-Effects of the electron-transfer on the 6 bacteriochlorin pigments. Biochim Biophys Acta 810: 49–61

Kirmaier C and Holten D (1993) Electron transfer and charge recombination reactions in wild-type and mutant bacterial reaction centers. In: Norris JR and Deisenhofer J (eds) Photosynthetic Reaction Centers, pp 49–70. Academic Press, San Diego

Kobayashi T, Straub KD and Rentzepis PM (1979) Energy relaxation mechanism in Ni(II), Pd(II), Pt(II) and Zn(II) porphyrins. Photochem Photobiol 29: 925–931

Kräutler B, Jaun B, Amrein W, Bortlik K, Schellenberg M and Matile P (1992) Breakdown of chlorophyll: Constitution of a secoporphinoid chlorophyll catabolite isolated from senescent barley leaves. Plant Physiol Biochem. 30: 333–346

Kühlbrandt W (1988) Structure of light-harvesting chlorophyll *a/b* protein complex from plant photosynthetic membranes at 7A resolution in projection. J Mol Biol 202: 849–864

Losev AP, Kryukshto ND, Kochubeeva KN and Solovev KN (1990) Triplet-singlet phosphorescence of complexes of bacteriopheophytin *a* with palladium and copper. Opt Spectrosc (USSR) 69: 97–110, Engl Transl 59–61

Lous EJ and Hoff AJ (1987) Exciton interactions in reaction centers of the photosynthetic bacterium *Rhodopseudomonas viridis* probed by optical triplet-minus-singlet polarization spectroscopy at 1.2 K monitored through absorbance-detected magnetic-resonance. Proc Natl Acad Sci USA 84: 6147–6151

Lubitz W (1991) EPR and ENDOR studies of chlorophyll cation and anion radicals. In: Scheer H (ed) Chlorophylls, pp 903–944. CRC-Press, Boca Raton

Lutz M and Mäntele W (1991) Vibrational spectroscopy of chlorophylls. In: Scheer H (ed) Chlorophylls, pp 855–902. CRC-Press, Boca Raton

Martin JL and Vos MH (1992) Femtosecond biology. Ann Rev Biophys Biomol Struct 21: 199–222

Meyer M (1992) Austauschversuche mit Bakteriophäophytin und Phäophytinen in die Bindungsstellen $H_{A,B}$ von Reaktionszentren aus *Rhodobacter sphaeroides* R26. Diplomarbeit, Universität München

Meyer M and Scheer H (1995) Reaction centers of *Rhodobacter sphaeroides* R26 containing c-3 acetyl- and vinyl(bacterio)-pheophytins at sites $H_{A,B}$. Photosynth Res, in press

Michel H and Deisenhofer J (1988) Relevance of the photosynthetic reaction center from purple bacteria to the structure of Photosystem II. Biochemistry 27: 1–7

Monger TB and Parson WW (1977) Singlet-triplet fusion in *Rhodopseudomonas sphaeroides* chromatophores. A probe of the organization of the photosynthetic apparatus. Biochim Biophys Acta 460: 393

Mühlecker W and Kräutler B (1993) Breakdown of chlorophyll: A tetrapyrrolic chlorophyll catabolite from senescent rape leaves. Helv Chim Acta 76: 2976–2980

Okamura MY, Isaacson RA and Feher G (1975) Primary acceptor in bacterial photosynthesis: Obligatory role of quinone in the photoactive reaction center of *Rhodopseudomonas sphaeroides*. Proc Natl Acad Sci USA 72: 3491–3495

Olson JM (1978) Bacteriochlorophyll *a* proteins from green bacteria. In: Clayton RK and Sistrom WR (eds) The Photosynthetic Bacteria, pp 161–178. Plenum Press, New York

Owens TG and Falkowski PG (1982) Enzymatic degradation of chlorophyll *a* by marine phytoplankton in vitro. Phytochemistry 21: 979–984

Parkes-Loach P S, Michalski T J, Bass W J, Smith U J and Loach PA (1990) Probing the bacteriochlorophyll binding site by reconstitution of the light-harvesting complex of *Rhodospirillum rubrum* with bacteriochlorophyll *a* analogues. Biochemistry 29: 2951–2960

Parkes-Loach P S, Sprinkle J.R. and Loach P A (1988) Reconstitution of the B873 light-harvesting complex of *Rhodospirillum rubrum* from the separately isolated α-and β-polypeptides and bacteriochlorophyll *a*. Biochemistry 27: 2718–2727

Parson WW and Monger TG (1976) Interrelationships among excited states in bacterial reaction centers. Brookhaven Symp Biol 28: 195–212

Paulsen H, Rümler U and Rüdiger W (1990) Reconstitution of pigment-containing complexes from light-harvesting chlorophyll *a/b*-binding protein overexpressed in *Escherichia coli*. Planta 181: 204–211

Paulsen H, Hobe S and Eisen C (1991) Reconstitution of LHCII in vitro from mutant LHCP and pigment analogs. In: Argyroudi-Akoyonoglu JH (ed) Regulation of Chlroplast Biogenesis, pp 343–348. Plenum Press, New York

Plumley F G and Schmidt G W (1987) Reconstitution of Chlorophyll *a/b* light-harvesting complexes—Xanthophyll-dependent assembly and energy-transfer. Biochim Biophys Acta 84: 146–150

Scharnagl C and Schneider S (1989) UV-visible absorption and circular dichroism spectra of the subunits of C-Phycocyanin. 1. Quantitative assessment of the effect of chromophore protein interaction in the α-subunit. J Photochem Photobiol B. 3: 603–614

Scharnagl C and Schneider S (1991) UV-visible absorption and circular dichroism spectra of the subunits of C-Phycocyanin. II. A quantitative discussion of the chromophore-protein and chromophore-chromophore interactions in the β-subunit. J Photochem Photobiol B 8: 129–157

Scheer H and Katz JJ (1975) Nuclear magnetic resonance spectroscopy of porphyrins and metalloporphyrins. In: Smith KM (ed) Porphyrins and Metalloporphyrins, pp 399–524. Elsevier, New York

Scheer H and Struck A (1993) Bacterial reaction centers with modified tetrapyrrole chromophores. In: Norris JR and Deisenhofer J (eds) Photosynthetic Reaction Centers, pp 157–191. Academic Press, San Diego

Scheer H, Meyer M and Katheder I (1992) Bacterial reaction centers with plant-type pheophytins. In: Breton J and Vermeglio A (eds) The Photosynthetic Bacterial Reaction Center: Structure, Spectroscopy and Dynamics, pp 49–57. Plenum Press, New York

Schenck CC, Mathis P and Lutz M (1984) Triplet formation and triplet decay in reaction centers from the photosynthetic bacterium *Rhodopseudomonas sphaeroides*. Photochem Photobiol 39: 407–417

Scherer POJ and Fischer SF (1991) Interpretation of optical reaction center spectra. In: Scheer H (ed) Chlorophylls, pp 1079–1096. CRC Press, Boca Raton

Scherz A, Rosenbach-Belkin V and Fisher JRE (1991) Chlorophyll aggregates in aqueous solution. In: Scheer H (ed) Chlorophylls, pp 237–268. CRC-Press, Boca Raton

Schirmer T, Bode W and Huber R (1987) Refined three-dimensional structures of two cyanobacterial C-phycocyanins at 2.1 and 2.5Å resolution—A common principle of phycobilin-protein interaction. J Mol Biol 196: 677–695

Senge MO, Gerzevske KR, Vicente MGH, Forsyth TP and Smith KM (1993) Modelle für das photosynthetische Reaktionszentrum: Synthese und Struktur von *cis* und *trans*-Ethen-verbrückten sowie gewinkelten Hydroxymethylen-verbrückten Porphyrindimeren. Angew Chem 105: 745–747

Shkurupatov A Y and Shuvalov V A (1993) Electron transfer in pheophytin *a*-modified reaction centers from *Rhodobacter sphaeroides* (R-26). FEBS Lett 322: 168–172

Shuvalov VA (1993) Time and frequency domain study of different electron transfer processes in bacterial reaction centers. In: Norris JR and Deisenhofer J (eds) Photosynthetic Reaction Centers, pp 89–104. Academic Press, San Diego

Struck A (1990) Chemisch modifizierte Bakteriochlorophylle und -pheophytine in den Bindungsstellen $B_{A,B}$ und $H_{A,B}$ von photosynthetischen Reaktionszentren aus *Rhodobacter sphaeroides* R26: Pigmentsynthese, Pigmentaustausch und Spektroskopie. Dissertation, Universität München

Struck A, Beese D, Cmiel E, Fischer M, Müller A, Schäfer W and Scheer H (1990) Modified bacterial reaction centers: 3. Chemically modified chromophores at sites B_A, B_B and H_A, H_B. In: Michel-Beyerle ME (ed) Reaction Centers of Photosynthetic Bacteria, pp 313–326. Springer, Berlin

Struck A, Müller A and Scheer H (1991) Modified bacterial reaction centers: 4. The borohydride treatment reinvestigated: comparison with selective exchange experiments at binding sites $B_{A,B}$ and $H_{A,B}$. Biochim Biophys Acta 1060: 262–270

Teuchner K, Stiel H, Leupold D, Katheder I and Scheer H (1994) From chlorophyll *a* towards bacteriochlorophyll *a*: Excited-state processes of modified pigments. J Luminesc 60161: 520–522

Vermeglio A, Breton J, Paillotin G and Cogdell R (1978) Orientation of chromophores in reaction centers of *Rhodopseudomonas sphaeroides:* A photoselection study. Biochim Biophys Acta 501: 514

Walter E, Schreiber J, Zass E and Eschenmoser A (1979) Bakteriochlorophyll a_{gg} und Bakteriophophytin a_p in den photosynthetischen Reaktionszentren von *Rhodospirillum rubrum* G9. Helv Chim Acta 62: 899–920

Watanabe T and Kobayashi M (1991) Electrochemistry of chlorophylls. In: Scheer H (ed) Chlorophylls, pp 287–316. CRC-Press, Boca Raton

Ziegler R, Blaheta A, Guha N and Schnegge B (1988) Enzymatic formation of pheophorbide and pyropheophorbide during chlorophyll degradation in a mutant of *Chlorella fusca* Shihira et Kraus. J Plant Physiol 132: 327–332

Zinth W and Kaiser W (1993) Time-resolved spectroscopy of the primary electron transfer in reaction centers of *Rhodobacter sphaeroides* and *Rhodopseudomonas viridis*. In: Norris JR and Deisenhofer J (eds) Photosynthetic Reaction Centers, pp 71–88. Academic Press, San Diego

Chapter 30

The Reaction Center from Green Sulfur Bacteria

Ute Feiler
Département de Biologie Cellulaire et Moléculaire, Section de Biophysique des Protéines et des Membranes, URA CNRS 1290, Centre d'Etudes de Saclay, F-91191 Gif sur Yvette Cedex, France

Günter Hauska
Institut für Pflanzenphysiologie und Zellbiologie, Universität Regensburg, Universitätsstr. 31, D-93040 Regensburg, Germany

R. E. Blankenship, M. T. Madigan and C. E. Bauer (eds): Anoxygenic Photosynthetic Bacteria, pp. 665–685.

Summary

Green sulfur bacterial reaction centers have been usually cited as counterparts of Photosystem I. However, they have remained poorly characterized for a long time, since the main interest was directed towards understanding the unique antenna system of green bacteria, the so-called chlorosomes. Many exciting developments in green sulfur bacterial reaction center research have occurred during the years 1990 to 1995. Important similarities between the reaction centers of green sulfur bacteria, heliobacteria, and Photosystem I have been discovered, such as the presence of the same set of electron acceptors and a similar polypeptide composition. Furthermore, similarities to the purple bacterial-type reaction centers have been described. These include the presence of a bacteriochlorophyll dimer as the primary donor and a reaction-center-associated cytochrome subunit as electron donor to P840. Most surprising, however, has been the discovery of only one gene coding for the dimeric, large reaction center subunits. This feature is in contrast, not only to Photosystem I, but also, to purple bacterial-type reaction centers, and it has raised interesting evolutionary questions which are discussed in this chapter. It has been shown that both types of reaction centers, i.e. purple bacterial- and Photosystem I -type reaction centers share common features, such as the charge separation processes, those pigments involved and a similar organization of the main polypeptides. All this led to the speculation that the different kinds of photosynthetic reaction centers have a common evolutionary origin. In this review we summarize the current state of knowledge about the green sulfur bacterial reaction center structure and function and compare these with those of other Photosystem I-type and purple bacterial reaction centers.

I. Introduction

The observation of a photosynthetic reaction center (RC) in green sulfur bacteria dates back to 1963, when Sybesma and Vredenberg reported photobleaching at 840 nm due to a photooxidizable primary donor, P840, for *Chloropseudomonas ethylicum* 2K. This 'organism' has since been identified as a syntrophic mixture of the sulfur-photooxidizing green sulfur bacterium *Prosthecochloris (Pc.) aestuarii* and the sulfur-respiring bacterium *Desulfovibrio* (Olson, 1978a). P840-photobleaching has subsequently been described for *Chlorobium (Cb.) limicola* (Olson et al., 1973), and other green sulfur bacteria, in pure culture.

The early work on green sulfur bacteria, however, was focused on two other components of their peculiar photosynthetic apparatus—the chlorosomes, which are huge antennae complexes (Cohen-Bazire et al., 1964; see Chapter 13 by Oelze and Golecki and 20 by Blankenship et al.), and a bacteriochlorophyll (BChl) *a*-protein (FMO-protein), which was the first chlorophyll protein structure to be resolved by X-ray crystallography (Mathews et al., 1979). Therefore little was known about the RC proper of green sulfur bacteria in the pertinent chapters of the comprehensive account on photosynthetic bacteria from 1978 (Gingras, 1978, Olson 1978b, Pierson and Castenholz, 1978). Two years later, Olson (1980) summarized the notion of the photosynthetic apparatus in green sulfur bacteria in more detail, and put forward a structural model for the participating BChl-proteins, which includes the studies by Swarthoff and Amesz (1979), and by Staehelin et al. (1980). Its essentials were: 1) The photosynthetic unit contains 900–4500 BChl *c*, *d*, or *e* molecules in the chlorosome, and 80 – 250 BChl *a* molecules per RC. 2) The BChl *a* molecules are bound in groups of seven on to 42 kDa-polypeptides, which build trimers, most of which form a baseplate between the chlorosome and the RC in the cytoplasmic membrane. 3) Each RC is attached to two BChl *a*-protein trimers. 4) The RC was speculated to be built of 5 BChl-proteins similar to the known 42 kDa BChl *a*-protein, and a core holding P840 and carotenoids.

Another central interest in the RC of green sulfur bacteria is its similarity to the RC of Photosystem (PS) I, which goes back to the finding that green sulfur bacteria are capable of pyridine nucleotide photoreduction without further energy requirement

Abbreviations: ADMR – absorption detected magnetic resonance; BChl – bacteriochlorophyll; BPh – bacteriopheophytin; *Cb.* – *Chlorobium*; Chl – chlorophyll; FeS – iron-sulfur; FMO – Fenna-Mathews-Olson; FT – Fourier transform; HPLC – high performance liquid chromatography; Li/SDS – lithium/sodium dodecyl sulphate; P700 – primary donor of Photosystem I; P800 – primary donor of heliobacterial reaction centers; P840 – primary donor of green sulfur bacterial reaction centers; *Pc.* – *Prosthecochloris*; PS – Photosystem; RC – reaction center

(Buchanan and Evans, 1969), which was substantiated by the identification of iron-sulfur (FeS)-centers as early electron acceptors (Knaff and Malkin, 1976; Jennings and Evans, 1977). After the stimulus that research on PS II obtained from the similarity to the well characterized RCs in purple photosynthetic bacteria (see Chapter 23 by Lancaster et al.), the prospect of elucidating PS I structure and function from its bacterial counterparts grew, inspiring intensive investigations (see below). The recent discovery that the RC of heliobacteria also resembles PS I adds to the value of this comparison (Nitschke et al., 1990b; Liebl et al., 1993; for recent reviews see Mathis, 1990; Nitschke and Rutherford, 1991; Nitschke and Lockau, 1993). Moreover, both green sulfur bacteria (Büttner et al., 1992a, Okkels et al., 1992, Xie et al., 1993) and heliobacteria (Liebl et al., 1993) recently became accessible to molecular biology. In the following we will attempt to compile these new developments, and to update our concept of the RC in green sulfur bacteria. For the similar RC of heliobacteria this will be done by J. Amesz in Chapter 31.

II. Composition of the Reaction Center

A. Different Reaction Center Preparations

The first attempts to purify the RC of Chlorobiaceae consisted in the mechanical separation of the chlorosomes from the cytoplasmic membranes, using the French press and sucrose density gradient centrifugation (Fowler et al., 1971; Olson et al., 1973; Schmidt, 1980). This is equivalent to the separation of the BChl *c*-, *d*- or *e*- from BChl *a*-containing fractions. The latter consisted of a photoactive BChl *a* RC and the 42 kDa BChl *a*-antenna protein. By treatment of such a preparation from *Cb. limicola* with guanidine-HCl, followed by gel filtration on Sepharose, the BChl *a*-protein could be removed, but photoactivity was diminished (Olson et al., 1976a). The residual RC fraction retained about half the BChl *a* and carotenoid, together with photooxidizable cytochrome *c* and photoreducible cytochrome *b*.

A more defined RC particle was subsequently obtained from *Pc. aestuarii* by means of Triton X-100 treatment and sucrose density gradient separation (Swarthoff and Amesz, 1979). It closely resembled the cytoplasmic membrane vesicles mentioned above, in pigment composition (about 75 BChl *a* molecules per RC) and photoactivity, and its size corresponded to about 600 kDa. Again by treatment with guanidine-HCl, keeping the incubation time short, a sub-particle of about 350 kDa and 35 BChl *a* /RC was prepared, which retained primary charge separation, however with a lower yield. A smaller sub-particle, of about 200 kDa, called the 'core complex' was later described, which had all the BChl *a*-antenna protein removed by treatment with lithium dodecyl sulphate (LiDS), but had lost its photochemical activity (Vasmel et al., 1983). The authors estimated the removal of about two 42 kDa BChl *a*-proteins (14 BChl *a*-molecules), about 20 BChl *a*-molecules remaining in the core complex.

Another extensively purified RC preparation from *Cb. limicola f thiosulphatophilum*, strain Tassajara was obtained with the combination of octylglucoside/cholate for solubilization (Hurt and Hauska, 1984). The BChl *a*-antenna, as well as cytochrome *b* were removed from this preparation. Photooxidation of P840 and of cytochrome *c* were still observable, but were unstable.

More recently, a procedure yielding more stable RC preparations from *Cb. limicola* and from *Cb. phaeobacteroides* was described, using dodecyl-maltoside treatment, and avoiding ammonium sulphate precipitation (Feiler et al., 1992). These preparations retained part of the BChl *a*-antenna protein, and showed light-induced reduction of FeS-centers F_A and F_B as measured by EPR (see III.B.3.). Following this method, the RC has also been purified from *Cb. vibrioforme* (Okkels et al., 1992), with emphasis on the characterization of an electron donating cytochrome *c*-551 (see III.C.).

Meanwhile two more photoactive RC preparations which reveal photoreduction of FeS-centers have been reported, one from *Cb. limicola f thiosulphatophilum*, strain Larsen, using octylglucoside/cholate in the presence of cysteine (Oh-oka et al., 1993), and one from *Cb. tepidum* using Triton X-100 in the presence of dithiothreitol (Kusumoto et al., 1994). Furthermore, the presence of photoreducible FeS centers in the RC preparation of *Cb. vibrioforme* (Okkels et al., 1992) has been observed (Kjaer et al., 1994).

B. Polypeptide Composition

The isolation procedure of Hurt and Hauska (1984) was optimized for obtaining an RC complex with the minimum polypeptide composition necessary for primary charge separation. Three polypeptides of 65, 32 and 24 kDa M_{app}'s in SDS-PAGE, together with a small amount of residual BChl *a*-antenna protein of 42 kDa were present. Actually the BChl *a*-antenna protein is partially resolved from the other polypeptides on sucrose density gradients, and fractions totally devoid of it can be obtained (Büttner et al. (1992a), demonstrating that it is not required for the primary charge separation. Nevertheless, the 42 kDa BChl *a*-antenna-protein seems to stabilize the RC, and is present in excess in the more stable, photoactive preparations (Feiler et al. 1992; Okkels et al., 1992; Oh-oka et al., 1993). Following the more detailed picture of Olson (1980) we would state, that in the membrane the RC is attached to two trimers of the 42 kDa BChl *a*-protein ($2 \times 3 \times 7 = 42$ BChl *a* molecules), which can be removed by guanidine-HCl (Swarthoff and Amesz, 1979), plus two additional BChl *a* proteins ($2 \times 7 = 14$ BChl*a* molecules), which can be removed by LiDS (Vasmel et al., 1983), or by octylglucoside/cholate (Hurt and Hauska, 1984) treatment.

The 65 kDa polypeptide (Hurt and Hauska, 1984), observed already in the above-mentioned 'core complex' (Vasmel et al., 1983), in the more recent RC preparations is seen at 50 kDa (Feiler et al., 1992), 68 kDa (Oh-oka et al., 1993) or 80 kDa (Okkels et al., 1992). This variation may be caused by different SDS-PAGE conditions. In all cases this polypeptide migrated as a diffuse band, characteristic of hydrophobic proteins, at a similar R_f-value to that of the large subunits of PS I (see Golbeck and Bryant, 1991). For the 65 kDa polypeptide, it was additionally shown that it retained BChl during LiDS-PAGE (Hurt and Hauska, 1984), further, it was not fully unfolded in SDS-PAGE, yielding a green band at about 110 kDa and was easily aggregated by boiling in SDS sample buffer (Hurt and Hauska, unpublished). All this is similar to what is observed for the PS I large subunits. The resemblance of the diffuse 65 kDa band to the large subunits of PS I, binding the primary donor and the early acceptors, has been confirmed by comparison of the genes (see IV.A.).

A heme-staining polypeptide was observed in most of the described RC preparations. The 24 kDa polypeptide (Hurt and Hauska, 1984) represents the photooxidizable cytochrome *c*-550.5, and corresponds to the 21 kDa polypeptide of Oh-oka et al. (1993). Probably, both resemble the 18 kDa cytochrome *c* in the preparation from *Cb. vibrioforme*, which has been characterized via the gene sequence (Okkels et al., 1992). A cytochrome *c*-553 has been separated from the RC in the purification procedure of Hurt and Hauska (1984). In contrast, the preparation by Feiler et al. (1992) keeps the 32 kDa cytochrome *c*-553 bound and removes the 24 kDa cytochrome *c*-550.5. In a recent study on the isolated 32 kDa cytochrome *c*-553 it was shown that it is a multihemic cytocrome (D. Albouy, W. Nitschke, B. Robert and U. Feiler, unpublished). A cytochrome *c*-553 was already described as the immediate electron donor to P840 in *Chloropseudomonas* by Fowler et al. (1971), and in *Chlorobium* by Prince and Olson (1976). Thus the cytochrome *c*-553 seems to be the electron donor and as to EPR studies on membranes it is possible that it corresponds to a tetraheme cytochrome (Feiler et al., 1989; see Section III.C. and Chapter 36 by Nitschke and Dracheva).

A subunit of about 32 kDa, formerly ascribed to the Rieske protein (Hurt and Hauska, 1984), has recently been identified by immunoblotting as the protein carrying the FeS centers F_A and F_B (Illinger et al., 1993). Therefore, all the RC preparations exhibiting photoreduction of the terminal acceptors, the FeS center F_A and F_B (see III.B.3.), should contain a respective subunit. A 32 kDa polypeptide is indeed present in the preparation described by Okkels et al. (1992) and Oh-oka et al. (1993), but in substoichiometric amounts. The heme-positive 32 kDa band on SDS-PAGE of the preparation of Feiler et al. (1992) might, then, either represent two proteins, cytochrome *c*-553 and the F_A/F_B binding protein or the F_A/F_B binding protein is migrating on SDS-PAGE at different apparent molecular weight.

Minor heme-negative polypeptides of 22, 15 and 12 kDa, or of 15, 9 and 6 kDa are present in the preparations of Feiler et al. (1992) and Okkels et al. (1992), respectively. The preparation by Oh-oka et al. (1993) contains only one small polypeptide of about 12 kDa, together with a number of larger polypeptides, which probably represent contaminations. The possible role of the small polypeptides remains to be established.

The minimal number of subunits in a stable and photoactive *Cb. limicola* RC preparation included the P840 protein (pscA), the 42 kDa BChl *a* antenna protein (FMO), the FeS protein (pscB) and a fourth subunit of 17kDa (pscD). The presence of PscC, the cytochrome *c*-551, was not required. The gene for PscD was sequenced and codes for a polar, basic

16.5 kDa protein, which showed no sequence homologies, but could still functinally resemble PsaD of PSI (C. Hager-Braun, D.-L. Xie, U. Jarosch, E. Herold, M. Büttner, R. Zimmermann, G. Hauska and N. Nelson, unpublished).

C. Pigment Composition

All green sulfur bacteria contain BChl *a* in their RC complex. Characteristic absorption bands of BChl*a* are the Q_y transition band at ca. 810 nm, the Q_x transition band at ca. 610 nm and the Soret band at ca. 370 nm (in vivo; Olson, 1980). In the case of the primary donor, which is most likely a BChl *a* dimer (see III.A.), the Q_y transition band is shifted to ca. 840 nm.

Figure 1 shows the absorption spectrum of the RC complex of *Cb. limicola f thiosulphatophilum* isolated following the procedure of Feiler et al. (1992). We can estimate the pigment content of the RC complex from this spectrum: At 340, 370, 603 and 810 nm, the characteristic absorption bands of BChl*a* and, additionally, the shoulder at 838 nm indicating the primary donor, are seen. The ratio of BChl *a* to P840 was about 40:1 in this preparation. Considering the model of Olson (1980) discussed above, this ratio might be interpreted as implying two BChl *a*-antenna proteins are attached to the RC complex, while the remaining BChl*a*' s are bound to the RC core.

In Fig. 1, the insert shows part of the absorption spectrum of a preparation enriched in the RC core which contained only about 25 BChl *a* per P840 (Vasmel et al., 1983; Hurt and Hauska, 1984). Characteristic for a RC core complex, from which the BChl *a*-antenna protein is separated, is the shift of the absorption maximum of the Q_y and Q_x transitions of BChl *a*. The absorption maximum of the Q_y transition is now at 814 nm and the shoulder arising from P840 is more pronounced. Additionally, the absorption maximum of the Q_x transition is at 597 nm and shows, like the Q_y transition, an asymmetry possibly reflecting the presence of a heterogeneous population of BChl *a*. The Q_y transition of BChl *a* shown in Fig. 1 are similar to those reported for *Pc. aestuarii* (Swarthoff and Amesz, 1979; Vasmel et al., 1983).

Low-temperature absorption spectra of purified membrane particles resolved several BChl *a* Qy bands arising from exciton interactions within BChl*a* aggregates. Peaks at 805, 816 and 823–824 nm were present in RC containing preparations as well as in the BChl *a*-antenna protein, while 1 or 2 peaks between 833 and 836 nm, due to the primary donor were additionally present in the former only (Swarthoff and Amesz, 1979; Olson, 1980; Miller et al., 1994). Likewise, the CD spectra of the same samples were comparable in both band sign and band position. From these similarities in absorption and CD spectra of the two BChl*a*-containing proteins, it was concluded that the exciton interactions and thus, presumably, the organization of the BChl *a*'s, present in the BChl *a*-antenna protein and in the RC-containing membranes, are similar (Olson, 1980).

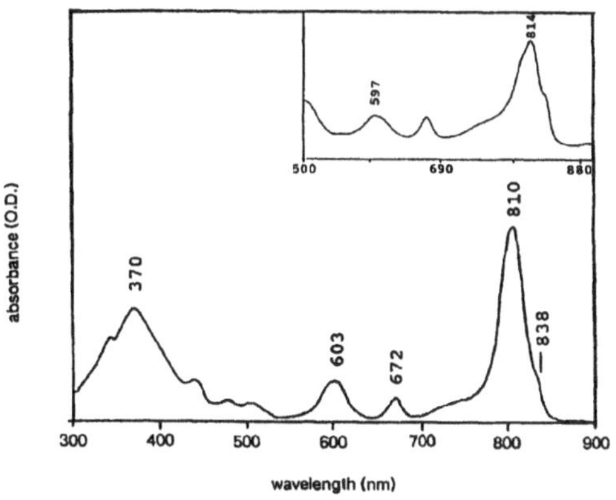

Fig. 1. Room temperature absorption spectrum of the photoactive reaction center complex of *Chlorobium limicola f thiosulphatophilum*. The isolation procedure of Feiler et al. (1992) was used. The insert shows the reaction center core-complex isolated following the procedure of Hurt and Hauska (1984).

The absorption band at 672 nm (see Fig. 1) arises, at least partially, from the primary acceptor pigment, called BChl-663, which is a Chl *a*-molecule (van de Meent et al., 1992; Feiler et al., 1994; see III.B.1.). No BChl *c* or bacteriopheophytin (BPh) *c* contributions could be detected by means of resonance Raman spectroscopy in the preparation of Feiler et al. (1992,1994). However, in earlier RC preparations of *Chlorobium* and *Prosthecochloris* (see II.A.) both BChl *c* and BPh *c* contributed to absorption bands at 670 nm. As detected by low-temperature absorption and CD spectroscopy (Olson, 1980), these contributions are most likely due to residual BChl *c* and BPh *c* presumably formed from chlorosomal BChl *c*.

Absorption bands in the wavelength range from 440 to 505 nm arise from the carotenoids' rhodopin and/or hydroxychlorobactene (Braumann et al., 1986). A ratio of ca. 2 carotenoids per P840 can be calculated for the spectrum in Fig. 1.

In the RC preparations, *c*-type hemes are present associated with membrane-bound cytochrome(s) (Hurt and Hauska, 1984; Feiler et al. 1992, Okkels et al., 1992; Oh-oka et al., 1993; see III.C.). However, the stoichiometry and the exact nature of these cytochromes associated with the RC is not yet clear (see III.C.). The described *b*-type cytochromes in membranes (Fowler, 1974) do not copurify with the RC.

III. Electron Transfer Within the Reaction Center

A. The Primary Electron Donor P840

1. Optical Properties

Photobleaching of BChl *a* at around 840 nm (P840) was first reported in whole cells of *Chloropseudomonas ethylicum* 2K (Sybesma and Vredenberg, 1963), and has since been observed in both membranes and isolated RC preparations of all green sulfur bacteria studied. This photobleaching is accompanied by additional bleachings at ca. 830 nm and, weaker, at ca. 790 nm (Fowler et al., 1971; Olson et al., 1973; Olson et al., 1976b; Swarthoff and Amesz, 1979). The origin of these additional signals is not yet clear. These bleachings have also been observed in chemically-induced oxidized-minus-reduced difference spectra (Fowler et al., 1971; Olson et al., 1973; Feiler et al., 1992; Miller et al., 1992; Oh-oka et al., 1993).

In light-minus-dark CD difference spectra, only the band at ca. 840 nm was visible (Olson et al., 1973). This might be interpreted as evidence that the two additional bleaching bands at 830 and 790 nm, seen in the absorption difference spectra, arise from monomeric pigments which have very little rotational strength.

Two bands at around 830 and 840 nm, were also observed in low-temperature triplet-minus-singlet absorbance difference spectra of RCs from *Pc. aestuarii* (Vasmel et al., 1984). It was concluded from LD-absorbance detected magnetic resonance (ADMR) spectroscopy experiments that these two transitions are approximately parallel. They, thus, cannot arise from an excitonically split transition, as the transition moments of such an exciton pair are mutually perpendicular. These two bands therefore do not belong to the same molecular species (e.g. a special pair). A possible explanation is, that only one of the bands, presumably the one at 840 nm, reflects the primary donor (the low energy transition of a strongly exciton coupled, CD-active dimer) while the second band at around 830 nm is arising from electrochromic changes that occur in neighbouring, monomeric pigments.

More specifically, this transition at 830 nm may arise from a special pigment, a so-called 'voyeur'-BChl, as found in purple bacterial RCs (for a review see Kirmaier and Holten, 1987). This is supported by observations of chemical-induced absorption difference spectra of *Chlorobium* RC complexes (Feiler et al., 1992; Okkels et al. 1992); no negative band at ca. 790 nm was observed in the purified RCs. The difference spectra in this region could be interpreted as arising from the shift of the transition of a pigment absorbing at ca. 815 nm resulting in a negative band at ca. 830 nm and a positive band at ca. 800 nm. This is reminiscent of the situation in the purple bacterial RC, where oxidation of the primary donor causes an electrochromic shift of the 'voyeur'-BChl absorption band near 800 nm.

An absorption band at 1157 nm, attributed to $P840^{+}$, was observed in both light-induced and chemical-induced absorption difference spectra (Olson et al., 1976b; U. Feiler, T. A. Mattioli and B. Robert, unpublished, respectively). Like in the purple bacterial RC (for a review see Kirmaier and Holten, 1987) the P^{+} band of green sulfur bacterial RC shows a large red-shift when compared to monomeric BChl $a^{+\cdot}$ which absorbs near 900 nm (Fajer et al., 1975). This suggests that in green sulfur bacteria the primary donor is a dimer of BChl *a*, or more precisely it keeps its dimeric character in the P^{+} state, as in the purple bacterial RCs.

2. Redox Potential

Measurement of the magnitude of light-induced bleaching of P840 as a function of applied redox potential yielded a titration curve well described by a one-electron Nernst curve with a midpoint potential of +240 mV (Fowler et al., 1971; Prince and Olson, 1976). This potential is ca. 150 mV lower than those of the primary donors in either purple bacteria (see Kirmaier and Holten, 1987) or PS I of plants and cyanobacteria (Golbeck and Bryant, 1991). This decrease in the redox potential is also observed in other redox carriers of Chlorobiaceae (Rieske-protein,

Knaff and Malkin, 1976; cytochrome *c*-555, Meyer et al., 1968) and seems to be an adaptation to the reducing conditions in which they live.

3. Structural Studies

Magnetic studies have given valuable insight into the delocalization of unpaired electrons over the π-system of P840$^+$ and hence the structure of the primary donor. The EPR lineshape of P840$^+$, studied by the second moment method in ^{13}C-enriched *Chlorobium*, demonstrated that the unpaired electron is delocalized over two macrocycles (Wasielewski et al., 1982). Furthermore, the 9–10 Gauss EPR linewidth of P840$^+$ (Olson et al., 1976b; Swarthoff et al., 1981a; Feiler, 1991) is narrowed compared to that of monomeric BChl a^+ (12.8 Gauss) by a factor of ca. $1/\sqrt{2}$, as in *Rhodobacter sphaeroides* (Norris et al. 1971), and has also been attributed to a delocalization of the unpaired electron over two BChl *a* molecules.

Additionally, the triplet state of P840 has been studied by both EPR and ADMR spectroscopy. The zero field splitting parameters of ^{T}P840 (Swarthoff et al., 1981a; Vasmel et al., 1984; Nitschke et al., 1990a) are similar to those of the primary donor triplet signal in purple bacterial RC (Dutton et al., 1971). Moreover, the orientation of the ^{T}P840 EPR signal (Nitschke et al., 1990a) is very similar to that found for ^{T}P in purple bacterial RCs (Hales and Gupta, 1979; Tiede and Dutton, 1981).

These studies together with the observation of the P840$^+$ band at 1157 nm of Olson et al. (1976) (see III.A.1.) have shown that the primary donor in green sulfur bacterial RCs is most likely a BChl *a* dimer with a similar geometry to that of the purple bacterial RC.

Analysis of the unpaired electron spin distribution of P840$^{+\bullet}$ using electron nuclear double resonance (ENDOR) and special TRIPLE spectroscopy was perfomed by Rigby et al. (1994). A highly symmetrical distribution of the electron spin density between the two BChl *a* molecules of the special pair was shown, indicating a symmetrical protein environment of the special pair. This is consistent with the proposal of a homodimeric photosynthetic RC (see IV).

Recently, near-infrared Fourier transform (FT) Raman spectroscopy (excited at 1064 nm) has been performed on the isolated RC (U. Feiler, T. A. Mattioli and B. Robert, unpublished). The FT resonance Raman spectrum of the *Chlorobium* RC in the P$^+$ state exhibits a band at 1707 cm^{-1} which is significantly lower than its counterpart in the P$^+$ spectrum from BChl *a*-containing purple bacterial RCs (ca. 1717 cm^{-1}). This 1707 cm^{-1} band arises from the C$_9$-keto carbonyl stretching mode of the P$^+$ species which is upshifted from 1690 cm^{-1}, observed in the P^o FT Raman spectrum of ascorbate-reduced *Chlorobium* RCs. This 17 cm^{-1} upshift of a free keto carbonyl stretching frequency has been interpreted as indicating a nearly complete and symmetric delocalization of the residual unpaired electron over both BChl *a* molecules in the P$^+$ state on the timescale of the resonance Raman effect (10^{-13}s). In the RC of *Rhodobacter sphaeroides* the corresponding upshift is 26 cm^{-1} which indicates that nearly the full +1 charge is carried by one of the two dimeric BChl *a* molecules (Mattioli et al., 1991). Therefore the delocalization of the +1 charge in the P$^+$ state over two BChl *a* molecules in the *Chlorobium* primary donor is markedly different from the situation in purple bacteria, and supports the suggested homodimeric structure of the RC, which was based on the observation of a sole RC gene (see IV).

B. The Electron Acceptors

1. The Primary Electron Acceptor

In light-minus-dark absorbance difference spectra Swarthoff and Amesz (1979) observed a negative band at 660 nm and a positive band at 678 nm which they proposed to be caused by an electrochromic red shift of a pigment absorbing at 670 nm, possibly a BPh *c*. Since, in the two different RC preparations used (see I.A.), this pigment was present in an approximately constant ratio to P840, they suggested that it might be a functional component of the RC complex. Later, flash-induced absorption changes at around 670 nm were attributed to reduction of a primary acceptor molecule, variously identified as BChl *c* or BPh *c* (van Bochove et al., 1984; Nuijs et al., 1985; Shuvalov et al., 1986).

Evidence of BPh *c* as primary acceptor was based on absorption and CD spectrophotometry data (Olson, 1981) and LD spectroscopy data (Vasmel et al., 1983). From comparison of the absorption spectra of both BChl *c* and BPh *c* in organic solvents with absorption spectra of different RC preparations it was concluded that the observed 670 nm pigment is mainly BPh *c*. CD data showed that only some of the BPh *c* molecules were optically active. Furthermore, a correlation between distorted CD and LD signals of the 670 nm pigment and the loss of photochemical activity of the RC was shown.

Thin layer chromatography showed, besides BChl *a* and carotenoids, small amounts of BPh *c* and an unidentified pigment (named P-665), with an absorption spectrum similar to that of BPh *c*, were present in *Pc. aestuarii* membranes (Swarthoff et al., 1982). Further analysis of the pigment composition of various pigment-protein complexes from the same bacterium by reversed-phase high-performance liquid chromatography (HPLC) (Braumann et al., 1986), demonstrated the presence of a special pigment with absorption maxima at 663 and 433 nm in these complexes. This pigment, labelled BChl-663, was more lipophilic than BChl *c* and had phytol as the esterified alcohol rather than the *trans-trans* farnesol of BChl *c*. A ratio of 10–15 molecules BChl-663 per RC was calculated. Since this pigment absorbs at 670 nm in vivo, a wavelength where absorption changes associated with the primary acceptor reduction were observed (see above and III.C.), it was proposed that BChl-663 is the primary acceptor. BPh *c* which may also absorb at 670 nm was absent from the pigment protein complexes studied. This work was extended by van de Meent et al. (1992) who showed, by normal-phase HPLC, that BChl-663 was present, at a ratio of 3–9 molecules/RC, in five species of green sulfur bacteria studied. By using plasma desorption mass spectroscopy, NMR and absorption spectroscopy they concluded that BChl-663 was a Chl *a*-isomer.

More direct evidence for the role of a Chl *a*-like pigment as the primary acceptor in situ was provided by a resonance Raman study of isolated, photoactive, RCs (Feiler et al., 1994). This work demonstrated not only the presence of a Chl *a*-like pigment in highly purified RCs but also that this molecule was bound to the RC protein in a manner very similar to the primary acceptors of both purple bacterial RCs and PS II (see V.A.).

Therefore, it appears that the primary acceptor in green sulfur bacterial RCs is a Chl *a*-like pigment, as it is likely in PS I (Golbeck and Bryant, 1991) and in the heliobacteria (see Chapter 31 by Amesz). We will, thus, refer to it as A_0 below.

2. The Secondary Electron Acceptor

The first evidence for the presence of a quinone in the RC of green sulfur bacteria was given by Nitschke et al. (1987) where a low-temperature EPR signal in the *Chlorobium* RC similar to that of A_1 in PS I (Mansfield et al., 1987; Mansfield and Evans, 1988) was observed. However, a controversy has since arisen as to the origin of the low-temperature EPR signal of A_1 in PS I (Ziegler et al., 1987; Barry et al., 1988; Heathcote et al., 1993). The experimental conditions necessary to induce the triplet EPR-signal, and thus to reduce the secondary acceptor, were similar in *Chlorobium* and PS I. This provides indirect evidence for the secondary acceptor behaving like a quinone in the *Chlorobium* RC (Nitschke et al. 1990a). In a preliminary study of electron spin polarized EPR spectroscopy performed on *Chlorobium* membranes, an emission/absorption signal was observed, resembling the spectra of purple bacterial RCs in which the rate of electron transfer from BPh (H_M) to quinone (Q_A) was reduced. This can be understood in terms of sequential electron transfer in green sulfur bacteria (Snyder and Thurnauer, 1993).

A preliminary study of the lipid composition of *Chlorobium* RCs by thin-layer-chromatography (Feiler and Hauska, unpublished) demonstrated the presence of a quinone-like component, probably 1'-hydroxy-menaquinone-7 which is one of the three menaquinones identified in whole cells (Powls and Redfearn, 1969). Furthermore, in the same study, a quinone component was observed by redox difference spectroscopy.

Although one could expect a quinone participating in the electron transport chain as in PS I (Hauska, 1988), involvement of such a quinone as secondary acceptor in the RC complex remains unclear and thus further experiments are necessary to clarify the identity of the secondary acceptor. In this context it is noteworthy that the electron transport in the RC of heliobacteria is not affected by removal of extractable menaquinone (Kleinherenbrink et al. 1993 and Chapter by 31 Amesz).

3. The Iron-Sulfur Centers as Electron Acceptors

Early EPR data on *Chlorobium* membranes suggested that FeS centers are involved in primary electron transfer processes as RC bound acceptors. However, the early results were inconsistent: Jennings and Evans (1977) observed a signal at g = 1.90 which was photoinduced at 4 K and its midpoint potential was estimated as –550 mV. They associated this light-induced signal with the low g-value component, previously reported by Knaff and Malkin (1976) who originally attributed it to a Rieske FeS center, and suggested that the +160 mV midpoint potential, determined by Knaff and Malkin, was artifactual. Re-examination (Knaff et al., 1979) confirmed the original assignment of the g = 1.90 signal to a Rieske FeS center with the E_m = +160 mV, and challenged the existence of the photoreducible signal.

Swarthoff et al., (1981b) reported the EPR spectrum of an FeS center with g-values of g_y = 1.94 g_x = 1.89, and determined the midpoint potential to be more positive than −420 mV. They attributed this signal to the terminal acceptor of the RC. A sample poised at −420 mV and frozen during illumination gave rise to an additional FeS center signal at g_y = 1.94 and g_x = 1.88. This observation is in line with earlier redox titration experiments (Knaff and Malkin, 1976) that described an FeS center with a g-value of g = 1.94 and a redox potential of about −550 mV. Swarthoff and co-workers suggested that the latter FeS center was an earlier electron acceptor with an E_m of ca. −560 mV functioning prior to the terminal acceptor of the RC (Swarthoff et al., 1981b).

Nitschke et al. (1987) showed that, at 200 K in the presence of dithionite at high pH an FeS center could be photoreduced. The EPR signal obtained in this study was similar to the photoinduced FeS signal of Swarthoff et al. (1981b). However, it was still not clear how many FeS centers are involved in photoinduced electron transport within the RC.

A clarifying re-examination by EPR spectroscopy of *Chlorobium* membranes (Nitschke et al., 1990a) finally showed the presence of three different photoreduced FeS centers involved in electron transfer processes in the RC of green sulfur bacteria. One FeS center ($g_{z,y,x}$–values: 2.07, 1.91, 1.86) was stably photoreduced at 4 K and is probably identical with the 1.90 signal reported by Jennings and Evans (1977). At 200 K an additional FeS center was photoaccumulated ($g_{z,y,x}$–values: 2.05, 1.94, 1.88) whose spectrum resembles that published by Swarthoff et al. (1981b) when the chemically reduced sample was illuminated during freezing. A further signal was obtained after prolonged illumination at 200 K, identified by a g_x = 1.78 and assigned to a third FeS center. Based on the similarities of their g-values and orientations to those of the PS I FeS centers they have been designated as F_B, F_A and F_X, respectively. If this assignment is warranted, the order of low temperature photoreduction of the FeS centers in green sulfur bacterial RCs (i.e. first F_B, second F_A) is reversed compared to PS I (first F_A, second F_B) under the same conditions. FeS centers in RCs of *Cb. limicola f thiosulphatophilum* are not chemically reducible under these conditions (i.e. 20 mM dithionite, 200 mM glycine pH 11) and thus should have lower redox potentials than ca. −600 mV. A recent report, however, indicates that in the thermophilic *Cb. tepidum* the redox potentials of the FeS centers (as well as that of P840) are slightly more positive, rendering them reducible by dithionite at high pH (Kusumoto et al., 1994).

Both the FeS centers F_A and F_B have been shown to be present in an isolated RC complex (Feiler et al., 1992), and since, the presence of photoreducible FeS centers in *Chlorobium* RCs has been confirmed by several groups (Miller et al., 1992; Oh-oka et al., 1993; Kusumoto et al., 1994; Kjaer et al., 1994).

Preliminary data using electron spin polarized EPR spectroscopy at 283 K have been obtained on *Chlorobium* membranes (U Feiler, W Nitschke, D Stehlik and C Bock, unpublished). An emission/absorption spectrum was observed. Based on the similarities of its g-values and emission/absorption characteristics to those in comparable spectra of PS I, it might be arising from the radical pair P^+FeS^-. However, a clear interpretation of these data is not possible yet.

The putative binding elements for the FeS center F_X are conserved in RCs of *Chlorobium*, PS I and heliobacteria (Golbeck and Bryant, 1991; Büttner et al., 1992a; Liebl et al., 1993). A gene related to the *psaC* gene of PS I, which codes for the protein binding the FeS centers F_A and F_B (Büttner et al., 1992a; see IV), as well as the corresponding protein (Illinger et al., 1993; see II.B.), is also found in *Cb. limicola f thiosulphatophilum*

4. Ferredoxin and Ferredoxin-NAD(P)-Reductase

Green sulfur bacteria are capable of direct photoreduction of $NAD(P)^+$ via ferredoxin without additional energy requirement (i.e. an electrochemical proton gradient) (Buchanan and Evans, 1969). The light-induced $NAD(P)^+$ reduction is insensitive to uncoupling chemicals added in order to collapse the proton gradient (Knaff, 1978). Both soluble ferredoxin and the flavoprotein ferredoxin-NAD(P)-reductase (FNR) have been isolated (Rao et al., 1969; Buchanan et al., 1969; Kusai and Yamanaka, 1973; Shioi et al., 1976). The amino acid sequence of *Chlorobium* ferredoxin (Tanaka et al., 1974) indicates that it is a protein containing two 4Fe4S-clusters, which is in contrast to the 2Fe2S ferredoxin in PS I. The *Chlorobium* ferredoxin is proposed to be a link between the ferredoxins of non-photosynthetic bacteria and those of photosynthetic bacteria of the *Chromatium*-type (Buchanan et al., 1969; Rao et al., 1969). The FNR catalyses the reduction of $NAD(P)^+$ by photoreduced ferredoxin (Kusai and Yamanaka, 1973; Shioi et al., 1976). Both proteins were proposed to be involved in transferring electrons from the bound FeS centers to $NAD(P)^+$ similar to the respective process in PS I (for a review see Knaff and Hirasawa, 1991).

C. The Electron Donor to P840

As early as 1971, a membrane-bound cytochrome *c* with an α-band at 553 nm was proposed to act as electron donor to the photooxidized primary electron donor P840 (Fowler et al., 1971). This was confirmed by light induced kinetic measurements on membrane fragments showing that $P840^+$ is rapidly (biphasically, $\tau < 7$ μs and 70 μs) rereduced by cytochrome *c*-553 (Prince and Olson, 1976). In whole cells, the rereduction of $P840^+$ by cytochrome *c*-553 is seen to be monophasic with $\tau = 6.9$ μs (U Feiler and W Nitschke, unpublished).

Recently, however, different RC preparations have been reported which have triggered a controversy as to the nature of the cytochrome serving as immediate electron donor to photooxidized P840. A cytochrome *c* with an α-band at 553 nm and a M_{app} of 32 kDa copurifies with the RC proteins in the RC preparation of Feiler et al. (1992). Based on EPR data on membranes and partially isolated cytochrome (Feiler et al., 1989; Feiler, 1991) this cytochrome *c*-553 was proposed to contain four heme groups like the tetraheme cytochromes associated with the RCs of purple bacteria and *Chloroflexus* (see Chapter 36 by Nitschke and Dracheva). A ratio of 4 hemes per RC was determined in solubilized membranes and the isolated RC complex still showed a stoichiometry of 2 hemes/RC (Feiler et al., 1992). Recently this cytochrome *c*-553 was isolated. Its characterization by EPR spectroscopy showed that this cytochrome is multihemic (D. Albouy, W. Nitschke, B. Robert and U. Feiler, unpublished). By contrast, in a preparation reported by Okkels et al. (1992), a photooxidizable 18 kDa monoheme cytochrome with an α–band at 551 nm was observed to copurify with the RC proteins and was proposed as electron donor to $P840^+$. Its primary sequence was determined and no significant homology to known cytochromes was found. This cytochrome might be similar to that observed in the preparation of Hurt and Hauska (1984), where a photooxidizable 24 kDa (M_{app}) cytochrome *c*-550.5 copurified with the large RC subunit. More recently, Oh-oka et al. (1993) proposed a cytochrome *c*-551 as electron donor to P^+ which exhibited a M_{app} of 21 kDa. However, in purified samples the rereduction of $P840^+$ via this cytochrome *c*-551 occurred with rather high time constants (130 μs and 560μs).

To resolve this controversy, further functional studies of the various cytochromes are needed. Further data on amino acid sequences of the several proposed cytochromes and deletions of their respective genes should be performed to obtain more information.

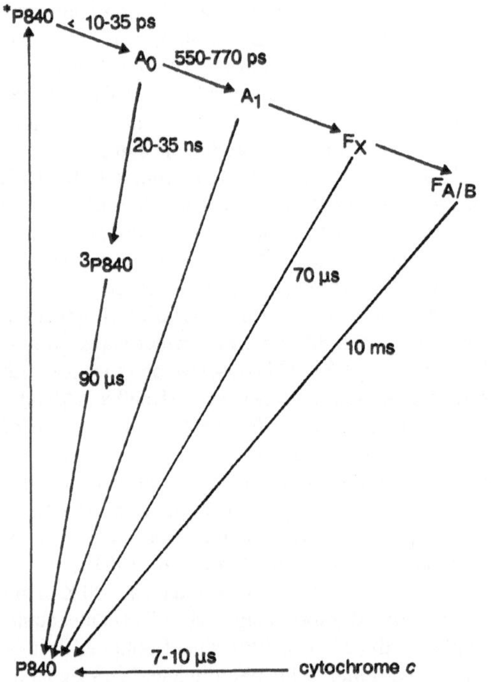

Fig. 2. Room temperature kinetics of electron flow in the reaction center complex of green sulfur bacteria. The electron-transfer rates are given, either as single values or, where discrepancies exist, as a range of values (described further in the text). Although the chain of electron carriers conforms to current thinking on the role of each component, uncertainties concerning the identity and, thus, the role of A_1 exist (described in the text).

D. Kinetics of Electron Transfer

Primary charge separation and further electron transfer steps in the RC of green sulfur bacteria have been studied by time-resolved optical spectroscopy. Experiments have been performed on different RC preparations with antenna sizes ranging from ca. 20 to ca. 100 BChl *a* per RC (see II.) at both low and room temperatures. In this paragraph, only the data obtained at room temperature are discussed (see Fig. 2).

The time constant (τ) for the primary charge separation was calculated to be <10 (Shuvalov et al., 1986) and 35 ps (Nuijs et al., 1985). The primary acceptor A_0 was identified by measuring a light-induced bleaching around 670 nm together with the primary donor bleaching. Both showed the same kinetics (Swarthoff and Amesz, 1979). In a further investigation by flash-induced optical spectroscopy on various kinds of RC preparations, P840 oxidation at 810–850 and 610 nm together with A_0 bleaching around 670 nm, was observed (Swarthoff et al., 1981c). Recovery of the A_0 neutral state was measured by nanosecond flash-induced spectroscopy (van Bochove et al., 1984). A decay time of 20–35 ns was measured for the $P^+A_0^-$ radical pair, most likely mainly to the triplet state of P840, because the preparation used was impaired in electron transfer between the Chl *a*-like acceptor and the secondary acceptor (Swarthoff et al., 1981c). In picosecond absorption spectroscopy the decay time of the A_0^- signal measured at 670 nm was 550 to 770 ps which was attributed to the electron transfer time to the secondary acceptor present in the more intact membrane systems of these studies (Nuijs et al., 1985 and Shuvalov et al., 1986, respectively). This secondary acceptor might be a quinone (see III.B.2.).

No kinetic data are available for forward transfer reactions from A_1 to the terminal acceptors and on recombination reactions of the secondary acceptors to P840.

A BChl *a*-triplet signal with a life time of 90 μs was interpreted as being due to the triplet-state of P840 (Swarthoff et al., 1981c).

Recombination between the FeS centers F_A and F_B and photooxidized P840 was measured and a time constant of 10 ms was observed (Miller et al., 1992). After removing the FeS centers F_A and F_B by treatment of the RC samples with chaotropic agents, faster rereduction kinetics (τ= 70μs) were introduced which were suggested by the same authors as resulting from a recombination between the FeS center F_X^- and $P840^+$.

The respective kinetic data in PS I occurred in the same order of magnitude (see Golbeck and Bryant, 1991), further supporting the similarities to PS I.

Further, rereduction of the primary donor was measured together with the photooxidation of cytochrome *c*. However, several somewhat inconsistent, results have been presented in the literature. Fast electron transfer kinetics (τ = 6.9 μs) between cytochrome *c* with its α-band at 553 nm and $P840^+$ was observed in whole cells (U Feiler and W Nitschke, unpublished; see III.B.5. and chapter by Nitschke and Dracheva in this volume). A similar value (τ< 7 μs) was measured in membrane fractions, however the kinetics showed also a slower phase with a time constant of 70 μs (Prince and Olson, 1976), which might indicate some damage on the donor side to $P840^+$. In an isolated RC complex, similar kinetics (time constants of 10μs and 70 μs) were observed. However, they were due to reaction of a cytochrome *c*-551 with $P840^+$ (Okkels et al., 1992). The slow kinetics have also been observed by Swarthoff et al. (1981c) (τ = 130 μs) and Miller et al. (1992) (τ = 70–120 μs), however, in both cases due to reduction of $P840^+$ by cytochrome *c*-553. Depending on the preparation used an even slower phase ($\tau \approx$ 40 ms) was seen, which was due to damage on the donor side during purification (Swarthoff et al. 1981c). A 130 μs and an additional 560 μs phase, due to oxidation of cytochrome *c*-551 were described by Oh-oka et al. (1993).

It seems clear that the immediate donor to $P840^+$ is a membrane bound cytochrome *c*, however, the exact nature still remains unclear. The 7–10 μs time constant of oxidation of cytochrome *c* likely indicates that this cytochrome is closely associated if not bound to the native RC and slowing down of the respective donation kinetics may indicate damage of the RC preparations.

IV. Genes Encoding Reaction Center Proteins

A transcription unit coding for two subunits of the RC of *Cb. limicola* has been cloned and sequenced by Büttner et al. (1992a). At about the same time the sequence of the gene and its interpretation for the electron donating cytochrome *c*-551 in a RC preparation from *Cb. vibrioforme* has been published (Okkels et al., 1992). Furthermore, a gene for the large subunit of the RC from *Heliobacillus mobilis* (Liebl et al. 1993; see Chapter 31 by Amesz in this volume) has been obtained. Thus the photosynthetic bacteria with PS I-like RCs have become accessible to molecular biology.

The first gene (*pscA*) in the transcription unit of *Cb. limicola* codes for the P840-protein (PscA) related to the P700-proteins PsaA and PsaB, the second (*pscB*) for the protein related to PsaC, which carries the FeS-centers F_A and F_B. Upstream of the transcription unit promoter and ribosome binding motifs are found, and a transcription terminating stem-loop structure occurs after the second gene. This suggests that the two genes are transcribed together into a bicistronic mRNA. See also Section II.B.

A. The Gene for the P840-Protein

The gene for the P840-protein (*pscA*) codes for 730 amino acids, with a calculated mass of 82.24 kDa. This is almost the same mass as the ones for the PsaA/B-proteins, which are 750 and 742 amino acids long, respectively (Fish et al., 1985; see Golbeck and Bryant, 1991). Several hydrophobic peptides long enough to span the membrane occur along the sequence. The 'positive-inside-rule' (von Heijne, 1986) favors eleven transmembrane helices (Büttner et al. 1992a,b). Eleven spans each have also been postulated for the PsaA/B-proteins (Fish et al., 1985), and have been detected in the gene (*pshA*) for the shorter P800-protein (608 amino acids) from *Heliobacillus mobilis* (Liebl et al., 1993). Most striking and valuable for an alignment of these RC proteins is the occurrence of a highly conserved dodecapeptide close to the hydrophobic span IX, as seen in Fig. 3. Of the 12 amino acids identical between PsaA and PsaB, 9 each are conserved in both the P840- and the P800-protein (Trost et al., 1992), which themselves have 11 in common. The PS I-dodecapeptides contain 2 cysteines each, which are suggested to bind the FeS-center F_X between PsaA and PsaB (see Golbeck and Bryant, 1991). From its position the alignment of the P840-protein with the PsaA/B-proteins was started (Büttner et al., 1992a).

The overall identity of the P840-protein to PsaA/B-proteins is only 15 and 14%, respectively. For the P800-protein it is about 17% to both PsaA and B (Liebl et al., 1993). The two large PS I-proteins have 45% of their amino acids in common, indicating that gene duplication occurred later than the separation of green sulfur bacteria and heliobacteria from cyanobacteria. The low overall identity between both P840- and P800-protein to PsaA and -B points to the significant residues, which may be valuable for suggesting site–directed mutations of PS I. These are now possible after successful deletion of the genes for PsaA or -B (Smart et al. 1991; Toelge et al. 1991).

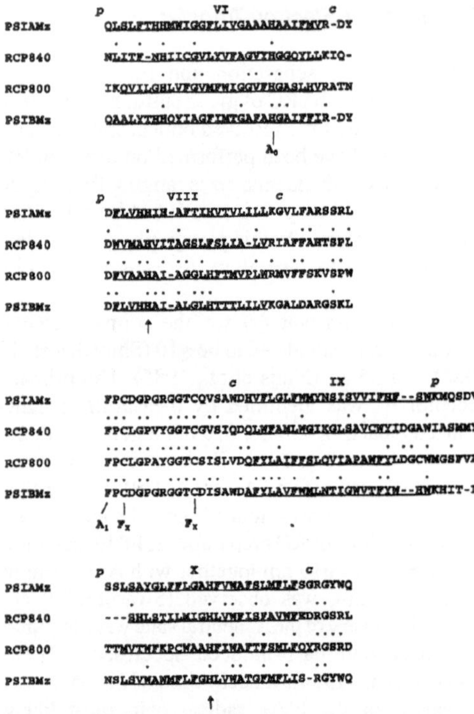

Fig. 3. Alignment of regions of the amino acid sequences for the P840 protein (RCP840) with the large reaction center subunit of heliobacteria (RCP800) and the two large subunits of PS I from maize (PS IAMz/PS IBMz). Identical residues are indicated by dots. The Roman numerals mark three of the putative membrane spanning helices, which are underscored. The putative binding sites of the primary donors (P700, P840, P800) are indicated by arrows, those of the electron acceptors A_0, A_1, and F_X are indicated below the alignment. Letter *c* indicates the cytoplasmic side and *p* the periplasmic side of the membrane. The RCP840 sequence is taken from Büttner et al. (1992a), the RCP800 sequence from Liebl et al. (1993) and the PS IA/BMz sequences from Fish et al. (1985).

The P840-protein contains only 20 histidines, compared to 42 and 37 in PsaA and -B, respectively. Of these 20 histidines only 7 are found at identical positions in the alignment with PsaA and -B (Büttner et al. 1992a). One of the two histidines conserved in hydrophobic span VI, more likely the one closer to the cytoplasmic surface (Fig. 3), may hold the primary acceptor A_0 (see III.B.1). The one in putative transmembrane helix VIII (Fig. 3) might bind P840, the 'special pair' of BChl *a*'s (see III.A). However, recently, site-directed mutagenesis experiments performed on chloroplast *psaB* gene of *Chlamydomonas reinhardtii* indicated that a conserved histidine in helix X might be the ligand of the primary donor P700 (Cui et al., 1995). Indeed, conserved histidine residues are found for PsaA and PsaB of maize as well as for the reaction center proteins of both *Chlorobium* and *Heliobacillus* (see Fig. 3). The aromatic residues conserved in span IX may contribute to the transmembrane charge separation. Furthermore, the phenylalanine in the conserved dodecapeptide may be involved in the function of A_1 (see III.B.2.).

In the alignment of Büttner et al. (1992a) helices I–III and VI–IX are found at corresponding locations in all RC proteins, while IV and V are not. Possibly helices IV and V are not in the membrane, as considered in a recent folding model (Xie et al. 1992). Indeed, a first structure of the RC of PS I from X-ray crystallography shows only 8 transmembrane helices per PsaA or -B (Krauss et al., 1993).

B. Evidence for a Reaction Center-Homodimer

Only one gene coding for a large, hydrophobic protein is present in the transcription unit as opposed to two genes, mostly in tandem, for PsaA and -B of PS I (Golbeck and Bryant, 1991). A second band in Southern blot analyses with 16 different restriction enzymes, even under low stringency, is not observed (Büttner et al. 1992b), although cross hybridization occurs between the genes for PsaA and PsaB. This again indicates the presence of only one gene, unless the two proteins in the RC are significantly less related than PsaA to PsaB. The most convincing evidence for only one gene, and thus for a homodimeric RC, is that it represents all 7 peptides obtained so far from the large 65 kDa-subunit by proteolytic degradation and amino acid sequencing (Büttner et al., 1992a,b and unpublished). This leaves a probability of 1/128 for a second protein in the 65 kDa-subunit, if equal accessibility to proteases is assumed. Very similar negative evidence for a respective second gene has been provided for the RC of heliobacteria (Liebl et al. 1993; see Chapter 31 by Amesz).

C. The Gene for the Protein Binding the FeS-Centers F_A and F_B

The second gene (*pscB*) in the transcription unit codes for a protein of 23.87 kDa with 232 amino acids. It contains 8 cysteines in the C-terminal part, which are spaced similar to the smaller PsaC-proteins (8,8 kDa) of PS I (see Golbeck and Bryant, 1991) and probably hold the two 4Fe4S-clusters of centers F_A and F_B. As mentioned above (see II.B.), the gene product has been identified as the 32 kDa-polypeptide (Illinger et al., 1993). The comparatively high apparent mass is probably caused by the positively charged N-terminal extension of 132 amino acids, with a repetitive sequence, which may function in docking the negatively charged ferredoxin (Büttner et al., 1992a), and which may thus substitute for subunit PsaD in PS I (see Golbeck and Bryant, 1991). A similar N-terminal extension is encoded in the gene for cytochrome *b* of *Chlorobium* (Schütz et al., 1994), and a shorter version of it also in the N-terminus of the P840-protein (Büttner et al., 1992a; see Fig. 4).

The redox potentials of FeS centers F_A and F_B are more negative and photoreduction order is turned around in *Chlorobium* RC compared to PS I. An explanation of the latter might be that center F_A in *Chlorobium* is more negative than center F_B (Nitschke et al., 1990a). On the basis of the structural similarities of the PsaC-protein to bacterial 2(4Fe4S) ferredoxins (Adman et al., 1973; Dunn and Gray, 1988; Oh-oka et al., 1988) it was suggested for the PsaC-protein that the first 3 and the last of the 8 cysteines bind one of the two 4Fe4S clusters, while the residual cysteines bind the second cluster. Two positive charges, likely responsible for stabilization of the reduced center F_A, are conserved between the 6th and the 7th cysteines in the PsaC-proteins, but missing from the respective *Chlorobium* sequence. This could explain a shift of the redox potential of F_A to a more negative value in *Chlorobium*. Therefore, center F_A is probably bound to cysteines 4 to 7, while center F_B is bound by the residual four cysteines (Büttner et al., 1992a). This

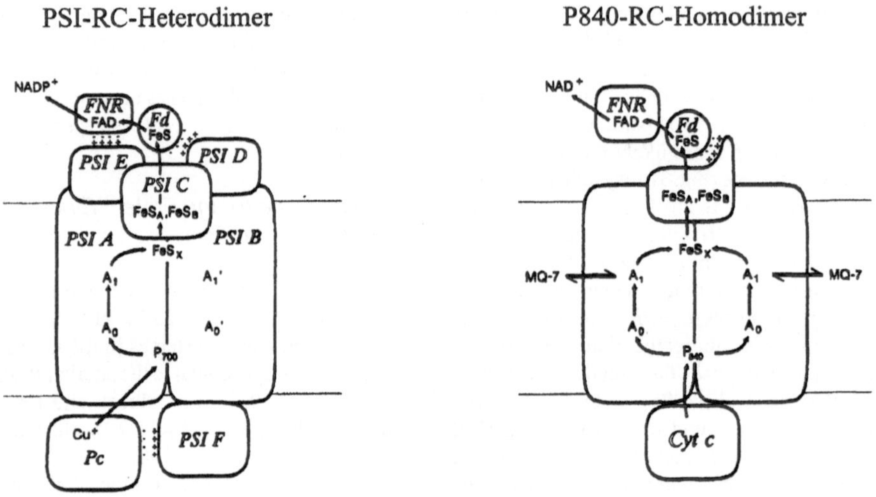

Fig. 4. Comparison of the organization of the heterodimeric reaction center of PS I and the homodimeric reaction center of green sulfur bacteria. Pc stands for plastocyanin, Fd for ferredoxin, FNR for ferredoxin-NAD(P)-oxidoreductase, PsaA-F for the six major subunits of PS I, and MQ-7 for menaquinone-7. The other abbreviations are explained in the text. (Redrawn from Xie et al., 1992)

assignment is in agreement with site directed mutagenesis of *psaC* (Zhao et al., 1992). However, the latter work rose also doubts about correlation of predominant photoreduction and relative redox potentials of F_A and F_B.

D. A Gene for an Electron Donating Cytochrome

Another gene, *pscC* (formerly called *cycA*, Bryant (1994)) was isolated and characterized from *Cb. vibrioforme* by Okkels et al. (1992). The gene codes for the 18 kDa (M_{app}) cytochrome *c*-551 that copurifies with the RC complex and was proposed to function as electron donor to P840. The 621 base pair open reading frame of *pscC* encodes an apoprotein of 22.858 kDa with a single heme binding site. Three N-terminal membrane spanning helices were predicted. Sequence comparison showed no significant similarities to other cytochromes. Northern blot analysis indicated a monocistronic transcript of *pscC* gene and presence of only one *pscC* gene was confirmed by Southern blot analysis. The unique membrane-bound monoheme cytochrome *c*-551 of *Cb. vibrioforme* was assigned to a new class of *c*-type cytochromes. Very recently the sequence of the 24 kDa (M_{app}) cytochrome *c*-550.5, copurifying with the RC proteins in the preparation of Hurt and Hauska (1984), was obtained. It showed a high degree of identity to the cytochrome *c*-551 sequence (D-L Xie, N Nelson and G Hauska, unpublished). This cytochrome *c*-550.5 was additionally observed to copurify with cytochrome *b*, being separated from the RC (Hurt and Hauska, 1984). Therefore, a function as an equivalent to cytochromes c_1/f in a cytochrome *bc*-complex (for review see Hauska (1986) and Chapter 35 by Gray and Daldal) cannot be excluded. This is even more likely since a transcription unit for a cytochrome *bc*-complex of *Cb. limicola* includes the genes for the Rieske FeS-protein and cytochrome *b*, but lacks a gene for cytochrome c_1/f (Schütz et al., 1994). Cytochrome *c*-550.5 thus reduced by the cytochrome *bc*-complex could then function as electron donor to either P840 directly, or to a tetraheme cytochrome *c*-553 (see III.C.).

V. Comparison with Other Reaction Centers

All photosynthetic RCs characterized so far can be classified into two types, the 'pheophytin-quinone' RCs of purple bacteria, Chloroflexaceae and PS II, and the 'FeS' RCs of green sulfur bacteria, heliobacteria and PS I (Blankenship, 1992).

The similarity of the Photosystem of green sulfur bacteria to PS I suggested from earlier studies (Buchanan and Evans, 1969) has been substantiated by more recent results on electron transport, polypeptides and their genes, as detailed in the previous sections. The respective similarities of the heliobacterial PS to PS I have been discovered more recently (Trost et al., 1992; Liebl et al., 1993; see Chapter 31 by Amesz).

Both types of RCs consist of a dimeric core structure containing two hydrophobic subunits, which bind the primary electron donor and the early electron acceptors (see above and Chapter 23 by Lancaster et al.). The primary electron donor is probably a 'special pair' of (bacterio)chlorophylls in all cases (see above), although this is still not clearly shown in PS II (van Mieghem and Rutherford, 1994). The primary electron acceptors are tetrapyrroles, either (bacterio)pheophytins, or Chl *a*-like molecules (see above and Chapters 23, 31 and 32 by Lancaster et al., Amesz, and Feick, respectively, and Golbeck and Bryant, 1991). Quinones universally seem to function as secondary electron acceptors (see above and Chapter 23 by Lancaster et al., however, also see Chapter 31 by Amesz). Furthermore it has been concluded from X-ray crystallography that the distance between the special pair and the Fe at the acceptor side of the 'pheophytin-quinone' RCs is similar to the distance between P700 and FeS center F_X in the PS I RC (Krauss et al., 1993).

The major differences between 'pheophytin-quinone' RCs and 'FeS' RCs are, that the core subunits of the latter are more than twice as large, carrying tens of antenna (bacterio)chlorophylls in the N-terminal part, and that FeS centers of low redox potentials enable the reduction of $NAD(P)^+$ via ferredoxin (see Golbeck and Bryant, 1991). The two subunits of the 'pheophytin-quinone' RCs with a M_{app} around 30 kDa span the membrane 5 times, while the corresponding subunits of the 'FeS' RCs contain up to 11 hydrophobic spans (Fish et al., 1985; Büttner et al., 1992a,b; Liebl et al., 1993).

Despite these differences, several similarities in the primary structure and the folding models of RC cores have been noted (Robert and Moënne-Loccoz, 1990; Margulies, 1991; Nitschke and Rutherford, 1991; Otsuka et al., 1992). As indicated by the shaded parts in Fig. 6 (see below) it has been suggested that the five transmembrane helices of the 'pheophytin-quinone' RC subunits correspond to the C-terminal part of the two larger 'FeS' RC subunits from transmembrane spans V to IX, based on both conserved residues and folding pattern (Büttner et al., 1992a,b; Liebl et al., 1993). By principal component analysis Otsuka et al. (1992) extended the structural similarity. According to their results, an additional similarity was proposed between the N-terminal half of the large 'FeS' RC subunits and the 'pheophytin-quinone' RC subunits. This would imply gene duplication and fusion of the small RC core subunits to construct the large RC during evolution.

A. Homodimeric and Heterodimeric Reaction Centers

All RCs exhibit a dimeric core structure which catalyzes transmembrane charge separation. The dimeric structure is apparently necessary for binding the special, electron donating pair of (bacterio)chlorophylls, and also to bind the Fe in the case of the 'pheophytin-quinone' RCs (see Chapter 23 by Lancaster et al.), or the FeS center F_X in the case of the 'FeS' RCs. While the 'pheophytin-quinone' RCs and the PS I RCs are heterodimers, built by two different, although related core subunits, a homodimeric structure is indicated for the RCs of both *Chlorobium* (Büttner et al., 1992a,b) and heliobacteria (Liebl et al., 1993).

The near symmetry of the RCs of purple bacteria not only relates the two subunits L and M, but also the two sets of pigments, one set being strongly favored for charge separation (see Chapters 23 by Lancaster et al. and 24 by Woodbury and Allen). Similar heterodimeric structures are envisaged for the RCs of PS II (Trebst, 1987; Michel et al., 1986) and PS I (see Golbeck and Bryant, 1991). The reasons for the presence of one electron transfer branch, and for the continued existence of the other remains unclear. If the RCs of green sulfur- and heliobacteria are indeed homodimeric, then it is possible their study may aid in the resolution of this question. Furthermore, it is intriguing to consider whether the existence of a homodimeric structure is related to their anaerobic life.

In Fig. 4 schemes for the organization and the electron transfer of the PS I RC heterodimer and of the simpler P840 RC homodimer are compared. It is conceived that the positively charged, N-terminal extension of PscB (equivalent to PsaC) in *Chlorobium* (see IV. C.) substitutes for PsaD in binding ferredoxin, and that electron transfer occurs in the two parallel, equally probable, branches of the homodimer. As a consequence, each of the two secondary acceptors A_1 should exchange with the menaquinol pool with equal ease. This might limit the occupancy of the A_1

site and lead to a faster back reaction to P840$^+$ from A_0^-. An increased back reaction rate might raise the vulnerability of this system to oxygen.

Possibly associated with parallel electron transfer in the RCs of green sulfur bacteria is the presence of the charged cluster, which in 'pheophytin-quinone' RCs is C-terminal to transmembrane helix II of subunit L or D1, or C-terminal to helix VI in the PsaA subunit of PS I. The glutamic acid D131 in PS II RC and aspartic acid PsaA414 in PS I have been proposed to bind the respective primary electron acceptors (Robert and Moënne-Loccoz, 1990). From resonance Raman data of *Chlorobium* RCs a single-type of binding site for the primary electron acceptor of *Chlorobium* was proposed (Feiler et al., 1994), supporting the presence of a homodimeric RC core. Additionally, a similar binding force as to those in PS II and purple bacterial RCs was estimated. If PscA is added to the sequence alignment mentioned above (see Fig. 5) it is clear that glutamine PscA397 could be the residue H-bonding the acceptor Chl *a* molecule in each half of the homodimer. Further evidence for the presence of a homodimer comes from ENDOR and special TRIPLE spectroscopy and from FT Raman spectroscopy in the near infrared, which reveals an even distribution of the positive charge in P840$^+$ over the special pair of bacteriochlorophylls in *Chlorobium*, as opposed to purple bacteria (see III.A.3.).

Without a three dimensional structure, it can not be excluded that the two identical core subunits of the P840 RC are rendered asymmetric by binding the protein carrying FeS centers F_A and F_B, as is seen in receptor dimerization by a single, asymmetrically binding growth factor molecule (de Vos et al., 1992). Further, post-translational modification of the polypeptides could lead to an asymmetry. Both possibilities could lead to two unequal electron transfer branches.

B. Evolutionary Aspects

The overall similarities of photosynthetic RCs suggest that all of them originate from one common ancestor (Nitschke and Rutherford, 1991; Blankenship, 1992), and it is rational to assume that the heterodimeric forms of the RC cores evolved from a homodimeric form as depicted in Fig. 6. Based on the parsimony analysis of the RC core subunit sequences at least two independent gene duplications have occurred, giving rise to the L/M and D1/D2 heterodimers, of

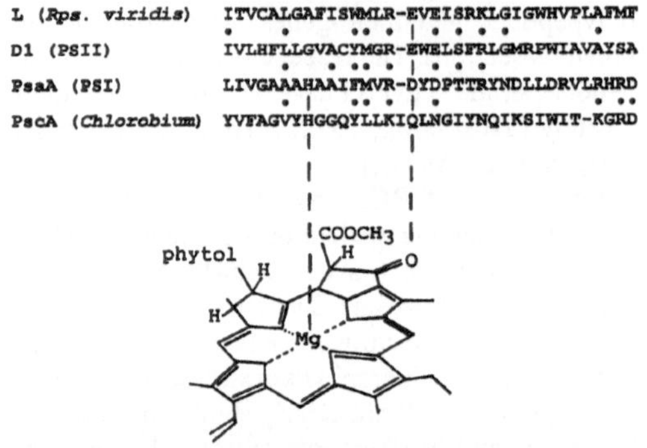

Fig. 5. Amino acid sequence comparison of regions of: the B-helix of the L-subunit from *Rhodopseudomonas viridis*; the helix IV of D1 (PS II) and, the helix VI of PsaA (PS I) from spinach; and the helix VI of P840-protein (PscA) from *Cb. limicola*. Conserved amino acids with functional similarities between L, D1, PsaA and/or PscA are indicated by points, or in the case of the amino acids presumed responsible for binding the primary acceptor, marked by lines. The L sequence was taken from Michel et al. (1986), D1 sequence from Zurawski et al. (1982), PsaA sequence from Fish et al. (1985) and PscA sequence from Büttner et al. (1992a).

the RCs of purple bacteria and *Chloroflexus*, and of PS II, respectively (Blankenship, 1992). Thus the divergence of the reaction centers in purple bacteria and those in PS II might have taken place before the gene duplications. Since PsaA and B, in turn, are substantially more closely related to each other, than to the RC subunits in either helio- or green sulfur bacteria (Liebl et al., 1993; Büttner et al., 1992a), gene duplication must have occurred after the divergence of the RC of PS I, and those of green sulfur bacteria and heliobacteria. Earlier, the 'pheophytin-quinone' RCs must have separated from the 'FeS' RCs, but it remains undecided which was first. One may make a case for the 'FeS' RCs, because they occur in strictly anaerobic bacteria, and the early atmosphere was anaerobic. The idea of an iron-sulfur world at early stages of evolution supports this view (Drobner et al., 1990). If the ancestral gene indeed coded for an 'FeS' RC, as shown in Fig. 6, was it small or large? According to the analysis of Otsuka et al. (1992), which suggests the evolution of the large subunits of the recent 'FeS' RCs by yet another gene duplication, a small ancestral gene is more likely. Whether the ancestral photosynthetic RC had a monomeric or dimeric structure remains unclear, although a monomeric form is favored for simplicity (Blankenship, 1992). Finally, recent

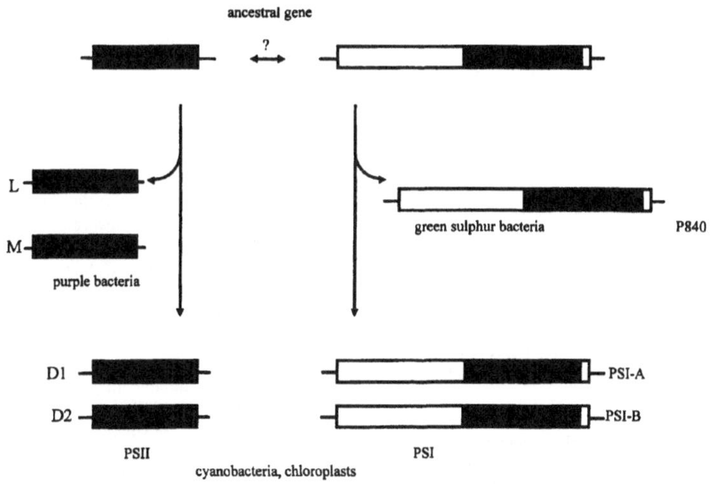

Fig. 6. Evolutionary concept of the reaction center dimers. L and M, and D1 and D2 stand for the subunits constituting the heterodimer of the reaction centers in purple bacteria and PS II, respectively, while *psaA* and *psaB* are the denotations of the genes coding for the two large subunits of PS I. Shaded areas of the bars indicate related regions in all reaction centers. (Redrawn from Büttner et al., 1992a)

oxygenic photosynthesis in cyanobacteria and chloroplasts probably evolved by fusion of two bacterial ancestors (Büttner et al., 1992a; Blankenship, 1992) as indicated in Fig. 6, although other alternatives are under discussion (Olson and Pierson, 1989).

VI. Conclusions

Summarizing the present state of our knowledge, it is now certain that the RC complex of green sulfur bacteria is a bacterial counterpart of PS I. This is based on both functional and structural features as well as genetic data. The latter established a convincing level of identity between the sequences of the PS I genes and that of *Chlorobium*. This permits the identification of regions of functional and structural importance in both systems and thus can be used for designing site-directed mutations. Such mutations will lead to a better understanding of the properties of the FeS-type RCs.

Of considerable interest, also, is the nature of the electron donor to P840 since two different candidates (a monoheme- or a tetraheme-containing cytochrome, see III.C. and Chapter 36 by Nitschke and Dracheva) are discussed. In either case, intriguing evolutionary questions should arise.

New and most striking is the possibility of a homodimeric organization of the green sulfur bacterial RC core, since only one gene coding for the two core subunits has as yet been found. However, it remains unclear whether electron transport occurs symmetrically in the two branches. A three dimensional structure of the RC complex could help answer this question.

Acknowledgments

The authors thank all the people having made essential contributions to the work described in this chapter (some of which was unpublished at this time): M Büttner, C Bock, J Golbeck, N Illinger, W Nitschke, M Schütz, D Stehlik, J Sturgis, M Thurnauer and D-L Xie. We also like to thank D Albouy, RE Blankenship, K Brettel, U Liebl, TA Mattioli, N Nelson, JM Olson, B Robert, AW Rutherford and W Vermaas for many stimulating discussions and TA Mattioli, B Robert and W Nitschke for critical reading of the manuscript. U.F. would like to thank in particular J Sturgis for his patience in translating her 'German' English into 'English' English. U.F. was supported by the EEC (Science program) and G.H. by the Deutsche Forschungsgemeinschaft (SFB43 C2).

References

Adman ET, Sieker LC and Jensen LH (1973) The structure of a bacterial ferredoxin. J Biol Chem 248: 3987–3996

Barry BA, Bender CJ, McIntosh L, Ferguson-Miller S and Babcock GT (1988) Photoaccumulation in Photosystem I does not produce a phylloquinone radical. Israel J Chemistry 28: 129–132

Blankenship RE (1992) Origin and early evolution of photosynthesis. Photosynth Res 33: 91–111

Braumann T, Vasmel H, Grimme LH and Amesz J (1986) Pigment composition of the photosynthetic membrane and reaction center of the green bacterium *Prosthecochloris aestuarii*. Biochim Biophys Acta 848: 83–91

Bryant DA (1994) Gene nomenclature recommendations for green photosynthetic bacteria and heliobacteria. Photosynth Res 41: 27–28

Buchanan BB and Evans MCW (1969) Photoreduction of ferredoxin and its use in $NAD(P)^+$ reduction by a subcellular preparation from the photosynthetic bacterium *Chlorobium thiosulphatophilum*. Biochim Biophys Acta 180: 123–129

Buchanan BB, Matsubara H and Evans MCW (1969) Ferredoxin from the photosynthetic bacterium, *Chlorobium thiosulphatophilum*. A link to ferredoxins from nonphotosynthetic bacteria. Biochim Biophys Acta 189: 46–53

Büttner M, Xie D-L, Nelson H, Pinther W, Hauska G and Nelson N (1992a) Photosynthetic reaction center genes in green sulfur bacteria and in Photosystem I are related. Proc Natl Acad Sci USA 89: 8135–8139

Büttner M, Xie D-L, Nelson H, Pinther W, Hauska G and Nelson N (1992b) The Photosystem I-like reaction center of green S-bacteria is a homodimer. Biochim Biophys Acta 1101: 154–156

Cohen-Bazire GN, Pfennig N and Kunisawa J (1964) The fine structure of green bacteria. J Cell Biol 22: 207–225

Cui L, Bingham SE, Kuhn M, Käß H, Lubitz W and Webber AN (1995) Site-directed mutagenesis of conserved histidines in the helix VIII domain of PsaB impairs assembly of the photosystem I reaction center without altering spectroscopic characteristic of P700. Biochemistry 34: 1549–1558

De Vos AM, Ultsch M and Kossiakoff AA (1992) Human growth hormone and extracellular domain of its receptor: Crystal structure of the complex. Science 255: 306–312

Drobner E, Huber H, Wächtershäuser G, Rose D and Stetter KO (1990) Pyrite formation linked with hydrogen evolution under anaerobic conditions. Nature 346: 742–744

Dunn PPJ and Gray JC (1988) Localization and nucleotide sequence of the gene for the 8 kDa subunit of Photosystem I in pea and wheat chloroplast DNA. Plant Mol Biol 11: 311–319

Dutton PL, Leigh JS and Seibert M (1971) Primary processes in photosynthesis: *In situ* ESR studies on the light induced oxidized and triplet state of reaction center bacteriochlorophyll. Biochem Biophys Res Commun 46: 406–413

Fajer J, Brune DC, Davis MS, Forman A and Spaulding LD (1975) Primary charge separation in bacterial photosynthesis: Oxidised chlorophylls and reduced pheophytin. Proc Nat Acad Sci USA 72: 4956–4960

Feiler U (1991) Untersuchungen am Photosystem von verschiedenen photosynthetischen Bakterien. Thesis, University Frankfurt

Feiler U, Nitschke W, Michel H and Rutherford AW (1989) The membrane-bound cytochrome hemes from the green sulfur bacterium *Chlorobium phaeobacteroides* studied by EPR spectroscopy. In: Symposium on Molecular Biology of Membrane-bound Complexes in Phototrophic Bacteria, Abstr. P-6, 19, Freiburg, Germany

Feiler U, Nitschke W and Michel H (1992) Characterisation of an improved reaction center preparation from the photosynthetic green sulfur bacterium *Chlorobium* containing the FeS centers F_A and F_B and a bound cytochrome subunit. Biochemistry 31: 2608–2614

Feiler U, Albouy D, Pourcet C. Mattioli T, Lutz M and Robert B (1994) Structure and binding site of the primary electron acceptor in the reaction center of *Chlorobium*. Biochemistry 33: 7594–7599

Fish LE, Kück U and Bogorad L (1985) Two partially homologous adjacent light-inducible maize chloroplast genes encoding polypeptides of the P700 chlorophyll *a*-protein complex of Photosystem I. J Biol Chem 260: 1413–1421

Fowler CF (1974) Evidence for a cytochrome *b* in green bacteria. Biochim Biophys Acta 357: 327–331

Fowler CF, Nugent NA and Fuller RC (1971) The isolation and characterization of a photochemically active complex from *Chloropseudomonas ethylica*. Proc Nat Acad Sci USA 68: 2278–2282

Gingras G (1978) A comparative review of photochemical reaction center preparations from photosynthetic bacteria. In: Clayton RK and Sistrom WR (eds) The Photosynthetic Bacteria, pp 119–132. Plenum Press, New York and London

Golbeck JH and Bryant DA (1991) Photosystem I. Curr Top Bioenerg 16: 83–177

Hales BJ and Gupta AD (1979) Orientation of the BChl triplet and the primary ubiquinone acceptor of *Rhodospirillum rubrum* in membrane multilayers determined by ESR spectroscopy (I). Biochim Biophys Acta 548: 276–286

Hauska G (1986) Composition and structure of cytochrome bc_1 and b_6f complexes. In: Staehelin LA and Arntzen CJ (eds) Encyclopedia of Plant Physiology, New Series Vol 19, pp 496–507. Springer Verlag, Berlin

Hauska G (1988) Phylloquinone in Photosystem I: are quinones the secondary electron acceptors in all types of photosynthetic reaction centers? Trends Biochem Sci 13: 414–416

Heathcote P, Hanley JA and Evans MCW (1993) Double-reduction of A_1 abolishes the EPR signal attributed to A_1^-: Evidence for C_2 symmetry in the Photosystem I reaction center. Biochim Biophys Acta 1144: 54–61

Hurt EC and Hauska G (1984) Purification of membrane-bound cytochromes and a photoactive P840 protein complex of the green sulfur bacterium *Chlorobium limicola f thiosulphatophilum*. FEBS Lett 168: 149–154

Illinger N, Xie D-L, Hauska G and Nelson N (1993) Identification of the subunit carrying FeS centers A and B in the P840-reaction center preparation of *Chlorobium limicola*. Photosynth Res 38: 111–114

Jennings JV and Evans MCW (1977) The irreversible photoreduction of a low potential component at low temperatures in a preparation of the green photosynthetic bacterium *Chlorobium limicola f thiosulphatophilum*. FEBS Lett 75: 33–36

Kirmaier C and Holten D (1987) Primary photochemistry of

reaction centers from the photosynthetic purple bacteria. Photosynth Res 13: 225–260
Kjaer B, Jung Y-S, Yu L, Golbeck JH and Scheller HV (1994) Iron sulfur centers in the photosynthetic reaction center complex from *Chlorobium vibrioforme*. Differences from and similarities to iron sulfur centers in Photosystem I. Photosynth Res 41: 105–114
Kleinherenbrink FAM, Ikegami I, Hiraishi A, Otte SCM and Amesz J (1993) Electron transfer in menaquinone-depleted membranes of *Heliobacterium chlorum*. Biochim Biophys Acta 1142: 69–73
Knaff DB (1978) Reducing potentials and the pathway of NAD^+ reduction. In: Clayton RK and Sistrom WR (eds) The Photosynthetic Bacteria, pp 629–640. Plenum Press, New York
Knaff DB and Hirasawa M (1991) Ferredoxin-dependent chloroplast enzymes. Biochim Biophys Acta 1056: 93–125
Knaff DB and Malkin R (1976) Iron-sulfur proteins of the green photosynthetic bacterium *Chlorobium*. Biochim Biophys Acta 430: 244–252
Knaff DB, Olson JM and Prince RC (1979) The light-reaction of the green photosynthetic bacterium *Chlorobium limicola f thiosulphatophilum* at cryogenic temperatures. FEBS Lett 98: 285–289
Krauss N, Hinrichs W, Witt I, Fromme P, Pritzkow W, Dauter Z, Betzel C, Wilson KS, Witt HT and Saenger W (1993) Three-dimensional structure of system I of photosynthesis at 6 Å resolution. Nature 361: 326–331
Kusumoto N, Inoue K, Nasu H, Takano H and Sakurai H (1994) Preparation of reaction center complex containing photo-reducible FeS centers and photooxidizable cytochrome *c* from green sulfur bacterium *Chlorobium tepidum*. Plant Cell Physiol 35: 17–25
Kusai A and Yamanaka T (1973) An NAD(P) reductase derived from *Chlorobium thiosulphatophilum*: purification and some properties. Biochim Biophys Acta 292: 621–633
Liebl U, Mockensturm-Wilson M, Trost JT, Brune DC, Blankenship RE and Vermaas W (1993) Single core polypeptide in the reaction center of the photosynthetic bacterium *Heliobacillus mobilis* - structural implications and relations to other Photosystems. Proc Natl Acad Sci USA 90: 7124–7128
Mathews BW, Fenna RE, Bolognesi MC, Schmid MF and Olson JM (1979) Structure of a bacteriochlorophyll *a* protein from the green photosynthetic bacterium *Prosthecochloris aestuarii*. J Mol Biol 131: 259–285
Mathis P (1990) Compared structure of plant and bacterial reaction centers. Evolutionary implications. Biochim Biophys Acta 1018: 163–167
Mansfield RW and Evans MCW (1988) EPR characteristics of the electron acceptors A_0, A_1, and (Fe-S)X in digitonin and triton X-100 solubilized pea Photosystem I. Israel J Chemistry 28: 97–102
Mansfield RW, Nugent JHA and Evans MCW (1987) ESR characteristics of Photosystem I in deuterium oxide: further evidence that electron acceptor A_1 is a quinone. Biochem Biophys Acta 894: 515–523
Margulies MM (1991) Sequence similarity between Photosystem I and II. Identification of a Photosystem I reaction center transmembrane helix that is similar to transmembrane helix IV of the D2 subunit of Photosystem II and the M subunit of the non-sulfur purple and flexible green bacteria. Photosynth Res 29: 133–147
Mattioli TA, Hoffmann A, Robert B, Schrader B and Lutz M (1991) Primary donor structure and interactions in bacterial reaction centers from near-infrared Fourier transform resonance Raman spectroscopy. Biochemistry 30: 4648–4654
Meyer TE, Bartsch RG, Cusanovich MA and Mathewson JH (1968) The cytochromes of *Chlorobium thiosulphatophilum*. Biochim Biophys Acta 153: 854–861
Michel H, Weyer KA, Gruenberg H, Dunger I, Oesterhelt D and Lottspeich F (1986) The 'light' and 'medium' subunits of the photosynthetic reaction center from *Rhodopseudomonas viridis*: isolation of the genes, nucleotide and amino acid sequence. EMBO J 5: 1149–1158
Miller M, Liu X, Snyder S, Thurnauer MC and Biggins J (1992) Photosynthetic electron-transfer reactions in the green sulfur bacterium *Chlorobium vibrioforme*: Evidence for the functional involvement of iron-sulfur redox centers on the acceptor side of the reaction center. Biochemistry 31: 4354–4363
Nitschke W and Lockau W (1993) Photosystem I and its bacterial counterparts. Physiologia Plantarum 88: 372–381
Nitschke W and Rutherford AW (1991) Are all the different types of photosynthetic reaction centers variations of a common structural theme? Trends Biochem Sci 16: 241–245
Nitschke W, Feiler U, Lockau W and Hauska G (1987) The Photosystem of the green sulfur bacterium *Chlorobium limicola* contains two early electron acceptors similar to Photosystem I. FEBS Lett 218: 283–286
Nitschke W, Feiler U and Rutherford AW (1990a) Photosynthetic reaction center of green sulfur bacteria studied by EPR. Biochemistry 29: 3834–3842
Nitschke W, Sétif P, Liebl U, Feiler U and Rutherford AW (1990b) Reaction center photochemistry of *Heliobacterium chlorum*. Biochemistry 29: 11079–11088
Norris JR, Uphaus RA, Crespi HG and Katz JJ (1971) Electron spin resonance of chlorophyll and the origin of signal I in photosynthesis. Proc Natl Acad Sci USA 68: 625–629
Nuijs AM, Vasmel H, Joppe HLP, Duysens LNM and Amesz J (1985) Excited states and primary charge separation in the pigment system of the green photosynthetic bacterium *Prosthecochloris aestuarii* as studied by picosecond absorbance difference spectroscopy. Biochim Biophys Acta 807: 24–34
Oh-oka H, Takahashi Y, Kuriyama K, Saeki K and Matsubara H (1988) The protein responsible for center A/B in spinach Photosystem I: Isolation with iron-sulfur cluster(s) and complete sequence analysis. J Biochem (Tokyo) 103: 962–968
Oh-oka H, Kakutani S, Matsubara H, Malkin R and Itoh S (1993) Isolation of the photoactive reaction center complex that contains three types of Fe-S centers and a cytochrome c subunit from the green sulfur bacterium *Chlorobium limicola f thiosulphatophilum*, strain Larsen. Plant Cell Physiol 34: 93–101
Okkels JS, Kjær B, Hansson Ö, Svendsen I, Lindberg-Møller B and Scheller HV (1992) A membrane-bound monoheme cytochrome c_{551} of a novel type is the immediate electron donor to P840 of the *Chlorobium vibrioforme* photosynthetic reaction center complexes. J Biol Chem 267: 21139–21145
Olson JM (1978a) Confused history of *Chloropseudomonas ethylica* 2K. Int J Syst Bacteriol 28: 128–129
Olson JM (1978b) Bacteriochlorophyll a-proteins from green

bacteria. In: Clayton RK and Sistrom WR (eds) The Photosynthetic Bacteria, pp 119–132. Plenum Press, New York and London

Olson JM (1980) Chlorophyll organization in green photosynthetic bacteria. Biochim Biophys Acta 594: 33–51

Olson JM (1981) Bacteriopheophytin *c* in reaction center complexes of green photosynthetic bacteria. Biochim Biophys Acta 637: 185–188

Olson JM, Philipson K and Sauer K (1973) Circular dichroism and absorption spectra of bacteriochlorophyll-protein and contains reaction center complexes from *Chlorobium thiosulphatophilum*. Biochim Biophys Acta 292: 206–217

Olson JM, Giddings TH and Shaw EJ (1976a) An enriched reaction center preparation from green photosynthetic bacteria. Biochim Biophys Acta 449: 197–208

Olson JM, Prince RC and Brune DC (1976b) Reaction center complexes from green bacteria. Brookhaven Symp Biol 28: 238–246

Otsuka J, Miyachi H and Horimoto K (1992) Structure model of core proteins in Photosystem I inferred from the comparison with those in Photosystem II and bacteria; an application of principal component analysis to detect the similar regions between distantly related families of proteins. Biochim Biophys Acta 1118: 194–210

Pierson BK and Castenholz RW (1978) Photosynthetic apparatus and cell membranes of the green bacteria. In: Clayton RK and Sistrom WR (eds) The Photosynthetic Bacteria, pp 119–132. Plenum Press, New York and London

Pierson BK and Olson JM (1989) Evolution of photosynthesis in anoxygenic photosynthetic procaryotes. In: Cohen Y and Rosenberg E (eds) Mirobial Mats: Physiological Ecology of Benthic Microbial Communities, pp 402–427. Am Soc Microbiol, Washington

Powls R and Redfearn RE (1969) Quinones of the Chlorobiaceae. Properties and possible function. Biochim Biophys Acta 172: 429–437

Prince RC and Olson JM (1976) Some thermodynamic and kinetic properties of the primary photochemical reactants in a complex from a green photosynthetic bacterium. Biochim Biophys Acta 423: 357–362

Rao KK, Matsubara H, Buchanan BB and Evans MCW (1969) Amino acid composition and terminal sequences of ferredoxins from two photosynthetic green bacteria. J Bact 100: 1411–1412

Rigby SEJ, Thapar R, Evans MCW and Heathcote P (1994) The electronic structure of $P840^{+\cdot}$. The primary donor of the *Chlorobium limicola* f. sp. *thiosulphatophilum* photosynthetic reaction center. FEBS Lett 350: 24–28

Robert B and Moënne-Loccoz P (1990) Is there a protein substructure common to all photosynthetic reaction centers? In: Baltscheffsky M (ed) Current research in photosynthesis, Vol 1, pp 65–68. Kluwer Academic Publishers, Dordrecht

Schmidt K (1980) A comparative study of the composition of chlorosomes (*Chlorobium* vesicles) and cytoplasmic membranes from *Chloroflexus aurantiacus* strain Ok-70-fl and *Chlorobium limicola* f. *thiosulphatophilum* strain 6230. Arch Microbiol 124: 21–31

Schütz M, Zirngibl S, le Coutre J, Büttner M, Xie D-L, Nelson N, Deutzmann R and Hauska G (1994) A transcription unit for the Rieske FeS-protein and cytochrome *b* in *Chlorobium limicola*. Photosynth Res 39: 167–174

Shioi Y, Takamiya K and Nishimura M (1976) Isolation and some properties of NAD^+ reductase of the green photosynthetic bacterium *Prosthecochloris aestuarii*. J Biochem 79: 361–371

Shuvalov VA, Amesz J and Duysens LNM (1986) Picosecond spectroscopy of isolated membranes of the photosynthetic green sulfur bacterium *Prosthecochloris aestuarii* upon selective excitation of the primary electron donor. Biochim Biophys Acta 851: 1–5

Smart LB, Anderson SL and McIntosh L (1991) Targeted genetic inactivation of the Photosystem I reaction center in the cyanobacterium *Synechocystis* sp PCC 6803. EMBO J. 10, 3289–3296

Snyder SW and Thurnauer MC (1993) Electron spin polarization in photosynthetic reaction centers. In: Norris JR and Deisenhofer J (eds) The Photosynthetic Reaction Centre, Vol. II, chapter 11, pp 285–330. Academic press, New York

Staehelin LA, Golecki JR and Drews G (1980) Supramolecular organization of chlorosomes (*Chlorobium* vesicles) and of their membrane attachment sites in *Chlorobium limicola*. Biochim Biophys Acta 589: 30–45

Swarthoff T and Amesz J (1979) Photochemically active pigment-protein complexes from the green photosynthetic bacterium *Prosthecochloris aestuarii*. Biochim Biophys Acta 548: 427–432

Swarthoff T, Gast P and Hoff AJ (1981a) Photooxidation and triplet formation of the primary electron donor of the green photosynthetic bacterium *Prosthecochloris aestuarii*, observed with ESR spectroscopy. FEBS Lett 127: 83–86

Swarthoff T, Gast P, Hoff AJ and Amesz J (1981b) An optical and ESR investigation on the acceptor side of the reaction center of the green photosynthetic bacterium *Prosthecochloris aestuarii*. FEBS Lett 130: 93–98

Swarthoff T, van der Veek-Horsley KM and Amesz J (1981c) The primary charge separation, cytochrome oxidation and triplet formation in preparations from the green photosynthetic bacterium *Prosthecochloris aestuarii*. Biochim Biophys Acta 635: 1–12

Swarthoff T, Kramer HJM and Amesz J (1982) Thin-layer chromatography of pigments of the green photosynthetic bacterium *Prosthecochloris aestuarii*. Biochim Biophys Acta 681: 354–358

Sybesma C and Vredenberg WJ (1963) Evidence for a reaction center P840 in the green photosynthetic bacterium *Chloropseudomonas ethylicum*. Biochim Biophys Acta 75: 439–441

Tanaka M, Haniu M, Yasunobu KT, Evans MCW and Rao KR (1974) Amino acid sequence of ferredoxin from a photosynthetic green bacterium, *Chlorobium limicola*. Biochemistry 13: 2953–2959

Tiede DM and Dutton PL (1981) Orientation of primary quinones of bacterial photosynthetic reaction centers contained in chromatophores and reconstituted membranes. Biochim Biophys Acta 637: 278–290

Toelge M, Ziegler K, Maldener I and Lockau W (1991) Directed mutagenesis of the *psaB* of Photosystem I of the cyanobacterium *Anabaena variabilis* ATCC 29413. Biochim Biophys Acta 1060: 233–236

Trebst A (1986) The topology of the plastoquinone and herbicide binding peptides of Photosystem II in the thylakoid membrane. Z Naturforsch 41c: 240–245

Trost JT, Brune DC and Blankenship RE (1992) Protein sequences and redox titrations indicate that the electron acceptors in the reaction centers from heliobacteria are similar to Photosystem I. Photosynth Res 32: 11–22

Van Bochove AC, Swarthoff T, Kingma H, Hof RM, van Grondelle R, Duysens LNM and Amesz J (1984) A study of the primary charge separation in green bacteria by means of flash spectroscopy. Biochim Biophys Acta 764: 343–346

Van de Meent EJ, Kobayashi M, Erkelens C, van Veelen PA, Otte SCM, Inoue K, Watanabe T and Amesz J (1992) The nature of the primary electron acceptor in green sulfur bacteria. Biochim Biophys Acta 1102: 371–378

Van Mieghem FJE and Rutherford AW (1993) Comparative spectroscopy of Photosystem II and purple bacterial reaction centers. Biochem Soc Trans 21: 986–991

Vasmel H, Swarthoff T, Kramer HJM and Amesz J (1983) Isolation and properties of a pigment-protein complex associated with the reaction center of the green photosynthetic sulfur bacterium *Prosthecochloris aestuarii*. Biochim Biophys Acta 725: 361–367

Vasmel H, den Blanken HJ, Dijkman JT, Hoff AJ and Amesz J (1984) Triplet-minus-singlet absorbance difference spectra of reaction centers and antenna pigments of the green photosynthetic bacterium *Prosthecochloris aestuarii*. Biochim Biophys Acta 767: 200–208

Von Heijne G (1986) The distribution of positively charged residues in bacterial inner membrane proteins correlates with the transmembrane topology. EMBO J 5: 3021–3027

Wasielewski MR, Smith UH and Norris JR (1982) ESR study of the primary electron donor in highly ^{13}C-enriched *Chlorobium limicola f thiosulphatophilum*. FEBS Lett 149: 138–140

Xie D-L, Büttner M, Nelson H, Chitnis P, Pinther W, Hauska G and Nelson N (1992) A transcription unit for the Photosystem 1-like reaction center of the green S-bacterium *Chlorobium limicola*. In: Murata N (ed) Research in Photosynthesis, Vol I, pp 513–520. Kluwer Academic Publishers, Dordrecht

Xie D-L, Lill H, Hauska G, Maeda M, Futai M and Nelson N (1993) The *atp*2 operon of the green bacterium *Chlorobium limicola*. Biochim Biophys Acta 1172: 267–273

Zhao J, Li N, Warren PV, Golbeck JH and Bryant DA (1992) Site-directed conversion of a cysteine to aspartate leads to the assembly of a 3Fe4S-cluster in *psaC* of Photosystem I1. The photoreduction of F_A is independent of F_B. Biochemistry 31: 5093–5099

Ziegler K, Lockau W and Nitschke W (1987) Bound electron acceptors of Photosystem I: Evidence against the identity of redox center A_1 with phylloquinone. FEBS Lett 217: 16–20

Zurawski G, Bohnert HJ, Whitfield PR and Bottomley W (1982) Nucleotide sequence of the gene for the M*r* 32,000 thylakoid membrane protein from *Spinacia oleracea* and *Nicotiana debneyi* predicts a totally conserved primary translation product of M*r* 38,950. Poc Natl Acad Sci USA 79: 7699–7703

Chapter 31

The Antenna-Reaction Center Complex of Heliobacteria

Jan Amesz
Department of Biophysics, Huygens Laboratory, University of Leiden, P.O. Box 9504, 2300 RA Leiden, The Netherlands

Summary

The heliobacteria are a recently discovered group of photosynthetic bacteria that are characterized by the presence of bacteriochlorophyll (BChl) *g* as major pigment. All of the pigments are contained in a single pigment-protein complex, which contains about 35 BChls *g* per reaction center. The primary electron donor, P798, is presumably a dimer of BChl *g* or of its epimer, BChl *g'*; the primary electron acceptor has been identified as hydroxy chlorophyll *a*. There is evidence that iron-sulfur centers act as secondary electron acceptors, while a *c*-type cytochrome, cytochrome *c*-553, acts as secondary electron donor to P798. Energy transfer from the antenna pigments to the reaction center appears to be distinctly faster than in purple bacteria. The major subunit of the antenna reaction center complex is a peptide of 68 kDa, containing 609 amino acid residues, which shows an about 20% homology to the PS I-A and PS I-B reaction center-core peptides of Photosystem I.

The above observations all indicate that the reaction center of heliobacteria, like that of green sulfur bacteria is related to that of Photosystem I of plants and, in fact, the heliobacteria are in many ways excellent material to study the basic processes of Photosystem I-type reaction centers. However, the reaction center is homodimeric, i.e., it contains two identical subunits that bind the antenna pigments and the reaction center components. This indicates that the ancestral lines of Photosystem I and heliobacteria separated before gene duplication occurred in the first one.

R. E. Blankenship, M. T. Madigan and C. E. Bauer (eds): Anoxygenic Photosynthetic Bacteria, pp. 687–697.

I. Introduction

The heliobacteria have been known for only about a decade. The first species, *Heliobacterium (Hb.) chlorum* was discovered by Gest and Favinger (1983) in a soil sample from the Indiana University campus (Gest, 1994), and soon thereafter several species were discovered in rice fields in Asia and Africa (Beer-Romero and Gest, 1987; Beer-Romero et al., 1988; Ormerod et al., 1990). More recently, various strains or species were also found in more natural habitats, like lakeshore muds (A. Hiraishi, personal communication) and even an alkaline hot spring (Starynin and Gorlenko, 1993). It is recognized now that the heliobacteria, at least the species that have been more than summarily investigated, are all very closely related, and it has been proposed to assign them all to one genus (Ormerod et al., 1990), although one species, *Heliobacillus (Hc.) mobilis* (Beer-Romero and Gest, 1987), is still known under a different genus name. In fact, the only three species that have been studied more or less extensively (*Hb. chlorum, Hc. mobilis and Hb. gestii*) are photosynthetically and spectroscopically virtually indistinguishable, although they are morphologically quite different. 16S rRNA sequence analyses have shown that the heliobacteria are related to the Gram-positive bacteria (Woese et al., 1985; Stackebrandt et al., 1988). All species investigated so far are able to fix nitrogen (Gest and Favinger, 1983; Kimble and Madigan, 1992), which may be of some ecological and economic importance.

For a discussion of the taxonomy and physiology of the heliobacteria refer to Chapter 2 by Madigan and Ormerod in this volume.

II. Pigments and Spectral Properties

The heliobacteria are distinguished from other photosynthetic organisms by the possession of the newly discovered bacteriochlorophyll (BChl) *g* (Brockmann and Lipinski, 1983)(Fig. 1). The pigment resembles BChl *b* with respect to its ethylidene group at C8 (ring II) but has a vinyl group at C3 (ring I), like chlorophyll (Chl) *a*. It is esterified with farnesol (Michalski et al., 1987). The only carotenoid present in *Hb. chlorum* and *Hc. mobilis* is neurosporene (Gest and Favinger, 1983; Kobayashi et al., 1991a). Ultrastructural studies indicate that these pigments are contained in the cytoplasmic membrane only; heliobacteria have no chlorosomes, and intracytoplasmic membranes, as in purple bacteria, appear to be absent too (Gest and Favinger, 1983; Miller et al., 1986; Starynin and Gorlenko, 1993).

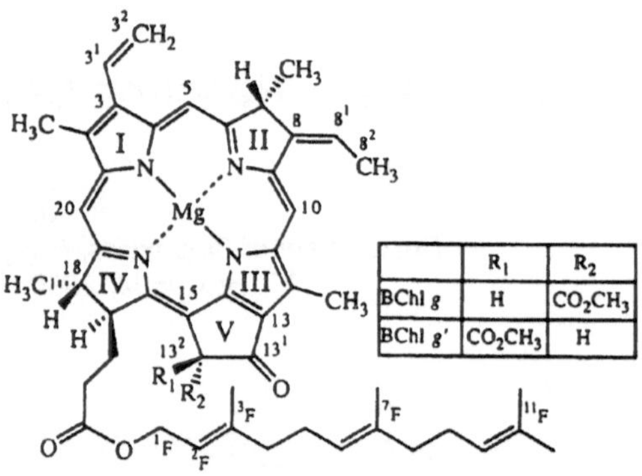

Fig. 1. Structure and partial carbon numbering of BChl *g* and BChl *g'*.

Figure 2 shows the absorption spectrum of BChl *g* in acetone. The Soret transitions of BChl *g* in acetone are located at 364.8 and 404.8 nm., the Q_x and Q_y transitions at 566.4 and 761.6 nm, respectively (Kobayashi et al., 1991a), with a specific extinction

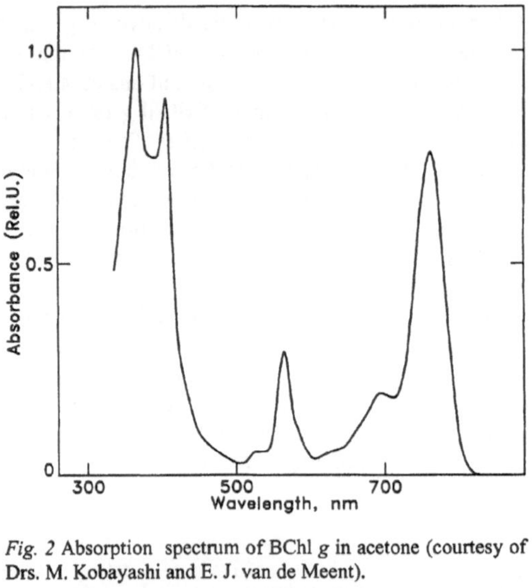

Fig. 2 Absorption spectrum of BChl *g* in acetone (courtesy of Drs. M. Kobayashi and E. J. van de Meent).

Abbreviations: A_0 – primary electron acceptor; BChl – bacteriochlorophyll; Chl – chlorophyll; *Hb.* – *Heliobacterium*; *Hc.* – *Heliobacillus*; P798 – primary electron donor coefficient of 76 ± 4 mM^{-1} cm^{-1} at 761.6 nm (van de Meent et al., 1991). Cells as well as isolated membranes show bands at 368, 409, 575 and 787 nm (Fig. 3 and Fuller et al., 1985). The amplitude of the Q_y band is higher than in solution, with a specific extinction coefficient of 110 mM^{-1} cm^{-1} (Kleinherenbrink and Amesz, 1993). At low temperature, the Q_y band is resolved into at least three components (Fig. 4), supposedly due to three spectral forms of BChl *g*, BChl 778, BChl 793 and BChl 808 (van Dorssen et al., 1985; Smit et al., 1989). The circular dichroism spectrum shows an approximately conservative signal centered at 793 nm, suggesting that BChl 793 may have a dimeric structure (van Dorssen et al., 1985). Linear dichroism measurements indicate that the Q_y transition of BChl 808 lies more or less in the plane of the membrane, while those of the short-wave forms make a larger angle with the membrane plane (van Dorssen et al., 1985). The Q_x transitions are more perpendicular. The orientation of BChl *g* thus is similar to that of antenna BChl *a* in the membrane of purple bacteria (Breton, 1974).

At room temperature, the fluorescence spectrum of *Hb. chlorum* shows a band at 813 nm. Upon cooling to 4 K, the band intensifies about 20-fold and shifts to 819 nm (van Dorssen et al., 1985) indicating that, at least at low temperature, all of the emission comes from BChl 808. Excitation spectra indicate a nearly 100% efficiency of energy transfer from BChl 778 and BChl 793 to BChl 808, while neurosporene transfers its excitation energy with about 70% efficiency.

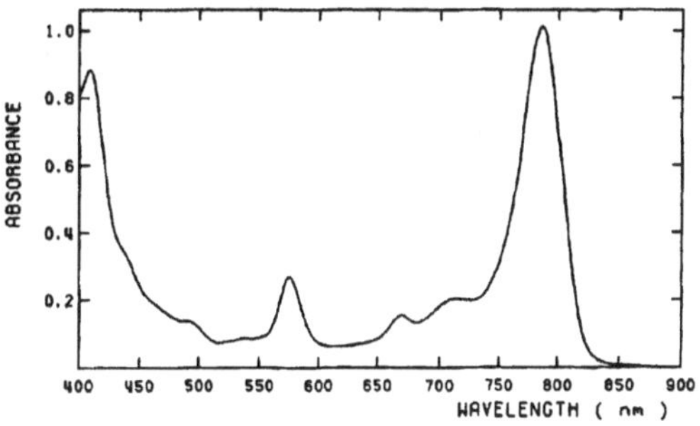

Fig. 3. Absorption spectrum of membranes of *Heliobacterium chlorum*.

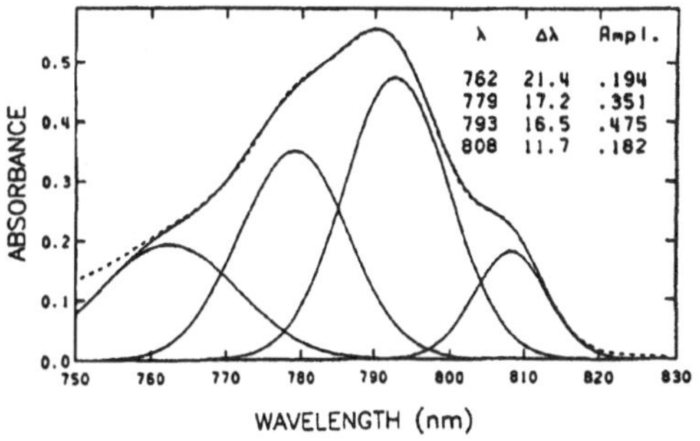

Fig. 4. Gaussian deconvolution of the 5 K absorption spectrum of membranes of *Heliobacterium chlorum*, showing the estimated contributions by the three spectral species, BChl 778, BChl 793 and BChl 808. Band centers, half-widths and amplitudes are listed in the figure. The band at 763 mm may be a composite of vibrational subbands (Smit et al., 1989).

III. The Primary Electron Donor and the Antenna-Reaction Center Complex

The primary electron donor of heliobacteria is called P798, after the wavelength of its maximum bleaching in the Q_y-region upon photooxidation (Fuller et al., 1985; Prince et al., 1985). Its redox potential is 225 – 250 mV and the EPR spectrum of the oxidized radical indicates a dimeric structure (Prince et al., 1985; Brok et al., 1986; Fischer, 1990). A rapid photooxidation of P798 was observed by flash spectroscopy in the ps region, at room temperature as well as at 15 K, confirming the notion that it is the primary electron donor (Nuijs et al., 1985a; van Kan et al., 1989; van Kan, 1991).

At room temperature, the maximum bleaching at 798 nm upon illumination as well as upon chemical oxidation is 1/20 to 1/22 of the antenna absorption maximum at 788 nm, both in *Hb. chlorum* and *Hc. mobilis* (Trost and Blankenship, 1989; van de Meent et al., 1990; Kobayashi et al., 1991a). Evidence to be discussed below indicates that the antenna of both species contains about 35 BChls *g* per P798, which means that the differential extinction coefficient of P798/P 798$^+$ is considerably higher than the specific extinction coefficient of the antenna (see previous section). In fact, a value of 180 mM^{-1} cm^{-1} has been determined recently (Kleinherenbrink and Amesz, 1993), higher than of any primary electron donor determined to date. An even higher number (225–240 mM^{-1} cm^{-1}) was reported by Kleinherenbrink et al. (1994a).

The reaction center and the antenna of heliobacteria appear to be bound to a single pigment-protein complex. Antenna-reaction center complexes have

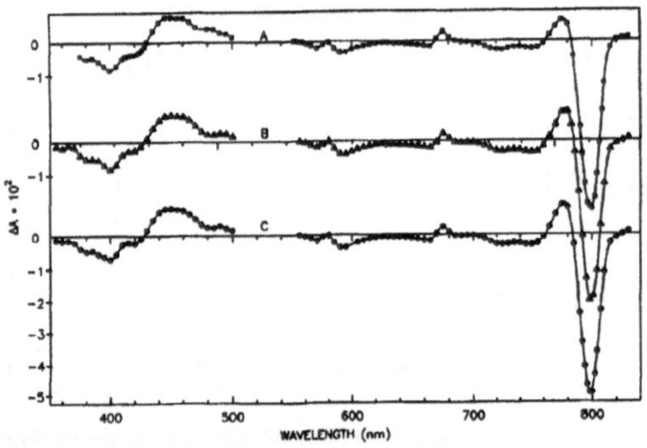

Fig. 5. Absorbance difference spectra caused by photooxidation of P798 obtained upon illumination of (A) membranes of *Heliobacterium chlorum*, (B) the isolated antenna-reaction center complex of *Hb. chlorum* and (C) the antenna-reaction center complex of *Heliobacillus mobilis*. The spectra are normalized to the same antenna absorption (van de Meent et al., 1990).

```
                                                          I
1    MATADAAFNP RAQVFEWFKD KVPATRGAVL KAHINHLGMV AGFVSFVLVH
                                                          II
51   HLSWLSDLVL FAPTPIFYAR LYQLGLDASA RSADALHVAR LHLPAAIIFH
                                                III
101  IIGHIKTPRE DEFLKNVTFG KTLVAQFHFL ALVATLWGMH MAYIGVRGAN
                                                IV
151  GGIVPTGLSF DMFGPITGAT LAGNHVAFGA LLFLGGVFHH FAGFNTKRFA
                                                           V
201  FFEKDWEAVL SVSAQVLAFH FATVVFAMII WNRPDQPILS FYFMQDYALS
                                               VI
251  NYAAPEIREI ASQNPGFLIK QVILGHLVFG VMFWIGGVFH GASLHVRATN
                                                     VII
301  DPKLAEALKD FKMLKRCYDH DFQKKFLALI MFGAFLPIFV SYGIATHNTI
                                                           VIII
351  SDLHHLAKAG MFANMTYINI GTPLHDAIFG SHGTVSDFVA AHAIAGGLHF
401  TMVPLWRMVF FSKVSPWTTK VGMKAKRDGE FPCLGPAYGG TCSISLVDQF
               IX
451  YLAIFFSLQV IAPAWFYLDG CWMGSFVATS SEVYKQAAEL FKANPIWFSL
                                                     X
501  HAVSNFTSEV TSATSSLKPL VCSNTTMVTH FKPCWAAHFI WAFTFSMLFQ
                                                           XI
551  YRGSRDEGAM VLKWAHEQVG LGFAGKVYNR ALSLKEGKAI GTFLFFKMTV
601  LCMWCLAMV*
```

Fig. 6 Amino acid sequence of the antenna-reaction center peptide of *Heliobacillus mobilis*. 11 Predicted membrane-spanning helices are underlined (Liebl et al., 1992).

been obtained by detergent solubilization of membranes of *Hb. chlorum* and *Hc. mobilis* (Trost and Blankenship, 1989; van de Meent et al., 1990) and from *Hb. gestii* (Trost et al., 1992). These complexes have the same spectroscopic properties as isolated membranes and they are fully active in P798 photooxidation (Fig. 5). The fluorescence excitation spectrum showed unimpaired efficiency for energy transfer from short wave absorbing BChls *g* to BChl 808. (van de Meent et al., 1990). These observations indicate that the isolated complexes contain the complete pigment complement of the cells, although some neurosporene may be lost during the isolation (Trost and Blankenship, 1989), and that the structural integrity of the complex is essentially conserved during its isolation.

SDS-PAGE of purified antenna-reaction center preparations showed only one major band on the electrophoresis gel (Trost and Blankenship, 1989; van de Meent et al., 1990). Attempts to sequence the polypeptide were at first hampered by blockage at the N-terminus, but after enzymatic and chemical cleavage, 15 peptide fragments were sequenced from the peptide of *Hc. mobilis* (Trost et al., 1992; Liebl et al., 1993). Subsequent identification and sequencing of the corresponding gene resulted in the amino acid sequence shown in Fig. 6 (Liebl et al., 1992, 1993). Inspection of the sequence reveals a distinct, but limited homology of about 20% with the PS I-A and PS I-B reaction center-core peptides of Photosystem I. Although the molecular mass (68 kDa) and the number of amino acid residues (609) are about 20% smaller than of PS I-A and PS I-B, the number of predicted membrane-spanning helices is the same (11). These helices contain 17 histidines as compared to 29 and 24, respectively, on the Photosystem I peptides (Golbeck and Bryant, 1991), in agreement with the notion that the antenna-reaction center complex of heliobacteria contains fewer (B)Chls than does the Photosystem I core. It may be noted, however, that the preliminary results of an X-ray analysis of the reaction center complex of Photosystem I revealed only 8 such helices (Krauss et al., 1993), suggesting that the hydropathy plots may not be fully reliable. The homology with the PS I-A and B peptides is mainly apparent in the putative reaction center binding areas. A homology of about 50% was found in the extramembranous loop between helices VIII and IX (Fig. 6), that contains two conserved cysteines. These residues are the presumed binding sites for iron in the analogon to F_x, the first iron-sulfur electron acceptor of Photosystem I.

There is no evidence for the existence of a second, related peptide in the antenna-reaction complex of *Hc. mobilis*, and the same applies to the recently sequenced core-complex of green sulfur bacteria, which contains a peptide of 730 amino acid residues and shows a homology pattern similar to that of heliobacteria (Büttner et al., 1992a,b; see also Chapter 30 by Feiler and Hauska). These observations strongly indicate that the reaction centers of heliobacteria and of green sulfur bacteria, in contrast to Photosystem I, are homodimers. The results clearly support the notion, based on studies of electron transfer (see below), that the three reaction centers are evolutionarily of the same origin, but one has to conclude that the evolutionary tree has branched off before gene duplication occurred in the ancestral line of Photosystem I (Liebl et al., 1992, 1993; Blankenship, 1992; Büttner et al., 1992b). Interestingly, the N-terminal part of the heliobacterial protein, including the first six major hydrophobic regions, shows a significant homology to CP47, suggesting an evolutionary relationship to this antenna protein of Photosystem II (Vermaas, 1994).

A quantitative pigment analysis of isolated antenna-reaction particles as well as membrane fragments and cells of *Hb. chlorum* and *Hc. mobilis* revealed the presence of neurosporene and BChl *g*, but essentially no bacteriopheophytin when material from rapidly growing cells was used (Kobayashi et al., 1991a, van de Meent et al., 1991). In addition, small amounts of the epimer of BChl *g*, BChl *g′* were found. The latter pigment differs from BChl *g* in the configuration of the carboxymethyl group at $C13^2$ (ring V), which has the opposite position with respect to the macrocycle (Fig. 1). The absorption spectrum of BChl *g′* in organic solvents resembles that of BChl *g*, but the circular dichroism spectrum is distinctly different (Kobayashi et al., 1991a). Control experiments indicated that BChl *g′* was not produced by epimerization during or after extraction; however, in old cells some artifactual epimerization appeared to occur (Kobayashi et al., 1991a,b).

The epimer of Chl *a*, Chl *a′*, has been implied to constitute the primary electron donor of Photosystem I, P700 (Maeda et al., 1992). By analogy, BChl *g′* might function as primary electron donor in heliobacteria. The amount of BChl *g′* was found to be one molecule per ca. 17 BChls *g* for both species investigated, and it was therefore proposed that P798 would be a dimer of BChl *g′*, giving a BChl *g* to reaction center ratio of ca. 35. It may be noted that such a number would be in fair agreement with the number of histidines in the membrane spanning regions of the reaction center peptides.

IV. The Primary Charge Separation

Measurements of flash-induced absorbance changes in membranes of *Hb. chlorum* at 800 nm (Nuijs et al, 1985a) showed a transient bleaching, which reversed in some tens of ps, and a 'permanent' one, which did not reverse on the timescale (several ns) of the experiment. When the reaction centers had been closed by a 'preflash', given a few ns before the actual measurement, only the transient bleaching was seen. Thus, the transient bleaching, which will be discussed in more detail in the next section, can be ascribed to excited antenna BChl *g*, whereas the 'permanent' one is caused by the oxidation of P798. A bleaching was also observed near 670 nm. This bleaching again was only observed with 'active' but not with closed reaction centers, indicating that it is caused by the photoreduction of the primary, or at least an early, electron acceptor, which we shall call A_0. This bleaching decayed in 500–800 ps (Nuijs et al., 1985a; van Kan et al., 1989; Lin et al., 1994a), presumably by electron transfer to the next acceptor, X_1. Because of energy transfer and absorbance changes occurring in the antenna, an accurate determination of the rate of charge separation has not yet been possible, but it must be faster than 10 ps (van Noort et al., 1992a). In fact, model calculations indicate a time constant of, at most, 1.2 ps for the primary charge separation (Lin et al., 1994a). Measurements at 15 K confirmed the above picture. Photooxidation of P798 could now be observed by the bleaching of a band centered at 793 nm and the reoxidation of A_0^- now occurred in 300 ps, i.e. 2–3 times faster than at room temperature (van Kan et al., 1989; van Kan, 1991). A similar effect has been observed for electron transfer from reduced bacteriopheophytin to the first acceptor quinone in purple bacteria (Kirmaier et al., 1985).

It thus appears that the minor band at 669 nm in the absorption spectrum of heliobacteria (Fig. 3) is, at least in part, due to the primary acceptor. This situation is reminiscent of that in green sulfur bacteria (Nuijs et al., 1985b, Shuvalov et al., 1986; see also Chapter 30 by Feiler and Hauska). Experiments of van Noort et al. (1992a) showed that the bleaching at 670 nm was brought about not only upon excitation at 532 nm, as in the experiments of Nuijs et al., but also upon excitation at 790 nm, excluding the possibility that it would be due to an excited state of the pigment. Supporting evidence came also from measurements of Kleinherenbrink et al. (1991, 1994b), who showed that the lifetime of A_0^- could be extended to 17 ns by applying strongly reducing conditions. The same time constant was now observed in the decay of P798$^+$, whereas P798$^+$ normally decays in several ms.

The radical pair recombination resulted in the formation of the triplet of P798, with a yield of 20–25%. At 6 K, the yield increased slightly, and radical pair recombination slowed down to 55 ns. The lifetimes of the triplet was 35 μs at room temperature and 500 μs at 6 K (Nitschke et al., 1990; Kleinherenbrink et al., 1991). By means of absorption detected magnetic resonance (ADMR) the zero-field splitting parameters (D and E values) of the reaction center triplet, as well as the decay times for the x, y and z sublevels at 1.2 K have been determined (Vrieze et al., 1992). The latter were found to be 270 ± 20, 360 ± 45 and 1150 ± 60 μs, respectively. Very similar decay times were observed for the triplet of BChl *g* in ethanol. Absorption difference spectra of reaction center triplet formation show a minimum at 794 nm, and have been obtained by flash spectroscopy (Kleinherenbrink et al., 1991) as well as by ADMR (Vrieze et al., 1992). The latter method gives more detailed spectra.

The chemical identity of the primary electron acceptor has recently been determined after purification by HPLC (van de Meent et al., 1991). A pigment was isolated from membranes of *Hb. chlorum* and *Hc. mobilis* which had essentially the same absorption spectrum as Chl *a* from plants, but showed a higher polarity. NMR and mass spectroscopy established its identity as 8^1-OH-Chl *a*, esterified with farnesol. This observation again points to the similarity with the reaction centers of Photosystem I and green sulfur bacteria, where the primary acceptors are Chl *a* (Golbeck and Bryant, 1991) and an isomer of Chl *a* (van de Meent et al., 1992), respectively. The amount of 8^1-OH-Chl *a* was found to be one molecule per 17 BChls *g*, the same as of BChl *g'* (see previous section). With a BChl *g* to reaction center ratio of about 35, this would indicate the presence of two molecules of OH-Chl *a* per reaction center. Since the reaction center is presumably homodimeric, and may be assumed to be inherently symmetrical it is likely that both molecules occupy equivalent positions and that there is no preferred pathway for electron transfer, as in purple bacteria and in Photosystem II. The Q_y transitions of these pigments make a large angle with the plane of the membrane (Francke et al., 1994). It is conceivable that OH-Chl *a* is formed by isomerization and oxidation of BChl *g* at ring II (see also Michalski et al., 1987); some additional OH-Chl *a* and other oxidation products of BChl *g* are formed in old cells (van de Meent et al., 1991).

V. Transfer and Trapping of Excitation Energy

At room temperature, the fluorescence lifetime of BChl *g* in membranes and isolated antenna-reaction center complexes of *Hc. mobilis* is ca. 25 ps (Trost and Blankenship, 1989), which explains, at least qualitatively, the low fluorescence yield (van Dorssen et al., 1985). Similar results were obtained in time-resolved absorption measurements on membranes of *Hb. chlorum* and *Hc. mobilis*, which showed a bleaching, centered at around 805 nm, which decayed in 20–30 ps (van Noort et al., 1992a; Lin et al., 1992, 1994a). The bleaching developed in less than about 1 ps (Lin et al., 1992, 1994a) indicating a very rapid equilibration of the excitations among the various spectral forms of BChl *g*, BChl 798, 793 and 808 (see section II). Its lifetime was essentially independent of the redox state of P798, in agreement with measurements of Deinum et al. (1991), who showed that, at room temperature, P798 and P798$^+$ are approximately equally effective in quenching antenna fluorescence. At 6 K, P798$^+$ is almost twice as effective a quencher as P798.

The above measurements thus lead to a simple model, where the excitations are equilibrated between BChl 778, BChl 793 and BChl 808 within a very short time, and energy transfer to P798 or P798$^+$ effectively takes 20–30 ps (van Noort et al., 1992a), but a model involving more rapid transfer to P798 has also been proposed (Lin et al., 1992, 1994a). Energy transfer to the reaction center thus is significantly faster than in purple bacteria (Sundström and van Grondelle, 1992), which may be explained by the binding of P798 and antenna pigments to the same protein complex. Because the equilibration is so fast, the results do not permit distinguishing between energy transfer from BChl 793 and BChl 808 to P798. The lifetime of excited P798 appears to be much shorter than 20 ps, since the bleaching (i.e. reduction) of A_0 was found to occur at the same rate as the disappearance of excited antenna BChl *g* (van Noort et al., 1992a; but see also Lin et al., 1994a).

The fluorescence and absorption kinetics at low temperature cannot be explained by a simple model, however. The action spectrum of P798 photooxidation indicates that, at least at low temperature energy transfer to the reaction center proceeds via BChl 808 (Kleinherenbrink et al., 1992) and steady state fluorescence (van Dorssen et al., 1985) as well as time resolved fluorescence and absorption measurements (van Noort et al., 1992b, 1994; Kleinherenbrink et al., 1993a) indicate very fast energy transfer from the other BChls to BChl 808, presumably within 1 ps. However, the decay of the excitations on BChl 808 shows a complicated pattern, with various time constants, ranging from about 4 to 900 ps. The 4 ps decay appears to reflect energy transfer to the (oxidized) reaction center, while the longer-lived components are associated with BChl 808 species with red-shifted absorption and fluorescence spectra (van Noort et al., 1992b, 1994; Kleinherenbrink et al., 1993a), suggesting the existence of different sites in the antenna which are only weakly coupled to each other and to the reaction center. Energy transfer to the open reaction center was found to occur in about 70 ps (Lin et al., 1994b).

In addition to the reaction center triplet, mentioned in the previous section, antenna triplets are also generated at liquid helium temperature. The overall yield is 1–2%. Most of these appear to be produced by intersystem crossing on long-wavelength forms of BChl 808 (Nitschke et al., 1990; Kleinherenbrink et al., 1991) but a triplet on a BChl *g* species absorbing at 798 nm was also observed (Vrieze et al., 1992). Three different triplets of BChl 808, with somewhat different zero field splitting parameters could be distinguished, confirming the heterogeneity of this antenna component (Vrieze et al., 1992). The lifetimes of the antenna triplets are not much different from those of the reaction center triplet (section IV).

VI. Secondary Electron Transport

There is various evidence, mainly based on low temperature EPR measurements, that iron-sulfur centers act as secondary electron acceptors (Prince et al., 1985 Brok et al., 1986; Smit et al., 1987; Nitschke et al., 1990; Fischer, 1990), as in Photosystem I (Golbeck, 1993). EPR signals with g-values of 2.07, 1.93 and 1.89 and of 2.05, 1.95 and 1.90 have been observed which have been attributed to centers which are, except for their relative midpoint potentials, analogous to F_B and F_A in Photosystem I (Nitschke et al., 1990). Since *rates* of reduction of iron sulfur centers are hard to measure, and since the EPR method cannot be applied at room temperature, direct evidence concerning the role of these iron sulfur centers is not available. Optical evidence for the role of an iron sulfur center analogous to F_x in Photosystem I has recently been reported (Kleinherenbrink et al., 1994a), but no corresponding EPR signal has been observed so far (Trost and Blankenship, 1989). Indirect evidence for the presence of such a center comes from the primary structure analysis (Trost et al., 1992; Liebl et al., 1992, 1993).

The isolated antenna-reaction center complex of *Hc. mobilis* contains about one molecule of menaquinone (vitamin K-2) (Trost and Blankenship, 1989) and 3–4 times larger amounts have been found in isolated membranes of *Hb. chlorum* and *Hc. mobilis*, the main component being menaquinone-9 (Hiraishi, 1989; Trost and Blankenship, 1989; Kleinherenbrink et al., 1993b) Photoaccumulation of reduced quinone was observed by EPR (Brok et al., 1986) and the double reduction of a quinone acceptor was proposed to explain the titration behavior of the recombination kinetics of $P798^+$ (Trost et al., 1992). These observations thus would indicate that menaquinone functions as a first secondary electron acceptor in heliobacteria in an analogous way as phylloquinone (vitamin K-1) does in Photosystem I (Golbeck and Bryant, 1991). However, recent experiments have shed doubt on this hypothesis. It was observed that by extraction with appropriate ether-water mixtures essentially all menaquinone could be removed from membranes while secondary electron transport and charge recombination were not affected (Kleinherenbrink et al., 1993b). In Photosystem I preparations such treatments have a major effect on electron transport (Ikegami et al., 1987). These results do not exclude a function for a very strongly bound quinone, but is should be noted that there is also evidence that at least some of the rates of electron transfer in heliobacteria are about an order of magnitude slower than for Photosystem I, including that of electron transfer from A_0 to the next electron acceptor (see section I and Fig. 7, and also Kleinherenbrink and Amesz, 1993), indicating that there may be rather fundamental differences between the electron acceptor chains of Photosystem I and heliobacteria.

The electron donor to P798 is a *c*-type cytochrome, cytochrome *c*-553. In intact cells of *Hb. chlorum* its oxidation is fairly rapid, with half times of 110 and

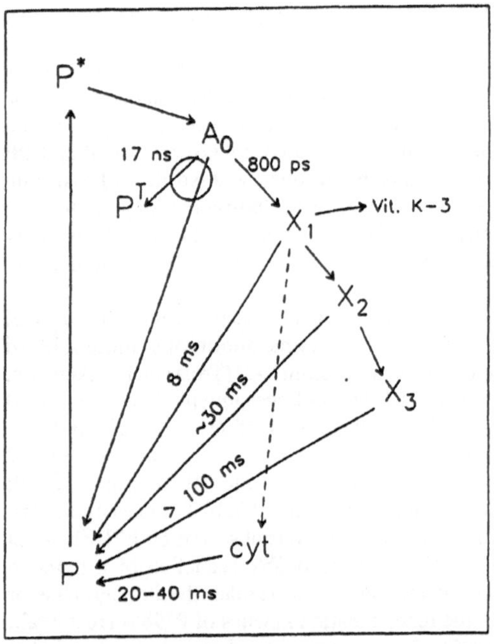

Fig. 7. Tentative scheme for electron transfer in membranes of heliobacteria at room temperature (Kleinherenbrink and Amesz, 1993). The existence of three electron acceptors, which may all be iron-sulfur centers, is in part proposed on basis of kinetic evidence. A_0 may represent two identical acceptor molecules acting in parallel (Section IV). Further details are discussed in the text.

700 μs (Vos et al., 1989), but in isolated membranes a much slower and incomplete photooxidation with a time constant of about 6 ms occurs (Prince et al., 1985; Smit et al., 1987). The cause of this effect is not known; it might be due to a change in binding constant. The cytochrome is removed by detergent treatment to solubilize the antenna-reaction center complex (Trost and Blankenship, 1989; van de Meent et al., 1990), indicating that it is not strongly bound to the reaction center. Recent experiments have shown that the 6 ms time constant is mainly determined by a concomitant back reaction of $P798^+$ with a reduced electron acceptor (Kleinherenbrink and Amesz, 1993). When the back reaction was prevented by the addition of vitamin K-3 (menadione) as an artificial electron acceptor, P798 and cytochrome *c*-553 reacted in stoichiometric amounts with a time constant of approximately 20 ms.

Although outside the scope of this chapter, it may be mentioned here that the cytoplasmic membrane of heliobacteria contains a cytochrome *bc* complex which may function in cyclic electron transport. The presence of a Rieske iron-sulfur center has been demonstrated (Liebl et al., 1990) and recently the isolation of a cytochrome *bc* complex by detergent solubilization of membranes of *Hc. mobilis* was reported (Nitschke and Liebl, 1992). Interestingly, in various aspects the complex resembles more the Cyt b_6f-type complex of oxygenic photosynthesis than the *bc*-type complex of purple bacteria and mitochondria (Liebl, 1992).

VII. Concluding Remarks

Hopefully, the foregoing sections have demonstrated that the heliobacteria, although they may occupy a tiny niche in the ecological spectrum of photosynthetic organisms, are of considerable scientific interest. Even to the reader who is not interested in the heliobacteria *per se* it should be clear that they present a paradigm of the Photosystem I-type reaction center which in many ways is more amenable to investigation than the reaction center of Photosystem I of plants or the related reaction center of green sulfur bacteria. Heliobacteria have a relatively small antenna, and for many experiments the preparation of membrane fragments suffices to obtain material with a higher reaction center to pigment ratio than has the core complex of Photosystem I, which can only be obtained by detergent solubilization and separation of the solubilized pigment protein complexes. A similar comparison applies with respect to the green sulfur bacteria, which have a very large antenna which is difficult to remove entirely. Moreover, the primary electron acceptor, in contrast to Photosystem I of plants, is chemically different from the bulk pigment, and absorbs at a wavelength (670 nm) far removed from the Q_y absorption of the antenna and the primary electron donor, which facilitates optical measurements on charge separation considerably.

Altogether, the heliobacteria provide excellent material for studies of the basic processes of photosynthesis, and although these have been performed in only a few laboratories to date, the heliobacteria are potentially as useful for such studies as the much more extensively investigated purple bacteria.

References

Beer-Romero P and Gest H (1987) *Heliobacillus mobilis*, a peritrichously flagellated anoxyphototroph containing bacteriochlorophyll *g*. FEMS Microbiol Lett 41: 109–114

Beer-Romero P, Favinger JL and Gest H (1988) Distinctive properties of bacilloform photosynthetic heliobacteria. FEMS Microbiol Lett 49: 541–454

Blankenship R.E. (1992) Origin and early evolution of photosynthesis. Photosynth Res 33: 91–111

Breton J (1974) The state of chlorophyll and carotenoid in vivo II. A linear dichroism study of pigment orientation in photosynthetic bacteria. Biochem Biophys Res Comm 95: 1011–1017

Brockmann H and Lipinski A (1983) Bacteriochlorophyll *g*. A new bacteriochlorophyll from *Heliobacterium chlorum*. Arch Microbiol 136: 17–19

Brok M, Vasmel H, Horikx JTG and Hoff AJ (1986) Electron transport components of *Heliobacterium chlorum* investigated by EPR spectroscopy at 9 and 35 GHz. FEBS Lett 194: 322–326

Büttner M, Xie D-L, Nelson H, Pinther W, Hauska G and Nelson N (1992a) The Photosystem I-like P840-reaction center of green S-bacteria is a homodimer. Biochim Biophys Acta 1101: 154–156

Büttner M, Xie D-L, Nelson H, Pinther W, Hauska G and Nelson N (1992b) Photosynthetic reaction center genes in green sulfur bacteria and in Photosystem I are related. Proc Natl Acad Sci USA 89: 8135–8139

Deinum G, Kramer H, Aartsma TJ, Kleinherenbrink FAM and Amesz J (1991) Fluorescence quenching in *Heliobacterium chlorum* by reaction centers in the charge separated state. Biochim Biophys Acta 1058: 339–344

Fischer MR (1990) Photosynthetic electron transfer in *Heliobacterium chlorum* studied by EPR spectroscopy. Biochim Biophys Acta 1015: 471–482

Francke C, Otte SCM, van der Heiden JC and Amesz J (1994) Spurious circular dichroism signals with intact cells of heliobacteria. Biochim Biophys Acta 1186: 75–80

Fuller RC, Sprague SG, Gest H and Blankenship RE (1985) A unique photosynthetic reaction center from *Heliobacterium chlorum*. FEBS Lett 182: 345–349

Gest H (1994) Discovery of the heliobacteria. Photosynth Res 41: 17–21

Gest H and Favinger JL (1983) *Heliobacterium chlorum*, an anoxygenic brownish-green photosynthetic bacterium containing a 'new' form of bacteriochlorophyll. Arch Microbiol 136: 11–16

Golbeck JH and Bryant DA (1991) Photosystem I. Curr Top Bioenerg 16: 83–177

Golbeck JH (1993) Shared thematic elements in photochemical reaction centers. Proc Natl Acad Sci USA 90: 1642–1646

Hiraishi A (1989) Occurrence of menaquinone as the sole isoprenoid quinone in the photosynthetic bacterium *Heliobacterium chlorum*. Arch Microbiol 151: 378–379

Ikegami I, Sétif P and Mathis P (1987) Absorption studies of Photosystem I photochemistry in the absence of vitamin K 1. Biochim Biophys Acta 894: 414–422

Kimble LK and Madigan MT (1992) Nitrogen fixation and nitrogen metabolism in heliobacteria. Arch Microbiol 158: 155–161

Kirmaier C, Holten D and Parson WW (1985) Temperature and detection-wavelength dependence of the picosecond electron-transfer kinetics measured in *Rhodopseudomonas sphaeroides* reaction centers. Resolution of a new spectral and kinetic component. Biochim Biophys Acta 810: 33–48

Kleinherenbrink FAM and Amesz J (1993) Stoichiometries and rates of electron transfer and charge recombination in *Heliobacterium chlorum*. Biochim Biophys Acta 1143: 77–83

Kleinherenbrink FAM, Aartsma TJ and Amesz J (1991) Charge separation and formation of bacteriochlorophyll triplets in *Heliobacterium chlorum*. Biochim Biophys Acta 1057: 346–352

Kleinherenbrink FAM, Deinum G, Otte SCM, Hoff AJ and Amesz J (1992) Energy transfer from long-wavelength absorbing antenna bacteriochlorophylls to the reaction center. Biochim Biophys Acta 1099: 175–181

Kleinherenbrink FAM, Cheng P, Amesz J and Blankenship RE (1993a) Lifetimes of bacteriochlorophyll fluorescence in *Rhodopseudomonas viridis* and *Heliobacterium chlorum* at low temperatures. Photochem Photobiol 57: 13–18

Kleinherenbrink FAM, Ikegami I, Hiraishi A, Otte SCM and Amesz J (1993b) Electron transfer in menaquinone-depleted membranes of *Heliobacterium chlorum*. Biochim Biophys Acta 1142: 69–73

Kleinherenbrink FAM, Chiou H-C, LoBrutto R and Blankenship RE (1994a) Spectroscopic evidence for the presence of an iron-sulfur center similar to F_X of Photosystem I in *Heliobacillus mobilis*. Photosynth Res 41: 115–123

Kleinherenbrink FAM, Hastings G, Wittmershaus BP and Blankenship RE (1994b) Delayed fluorescence from Fe-S type photosynthetic reaction centers at low redox potential. Biochemistry 33: 3096–3105

Kobayashi M, van de Meent EJ, Erkelens C, Amesz J, Ikegami I and Watanabe T (1991a) Bacteriochlorophyll *g* epimer as a possible reaction center component of heliobacteria. Biochim Biophys Acta 1057: 89–96

Kobayashi M, Watanabe T, Ikegami I, van de Meent EJ and Amesz J (1991b) Enrichment of bacteriochlorophyll *g'* in membranes of *Heliobacterium chlorum* by ether extraction. Unequivocal evidence for its existence in vivo. FEBS Lett 284: 129–131

Krauss N, Hinrichs W, Witt I, Fromme P, Pritzkow W, Dauter Z, Betzel C, Wilson KS, Witt HT and Saenger W (1993) Three-dimensional structure of system I of photosynthesis at 6 Å resolution. Nature 361: 326–331

Liebl U (1992) Components of electron transport in heliobacteria. Doctoral Thesis, University of Regensburg

Liebl U, Rutherford AW and Nitschke W (1990) Evidence for a unique Rieske iron-sulfur centre in *Heliobacterium chlorum*. FEBS Lett 261: 427–430

Liebl U, Mockensturm-Wilson M, Trost JT, Brune DC, Blankenship RE and Vermaas WFJ (1992) The reaction center core polypeptide in the photosynthetic bacterium *Heliobacillus mobilis*. In: Murata N (ed) Research in Photosynthesis, Vol II, pp 595–598. Kluwer Academic Publishers, Dordrecht

Liebl U, Mockensturm-Wilson M, Trost J, Brune D, Blankenship R and Vermaas W (1993) Single core polypeptide in the reaction center of the photosynthetic bacterium *Heliobacillus mobilis*. Structural implications and relations to other photosystems. Proc Natl Acad Sci USA 90: 7134–7128

Lin S, Chiou H-C and Blankenship RE (1992) Energy transfer and photochemistry in *Heliobacillus mobilis*. In: Murata N (ed) Research in Photosynthesis Vol I, pp 417–420. Kluwer Academic Publishers, Dordrecht

Lin S, Chiou H-C, Kleinherenbrink FAM and Blankenship RE (1994a) Time-resolved spectroscopy of energy and electron transfer processes in the photosynthetic bacterium *Heliobacillus mobilis*. Biophys J 66: 437–445

Lin S, Kleinherenbrink FAM, Chiou H-C and Blankenship RE (1994b) Spectral heterogeneity and time-resolved spectroscopy of excitation energy transfer in membranes of *Heliobacillus mobilis* at low temperatures. Biophys J 67: 2479–2489

Maeda H, Watanabe T, Kobayashi M and Ikegami I (1992) Presence of two chlorophyll *a'* molecules at the core of Photosystem I. Biochim Biophys Acta 1099: 343–346

Michalski TJ, Hunt JE, Bowman MK, Smith U, Bardeen K, Gest H, Norris JR and Katz JJ (1987) Bacteriopheophytin *g*: Properties and some speculations on a possible primary role for bacteriochlorophylls *b* and *g* in the biosynthesis of chlorophylls. Proc Natl Acad Sci USA 84: 2570–2574

Miller KR, Jacob JS, Smith U, Kolaczkowski S and Bowman MK (1986) *Heliobacterium chlorum:* Cell organization and structure. Arch Microbiol 146: 111–114

Nitschke W and Liebl U (1992) The cytochromes in *Heliobacterium mobilis*. In: Murata N (ed) Research in Photosynthesis Vol III, pp 507–510. Kluwer Academic Publishers, Dordrecht

Nitschke W, Sétif P, Liebl U, Feiler U and Rutherford AW (1990) Reaction center photochemistry of *Heliobacterium chlorum*. Biochemistry 29: 11079–11088

Nuijs AM, van Dorssen RJ, Duysens LNM and Amesz J (1985a) Excited states and primary photochemical reactions in the photosynthetic bacterium *Heliobacterium chlorum*. Proc Natl Acad Sci USA 82: 6865–6868

Nuijs AM, Vasmel H, Joppe HLP, Duysens LNM and Amesz J (1985b) Excited states and primary charge separation in the pigment system of the green photosynthetic bacterium *Prosthecochloris aestuarii* as studied by picosecond absorbance difference spectroscopy. Biochim Biophys Acta 807: 24–34

Ormerod J, Nesbakken T and Torgersen Y (1990) Phototrophic bacteria that form heat resistant endospores. In: M. Baltscheffsky (ed) Current Research in Photosynthesis Vol IV, pp 935–938. Kluwer Academic Publishers, Dordrecht

Prince RC, Gest H and Blankenship RE (1985) Thermodynamic properties of the photochemical reaction center of *Heliobacterium chlorum*. Biochim Biophys Acta 810: 377–384

Shuvalov VA, Amesz J and Duysens LNM (1986) Picosecond spectroscopy of isolated membranes of the photosynthetic green sulfur bacterium *Prosthecochloris aestuarii* upon selective excitation of the primary electron donor. Biochim Biophys Acta 851: 1–5

Smit HWJ, Amesz J and van der Hoeven MFR (1987) Electron transport and triplet formation in membranes of the photosynthetic bacterium *Heliobacterium chlorum* at low temperature. Biochim Biophys Acta 893: 232–240

Smit HWJ, van Dorssen RJ and Amesz J (1989) Charge separation and trapping efficiency in membranes of *Heliobacterium chlorum* at low temperature. Biochim Biophys Acta 973: 212–219

Stackebrandt E, Embley M and Weckesser J (1988) Phylogenetic, evolutionary and taxonomic aspects of phototrophic eubacteria. In: Olson JM, Ormerod JG, Amesz J, Stackebrandt E and Trüper HG (eds) Green Photosynthetic Bacteria, pp 201–215. Plenum Press, New York

Starynin DA and Gorlenko VM (1993) Sulphide oxidizing spore forming heliobacteria isolated from a thermal hot spring. Microbiology (Engl transl) 62: 99–104

Sundström V and van Grondelle R (1992) Ultrafast dynamics of excitation energy transfer and trapping in BChl *a* and BChl *b*-containing photosynthetic bacteria. J Photochem Photobiol B: Biol 15: 141–150

Trost JT and Blankenship RE (1989) Isolation of a photoactive photosynthetic reaction center-core antenna complex from *Heliobacillus mobilis*. Biochemistry 28: 9898–9904

Trost JT, Brune DC and Blankenship RE (1992) Protein sequences and redox titrations indicate that the electron acceptors in reaction centers from heliobacteria are similar to Photosystem I. Photosynth Res 32: 11–22

van de Meent EJ, Kleinherenbrink FAM and Amesz J (1990) Purification and properties of an antenna-reaction center complex from heliobacteria. Biochim Biophys Acta 1015: 223–230

van de Meent EJ, Kobayashi M, Erkelens C, van Veelen PA and Amesz J (1991) Identification of 8^1-hydroxychlorophyll *a* as a functional reaction center pigment in heliobacteria. Biochim Biophys Acta 1058: 356–362

van de Meent EJ, Kobayashi M, Erkelens C, van Veelen PA, Otte SCM, Inoue K, Watanabe T and Amesz J (1992) The nature of the primary electron acceptor in green sulfur bacteria. Biochim Biophys Acta 1102: 371–378

van Dorssen RJ, Vasmel H and Amesz J (1985) Antenna organization and energy transfer in membranes of *Heliobacterium chlorum*. Biochim Biophys Acta 809:, 199–203

van Kan PJM (1991) Energy transfer and charge separation in photosynthetic systems at low temperature. Doctoral thesis, University of Leiden

van Kan PJM, Aartsma TJ and Amesz J (1989) Primary processes in *Heliobacterium chlorum* at 15 K. Photosynth Res 22: 61–68

van Noort PI, Gormin DA, Aartsma TJ and Amesz J (1992a) Energy transfer and primary charge separation in *Heliobacterium chlorum* studied by picosecond time-resolved transient absorption spectroscopy. Biochim Biophys Acta 1140: 15–21

van Noort PI, Aartsma TJ and Amesz J (1992b) Energy transfer in *Heliobacterium chlorum* at room temperature and at 15 K. In: Murata N (ed) Research in Photosynthesis Vol II, pp 595–598. Kluwer Academic Publishers, Dordrecht

van Noort PI, Aartsma TJ and Amesz J (1994) Energy transfer and trapping of excitations in membranes of *Heliobacterium chlorum* at 15 K. Biochim Biophys Acta 1184: 21–27

Vermaas WFJ (1994) Evolution of heliobacteria: Implications for photosynthetic reaction center complexes. Photosynth Res 41: 285–294

Vos MH, Klaassen HE and van Gorkom HF (1989) Electron transport in *Heliobacterium chlorum* whole cells studied by electroluminescence and absorbance difference spectroscopy. Biochim Biophys Acta 973: 163–169

Vrieze J, van de Meent EJ and Hoff AJ (1992) Triplet-minus-singlet absorbance difference spectroscopy of *Heliobacterium chlorum* monitored with absorbance detected magnetic resonance. In: Breton J and Verméglio A (eds) The Photosynthetic Bacterial Reaction Center II: Structure, Spectroscopy and Dynamics, pp 67–78. Plenum Press, New York, London

Woese CR, Debrunner-Vossbrinck BA, Oyaizu H, Stackebrandt E and Ludwig W (1985) Gram-positive bacteria: Possible photosynthetic ancestry. Science 229: 762–765

Chapter 32

Biochemical and Spectroscopic Properties of the Reaction Center of the Green Filamentous Bacterium, *Chloroflexus aurantiacus*

Reiner Feick, Judith A. Shiozawa and Angelika Ertlmaier
Max Planck Institut für Biochemie, Am Klopferspitz 18a, 82152 Martinsried, Germany

Summary

The reaction center of the thermophilic green bacterium *Chloroflexus (Cf.) aurantiacus* has a bacteriochlorophyll donor-bacteriopheophytin-quinone acceptor system which is characteristic for purple bacteria and not for Chlorobiaceae reaction centers. It is comprised of 3 BChl *a*-, 3 BPheo *a*-, 1-2 menaquinone molecules and only two protein subunits with almost identical molecular masses of approximately 35,000 Da. The RC complex possesses thermal stability. Despite the absence of the so called H-protein subunit, electron transfer steps are not impaired although the kinetics are somewhat slower than that in the three protein subunit RC of purple bacteria. Hence, the RC of *Cf. aurantiacus* is the smallest functional RC isolated thus far. The two polypeptides possess a moderate identity of 40% each to the L- and M proteins of *Rhodobacter sphaeroides* and *Rhodopseudomonas viridis* RC, respectively. A characteristic feature of the N-terminus of the *Cf. aurantiacus* RC is the extended hydrophilic region at the N-terminus. When compared to purple bacteria RC, a few functionally important amino acids are not conserved; these differences help to explain some of the spectral and kinetic features of the *Cf. aurantiacus* RC.

R. E. Blankenship, M. T. Madigan and C. E. Bauer (eds): Anoxygenic Photosynthetic Bacteria, pp. 699–708.

I. Introduction

Over the last twenty years reaction centers of two members of the Rhodospirillaceae, namely *Rhodobacter (Rb.) sphaeroides* and *Rhodopseudomonas (Rp.) viridis* have emerged as the spectroscopically and biochemically best characterized photoactive pigment-protein complexes for two reasons: the RC of *Rb. sphaeroides* was the first to have been isolated, and the reaction center (RC) of *Rp. viridis* was the first to be crystallized and the first to have its structure elucidated. While RC preparations from other purple bacteria have been described more recently such as that of *Rhodobacter capsulatus*, *Rhodospirillum (Rs.) rubrum* and *Chromatium (Cm.) vinosum*, the reaction center of *Cf. aurantiacus* has emerged over the last decade as perhaps the third best characterized RC. The thermophilic nature of *Cf. aurantiacus* probably contributes to the unusual stability of its RC (Pierson et al., 1983; Nozawa and Madigan, 1991) and makes this RC particularly well suited to spectroscopic analysis. This progress was certainly influenced by some inherent differences between the purple nonsulfur bacteria and the green filamentous bacteria RC.

II. Biochemical Composition of the *Chloroflexus aurantiacus* Reaction Center

A. Cofactors

1. Pigments

Although the preparation of the first *Cf. aurantiacus* RC enriched fraction dates back to 1977 (Fuller and Feick, unpublished), the first isolation procedure yielding RC free of antenna pigment was published by Pierson and Thornber (1983). Purified RC exhibited in the near infrared absorption bands at 865, 813, 756 and a shoulder at around 792 nm in a characteristic ratio of 0.77 : 1 : 0.95 (Fig. 1) which were attributed to BChl, BChl, BPheo and BChl, respectively (Shiozawa et al., 1987). The light-induced reduced minus oxidized difference spectrum in general resembled that of purple nonsulfur RC with a few minor differences in the 800 nm region.

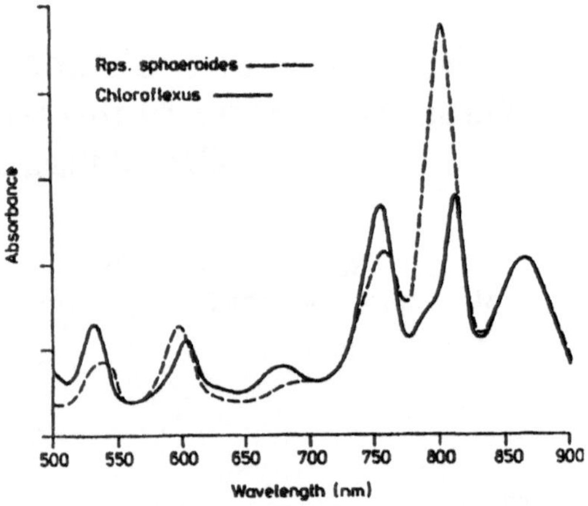

Fig. 1. Absorption spectra of RC, normalized at 865 nm, from *Cf. aurantiacus* (solid line) and *Rhodobacter sphaeroides* (dashed line). Reprinted from Blankenship et al. (1983) with permission of the publishers.

Based on a spectral comparison to the *Rb. sphaeroides* RC in which the absorbances at 865 nm of both reaction center preparations were normalized, Pierson and Thornber (1983) concluded that the BChl *a* to BPheo *a* stoichiometry is 3:3 in the *Cf. aurantiacus* RC instead of the 4:2 commonly present in purple nonsulfur bacteria RC. No carotenoid was found in the isolated *Cf. aurantiacus* RC despite the fact that this organism contains large amounts of carotenoids. A Gaussian deconvolution of low temperature absorption spectra led Vasmel et al., (1983) to assume a pigment composition of 4 BChl *a* to 3 BPheo *a* molecules per *Cf. aurantiacus* RC. This, however, was subsequently refuted as the following experiments show.

Quantitative pigment extraction experiments (Blankenship et al., 1983) and, in addition, comparative optical and ESR experiments (Blankenship et al., 1984; Mancino et al., 1985) not only confirmed the 3:3 stoichiometry but, furthermore, established extinction coefficients (135 $mM^{-1}cm^{-1}$ at 865 nm and 175 $mM^{-1}cm^{-1}$ at 813 nm) and identified the photobleachable primary electron donor as a BChl *a* dimer which absorbs maximally at 865 nm (P865) in analogy to purple nonsulfur bacteria RC.

Abbreviations: B_A – bacteriochlorophyll$_L$; B_B – bacteriochlorophyll$_M$; BChl – bacteriochlorophyll; BPheo – bacteriopheophytin; *Cf. – Chloroflexus; Cm. – Chromatium;* H_A – bacteriopheophytin$_L$; H_B – bacteriopheophytin$_M$; MQ – menaquinone; P – the special BChl dimer pair; PAGE – polyacrylamide gel electrophoresis; Q_A – primary quinone acceptor; Q_B – secondary quinone acceptor; *Rb. – Rhodobacter*; RC – reaction center; *Rp. – Rhodopseudomonas; Rs. – Rhodospirillum;* SDS – Sodium dodecyl sulfate; UQ – ubiquinone

This pigment composition and spectral characteristics was further supported from theoretical calculations of the *Cf. aurantiacus RC* absorption spectra with the aid of exciton theory (Vasmel et al., 1986; Scherer and Fischer, 1987); however, it was assumed that the pigment geometry was the same as in the structurally known *Rp. viridis* RC.

Initial studies on the primary charge separation in *Cf. aurantiacus* RC clearly established that BPheo is an intermediate electron acceptor of the photochemistry (Kirmaier et al., 1983, 1986; Nuijs et al., 1986; Shuvalov et al., 1986b). Photoselection experiments revealed that the photoreducible BPheo(s) is oriented perpendicular to the special pair P865 transition dipole. The 'third' BChl absorbing at 813 nm possesses roughly the same orientation with an angle of 35–65° with respect to the tetrapyrrole plane of P865 (Kirmaier et al., 1984; Parot et al., 1985).

The midpoint potential for P865 in purified membranes has been determined to be $E_{m,8.2}$ = +362 mV, n = 1 (Bruce et al., 1982) and in isolated RC to be $E_{m,8.0}$ = +386 mV (Shuvalov et al., 1986a) or +420 mV (Venturoli and Zannoni, 1988).

2. Primary and Secondary Electron Acceptors

The primary electron acceptor has been identified through a combination of spectroscopy on isolated RC and by quinone analysis of *Cf. aurantiacus* cells. Time resolved light-induced difference spectroscopy (in the presence of N-methylphenazonium methosulfate) under conditions which prevent the accumulation of oxidized P865 yielded a difference spectrum characteristic of menaquinone (vitamin K_2) anion (Vasmel and Amesz, 1983). After isolated *Cf. aurantiacus* RC have been exposed to a 1–5 μs light pulse, charge recombination (electron transfer back to $P865^+$) with a monoexponential decay of = 60 ms occurred (Blankenship et al., 1983; Venturoli et al., 1991). Addition of excess quinone analogues slowed the charge recombination by almost 10-fold (= 550 ms), if Vit K_1 was used (Blankenship et al., 1983) or by a factor of 100 (= 6 s) if UQ_3 was added (Venturoli et al., 1991). This recombination depended to a certain degree on the type of quinone used (Venturoli et al., 1991). If a suitable donor was added to such a quinone-reconstituted RC, so that P865 was maintained in a reduced state, a binary oscillation in semiquinone absorbance was observed which is characteristic of a two electron gate or a two quinone acceptor system typically observed in purple nonsulfur bacteria RC (Blankenship et al., 1988). Once again o-phenanthroline abolished this spectroscopic behavior while, in contrast to purple nonsulfur bacteria RC, herbicides such as terbutryn and atrazine had no effect (Blankenship, 1984). These experiments clearly showed that the *Cf. aurantiacus* RC has i) a two quinone acceptor system (Q_A, Q_B), ii) the secondary quinone (Q_B) is usually lost during RC isolation, iii) MQ functions both as Q_A and Q_B. Since functional reconstitution of Q_B can be achieved with different quinone types, either UQ or MQ, the identification of Q_B as a MQ rests solely on the quinone analysis of *Cf. aurantiacus* cells (Hale et al., 1983). The absence or noninvolvement of ferredoxin as a primary electron acceptor, commonly found in green sulfur bacteria RC can be indirectly surmised from the amino acid sequence of the *Cf. aurantiacus* RC protein entity.

Redox titrations with membranes yielded a midpoint potential for Q_A of $E_{m,8.1}$ = –50 mV (Bruce et al., 1982) or $E_{m,8.0}$ = –210 mV (with a pH dependence of 60 mV/pH unit up to pH 9.3 (Venturoli and Zannoni, 1988). Removal of the primary acceptor of *Cf. aurantiacus* RC does not require as harsh of a treatment as for *Rb. sphaeroides* RC. It can be readily removed by a 1% Triton X-100 wash (see below).

B. The Polypeptide Backbone

In contrast to the initial pigment characterization and spectroscopy, the polypeptide composition was at first ambiguous. Using SDS-PAGE as an analytical tool, Pierson et al. (1983) found, in addition to some contaminating polypeptides, two dominant proteins with M_r of 30,000 and 28,000. Feick and Fuller (1984) reported only one major polypeptide with M_r = 24,000–26,000. There was, however, agreement that when the RC preparation was subjected to less harsh denaturation conditions, a photochemically active RC pigment-protein band having M_r = 52,000 was observed and was present in relatively high amounts (Blankenship and Fuller, 1986; see also Fig. 2B lane 2).

Complete denaturation and re-electrophoresis of the extracted pigment-protein band (M_r 52,000 – 55,000) revealed the presence of just two polypeptides with very similar mobilities of approximately M_r of 24,000 (Fig. 2). Subsequently, the isolation of a highly purified *Cf. aurantiacus* reaction center preparation having an A_{280}/A_{813} of 1.3 to 1.4 was

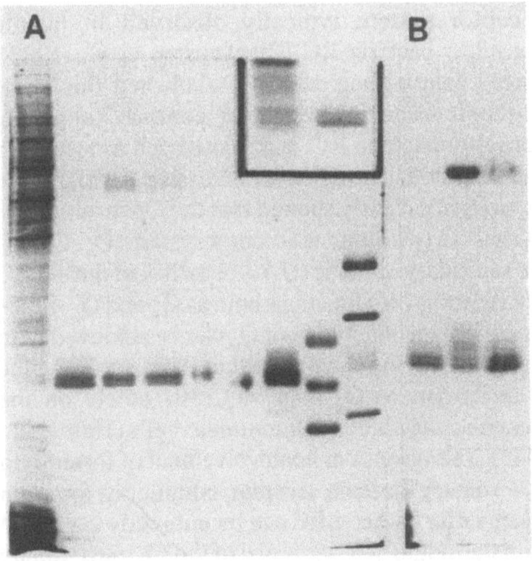

Fig. 2. SDS-PAGE. (A) Various fractions during the purification of the *Cf. aurantiacus* RC. Lane 1, whole membrane fraction; Lane 2, RC after DEAE-cellulose chromatography. Lanes 3 and 4, two different final RC-preparations; Lane 5 and 6; electrophoretically isolated subunits of RC; Lane 7, *Cf. aurantiacus* RC; Lane 8, *Rb. sphaeroides* RC; Lane 9, *Rp. viridis* RC. Inset enlargement of isolated *Cf. aurantiacus* RC and the corresponding part of whole membrane fraction. (B) Lane 1, heat denatured *Cf. aurantiacus* RC; Lane 2, RC incubated at 30 °C for 30 min in Laemmli's solubilization buffer and then applied onto the gel. Lane 3, excised and heat denatured pigmented RC band from lane 2. Reprinted from Shiozawa et al. (1987) with permission from the publishers.

described. After denaturation and electrophoresis on a SDS-PAGE having an optimized composition, only two polypeptides with similar M_r, 24,500 and 24,000, in an 1:1 stoichiometry were observed (Fig. 2, lanes 2–5). These data confirmed that the M_r 30,000 constituent in the Pierson and Thornber (1983) preparation was a proteinaceous contaminant. A third polypeptide corresponding to the H subunit of the purple nonsulfur bacteria reaction center was not found in *Cf. aurantiacus* reaction center preparations. Purified *Cf. aurantiacus* membranes showed no cross reaction with polyclonal antibodies against the H subunit of *Rb. sphaeroides* RC when analyzed by western blot experiments (Shiozawa and Feick, unpublished). Electron transfer and charge recombination kinetics from Q_A and Q_B in isolated and reconstituted *Cf. aurantiacus* RC suggest that a third protein subunit is not present in the reaction center of this bacterial species (Venturoli et al., 1991). Thus, the *Cf. aurantiacus* reaction center is the smallest functional reaction center isolated thus far.

A comparison of the RC constituents from *Cf. aurantiacus*, *Rb. sphaeroides* and *Rp. viridis* is shown in Fig. 3.

III. Structure-Function Relationships of the RC Complex

Protein chemical sequencing of the proteins and nucleotide sequencing of the genes encoding the reaction center polypeptides led to the elucidation of their primary structures (Ovchinnikov et al., 1988a, b, Shiozawa et al., 1989). These studies clearly showed a LM subunit analogy to the purple nonsulfur bacteria reaction center. The M_r 24,000 protein has 40% sequence identity with the L subunit of *Rb. sphaeroides*, is 310 amino acids long and has a M_m of 35,010 Da. The 24,500 polypeptide is 42% identical to the *Rb. sphaeroides* M subunit and possesses a M_m of 34,948 Da. The N-terminal amino acid of the M subunit is modified by a group having a M_m of 60 Da. A characteristic feature of the *Cf. aurantiacus* L subunit is its extended (31 amino acid residues long) hydrophilic (10 of 31 amino acid residues) N-terminus (see Fig. 4). Due to the low degree of sequence identity to the purple nonsulfur bacteria H subunit, it is doubtful that this region could substitute for the missing H subunit. However, it could be involved in stabilizing or anchoring the chlorosome attachment to the cytoplasmic membrane. Also it is not known with certainty whether reaction centers are located exclusively under chlorosomes.

The Kyte and Doolittle hydropathic profiles indicate that each polypeptide has five transmembrane helices. From the overall shape of these profiles, it is assumed that the location of the N- and C-termini correspond to that of purple nonsulfur bacteria reaction centers.

Comparing the primary structure and results of spectroscopic measurements on the *Cf. aurantiacus* RC with that of the two structurally known purple nonsulfur bacteria RC, the arrangement of the cofactors, eg. 3 BChl, 3 BPheo and 2 menaquinones (Q_A and Q_B), should be analogous to that in the *Rb. sphaeroides* and *Rp. viridis* RC where they are arranged in two transmembrane branches. These two branches will be denoted as A (L) and B (M) according to their relative proximity to the corresponding protein subunit (Fig. 5). In the *Cf. aurantiacus* RC, the $BChl_B$ (B_B) is replaced by a BPheo, hence, the B-branch contains two BPheo molecules, H_{1B} and H_{2B}.

Comparison of Reaction Centers

component	*Chloroflexus aurantiacus*	*Rb. sphaeroides*	*Rps. viridis*
Cytochrome(1)	--	--	41 000
H (1)	--	28 000	28 000
M (1)	35 000	35 000	36 000
L (1)	35 000	31 000	31 000
Bchl	3	4	4
Bpheo	3	2	2
Q_A ; Q_B	MQ ; MQ	UQ ; UQ	MQ ; UQ
Electron Transfer + Herbizide	'Resistant'	Sensitive	Sensitive
Sequence Similarities Cflx-L / Cflx-M		40 /42 L M	41 /40 L M
Transmembrane Helices, L + M	5 + 5	5 + 5	5 + 5

Fig. 3. Comparison of constituents of RC from *Cf. aurantiacus*, *Rb. sphaeroides* and *Rp. viridis*

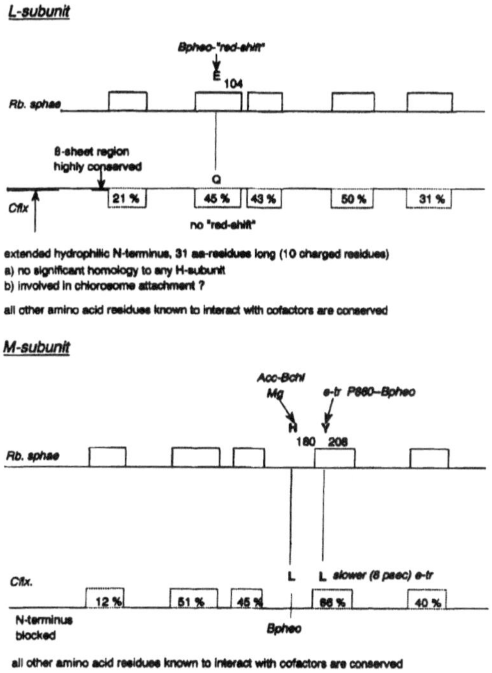

Fig. 4. A schematic comparison of the primary structures of the L- and M subunits of *Cf. aurantiacus* and *Rhodobacter sphaeroides* RC. The lines are drawn in proportion to their M_m. The boxes represent transmembrane helices as revealed by the three-dimensional structure of *Rp. viridis* RC and of putative transmembrane helices for *Cf. aurantiacus* RC as indicated from hydropathy plots. The degree of identity is given in percent. A complete comparison of amino acid sequences is given in Shiozawa et al. (1989)

The regions of highest sequence identity between *Cf. aurantiacus* and purple bacteria RC are observed where amino acid residues are known to interact with the special pair BChl (P865 or P) and accessory BChl molecules in the latter RC. A comparison of the functionally important amino acid residues which are involved in pigment binding and electron transfer (Michel et al., 1986; Allen et al., 1988; El-Kabbani et al., 1991) reveals some similarity, though significant differences exist (Fig. 4) and deserve some attention.

His^{M180} which acts as a ligand for the accessory BChl (B_B) in *Rp. viridis* RC is replaced by a Leu in *Cf. aurantiacus* and, hence, might be the reason for the replacement of BChl by the third BPheo in *Cf. aurantiacus* reaction center.

Tyr^{M210} located in the vicinity of P865, $BChl_L$ (B_A) and $BPheo_L$ (H_A) is conserved in all purple nonsulfur bacteria RC and it is proposed to facilitate or influence the primary electron transfer step (Allen et al., 1988) by lowering the energy of the state $P^+B_A^-$ (Parson et al., 1990). A replacement of this residue by Leu could explain the slower photochemical charge separation in *Cf. aurantiacus* RC. The functional role of this amino acid residue could indeed be confirmed by site-directed mutagenesis of *Rb. sphaeroides RC* where a $Tyr^{M210} \rightarrow$ Leu change drastically changed the time constant from 3.5 to 22 ps (Gray et al., 1990; Finkele et al., 1990; Nagarajan et al., 1990).

Glu^{L104} is located near $BPheo_A$ (H_A) and is hydrogen bonded with the carbonyl group of ring V of this pigment. This interaction is thought to be responsible for a red shift of the Q_x transition of this BPheo and, hence, allowed a spectroscopic discrimination between $BPheo_L$ (H_A) and $BPheo_M$ (H_B) (Zinth et al., 1983; Zinth et al., 1985). Exchanging this amino acid for Gln, which occurs in *Cf. aurantiacus* RC, significantly diminished this red shift and caused an absorbance in the BPheo Q_x region which is reminiscent of that of *Cf. aurantiacus* RC (Bylina et al., 1988).

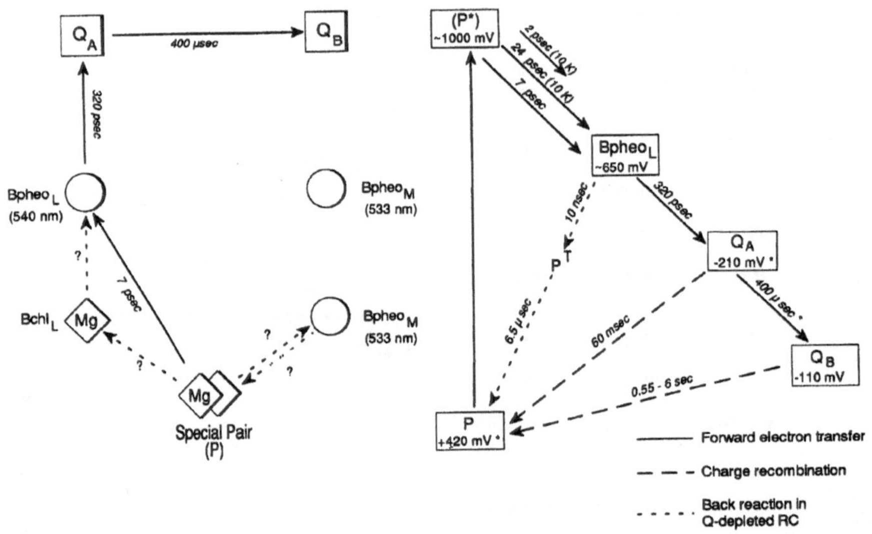

Fig. 5. Pigment organization and energetics of charge transfer states in *Cf. aurantiacus* RC. Left side: Hypothetical pigment and quinone organization. Right side: Time constants (τ, 1/e times) for electron transfer steps and approximate midpoint potentials of charge transfer states. * Slightly different values have been reported by Blankenship et al. (1983, 1988)

IV. The Primary Photochemistry

The early events of photochemistry originate from P^{*}· which is directly formed after absorption of a light quantum by the special pair (P). The initial electron transfer steps which occur well within 1 ns after light excitation have been studied by transient absorption and, more recently, by fluorescence emission spectroscopy with pico- and femtosecond time resolution.

A. The Early Event(s): Before 15 ps —The Formation of $P^{+}H_A^{-}$

Initial studies with *Cf. aurantiacus* RC revealed the involvement of BPheo as an intermediate electron acceptor which is reduced within 13 ps. An electron is subsequently transferred within 280–400 ps from the reduced BPheo to Q_A (Kirmaier et al., 1983, 1984, 1986; Nuijs et al., 1986). Despite an improved spectral resolution in the near infrared absorption range compared to purple bacteria RC, the time constants of the kinetics differed somewhat with the detection wavelength. In similar experiments, in which P was directly excited with a 32 ps light pulse, Shuvalov et al. (1986b) interpreted their results as a two step electron transfer, first from P^{*} to B_A with $\tau=$ 10 ps (1/e) and, subsequently, to H_A within 3 ps; these time constants were well within the time resolution of the instrument.

Subpico- and femtosecond absorbance spectroscopy with *Rb. sphaeroides* and *Rp. viridis* RC contradicted this electron transfer sequence (Woodbury et al., 1985; Martin et al., 1986; Breton et al., 1988; Kirmaier and Holten, 1991). No evidence for a state B_A^{-} was found. According to their data evaluation, charge transfer occurred from P^{*} directly to H_A within less than 2.2–2.8 ps via a superexchange mechanism (Bixon et al., 1988). See Chapter 24 by Woodbury and Allen for a discussion of this issue.

In contrast, Zinth and coworkers' assessment of their data from femtosecond absorption and fluorescence spectroscopy was fully consistent with a sequential two step electron transfer from P^{*} to B_A in 3 ps and then to H_A in 0.9 ps (Holzapfel et al., 1990; Hamm et al., 1993; Arlt et al., 1993).

This controversy might be partially related to the mathematical treatment of kinetics (curve fitting) and to the very complex behavior of absorbance changes and their kinetics which are dependent upon i) the temperature, ii) the wavelength of the probing beam and iii) the occurrence of mono- and non-monoexponential decay patterns.

For *Cf. aurantiacus* RC, the quantum yield of

charge separation (to $P^+Q_A^-$) has been initially reported to be 0.6 (Shuvalov et al., 1986a) which is considerably lower than the 1.0 reported for *Rb. sphaeroides* RC (Wraight and Clayton, 1973; Cho et al., 1984). Since such a low yield could be indicative of certain unusual electron transfer characteristics and/or might simply reflect a partial loss of Q_A. This measurement was repeated on a ns time scale, thus making it insensitive to Q-depletion. The value obtained was 1.0 at 280 K (Volk et al., 1991). Furthermore, it was shown that standard *Cf. aurantiacus* RC preparations contained usually 90% Q_A if LDAO, but only 20% Q_A occupancy if Triton X-100 was used as detergent.

Subpicosecond absorption changes associated with the decay of P^*, mirroring the earliest electron transfer step, revealed monoexponential kinetics with a $\tau = 7$ ps at 296 K and $\tau = 9$ ps at 320 K in *Cf. aurantiacus* RC (Becker et al., 1991). At lower temperatures the kinetics became markedly biphasic and could be best fitted with two exponential functions. The time constants for P^* decay were 2 and 24 ps at 10 K and 2 and 12 ps at 240 K with amplitudes of 1:1 at both temperatures (Martin et al., 1990; Feick et al., 1990).

With fluorescence emission studies several (3, 7, 18 ps, Müller et al., 1991) or two (2.7 and 9 ps, Hamm et al., 1993) short lived components could be detected. The 7–9 ps constituent agrees nicely with the $\tau = 7$ ps observed by Becker et al. (1991) and should clearly reflect the formation of P^+ ($P^+B_A^-$ or $P^+H_A^-$). The 2–3 ps kinetics was attributed to energy transfer from the excited B_A to P within the RC (Müller et al., 1991). However, ps absorption measurements with 812 nm excitation failed to confirm this assignment (Hastings et al., 1993). This 2–3 ps energy transfer component could not be the 2 ps component observed by Martin et al. (1990) because P was directly excited in their experiments and uphill energy transfer can be ruled out on such a time scale. The 18 ps fluorescence kinetics was interpreted to originate from charge transfer from P in closed (PQ^-) RC which the authors contributed to a open/closed RC sample heterogeneity. In light of the results of Volk et al. (1991) where standard RC preparations contain about 10% Q-depleted centers, this type of heterogeneity cannot formally be ruled out. Heterogeneous populations of *Cf. aurantiacus* RC which could be caused by differently charged complexes have been reported earlier (Feick et al., 1990). Other interpretations for the origin of the complex decay kinetics of P^* are discussed in detail by Becker et al. (1991), Müller et al. (1991), Kirmaier and Holten (1991), Hamm et al. (1993), Arlt et al. (1993).

B. Events After 15 ps: Electron Transfer from $P^+H_A^-$ to $P^+Q_A^-$

Regardless of whether the reduction of BPheo (H_A^- within 10 ps) occurs via a one or two step mechanism, the subsequent forward electron transfer from H_A^- results in the formation of the state $P^+Q_A^-$ with $\tau =$ 320 ps at 280 K (Kirmaier et al., 1986). The authors observed a dependence of τ on the detection wavelength used. The observed value of τ ranged from 365 ps in the 530–545 nm region to 281 ps at 810 nm. With the knowledge of the *Rb. sphaeroides* and *Rp. viridis* RC crystal structures, a spectroscopic discrimination between the two functionally nonequivalent BPheo-molecules was possible (Zinth et al., 1983, 1985). Subsequently, with the help of a high resolution spectrophotometer, this distinction could also be made on a picosecond time scale (Eberl, 1988; Aumeier, 1990; Aumeier et al., 1990). It was found that, after a 25 ps flash, differential absorption changes at 540 and 533 nm mirrored forward electron transfer from H_A^- or from (if it occurs at all) H_B^- to Q_A. In *Rb. sphaeroides RC* (the equivalent wavelengths were 547 and 530 nm). Since only the 547 nm showed a differential absorption change of $\tau = 200$ ps, the 530 nm feature remained constant over a ms time scale. Thus, it was possible to selectively probe electron transfer along the A- and B-pigment branch. In *Cf. aurantiacus* RC, however, the changes of the 540 nm (due to the disappearance of H_A^-) and 533 nm band with a amplitude ratio of 6:1 disappeared with different kinetics $\tau_{540\ nm} = 320$ ps, $\tau_{533\ nm} = 280$–450 ps. This behavior was consistent with the interpretation of a significant charge transfer into the B-branch of *Cf. aurantiacus* RC (Eberl, 1988). Later measurements and, also, the high quantum yield of 1.0 were no longer compatible with this notion. Hence, unidirectionality of electron transport, i.e. charge transfer occurs via the A-branch only, also appeared to be a salient feature of *Cf. aurantiacus* RC (Aumeier et al., 1990). However, according to recent picosecond fluorescence measurements under Q_A reduced conditions, a reversible electron transfer into the B-branch, from P^* to H_{IB}, had to be considered as one alternative to explain the observed results (Schweitzer et al., 1992).

C. Electron Transfer from Q_A to Q_B and Charge Recombination

Very few data exist on the forward electron transfer from Q_A to Q_B. This might in part be related due to the fact that isolated *Cf. aurantiacus* RC are Q_B depleted and, hence, require Q-reconstitution prior to measurement. Time constants of 1.3 ms (Blankenship et al., 1988) and 400 μs (Venturoli and Zannoni, personal communication) for electron transfer from Q_A to Q_B have been determined. The difference could be due to the type of quinone (UQ and MQ analogues) with which the RC have been reconstituted. An even more pronounced quinone dependence has been observed for charge recombination (back electron transfer to the state PHQ_AQ_B) which varied from τ = 140 ms in the presence of MQ-6 and τ = 5.5 s in the presence of UQ-3. Further spectroscopic characteristics of Q_A and Q_B have already been addressed above. A summary of the redox potentials and time constants of all photochemical processes occurring in *Cf. aurantiacus* RC are presented in Fig. 5

V. Summary and Conclusions

The photosynthetic apparatus of the thermophilic *Cf. aurantiacus* has many distinctive features which are characteristic for green bacteria. Its RC and photochemical processes are reminiscent of purple nonsulfur bacteria RC. However, significant differences exist: i) a modified pigment composition of 3 BChl:3 BPheo molecules per RC (compared to a 4:2 stoichiometry in purple nonsulfur bacteria RC, ii) it is composed of only two protein subunits with almost identical molecular masses of approximately 35,000 Da and iii) increased thermal stability of the complex. Due to the absence of the so called H-protein subunit, the *Cf. aurantiacus* RC is the smallest functional RC isolated thus far. The two polypeptides possess a moderate identity of 40% each to the L- and M polypeptides of *Rb. sphaeroides*- and *Rp. viridis* RC, respectively. A characteristic feature of the N-terminus of the L-subunit of the *Cf. aurantiacus* RC is the extended hydrophilic region at the N-terminus. This region could be involved in stabilizing or anchoring the chlorosome attachment to the cytoplasmic membrane. A three-dimensional structure is not yet available; all evidence points to a two branch organization of cofactors similar to the two structurally known RC. A few functionally important amino acids are not conserved and help to explain certain spectroscopic features such as the presence of a BPheo instead of a BChl and slower photochemical reactions. These characteristics and a better resolved absorption spectrum in the near infrared should, in theory, make the *Cf. aurantiacus* RC particularly well suited for ultrafast spectroscopy to complement the existing knowledge of primary photochemical charge separation.

Acknowledgments

We would like to thank Professors M.E. Michel-Beyerle, D. Oesterhelt and H. Scheer for support and Mr. G. Niedermeier for helpful discussions. This work was supported by the Deutsche Forschungsgemeinschaft (SFB 143).

References

Allen JP, Feher G, Yeates TO, Komiya H and Rees DC (1988) Structure of the reaction center from *Rhodobacter sphaeroides* R-26: Protein-cofactor (quinones and Fe^{+2}) interactions. Proc Natl Sci USA 85: 8487–8491

Arlt T, Schmidt S, Kaiser W, Lauterwasser C, Meyer M, Scheer H and Zinth W (1993) The accessory bacteriochlorophyll: A real electron carrier in primary photosynthesis. Proc Natl Acad Sci USA 90:11757–11761

Aumeier W (1990) Pikosekunden-zeitaufgelöste Absorptionsmessungen an Reaktionszentren von Purpurbakterien, sowie an *Chloroflexus aurantiacus*. Ph. D. Thesis, Technische Universität München

Aumeier W, Eberl U, Ogrodnik A, Volk M, Scheidel G, Feick R, Plato M and Michel-Beyerle ME (1990) Unidirectionality of charge separation in reaction centers of *Rb. sphaeroides* and *Chloroflexus aurantiacus*. In: Baltscheffsky M (ed) Current Research in Photosynthesis, Vol 1, pp 133–136. Kluwer Academic Publishers

Becker M, Nagarajan V, Middendorf D, Parson WW, Martin JE and Blankenship RE (1991) Temperature dependence of the initial electron-transfer kinetics in photosynthetic reaction centers of *Chloroflexus aurantiacus*. Biochim Biophys Acta 1057: 299–312

Bixon M, Jortner J, Plato M and Michel-Beyerle ME (1988) Mechanism of the primary charge separation in bacterial photosynthetic reaction centers. In: Breton J and Vermeglio A (eds) The Photosynthetic Bacterial Reaction Center–Structure and Dynamics, pp 399–419. Plenum Press, New York

Blankenship RE (1984) Primary photochemistry in green photosynthetic bacteria. Photochem and Photobiol 40: 801–806

Blankenship RE and Fuller RC (1986) Membrane topology and photochemistry of the green photosynthetic bacterium *Chloroflexus aurantiacus*. In: Staehelin LA and Arntzen CJ (eds) Encyclopedia of Plant Physiology, Vol 19, pp 390–806. Springer Verlag, Berlin

Blankenship RE, Feick R, Bruce BD, Kirmaier C, Holten D and Fuller RC (1983) Primary photochemistry in the facultative green photosynthetic bacterium *Chloroflexus aurantiacus*. J Cell Biochem 22: 251–261

Blankenship RE, Mancino LJ, Feick R, Fuller RC, Machnicki J, Frank HA, Kirmaier C and Holten D (1984) Primary photochemistry and pigment composition of reaction centers isolated from the green photosynthetic bacterium *Chloroflexus aurantiacus*. In: Sybesma C (ed) Advances in Photosynthesis Research, Vol I, Proc VI Inter Cong Photosyn, pp 203–206. Dr W Junk, The Hague

Blankenship RE, Trost JT and Mancino LJ (1988) Properties of reaction centers from the green photosynthetic bacterium *Chloroflexus aurantiacus*. In: Breton J and Vermeglio A (eds) Structure of Bacterial Reaction Centers, pp 119–127. Plenum Press, New York

Breton J, Martin J-L, Fleming GR and Lambry J-C (1988) Low Temperature femtosecond spectroscopy of the initial step of electron transfer in reaction centers from photosynthetic purple bacteria. Biochemistry 27: 8276–8284

Bruce BD, Fuller RC and Blankenship RE (1982) Primary photochemistry in the facultatively aerobic green photosynthetic bacterium *Chloroflexus aurantiacus*. Proc Natl Acad Sci USA 79: 6532–6536

Bylina EJ, Kirmaier C, McDowell L, Holten D and Youvan D (1988) Influence of an amino-acid residue on the optical properties and electron transfer dynamics of a photosynthetic reaction centre complex. Nature 336: 182–184

Cho, HM, Mancino LJ, and Blankenship RE (1984) Light saturation curves and quantum yields in reaction centers from photosynthetic bacteria. Biophys J 45: 455–461

Den Blanken HJ, Vasmel H, Jongenelis APLM, Hoff AJ and Amesz J (1983) The triplet state of the primary donor of the green photosynthetic bacterium *Chloroflexus aurantiacus*. FEBS Lett 161: 185–189

Eberl U (1988) Temperaturabhängige Pikosekunden-absorptions-spektroskopie an bakteriellen Reaktionszentren von *Rhodobacter sphaeroides* und *Chloroflexus aurantiacus*. Diplomarbeit, Technische Universität München

El-Kabbani O, Chang C-H, Tiede D, Norris J and Schiffer M (1991) Comparison of reaction centers from *Rhodobacter sphaeroides* and *Rhodopseudomonas viridis*: Overall architecture and protein-pigment interactions. Biochemistry 30: 5361–5369

Feick R and Fuller RC (1984) Topography of the photosynthetic apparatus of *Chloroflexus aurantiacus*. Biochemistry 23: 3693–3700

Feick R, Martin JL, Breton J, Volk M, Scheidel G, Langenbacher T, Urbano C, Ogrodnik A and Michel-Beyerle ME (1990) Biexponential charge separation and monoexponential decay of P^+H^- in reaction centers of *Chloroflexus aurantiacus*. In: Michel-Beyerle (ed) Reaction Centers of Photosynthetic Bacteria (Feldafing II Meeting), pp 181–188. Springer Verlag, Berlin

Finkele U, Lauterwasser C, Zinth W, Gray KA and Oesterhelt D (1990) Role of tyrosine M210 in the initial charge separation of reaction centers of Rhodobacter sphaeroides. Biochemistry 29: 8517–8521

Gray KA, Farchaus JW, Wachtveitl J, Breton J and Oesterhelt D (1990) Initial characterization of site-directed mutants of tyrosine M210 in the reaction centre of *Rhodobacter sphaeroides*. EMBO J 9: 2061–2070

Hale MB, Blankenship RE and Fuller RC (1983) Menaquinone is the sole quinone in the facultatively aerobic green photosynthetic bacterium *Chloroflexus aurantiacus*. Biochim Biophys Acta 723: 376–382

Hamm P, Gray KA, Oesterhelt D, Feick R, Scheer H and Zinth W (1993) Subpicosecond emission studies of bacterial reaction centers. Biochim Biophys Acta 1142: 99–105

Hastings G, Lin S, Zhou W and Blankenship RE (1993) Femtosecond spectroscopy of *Chloroflexus aurantiacus* reaction centers. Photochem Photobiol 57: 65S

Holzapfel W, Finkele U, Kaiser W, Oesterhelt D, Scheer HU and Zinth W (1990) Initial electron-transfer in the reaction center from *Rhodobacter sphaeroides*. Proc Natl Acad Sci USA 87: 5168–5172

Kirmaier C and Holten D (1991) An assessment of the mechanism of initial electron transfer in bacterial reaction centers. Biochemistry 30: 609–613

Kirmaier C, Holten D, Feick R and Blankenship RE (1983) Picosecond measurements of the primary photochemical events in reaction centers isolated from the facultative green photosynthetic bacterium *Chloroflexus aurantiacus*. FEBS Lett 158: 73–78

Kirmaier C, Holten D, Mancino LJ and Blankenship RE (1984) Picosecond photodichroism studies on reaction centers from the green photosynthetic bacterium *Chloroflexus aurantiacus*. Biochim Biophys Acta 765: 138–146

Kirmaier C, Blankenship RE and Holten D (1986) Formation and decay of radical-pair state P^+I^- in *Chloroflexus aurantiacus* reaction centers. Biochim Biophys Acta 850: 275–285

Lang E, Lersch W, Tappermann P, Coleman WJ, Youvan DC, Feick R and Michel-Beyerle ME (1990) High power RYDMR spectra of P^+H^- in reaction centers of photosynthetic bacteria. In: Baltscheffsky M (ed) Current Research in Photosynthesis, Vol 1, pp 137–140. Kluwer Academic Publishers

Mancino LJ, Hansen PL, Stark RE and Blankenship RE (1985) Chemical composition and photochemical properties of reaction centers from the photosynthetic bacterium *Chloroflexus aurantiacus*. Biophys J 47: 2a

Martin J-L, Breton J, Hoff AJ, Migus A and Antonetti A (1986) Femtosecond spectroscopy of electron transfer in the reaction center of the photosynthetic bacterium *Rhodopseudomonas sphaeroides* R-26: Direct electron transfer from the dimeric bacteriochlorophyll primary donor to the bacteriopheophytin acceptor with a time constant of 2.8 ± 0.2 ps. Proc Natl Acad Sci USA 83: 957–961

Martin J-L, Lambry JC, Ashokkumar M, Michel-Beyerle ME, Feick R and Breton J (1990) In: Harris CB, Ippen EB, Mourou GA and Zwail AH (eds) Ultrafast Phenomena VIII, Vol 53, pp. 524–528. Springer Series Chem Phys, Springer Verlag, Berlin

Michel H, Epp O and Deisenhofer J (1986) Pigment-protein interactions in the photosynthetic reaction centre from *Rhodopseudomonas viridis*. EMBO J 5: 2445–2451

Müller MG, Griebenow K and Holzwarth AR (1991) Primary processes in isolated photosynthetic bacterial reaction centres from *Chloroflexus aurantiacus* studied by picosecond fluorescence spectroscopy. Biochim Biophys Acta 1098: 1–12

Nagarajan V, Parson WW, Gaul D and Schenck CC (1990) Effect of specific mutations of tyrosine-(M)210 on the primary photosynthetic electron-transfer process in *Rhodobacter sphaeroides*. Proc Natl Acad Sci USA 87: 7888–7892

Nozawa T and Madigan MT (1991) Temperature and solvent effects on reaction centers from *Chloroflexus aurantiacus* and *Chromatium tepidum*. J Biochem 110: 588–594

Nuijs AM, Vasmel H, Duysens LNM and Amesz J (1986) Antenna and reaction-center processes upon picosecond-flash excitation of membranes of the green photosynthetic bacterium *Chloroflexus aurantiacus*. Biochim Biophys Acta 849: 316–324

Ovchinnikov YA, Abdulaev NG, Zolotarev AS, Shmukler BE, Zargarov AA, Kutuzov MA, Telezhinskaya IN and Levina NB (1988a) Photosynthetic reaction centre of *Chloroflexus aurantiacus*. I. Primary structure of L-subunit. FEBS Lett 231: 237–242

Ovchinnikov YA, Abdulaev NG, Shmukler BE, Zargarov AA, Kutuzov MA, Telezhinskaya IN, Levina NB and Zolotarev AS (1988b) Photosynthetic reaction centre of *Chloroflexus aurantiacus* - Primary structure of M-subunit. FEBS Lett 232: 364–368

Parot P, Delmas N, Garcia D and Vermeglio A (1985) Structure of *Chloroflexus aurantiacus* reaction center: Photoselection at low temperature. Biochim Biophys Acta 809: 137–140

Parson WW, Nagarajan V, Gaul D, Schenck CC, Chu Z-T and Warshel A (1990) Electrostatic effects on the speed and directionality of electron transfer in bacterial reaction centers: The special role of tyrosine M-208. In: Michel-Beyerle ME (ed) Reaction Centers of Photosynthetic Bacteria (Feldafing II Meeting), pp 239–249. Springer Verlag, Berlin

Pierson BK and Thornber JP (1983) Isolation and spectral characterization of photochemical reaction centers from the thermophilic bacterium *Chloroflexus aurantiacus* strain J-10-fl. Proc Natl Acad Sci USA 80: 80–84

Pierson BK, Thornber JP and Seftor REB (1983) Partial purification, subunit structure and thermal stability of the photochemical reaction center of the thermophilic green bacterium *Chloroflexus aurantiacus*. Biochim Biophys Acta 723: 322–326

Scherer POJ and Fischer SF (1987) Model studies to low-temperature optical transitions of photosynthetic reaction centers. II. *Rhodobacter sphaeroides* and *Chloroflexus aurantiacus*. Biochim Biophys Acta 891: 157–164

Schweitzer G, Hucke M, Griebenow K, Müller MG and Holzwarth AR (1992) Charge separation kinetics in isolated photosynthetic reaction centers of *Chloroflexus aurantiacus* (with Q_A reduced) at low temperatures. Chem Phys Lett 190: 149–154

Shiozawa JA, Lottspeich F and Feick R (1987) The photochemical reaction center of *Chloroflexus aurantiacus* is composed of two structurally similar polypeptides. Eur J Biochem 167: 595–600

Shiozawa JA, Lottspeich F, Oesterhelt D and Feick R (1989) The primary structure of the *Chloroflexus aurantiacus* reaction-center polypeptides. Eur J Biochem 180: 75–84

Shuvalov VA, Shkuropatova YA, Kulkkova SM, Ismailiov MA and Shkuropatov VA (1986a) Photoreactions of bacteriopheophytins and bacteriochlorophylls in reaction centers of *Rhodopseudomonas sphaeroides* and *Chloroflexus aurantiacus*. Biochim Biophys Acta 849: 337–346

Shuvalov VA, Vasmel H, Amesz J and Duysens LNM (1986b) Picosecond spectroscopy of the charge separation in reaction centers of *Chloroflexus aurantiacus* with selective excitation of the primary electron donor. Biochim Biophys Acta 851: 361–368

Vasmel H and Amesz J (1983) Photoreduction of menaquinone in the reaction center of the green photosynthetic bacterium *Chloroflexus aurantiacus*. Biochim Biophys Acta 724: 118–122

Vasmel H, Meiburg RF, Kramer HJM, De Vos LJ and Amesz J (1983) Optical properties of the photosynthetic reaction center of *Chloroflexus aurantiacus* at low temperature. Biochim Biophys Acta 724: 333–339

Vasmel H, Amesz J and Hoff AJ (1986) Analysis by exciton theory of the optical properties of the *Chloroflexus aurantiacus* reaction center. Biochim Biophys Acta 852: 159–168

Venturoli G and Zannoni D (1988) Oxidation-reduction thermodynamics of the acceptor quinone complex in whole-membrane fragments from *Chloroflexus aurantiacus*. Eur J Biochem 178: 503–509

Venturoli G, Feick R, Trotta M and Zannoni D (1990) Thermodynamic and kinetic features of the redox carriers operating in the photosynthetic electron transport of *Chloroflexus aurantiacus*. In: Drews G and Dawes EA (eds) Molecular Biology of Membrane-bound Complexes in Phototrophic Bacteria, pp 425–432. Plenum Press, New York

Venturoli G, Trotta M, Feick R, Melandri BA and Zannoni D (1991) Temperature dependence of charge recombination from the $P^+Q_A^-$ and $P^+Q_B^-$ states in photosynthetic reaction centers isolated from the thermophilic bacterium *Chloroflexus aurantiacus*. Eur J Biochem 202: 625–634

Volk M, Scheidel G, Ogrodnik A, Feick R and Michel-Beyerle ME (1991) High quantum yield of charge separation in reaction centers of *Chloroflexus aurantiacus*. Biochim Biophys Acta 1058: 217–224

Volk M, Aumeier G, Häberle T, Ogrodnik A, Feick R and Michel-Beyerle ME (1992) Sensitive analysis of the occupancy of the quinone binding site at the active branch of photosynthetic reaction centers. Biochim Biophys Acta 1102: 253–259

Woodbury NW, Becker M, Middendorf D and Parson WW (1985) Picosecond kinetics of the initial photochemical electron-transfer reaction in bacterial photosynthetic reaction centers. Biochemistry 24: 7516–7521

Wraight CA and Clayton RK (1973) The absolute quantum efficiency of bacteriochlorophyll photooxidation in reaction centers of *Rhodopseudomonas sphaeroides*. Biochim Biophys Acta 333: 246–260

Zinth W, Kaiser W and Michel H (1983) Efficient photochemical activity and strong dichroism of single crystals of reaction centers from *Rhodopseudomonas viridis*. Biochim Biophys Acta 723: 128–131

Zinth W, Sander M, Dobler J, Kaiser W and Michel M (1985) Single crystals from reaction centers of *Rhodopseudomonas viridis* studied by polarized light. In: Michel-Beyerle M (ed) Antennas and Reaction Centers of Photosynthetic Bacteria, pp 97–102. Springer Verlag, Berlin

Anoxygenic Photosynthetic Bacteria

Advances in Photosynthesis

VOLUME 2

Series Editor:

GOVINDJEE

Department of Plant Biology

University of Illinois, Urbana, Illinois, U.S.A.

Consulting Editors:

Jan AMESZ, *Leiden, The Netherlands*

James BARBER, *London, United Kingdom*

Robert E. BLANKENSHIP, *Tempe, Arizona, U.S.A.*

Norio MURATA, *Nagoya, Japan*

Donald R. ORT, *Urbana, Illinois, U.S.A.*

Advances in Photosynthesis provides an up-to-date account of research on all aspects of photosynthesis, the most fundamental life process on earth. *Photosynthesis* is an area that requires, for its understanding, a multidisciplinary (biochemical, biophysical, molecular biological, and physiological) approach. Its content spans from physics to agronomy, from femtosecond reactions to those that require an entire season, from photophysics of reaction centers to the physiology of the whole plant, and from X-ray crystallography to field measurements. The aim of this series of publications is to present to beginning researchers, advanced graduate students and even specialists a comprehensive current picture of the advances in the various aspects of photosynthesis research. Each volume focusses on a specific area in depth.

The titles to be published in this series are listed on the backcover of this volume.

Anoxygenic Photosynthetic Bacteria

Edited by

Robert E. Blankenship
Department of Chemistry and Biochemistry,
Arizona State University,
Tempe, Arizona, U.S.A.

Michael T. Madigan
Department of Microbiology,
Southern Illinois University,
Carbondale, Illinois, U.S.A.

and

Carl E. Bauer
Department of Biology,
Indiana University,
Bloomington, Indiana, U.S.A.

KLUWER ACADEMIC PUBLISHERS
DORDRECHT / BOSTON / LONDON

A C.I.P. Catalogue record for this book is available from the Library of Congress

ISBN 0-7923-3681-X

Published by Kluwer Academic Publishers,
P.O. Box 17, 3300 AA Dordrecht, The Netherlands.

Kluwer Academic Publishers incorporates
the publishing programmes of
D. Reidel, Martinus Nijhoff, Dr W. Junk and MTP Press.

Sold and distributed in the U.S.A. and Canada
by Kluwer Academic Publishers,
101 Philip Drive, Norwell, MA 02061, U.S.A.

In all other countries, sold and distributed
by Kluwer Academic Publishers Group,
P.O. Box 322, 3300 AH Dordrecht, The Netherlands.

The camera ready text was prepared
by Lawrence A. Orr, Center for the Study of Early Events in Photosynthesis
Arizona State University, Tempe, Arizona 85287-1604, U.S.A.

Printed on acid-free paper

Printed in the Netherlands

Contents

Part II: Molecular Structure and Biosynthesis of Pigments and Cofactors

Part III. Membrane and Cell Wall Architecture and Organization

Part V: Reaction Center Structure, Electron and Proton Transfer Pathways

Part VI: Cyclic Electron Transfer Components and Energy Coupling Reactions

Part VIII: Genetics and Genetic Manipulations

Part IX: Regulation of Gene Expression

Part X: Applications

Preface

The editors are proud to present *Anoxygenic Photosynthetic Bacteria* to students and researchers in the field of photosynthesis. We feel that this book will be the definitive volume on non-oxygen evolving photosynthetic bacteria for years to come, as it is literally loaded with the most recent information available in this field. Contributors were given the freedom to develop topics in depth, and although this has lead to a lengthy volume, readers can be assured that nothing of significance in the field of anoxygenic photosynthetic bacteria has been neglected.

We have organized the book along ten major themes: (1) Taxonomy, physiology and ecology, (2) Molecular structure of pigments and cofactors, (3)Membrane and cell wall structure, (4) Antenna structure and function, (5) Reaction center structure and electron/proton pathways, (6) Cyclic electron transfer, (7) Metabolic processes, (8) Genetics, (9), Regulation of gene expression, and (10) Applications. For each theme, several chapters, written by leading experts in the field, combine to paint a detailed picture of the current state of affairs of research in that area. The editors have also tried to provide ample cross-references between chapters to help guide the reader to all of the coverage on a particular topic.

The last time a comprehensive volume on anoxygenic photosynthetic prokaryotes was published was in 1978. This was the landmark volume *The Photosynthetic Bacteria*, edited by R. K. Clayton and W. R. Sistrom, and published by Plenum Press in New York. That book brought together in one place the state of current knowledge on the ecology, physiology, taxonomy, ultrastructure, biochemistry, biophysics and genetics of this diverse group of bacteria. As a testament to its value as a reference book, every copy of *The Photosynthetic Bacteria* that any of the three editors of the present volume have ever seen was tattered from years of almost constant use and stained with coffee and cell extracts of various colors. *The Photosynthetic Bacteria* had a tremendous impact on a generation of scientists in many disciplines and will remain a classic for many years to come.

Thus it was with some trepidation that we undertook what was in essence the same task that faced Clayton and Sistrom 17 years earlier. The field has changed dramatically in that time in nearly every one of the areas mentioned above, perhaps most noticeably in the area of genetics and related topics. A single chapter on this subject sufficed in 1978, whereas 14 of the 62 chapters in the present volume have 'gene' or 'genetic' in the title and nearly every chapter makes use of information obtained using mutants.

Another area of bacterial photosynthesis that has seen tremendous progress in 17 years is structural studies. There is only one protein structure determined using X-ray diffraction that appears as a figure in the 1978 volume, whereas dozens of structural figures are included in this volume, including reaction centers, antenna complexes, electron transfer proteins and porins. An even more dramatic difference is that there is only a single, partial amino acid sequence in the 1978 volume, while there are literally hundreds of sequences in this book. The emphasis in every area covered in this book, from taxonomy and ecology all the way to biophysics has continually shifted more and more toward analysis at the molecular level.

This is the second volume in a series titled 'Advances in Photosynthesis,' with Govindjee as the series editor. The series provides an up to date account of all aspects of the process of photosynthesis. A statement of the scope of the series and a list of previous and upcoming titles can be found on the back cover.

We would like to dedicate this volume to Howard Gest, who is unquestionably the father of modern research involving photosynthetic bacteria. He has also greatly influenced the careers of each of the editors as either a collaborator, mentor or colleague and to all of us as a friend.

Special thanks go to Larry Orr, without whose

extraordinary talents and hard work this effort could never have come to fruition. Larry prepared all the camera ready copy and is primarily responsible for the layout and 'look' of the book.

Finally, we thank our wives, Liz, Nancy and Chris and children, Larissa, Sam, Scott and Kevin for their patience and understanding of the time needed to complete this volume.

Robert E. Blankenship
Michael T. Madigan
Carl E. Bauer

Chapter 33

Cytochrome Biogenesis

Robert G. Kranz and Diana L. Beckman
Department of Biology, Box 1137, Washington University, One Brookings Drive, St. Louis, MO 63130

Summary

A variety of cytochromes are present in anoxygenic photosynthetic bacteria including *c*-type, *b*-type and *a*-type. The properties that biosynthetically distinguish these types include: the type of heme group, the nature of the heme linkage to the apoprotein and the topological characteristics within the cell, which are in part determined by the secretion and/or assembly pathway. For example, cytochromes *a* contain heme *a* that is enzymatically modified with formyl and farnesylhydroxymethyl hydrophobic side chains. Although the *b*- and *c*-type cytochromes both have heme *b*, the heme in *c*-type cytochromes is covalently bound through thioether linkages to CysXxxYyyCysHis of the apoprotein. Enzyme(s) must be present in the cell for this ligation reaction. Topologically, the *c*-type cytochromes of Gram-negative bacteria are present in the periplasmic space (or are oriented toward it) and the apocytochromes *c* are synthesized with typical bacterial signal sequences for secretion. In many cases, the *b*- and *a*-type cytochromes are membrane bound and contain numerous transmembrane helices. The topological location of the heme in these cases may vary, thus presenting a further

R. E. Blankenship, M. T. Madigan and C. E. Bauer (eds): Anoxygenic Photosynthetic Bacteria, pp. 709–723.

complication from a biosynthetic standpoint. This chapter discusses the cytochrome assembly pathways and assembly factors that have been discovered in anoxygenic photosynthetic bacteria and their relationship to those in other organisms. Biogenesis pathways of macromolecular membrane complexes like cytochrome bc_1 and the cytochrome aa_3 oxidase are complicated, with some of the assembly rules and potential assembly factors or proteins just beginning to be elucidated. There appears to be at least seven assembly proteins required for the biogenesis of all *c*-type cytochromes in the anoxygenic photosynthetic bacterium *Rhodobacter capsulatus*. These assembly proteins (called Hel and Ccl) are discussed with respect to their properties and putative roles within a specific cytochrome(s) *c* biogenesis pathway. The recent discovery of *hel*- and *ccl*-like genes in mitochondrial genomes underscores the idea that related biogenesis pathways may exist in eukaryotes. Their predicted close phylogenetic relationship to the ancestral mitochondrial endosymbiont make *Rhodobacter* and organisms like it ideal systems to study the biogenesis of cytochromes.

I. Introduction: Topological and Biosynthetic Distinctions Among Cytochromes

Cytochromes are heme proteins that function in electron-transfer systems of the cell. The physical and functional properties that distinguish one cytochrome from another include: reduction/ oxidation potential, interaction with specific substrates (i.e., electron donors or acceptors), type of heme group and the nature of its binding to the apocytochrome, and the topological location of the cytochrome within the cell. Five types of cytochromes have been classified based on heme characteristics: *a, b, c, o* and *d* types (Palmer and Reedijk, 1992). Thus, *a*- and *d*-type cytochromes are distinguished by their heme prosthetic group: hemes *o* and *a* contain a farnesylhydroxymethyl group, heme *a* an additional formyl group while cytochromes *d* contain heme *d* (also known as heme a_2). Heme *d* contains two hydroxyl groups on pyrrole ring III. In contrast, both *b*- and *c*-type cytochromes contain heme *b* and are distinguished from each other in that only *c*-type cytochromes have this heme covalently attached by thioether linkages to Cys XxxYyy Cys His residues of the apocytochrome. Cytochromes *c* also differ from most *b*-type cytochromes in their subcellular location; they are either soluble in the periplasm (e.g., cytochrome c_2) or periplasmically orientated if membrane bound (e.g., cytochrome c_1 of the bc_1 complex). (For a detailed discussion of the structure and function of various cytochromes see Chapter 34 by Meyer and Donohue, Chapter 35 by Gray and Daldal and Chapter 36 by Nitschke and Dracheva, and of the structure and biosynthesis of heme see Chapter 9 by Beale.)

A variety of heme proteins have been reported to be present in anoxygenic photosynthetic bacteria, including *a*-, *b*- and *c*-type cytochromes. In fact, it is not uncommon to find these various types of cytochromes present in a single complex (e.g., cytochrome bc_1 or cytochrome *cb* oxidases in phototrophic bacteria or as *bd* oxidases or caa_3 oxidases in other bacteria). Understanding the assembly of these diverse macromolecular complexes is clearly at a very early stage. Nevertheless, it is clear that specific assembly proteins, either transiently or permanently bound to such complexes as the oxidases, may aid in the folding and/or insertion of these polypeptides into the membrane and/or their stability (see IIB below for an example).

Putative assembly factors have been reported for the cytochrome aa_3 oxidase in *Rhodobacter sphaeroides* (Cao et al., 1992; see IIB). Some extragenic (i.e., not apocytochrome structural gene) factors have also been discovered in *Rhodobacter (Rb.) capsulatus* that are specifically required for the synthesis of *c*-type cytochromes. These factors will be described here and we will evaluate their proposed functions with respect to the assembly of non-*c*-type cytochromes. Chapter 34 by Meyer and Donohue will address the structural requirements of soluble cytochromes *c* with respect to heme attachment in *Rb. sphaeroides*. The majority of studies on cytochrome *c* biogenesis have been conducted in *Rb. capsulatus*. However, the significant homologies observed between many of these *Rb. capsulatus* components and proteins encoded by mitochondrial and chloroplast genes underscores the idea that similar assembly processes may occur in plants and animals. These conservations also reveal the close ancestral

Abbreviations: CCHL – cytochrome *c* heme lyase; *Rb. – Rhodobacter*; TMPD – tetramethyl *p*-phenylamindiamine; *S. – Saccharomyces*; *E. – Escherichia*; *B. – Bradyrhizobium*; *P. – Paracoccus*; *N. – Neurospora*

relationship between anoxygenic non-sulfur purple bacteria like *Rb. capsulatus* and mitochondria (Woese, 1987). Clearly, *Rb. capsulatus* and bacteria like it can be used as model systems for the biogenesis of cytochromes in general. A working model for cytochromes *c* biogenesis in *Rb. capsulatus* is presented in Fig. 1 to serve as an aid in understanding the properties and putative functions of these components.

II. Transport of Apocytochromes to the Appropriate Subcellular Locations

A. c-Type Cytochromes

The structural genes encoding many *c*-type cytochromes from photosynthetic and non-photosynthetic bacteria have been cloned and sequenced. From these studies, it is clear that these proteins are synthesized with amino-terminal extensions that are not present in the purified (or crystallized) *c*-type cytochrome. These extensions have the properties of typical bacterial signal sequences: 15–30 amino acid (aa) residues in length with a net positively charged region of 5–8 aa long, followed by a 8–12 aa central hydrophobic core, a predicted 'turn' (e.g., proline) and a polar region of 5–7 aa that terminates in a site for signal cleavage (von Heijne, 1985). In recent years *Escherichia coli* alkaline phosphatase gene (*phoA*) fusions have been used as reporters of secretion since alkaline phosphatase is active only in the periplasm (see Manoil et al., 1990 for review). Gene fusions between *phoA* (without its signal sequence) and the N-termini of specific cytochrome *c* genes have been used to verify the use of these signal sequences in both the photosynthetic bacterium and in the heterologous *E. coli* system (e.g., Varga and Kaplan, 1989). Interestingly, the heterologous secretion of these cytochrome *c*-alkaline phosphatase fusion proteins implies that the general secretory components used in *E. coli* (e.g., SecA, Y, E, D, F) are present in anoxygenic phototrophs. This is supported by the discovery of SecD and SecF homologs in *Rb. capsulatus* (Beckman and Kranz, 1993). Certain indirect evidence (presented in Section IIIA) strongly suggests that the apoproteins and not the holoproteins are secreted across the cytoplasmic

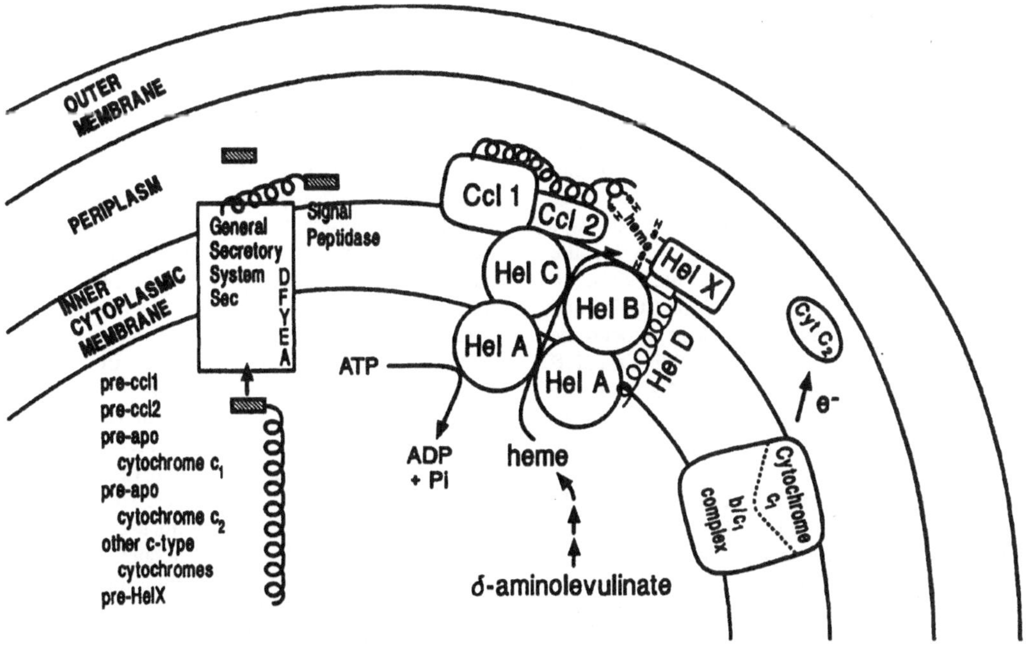

Fig. 1. Working model of cytochromes *c* biogenesis in *Rb. capsulatus* (see text for details).

membrane, probably in an unfolded state. Accordingly, heme and final assembly of the holocytochrome occurs in the periplasmic compartment or on the outer leaflet of the cytoplasmic membrane (see III below). This suggestion is consistent with the hypothesis that during signal sequence-dependent transport, a protein must be in the unfolded state for efficient secretion.

B. Other Cytochromes

Specific modification enzymes are needed to form heme *a* or heme *o* for these types of cytochromes (e.g., see review by von Wachenfeldt and Hederstedt, 1992). Although these enzymes or their genes have yet to be discovered in anoxygenic photosynthetic bacteria, they have been characterized in other bacteria (eg. Saiki et al., 1993; Svensson et al., 1993). Undoubtedly, similar enzymes and genes will be discovered in anoxygenic phototrophs.

Topological studies on cytochrome *b* of the cytochrome bc_1 complex in *Rb. sphaeroides* have indicated that some domains are oriented toward the periplasm and others towards the cytoplasm (Yun et al., 1991; see Chapter 35 by Gray and Daldal). A number of important assembly questions remain for the bc_1 complex. Does the *b* apoprotein naturally insert within the membrane via specific hydrophobic rules dictated by the cytochrome *b* primary amino acid sequence? Is heme bound before, during or after this folding pathway? Do other polypeptides within the bc_1 complex play a role in directing this assembly pathway? If heme is oriented towards the periplasm, how does heme get to this location? Are there assembly factors required for this folding pathway?

Putative factors have been proposed for the assembly of specific oxidases (Cao et al., 1992). For example, the *Saccharomyces cerevisiae* COX10 and COX11 mutations result in a block in cytochrome aa_3 oxidase assembly (Nobrega et al., 1990; Tzagoloff et al., 1990). Proteins homologous to COX10 that are encoded by genes adjacent to cytochrome oxidase subunit genes have been discovered in Gram-negative bacteria (e.g., Chepuri et al., 1990) and Gram-positive bacteria (see von Wachenfeldt and Hederstedt, 1992). Very recently, two *Rb. sphaeroides* genes that are linked to cytochrome aa_3 oxidase subunit genes have been shown to be required for proper cytochrome aa_3 oxidase assembly and the proteins encoded by these genes are related to the COX10 and COX11 genes (Cao et al. 1992). These authors (Cao et al., 1992) note that the high degree of identity between the eukaryotic aa_3 oxidase and that of *Rb. sphaeroides* establish this organism as an excellent model for function, structure and assembly of this important oxidase.

It should be noted that *b*-type cytochromes can be periplasmically located in Gram-negative bacteria. Cytochrome *b*-562, which has a crystal structure surprisingly similar to some soluble *c*-type cytochromes (Lederer et al., 1981), is present in the periplasm of certain *E. coli* strains. Although its function is unknown, cytochrome *b*-562 is synthesized with a typical signal sequence (Nikkila et al., 1991). Accordingly, the question of how heme gets to apocytochrome *b*-562 is particularly pertinent: Could it involve similar mechanisms employed for cytochrome *c* assembly? Some of these generic questions concerning biogenesis of cytochromes will be addressed in this chapter when considering the specific assembly pathway of *c*-type cytochromes (e.g., see IIIA below).

III. Components Required Specifically for Cytochrome *c* Biogenesis

Rb. capsulatus is the only anoxygenic photosynthetic bacterium for which genes involved specifically in the synthesis of *c*-type cytochromes have been characterized. *Rb. capsulatus* contains a *b*-type ubiquinol oxidase in addition to a cytochrome *c* oxidase. Previous studies (La Monica and Marrs, 1976; Hüdig and Drews, 1982) have demonstrated that because either oxidase is sufficient for aerobic growth, it is possible to isolate mutants specifically deficient in *c*-type cytochromes. This property, along with the amenable genetics and gene transfer systems, make *Rb. capsulatus* an excellent model for studying cytochrome *c* biogenesis. A *Rb. capsulatus* mutant that cannot make *c*-type cytochromes will be unable to grow anaerobically since the bc_1 complex is required for photosynthetic growth (Prince and Daldal, 1987; Daldal et al., 1987) and it has been shown that a *c*-type cytochrome may be involved in dark anaerobic growth (Zsebo and Hearst, 1984; Kranz, 1989). In fact, cytochromes *c* biogenesis mutants (Davidson et al., 1987; Kranz, 1989; Biel and Biel, 1990; Beckman et al., 1992; Beckman and Kranz, 1993) have such a phenotype, i.e., they can only grow aerobically. Whether it is possible to isolate null mutants of this pathway in other anoxygenic photosynthetic bacteria will depend on the particular physiology of the microorganism in that it must have a non *c*-type ubiquinol oxidase or other growth mode that is independent of *c*-type cytochromes.

Analyses of *Rb. capsulatus* mutants and genes have defined seven genes at two loci that are required for cytochromes *c* biogenesis (Kranz, 1989; Biel and Biel, 1990; Beckman et al., 1992; Beckman and Kranz, 1993). Physical maps of these genes are shown in Fig. 2. The *hel* locus contains five genes (*helABCDX*) and the second locus contains two genes (*ccl12*) that are required for cytochromes *c* biogenesis. In addition to the growth phenotype described above, mutants in each of these seven genes have the following properties:

(1) Negative for tetramethyl *P* phenylenediamine (TMPD) oxidase activity.

(2) Complete deficiency of all *c*-type cytochromes, as detected by (a) heme stained polypeptides separated on sodium dodecyl sulfate gels and (b) spectroscopically.

(3) Wild-type levels of *b*-type cytochromes, except for the cytochrome *b* of the cytochrome bc_1 complex which is spectroscopically and antigenically missing. (It should be noted that mutants lacking cytochrome c_1 have greatly reduced levels of the cytochrome *b* subunit of the bc_1 complex; thus, it is predicted that cytochrome c_1 is required for proper assembly of the bc_1 complex and without cytochrome c_1, the bc_1 polypeptides are rapidly degraded (Konishi et al., 1991; Davidson et al., 1992; see Chapter 35by Gray and Daldal).

(4) Secrete extracellular protoporphyrin and coprotoporphyrin. This property appears to be a response to a deficiency in *c*-type cytochromes (or electron transport catalyzed by such) and will be discussed in Chapter 52 by Biel.

(5) All mutants have the ability to synthesize and

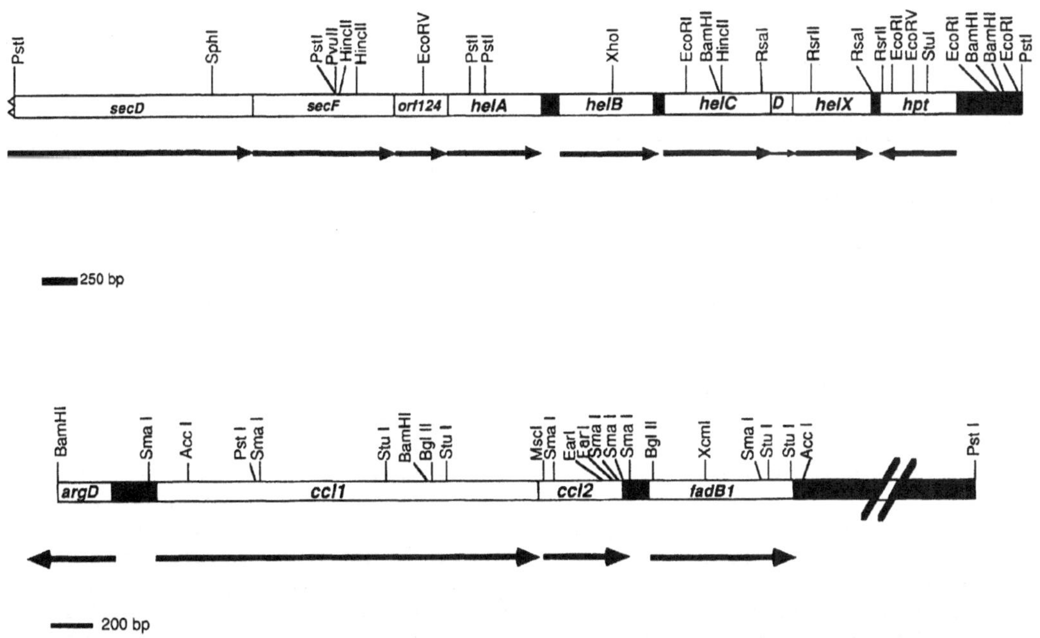

Fig. 2. Restriction maps of the *hel* and *ccl* loci of *Rb. capsulatus*, including upstream and downstream regions. At the *hel* locus, *helA, helB, helC, helD,* and *helX* are required for cytochromes *c* biogenesis while 2 genes at the *ccl* locus, *ccl1* and *ccl2*, are required. The predicted gene product of *orf124* exhibits no known homology. Since the predicted gene products of the other ORFs at these loci exhibit significant homology to known proteins, they were thus named accordingly (see Sections IIIA and IIIB).

secrete an apocytochrome c_2-alkaline phosphatase reporter into the periplasm (Beckman et al., 1992; Beckman and Kranz, 1993 and unpublished). Thus, these mutants still transcribe, translate and transport the apocytochrome c_2 molecule, presumably through the *sec*-dependent pathway. Accordingly, it is probable that each of the proteins encoded by these seven genes plays a role in the biogenesis pathway at a step after the secretion of the apocytochrome molecule to the periplasm.

Following is a discussion of each of the gene products. They will be discussed separately with respect to their properties and putative functions.

A. HelABC Proteins: A Putative Heme Exporter that Fits the Paradigm of an ATP-Dependent Transporter

Results with the *Rb. capsulatus* cytochrome c_2-alkaline phosphatase reporter, as described above, suggest that apocytochrome c_2 is transported either independently of or concurrently with the covalent ligation of heme. Secretion is therefore predicted to occur prior to the assembly of periplasmic *c*-type cytochromes. This is also the conclusion of studies carried out with cytochromes *c* in *Paracoccus denitrificans;* mutants in *P. denitrificans* that do not form the final holocytochromes *c* are still able to synthesize and secrete the apocytochrome to the periplasm (Page and Ferguson, 1989, 1990). Moreover, three separate proteins (i.e., Ccl1, Ccl2, HelX) in *Rb. capsulatus* have been shown to be required for cytochromes *c* biogenesis and present in the periplasm (see below; Beckman et al., 1992; Beckman and Kranz, 1993), a result that is consistent with the post-secretion assembly hypothesis. It is therefore necessary to understand how heme and other assembly factors are transported to the periplasmic compartment.

1. The hel Locus in Rb. capsulatus

One class of *Rb. capsulatus* mutations and genes involved in cytochromes *c* biogenesis were mapped to a single locus called *hel* (for *he*me for *l*igation). Genetic analyses determined that 5 genes at this locus (called *helABCDX*) are required for cytochromes *c* biogenesis (Kranz, 1989; Beckman et al., 1992; Beckman and Kranz, 1993). Each *hel* gene on the chromosome was inactivated and complemented with a DNA fragment containing only that *hel* gene. Also present at this locus are genes which are highly homologous to the *E. coli secD* and *secF* genes (Gardel et al., 1990); a gene called *orf124* that encodes an open reading frame (ORF) of unknown function that is not required for cytochromes *c* biogenesis; and the *hpt* gene, which is homologous to the human HPRT gene (Beckman and Kranz, 1991a), is also not required for cytochromes *c* biogenesis. Considering transcription at this locus, since the stop of *secF* overlaps the predicted *orf124* translational start site, it is thought that these may comprise a single transcription unit (or a superoperon with multiple promoters). Because it was not possible to inactivate the chromosomal *secD* or *secF* by insertional mutagenesis with growth on minimal media it is assumed that, as in *E. coli*, these genes are required for optimal protein secretion in *Rb. capsulatus*. A mini-*Mu* insertion in the chromosomal *orf124* was obtained, ruling out that *orf124* is required for secretion. Additionally, this insert showed polarity on *helA*, resulting in a cytochromes *c* biogenesis defect. This polarity was confirmed by complementation of this insertion strain using a plasmid with only *helA* and a vector promoter. Similar complementation studies in *helB*, *helC*, *helD* and *helX* indicate that multiple promoters exist, representing a potential *secD - secF - orf124 - helA - helB - helC - helD - helX* superoperon. It will be interesting to determine if the products of all these genes also interact in a direct fashion to facilitate cytochromes *c* biogenesis (see model in Fig. 1). It should be noted that although the exact functions of SecD and SecF in the *E. coli* secretory pathway are unknown, SecD and SecF are membrane-bound proteins which are oriented towards the periplasm (Gardel et al., 1990). A role at a later stage in the secretory pathway is envisioned.

2. The Hel ABCD Proteins

The *helA* gene encodes a product that is a member of a larger class of proteins recently termed ABC transporters (for *A*TP-*b*inding *c*assette) or traffic ATPases. Recent reviews on ABC transporters in general, including bacterial and eukaryotic systems, have been published (e.g., Ames, 1992; Higgins, 1992; Reizer et al., 1992) Additionally, a comprehensive review of putative exporters of the ABC protein family, including the *hel* system, has been published (Fath and Kolter, 1993). Substrates for such exporters include peptides, proteins, polysaccharides and certain antibiotics and anti-tumor agents. An ABC transporter is typically composed of a hydrophilic polypeptide containing a consensus ATP-binding site (i.e., Walker motif; Walker et al., 1982). Additional C-terminal amino acid residues are conserved. In the bacterial systems, the ATP-subunit of the ABC transporter is thought to function either as a homodimer or as a heterodimer of two ATP-binding gene products (see Fig. 1). Members of the ABC transporter family include HisP (for histidine import in *E. coli*), MalK (for maltose import), and HlyA (for hemolysin export in *E. coli*) (see reviews for extended references). Associated with this ATP-binding component of the ABC transporter is one (e.g., hemolysin transporter) or multiple (e.g., histidine transporter) polypeptides containing transmembrane domains. Both HelB and HelC are predicted to have 6 transmembrane helices and they are predicted to associate with a HelA homodimer (see Fig. 1). Accordingly, HelB and HelC are predicted to confer the substrate specificity to this transport system. Recently, it has been demonstrated that the HelA polypeptide is unstable and degraded in *Rb. capsulatus helB* and *helC* mutants, results that support the hypothesis that HelABC forms a complex (Goldman and Kranz, unpublished). In contrast, the eukaryotic ABC transporters appear to be composed of a single large polypeptide chain with at least 12 transmembrane helices, 6 each adjoining 2 individual ATP-binding sections. Examples of the eukaryotic ABC transporters include the cystic fibrosis transmembrane conductance regulator protein (chloride transport) and the multidrug-resistance protein (also known as MDR or P-glycoprotein). A notable exception to the eukaryotic 'single polypeptide rule' may be the HelB- and HelC-like proteins observed in mitochondria (see below).

The *helD* gene encodes a 52 aa protein with one potential transmembrane domain in the N-terminus and a highly basic C-terminus. Although this gene has been shown to be essential for cytochrome *c* biogenesis, it is unknown whether it has a role within the ABC transporter model since it does not fit within the current paradigm of ABC transporters. Recent studies indicate that HelD may anchor HelX to the cytoplasmic membrane within the periplasmic space (Beckman and Kranz, unpublished).

The discovery of a specific ABC transporter required for cytochromes *c* biogenesis initially led to a number of theories on its exact function. Substrates for the Hel transporter included three likely candidates: (1) apocytochromes *c*, (2) heme, or (3) other assembly component(s). An additional possibility was that the *hel* genes encoded an iron importer or exporter. This possibility was rejected as unlikely for a number of reasons. These include the specific nature of the defects, the fact that alternative iron sources do not correct the *hel* defects, the *hel* mutants are as sensitive as wild-type to the iron chelator dipyridyl, and because the Hel components appear to be well conserved among bacterial and eukaryotic organisms that synthesize *c*-type cytochromes. Iron import systems, on the other hand, are diverse with multiple mechanisms usually operating and available (e.g., Angerer et al., 1992).

That apocytochromes *c* is a substrate of the Hel transporter was ruled out by the previously described apocytochromes *c* - alkaline phosphatase gene fusion studies. Possibility (3) seems unlikely since the currently known periplasmic assembly components (Ccl1, Ccl2 and HelX) all have signal sequences and are presumably exported through a *sec*-dependent pathway. Likewise, Ccl1-, Ccl2- and HelX-PhoA fusion proteins are transported to the periplasm in an *hel*-independent manner. Accordingly, the most likely substrate for the *hel*-encoded transporter is heme. This hypothesis is consistent with the suggestion that assembly takes place within the periplasm. Furthermore, the transport function is corroborated by the discovery of Hel-like components in mitochondria, the site of heme biosynthesis in eukaryotes (see below). Biochemical analysis of heme transport will be required to confirm this hypothesis.

B. Periplasmic Proteins

1. Ccl1

A *Rb. capsulatus* mutant in *ccl* was first isolated by Biel and Biel (1990). This mutant, called AJB530, was shown to possess the cytochrome defects previously described for *hel* mutants, yet the DNA fragment that complemented AJB530 was different than the *hel* DNA fragment. Subsequently, Beckman et al., (1992), showed that two genes at this locus were required for cytochrome *c* biogenesis, called *ccl1* and *ccl 2* (for cytochrome *c* ligation). Sequence and genetic analysis of this locus indicated that *ccl1* and *ccl2* are located between two other genes not involved in cytochromes *c* biogenesis: *argD* encodes a protein related to *E. coli* acetylornithine aminotransferase and *fadB1* (Beckman and Kranz, 1991b), encodes a protein related to the mitochondrial enoyl CoA hydratase, an enzyme involved in fatty acid degradation. The *Rb. capsulatus ccl1* gene is predicted to encode a 653 aa protein that has a typical bacterial signal sequence. This observation and the high alkaline phosphatase activity exhibited by a *ccl1- phoA* fusion in *Rb. capsulatus* suggest that Ccl1 is oriented toward the periplasm. However, the Ccl1 protein is also predicted to have as many as 12 potential transmembrane helices, indicating that Ccl1 may be integrally associated with the cytoplasmic membrane. While the defects in *Rb. capsulatus* Ccl1 strains clearly indicate a role in the post-secretory assembly of cytochromes *c*, the exact function can only be speculated. The Ccl1 protein is not homologous to known eukaryotic cytochromes *c* heme lyases (CCHL), the enzyme that ligates heme to apocytochromes *c* in yeast (see below). Moreover, the discovery of mitochondrial ORFs homologous to Ccl1 leads to the premise that the Ccl1 class of proteins have a conserved function that remains to be defined. Possibly, the Ccl1 protein acts as a specific chaperone or scaffold used to concentrate the substrates (i.e., heme and apocytochromes *c*). Curiously, the Ccl1 protein shows limited but significant homology to the HelC protein (in a region called the WWD domain because of the highly conserved tryptophan character of this domain across kingdoms; see Fig. 3). It is conceivable that this WWD domain represents a specific heme and/or apocytochrome *c* binding site for the respective functions of HelC and Ccl1. In fact, a similar tryptophan-rich motif (including WFWD) is conserved in the hemopexin protein from rabbit, rat and human (eg. Morgan et al., 1993). Hemopexin is present in the plasma and carries heme to the liver. Recently, it has been shown using *pho* fusion analysis that the WWD domain of Ccl1 is exposed to the periplasmic space (Beckman and Kranz, unpublished).

2. Ccl2

The *ccl2* gene is located downstream of *ccl1* in a potential *ccl12* operon. *ccl2* is predicted to encode a 149 aa polypeptide with a typical bacterial signal sequence and, with the exception of the hydrophobic domain of its signal sequence and aa residues 105 to 122, is predicted to be hydrophilic in character. Recently, the amino two thirds of Cc12 (up to residue 105) has been overproduced and purified from the periplasmic fraction in *E. coli*; the signal sequence of this protein has been cleaved off, proving the presence and function of this signal (Monika and Kranz, unpublished). Ccl2 contains a sequence motif within its N-terminus (RCPVCQGEN) that is similar to a motif found in the *S. cerevisiae* CCHL (CPVMQGDN) (Dumont et al., 1987). However, there is no overall similarity between Ccl2 and CCHL. Additionally, the sequence motifs RCVDH and RCPVL have been identified, respectively, as heme-binding motifs in the HAP1 regulatory protein from *S. cerevisiae* (Creusot et al., 1988; Pfeifer et al., 1989) and erythroid aminolevulinate synthase (Lathrop and Timko, 1993). Ccl 2 may play a role in the heme ligation process itself, although definitive proof awaits biochemical analyses of this aspect of the assembly pathway.

3. HelX, a Thioredoxin-Like Protein

Using deletion mutagenesis, the *Rb. capsulatus helX* gene, downstream of *helD,* was shown to be specifically required for cytochromes *c* biogenesis (Beckman and Kranz 1993). The *helX* gene is predicted to encode a hydrophilic protein of 176 aa residues with a typical N-terminal signal sequence. Using *helX-phoA* as a secretion reporter, it was shown that HelX is a periplasmic protein. HelX shares significant overall homology to *E. coli* thioredoxin (49% similarity and 23% identity) as well as to other thioredoxins (Beckman and Kranz, 1993). Thioredoxins are enzymes involved in reducing protein disulfide bonds and they are characterized by their active site cysteines (WCXPC) that are involved in the redox reactions. The *Rb. capsulatus* HelX protein has this conserved active site. Site-directed mutagenesis of each cysteine residue of HelX, when replaced with a serine residue, results in an inactive protein, unable to complement the *helX* deletion strain (Monika and Kranz, unpublished). This result is taken as evidence that HelX is involved in a red-ox reaction at its active site. It is noteworthy that another class of periplasmic oxido-reductases has recently been characterized. Represented by its first member, the *E. coli* DsbA (Bardwell et al., 1991; Kamitani et al., 1992) protein has been shown to be required for oxidizing cysteine residues of specific secreted proteins (presumably for the latter's proper folding pathway)(Bardwell et al., 1991). The DsbA-class of periplasmic oxido-reductases do not have significant overall similarity to thioredoxins but do contain the active site cysteine residues. HelX is not homologous to DsbA overall. Moreover, HelX cannot be DsbA since high levels of alkaline phosphatase activities were observed in vivo in a ΔhelX mutant (Beckman and Kranz, 1993). In contrast, an *E. coli* DsbA mutant strain shows significantly reduced levels of alkaline phosphatase activity because its cysteine residues are not oxidized, consequently altering the folding pathway of this enzyme (Bardwell et al., 1991). Thus, the HelX protein is proposed to be involved not in the oxidation of cysteine, but in maintenance of reduced cysteine residues. Alternatively, HelX may be required for the maintenance of heme in the reduced state. Each hypothesis is consistent with the view that both heme (Nicholson and Neupert, 1989) and apocytochrome cysteine residues (Enosawa and Ohashi, 1987) must be in the reduced state prior to ligation. Many questions remain, including the ultimate need or supply of reducing equivalents for such reactions to occur within the periplasmic compartment. Obviously, some reactions within this compartment are reducing and some are oxidizing; specific rather than general redox mediating enzymes are therefore likely.

C. Other Proteins

It is important to remember that it remains to be determined whether there exists only one generic CCHL (or one for each *c*-type cytochrome in the cell). The phenotype of mutants in all seven *Rb. capsulatus* genes described above (*helABCDX* and *ccl 12*) are clear—they are totally deficient in all *c*-type cytochromes. Thus, if Ccl2 represents a CCHL, then it is probably a CCHL for all *c*-type cytochromes. Nevertheless, it is possible that a CCHL exists for each (or for soluble versus membrane-bound) *c*-type cytochrome. In this respect, it is interesting to note that a recently described mutant from *Bradyrhizobium japonicum* is deficient in soluble cytochrome *c* but not membrane-bound *c*-type cytochromes (Ritz et al. 1993). The gene complementing this mutant, *cycH*, encodes a periplasmically oriented protein that is not homologous to Hel, Ccl, or eukaryotic CCHL proteins. The function of CycH awaits further investigation. Some results seem to imply that bacterial lyase systems are not highly specific. A number of reports have been made that indicate synthesis and heme attachment of certain *c*-type cytochromes can occur in heterologous *E. coli* expression systems (e.g., Sanbongi et al., 1991; von Wachenfeldt and Hederstedt, 1990).

IV. Synthesis of *c*-Type Cytochromes in Other Bacteria and in Eukaryotes: Analogous Components

A. Other Gram-Negative Bacteria

Genes that are homologous to the *Rb. capsulatus helABCDX* have been reported in the nonphotosynthetic bacterium *B. japonicum* (Ramseier et al., 1991). *B. japonicum* mutants in *cycV* (*helA*-like) *cycW* (*helB*-like) and *cycX* (*helD*-like) were isolated using Tn5. These mutations resulted in strains that are specifically deficient in *c*-type cytochromes. Surprisingly, inactivation of *orf263*, which is located between *cycW* and *cycX* and homologous to *helC*, resulted in the wild-type *B. japonicum* phenotype. Why this gene is not required is unknown. Another gene product in *B. japonicum* called Orf132 is related to the *Rb. capsulatus* HelX. However, in contrast to *helX*, a deletion or insertion in *orf132* could not be constructed and the *orf132* gene product is neither reported to have a signal sequence, nor does its N-terminus show resemblance to HelX. However, recent results indicate that a cosmid containing the *B. japonicum orf132* (provided by Dr. Mark O'Brian, University of New York, Buffalo) complements the *Rb. capsulatus helX* deletion strain to photosynthetic growth (Beckman and Kranz, unpublished). Accordingly, similar functions are envisioned for *B. japonicum* Orf132 and HelX.

Very recently, mutants deficient in cytochrome *c* biogenesis in *Methylobacterium extorquens* AM1 have been reported (Oozeer et al., 1993). DNA fragments complementing these mutants mapped to two locations and the authors speculate that these may be *ccl*- and *hel*-like genes.

Clearly, biogenesis components similar to those of *Rb. capsulatus* are present in non-photosynthetic Gram-negative bacteria.

B. Gram-Positive Bacteria

Because of the different structural limitations placed on the cell envelope compartments within Gram-positive bacteria, *c*-type cytochromes are of necessity bound to the cytoplasmic membrane. Some of the genes encoding *c*-type cytochromes in *Bacillus* have

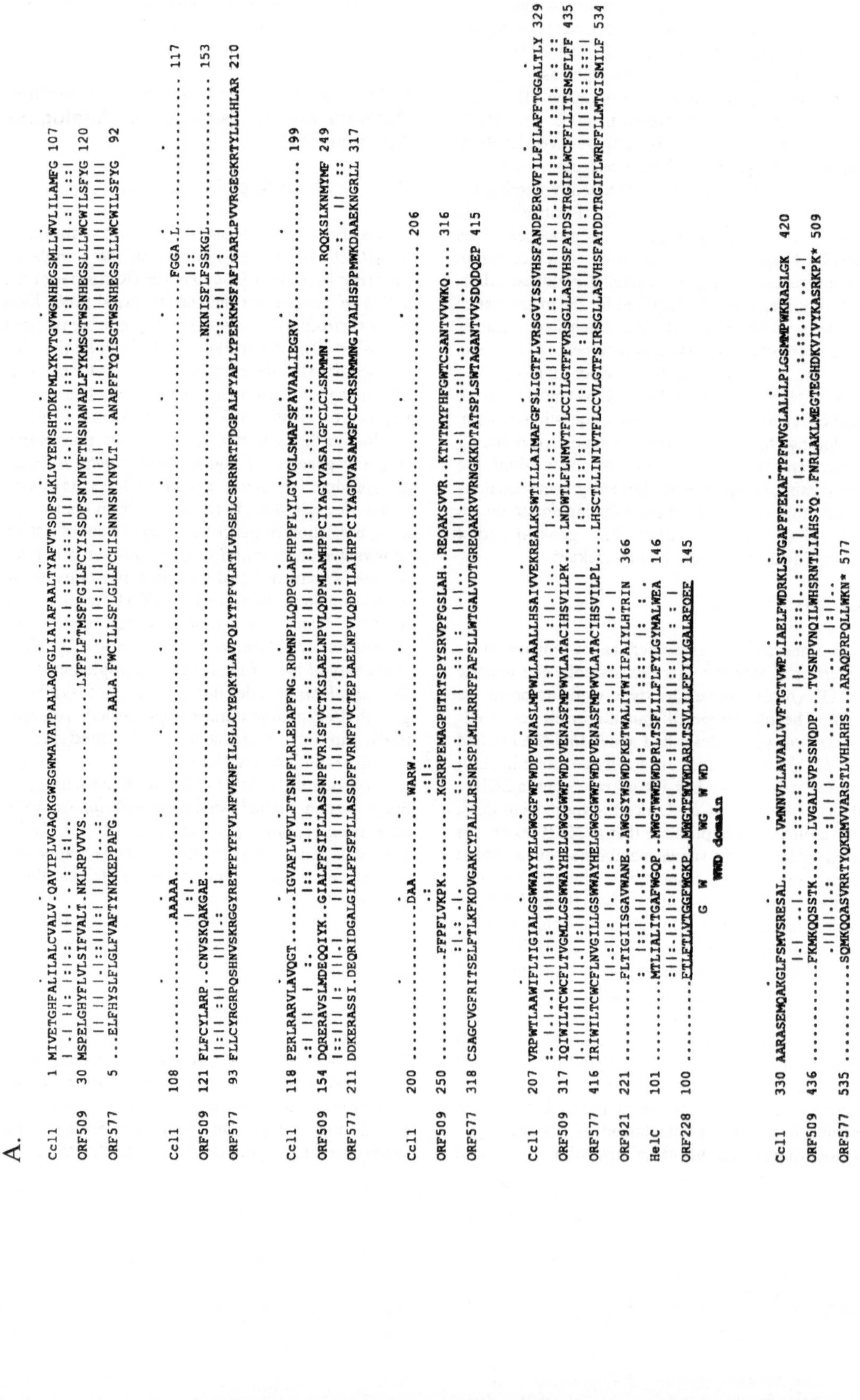

B.

```
                  .         .         .         .         .         .         .         .         .         .
HelC      5 EYANPVKFMQTSGRLLPWVVAATVLTLLPGLVWGFFFTPVAAEFGATVKVIYVHVPAATLAINIWVMMLVASLIWLIRRHHVSALAAKAAAPIGMVMTLI 104
            .. .|. || .| |. . ::: . :     :: ::::..| . : |.. ::|||||||| :.: |:: | :.|:::|: :|.: .| .|.||.:| ::||:
ORF228    4 PLLRPFFFMCCSFRYAQILIGFCWFLTAMAIYLSIWVAPSDFQQGENYRIIYVHVPAAWMSLLIYIAMAISSVLFLLTKHPLFQLFSKTAAKMGALFTLF 103

                  .         .         .         .         .         .         .         .         .         .
HelC    105 ALITGAFWGQPMWGTWWEWDPRLTSFLILFLFYLGYMALWEAIENPDTAADLTGVLCLVGSVFAVLSRYAAIFWNQGLHQGSTLS.LDKEEHIADVYWQP 203
            .|:||:|||.|||||:| ||:||||.||||::||| : :        .: .||:.:::  :| :  .: ::.. :|| .|||.|.:| ::.. ||. :. |
ORF228  104 TLVTGGFWGKPMWGTFWVWDARLTSVLILFFIYLGALRF......QEFSADVASIFMCMGLINMPMIKFSVNWWN.TLHQPSSISQFGTSMHISMLI..P 194

                  .         .         .
HelC    204 LVLSIAGFGMLFVALLLLRTRTEIRARRLKALEQ  237
            :.| :|:| :| . :::| ||  | .  :.  .|
ORF228  195 IFLIFASFFFLTGIFFILETRQMILSFYFQRKSQ* 228
```

Fig. 3. Ccl1 homology to other proteins. A. Pairwise alignment of Ccl1 to ORF509 from the mitochondrial genome of *Marchantia polymorpha* (liverwort) (Oda et al., 1992) and to ORF577 from the mitochondrial genome of *Oenothera* (Schuster et al., 1993). Using the Wisconsin GCG version 7.2 BESTFIT analysis (Devereux et al., 1984), Ccl1 shows 38% identity and 64% similarity to ORF509 and 22% identity and 48% similarity to ORF577. Using the BLASTP program (Altschul et al., 1990), overall homologies to Ccl which are not shown in this figure include the carrot mitochondrial ORF579 (unpublished; GenBank accession number X69554); the *Paramecium* mitochondrial ORF238 (Pritchard et al., 1990; and see Beckman et al., 1992 figure 9 legend) the mitochondrial liverwort ORF322 (to CCl1 aa residues 492-630) and ORF169 (to Ccl1 aa residues 353-405) (Oda et al., 1992). Significant homology is also obtained to other proteins in the WWD region. Shown is the WWD region of the red alga *Cyanidium caldarium* chloroplast ORF921 (a 306 aa protein; Valentin et al., 1992), HelC of *Rb. capsulatus* (Beckman et al., 1992) and the HelC-like mitochondrial ORF228 from liverwort (Oda et al., 1992). Other ORFs with significant homology to Ccl1 include the *Cryptomonas* plastid ORF301 (unpublished; GenBank accession number X52159); liverwort chloroplast ORF320 (Ohyama et al., 1988); tobacco chloroplast ORF313 (Shinozaki et al., 1986); and the rice chloroplast ORF321 (Hiratsuka et al., 1989). See text, section III, for a description. B. Wisconsin GCG version 7.2 BESTFIT alignment (Devereux et al., 1984) of *Rb. capsulatus* HelC to the liverwort mitochondrial ORF228 (Oda et al., 1992) showing 33% identity and 59% similarity over this region. In addition to ORF228, BLASTP (Altschul et al., 1990) significant homologies to HelC include *B. japonicum* ORF263 (Ramseier et al., 1991), the liverwort chloroplast ORF320 (Ohyama et al., 1988); the *Cryptomonas* chloroplast ORF301 (unpublished; Genbank accession number X52159); the *Cyanidium caldarium* chloroplast ORF921 (Valentin et al., 1992); the tobacco chloroplast ORF313 (Shinozaki et al., 1986) and the rice chloroplast ORF321 (Hiratsuka et al., 1989). (The WWD region of ORF509 has a Poisson Probability score of 0.9998) been characterized and discussed in a recent review (von Wachenfeldt and Hederstedt, 1992). Briefly, these cytochromes are proposed to be synthesized with a typical prokaryotic signal sequence. Some of these signals are apparently not cleaved but serve to anchor the cytochrome to the membrane. With regards to biogenesis, little is known about the involvement of extragenic factors or the location and timing of assembly. Whether the biogenesis components described for *Rb. capsulatus* (and for *B. japonicum* or eukaryotic organelles — see below) are also present in *Bacillus* and other Gram-positive bacteria remains to be determined.

C. Eukaryotes

1. Mitochondria

The site of eukaryotic ATP synthesis is the mitochondrion. Mitochondrial electron transport pathways are similar to that of some anoxygenic photosynthetic bacteria. These aerobic electron transfer chains include a membrane-bound cytochrome bc_1 complex and soluble cytochrome *c* that are structurally and functionally analogous to the bacterial types (see Chapters 35 by Gray and Daldal and 34 by Meyer and Donohue). Topologically, these cytochromes are also similar to those in procaryotes. Cytochrome c_1 is membrane-bound and oriented toward the intermembrane space while cytochrome *c* is in the mitochondrial intermembrane space. The intermembrane space, located between the mitochondrial inner membrane and outer membrane, is topologically equivalent to the periplasmic space of a Gram-negative bacterium. Transport of the apocytochrome *c* and c_1 of *S. cerevisiae* and *Neurospora crassa* has been the subject of many exceptional experiments during the last 20 years. These transport pathways have been reviewed (Gonzales and Neupert, 1990). Briefly, both apocytochrome *c* and c_1 are synthesized in the cytoplasm and posttranslationally transported to the intermembrane space. Apocytochrome *c* is synthesized without a signal sequence and is brought directly into the intermembrane space by a reaction that is proposed to depend on the ligation of heme to the apocytochrome *c* within the intermembrane space (see below). The apocytochrome c_1 is synthesized with a bipartite signal sequence which is initially used to enter the mitochondrial matrix where cleavage of the first part of the bipartite signal occurs. The remaining signal directs the apocytochrome c_1 back across the inner membrane. Heme ligation to apocytochrome c_1, although it may occur prior to cleavage of the second presequence, is not obligatory for this translocation process.

S. cerevisiae and *N. crassa* mutants in CCHL and cytochrome c_1 heme lyase (CC_1HL) have been characterized and the genes encoding these two lyases have been cloned and sequenced (Dumont et al., 1987; Drygas et al., 1989; Zollner et al., 1992). Accordingly, it is well documented that two separate heme lyases, both of which are present in the intermembrane space, are required for formation of holocytochrome *c* and c_1 (Nargang et al., 1988; Dumont et al., 1988). Recently, transport of CCHL to the intermembrane space has been characterized (Lill et al., 1992). It is thought that a signal present in the mature CCHL is sufficient for translocation across the outer membrane directly to the intermembrane space.

Certain questions concerning the heme ligation reaction in mitochondria have been answered. In contrast to bacterial systems in which in vitro heme lyase activity has not been reported, this activity has been detected in vitro in the two eukaryotic systems described above. Although the CCHL is significantly homologous to the CC_1 HL and both appear to be membrane bound intermembrane proteins (Zollner et al., 1992; Lill et al., 1992; Dumont et al., 1991; Nicholson et al., 1989), CC_1HL cannot use apocytochrome *c* as a substrate (Stuart et al., 1990). Interestingly, both CCHL- and CC_1HL- catalyzed ligation reactions require that heme is in the reduced state (Nicholson and Neupert, 1989; Nicholson et al., 1989). This reduction was accomplished by the addition of NADH and FMN in the in vitro assays (Nicholson and Neupert, 1989). However, sodium dithionite can replace these reductants (Nicholson and Neupert, 1989) and it is likely that the cysteine residues of apocytochromes *c* need to be reduced for the ligation reaction to proceed (Enosawa and Ohashi, 1987). The mechanism(s) by which heme and apocytochromes are reduced in vivo in the mitochondria is unknown.

Furthermore, the involvement of components besides CCHL and CC_1HL in eukaryote cytochromes *c* biogenesis remains an open question. The last step of heme biosynthesis occurs in the mitochondrial matrix. How does heme get to the mitochondrial intermembrane space? Recently, searches of the protein and DNA databases with the

Rb. capsulatus Hel and Ccl proteins have resulted in some striking comparisons. A number of mitochondrial genomes have recently been sequenced; these genomes show ORFs that have remarkable similarities to the *Rb. capsulatus* HelB, HelC and Ccl1 proteins. In fact, the *helB-*,*helC-*, and *ccl1*-like genes are adjacent to each other in the liverwort mitochondrial genome (Oda et al., 1992). Examples of these comparisons and references are shown in Fig. 3. The HelC and Ccl1 comparisons to the eukaryotic proteins are particularly identical in a domain, called the 'WWD' domain. These mitochondrial analogies raise obvious questions concerning function. Could the Hel ABC transport system be required for cytochromes *c* biogenesis (and possibly heme transport) in eukaryotes? Could the Ccl1 protein be required for cytochrome *c* biogenesis in eukaryotes? Two very recent reports note the striking similarity between an *Oenothera* (primrose) mitochondrial Orf577 (Schuster et al. 1993) or a wheat mitochondrial Orf589 (Gonzales et al., 1993) and the *Rb. capsulatus* Ccl1. Both groups show that mRNA editing occurs within these Orfs, especially within the WWD domain, to yield even more homology. Based on these results, the authors conclude that these Orfs may also be involved in cytochromes *c* biogenesis in plant mitochondria.

2. Chloroplasts

Chloroplasts possess a *c*-type cytochrome called cytochrome *f* which is a membrane component of the cytochrome b_6f complex, and it is the chloroplast equivalent of the mitochondrial cytochrome c_1 (for recent reviews on cytochrome *f* see Gray, 1992; for cytochrome *bf* see Hope, 1993 and Anderson, 1992). Proteins encoded by chloroplast genomes from a number of plant species have previously been shown to be homologous to *Rb. capsulatus* Ccl1 (see Fig. 3 for example). Moreover, certain mutations which map to the chloroplast genome of *Chlamydomonas* yield cells that are pleiotrophically deficient in chloroplast *c*-type cytochromes (Howe and Merchant, 1992). Could these mutations be located in the Ccl1-like ORFs encoded in the chloroplast?

V. Future Directions

Much work is required on the folding and assembly pathways of the membrane-bound cytochrome complexes. What accessory factors are required and what are their functions within the biogenesis pathway for macromolecular cytochrome complexes? When and how is heme incorporated into the complexes? Questions concerning cytochrome *c* biogenesis include what are the exact functions of Ccl and Hel proteins? Are other accessory factors required (e.g., individual heme lyases)? Clearly, multifaceted approaches that include genetics, molecular genetics and biochemistry will be required to answer these and other important questions. The discovery of eukaryotic counterparts to the *Rhodobacter* components underscore the significance of these factors and the use of anoxygenic photosynthetic bacteria as models for these biogenesis pathways.

Acknowledgments

We thank Dr. Barry Goldman for critiquing this chapter. RGK is supported by NSF and NIH.

References

Altschul SF, Gish W, Miller W, Myers EW and Lipman DJ (1990) Basic local alignment search tool. J Mol Biol 215: 403–410

Ames GF-L (1992) Bacterial periplasmic permeases as model systems for the superfamily of traffic ATPases, including the multidrug resistance protein and the cystic fibrosis transmembrane conductance regulator. Intl Rev Cytology 137A: 1–35

Anderson JM (1992) Cytochrome b_6f complex: Dynamic molecular organization, function and acclimation. Photosynth Research 34: 341–357

Angerer A, Klupp B and Braun V (1992) Iron transport systems of *Serratia marcescens*. J Bacteriol 174: 1378–1387

Bardwell JCA, McGovern K and Beckwith J (1991) Identification of a protein required for disulfide bond formation in vivo. Cell 67: 581–589

Beckman DL and Kranz RG (1991a) A bacterial homolog to HPRT. Biochim Biophys Acta 1129: 112–114

Beckman DL and Kranz RG (1991b) A bacterial homolog to the mitochondrial enoyl-CoA hydratase. Gene 107: 171–172

Beckman DL, Trawick DR and Kranz RG (1992) Bacterial cytochromes *c* biogenesis. Genes Dev 6: 268–283

Beckman DL and Kranz RG (1993) Cytochromes *c* biogenesis in a photosynthetic bacterium requires a periplasmic thioredoxin-like protein. Proc Natl Acad Sci USA 90: 2179–2183

Biel SW and Biel AJ (1990) Isolation of a *Rhodobacter capsulatus* mutant that lacks *c*-type cytochromes and excretes porphyrins. J Bacteriol 172: 1321–1326

Cao J, Hosler J, Shapleigh J, Revzin A and Ferguson-Miller S (1992) Cytochrome aa_3 of *Rhodobacter sphaeroides* as a model for mitochondrial cytochrome *c* oxidase. J Biol Chem 267: 24273–24278

Chepuri V, Lemieux L, Au DC-T and Gennis RB (1990) The sequence of the *cyo* operon indicates substantial structural similarities between the cytochrome *c* ubiquinol oxidase of *Escherichia coli* and the aa_3-type family of cytochrome *c* oxidases. J Biol Chem 265: 11185–11192

Creusot F, Verdiere J, Gaisne M and Slonimski PP (1988) CYP1 (HAP1) regulator of oxygen-dependent gene expression in yeast. Overall organization of the protein sequence displays several novel structural domains. J Mol Biol 204: 263–276

Daldal F, Davidson E and Cheng S (1987) Isolation of the structural genes for the Rieske Fe-S protein, cytochrome *b*, and cytochrome c_1. All components of the ubiquinol:cytochrome c_2 oxidoreductase complex of *Rhodopseudomonas capsulata*. J Mol Biol 195: 1–12

Davidson E, Prince RC, Daldal F, Hauska G, and Marrs BL (1987) *Rhodobacter capsulatus* MT113: A single mutation results in the absence of *c-type* cytochromes and in the absence of the cytochrome bc_1 complex. Biochim Biophys Acta 890: 292–301

Davidson E, Ohnishi T, Tokito M and Daldal F (1992) *Rhodobacter capsulatus* mutants lacking the Rieske FeS protein form a stable cytochrome bc_1 subcomplex with an intact quinone reduction site. Biochemistry 31: 3351–3358

Devereux J, Haeberli P and Smithies O (1984) A comprehensive set of sequence analysis programs for the VAX. Nuc Acids Res 12: 387–395

Drygas ME, Lambowitz AM and Nargang FE (1989) Cloning and analysis of the *Neurospora crassa* gene for cytochrome *c* heme lyase. J Biol Chem 264: 17897–17906

Dumont ME, Ernst JF, Hampsey DM and Sherman F (1987) Identification and sequence of the gene encoding cytochrome *c* heme lyase in the yeast *Saccharomyces cerevisiae*. EMBO J 6: 235–241

Dumont ME, Ernst JF and Sherman F (1988) Coupling of heme attachment to import of cytochrome *c* into yeast mitochondria. Studies with heme lyase-deficient mitochondria and altered apocytochromes *c*. J Biol Chem 263: 15928–15937

Dumont ME, Cardillo TS, Hayes MK and Sherman F (1991) Role of cytochrome *c* heme lyase in mitochondrial import and accumulation of cytochrome *c* in *Saccharomyces cerevisiae*. Mol Cell Biol 11: 5487–5496

Enosawa S and Ohashi A (1987) A simple and rapid assay for heme attachment to apocytochrome *c*. Anal Biochem 160: 211–216

Fath MJ and Kolter R (1993) ABC transporters: The bacterial exporters. Microbiol Reviews 57: 995–1017

Gardel C, Johnson K, Jacq A and Beckwith J (1990) The *secD* locus of *E. coli* codes for two membrane proteins required for protein export. EMBO J 9: 3209–3216

Gonzales DH and Neupert W (1990) Biogenesis of mitochondrial *c*-type cytochromes. J Bioenerg. Biomembr. 22: 753–768

Gonzalez DH, Bonnard G and Grienenberger J-M (1993) A gene involved in the biogenesis of cytochromes is co-transcribed with a ribosomal protein gene in wheat mitochrondria. Curr Genet 24: 248–255

Gray JC (1992) Cytochrome *f*: Structure, function and biosynthesis. Photosynth Res 34: 359–374

Higgins CF (1992) ABC transporters: From microorganisms to man. Annu Rev Cell Biol 8: 67–113

Hiratsuka J, Shimada H, Whittier R, Ishibashi T, Sakamoto M, Mori M, Kondo C, Honji Y, Sun CR, Meng BY, Li YQ, Kanno A, Nishizawa Y, Hirai A, Shinozaki K and Sugiura M (1989) The complete sequence of the rice (Oryza sativa) chloroplast genome: Intermolecular recombination between distinct tRNA genes accounts for a major plastid DNA inversion during the evolution of the cereals. Mol Gen Genet 217: 185–194

Hope AB (1993) The chloroplast cytochrome *bf* complex: A critical focus on function. Biochim Biophys Acta 1143: 1–22

Howe G and Merchant S (1992) The biosynthesis of membrane and soluble plastidic *c*-type cytochromes of *Chlamydomonas reinhardtii* is dependent on multiple common gene products. EMBO J 11: 2789–2801

Hüdig H and Drews G (1982) Isolation of a *b*-type cytochrome oxidase from membranes of the phototrophic bacterium *Rhodopseudomonas capsulata*. Z Naturforsch 37: 193–198

Kamitani S, Akiyama Y and Ito K (1992) Identification and characterization of an *Escherichia coli* gene required for the formation of correctly folded alkaline phosphatase, a periplasmic enzyme. EMBO J 11: 57–62

Konishi K, VanDoren SR, Kramer DM, Crofts AR and Gennis RB (1991) Preparation and characterization of the water-soluble heme-binding domain of cytochrome c_1 from the *Rhodobacter sphaeroides* bc_1 complex. J Biol Chem 266: 14270–14276

Kranz RG (1989) Isolation of mutants and genes involved in cytochromes *c* biosynthesis in *Rhodobacter capsulatus*. J Bacteriol 171: 456–464

La Monica RF and Marrs BL (1976) The branched respiratory system of photosynthetically grown *Rhodopseudomonas capsulata*. Biochim Biophys Acta 423: 431–439

Lathrop JT and Timko MP (1993) Regulation by heme of a mitochondrial protein transport through a conserved amino acid motif. Science 259: 522–525

Lederer F, Glatigny A, Bethge PH, Bellamy HD and Mathews FS (1981) Improvement of the 2.5 Å Resolution model of cytochrome *b*-562 by redetermining the primary structure and using molecular graphics. J Mol Biol 148: 427–448

Lill R, Stuart RA, Drygas ME, Nargang FE and Neupert W (1992) Import of cytochrome *c* heme lyase into mitochondria: A novel pathway into the intermembrane space. EMBO J 11: 449–456

Manoil C, Mekalanos, JJ and Beckwith J (1990) Alkaline phosphatase fusions: Sensors of subcellular location. J Bacteriol 172: 515–518

Morgan W, Muster P, Tatum F, Kao S, Alam J and Smith A (1993) Identification of the histidine residues of hemopexin that coordinate with heme-iron and of a receptor-binding region. J Biol Chem 268: 6256–6262

Nargang FE, Drygas ME, Kwong PL, Nicholson DW and Neupert W (1988) A mutant of *Neurospora crassa* deficient in cytochrome *c* heme lyase activity cannot import cytochrome *c* into mitochondria. J Biol Chem *263*, 9388–9394

Nicholson, DW and Neupert W (1989) Import of cytochrome *c* into mitochondria: Reduction of heme, mediated by NADH and flavin nucleotides, is obligatory for its covalent linkage to apocytochrome *c*. Proc Natl Acad Sci USA 86: 4340–4344

Nicholson DW, Stuart RA and Neupert W (1989) Biogenesis of cytochrome c_1. Role of cytochrome c_1 heme lyase and of the two proteolytic processing steps during import into mitochondria. J Biol Chem 264: 10156–10168

Nikkila H, Gennis RB and Sligar SG (1991) Cloning and expression of the gene encoding the soluble cytochrome *b*-562 of *E. coli*. Eur. J Biochem 202: 309–313

Nobrega MP, Nobrega FG and Tzagoloff A (1990) *COX10* codes for a protein homologous to the ORF1 product of *Paracoccus denitrificans* and is required for the synthesis of yeast cytochrome oxidase. J Biol Chem 265: 14220–14226

Oda K, Yamato K, Ohta E, Nakamura Y, Takemura M, Nozato N, Kohchi T, Ogura Y, Kanegae T, Akashi K and Ohyama K (1992) Gene organization deduced from the complete sequence of liverwort *Marchantia polymorpha* mitochondrial DNA. J Mol Biol 223: 1–7

Ohyama K, Fukuzawa H, Kohchi T, Sano T, Sano S and Shirai H (1988) Structure and organization of *Marchantia polymorpha* chloroplast genome. J Mol Biol 203: 281–298

Oozeer F, Page MD, Ferguson SJ and Goodwin PM (1993) Phenotypic characterization of *c*-type cytochrome-deficient mutants of *Methylobacterium extorquens* AM1 and identification of two chromosomal regions essential for the production of *c*-type cytochromes. J General Microbiology 139: 11–19

Page MD and Ferguson SJ (1989) A bacterial *c*-type cytochrome can be translocated to the periplasm as an apo form; the biosynthesis of cytochrome cd_1 (nitrite reductase) from *Paracoccus denitrificans*. Mol Microbiol 3: 653–661.

Page MD and Ferguson SJ (1990). Apo forms of cytochrome *c*-550 and cytochrome cd_1 are translocated to the periplasm of *Paracoccus denitrificans* in the absence of haem incorporation caused by either mutation or inhibition of haem synthesis. Mol Microbiol 4: 1181–1192.

Palmer G and Reedijk J (1992) Nomenclature of electron-transfer proteins. J Biol Chem 267: 665–677

Pfeifer K, Kim K-S, Kogan S and Guarente L (1989) Functional dissection and sequence of yeast HAP1 activator. Cell 56: 291–301

Prince RG and Daldal F (1987) Physiological electron donors to the photochemical reaction center of *Rhodobacter capsulatus*. Biochim Biophys Acta 894: 370–378

Pritchard AE, Seilhamer JJ, Mahalingam R, Sable CL, Venuti SE and Cummings DJ (1990) Nucleotide sequence of the mitochondrial genome of *Paramecium*. Nuc Acids Res 18: 173–180

Ramseier TM, Winteler HV, Hennecke H (1991) Discovery and sequence analysis of bacterial genes involved in the biogenesis of *c*-type cytochromes. J Biol Chem 266: 7793–7803

Reizer J, Reizer A and Saier MH Jr (1992) A new subfamily of bacterial ABC-type transport systems catalyzing export of drugs and carbohydrates. Protein Science 1: 1326–1332

Ritz D, Bott M and Hennecke H (1993) Formation of several bacterial *c*-type cytochromes requires a novel membrane-anchored protein that faces the periplasm. Mol Microbiol 9: 729–740

Saiki K, Mogi T, Ogura K and Anraku Y (1993) In vitro heme *o* synthesis by the *cyoE* gene product form *Escherichia coli*. J Biol Chem 268: 26041–26045

Sanbongi Y, Yang J-H, Igarashi Y and Kodama T (1991) Cloning, nucleotide sequence and expression of the cytochrome *c*-552 gene from *Hydrogenobacter thermophilus*. Eur J Biochem 198: 7–12

Schuster W, Combettes B, Flieger K and Brennicke A (1993) A plant mitochondrial gene encodes a protein involved in cytochrome *c* biogenesis. Mol Gen Genet 239: 49–57

Shinozaki K, Ohme M, Tanaka M, Wakasugi T, Hayashida N, Matsubayashi T, Zaita N, Chunwongse J, Obokata J, Yamaguchi-Shinozaki K, Ohto C, Torazawa K, Meng BY, Sugita M, Deno H, Kamogashira T, Yamada K, Kusuda J, Takaiwa F, Kato A, Tohdoh N, Shimada H and Sugiura M (1986) The complete nucleotide sequence of the tobacco chloroplast genome: Its gene organization and expression. EMBO J 5: 2043–2049

Stuart RA, Nicholson DW, Wienhues U and Neupert W (1990) Import of apocytochrome *c* into the mitochondrial intermembrane space along a cytochrome c_1 sorting pathway. J Biol Chem 265: 20210–20219

Svensson B, Lübben M and Hederstedt L (1993) *Bacillus subtilis* CtaA and CtaB function in haem *a* biosynthesis. Mol Microbiol 10: 193–201

Tzagoloff A., Capitanio N, Nobrega MP and Gatti D (1990) Cytochrome oxidase assembly in yeast requires the product of *COX11*, a homolog of the *P. denitrificans* protein encoded by *ORF3*. EMBO J 9: 2759–2764

Valentin K, Maid U, Emich A and Zetsche K (1992) Organization and expression of a phycobiliprotein gene cluster from the unicellular red alga *Cyanidium caldarium*. Plant Mol Biol 20: 267–276

Varga AR and Kaplan S (1989) Construction, expression, and localization of a CycA::PhoA fusion protein in *Rhodobacter sphaeroides* and *Escherichia coli*. J Bacteriol 171: 5830–5839

von Heijne G (1985) Signal sequences: The limits of variation. J Mol Biol 184: 99–105

von Wachenfeldt C and Hederstedt L (1990) *Bacillus subtilis* holo-cytochrome *c*-550 can be synthesized in aerobic *Escherichia coli*. FEBS Lett 270: 147–151

von Wachenfeldt C and Hederstedt L (1992) Molecular biology of *Bacillus subtilis* cytochromes. FEMS Microbiol Lett 100: 91–100

Walker JE, Saraste M, Runswick MJ and Gay NJ (1982) Distantly related sequences in the alpha- and beta-subunits of ATP synthase, myosin, kinases and other ATP-requiring enzymes and a common nucleotide binding fold. EMBO J 8: 945–951

Woese CR (1987) Bacterial evolution. Microbiol Rev 51: 221–271

Yun C-H, Van Doren S, Crofts AR and Gennis RB (1991) The use of gene fusions to examine the membrane topology of the L-subunit of the photosynthetic reaction center and of the cytochrome *b* subunit of the bc_1 complex from *Rhodobacter sphaeroides*. J Biol Chem 266: 10967–10973

Zollner A, Rodel G and Haid A (1992) Molecular cloning and characterization of the *Saccharomyces cerevisiae cyt2* gene encoding cytochrome-c_1-heme lyase. Eur J Biochem 207: 1093–1100

Zsebo KM and Hearst JE (1984) Genetic-physical mapping of a photosynthetic gene cluster from *Rb. capsulata*. Cell 37: 937–947

Chapter 34

Cytochromes, Iron-Sulfur, and Copper Proteins Mediating Electron Transfer from the Cyt bc_1 Complex to Photosynthetic Reaction Center Complexes

T. E. Meyer
Department of Biochemistry, University of Arizona, Tucson, AZ 85721, USA

Timothy J. Donohue
Department of Bacteriology, University of Wisconsin, 1550 Linden Drive, Madison, WI 53706, USA

R. E. Blankenship, M. T. Madigan and C. E. Bauer (eds): Anoxygenic Photosynthetic Bacteria, pp. 725–745.

Summary

Electron transfer between the photosynthetic reaction center and the cytochrome bc_1 complexes is often mediated by a high redox potential soluble cytochrome. In purple non-sulfur bacteria, this electron donor is usually cytochrome c_2 (Cyt c_2), while cyanobacteria and green algae can use the distantly related cytochrome c_6 protein. Genetic and biochemical support for the role of these soluble mediators has been obtained in purple bacteria *Rhodobacter sphaeroides* and *Rb. capsulatus*, and in several cyanobacteria and green algae. However, recent experiments indicate that photosynthetic organisms often contain redundant electron donors to light-oxidized reaction center complexes. For example, soluble or membrane-bound cytochromes can substitute for the normal electron donor, Cyt c_2, in purple non-sulfur bacteria. In addition, the soluble copper protein, plastocyanin, is interchangeable with cytochrome c_6 in cyanobacteria and green algae even though it is apparently the sole electron donor in green plants.

In contrast, the soluble electron donor is either unknown or presumed to be different in photosynthetic bacteria that lack soluble high potential proteins in the Cyt c_2 family. Spectroscopic analysis indicates this donor could be the high redox potential ferredoxin, HiPIP, in *Rhodocyclus sp.* and members of the *Chromatiaceae* and *Ectothiorhodospiraceae* families. In addition, cytochromes other than Cyt c_2 may reduce reaction center complexes in some photosynthetic bacteria. For example, Cyt *c*-551 may be the donor in *Ectothiorhodospira* sp., while related cytochromes could function in *Rc. purpureus* and green bacteria. Finally, the membrane bound copper protein, auracyanin, from the green bacterium *Chloroflexus aurantiacus* could also be responsible for reaction center reduction. In these latter species, the combination of biochemical and genetic analyses have not been used to identify the actual electron donor(s) to reaction center complexes.

I. Introduction

In the broadest sense, photosynthesis may be defined as the conversion of light into chemical energy. In photosynthetic (PS) prokaryotes, this conversion is accomplished by coupling light-driven oxidation of specific chlorophylls or bacteriochlorophylls in a reaction center (RC) complex to subsequent electron transfer through membrane-bound redox components. The resulting formation of a proton gradient by PS membrane redox complexes is used to drive ATP synthesis. In its simplest form, there are two essential membrane-bound components of PS electron transport pathways, the RC and the cytochrome bc_1 (Cyt bc_1) complexes. Function of these two integral membrane complexes is linked via membrane-diffusible quinones and soluble electron carrier proteins. This chapter discusses the soluble redox proteins which are capable of mediating electron transfer from the Cyt bc_1 complex to a light-oxidized RC complex.

A. Why Do Photosynthetic Cells Use Soluble Electron Carriers?

As described elsewhere in this volume, light-induced electron transfer in the bacterial RC results in the rapid oxidation of a special pair of pigment molecules at the outer or periplasmic side of the PS membrane and the reduction of a quinone on the inner or cytoplasmic side of this membrane. This so-called special pair of pigment molecules can be either bacteriochlorophyll or chlorophyll in different PS prokaryotes. The role of soluble redox proteins in PS electron flow is to reduce the photo-oxidized special pair (Fig. 1) to a ground state. In order for photosynthesis to generate net biological energy, special pair reduction must occur before the return of an electron from the RC quinone (the so-called 'back reaction'). Since full quinone reduction requires the addition of two electrons, it is generally considered essential that the singly-reduced semiquinone remain bound to the RC until it is fully reduced to quinol by a second electron. Either semiquinone or quinol would be capable of directly reducing the special pair via this back reaction if soluble electron donors were not capable of efficient RC reduction. Once quinone is fully reduced to quinol, it diffuses from the RC through the membrane, where it is oxidized by the Cyt bc_1 complex, which in turn reduces the soluble redox protein that ultimately donates its electron to the special pair. Presumably, the inability to position the respective redox centers of the RC and Cyt bc_1 complexes in close enough proximity make direct electron transfer between the these two supramolecular aggregates too slow to compete with the non-productive return of an electron from a reduced quinone species to the oxidized special pair.

The current understanding of bacterial photosynthesis indicates that the RC electron donor must have several characteristics. First, the RC electron donor must be on the proper side of the membrane to reduce the special pair and accept electrons from Cyt c_1 of the Cyt bc_1 complex. In bacterial systems, these donors are found either in the periplasm of Gram-negative bacteria or associated with the outer surface of the cytoplasmic membrane of Gram-positive species. Thus, their genes commonly contain signal sequences to presumably help direct the respective precursor proteins to their proper cellular location (Daldal et al., 1986; Donohue et al., 1986; Brandner and Donohue, 1994; see Chapter 33 by Kranz and Beckman). In addition, these donors should have midpoint potentials between Cyt c_1 and the special pair. Cyt c_1 has a midpoint potential of 265–318 mV in purple bacteria (Gabellini et al., 1982; Guner et al., 1991; Gray et al., 1992; Hacker et al., 1993), while the midpoint potential of the RC special pair is ~500 mV (Case and Parson, 1973; Gao et al., 1990; Murchison et al., 1993). Thus, the theoretical range of midpoint potentials for soluble electron donors is ~300–500 mV for cells in which these soluble proteins are directly responsible for special pair reduction. In cells that have a membrane-bound tetraheme cytochrome *c* in the RC (see below), the soluble donors should have midpoint potentials (~300–350 mV) which approximate that of the cytochrome *c* heme, which is the immediate electron acceptor. Significant deviations from these midpoint potentials for the soluble electron donor are likely to be reflected in the corresponding values for the Cyt bc_1 complex and the RC. In addition, these diffusible proteins often contain a so-called interaction domain to reversibly dock with their membrane-bound electron donor (Cyt bc_1 complex) or acceptor (RC complex). For membrane bound electron donors, there presumably must be a membrane anchor, a 'soluble' or extra-membrane portion, and a flexible arm which permits the interaction domain to flip back and forth from one complex to the other. Finally, to sustain the rate of electron transport necessary to generate a sufficient proton gradient for supporting PS growth, the cellular abundance of these electron donors should approximate that of the RC.

This chapter summarizes what is currently known about proteins which can mediate electron transfer between the Cyt bc_1 complex and the RC special pair. Historically, the ability to monitor light-dependent electron transport reactions in either whole cells, isolated membranes or with purified components provided investigators with a relatively simple way to identify and characterize proteins capable of catalyzing RC reduction. The first part of this chapter summarizes what is known about soluble proteins that are capable of performing this reaction. Next, the mechanisms by which these soluble proteins might interact with their membrane-bound redox partners is discussed. Finally, we summarize how recent molecular genetic of several PS bacterial species has provided the means to test whether those proteins which can participate in RC reduction in vitro do, in fact, perform this function in vivo. This

Abbreviations: AdhI – Zn-dependent alcohol dehydrogenase; *Cf.* – *Chloroflexus;* ChrR – Cohemin-resistance regulatory protein; Cyt b_6f – cytochrome b_6f; Cyt bc_1 – ubiquinol Cyt c_2-oxidoreductase; Cyt c_2 – cytochrome c_2; Cyt c_6 – cytochrome c_6; *Ec.* – *Ectothiorhodospira;* HiPIP – high potential ferredoxin; isocyt c_2 – isocytochrome c_2; NAD – nicotinamide adenine dinucleotide; PC – plastocyanin; PrrA – photosynthesis regulatory protein; PS II – Photosystem II; PS – photosynthetic; PS I – Photosystem I; *Rb.* – *Rhodobacter;* RC – reaction center; *Rc.* – *Rhodocyclus; Rh.* – *Rhodophila; Rp.* – *Rhodopseudomonas; Rs.* – *Rhodospirillum* interdisciplinary approach has shown that cells often contain more than one protein capable of participating in light-driven electron transport in vivo. Therefore, this chapter seeks to summarize how genetic and biochemical analysis of PS bacteria has provided new and important insights into the numbers and types of soluble proteins that can reduce RC complexes.

B. The Two Major Classes of Reaction Center Complexes Found in Photosynthetic Cells

To understand how and why PS cells use a variety of soluble proteins to reduce RC complexes, one must remember there are two very different kinds of PS RC complexes (Golbeck, 1993; Chapter 30 by Feiler and Hauska). Thus, the interaction of an electron donor with the RC can vary significantly depending on the type of membrane-bound complexes present.

1. Purple Bacterial Reaction Center Complexes Are Similar to Photosystem II

The purple bacterial RC is the best understood complex from both structural and functional standpoints (Fig. 1, Panel A). At a minimum, it contains similar 30 kDa core membrane-spanning polypeptides (the L and M subunits) which bind the bacteriochlorophyll special pair and the electron acceptor quinones (Deisenhofer et al., 1985). The overall architecture of the core subunits of cyanobacterial and plant PS II complexes appears similar to that of the purple bacterial RC (Michel and Deisenhofer, 1988; Komiya et al., 1988). However, PS II complexes do not interact with soluble cytochromes because they utilize light energy to oxidize water with concomitant quinone reduction. Rather the membrane-bound subunits for water oxidation generate both electrons for PS II reduction and hydrogen ions to contribute to the proton gradient.

There appears to be considerable similarity in the core structure of the RC complex among different bacterial species. For example, the three-dimensional structures of RC complexes from *Rhodopseudomonas (Rp.) viridis* (Deisenhofer et al., 1985) and *Rhodobacter (Rb.) sphaeroides* (Allen et al., 1987; Chang et al., 1991) show that the sequence and organization of these core subunits are similar even though the *Rp. viridis* RC contains an additional 45 kDa membrane-bound tetraheme cytochrome *c* subunit. In addition to *Rb. sphaeroides*, there are only a few other species which lack a membrane-bound tetraheme cytochrome *c* subunit in the RC; the best characterized are those found in *Rb. capsulatus*, *Rhodospirillum (Rs.) rubrum* and *Rp. palustris* (Rickle and Cusanovich, 1979; Matsuura and Shimada, 1986; Hall et al., 1987; Jones et al., 1990; Bartsch, 1991).

2. Bacterial Reaction Center Complexes That Are Similar to Photosystem I

The cyanobacterial PS I has a pair of non-identical membrane spanning 82 kDa core polypeptides that bind the special pair chlorophylls and a 4-Fe-S cluster that serves as the immediate electron acceptor (Krauss et al., 1993). The green bacterial RC (Fig. 1, Panel B) appears homologous to cyanobacterial and plant PS I complexes except that the green bacterial core subunits are identical to one another (Büttner et al., 1992). Instead of catalyzing light-dependent quinone reduction, this type of RC transfers electrons from the 4-Fe-S cluster to a bound low potential 8-Fe-S ferredoxin subunit (the F_AF_B protein). Ultimately, pyridine nucleotide is reduced via soluble 2-Fe-S or 8-Fe-S ferredoxins and a flavoprotein (FNR). In this case, the generation of a proton gradient may involve additional membrane-bound components such as NADH dehydrogenase (Mi et al., 1992) as well as the Cyt bc_1 or Cyt b_6f complexes.

3. Why Might Bacterial Reaction Center Complexes Contain a Membrane-Bound Tetraheme Cytochrome c?

In species that contain a membrane-bound cytochrome *c* subunit, this cytochrome serves the same function as the soluble electron donors in species that contain only core RC subunits; namely to be the immediate electron donor to the special pair. From current analysis, purple bacterial RC complexes that contain membrane-bound cytochrome *c* subunits appear more common than those where the soluble electron donor interacts directly with the core subunits (Bartsch, 1991). This could suggest that the membrane-bound cytochrome *c* is generally more efficient in reducing the special pair than the soluble proteins present in these species.

Based on the available examples, it seems generally true that the membrane-bound tetraheme cytochrome *c* is missing in species that contain Cyt c_2. This might suggest that members of the Cyt c_2 protein family are

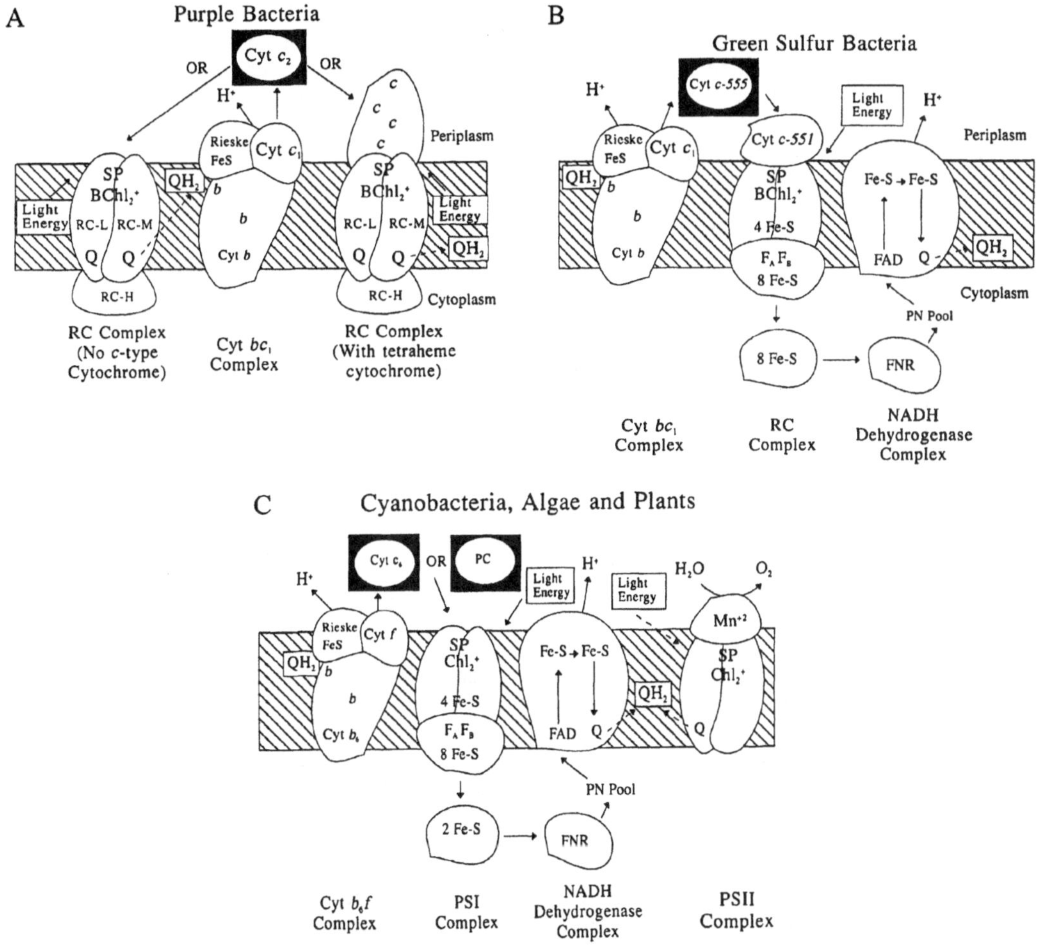

Fig. 1. Different PS electron transfer pathways. In all panels, the PS membrane is represented by diagonal shading.
Panel A: The function of Cyt c_2 in RC reduction by purple bacteria. In these organisms, soluble members of the Cyt c_2 family accept electrons from a membrane bound Cyt bc_1 complex (center). Reduced Cyt c_2 then donates electrons to either core subunits of the RC complex (left) or to a specific heme *c* (the so-called heme 2) within a RC complex containing a tetraheme cytochrome *c* subunit (right). Abbreviations: RC-H, RC-M, and RC-L: heavy, medium and light subunits of the RC complex; SP $BChl_2^+$: special pair bacteriochlorophyll dimer which is the RC electron acceptor from Cyt c_2; Q: quinone; QH_2: quinol; Rieske FeS: Rieske 2-Fe-S subunit of the Cyt bc_1 complex; Cyt c_1: Cyt c_1 subunit of the Cyt bc_1 complex; Cyt *b*: diheme membrane spanning Cyt *b* subunit of the Cyt bc_1 complex.
Panel B: The function of soluble Cyt *c*-555 in RC reduction by green sulfur bacteria. In these organisms, soluble Cyt *c*-555 uses electrons from the Cyt bc_1 complex to reduce light oxidized RC complexes. Abbreviations: Cyt *c*-551: monoheme cytochrome *c* subunit of the green sulfur bacterial RC complex; 4 Fe-S: electron acceptor 4-Fe-S center of PS I complex; F_A F_B: 8-Fe-S ferredoxin subunit of PS I complex F_A F_B; 8 Fe-S: ferredoxin mediator between F_A F_B and FNR; FNR: soluble ferredoxin $NADP^+$ reductase; PN: pyridine nucleotide; FAD: flavin adenine dinucleotide; Fe-S: iron sulfur centers of NADH dehydrogenase complex. All other abbreviations are as in Panel A.
Panel C: The function of cytochrome c_6 (Cyt c_6) or plastocyanin (PC) as interchangeable electron donors to PS I in cyanobacteria, green algae and plants. In these organisms, Cyt c_6 or PC use electrons from the Cyt b_6f complex to donate light-oxidized PS I complexes. Abbreviations: PS I: pyridine nucleotide reducing photosystem I complex; 4 Fe-S: electron acceptor 4-Fe-S center of PS I complex; F_A F_B: 8 Fe-S ferredoxin subunit of PS I complex; 2 Fe-S: soluble 2 Fe-S ferredoxin mediator; Cyt b_6: diheme membrane spanning Cyt b_6 subunit of the Cyt b_6f complex; Rieske FeS: Rieske Fe-S subunit of the Cyt b_6f complex; Cyt *f*: Cyt *f* subunit of the Cyt b_6f complex; PS II: oxygen evolving photosystem II; Mn^{+2}: water-splitting activity of PS II. All other abbreviations are as in Panels A and B.

at least as effective in reducing the RC as the membrane-bound tetraheme cytochrome *c* subunit. However, there are some exceptions to this generalization. The most well-studied is *Rp. viridis*, which contains both a RC tetraheme cytochrome *c* subunit and soluble Cyt c_2. Additional exceptions include *Roseobacter denitrificans*, whose Cyt c_2 is closely related to that of *Rb. capsulatus* (Okamura et al., 1987), which has a tetraheme RC cytochrome *c* (Liebetanz et al., 1991). On the other hand, *Rhodomicrobium vannielli*, which contains a Cyt c_2 closely related to that of *Rp. viridis* (Ambler et al., 1979a), appears to lack a membrane-bound cytochrome *c* in its RC (Bartsch, 1991). Furthermore, the *Rp. palustris* type strain appears to lack the tetraheme RC cytochrome *c*, but *Rs. centenum* (which has a closely-related Cyt c_2) appears to have this additional subunit (Bartsch, 1991). From this developing analysis, we might also conclude that species that rely on mediators that are not related to Cyt c_2 (i.e., auracyanin, HiPIP, or other c-type cytochromes; see below) contain a membrane-bound cytochrome *c* because these electron donors are not very efficient. Thus, these species may be more primitive in PS electron transport than those which utilize Cyt c_2 for direct and rapid reduction of the RC special pair.

II. What Proteins Participate in Light-Dependent Reaction Center Reduction?

The central problem common to all RC complexes is the necessity to rapidly reduce the photo-oxidized special pair to prevent non-productive back electron transfer from either reduced quinone or the 4-Fe-S cluster. In most cases, productive reduction of the special pair is accomplished by a soluble cytochrome. However, as described below, this function may be assumed by a variety of proteins in different PS bacteria.

The electron donor to RC complexes has often been inferred by spectroscopic examination of characteristic light-dependent absorbance changes after excitation of either whole cells, membranes or isolated components. It should be noted that cytochromes are easy to characterize in this manner, but iron-sulfur proteins and copper proteins have broad absorption peaks of low absorbency, which make their oxidation difficult to monitor in vivo. Therefore, in some species potential electron donors have been identified solely from analyzing their redox properties and cellular abundance. Although each system has its limitations, the following sections describe what proteins are believed to function as electron donors to light-dependent RC complexes in different classes of PS bacteria.

A. Electron Donors From Purple Bacteria

1. Rhodospirillaceae—*Species Which Use Members of the Cyt* c$_2$ *Family*

In most of these bacteria, the soluble electron carrier to the RC special pair is a protein in the Cyt c_2 family. Members of this family are defined as the nearest bacterial homologue to mitochondrial cytochrome *c*. In *Rb. sphaeroides* (Overfield and Wraight, 1980; Rosen et al., 1983; Allen et al., 1987; Tiede and Chang, 1988; Moser and Dutton, 1988; Caffrey et al., 1992), Cyt c_2 binds transiently and reversibly to the RC core subunits. However, electron transfer occurs between Cyt c_2 and heme 2 of the membrane-bound tetraheme cytochrome *c* in *Rp. viridis* (Knaff et al., 1991; Meyer et al., 1993). Even though these soluble cytochromes can interact with dissimilar RC subunits, the ability to monitor light-induced electron transport with mixtures of Cyt c_2 proteins and RC complexes from different sources suggests that at least some of the interactions used for binding and formation of a productive complex are conserved (see Section III).

An ~1:2 ratio of Cyt c_2 to RC complexes is sufficient to support PS growth of *Rb. sphaeroides* (Crofts, et al., 1983) and *Rb. capsulatus* (Prince et al., 1986), but the minimal amount of electron donor necessary to support wild type growth rates is not yet clear. Recent genetic analysis has confirmed the role of either soluble or membrane-bound members of the Cyt c_2 family in light-induced electron transport in these two species (Daldal et al., 1986; Donohue et al., 1986; Prince et al., 1986; Donohue et al., 1988; Fitch et al., 1989; Rott and Donohue, 1990; Rott et al., 1993; Jenney and Daldal, 1993; see below). However, other soluble or membrane bound mediators must be used in some purple bacteria because not all of them contain measurable quantities of soluble Cyt c_2 (Bartsch, 1991). As described below, *Rhodocyclus* (Rc.), *Chromatium*, and *Ectothiorhodospira (Ec.)* sp. generally lack soluble members of the Cyt c_2 family.

2. Chromatium vinosum*—Are Cytochrome* c-*551 or HiPIP Electron Donors to Reaction Center Complexes?*

The purple bacterium, *Chromatium vinosum*, has also been well studied with respect to light-induced absorbance changes and the involvement of cytochromes in RC reduction (Van Grondelle et al., 1977; Knaff et al., 1980; Coremans et al., 1985). This RC contains a membrane-bound 45 kDa tetraheme cytochrome *c* similar to that of *Rp. viridis* with both high and low potential hemes (Kennel and Kamen, 1971; Case and Parson, 1973; Itoh, 1980). However, this bacterium does not grow aerobically and genetic techniques are lacking to test the role of soluble proteins in RC reduction. Consequently, the identity of soluble proteins that presumably function as reductants for the RC Cyt *c* subunit has rested solely on spectroscopic studies.

Spectroscopic analysis suggests that a high potential soluble, Cyt *c*-551 (E_m ~240–260 mV) can donate electrons to the RC (Van Grondelle et al., 1977; Tomiyama et al., 1983; Gray et al., 1983). However, the midpoint potential of purple bacterial Cyt c_1 proteins (265–318 mV, Gabellini et al., 1982; Guner et al., 1991; Gray et al., 1992; Hacker et al., 1993) suggest that the redox potential of Cyt *c*-551 may be too low unless this bacterium's Cyt c_1 and Rieske iron sulfur proteins have correspondingly lower potentials. Furthermore, the relationship of Cyt *c*-551 to other *c*-type cytochromes is unknown because it has never been purified or sequenced. For example, Cyt *c*-551 may not be a member of the Cyt c_2 family or it could lack the proper charge at the interaction domain. Finally, even if its properties were optimal, the concentration of Cyt *c*-551 may be below that needed to allow rates of RC reduction required for PS growth. The recent analysis of *Rb. sphaeroides* Spd mutants provides an example of how PS growth can be limited by low levels of a functional electron donor (see Section IV).

Another possible soluble electron donor in *Chromatium* is the high redox potential ferredoxin, HiPIP. HiPIP is relatively abundant and has a midpoint potential (+330 mV) close to that of the membrane-bound RC cytochrome *c* subunit (Case and Parson, 1973). HiPIP can transfer electrons to the isolated *Chromatium* RC (Kennel et al., 1972), even though its small negative charge at the interaction domain makes this protein a rather poor donor to the purified *Rp. viridis* RC complex (Meyer et al., 1993a). On the other hand, Knaff et al. (1980) found that HiPIP was not very effective in *Chromatium* RC reduction.

A final candidate for an electron donor to the *Chromatium* RC is a minor high potential cytochrome (Cusanovich and Bartsch, 1969) that has been identified as a diheme cytochrome c_4 (Van Beeumen, 1991). Unfortunately, this protein is also not very efficient in electron transfer to RC complexes (Knaff et al., 1980). Therefore, in spite of this extensive search, additional experiments are necessary to identify the actual electron donor to *Chromatium* RC complexes.

3. Rhodocyclus *sp.—Are Cyt* c-*553 or HiPIP Electron Donors to Reaction Center Complexes?*

The purple non-sulfur bacteria in the genus *Rhodocyclus* also contain a RC with a membrane-bound tetraheme cytochrome *c*. The tetraheme cytochrome *c* in *Rc. gelatinosus* RC complexes is easily separated from the other subunits (Clayton and Clayton, 1978; Fukushima et al., 1988; Agalidis and Reiss-Husson, 1992) and the midpoint potentials of the four hemes in the RC cytochrome are comparable to those of *Rp. viridis* (Nitschke et al., 1992). This suggests that a relatively high potential soluble electron donor should exist in this species. However, both *Rhodocyclus* sp. (Bartsch, 1991) and *Rs. salinarum* (Meyer et al., 1990a) appear to lack soluble proteins related to Cyt c_2. In fact, *Rc. gelatinosus* and *Rs. salinarum* have no measurable soluble high redox potential cytochromes. These observations suggest that either a soluble mediator is not required or that proteins other than cytochromes must be considered for this role. Notwithstanding these observations, Matsuura et al. (1988) reported that a soluble cytochrome from a strain of *Rc. gelatinosus* was as efficient as horse Cyt *c* in reducing RC preparations that either contained or lacked its membrane-bound tetraheme cytochrome *c* subunit. It is important to note that these two seemingly contradictory sets of results came from studies performed with different strains. Thus, it is possible that some isolates produce different bacterial cytochromes than the type strain (for which the sequence and redox potential of Cyt *c*-551 were determined) or that some isolates may not indeed be *Rc. gelatinosus*. Until these discrepancies are resolved, the nature of the soluble mediator in *Rc. gelatinosus* will remain an open question.

Rc. gelatinosus, *Rc. tenuis*, and *Rs. salinarum* all

contain abundant levels of HiPIP, that might function as a soluble electron donor to the RC. *Rc. gelatinosus* will grow aerobically, so it may be amenable to the types of genetic analyses necessary to identify the electron donor to the RC (see below).

On the other hand, the soluble Cyt *c*-551 of *Rc. gelatinosus* has a redox potential of only ~28 Mv (Meyer et al., 1983). It is a minor component, and is related to *Pseudomonas* Cyt *c*-551 (Ambler et al., 1979b). Thus, it would not be able to efficiently couple electron flow from the Cyt bc_1 complex to the RC if the Cyt c_1 has a potential near 300 mV as found in other species. *Rc. tenuis* contains both HiPIP and an abundant soluble high redox potential, Cyt *c*-553, that is related to *Pseudomonas* Cyt *c*-551 (Ambler et al., 1979b). From their properties, it is possible that either HiPIP or Cyt *c*-553 function interchangeably to reduce RC complexes in this bacterium. Both are positively charged and should interact favorably with the RC, although neither was effective in electron transfer to complexes from *Rp. viridis* (Meyer et al., 1993a). In contrast *Rc. purpureus* does not have HiPIP, but it has a Cyt *c*-553 similar to that found in *Rc. tenuis*. Thus, it is likely that Cyt *c*-553 from *Rc. purpureus* is the sole soluble electron donor to the RC in this species.

4. Rhodophila globiformis*—Cyt* c_2 *and HiPIP Are Potential Donors to Reaction Center Complexes*

The purple bacterium, *Rh. globiformis*, is a relatively interesting example since it contains a RC complex with a membrane-bound tetraheme cytochrome *c* and both Cyt c_2 and HiPIP as potential soluble electron donors (Bartsch, 1991; Ambler et al., 1987). *Rp. marina* (Meyer et al., 1990b) and *Rhodomicrobium vannielli* also share these characteristics, although HiPIP is only a minor constituent in the latter species. Presumably, Cyt c_2 would be more reactive with the RC than HiPIP and would be the primary electron donor in these species. However, HiPIP could be of greater significance if growth conditions alter its cellular level relative to Cyt c_2. In this regard, *Rp. marina* will grow aerobically, so it has the potential for the types of genetic analyses necessary to show what proteins are electron donors in vivo. Finally, both Cyt c_2 and HiPIP from *Rh. globiformis* have redox potentials >100 mV higher than the average (450 mV for HiPIP and 470 mV for Cyt c_2), so it is possible that the membrane bound tetraheme RC cytochrome *c* and the special pair will have correspondingly higher redox potentials.

5. Ectothiorhodospira *sp.—Are Cyt c-551 and HiPIP Electron Donors to Reaction Center Complexes?*

The halophilic purple bacteria in the family Ectothiorhodospiraceae all have a membrane-bound tetraheme cytochrome *c* RC subunit (Lefèbvre et al., 1984; Engelhardt et al., 1986; Bartsch, 1991) although it appears that the cytochrome subunit might be easily dissociated (Leguijt and Hellingwerf, 1991). An apparent characteristic of phototrophic halophiles is that the soluble electron transfer proteins have lower than average midpoint potentials. For example, *Ec. halophila* (Meyer, 1985), *Ec. shaposhnikovii* (Kusche and Trüper, 1984), and *Ec. vacuolata* (unpublished) all have equally abundant isozymes of HiPIP with the lowest redox potentials of proteins in their class (50–250 mV). *Ec. halophila* (Meyer, 1985), *Ec. halochloris* (unpublished), and *Ec. abdelmalekii* (Then and Trüper, 1983) also have a Cyt *c*-551 related to *Pseudomonas* Cyt c_5 (Ambler et al., 1993) with a low midpoint potential (–7 to +58 mV) similar to those of HiPIP isozymes (Meyer et al., 1983; Then and Trüper, 1983). Finally, *Ec. shaposhnikovii* has a soluble Cyt *c*-553(549) with a midpoint potential of ~248 mV (Kusche and Trüper, 1984). Thus, if any of these proteins function as electron donors to RC complexes, then it seems likely that both the Cyt bc_1 complex and tetraheme RC cytochrome *c* subunit from these bacteria also have lower than average midpoint potentials.

B. Electron Donors from Green Sulfur Bacteria

1. Chlorobium*—Does Cytochrome c-555 Reduce Reaction Center Complexes?*

The green bacterial RC found in *Chlorobium* is quite different from that of purple bacteria (Fig. 1, Panel B). In this bacterium, both the bacteriochlorophyll special pair and the electron acceptors have much lower redox potentials (in the vicinity of +250 mV and –600 mV, respectively) than either PS I or purple RC complexes (Fowler et al., 1971; Prince and Olson, 1976; Hurt and Hauska, 1984; Nitschke et al., 1990a; Feiler et al., 1992; Miller et al., 1992). This suggests that both the hypothetical cytochrome bc_1 complex and soluble electron donors also have lower midpoint potentials than the proteins typically found in purple bacteria. The gene sequence of the green bacterial RC shows that it is homologous to PS I and an adjacent gene encodes a protein resembling an 8 Fe-S F_AF_B protein (Büttner et al., 1992). There is a bound cytochrome *c* subunit in the green bacterial RC (redox potential of 165–220 mV). However, the sequence of this structural gene indicates only a single heme binding site on this ~23 kDa protein (Okkels et al., 1992), unlike the 45 kDa *Rp. viridis* tetraheme RC cytochrome *c* subunit.

All species of green sulfur bacteria contain an abundant soluble Cyt *c*-555 (redox potential of 80–160 mV; Fischer, 1988), which could be the soluble electron donor to RC complexes. This protein is not closely related to Cyt c_2 (Van Beeumen et al., 1976). Rather, Cyt *c*-555 is similar to *Ec. halophila* Cyt *c*-551 and cyanobacterial Cyt c_6 (Ambler et al., 1993).

2. Chloroflexus—*Copper proteins Reduce Reaction Center Complexes*

Chloroflexus aurantiacus has properties of both green and purple bacteria. This moderate thermophile is a facultative aerobe but the presence of light harvesting bacteriochlorophylls and chlorosomes similar to those in *Chlorobium* (Zuber, 1988) led to its classification as a green photosynthetic bacterium, even though all other green bacteria are strict anaerobes. In contrast, *Cf. aurantiacus* contains a RC which is closely related to those of purple bacteria (Bruce et al., 1982; Pierson and Thornber, 1983; Shiozawa et al., 1989; Van Vliet et al., 1991). This RC is surrounded by typical purple bacterial antenna proteins (Zuber, 1988) and it contains a membrane-bound tetraheme cytochrome *c* (Freeman and Blankenship, 1990; Dracheva et al., 1991). A simple explanation for the occurrence of both green and purple bacterial characteristics is that *Chloroflexus* is a hybrid resulting from gene transfer (Blankenship, 1992).

For this chapter, perhaps one of the most interesting characteristics of *Chloroflexus* is that it appears to be gram positive (Meissner et al., 1988) and it lacks detectable soluble cytochromes. Instead, membrane-bound copper proteins, known as auracyanins (McManus et al., 1992), which are easily washed from the membrane, might fill the role usually reserved for soluble cytochromes in electron transfer between the presumed cytochrome bc_1 complex and the RC. Auracyanins are most closely related to the azurin type of copper proteins found in *Pseudomonas*. Because *Chloroflexus* can grow aerobically, there is a possibility for the types of genetic manipulation necessary to test the role of these copper proteins in RC reduction.

C. Heliobacterium *Contains No Known Soluble Electron Donors to the Reaction Center Complex*

Heliobacterium appears to have the green bacterial type of RC (Fuller et al., 1985; Prince et al., 1985; Smit et al., 1987; Vos et al., 1989; Nitschke et al., 1990b; Trost et al., 1992) and a RC peptide from this bacterium is homologous to those of plant PS I and green bacterial subunits (Trost et al., 1992). The primary donor has a midpoint potential of +225 mV and the acceptor has a potential of –510 mV (Prince et al., 1985).

There is no evidence for soluble electron donors to RC complexes in *Heliobacterium*. This suggests the existence of a membrane-bound electron donor similar to the apparent situation in *Chloroflexus*. *Heliobacterium* contains several membrane-bound *c*-type cytochromes of ~50, 33, and 20 kDa (Fuller et al., 1985). In addition, a membrane-bound cytochrome *c*-553 (redox potential ~170 mV) is photooxidized (Prince et al., 1985). Thus, it is possible that a bound RC cytochrome *c* similar to that found in *Chlorobium* may be present. However, the immediate electron donor to this membrane-bound cytochrome is still unknown.

D. Cyanobacteria, Green Algae, and Chloroplasts Use Cyt c_6 *and Plastocyanin as Interchangeable Electron Donors to Photosystem I*

Cyanobacteria, algae and plants contain two photosystems that are homologous to either purple bacterial (PS II) or green bacterial (PS I) RC complexes (Fig. 1, Panel C). Because PS II oxidizes water via a non-cyclic electron transport pathway, a soluble electron donor such as Cyt c_2 of purple bacteria is not necessary. The PS I RC, unlike those found in green sulfur bacteria and *Heliobacterium*, does not contain a membrane-bound cytochrome *c* subunit. In addition to reducing pyridine nucleotides in a non-cyclic pathway, PS I is involved in cyclic electron transfer between NADH dehydrogenase and Cyt b_6f complexes (Mi et al., 1992). In this pathway, PS I complexes interact with the high potential end of the b_6f complex (the plant and cyanobacterial version of the Cyt bc_1 complex) via soluble protein mediators. There is overwhelming evidence for a soluble electron donor to light-oxidized PS I complexes.

Cyt c_6 (also called Cyt c-553) or the copper protein plastocyanin (PC) transfers electrons between the b_6f and PS I complexes in many cyanobacteria and green algae (Wood, 1978; Lochau, 1981; Ho and Krogmann, 1984; Sandmann, 1986). The interchangeable function of PC and Cyt c_6 has been confirmed by spectroscopic, genetic and physiological experiments in the cyanobacteria and green algae that are amenable to such interdisciplinary approaches (Merchant and Bogorad, 1987; Briggs et al., 1990; Hill et al., 1991; Nakamura et al., 1992; Bovy et al., 1992; Zhang et al., 1992). However, some species utilize only Cyt c_6, whereas others (including all green plants) use only PC as an electron donor to PS I.

PC is homologous to other electron transfer copper proteins including auracyanin from *Chloroflexus*, although the latter protein appears closer to the azurins of *Pseudomonas*. There is not likely to be a close evolutionary connection between these copper protein classes because they interact with widely different RC complexes. In contrast, Cyt c_6 appears most closely related to *Chlorobium* Cyt c-555 and *Ectothiorhodospira* Cyt c-551 (Ambler et al., 1993), suggesting a more direct evolutionary link between green sulfur bacteria and cyanobacteria. Cyanobacterial PS I complexes do not have a membrane-bound tetraheme cytochrome *c* subunit similar to that found in green bacteria. However, it appears that the *psaF* gene product is necessary for binding and efficient electron transfer from PC (Hippler et al., 1989; Wynn et al., 1989).

III. How Do Electron Donors Interact with Membrane-Bound Redox Complexes?

To support PS growth, soluble redox proteins must reversibly interact with membrane bound electron donors (Cyt bc_1 or Cyt b_6f complexes) and acceptors (RC complexes). This suggests that transient binding of these soluble proteins requires a specific interaction with their membrane-bound redox partners. In addition, the ability of these diffusible electron carriers to interact with more than one membrane-bound redox partner raises the question of whether their binding to each complex is mediated by the same interaction domain.

A. Interaction of Cyt c_2 with Reaction Center Subunits

Among PS systems, the most detailed information is available on the interaction between Cyt c_2 and those RC complexes that lack a membrane-bound cytochrome *c* subunit. Proteins in the Cyt c_2 family have relatively high affinities for these RC complexes ($K_D \approx 1\ \mu M$; Rickle and Cusanovich, 1979; Overfield and Wraight, 1986; Moser and Dutton, 1988; Venturoli et al., 1990, Tiede et al., 1993). The salt dependence of the rate constant for RC reduction suggests that electrostatic interactions make a significant contribution to the productive interaction between these proteins (Caffrey et al., 1992). Support for the potential contribution of electrostatic interactions also comes from the existence of a net positive charge at the Cyt c_2 interaction domain with either small molecule reactants (Meyer et al., 1984) or proteins (Tollin et al., 1984), the location of lysines on the interaction surface in the available crystal structure of several proteins in this family (Benning et al., 1991), and the expected decrease in either affinity for the cytochrome or its rate of electron transfer when individual lysines are modified chemically (Hall et al., 1987; Long et al., 1989) or genetically (Caffrey et al., 1992). Although some of the highly conserved surface lysines appear to be more important than others, multiple electrostatic interactions must aid formation of a productive complex since no chemical or genetic modification at a single lysine totally abolishes binding or electron transfer to RC complexes.

If electrostatic interactions play a significant role in formation of productive complexes between Cyt c_2 and RC complexes, then one might expect its membrane-bound redox partners to contain a complementary set of charged residues at their interaction domains. The crystal structure of the *Rb. sphaeroides* RC complex indicates the existence of several negatively charged residues on the periplasmic surface (Allen et al., 1987; Chang et al., 1991). Although structural representations of Cyt c_2 docked on a RC complex suggest these residues could represent sites of electrostatic contact (Allen et al., 1987; Tiede and Chang, 1988; Caffrey et al., 1992; Tiede et al., 1993), the orientation of the bound cytochrome differs significantly in individual models. Therefore, the contribution of individual charge pairs predicted by different models to the formation of a productive complex remains to be proven. Indeed, while considerable attention has been devoted to defining the importance of electrostatics, it is important to note that the binding or rate of electron transfer by various members of this cytochrome family does not correlate directly with the magnitude of the charge difference at the interaction domain. Therefore, forces other than electrostatics (i.e., hydrophobic and van der Waals interactions) are likely to contribute significantly to formation of a productive complex between these redox partners. Currently, *Rb. sphaeroides* is the only genetically-manipulatable PS organism for which a crystal structure for both Cyt c_2 (Allen, 1988; Axelrod et al., 1994) and its membrane-bound redox partner (Allen et al., 1987; Chang et al., 1991) is available. In the absence of co-crystals, this type of detailed information should make *Rb. sphaeroides* an extremely attractive model for future genetic and biochemical analysis of the determinants required for Cyt c_2 docking and electron transfer to RC complexes.

Studies of electron transfer between Cyt c_2 and the *Rp. viridis* RC complex suggests that Cyt c_2 directly binds and donates electrons to the membrane-bound tetraheme Cyt *c* subunit (Knaff et al., 1991; Meyer et al., 1993). In this case, a negatively charged region of the tetraheme cytochrome *c* subunit appears to be involved. The structure of the *Rp. viridis* RC cytochrome *c* subunit indicates that a negatively charged surface exists in the vicinity of the ~330 mV *c*-556 (often called heme 2) that serves as the immediate electron acceptor from Cyt c_2. However, as was the case for Cyt c_2 electron transfer to simpler RC complexes (see above), electrostatic interactions appear to be only one part of the driving force for the formation of a productive complex between these redox partners.

B. Interaction of Cyt c_2 with Other Photosynthetic Membrane Complexes

The interaction domain of Cyt c_2 used for RC binding appears to be similar to that used for interaction with Cyt bc_1 complexes, since the rates of electron transfer from this donor are often influenced by the same alterations that affect the rates of RC reduction (Hall et al., 1989; Guner et al., 1993). Although these results suggest that electrostatic interactions contribute significantly to the formation of a productive complex between Cyt c_2 and Cyt bc_1 complexes, structural information on the Cyt c_1 subunit (the immediate electron donor to Cyt c_2) is required to begin identifying the potential site of this interaction.

C. Interaction of Other Soluble Electron Donors with Photosynthetic Membrane Redox Partners

Distantly related cytochromes, HiPIP and azurins are relatively poor electron donors to the *Rp. viridis* RC (Meyer et al., 1993a). In spite of significant differences in their midpoint potentials or net charge at the interaction domain (positive or negative), these soluble electron donors all appear to interact with a similar negatively charged region surrounding heme 2 of the RC tetraheme cytochrome *c* subunit. These results may suggest that hydrophobic, van der Waals, and electrostatic interactions all contribute to formation of a productive complex. In addition, the lack of a correlation between the redox potential difference of the reactants and the rate constants for electron transfer implies that factors such as distance and orientation between electron donors and acceptors may also contribute to formation of productive complexes and the reaction rate. Considerably less is known about the interaction of PC with either PS I RC complexes or Cyt *f* subunits of the Cyt b_6f complex. Chemical modification of PS I shows that the requirement for divalent cations, such as magnesium, is to neutralize negative charges on PS I and optimize binding and reaction with PC, which is also negatively charged at its interaction domain (Burkey and Gross, 1981). The kinetics of electron transfer in the complex between Cyt *f* and PC indicate that some rearrangement of reactants is necessary for rapid electron transfer (Qin and Kostic, 1992; Meyer et al., 1993b). Chemical cross-linking has implicated the interaction of a negatively charged region of spinach PC with both PS I and Cyt b_6f complexes. Specific cross-linking of spinach PC to lysine residues of turnip Cyt *f* has been found (Morand et al., 1989), while the basic residues of a positively charged region within spinach PsaF have yet to be cataloged (Wynn et al., 1989; Hippler et al., 1989). From this initial analysis, electrostatic interaction between a negatively-charged domain of PC and a positively charged region on the membrane-bound redox partner has been proposed. The contribution of this type of interaction is likely to be conserved because *Euglena* Cyt c_6 could also be covalently linked to turnip Cyt *f* (Morand et al., 1989). However, a variety of cyanobacterial Cyt c_6 proteins and a cyanobacterial PC could not be cross-linked to turnip Cyt *f*. This lack of cross-linking could reflect the lack of an acidic domain in the cyanobacterial proteins or a slightly different arrangement of acidic domains relative to their eukaryotic counterparts. In this regard, several labs have also noted significant sequence differences between the cyanobacterial and plant Cyt *f* subunits in the regions identified by cross-linking with PC (Ho and Krogmann, 1984; Widger, 1991). Therefore, it is also possible that cyanobacterial and higher plant PC may use a different series of interactions to form productive complexes with their physiological redox partners. In support of this notion, the net charge at the interaction domain in Cyt c_6 and PC was found to be highly variable in different species (Meyer et al., 1987).

IV. What Insights Has Genetics Provided About Electron Donors to Reaction Center Complexes?

Among PS bacteria, Cyt c_2 is the only soluble redox protein whose physiological function has been confirmed both biochemically and genetically (Daldal et al., 1986; Donohue et al., 1988; Rott et al., 1993; Jenney and Daldal, 1993). Genetic analysis of Cyt c_2 function was facilitated by the fact that some PS bacteria grow under conditions where this protein is dispensable. Although this analysis has only been performed to date in *Rb. sphaeroides* and *Rb. capsulatus*, biochemical and genetic (Self et al., 1990; Liang et al., 1991; Donohue and Kaplan, 1991) advances in *Rs. rubrum* suggest this bacterium is currently amenable to similar experiments.

Previous genetic analysis of Cyt c_2 function took advantage of the existing protein sequence to clone the respective structural gene (*cycA*) and construct defined null mutants by marker exchange (Daldal et al., 1986; Donohue et al., 1988; Donohue and Kaplan 1991). In retrospect, there are several reasons why this approach was preferable to random genetic screens for mutant cells lacking Cyt c_2. First, the relatively small size of the *cycA* locus (<1 kb) predicts that such mutants would represent only a small fraction of any given population. For example, a screen for *Rb. sphaeroides* strains lacking isocyt c_2 identified only one mutant in over 20,000 independent isolates. Even in this case, the mutation did not directly inactive this structural gene (*cycI*); rather the loss of isocyt c_2 was due to polarity of the mutation on *cycI* expression (Rott et al., 1993). More importantly, the existence of functionally redundant electron donors to the RC dramatically underscores why *cycA* mutants might not have been isolated in some PS bacteria (Rott et al., 1993; Jenney and Daldal, 1993) and cyanobacteria (Laudenbach et al., 1990) if photosynthetically-incompetent (PS^-) cells were directly screened for a loss of proteins in the Cyt c_2 family. The following sections briefly describe what new insights into PS electron transport have been obtained by genetically analyzing PC electron donors.

A. Purple Bacteria That Require Members of the Cyt c_2 Family

The inability of *Rb. sphaeroides* Cyt c_2 null mutants to grow under PS conditions and reduce light-oxidized RC complexes (Donohue et al., 1988) confirmed this protein's presumed function in this species. Although analogous *Rb. capsulatus* Cyt c_2 null mutants are PS^+ (Daldal, et al., 1986), they lack the rapid and complete RC reduction that is diagnostic of electron transfer from Cyt c_2 (Prince et al., 1986). Initially it was proposed that PS growth of *Rb. capsulatus* Cyt c_2 null mutants resulted from direct electron transfer between the Cyt bc_1 and RC complexes (Prince and Daldal, 1987; Prince et al., 1988). However, subsequent biochemical (Jones et al., 1990; Zannoni et al., 1992) and genetic (Jenney and Daldal, 1993) experiments demonstrated that both *Rb. sphaeroides* and *Rb. capsulatus* use different isoforms of Cyt c_2 to support PS growth when *cycA* is inactivated (see below). This is one clear instance where a genetic approach to identifying electron donors to the RC allowed PS bacteria to reveal aspects of their physiology that remained hidden during previous biochemical analysis of wild type cells.

Both *Rb. sphaeroides* and *Rb. capsulatus* grow with virtually wild type growth rates under PS conditions at saturating light intensities when they use an alternative electron donor to the RC. This is true even though the magnitude of RC reduction is considerably lower than that seen in wild type cells (Daldal et al., 1986; Rott and Donohue 1990). This indicates that the rapid RC reduction seen in wild type purple bacteria is not limiting for PS growth rates at saturating light intensities. In this regard, it may be more significant that the apparent rate of proton pumping through the Cyt bc_1 complex (as estimated by the slow phase of the carotenoid bandshift) is similar in both wild type and mutant cells. Electron transport through the Cyt bc_1 complex is a primary source of the membrane potential under PS conditions, so a simple explanation for the virtually wild type growth rates is that the lesions do not dramatically alter the rate at which these mutants generate a proton gradient from light.

1. Rb. sphaeroides *Uses a Soluble Isoform of Cyt* c_2 *as an Alternative Electron Donor to Reaction Center Complexes*

In *Rb. sphaeroides*, *spd* mutations allow PS growth of *cycA* null mutants because they increase the cellular abundance of a soluble Cyt c_2 isoform (isocyt c_2; Fitch et al., 1989; Rott and Donohue, 1990; Rott et al., 1992). A simple explanation for the inability of isocyt c_2 to support PS growth of *Rb. sphaeroides cycA* null mutants in the absence of an *spd* mutation is that these cells contain only ~1 molecule of this protein for every 100 RC complexes. However, the ~25-fold increase in isocyt c_2 levels under PS conditions brought about by the *spd* mutation brings the cellular abundance of this protein close to that of RC complexes (Rott et al., 1992).

Genetic and biochemical experiments confirmed that isocyt c_2 reduces RC complexes in *Rb. sphaeroides* Spd mutants. The accumulation of isocyt c_2 in *Rb. sphaeroides* Spd mutants initially suggested it was responsible for RC reduction in the absence of Cyt c_2 (Rott and Donohue, 1990). The ability of purified isocyt c_2 to reduce light-oxidized RC complexes in vitro supported this notion (V. C. Witthuhn, M. A. Rott, C. A. Wraight and T. J. Donohue, unpublished). In spite of the fact that isocyt c_2 is ~45% identical to *Rb. sphaeroides* Cyt c_2, these in vitro experiments demonstrated that saturating concentrations of isocyt c_2 reduce RC complexes 50–100 fold slower than Cyt c_2 at low ionic strength. This reduced rate of RC reduction by isocyt c_2 apparently is not due to an altered affinity of RC complexes for this protein. In addition, the salt dependence of RC reduction by isocyt c_2 suggests that it interacts with RC complexes via similar electrostatic contacts with a positive interaction domain. This slow rate of RC reduction by isocyt c_2 appears to be a major reason for the relatively slow kinetics of whole cell PS electron transport seen in Spd mutants (Donohue et al., 1988; V. C. Witthuhn et al., unpublished).

If isocyt c_2 was responsible for PS growth of Spd mutants, then the ability to grow photosynthetically should be lost in a strain lacking this protein. The isolation of a mutant lacking both Cyt c_2 and isocyt c_2 (Rott et al., 1993) confirmed the function of isocyt c_2 in PS electron transport. In addition, it showed that PS growth of *Rb. sphaeroides* Cyt c_2 null mutants required increased cellular levels of isocyt c_2. All of the data indicate that the failure of previous investigators to discover isocyt c_2 reflects the fact that synthesis of this protein is repressed in wild type cells (Rott et al., 1992). In contrast, isocyt c_2 levels necessary to support PS growth can be obtained either by an *spd* mutation or by placing the isocyt c_2 structural gene under control of a promoter that bypasses negative regulation (Rott et al., 1993).

2. Rb. capsulatus *Uses a Membrane-Bound Member of the Cyt* c_2 *Family as an Alternative Electron Donor to Reaction Center Complexes*

Although the initial analysis of the *Rb. capsulatus* Cyt c_2 mutant generated by Daldal and co-workers suggested that light-oxidized RC complexes might be directly reduced by the Cyt bc_1 complex (Prince et al., 1986; Prince and Daldal, 1987), subsequent experiments by several groups implicated a membrane-bound *c*-type cytochrome in Cyt c_2-independent PS growth of this bacterium (Jones et al., 1990; Zannoni et al., 1992). To test this possibility, Jenney and Daldal (1993) used the PS^- phenotype of an *Rb. sphaeroides* Cyt c_2 null mutant to screen *Rb. capsulatus* DNA for sequences that allowed PS growth. This screen identified a gene that apparently encodes a membrane-bound member of the Cyt c_2 family (this protein has been called Cyt c_y, but it actually appears to be a membrane-associated isozyme of Cyt c_2 similar to those found in other prokaryotes; Bott et al., 1991). This membrane-bound isoform of Cyt c_2 appears responsible for PS growth of the *Rb. capsulatus* Cyt c_2 null mutant isolated by Daldal and co-workers because a strain lacking both it and Cyt c_2 is unable to grow via photosynthesis (Jenney and Daldal, 1993). In addition, mutants lacking either Cyt c_2 or this membrane-bound isoform of Cyt c_2 are PS^+, so it appears that either protein is capable of supporting rates of RC reduction necessary to support PS growth. Estimates of the cellular abundance of this membrane-bound isoform of Cyt c_2 are not yet available in different *Rb. capsulatus* genetic backgrounds. Thus, it is not clear why previous studies of light-induced electron transport in *Rb. capsulatus* failed to detect significant RC reduction by this membrane-bound protein. In this regard, it is possible that electron transport previously ascribed to tightly bound Cyt c_2 in *Rb. capsulatus* spheroplasts was actually due to the membrane-bound isoform of Cyt c_2 (Prince et al., 1975). It has also been suggested that the cellular abundance of the membrane-bound isoform of Cyt c_2 is dependent on nutrient supply, so it is possible that many of the previous experiments used cells grown under conditions where low levels of this protein obscured its contribution to RC reduction (Jones et al., 1990; Zannoni et al., 1992). Finally, Cyt c_2 mutants from another wild type *Rb. capsulatus* strain were PS$^-$ (Hudig, et al., 1986) so it is possible that strain differences contributed to the previous failure to detect the function of this membrane-bound isoform of Cyt c_2 in RC reduction.

B. How is Synthesis of Reaction Center Electron Donors Regulated?

In facultative species, chlorophyll-protein complexes are often accumulated only under conditions of low oxygen tension that induce PS membrane synthesis (Kiley and Kaplan, 1988). If the RC electron donor is also not accumulated under non-PS conditions, then its cellular abundance will probably parallel that of chlorophyll-protein complexes. However, when considering how synthesis of RC electron donors might be regulated it is important to remember that these proteins might serve an important physiological function under non-PS conditions. Indeed, the apparent function of *Rb. sphaeroides* Cyt c_2 in aerobic respiration (Gennis, et al., 1982) means that significant levels of this protein are present in cells that lack detectable quantities of PS membrane pigment-protein complexes (Chory et al., 1984).

In many systems, the cellular level of PS membrane complexes is also controlled by changes in light availability (Kiley and Kaplan, 1988). However, in *Rb. sphaeroides* neither Cyt c_2 levels nor the cellular abundance of mRNA specific for this RC electron donor appear to be significantly altered by changes in light intensity (Donohue et al., 1986). These results appear to support the notion that synthesis of RC electron donors need not be coupled to that of the chlorophyll-protein complexes within the PS apparatus.

From recent studies, some general themes are beginning to emerge on how synthesis of RC electron donors is regulated. The following sections briefly illustrate potential strategies that different PS bacteria use to regulate the synthesis of electron donors to the RC complex. For example, ligand availability can influence the cellular level of RC electron donor. In several cyanobacteria and green algae, the choice between PC and Cyt c_6 as an electron donor to PS I is governed by copper availability. In addition, recent work indicates that *Rb. sphaeroides* couples Cyt c_2 synthesis to the accessibility of this protein's heme ligand. The recent cloning of alternative electron donors to RC complexes from purple bacteria has provided investigators with some insights into how synthesis of these proteins might be regulated.

1. Cyt c_2 Synthesis in Purple Bacteria

In *Rb. sphaeroides*, the normal RC electron donor, Cyt c_2, also participates in a branch of the aerobic respiratory chain (Gennis et al., 1982). Because of this additional function, *Rb. sphaeroides* Cyt c_2 levels only increase ~2–4 fold under PS conditions (Brandner et al., 1989). Indeed, when one compares the induction of Cyt c_2 to that of chlorophyll-protein complexes during a shift from aerobic to PS growth conditions, the cellular levels of this electron donor do not increase until after chlorophyll-protein complexes are assembled and growth resumes (Chory et al., 1984). From these and other experiments, it appears that the pool of Cyt c_2 present in aerobically grown cells is sufficient to support RC reduction during induction of the PS membrane.

Analysis of Cyt c_2 synthesis in *Rb. sphaeroides* indicates that increased *cycA* transcription under PS conditions contributes significantly to the increased levels of this protein. Several lines of evidence indicate that the Cyt c_2 structural gene (*cycA*) contains several promoters and that transcription from at least one upstream promoter region is increased under conditions that induce synthesis of the PS apparatus (MacGregor and Donohue, 1991; Schilke and Donohue 1992). To test whether this increase in *cycA* transcription could be coupled to changes in heme availability that accompany induction of the PS apparatus (see Lascelles, 1978 for review), β-galactosidase levels from a series of *cycA′:lacZ* operon fusions were assayed after tetrapyrrole intermediates were added to cells. From the results of these experiments, it appeared that a negative feedback loop links *cycA* transcription to heme availability because the addition of δ-aminolevulinic acid (ALA), the first committed intermediate in the tetrapyrrole biosynthetic pathway, effectively repressed *cycA* transcription (Schilke and Donohue, 1992).

The existence of this negative feedback loop predicts that mutants in which *cycA* transcription is insensitive to exogenous ALA should have increased expression of this gene. Such regulatory mutants (Chr4) have been isolated (Schilke and Donohue, 1992) and recent data indicates that the ChrR gene product is an activator of *cycA* transcription whose function is poisoned by the addition of exogenous ALA (Schilke and Donohue, 1995). If this were the case, then the ChrR protein in mutants like Chr4 is locked in an active state and can not respond to the signal generated by the addition of ALA. In this regard it is exciting to note that the mutant gene product in Chr4 contains a Cys to Arg substitution in a Cys-X-Y-Cys motif reminiscent of heme binding sites in cytochromes (Schilke and Donohue, 1995). While it is tempting to speculate that this motif represents a target site for the interaction of ChrR with heme or some other tetrapyrrole intermediate produced from ALA, final conclusions await additional physiological and genetic analysis. It will also be interesting to see if ChrR regulates transcription of additional cytochromes or chlorophyll-binding proteins within the PS membrane which are known to respond to the presence of exogenous ALA (Neidle and Kaplan, 1993). Regardless of the number of gene products controlled by ChrR, this negative feedback loop serves a similar function to the systems used by cyanobacteria and green algae that insure that Cyt c_6 is only synthesized when cells are limited for the copper ligand required for PC formation.

Recent genetic experiments suggest that *Rb. sphaeroides cycA* transcription requires a second gene product (PrrA) which positively controls transcription of many PS membrane gene products (Eraso and Kaplan, 1994). The observation that *cycA* transcription is still induced under PS conditions in ChrR null mutants might suggest that the PrrA pathway is still operative (Schilke and Donohue, 1995). However, additional experiments are required to test whether these two regulatory networks independently regulate *cycA* transcription.

2. Do Purple Bacteria Control Synthesis of Cyt c_2 Isoforms?

In the few bacterial systems known to contain alternate electron donors to the RC complex, little is known about whether individual proteins are used under specific physiological conditions. In addition, information is lacking about the molecular cues which might regulate synthesis of an individual electron donor to the RC complex.

In *Rb. capsulatus*, levels of the membrane-bound isoform of Cyt c_2 are believed to respond to nutrient conditions (Zannoni et al., 1992). However, the signals mediating this response and the magnitude of the changes in the abundance of this protein have not been elucidated. The ability of *Rb. capsulatus* Cyt c_2 null mutants to grow photosynthetically (Daldal et al., 1986) suggests that levels of this alternative electron donor are sufficiently high to support PS growth in the absence of any other mutation or specific physiological stimulus.

In contrast, *Rb. sphaeroides* Cyt c_2 mutants do not grow photosynthetically because isocyt c_2 synthesis is repressed when cells are grown in a succinate-based minimal medium (Rott et al., 1992). However, *spd* mutations bypass negative regulation of isocyt c_2 synthesis in this medium and allow this protein to accumulate to levels necessary to support PS growth. While sequencing the isocyt c_2 structural gene (*cycI*), it was noted that *cycI* was apparently part of an operon with an open reading frame encoding a typical member of the Zn-dependent alcohol dehydrogenase enzyme family (Rott et al., 1993; R. Barber, M.A. Roth and T. J. Donohue, unpublished). Soluble members of the AdhI family normally function in the oxidation of primary alcohols (Jornvall, et al., 1987). However, recent experiments confirmed that this protein is actually a member of the glutothione-dependent formaldehyde dehydrogenase family of alcohol dehydrogenases (R. Barber et al., unpublished). Since this protein apparently lacks a signal peptide, it seems unlikely that it functions as a direct electron donor to the periplasmic isocyt c_2 protein in a fashion similar to the membrane-bound alcohol dehydrogenase of *Acetobacter pasteurians* (Takemura et al., 1991) and *Gluconobacter suboxydans* (Takeda and Shimizu, 1992). Additional experiments are also required to determine the potential relationship of AdhI to other broad specificity alcohol dehydrogenases from Rhodospirillaceae (Bamforth and Quayle, 1978).

While there are reports of methanol utilization by some purple bacteria (Quayle and Pfennig, 1975), *Rb. sphaeroides* is one of the few Rhodospirillaceae that can utilize ethanol as sole carbon and energy sources (Imhoff and Trüper, 1984). Although it might be tempting to speculate that isocyt c_2 synthesis might be inducible by the addition of alternate carbon sources to wild type cells, additional experiments are required to determine the potential function of isocyt c_2. In this regard, cloned genes and defined mutants should facilitate the experiments necessary to determine whether growth in the presence of alternate carbon sources allows increased isocyt c_2 synthesis or whether this protein could contribute to PS electron transport under other physiological conditions in wild type cells. Therefore, in spite of the potential physiological function of isocyt c_2 suggested by the genetic linkage of these two genes, experimental support is currently lacking for a role of isocyt c_2 in any other metabolic pathway.

3. Copper Availability Dictates the Use of Plastocyanin or Cyt c_6

Many oxygenic phototrophs contain a regulatory network to control synthesis of functionally redundant proteins like Cyt c_6 and PC. For example, one third of the cyanobacteria contain Cyt c_6 alone, whereas two thirds contained both proteins (Sandmann, 1986). In contrast, three quarters of the green algae have both proteins while one quarter have only PC. Therefore, it will be interesting to see if bacteria that use copper proteins as electron donors to RC complexes contain a similar regulatory network.

In cyanobacteria and green algae that are able to make both proteins, PC seems to be the preferred electron donor because only copper-limited cells accumulate significant levels of Cyt c_6. Thus, cells containing both genes only use Cyt c_6 as an electron donor to PS I when they lack the copper ligand required for assembly of an active PC molecule (Merchant and Bogorad, 1987a and 1987b, Bovy et al., 1992). In all of these species tested, transcription of the Cyt c_6 gene is repressed by copper, whereas regulated transcription of the PC gene by copper is found only in some species (Briggs et al., 1990, Hill et al., 1991; Nakamura et al., 1992; Bovy et al., 1992; Zhang et al., 1992). The behavior of this regulatory network suggests that PC is generally a more efficient electron donor, although the molecular basis for this difference must await completion of the three-dimensional structure of PS I and the appropriate kinetic studies.

V. Future Perspectives

This chapter has attempted to provide a current understanding of the soluble proteins capable of electron transfer to light-oxidized RC complexes. The recent biochemical, structural and genetic analysis of these electron transfer proteins has demonstrated the power and versatility of such an interdisciplinary approach to this important biological question. In addition to providing new ways to address old biological questions, this interdisciplinary approach has uncovered new and exciting areas of inquiry for future research. Hopefully, like most areas of research, these recent advances have demonstrated that model building or the assignment of a function to a given protein from a single type of analysis can be misleading. Rather, we hope this chapter has clearly illustrated that the ultimate confirmation of a protein's function in PS electron transport requires the complementary information gleaned from biochemical, physiological and genetic experiments. Therefore, one challenge for future investigators working with new and exciting species of PS bacteria is to develop the necessary ensemble of structural, biochemical and genetic techniques to unambiguously assign a function to a given protein.

When one considers the explosion of structural, biochemical and genetic information that has occurred since the last review of soluble electron transfer proteins from PS bacteria (Bartsch, 1978), it is clear that remarkable progress has been made on many fronts. Having accumulated this plethora of information, it is also obvious that many important questions remain to be answered about the synthesis, structure and function of proteins that transfer electrons to light-oxidized RC complexes. A further challenge for future researchers is the development of the techniques and experimental systems to answer these important biological questions.

Finally, many of the questions we posed regarding soluble electron donors to the RC have parallels in the function of redox proteins in virtually all energy generating pathways. The unequaled wealth of biophysical data on bacterial PS electron transport, the RC crystal structures, and the rapidly expanding array of genetic techniques that can be brought to bear on soluble proteins and their membrane-bound redox partners all promise to make PS bacteria a unique system for future analysis of photosynthesis and biological energy generation.

Acknowledgments

TEM was supported in part by a grant from the National Institutes of Health (GM 21277). TJD was supported by grants from the National Institutes of Health (GM37509), the USDA (90-37262-5588) and University of Wisconsin-Madison Hatch project WIS3384.

References

Agalidis I and Reiss-Hutton F (1992) Purification and characterization of *Rhodocyclus gelatinosus* photochemical reaction center. Biochim Biophys Acta 1098: 201–208

Allen JP (1988) Crystallization and preliminary X-ray diffraction analysis of cytochrome c_2 from *Rhodobacter sphaeroides*. J Mol Biol 204: 495–496

Allen JP, Feher G, Yeates TO, Komiya H and Rees DC. (1987) Structure of the reaction center from *Rhodobacter sphaeroides* R26: The protein subunits. Proc Natl Acad Sci USA 84: 6162–6166

Ambler RP, Daniel M, Hermoso J, Meyer TE, Bartsch RG and Kamen MD (1979a) Cytochrome c_2 sequence variation among the recognised species of purple nonsulphur photosynthetic bacteria. Nature 278: 659–660

Ambler RP, Meyer TE and Kamen MD (1979b) Anomalies in amino acid sequences of small cytochromes *c* and cytochromes *c'* from two species of purple photosynthetic bacteria. Nature 278: 661–662

Ambler RP, Meyer TE, Cusanovich MA and Kamen MD (1987) The amino acid sequence of the cytochrome c_2 from the phototrophic bacterium *Rhodopseudomonas globiformis*. Biochem J 246: 115–120

Ambler RP, Meyer TE and Kamen MD (1993) Amino acids sequences of cytochromes *c*-551 from the halophilic purple phototrophic bacteria, *Ectothiorhodospira halophila* and *Ec. halochloris*. Arch Biochem Biophys 306: 83–93

Axelrod HL, Feher G, Chirino A, Day M, Hsu BT and Rees DC (1994) Crystallization and X-ray structure determination of cytochrome c_2 from *Rhodobacter sphaeroides* in three crystal forms. Acta Cryst D50: 596–602

Bamforth CW and Quayle JR (1978) The dye-linked alcohol dehydrogenases of *Rhodopseudomonas acidophila*. Biochem J 169: 677–686

Bartsch RG, (1978) Cytochromes. In Clayton RK and Sistrom WR (eds) The Photosynthetic Bacteria, pp 249–279. Plenum Publishing Corp., New York

Bartsch RG (1991) The distribution of soluble metallo-redox proteins in purple phototrophic bacteria. Biochim Biophys Acta 1058: 28–30

Benning MM, Wesenberg G, Caffrey MS, Bartsch RG, Meyer TE, Cusanovich MA, Rayment I and Holden HM (1991) Molecular structure of cytochrome c_2 isolated from *Rhodobacter capsulatus* determined at 2.5Å resolution. J Mol Biol 220: 673–685

Blankenship RE (1992) Origin and early evolution of photosynthesis. Photosyn Res 33: 91–111

Bott M, Ritz D and Hennecke H (1991) The *Bradyrhizobium japonicum cycM* gene encodes a membrane-anchored homolog of mitochondrial cytochrome *c*. J Bacteriol 173: 6766–6772

Bovy A, DeVrieze G, Borrias M and Weisbeek P (1992) Transcriptional regulation of the plastocyanin and cytochrome c_{553} genes from the cyanobacterium *Anabaena* species PCC7937. Mol Microbiol 6: 1507–1513

Brandner JP and Donohue TJ (1994) The *Rhodobacter sphaeroides* cytochrome c_2 signal peptide is not necessary for export and heme attachment. J Bacteriol 176: 602–609

Brandner JP, McEwan AG, Kaplan S and Donohue TJ (1989) Expression of the *Rhodobacter sphaeroides* cytochrome c_2 structural gene. J Bacteriol 171: 360–368

Brandner JP, Stabb EV, Temme R and Donohue TJ (1991) Regions of *Rhodobacter sphaeroides* cytochrome c_2 required for export, heme attachment, and function. J Bacteriol 173: 3958–3965

Briggs LM, Pecoraro VL and McIntosh L (1990) Copper-induced expression, cloning and regulatory studies of the plastocyanin gene from the cyanobacterium *Synechocystis* sp. PCC6803. Plant Mol Biol 15: 633–642

Bruce BD, Fuller RC and Blankenship RE (1982) Primary photochemistry in the facultatively aerobic green photosynthetic bacterium *Chloroflexus aurantiacus*. Proc Natl Acad Sci USA 79: 6532–6536

Büttner M, Xie DL, Nelson H, Pinther W, Hauska G and Nelson N (1992) Photosynthetic reaction center genes in green sulfur bacteria and photosystem I are related. Proc Natl Acad Sci USA 89: 8135–8139

Burkey KO and Gross EL (1981) Chemical modification of spinach plastocyanin: Separation and characterization of four different forms. Biochemistry 20: 2961–2967

Caffrey MS, Bartsch RG and Cusanovich MA (1992) Study of cytochrome c_2-reaction center interaction by site-directed mutagenesis. J Biol Chem 267: 6317–6321

Case GD and Parson WW (1973) Redistribution of electric charge accompanying photosynthetic electron transport in *Chromatium*. Biochim Biophys Acta 292: 677–684

Chang CH, El-Kabbani O, Tiede D, Norris J and Schiffer M (1991) Structure of the membrane-bound protein photosynthetic reaction center from *Rhodobacter sphaeroides*. Biochemistry 30: 5352–5360

Chory J, Donohue TJ, Varga AR, Staehelin LA and Kaplan S (1984) Induction of the photosynthetic membrane of *Rhodopseudomonas sphaeroides*: Biochemical and morphological studies. J Bacteriol 159: 540–554

Clayton RK and Clayton BJ (1978) Properties of photochemical reaction centers purified from *Rhodopseudomonas gelatinosa*. Biochim Biophys Acta 501: 470–477

Coremans JMCC, Van der Wal HN, Van Grondelle R, Amesz J and Knaff DB (1985) The pathway of cyclic electron transport in chromatophores of *Chromatium vinosum*. Evidence for a Q-cycle mechanism. Biochim Biophys Acta 807: 134–142

Crofts AR, Meinhardt SW, Jones KR and Snozzi M (1985) The

role of the quinone pool in the cyclic electron transfer chain of *Rhodopseudomonas sphaeroides*: A modified Q-cycle mechanism. Biochim Biophys Acta 723: 202–218

Cusanovich MA and Bartsch RG (1969) A high potential cytochrome *c* from *Chromatium* chromatophores. Biochim Biophys Acta 189: 245–255

Daldal F, Cheng S, Applebaum J, Davidson E and Prince RC (1986) Cytochrome c_2 is not essential for photosynthetic growth of *Rhodopseudomonas capsulata*. Proc Natl Acad Sci USA 83: 2012–2016

Deisenhofer J, Epp O, Miki K, Huber R and Michel H (1985) Structure of the protein subunits in the photosynthetic reaction center of *Rhodopseudomonas viridis* at 3Å resolution. Nature 318: 618–624

Donohue TJ and Kaplan S (1991) Genetic techniques in the *Rhodospirillaceae*. Methods in Enzymol 204: 459–485

Donohue TJ, McEwan AG and Kaplan S (1986) DNA sequence and expression of the *Rhodobacter sphaeroides* cytochrome c_2 structural gene. J Bacteriol 168: 962–972

Donohue TJ, McEwan AG, Van Doren S, Crofts AR and Kaplan S (1988) Phenotypic and genetic characterization of cytochrome c_2-deficient mutants of *Rhodobacter sphaeroides*. Biochemistry 27: 1918–1925

Dracheva S, Williams JC, Van Driessche G, Van Beeumen JJ and Blankenship RE (1991) The primary structure of cytochrome *c*-554 from the green photosynthetic bacterium *Chloroflexus aurantiacus*. Biochemistry 30: 11451–11458

Engelhardt H, Baumeister W and Engel A (1986) Stoichiometric model of the photosynthetic unit of *Ectothiorhodospira halochloris*. Proc Natl Acad Sci USA 83: 8972–8976

Eraso J and Kaplan S (1994) *prr A*, a putative response regulator involved in oxygen regulation of photosynthetic gene expression in *Rhodobacter sphaeroides*. J Bacteriol 176: 32–43

Feiler U, Nitschke W and Michel H (1992) Characterization of an improved reaction center preparation from the photosynthetic green sulfur bacterium *Chlorobium* containing the iron sulfur centers F-A and F-B and a bound cytochrome subunit. Biochemistry 31: 2608–2614

Fischer U (1988) Soluble electron transport proteins of *Chlorobiaceae*. In Olson JM, Ormerod JG, Amesz J, Stackebrandt E and Trüper HG (eds) Green Photosynthetic bacteria, pp 127–131. Plenum Press, NY

Fitch J, Cannac V, Meyer TE, Cusanovich MA, Tollin G, Van Beeumen J, Rott MA and Donohue, TJ (1989) Expression of a cytochrome c_2 isozyme restores photosynthetic growth of *Rhodobacter sphaeroides* mutants lacking the wild type cytochrome c_2 gene. Arch Biochem Biophys 271: 502–507

Fowler CF, Nugent NA and Fuller RC (1971) The isolation and characterization of a photochemically active complex from *Chloropseudomonas ethylica*. Proc Natl Acad Sci USA 68: 2278–2282

Freeman JC and Blankenship RE (1990) Isolation and characterization of the membrane-bound cytochrome *c*-554 from the thermophilic green photosynthetic bacterium *Chloroflexus aurantiacus*. Photosyn Res 23: 29–38

Fuller RC, Sprague SG, Gest H and Blankenship RE (1985) A unique photosynthetic reaction center from *Heliobacterium chlorum*. FEBS Letts 182: 345–349

Gabellini N, Bowyer JR, Hurt E, Melandri BA and Hauska G. (1982) A cytochrome bc_1 complex with ubiquinol-cytochrome c_2 oxidoreductase activity from *Rhodopseudomonas sphaeroides* GA. Eur J Biochem 126: 105–111

Gao JL, Shopes RJ and Wraight CA (1990) Charge recombination between the oxidized high-potential *c*-type cytochrome and Q-A negative in reaction centers from *Rhodopseudomonas viridis*. Biochim Biophys Acta 1015: 96–108

Gennis RB, Casey RP, Azzi A and Ludwig B (1982) Purification and characterization of the cytochrome *c* oxidase from *Rhodopseudomonas sphaeroides*. Eur. J. Biochem. 125: 189–195.

Golbeck JH (1993) Shared thematic elements in photochemical reaction centers. Proc Natl Acad Sci USA 90: 1642–1646

Gray GO, Gaul DF and Knaff DB (1983) Partial purification and characterization of two soluble *c*-type cytochromes from *Chromatium vinosum*. Arch Biochem Biophys 222: 78–86

Gray KA, Davidson E and Daldal F (1992) Mutagenesis of methionine-183 drastically affects the physicochemical properties of cytochrome c_1 of the bc_1 complex of *Rhodobacter capsulatus*. Biochemistry 31: 11864–11873

Guner S, Robertson DE, Yu L, Qiu ZH, Yu CA and Knaff DB (1991) The *Rhodospirillum rubrum* cytochrome bc_1 complex: Redox properties, inhibitor sensitivity and proton pumping. Biochim Biophys Acta 1058: 269–279

Guner S, Willie A, Millett F, Caffrey MS, Cusanovich MA, Robertson DE and Knaff DB (1993) The interaction between cytochrome c_2 and the cytochrome bc_1 complex in the photosynthetic purple bacteria *Rhodobacter capsulatus* and *Rhodopseudomonas viridis*. Biochemistry 32: 4793–4800

Hacker B, Barquera B, Crofts AR and Gennis RB (1993) Characterization of mutations in the cytochrome *b* subunit of the bc_1 complex of *Rhodobacter sphaeroides* that affect the quinone reductase site (Q_c). Biochemistry 32: 4403–4410

Hall J, Ayres M, Zha X, O'Brien P, Durham B, Knaff DB and Millett F (1987) The reaction of cytochromes *c* and c_2 with the *Rhodospirillum rubrum* reaction center involves the heme crevice domain. J Biol Chem 262: 11046–11051

Hall J, Zha X, Yu L, Yu CA and Millett F (1989) Role of specific lysine residues in the reaction of *Rhodobacter sphaeroides* cytochrome c_2 with the cytochrome bc_1. Biochemistry 28: 2568–2571

Hill KL, Li HH, Singer J and Merchant S (1991) Isolation and structural characterization of the *Chlamydomonas reinhardtii* gene for cytochrome *c-6*; analysis of the kinetics and metal sensitivity of its copper-responsive expression. J Biol Chem 266: 15060–15067

Hippler M, Haehnel W and Ratajczak R (1989) Identification of the plastocyanin binding subunit of photosystem I. FEBS Letts 250: 280–284

Ho KK and Krogmann DW (1984) Electron donors to P700 in cyanobacteria and algae: An instance of unusual genetic variability. Biochim Biophys Acta 766: 310–316

Hudig H, Kaufmann N and Drews G (1986) Respiratory deficient mutants of *Rhodobacter capsulatus*. Arch Microbiol 145: 378–385

Hurt EC and Hauska G (1984) Purification of membrane bound cytochromes and a photoactive P840 protein complex of the green sulfur bacterium *Chlorobium limicola* f. *thiosulfatophilum*. FEBS Letts 168: 149–154

Itoh S (1980) Effects of surface potential and membrane potential on the midpoint potential of cytochrome *c*-555 bound to the chromatophore membrane of *Chromatium vinosum*. Biochim Biophys Acta 591: 346–355

Imhoff JF and Trüper HG (1984) Purple nonsulfur bacteria. In Hensyl WR (ed) Bergy's Manual of Systematic Bacteriology, Vol. 3, pp 1658–1682 Williams and Wilkins, Baltimore Md

Jenney FE and Daldal F (1993) A novel membrane-associated *c*-type cytochrome, Cyt c_y, can mediate the photosynthetic growth of *Rhodobacter capsulatus* and *Rhodobacter sphaeroides*. EMBO J 12: 1283–1292

Jones MR, McEwan AG and Jackson JB (1990) The role of *c*-type cytochromes in the photosynthetic electron transport pathway of *Rhodobacter capsulatus*. Biochim Biophys Acta 1019: 59–66

Jornvall H, Persson B and Jeffery J (1987) Characteristics of alcohol/polyol dehydrogenases. The zinc-containing long chain alcohol dehydrogenases. Eur J Biochem 167: 195–201

Kennel SJ and Kamen MD (1971a) Iron containing proteins in *Chromatium*. I. Solubilization of membrane-bound cytochrome. Biochim Biophys Acta 234: 458–467

Kennel SJ and Kamen MD (1971b) Iron containing proteins in *Chromatium*. II. Purification and properties of cholate-solubilized cytochrome complex. Biochim Biophys Acta 253: 153–166

Kennel SJ, Bartsch RG and Kamen MD (1972) Observations on light-induced oxidation reactions in the electron transport system of *Chromatium*. Biophys J 12: 882–896

Kiley PJ and Kaplan S (1988) Molecular genetics of photosynthetic membrane biosynthesis in *Rhodobacter sphaeroides*. Microbiol Rev 52: 50–69

Knaff DB, Whetstone R and Carr JW (1980) The role of soluble cytochrome *c*-551 in cyclic electron flow-driven active transport in *Chromatium vinosum*. Biochim Biophys Acta 590: 50–58

Knaff DB, Willie A, Long JE, Kriauciunas A, Durham B and Millett F (1991) Reaction of cytochrome c_2 with photosynthetic reaction centers from *Rhodopseudomonas viridis*. Biochemistry 30: 1303–1310

Komiya H, Yeates TO, Rees DC, Allen JP and Feher G (1988) Structure of the reaction center from *Rhodobacter sphaeroides* R-26 and 2.4.1; symmetry relations and sequence comparisons between different species. Proc Natl Acad Sci USA 85: 9012–9016

Krauss N, Hinrichs W, Witt I, Fromme P, Pritzkow W, Dauter Z, Betzel C, Wilson KS, Witt HT and Saenger W (1993) 3-dimensional structure of system I of photosynthesis at 6 angstrom resolution. Nature 361: 326–331

Kukushima A, Matsuura K, Shimada K and Satoh T (1988) Reaction center-B870 pigment protein complexes with bound cytochrome *c*-555 and cytochrome *c*-551 from *Rhodocyclus gelatinosus*. Biochim Biophys Acta 933: 399–405

Kusche WH and Trüper HG (1984) Cytochromes of the purple sulfur bacterium *Ectothiorhodospira shaposhnikovii*. Z Naturforsch. 39c: 894–901

Lascelles J (1978) Regulation of pyrrole synthesis. In Clayton RK and Sistrom WR (eds) The Photosynthetic Bacteria, pp 795–808. Plenum Publishing Corp., NY

Laudenbach DE, Hebert SK, McDowell C, Fork DC, Grossman AR and Strauss NA (1990) Cytochrome *c*-553 is not required for photosynthetic activity in the cyanobacterium *Synechococcus*. Plant Cell 2: 913–924

Lefèbvre S, Picorel R, Cloutier Y and Gingras G (1984) Photoreaction center of *Ectothiorhodospira* sp. Pigment, heme, quinone and polypeptide composition. Biochemistry 23: 5279–5288

Leguijt T and Hellingwerf KJ (1991) Characterization of reaction center/antenna complexes from bacteriochlorophyll *a* containing *Ectothiorhodospiraceae*. Biochim Biophys Acta 1057: 353–360

Liang JH, Nielsen GM, Lies DP, Burriss RH, Roberts GP and Ludden PW (1991) Isolation and physiological characterization of mutations in the genes encoding the reversible ADP-ribosylation system of *Rhodospirillum rubrum*. J Bacteriol 173: 6903–6909

Liebetanz R, Hornberger U and Drews G (1991) Organization of the genes for the reaction center L and M subunits and B870 antenna polypeptides alpha and beta from the aerobic photosynthetic bacterium *Erythrobacter* sp. OCH114. Mol Microbiol 5: 1459–1468

Lochau W (1981) Evidence for a dual role of cytochrome *c*-553 and plastocyanin in photosynthesis and respiration of the cyanobacterium *Anabaena variabilis*. Arch Microbiol 128: 336–340

Long JE, Durham B, Okamura M and Millett F (1989) Role of specific lysine residues in binding cytochrome c_2 to the *Rhodobacter sphaeroides* reaction center in optimal orientation for rapid electron transfer. Biochemistry 28: 6970–6974

MacGregor BJ and Donohue TJ (1991) Evidence for two promoters for the *Rhodobacter sphaeroides* cytochrome c_2 gene. J Bacteriol 173: 3949–3957

Matsuura K and Shimada K (1986) Cytochromes functionally associated to photochemical reaction centers in *Rhodopseudomonas palustris* and *Rhodopseudomonas acidophila*. Biochim Biophys Acta 852: 9–18

Matsuura K, Fukushima A, Shimada K and Satoh T (1988) Direct and indirect electron transfer from cytochrome *c* and cytochrome c_2 to the photosynthetic reaction center in pigment-protein complexes isolated from *Rhodocyclus gelatinosus*. FEBS Letts 237: 21–25

McManus JD, Brune DC, Han J, Sanders-Loehr J, Meyer TE, Cusanovich MA, Tollin G and Blankenship RE (1992) Isolation, characterization and amino acid sequences of auracyanins, blue copper proteins from the green photosynthetic bacterium *Chloroflexus aurantiacus*. J Biol Chem 267: 6531–6540

Meissner J, Krauss JH, Jurgens UJ and Weckesser J (1988) Chemical analysis of peptidoglycan from species of *Chromatiaceae* and *Ectothiorhodospiraceae*. J Bacteriol 170: 3213–3216

Merchant S and Bogorad L (1987a) The Cu(II)-repressible plastidic cytochrome *c*. J Biol Chem 262: 9062–9067

Merchant S and Bogorad L (1987b) Metal ion regulated gene expression: Use of a plastocyanin-less mutant of *Chlamydomonas reinhardtii* to study the Cu(II)-dependent expression of cytochrome *c*-552. EMBO J 6: 2531–2535

Meyer TE (1985) Isolation and characterization of soluble cytochromes, ferrodoxins and other chromophoric proteins from the halophilic phototrophic bacterium *Ectothiorhodospira halophila*. Biochim Biophys Acta 806: 175–183

Meyer TE, Przysiecki CT, Watkins JA, Bhattacharyya A, Simondsen RP, Cusanovich MA and Tollin G (1983) Correlation between rate constant for reduction and redox potential as a basis for systematic investigation of reaction mechanism of electron transfer proteins. Proc Natl Acad Sci USA 80: 6740–6744

Meyer TE, Watkins JA, Przysiecki CT, Tollin G and Cusanovich MA (1984) Electron-transfer reactions of photoreduced flavin analogues with *c*-type cytochromes: Quantitation of steric and electrostatic factors. Biochemistry 23: 4761–4767

Meyer TE, Cusanovich MA, Krogmann DW, Bartsch RG and Tollin G (1987) Kinetics of reduction by free flavin semiquinones of algal cytochromes and plastocyanin. Arch Biochem Biophys 258: 307–314

Meyer TE, Fitch JC, Bartsch RG, Tollin D and Cusanovich MA (1990a) Unusual high redox potential ferrodoxins and soluble cytochromes from the moderately halophilic purple phototrophic bacterium *Rhodospirillum salinarum*. Biochim Biophys Acta 1017: 118–124

Meyer TE, Cannac V, Fitch J, Bartsch RG, Tollin D, Tollin G and Cusanovich MA (1990b) Soluble cytochromes and ferrodoxins from the marine purple phototrophic bacterium *Rhodopseudomonas marina*. Biochim Biophys Acta 1017: 125–138

Meyer TE, Bartsch RG, Cusanovich MA and Tollin G (1993a) Kinetics of photooxidation of soluble cytochromes, HiPIP, and azurin by the photosynthetic reaction center of the purple phototrophic bacterium *Rhodopseudomonas viridis*. Biochemistry 32: 4719–4726

Meyer TE, Zhao ZG, Cusanovich MA and Tollin G (1993b) Transient kinetics of electron transfer from a variety of *c*-type cytochromes to plastocyanin. Biochemistry 32: 4552–4559

Mi H, Endo T, Schreiber U, Ogawa T and Osada K (1992) Electron donation from cyclic and respiratory flows to the photosynthetic intersystem chain is mediated by pyridine nucleotide dehydrogenase in the cyanobacterium *Synechocystis* PCC6803. Plant Cell Physiol 33: 1233–1237

Michel H and Deisenhofer J (1988) Relevance of the photosynthetic reaction center from purple bacteria to the structure of photosystem II. Biochemistry 27: 1–7

Miller M, Liu X, Snyder SW, Thurnauer MC and Biggins J (1992) Photosynthetic electron transfer reactions in the green sulfur bacterium *Chlorobium vibrioforme*: Evidence for the functional involvement of iron-sulfur redox centers on the acceptor side of the reaction center. Biochemistry 31: 4354–4363

Morand LZ, Frame MK, Colvert KK, Johnson DA, Krogmann DW and Davis DJ (1989) Plastocyanin cytochrome *f* interaction. Biochemistry 28: 8039–8047

Moser CC and Dutton PL (1988) Cytochrome *c* and c_2 binding dynamics and electron transfer with photosynthetic reaction center protein and other integral membrane redox proteins. Biochemistry 27: 2450–2461

Murchison HA, Alden RG, Allen JP, Peloquin JM, Taguchi AKW, Woodbury NW and Williams JC (1993) Mutations designed to modify the environment of the primary electron donor of the reaction center from *Rhodobacter sphaeroides*: Phenylalanine to leucine at position L167 and histidine to phenylalanine at L168. Biochemistry 32: 3498–3505

Nakamura M, Yamagishi M, Yoshizaki F and Sugimura Y (1992) Synthesis of plastocyanin and cytochrome *c*-553 are regulated by copper at the pre-translation level in a green alga, *Pediastrum borganum*. J. Biochem. 111: 219–224

Neidle EL and Kaplan S (1993) 5-aminolevulinic acid availability and control of spectral complex formation in HemA and HemT mutants of *Rhodobacter sphaeroides*. J Bacteriol 175: 2304–2313.

Nitschke W, Feiler U and Rutherford AW (1990a) Photosynthetic reaction center of green sulfur bacteria studied by EPR. Biochemistry 29: 3834–3842

Nitschke W, Setif P, Liebl U, Feiler U and Rutherford AW (1990b) Reaction center photochemistry of *Heliobacterium chlorum*. Biochemistry 29: 11079–11088

Okamura K, Miyata T, Iwanaga S, Takamiya K and Nishimura M (1987) Complete amino acid sequence of cytochrome *c*-551 from *Erythrobacter* species strain OCH-114. J. Biochem. 101: 957–966

Okkels JS, Kjaer B, Hansson O, Svendsen I, Moller BL and Scheller HV (1992) A membrane-bound monoheme cytochrome *c*-551 of a novel type is the immediate electron donor to P840 of the *Chlorobium vibrioforme* photosynthetic reaction complex. J Biol Chem 267: 21139–21145

Overfield RE and Wraight CA (1980) Oxidation of cytochromes *c* and c_2 by bacterial photosynthetic reaction centers in phospholipid vesicles. I. Studies with neutral membranes. Biochemistry 19: 3322–3327

Pierson BK and Thornber JP (1983) Isolation and spectral characterization of photochemical reaction centers from the thermophilic green bacterium *Chloroflexus aurantiacus* strain J-10-fl. Proc Natl Acad Sci USA 80: 80–84

Prince RC and Daldal F (1987) Physiological electron donors to the photochemical reaction center of *Rhodobacter capsulatus*. Biochim Biophys Acta 894: 370–378

Prince RC and Olson JM (1976) Some thermodynamic and kinetic properties of the primary photochemical reactants in a complex from a green photosynthetic bacterium. Biochim Biophys Acta 423: 357–362

Prince RC, Baccarini-Melandri A, Hauska GA, Melandri BA and Crofts AR (1975) Asymmetry of an energy transducing membrane: The localization of cytochrome c_2 in *Rhodopseudomonas sphaeroides* and *Rhodopseudomonas capsulata*. Biochim Biophys Acta 387: 212–227

Prince RC, Gest H and Blankenship RE (1985) Thermodynamic properties of the photochemical reaction center of *Heliobacterium chlorum*. Biochim Biophys Acta 810: 377–384

Prince RC, Davidson E, Haith CE and Daldal F (1986) Photosynthetic electron transfer in the absence of cytochrome c_2 in *Rhodopseudomonas capsulata*: Cytochrome c_2 is not essential for electron flow from the Cyt bc_1 complex to the reaction center. Biochemistry 25: 5208–5214

Qin L and Kostic NM (1992) Electron transfer reactions of cytochrome *f* with flavin semiquinones and with plastocyanin: Importance of protein-protein electrostatic interactions and of donor-acceptor coupling. Biochemistry 31: 5145–5150

Quayle JC and Pfennig N (1975) Methanol utilization by *Rhodospirillaceae*. Arch Microbiol 102: 193–198

Rickle GK and Cusanovich MA (1979) The kinetics of photooxidation of *c*-type cytochromes by *Rhodospirillum rubrum* reaction centers. Arch Biochem Biophys 197: 589–598

Rosen D, Okamura MY, Abresch EC, Valkirs GE and Feher G (1983) Interaction of cytochrome *c* with reaction centers of *Rhodopseudomonas sphaeroides* R-26: Localization of the binding site by chemical cross-linking and immunological studies. Biochemistry 22: 335–341

Rott MA and Donohue TJ (1990) *Rhodobacter sphaeroides spd*

mutations allow cytochrome c_2-independent photosynthetic growth. J Bacteriol 172: 1954–1961

Rott MA, Fitch J, Meyer TE and Donohue TJ (1992) Regulation of a cytochrome c_2 isoform in wild-type and cytochrome c_2 mutant strains of *Rhodobacter sphaeroides*. Arch Biochem Biophys 292: 576–582

Rott MA, Witthuhn VC, Schilke BA, Soranno M, Ali A and Donohue TJ (1993) Genetic evidence for the role of isocytochrome c_2 in photosynthetic growth of *Rhodobacter sphaeroides* Spd mutants. J Bacteriol 175: 358–366

Sandmann G (1986) Formation of plastocyanin and cytochrome *c*-553 in different species of blue green algae. Arch Microbiol 145: 76–79

Schilke BA and Donohue TJ (1992) δ-Aminolevulinic acid couples *cycA* transcription to changes in heme availability in *Rhodobacter sphaeroides*. J Mol Biol 226: 101–115

Schilke BA and Donohue TJ (1995) ChrR positively regulates transcription of the *Rhodobacter sphaeroides* cytochrome c_2 gene. J Bacteriol 177: 1929–1937

Self SJ, Hunter CN, and Leatherbarrow RJ (1990) Molecular cloning, sequencing and expression of cytochrome c_2 from *Rhodospirillum rubrum*. Biochem J 265: 599–604

Shiozawa JA, Lottspeich F, Oesterhelt D and Feick R (1989) The primary structure of the *Chloroflexus aurantiacus* reaction-center polypeptides. Eur J Biochem 180: 75–84

Smit HWJ, Amesz J and Van der Hoeven MFR (1987) Electron transport and triplet formation in membranes of the photosynthetic bacterium *Heliobacterium chlorum*. Biochim Biophys Acta 893: 232–240

Takeda Y and Shimizu T (1992) Expression of a cytochrome *c*-553(CO) gene that complements the second subunit deficiency of membrane bound alcohol dehydrogenase in *Gluconobacter suboxydans* subsp. a. J Ferm Bioeng 73: 89–93

Takemura H, Horinouchi S and Beppu T (1991) Novel insertion sequence IS1380 from *Aceotbacter pasteurianus* is involved in loss of ethanol-oxidizing ability. J Bacteriol 173: 7070–7076

Then J and Trüper HG (1983) Sulfide oxidation in *Ectothiorhodospira abdelmalekii*. Evidence for the catalytic role of cytochrome *c*-551. Arch Microbiol 135: 254–258

Tiede DM and Chang CH (1988) The cytochrome *c* binding surface of reaction centers from *Rhodobacter sphaeroides* Israel J Chem 28: 183–191

Tiede DM, Vashishta AC and Gunner MR (1993) Electron-transfer kinetics and electrostatic properties of the *Rhodobacter sphaeroides* reaction center and soluble *c*-cytochromes. Biochemistry 32: 4515–4531

Tollin G, Cheddar G, Watkins JA, Meyer TE and Cusanovich MA (1984) Electron transfer between flavodoxin semiquinone and *c*-type cytochromes: Correlations between electrostatically corrected rate constants, redox potentials, and surface topologies. Biochemistry 23: 6345–6349

Tomiyama Y, Doi M, Takamiya K and Nishimura M (1983) Isolation, purification and some properties of cytochrome *c*-551 from *Chromatium vinosum*. Plant Cell Physiol 24: 11–16

Trost JT and Blankenship RE (1989) Isolation of a photoactive photosynthetic reaction center-core antenna complex from *Heliobacillus mobilis*. Biochemistry 28: 9898–9904

Trost JT, McManus JD, Freeman JC, Ramakrishna BL and Blankenship RE (1988) Auracyanin, a blue copper protein from the green photosynthetic bacterium *Chloroflexus aurantiacus*. Biochemistry 27: 7858–7863

Trost JT, Brune DC and Blankenship RE (1992) Protein sequences and redox titrations indicate that the electron acceptors in reaction centers from *Heliobacteria* are similar to photosystem I. Photosyn Res 32: 11–22

Van Beeumen JJ (1991) Primary structure of procaryotic diheme cytochromes *c*. Biochim Biophys Acta 1058: 56–60

Van Beeumen J, Ambler RP, Meyer TE, Kamen MD, Olson JM and Shaw EK (1976) The amino acids sequences of cytochromes *c*-555 from two green-sulphur bacteria of the genus *Chlorobium*. Biochem J 159: 757–774

Van Grondelle R, Duysens LNM, Van der Wel JA and Van der Wal HN (1977) Function of three cytochromes in photosynthesis of whole cells of *Rhodospirillum rubrum* as studied by flash spectroscopy. Biochim Biophys Acta 461: 188–201

Van Vliet P, Zannoni D, Nitschke W, and Rutherford AW (1991) Membrane-bound cytochromes in *Chloroflexus aurantiacus* studied by EPR. Eur J Biochem 199: 317–323

Vos MH, Klaasen HE and Van Gorkom HJ (1989) Electron transport in *Heliobacterium chlorum* whole cells studied by electroluminescence and absorbance difference spectroscopy. Biochim Biophys Acta 973: 163–169

Widger WR (1991) The cloning and sequencing of *Synechococcus* sp PCC7002 *petCA* operon: Implications for cytochrome *c*-553 binding domain of cytochrome *f*. Photosyn Res 30: 71–84

Wood PM (1978) Interchangeable copper and iron proteins in algal photosynthesis. Studies on plastocyanin and cytochrome *c*-553 in *Chlamydomonas*. Eur J Biochem 87: 9–19

Wynn RM, Luong C and Malkin R (1989) Maize photosystem I: Identification of the subunit which binds plastocyanin. Plant Physiol 91: 445–449

Zannoni D, Venturoli G and Daldal F (1992) The role of the membrane bound cytochromes of *b*-and *c*-type in the electron transport chain of *Rhodobacter capsulatus*. Arch Microbiol 157: 367–374

Zhang L, McSpadden B, Pakrasi HB and Whitmarsh J (1992) Copper-mediated regulation of cytochrome *c*-553 and plastocyanin in the cyanobacterium *Synechocystis* 6803. J Biol Chem 267: 19054–19059

Zuber H (1988) Structural studies on the antenna complexes and polypeptides of *Chloroflexus aurantiacus* and other green photosynthetic bacteria. In Olson JM, Ormerod JG, Amesz J, Stackebrandt E and Trüper HG (eds) Green Photosynthetic Bacteria, pp 53–55. Plenum Press, New York

Chapter 35

Mutational Studies of the Cytochrome bc_1 Complexes

Kevin A. Gray and Fevzi Daldal
Department of Biology, Plant Science Institute, University of Pennsylvania, Philadelphia, PA 19104, USA

Summary

The ubihydroquinone:cytochrome *c* oxidoreductase (Cyt bc_1 complex) is a multisubunit integral membrane protein present in a wide range of organisms. In the photosynthetic bacteria the Cyt bc_1 complex is involved in both light driven cyclic electron transfer as well as respiratory electron transfer. It oxidizes ubihydroquinone generated either by the reaction center or the respiratory dehydrogenases and reduces Cyt c_2 (or Cyt c_y) which then either re-reduces the reaction center or donates electrons to a terminal cytochrome *c* oxidase. During electron transfer, protons are translocated from the cytoplasm to the periplasm and contribute to the formation of an electrochemical gradient used for ATP synthesis. A large body of spectroscopic data supports a Q-cycle

R. E. Blankenship, M. T. Madigan and C. E. Bauer (eds): Anoxygenic Photosynthetic Bacteria, pp. 747–774.

mechanism of electron transfer through the Cyt bc_1 complex. The enzyme is composed of three metal containing subunits, Cyt *b*, Cyt c_1 and the 'Rieske' iron sulfur protein. The structural genes encoding these three core subunits (designated *petA*, *B* and *C* or *fbcF*, *B* and *C*) have now been cloned and sequenced from several photosynthetic bacteria. Two of them, *Rhodobacter capsulatus* and *Rhodobacter sphaeroides*, have been genetically modified to allow site-directed mutagenesis, and a number of single site mutants have been obtained in each of the three subunits. In recent years, the analyses of both site-directed mutants and spontaneous inhibitor resistant mutants have led to the identification of the ligands to the metal clusters, the general location of the sites of quinone interaction and more accurate folding models of the individual subunits of the complex. This chapter reviews recent biochemical and genetic studies (inclusive of 1994) that have greatly increased our understanding of how the Cyt bc_1 complex functions at the molecular level.

I. Cytochrome bc_1 Complex

A. Physiological Role in Photosynthesis and Respiration

Ubihydroquinone-cytochrome *c* oxidoreductase (or the Cyt bc_1 complex) is an integral membrane protein involved in energy transduction in a wide range of organisms (for recent reviews see Prince, 1990; Trumpower, 1990; Cramer and Knaff, 1990; Knaff, 1993; Gennis et al., 1993; Brandt and Trumpower, 1994). The enzyme is a major component of the mitochondrial electron transfer chain and performs a similar function in aerobic prokaryotes (Fig. 1). Anoxygenic photosynthetic prokaryotes contain a Cyt bc_1 complex that is utilized for both light driven electron transfer and dark respiration (Crofts and Wraight, 1983; Dutton, 1986; Prince, 1990; Knaff, 1993). Oxygenic photosynthetic organisms contain a similar redox complex (called the Cyt b_6f complex) which functions between Photosystems I and II (Hauska, 1986; Cramer et al., 1987; Malkin, 1992). In all cases the complexes oxidize and reduce the lipid soluble two electron carrier (plasto- , ubi- or mena-hydroquinone) in the membrane phase, and reduce a one electron carrier (the water soluble Cyt *c* or plastocyanin or the newly discovered membrane-associated Cyt c_y [Jenney and Daldal, 1993; Jenney et al., 1994]) at the membrane aqueous interface. During respiration ferrocyt *c* then donates electrons to Cyt *c* oxidase (Gray et al., 1994) which catalyzes the four electron reduction of dioxygen to water. On the other hand, during anoxygenic photosynthesis, a predominantly cyclic process, ferrocyt *c* re-reduces the photooxidized BChl dimer of the reaction center, resulting in no net redox change (Fig. 1). During electron transfer, charges (electrons and protons) are vectorially transported across the membrane, generating an electrochemical gradient ($\Delta\mu_{H^+}$) used to drive energy requiring processes such as ATP synthesis, active transport and taxis (Dutton, 1986; Prince, 1990).

Abbreviations: B. – Bacillus; Br. – Bradyrhizobium; Cb. – Chlorobium; Cyt – cytochrome; *E. – Escherichia; N. – Neurospora; P. – Paracoccus; Rb. – Rhodobacter; Rp. – Rhodopseudomonas; Rs. – Rhodospirillum;* UHDBT – undecylhodroxydioxobenzothiazole; UNHQ – undecylhydroxynapthoquinone

B. Experimental Advantages of Phototrophs in Studying the Cyt bc$_1$ Complex

Photosynthetic bacteria, in general, and *Rhodobacter (Rb.) capsulatus* and *Rhodobacter sphaeroides*, in particular, provide important experimental advantages over mitochondria and chloroplasts for studying the structure and function of the Cyt bc_1 complex. First, the bacterial complexes are structurally much simpler, and yet functionally very similar, when compared with those from mitochondria. The bacterial enzymes contain only three or four subunits, while mitochondrial complexes contain several additional subunits of undefined function. Second, in phototrophic organisms the reactions catalyzed by the Cyt bc_1 complex can be measured in situ by flash activation of the photochemical reaction center using either chromatophore membranes or whole cells. The initial events that take place in the reaction center are extremely fast, and this photoactivation provides an oxidant (ferricyt *c*) and a reductant (UQH_2) for the Cyt bc_1 complex allowing the use of time-resolved optical spectroscopy to monitor the internal reactions of the complex. Examples of flash-induced electron transfer kinetics and the corresponding specific reactions are shown in Fig. 2. Further, in response to the change in the electric field across the low dielectric of the membrane (i.e. the membrane potential) caused by charge transfer the absorbance

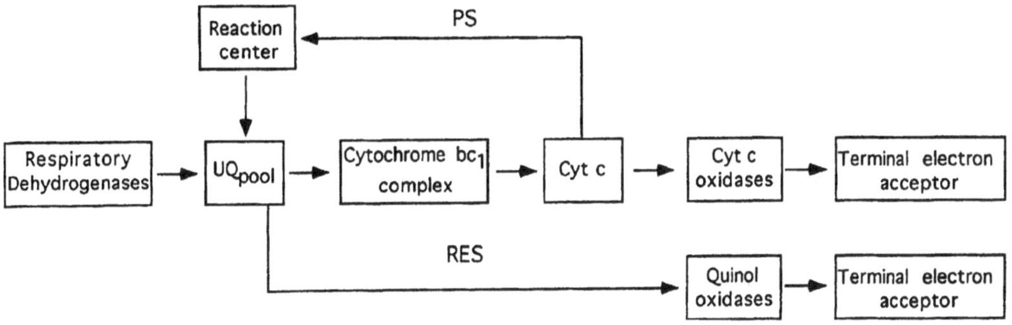

Fig. 1. Schematic representation of energy transduction in bacterial membranes indicating the route of electrons through the various membrane associated and soluble components. In purple bacteria, photosynthetic electron transfer cycles around the reaction center, UQ_{pool}, Cyt bc_1 complex and Cyt c (c_2 and c_y). Respiratory electron transfer can either be routed to a Cyt c oxidase via the Cyt bc_1 complex and Cyt c or directly to a terminal electron acceptor via a quinol oxidase bypassing the Cyt bc_1 complex.

spectra of the membrane-associated carotenoids are shifted (a Stark effect). Thus the electrogenic steps of the internal reactions can be followed spectroscopically on a fast time scale (see for example Matsuura et al., 1983b; Glaser and Crofts, 1984; Robertson and Dutton, 1988). Third, photosynthetic growth of these bacteria is absolutely dependent upon the presence of a functional Cyt bc_1 complex, thus mutations affecting the complex can be detected by simply monitoring the growth ability of the mutants (Fig. 1) (Daldal et al., 1987). Yet, non-functional Cyt bc_1 complex mutants of facultative phototrophic bacteria are still viable due to the availability of other growth modes independent of the complex (Marrs and Gest, 1973; Zannoni et al., 1976; LaMonica and Marrs, 1976). Appropriate growth conditions (low O_2 in the dark) can be chosen so that the photosynthetic apparatus is induced, thus non-functional mutants can be analyzed in detail. Fourth, for both *Rb. capsulatus* and *Rb. sphaeroides* molecular and genetic tools are available for sophisticated genetic analyses such as 'gene tagging' and suppressor analyses. Furthermore in *Rb. capsulatus* the genetic system to be used can be chosen so that a mutant complex is either over–produced to facilitate its physicochemical analysis, or kept at the wild type level to analyze its physiological impact. The overproduction also eases the purification and characterization of the wild type as well as mutant complexes, and allows the detection of very low activity of some mutants which may be otherwise undetected. Finally, the structural genes of most bacterial Cyt bc_1 complexes are genetically organized as an operon (with a few exceptions) so that both classical and reverse genetic approaches are often facilitated. Thus, multidisciplinary approaches encompassing biochemistry, biophysics, molecular biology and genetics can be more easily applied in the bacterial systems. Similar approaches are either not yet available or more difficult to perform in mitochondrial and chloroplast systems.

In this chapter we review the salient features of the Cyt bc_1 complex and describe recent mutagenesis work being performed on the complex. Although we mainly focus on the work accomplished using the purple non-sulfur photosynthetic bacteria, which is currently the most advanced, similar approaches are also being pursued in other bacterial systems such as *Paracoccus (P.) denitrificans*, *Bradyrhizobium (Br.) japonicum* and cyanobacteria, as well as yeast and other higher eukaryotes (see for example Graham and Trumpower, 1991).

C. Structure

1. Protein Subunits

While the subunit content of the Cyt bc_1 complexes is variable from organism to organism the smallest number is in general found in bacteria. Those complexes from the purple non-sulfur, photosynthetic bacteria *Rb. capsulatus* (Robertson et al., 1993), *Rhodospirillum (Rs.) rubrum* (Wynn et al., 1986; Kriauciunas et al., 1989; Majewski and Trebst, 1990; Purvis et al., 1990; Güner et al., 1991) and *Rhodopseudomonas (Rp.) viridis* (Wynn et al, 1986) contain three polypeptides while *Rhodobacter sphaeroides* contains a fourth subunit of unknown

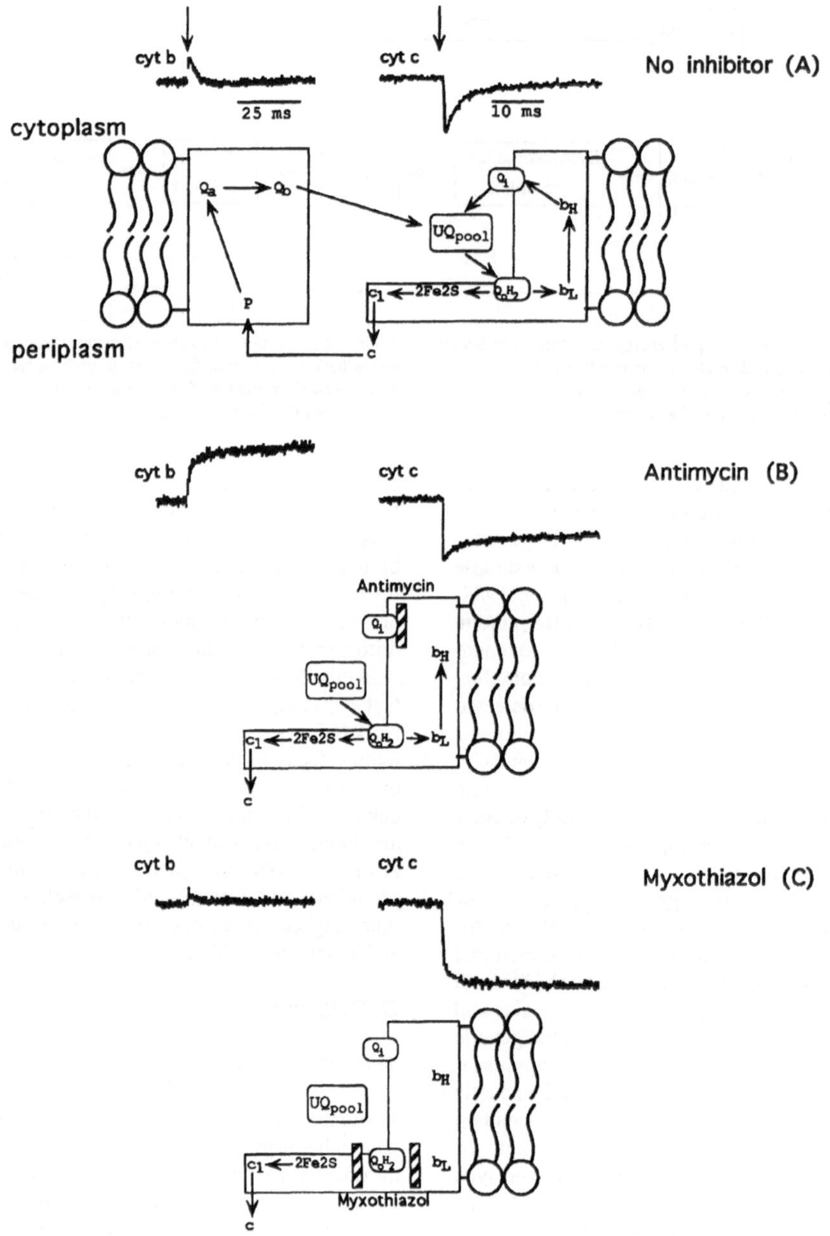

Fig. 2. Kinetic traces of flash induced electron transfer through the Cyt bc_1 complex in the absence (A) and presence (B and C) of specific inhibitors (10 μM antimycin and 5 μM myxothiazol). *Rhodobacter capsulatus* chromatophores were poised at an ambient redox potential of 100 mV in 50 mM MOPS, pH 7 containing 100 mM KCl, 6 μM valinomycin and redox mediators. Below each set of traces is shown a schematic representation of the pathway of electron transfer under the specified conditions. Electron transfer through the reaction center is only shown once in a highly simplified form. The flash point is indicated by an arrow, redox changes in total Cyt *b* are followed by monitoring the wavelength pair 560 nm minus 570 nm while redox changes in total Cyt *c* are followed by monitoring the wavelength pair 550 nm minus 540 nm.

function (Yu et al., 1984; Ljungdahl et al., 1987; Andrews et al., 1990; Purvis et al., 1990; Usui et al., 1990, 1991). Recent immunological evidence suggests that this fourth subunit appears to be a bona fide component of the complex, possibly interacting with quinone and antimycin (Wilson et al., 1985; Yu and Yu, 1987, 1991; Usui et al., 1991; Chen et al., 1994). The Cyt bc_1 complexes from the purple sulfur bacteria, *Ectothiorhodospira mobilus* (Leguijt et al., 1993) and *Chromatium vinosum* (Tan et al., 1993) also appear very similar to those from the purple non-sulfur bacteria. It now appears that the bacteria utilizing menaquinone in the Q_{pool} (for example heliobacteria, bacilli and *Chlorobium*-type green bacteria) contain Cyt *bc* complexes with somewhat different properties (Knaff and Malkin, 1976; Kutoh and Sone, 1988; Liebl et al., 1990, 1992; Riedel et al., 1993), perhaps to better to accommodate the approximately 150 mV lower redox midpoint potential of menaquinone in comparison to that of ubiquinone.

Regardless of the total subunit content, all Cyt bc_1 complexes contain three core polypeptides, called Cyt *b*, Cyt c_1 and the 'Rieske' iron-sulfur protein (FeS), that bind the metal clusters participating in charge separation through the complex. Cyt b_6f complexes actually contain another subunit referred to as subunit IV, however it is highly homologous to the C-terminal region of Cyt *b* while Cyt b_6 is highly homologous to the N-terminal region of Cyt *b*. Thus, Cyt b_6 and subunit IV of the chloroplast complex could be considered a 'split' Cyt *b*. A similar situation probably exists in some Gram positive bacteria, such as *Bacillus* PS3 (Kutoh and Sone, 1988 and N. Sone, personal communication) and *Bacillus subtilus* (Sorokine and Erlich, personal communication).

2. Prosthetic Groups and Quinone Binding Sites

The Cyt *b* polypeptide (typically about 45 kDa) is an integral membrane protein with multiple transmembrane helices that contains two spectrochemically distinct molecules of protoheme, Cyt b_L and b_H, referring to the relatively 'low' (approximately −100 mV in *Rb. capsulatus* and *Rb. sphaeroides*) and 'high' (approximately 50 mV in *Rb. capsulatus* and *Rb. sphaeroides*) redox midpoint potentials (E_{m7}) of the two hemes. Spectroscopic analyses have suggested that both *b* hemes have bis-imidazole ligation (Carter et al., 1981; Salerno et al., 1986; Simpkin et al., 1989; McCurley et al., 1990; Hobbs et al., 1990), and current folding models of this protein with eight transmembrane helices (see below) use the four universally conserved histidine residues located on the same pair of helices (two in B and two in D) as their ligands.

Cyt c_1 (typically about 30 kDa, except in *P. denitrificans* where it is approximately 65 kDa), covalently binds one molecule of heme ($E_{m7} \approx 290$ mV) with the most probable axial ligation of histidine and methionine (Simpkin et al., 1989). The consensus heme binding motif (-Cys-Xaa-Yaa-Cys-His-) is located close to the amino terminus of both Cyt c_1 and Cyt *f* of the Cyt b_6f complex, and these two cysteines covalently link the heme macrocycle to the protein via thioether linkages. Magnetic circular dichroism (MCD) and resonance Raman (RR) measurements suggested that the sixth axial ligand to the heme Fe of Cyt *f* is not methionine but an amine group perhaps from a lysine (Rigby et al., 1988; Davis et al., 1988; Simpkin et al., 1989). Recent 3-D structural analysis of isolated Cyt *f* from turnip has shown that its sixth axial ligand is surprisingly the amino terminus of the protein (Martinez et al., 1992, 1994).

The FeS protein (typically about 20 kDa) has a $[Fe_2S_2]$ cluster with an unusual EPR spectrum and a relatively high E_m value (approximately 260 mV). Spectroscopic analyses have suggested that two histidines and two cysteines are the ligands to the cluster in contrast to the more common four cysteines seen in the ferredoxins (Telser et al., 1987; Powers et al., 1989; Britt et al., 1991; Miki et al., 1991; Gurbiel et al., 1991). The $[Fe_2S_2]$ cluster interacts with the site of hydroquinone oxidation as evidenced by the sensitivity of the EPR spectrum of the cluster to the redox state of the UQ_{pool} and to the occupancy of the Q_o site of the complex (Matsuura et al., 1983a; Robertson et al., 1986; Robertson et al., 1990; Ding et al., 1992).

In addition to these metal-containing cofactors the enzyme has interaction sites for its reaction partners, Cyt c_2, Cyt c_y (in *Rb. capsulatus)* and ubi(hydro)-quinone. The two quinone binding sites, called Q_o and Q_i, are at least partially located on the Cyt *b* polypeptide and have different catalytic and thermodynamic properties (see the section on the Q-cycle below). The interaction between Cyt c_1 and c_2 is electrostatic and involves lysine residues on the 'front' face (i.e. the side of the exposed heme edge) of Cyt c_2 (Margoliash and Bosshard, 1983; Wynn et al., 1986; Bosshard et al., 1987; Güner et al., 1993). Differential chemical modification (Hall et al., 1987a,b,c; Hall et al., 1989) and later, site-directed mutagenesis (Caffery et al., 1992; Güner et al., 1993), of these lysines resulted in a perturbation of the Cyt c_1/c_2 interaction and the kinetics of electron transfer. The positions of Cyt c_1 providing the counter charges are unknown but the effective negative charges supplied by Cyt c_1 in the interaction appear approximately the same in *Rb. capsulatus* and *Rp. viridis* (Güner et al., 1993). There are two patches of highly conserved acidic residues on Cyt c_1 in *Rb. capsulatus* (G88 to D107 and V209 to G218) which could be responsible for these interactions (Davidson and Daldal, 1987b). At present the interaction between Cyt c_1 and Cyt c_y of *Rb. capsulatus* is entirely unknown (Jenney and Daldal, 1993).

D. Q-cycle Mechanism of Charge Separation

1. Inner Reactions

Light-induced electron transfer through the Cyt bc_1 complex is initiated by the very rapid (microsecond) oxidation of Cyt *c* in the periplasmic space by the photo-oxidized bacteriochlorophyll dimer (P^+) of the reaction center (Fig. 2). Ferricyt *c* triggers a cascade of events beginning with the oxidation of Cyt c_1. The latter then oxidizes the $[Fe_2S_2]$ cluster, which in turn is reduced by ubihydroquinone bound at a site (Q_o) formed by Cyt *b* and the FeS protein. The transient semiquinone generated by the one electron oxidation of UQH_2 is unstable and quickly reduces the low potential Cyt b_L. The observation of reduction of Cyt *b* following the addition of an 'oxidant' (the oxidized $[Fe_2S_2]$) led to the proposal of a Q-cycle mechanism of electron transfer through the complex (Wikström and Berden, 1972; Mitchell, 1976). Later, modifications to the Q-cycle were introduced to better describe the pathways of charge separation (Crofts et al., 1983; Ding et al., 1992). Inherent to these models are two separate, independent sites of (hydro)quinone interaction on the enzyme. One site, that of UQH_2 oxidation (Q_o, for proton 'output'), is near the periplasmic surface of the membrane and undergoes a concerted two-electron oxidation by the $[Fe_2S_2]$ and Cyt b_L. Ding et al. (1992) have recently analyzed the EPR spectra of ubiquinone extracted- and reconstituted-chromatophores and proposed that two molecules of UQH_2 may occupy the Q_o site. These two sites are called Q_{ow} (Q_o 'weak') and Q_{os} (Q_o 'strong') in reference to their strength for binding UQH_2.

After the reduction of Cyt b_L, electron transfer occurs in an electrogenic step from Cyt b_L to Cyt b_H. From carotenoid electrochromic shift measurements the distance from Cyt b_L to Cyt b_H is estimated to span more than half of the membrane dielectric (Glaser and Crofts, 1984; Robertson et al., 1988). Cyt b_H is then oxidized by a quinone bound at a second site, called the Q_i site (for proton 'input') forming an antimycin sensitive EPR detectable semiquinone ($Q^{\cdot -}$) species (Ohnishi and Trumpower, 1980; Robertson et al., 1984). Carotenoid bandshift measurements have estimated that the distance between quinone bound at the Q_i site and Cyt b_H represents the remainder of the membrane dielectric. Recently it has been calculated from spin relaxation enhancement of the EPR signal of $Q^{\cdot -}$ by an external spin probe that the semiquinone anion is 6 to 10 Å from the protein surface (Meinhardt and Ohnishi, 1992). Note that a second turnover of the Q_o site is needed for a complete turnover of the Q_i site, and the presence of one Q_{ow} and one Q_{os} may allow two turnovers of the Q_o site without the need of replenishment from the Q_{pool}. Each oxidation of UQH_2 at the Q_o site releases two H^+ into the periplasmic space while reduction of UQ to UQH_2 at the Q_i site results in the consumption of two H^+ from the cytoplasmic side of the membrane. Thus the oxidation and reduction of both an electron and proton carrier on opposite sides of the membrane forms a classical redox loop (Mitchell, 1966), and explains how turnover of the Cyt bc_1 complex contributes to the electrochemical potential difference between the two sides of the membrane.

2. Inhibitors

The use of inhibitors has aided greatly in the elucidation of the pathways of electron transfer through the Cyt bc_1 complex. The most common inhibitors of the Cyt bc_1 complex predominantly affect either the Q_o or Q_i-sites specifically (reviewed in von Jagow and Link, 1986) although they may have additional effects elsewhere (Tsai and Palmer, 1982; Rich et al., 1990; Howell and Robertson, 1993). The Q_o site inhibitors can be subdivided into the methoxyacrylates, which include myxothiazol and mucidin, and the chromones like stigmatellin. The methoxyacrylates inhibit reduction of both Cyt

b_L and the $[Fe_2S_2]$ cluster via the Q_o site but allow turnover of the Q_i site (see Fig. 2 for the effects of the inhibitors on the electron transfer kinetics). Binding of myxothiazol perturbs the EPR spectrum of the $[Fe_2S_2]$ cluster, which may be a consequence of displacing Q from the Q_o site (Ding et al., 1992 and references therein). The hydroxyquinones, which include UHDBT (undecylhydroxydioxobenzothiazole) and UHNQ (undecylhydroxynaphthoquinone), act between the $[Fe_2S_2]$ cluster and Cyt c_1. The mode of action of these quinone analogs appears different from the methoxyacrylates, in that they increase the E_m value of the $[Fe_2S_2]$ cluster by approximately 50 mV and inhibit oxidation of the cluster by Cyt c_1 (Matsuura et al., 1983a). The inhibitory effect of stigmatellin also seems to be centered on the $[Fe_2S_2]$ cluster since its binding to the Cyt bc_1 complex induces a large (approximately 250 mV) increase in the E_m value of this cluster, strongly stabilizing the reduced state and inhibiting oxidation by Cyt c_1 (von Jagow and Ohnishi, 1985). Interestingly, this effect is not observed in menaquinone utilizing Cyt bc_1 complexes such as *Heliobacterium chlorum* (Liebl et al., 1990, 1992), *Bacillus (B.)* PS3 and *B. firmus* OF4 (Riedel et al., 1993).

Examples of Q_i-specific inhibitors are antimycin, HQNO (heptylhydroxy-quinoline-N-oxide), diuron and funiculosin. All inhibit the oxidation of Cyt b_H by displacing the semiquinone bound at the Q_i site without affecting the oxidation of the first QH_2 at the Q_o site (Fig. 2, Ohnishi and Trumpower, 1980). Antimycin also induces a red shift in the absorption spectrum of Cyt b_H suggesting that its binding site is close to that heme. Funiculosin causes an increase in the E_m value of Cyt b_H (Howell and Robertson, 1993) and the $[Fe_2S_2]$ cluster in mitochondria (Tsai and Palmer, 1982) but the photosynthetic bacteria appear to be immune to the effects of this latter inhibitor both in vivo and in vitro (M. K. Tokito and F. Daldal, unpublished data).

E. Genetic Organization of the Structural Genes of the Cyt bc$_1$ *Complex*

1. Cloning of the fbc/pet *Operons*

The first structural gene of a Cyt bc_1 complex from a photosynthetic bacterium initially thought to be *Rb. sphaeroides* was isolated by Gabellini et al. (1985) using the Rieske iron sulfur gene of *Neurospora (N.) crassa* as a heterologous probe (Harnisch et al., 1985). The determination of the nucleotide sequence of this cloned region indicated that it contained not only the gene for the FeS protein (*fbcF*) but also the structural genes for the Cyt *b* (*fbcB*) and Cyt c_1 (*fbcC*) subunits (Gabellini and Sebald, 1986). S1 mapping and Northern blot analysis indicated that the corresponding mRNA was approximately 2.9 kb long, and proved that these three genes were cotranscribed and constitute an operon (Gabellini et al., 1985; Gabellini and Sebald, 1986).

An independent approach to clone the structural genes of the Cyt bc_1 complex was used by Daldal and co-workers with a different purple non-sulfur photosynthetic bacterium, *Rb. capsulatus* (Daldal et al. 1987, Davidson and Daldal, 1987a,b). This organism was chosen due to its amenability to genetic approaches and to the availability of mutants related to bacterial energy transduction isolated by Marrs and co-workers (for a review see Scolnik and Marrs, 1987). One of these mutants, R126, was initially described by Zannoni et al. (1981) as a non-photosynthetic mutant defective in the electron transfer pathways around the Cyt bc_1 complex.

Genetic complementation of R126 with the *Rb. capsulatus* chromosomal library of Scolnik and Haselkorn (1984) yielded a plasmid containing the structural genes (*petABC*) for the Cyt bc_1 complex. As in the case of Gabellini et al. (1985) this study also indicated that the three structural genes *petA*, *petB* and *petC* corresponding to the FeS, Cyt *b* and Cyt c_1 polypeptides respectively, are organized in the 5' to 3' order. This work also revealed unexpectedly that the sequence obtained by Gabellini and Sebald (1986) was almost identical to the sequence obtained by Davidson and Daldal (1987a), and that the strain used by Gabellini and coworkers was not *Rb. sphaeroides* but was rather a photosynthetic bacterium closely related to *Rb. capsulatus* (Davidson and Daldal, 1987b). The first partial *fbc/pet* sequence of a bona fide *Rb. sphaeroides* strain was reported by Davidson and Daldal (1987b), and the entire sequence from *Rb. sphaeroides* was later completed by Yun et al. (1990).

The isolation of the structural genes led to the construction of Cyt bc_1-minus mutants of *Rb. capsulatus* and *Rb. sphaeroides*, and to the demonstration that a functional Cyt bc_1 complex is absolutely essential for photosynthetic growth of these species (Daldal et al., 1987; Yun et al., 1990).

Insertional mutagenesis of these genes further confirmed that they constitute an operon since insertions in the proximal genes are polar to the expression of the distal genes (Daldal et al., 1987). While no other gene related to the Cyt bc_1 complex located at the 3' end of the operon was found insertions near its 5' end negatively affected the growth of *Rb. capsulatus*. Subsequent analysis of this latter region revealed the presence of two additional genes, *petP* and *petR*, whose activities are essential for both photosynthetic and respiratory growth of *Rb. capsulatus* (Tokito and Daldal, 1992). While one of these bc_1-unrelated genes (*petR*) shows strong homology to *ompR*, the well studied response regulator involved in osmoregulation in *E. coli* (Stock et al., 1989), neither its role in *Rb. capsulatus* nor the role of the preceding gene (*petP*) are known. The strong homology observed infers that they may be regulatory genes involved in sensing some unknown environmental signal(s).

The structural genes of the Cyt bc_1 complex have also been obtained from other purple non sulfur photosynthetic bacteria, *Rp. viridis* (Verbist et al., 1989) and *Rs. rubrum* (Majewski and Trebst, 1990; Shankar et al., 1992) using heterologous probes and were found to form an operon as well. These genes are also available from the non-phototrophs, *P. denitrificans* (Kurowski and Ludwig, 1987), *B. japonicum* (Thöny-Meyer et al., 1988), and the green sulfur photosynthetic bacterium *Chlorobium (Cb.) limicola* (Schütz et al., 1993) with the exception of Cyt c_1 gene in this latter case. Partial or preliminary genetic data have recently been obtained for the strict aerobes *Bacillus* PS3 and *Bacillus subtilis* (N. Sone, and Sorokin and Ehrlich, respectively, personal communication). In addition, the structural genes have also been cloned and sequenced from the cyanobacterial species *Nostoc* PCC 7906 (Kallas et al., 1988a) and *Synechococcus* PCC 7002 (Widger, 1991; Brand et al., 1992). Hopefully sequence data from two other important branches of the phototrophs, namely the purple sulfur bacteria such as *Chromatium* and the green gliding bacteria such as *Chloroflexus*, will soon be added to this growing database and provide further insight into the structure and function as well as evolution of the Cyt bc_1 complex.

2. Primary Sequences of the Subunits from Different Species

While the operonal organization of the *fbcF* (*petA*), *fbcB* (*petB*) and *fbcC* (*petC*) genes for the three essential subunits of the Cyt bc_1 complex is generally conserved, some interesting deviations have also been observed (Fig. 3). For example, in *Cb. limicola fbcF* and *fbcB* are next to each other, but *fbcC* is not adjacent to this cluster (Schütz et al., 1993). In cyanobacteria *fbcF* and *fbcC* analogues have remained together while the equivalent of the *fbcB* gene was found elsewhere in a split form (Kallas et al., 1988b; Widger, 1991; Brand et al., 1992). In addition, the Cyt *b* and Cyt c_1 genes of *Br. japonicum* are fused to form a single gene (referred to as *fbcH*) whose product is apparently post-translationally processed to yield a complex with three polypeptide subunits (Thöny-Meyer et al., 1991). Finally, there are approximately 150 additional residues of unknown role at the N-terminal end of Cyt c_1 from *P. denitrificans* (Kurowski and Ludwig, 1987) that are not present in other known species, and whose deletion does not affect drastically either the assembly or the function of the Cyt bc_1 complex (B. Ludwig, personal communication).

The currently available sequences of Cyt *b* from the photosynthetic bacteria can be divided into two distinct groups based upon the presence or absence of specific deletions (Fig. 4a). Both *Rp. viridis* and *Rs. rubrum* lack the residues from 231 to 237 in the Q_iII region and between 309 to 326 in the Q_oII region (Shankar et al, 1992). The deletion of either of these regions from *Rb. capsulatus* Cyt *b* has no drastic effect on the Cyt bc_1 complex (K. A. Gray and F. Daldal, unpublished). Also, helix H present in other bacterial Cyt *b* is absent from that of *Cb. limicola*. The alignment of the amino acid sequences of the three subunits from photosynthetic species (Fig. 4a, b, c) reveals that while the Cyt *b* subunit is extremely well-conserved between various species, such pronounced homologies in the other subunits are limited to the metal binding domains of the FeS protein (the C-terminal region) and Cyt c_1 (the N-terminal region) (see also Hauska et al., 1988). An extensive comparison of the primary sequence of Cyt *b* subunit of the Cyt bc_1 complex from many phylogenetically distinct species has recently been reported (Degli Esposti et al., 1993).

3. Folding Models for the Different Subunits

The determination of the primary sequences of the three subunits of the Cyt bc_1 complex has led to predictions of their secondary structures using various computer programs (Rao and Argos, 1986; Crofts et al., 1987; Brasseur, 1988). These analyses indicated

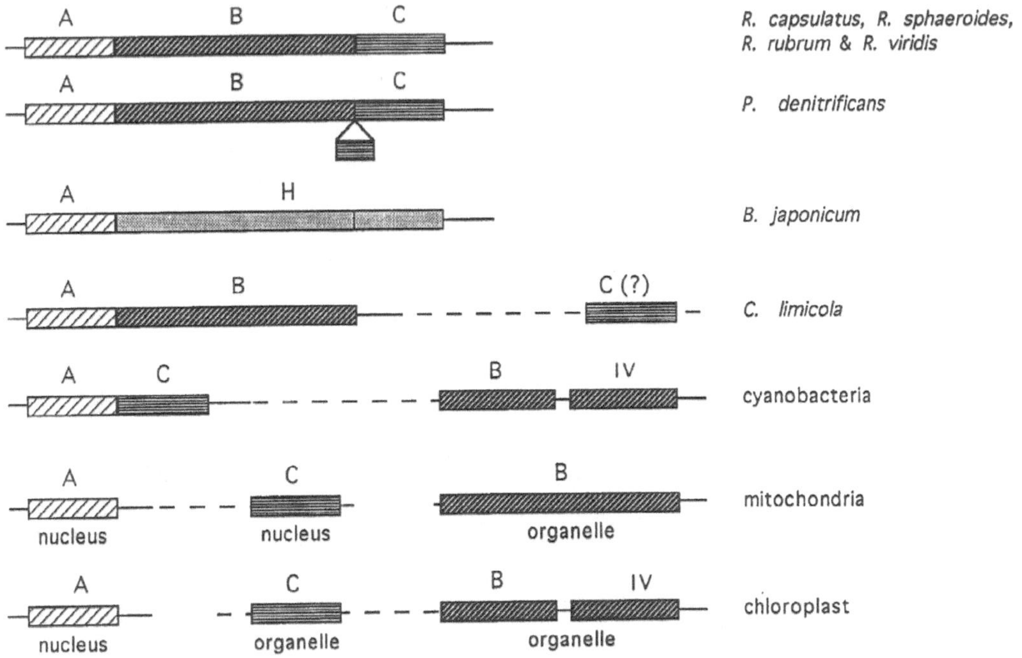

Fig.3. Genetic organization of the *fbc/pet* operons in various organisms. The organization of the structural genes for the three core subunits of the Cyt bc_1 complex are shown schematically. A, B, C and IV represent the genes for the FeS protein, Cyt *b*, Cyt c_1 (Cyt *f*) and Cyt b_6 subunit IV respectively.

that the Rieske FeS protein is mainly hydrophilic with its two mostly hydrophobic regions being located near the N-terminus and around the $[Fe_2S_2]$ cluster binding region near the C-terminal end of the protein (Fig. 5). While there is no cleaved signal sequence in the FeS protein the bulk of the protein appears to be on the periplasmic face of the membrane. Its amino terminal region has been implicated in binding it to the membrane, although the exact nature of this association is unclear (Li et al., 1981b; Schägger et al., 1987; Gonzalez-Halphen et al., 1988, 1991; Cocco et al., 1991; Theiler and Niederman, 1991; Van Doren et al., 1993b).

Cyt c_1 is also mainly a hydrophilic protein but with two hydrophobic helical regions (Fig. 6). The first region corresponds to its membrane translocation signal sequence located at the N-terminus. This sequence is processed during secretion across the cytoplasmic membrane into the periplasm and is not present in the mature protein found in the complex (Gabellini and Sebald, 1986; Yun et al., 1990). The second hydrophobic domain of Cyt c_1 (about 15 amino acids delimited by charged residues) located close to its C-terminal end is thought to anchor the subunit to the cytoplasmic membrane. Limited proteolysis experiments which release the subunit from the membrane in *Neurospora crassa* (Li et al., 1981a) as well as genetic approaches involving the introduction of a stop codon at an appropriate position to prematurely terminate the subunit (Konishi et al., 1991, Gray et al., 1992) are consistent with this topology of Cyt c_1 (Fig. 6).

Secondary structure analyses predicted that Cyt *b* is an integral membrane protein with multiple transmembrane helical domains (Fig. 7). While initially it was proposed that it contained nine transmembrane helices (Saraste, 1984; Widger et al., 1984) more refined analyses using several different secondary structure prediction programs (Crofts et al., 1987) revealed that the helix 4 of the initial model is more amphipathic and probably not transmembranous. This extramembranous loop (referred to as helix *cd*) is reminiscent of the amphipathic helix in the purple bacterial photosynthetic reaction center which forms part of the Q_b binding pocket (Deisenhofer et al., 1985; Allen at al., 1988).

```
          1                                                 49
R.c.      MSGIPHDHYE PKTGIEKWLH DRLPIVGLVY DTIMI.PTPK NLNWWWIWGI
R.s.      ---------- -r-------- s-----a-a- -----.---r ----m----v
R.v.      ---..-ss-q -s----r--d t-----rmm- d-fvaf-v-- -i-yayaf-a
R.r.      -yt..pprwn n-a..l--fd e---vltvah kelvvy-a-r -i-yf-nf-s
C.l.      npfkdskrda vagwfqerfy vln--idylk hkev..-k.h a-sf-yyf-g

          50                                                99
R.c.      VLAFTLVLQI VTGIVLAMHY TPHVDLAFAS VEHIMRDVNG GWAMRYIHAN
R.s.      ----c----- ---------- ---------- ------n--- -fml--l---
R.v.      i--vf-ii-- -s-v------ vaqdt----- i--------y --li----n-
R.r.      lagiamiim- a---f---s- -a---h--d- --r------y --l---m---
C.l.      lgl-ff-i-- l--ll-lqy- k-tetd---- flf-qge-pf --ll-q---w
                                                         *

          100                                               148
R.c.      GASLFFLAVY IHIFRGLYYG SY.KAPREIT WIVGMVIYLL MMGTAFMGYV
R.s.      ------i--- l--------- --.-----v- -----l---a --a-------
R.v.      ---f--f--- a-t---m--- --.-e---vl --l-vi-ii- --a-------
R.r.      ---m--iv-- v-m------- --.-p---vl -wl-l--ll- --a-------
C.l.      s-n-mimmlf --m-stffmk --r-.---lm -vs-f-ll-- sl-fg-t--l
            *
          149                                               198
R.c.      LPWGQMSFWG ATVITGLFGA IPGIGPSIQA WLLGGPAVDN ATLNRFFSLH
R.s.      ---------- ---------- -----h---t ---------- ----------
R.v.      ---------- -----n--s- --y--dp-vt --w--ys-g- p--t--y---
R.r.      ---------- -----n--s- --vv-dd-vt l-w--fs--- p---------
C.l.      ---nela-fa tq-g-evpkv a--gaflvei lrg-pevgge tl-t-m----
                                                                *
          199                                               247
R.c.      YLL.PFVIAA LVAIHIWAFH TTGNNNPTGV EVRRTSKADA EKDTLPFWPY
R.s.      ---.------ ---------- s--------- --------e- q---v-----
R.v.      ---.----vg v-ml-v--l- v--qt----- --k......s ----vr-t-f
R.r.      --f.-mllf- v-fl-m--l- vkks---l-i dak......g pf--i--h--
C.l.      vv-l-glvml vl-a-ltlvq il-tsa-i-y ..keaglikg ydkff-tfla
          248   *                                           297
R.c.      FVIKDLFALA LVLLGFFAVV AYMPNYLGHP DNYVQANPLS TPAHIVPEWY
R.s.      -i---v---- v---v---i- gf-------- ---ie----- ----------
R.v.      alt--av--g vcfia-awf- ffv------a ---ip---gv ----------
R.r.      ytv--a-g-g if-mv-cff- ffa--ff-e- ---ip---mv --t-------
C.l.      ...--gigwl igfalliyla vmf-wei-vk a-plspa-lg ik....----

          298                                               347
R.c.      FLPFYAILRA FAADVWVVIL VDGLTFGIVD AKFFGVIAMF GAIAVMALAP
R.s.      ---------- -t------qi anfis---i- ------l--- ---l----v-
R.v.      l--------- .......... ........-p d-lg------ --lv-llfl-
R.r.      ---------- .......... ........-p d-lg--l--- ---lilfvl-
C.l.      -waqfql-kd -kfeggella iilfti-g-v wllvpf-drq aseekkspif

          348                                               397
R.c.      WLDTSKVRSG AYRPKFRMWF WFLVLDFVVL TWVGAMPTEY PYDWISLIAS
R.s.      -----p---- r---m-kiy- -l-aa---i- -----qq-tf ----------
R.v.      ---g-----a ----ly-if- -lf-a-cif- g-l----a-g i-ptl-qvgt
R.r.      ---------a tf--v-kgf- -vfla-cll- gyl----a-e --vt-tql-t
C.l.      tifgil-laf llinty-vya eysm-k

          398                                    437
R.c.      TYWFAYFLVI LPLLGATEKP EPIPASIEED FNSHYGNPAE .......
R.s.      a--------- --i---i--- vap--t---- --a--spatg gtktvvae
R.v.      vwy--h---- v-a--yf--- k-l-e--saa vleahsr-sl larlinr
R.r.      iyy-lh---- t-lv-wf--- k-l-v--ssp vttqa..... .......
```

Fig.4. (a) Sequence comparison of Cyt *b* from the photosynthetic bacteria. The four boxed histidines are the ligands for heme b_H and b_L. (b) Sequence comparison of Cyt c_1 from the photosynthetic bacteria. The boxed sequence (CxxCH) covalently attaches the heme. (c) Sequence comparison of Rieske FeS protein from the photosynthetic bacteria. The boxed sequences contain the histidines and cysteines which are thought to coordinate the $[Fe_2S_2]$ cluster. R. c. : *Rb. capsulatus*; R. s. : *Rb. sphaeroides*; R. v. : *Rp. viridis*; R. r. : *Rs. rubrum*, *C. l.* : *Cb. limicola*. Only the complete sequence is shown from *Rb. capsulatus* and identities are indicated with dashes in the other sequences while non-identities are given in lower case letters. Dots correspond to gaps between different species.

```
         1                                                 41
R.c.     ....MKKLLI SAVSALVLGS GAAFANSNVP D.....HAFS FEGIFGKYDQ
R.s.     ....-irk-t ltaatalal- -gaamaaggg .hvedvp.-- ---p--tf--
R.v.     .mtiklrfva -lalvfg-aa asvp-qasgg -tp.hlqsw- -a-p--q--k
R.r.     mttiv-ralv a-gmvlai-g a...-qa-eg gvslhkqdw- wk----r--q

         42                                               90
R.c.     AQLRRGFQVY NEVCSACHGM KFVPIRTLAD DGGPQLDPTF VREYAAGLD.
R.s.     h--q--l--- t---aa---- -------s-se p---e-pedq --a--tqftv
R.v.     ---------f qn--vs---l enggf-n-ps raa-nwplde --ql--swpv
R.r.     p--q-----f h----t---- ---ay-n-.. .salgfsedg ik-l--ekef
                       *   **

         91                                                127
R.c.     ...TIIDKDS GEERDRKETD MFPTRV.... ......GDGM GPDLSVMAKA
R.s.     .....t-eet --d-eg-p-- h--hsa.... .......len aa---l----
R.v.     qvkd-n--gd pmq-ap-lp- ri-sqyanea aariihngav p-----i---
R.r.     pa.gpd-ngd mft-pgtpa- hi-spfandk aaaaan-gaa p----ll---

         128                                               175
R.c.     RAGFSGPAGS GMNQLFK.GM GGPEYIYNYV IGF.EENPEC APEGIDG.YY
R.s.     ----h--m-t -is---n.-i -------svl t--p--p-k- -eghepdgf-
R.v.     -tfqr-fpww vtdiftqyne n-vd--vall n-y.....-d p--......r
R.r.     -p........ .......... ---n---sll e-yasds-ge pa-......w

         176                                               224
R.c.     YNKTFQIGGV PDTCKDAAGV KITHGSWA.R MPPPLVDDQV TYEDGTPATV
R.s.     --ra--n-s- -------n-- -t-a---i.a -----m--l- e-a--hd-s-
R.v.     f......... ..kvp-gsfy nkyfpghiig -t--ia-gl- --g----e-q
R.r.     wv-qqqek-l evafne-kyf ndyfpghais -----m--li ------a--k

         225                                               274
R.c.     DQMAQDVSAF LMWAAEPKLV ARKQMGLVAM VMLGLLSVML YLTNKRLWAP
R.s.     ha--e----- ---------m ----a-ft-v mf-tv---l- ---------g
R.v.     l-ysk--a-- -------t-d v--ri-wwvl gf-viftgl- va-kivv-r-
R.r.     -------v-y -n-----e-d ---sl--kvl lf--v-ta-- lalklai-rd

         275
R.c.     YKGHKA
R.s.     v-gk-ktnv
R.v.     v-kgl-
R.r.     v-h...
```

Fig. 4b. Continued

Furthermore, a large number of spontaneous Q_o and Q_i site inhibitor resistant mutants have been mapped to Cyt *b*, and the observed nature and distribution of the sites of these mutations were inconsistent with the nine transmembrane model (Fig. 7a). These findings led several groups to propose a newer model of Cyt *b* with eight transmembrane helices (A to H) by pulling the old helix 4 out of the membrane and reversing the orientation of the following five helices (D to H). The distribution of charged residues of Cyt *b* (inside more positive than the outside, von Heijne et al., 1989) also supports the latter model. There is another large loop on the periplasmic side of the membrane (between the helices E and F) and one large cytoplasmic loop (between helices D and E) which may also contain amphipathic domains.

Indirect experimental proof confirming the topology of the Cyt *b* polypeptide has also been obtained using alkaline phosphatase (*phoA*) fusions in *Rb. sphaeroides* (Yun et al., 1991b). This frequently used technique has been shown to be reliable in predicting the topology of several *E. coli* membrane proteins (Manoil and Beckwith, 1986), and also to correctly predict the overall topology of the *Rb. sphaeroides* reaction center subunits L and M (Yun et al., 1991b) for which a high resolution structure is available. The data on Cyt *b* clearly indicate that the inside/outside orientation of its first six helices is consistent with the eight helix model predicted from the distribution of inhibitor resistant mutations. However, there is still little information pertinent to the last two helices G and H of Cyt *b* in bacteria. The recent determination of the primary structure of Cyt *b* from *Cb. limicola* indicates that the last helix (H)

```
                                                                1    6
R.c.    .......... .......... .......... .......... ....MSHAED
R.s.    .......... .......... .......... .......... ....--n---
R.r.    .......... .......... .......... .......... .....ae-eh
R.v.    miisifnqlh ltensslmas ftlssatpsq lcssknqmfa pslalak-gr
C.l.    .......... .......... .......... .......... maqtgnfksp

        7                                              42
R.c.    NAGT...... R.RDFLYHAT AATGVVVTGA AVWPLINQMN ASA.......
R.s.    h---...... -.-----y-- -ga-a-a--- ---------- p--.......
R.r.    t-s-pggess -.----iyg- t-v-a-gval ----f-df-- pa-.......
R.v.    vnvliskeri -gmkltcqa- sipadn-pdm qkretl-lll lg-lslptgy
C.l.    arms...... slgqgaap-s sgavtggkpr egglkgvdfe rrg......f

        43                                                92
R.c.    DVKAMASIFV DVSAVEVGTQ LTVKWRGKPV FIRRRDEKDI ELARSVPLGA
R.s.    --q-l----- ---s--p-v- ----fl---i -----t-a-- --g---q--q
R.r.    -tl-l--te- ----iae-qa i--t------ -v-h-tq-e- vv--a-dpas
R.v.    mllpy--f-- ppgg.ga--g g-ia-da... .......... .-gnd-iaae
C.l.    lh-ivggvga v-avstlypv vkyiippar. ....kiknvd --tvgkasev

        93                                               142
R.c.    LRDTSAENAN KPGAEATDEN RTLPAFDGTN TGEWLVMLGV CTHLGCVPMG
R.s.    -v--n-r--- idagaeatdq nrtlldeage -..----w-- --------i-
R.r.    ---pqtde-r vqq-...... .......... ..q----v-- ------i-l-
R.v.    wlk-h-p.gd rtltqglkgd p-ylvvesdk -latfgina- -------vpf
C.l.    pdgk-kif.. .....qfn-d kvivvnkgga lt...avsa- ------lvnw
                                                   * *  *

        143                                              188
R.c.    DKS....GDF GGWFCPCHGS HYDSAGRIRK GPAPRNLDIP VAAFVDETTI
R.s.    gv-....--- ---------- ---------- ----e--p-- l-k-i-----
R.r.    q-agdpk--- dg-------- ---------- ----l--pv- py--t-d--v
R.v.    naae...... nkfi------ q-nnq--vvr ----ls-ala hcdvd-gkvv
C.l.    v.....dad nqy------A k-klt-eiis --q-lp-kqy k-riegdsii
                       * **

        189
R.c.    KLG....... .........
R.s.    ql-....... .........
R.r.    li-....... .........
R.v.    fvpwtetdfr tgeapwwsa
C.l.    iska...... .........
```

Fig. 4c. Continued

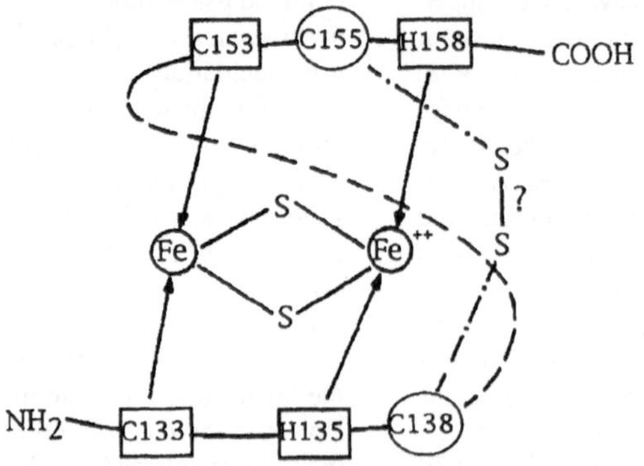

Fig. 5. Model of the C-terminal $[Fe_2S_2]$ binding region of the 'Rieske' FeS protein subunit of the Cyt bc_1 complex. Adapted from Davidson et al., 1992a. The boxed residues correspond to the putative ligands of the metal cluster and the circled residues may form a disulfide bridge.

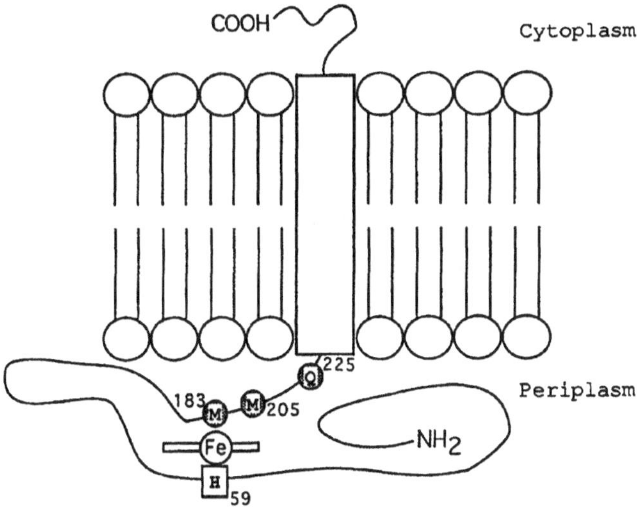

Fig. 6. Folding model of the Cyt c_1 polypeptide showing the putative single transmembrane α-helix and the sites of site specific mutations constructed in both *Rb. capsulatus* and *Rb. sphaeroides*. Histidine 59 shown in the box is the fifth axial ligand to the heme Fe, and the circled residue are positions that have been altered by mutagenesis.

may not be present in that polypeptide (Schütz et al., 1993), and it has even been proposed that this helix, in those cases where it exists, may not be entirely transmembranous (Cramer and Trebst, 1991).

It should be noted that although several speculative models for specific portions of each of the subunits have been successfully developed in recent years no crystallography-based direct structural data are yet available for any of them with the exception of the hydrophilic portion of Cyt *f* of turnip chloroplast Cyt b_6f complex (Martinez et al., 1992, 1994). Unfortunately, the poor homology between Cyt *f* and Cyt c_1 may limit the direct application of this important information to the analogous subunits of bacterial and mitochondrial complexes.

II. Mutational Approaches

A combination of molecular genetics to obtain desired mutants designed to test specific hypotheses followed by in-depth biochemical and biophysical analyses is a powerful experimental approach to obtain structural and functional information about membrane-associated complexes. This approach is being used extensively in both *Rb. capsulatus* and *Rb. sphaeroides* to address a number of issues related to the Cyt bc_1 complex including the steps of electron transfer and proton translocation processes, the physicochemical properties and the assembly of the subunits, the active sites of the complex and the validity of current folding models.

A. Spontaneous Mutants

The isolation and molecular genetic characterization of spontaneous mutants has been invaluable to further our knowledge on how the Cyt bc_1 complex functions. R126 was one of the earliest *Rb. capsulatus* mutants known to affect electron transfer between Cyt *b* and Cyt c_2 (Zannoni and Marrs, 1981). A detailed biochemical analysis by Robertson et al. (1986) indicated that this mutant was kinetically impaired at the Q_o site of the Cyt bc_1 complex while its Q_i site function was retained. The mutation in R126 was later determined to be a single amino acid substitution, G158D, located in the transversal helix *cd* of the Cyt *b* polypeptide (Daldal et al., 1989) (Fig. 7).

Another set of spontaneous mutants that has been highly informative for the location of the Q_o site are *Rb. capsulatus* mutants resistant to the Q_o site inhibitors stigmatellin, myxothiazol and mucidin. Several of these have been isolated, and the genetic lesions conferring inhibitor resistance have been determined (Daldal et al., 1989). Similar Cyt *b* mutations have also been obtained in yeast (di Rago

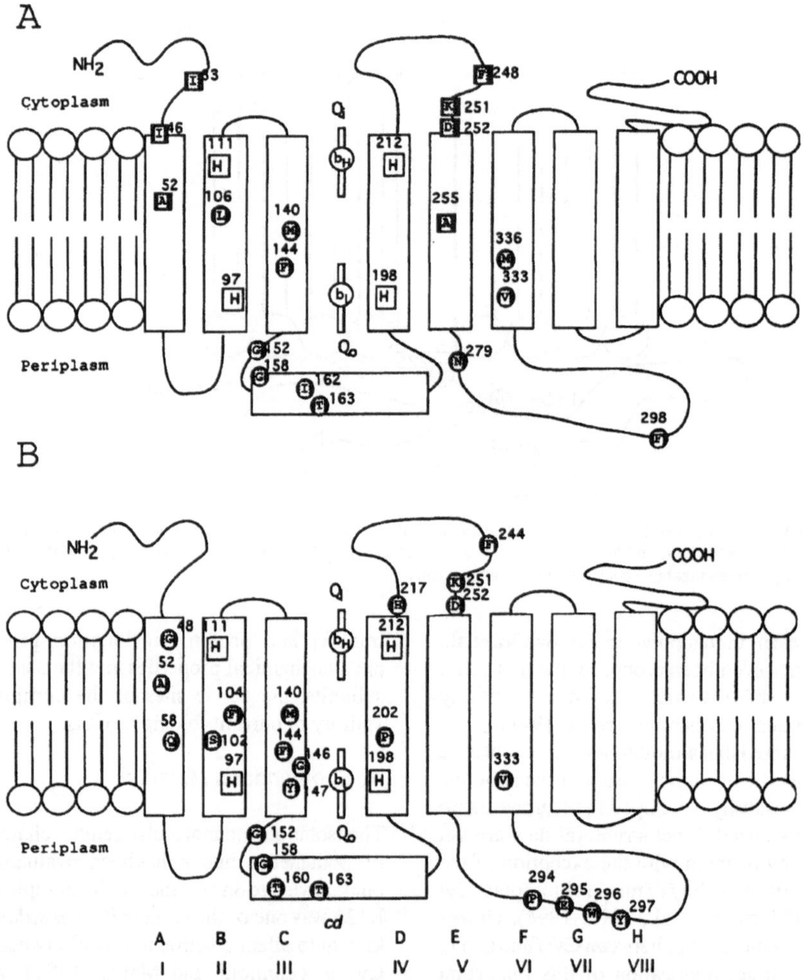

Fig. 7. (a) Eight transmembrane helix folding model of the Cyt *b* polypeptide from *Rhodobacter capsulatus*. Helices have been designated both by letter (A-H) and number (I - VIII). The four histidines (H97, H111, H198 and H212) which provide the axial ligands to Cyt b_L and b_H are in empty boxes. The filled-in boxes are the locations of spontaneous Q_i site inhibitor resistant loci from yeast and mouse mitochondria and *Rs. rubrum*. The filled-in circles are the locations of spontaneous Q_o site inhibitor resistant loci from *Rb. capsulatus* and yeast and mouse mitochondria. (b) Same as (a) except the highlighted positions are the locations of site specific mutations constructed in both *Rb. capsulatus* and *Rb. sphaeroides*.

et al., 1989) and mouse (Howell and Gilbert, 1988) mitochondria. The great majority of these mutations have been mapped to the Cyt *b* polypeptide and are distributed nonrandomly (Fig. 7a). They are clustered in two distinct domains referred to as Q_oI and Q_oII. The Q_oI region is a very small segment of the Cyt *b* polypeptide, spanning the residues 140 to 163 (in the numbering of *Rb. capsulatus*) containing part of helix C and most of the amphipathic transversal helix *cd* while the Q_oII region is located towards the C-terminal end of Cyt *b*. In addition, one substitution (L106P) was found located on helix B between the histidines that coordinate heme b_H and b_L. Robertson et al. (1990) observed that not only was the affinity for inhibitor decreased in these mutants but there was also a weakening of the binding of the physiological substrate, ubiquinol, at the Q_o site suggesting that the inhibitor and substrate interaction sites share common structural features. The study of the locations and the nature of the inhibitor resistant mutants has led to a speculative model for the Cyt *b* portion of the Q_o site based on the three dimensional structure of the Q_a and Q_b sites of the reaction center (Robertson et al., 1990). In addition, this region of Cyt *b* polypeptide has now been shown to bind ubiquinone by photoaffinity labeling using an azido-Q derivative (He et al., 1994).

Interestingly, not all Cyt bc_1 inhibitors are active in the photosynthetic bacteria. Since photosynthetic growth of these bacteria is absolutely dependent on the presence of a functional Cyt bc_1 complex (Daldal et al., 1987) it could be expected that most electron transfer inhibitors also inhibit growth. While this is true for the Q_o site inhibitors it is not the case in *Rb. capsulatus* and *Rb. sphaeroides* for the Q_i site inhibitors. This natural resistance to Q_i site inhibitors in vivo precluded the isolation of spontaneous Q_i inhibitor resistant mutants in *Rb. capsulatus* and *Rb. sphaeroides*. However it has been shown recently that the photosynthetic growth of the purple non-sulfur bacterium *Rs. rubrum* is sensitive to antimycin (Shankar et al., 1992; Uhrig et al., 1994). Antimycin-resistant mutants were isolated using this latter species and lesions responsible for the resistance were identified as the substitutions D252E and D252H (Park and Daldal, 1992; Uhrig et al., 1994). D252 is located near the beginning of helix E in close proximity to mutations conferring resistance to antimycin and diuron in yeast and mouse mitochondrial Cyt bc_1 complex (di Rago et al., 1986; Howell et al., 1987; di Rago and Colson, 1988; Weber and Wolf, 1988; di Rago et al., 1990). The many Q_i mutations available in mitochondrial Cyt *b* are also clustered in two distinct regions, which could be called by analogy Q_iII, consisting of the loop extending into the cytoplasm connecting helix D and E, and Q_iI consisting of a stretch close to the amino terminus of the protein also extending into the cytoplasm (Fig. 7a).

B. Site-Directed Mutants

The site directed mutagenesis systems that are developed in both *Rb. capsulatus* and *Rb. sphaeroides* have two components. The first is a deletion background in which the chromosomal copy of the wild type structural genes has been eliminated using either interposon mutagenesis as in *Rb. capsulatus* (Atta-Asafo-Adjei and Daldal, 1991), or non-replicating suicide plasmids as in *Rb. sphaeroides* (Yun et al., 1990). The use of a deletion background avoids unwanted homologous recombination events that otherwise may yield mixtures of wild type and mutant complexes and complicate the biochemical/biophysical analyses.

The second component of the genetic system is a broad host range plasmid into which the wild type structural genes have been cloned. The mutations are often generated by in vitro mutagenesis techniques and exchanged with their wild type counterparts on this plasmid. Any desired mutation can be efficiently constructed using these systems, and the growing lists of mutants of *Rb. capsulatus* (Tables 1, 2 and 3) and *Rb. sphaeroides* (Gennis et al., 1993) are rather impressive. Other phototrophic species are also amenable to similar approaches, and an example is *Rp. viridis* for which a similar system to study the reaction center has recently been described (Laussermaier et al., 1992). Similar deletion and conjugation procedures also work well for the Cyt bc_1 complex of *Rs. rubrum* (S. Y. Park and F. Daldal, unpublished results) for which genetic systems are also developed to study the reaction center (Hessner et al., 1991), CO dehydrogenase (Kerby et al., 1992) and ribulose bisphosphate carboxylase (Hartman et al., 1987).

1. Cytochrome b Mutants

Figure 7b and Table 1 show the locations and effects of site directed mutations that are currently available in the Cyt *b* polypeptide, and are grouped into different classes as described below.

a. Q_o site

The majority of the initial work on the Q_o site has been geared towards defining the regions of Cyt *b* affecting UQH_2 oxidation. Clues to the approximate location of the site have come from earlier analyses of spontaneous inhibitor resistant mutants (see above) and the non-photosynthetic mutant of *Rb. capsulatus* R126. One position chosen for detailed study was G158. This site yields a Cyt bc_1 complex defective in the Q_o site when substituted with aspartate in bacteria (Robertson et al., 1986), or a myxothiazol resistant complex when substituted with alanine in mouse mitochondria (Howell et al., 1988). To define the role

Table 1. Selected Cytochrome *b* Mutants of *Rhodobacter capsulatus*[1]

Residue	Growth[2]	Rev[3]	Inh[4]	Assembly[5]	$QH_2 \rightarrow$ Cyt c[6]	Biophysical/biochemical Properties
Wild type	Ps^+	na	My^S, St^S	$(FBC)^+$	100.0	Fully functional
Spontaneous mutants						
FbcB : L106P	Ps^+	na	My^R, St^S	$(FBC)^+$	nd	No detected Q/QH_2 binding change by EPR g_x signal of FeS; little effect on first QH_2 oxidized; full turnover impaired.
FbcB : M140I	Ps^+	na	My^R, St^R	$(FBC)^+$	nd	No detected Q/QH_2 binding change by EPR g_x signal; no effect on QH_2 oxidation.
FbcB : F144S	Ps^+	na	My^R, St^S	$(FBC)^+$	nd	Q/QH_2 binding weakened; QH_2 oxidation slowed.
FbcB : F144L	Ps^+	na	My^R, St^R	$(FBC)^+$	nd	Q/QH_2 binding weakened; QH_2 oxidation slowed drastically.
FbcB : G152S	Ps^+	na	My^R, St^R	$(FBC)^+$	nd	No detected Q/QH_2 binding change by EPR g_x signal of FeS; no or little slowing in QH_2 oxidation.
FbcB : G158D	Ps^-	10^{-7}	na	$(FBC)^+$	nd	Q/QH_2 binding weakened; no QH_2 oxidation detected at a fast time scale.
FbcB : T163A	Ps^+	na	My^S, St^R	$(FBC)^+$	nd	No detected Q/QH_2 binding change by EPR g_x signal; no effect on QH_2 oxidation.
FbcB : V333A	Ps^+	na	My^S, St^R	$(FBC)^+$	nd	No detected Q/QH_2 binding change by EPR g_x signal; no effect on QH_2 oxidation.
Site directed non functional mutants						
FbcB : M140R[8]	Ps^-	10^{-7}	na	$(FBC)^+$	< 5	Q_o site fully occupied based on the EPR g_x signal for FeS, but no or litte QH_2 oxidation
FbcB : F144R[8]	Ps^-	10^{-7}	na	$(FBC)^+$	< 5	Q_o site occupied, a novel EPR g_x signal for FeS seen; no or litte QH_2 oxidation.
FbcB : G152P[8]	Ps^-	10^{-7}	na	$(FBC)^+$	< 5	No detected Q/QH_2 binding change by EPR g_x signal of FeS; no or little QH_2 oxidation.
FbcB : G158X[7]	Ps^-	10^{-7}	na	$(FBC)^+$	< 10	Q_o site occupancy is defective; No or little QH_2 oxidation.
FbcB : T163F, P[8]	Ps^-	10^{-4}–10^{-5}	na	$(FBC)^-$	0	No subunit of the complex detected in the membrane under steady state growth.

[1] Data from Daldal et al., 1989; Robertson et al., 1990; Ding et al., 1992; Tokito & Daldal, 1993 and Ding et al., unpublished.
[2] Ps^+ and Ps^- correspond to the photosynthetic growth ability of the mutants on rich medium.
[3] Rev indicate the approximate reversion frequency of the mutants to Ps^+ phenotype; na, not applicable.
[4] Inh, inhibitor resistance pattern; S and R correspond to Ps^+ growth ability of the mutants in the presence of My (myzothiazol, 5×10^{-6} M) and St (stigmatellin, 4×10^{-6} M).
[5] Steady state presence (+) or absence (–) of the FeS protein (F), Cyt *b* (B) and Cyt c_1 (C) in chromatophore membranes probed immunologically.
[6] DBH: Cyt *c* reductase activity of chromatophore membranes indicated as a % of that of a wild type strain overproducing Cyt bc_1 complex (typically between 3,800 to 5,400 nmol of Cyt c reduced/min/mg of protein), nd, not done.
[7] See Atta-Asafo-Adjei and Daldal (1991) for a list of the G158x substitutions.
[8] See Tokito and Daldal (1993) for other substitutions in these positions.

of position 158 Atta-Asafo-Adjei and Daldal (1991) substituted it with fifteen different amino acid residues, and observed that all but the A and S substitutions had lower Cyt *c* reductase activities and failed to support photosynthetic growth. While a glycine in this position allows free access of both ubiquinol and myxothiazol into the Q_o site, substitutions with side chains like $-CH_3$ (A) or $-CH_2OH$ (S) hinders the access of myxothiazol and confers inhibitor resistance. Based on these facts it was proposed that side chains larger than methyl and hydroxymethyl at this position exclude the substrate and lead to nonfunctional mutants. More recent data obtained by EPR spectroscopy, using the $[Fe_2S_2]$ cluster g_x signal to monitor the occupancy of the Q_o site in these mutants, further confirmed this finding (Ding et al., 1992; Ding et al., unpublished).

Other positions in Cyt *b* conferring resistance to Q_o specific inhibitors were also mutagenized extensively (Tokito and Daldal, 1993). These include M140, F144, G152, T163 and V333. All, except for V333, are located in the Q_oI region either in transmembrane helix C or in the *cd* amphipathic helix. In contrast to G158 mutants, most of these mutations yielded photosynthetically competent strains with, however, perturbed rates of UQH_2 oxidation and modified inhibitor recognition. At position 140 the presence of a methyl group on the β carbon atom correlated with myxothiazol and stigmatellin resistance, while position 144 required an aromatic residue to yield a myxothiazol-sensitive Cyt bc_1 complex with a high turnover rate (Tokito and Daldal, 1993). An extension of this work has now shown that mutations at positions 158 and 144 perturb the binding of substrate to both the Q_{ow} and Q_{os} domains (Ding et al., unpublished), and the study of these mutants has led to a correlation between the occupancy of the Q_o site (as determined from EPR of the $[Fe_2S_2]$ cluster) and the rate of turnover of the site. On the other hand positions 163 and 333 do not appear to define strong interaction points with myxothiazol though they affect stigmatellin resistance.

Furthermore some mutations at T163 affect the assembly of the complex and in particular the T163F and P mutants are devoid of the three subunits of the Cyt bc_1 complex (Tokito and Daldal, 1993). It was proposed that T163 provides a contact point between the subunits based on the fact that suppressors of T163F were found in the other subunits of the complex (see below). This finding was used to orient the transversal helix *cd* in such a way that the face containing T163 is turned away from the protein interior, thus positioning G158 toward the protein interior of Cyt *b* as proposed earlier (Robertson et al., 1990). Crofts et al. (1992) have also predicted a similar orientation for helix *cd* in a computer generated model of the topology and folding of Cyt *b*. Thus the steric restraints on the residues occupying positions 152 and 158 can be understood since the side chains would project into the UQH_2 binding pocket. More recent work on positions Y147 and T160 are also consistent with this orientation (S. Saribas and F. Daldal, unpublished results).

The loop connecting helix E and F contains the highly conserved -PEWY- span (Hauska et al., 1988). Mutations at these sites in *Rb. sphaeroides* cause a decrease in the rate of UQH_2 oxidation at the Q_o site (Crofts et al., 1992; Table 1). For example the substitution E295Q decreased the rate of Q_o turnover by a factor of approximately 50, but did not prevent photosynthetic growth under laboratory conditions. Considering that the midpoint potential of Cyt b_L is more positive in most of the -PEWY- mutants and that some substitutions also confer resistance to Q_o specific inhibitors, it was proposed that this portion of Cyt *b* is also involved in the interaction with UQH_2 at the Q_o site.

b. Ligands of the Cyt b *hemes*

Mutagenesis of Cyt *b* in *Rb. capsulatus* and *Rb. sphaeroides* has also involved the residues H97, H111, H198 and H212 (Fig. 7), proposed to be the ligands of Cyt b_H and b_L hemes (Widger et al., 1984; Saraste et al., 1984). These residues were altered in *Rb. sphaeroides* to amino acids unable to coordinate metal ions, and it was shown that all are absolutely essential for photosynthetic growth (Yun et al., 1991a). The substitutions of H111 and H212 led to the selective loss of Cyt b_H while those of H97 and H198 resulted in the loss of both hemes. Similar data were also obtained for H212 mutants in *Rb. capsulatus* (F. Daldal, unpublished data). These findings indicated that H212 and H111 are ligands to Cyt b_H and H97 and H198 are ligands to Cyt b_L. This assignment is totally consistent with the eight transmembrane helices model of Cyt *b*, and indicated that the Q_o site stays fully functional even when Cyt b_H is selectively destroyed (Yun et al., 1991a). The opposite of this finding has been seen with R126 where the G158D mutation selectively inactivates the Q_o site while leaving the Q_i site intact (Robertson et al., 1986). These experiments thus provide clear support for the remarkable functional independence of the quinone interaction sites.

c. Q_i site

Several highly conserved residues thought to be in the vicinity of the Q_i site have been altered (Yun et al., 1992; Crofts et al., 1992; Hacker et al., 1993; Gray et al., 1994). These positions were chosen based on the known location of spontaneous inhibitor (antimycin, diuron and funiculosin) resistant mutations in mitochondrial Cyt *b*, and on their universal conservation in Cyt *b* polypeptides from many phylogenetically different species (Degli-Esposti et al., 1993). The high resolution structure of the quinone binding sites from the *Rb. sphaeroides* photosynthetic reaction center also offers clues to deciphering analogous interactions that may occur between the quinone moiety and the Cyt *b* polypeptide (Allen et al., 1988). In this respect, the Q_b site of the reaction center appears to be a good model for the Q_i site since both sites catalyze similar redox chemistry. Of particular interest is the finding that in the reaction center H190 and S223 of the L-subunit form hydrogen bonds with the quinone carbonyl groups at the Q_b site.

The main effect of the mutations in this region of Cyt *b* is an inhibition of re-oxidation of Cyt b_H following flash activation. This kinetic defect accounts for the observation that these mutants are impaired to various degrees for their photosynthetic growth (from no growth to slow growth). Further, some of these mutations also affect the amount of Cyt b_{150} which is postulated to be a high potential form of Cyt b_H due to a redox interaction between the heme and the quinone bound at the Q_i site (de la Rosa and Palmer, 1983; Salerno et al., 1989; Rich et al., 1990). The increased amount of Cyt b_{150} in those mutants suggested that the interaction of the protein with the quinone bound at the site has been perturbed. Redox titrations of the antimycin sensitive semiquinone have shown that mutations at H217 either highly destabilized (L and D substitutions) or over-stabilized (R substitution) the semiquinone with essentially the same phenotypic effect (Gray et al., 1994). Recent ENDOR data have suggested that the semiquinone at Q_i interacts with the protein via hydrogen bonds, and that the protons of these bonds are exchangeable with the medium (Salerno et al., 1990). Further, these authors suggest that these interactions could be strong enough to account for the stabilization of the semiquinone, thus modulating the thermodynamic properties of the bound quinone. It was therefore proposed that H217 may, in a manner analogous to H190 of the L-subunit of the reaction center, form a hydrogen bond with the quinone at the Q_i site (Gray et al., 1994). Obviously, a better picture of the quinone binding pocket as well as a detailed kinetic model of quinone catalysis at the Q_i site will emerge as more work is performed in this region of Cyt *b*.

d. Other Regions of Cyt b

Some of the other highly conserved Cyt *b* residues have also been altered with the hope of discovering their basis of conservation (Yun et al., 1992). However none of the positions tested thus far appears to be absolutely essential for a functional Cyt bc_1 complex. For example, F104 is located in helix B between the two sets of axial ligands to the two hemes and is often modeled to project between them. It is conserved in most Cyt *b* sequences, with the exception of Cyt b_6 from chloroplasts, and Cyt *b* from some protozoans. It had been suggested that the rate of electron transfer between Cyt b_L and b_H may be enhanced by the presence of this aromatic residue (Widger et al., 1984). However, Yun et al. (1992) replaced F104 with I and observed minimal effects on the rates of electron transfer within the Cyt bc_1 complex, suggesting that F104 does not function in the rate-determining step of electron transfer.

Another example is P202, which is the only conserved proline within a transmembrane helix (D) in the Cyt *b* polypeptide. The presence of a proline would be expected to change the hydrogen bonding patterns in the helix and may participate in proton translocation via a *cis-trans* isomerization (Brandl and Deber, 1986). However replacement of P202 with the helix-favoring amino acid leucine had only minor effects on the properties of the Cyt bc_1 complex (Yun et al., 1992). These examples suggest that conservation of a particular amino acid alone is not sufficient to indicate that it has a specialized role for the function of the Cyt bc_1 complex.

2. Rieske Iron Sulfur Protein

On the basis of proteolytic cleavage, the C-terminal

region of the Rieske protein has been implicated in the binding of its [Fe_2S_2] cluster (Li et al., 1981b). This region of the protein contains two groups of highly conserved residues, called box I (CTHLGC) and box II (CPCHGS). In *Rb. capsulatus* these correspond to residues C133 to C138 and C153 to S158 (Fig. 4c). Each box contains two cysteines and one histidine, residues thought to coordinate the cluster. However, considering that the [Fe_2S_2] cluster requires only two N and two S ligands (Gurbiel et al., 1991; Britt et al., 1991) the number of conserved potential ligands exceeds the number needed. Each potential ligand in box I and box II (as well as a non-conserved H159 next to box II and the G137 in box I) have been altered by site directed mutagenesis (Table 2; Davidson et al., 1992a; van Doren et al., 1993a). Mutations of H159 did not affect the properties of the cluster or the assembly of the complex, and it was concluded that H159 is not a ligand (Davidson et al., 1992a). Substitution of G133 in *Rb. sphaeroides* (G137 in *Rb. capsulatus*) to aspartate yielded a photosynthetically competent strain, but interestingly, with a decreased turnover rate of the Q_o site (van Doren et al., 1993a). On the other hand mutations of either H135 or H156 in *Rb. capsulatus* (and the corresponding residues in *Rb. sphaeroides*) resulted in photosynthetically incompetent strains, and led to the loss of the cluster and degradation of the mutant FeS subunit. It was concluded from these data that the imidazole rings of

Table 2. Rieske FeS Protein Mutants of *Rhodobacter capsulatus*[1]

Residue	Growth[2]	Rev[3]	Inh[4]	Assembly[5]	$QH_2 \rightarrow$ Cyt c[6]	Biophysical/biochemical Properties
Wild type	Ps^+	na	My^S, St^S	$(FBC)^+$	100.0	Fully functional
FbcF : C133R	Ps^-	10^{-5}	na	$F^-(BC)^+$	0.0	Like H135L
FbcF : C133S	Ps^-	10^{-9}	na	$F^-(BC)^+$	0.2	Like H135L
FbcF : H135P	Ps^-	nd	na	$F^-(BC)^+$	0.1	Like H135L
FbcF : H135L	Ps^-	10^{-8}	na	$F^-(BC)^+$	nd	No [2Fe2S] cluster; Q_o site defective; spectra and E_m of b_L, b_H and c_1 unchanged; $Q_i^{\cdot-}$ present; Q_i site functional
FbcF : C138R	Ps^-	3×10^{-8}	na	$F^-(BC)^+$	nd	Like H135L
FbcF : C138S	Ps^-	6×10^{-6}	na	$F^-(BC)^+$	nd	Like H135L, except tiny amount of [2Fe2S] cluster with $g_x = 1.89$ and lower E_m
FbcF : C138F	Ps^-	nd	na	$F^-(BC)^+$	nd	Like H135L
FbcF : C153R	Ps^-	2×10^{-8}	na	$F^-(BC)^+$	nd	Like H135L
FbcF : C153S	Ps^-	nd	na	$F^-(BC)^+$	0.0	Like H135L
FbcF : C155D	Ps^-	10^{-10}	na	$F^-(BC)^+$	nd	Like H135L
FbcF : C155G	Ps^-	nd	na	$F^-(BC)^+$	nd	Like H135L
FbcF : C155S	Ps^-	6×10^{-7}	na	$F^-(BC)^+$	0.0	Like H135L, except tiny amount of [2Fe2S] cluster with g_x=1.90 and E_m=160mV and does not respond to redox state of Q_{pool} and to St.
FbcF : H156L	Ps^-	nd	na	$F^-(BC)^+$	0.2	Like H135L
FbcF : H156P	Ps^-	3×10^{-6}	na	$F^-(BC)^+$	0.0	Like H135L
FbcF : H156F	Ps^-	10^{-10}	na	$F^-(BC)^+$	0.0	Like H135L
FbcF : H156T	Ps^-	10^{-10}	na	$F^-(BC)^+$	0.0	Like H135L
FbcF : H156Y	Ps^-	10^{-6}	na	$F^-(BC)^+$	0.0	Like H135L
FbcF : H159A	Ps^+	na	My^S, St^S	$(FBC)^+$	100.0	wild type-like, E_m of FeS=317 mV; responds to redox state of Q_{pool} and to St.
FbcF : H159S	Ps^+	na	My^S, St^S	$(FBC)^+$	100.0	wild type-like with E_m of FeS=287 mV.

[1] data from Davidson et al., 1992a,b.
[2,3,4,5 and 6] See Table 1 footnotes.

H135 of box I and H156 of box II provide the nitrogenous ligands to the cluster (Fig. 5).

Of the four cysteines mutagenized only C138S (Ohnishi et al., 1993) and C155S (Davidson et al., 1992a) contained a very small amount of a perturbed $[Fe_2S_2]$ cluster detectable by EPR, suggesting that these residues are not ligands. Substitutions of the remaining two cysteines (C133 and C153) resulted in a loss of the cluster and very small amounts of detectable apoprotein (Table 2). Sequence comparisons of the *Rb. capsulatus* FeS protein with 'Rieske-type' $[Fe_2S_2]$ cluster-containing dioxygenase proteins also suggested that the non-conserved C138 and C155 residues are not the ligands to the $[Fe_2S_2]$ cluster (Davidson et al., 1992a; van Doren et al., 1993a), thus leaving C133 and C153 as possible ligands. Considering that substitutions at C155 and C138 highly perturbed the assembly of the complex it was further suggested that they may form a disulfide bond which is essential for the structure of the wild type protein. The overall data has led to a folding model of the binding pocket of the $[Fe_2S_2]$ cluster in *Rb. capsulatus* (Fig. 5) (Davidson et al., 1992a). On the other hand, it should be noted that other residues were also proposed as ligands in other systems (Graham and Trumpower, 1991), and the exact nature of the ligands of the $[Fe_2S_2]$ cluster has not yet been unequivocally established.

3. Cytochrome c$_1$ *Mutants*

Mutagenesis of Cyt c_1 has been geared towards two main goals. One was the identification of the amino acid residue that provides the sixth ligand to the heme Fe. Data from magnetic circular dichroism of Cyt c_1 from mitochondria are consistent with histidine-methionine ligation (Simpkin et al., 1989). Further, sequence alignments of Cyt c_1 from different species reveal the presence of two highly conserved methionines (M183 and M205 in the numbering of the mature *Rb. capsulatus* protein) near the C-terminal portion of the protein (Fig. 6). Both M183 and M205 have been altered to non-liganding aliphatic residues (M183L and M205V) in Cyt c_1 from *Rb. capsulatus* (Table 3; Gray et al., 1992). The substitution M205V caused no growth defect, and had very minor effects on the enzymatic activity of the Cyt bc_1 complex whereas the substitution M183L had quite drastic effects. The strain containing this mutation was Ps^- while the entire complex was still present in the membrane. Redox titrations of purified complex indicated that the E_{m7} value of the mutant Cyt c_1 was decreased by approximately 390 mV (to –70 mV) as compared to wild type ($E_{m7} \approx 315$ mV). Interestingly while the isolated Cyt c_1 could bind CO its absorption spectrum indicated that it was still at least partially low spin. This suggests that if M183 is indeed the ligand in the wild type protein yet another residue that can be displaced by CO could replace it in the ligation sphere. These results suggest that M183 is more likely than M205 to be the sixth ligand of the heme Fe, and a similar analysis in yeast (Nakai et al., 1990) has also come to the same conclusion. However, this assignment needs to be confirmed by direct structural analyses, especially in light of the recent structure of Cyt *f* which indicated that the N-terminal amino group of this protein is the sixth ligand of its heme Fe (Martinez et al., 1992, 1994) .

A second focus of mutagenesis in Cyt c_1 has been to isolate a truncated, non-membrane anchored form which may be more readily crystallized (Table 3). For this purpose, Konishi et al (1991) engineered a stop codon at position 228 (Q) in *Rb. sphaeroides* Cyt c_1 right before the putative transmembrane anchor,

Table 3. Cytochrome c_1 Mutants of *Rhodobacter capsulatus*[1]

Residue	Growth[2]	Rev[3]	Inh[4]	Assembly[5]	$QH_2 \rightarrow$ Cyt c[6]	Biophysical/biochemical Properties
Wild type	Ps^+	na	My^S, St^S	$(FBC)^+$	100.0	Fully functional
FbcC : 183L	Ps^-	10^{-10}	na	$(FBC)^+$	0.3	$E_m = -74$ mV, 390 mV lower than wt; low spin heme but binds CO
FbcC : 205V	Ps^+	na	nd	$(FBC)^+$	83	wild type-like
FbcC : 205***[7]	Ps^-	10^{-6}	na	$(FBC)^-$	0.0	No truncated Cyt c_1; No Cyt bc_1 complex in the membrane
FbcC : del32–159		Ps^-	na	na	$(FBC)^-$	0.0 Like FbcC : 205***.

[1] data from Gray et al., 1992.

[2,3,4,5 and 6] See Table 1 footnotes.

[7] *** indicates the presence of a stop codon at position 205.

and were able to detect traces of a soluble form of Cyt c_1 which could still donate electrons to Cyt c_2. In this mutant, a small amount of a crippled Cyt bc_1 complex with relatively unperturbed $[Fe_2S_2]$ and Cyt b_H was also detected, suggesting that the absence of Cyt c_1 does not entirely destroy the essential architecture of these subunits. On the other hand a similar mutation in *Rb. capsulatus* Cyt c_1 as well as an internal deletion eliminating its heme binding site (from residue 32 to 159) has not led to the production of any truncated derivative of Cyt c_1 (Gray et al., 1992). This finding suggests that both the signal sequence and bound heme appear necessary for effective translocation of Cyt c_1 across the membrane and its insertion into the lipid bilayer.

4. Assembly/Subunit Interaction Mutants

a. Partial Cyt bc$_1$ Complexes

During the characterization of FeS protein mutants it was observed that the cytochromes *b* and c_1 were fully present in the intracytoplasmic membranes of *Rb. capsulatus* mutants containing little, or no FeS apoprotein (Davidson et al., 1992a). Further characterization of these mutants have shown that a stable subcomplex with a functional Q_i site could be formed in the absence of a Q_0 site (Davidson et al., 1992b). To further examine the stabilizing interactions among the three subunits of the Cyt bc_1 complex in the membrane, additional strains which contained deletions of the *fbcFBC* genes in various combinations were obtained. These mutants contained both Cyt *b* and c_1 in the absence of the FeS protein in the membrane, consistent with the point mutations affecting box I and box II of the latter subunit. However in the absence of Cyt c_1 very little of the remaining subunits were detected, leading to the suggestion that Cyt c_1 and Cyt *b* must somehow interact and protect each other from degradation (Davidson et al., 1992b).

Recently Van Doren et al. (1993b) have succeeded in expressing low amounts of the FeS protein in *E. coli* and *Rb. sphaeroides* in the absence of the other two subunits. When the subunit was expressed alone in either *Rb. sphaeroides* or *E. coli* it was located in the cytoplasmic membrane and contained a Rieske-like cluster but with aberrant EPR properties. Further, stigmatellin did not affect the EPR spectrum of the cluster in the absence of the other two subunits. Finally, truncation of the putative transmembrane anchor by limited proteolysis resulted in a water soluble form of the FeS protein retaining the $[Fe_2S_2]$ cluster. These studies suggested that the Rieske protein is anchored to the membrane via a single transmembrane helix which may interact with helices of the other subunits.

b. Suppressors of a Cyt b Mutant

During a systematic analysis of the role on inhibitor recognition and quinol catalysis of the previously defined Q_0-inhibitor positions, two mutations in the Q_0I region of Cyt *b*, T163F and T163P, have been found to affect drastically the overall assembly and in vivo stability of the Cyt bc_1 complex (Tokito and Daldal, 1993). Chromatophore membranes of the strains carrying these mutations were devoid of any of the three subunits of the complex. Further, the Ps^- phenotype of these mutants was rather unstable, and reverted to a Ps^+ phenotype very frequently. The basis of the severe effect of these mutations on the assembly and in vivo stability of the Cyt bc_1 complex is not clear. However, considering that the 'Rieske-less' mutants of *Rb. capsulatus* can still form a two subunit Cyt bc_1 subcomplex the data suggest that these mutations must severely disrupt subunit interactions around the Q_0 site. In fact analyses of the Ps^+ revertants of the T163F mutation indicated that there are at least three genetically different classes of suppressors overcoming its deleterious assembly effect (Tokito and Daldal, in preparation). While one class affects another Cyt *b* residue close to T163 (G182T mutation) two other classes are located in Cyt c_1 (R46C) and in the FeS protein (A46T), close to their respective N-terminal regions. If the effect of these suppressor mutations is direct then these findings suggest that the amino terminal domains of these two latter subunits must be close to the Q_0I region of Cyt *b* and contribute to the formation of the Q_0 site. The isolation and analysis of more mutations with similar effects are needed to further define the subunit interactions in the Cyt bc_1 complex.

III. Conclusions and Perspectives

The Cyt bc_1 complex is a paradigm for redox-driven proton translocation and the combination of multidisciplinary approaches including molecular genetic, biochemical and biophysical studies have led in recent years to substantial progress in our understanding of the structure, function, genetic organization and physiological importance of this enzyme. In particular, our knowledge has been greatly increased by the ever growing set of structural genes from many different organisms. This database now allows meaningful structural and functional comparisons and also points to the possible steps during the evolution of this complex. The primary sequences of the subunits have led to more accurate secondary structural models and to the identification of the ligands for the metal cofactors that have been probed by mutagenesis. Currently while the ligands to Cyt b_L and b_H are known with some confidence strong contenders for the sixth axial ligand of Cyt c_1 and the possible ligands for the $[Fe_2S_2]$ cluster of the FeS protein have also been found. The tentative identification of these ligands now raises several other crucial questions. For example, how the protein environment modulates the redox and spectroscopic properties of these metal clusters.

The extensive work using both spontaneous and site directed inhibitor resistant mutants has been very informative in globally defining the regions involved in the interaction with UQH_2 at the Q_o site as well as in establishing the topology of the transmembrane helices of Cyt *b*. A better picture of the architecture of the Q_i site and the role of the Cyt *b* residues stabilizing the semiquinone is also beginning to emerge. Working models for the Q_o and Q_i sites are being developed using analogies with the quinone binding sites of the photosynthetic reaction center. Yet, the roles of amino acid residues located in the vicinity of these sites need to be analyzed in depth to define their contributions to the different steps of the catalytic cycle. Although there are a number of mutants now available it should be noted that only 27 positions out of 437 of Cyt *b* have been probed by mutagenesis with often only one or just a few substitutions at each position. Further, several other areas of Cyt *b* still need to be explored, in particular very little information is available about the last two transmembrane helices G and H. To date, mutagenesis has been limited to single site changes, and undoubtedly, future experiments could use a judicious combination of multiple, and perhaps large scale, substitutions similar to those currently being carried out in other systems. Finally, the characterization of second site revertants of Ps^- mutants has only been initiated very recently, and has already been extremely informative about the crucial subunit interactions in the complex. Until now while most of the work has been focused on the Cyt *b* subunit, comparatively little effort has been devoted to the Rieske FeS protein and Cyt c_1.

Currently our knowledge about the biogenesis and assembly of the Cyt bc_1 complex is rudimentary at best. In the bacterial case, while we now realize the importance of the in vivo stabilizing interactions between some subunits we are still in the dark about the temporal order of assembly of the subunits, their translocation across the cytoplasmic membrane and the incorporation of their prosthetic groups during biogenesis and maturation. Another important direction for future work could be the issue of unifying the different Cyt bc_1/Cyt b_6f variants of the oxidoreductases encountered in nature. While there are some distinct differences between these complexes, they catalyze the same overall charge separation processes across the lipid bilayer. Systematic attempts to convert a Cyt bc_1 structure to a Cyt b_6f structure by stepwise molecular engineering, and to analyze the implications of these cumulative mutations would be extremely informative, and attempts in this direction are already underway. Without question these future studies will be greatly enhanced with the availability of a high resolution structure for a Cyt bc_1 complex, and recently several groups have reported promising advances in this direction (Kubota et al., 1991; Yue et al., 1991; Berry et al., 1992; Yu and Yu, 1993; Kawamoto et al., 1994).

While there are so many unanswered questions related to the structure and function of the Cyt bc_1 complex, the well developed molecular genetic and biochemical/biophysical tools available in photosynthetic bacteria will certainly further our understanding of how a Cyt bc_1 complex works as a redox-driven proton pump. Further, the Cyt bc_1 complex will also continue to contribute as a model system to issues with broader implications, such as the investigation of the general phenomenon of membrane protein targeting and insertion, and the coordinated biogenesis and assembly of the multisubunit membrane complexes.

Acknowledgments

We thank our many colleagues who have graciously allowed us to refer to their unpublished work in this review and Dr. Dan Robertson for critical reading of the manuscript. Work in this laboratory is supported by NIH grant GM 38237.

References

Allen JP, Feher G, Yeates TO, Komiya H and Rees DC (1988) Structure of the reaction center from *Rhodobacter sphaeroides* R-26: protein-cofactor (quinones and Fe^{2+}) interactions. Proc Natl Acad Sci USA 85: 8487–8491

Andrews KM, Crofts AR and Gennis RB (1990) Large-scale purification and characterization of a highly active four subunit cytochrome bc_1 complex from *Rhodobacter sphaeroides*. Biochemistry 29: 2645–2651

Atta-Asafo-Adjei E and Daldal F (1991) Size of the amino acid side chain at position 158 of cytochrome *b* is critical for an active cytochrome bc_1 complex and for photosynthetic growth of *Rhodobacter capsulatus*. Proc Natl Acad Sci USA 88: 492–496

Berry EA, Huang LS, Earnest TN and Jap BK (1992) X-ray diffraction by crystals of beef heart ubiquinol:cytochrome *c* oxidoreductase. J Mol Biol 224: 1161–1166

Bosshard HR, Wynn RM and Knaff DB (1987) Binding site on *Rhodospirillum rubrum* cytochrome c_2 for the *Rhodospirillum rubrum* cytochrome bc_1 complex. Biochemistry 26: 7688–7693

Brand SN, Tan X and Widger WR (1992) Cloning and sequencing of the *petBD* operon from the cyanobacterium *Synechococcus* sp. PCC 7002. Plant Molec Biol 20: 481–491

Brandl CJ and Deber CM (1986). Hypothesis about the function of membrane-buried proline residues in transport protein. Proc Natl Acad Sci USA 83: 917–921

Brandt U and Trumpower BL (1994) The protonmotive Q cycle in mitochondria and bacteria. Crit Rev Biochem Molec Biol 29: 165–197

Brasseur R (1988) Calculation of the three-dimensional structure of *Saccharomyces cerevisiae* cytochrome *b* inserted in a lipid matrix. J Biol Chem 263: 12571–12575

Britt RD, Sauer K, Klein MP, Knaff DB, Kriauciunas A, Yu C-A, Yu L and Malkin R (1991) Electron spin echo envelope modulation spectroscopy supports the suggested coordination of two histidine ligands to the Rieske Fe-S centers of the cytochrome b_6f complex of spinach and the cytochrome bc_1 complexes of *Rhodospirillum rubrum, Rhodobacter sphaeroides* R-26, and bovine heart. Biochemistry 30: 1892–1901

Caffrey M, Davidson E, Cusanovich M and Daldal F (1992) Cytochrome c_2 mutants of *Rhodobacter capsulatus*. Arch Biochem Biophys 292: 419–426

Carter K, Tsai A and Palmer G (1981) The coordination environment of mitochondrial cytochromes *b*. FEBS Lett 132: 243–246

Chen YR, Usui S, Yu CA and Yu L (1994) Role of subunit IV in the cytochrome bc_1 complex from *Rhodobacter sphaeroides*. Biochemistry 33: 10207–10214

Cocco T, Lorusso M, Sardanelli AM, Minuto M, Ronichi S and Tedeschi G (1991) Structural and functional characteristics of polypeptide subunits of the bovine heart ubiquinol-cytochrome-*c* reductase complex. Eur J Biochem 195: 731–734

Cramer WA and Knaff DB (1990) Energy Transduction in Biological Membranes. Springer-Verlag, New York

Cramer WA and Trebst A (1991) Membrane protein structure prediction: cytochrome *b*. Trends Biochem Sci 16: 207

Cramer WA, Black MT, Widger WR and Girvon ME (1987) In: Barber J (ed) The Light Reactions, pp 447–493. Elsevier Science Publishers, Amsterdam

Crofts AR and Wraight CA (1983) The electrochemical domain of photosynthesis. Biochim Biophys Acta 726: 149–185

Crofts AR, Meinhardt SW, Jones KR and Snozzi M (1983) The role of the quinone pool in the cyclic electron transfer chain of *Rhodopseudomonas sphaeroides*. A modified Q-cycle mechanism. Biochim Biophys Acta 723: 202–218

Crofts AR, Robinson HH, Andrews K, van Doren S and Berry E (1987) Catalytic sites for reduction and oxidation of quinone. In: Papa S, Chance B and Ernster L (eds) Cytochrome Systems: Molecular Biology and Bioenergetics, pp 617–214. Plenum, New York

Crofts AR, Hacker B, Barquera B, Yun C-H and Gennis R (1992) Structure and function of the *bc*-complex of *Rhodobacter sphaeroides*. Biochim Biophys Acta 1101: 162–165

Daldal F, Davidson E and Cheng S (1987) Isolation of the Structural Genes for the Rieske Fe-S Protein, Cytochrome *b* and cytochrome c_1, all components of the ubiquinol:cytochrome c_2 oxidoreductase complex of *Rhodopseudomonas capsulata*. J Molec Biol 195: 1–12

Daldal F, Tokito M, Davidson E and Faham M (1989) Mutations conferring resistance to quinol oxidation (Q_z) inhibitors of the Cyt bc_1 complex of *Rhodobacter capsulatus*. EMBO J 8: 3951–3961

Davidson E and Daldal F (1987a) Primary structure of the bc_1 complex of *Rhodopseudomonas capsulata*. Nucleotide sequence of the *pet* operon encoding the Rieske, cytochrome *b*, and cytochrome c_1 apoproteins. J Mol Biol 195: 13–24

Davidson E and Daldal F (1987b) *fbc* Operon, encoding the Rieske Fe-S protein, cytochrome *b*, and cytochrome c_1 apoproteins previously described from *Rhodopseudomonas sphaeroides*, is from *Rhodopseudomonas capsulata*. J Mol Biol 195: 25–29

Davidson E, Ohnishi T, Atta-Asafo-Adjei E and Daldal F (1992a) Potential Ligands to the [2Fe-2S] Cluster of the cytochrome bc_1 complex of *Rhodobacter capsulatus* probed by site-directed mutagenesis. Biochemistry 31: 3342–3351

Davidson E, Ohnishi T, Tokito M and Daldal F (1992b) *Rhodobacter capsulatus* mutants lacking the Rieske FeS protein form a stable cytochrome bc_1 subcomplex with an intact quinone reduction site. Biochemistry 31: 3351–3358

Davis DJ, Frame MK and Johnson DA (1988) Resonance Raman spectroscopy indicates a lysine as the sixth iron ligand in cytochrome *f*. Biochim Biophys Acta 936: 61–66

de la Rosa FF and Palmer G (1983) Reductive titration of CoQ-depleted Complex III from Baker's yeast. Evidence for an exchange-coupled complex between QH· and low-spin ferricytochrome *b*. FEBS Lett 163: 140–143

Degli Esposti M, De Vries S, Crimi M, Ghelli A, Patarnello T and Meyer A (1993) Mitochondrial cytochrome *b*: relation between natural variations in its primary structure and its properties. Biochim. Biophys Acta 1143: 243–271

Deisenhofer J, Epp O, Miki K, Huber R and Michel H (1985) Structure of the protein subunits in the photosynthetic reaction centre of *Rhodopseudomonas viridis* at 3Å resolution. Nature 318: 618–622

di Rago J-P and Colson AM (1988) Molecular basis for resistance to antimycin and diuron, Q-Cycle inhibitors acting at the Q_i site in the mitochondrial ubiquinol-cytochrome *c* reductase in *Saccharomyces cerevisiae*. J Biol Chem 263: 12564–12570

di Rago J-P, Perea J and Colson AM (1986) DNA Sequence analysis of diuron-resistant mutations in the mitochondrial cytochrome *b* gene of *Saccharomyces cerevisiae.* FEBS Lett 208: 208–210

di Rago J-P, Coppée J-Y and Colson AM (1989) Molecular basis for resistance to myxothiazol, mucidin (strobilurin A), and stigmatellin. Cytochrome *b* inhibitors acting at the center o of the mitochondrial ubiquinol-cytochrome *c* reductase in *Saccharamyces cerevisiae.* J Biol Chem 264: 14543–14548

di Rago JP, Perea J and Colson AM (1990) Isolation and RNA sequence analysis of cytochrome *b* mutants resistant to funiculosin, a center i inhibitor of the mitochondrial ubiquinol-cytochrome *c* reductase in *Saccharomyces cerevisiae.* FEBS Lett 263: 93–98

Ding H, Robertson DE, Daldal F and Dutton PL (1992) Cytochrome bc_1 complex [2Fe-2S] cluster and its interaction with ubiquinone and ubihydroquinone at the Q_o site: A double-occupancy Q_o site model. Biochemistry 31: 3144–3158

Dutton PL (1986) Energy transduction in anoxygenic photosynthesis. In: Staehelin LA and Arntzen CJ (eds) Encyclopedia of Plant Physiology, Photosynthesis III, Vol 19, pp 197–237. Springer-Verlag, Berlin

Gabellini N and Sebald W (1986) Nucleotide sequence and transcription of the *fbc* operon from *Rhodopseudomonas sphaeroides.* Eur J Biochem 154: 569–579

Gabellini N, Harnisch U, McCarthy JED, Hauska G and Sebald W (1985) Cloning and expression of the *fbc* operon encoding the FeS protein, cytochrome *b* and cytochrome c_1 from the *Rhodopseudomonas sphaeroides* b/c_1 complex. EMBO J 4: 549–533

Gennis RB, Barquera B, Hacker B, Van Doren SR, Arnaud S, Crofts AR, Davidson E, Gray KA and Daldal F (1993) The bc_1 complexes of *Rhodobacter sphaeroides* and *Rhodobacter capsulatus.* J Bioener Biomem 25: 195–210

Glaser E and Crofts AR (1984) A new electrogenic step in the ubiquinol: cytochrome c_2 oxidoreductase complex of *Rhodopseudomonas sphaeroides.* Biochim Biophys Acta 766: 322–333

González-Halphen D, Lindorfer MA and Capaldi RA (1988) Subunit arrangement in beef heart complex III. Biochemistry 27: 7021–7031

González-Halphen D, Vázquez-Acevedo M and Garcia-Ponce B (1991) On the interaction of mitochondrial complex III with the Rieske iron-sulfur protein (Subunit V). J Biol Chem 266: 3870–3876

Graham LA and Trumpower BL (1991) Mutational analysis of the mitochondrial Rieske iron-sulfur protein of *Saccharomyces cerevisiae.* J Biol Chem 266: 22485–22492

Gray KA, Davidson E and Daldal F (1992) Mutagenesis of methionine-183 drastically affects the physicochemical properties of cytochrome c_1 of the bc_1 complex of *Rhodobacter capsulatus.* Biochemistry 31: 11864–11873

Gray KA, Dutton PL and Daldal F (1994) The requirement of histidine 217 for ubiquinone reductase activity (Q_i Site) in the cytochrome bc_1 complex. Biochemistry 33: 723–733

Gray KA, Grooms M, Myllykallio H, Moomaw C, Slaughter C and Daldal F (1994) *Rhodobacter capsulatus* contains a novel *cb*-type cytochrome *c* oxidase without a Cu_A center. Biochemistry 33: 3120–3127

Güner S, Robertson DE, Yu L, Qiu ZH, Yu CA and Knaff DB (1991) The *Rhodospirillum rubrum* cytochrome bc_1 complex: redox properties, inhibitor sensitivity and proton pumping. Biochim Biophys Acta 1058: 269–279

Güner S, Willie A, Millett F, Caffrey MS, Cusanovich MA, Robertson DE and Knaff DB (1993) The interaction between cytochrome c_2 and the cytochrome bc_1 complex in the photosynthetic purple bacteria *Rhodobacter capsulatus* and *Rhodopseudomonas viridis.* Biochemistry 32: 4793–4800

Gurbiel RJ, Ohnishi T, Robertson DE, Daldal F and Hoffman BM (1991) Q-band ENDOR spectra of the Rieske protein from *Rhodobacter capsulatus* ubiquinol-cytochrome *c* oxido-reductase show two histidines coordinated to the [2Fe-2S] cluster. Biochemistry 30: 11579–11584

Hacker B, Barquera B, Crofts AR and Gennis RB (1993) Characterization of mutations in the cytochrome *b* subunit of the bc_1 complex of *Rhodobacter sphaeroides* that affect the quinone reductase site (Q_c). Biochemistry 32: 4403–4410

Hall J, Ayres M, Zha X, O'Bien P, Durham B and Knaff D (1987a) The reaction of cytochrome *c* and c_2 with the *Rhodospirillum rubrum* reaction center involves the heme crevice domain. J Biol Chem 262: 11046–11051

Hall J, Kriauciunas A, Knaff D and Millett F (1987b) The reaction domain on *Rhodospirillum rubrum* cytochrome c_2 and horse cytochrome *c* for the *Rhodospirillum rubrum* cytochrome bc_1 complex. J Biol Chem 262: 14005–14009

Hall J, Zha X, Yu L, Yu C-A and Millett F (1987c) The binding domain on horse cytochrome *c* and *Rhodobacter sphaeroides* cytochrome bc_1 complex. Biochemistry 26: 4501–4504

Hall J, Zha X, Yu L, Yu C-A and Millett F (1989) Role of specific lysine residues in the reaction of *Rhodobacter sphaeroides* cytochrome c_2 with the cytochrome bc_1 complex. Biochemistry 28: 2568–2571

Harnisch U, Weiss H and Sebald W (1985) The primary structure of the iron-sulfur subunit of ubiquinol-cytochrome *c* reductase from *Neurospora,* determined by cDNA and gene expression sequencing. Eur J Biochem 149: 95–99

Hartman FC, Soper TS, Niyogi SK, Mural RJ, Foote RS, Mitra S, Lee EH, Machanoff R and Larimer FW (1987) Function of Lys-166 of *Rhodospirillum rubrum* ribulose bisphosphate carboxylase/oxygenase as examined by site-directed mutagenesis. J Biol Chem 262: 3496–3501

Hauska G (1986) Composition and structure of cytochrome bc_1 and b_6f complexes. In: Staehelin AL and Arntzen CJ (eds) Encyclopedia of Plant Physiology, Photosynthesis III, Vol 19, pp 496–507. Springer-Verlag, Berlin

Hauska G, Nitschke W and Herrmann RG (1988) Amino acid identities in the three redox center-carrying polypeptides of cytochrome bc_1/b_6f Complexes. J Bioenerg Biomem 20: 211–228

He D-Y, Yu L and Yu CA (1994) Ubiquinone binding domains in bovine heart mitochondrial cytochrome *b.* J Biol Chem 269: 2292–2298

Hessner MJ, Wejksnora PJ, Perille Collins ML (1991) Construction, characterization and complementation of *Rhodospirillum rubrum puf* region mutants. J Bacteriol 173: 5712–5722

Hobbs DD, Kriauciunas A, Güner S, Knaff DB and Ondrias MR (1990) Resonance Raman spectroscopy of cytochrome bc_1 complexes from *Rhodospirillum rubrum*: Initial characterization and reductive titrations. Biochim Biophys Acta 1018: 47–54

Howell N and Gilbert K (1988) Mutational analysis of the mouse mitochondrial cytochrome *b* gene. J Mol Biol 203: 607–618

Howell N and Robertson DE (1993) Electrochemical and spectral analysis of the long range interactions between the Q_o and Q_i sites and the heme prosthetic groups in ubiquinol-cytochrome c oxidoreductase. Biochemistry 32: 11162–11172

Howell N, Appel J, Cook JP, Howell B and Hausworth WW (1987) The molecular basis of inhibitor resistance in a mammalian mitochondrial cytochrome *b* mutant. J Biol Chem 262: 2411–2414

Jenney FE and Daldal F (1993) A novel membrane-associated *c*-type cytochrome, Cyt c_y, can mediate the photosynthetic growth of *Rhodobacter capsulatus* and *Rhodobacter sphaeroides*. EMBO J 12: 1283–1292

Jenney FE, Prince RC and Daldal F (1994) Roles of the soluble cytochrome c_2 and membrane-associated cytochrome c_y of *Rhodobacter capsulatus* in photosynthetic electron transfer. Biochemistry 33: 2496–2502

Kallas T, Spiller S and Malkin R (1988a) Characterization of two operons encoding the cytochrome b_6-f complex of the cyanobacterium *Nostoc* PCC 7906. J Biol Chem 263: 14334–14342

Kallas T, Spiller S and Malkin R (1988b) Primary structure of cotranscribed genes encoding the Rieske Fe-S and cytochrome *f* proteins of the cyanobacterium *Nostoc* PCC 7906. Proc Natl Acad Sci USA 85: 5794–5798

Kawamoto M, Kubota T, Matsuraga T, Fukuyama K, Matsubara H, Shinzawaitoh K and Yoshikawa S (1994) New crystal forms and preliminary X-ray diffraction studies of mitochondrial cytochrome bc_1 complex from bovine heart. J Mol Biol 244: 238–241

Kerby RL, Hong SS, Ensign SA, Coppoc LJ, Ludden PW and Roberts GP (1992) Genetic and physiological characterization of the *Rhodospirillum rubrum* carbon monoxide dehydrogenase system. J Bacteriol 174: 5284–5294

Knaff DB (1993) The cytochrome bc_1 complexes of photosynthetic bacteria. Photosyn Res 35: 117–133

Knaff DB and Malkin R (1976) Iron sulfur proteins of the green photosynthetic bacterium *Chlorobium*. Biochim Biophys Acta 430: 244–252

Konishi K, Van Doren SR, Kramer DM, Crofts AR and Gennis RB (1991) Preparation and characterization of the water-soluble heme-binding domain of cytochrome c_1 from the *Rhodobacter sphaeroides* bc_1 complex. J Biol Chem 266: 14270–14276

Kriauciunas A, Yu L, Yu C-A, Wynn RM and Knaff DB (1989) The *Rhodospirillum rubrum* cytochrome bc_1 complex: Peptide composition, prosthetic group content and quinone binding. Biochim Biophys Acta 976: 70–76

Kubota T, Kawamoto M, Fukuyama K, Shizawa-Itoh K, Yoshikawa S and Matsubara H (1991) Crystallization and preliminary X-ray crystallographic studies of bovine heart mitochondrial cytochrome bc_1 complex. J Mol Biol 221: 379–382

Kurowski B and Ludwig B (1987) The genes of the *Paracoccus denitrificans* bc_1 complex. J Biol Chem 262: 13805–13811

Kutoh E and Sone N (1988) Quinol-cytochrome *c* oxidoreductase from the thermophilic bacterium PS3. J Biol Chem 263: 9020–9026

LaMonica R and Marrs BL (1976) The branched respiratory system of photosynthetically grown *Rhodopseudomonas capsulata*. Biochim Biophys Acta 423: 431–439

Laussermair E and Oesterhelt D (1992) A system for site-specific mutagenesis of the photosynthetic reaction center in *Rhodopseudomonas viridis*. EMBO J 11: 777–783

Leguijt T, Engels PW, Crielaard W, Albracht SPJ and Hellingwerf KJ (1993) Abundance, subunit composition, redox properties, and catalytic activity of the cytochrome bc_1 complex from the alkalophilic and halophilic, photosynthetic members of the family *Ectothiorhodospiraceae*. J Bacteriol 175: 1629–1636

Li Y, De Vries S, Leonard K and Weiss H (1981b) Topography of the iron-sulfur subunit in mitochondrial ubiquinol-cytochrome *c* reductase. FEBS Lett 135: 277–280

Li Y, Leonard K and Weiss H (1981a) Membrane-bound and water soluble cytochrome c_1 from *Neurospora* mitochondria. Eur J Biochem 116: 199–295

Liebl U, Rutherford AW and Nitschke W (1990) Evidence for a unique Rieske iron-sulphur centre in *Heliobacterium chlorum*. FEBS Lett 261: 427–430

Liebl U, Pezennec S, Riedel A, Kellner E and Nitschke W (1992) The Rieske FeS center from the Gram-positive bacterium PS3 and its interaction with the menaquinone pool studied by EPR. J Biol Chem 267: 14068–14072

Ljungdahl PO, Pennoyer JD, Robertson DE and Trumpower BL (1987) Purification of highly active cytochrome bc_1 complexes from phylogenetically diverse species by a single chromatographic procedure. Biochim Biophys Acta 891: 227–241

Majewski C and Trebst A (1990) The *pet* genes of *Rhodospirillum rubrum*: cloning and sequencing of the genes for the cytochrome bc_1-complex. Molec Gen Gen 224: 373–382

Malkin R (1992) Cytochrome bc_1 and b_6f complexes of photosynthetic membranes. Photosynth Res 33: 121–136

Manoil C and Beckwith J (1986) A genetic approach to analyzing membrane protein topology. Science 233: 1403–1408

Margoliash E and Bosshard HR (1983) Guided by electrostatics, a textbook protein comes of age. Trends Biochem Sci 8: 316–320

Marrs B and Gest H (1973) Genetic mutations affecting the respiratory electron-transport system of the photosynthetic bacterium *Rhodopseudomonas capsulata*. J Bacteriol 114: 1045–1051

Martinez SE, Smith JL, Huang D, Szczepaniak A and Cramer WA (1992) Crystallographic studies of the lumen side domain of turnip cytochrome *f*. In: Murata N (ed) Research in Photosynthesis, Vol II, pp 495–498. Kluwer Academic Press, Dordrecht

Martinez SE, Huang D, Szczepaniak, Cramer, WA and Smith JL (1994) Crystal structure of chloroplast cytochrome *f* reveals a novel cytochrome fold and unexpected heme ligation. Sturcutre 2: 95–105

Matsuura K, Bowyer JR, Ohnishi T and Dutton PL (1983a) Inhibition of electron transfer by 3-alkyl-2hydroxy-1,4-naphthoquinone in the ubiquinol-cytochrome *c* oxidoreductases of *Rhodopseudomonas sphaeroides* and mammalian mitochondria. J Biol Chem 258: 1571–1579

Matsuura K, O'Keefe DP and Dutton PL (1983b) A re-evaluation of the events leading to the electrogenic reaction and proton translocation in ubiquinol-cytochrome *c* oxidoreductase of *Rhodopseudomonas sphaeroides*. Biochim Biophys Acta 722: 12–22

McCurley JP, Miki T, Yu L and Yu C-A (1990) EPR characterization of the cytochrome bc_1 complex from *Rhodobacter sphaeroides*. Biochim Biophys Acta 1020: 176–186

Meinhardt SW and Ohnishi T (1992) Determination of the

position of the $Q_i^{\cdot-}$ quinone binding site from the protein surface of the cytochrome bc_1 complex in *Rhodobacter capsulatus* chromatophores. Biochim Biophys Acta 1100: 67–74

Miki T, Yu L and Yu C-A (1991) Hematoporphyrin-promoted photoinactivation of mitochondrial ubiquinol-cytochrome *c* reductase: selective destruction of the histidine ligands of the iron-sulfur cluster and protective effect of ubiquinone. Biochemistry 30: 230–238

Mitchell P (1966) Chemiosmotic coupling in oxidative and photosynthetic phosphorylation. Biol Rev 41: 445–502

Mitchell P (1976) Possible molecular mechanisms of the protonmotive function of cytochrome systems. J Theor Biol 62: 327–367

Nakai M, Ishiwatari H, Asada A, Bogaki M, Kawai K, Tanaka Y and Matsubara H (1990) Replacement of putative axial ligands of heme iron in yeast cytochrome c_1 by site-directed mutagenesis. J Biochem (Tokyo) 108: 798–803

Ohnishi T and Trumpower BL (1980) Differential effects of antimycin on ubisemiquinone bound in different environments in isolated succinate-cytochrome reductase complex. J Biol Chem 255: 3278–3284

Ohnishi T, Sled V, Meinhardt SW, Yagi T, Hatefi Y, Saribas S and Daldal F (1994) Studies of the topographical distribution of the redox centers and the 'Q_p' site in ubiquinol-cytochrome c oxidoreductase (Complex III) and of the ligand structure of the Rieske iron-sulfur cluster. Biochemical Society Transactions 22: 191–197

Park SY and Daldal F (1992) The Qi site of cytochrome bc_1 complex of purple non-sulfur bacterium *Rhodospirillum rubrum*. In: Murata N (ed) Research in Photosynthesis, Vol II, pp 761–764. Kluwer Academic Publishers, Dordrecht

Powers L, Schägger H, von Jagow G, Smith J, Chance B and Ohnishi T (1989) EXAFS studies of the isolated bovine heart Rieske [2Fe-2S] 1+(1+, 2+) cluster. Biochim Biophys Acta 975: 293–298

Prince RC (1990) Bacterial photosynthesis: from photons to Δp. In: Krulwich T (ed) The Bacteria, Vol XII, pp 111–149. Academic Press, New York

Purvis D, Theiler R and Niederman RA (1990) Chromatographic and protein chemical analysis of the ubiquinol-cytochrome c_2 oxidoreductase isolated from *Rhodobacter sphaeroides*. J Biol Chem 265: 1208–1215

Rao JKM and Argos P (1986) A conformational preference parameter to predict helices in integral membrane proteins. Biochim Biophys Acta 869: 197–214

Rich PR, Jeal AE, Madgwick SA and Moody AJ (1990) Inhibitor effects on redox-linked protonations of the *b* heams of the mitochondrial bc_1 complex. Biochim Biophys Acta 1018: 29–40

Riedel A, Kellner E, Grodzitzki D, Liebl U, Hauska G, Müller A, Rutherford AW and Nitschke W (1993) The 2Fe2S centre of the cytochrome *bc* complex in *Bacillus firmus* OF4 in EPR: An example of a menaquinol oxidizing Rieske centre. Biochim Biophys Acta 1183: 263–268

Rigby SEJ, Moore GR, Gray JC, Gadsby PMA, George SJ and Thomson AJ (1988) N.M.R., e.p.r. and magnetic-c.d. studies of cytochrome *f*. Identity of the heam axial ligands. Biochem J 256: 571–577

Robertson DE and Dutton PL (1988) The nature and magnitude of the charge-separation reactions of ubiquinol cytochrome c_2 oxidoreductase. Biochim Biophys Acta 935: 273–291

Robertson DE, Prince RC, Bowyer JR, Matsuura K, Dutton PL and Ohnishi T (1984) Thermodynamic properties of the semiquinone and its binding site in the ubiquinol-cytochrome c (c_2) oxidoreductase of respiratory and photosynthetic systems. J Biol Chem 259: 1758–1763

Robertson DE, Davidson E, Prince RC, van der Berg WH, Marrs BL and Dutton PL (1986) Discrete catalytic sites for quinone in the ubiquinol-cytochrome c_2 oxidoreductase of *Rhodopseudomonas capsulata*. J Biol Chem 261: 584–591

Robertson DE, Daldal F and Dutton PL (1990) Mutants of ubiquinol-cytochrome c_2 oxidoreductase resistant to Q_o site inhibitors: Consequences for ubiquinone and ubiquinol affinity and catalysis. Biochemistry 29: 11249–11260

Robertson DE, Ding H, Chelminski PR, Slaughter C, Hsu J, Moomaw C, Tokito M, Daldal F and Dutton PL (1993) Hydroubiquinone-cytochrome c_2 oxidoreductase from *Rhodobacter capsulatus*: Definition of a minimal, functional isolated preparation. Biochemistry 32: 1310–1317

Salerno JC, McCurley JP, Dong J-H, Doyle MF, Yu L and Yu C-A (1986) The EPR spectra of the cytochrome bc_1 complex of *Rhodopseudomonas sphaeroides*. Biochem Biophys Res Comm 136: 616–621

Salerno JC, Xu Y, Osgood MP, Kim CH and King TE (1989) Thermodynamic and spectroscopic characteristics of the cytochrome bc_1 complex. J Biol Chem 264: 15398–15403

Salerno JC, Osgood M, Liu Y, Taylor H and Scholes CP (1990) Electron Nuclear Double Resonance (ENDOR) of the $Q_c^{\cdot-}$ ubisemiquinone radical in the mitochondrial electron transport chain. Biochemistry 29: 6987–6993

Saraste M (1984) Location of haem-binding sites in the mitochondrial cytochrome *b*. FEBS Lett 166: 367–372

Schägger H, Borchart U, Machleidt W, Link TA and von Jagow G (1987) Isolation and amino acid sequence of the 'Rieske' iron sulfur protein of beef heart ubiquinol:cytochrome c reductase. FEBS Lett 219: 161–168

Schütz M, Zirngibl S, le Coutre J, Büttner M, Xie D-L, Nelson N, Deutzmann R and Hauska G (1994) A transcription unit for the Rieske FeS-protein and cytochrome *b* in *Chlorobium limicola*. Photosynth Res 39: 163–174

Scolnik PA and Haselkorn R (1984) Activation of extra copies of genes coding for nitrogenase in *Rhodopseudomonas capsulata*. Nature 307: 289–292

Scolnik PA and Marrs BL (1987) Genetic research with photosynthetic bacteria. Ann Rev Microbiol 41: 703–726

Shankar S, Güner S, Daldal F, Tokito MK, Moomaw C, Knaff DB and Harman JG (1992) Characterization of the *pet* operon of *Rhodospirillum rubrum*. Photosyn Res 32: 79–94

Simpkin D, Palmer G, Devlin FJ, McKenna MC, Jensen GM and Stephens PJ (1989) The axial ligands of heme in cytochromes: A near-infrared magnetic circular dichroism study of yeast cytochromes *c*, c_1, and *b* and spinach cytochrome *f*. Biochemistry 28: 8033–8039

Stock JB, Ninfa, A and Stock AM (1989) Protein phosphorylation and regulation of adaptive responses in bacteria. Microbiol Rev 53: 450–490

Tan J, Corson GE, Chen YL, Garcia MC, Güner S and Knaff DB (1993) The ubiquinol:cytochrome c_2/c oxidoreductase of *Chromatium vinosum*. Biochim Biophys Acta 1144: 69–76

Telser J, Hoffman BM, LoBrutto R, Ohnishi T, Tsai A-L, Simpkin D and Palmer G (1987) Evidence for N coordination to Fe in the [2Fe-2S] center in yeast mitochondrial complex III. FEBS Lett 214: 117–121

Theiler R and Niederman RA (1991) Localization of chromatophore proteins of *Rhodobacter sphaeroides*. II. Topography of cytochrome c_1 and the Rieske iron-sulfur protein as determined by proteolytic digestion of the outer and luminal membrane surfaces. J Biol Chem 266: 23163–23168

Thöny-Meyer L, Stax D and Hennecke H (1989) An unusual gene cluster for the cytochrome bc_1 complex in *Bradyrhizobium japonicum* and its requirement for effective root nodule symbiosis. Cell 57: 683–697

Thöny-Meyer L, James P and Hennecke H (1991) From one gene to two proteins: The biogenesis of cytochromes *b* and c_1 in *Bradyrhizobium japonicum*. Proc Natl Acad Sci, USA 88: 5001–5005

Tokito MK and Daldal F (1992) *pet*R, located upstream of the *fbc*FBC operon encoding the cytochrome bc_1 complex, is homologous to bacterial response regulators and necessary for photosynthetic and respiratory growth of *Rhodobacter capsulatus*. Mol Microbiol 6: 1645–1654

Tokito MK and Daldal F (1993) Roles in inhibitor recognition and quinol oxidation of the amino acid side chains at positions of Cyt *b* providing resistance to Q_o-inhibitors of the Cyt bc_1 complex from *Rhodobacter capsulatus*. Mol Microbiol 9: 965–978

Trumpower BL (1990) Cytochrome bc_1 complexes of microorganisms. Microbiol Rev 54: 101–129

Tsai AH and Palmer G (1982) Purification and characterization of highly purified cytochrome *b* from complex III of Baker's yeast. Biochim Biophys Acta 681: 484–495

Uhrig J, Jakobs C, Majewski C and Trebst (1994) Molecular characterization of two spontaneous antimycin A resistant mutants of *Rhodospirillum rubrum*. Biochim Biophys Acta 1187: 347–353

Usui S, Yu L and Yu C-A (1990) The small molecular mass ubiquinone-binding protein (QPc-9.5 kDa) in mitochondrial ubiquinol-cytochrome *c* reductase: isolation, ubiquinone binding domain, and immunoinhibition. Biochemistry 29: 4618–4626

Usui S, Yu L, Harmon J and Yu CA (1991) Immunochemical study of subunit VI (M_r 13,400) of mitochondrial ubiquinol-cytochrome *c* reductase. Arch Biochem Biophys 289: 109–117

Van Doren SR, Gennis RB, Barquera B and Crofts AR (1993a) Site-directed mutations of conserved residues of the Rieske iron-sulfur subunit of the cytochrome bc_1 complex of *Rhodobacter sphaeroides* blocking or impairing quinol oxidation. Biochemistry 32: 8083–8091

Van Doren SR, Yun C-H, Crofts AR and Gennis RB (1993b) Assembly of the Rieske iron-sulfur subunit of the cytochrome bc_1 complex in the *Escherichia coli* and *Rhodobacter sphaeroides* membranes independent of the cytochrome *b* and c_1 subunits. Biochemistry 32: 628–636

Verbist J, Lang F, Gabellini N and Oesterhelt D (1989) Cloning and sequencing of the *fbcF, B* and *C* genes encoding the cytochrome b/c_1 complex from *Rhodopseudomonas viridis*. Molec Gen Genet 219: 445–452

von Heijne G (1989) Control of topology and mode of assembly of a polytopic membrane protein by positively charged residues. Nature 341: 456–458

von Jagow G and Ohnishi T (1985) The chromone inhibitor stigmatellin – binding to the ubiquinol oxidation center at the C-side of the mitochondrial membrane. FEBS Lett 185: 311–315

von Jagow G and Link TA (1986) Use of specific inhibitors on the mitochondrial bc_1 complex. Methods Enzymol. 126: 253–271

Weber S and Wolf K (1988) Two changes of the same nucleotide confer resistance to diuron and antimycin in the mitochondrial cytochrome *b* gene of *Schizosaccharomyces pombe*. FEBS Lett 237: 31–34

Widger WR (1991) The cloning and sequencing of *Synechococcus* sp. PCC 7002 *pet*CA operon: Implications for the cytochrome *c*-553 binding domain of cytochrome *f*. Photosyn Res 30: 71–84

Widger WR, Cramer WA, Hermann RG and Trebst A (1984) Sequence homology and structural similarity between cytochrome *b* of mitochondrial complex III and the chloroplast b_6-*f* complex: Position of the cytochrome *b* hemes in the membrane. Proc Natl Acad Sci USA 81: 674–678

Wikström MKF and Berden JA (1972) Oxidoreduction of cytochrome *b* in the presence of antimycin. Biochim Biophys Acta 283: 403–420

Wilson E, Farley TM and Takemoto JY (1985) Photoaffinity labeling of an antimycin-binding site in *Rhodopseudomonas sphaeroides*. J Biol Chem 260: 10288–10292

Wynn RM, Gaul DF, Choi W-K, Shaw RW and Knaff DB (1986) Isolation of cytochrome bc_1 complexes from the photosynthetic bacteria *Rhodopseudomonas viridis* and *Rhodospirillum rubrum*. Photosyn Res 9: 181–195

Yu C-A and Yu L (1993) Mitochondrial ubiquinol-cytochrome c reductase complex: Crystallization and protein:ubiquinone interaction. J Bioenerg Biomem 23: 259–273

Yu L and Yu C-A (1987) Identification of cytochrome *b* and a molecular weight 12K protein as the ubiquinone-binding proteins in the cytochrome bc_1 complex of a photosynthetic bacterium *Rhodobacter sphaeroides* R-26. Biochemistry 26: 3658–3664

Yu L and Yu C-A (1991) Essentiality of the molecular weight 15,000 protein (Subunit IV) in the cytochrome bc_1 complex of *Rhodobacter sphaeroides*. Biochemistry 30: 4934–4939

Yu L, Mei Q and Yu C (1984) Characterization of purified cytochrome bc_1 complex from *Rhodopseudomonas sphaeroides* R-26. J Biol Chem 259: 5752–5760

Yue W-H, Zou Y-P, Yu L and Yu C-A (1991) Crystallization of mitochondrial ubiquinol-cytochrome *c* reductase. Biochemistry 30: 2303–2306

Yun C-H, Beci R, Crofts AR, Kaplan S and Gennis RB (1990) Cloning and DNA sequencing of the *fbc* operon encoding the cytochrome bc_1 complex from *Rhodobacter sphaeroides*. Characterization of *fbc* deletion mutants and complementation by a site-specific mutational variant. Eur J Biochem 194: 399–411

Yun C-H, Crofts A R and Gennis RB (1991a) Assignment of the histidine axial ligands to the cytochrome b_H and cytochrome b_L components of the bc_1 complex from *Rhodobacter sphaeroides* by site-directed mutagenesis. Biochemistry 30: 6747–6754

Yun C-H, Van Doren SR, Crofts AR and Gennis RB (1991b) The use of gene fusions to examine the membrane topology of the L-subunit of the photosynthetic reaction center and of the cytochrome *b* subunit of the bc_1 complex from *Rhodobacter*

sphaeroides. J Biol Chem 266: 10967–10973
Yun C-H, Wang Z, Crofts AR and Gennis RB (1992) Examination of the functional roles of 5 highly conserved residues in the cytochrome *b* subunit of the bc_1 complex of *Rhodobacter sphaeroides*. J Biol Chem 267: 5901–5909
Zannoni D and Marrs BL (1981) Redox chain and energy transduction in chromatophores from *Rhodopseudomonas capsulata* cells grown anaerobically in the dark on glucose and dimethyl sulfoxide. Biochim Biophys Acta 637: 96–106
Zannoni D, Melandri BA and Baccarini-Melandri A (1976) Energy transduction in photosynthetic bacteria. X. Composition and function of the branched oxidase system in wild type and respiration deficient mutants of *Rhodopseudomonas capsulata*. Biochim Biophys Acta 423: 413–430

Chapter 36

Reaction Center Associated Cytochromes

Wolfgang Nitschke
Biologisches Institut II, University of Freiburg, 79104 Freiburg, FRG

Stella M. Dracheva
Boehringer Ingelheim Pharmaceuticals, Inc., Department of Molecular Biology, Ridgefield, CT 06877, USA

R. E. Blankenship, M. T. Madigan and C. E. Bauer (eds): Anoxygenic Photosynthetic Bacteria, pp. 775–805.

Summary

The X-ray structure of the *Rhodopseudomonas viridis* reaction center has for the first time provided a detailed picture of a reaction center associated cytochrome. Guided by the domain structure visible in this atomic model of the protein, the amino acid sequence of the *Rp. viridis* subunit can similarly be divided up into the respective structurally important segments. A comparison of presently available amino acid sequences for reaction center associated tetraheme subunits in the light of this segmented structure strongly suggests that all these proteins have a common global structure.

Many experimental results obtained on a variety of tetraheme subunits prior to the *Rp. viridis* crystal structure, however, were interpreted to yield a model that looks rather different from this structure. A compilation of these results is presented and alternative interpretations are proposed. It is shown that all experimental data reported so far on tetraheme subunits can be rationalized assuming structural arrangements comparable to that of the crystal structure.

Presently available data on electron transfer properties of cytochrome subunits are summarized.

In addition to the description of tetraheme subunits found in purple and green filamentous bacteria, the possible nature of the respective reaction center associated cytochromes in green sulfur and heliobacteria is discussed.

Present ideas on evolutionary relationships (or lack thereof) among these cytochrome subunits are presented.

I. Introduction

A hypothetical young researcher who entered the subject later than in 1985 probably would have become aware of the presence of such things as reaction center (RC) associated cytochrome subunits while gazing at the glossy, multicolored pictures of the X-ray structure of the *Rhodopseudomonas (Rp.) viridis* RC (e.g. in Deisenhofer et al., 1985). Upon reading these articles, our young researcher might almost be led to believe that the RC associated cytochromes were discovered via the *Rp. viridis* crystal structure. Only occasionally will she or he come across references to older work on this subunit in *Rp. viridis* or in other photosynthetic bacteria. However, having found a way into the past along such cross references, our young scientist is likely to soon become lost in a vast labyrinth of data on RC associated cytochromes (close to a third of the articles in the reference list appeared before 1978).

Abbreviations: Cb. – Chlorobium; Cf. – Chloroflexus; Cm. – Chromatium; Ec. – Ectothiorhodospira; E_m – redox midpoint potential; EPR – electron paramagnetic resonance; LD – linear dichroism; *Rb. – Rhodobacter;* RC – reaction center; *Rc. – Rhodocyclus; Ro. – Roseobacter; Rp. – Rhodopseudomonas; Rs. – Rhodospirillum; Tc. – Thiocapsa*

The resulting confusion of our hypothetical young researcher reflects the trouble with the present state of the art on RC associated cytochromes: Since 1984, a high resolution crystal structure of the cytochrome subunit attached to the *Rp. viridis* RC is at hand and (mainly based on this structure) a wealth of data with respect to electron transfer between and properties of the four hemes within this subunit has been obtained. Nevertheless, no convincing model explaining how the cytochrome subunit functions and why it is there has yet been put forward. On the contrary, many of the results obtained on the *Rp. viridis* cytochrome actually added to the 'mysteriousness' of this protein.

However, this is only one part of the problem. The surprise of the 3-D structure of the tetraheme subunit in *Rp. viridis* was the fact that it provided a picture that looked rather different from what people had anticipated based on previously obtained data on the respective cytochrome subunits from various purple bacteria. Many of the data reported within 25 years of research on RC associated cytochromes actually seemed to contradict the structure. The incomprehensibility of these data as looked at against the background of the *Rp. viridis* structure caused a tendency to willfully ignore these discomforting results. A wealth of information therefore lies dormant that might well contribute important clues for solving some of the riddles of the cytochrome subunits. One aim of this article is therefore to reexamine and, when necessary reinterpret these data in the light of our present knowledge based on the X-ray structure.

One might be tempted to say: 'Why bother with a superfluous appendix on an RC from a bacterium most people hadn't even noticed before it became the shooting star of photosynthesis due to the crystal structure.' It is certainly true that a few species of purple bacteria grow perfectly well without such a cytochrome subunit. However, in the world of purple bacteria presently known to the microbiologists, these are rather the exceptions than the rule (see for example Matsuura and Shimada, 1990). Even more strikingly, in all eubacterial phyla containing photosynthetic organisms (with the only exception of the cyanobacteria) the great majority of species seem to use an RC associated cytochrome as immediate reductant for the photooxidized primary donor P^+. Several of these subunits have been proposed to be tetraheme cytochromes, evolutionarily related to each other, while the precise nature of others is under debate. Thus, the advantage conveyed by the presence of a cytochrome subunit (whatever this advantage will turn out to be) seems to be widely appreciated by photosynthetic eubacteria and the molecular evolution of the cytochrome subunits might even be related to the evolution of photosynthetic reaction centers. A second goal of this article is therefore to present an overview of presently available data concerning distribution among species, nature of the cytochromes and evolutionary relatedness.

II. A Brief History of Reaction Center Associated Cytochromes

A. RC Associated Cytochromes: A Segregating Factor in the World of Photosynthetic Bacteria

In whole cells of photosynthetic eubacteria, the photooxidizable pigment (presumably a dimer of BChls in all species) is rereduced by a secondary electron donor in the time range between 100 ns and several hundreds of μs depending on species (see also Chapter 34 by Meyer and Donohue). This secondary electron donator can either be a soluble redox protein (mostly a small heme protein called cytochrome c_2) or it can be a membrane associated heme protein which is additionally interpositioned between the soluble components and the photooxidizable pigment.

The phenomenon that some species only contain soluble electron carriers whereas others additionally have membrane-bound electron donors has been well studied already in the 1970s for a range of different organisms (for reviews see Dutton and Prince, 1978 and Dutton et al., 1978).

The degree of membrane-association of the nonsoluble cytochromes, however, varies significantly between species. Whereas, for some species this cytochrome was retained even in highly purified RC samples (e.g. Clayton and Clayton, 1978 for *Rp. viridis*), in most species examined so far it is easily released into the aqueous phase during purification procedures (e.g. in *Rhodocyclus (Rc.) gelatinosus*, Prince et al., 1978; for a detailed study of such loss of the tetraheme cytochrome see Fukushima et al., 1988). An experimental distinction between organisms that contain and organisms that lack the RC associated cytochrome is therefore not always a trivial task (see for example the conflicting data reported for *Rp. palustris* by Kihara and Chance, 1969, Kihara and Dutton, 1979 and Matsuura and Shimada, 1986).

B. Distribution Among Species

As already mentioned in the Introduction, the occurrence of membrane-bound cytochromes serving as immediate electron donors to photooxidized P^+ is not confined to purple bacteria. In addition to the proteobacterial phylum (containing the purple bacteria, Stackebrandt et al., 1988), examples for such a situation were found in green filamentous (Bruce et al., 1982), green sulfur (Prince and Olson, 1976) and heliobacteria (Prince et al., 1985) (for a description of the classification of species presently in use see Woese, 1987; also see Chapter 1 by Imhoff).

A compilation of data concerning RC associated cytochromes reported so far in a variety of photosynthetic bacteria is given in Table 1.

C. The Tetraheme Subunits of Purple Bacterial Reaction Centers

It was observed for a number of purple bacterial species that depending on redox conditions (i.e. anaerobiosis or aerobiosis in the case of whole cells)

Table 1. Compilation of presently available data on RC-associated cytochromes

Species	Phylogenetic classification	MW_{app} [MW_{calc}] (kDa)	E_m values of soluble donor (mV)	E_m values of the individual hemes in cytochrome subunits (mV)		Orientations of hemes with respect to the membrane	E_m values of the primary donor (mV)	Halftimes for electron donation to P+ (μs)
				equilibrium titrations	flash titrations			
Chromatium vinosum	P γ	45[1]	240[67]	360[3]–370[4] 330[3] 2 × 10[3,5,6]	350[5] 320[5] 30[7] −12[7]	30[3] 90 0 or 40 40 or 0	470[5] 490[6,8]	2–3 (>50mV)[9,10] 1 (< 0mV)[9,10] 2000 (< 0 mV, 77K)[11,12]
Chromatium tepidum	P γ	36[13]						
Chromatium minutissimum	P γ			380[14] 270[14]			490[15]	3 (RT–200K, E_h=300mV) 1 (RT, E_h < 0 mV)[14] 35 (200K,E_h < 0 mV)
Ectothiorhodo-spira halochloris	P γ	41[60]						
Ectothiorhodo-spira mobilis	P γ	42[17]		297[17] 25				0.5–3 (50%) 200–400 (50%)[66] <5 (100%) sulfide present
Ectothio-rhodospira shaposhnikovii	P γ	39[16]		290 0–10[18]			400[20]	2.5 (RT–120K, E_h>150mV)[18,20] 1 (RT, E_h<−50mV)[20]
Ectothio-rhodospira sp.	P γ	36[59]						
Rhodoferax fermentans	P ?		287[70]	358 296 78[70] −1			471[70]	
Rhodopseudomonas acidophila	P α		370[24]	360[21] 110	385 130			
Rhodopseudomonas palustris	P α							1.8 (RT)[22] 12000 (77K)

Table 1. Continued

Species	Phylogenetic classification	MW_{app} [MW_{calc}] (kDa)	E_m values of soluble donor (mV)	E_m values of the individual hemes in cytochrome subunits (mV): equilibrium titrations	E_m values of the individual hemes in cytochrome subunits (mV): flash titrations	Orientations of hemes with respect to the membrane	E_m values of the primary donor (mV)	Halftimes for electron donation to P+ (μs)
Rhodopseudomonas	P	38		360 to 400	370[25]	80	500[25]	0.23 (RT, E_h = 380 mV)[26]
viridis	α	[40.5]	290[24]	300 to 320	320[25]	55		0.19 (RT E_h = 250 mV)[27,28,29,26]
		23		10 to 50	25[68]	60		0.115 (RT, E_h = –20 mV)[26,30]
				–80 to –50	–70[25]	85		1100 (<50K, E_h = –20 mV)[26,31]
Rhodomicrobium	P	38[32]						
vannielii	α							
Rhodospirillum	P		395[24]	390[33]			550[33]	
molischianum	α		300	3x<100				
Roseobacter	P	45[35]	250[36]					
denitrificans	α	(partial						
(*Erythrobacter*		seq.[34])						
sp.OCH 114)								
Rubrivivax	P	43[38,39]		320[40,41]		0[41]	400[5]	0.7 (RT, anaerobic)[42]
gelatinosus	β	[39.8][37]		300		90		4.6 (80K, anaerobic)[42]
(*Rhodocyclus*	α[37]			130	130[5]	90		
gelatinosus)				70	70[5]	0		
Thiocapsa	P	40[43]		340[44]	360		490[44]	
pfennigii	γ			0	20			
Chloroflexus	gfb	43[45]	240[58]	280	295[19]	30[47]		1.4 (one heme reduced)[63]
aurantiacus		[45.4][46]		220/95[45, 47]	220[19]	40–50	360[48,49]	0.7 (two hemes reduced)
				150	140[19]	90	420[50]	0.4 (three hemes reduced)
				0	20[19]	45		0.3 (four hemes reduced)
Chlorobium	gsb	32[51]	145[57]		180[2]		250[2]	5[2] 50[2]
limicola f. sp.		24[52]			150			90[56] 390[56]
thiosulfatophilum		21[56]						

Table 1. Continued

Species	Phylogenetic classification	MW_{app} [MW_{calc}] (kDa)	E_m values of soluble donor (mV)	E_m values of the individual hemes in cytochrome subunits (mV): equilibrium titrations	E_m values of the individual hemes in cytochrome subunits (mV): flash titrations	Orientations of hemes with respect to the membrane	E_m values of the primary donor (mV)	Halftimes for electron donation to P+ (μs)
Chlorobium phaeobacteroides	gsb	32[54]		190[55] 165 (hemes at 110/0 mV present in membr.)				
Chlorobium vibrioforme	gsb	18[53] [22.8]						5 (30%) 50 (70%)[53]
Chlorobium tepidum	gsb	23[64]						
Heliobacterium chlorum	F				170[62,69] 150		220[69]	110 (50%) 700(50%)[65]
Heliobacillus mobilis	F				190[62] 160		240[62]	110 (50%) 400–600 (50%)[62]

The following abbreviations are used: P, Proteobacteria; gfb, green filamentous bacteria; gsb, green sulfur bacteria; F, Firmicutes.

[1]Kennel & Kamen (1971); [2]Prince & Olson (1976); [3]Nitschke et al. (1993); [4]Alegria & Dutton (1990); [5]Dutton (1971); [6]Case & Parson (1971); [7]Case & Parson (1973); [8]Cusanovich et al. (1968); [9]Seibert & DeVault (1970); [10]Case et al. (1970); [11]DeVault & Chance (1966); [12]Dutton et al. (1971); [13]Garcia et al. (1987); [14]Rubin et al. (1989); [15]Shinkarev et al. (1991); [16]Levebvre et al. (1984); [17]Leguijt & Hellingwerf (1991); [18]Chamorovsky et al. (1980); [19]Zannoni & Venturoli (1988); [20]Chamorovsky et al. (1985); [21]Matsuura & Shimada (1986); [22]Kihara & Dutton (1970); [23]Weyer et al. (1987a); [24]Pettigrew et al. (1978); [25]Prince et al. (1976); [26]Ortega & Mathis (1993); [27]Holten et al. (1978); [28]Dracheva et al. (1986); [29]Shopes et al. (1987); [30]Dracheva et al. (1988); [31]Ortega & Mathis (1992); [32]Kelly & Dow (1985); [33]Nagashima et al. (1993a); [34]Liebetanz et al. (1991); [35]Shimada et al. (1985); [36]Okamura et al. (1984); [37]Nagashima et al. (1993b); [38]Fukushima et al. (1988); [39]Agalidis et al. (1990); [40]Matsuura et al. (1988); [41]Nitschke et al. (1992); [42]Kihara & McCray (1973); [43]Seftor & Thornber (1984); [44]Prince (1978); [45]Freeman & Blankenship (1990); [46]Dracheva et al. (1991); [47]Van Vliet et al. (1991); [48]Bruce et al. (1982); [49]Blankenship et al. (1983); [50]Venturoli & Zannoni (1988); [51]Hurt & Hauska (1984); [52]Büttner (1993); [53]Okkels et al. (1992); [54]Feiler et al. (1992); [55]Feiler et al. (1989); [56]Oh-oka et al. (1993); [57]Meyer et al. (1968); [58]Trost et al. (1988); [59]Levebvre et al. (1989); [60]Engelhardt et al. (1986);[61]Meyer et al. (1973); [62]Nitschke & Liebl (unpublished); [63]Mulkidjanian et al. (1993); [64]Kusumoto et al. (1993); [65]Vos et al. (1989); [66]Leguijt, Verméglio, Crielaard & Hellingwerf pers. communication; [67]Tomiyama et al. (1983); [68]Cogdell & Crofts (1972); [69]Prince et al. (1985); [70]Hochkoeppler et al. (1993); heme orientations for *R. viridis* were obtained from the crystal structure. Spectral data on *R. photometricum*, *R. fulvum*, *R. globiformis*, *Protomonas* sp., *R. tenuis* were furthermore reported by Matsuura & Shimada (1990).

different hemes were oxidized upon illumination. This was interpreted to reflect two different populations of hemes which are both able to reduce P^+ (Parson and Case, 1970). The hemes which were in the reduced state only under anaerobic conditions (the so-called 'low potential' hemes) were able to rereduce P^+ faster than the hemes which had a higher redox potential (i.e. which were in the reduced state already under aerobic conditions). Therefore, these low potential hemes were considered to be closer to P than the high potential ones. The assumed shorter distance between the low potential hemes and the special pair also served to rationalize the phenomenon that photoinduced oxidation of the low potential hemes persisted down to the lowest temperatures attained (2K), whereas there was no stable low temperature photooxidation when only the high potential hemes were reduced prior to illumination (Vredenberg and Duysens, 1964). Below a certain temperature the kinetics of this reaction became virtually temperature-independent leading to the proposal that this low-temperature photooxidation of hemes was the first biological example of electron tunneling (DeVault and Chance, 1966; see also III.C.1.b.).

After a short period of time when numbers up to 9 hemes per RC were reported (Thornber, 1970; Thornber and Olson, 1971; Seibert and DeVault, 1970), the values obtained both spectroscopically and biochemically converged towards four hemes/P, two of which had a high (> 250 mV) and two a low (< 150 mV) redox midpoint potential (Case and Parson, 1971; Clayton and Clayton, 1978). Kennel and Kamen (1971b) demonstrated that in *Chromatium (Cm.) vinosum* the low and the high potential hemes were located on one polypeptide, thereby preparing the ground for the notion of the 'tetraheme cytochrome subunit.' No conspicuous distinguishing characteristics were observed within the pair of low potential and within the pair of high potential hemes. The four hemes were therefore considered to consist of two twins of hemes suggesting an arrangement displaying some two-fold symmetry with respect to the RC. In such an arrangement, all four hemes were considered to be able to directly reduce P^+ and it was only due to their faster donation reactions (supposedly due to their closer proximity) that the low potential hemes dominated the rereduction process once they were reduced.

This rough picture was refined into a detailed structural model based on data obtained on *Cm. vinosum* (Tiede et al., 1978). The respective model was characterised by two symmetric pairs of hemes, i.e. one pair of low potential hemes oriented parallel with respect to the membrane and closer to P than the pair of high potential hemes considered to be oriented perpendicular to the membrane plane (see Tiede et al., 1978).

D. The 'New Age'

The scientific community was apparently satisfied with this model (or had possibly become tired of RC associated cytochromes), since from 1978 to 1985 only very few studies were published on this specific subject.

It was not before the mid-80's that the RC associated cytochromes had a glamorous comeback, promoted by major breakthroughs in two rather different fields.

1. The Impact of the Crystal Structure

One breakthrough was the crystallization and subsequent structure determination of the photosynthetic reaction center from *Rp. viridis*, a species that contained an RC associated tetraheme cytochrome (Deisenhofer et al., 1985). This structure showed that the model arrived at in 1978 for the *Chromatium* subunit had hardly anything more in common with the *Rp. viridis* structure than the total of four for the number of hemes (see section III.A.1.). This led to the question whether the apparent inconsistencies between previous models and the structure of the *Rp. viridis* subunit were due to misleading interpretations of data or to real interspecies differences. Amino acid sequence information appeared to be a first step to address the question of relatedness between the tetraheme subunits from different organisms. Primary sequences for *Chloroflexus aurantiacus*, *Roseobacter denitrificans* and *Rubrivivax gelatinosus* have been obtained providing evidence that, at least for the tetraheme subunits from these three organisms, the global structure elements are comparable to those of the *Rp. viridis* subunit (see III.B.4.).

Furthermore, several of the species for which reported data were in conflict with the *Rp. viridis* crystal structure were reexamined yielding explanations for at least part of the obvious incompatibilities. Most conspicuously, the model featuring two twins of identical hemes was in blatant conflict with the structure of the *Rp. viridis* cytochrome subunit showing the presence of four individual, *non-equivalent* hemes. Only when they saw the structure, spectroscopists and electrochemists went back to their experimental setups and duly found what this structure had ordered them to find, i.e. differences in spectral parameters, redox midpoint potentials and orientations between the individual hemes of each 'pair' both in *Rp. viridis* and in other species containing a tetraheme subunit (see III.A.2.a and III.B.1./2.).

Whereas in these cases, improved experimental data were required for a reconciliation with the crystal structure, in other cases a mere reinterpretation of published work might be sufficient to this end (see III.B.3.).

Independently, the advantage conveyed by the sound basis of a high resolution crystal structure attracted many research groups to perform in-depth studies of the cytochrome subunit from *Rp. viridis* with respect to spectroscopic, electrochemical and electron transfer parameters (see III.A.2. and III.C.).

2. The New Phylogenetic Landscape of the Eubacteria

The second important re-launching of research efforts on membrane associated cytochrome donors was triggered by profound changes in our conceptions of bacterial taxonomy and phylogeny. Application of approaches based on molecular biology (rRNA catalogues) and phylogenetic tree building algorithms resulted in extensive rearrangements of bacterial taxa that formerly were mainly based on morphology and ecology (Woese, 1987). Since it has been recognized during recent years that RC associated cytochromes are abundant in the world of photosynthetic organisms, these proteins have become important markers in tracing the evolutionary relationships between species and between types of reaction centers (see IV. and V.).

III. Tetraheme Cytochrome Subunits

A. The Special Case: Rp. viridis

1. Structure

a. The Global Structure

Since the high resolution structure of the *Rp. viridis* reaction center including the cytochrome subunit is the topic of an entire chapter in this volume, the reader is referred to this contribution and figures therein for a general outline of the structure (see Chapter 23 by Lancaster et al.). Only features of the X-ray structure dealing specifically with the cytochrome subunit will be discussed in more detail in the following sections.

The tetraheme subunit in *Rp. viridis* consists of 336 amino acid residues, four covalently linked hemes and a covalently attached lipid molecule adding up to a molecular weight of 40.5 kDa. Whereas in the models developed prior to the *Rp. viridis* crystal structure, the cytochrome subunits were thought to be largely membrane-integral (Tiede et al., 1978; Dutton and Prince, 1978; Prince, 1978), the cytochrome seen in the crystal structure appears as an oblong soluble protein which was stuck at one of its ends to the periplasmic surface of the membrane-integral RC core (i.e. the L- and M-subunits, Fig. 1). The interaction surface of the cytochrome subunit with the RC-core amounts to approximately 15 % of the cytochrome's total surface, the remainder being exposed to the aqueous phase of the periplasmic space. The oblong protein harbors the four heme groups which are arranged in a roughly linear row spanning a large part of the periplasmic space and pointing from the far end of the cytochrome subunit right towards the special pair (for the arrangement of the cofactors in the RC, see Chapter 23 by Lancaster et al.). This again is a major difference to the previously assumed two similar, parallel pathways each containing one high and one low potential heme.

The row of hemes breaks the C_2-symmetry of the RC core's redox centers being inclined by about 30° with respect to the membrane normal. Parts of the cytochrome subunit, however, display an internal C_2-symmetry with respect to an axis not far from parallel to the membrane plane (see below).

The four hemes in the subunit can be identified either by order of distance from the RC, i.e. heme 1 being closest to the special pair etc., or by the position of the respective heme binding motifs (CXXCH) in the primary sequence. Both conventions are currently in use. To avoid confusion, Roman numerals will be used in the following sections to denote the sequence based, whereas Arabic numerals denote the structure-based nomenclature. Fig. 1a defines the relationship between both systems.

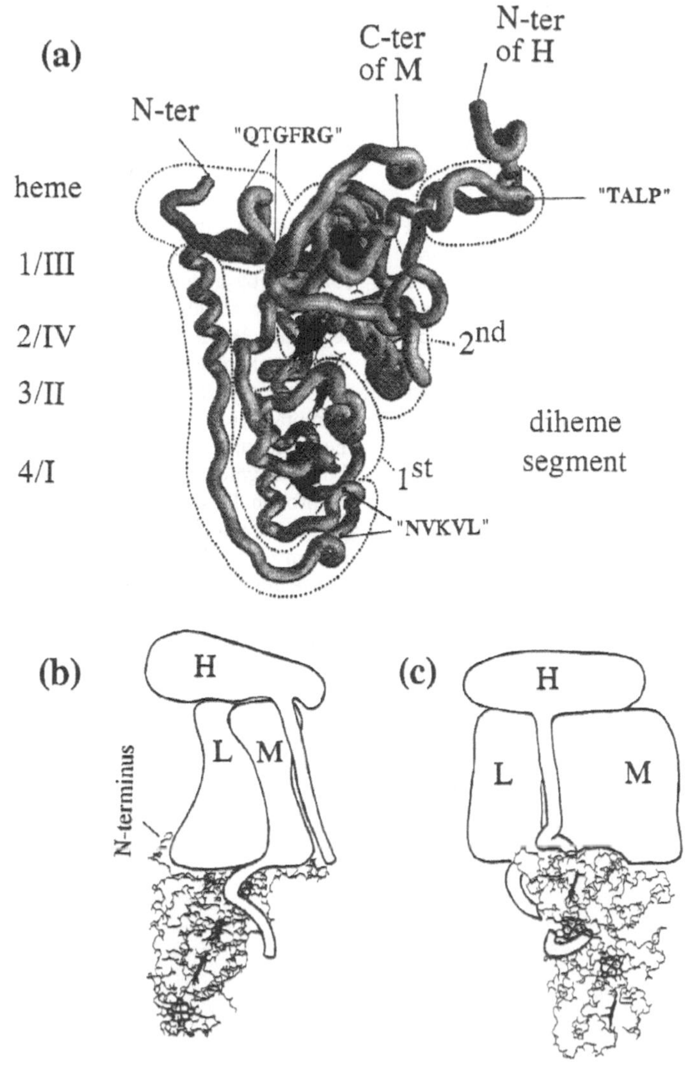

Fig. 1: Molecular models of the tetraheme subunit associated to the *Rp. viridis* reaction center. (a) defines the structure elements referred to in the text and used for the layout of the sequence comparison in Fig. 4. Only the α-carbon backbone and the four heme groups are included. Structurally important (and conserved) peptides are shown in between quotation marks. Arabic numerals refer to increasing distance from the primary donor and Roman numerals follow the heme order within the amino acid sequence as referred to in the text. Note that the pieces of protein backbone labeled 'C-ter of M' and 'N-ter of H' are clipped-off parts of the M- and H- subunits, respectively. (b) and (c) depict the structural association of the cytochrome subunit with the other subunits in the reaction center, viewed from two directions parallel to the plane of the membrane and perpendicular to each other. (a) was produced by S. Grzybek (Freiburg) using the program GRASP developed by A. Nicholls and B. Honig, whereas (b) and (c) were constructed on a PC using the pdViewer software by H.H. Robinson and A.R. Crofts.

b. Breakdown into Structure Elements

As has been pointed out by Deisenhofer et al. (1985), parts of the tetraheme subunit can be superpositioned onto one another via a C_2-symmetry operation. Hemes 4/I and 3/II superimpose on hemes 1/III and 2/IV, respectively, when the lower half of the cytochrome subunit is turned up by 180° about the C_2-symmetry axis. Equally, the α-helices providing ligands to hemes 4/I and 3/II can be mapped onto the respective parts of the protein surrounding hemes 1/III and 2/IV, respectively. Thus, it appears as if the central part of the tetraheme subunit was made up from two similar building blocks (each of which contains two hemes), arranged in C_2-symmetry with respect to each other. This symmetry seen on the structural level finds its expression also on the level of the primary sequence of the *Rp. viridis* tetraheme subunit (Weyer et al., 1987a), since higher sequence homologies have been found between the heme-binding regions of the pairs I/III and II/IV than between all other possible pairs.

In addition to the C_2-symmetric parts liganding the heme groups, the following structural elements (Fig. 1a) can be distinguished (note that in the following discussion of structurally important sequence stretches, the numbering for the protein including the presequence will be used; in this numbering, the mature protein starts at residue number 21).

At the N-terminal end of the protein, the first 24 amino acid residues (residues 21 to 44) run roughly parallel (containing a β-sheet region) to the membrane plane in contact with the L-subunit. Between residue 45 and residue 65, the chain protrudes vertically away from the membrane towards the far end of the subunit. After winding around the bottom of the protein between residues 66 and 82, the chain enters the C_2-symmetric part discussed above, liganding hemes 4/I and 3/II. This first 'diheme segment' is followed by a long stretch that is not part of the symmetric structure but links the two symmetric segments. Within this sequence stretch, the chain winds back to the periplasmic surface of the RC core and forms a large loop along the plane of the membrane (residues 222–244) in close contact with the M-subunit. From residues 245 on, the second diheme segment is formed and the C-terminal end of the protein disappears in mystery since the last four residues are apparently disordered and do not yield attributable electron density in the structure (Deisenhofer and Michel, 1989).

However, the C_2-symmetry between the two diheme segments is broken when it comes to the 6th ligand of

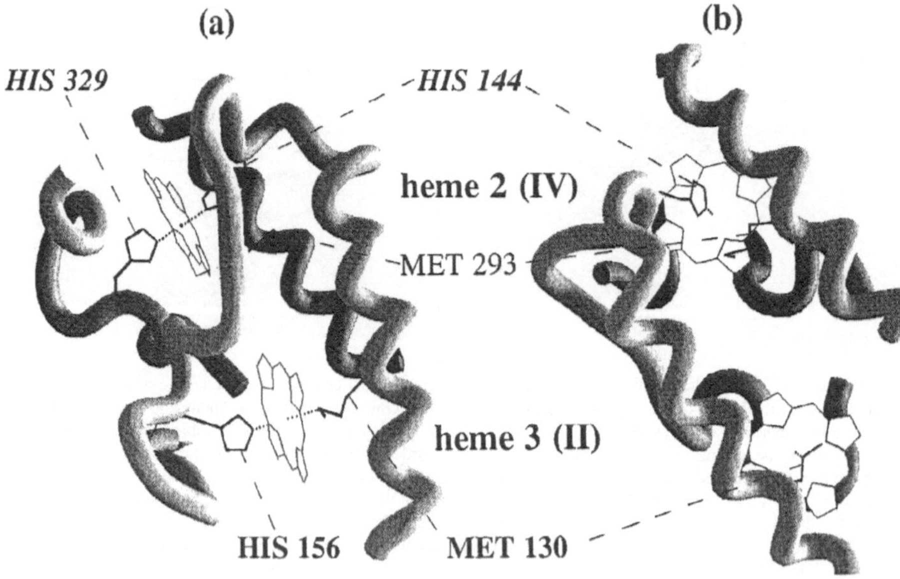

Fig. 2: Molecular models of the contact region between the two individual diheme segments in the cytochrome subunit from *Rp. viridis* viewed from directions approximately parallel to the plane of the membrane and mutually perpendicular. Only a few relevant residues (as referred to in the text) have been added to the backbone model. Dotted lines represent bonds from 5th and 6th ligands to the heme irons. The models were produced as detailed for Fig. 1a.

heme 2/IV. An α-helix providing the 6th ligand, a methionine, in the case of hemes I, II and III, is also present in the equivalent position close to heme 2/IV, however, the only methionine (Met 293) in this helix is one helix-turn too far to serve as a ligand for the iron (Fig. 2a,b). Instead, this role is taken over by a histidine residue (His 144) situated on the first diheme segment and within the sequence stretch liganding heme 3/II, i.e. the neighboring heme in the direction of the far end of the subunit. This altered ligation pattern for heme 2/IV has important consequences for the functional properties of the *Rp. viridis* subunit, as will be detailed below.

c. Attachment to and Interactions with the Core of the RC

A rather unusual structural element participating in the attachment of the cytochrome subunit to the membrane is the so-called lipid-anchor. Weyer et al. (1987b) demonstrated that the N-terminal cysteine residue of the mature protein was covalently linked to a diglyceride. Furthermore, the precursor protein (the sequence of which was deduced from the DNA-sequence of the gene; Weyer et al, 1987a) contained a stretch of amino acids around the N-terminus of the protein (residues 17 to 21) which is known as a recognition site for signal peptidase II, a protease that specifically cleaves presequences before lipid-modified cysteines in many bacterial lipoproteins (Weyer et al., 1987a). As can be seen in Figs. 1a and 1b, the N-terminus is close to the plane of the membrane. Since this N-terminus furthermore contacts the membrane at the very periphery of the core of the RC (Fig. 1b), the fatty acids attached to the terminal cysteine seem to be inserted into the membrane to become part of the lipid bilayer. Accordingly, no electron density attributable to the lipid anchor is seen in the X-ray structure (Deisenhofer and Michel, 1989), indicating that the two fatty acids are disordered in the membrane. The lack of specific interaction of the lipid anchor with other parts of the RC makes it rather unlikely that it contributes significantly to the well-determined geometry of the cytochrome-RC complex.

The N-terminal segment consisting of 25 residues is mainly in contact with the L-subunit of the reaction center, whereas the loop region (residues 222–244) connecting the two diheme segments touches the M-subunit but also reaches out of the circumference of the RC core at the periplasmic surface of the membrane contacting (with the 'TALP'-sequence) the H-subunit at its N-terminal residues (Fig. 1a,b).

In contrast to this very local interaction between the cytochrome and the H-subunit, a rather extended interaction surface exists between the cytochrome and the C-terminal 20 residues of the M-subunit which run along one side of the cytochrome away from the membrane plane down to a level between the 2nd and the 3rd heme. The sequence segment between residues 276 and 286 of the cytochrome subunit (i.e. the region between the heme binding cysteines of heme 1/III and the flanking α-helix of heme 2/IV is virtually clamped to the membrane via these 20 N-terminal residues of the M-subunit (Fig. 1a–c).

It is of note, that in *Rhodobacter (Rb.) sphaeroides*, *Rhodospirillum (Rs.) rubrum* and *Rb. capsulatus*, that are all species lacking an RC associated cytochrome, the polypeptide chain of the M-subunit terminates about 20 residues earlier and therefore the respective stretch reaching out into the periplasmic space to grab the cytochrome is missing. By contrast, all M-subunits from purple bacteria that contain the cytochrome subunit and that have been sequenced so far, possess the C-terminal extension. Thus, the (admittedly small) statistical sample of some half a dozen examples presently indicate that the presence of the RC-attached cytochrome subunit and the M-terminal extension are correlated in purple bacteria.

2. Redox Properties

a. The Equilibrium Redox Midpoint Potentials of the Four Hemes

By 1976, it had become clear that there were four hemes functionally associated to the *Rp. viridis* reaction center (Case et al., 1970; Thornber and Olson, 1971; Cogdell and Crofts, 1972; Prince et al. 1976). The high and the low potential hemes, designated Cyt c_{558} and c_{553}, were found to titrate as identical pairs at E_m-values of about +340 mV and −10 mV, respectively, in dark equilibrium redox titrations (Thornber and Olson, 1971). The arrangement of the hemes seen in the X-ray structure (Fig. 1), however, made such a situation hard to believe, since no obvious pairs of identical hemes could be seen in the cytochrome. A subsequent reexamination of the spectral and electrochemical properties of the cytochrome subunit from *Rp. viridis* yielded distinct α-band absorption wavelengths, vibrational spectra and E_m-values for each of the hemes of the former two 'pairs' (see Table 1; Dracheva et al. 1986; Shopes et al. 1987; Dracheva et al. 1988; Nabedryk et al. 1991). These individual E_m-values have since been confirmed by several independent studies using various different methods (Nitschke and Rutherford 1989; Fritzsch et al. 1989; Alegria and Dutton 1991b; Fritz et al. 1992). Values obtained in these studies are within ± 20 mV of the midpoint potentials first obtained by Dracheva et al. (1988). This range of measured values, however, seems to be at least in part due to using differently treated materials rather than reflecting experimental error. An investigation of the influence of detergents on the E_m-values of the hemes has been presented by Alegria and Dutton (1991a).

In a next step, these four individual, spectrally and electrochemically distinguishable hemes had to be related to the hemes seen in the X-ray structure. The most obvious choice was to arrange the hemes by order of their redox midpoint potentials, i.e. –60 mV → +20 mV → +320 mV → +380 mV → special pair (Michel and Deisenhofer, 1986). However, the finding that heme 2/IV had histidines both as 5th and 6th ligands (usually resulting in rather low redox midpoint potentials), whereas all other hemes had histidine-methionine ligation (Weyer et al. 1987), together with comparisons of data obtained on *Rp. viridis* with those previously gained on *Cm. vinosum* quickly raised doubts about this attribution (Dracheva et al. 1988).

Indeed, the consensus of subsequent studies addressing this specific question was the following order of hemes: –60 mV → +320 mV → +20 mV → +380 mV → special pair (Nitschke and Rutherford 1989, Fritzsch et al. 1989; Alegria and Dutton 1991b).

From what is known for the organization of typical electron transfer chains, the interposition of a low potential heme in between the two high potential hemes would certainly not have been anticipated. The role of this very heme as well as the 'raison d'être' of the low potential hemes in general, are part of the enigmatic features of the tetraheme subunits (see III.C.3.).

b. Factors Influencing the Redox Potentials of the Hemes

In the case of heme 2/IV with its histidine/histidine-ligation, the correlation between E_m (low, +20 mV) and structure seems rather straightforward. Heme 4/I, however, with its even lower E_m (–60 mV), has histidine/methionine ligands just as the two high potential hemes do, nevertheless being more negative by almost 400 mV. Therefore, other factors must exert a significant impact on the electrochemical properties of the hemes.

Recent progress in computational treatment of continuum electrostatics (Sharp and Honig, 1990), applied to the high resolution crystal structure of the cytochrome subunit provided the first quantitative estimations of factors contributing to the equilibrium E_m-values of the four hemes (Gunner and Honig, 1991). These calculations were able to reproduce the alternating sequence of redox midpoint potentials yielding E_m-values comparable to the experimentally determined potentials. The following conclusions were drawn from these calculations (Gunner and Honig, 1991):

(a) Specific residues and parts of the cofactors (i.e. propionic acid side chains on hemes) were predicted to strongly modulate E_m-values of the hemes. These predictions can serve to suggest promising targets for site-directed mutagenesis. They furthermore can help to rationalize inter-species differences once the respective amino acid sequences are available (see sequence comparisons in III.C.4.).

(b) Taking only the protein into account and neglecting electrostatic interaction between hemes yielded the correct division into low and high potential hemes. In absolute values, however, the experimental data could only be reproduced when heme-heme interactions between the individual hemes were included in the calculations. Electrostatic interaction energies of up to 100 mV were found (for example between heme 3 and the two neighbouring low potential hemes). The important take-home message from this result was that the E_m values measured in equilibrium titrations need not apply to all experimental conditions. For example, using the values given by Gunner and Honig (1991), it can be calculated that the potential of heme 1, i.e. the closest one to the special pair, drops from its experimental E_m of +380 mV (when all other hemes are oxidized) to +289 mV in a state where hemes 1, 2 and 3 are reduced (see III.C.1. for a further discussion of these effects).

(c) Based on these calculations of electrostatic potentials, effective dielectric constants (ϵ_{eff}) were predicted for all possible pairs of charges within the tetraheme cytochrome. As shown by Gunner and Honig (1990), ϵ_{eff}-values are not identical for all pairs of hemes but rather increase dramatically with increasing distance between the members of the respective pair. This dependence of ϵ_{eff} on distance can be intuitively understood considering the almost rod-shaped geometry of the cytochrome subunit which is surrounded by a medium with high dielectric constant, i.e. water ($\epsilon = 81$). Basic electrostatics predict for such a geometry that electric field lines between two charges with opposite sign will bend out of the low dielectric protein interior (ϵ probably close to 4, see Gunner and Honig, 1991) into the surrounding phase, thereby significantly increasing the effective dielectric constant between the two charges. Accordingly, an ϵ_{eff} of 13 was calculated for the neighboring hemes 1 and 2, whereas this value is as high as 81 for the pair heme 1-heme 4, i.e. for the hemes at the extremes of the oblong cytochrome subunit (see also Moore and Rogers, 1985).

The resulting inhomogeneity of the dielectric medium might help to rationalize apparently incomprehensible data with respect to the contribution of the cytochrome to the buildup of membrane potential (see III.B.3.).

B. The 'Tetraheme Family'

1. The 'Pairs of Hemes' (a Tentative Reinterpretation of Previous Data)

The dogma of pairwise identical high and low potential hemes was somewhat in conflict with experimental data already in the 1970's. Redox titrations of multiple-flash induced cytochrome oxidations showed lower E_m-values for the hemes oxidized in the second flash, compared to what was obtained for the first flash oxidations. These '1st and 2nd flash E_m-values' usually were found at 20 mV above and below the E_m-value of the 'pair' as determined in dark titrations. This discrepancy, however, was resolved by postulating a number n of identical hemes all of which were competent in rereduction of photooxidized P^+. Apparent E_m values to be expected in such a situation explained the data sufficiently well (Case and Parson, 1971; Jackson and Dutton, 1973). These considerations made perfect sense as long as their basic assumptions were justified.

We now know that this is not the case for the RC of *Rp. viridis* since only heme 1 seems to be well-placed for efficient electron transfer to the special pair. However, taking the multiple flash titrations curves obtained for *Rp. viridis* at face value instead of passing them through the 'identical hemes algorithm', yields E_m-values in perfect agreement with the midpoint potentials of the individual hemes (see Table I).

Recently, distinct E_m-values for the four hemes have been obtained by EPR for the tetraheme systems from *Cm. vinosum* (Nitschke et al., 1993) and *Rc. gelatinosus* (Nitschke et al., 1992). In both cases, the redox midpoint potentials seen in the EPR titrations correspond well to the E_m-values obtained by multiple-flash-induced cytochrome oxidations and electrochromic changes (Table I). This is also true for the tetraheme subunit of *Chloroflexus (Cf.) aurantiacus* (n.b.: *Cf. aurantiacus* is not a purple bacterium but belongs to the green filamentous bacteria) where equilibrium redox titrations (Freeman and Blankenship, 1990; Van Vliet et al., 1991) could be reconciled with multiple flash experiments (Zannoni and Venturoli, 1988).

In short, whereas in the older literature it was assumed that the equilibrium titrations gave correct values and the multiple flash data were 'distorted', the opposite appears to be the case, i.e. multiple flash experiments monitored the 'true' E_m-values of the individual hemes whereas the equilibrium dark titrations actually yielded non-resolved curves.

We assume that this holds true also for other tetraheme systems for which multiple flash data are available but no differentiation in equilibrium titrations has been achieved yet, such as *Thiocapsa (Tc.) pfennigii* (Prince, 1978), *Rp. acidophila* (Matsuura and Shimada, 1980) or *Ectothiorhodospira (Ec.) shaposhnikovii* (Chamorovsky et al., 1985).

Therefore, all tetraheme cytochromes are likely to contain four individual, distinguishable hemes rather than two twins of identical hemes.

2. Heme Orientations in Tetraheme Subunits

Both optical LD- (Vermeglio et al., 1989; Fritsch et al., 1989; Alegria and Dutton, 1990, 1991a,b) and EPR spectroscopy (Tiede et al., 1978; Nitschke and Rutherford, 1989; Van Vliet et al., 1991; Nitschke et al., 1992, 1993) have been employed in order to determine orientations of the hemes with respect to the membrane in RC associated cytochromes.

When applied to the case of *Rp. viridis*, these techniques yielded roughly correct orientations (±15°) as judged by comparisons with the crystal structure showing that they have the potential to yield structural information in species for which no crystal structure is available as yet.

Presently, such orientation data are available for *Cm. vinosum*, *Rhodocyclus (Rc.) gelatinosus* and *Cf. aurantiacus* (see Table 1).

The heme orientations (and redox midpoint potentials) determined by EPR for the cytochrome subunit from *Cm. vinosum* (Dutton and Leigh, 1973; Tiede et al., 1978) had contributed significantly to the formulation of the 'pairs-of-hemes'-dogma. Reexaminations of this cytochrome both by LD (Alegria and Dutton, 1990) and by EPR (Nitschke et al., 1993), however, yielded results supporting the presence of four distinguishable hemes and therefore in disagreement with the previously proposed model.

Surprisingly, none of the sets of heme orientations determined in the above mentioned species fitted the pattern of the *Rp. viridis* cytochrome. Even more surprisingly, they all seem rather different from one another.

A tentative model rationalizing these interspecies differences in the light of presently available primary sequences will be presented below (III.B.4.c.).

3. The $\Delta\Psi$ Paradox

The contribution of the RC associated cytochrome to the light-induced generation of membrane potential has been studied in *Cm. vinosum* (Dutton, 1971; Case and Parson, 1973; Itoh, 1980), *Rc. gelatinosus* (Dutton, 1971), *Tc. pfennigii* (Prince, 1978) and *Rs. molischianum* (Nagashima et al., 1993a) by monitoring electrochromic changes of carotenoids and bacteriochlorophylls. Chemical reduction of the high potential hemes prior to the actinic flashes resulted in an additional electrochromic component the extent of which ranged between 0.5 and 1.5 of the change due to charge separation between P and Q_A. The migration of the positive charge from P^+ to the cytochrome therefore seemed to roughly double the dielectric distance between the separated pair of charges. However, at ambient redox potentials where the low potential hemes became reduced, the electrochromic response diminished to a level even below that of the P^+Q^- state (about 60% of this state). Assuming a homogeneous dielectric medium, this was interpreted to indicate that the low potential hemes were much closer to the special pair than were the high potential hemes (Dutton and Prince, 1978; Tiede et al., 1976; Itoh, 1980; Chamorovsky et al., 1985) or even located in between P and Q_A (Prince, 1978). However, Case and Parson (1973) suggested that the assumption of a dielectrically homogeneous phase harboring the redox centers might be misleading.

Such caution turned out to be fully justified when the crystal structure showed that the *Rp. viridis* cytochrome subunit was not membrane-integral at all but reached far out into the aqueous periplasmic phase. Thus, the relationship between dielectric and spatial distance is predicted to increasingly deviate from linearity the further the positive charge moves out within the tetraheme system into the periplasmic space (cf. III.A.2.b.).

Due to the absence of significant electrochromic shifts in *Rp. viridis*, data with respect to the buildup of membrane potential had to be collected by direct electrometric measurements on proteoliposome-collodion films (Skulachev et al., 1987; Dracheva et al., 1988).

Figure 3 relates the dielectric distances obtained in these studies for the high potential hemes to the distance of charge separation perpendicular to the membrane plane (arrows and values left to the RC in Fig. 3). Assuming a dielectric constant of about 4 for the protein inserted in the membrane (i.e. the RC core), effective ϵ-values of about 6 and about 8 would result from the positions of heme 1 and heme 3, respectively.

Furthermore, using data obtained on *Cm. vinosum* or *Tc. pfennigii* for electron donation from the low potential hemes to P^+ and assuming that the respective electron comes (like it does in *Rp. viridis*) ultimately (i.e. at very low potentials) from the outermost heme 4, would require an ϵ_{eff} of about 20 (note that this and the following considerations rely on the fact that the 2nd heme from P is a low potential heme in *Cm. vinosum* and *Tc. pfennigii*, just as it is in *Rp. viridis*; see also III.B.4.d.).

Published data on the contribution of heme 2 (i.e. the less reducing of the low potential hemes in the case of *Rp. viridis*) are ambiguous. In the study on *Cm. vinosum* by Case and Parson (1973), the pronounced decrease of the electrochromic response seems to titrate with the lowest potential heme, suggesting that donation from the second low

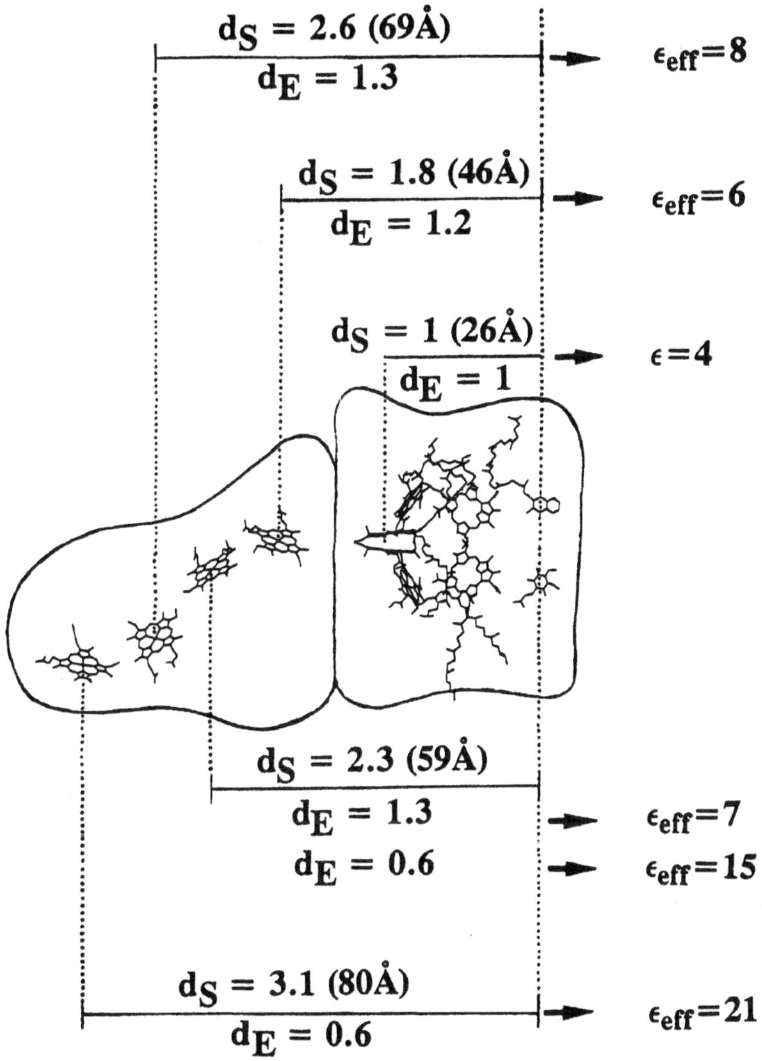

Fig. 3: Schematic model of the *Rp. viridis* reaction center embedded in the membrane (showing only chromophores and the general outline of the protein; the H-subunit is omitted) correlating spatial and dielectric distances for some selected charge separated states. Values for dielectric distances shown to the right of the RC were calculated for electron donation from the high potential hemes in *Rp. viridis* as reported by Dracheva et al.(1988) whereas the values to the left of the RC are averages over data from several species studied with respect to electron donation from the low potential hemes (see text). All spatial distances were taken from the crystal structure of the *Rp. viridis* RC.

potential heme is still approximately as electrogenic as is donation from the high potential hemes. This would predict an ϵ_{eff} of 7, which is just intermediate between those for hemes 1 and 3, i.e. suggesting that the magnitude of ϵ_{eff} is dominated by the distance from the membrane. In Itoh's work on *Cm. vinosum* (1980) and Prince's study on *Tc. pfennigii* (1978), however, the membrane potential titrates rather with the higher component of the low potential hemes. In that case, ϵ_{eff} must be as high as 15. Since hemes 2 and 4 are rather exposed to solvent (Deisenhofer and Michel, 1989; Gunner and Honig, 1990), the high and low dielectrics would in this case predominantly depend on solvation energy.

Alternatively, it is possible that in the respective species, heme 4 has a higher E_m than heme 2 (Nitschke and Rutherford, 1994). In that case, the electrochromic response should increase on reduction of the lowest potential heme, an effect which might be rendered difficult to observe by a compensating decrease in efficiency of charge separation due to chemical reduction of Q_A (for indications of such an effect see Nagashima et al., 1993a).

Thus, the presence of a drastically inhomogeneous dielectric medium in the region of the cytochrome subunit as suggested by electrostatic calculations (Gunner and Honig, 1991) can explain why previous interpretations of $\Delta\Psi$ measurements (based on constant ϵ_{eff}-values) yielded models in conflict with the crystal structure.

Once the contributions of the individual hemes to $\Delta\Psi$ formation (e.g. via time-resolved experiments) will be resolved in tetraheme systems for which no crystal structure is available, this method can become a powerful tool for investigating the order of redox potentials in these systems.

4. Sequence Comparisons

a. The 'Building Blocks' of a Tetraheme Cytochrome

By the time of completion of this contribution, the primary sequences of the RC associated cytochromes from *Rp. viridis* (Weyer et al., 1987a), *Cf. aurantiacus* (Dracheva et al., 1991), *Roseobacter (Ro.) denitrificans* (formerly called *Erythrobacter* sp. OCH 114; only a partial sequence is available so far; Liebetanz et al., 1991) and *Rc. gelatinosus* (Nagashima et al., 1993b, 1994) have been published. All four amino acid sequences are indicative of four heme groups bound to the protein by sequence motifs very similar to what is seen in the crystal structure of the *Rp. viridis* cytochrome subunit (see Fig. 4). The amino acid compositions determined for the cytochromes from *Cm. vinosum* (Kennel and Kamen, 1971b) and *Ectothiorhodospira* sp. (Lefèbvre et al., 1984, 1989) agree well with those calculated from the two complete purple bacterial sequences providing additional evidence for structural relatedness of all these proteins.

An alignment of the four sequences is shown in Fig. 4. This alignment is similar to those proposed by Dracheva et al. (1991) and by Nagashima et al. (1993b, 1994) with one exception: the stretch proposed to form an α-helix parallel to heme IV in *Cf. aurantiacus* is shifted by three residues (i.e. one helix turn) with respect to the other sequences (see below).

Note that the first three sequences are all from purple bacteria, i.e. from species belonging to the proteobacterial phylum, from which the organism shown in the last line, i.e. *Cf. aurantiacus*, is evolutionarily rather distant, according to 16 Sr-RNA (Woese, 1987).

The sequence stretches corresponding to the structural elements of the *Rp. viridis* subunit as discussed in section III.A.1 are indicated above the amino acid sequence of the *Rp. viridis* cytochrome (uppermost line). Obviously, the amount of detail permitted in conclusions concerning such structure elements in the other species depends on the quality of the alignment. Whereas this alignment seems to be quite reliable within each of the diheme segments due to the possibility to align heme binding motifs, the conservation of amino acids in other parts of the sequence is rather poor, resulting in a large number of possible alignments with comparable probabilities. To facilitate comparisons with the original articles, the previously proposed alignments for the sequence stretches of low conservation have been adopted. It is of note, however, that these alignments based on homology are not always in line with independent pieces of information. For example, the α-helical part of the segment going down from the N-terminus to the bottom of the cytochrome (see Fig. 1) is correctly predicted (as a Chou-Fasman α-helix) between residues 45 and 56, i.e. PAT...RDA in *Rp. viridis*. The respective stretches with high probability for α-helices align well for the case of *Cf. aurantiacus* (residues 41–51, RAQ...AGA) but badly for that of *Ro. denitrificans* (predicted between residues 59 and 67, HVA...VTR). Secondary structure prediction would therefore suggest that the *Ro. denitrificans* sequence should be shifted by 7 residues with respect to the other species. No α-helix is predicted for *Rc. gelatinosus* in this region of the protein.

The global alignment of the four sequences is nevertheless validated by numerous spots of significant homology throughout the entire sequence and it is therefore likely that the arrangement of structural elements seen in the *Rp. viridis* crystal structure is conserved in the other three species

```
                           Presequences                      Interaction surface with the L-subunit
R. vir.    1                  MKQLIVN-S---VA-T-VALASLVAG 20    21 CFEPPP-ATTTQTGFRGLSHGE-VLH          44
R. gel.    1                  MA-LAVRISTLTVAVT-AA-A-LLAG 22    23 CERPPVDAV--QRGYRGTGMQHIVN-         45
R. den.    1 MFPKWFDKWNADNPTNIFGPAILIGVLGVAVFGAA-A-IVSI 40     41 GNPAQTASM--QTGPRGTGMHVAEFN          64
C. aur.    1           MQSSRPSDRQLAI------VV-S-VAVGIVVA- 24    25 -------VITTATFWWVYDLT---LG          40

        Chain spanning the distance between the membrane and the far end of the cytochrome
R. vir.   45 PATVKAKKERDAQYPPALA---A--VKAE--GPPV-----------SQVYK-----------------NVKVL            82
R. gel.   46 PRTL-A--E---QIPTQQA-PVATPV-ADNSGPRA-----------NQVFQ-----------------NVKVL            82
R. den.   65 VTRF-A--P---DPTIEEY-YTEAPY-IPEGGEEL-----------AKDIY----------------ENVQVL          102
C. aur.   41 RAQREAAQTAGARWSPSDG--IK--VITS--SPPVTPTDGRQNWMGTQAWNEGVQAGQAWIQQYPNTVNVQVL          107

                                          1st diheme segment
                       *                                                      *
R. vir.   83    GNLTEAEFLRTHTAITEWVSP             --             QEGCTYCHDE---NNLASEA        120
R. gel.   83    GHLSVAEFTRQMAAINEWVAP             --             TEGCNYCHTE---N-LADDS        119
R. den.  103    GDLTDDNFNRVMTAMTQWIAP             --             EEGCVYCHGEGDLETYGEDN        143
C. aur.  108    IGMSSAQIWTYM---QQYVSG             AL             GVGCQYCHNI---NNFASDE        144
                                                  +
R. vir.  121     KYPYVVARRMLEMTRAINTNWT          QHVA     Q---TGVTCYTCHRGTPLP               162
R. gel.  120     KYQKVVSRRMLEMTQKVNTQWT          HHVA     A---TGVTCYTCHRGNPVP               161
R. den.  144     LYTKVVARRMIHMTQNINENWD          GHVN     ANAEVGVNCYTCHRGEHV···             187
C. aur.  145     YPQKIAARNMLRLVRDVNAEFI          VNLP     NWQGNYVQCATCHNN--AP               187
                          *                                              *
                                 Chain connecting the two diheme segments
R. vir.  163    ---------------PYVRYLEP----TLPLNNRETPTHVERVETRSGYVVRLAKYTAYSALNYDPFTMFLANDKRQ--V   221
R. gel.  162    ---------------KEIWFTAV----PQNKRADFIGNLD--GQNQAAKVVGL---T--S-LPYDPFTTFL---KEETNV  211
C. aur.  188    NNLEGFGAQFINSVPPIKVTVDP-----LDANGMAILDPAQKPEAIREPVL-LKDAILFYIYNYQVWKPFDPNDPES--G   259

                Loop in contact with the membrane and the N-terminus of the H-subunit
R. vir.  222    RVVPQTALPLVGVSRGKERRPLS                                                          244
R. gel.  212    RVYGTTALPT-GTS--KA-----                                                          226
C. aur.  260    R----GSLALT-YDGGR---TQD                                                          274

                       *                2nd diheme segment                    *
R. vir.  245    DAY-A--TFALMMSISDSL                                 GTNCTFCHNAQTFESWGKKS--------  280
R. gel.  227    DIKQAEKTYGLMMHFSGAL                                 GVNCTYCHNTNGFGSWDNAA--------  265
C. aur.  275    QVT-I--NQNVMNYQAWSL                                 GVGCTFCHNSRNFVAYELNPAGDNVLNP  318

R. vir.  281    --TPQRAIAWWGIRMVRDLNMNYLAP--------LNASLPASRLG--RQGEAP---QADCRTCHQGVTKPLF         337
R. gel.  266    ---PQRATAWYGIRMARDLNNNFMEG--------LTKTFPAHRLG--PTGDVA---KINCSTCHQGAYKPLY         321
C. aur.  319    LYAYNKLKAQRMLLLTTWLAENWPRYGAIAKPEIPTGSGAASRYSYQRLGDGQIYNVPGCYTCHQGNNIPLA         390
                              *                                                  *
                C-terminal stretch
R. vir.  338    GASRLKDYPELGPIKAAAK                                                              356
R. gel.  322    GAQMAKDYPGLKPAPAAAAASAVEAAPVDAAASAAPVATVATAAK                                    366
C. aur.  391    SINQANIPSGDAGIVVLPPQIRGR                                                         414
```

Fig. 4: Sequence comparison for the tetraheme subunits (including putative precursor peptides) from the purple bacteria *Rp. viridis*, *Rc. gelatinosus* and *Ro. denitrificans* as well as from the green filamentous bacterium *Cf. aurantiacus*. Residues conserved among the purple bacteria and among all four species are marked as bold letters and by bars above the purple bacterial and below the *Cf. aurantiacus* sequences, respectively. The positions of the 'usual' 5th and 6th ligands are indicated by '*' whereas the position of the 'unusual' 6th histidine ligand to heme 2/IV is denoted by '+'. The layout of the representation reflects the structure elements as defined in Fig. 1a and in the text. Only a partial primary sequence is available for *Ro. denitrificans*.

sequenced so far: An N-terminal contact region is followed by a segment protruding away from the membrane towards the far end of the subunit (containing an α-helical part at the beginning and the highly conserved stretch NVXVL at the end), the 1st and the 2nd diheme segments connected by a long stretch of low homology and by a C-terminal patch of variable length with essentially no homology.

The layout of the sequence comparison shown in Fig. 4 is meant to emphasize these structural elements. Sequence stretches within the two diheme segments have been arranged in order to locate 5th and 6th ligands (i.e. α-helices and CXXCH-patches) at equivalent positions.

In previous alignments Met330 of *Cf. aurantiacus* has been aligned with Met293 of *Rp. viridis*, because Met330 most probably serves as 6th ligand for heme IV in *Cf. aurantiacus* (Dracheva et al., 1991; also see below). An inspection of the crystal structure, however, shows that Gly290 rather than Met293 would be at the correct position for liganding heme IV. Moreover, analysis of internal homology between different heme binding sites in the tetraheme subunit of *Rp. viridis* (Weyer et al., 1987a) also favors alignment of Gly290 in the segment binding heme IV with the 6th ligand Met in the segments binding hemes I–III. The stretch 317–344 in the *Cf. aurantiacus* sequence has been realigned accordingly, i.e. placing Met330 (*Cf. aurantiacus*) at the position of Gly290 (*Rp. viridis*).

b. Attachment to the Reaction Center

Within the N-terminal part interacting with the L-subunit in *Rp. viridis*, a rather conserved patch is found in the purple bacterial cytochromes between residues 31 and 36 (*Rp. viridis* numbering). This conservation of amino acid residues does not extend to the green filamentous bacterium *Cf. aurantiacus*. Moreover, in *Cf. aurantiacus*, the predicted respective stretch of residues is significantly shorter than in the purple bacteria. The recognition sequence for signal peptidase II and an N-terminal cysteine residue are only present in *Rp. viridis* and *Rc. gelatinosus*. Therefore, it seems likely that both *Ro. denitrificans* and *Cf. aurantiacus* do not have the lipid anchor.

Another important piece of sequence contributing to interaction of the cytochrome with the RC is the loop between residues 222 and 244 (*Rp. viridis*). The residues within this loop that are in contact with the N-terminus of the H-subunit (see Figs. 1 and 4) are fully conserved in the purple bacteria (residues 227-230, 'TALP'). This homology is absent in *Cf. aurantiacus* in line with the finding that the *Chloroflexus* RC does not contain the H-subunit (Blankenship et al., 1983).

For the short loop clamped to the membrane by the C-terminal 'arm' of the M-subunit (see Figs. 1 and 4), i.e. residues 276–280 in *Rp. viridis*, no homologous residues are seen in the other species. However, the *Chloroflexus* sequence has an insertion of 10 amino acid residues at precisely this position, indicating that the respective loop might be more extended than in purple bacteria. It is tempting to correlate such a larger loop with the finding that the C-terminal arm of the M-subunit in *Cf. aurantiacus* is only half as long as it is in purple bacteria (Nagashima et al., 1993b, 1994).

c. Primary Sequence and Heme Orientation

The non-conservation of heme orientations between species has been tentatively attributed (Nitschke et al., 1993) to the low homology in the regions of the cytochrome in contact with the RC and in the stretch that runs along the protein from the N-terminal contact area with the membrane to the far end near heme I. Low conservation in this latter part is certainly also due to its being largely exposed to the aqueous phase resulting in low constraints on mutability. However, the almost twice as long respective stretch in *Cf. aurantiacus* as compared to the purple bacterial sequences is likely to produce an altered tertiary structure in this region of the protein which might entail a tilt angle of the whole subunit different from that in *Rp. viridis*. In that respect it is noteworthy that the possible absence of an α-helical part in this segment in *Rc. gelatinosus* would also result in an altered 'spatial' length of the segment, even if the 'sequence' length is roughly conserved. For the third species studied with respect to heme orientation, i.e. *Cm. vinosum*, no sequence data are available as yet.

As suggested by the model proposed in Fig. 5, the above discussed local alterations of structural elements could result in heme orientations significantly differing between species while still retaining a globally similar structure of the various tetraheme subunits.

The examination of further species both with respect to structural and sequence data, or site-directed mutagenesis in the respective sequence segments will provide a more solid basis for these

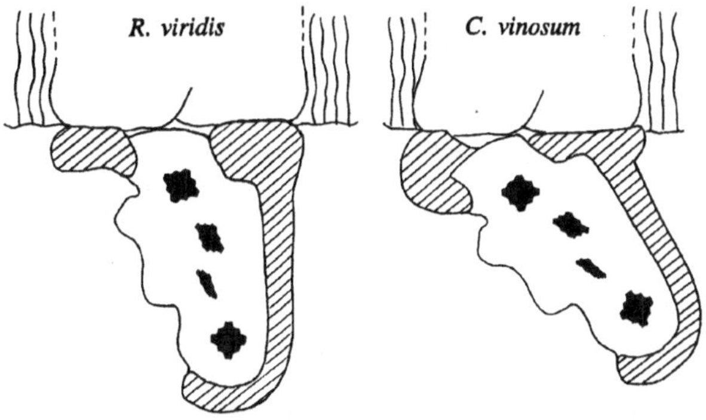

Fig. 5. Comparison of the attachment of the tetraheme subunit from *Rp. viridis* to the RC with a hypothetical structure for this attachment based on data obtained for *Cm. vinosum*. The shaded areas denote the structure elements corresponding to residues 21 to 82 and 163 to 244 in the sequence of the *Rp. viridis* subunit (see Fig. 4). The figure was reprinted with permission from Nitschke et al., 1993).

kinds of considerations.

d. Primary Sequence and Electrochemistry

The 6th ligand of heme IV being histidine instead of methionine in *Rp. viridis* contributes significantly to the fact that this heme has a low redox potential. This histidine is found to be conserved in all three purple bacterial sequences whereas the 'legitimate' 6th ligand methionine is replaced by a conserved glycine (Gly 290 in *Rp. viridis*) the small size of which might be required to facilitate access of the histidine to the heme without major displacements of the α-helix (see Fig. 2). This pattern of conservation indicates that heme 2 is probably a low potential heme in *all* these species. It therefore seems likely that the alternating sequence of redox midpoint potentials is common to these species. However, it cannot be ruled out that in some of these species the order might be high-low-low-high-special pair.

Whereas histidine as 6th ligand to heme IV is conserved in purple bacterial sequences, it is absent in the sequence of the *Cf. aurantiacus* cytochrome. Since no histidines besides the four His residues contained in the CXXCH-stretches are found in this sequence, the 6th ligand to heme IV must be some other residue, most probably (based on the alignment) Met330. It appears tempting to suggest that this is the reason why no second heme titrating around 0 mV is found in *Cf. aurantiacus* (Zannoni and Venturoli, 1988; Freeman, 1989; Freeman and Blankenship, 1990; Van Vliet et al., 1991). However, already the highest potential heme in the *Cf. aurantiacus* tetraheme subunit has an E_m that is about 100 mV lower than that of the respective heme in *Rp. viridis*. Therefore, the influence exerted by the protein is obviously too dominant to allow such offhand conclusions. More structural and/or sequence information is required to assay the contributions of various factors on redox midpoint potentials of the four hemes. In this context it is of note that five out of seven residues pointed out by Gunner and Honig (1991) as significantly influencing the E_m-values of the hemes in *Rp. viridis*, are conserved in the purple bacterial sequences, i.e. the arginines 128, 157, 222, 284 and 292. Important pieces of information with respect to this question can also be expected from the primary sequence of the cytochrome subunit in *Rs. molischianum* where three low potential hemes and only one high potential heme have been found (Nagashima et al, 1993a).

C. Electron Transfer Reactions Involving Tetraheme Cytochromes

1. Electron Transfer Involving the Tetraheme Subunit and the RC

a. At Room Temperature

Kinetic parameters of the electron donation from the RC associated cytochromes to the photooxidized special pair have been determined by optical spectroscopy as well as by direct electrical measurements in *Cm. vinosum* (Chance and Nishimura, 1960; DeVault and Chance, 1966; Parson, 1969; Kihara and Chance, 1969; Parson and Case, 1970; Seibert and DeVault, 1970; Case et al., 1970;

Dutton et al., 1971; Case and Parson, 1971; Kihara and McCray, 1973), *Cm. minutissimum* (Rubin et al., 1989; Shinkarev et al., 1991), *Ec. shaposhnikovii* (Chamorovsky et al. 1985), *Rc. gelatinosus* (Dutton et al., 1970) and *Rp. viridis* (Case et al., 1970; Holten et al., 1978; Shopes et al., 1987; Dracheva et al., 1986; Skulachev et al., 1987; Dracheva et al., 1988; Gao et al., 1990; Ortega and Mathis, 1992; Ortega and Mathis, 1993).

Whereas in previous publications, kinetics of P^+-rereduction by the cytochrome in *Rp. viridis* were interpreted to be monoexponential, Ortega and Mathis (1993) recently claimed to be able to distinguish multiple (up to three) kinetic components. Under all experimental conditions, however, there was one phase which was strongly dominant, the rate constants of which being comparable to those reported earlier.

In *Rp. viridis*, when only the highest potential heme (heme 1) is reduced prior to the flash, photooxidized P^+ is rereduced by this heme with a dominant phase of $t_{1/2}$ = 230 ns and a minor phase (about 10%) having $t_{1/2}$ = 2.1 μs (Ortega and Mathis, 1993).

When the potential is lowered so that both high potential hemes are reduced (i.e. heme 1 and heme 3), both donation phases from heme 1 to P^+ are sped up to $t_{1/2}$ = 190 ns (Holten et al., 1978; Dracheva et al, 1986; Skulachev et al., 1987; Shopes et al., 1987; Dracheva et al., 1988; Ortega and Mathis, 1993) and 1.5 μs (10%; Ortega and Mathis, 1993). Heme 1 is subsequently rereduced by heme 3 with $t_{1/2}$ of 1.7μs (Shopes et al., 1987; Dracheva et al., 1988; Ortega and Mathis, 1993).

At an E_h of –20 mV (i.e. hemes 1, 2 and 3 are reduced), the rereduction of P^+ becomes even faster ($t_{1/2}$ = 115 ns, 85% and $t_{1/2}$ = 670 ns; Dracheva et al., 1988; Ortega and Mathis, 1993), however, the heme which is seen to be oxidized with the same kinetics is now heme 2. This can be interpreted in two ways. (a) Once the low potential heme 2 is reduced, it donates its electron directly (without passing via heme 1) to P^+. (b) The kinetics seen for P^+ rereduction corresponds to the oxidation kinetics of heme 1. This oxidation of heme 1, however, is masked by a much faster rereduction of heme 1 by heme 2 (<40 ns; Ortega and Mathis, 1993). (a) seems to be in conflict with considerations based on electron transfer theory (Moser et al., 1992) predicting rates for the direct electron transfer from heme 2 to P^+ that are significantly slower than what is observed (Ortega and Mathis, 1993).

The speed-up of the dominant phase from 230 ns through 190 ns to 115 ns when one, two or three hemes are reduced, can be rationalized by an electrostatic effect of the electrons on hemes 2 and 3 on the driving force for electron transfer from heme 1 to P^+. Neglecting electrostatic influences on the E_m of P/P^+, driving forces of 120 mV, 134 mV and 211 mV are predicted by electrostatic calculations (Gunner and Honig, 1991; Ortega and Mathis, 1993) for the respective three states. Assuming these values for ΔG, the measured electron transfer rates agree well with the free-energy-rate-relationship proposed by Moser et al. (1992) for this reaction.

In the other species studied so far, the electron donation seems to proceed slightly slower with values for $t_{1/2}$ varying between 800 ns and 3.5 μs in different species (see Table I). For a structure identical to that of the *Rp. viridis* RC, large variations in driving force would be necessary to explain the altered rate constants. Such variations are not observed [see Table I, E_m (cytochrome) vs. $E_m(P/P^+)$]. Slight differences in structure resulting in altered distances for electron transfer, however, could rationalize the observed differences in rate constants. Such variability in structure has already been invoked based on the lack of conservation of heme orientation between species (see III.B.4.c.). Rearrangements on the order of 0.2 Å would be sufficient to yield the observed differences in rate constants (Moser et al., 1992).

Apart from the absolute value of $t_{1/2}$, the electron donation reactions observed in the other species show most of the characteristic features seen in *Rp. viridis*. Electron donation to P^+ is sped up by a factor of two to four upon additional reduction of the low potential hemes in *Cm. vinosum*, *Cm. minutissimum*, *Ec. shaposhnikovii* (Seibert and DeVault, 1979; Rubin et al., 1989; Chamorovsky et al., 1985) and *Cf. aurantiacus* (Mulkidjanian et al., 1993).

Furthermore, a somewhat slower electron transfer from the cytochrome to P^+ at potentials where only one high potential heme is reduced as compared to the two-hemes-reduced state as reported for *Rp. viridis* by Ortega and Mathis (1993) has been observed in *Ec. shaposhnikovii* (Chamorovsky et a., 1985) and *Cf. aurantiacus* (Mulkidjanian et al., 1994).

The presence of multiple kinetic phases reported by Ortega and Mathis (1992, 1993) was taken as evidence for heterogeneity in the samples and attributed to conformational substates of the RC.

b. At Cryogenic Temperatures

The following description of the behavior of photooxidation of hemes at low temperatures is presented in an order intended to facilitate a global overview of the data. This order of presentation, however, deviates significantly from the chronological order of discoveries in this field. These discoveries constitute a fascinating period of research in photosynthesis performed by remarkable scientists. Interested readers are referred to more 'historical' presentations, e.g. Don DeVault (1989).

In a state, where one or both of the high potential hemes are reduced but the two low potential hemes are oxidized, lowering the temperature results in a relatively small increase in half-times of cytochrome oxidation in *Ec. shaposhnikovii* (Chamorovsky et al., 1980, 1986), *Cm. vinosum* (Sarai and DeVault, 1984), *Cm. minutissimum* (Rubin et al., 1989) and *Rp. viridis* (Ortega and Mathis, 1993) yielding activation energies in the range of 4–8 $kJmol^{-1}$. For *Rp. viridis*, an activation energy of 6–8 $kJmol^{-1}$ was found for the state with only one heme reduced whereas this E_a dropped to 3–5 $kJmol^{-1}$ when both high potential hemes were reduced prior to the flash (Ortega and Mathis, 1993).

By contrast, the extent of heme oxidized by P^+ decreases drastically from close to 100% at room temperature to relatively low values below 100 K, i.e. to < 10% in *Cm. vinosum* (Vredenberg and Duysens, 1964; Sarai and DeVault, 1984), 10–20% in *Ec. shaposhnikovii* (Chamorovsky et al., 1980) and roughly 0% in *Cm. minutissimum* (Rubin et al., 1989). For *Rp. viridis*, a broad range of values has been reported by different groups (20% Kaminskaya et al., 1990; 4% by Ortega and Mathis, 1993; close to 0% by Gao et al., 1990 and Nitschke and Rutherford, 1989; for a conflicting EPR result, however, see Hubbard and Evans, 1989). The transition from full to only fractional oxidation of heme occurs in a narrow temperature range (50–70K) in all studies published so far. Again, the absolute value of this 'transition temperature' varies by as much as 80K for the same species (*Rp. viridis*) between groups. From the work of Ortega and Mathis (1993), though, it seems likely that the transition temperature falls in correlation with increasing number of hemes reduced (being 250K with one heme reduced and 210K with two hemes reduced) prior to the flash. This again is indicative of an increase in driving force for electron donation to P^+ due to electrostatic interactions when an increasing number of the outer hemes are reduced (see III.A.2.b.).

Reports observing blockage of electron transfer from heme to P^+ already at room temperatures in the presence of cryoprotectants above a well-defined threshold concentration in *Cm. vinosum* (Kihara and McCray, 1973) or induced by partial dehydration in *Ec. shaposhnikovii* (Chamorovsky et al., 1986) and in *Cm. vinosum* (Kihara and McCray, 1973) might actually observe related effects.

The charge separated ($heme^{ox} Q_A^-$) state recombines with halftimes of 500 ms (recombination between heme 1 and Q_A) and 7s (recombination between heme 3 and Q_A) below 200K in *Rp. viridis* (Gao et al., 1990; Kaminskaya et al., 1990). An in-depth study of kinetic and thermodynamic parameters of these recombination reactions has been presented by Gao et al. (1990). A recombination under these conditions has also been observed in *Cm. vinosum*, its kinetics, however, have not been studied (Sarai and DeVault, 1984). Considerable disagreement prevails with respect to the amount of oxidized high potential heme that does not recombine with the electron on Q_A but becomes stably photooxidized below 100K. Values ranging from 40% (Hubbard and Evans, 1989) through 10% (Kaminskaya et al., 1990) down to values below detection (Nitschke and Rutherford, 1989; Gao et al., 1990) have been reported for the case of *Rp. viridis*.

In contrast to what is seen for *Rp. viridis*, *Cm. vinosum*, *Cm. minutissimum* and *Ec. shaposhnikovii*, significant amounts of stable photooxidation of high potential hemes can be observed below 100 K in *Rc. gelatinosus* (Dutton et al., 1971; Nitschke et al., 1992) and in *Cf. aurantiacus*. In *Cf. aurantiacus*, three of the four hemes were seen to undergo stable photooxidation at 4 K (Van Vliet et al., 1991).

The situation changes drastically when one or both of the low potential hemes are additionally reduced prior to illumination. Under these conditions, the major part of the photoinduced charge dipole of heme $2(4)^{ox} - Q_A^-$ does not recombine any more at low temperatures in all species containing a tetraheme subunit, i.e. low temperature photooxidation of the low potential hemes is an *irreversible* process.

In 1960, Chance and Nishimura have shown that in whole cells of *Cm. vinosum*, electron donation rates from the low potential hemes to P were essentially independent of temperature below 100 K. A subsequent study by DeVault and Chance (1966) established the dependence of donation rates in the temperature range from 300 K down to 30 K, showing that above 100 K, the reaction was described by an activation energy of about 14 $kJmol^{-1}$ whereas it became almost activationless below 100 K. Further experiments on chromatophores and subchromatophore particles (i.e. partially detergent-solubilized material) confirmed the results obtain on whole cells (Dutton et al., 1971). Since in the structural model in use prior to the *Rp. viridis* crystal structure the hemes closest to the special pair were considered to be low potential, a direct donation from these low potential hemes without implication of the high potential hemes was assumed. Electron tunneling from these low potential hemes has been invoked to explain the apparently temperature independent donation reaction at low temperatures (DeVault and Chance, 1966; Hopfield, 1974; Hales, 1976). After the *Rp. viridis* structure was published, alternative explanations were put forward (Bixon and Jortner, 1986, 1989), based on activationless electron transfer rather than tunneling, however, still keeping the idea of direct electron transfer from heme 2 to P^+.

In contrast to the theories based on direct donation from the low potential hemes, other models are based on the assumption that electron donation to P^+ is likely to proceed via heme 1 under all conditions. In these models, it is proposed that the redox equilibrium between heme 1 and P is temperature dependent, i.e. that below a certain threshold temperature the equilibrium shifts from heme 1^{ox}-P^{red} to heme 1^{red}-P^{ox}. This would explain the observed sharp drop in efficiency of the donation reaction.

Additional reduction of the low potential heme 2 would in this model result in (a) a lower transition temperature due to a higher driving force and (b) a rapid rereduction of heme 1 during the time interval when the electron resides transiently on P, thereby rendering the donation event irreversible. In this model, the long (ms) donation times under these conditions would reflect the time of equilibration between all states rather than being a true electron transfer rate.

To rationalize the required temperature dependent equilibrium, Kaminskaya et al. (1990) proposed the existence of an excited intermediate state that has to relax via medium reorganization to the thermodynamic equilibrium. At low temperatures, medium reorganization is proposed to become frozen out making the free energy of the transition state the determining factor for the equilibrium between heme 1 and the special pair.

Although more work is required to decide between the various models proposed, the experimentally similar behavior in the electron transfer reactions described above for all species examined, provides further circumstantial evidence for a rather similar structural and electrochemical arrangement of the four hemes in all these tetraheme cytochromes.

2. Rereduction of the Tetraheme Subunit

The rereduction of photooxidized hemes in RC associated cytochromes has been studied in vivo in *Cm. vinosum* (van Grondelle et al., 1977), *Ec. mobilis* (Leguijt, 1993) and *Rp. viridis* (Olson and Nadler, 1965; Garcia et al., 1993). Both in *Cm. vinosum* and in *Rp. viridis* under most physiological conditions, this rereduction is probably mediated by a soluble cytochrome (Cyt c_2). Cytochrome c_2 was found to be in turn rereduced by the cytochrome bc_1 complex (Shill and Wood, 1984; Coremans et al., 1985; Tan et al., 1993; Garcia et al., 1993). Halftimes of the dominant phases of electron transfer from cytochrome c_2 to the tetraheme subunits were determined to be 300 μs (Coremans et al., 1985) in the case of *Cm. vinosum* and to be 110 μs or 40 μs (Garcia et al., 1993) for aerobic or anaerobic conditions, respectively, in the case of *Rp. viridis*. In both species, these halftimes were independent of the intensities of subsaturating flashes, indicating that they may not correspond to diffusion controlled, second order processes (van Grondelle et al., 1977; Garcia et al., 1993). The formation of an electron transfer complex between cytochrome c_2 and the tetraheme cytochrome prior to the flash was therefore proposed (Garcia et al., 1993). In *Rp. viridis*, even under anaerobic conditions, no involvement of the low potential hemes was seen. It was therefore argued that under most physiological culture conditions, the low potential hemes do not participate directly in rereduction of the photooxidized special pair (Garcia et al., 1993).

The existence of electron transfer complexes has also been reported using purified reaction center and cytochrome c_2 of *Rp. viridis* (Knaff et al., 1991; Meyer et al., 1993). Complex formation as evidenced by the presence of first order rate constants was only observed at low ionic strengths indicative for an electrostatic nature of this complex. The halftimes for the intra-complex electron transfer reaction (2600 μs according to Knaff et al., 1991 and 530 μs according to Meyer et al., 1993) are somewhat longer that those observed in vivo. Studies comparing binding of cytochrome c_2 with either the cytochrome bc_1 complex or the tetraheme cytochrome indicate substantially different docking modes for these two electron transfer complexes (Güner et al, 1993).

Since in *Rp. viridis* the in vivo halftime of donation from cytochrome c_2 to the high potential hemes is an order of magnitude slower than electron transfer from heme 3 to heme 1 within the tetraheme subunit, it was concluded that, if the binding site for cytochrome c_2 was near to heme 1, donation from cytochrome c_2 would always be outcompeted by intrasubunit heme-heme electron transfer. An obligatory donation into heme 3, i.e. the heme with the second highest potential, and consequently a binding site close to this heme was therefore proposed (Meyer et al., 1993) to explain the observed electron transfer events. Molecular modeling studies, however, were interpreted to indicate possible binding sites also near heme 4 (Knaff et al., 1991) and heme 1 (Sogabe et al., 1992).

In the case of a specific strain of *Rc. gelatinosus*, a soluble high potential iron sulfur protein (HiPIP) was recently identified as the physiological electron donor to the RC-associated tetraheme subunit (Schoepp et al., 1994; Schoepp, 1994). The characteristics of inter-protein electron transfer between the HiPIP and the tetraheme subunit were found to resemble those observed for cytochrome c_2 and the cytochrome subunit.

3. What Are Those Tetraheme Subunits Good For?

In the years before the *Rp. viridis* crystal structure, the RC associated cytochromes of purple bacteria were enigmatic on three different levels. (1) Why did RC associated cytochromes exist at all when species like *Rb. sphaeroides* proved that they were dispensable, (2) why was one heme not enough and (3) why were there hemes with redox potentials too low to fit into the cyclic electron transfer scheme. A tentative explanation for (2) and (3) was proposed by the hypothesis that the low potential hemes might serve as entry points for electrons coming ultimately from low potential substrates (for a discussion see Dutton and Prince, 1978), a hypothesis that was as elegant and appealing as it was difficult to check experimentally. Eventually, the crystal structure of the *Rp. viridis* RC and subsequent studies based on this structure provided a wealth of data and another level of mystery, i.e. the question (4) why the four hemes were arranged in an alternating sequence of redox potentials. The interposition of a low potential heme in between two high potential hemes would not seem to be required for this heme to act as an acceptor for electrons coming from low potential substrates but rather points to a more complicated role of this arrangement, especially since it seems to be conserved at least among the purple bacterial tetraheme subunits (III.B.4.d.). For example, electrostatic interaction energies between two adjacent high potential hemes might be too high to allow efficient electron donation to P^+ under all redox conditions (see III.A.2.b.). Although the *Rp. viridis* crystal structure did not solve these questions, answers seem nevertheless likely to be provided by *Rp. viridis*. The recently described transformation system for this bacterium (Laußermair and Oesterhelt, 1992) allows site-directed mutations to be introduced into the tetraheme subunit, providing a tool to study the importance of the individual hemes and their redox midpoint potentials for in vivo metabolic processes. Complete deletion of the cytochrome subunit might furthermore give hints to the kind of advantage provided by these additional, RC associated cytochrome electron donors.

IV. Nature of RC Associated Cytochromes in Photosynthetic Eubacteria

A. Proteobacteria

The evidence that the RC associated cytochromes in purple bacteria all belong to the 'family' of tetraheme cytochromes appeared to be rather compelling until recently. Conflicting data have been presented for the case of the *Ectothiorhodospira* (Leguijt and Hellingwerf, 1991), where the purification of a protein containing only the low potential hemes was reported. Such a situation would certainly raise interesting questions with respect to the evolution of RC associated cytochromes (Leguijt, 1993). Even so, some caution seems to be appropriate since similar observations have been made in *Cm. vinosum* (Doi et al., 1983; Knaff and Kraichoke, 1983) and in *Tc. pfennigii* (Meyer et al., 1973), the RC associated cytochromes of which are now generally accepted to be members of the tetraheme family (see also III.B.4.a.). Alterations of the physicochemical properties of the hemes induced by detergent treatments have been invoked to explain the observed phenomena (Meyer et al., 1973). Moreover, the amino acid composition of the cytochrome copurifying with the RC in *Ectothiorhodospira* sp. agrees well with that of other tetraheme cytochromes (Lefèbvre et al., 1984, 1989). Further circumstantial evidence for the cytochrome associated to the *Ectothiorhodospira* RCs being a true tetraheme cytochrome comes from its photochemical behavior (Chamorovsky et al., 1980, 1985, 1986) which is strikingly similar to that observed in *Rp. viridis* (see III.C.1.).

B. Green Filamentous Bacteria (Chloroflexaceae)

The cytochrome subunit of *Cf. aurantiacus* has been purified and characterized (Freeman and Blankenship, 1990; Meyer et al., 1989; Heibel et al., 1991). A partial amino acid sequence has been obtained by Edman degradation (Freeman, 1989) and subsequently the complete primary sequence was determined from the gene (Dracheva et al., 1991). All these data demonstrate that the *Chloroflexus* cytochrome is a member of the tetraheme family.

C. Green Sulfur Bacteria (Chlorobiaceae)

The nature of the immediate electron donor to P^+ in green sulfur bacteria is presently a matter of debate and both mono- and tetraheme cytochromes have been proposed to fulfill this role.

Four hemes in approximately equal amounts were observed by EPR in membrane fragments of *Chlorobium phaeobacteroides* (Feiler et al., 1989). This was interpreted as evidence for the presence of a tetraheme cytochrome donor in green sulfur bacteria (Nitschke and Rutherford, 1991). This putative tetraheme cytochrome was tentatively identified with a 32 kDa cytochrome *c*-553 found to copurify with the RC (Feiler et al., 1992). The recent purification and characterization of the 32 kDa cytochrome *c*-553 demonstrated, in fact, that this cytochrome is a multiheme protein (Albouy et al., 1994). By contrast, in a preparation of the RC from *Cb. vibrioforme* (Okkels et al., 1992), an 18 kDa cytochrome *c*-551 was observed. Its sequence was determined from the respective gene (named *pscC*, formerly called *cycA*), demonstrating that it contained only one heme binding motif (Okkels et al., 1992). No sequence homology to tetraheme cytochromes was seen and a (weak) resemblance was only found to the membrane-anchored *cycM* gene product studied in *Bradyrhizobium japonicum* (Bott et al., 1981). Deletion of the *pscC* gene seemed to be lethal for the organism (Kjærulff et al., 1993) suggesting an indispensable metabolic function of the gene product. Possibly similar cytochromes were reported in RCs of *Cb. limicola* by Hurt and Hauska (1984; 24 kDa, Cyt *c*-550.5) and Oh-oka et al. (1993; 21 kDa, Cyt *c*-551). The α-band wavelength of 550–551 nm is significantly different from the wavelength (553 nm) of the heme photooxidized in membrane fragments (Olson and Sybesma, 1963; Sybesma and Vredenberg, 1964; Sybesma, 1967; Fowler et al., 1971; Prince and Olson, 1976; Swarthoff and Amesz, 1979) and in whole cells (Feiler and Nitschke, unpublished) or the heme components found to be abundant in *Chlorobium* (Gibson, 1961; Fowler et al., 1971), but this might again be due to detergent-induced alterations of spectral properties.

It is of note, however, that in double flash experiments on *Cb. limicola*, Prince and Olson (1976) obtained differing E_m-values for the 1st and 2nd flash-induced cytochrome oxidation reminiscent of the situation found with purple bacterial tetraheme subunits (see III.B.1.). Two of the four hemes observed in EPR equilibrium titrations (Feiler et al., 1989) had E_m values comparable to those determined in the double flash experiment (see Table I) which was considered as further support for the cytochrome associated to the *Chlorobium* RC being a tetraheme system (Feiler et al., 1989; Nitschke and Rutherford, 1991). Similarly different titration curves for 1st and 2nd flash induced cytochrome oxidation in *Cb. tepidum* were recently interpreted assuming two identical monoheme cytochromes associated with one RC (Okumura et al., 1994). If this model should turn out to be correct, then *Chlorobium* might provide an example where the rationalizing of multiple flash data by assuming two or more identical electron donating cytochromes would eventually be applicable (see III.B.1.). Both the 18 kDa cytochrome *c*-551 and the 32 kDa cytochrome *c*-553 were shown to contain hemes with E_m-vlaues in the range of those of the photooxidized heme(s) (Kjaer et al., 1994; Albouy et al., 1994).

No stable photooxidation of hemes at low temperatures has been observed in green sulfur bacteria.

For a more detailed description of the green sulfur bacterial reaction center, the reader is referred to the contribution by Hauska and Feiler in this volume.

D. Heliobacteria

In Firmicutes (formerly called Gram-positive phylum; see Wayne et al., 1987), the last phylum containing photosynthetic representatives (i.e. the Heliobacteria), data with respect to the immediate electron donor to the RC are scarce. A membrane-bound cytochrome *c*-553 can donate to the photooxidized primary donor with $t_{1/2}$ in the ms-range in membrane fragments (Prince et al., 1985; Smit et al., 1987). The E_m of this heme photooxidized in membranes has been determined to +170 mV (Prince et al., 1985). In whole cells, electron transfer from cytochrome *c*-553 to P^+ is significantly faster and biphasic (Vos et al., 1989). Promising candidates on the level of proteins for being cytochrome *c*-553 are a 50 kDa and a 17 kDa protein which were found in membrane samples (Fuller et al., 1985; Trost and Blankenship, 1990; Nitschke and Liebl, 1992). All purification procedures reported so far yielded photosynthetic reaction centers devoid of cytochrome (Trost and Blankenship, 1989; Van de Meent et al., 1990). From optical spectra, no differentiation into multiple heme components has been achieved so far. In EPR, however, three to four inequivalent hemes were observed in membranes, two of which titrated around 170 mV whereas the other one or two had potentials somewhat below –50 mV (Liebl, Rutherford and Nitschke, unpublished). Multiple flash experiments on whole cells also yielded distinguishable titration curves (Nitschke, unpublished), reminiscent of what is described in III.B.1. for the case of the tetraheme cytochromes from purple bacteria (however, see IV.C.). Indications for more than one photooxidizable heme in a fraction of RCs have also been reported by Kleinherenbrink and Amesz (1993).

Just as in green sulfur bacteria, no (Nitschke and Rutherford, unpublished) or very low (4%; Smit et al., 1989) photooxidation of hemes at low temperatures (< 100K) has been observed in Heliobacteria (see also Chapter 31 by Amesz on the heliobacterial RC).

V. Some Considerations with Respect to Evolution

Efforts in the fields of structure determination, protein/gene sequencing and functional studies have transformed the subject 'evolution of photosynthetic reaction centers' from an intellectual pastime discussed in scientific 'salons' into experimental research performed on lab-benches and in CPUs (for a review see Blankenship, 1992). There is some reason to believe that the two types of reaction centers, i.e. RCI (as represented by the green sulfur and heliobacterial RCs as well as PS I) and RCII (as represented by the RCs from purple and green filamentous bacteria as well as PS II) have evolved from a common origin.

In phylogenetic trees based on 16 S r-RNA (Woese, 1987), green filamentous bacteria are evolutionarily the earliest photosynthetic phylum to branch off from the tree. Since these organisms were shown to possess a tetraheme cytochrome subunit related to that of purple bacteria (IV.B.) and since some data suggested that the cytochrome subunit in green sulfur and heliobacteria might also be a tetraheme cytochrome (IV.C./D.), it was inferred that the principle of the tetraheme subunit was ancient and that such cytochromes already served as electron donors to the common ancestor of all extant types of reaction centers (Nitschke and Rutherford, 1991).

However, based on inconsistencies of phylogenetic trees constructed from 16S r-RNA and from sequences of the RC-subunits L and M, the conclusion was drawn that the photosynthetic apparatus of *Chloroflexus* might have been imported via lateral gene transfer from an ancestor of the Proteobacteria (Blankenship, 1992). This would mean that the photosynthetic apparatus of *Cf. aurantiacus* cannot be related to the position of the organism within the phylogenetic tree of the eubacteria. This would weaken the significance of the *Chloroflexus* cytochrome subunit being a tetraheme protein with respect to evolutionary scenarios.

Furthermore, recent results suggest that the cytochrome subunit in green sulfur bacteria might be a monoheme cytochrome (IV.C.). On the basis of this finding, an alternative model for the evolution of RC associated cytochromes has been put forward (Kjær et al., 1992). In this model, a monoheme would have been attached to the ancestral reaction center (and would still be present in extant RCI-type reaction centers), subsequently evolving (via two gene duplications) into the tetraheme subunits of purple and green filamentous bacteria.

As pointed out by Okkels et al. (1992), if the green sulfur bacterial subunit was a monoheme cytochrome, then it would also be possible that the ancestral RC was devoid of bound secondary electron donors and mono- or tetraheme cytochromes were captured only later in evolution by the different descendants of the primordial RC. If the *pscC* gene product should turn out to be the immediate secondary donor to P^+ in green sulfur bacteria, then the latter hypothesis actually seems more likely than an evolution from a mono- through a di- into a tetraheme cytochrome. From the amino acid sequence of *pscC* it seems clear that not only sequence but also structural homologies are absent between this monoheme and the tetraheme cytochromes. The proposed 6th ligand to the heme, methionine, in *pscC* is located about 30 residues further down towards the C-terminal end of the protein than is the CXXCH-heme binding motif. A heme binding structure element comparable to that seen for all four hemes in the tetraheme subunit (see III.A.1.b.) would therefore be excluded in *pscC*.

Thus, a number of rather different scenarios for the evolution of the RC associated cytochromes are presently being proposed. This might seem reminiscent of the situation way back in the days of the evolutionary 'salons'. The present situation is nevertheless different in that further data can be expected to gradually exclude some of these scenarios.

Both the hypothesis on the importance of lateral gene transfer (see for example Nagashima et al., 1993b reporting further inconsistencies between phylogenetic trees based on 16S r-RNA and protein sequences) and the controversy on the nature of the RC associated cytochrome in green sulfur bacteria (see Kjærulff et al., 1993, presenting a transformation system for *Chlorobium* which can ultimately allow site-directed mutagenesis to be performed on this organism) are presently under investigation.

Therefore, a much clearer picture for these evolutionary considerations can be expected in the near future and the authors of a review article on RC associated cytochromes to be written in 10 years time will probably shake their heads in wonder about what people could have been imagining in 1993.

Acknowledgments

We thank Drs. J. Amesz, R.E. Blankenship, M.A. Cusanovich, G. Hauska, B. Kjær, F.A.M. Kleinherenbrink, T. Leguijt, P. Mathis, K. Matsuura, A.Ya. Mulkidjanian, K.V.P. Nagashima, N. Okumura, R.C. Prince, H.V. Scheller, J. Shiozawa, D.M. Tiede and A. Verméglio for communicating results prior to publication. Thanks are furthermore due to Drs. A.W. Rutherford, M.R. Gunner, I. Agalidis, D. Albouy, J. Breton, S. Creuzet, F. Drepper, G. Drews, P.L. Dutton, U. Feiler, M. Hippler, A. Hochkoeppler, P. Joliot, D.B. Knaff, U. Liebl, B. Schoep, J.T. Trost, J. Vachtweitl, P. van Vliet and D. Zannoni for stimulating discussions.

Molecular models were produced on a SG computer by S. Grzybek using the GRASP-software (developed by A. Nicholls and B. Honig) and on a PC (with skillful assistance by J. Reichert) using the pdViewer software developed by H.H. Robinson and A.R. Crofts. Coordinates were obtained from the Brookhaven Data Bank (entry 1PRC).

Access to computer facilities was kindly provided by W. Mäntele and M. Wissen.

References

Agalidis I, Rivas E and Reiss-Husson (1990) Reaction center light harvesting B875 complexes from *Rhodocyclus gelatinosus*: Characterization and identification of quinones. Photosynth Res 23: 249–255

Albouy D, Feiler U, Sturgis J, Nitschke W and Robert B (1994) The cytochromes associated with the cytoplasmic membrane in *Chlorobium limicola* f. *thiosulfatophilum*. Abstracts of the VIII Intl Symposium on Phototrophic Prokaryotes, p 168B. Urbino, Italy

Alegria G and Dutton PL (1990) Spectroscopic and electrochemical resolution of the high potential hemes in the reaction center of *Chromatium vinosum* using oriented Langmuir-Blodgett films. Biophys J 57: W–Pos607

Alegria G and Dutton PL (1991a) Langmuir-Blodgett monolayer films of bacterial photosynthetic membranes and isolated reaction centers: Preparation, spectrophotometric and electrochemical characterization. Biochim Biophys Acta 1057: 239–257

Alegria G and Dutton PL (1991b) Langmuir-Blodgett monolayer films of the *Rhodopseudomonas viridis* reaction center: determination of the order of the hemes in the cytochrome *c* subunit. Biochim Biophys Acta 1057: 258–272

Bixon M and Jortner J (1986) On the mechanism of cytochrome oxidation in bacterial photosynthesis. FEBS Lett 200: 303–308

Bixon M and Jortner J (1989) Cytochrome oxidation in bacterial photosynthesis. Photosynth Res 22: 29–37

Blankenship RE (1992) Origin and early evolution of photosynthesis. Photosynth Res 33: 91–111

Blankenship RE, Feick R, Bruce BD, Kirmaier C, Holten D and Fuller RC (1983) Primary photochemistry in the facultative green photosynthetic bacterium *Chloroflexus aurantiacus*. J Cell Biochem 22: 251–261

Bott M, Ritz D and Hennecke H (1991) The *Bradyrhizobium japonicum cyc M* gene encodes a membrane-anchored homolog of mitochondrial cytochrome *c*. J Bacteriol 173: 6766–6772

Bruce BD, Fuller RC and Blankenship RE (1982) Primary

photochemistry in the facultative aerobic green photosynthetic bacterium *Chloroflexus aurantiacus*. Proc Natl Acad Sci USA 79: 6532–6536

Büttner M (1993) Molekularbiologische charakterisierung des P840-reaktionszentrums von *Chlorobium limicola* f. sp. *thiosulfatophilum*. PhD thesis, University of Regensburg, Germany

Case GD and Parson WW (1971) Thermodynamics of the primary and secondary photochemical reactions in *Chromatium*. Biochim Biophys Acta 253: 187–202

Case GD and Parson WW (1973) Shifts of bacteriochlorophyll and carotenoid absorption bands linked to cytochrome *c*-555 photooxidation in *Chromatium*. Biochim Biophys Acta 325: 441–453

Case GD, Parson WW and Thornber JP (1970) Photo-oxidation of cytochromes in reaction center preparations from *Chromatium* and *Rhodopseudomonas viridis*. Biochim Biophys Acta 223: 122–128

Chamorovsky SK, Kononenko AA, Remennikov SM and Rubin AB (1980) The oxidation rate of high-potential *c*-type cytochrome in the photochemical reaction centre is temperature-independent. Biochim Biophys Acta 589: 151–155

Chamorovsky SK, Drachev AL, Drachev LA, Karagul'yan AK, Kononenko AA, Rubin AB, Semenov AY and Skulachev VP (1985) Fast phases of generation of the transmembrane electric potential in chromatophores of the photosynthetic bacterium *Ectothiorhodospira shaposhnikovii*. Biochim Biophys Acta 808: 201–208

Chamorovsky SK, Kononenko AA, Petrov EG, Pottosin II and Rubin AB (1986) Effects of dehydration and low temperatures on the oxidation of high-potential cytochrome *c* by photosynthetic reaction centers in *Ectothiorhodospira shaposhnikovii*. Biochim Biophys Acta 848: 402–410

Chance B and Nishimura M (1960) On the mechanism of chlorophyll-cytochrome interaction: The temperature insensitivity of light induced cytochrome oxidation in *Cm. vinosum*. Proc Natl Acad Sci USA 46. 19–24

Clayton RK and Clayton BJ (1978) Molar extinction coefficients and other properties of an improved reaction center preparation from *Rhodopseudomonas viridis*. Biochim Biophys Acta 501: 478–487

Cogdell RJ and Crofts AR (1972) Some observations on the primary acceptor of *Rps. viridis*. FEBS Lett 27: 176–178

Coremans JMCC, van der Wal HN, van Grondelle R, Amesz J and Knaff DB (1985) The pathway of cyclic electron transport in chromatophores of *Chromatium vinosum*. Evidence for a Q-cycle mechanism. Biochim Biophys Acta 807: 134–142

Cusanovich MA, Bartsch RG and Kamen MD (1968) Light-induced electron transfer in *Chromatium* strain D. II. Light-induced absorbance changes in *Chromatium* chromatophores. Biochim Biophys Acta 153: 397–417

Deisenhofer J and Michel H (1989) The photosynthetic reaction centre from the purple bacterium *Rhodopseudomonas viridis*. EMBO J 8: 2149–2169

Deisenhofer J, Epp O, Miki K, Huber R and Michel H (1985) Structure of the protein subunits in the photosynthetic reaction centre of *Rhodopseudomonas viridis* at 3 Å resolution. Nature (Lond.) 318: 618–624

DeVault D (1989) Tunneling enters biology. Photosynth Res 22: 5–10

DeVault D and Chance B (1966) Studies of photosynthesis using a pulsed laser. I. Temperature dependence of cytochrome oxidation rate in *Chromatium*. Evidence for tunneling. Biophys J 6: 825–847

Dracheva SM, Drachev LA, Zaberezhnaya SM, Konstantinov AA, Semenov AY and Skulachev VP (1986) Spectral, redox and kinetic characteristics of high-potential cytochrome c hemes in *Rhodopseudomonas viridis* reaction center. FEBS Lett 205: 41–46

Dracheva SM, Drachev LA, Konstantinov AA, Semenov AY, Skulachev VP, Arutjunjan AM, Shuvalov VA and Zaberezhnaya SM (1988) Electrogenic steps in the redox reactions catalyzed by photosynthetic reaction-centre complex from *Rhodopseudomonas viridis*. Eur J Biochem 171: 253–264

Dracheva S, Williams JA, Van Driessche G, Van Beeumen JJ and Blankenship RE (1991) The primary structure of cytochrome *c*-554 from the green photosynthetic bacterium *Chloroflexus aurantiacus*. Biochemistry 30: 11451–11458

Dutton PL (1971) Oxidation-reduction potential dependence of the interaction of cytochromes, bacteriochlorophyll and carotenoids at 77 K in chromatophores of *Chromatium* D and *Rhodopseudomonas gelatinosa*. Biochim Biophys Acta 226: 63–80

Dutton PL and Leigh JS (1973) Electron spin resonance characterization of *Chromatium* D hemes, non-heme irons and the components involved in primary photochemistry. Biochim Biophys Acta 314: 178–190

Dutton PL and Prince RC (1978) Reaction-center-driven cytochrome interactions in electron and proton translocation and energy coupling. In: Clayton RK and Sistrom WR (eds) The Photosynthetic Bacteria, pp 525–570. Plenum Press, New York

Dutton PL, Kihara T and Chance B (1970) Early reactions in photosynthetic energy conservation: The photooxidation at liquid nitrogen temperatures of two cytochromes in chromatophores of *Rhodopseudomonas gelatinosa*. Arch Biochem Biophys 139: 236–240

Dutton PL, Kihara T, McCray JA and Thornber JP (1971) Cytochrome *c*-553 and bacteriochlorophyll interaction at 77 K in chromatophores and a subchromatophore preparation from *Chromatium* D. Biochim Biophys Acta 226: 81–87

Dutton PL, Prince RC and Tiede DM (1978) The reaction center of photosynthetic bacteria. Photochem Photobiol 28: 939–949

Engelhardt H, Engel A, Hart B and Baumeister W (1986) Stoichiometric model of the photosynthetic unit of *Ectothiorhodospira halochloris*. Proc Natl Acad Sci USA 83: 8972–8976

Feiler U, Nitschke W, Michel H and Rutherford, AW (1989) *Title* Symposium on molecular biology of membrane bound complexes in phototrophic bacteria, Freiburg FRG, Abstract P–6,19

Feiler U, Nitschke W and Michel H (1992) Characterization of an improved reaction center preparation from the photosynthetic green sulfur bacterium *Chlorobium* containing the FeS centers F_A and F_B and a bound cytochrome subunit. Biochemistry 31: 2608–2614

Fowler CF, Nugent NA and Fuller RC (1971) The isolation and characterization of a photochemically active complex from *Chloropseudomonas ethylica*. Proc Natl Acad Sci USA 68, 2278–2282

Freeman JC (1989) Characterization of the membrane-bound cytochrome *c*-554 from the thermophilic green photosynthetic bacterium *Chloroflexus aurantiacus*. PhD thesis, Arizona State University, Tempe

Freeman JC and Blankenship RE (1990) Isolation and characterization of the membrane-bound cytochrome c-554 from the thermophilic green photosynthetic bacterium *Chloroflexus aurantiacus*. Photosynth Res 23: 29–38

Fritz F, Moss DA and Mäntele W (1992) Electrochemical titration of the cytochrome hemes in the *Rhodopseudomonas viridis* reaction center. Cyclic equilibrium titrations yield midpoint potentials without evidence for heme cooperativity. FEBS Lett 297: 167–170

Fritzsch G, Buchanan S and Michel H (1989) Assignment of cytochrome hemes in crystallized reaction center from *Rhodopseudomonas viridis*. Biochim Biophys Acta 977: 157–162

Fuller RC, Sprague SG, Gest H and Blankenship RE (1985) A unique photosynthetic reaction center from *Heliobacterium chlorum*. FEBS Lett 182: 345–349

Fukushima A, Matsuura K, Shimada K and Satoh T (1988) Reaction center-B870 pigment protein complexes with bound cytochrome *c*-555 and *c*-551 from *Rhodocyclus gelatinosus*. Biochim Biophys Acta 933: 399–405

Gao J-L, Shopes RJ and Wraight CA (1990) Charge recombination between the oxidized high-potential *c*-type cytochromes and Q_A^- in reaction centers from *Rhodopseudomonas viridis*. Biochim Biophys Acta 1015: 96–108

Garcia A, Vernon LP and Mollenhauer H (1966) Properties of *Chromatium* subchromatophore particles obtained by treatment with Triton X-100. Biochemistry 5: 3299–2407

Garcia D, Parot P and Verméglio A (1987) Purification and characterization of the photochemical reaction center of the thermophilic purple sulfur bacterium *Chromatium tepidum*. Biochim Biophys Acta 894: 379–385

Garcia D, Richaud P and Verméglio A (1993) The photoinduced cyclic electron transfer in whole cells of *Rhodopseudomonas viridis*. Biochim Biophys Acta 1144: 295–301

Gibson J (1961) Cytochrome pigments from the green photosynthetic bacterium *Chlorobium thiosulphatophilum*. Biochem J 79: 151–158

Güner S, Willie A, Millet F, Caffrey MS, Cusanovich MA, Robertson DE and Knaff DB (1993) The interaction between cytochrome c_2 and the cytochrome bc_1 complex in the photosynthetic purple bacteria *Rhodobacter capsulatus* and *Rhodopseudomonas viridis*. Biochemistry 32: 4793–4800

Gunner MR and Honig B (1990a) Electrostatic analysis of the midpoints of the cofactors in the reaction center protein of *Rp. viridis*. In: Baltscheffsky M (ed) Current Research in Photosynthesis, Vol I, pp 47–52. Kluwer Academic Publishers, Dordrecht

Gunner MR and Honig B (1990b) Electrostatic analysis of the midpoints of the four hemes in the bound cytochrome of the reaction center of *Rp. viridis*. In: Jortner J and Pullman B (eds) Perspectives in Photosynthesis: Proceedings of the 22 Jerusalem Symposium on Quantum Chemistry and Biochemistry, pp 53–60. Kluwer Academic Publishers, Dordrecht

Gunner MR and Honig B (1991) Electrostatic control of midpoint potentials in the cytochrome subunit of the *Rhodopseudomonas viridis* reaction center. Proc Natl Acad Sci USA 88: 9151–9155

Hales BJ (1976) Temperature dependence of the rate of electron transport as a monitor of protein motion. Biophys J 594: 193–230

Heathcote P and Rutherford AW (1987) An EPR signal arising from Q_B^- Fe in *Chromatium vinosum* strain D. In: Biggins J (ed) Progress in Photosynthesis Research, Vol. I, pp 2.201–2.204. Martinus Nijhoff Publishers, Dordrecht

Heibel G, Griebenow K and Hildebrandt P (1991) Structural studies of cytochrome *c*-554 from *Chloroflexus aurantiacus* by resonance Raman spectroscopy techniques. Biochim Biophys Acta 1060: 196–202

Hochkoeppler A, Venturoli G and Zannoni D (1993) The electron transport chain of *Rhodoferax fermentans*. Biol Chem Hoppe-Seyler 374: 831

Holton D, Windsor MW, Parson WW and Thornber JP (1978) Primary photochemical processes in isolated reaction centers of *Rhodopseudomonas viridis*. Biochim Biophys Acta 501: 112–126

Hopfield JJ (1974) Electron transfer between biological molecules by thermally activated tunneling. Proc Natl Acad Sci USA 71: 3640–3644

Hubbard JAM and Evans MCW (1989) Electron donation by the high-potential haems in *Rhodopseudomonas viridis* reaction centres at low temperatures. FEBS Lett 244: 71–75

Hurt EC and Hauska G (1984) Purification of membrane-bound cytochromes and a photoactive P840 protein complex of the green sulfur bacterium *Chlorobium limicola* f. *thiosulfatophilum*. FEBS Lett 168: 149–154

Itoh S (1980) Study of electrogenic electron transfer steps in chromatophore membrane of *Chromatium vinosum* by the response of merocyanin dye. Biochim Biophys Acta 593: 212–223

Kaminskaya O, Konstantinov AA and Shuvalov VA (1990) Low-temperature photooxidation of cytochrome *c* in reaction centre complexes from *Rhodopseudomonas viridis*. Biochim Biophys Acta 1016: 153–164

Kelly DJ and Dow CS (1985) Isolation, characterization and topographical relationships of pigment-protein complexes from membranes of *Rhodomicrobium vannielii*. J Gen Microbiol 131: 2941–2952

Kennel SJ and Kamen MD (1971a) Iron containing proteins in *Chromatium*. I. Solubilization of membrane-bound cytochrome. Biochim Biophys Acta 234: 458–467

Kennel SJ and Kamen MD (1971b) Iron containing proteins in *Chromatium*. II. Purification and properties of cholate-solubilized cytochrome complex. Biochim Biophys Acta 253: 153–166

Kihara T and Chance B (1969) Cytochrome photooxidation at liquid nitrogen temperatures in photosynthetic bacteria. Biochim Biophys Acta 189: 116–124

Kihara T and Dutton PL (1970) Light-induced reactions of photosynthetic bacteria. I. Reactions in whole cells and in cell-free extracts at liquid nitrogen temperatures. Biochim Biophys Acta 205: 196–204

Kihara T and McCray JA (1973) Water and cytochrome oxidation-reduction reactions. Biochim Biophys Acta 292: 297–309

Kjær B, Okkels JS, Moeller BL and Scheller HV (1992) The cytochrome of the photosynthetic reaction center complex from *Chlorobium vibrioforme*. In: Murata N (ed) Research in Photosynthesis, Vol. I, pp 3.465–3.469. Kluwer Academic Publishers, Dordrecht

Kjær B, Golbeck JH and Scheller HV (1994) A reaction center complex from the green sulfur bacterium *Chlorobium vibrioforme* active in photoreduction of $NADP^+$. Abstracts of the VIII Intl Symposium on Phototrophic Prokaryotes, p 181. Urbino, Italy

Kjærulff S, Okkels JS, Kjær B, Ormerod JG and Scheller HV (1993) Electro-transformation of the green sulfur bacteria *Chlorobium vibrioforme*. In: Olson J (ed) Abstracts of the EMBO Workshop on Green and Heliobacteria, p 39

Kleinherenbrink FAM and Amesz J (1993) Stoichiometries and rates of electron transfer and charge recombination in *Heliobacterium chlorum*. Biochim Biophys Acta 1143: 77–83

Knaff DB and Kraichoke S (1983) Oxidation-reduction and EPR properties of a cytochrome complex from *Chromatium vinosum*. Photochem Photobiol 37: 243–246

Knaff DB, Willie A, Long JE, Kriauciunas A, Durham B and Millet F (1991) Reaction of cytochrome c_2 with photosynthetic reaction centers from *Rhodopseudomonas viridis*. Biochemistry 30: 1304–1310

Kusumoto N, Inoue K, Nasu H and Sakurai H (1993) Some properties of a photoactive reaction center complex containing photoreducible Fe-S centers and photooxidizable cytochrome *c* from the green sulfur bacterium *Chlorobium tepidum*. In: Olson J (ed) Abstracts of the EMBO Workshop on Green and Heliobacteria, p 27

Laußermair E and Oesterhelt D (1992) A system for site-specific mutagenesis of the photosynthetic reaction center in *Rhodopseudomonas viridis*. EMBO J 11: 777–783

Lefèbvre S, Picorel R, Cloutier Y and Gingras G (1984) Photoreaction center of *Ectothiorhodospira* sp. Pigment, heme, quinone, and polypeptide composition. Biochemistry 23: 5279–5288

Lefèbvre S, Picorel R and Gingras G (1989) Further characterization of the photoreaction center from *Ectothiorhodospira* sp. Detection of the H subunit by monoclonal antibodies. FEMS Microbiol Lett 65: 247–252

Leguijt T (1993) Photosynthetic electron transfer in Ectothiorhodospira. PhD thesis, University of Amsterdam, The Netherlands

Leguijt T and Hellingwerf KJ (1991) Characterization of reaction center/antenna complexes from bacteriochlorophyll *a* containing Ectothiorhodospiraceae. Biochim Biophys Acta 1057: 353–360

Liebetanz R, Hornberger U and Drews G (1991) Organization of the genes coding for the reaction-centre L and M subunits and B870 antenna polypeptides α and β from the aerobic photosynthetic bacterium *Erythrobacter* species OCH114. Molecular Microbiology 5: 1459–1468

Matsuura K and Shimada K (1986) Cytochromes functionally associated to photochemical reaction centers in *Rhodopseudomonas palustris* and *Rhodopseudomonas acidophila*. Biochim Biophys Acta 852: 9–18

Matsuura K and Shimada K (1990) Evolutionary relationships between reaction center complexes with and without cytochrome *c* subunits in purple bacteria. In: Baltscheffsky M (ed) Current Research in Photosynthesis, Vol. I, pp 193–196. Kluwer Academic Publishers, Dordrecht

Matsuura K, Fukushima A, Shimada K and Satoh T. (1988) Direct and indirect electron transfer from cytochromes *c* and c_2 to the photosynthetic reaction center in pigment-protein complexes isolated from *Rhodocyclus gelatinosus*. FEBS Lett 237: 21–25

Meyer TE, Bartsch RG, Cusanovich MA and Mathewson JH (1968) The cytochromes of *Chlorobium thiosulfatophilum*. Biochim Biophys Acta 153: 854–861

Meyer TE, Kennel SJ, Tedro SM and Kamen MD (1973) Iron protein content of *Thiocapsa pfennigii*, a purple sulfur bacterium of atypical chlorophyll composition. Biochim Biophys Acta 292: 534–643

Meyer TE, Tollin G, Cusanovich MA, Freeman JC and Blankenship RE (1989) In vitro kinetics of reduction of cytochrome c-554 isolated from the reaction center of the green phototrophic bacterium *Chloroflexus aurantiacus*. Arch Biochem Biophys 272: 254–261

Meyer TE, Bartsch RG, Cusanovich MA and Tollin G (1993) Kinetics of photooxidation of soluble cytochromes, HiPIP and azurin by the photosynthetic reaction center of the purple phototrophic bacterium *Rhodopseudomonas viridis*. Biochemistry 32: 4719–4726

Michel H and Deisenhofer J (1986) In: Staehelin LA and Arntzen CJ (eds) Encyclopedia of Plant Physiology, New Series, Photosynthesis III, Vol. 19, pp 371–381. Springer, Berlin

Moore GR and Rogers NK (1985) The influence of electrostatic interactions between buried charges on the properties of membrane proteins. J Inorg Biochem 23: 219–226

Moser CC, Keske JM, Warncke K, Farid RS and Dutton PL (1992) Nature of biological electron transfer. Nature 355: 796–802

Mulkidjanian A, Venturoli G, Hochkoeppler A, Zannoni D, Melandri BA and Drachev LA (1994) Photosynthetic electrogenic events in native membranes of *Chloroflexus aurantiacus*. Flash-induced charge displacements within the reaction center-cytochrome *c*-554 complex. Photosynth Res 41: 135–143

Nabedryk E, Berthomieu C, Verméglio A and Breton J (1991) Photooxidation of high-potential (*c*-559, *c*-556) and low-potential (*c*-552) hemes in the cytochrome subunit of *Rhodopseudomonas viridis* reaction center. Characterization by FTIR spectroscopy. FEBS Lett 293: 53–58

Nagashima KVP, Itoh S, Shimada K and Matsuura K (1993a) Photo-oxidation of reaction center-bound cytochrome *c* and generation of membrane potential determined by carotenoid band shift in the purple photosynthetic bacterium, *Rhodospirillum molischianum*. Biochim Biophys Acta 1140: 297–303

Nagashima KVP, Shimada K and Matsuura K (1993b) Phylogenetic analysis of photosynthetic genes of *Rhodocyclus gelatinosus*: possibility of horizontal gene transfer in purple bacteria. Photosynth Res 36: 185–191

Nagashima KVP, Matsuura K, Ohyama S and Shimada K (1994) Primary structure and transcription of genes encoding B870 and photosynthetic reaction centre apoproteins from *Rubrivivax gelatinosus*. J Biol Chem 269: 2477–2484

Nitschke W and Liebl U (1992) The cytochromes of *Heliobacillus mobilis*. In: Murata N (ed) Research in Photosynthesis, Vol. III, pp 507–510. Kluwer Academic Publishers, Dordrecht

Nitschke W and Rutherford AW (1989) Tetraheme cytochrome *c* subunit of *Rhodopseudomonas viridis* characterized by EPR. Biochemistry 28: 3161–3168.

Nitschke W and Rutherford AW (1991) Photosynthetic reaction

centres: Variations on a common structural theme? Trends Biochem Sci 16: 241–245

Nitschke W and Rutherford AW (1994) The tetrahaem cytochromes associated with photosynthetic reaction centres: A model system for intraprotein redox centre interactions. Biochem Soc Trans 22: 692–697

Nitschke W, Agalidis I and Rutherford AW (1992) The reaction-centre associated cytochrome subunit of the purple bacterium *Rhodocyclus gelatinosus*. Biochim Biophys Acta 1100: 49–57

Nitschke W, Jubault-Bregler M and Rutherford AW (1993) The reaction centre associated tetraheme cytochrome subunit from *Chromatium vinosum* revisited: A reexamination of its EPR properties. Biochemistry 32: 8871–8879

Oh-oka H, Kakutani S, Matsubara H, Malkin R and Itoh S (1993) Isolation of the photoactive reaction center complex that contains three types of Fe-S centers and a cytochrome *c* subunit from the green sulfur bacterium *Chlorobium limicola* f. *thiosulfatophilum*, strain Larsen. Plant Cell Physiol 34: 93–101

Okamura K, Kisaichi K, Takamiya K and Nishimura M (1984) Purification and some properties of a soluble cytochrome (Cyt *c*-551) from *Erythrobacter* species strain OCh 114. Arch Microbiol 139: 143–146

Okkels JS, Kjær B, Hansson Ö, Svendsen I, Moeller BL and Scheller HV (1992) A membrane-bound monoheme cytochrome *c*-551 of a novel type is the immediate electron donor to P840 of the *Chlorobium vibrioforme* photosynthetic reaction center complex. J Biol Chem 267: 21139–21145

Okumura N, Shimada K and Matsuura K (1994) Photo-oxidation of membrane-bound and soluble cytochrome *c* in the green sulfur bacterium *Chlorobium tepidum*. Photosynth Res 41: 125–134

Olson JM and Nadler KD (1965) Energy transfer and cytochrome function in a new type of photosynthetic bacterium. Photochem Photobiol 4: 783–791

Olson JM and Sybesma C (1963) Energy transfer and cytochrome oxidation in green bacteria. In: Gest H, San Pietro A and Vernon LP (eds) Bacterial Photosynthesis, pp 413–422, Antioch Press, Yellow Springs

Ortega JM and Mathis P (1992) Effect of temperature on the kinetics of electron transfer from the tetraheme cytochrome to the primary donor in *Rhodopseudomonas viridis*. FEBS Lett 301: 45–48

Ortega JM and Mathis P (1993) Electron transfer from the tetraheme cytochrome to the special pair in isolated reaction centers of *Rhodopseudomonas viridis*. Biochemistry 32: 1141–1151

Parson WW (1969) Cytochrome photooxidations in *Chromatium* chromatophores. Each P870 oxidizes two cytochrome *c*-422 hemes. Biochim Biophys Acta 189: 397–403

Parson WW and Case GD (1970) In *Chromatium*, a single photochemical reaction center oxidizes both cytochrome *c*-552 and cytochrome *c*-555. Biochim Biophys Acta 205: 232–245

Pettigrew G, Bartsch RG, Meyer TE and Kamen M (1978) Redox potentials of the photosynthetic bacterial cytochromes c_2 and the structural bases for variability. Biochim Biophys Acta 503: 509–523

Prince RC (1978) The reaction center and associated cytochromes of *Thiocapsa pfennigii*: their thermodynamic and spectroscopic properties, and their possible location within the photosynthetic membrane. Biochim Biophys Acta 501: 195–207

Prince RC and Olson JM (1976) Some thermodynamic and kinetic properties of the primary photochemical reactants in a complex from a green photosynthetic bacterium. Biochim Biophys Acta 423: 357–362

Prince RC, Leigh JS Jr. and Dutton PL (1976) Thermodynamic properties of the reaction center of *Rhodopseudomonas viridis*. In vivo measurement of the reaction center bacteriochlorophyll-primary acceptor intermediary electron carrier. Biochim Biophys Acta 440: 622–636

Prince RC, Dutton PL, Clayton BJ and Clayton RK (1978) EPR properties of the reaction center of *Rhodopseudomonas gelatinosa* in situ and in a detergent-solubilized form. Biochim Biophys Acta 502: 354–358

Prince RC, Gest H and Blankenship RE (1985) Thermodynamic properties of the photochemical reaction center of *Heliobacterium chlorum*. Biochim Biophys Acta 810: 377–384

Rubin AB, Shaitan KV, Kononenko AA and Chamorovsky SK (1989) Temperature dependence of cytochrome photooxidation and conformational dynamics of *Chromatium* reaction center complexes. Photosynth Res 22: 219–231

Sarai A and DeVault D (1983) Temperature dependence of high-potential cytochrome photo-oxidation in *Cm. vinosum*. In: Sybesma C (ed) Advances in Photosynthesis Research, Vol. I, pp 5.653–5.656. Martinus Nijhoff, The Hague

Schoepp B (1994) Aspects fonctionnels et structuraux du complexe *bc* de micro-organismes photosynthétiques. PhD thesis, Institut National Agronomique, Paris-Grignon, France

Schoepp B, Parot P, Richard P and Verméglio A (1994) Electron transfer pathway and transmembrane electrical potential of *Rubrivivax gelatinosus*. Abstracts, VIII Intl Symposium Phototrophic Prokaryotes, p 5B. Urbino, Italy

Seftor REB and Thornber JP (1984) The photochemical reaction center of the bacteriochlorophyll *b*-containing organism *Thiocapsa pfennigii*. Biochim Biophys Acta 764: 148–159

Seibert M and DeVault D (1970) Relationship between the laser-induced oxidation of the high and low potential cytochromes of *Chromatium* D. Biochim Biophys Acta 205: 222–231

Sharp KA and Honig B (1990) Electrostatic interactions in macromolecules. Theory and Applications. Annu Rev Biophys Biophys Chem 19: 301–332

Shill DA and Wood PM (1984) A role for cytochrome c_2 in *Rhodopseudomonas viridis*. Biochim Biophys Acta 764: 1–7

Shimada K, Hayashi H and Tasumi M (1985) Bacteriochlorophyll-protein complexes of aerobic bacteria, *Erythrobacter longus* and *Erythrobacter* species OCh 114. Arch Microbiol 143: 244–247

Shinkarev VP, Verkhovsky MI, Sabo J, Zakharova NI and Kononenko AA (1991) Properties of photosynthetic reaction centers isolated from chromatophores of *Chromatium minutissimum*. Biochim Biophys Acta 1098: 117–126

Shopes RJ, Levine LMA, Holten D and Wraight CA (1987) Kinetics of oxidation of the bound cytochromes in reaction centers from *Rhodopseudomonas viridis*. Photosynth Res 12: 165–180

Skulachev VP, Drachev LA, Dracheva SM, Konstantinov AA, Semenov AY and Shuvalov VA (1987) Structure-function relationships in the *Rhodopseudomonas viridis* reaction center complex: electrogenic steps contributing to $\Delta\Psi$ formation. In: Papa S, Chance B and Ernster L (eds) Cytochrome Systems, pp 609–616. Plenum Press, New York

Smit HWJ, Amesz J and van der Hoeven MFR (1987) Electron transport and triplet formation in membranes of the photosynthetic bacterium *Heliobacterium chlorum*. Biochim Biophys Acta 893: 232–240

Smit HWJ, van Dorssen RJ and Amesz J (1989) Charge separation and trapping efficiency in membranes of *Heliobacterium chlorum* at low temperature. Biochim Biophys Acta 973: 212–219

Sogabe S, Saeda M, Uno A, Ezoe T, Miki M, Matsuura Y, Matsuura K and Miki K (1992) Crystal structure of cytochrome c_2 from *Rhodopseudomonas viridis* at 3.0 Å resolution. In: Murata N (ed) Research in Photosynthesis, Vol. II, pp 531–534. Kluwer Academic Publishers, Dordrecht

Stackebrandt E, Murray RGE and Trüper HG (1988) *Proteobacteria* classis nov., a name for the phylogenetic taxon that includes the 'purple bacteria and their relatives.' Int J Sys Bact 38: 321–325

Sybesma C (1967) Light-induced cytochrome reactions in the green photosynthetic bacterium *Chloropseudomonas ethylicum*. Photochem Photobiol 6: 261–267

Sybesma C and Vredenberg WJ (1964) Kinetics of light-induced cytochrome oxidation and P840 bleaching in green photosynthetic bacteria under various conditions. Biochim Biophys Acta 88: 205–207

Tan J, Corson GE, Chen YL, Garcia MC, Güner S and Knaff DB (1993) The ubiquinol:cytochrome c_2/c oxidoreductase of *Cm. vinosum*. Biochim Biophys Acta 1144: 69–76

Thornber JP (1970) Photochemical reactions of purple bacteria as revealed by studies of three spectrally different carotenobacteriochlorophyll-protein complexes isolated from *Chromatium*, strain D. Biochemistry 9: 2688–2698

Thornber JP and Olson JM (1971) Chlorophyll proteins and reaction center preparations from photosynthetic bacteria, algae and higher plants. Photochem Photobiol 14: 329–341

Tiede DM, Leigh JS and Dutton PL (1978) Structural organisation of the *Chromatium vinosum* reaction center associated *c*-cytochromes. Biochim Biophys Acta 503: 524–544

Tiede DM, Prince RC and Dutton PL (1976) EPR and optical spectroscopic properties of the electron carrier intermediate between the reaction center bacteriochlorophylls and the primary acceptor in *Chromatium vinosum*. Biochim Biophys Acta 449: 447–467

Tomiyama Y, Doi M, Takamiya K-i and Nishimura M (1983) Isolation, purification and some properties of cytochrome *c*-551 from *Chromatium vinosum*. Plant Cell Physiol 23: 11–16

Trost JT and Blankenship RE (1989) Isolation of a photoactive photosynthetic reaction center-core antenna complex from *Heliobacillus mobilis*. Biochemistry 28: 9898–9904

Trost JT and Blankenship RE (1990) Isolation of a reaction center particle and a small *c*-type cytochrome from *Heliobacillus mobilis*. In: Baltscheffsky M (ed) Current Research in Photosynthesis, Vol. II, pp 703–706. Kluwer Academic Publishers, Dordrecht

Trost JT, McManus JD, Freeman JC, Ramakrishna BL and Blankenship RE (1988) Auracyanin, a blue copper protein from the green photosynthetic bacterium *Chloroflexus aurantiacus*. Biochemistry 27: 7858–7863

Van de Meent EJ, Kleinherenbrink FAM and Amesz J (1990) Purification and properties of an antenna-reaction center complex from heliobacteria. Biochim Biophys Acta 1015: 223–230

Van Grondelle R, Duysens LNM, Van der Wel JA and Van der Wal HN (1977) Function and properties of a soluble *c*-type cytochrome *c*-551 in secondary electron transport in whole cells of *Chromatium vinosum* as studied with flash spectroscopy. Biochim Biophys Acta 461: 188–201

Van Vliet P, Zannoni D, Nitschke W and Rutherford AW (1991) Membrane-bound cytochromes in *Chloroflexus aurantiacus* studied by EPR. Eur J Biochem 199: 317–323

Venturoli G and Zannoni D (1988) Oxidation-reduction thermodynamics of the acceptor quinone complex in whole-membrane fragments from *Chloroflexus aurantiacus*. Eur J Biochem 178: 503–509

Vermeglio A, Richaud P and Breton J (1989) Orientation and assignment of the four cytochrome hemes in *Rhodopseudomonas viridis* reaction centers. FEBS Lett 243: 259–263

Vos MH, Klaassen HE and van Gorkom HJ (1989) Electron transport in *Heliobacterium chlorum* whole cells studied by electroluminescence and absorbance difference spectroscopy. Biochim Biophys Acta 973: 163–169

Vredenberg WJ and Duysens LNM (1964) Light-induced oxidation of cytochromes in photosynthetic bacteria between 20 and −170°. Biochim Biophys Acta 79: 456–463

Wayne LG, Brenner DJ, Colwell RR, Grimont PAD, Kandler O, Krichevsky MI, Moore LH, Moore WEC, Murray RGE, Stackebrandt E, Starr MP and Trüper HG (1987) Report on the ad hoc committee on reconciliation of approaches to bacterial systematics. Int J Sys Bact 37: 463–464

Weyer KA, Lottspeich F, Gruenberg H, Lang F, Oesterhelt D and Michel H (1987a) Amino acid sequence of the cytochrome subunit of the photosynthetic reaction centre from the purple bacterium *Rhodopseudomonas viridis*. EMBO J 6: 2197–2202

Weyer KA, Schäfer W, Lottspeich F and Michel H (1987b) The cytochrome subunit of the photosynthetic reaction center from *Rhodopseudomonas viridis* is a lipoprotein. Biochemistry 26: 2909–2914

Woese CR (1987) Bacterial evolution. Microbiol Rev 51: 221–271

Zannoni D and Venturoli G (1988) The mechanism of photosynthetic electron transport and energy transduction by membrane fragments from *Chloroflexus aurantiacus*. In: Olson JM, Ormerod JG, Amesz J, Stackebrandt E and Trüper HG (eds) Green Photosynthetic Bacteria, pp 135–143. Plenum Press, New York

Chapter 37

The Proton-Translocating F_0F_1 ATP Synthase-ATPase Complex

Zippora Gromet-Elhanan
Department of Biochemistry, The Weizmann Institute of Science, Rehovot 76100, Israel

Summary

The F_0F_1 ATP synthase is responsible for electron-transport coupled ATP synthesis in every living cell, and functions also as a reversible ATPase. It is composed of an integral membrane sector, F_0, containing four subunits in a stoichiometry of $a_1b_1b'_1c_{6-12}$, and an extrinsic sector, F_1, containing five subunits in a stoichiometry of $\alpha_3\beta_3\gamma_1\delta_1\varepsilon_1$. The detailed structure of the catalytic site and mechanism of action of this very complex enzyme are still unknown. Work by many research groups led to isolation of the whole F_0F_1 complex and the F_1-ATPase from many bacteria. From *Rhodospirillum rubrum* chromatophores the catalytic RrF_1-$\alpha\beta$-core complex and the $RrF_1\beta$ subunit have also been isolated. Removal of *all* $RrF_1\beta$ from the membrane enabled the separation of inactive, but fully reconstitutable β-less *Rs. rubrum* chromatophores. The $RrF_1\gamma$ subunit could be sequentially removed from these chromatophores. All isolated whole and partial complexes and individual subunits have been purified and characterized. Most important results include: 1) Demonstration of a low but continuous light-driven ATP synthesis by purified RcF_0F_1 reconstituted into phospholipid vesicles together with reaction

R. E. Blankenship, M. T. Madigan and C. E. Bauer (eds): Anoxygenic Photosynthetic Bacteria, pp. 807–830.

centers and a cytochrome bc_1 complex purified from the same bacteria. This is a first step towards reconstitution of a functional photosynthetic membrane. 2) Formation and characterization of active hybrid membrane-bound F_1-ATPases by reconstituting β-less *Rs. rubrum* chromatophores with $F_1\beta$ subunits isolated from *E. coli* EcF_1 and spinach CF_1. The restored ATPase activity demonstrated the functional homology of all $F_1\beta$ subunits, but their different response to various known F_1 effectors. 3) Characterization of two binding sites for ATP and ADP on the purified $RrF_1\beta$. One of them, which does also bind P_i, appears to be the catalytic site of the F_1-ATPase. Recent successful attempts at cloning and functional expression of this $RrF_1\beta$ subunit open up exciting possibilities for future research aimed at elucidating the structure of this catalytic site and identifying amino acid residues essential for assembly of the $F_1\beta$ subunit into an active F_1-ATPase.

A concerted effort involving biochemical, genetic, electron microscopic and crystallographic techniques will hopefully lead to resolution of the as yet enigmatic mechanism of action of this most important enzyme complex.

I. Introduction

Energy transducing membranes of bacteria, chloroplasts and mitochondria contain F_0F_1-type proton-translocating ATP synthases that catalyze electron-transport coupled phosphorylation. The coupling mechanism involves utilization of an electrochemical gradient of protons, generated during respiratory or photosynthetic electron-transport, as driving force for the respective oxidative and photosynthetic phosphorylation (Mitchell, 1966). All membrane embedded F_0F_1 ATP synthases can also work in the reverse direction as proton-translocating ATPases, but their major biological function is the synthesis of ATP. This unique metabolic role is reflected in the very unusual structural complexity of the ATP synthases in comparison to other ATPases, as well as in the high degree of conservation of this structure in enzymes from different sources. All F_0F_1 ATP synthases consist of two distinct sectors, termed F_0 and F_1. F_0 is an integral membrane complex, which mediates transmembrane proton conduction, and F_1 is an extrinsic membrane complex, which when removed from the membrane functions as a water soluble ATPase. However, proton-transport coupled ATP synthesis and hydrolysis occurs only when the F_1 sector is combined with the membrane-bound F_0 sector, hence the widely used term of coupling factor (Racker, 1976).

Each of the two sectors is a complex structure by itself, composed of a number of different subunits with varying stoichiometries. Although the overall function and shape of all F_0F_1-type enzymes is similar, there are some important differences, especially in regulation of their ATPase activity and in the subunit composition of various F_0 sectors. The enzyme from respiratory bacteria is the least complex, containing three types of F_0 subunits in addition to the conserved structure of five types of F_1 subunits assembled with a stoichiometry of $\alpha_3\beta_3\gamma\delta\epsilon$. The mitochondrial enzyme is the most complex. It contains, besides the five F_1 subunits, up to nine additional types of subunits (Collinson et al., 1994), which form the F_0 sector and link it to the F_1 sector. The photosynthetic F_0F_1 complexes, including those of anoxygenic photosynthetic bacteria, form a separate group with four distinct F_0 subunits (see II. A.1).

A vast amount of structural and functional information obtained through genetic, biochemical and electron-microscopic investigations of various types of F_0F_1 ATP synthases, has led to the formulation of plausible models for the mechanism of action of this group of enzymes (see Section IV). However for elucidation of the exact molecular mechanism of proton-coupled ATP synthesis/hydrolysis much more detailed direct structural information is required. The first published 3.6 Å X-ray crystallographic structure of an F_1 complex was that of rat-liver MF_1

Abbreviations: AMP-PNP – Adenylyl-β,γ-imidodiphosphate; BChl – bacteriochlorophyll; *Cb.* – *Chlorobium; Cf.* – *Chloroflexus;* CF_0 – chloroplast F_0; CF_0F_1 – chloroplast F_0F_1 ; CF_1 – chloroplast F_1 ; $CF_1\alpha$ – chloroplast F_1 α subunit; $CF_1\beta$ – chloroplast F_1 β subunit; *Cm.* – *Chromatium;* CvF_1 – *Cm. vinosum* F_1 ; DCCD – N,N-dicyclohexyl carbodiimide; DEPC – diethylpyrocarbonate; DTT – dithiothreitol; *E.* – *Escherichia;* EcF_0 – *E. coli* F_0 ; EcF_0F_1 – *E. coli* F_0F_1 ; EcF_1 – *E. coli* F_1 ; $EcF_1\beta$ – *E. coli* $F_1\beta$ subunit; EDTA – ethylenediaminetetraacetic acid; MF_0F_1 – mitochondrial F_0F_1 ; MF_1 – mitochondrial F_1 ; NBf-Cl – 4-chloro-7-nitrobenzofurazan; PMS – N-methylphenazonium methosulfate; *Rb.* – *Rhodobacter;* RcF_0F_1 – *Rb. capsulatus* F_0F_1 ; RcF_1 – *Rb. capsulatus* F_1 ; *Rp.* – *Rhodopseudomonas*; RrF_0F_1 – *Rs. rubrum* F_0F_1 ; RrF_1 – *Rs. rubrum* F_1 ; $RrF_1\beta$ – *Rs. rubrum* $F_1\beta$ subunit; $RrF_1\alpha$ – *Rs. rubrum* F_1 α subunit; $RrF_1\gamma$ – *Rs. rubrum* $F_1\gamma$ subunit; *Rs.* – *Rhodospirillum;* RsF_1 – *Rb. sphaeroides* F_1 ; SDS-PAGE – sodium dodecyl sulfate - polyacrylamide gel electrophoresis; TF_1 – Thermophilic bacterium PS3 F_1 ; WRK – Woodward's reagent K

(Bianchet et al., 1991), and recently the structrure of bovine heart MF_1, at 2.8 Å resolution, has also been reported (Abrahams et al, 1994).

This chapter will review the current knowledge about the structure and function of complete and partial F_0F_1 ATP synthase complexes isolated from anoxygenic photosynthetic bacteria, and compare their properties with those of other types of photosynthetic F_0F_1 enzyme complexes. Due to space limitations chromatophore photophosphorylation and proton-coupled ATP hydrolysis will not be discussed, except for a short summary of membrane-bound F_0F_1 activities, that are useful for comparison with the isolated enzyme. Recent comprehensive reviews of the presently available literature on the structure and mechanism of action of F_0F_1 ATP synthases can be found in: Futai et al. (1989); Fillingame (1990); Penefsky and Cross (1991); Cox et al. (1992); Issartel et al. (1992); Hatefi (1993); Boyer (1993). Specific reviews of the chloroplast enzyme include: Strotmann and Bickel-Sandkotter (1984); McCarty and Moroney (1985); Nalin and Nelson (1987); Graber et al. (1990); Jagendorf et al. (1991) and Glaser and Norling (1991). There is only one earlier review on the photosynthetic bacterial enzyme (Baccarini-Melandri and Melandri, 1978).

II. The F_0F_1 Complex

A. Overall Structural Features

1. Organization of Gene Clusters and Gene Sequence Homology

F_0F_1 complexes of anoxygenic photosynthetic bacteria have first been isolated from chromatophores of the purple non-sulfur bacterium *Rhodospirillum rubrum* (Oren and Gromet-Elhanan, 1977; Schneider et al., 1978). As has been observed with the chloroplast CF_0F_1 (see Glaser and Norling, 1991 and references therein), the best purified RrF_0F_1 preparations showed on SDS-PAGE either eight (Bengis-Garber and Gromet-Elhanan, 1979) or at least nine (Muller and Baltscheffsky, 1979) different polypeptide subunits. Nine subunits have also been observed in the RcF_0F_1 complex purified from another purple non-sulfur bacterium *Rhodobacter capsulatus* (Gabellini et al., 1988). These earlier discrepancies in the reported number of subunits in various photosynthetic F_0F_1 complexes have been resolved via gene sequence analysis, which revealed that RrF_0F_1, CF_0F_1 and all other examined photosynthetic F_0F_1 complexes, contain nine distinct polypeptide subunits rather than the eight found in the *Escherichia coli* EcF_0F_1.

Figure 1 compares the organization of genes encoding F_0F_1 subunits in two anoxygenic photosynthetic bacteria with those of two oxygenic cyanobacteria, plant chloroplasts and *E. coli*. All the genes for the eight EcF_0F_1 subunits form a single operon organized in two sub-clusters encoding the three F_0 a, c and b subunits followed by the five F_1 subunits in the order of $\delta\,\alpha\,\gamma\beta$ and ε (see Futai et al., 1989 and references therein). Upstream of the F_0 subunit a there is a ninth gene, designated I, encoding a small hydrophobic protein of unknown function that is not assembled into the F_0F_1 complex. In *Rs. rubrum* the genes for the F_0 and F_1 sectors are organized in separate operons. The F_0 operon consists, as in EcF_0F_1, of genes for I, a, c, b and an additional related but different gene for b' (Falk and Walker, 1988). The second operon consists of the five F_1 genes in the same order as in the *E. coli* operon (Falk et al., 1985). In a related purple non-sulfur bacterium *Rhodopseudomonas blastica* only the F_1 operon has been identified. It contains six genes arranged in the same order as the F_1 genes in *E. coli* and *Rs. rubrum*, except for the presence of an additional unknown gene, designated x, inserted between the γ and β ones (Tybulewicz et al., 1984).

A different organization of two or even three operons encoding nine F_0F_1 subunits has been reported in two closely related cyanobacteria (Fig. 1). In both cases the F_0 operon, as the *Rs. rubrum* F_0 operon, contains two genes for b and b' but in *Synechococcus* 6301 (Cozens and Walker, 1987), it carries also the genes for F_1 subunits δ, α, and γ, whereas in *Synechococcus* 6716 (Van Walraven et al., 1993) it carries only the genes for F_1 δ and α. In this cyanobacterium the gene for γ appears on a completely separate cluster. The genes for $F_1\beta$ and ε form a separate far apart operon in both cyanobacteria as well as in higher plant chloroplasts. However, in plants and green algae the organization of genes encoding the remaining seven ATP synthase subunits is different, because CF_1 subunits γ, δ and CF_0 subunit II, which is equivalent to b' (Hermann et al., 1993), are encoded in the nucleus (see Jagendorf et al., 1991). So the plastid F_0 operon carries only genes for F_0 subunits IV (a), III (c), I (b) and F_1 subunit α.

The presence of a separate operon encoding the F_1 β and ε subunits has recently been reported in the

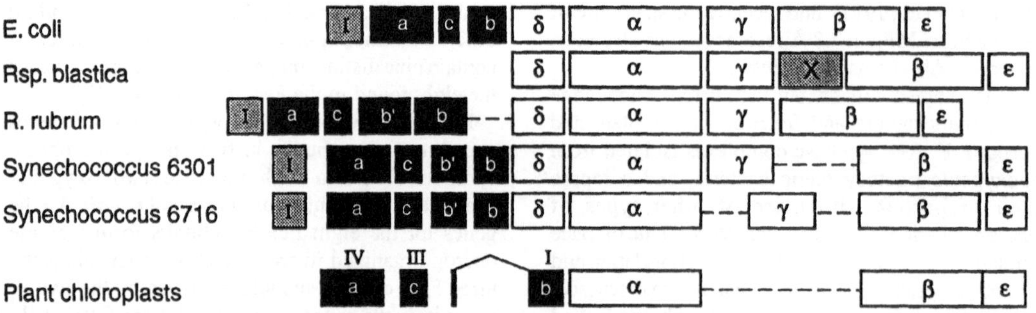

Fig. 1. Diagrammatic representation of genes encoding F_0F_1 subunits in *E. coli*, anoxygenic photosynthetic bacteria, cyanobacteria and plastid genomes. Modified from Falk et al. (1985), Falk and Walker (1988) and Van Walraven et al., (1993). Separate operons are indicated by dashed lines. White boxes denote F_1 genes; black boxes - F_0 genes and shaded boxes denote genes encoding proteins that do not appear in the F_0F_1 complex.

green sulfur bacterium *Chlorobium limicola*, but the organization of the other genes coding for F_0F_1 subunits is as yet unknown (Xie et al., 1993). The green photosynthetic bacteria, unlike the purple bacteria, utilize a reaction center related to Photosystem I of cyanobacteria and chloroplasts (see chapter by Feiler and Hauska). It would therefore be very interesting to find out whether the organization of the genes encoding other F_0F_1 subunits is also similar in *Cb. limicola* and in cyanobacteria.

Some general features observed in all examined species include the highly significant homology between their β subunits (>40% sequence identity in 20 different β subunits) and to a lesser extent α subunits (Futai et al., 1989). A functional homology has also been demonstrated among β subunits isolated from various sources (see III.B.2). A comparison of the sequences of F_0F_1 subunits from all photosynthetic sources with available sequences of plant and bovine mitochondria (Cozens and Walker, 1987; Falk et al., 1988) indicates rather clearly that both mitochondrial enzymes are more closely related to the enzyme from the anoxygenic photosynthetic bacteria than to any other group. They show >70% sequence identity in their α and β subunits and >40% in their F_0 c subunits. The chloroplast enzyme on the other hand is more closely related to the cyanobacterial one, showing >70% sequence identity in their α subunits and >80% in their β and F_0III (c) subunits.

2. Subunit Stoichiometry, Structure and Topology

The stoichiometry of the F_0F_1 subunits has been investigated in detail in *E. coli* and chloroplast complexes. This involved careful determination of molecular weights of each individual subunit and of purified F_0F_1, F_1, and F_0 complexes, as well as estimation of relative amounts of each subunit in radioactively labeled complexes (McCarty and Moroney, 1985; Futai et al., 1989; Glaser and Norling, 1991 and references therein). An identical pattern of F_1 subunit stoichiometry of $\alpha_3\beta_3\gamma\delta\varepsilon$ has been established for both bacterial and chloroplast complexes. The subunit stoichiometry of the respective F_0 complexes is less certain, especially for the F_0 c subunit. The presently proposed ratios are $a_1b_2c_{6-12}$ for EcF_0, and $a_1b_1b'_1c_{6-12}$ for CF_0. The main difference between both types of enzymes is in the assembly per complex of two identical b subunits in EcF_0 as compared to one of each of the non-identical homologues b and b' in CF_0. The subunit stoichiometry of F_1 and F_0F_1 complexes isolated from various photosynthetic bacteria has not been examined in detail. At present the identical numbers obtained for EcF_0F_1 and CF_0F_1 suggest that the basic overall subunit stoichiometry and structure of at least all bacterial and chloroplast F_0F_1 complexes is very conserved.

Low resolution information on the overall shape of what we recognize now as the membrane-bound RrF_0F_1 complex, has been obtained already in 1965 by electron microscopic examination of negatively stained *Rs. rubrum* chromatophores (Low and Afzelius, 1965). Their pictures suggested the presence of protruding knobs of 120 Å diameter, that are located on the outside surface of the chromatophores. These structures have been observed on all examined energy-transducing membranes, and their release upon removal of the membrane-bound ATPase activity by washing with EDTA, suggested that they are morphological representations of the coupling factor-F_1 sector of the membrane-bound F_0F_1 ATP synthase complex (see Racker, 1976). This suggestion was ascertained by experiments with antibodies raised against the purified coupling factor F_1-ATPase isolated from *Rs. rubrum* (Berzhorn et al., 1975) and from *Rb. capsulatus* (Prince et al., 1975). The antibodies agglutinated chromatophores but not spheroplasts, thus demonstrating that the F_1 sector of the F_0F_1-ATPase is located exclusively outside the chromatophore membrane. Further immunological studies compared the effect of antibodies prepared against the RrF_1 and its purified native β subunit (Philosoph and Gromet-Elhanan, 1981a, see also III.B.1). Both types of antibodies agglutinated the *Rs. rubrum* chromatophores, suggesting that the F_1 β subunit is located on the external part of RrF_1 (see Fig. 2).

No further structural work has been reported on the F_0F_1 from any photosynthetic bacterium. But electron microscopic examinations of detergent-solubilized F_0F_1 ATP synthases from *E. coli*, chloroplasts and mitochondria led to formulation of two-dimensional structural models for the F_0F_1 (Gogol et al., 1987; Graber et al., 1990; Walker et al., 1990; Pedersen and Amzel, 1993). These models propose a tripartite structure containing a headpiece, identified as the F_1 sector, a short stalk and a basepiece, identified as the F_0 sector. The stalk is probably composed of subunits or parts of subunits that connect the F_1 to the F_0 (see Fig. 2). Electron microscopy has also succeeded in defining the gross arrangement of subunits within the F_1 complex, revealing a hexagonal arrangement of the large subunits enclosing a mass suggested to be one or more of the small γ, δ, and ε subunits. Studies with CF_1 labeled with monoclonal antibodies against the CF_1 α subunit, demonstrated the binding of three antibodies at an angle of 120 degrees per CF_1 molecule. This observation verified the stoichiometry of $\alpha_3\beta_3$, and indicated that the α and β subunits are arranged in alternating positions (see Graber et al., 1990).

The X-ray crystallographic structure at 3·6Å resolution of rat liver MF_1 gives a more detailed picture of the organization of the α and β subunits (Bianchet et al., 1991). They are arranged in two interdigitated layers, each containing a trimeric array of one type of subunit. The α trimer starts below the β trimer, which in turn protrudes about 15Å above the α one, and both α and β subunits have similar ellipsoidal shapes. A different picture has recently been reported for the bovine heart MF_1 structure at 2.8 Å resolution (Abrahams et al., 1994). It is clear that a higher resolution crystallographic structure of these, as well as additional types of F_1 complexes, including the photosynthetic ones, is needed.

We know less about the overall shape of the F_0 sector. But from the primary structure of the F_0 subunits and their hydrophobic profiles, including those reported for the *Rs. rubrum* F_0 subunits (Hoppe and Sebald, 1984; Falk and Walker, 1988), secondary structural models for their orientation in the membrane have been deduced (see Fillingame, 1990; Graber et al., 1990; Glaser and Norling, 1991; Deckers-Herbestreit and Altendorf, 1992). Subunit c has two hydrophobic stretches, thought to span the membrane in a hairpin-like structure, linked by a polar loop projecting towards the F_1 sector. A conserved acidic residue, glutamic acid in *Rs. rubrum* (Hoppe and Sebald, 1984), within the C-terminal hydrophobic region is the site of DCCD-binding, that results in inhibition of proton translocation and thus of coupled ATP synthesis and hydrolysis. Subunit b, and the related b' in RrF_0F_1 and other examined photosynthetic enzymes, have a hydrophobic N-terminal domain, while the rest of the protein is hydrophilic (Falk and Walker, 1988). This structure has been interpreted as indicating that the two copies of b in *E. coli* (Deckers-Hebestreit and Altendorf, 1992) or the two homologous b and b' in the photosynthetic enzymes (Otto and Berzborn, 1989) are anchored in the membrane via their two N-terminal parts in association with the transmembrane segments of subunits c and a. The remaining hydrophilic part is proposed to protrude into the cytoplasm and interact with F_1 subunits. Subunit a is the largest F_0 subunit, and the least resolved in structure. It is extremely hydrophobic and various reports suggest that it inserts in the membrane with 5 to 8 membrane spanning α-helices. A schematic model of the overall F_0F_1 shape with a more detailed presentation of the organization of F_1 α and β subunits and the suggested arrangement of the F_0 subunits is given in Fig. 2.

B. Catalytic Properties of the Membrane-Bound F_0F_1

Chromatophores of most photosynthetic bacteria effectively catalyze both ATP synthesis and hydrolysis. Rates of ATP synthesis in absence of any

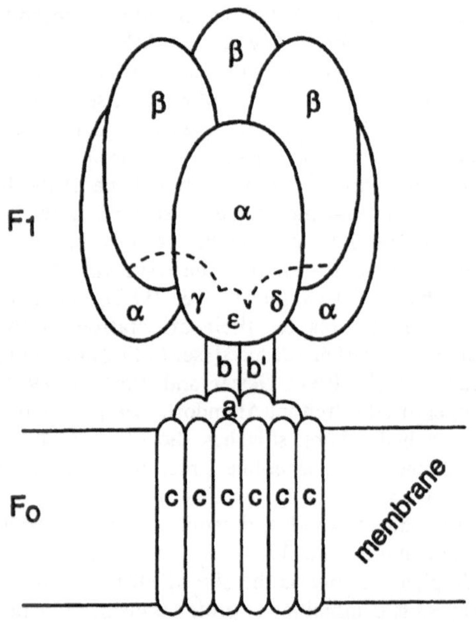

Fig. 2. A schematic model of the subunit organization in photosynthetic F_0F_1 ATP synthases. See text for explanations.

added electron carriers vary between 80–200 μmol/hr per mg BChl for both *Rs. rubrum* and *Rb. capsulatus*. Addition of PMS increases these rates to 800–1200 in the first and 400–500 in the second (Gromet-Elhanan and Gest, 1978; Baccarini-Melandri et al., 1979; Gromet-Elhanan and Weiss, 1989). In contrast to chloroplasts, chromatophores also hydrolyze ATP in the dark at relatively high rates in the presence of either Mg^{2+} or Ca^{2+}. These rates reach 30 and 10%, respectively of the rates of photophosphorylation (Baccarini-Melandri and Melandri, 1978; Gromet-Elhanan and Weiss, 1989). Most chromatophore membrane-bound ATPase activities do not respond to the activation treatments employed in chloroplasts, such as tryptic digestion or illumination. Only in special cases, when the dark rates are very low, as in *Chromatium vinosum* chromatophores (Gepshtein and Carmeli, 1974), has tryptic digestion been shown to cause a large stimulation.

All chromatophore bound ATP synthesis and hydrolysis activities are inhibited by energy-transfer inhibitors that were shown to block proton-translocation through the F_0 sector (Hoppe and Sebald, 1984). DCCD is a general inhibitor of all membrane-bound F_0F_1, including chromatophore F_0F_1 complexes. It does also inhibit the isolated F_1-ATPase, but at much higher concentrations and different conditions (see III.A.3). Oligomycin is a much more selective inhibitor. It inhibits the mitochondrial membrane-bound and solubilized MF_0F_1, but not MF_1, and is completely ineffective towards the chloroplast CF_0F_1. Oligomycin does inhibit the membrane-bound ATP synthesis and hydrolysis activities of RrF_0F_1 as effectively as the mitochondrial MF_0F_1 activities (Johansson et al., 1971). It is somewhat less effective in various *Rhodobacter* chromatophores (Melandri and Baccarini-Melandri, 1971; Müller et al., 1982), and completely ineffective in *Chromatium* chromatophores (Gepshtein and Carmeli; 1974).

Uncouplers exert a different effect. They stimulate the Mg^{2+}-ATPase activity, while inhibiting photophosphorylation in all chromatophore membranes. Illumination in presence of uncouplers causes further stimulation of the Mg^{2+}-ATPase activity, especially at very high ratios of Mg^{2+}/ATP (Edwards and Jackson, 1976; Slooten and Nuyten, 1981). Recently a transient stimulation of 10 to 15 fold was reported, when uncouplers were added immediately after the illumination. The rate of the transient Mg^{2+}-ATPase observed in this case was two fold higher than the rate of ATP synthesis (Turina et al., 1992).

The Ca^{2+}-ATPase activity is much less responsive to uncoupler stimulation than the Mg^{2+}-ATPase. This difference between Ca^{2+} and Mg^{2+} has been traced to their effect on H^+-translocation by the membrane-bound RrF_0F_1 complex (Casadio, 1988; Gromet-Elhanan and Weiss, 1989). In the presence of Mg^{2+}, and also Mn^{2+}, the RrF_0F_1 complex catalyzes both ATP synthesis and hydrolysis as well as ATP-induced H^+-translocation, but in the presence of Ca^{2+} it catalyzes only ATP hydrolysis, which is *not* coupled to H^+-translocation. Further studies have indicated that the inability of Ca^{2+} to maintain the coupling process is not due to opening of a proton leak in its presence, but rather to the inability of bound CaATP to lead to a conformational change that opens up the pathway for proton translocation through the F_0F_1 complex during catalysis (Gromet-Elhanan and Weiss, 1989; Strid and Nyren, 1989). These findings direct attention to the earlier kinetic analysis of the membrane bound F_0F_1-ATPase activity in *Chromatium* (Gepshtein and Carmeli, 1977) as well as in *Rs.*

rubrum (Oren and Gromet-Elhanan, 1979). The cation-ATP complex was shown to be the substrate for ATP hydrolysis in both bacteria and free Mg^{2+}, but not free Ca^{2+}, was a very efficient competitive inhibitor. It is thus clear that Mg^{2+}-ions play an important regulatory role in proton-coupled ATP hydrolysis.

The chromatophore membrane-bound RrF_0F_1 has been shown to synthesize ATP in the dark at the expense of an artificially imposed electrochemical proton gradient composed of both a ΔpH and a $\Delta\psi$ (Leiser and Gromet-Elhanan, 1974). This capacity, first observed in chloroplasts (Jagendorf and Uribe, 1966), helped confirm the basic concepts of the chemiosmotic theory (Mitchell, 1966). It has also been very useful in establishing the capacity of the detergent-solubilized purified F_0F_1 preparations for $\Delta\bar{\mu}H^+$-driven ATP synthesis after their reconstitution into phospholipid vesicles (see II.C.2).

C. Resolution and functional reconstitution

1. Purification and properties of isolated F_0F_1-ATPases

Highly purified F_0F_1 complexes have been obtained from *Rs. rubrum* (Bengis-Garber and Gromet-Elhanan, 1979; Müller and Baltscheffsky, 1979, and Slooten and Vanderbranden, 1989a), *Rb. capsulatus* (Gabellini et al., 1988) and recently also from the green non-sulfur bacterium *Chloroflexus aurantiacus* (Yanyushin, 1991). The procedures used for solubilization and purification have been adapted from those developed for the detergent solubilized mitochondrial and chloroplast enzymes. The earlier preparations employed the mitochondrial procedure (Tzagoloff and Meagher, 1971) of solubilization by Triton X-100, and purification of the concentrated extract by glycerol gradient centrifugation. Additional steps of ion-exchange chromatography or gel filtration (Bengis-Garber and Gromet-Elhanan, 1979; Schneider et al., 1979) yielded the most active purified F_0F_1 preparations reported to date in photosynthetic bacteria (Table 1). More recent preparations used the procedure introduced by Pick and Racker (1979) for extraction and purification of CF_0F_1, in which the F_0F_1 is solubilized by a combination of octylglucoside and cholate (or deoxycholate) and purified by ammonium sulfate fractionation and sucrose gradient centrifugation. All isolated and purified bacterial

Table 1. Summary of ATPase activities of the *Rs. rubrum* membrane-bound and detergent solubilized RrF_0F_1, the purified water soluble five subunit RrF_1, and the RrF_1-$\alpha\beta$-dimer

Type of assayed enzyme[1]	ATPase activity assayed with[2] (U/mg of protein)		Inhibition (%) of Mg^{2+}-ATPase by oligomycin
	Mg^{2+}	Ca^{2+}	
RrF_0F_1:			
membrane-bound	0.2	0.1	90
purified	1.8	3.3	70
Purified RrF_1[3]:			
assayed by itself	1.3	8.6	<5
assayed with OG	8.2	1.8	ND
RrF_1-$\alpha\beta$	0.09	0.08	ND

[1] Data for RrF_0F_1 are modified from Bengis-Garber and Gromet-Elhanan (1979), for purified RrF_1 from Weiss et al. (1994), and for the RrF_1-$\alpha\beta$-dimer from Andralojc and Harris (1993).
[2] ATPase activity was assayed with 4 mM ATP and either 2 mM $MgCl_2$ or 4 to 8 mM $CaCl_2$, except for RrF_1-$\alpha\beta$ where 200 μM ATP and 5 mM $MgCl_2$ or $CaCl_2$ were used. When indicated 5×10^{-5}M oligomycin was added. One unit (U) of activity is defined as the number of μ mol of P_i released per min at 35 °C. ND – not determined.
[3] The activity of purified RrF_1 was determined either in absence or presence of 16 mM octylglucoside (OG) in the assay medium.

F_0F_1 complexes were shown to retain the characteristic DCCD-sensitive Mg^{2+}-ATPase activities of their parent chromatophore membrane-bound enzymes (see II.B).

An extensive characterization of the catalytic properties of the isolated, as compared to the membrane-bound F_0F_1-ATPase, has been carried out with the RrF_0F_1. As is illustrated in Table 1 the membrane-bound enzyme is both a Mg^{2+}- and Ca^{2+}-dependent ATPase, its Mg^{2+}-ATPase activity being two to three fold faster than the Ca^{2+}-ATPase one (Johansson et al., 1971; Bengis-Garber and Gromet-Elhanan, 1979 and see II.B). Both activities are retained also in the detergent solubilized enzyme, but the Ca^{2+}-ATPase is purified to a much greater extent than the Mg^{2+}-ATPase activity, so that the finally purified complex shows a two fold higher Ca^{2+}-ATPase activity. Table 1 shows also that the water soluble RrF_1-ATPase is predominantly a Ca^{2+}-ATPase (see III.A.2). In *Rb. capsulatus* only the Mg^{2+}-ATPase activity of the purified F_0F_1 complex has been followed, and found to be five to ten fold slower than that of purified RrF_0F_1 (Gabellini et al., 1988).

Optimal ratios of 0.2–0.5 Mg^{2+}/ATP and 1–3 Ca^{2+}/ATP were found for both membrane-bound and solubilized RrF_0F_1 (Oren and Gromet-Elhanan, 1979) as well as for RrF_1 (see III.A.2.). The results summarized in Table 1 have been obtained with these ratios. The lower optimal ratio of Mg^{2+}/ATP reflects the fact that free Mg^{2+}-ions are, whereas Ca^{2+}-ions are not potent competitive inhibitors of the membrane-bound (see II.B) as well as the solubilized RrF_0F_1. Here the K_i for free Mg^{2+}-ions is 7 μM, whereas free Ca^{2+}-ions do not inhibit even at 10mM (Oren and Gromet-Elhanan, 1979).

The nucleotide specificity of soluble RrF_0F_1 and of the membrane-bound complex are very similar. The preferential substrate is ATP, but GTP and, to a lesser extent ITP, are also hydrolysed (Schneider et al., 1979). Table 1 illustrates that both bound and soluble RrF_0F_1-ATPase activities are very sensitive to the energy-transfer inhibitor oligomycin, that does not inhibit the water soluble RrF_1-ATPase (Oren and Gromet-Elhanan, 1977 and 1979; Schneider et al., 1978). The RrF_0F_1-ATPase is the only F_0F_1-ATPase in both groups of bacterial and photosynthetic enzymes that is as sensitive to oligomycin and rutamycin as the mitochondrial MF_0F_1 complex (see II. B). The RrF_0F_1 Ca^{2+}-ATPase is somewhat less sensitive than the Mg^{2+}-ATPase in both membrane-bound (Johansson et al., 1971) and solubilized states (Oren and Gromet-Elhanan, 1977).

2. Reconstitution of Proton-Coupled ATP-Linked Activities

F_0F_1 complexes isolated and purified from various photosynthetic bacteria have been reconstituted into phospholipid vesicles and found to catalyze, as the membrane-bound enzyme, a number of proton-coupled ATP-linked reactions. These range from ATP-dependent quenching of acridine dye fluorescence (Schneider et al., 1980; Slooten and Vanderbranden, 1989a), to $^{32}P_i$-ATP exchange, and finally to $\Delta\bar{\mu}H^+$-dependent net ATP synthesis (see II. B). All these reactions were inhibited by the energy-transfer inhibitors oligomycin and DCCD as well as by uncouplers. Reconstitution of isolated RrF_0F_1 (Bengis-Garber and Gromet-Elhanan, 1979) and RcF_0F_1 (Gabellini et al., 1988) into phospholipid vesicles, enabled catalysis of $^{32}P_i$-ATP exchange at reproducible but low rates, amounting to 23 and 5 nmol ATP formed/min per mg of RrF_0F_1 and RcF_0F_1 respectively. Ten-fold higher rates, similar to those observed with reconstituted CF_0F_1 (Pick and Racker, 1979), have recently been obtained with the *Chloroflexus* enzyme, upon its reconstitution into liposomes prepared with intrinsic lipids isolated from the same bacterium (Yanyushin, 1991). The use of lipids prepared from the same bacterium was earlier found to yield higher $^{32}P_i$-ATP exchange activities as well as $\Delta\bar{\mu}H^+$-driven ATP synthesis in a reconstituted cyanobacterial F_0F_1 complex (Van Walraven et al., 1983; 1986). Purified RrF_0F_1 reconstituted into phospholipid vesicles is also capable of ATP synthesis at the expense of artificially applied ΔpH plus $\Delta\psi$ (Slooten and Vandenbranden, 1989b). The amount of ATP formed in this discontinuous system, 80 mol/mol enzyme, is similar to that obtained with reconstituted CF_0F_1 (Pick and Racker, 1979).

Continuous net ATP synthesis was first obtained with a purified RrF_0F_1 reconstituted together with purple membrane bacteriorhodopsin into phospholipid vesicles (Oren et al., 1980). The light-induced ATP synthesis was proportional to the amount of RrF_0F_1 and bacteriorhodopsin (at a 1:1 ratio) and linear for at least 20 min of illumination . The maximal rate of ATP synthesis of 8·4 nmol/min per mg of protein obtained in this system is low, but 10 to 30 fold higher rates have been obtained when RrF_0F_1 was reconstituted into soybean liposomes together with monomeric bacteriorhodopsin (Wagner et al., 1987). Concomitantly measured rates of light-driven proton translocation enabled the calculation of H^+/ATP stoichiometries in this reconstituted system. The constant ratio of H^+/ATP of 5·2, that was obtained under saturating light intensities, is lower than the value of H^+/ATP of 3 obtained in native membranes (Ferguson and Sorgato, 1982). Net continuous ATP synthesis has also been obtained upon reconstitution of the purified RrF_0F_1 together with an energy-linked pyrophosphatase, partially purified from the same bacterium (Nyren and Baltscheffsky, 1983). The maximal rate of this PP_i-driven ATP synthesis was 25 nmol of ATP/min per mg of protein and it was sensitive to a variety of uncouplers and inhibitors of both the RrF_0F_1-ATPase and the pyrophosphatase.

A more ambitious system for coreconstitution of cyclic electron transport and photophosphorylation has been developed by Gabellini et al. (1989). They have coincorporated into phospholipid vesicles purified preparations of the *Rb. capsulatus* reaction center, cytochrome bc_1 complex, and RcF_0F_1, and obtained light-driven ATP synthesis at a rate of 36 nmol ATP/min per mg of RcF_0F_1. A rate of up to 55 nmol/min per mg of protein has been reported in an earlier developed system for coreconstitution of chloroplast Photosystem I reaction centers and CF_0F_1, where the cyclic electron transport was completed with the artificial redox component PMS (Hauska et al., 1980). Although the reported maximal rates amount to only about 0.3–3% of the rates of photophosphorylation observed in native chromatophores (see II. B), when calculated by using the two different published ATP synthase contents of coupled chromatophores (see III.A.1), this is a first step towards reconstitution of a functional photosynthetic membrane.

III. The F_1-ATPase

A. General Features of the Whole F_1 Complex

1. Isolation and Purification

F_1-type ATPases have been isolated from *Rb. capsulatus* (Baccarini-Melandri et al., 1970), *Cm. vinosum* (Hochman and Carmeli, 1971), *Rs. rubrum* (Johansson, 1972), *Rb. sphaeroides* (Reed and Raveed, 1972), *Rp. palustris* (Müller et al., 1982), and *Cf. aurantiacus*, (Yanyushin, 1988). The methods employed for complete detachment of the F_1 complex from the chromatophore membrane vary depending on the tightness of its attachment to the membrane-bound F_0 sector. A 10 min incubation in a low ionic strength buffer was sufficient for elimination of practically all the photophosphorylation from *Cm. vinosum* chromatophores, enabling the separation of fully resolved chromatophores and a supernatant from which the solubilized CvF_1-ATPase has been purified (Gepshtein et al., 1978). In *Rb. capsulatus* a similar incubation, even in presence of 1 mM EDTA, decreased phosphorylation by only 50%, and for an 85% decrease a 90 sec sonication in presence of EDTA was required. In *Rs. rubrum* chromatophores this harsher treatment decreased the endogenous phosphorylation by 80% (Johansson, 1972), but the PMS-supported phosphorylation by less than 50% (Gromet-Elhanan, 1974a). It is thus clear that sonication in a low ionic strength buffer in presence of EDTA releases in *Rs. rubrum* not only the F_1-ATPase but also electron transport components that are bypassed in the presence of PMS (Gromet-Elhanan, 1977).

Up to now no satisfactory procedure has been developed for preparing *Rs. rubrum* chromatophores that are completely devoid of the whole F_1 complex and show no extensive damage of their electron transport chain. In the absence of such F_1-depleted chromatophores it is impossible to determine the F_1 content of coupled chromatophores by the method developed in chloroplasts (Strotmann et al., 1973) of assaying their ATPase activities before and after removal of the F_1-ATPase, as compared to the activity of the removed F_1. Calculations based on other methods yielded for *Rs. rubrum* chromatophores values of 9.1 mmol of RrF_1/mol BChl (Norling et al., 1989) and 0.9 mmol RrF_1/mol BChl (Andralojc and Harris, 1993). Differences in the light intensity at which the *Rs. rubrum* cultures were grown were suggested as a possible reason for this 10 fold difference in the calculated RrF_1 content. But the relevant light intensities were not stated for any one of the published estimates. So a reliable determination of the RrF_1 content will require a careful examination of light intensity dependent changes in the ratios of total protein and RrF_1-ATPase protein/BChl.

RrF_1 as well as other bacterial F_1-ATPases were usually removed from their parent chromatophores by drastic treatments that did not enable isolation of resolved chromatophores. These involve either aqueous extraction of acetone powders of coupled chromatophores (Baccarini-Melandri and Melandri, 1971) or the chloroform extraction technique developed by Beechey et al (1975) for removal of the MF_1-ATPase, and introduced by Webster and Jackson (1978) for isolation of RrF_1. In many cases the drastic extraction procedures and even some purification procedures yielded F_1 complexes containing only four, or even three, subunits. Thus, in *Cm. vinosum* the δ subunit was present in the crude extract, but only a small amount of it remained in the purified CvF_1-ATPase (Gepshtein et al., 1978). The purified *Cf. aurantiacus* F_1-ATPase is also depleted of the δ subunit (Yanyushin, 1988). The δ subunit is relatively easily dissociated from various CF_1 and EcF_1 complexes during extraction and/or purification, and in its absence the F_1-ATPase can bind to resolved membranes, but does not restore phosphorylation (McCarty and Moroney, 1985; Glaser and Norling, 1991; Futai et al, 1989). Indeed the crude CvF_1 containing the δ subunit could, whereas the purified δ-less CvF_1 could not, restore phosphorylation to resolved *Cm. vinosum* chromatophores (Gepshten and Carmeli, 1974; Gepshtein et al., 1978). A five subunit F_1 has not, as yet, been purified from these two bacteria.

Conflicting data earlier reported with the RrF_1 seem in retrospect to reflect a similar case of loss of one or more of the small single copy subunits upon extraction and purification. Thus, RrF_1 purified from an acetone powder extract (Johansson et al., 1973) was found to contain five subunits with molecular weights of α-54000; β-50,000; γ-32,000; δ-13,000 and ε-7,500 (Johansson and Baltscheffsky, 1975). Identical numbers were also reported by Soe et al. (1978). The molecular weights of the first three subunits correspond to those calculated by Falk et al. (1985) from predicted protein sequences, but the last two subunits show much lower molecular weights than the calculated δ-19,500 and ε-14,300. These purified RrF_1-ATPases were not assayed for restoration of phosphorylation activity to resolved chromatophores, so it was not realized that their content of δ and ε could be questionable. Later, Müller et al. (1979) reported that they could purify only a four subunit RrF_1-ATPase complex, composed of α, β, γ, and δ, from a similar acetone powder extract. They could, however, purify a five subunit RrF_1 from a chloroform extract, and the molecular weights of these subunits did correspond to those calculated by Falk et al. (1985). In light of these results, it seems possible that the enzyme purified and assayed by Johansson and Baltscheffsky (1975) and Soe et al. (1978) could be composed of α, β, γ, and either ε or a split δ. F_1 complexes purified from acetone extracts of *Rp. palustris* and *Rb. sphaeroides* (Müller et al., 1982) show also molecular weights of 13,000 for δ and 8,000 for ε, that do not correspond to those calculated from the predicted protein sequence of their close relative *Rp. blastica* (Tybulewicz et al., 1984).

The chloroform extraction does also yield variable results. Webster and Jackson (1978) have reported that their crude chloroform extract could *not* restore phosphorylation to partially resolved EDTA-treated *Rs. rubrum* chromatophores and purified from it only a three subunit RrF_1 containing α, β, and γ. Müller et al. (1979) state on the other hand that the chloroform extract from which they have purified the five subunit RrF_1 was prepared as described by Webster and Jackson. It is thus clear that both the acetone and chloroform extraction procedures used in the earlier reported attempts at solubilization and purification are not ideal.

A milder technique of extraction, using a much lower amount of chloroform than that employed by Beechey et al (1975), has been introduced by Fisher et al. (1981). This technique was employed for extraction and purification of a five subunit RrF_1 (Norling et al., 1988). Protein sequencing of the N-terminii of all five subunits has established that they are products of the earlier sequenced *Rs. rubrum* genes. So this extraction technique seems to provide at present the best available procedure for obtaining purified five subunit F_1 complexes from photosynthetic bacteria. It should however be emphasized that up to now none of the extensively purified five subunit F_1 complexes from any photosynthetic bacterium have been assayed for their capacity to restore photophosphorylation in resolved chromatophores.

F_1-ATPases have also been prepared from a number of cyanobacteria, including *Mastigocladus laminosus*(Binder and Bachofen, 1979); *Synechococcus* 6716 (Lubberding et al., 1981); *Spirulina platensis* (Hicks and Yocum, 1986a) and *Anacystis nidulans* (Nemoto et al., 1990). Most cyanobacterial complexes could be removed from the membrane by incubation in low ionic strength buffers without or with 1 mM EDTA, but the usual Beechey-type chloroform extract has also been used. Both extraction methods enabled the purification of five subunit F_1 complexes for *Spirulina* and *Anacystis*, but from the other cyanobacteria only four subunit F_1-ATPases, lacking the δ subunit, have been purified.

2. Catalytic Properties of Isolated F_1-ATPases

The F_1 complexes isolated from various photosynthetic bacteria exhibit a rather varying spectrum of properties. Thus the partially purified RcF_1 has a very weak Mg^{2+}-ATPase activity of 0.16 μmol/min = (u)/mg of protein, and no measurable Ca^{2+}-ATPase activity (Melandri and Baccarini-Melandri, 1971). Its inactivity as Ca^{2+}-ATPase seems to be due to loss of this activity rather than to latency, as in CF_1, because a) it could not be stimulated by treatments, such as heat or trypsin, that enhance the Ca^{2+}-ATPase activity of CF_1, and b) unlike in chloroplasts, the membrane-bound RcF_0F_1 is rather an active Mg^{2+} and Ca^{2+}-dependent ATPase (see II.B). So although this partially purified RcF_1 could restore some photophosphorylation to resolved *Rb. capsulatus* chromatophores (Baccarini-Melandri and Melandri, 1971) it has not been subjected to any further purification or characterization. The partially purified CvF_1 exhibits low but reproducible Ca^{2+}- and Mg^{2+}-

dependent ATPase activities that are stimulated five-fold on addition of trypsin to the assay medium. The Ca^{2+}-ATPase activity is 3 fold higher than the Mg^{2+}-ATPase activity (Gepshtein and Carmeli, 1977), and amounts to 9U/mg of protein in the highly purified four subunit CvF_1 (Gepshtein et al., 1978).

All purified RrF_1 preparations show, unlike RcF_1 and CvF_1, relatively high native Ca^{2+}-ATPase activities in absence of trypsin (Soe et al., 1978; Müller et al., 1979, and Table 1), but their Mg^{2+}-ATPase activities vary depending on the extraction procedure. The acetone extracted enzymes show no Mg^{2+}-ATPase activity (Johansson et al., 1973; Soe et al., 1978; Müller et al., 1979), whereas the various chloroform extracted and purified five subunit RrF_1 complexes, exhibit some Mg^{2+}-ATPase activity (Müller et al., 1979; Norling et al., 1988). When assayed at an optimal ratio of Mg/ATP of 0.2 to 0.5 it amounts to about 15% of the Ca^{2+}-ATPase activity (Müller et al., 1979; Weiss et al., 1994, see Table 1). This range of Ca^{2+}- and Mg^{2+}-ATPase activities is very similar to those of activated CF_1 from which the ε subunit has been removed (McCarty and Moroney, 1985).

In both activated ε-depleted CF_1 and native RrF_1 the low Mg^{2+}-ATPase activity has been stimulated by addition of certain anions (Nelson et al., 1972; Webster et al., 1977) or detergents (Soe et al., 1978 and 1980; Pick and Bassilian, 1982) to the assay medium . With anions, such as malate or sulfite, the Mg^{2+}-ATPase is highly stimulated while the Ca^{2+}-ATPase is either unaffected (Nelson et al., 1972), or even somewhat stimulated (Webster et al., 1977). Detergents, on the other hand, inhibit the Ca^{2+}-ATPase activity while stimulating the Mg^{2+}-ATPase activity. Recent examinations of these two opposite effects, using octylglucoside as the detergent, revealed that the stimulation is dependent on the ratio of Mg^{2+}/ATP. At high Mg^{2+}/ATP ratios higher concentrations of octylglucoside are required for stimulation of the Mg^{2+}-ATPase activity than for inhibition of the Ca^{2+}-ATPase one (Norling et al., 1988). But at optimal Mg^{2+}/ATP ratios of 0.2 to 0.5 both inhibition and stimulation occur at identical and very low concentrations of octylglucoside. A 50% effect is obtained at 5 mM octylglucoside, a concentration well below its critical micellar concentration (Weiss et al., 1994, see Table 1). These results indicate that the effect of octylglucoside cannot be restricted to relief of inhibition by free Mg^{2+}-ions. It must also have a more direct effect on the affinity of the RrF_1 cation binding site towards Ca^{2+} and Mg^{2+}. A similar almost complete conversion of the RrF_0F_1 and RrF_1 from a mostly Ca^{2+}-dependent into a mainly Mg^{2+}-dependent enzyme, has been obtained when ATP was replaced by 1,N^6-etheno ATP (Schafer et al., 1980).

The ATPase activity of the purified *Rp. palustris*, *Rb. sphaeroides* and *Cf. aurantiacus* enzymes has also been extensively characterized (Müller et al., 1982 and 1983; Yanyushin, 1988). All complexes are active Ca^{2+}-ATPases, and exhibit a Mg^{2+}-ATPase activity, amounting to about 40% of their Ca^{2+}-ATPase activities, that can be stimulated by sulfite and octylglucoside. None of these activities is stimulated by trypsin. The nucleotide specificity of these enzymes is similar; the preferential substrate is ATP, but GTP, UTP, and to a lesser extent CTP, are also hydrolyzed, especially in presence of Ca^{2+}-ions.

A kinetic analysis of the partially purified CvF_1 (Gepshtein and Carmeli, 1977) and the completely purified RrF_1 (Müller et al., 1979) and RsF_1 (Müller et al., 1983) revealed that in all complexes, as in the membrane-bound and isolated RrF_0F_1 (see II.B. and II.C.1), the cation-ATP complex is the functional substrate for ATP hydrolysis. The kinetic parameters obtained for $(CaATP)^{2-}$ are very similar in all examined complexes. The K_m for $(CaATP)^{2-}$ is between 0.2–0.5 mM, the K_i for free ATP is 0.5 to 1.0 mM, and the K_i for free Ca^{2+}-ions is 5 to 8 mM. The kinetic parameters for $(MgATP)^{2-}$ show however large differences. Thus, the K_m for $(MgATP)^{2-}$ is 1 mM in CvF_1 and RsF_1, but only 30 μM in RrF_1. The K_i for free ATP is similar in all complexes - about 1 mM, but the K_i for free Mg^{2+}-ions is again very different, although in this case it is around 10 to 20 μM in both CvF_1 and RrF_1, but 0.65 mM in RsF_1 . The reason for these differences has not been clarified.

Some general catalytic properties of these isolated bacterial F_1-ATPases do, however, emerge from the above summary. They are not latent ATPases and do not respond to the CF_1 activation treatments, except for CvF_1, which is stimulated by trypsin. They are also both Ca^{2+}- and Mg^{2+}-ATPases with a higher Ca^{2+}-dependent activity, except for RcF_1 whose activities are not clear. These enzymes thus form a separate group between the very active EcF_1-ATPase, that has a much higher Mg^{2+}- than Ca^{2+}-ATPase activity (Dunn and Heppel, 1981), and the tightly regulated latent CF_1 (McCarty and Moroney, 1985).

All isolated cyanobacterial F_1-ATPases, unlike most photosynthetic bacterial ones, are rather latent enzymes. Their Ca^{2+}-ATPases, as the chloroplast

CF_1-ATPase, is stimulated 5 to 10 fold by trypsin. They show also a very low Mg^{2+}-ATPase activity that can be stimulated 10 to 20 fold by methanol (Binder and Bachofen, 1979; Lubberding et al., 1981; Hicks and Yocum, 1986b; Nemoto et al., 1990). But except for the *Spirulina* enzyme, the cyanobacterial F_1-ATPases do not respond to additional treatments, such as heat and DTT, that elicit ATPase activity in latent chloroplast CF_1. The molecular basis of thiol activation involves the reduction of a disulfide bond between cysteines 199 and 205 in the spinach CF_1 γ subunit (McCarty and Moroney, 1985; Glaser and Norling, 1991). These two cysteines are rather specific to plant chloroplast γ subunits, since all other $F_1\gamma$ subunit sequences reported to date, including those of various *Synechococcus* strains, as well as of *Rs. rubrum* and *Rp. blastica*, do not contain them (see Van Walraven et al., 1993 and references therein). In light of the positive response of the *Spirulina* enzyme to DTT (Hicks and Yocum, 1986b) it would be most interesting to find out whether the *Spirulina* F_1 γ subunit sequence contains these two cysteines.

3. Nucleotide Binding and Chemical Modifications

The determination of the number of nucleotide binding sites in various purified F_1-ATPases and their characterization played a major role in formulating the presently prevailing hypothesis for the mechanism of action of the F_0F_1 ATP synthase-ATPase complex (Boyer, 1987 and 1993). The maximal number of nucleotide binding sites retained on MF_1 and EcF_1 after passage through gel filtration centrifuge columns is six (Futai et al., 1989; Penefsky and Cross, 1991 and references therein). In CF_1 such nonequilibrium binding measurements revealed the presence of four sites (Xue et al., 1987; Shapiro et al., 1991a) and only under strict equilibrium conditions have six sites been recently observed (Girault et al., 1988; Shapiro et al., 1991b). Three of the sites can readily exchange bound nucleotides for medium nucleotides and are suggested to have a catalytic role, whereas the three less exchangeable sites are referred to as noncatalytic (Penefsky and Cross, 1991).

The number and properties of nucleotide binding sites on RrF_1 and CF_1 have recently been compared with enzymes depleted of all loosely bound nucleotides by passage through three successive centrifuge columns (Weiss et al., 1994). They still contain 1.6 mol of ADP plus 0.1 mol of ATP/mol of CF_1, and 1.5 mol of ATP plus 1.1 mol of ADP/mol of RrF_1. After incubation with saturating concentrations of Mg^{2+}-AMP-PNP, followed by removal of all excess and loosely bound nucleotides, both enzymes have been found to retain up to four mol of tightly bound nucleotides/mol F_1. In RrF_1 they include 2.3 mol of AMP-PNP, practically no ADP, and all the 1.5 mol of ATP/mol RrF_1 that were retained on the depleted RrF_1. So in RrF_1 about two of the observed four nucleotide binding sites do exchange with medium AMP-PNP and two are occupied by nonexchangeable ATP molecules. The lower maximal number of four occupied nucleotide binding sites retained after column centrifugation on the photosynthetic CF_1 and RrF_1, as compared to six in the respiratory MF_1 and EcF_1, is probably due to faster K_{off} constants of two of the Mg-dependent nucleotide binding sites on RrF_1 and CF_1. They might lead to the much lower Mg^{2+}-ATPase activities of the photosynthetic F_1-ATPases.

The location of nucleotide binding sites has been studied in respiratory and chloroplast F_1-ATPases with labeled nucleotide affinity probes or photo-reactive nucleotide analogs and, where available, with isolated F_1 subunits. These studies have revealed that all nucleotide binding sites reside exclusively on the F_1 α and β subunits (Vignais and Lumardi, 1985; Futai et al., 1989; Penefsky and Cross, 1991). Some of these reagents that inhibit F_1-ATPase activity are also known to modify specific amino acid residues. The effect of such reagents as well as of other inhibitors of F_1 -ATPase activity have been studied mainly with RrF_1.

DCCD is a hydrophobic carboxyl group reagent, which inhibits all types of F_0F_1-ATPases by binding specifically to the conserved carboxyl residue within the C-terminal hydrophobic segment in the F_0 c subunit (Hoppe and Sebald, 1984). Gepshtein et al., (1978) were the first to observe that DCCD can also inhibit the purified CvF_1 Ca^{2+}-ATPase activity. Further experiments with [^{14}C] DCCD have established that it does indeed bind to and inactivates all examined F_1-ATPases (Vignais and Lunardy, 1985). Khananshvili and Gromet-Elhanan (1983b) have shown that DCCD inactivates purified RrF_1 in a time- concentration- and pH-dependent manner. Complete inhibition requires binding of 1 mol of [^{14}C]DCCD/mol of RrF_1, suggesting that inactivation is associated with modification of one essential carboxyl group in the RrF_1 complex. The inhibition of both membrane-bound and isolated RrF_1 by another, hydrophilic carboxyl group modifier, WRK, has also been proposed to be due to modification of one essential carboxyl group on RrF_1 (Ceccarelli and Vallejos, 1983). But since a previous modification with WRK did not prevent binding of [^{14}C] DCCD, these two reagents seem to modify two different carboxyl groups on RrF_1. This suggestion has indeed been corroborated by studies with the isolated RrF_1 β subunit (see III.B.3).

The adenine analogue NBf-Cl does also inhibit both membrane-bound and soluble F_1-ATPase from various sources (Vignais and Lunardy, 1985). In *Rs. rubrum* it inhibits photophosphorylation and the membrane-bound Mg^{2+}-ATPase activity as well as the Ca^{2+}-ATPase activity of soluble RrF_1, in a DTT-reversible manner (Khananshvili and Gromet-Elhanan, 1983a). Isolated RrF_1 binds covalently [^{14}C] NBf-Cl with an accompanying increase in absorbance at 385 nm, indicative of a tyrosyl-0-NBf bond. Full protection of the inhibition of the RrF_1 Ca^{2+}-ATPase by NBf-Cl has been obtained in presence of 5 mM P_i (Cortez et al., 1983). Another reagent that effectively inhibits the RrF_1 Ca^{2+}-ATPase activity is the histidine reagent DEPC. Complete inactivation is obtained on modification of 2 to 3 histidine residues per molecule of RrF_1 (Khananshvili and Gromet-Elhanan, 1983c).

Various other known inhibitors of F_1-ATPases have also been reported to inhibit the RrF_1 Ca^{2+}-ATPase. These include azide (Johansson et al., 1973), aurovertin (Gromet-Elhanan, 1974b; Ravizzini et al., 1975), efrapeptin (Webster et al., 1977) and quercetin (Weiss et al., 1994). Some inhibitors seem to exert a similar effect on all types of F_1-ATPases, while others, such as aurovertin, efrapeptin and tentoxin, are more specific. Aurovertin is a fluorescent compound which inhibits MF_1 and EcF_1 ATPase activities. Its fluorescence is enhanced when it binds to these enzymes, and undergoes marked changes on addition of nucleotides, that have been used to monitor nucleotide binding sites on the sensitive F_1-ATPases (Vignais and Lunardy, 1985). Aurovertin has been found to inhibit the RrF_1-ATPase, but no accompanying changes in its fluorescence could be observed (Gromet-Elhanan, 1974b).

The effects of efrapeptin and tentoxin on the Ca^{2+}- and Mg^{2+}-ATPase activities of RrF_1 and activated CF_1 have recently been compared under identical assay conditions (Weiss et al., 1994). Efrapeptin was found to be a much more effective inhibitor of RrF_1 than CF_1. Half maximal inhibition of both Ca^{2+}- and Mg^{2+}-dependent RrF_1-ATPase activities is obtained with a ratio of 1.6 mol of efrapeptin/mol of RrF_1, whereas for CF_1 a thirty fold higher ratio is required. RrF_1 is thus as sensitive to efrapeptin as the mitochondrial F_1-ATPase (Cross and Kohlbrenner, 1978), while the lower sensitivity of CF_1 is similar to that reported for the *E. coli* enzyme (Wise et al., 1983). Tentoxin, on the other hand, is a very potent species specific inhibitor of CF_1 (Steele et al., 1976; Selman and Durbin, 1978). Half maximal inhibition of the spinach CF_1-ATPase activities are obtained at a concentration below 0.1 μM tentoxin. All RrF_1 ATPase activities are however completely resistant to tentoxin concentrations of up to 300 μM under all assay conditions used (Weiss et al., 1994; see also III.B.2). The membrane-bound and isolated F_1-ATPase of the cyanobacterium *Anacystis nidulans* has recently been reported to be as sensitive to tentoxin as spinach CF_1 (Ohta et al., 1993).

B. Individual F_1 Subunits and Partial Complexes

1. Isolation and Reconstitution of Single Subunits

The complex structure of the F_0F_1 ATP synthase and its soluble F_1-ATPase sector (see Fig. 2) impedes studies aimed at elucidating the structure of their catalytic sites and mechanism of action. A precise determination of the function of each subunit is a prerequisite for achieving this goal. Two main methods have been developed for the isolation of single F_1 subunits in a native, active state. The first involves partial or complete dissociation of the F_1 complex, followed by separation and purification of the dissociated subunits. Complete dissociation of all five F_1 subunits was reported for F_1 from various respiratory bacteria, and reassociation of mixtures of such isolated subunits led to isolation of partial active complexes (Futai and Kanazawa, 1983, and references therein). There are no reported attempts to dissociate F_1-ATPases isolated from photosynthetic bacteria. With CF_1 only a partial dissociation was achieved, enabling up to now isolation and purification of $CF_1\beta$ containing about 5% of $CF_1\alpha$ (Richter et al., 1986).

The second method developed in *Rs. rubrum* chromatophores involves selective removal of individual subunits or partial complexes from the membrane-bound F_1-ATPase. Incubation of chromatophores with 2M LiCl followed by 2M LiBr led to sequential removal of all their $RrF_1\beta$ and γ subunits, leaving inactive, but fully reconstitutable β-less and β, γ-less chromatophore membranes (Philosoph et al., 1977; Khananshvili and Gromet-Elhanan, 1982). The thoroughly washed LiCl-treated chromatophores, retain about 5% of their ATP synthesis and hydrolysis activities, but over 80% of their capacity to take up protons during light-induced electron transport. Their lost ATP-linked activities could be restored by more than 80% after reconstitution with three volumes of the dialyzed LiCl-extract (Binder and Gromet-Elhanan, 1974).

The $RrF_1\beta$ removed by extraction of *Rs. rubrum* chromatophores with LiCl and purified by ammonium sulfate fractionation and gel filtration (Philosoph et al., 1977) or ion-exchange chromatography (Gromet-Elhanan and Khananshvili, 1986) could rebind to the LiCl-treated chromatophores and restore their lost ATP-linked activities, but showed no measurable ATPase activity. A previous claim for a significant ATPase activity in a similarly prepared $RrF_1\beta$ (Harris et al., 1985) has recently been explained as due to contamination by the RrF_1 $\alpha\beta$-ATPase (Andralojc and Harris, 1992; see III.B.5). Antibodies prepared against the purified $RrF_1\beta$ did agglutinate control active *Rs. rubrum* chromatophores, but not the washed LiCl-treated inactive ones (Philosoph and Gromet-Elhanan, 1981b). These antibodies could also form precipitin lines with whole native RrF_1, but not with a four subunit RrF_1, composed of α, γ, δ, and ε, that has been solubilized from the washed LiCl-treated chromatophores (Philosoph and Gromet-Elhanan, 1981b). These results indicate that *all* the $RrF_1\beta$ subunit has indeed been removed from the thoroughly washed LiCl-treated β-less *Rs. rubrum* chromatophores.

Optimal conditions for rebinding of purified $RrF_1\beta$ to β-less chromatophores, which lead to restoration of 70–90% of their lost ATP synthesis and hydrolysis activities, require the presence of at least 2 mM ATP and 20 mM $MgCl_2$ during the long reconstitution at 35 °C, but much lower concentration of $MgCl_2$ for assays of restored activity (Gromet-Elhanan, 1974a; Philosoph et al., 1977). Under these conditions the reconstitution of a fixed amount of β-less chromatophores with increasing amounts of $RrF_1\beta$ follows a simple saturation curve, and results in a similar degree of restoration of photophosphorylation and Ca^{2+}- and Mg^{2+}-ATPase activities (Khananshvili and Gromet-Elhanan, 1982). For restoration of 50% of these activities a ratio of 1.0 μg of $RrF_1\beta/\mu$g BChl is required.

Further extraction of β-less *Rs. rubrum* chromatophores by LiBr removed one additional subunit, $RrF_1\gamma$, that was purified by ammonium sulfate fractionation and hydroxyapatite chromatography (Philosoph et al., 1981). The resulting doubly extracted chromatophores have, of course, no ATP-linked activities, but can be restored to about 65% by reconstitution with both purified $RrF_1\beta$ and $RrF_1\gamma$, although not with either one alone (Khananshvili and Gromet-Elhanan, 1982). These LiCl-LiBr doubly extracted chromatophores are therefore defined as β, γ-less chromatophores. Reconstitution of β,γ-less chromatophores by the missing subunits was optimal at pH 6.2, and required the presence of 20 mM $MgCl_2$ (Philosoph et al., 1981; Gromet-Elhanan et al., 1981). Because the reconstitution of β-less chromatophores with $RrF_1\beta$ was similar between pH 6.2 and 8.0, both systems could be compared under identical conditions. All assays of restored activity were, however, conducted at their optimal pH 8.0.

Incubation of chloroplast thylakoid membranes from various plants with LiCl resulted in release of CF_1 β together with varying, but large amounts of $CF_1\alpha$ and traces of the single copy γ, δ, and ε subunits. The ratio of released $CF_1\alpha/CF_1\beta$ varied from 0.7–0.8 in spinach (Avital and Gromet-Elhanan, 1990, 1991), to 0.5 in lettuce and 0.2–0.3 in tobacco (Avni et al., 1991). Fractionation of the spinach LiCl-extract led to isolation of pure CF_1 β, a $CF_1(\alpha\beta)$ complex, and small amounts of various larger partial CF_1 complexes, but no pure $CF_1\alpha$ (Avital and Gromet-Elhanan, 1991; Sokolov et al., 1992; and see III.B.2 and III.B.5). Washed LiCl-treated spinach thylakoids, as the *Rs. rubrum* LiCl-treated chromatophores, lost >95% of their photophosphorylation capacity. But, because of the more drastic removal of CF_1 subunits, the reconstitution of the treated thylakoids with their LiCl-extract restored at the most 20% of the lost activity (Avital and Gromet-Elhanan, 1991).

In light of the observed release of large amounts of $CF_1\alpha$ from LiCl-treated chloroplasts, we have reexamined the *Rs. rubrum* LiCl-extracts for presence of $RrF_1\alpha$ by Western immunoblotting with mixtures of anti $RrF_1\alpha$ and $RrF_1\beta$. The ratio of $RrF_1\alpha/RrF_1\beta$ appearing in these extracts varied between 0.03–0.15 (Weiss and Gromet-Elhanan, unpublished). The small amount of $RrF_1\alpha$ extracted by LiCl together with *all* the $RrF_1\beta$, is partly associated in an $RrF_1(\alpha\beta)$ complex (Androlojc and Harris, 1993; and see

III.B.5). It is also the source of the ~5% $RrF_1\alpha$ found to accompany the purified $RrF_1\beta$ preparations which can by themselves restore activity to the β-less chromatophores (Weiss and Gromet-Elhanan, unpublished). Purified preparations of $RrF_1\beta$ containing <1% of $RrF_1\alpha$, as well as purified expressed $RrF_1\beta$, that has no trace of the α subunit can restore the β-less RrF_0F_1 activity only when supplemented with ~5% of $RrF_1\alpha$ (see also III.B.2 and 4).

2. Hybrid F_1-ATPases: Formation and Characterization

The inactive, but fully reconstitutable, β-less *Rs. rubrum* chromatophores from which all $RrF_1\beta$ has been removed provide a uniquely suitable system for testing the capacity of $F_1\beta$ subunits, isolated from various other sources, for reconstitution of active hybrid membrane-bound F_1 ATPases. Two such active hybrids have been formed with isolated $EcF_1\beta$ (Gromet-Elhanan et al., 1985) and spinach $CF_1\beta$ that contained about 5% of $CF_1\alpha$ (Richter et al., 1986). Pure $CF_1\beta$ containing no trace of $CF_1\alpha$ could not form an active hybrid when reconstituted by itself into β-less *Rs. rubrum* chromatophores (Avital and Gromet-Elhanan, 1991; Avni et al., 1991). It could, however, do so when supplemented by trace amounts of a pure $CF_1(\alpha\beta)$ complex, at a ratio of 0.1 μg of $CF_1(\alpha\beta)$/1 μg of BChl (Avni et al., 1991). These traces of $CF_1(\alpha\beta)$ were inactive by themselves. It was therefore suggested that the $CF_1\alpha$ component of the added $CF_1(\alpha\beta)$ exerts a chaperonin-like function of keeping $CF_1\beta$ in the correct folding during the reconstitution at 35 °C with 20 mM $MgCl_2$ (see III.B1).

All types of $EcF_1\beta/RrF_1$ and CF_1/RrF_1 hybrids could restore 50 to 140% of the Mg^{2+}-ATPase activity, but less than 10% of the photophosphorylation activity obtained in β-less chromatophores reconstituted with their homologous $RrF_1\beta$. Moreover in the hybrids, unlike in the homologous enzyme, the restored Mg^{2+}-ATPase activity was not coupled to H^+-translocation. In this respect the hybrid Mg^{2+}-ATPase activities are rather similar to the Ca^{2+}-ATPase activity of the control membrane-bound RrF_0F_1, which has been shown to be completely disconnected from H^+-translocation (see II.B). These results indicate that in the hybrid enzymes the RrF_1 catalytic site is restored, but its connection with others subunits is not tight enough to trigger proton translocation through the membrane-bound RrF_0 sector. The relatively high Mg^{2+}-ATPase activity of the hybrid enzymes thus suggests that they might enable leakage of the protons released during ATP hydrolysis to the outside medium, rather than translocate them inside through the F_0 sector. Experimental evidence for this suggestion is provided by the finding that both hybrid Mg^{2+}-ATPases, unlike the homologous RrF_1 Mg^{2+}-ATPase, are not stimulated by uncouplers (Richter et al., 1986; Gromet-Elhanan, 1988).

The hybrid enzymes retain specific properties of their parent F_1-ATPases that enable clear identification of the activity of each one of them. Thus only the EcF_1/RrF_1 hybrid Mg^{2+}-ATPase is inhibited by anti EcF_1 antibodies and insensitive to sulfite, and only the CF_1/RrF_1 hybrid is inhibited by tentoxin. Indeed, the $CF_1\beta/RrF_1$ hybrid is as sensitive to tentoxin as the CF_1-ATPase, being fully inhibited by 0.1 μM, whereas the native RrF_1-ATPase is resistant even to 10^4-fold higher concentrations (see III.A.3). So the complete elimination of restored ATPase activity in the $EcF_1\beta/RrF_1$ and $CF_1\beta/RrF_1$ hybrids in presence of anti EcF_1 and tentoxin, respectively demonstrate the absence of any remaining homologous RrF_1 activity. These findings reinforce an earlier conclusion that *all* the $RrF_1\beta$ has been removed from the β-less *Rs. rubrum* chromatophores (see III.B.1).

3. Nucleotide Binding Sites on Isolated $F_1\beta$ Subunits

The finding of six nucleotide binding sites on various F_1-ATPases that are located exclusively on the F_1 α and β subunits (Futai et al., 1989; Penefsky and Cross, 1991), together with the conserved stoichiometry of $\alpha_3\beta_3$ (see II.A.2), led to the suggestion that each of these subunits contains a single nucleotide binding site. These sites were designated as catalytic or noncatalytic according to their exchangeability with medium nucleotides, and since nucleotide affinity probes were shown to label the β subunits of many F_1-ATPases under conditions that led to loss of activity, the catalytic sites were assumed to be located on the β subunits.

Direct binding studies with labeled ATP or ADP have indeed revealed the presence of a single nucleotide binding site on the F_1 α and β subunits isolated from two respiratory bacteria (Futai et al. 1989). In isolated $RrF_1\beta$ such direct binding studies have however identified *two* different binding sites for either ATP (Gromet-Elhanan and Khananshvili,

1984) or ADP (Khananshvili and Gromet-Elhanan, 1984), one being Mg-independent and the other Mg-dependent (Table 2). These results questioned the validity of the proposed model of a single binding site on each α and β subunit. To accommodate these findings with the overall number of six nucleotide binding sites per F_1 molecule, one of the two nucleotide binding sites on the β subunit, the Mg-independent site, has been suggested to reside at an interface between the α and β subunits (Gromet-Elhanan and Khananshvili, 1984). These binding properties have also been observed in pure expressed $RrF_1\beta$ that has no trace of the α subunit (see III.B.4).

Further evidence for this suggestion came from $^{32}P_i$ binding tests, which revealed only one Mg-dependent P_i binding site on $RrF_1\beta$ (Khananshvili and Gromet-Elhanan, 1985a, see Table 2), and from the effect of chemical modification of $RrF_1\beta$ by DEPC and WRK (Khananshvili and Gromet-Elhanan, 1985b). Table 2 illustrates that DEPC did not affect the binding of ADP to both sites on $RrF_1\beta$, nor its capacity to rebind to β-less chromatophores, but inhibited completely the Mg-independent binding of P_i, the binding of ATP to its Mg-dependent site, and the capacity of the modified rebound $RrF_1\beta$ to restore ATP synthesis in the reconstituted chromatophores. These results have thus identified the Mg-dependent ATP and P_i binding site on $RrF_1\beta$ as the catalytic site and the Mg-independent site as the noncatalytic one (Khananshvili and Gromet-Elhanan, 1985b). More recent studies on the photolysis of 2-azidoadenine nucleotides bound at either catalytic or noncatalytic sites on MF_1, EcF_1, and CF_1 have indicated that the β subunit was labeled in all cases (Boyer, 1987). These findings suggested the presence of at least parts of the adenosine moiety of both nucleotide binding sites on the $F_1\beta$ subunit, and the possible sharing of one (Penefsky and Cross, 1991) or even both sites (Gromet-Elhanan, 1992) by the α and β subunits.

Occupation of the nucleotide binding sites on $Rr\beta$ by ADP, ATP and P_i was found to induce large conformational changes, that could be followed by their effect on trypsin sensitivity of the isolated subunit (Khananshvili and Gromet-Elhanan, 1986). Two distinct types of changes were obtained: one on occupation of the Mg-independent nucleotide binding site, that is further stabilized by MgADP, the other on binding of MgATP or MgP_i, suggesting that it is induced by occupation of the γ-phosphoryl subsite in the Mg-dependent catalytic site on the β subunit.

Table 2. Nucleotide binding to and restored ATP synthesis by native and diethylpyrocarbonate modified $RrF_1\beta$ [1]

Assayed activity	Type of assayed RrF_1 β	
	native	modified
Substrate binding[2]		
[^{3}H]ADP	1.88	1.76
[^{3}H]ATP	1.84	0.78
$^{32}P_i$	0.93	0.01
Restored ATP synthesis[3]	647	37

[1] Modified from Khananshvili and Gromet-Elhanan (1985b).
[2] Binding stoichiometry in mol/mol of $RrF_1\beta$, assayed in presence of $MgCl_2$.
[3] ATP synthesis in μmol/hr per mg of BChl, recorded in β-less chromatophores reconstituted with native or DEPC modified $RrF_1\beta$.

Purified $RrF_1\beta$ was found to bind [^{14}C] NBf-Cl and [^{14}C] DCCD, with a binding stoichiometry of 1 mol of either reagent/mol of $RrF_1\beta$ (Khananshvili and Gromet-Elhanan, 1983a,b). The NBf-modified $RrF_1\beta$, unlike the DEPC or WRK modified subunit, did rebind to the β-less chromatophores and restored all their lost ATP-linked activities as efficiently as native unmodified $RrF_1\beta$, whereas the DCCD-modified subunit lost its capacity to rebind to the β-less chromatophores. Moreover, modification of $RrF_1\beta$ by either DCCD or NBf-Cl had little effect on its capacity to bind ADP, ATP and P_i (Khananshvili and Gromet-Elhanan, 1985b). These results indicate that DCCD and WRK modify two different carboxyl groups on isolated $RrF_1\beta$ (see III.A.3). The WRK modified carboxyl group is located within the catalytic site, while the DCCD modified carboxyl group is essential for rebinding of the isolated $RrF_1\beta$ to the β-less chromatophores (see III.B.4).

4. Cloning and Expression of F_1 β Subunits

The large number of direct assays for binding of nucleotides, P_i and various modifying reagents by isolated $RrF_1\beta$, together with its capacity to rebind to β-less chromatophores and restore their ATP-linked activities, make $RrF_1\beta$ an especially interesting candidate for studies aimed at identifying the amino acid residues essential for these activities. A most promising approach towards this goal is the use of site directed mutagenesis, which requires cloning and expression of the $RrF_1\beta$ gene. This technique can also ensure the absence of contamination by other F_1 subunits in the expressed protein.

Baltscheffsky et al., (1992) have recently reported on the cloning of the $RrF_1\beta$ gene, and its expression as a soluble fusion protein in *E. coli*. Activity tests have not been reported yet for this expressed $RrF_1\beta$. In another attempt (Nathanson and Gromet-Elhanan, 1994), cloning and functional expression of wild type and Glu 195 mutant $RrF_1\beta$ has been achieved. The expressed β subunits, that are free of any contaminating $RrF_1\alpha$, could bind ATP as the native $RrF_1\beta$ (III.B.3), but could not restore ATP-linked activity to β-less *Rs. rubrum* chromatophores. They could, however, do so when supplemented with traces of about 5% of $RrF_1\alpha$, as has been shown for native $RrF_1\beta$ containing <1% of $RrF_1\alpha$ (III.B.1) and for pure $CF_1\beta$ containing no trace of $CF_1\alpha$ (III.B.2). The $RrF_1\beta$ Glu 195 $\rightarrow$ Gly mutant, unlike the wild type native or expressed $RrF_1\beta$, could form only an unstable assembled RrF_0F_1, that dissociated upon centrifugation (Nathanson and Gromet-Elhanan, 1994). Glu 195 in $RrF_1\beta$ is identical to Glu 192 in $EcF_1\beta$ and Glu 199 in $MF_1\beta$ that are modified by DCCD (Yoshida et al., 1982). The findings that both the Glu 195 $\rightarrow$ Gly mutant and the DCCD modified $RrF_1\beta$ (see III.B.3) form unstable assembled F_0F_1 complexes indicate that the fully conserved Glu 195 in $RrF_1\beta$ does play an important role in the assembly of the β subunit into a stable F_1 complex.

Three recent reports describe the cloning and expression of chloroplast $CF_1\beta$ genes from *Chlamydomonas reinhardtii* (Blumenstein et al., 1990), Spinach (Chen et al., 1992) and maize (Wang et al., 1992). The expressed CF_1 β polypeptide, unlike the expressed $RrF_1\beta$, appeared in insoluble inclusion bodies. It was solubilized in all cases by 4M urea. Slow removal of the urea by stepwise dialysis restored a functional $CF_1\beta$, that could bind the nucleotide analog TNP-ATP (Chen et al., 1992).

5. Isolation, Structure and Function of minimal catalytic F_1 Complexes

$F_1(\alpha\beta)$ complexes, containing equimolar ratios of the α and β subunits, have been shown to function as active ATPases, with rates varying between 1 to 10% of those found in their parent F_1-ATPases. The individually isolated α and β subunits show about 100 fold lower rates than those of the $F_1(\alpha\beta)$ complexes (Gromet-Elhanan, 1992 and references therein). Such complexes have either been assembled from a 1:1 mixture of cloned and expressed $TF_1\alpha$ and β subunits (Kagawa et al., 1989; Miwa and Yoshida, 1989) or isolated from LiCl extracts of spinach (Avital and Gromet-Elhanan, 1991) and lettuce chloroplasts (Avni et al., 1991), and *Rs. rubrum* chromatophores (Andralojc and Harris, 1992).

The TF_1 α and β subunits were found to assemble into $\alpha_3\beta_3$-hexamers in absence of MgATP, but the hexamers dissociated into $\alpha_1\beta_1$-dimers in presence of MgATP (Harada et al., 1991). So in an ATPase assay medium a varying mixture of both structures exists. The $RrF_1(\alpha\beta)$ complex appeared as an $\alpha_1\beta_1$-dimer on size exclusion HPLC in presence of MgATP and removal of Mg^{2+}-ions caused complete dissociation of this dimer into the monomeric α and β subunits (Andralojc and Harris, 1992). The CF_1 $(\alpha\beta)$ complex is very unstable and dissociates into monomers upon dilution and/or removal of ATP. A stabilized CF_1-$\alpha_3\beta_3$-hexamer has recently been isolated by gel filtration chromatography (M. Sokolov and Z. Gromet-Elhanan, unpublished).

All types of isolated F_1 $(\alpha\beta)$ complexes are active ATPases. They exhibit however lower rates of catalysis than their parent F_1-ATPases (see Table 1) and are resistant to inhibition by azide. The $CF_1(\alpha\beta)$ is also resistant to inhibition by tentoxin (Gromet-Elhanan and Avital, 1992). These properties suggest that the single copy F_1 subunits are not required for assembly of these F_1 $(\alpha\beta)$ core complexes nor for their activity. They are however responsible for the 10 to 100 fold increased rates and for the different properties of the whole F_1-ATPases. The F_1 $(\alpha\beta)$ complexes can therefore provide an important tool for elucidating the role of the single copy subunits in the structure and function of the F_1-ATPase.

An interesting result obtained with the CF_1 $(\alpha\beta)$ and RrF_1 $(\alpha\beta)$ complexes is their capacity to bind to the inactive LiCl-treated *Rs. rubrum* chromatophores and restore their activity (Avital and Gromet-Elhanan, 1991; Andralojc and Harris, 1992, 1993). The activity restored by each complex is identical to that restored by the respective isolated $F_1\beta$ subunits, which contained about 5% $F_1\alpha$ (see III.B.1 and 2). Thus the homologous $RrF_1(\alpha\beta)$ restored both ATP synthesis and hydrolysis, whereas the $CF_1(\alpha\beta)$ restored only ATP hydrolysis, that is fully sensitive to inhibition by tentoxin (Avital and Gromet-Elhanan, 1991; Gromet-Elhanan and Avital, 1992). It is not clear as yet whether these $F_1(\alpha\beta)$ complexes bind as a complex to the depleted membranes or dissociate during the reconstitution step which is conducted in 20 mM $MgCl_2$ at 35 °C (see III.B.1).

One reported difference between $CF_1(\alpha\beta)$ and $RrF_1(\alpha\beta)$ is the 10-fold lower ratio of 0.1 μg $RrF_1\alpha\beta$/ μg BChl as compared to 1 μg $CF_1(\alpha\beta)$/μg BChl required for 50% restoration of activity to the LiCl-treated chromatophores (Andralojc and Harris, 1993; Avital and Gromet-Elhanan, 1991). The reason for this difference can be traced to the different types of LiCl-treated chromatophores used in both cases. Andralojc and Harris (1993) studied the rebinding of RrF_1-$\alpha_1\beta_1$ to 'LiCl-washed' chromatophores in which they found, according to their nucleotide binding properties, more than 50% of the β subunits still attached to the chromatophore membrane. Avital and Gromet-Elhanan (1991) used, on the other hand, thoroughly washed β-less chromatophores from which *all* the β subunit has been removed (see III.B.1). When pure $CF_1\beta$, containing no trace of $CF_1\alpha$, was added to these β-less chromatophores it could rebind, but could not restore any ATPase activity (see III.B.2). These β-less chromatophores containing bound inactive $CF_1\beta$ are similar to the 'LiCl-washed' chromatophores, containing large amounts of attached but inactive $RrF_1\beta$, that have been used by Andralojc and Harris (1993). The ATPase activity of these $CF_1\beta$ containing β-less chromatophores could indeed be restored upon addition of a low ratio of 0.1 μg $CF_1(\alpha\beta)$/μg BChl (Avni et al., 1991).

The above described results indicate that the attached $RrF_1\beta$ present in the 'LiCl-washed' membranes (Andralojc and Harris, 1993), as the pure $CF_1\beta$, added to the β-less chromatophores (Avni et al., 1991), are activated by added traces of the respective $F_1(\alpha\beta)$ complexes. These activated $RrF_1\beta$ and $CF_1\beta$ subunits restore the same level of activity to β-less chromatophores as the $RrF_1\beta$ and $CF_1\beta$ containing about 5% of their respective $F_1\alpha$ subunits. It is, therefore, clear that restoration of activity in all LiCl-treated chromatophores does *not* require stoichiometric amounts of α and β subunits. For an accurate determination of whether this activation reflects a chaperonin-like function of $F_1\alpha$ (Avni et al., 1991) labeled $CF_1(\alpha\beta)$ and $RrF_1(\alpha\beta)$ complexes will be required.

IV. Mechanism of Action of the F_0F_1 ATP Synthase

A complete resolution of the mechanism of action of the ATP synthase requires the presence of detailed information on its structure, as well as on the pathway of proton-translocation through the whole enzyme complex and on the specific mode of coupling of this proton-transport to ATP synthesis. Although the general features of the F_0F_1 structure have been elucidated (see II.A.2 and Fig. 2), the detailed organization of its catalytic sites is as yet unknown. Also, the role of F_0 in proton-translocation through the membrane is clear, and the detailed pathway for their translocation through this sector is being elucidated (Fillingame, 1990). But there is almost no information on if and how protons move from the F_0 sector to the F_1 sector, or on how they affect the catalytic energy-linked steps. All proposed models for the mechanism of action of the ATP synthase are therefore still at a rather speculative stage.

The hypothesis for which considerable experimental evidence exists, and that is currently the most widely used, is the binding change mechanism (see Boyer, 1987 and 1993 and Penefsky and Cross, 1991). It is based on the existing information on the presence of three catalytic sites in the F_0F_1 ATP synthase (see III.A.3) and their location mainly on the F_1 β subunits (see III.B.3). The basic features of this model have evolved from a large body of data collected from studies with MF_0F_1, CF_0F_1, and some respiratory bacterial enzymes. Very few mechanistic studies have been conducted with the photosynthetic bacterial enzymes, so only the overall features will be presented below. They are as follows:

1. The $\Delta\tilde{\mu}H^+$ driving force is not required for the final step of ATP synthesis at each catalytic site.

2. Energy is required to promote the release of product, ATP, from the site of synthesis and to enable the binding of substrate, ADP and P_i, to another site.

3. The energy-linked substrate binding and product release occur simultaneously at separate catalytic sites, that must therefore interact during the catalytic event.

This hypothesis thus suggests that the newly formed product is tightly bound to the catalytic site. The energy driving its loosening and final release does also tighten the binding of a new substrate at an alternate site, where product is formed with no additional energy input. Thus, during catalysis each site changes its ligand affinity properties, hence the definition of binding change mechanism (Boyer, 1987 and 1993). The involvement of protons in this mechanism is linked with a proposed rotation of an aggregate of subunits, possibly the internally located

$\gamma\delta\epsilon$abb', against a complex of the remaining $\alpha_3\beta_3c_{6-12}$ subunits. There is however no direct evidence for this rotation mechanism.

The recently isolated $F_1(\alpha\beta)$ catalytic core complexes provide important tools for further investigation of the above proposed mechanism of action. A kinetic investigation of stable, nondissociating $\alpha_3\beta_3$-hexamers could establish whether the specific alternation of sites is dependent on the presence of the single copy subunits.

Acknowledgments

Recent research from the author's laboratory was supported by grants from the Basic Research Foundation administered by the Israel Academy of Sciences and Humanities, the United States-Israel Binational Science Foundation (BSF), Jerusalem, Israel and the Minerva Willstatter Center for Research in Photosynthesis, Rehovot, Israel.

References

Abrahams JP, Leslie AGW, Lutter R and Walker JE (1994) Structure at 2.8 Å resolution of F_1-ATPase from bovine heart mitochrondria. Nature 370: 621–628

Andralojc PJ and Harris DA (1992) Isolation and characterisation of a functional $\alpha\beta$ heterodimer from ATP synthase of *Rhodospirillum rubrum*. FEBS Lett310: 187–192

Andralojc PJ and Harris DA (1993) Preparation and characterisation of an $\alpha\beta$ heterodimer from the ATP synthase of *Rhodospirillum rubrum*. Biochim Biophys Acta 1143: 51–66

Avital S and Gromet-Elhanan Z (1990) Isolation of an active β subunit from chloroplast CF_0F_1-ATP synthase. In: Baltscheffsky M (ed) Current Research in Photosynthesis, Vol III, pp 45–48. Kluwer Academic Publishers, Dordrecht

Avital S and Gromet-Elhanan Z (1991) Extraction and purification of the β subunit and an active $\alpha\beta$-core complex from the spinach chloroplast CF_0F_1-ATP synthase. J Biol Chem 266: 7067–7072

Avni A, Avital S and Gromet-Elhanan Z (1991) Reactivation of the chloroplast CF_1-ATPase β subunit by trace amounts of the CF_1 α subunit suggests a chaperonin-like activity for $CF_1\alpha$. J Biol Chem 266: 7317–7320

Baccarini-Melandri A and Melandri BA (1971) Partial resolution of the photophosphorylating system of *Rhodopseudomonas capsulata*. Methods Enzymol 23: 556–561

Baccarini-Melandri A and Melandri BA (1978) Coupling factors. In: Clayton RK and Sistrom WR (eds) The Photosynthetic Bacteria, pp 615–628. Plenum Press, New York

Baccarini-Melandri A, Gest H and San Pietro A (1970) A coupling factor in bacterial photophosphorylation. J Biol Chem 245: 1224–1226

Baccarini-Melandri A, Melandri BA and Hauska G (1979) The stimulation of photophosphorylation and ATPase by artificial redox mediators in chromatophores of *Rhodopseudomonas capsulata* at different redox potentials. J Bioenerg Biomembr 11: 1–16

Baltscheffsky M, Nadanaciva S and Harris DA (1992) Cloning and expression in *Escherichia coli* of the β subunit from *Rhodospirillum rubrum* F_1-ATPase. In: Murata N (ed) Research in Photosynthesis, Vol III, pp 385–388. Kluwer Academic Publishers, Dordrecht

Beechey RB, Hubbard SA, Linnett PE, Mitchell AD and Munn EA (1975) A simple and rapid method for the preparation of adenosine triphosphatase from submitochondrial particles. Biochem J 148: 533–537

Bengis-Garber C and Gromet-Elhanan Z (1979) Purification of the energy-transducing adenosine triphosphatase complex from *Rhodospirillum rubrum*. Biochemistry 18: 3577–3581

Berzborn RJ, Johansson BC and Baltscheffsky M (1975) Immunological and fluorescence studies with the coupling factor ATPase from *Rhodospirillum rubrum*. Biochim Biophys Acta 396: 360–370

Bianchet M, Ysern X, Hullihen J, Pedersen PL and Amzel LM (1991) Mitochondrial ATP synthase. Quaternary structure of the F_1 moiety at 3·6Å determined by X-ray diffraction analysis. J Biol Chem 266: 197–212

Binder A and Bachofen R (1979) Isolation and characterization of a coupling factor I ATPase of the thermophilic blue-green alga (cyanobacterium) *Mastigocladus laminosus*. FEBS Lett 104:66–70

Binder A and Gromet-Elhanan Z (1974) Depletion and reconstitution of photophosphorylation in chromatophore membranes of *Rhodospirillum rubrum*. In: Avron M (ed) Proceedings of the Third International Congress on Photosynthesis, pp 1163–1170. Elsevier Scientific Publishing Co., Amsterdam

Blumenstein S, Leu S, Abu-Much E, Bar-Zvi D, Shavit N and Michaels A (1990) Expression of the chloroplast *atp*B gene of *Chlamydomonas reinhardtii* in *E. coli*. In: Baltscheffsky M (ed) Current Research in Photosynthesis, Vol III, pp 193–196. Kluwer Academic Publishers, Dordrecht

Boyer PD (1987) The unusual enzymology of ATP synthase. Biochemistry 26: 8503–8507

Boyer PD (1993) The binding change mechanism for ATP synthase-Some probabilities and possibilities. Biochim Biophys Acta 1140: 215–250

Casadio R (1988) The oligomycin-sensitive Ca-ATPase of chromatophores from photosynthetic bacteria is not coupled to $\Delta\mu_H^+$ generation. In: Stein W (ed) The Ion Pumps: Structure, Function and Regulation, pp 201–206. Alan R. Liss, Inc., New York

Ceccarelli E and Vallejos RH (1983) Two types of essential carboxyl groups in *Rhodospirillum rubrum*. Arch Biochem Biophys 224: 382–388

Chen Z, Wu I, Richter ML and Gegenheimer P (1992) Over-expression and refolding of β-subunit from the chloroplast ATP synthase. FEBS Lett 298: 69–73

Collinson IR, Runswick MJ, Buchanan SK, Fearnly IM, Skehel JM, Van Raaij Mj, Griffiths DE and Walker JE (1994) F_0 membrane domain of ATP synthase from bovine heart mitochondria: Purification, subunit composition, and reconstitution with F_1-ATPase. Biochemistry 33: 7971–7978

Cortez N, Lucero HA and Vallejos RH (1983) Inactivation of

Rhodospirillum rubrum coupling factor by 7-chloro-4-nitrobenzo-2-oxa-1,3-diazole. Biochim Biophys Acta 724: 396–403

Cox GB, Devenish RJ, Gibson F, Howitt SM and Nagley P (1992) The structure and assembly of ATP synthase. In: Ernster L (ed) Molecular Mechanisms in Bioenergetics, pp 283–315. Elsevier

Cozens AL and Walker JE (1987) The organization and sequence of the genes for ATP synthase subunits in the cyanobacterium *Synechococcus* 6301. Support for an endosymbiotic origin of chloroplasts. J Mol Biol 194: 359–383

Cross RL and Kohlbrenner WE (1978) The mode of inhibition of oxidative phosphorylation by efrapeptin (A23871). J Biol Chem 253: 4865–4873

Deckers-Hebestreit G and Altendorf K (1992) The F_0 complex of the proton-translocating F-type ATPase of *Escherichia coli*. J Exp Biol 172: 451–459

Dunn DF and Heppel LA (1981) Properties and functions of the subunits of the *Escherichia coli* coupling factor ATPase. Arch Biochem Biophys 210: 421–436

Edwards PA and Jackson JB (1976) The control of the adenosine triphosphatase of *Rhodospirillum rubrum* chromatophores by divalent cations and the membrane high energy state. Eur J Biochem 62: 7–14

Falk G and Walker E (1988) DNA sequence of a gene cluster coding for subunits of the F_0 membrane sector of ATP synthase in *Rhodospirillum rubrum*. Biochem J 254: 109–122

Falk G, Hampe A and Walker JE (1985) Nucleotide sequence of the *Rhodospirillum rubrum atp* operon. Biochem J 228: 391–407

Ferguson SJ and Sorgato MC (1982) Protein electrochemical gradients and energy-transduction processes. Ann Rev Biochem 51: 185–217

Fillingame RH (1990) Molecular mechanics of ATP synthesis by F_0F_1-type H^+-transporting ATP synthases. In: Krulwich TA (ed) The Bacteria, Vol 12, pp 345–391. Academic Press, San Diego

Fisher RJ, Liang AM and Sundström GC (1981) Selective disaggregation of the H^+-translocating ATPase. J Biol Chem 256: 707–715

Futai M and Kanazawa H (1983) Structure and function of proton-translocating adenosine triphosphatase (F_0F_1): Biochemical and molecular biological approaches. Microbiol Rev 47: 285–312

Futai M, Noumi T and Maeda M (1989) ATP synthase (H^+-ATPase): Results by combined biochemical and molecular biological approaches. Annu Rev Biochem 58: 111–136

Gabellini N, Gao Z, Eckerskorn C, Lottspeich F and Oesterhelt D (1988) Purification of the H^+-ATPase from *Rhodobacter capsulatus*, identification of thc F_0F_1 components and reconstitution of the active enzyme. Biochim Biophys Acta 934: 227–234

Gabellini N, Gao Z, Oesterhelt D, Venturoli G and Melandri BA (1989) Reconstitution of cyclic electron transport and photophosphorylation by incorporation of the reaction center, cytochrome bc_1 complex and ATPsynthase from *Rhodobacter capsulatus* into ubiquinine-10/phospholipid vesicles. Biochim Biophys Acta 974: 202–210

Gepshtein A and Carmeli C (1974) Properties of adenosinetriphosphatase in chromatophores and in coupling factor from the photosynthetic bacteria *Chromatium* strain D. Eur J Biochem 44: 593–602

Gepshtein A and Carmeli C (1977) Properties of ATPase activity in coupling factor from *Chromatium* strain D chromatophores. Eur J Biochem 74: 463–469

Gepshtein A, Carmeli C and Nelson N (1978) Purification and properties of adenosine triphosphatase from *Chromatium vinosum* chromatophores. FEBS Lett 85: 219–223

Girault G, Berger G, Galmisch JM and Andre F (1988) Characterization of six nucleotide-binding sites on chloroplast coupling factor 1 and one site on its purified β subunit. J Biol Chem 263: 14690–14695

Glaser E and Norling B (1991) Chloroplast and plant mitochondrial ATP synthases. In: Lee CP (ed) Current Topics in Bioenergetics, Vol 16, pp 223–263. Academic Press, San Diego

Gogol EP, Lucken V and Capaldi RA (1987) The stalk connecting the F_1 and F_0 domains of ATP synthase visualized by electron microscopy of unstained specimens. FEBS Lett 219: 274–278

Graber P, Böttcher B and Boekema E (1990) The structure of the ATP synthase from chloroplasts. In: Milazzo G and Blank M (eds) Bioelectrochemistry, Vol III, pp 247–276. Plenum Press, New York

Gromet-Elhanan Z (1974a) Role of photophosphorylation coupling factor in energy conversion by depleted chromatophores of *Rhodospirillum rubrum*. J Biol Chem 249: 2522–2527

Gromet-Elhanan Z (1974b) Effect of aurovertin on energy conversion reactions in *Rhodospirillum rubrum* chromatophores. In: Avron M (ed) Proceedings of the Third International Congress on Photosynthesis, pp 791–797. Elsevier, Amsterdam

Gromet-Elhanan Z (1977) Electron transport and photophosphorylation in photosynthetic bacteria. In: Trebst A and Avron M (eds) Encyclopedia of Plant Physiology, New Series Vol 5, pp 637–662. Springer-Verlag, Berlin

Gromet-Elhanan Z (1988) Determination of the functional homology of β subunits isolated from various F_1-ATPase complexes: Their role in catalysis and coupled proton-translocation. In: Stein W. (ed) The Ion Pumps: Structure, Function and Regulation, pp 299–306. Alan R. Liss, Inc. New York

Gromet-Elhanan Z (1992) Identification of subunits required for the catalytic activity of the F_1-ATPase. J Bioenerg Biomembr 24: 447–452

Gromet-Elhanan Z and Avital S (1992) Properties of the catalytic ($\alpha\beta$)-core complex of chloroplast CF_1-ATPase. Biochim Biophys Acta 1102: 379–385

Gromet-Elhanan Z and Gest H (1978) A comparison of electron transport and photophosphorylation systems of *Rhodopseudomonas capsulata* and *Rhodospirillum rubrum*. Arch Microbiol 116: 29–34

Gromet-Elhanan Z and Khananshvili D (1984) Characterization of two nucleotide binding sites on the isolated, reconstitutively active β subunit of the F_0F_1 ATP synthase. Biochemistry 23: 1022–1028

Gromet-Elhanan Z and Khananshvili D (1986) Selective extraction and reconstitution of F_1 subunits from *Rhodospirillum rubrum* chromatophores. Methods Enzymol 126: 528–538

Gromet-Elhanan Z and Weiss S (1989) Regulation of $\Delta\bar{\mu}H^+$-coupled ATP synthesis and hydrolysis: role of divalent cations and of the F_0F_1-β subunit. Biochemistry 28: 3645–3650

Gromet-Elhanan Z, Philosoph S and Khananshvili D (1981) Sequential removal and reconstruction of subunits β and γ of the *Rhodospirillum rubrum* membrane-bound ATP synthase. In: Selman B and Selman-Reimer S (eds) Energy Coupling in Photosynthesis, pp 323–332. Elsevier, North Holland

Gromet-Elhanan Z, Khananshvili D, Weiss S, Kanazawa H and Futai M (1985) ATP synthesis and hydrolysis by a hybrid system reconstituted from the β-subunit of *Escherichia coli* F_1-ATPase and β-less chromatophores of *Rhodospirillum rubrum*. J Biol Chem 260: 12635–12640

Harada M, Ohta S, Sato M, Ito Y, Kobayashi Y, Sone N, Ohta T and Kagawa Y (1991) The $\alpha_1\beta_1$ heterodimer, the unit of ATP synthase. Biochim Biophys Acta 1056: 279–284

Harris DA, Boork J and Baltscheffsky M (1985) Hydrolysis of adenosine 5'-triphosphate by the isolated catalytic subunit of the coupling ATPase from *Rhodospirillum rubrum*. Biochemistry 24: 3876–3883

Hatefi Y (1993) ATP synthesis in mitochondria. Eur J Biochem 218: 759–767

Hauska G, Samoray D, Orlich G and Nelson N (1980) Reconstitution of photosynthetic energy conservation. II. Photophosphorylation in liposomes containing Photosystem-I reaction center and chloroplast coupling-factor complex. Eur J Biochem 111: 535–543

Hermann RG, Steppuhn J, Herrmann GS and Nelson N (1993) The nuclear-encoded polypeptide CF_0 -II from spinach is a real, ninth subunit of chloroplast ATP synthase. FEBS Lett 326: 192–198

Hicks DB and Yocum CF (1986a) Properties of the cyanobacterial coupling factor ATPase from *Spirulina platensis*. Arch Biochem Biophys 245: 220–229

Hicks DB and Yocum CF (1986b) Properties of the cyanobacterial coupling factor ATPase from *Spirulina platensis*. II. Activity of the purified and membrane-bound enzymes. Arch Biochem Biophys 245: 230–237

Hochman A and Carmeli C (1971) A coupling factor from *Chromatium* strain D chromatophores. FEBS Lett 13: 36–40

Hoppe J and Sebald W (1984) The proton conducting F_0-part of bacterial ATP synthases. Biochim Biophys Acta 768: 1–27

Issartel JP, Dupuis A, Garin J, Lunardi J, Michel L and Vignais PV (1992) The ATP synthase (F_0F_1) complex in oxidative phosphorylation. Experientia 48: 351–362

Jagendorf AT and Uribe E (1966) ATP formation caused by acid-base transition of spinach chloroplasts. Proc Natl Acad Sci USA 55: 170–177

Jagendorf AT, McCarty RE and Robertson D (1991) Coupling factor components: Structure and function. In: Bogorad L and Vasil IK (eds) Cell Culture and Somatic Cell Genetics of Plants, Vol 7B, pp 225–254. Academic Press, San Diego

Johansson BC (1972) A coupling factor from *Rhodospirillum rubrum* chromatophores. FEBS Lett 20: 339–340

Johansson BC and Baltscheffsky M (1975) On the subunit composition of the coupling factor (ATPase) from *Rhodospirillum rubrum*. FEBS Lett 53: 221–224

Johansson BC, Baltscheffsky M and Baltscheffsky H (1971) Coupling factor capabilities with chromatophore fragments from *Rhodospirillum rubrum*. In: Forti G, Avron M and Melandri A (eds) Proceedings of the IInd International Congress on Photosynthesis Research, Vol 2, pp 1203–1209. Dr. W Junk N.V., The Hague

Johansson BC, Baltscheffsky M and Baltscheffsky H (1973) Purification and properties of a coupling factor (Ca^{2+}-dependent adenosine triphosphatase) from *Rhodospirillum rubrum*. Eur J Biochem 40: 109–117

Kagawa Y, Ohta S and Otawara-Hamamoto Y (1989) $\alpha_3\beta_3$ complex of thermophilic ATP synthase: Catalysis without the γ-subunit. FEBS Lett 249: 67–69

Khananshvili D and Gromet-Elhanan Z (1982) Isolation and purification of an active γ-subunit of the F_0F_1-ATP synthase from chromatophore membranes of *Rhodospirillum rubrum*. J Biol Chem 257: 11377–11383

Khananshvili D and Gromet-Elhanan Z (1983a) The interaction of 4-Chloro-7-nitrobenzofurazan with *Rhodospirillum rubrum* chromatophores, their soluble F_1-ATPase, and the isolated purified β-subunit. J Biol Chem 258: 3714–3719

Khananshvili D and Gromet-Elhanan Z (1983b) The interaction of carboxyl group reagents with the *Rhodospirillum rubrum* F_1-ATPase and its isolated β-subunit. J Biol Chem 258: 3720–3725

Khananshvili D and Gromet-Elhanan Z (1983c) Modification of histidine residues by diethyl pyrocarbonate leads to inactivation of the *Rhodospirillum rubrum* RrF_1-ATPase. FEBS Lett 159: 271–274

Khananshvili D and Gromet-Elhanan Z (1984) Demonstration of two binding sites for ADP on the isolated β-subunit of the *Rhodospirillum rubrum* RrF_0F_1-ATP synthase. FEBS Lett 178: 10–14.

Khananshvili D and Gromet-Elhanan Z (1985a) Characterization of an inorganic phosphate binding site on the isolated, reconstitutively active β subunit of F_0F_1 ATP synthase. Biochemistry 24: 2482–2487

Khananshvili D and Gromet-Elhanan Z (1985b) Evidence that the Mg-dependent low-affinity binding site for ATP and P_i demonstrated on the isolated β subunit of the F_0F_1 ATP synthase is a catalytic site. Proc Natl Acad Sci USA 82: 1886–1890

Khananshvili D and Gromet-Elhanan Z (1986) Partial proteolysis as a probe for ligand-induced conformational changes in the isolated β subunit of the H^+-translocating F_0F_1 ATP synthase. Biochemistry 25: 6139–6144

Leiser M and Gromet-Elhanan Z (1974) Demonstration of acid-base phosphorylation in chromatophores in the presence of a K^+ diffusion potential. FEBS Lett 43: 267-270

Low H and Afzelius BA (1965) Subunits of the chromatophore membranes in *Rhodospirillum rubrum*. Exp Cell Biol 35: 431–434

Lubberding HJ, Offerijns F, Vel WAC and de Vries PJR (1981) Characterization of the ATPase of the thermophilic cyanobacterium *Synechococcus lividus*. In: A Koyanoglou G (ed) Photosynthesis II. Electron Transport and Photophosphorylation, pp 779–788. Balaban Int. Sci. Services, Philadelphia

McCarty RE and Moroney JV (1985) Structure and function of chloroplast coupling factor 1. In: Martonosi A (ed) The Enzymes of Biological Membranes, pp 383–413. Plenum Press, New York

Melandri BA and Baccarini-Melandri A (1971) Energy-transduction in photosynthetic bacteria. I. Properties of solubilized and reconstituted ATPase in *Rhodopseudomonas capsulata* photosynthetic membranes. In: Forti G, Avron M and Melandri A (eds) Proceedings of the IInd International

Congress on Photosynthesis Research, Vol 2, pp 1169–1192. Dr. W Junk N.V., The Hague

Mitchell P (1966) Chemiosmotic Coupling in Oxidative and Photosynthetic Phosphorylation. Glynn Research Ltd., Bodmin

Miwa K and Yoshida M (1989) The $\alpha_3\beta_3$ complex, the catalytic core of F_1-ATPase. Proc Natl Acad Sci USA 86: 6484–6487

Müller H, Neufang H and Knobloch K (1982) Purification and properties of the coupling-factor ATPases F_1 from *Rhodopseudomonas palustris* and *Rhodopseudomonas sphaeroides*. Eur J Biochem 127: 559–566

Müller H, Neufang H and Knobloch K (1983) Kinetic studies on the membrane-bound and the purified coupling factor ATPase from *Rhodopseudomonas sphaeroides*. Arch Biochem Biophys 224: 283–289

Müller HW and Baltscheffsky M (1979) On the oligomycin-sensitivity and subunit composition of the ATPase complex from *Rhodospirillum rubrum*. Z Naturforsch 34c: 229–232

Müller HW, Schwuléra U, Salzer M and Dose K (1979) Purification, subunit structure, and kinetics of the chloroform-released F_1ATPase complex from *Rhodospirillum rubrum* and its comparison with F_1 ATPase forms isolated by other methods. Z Naturforsch 34: 38–45.

Nalin CM and Nelson N (1987) Structure and biogenesis of the chloroplast coupling factor CF_0CF_1-ATPase. In: Lee CP (ed) Current Topics in Bioenergetics, Vol 15, pp 273–294. Academic Press, San Diego

Nathanson L and Gromet-Elhanan Z (1994) Cloning and functional expression of wild type and mutant β subunits of the *Rhodospirillum rubrum* F_0F_1 ATP synthase. In: 8th European Bioenergetics Conferece (EBEC) Short Reports, Vol 8, p 19. Elsevier, Amsterdam

Nelson N, Nelson H and Racker E (1972) Partial resolution of the enzymes catalyzing photophosphorylation. XI. Magnesium adenosine triphosphatase properties of heat-activated coupling factor I from chloroplasts. J Biol Chem 247: 6506–6510

Nemoto H, Ohta Y, Hisobori T, Shinohara K and Sakurai H (1990) Isolation, purification and characterization of coupling factor ATPase from *Anacystis nidulans*. In: Baltscheffsky M (ed) Current Research in Photosynthesis, Vol III, pp 169–172. Kluwer Academic Publishers, Dordrecht

Norling B, Strid Å and Nyrén P (1988) Conversion of coupling factor 1 of *Rhodospirillum rubrum* from a Ca^{2+}-ATPase into a Mg^{2+}-ATPase. Biochim Biophys Acta 935: 123–129

Norling B, Strid Å Tourikas C and Nyrén P (1989) Amount and turnover rate of the F_0F_1-ATPase and the stoichiometry of its inhibition by oligomycin in *Rhodospirillum rubrum* chromatophores. Eur J Biochem 186: 333–337

Nyrén P and Baltscheffsky M (1983) Inorganic pyrophosphate-driven ATP-synthesis in liposomes containing membrane-bound inorganic pyrophosphatase and F_0-F_1 complex from *Rhodospirillum rubrum*. FEBS Lett 155: 125–130

Ohta Y, Yoshioka T, Mochimaru M, Hisabori T and Sakurai H (1993) Tentoxin inhibits both photophosphorylation in thylakoids and the ATPase activity of isolated coupling factor F_1 from the cyanobacterium *Anacystis nidulans*. Plant Cell Physiol 34: 523–529

Oren R and Gromet-Elhanan Z (1977) Coupling factor adenosine triphosphatase-complex of *Rhodospirillum rubrum*. Isolation of an oligomycin-sensitive Ca^{2+}, Mg^{2+}-ATPase. FEBS Lett 79: 147–150

Oren R and Gromet-Elhanan Z (1979) Coupling factor ATPase complex of *Rhodospirillum rubrum*. Purification and characterization of an oligomycin and N-N'-dicyclohexyl-carbodiimide- sensitive (Ca^{2+} + Mg^{2+})-ATPase. Biochim Biophys Acta 548: 106–118

Oren R, Weiss S, Garty H, Caplan SR and Gromet-Elhanan Z (1980) ATP synthesis catalyzed by the ATPase complex from *Rhodospirillum rubrum* reconstituted into phospholipid vesicles together with bacteriorhodopsin. Arch Biochem Biophys 205: 503–509

Otto J and Berzborn RJ (1989) Quantitative immunochemical evidence for identical topography of subunits CF_0II and CF_0I within the photosynthetic ATP-synthase of spinach chloroplasts. FEBS Lett 250: 625–628

Pedersen PL and Amzel LM (1993) ATP synthases. Structure, reaction center, mechanism, and regulation of one of nature's most unique machines. J Biol Chem 268: 9937–9940

Penefsky HS and Cross RL (1991) Structure and mechanism of F_0F_1-type ATP synthases and ATPases. Adv Enzymol 64: 173–214

Philosoph S and Gromet-Elhanan Z (1981a) Antibodies to the F_1-ATPases of *Rhodospirillum rubrum* and its purified native β subunit: Inhibition of ATP-linked activities in *Rs. rubrum* and in lettuce. Eur J Biochem 119: 107–113

Philosoph S and Gromet-Elhanan Z (1981b) Sequential removal of specific subunits from the ATPase complex of *Rhodospirillum rubrum*. In: Akoyunoglou G (ed) Photosynthesis II. Electron Transport and Photophosphorylation, pp 741–751. Balaban Intl Sci Services, Philadelphia

Philosoph S, Binder A and Gromet-Elhanan Z (1977) Coupling factor ATPase complex of *Rhodospirillum rubrum*. Purification and properties of a reconstitutively active single subunit. J Biol Chem 252: 8747–8752

Philosoph S, Khananshvili D and Gromet-Elhanan Z (1981) Sequential removal and reconstitution of subunits β and γ from a membrane-bound F_0F_1-ATP synthase. Biochem Biophys Res Commun 101: 384–389

Pick U and Bassilian S (1982) Activation of magnesium ion specific adenosinetriphosphatase in chloroplast coupling factor 1 by octyl glucoside. Biochemistry 24: 6144–6152

Pick U and Racker E (1979) Purification and reconstitution of the *N,N'*-dicyclohexylcarbo-diimide-sensitive ATPase complex from spinach chloroplasts. J Biol Chem 254: 2793–2799

Prince RC, Baccarini-Melandri A, Hauska GA, Melandri BA and Crofts AR (1975) Asymmetry of an energy transducing membrane: The location of cytochrome C_2 in *Rhodopseudomonas spheroides* and *Rhodopseudomonas capsulata*. Biochim Biophys Acta 387: 212–227

Racker E (1976) A New Look at Mechanism in Bioenergetics, Academic Press, New York

Ravizzini RA, Lescano WIM and Vallejos RH (1975) Effect of aurovertin on energy transfer reactions in *Rhodospirillum rubrum* chromatophores. FEBS Lett 58: 285–288

Reed DW and Raveed D (1972) Some properties of the ATPase from chromatophores of *Rhodopseudomonas sphaeroides* and its structural relationship to bacteriochlorophyll proteins. Biochim Biophys Acta 283: 79–91

Richter ML, Gromet-Elhanan Z and McCarty RE (1986) Reconstitution of the H^+-ATPase complex of *Rhodospirillum rubrum* by the β subunit of the chloroplast coupling factor 1. J Biol Chem 261: 12109–12113

Schafer HJ, Müller HW and Dose K (1980) Conversion of the

Ca^{2+}-ATPase from *Rhodospirillum rubrum* into a Mg^{2+}-dependent enzyme by 1,N^6-etheno ATP. Biochem Biophys Res Commun 95: 1113–1118

Schneider E, Schwuléra U, Müller HW and Dose K (1978) Solubilization of an oligomycin-sensitive ATPase complex from *Rhodospirillum rubrum* chromatophores and its inhibition by various antibiotics. FEBS Lett 87: 257–260

Schneider E, Müller HW, Rittinghaus K, Thiele V, Schwulera U and Dose K (1979) Properties of the F_0F_1 ATPase complex from *Rhodospirillum rubrum* chromatophores, solubilized by Triton X-100. Eur J Biochem 97: 511–517

Schneider E, Friedl P, Schwuléra U and Dose K (1980) Energy-linked reactions catalyzed by the purified ATPase complex (F_0F_1) from *Rhodospirillum rubrum* Chromatophores. Eur J Biochem 108: 331–336

Selman BR and Durbin RD (1978) Evidence for a catalytic function of the coupling factor 1 protein reconstituted with chloroplast thylakoid membranes. Biochim Biophys Acta 502: 29–37

Shapiro AB, Huber AH and McCarty RE (1991a) Four tight nucleotide binding sites of chloroplast coupling factor 1. J Biol Chem 266: 4194–4200

Shapiro AB, Gibson KD, Scheraga HA and McCarty RE (1991b) Fluorescence resonance energy transfer, mapping of the fourth of six nucleotide binding sites of chloroplast coupling factor 1. J Biol Chem 266: 12276–12280

Slooten L and Nuyten A (1981) Activation-deactivation reactions in the ATP-ase enzyme in *Rhodospirillum rubrum* chromatophores. Biochim Biophys Acta 638: 305–312

Slooten L and Vandenbranden S (1989a) Isolation of the proton-translocating F_0F_1-ATPase from *Rhodospirillum rubrum* chromatophores, and its functional reconstitution into proteoliposomes. Biochim Biophys Acta 975: 148–157

Slooten L and Vandenbranden S (1989b) ATP-synthesis by proteoliposomes incorporating *Rhodospirillum rubrum* F_0F_1 as measured with firefly luciferase: dependence on $\Delta\psi$ and ΔpH. Biochim Biophys Acta 976: 150–160

Soe G, Nishi N, Kakuno T and Yamashita J (1978) Reversible conversion from Ca^{2+}-ATPase activity to Mg^{2+} and Mn^{2+}-ATPase activities of coupling factor purified from acetone powder of *Rhodospirillum rubrum* chromatophores. J Biochem 84: 805–814

Soe G, Nishi N, Kakuno T, Yamashita J and Horio T (1980) Purification and identification of the factor capable of converting Ca^{2+}-ATPase into Mg^{2+}-ATPase present in *Rhodospirillum rubrum* chromatophores. J Biochem 87: 473–481

Sokolov M, Avital S and Gromet-Elhanan Z (1992) Structure and function of active partial complexes isolated from the chloroplast CF_1-ATPase. In: Murata N (ed) Research in Photosynthesis, Vol II, p 653–659. Kluwer Academic Publishers, Dordrecht

Steele JA, Uchytil TF, Durbin RD, Bhatnagar P and Rich DH (1976) Chloroplast coupling factor 1: A species-specific receptor for tentoxin. Proc Natl Acad Sci USA 73: 2245–2248

Strid A and Nyren P (1989) Division of divalent cations into two groups in relation to their effect on the coupling of the F_0F_1-ATPase of *Rhodospirillum rubrum* to the proton motive force. Biochemistry 28: 9718–9724

Strotmann H and Bickel-Sandkötter S (1984) Structure, function, and regulation of chloroplast ATPase. Annu Rev Plant Physiol 35: 97–120

Strotmann H, Hesse H and Edelmann K (1973) Quantitative determination of coupling factor CF_1 of chloroplasts. Biochim Biophys Acta 314: 202–210

Turina P, Rumberg B, Melandri A and Gräber P (1992) Activation of the H^+-ATP synthase in the photosynthetic bacterium *Rhodobacter capsulatus*. J Biol Chem 267: 11057–11063

Tybulewicz VLJ, Falk G and Walker JE (1984) *Rhodopseudomonas blastica atp* operon. Nucleotide sequence and transcription. J Mol Biol 179: 185–214

Tzagoloff A and Meagher P (1971) Assembly of the mitochondrial membrane system. V. Properties of a dispersed preparation of the rutamycin-sensitive adenosine triphosphatase of yeast mitochondria. J Biol Chem 246: 7328–7336

Van Walraven HS, Lubberding HJ, Marvin HJP and Kraayenhof R (1983) Characterization of reconstituted ATPase complex proteoliposomes prepared from the thermophilic cyanobacterium *Synechococcus* 6716. Eur J Biochem 137: 101–106

Van Walraven HS, Van der Bend RL, Hagendoorn MJM, Haak NP, Oskam A, Oostdam A, Krab K and Kraayenhof R (1986) Comparison of ATP synthesis efficiencies in ATPase proteoliposomes of different complexities. Bioelectrochem Bioener 16: 167–180

Van Walraven HS, Lutter R and Walker JE (1993) Organization and sequences of genes for the subunits of ATP synthase in the thermophilic cyanobacterium *Synechococcus* 6716. Biochem J 294: 239–251

Vignais PV and Lunardi J (1985) Chemical probes of the mitochondrial ATP synthesis and translocation. Ann Rev Biochem 54: 977–1014

Wagner N, Gutweiler M, Pabst R and Dose K (1987) Coreconstitution of bacterial ATP synthase with monomeric bacteriorhodopsin into liposomes. A comparison between the efficiency of monomeric bacteriorhodopsin and purple membrane patches in coreconstitution experiments. Eur J Biochem 165: 177–183

Walker JE, Fearnley IM, Lutter R, Todd RJ and Runswick MJ (1990) Structural aspects of proton-pumping ATPases. Phil Trans R Soc Lond B 326: 367–378

Wang ZG, Wei JM, Wu YQ and Shen YK (1992) Expression of fusion proteins of maize chloroplast CF_1 beta subunit in E. coli. In: Murata N (ed) Research in Photosynthesis, Vol III, pp 243–246. Kluwer Academic Publishers, Dordrecht

Webster GD and Jackson JB (1978) Affinity chromatography of H^+-translocating adenosine triphosphatase isolated by chloroform extraction of *Rhodospirillum rubrum* chromatophores. Modification of binding affinity by divalent cations and activating anions. Biochim Biophys Acta 503: 135–154

Webster GD, Edwards PA and Jackson JB (1977) Interconversion of two kinetically distinct states of the membrane-bound and solubilized H^+-translocating ATPase from *Rhodospirillum rubrum*. FEBS Lett 76: 29–35

Weiss S, McCarty RE and Gromet-Elhanan Z (1994) Tight nucleotide binding sites and ATPase activities of the *Rhodospirillum rubrum* RrF_1-ATPase as compared to spinach chloroplast CF_1-ATPase. J Bioengr Biomembr 26: 573–580

Wise JG, Duncan TM, Richardson-Latchney L, Cox DN and Senior AE (1983) Properties of F_1-ATPase from the *uncD412* mutant of *Escherichia coli*. Biochem J 215: 343–350

Xie OL, Lill H, Hauska G, Maeda M, Futai M and Nelson N (1993) The *atp2* operon of the green bacterium *Chlorobium limicola*. Biochim Biophys Acta 1172: 267–273

Xue Z, Zhou JM, Melese T, Cross RL and Boyer PD (1987) Chloroplast F_1 ATPase has more than three nucleotide binding sites, and 2-azido-ADP or 2-azido-ATP at both catalytic and noncatalytic sites labels the β subunit. Biochem 26: 3749–3753

Yanyushin MF (1988) Isolation and characterization of F_1-ATPase from the green nonsulfur photosynthetic bacterium *Chloroflexus aurantiacus*. Biokhumiya 53: 1288–1295

Yanyushin MF (1991) ATP synthase of the green nonsulfur photosynthetic bacterium *Chloroflexus aurantiacus*. Biokhimiya 56: 1131–1139

Yoshida M, Allison WS, Esch FS and Futai M (1982) The specificity of carboxyl group modification during the inactivation of the *Escherichia coli* F_1-ATPase with dicyclohexyl-[^{14}C] carbodiimide. J Biol Chem 257: 10033–10037

Chapter 38

Proton-Translocating Transhydrogenase and NADH Dehydrogenase in Anoxygenic Photosynthetic Bacteria

J. Baz Jackson
School of Biochemistry, The University of Birmingham, Edgbaston, Birmingham B15 2TT, U.K.

Summary

Some species of anoxygenic photosynthetic bacteria possess transhydrogenase and NADH dehydrogenase enzymes that are closely related to equivalent enzymes from animal mitochondria. Both are membrane proteins which couple redox reactions of the nicotinamide nucleotides to the translocation of hydrogen ions. Transhydrogenase catalyses the reversible transfer of reducing equivalents between NAD(H) and NADP(H); it has a low proton pumping stoichiometry. Probably under most conditions NADH dehydrogenase serves to transfer reducing equivalents from NADH to ubiquinone; it has a high proton pumping stoichiometry. Both enzymes are very active in chromatophore membranes isolated from cells of *Rhodobacter* and *Rhodospirillum* species grown anaerobically in the light, but their physiological function during phototrophic growth is not clear. Transhydrogenase may be involved in the transfer of reducing power between the cytoplasmic NAD(H) and NADP(H) pools, governed by the transmembrane proton electrochemical gradient (Δp), and in response to the metabolic needs of the cell. It might have a lesser role in the generation or regulation of Δp. During aerobic growth in the dark, NADH dehydrogenase undoubtedly serves as a major generator of Δp; during phototrophic growth it might be essential for the oxidation of highly reduced carbon substrates by auxiliary oxidants such as nitrate or dimethylsulfoxide, or it might be expressed in 'readiness' for the return to dark aerobic growth.

Interest in NADH dehydrogenase in photosynthetic bacteria will increase as gene sequences become available and, perhaps, in response to the need to elucidate the role of the related protein in chloroplasts from higher plants. Present studies are limited by the lack of a purification procedure for the enzyme. Transhydrogenase,

R. E. Blankenship, M. T. Madigan and C. E. Bauer (eds): Anoxygenic Photosynthetic Bacteria, pp. 831–845.

on the other hand, is amenable to study. In many ways it presents an excellent model for the study of other ion-translocating membrane proteins, particularly in photosynthetic membranes where its activity can be measured by spectroscopy with good time resolution.

I. Introduction

A. The Reactions Catalysed by Transhydrogenase and NADH Dehydrogenase

Proton-translocating transhydrogenase (H^+-Thase) catalyses the transfer of reducing equivalents (hydride ion equivalents) between NAD(H) and NADP(H) coupled to the translocation of protons across a membrane,

$$\mathrm{NADH + NADP^+ + xH^+_{out} \rightleftharpoons NAD^+ + NADPH + xH^+_{in}} \quad (1)$$

where x is the stoichiometry of H^+ translocated per H^- equivalent transferred between the nucleotides. There are no scalar protons associated with the net reaction. Recent reviews on transhydrogenase have been published by Rydstrom et al. (1987), Jackson (1991) and Yamaguchi and Hatefi (1992).

NADH dehydrogenase (NADH-DH), more properly called NADH:ubiquinone oxidoreductase and sometimes, especially in mitochondria, referred to as 'complex I', catalyses the transfer of reducing equivalents from NADH to ubiquinone coupled to the translocation of protons across a membrane,

$$\mathrm{NADH + UQ + H^+_{scalar} + nH^+_{in} \rightleftharpoons NAD^+ + UQH_2 + nH^+_{out}} \quad (2)$$

where n is the stoichiometry of H^+ translocated per $2e^-$ transferred from NADH to UQ. A scalar proton is associated with the net reaction. Recent reviews on NADH-DH can be found (Ragan, 1987; Weiss and Friedich, 1991; Weiss et al., 1991; Yagi, 1991, 1993; Fearnley and Walker, 1992; Walker, 1992).

Both enzymes are found in the inner membranes of animal and plant mitochondria and the cytoplasmic membranes of many bacteria. In the anoxygenic photosynthetic bacteria, H^+-Thase has been studied in some detail in *Rhodobacter (Rb.) capsulatus* and *Rhodospirillum (Rs.) rubrum* but, in other organisms from the group, the available information is sketchy. There have been few investigations into the properties of NADH-DH from photosynthetic bacteria but this situation might change in view of the resurgence of interest in the enzyme from mitochondria and the finding that there may be homologues of some subunits of mitochondrial complex 1 in the chloroplast genome. Unfortunately, the enzyme from photosynthetic and other bacteria has been difficult to isolate in a form that can actively transport protons (Berks and Ferguson, 1991). Remarkably, the functions of H^+-Thase and of NADH-DH during phototropic growth are not well understood (Jackson and McEwan, 1993). However, both are present at high activities during chemoheterotrophic and photoheterotrophic growth. They are both directly coupled to the proton electrochemical gradient (Δp), the primary energetic intermediate of the bacterial cell, and to the redox state of the nicotinamide nucleotide pools. Some speculations as to the function of H^+-Thase and NADH-DH will be offered at the end of the chapter.

B. Historical Perspective

The discovery of H^+-Thase and NADH-DH in photosynthetic bacteria followed on rapidly from research into the mechanism of energy coupling in animal mitochondria during the 1960's.

Keister and Yike (1966, 1967b) and Orlando et al. (1966) described 'energy-linked' transhydrogenation from NADH to $NADP^+$ in chromatophores from a number of species of photosynthetic bacteria. The reaction proceeded at a low rate in darkened suspensions of chromatophores but was substantially increased by ATP, PPi or light. The acceleration was blocked by uncoupling agents. From these observations and from the effects of specific inhibitors, for example antimycin, on the light-driven reaction and oligomycin on the ATP-driven reaction, it was concluded that transhydrogenation was coupled to the consumption of 'the high energy intermediate,

Abbreviations: $AcPdAD^+$ – acetyl pyridine adenine dinucleotide; *E.* – *Escherichia*; H^+-Thase – proton-translocating nicotinamide dinucleotide transhydrogenase; *N.* – *Neurospora*; NADH-DH – NADH dehydrogenase; *Rb.* – *Rhodobacter*; *Rs.* – *Rhodospirillum*; Δp – proton electrochemical gradient

X~I'. Subsequent appreciation that the high energy intermediate is in fact a proton electrochemical gradient led to the recognition that transhydrogenase is a proton pump, confirmed by Rydstrom (1979) and Earle and Fisher (1980) with the purified mitochondrial enzyme reconstituted into liposomes. An equivalent experiment with purified transhydrogenase from photosynthetic bacteria was not reported until recently (Jackson et al., 1991).

It has long been known that several species of the anoxygenic photosynthetic bacteria are capable of oxidative growth in the dark under aerobic conditions. It was realized in the 1950's and 1960's that chromatophores, for example from *Rs. rubrum*, *Rb. sphaeroides* and *Rb. capsulatus*, had respiratory activity with NADH as an electron donor. The activity was shown to be inhibited by rotenone and piericidin A and might thus be related to NADH-DH from mitochondria. The existence of an energy coupling site at the level of NADH-DH in chromatophores was indicated by the observation of 'energy-linked reversed electron flow' first, in *Rs. rubrum* (Keister and Yike, 1967a) and then in *Rb. capsulatus* (Klemme, 1969): reduction of NAD^+ by succinate was observed only if chromatophores were supplied with ATP, PPi or light; uncoupling agents inhibited the reaction. Although the clear implication of these findings would now be that chromatophore NADH-DH is a proton pump, a direct demonstration of this, either in chromatophores, or in a reconstituted system is still required.

II. The Structure of Transhydrogenase

Until recently most work on the structure of H^+-Thase had been carried out with the enzymes from mitochondria and *Escherichia coli* but information is now emerging on the somewhat different structure of the *Rs. rubrum* protein.

A. The Structure of Transhydrogenase from Mitochondria and E. coli *as a Paradigm for the Enzyme from Photosynthetic Bacteria?*

The amino acid sequences for H^+-Thase from *E. coli* and mitochondria have been predicted from the nucleotide sequences of the genes and the cDNA, respectively (Clarke et al., 1986; Yamaguchi et al., 1988; Ahmad et al., 1992). It is established that the *E. coli* enzyme comprises two polypeptides (α, Mr 54000 and β, Mr 49000) which in the active enzyme probably form an $\alpha_2\beta_2$ structure (Clarke and Bragg, 1985; Hou et al., 1990). The mitochondrial enzyme is a dimer of two identical polypeptides (Anderson and Fisher, 1981; Persson et al., 1987). When the α and β polypeptides of H^+-Thase from *E. coli* are arranged contiguously there is a clear similarity in sequence with the mitochondrial enzyme. Hydropathy profiles indicate that both enzymes might be organised in three large domains—see Fig. 1 (Clarke et al., 1986; Yamaguchi et al., 1988). In mitochondria, domain I (approximately 400 amino acids at the N-terminus) and domain III (approximately 200 amino acids at the C-terminus) are relatively hydrophilic and are thought to protrude from the membrane whereas domain II (approximately 400 amino acids at the center of the polypeptide) is strongly hydrophobic and might comprise up to 14 trans-membrane helices. In *E. coli* the structure is similar except that there is a 'break' in the polypeptide chain within the hydrophobic domain II. Experiments with proteases and with specific antibodies raised to segments of the protein indicate that domains I and

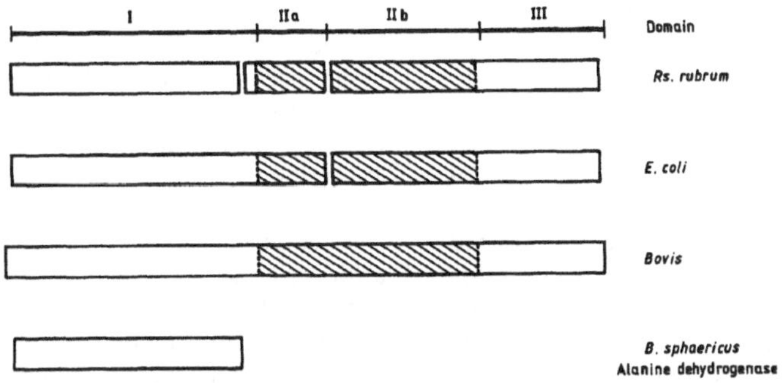

Fig. 1. The domain structure of transhydrogenases from mitochondria, *E. coli* and *Rs. rubrum* and homology with alanine dehydrogenase.

III protrude from the same side of the membrane - the matrix side in mitochondria (Yamaguchi and Hatefi, 1991) and the cytoplasmic side in *E. coli* (Tong et al., 1991). There is no evidence for the existence of prosthetic groups in H^+-Thase from any source.

There is clear evidence from analysis of steady-state kinetics (Hanson, 1979; Homyk and Bragg, 1979; Enander and Rydstrom, 1982; Lever et al., 1991) that there are separate sites on H^+-Thase for NAD^+ or NADH and for $NADP^+$ or NADPH. Substrates bound at the NAD(H) site donate or receive H^- equivalents to or from the 4A position of the nicotinamide ring whereas substrates bound at the NADP(H) site donate or receive H^- equivalents to or from the 4B position (Kawasaki et al., 1964; Griffiths and Robertson, 1966; Lee et al., 1965).

There is presently considerable interest in the indentification of the two nucleotide binding sites in H^+-Thase. There is a clearly recognizable $\beta\alpha\beta$ 'Rossman fold' in domain I (Clarke et al., 1986; Yamaguchi et al., 1988) which has been predicted to be the NAD(H) site on the basis of a G-G--G motif. However, recent findings (Lilly et al., 1991) suggest that there may be exceptions to what was thought to be a general rule for the specificity of nucleotide binding. Some experiments with radiolabeled dicyclohexyl-carbodiimide and 5'-[p-(fluorosulfonyl)benzoyl] adenosine were taken to support the view that the NAD(H) site lies within domain I and also to indicate that the NADP(H) site is in domain III (Wakabayashi and Hatefi, 1987a,b). However, other studies with radiolabelled 8-azido-AMP are less conclusive (Hu et al., 1992).

B. The Unique Structure of Transhydrogenase in Rhodospirillum rubrum

Consistent with the existence of stretches of very hydrophobic amino acids in their sequences, the H^+-Thases from mitochondria and from *E. coli* require detergent for solubilization. The enzyme from *Rb. capsulatus* has a polypeptide composition that resembles the one from *E. coli* and it too has the character of a membrane protein (Lever et al., 1991). However, in an elegant series of papers, Fisher and colleagues showed that the structure of transhydrogenase in *Rs. rubrum* is different. When chromatophores from this organism were washed by centrifugation through low ionic strength buffer, light-dependent transhydrogenation was abolished. The supernatant fraction, though inactive itself, was fully able to restore the reaction to depleted chromatophores (Fisher and Guillory, 1969a,b). In the intact chromatophore membranes, the 4A to 4B stereospecific transfer of H^- from NADH to $NADP^+$ was similar to that found for the mitochondrial enzyme, indicating that it is the same type of transhydrogenation reaction (Fisher and Guillory, 1971a). It was supposed that H^+-Thase in *Rs. rubrum* is composed of a water soluble protein, Th_s, and a membrane-bound component, Th_m, which can be reversibly associated. The recombination of Th_s with Th_m was dependent on the presence of low concentrations of nicotinamide nucleotides (Fisher and Guillory, 1971b). The potential of the system for the study of H^+-Thase was exemplified by experiments in which Th_s and washed, depleted chromatophores were separately treated with protein modification reagents and then reconstituted with the untreated complementary component (Fisher et al., 1975; McFadden and Fisher, 1978; Jacobs and Fisher, 1979; Palmer et al., 1993).

As the primary amino acid sequences for H^+-Thases from *E. coli* and mitochondria became available and, with them, predictions of the domain structure of the enzymes (Clarke et al., 1986; Yamaguchi et al., 1988), it became of interest to re-evaluate the relationship with the transhydrogenase from *Rs. rubrum*. It was conceivable that Th_s could correspond to either of the predominantly hydrophilic domains I or III, or to an activating protein factor for transhydrogenase unique to this bacterium. Purification of Th_s from *Rs. rubrum* and analysis of the nucleotide sequence of the gene have led to a solution of this problem.

In earlier work, only partial purification of Th_s and Th_m had been achieved (Konings and Guillory, 1972; Jacobs et al., 1977) but we recently described a procedure that routinely leads to small quantities of highly-purified material (Cunningham et al., 1992b). It makes use of the fact that other peripheral proteins can be displaced from *Rs. rubrum* chromatophores by washing in the presence of $NADP^+$ and then Th_s can be specifically removed by centrifugation in the absence of nucleotide. Further purification is then achieved by ion-exchange and affinity dye chromatography. The protein is assayed on the basis of its ability to restore light-driven transhydrogenase to depleted chromatophores. It runs predominantly as a single band, Mr 43000 Da, on SDS-PAGE but appears to be dimeric under non-denaturing conditions. The complete purification of Th_m still remains to be achieved. The solubilization procedure is particularly demanding, with only two detergents being successful of more than thirty tried (Palmer et al., 1993). Some purification was achieved by ion exchange and affinity dye chromatography but the small quantity of available material limited further success.

The N-terminal amino acid sequence of the 43 kDa polypeptide of Th_s (Cunningham et al., 1992b), and of proteolytic fragments derived therefrom, show clearly that it is related to domain I of the mitochondrial and *E. coli* proteins. The nucleotide sequence of the gene (Williams et al., 1994) leads us to expect that the relationship of the *Rs. rubrum* protein to other members of the family is as shown in Fig. 1. The essential difference with *E. coli* transhydrogenase is that the *Rs. rubrum* protein has three polypeptides. In addition to that, in the strongly hydrophobic domain II, there is an additional 'break' between domain II and the relatively hydrophilic domain I. This is nicely consistent with the view that domain I protrudes from the cytoplasmic side of the membrane. Evidently, under physiological conditions the domain I polypeptide of *Rs. rubrum* is bound to domains II and III by interactions that depend on the occupation of the nicotinamide nucleotide sites. In the absence of nucleotides, the domain I protein, Th_s, dissociates from domains II and III. In the presence of nucleotides, it can be rebound (Cunningham et al., 1992a).

Although the predicted amino acid sequences of H^+-Thase from *E. coli* and mitochondria have been available for some time, there had been no reported sequence similarities with other proteins except in the region of the nucleotide binding region. However, it transpires (Cunningham et al., 1992) that domain I of H^+-Thase has close similarities with the recently-published sequences of alanine dehydrogenase from *Bacillus* sp. (Kuroda at el., 1990). Alanine dehydrogenase catalyses the reversible reductive amination of pyruvate to alanine with NAD(P)H. It is a soluble enzyme, widespread in bacteria, and is generally thought to be cytoplasmic, although alanine dehydrogenase from *Mycobacterium tuberculosis* (for which amino acid sequence is now available, Anderson et al., 1992) is secreted from the cells during growth. The implications of these observations for the structure, mechanisms and evolution of alanine dehydrogenase and H^+-Thase are extremely interesting. Possibly H^+-Thase evolved from interaction between a soluble alanine dehydrogenase-like protein and an intrinsic membrane protein with a capacity for proton translocation.

III. Bioenergetics and Kinetics of Transhydrogenase

A. Transhydrogenase is a Simple Proton Pump

The primary energy-conserving process in phototropic bacteria is the chemiosmotic proton circuit utilising the photosynthetic electron transport chain as a generator and the H^+-ATP synthase as a consumer of Δp. However, as an experimental system this can sometimes be complex and, in contrast, electron transport $\rightarrow \Delta p \rightarrow H^+$-Thase, is relatively simple. For example, the dependence of the rate of ATP synthesis in chromatophores on the value of Δp indicates that the activity of the H^+-ATP synthase may be modulated by interactions with the electron transport system (Baccarini-Melandri et al., 1977; Venturoli and Melandri, 1982) but the reaction catalysed by H^+-Thase, operating in the 'forward' direction (Eq. (1)), has a unique dependence on Δp regardless of whether Δp is varied with uncoupling agents or with inhibitors of electron transport (Cotton et al., 1987). Thus, H^+-Thase can be treated as a simple and independent proton pump.

Furthermore, as in other membrane systems, the chromatophore H^+-ATP synthase is not only driven by, it is also regulated by, Δp: increase in Δp leads to a conformationally-activated enzyme. This is most clearly demonstrated by experiments in which an increase in Δp upon illumination causes an accelerated rate of ATP hydrolysis (Melandri et al., 1972). In contrast, when conditions were chosen greatly to favor the 'reverse' transhydrogenase reaction (Eq. (1)), even up to substantial values of Δp there was no increase in the rate of reaction for any incremental increase in Δp (Palmer and Jackson, 1992). Thus, there is no evidence that transhydrogenase can be switched between relatively active and inactive states.

B. Steady-State and Transient-State Kinetics of Transhydrogenase

The kinetic mechanism for the reduction, by NADPH,

of AcPdAD$^+$ (an analogue of NAD$^+$, whose reduced form, AcPdADH, has a characteristic absorbance band at 363 nm), has been determined for membrane-bound and solubilized preparations of H$^+$-Thase from *E. coli*, mitochondria and *Rb. capsulatus* (Hanson, 1979; Homyk and Bragg, 1979; Enander and Rydstrom, 1982; Lever et al., 1991). The reaction proceeds by random addition of substrates to give a ternary complex with the enzyme. The substrate binding reactions appear to reach equilibrium. Thus, despite some speculations to the contrary (see Fisher and Earle, 1982), there is little evidence to support a substituted enzyme ('ping-pong') mechanism or for the existence, in the absence of substrates, of a reduced enzyme intermediate. The transfer of H$^-$ equivalents between the bound nucleotide substrates on the ternary complexes might be direct - at least it is clear that the transferred H$^-$ does not exchange with water protons (Lee et al., 1965).

The relationship between the rate of transhydrogenation, from NADH to thio-NADP$^+$ (an analogue of NADP$^+$), and Δp in steady-state, when the nucleotides were in excess, has been examined in chromatophores of *Rb. capsulatus* (Cotton et al., 1989; Jackson, 1991). The data were consistent with a simple scheme in which an increase in $\Delta\psi$ (= Δp in the conditions of the experiment) gives rise to an increase in a single rate constant in the reaction sequence, i.e. there is only a single step responsible for energy transduction. At high values of $\Delta\psi$ this step is saturated and a reaction component unaffected by $\Delta\psi$ becomes rate-limiting. The dependence of the relationship on pH was complex. There was little to choose between two models of energy transduction: (i) $\Delta\psi$ is transformed by a 'proton well' within the enzyme into a ΔpH (Mitchell, 1969). The elevated concentration of H$^+$ at catalytic center, at the bottom of the well, would lead to an accelerated rate of a single component in the reaction. (ii) $\Delta\psi$ increases the rate constant for the movement of charge within the enzyme across the membrane dielectric. The charge displacement would be coupled to a single component in the catalytic reaction.

To date there have been no reports of any transient-state kinetic analyses with solubilised H$^+$-Thase from any source but these would assuredly be of value in understanding the mechanism of the enzyme. An experiment with chromatophores from *Rb. capsulatus* was described in which transhydrogenation from NADH to thio-NADP$^+$ was measured at the onset of illumination (Palmer and Jackson, 1990). Because, under these conditions, Δp rose rapidly (within a few milliseconds) to a high value, resolution of a hitherto undetected component in the transhydrogenase reaction was possible: a burst ($t_{1/2}$ ~ 6 ms) of thio-NADPH formation was recorded prior to the establishment of the steady-state rate. On the assumption that, under de-energized conditions before illumination, the nucleotide substrates reached binding equilibrium on the enzyme, the existence of the burst was simply explained. The rapid increase in Δp upon illumination caused acceleration of a single rate constant (see above) and hence the rapid formation of thio-NADPH *on the enzyme* during its first turnover. Release of product (thio-NADPH or NAD$^+$) was rate-limiting in the steady-state reaction during subsequent turnovers. On this basis, it was calculated that H$^+$-Thase must be in excess of 0.12 mol/mol of photosynthetic reaction centers (about 1.2 enzymes per chromatophore) and, thus, k_{cat} was in the region of 20s^{-1}. This figure is consistent with an estimation of the turnover number of H$^+$-Thase based on the yield of transhydrogenation by chromatophores during trains of short flashes (Palmer et al., 1991).

C. The H$^+$/H$^-$ Ratio of Transhydrogenase

A consensus was reached that the H$^+$/H$^-$ ratio (the ratio of the number of protons translocated across the membrane per hydride ion equivalent transferred between the nucleotide substrates) was 1.0 for mitochondrial transhydrogenase (Fisher and Earle, 1982; Rydstrom et al., 1987). A ratio of 1.0 can be easily accommodated by several types of reaction mechanism (Jackson, 1991). Chromatophores provide a good system for the measurement of H$^+$/H$^-$ ratios for transhydrogenase since they can be energized conveniently with light and because their electrochromic absorbance changes provide a rapid measure of membrane potential and of membrane ion flux. Published experiments have led to ratios in the region of 0.4–0.7 (Cotton et al., 1989; Jackson et al., 1990; Palmer and Jackson, 1992; Bizouarn and Jackson, 1993). In recent work using chromatophores from a strain of *Rs. rubrum* that over-expresses transhydorgenase, and a rapid mixing procedure that permitted initial rate measurements of proton translocation with negligible contamination by 'leaks,' it emerged that the true ratio is probably 1.0 (T. Bizouarn, L. A. Sazanov, S. Aubourg and J. B. Jackson, unpublished).

D. Comment on the Mechanism of Transhydrogenase

Many authors have favored mechanisms of action of transhydrogenase in which energy coupling between the proton electrochemical gradient and the chemical reaction of hydride transfer is mediated by conformational changes in the protein (Rydstrom et al., 1981; Fisher and Earle, 1982; Yamaguchi and Hatefi, 1989). Specifically, it has been suggested that energy-dependent changes in the binding of the nucleotide substrates are critical in the energy-transduction process (see Fig. 2). The enzyme may function as a dimer (Anderson and Fisher, 1981; Persson et al., 1987; Hou et al., 1990) and interaction between the monomeric units might be indicated by observations that transhydrogenase displays 'half of the sites' reactivity (Phelps and Hatefi, 1984, 1985). Studies on the cyclic reduction of $AcPdAD^+$ by NADH in the presence of either $NADP^+$ or NADPH in purified solubilized transhydrogenase from *Rb. capsulatus* (Palmer et al., 1993) and *E. coli* (Hutton et al., 1994) have revealed that, at low pH, or after inhibition by dicyclohexyl carbodiimide, NADP(H) is very tightly bound by the enzyme. The pH dependence of the cyclic reactions suggest that binding/release of NADP(H) to/from the protein are accompanied by the binding/release of a proton. It was proposed that these proton binding and release reactions might, in the membrane-bound enzyme be components of proton translocation.

IV. Structure of NADH Dehydrogenase

As a guide to the structure of NADH dehydrogenase in photosynthetic bacteria, the enzyme from other organisms will be briefly considered. It is the most complex of the respiratory proteins in animal mitochondria. It comprises in the region of 40 different polypeptides, all of which have now been sequenced (Arizmendi et al., 1992). The enzyme from *Neurospora crassa* contains more than 30 polypeptides, some of which have been sequenced (Weiss et al., 1991). Clear similarities between the sequences of some of the polypeptides are emerging. In both enzymes most of the polypeptides are nuclear encoded. A minority, mostly hydrophobic polypeptides, are encoded in the mitochondrial DNA.

NADH-DH from both sources has one FMN which might be the immediate electron acceptor for NADH and at least four iron-sulfur centers, N-1 to N-4 (Ingledew and Ohnishi, 1980; Beinert and Albracht, 1982; Wang et al., 1991). Centers N-1, N-3 and N-4 have low redox potentials (E_{m7} in the region of −250 mV), whereas N-2, with a midpoint potential around − 100 mV, may be an electron donor to ubiquinone. The location of the Fe-S clusters on individual peptides of the bovine complex has been predicted on the basis of sequence motifs (Walker, 1992). The complex from beef heart might also contain a single, tightly-bound ubiquinone (Burbaev et al., 1989).

NADH-DH can be resolved into smaller complexes by a variety of means. After treatment of the enzyme from beef-heart with chaotropic reagents, it can be fractionated into a 'flavoprotein' fragment and an 'iron-protein' fragment, which are both water-soluble, and a 'hydrophobic protein' (Hatefi, 1985; Ragan, 1987). The bovine enzyme can alternatively be split by detergent treatment into two subcomplexes, 'Iα' and 'Iβ', which are thought to correspond to the peripheral and the intrinsic membrane domains (Finel et al., 1992). When mitochondrial protein synthesis is inhibited in *N. crassa*, a 'small form' of NADH-DH is produced (Friedrich et al., 1989) which appears to correspond to the combined flavoprotein fragment and iron-protein fragment of the beef-heart enzyme. Alternatively, when the organism is grown under Mn limitation, the membrane component is accumulated (Schmidt et al., 1992). Three-dimensional image reconstruction of electron micrographs of the intact

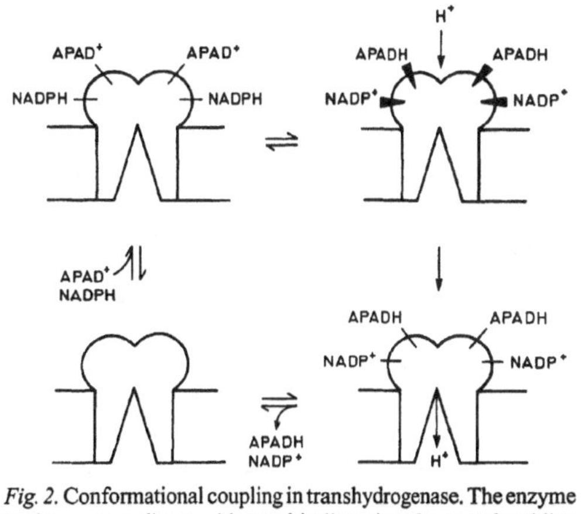

Fig. 2. Conformational coupling in transhydrogenase. The enzyme is shown as a dimer with two binding sites for acetylpyridine adenine dinucleotide ($APAD^+$/APADH) and for $NADP^+$/NADPH. The nucleotides may be weakly bound (–) or strongly bound (▶) to the enzyme. Proton translocation is shown to accompany the change in nucleotide binding affinity.

and small form of *N. crassa* NADH-DH indicate that the enzyme is L-shaped, with the hydrophobic part of the protein within the membrane corresponding to one arm, and the extrinsic peripheral arm protruding into the mitochondrial matrix (Hofhaus et al., 1991). All the known redox centers of NADH-DH are located in the peripheral polypeptides of the complex (Weiss and Friedrich, 1991; Schmidt et al., 1992; Walker, 1992).

Sequence comparisons of the polypeptides of NADH-DH have revealed some interesting evolutionary relationships with bacterial redox enzymes. A part of the water-soluble NAD^+-reducing hydrogenase from *Alcaligenes eutrophus* (actually the $\alpha\gamma$ dimer, which binds FMN and two [4Fe-4S] clusters) is related to subunits in the small form of NADH-DH from *N. crassa* and the flavoprotein and iron-protein fragments of the beef heart enzyme (Pilkington et al., 1991). In other studies it has been shown that there may be sequence homologies between some of the hydrophobic polypeptides of NADH-DH and the membrane-spanning regions of glucose dehydrogenase, a ubiquinone-reducing enzyme from *Acinetobacter calcoaceticus*, and formate hydrogen lyase from *E. coli* (Weiss and Friedrich, 1991).

The difficulties of isolating NADH-DH from photosynthetic and indeed other bacteria have been the main reason for the lack of progress in understanding the structure of the bacterial enzyme (Berks and Ferguson, 1991). An NADH dehydrogenase, purified from membranes of *Rb. capsulatus* by Ohshima and Drews (1981), seems structurally not to resemble the mitochondrial type of enzyme (class NDH-1, in the terminology of Yagi, 1991). However, the inhibitor sensitivity of NADH-dependent respiration in chromatophores of *Rb. capsulatus* and its coupling to Δp (Klemme, 1969; Berks and Ferguson, 1991), the presence of Fe-S centers N1-N4 (Zannoni and Ingledew, 1983; Meinhardt et al., 1989) and, recently, the identification of two genes of *Rb. capsulatus* which code for proteins that are homologous to polypeptides from mitochondrial complex I (Dupuis, 1992) all testify to the existence of an NDH-1 enzyme in this bacterium. It seems very likely that the next advances in our understanding the structure and mechanism of NADH-DH in photosynthetic bacteria will come from molecular genetics (eg. Dupuis, 1992) but in the longer term it will be necessary to develop a protocol for solubilizing and purifying the protein. Yagi (1986) and George and Ferguson (1984) have made some progress with the NADH-DH from the related non-photosynthetic bacterium *Paracoccus denitrificans* and, very recently, a preparation of the enzyme from *E. coli* has been described (Leif et al., 1993). The impetus for further research is that precedents with other electron transport systems suggest that the bacterial complexes may be simpler and easier to study than their mitochondrial counterparts and this is born out by the fact that the NADH-DH from *E. coli* has only 14 polypeptides.

V. Integration Of NADH Dehydrogenase and Transhydrogenase in Metabolism

A remarkable feature of many species of photosynthetic bacteria is their versatility (Madigan and Gest, 1979). However, information about their metabolic profiles is often very limited. Not even during photoheterotrophic growth in the species commonly used for studies of photosynthetic electron transport, are the central metabolic pathways clearly established. In that they are coupled to Δp and can also affect the redox state of the nicotinamide nucleotide pools, NADH-DH and H^+-Thase can, in principle, have profound effects on metabolism but, because of the lack of precise definition of the ongoing pathways, their exact function is not clear. The following discussion is intended to indicate what possible roles these enzymes might play and in what way their activities must be regulated.

A. The Function of NADH Dehydrogenase

During chemoheterotrophic growth of organisms such as *Rb. sphaeroides, Rb. capsulatus* and *Rs. rubrum* under aerobic conditions in the dark, it seems likely that NADH-DH serves to input electrons from NADH to the respiratory chain and contribute to the generation of Δp across the cytoplasmic membrane (Fig. 3). Consistent with this, *Rb. capsulatus* M-1, a mutant with defective NADH-DH, is unable to grow under these conditions (Marrs and Gest, 1973). It also seems likely that the enzyme functions in an analogous manner when these organisms grow in the dark and obtain their energy by anaerobic respiration, for example with nitrous oxide, trimethylamine-N-oxide or dimethylsulphoxide as electron acceptors (Yen and Marrs, 1977; Madigan and Gest, 1979; Schulz and Weaver, 1982; McEwan et al., 1985).

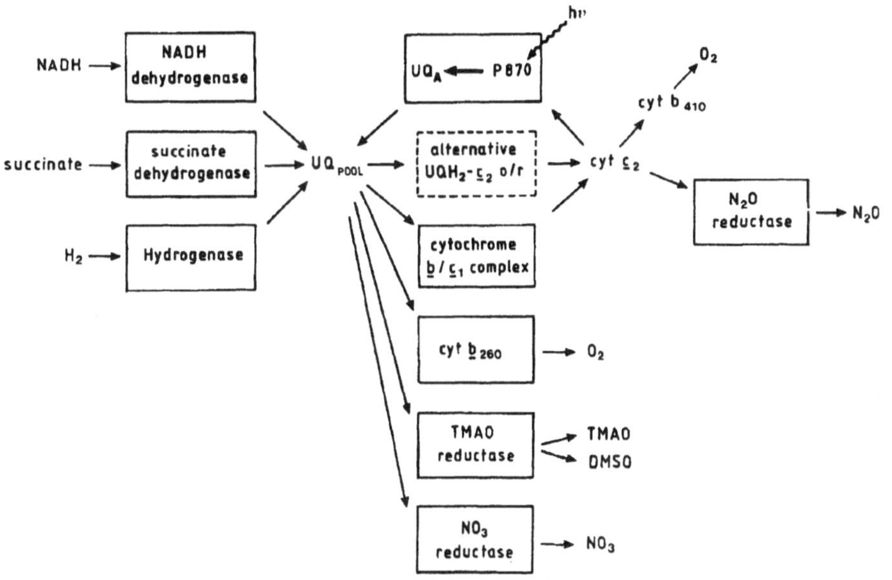

Fig. 3. The electron transport pathways of *Rb. capsulatus*. See Ferguson et al., 1987 and Richardson et al., 1991.

The function of NADH-DH during photoheterotrophic growth is not known. It has a high activity in chromatophores from photoheterotrophically-grown bacteria but the M-1 mutant of *Rb. capsulatus* grows rapidly with malate as carbon source during photosynthetic conditions. Malate is presumably metabolized by the pathways shown in Fig. 4: NADH generated in the first step of malate metabolism (Tayeh and Madigan, 1987) could be reoxidized (a) during carbohydrate synthesis, (b) during re-fixation of CO_2 by the Calvin cycle (note that triosephosphate dehydrogenase is NADH-dependent in *Rb. sphaeroides*, Lascelles, 1960), (c) during production of storage polymers, or (d) by H^+-Thase (see below). Thus, it is understandable why, on malate, NADH-DH is not essential to maintain the balance of the NAD^+-NADH pools. It might have been expected that NADH-DH would be necessary during photoheterotrophic growth on succinate, to reoxidize the bound $FADH_2$ of succinate dehydrogenase by energy-linked reversed electron transport via the ubiquinone pool, although the M-1 mutant is reported to grow normally under these conditions (Marrs and Gest, 1973). When anaerobic, phototrophic growth on highly-reduced substrates, such as propionate and butyrate is facilitated by auxiliary electron acceptors such as trimethylamine-N-oxide, dimethylsulphoxide, nitrate or nitrous oxide (Richardson et al., 1988), then NADH-DH is probably involved in the transfer of reducing equivalents (Fig. 3). It is therefore predicted that the M-1 mutant of *Rb. capsulatus* would not grow phototrophically under anaerobic conditions with propionate (or butyrate) and auxiliary oxidant, but this has not been tested.

Because NADH-DH links the cytoplasmic NAD^+/NADH pools with the membrane phase UQ/UQH_2 pools, and since the redox state of the latter can critically influence the rate of cyclic electron transport (Venturoli et al., 1986), the enzyme could, in principle, be involved in the regulation of photosynthesis. It might supply or withdraw reducing equivalents to or from the quinone pool, under the control of Δp, and thus modulate the cyclic electron transport rate. Indeed, unless the NADH-DH is itself subject to regulation, or unless the nicotinamide nucleotide pools and the quinone pools are somehow compartmentalized, then it is difficult to escape the conclusion that there will be a tendency towards redox equilibration. If the $H^+/2e^-$ ratio of NADH-DH is 4.0 (see Weiss and Friedrich, 1991) and Δp is 250 mV, then the redox potential difference between NAD^+/NADH and UQ/UQH_2 at equilibrium would be 500 mV. The advantage to the bacterial cell could be

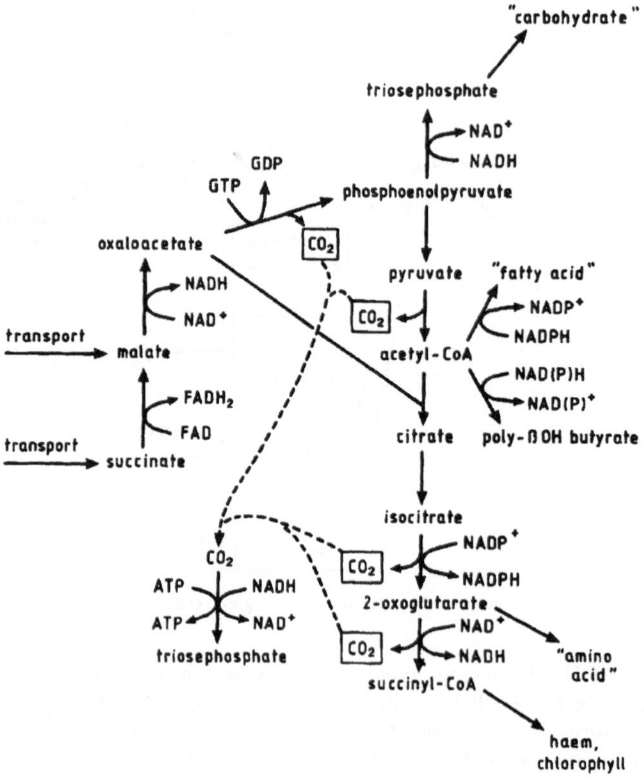

Fig. 4. The balance of reducing power in intermediary metabolism during anaerobic phototrophic growth on malate or succinate (a plausible scheme).

that this provides a vehicle by which carbon metabolism, via the redox state of NAD^+/NADH, might exert control on photosynthetic electron transport. At low values of Δp (eg. when bacteria are exposed to only low light intensities) and particularly with relatively reduced carbon sources, there is evidence (Jones et al., 1990) that the resulting low potential of the UQ pool restricts the rate of light-driven cyclic electron transport. Oxidation of the UQ pool with the auxiliary oxidants, nitrate, nitrous oxide or trimethylamine-N-oxide under these conditions increases the cyclic electron transport rate.

B. The Function of Transhydrogenase

The function of H^+-Thase in photosynthetic bacteria (or indeed in other bacteria) is not clear. However, in *Rb. capsulatus* the enzyme is expressed constitutively at high levels during chemotrophic and phototrophic growth on a variety of carbon and nitrogen sources (Jackson and Cotton, unpublished) and, in view of its interaction with Δp and the redox state of the nicotinamide nucleotide pools, it seems likely that it has a key role in metabolism. It is probably involved in balancing the supply of NADH and NADPH for biosynthesis and detoxification in the cell, dependent on the supply of reduced nucleotides from catabolic dehydrogenases and on the value of Δp (Jackson and McEwan, 1993).

Although experimental proof for bacteria is unavailable, it is widely thought that NAD(H) is maintained in the oxidized state in the cell and NADP(H) in the reduced state. The action of H^+-Thase would contribute to this separation using the energy of Δp. For an H^+/H^- ratio of 1.0 and a Δp of 200 mV, the redox potential difference at equilibrium would be about 100 mV, in which case the ratio of

$[NAD^+]/[NADH]$ would be about 2300 times greater than the ratio of $[NADP^+]/[NADPH]$. This situation favors the action of the catabolic dehydrogenases, which are mainly (though not universally) NAD^+-linked, and reductive biosynthetic reactions and detoxification reactions, which generally require NADPH. A role for H^+-Thase in this context might be particularly acute during anaerobic photosynthetic growth (as distinct from growth under aerobic conditions) since NADH cannot be oxidized by the respiratory chain. Even so, this function of H^+-Thase might not be essential, since NADH might also be reoxidized by the NAD^+-linked triose phosphate dehydrogenase during refixation of CO_2 (Ormerod and Gest, 1962) or during the synthesis of energy storage polymers (Eidels and Preiss, 1970). Furthermore, H^+-Thase is not the only enzyme capable of generating NADPH at substantial rates; $NADP^+$-linked dehydrogenases, for example isocitrate dehydrogenase (Beatty and Gest, 1981; Leyland and Kelly, 1991), can also fulfill this role. In *E. coli*, there are clear indications that there are indeed several metabolic sources of cytoplasmic NADPH, including H^+-Thase (Hanson and Rose, 1980). The importance of the transhydrogenase pathway in this context might be to provide a regulatory link between Δp and biosynthesis (Fig. 4).

It is also possible that H^+-Thase could serve in the opposite sense, to transfer reducing equivalents from NADPH (generated by $NADP^+$-specific dehydrogenases) to NADH, for example if the supply of the N-source were limiting and it therefore became desirable to divert carbon towards the production of storage compounds such as glycogen or beta-hydroxybutyrate. At lease in some bacteria the enzymes catalyzing these reactions prefer NADH to NADPH (Haywood et al., 1988). Of course this reaction of H^+-Thase would be depressed at high values of Δp. Under these conditions of N-limitation, and in order to permit continued metabolic turnover, NADH is also re-oxidized by H^+ in the production of H_2 by nitrogenase (Hillmer and Gest, 1977).

Transhydrogenase might also participate in the generation or utilization of Δp (Jackson and McEwan, 1993). However, as there is increasing evidence that the H^+/H^- ratio for the enzyme is only 1.0 (Section III.C), it would seem that transhydrogenase is not a very effective Δp generator or consumer; its major function is more likely to be in the distribution of reducing equivalents between the NAD(H) and NADP(H) pools, *biased* by Δp (see above). Perhaps in some circumstances its proton-translocating properties are directly important. Conceivably during short periods of energy deprivation it might operate in reverse (cf. Eq. (1)) and pump H^+ outwards across the cytoplasmic membrane to maintain Δp. Because of the non-Ohmic conductance of the bacterial membrane (Golby et al., 1990), H^+-Thase could support a low but substantial Δp, enough to support some essential maintenance functions, such as the cellular osmotic balance, even despite a low H^+/H^- ratio. Another possibility (Jackson and McEwan, 1993) is that pairs of dehydrogenases operating cyclically would lead to continued generation of NADH and $NADP^+$ as substrates for H^+-Thase in the forward direction. This would represent a 'futile' or 'substrate' cycle that might be important in the control of metabolic flux or even as a 'valve' for the dissipation of Δp in a process that would contribute to 'metabolic uncoupling' or 'overflow metabolism' (Tempest and Neijssel, 1984).

VI. Concluding Remarks

NADH-DH and H^+-Thase are both involved in the coupling of the redox chemistry of the nicotinamide nucleotides to proton translocation across the cytoplasmic membrane of at least some species of anoxygenic photosynthetic bacteria. The metabolic role of neither of these enzymes is known with certainty but it is likely that they function in different capacities in different metabolic states. They are both clearly related to the equivalent enzymes in mitochondria. Difficulties in the development of a good solubilization and purification procedure for NADH-DH have limited studies on this enzyme. H^+-Thase, on the other hand, is rather easy to purify from the photosynthetic bacteria and indeed its unique properties in *Rs. rubrum* make it a good system for study.

References

Ahmad S, Glavas NA and Bragg PD (1992) A mutation at Gly314 of the beta subunit of the *Escherichia coli* pyridine nucleotide transhydrogenase abolishes activity and affects the NADP(H)-induced conformational change. Eur J Biochem 207: 733–739

Andersen AB, Andersen P and Ljungqvist L (1992) Structure and function of a 40000-molecular weight protein antigen of *Mycobacterium tuberculosis*. Infect Immun 60: 2317–2323

Anderson WM and Fisher RR (1981) The subunit structure of

bovine heart mitochondrial transhydrogenase. Biochim Biophys Acta 635: 194–199

Arizmendi JM, Skehel JM, Runswick MJ, Fearnley IM and Walker JE (1992) Complementary DNA sequences of two 14.4 kDa subunits of NADH:ubiquinone oxidoreductase from bovine heart mitochondria. FEBS Lett 313: 80–84

Baccarini-Melandri A, Casadio R and Melandri BA (1977) Thermodynamics and kinetics of photophosphorylation in bacterial chromatophores and their relation with the transmembrane electrochemical potential difference of protons. Eur J Biochem 78: 389–402

Beatty JT and Gest H (1981) Biosynthetic and bioenergetic functions of citric acid cycle reactions in *Rhodopseudomonas capsulata*. J Bacteriol 148: 584–593

Beinert H and Albracht SPJ (1982) New insights, ideas and unanswered questions concerning iron-sulphur clusters in mitochondria. Biochim Biophys Acta 683: 245–277

Berks BB and Ferguson SJ (1991) Simplicity and complexity of electron transfer between NADH and *c*-type cytochromes in bacteria. Biochem Soc Trans 19: 581–588

Bizouarn T and Jackson JB (1993) The ratio of protons translocated per hydride ion equivalent transferred by nicotinamide nucleotide transhydrogenase in chromatophores from *Rhodospirillum rubrum*. Eur J Biochem 217: 763–770

Burbaev DSH, Moroz IA, Kotlyar AB, Sled VD and Vinogradov AD (1989) Ubisemiquinone in the NADH-ubiquinone reductase region of the mitochondrial respiratory chain. FEBS Lett 254: 47–51

Clarke DM and Bragg PD (1985) Purification and properties of reconstitutively active nicotinamide nucleotide transhydrogenase of *Escherichia coli*. Eur J Biochem 149: 517–523

Clarke DM, Loo TW, Gillam S and Bragg PD (1986) Nucleotide sequence of the *pntA* and *pntB* genes encoding the pyridine nucleotide transhydrogenase of *Escherichia coli*. Eur J Biochem 158: 647–653

Cotton NPJ, Myatt JF and Jackson JB (1987) The dependence of the rate of transhydrogenase on the value of the protonmotive force in chromatophores from photosynthetic bacteria. FEBS Lett 219: 88–92

Cotton NPJ, Lever TM, Nore BF, Jones MR and Jackson JB (1989) The coupling between protonmotive force and the NAD(P) transhydrogenase in chromatophores from photosynthetic bacteria. Eur J Biochem 182: 593–603

Cunningham IJ, Baker JA and Jackson JB (1992a) Reaction between the soluble and membrane-associated proteins of the transhydrogenase of *Rhodospirillum rubrum*. Biochim Biophys Acta 1101: 345–352

Cunningham IJ, Williams R, Palmer T, Thomas CM and Jackson JB (1992b) The relation between the soluble factor associated with H+-transhydrogenase of *Rhodospirillum rubrum* and the enzyme from mitochondria and *Escherichia coli*. Biochim Biophys Acta 1100: 332–338

Dupuis A (1992) Identification of two genes of *Rhodobacter capsulatus* coding for proteins homologous to the ND1 and 23 kDa subunits of the mitochondrial complex 1. FEBS Lett 301: 215–218

Earle SR and Fisher RR (1980) A direct demonstration of proton translocation coupled to transhydrogenation in reconstituted vesicles. J Biol Chem 255: 827–830

Eidels L and Preiss J (1970) Carbohydrate metabolism in *Rhodopseudomonas capsulata*: Enzyme titers, glucose metabolism and polyglucose polymer synthesis. Arch Biochem Biophys 140: 75–89

Enander K and Rydstrom J (1982) Energy-linked nicotinamide nucleotide transhydrogenase- Kinetics and regulation of purified and reconstituted transhydrogenase from beef heart mitochondria. J Biol Chem 257: 14760–14766

Fearnley IM and Walker JE (1992) Conservation of sequences of subunits of mitochondrial complex I and their relationships with other proteins. Biochim Biophys Acta 1140: 105–134

Ferguson SJ, Jackson JB and McEwan AG (1987) Anaerobic respiration in the Rhodospirillaceae: Characterisation of pathways and evaluation of roles in redox balancing during photosynthesis. FEMS Microbiol Revs 46: 117–143

Finel M, Skehel JM, Albracht SPJ, Fearnley IM and Walker JE (1992) Resolution of NADH:ubiquinone oxidoreductase from bovine mitochondria into two subcomplexes one of which contains the redox centers of the enzyme. Biochemistry 31: 11425–11434

Fisher RR and Earle SR (1982) Membrane-bound pyridine dinucleotide transhydrogenases. In: Everse J, Anderson BM, You KS (eds) The Pyridine Nucleotide Coenzymes, pp 279–324. Academic Press, New York

Fisher RR and Guillory RJ (1969a) A soluble factor related to the energy-linked transhydrogenase reaction of *Rhodospirillum rubrum* chromatophores. J Biol Chem 244: 1078–1079

Fisher RR and Guillory RJ (1969b) Partial resolution of energy-linked reactions in *Rhodospirillum rubrum chromatophores. FEBS Letters* 3: 27–30

Fisher RR and Guillory RJ (1971a) Resolution of enzymes catalyzing energy-linked transhydrogenation—Interaction of transhydrogenase factor with the *Rhodospirillum rubrum* chromatophore membrane. J Biol Chem 246: 4679–4686

Fisher RR and Guillory RJ (1971b) Resolution of enzymes catalyzing energy-linked transhydrogenation—Preparation and properties of *Rhodospirillum rubrum* transhydrogenase factor. J Biol Chem 246: 4687–4693

Fisher RR, Rampey SA, Sadighi A and Fisher K (1975) Resolution and reconstitution of *Rhodospirillum rubrum* pyridine dinucleotide transhydrogenase—Proteolytic and thermal inactivation of the membrane component. J Biol Chem 250: 819–825

Friedrich T, Hofhaus G, Ise W, Nehls U, Schmitz B and Weiss H (1989) A small isoform of NADH:ubiquinone oxidoreductase (complex I) without mitochondrially-synthesised subunits is made in chloramphenicol-treated *Neurospora crassa*. Eur J Biochem 180: 173–180

George CL and Ferguson SJ (1984) Immunochemical identification of a two subunit NADH-ubiquinone oxidoreductase from *Paracoccus denitrificans*. Eur J Biochem 143: 567–573

Golby P, Carver M and Jackson JB (1990) Membrane ionic currents in *Rhodobacter capsulatus*—Evidence for electrophoretic transport of K^+,Rb^+ and NH_4^+. Eur J Biochem 187: 589–597

Griffiths DE and Robertson AM (1966) Energy-linked reactions in mitochondria: Studies on the mechanism of the energy-linked transhydrogenase reaction. Biochim Biophys Acta 118: 453–464

Hanson RL (1979) Kinetic mechanism of pyridine nucleotide transhydrogenase from *Escherichia coli*. J Biol Chem 254: 888–893

Hanson RL and Rose C (1980) Effects of an insertion mutation in a locus affecting pyridine nucleotide transhydrogenase (pnt::Tn5) on the growth of *Escherichia coli*. J Bacteriol 141: 401–404

Hatefi Y (1985) The mitochondrial electron transport and oxidative phosphorylation system. Ann Rev Biochem 54: 1676–1680

Hatefi Y and Yamaguchi M (1992) Energy-transducing nicotinamide nucleotide transhydrogenase. In: Ernster L, (ed) Molecular Mechanisms in Bioenergetics, pp 265–281. Elsevier, Amsterdam

Hillmer P and Gest H (1977) Hydrogen metabolism in the photosynthetic bacterium *Rhodopseudomonas capsulatus*; production and utilisation of hydrogen by resting cells. J Bacteriol 124: 732–739

Hofhaus G, Weiss H and Leonard K (1991) Electron microscopic analysis of the peripheral and membrane parts of NADH dehydrogenase (complex I) J Mol Biol 221: 1027–1043

Homyk M and Bragg PD (1979) Steady-state kinetics and the inactivation by 2,3-butandione of the energy independent transhydrogenase of *Escherichia coli* cell membranes. Biochim Biophys Acta 571: 201–217

Hou C, Potier M and Bragg PD (1990) Crosslinking and radiation inactivation analysis of the subunit structure of the pyridine nucleotide transhydrogenase of *Escherichia coli*. Biochim Biophys Acta 1018: 61–66

Hu PS, Persson B, Hoog JO, Jornvall H, Hartog AF, Berden JA, Holmberg E and Rydstrom J (1992) Energy-linked transhydrogenase. Characterisation of a nucleotide-binding sequence in nicotinamide nucleotide transhydrogenase from beef heart. Biochim Biophys Acta 1102: 19–29

Hutton M, Day JM, Bizouarn T and Jackson JB (1994) Kinetic resolution of the reaction catalysed by proton-translocating transhydrogenase from *Escherichia coli* as revealed by experiments with analogues of the nucleotide substrates. Eur J Biochem 219: 1041–1051

Ingledew WJ and Ohnishi T (1980) An analysis of some thermodynamic properties of iron-sulphur centers in site I of mitochondria. Biochem J 186: 111–117

Jackson JB (1991) The Proton-Translocating Nicotinamide Adenine Dinucleotide Transhydrogenase. J Bioenerget Biomembranes 23: 715–741

Jackson JB and McEwan AG (1993) NAD(P)H transhydrogenase and NADH dehydrogenase in photosynthetic membranes. In: Barber J (ed) Molecular Processes in Photosynthesis, Jai Press, Greenwich

Jackson JB, Cotton NPJ, Lever TM, Cunningham IJ, Palmer T and Jones MR (1990) Nicotinamide Nucleotide Transhydrogenase in Photosynthetic Bacteria. In: Drews G and Dawes EA (eds) Molecular Biology of Membrane-Bound Complexes in Phototrophic Bacteria, pp 415–424. Plenum Press, New York

Jackson JB, Lever TM, Rydstrom J, Persson B and Carlenor E (1991) Proton-translocating transhydrogenase from photosynthetic bacteria. Biochem Soc Trans 19: 573–575

Jacobs E and Fisher RR (1979) Resolution and reconstitution of *Rhodospirillum rubrum* pyridine nucleotide transhydrogenase. Modification with N-ethyl maleimide and 2,4-pentanedione. Biochemistry 18: 4315–4322

Jacobs E, Heriot K and Fisher RR (1977) Resolution and reconstitution of *Rhodospirillum rubrum* pyridine nucleotide transhydrogenase II solubilisation of the membrane-bound component. Arch Microbiol 115: 151–156

Jones MR and Jackson JB (1989) Proton release by the quinol oxidase site of the cytochrome b/c_1 complex following single turnover flash excitation of intact cells of Rhodobacter capsulatus. Biochim Biophys Acta 975: 34–43

Kawasaki T, Satoh K and Kaplan NO (1964) The involvement of pyridine nucleotide transhydrogenase in ATP-linked TPN reduction by DNPH. Biochem Biophys Res Commun 17: 648–654

Keister DL and Yike NJ (1966) Studies on an energy-linked pyridine nucleotide transhydrogenase in photosynthetic bacteria I—Demonstration of the reaction in *Rhodospirillum rubrum*. Biochem Biophys Res Commun 24: 519–525

Keister DL and Yike NJ (1967a) Energy-linked reactions in photosynthetic bacteria I. Succinate-linked ATP-driven NAD reduction by *Rhodospirillum rubrum* chromatophores. Arch Biochim Biophys 121: 415–422

Keister DL and Yike NJ (1967b) Energy-linked reactions in photosynthetic bacteria II -The energy-dependent reduction of oxidized nicotinamide-adenine dinucleotide phosphate by reduced nicotinamide-adenine dinucleotide in chromatophores of *Rhodospirillum rubrum*. Biochemistry 6: 3847–3857

Klemme JH (1969) Studies on the mechanism of NAD-photoreduction by chromatophores of the facultative phototroph, *Rhodopseudomonas capsulata*. Z Naturforsch 24b: 67–76

Konings AWT and Guillory RJ (1972) Specificity of the transhydrogenase factor for chromatophores of *Rhodopseudomonas spheroides* and *Rhodospirillum rubrum*. Biochim Biophys Acta 283: 334–338

Kuroda S, Tanizawa K, Sakamoto Y, Tanaka H and Soda K (1990) Alanine dehydrogenases from two *Bacillus* species with distinct thermostabilities: Molecular cloning, DNA and protein sequence determination and structural comparison with other NAD(P)-dependent dehydrogenases. Biochemistry 29: 1009–1015

Lascelles J (1960) The formation of ribulose 1:5-diphosphate carboxylase by growing cultures of Athiorhodaceae. J Gen Microbiol 23: 499–510

Lee CP, Simard-Duquesne N, Ernster L and Hoberman HD (1965) Stereochemistry of hydrogen transfer in the energy-linked transhydrogenase and related reactions. Biochim Biophys Acta 105: 397–409

Lever TM, Palmer T, Cunningham IJ, Cotton NPJ and Jackson JB (1991) Purification and properties of the H^+-nicotinamide nucleotide transhydrogenase from *Rhodobacter capsulatus*. Eur J Biochem 197: 247–255

Leyland ML and Kelly DJ (1991) Purification and characterisation of a monomeric isocitrate dehydrogenase with dual coenzyme specificity from the photosynthetic bacterium *Rhodomicrobium vannielii*. Eur J Biochem 202: 85–93

Lilley KS, Baker PJ, Britton KL, Stillman TJ, Brown PE, Moir AJG, Engel PC, Rice DW, Bell JE and Bell E (1991) The partial amino acid sequence of the NAD-dependent glutamate dehydrogenase of *Chlostridium symbiosum*: Implications for the evolution and structural basis of coenzyme specificity. Biochim Biophys Acta 1080: 191–197

Madigan MT and Gest H (1979) Growth of photosynthetic bacterium *Rhodopseudomonas capsulata* chemoautotrophically in darkness and H_2 as the energy source. J Bacteriol 137: 524–530

Marrs B and Gest H (1973) Genetic mutations affecting the respiratory electron transport system of the photosynthetic bacterium *Rhodopseudomonas capsulata*. J Bacteriol 114: 1045–1051

McEwan AG, Cotton NPJ, Ferguson SJ and Jackson JB (1985) The role of auxiliary oxidants in the maintenance of a balanced redox poise for photosynthesis in bacteria. Biochim Biophys Acta 810: 140–147

McFadden BJ and Fisher RR (1978) Resolution and reconstitution of *Rhodospirillum rubrum* pyridine dinucleotide transhydrogenase: Localisation of substrate binding sites. Arch Biochem Biophys 190: 820–828

Meinhardt SW, Matsushita K, Kaback HR and Ohnishi T (1989) EPR characterisation of the iron-sulphur-containing NADH-ubiquinone oxidoreductase of the *Escherichia coli* aerobic respiratory chain. Biochemistry 28: 2153–2160

Melandri BA, Baccarini-Melandri A and Fabbri E (1972) Energy transduction in photosynthetic bacteria. IV Light-dependent ATPase in photosynthetic membranes from *Rhodopseudomonas capsulata*. Biochim Biophys Acta 275: 383–394

Mitchell P (1969) Chemiosmotic coupling and energy transduction. Theoret Exp Biophys 2: 159–215

Ohshima T and Drews G (1981) Isolation and partial characterisation of the membrane-bound NADH dehydrogenase from the phototrophic bacterium *Rhodopseudomonas capsulata*. Z Naturforsch 36c: 400–406

Orlando JA, Sabo A and Curnym C (1966) Photoreduction of pyridine nucleotide by subcellular preparations from *Rhodopseudomonas sphaeroides*. Plant Physiol 41: 937–945

Ormerod JG and Gest H (1962) Hydrogen photosynthesis and alternative metabolic pathways in photosynthetic bacteria. Bact Revs 26: 51–66

Palmer T and Jackson JB (1990) A rapid burst preceding the steady-state rate of H^+-transhydrogenase during illumination of chromatophores of *Rhodobacter capsulatus* : Implications for the mechanism of interaction between protonmotive force and enzyme. FEBS Letters 277: 45–48

Palmer T and Jackson JB (1992) Nicotinamide nucleotide transhydrogenase from *Rhodobacter capsulatus*; the H+/H- ratio and the activation state of the enzyme during reduction of acetyl pyridine adenine dinucleotide. Biochim Biophys Acta 1099: 157–162

Palmer T, Cotton NPJ and Jackson JB (1991) H+-transhydrogenase in chromatophores from *Rhodobacter capsulatus* after periods of continuous illumination and short flash excitation. Biochim Biophys Acta 1098: 21–26

Palmer T, Williams R, Cotton NPJ, Thomas CM and Jackson JB (1993) Inhibition of proton-translocating transhydrogenase from photosynthetic bacteria by N,N-dicyclohexylcarbodiimide. Eur J Biochem 211: 663–669

Persson B, Ahnstrom G and Rydstrom J (1987) Energy-linked nicotinamide transhydrogenase: Hydrodynamic properties and active form of purified and membrane bound mitochondrial transhydrogenase from beef heart. Arch Biochem Biophys 259: 341–349

Phelps DC and Hatefi Y (1984) Interaction of purified nicotinamidenucleotide transhydrogenase with dicyclohexylcarbodiimide. Biochemistry 23: 4475–4480

Phelps DC and Hatefi Y (1985) Mitochondrial nicotinamide nucleotide transhydrogenase: Active site modification by 5'-(p-(fluorosulfonyl)benzoyl)adenosine. Biochemistry 24: 3503–3507

Pilkington SJ, Skehel JM, Gennis RB and Walker JE (1991) Relationship between mitochondrial NADH-ubiquinone reductase and a NAD+reducing dehydrogenase. Biochemistry 30: 2166–2175

Ragan CI (1987) The structure of NADH-ubiquinone reductase (complex I) Curr Topics Bioenergetics 15: 1–36

Richardson RJ, King GF, Kelly DJ, McEwan AG, Ferguson SJ and Jackson JB (1988) The role of auxiliary oxidants in maintaining redox balance during phototrophic growth of *Rhodobacter capsulatus*. Arch Microbiol 150: 131–137

Richardson RJ, Bell LC, McEwan AG, Jackson JB and Ferguson SJ (1991) Cytochrome c_2 is essential for electron transfer to nitrous oxide reductase from physiological substrates in *Rhodobacter capsulatus* and can act as an electron donor to the reductase in vitro. Eur J Biochem 199: 677–683

Rydstrom J (1979) Energy linked nicotinamide nucleotide transhydrogenase—Properties of proton translocating and ATP driven transhydrogenase reconstituted from synthetic phospholipids and purified transhydrogenase from beef heart mitochondria. J Biol Chem 254: 8611–8619

Rydstrom J, Lee CP and Ernster L (1981) Energy-linked nicotinamide nucleotide transhydrogenase. In: Skulachev VP and Hinkle PC (eds) Chemiosmotic Proton Circuits in Biological Membranes, pp 483–503 Addison-Wesley, Reading

Rydstrom J, Persson B and Carlenor E (1987) Transhydrogenase linked to pyridine nucleotides. In: Dolphin D, Poulson R and Avramovic O (eds) Pyridine Nucleotide Coenzymes: Chemical, Biochemical, and Medical Aspects, Vol. 2B, pp 433–460 John Wiley and Sons, New York

Schmidt M, Friedrich T, Wallrath J, Ohnishi T and Weiss H (1992) Accumulation of the pre-assembled membrane arm of NADH:ubiquinone oxidoreductase in mitochondria of manganese-limited grown *Neurospora crassa*. FEBS Lett 313: 8–11

Schulz JE and Weaver PF (1982) Fermentation and anaerobic respiration by *Rhodospirillum rubrum* and *Rhodopseudomonas capsulata*. J Bacteriol 149: 181–190

Tayeh MA and Madigan MT (1987) Malate dehydrogenase in phototrophic purple bacteria: Purification, molecular weight and quaternary structure. J Bacteriol 169: 4196–4202

Tempest DW and Neijssel OM (1984) The status of Y_{ATP} and maintenance energy as biologically interpretable phenomena. Ann Rev Microbiol 38: 459–486

Tong RCW, Glavas NA and Bragg PD (1991) Topological analysis of the pyridine nucleotide transhydrogenase of *E. coli* using proteolytic enzymes. Biochim Biophys Acta 1080: 19–28

Venturoli G and Melandri BA (1982) The localised coupling of bacterial photophosphorylation: Effect of antimycin A and dicyclohexylcarbodiimide in chromatophores from *Rhodopseudomonas sphaeroides* studied by single turnover analysis. Biochim Biophys Acta 680: 8–16

Venturoli G, Fenandez-Velasco JG, Crofts AR and Melandri BA (1986) Demonstration of a collisional interaction of ubiquinol with the ubiquinol-cytochrome c_2 oxidoreductase complex in chromatophores from *Rhodobacter sphaeroides*. Biochim Biophys Acta 851: 340–352

Wakabayashi S and Hatefi Y (1987a) Amino acid sequence of the NAD(H)-binding region of the mitochondrial nicotinamide nucleotide transhydrogenase modified by N,N'-dicyclohexylcarbodiimide. Biochem Internat 15: 667–675

Wakabayashi S and Hatefi Y (1987b) Characterization of the

substrate-binding sites of the mitochondrial nicotinamide nucleotide transhydrogenase. Biochem Internat 15: 915–924
Walker JE (1992) The NADH:ubiquinone oxidoreductase (complex I) of respiratory chains. Quart Rev Biophys 25: 253–324
Wang DC, Meinhardt SW, Sackmann U, Weiss H and Ohnishi T (1991) The iron-sulphur clusters in the two related forms of mitochondrial NADH:ubiquinone oxidoreductase made by *Neurospora crassa*. Eur J Biochem 197: 257–264
Weiss H and Friedrich T (1991) Redox-linked proton translocation by NADH-ubiquinone reductase (complex I) J Bioenerget Biomembranes 23: 743–754
Weiss H, Freidrich T, Hofhaus G and Preis D (1991) The respiratory chain NADH dehydrogenase (complex I) of mitochondria. Eur J Biochem 197: 563–576
Williams R, Cotton NPJ, Thomas CM and Jackson JB (1994) Cloning and sequencing of the genes for the proton-translocating nicotinamide nucleotide transhydrogenase from *Rhodospirillum rubrum* and the implications for the domain structure of the enzyme. Microbiology 140: 1595–1604
Yagi T (1986) Purification and characterisation of NADH dehydrogenase complex from *Paracoccus denitrificans*. Arch Biochem Biophys 250: 302–311
Yagi Y (1993) The bacterial energy-transducing NADH-quinone oxidoreductase. Biochim Biophys Acta 141: 1–17
Yamaguchi M and Hatefi Y (1989) Mitochondrial nicotinamide nucleotide transhydrogenase: NADPH binding increases and NADP binding decreases the acidity and susceptibility to modification of cysteine-893. Biochemistry 28: 6050–6056
Yamaguchi M and Hatefi Y (1991) Mitochondrial energy linked nicotinamide nucleotide transhydrogenase. Membrane topography of the bovine enzyme. J Biol Chem 266: 5728–5735
Yamaguchi M, Hatefi Y, Trach K and Hoch JA (1988) The primary structure of the mitochondrial energy-linked nicotinamide nucleotide transhydrogenase deduced from the sequence of cDNA clones. J Biol Chem 263: 2761–2767
Yen HC and Marrs B (1977) Growth of *Rhodopseudomonas capsulata* under anaerobic dark conditions with dimethyl sulphoxide . Arch Biochem Biophys 181: 411–418
Zannoni D and Ingledew WJ (1983) *Rhodopseudomonas capsulata* respiratory dehydrogenase mutants: An electron paramagnetic resonance study. FEMS Microbiol Lett 17: 331–334

Chapter 39

Sulfur Compounds as Photosynthetic Electron Donors

Daniel C. Brune
Department of Chemistry and Biochemistry, and
Center for the Study of Early Events in Photosynthesis,
Arizona State University, Tempe, AZ 85287-1604, USA

Summary

Most photosynthetic bacteria can grow photoautotrophically using inorganic sulfur compounds (i. e. sulfide, sulfur, polysulfides, thiosulfate, or sulfite) as electron donors for CO_2 fixation. The different types of phototrophs that use sulfur compounds as electron donors, and their varying sulfur-oxidizing capabilities are briefly described. Several species of purple sulfur bacteria can also grow aerobically or microaerophilically as chemolithotrophs, oxidizing sulfur compounds to obtain energy as well as electrons for CO_2 reduction. They share this ability with nonphotosynthetic sulfur bacteria such as thiobacilli. Although thiobacilli are not

R. E. Blankenship, M. T. Madigan and C. E. Bauer (eds): Anoxygenic Photosynthetic Bacteria, pp. 847–870.

particularly closely related to purple sulfur bacteria, studies on sulfur oxidation by thiobacilli have yielded information that may be relevant to sulfur oxidation by photosynthetic bacteria. A variety of enzymes catalyzing sulfur oxidation reactions have been isolated from photosynthetic bacteria, and possible pathways for sulfur oxidation involving those enzymes are discussed. Except for flavocytochrome *c* and sulfide-quinone reductase, which catalyze electron transfer from sulfide to cytochrome *c* and quinone, respectively, the in vivo electron acceptors used by the sulfur-oxidizing enzymes are generally unknown. So far, no enzyme has been isolated that catalyzes oxidation of elemental sulfur, and some new possibilities for how elemental sulfur is oxidized are considered. Finally, some suggestions for future research are made that use metabolically versatile purple bacteria to examine donation of electrons by sulfur compounds to the electron transport chain, active transport of ionic sulfur compounds, and the molecular genetics of sulfur-oxidizing enzymes.

I. Introduction

Photosynthetic bacteria, other than the chlorophyll *a*-containing cyanobacteria, do not photochemically generate a sufficiently strong oxidant to use water as an electron donor. Thus many of them grow simply as photoheterotrophs, using light as the source of energy for converting organic nutrients to metabolic products (Gest, 1993). Others, however, find environmental niches in which sulfide, polysulfide or elemental sulfur, or thiosulfate is available. Sulfide is typically generated by sulfate-reducing bacteria, which reduce sulfate in oxygen-depleted environments such as sediments or stagnant lower levels in stratified lakes and lagoons (Postgate, 1984). Sulfide spontaneously reduces oxygen as it diffuses upward, thus expanding the anaerobic zone. When light reaches such habitats, photoautotrophic sulfur bacteria proliferate.

Inorganic sulfur compounds are concentrated sources of electrons at a low redox potential (see Table 1). Sulfide, for example, donates 8 electrons while being oxidized to sulfate at an overall midpoint redox potential (E_0') of –215 mV at pH 7. For thiosulfate, E_0' = –245 mV for its 8-electron oxidation to sulfate. A variety of enzymes participate in the oxidation of these electron donors, with sulfur compounds of intermediate redox states being formed in the process. This chapter will review what is known about the sulfur-oxidizing capabilities of different bacterial species and the enzymes mediating their sulfur oxidation reactions. As will become apparent, much remains to be learned about the biochemistry of sulfur oxidation, and some thoughts about possible future experiments in this area will also be presented.

The low redox potentials for oxidation of sulfide and thiosulfate imply that a substantial amount of energy can also be obtained through aerobic oxidation of these electron donors. In fact thiotrophy (i. e. chemolithotrophic growth using energy derived from oxidation of inorganic sulfur compounds) is fairly widespread among the prokaryotes, occurring both among the Archaea (*Sulfolobus,* for example) and the Bacteria. This wide distribution of thiotrophy may indicate that it appeared at an early stage in the evolution of life, but comparative information about the structures of the enzymes involved is needed to decide between that possibility and the alternative

Table 1. Redox potentials of sulfur compounds

Redox couple	E_0' (mV)
SO_4^{2-}/SO_3^-	–515
$S_2O_3^{2-}/HS^- + HSO_3^-$	–400
S^0/HS^-	–260
$2SO_4^{2-}/S_2O_3^{2-}$	–245
SO_4^{2-}/HS^-	–215
SO_4^{2-}/S^0	–200
HSO_3^-/HS^-	–115
APS/AMP + HSO_3^-	–60
HSO_3^-/S^0	–40
$S_4O_6^{2-}/2S_2O_3^{2-}$	+25

Tabulated values are from Brune (1989) and from Wood (1988). The latter also has a useful discussion of redox potentials for reactions involving elemental sulfur as a function of its activation state in sulfur globules.

Abbreviations: APS – adenosine-5'-phosphosulfate; BChl – bacteriochlorophyll; *Cm. – Chromatium; Cb. – Chlorobium; Dv. – Desulfovibrio; Ec. – Ectothiorhodospira*; FAD – flavin adenine dinucleotide; FMN – flavin mononucleotide; GSH – glutathione; HiPIP – high potential iron-sulfur protein; HQNO – 2-heptyl-4-hydroxyquinoline-N-oxide; NAD^+ and NADH – oxidized and reduced forms of nicotinamide adenine dinucleotide; NEM – N-ethylmaleimide; NQNO – 2-nonyl-4-hydroxyquinoline-N-oxide; *Rb. – Rhodobacter; Rp. – Rhodopseudomonas; Rs. – Rhodospirillum*; SDS-PAGE – sodium dodecyl sulfate-polyacrylamide gel electrophoresis; SQR – sulfide-quinone reductase; *Tb. – Thiobacillus; Tc. – Thiocapsa* that the requisite enzymes of (thiotrophic) Archaea and Bacteria evolved independently. Among the Bacteria, chemolithotrophy via aerobic oxidation of sulfide, elemental sulfur, or thiosulfate is characteristic of the thiobacilli, a diverse group of which some are economically important because of their ability to oxidize and solubilize mineral sulfide ores and because of the damage to concrete and limestone or marble structures caused by the acidity produced during their sulfur oxidation reactions (Fischer, 1988). Interestingly, several species of purple sulfur bacteria resemble the thiobacilli in that they can also grow on sulfur as chemolithotrophs.

II. Types of Anoxygenic Phototrophs and Their Sulfur-Oxidizing Capabilities

Figure 1 shows a current scheme for the phylogeny of bacteria based on 16S ribosomal RNA sequences. Taxonomy/phylogeny of the anoxygenic phototrophs is discussed in detail by Imhoff (Chapter 1) and is discussed here only as it relates to their sulfur-oxidizing capabilities.

A. Purple Photosynthetic Bacteria

The best known anoxygenic phototrophs belong to a large and diverse bacterial phylum called the Proteobacteria (previously referred to as the purple bacteria) (Stackebrandt et al., 1988). The Proteobacteria have been subdivided into four groups, designated α, β, γ, and δ (Woese, 1987), although recent RNA analyses (Lane et al., 1992) suggest that the genera *Campylobacter* and *Thiovulum* belong to a fifth (ε) subdivision branching off near the root of the δ subdivision (Fig. 1). The classical purple sulfur bacteria (family Chromatiaceae), and their close relatives the Ectothiorhodospiraceae, belong to the γ subdivision. Both families form coherent groups on the basis of 16S rRNA sequences as well as the classical taxonomic criteria of cellular morphology and physiology.

1. Chromatiaceae

The Chromatiaceae are easily distinguished by the presence of microscopically observable, highly refractile globules of elemental sulfur that form

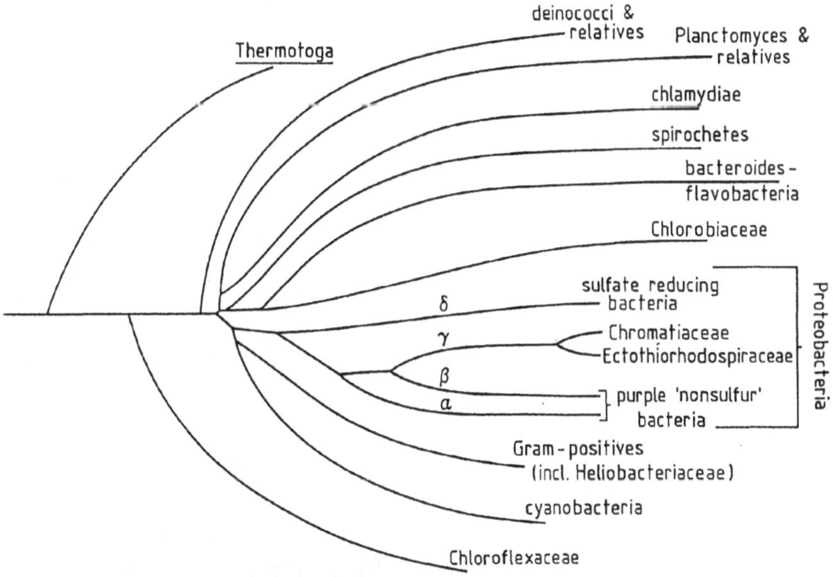

Fig. 1. Phylogenetic scheme for the domain Bacteria showing the positions of the different types of photosynthetic bacteria within the different bacterial phyla. Subdivisions are indicated within the Proteobacteria to indicate the relative positions of the Chromatiaceae, Ectothiorhodospiraceae, and the purple 'nonsulfur' bacteria. (Slightly modified from Brune, 1989).

intracellularly when they are grown on sulfide or thiosulfate. The sulfur in these globules exists in a metastable liquid state at room temperature, possibly due to the presence of polysulfides which give the sulfur a somewhat hydrophilic character and prevent its crystallization (Steudel et al., 1990). Sulfur globules isolated from *Chromatium vinosum* contain about 5% protein on a dry weight basis, and evidence was presented that this protein forms a monolayer enclosing the sulfur globule (Schmidt et al., 1971; Nicolson and Schmidt, 1971). A recent analysis of *Cm. vinosum* sulfur globules showed that three major proteins could be purified by high pressure liquid chromatography, two with molecular masses of about 10.5 kDa and one with a mass of about 8.6 kDa (Brune, in preparation). Similar proteins were isolated from sulfur globules of *Thiocapsa roseopersicina*, and N-terminal sequencing showed them to be homologous to the *Cm. vinosum* proteins. These proteins are rich in glycine and tyrosine and have weakly repetitive sequences, making it likely that they are structural rather than enzymatic.

Sulfur globules in the Chromatiaceae are generally considered to be cytoplasmic. However, electron micrographs by Remsen (1978) show chromatophores opening into the space enclosing sulfur globules. Since the interior of chromatophores of purple photosynthetic bacteria is thought to be continuous with the periplasm, this implies a periplasmic location for the sulfur globules, at least at some stage of their formation. Remsen (1978) suggested that sulfur-filled chromatophores coalesce followed by disintegration of their lipid bilayer membranes to eventually form sulfur globules that are separated from the cytoplasm only by the surrounding proteinaceous membrane.

The sulfur-oxidizing capabilities of the Chromatiaceae are summarized in Table 2. Most of these data were discussed previously (Brune, 1989), so only some general observations and new information will be discussed here.

On the basis of their sulfur metabolism, the Chromatiaceae can be divided into two groups (Imhoff, 1992a). One, consisting of the large-celled *Chromatium* species (*buderi, okenii, warmingii, weissei*), *Cm. tepidum*, *Thiospirillum jenense, Thiocapsa pfennigii, Lamprobacter modestohalophilus*, and the *Thiodictyon* species, requires sulfide or elemental sulfur (S^0) both as a photosynthetic electron donor and as a source of sulfur for biosynthesis. These do not use thiosulfate as an electron donor, and cannot grow aerobically. The other group, which includes most of the small-celled *Chromatium* species and *Thiocapsa roseopersicina,* is more versatile. Most use thiosulfate, and some also use sulfite as an electron donor. Assimilatory sulfate reduction is common in this group (although it is absent in *Cm. minus, Cm. salexigens*, the *Amoebobacter* species, and *Thiospirillum winogradskyi*), allowing them to grow photoheterotrophically with sulfate as the sulfur source. Generally they can grow as chemolithotrophs using aerobic oxidation of reduced sulfur compounds as their source of energy and electrons for CO_2 fixation (Kondratieva et al., 1976, Kämpf and Pfennig, 1980, 1986; de Wit and van Gemerden, 1990, Overmann and Pfennig, 1992). Some can even grow as chemoorganotrophs, although they may require sulfide or thiosulfate due to a suppression of sulfate assimilation by aerobic conditions (Kondratieva et al., 1981).

Although not included in Table 2 because the small number of investigations of its use do not represent a comprehensive survey, two species of Chromatiaceae, namely *Cm. vinosum* and *Thiocapsa roseopersicina*, have been shown to use polysulfide (average chain length 3–4 S atoms) as a photosynthetic electron donor (van Gemerden, 1987, Steudel et al., 1990, Visscher et al., 1990). Cultures grown on sulfide oxidize polysulfide to sulfate (with intermediate accumulation of elemental sulfur globules) without any lag and show a high affinity for polysulfide. A possible role of polysulfides as intermediates during oxidation of sulfide to elemental sulfur was discussed previously (Brune, 1989).

Recently some organic sulfur compounds have also been shown to function as electron donors for photosynthetic growth of some Chromatiaceae. Visscher and Taylor (1993) found that *Tc. roseopersicina* could decompose mercaptomalate to fumarate + H_2S and mercaptopropionate to acrylate + H_2S and then use the liberated sulfide as an electron donor. This ability was shared by an as yet unclassified *Rhodopseudomonas* species that, unlike *Tc. roseopersicina*, could also assimilate the fumarate or acrylate formed in the decomposition reaction. Previously, Visscher and van Gemerden (1991) reported that dimethylsulfide functioned as an electron donor for photosynthetic growth of *Tc. roseopersicina*, being oxidized to dimethyl sulfoxide, which is not further metabolized.

Table 2. Sulfur oxidizing capabilities of the anoxygenic phototrophs

I. Proteobacteria

- A. Chromatiaceae
  - 1. Oxidize H_2S and S^0 to SO_4^{2-}.
    Chromatium buderi, Cm. okenii, Cm. tepidum, Cm. warmingii, Cm. weissei, Lamprocystis roseopersicina, Thiocapsa pfennigii, Thiocystis gelatinosa, Thiocystis violacea, Thiodictyon bacillosum, Thiodictyon elegans, Thiopedia rosea, Thiospirillum jenense, Thiospirillum winogradskyi [†,(a)].
  - 2. Oxidize H_2S, S^0 and $S_2O_3^{2-}$ to SO_4^{2-}.
    *Amoebobacter pedioformis, Amoebobacter purpureus, Cm. gracile, Cm. minus, Cm. minutissimum, Cm. purpuratum, Cm. violascens, Thiocapsa roseopersicina**.
  - 3. Oxidize H_2S, S^0, $S_2O_3^{2-}$, and SO_3^{2-} to SO_4^{2-}.
    Amoebobacter pendens, Amoebobacter roseus, Cm. salexigens[(b)], *Cm. vinosum, Thiocapsa halophila*[(c)].
- B. Ectothiorhodospiraceae
  - 1. Oxidize H_2S to S^0.
    Ectothiorhodospira halochloris, Ec. abdelmalekii.
  - 2. Oxidize H_2S and S^0 to SO_4^{2-}.
    Ec. marismortui[(d)].
  - 3. Oxidize H_2S, S^0, and $S_2O_3^{2-}$ to SO_4^{2-}.
    Ec. halophila, Ec. shaposhnikovii, Ec. vacuolata.
  - 4. Oxidize H_2S, S^0, $S_2O_3^{2-}$, and SO_3^{2-} to SO_4^{2-}.
    Ec. mobilis.
- C. Purple 'nonsulfur' bacteria.
  - 1. Oxidize H_2S to S^0.
    Rhodobacter capsulatus, Rb. sphaeroides, Rhodospirillum mediosalinum[(e)], *Rs. rubrum, Rp. marina*#.
  - 2. Oxidize H_2S to $S_4O_6^{2-}$.
    Rhodomicrobium vanneilii
  - 3. Oxidize $S_2O_3^{2-}$ to $S_4O_6^{2-}$.
    Rhodopila globiformis[+]
  - 4. Oxidize H_2S and $S_2O_3^{2-}$ to SO_4^{2-}.
    Rb. sulfidophilus, Rp. palustris.
  - 5. Oxidize H_2S, S^0, and $S_2O_3^{2-}$ to SO_4^{2-}.
    Rb. adriaticus, Rb. euryhalinus[(f)], *Rb. veldkampii, Rp. sulfoviridis.*

II. Green Sulfur Bacteria

- A. Chlorobiaceae
  - 1. Oxidize H_2S and S^0 to SO_4^{2-}.
    Ancalochloris perfilievii, Chlorobium limicola, Cb. chlorovibrioides, Cb. phaeobacteroides, Cb. vibrioforme, Chloroherpeton thalassium, Pelodictyon clathratiforme, Pelodictyon luteolum, Pelodictyon phaeum, Prosthecochloris aestuarii, Prosthecochloris phaeoasteroides.
  - 2. Oxidize H_2S, S^0, and $S_2O_3^{2-}$ to SO_4^{2-}.
    Cb. limicola f. thiosulfatophilum‡, *Cb. tepidum*[(g)], *Cb. vibrioforme f. thiosulfatophilum*‡, *Pelodictyon phaeoclathratiforme*[(h)].

III. Green Gliding Bacteria

- A. Chloroflexaceae[(i)]
  - 1. Oxidize H_2S to S^0.
    Chloroflexus aurantiacus, Oscillochloris trichoides..

Tabulated data are taken from Brune (1989) supplemented with information from the following: (a) Overmann et al., 1992; (b) Caumette et al., 1988; (c) Caumette et al., 1991; (d) Oren et al., 1989; (e) Kompantseva and Gorlenko, 1984; (f) Kompantseva, 1985; (g) Wahlund et al., 1991; (h) Overmann and Pfennig, 1989; (i) Madigan and Brock, 1975, and see text.

† One strain also oxidizes thiosulfate.

* Some strains also oxidize sulfite.

Thiosulfate also formed during sulfide oxidation.

[+] Oxidation reaction observed during photomixotrophic growth.

‡ These strains also oxidize tetrathionate to sulfate.

2. *Ectothiorhodospiraceae*

As the name implies, species of the family Ectothiorhodospiraceae deposit elemental sulfur extracellularly during sulfide oxidation. The Ectothiorhodospiraceae are halophilic and alkalophilic phototrophs, typically being found in salt lakes or marine tidal pools (Imhoff, 1992b). Polysulfides, which are stabilized by alkaline conditions, have been found to accumulate during sulfide oxidation and are attributed to a chemical reaction between sulfide and sulfur. Alternatively, polysulfides may be intermediates in the oxidation of sulfide to sulfur.

16S rRNA sequences indicate that the Ectothiorhodospiraceae can be divided into two subgroups, one consisting of the extreme halophiles *Ec. halochloris*, *Ec. abdelmalekii* (both of which contain BChl *b*), and *Ec. halophila*, and the other containing the remainder of the *Ectothiorhodospira* species (Trüper, 1987). The two BChl *b*-containing species have the most limited sulfur-oxidizing capabilities, oxidizing sulfide only to elemental sulfur while growing photomixotrophically (i. e. with both CO_2 and an organic carbon source). They are incapable of photoautotrophic growth (Then and Trüper, 1984). The remaining species oxidize sulfide and sulfur to sulfate, and except for *Ec. marismortui*, also use thiosulfate as an electron donor. Kusche (1985) found that elemental sulfur is not an intermediate during thiosulfate oxidation by *Ec. shaposhnikovii*, an observation that probably can be generalized to other thiosulfate-oxidizing Ectothiorhodospiraceae.

None of the species of Ectothiorhodospiraceae are able to grow as chemolithotrophs, although two species, *Ec. mobilis* and *Ec. shaposhnikovii*, can grow as chemoorganotrophs if sulfide or thiosulfate is supplied as a sulfur source (Kondratieva et al., 1981). The same two species are the only Ectothiorhodospiraceae able to assimilate sulfate when growing anaerobically as photoheterotrophs.

3. *Purple 'Nonsulfur' Bacteria*

Purple photosynthetic bacteria belonging to the α and β subdivisions of the Proteobacteria were formerly placed together in the family Rhodospirillaceae (also called the purple nonsulfur bacteria because of their predominantly photoheterotrophic growth mode). Ribosomal RNA sequence comparisons show that many species of 'Rhodospirillaceae' are more closely related to non-photosynthetic bacteria than to each other. Family designations are not used in the latest edition of *The Prokaryotes*, which discusses the α division phototrophs (*'Rhodospirillum* and related genera') and the β division phototrophs (the genera *Rhodocyclus* and *Rubrivivax*) in separate chapters (Imhoff and Trüper, 1992; Trüper and Imhoff, 1992). *Rhodocyclus* and *Rubrivivax* apparently are incapable of using reduced sulfur compounds as photosynthetic electron donors and will not be discussed further here.

Generally speaking, the α division photosynthetic Proteobacteria (here called α purple phototrophs) are photoheterotrophs, and with a few exceptions, were generally considered incapable of growing photoautotrophically with reduced sulfur compounds as electron donors. That perception changed when Hansen and van Gemerden (1972) demonstrated that *Rhodopseudomonas palustris, Rhodospirillum rubrum, Rhodobacter sphaeroides* and *Rb. capsulatus* could grow photoautotrophically in a chemostat when sulfide was supplied continuously at a low concentration (below 0.5 mM for the first three and 2 mM for *Rb. capsulatus*). The higher sulfide concentrations typically used in the culture media of purple and green sulfur bacteria prevent the growth of these species. It is now known that all species of *Rhodobacter* can grow photosynthetically with sulfide as the electron donor, and one species, *Rb. sulfidophilus*, resembles the true sulfur bacteria in its tolerance of high sulfide concentrations.

The diverse sulfur-oxidizing capabilities of the α purple phototrophs are summarized in Table 2. Most of these data have been discussed previously (Fischer, 1988; Brune, 1989), although the ability of *Rs. rubrum* and *Rb. sphaeroides* to grow photolithotrophically on thiosulfate was reported only recently by Wang et al. (1993), who noted an increase in turbidity of the bacterial culture which they attributed to formation of elemental sulfur. This suggests that the products of thiosulfate oxidation are S^0 and SO_4^{2-}. An interesting aspect of the work by Wang et al. (1993) is that CO_2 fixation in *Rs. rubrum* and *Rb. sphaeroides* does not require rubulose-1,5-bisphosphate carboxylase and thus does not proceed via the Calvin cycle, which is usually used for purple bacterial CO_2 fixation.

4. *Relationship of Photosynthetic to Nonphotosynthetic Sulfur Bacteria*

Included with the purple photosynthetic bacteria in the phylum Proteobacteria are such nonphotosynthetic sulfur bacteria as the thiobacilli, beggiatoas, and a group of so-called 'morphologically conspicuous' sulfur bacteria, of which only *Thiovulum* has been obtained in pure culture (La Rivière and Schmidt, 1992). The thiobacilli were originally grouped together solely on the basis of their ability to grow chemolithotrophically with reduced inorganic sulfur compounds as respiratory substrates. Physiological and biochemical characterizations (Kelly, 1989) as well as ribosomal RNA sequences (Lane et al., 1992) indicate that the thiobacilli do not form a coherent taxonomic group. Most of them belong to the β branch of the Proteobacteria, where the relatively well studied species *Thiobacillus ferrooxidans, Tb. thiooxidans, Tb. neapolitans,* and *Tb. tepidarius* cluster together in one group. RNA sequences indicate that *Tb. thioparus* is somewhat related to *Rhodocyclus purpureus*, while *Tb. perometabolis* is related to *Rubrivivax gelatinosus*. However, as noted above, neither of those β branch phototrophs can oxidize reduced sulfur compounds.

Tb. versutus (and two marine thiobacilli isolated from hydrothermal vent environments) appear to be fairly closely related to *Rb. capsulatus* (Lane et al., 1992) (and thus presumably also to *Rb. sulfidophilus*, which has fairly well developed sulfur-oxidizing capabilities). A periplasmic enzyme complex has been isolated from *Tb. versutus* that oxidizes thiosulfate (and sulfite) to sulfate with a *c*-type cytochrome as the electron acceptor (Kelly, 1989). Too little is known about the biochemistry of thiosulfate oxidation in those species of *Rhodobacter* able to use it to know if they contain enzymes like those in *Tb. versutus*. Unlike *Tb. versutus*, none of the *Rhodobacter* species grow chemolithotrophically on thiosulfate (Kondratieva, 1989).

Considering the ability of some species of Chromatiaceae to grow aerobically or micro-aerophilically in the dark on thiosulfate or sulfide, it is surprising that they have no close relatives that, having lost the ability to carry out photosynthesis, grow as chemolithotrophs. The nonphotosynthetic sulfur bacteria in the γ branch of Proteobacteria include *Beggiatoa*, a genus of filamentous bacteria that characteristically form periplasmic intracellular sulfur globules when grown in the presence of sulfide and oxygen. Many of these, however, cannot actually grow autotrophically on sulfide (Nelson, 1989). Also included in the γ branch are several sulfur bacteria that occur in symbiotic associations with marine invertebrates living near hydrothermal vents and *Thiothrix*. They appear to be more closely related to the fluorescent pseudomonads than to the purple sulfur bacteria (Lane et al., 1992).

The lack of a close relationship between the purple sulfur bacteria and the thiobacilli indicated by ribosomal RNA sequence comparisons is consistent with the differences in their sulfur oxidation pathways. Unlike the purple sulfur bacteria, most thiobacilli (*Tb. versutus* is an exception) oxidize thiosulfate to tetrathionate en route to sulfate, and can grow with tetrathionate as the electron donor (Kelly, 1989). This difference is revealed in the different biphasic oxygen uptake curves for thiobacilli and Chromatiaceae growing on thiosulfate. *Thiocystis violacea* and *Cm. vinosum* consume one molecule of oxygen per two molecules of thiosulfate in a rapid, initial phase of oxygen uptake in which 2 S° and 2 SO_4^{2-} are formed (Overmann and Pfennig, 1992). In typical thiobacilli (e.g. *Tb. acidophilus, Tb. ferrooxidans, Tb. thiooxidans, Tb. neapolitanus,* and *Tb. tepidarius*), oxygen uptake is also biphasic, but 4 $S_2O_3^{2-}$ are consumed per O_2 with 2 $S_4O_6^{2-}$ being the product of their oxidation in the rapid initial phase of oxygen consumption (Lu and Kelly, 1988a, Meulenberg et al., 1992; Pronk et al., 1990). Similar experiments indicate that the acidophilic archaean *Sulfolobus* has a thiosulfate oxidation pathway similar to that of the thiobacilli (Nixon and Norris, 1992).

Several species of sulfate-reducing bacteria were recently found to oxidize sulfide and thiosulfate under aerobic conditions (Dannenberg et al., 1992). Like the purple sulfur bacteria, the thiosulfate-oxidizing species *Desulfovibrio desulfuricans* does not form tetrathionate as an intermediate and is thought to reductively cleave thiosulfate initially to sulfide and sulfite. The sulfate reducing bacteria belong to the δ group of the Proteobacteria (Woese, 1987).

B. Green Sulfur Bacteria

The green sulfur bacteria, or Chlorobiaceae, form a coherent taxonomic group (resembling in this respect the Chromatiaceae and the Ectothiorhodospiraceae) and are the characteristic members of a separate bacterial phylum (Woese, 1987). They differ from the purple photosynthetic bacteria both in their primary photochemical processes and in their carbon metabolism (Amesz, 1991). All are photoautotrophs, although acetate and a few other simple organic substrates also can be assimilated when CO_2 is present. Assimilatory sulfate reduction is lacking, so reduced sulfur compounds are required as sulfur sources as well as photosynthetic electron donors.

The sulfur-oxidizing capabilities of the Chlorobiaceae are summarized in Table 2. Until recently, only two strains, namely *Chlorobium limicola* f. *thiosulfatophilum* and *Cb. vibrioforme* f. *thiosulfatophilum* were known to oxidize thiosulfate. These strains are also the only phototrophs known to use tetrathionate as an electron donor and to be able to photochemically disproportionate elemental sulfur into H_2S and $S_2O_3^{2-}$, with the latter product probably being formed via reaction of a SO_3^{2-} intermediate with elemental sulfur (Trüper et al., 1988). Recently two other green sulfur bacteria that use thiosulfate as an electron donor have been discovered, namely *Pelodictyon phaeoclathratiforme* (Overmann and Pfennig, 1989) and *Cb. tepidum* (Wahlund et al., 1991). So far, the tetrathionate-utilizing and sulfur-disproportionating capabilities of those species have not been reported. None of the Chlorobiaceae are known to use SO_3^{2-} as a photosynthetic electron donor.

C. Green Gliding Bacteria

The family Chloroflexaceae, or green gliding (green 'nonsulfur') bacteria, belongs to a phylum which, according to 16S rRNA sequence data, branched off early in the evolutionary history of Bacteria. The only species of that family that has been obtained in pure culture and characterized biochemically is *Chloroflexus aurantiacus*, a thermophile capable of slow photoautotrophic growth while oxidizing sulfide to elemental sulfur (Madigan and Brock, 1975; Pierson and Castenholz, 1992). Several other species, particularly in the genera *Heliothrix, Oscillochloris,* and *Chloronema* are often found in sulfide-containing environments, and Gorlenko (1988) has reported that *Oscillochloris trichoides* grows photoautotrophically while oxidizing sulfide to sulfur. *Cf. aurantiacus* resembles the purple phototrophs in its primary photochemistry and reaction center structure (Blankenship, 1992).

III. Electron Transport in Purple and Green Sulfur Bacteria

A. Purple Bacteria

Electron transport in the photosynthetic Proteobacteria (purple bacteria) is basically cyclic (Fig. 2A). A photochemical reaction center complex integral to the photosynthetic membrane uses light energy to transfer electrons from a periplasmic *c*-type cytochrome (and perhaps a high potential iron-sulfur protein [HiPIP] in some species, particularly in the family Ectothiorhodospiraceae) (Bartsch, 1991; Leguijt, 1993) to a quinone within the membrane. Quinone reduction occurs near the cytoplasmic membrane surface and is accompanied by uptake of protons (one per electron) from the cytoplasm (see chapter by Okamura and Feher). The cycle is completed when electrons are transferred to a periplasmic cytochrome (or HiPIP) via a membrane-spanning cytochrome bc_1 complex (Trumpower, 1990). Quinol oxidation releases protons into the periplasmic space.

Due to a cycle of quinone reduction as well as oxidation by the cytochrome bc_1 complex, 2 H^+ per electron are translocated from the cytoplasm to the periplasm, creating a gradient in the chemical potential of hydrogen ions ($\Delta\mu H^+$) across the cytoplasmic membrane. This H^+ ion gradient drives ATP synthesis (photophosphorylation) and reduction of NAD^+ to NADH with quinol as the reductant. NADH and ATP are used to reduce CO_2 to carbohydrate via the reductive pentose phosphate pathway (Calvin cycle).

Electrons drained from the reduced quinone pool to reduce NADH (and ultimately CO_2) are replaced in the photosynthetic sulfur bacteria by oxidation of inorganic sulfur compounds. Given the redox potentials of the sulfur compounds used as photosynthetic electron donors, these electrons can be transferred initially either to a periplasmic cytochrome *c* or directly to the quinone pool within the membrane. In the former case, subsequent quinone reduction would occur photochemically, resulting in charge transfer through the membrane and contributing to the transmembranous $\Delta\mu H^+$. The energetic differences between these two cases have been discussed previously (Brune, 1989).

Electron transport in the Chloroflexaceae is basically similar to that in purple bacteria (Amesz, 1991), but CO_2 fixation follows a different pathway (see Chapter 40 by Sirevåg).

B. Green Sulfur Bacteria

Electron transport in the green sulfur bacteria is fundamentally different from that in the purple bacteria in that the membrane-spanning photochemical reaction center transfers electrons completely through the membrane from a periplasmic

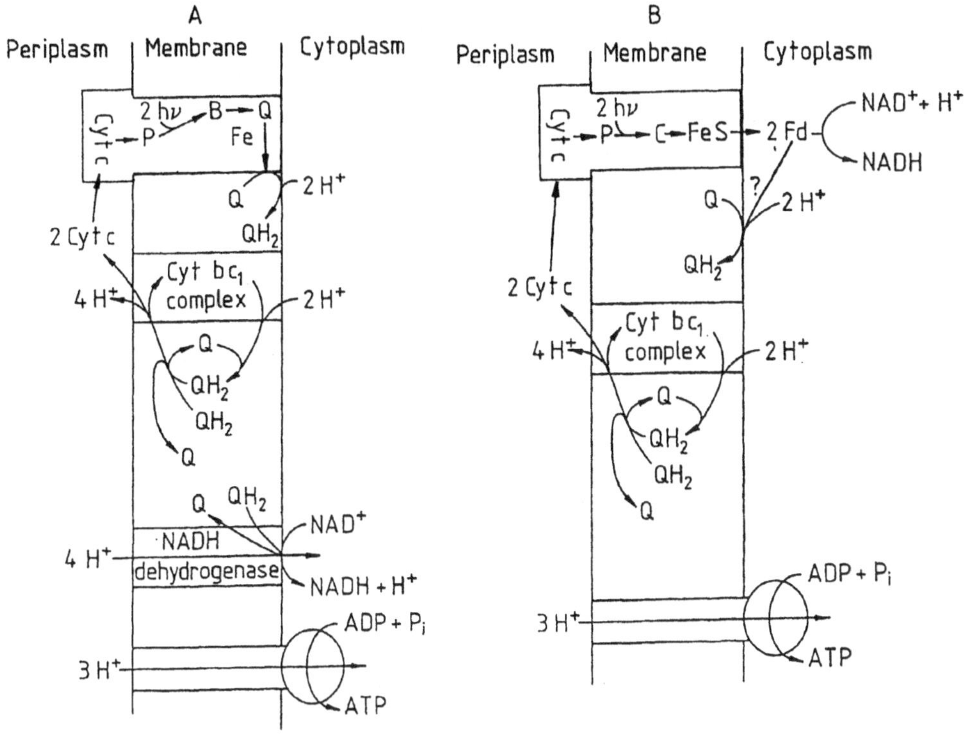

Fig. 2. Electron transport pathways in purple bacteria (A) and in green sulfur bacteria (B). Abbreviations: B, bacteriopheophytin molecule acting as initial electron acceptor in purple bacteria; C, bacteriochlorophyll *c*-like initial electron acceptor in green sulfur bacteria; Cyt, cytochrome, Fd, ferredoxin; FeS, iron-sulfur cluster acting as early electron acceptor; hν, photon of light absorbed to drive the initial electron transfer reaction; P, bacteriochlorophyll dimer acting as the primary electron donor; Q, quinone. A question mark next to the arrow from Fd to Q in (B) indicates that this electron transfer reaction has not been established, although it is likely to occur during cyclic electron transport. Because quinones and NAD^+ require two electrons for reduction, the electron transfer events shown here are those resulting from two sequential excitations of the reaction center pigments. Both green and purple bacteria are shown as having a bound cytochrome associated with the reaction center complex, although in some purple bacterial species, this cytochrome is missing. Note that logical sites for entry of electrons from inorganic sulfur compounds into the photosynthetic electron transport chain are the pool of quinones within the membrane or cytochrome *c* in the periplasm.

cytochrome *c* to a cytoplasmic ferredoxin molecule (Fig. 2B). Ferredoxin, a low-potential iron-sulfur protein, reduces pyridine nucleotides in a reaction catalyzed by ferredoxin-NAD^+ reductase (Kusai and Yamanaka, 1973; Knaff and Buchanan, 1975). CO_2 reduction in the green sulfur bacteria uses reduced ferredoxin, NADH, and NADPH as electron donors and occurs via a reductive tricarboxylic acid cycle (see Chapter 40 by Sirevåg for further details).

Like the purple bacteria, the green sulfur bacteria contain a cytochrome bc_1 complex (Hurt and Hauska, 1984) and quinones (menaquinone and chlorobium quinone) that can participate in cyclic electron transport. Electrons from inorganic sulfur compounds can enter the electron transport chain either via a periplasmic cytochrome *c* or via the quinone pool within the photosynthetic membrane. In the latter case, electron transfer would be via the cytochrome bc_1 complex to a periplasmic cytochrome *c* and then via the reaction center to ferredoxin and NAD(P). A similar pathway for electron transport in green bacteria was presented by Okumura et al. (1994) who examined the kinetics of cytochrome *c* oxidation and rereduction during cyclic and noncyclic electron transport in the thermophilic green sulfur bacterium, *Chlorobium tepidum*.

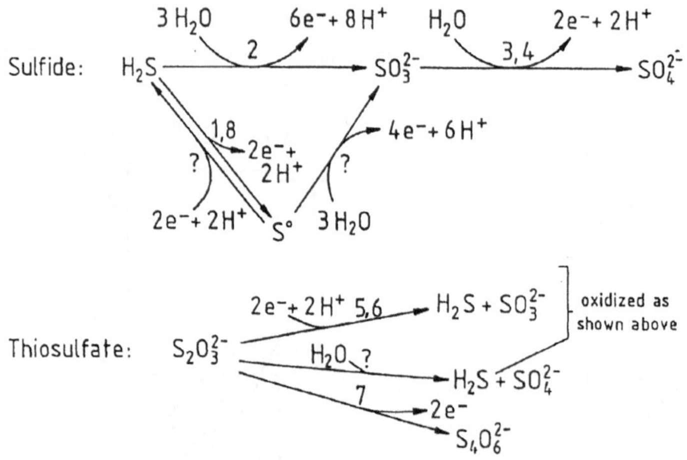

Fig. 3. Sulfide and thiosulfate oxidation pathways. Catalytic enzymes are indicated by numbers next to the reactions that they catalyze. 1 flavocytochrome *c*; 2 sulfite reductase; 3 APS reductase + ADP- or ATP-sulfurylase; 4 sulfite:acceptor oxidoreductase; 5 rhodanese; 6 thiosulfate reductase; 7 thiosulfate:acceptor oxidoreductase; 8 sulfide-quinone reductase. Question marks indicate reactions for which a catalytic enzyme is not yet known. (Slightly modified from Brune, 1989).

IV. Pathways of Sulfide and Thiosulfate Oxidation and Catalytic Enzymes

Enzymes able to catalyze sulfur oxidation reactions have been isolated from various species of photosynthetic bacteria and the reactions they catalyze arranged into possible sulfur oxidation pathways, bearing in mind the intermediates observed in bacterial cultures and what intermediates might be stable en route to sulfate formation (for reviews, see Trüper and Fischer, 1982; Fischer, 1988, 1989; Trüper, 1989; and Brune, 1989). One such pathway is shown in Fig. 3.

Only two intermediates are shown during sulfide oxidation, namely elemental sulfur, which accumulates as sulfur globules during sulfide oxidation by all species of green and purple sulfur bacteria (but not in some α purple phototrophs), and sulfite. Although sulfite formation has been observed only in *Rb. sulfidophilus* (Neutzling et al., 1985), enzymes that can catalyze its formation and consumption have been isolated from a variety of photosynthetic bacterial species. Possibly sulfite is not observed in other species because it is formed in the cytoplasm and is further oxidized to sulfate before it can escape to the external medium. The inability of most phototrophs to use sulfite as a photosynthetic electron donor could also be explained by its consumption in the cytoplasm together with impermeability of the cell membrane to it.

Thiosulfate oxidation is thought to proceed through the same intermediates as are formed during sulfide oxidation, except in special cases in which tetrathionate is the end product of its oxidation. The enzymes catalyzing the sulfur transformations shown in Fig. 3 are discussed below.

A. Oxidation of Sulfide to Elemental Sulfur

1. Flavocytochrome c

The distribution of flavocytochrome *c* among the photosynthetic bacteria and its chemical and catalytic properties were discussed in detail in a recent review (Brune, 1989; see also Meyer and Cusanovich, 1989, and Bartsch, 1991). Recently it was also shown to be present in *Amoebobacter pedioformis* (Fischer and Ufken, 1991). What is important for the present discussion is that flavocytochromes *c* can catalyze electron transfer from sulfide to a variety of small *c*-type cytochromes, including cytochrome *c*-555 from *Cb. limicola* f. *thiosulfatophilum* (Yamanaka and Kusai, 1976) and cytochrome *c*-550 from *Cm. vinosum* (Davidson et al., 1985) which are thought to donate electrons to the reaction centers in those species.

The best-studied flavocytochromes are those from *Cb. limicola* f. *thiosulfatophilum* and from *Cm. vinosum*. Both contain a 46–47 kDa flavoprotein subunit and a smaller hemoprotein subunit. The hemoprotein subunit of the *Cb. limicola* flavocytochrome contains a single heme group and has a molecular mass of 10 kDa (van Beeumen et al., 1990), while in the *Cm. vinosum* flavocytochrome it is larger (21 kDa) and contains two heme groups (van Beeumen et al., 1991). Electron transfer from sulfide is initially to the flavin group and is followed by intramolecular electron transfer to the heme group(s) (Tollin et al., 1982; Cusanovich et al., 1985), which then reduce various cytochromes *c*. A recent crystal structure of flavocytochrome *c* from *Cm. vinosum* (Chen et al., 1994) has provided new insights into electron transfer between the flavin and heme cofactors, and also revealed a disulfide bond near the flavin that may participate in sulfide oxidation. Knaff and coworkers (Davidson et al., 1985, 1986) have shown that interaction between the flavocytochrome heme subunits and cytochrome *c* is electrostatic, and is inhibited at high ionic strength with concomitant inhibition of flavocytochrome-catalyzed cytochrome *c* reduction.

Van Beeumen and coworkers (1990, 1991) have obtained complete amino acid sequences for the hemoprotein subunits of the *Chlorobium* and the *Cm. vinosum* flavocytochromes as well as N-terminal sequences for the flavoprotein subunits. The *Cm. vinosum* hemoprotein apparently evolved by fusion of the genes for two different ≈10 kDa hemoproteins, rather than by gene duplication, since the two halves of that hemoprotein exhibit only 7% sequence identity. The N-terminal half of the *Cm. vinosum* hemoprotein appears to be related to the *Chlorobium* hemoprotein, with which it has 39% sequence identity. The flavoprotein subunits are also homologous, with their N-terminal sequences being about 60% identical. Recently the protein sequence data were used to isolate and clone the genes encoding the *Cm. vinosum* hemoprotein and about 25% of the flavoprotein (Dolata et al., 1993). The translated gene sequences revealed that both flavocytochrome subunits have signal peptide leader sequences that are absent in the mature proteins, implying a periplasmic location.

Not all bacteria that oxidize sulfide to elemental sulfur have a flavocytochrome *c*. For example flavocytochromes have not been found in any of the Ectothiorhodospiraceae or in *Tc. pfennigii* (Bartsch, 1991) or in *Cm. warmingii* (Wermter and Fischer, 1983). The green sulfur bacteria *Cb. vibrioforme* (the non-thiosulfate-utilizing strain) and *Pelodictyon luteolum* also lack flavocytochromes (Steinmetz et al., 1983).

In spite of these indications that flavocytochrome *c* is not necessary for sulfide oxidation, it is difficult to assign another function to it in those species in which it occurs. The low midpoint redox potentials of its flavin and heme prosthetic groups (0 to +100 mV) (Tollin et al., 1982) and its ability to mediate electron transfer to small periplasmic cytochromes *c* are most consistent with a role in transferring electrons from a periplasmic substrate to the reaction center. Although flavocytochromes *c* form complexes with sulfite and thiosulfate (Meyer et al., 1988, 1991), neither appears to be oxidized by the flavocytochrome, leaving sulfide as the most likely of the normally encountered substrates. Possibly there are parallel sulfide oxidation pathways. The large redox potential drop when electrons are transferred via flavocytochrome *c* to the reaction center would keep flavocytochrome *c* in a highly oxidized state, creating a high affinity system for sulfide oxidation, especially at low sulfide concentrations. This might be important in environments in which bacteria have to compete for sulfide. The fact that flavocytochrome *c* is constitutive (at least in *Cm. vinosum*) (Bartsch, 1978) would keep sulfide-starved cells ready to oxidize sulfide as soon as it was encountered. This high affinity system might supplement an energetically more efficient system involving electron transfer from sulfide to quinone via a sulfide-quinone reductase (see below).

Cytochromes without flavin prosthetic groups have also been proposed to mediate electron transfer from sulfide to the reaction center (see Fischer, 1984 and Brune, 1989, for discussion of the older literature). Recently Leguijt and coworkers (Leguijt, 1993) found that a low potential (E_m = +25 mV) membrane bound cytochrome *c* in *Ectothiorhodospira mobilis* is reduced by sulfide with a half-time of 465 ms. They proposed that this cytochrome, which is in turn oxidized by the reaction center in less than 5 μs, mediates sulfide oxidation in that species. As noted above, periplasmic sulfide oxidation by such a mechanism could contribute to the transmembranous $\Delta\mu H^+$ required for NAD^+ reduction and ATP synthesis. In fact, sulfide stimulated formation of the transmembranous $\Delta\mu H^+$ when *Ectothiorhodospira mobilis* cells were illuminated (Leguijt, 1993), although the sensitivity of this stimulation to inhibitors of the cytochrome bc_1 complex present in *Ec. mobilis* (Leguijt et al., 1993) was not tested.

2. Sulfide-Quinone Reductase

Recently Shahak and coworkers provided a new

perspective on photosynthetic bacterial sulfide oxidation when they developed an assay for and then isolated an enzyme catalyzing electron transfer from sulfide to plastoquinone in the facultatively anoxygenic cyanobacterium *Oscillatoria limnetica* (Arieli et al., 1991; Shahak et al., 1992b,c). They named this enzyme sulfide-quinone reductase, or SQR. Subsequent work showed that analogous enzymes are present in the green sulfur bacterium *Cb. limicola* f. *thiosulfatophilum* and in *Rb. capsulatus* (Shahak et al., 1992a,b,c). Earlier experiments by Brune and Trüper (1986) showing nonphotochemical reduction of ubiquinone by sulfide in *Rb. sulfidophilus* can now be explained by assuming that SQR is also present in that species. The *O. limnetica* enzyme has a molecular weight of 57,000 (from SDS-PAGE) (Shahak et al., 1992b,c) and contains a flavin prosthetic group that probably mediates electron transfer from sulfide to quinone (Y. Shahak, personal communication). Its K_m for sulfide is 4 μM (Arieli et al., 1991), a value comparable to that of flavocytochrome *c*-552 from *Cm. vinosum* (Fukumori and Yamanaka, 1979). Preliminary N-terminal protein sequence data on the *O. limnetica* and *Rb. capsulatus* enzymes indicate that they are homologous proteins with FAD-binding domains near their N-terminal ends (Y. Shahak, personal communication).

The *Oscillatoria, Chlorobium,* and *Rhodobacter* enzymes exhibit different sensitivities to various quinone analog inhibitors (Shahak et al., 1992a,b), a result consistent with the occurrence of different quinone acceptors in these species. The sensitivity of SQR in *Cb. limicola* to antimycin A and NQNO probably accounts at least partly for the previously observed sensitivity of photochemical electron transport from sulfide to those inhibitors in membranes isolated from that species (Knaff and Buchannan, 1975; Shill and Wood, 1985).

The experimental evidence for SQR in such different species as *O. limnetica, Cb. limicola,* and *Rb. capsulatus* raises the possibility that this enzyme is present in all sulfide-oxidizing phototrophs. Electron transfer from sulfide to quinone in the Chromatiaceae was previously suggested based on the energetics of chemolithotrophic growth of *Tc. roseopersicina* and *Cm. vinosum* (Brune, 1989).

B. Oxidation of Sulfide and Sulfur to Sulfite

Oxidation of sulfur is a problem in bacterial photosynthesis in that no enzyme has been found that catalyzes this reaction. Some recent observations on sulfur oxidation by thiobacilli may be relevant to the reaction in phototrophs. A variety of thiobacilli form elemental sulfur as an intermediate during sulfide and tetrathionate oxidation (Javor et al., 1990; Meulenberg et al., 1992; Pronk et al., 1990). Subsequent sulfur oxidation is completely inhibited by the sulfhydryl reagent N-ethyl maleimide (NEM). The effect of NEM on sulfur oxidation is specific; oxidations of thiosulfate, trithionate, and even tetrathionate (as measured by oxygen uptake rates) were not affected. Analogous experiments on photosynthetic bacteria have not yet been carried out.

Bacon and Ingledew (1989) found that *Tb. ferrooxidans* cells formed H_2S from added sulfur (and, less efficiently, H_2Se from red selenium). They proposed that the S_8 rings of elemental sulfur were activated by a two electron reduction to give an enzyme-bound polysulfide, i. e. $Enz + S_8 + 2e^- \rightarrow Enz\text{-}^-S\text{-}S_6\text{-}S^-$. The enzyme-bound polysulfide is subsequently oxidized to eight sulfate ions in a manner not specified. The sulfide detected in their experiments was thought to be a product formed from a side reaction of the enzyme-bound polysulfide. Sulfhydryl groups are strong nucleophiles that readily attack elemental sulfur to form polysulfides (Roy and Trudinger, 1970), and if the enzyme reacts with sulfur via one or more cysteine SH groups, the inhibition by NEM would be expected. Analogous schemes for activation of elemental sulfur by reaction with glutathione have been proposed previously (Oh and Suzuki, 1980).

Sulfur oxidation in thiobacilli is also inhibited by uncouplers and by cytochrome bc_1 complex inhibitors such as myxothiazole and HQNO (Beffa et al., 1992a,b; Meulenberg et al.., 1992; Pronk et al., 1990). Inhibition by uncouplers may indicate an energy requirement for sulfur activation, while the inhibitor data could indicate either that electron transfer from sulfur to oxygen occurs via the cytochrome bc_1 complex or that coupled electron transfer through the complex is needed to provide energy for activation. Analogous experiments on sulfur oxidation by purple sulfur bacteria that can grow as chemolithotrophs have not yet been done.

Even if elemental sulfur is reductively activated (which, from a redox point of view, is a step backward), this information per se provides little insight into how it is further oxidized to sulfite or sulfate. An alternative possibility might be that

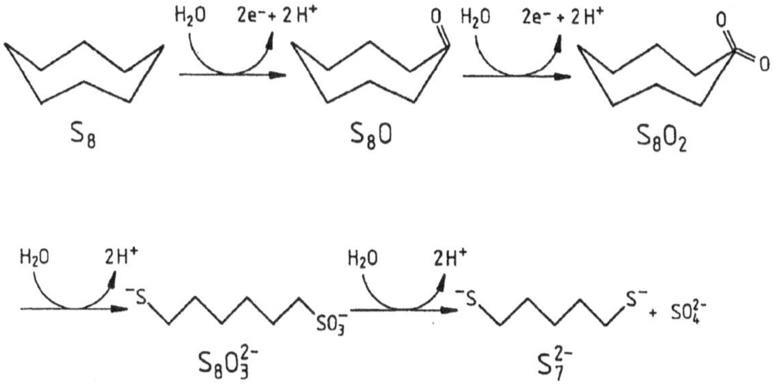

Fig. 4. Possible initial steps for oxidation of elemental sulfur. So far, the enzymes catalyzing the reactions shown here have not been isolated, although there is evidence for polysulfide formation during sulfur oxidation in some sulfur-oxidizing bacteria.

polysulfides are formed by initial oxidation of elemental sulfur followed by ring opening and hydrolysis of the resulting sulfane monosulfonate (Fig. 4). Oxidation of S_8 to S_8O might be catalyzed by an enzyme analogous to that catalyzing the oxidation of dimethyl sulfide to dimethyl sulfoxide that occurs in some purple sulfur bacteria (Visscher and Van Gemerden, 1991). A second oxidation, yielding S_8O_2 followed by hydrolytic ring opening to form $^-S\text{-}S_6\text{-}SO_3^{2-}$ and subsequent hydrolysis to $^-S\text{-}S_5\text{-}S^- + SO_4^{2-}$ could generate a polysulfide without an initial reduction reaction. The intermediate S_8O proposed in this scheme has been synthesized and characterized, while S_8O_2 is unstable, decomposing to SO_2 and S_7 (Steudel, 1984). Spontaneous hydrolysis of sulfane monosulfonates at room temperature is slow (Steudel et al., 1988) and enzymes catalyzing this reaction are unknown. However, enzymes catalyzing the analogous hydrolyses of trithionate and tetrathionate have been found in thiobacilli (Meulenberg et al., 1992; Lu and Kelly, 1988b). A similar pathway for sulfur oxidation was suggested earlier (Kodama and Mori, 1968).

1. Sulfite Reductase

Sulfite reductase is an enzyme which, as its name implies, can catalyze reduction of sulfite to sulfide. It occurs in bacteria and in some algae where it participates in assimilatory reduction of sulfate for incorporation into sulfur-containing amino acids, coenzyme A, etc. (Schiff and Fankhauser, 1981), and in sulfate-reducing bacteria, where it participates in dissimilatory sulfate reduction (Thauer, 1989). Kobayashi et al. (1978) and Schedel et al. (1979) found that sulfite reductase was also present in extracts of *Cm. vinosum* cells grown photoautotrophically on sulfide and CO_2, and the latter obtained the enzyme in about 80% purity. Schedel et al. (1979) suggested that the enzyme catalyzed sulfide oxidation, rather than sulfite reduction, and noted that it was undetectable in cells grown photoheterotrophically on malate with sulfate as the sulfur source. There is no information about the electron acceptor for sulfide oxidation via sulfite reductase, and in fact enzyme activity was measured as sulfite reduction.

Analysis of the isolated enzyme indicated that it contained siroheme and Fe-S prosthetic groups as is characteristic of sulfite reductases. SDS-PAGE revealed approximately equal amounts of two subunits, with molecular masses of 37 and 43 kDa. Schedel et al. (1979) suggested an $\alpha_4\beta_4$ structure for the enzyme in approximate agreement with a molecular weight of 280,000 determined by gel filtration.

Much has been learned about the mechanism of action of sulfite reductases from extensive studies done on the assimilatory enzyme from *E. coli*. Although it is beyond the scope of this review to discuss that work in depth, a few conclusions from it are relevant. The *E. coli* enzyme is a large (nearly 700 kDa) $\alpha_8\beta_4$ complex in which the 8 α subunits contain FAD and FMN prosthetic groups that transfer electrons from NADPH to the siroheme-containing β subunits (Siegel et al., 1982). The sulfite-binding β subunit can be isolated using chaotropic agents, and its crystal structure has been determined (McRee et al., 1986). The crystal structure indicated that the siroheme and Fe_4S_4 prosthetic groups are only 4.4 Å apart, and probably share a cysteine sulfur ligand. The side of the siroheme group away from the Fe_4S_4 cluster is solvent exposed, and is thought to bind sulfite via ligation of its sulfur atom to the siroheme iron (Young and Siegel, 1988). A similar arrangement of the siroheme and Fe_4S_4 prosthetic groups and for substrate binding has been proposed for a smaller (27 kDa) sulfite reductase isolated from *Desulfovibrio vulgaris*, although in that case the bridging ligand between the siroheme and Fe_4S_4 prosthetic groups appears to be a sulfide ion (Tan and Cowan, 1991). This configuration allows concerted transfers of electron pairs between the enzyme and bound substrate molecules, and is probably a general feature of both assimilatory and dissimilatory sulfite reductases.

Trüper (1984) has suggested that sulfite reductase can catalyze oxidation of elemental sulfur and of polysulfides as well as sulfide. Zero-valent sulfur is a likely intermediate between sulfite and sulfide in the reaction catalyzed by sulfite reductase and conceivably could be transferred to that enzyme from a polysulfide chain by a sulfane sulfur transferase. Alternatively, a cycle (see below) can be written in which sulfite produced by sulfite reductase (4) is used to activate elemental sulfur via nucleophilic attack (1) to form a transient sulfane monosulfonic acid that, as noted in the discussion of Fig. 4, could be hydrolyzed (2) to form sulfate + a polysulfide. Disproportionation of polysulfides to elemental sulfur and sulfide (3) is a thermodynamically favored reaction, and could regenerate the sulfide substrate for a second round of sulfite formation via sulfite reductase (4). The series of reactions in this cycle is as follows:

$$HSO_3^- + S_8 \rightarrow {}^-S\text{-}S_6\text{-}S\text{-}SO_3^- + H^+ \quad (1)$$

$$^-S\text{-}S_6\text{-}S\text{-}SO_3^- + H_2O \rightarrow {}^-S\text{-}S_6\text{-}S^- + SO_4^{2-} \quad (2)$$

$$^-S\text{-}S_6\text{-}S^- + H^+ \rightarrow {}^7\!/\!_8\, S_8 + HS^- \quad (3)$$

$$HS^- + 3\, H_2O \rightarrow HSO_3^- + 6\, H^+. \quad (4)$$

With some modifications, this scheme might also apply to the polythionate-utilizing thiobacilli, and in fact Schedel and Trüper (1979, 1980) isolated and characterized a sulfite reductase from *Tb. denitrificans* which they suggested was also involved in sulfur oxidation. In the acidophilic thiobacilli, disproportionation of the sulfane monosulfonic acid formed in reaction (2) to form polythionates + polysulfides would also have to be included, as well as hydrolysis reactions involving the polythionates. It is known that polythionates are most stable at acidic pH values (Le Faou et al., 1990), which may account for their occurrence in the sulfur globules of acidophilic thiobacilli such as *Tb. ferrooxidans*. (Steudel et al., 1987, 1988; Steudel, 1989). Polythionates have also been detected in sulfur globules formed by the thiosulfate-oxidizing *Chlorobium* species (Steudel et al., 1988).

A key feature of the proposed cycle is that sulfite acts as the reductant for activation of sulfur to a polysulfide [reactions (1) and (2)], with sulfite being oxidized to sulfate in the process. Reaction (1) is well known in sulfur chemistry (Roy and Trudinger, 1970). In the presence of excess sulfite, subsequent reactions yield thiosulfate, which could account for the observed formation of small amounts of thiosulfate as a side product during sulfide oxidation by *Cm. vinosum* (Steudel et al., 1990) and by the thiosulfate-utilizing species of *Chlorobium* (Fischer, 1984). Oxidation of sulfite by sulfur might also account for the inability of many photosynthetic sulfur bacteria to grow using sulfite alone as the electron donor.

2. Sulfite Formation by Other Enzymes

A sulfite reductase like the *Cm. vinosum* enzyme has not yet been found in any other bacterial species, making it questionable whether or not this enzyme has any general role in photosynthetic bacterial sulfide (or sulfur) oxidation. Sugio et al. (1987,1989) have isolated a very different enzyme from *Tb. ferrooxidans* that also catalyzes oxidation of sulfide to sulfite with Fe^{3+} as the electron acceptor. Best results were obtained with sulfide produced at constant low levels by the reaction of elemental sulfur with glutathione (Sugio et al., 1989). No redox active prosthetic group has yet been identified for that 46 kDa enzyme, which contains two identical 23 kDa subunits. Although ferric ions are not plausible electron acceptors in phototrophs, the midpoint potential for oxidation of sulfide to sulfite (–115 mV at pH 7) is sufficiently low that a variety of other electron acceptors might be used by an analogous photosynthetic bacterial enzyme, if one exists.

Another possibility that perhaps should be considered is oxidation of sulfur via persulfide (R-S-

S-H) and S-sulfo-, or Bunte salt, (R-S-SO_3^{2-}) intermediates. This would involve redox transformations that are essentially the reverse of those occurring during assimilatory sulfite reduction in *Chlorella* and in higher plants (Schiff and Fankhauser, 1981; Fischer, 1988). Westley and Westley (1991) have noted that the pool of sulfane sulfur, including elemental sulfur and persulfides, in animal tissues is in rapid equilibrium in the presence of glutathione and can be assayed via rhodanese-catalyzed thiocyanate formation (see below for further discussion of rhodanese.) Such an equilibrium might possibly supply persulfides for further oxidation. Currently, however, there is no experimental evidence for this kind of sulfur oxidation pathway in any bacterial species.

C. Oxidation of Sulfite to Sulfate

Although hypothetical mechanisms for sulfite oxidation have been mentioned in the previous section, other sulfite oxidation reactions have been suggested, and enzymes that catalyze them have been isolated and characterized. These enzymes will be considered briefly in the sections below.

1. APS Reductase

APS reductases occur in sulfate reducing-bacteria where they catalyze the reductive cleavage of adenosine-5'-phosphosulfate, or APS, to form AMP (adenosine-5'-monophosphate) and sulfite (Thauer, 1989). They have also been found in some thiobacilli (Takakuwa, 1992) and in several species of purple and green sulfur bacteria (Trüper and Fischer, 1982; Fischer, 1989) where their function is assumed to be an oxidative binding of sulfite to AMP, giving APS as the product. Typically enzyme activity is measured as AMP-dependent sulfite oxidation with ferricyanide as the electron acceptor. With most of the APS reductases, *c*-type cytochromes can also be used as electron acceptors. The in vivo electron acceptor is unknown. Although APS is also an intermediate in assimilatory sulfate reduction, it is not reduced by APS reductase in that process; thus APS reductases are generally regarded as dissimilatory enzymes.

Sulfate can be liberated from APS in a subsequent reaction with inorganic phosphate in which ADP is formed (Trüper, 1989, Fischer, 1989). That reaction is catalyzed by the enzyme ADP sulfurylase. Alternatively, APS may react with pyrophosphate in an ATP sulfurylase-catalyzed reaction giving ATP and sulfate as the products. In either case, part of the free energy from hydrolysis of the phosphosulfate anhydride bond ($\Delta G_0' = -21$ kcal/mol) (Thauer et al., 1977) is conserved by formation of a phosphate anhydride bond. When ADP is the product, AMP can be regenerated in the reaction 2ADP $\rightarrow$ AMP + ATP, which is catalyzed by adenylate kinase.

APS reductases have been found in several species of Chromatiaceae, in which they are usually membrane bound (Trüper and Fischer, 1982; Schwenn and Biere, 1979), a factor which has inhibited their isolation and purification. The enzyme from *Tc. roseopersicina* strain 6311 is an exception, and was purified by Trüper and Rogers (1971). Recently, Dahl and Trüper (1989) also described an isolation procedure for the membrane-bound enzyme from *Tc. roseopersicina* strain M1. APS reductases have also been isolated from two green sulfur bacteria, namely *Cb. limicola* f. *thiosulfatophilum* (Kirchoff and Trüper, 1974) and *Cb. vibrioforme* f. *thiosulfatophilum* (Khanna and Nicholas, 1983). Both were obtained from the soluble fraction from broken cells. APS reductases apparently do not occur in any of the α purple phototrophs and are either absent or present at very low levels in the Ectothiorhodospiraceae (Trüper and Fischer, 1982; Fischer, 1989). They also were reported not to occur in *Cm. gracile* or *Cm. purpuratum* (Trüper and Fischer, 1982).

The photosynthetic bacterial enzymes are generally similar to those from thiobacilli and from sulfate-reducing bacteria (Takakuwa, 1992). All contain FAD and Fe-S prosthetic groups and have molecular masses near 200 kDa. The enzyme from *Tc. roseopersicina* was also reported to contain two heme groups per molecule (Trüper and Rogers, 1971). Recently, however, Wang (1991) presented the following arguments that the heme groups belong to a contaminating protein that copurifies with the *Tc. roseopersicina* enzyme: (1) Adding sulfite + AMP to the enzyme gave spectral changes indicating flavin and Fe-S reduction, but did not affect the oxidized heme absorbance, and (2) The heme absorbance in different fractions from the APS reductase band collected during chromatography on hydroxyapatite was not proportional to enzyme activity.

Very little is known about the subunit structure of the APS reductases from either photosynthetic bacteria or thiobacilli. Bramlett and Peck (1975) found that the 220 kDa enzyme from the sulfate-reducing bacterium *Desulfovibrio* (*Dv.*) *vulgaris* consisted of 70 and 20 kDa subunits and proposed a 3 to 1 stoichiometry in which each 70 kDa subunit bound a Fe_4S_4 group while the single 20 kDa subunit bound FAD. This composition was consistent with the staining intensities of the protein bands on polyacrylamide gels and with the iron, sulfur, and FAD analyses. Wang (1991) also found 70 and 20 kDa subunits for the *Tc. roseopersicina* enzyme, possibly suggesting a subunit structure similar to that of the *Dv. vulgaris* enzyme. However, nonheme iron and acid labile sulfur contents of APS reductases from photosynthetic bacteria (Trüper and Rogers, 1971; Khanna and Nicholas, 1983) and thiobacilli (see Takakuwa, 1992 for a table) are sufficient only for the presence of one or two Fe_4S_4 clusters per FAD group in those enzymes. Two would be reasonable to accept the two electrons released during sulfite oxidation.

On the basis of EPR and spectroscopic data, Peck and Bramlett (1982) proposed a catalytic mechanism for APS reductase in which SO_3^{2-} initially formed a complex with the flavin. This then reacts with AMP to yield APS, releasing two electrons that are transferred via the flavin to the Fe_4S_4 centers. A kinetic analysis of AMP-dependent ferricyanide reduction catalyzed by the *Tc. roseopersicina* enzyme (Wang, 1991) indicated a ping-pong type mechanism in which AMP and sulfite first reacted with the enzyme (probably to reduce its prosthetic groups) to form an intermediate which then reacted with ferricyanide. This mechanism is consistent with that proposed for the *Dv. vulgaris* enzyme by Peck and Bramlett (1982).

2. Sulfite: Acceptor Oxidoreductase

Often extracts from photosynthetic bacteria have been observed to catalyze an AMP-independent reduction of ferricyanide as well as (or instead of) the APS reductase catalyzed reaction. This reaction may be catalyzed by an enzyme called sulfite: acceptor oxidoreductase, or sulfite oxidoreductase for short. Because APS reductase is not found in the Ectothiorhodospiraceae or in the α purple phototrophs, it has been suggested that sulfite oxidation in those groups is catalyzed by this enzyme.

Sulfite is a strong reducing agent ($E_0' = -515$ mV for the sulfate/sulfite couple), and nonenzymatic artifacts are a problem when assaying ferricyanide reduction. For example, Dahl and Trüper (1989) found a membrane-associated activity in two *Tc. roseopersicina* strains that was not destroyed by boiling, indicating that it was not enzymatic.

Sulfite oxidoreductases have been found in almost all photosynthetic bacteria examined (Fischer, 1989), but in some cases it is not clear what controls were performed. Dahl and Trüper (1989) reported that only one of the three strains of *Tc. roseopersicina* that they examined contained a true heat sensitive sulfite oxidoreductase. That enzyme, which was present at only low levels, used cytochrome *c* as the electron acceptor and gave no activity with ferricyanide. In isotope fractionation experiments on *Cm. vinosum*, Fry et al. (1985) noted a change from a preferential formation of $^{34}SO_4^{2-}$ (an inverse isotope effect) to preferential formation of $^{32}SO_4^{2-}$ during sulfite oxidation. They interpreted this result as indicating competition for sulfite between two different enzymes, possibly APS reductase and sulfite oxidoreductase. *Cm. vinosum* was previously reported to contain a 68 kDa protein with sulfite oxidoreductase activity (see Neutzling et al., 1985). So far, nothing is known about the prosthetic groups or the in vivo electron acceptors of any of the photosynthetic bacterial sulfite oxidoreductases.

Several species of thiobacilli have been shown to contain sulfite: acceptor oxidoreductases in which the preferred electron acceptor is cytochrome *c* (Takakuwa, 1992). These are molybdoproteins in which molybdenum is complexed with an organic cofactor called molybdopterin (see Takakuwa, 1992, for a proposed structure). The *Tb. versutus* enzyme appears to be part of a larger periplasmic enzyme complex that catalyzes oxidation of thiosulfate to sulfate, and it is not clear that sulfite oxidation is actually the function of sulfite oxidase in that complex (Lu and Kelly, 1984). *Tb. versutus* and *Tb. pantophora* require molybdenum to grow on thiosulfate (Friedrich et al., 1986), and transposon insertional mutants of the latter species that are unable to synthesize molybdopterin are also unable to grow on thiosulfate (Chandra and Friedrich, 1986; Mittenhuber et al., 1991). Sulfide oxidation is also abolished in these mutants. Whether or not any phototrophs have an analogous molybdenum requirement is not known.

D. Thiosulfate Oxidation

Basically, three different initial reactions have been considered for oxidation of thiosulfate in photosynthetic bacteria. These are (1) oxidation to tetrathionate by a thiosulfate: acceptor oxidoreductase, (2) reduction to sulfite and sulfide by a rhodanese or thiosulfate reductase, and (3) hydrolysis to sulfide and sulfate. These will now be discussed briefly in that order.

1. Thiosulfate: Acceptor Oxidoreductases

In 1966, Smith reported that *Cm. vinosum* oxidized thiosulfate to tetrathionate in a mildly acidic culture medium (pH = 6.25). Similarly, *Rp. palustris* was found to oxidize thiosulfate to tetrathionate when its concentration in the culture medium exceeded 10 mM (Trüper, 1984). Neither these nor any other purple bacteria have been found to oxidize tetrathionate further, however, making it unlikely that this oxidation reaction is on the main pathway of thiosulfate oxidation. In fact, Smith (1966) reported that tetrathionate inhibited the normal oxidation of thiosulfate to sulfate as well as growth of *Cm. vinosum*.

Enzymes that catalyze reduction of various electron acceptors (i. e. ferricyanide, various cytochromes, and a high potential iron-sulfur protein, or HiPIP) by thiosulfate have been isolated from *Cm. vinosum* and from *Rp. palustris*. That work was reviewed previously (Brune, 1989), and that review can be consulted for further information. If tetrathionate is in fact the product of thiosulfate oxidation in these reactions, then they are unlikely to play any significant role in the complete oxidation of thiosulfate to sulfate. Therefore they will not be considered further here.

Two strains of *Chlorobium* are able to use tetrathionate as an electron donor, but there is no evidence that it is an intermediate in thiosulfate oxidation. Experiments with thiosulfate labelled with ^{35}S either in the sulfane ($^{35}SSO_3^{2-}$) or sulfone ($S^{35}SO_3^{2-}$) position indicated that *Cb. vibrioforme* f. *thiosulfatophilum* rapidly released the sulfone sulfur as sulfate, while the sulfane sulfur was retained (probably as elemental sulfur) with the bacterial cells upon filtration (Khanna and Nicholas, 1982). Acidophilic thiobacilli apparently contain enzymes able to hydrolyze tetrathionate to thiosulfate, sulfur, and sulfate (Meulenberg et al., 1992), and it may be that a similar enzyme is present in these *Chlorobium* strains. In this case, tetrathionate would be metabolized via thiosulfate, rather than *vice versa*.

2. Rhodanese and Thiosulfate Reductase

Rhodaneses are thiosulfate: cyanide sulfur transferases, meaning that they transfer the sulfane sulfur atom of thiosulfate to CN^-, generating SO_3^{2-} and SCN^- as the products (Westley, 1988). The best-studied rhodanese is the 33 kDa enzyme from bovine liver, for which a high-resolution crystal structure has been obtained (Hol et al., 1983), providing useful insights into the catalytic mechanism of rhodanese. The initial reaction is between thiosulfate and a cysteine at the active site of the enzyme to form a cysteine persulfide (Cys-SSH) and sulfite, which is released into the medium. The persulfide form of the enzyme is rather stable, and was in fact the form crystallized. The enzyme persulfide can transfer the bound sulfane sulfur to a variety of thiophilic acceptors such as cyanide, various thiols, and, by reversal of the initial reaction, sulfite. The preferred substrate for the mammalian enzyme is cyanide, and thiocyanate formation catalyzed by rhodanese is important as a means of cyanide detoxification (Westley, 1988). Cerletti (1986) has argued that another important function of rhodaneses is to generate sulfide for assembly of Fe-S centers in iron-sulfur proteins. Reaction with a thiol generates a persulfide from which sulfide may be released by reaction with a second thiol; i. e. $RSSH + R'SH \rightarrow RSSR' + H_2S$. An intramolecular disulfide is formed if the sulfane sulfur is transferred to a dithiol, such as dihydrolipoic acid.

Thiosulfate reductases are similar to rhodaneses in their mode of action, but do not catalyze reactions of the persulfide with cyanide. The best characterized of these is a 17 kDa enzyme from yeast for which a useful substrate is glutathione (Chauncey et al., 1987). It first binds the thiol (e.g. GSH) and then catalyzes its reaction with thiosulfate to form a persulfide (e.g. GSSH) and sulfite, both of which then dissociate from the enzyme. Formation of sulfide can then occur in an uncatalyzed reaction with a second thiol, resulting in a net reduction of thiosulfate to sulfite + sulfide. Alternatively the persulfide can react with cyanide in an uncatalyzed reaction, causing thiosulfate reductases to have a low apparent rhodanese activity (Westley, 1988).

The overall reaction for thiosulfate reduction catalyzed by either rhodanese or thiosulfate reductase is as follows:

$$S_2O_3^{2-} + 2\,RSH \rightarrow HS^- + HSO_3^- + RSSR \qquad (5)$$

Trüper and Pfennig (1966) demonstrated that *Tc. roseopersicina* could form sulfide from thiosulfate in the dark by flushing the H_2S produced from the cell suspension into a zinc acetate trap with a stream of argon, but were unable to detect sulfite. Sulfide was not produced by broken cells, possibly because of a need for cellular metabolism to produce the reduced thiol needed for thiosulfate reduction. Subsequently Khanna and Nicholas (1982) detected formation of both sulfite and sulfide from thiosulfate in a suspension of *Cb. vibrioforme* f. *thiosulfatophilum* cells by trapping them as complexes with N-ethylmaleimide.

Smith and Lascelles (1966) isolated a rhodanese from *Cm. vinosum* and reported that it had a K_m for thiosulfate of 0.6 mM and a K_m for cyanide of 20 mM. Hashwa (1975) later partially purified both a rhodanese and a thiosulfate reductase from the same species, and determined the molecular weight of the rhodanese to be 45,000 while that of the thiosulfate reductase was 90,000. The thiosulfate reductase had a K_m for dihydrolipoic acid of 1.25 mM and an apparent K_m for cyanide of 3.3 mM. Steinmetz and Fischer (1985) isolated two rhodaneses from *Cb. limicola* f. *thiosulfatophilum*, the main one of which was a 39 kDa protein with a K_m for thiosulfate of 0.25 mM and a K_m for cyanide of 5 mM. Previously, Trüper and Fischer (1982) had reported that neither rhodanese nor thiosulfate reductase occurred in the non-thiosulfate utilizing strains of *Cb. limicola* and *Cb. vibrioforme*, while both enzyme were present in the strains able to grow on thiosulfate (See also Trüper et al., 1988). This observation suggests a specific role for the enzymes isolated by Steinmetz and Fischer (1985) in thiosulfate oxidation.

It has sometimes been suggested that an enzyme like rhodanese might also mediate dissociation of thiosulfate into sulfur and sulfite. While it is possible that the sulfane sulfur of a persulfide could be transferred to a sulfur globule, Wood (1988) has pointed out that the equilibrium constant for the overall process strongly favors thiosulfate formation rather than dissociation. Consequently, dissociation of thiosulfate into sulfite and sulfur is unlikely, unless there is an input of energy.

It should be emphasized that rhodaneses (and thiosulfate reductases) are widely distributed enzymes, whose occurrence is not limited to bacteria that oxidize thiosulfate. They are versatile enzymes, and may be employed by the organisms that have them for a variety of purposes (Cerletti, 1986). Utilization of thiosulfate as a photosynthetic electron donor may well be one of these purposes. If they really are key enzymes in thiosulfate utilization, it is then necessary to study them further to see what distinguishes the enzymes (or their biochemical environment) in phototrophs able to oxidize thiosulfate from those in species unable to do so.

3. Thiosulfate Hydrolysis

Hydrolysis of thiosulfate to form sulfate and sulfide directly occurs with a $\Delta G_0'$ of −5.2 kcal/mol (Bak and Pfennig, 1987) and thus is thermodynamically favorable. Nearly 30 years ago, Trüper and Pfennig (1966) considered the possibility that an enzyme catalyzing this reaction might be involved in thiosulfate utilization by photosynthetic bacteria. However, an enzyme able to catalyze that reaction has never been isolated, and the possibility has received little further attention. The discovery of enzymes in thiobacilli able to hydrolyze S-SO_3^{2-} bonds in trithionate and tetrathionate (Meulenberg et al., 1992) is perhaps indirect evidence that a thiosulfate hydrolyzing enzyme in phototrophs should not yet be completely ruled out.

V. Considerations for Future Research

The metabolic versatility of some species of photosynthetic bacteria is a feature that makes them uniquely useful for studies on oxidative sulfur metabolism. It is also a feature that in my opinion has been underexploited in the work done so far. Several species of purple sulfur bacteria, notably *Tc. roseopersicina*, but also several others (see Overmann and Pfennig, 1992, for example) can grow as photoheterotrophs, photoautotrophs, or as chemolithotrophs. When growing as chemolithotrophs, these bacteria presumably utilize a linear electron transport chain from the inorganic sulfur compounds acting as electron donors to oxygen. The site of electron donation to this electron transport chain might logically be at either the ubiquinone or the cytochrome *c* level. In principle, inhibitors of electron transfer through cytochrome bc_1 complexes, which mediate electron transfer from ubiquinone to cytochrome *c*, could distinguish between these possibilities. Knaff and coworkers have partially characterized the components of the electron transport chain in microaerobically grown *Cm. vinosum* cells (Wynn et al., 1985) as well as the cytochrome bc_1 complex in phototrophically grown cells (Tan et al., 1993).

Measurements of oxygen uptake rates in the presence and absence of cytochrome bc_1 inhibitors have been done with thiobacilli (e.g. Lu and Kelly, 1988a; Beffa et al., 1992b), but not yet with chemolithotrophic purple sulfur bacteria.

In thiobacilli, energy generation and sulfur oxidation are coupled processes. This is a disadvantage for studying active transport of the oxidizable sulfur substrates. Phototrophs, on the other hand, obtain energy from light which is converted via cyclic electron transport to a transmembranous H^+ ion gradient and eventually ATP. This energy can be used for active transport of substrates like thiosulfate and sulfite in the absence of oxidation reactions if CO_2, the terminal electron acceptor in photosynthesis, is omitted from the cell suspension. (H_2S, at pH values near its pK_1 of 7.04, is uncharged and probably can freely diffuse through the membrane with slight accumulation on the more alkaline side of the membrane.) So far, nothing is known about transport of sulfur compounds into photosynthetic bacterial cells. Such information is of course relevant to the locations of the enzymes that oxidize these substrates.

The metabolic versatility of many purple sulfur bacteria (and of the α purple phototrophs able to grow photoautotrophically on sulfide and thiosulfate) makes dissimilatory sulfur metabolism optional. Thus mutants of these species lacking enzymes required for sulfur oxidation could grow as photoheterotrophs (or chemoorganotrophs, in some cases). Transposon mutagenesis, which was used to generate mutants of *Thiosphaera pantotropha* unable to oxidize thiosulfate and sulfide (Chandra and Friedrich, 1986; Mittenhuber et al., 1991) should also be applicable to sulfur-oxidizing purple bacteria. This approach might also be useful as a first step toward developing procedures for gene transfer and recombination in the purple sulfur bacteria.

Molecular biological methods can also be used to isolate and clone genes encoding known proteins thought to have key roles in sulfur metabolism. Examples are APS reductase from *Tc. roseopersicina* and sulfite reductase from *Cm. vinosum*. Partial protein sequences of purified subunits of these enzymes can be used to design oligonucleotide probes that can in turn be used to isolate and clone the genes encoding these enzymes. Complete protein sequences for the enzymes can be derived from the gene sequences. The cloned genes can also be used as probes in Northern blots to study regulation at the transcriptional level as well as for gene mapping. An interesting possibility is that many of the genes for enzymes with specific functions in sulfur oxidation may be clustered together to facilitate their joint regulation.

Eventually it should be possible to use the cloned genes to generate mutants lacking these enzymes by homologous recombination events (Saunders, 1992; Hunter and Mann, 1992). This might be done by interrupting the cloned gene by inserting an antibiotic resistance cartridge, using this construct to transform wild-type cells, selecting transformants for antibiotic resistance, and testing the colonies obtained for the presence of the enzyme in question by Western blotting (assuming that an antibody specific for the enzyme has been prepared). Sulfur oxidation pathways in bacteria are complex, and some reactions may be catalyzed by more than one enzyme. *Cb. limicola* f. *thiosulfatophilum* contains both flavocytochrome *c* and sulfide quinone reductase, for example. In such cases sulfur oxidation may still occur even when an enzyme normally involved in the process has been eliminated. If so, examining the physiological and biochemical consequences of the mutation may reveal what benefit is derived by the organism from the presence of parallel oxidation reactions. The situation is of course more straightforward when alternative reactions do not exist and mutagenesis eliminates the ability to oxidize a particular sulfur substrate.

It should be apparent from the preceding discussion that much remains to be learned about dissimilatory sulfur metabolism in photosynthetic bacteria. Nevertheless, new techniques in molecular genetics as well as in biochemistry and biophysics make it likely that future research in this field will be rewarded with definitive new results.

Acknowledgments

I thank Tina Leguijt and Yosepha Shahak for providing information prior to publication. This is publication number 196 from the Arizona State University Center for the Study of Early Events in Photosynthesis.

References

Amesz J (1991) Green photosynthetic bacteria and heliobacteria. In: Shively JM and Barton LL (eds) Variations in Autotrophic Life, pp 99–119. Academic Press, London

Arieli B Padan E and Shahak Y (1991) Sulfide-induced sulfide-quinone reductase activity in thylakoids of *Oscillatoria limnetica*. J Biol Chem 286: 104–111

Bacon M and Ingledew WJ (1989) The reductive reactions of *Thiobacillus ferrooxidans* on sulphur and selenium. FEMS Microbiol Lett 58: 189–194

Bak F and Pfennig N (1987) Chemolithotrophic growth of *Desulfovibrio sulfodismutans* sp. nov. by disproportionation of inorganic sulfur compounds. Arch Microbiol 147: 184–189

Bartsch RG (1978) Cytochromes. In: Clayton RK and Sistrom WR (eds) The Photosynthetic Bacteria, pp 249–279. Plenum Press, New York

Bartsch RG (1991) The distribution of soluble metallo-redox proteins in purple phototrophic bacteria. Biochim Biophys Acta 1058: 28–30

Beffa T, Berczy M and Aragno M (1992a) Inhibition of respiratory oxidation of elemental sulfur (S^0) and thiosulfate in *Thiobacillus versutus* and another sulfur-oxidizing bacterium. FEMS Microbiol Lett 90: 123–128

Beffa T, Fischer C and Aragno M (1992b) Respiratory oxidation of reduced sulfur compounds by intact cells of *Thiobacillus tepidarius* (type strain). Arch Microbiol 158: 456–458

Blankenship RE (1992) Origin and early evolution of photosynthesis. Photosynth Res 33: 91–111

Bramlett RN and Peck HD Jr. (1975) Some physical and kinetic properties of adenylyl sulfate reductase from *Desulfovibrio vulgaris*. J Biol Chem 250: 2979–2986

Brune DC (1989) Sulfur oxidation by phototrophic bacteria. Biochim. Biophys Acta 975: 189–221

Brune DC and Trüper HG (1986) Noncyclic electron transport in chromatophores from photolithotrophically grown *Rhodobacter sulfidophilus*. Arch Microbiol 145: 295–301

Caumette P, Baulaigue R and Matheron R (1988) Characterization of *Chromatium salexigens* sp. nov., a halophilic *Chromatiaceae* isolated from Mediterranean salinas. Syst Appl Microbiol 10: 284–292

Caumette P, Baulaigue R and Matheron R (1991) *Thiocapsa halophila* sp. nov., a new halophilic phototrophic purple sulfur bacterium. Arch Microbiol 155: 170–176

Cerletti P (1986) Seeking a better job for an under-employed enzyme: rhodanese. Trends Biochem Sci 11: 369–372

Chandra TS and Friedrich CG (1986) Tn5-induced mutations affecting sulfur-oxidizing ability (Sox) of *Thiosphaera pantophora*. J Bacteriol 166: 446–452

Chauncey TR, Uhteg LC and Westley J (1987) Thiosulfate reductase. Meth Enzymol 143: 350–354

Chen Z, Koh M, Van Dreissche G, Van Beeumen JJ, Bartsch RG, Meyer TE, Cusanovich MA and Mathews FS (1994) The structure of flavocytochrome *c* sulfide dehydrogenase from a purple phototrophic bacterium. Science 266: 430–432

Cusanovich MA, Meyer TE and Tollin G (1985) Flavo-cytochromes *c*: Transient kinetics of photoreduction by flavin analogues. Biochemistry 24: 1281–1287

Dahl C and Trüper HG (1989) Comparative enzymology of sulfite oxidation in *Thiocapsa roseopersicina* strains 6311, M1, and BBS under chemotrophic and phototrophic conditions. Z Naturforsch 44c: 617–622

Dannenberg S, Kroder M, Dillling W and Cypionka H (1992) Oxidation of H_2, organic compounds and inorganic sulfur compounds coupled to reduction of O_2 or nitrate by sulfate-reducing bacteria. Arch Microbiol 158: 93–99

Davidson MW, Gray GO and Knaff DB (1985) Interaction of *Chromatium vinosum* flavocytochrome *c*-552 with cytochromes *c* studied by affinity chromatography. FEBS Lett 187: 155–159

De Wit R and van Gemerden H (1990) Growth of the phototrophic purple sulfur bacterium *Thiocapsa roseopersicina* under oxic/anoxic regimens in the light. FEMS Microbiol Ecol 73: 69–76

Dolata MM, van Beeumen JJ, Ambler RP, Meyer TE and Cusanovich MA (1993) Nucleotide sequence of the heme subunit of flavocytochrome *c* from the purple photosynthetic bacterium *Chromatium vinosum*. J Biol Chem 268: 14426–14431

Fischer U (1984) Cytochromes and iron sulfur proteins in sulfur metabolism of phototrophic sulfur bacteria. In: Müller A and Krebs B (eds) Sulfur, Its Significance for Chemistry, for the Geo-, Bio- and Cosmosphere and Technology, pp 383–407. Elsevier, Amsterdam

Fischer U (1988) Sulfur in biotechnology. In: Rehm HJ and Reed G (eds) Biotechnology, Vol 6b pp 463–496. VCH Verlagsgesellschaft, Weinheim

Fischer U (1989) Enzymatic steps and dissimilatory sulfur metabolism by whole cells of anoxyphotobacteria. In: Saltzman ES and Cooper WJ (eds) Biogenic Sulfur in the Environment pp 262–279. American Chemical Society, Washington, DC

Fischer U and Ufken H (1991) Soluble electron transfer proteins of *Amoebobacter pedioformis*. Abstract 68A, VII International Symposium on Photosynthetic Prokaryotes, Amherst, MA

Friedrich CG, Meyer O and Chandra TS (1986) Molybdenum-dependent sulfur oxidation in facultatively lithoautotrophic thiobacteria. FEMS Microbiol Lett 37: 105–108

Fry B, Gest H and Hayes JM (1985) Isotope effects associated with the anaerobic oxidation of sulfite and thiosulfate by the photosynthetic bacterium, *Chromatium vinosum*. FEMS Microbiol Lett 27: 227–232

Fukumori Y and Yamanaka T (1979) Flavocytochrome *c* of *Chromatium vinosum*. Some enzymatic properties and subunit structure. J Biochem 85: 1405–1414

Gest H (1993) History of concepts of the comparative biochemistry of oxygenic and anoxygenic photosynthesis. Photosynth Res 35: 87–96

Gorlenko VM (1988) Ecological niches of green sulfur and green gliding bacteria. In: Olson JM, Ormerod JG, Amesz J, Stackebrandt E and Trüper HG (eds) Green Photosynthetic Bacteria, pp 257–267. Plenum Press, New York

Hansen TA and van Gemerden H (1972) Sulfide utilization by purple nonsulfur bacteria. Arch Microbiol 86: 49–56

Hashwa F (1975) Thiosulfate metabolism in some red phototrophic bacteria. Plant Soil 43: 41–47

Hol WGJ, Lijk LJ and Kalk KH (1983) The high resolution three-dimensional structure of bovine liver rhodanese. Fundam Appl Toxicol 3: 370–376

Hunter CN and Mann NH (1992) Genetic manipulation of photosynthetic prokaryotes. In: Mann NH and Carr NG (eds) Photosynthetic Prokaryotes, pp 153–179. Plenum Press, New York

Hurt EC and Hauska G (1984) Purification of membrane-bound cytochromes and a photoactive P840 protein complex of the green sulfur bacterium *Chlorobium limicola f. thiosulfatophilum*. FEBS Lett 168: 149–154

Imhoff JF (1992a) Taxonomy, physiology, and general ecology of anoxygenic phototrophic bacteria. In: Mann NH and Carr

NG (eds) Photosynthetic Prokaryotes, pp 53–92. Plenum Press, New York

Imhoff JF (1992b) The family Ectothiorhodospiraceae. In: Balows A, Trüper HG, Dworkin M, Harder W and Schleifer KH (eds) The Prokaryotes, Second Ed, pp 3222–3229. Springer-Verlag, New York

Imhoff JF and Trüper HG (1992) The genus *Rhodospirillum* and related genera. In: Balows A, Trüper HG, Dworkin M, Harder W and Schleifer KH (eds) The Prokaryotes, Second Ed, pp 2141–2155. Springer-Verlag, New York

Javor BJ, Wilmot DB and Vetter RD (1990) pH-Dependent metabolism of thiosulfate and sulfur globules in the chemolithotrophic marine bacterium *Thiomicrospira crunogena*. Arch Microbiol 154: 231–238

Kämpf C and Pfennig N (1980) Capacity of Chromatiaceae for chemotrophic growth. Specific respiration rates of *Thiocystis violacea* and *Chromatium vinosum*. Arch Microbiol 127: 125–135

Kämpf C and Pfennig N (1986) Chemoautotrophic growth of *Thiocystis violacea, Chromatium gracile* and *Cm. vinosum* in the dark at various oxygen concentrations. J Basic Microbiol 26: 517–531

Kelly DP (1989) Physiology and biochemistry of unicellular sulfur bacteria. In: Schlegel HG and Bowien B (eds) Autotrophic Bacteria, pp 193–217. Science Tech Publishers, Madison, WI

Khanna S and Nicholas DJD (1982) Utilization of tetrathionate and ^{35}S-labelled thiosulfate by washed cells of *Chlorobium vibrioforme* f. sp. *thiosulfatophilum*. J Gen Microbiol 128: 1027–1034

Khanna S and Nicholas DJD (1983) Substrate phosphorylation in *Chlorobium vibrioforme* f. sp. *thiosulfatophilum*. J Gen Microbiol 129: 1365–1370

Kirchoff J and Trüper HG (1974) Adenylyl sulfate reductase of *Chlorobium limicola*. Arch Microbiol 100: 115–120

Knaff DB and Buchanan BB (1975) Cytochrome *b* and sulfur bacteria. Biochim Biophys Acta 376: 549–560

Kobayashi K, Katsura E, Kondo T and Ishimoto M (1978) *Chromatium* sulfite reductase. I. Characterization of thiosulfate-forming activity at the cell extract level. J Biochem 84: 1209–1215

Kodama A and Mori T (1968) Studies on the metabolism of a sulfur-oxidizing bacterium. IV. Growth and oxidation of sulfur compounds in *Thiobacillus thiooxidans*. Plant Cell Physiol 9: 709–723

Kompantseva EJ (1985) *Rhodobacter euryhalinus* sp. nov., a new halophilic bacterial species. Mikrobiologiya 54: 974–982

Kompantseva EJ and Gorlenko VM (1984) A new species of moderately halophilic purple bacterium *Rhodospirillum mediosalinum* sp. nov. Microbiologiya 53: 775–783

Kondratieva EN (1989) Chemolithotrophy of phototrophic bacteria. In: Schlegel HG and Bowien B (eds) Autotrophic Bacteria, pp 283–287. Science Tech Publishers, Madison, WI

Kondratieva EN, Zhukov VG, Ivanovsky RN, Petushkova, Yu P and Monosov EZ (1976) The capacity of phototrophic sulfur bacterium *Thiocapsa roseopersicina* for chemosynthesis. Arch Microbiol 108: 287–292

Kondratieva EN, Ivanovsky RN and Krasilnikova EN (1981) Light and dark metabolism in purple sulfur bacteria. In: Skulachev VP (ed) Soviet Science Reviews Section D, Biology Reviews, pp 325–364.

Kusai A and Yamanaka T (1973) An NAD(P) reductase derived from *Chlorobium thiosulfatophilum*: purification and some properties. Biochim Biophys Acta 292: 621–633

Kusche WH (1985) Untersuchungen an Elektronentransportproteinen und zum Schweffelstoffwechsel in Ectothiorhodospiraceae. PhD thesis, University of Bonn

Lane DJ, Harrison Jr AP, Stahl D, Pace B, Giovannoni SJ, Olsen GJ and Pace NR (1992) Evolutionary relationships among sulfur- and iron-oxidizing eubacteria. J Bacteriol 174: 269–278

La Rivière JMW and Schmidt K (1992) Morphologically conspicuous sulfur-oxidizing eubacteria. In: Balows A, Trüper HG, Dworkin M, Harder W and Schleifer KH (eds) The Prokaryotes, Second Ed, pp 3934–3947. Springer-Verlag, New York

Le Faou A., Rajagopal BS, Daniels L and Fauque G (1990) Thiosulfate, polythionates and elemental sulfur assimilation and reduction in the bacterial world. FEMS Microbiol Rev 75: 351–382

Leguijt T (1993) Photosynthetic Electron Transfer in *Ectothiorhodospira*. PhD thesis, University of Amsterdam

Leguijt T, Engels PW, Crielaard W, Albracht SPJ and Hellingwerf KJ (1993) Abundance, subunit composition, redox properties and catalytic activity of the cytochrome bc_1 complex from alkalophilic and halophilic, photosynthetic members of the family *Ectothiorhodospiraceae*. J Bacteriol 175: 1629–1636

Lu WP and Kelly DP (1984) Properties and role of sulphite: cytochrome *c* oxidoreductase purified from *Thiobacillus versutus* (A2). J Gen Microbiol 130: 1683–1692

Lu WP and Kelly DP (1988a) Kinetic and energetic aspects of inorganic sulfur compound oxidation by *Thiobacillus tepidarius*. J Gen Microbiol 134: 865–876

Lu WP and Kelly DP (1988b) Cellular location and partial purification of the 'thiosulfate-oxidizing enzyme' and 'trithionate hydrolase' from *Thiobacillus tepidarius*. J Gen Microbiol 134: 877–885

Madigan MT and Brock TD (1975) Photosynthetic sulfide oxidation by *Chloroflexus aurantiacus*, a filamentous photosynthetic gliding bacterium. J Bacteriol 122: 782–784

McRee DE, Richardson DC, Richardson JS and Siegel LM (1986) The heme and Fe_4S_4 cluster in the crystallographic structure of *Escherichia coli* sulfite reductase. J Biol Chem 261: 10277–10281

Meulenberg R, Pronk JT, Hazeu W, Bos P and Kuenen JG (1992) Oxidation of reduced sulfur compounds by intact cells of *Thiobacillus acidophilus*. Arch Microbiol 157: 161–168

Meyer TE and Cusanovich MA (1989) Structure, function and distribution of soluble bacterial redox proteins. Biochim Biophys Acta 973: 1–28

Meyer TE, van Beeumen J, Holden HM, Rayment I, Bartsch RG and Cusanovich MA (1988) Kinetics of ligand binding, amino acid sequence, and crystallization of *Chlorobium* and *Chromatium* flavocytochrome *c*. In: Edmondson DE and McCormick D (eds) Flavins and Flavoproteins, pp 365–369. Walter de Gruyter, New York

Meyer TE, Bartsch RG and Cusanovich MA (1991) Adduct formation between sulfite and the flavin of phototrophic bacterial flavocytochromes *c*. Kinetics of sequential bleach, recolor, and rebleach of flavin as a function of pH. Biochemistry 30: 8840–8845

Mittenhuber G, Sonomoto K, Egert M and Friedrich CG (1991) Identification of the DNA region responsible for sulfur-

oxidizing ability of *Thiosphaera pantophora*. J Bacteriol 173: 7340–7344
Nelson DC (1989) Physiology and biochemistry of filamentous sulfur bacteria. In: Schlegel HG and Bowien B (eds) Autotrophic Bacteria, pp 219–238. Science Tech Publishers, Madison, WI
Neutzling O, Pfleiderer C and Trüper HG (1985) Dissimilatory sulphur metabolism in phototrophic 'non-sulphur' bacteria. J Gen Microbiol 131: 791–798
Nicolson GL and Schmidt GL (1971) Structure of the *Chromatium* sulfur particle and its protein membrane. J Bacteriol 105: 1142–1148
Nixon A and Norris PR (1992) Autotrophic growth and inorganic sulfur compound oxidation by *Sulfolobus* sp. in chemostat culture. Arch Microbiol 157: 155–160
Oh JK and Suzuki I (1980) Respiration in chemolithotrophs oxidizing sulfur compounds. In: Knowles CJ (ed) Diversity of Bacterial Respiratory Systems, Vol 2, pp 113–137. CRC Press, Boca Raton
Okumura N, Shimada K and Matsuura K (1994) Photo-oxidation and rereduction of membrane-bound and soluble cytochrome *c* in the green sulfur bacterium *Chlorobium tepidum*. Photosynth Res 41: 125–134
Oren A, Kessel M and Stackebrandt E (1989) *Ectothiorhodospira marismortui* sp. nov., an obligately anaerobic, moderately halophilic purple sulfur bacterium from a hypersaline sulfur spring on the shore of the Dead Sea. Arch Microbiol 151: 524–529
Overmann J and Pfennig N (1989) *Pelodictyon phaeoclathratiforme* sp. nov., a new brown-colored member of the Chlorobiaceae forming net-like colonies. Arch Microbiol 152: 401–406
Overmann J and Pfennig N (1992) Continuous chemotrophic growth and respiration of Chromatiaceae species at low oxygen concentrations. Arch Microbiol 158: 59–67
Peck Jr HD and Bramlett RN (1982) Flavoproteins in sulfur metabolism. In: Massey V and Williams CH (eds) Flavins and Flavoproteins, pp 851–858. Elsevier, Amsterdam
Pierson BK and Castenholz RW (1992) The family Chloroflexaceae. In: Balows A, Trüper HG, Dworkin M, Harder W and Schleifer KH (eds) The Prokaryotes, Second Ed, pp 3754–3774. Springer-Verlag, New York
Postgate JR (1984) The Sulfate Reducing Bacteria. Cambridge University Press, Cambridge
Pronk JT, Meulenberg R, Hazeu W, Bos P and Kuenen JG (1990) Oxidation of reduced inorganic sulfur compounds by acidophilic thiobacilli. FEMS Microbiol Rev 75: 293–306
Remsen CC (1978) Comparative subcellular architecture of photosynthetic bacteria. In: Clayton RK and Sistrom WR (eds) The Photosynthetic Bacteria, pp 31–60. Plenum Press, New York
Roy AB and Trudinger PA (1970) The Biochemistry of Inorganic Compounds of Sulfur. Cambridge University Press, Cambridge
Saunders VA (1992) Genetics of the photosynthetic prokaryotes. In: Mann NH and Carr NG (eds) Photosynthetic Prokaryotes, pp 121–152. Plenum Press, New York
Schedel M and Trüper HG (1979) Purification of *Thiobacillus denitrificans* siroheme sulfite reductase and investigation of some molecular and catalytic properties. Biochim Biophys Acta 568: 454–467
Schedel M and Trüper HG (1980) Anaerobic oxidation of thiosulfate and elemental sulfur in *Thiobacillus denitrificans*. Arch Microbiol 124: 205–210
Schedel M, Vanselow M and Trüper HG (1979) Siroheme sulfite reductase isolated from *Chromatium vinosum*. Purification and investigation of some of its molecular and catalytic properties. Arch Microbiol 121: 29–36
Schiff JA and Fankhauser H (1981) Assimilatory sulfate reduction. In: Bothe H and Trebst A (eds) Biology of Inorganic Nitrogen and Sulfur, pp 153–168. Springer-Verlag, Berlin
Schmidt GL, Nicolson GL and Kamen MD (1971) Composition of the sulfur particle of *Chromatium vinosum* strain D. J Bacteriol 105: 1137–1141
Schwenn JD and Biere M (1979) APS reductase activity in chromatophores of *Chromatium vinosum* strain D. FEMS Microbiol Lett 6: 19–22
Shahak Y, Arieli B, Padan E and Hauska G (1992a) Sulfide quinone reductase (SQR) activity in *Chlorobium*. FEBS Lett 299: 127–130
Shahak Y, Arieli B, Hauska G, Herrmann I and Padan E (1992b) Isolation of sulfide-quinone reductase (SQR) from prokaryotes. Phyton 32: 133–137
Shahak Y, Hauska G, Herrmann I, Arieli B, Taglicht D and Padan E (1992c) Sulfide-quinone reductase (SQR) drives anoxygenic photosynthesis in prokaryotes. In: Murata N (ed) Research in Photosynthesis, Vol 2, pp 483–486. Kluwer Academic Publishers, Dordrecht
Shill DA and Wood PM (1985) Light-driven reduction of oxygen as a method for studying electron transport in the green photosynthetic bacterium *Chlorobium limicola*. Arch Microbiol 143: 82–87
Siegel LM, Rueger DC, Barber MJ, Krueger RJ, Orme-Johnson NR and Orme-Johnson WH (1982) *Escherichia coli* sulfite reductase hemoprotein subunit. Prosthetic groups, catalytic parameters, and ligand complexes. J Biol Chem 257: 6343–6350
Smith AJ (1966) The role of tetrathionate in the oxidation of thiosulfate by *Chromatium* sp. strain D. J Gen Microbiol 42: 371–380
Smith AJ and Lascelles J (1966) Thiosulfate metabolism and rhodanese in *Chromatium* sp. strain D. J Gen Microbiol 42: 357–370
Stackebrandt E, Murray RGE and Trüper HG (1988) *Proteobacteria* classis nov., a name for the phylogenetic taxon that included the 'purple bacteria and their relatives'. Int J Syst Bacteriol 38: 321–325
Steinmetz MA and Fischer U (1985) Thiosulfate sulfur transferases (rhodaneses) of *Chlorobium vibrioforme* f. *thiosulfatophilum*. Arch Microbiol 142: 253–258
Steinmetz MA, Trüper HG and Fischer U (1983) Cytochrome c-555 and iron-sulfur proteins of the non-thiosulfate-utilizing green sulfur bacterium *Chlorobium vibrioforme*. Arch Microbiol 135: 186–190
Steudel R (1984) Elemental sulfur and related homocyclic compounds and ions. In: Müller A and Krebs B (eds) Sulfur: Its Significance for Chemistry, for the Geo-, Bio-, and Cosmosphere and Technology, pp 3–37. Elsevier, Amsterdam
Steudel R (1989) On the nature of the 'elemental sulfur' (S^0) produced by sulfur-oxidizing bacteria - a model for S^0 globules. In: Schlegel HG and Bowien B (eds) Autotrophic Bacteria, pp 289–303. Science Tech Publishers, Madison, WI

Steudel R, Holdt G, Göbel T and Hazeu W (1987) Chromatographic separation of higher polythionates $S_nO_6^{2-}$ (n=3···22) and their detection in cultures of *Thiobacillus ferrooxidans*; molecular composition of bacterial sulfur secretions. Angew Chem Int Ed Engl 26: 151–153

Steudel R, Göbel T and Holdt G (1988) The molecular composition of hydrophilic sulfur sols prepared by acid decomposition of thiosulfate. Z Naturforsch 43b: 203–218

Steudel R, Holdt G, Visscher PT and van Gemerden H (1990) Search for polythionates in cultures of *Chromatium vinosum* after sulfide incubation. Arch Microbiol 153: 432–437

Sugio T, Mizunashi W, Inagaki K and Tano T (1987) Purification and some properties of sulfur: ferric ion oxidoreductase from *Thiobacillus ferrooxidans*. J Bacteriol 169: 4916–4922

Sugio T, Katagiri T, Inagaki K and Tano T (1989) Actual substrate for elemental sulfur oxidation by sulfur: ferric ion oxidoreductase purified from *Thiobacillus ferrooxidans*. Biochim Biophys Acta 973: 250–256

Takakuwa S (1992) Biochemical aspects of microbial oxidation of inorganic sulfur compounds. In: Oae S and Okuyama T (eds) Organic Sulfur Chemistry: Biochemical Aspects, pp 1–43. CRC Press, Boca Raton

Tan J and Cowan JA (1991) Enzymatic redox chemistry: A proposed reaction pathway for the six-electron reduction of SO_3^{2-} to S^{2-} by the assimilatory-type sulfite reductase from *Desulfovibrio vulgaris* (Hildenborough). Biochemistry 30: 8910–8917

Tan J, Corson GE, Chen YL, Garcia MC, Güner S and Knaff DB (1993) The ubiquinol: cytochrome c_2/c oxidoreductase of *Chromatium vinosum*. Biochim Biophys Acta 1144: 69–76

Thauer RK (1989) Energy metabolism of sulfate-reducing bacteria. In: Schlegel HG and Bowien B (eds) Autotrophic Bacteria, pp 397–413. Science Tech Publishers, Madison, WI

Thauer RK, Jungermann K and Decker K (1977) Energy conservation in chemotrophic anaerobic bacteria. Bacteriol Rev 41: 100–180

Then J and Trüper HG (1984) Utilization of sulfide and elemental sulfur by *Ectothiorhodospira halochloris*. Arch Microbiol 139: 295–298

Tollin G, Meyer TE and Cusanovich MA (1982) Intramolecular electron transfer in *Chlorobium thiosulfatophilum* flavocytochrome *c*. Biochemistry 21: 3849–3856

Trumpower BL (1990) Cytochrome bc_1 complexes of microorganisms. Microbiol Rev 54: 101–129

Trüper HG (1984) Phototrophic bacteria and their sulfur metabolism. In: Müller A and Krebs B (eds) Sulfur, Its Significance for Chemistry, for the Geo-, Bio-, and Cosmosphere and Technology, pp 367–382, Elsevier, Amsterdam

Trüper HG (1987) Phototrophic bacteria (an incoherent group of prokaryotes). A taxonomic versus phylogenetic survey. Microbiologia SEM 3: 71–89

Trüper HG (1989) Physiology and biochemistry of phototrophic bacteria. In: Schlegel HG and Bowien B (eds) Autotrophic Bacteria, pp 267–281. Science Tech Publishers, Madison, WI

Trüper HG and Fischer U (1982) Anaerobic oxidation of sulfur compounds as electron donors for bacterial photosynthesis. Phil Trans R Soc Lond B 298: 529–542

Trüper HG and Imhoff JF (1992) The genera *Rhodocyclus* and *Rubrivivax*. In: Balows A, Trüper HG, Dworkin M, Harder W and Schleifer KH (eds) The Prokaryotes, Second Ed, pp 2556–2561. Springer-Verlag, New York

Trüper HG and Pfennig N (1966) Sulphur metabolism in Thiorhodaceae. III. Storage and turnover of thiosulphate sulphur in *Thiocapsa floridana* and *Chromatium* species. Antonie van Leeuwenhoek 32: 261–276

Trüper HG and Rogers LA (1971) Purification and properties of adenylyl sulfate reductase from the phototrophic sulfur bacterium, *Thiocapsa roseopersicina*. J Bacteriol 108: 1112–1121

Trüper HG, Lorenz C, Schedel M and Steinmetz M (1988) Metabolism of thiosulfate in *Chlorobium*. In: Olson JM, Ormerod JG, Amesz J, Stackebrandt E and Trüper HG (eds) Green Photosynthetic Bacteria, pp 189–200. Plenum Press, New York

Van Beeumen J, van Bun S, Meyer TE, Bartsch RG and Cusanovich MA (1990) Complete amino acid sequence of the cytochrome subunit and amino-terminal sequence of the flavin subunit of flavocytochrome *c* (sulfide dehydrogenase) from *Chlorobium thiosulfatophilum*. J Biol Chem 17: 9793–9799

Van Beeumen JJ, Demol H, Samyn B, Bartsch RG, Meyer TE, Dolata MM and Cusanovich MA (1991) Covalent structure of the diheme cytochrome subunit and amino terminal sequence of the flavoprotein subunit of flavocytochrome *c* from *Chromatium vinosum*. J Biol Chem 266: 12921–12931

Van Gemerden H (1987) Competition between purple sulfur bacteria and green sulfur bacteria: Role of sulfide, sulfur and polysulfides. Acta Academiae Aboensis 47: 13–27

Visscher PT and Taylor BF (1993) Organic thiols as organolithotrophic substrates for growth of phototrophic bacteria. Appl Environ Microbiol 59: 93–96

Visscher PT and Van Gemerden H (1991) Photoautotrophic growth of *Thiocapsa roseopersicina* on dimethyl sulfide. FEMS Microbiol Lett 81: 247–250

Visscher PT, Nijburg JW and van Gemerden H (1990) Polysulfide utilization by *Thiocapsa roseopersicina*. Arch Microbiol 155: 75–81

Wahlund TM, Woese CR, Castenholz RW and Madigan MT (1991) A thermophilic green sulfur bacterium from New Zealand hot springs, *Chlorobium tepidum* sp. nov. Arch Microbiol 156: 81–90

Wang S (1991) Purification and characterization of APS reductase from the photosynthetic bacterium *Thiocapsa roseopersicina*. MS Thesis, Arizona State University

Wang X, Modak HV and Tabita FR (1993) Photolithoautotrophic growth and control of CO_2 fixation in *Rhodobacter sphaeroides* and *Rhodospirillum rubrum* in the absence of ribulose bisphosphate carboxylase-oxygenase. J Bacteriol 175: 7109–7114

Wermter U and Fischer U (1983) Cytochromes and anaerobic sulfide oxidation in the purple sulfur bacterium *Chromatium warmingii*. Z Naturforsch 38c: 960–967

Westley AM and Westley J (1991) Biological sulfane sulfur. Anal Biochem 195: 63–67

Westley J (1988) Mammalian cyanide detoxification with sulfane sulfur. In: Evered D and Harnett S (eds) Cyanide Compounds in Biology, pp 201–218. John Wiley and Sons, Chichester

Woese CR (1987) Bacterial evolution. Microbiol Rev 51: 221–271

Wood PM (1988) Chemolithotrophy. In Anthony C. (ed) Bacterial

Energy Transduction, pp 183–230. Academic Press, London
Wynn RM, Kämpf C, Gaul DF, Choi WK, Shaw RW and Knaff DB (1985) The membrane-bound electron-transfer components of aerobically grown *Chromatium vinosum*. Biochim Biophys Acta 808: 85–93
Yamanaka T and Kusai A (1976) The function and some molecular features of cytochrome *c*-553 derived from *Chlorobium thiosulfatophilum*. In: Singer TP (ed) Flavins and Flavoproteins, pp 292–301. Elsevier, Amsterdam
Young LJ and Siegel LM (1988) Activated conformers of *Escherichia coli* sulfite reductase heme protein subunit. Biochemistry 27: 4991–4999

Chapter 40

Carbon Metabolism in Green Bacteria

Reidun Sirevåg
Department of Biology, Division of Molecular Cell Biology, University of Oslo, Box 1050, Blindern, 0316 Oslo, Norway

Summary

The only common feature of the two families of green bacteria, Chlorobiaceae (green sulfur bacteria) and Chloroflexaceae (green gliding bacteria), is their type of light harvesting chlorophyll and the organization of these pigments into chlorosomes. In most other respects, including metabolism, photosynthetic apparatus and phylogeny, they are very different and far apart. Each of the two most studied genera possesses a unique pathway for autotrophic fixation of CO_2: the reductive tricarboxylic acid cycle used by *Chlorobium* and the newly discovered 3-hydroxypropionate cycle used by *Chloroflexus*.

The two pathways are described in detail and the experimental evidence for their existence is given. Likewise are the experimental approaches used to elucidate these pathways discussed and evaluated. The biochemical reactions participating in the pathways are considered in relation to the nature of the photosynthetic apparatus of the organisms and their ecology.

The current knowledge of the metabolism of organic compounds in *Chlorobium* and *Chloroflexus* is reviewed, and the accumulation of glycogen and its breakdown in *Chlorobium* is discussed in detail.

Assuming that life originated in a primordial sea of abiotically produced organic molecules, and taking into account phylogenetic relationships based on 16S rRNA analysis, the possible development by retroevolution of the metabolic pathways used by the green sulfur bacteria and the green gliding bacteria is discussed.

R. E. Blankenship, M. T. Madigan and C. E. Bauer (eds): Anoxygenic Photosynthetic Bacteria, pp. 871–883.

I. Introduction

The members of the two families of green bacteria, the green sulfur bacteria (*Chlorobiaceae*) as represented by *Chlorobium*, and the green gliding bacteria (*Chloroflexaceae*) as represented by *Chloroflexus*, can grow photoautotrophically with CO_2 as the sole carbon source. In addition they are able to utilize organic compounds for growth. Whereas *Chlorobium* is able to assimilate organic compounds to a limited degree and only in the presence of CO_2 and an inorganic electron donor, *Chloroflexus* grows on a variety of carbon sources both anaerobically in the light and aerobically in the dark.

A profound difference between *Chlorobium* and *Chloroflexus* is the nature of their photosynthetic apparatus. In spite of the fact that they both contain chlorosomes and have the same types of bacteriochlorophyll, *Chlorobium* possesses a PSI type reaction center, whereas that of *Chloroflexus* is of the PS II type. Because of the very low (~ –0. 9 V) redox potential of the primary acceptor in *Chlorobium*, the reaction center is able to photoreduce ferredoxin and therefore reducing power for the reduction of pyridine dinucleotides is easily available. In *Chloroflexus* the redox potential of the primary acceptor is less negative (~ –0. 5V), and as a result *Chloroflexus* has to reduce pyridine dinucleotides by reverse electron transport in a similar way as the purple bacteria. A detailed discussion of the reaction centers of green sulfur bacteria and purple bacteria, to which the *Chloroflexus* reaction center is most clearly related, appears in Chapters 30 and 23 by Feiler and Hauska, and Lancaster et al., respectively.

The nature and availability of reducing power determine to a large extent the biochemical reactions which are possible in a cell. Thus the differences in the photosynthetic apparatus of *Chlorobium* and *Chloroflexus* are reflected not only in their carbon metabolism but also in their ecology and evolution.

II. Mechanisms of Autotrophic Carbon Dioxide Assimilation

The formation of organic material from carbon dioxide by autotrophic organisms can be expressed in the following way:

$$CO_2 \xrightarrow{4(H) + ATP + enzymes} (CH_2O) + H_2O$$

where (CH_2O) is a rough expression of cell material.

The ability to grow autotrophically is widespread in nature and is the common feature of higher plants and algae. This mode of life is also found among members of several groups of prokaryotes, notably the phototrophic prokaryotes.

In phototrophic organisms, the ATP and reducing power needed to form organic material from carbon dioxide are produced as a result of electron transport initiated by light reactions. In the case of higher plants, algae and cyanobacteria, the ultimate H-donor in photosynthesis is H_2O, whereas reduced sulfur compounds or molecular hydrogen serve as reductants for green and purple bacteria.

During evolution of autotrophic organisms, several routes of assimilation of CO_2 have developed, each with its own biochemical reactions requiring its own enzymes and reducing power of a specific nature. The most widespread mechanism is the reductive pentose phosphate cycle or Calvin cycle (Bassham 1979) which is found in higher plants, algae and most known groups of autotrophic prokaryotes. In green bacteria, however, two alternative cyclic mechanisms of CO_2 assimilation have evolved: the reductive tricarboxylic acid (TCA) cycle in *Chlorobium,* first proposed by Evans et al. (1966), and the 3-hydroxypropionate pathway suggested by Holo (1989) for *Chloroflexus*. Recently the latter pathway has been supported by the biochemical studies in the laboratory of Georg Fuchs (Eisenreich et al., 1993; Strauss et al., 1993; Strauss and Fuchs, 1993), but has also been contested by Ivanovsky et al. (1993).

A. Experimental Approaches Employed in Studying the Path of Carbon Dioxide in Autotrophic Organisms

Proof for the operation of a cyclic biochemical mechanism should be based on several lines of evidence. Among these are (i) demonstration of activity of the necessary enzymes in cell-free extracts in amounts consistent with the rate of corresponding metabolism in intact cells; (ii) the appearance of radioactivity in intermediate compounds in a manner consistent with the operation of the cycle when

Abbreviations: A. – Anacystis; Cb. – Chlorobium; Cf. – Chloroflexus; Fd – ferredoxin; NMR – nuclear magnetic resonance; 3-PGA – 3-phosphoglyceric acid; PS – Photosystem; RuBisCO – ribulose 1,5-bisphosphate carboxylase; TCA – tricarboxylic acid radioactive substrate is added to the cells; (iii) inhibitors specific for enzymes of the cycle should have the expected effect in terms of lowered overall rates and accumulation of intermediates.

In addition to *enzymic studies* and the effect of specific *inhibitors* on known reactions, *short-term labeling* experiments are useful in studying the path of carbon in autotrophic organisms. Such experiments were originally employed by Calvin and coworkers (see Bassham, 1979) when they elucidated the path of carbon in algae by exposing cells to $^{14}CO_2$ for various periods of time before they were killed and analyzed for stable, radioactive products. Cells which have a functional reductive pentose phosphate cycle, give a characteristic pattern of distribution of the radioisotope. After very short exposures (1s) to $^{14}CO_2$, most of the radioactivity is recovered in the first stable product of CO_2-fixation, 3-PGA. The relative amount of radioactivity in this compound decreases with time as other intermediates in the cycle are formed from it. Thus, when the percentage of total activity in 3-PGA is plotted against time, a characteristic negative slope is obtained after the initial rise. However, it should be pointed out that if an organism has more than one CO_2 fixation reaction, as is the case for green bacteria, the picture obtained from such experiments is much more complex and difficult to interpret.

Long-term labeling experiments, combined with subsequent degradation of individual intermediates, have also proven a valuable experimental tool to elucidate the mechanism of CO_2 fixation in *Chlorobium* as well as in *Chloroflexus*.

Carbon isotope fractionation studies are based on the fact that some enzymes which use CO_2 as substrate, i. e. ribulose 1, 5-bisphosphate carboxylase (RuBisCO) show a much greater preference for $^{12}CO_2$ than for the slightly heavier $^{13}CO_2$ (Whelan et al., 1973). Thus organisms which use different carboxylation reactions for initial CO_2-fixation can be distinguished by comparing their carbon isotope composition after autotrophic growth.

The fact that the genes coding for RuBisCO have been isolated from various organisms and appear to be conserved in evolution, makes these genes and their transcripts valuable markers in the identification of carboxylation reactions in various organisms.

B. Carbon Dioxide Fixation in Chlorobium

It was long assumed that the mechanism of CO_2 fixation in *Chlorobium* was the reductive pentose phosphate cycle. This assumption was based on reports that RuBisCO was present in cell free extracts of *Cb. limicola* (Smillie et al., 1962; Tabita et al., 1974). However, it is now firmly established that the mechanism of CO_2 fixation in *Chlorobium* is the reductive tricarboxylic acid (TCA) cycle, which is depicted in Fig. 1. This cycle was proposed by Evans et al. (1966) on the basis of the discovery in phototrophic bacteria and other anaerobes of two new ferredoxin dependent carboxylation reactions:

acetyl-CoA + CO_2 + Fd_{red} $\rightarrow$ pyruvate + CoA + Fd_{ox}

succinyl-CoA + CO_2 + Fd_{red} $\rightarrow$ α-oxoglutarate + CoA + Fd_{ox}

catalyzed by pyruvate synthase and α-oxoglutarate synthase, respectively (Evans and Buchanan, 1965; Buchanan and Evans, 1965). These two reactions make possible a reversal of the reactions of the TCA cycle and result in the net synthesis of one molecule of acetyl-CoA from two molecules of carbon dioxide. Of the acetyl-CoA formed, some is removed for biosynthesis while some reenters the cycle.

One major obstacle to the proposed reductive TCA cycle was the report that a key enzyme, citrate lyase, could not be detected in cell free extracts of *Chlorobium* (Beuscher and Gottschalk, 1972). However, the finding of an ATP-linked citrate lyase in *Chlorobium* (Ivanovsky et al. 1980) and evidence from long-term labeling experiments combined with degradation of radioactive intermediates (Fuchs et al., 1980a,b), strongly supports its presence.

Originally the reductive TCA cycle was thought to be an additional mechanism to the reductive pentose phosphate cycle in *Chlorobium*, and it was suggested that its main function was to form precursors for the synthesis of amino acids, lipids and porphyrins while the reductive pentose phosphate cycle accounted for carbohydrate formation (Evans et al., 1966). However, inconsistency of data regarding the presence of RuBisCO (Tabita et al., 1974; Buchanan and Sirevåg, 1976; Sirevåg et al., 1977; Takabe and Akazawa, 1977) made the role of the reductive pentose phosphate cycle questionable. Several lines of evidence, including short-term labeling experiments (Sirevåg, 1974), use of washed cells and inhibitors (Sirevåg and Ormerod, 1970a,b), long-term labeling experiments (Fuchs et al., 1980b) and isotope fractionation studies (Quandt et al., 1977, 1978;

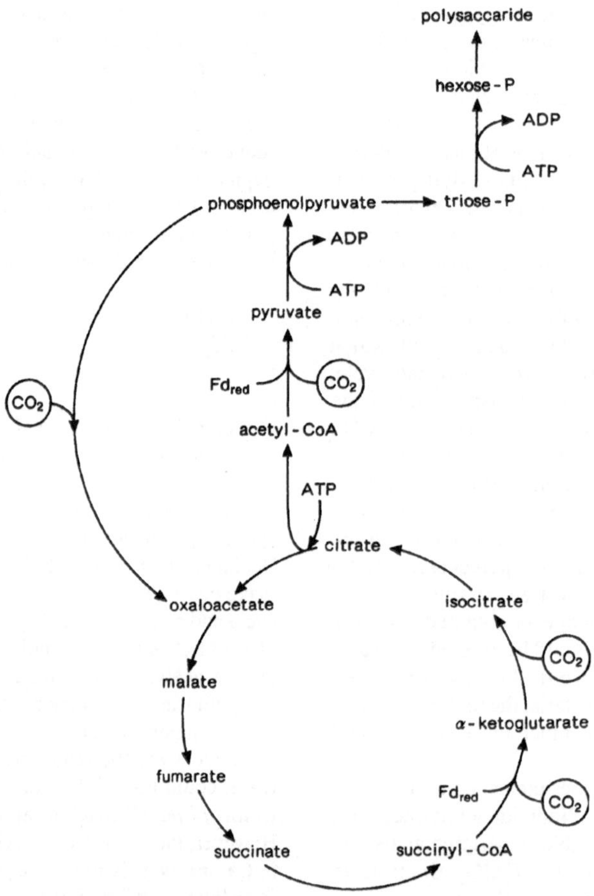

Fig. 1. The reductive tricarboxylic acid cycle of *Chlorobium* as proposed by Evans et al. (1966).

Sirevåg et al., 1977), indicate that the reductive TCA cycle is the sole mechanism of CO_2 fixation in this organism and that the product of the cycle, acetyl-CoA, can be directly converted to carbohydrate (Sirevåg, 1975). Furthermore, when the gene for RuBisCO from *Rhodospirillum rubrum* was used as a probe, no hybridization with *Chlorobium* DNA was detected (Shively et al. 1986). The same negative results were obtained when the RuBisCO genes from *Anacystis nidulans* were used.

C. Carbon Dioxide Fixation in Chloroflexus

Like most phototrophic bacteria, *Chloroflexus* is able to grow photoautotrophically with CO_2 as the sole carbon source. Slow photoautotrophic growth on sulfide was first reported by Madigan et al. (1974), and later sulfide was shown to be an electron donor in photosynthesis of *Chloroflexus* (Madigan and Brock, 1975, 1977). *Cf. aurantiacus* is also able to use molecular hydrogen for reduction of CO_2 and grows well photoautotrophically on H_2 and CO_2 (Sirevåg and Castenholz, 1979, Holo and Sirevåg, 1986).

Stimulation by ribulose-1, 5-bisphosphate of $^{14}CO_2$-uptake has been reported using cell-free extracts of photoheterotrophically grown *Cf. aurantiacus* (Madigan and Brock, 1977; Sirevåg and Castenholz, 1979). However, when cell-free extracts of cells grown photoautotrophically with H_2 as an electron donor were tested for RuBisCO activity, this enzyme was not detected. Furthermore, short-term labeling experiments with $^{14}CO_2$, use of inhibitors in combination with various ^{14}C-labeled substrates, and $^{13}C/^{12}C$ isotope fractionation studies indicate that the Calvin cycle does not operate in *Cf. aurantiacus* (Holo and Sirevåg, 1986; Holo and Grace, 1987; Holo, 1989). This is in agreement with the findings that *Cf. aurantiacus*-DNA did not hybridize with RuBisCO genes from either *Rs. rubrum* or *A. nidulans* (Shively et al., 1986)

The possibility that *Chloroflexus* might have a reductive TCA cycle similar to that of *Chlorobium* for fixation of CO_2 has been examined. It was found that one of the key enzymes of this cycle, pyruvate synthase, which catalyzes the formation of pyruvate from acetyl-CoA and CO_2 (see Fig. 1) is present in *Cf. aurantiacus*, whereas the other enzymes specific for the complete reductive TCA cycle were lacking (Holo and Sirevåg, 1986). Thus in *Cf. aurantiacus* acetyl-CoA is synthesized from CO_2 by a mechanism different from that of *Chlorobium*.

In order to investigate whether acetyl-CoA was formed from CO_2 by a $C_1 + C_2$ mechanism or by a cyclic pathway in *Cf. aurantiacus*, long-term labeling experiments were performed in which autotrophically grown cells were allowed to metabolize ^{13}C-labeled acetate under conditions where metabolism of acetate by the tricarboxylic acid cycle and the glyoxylate cycle was inhibited. When the labeling pattern was analyzed by ^{13}C-NMR, the results indicated a novel metabolic cycle in *Cf. aurantiacus* in which acetyl-CoA acted both as a product and an intermediate (Holo and Grace, 1987). Later Holo (1989) found that 3-hydroxypropionate was excreted by cells grown photoautotrophically under conditions which favored accumulation of metabolic intermediates. On this basis he suggested that under photoautotrophic conditions, *Cf. aurantiacus* converts acetyl-CoA to 3-hydroxypropionate which might serve as an intermediate in a CO_2 fixation pathway where it gives rise to malate and succinate.

Support for this hypothesis was obtained by Strauss et al. (1992) who observed that autotrophically grown cells of *Chloroflexus* excreted succinate and large amounts of 3-hydroxypropionate in the late exponential to stationary phase. When these workers labeled growing cultures of *Cf. aurantiacus* with [1, 4–^{13}C]succinate and analyzed the various cell constituents by ^{13}C-NMR spectroscopy to determine the distribution of ^{13}C in various cell constituents, their results were consistent with the role of 3-hydroxypropionate as an intermediate metabolite in a CO_2 fixation mechanism.

The operation in *Cf. aurantiacus* of a cyclic CO_2 fixation mechanism involving 3-hydroxypropionate has recently been contested by Russian workers (Kondratieva et al., 1992; Ivanovsky et al., 1993) who have proposed an alternative mechanism shown in Fig. 2. The key enzyme of this cycle is an ATP-dependent malate lyase (EC 4. 1. 3. 24) and the net product of the cycle is glyoxylate. These authors explain the finding by Holo (1989) that 3-hydroxypropionate is excreted by *Cf. aurantiacus* by suggesting that this compound is a dead-end product synthesized from acetate by a side path. However, a serious objection to the mechanism proposed by Ivanovsky et al. (1992) is that it is not in accord with the labeling pattern obtained previously (Holo and Sirevåg, 1986; Holo and Grace, 1987; Holo, 1989) and more recently (Strauss et al., 1992; Eisenreich et al., 1993; Strauss and Fuchs, 1993).

The possibility that 3-hydroxypropionate might be an intermediate in a cyclic CO_2 fixation mechanism has very recently been tested further by Fuchs and coworkers by long-term labeling experiments (Eisenreich et al., 1993). The relative ^{13}C content of individual carbon atoms in various amino acids and nucleosides was determined by ^{13}C-NMR spectroscopy after growing *Cf. aurantiacus* in the presence of [1–^{13}C]3-hydroxypropionate, [1–^{13}C]acetate, or [2–^{13}C]acetate in addition to $^{12}CO_2$. From the labeling patterns obtained, the ^{13}C labeling of central metabolic intermediates like trioses and dicarboxylic acids were deduced. These experiments showed clearly that when growing cells of *Cf. aurantiacus* were fed [1–^{13}C]3-hydroxypropionate over several generations, this compound served as a precursor for all cell compounds, a finding which can only be explained if 3-hydroxypropionate functions as an intermediate in the central metabolism of the organism. Their data obtained with labeled acetate also confirmed the possible role of 3-hydroxypropionate as an intermediate in the novel cyclic mechanism of CO_2 fixation which is shown in Fig. 3.

To test the operation of this cycle further, Strauss and Fuchs (1993) undertook enzymic studies and were able to demonstrate activity of all the enzymes necessary for the operation of the novel CO_2 fixation cycle. In this 3-hydroxypropionate cycle, acetyl-CoA is carboxylated to malonyl-CoA and then reductively converted via 3-hydroxypropionate to propionyl-CoA. This in turn is carboxylated and

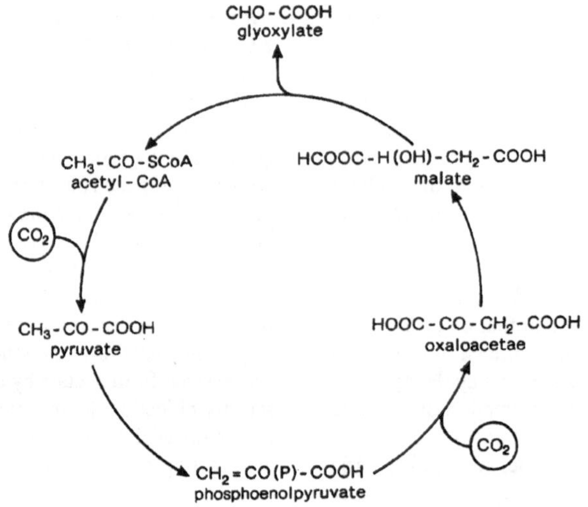

Fig. 2. The pathway for autotrophic CO_2 assimilation in *Chloroflexus aurantiacus* proposed by Ivanovsky et al. (1993)

converted via succinyl-CoA to malyl-CoA, which is cleaved to acetyl-CoA and glyoxylate in a reaction catalyzed by malyl-CoA lyase (EC 4. 1. 3. 24). Acetyl-CoA reenters the cycle whereas glyoxylate is its net fixation product. Thus, the two different autotrophic CO_2 fixation mechanisms which have been proposed for *Cf. aurantiacus* both have the same net product, glyoxylate (Figs. 2 and 3).

How glyoxylate might be further metabolized into important central metabolic intermediates has not yet been fully elucidated. However, on the basis of enzymic studies, Ivanovsky et al. (1993) have suggested that glyoxylate is converted into 3-PGA in reactions involving the serine and the glycine pathway.

III. The Assimilation of Organic Carbon

In chemoorganotrophic organisms, organic substrates have a dual role in that they serve both as an energy source and as a carbon source. This is not necessary in phototrophic bacteria where sufficient ATP as well as reducing power are provided by the reactions of their photosynthetic apparatus. Thus when exogenous organic compounds enter such cells, they can be assimilated directly into cell material resulting in both an increased rate of growth and a high yield of cell mass.

We have seen that the nature of the photosynthetic apparatus through the type of reducing power it makes available to the organism, determines the mechanism by which CO_2 is fixed in green bacteria. In its turn, as will be seen, the nature of these mechanisms has a bearing on the metabolism of organic compounds in these bacteria.

A. Metabolism of Organic Compounds in Chlorobium

The green sulfur bacteria are characterized as being obligately anaerobic and absolutely dependent on light, carbon dioxide and oxidizable sulfur or hydrogen for growth. Larsen (1953) investigated the possible participation of some simple organic compounds in the photometabolism of *Cb. thiosulfatophilum* and found that they did not support growth. This was long before it was understood on the biochemical level. Later, however, it was recognized by Sadler and Stanier (1960) that *Chlorobium* was able to assimilate some simple organic compounds in the light, provided that CO_2 and a source of reducing power were present. Today our knowledge of the reductive TCA cycle makes it possible to explain biochemically the inability of *Chlorobium* to grow photoheterotrophically. In addition to functioning as a mere mechanism of CO_2 fixation, the intermediates of this cycle (Fig. 1) also provide the cells with the necessary organic precursors for synthesis of fatty acids (from acetyl-CoA), amino acids (from pyruvate, α-oxoglutarate and oxalo-

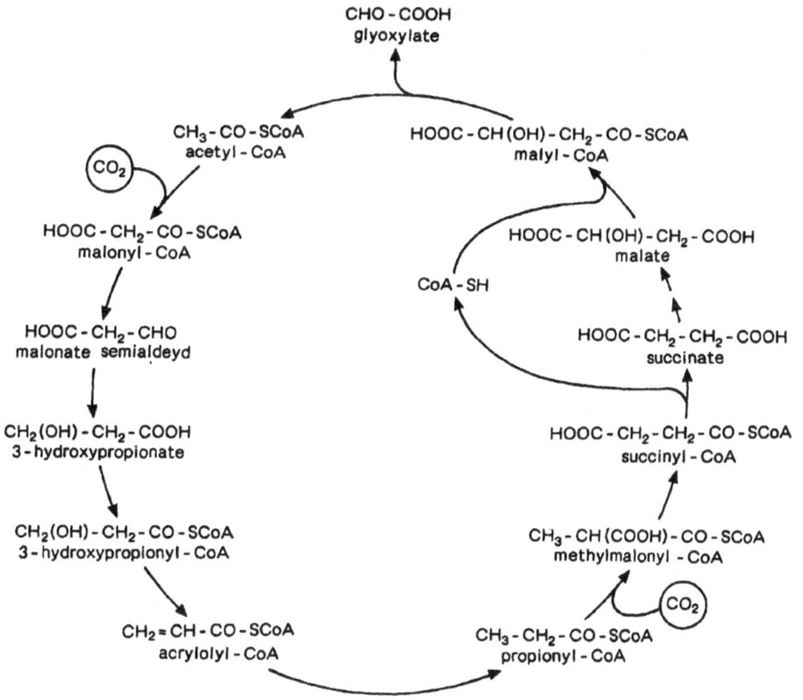

Fig. 3. The 3-hydroxypropionate cycle suggested by Holo (1989) for autotrophic CO_2 assimilation in *Chloroflexus aurantiacus* as presented by Strauss and Fuchs (1993).

acetate) and carbohydrates (from pyruvate). However, since α-oxoglutarate dehydrogenase is not present in *Chlorobium* (Beatty and Gest, 1981), this cycle can only function in the reductive direction and therefore organic compounds cannot be oxidized to provide CO_2 or reducing power.

Among the various organic compounds which can be assimilated by green sulfur bacteria are acetate, propionate, pyruvate, lactate, glutamate and glucose. Of these, acetate is by far the best substrate for growth. When this compound is added to the inorganic medium normally used for photoautotrophic growth, the growth rate as well as the yield are almost doubled. In view of the fact that acetyl-CoA is the immediate product of the reductive tricarboxylic acid cycle, this is not surprising. Presumably acetate in the medium is taken up by the cells and immediately converted to acetyl-CoA (Sirevåg and Ormerod, 1970b).

Evidence indicating that exogenous acetate is metabolized in this fashion was in fact provided by Hoare and Gibson (1964) before the reductive TCA cycle was proposed as a mechanism of CO_2 fixation in green sulfur bacteria. These workers grew *Chlorobium* in a medium containing ^{14}C-acetate and CO_2. The labeling pattern obtained when amino acids from protein of such cells were isolated and degraded, are in keeping with the assimilation of acetate *via* the now established reductive TCA cycle.

The photometabolism of propionate was investigated in detail by Larsen (1951), who discovered that *Chlorobium* is able to form succinyl-CoA from propionate. Thus after carboxylation and activation, the carbon atoms of propionate participate in the reductive TCA cycle. Later this knowledge proved valuable when radioactively labelled propionate was used in long term labeling experiments to gain evidence for the operation of the reductive TCA cycle in *Chlorobium* (Fuchs et al. 1980b).

The growth response of green sulfur bacteria to organic nutrients in the environment was investigated in detail by Kelly (1974). In addition to numerous amino acids, formate and glucose were readily assimilated in the presence of light, CO_2 and thiosulphate. The labeling patterns among cell constituents after assimilation of ^{14}C-labelled substrates indicated that the pathway of intermediary metabolism for these compounds in *Chlorobium* are similar to those known in many other organisms.

B. Metabolism of Organic Compounds in Chloroflexus

In contrast to green sulfur bacteria, the green gliding bacteria are able to utilize a variety of organic compounds for growth both under anaerobic conditions in the light and aerobically in the dark. These organic compounds include acetate, pyruvate, lactate, butyrate in addition to C_4 dicarboxylic acids, some alcohols, sugars and amino acids (Madigan et al. 1974). The natural habitat of the *Chloroflexaceae* is hot springs where they live in close association with other microorganisms which liberate a variety of organic compounds (Bauld and Brock, 1974). Thus the nutritional versatility of the green gliding bacteria reflects their ecology.

In order to examine whether the oxidative TCA cycle plays any role in the metabolism of *Chloroflexus*, fluoroacetate was used. In aerobic cells with a functional TCA cycle, this compound is converted to fluorocitrate which in turn inhibits aconitase and thus causes accumulation of citrate (Morrison and Peters, 1954). The results from such experiments indicated that this cycle operates in *Chloroflexus* under anaerobic as well as aerobic conditions (Sirevåg and Castenholz, 1979).

The metabolism of acetate in *Chloroflexus* was studied in detail in cells grown photoheterotrophically with acetate as the sole organic carbon source. In cell-free extracts from such cells, the specific activity of the key enzymes of the glyoxylate cycle, isocitrate lyase and malate synthase, was 5-fold higher than in extracts from cells grown in a complex medium. These findings together with results from experiments with inhibitors, indicated that operation of the glyoxylate cycle was necessary for acetate utilization in *Chloroflexus* (Løken and Sirevåg, 1982).

IV. Carbon Containing Reserve Materials and Endogenous Metabolism

In nature, phototrophic organisms are always exposed to regular variations in the supply of energy. Such variations are the case not only during the 24 hours (day/night), but also during the year (summer/winter). In the case of *Chlorobium*, the latter situation is dramatic since this organism frequently lives deep down (6 m or greater) in meromictic lakes, which might be covered by ice and snow for several months during the year (in Norway). It is therefore not surprising that phototrophic organisms have the capacity under favorable conditions to convert their carbon source into reserve materials, which can serve as maintenance under less favorable conditions. During balanced growth, the phototrophic bacteria usually form only small amounts of reserve materials, whereas such compounds usually accumulate in bacterial cells when carbon and energy are in excess, as under conditions of nitrogen starvation.

Among the phototrophic bacteria there are two carbonaceous reserve materials formed; glycogen and poly-β-hydroxybutyrate. Green sulfur bacteria such as *Chlorobium* are able to synthesize only glycogen, whereas the green gliding bacteria such as *Chloroflexus* appear to have the capacity to synthesize both. The type of storage material formed in the latter appears to be dependent on the conditions in the environment (Sirevåg and Castenholz, 1979).

A. Glycogen

Glycogen is a polymer in which numerous glucose residues are joined together by α-(1, 4)-glucosidic linkages into long chains which are branched, with α-(1, 6) linkages at the branch points.

The biosynthesis of glycogen in bacteria occurs via several reactions. First ADP-glucose is formed from ATP and α-glucose-1-P in a reaction catalyzed by ADP-glucose pyrophosphorylase (EC 2. 7. 7. 27) (in eukaryotic cells the corresponding reaction involves UTP). Two additional enzymes are further required; bacterial glycogen synthase (ADP-glucose:1,4-glucan transferase, EC 2. 4. 1. 21) which catalyzes the transfer of ADP-glucose to an α-glucan primer to form the α-(1, 4) linkages in the polymer. Finally, the 'branching enzyme' (α-1, 4-glucan-6-glycosyl transferase, EC 2. 4. 1. 18) forms the α-(1, 6) linkages by transferring part of an outer chain into the 6-position of a glucose residue further into the molecule.

It has been found that glycogen biosynthesis is regulated at the step where ADP-glucose is synthesized, and the properties and regulation of ADP-glucose pyrophosphorylase has been studied in detail in numerous bacteria by Preiss and coworkers (Preiss, 1984; Preiss and Romeo, 1989). Based on the nature of activators of the enzyme from various sources, seven groups of ADP-glucose pyrophosphorylases are recognized of which ADP-glucose pyrophosphorylase from *Chlorobium* belongs to the group activated by pyruvate and fructose-6-P (Preiss, 1984). That these specific compounds are activators of ADP-glucose phosphorylase in *Chlorobium* is not surprising in view of the fact that in this green sulfur bacterium pyruvate is a central metabolite which is directly converted to glucose (Sirevåg and Ormerod, 1970a). Furthermore, it has been shown that in *Chlorobium*, glucose is formed by the reactions of gluconeogenesis where fructose-6-P is an intermediate (Sirevåg, 1975; Paalme et al., 1982)

In prokaryotic as well as in eukaryotic cells, glycogen is deposited within the cells in the form of rosette-like non membrane-bound granules made up of smaller particles (Candy, 1980). When the deposition of glycogen in cells of *Chlorobium* was examined by electron microscopy of thin sections stained specifically for carbohydrate, results corresponding to this general picture were obtained (Sirevåg and Ormerod, 1977). The size of the granules in *Chlorobium* was limited to 30 nm and as the glycogen content of the cells increased, the number, rather than the their size, increased. When cells of *Chlorobium* containing polyglucose are incubated in the dark, the polymer is broken down. Experiments using electron microscopy and labeling with ^{14}C-acetate revealed that degradation of polyglucose occurs in such a way that all granules are subject to degradation simultaneously, and the polyglucose which has been formed most recently in the light becomes the first to be metabolized in the dark (Thorud and Sirevåg, 1982).

In his pioneering work on the green sulfur bacteria, Larsen (1953) investigated the endogenous dark metabolism of *Chlorobium* and found that washed cells evolved CO_2 and H_2S and produced acid in the dark. Later it was shown that resting cells of *Chlorobium* form glycogen in the light when incubated with CO_2 and a suitable hydrogen donor, and that the formation is increased by the addition of acetate (Sirevåg and Ormerod, 1970b). When washed cell suspensions enriched in glycogen are incubated in the dark, the glycogen decreases with time and organic acids such as acetate, propionate, caproate, and succinate are excreted into the medium. When glycogen-rich cells were incubated without a hydrogen donor, the level of glycogen decreased regardless of whether the cells were incubated in the dark or in the light. Since the products formed from polyglucose under the two different conditions are not the same, it has been suggested that glycogen serves as an energy source in the dark and as a source of reducing power in the light (Sirevåg and Ormerod 1977).

Although no detailed investigation of biosynthesis and storage of glycogen in *Chloroflexus* has been performed, it has been shown that washed cells of *Cf. aurantiacus* make glycogen as a reserve material from acetate and several other organic compounds (Sirevåg and Castenholz, 1979; Holo and Sirevåg, 1989).

B. Poly-β-Hydroxybutyrate

Poly-*β*-hydroxybutyrate is a linear polymer of D-*β*-hydroxybutyric acid and its molecular weight may be as high as several hundred thousand (Lundgren *et* al. 1965). In *Rs. rubrum* this polymer is synthesized from acetylCoA via *β*-hydroxybutyrate without the involvement of acyl carrier protein (Moskowitz and Merrick, 1969), and is deposited as large (up to 400 nm) intracellular granules which are surrounded by a thin, non-unit membrane (Boatman, 1964).

Results which indicated that *Chloroflexus* produced poly-*β*-hydroxybutyrate as reserve material during photoheterotrophic growth was obtained by chemical analysis as well as by electron microscopy by Pierson and Castenholz (1974). In experiments with washed cells of *Cf. aurantiacus* it was shown that this organism forms significant amounts of poly-*β*-hydroxybutyrate from acetate provided an inorganic hydrogen donor, H_2, was present. In the absence of the hydrogen donor, polyglucose was formed instead (Sirevåg and Castenholz. 1979).

VI. Evolution of Metabolic Pathways Involved in Carbon Metabolism of Green Bacteria

Three fundamentally different mechanisms exist for the fixation of CO_2 under autotrophic conditions in phototrophic prokaryotes: the Calvin cycle (see Chapter 41 by Tabita) used by purple bacteria and cyanobacteria, the reductive TCA cycle (Fig. 1) used by green sulfur bacteria, and the very recently discovered 3-hydroxypropionate cycle (Fig. 3) used by green gliding bacteria. In some nonphototrophic bacteria an additional autotrophic mechanism of CO_2 fixation occurs, the acetyl-CoA pathway, which is noncyclic and used by homoacetogens, methanogens, and most autotrophic sulfate reducers (Fuchs, 1989). The fact that completely different reactions participate in these four CO_2 fixation pathways makes it unlikely that they are derived from a common ancient mechanism. Instead they have probably evolved independently at different times during evolution to meet various needs in different organisms. Thus, when discussing evolution of the CO_2 fixation pathways in green bacteria one has to take into account what is known about the evolution of the organisms themselves and their phylogenetic relationship.

Based on sequence similarities in 16S ribosomal RNA from eubacteria, Woese (1987) introduced a now widely accepted classification scheme which contains 10 phyla. Results from this type of analysis show a deep divergence between *Chlorobium* and *Chloroflexus*, indicating that they are phylogenetically far apart (Gibson et. al 1985). Accordingly, on the evolutionary tree based on 16S rRNA analysis *Chloroflexus* and *Chlorobium* are separated into two different phyla, of which that containing *Chloroflexus* branches off very early. In this connection it should be pointed out that Blankenship (1993) recently presented an evolutionary tree based on information from reaction center sequence data, where the position of *Chloroflexus* is much closer to the purple bacteria than in the well known evolutionary tree of Woese (1987). However, regardless of their individual phylogenetic positions, it is clear that most evidence so far indicate that the only major character common to the Chlorobiaceae and the Chloroflexaceae is their type of light-harvesting bacteriochlorophylls and the organization of these pigments into chlorosomes. That these organisms share such important properties can however be explained by lateral gene transfer as was pointed out by Blankenship (1993). On this background it is therefore not surprising that two separate pathways for the fixation of CO_2 have evolved in the green sulfur bacteria and the green gliding bacteria.

The widely accepted idea that life originated in a primeval sea, containing a wide variety of abiotically produced organic compounds was put forward independently by Oparin in 1924 (Oparin 1953) and Haldane in 1929. It has also been widely accepted that no molecular oxygen was present in the atmosphere surrounding the primeval sea, and accordingly the first organisms must have been anaerobes. Presumably they were chemoorganotrophs, dependent on abiotically produced organic molecules as their source of carbon and energy. Since no inorganic electron acceptors were believed to be present, it is likely that these first organisms were fermentative.

Recently, the theory of a chemoorganotrophic origin of life has been challenged by Wachtershauser (1992) who has proposed a chemolithotrophic origin of life as an alternative. According to this theory, the source of reducing power for a chemolithotrophic origin as well as for early evolution is the oxidative formation of pyrite (FeS_2) from hydrogen sulfide and either ferrous ions or iron sulfide: $FeS + HS^- \rightarrow FeS_2 + H^+ + 2e^-$

Two different views have been put forward to explain the evolution of biosynthetic pathways. Horowitz (1945) speculated that metabolic pathways may have been constructed in a backwards manner by so called retroevolution. In line with this hypothesis, the activities of primitive organisms deprived the primeval seas of organic nutrients, resulting in a changed environment. As biochemically important substances were depleted, it became essential for survival of the organisms to develop the ability to convert other organic compounds into the depleted ones. Thus, selection would favor those organisms which acquired the ability to affect such conversions. On this background it is not difficult to envisage that the autotrophic way of life emerged early in evolution when organic nutrients became limiting. The other point of view, put forward by Granick (1965), implies that the intermediates in the biosynthetic pathway of a complex molecule like chlorophyll are thought to be the final products of biosynthetic pathways of earlier evolutionary forms. As was pointed out by Blankenship (1992) both of these concepts may in part be valid.

Founded on the ideas of a chemoorganotrophic origin of life and retroevolution in addition to evidence from paleomicrobiology and biogeochemistry, molecular phylogeny and comparative biochemistry, the generally accepted time sequence of metabolic evolution is the following:

fermentation
↓
anaerobic photosynthesis
↓
aerobic photosynthesis
↓
oxidative respiration

In the context of this scheme and assuming that the basic reactions of intermediary carbon metabolism in the ancient fermentative organisms were anaerobic, Weitzman (1985) has developed a possible pattern for the evolution of the tricarboxylic acid cycle. This work was to a large extent based on the speculations by Dillon (1981) who proposed a strictly biosynthetic evolutionary route by stepwise retroevolution. Starting from three key amino acids; glycine, aspartate and glutamate, the possible evolution of the TCA cycle and the glyoxylate cycles are described. As the environment became depleted of these amino acids, organisms evolved which could produce these compounds by amination of the corresponding α-oxoacids; glyoxylate, oxaloacetate and α-oxoglutarate. In order to have a supply of oxaloacetate, malate was converted to oxaloacetate in the next step in the scheme of Dillon (1981), whereas malate in turn was formed from glyoxylate and an activated acetyl group. Since *Chloroflexus* uses the reactions of the tricarboxylic acid as well as the glyoxylate cycles under anaerobic conditions in the light (Sirevåg and Castenholz, 1979; Løken and Sirevåg, 1982), these retroevolutionary speculations are very interesting in connection with the proposal that *Chloroflexus* might have branched off very early in evolution (Woese, 1987). The central role of glyoxylate in the scheme of Dillon (1981) is especially intriguing in view of the fact that glyoxylate is the product of the newly discovered 3 hydroxypropionate cycle (Fig. 3) for autotrophic CO_2 fixation in *Chloroflexus* (Holo, 1989; Strauss and Fuchs, 1993).

Taken together, our present knowledge of carbon metabolism in *Chloroflexus* is in accord with the hypothesis that the green gliding bacteria branched off very early in evolution.

If the TCA cycle and the glyoxylate cycle first evolved in an organism like *Chloroflexus*, it is not difficult to envisage how the reductive TCA cycle might have evolved from the TCA cycle. It was pointed out by Weitzman (1985) that once a photosynthetic mechanism evolved which could photoreduce ferredoxin, the possibility existed to produce pyruvate and α-oxoglutarate by the synthase reactions described earlier herein. Thus, the TCA cycle could be run reductively in reverse and function not only as a mechanism for the production of important organic metabolites, but also as a mechanism for the fixation of CO_2 as well. This is exactly what is the case in *Chlorobium*.

To investigate the possible evolutionary relationship of carbon metabolism in various phototrophic bacteria, we have studied the molecular properties of the enzyme malate dehydrogenase (MDH). In addition to its central role in the intermediary metabolism of most if not all organisms, this enzyme is also necessary in the CO_2 fixation pathway of *Chlorobium*. Analysis of the amino acid sequence of the N-terminal end of MDH from *Cb. vibrioforme*, *Cb. tepidum*, *Cf. aurantiacus*, and *Hb. gestii* showed an identical glycine motif (GXGXXG) in the dinucleotide binding domain of the enzyme from each organism (Rolstad *et* al. 1988, Charnock *et* al. 1992). Recent unpublished results show the same pattern for MDH from *Rs. rubrum*. This specific motif is different from that of all other MDHs sequenced to date, but is characteristic for other types of dehydrogenases and particularly for lactate dehydrogenases (LDH). Sequencing of the *mdh* structural genes of green sulfur bacteria and green gliding bacteria further indicate that in all other properties, except for the substrate binding domain, these MDHs resemble LDHs (Sirevåg, unpublished results). However the difference is small, since only three specific amino acids in the substrate binding domain determines the specificity of dehydrogenases (Wilks et al. 1988). Since lactic acid is a typical fermentation product, it is reasonable to assume that LDH appeared earlier in evolution than MDH. On this basis it is not difficult to envisage that the MDHs from green bacteria and possibly purple bacteria and heliobacteria are in principle LDHs which have evolved into MDHs by substituting specific amino acids in the substrate binding domain.

References

Bassham JA (1979) The reductive pentose phosphate cycle. Encycl Plant Physiol New Ser 6: 9–30

Bauld J and Brock TD (1974) Algal excretion and bacterial assimilation in hot spring algal mats. J Phycol 10: 101–106

Beatty JT and Gest H (1981) Generation of succinyl-coenzyme A in photosynthetic bacteria. Arch Microbiol 129: 335–40

Beuscher N and Gottschalk G (1972) Lack of citrate lyase—the key enzyme of the reductive carboxylic acid cycle—in *Chlorobium thiosulfatophilum* and *Rhodospirillum rubrum*. Z Naturforsch 27b: 967–73

Blankenship RE (1992) Origin and early evolution of photosynthesis. Photosynth Res 33: 91–111

Boatman E (1964) Observations on the fine structure of sphaeroplasts of *Rhodospirillum rubrum*. J Cell Biol 20: 297–311

Buchanan BB and Evans MCW (1965) The synthesis of alpha-ketoglutarate from succinate and carbon dioxide by a subcellular preparation of a photosynthetic bacterium. Proc Natl Acad Sci USA 54:1212–1218

Buchanan BB and Sirevåg R (1976) Ribulose 1, 5-diphosphate carboxylase and *Chlorobium thiosulfatophilum*. Arch Microbiol 109: 15–19

Candy DS (1980) Biological Function of Carbohydrates. Blackie, Glasgow and London

Charnock C, Refseth UH and Sirevåg R (1992) Malate dehydrogenase from *Chlorobium vibrioforme*, *Chlorobium tepidum*, and *Heliobacterium gestii*: Purification, characterization, and investigation of dinucleotide binding by dehydrogenases by use of empirical methods of protein sequence analysis. J Bacteriol 174: 1307–1313

Dillon LS (1981) Ultrastructure, Macromolecules and Evolution. Plenum Press, New York

Eisenreich W, Strauss G, Werz U, Fuchs G and Bacher A (1993) Retrobiosynthetic analysis of carbon fixation in the phototrophic eubacterium *Chloroflexus aurantiacus*. Eur J Biochem 215:619–632

Evans MCW and Buchanan BB (1965) Photoreduction of ferredoxin and its use in carbon dioxide fixation by a subcellular system from a photosynthetic bacterium. Proc Natl Acad Sci USA 53:1420–1425

Evans MCW, Buchanan BB and Arnon DI (1966) A new ferredoxin dependent carbon reduction cycle in a photosynthetic bacterium. Proc Natl Acad Sci USA 55: 928–34

Fuchs G (1989) Alternative pathways of autotrophic CO_2 fixation. In: Schlegel HG and Bowien B (eds) Autotrophic Bacteria. Science Tech Publishers, Madison

Fuchs G, Stupperich E and Jaenchen R (1980a) Autotrophic CO_2 fixation in *Chlorobium limicola*. Evidence against the operation of the Calvin cycle in growing cells. Arch Microbiol 128: 56–63

Fuchs G, Stupperich E and Eden G (1980b) Autotrophic CO_2 fixation in *Chlorobium limicola*. Evidence for the operation of a reductive tricarboxylic acid cycle in growing cells. Arch Microbiol 128: 64–71

Gibson J, Ludwig W, Stackebrandt E and Woese CR (1985) The phylogeny of the green photosynthetic bacteria: absence of a close relationship between *Chlorobium* and *Chloroflexus*. Syst Appl Microbiol 6: 152

Granick S (1965) Evolution of heme and chlorophyll. In: Bryson V, Vogel, H (eds) Evolving Genes and Proteins, pp 67–88. Academic Press, New York

Haldane JBS (1929) The Origin of Life. The Rationalist Annual: 3–10

Hoare DS and Gibson J (1964) Photoassimilation of acetate and the biosynthesis of amino acids by *Chlorobium thiosulfatophilum*. Biochem J 91: 546–59

Holo H (1989) *Chloroflexus aurantiacus* secretes 3-hydroxypropionate, a possible intermediate in the assimilation of CO_2 and acetate. Arch Microbiol 151: 252–56

Holo H and Grace D (1987) Polyglucose synthesis in *Chloroflexus aurantiacus* studied by ^{13}C-NMR. Evidence for acetate metabolism by a new metabolic pathway in autotrophically grown cells. Arch Microbiol 148: 292–97

Holo H and Sirevåg R (1986) Autotrophic growth and CO_2 fixation of *Chloroflexus aurantiacus*. Arch Microbiol 145: 173–80

Horowitz NH (1945) On the evolution of biochemical synthesis. Proc Natl Acad Sci USA 31: 153–157

Ivanovsky RN, Sinstov NV and Kondratieva EN (1980) ATP-linked citrate lyase activity in the green sulfur bacterium *Chlorobium limicola* forma *thiosulfatophilum*. Arch Microbiol 128: 239–41

Ivanovsky RN, Krasilnikova EN and Fal YI (1993) A pathway of the autotrophic CO_2 fixation in *Chloroflexus aurantiacus*. Arch Microbiol 159: 257–64

Kelly DP (1974) Growth and metabolism of the obligate photolithotroph *Chlorobium thiosulfatophilum* in the presence of added organic nutrients. Arch Microbiol 100:163–178

Kondratieva EN, Ivanovsky RN and Krasilnikova EN (1992) Carbon metabolism in *Chloroflexus aurantiacus*. FEMS Microbiol Lett, 100: 269–72

Larsen H (1951) Photosynthesis of succinic acid by *Chlorobium thiosulfatophilum*. J Biol Chem 193: 167–73

Larsen H (1953) On the microbiology and biochemistry of the photosynthetic green sulfur bacteria. K. norske vidensk. selsk. Skr. 1

Lundgren DG, Alper R, Schneitman C and Marchessault RH (1965) Characterization of poly-beta-hydroxybutyrate extracted from different bacteria. J Bacteriol 89:245–251

Løken Ø and Sirevåg R (1982) Evidence for the presence of the glyoxylate cycle in *Chloroflexus*. Arch Microbiol 132: 276–79

Madigan MT and Brock TD (1975) Photosynthetic sulfide oxidation by *Chloroflexus aurantiacus*, a filamentous, photosynthetic gliding bacterium. J Bacteriol 122: 782–784

Madigan MT and Brock TD (1977) CO_2 fixation in photosynthetically-grown *Chloroflexus aurantiacus*. FEMS Microbiol Lett, 1: 301–304

Madigan MT, Petersen SR and Brock TD (1974) Nutritional studies on *Chloroflexus*, a filamentous photosynthetic gliding bacterium. Arch Microbiol 100: 97–103

Morrison JF and Peters RA (1954) The inhibition of aconitase by fluorocitrate. Biochem J 56: XXXVI

Moscowitz GJ and Merrick JM (1969) Metabolism of poly-beta-hydroxybutyrate. II. Enzymatic synthesis of D(-)-beta-hydroxybutyryl coenzyme A by an enoyl hydrase from *Rhodospirillum rubrum*. Biochemistry 8: 2748–2755

Oparin A (1953) Origin of Life. Dover Publications, Inc. , New York

Paalme T, Olivson A and Vilu R (1982) ^{13}C -NMR study of the glucose synthesis pathways in the bacterium *Chlorobium thiosulfatophilum*. Biochim Biophys Acta 720: 303–10

Pierson BK and Castenholz RW (1974) A phototrophic gliding filamentous bacterium of hot springs, *Chloroflexus aurantiacus*, gen and sp. nov. Arch Microbiol 100:5–24

Preiss J (1984)Bacterial glycogen synthesis and its regulation. Ann Rev Microbiol 38: 419–458

Preiss J and Romeo T (1989) Physiology, biochemistry and genetics of bacterial glycogen synthesis. In: Rose AR and Tempest DH (eds) Advances in Microbial Physiology, Vol 30, pp 183–238. Academic Press, New York

Quandt L, Gottschalk G, Ziegler H and Stichler W (1977) Isotope discrimination by photosynthetic bacteria. FEMS Microbiol Lett 1: 125–28

Quandt L, Pfennig N and Gottschalk G (1978) Evidence for the key position of pyruvate synthase in the assimilation of CO_2 by *Chlorobium*. FEMS Microbiol Lett 3: 227–30

Rolstad AK, Howland, E and Sirevåg R (1988) Malate dehydrogenase from the thermophilic green bacterium

Chloroflexus aurantiacus: purification, molecular weight, amino acid composition, and partial amino acid sequence. J Bacteriol 170:2947–2953

Sadler WR and Stanier RY (1960) The function of acetate in photosynthesis by green bacteria. Proc Natl Acad Sci USA 46: 1328–34

Shively JM, Devore W, Stratford L, Porter L, Stevens SE (1986) Molecular evolution of the large subunit of ribulose-1, 5-bisphosphate carboxylase /oxygenase (Rubisco). FEMS Microbiol Lett 37:251–257

Sirevåg R (1974) Further studies on carbon dioxide fixation in *Chlorobium*. Arch Microbiol 93: 3–18

Sirevåg R (1975) Photoassimilation of acetate and metabolism of carbohydrate in *Chlorobium thiosulfatophilum*. Arch Microbiol 104: 105–111

Sirevåg R and Castenholz RW (1979) Aspects of carbon metabolism in *Chloroflexus*. Arch Microbiol 120: 151–3

Sirevåg R and Ormerod JG (1970a) Carbon dioxide fixation in green sulphur bacteria. Biochem J 120: 399–408

Sirevåg R and Ormerod JG (1970b) Carbon dioxide fixation in photosynthetic green sulphur bacteria. Science 169: 186–8

Sirevåg R and Ormerod JG (1977) Synthesis, storage and degradation of polyglucose in *Chlorobium thiosulfatophilum*. Arch Microbiol, 111: 239–44

Sirevåg R, Buchanan BB, Berry JA and Troughton JH (1977) Mechanisms of CO_2 fixation in bacterial photosynthesis studied by the carbon isotope fractionation technique. Arch Microbiol 112: 35–38

Smillie RM, Rigopoulos N and Kelly H (1962) Enzymes of reductive pentose phosphate cycle in the purple and the green photosynthetic sulphur bacteria. Biochim Biophys Acta 56: 612–14

Strauss G and Fuchs G (1993) Enzymes of a novel autotrophic CO_2 fixation pathway in the phototrophic bacterium *Chloroflexus aurantiacus*, the 3-hydroxypropionate cycle. Eur J Biochem 215:633–643

Strauss G, Eisenreich W, Bacher A and Fuchs G (1992) ^{13}C-NMR study of autotrophic CO_2 fixation pathways in the sulfur-reducing Archaebacterium *Thermoproteus neutrophilus* and in the phototrophic Eubacterium *Chloroflexus aurantiacus*. Eur J Biochem 205: 853–66

Tabita RF, McFadden B and Pfennig N (1974) D-ribulose-1, 5-bisphosphate carboxylase in *Chlorobium thiosulfatophilum*. Biochim Biophys Acta 341:187–194

Takabe T and Akazawa T (1977) A comparative study on the effect of O_2 on photosynthetic carbon metabolism by *Chlorobium thiosulfatophilum* and *Chromatium vinosum*. Plant and Cell Physiol 18: 753–756

Thorud M and Sirevåg R (1982) Kinetics of polyglucose breakdown in *Chlorobium*. Arch Microbiol 133:114–117

Wachterhauser G (1992) Groundworks for an evolutionary biochemistry: the iron-sulphur world. Prog Biophys Molec Biol 58:85–201

Weitzman PD (1985) Evolution of the citric acid cycle. In: Schleifer KH, Stackebrandt E (eds) Evolution of Prokaryotes, FEMS Symp. No 329, pp 253–275. Academic Press, London

Whelan T, Sackett WM and Benedict RC (1973) Enzymatic fractionation of carbon isotopes by phosphoenolpyruvate carboxylase from C4 plants. Plant Physiol 51:1051–1054

Wilks HM, Hart KW, Feeney RC, Dunn CR, Muirhead H, Chia WN, Barstow DA, Atkinson T, Clarke AR and Holbrook JJ (1988) A specific highly active malate dehydrogenase by redesign of a lactate dehydrogenase framework. Science 242: 1541–1544

Woese CR (1987) Bacterial evolution. Microbiol Rev 51:221–271

Chapter 41

The Biochemistry and Metabolic Regulation of Carbon Metabolism and CO_2 Fixation in Purple Bacteria

F. Robert Tabita
Department of Microbiology and The Biotechnology Center, The Ohio State University, 484 West 12th Avenue, Columbus, Ohio 43210-1192, USA

Summary

Purple photosynthetic bacteria exhibit great diversity in the metabolism of simple carbon compounds. In this chapter, the reactions and metabolic schemes that the organisms, particularly purple nonsulfur bacteria, employ to break down and/or assimilate one-carbon, two-carbon, three-carbon, and four-carbon compounds and sugars is examined. Knowledge of the biochemistry and physiology of carbon metabolism and its molecular control has benefited somewhat from the application of recombinant DNA approaches, yet there are significant gaps in our understanding of how catabolic and anabolic reaction sequences are integrated. By contrast, great advances have been made relative to the biochemistry of CO_2 fixation, due primarily to recent enzymological, molecular, and structural studies of the key enzyme, RubisCO. Indeed the enzyme from *Rhodospirillum rubrum* has become the paradigm for such work. The ability to prepare recombinant enzymes that catalyze additional key steps of CO_2 fixation should result in similar advances concerning these proteins in the future. Facile genetic manipulation of mutant *Rhodobacter* and *Rhodospirillum* strains has already resulted in the uncovering of alternative CO_2 assimilatory routes that replace the Calvin cycle, and prospects for gaining an understanding of the biochemistry and molecular control of both schemes should follow. The latter studies probably would not

R. E. Blankenship, M. T. Madigan and C. E. Bauer (eds): Anoxygenic Photosynthetic Bacteria, pp. 885–914.

be possible with other organisms, further attesting to the versatility of the purple nonsulfur bacteria for investigating basic metabolic processes. This chapter considers the current state of our knowledge of carbon dioxide fixation and carbon metabolism in purple bacteria.

I. Introduction

Purple photosynthetic bacteria, including the purple nonsulfur (PNS) and purple sulfur (PS) bacteria, represent a diverse group of organisms which exhibit great versatility in their carbon metabolism. Indeed, this metabolic versatility and the ability to grow under a variety of conditions present wonderful opportunities to study the biochemical and molecular basis for metabolic control. The PNS bacteria are particularly noteworthy in this regard, as these organisms are probably the most metabolically diverse organisms found in nature (Madigan and Gest, 1979). In this chapter, an overview of both heterotrophic and autotrophic carbon metabolism of purple bacteria is presented; this review focuses on developments since this topic was last reviewed (Fuller, 1978; Kondratieva, 1979) and stresses aspects of control and metabolic regulation in PNS bacteria.

A. Redox Balance and Carbon Metabolism

Under phototrophic growth conditions, purple photosynthetic bacteria must be supplied with an exogenous electron donor and an electron acceptor. Indeed, the organisms are exquisitely sensitive to the redox potential of the electron donor. The use of highly reduced electron donors; e.g., H_2, propionate, or butyrate by purple nonsulfur bacteria has a profound effect on the flow of carbon in these cells since under such conditions CO_2 may serve as both carbon source and electron acceptor. Thus, for purposes of this discussion it must be stressed that metabolic control of carbon flow is inexorably linked to the need for these organisms to balance or maintain redox poise.

Abbreviations: αKDGH – α-ketoglutarate dehydrogenase; AMP – adenosine monophosphate; *cbb* – Calvin, Benson, Bassham; *Cm.* – *Chromatium*; Cpn – chaperonin; DMSO – dimethylsulfoxide; ED – Entner-Doudoroff pathway; EMP – Embden-Meyerhof-Parnas pathway; FBP – fructose 1,6-bisphosphate; GPD – glyceraldehyde phosphate dehydrogenase; GS – glutamine synthetase; ICL – isocitrate lyase; KDG – 2-keto-3-deoxygluconate; MSX – methionine sulfoximine; PC – pyruvate carboxylase; PEP – phosphoenolpyruvate; PEPCK – PEP carboxykinase; 3-PGA – 3-phosphoglyceric acid; 3-PGAL – 3-phosphoglyceraldehyde; PGK – phosphoglyceric acid kinase; PK – pyruvate kinase; PNS – purple nonsulfur; PRK – phosphoribulokinase; PS – purple sulfur; *Rb.* – *Rhodobacter*; *Rp.* – *Rhodopseudomonas*; *Rs.* – *Rhodospirillum*; *Rv.* – *Rubrivivax*; RubisCO – RuBP carboxylase/oxygenase; Ru-5-P – ribulose-5-phosphate; RuBP – ribulose 1,5-bisphosphate; SBP – sedoheptulose 1,7-bisphosphate; TCA – tricarboxylic acid

II. Central Pathways for Organic Carbon Metabolism

In addition to CO_2, a number of organic carbon sources will support the growth of purple bacteria. For purple nonsulfur bacteria, such carbon compounds may support both aerobic chemotrophic or anaerobic phototrophic metabolism; the purple sulfur bacteria generally favor strictly anoxygenic growth conditions, but there are well documented reports where some strains have been shown to grow under aerobic conditions (Kondratieva et al., 1976).

A. Formate, Methanol, and CO Metabolism

There are a number of instances where one-carbon compounds, other than CO_2, are actively metabolized. Certainly, formate may be used as a growth substrate by a number of aerobic chemolithotrophic bacteria. In such organisms, formate dehydrogenase catalyses the oxidation of formate to CO_2, which is then assimilated via the Calvin cycle (Quayle and Keech, 1959). Amongst the purple nonsulfur bacteria, *Rhodopseudomonas palustris* has been classically considered as the one strain which may actively metabolize and grow on formate under photosynthetic conditions (Qadri and Hoare, 1968). Strains of *Rp. palustris* apparently contain an inducible soluble formate dehydrogenase (Qadri and Hoare, 1968), the activity of which is enhanced by either FAD or FMN (Yoch and Lindstrom, 1968). An inducible particulate hydrogenase is also synthesized to assist in oxidizing hydrogen formed from formate (Qadri and Hoare, 1968).

Another one carbon substrate that may be photoassimilated by some strains is methanol. Enrichment in methanol-bicarbonate medium resulted primarily in the isolation of strains of

Rhodocyclus gelatinosa (now *Rubrivivax gelatinosus*) and *Rp. acidophila* (Quayle and Pfennig, 1975); in all cases the pH optimum for growth on methanol was considerably higher than for other carbon sources. Bicarbonate (or CO_2) is absolutely required as an electron acceptor for growth since methanol is considerably more reduced than cell material. Several other Rhodospirillaceae were found to grow in a methanol-bicarbonate medium but *Rp. acidophila* appeared the most active. Interestingly, formate also supported the growth of *Rp. acidophila*, however other C_1 compounds such as methylamine or formaldehyde did not. Formate was also able to substitute for bicarbonate as the required electron acceptor. Further analyses showed that one molecule of CO_2 is required for every 2 molecules of methanol assimilated to cellular carbon (CH_2O):

$$2\ CH_3OH + CO_2 \rightarrow 3(CH_2O) + H_2O \qquad (1)$$

Under these growth conditions, ribulose bisphosphate carboxylase/oxygenase (RubisCO) is increased some 6-fold over cells grown with succinate. Subsequent studies showed that methanol is anaerobically oxidized to formaldehyde, then to formate, and eventually to CO_2, with the Calvin cycle used for the assimilation of CO_2 (Sahm et al., 1976). In this case, the Calvin cycle is also used to facilitate the transfer of excess reducing power from methanol to CO_2. As will be discussed later, RubisCO thus serves both as a key enzyme for carbon assimilation as well as an important catalyst that allows CO_2 to function as an electron acceptor.

An extremely interesting system for the oxidation of carbon monoxide to CO_2 has been described in *Rhodospirillum rubrum* (Bonam et al., 1984). Previous studies had shown that various photosynthetic bacteria grow in the presence of CO (Hirsch, 1968) and *Rhodocyclus (Rubrivivax) gelatinosus* grows anaerobically in the dark on CO as sole carbon and energy source (Uffen, 1976). The *Rv. gelatinosus* CO-oxidizing system is membrane bound and yields both CO_2 and H_2 (Uffen, 1983); presumably CO_2 is assimilated via RubisCO and the Calvin cycle but this has not been conclusively established as yet. It is interesting that light-grown *Rv. gelatinosus* does not oxidize or assimilate CO. By contrast, *Rs. rubrum* synthesizes both CO dehydrogenase and a CO-insensitive hydrogenase in the light when cells are exposed to CO (Bonam et al., 1989). This inducible CO dehydrogenase has been thoroughly characterized and found to be a nickel-containing monomeric enzyme often associated with a small Fe/S protein (Bonam and Ludden, 1987; Ensign and Ludden, 1991). Moreover, the structural genes, *cooF* and *cooS*, encoding the Fe/S protein and CO dehydrogenase, respectively, are present in a single transcriptional unit with part of the *cooH* gene (presumably encoding the CO-induced hydrogenase) upstream from *cooF*, which is found in a different transcriptional unit. There is also evidence for a fourth open reading frame whose partial deduced amino acid sequence suggests a protein with an ATP-binding site motif (Ensign and Ludden, 1991). Together, the biochemical results and the *cooHFS* genetic organization suggested an analogy to the *E. coli* formate hydrogenlyase system (Kerby et al., 1992), since *Rs. rubrum* is known to couple H_2 synthesis to formate oxidation (Gorrell and Uffen, 1977). Further experiments of this interesting system should establish whether the *Rs. rubrum* enzyme and associated proteins may catalyze more elaborate functions, such as the well-described acetyl CoA synthase reaction of CO dehydrogenase from homoacetogenic and methanogenic bacteria (Wood, 1991).

B. Acetate and 2-Carbon Metabolism

The pathways and reactions that purple bacteria, particularly PNS bacteria, use to metabolize acetate are clouded with uncertainty and controversy. All, or virtually all, PS and PNS bacteria grow on acetate (Thiele, 1968; Madigan, 1988). Exactly how this occurs, however, may vary in different strains. In *Chromatium*, a modified glyoxylate cycle lacking malate dehydrogenase is present (Fuller et al., 1961). This organism decarboxylates malate to form pyruvate via an $NADP^+$-linked malic enzyme; pyruvate may then be used for carbohydrate synthesis through the mediation of an acetyl-CoA dependent pyruvate carboxylase and the enzyme phosphoenolpyruvate (PEP) carboxykinase. In the PNS bacteria, the situation is much more complicated. In 1960, Kornberg and Lascelles (1960) reported that isocitrate lyase (ICL) was present in extracts of acetate or butyrate-grown *Rp. palustris* or *Rb. capsulatus* cultured under both phototrophic or chemotrophic growth conditions. However, both *Rb. sphaeroides* and *Rs. rubrum* contained very low levels of this key glyoxylate cycle enzyme, yet the two organisms grew well with acetate. A later report indicated that *Rb. capsulatus* lacked ICL (Albers and Gottschalk,

1976). Moreover, Payne and Morris (1969a) state that acetate-grown aerobic cultures of *Rb. sphaeroides* failed to incorporate [2-^{14}C] acetate into glyoxylate, glycolate, or glycine, suggesting that this organism uses a novel pathway to metabolize acetate. The early work of Hoare (1963), which showed that *Rs. rubrum* incorporated acetate into glutamate, would also tend to discount the glyoxylate cycle. Subsequent studies verified that three different strains of *Rb. sphaeroides* were ICL$^-$, while four *Rb. capsulatus* strains were shown to be ICL$^+$ (Willison, 1988). As pointed out by Willison (1988), the ICL$^-$ phenotype appears to be reminiscent of the situation observed with various methylotrophic bacteria and *Thiobacillus versutus*, which is both ICL$^+$ and ICL$^-$ depending on whether the organism is cultured anaerobically or aerobically, respectively (Claassen and Zehnder, 1986). Complicating this situation further, using an assay based on the colorimetric determination of the dinitrophenylhydrazone of glyoxylate, Blasco et al. (1989) were able to detect significant levels of ICL activity in *Rb. sphaeroides*, *Rs. rubrum*, as well as *Rb. capsulatus*. These authors contend that the continuous spectrophotometric assays previously employed may have led to difficulties in accurately measuring this enzyme. Perhaps the variance in ICL activity measured in different laboratories (with different strains) may be related to a faster adaptation by some strains to acetate, which seems in two cases to be related to acetate-dependent growth (Blasco et al., 1989; Nielsen and Sojka, 1979). When an acetate-adapted mutant became the dominant strain in cultures of *Rb. capsulatus* St. Louis, the levels of ICL also increased dramatically (Nielsen and Sojka, 1979). Furthermore, unlike ICL, malate synthase was shown to be constitutive in *Rb. capsulatus* E1F1 (Blasco et al., 1989; 1991). Exhaustion of acetate from cultures resulted in a rapid loss of ICL activity which was dependent on active photosynthesis. Restoration of activity required the addition of acetate and de novo protein synthesis (Nielsen et al., 1979). Taking these studies further, Blasco et al., (1991) found that there was a 3-fold increase in ICL activity after washed acetate-grown logarithmic *Rb. capsulatus* E1F1 cells were resuspended in fresh media containing 1 mM acetate. As described for the St. Louis strain, cultures which became depleted for acetate rapidly lost ICL activity. Perhaps such phenomena are related to the divergent reports of ICL in PNS bacteria.

C. Pyruvate and 3-Carbon Metabolism

Virtually all purple photosynthetic bacteria are able to grow on pyruvate under photoheterotrophic growth conditions. As an important hub in microbial metabolism (Gest, 1981), the reactions leading to and from pyruvate have attracted some interest in PNS bacteria, particularly since these organisms readily grow both aerobically and anaerobically. There are several ways for PNS bacteria to metabolize pyruvate. Several species have been shown to degrade pyruvate under phototrophic conditions via pyruvate decarboxylase to form CO_2, acetaldehyde and acetoin (Qadri and Hoare, 1967) and most, if not all, species contain pyruvate dehydrogenase (Willison, 1988). *Rs. rubrum* uses both PEP synthase and PEP carboxylase to convert pyruvate to oxaloacetate and these reactions are important anaplerotic enzymes of the reductive tricarboxylic acid cycle, which are present in this organism (Buchanan et al., 1967). However, *Rb. sphaeroides* and *Rb. capsulatus* contain no PEP carboxylase or PEP synthase (Payne and Morris, 1969b; Willison, 1988) and pyruvate phosphate dikinase was not found in *Rb. capsulatus* (Willison, 1988). In *Rb. sphaeroides* and *Rb. capsulatus*, pyruvate carboxylase (PC), which is absolutely dependent on acetyl CoA for activity, is found at significant levels in both phototropically and chemotrophically grown cells. Furthermore, PC negative mutants of both species were unable to grow on pyruvate or other compounds metabolized through pyruvate. Thus, even though both *Rb. sphaeroides* and *Rb. capsulatus* contained PEP carboxykinase (PEPCK) (Payne and Morris, 1969b; Willison, 1988), it is apparent that these organisms cannot use PEPCK to produce oxaloacetate. Rather, this enzyme, in vivo, must function to produce PEP and, as pointed out by Payne and Morris (1969b), *Rb. sphaeroides* (and *Rb. capsulatus*) is '...wholly reliant upon its pyruvate carboxylase to supply the C_4-dicarboxylic acids required during growth on carbon sources which only yield phosphoenolpyruvate and/or pyruvate.' Thus, it is apparent that PC is an important enzyme in *Rb. sphaeroides* and *Rb. capsulatus*; the enzyme has recently been purified to homogeneity from *Rb. capsulatus* and its gene cloned (the first prokaryotic PC gene to be isolated) (H.V. Modak and D.J. Kelly, personal communication). If the PEPCK reaction runs solely or predominately in the direction of PEP, there must be other enzyme(s) in the cell whose activity pulls the PEPCK reaction towards PEP. An excellent candidate is pyruvate kinase (PK),which has been detected and partially purified from *Rs. rubrum* (Klemme, 1973), *Rb. sphaeroides* (Klemme, 1974), and *Rb. capsulatus* (Willison, 1988; Schedel et al., 1975). The enzymes from the three sources have surprisingly different properties although they each show classic allosteric regulation. For example, the *Rs. rubrum* enzyme is highly sensitive to inorganic phosphate, which is an allosteric inhibitor at fairly low levels ($K_i = 50\ \mu M$) (Klemme, 1973). This sensitivity to phosphate is not exhibited by the *Rb. capsulatus* enzyme where the K_i was calculated to be 6.5 mM (Klemme, 1974). However, fumarate strongly inhibits the *Rb. capsulatus* PK and enhances the sigmoidal behavior at varying concentrations of PEP. The *Rb. sphaeroides* enzyme closely resembles the *Rb. capsulatus* enzyme with respect to its kinetic properties (Schedel et al., 1975). The most interesting and intriguing property of the *Rb. sphaeroides* and *Rb. capsulatus* PK enzymes is their cold lability, which appears to be due to a rapid, reversible and a slow, irreversible process that presumably involves dissociation-reassociation of enzyme protomers (Schedel et al., 1975). This important enzyme deserves to be studied in greater detail since it functions at the crossroads of important metabolic pathways of PNS bacteria. Other reactions involving pyruvate that have been demonstrated in PNS bacteria are malic enzyme (Fuller et al., 1961), pyruvate synthase (Buchanan et al., 1967), and pyruvate flavodoxin reductase (Brostedt and Nordlund, 1991). Finally, as previously discussed, *Rs. rubrum* shows the interesting capacity to grow fermentatively on pyruvate in the dark (Gorrell and Uffen, 1977) to form propionate (Kohlmiller and Gest, 1951), H_2 (Uffen and Wolfe, 1970), and formate (Schön and Voelskow, 1976), with formate subsequently metabolized to CO_2 and H_2. Pyruvate formate lyase appeared to be the key enzyme of the fermentation and formate hydrogen lyase is required for the generation of H_2; the appearance of different fermentation end products seems to be dependent on the growth conditions (Schön and Voelskow, 1976).

PNS bacteria show differential abilities to photometabolize propionate, as 33 strains of *Rb. capsulatus* grew in a propionate-CO_2 medium while *Rb. sphaeroides* did not (Weaver et al., 1975). CO_2 is absolutely required for growth on propionate, as it is for all reduced electron donors. In *Rs. rubrum* the metabolism of propionate has been studied in great detail; this 3-carbon compound is assimilated to succinate via a propionyl CoA carboxylase whose activity and/or synthesis is shut down in cells grown in a malate medium (Knight, 1962).

Like propionate, another 3-carbon substrate, glycerol, supports the growth of some PNS bacteria but not others. Thus, Van Niel found that *Rb. sphaeroides* and *Rp. palustris* could grown on glycerol while *Rb. capsulatus* and *Rhodocyclus (Rubrivivax) gelatinosus* could not (van Niel, 1944). Some years later, Lueking et al. (1976) isolated a spontaneous mutant of *Rb. capsulatus* after several rounds of selection that had 'gained' the ability to grow on glycerol. Under photosynthetic conditions, CO_2 is absolutely required, as might be expected for bacteria grown on such a reduced substrate. In this strain, glycerokinase and pyridine nucleotide independent L-α-glycerophosphate dehydrogenase activities were found to be elevated several fold over the wild-type strain. The mutant, strain L_1, contained constitutive levels of these enzymes, as the difference in activity levels of cells grown on malate compared to cells grown on glycerol-CO_2 was not great (Lueking et al., 1976). A subsequent study has shown that gain-of-function glycerol-catabolizing mutants of *Rb. capsulatus* can easily be selected by subculturing wild-type strains on media containing mixtures of malate and glycerol (Spear and Sojka, 1984).

By contrast, *Rb. sphaeroides*, which naturally grows on glycerol, followed a diauxic growth response in a malate-glycerol medium and exhibited a typical repression/induction response relative to the glycerol catabolizing enzymes in a malate and glycerol medium, respectively (Pike and Sojka, 1975). Since the authors indicate that there is no evidence for a specific glycerol transport system in *Rb. sphaeroides* and *Rb. capsulatus*, it would appear that strain L_1 is a mutant that has lost the capacity to control the expression of the glycerol catabolizing enzymes (Lueking et al., 1976). As will be discussed later, this scenario is similar to the gain in the ability of *Rb. sphaeroides* strain 16 to use CO_2 as an electron acceptor when malate is the carbon source (Wang et al., 1993a).

Lastly, strains of *Rb. capsulatus* and *Rhodomicrobium vannieli* were found to actively photometabolize and grow on acetone; again, as expected, growth was CO_2-dependent due to the reduction level of acetone (Madigan, 1990). No growth was obtained under dark aerobic or dark anaerobic conditions. Earlier studies described work with other interesting isolates, since lost, capable of quantitative conversion of isopropanol to acetone and CO_2. Unpublished results from our own laboratory had previously indicated that high levels of RubisCO may be obtained in *Rb. capsulatus* cells grown with a variety of reduced substrates and alcohols, including acetone.

D. Butyrate and the Metabolism of Other Reduced Fatty Acids

In previous sections, the absolute requirement of exogenous CO_2 for the growth of PNS bacteria on reduced organic compounds such as methanol, propionate, glycerol, and acetone was stressed. However, it is well known that PNS bacteria grow better, with faster doubling times and to higher density, when C_4 dicarboxylic acids such as malate or succinate are used as the electron donor. Such compounds, along with pyruvate and lactate, and other common growth substrates are more oxidized than the intracellular milieu; the cells thus do not require an exogenous electron acceptor. Indeed, in an insightful study over 60 years ago, Muller found that organisms growing at the expense of oxidized organic acids, such as acetate or malate, showed a significant net release of CO_2 into the growth medium. For acetate and malate, 0.17 and 1.2 moles of CO_2 were produced per mole of substrate consumed, respectively; for succinate, somewhat more reduced than malate, the value dropped to 0.7 (Muller, 1933). When reduced electron donors, such as butyrate, were used, addition of CO_2 to the growth medium was required and 0.7 mole of CO_2 was utilized per mole of butyrate consumed. Thus, even though acetate and butyrate are rapidly assimilated under phototrophic growth conditions, presumably to form large amounts of poly-β-hydroxybutyrate (PHB) and other cellular constituents, metabolism of the two growth substrates obviously cause profound physiological changes. As discussed by Stanier et al., (1976), the synthesis of PHB from acetate is a reductive process involving the metabolic sequences of the TCA cycle:

$$9n\ CH_3COOH \rightarrow 4(CH_3CH_2CHCOO^-)_n + 2nCO_2 + 6_nH_2O \quad (2)$$

However, synthesis of PHB from butyrate is an oxidative process:

$$nCH_3CH_2CH_2COOH \rightarrow (CH_3CH_2CHCOO^-)_n + 2nH \quad (3)$$

Under anaerobic conditions, there must be a readily available hydrogen acceptor and, obviously, this must be supplied exogenously because the cells will not grow without the addition of CO_2 in the form of bicarbonate, with CO_2 subsequently reductively assimilated to carbohydrate $(CH_2O)_n$. Thus, butyrate photoassimilation is coupled to CO_2 fixation:

$$2nCH_3CH_2CH_2COOH + nCO_2 \rightarrow 2(CH_3CH_2CHCOO^-)_n + (CH_2O)_n + nH_2O \quad (4)$$

This last equation readily explains how PNS bacteria are able to grow on reduced fatty acids, and probably relates to the CO_2 requirement of bacteria cultured on other reduced carbon sources as well. At one time, it was thought that organic carbon served mainly as a H-donor for CO_2 fixation (van Niel, 1941). However, the discovery (Gest and Kamen, 1949) that PNS bacteria may evolve copious quantities of H_2 when grown on organic carbon caused a reevaluation of Van Niel's hypothesis on the role of organic carbon. Indeed as stated by Gest (1951): "The accessory (organic) 'hydrogen donor' required in bacterial photosynthesis ordinarily does not undergo a simple one-step oxidation; it generally supplies carbon intermediates, other than CO_2, which are directly used by the cell for synthetic purposes. This is apparently true for practically all classes of organic 'hydrogen donors.' " Indeed, it should be stressed that CO_2 need not be a major carbon source at all under such growth conditions since it is easily replaced by other exogenous electron acceptors that have no ability to function directly in carbon assimilation (Richardson et al., 1988). The importance of RubisCO and the Calvin cycle to this whole question of balancing the redox state of the cell is discussed in Sections III.D and IV.A.

E. Sugar Metabolism

Most, if not all, PNS bacteria are able to grow on one or more of the common sugars; in some cases the ability or inability to use fructose is characteristic of the species (Madigan, 1988). PS bacteria are much

$$\text{Glucose} \xrightarrow{\text{dehydrogenase}} \text{Gluconate} \xrightarrow{\text{dehydratase}} \text{KDG} \xrightarrow{\text{aldolase}} \begin{cases} \text{Glyceraldehyde} \\ \text{Pyruvate} \end{cases} \quad (5)$$

more restrictive in their ability to metabolize sugars (Thiele, 1968). For purposes of this discussion, the metabolism of glucose and fructose by *Rb. sphaeroides* and *Rb. capsulatus* has been the best described. Both organisms transport these sugars via the phosphotransferase system (Saier et al., 1971). Early studies by Szymona and Doudoroff (1960) had indicated that *Rb. sphaeroides* uses a modified version of the classic Entner-Doudoroff pathway to metabolize glucose. In this scheme, no phosphorylated intermediates are employed (Conway, 1992) and this 2-keto-3-deoxygluconate (KDG) bypass proceeds as shown in Eq. (5).

In *Rb. capsulatus*, glucose is degraded by the Entner-Doudoroff pathway but fructose metabolism proceeds through the Embden-Meyerhof pathway (Eidels and Preiss, 1970; Conrad and Schlegel, 1974; 1977a,b). In the latter instance, a specific 1-phosphofructokinase is substantially induced under both phototrophic and aerobic growth modes in the presence of fructose (Conrad and Schlegel, 1974). This enzyme is encoded by the *fruK* gene (Wu et al., 1991) which is part of the *fru* operon, composed of other genes that specify enzymes of fructose metabolism (Wu and Saier, 1990; Wu et al., 1990). Interestingly, some homology of an internal portion of FruK to a region of RubisCO large subunits containing residues involved in ribulose 1,5-bisphosphate (RuBP) binding was noted. In *Rb. sphaeroides*, fructose is metabolized via both the Entner-Doudoroff (ED) and Embden-Meyerhof-Parnas (EMP) pathways; aerobically, the cells predominantly use the ED pathway, while under phototrophic conditions, fructose was metabolized by the EMP pathway (Conrad and Schlegel, 1977a). Glucose-grown cells employed only the ED pathway under both phototrophic and aerobic chemotrophic growth conditions (Conrad and Schlegel (1977a). These results are quite fascinating, especially in *Rb. sphaeroides*, as they point to sophisticated molecular controls over the expression of enzymes of sugar dissimilation that are dependent on both the growth substrate and the levels of oxygen. Conditions which lead to the induced metabolism of either glucose or fructose should result in the repression of enzymes of the Calvin cycle, which is involved in carbohydrate biosynthesis. More than likely when the Calvin cycle (*cbb*) operons are switched on, the sugar dissimilatory system will be off. A fruitful area of investigation would be to isolate regulatory elements which control or influence the molecular 'switching' between carbohydrate synthesis and catabolism. Such studies might also relate to the control of oxidant-dependent sugar fermentation catalyzed by both *Rb. sphaeroides* and *Rb. capsulatus* (Madigan et al., 1980). This special type of energy conversion is discussed in the Chapter by Zannoni of this volume.

Finally, any discussion of heterotrophic metabolism (summarized in Fig. 1), under both aerobic or phototrophic growth conditions, must consider that the TCA cycle plays an essential role in the metabolism of all PNS bacteria. In *Rb. capsulatus*, it was found that molecular oxygen regulates through a rather specific effect on the synthesis of the α-ketoglutarate dehydrogenase (αKDGH) enzyme complex; low basal levels of αKGDH found under phototrophic growth conditions are presumably required to generate sufficient succinyl CoA for the biosynthesis of metalloporphyrins required for photosynthesis (Beatty and Gest, 1981a). Aerobic metabolism, however, exacts a high requirement for reduced pyridine nucleotide for ATP biosynthesis. Thus, αKGDH is induced under aerobic conditions and the levels of isocitrate dehydrogenase and citrate synthase are also elevated (Beatty and Gest, 1981b).

III. Calvin Reductive Pentose Phosphate Pathway

Classic label incorporation experiments performed many years ago (reviewed in Tabita, 1988) established that, much like green plants, PS and PNS bacteria assimilate $^{14}CO_2$ primarily into 3-phosphoglyceric acid (3-PGA). The kinetics of incorporation, plus the identification of additional products of CO_2 fixation unequivocally established that the Calvin-Benson-Bassham reductive pentose phosphate cycle is the major route for CO_2 reduction (Glover et al., 1952; Stoppani et al., 1955). With regard to the PNS bacteria it is interesting that the conditions under which the cells were grown (in a medium containing malate as

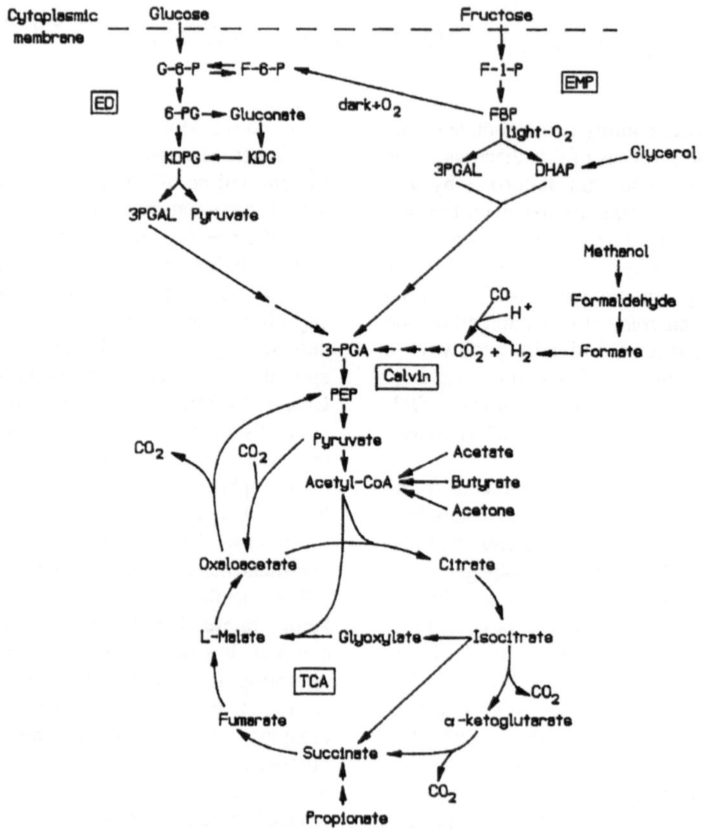

Fig. 1. Composite reactions of central carbon metabolism in purple nonsulfur bacteria. ED, Entner Doudoroff pathway; EMP, Embden-Meyerhof pathway; Calvin, Calvin reductive pentose pathway; TCA, tricarboxylic acid cycle. Modified from Conrad and Schlegel (1977a) and Willison (1988).

electron donor and primary carbon source) result in a net release of CO_2 into the growth medium (section II C). Thus, at least part of the CO_2 produced as a result of the metabolism of the oxidized carbon source is assimilated through the Calvin pathway. Presumably under these conditions the Calvin cycle is used more as an anaplerotic carbon assimilatory route. These studies were followed by the actual demonstration of RubisCO activity in extracts of *Rs. rubrum* and *Chromatium* sp. (Fuller and Gibbs, 1959). Based on the time-dependent isolation of labeled intermediates of the pathway, as well as measurement of several enzyme activities, a cyclic pathway was established (Fig. 2). Aside from the pyridine nucleotide specificity of the glyceraldehyde phosphate dehydrogenase reaction, the scheme is the same in prokaryotes and eukaryotes. There are two unique reactions, those catalyzed by RubisCO, the actual CO_2 assimilatory enzyme, and phosphoribulokinase, which catalyses the ATP-dependent phosphorylation of ribulose 5-phosphate (Ru-5-P) to form RuBP. In PNS bacteria, one enzyme, fructose 1,6-/sedoheptulose 1,7- bisphosphatase, catalyzes the removal of inorganic phosphate from fructose 1,6-bisphosphate (FBP) and sedoheptulose 1,7-bisphosphate (SBP), respectively. Examination of the reactions involved in the metabolism and reduction of 3-PGA indicate many familiar enzymes of intermediary metabolism. These reactions are devoted to the regeneration of the CO_2 acceptor, RuBP, with the net result that one molecule of triose phosphate is produced from three molecules of CO_2. In the following discussion, the properties of each of the important enzymes are reviewed and considered in conjunction with the

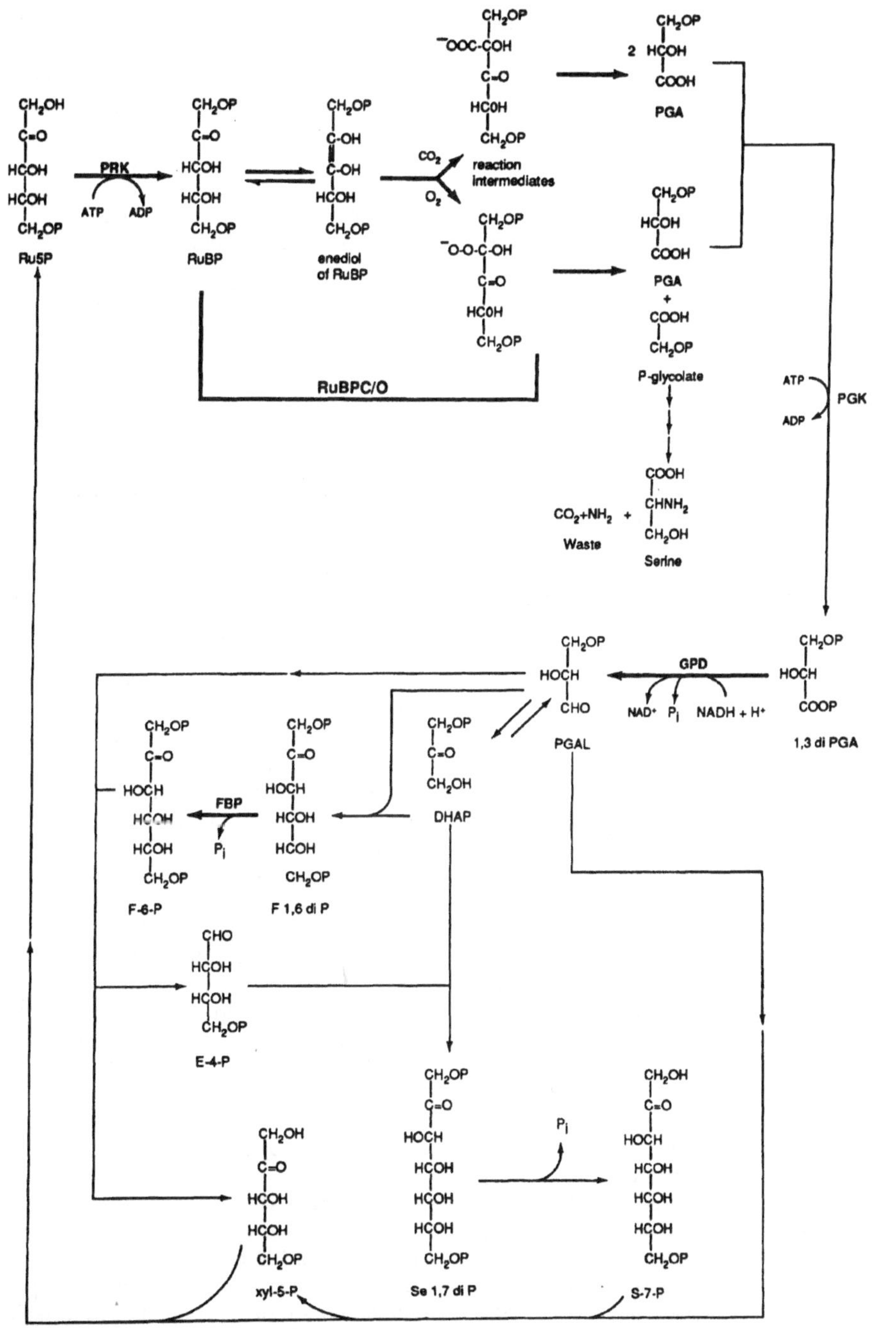

Fig. 2. Calvin reductive pentose phosphate pathway, highlighting the two reactions catalyzed by RubisCO (RuBPC/O). Other key enzymes including phosphoribulokinase (PRK), phosphoglyceric acid kinase (PGK), glyceraldehyde phosphate dehydrogenase (GPD), aldolase (FBA), and fructose bisphosphatase (FBP) are highlighted. From Tabita (1988), with permission.

regulation of CO_2 assimilation by purple photosynthetic bacteria.

A. RubisCO

This important catalyst has generated considerable interest over the years, mainly due to its capacity to catalyze the first reaction of two competing metabolic pathways, the Calvin reductive CO_2 assimilatory pathway and an oxidative pathway involving the oxygenolysis of RuBP and subsequent oxidative metabolism of phosphoglycolate, the product of RuBP oxygenolysis. The two major physiological reactions catalyzed by RubisCO (Fig. 2), carboxylation and oxygenolysis, are catalyzed by a single protein at a single active site; the overall significance of both reactions to agricultural productivity has resulted in this enzyme becoming one of the most intensely studied of all proteins. RubisCO is the most abundant protein found in nature (Ellis, 1979) and is an extremely poor catalyst with a k_{cat} of about 3 sec^{-1}, undoubtedly a contributing factor to its massive synthesis in organisms that employ the Calvin cycle to grow with CO_2 as sole carbon source. The enzyme itself may be isolated as a complex hexadecameric protein (type I or form I RubisCO) comprised of eight large (catalytic) subunits (M_r of about 55,000) and eight small subunits (M_r about 15,000). The type I enzyme thus has an L_8S_8 arrangement of subunits. Type II (form II) RubisCO, found in several PNS bacteria (Tabita, 1988), is comprised of only large subunits (either L_2, L_4, or L_8). The large subunits from both types are unrelated, a fact that was first noted when the deduced amino acid sequence of the type II enzyme of *Rs. rubrum* was compared to various type I large subunits (Nargang et al., 1984) and when peptide maps of the large subunits of form I and form II proteins of *Rb. sphaeroides* were compared (Gibson and Tabita, 1985). Subsequently the genes of the two distinct *Rb. sphaeroides* proteins were sequenced (see Chapter by Gibson) and the deduced amino acid sequences clearly showed that there was only about 25% identity. Moreover, the form I and form II RubisCO genes are found on separate chromosomal genetic elements (Suwanto and Kaplan, 1989). A recent review traces the history of the discovery of the different structural forms of RubisCO in photosynthetic bacteria (Tabita, 1988).

The type II RubisCO from *Rs. rubrum*, which is a simple homodimer of large subunits (Tabita and McFadden, 1974b), has become the paradigm for structure-function studies of this enzyme (Schneider et al., 1990). Indeed one of the more interesting findings has been the universality of the dimer as a basic structural motif, even in more complex type I RubisCO enzymes. Thus in the type I enzyme, there are essentially four *Rs. rubrum*-like dimers that form the octameric core of the protein, most accurately depicted as an $(L_2)_4$ arrangement (Knight et al., 1990). Decorating both the top and bottom of the octameric core are 4 small subunits, such that the L_8S_8 structure in reality is an $(L_2)_4(S_2)_4$ type structure. The active site is formed from the interface between the N-terminal domain of one subunit and the α/β barrel of the C-terminal domain of the second subunit of the dimer with at least two subunits required to form the complete active site (Schneider et al., 1990; Knight et al., 1990; Schreuder et al., 1993). Like most α-β barrel enzymes, various loops appear to be extremely important for catalysis; for RubisCO loop 6 is particularly cogent since this 11-amino acid region appears to be involved with stabilizing the carboxy and peroxy intermediates that form during the carboxylase and oxygenase reactions, respectively. Of particular importance in this region is Lys-329 of the *Rs. rubrum* enzyme (Lys-334 of type I RubisCO), which is absolutely required for activity (Hartman and Lee, 1989) and whose function appears to be influenced by other residues found in loop 6 or which contact loop 6 (Spreitzer, 1993). Although not within the purview of this review, various aspects of the structure of RubisCO that relate to its function have greatly influenced the experimental approaches that have been taken by various groups to elucidate the catalytic mechanism and the basis for CO_2/O_2 specificity. For more information on this interesting topic, the reader is directed to several papers that relate to the X-ray structure of RubisCO (Schneider et al., 1990; Knight et al., 1990; Schreuder et al., 1993).

1. Biochemical Properties of RubisCO from Purple Photosynthetic Bacteria

Before catalysis may proceed, RubisCO must first be activated. This involves carbamylation of the ε-amino group of Lys-191/Lys-201 with 'activator' CO_2 in a slow, rate-determining step, followed by the rapid addition of divalent cation to form a stable ternary complex, which then may catalyze either the carboxylation or oxygenolysis of RuBP (see Eq. (6)) where E represents RubisCO; C is CO_2; M is divalent

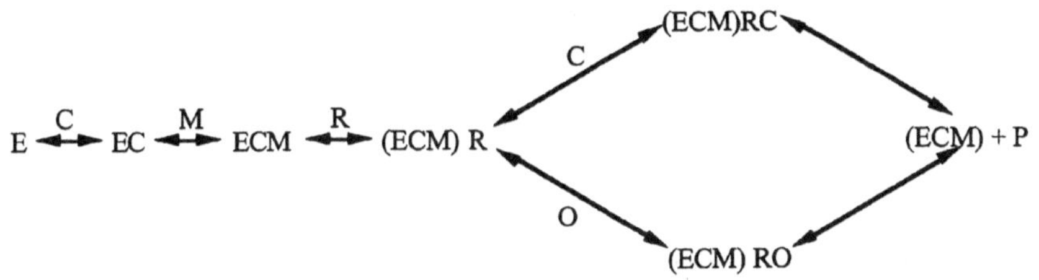

(6)

cation; R is RuBP; O is oxygen; and P is product (Miziorko and Lorimer, 1983). This basic catalytic pathway is obligatory for all sources of RubisCO and readily illustrates the well described competition of CO_2 and O_2 for the enzyme bound enediolate of RuBP. It should also be noted that in many instances RuBP prevents activation by forming a tight E-R complex, basically locking the enzyme in a conformation that prevents the obligatory carbamylation (activation). The capacity to form the E-R complex is a property which accentuates differences between type II enzymes and most type I enzymes. For the *Rs. rubrum* type II enzyme, early studies indicated that preincubation with RuBP does not drastically inhibit activity (Tabita and McFadden, 1974a; Christeller and Laing, 1978; Whitman et al., 1979). These differences are readily illustrated by comparing the responses of the form I and form II *Rb. sphaeroides* enzymes (Fig. 3). The only known type I enzymes that do not show RuBP-mediated inhibition are the enzymes isolated from cyanobacteria (Tabita, 1994). In addition to differences in the response to RuBP, type I and II enzymes show pronounced differences in their sensitivity to phosphorylated metabolites such as 6-phosphogluconate (Tabita and McFadden, 1972; Gibson and Tabita, 1977; 1979; Whitman et al., 1979). Perhaps the most impressive differences in catalytic properties involve comparisons of the Michaelis constants, particularly for CO_2 and O_2, and the $V_{CO_2}K_{O_2}/V_{O_2}K_{CO_2}$ ratio or substrate specificity factor (τ). Jordan and Ogren (1981) compared the substrate specificity factor and kinetic constants of RubisCO from a number of photosynthetic organisms. This ratio (τ) contrasts the relative rates of the carboxylase and oxygenase reactions at any given CO_2 and O_2 concentration and is derived from the equation (Laing et al., 1974):

$$v_{CO_2}/v_{O_2} = (V_{CO_2}K_{O_2}/V_{O_2}K_{CO_2})([CO_2]/[O_2])$$
$$= \tau([CO_2]/[O_2]) \quad (7)$$

where V_{CO_2} and K_{CO_2} represent the maximal velocity and Michaelis constant, respectively, for CO_2; V_{O_2} and K_{O_2} are the maximal velocity and Michaelis constant for O_2; and v_{CO_2} and v_{O_2} are the initial velocities for the carboxylase and oxygenase reactions, respectively. Using a simultaneous and sensitive dual label assay to measure the two activities,

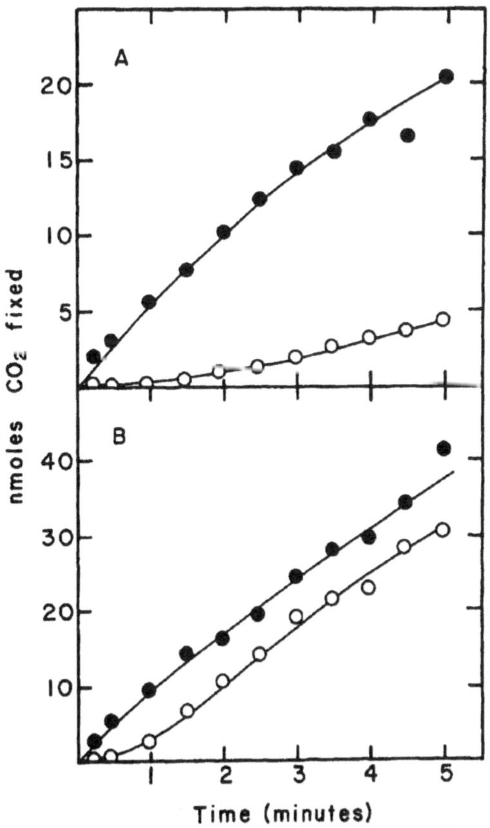

Fig. 3. Time course of *Rb. sphaeroides* form I (A) or form II (B) RubisCO activity after preincubation in the presence of 0.4 mM RuBP (○) or 10 mM Mg^{2+} and 20 mM HCO_3^- (●). From Gibson and Tabita (1979), with permission.

it was found that the type II enzymes from *Rs. rubrum* and *Rb. sphaeroides* have considerably lower τ values than the enzyme from higher plants. Indeed there seems to be a gradation in the τ values from higher plants to type II enzymes. Interestingly the τ value for the form I enzyme of *Rb. sphaeroides* resembled values obtained for eukaryotic green algal enzymes, but was somewhat lower than the plant enzyme. The major reason for the low τ value of the two type II enzymes appears to be primarily due to the high K_{CO_2} for these enzymes, which also appears to be true of the low specificity factor cyanobacterial, vent symbiont (Stein and Felbeck, 1993), and *Chromatium* (Jordan and Chollet, 1985) enzymes. More recently the substrate specificity value for several additional microbial enzymes have been determined and these data, along with results summarized from earlier studies are compared (Table 1). In all cases, the spinach enzyme has been used as an internal standard and all τ values are based on the double label assay; the type I enzymes have been grouped according to their amino acid sequence relatedness. Obviously, there is a great need for additional determinations amongst the bacterial type I RubisCO subclasses, particularly for additional PS photosynthetic bacterial and sulfur bacterial enzymes of Type IA and for PNS bacterial and chemolithotrophic bacterial enzymes of Type IC. Now that additional Type II enzymes have been isolated (English et al., 1992; Chung et al., 1993), it will be important to determine their sequence and the relevant kinetic constants so that comparisons may be made. The most important question facing RubisCO biochemists continues to involve elucidation of the determinants which influence CO_2 and O_2 substrate specificity. As noted earlier, the importance of loop 6 is well appreciated. However, it is also apparent from site-directed mutagenesis and genetic studies (Hartman

Table 1. Classification of diverse RubisCO enzymes[a]

RubisCO type	Organism	$V_{CO_2}K_{O_2}/V_{O_2}K_{CO_2}(\tau)$[b]	K_{CO_2} (μM)
Type IA	*Thiobacillus ferrooxidans*	N.D.[c]	N.D.
	Vent symbiont	30	80
	Chromatium vinosum	40	35
Type IB	Cyanobacteria		
	Anabaena 7120	35	150
	Synechococcus	40	150
	Green algae		
	Chlamydomonas reinhardtii	60	30
	Plants—many species	80	10–30
Type IC	Purple bacteria class		
	Rhodobacter sphaeroides	60	35
	Alcaligenes eutrophus	75	N.D.
	Bradyrhizobium japonicum	75	N.D.
	Xanthobacter flavus	40	N.D.
Type ID	Marine nongreen algae		
	Cylindrotheca sp. strain N1	105	30
	Olisthodiscus luteus	100	60
	Porphyridium cruenteum	130	20
	Cylindrotheca fusiformis	110	35
Type II	*Rhodospirillum rubrum*	15	100
	Rhodobacter sphaeroides II	10	100
	Thibacillus denitrificans II	10	250
	Hydrogenovibrio marinus	N.D.	N.D

[a] RubisCO enzymes have been classified according to their sequence relatedness; only those enzymes for which sequence information and/or kinetic data are available are presented.

[b] Values (rounded and averaged) obtained from the dual label assay except the *Anabaena* and vent symbiont enzmes. From Jordan and Ogren (1981); Jordan and Chollet (1985); Lee et al. (1991); Read and Tabita (1992 a,b; 1994); Chung et al. (1993); Larimer and Soper (1993); Stein and Felbeck (1993); Horken and Tabita (unpublished); Hernandez and Tabita (unpublished).

[c] Not determined

and Harpel, 1993, 1994; Spreitzer, 1993) that other unknown determinants remain to be discovered. A good representative sample of τ values for additional bacterial enzymes, in which the sequence is known, should give some idea as to what regions have been altered. Indeed, recent studies and comparisons of this type have pointed to several distinct changes in residues that may influence loop 6 of several nongreen algal RubisCOs, enzymes which all have τ values higher than plant RubisCO (Read and Tabita, 1992a,b; 1994). These comparisons, in conjunction with X-ray modeling studies and computer programs that can simulate perturbations or influences in the active site, will be most important for designing future enzyme engineering approaches. Obviously, if one could design the perfect RubisCO, it would be highly active and possess a high substrate specificity. At present, genetic engineering approaches have resulted in the isolation of RubisCO enzymes with altered, and even increased substrate specificity, however the enzymes uniformly exhibit low k_{cat} values (reviewed in Spreitzer, 1993; Hartman and Harpel, 1993, 1994).

2. Biosynthesis and Assembly of RubisCO

It is apparent that the complexity of RubisCO from photosynthetic bacteria ranges from the simple dimeric structure of the *Rs. rubrum* type II enzyme to the hexadecameric type I structure found in most purple photosynthetic bacteria. At this time, only *Rs. rubrum* contains the simple dimer; this structural form has not been isolated from any other photosynthetic bacterium, not even from other species purported to be members of the genus *Rhodospirillum* (Tabita, 1988). The fact that all other type II enzymes from PNS bacteria that have been described are found as higher aggregates, implies that the *Rs. rubrum* dimer is unusual and lacks some key assembly determinant(s) that prevent assembly into higher aggregated forms. Inasmuch as the recombinant *Rs. rubrum* enzyme is assembled into an active dimer in *Escherichia coli* (Sommerville and Sommerville, 1984), the assembly determinants must be related to the specific sequence. The same may be said for the *Rb. sphaeroides* form II enzyme, in which aggregates ranging from dimers to structures greater than L_8 may be found (Gibson and Tabita, 1977) as well as the apparently native tetrameric/hexameric structure formed in vivo by both the parent organism and *E. coli* (Tabita et al., 1986). The preparation of truncated recombinant *Rs. rubrum* enzymes by site-directed mutagenesis indicated that removal of the C-terminal tail section of the protein resulted in its assembly to dimers, octamers, or both, depending on the site of truncation (Ranty et al., 1990). The L_8 mutant proteins exhibited very low levels of activity and the removal of the C-terminal helical regions appeared to expose new hydrophobic surfaces to solvent. Thus, the authors postulate that the newly exposed hydrophobic surfaces interact to form the higher aggregate. This hypothesis seemed valid since the introduction of a polar residue within the new hydrophobic region resulted in the formation of dimers. Analogous work yielded an 18-residue truncated protein (Morell et al., 1990) whose properties were totally different from the 17-residue truncation prepared by Ranty et al., (1990). In the 18-residue truncated mutant enzyme, activity was quite similar to the wild-type protein and formed the usual dimer structure. The 18-residue truncation contains a new C-terminal Leu which replaces the Tyr-Pro C-terminus of the 17-residue truncation. However, as pointed out by the authors, it is difficult to understand how these sequence differences yield such profound structural and catalytic differences (Morell et al., 1990). In any case, these studies illustrate that localized structural determinants of the large subunits directly influence the final assembly product.

Previous studies on the biosynthesis of RubisCO from *Chromatium vinosum* have shown that the synthesis of both large and small subunits is synchronized and maximized under autotrophic growth conditions in the presence of reduced sulfur compounds (Kobayashi and Akazawa, 1982a,b). A constant ratio of large and small subunits was achieved throughout the induction. Taken together, these results were interpreted to indicate that the synthesis of the two subunits is coordinately regulated. Although similar studies have not been performed for other type I bacterial enzymes, it would be fairly safe to predict that similar results would be obtained. Indeed, now that it is known that the large and small subunit genes of all type I RubisCO enzymes (*cbbL* and *cbbS*) (Tabita et al., 1992) are cotranscribed as part of an operon in all bacteria (Tabita, 1988), the molecular basis for the *Chromatium* findings is readily apparent.

Despite the obvious requirement for specific residues on the large subunits, research on the assembly of RubisCO has taken on a whole new, more complicated and more interesting dimension. These studies have primarily focused on the simple dimeric enzyme from *Rs. rubrum*, where it has been shown that accessory proteins or 'molecular chaperones' (Ellis, 1990) are required to assist the assembly of unfolded monomers or nascent polypeptide chains along a pathway that favors the formation of the correctly folded and properly arranged active dimer (Goloubinoff et al., 1989a,b). Complexes of chaperones and newly synthesized proteins are often found to be associated in the cell. These can then dissociate, often through the mediation of other proteins and the energy obtained from ATP hydrolysis, to form the final functional oligomeric protein and free chaperone protein. For RubisCO, there is an interesting history relative to the discovery of the 'RubisCO-binding protein' (Barraclough and Ellis, 1980; Roy et al., 1982) and its remarkable homology to the product of the *groEL* gene of *E. coli* (Hemmingsen et al., 1988), which along with *groES*, comprise the *groEL* heat shock operon (Zeilstra-Ryalls et al., 1991). Both the *Rs. rubrum* and cyanobacterial RubisCO genes, *cbbM* and *rbcLrbcS*, respectively, when expressed in *E. coli*, were shown to require the GroEL (cpn60) and GroES (cpn10) proteins for proper assembly in vivo (Goloubinoff et al., 1989a), and for the *Rs. rubrum* protein, in vitro as well (Goloubinoff et al., 1989b; Viitanen et al., 1990). Homologs of cpn60 have been isolated from *Chromatium vinosum* (McFadden et al., 1989; Torres-Ruiz, 1989, 1992) and *Rb. sphaeroides* (Watson et al., 1990; Terlesky and Tabita, 1991) and the *groE* genes from both organisms have been isolated and sequenced (Ferreyra et al., 1993; Lee, Terlesky, and Tabita, unpublished). In each case, the cpn60 levels were shown to vary according to the growth conditions, following the same pattern exhibited for RubisCO synthesis (McFadden et al., 1989; Terlesky and Tabita, 1991). The *Cm. vinosum groE* genes were functional in *E. coli*; indeed, in a very interesting experiment, lambda clones containing the *Cm. vinosum* genes were shown to form plaques in a *groE* host strain of *E. coli*, indicating that the *Chromatium* genes could replace the well-known requirement of the *E. coli* genes for bacteriophage assembly (Zeilstra-Ryalls et al., 1991). Moreover, expression of the *Cm. vinosum* genes in *E. coli* was enhanced by heat-shock and a putative heat shock promoter was identified. The *Cm. vinosum* cpn60 and cpn10 proteins, or mixtures of these proteins and *E. coli* chaperonin proteins, were effective in enhancing productive assembly of the *Rs. rubrum*, *Synechococcus*, and *Cm. vinosum* RubisCO enzymes in *E. coli* (Ferreyra et al., 1993). Since the L_2 dimer is the basic structural motif (Andersson et al., 1989), a model for chaperonin-mediated RubisCO assembly for the *Rs. rubrum* dimer (type II) and type I cyanobacterial enzymes has been proposed (Goloubinoff et al., 1989a). For *Rb. sphaeroides*, containing a more aggregated type II enzyme, as well as the conventional L_8S_8 type I enzyme, a similar assembly pathway is envisioned (Fig. 4) that shows the involvement of the products of the form I and form II operons, and the *groE* operon for RubisCO assembly in this organism. Although the form I and form II operons are located on the separate chromosomal elements of *Rb. sphaeroides* (Suwanto et al., 1989), it is not known at this juncture where the *groE* operon maps. As discussed above, the levels of cpn60 are maximized under growth conditions that favor maximum RubisCO synthesis (Terlesky and Tabita, 1991), however, the molecular basis for this observation has not been elucidated. Finally, *Cm. vinosum* RubisCO, presumably encoded by one of the two sets of type I RubisCO genes found in this organism (Viale et al., 1989), was isolated in both the L_8S_8 and L_8 forms, with the L_8 form presumably derived from the L_8S_8 enzyme and fully active in the absence of small subunits (Torres-Ruiz and McFadden, 1985, 1987). Although the N-terminal amino acid sequence of the large subunit of these two enzymes is identical, it differs somewhat from the amino acid sequences deduced from the two type I large subunits genes of this organism (Kobayashi et al., 1991). This potentially interesting system has not been studied further. If it is verified that L_8 subunits are fully active, current perceptions as to the requirements of small subunits for catalysis (Andrews, 1988; Lee and Tabita, 1990; Smrcka et al., 1991) will need to be revised as will the ability of type I large subunits to form assembly domains that favor activity.

3. In vivo Regulation of RubisCO Activity

As a key enzyme in CO_2 assimilation, there have been several attempts to relate physiological responses to the activity of RubisCO. Perhaps the first indication that the activity of the enzyme might be regulated in vivo came from studies with *Chromatium* and *Thiopedia (Thiocapsa)* spp. (Hurlbert and Lascelles, 1963). When thiosulfate-bicarbonate grown *Chromatium* was transferred to a medium containing pyruvate,

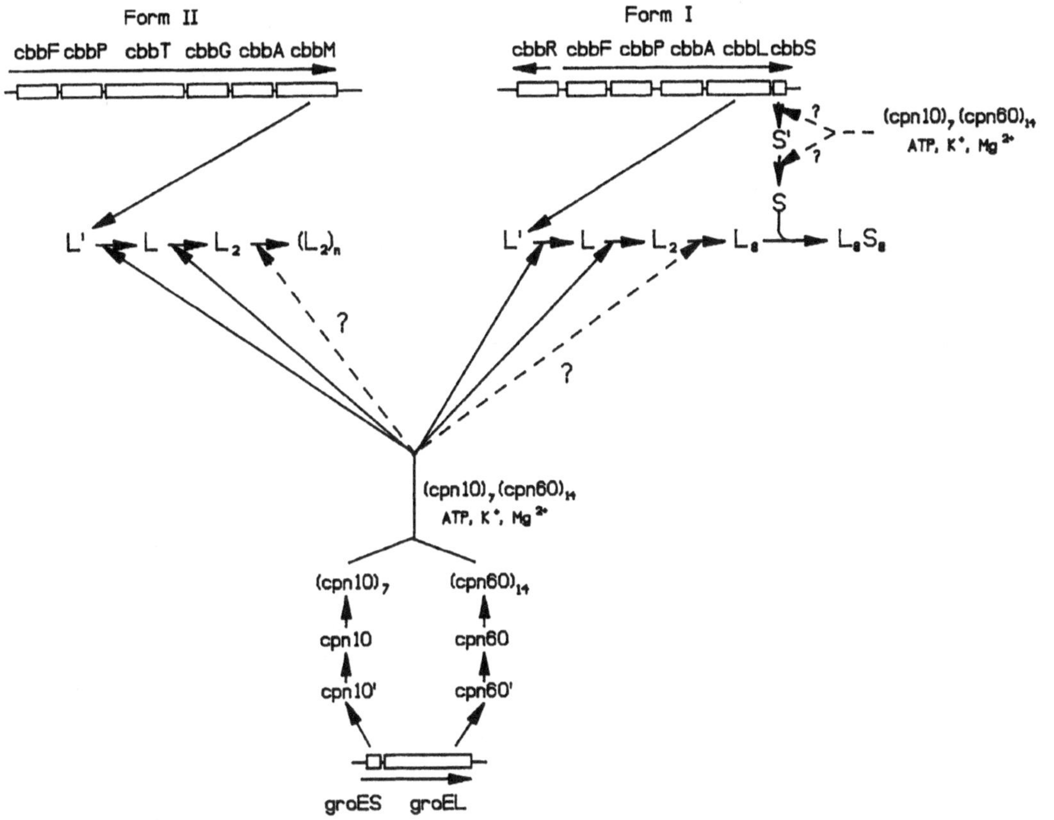

Fig. 4. Integration of form I and form II RubisCO expression and assembly with *groE* transcription and synthesis of chaperonin proteins. Based on the model of Goloubinoff et al. (1989a). L', S', cpn10', and cpn60' refer to the unfolded forms of these proteins after they are produced as nascent proteins. The dashed arrows and question mark (?) refer to assembly steps that have not been shown to be chaperonin-associated.

a rapid decrease in RubisCO activity was obtained that was dependent on protein synthesis. It was further shown that cells placed in a thiosulfate-pyruvate (mixotrophic) growth medium blocked the normal increase in RubisCO activity characteristic of cells growing solely in a thiosulfate-bicarbonate medium. So long as there was oxidizable thiosulfate in the mixotrophic growth medium, the level of activity remained constant. Depletion of thiosulfate led to a rapid decrease in RubisCO activity. Excess CO_2, as well as pyruvate, was subsequently shown to affect the level of activity (Kobayashi and Akazawa, 1982a). These results clearly indicate that some alteration of the enzyme occurs in the cell, either through the binding of some inhibitor or through some modification in the structure of the enzyme. Various phosphorylated metabolites are known to modulate the activity of RubisCO (Tabita, 1988), and the intracellular level of many of these metabolites correlates with enzyme activity in two different cyanobacteria (Pelroy and Bassham, 1972, 1976). Most interesting, these effectors appeared to be able to regulate the activity of the enzyme in the intracellular milieu as opposed to in vitro (Tabita and Colletti, 1979). In *Rs. rubrum*, in situ assays of toluene-permeabilized cells indicated that the affinity of the enzyme for RuBP was lower than the enzyme assayed in vitro (Störro and McFadden, 1983). Taken together, these in situ activity measurements indicated that the enzyme microenvironment, obviously different in situ and in vitro, influences RubisCO activity. Certainly in plants, several additional factors contribute to the modulation of RubisCO activity in vivo, including light-dependent synthesis of tightly-binding specific inhibitors (Gutteridge et al., 1986; Berry et al., 1987) and the enzyme RubisCO activase (Portis, Jr. 1992). Whether these rather specific inhibitory compounds and/or RubisCO activase play any role in the control of RubisCO activity in purple photosynthetic bacteria, remains to be determined.

In *Rb. sphaeroides*, the inactivation/reactivation of RubisCO in vivo appears to play an important role in modulating activity. Much like *Cm. vinosum* (Hurlbert and Lascelles, 1963), the addition of organic compounds, particularly pyruvate, to photolithoautotrophically-grown (1.5% CO_2/98.5% H_2) cultures, results in a drop in RubisCO activity. This effect was mimicked in autotrophic cultures after the concentration of CO_2 was raised from the usual 1.5% to 6% (Jouanneau and Tabita, 1987). Immunological assays indicated that the levels of both form I and form II RubisCO protein remained constant during the time activity was lost, suggestive of some modulation of enzyme activity not related to protein degradation. Isolation and separation of inactivated from active enzyme further indicated that only the form I RubisCO was affected. Complicating things, the inactivated form I enzyme exhibited a marked propensity to become reactivated in vitro (Jouanneau and Tabita, 1987). Subsequent studies showed that inactivation occurred by some process that was reversible in vivo, since activity recovered soon after the cells consumed pyruvate or α-ketoglutarate from the growth medium in the presence of protein synthesis inhibitors (Wang and Tabita, 1992a). Reversibility of inactivation, however, occurred only when the culture medium contained saturating levels of ammonia; if the cells depleted the available ammonia during the time they metabolized added pyruvate, recovery did not occur. Subsequent studies showed that blockage of the major ammonia assimilatory enzyme, glutamine synthetase (GS), with the specific inhibitor methionine sulfoximine (MSX), prevented the recovery of pyruvate-mediated inactivation (Wang and Tabita, 1992b). Indeed, recovery occurred only at concentrations of MSX that failed to completely inhibit in situ GS activity. These results were interpreted to indicate that the intracellular ratio of glutamine/α-ketoglutarate, which effectively defines the nitrogen status of bacteria and initiates a complex regulatory cascade in many organisms including *Rb. sphaeroides* (Wang and Song, 1989), may contribute to the regulation of both the carbon and nitrogen assimilatory systems of this organism. Studies with a RubisCO deletion mutant of *Rb. sphaeroides* showed that there was specificity of the inactivation system for *Rb. sphaeroides* form I RubisCO. This was determined after expressing several diverse type I RubisCO genes in the *Rb. sphaeroides* RubisCO deletion strain (Wang and Tabita, 1992a). Moreover, expression of the form I *Rb. sphaeroides* genes in a *Rs. rubrum* RubisCO deletion strain did not result in pyruvate-mediated inhibition of form I RubisCO in this environment, suggesting that this inactivation system may be unique to *Rb. sphaeroides*, although other interpretations of these results are possible (Wang and Tabita, 1992a). Finally, the inactivated form I RubisCO was purified to homogeneity and the propensity of the purified enzyme to become reactivated in vitro found to closely parallel what had been found with partially purified preparations. Interestingly, the purified inactivated enzyme was shown to exhibit mobility on both nondenaturing and sodium dodecyl sulfate gels different from that of the active enzyme prepared from cells not treated with organic acids. However, the Michaelis constants for RuBP, CO_2, or O_2 were not significantly altered. Several factors were found to lead to activation in vitro, including increases in temperature, the levels of Mg^{2+}, and the levels of ATP; the latter two compounds were shown to mediate reactivation in a time-dependent and concentration-dependent process which appeared to be reversible (Wang and Tabita, 1992a,c). Concomitant with the recovery of enzymatic activity, the migration of the inactivated form I RubisCO on gels changed from a pattern that was characteristic of inactivated enzyme to a pattern that was identical to that of the active protein. It was further found that the form I enzyme and cpn60 formed a complex in the presence of ATP making it apparent that the form I enzyme contained a specific ATP-binding site that might contribute to both regulation of activity as well as to the assembly of active enzyme. These results emphasize that some powerful negative effector must be produced in vivo to maintain the enzyme in the inactivated state during the time that the form I RubisCO becomes inactivated after the addition of pyruvate to cultures; otherwise the enzyme would become activated and remain activated by the high intracellular nucleotide pool (Wang and Tabita, 1992c).

To further understand in vivo inactivation of form I RubisCO, it is necessary that the nature and mechanism of the apparent modification be characterized further. Moreover, it is necessary that some genetic component of the system become identified. The demonstration that inactivation is reversible in vivo (and in vitro) and depends on important physiological parameters such as the nitrogen status of the cells, lends credence to the assumption that this is an important physiological mechanism to regulate RubisCO activity in *Rb. sphaeroides*.

Another system for posttranslational modification of RubisCO activity was demonstrated to occur in *Rs. rubrum*. Not surprisingly, O_2 readily inhibits the photoassimilation of CO_2 in PNS bacteria (Khanna et al., 1981), and *Chromatium* and *Rs. rubrum* will excrete glycolate (derived from the oxygenase activity of RubisCO) in the presence of O_2 in the light by a process that is inhibited by CO_2 (Asami and Akazawa, 1974; Störro and McFadden, 1981). Beyond competition for enzyme-bound enediolate, it seems likely that O_2 control might involve RubisCO directly. Thus it was not surprising that from 70–80% of the RubisCO activity was lost when *Rs. rubrum* cells were exposed to atmospheric levels of oxygen over a 12–24 h period (Cook and Tabita, 1988). The quantity of RubisCO protein, determined immunologically and on SDS gels, did not decrease during the time activity was lost. Only after prolonged periods of incubation with oxygen did enzyme degradation occur, suggesting that the inactivated enzyme becomes 'marked' for proteolytic processing, much like *E. coli* GS (Levine et al., 1981). Unlike the organic acid-mediated response noted earlier, oxygen-mediated inactivation of RubisCO appears to be due to some form of irreversible modification mediated by some gene product(s) found in *Rs. rubrum* (Cook and Tabita, 1988). An artificial model system with purified *Rs. rubrum* RubisCO was established to mimic the oxidative modification of the enzyme. This ascorbate-$FeSO_4$ system established that a reactive oxygen species is generated which interacts at a specific site on the noncarbamylated enzyme from several sources; carbamylation protects against inactivation (Cook et al., 1988). In the PS bacterium *Thiocapsa roseopersicina*, it was found that thiosulfate and sulfide were oxidized under both anaerobic photolithoautotrophic and aerobic chemolithoautotrophic growth conditions, in each case using RubisCO and the Calvin cycle to assimilate CO_2 (Kondratieva et al., 1976). Thus, it is apparent that in this organism, a similar O_2-mediated inactivation process does not occur.

Finally, Turner and Mann (1986) found that the major phosphoprotein detected in *Rhodomicobium vannielii* had a M_r of 55,000; this was subsequently proposed to be the large subunit of type I RubisCO based on immunoprecipitation and sucrose density gradient centrifugation experiments (Mann and Turner, 1988). Phosphotyrosine was the modified amino acid identified, however at this time there is no indication that the phosphorylated tyrosine is part of a discrete peptide that coincides with a known RubisCO sequence or that the modified enzyme has been affected in its catalytic activity. The authors provide evidence that the large subunit is phosphorylated and present in a large molecular weight complex prior to its assembly into the mature holoenzyme (Mann and Turner, 1988). Perhaps these results relate to the complexation of cpn60 and form I RubisCO in *Rb. sphaeroides* (Wang and Tabita, 1992c). In any case, further progress on this potentially interesting posttranslational modification of the *Rm. vannielii* enzyme is awaited with great interest.

4. Enzyme Engineering of RubisCO and Mutant Selection

Thus far, type I (*cbbLcbbS*) and type II (*cbbM*) RubisCO genes from *Rs. rubrum*, *Rb. sphaeroides*, and *Cm. vinosum* have been expressed in *E. coli* and purified recombinant proteins have been prepared in each instance (Somerville and Sommerville, 1984; Viale et al., 1985; Gibson and Tabita, 1986; Tabita et al., 1986; Viale et al., 1990). Obviously, this is an obligatory first step before studies may be undertaken to modify or engineer the properties of RubisCO via site-directed mutagenesis. A thermostable RubisCO was prepared from *Chromatium tepidum* (Heda and Madigan, 1988, 1989) and this system might be effectively employed to elucidate the molecular basis for thermal stability or for determining how τ is affected by temperature. To date, only the *Rs. rubrum cbbM* gene has been subject to substantial modification, with the result that an impressive amount of work has been produced which has greatly improved our understanding of the mechanism of catalysis (reviewed in Hartman, 1992; Hartman and harpel, 1993). However, the *Rb. sphaeroides* and *Cm. vinosum cbbLcbbS* genes are excellent potential systems for genetic engineering because they encode enzymes that are fairly well characterized, yet they have substantially different properties compared to the more heavily studied cyanobacterial type I system (Tabita, 1994), including different substrate specificity values (Table 1) and the capacity to form

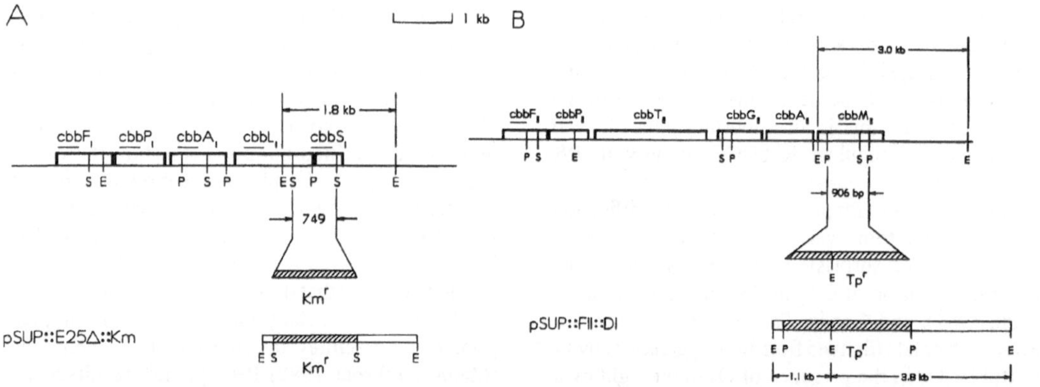

Fig. 5. Construction of a RubisCO double deletion mutant of *Rb. sphaeroides.* (A) The form I gene cluster; (B), the form II gene cluster. The portions of sequence removed within the RubisCO structural genes and the deletion-insertion derivative plasmids used for construction of the deletion mutants are indicated below each cluster. Plasmid diagrams below the maps are drawn to scale to indicate the approximate amount of homologous DNA on either end of the plasmid resistance cartridge insert (hashed lines). From Falcone and Tabita (1991), with permission.

inhibitory E-R complexes. The *Rb. sphaeroides* RubisCO deletion strain, prepared by inserting antibiotic resistant cartridges within the remaining RubisCO gene coding sequences (Fig. 5), has also proven to be a valuable host for the expression of foreign RubisCO genes, using either a specific promoter-vector system or the promoter specific to the RubisCO genes being expressed (Falcone and Tabita, 1991; Falcone and Tabita, 1993a). The value of RubisCO deletion strains of *Rb. sphaeroides* (Falcone and Tabita, 1991) and *Rs. rubrum* (Falcone and Tabita, 1993b), strains 16 and I-19, respectively, lies in the fact that these strains may be complemented to photolithoautotrophic growth via the expression of different RubisCO genes, such that the growth properties of the complemented strain reflects the idiosyncrasies of the expressed RubisCO. For example, *Rb. sphaeroides* strain 16, when it synthesizes the *Synechococcus* 6301 enzyme, does not grow at low CO_2 levels, due to the inherently poor Michaelis constant for CO_2 of the cyanobacterial enzyme (Falcone and Tabita, 1991). In addition, various cyanobacterial mutant and hybrid enzymes that have weak catalytic activity, exhibit very poor growth rates (Hernandez and Tabita, unpublished). Such results suggest that methods to select additional mutant enzymes may be developed by randomly mutagenizing the RubisCO genes of interest, followed by transformation and complementation of strains 16 or I-19 with the mutagenized DNA. Subsequent growth, under environmental conditions that favor the trait desired, should result in the selection of novel mutant enzyme forms. It is anticipated that biological selection of RubisCO mutations using photosynthetic bacterial hosts will result in the isolation of a number of interesting mutant forms of RubisCO that should greatly supplement those that may be isolated using the *Chlamydomonas* system (Spreitzer, 1993). Moreover the bacterial system provides a facile approach to screen the in vivo consequences of changes in RubisCO substrate specificity made by in vitro genetic engineering.

B. Phosphoribulokinase (PRK)

PRK, the other unique enzyme of the Calvin cycle, has recently received increased scrutiny. The enzyme from *Rb. capsulatus* was the first preparation of electrophoretically homogenous enzyme to be isolated from any source (Tabita, 1980). Early studies with partially purified *Cm. vinosum* PRK had indicated that the enzyme exhibited complex properties, including sigmoidal substrate saturation curves indicative of allosteric behavior (Hart and Gibson, 1971). However, unlike the enzyme from PNS bacteria (Rindt and Ohmann, 1969; Tabita, 1980; Rippel and Bowien, 1984; Tabita, 1988), the *Cm. vinosum* enzyme does not require NADH for activity; in this respect the *Cm. vinosum* enzyme resembles the enzyme of oxygen-evolving photosynthetic organisms (Tabita, 1980; 1987). Furthermore, the activity of enzymes isolated from all purple bacteria is negatively regulated by compounds such as AMP, 3-PGA, 3-PGAL, and PEP (Tabita, 1988).

Beyond these observations, thorough kinetic characterization of homogenous bacterial PRK has not appeared in the literature as yet. With respect to the PNS bacteria, PRK genes have now been cloned from *Rb. sphaeroides* (Gibson and Tabita, 1987; Hallenbeck and Kaplan, 1987), *Rs. rubrum* (Falcone and Tabita, 1993b), and *Rb. capsulatus* (G.C. Paoli, J.S. Shively, and F.R. Tabita, unpublished). Interestingly, there are two PRK enzymes in *Rb. sphaeroides*, the products of the $cbbP_I$ and $cbbP_{II}$ genes (formerly *prkA* and *prkB*), which are found on Calvin cycle operons I and II; recombinant PRKI and PRKII have each been purified to homogeneity after expression of the respective genes in *E. coli*. In the host organism, the enzyme is isolated as a tight complex of the individual PRKI and PRKII subunits (Gibson and Tabita, 1987), yet each individual PRK enzyme appears to be functional in *Rb. sphaeroides* since mutations in $cbbP_I$ or $cbbP_{II}$ allow for phototrophic growth with CO_2 as carbon source and/or electron acceptor (Gibson et al., 1990; Hallenbeck et al., 1990; Gibson et al., 1991). Analysis of the deduced amino acid sequence of bacterial *cbbP* genes from *Rb. sphaeroides* (Gibson et al., 1990; Gibson et al., 1991), *Alcaligenes eutrophus* (Kossmann et al., 1989), and *Xanthobacter flavus* (Meijer et al., 1990), indicates very close homology. Interestingly, however, only slight sequence homology of bacterial PRKs towards the plant (Milanez and Mural, 1988; Lloyd et al., 1991), algal (Roesler and Ogren, 1990), or cyanobacterial (Su and Bogorad, 1991) enzymes is apparent; for example the plant enzymes share only

A

Source	Residues	●○○○○ ●● ● ○○
RsI	8-20	ISVTGSSGAGTST
RsII	8-20	ISVVGSSGAGTST
Ae	8-20	IAITGSSGAGTTS
Xf	8-20	IVVTGSSGAGTTS
Sp	8-20	IGLAADSGCTKST
Wh	8-20	IGLAADSGCGKST
Ch	8-20	IGLAADSGCGKST
Syn	11-23	IGVAGDSGCGKST

B

		● ○● ○● ○
RsI	42-53	DAFHRFNRADRK
RsII	42-53	DAFHRFNRADMK
Ae	42-53	DSFHRYDRAEMK
Xf	42-53	DSFHRYDRYEMR
Sp	57-58	DDFHSLDRNGRK
Wh	57-68	DDYHSLDRTGRK
Ch	57-68	DDYHCLDRNGRK
Syn	43-54	DDYHSLDRQGRK

C

		○ ●●●●
RsI	127-134	LLFYEGLH
RsII	127-134	LLFYEGLH
Ae	127-134	LLFYEGLH
Xf	127-134	ILFYEGLH
Sp	121-128	ILVIEGLH
Wh	121-128	IFVIEGLH
Ch	121-128	ILVIEGLH
Syn	107-114	VVVIEGLH

D

		●○ ●● ●● ○ ● ○ ● ● ○ ● ●
RsI	165-198	KIHRDRATRGYTTEAVTDVILRRMHAYVHCIVPQ
RsII	165-198	KIHRDRATRGYTTEAVTDVILRRMYAYVHCIVPQ
Ae	165-198	KLWRDKKQRGYSTEAVTDTILRRMPDYVNYICPQ
Xf	165-198	KIHRDKATRGYTTEDVTDTIMRRMPDYVRYICPQ
Sp	156-189	KIQRDMKERGHSLESIKASIESRKPDFDAYIDPQ
Wh	156-189	KIQRDMAERGHSLESIKASIEARKPDFDAFIDRQ
Ch	156-189	KIQRDMAERGHSLESIKSSIAARKPDFDAYIDPQ
Syn	142-175	KIQRDMAERGHTYEDILASINARKPDFTAYIEPQ

Fig. 6. Conserved amino acid sequence regions (A,B,C, and D) in PRK from various sources; RsI and RsII, *Rb. sphaeroides* form I and form II PRK (Gibson et al., 1990; 1991); Ae, *A. eutrophus* (Kossmann et al., 1989); Xf, *X. flavus* (Meijer et al., 1990); Sp, spinach (Milanez and Mural, 1988); Wh, wheat (Lloyd et al., 1991); Ch, *Chlamydomonas reinhardtii* (Roesler and Ogren, 1990); Syn, *Synechocystis* PCC6803 (Su and Bogorad, 1991). Identical residues (●) and conservative amino acid replacements (○) are shown.

13% identity towards the *Rb. sphaeroides* proteins. However, now that four different bacterial genes have been sequenced, along with four different PRK genes from oxygen-evolving organisms, it is apparent there are regions which do show some similarity (Fig. 6). Of particular interest are regions A and B of Fig. 6, which have been implicated in ATP-binding and Ru-5-P binding, respectively (Krieger et al., 1987; Roessler et al., 1992; Sandbaken et al., 1992). Asp-42 and Asp-169 appear to be excellent candidates for the active site base or divalent cation ligand, respectively (Charlier et al., 1994). The tetrapeptide Glu-Gly-Leu-His is conserved in all known PRK enzymes (region C) but thus far its significance is not apparent, however a recent study indicated that Glu-131 of this conserved tetrapeptide may be involved with allosteric NADH activation (Charlier et al., 1994). The interesting F227S mutation in the *Synechocystis* PRK, which renders this cyanobacterium light sensitive (Su and Bogorad, 1991), is not conserved in bacterial enzymes. Finally, PRK was isolated as a complex with FBPase (*cbbF* gene) from extracts of autotrophically-grown *Rs. rubrum* (Joint et al., 1972); the close juxtaposition of the structural genes for these two enzymes in *Rs. rubrum* (Falcone and Tabita, 1993b) and *Rb. sphaeroides* (6 bases separate $cbbP_I$ and $cbbF_I$ and three bases separate $cbbP_{II}$ and $cbbF_{II}$) and their cotranscription (Gibson et al., 1990, 1991) suggests that such complexes may form after these proteins are synthesized on the ribosome.

C. Other Enzymes of the Calvin Cycle

The organization of Calvin cycle structural genes into discrete operons in *Rb. sphaeroides* and *Rs. rubrum* has led to the isolation, cloning and sequencing of many of the relevant genes (Gibson and Tabita, 1987; Hallenbeck and Kaplan, 1987; Gibson and Tabita, 1988; Gibson et al., 1990; Gibson et al., 1991; Falcone and Tabita, 1993b). Moreover, recombinant FBPase, transketolase, and glyceraldehyde phosphate dehydrogenase have been prepared from *E. coli*, and the ability to produce recombinant ribulose 5-phosphate-3-epimerase and aldolase is apparent as well. Thus, these enzymes might now be studied in depth, as there is a paucity of information relative to the properties of these important catalysts (however, see Springgate and Stachow, 1972a,b). Future developments are awaited with great interest.

D. Metabolic Control of Calvin Cycle Enzyme Synthesis

The first indication that the synthesis of RubisCO is regulated came from the work of Lascelles (1960), who showed that synthesis immediately stopped when phototrophically grown *Rb. sphaeroides* cells were incubated under aerobic conditions. These results led to the hypothesis that the oxidation state of some intracellular metabolite might regulate enzyme synthesis. This insightful analysis of the physiological results have become even more pertinent now that there is some understanding of the molecular biology of the process, since it is apparent that the promoters

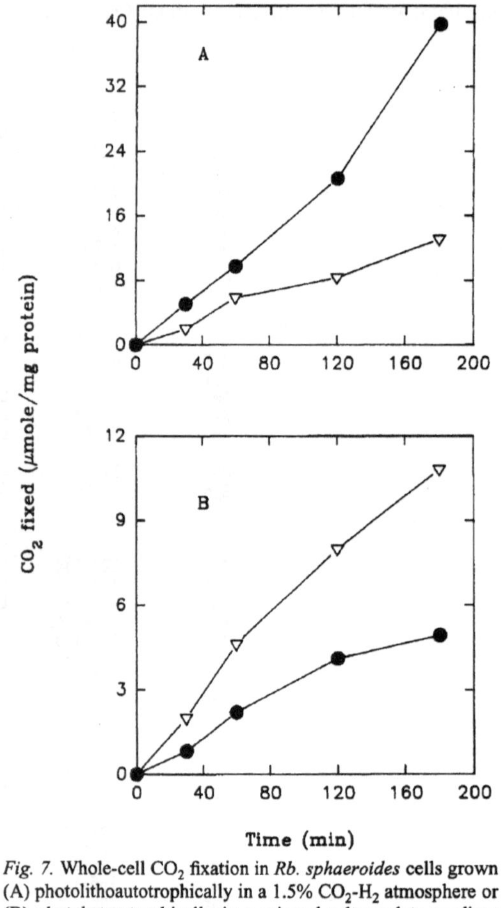

Fig. 7. Whole-cell CO_2 fixation in *Rb. sphaeroides* cells grown (A) photolithoautotrophically in a 1.5% CO_2-H_2 atmosphere or (B) photoheterotrophically in a mineral salts malate medium with CO_2 as the electron acceptor. CO_2 fixation assays were performed with hydrogen (●) or malate (▽) as the electron donor. From Wang et al. (1993a), with permission.

that control gene expression are oxygen regulated (see Chapter by Bauer). These initial results have been confirmed and extended to other PNS bacteria (Anderson and Fuller, 1967a,b; Hicughi and Kikuchi, 1969). As discussed earlier, there is an absolute requirement for HCO_3^- (CO_2) when PNS bacteria are grown with reduced carbon sources/electron donors. Not unexpectedly, this is reflected by an increase in whole-cell CO_2 fixation by resting cells. Interestingly, when *Rb. sphaeroides* was grown under photoheterotrophic conditions, the photoheterotrophic growth substrate (malate) was the preferred electron donor in whole-cell CO_2 fixation assays (Fig. 7). However, photolithoautotrophically grown cells preferentially use H_2 as electron donor. These results indicate that the metabolism and/or utilization of malate and H_2 are subject to induction/repression. From the foregoing, it is not surprising that RubisCO synthesis is markedly stimulated when PNS bacteria are cultured with reduced electron donors. This is illustrated by the data compiled in Table 2, where RubisCO specific activity and protein levels were compared in several strains using representative oxidized and reduced electron donors. The ability of these organisms to derepress synthesis is particularly evident in *Rs. rubrum*, where up to 50% of the soluble protein is RubisCO. Presumably the extremely high levels of RubisCO, far beyond what is required to support either the observed growth rate or the observed in vivo CO_2 fixation rates (Sarles and Tabita, 1983; Wang et al., 1993b), may be important to remove excess reducing power (Wang et al., 1993a) or possibly for scavenging CO_2 from the media since the K_{CO_2} of RubisCO is relatively high (Table 1). Both *Rs. rubrum* and *Rb. sphaeroides* were recently shown to grow photolithoautotrophically using either thiosulfate or sulfide as electron donors. Under these growth conditions, both organisms drastically reduce the levels of RubisCO protein (Table 2), and these results, combined with the fact that RubisCO negative strains are able to grow using either thiosulfate or sulfide in the complete absence of organic carbon, suggested that these organisms may use some alternative means to reduce CO_2 (Wang et al., 1993b). It is also apparent that *Rb. sphaeroides* (Jouanneau and Tabita, 1986) and *Rb. capsulatus* (Shively et al., 1984) preferentially synthesize form I RubisCO at low levels of CO_2. In *Rp. blastica*, specific synthesis of form I RubisCO occurred at the late logarithmic growth phase in cells grown in a butyrate-bicarbonate growth medium (Dow, 1987). Further, [^{35}S] methionine incorporation studies indicated that the form II *Rp. blastica* enzyme was expressed only at high levels of CO_2. Differential regulation of the synthesis of form I and form II RubisCO in *Rb. sphaeroides*, *Rb. capsulatus*, and *Rp. blastica* clearly indicates that there are separate molecular signals that influence the synthesis of these distinct proteins.

Table 2. Growth substrate dependent synthesis of RubisCO in purple nonsulfur photosynthetic bacteria

Organism	Growth substrate	Sp act (Umg^{-1} protein)	RubisCO (% sol. prot.)		Reference
Rs. rubrum	Malate	0.017	0.4		Sarles and Tabita (1983)
	Butyrate-HCO_3^-	0.600	13.9		
	6.4% CO_2 in H_2	0.177	3.9		
	1.5% CO_2 in H_2	1.680	50.2		
	$S_2O_3^{2-}$-HCO_3^-	0.007	0.8		Wang et al. (1993b)
			form I	form II	
Rb. sphaeroides	Malate	0.032	0.5	0.9	Jouanneau and Tabita (1986)
	1.5% CO_2 in H_2	0.370	8.4	4.0	
	$S_2O_3^{2-}$-HCO_3^-	0.008	N.D.[a]	0.4	Wang et al. (1993b)
Rp. acidophila	Succinate	0.008			Quayle and Pfennig (1975)
	Methanol-HCO_3^-	0.050			
Rp. blastica	Malate	0.010			Dow (1987)
	Butyrate-HCO_3^-	0.176			
Rb. capsulatus[b]	Malate	0.054			
	1.5 CO_2 in H_2	0.488			

[a] Not capable of being quantitated
[b] G.C. Paoli and FR Tabita, unpublished

Complicating this further, under certain conditions, form I and form II RubisCO synthesis in *Rb. sphaeroides* is under the influence of some common regulatory locus, since mutations that affect the synthesis of one enzyme are compensated by the overproduction of the other enzyme (Falcone and Tabita, 1988; 1991; Hallenbeck et al., 1990; Gibson et al., 1991). The isolation of the *cbbR* gene, a transcriptional activator gene which controls both form I and form II RubisCO synthesis in *Rb. sphaeroides* (Gibson and Tabita, 1993), promises to provide answers relative to the molecular basis of this interesting interdependent regulation. By the same token, it is apparent that other genes must separately control form I and form II RubisCO synthesis. Other key enzymes of the Calvin cycle, particularly PRK and FBPase, are also found at higher levels in cells grown under conditions which favor maximum RubisCO synthesis (Anderson and Fuller, 1967a; Gibson et al., 1991).

Two separate RubisCO negative strains were isolated in *Rb. sphaeroides* (Hallenbeck et al., 1990; Falcone and Tabita, 1991). Each strain was incapable of growing using CO_2 as electron acceptor when either organic carbon (i.e., malate or succinate) or molecular hydrogen were used as electron donors. However, both mutants could grow under photoheterotrophic conditions when dimethylsulfoxide (DMSO) was added to the growth media and used as the electron acceptor. Presumably, the inability of the RubisCO negative strains to grow photoheterotrophically is related to the capacity of RubisCO and the Calvin cycle to function as an electron sink to maintain a proper redox balance under these growth conditions (also see Richardson et al., 1988). Moreover, even reduced electron donors, such as butyrate, support growth of the RubisCO deficient strains when DMSO is used as electron acceptor, further substantiating the proposed redox balancing role of the Calvin cycle. However, it is not clear whether this suggested function for the Calvin cycle applies under photolithoautotrophic growth conditions where H_2 is the electron donor. Recent experiments (Wang et al., 1993a) addressed this issue when the effects of DMSO on whole cell CO_2 fixation, RubisCO activity, and RubisCO synthesis were examined in form I and form II RubisCO deletion mutants of *Rb. sphaeroides* grown in a H_2/CO_2 atmosphere. Both strains were influenced by the addition of DMSO, but form II RubisCO activity (in the form I deletion mutant) was influenced considerably, such that there was a loss of RubisCO activity which greatly exceeded a slight decrease in RubisCO protein. In the form II deletion strain, there was a much diminished effect on CO_2 fixation and form I RubisCO activity. These experiments are entirely consistent with the proposal that considerable amounts of RubisCO (particularly form II RubisCO) may be utilized to maintain the redox balance of the cell when the organism uses highly reduced electron donors, such as molecular hydrogen. Finally, as noted previously, the PS bacteria *Chromatium* and *Thiopedia* maximize RubisCO synthesis under autotrophic conditions as well (Hurlbert and Lascelles, 1963; Kobayashi and Akazawa, 1982a,b).

IV. Alternative CO_2 Assimilatory Routes

The Calvin cycle is obviously the predominant CO_2 assimilatory path employed by phototrophic and chemolithotrophic organisms in the biosphere. However, there are many notable exceptions to the Calvin scheme in several diverse bacteria, particularly in the photosynthetic green sulfur bacterium, *Chlorobium*, and diverse chemolithoautotrophic bacteria (see Chapter 40 by Sirevåg). In the purple photosynthetic bacteria (i.e., *Chromatium* and *Rs. rubrum*), several early studies, including determinations of the initial products of CO_2 fixation and several enzyme measurements, had indicated that these organisms assimilate significant amounts of CO_2 via alternative (nonCalvin) routes, specifically the reductive tricarboxylic acid (TCA) cycle (Buchanan et al., 1964; 1967). Unlike the green bacteria, PS and PNS bacteria also employ the Calvin cycle, which implies that these organisms may be able to regulate the use of each of these pathways under certain conditions; it was also suggested that the reductive TCA cycle might serve an auxiliary role in providing carbon skeletons in *Rs. rubrum* (Tabita, 1988). More recent studies with RubisCO deletion strains of *Rs. rubrum* and *Rb. sphaeroides* (strains 16 and 16PHC) extend the early biochemical studies and confirm the importance of alternative CO_2 fixation pathway(s). From strain 16, which cannot grow phototrophically using malate or H_2 as electron donors when CO_2 is the electron acceptor (Falcone and Tabita, 1991), a strain was selected that maintained the RubisCO negative phenotype yet was capable of photoheterotrophic growth on malate with CO_2 as electron acceptor. This strain, 16PHC, had the exact same phenotype as *Rs. rubrum* strain I-19. Strain 16PHC exhibited significant rates of whole-cell CO_2 fixation; moreover wild-type *Rb. sphaeroides* cells exhibited rates of whole-cell CO_2 fixation that greatly exceeded whole-cell RubisCO activity. These observations with the mutant and wild-type strains thus suggested that considerable CO_2 fixation occurs in the absence of a functional Calvin cycle (Wang et al., 1993a). Furthermore, the CO_2 fixation capacity (about 30% of the wild-type) is repressed by the addition of the alternate electron acceptor DMSO and is induced upon DMSO removal. Strain 16, although not capable of photoheterotrophic growth on C_4 acids with CO_2 as electron acceptor, could grow with CO_2 as electron acceptor when pyruvate or acetate is used as the electron donor (Y. Qian and F.R. Tabita, unpublished). Since C_4 cleavage reactions have recently been postulated to be essential to generate CO_2 acceptor molecules in the green bacterium *Chloroflexus* (Strauss et al., 1992; Ivanovsky et al., 1993), it seems plausible that strain 16PHC has somehow overcome this defect. Recent studies indicate that the 16PHC phenotype is accompanied by major changes in its protein profile, suggestive of some alternation of a major regulatory system under conditions of active CO_2 fixation (H. Modak and F.R. Tabita, unpublished). The importance of alternative CO_2 fixation reactions was greatly strengthened by the finding that all the RubisCO negative mutants of *Rb. sphaeroides* and *Rs. rubrum* (strains 16, 16PHC, and I-19) could grow photolithoautotrophically, using CO_2 as the sole carbon source, in a medium containing either thiosulfate or sulfide as electron donor (Wang et al., 1993b). It is interesting that photolithoautotrophic growth of the RubisCO deficient strains is possible only with electron donors of higher redox potential than hydrogen. The recent demonstration of photolithoautotrophic growth by PNS bacteria using reduced iron compounds as the source of reducing power (Widdell et al., 1993) suggests that compounds of intermediate redox potential might be important to support growth and CO_2 fixation in environments where hydrogen and organic substrates are unavailable. It would appear that only RubisCO is capable of disposing of the excess reducing power generated by hydrogen-dependent growth, suggesting that the enormously high levels of RubisCO synthesized under these growth conditions might be exceedingly important for this function. As noted earlier, the levels of RubisCO synthesized in a 1.5% CO_2/98.5 H_2 atmosphere are far in excess of what both *Rs. rubrum* and *Rb. sphaeroides* require for carbon assimilation and growth. Both the Calvin cycle and the alternative CO_2 assimilatory pathway must be under metabolic control in the two organisms, since the RubisCO activity levels and the amount of RubisCO protein are significantly reduced in thiosulfate-grown wild-type cells (Table 2). Thus, it is not surprising that conditions which allow for CO_2-dependent growth of the RubisCO deficient strains result in repression of RubisCO synthesis in the wild-type. Beyond the fundamental questions of metabolic control of the alternative and Calvin CO_2 fixation pathways, the finding that some alternative CO_2 fixation scheme could take the place of the Calvin pathway and support photolithoautotrophic growth raises interesting questions relative to the evolution of CO_2 assimilatory schemes. Perhaps the PNS bacteria represent an evolutionary link between bacteria which use the reductive TCA cycle (Evans et al., 1966), the hydroxypropionate pathway (Holo, 1989; Strauss et al., 1992; Ivanovsky et al., 1993;) or the non-cyclic reductive acetyl CoA/carbon monoxide dehydrogenase pathway (Fuchs, 1989) and more evolutionary advanced organisms. Indeed, the chloroplasts of the green alga *Chlamydomonas reinhardtii* were recently shown to contain the ferredoxin-linked enzymes of the reductive tricarboxylic acid cycle (Chen and Gibbs, 1992). In this context, it would be interesting to determine whether cyanobacteria which are capable of anoxygenic photosynthesis (Padan, 1979) employ the reductive TCA cycle or alternative reactions to fix CO_2 under these growth conditions. In any case, it would appear that the original hypothesis of Lascelles (1960), that RubisCO and the Calvin cycle play an important role in providing redox balance for the cell, has gained added significance. The development of the Calvin reductive pentose phosphate pathway and the acquisition of RubisCO could be a very important evolutionary event since this would enable organisms to thrive under a much wider spectrum of environmental growth conditions than organisms that fix CO_2 solely by using other pathways. The capacity for photosynthetic metabolism would thus be much enhanced and substrates other than reduced iron and sulfur compounds could be employed as reductants to support CO_2 assimilation. Although it is recognized that the purple nonsulfur photosynthetic bacteria are probably the most metabolically versatile organisms found in nature (Madigan and Gest, 1979; Madigan,

1988), recent studies provide additional support for the great adaptability of these bacteria.

V. Conclusions

The purple photosynthetic bacteria are capable of diverse modes of carbon catabolism and anabolism. This is particularly true of the PNS bacteria, which provide excellent systems for the study of fundamental aspects of carbon metabolism and its regulation at the biochemical and molecular levels. Indeed, recent studies indicate that common regulatory elements may serve to control biosynthetic CO_2 assimilation and the degradation of fixed carbon compounds and it is anticipated that this will become a fruitful area for future investigation. The enzymes of carbon metabolism have been studied to some extent, but only RubisCO has attracted the attention of a large number of enzyme chemists, resulting in a deep understanding of how this enzyme functions. Nevertheless, key questions remain concerning the basis for gas substrate specificity and the mechanisms PNS bacteria use to regulate activity in vivo. Other key catalysts deserve future scrutiny, and with the cloning, sequencing, and expression of virtually all the structural genes of the Calvin cycle, it is anticipated that these enzymes will be the focus of considerable future in depth study. The role of chaperonin proteins in enzyme assembly is also a burgeoning field and many questions remain unanswered relative to how purple bacteria use these accessory proteins for placing complex oligomeric proteins in the proper conformation for efficient catalysis. Finally, it has become apparent that PNS bacteria may use alternative means to fix CO_2; these pathway(s) may support autotrophic growth in the absence of a functional Calvin cycle. Elucidation of the nature of these alternative reactions and their biochemistry is in progress, with the expectation that an understanding of the metabolic control of both CO_2 assimilatory paths will be forthcoming.

Acknowledgments

Support from the National Institutes of Health, Department of Energy, and the Department of Agriculture is gratefully acknowledged. I particularly wish to thank my students, postdoctoral associates, and colleagues for their inspiration and many contributions.

References

Albers H and Gottschalk G (1976) Acetate metabolism in *Rhodopseudomonas gelatinosa* and several other *Rhodospirillaceae*. Arch Microbiol 111: 45–49

Anderson L and Fuller RC (1967a) Photosynthesis in *Rhodospirillum rubrum*. II. Photoheterotrophic carbon dioxide fixation. Plant Physiol 42: 491–496

Anderson L and Fuller RC (1967b) Photosynthesis in *Rhodospirillum rubrum*. III. Metabolic control of reductive pentose phosphate and tricarboxylic acid cycle enzymes. Plant Physiol 42: 497–502

Andersson I, Knight S, Schneider G, Lindqvist Y, Lundqvist T, Branden C-I and Lorimer GH (1989) Crystal structure of the active site of ribulose-bisphosphate carboxylase. Nature 337: 229–234

Andrews TJ (1988) Catalysis by cyanobacterial ribulose-bisphosphate carboxylase large subunits in the complete absence of small subunits. J Biol Chem 263: 12213–12219

Asami S and Akazawa T (1974) Oxidative formation of glycolic acid in photosynthesizing cells of *Chromatium*. Plant Cell Physiol 15: 571–576

Barraclough R and Ellis RJ (1980) Protein synthesis in chloroplasts. IX. Assembly of newly-synthesized large subunits into ribulose bisphosphate carboxylase in isolated intact pea chloroplasts. Biochim Biophys Acta 608: 19–31

Beatty JT and Gest H (1981a) Generation of succinyl-coenzyme A in photosynthetic bacteria. Arch Microbiol 129: 335–340

Beatty JT and Gest H (1981b) Biosynthetic and bioenergetic functions of citric acid cycle reactions in *Rhodopseudomonas capsulata*. J Bacteriol 148: 584–593

Berry JA, Lorimer GH, Pierce J, Seemann JR, Meek J and Freas S (1987) Isolation, identification, and synthesis of 2-carboxyarabinitol 1-phosphate, a diurnal regulator of ribulose-bisphosphate carboxylase activity. Proc Natl Acad Sci USA 84: 734–738

Blasco R, Cardenas J and Castillo F (1989) Acetate metabolism in purple non-sulfur bacteria. FEMS Microbiol Lett 58: 129–132

Blasco R, Cardenas J and Castillo F (1991) Regulation of isocitrate lyase in *Rhodobacter capsulatus* E1F1. Current Microbiol 22: 73–76

Bonam D and Ludden PW (1987) Purification and characterization of carbon monoxide dehydrogenase, a nickel, zinc, iron-sulfur protein, from *Rhodospirillum rubrum*. J Biol Chem 262: 2980–2987

Bonam D, Murrell SA and Ludden PW (1984) Carbon monoxide dehydrogenase from *Rhodospirillum rubrum*. J Bacteriol 159: 693–699

Bonam D, Lehman L, Roberts GP and Ludden PW (1989) Regulation of carbon monoxide dehydrogenase and hydrogenase in *Rhodospirillum rubrum*: Effects of CO and oxygen on synthesis and activity. J Bacteriol 171: 3102–3107

Brostedt E and Nordlund S (1991) Purification and partial characterization of a pyruvate oxidoreductase from the

photosynthetic bacterium *Rhodospirillum rubrum* grown under nitrogen-fixing conditions. Biochem J 279: 155–158
Buchanan BB, Bachofen R and Arnon DI (1964) Role of ferredoxin in the reductive assimilation of CO_2 and acetate by extracts of the photosynthetic bacterium, *Chromatium*. Proc Natl Acad Sci USA 52: 839–847
Buchanan BB, Evans MCW and Arnon DI (1967) Ferredoxin-dependent carbon assimilation in *Rhodospirillum rubrum*. Arch Mikrobiol 59: 32–40
Charlier HA Jr, Runquist JA and Miziorko HM (1994) Evidence supporting catalytic roles for aspartate residues in phosphoribulokinase. Biochemistry 33: 9343–9350
Chen C and Gibbs M (1992) Some enzymes and properties of the reductive carboxylic acid cycle are present in the green alga *Chlamydomonas reinhardtii* F-60. Plant Physiol 98: 535–539
Christeller JT and Laing WA (1978) A kinetic study of ribulose bisphosphate carboxylase from the photosynthetic bacterium *Rhodospirillum rubrum*. Biochem J 173: 467–473
Chung SY, Yaguchi T, Nishihara H, Igarashi Y, Kodama T (1993) Purification of form L_2 RubisCO from a marine obligate autotrophic hydrogen-oxidizing bacterium, *Hydrogenovibrio marinus* strain MH-110. FEMS Microbiol Lett 109: 49–54
Claassen PAM Zehnder AJB (1986) Isocitrate lyase activity in *Thiobacillus versutus* grown anaerobically on acetate and nitrate. J Gen Microbiol 132: 3179–3185
Conrad R and Schlegel HG (1974) Different pathways for fructose and glucose utilization in *Rhodopseudomonas capsulata* and demonstration of 1-phosphofructokinase in phototrophic bacteria. Biochim Biophys Acta 358: 221–225
Conrad R and Schlegel HG (1977a) Influence of aerobic and phototrophic growth conditions on the distribution of glucose and fructose carbon into the Entner-Doudoroff and Embden-Meyerhof pathways in *Rhodopseudomonas sphaeroides*. J Gen Microbiol 101: 277–290
Conrad R Schlegel HG (1977b) Different degradation pathways for glucose and fructose in *Rhodopseudomonas capsulatus*. Arch Microbiol 112: 39–48
Conway T (1992) The Entner-Doudoroff pathway: History, physiology and molecular biology. FEMS Microbiol Rev 103: 1–27
Cook LS and Tabita FR (1988) Oxygen regulation of ribulose 1,5-bisphosphate carboxylase activity in *Rhodospirillum rubrum*. J Bacteriol 170: 5468–5472
Cook LS, Im H and Tabita FR (1988) Oxygen-dependent inactivation of ribulose 1,5-bisphosphate carboxylase/oxygenase in crude extracts of *Rhodospirillum rubrum* and establishment of a model inactivation system with purified enzyme. J Bacteriol 170: 5473–5478
Dow CS (1987) CO_2 fixation in *Rhodopseudomonas blastica*. In: Van Verseveld HW and Duine J (eds) Microbial growth on C1 compounds, pp 28–37. Martinus Nijhoff Publishers, Dordrecht
Eidels L and Preiss J (1970) Carbohydrate metabolism in *Rhodopseudomonas capsulata*: Enzyme titers, glucose metabolism, and polyglucose polymer synthesis. Arch Biochem Biophys 140: 75–89
Ellis RJ (1979) The most abundant protein in the world. Trends Biochem Sci 4: 241–244
Ellis RJ (1990) Molecular chaperones: The plant connection. Science 250: 954–959
English RS, Williams CA, Lorbach SC and Shively JM (1992) Two forms of ribulose-1,5-bisphosphate carboxylase/oxygenase from *Thiobacillus denitrificans*. FEMS Microbiol Lett 94: 111–120
Ensign SA and Ludden PW (1991) Characterization of the CO oxidation/H_2 evolution of *Rhodospirillum rubrum*: Role of a 22 kDa iron-sulfur protein in mediating electron transfer between carbon monoxide dehydrogenase and hydrogenase. J Biol Chem 266: 18395–18403
Evans MCW, Buchanan BB and Arnon DI (1966) A new ferredoxin dependent carbon reduction cycle in a photosynthetic bacterium. Proc Natl Acad Sci USA 55: 928–934
Falcone DL and Tabita FR (1991) Expression of endogenous and foreign ribulose 1,5-bisphosphate carboxylase-oxygenase (RubisCO) genes in a RubisCO deletion mutant of *Rhodobacter sphaeroides*. J Bacteriol 173:2099–2108
Falcone DL and Tabita FR (1993a) Expression and regulation of *Bradyrhizobium japonicum* and *Xanthobacter flavus* CO_2 fixation genes in a photosynthetic bacterial host. J Bacteriol 175: 866–869
Falcone DL and Tabita FR (1993b) Complementation analysis and regulation of CO_2 fixation gene expression in a ribulose 1,5-bisphosphate carboxylase-oxygenase deletion strain of *Rhodospirillum rubrum*. J Bacteriol 175: 5066–5077
Falcone DL, Quivey Jr, RG and Tabita FR (1988) Transposon mutagenesis and physiological analysis of strains containing inactivated form I and form II ribulose bisphosphate carboxylase/oxygenase genes in *Rhodobacter sphaeroides*. J Bacteriol 170: 5–11
Ferreyra RG, Soncini FC and Viale AM (1993) Cloning, characterization, and functional expression in *Escherichia coli* of chaperonin (groESL) genes from the phototrophic sulfur bacterium *Chromatium vinosum*. J Bacteriol 175:1514–1423
Fuchs G (1989) Alternative pathways of autotrophic CO_2 fixation. In: Schlegel HG and Bowien B (eds) Autotrophic Bacteria, pp 365–382. Science Tech Publishers, Madison
Fuller RC (1978) Photosynthetic carbon metabolism in the green and purple bacteria. In: Clayton RK and Sistrom WR (eds) The Photosynthetic Bacteria, pp 691–705. Plenum Press, New York
Fuller RC and Gibbs M (1959) Intracellular and phylogenetic distribution of ribulose 1,5-diphosphate carboxylase and D-glyceraldehyde 3-phosphate dehydrogenases. Plant Physiol 34: 324–329
Fuller RC, Smillie RM, Sisler EC and Kornberg HL (1961) Carbon metabolism in *Chromatium*. J. Biol. Chem. 236: 2140–2149
Gest H (1951) Metabolic patterns in photosynthetic bacteria. Bacteriol Rev 15: 183–210
Gest H (1981) Evolution of the citric acid cycle and respiratory energy conversion in prokaryotes. FEMS Microbiol Lett 12: 209–215
Gest H and Kamen MD (1949) Photoproduction of molecular hydrogen by *Rhodospirillum rubrum*. Science 109: 558–559
Gibson JL and Tabita FR (1977) Different molecular forms of D-ribulose-1,5-bisphosphate carboxylase from *Rhodopseudomonas sphaeroides*. J Biol Chem 252: 943–949
Gibson JL and Tabita FR (1979) Activation of ribulose 1,5-bisphosphate carboxylase from *Rhodopseudomonas sphaeroides*: Probable role of the small subunit. J Bacteriol 140: 1023–1027

Gibson JL and Tabita FR (1985) Structural differences in the catalytic subunits of form I and form II ribulose 1,5-bisphosphate carboxylase/oxygenase from *Rhodopseudomonas sphaeroides*. J Bacteriol 164: 1188–1193

Gibson JL and Tabita FR (1986) Isolation of the *Rhodopseudomonas sphaeroides* form I ribulose 1,5-bisphosphate carboxylase/oxygenase large and small subunit genes and expression of the active hexadecameric enzyme in *Escherichia coli*. Gene 44:271–278

Gibson JL and Tabita FR (1987) Organization of phosphoribulokinase and ribulose bisphosphate carboxylase/oxygenase genes in *Rhodopseudomonas (Rhodobacter) sphaeroides*. J Bacteriol 169: 3685–3690

Gibson JL and Tabita FR (1988) Localization and mapping of CO_2 fixation genes within two gene clusters in *Rhodobacter sphaeroides*. J Bacteriol 170: 2153–2158

Gibson JL and Tabita FR (1993) Nucleotide sequence and functional analysis of *cbbR*, a positive regulator of the Calvin cycle operons of *Rhodobacter sphaeroides*. J Bacteriol 175:5778–5784

Gibson JL, Chen C-H, Tower PA and Tabita FR (1990) The form II fructose 1,6-bisphosphatase and phosphoribulokinase genes form part of a large operon in *Rhodobacter sphaeroides*: Primary structure and insertional mutagenesis analysis. Biochemistry 29: 8085–8093

Gibson JL, Falcone DL and Tabita FR (1991) Nucleotide sequence, transcriptional analysis, and expression of genes encoded within the Form I CO_2 fixation operon of *Rhodobacter sphaeroides*. J Biol Chem 266: 14646–14653

Glover J, Kamen MD and Van Genderen H (1952) Studies on the metabolism of photosynthetic bacteria. XII. Comparative light and dark metabolism of acetate and carbonate by *Rhodospirillum rubrum*. Arch Biochem Biophys 35: 343–408

Goloubinoff P, Gatenby AA and Lorimer GH (1989a) GroE heat-shock proteins promote assembly of foreign prokaryotic ribulose bisphosphate carboxylase oligomers in *Escherichia coli*. Nature 337: 44–47

Goloubinoff P, Christeller JT, Gatenby AA and Lorimer GH (1989b) Reconstitution of active dimeric ribulose bisphosphate carboxylase from an unfolded state depends on two chaperonin proteins and Mg-ATP. Nature 342: 884–889

Gorrell TE and Uffen RL (1977) Fermentative metabolism of pyruvate by *Rhodospirillum rubrum* after anaerobic growth in darkness. J Bacteriol 131: 533–543

Gutteridge S, Parry MAJ, Burton S, Keys AJ, Mudd A, Feeney J, Servaites JC and Pierce JA (1986) A nocturnal inhibitor of carboxylation in leaves. Nature 324: 274–276

Hallenbeck P and Kaplan S (1987) Cloning of the gene for phosphoribulokinase activity from *Rhodobacter sphaeroides* and its expression in *Escherichia coli*. J Bacteriol 169: 3669–3678

Hallenbeck PL, Lerchen R, Hessler P, and Kaplan S (1990) Phosphoribulokinase activity and regulation of CO_2 fixation critical for photosynthetic growth of *Rhodobacter sphaeroides*. J Bacteriol 172: 1749–1761

Hart BA and Gibson J (1971) Ribulose-5-phosphate kinase from *Chromatium* sp. strain D. Arch Biochem Biophys 144: 308–321

Hartman FC (1992) Structure-function relationships of ribulose bisphosphate carboxylase/oxygenase as suggested by site-directed mutagenesis. In: Shewry PR and Gutteridge S (eds) Plant Protein Engineering, Vol 1, pp 61–92. Cambridge University Press, London

Hartman FC and Harpel MR (1993) Chemical and genetic probes of the active site of D-ribulose-1,5-bisphosphate carboxylase/oxygenase: A retrospecive based on the three-dimensional structure. Adv Enzmol 67: 1–75

Hartman FC and Harpel MR (1994) Structure, function, regulation, and assembly of D-ribulose 1,5-bisphosphate carboxylase/oxygenase. Ann Rev Biochem 63: 197–234

Hartman FC and Lee EH (1989) Examination of the function of active site lysine 329 of ribulose-bisphosphate carboxylase/oxygenase as revealed by the proton exchange reaction. J Biol Chem 246: 11784–11789

Heda GD and Madigan MT (1988) Thermal properties and oxygenase activity of ribulose 1,5-bisphosphate carboxylase from the thermophilic purple bacterium *Chromatium tepidum*. FEMS Microbiol Lett 51: 45–50

Heda DG and Madigan MT (1989) Purification and characterization of the thermostable ribulose-1,5-bisphosphate carboxylase/oxygenase from the thermophilic purple bacterium *Chromatium tepidum*. Eur J Biochem 184: 313–319

Hemmingsen SM, Woolford C, van der Vies SM, Tilly K, Dennis DT, Georgopoulos CP, Hendrix RW and Ellis RJ (1988) Homologous plant and bacterial proteins chaperone oligomeric protein assembly. Nature 333: 330–334

Hicughi M and Kikuchi G (1969) Induced formation of ribulose 1,5-diphosphate carboxylase in *Rhodopseudomonas spheroides* with particular concern to its relation with chromatophore formation. Plant Cell Physiol 10: 149–160

Hirsch P (1968) Photosynthetic bacterium growing under carbon monoxide. Nature 217: 555–556

Hoare DS (1963) The photo-assimilation of acetate by *Rhodospirillum rubrum*. Biochem J 87: 284–301

Holo H (1989) *Chloroflexus aurantiacus* secretes 3-hydroxypropionate, a possible intermediate in the assimilation of CO_2 fixation and acetate. Arch Microbiol 151: 252–256

Hurlbert RE and Lascelles J (1963) Ribulose bisphosphate carboxylase in Thiorhodaceae. J Gen Microbiol 33: 445–458

Ivanovsky RN, Krasilnikova EN and Fal YI (1993) A pathway of the autotrophic CO_2 fixation in *Chloroflexus aurantiacus*. Arch Microbiol 159: 257–264

Joint IR, Morris I and Fuller RC (1972) Purification of a complex of alkaline fructose 1,6-bisphosphatase and phosphoribulokinase from *Rhodospirillum rubrum*. J Biol Chem 247: 4833–4838

Jordan DB and Chollet R (1985) Subunit dissociation and reconstitution of ribulose-1,5-bisphosphate carboxylase from *Chromatium vinosum*. Arch Biochem Biophys 236: 487–496

Jordan DB and Ogren WL (1981) Species variation in the specificity of ribulose bisphosphate carboxylase/oxygenase. Nature 291: 513–515

Jouanneau Y and Tabita FR (1986) Independent regulation of form I and form II ribulose bisphosphate carboxylase-oxygenase in *Rhodopseudomonas sphaeroides*. J Bacteriol 165: 620–624

Jouanneau Y and Tabita FR (1987) In vivo regulation of form I ribulose 1,5-bisphosphate carboxylase/oxygenase from *Rhodopseudomonas sphaeroides*. Arch Biochem Biophys 254: 290–303

Kerby RL, Hong SS, Ensign SA, Coppoc LJ, Ludden PW and

Roberts GP (1992) Genetic and physiological characterization of the *Rhodospirillum rubrum* carbon monoxide dehydrogenase system. J Bacteriol 174: 5284–5294

Khanna S, Kelley BC and Nicholas DJD (1981) Oxygen inhibition of the photoassimilation of CO_2 in *Rhodopseudomonas capsulata*. Arch Microbiol 128: 421–423

Klemme J-H (1973) Allosterische kontrolle der pyruvatkinase aus *Rhodospirillum rubrum* durch anorganisches phosphat und zucherphosphatester. Arch Mikrobiol 90: 305–322

Klemme J-H (1974) Modulation by fumarate of a Pi-insensitive pyruvate kinase from *Rhodopseudomonas capsulata*. Arch Microbiol 100: 57–63

Knight M (1962) The photometabolism of propionate by *Rhodospirillum rubrum*. Biochem J 84: 170–185

Knight S, Andersson I, and Branden C-I (1990) Crystallographic analysis of ribulose 1,5-bisphosphate carboxylase from spinach at 2.4 A resolution. J Mol Biol 215: 113–160

Kobayashi H and Akazawa T (1982a) Biosynthetic mechanism of ribulose 1,5-bisphosphate carboxylase in the purple photosynthetic bacterium *Chromatium vinosum*. I. Inducible formation. Arch Biochem Biophys 214: 531–539

Kobayashi H and Akazawa T (1982b) Biosynthetic mechanism of ribulose-1,5-bisphosphate carboxylase in the purple photosynthetic bacterium, *Chromatium vinosum*. II. Biosynthesis of constituent subunits. Arch Biochem Biophys 214: 540–549

Kobayashi H, Viale AM, Takabe T, Akazawa T, Wada K, Shinozaki K, Kobayashi K and Sugiura M (1991) Sequence and expression of genes encoding the large and small subunits of ribulose 1,5-bisphosphate carboxylase/oxygenase from *Chromatium vinosum*. Gene 97: 55–62

Kohlmiller EF and Gest H (1951) A comparative study of light and dark fermentations of organic acids by *Rhodospirillum rubrum*. J Bacteriol 61: 269–282

Kondratievea EN (1979) Interrelation between modes of carbon assimilation and energy production in phototrophic purple and green bacteria. In: Quayle JR (ed) International Review of Biochemistry, Microbial Biochemistry, Vol 21, pp 117–175. University Press, Baltimore

Kondratieva EN, Zhukov VG, Ivanovsky RN, Petushkova, YP and Monosov EZ (1976) The capacity of phototrophic sulfur bacterium *Thiocapsa roseopersicina* for chemosynthesis. Arch Microbiol 108: 287–292

Kornberg HL and Lascelles J. (1960) The formation of isocitratase by the Athiorhodaceae. J Gen Microbiol 23: 511–517

Kossmann J, Klintworth R and Bowien B (1989) Sequence analysis of the chromosomal and plasmid genes encoding phosphoribulokinase from *Alcaligenes eutrophus*. Gene 85: 247–252

Krieger TJ, Mende-Muller L Miziorko HM (1987) Phosphoribulokinase: Isolation and sequence determination of the cysteine-containing active-site peptide modified by 5_-*p*-fluorosulfonylbenzoyladenosine. Biochim Biophys Acta 915:112–119

Laing WA, Ogren WL and Hageman RH (1974) Regulation of soybean net photosynthetic CO_2 fixation by the interaction of CO_2, O_2, and ribulose 1,5-diphosphate carboxylase. Plant Physiol 54: 678–685

Larimer FW and Soper TS (1993) Overproduction of *Anabaena* 7120 ribulose-bisphosphate carboxylase/oxygenase in *Escherichia coli*. Gene 126:85–92

Lascelles J (1960) The formation of ribulose 1:5-diphosphate carboxylase by growing cultures of *Athiorhodaceae*. J Gen Microbiol 23: 499–510

Lee B and Tabita FR (1990) Purification of recombinant ribulose-1,5-bisphosphate carboxylase/oxygenase large subunits suitable for reconstitution and assembly of active L_8S_8 enzyme. Biochemistry 29: 9352–9357

Lee B, Read BA and Tabita FR (1991) Catalytic properties of recombinant octameric, hexadecameric, and heterologous cyanobacterial/bacterial ribulose-1,5-bisphosphate carboxylase/oxygenase. Arch Biochem Biophys 291:263–269

Levine RL, Oliver CN, Fulks RM and Stadtman ER (1981) Turnover of bacterial glutamine synthetase: Oxidative inactivation precedes proteolysis. Proc Natl Acad Sci USA 78: 2120–2124

Lloyd JC, Horsnell PR, Dyer TA and Raines CA (1991) Structure and sequence of wheat phosphoribulokinase gene. Plant Mol Biol 17: 167–168

Lueking D, Pike L and Sojka G (1976) Glycerol utilization by a mutant of *Rhodopseudomonas capsulata*. J Bacteriol 125: 750–752

Madigan MT (1988) Microbiology, physiology, and ecology of phototrophic bacteria. In: Zehnder AJB (ed) Biology of anaerobic microorganisms, pp 39–111. John Wiley and Sons, Inc, New York

Madigan MT (1990) Photocatabolism of acetone by nonsulfur purple bacteria. FEMS Microbiol Lett 71: 281-286

Madigan MT and Gest H (1979) Growth of the photosynthetic bacterium *Rhodopseudomonas capsulata* chemolithotrophically in darkness with H_2 as the energy source. J Bacteriol. 137: 524–530

Madigan MT, Cox JC and Gest H (1980) Physiology of dark fermentative growth of *Rhodopseudomonas capsulata*. J Bacteriol 142: 908–915

Mann NH and Turner AM (1988) Covalent modification of ribulose 1,5-bisphosphate carboxylase/ oxygenase in *Rhodomicrobium vannielii*. Molecular Microbiology 2: 427–432

McFadden BA, Torres-Ruiz JA and Franceschi VR (1989) Localization of ribulose-bisphosphate carboxylase-oxygenase and its putative binding protein in the cell envelope of *Chromatium vinosum*. Planta 178: 297–302

Meijer WG, Enequist HG, Terpstra P and Dijkhuizen L (1990) Nucleotide sequences of the genes encoding fructose-bisphosphatase and phosphoribulokinase from *Xanthobacter flavus* H4-14. J Gen Microbiol 136: 2225–2230

Milanez S and Mural RJ (1988) Cloning and sequencing of cDNA encoding the mature form of phosphoribulokinase from spinach. Gene 66: 55–63

Miziorko HM and Lorimer GH (1983) Ribulose-1,5-bisphosphate carboxylase-oxygenase. Ann Rev Biochem 52: 507–535

Morell MK, Kane HJ and Andrews TJ (1990) Carboxyterminal deletion mutants of ribulosebisphosphate carboxylase from *Rhodospirillum rubrum*. FEBS Lett 265: 41–45

Muller FM (1933) On the metabolism of the purple sulfur bacteria in organic media. Arch Mikrobiol 4: 131–166.

Nargang F, McIntosh L and Somerville C (1984) Nucleotide sequence of the ribulosebisphosphate carboxylase gene from *Rhodospirillum rubrum*. Mol Gen Genet 193: 220–224

Nielsen AM and Sojka GA (1979) Photoheterotrophic utilization of acetate by the wild type and an acetate-adapted mutant of

Rhodopseudomonas capsulata. Arch Microbiol 120: 39–42

Nielsen AM, Rampsch BJ and Sojka GA (1979) Regulation of isocitrate lyase in a mutant of *Rhodopseudomonas capsulata* adapted to growth on acetate. Arch Microbiol 120: 43–46

Padan E (1979) Facultative anoxygenic photosynthesis in cyanobacteria. Ann Rev Plant Physiol 30: 27–40

Payne J and Morris JG (1969a) Acetate utilisation by *Rhodopseudomonas spheroides*. FEBS Lett. 4:52–54

Payne J and Morris JG (1969b) Pyruvate carboxylase in *Rhodopseudomonas spheroides*. J Gen Microbiol 59: 97–101

Pelroy RA and Bassham JA (1972) Photosynthetic and dark carbon metabolism in unicellular blue-green algae. Arch Mikrobiol 86: 25–38

Pelroy RA and Bassham JA. (1976) Kinetics of light-dark CO_2 fixation and glucose assimilation by *Aphanocapsa*. J Bacteriol 128: 633–643

Pike L and Sojka GA (1975) Glycerol dissimilation in *Rhodopseudomonas sphaeroides*. J Bacteriol 124: 1101–1105

Portis Jr AR (1992) Regulation of ribulose 1,5-bisphosphate carboxylase/oxygenase activity. Ann Rev Plant Physiol 43: 415–437

Qadri SMH and Hoare DS (1967) Pyruvate decarboxylase in photosynthetic bacteria. Biochim Biophys Acta 148: 304–306

Qadri SMH and Hoare DS (1968) Formic hydrogenlyase and the photoassimilation of formate by a strain of *Rhodopseudomonas palustris*. J Bacteriol 95: 2344–2357

Quayle, JR and Keech DB (1959) Carboxydismutase activity in formate- and oxalate-grown *Pseudomonas oxalaticus* (strain OX1). Biochim Biophys Acta 31: 587–588

Quayle JR and Pfennig N (1975) Utilization of methanol by Rhodospirillaceae. Arch Microbiol 102: 193–198

Ranty B, Lundqvist T, Schneider G, Madden M, Howard R and Lorimer G (1990) Truncation of ribulose-1,5-bisphosphate carboxylase/oxygenase (Rubisco) from *Rhodospirillum rubrum* affects the holoenzyme assembly and activity. EMBO J 9:1365–1373

Read BA and Tabita FR (1992a) A hybrid ribulosebisphosphate carboxylase/oxygenase enzyme exhibiting a substantial increase in substrate specificity factor. Biochemistry 31: 5553–5560

Read BA and Tabita FR (1992b) Catalytic properties of a hybrid ribulose bisphosphate carboxylase/oxygenase enzyme containing cyanobacterial large subunits and diatom small subunits. FASEB J 6, A209

Read BA and Tabita FR (1994) High substrate factor ribulose bisphosphate carboxylase/oxygenase from eukaryotic marine algae and properties of recombinant cyanobacterial rubisco cotaining 'algal' residue modifications. Arch Biochem Biophys 312: 210–218

Richardson DJ, King GF, Kelly DJ, McEwan AG, Ferguson SJ and Jackson JB (1988) The role of auxiliary oxidants in maintaining redox balance during phototrophic growth of *Rhodobacter capsulatus* on propionate or butyrate. Arch Microbiol 150: 131–137

Rindt K-P and Ohmann E (1969) NADH and AMP as allosteric effectors of ribulose-5-phosphate kinase in *Rhodopseudomonas sphaeroides*. Biochem Biophys Res Commun 36: 357–364

Rippel S and Bowien B (1984) Phosphoribulokinase from *Rhodopseudomonas acidophila*. Arch Microbiol 139: 207–212

Roesler KR and Ogren WL (1990) *Chlamydomonas reinhardtii* phosphoribulokinase. Plant Physiol 93: 188–193.

Roy H, Bloom M, Milos P and Monroe M (1982) Studies on the assembly of large subunits of ribulose bisphosphate carboxylase in isolated pea chloroplasts. J Cell Biol 94: 20–27

Roesler KR, Marcotte BL and Ogren WL (1992) Functional importance of arginine 64 in *Chlamydomonas reinhardtii* phosphoribulokinase. Plant Physiol. 98: 1285–1289

Sahm H, Cox RB and Quayle JR (1976) Metabolism of methanol by *Rhodopseudomonas acidophila*. J Gen Microbiol 94: 313–322

Saier, Jr MH, Feucht BU and Roseman S (1971) Phosphoenolpyruvate-dependent fructose phosphorylation in photosynthetic bacteria. J Biol Chem 246: 7819–7821

Sandbaken MG, Runquist JA, Barbieri JT and Miziorko HM (1992) Identification of the phosphoribulokinase sugar phosphate binding domain. Biochemistry 31: 3715–3719

Sarles LS and Tabita FR (1983) Derepression of the synthesis of D-ribulose 1,5-bisphosphate carboxylase/oxygenase from *Rhodospirillum rubrum*. J Bacteriol 153: 458–464

Schedel M, Klemme J-H and Schlegel HG (1975) Regulation of C_3-enzymes in facultative phototrophic bacteria. The cold-labile pyruvate kinase of *Rhodopseudomonas sphaeroides*. Arch Microbiol 103: 237–245

Schneider G, Lindqvist Y and Lundqvist T (1990) Crystallographic refinement and structure of ribulose-15,-bisphosphate carboxylase from *Rhodospirillum rubrum* at 1.7 A resolution. J Mol Biol 211: 989–1008

Schön G and Voelskow H (1976) Pyruvate fermentation in *Rhodospirillum rubrum* after transfer from aerobic to anaerobic conditions in the dark. Arch Microbiol 107: 87–92

Schreuder HC, Knight S, Curmi PMG, Andersson I, Cascio D, Sweet RM, Bradnon C-I, and Eisenberg D (1993) Crystal structure of activated tobacco RubisCO complexed with the reaction-intermediate analog 2-carboxy-arabinitol 1,5-bisphosphate. Protein Science 2, 1136–1146

Shively JM, Davidson E and Marrs BL (1984) Derepression of the synthesis of the intermediate and large forms of ribulose-1,5-bisphosphate carboxylase/oxygenase in *Rhodopseudomonas capsulata*. Arch Microbiol 138: 233–236

Smrcka AV, Bohnert HJ and Jensen RG (1991) Modulation of the tight binding of carboxyarabinitol 1,5-bisphosphate to the large subunit of ribulose 1,5-bisphosphate carboxylase/oxygenase. Arch Biochem Biophys 286: 14–19

Somerville CR and Somerville SC (1984) Cloning and expression of the *Rhodospirillum rubrum* ribulosebisphosphate carboxylase gene in *E. coli*. Mol Gen Genet 193: 214–219

Spear N and Sojka G (1984) Conversion of two distinct *Rhodopseudomonas capsulata* isolates to the glycerol-utilizing phenotype. FEMS Microbiol Letts 22: 259–263

Spreitzer RJ (1993) Genetic dissection of RubisCO structure and function. Plant Mol Biol 44: 411–434

Springgate CF and Stachow CS (1972a) Fructose 1,6-diphosphatase from *Rhodopseudomonas palustris*. I. Purification and properties. Arch Biochem Biophys 152: 1–12

Springgate CF and Stachow CS (1972b) Fructose 1,6-diphosphatase from *Rhodopseudomonas palustris*. II. Regulatory properties. Arch Biochem Biophys 152: 13–20

Stanier RY, Adelberg EA, and Ingraham JL (1976) The Microbial World, Fourth Edition pp 546–548 Englewood Cliffs, New Jersey

Stein JL and Felbeck H (1993) Kinetic and physical properites of a recombinant RubisCO from a chemoautotrophic endosymbiont. Molec Mar Biol Biotech 2: 280–290

Stoppani AOM, Fuller RC and Calvin M (1955) Carbon dioxide

fixation by *Rhodopseudomonas capsulatus*. J Bacteriol 69: 491–501

Störro I and McFadden BA (1981) Glycolate excretion by *Rhodospirillum rubrum*. Arch Microbiol 129: 317–320

Störro I and McFadden BA (1983) Ribulose bisphosphate carboxylase/oxygenase in toluene-permeabilized *Rhodospirillum rubrum*. Biochem J 212: 45–54

Strauss G, Eisenreich W, Bacher A and Fuchs G (1992) ^{13}C-NMR study of autotrophic CO_2 fixation pathways in the sulfur-reducing archaebacterium *Thermoproteus neutrophilus* and in the phototrophic eubacterium *Chloroflexus aurantiacus*. Eur J Biochem 205: 853–866

Su X and Bogorad L (1991) A residue substitution in phosphoribulokinase of *Synechocystis* PCC 6803 renders the mutant light-sensitive. J Biol Chem 266: 23698–23705

Suwanto A and Kaplan S (1989) Physical and genetic mapping of the *Rhodobacter sphaeroides* 2.4.1 genome: Genome size, fragment identification, and gene localization. J Bacteriol 171: 5840–584

Szymona M Doudoroff M (1960) Carbohydrate metabolism in *Rhodopseudomonas spheroides* J Gen Microbiol 22: 167–183

Tabita FR (1980) Pyridine nucleotide control and subunit structure of phosphoribulokinase from photosynthetic bacteria. J Bacteriol 143: 1275–1280

Tabita FR (1987) Carbon dioxide fixation and its regulation in cyanobacteria. In: Fay P and Van Baalen C (eds) The Cyanobacteria, pp 96–117. Elsevier, Amsterdam

Tabita FR (1988) Molecular and cellular regulation of autotrophic carbon dioxide fixation in microorganisms. Microbiol Rev 52: 155–189

Tabita FR (1994) The biochemistry and molecular regulation of carbon dioxide metabolism in cyanobacteria. In: Bryant DA (ed) The Molecular Biology of Cyanobacteria, pp 437–467. Kluwer Academic Publishers, Dordrecht

Tabita FR and Colletti C (1979) Carbon dioxide assimilation in cyanobacteria: Regulation of ribulose 1,5-bisphosphate carboxylase. J Bacteriol 140: 452–458

Tabita FR and McFadden BA (1972) Regulation of ribulose-1,5-diphosphate carboxylase by 6-phospho-D-gluconate. Biochem Biophys Res Commun 48: 1153–1159

Tabita FR and McFadden BA (1974a) D-Ribulose 1,5-diphosphate carboxylase from *Rhodospirillum rubrum*. I. Levels, purification, and effect of metallic ions. J Biol Chem 249: 3453–3458

Tabita FR and McFadden BA (1974b) D-Ribulose 1,5-diphosphate carboxylase from *Rhodospirillum rubrum*. II. Quaternary structure, composition, catalytic and immunological properties. J Biol Chem 249: 3459–3464

Tabita FR, Gibson JL, Mandy WJ and Quivey, Jr RG (1986) Synthesis and assembly of a novel recombinant ribulose bisphosphate carboxylase/oxygenase. Bio/Technology 4:138–141

Tabita FR, Gibson JL, Bowien B, Dijkhuizen L and Meijer WG (1992) Uniform designation for genes of the Calvin-Benson-Bassham reductive pentose phosphate pathway of bacteria. FEMS Microbiol Lett 99: 107–110

Terlesky KC and Tabita FR (1991) Purification and characterization of the chaperonin 10 and chaperonin 60 proteins from *Rhodobacter sphaeroides*. Biochemistry 30: 8181–8186

Thiele HH (1968) Die verwertung einfacher organischer substrate durch Thiorhodaceae. Arch Mikrobiol 60:124–138

Torres-Ruiz JA and McFadden BA (1985) Isolation of L_8 and L_8S_8 forms of ribulose bisphosphate carboxylase/oxygenase from *Chromatium vinosum*. Arch Microbiol 142: 55–60

Torres-Ruiz J and McFadden BA (1987) The nature of L_8 and L_8S_8 forms of ribulose bisphosphate carboxylase/oxygenase from *Chromatium vinosum*. Arch Biochem Biophys 254: 63–68

Torres-Ruiz JA and McFadden BA (1989) A homolog of ribulose bisphosphate carboxylase/oxygenase-binding protein in *Chromatium vinosum*. Arch Biochem Biophys 261:196–204

Torres-Ruiz JA and McFadden BA (1992) Purification and characterization of chaperonin 10 from *Chromatium vinosum*. Arch Biochem Biophys 295:172–179

Turner AM and Mann NH (1986) Protein phosphorylation in *Rhodomicrobium vannielii*. J Gen Microbiol 132: 3433–3440

Uffen RL (1976) Anaerobic growth of a *Rhodopseudomonas* species in the dark with carbon monoxide as sole source of carbon and energy source. Proc Natl Acad Sci USA 73: 3298–3302

Uffen RL (1983) Metabolism of carbon monoxide by *Rhodopseudomonas gelatinosis*: Cell growth and properties of the oxidation system. J Bacteriol 155: 956–965

Uffen RL and Wolfe RS (1970) Anaerobic growth of purple non-sulfur bacteria under dark conditions. J Bacteriol 104: 462–472

van Niel CB (1941) The bacterial photosyntheses and their importance for the general problem of photosynthesis. Adv Enzymol 1: 263–328

van Niel CB (1944) The culture, general physiology, morphology, and classification of the non-sulfur purple and brown bacteria. Bacteriol Rev 8: 1–118

Viale AM, Kobayashi H, Takabe T and Akazawa T (1985) Expression of genes for subunits of plant-type RubisCO from *Chromatium* and production of the enzymatically active molecule in *Escherichia coli*. FEBS Lett 192: 283–288

Viale AM, Kobayashi H and Akazawa T (1989) Expressed genes for plant-type ribulose 1,5-bisphosphate carboxylase/oxygenase in the photosynthetic bacterium *Chromatium vinosum*, which possess two complete sets of the genes. J Bacteriol 171: 2391–2400

Viale AM, Kobayashi H, and Akazawa T (1990) Distinct properties of *Escherichia coli* products of plant-type ribulose-1,5-bisphosphate carboxylase/oxygenase directed by two sets of genes from the photosynthetic bacterium *Chromatium vinosum*. J Biol Chem 265: 18383–18392

Viitanen PV, Lubben TH, Reed J, Goloubinoff P, O'Keefe DP and Lorimer GH (1990) Chaperonin-facilitated refolding of ribulosebisphosphate carboxylase and ATP hydrolysis by chaperonin 60 (groEL) are K^+ dependent. Biochemistry 29: 5665–5671

Wang X and Song H (1989) Allosteric regulation of the state of adenylylation of glutamine synthetase in permeabilized preparations of *Rhodopseudomonas sphaeroides*. Sci China Ser B 32: 960–969

Wang X and Tabita FR (1992a) Reversible inactivation and characterization of purified inactivated form I ribulose 1,5-bisphosphate carboxylase/oxygenase of *Rhodobacter sphaeroides*. J Bacteriol 174: 3593–3600

Wang X and Tabita FR (1992b) Interaction between ribulose 1,5-bisphosphate carboxylase/oxygenase activity and the ammonia assimilatory system of *Rhodobacter sphaeroides*. J Bacteriol 174: 3601–3606

Wang X and Tabita FR (1992c) Interaction of inactivated and

active ribulose 1,5-bisphosphate carboxylase/oxygenase of *Rhodobacter sphaeroides* with nucleotides and the chaperonin 60 (GroEL) protein. J Bacteriol 174: 3607–3611

Wang X, Falcone DL and Tabita FR (1993a) Reductive pentose phosphate-independent CO_2 fixation in *Rhodobacter sphaeroides* and evidence that RubisCO activity serves to maintain the redox balance of the cell. J Bacteriol 175: 3372–3379

Wang X, Modak HV, and Tabita FR (1993b) Photolithoautotrophic growth and control of CO_2 fixation in *Rhodobacter sphaeroides* and *Rhodospirillum rubrum* in the absence of ribulose bisphosphate carboxylase/oxygenase. J Bacteriol 175: 5066–5077

Watson GMF, Mann NH, MacDonald GA and Dunbar B (1990) Identification and characterization of a GroEL homologue in *Rhodobacter sphaeroides*. FEMS Microbiol Lett 72: 349–354

Weaver PF, Wall JD and Gest H (1975) Characterization of *Rhodopseudomonas capsulata*. Arch Microbiol 105: 207–216

Whitman WB, Martin MN and Tabita FR (1979) Activation and regulation of ribulose bisphosphate carboxylase-oxygenase in the absence of small subunits. J Biol Chem 254: 10184–10189

Widdell F, Schnell S, Heising S, Ehrenreich A, Assmus B and Schink B (1993) Ferrous iron oxidation by anoxygenic phototrophic bacteria. Nature 362: 834–836

Willison JC (1988) Pyruvate and acetate metabolism in the photosynthetic bacterium *Rhodobacter capsulatus*. J Gen Microbiol 134: 2429–2439

Wood HG (1991) Life with CO or CO_2 and H_2 as a source of carbon and energy. FASEB J. 5: 156–163

Wu L-F, Reizer A, Reizer J, Cai B, Tomich JM and Saier, MH Jr (1991) Nucleotide sequence of the *Rhodobacter capsulatus fruK* gene, which encodes fructose-1-phosphate kinase: Evidence for a kinase superfamily including both phosphofructokinases of *Escherichia coli*. J Bacteriol. 173: 3117–3127

Wu L-F and Saier MH Jr (1990) Nucleotide sequence of the *fruA* gene, encoding the fructose permease of the *Rhodobacter capsulatus* phosphotransferase system, and analyses of the deduced protein sequence. J Bacteriol 172: 7167–7178

Wu L-F, Tonich JM, and Saier MH Jr (1990) Structure and evolution of a multidomain, multiphosphoryl transfer protein: Nucleotide sequence of the *fruB(HI)* gene in *Rhodobacter capsulatus* and comparisons with homologous genes from other organisms. J Mol Biol 213: 687–703

Yoch DC and Lindstrom ES (1968) Nicotinamide adenine dinucleotide-dependent formate dehydrogenase from *Rhodopseudomonas palustris*. Arch Mikrobiol 67: 182–186

Zeilstra-Ryalls J, Fayet O and Georgopoulos C. (1991) The universally conserved GroE (Hsp60) chaperonins. Ann Rev Microbiol 45: 301–325

Chapter 42

Microbiology of Nitrogen Fixation by Anoxygenic Photosynthetic Bacteria

Michael T. Madigan
Department of Microbiology, Southern Illinois University, Carbondale, IL 62901-6508, USA

Summary

Anoxygenic phototrophic bacteria conduct many specialized metabolic processes but one of the most important is nitrogen fixation, the reduction of N_2 to NH_3. Nitrogen fixation is catalyzed by the enzyme nitrogenase which is widely distributed among anoxygenic phototrophs. The most comprehensive understanding of N_2 fixation in photosynthetic bacteria is among the nonsulfur purple bacteria, where the capacity to fix N_2 is nearly universal. In this group N_2 fixation occurs both in the light and in darkness and nitrogenase expression and activity is highly regulated. Available evidence indicates that N_2 fixation is fairly widespread among purple and green sulfur bacteria and probably universal among heliobacteria. By contrast, the thermophilic phototroph *Chloroflexus aurantiacus*, a very few purple nonsulfur bacteria and green sulfur bacteria, and the 'aerobic phototrophs' are not diazotrophic. The recently discovered bacteriochlorophyll-containing rhizobia are also nitrogen fixers but their status as anoxygenic phototrophs is at present unclear. Although the ecological significance of N_2 fixation by anoxygenic phototrophs is unknown, the widespread distribution of these organisms in nature and the fact that most species are diazotrophic, suggests that they may well be significant, particularly in specialized environments such as microbial mats and paddy soils.

R. E. Blankenship, M. T. Madigan and C. E. Bauer (eds): Anoxygenic Photosynthetic Bacteria, pp. 915–928.

I. Background

The biological utilization of dinitrogen (N_2) as a source of cell nitrogen is a process called *nitrogen fixation* and is a property of only certain prokaryotes. In the fixation process, N_2 is reduced to NH_3 and the latter converted into organic form. Nitrogen fixation is catalyzed by the enzyme complex *nitrogenase* (Burris and Roberts, 1993). The biochemistry and genetics of nitrogen fixation in anoxygenic photosynthetic bacteria is discussed in detail in Roberts and Ludden (1992) and in Chapter 43 of this volume by the same authors. The present chapter documents the occurrence of nitrogen fixation in anoxygenic phototrophs, compares the efficacy of this process in different species where possible, and discusses the potential ecological ramifications of N_2 fixation by photosynthetic bacteria as it relates to their competitiveness in aquatic and terrestrial habitats. For a general treatment of the microbiology of N_2 fixation the reader is referred to the excellent reviews by Eady (1992) and Young (1992).

A. Discovery of N_2 Fixation in Anoxygenic Photosynthetic Bacteria

The first hint that anoxygenic photosynthetic bacteria could fix molecular nitrogen emerged 45 years ago during an investigation of light-dependent H_2 production in the purple nonsulfur bacterium *Rhodospirillum rubrum* (Gest and Kamen, 1949). Cells growing photosynthetically on malate with glutamate as nitrogen source evolved large quantities of H_2 (now known to be a result of nitrogenase activity, see Chapter 43 by Ludden and Roberts) while parallel cultures grown on ammonia did not. In experiments with resting cells, glutamate-grown cells also photoproduced H_2 and this process was strongly inhibited by N_2; the latter was a major clue that *Rs. rubrum* was capable of nitrogen fixation (Kamen and Gest, 1949). Confirmation of this was made by measurement of the fixation of $^{15}N_2$ into cells of *Rs. rubrum* and other species of purple bacteria (Lindstrom et al., 1949), and several papers appearing shortly afterwards suggested that the capacity to fix nitrogen was widely distributed among purple and green bacteria (Lindstrom et al., 1950, 1951; Newton and Wilson, 1953).

Abbreviations: Cf. – Chloroflexus; Cm. – Chromatium; Rb. – Rhodobacter; Rc. Rhodocyclus; Rs. – Rhodospirillum; Tc. – Thiocapsa

Since the early 1950s studies of particular species of anoxygenic phototrophs and one major survey (Madigan et al., 1984) of nitrogen fixation in purple nonsulfur bacteria have indicated that most but not all species can fix N_2 and that in addition, some species are naturally 'strong' nitrogen-fixers (i.e., grow rapidly on N_2) while other species are more feeble in this connection (see e.g., Madigan et al., 1984). Such differences in efficacy of N_2 fixation may reflect the ecological importance of N_2 fixation to particular species but is undoubtedly also a reflection of genetic and metabolic differences between species of photosynthetic bacteria as well.

B. Nitrogenases

For many years it was thought that molybdenum was absolutely required for N_2 fixation because of its central importance as a metal cofactor of nitrogenase (Bishop and Premakumar, 1992; Burris and Roberts, 1993). However it is now clear that biological N_2 fixation can proceed in the absence of molybdenum through the activity of *alternative nitrogenases* that lack this element (see Bishop and Premakumar, 1992 for a recent review).

All nitrogenases consist of two proteins, dinitrogenase and dinitrogenase reductase. Iron is present along with molybdenum in classical dinitrogenase (referred to as nitrogenase-1). In addition to nitrogenase-1, two non-molybdenum nitrogenases have been described, including one in which vanadium substitutes for molybdenum in dinitrogenase (called nitrogenase-2) and another in which iron only is present in dinitrogenase (nitrogenase-3). Dinitrogenase reductases from all forms of nitrogenase contain iron as a metal cofactor. Each form of nitrogenase is coded for by its own distinct gene complex but significant amino acid sequence homology exists among the three forms of nitrogenase, especially among dinitrogenase reductases (Bishop and Premakumar, 1992).

Among anoxygenic photosynthetic bacteria alternative nitrogenases have been documented in only four species. Iron-only (nitrogenase-3) type nitrogenases have been characterized from the purple nonsulfur bacteria *Rhodobacter capsulatus* (Schneider et al., 1991; Shüddekopt, et al., 1993) and from *Rhodospirillum rubrum* (Lehman and Roberts, 1991). Physiological evidence for alternative nitrogenase systems, presumably of the nitrogenase-3 type but not proven unequivocally as such, has also emerged from studies of *Rhodospirillum fulvum* (Gogotov, et al., 1991) and *Heliobacterium gestii* (Kimble and Madigan, 1992b). Curiously, experiments with *Rhodobacter sphaeroides*, a close relative of *Rb. capsulatus* (Woese et al., 1984a), indicated that this species lacked an alternative nitrogenase system (Gogotov et al., 1991). In addition, to the author's knowledge no reports of alternative nitrogenases from purple or green sulfur bacteria have been published. It can thus be concluded that alternative nitrogenases are present in at least some anoxygenic phototrophs and are of the nitrogenase-3 type. As in nonphotosynthetic bacteria, alternative nitrogenases presumably allow species of photosynthetic bacteria to continue diazotrophic growth in environments limiting in molybdenum but containing sufficient iron to otherwise satisfy the needs of nitrogen fixation.

C. Physiological Prerequisites for N_2 Fixation by Photosynthetic Bacteria

Photosynthetic purple bacteria show remarkable metabolic diversity and are capable of growing both phototrophically (anaerobic/light) and in darkness by both respiration and fermentation (Madigan and Gest, 1979; Madigan, 1988; see also Chapter 41 by Tabita). *Rhodobacter capsulatus* shows particularly well-developed abilities in this connection, being capable of growth under five distinctly different growth modes: photoautotrophic, photoheterotrophic, chemoorganotrophic by aerobic respiration, chemoorganotrophic by fermentation, and chemolithotrophic with H_2 as electron donor (O_2 as electron acceptor, Madigan and Gest, 1979). Diazotrophic growth by *Rb. capsulatus* is possible under at least the first four of these conditions.

1. The Effect of Light

In the original discovery of N_2 fixation in photosynthetic bacteria (Gest and Kamen, 1949; Kamen and Gest, 1949), photoheterotrophic growth conditions were employed. All purple nonsulfur bacteria capable of N_2 fixation (see Section II) grow best diazotrophically under photoheterotrophic conditions (Madigan et al, 1984). However, many purple bacteria can also grow photoautotrophically with H_2, $S_2O_3^{2-}$, H_2S, Fe^0, or some other inorganic substance as electron donor (Madigan, 1988); N_2 fixation is also possible under these conditions.

Dark *aerobic* growth of purple nonsulfur bacteria has been known since before the classic study of van Niel (1944), but dark *anaerobic* growth, either by fermentation (Uffen and Wolfe, 1970; Madigan and Gest, 1978) or by various anaerobic respiratory means (see Chapter 44 by Zannoni) was only discovered more recently. Once dark anaerobic growth of purple bacteria was established (Madigan and Gest, 1978) it was possible to test hypotheses (Meyer et al., 1978a,b) suggesting an obligatory link between photosynthesis and N_2 fixation in these organisms. The lack of any such connection was clearly shown in experiments where *Rb. capsulatus* was grown anaerobically in darkness with fructose as sole carbon and energy source and N_2 as sole nitrogen source (growth under these conditions also requires the presence of an 'accessory oxidant', such as dimethyl sulfoxide (DMSO), to allow for catabolism of the fructose, see Madigan and Gest, 1978; Madigan et al., 1980, and Chapter 44 of this volume by Zannoni). Under dark anaerobic conditions diazotrophic growth of *Rb. capsulatus* occurred and cell suspensions readily reduced acetylene to ethylene (a common assay of nitrogenase activity) in darkness (Fig. 1 and Madigan et al., 1979). Dark anaerobic nitrogen fixation has also been reported in the heliobacteria. Anaerobic cultures of all species of heliobacteria continue to fix N_2 and grow slowly in darkness (Kimble and Madigan, 1992a) via fermentation of pyruvate, yielding acetate and in some species H_2 (Kimble et al., 1994).

To the author's knowledge, no report of dark anaerobic N_2 fixation by purple *sulfur* bacteria has been published, although dark microaerobic growth (on ammonia) of some species is possible (Kämpf and Pfennig, 1980) and several of these also fix N_2 microaerobically in darkness (Jouanneau et al., 1980; Kämpf and Pfennig, 1986). The ability of green sulfur bacteria to fix N_2 in darkness is unknown; to date, no one has successfully grown these organisms in darkness under *any* growth conditions. Photosynthetically-grown cell suspensions of *Chlorobium* species incubated in darkness showed essentially no nitrogenase activity (Heda and Madigan, 1986*a*). However, should conditions for dark growth of green sulfur bacteria be discovered, N_2 fixation will probably be possible here as well.

2. The Effect of Oxygen

Although nitrogenase is obviously protected from oxygen inactivation during *anaerobic* growth photosynthetically or in darkness (see Fig. 1), *aerobic*

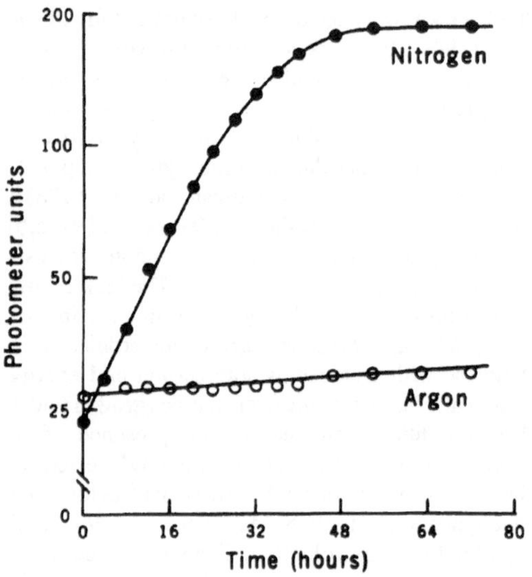

Fig. 1. Anaerobic growth of *Rhodobacter capsulatus* strain B10 in darkness with N_2 as sole nitrogen source. The mineral salts growth medium contained fructose as sole carbon and energy source and dimethyl sulfoxide (DMSO) as electron acceptor. Gas phases consisted of N_2:CO_2 (95:5) (●) or argon (○). Cultures in side-arm flasks were incubated in total darkness at 33 °C; 200 Klett units are equal to approximately 650 μg of bacterial dry weight per milliliter. Suspensions of cells grown in this fashion readily reduced acetylene to ethylene. (From Madigan et al., 1979.)

dark growth on N_2 is problematic. Relatively little O_2 is required to inactivate preexisting nitrogenase in *Rb. capsulatus* (Hochman and Burris, 1981) and other purple bacteria (Madigan et al, 1984), and O_2 is also a potent repressor of nitrogenase synthesis in facultatively aerobic bacteria including purple bacteria (Burris and Roberts, 1993). Thus, to achieve aerobic dark diazotrophic growth, photosynthetic bacteria must receive sufficient O_2 to carry out respiration (for ATP production) while at the same time prevent inactivation and/or repression of synthesis of nitrogenase.

Conditions for aerobic diazotrophic growth of anoxygenic phototrophs were first described by Siefert and Pfennig (1980) using auxanographic techniques. When molten agar media is seeded with a culture of a photosynthetic bacterium and then allowed to solidify, O_2 and N_2 diffuse downward and respiratory growth on N_2 is observed as a tight band of cells a defined distance from the agar surface; the band represents the only zone in the agar tube where O_2 levels are compatible with both diazotrophy and respiration (Fig. 2). Interestingly, however, because of differences in respiratory potential among species of purple bacteria, the distance from the agar surface that the growth band appears varies considerably. Species such as *Rb. sphaeroides*, which grow well at full oxygen tensions, form growth bands much nearer the surface than do extremely microaerophilic species, such as *Rs. molischianum* (Fig. 2). Using auxanographic methods, dark microaerobic N_2 fixation occurred in all diazotrophic purple nonsulfur bacteria tested except for one strain of *Rs. fulvum* (Madigan et al., 1984) and also occurred among certain species of Chromatiaceae (Kämpf and Pfennig, 1986).

II. Nitrogen Fixation by Purple and Green Bacteria

This section is divided into four parts, each developed around a summarizing table listing the species of photosynthetic bacteria in which N_2 fixation has been documented. In addition, where data are available, an indication of the relative efficacy of the N_2 fixation process as measured by in vivo nitrogenase activities in each species is given; such data identify the 'stronger' and 'weaker' N_2-fixing species of photosynthetic bacteria (see Madigan et al. 1984).

A. Purple Nonsulfur Bacteria

In a major survey of 18 species of purple nonsulfur bacteria performed in my laboratory over 10 years ago all but one species, *Rhodocyclus purpureus*, were found capable of N_2 fixation (Madigan et al., 1984). Species of *Rhodobacter*, such as *Rb. capsulatus* and *Rb. sphaeroides*, expressed high levels of nitrogenase in vivo and grew most rapidly on N_2, while species like *Rhodopseudomonas palustris* and *Rhodopila globiformis* expressed comparatively low nitrogenase levels and grew only very slowly on N_2 (Table 1). A correlation between in vivo specific nitrogenase contents and growth rate was observed in this study, and strains of *Rb. capsulatus* consistently showed the highest specific nitrogenase activities and shortest generation times for photoheterotrophic growth on N_2 (Madigan et al., 1984). This finding supports the observation that enrichment cultures for purple nonsulfur bacteria using the ability to fix N_2 as selective agent commonly result in *Rhodobacter* species, especially *Rb. capsulatus* (Gest et al., 1985). Since the study of Madigan et al. (1984) was

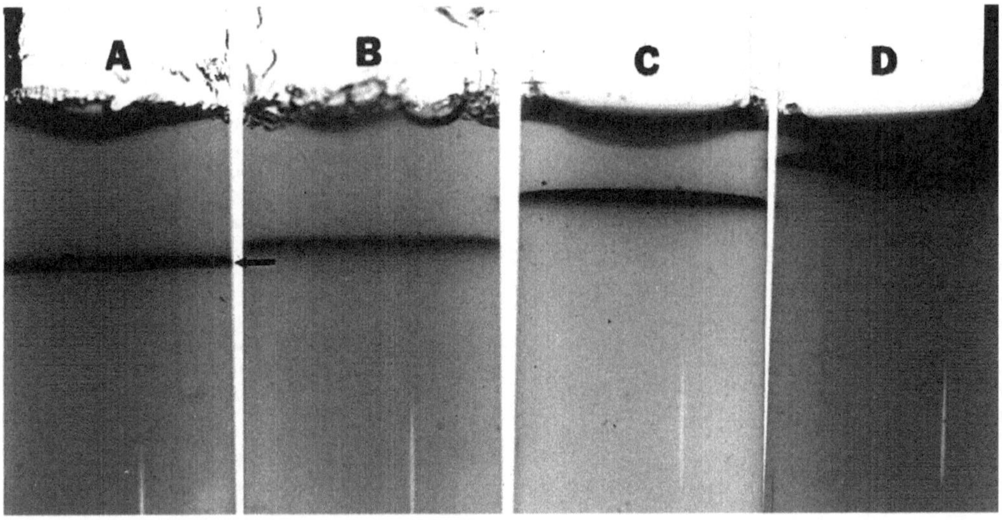

Fig. 2. Photograph of microaerobic dark growth of various species of anoxygenic phototrophic bacteria with N_2 as the nitrogen source. Cultures were seeded into molten agar tubes of a mineral salts/malate medium lacking combined nitrogen, quickly cooled to solidify the agar, and incubated in air and total darkness. Growth is indicated by the pigmented band of cells (e.g., see arrow in A) located at various distances from the agar surface. Organisms are: A, *Rhodospirillum molischianum* ATCC 14031; B, *Rhodopseudomonas viridis* DSM 133; C, *Rhodospirillum fulvum* DSM 2860; D, *Rhodobacter sphaeroides* 2.4.1.

undertaken, several new species of purple nonsulfur bacteria have been described and their nitrogen-fixing status, if known, is included in Table 1.

Only two species of purple nonsulfur bacteria have been shown by growth and acetylene reduction experiments to be incapable of N_2 fixation. These include *Rc. purpureus* (Masters and Madigan, 1983; Madigan et al., 1984), and the Dead Sea halophile, *Rhodospirillum sodomense* (Mack et al., 1993). Subsequent probing of DNA from both of these species with a *nifH* gene probe (*nifH* codes for dinitrogenase reductase) that hybridizes strongly with DNA from other species of purple bacteria confirmed that *Rc. purpureus* and *Rs. sodomense* lacked *nif* genes, thus explaining their failure to fix N_2 (PW Ludden, GP Roberts and MT Madigan, unpublished results). It is possible that the unusual habitats of *Rc. purpureus* and *Rs. sodomense*, a swine waste lagoon rich in ammonia and amines (Pfennig, 1978) and the Dead Sea (Mack et al., 1993), respectively, have for some reason(s) selected for an inability to fix N_2 in these phototrophs.

An interesting situation exists in regard to N_2 fixation in the moderately halophilic species *Rhodospirillum salexigens*. This organism was originally described as a glutamate auxotroph (Drews, 1981), however experiments in my laboratory showed this to be unfounded (Rubin and Madigan, 1986). Although *Rs. salexigens* grows well either phototrophically or chemotrophically (aerobic/dark) in media containing glutamate as nitrogen source, utilization of ammonia or N_2 as sole nitrogen source is also possible. For unknown reasons substitution of NH_4Cl for glutamate in mineral media containing acetate (the preferred carbon source for *Rs. salexigens*, Drews, 1981) leads to a rapid rise in pH and cessation of growth. However, modifications to the medium, particularly the addition of pyruvate in addition to acetate, stabilized pH and supported good growth of *Rs. salexigens* on either ammonia or on N_2 (Rubin and Madigan, 1986).

Table 1 summarizes the microbiology of N_2 fixation in purple nonsulfur bacteria.

B. Purple Sulfur Bacteria (Chromatiaceae and Ectothiorhodospiraceae)

Documentation of N_2 fixation in purple sulfur bacteria is sketchy outside of a few species of the genus *Chromatium*. *Chromatium vinosum* has been well

Table 1. Occurrence and rates of nitrogen fixation in purple nonsulfur bacteria[a]

Genus/Species	N_2 Fixation[b]	Nitrogenase Activity[c]	Reference
Rhodobacter			
adriaticus	Y	ND	Neutzling et al., 1984
capsulatus	Y	3000	Lindstrom et al., 1951; Madigan et al., 1984
sphaeroides	Y	1800	Madigan et al., 1984 Lindstrom et al., 1951
sulfidophilus	Y	650	Hansen and Veldkamp, 1973; Kelley et al., 1979; Madigan et al., 1984
veldkampii	Y	ND	Hansen and Imhoff, 1985
Rhodopseudomonas			
acidophila	Y	700	Madigan et al., 1984
blastica	Y	1200	Eckersley and Dow, 1980; Madigan et al., 1984
cryptolactis	Y	ND	Stadtwald-Demchick et al., 1990
marina (agilis)	Y	1100	Imhoff 1983; Mangels et al., 1986
palustris	Y	500	Lindstrom et al., 1951; Madigan et al., 1984
rutila (palustris)	Y	ND	Akiba et al., 1983; Hiraishi et al., 1992
sulfoviridis	Y	500	Madigan et al., 1984
viridis	Y	2300	Gogotov and Glinskii, 1973; Howard et al, 1983; Madigan et al., 1984
Rhodospirillum			
centenum	Y	500–1000	Favinger et al., 1989
fulvum	Y	700	Madigan et al., 1984
molischianum	Y	1000	Madigan et al., 1984
photometricum	Y	750	Madigan et al., 1984
rubrum	Y	1000	Gest and Kamen, 1949; Kamen and Gest, 1949; Madigan et al., 1984; Schick, 1971
salexigens	Y	650	Madigan et al., 1984; Rubin and Madigan, 1986
sodomense	N	-	Mack et al., 1993
Rhodocyclus			
purpureus	N	-	Masters and Madigan, 1983; Madigan et al., 1984
tenuis	Y	800	Masters and Madigan, 1983; Madigan et al., 1984
Rhodopila			
globiformis	Y	500	Madigan and Cox, 1982; Madigan et al., 1984
Rhodomicrobium			
vannielii	Y	1000	Lindstrom et al., 1951; Gogotov and Glinskii, 1973; Madigan et al., 1984
Rhodoferax			
fermentans	Y	ND	Hiraishi et al., 1991
Rubrivivax			
gelatinosus	Y	1300	Lindstrom et al., 1951; Madigan et al., 1984

[a] Only those species in which N_2 fixation could be documented are included in this table.
[b] Y, yes (N_2 fixation documented); N, no (N_2 fixation absent)
[c] Average nitrogenase activity of suspensions of intact cells (grown on N_2 as sole nitrogen source) in nmoles $C_2H_4 \cdot h^{-1} \cdot mg$ cell dry weight^{-1}. In some cases values are the average of experiments with several strains of one species while in other cases only a single strain may have been tested. ND, not determined. Most values from Madigan et al., 1984.

studied as to its N_2-fixing capabilities and was the first phototrophic bacterium in which consistently active nitrogen-fixing cell-free extracts were obtained (Winter and Arnon, 1970). In vivo studies of nitrogenase in *Cm. vinosum* have also shown that the organism employs the ammonia 'switch-off' effect to control nitrogenase activity, as in purple nonsulfur bacteria (Gotto and Yoch, 1985). In addition to *Cm. vinosum*, dinitrogen fixation has been documented in six other species of *Chromatium*, four species of *Ectothiorhodospira*, two species of *Thiocapsa*, and in a single species each of the genera *Amoebobacter*, *Thiocystis*, and *Lamprobacter* (Zakhvataeva et al., 1970; Postgate, 1982; Caumette et al., 1985; Ventura

Table 2. Species of Chromatiaceae in which N_2 fixation has been demonstrated[a]

Genus/Species	Reference
Chromatium	
gracile[b]	Matheron and Baulaigue, 1983; Pfennig and Trüper, 1989
minus	Pfennig and Trüper, 1989
minutissimum	Pfennig and Trüper, 1989
vinosum[b]	Lindstrom et al., 1949; 1950; Newton and Wilson, 1953
violascens	Pfennig and Trüper, 1989
warmingii	Pfennig and Trüper, 1989
weissei	Pfennig and Trüper, 1989
Thiocapsa	
pfennigii	Pfennig and Trüper, 1989
roseopersicina[b]	Gogotov and Glinskii, 1973
Thiocystis	
violacea[b]	Matheron and Baulaigue, 1983; Pfennig and Trüper, 1989
Amoebobacter roseus	Postgate 1982; Stewart et al., 1980
Lamprobacter	
modesohalophilus	Gorlenko et al., 1979
Ectothiorhodospira	
marismortui	Oren et al., 1989
mobilis	Bast, 1977; Bognar et al., 1982; Madigan, unpublished
shaposnikovii	Zakhvataeva et al., 1971; Gogotov and Glinskii, 1973
vacuolata	Imhoff 1989
species, strains EST4/ EST8	Ventura et al., 1988

[a] See footnote a in Table 1. All N_2 fixation reported from photosynthetically-grown cultures.
b Capable of nitrogen fixation microaerobically in darkness, see Kämpf and Pfennig (1986)

et al., 1988; Pfennig and Trüper, 1989; Young, 1992). Of these organisms, *Tc. roseopersicina* is probably the most widespread in nature and is frequently found as a component of microbial mats (see Chapter 4 by van Gemerden and Mas) where it may be a major nitrogen-fixing species.

The capacity for N_2 fixation has not been reported for species of *Lamprocystis, Thiodictyon, Thiopedia, Thiorhodovibrio and Thiospirillum.* Thus the family Chromatiaceae obviously needs more study as regards nitrogen fixation. The current status of N_2 fixation in purple sulfur bacteria is summarized in Table 2.

C. Green Sulfur Bacteria (Chlorobiaceae)

Our knowledge of nitrogen fixation in green sulfur bacteria is somewhat more developed than that of purple sulfur bacteria, although the story is far from complete. Clear evidence for N_2 fixation has been obtained with several species of *Chlorobium* and with single species each of the genera *Pelodictyon, Prosthecochloris* and *Chloroherpeton* (see Table 3 and Bergstein et al., 1981; Heda and Madigan, 1986a,b; 1988; Rodionov et al., 1986; Gibson et al., 1984). Interestingly, like purple bacteria, nitrogen-fixing cultures of *Chlorobium* species showed ammonia 'switch-off' of nitrogenase activity (Heda and Madigan, 1986a; Rodionov et al., 1986; Wahlund and Madigan, 1993), indicating that this form of regulation, apparently universal among nitrogen-fixing purple bacteria (see Chapter 43 by Ludden and Roberts), extends also to the green sulfur bacteria, a phylogenetically distinct lineage from that of the purple bacteria (Gibson et al., 1985; Oyaizu et al., 1987).

The moderately thermophilic green sulfur bacterium *Chlorobium tepidum* (Wahlund et al., 1991) is an excellent nitrogen-fixing bacterium. This organism, which has a generation time of as little as 2 hr on ammonia, also grows rapidly on N_2 and intact cells contain extremely high levels of nitrogenase (Wahlund and Madigan, 1993). In a study of the mesophilic green bacterium *Cb. limicola* Heda and Madigan (1986a) also observed high levels of nitrogenase in vivo. These facts, coupled with the observations that *Cb. limicola* strongly derepressed nitrogenase when grown on glutamate as nitrogen source yet did not photoproduce H_2 (Heda and

Table 3. Occurrence and rates of N_2 Fixation in Chlorobiaceae (green sulfur) and Chloroflexaceae (green nonsulfur) bacteria[a]

Genus/Species	N_2 Fixation Established[b]	Nitrogenase Activity[c]	Reference(s)
Green Sulfur Bacteria			
Chlorobium			
limicola f. sp. *thiosulfatophilum*	Y	1550[d]	Lindstrom et al., 1949; 1950; Keppen et al., 1985; Heda and Madigan, 1986a
		8500[e]	Heda and Madigan, 1986a
phaeobacteroides	Y	ND	Bergstein et al., 1981; Kimble and Madigan, unpublished
tepidum	Y	6300	Wahlund et al., 1991; Wahlund and Madigan, 1993
Pelodictyon			
luteolum	Y	1200[f]	Heda and Madigan (1988)
phaeoclathratiforme	N	–	Overmann and Pfennig, 1989
Prosthecochloris aestaurii	Y	ND	Stewart et al., 1980
Chloroherpeton			
thalassium	Y	ND	Gibson et al., 1984
Green Nonsulfur Bacteria			
Chloroflexus			
aurantiacus	N	–	Heda and Madigan, 1986b
Oscillochloris			
trichoides	Y	180[f]	Keppen et al., 1989

[a,b] See footnotes a and b in Table 1

[c] Nitrogenase activity expressed as nmoles C_2H_4 produced·hr^{-1}·mg cell dry $weight^{-1}$.

[d] Average of two strains, strain 8327 and Tassajara, Heda and Madigan 1986*a*.

[e] Result with one strain, strain 8327, cells grown on 10mM glutamate instead of N_2 as sole nitrogen source.

[f] Activity expressed as nmoles C_2H_4·h^{-1}·mg cell $protein^{-1}$

Madigan, 1986a) and that nitrogenase from this organism did not cross-react immunologically with that of other diazotrophs (Heda and Madigan, 1988), warrants more thorough characterization of the nitrogenase system of green sulfur bacteria.

Interestingly, despite the good nitrogen-fixing properties of *Chlorobium tepidum*, its thermophilic counterpart among purple sulfur bacteria, *Chromatium tepidum* (Madigan, 1984, 1986), was found incapable of N_2 fixation (Madigan, 1986) despite its apparent possession of *nif* genes (PW Ludden, GP Roberts and MT Madigan, unpublished results). The reason for the lack of growth on N_2 by *Chromatium tepidum* remains unknown. Although N_2 fixation in *Chlorobium tepidum* was found to require elevated levels of sulfide (which was otherwise not required for growth of this thiosulfate-utilizing green bacterium, Wahlund et al., 1991; Wahlund and Madigan, 1993), elevated levels of sulfide do not support N_2 fixation in *Chromatium tepidum*. *Chlorobium tepidum* grows on N_2 up to its growth temperature maximum of 52 °C (Wahlund and Madigan, 1993) and this is the highest temperature reported for N_2 fixation by any green sulfur bacterium.

The current status of N_2 fixation in green sulfur bacteria is summarized in Table 3.

D. Green Nonsulfur Bacteria (Chloroflexaceae)

Although phenotypically resembling green sulfur bacteria in terms of pigments and the production of chlorosomes, green nonsulfur bacteria (the *Chloroflexus* group, see Chapter 3 by Pierson and Castenholz) are phylogenetically unrelated to green sulfur bacteria (Gibson et al., 1985; Oyaizu et al., 1987). Study of the N_2-fixing potential of four strains of *Chloroflexus aurantiacus* showed the organism to be totally unable to fix N_2 (Heda and Madigan, 1986b). This result was supported by the complete absence of hybridization of *Cf. aurantiacus* DNA to a *nifH* probe (PW Ludden, GP Roberts and MT Madigan, unpublished results). *Cf. aurantiacus* grows well on any of several amino acids (Heda and Madigan, 1986b) including ones that normally allow for derepression of nitrogenase in purple bacteria (Arp and Zumft, 1983). Despite this, however, cultures of *Cf. aurantiacus* grown on amino acids or on growth-limiting levels of ammonia failed to reduce acetylene or photoproduce H_2 (Heda and Madigan, 1986b). One published report of growth of *Cf. aurantiacus* in a medium lacking combined nitrogen, presumably at the expense of N_2 (Gallon and Chaplin, 1987), has not been confirmed in my laboratory, and it is possible that the slight growth observed was due to trace levels of fixed nitrogen compounds in the medium (J.R. Gallon, personal communication).

Weak but detectable N_2 fixation has been reported (Keppen et al., 1989) in the newly described green filamentous bacterium *Oscillochloris trichoides*, which is phylogenetically related to *Cf. aurantiacus* (Keppen et al., 1994). Thus it is possible that N_2-fixing *Cf. aurantiacus* strains exist but have thus far not been cultured. Temperature is obviously not a barrier here because fixation in thermophilic green sulfur bacteria (Wahlund and Madigan 1993) and thermophilic heliobacteria (Kimble et al., 1995) occurs within the temperature range of optimal growth of *Chloroflexus* (50–55 °C). Because of the more extreme thermophilic character (growth up to 72 °C, see Chapter 3 by Pierson and Castenholz) and unique phylogeny (Gibson et al., 1985; Oyaizu et al., 1987) of *Chloroflexus*, the properties of N_2-fixing species, if they exist, should be very interesting.

E. Heliobacteria (Heliobacteriaceae)

All known species of heliobacteria fix nitrogen (Kimble and Madigan, 1992a; Kimble et al., 1995). *Heliobacillus mobilis* appears to be the best N_2-fixing species presently in culture, both in terms of its ability to grow rapidly on N_2 and in expression of nitrogenase (Kimble and Madigan, 1992a). Even thermophilic heliobacteria are diazotrophic and fix N_2 up to 55°C (Kimble et al., 1995). Thus nitrogen fixation may be of strong selective advantage to heliobacteria in their soil, and especially rice soil, habitat (see Chapter 2 by Madigan and Ormerod).

All species of heliobacteria examined by Kimble and Madigan (1992a) and Kimble et al. (1995) were subject to ammonia 'switch-off' of nitrogenase activity. This finding in the heliobacteria, which phylogenetically are Gram-positive bacteria (Woese et al., 1985), suggests that this form of enzyme regulation is of such strong survival value for anoxygenic phototrophic bacteria that it has evolved across deep evolutionary lines from Gram-negative to Gram-positive phototrophs (ammonia 'switch-off' in heliobacteria is the first such instance reported from Gram-positive bacteria, Kimble and Madigan 1992a). Finally, as previously discussed (see Section IB), alternative (non-molybdenum) nitrogenases are present in certain heliobacteria, the best example being *Heliobacterium gestii* (Kimble and Madigan, 1992b).

A summary of nitrogen-fixation in heliobacteria is given in Table 4.

III. Phototrophic Rhizobia and Other Bacteriochlorophyll *a*-Containing Species

A number of *Rhizobium* species have been discovered that contain bacteriochlorophyll *a*. These, along with the 'aerobic phototrophs' (see below) have been termed 'quasi-photosynthetic bacteria' by Gest (1993) because they share certain properties with true anoxygenic phototrophs but lack many other characteristic traits (e.g. the ability to grow anaerobically, see Chapters 6 by Shimada and 7 by Fleischman et al. for a description of these organisms).

Bacteriochlorophyll *a*-containing *Rhizobium* species (now referred to as *Photorhizobium* species, see Chapter 7 by Fleischman et al.) nodulate the stems of several different, primarily tropical, leguminous plants. As is well-known, a prime physiological characteristic of *Rhizobium* is its ability to fix N_2, and evidence exists that *Photorhizobium* fixes N_2 both in association with host plants in stem nodules and *ex planta* in pure culture in the laboratory. It is hypothesized that photophosphorylation by *Photorhizobium* at least partially supports the energy needs of N_2 fixation in stem nodules; this in turn would reduce the energy drain on the plant leading to more efficient symbiotic nitrogen fixation.

Several aerobic BChl *a*-containing bacteria have been isolated by Japanese workers and others in the last 15 years. These organisms, referred to as 'aerobic phototrophs', are mainly marine and grow best under

Table 4. Occurrence and rates of N_2 fixation in species of heliobacteria (Heliobacteriaceae)[a]

Genus/Species	N_2 Fixation[b]	Nitrogenase Activity[c]	Reference(s)
Heliobacterium			
chlorum	Y	1700(P); 2900 (L)	Kimble and Madigan, 1992a
fasciatum[d]	Y	1450 (P)	Kimble and Madigan, unpublished
gestii	Y		
strain Chainat		1800 (P); 1300 (L)	Kimble and Madigan, 1992a
strain HD7 (Pfennig)		2600 (P)	Kimble and Madigan, unpublished
strain T15–1		2000 (P)	Stevenson and Madigan, unpublished
modesticaldum	Y		
strain Ice 1		3800 (P)	Kimble et al., 1995
strain YS6		2180 (P)	Kimble et al., 1995
Heliobacillus			
mobilis	Y	4400(P); 3800(L)	Kimble and Madigan, 1992a

[a,b] See footnotes to Table 1

[c] Nitrogenase activities expressed as nmoles C_2H_4 produced·h^{-1}·mg dry weight of $cells^{-1}$. P, pyruvate grown cells; L, lactate grown cells

[d] This organism is to be renamed as a new genus, *Heliophilum fasciatum* (Ormerod, Kimble and Madigan, unpublished)

fully aerobic conditions (Shiba and Harashima, 1986, and see also Chapter 6 by Shimada). However, no evidence for N_2 fixation by aerobic phototrophs has emerged and gene probe experiments (using a *nifH* probe) carried out on two species failed to show any hybridization of the probe to their chromosomal DNA (PW Ludden, GP Roberts and MT Madigan, unpublished results). Thus, all available evidence suggests that the 'aerobic phototrophs' are nondiazotrophic.

IV. Ecological Aspects of N_2 Fixation by Anoxygenic Photosynthetic Bacteria

With the exception of a few well studied lake ecosystems where mass developments of anoxygenic phototrophic bacteria occur (see review of Madigan, 1988 and Chapter 4 of this volume by van Gemerden and Mas), few data are available on the contributions of photosynthetic bacteria to the carbon balance of aquatic or terrestrial habitats. Much the same can be said about nitrogen fixation. A few reports of in situ nitrogen fixation by photosynthetic bacteria have been published (see e.g. Kobayashi et al., 1967; Habte and Alexander, 1980; Bergstein et al., 1981), but too little data exist to generalize on a global basis.

The nitrogen-fixing activities of photosynthetic bacteria in specialized habitats such as paddy soils or microbial mats, however, may be very ecologically significant (Kobayashi and Haque, 1971). In a detailed report on the distribution of nitrogen-fixing microorganisms in paddy soils of Southeast Asia, Kobayashi et al. (1967) clearly implicated anoxygenic phototrophs, particularly purple nonsulfur bacteria, as contributors to the nitrogen economy of these soils. A later study of N_2 fixation in rice soils by Habte and Alexander (1980) employed acetylene reduction methods and showed more directly the importance of anoxygenic photosynthetic bacteria to the fertility of paddy soils. Presumably photosynthetic bacteria growing in such environments utilize organic substrates leached from the rice plants or H_2S as electron donors for reduction of N_2 to ammonia. In addition, microbial mats (see Chapter 4 by van Gemerden and Mas) can be major sources of N_2 fixation by anoxygenic phototrophs; the common occurrence of the diazotroph *Thiocapsa roseopersicina*, a purple sulfur bacterium with well developed aerobic/dark metabolic capacities (Jouanneau et al., 1980) in such habitats suggests that this organism is a major contributor to fixed nitrogen inputs therein.

Another link between the paddy soil environment and photosynthetic bacteria is the recent discovery of the heliobacteria. These anoxygenic phototrophs are terrestrial organisms and are particularly abundant in rice soils (see Chapter 2 by Madigan and Ormerod for a discussion of the heliobacteria). Because heliobacteria are active nitrogen-fixers (Kimble and Madigan, 1992a,b; Kimble et al., 1995) and also capable of dark growth (Kimble et al., 1994), it is likely that these organisms contribute fixed nitrogen to paddy soil environments (see also Madigan, 1992 for a discussion of the ecology of heliobacteria).

The ability of photosynthetic bacteria to fix N_2 can be used to advantage in their enrichment. This is particularly true of purple nonsulfur bacteria (Gest et al., 1985) and heliobacteria (Madigan, 1992), however experiments in this laboratory have shown that diazotrophic enrichments also work for the isolation of purple and green sulfur bacteria (unpublished results and Heda and Madigan, 1986a). Combining these observations with the fact that results thus far have shown that most photosynthetic bacteria can fix N_2 (see Section II), it stands to reason that N_2 fixation is of ecological importance to anoxygenic phototrophs, at least under certain environmental conditions, and that their diazotrophic capacity allows them to compete better in their aquatic and terrestrial habitats. However, now that phylogenetic probes, biomarker analyses, and other highly specific analytical methods are available for characterizing the microbiology of virtually any habitat, it would be worthwhile to investigate the N_2-fixing activities of photosynthetic bacteria in situ using these modern techniques. Only from well controlled field studies that specifically target a group or groups of anoxygenic phototrophs will the real contributions of these organisms to the nitrogen economy of soils and waters be known.

V. Concluding Remarks

In conclusion it can be stated that N_2 fixation is a universal property of heliobacteria and nearly universal among purple nonsulfur bacteria; however, much is yet to be learned about this major physiological process in other groups of anoxyphototrophs. It will be of interest in future years to test more isolates of photosynthetic bacteria for the capacity to fix N_2, especially new species isolated from extreme environments or from marine environments (the latter of which have only been minimally explored, Imhoff 1983; Mangels et al., 1986; Wynn-Williams and Rhodes 1974), and to better understand the nature and distribution of alternative nitrogenases among anoxygenic phototrophs. Such studies will yield a more complete picture of the diazotrophic systems of photosynthetic bacteria and should complement our new evolutionary insight into these organisms, born with the advent of ribosomal RNA sequencing (Woese, 1987; 1992). The prominence of purple bacteria as likely ancestors of the majority of known Gram-negative bacteria (see Woese et al 1984a, b, and reviews of Woese 1987, 1992) and the heliobacteria of Gram-positive bacteria (Woese et al., 1985) mandates that we understand important metabolic processes such as N_2 fixation in anoxygenic phototrophs in considerably more detail.

Acknowledgments

Work on N_2 fixation in photosynthetic bacteria in my laboratory has been supported by the United States Department of Agriculture and the National Science Foundation. I thank Linda Kimble for unpublished results.

References

Akiba T, Usami R and Horikoshi K (1983) *Rhodopseudomonas rutila*, a new species of nonsulfur purple photosynthetic bacteria. Intl J Syst Bacteriol 33: 551–556

Arp DJ and Zumft WG (1983) Overproduction of nitrogenase by nitrogen-limited cultures of *Rhodopseudomonas palustris*. J Bacteriol 153: 1322–1330

Bast, E (1977) Utilization of nitrogen compounds and ammonia assimilation by *Chromatiaceae*. Arch Microbiol 113: 91–94

Bergstein T, Henis Y and BZ Cavari (1981) Nitrogen fixation by the photosynthetic sulfur bacterium *Chlorobium phaeobacteroides* from Lake Kinneret. Appl Environ Microbiol 41: 542–544

Bishop PE and Premakumar R (1992) Alternative nitrogen fixation systems. In Stacey G, Burris RH, and Evans HJ (eds), Biological Nitrogen Fixation, pp 736–762. Chapman and Hall, New York

Bognar A, Desrosiers L, Libman M and Newman EB (1982) Control of nitrogenase in the photosynthetic autotrophic bacterium, *Ectothiorhodospira* sp. J Bacteriol 152: 706–713

Burris RH and Roberts GP (1993) Biological nitrogen fixation. Annu Rev Nutr 13: 317–335

Caumette P, Schmidt K, Biebl H and Pfennig N (1985) Characterization of a *Thiocapsa* strain containing okenone as major carotenoid. System Appl Microbiol 6: 132–136

Drews G (1981) *Rhodospirillum salexigens*, spec. nov., an obligatory halophilic phototrophic bacterium. Arch Microbiol 130: 325–327

Eady RE (1992) The dinitrogen-fixing bacteria. In Balows A, Trüper HG, Dworkin M, Harder W and H-K Schliefer (eds), The Prokaryotes, 2nd edition, pp 534–553. Springer-Verlag, New York

Eckersley K and Dow CS (1980) *Rhodopseudomonas blastica* sp. nov.: A member of the Rhodospirillaceae. J Gen Microbiol 119: 465–473

Favinger J, Stadtwald R and Gest H (1989) *Rhodospirillum centenum*, sp. nov. a thermotolerant cyst-forming anoxygenic photosynthetic bacterium. Antonie van Leewenh 55: 291–296

Gallon JR and Chaplin AE (1987) An Introduction to Nitrogen Fixation. Cassell, London

Gest H (1993) Photosynthetic and quasi-photosynthetic bacteria. FEMS Microbiol Letts 112: 1–6

Gest, H and Kamen MD (1949) Photoproduction of molecular hydrogen by *Rhodospirillum rubrum*. Science 109: 558

Gest H, Favinger JL and Madigan MT (1985) Exploitation of N_2-fixation capacity for enrichment of anoxygenic photosynthetic bacteria in ecological studies. FEMS Microbiol Ecol 31: 317–322

Gibson J, Pfennig N and Waterbury JB (1984) *Chloroherpeton thalassium* gen. nov., et spec. nov., a nonfilamentous, flexing, and gliding green sulfur bacterium. Arch. Microbiol 138: 96–101

Gibson J, Ludwig W, Stackebrandt E and Woese CR (1985) The phylogeny of the green photosynthetic bacteria: Absence of a close relationship between *Chlorobium* and *Chloroflexus*. System Appl Microbiol 6: 152–156

Gogotov IN and Glinskii VP (1973) A comparative study of nitrogen fixation in the purple bacteria. Microbiology (English translation of Mikrobiologiya) 42: 877–880

Gogotov IN, Yakunin AF, Tsygankov AA and Denisova EN (1991) Ability of phototrophic microorganisms to synthesize the alternative nitrogenase(s). Abstr. VII Intl Symp Photosyn Prokary, Amherst, MA

Gorlenko VM, Krasil'nikova EN, Kikina OG and Tatarinova NY (1979) The new motile purple sulfur bacterium *Lamprobacter modestohalophilus* nov. gen., nov. sp., with gas vacuoles. Izv Akad Nauk SSSR Ser Biol 5: 755–767 (in Russian)

Gotto JW and Yoch DC (1985) Regulation of nitrogenase activity by covalent modification in *Chromatium vinosum*. Arch Microbiol 141: 40–43

Habte M and Alexander M (1980) Nitrogen fixation by photosynthetic bacteria in lowland rice culture. Appl Environ Microbiol 39: 342–347

Hansen TA and Imhoff JF (1985) *Rhodobacter veldkampii* sp. nov., a new species of the phototrophic purple bacteria. Intl J Syst Bacteriol 35: 115–116

Hansen TA and Veldkamp H (1973) *Rhodopseudomonas sulfidophila*, nov. spec. a new species of the purple nonsulfur bacteria. Arch Mikrobiol 92: 45–58

Heda GD and Madigan MT (1986a) Aspects of nitrogen fixation in *Chlorobium*. Arch Microbiol 143: 330–336

Heda GD and Madigan MT (1986b) Utilization of amino acids and lack of diazotrophy in the thermophilic anoxygenic phototroph *Chloroflexus aurantiacus*. J Gen Microbiol 132: 2469–2473

Heda GD and Madigan MT (1988) Nitrogen metabolism and N_2 fixation in phototrophic green bacteria. In Olson JM, Ormerod JG, Amesz J, Stackebrandt E and Trüper HG (eds), Green Photosynthetic Bacteria, pp 175–187. Plenum Press, New York

Hiraishi A, Santos TS, Sugiyana J and Komagata K (1992) *Rhodopseudomonas rutila* is a later subjective symonym of *Rhodopseudomonas palustris* Intl J Syst Bacteriol 42: 186–188

Hochman A and Burris RH (1981) Effect of oxygen on acetylene reduction by photosynthetic bacteria. J Bacteriol 147: 492–499

Howard KS, Hales BJ and Socolofsky MD (1983) Nitrogen fixation and ammonia switch-off in the photosynthetic bacterium *Rhodopseudomonas viridis*. J Bacteriol 155: 107–112

Imhoff JH (1983) *Rhodopseudomonas marina* sp. nov., a new marine phototrophic purple bacterium. Syst Appl Microbiol 4: 512–521

Imhoff JF (1989) Family II. Ectothiorhodospiraceae Imhoff 1984, 33. In Staley JT, Bryant NP Pfennig N and Holt G (eds), Bergey's Manual of Systematic Bacteriology, Vol 3, pp 1654–1658. Williams and Wilkins, Baltimore

Jouanneau Y, Siefert E and Pfennig N (1980) Microaerobic nitrogenase activity in *Thiocapsa* sp. strain 5811. FEMS Microbiol Letts 9: 89–93

Kamen MD and Gest H (1949) Evidence for a nitrogenase system in the photosynthetic bacterium *Rhodospirillum rubrum*. Science 109: 560

Kämpf C and Pfennig N (1980) Capacity of Chromatiaceae for chemotrophic growth. Specific respiration rates of *Thiocystis violacea* and *Chromatium vinosum*. Arch Microbiol 127: 125–135

Kämpf C and Pfennig N (1986) Isolation and characterization of some chemoautotrophic Chromatiaceae. J Basic Microbiol 26: 507–515

Kelley BC, Jouanneau Y and Vignais PM (1979) Nitrogenase activity in *Rhodopseudomonas sulfidophila*. Arch Microbiol 122: 145–152

Keppen OI, Lebedeva NV, Petukhov SA and Rodionov YV (1985) Nitrogenase activity in the green bacterium *Chlorobium limicola*. Microbiology (English translation of Mikrobiologiya) 54: 28–32

Keppen OI, Lebedeva NV, Troshina OY and Rodionov YV (1989) The nitrogenase activity of filamentous phototrophic green bacterium. Mikrobiologiya 58: 520–521

Keppen OI, Baulina OI and Kondratieva EN (1994) *Oscillochloris trichoides* neotype strain DG-6. Photosyn Res 41: 29–33

Kimble LK and Madigan MT (1992a) Nitrogen fixation and nitrogen metabolism in heliobacteria. Arch Microbiol 158: 155–161

Kimble LK and Madigan MT (1992b) Evidence for an alternative nitrogenase in *Heliobacterium gestii*. FEMS Microbiol Letts 100: 255–260

Kimble LK, Stevenson AK and Madigan MT (1994) Chemotrophic growth of heliobacteria in darkness. FEMS Microbiol Letts 115: 51–56

Kimble LK, Mandelco L, Woese CR and Madigan MT (1995) *Heliobacterium modesticaldum* sp. nov., a thermophilic heliobacterium of hot springs and volcanic soils. Arch Microbiol 163: 259–267

Kobayashi M and Haque MZ (1971) Contribution to nitrogen fixation and soil fertility by photosynthetic bacteria. Plant Soil Spec. Vol. pp. 443–456

Kobayashi M, Takahashi E and Kawaguchi K (1967) Distribution of nitrogen–fixing microorganisms in paddy soils of southeast Asia. Soil Science 104: 113–118

Lehman LJ and Roberts GP (1991) Identification of an alternative nitrogenase system in *Rhodospirillum rubrum*. J Bacteriol 173: 5705–5711

Lindstrom ES, Burris RH and Wilson PW (1949) Nitrogen fixation by photosynthetic bacteria. J Bacteriol 58: 313–316

Lindstrom ES, Tove SR and Wilson PW (1950) Nitrogen fixation by the green and purple sulfur bacteria. Science 112: 197–198

Lindstrom ES, Lewis SM and Pinsky MJ (1951) Nitrogen fixation and hydrogenase in various bacterial species. J Bacteriol 61: 481–487

Mack EE, Mandelco L, Woese CR and Madigan MT (1993) *Rhodospirillum sodomense*, sp. n., a Dead Sea *Rhodospirillum* species. Arch Microbiol 160: 363–367

Madigan MT (1984) A novel photosynthetic purple bacterium isolated from a Yellowstone hot spring. Science 225: 313–315

Madigan MT (1986) *Chromatium tepidum* sp. n., a thermophilic photosynthetic bacterium of the family Chromatiaceae. Intl J Syst Bacteriol 36: 222–229

Madigan MT (1988) Microbiology, physiology and ecology of phototrophic bacteria. In Zehnder AJB (ed), Biology of Anaerobic Microorganisms, pp 39–111. John Wiley and Sons, New York

Madigan MT (1992) The family Heliobacteriaceae. In: Balows A. Trüper, HG, Dworkin M, Harder W and Schleifer K-H (eds). The Prokaryotes, 2nd ed., pp 1981–1992. Springer-Verlag, New York

Madigan MT and Cox SS (1982) Nitrogen metabolism in *Rhodopseudomonas globiformis*. Arch Microbiol 133: 6–10

Madigan MT and Gest H (1978) Growth of a photosynthetic bacterium anaerobically in darkness, supported by 'oxidant-dependent' sugar fermentation. Arch Microbiol 117: 119–122

Madigan MT and Gest H (1979) Growth of the photosynthetic bacterium *Rhodopseudomonas capsulata* chemoautotrophically in darkness with H_2 as the energy source. J Bacteriol 137: 524–530

Madigan MT, Wall JD and Gest H (1979) Dark anaerobic dinitrogen fixation by a photosynthetic microorganism. Science 204: 1429–1430

Madigan MT, Cox JC and Gest H (1980) Physiology of dark fermentative growth of *Rhodopseudomonas capsulata*. J Bacteriol 142: 908–915

Madigan MT, Cox SS and Stegeman RA (1984) Nitrogen fixation and nitrogenase activities in members of the family *Rhodospirillaceae*. J Bacteriol 157: 73–78

Mangels LA, Favinger JL, Madigan MT and Gest H (1986) Isolation and characterization of the N_2-fixing marine photosynthetic bacterium *Rhodopseudomonas marina*, variety *agilis*. FEMS Microbiol Letts 36: 99–104

Masters RA, and Madigan MT (1983) Nitrogen metabolism in the phototrophic bacteria *Rhodocyclus purpureus* and *Rhodospirillum tenue*. J Bacteriol 155: 222–227

Matheron R and Baulaigue R (1983) Photoproduction d'hydrogène sur soufre et sulfure par des Chromatiaceae. Arch Microbiol 135: 211–214

Meyer J, Kelley BC and Vignais PM (1978a) Nitrogen fixation and hydrogen metabolism in photosynthetic bacteria. Biochimie 60: 245–260

Meyer J, Kelley BC and Vignais PM (1978b) Effect of light on nitrogenase function and synthesis in *Rhodopseudomonas capsulata*. J Bacteriol 136: 201–208

Neutzling O, Imhoff JF and Trüper HG (1984) *Rhodopseudomonas adriatica* sp. nov., a new species of the Rhodospirillaceae, dependent on reduced sulfur compounds. Arch Microbiol 137: 256–261

Newton JW and Wilson PW (1953) Nitrogen fixation and photoproduction of molecular hydrogen by *Thiorhodaceae*. Antonie van Leeuwenh 19: 71–77

Oren A, Kessel M and Stackebrandt E (1989) *Ectothiorhodospira marismortui* sp. nov., an obligately anaerobic, moderately halophilic purple sulfur bacterium from a hypersaline sulfur spring on the shore of the Dead Sea. Arch Microbiol 151: 524–529.

Overmann J and Pfennig N (1989) *Pelodictyon clathratiforme* sp. nov., a new brown-colored member of the Chlorobiaceae forming net-like colonies. Arch Microbiol 152: 401–406

Oyaizu H, Debrunner-Vossbrinck B, Mandelco L, Studier JA, and Woese CR (1987) The green non-sulfur bacteria: A deep branching in the eubacterial line of descent. System Appl Microbiol 9: 47–53

Pfennig N (1978) *Rhodocyclus purpureus* gen. nov., and sp. nov., a ring-shaped, vitamin B_{12}-requiring member of the family Rhodospirillaceae. Intl J Syst Bacteriol 28: 283–288

Pfennig N and Trüper HG (1989) Family I. Chromatiaceae Bavendamm 1924, 125, AL* emended description Imhoff 1984, 339. In Staley JT, Bryant MP, Pfennig N and Holt JG (eds), Bergey's Manual of Systematic Bacteriology, Vol 3, pp 1637–1653. Williams and Wilkins, Baltimore

Postgate JR (1982) The Fundamentals of Nitrogen Fixation. Cambridge University Press, Cambridge

Rodionov YV, Lebedeva NV and Kondratieva EN (1986) Ammonia inhibition of nitrogenase activity in purple and green bacteria. Arch Microbiol 143: 345–347

Roberts GP and Ludden PW (1992) Nitrogen fixation by photosynthetic bacteria. In Stacey G, Burris RH, and Evans HJ (eds) Biological Nitrogen Fixation, pp 135–165. Chapman and Hall, New York

Rubin CA and Madigan MT (1986) Amino acid and ammonia metabolism by the halophilic nonsulfur purple bacterium *Rhodospirillum salexigens*. FEMS Microbiol Letts 34: 73–77

Schick HJ (1971) Substrate and light dependent fixation of molecular nitrogen in *Rhodospirillum rubrum*. Arch Mikrobiol 75: 89–101

Schneider K, Müller A, Schramm U and Klipp W (1991) Demonstration of a molybdenum and vanadium-independent nitrogenase in a *nifHDK*-deletion mutant of *Rhodobacter capsulatus*. Eur J Biochem 195: 653–661

Schüddekopf K, Hennecke S, Liese U, Kutsche M and Klipp W (1993) Characterization of *anf* genes specific for the alternative nitrogenase and identification of *nif* genes required for both nitrogenases in *Rhodobacter capsulatus*. Mol Micro 8: 673–684

Shiba T and Harashima K (1986) Aerobic photosynthetic bacteria. Microbiol Sci 3: 376–378

Siefert E and Pfennig N (1980) Diazotrophic growth of *Rhodopseudomonas acidophila* and *Rhodopseudomonas capsulata* under microaerobic conditions in the dark. Arch Microbiol 125: 73–77

Stadtwald-Demchick R, Turner FR and Gest H (1990) *Rhodopseudomonas cryptolactis*, sp. nov., a new thermotolerant species of budding phototrophic purple bacteria. FEMS Microbiol Letts 71: 117–122

Stewart WDP, Rowell P and Rai AN (1980) Symbiotic nitrogen-fixing cyanobacteria. In Stewart WDP and Gallon JR (eds), Nitrogen Fixation, pp 329–277. Academic Press, London

Uffen RL and Wolfe RS (1970) Anaerobic growth of purple

nonsulfur bacteria under dark conditions. J Bacteriol 104: 462–472

Van Niel CB (1944) The culture, general physiology, morphology and classification of the non-sulfur purple and brown bacteria. Bacteriol Revs 8: 1–118

Ventura S, De Philippis R, Materassi R and Balloni W (1988) Two halophilic *Ectothiorhodospira* strains with unusual morphological, physiological and biochemical characters. Arch Microbiol 149: 273–279

Wahlund TM and Madigan MT (1993) Nitrogen-fixation by the thermophilic green sulfur bacterium *Chlorobium tepidum*. J Bacteriol 175: 474–478

Wahlund TM, Woese CR, Castenholz RW and Madigan MT (1991) A thermophilic green sulfur bacterium from New Zealand hot springs, *Chlorobium tepidum* sp. nov. Arch Microbiol 159: 81–90

Wall JS, Wagenkneckt AC, Newton JW and Burris RH (1952) Comparison of the metabolism of ammonia and molecular nitrogen in photosynthesizing bacteria. J Bacteriol 63: 563–573

Winter HC and Arnon DI (1970) The nitrogen fixation system of photosynthetic bacteria. I. Preparation and properties of a cell-free extract from *Chromatium*. Biochim Biophys Acta 197: 170–79

Woese CR (1987) Bacterial evolution. Microbiol Rev 51: 221–271

Woese CR (1992) Prokaryote systematics: The evolution of a science. In Balows A, Trüper HG, Dworkin M, Harder W and Schliefer K-H (eds) The Prokaryotes, 2nd ed., pp 3–18. Springer-Verlag, New York

Woese CR, Stackebrandt E, Weisburg WG, Paster PJ, Madigan MT, Fowler VJ, Hahn CM, Blanz P, Gupta R and Fox GE (1984a) The phylogeny of purple bacteria: The alpha subdivision. System Appl Microbiol 5: 315–326

Woese CR, Weisberg, WG, Paster BJ, Hahn CM, Tanner RS, Krieg NR, Koops HP, Harms H and Stackebrandt E (1984b) The phylogeny of purple bacteria: The beta subdivision. System Appl Microbiol 5: 327–336

Woese CR, Debrunner-Vossbrinck BA, Oyaizu H, Stackebrandt E and Ludwig W (1985) Gram-positive bacteria: Possible photosynthetic ancestry. Science. 229: 762–765

Wynn-Williams DD and Rhodes ME (1974) Nitrogen fixation by marine photosynthetic bacteria. J Appl Bacteriol 37: 217–224

Young JPW (1992) Phylogenetic classification of nitrogen-fixing organisms. In Stacy G, Burris RH, and Evans HJ (eds) Biological Nitrogen Fixation, pp 43–86. Chapman and Hall, New York

Zakhvataeva NV, Malofeeva IV and Kondrat′eva EN (1971) Nitrogen fixation capacity of photosynthesizing bacteria. Microbiology (English translation of Mikrobiologia) 39: 661–666

Chapter 43

The Biochemistry and Genetics of Nitrogen Fixation by Photosynthetic Bacteria

Paul W. Ludden[†] and Gary P. Roberts[‡]
Departments of [†] Biochemistry and [‡] Bacteriology,
University of Wisconsin-Madison, Madison, WI 53706

Summary

Most photosynthetic bacteria are capable of nitrogen fixation and the biochemistry and genetics of nitrogen fixation have been studied extensively in a few of these. Both the traditional, molybdenum-dependent, nitrogenase and the third alternative form of nitrogenase have been found in anoxygenic phototrophs and the properties of these are similar to those in heterotrophic nitrogen fixers. The *nif* genes for the molybdenum nitrogenase system and the *anf* genes for the alternative nitrogenase system have been isolated. Nitrogenase activity in some photosynthetic bacteria is regulated by reversible ADP-ribosylation at arg101 of the dinitrogenase reductase protein, which prevents it from transferring electrons to the dinitrogenase protein. ADP-ribosylation of dinitrogenase is catalyzed by dinitrogenase reductase ADP-ribosyl transferase (DRAT) and the removal of ADP-ribose is performed by dinitrogenase reductase activating glycohydrolase (DRAG). These are encoded by the *draT* and *draG* genes, respectively.

R. E. Blankenship, M. T. Madigan and C. E. Bauer (eds): Anoxygenic Photosynthetic Bacteria, pp. 929–947.

I. Background and Historical Introduction

A. Discovery of Nitrogen Fixation by Photosynthetic Bacteria

Nitrogen fixation by photosynthetic bacteria was discovered by Gest and Kamen during their investigation of H_2 evolution by *Rhodospirillum rubrum* (Gest and Kamen, 1949; Kamen and Gest, 1949) and Kamen provides a dramatic recounting of the discovery in his autobiography (Kamen, 1986). They found that N_2 was not an inert gas when used as the gas phase over cells of *Rs. rubrum* grown on glutamate as the nitrogen source, and subsequent studies confirmed the ability of *Rs. rubrum* and other photosynthetic bacteria to utilize N_2 as the sole nitrogen source (Gest et al., 1950). As a final confirmation, a culture of *Rs. rubrum* was brought to the laboratory of R. H. Burris in Madison for a mass spectrometric demonstration that $^{15}N_2$ was incorporated into cellular nitrogen (Lindstrom et al., 1949).

Gest and Kamen also observed that the uptake of N_2 by actively fixing *Rs. rubrum* cells was inhibited by NH_4^+(Gest, Kamen et al., 1950). While this observation was not explained at the time, it led to studies by others that eventually demonstrated the regulation of nitrogenase activity by reversible ADP-ribosylation (reviewed here and in Ludden and Roberts, 1989). Although Gest and Kamen did not realize that the H_2 evolution they observed was, in fact, mediated by nitrogenase, Gest was the first to speculate that nitrogenase might function as an H_2-evolving enzyme (Ormerod and Gest, 1962). Gest's hypothesis was confirmed by Burns and Bulen (1965), who demonstrated H_2 evolution by the isolated nitrogenase enzyme. The discovery of nitrogen fixation by *Rs. rubrum* led to the description of growth on N_2 in many photosynthetic bacteria (Lindstrom et al., 1949, 1950). While this will be discussed in more detail in this volume in Chapter 42 by Madigan, it is worth noting here that the ability to fix nitrogen is the rule rather than the exception among photosynthetic bacteria, in contrast to the distribution of the ability to fix nitrogen among other genera of prokaryotes.

Abbreviations: ADP-ribose, ; ADPR – Adenosinediphosphoribose; *anf* – designation for genes involved in nitrogen fixation by alternative nitrogenase; DRAG – Dinitrogenase reductase activating glycohydrolase; DRAT – Dinitrogenase reductase ADP-ribosyl transferase; FeMo-co – Iron Molybdenum-cofactor of nitrogenase; *nif* – designation for genes involved in nitrogen fixation by molybdenum nitrogenase

Following Gest and Kamen's work, Schick (1971) carried out extensive investigations of N_2 fixation by *Rs. rubrum* using manometric methods, as his experiments predated the acetylene reduction technique. Schick confirmed the inhibition of N_2 uptake by ammonium in *Rs. rubrum* and also showed that N_2 uptake is typically light-dependent.

B. Early Work on Isolated Nitrogenases From Photosynthetic Bacteria

Schneider et al. (1960) adapted the method of Carnahan et al. (1960) to achieve the first success at detecting nitrogenase activity in extracts of a photosynthetic bacterium. However, activity in these *Rs. rubrum* extracts was low and unpredictable. Consistently active preparations were obtained by Burns and Bulen (1966) when they removed the membrane fraction from their extracts. The usefulness of the removal of membranes is surprising in light of current knowledge of the regulation of nitrogenase activity (section IIC below) in which an enzyme involved in activation of nitrogenase is found in this fraction. However, when this fraction is exposed to air, it becomes inhibitory and it is possible that methods for excluding air were less refined in earlier studies than those available now. They also found that inclusion of Mg^{2+} in 2.5-fold excess over ATP in the assay mixture greatly stimulated activity (Burns and Bulen, 1966) and it was subsequently discovered that addition of 0.5 mM Mn^{2+} along with high Mg^{2+} in the assay amplified this stimulation (Ludden and Burris, 1976).

Munson and Burris (1969) then found that the most consistently active preparations were obtained from cells grown in an N-limited chemostat under an atmosphere of H_2. In these studies, it was also discovered that the activity in extracts was non-linear, with increased activity observed at later time points (Munson and Burris, 1969). This was interpreted as indicating that some inhibition of nitrogenase activity was being overcome. Now it is known that ADP-ribose is being removed from dinitrogenase reductase during the assay (see below) and this results in activation of the enzyme. All attempts at purification of nitrogenase from these extracts resulted in the loss of activity.

While purification of nitrogenases from other organisms was achieved, there was little or no progress with nitrogenases from phototrophs. Most of the effort went toward the purification of nitrogenases from *Rs. rubrum* and the purple sulfur bacterium,

Chromatium vinosum. Winter and Arnon (1970) obtained consistently active preparations from *Cm. vinosum* and Winter and Ober (1973) found that a particulate form of *Cm. vinosum* nitrogenase could be sedimented by high-speed centrifugation. This particulate nitrogenase was reminiscent of the particulate nitrogenase from *Azotobacter vinelandii* isolated by Bulen and coworkers (Silverstein and Bulen, 1970), but the *Cm. vinosum* enzyme lost activity upon solubilization. Unfortunately, this particulate nitrogenase has not been investigated further.

The reason for the problem of nitrogenase purification from *Rs. rubrum* was clarified when a third factor for the nitrogenase system was discovered. This 'Activating Factor' as it was called by the Madison group (Ludden and Burris, 1976), or 'Membrane Component' as it was called by the Stockholm group (Nordlund et al., 1977), was associated with the chromatophore fraction of extracts and could be released by treatment of the pelleted fraction with 0.5 M NaCl. This fraction is now called Dinitrogenase Reductase Activating Glycohydrolase (DRAG) (Pope et al., 1986) and is discussed in more detail in section IIC. DRAG activity requires a free divalent metal, with Mn^{2+} being most effective, thus explaining the results with metal additions to the nitrogenase assays.

C. Biochemistry and Genetics of Nitrogenases

1. Molybdenum Nitrogenases

a. Enzymology

Nitrogenases catalyze the MgATP and reductant-dependent reduction of N_2 to ammonium as shown in Eq. (1), reviewed in (Burris, 1991; Burris and Roberts, 1993).

$$N_2 + nMgATP + 8e^- + 10\,H^+ \rightarrow 2\,NH_4^+ + nMgATP + nPi + H_2 \quad (1)$$

In this reaction, n is $\geq 2/e^-$. H_2 is produced as an obligate product of the reaction in which N_2 is reduced and thus the requirement for 8 electrons (Simpson and Burris, 1984).

Nitrogenases consist of two proteins, the dinitrogenase protein (also known as the MoFe protein or component 1 and abbreviated by the initials of the organism from which it is derived and the number one; i.e. Rr1 is the dinitrogenase from *Rs. rubrum*) and the dinitrogenase reductase protein (also known as the Fe protein or component 2 and abbreviated as, for example, Rr2). The dinitrogenase protein is the product of the *nifK* and *D* genes (Roberts et al., 1978) and exists as an $\alpha_2\beta_2$ tetramer containing 2 molecules of the iron-molybdenum cofactor (FeMo-co) (Shah and Brill, 1977) and two Fe_8S_8 clusters known as P-clusters (Bolin et al., 1993; Dean et al., 1993). The structures of these prosthetic groups have recently become known as a result of the solution of the crystal structure of the dinitrogenase protein from *Azotobacter vinelandii* (Kim and Rees, 1992). The composition of FeMo-co is $MoFe_7S_9$Homocitrate (Shah and Brill, 1977; Hoover et al., 1987) and its structure is shown in Fig. 1. FeMo-co is liganded to the α(NIFD) subunit of the dinitrogenase protein via his441 as a sixth ligand to Mo and cys275 as a 4th ligand to the Fe atom most distant from the Mo atom of the molecule. The P clusters were originally thought to exist as unusually-liganded Fe_4S_4 clusters, but are now known to be Fe_8S_8 clusters. The P clusters bridge the α and β subunits of the enzyme and are liganded by cysα62,88,154 and cysβ70,95,153. Each molecule of FeMo-co is approximately 16 Å from one P cluster and 69 Å from the other molecule of FeMo-co in the protein and, thus, the dinitrogenase tetramer is thought of as two $\alpha\beta$ pairs functioning independently.

FeMo-co is the site of substrate binding and reduction, although neither process is yet understood. The general role of FeMo-co is based on the observation by Smith and coworkers that the dinitrogenase from *nifV* mutants shows aberrant substrate reduction properties and those aberrant properties can be transferred to another molecule of dinitrogenase by the transfer of the FeMo-co center alone (McLean and Dixon, 1981; Hawkes et al., 1984). It was later shown that homocitrate is produced for FeMo-co synthesis in cells that have a functional *nifV* gene and that the unusual properties arise from substitution of other organic acids for homocitrate in FeMo-co in *nifV* mutants (Hoover et al., 1989; Imperial et al., 1989). When various organic acids were substituted for homocitrate in FeMo-co in vitro, dinitrogenases with altered properties were also obtained.

Dinitrogenase reductase is an α_2 dimer of the *nifH* product (St. John et al., 1975). This protein has several roles in the overall nitrogen fixation process, including the specific transfer of electrons to the dinitrogenase protein (Ljones and Burris, 1972; Orme-Johnson et al., 1972). Dinitrogenase reductase

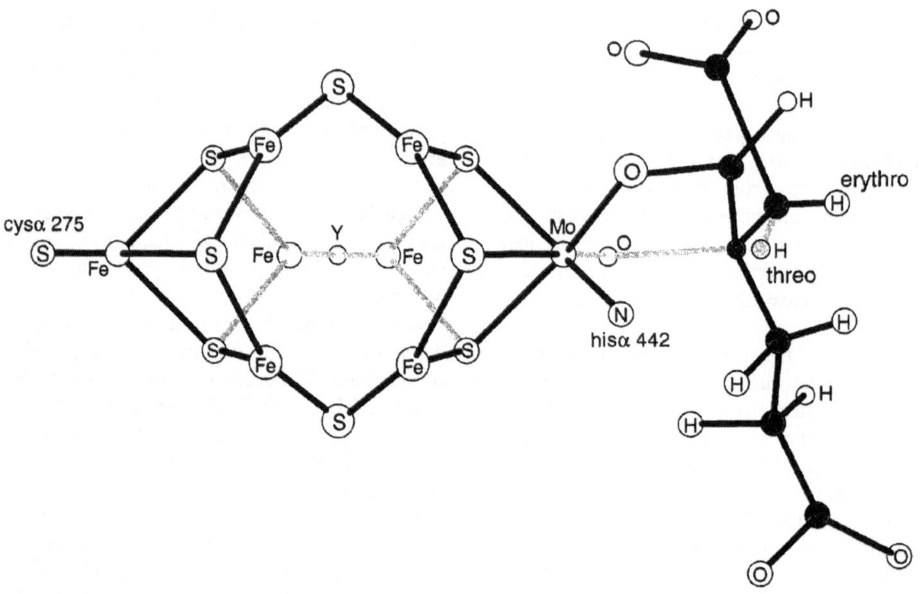

Fig. 1. The structure of the iron molybdenum-cofactor (after Kim and Rees, 1992).

contains a single Fe_4S_4 cluster and bridges the two subunits of the protein through its ligation by cys98 and cys132 of each subunit (Hausinger and Howard, 1983). The biosynthesis of the Fe_4S_4 cluster of dinitrogenase reductase has recently been shown to involve the *nifS* gene product (Zheng et al., 1993; Zheng and Dean, 1994). Dinitrogenase reductase has binding sites for two molecules of MgATP and one molecule of MgADP (Tso and Burris, 1973) and these sites are distal from the Fe_4S_4 cluster of the enzyme (Georgiadis et al., 1992). The MgADP binding site is competitive with one MgATP binding site. The crystal structure of the dinitrogenase reductase protein from *Azotobacter vinelandii* has recently been determined and the structure reveals relatively little interaction of the two subunits other than at the region surrounding the site of Fe_4S_4 cluster ligation (Georgiadis et al., 1992).

Both nitrogenase proteins are extremely oxygen-labile and organisms have devised various physiological strategies to overcome this problem. In the case of the anoxygenic photosynthetic bacteria, nitrogenase synthesis is strongly repressed in the presence of oxygen. Organisms with high respiratory rates, such as *Rb. capsulatus,* are able to tolerate higher levels of oxygen coincident with active nitrogen fixation (Meyer et al., 1987), but continued exposure of *Rb. capsulatus* cells to high levels of oxygen results in oxidative destruction of the nitrogenase proteins.

Nitrogenases will reduce a variety of triply- and doubly-bonded molecules in addition to the physiologically relevant substrate, N_2 (Rivera-Ortiz and Burris, 1975). In the absence of any other substrate, the enzyme will reduce protons to H_2 gas at a rate of electron flow through the enzyme equivalent to the rate when the enzyme is saturated with N_2. This ATP-dependent hydrogenase activity creates a physiological problem for nitrogen-fixing organisms because ATP and reductant are consumed without nitrogen fixation. This activity also creates experimental problems, because it is never possible to isolate and study the fully reduced form of the dinitrogenase protein—as soon as it is reduced, it will transfer electrons to protons. The reduction of acetylene to ethylene is a 2 electron reaction and is the basis of an extremely useful assay for nitrogenase because the product, ethylene, can be easily quantitated by gas chromatography (Stewart et al., 1967). Unlike N_2, acetylene can completely suppress H_2 evolution by nitrogenase and this should be taken into account when using the results of the acetylene reduction assay to estimate the amount of N_2 fixed by nitrogen-fixing cultures. CN^-, N_3^-, N_2O, CH_3CN, cyclopropene and diazerene are also substrates for nitrogenase (Rivera-Ortiz and Burris, 1975; McKenna et al., 1979). H_2 is a specific inhibitor of the N_2 reduction assay (Burris, 1985) and CO inhibits the reduction of all substrates except protons (Rivera-Ortiz and Burris, 1975).

The dinitrogenase protein exhibits a nondescript UV-visible spectrum from < 300 nm to beyond 800 nm; the difference spectrum between the oxidized and reduced forms of the protein is not significant (Shah and Brill, 1973). The dinitrogenase reductase protein also has a broad spectrum, but it has a distinct shoulder that shows a significant difference between the oxidized and reduced forms at 430 nm, and changes in the spectrum have been used to follow the oxidation state of the protein (Ljones and Burris, 1972). The dinitrogenase protein has a unique EPR spectrum with g values at 4.3, 3.65 and 2.01 (Eady et al., 1972; Orme-Johnson et al., 1972; Zumft et al., 1974). The signal at 3.65 is unique in nature and can even be detected in vivo (Shah et al., 1973). This signal is referred to as the M center as it arises from the magnetic center of the protein, the FeMo-co. Upon reduction of dinitrogenase by dinitrogenase reductase, the signals at g 4.3 and 3.65 disappear. The P clusters are not observed in EPR under normal conditions but can be seen in oxidized forms of the enzyme. Dinitrogenase reductase has an EPR spectrum in the g = 2 region (Eady et al., 1972; Orme-Johnson et al., 1972; Zumft et al., 1974). The spectrum is axial and becomes more rhombic upon binding of MgATP to the protein. EPR spectroscopy was employed to establish the sequence of electron transfer between nitrogenase proteins. Both dinitrogenase and dinitrogenase reductase have been investigated with Mössbauer and ENDOR spectroscopy (Münck et al., 1975; Hoffman et al., 1982; Ventures et al., 1986).

The features described above for molybdenum nitrogenases have been determined primarily using non-photosynthetic organisms, especially *Azotobacter vinelandii, Clostridium pasteurianum* and *Klebsiella pneumoniae*, and apparently these features are retained in the properties of photosynthetic species. Nitrogenases from *Rs. rubrum* (Ludden and Burris, 1978), *Rb. capsulatus* (Hallenbeck et al., 1982b) and *Cm. vinosum* (Albrecht and Evans, 1973) have been purified and characterized, and crude extracts with nitrogenase activity or partially purified nitrogenases have been prepared from a number of other phototrophs. Purification of nitrogenases from phototrophs is performed under anaerobic conditions using methods developed for heterotrophic diazotrophs but the large amount and small size of the membrane material requires additional, high speed centrifugation of the crude extract. Photosynthetic bacteria grown under nitrogen-fixing conditions also store large amounts of poly-β hydroxybutyric acid that is hydrolyzed upon breakage of the cells. The ionization of the hydroxybutyric acid monomers can result in a drastic lowering of the pH of the extract and high buffer concentrations (100–200 mM) are required to maintain the pH of extracts.

The physiological electron donor to nitrogenase has been unambiguously established only in *K. pneumoniae* where the *nifJ* gene product, pyruvate: flavodoxin oxidoreductase, and the *nifF* product, flavodoxin, perform the task of supplying low potential electrons to nitrogenase (Nieva-Gomez et al., 1980; Shah et al., 1983). A pyruvate:ferredoxin oxidoreductase has been isolated from *Rs. rubrum* and a gene with significant sequence similarity to the *K. pneumoniae nifJ* gene has been isolated (Brostedt and Nordlund, 1991). However, mutations in the *nifJ* gene of *Rs. rubrum* do not render the organism Nif^-. Similarly, several ferredoxins that are capable of donating electrons to nitrogenase in vitro have been isolated from *Rs. rubrum* and from *Rb. capsulatus*, but none of these has been demonstrated to be the in vivo electron donor (Hallenbeck et al., 1982a; Meyer et al., 1987). Recently a *nif*-specific flavodoxin has been identified in *Rb. capsulatus* and shown to have the ability to transfer electrons to dinitrogenase reductase (Yakunin et al., 1993). A flavodoxin that accumulates under conditions of iron starvation has been isolated from *Rs. rubrum*, but it is not known if this protein plays any role in nitrogen fixation (Cusanovich and Edmondson, 1971).

It is possible that the physiological electron donor to nitrogenase in phototrophs may vary depending on the conditions of growth. For example, *Rs. rubrum* is able to grow and fix nitrogen under fermentative conditions with pyruvate as the carbon and energy source (Uffen and Wolfe, 1970; Madigan et al., 1979; Schultz et al., 1985). Under these conditions, a system like that in *K. pneumoniae* might be expected, whereas a light-dependent system might operate under photosynthetic growth conditions. As noted below, Schmel et al. (1993) have recently described a set of genes in *Rb. capsulatus* whose encoded products apparently function as electron donors to nitrogenase. Further biochemical analysis of the gene products should clarify this assignment.

The interaction of the dinitrogenase reductase protein with its electron donor has been an unstudied area, but one of potential importance to the phototrophs. The site for regulation of the dinitrogenase reductase protein by reversible ADP-ribosylation (see IIA below) is near the Fe_4S_4 cluster of the enzyme. It is reasonable to suggest that the electron-donating proteins, whatever they be, bind at or near the same site. It will be interesting to determine if ADP-ribosylation disrupts the ability of the dinitrogenase protein to be reduced by ferredoxins and flavodoxins; it is known that the modified protein can be reduced by low molecular weight reductants such as dithionite (Ludden and Burris, 1979).

b. Gene Organization for the Mo-Nitrogenase

The genes for the Mo-nitrogenase system of a phototroph have only been well-described in the case of *Rb. capsulatus,* largely through the work of the laboratories of Klipp (Moreno-Vivian et al., 1988), Haselkorn (Ahombo et al., 1986) and Vignais (Ahombo et al., 1986) [reviewed in Dean and Jacobson (1993) and Haselkorn (1986)]. The genes that are necessary for the expression and function of the Mo-nitrogenase fall into three non-contiguous gene clusters, termed A, B, and C.

Region A contains 26 open reading frames (ORFs) that are apparently organized into six transcriptional units (Klipp et al., 1988; Fonstein et al., 1992). Eleven of these have homologs in the *nif* cluster of *K. pneumoniae* and three others are similar to ORFs found in the *nif* region of *Azotobacter vinelandii.*

Immediately adjacent to these *nif* gene homologs in Region A is a set of genes that have been termed *rnfA-F* (for Rhodobacter nitrogen fixation) (Schmehl et al., 1993). While these genes are not homologous to any sequences stored in available data banks, their sequences suggest that many encode either membrane proteins or contain cysteine motifs typical of iron-sulfur proteins. Both mutational analyses and the effects of metronidazole suggest that the encoded proteins are involved in electron transport to nitrogenase. The authors have speculated that the encoded proteins form a complex and that it might either serve as an oxido-reductase for an unknown substrate or utilize membrane potential as a source of reductant for nitrogenase (Schmehl et al., 1993).

Region B contains the *nifHDK* genes, which encode dinitrogenase and dinitrogenase reductase, and also 'second' copies of *nifA*, *nifB*, and *nifU* that duplicate the same functions found in region A. Finally, this region contains *nifR4*, which encodes a *nif*-specific sigma factor (Jones and Haselkorn, 1989).

Region C contains *nif*-specific homologs of *ntrBC*, termed *nifR2* and *nifR1*. This regulatory system is discussed in detail in Chapter 56 in this volume by Kranz and Cullen on the regulation of nitrogen fixation.

Immediately adjacent to Region B is a cluster of six ORFs, designated *modABCD* and *mopAB*, whose products appear to be involved in molybdenum transport and perhaps other aspects of molybdenum processing (Wang et al., 1993). While the precise roles of the different gene products are unknown and the phenotypes of mutants are not identical, in general the presence of these genes is important for both molybdenum uptake and for the molybdenum repression of the alternative nitrogenase in *Rb. capsulatus*.

2. Alternative Nitrogenase Systems

a. Enzymology

Following the description of the alternative nitrogenases in *Azotobacter vinelandii* by Bishop (Bishop et al., 1980; Premakumar et al., 1989; Bishop and Premakumar, 1992) non-molybdenum nitrogenases have been described in a number of organsims, including the photosynthetic bacteria. To date, only the iron-only (third alternative) nitrogenases have been described in the photosynthetic bacteria and these have been purified from *Rb. capsulatus* (Schneider et al., 1991) and *Rs. rubrum* (Davis and Ludden, unpublished results). Like the alternative nitrogenases from *Azotobacter vinelandii* and *A. chroococcum*, the dinitrogenase proteins from *Rs. rubrum* and *Rb. capsulatus* contain a third (δ, the product of *anfG*) subunit with a molecular weight of approximately 20 kD (Schneider et al., 1991). The alternative nitrogenase systems from *Rb. capsulatus* and *Rs. rubrum* reduce protons well, N_2 at a rate sufficient to sustain growth, and acetylene very poorly. The maximal rate of N_2 reduction by the alternative system is 10% of the rate observed with the molybdenum nitrogenase. The enzymes are extremely labile and purification is difficult. The alternative dinitrogenase proteins from photosynthetic bacteria do not accept electrons from the dinitrogenase reductases from the molybdenum systems. The alternative dinitrogenase reductases are effective in transferring electrons to the molybdenum dinitrogenase. Like the alternative nitrogenases from other systems (Dilworth et al., 1987), the alternative nitrogenases of phototrophs reduce acetylene to ethylene and ethane. Approximately 2–5% of the product of acetylene reduction is ethane.

In both *Rb. capsulatus* and *Rs. rubrum*, the alternative dinitrogenase reductases are substrates for regulation by the reversible ADP-ribosylation systems described below (Lehman and Roberts, 1991; Masepohl et al., 1993). The proteins can be modified in vivo and in vitro. Thus, the alternative nitrogenase does not serve as an unregulated back-up to the molybdenum nitrogenase. This result predicts that the dinitrogenase reductase of each system contains an arginine residue at position 101.

The UV-visible and EPR spectra of the alternative dinitrogenase reductases are similar to those of the dinitrogenase reductases of the molybdenum system. The $g = 3.65$ EPR signal attributed to FeMo-co in the molybdenum system is lacking in the alternative system. The *Rs. rubrum* alternative dinitrogenase exhibits a signal in the $g = 2$ region with no apparent higher field signals. The spectra of the *Rb. capsulatus* protein are reported to be complex and include a signal like that of the *Rs. rubrum* alternative dinitrogenase as well as additional signals in the $g = 4$ region (Schneider et al., 1991).

A cofactor analogous to FeMo-co can be isolated from the *Rs. rubrum* alternative dinitrogenase protein (Ludden, Davis and Shah, unpublished results). This cofactor contains no molybdenum, but will replace FeMo-co in the activation of apo-dinitrogenase (lacking FeMo-co) of the molybdenum nitrogenase. The *Rs. rubrum* alternative dinitrogenase reductase also contains homocitrate, presumably in its cofactor. The reported iron and acid-labile sulfur contents of the alternative dinitrogenase proteins are significantly less than those of the molybdenum dinitrogenase proteins and thus it is possible that they contain fewer or different clusters than the molybdenum dinitrogenase. Alternatively, it is possible that the fully active proteins contain a full complement of metal clusters, but that these clusters are more labile and thus lost during purification.

b. Gene Organization for the Alternate Nitrogenase

Among phototrophs, alternate nitrogenase systems have been genetically well-characterized only in *Rb. capsulatus*. In this organism, Klipp and coworkers (Schüddekopf et al., 1993) have used mutational analysis to identify the genes for the alternate dinitrogenase and dinitrogenase reductase as well as a number of other loci important for their function.

The region encoding the components of the alternate nitrogenase (*anfHDGK*) has been cloned and sequenced and display the greatest similarity to the *anf* genes of *Azotobacter vinelandii*. Immediately upstream of this region appears to be a homolog of *anfA* of *Azotobacter vinelandii*, suggesting that it encodes a regulator for the *anf* region. In addition to this locus, random transposon mutagenesis identified at least ten additional restriction fragments wherein insertions appear to specifically result in an Anf^- phenotype, although the nature of the encoded functions is currently unknown.

Finally, the Klipp lab has performed a rather thorough analysis of the 32 genes and open reading frames that fall in Regions A, B and C of the *nif* gene clusters noted in section IC1b, with the goal of identifying those necessary for functionality of the alternate system (Schüddekopf et al., 1993). The results indicate that the products of *nifB* and *nifV* are critical for both systems, as are those of the regulatory genes, *nifR1*, *nifR2*, and *nifR4*. Surprisingly, at least four other ORFs, scattered among the *nif*-specific genes, are also necessary for both systems, and predicted amino acid sequences suggest that the products might form a membrane complex involved in electron transport to both nitrogenase systems (Schmehl et al., 1993).

The genes for the alternate nitrogenase of *Rs. rubrum* have also been cloned, although sequence and mutational analysis have not been performed (Lehman and Roberts, 1991).

II. Regulation of Nitrogenase by Reversible ADP-ribosylation

A. Model for Regulation of Nitrogenase by Reversible ADP-ribosylation

The current understanding of the reversible ADP-ribosylation of dinitrogenase reductase in *Rs. rubrum*

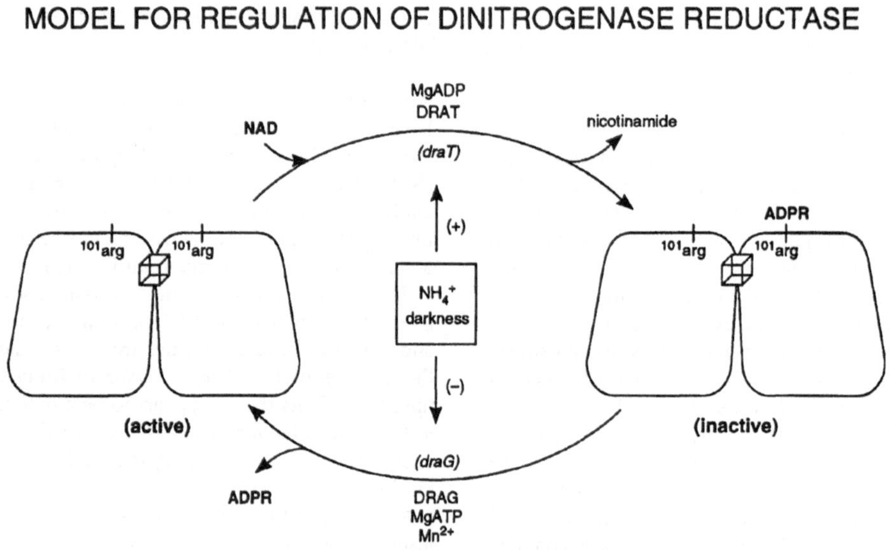

Fig. 2. Model for the reversible ADP-ribosylation of dinitrogenase reductase in *Rhodospirillum rubrum.*

and related organisms is shown in Fig. 2. In this model, active dinitrogenase reductase is inactivated by ADP-ribosylation at arg101 through the action of dinitrogenase reductase ADP-ribosyl transferase (DRAT) in an NAD- and MgADP-dependent reaction in response to darkness or a source of fixed nitrogen. Only one of the two potential sites for ADP-ribosylation is modified at one time. When the culture is re-illuminated or the source of fixed nitrogen is depleted, ADP-ribose is removed by hydrolysis from the protein by dinitrogenase reductase activating glycohydrolase (DRAG) in a MgATP- and M^{2+}-dependent reaction. Dinitrogenase reductase is activated by the removal of ADP-ribose.

B. Isolation and Identification of ADP-ribose and Effects on Dinitrogenase Reductase

Once DRAG had been discovered (Section IB above), it was possible to purify the nitrogenase from *Rs. rubrum* and to follow the purification by monitoring activity of the dinitrogenase reductase after activation. The protein was purified by methods similar to those employed for other, active dinitrogenase reductases and it had several properties and the characteristic brown color of other dinitrogenase reductases. Thus it appeared that the inactive form of dinitrogenase reductase from *Rs. rubrum* was normal but required activation, and the hypothesis that it existed in a modified form was developed. Phosphatase-resistant phosphate was detected on the inactive protein and an analysis of the UV-visible spectrum of the protein revealed a shoulder at 265 nm that could not be explained by the known amino acid content of the enzyme (dinitrogenase reductases contain no tryptophan). Tests for adenine nucleotide and for pentoses were positive and it was concluded that the enzyme was modified by a nucleotide. Originally it was thought that the adenine moiety alone was removed upon activation (Ludden and Burris, 1979), but further studies showed that the entire nucleotide was removed upon activation (Pope et al., 1986).

The inactive form of the protein was found to migrate as two bands of equal staining intensity on SDS gels and the conversion of the slower migrating band to the single band observed for active enzyme was correlated with activation (Preston and Ludden, 1982). It was found that the band conversion could also be affected by heating the protein at pH above 8.0 (Dowling et al., 1982). This observation was applied to the isolation of the nucleotide which could be thermally released from the protein and purified by affinity chromatography on boronate affinity resin by virtue of the binding of the cis hydroxyls of the compound to the resin (Pope et al., 1985b). Alternatively, the modified dinitrogenase reductase was treated with proteases and the peptidyl nucleotide was purified on the boronate resin. In each case, the nucleotide was further purified by HPLC.

The identity of the compound as ADP-ribose was established by mass spectrometric and NMR analyses (Pope et al., 1985a). The peptide with bound nucleotide was sequenced and found to have the sequence gly-X-gly-val-ile-thr and the amino acid analysis of the material indicated the presence of a single arginine, presumably the X component of the sequence. This sequence is highly conserved in dinitrogenase reductases and is found uniquely at positions 100–105 of the sequence (Pretorius et al., 1987). It was of immediate interest that arg101 was only 3 amino acids removed from cys98 which had been proposed as one site of ligation of the Fe_4S_4 cluster to the protein by Hausinger and Howard (1983) (cys132 being the other site of ligation). It was originally confusing why only one of the two subunits became modified, but the recent crystal structure by Georgiadis et al. (1992) shows that the two arg101 residues are close to each other near the site of interaction of the two subunits. This suggests that the reason only one subunit becomes modified is that once one site is modified, the other site is not accessible to the modifying enzyme.

The attachment of the ADP-ribose group was established as an N-glycosidic linkage to the guanidinium N of arginine by mass spectrometry (Pope et al., 1985a) and it was determined that the ADP-ribose-arginine bond in the α configuration may be cleaved by DRAG (Pope et al., 1986). However, there is mutarotation around the N-glycosidic bond and it becomes a 60:40 mixture of α:β with a $t_{1/2}$ of approximately 6 hours (Pope et al., 1987). The β anomer of the glycosidic bond is inaccessible to DRAG.

ADP-ribosylation of dinitrogenase reductase inhibits its ability to transfer electrons to dinitrogenase. The modified protein is able to bind MgATP and undergoes the conformational shift evidenced by the MgATP-dependent loss of the Fe_4S_4 center to chelators (Ludden and Burris, 1979; Ludden et al., 1982). The EPR signal of the modified protein is identical to that of the unmodified enzyme and thus, ADP-ribosylation has no observable direct effect on the Fe_4S_4 center (Ludden and Burris, 1979). The modified protein does not inhibit the ability of unmodified protein to transfer electrons to dinitrogenase, and thus it is suggested that modification renders the protein unable to bind to dinitrogenase. Evidence in support of this was obtained when it was shown that modification of the dinitrogenase reductase from *Clostridium pasteurianum* prevented that protein from forming the tight, inhibitory complex with dinitrogenase from *Azotobacter vinelandii* that had been reported by Emerich and Burris (1976; Murrell et al., 1988). Removal of the ADP-ribose with DRAG restored the inhibition. ADP-ribosylation does not appear to block its role in FeMo-co synthesis or FeMo-co insertion, as both of these activities can be observed in vitro with the ADP-ribosylated form (Shah et al., 1988); it should be noted that it is difficult to determine if absolutely all of the dinitrogenase reductase has been ADP-ribosylated and thus it is possible that the FeMo-co synthesis and insertion activities are supported by a small amount of unmodified protein in the mixture. ADP-ribosylation of dinitrogenase reductase does not make the protein less sensitive to oxygen.

C. Isolation and Characterization of DRAG and DRAT

1. Isolation of DRAG

DRAG was the first component of the regulatory system to be identified and it was purified based on its ability to activate dinitrogenase reductase even though the mechanism of this activation was not known (Saari et al., 1984). The predominant amount of DRAG was found in the pellet fraction after high speed centrifugation of the crude extract. It has been assumed that DRAG is associated with the chromatophore membranes of the extract and this is likely the case, but it remains a possibility that DRAG is bound to some other macromolecular complex in the extract, as no further characterization of the binding has been performed. DRAG can be eluted from the pellet fraction with 0.5 M NaCl in buffer and remains stable in its soluble form (Ludden and Burris, 1976). It should not be assumed that DRAG is bound to the photosynthetic membrane in the living cell, as the ionic strength of the cell cytoplasm is likely to be sufficiently high to elute the enzyme.

The ability to elute the enzyme from the membrane with salt was employed in the purification of DRAG. Repeated washing of the enzyme with buffer removed many contaminating proteins, allowing DRAG to be eluted rather specifically and in good yield with a 30-

fold purification. Following desalting, the enzyme was further purified by chromatography on DEAE-cellulose, hydroxylapatite, reactive red agarose and finally by gel filtration on G-75 Sephadex. The enzyme was purified 12,000-fold to homogeneity and runs as a single band on SDS gels with a molecular weight of 32,000 (Saari et al., 1984). The protein migrates anomolously on gel filtration columns and early estimates gave the molecular weight of the enzyme as approximately 20 kD (Gotto and Yoch, 1982a). The correct molecular weight of the enzyme was established when the sequence of the protein was deduced from the sequence of its gene (Fitzmaurice et al., 1989); this value confirmed that obtained by SDS PAGE analysis of the purified enzyme (Saari et al., 1984).

DRAG was originally found to be very oxygen-labile and was always purified under strict anaerobic conditions, including the presence of the oxygen scavenger sodium dithionite in all buffers (Ludden and Burris, 1976; Nordlund et al., 1977; Zumft and Nordlund, 1981). More recently, it has been discovered that the enzyme is stable to oxygen provided that it is purified in the absence of sodium dithionite (Neilsen et al., 1994). Upon exposure of the dithionite-containing solution to oxygen, hydrogen peroxide is formed, to which DRAG is particularly sensitive, especially in the presence of Mn^{2+}, its metal cofactor. ADP-ribose protects the enzyme from hydrogen peroxide attack, suggesting that the damage by hydrogen peroxide occurs near the active site of the enzyme. Dithiothreitol is required for stability of the enzyme, but does not protect against the loss of DRAG activity in the presence of hydrogen peroxide and Mn^{2+}. Under anaerobic conditions, Mn^{2+} was found to increase the stability of the enzyme and thus is included in the buffers used during purification. The preferred protocol for DRAG purification calls for use of cold, degassed buffer containing Mn^{2+} but lacking dithionite.

Although ADP-ribosylated dinitrogenase reductase and DRAG are both present in crude extracts, the dinitrogenase reductase remains ADP-ribosylated because the activation reaction requires MgATP and divalent metal. The binding of Mn^{2+} to DRAG is not very tight and, upon extraction, the divalent metal is diluted away from the enzyme. Similarly, the nucleotide concentration in crude extracts is much lower than needed to lead to significant activation. Munson (1969) first noted the non-linear time course of nitrogenase assays with crude extracts of *Rs. rubrum,* and this can now be explained as a result of the inclusion of MgATP in the assay mix. The MgATP is required for nitrogenase activity and, in the case of *Rs. rubrum* extracts, also allows activation of the enzyme.

The nucleotide involved in activation of dinitrogenase reductase by DRAG is thought to bind to the ADP-ribosylated form of dinitrogenase reductase. The interaction of MgATP with ADP-ribosylated dinitrogenase reductase was demonstrated using Walker and Mortenson's assay to follow the MgATP-dependent release of the iron-sulfur center to iron chelators (Ludden and Burris, 1979; Gotto and Yoch, 1982b). MgATP is not hydrolyzed in stoichiometric amounts during the activation of dinitrogenase reductase. MgATP is not required for the removal of ADP-ribose from the oxygen-denatured form of dinitrogenase reductase (Ludden and Burris, 1979) or from low molecular weight substrates such as the ADP-ribosylated hexapeptide obtained by protease treatment of ADP-ribosylated dinitrogenase reductase (Pope et al., 1986; Saari et al., 1986). Thus it is thought that the nucleotide exerts its effect on the process by binding to the substrate (dinitrogenase reductase) rather than to DRAG. The available evidence indicates that MgATP does not bind to DRAG, although these studies were hindered by the lack of appreciable quantities of the purified enzyme. MgADP is known to inhibit the activity of dinitrogenase reductase and it also inhibits the activation of dinitrogenase reductase by DRAG.

The free divalent metal is thought to bind directly to DRAG and play a catalytic or regulatory role in the DRAG reaction. Burns and Bulen (1966) first noted the stimulatory effect of high Mg^{2+} on the activity of crude extracts of *Rs. rubrum* and then it was discovered that low levels of Mn^{2+} provided even greater stimulation (Ludden and Burris, 1976). Guth and Burris (1983) performed a careful analysis of the stimulation of DRAG by Mg^{2+} and Mn^{2+} and came to the conclusion that the role of Mg^{2+} was to release any Mn^{2+} present in the reaction mixture for DRAG activity. Nordlund and Noren (1984) tested various metals for activity with DRAG and found that Mn^{2+} was the most effective, followed by Fe^{2+} and then Mg^{2+}. It has not been established which of these metal ions functions in vivo, and experiments with Mn-depleted medium have given ambiguous results. One mM Mn^{2+} in excess over the ATP concentration in the assay will saturate the divalent metal requirement for DRAG (Saari et al., 1986).

The substrate specificity of DRAG has been investigated in detail. The enzyme will carry out the hydrolysis of N-glycosidic bonds and is specific for the α anomer (Pope et al., 1986). DRAG does not perform the lyase reaction reported for the enzyme that removes the terminal ADP-ribose moiety of polyADP-ribosylated histones (Oka et al., 1984). As noted above, DRAG will cleave ADP-ribose from tri- and hexapeptides purified from the protease hydrolysate of ADP-ribosylated dinitrogenase reductase with Kms of 11–12 μM and a Vmax equivalent to that for ADP-ribosylated dinitrogenase reductase (Pope et al., 1986; Saari et al., 1986). ADP-ribosylarginine is a very poor substrate for DRAG with a Km of 2 mM, whereas Nα-dansyl-NωADP-ribosylarginine is a very good substrate with a Km of 12 μM. Blocking the amino group of the arginine molecule is essential for good activity of DRAG while blocking the carboxyl group has little effect. The enzyme is non-specific with respect to the nucleotide and hydrolyzes dansylarginine-guanosine diphosphoribose as well as the ADP-ribosylated compound. However, DRAG will not cleave the N-glycosidic bonds of compounds that lack the adenosine group or analogs such as the dansylated forms of phosphoribosylarginine or ribosylarginine. The fluorescent etheno analog of ADP-ribose attached to the dansylarginine methylester (ε-ADPR-DAME) is an effective substrate that also provides a sensitive assay for the enzyme (Pope et al., 1987). The fluorescence of the ε-ADPR-DAME molecule is quenched while both the ethenoadenine and the dansyl groups are attached to the same molecule, but become highly fluorescent upon cleavage. DRAG will also cleave ADP-ribose from cholera toxin-modified histones, suggesting that it will recognize ADP-ribose attached to proteins (presumably at arginine) at a variety of sequences, since the gly-arg-gly-val-ile-thr motif is not present in histones.

The Km for the natural substrate (74 μM) of DRAG is higher than that determined for a number of the low molecular weight substrates (11–15 μM). The true Km for ADP-ribosylated dinitrogenase reductase may be lower by one-half because the N-glycosidic bond between arginine and ADP-ribose is subject to mutarotation and, during purification, there is sufficient time for the bond to mutarotate to an approximately 60:40 mix of α and β anomers; DRAG is specific for the α linkage. A Km of 40 μM for ADP-ribosylated dinitrogenase reductase would be in the physiological range, as dinitrogenase reductase is estimated to be present in the cell at a concentration approaching 100 μM.

2. Isolation of DRAT

Once the modifying group for inactive dinitrogenase was identified as ADP-ribose, an NAD-dependent reaction analogous to the cholera toxin-catalyzed modification of G proteins was sought. Lowery et al. (1986) identified an activity that would transfer label from α^{32}P-NAD to dinitrogenase reductase. This label was susceptible to removal by DRAG, and peptide analysis revealed that arg101 was the sole and unique site of modification. Antibodies against DRAG were unreactive with the ADP-ribosyl transferase activity and it was concluded that the two activities resided on distinct gene products (Lowery and Ludden, 1988). This conclusion was confirmed biochemically when the amino terminal amino acid sequences of the DRAT and DRAG proteins were shown to be different, and genetically when the genes for the two enzymes were isolated (Fitzmaurice et al., 1989). The new enzyme was named Dinitrogenase Reductase ADP-ribosyl Transferase (DRAT) and its gene designated *draT*.

DRAT activity was found in the soluble fraction of extracts and it was found to be stable to oxygen although otherwise less stable than DRAG activity. The enzyme was purified by sequential ion exchange (DEAE cellulose), dye matrix affinity (reactive red agarose), gel filtration chromatography steps followed by ion exchange HPLC (Lowery and Ludden, 1988). Like DRAG, the enzyme is very non-abundant and a 20,000-fold purification of the enzyme was required.

DRAT uses β-NAD as the donor for ADP-ribose and releases stoichiometric amounts of nicotinamide (Lowery and Ludden, 1988). The enzyme will not transfer the ADP-ribose moiety of NADH or NADPH but will utilize etheno-NAD and nicotin-amide guanine dinucleotide as donors. The only known acceptors of ADP-ribose from DRAT are the dinitrogenase reductases. The sequences of dinitro-genase reductases are remarkably conserved in nature, and the region surrounding the site of ADP-ribosylation is nearly constant. All dinitrogenase reductases tested to date have arginine at the position analogous to arg101 in the *Rs. rubrum nifH* product and are substrates for DRAT, with the exception of the dinitrogenase reductases from *Azospirillum amozonense* (for which the sequence is not known) (Lowery and Ludden, 1988) and from mutants with changes at arg101 or immediately adjacent residues (Chang et al., 1988; Lowery et al., 1989; Lehman and Roberts, 1991; Pierrard, 1993). The alternative dinitrogenase reductase (the *anfH* gene product) from *Rs. rubrum* is reversibly modified in vivo and thus is presumed to be modified by DRAT (Lehman and Roberts, 1991). The specificity of DRAT is in marked contrast to the ability of the arginine-specific toxins to modify a wide range of proteins. DRAT exhibits no detectable NADase activity, also in contrast to the toxin ADP-ribosyl transferases and the turkey erythrocyte ADP-ribosyl transferase isolated by Moss (Watkins et al., 1987). Small molecules, including the peptide gly-arg-gly-val-ile-thr do not accept NAD from DRAT (Lowery and Ludden, 1988). The products of the reactions of DRAG, with its small molecule substrates, do not serve as acceptors. DRAT is unable to recognize oxygen-denatured dinitrogenase reductase, suggesting that a native conformation of the substrate protein is essential.

The modification of *Rs. rubrum* dinitrogenase reductase is highly dependent on the presence of ADP in the reaction mixture (Lowery and Ludden, 1989). DRAT-dependent incorporation of label from ^{32}P-NAD into dinitrogenase reductase can be observed, but only a small amount of the total dinitrogenase reductase protein becomes ADP-ribosylated in the absence of ADP. In the presence of ADP, it is possible to quantitatively convert dinitrogenase reductase into its ADP-ribosylated form. The measured Km for NAD in the presence of ADP and saturating amounts of dinitrogenase reductase is approximately 2 mM, while the measured Km for NAD under the same conditions but in the absence of ADP is at least 20-fold higher. ADP is thought to exert its effect by binding to dinitrogenase reductase rather than DRAT because DRAT will effectively modify the dinitrogenase reductase protein from *Azotobacter vinelandii* in the absence of ADP. In this case, the measured Km for NAD in the absence of ADP is approximately 1 mM and 30–50 μM in the presence of ADP.

The nucleotide requirements for DRAT and DRAG suggest a role for these nucleotides in the regulation of the DRAT/DRAG system. However, no consistent change in ATP/ADP ratio has been shown to correlate with changes in the modification status of dinitrogenase reductase in vivo (Paul and Ludden, 1984; Nordlund and Hoglund, 1986). Based on in vitro studies, the ATP/ADP ratio that is thought to occur in vivo would seem to predict: 1) a continuous turnover of ADP-ribose on the dinitrogenase reductase protein in vivo, and 2) significant inhibition of nitrogenase activity. Neither of these predictions can be verified; in fact, the evidence shows the opposite. Perhaps there is a level of compartmentation or organization of the nitrogenase proteins and their regulators that has not yet been detected.

D. Genetic Analysis of the dra *System of Regulation in* Rs. rubrum *and* Rb. capsulatus

By far the most complete analysis of the *dra* system has been in *Rs. rubrum* (Fitzmaurice et al., 1989), although work in *Rb. capsulatus* has demonstrated a substantially similar system (Masepohl et al., 1993). In *Rs. rubrum*, the *dra* genes were identified using degenerate oligonucleotides based on the sequences of DRAG and DRAT. The *draT* and *draG* genes are adjacent to each other and are probably cotranscribed. These genes are immediately adjacent to the *nifHDK* genes, which are transcribed in the opposite direction. The sequences of *draT* and *draG* are unremarkable and possess no obvious similarity to genes in the data banks. Immediately adjacent to *draTG* is another ORF that does possess similarity to ORF-14 in *Azotobacter vinelandii* (Dean and Jacobson, 1993)—ORF-14 has been recently termed *nifO* by Rodriguez-Quinones et al. (1993)—and similarity to the *arsC* of *E. coli* (Rosen et al., 1991), but the significance of these similarities is unclear.

The cloned *dra* region of *Rs. rubrum* has also been moved into other diazotrophs, including the non-phototrophic microaerophile *Azospirillum brasilense*. This organism itself possesses a *dra* system that ADP-ribosylates dinitrogenase reductase in response to fixed nitrogen or anaerobiosis, but not darkness (Fu et al., 1989; Zhang et al., 1992). A *draT*$^-$ mutant of *A. brasilense*, carrying the *draTG* genes from *Rs. rubrum*, responds in a fashion similar to the wild-type, indicating proper regulation of the heterologous DRAG and DRAT. Similarly, introduction of *draTG* from either *Rs. rubrum* or *A. brasilense* into *Klebsiella pneumoniae*, which normally lacks this system, results in regulated ADP-ribosylation of dinitrogenase reductase in response to added fixed nitrogen (Fu et al., 1990). These two sets of results strongly suggest that the signal transduction pathway, whether composed of proteins or small molecules, is sufficiently similar in the diverse organisms to effect apparently proper regulation of DRAG and DRAT.

The *draTG* genes have also been identified in *Rb. capsulatus*, using synthetic oligo probes (Masepohl et al., 1993). In this organism, the *dra* region is not located near the *nif* genes, nor is there a downstream ORF similar to that noted in the above organisms. An insertion in *draT* of *Rb. capsulatus* not only eliminates the reversible ADP-ribosylation of dinitrogenase reductase in response to both darkness and fixed nitrogen, it also apparently eliminates a second regulatory mechanism, which is described further in section IIF below. If this result is confirmed, it indicates that there is another role for DRAT besides ADP-ribosylation of dinitrogenase reductase; a possibility might be the ADP-ribosylation of another protein necessary for nitrogenase activity. DRAT also ADP-ribosylates the alternate dinitrogenase reductase in this organism in response to darkness or ammonia, but does not in response to molybdenum addition.

E. Regulation of DRAG/T Activities

In a regulatory cascade such as the DRAT/DRAG system, it would be possible to achieve control by regulating one or both enzymes. Furthermore, the regulatory enzymes might have either static states (i.e. on or off) or a range of activities. A range of activities can be achieved by partial activation of each enzyme molecule or by controlling a population of enzyme molecules so that some enzyme molecules are active while others are inactive. In the case of DRAT/DRAG, it appears that both enzyme molecules are posttranslationally regulated and that the regulation is on/off.

Demonstration of the reversible modification of dinitrogenase reductase was performed by Kanemoto (1984), who showed that the nitrogenase system could be subjected to repeated cycles of switch-off/switch-on. The modified subunit of dinitrogenase reductase can be separated from the unmodified subunit by electrophoresis on SDS gels, and this allows the analysis of the modification status of the protein. Modification of one subunit of the dimer renders dinitrogenase reductase completely inactive, and it is not possible to modify both subunits of the dimer at one time. The recent crystal structure determination by Georgiadis et al. (1992) shows that the two sites for modification are very close to each other at the surface of the protein and suggests that modification of one subunit blocks the modification of the other. The acetylene reduction assay allows an estimation of the in vivo nitrogenase activity. A comparison of modification status of dinitrogenase reductase with the whole-cell nitrogenase activities of samples taken during dark/light cycles allowed Kanemoto to conclude that the modification was reversible. Because three cycles of modification/demodification could be observed, it was concluded that sites for modification were being regenerated on the dimeric dinitrogenase reductase protein in vivo.

The conclusion that ADP-ribosylation is responsible for the loss of activity in vivo was confirmed when mutants lacking the system were prepared (see discussion below). Yakunin and Gogotov (1984) proposed that the covalent ADP-ribosylation of dinitrogenase did not affect the ability of the *Rb. capsulatus* enzyme to donate electrons to dinitrogenase if ferredoxin served as the electron donor, and that the loss of activity in vitro was an artifact of the use of dithionite as the electron donor. It is difficult to accept this proposal in light of the overwhelming biochemical and genetic evidence, and Gotto and Yoch (1985a) have directly refuted this hypothesis.

Evidence that DRAG is regulated comes from a pulse-chase experiment in which the turnover of ^{32}P of ADP-ribose bound to dinitrogenase reductase during switch-off conditions was monitored (Kanemoto and Ludden, 1984). *Rs. rubrum* cells were grown under nitrogenase derepressing conditions in the presence of $^{32}PO_4^{2-}$ and then treated with ammonium to initiate the modification of dinitrogenase reductase. The cell culture was then treated with an excess of $^{31}PO_4^{2-}$ to chase the label out of ^{32}P-labelled molecules in the cell. The amount of ^{32}P label remaining in ADP-ribose attached to dinitrogenase reductase was monitored and found to have a half-life 3–4 times longer than the half-life of the total cell P, indicating that under the conditions tested, DRAG was not functioning. Under conditions where DRAG is reactivating dinitrogenase reductase, ADP-ribose is removed with a half-life of 4 minutes.

Evidence that DRAT is regulated comes from experiments with *draG* mutants (Liang et al., 1991). Such mutants express *draT* and accumulate DRAT activity. When *draG*$^-$ cells are grown under nitrogenase derepressing conditions, the significant majority of dinitrogenase reductase accumulates in the active form until the signal for modification (darkness or ammonium) is given to the culture. Upon signaling, dinitrogenase reductase is irreversibly ADP-ribosylated in these cells. The interpretation of this result is that DRAT is regulated in an on/off manner in the cell because dinitrogenase reductase was not modified until the signal to modify was given, thus DRAT was off. Analysis of the dinitrogenase reductase present in the cell indicated that, prior to the negative stimulus, very little of the dinitrogenase reductase was modified, indicating that only trace DRAT activity was present.

DRAT and DRAG have been overexpressed approximately 200-fold in *Rs. rubrum* and the effects of overexpression on the regulation of DRAT and DRAG activities have been studied (Grunwald et al., 1995). In a DRAG overexpresser (DRAT at normal levels) the regulatory system is overcome, and at any time after signaling with ammonium or darkness, only a small percentage of the dinitrogenase reductase is in the modified form. In a DRAT overexpresser (DRAG at normal levels) the regulatory cycle is maintained. Surprisingly, a small amount of DRAG is able to counter the activity of a large excess of DRAT; apparently, the regulation of DRAT is maintained in the presence of a large amount of DRAT, indicating that its regulation is not dependent on stoichiometric amounts of a gene product or small molecule that is present in amounts equivalent to the concentration of DRAT in the wild-type. However, the whole cell nitrogenase activity (acetylene reduction) in a DRAT overexpresser is only 5–10% that of wild-type cells, although the nitrogenase proteins accumulate to at least 50% of the levels observed in wild-type. One explanation for this might be that DRAT binds to dinitrogenase without ADP-ribosylating the protein. At its normal levels, this is not a problem as it ties up less than a percent or two of the total dinitrogenase reductase. However, when DRAT is overexpressed 200-fold, it might tie up a significant amount of the dinitrogenase reductase protein, inhibiting activity. This model has yet to be tested.

Very little is known about proteins that might be involved in the regulation of DRAG and DRAT activities. Mutations downstream of the *dra* region have been isolated and analyzed (Fitzmaurice et al., 1989). ORF mutants show a faster time course of modification in vivo than the wild-type, but show normal kinetics of de-modification of dinitrogenase reductase (switch-on) (Liang et al., 1991). The significance of this relatively small effect on ADP-ribosylation is unclear. In *Rb. capsulatus,* mutants have been isolated with constitutive *nif* expression, but with apparently normal ADP-ribosylation; *glnB* was one of the affected loci (Hallenbeck, 1992). This result indicates that there is at least some independence between the systems that regulate the transcription of *nif* genes and those that posttranslationally regulate *nif* gene products. Analysis of *ntrBC* mutants of *A. brasilense* suggest that the ADP-ribosylation of dinitrogenase reductase in response to ammonium or anaerobiosis occurs through independent signal transduction pathways (Zhang et al., 1994).

A number of metabolites have been analyzed as potential regulatory agents in the DRAT/DRAG system. Ammonium ion, glutamine, and other nitrogen sources have no observable effect on DRAT or DRAG activity in vitro (Saari et al., 1986; Lowery and Ludden, 1988). The relative in vivo concentrations of ATP, ADP and AMP during switch-off and switch-on have been estimated in several studies (Paul and Ludden, 1984; Nordlund and Hoglund, 1986). In these studies, either darkness, ammonium ion or other effectors leading to switch-off have been employed to perturb the system. In no case has a consistent correlation with an increase in nucleotide concentration been observed. The in vivo glutamine level has been reported to correlate with the onset of switch-off, but other studies have shown that neither glutamine nor any other amino acid level is obligately correlated with switch-off. The pools of pyridine nucleotides have been investigated as well (Nordlund and Hoglund, 1986), and no change in their pools has been tied to changes in the modification status of dinitrogenase reductase. However, Soliman and Nordlund (1992) have provided interesting evidence that exogenously-supplied NAD can stimulate switch-off, and they have shown that label from ^{32}P-NAD supplied to the cells can be incorporated into dinitrogenase reductase.

Because a divalent metal ion such as Mn^{2+} is required for DRAG activity, fluxes in divalent ion pools have been proposed to play a role in regulation of the DRAT/DRAG system. Yoch (1979) first studied Mn^{2+}-starved cells for the ability to regulate nitrogenase activity. His results showed that the cells grew with glutamate as the N source but produced only low levels of nitrogenase activity. These cells were able to switch-off nitrogenase activity. Nordlund and coworkers (Cadez and Nordlund, 1991) have investigated the Ca^{2+} concentration in the medium of *Rs. rubrum* cells during switch-off/switch-on and have observed transients in the concentration of this ion, suggesting a role for this ion in the regulatory scheme.

Nutritional conditions are important in determining the ability of *Rs. rubrum* to regulate nitrogenase. Munson and Burris (1969) noted that the best activity was obtained in extracts of cells grown under nitrogen starvation. Yoch and Cantu (1980) provided an explanation for this observation when they showed that nitrogen-starved cells were incapable of switch-off. The reason for the inability of nitrogen-starved cells to switch-off is not known, as both DRAT and DRAG are present in these cells and active in extracts (Lowery et al., 1986).

While most experiments on the regulation of nitrogenase have been performed in light-grown cells, *Rs. rubrum* and certain other phototrophs are capable of dark-anaerobic fermentation of pyruvate and nitrogen fixation under these conditions (Uffen and Wolfe, 1970; Madigan et al., 1979; Schultz et al., 1985). Switch-off of nitrogenase occurs in dark pyruvate-grown cells in response to ammonium (Schultz et al., 1985).

While much of the work on regulation of nitrogenase by reversible ADP-ribosylation has been performed in *Rs. rubrum*, the system has also been demonstrated in *Cm. vinosum* (Gotto and Yoch, 1985b), *Rb. capsulatus* (Hallenbeck et al., 1982c; Gotto and Yoch, 1985a; Jouanneau et al., 1989), and *Rhodopseudomonas palustris* (Alef et al., 1981). Evidence against the presence of the system in *Rhodopseudomonas spheroides* has been presented (Jones and Monty, 1979). The system has also been extensively studied in the non-photosynthetic bacteria of the genus *Azospirillum* (Hartmann et al., 1986; Fu et al., 1989; Burris et al., 1991; Zhang et al., 1993).

F. Non-ADPR Regulatory Mechanisms

Although the original observations of Gest and Kamen (1949) can be explained by the action of the reversible ADP-ribosylation system described above, it has become apparent that there are additional complexities in the regulation of nitrogenase activity in phototrophs. The presence of the phenomenon described below not only suggests such a case but serves to remind us that the presence of one regulatory system by no means precludes organisms from elaborating additional, even redundant, mechanisms for regulation.

Two lines of evidence indicate that *Rb. capsulatus* responds to the addition of fixed nitrogen by a mechanism in addition to the ADP-ribosylation of dinitrogenase reductase (Pierrard et al., 1993). First, following addition of fixed nitrogen to wild-type, acetylene reduction activity falls much faster and to a greater extent than explained by a direct measurement of ADP-ribosylated dinitrogenase reductase. Second, partially active *nifH* mutants, in which the target arginine is replaced by another amino acid, are unable to be ADP-ribosylated but continue to show a rapid loss of nitrogenase activity in response to added fixed nitrogen. In contrast to ADP-ribosylation, this regulatory response does not occur in response to darkness.

The biochemical basis for this effect is unknown, but recent results suggest that it does not occur in a *draT* mutant (Masepohl et al., 1993). This result is consistent with several models: (1) In this organism, DRAT has another substrate in the cell, whose ADP-ribosylation is more rapid than that of dinitrogenase reductase and leads to a loss of nitrogenase activity. Because there is no detectable effect on the electrophoretic mobility of dinitrogenase itself in these strains, a reasonable candidate might be a protein involved in electron transfer to dinitrogenase reductase. Presumably, if it were present at a lower concentration than dinitrogenase reductase (as is the case with flavodoxin in *Klebsiella pneumoniae*), its ADP-ribosylation might be both rapid and complete, relative to that of dinitrogenase reductase. (2) DRAT might physically block activity of dinitrogenase reductase by other mechanisms, though there would appear to be too little DRAT to accomplish this task. The explanation of this surprising result will await further experimentation, but serves as a reminder that the presence of any single regulatory mechanism does not preclude the presence of redundant ones in the same organism. There is currently no evidence for the above response in *Rs. rubrum*.

Acknowledgments

Results from the authors' laboratories reported here have been generously supported by NSF and by the USDA/NRI Competitive Grants Program. The authors thank Gary Nielsen, Doug Lies, Yaoping Zhang, Sandy Grunwald, Stefan Nordlund and Werner Klipp for sharing results of their work before publication. The authors thank Ms. Gail Stirr for her help in editing and preparation of the manuscript.

References

Ahombo G, Willison JC and Vignais PM (1986) The *nifHDK* genes are contiguous with a *nifA*-like regulatory gene in *Rhodobacter capsulatus*. Mol Gen Genet 205: 442–445

Albrecht SL and Evans MCW (1973) Measurement of the oxidation reduction potential of the EPR detectable active centre of the molybdenum iron protein of *Chromatium* nitrogenase. Biochem Biophys Res Comm. 55: 1009–1014

Alef K, Arp DJ and Zumft WG (1981) Nitrogenase switch-off by ammonia in *Rhodopseudomonas palustris*: Loss under nitrogen deficiency and independence from the adenylylation state of glutamine synthetase. Arch Microbiol 130: 138–142

Bishop PE and Premakumar R (1992) Alternative Nitrogen Fixation Systems. In: Stacey G, Burris RH and Evans JH (eds) Biological Nitrogen Fixation. New York, Chapman and Hall. 736–762

Bishop PE, Jarlenski DML and Hetherington DR (1980) Evidence for an alternative nitrogen fixation system in *Azotobacter vinelandii*. Proc Natl Acad Sci USA 77: 7342–7346

Bolin JT, Ronco AE, Morgan TV, Mortenson LE and Xuong N-H (1993) The unusual metal clusters of nitrogenase: Structural features revealed by x-ray anomalous diffraction studies of the MoFe protein from *Clostridium pasteurianum*. Proc Natl Acad Sci USA 90: 1078–1082

Brostedt E and Nordlund S (1991) Purification and partial characterization of a pyruvate oxidoreductase from the photosynthetic bacterium *Rhodospirillum rubrum* grown under nitrogen-fixing conditions. Biochem J 279: 155–158

Burns RC and Bulen WA (1965) ATP-dependent hydrogen evolution by cell-free preparations of *Azotobacter vinelandii*. Biochim Biophys Acta 105: 437–445

Burns RC and Bulen WA (1966) A procedure for the preparation of extracts from *Rhodospirillum rubrum* catalyzing N_2 reduction and ATP-dependent H_2 evolution. Arch Biochem Biophys 113: 461–463

Burris RH (1985) H_2 as an inhibitor of N_2 fixation. Physiol Veg 23: 843–848

Burris RH (1991) Nitrogenases. J Biol Chem 266: 9339–9342

Burris RH and Roberts GP (1993) Biological Nitrogen Fixation. Annu Rev Nutr 13: 317–335

Burris RH, Hartmann A, Zhang Y and Fu H (1991) Control of nitrogenase in *Azospirillum* sp. Plant and Soil 137: 127–134

Cadez P and Nordlund S (1991) The requirement for Mn^{2+} and Ca^{2+} in nitrogen fixation by the photosynthetic bacterium *Rhodospirillum rubrum*. FEMS Microbiol Lett 81: 279–282

Carnahan JE, Mortenson LE, Mower HF and Castle JE (1960) Nitrogen fixation in cell-free extracts of *Clostridium pasteurianum*. Biochim Biophys Acta 44: 520–535

Chang CL, Davis LC, Rider M and Takemoto DJ (1988) Characterization of *nifH* mutations of *Klebsiella pneumoniae*. J Bacteriol 170: 4015–4022

Cusanovich MA and Edmondson DE (1971) The isolation and characterization of *Rhodospirillum rubrum* flavodoxin. Biochem Biophys Res Comm 45: 327

Dean DR and Jacobson MR (1993) Biochemical genetics of nitrogenase. In: Stacey G, Burris RH and Evans JH (eds) Biological Nitrogen Fixation. New York, Chapman and Hall. 763–834

Dean DR, Bolin JT and Zheng L (1993) Nitrogenase metalloclusters: Structures, organization, and synthesis. J Bacteriol 175: 6737–6744

Dilworth MJ, Eady RR, Robson RL and Miller RW (1987) Ethane formation from acetylene as a potential test for vanadium nitrogenase *in vivo*. Nature 327: 167–168

Dowling TE, Preston GG and Ludden PW (1982) Heat activation of the Fe protein of nitrogenase from *Rhodospirillum rubrum*. J Biol Chem 257: 13987–13992

Eady RR, Smith BE, Cook KA and Postgate JR (1972) Nitrogenase of *Klebsiella pneumoniae*: Purification and properties of the component proteins. Biochem J 128: 665–675

Emerich DW and Burris RH (1976) Interactions of heterologous nitrogenase components that generate catalytically inactive complexes. Proc Natl Acad Sci USA 73: 4369–4373

Fitzmaurice WP, Saari LL, Lowery RG, Ludden PW and Roberts GP (1989) Genes coding for the reversible ADP-ribosylation system of dinitrogenase reductase from *Rhodospirillum rubrum*. Mol Gen Genet 218: 340–347

Fu H, Burris RH and Roberts GP (1990) Reversible ADP-ribosylation is demonstrated to be a regulatory mechanism in prokaryotes by heterologous expression. Proc Natl Acad Sci USA 87: 1720–1724

Fu H-A, Hartmann A, Lowery RG, Fitzmaurice WP, Roberts GP and Burris RH (1989) Posttranslational regulatory system for nitrogenase activity in *Azospirillum* spp. J Bacteriol 171(9): 4679–4685

Georgiadis MM, Komiya H, Chakrabarti P, Woo D, Kornuc JJ and Rees DC (1992) Crystallographic structure of the nitrogenase iron protein from *Azotobacter vinelandii*. Science 257: 1653–1659

Gest H and Kamen MD (1949) Photoproduction of molecular hydrogen by *Rhodospirillum rubrum*. Science 109: 558–559

Gest H, Kamen MD and Bregoff HM (1950) Studies on the metabolism of photosynthetic bacteria. V. Photoproduction of hydrogen and nitrogen fixation by *Rhodospirillum rubrum*. J Biol Chem 182: 153–170

Gotto JW and Yoch DC (1982a) Purification and Mn^{2+} activation of *Rhodospirillum rubrum* nitrogenase activating enzyme. J Bacteriol 152: 714–721

Gotto JW and Yoch DC (1982b) Regulation of *Rhodospirillum rubrum* nitrogenase activity. J Biol Chem 257: 2868–2873

Gotto JW and Yoch DC (1985a) The regulation of ferredoxin-dependent nitrogenase activity in *Rhodospirillum rubrum* and *Rhodopseudomonas capsulata*. FEMS Microbiol Lett 28: 107–111

Gotto JW and Yoch DC (1985b) Regulation of nitrogenase activity by covalent modification in *Chromatium vinosum*. Arch Microbiol 141: 40–43

Grunwald SK, Lies DP, Roberts GP and Ludden PW (1995) Posttranslational regulation of nitrogenase in *Rhodospirillum rubrum* strains overexpressing the regulatory enzymes dinitrogenase reductase ADP-ribosyl transferase and dinitrogenase reductase activating glycohydrolase. J Bacteriol 177: 628–635

Guth JH and Burris RH (1983) Comparative study of the active and inactive forms of dinitrogenase reductase from *Rhodospirillum rubrum*. Biochim Biophys Acta 749: 91–100

Hallenbeck PC (1992) Mutations affecting nitrogenase switch-off in *Rhodobacter capsulatus*. Biochim Biophys Acta 1118: 161–168

Hallenbeck PC, Jouanneau Y and Vignais PM (1982a) Purification and molecular properties of a soluble ferredoxin from *Rhodopseudomonas capsulata*. Biochim Biophys Acta 681: 168–176

Hallenbeck PC, Meyer CM and Vignais PM (1982b) Nitrogenase from the photosynthetic bacterium *Rhodopseudomonas capsulata*: purification and molecular properties. J Bacteriol 149: 708–717

Hallenbeck PC, Meyer CM and Vignais PM (1982c) Regulation of nitrogenase in the photosynthetic bacterium *Rhodopseudomonas capsulata* as studied by two-dimensional gel electrophoresis. J Bacteriol 151: 1612–1616

Hartmann A, Fu H and Burris RH (1986) Regulation of nitrogenase activity by ammonium chloride in *Azospirillum* sp. J Bacteriol 165: 864–870

Haselkorn R (1986) Organization of the genes for nitrogen fixation in photosynthetic bacteria and cyanobacteria. Ann Rev Microbiol 40: 525–547

Hausinger RP and Howard J (1983) Thiol reactivity of the nitrogenase Fe-protein from *Azotobacter vinelandii*. J Biol Chem 258: 13486–13492

Hawkes TR, McLean PA and Smith BE (1984) Nitrogenase from *nifV* mutants of *Klebsiella pneumoniae* contains an altered form of the iron-molybdenum cofactor. Biochem J 217: 317–321

Hoffman BM, Ventures RA, Roberts JE, Nelson MJ and Orme-Johnson WH (1982) ^{57}Fe ENDOR of the nitrogenase MoFe protein. J Am Chem Soc 104: 4711

Hoover TR, Robertson AD, Cerny RL, Hayes RN, Imperial J, Shah VK and Ludden PW (1987) Identification of the V factor needed for synthesis of the iron-molybdenum cofactor of nitrogenase as homocitrate. Nature 329: 855–857

Hoover TR, Imperial J, Ludden PW and Shah VK (1989) Homocitrate is a component of the iron-molybdenum cofactor of nitrogenase. Biochemistry 28: 2768–2771

Imperial J, Hoover TR, Madden MS, Ludden PW and Shah VK (1989) Substrate reduction properties of dinitrogenase activated in vitro are dependent upon the presence of homocitrate or its analogues during iron-molybdenum cofactor synthesis. Biochemistry 28(19): 7796–7799

Jones BL and Monty KJ (1979) Glutamine as a feedback inhibitor of the *Rhodopseudomonas sphaeroides* nitrogenase system. J Bacteriol 139: 1007–1013

Jouanneau Y, Roby C, Meyer CM and Vignais PM (1989) ADP-ribosylation of dinitrogenase reductase in *Rhodobacter capsulatus*. Biochemistry 28: 6524–6530

Kamen MD (1986) Radiant Science, Dark Politics. Berkeley, CA, University of California Press

Kamen MD and Gest H (1949) Evidence for a nitrogenase system in the photosynthetic bacterium *Rhodospirillum rubrum*. Science 109: 560

Kanemoto RH and Ludden PW (1984) Effect of ammonia, darkness, and phenazine methosulfate on whole-cell nitrogenase activity and Fe protein modification in *Rhodospirillum rubrum*. J Bacteriol 158: 713–720

Kim J and Rees DC (1992) Crystallographic structure and functional implications of the nitrogenase molybdenum-iron protein from *Azotobacter vinelandii*. Nature 360: 553–560

Lehman LJ and Roberts GP (1991) Glycine 100 in the dinitrogenase reductase of *Rhodospirillum rubrum* is required for nitrogen fixation but not for ADP-ribosylation. J Bacteriol 173: 6159–6161

Lehman LJ and Roberts GP (1991) Identification of an alternative nitrogenase system in *Rhodospirillum rubrum*. J Bacteriol 173: 5705–5711

Liang J, Nielsen GM, Lies DP, Burris RH, Roberts GP and Ludden PW (1991) Mutations in the *draT* and *draG* genes of *Rhodospirillum rubrum* result in loss of regulation of nitrogenase by reversible ADP-ribosylation. J Bacteriol 173: 6903–6909

Lindstrom ES, Burris RH and Wilson PW (1949) Nitrogen fixation by photosynthetic bacteria. J Bacteriol 58: 313–316

Lindstrom ES, Tove SR and Wilson PW (1950) Nitrogen fixation by the green and purple sulfur bacteria. Science 112: 197–198

Ljones T and Burris RH (1972) ATP hydrolysis and electron transfer in the nitrogenase reaction with different combinations of the iron protein and the molybdenum-iron protein. Biochim Biophys Acta 275: 93–101

Lowery RG and Ludden PW (1988) Purification and properties of the dinitrogenase reductase inactivating ADP-ribosyltransferase from *Rhodospirillum rubrum*. J Biol Chem 263: 16714–16719

Lowery RG and Ludden PW (1989) Effect of nucleotides on the activity of dinitrogenase reductase ADP-ribosyltransferase from *Rhodospirillum rubrum*. Biochemistry 28: 4956–4961

Lowery RG, Saari LL and Ludden PW (1986) Reversible regulation of the iron protein of nitrogenase from *Rhodospirillum rubrum* by ADP-ribosylation *in vitro*. J Bacteriol 166: 513–518

Lowery RG, Chang CL, Davis LC, McKenna MC, Stephens PJ and Ludden PW (1989) Substitution of histidine for arginine-101 of dinitrogenase reductase disrupts electron transfer to dinitrogenase. Biochemistry 28: 1206–1212

Ludden PW and Burris RH (1976) Activating factor for the iron protein of nitrogenase from *Rhodospirillum rubrum*. Science 194: 424–426

Ludden PW and Burris RH (1978) Purification and properties of nitrogenase from *Rhodospirillum rubrum* and evidence for phosphate, ribose and an adenine-like unit covalently bound to the iron protein. Biochem J 175: 251–259

Ludden PW and Burris RH (1979) Removal of an adenine-like molecule during activation of dinitrogenase reductase from *Rhodospirillum rubrum*. Proc Natl Acad Sci USA 76: 6201–6205

Ludden PW and Roberts GP (1989) Regulation of nitrogenase activity by reversible ADP-ribosylation. Current Topics in Cellular Regulation. Orlando, Academic Press Inc. 23–55

Ludden PW, Preston GG and Dowling TE (1982) Comparison of active and inactive forms of iron protein from *Rhodospirillum rubrum*. Biochem J 203: 663–668

Madigan MT, Wall JD and Gest H (1979) Dark anaerobic dinitrogen fixation by a photosynthetic microorganism. Science 204: 1430

Masepohl B, Krey R and Klipp W (1993) The *draTG* gene region of *Rhodobacter capsulatus* is required for posttranslational regulation of both the molybdenum and the alternative nitrogenase. J Gen Microbiol 139: 2667–2675

McKenna CE, McKenna M-C and Huang CW (1979) Low stereoselectivity in methylacetylene and cycolpropene reductions by nitrogenase. Proc Natl Acad Sci USA 76: 4773–4777

McLean PA and Dixon RA (1981) Requirement of *nifV* gene for production of wild-type nitrogenase enzyme in *Klebsiella pneumoniae*. Nature 292: 655–656

Meyer J, Kelley BC and Vignais PM (1978) Aerobic nitrogen fixation by *Rhodopseudomonas capsulata*. FEBS Lett 85: 224–228

Moreno-Vivian C, Masepohl B, Schmehl M, Klipp W and Puhler A (1988) Nucleotide sequence of *Rhodobacter capsulatus nif* gene regions carrying homologous genes to *Klebsiella pneumoniae, nifE, nifN, nifX, nifQ, nifS* and *nifV*. In: Bothe H, de Bruihn FJ and Newton FE (eds) Nitrogen Fixation: Hundred Years After, p 177. Gustav Fischer, Stuttgart

Munck E, Rhodes H, Orme-Johnson WH, Davis LC, Brill WJ and Shah VK (1975) Nitrogenase. VIII. Mossbauer and EPR spectroscopy. The MoFe protein component from *Azotobacter vinelandii*. Biochim Biophys Acta 400: 32–53

Munson TO and Burris RH (1969) Nitrogen fixation by *Rhodospirillum rubrum* grown in nitrogen limited continuous culture. J Bacteriol 97: 1093–1098

Murrell SA, Lowery RG and Ludden PW (1988) ADP-ribosylation of dinitrogenase reductase from *Clostridium pasteurianum* prevents its inhibition of nitrogenase from *Azotobacter vinelandii*. Biochem J 251: 609–612

Nielsen GM, Bao Y, Roberts GP and Ludden PW (1994) Purification and characterization of an oxygen-stable form of dinitrogenase reductase-activating glycohydrolase from *Rhodospirillum rubrum*. Biochem J 302: 801–806

Nieva-Gomez E, Roberts GP, Klevickis S and Brill WJ (1980) Electron transport to nitrogenase in *Klebsiella pneumoniae*. Proc Natl Acad Sci USA 77: 2555–2558

Nordlund S and Hoglund L (1986) Studies of the adenylate and pyridine nucleotide pools during nitrogenase 'switch-off' in *Rhodospirillum rubrum*. Plant Soil 90: 203–209

Nordlund S and Noren A (1984) Dependence on divalent cations of the activation of inactive Fe-protein of nitrogenase from *Rhodospirillum rubrum*. Biochim Biophys Acta 791: 21–27

Nordlund S, Eriksson U and Baltscheffsky H (1977) Necessity of a membrane component for nitrogenase activity in *Rhodospirillum rubrum*. Biochim Biophys Acta 462: 187–195

Oka J, Kunihiro U, Hayaishi O, Komura H and Nakanishi K (1984) ADP-ribosyl protein lyase. Purification, properties and identification of the product. J Biol Chem 259: 986–995

Orme-Johnson WH, Hamilton WD, Ljones T, Tso M-Y, Burris RH, Shah VK and Brill WJ (1972) Electron paramagnetic resonance of nitrogenase and nitrogenase components from *Clostridium pasteurianum* W5 and *Azotobacter vinelandii* OP. Proc Natl Acad Sci USA 69: 3142–3145

Ormerod JG and Gest H (1962) Symposium on metabolism of inorganic compounds: Hydrogen photosynthesis and alternative metabolic pathways in photosynthetic bacteria. Bacteriol Rev 26: 51–66

Paul TD and Ludden PW (1984) Adenine nucleotide levels in *Rhodospirillum rubrum* during switch-off of whole-cell nitrogenase activity. Biochem J 224: 961–969

Pierrard J, Ludden PW and Roberts GP (1993) Posttranslational regulation of nitrogenase in *Rhodobacter capsulatus*: Existence of two independent regulatory effects of ammonium. J Bacteriol 175: 1358–1366

Pope MR, Murrell SA and Ludden PW (1985a) Covalent modification of the iron protein of nitrogenase from *Rhodospirillum rubrum* by adenosine diphosphoribosylation of a specific arginyl residue. Proc Natl Acad Sci USA 82: 3173–3177

Pope MR, Murrell SA and Ludden PW (1985b) Purification and properties of the heat released nucleotide modifying group from the inactive Fe protein of nitrogenase from *Rhodospirillum rubrum*. Biochemistry 24: 2374–2380

Pope MR, Saari LL and Ludden PW (1986) *N*-glycohydrolysis of adenosine diphosphoribosyl arginine linkages by dinitrogenase reductase activating glycohydrolase (activating enzyme) from *Rhodospirillum rubrum*. J Biol Chem 261: 10104–10111

Pope MR, Saari LL and Ludden PW (1987) Fluorometric assay for ADP-ribosylarginine cleavage enzymes. Anal Biochem 160: 68–77

Premakumar R, Chisnell JR and Bishop PE (1989) A comparison of the three dinitrogenase reductases expressed by *Azotobacter vinelandii*. Can J Microbiol 35: 344–348

Preston GG and Ludden PW (1982) Change in subunit composition of the iron protein of nitrogenase from *Rhodospirillum rubrum* during activation and inactivation of iron protein. Biochem J 205: 489–494

Pretorius I-M, Rawlings DE, O'Neill EG, Jones WA, Kirby R and Woods DR (1987) Nucleotide sequence of the gene encoding the nitrogenase iron protein of *Thiobacillus ferrooxidans*. J Bacteriol 169: 367–370

Rivera-Ortiz JH and Burris RH (1975) Interactions among substrates and inhibitors of nitrogenase. J Bacteriol 123: 537–545

Roberts GP, MacNeil T, MacNeil D and Brill WJ (1978) Regulation and characterization of protein products coded by the *nif* (nitrogen fixation) genes of *Klebsiella pneumoniae*. J Bacteriol 136: 267–279

Rodriguez-Quinones F, Bosch R and Imperial J (1993) Expression of the *nifBfdxNnifOQ* region of *Azotobacter vinelandii* and its role in nitrogenase activity. J Bacteriol 175: 2926–2935

Rosen BP, Weigel U, Monticello RA and Edwards BPF (1991) Molecular analysis of an anion pump: purification of the ArsC protein. Biochim Biophys Acta 284: 381–385

Saari LL, Triplett EW and Ludden PW (1984) Purification and properties of the activating enzyme for iron protein of nitrogenase from the photosynthetic bacterium *Rhodospirillum rubrum*. J Biol Chem 259: 15502–15508

Saari LL, Pope MR, Murrell SA and Ludden PW (1986) Studies on the activating enzyme for iron protein of nitrogenase from *Rhodospirillum rubrum*. J Biol Chem 261: 4973–4977

Schick H-J (1971) Substrate and light dependent fixation of molecular nitrogen in *Rhodospirillum rubrum*. Arch Mikrobiol 75: 89–101

Schmehl M, Jahn A, Meyer zu Vilsendorf A, Hennecke S, Masepohl B, Schuppler M, Marxer M, Oelze J and Klipp W (1993) Identification of a new class of nitrogen fixation genes in *Rhodobacter capsulatus*: a putative membrane complex is involved in electron transport to nitrogenase. Mol Gen Genet 241: 602–615

Schneider K, Muller A, Schramm U and Klipp W (1991) Demonstration of a molybdenum- and vanadium-independent nitrogenase in a *nifHDK*-deletion mutant of *Rhodobacter capsulatus*. Eur J Biochem 195: 653–661

Schneider KC, Bradbeer C, Singh RN, Wang LC, Wilson PW and Burris RH (1960) Nitrogen fixation by cell-free preparations from microorganisms. Proc Natl Acad Sci USA 46: 726–733

Schüddekopf K, Hennecke S, Liese U, Kutsche M and Klipp W (1993) Characterization of *anf* genes specific for the alternative nitrogenase and identification of *nif* genes required for both nitrogenases in *Rhodobacter capsulatus*. Mol Micro 8: 673–684

Schultz JE, Gotto JW, Weaver PF and Yoch DC (1985) Regulation of nitrogen fixation in *Rhodospirillum rubrum* grown under dark, fermentative conditions. J Bacteriol 162: 1322–1324

Shah VK and Brill WJ (1973) Nitrogenase. IV. Simple method of purification to homogeneity of nitrogenase components from *Azotobacter vinelandii*. Biochim Biophys Acta 305: 445–54

Shah VK and Brill WJ (1977) Isolation of an iron-molybdenum cofactor (FeMo-co) from nitrogenase. Proc Natl Acad Sci USA 74: 3249–3253

Shah VK, Davis LC, Gordon JK, Orme-Johnson WH and Brill WJ (1973) Nitrogenase. III. Nitrogenaseless mutants of *Azotobacter vinelandii*: activities, cross reactions and EPR spectra. Biochim Biophys Acta 292: 246–55

Shah VK, Stacey G and Brill WJ (1983) Electron transport to nitrogenase—Purification and characterization of pyruvate-flavodoxin oxidoreductase, the *nifJ* gene product. J Biol Chem 258: 12064–12068

Shah VK, Hoover TR, Imperial J, Paustian TD, Roberts GP and Ludden PW (1988) Role of *nif* gene products and homocitrate in the biosynthesis of the iron-molybdenum cofactor. In: Bothe H, deBruijn FJ and Newton WE (eds) Nitrogen Fixation: Hundred Years After, pp 115–120. Gustav Fischer, Stuttgart

Silverstein R and Bulen WA (1970) Kinetic studies of the nitrogenase-catalyzed hydrogen evolution and nitrogen reduction reactions. Biochemistry 9: 3809–3815

Simpson FB and Burris RH (1984) A nitrogen pressure of 50 atmospheres does not prevent evolution of hydrogen by nitrogenase. Science 224: 1095–1097

Soliman A and Nordlund S (1992) Studies on the effect of NAD(H) on nitrogenase activity in *Rhodospirillum rubrum*. Arch Microbiol 157: 431–435

St. John RT, Johnston HM, Seidman C, Garfinkel D, Gordon JK, Shah VK and Brill WJ (1975) Biochemistry and genetics of *Klebsiella pneumoniae* mutant strains unable to fix N_2. J Bacteriol 121: 759–765

Stewart WPD, Fitzgerald GP and Burris RH (1967) *In situ* studies on N_2 fixation using the acetylene reduction technique. Proc Natl Acad Sci USA 58: 2071–2078

Tso MYW and Burris RH (1973) The binding of ATP and ADP by nitrogenase components from *Clostridium pasteurianum*. Biochim Biophys Acta 309: 263–70

Uffen RL and Wolfe RS (1970) Anaerobic growth of purple nonsulfur bacteria under dark conditions. J Bacteriol 104: 462–472

Ventures RA, Nelson MJ, McLean PA, True AE, Levy MA, Hoffman BM and Orme-Johnson WH (1986) ENDOR of the resting state of nitrogenase molybdenum iron proteins from *Azotobacter vinelandii*, *Klebsiella pneumoniae* and *Clostridium pasteurianum*. ^{1}H, ^{57}Fe, ^{95}Mo and ^{33}S studies. J Am Chem Soc 108: 3487–3498

Watkins PA, Kanaho Y and Moss J (1987) Inhibition of the GTPase activity of transducin by an NAD^+: arginine ADP-ribosyltransferase from turkey erythrocytes. Biochem J 248: 749–754

Winter HC and Arnon DI (1970) The nitrogen fixation system of photosynthetic bacteria. I. Preparation and properties of a cell-free extract from *Chromatium*. Biochim Biophys Acta 197: 170–79

Winter HC and Ober JA (1973) Isolation of particulate nitrogenase from *Chromatium* strain D. Plant Cell Physiol 14: 769–773

Yakunin AF, Gennaro G and Hallenbeck PC (1993) Purification and properties of a *nif*-specific flavodoxin from the photosynthetic bacterium *Rhodobacter capsulatus*. J Bacteriol 175: 6775–6780

Yakunin AF and Gogotov IN (1984) The activity of two forms of nitrogenase from *Rhodopseudomonas capsulata* in the presence of different electron donors. FEMS Microbiol Lett 23: 217–220

Yoch DC (1979) Manganese, an essential trace element for N_2 fixation by *Rhodospirillum rubrum* and *Rhodopseudomonas capsulata*: role in nitrogenase regulation. J Bacteriol 140: 987–995

Yoch DC and Cantu M (1980) Changes in the regulatory form of *Rhodospirillum rubrum* nitrogenase as influenced by nutritional and environmental factors. J Bacteriol 142: 899–907

Zhang Y, Burris RH and Roberts GP (1992) Cloning, sequencing, mutagenesis, and functional characterization of *draT* and *draG* genes from *Azospirillum brasilense*. J Bacteriol 174: 3364–3369

Zhang Y, Burris RH, Ludden PW and Roberts GP (1993) Posttranslational regulation of nitrogenase activity by anaerobiosis and ammonium in *Azospirillum brasilense*. J Bacteriol 175: 6781–6788

Zhang Y, Burris RH, Ludden PW and Roberts GP (1994) Posttranslational regulation of nitrogenase activity in *Azospirillum brasilense* ntrBC mutants: Ammonium and anaerobic switch-off occurs through independent signal transduction pathways. J Bacteriol 176: 5780–5787

Zheng L and Dean DR (1994) Catalytic formation of a nitrogenase iron-sulfur cluster. J Biol Chem 269: 18723–18726

Zheng L, White RH, Cash VL, Jack RL and Dean DR (1993) Cysteine desulfurase activity indicates a role for NIFS in metallocluster biosynthesis. Proc Natl Acad Sci USA 90: 2754–2758

Zumft WG and Nordlund S (1981) Stabilization and partial characterization of the activating enzyme for dinitrogenase reductase (Fe-protein) from *Rhodospirillum rubrum*. FEBS Lett 127: 79–82

Zumft WG, Mortenson LE and Palmer G (1974) Electron-paramagnetic-resonance studies on nitrogenase. Investigation of the oxidation-reduction behaviour of azoferredoxin and molybdoferredoxin and potentiometric and rapid-freeze techniques. Eur J Biochem 46: 525–535

Chapter 44

Aerobic and Anaerobic Electron Transport Chains in Anoxygenic Phototrophic Bacteria

Davide Zannoni
Department of Biology, University of Bologna, 42 Irnerio, 40126 Bologna, Italy

Summary

Aerobic and anaerobic electron transport chains of facultative phototrophs have been of increasing interest because of their diverse organization of redox carriers and their adaptive regulatory mechanisms of gene expression. During the last decade, studies on the biochemistry of bacterial redox complexes such as NADH-deh and bc_1 from *Rhodobacter* species, and Cyt *c*-oxidases of aa_3 type from *Rb. sphaeroides* and *Chloroflexus aurantiacus,* have revealed the presence of fewer subunits than corresponding eukaryotic enzymes. This evidence has provided new insights into the biochemical evolution of respiration and also useful indications on structure/function relationships. Recent advances in studying the aerobic and anaerobic respiratory pathways of facultative phototrophs have taken advantage of modern molecular genetics. In particular, the role of soluble cytochrome c_2, until recent years considered to be essential for electron transport in the two closely related species *Rb. capsulatus* and *Rb. sphaeroides,* has been better defined. Indeed, it is now clear that two different classes of alternative electron carriers (soluble Cyt iso-c_2 and membrane-bound Cyt c_y) can operate between the membrane-bound redox complexes instead of, or along with, the Cyt c_2. The presence of multiple electron carriers between redox complexes suggests that Cyt c_y-like components might be more widely spread among those photosynthetic bacteria where photooxidizable soluble *c*-type hemes are not readily detected, e.g. *Cf. aurantiacus*. The outstanding metabolic versatility of *Rb. capsulatus* made also possible the use of mutants defective in redox carriers of aerobic respiration for the analysis of anaerobic electron transport pathways. Thus, if the role of Cyt c_2 in anaerobic light-driven electron flow has partially been reshuffled, Cyt c_2 seems to

R. E. Blankenship, M. T. Madigan and C. E. Bauer (eds): Anoxygenic Photosynthetic Bacteria, pp. 949–971.

play a key role in the dark anaerobic pathways leading to NO_2 and N_2O reduction. The use of Cyt *c*-deficient mutants also demonstrated that the ubiquinol/Cyt *c* oxidoreductase is not required for growth with DMSO or TMAO as electron acceptors. These dark anaerobic processes, however, cannot sustain a 'consistent' cell growth in the presence of non fermentable substrates; thus, they must be regarded as advantageous metabolic systems facilitating anaerobic growth in the dark and/or in the light.

I. Introduction

The anoxygenic phototrophic bacteria represent, in terms of morphological, biochemical, and physiological features, a quite heterogeneous group which is unified by the common property of growing by means of an anoxygenic type of photosynthesis. Under natural conditions, however, phototrophic bacteria receive light during only part of the solar day, and even if their growth is restricted to this period, it is likely that they display some alternative metabolic capacity during dark periods. In this respect, several species of purple and green phototrophic bacteria are therefore regarded as 'facultative' in that they can also obtain energy from aerobic and anaerobic dark growth.

This chapter will first address some general remarks on the physiology of facultative phototrophs, and then discuss in detail the biochemical processes that lead to the use of oxygen and other electron acceptors, e.g. nitrate or dimethyl sulfoxide, to convert the oxidation of various substrates to a form immediately available to growth and development.

A. Physiological Aspects: Some General Remarks

Facultative phototrophs are widely distributed in nature, although rarely in high concentrations (Pfennig, 1978; Gest and Favinger, 1983) (see Chapters 1, 4 and 41 by Imhoff, van Gemerden and Mas, and Tabita, respectively, for coverage of taxonomy, physiology, and ecology of anoxygenic phototrophs). The metabolic options available to purple nonsulfur bacteria, e.g. *Rhodobacter*, *Rhodospirillum*, *Rhodopseudomonas*, put them in a position to survive in quite different habitats and many species can use a variety of carbon sources (in the light or in the dark) such as organic acids or fatty acids (heterotrophy) but also CO_2 as sole carbon source and H_2 as sole electron donor (autotrophy) (Ormerod and Gest, 1962). Most species studied to date are mesophilic, but *Rhodospirillum centenum* has been isolated from a hot spring (Favinger et al., 1989). Typically, sulphide is a growth inhibitor to most species of the Rhodospirillaceae (Hansen and van Gemerden, 1972) unless it is kept at constant and very low levels in continuous cultures (see Chapter 4 by van Gemerden and Mas). In general, purple sulfur bacteria are rather strict anaerobes that can be found just below the oxic-anoxic interface of lakes. However certain members of the Chromatiaceae (*Chromatium, Thiocystis, Amoebobacter, Thiocapsa*) are capable of heterotrophic and/or lithotrophic growth under dark microaerobic conditions (Kämpf and Pfennig, 1980). This growth mode is generally considered a mechanism for temporary survival in transiently oxygenated environments or as a means for maintaining the 'energy-balance' at night (Hashwa and Trüper, 1978).

The green nonsulfur- or green gliding-bacteria, the Chloroflexaceae, are exemplified by the thermophilic genus *Chloroflexus aurantiacus* (Pierson and Castenholz, 1974; Madigan et al., 1974) (see Chapter 3 by Pierson and Castenholz). They usually occur as thick mats in alkaline hot springs, the maximum standing crop occurring at a temperature of 50–55 °C (Bauld and Brock, 1973). *Chloroflexus*, which shares physiological, biochemical, and ultrastructural relationships with both the green and purple bacteria, usually coexists with the cyanobacterium which is the primary producer for photoheterotrophic growth of *Chloroflexus* (Ward et al., 1989). In contrast to green sulfur bacteria, *Chloroflexus aurantiacus* is facultative and indeed grows well as a chemoorganotroph in the dark at full oxygen tension (Pierson, 1985; Zannoni, 1986; Wynn et al., 1987; Zannoni and Fuller, 1988).

In terms of plasma membrane arrangement, a clear distinction exists between purple and green bacteria and, among the same bacterial species, between light- or dark-grown cells (Oelze and Drews, 1972; Oelze and Drews, 1981; Drews, 1985,1986)

Abbreviations: Cf. – Chloroflexus; Cm. – Chromatium; DAD – 2,6-Diaminodurene; deh – dehydrogenase; DMSO – Dimethylsulphoxide; HQNO – 2-*n*-heptyl-4-hydroxyquinoline-N-oxide; MQ – Menaquinone; *Rb. – Rhodobacter; Rp. – Rhodopseudomonas;* RQ – Rhodoquinone; *Rs. – Rhodospirillum;* TMAO – Trimethylamine-N-oxide; UHDBT – Undecylhydroxydioxobenzothiazole (see also Chapters 12 and 13 by Drews and Golecki and Oelze and Golecki, respectively). Indeed, under phototrophic growth conditions, all members of the Chromatiaceae and Rhodospirillaceae develop an extensive intracytoplasmic membrane system (ICM) which forms species-specific vesicular, tubular or lamellar structures, often connected to each other and/or to the cytoplasmic membrane (CM) (Oelze and Drews, 1972). These lipoprotein membranes contain all the components involved in electron transport and energy transduction. The formation of the ICM system is strongly inhibited by increasing the light intensity (and/or the oxygen partial pressure) and this facilitates spheroplast disruption by osmotic shock. As will be reported later in this chapter, most of the available literature concerning electron transport and energy transduction in facultative phototrophs is based on results obtained with isolated membrane vesicles. For an extensive coverage of the plasma membrane formation and modulation see the excellent reviews by Drews and Oelze, 1981; Oelze and Drews, 1981; Drews 1985, 1986; Kiley and Kaplan, 1988; and Chapters 12 and 13 by Drews and Golecki and Oelze and Golecki, respectively.

II. Dark Aerobic Electron Transport Chains

A. Redox Complexes and Electron Transport Components

The respiratory systems of facultative phototrophs have greater diversity of electron transfer pathways than mitochondrial respiratory chains, and have exploited unique terminal oxidases, depending on their natural habitats and modes of aerobic growth. In most species there is more than one terminal oxidase, so that the respiratory chain is branched, generally, at the dehydrogenase(s) donor sites. The respiratory electron-transport components include quinones (ubi-, rhodo- and mena-quinones), iron-sulfur proteins, and *a*, *b*, and *c*-type cytochromes (Bartsch, 1978; Baccarini Melandri and Zannoni, 1978; Cramer and Crofts, 1982; Amesz and Knaff, 1988; Anraku, 1988; Jackson, 1988; Prince, 1990; Zannoni and Daldal, 1993). Most of these redox carriers are arranged as membrane-bound, intrinsic complexes. However, some components such as cytochrome c_2 are soluble and thought to diffuse along the membrane surface (van Grondelle, 1976,1977) (see however Chapter 34 by Meyer and Donohue). Quinone molecules are always present as a pool in excess over the other secondary electron transport carriers and in some cases, more than one quinone isoprenolog, e.g. UQ-8 and UQ-10, and species, e.g. UQ and RQ, is likely to be involved in respiration (Ramirez-Ponce et al., 1980; Zannoni and Melandri, 1985). This section describes the present status of information on aerobic respiratory chains of purple and green nonsulfur bacteria, with special emphasis on terminal oxidases. NADH and succinic dehydrogenases are scarcely discussed because of limitation of data, whereas subjects such as H_2-uptake hydrogenases and ubiquinol/Cyt *c* oxidoreductases have been extensively reviewed in recent years by others (see also Chapter 14 by Vermeglio et al. and Chapter 35 by Gray and Daldal).

1. H_2-Uptake Hydrogenase

Several purple nonsulfur bacteria are able to grow chemo-lithotrophically in the dark. Under these growth conditions, H_2 serves as the source of energy and reducing power, O_2 as the terminal electron acceptor, and CO_2 as the sole carbon source (Madigan and Gest, 1979; Kelley et al., 1979; Seifert and Pfennig, 1979; Colbeau et al., 1980). In species such as *Rb. capsulatus,* the oxidation of hydrogen is catalyzed by a membrane-bound enzyme (Colbeau and Vignais, 1981) which protrudes on the cytoplasmic side of the membrane (Colbeau et al., 1983; Lissolo et al., 1993); conversely, the hydrogenases of *Thiocapsa roseopersicina* and *Chromatium* are less tightly bound to the membrane (Kovacs et al., 1983; van Heerikhuizen et al., 1981) and, in the case of *T. roseopersicina*, it is oriented towards the periplasmic compartment (Kovacs et al., 1983). The majority of hydrogenases are hetero-dimeric (subunits *ca.* 34 and 65 kDa) (Adams and Hall, 1979; Colbeau et al., 1983). Nickel is required for the synthesis and activity of the enzyme (Takakuwa and Wall, 1981; Colbeau and Vignais, 1983) and the amount of Fe and labile sulfur seen by spectroscopic techniques is consistent with the presence of a 4Fe-4S cluster (Strekas et al., 1980; Albracht et al., 1983). Among the photosynthetic bacteria, the regulation of hydrogenase synthesis has been studied only in *Rb. capsulatus* (Colbeau and Vignais, 1983); however, the factors involved in the regulation of the enzyme are still poorly understood (see Vignais et al., 1985).

In cells of *Rb. capsulatus,* H_2 is rapidly oxidized either in the light or in the dark (Meyer et al., 1978) with a K_m for H_2 (pH-dependent between 7.5 and 9) of 0.25 μM (Colbeau and Vignais, 1981). In *Rb. capsulatus,* where the respiratory chain is branched and comprises two terminal oxidases having different sensitivities to CN^- (Marrs and Gest, 1973; La Monica and Marrs, 1976; Zannoni et al., 1974, 1976a,b), the hydrogenase delivers electrons to the ubiquinone pool. Indeed, inhibition of H_2 oxidation by CN^- shows a biphasic pattern (Paul et al., 1979) and mutants of *Rb. capsulatus* defective in either one of the two terminal oxidases or of a functional bc_1 complex are capable of H_2 consumption and chemolithotrophic growth (Madigan and Gest, 1979; Melandri et al., 1982).

In membranes from *Rb. capsulatus,* the oxidation of H_2 is coupled to ATP synthesis and the P/O ratios are generally as high as for NADH oxidation (O.4–O.5) and twice those reported for succinate oxidation (O.2–O.3) (Paul et al., 1979; Melandri et al., 1983). Thus, the pathway from H_2 must involve at least one extra energy transducing site compared to the pathway from succinate; this is consistent with the recent observation that protons resulting from the oxidation of H_2 by methylene blue are released into the periplasmic compartment (scalar protons) in a mutant strain BCX of *Rb. capsulatus* which lacks the *hupC* gene product involved in the transfer of electrons from the hydrogenase to the rest of the chain (Cauvin et al., 1991; Lissolo et al., 1993).

2. NAD(P)H Oxidation and NAD(P)$^+$ Reduction

A common feature of both NADH oxidation and NAD^+ reduction in purple bacteria is the inhibition by inhibitors of mitochondrial NADH dehydrogenase (complex I) such as rotenone, amytal, and piericidin A (Keister and Yike, 1967; Gromet-Elhanan, 1969; Klemme, 1969; Baccarini-Melandri et al., 1973; Malkin et al., 1981; Zannoni and Ingledew, 1983a,b). This suggests that the bacterial and the mitochondrial enzymes are related but also that NADH oxidation and NAD^+ reduction are catalyzed by the same enzyme operating in opposite directions. Further evidence for similarities between the two enzymes in bacteria and mitochondria comes from the detection of iron-sulfur centers in bacterial membranes (Dutton and Leigh, 1973; Prince and Ingledew, 1977; Malkin et al., 1981; Zannoni and Ingledew, 1983a,b; Sled' et al., 1993) with EPR spectra and Em values similar to most of the iron-sulfur centers of mitochondrial complex I (NADH-ubiquinone reductase) (Weiss and Friedrich, 1991; Friedrich et al., 1993).

The present state of knowledge of the NADH-Q oxidoreductases in bacterial respiratory systems is suggestive of two types of NADH-dehydrogenases (Yagi, 1991; Yagi, 1993). One type includes those enzymes (designated NDH-1) that bear an energy-transducing site whereas the enzymes of the second type (designated NDH-2) are not linked to proton translocation. The existence of NDH-1 has been reported in several purple bacteria. These include *Rs. rubrum* (Keister and Yike, 1967; Jones and Vernon, 1969), *Rb. capsulatus* (Klemme, 1969; Baccarini-Melandri et al., 1973; Zannoni and Ingledew, 1983a,b), *Rp. viridis* (Jones and Saunders, 1972) and *Cm. vinosum* (Malkin et al., 1981). Indeed, as the photochemical primary quinone acceptors (Q_A) in these bacteria have $Em_{7.0}$ values that range from 0 to −100 mV (Prince and Dutton, 1978; Knaff, 1978) reduction of NAD^+ ($Em_{7.0}$ of NADH/NAD^+ = −320 mV) could not occur without additional input of energy. The mechanism of such energy-dependent reversal of the electron flow requires therefore the use of an electric and proton potential difference across the membrane ($\Delta\tilde{\mu}_{H^+}$) to 'pump' electrons uphill from Q_A to NAD^+. It is apparent that this type of NAD^+ reduction involves the presence of a membrane-bound NADH-quinone oxidoreductase driving reducing equivalents to NAD^+. Thus, although the direct detection of a energy coupling linked to NADH-Q oxidoreduction has been reported only in a very few species of purple bacteria, e.g. *Rb. capsulatus* (Baccarini-Melandri et al., 1973), the presence of NDH-1 can be indirectly deduced from the above reported considerations.

In membranes from light-grown cells of *Rb. capsulatus,* 4 ferredoxin-like centers (g_y = 1.94, $Em_{7.0}$ of +120, −115, −280 and −370 mV) have been resolved by EPR spectroscopy (Zannoni and Ingledew, 1983b). In contrast, *Rb. capsulatus* M-1, a mutant deficient in both NADH dehydrogenase (Marrs and Gest, 1973) and energy-dependent NAD^+ reduction (Melandri et al., 1983), contain only those centers with $Em_{7.0}$ of +120 and −280 mV (Zannoni and Ingledew, 1983a). Thus, the inability of membranes from strain M-1 to perform $NADH^-$ dependent respiration is clearly related to the lack of centers with $Em_{7.0}$ at −370 and −115 mV (Zannoni and Ingledew, 1983a). EPR spectra of *Rb. sphaeroides* chromatophores demonstrated the existence of iron-sulfur clusters with spectral and thermodynamic features close to those of the mitochondrial centers N1a, N1b, N2, N3, N4, and, probably, N5 (Sled' et al., 1993). An NADH-1 preparation from *Rb. sphaeroides* strain GA cells has been reported recently (Sled' et al., 1993). EPR analysis has revealed that this NADH-1 preparation contains relatively intact clusters N1, N3 and N4, while cluster N2 is considerably altered.

Recently, a cluster of 14 genes (designated as *nuo* genes) coding for the subunits of the NADH-Q oxidoreductase of *Rb. capsulatus* has been cloned (Dupuis, 1992). These genes are gathered in a cluster similar to the NQO/*nuo* genes identified in *Paracoccus denitrificans* (Yagi, 1993) and in *E. coli* (Friedrich et al., 1993). The order of these genes is identical in the three species although some open reading frames unrelated to complex I of mitochondria and of *P. denitrificans* have been identified in *Rb. capsulatus* (Dupuis et al., 1993).

Enzymes designated as NDH-1, oxidize not only NADH but also deamino-NADH (Yagi, 1991; Yagi, 1993; Sled' et al., 1993). Conversely, NADPH does not act as a substrate because the NADPH oxidation rate is normally less than 2% of that using NADH. However, the rate of NADPH oxidation by aerobic membranes from *Rb. capsulatus* (MT1131 strain) is approximately 15–20% of that with NADH (D. Zannoni, unpublished).

To our knowledge, there are very few reports on the presence of NDH-2 enzymes in facultative phototrophs (Sled' et al., 1993), although it is clear that dehydrogenases of this type are widely distributed in bacteria (Yagi, 1991, 1993). In general, the membrane oxidation of deamino-NADH is taken as a selective assay for NDH-1 activity in the presence of NDH-2, the latter being insensitive to rotenone, capsaicin or DCCD (Yagi, 1991). To date, very few inhibitors of the NDH-2 type enzymes have been reported, i.e. flavone and platanetin. These compounds, when used at micromolar concentrations, act as both uncoupling agents and inhibitors for the external and internal rotenone-insensitive NAD(P)H dehydrogenases of yeast and plant mitochondria (Cook and Cammak, 1984; de Vries and Grivell, 1988; Rugolo and Zannoni, 1992). Notably, oxidation of NADH by membranes of *Rb. capsulatus* is strongly affected by the NDH-2 type inhibitor, platanetin (D. Zannoni, unpublished). Furthermore, the NADH-Q oxidoreductase activity of semi-aerobically light-grown *Rb. capsulatus* Kb1, is not linked to energy transduction (see also II.4 and III.5) (Zannoni et al., 1978; Melandri et al., 1983b). These results, along with evidence that the K_m^{app} value for UQ_1 of NADH-UQ_1 reductase activity varies from 8–10 μM to 30–40 μM when cells are shifted from dark-aerobic growth to semiaerobic light-growth, indicate the presence in *Rb. capsulatus* of both NDH-1 and NDH-2 enzymes. However, it must be noticed that the external and the internal rotenone-insensitive NAD(P)H dehydrogenases from plant mitochondria differ from bacterial NDH-2 dehydrogenases (Yagi, 1991,1993) in several ways: (1) the internal one, and probably also the external, is smaller; (2) they are loosely membrane-bound and are therefore extrinsic proteins; (3) they both oxidize NAD(P)H; (4) Ca^{2+} regulates their activity, at least in situ (Møller et al., 1993).

3. Succinic Dehydrogenase

The enzyme responsible for oxidation of succinate has been characterized in plasma membranes from several species of purple bacteria (Hatefi et al., 1972; Carithers et al., 1977; Prince and Ingledew, 1977; Zannoni and Ingledew, 1983a,b). This enzyme, like mitochondrial succinic dehydrogenase, contain two ferredoxin-type iron-sulfur centers with EPR signals at g = 1.93 to 1.94 (reduced form) plus one iron-sulfur center (HiPIP) that is paramagnetic in the oxidized form, having an EPR signal at g = 2.01 to 2.02. This latter center, which is reduced by succinate under steady-state respiration in the presence of TTFA (Zannoni and Ingledew, 1983b), is likely to be analogous to center S 3 of the mitochondrial complex (Ohnishi and Salerno, 1982). The S-3 center has $Em_{7.0}$ = +50 mV in *Cm. vinosum* (Malkin et al., 1981), +60 mV in *Cf. aurantiacus* (Zannoni and Ingledew, 1985), +60 mV in *Rb. capsulatus* (Zannoni and Ingledew, 1983b), and $Em_{7.0}$ = +80 mV in *Rb. sphaeroides* (Prince and Ingledew, 1977). The other two ferredoxin-type centers have $Em_{7.0}$ values of +120 and –280 mV in *Rb. capsulatus* (Zannoni and Ingledew, 1983b) and $Em_{7.0}$ = +90 mV and –200 mV in *Rb. sphaeroides* (Prince and Ingledew, 1977). Like the mitochondrial S-1 centers, only the more electropositive bacterial centers, i.e. +120 and +90 mV, are reducible by succinate. These similarities support the close correspondence between mitochondrial and bacterial succinic dehydrogenases.

4. Quinol/Cyt c Oxidoreductases and Quinol Oxidases

All photosynthetic bacteria examined to date have

been shown to contain a multiprotein complex which is functionally and structurally similar to the mitochondrial bc_1 complex (complex III) or ubiquinol/Cyt *c* oxidoreductase. Notably, the possibility of driving cyclic electron flow of photosynthetic bacteria by a single photochemical event, led to the result that most of the data concerning the mechanism and the molecular nature of the bc_1 complex of phototrophs have been obtained by examining phototrophically grown cells. However, considering that kinetic, structural and genetic approaches indicate that the bc_1 complex of aerobically grown cells is the same as the complex in light-grown grown cells, makes it likely that data available about the 'photosynthetic' bc_1 complex are also valid for the 'aerobic' quinol/Cyt *c* oxidoreductase (Jones and Plewis, 1974; Takamiya et al., 1982; Zannoni, 1982; Wynn et al., 1986; Davidson and Daldal, 1987; Daldal, 1988a; Donohue et al., 1988; Daldal et al., 1989; Lavorel et al., 1989; Venturoli et al., 1990; Zannoni and Moore, 1990). Thus, because of limitation of space, the 'aerobic' bc_1 complex will not be discussed here. For this subject the readers should refer to comprehensive reviews that have appeared in recent years (Dutton, 1986; Hauska, 1986; Amesz and Knaff, 1988; Daldal, 1988a; Jackson, 1988; Prince, 1990; Gennis et al., 1993) and also to Chapters 14 and 35 by Vermeglio et al. and Gray and Daldal, respectively, of this book.

The role of quinone molecules (RQ, UQ and MQ) as physiological electron donors to both the bc_1 complex and the cyanide-resistant oxidase of aerobically dark-grown facultative phototrophs, although largely established (Zannoni and Melandri, 1985), has been examined in some detail only in *Rp. palustris* (King and Drews, 1973) and *Rb. capsulatus* (Zannoni et al., 1976b; Zannoni and Moore, 1990). The recent use of the voltametric technique to detect the level of reduction of the quinone-pool in vivo (Moore et al., 1988; Dry et al., 1989; Moore et al., 1990) allowed the rapid and continuous monitoring of the Q redox state in aerobic membranes of *Rb. capsulatus* (Zannoni and Moore, 1990). The results obtained indicated that in the absence of inhibitors (Fig. 1), the relationship between electron flow and Q redox state approximates linearity over the full range of Q reduction level (from 4–5% to 30%). Conversely, the relationship becomes nonlinear when the cytochrome *c* oxidase dependent flow is blocked by CN^- and/or AA (Fig. 1) and the Q reduction level at which a significant rate of electron flux is seen, is shifted from 4–5% to 25%. This means that electron flow through the quinol oxidase pathway of *Rb. capsulatus* is strongly limited until the Q-pool reduction level reaches 25%. These results suggested an apparent K_m of QH_2 at the Q_O site of the bc_1 complex of 2.4–3 Q per RC and a K_m at the 'quinol oxidase pathway' site (Q_{OX}) of 15 Q per RC. Notably, in photosynthetic membranes of *Rb. sphaeroides* the apparent K_m of QH_2 at the Q_{OX}-site was calculated to be approximately 3.5 Q per RC (Venturoli et al., 1986). In plant mitochondria, the non-linearity of the relationship between the rate of electron flow catalyzed by the CN^--resistant pathway and the reduction level of the Q-pool, has been suggested to limit the extent of 'energetically wasteful' respiration (Dry et al., 1989). This proposal is not applicable to *Rb. capsulatus* because observations made with the respiratory mutant R126, lacking a functional bc_1 complex (Robertson et al., 1986), indicate that the quinol oxidase is linked to energy transduction (Zannoni, 1982). In this respect it is noteworthy that light-driven oxygen uptake (noncyclic electron flow) by membranes of *Rb. capsulatus* requires the quinol oxidase pathway (Zannoni et al., 1978; Zannoni et al., 1986). Results of Fig. 2 indicate that a transmembrane ΔpH is formed by membranes of *Rb. capsulatus* R126 as a result of quinol oxidase activity (probably a *bo*-type oxidase, see below). In proteoliposomes containing the Cyt *o* complex of *E. coli*, quinol oxidation is accompanied by H^+ release on the external surface and consumption on the internal surface with a ratio of $1H^+/e^-$ (Matsushita et al., 1984). However, oxidation of reduced TMPD resulted in little change in external pH, suggesting that Cyt *o* in *E. coli* generates a pH gradient from scalar (non-vectorial) proton-consuming and releasing reactions on opposite sides of the membrane (Hamamoto et al., 1985). In contrast, studies in spheroplasts yielded a ratio of $2H^+/e^-$ suggesting an additional vectorial mechanism similar to that observed in the aa_3-type cytochrome *c* oxidases (Puustinen et al., 1991; Larsen et al., 1992). Unfortunately, the mechanism of ΔpH formation and the contribution of the quinol oxidase to the overall efficiency of the respiratory chain of *Rb. capsulatus,* is not known. In this respect, Richaud et al. (1986) have demonstrated that restriction of the respiratory flux by light occurs at the NADH dehydrogenase level (NDH-1). This means that under aerobic conditions in the light the quinol oxidase would be

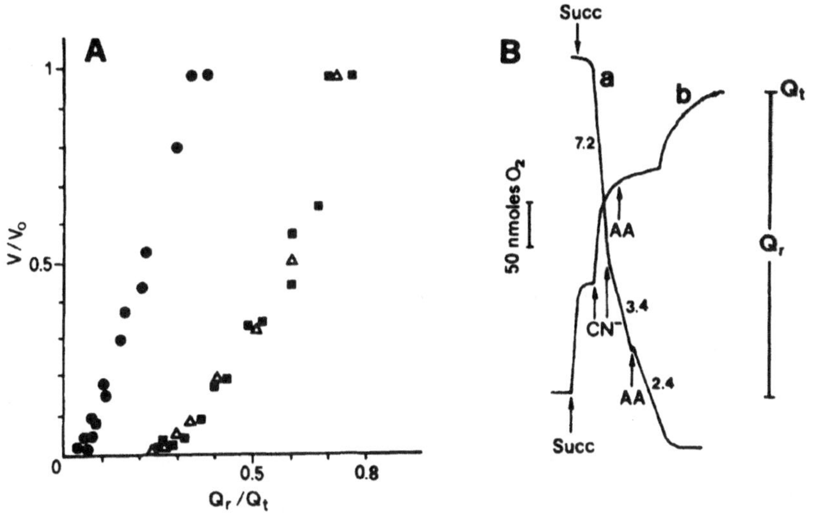

Fig. 1. In panel A, the dependence of respiratory flux on quinone redox state in membranes from *Rb. capsulatus* MT1131 either in the absence (●) or in the presence of 50 μM CN (■) and 50 μM CN⁻ plus 5 μM AA (△), is shown. The data are presented as the ratio v/V_0 (where v is the initial rate of respiration in the presence of inhibitor of electron input and V_0 is the uninhibited rate) plotted vs. the proportion of Q in the reduced state (Q_r/Q_t). In panel B, the traces of the simultaneous measurement of O_2 consumption and steady-state reduction level of Q as determined by the voltametric technique (see Zannoni and Moore, 1990), are shown. Non standard abbreviations: AA, antimycin A, CN^-, potassium cyanide; Succ, succinate. The results are from Zannoni and Moore (1990).

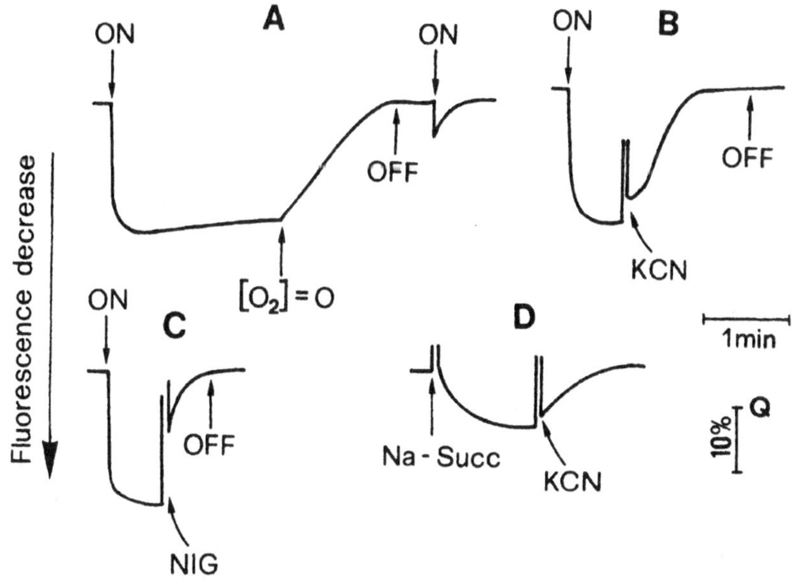

Fig. 2. Formation of an electrochemical potential of protons coupled to O_2 consumption through the quinol-oxidase pathway of *Rb. capsulatus* R126 (Q_0^- mutant), as monitored by the quenching of the atebrine fluorescence. Traces in A, B and C indicate the atebrine fluorescence quenching induced by light-dependent O_2 uptake with ascorbate (2 mM)/TMPD (0.1 mM) as donors (see text for details). In D, the transmembrane ΔpH is induced by succinate oxidation. Additions and abbreviations: Na-Succ, sodium succinate (3 mM); KCN, cyanide (5 mM); Nig, nigericin (2 μg/ml); ON,OFF, light on and off. Adapted from Zannoni (1982).

poorly engaged because of a restriction of electron input (Q_r/Q_t lower than 25%, see Fig. 1).

Quinol oxidases have been reported in several aerobically grown photosynthetic bacteria; however, their actual molecular nature is far from clear (Zannoni and Baccarini-Melandri, 1980). In general, the quinol oxidase activity is relatively insensitive to CN^- (≥ 2–3 mM) but inhibited by CO; these features have been considered indicative of the presence of cytochromes of the *o*-type in the catalytic activity of the enzyme (Zannoni et al., 1976a,b; Venturoli et al., 1987; Schrattenholz et al., 1989; Varela and Ramirez, 1990). Other inhibitors of the quinol oxidase pathway are HQNO, UHDBT and mefloquine, since they act, most likely, at the quinol oxidizing site (Zannoni, 1985; Anraku and Gennis, 1987; Venturoli et al., 1987; Schrattenholz et al., 1989). A cytochrome-*o* complex, claimed to be a quinol oxidase, has been isolated from phototrophically grown *Rhodospirillum rubrum* (Schrattenholz et al., 1989). The complex, composed of 4 subunits (MW of 66, 39, 20 and 11 kDa), contains 2 types of cytochromes (αmax at 553 and 561 nm), one of them a cytochrome *o*, and 2 Cu atoms. Recent evidence indicated that heme *o* is a modification of heme *b*, with a hydroxyfarnesyl side-chain addition (Puustinen and Wikstrom, 1991). The heme *b* of the cytochrome *bo* complex of *E. coli* is present as the low-spin six-coordinate Cyt *b*-562 component responsible for the visible absorption of the oxidase and has a split α band at 555 and 563.5 nm (Puustinen et al., 1991; Minghetti et al., 1992). Possibly, the low-spin Cyt *b*-562 corresponds to the *b*-type heme titrated at +250 mV in native membranes of *Rs. rubrum* (Venturoli et al., 1987), *Rb. sphaeroides* (Saunders and Jones, 1975) and *Rb. capsulatus* (Zannoni et al., 1976b; Zannoni et al., 1992). In this respect it might be important to note that most members of the heme-copper oxidase superfamily appear to have in common three core subunits that correspond to the three mitochondrially encoded subunits of the aa_3-type oxidase (Hosler et al., 1993). The entire operon of the *bo*-type quinol oxidase of *E. coli* contains five genes corresponding to the five subunits of the purified oxidase (Chepuri et al., 1990). The *bo*-type oxidase does not contain Cu_A which is consistent with the fact that the amino acid residues that are probable ligands of Cu_A in the cytochrome *c* oxidase are not conserved in the sequence of subunit II of the quinol oxidase (van der Oost et al., 1992). Unfortunately there are no data about the possible homology of the quinol oxidizing sites of the cytochrome *b* subunit of the bc_1 complex and that of subunit I of the *bo* complex (Hosler et al., 1993; Gennis et al., 1993).

5. Cytochromes of c *Type and Cyt* c *Oxidases*

Membranes isolated from aerobically grown cells of *Rp. viridis*, *Cm. vinosum* and *Cf. aurantiacus* lack a photochemical reaction center (RC) and the multiheme Cyt *c* that operates as immediate donor to the RC in light-grown cells (Wynn et al., 1985,1987; Kämpf et al., 1987; Zannoni and Venturoli, 1988; Zannoni and Fuller, 1988) (see also Chapters 14 and 35 by Vermeglio et al. and Gray and Daldal, respectively). Although lacking a tetrahaem Cyt *c*-554, aerobic-membranes from *Cf. aurantiacus* contain several membrane bound *c*-type cytochromes with $Em_{7.0}$ of +270, +130 and 0 mV (Zannoni, 1986; Wynn et al., 1987; Zannoni and Fuller, 1988). Inhibitor studies suggested that both Cyts c_{130} (splitted αmax at 550/553 nm) and c_{270} (αmax at 549 nm) are involved in respiratory electron flow through the Cyt *c* oxidase containing pathway whereas Cyt c_0 is not reducible by NADH and succinate (Zannoni and Fuller, 1988). In membranes from cells grown under oxygen saturated conditions no EPR signal characteristic of a reduced Rieske iron sulfur center could be detected (Wynn et al., 1987); conversely, phototrophically grown *Cf. aurantiacus* contains a Rieske-type center with an unusually low $Em_{7.0}$ (+100 mV) (Zannoni and Ingledew, 1985). Although the isolation of the bc_1 complex of *Cf. aurantiacus* has not yet been reported, thermodynamic considerations (Amesz and Knaff, 1988) suggest that Cyt c_{130} of this organism might be homologous to Cyt c_1 of the bc_1 complex, whereas Cyt c_{270} might be involved in Cyt *c* oxidase activity (see model scheme in Fig. 3). Perhaps surprisingly, *Cf. aurantiacus* seems to lack soluble *c*-type cytochromes (Bartsch, 1978). This unusual membrane arrangement has been suggested to be characteristic of Gram negative thermophiles (Wood, 1983). However the mesophilic facultative phototrophs, *Rhodoferax fermentans* and *Cm. vinosum*, seem to lack soluble *c*-type cytochrome(s) when grown aerobically in the dark (Kämpf et al., 1987; Hochkoeppler et al., 1993). Notably, the soluble high-potential iron-sulfur protein (HiPIP) isolated from *Rhodoferax fermentans* has recently been demonstrated to be competent in photosynthetic electron transfer (Hochkoeppler et al., 1995a). Coversely, light-grown *Cf. aurantiacus* contain a

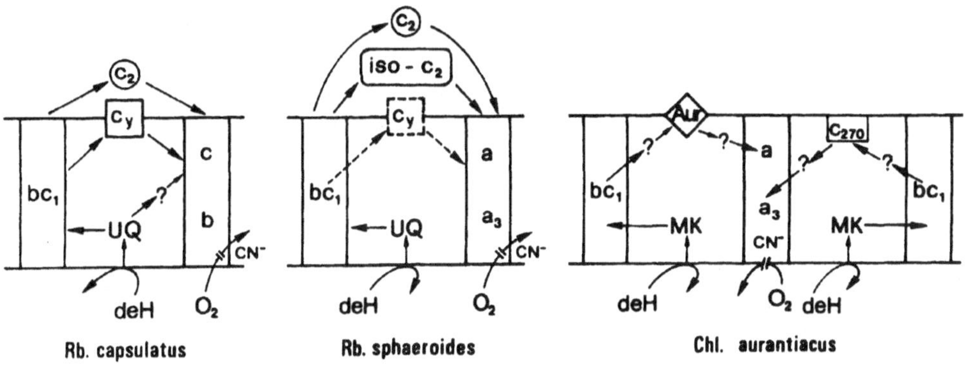

Fig. 3. Models for respiratory electron transport between a non specific dehydrogenase(s) (deH) and the Cyt *c* oxidases from 2 representative members of the 'purple' nonsulfur bacteria (*Rb. sphaeroides, Rb. capsulatus*) which contain the periplasmic soluble Cyt c_2, and the 'green' nonsulfur bacterium *Cf. aurantiacus* which does not. As discussed in the text, our knowledge of the respiratory systems in *Cf. aurantiacus* is far less complete than those of *Rb. capsulatus* and *Rb. sphaeroides*. The second Cyt *c* oxidase (*cb*-type) of *Rb. sphaeroides* is not shown. Solid lines symbolize components and pathways present in wild type cells, and dashed lines are those that require either a mutation (iso-Cyt c_2) or trans-complementation (Cyt c_y of *Rb. sphaeroides*) for function. The Cyt *c* oxidases are tentatively arranged as to transmembrane redox complexes containing aa_3 and *cb*-type hemes. Dotted lines, interrupted by question marks, suggest hypothetical intermolecular redox exchanges. Abbreviations: bc_1, the bc_1 complex; UQ, ubiquinone pool; MQ, menaquinone pool; Aur, auracyanin; c_2, iso-c_2 and c_y are the cytochromes c_2, iso-c_2, and c_y. See text for details.

small copper protein (MW = 12.8 kDa), auracyanin, with thermodynamic properties (E_m = +240 mV) compatible with a 'c_2-like' electron transport function (Trost et al., 1988). However, auracyanin behaves as a peripheral membrane protein that can be released by membrane fragments only by salt washing or detergent (LDAO) treatment (Trost et al., 1988). Auracyanin is therefore arranged in a way which appears as intermediate to the two extremes of membrane/bound and periplasmic/soluble *c*-type cytochromes (see below, c_y and c_2 Cyts) (see Fig. 3).

Considerable evidence has been obtained for the presence of a membrane bound cytochrome oxidase of the aa_3-type in chemotrophic cultures of *Cf. aurantiacus* in which the O_2 tension during growth was maintained below the saturation level (termed 'less-aerated cells' in Zannoni and Fuller, 1988; Zannoni, 1986). The two haems ($Em_{7.0}$ of +190 and +330 mV) are present in a 1: 1 ratio and they show spectroscopic properties described for bacterial and mitochondrial oxidases of the aa_3-type (Poole, 1983) (see also Table 1). This conclusion is further supported by evidence that a polyclonal antibody raised against the Cyt aa_3 of beef-heart mitochondria cross-reacts with a 31 kDa subunit of a Cyt aa_3-enriched fraction (solubilization with Triton X-100) from membranes of aerobically dark-grown *Cf. aurantiacus* (R. Bisson and G. Schiavo, personal communication). Aerobic membranes of *Cf. aurantiacus* can rapidly oxidize ascorbate-TMPD, [$K_{i\,(CN^{-}50\%)}$ = 8 μM], a redox couple which is thought to interact with cytochromes of the *c*-type. Unfortunately, the actual physiological donor to the oxidase is presently unknown (see however the tentative model in Fig. 3). Membranes isolated from cells of *Cf. aurantiacus* chemotrophically grown under oxygen-saturated conditions are endowed with a Cyt *c* oxidase having properties of a-type bacterial cytochromes, classified as a_1 (Wynn et al., 1987). This indicates that oxygen tension strongly affects the terminal oxidase phenotype, this phenomenon being largely present in other bacterial species such as *Escherichia coli* (Haddock and Jones, 1977) and *Rb. sphaeroides* (Sasaki et al., 1970) (see also below).

Aerobically grown cells of *Cm. vinosum* contain, similarly to *Cf. aurantiacus*, three membrane bound *c*-type cytochromes; however, in contrast to *Cf. aurantiacus*, they all have the α band maxima at 551–552 nm and $Em_{7.0}$ values shifted to more positive potentials (+340, +180 and +40 mV) (Wynn et al., 1985). In addition, a soluble *c*-type haem (c-550, α band at 550 nm) is present (Knaff et al., 1980). Although the role of the membrane bound cytochromes is presently obscure, it has been speculated that Cyt c_{180} might be related to Cyt c_1; this suggestion, however, is unlikely because the estimated Em for

Table 1. Some structural, spectroscopic and thermodynamic properties of Cyt *c* oxidases in a few representative species of facultative phototrophs

Bacterial species	Heme type	α-band (reduced, nm)	Subunits MW (kDa)	Heme-$Em_{7.0}$ (mV)	Ref
Rb. sphaeroides	aa_3	606	62, 33, 30	(a_{200}, a_{375})	1–3
	cb	552, 561	23, 16, 6	$c_{n.d.}, (b_{153})$	4–6
Cf. aurantiacus	(aa_3)	602	(31)*	(a_{190}, a_{330})	7
	(a_1)	n.d.	n.d.	≥ +250	8
Rb. capsulatus	*ccb*	550, 560	33, 29, 44	$c_{265}, c_{320}, b_{415}$	9
Rp. palustris	*co*	550, 560	20.5, 25.5, 12, 9.5	n.d.	10

Data between brackets are from intact membranes. * Detected by immunoblotting (see text for details); n.d., not detected.
References: 1. Hosler et al., 1992; 2. Gennis et al., 1982; 3. Saunders and Jones, 1974; 4. Sasaki et al., 1970; 5. Hosler et al., 1993; 6. Takamyia and Tanaka, 1983; 7. Zannoni, 1986; 8. Wynn et al., 1987; 9. Gray et al., 1994; 10. King and Drews, 1976.

the Rieske type iron sulfur center of *Cm. vinosum* is approximately +245 mV (Gaul and Knaff, 1983). A possible involvement of Cyt c_{340} in the Cyt *c* oxidase activity of *Cm. vinosum* has also been suggested (Wynn et al., 1985). The question of whether the soluble *c*-550 (Em = +240 mV) is involved in oxidative electron transport is still not solved. This soluble and periplasmic *c*-type species (Knaff et al., 1980) (previously referred to as *c*-551 by van Grondelle et al., 1977) has a MW of 15 kDa and is related to the family of *c*-type haems that includes the mitochondrial Cyt *c* and Cyt c_2 (Gray et al., 1983). Notably, a similar c_2-like cytochrome is also present in *Rp. viridis* and is thought to be the electron donor to the tetraheme subunit of the RC (Shill and Wood, 1984). However, while Cyt *c*-550 is present in aerobically grown *Cm. vinosum* (Wynn et al., 1985), no analogous soluble *c*-type cytochrome could be detected in chemotrophically grown *Rp. viridis* (Kämpf et al., 1987). A soluble *c*-type cytochrome (α max at 552 nm, Em = +287 mV) might be involved in photocyclic electron flow of *Rf. fermentans* but not in respiration of this facultative phototroph (Hochkoeppler et al., 1993).

Membranes isolated from chemotrophically grown *Rp. viridis* contain at least two putative terminal oxidases of the *d*- and *o*-type (Kämpf et al., 1987). How these two oxidases interact with the rest of the respiratory chain is unknown (Kämpf et al., 1987). Equally uncertain is the nature of the cytochrome oxidase system of aerobically grown *Cm. vinosum*, although the lack of heme *a* and *d* suggest that the oxidase(s) is neither a Cyt aa_3 type nor a Cyt *d* type (Wynn et al., 1985).

The role of the periplasmic Cyt c_2 (13090 MW, $Em_{7.0}$ = +350 mV) in carrying electrons between the bc_1 complex and the Cyt *c* oxidase complex (see below) during respiration, and between the bc_1 complex and the oxidized RC during photosynthesis, has long been recognized (Prince et al., 1975; Baccarini-Melandri and Zannoni, 1978; Michels and Haddock, 1980; Zannoni et al., 1980; Hudig et al., 1986; Prince, 1990; Gooley et al., 1990; Caffrey et al., 1992). In recent years, genetic approaches have better defined the role of Cyt c_2 in the two closely related species *Rb. capsulatus* and *Rb. sphaeroides*, and revealed that different classes of alternative electron carriers can operate between the membrane-bound redox complexes instead of, or along with, the periplasmic Cyt c_2 of these facultative phototrophs (Daldal et al., 1986; Prince et al., 1986; Donohue et al., 1986; Prince and Daldal, 1987) (see Fig. 3). Extensive analyses of *Rb. sphaeroides* and *Rb. capsulatus* Cyt c_2^- mutants for their ability to grow phototrophically have also revealed that although these purple nonsulfur species are closely related to each other, their Cyt c_2^--derivatives had different photosynthetic growth phenotypes. Indeed, a *Rb. sphaeroides* Cyt c_2^- mutant is unable to grow by photosynthesis (Donohue et al., 1986) whereas its *Rb. capsulatus* counterpart is capable of photosynthetic growth (Daldal et al., 1986). Notably, *Rb. sphaeroides* Cyt c_2^- mutants can give rise to phototrophic competent pseudorevertants (Rott and Donohue, 1990). These pseudorevertants overproduced a new soluble periplasmic Cyt *c*, named iso-Cyt c_2, which shows a sequence similarity to Cyt c_2 (Fitch et al., 1989) but different biochemical characteristics (α band at 552 nm instead of 550 nm; $Em_{7.0}$ of +300 mV instead of 350 mV; MW of 15 kDa instead of 13.5 kDa). Genetic analysis revealed that the redox component permitting phototrophic growth of *Rb. capsulatus* Cyt c_2^- mutants is a membrane-associated *c*-type monoheme cytochrome encoded by a gene called *cycY* (Jenney and Daldal, 1993). This membrane bound Cyt *c*, called c_y, has a redox potential of +350 mV (see Zannoni and Daldal, 1993) and it is likely to be homologous to Cyt c_x (Em = +354 mV) detected by spectroscopic methods in chromatophores of a Cyt c_2^-, c_1^- double mutant (MT-GS18) of *Rb. capsulatus* (Jones et al., 1990; Zannoni et al., 1992). Notably, a functional homolog Cyt c is not present in *Rb. sphaeroides* but it can be introduced into this species to complement the photosynthetic growth of a mutant lacking Cyt c_2 (Jenney and Daldal, 1993).

Similarly to photosynthesis, respiratory electron transport through the 'branch' involving the bc_1 complex and the cytochrome oxidase of *Rb. capsulatus* does not depend on the presence of soluble periplasmic Cyt c_2 (Daldal, 1988b) (Table 2). Indeed, the respiratory growth of *Rb. capsulatus* mutants (M6G-G4/S4) lacking both Cyt c_2 and a functional quinol oxidase (Qox^-) (Marrs and Gest, 1973; La Monica and Marrs, 1976; Zannoni et al., 1976a,b) is not arrested, and this growth was sensitive to inhibitors of the bc_1 complex and of the Cyt *c* oxidase (Daldal, 1988b). Recently, the kinetic competence of Cyt c_y in respiratory electron flow has been demonstrated (Hochkoeppler et al., 1995b). Thus, Cyt c_y might substitute for Cyt c_2 in both photosynthesis and respiration. Furthermore, Cyt c_y could also be operative in other redox reactions (see III, and see Fig. 4).

The terminal oxidase systems of aerobically dark grown *Rb. capsulatus* and *Rb. sphaeroides* differ markedly. In the latter species aerobic growth leads to the development of a Cyt *c* oxidase similar to the mitochondrial aa_3 complex (Saunders and Jones, 1974; Gennis et al., 1982; Hosler et al., 1993) whereas *Rb. capsulatus* does not synthesize haems of the *a*-type (Zannoni and Baccarini-Melandri, 1980). However, in membranes from cells of *Rb. sphaeroides* grown in the dark with low oxygen concentration or anaerobically in the light, a *cb*-type Cyt *c* oxidase is present (Wale and Jones, 1970; Sasaki et al., 1970; Hosler et al., 1992). This oxidase, which contains at least one *b*- and one *c*-type heme, was reported to be composed of at least 3 subunits (MW of 23,19 and 6 kDa) (Takamyia, 1983; Takamyia and Tanaka, 1983) and to be capable of oxidizing quite efficiently TMPD in a strain (JS100) from which aa_3 was genetically deleted (Hosler et al., 1992) (see Table 1). The interaction of the two Cyt *c* oxidases of *Rb. sphaeroides* with the rest of the respiratory chain is presently unknown.

Table 2. Respiratory growth rates of various *Rhodobacter capsulatus* wild type and mutant strains

Strain	Phenotype	Doubling time (min)
MT1131	wild type	122
MT-G4/S4	c_2^-	120
M6G	Qox_{260}^-	188
M7G	Cox_{410}^-	208
M6G-G4/S4	c_2^-, Qox_{260}^-	170
M7G-G4/S4	c_2^-, Cox_{410}^-	188
MT-CBC1	bc_1^-	146
MT-GS18	c_2^-, bc_1^-	148

Data from Daldal, 1988b.

The Cyt *c* oxidases of *Rb. capsulatus* and *Rs. rubrum* contain an unusual *b*-type cytochrome that is distinguished from classical cytochrome *o* by its inability to bind CO and by its high $Em_{7.0}$ (Zannoni et al., 1974; Venturoli et al., 1987); this is +410±5 mV in membranes (Zannoni et al., 1974) and +385 mV in a solubilized Cyt *c* oxidase complex from *Rb. capsulatus* (Hudig and Drews (1982a,b). This Cyt *b*-complex shows a DCCD-proton pumping activity of 0.9 H^+/e^- when incorporated into proteoliposomes (Hudig and Drews, 1984). A CO-binding cytochrome *c* (MW of 12600, $Em_{7.0}$ of +234 mV) was claimed to function as donor to the high potential *b*-type haem of the oxidase (Hudig and Drews, 1983). This hypothesis, however, was considered in contrast with the current accepted models of redox chains in which Cyt c_2 ($Em_{7.0}$ = +350 mV) lies on the pathway to the high potential oxidase (Wood, 1984) (see Fig. 3 and previous discussion on the role of Cyt c_2).

A Cyt *c* oxidase complex has recently been isolated from *Rb. capsulatus* MT-GS18 (c_2^-, c_1^-) (see also Table 2) (Gray et al., 1994). This complex contains 3 subunits (molecular masses of 44, 33 and 29 kDa in SDS-PAGE) with haem cofactors of *c*- and *b*-type (2 hemes *c*/2 hemes *b*). The purified complex oxidizes equine Cyt *c* and it is inhibited by CN^- (25 μM). Equilibrium redox titrations could be fitted into at least 3 components with $Em_{7.0}$ of +415 (Cyt *b*, 44 kDa), +320 (Cyt *c*, 29 kDa) and +265 mV (Cyt *c*, 33 kDa) so to suggest a *ccb* type oxidase. By comparing these results with those obtained in a similar preparation from *Rb. capsulatus* M7G-G4/S4 (c_2^-,

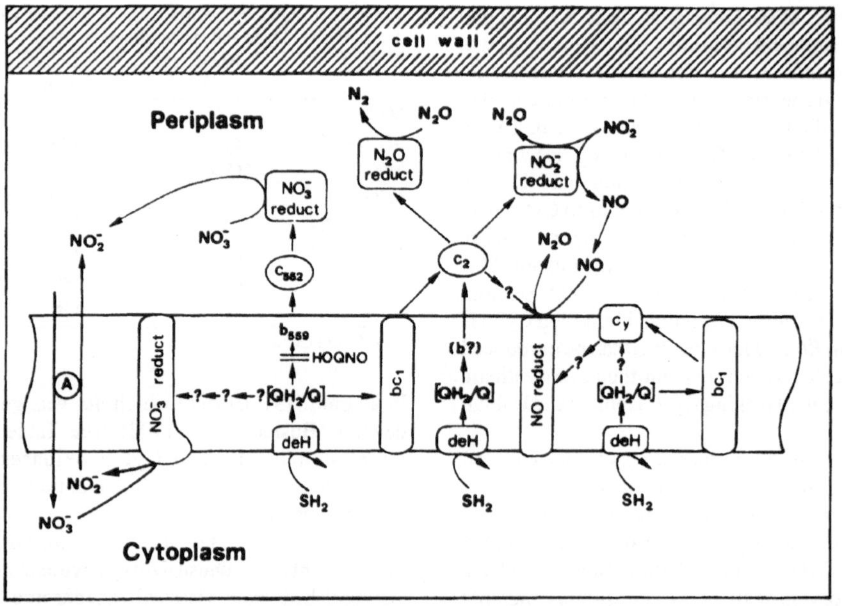

Fig. 4. The postulated organization of the electron transport pathways from ubiquinol to nitrate (NO_3^-), nitrite (NO_2^-), nitric oxide (NO) and nitrous oxide (N_2O) reductases in *Rb. capsulatus*. Two enzymes for NO_3^- reduction are shown, this conclusion based on results with different wild type strains (see text). Two routes for oxidation of ubiquinol and reduction of NO_2^-, NO and N_2O are shown (see text for results in *Rb. capsulatus* MT-CBC1, MT-G4/S4, MT-GS18 and H131 mutants). Whereas a role for Cyt c_2 in N_2O respiration is established (solid arrow) its participation in NO reduction is only by inference, hence the '?' signs (see text). Cyt c_y is suggested to mediate between the NO reductase and either the bc_1 complex or the ubiquinol bc_1-independent oxidation route. The lack of data about electron transfer from ubiquinol to the membrane bound NO_3^- reductase is symbolized by a dashed line, interrupted by question marks. A nitrate/nitrite antiporter (A) is shown tentatively. Non standard abbreviations: reduct, reductases; b_{559}, Cyt *b*-559; (b ?), undefined Cyt *b*; c_y, Cyt c_y; c_2, Cyt c_2; c_{552}, Cyt *c*-552; deH, undefined dehydrogenase; bc_1, Cyt bc_1 complex; SH_2, undefined electron donor; QH_2/Q, ubiquinol/ubiquinone pool.

Cyt c_{ox}^-) it was confirmed that the M7 mutant, previously characterized by Zannoni et al. (1974) as defective in Cyt b_{415}, lacks not only Cyt b_{415} but also another cytochrome of the *c*-type, most probably c_{270} (Gray et al., 1994). In this respect, it has recently been reported that mutations in the *fix*NOQP gene cluster of *Bradyrhizobium japonicum*, a symbiotic rhizobium, resulted in defective soybean root-nodule bacteroid development and nitrogen fixation (Presig et al., 1993). The predicted DNA-derived protein sequences suggested that the *fix*NOQP gene products are induced at low oxygen tension and constitute a heme/copper Cyt *c* oxidase which is composed of several *c*-type hemes and one heme *b* (possibly a *ccb* or a *cccb*-type oxidase). Preliminary data indicate that *fix*NOQP-homologous genes are also present in *Rb. capsulatus;* in addition, it has been reported that the *Cyc*M gene product in *Br. japonicum* is homologous to that of *cyc*Y (Cyt c_y) in *Rb. capsulatus* (Bott et al., 1991; Jenney and Daldal, 1993). It is evident how these recent acquisitions might shed a new light over the evolution of nitrogen-fixing organisms and also raise new questions about the structure and function of bacterial oxidases.

III. Dark Anaerobic Electron Transport Chains

A. Generalities

Growth of several facultative phototrophs in an anaerobic environment may proceed by fermentation or by anaerobic respiration (Uffen and Wolfe, 1979; Schultz and Weaver, 1982; McEwan et al., 1982; Richardson et al., 1986). These two modes of growth are not always wholly distinct; fermentation processes produce their own oxidants to balance NADH production and consumption, and in some cases these can be linked to the redox chain (Madigan and Gest, 1978; Yen and Marrs, 1977). Given an appropriate electron acceptor, the oxidation of fermentation products by the redox chain may also support ATP synthesis. This energy transduction, however, is generally inadequate to sustain a 'significant' bacterial growth so that the electron acceptor must be regarded as an 'accessory oxidant' which allows the dark anaerobic utilization of sugars (see Madigan and Gest, 1978; Yen and Marrs, 1977). Thus, these dark-anaerobic electron transport mechanisms are metabolically different from a true 'anaerobic respiratory process' because the latter implicates a 'consistent' cell growth driven by the oxidation of non fermentable substrates through the respiratory chain (see III.A.4). In summary, although different types of anaerobic electron transport are present in several species of purple non-sulfur bacteria, these respiratory processes are mainly advantageous metabolic systems facilitating anaerobic growth in the dark and in the light (Ferguson et al., 1987). A tentative model summarizing the present state of knowledge of the electron transport pathways from ubiquinol to nitrate, nitrite, nitric oxide and nitrous oxide reductases of *Rb. capsulatus* is depicted in Fig. 4.

1. Reduction of Fumarate

The available structural and functional information about the utilization of fumarate as a terminal oxidant for anaerobic respiration (fumarate reductase) concerns mainly the two anaerobes *Wolinella succinogenes* and *E. coli* (Ingledew and Poole, 1984; Hamilton, 1988). Although some scattered reports that refer to the presence of membrane-bound fumarate reductase activity in several species of Rhodospirillaceae can be found in the literature, the facultative phototrophs have received little attention with regard to their capability of fumarate reduction (Bose and Gest, 1962; Baccarini-Melandri et al., 1973; Ramirez-Ponce at al., 1980; Beatty and Gest, 1981; Hiraishi, 1988; Hiraishi et al., 1991).

In principle, under strict anaerobic conditions, fumarate can be used as a terminal electron acceptor and thus energy conserved via fumarate respiration; this hypothetical route would involve NADH dehydrogenase, quinone and the membrane bound fumarate reductase (Hamilton, 1988). However, the relatively high redox potential of the fumarate/succinate couple ($Em_{7.0}$ = +30 mV) precludes any significant energy conservation during reduction of fumarate by quinol and it also dictates that the donor is menaquinol (MQ, $Em_{7.0}$ = –74 mV) or rhodoquinol (RQ, $Em_{7.0}$ = –30 mV) rather than the less strongly reducing ubiquinol ($Em_{7.0}$ = +80 mV). Fumarate respiration is therefore restricted to MQ-containing species such as for example, *Rs. fulvum*, *Rs. molischianum*, *Rp. viridis*, *Rp. acidophila* and the *Rhodocyclus* group or RQ-containing species such as *Rs. rubrum*, *Rhodomicrobium vannielii* and *Rhodoferax fermentans* (Imhoff, 1984; Zannoni and Melandri, 1985; Hiraishi, 1988; Hiraishi et al., 1991). In this respect, it is interesting to notice that an early study on a RQ-deficient mutant of *Rs. rubrum* demonstrated that RQ was necessary for both photosynthetic growth and NADH-fumarate oxido-reductase activity (Giménez-Gallego et al., 1976; Ramìrez-Ponce et al., 1980). This suggests that RQ may be necessary in maintaining the redox balance during phototrophic growth of *Rs. rubrum* through the use of a membrane-bound fumarate reductase. A similar redox system might also be operative during fermentative growth of the purple nonsulfur bacterium *Rhodoferax fermentans* (Hiraishi, 1988). At present, however, the role of fumarate as 'auxiliary' oxidant requires a more accurate verification.

2. Reduction of Nitrate

Studies on the respiration of nitrate (NO_3^-, $Em_{7.0}{}^{NO_3^-/NO_2^-}$ = +420 mV) by phototrophically grown cells of *Rb. capsulatus* have identified two distinct types of NO_3^- reductase amongst different strains: (1) *Rb. capsulatus* strains AD2 and N22DNAR$^+$ contain a periplasmic soluble enzyme (McEwan et al., 1984; McEwan et al., 1987) while (2) *Rb. capsulatus* strain BK5 contains a membrane-bound enzyme which is functionally and structurally related to the enzyme found in *Paracoccus denitrificans* and *E. coli* (Ballard et al., 1990). In cells of *Rb. capsulatus* in which the periplasmic NO_3^- reductase is expressed, reduction of NO_3^- is coupled to generation of a membrane potential ($\Delta\psi$) (McEwan et al., 1984; McEwan et al., 1987). Reduction of NO_3^- cannot, however, support anaerobic dark growth of *Rb. capsulatus* on non-fermentable carbon sources, but it has been speculated that it can facilitate phototrophic growth of this organism in the presence of highly reduced carbon sources such as propionate or butyrate (Richardson et al., 1990). In contrast, several strains of *Rhodopseudomonas palustris* have been shown to grow anaerobically in the dark in the presence of nitrate (Klemme et al., 1980). Several lines of evidence indicate that the bc_1 complex is not involved in electron transfer to the periplasmic enzyme of *Rb. capsulatus* . The current model is that electrons destined for NO_3^- reduction involve (a) a soluble *c*-type cytochrome (c-552, $Em_{7.0}$ = +165 mV), that is released into the periplasm during cell fractionation and which co-fractionates with NO_3^- reductase, and (b) a membrane-bound *b*-type cytochrome (α band at 559 nm). An electron transport pathway that proceeds, $UQH_2 \rightarrow$ Cyt *b*-559$\rightarrow$ Cyt c-552$\rightarrow$ NO_3^- reductase has therefore been proposed (Richardson et al., 1990). Neither low concentrations of CN^- ($\leq 5\ \mu M$), nor similar amounts of HOQNO (shown to inhibit NO_3^- reduction) prevented the oxidation of the cytochromes by NO_3^-; this suggests that the site of inhibition is on the reducing side of Cyt b-559 and *c*-552 (Richardson et al., 1990). The periplasmic enzyme from *Rb. capsulatus* strain N22DNAR$^+$ has been purified as a single polypeptide chain of MW 90,000. It possesses the molybdenum cofactor (McEwan et al. 1987) that is characteristic of NO_3^- reductases in general, but differs from the membrane bound NO_3^- reductases of *E. coli* and *Paracoccus denitrificans* not only in terms of subunit composition (Craske and Ferguson, 1986) and cellular location but also in being relatively insensitive to azide and unable to use chlorate (ClO_3^-) as an alternative substrate (McEwan et al., 1984).

A soluble NO_3^- reductase with an associated *c*-type cytochrome has been isolated by Satoh (1981) from the periplasm of *Rb. sphaeroides* f. sp. *denitrificans* (Sawada and Satoh, 1980). Satoh (1981) suggested that the catalytic activity was due to a 60 kDa polypeptide, a value which is not consistent with that estimated by McEwan et al. (1987) for the analogous enzyme from *Rb. capsulatus*. Another major difference is that the *c*-type haem associated with the enzyme from *Rb. sphaeroides* had a higher $Em_{7.0}$ value (+251 mV) (Yokota et al., 1984) than that found for Cyt *c*-552 of *Rb. capsulatus* . Notably, in other studies (Michalski and Nicholas, 1985) a polypeptide with an apparent molecular mass of 75 kDa, which was induced by NO_3^-, was identified in soluble fractions. However, a later study by Byrne and Nicholas (1987) concluded that the NO_3^- reductase of *Rb. sphaeroides* f. sp. *denitrificans* was membrane-bound and not periplasmic, an enzyme presumably similar to that identified in *Rb. capsulatus* BK5 (Ballard et al., 1990). In view of the observations discussed above it seems more likely that *Rb. sphaeroides* f. sp. *denitrificans* can express both the periplasmic and membrane-bound types of respiratory NO_3^- reductase.

3. Nitrite and Nitrous Oxide Reductases

In general, the enzymes responsible for reduction of nitrite (NO_2^-) and nitrous oxide (N_2O) are located in the periplasm of the genera *Rhodobacter* and *Rhodospirillum* (see Hamilton, 1988). There are conflicting results about the nitrite reductase purified from *Rb. sphaeroides* f. sp. *denitrificans*. The enzyme isolated by Sawada et al. (1978) had a mass of 80 kDa, was composed of two subunits (39 kDa), and contained 1 Cu atom per subunit. The pathway of reduction was proposed to proceed from ubiquinol through the bc_1 complex and Cyt c_2, which is the immediate electron donor for the enzyme (Urata and Satoh, 1984). Michalski and Nicholas (1985) described the purification from the same bacterial species of a protein with two nonidentical subunits of 37.5 and 39.5 kDa, with different isoelectric points. The EPR spectrum was characteristic of types I and II Cu, respectively. Changes in the EPR spectrum observed in the presence of nitrite indicated that the type I Cu is involved in NO_2^- binding and, for topological considerations, the formation of the N-N bond. The type II Cu is thought to act as an electron acceptor.

Rhodobacter capsulatus (strains N22DNAR$^+$, MT1131 and AD2) contains a periplasmic nitrous oxide reductase when grown under anaerobic conditions (McEwan et al., 1985a; Richardson et al., 1991). Several lines of evidence which include the use of the Cyt c_2 deficient mutant of *Rb. capsulatus,* MTG4/S4, indicate that in this bacterial species, Cyt c_2 is necessary for electron transport to N_2O (Richardson et al., 1991). It is worth noting that this is the first respiratory pathway in *Rb. capsulatus* for which Cyt c_2 seems to be an essential component (see II.5). A role for the Cyt bc_1 complex in the electron transport pathway to N_2O has also been established (Richardson et al., 1991). However, the presence of N_2O reduction in a mutant of *Rb. capsulatus* (MT-CBC1) deficient in a functional Cyt bc_1 complex (Daldal et al., 1986) (see Table 2) revealed that the bc_1 complex can be substituted by at least one other route of electron flow. This second route involves at least a *b*-type cytochrome as judged by N_2O-oxidized minus reduced spectra of cells of the MT-

CBC1 mutant (Richardson et al., 1989). It should also be noted that in *Rb. sphaeroides* f. sp. *denitrificans* the partial inhibition of the AA-insensitive N_2O-reduction by myxothiazol is interpreted as meaning that in this organism there are also at least two pathways to N_2O (McEwan et al., 1985a).

4. Nitric Oxide Reductase

The evidence that nitric oxide (NO) is an intermediate between nitrite and nitrous oxide in the denitrification process remains to be established (Averill and Tiedje, 1982; Goretski and Hallocher, 1988; Zafiriou et al., 1989). On the other hand, the toxicity of nitric oxide along with its extensive presence in some environments (Rende et al., 1989), suggest that it may be advantageous to possess nitric oxide reductase activity although the extent to which this activity is widespread amongst bacteria in general remains to be seen. There is evidence that many strains of *Rb. capsulatus*, including strain 37b4 which lacks any other reductase activity towards oxy-species of nitrogen (see Richardson et al., 1991), catalyze reduction of NO (Bell et al., 1992). The finding of a discrete NO reductase activity in a denitrifying strain of *Rb. sphaeroides* (Itoh et al., 1989) and the close relationship of *Rb. capsulatus* and *Rb. sphaeroides* suggests that this activity may also be common to many non-denitrifying strains of the latter species. The retention of nitric oxide reductase activity by *Rb. capsulatus* mutants MT-CBC1, MT-GS4/S4, MT-GS18 and H123, deficient in Cyt bc_1, Cyt c_2, Cyt [bc_1 + c_2] and Cyt c', respectively (see also Table 2), establishes that neither of these redox carriers is obligatory for electron transport to NO (Bell et al., 1992).On the other hand, the partial inhibition by myxothiazol of NO reduction in both wild type and MT-G4/S4 mutant cells show that the Cyt bc_1 complex is involved in electron transfer to nitric oxide (Bell et al., 1992). This might be either a direct transfer of electrons from the bc_1 complex to the nitric oxide reductase or, which is more likely, an indirect electron transfer mediated by the recently described membrane-bound Cyt c_y (Zannoni et al., 1992; Jenney and Daldal, 1993) (see II.5). Evidence for including Cyt c_2 on a path to nitric oxide reductase of *Rb. sphaeroides* f. sp. *denitrificans* has been presented (Itoh et al., 1989). It is also noteworthy that antimycin A fails to inhibit completely the reduction of nitric oxide by the latter organism (Shapleigh and Payne, 1985) suggesting that the Cyt bc_1 complex may be bypassed by electrons destined for the NO reductase as shown for *Rb. capsulatus* (Bell et al., 1992). It has been shown that reduction of NO by $DADH_2$ does not generate a membrane potential in *Rb. capsulatus*, whereas electron flow from either NADH or succinate does so (Bell et al., 1992); this suggests that NO is reduced at the periplasmic side of the membrane where Cyt c_2 generally operates as electron acceptor from DAD. Thus if nitric oxide reductase is an integral membrane protein in *Rb. capsulatus*, as in *P. stutzeri* (Heiss et al., 1989) and *P. denitrificans* (Carr and Ferguson, 1990), it is likely to presume that protons required for the reduction of NO are taken from the periplasmic side of the plasma membrane.

5. DMSO and TMAO Reductases

Several Rhodospirillaceae can grow anaerobically in darkness in synthetic media containing glucose as sole carbon and energy source with dimethylsulfoxide (DMSO) or trimethylamine-N-oxide (TMAO) as electron acceptors (Yen and Marrs, 1977; Madigan and Gest, 1978; Cox et al., 1980; Madigan et al., 1980). In dark anaerobic batch cultures in fructose plus TMAO medium, reduction of TMAO to trimethylamine (TMA) paralleled growth, and typical fermentation products accumulated, namely, CO_2 and formic, acetic, and lactic acids (Madigan et al., 1980). Notably, pyruvate and TMAO competed for electrons produced in glycolysis; limiting TMAO resulted in far more lactate than was produced when fructose was limited (Madigan et al., 1980). On the other hand, Schultz and Weaver (1982) reported that *Rb. capsulatus* and *Rs. rubrum* both grew on fructose in the absence of an 'accessory oxidant' if the medium contained 0.1% sodium bicarbonate; in addition they found that TMAO and DMSO both supported growth on nonfermentable substrates, including malate, succinate, and acetate, although both rate and extent of growth was very limited. This latter observation led to consideration of the possibility that reduction of DMSO and TMAO might be a further example of an anaerobic respiratory process (Mc Ewan et al., 1983). In this, respect McEwan et al. (1983) demonstrated that the addition of TMAO or DMSO to cell suspensions of *Rb. capsulatus*, which had been grown under anaerobic, dark conditions, led to the generation of a rotenone- and uncoupler sensitive membrane potential ($\Delta\psi$) (except strain BK5, see Ballard et al., 1990). This observation could, in principle, be explained solely on the basis of proton translocation by the NADH-ubiquinone oxidoreductase (complex I), although the relative redox potentials of the electron acceptors ($Em_{7.0}$ (TMAO/TMA) = +130 mV; $Em_{7.0}$ (DMSO/DMS) = +160 mV) might contribute to generate a larger $\Delta\psi$ by a hypothetical ubiquinol-TMAO/DMSO oxidoreductase (see Ferguson 1988). On the other hand, it has also been shown by Yen and Marrs (1977) that a NADH dehydrogenase mutant of *Rb. capsulatus,* M1 strain (see Zannoni and Ingledew, 1983a), grew as well as the wild type strain in anaerobic dark conditions with DMSO so to exclude the requirement of complex I (NDH-1, see II.2) for growth on this oxidant. Moreover, Zannoni and Marrs (1981) demonstrated that membranes of *Rb. capsulatus* R126 and MT113, two mutants lacking of a functional bc_1 complex and of *c*-type cytochromes, respectively, contained a NDH-2 type dehydrogenase. The results with *Rb. capsulatus* M1, R126 and MT113 mutants suggest therefore that the complete energy transducing redox chain is not required for anaerobic/dark growth of this bacterial species. However, Yen and Marrs (1977) reported that an aminolevulinic acid auxotrophic mutant of *Rb. capsulatus,* which presumably cannot synthesize any cytochromes, was not able to grow under dark/anaerobic conditions with DMSO (supplemented with B_{12}). This suggests that certain cytochromes may be necessary for reduction of DMSO. Taken together these observations can be rationalized in the following way. Reduction of TMAO/DMSO may, in principle, generate a ($\Delta\tilde{\mu}_{H^+}$) that is inadequate to sustain a significant growth on a carbon source from which ATP derives solely by respiration. This explanation could account for superior growth of *Rb. capsulatus* on TMAO/DMSO with a sugar than with a non-fermentable carbon source. Clearly these aspects remain to be clarified by further studies.

A single periplasmic enzyme is responsible for reduction of both DMSO and TMAO (McEwan et al., 1985b, 1987; Kelly et al., 1988) which can be conveniently assayed by NMR (King et al., 1987). The enzyme has been purified as a single polypeptide with a MW of 82,000. It behaves as a monomer on gel filtration and EPR studies have shown it to be a pterin-type molybdenum protein (McEwan et al., 1991). A *c*-type cytochrome (α band at 556 nm; $Em_{7.6}$ = +105 mV; MW = 13,000) co-purifies with the reductase and is oxidised by TMAO or DMSO in the presence of the enzyme (Mc Ewan et al., 1989). In addition to the above redox components a membrane bound *b*-type heme ($Em_{7.0}$ = 0 mV; α band at 559 nm) clearly distinct from the *b*-type hemes of the bc_1 complex is involved in electron transport to DMSO in *Rb. capsulatus* (Zannoni and Marrs, 1981; McEwan et al., 1989). Notably, another *c*-type cytochrome of apparent molecular mass of 44 kDa has been identified by haem-staining of membrane components of *Rb. capsulatus* after SDS-PAGE (Hudig and Drews, 1986; Zsebo and Hearst, 1984). This 44 kDa-cytochrome is induced during phototrophic growth of *Rb. capsulatus* in the presence of DMSO but not TMAO and has been identified as a cytochrome *c* peroxidase (Hanlon et al., 1992). Its induction during growth on DMSO but not on TMAO has tentatively been related with the production of toxic dimethylsulfide (DMS) which under certain conditions can be photochemically oxidised to generate free radicals (Brimblecombe and Shooter, 1984; Hanlon et al., 1992).

Acknowledgments

I would like to thank the following colleagues and friends: A. McEwan, S. Ferguson and F. Daldal for supply of published and unpublished material. This work was supported by CNR of Italy (Progetto Finalizzato BTBS).

References

Albracht SPJ, Kalman ML and Slater EC (1983) Magnetic interaction of nickel (III) and the iron-sulfur cluster in hydrogenase from *Chromatium vinosum*. Biochim Biophys Acta 724: 309–316

Adams MWW and Hall DO (1979) Properties of the solubilized membrane-bound hydrogenase from the photosynthetic bacterium *Rhodospirillum rubrum*. Arch Biochem Biophys 195: 288–299

Amesz J and Knaff DB (1988) Molecular mechanisms of bacterial photosynthesis. In: Zehnder AJB (ed) Biology of Anaerobic Microorganisms, pp 113–178. John Wiley & Sons, New York

Anraku Y (1988) Bacterial electron transport chains. Ann Rev Biochem 57: 101–132

Anraku Y and Gennis RB (1987) The aerobic respiratory chain of *Escherichia coli*. Trends Biochem Sci 12: 262–266

Averill BA and Tiedje JM (1982) The chemical mechanism of microbial denitrification. FEBS Lett 138: 8–12

Baccarini-Melandri A, Zannoni D and Melandri BA (1973) Energy transduction in photosynthetic bacteria. VI. Respiratory sites and energy conservation in membranes from dark-grown cells of *Rhodopseudomonas capsulata*. Biochim Biophys Acta 314: 298–313

Baccarini-Melandri A and Zannoni D (1978) Photosynthetic and respiratory flow in the dual functional membrane of facultative photosynthetic bacteria. J Bioenerg Biomembr 10: 109–138

Ballard AL, McEwan AG, Richardson DJ, Jackson JB and Ferguson SJ (1990) *Rhodobacter capsulatus* strain BK5 possesses a membrane bound respiratory nitrate reductase rather than the periplasmic enzyme found in other strains. Arch Microbiol 154: 301–303

Bartsch RG (1978) Cytochromes. In: Clayton RK, Sistrom WR (eds) The Photosynthetic Bacteria. pp 249–279 Plenum Press, New York

Bauld J and Brock TD (1973) Ecological studies of *Chloroflexus*, a gliding photosynthetic bacterium. Arch Mikrobiol 92: 267–284

Beatty T and Gest H (1981) Generation of succinyl-coenzyme A in photosynthetic bacteria. Arch Microbiol 129: 335–340

Bell LC, Richardson DJ and Ferguson SJ (1992) Identification of nitric oxide reductase activity in *Rhodobacter capsulatus*: the electron transport pathway can either use or bypass both cytochrome c_2 and the bc_1 complex. J Gen Microbiol 138: 437–443

Bose SK and Gest H (1962) Hydrogenase and light-stimulated electron transfer reactions in photosynthetic bacteria. Nature 195: 1168–1171

Bott M, Ritz D and Henneke H (1991) The *Bradyrhizobium japonicum cycM* gene encodes a membrane anchored homolog of mitochondrial *c*. J Bacteriol 173: 6766–6772

Brimblecombe P and Shooter D (1984) Photooxidation of demethylsulphide in aqueous solutions. Marine Chem 19: 343–353

Byrne MD and Nicholas DJD (1987) A membrane bound dissimilatory nitrate reductase from *Rhodobacter sphaeroides* f. sp. *denitrificans*. Biochim Biophys Acta 915: 120–124

Caffrey M, Davidson E, Cusanovich M and Daldal F (1992) Mutants of *Rhodobacter capsulatus* cytochrome c_2. Arch Biochem Biophys 292: 419–426

Carr GJ and Ferguson SJ (1990) The nitric oxide reductase of *Paracoccus denitrificans*. Biochem J 269: 423–429

Carrithers RP, Yoch DC and Arnon DI (1977) Isolation and characterization of bound iron sulfur proteins from bacterial photosynthetic membranes. II. Succinate dehydrogenase from *Rhodospirillum rubrum* chromatophores. J Biol Chem 252: 7461–7467

Cauvin B, Colbeau A and Vignais P (1991) The hydrogenase structural operon in *Rhodobacter capsulatus* contain a third gene, *hup*M, necessary for the formation of a physiological competent hydrogenase. Mol Microbiol 5: 2519–2527

Chepuri V, Lemiuex LJ, Au D C-T and Gennis RB (1990) The sequence of the *cyo* operon indicates substantial structural similarities between the cytochrome *o* ubiquinone oxidase of *Escherichia coli* and the aa_3-type family of cytochrome *c* oxidases. J biol Chem 265: 11185–11192

Colbeau A and Vignais PM (1981) The membrane-bound hydrogenase of *Rhodopseudomonas capsulata*. Stability and catalytic properties. Biochim Biophys Acta 662: 271–284

Colbeau A and Vignais PM (1983) The membrane-bound hydrogenase of *Rhodopseudomonas capsulata* is inducible and contains nickel. Biochim Biophys Acta 748: 128–138

Colbeau A, Kelley BC and Vignais PM (1980) Hydrogenase activity in *Rhodopseudomonas capsulata*: relationship with nitrogenase activity. J Bacteriol 144: 141–148

Colbeau A, Chabert J and Vignais PM (1983) Purification, molecular properties and localization in the membrane of the hydrogenase of *Rhodopseudomonas capsulata*. Biochim Biophys Acta 748: 116–127

Cook ND and Cammack R (1984) Purification and characterization of the rotenone-insensitive NADH dehydrogenase of mitochondria from *Arum maculatum*. Eur J Biochem 141: 573–577

Cox JC, Madigan MT, Favinger JL and Gest H (1980) Redox mechanisms in 'oxidant dependent' hexose fermentation by *Rhodopseudomonas capsulata*. Arch Biochem Biophys 204: 10–17

Cramer WA and Crofts AR (1982) Electron and proton transport. In: Govindjee (ed) Photosynthesis. Energy Conversion by Plants and Bacteria. Vol 1, pp 387–467. Academic Press, New York

Craske AL and Ferguson SJ (1986) The respiratory nitrate reductase from *Paracoccus denitrificans*. Molecular characterisation and kinetic properties. Eur J Biochem 158: 429–436

Daldal F (1988a) Genetic approaches to study bacterial Cyt bc_1 complexes. In: Stevens SE and Bryant DA (eds), Light Energy Transduction in Photosynthesis: Higher Plants and Bacterial Models pp 259–273, American Society of Plant Physiologists, Washington D.C

Daldal F (1988b) Cytochrome c_2-independent respiratory growth of *Rhodobacter capsulatus*. J Bacteriol 170: 2388–2391

Daldal F, Cheng S, Applebaum J, Davidson E and Prince RC (1986) Cytochrome c_2 is not essential for photosynthetic growth of *Rhodopseudomonas capsulata*. Proc Natl Acad Sci 83: 2012–2016

Daldal F, Tokito MK, Davidson E and Faham M (1989) Mutations conferring resistance to quinol oxidation (Q_z)-inhibitors of the Cyt bc_1 complex of *Rhodobacter capsulatus*. EMBO J 8: 3951–3961

Davidson E and Daldal F (1987) Primary structure of the bc_1 complex of *Rhodobacter capsulatus*. Nucleotide sequence of the *pet* operon encoding the Rieske, Cyt *b* and Cyt c_1 apoproteins. J Mol Biol 195: 13–24

de Vries S and Grivel LA (1988) Purification and characterization of a rotenone-insensitive NADH-Q6 oxidoreductase from mitochondria of *Saccharomyces cerevisiae*. Eur J Biochem 176: 377–384

Donohue TJ, Mc Ewan AG and Kaplan S (1986) Cloning, DNA sequence and expression of the *Rhodopseudomonas sphaeroides* Cyt c_2 gene. J Bacteriol 168: 962–972

Donohue TJ, Kiley PJ and Kaplan S (1988) The *puf* operon region of *Rhodobacter sphaeroides*. Photosynth Res 19: 39–61

Drews G (1985) Structure and functional organization of light harvesting complexes and photochemical reaction centers in membranes of phototrophic bacteria. Microbiol Rev 49: 59–70

Drews G (1986) Adaptation of the bacterial photosynthetic apparatus to different light intensities. Trends Biochem Sci 11: 255–257

Drews G and Oelze J (1981) Organization and differentiation of membranes of phototrophic bacteria. Adv Microb Physiol 22: 1–92

Dry IB, Moore AL, Day DA and Wiskich JT (1989) Regulation of alternative pathway activity in plant mitochondria: nonlinear relationship between electron flux and the redox poise of the quinone pool. Arch Biochem Biophys 273: 148–157

Dupuis A (1992) Identification of two genes of *Rhodobacter capsulatus* coding for proteins homologous to the ND1 and 23 kDa subunits of the mitochondrial Complex I. FEBS Lett 301: 215–218

Dupuis A, Issartel JP, Lunardi J and Peinnequin A (1993) Study of the type I NADH-CoQ reductase of the purple bacterium *Rhodobacter capsulatus*. Biol Chem Hoppe-Seyler 374: 817

Dutton PL (1986) Energy transduction in anoxygenic photosynthesis. Encycl Plant Physiol New Ser 19: 197–237

Dutton PL and Leigh JS (1973) Electron spin resonance characterization of *Chromatium* D hemes. Non-heme irons and components involved in primary photochemistry. Biochim Biophys Acta 314: 178–190

Favinger J, Stadtwald R and Gest H (1989) *Rhodospirillum centenum*, sp. nov., a thermotolerant cyst-forming anoxygenic photosynthetic bacterium. Antonie van Leeuwenhoek 55: 291–296

Ferguson SJ (1988) Periplasmic electron transport reactions. In: Anthony C (eds) Bacterial Energy Transduction, pp 151–182. Academic Press, London

Ferguson SJ, Jackson JB and McEwan AG (1987) Anaerobic respiration in the *Rhodospirillaceae*: Characterization of pathways and evaluation of roles in redox balancing during photosynthesis. FEMS Microbiol Revs 46: 117–143

Fitch J, Meyer T, Cusanovich M, Tollin G, van Beeumen J, Rott M and Donohue TJ (1989) Expression of a Cyt c_2 isozyme restores photosynthetic growth of *Rhodobacter sphaeroides* mutants lacking the Cyt c_2 gene. Arch Biochem Biophys 271: 502–507

Friedrich T, Weidner U, Nehls U, Fecke W, Schneider R and Weiss H (1993) Attempts to define distinct parts of NADH:ubiquinone oxidoreductase (Complex I). J Bioenerg Biomembr 25: 331–337

Gaul DF and Knaff DB (1983) The presence of cytochrome c_1 in the purple sulfur bacterium *Chromatium vinosum*. FEBS Lett 162: 69–75

Gennis RB, Casey RP, Azzi A and Ludwig B (1982) Purification and characterization of the Cyt *c* oxidase from *Rhodopseudomonas sphaeroides*. Eur J Biochem 125: 189–195

Gennis RB, Barquera B, Hacker B, van Doren SR, Arnaud S, Crofts AR, Davidson E, Gray KA and Daldal F (1993) The bc_1 complexes of *Rhodobacter sphaeroides* and *Rhodobacter capsulatus*. J Bioenerg Biomembr 25: 195–209

Gest H and Favinger JL (1983) *Heliobacterium chlorum*, an anoxygenic brownish-green photosynthetic bacterium containing a 'new' form of bacteriochlorophyll. Arch Microbiol 136: 11–16

Giménez-Gallego G, del Valle-Tascòn S and Ramìrez JM (1976) A possible physiological function of the oxygen photoreducing system of *Rhodospirillum rubrum*. Arch Microbiol 109: 119–125

Gooley P, Caffrey M, Cusanovich M and McKenzie N (1990) Assignment of the 1H and 15N NMR spectra of the *Rb. capsulatus* ferrocytochrome c_2. Biochemistry 29: 2278–2290

Goretski J and Hallocher TC (1988) Trapping of nitric oxide produced during denitrification by extracellular haemoglobin. J Biol Chem 263: 2316–2323

Gray GO, Gaul DF and Knaff D (1983) Partial purification and characterization of two soluble *c*-type cytochromes from *Chromatium vinosum*. Arch Biochem Biophys 222: 78–86

Gray KA, Grooms M, Myllykallio H, Moomaw C, Sloughter C and Daldal F (1994) *Rhodobacter capsulatus* contains a novel cb-type cytochrome *c* oxidase without a Cu_A center. Biochemistry 33: 3120–3127

Gromet-Elhanan Z (1969) Inhibitors of photophosphorylation and photoreduction by chromatophores from *Rhodospirillum rubrum*. Arch Biochem Biophys 131: 299–315

Haddock BA and Jones CW (1977) Bacterial respiration. Bacteriol Rev 41: 47–99

Hamamoto T, Carrasco N, Matsushita K, Kaback HR and Montal M (1985) Direct measurement of the electrogenic activity of *O*-type cytochrome oxidase from *Escherichia coli* reconstituted into planar lipid bilayers. Proc Natl Acad Sci USA 82: 2570–2573

Hamilton WA (1988) Energy transduction in anaerobic bacteria. In: Anthony C (eds) Bacterial Energy Transduction, pp 84–1149. Academic Press, London

Hanlon SP, Holt RA and McEwan AG (1992) The 44-kDa c-type cytochrome induced in *Rhodobacter capsulatus* during growth with dimethylsulphoxide as an electron acceptor is a cytochrome *c* peroxidase. FEMS Microbiol Lett 97: 283–288

Hansen TA and van Gemerden H (1972) Sulfide utilization by purple nonsulfur bacteria. Arch Microbiol 86: 49–56

Hashwa FA and Trüper HG (1978) Viable phototrophic sulfur bacteria from the Black Sea bottom. Helgol Wiss Meeresunters 31: 249–253

Hauska G (1986) Composition and structure of cytochrome bc_1 and b_6f complexes. Encycl Plant Physiol New Ser 19: 494–507

Hatefi Y, Davis KA, Baltscheffsky H, Baltscheffsky M and Johansson BC (1972) Isolation and properties of succinate dehydrogenase from *Rhodospirillum rubrum*. Arch Biochem Biophys 152: 613–618

Heiss B, Frunzke K and Zumft WG (1989) Formation of the N-N bond from nitric oxide by a membrane bound cytochrome *bc* complex of nitrate respiring (denitrifying) *Pseudomonas stutzeri*. J Bacteriol 171: 3288–3297

Hiraishi A (1988) Fumarate reduction systems in members of the family *Rhodospirillaceae* with different quinone types. Arch Microbiol 150: 56–60

Hiraishi A, Hoshino Y and Satoh T (1991) *Rhodoferax fermentans* gen. no., sp. nov., a phototrophic purple nonsulfur bacterium previously referred to as the '*Rhodocyclus gelatinosus*-like' group. Arch Microbiol 155: 330–336

Hochkoeppler A, Venturoli G and Zannoni D (1993) The electron transport chain of *Rhodoferax fermentans*. Biol Chem Hoppe-Seyler 374: 831

Hochkoeppler A, Ciurli S, Venturoli g and Zannoni D (1995a) The high potential iron-sulfur protein (HiPIP) from *Rhodoferax fermentans* is competent in photosynthetic electron transfer. FEBS Lett 357: 70–74

Hochkoeppler A, Jenney FE, Lang SE, Zannoni D and Daldal F (1995b) Membrane-associated cytochrome c_y of *Rhodobacter capsulatus* is an electron carrier from cytochrome bc_1 complex to the cytochrome *c* oxidase during respiration. J Bacteriol 177: 608–613

Hosler JP, Ftter J, Tecklenburg MMJ, Espe M, Lerma C and Ferguson-Miller S (1992) Cytochrome aa_3 of *Rhodobacter sphaeroides* as a model for mitochondrial cytochrome *c* oxidase. J Biol Chem 34: 24264–24272

Hosler JP, Ferguson-Miller S, Calhoun MW, Thomas JW, Hill J, Lemieux L, Ma J, Georgiou C, Fetter J, Shapleigh J,

Tecklenburg MJ, Babcock GT and Gennis RB (1993) Insight into the active-site structure and function of cytochrome oxidase by analysis of site-directed mutants of bacterial cytochrome aa_3 and cytochrome *bo*. J Bioenerg Biomembr 25: 121–136

Hüdig H and Drews G (1982a) Characterization of a *b*-type cytochrome *c* oxidase of *Rhodopseudomonas capsulata*. FEBS Lett 146: 389–392

Hüdig H and Drews G (1982b) Isolation of a *b*-type cytochrome oxidase from membranes of the phototrophic bacterium *Rhodopseudomonas capsulata*. Z Naturforsch 37: 193–198

Hüdig H and Drews G (1983) Characterization of a new membrane bound cytochrome *c* of *Rhodopseudomonas capsulata*. FEBS Lett 152: 251–255

Hüdig H and Drews G (1984) Reconstitution of *b*-type cytochrome oxidase from *Rhodopseudomonas capsulata* in liposomes and turnover studies of proton translocation. Biochim Biophys Acta 765: 171–177

Hüdig H, Kaufmann N and Drews G (1986) Respiratory deficient mutants of *Rhodopseudomonas capsulata*. Arch Microbiol 145: 378–385

Imhoff J (1984) Quinones of phototrophic purple bacteria. FEMS Microbiol Lett 25: 85–89

Ingledew JW and Poole RK (1984) The respiratory chains of *Escherichia coli*. Microbiol Revs 48: 22–271

Itoh M, Mizukami S, Matsuura K and Satoh T (1989) Involvement of cytochrome bc_1 complex and cytochrome c_2 in the electron transfer pathway for NO reduction in a photodenitrifier *Rhodobacter sphaeroides* f. sp. *denitrificans*. FEBS Lett 24: 81–84

Jackson JB (1988) Bacterial photosynthesis. In: Anthony C (ed) Bacterial Energy Transduction pp 317–375 Academic Press, London

Jenney FE and Daldal F (1993) A novel membrane associated *c*-type cytochrome, Cyt c_y, can mediate the photosynthetic growth of *Rhodobacter capsulatus* and *Rhodobacter sphaeroides*. EMBO J 12: 1283–1293

Jones CW and Vernon LP (1969) NAD photoreduction in *Rhodospirillum rubrum* chromatophores. Biochim Biophys Acta 180: 149–161

Jones M, Mc Ewan A and Jackson B (1990) The role of *c*-type cytochromes in the photosynthetic electron transport pathway of *Rhodobacter capsulatus*. Biochim Biophys Acta 1019: 59–66

Jones OTG and Plewis KM (1974) Reconstitution of light-dependent electron transport in membranes from a bacterio-chlorophyll-less mutant of *Rhodopseudomonas sphaeroides*. Biochim Biophys Acta 357: 204–214

Jones OTG and Saunders VA (1972) Energy-linked electron transfer reactions in *Rhodopseudomonas viridis*. Biochim Biophys Acta 275: 427–436

Kämpf C and Pfennig N (1980) Capacity of Chromatiaceae for chemotrophic growth. Specific respiration rates of *Thiocystis violacea* and *Chromatium vinosum*. Arch Microbiol 127: 125–135

Kämpf C, Wynn RM, Shaw RW and Knaff DB (1987) The electron transfer chain of aerobically grown *Rhodopseudomonas viridis*. Biochim Biophys Acta 894: 228–238

Keister D and Yike JJ (1967) Energy-linked reactions in photosynthetic bacteria. I. Succinate-linked ATP-driven NAD^+ reduction by *Rhodospirillum rubrum* chromatophores. Arch Biochem Biophys 121: 415–422

Kelley BC, Jouanneau Y and Vignais PM (1979) Nitrogenase activity in *Rhodopseudomonas sulfidophila*. Arch Microbiol 122: 145–152

Kelly DJ, Richardson DJ, Ferguson SJ and Jackson JB (1988) Isolation of transposon Tn5 insertion mutants of *Rhodobacter capsulatus* unable to reduce trimethylamine-N-oxide and dimethylsulphoxide. Arch Microbiol 150: 138–144

Kiley P and Kaplan S (1988) Molecular genetics of photosynthetic membrane biosynthesis in *Rhodobacter sphaeroides*. Microbiol Rev 52: 50–69

King MT and Drews G (1973) The function and localization of ubiquinone in the NADH and succinate oxidase systems of *Rhodopseudomonas palustris*. Biochim Biophys Acta 305: 230–248

King MT and Drews G (1976) Isolation and partial purification of the cytochrome c oxidase from *Rhodopseudomonas palustris*. Eur J Biochem 68: 5–12

King GF, Richardson DJ, Jackson JB and Ferguson SJ (1987) Dimethylsulphoxide and trimethylamine-N-oxide as bacterial electron acceptors: Use of NMR to assay and characterise the reductase system in *Rhodobacter capsulatus*. Arch Microbiol 149: 47–51

Klemme JH (1969) Studies on the mechanism of NAD-photoreduction by chromatophores of the facultative phototroph, *Rhodopseudomonas capsulata*. Z Naturforsch B24: 67–76

Klemme JH, Chyla I and Preuss M (1980) Dissimilatory nitrate reduction by strains of the facultative phototrophic bacterium *Rhodopseudomonas palustris*. FEMS Microbiol Lett 9: 137–140

Knaff DB (1978) Reducing potentials and the pathway of NAD reduction. In: Clayton RK and Sistron WR (eds), The Photosynthetic Bacteria, pp 629–640. Plenum Press, New York

Knaff DB, Whestone R and Carr JW (1980) The role of soluble cytochrome c_{551} in cyclic electron flow driven active transport in *Chromatium vinosum*. Biochim Biophys Acta 590: 50–58

Kovacs KL, Bagyinka CS and Serebryakova LT (1983) Distribution and orientation of hydrogenase in various photosynthetic bacteria. Current Microbiol 9: 215–218

La Monica RF and Marrs BL (1976) The branched respiratory system of photosynthetically grown *Rhodopseudomonas capsulata*. Biochim Biophys Acta 423: 431–439

Larsen RW, Pan L-P, Musser SM, Li Z and Chan SI (1992) Could Cu_B be the site of redox linkage in cytochrome *c* oxidase?. Proc Natl Acad Sci USA 89: 723–727

Lavorel J, Richaud P and Vermeglio A (1989) Interaction of photosynthesis and respiration in *Rhodospirillaceae*: evidence of two functionally distinct bc_1 complex fractions. Biochim Biophys Acta 973: 290–295

Lissolo T, Colbeau A, Magnani P, Kovacs KL and Vignais PM (1993) The membrane-bound (Ni-Fe)hydrogenase of the photosynthetic bacteria *Rhodobacter capsulatus* and *Thiocapsa roseopersicina*. Biol Chem Hoppe-Seyler 374: 825

Madigan MT and Gest H (1978) Growth of a photosynthetic bacterium anaerobically in darkness, supported by oxidant-dependent sugar fermentation. Arch Microbiol 117: 119–122

Madigan MT and Gest H (1979) Growth of the photosynthetic bacterium *Rhodopseudomonas capsulata* chemoautotrophically in darkness with H_2 as the energy source. J Bacteriol 137: 524–530

Madigan MT, Petersen SR and Brock TD (1974) Nutritional studies on *Chloroflexus*, a filamentous, photosynthetic, gliding bacterium. Arch Microbiol 100: 97–103

Madigan MT, Cox JC and Gest H (1980) Physiology of dark fermentative growth of *Rhodopseudomonas capsulata*. J Bacteriol 142: 908–915

Malkin R, Chain RK, Kraichoke S and Knaff DB (1981) Studies on the function of the membrane-bound iron-sulfur centers of the photosynthetic bacterium *Chromatium vinosum*. Biochim Biophys Acta 637: 88–95

Marrs BL and Gest H (1973) Genetic mutations affecting the respiratory electron transport system of the photosynthetic bacterium *Rhodopseudomonas capsulata*. J Bacteriol 114: 1045–1051

Matsushita K, Patel L and Kaback HR (1984) Cytochrome *O*-type oxidase from *Escherichia coli*. Characterization of the enzyme and mechanism of electrochemical proton gradient generation. Biochemistry 23: 4703–4714

McEwan AG, George CL, Ferguson SJ and Jackson JB (1982) A nitrate reductase activity in *Rhodopseudomonas capsulatus* linked to electron transfer and generation of a membrane potential. FEBS Lett 150: 277–280

McEwan AG, Ferguson SJ and Jackson JB (1983) Electron flow to dimethylsulphoxide or trimethylamine-N-oxide generates a membrane potential in *Rhodopseudomonas capsulata*. Arch Microbiol 136: 300–305

McEwan AG, Jackson JB and Ferguson SJ (1984) Rationalization of properties of nitrate reductases in *Rhodopseudomonas capsulata*. Arch Microbiol 137: 344–349

McEwan AG, Greenfield AJ, Wetzstein HG, Jackson JB and Ferguson SJ (1985a) Nitrous oxide reduction by members of the family *Rhodospirillaceae* and the nitrous oxide reductase of *Rhodopseudomonas capsulata*. J Bacteriol 164: 823–830

McEwan AG, Wetzstein HG, Jackson JB and Ferguson SJ (1985b) Periplasmic location of the terminal reductase in trimethylamine-N-oxide and dimethylsulphoxide respiration in the photosynthetic bacterium *Rhodopseudomonas capsulata*. Biochim Biophys Acta 806: 410–417

McEwan AG, Wetstein HG, Meyer O, Jackson JB and Ferguson SJ (1987) The periplasmic nitrate reductase of *Rhodobacter capsulatus*; purification, characterization and distinction from a single reductase from trimethylamine-N-oxide, dimethylsulphoxide and chlorate. Arch Microbiol 147: 340–345

McEwan AG, Richardson DJ, Hudig H, Ferguson SJ and Jackson BJ (1989) Identification of cytochromes involved in electron transport to trimethylamine N-oxide/dimethylsulphoxide reductase in *Rhodobacter capsulatus*. Biochim Biophys Acta 973: 308–314

McEwan AG, Ferguson SJ and Jackson JB (1991) Purification and properties of dimethylsulphoxide reductase from *Rhodobacter capsulatus*. A periplasmic molybdoenzyme. Biochem J 274: 305–307

Meyer J, Kelley BC and Vignais PM (1978) Nitrogen fixation and hydrogen metabolism in photosynthetic bacteria. Biochemie 60: 245–360

Melandri BA, Zannoni D, De Santis A and Casadio R (1982) Hydrogen photometabolism in *Rhodopseudomonas capsulata*. In: Hall DO and Palz W (eds), Photochemical, Photoelectrochemical and Photobiological Processes, Vol 1-D, pp 164–164. D Riedel Publ, Dordrecht

Melandri BA, Zannoni D, De Santis A and Casadio R (1983) Hydrogen photometabolism in *Rhodopseudomonas capsulata*. In: Hall DO and Palz W (eds), Photochemical, Photoelectrochemical and Photobiological Processes, Vol 2-D, pp 208–213. D Riedel Publ, Dordrecht

Michalski W and Nicholas DJD (1985) Molecular characterization of a copper containing nitrite reductase from *Rhodopseudomonas sphaeroides* f. sp. *denitrificans*. Biochim Biophys Acta 828: 130–137

Michels PA and Haddock BA (1980) Cytochrome c deficient mutants of *Rhodopseudomonas capsulata*. EBEC Reports 1: 77–78

Minghetti KC, Goswitz VC, Gabriel NE, Hill JJ, Barassi C, Georgiou CD, Chan SI and Gennis RB (1992) Modified, large-scale purification of the cytochrome *o* complex (*bo*-type oxidase) of *Escherichia coli* yields a two heme/one copper terminal oxidase with high specific activity. Biochemistry 31: 6917–6924

Møller IM, Rasmusson AG and Fredlund KM (1993) NAD(P)H-Ubiquinone oxidoreductases in plant mitochondria. J Bioenerg Biomembr 25: 377–384

Moore AL, Dry IB and Wiskich JT (1988) Measurement of the redox state of the ubiquinone pool in plant mitochondria. FEBS Lett 235: 76–80

Moore AL, Day DA, Dry IB and Wiskich JT (1990) Regulation of electron transport activity in plant mitochondria by the redox poise of the quinol pool. In: Lenaz G, Barnabei O, Rabbi A and Battino M (eds), Highlights in Ubiquinone Research, pp 170–174. Taylor and Francis, London

Oelze J and Drews G (1972) Membranes of photosynthetic bacteria. Biochim Biophys Acta 265: 209–239

Oelze J and Drews G (1981) Membranes of phototrophic bacteria. In: Ghosh BK (eds), Organization of prokaryotic membranes, Vol 2, pp 131–195. CRC Press, Boca Raton

Ohnishi T and Salerno JC (1982) Iron sulfur clusters in the mitochondrial electron transport chain. In: Spiro TG (ed), Iron-Sulfur Proteins, Vol 4, pp 285–327. John Wiley and Sons, Chichester

Ormerod JG and Gest H (1962) Symposium on metabolism of inorganic compounds. IV. Hydrogen photosynthesis and alternative metabolic pathways in photosynthetic bacteria. Bacteriol Rev 26: 51–66

Paul F, Colbeau A and Vignais PM (1979) Phosphorylation coupled to H_2 oxidation by chromatophores from *Rhodopseudomonas capsulata*. FEBS Lett 106: 29–33

Pfennig N (1978) General physiology and ecology of photosynthetic bacteria. In: Clayton RK and Sistrom WR (eds) The Photosynthetic Bacteria pp 3–18 Plenum Press, New York

Pierson BK (1985) Cytochromes in *Chloroflexus aurantiacus* grown with and without oxygen. Arch Microbiol 143: 260–265.

Pierson BK and Castenholz RW (1974) A phototrophic, gliding filamentous bacterium of hot springs, *Chloroflexus aurantiacus*, gen. and sp. nov. Arch Microbiol 100: 5–24

Poole RK (1983) Bacterial cytochrome oxidases. A structural and functionally diverse group of electron transfer proteins. Biochim Biophys Acta 726: 205–243

Presig O, Anthamatten D, Thony-Meyer L, Beck C, Zufferey R and Henneke H (1993) Genetic and preliminary biochemical characterization of a novel member of the bacterial heme/copper cytochrome oxidase superfamily. Biol Chem Hoppe-Seyler 374: 821

Prince RC (1990) Bacterial photosynthesis: from photons to $\Delta\psi$. In: Krulwich TA (ed), The Bacteria, Vol XII, pp 111–149. Academic Press, New York

Prince RC and Daldal F (1987) Physiological electron donors to the photochemical reaction center of *Rhodobacter capsulatus*. Biochim Biophys Acta 894: 370–378

Prince RC and Dutton PL (1978) Protonation and the reducing potential of the primary electron acceptor. In: Clayton RK and Sistrom WR (eds), The Photosynthetic Bacteria, pp 439–453. Plenum Press, New York

Prince RC and Ingledew JW (1977) Thermodynamic resolution of the iron-sulfur centers of the succinic dehydrogenase of *Rhodopseudomonas sphaeroides*. Arch Biochem Biophys 178: 303–307

Prince RC, Baccarini-Melandri A, Crofts AR, Hauska GA and Melandri BA (1975) Asymmetry of an energy transducing membrane. Location of cytochrome c_2 in *Rhodopseudomonas sphaeroides* and *Rhodopseudomonas capsulata*. Biochim Biophys Acta 387: 212–227

Prince RC, Davidson E, Haith C and Daldal F (1986) Photosynthetic electron transfer in the absence of Cyt c_2 in *Rhodopseudomonas capsulata*: Cyt c_2 is not essential for electron flow from the bc_1 complex to the photochemical reaction center. Biochemistry 25: 5208–5212

Puustinen A and Wikström M (1991) The heme groups of the cytochrome *O* from *Escherichia coli*. Proc Natl Acad Sci USA 88: 6122–6126

Puustinen A, Finel M, Haltia T, Gennis RB and Wikström M (1991) Properties of the two terminal oxidases of *E. coli*. Biochemistry 30: 3936–3942

Ramìrez-Ponce MP, Ramìrez JM and Giménez-Gallego G (1980) Rhodoquinone as a constituent of the dark electron-transfer system of *Rhodospirillum rubrum*. FEBS Lett 119: 137–140

Rende A, Slemer F and Conrad R (1989) Microbial production and uptake of nitric oxide in soil. FEMS Microbiol Ecol 62: 221–230

Richardson DJ, Kelly DJ, Jackson, Ferguson SJ and Alef K (1986) Inhibitory effects of myxothiazol and 2-n-heptyl-4-hydroxy quinoline-N-oxide on the auxiliary electron transport pathways of *Rhodobacter capsulatus*. Arch Microbiol 146: 159–165

Richardson DJ, McEwan AG, Jackson JB and Ferguson SJ (1989) Electron transport pathways to nitrous oxide in *Rhodobacter* species. Eur J Biochem 185: 659–669

Richardson DJ, McEwan AG, Page MD, Jackson JB and Ferguson SJ (1990) The identification of cytochromes involved in the transfer of electrons to the periplasmic NO_3^- reductase of *Rhodobacter capsulatus* and resolution of a soluble NO_3^--reductase-cytochrome-c_{552} redox complex. Eur J Biochem 194: 263–270

Richardson DJ, Bell LC, McEwan AG, Jackson JB and Ferguson SJ (1991) Cytochrome c_2 is essential for electron transfer to nitrous oxide reductase from physiological substrates in *Rhodobacter capsulatus* and act as an electron donor to the reductase in vivo. Correlation with photoinhibition studies. Eur J Biochem 199: 677–683

Richaud R, Marrs BL and Vermeglio A (1986) Two modes of interaction between photosynthetic and respiratory chains in whole cells of *Rhodopseudomonas capsulata*. Biochim Biophys Acta 850: 256–263

Robertson DE, Davidson E, Prince RC, Van der Berg WH, Marrs BL and Dutton PL (1986) Discrete catalytic sites for quinone in the ubiquinol-cytochrome *c* oxidoreductase of *Rhodopseudomonas capsulata*. Evidence from a mutant defective in ubiquinol oxidation. J Biol Chem 261: 584–591

Rott MA and Donohue TJ (1990) *Rhodobacter sphaeroides* spd mutations allow Cyt c_2-independent photosynthetic growth. J Bacteriol 172: 1954-1961

Rugolo M and Zannoni D (1992) Oxidation of external NAD(P)H by Jerusalem artichoke (*Helianthus tuberosus*) mitochondria. Plant Physiol 99: 1037–1043

Sasaki T, Motokawa Y and Kikuchi G (1970) Occurrence of both *a*-type and *o*-type cytochromes as the functional terminal oxidases in *Rhodopseudomonas sphaeroides*. Biochim Biophys Acta 197: 284–291

Satoh T (1981) Soluble dissimilatory nitrate reductase containing cytochrome c from a photodenitrifier *Rhodopseudomonas sphaeroides* f. sp. *denitrificans*. Plant Cell Physiol 22: 423–432

Saunders VA and Jones OTG (1974) Properties of the cytochrome *a*-like material developed in the photosynthetic bacterium *Rhodopseudomonas sphaeroides*. Biochim Biophys Acta 333: 439–445

Saunders VA and Jones OTG (1975) Detection of two further *b*-type cytochromes in *Rhodopseudomonas sphaeroides*. Biochim Biophys Acta 396: 220–228

Sawada E and Satoh T (1980) Periplasmic location of dissimilatory nitrate and nitrite reductases in a denitrifying phototrophic bacterium *Rhodopseudomonas sphaeroides* f. sp. *denitrificans*. Plant Cell Physiol 24: 501–508

Sawada E, Satoh T and Kitamura H (1978) Purification and properties of a dissimilatory nitrite reductase of a denitrifying phototrophic bacterium. Plant Cell Physiol 19: 1339–1351

Schrattenholz AS, Nawroth T and Dose K (1989) Isolation and partial characterization of a cytochrome-*o* complex from chromatophores of the photosynthetic bacterium *Rhodospirillum rubrum* FR1. Eur J Biochem 181: 689–694

Schultz JE and Weaver PF (1982) Fermentation and anaerobic respiration by *Rhodospirillum rubrum* and *Rhodopseudomonas capsulata*. J Bacteriol 149: 181–190

Shapleigh JP and Payne WJ (1985) Nitric oxide dependent proton translocation in various denitrifiers. J Bacteriol 163: 837–840

Shill DA and Wood PM (1984) A role for cytochrome c_2 in *Rhodopseudomonas viridis*. Biochim Biophys Acta 764: 1–7

Seifert E and Pfennig N (1979) Chemotrophic growth of *Rhodopseudomonas* species with H_2 and chemotrophic utilization of methanol and formate. Arch Microbiol 122: 177–182

Sled' VD, Friedrich T, Leif H, Weiss H, Meinhardt SW, Fukumori Y, Calhoun MW, Gennis RB and Ohnishi T (1993) Bacterial NADH-quinone oxidoreductases: iron-sulfur clusters and related problems. J Bioenerg Biomembr 25: 347–356

Strekas T, Antanaitis BC and Krasna AI (1980) Characterization and stability of hydrogenase from *Chromatium*. Biochim Biophys Acta 616: 1–9

Takakuwa S and Wall JD (1981) Enhancement of hydrogenase activity in *Rhodopseudomonas capsulata* by nickel. FEMS Microbiol Lett 12: 359–363

Takamiya K (1983) Properties of the cytochrome c oxidase

activity of Cyt b_{561} from photoanaerobically grown *Rhodopseudomonas sphaeroides*. Plant Cell Physiol 24: 1457–1462

Takamiya K and Tanaka H (1983) Isolation and purification of Cyt b_{561} from a photosynthetic bacterium *Rhodopseudomonas sphaeroides*. Plant Cell Physiol 24: 1449–1455

Takamiya KI, Doi M and Okimatsu H (1982) Isolation and purification of a ubiquinone-Cyt bc_1 complex from a photosynthetic bacterium, *Rhodopseudomonas sphaeroides*. Plant Cell Physiol 23: 987–997

Trost JT, McManus JD, Freeman JC, Ramakrishna BL and Blankenship RE (1988) Auracyanin, a blue copper protein from the green photosynthetic bacterium *Chloroflexus aurantiacus*. Biochemistry 27: 7858–7863

Uffen RL and Wolfe RS (1979) Anaerobic growth of purple nonsulfur bacteria under dark conditions. J Bacteriol 104: 462–472

Urata K and Satoh T (1984) Evidence for cytochrome bc_1 complex involvement in nitrite reduction in a photodenitrifier, *Rhodopseudomonas sphaeroides* f. sp. *denitrificans*. FEBS Lett 172: 205–208

van der Oost J, Pappalainen P, Musacchio A, Warne A, Lemieux L, Rumbley J, Gennis RB, Aasa R, Pascher T, Malmstrom BG and Saraste M (1992) Restoration of a lost metal-binding site: construction of two different copper sites into a subunit of the *E. coli* cytochrome *O* quinol complex. EMBO J 11: 3209–3217

van Grondelle R, Duysens LNM and van der Wal HN (1976) Function of three cytochromes in photosynthesis of whole cells of *Rhodospirillum rubrum* as studied by flash spectroscopy. Biochim Biophys Acta 441: 169–187

van Grondelle R, Duysens LNM and van der Wal HN (1977) Function and properties of a soluble *c*-type cytochrome *c*-551 in secondary photosynthetic electron transport in whole cells of *Chromatium vinosum* as studied with flash spectroscopy. Biochim Biophys Acta 461: 188–201

van Heerikhuizen H, Albracht SPJ, Slater EC and van Rheenen PS (1981) Purification and some properties of the soluble hydrogenase from *Chromatium vinosum*. Biochim Biophys Acta 657: 26–39

Varela J and Ramires JM (1990) Oxygen-linked electron transfer and energy conversion in *Rhodospirillum rubrum*. In: Drews G and Dawes EA (eds) Molecular Biology of Membrane-Bound Complexes in Phototrophic Bacteria, pp 443–452. Plenum Press, New York

Venturoli G, Fenoll C and Zannoni D (1987) On the mechanism of respiratory and photosynthetic electron transfer in *Rhodospirillum rubrum*. Biochim Biophys Acta 892: 172–184

Venturoli G, Fernandez-Velasco JG, Crofts AR and Melandri BA (1986) Demonstration of a collisional interaction of ubiquinol with the ubiquinol-Cyt c_2 oxidoreductase complex in chromatophores from *Rhodopseudomonas sphaeroides*. *Biochim Biophys* Acta 851: 340–352

Venturoli G, Feick, Trotta M and Zannoni D (1990) Thermodynamic and kinetic features of the redox carriers operating in the photosynthetic electron transport of *Chloroflexus aurantiacus*. In: Drews G and Dawes EA (eds) Molecular Biology of Membrane-Bound Complexes in Phototrophic Bacteria, pp 425–432. Plenum Press, New York

Vignais PM, Colbeau A, Willison JC and Jouanneau Y (1985) Hydrogenase, nitrogenase and hydrogen metabolism in the photosynthetic bacteria. Adv Microbial Physiol 26: 155–234

Wale FR and Jones OTG (1970) The cytochrome system of heterotrophically grown *Rhodopseudomonas sphaeroides*. Biochim Biophys Acta 223: 146–157

Ward DM, Weller R, Shiea J, Castenholz RW and Cohen Y (1989) Hot spring microbial mats: anoxygenic and oxygenic mats of possible evolutionary significance. In: Cohen Y and Rosenberg E (eds) Microbial Mats, Physiological Ecology of Benthic Microbial Communities pp 3–15 Amer Soc Microbiol Washington D.C.

Weiss H and Friedrich T (1991) Redox-linked proton translocation by NADH-ubiquinone reductase (complex I). J Bioenerg Biomembr 23: 743–754

Wood PM (1983) Why do *c*-type cytochromes exists?. FEBS Lett 164: 223–226

Wood PM (1984) Bacterial proteins with CO-binding *b*- or *c*-type haem functions and absorption spectroscopy. Biochim Biophys Acta 768: 293–317

Wynn RM, Kämpf C, Gaul DF, Choi W-K, Shaw RW and Knaff DB (1985) The membrane bound electron transfer components of aerobically grown *Chromatium vinosum*. Biochim Biophys Acta 808: 85–93

Wynn RM, Gaul DF, Choi WK, Shaw RW and Knaff DB (1986) Isolation of cytochrome bc_1 complexes from the photosynthetic bacteria *Rhodopseudomonas viridis* and *Rhodospirillum rubrum*. Photosynth Res 9: 181–195

Wynn RM, Redlinger TE, Foster JM, Blankenship RE, Fuller RC, Shaw RW and Knaff DB (1987) Electron transport chains of phototrophically and chemotrophically grown *Chloroflexus aurantiacus*. Biochim Biophys Acta 981: 216–226

Yagi T (1991) Bacterial NADH-quinone oxidoreductases. J Bioenerg Biomembr 23: 211–225

Yagi T (1993) The bacterial energy-transducing NADH-quinone oxidoreductases. Biochim Biophys Acta 1141: 1–17

Yen HC and Marrs BL (1977) Growth of *Rhodopseudomonas capsulata* under anaerobic dark conditions with dimethyl sulphoxide. Arch Biochem Biophys 181: 411–418

Yokota S, Urata K and Satoh T (1984) Redox properties of membrane bound *b*-type cytochromes and a soluble *c*-type cytochrome of nitrate reductase in a photodenitrifier *Rhodopseudomonas sphaeroides* f. sp. *denitrificans*. J Biochem (Tokyo) 95: 1535–1541

Zafiriou OC, Hanley QS and Snyder G (1989) Nitric oxide and nitrous oxide production and cycling during dissimilatory nitrate reduction by *Pseudomonas perfectormarina*. J Biol Chem 264: 5694–5699

Zannoni D (1982) ATP synthesis coupled to light-dependent non-cyclic electron flow in chromatophores of *Rhodopseudomonas capsulata*. Biochim Biophys Acta 680: 1–7

Zannoni D (1985) Mefloquine: an antimalarial drug interacting with the *b/c* region of bacterial respiratory chains. FEBS Lett 183: 340–344

Zannoni D (1986) The branched respiratory chain of heterotrophically dark-grown *Chloroflexus aurantiacus*. FEBS Lett 198: 119–124

Zannoni D and Baccarini-Melandri A (1980) Respiratory electron flow in facultative photosynthetic bacteria. In: Knowles KJ (ed), Diversity of Bacterial Respiratory Systems, Vol II, pp183–202. CRC Press, Boca Raton

Zannoni D and Daldal F (1993) The role of *c* type cytochromes in catalyzing oxidative and photosynthetic electron transport in the dual functional plasma membrane of facultative phototrophs. Arch Microbiol Minirevs Ser (in press)

Zannoni D and Fuller RC (1988) Functional and spectral characterization of the respiratory chain of *Chloroflexus aurantiacus* grown in the dark under oxygen-saturated conditions. Arch Microbiol 150: 368–373

Zannoni D and Ingledew JW (1983a) *Rhodopseudomonas capsulata* respiratory dehydrogenase mutants: an electron paramagnetic resonance study. FEMS Microbiol Lett 17: 331–334

Zannoni D and Ingledew JW (1983b) A functional characterization of the membrane bound iron sulfur centres of *Rhodopseudomonas capsulata*. Arch Microbiol 135: 176–181

Zannoni D and Ingledew JW (1985) A thermodynamic analysis of the plasma membrane electron transport components in phototrophically grown cells of *Chloroflexus aurantiacus*. An optical and electron paramagnetic resonance study. FEBS Lett 193: 93–98

Zannoni D and Marrs BL (1981) Redox chain and energy transduction in chromatophores from *Rhodopseudomonas capsulata* cells grown anaerobically in the dark on glucose and dimethylsulphoxide. Biochim Biophys Acta 637: 96–106

Zannoni D and Melandri BA (1985) Function of ubiquinone in bacteria. In: Lenaz G (eds) Coenzyme Q. Biochemistry, Bioenergetics and Clinical Applications of Ubiquinone, pp 235–256 John Wiley and Sons Ltd, Chichester

Zannoni D and Moore AL (1990) Measurement of the redox state of the ubiquinone pool in *Rhodobacter capsulatus* membrane fragments. FEBS Lett 271: 123–127

Zannoni D and Venturoli G (1988) The mechanism of photosynthetic electron transport and energy transduction by membrane fragments from *Chloroflexus aurantiacus*. In: Olson JM, Ormerod JG, Amesz J, Stackebrandt E, Truper HG (eds) Green Photosynthetic Bacteria, pp 135–143. Plenum Press, New York

Zannoni D, Baccarini-Melandri A, Melandri BA, Evans EH, Prince RC and Crofts AR (1974) The nature of the cytochrome *c* oxidase in the respiratory chain of *Rhodopseudomonas capsulata*. FEBS Lett 48: 152–155

Zannoni D, Melandri BA and Baccarini-Melandri A (1976a) Composition and function of the branched oxidase system in wild type and respiratory mutants of *Rhodopseudomonas capsulata*. Biochim Biophys Acta 423: 413–430

Zannoni D, Melandri BA and Baccarini-Melandri A (1976b) Further resolution of the cytochrome of *b* type and the nature of the CO-sensitive oxidase present in the respiratory chain of *Rhodopseudomonas capsulata*. Biochim Biophys Acta 449: 386–400

Zannoni D, Melandri BA and Baccarini-Melandri A (1978) The branched respiratory system of the facultative photosynthetic bacterium *Rhodopseudomonas capsulata*. In: Degn H, Lloyd D and Hill GC (eds) Functions of Alternative Terminal Oxidases, pp 169–177. Pergamon Press, Oxford

Zannoni D, Prince RC, Dutton PL and Marrs BL (1980) Isolation and characterization of a cytochrome c_2 deficient mutant of *Rhodopseudomonas capsulata*. FEBS Lett 113: 289–293

Zannoni D, Peterson S and Marrs BL (1986) Recovery of the alternative oxidase dependent electron flow by fusion of membrane vesicles from *Rhodobacter capsulatus* mutant strains. Arch Microbiol 144: 375–380

Zannoni D, Venturoli G and Daldal F (1992) The role of the membrane bound cytochromes of *b*- and *c*-type in the electron transport chain of *Rhodobacter capsulatus*. Arch Microbiol 157: 367–374

Zsebo KM and Hearst J (1984) Genetic physical mapping of a photosynthetic gene cluster from *Rhodopseudomonas capsulata*. Cell 37: 937–947

Chapter 45

Storage Products in Purple and Green Sulfur Bacteria

Jordi Mas
Department of Genetics and Microbiology, Autonomous University of Barcelona, 08193 Bellaterra, Spain

Hans Van Gemerden
Department of Microbiology, University of Groningen, Kerklaan 30, 9751 NN Haren, The Netherlands

Summary

Synthesis and accumulation of storage materials occurs in phototrophic sulfur bacteria under virtually all environmental conditions. After summarizing the general characteristics of the different storage products and the maximum contents at which they are found in different organisms, attention is paid to the environmental conditions which affect their accumulation. Although the specific contents are usually high when resources are present in excess, deposition also occurs under conditions of limitation, thus suggesting a strategy which maximizes long term benefit in a fluctuating environment, rather than instantaneous growth. The physiological role of storage products is illustrated through several laboratory experiments, some were fluctuations play a major role, and later on, in a section which describes field situations which actually confirm several of the previous experimental results. The chapter ends with a section which emphasizes the physical effects of the deposition of storage products specially in organisms with a planktonic way of life. Changes in cell size and density derived from the presence of storage structures are discussed, specially in relation to their impact on buoyancy regulation and sinking rates.

R. E. Blankenship, M. T. Madigan and C. E. Bauer (eds): Anoxygenic Photosynthetic Bacteria, pp. 973–990.

I. General Characteristics of the Storage Inclusions of Phototrophic Sulfur Bacteria

The ability to store surplus substrates during unbalanced growth constitutes a widely spread strategy which enables microorganisms to cope with an environment in which fluctuations are the rule rather than the exception. The capacity to accumulate a certain compound during conditions of surplus, clearly becomes a selective advantage when later on this same compound may become limiting.

Phototrophic sulfur bacteria constitute almost a paradigm of this kind of behavior. Depending on the conditions in which they are growing, they can accumulate zero-valence sulfur (S°), glycogen, poly(3-hydroxyalkanoates) (PHA) and polyphosphate. These compounds are stored intracellularly (and therefore are inclusions) except for sulfur which, in the green sulfur bacteria and in the purple Ectothiorhodospiraceae, is stored extracellularly. However, even in these cases, the fact that extracellular sulfur (at least in the green sulfur bacteria) is not always available to other organisms (Van Gemerden, 1986) makes it a storage compound which behaves, for most purposes, as an intracellular inclusion.

The section which follows will provide some information regarding both the general characteristics of these products, as well as the extent to which they are found in phototrophic sulfur bacteria. Several reviews already exist which deal with different aspects of the metabolism and structure of storage compounds (Dawes and Senior, 1973; Shively, 1974; Merrick, 1978), however, for an in depth description of the macromolecular structure and the metabolism of these compounds the reader is referred to the excellent reviews by Shively et al. (1989) an Dawes (1992).

A. Sulfur

Sulfur globules accumulate during growth on reduced sulfur compounds. They occur extracellularly in green sulfur bacteria as well as in purple sulfur bacteria of the family Ectothiorhodospiraceae, and intracellularly in members of the family Chromatiaceae.

The amounts of sulfur stored depend largely on the organisms and especially on the growth conditions (see section II). Table 1 summarizes data on the maximum specific content of sulfur expressed as a percent of the dry weight in several phototrophic bacteria growing under different conditions. It seems that sulfur storage rarely exceeds 30-35% of the dry weight except in green sulfur bacteria, in which sulfur is formed extracellularly, and in the purple sulfur bacterium *Thiocapsa roseopersicina* in which the content of sulfur reaches values of 54-55%, well above the average. The physiological or ecological meaning of this increased storage remains to be elucidated.

Electron microscopy studies reveal that intracellular sulfur globules are spherical structures with a diameter which occasionally can reach 1 μm (Remsen, 1978), surrounded by a single layer protein membrane (Nicolson and Schmidt, 1971). They are found in the cytoplasm with no further membrane envelopes, although it has been suggested, taking as a basis the observations by Eimhjellen et al. (1967) that sulfur inclusions in *Thiocapsa* could be deposited surrounded by a protein membrane, inside invaginations of the plasma membrane (Shively et al., 1988). Statistical analyses of several samples from a natural population of *Chromatium minus* (Esteve et al., 1990) show rather homogeneous size distributions with average diameters between 317 and 353 nm.

Sulfur globules are evenly distributed in the cytoplasm except in *Chromatium warmingii* in which deposition occurs preferentially at the extremes of the cell (Pfennig and Trüper, 1989). Chemical analysis of the sulfur globules in *Chromatium vinosum* indicate a rather homogeneous composition with 93.5% sulfur, 5.2% protein, 0.55% lipid and 0.047% BChl*a* (Schmidt et al., 1971), the last two fractions being attributed by the authors to contamination from chromatophores. Early studies on the protein fraction of the sulfur globule found a single protein. This protein had a molecular weight of 13500 Da in the case of *Chromatium vinosum* (Schmidt et al., 1971) and 18500 Da in the case of *Chromatium buderi* (Remsen, 1978).

More recent studies carried out in *Chromatium vinosum* and *Thiocapsa roseopersicina* indicate the existence of three proteins with molecular weights of 10498, 10651, 8480 and 10619, 10661 and 8759 Da, respectively (D. Brune, personal communication). The protein sequence indicates that the two large proteins from *Chromatium vinosum* are homologous to each other and to the large proteins found in *Thiocapsa*. The small proteins display also a high sequence similarity. Whether these proteins play a

Abbreviations: BChl *a* – bacteriochlorophyll *a*; D – dilution rate; PHA – poly (3-hydroxyalkanoates); PHB – poly (3-hydroxybutyrate); S_R – reservoir concentration of limiting substrate; μ_{max} – maximum specific growth rate

Table 1. Glycogen and sulfur content of various phototrophic sulfur bacteria expressed as a percentage of the dry weight[a]

Organism	Strain	Conditions	Glycogen (%DW)	Sulfur (%DW)	Reference
Chromatium vinosum	DSM 185	Batch culture, S^{2-}/CO_2	16.0	37.0	Van Gemerden and Beeftink 1978
Chromatium vinosum	DSM 185	Continuous culture $D = 0.029\ h^{-1}$, S^{2-} limitation	9.2	0.0	Mas and Van Gemerden 1987
"		Continuous culture $D = 0.059\ h^{-1}$, S^{2-} limitation	5.7	3.5	"
"		Continuous culture $D = 0.086\ h^{-1}$, S^{2-} limitation	5.9	25.4	"
Chromatium vinosum	DSM 185	Batch culture, S^{2-}/CO_2	–	10.7	Van Gemerden 1974
Chromatium vinosum	DSM 185	Continuous culture $D = 0.071\ h^{-1}$, PO_4^{3-} limitation	–	25.1	Mas and Van Gemerden 1992
Chromatium vinosum	DSM 185	Batch culture, S^{2-}/CO_2	–	31.0	Van Gemerden 1968a
Chromatium vinosum	DSM 185	Continuous culture $D = 0.090\ h^{-1}$, S^{2-} limitation	–	27.6	Van Gemerden 1986
Chromatium vinosum	DSM 185	Continuous culture $D = 0.072\ h^{-1}$, light limitation	1.4	21.1	Van Gemerden et al. 1990
Chromatium vinosum D		Batch culture, S^{2-}/CO_2	–	34.8	Van Niel 1936
Chromatium vinosum D		Batch culture, S^{2-}/CO_2	–	15.0	Hara et al. 1973
Chromatium vinosum D		Batch culture, S^{2-}/CO_2	0.7	11.7	Schmidt and Kamen 1970
Chromatium minutissimum		Batch culture, S^{2-}/CO_2	40.0	–	Zaitseva et al. 1965
"		Batch culture, S^{2-}/CO_2/acetate	30.0	–	"
Chromatium okenii	Ostrau	Batch culture, S^{2-}/CO_2	–	30.5	Trüper and Schlegel 1964
Chromatium weissei	DSM 171	Batch culture, S^{2-}/CO_2	–	25.1	Van Gemerden 1974
Chromatium minus	UA 6001	Batch culture, S^{2-}/CO_2	–	36.0	Montesinos 1987
Thiocapsa roseopersicina	M1	Cont. culture $D = 0.045\ h^{-1}$, S^{2-} lim., chemotr. growth	5.1	54.9	De Wit and Van Gemerden 1987
Thiocapsa roseopersicina	M1	Continuous culture $D = 0.055\ h^{-1}$, S_3^{2-} limitation	7.3	47.3	Visscher et al. 1990
"		Continuous culture $D = 0.080\ h^{-1}$, S^{2-} limitation	9.0	23.5	"
Thiocapsa sp.	EP 2202	Batch culture, S^{2-}/CO_2	13.0	54.5	E. Ortega personal communication
Thiocapsa roseopersicina	DSM 5653	Batch culture, S^{2-}/CO_2	9.2	42.5	"
Thiocapsa sp.	EP 2208	Batch culture, S^{2-}/CO_2	8.0	55.5	"
Pelodictyon phaeoclathratiforme	BU1	Batch culture, $S_2O_3^{2-}/CO_2$/acetate	35.0	–	Overmann et al. 1991
Chlorobium limicola	DSM 249	Continuous culture $D = 0.090\ h^{-1}$, S^{2-} limitation	–	63.0	Van Gemerden 1986
Chlorobium phaeobacteroides	K1	Batch culture, S^{2-}/CO_2/acetate	20.6	21.7	Hofman et al. 1985

[a] Some of the data were originally expressed per weight of protein. In these cases conversion to a % dry weight was carried out assuming that protein accounted for 50% of the structural dry weight.

catalytic role in the deposition and oxidation of elemental sulfur or serve merely a structural function, has not yet been determined. Additional studies on the nature and functional characteristics of these proteins might yield a clue not only to the mechanism of sulfur formation and sulfur oxidation, but also to the actual state of the sulfur contained in the globules.

The molecular structure of the sulfur inclusions has been a matter of speculation for some time. As early as 1887 Winogradsky referred to these structures as 'semi-liquid sulfur'. He based his conclusions mainly on the fact that high temperatures (70 °C) induced the intracellular globules to coalesce into a single large globule. His observations, carried out in colorless sulfur bacteria, were later confirmed in purple sulfur bacteria which displayed a similar phenomenon (Winogradsky, 1945).

Although X-ray diffraction studies of dry preparations of sulfur globules show the presence of crystalline orthorhombic sulfur (Hageage et al., 1970; Trüper and Hathaway, 1967), in vivo studies indicate otherwise. After carrying out polarizing microscopy and X-ray diffraction studies on fresh sulfur globules isolated from *Chromatium*, Hageage et al (1970) concluded that sulfur in the inclusions existed in a state resembling that of liquid sulfur and suggested that the globules were symmetrical aggregates of radially arranged arrays of S_8 molecules. Later studies on the effect of sulfur inclusions in cell size and buoyant density provided estimates of the in vivo density of these inclusions. The values found were 1.22-1.31 pg.μm^{-3} for the sulfur globules of *Chromatium vinosum* (Guerrero et al., 1984; Mas and Van Gemerden, 1987) and 1.13 for the globules in *Chromatium warmingii* (Guerrero et al., 1984; Mas et al., 1985). These values were much lower than the density of elemental sulfur (1.92-2.07, Weast, 1986) and, thus, it was suggested that the sulfur in sulfur globules was hydrated.

Elemental sulfur is insoluble in water (Weast, 1986), however the presence of long chain sulfanes with hydrophilic termini could provide a certain degree of hydration. Although this hypothesis seems to be correct in the case of extracellularly deposited sulfur in *Thiobacillus* (Steudel, 1989), analyses of the sulfur species present in *Chromatium vinosum* indicate the existence of only S_6, S_7 and S_8 rings (Steudel et al., 1990) and fail to show the presence of any long chain polysulfur compounds. The authors suggest that the sulfur chains responsible for the hydration might be attached to the protein envelope at the prosthetic groups of the enzymes responsible for the deposition and breakdown of sulfur.

B. Glycogen

Accumulation of storage carbohydrate is observed simultaneously with the production of elemental sulfur during phototrophic growth on sulfide. Despite claims by some authors of specific contents as high as 40% of the dry weight, the specific content of storage carbohydrates in purple sulfur bacteria rarely exceeds 10% (Table 1). In the few cases in which green sulfur bacteria have been analyzed, the contents have been systematically higher (21 and 35% of the dry weight, see Table 1). Analysis of the structure of the storage carbohydrate in *Chromatium vinosum* (Hara et al., 1973) shows a glycogen structure, with chains of α-1,4 linked α-D-glucose with a variable degree of branching provided by α-1,6-glucosyl linkages. It displays a coefficient of sedimentation of 90S and an average chain length of 11, comparable to values of 10-13 found in other bacteria (Dawes, 1992). In *Chlorobium*, glycogen is deposited as rosette-like granules with a diameter of approximately 30 nm. The granules have been found to increase in number but not in size when the specific content of glycogen increases (Sirevåg and Ormerod, 1970).

C. Poly(3-hydroxyalkanoates)

Poly(3-hydroxyalkanoates) (PHA) are accumulated intracellularly by purple sulfur bacteria when growing in the light in the presence of organic acids, or in the dark, as a byproduct of the degradation of intracellularly stored glycogen. So far, PHA have not been detected in green sulfur bacteria.

These molecules are linear polymers with the general structure:

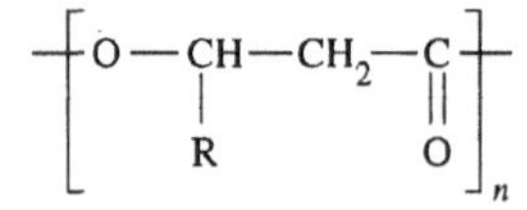

where R has a length between 1 and 10. Polymers constituted by monomers with a homogeneous length of the R side chain are referred to as *homopolymers*. When the monomer composition is heterogeneous, the term *copolymer* is used. Extensive analysis of the amount and composition of PHA deposited intracellularly by phototrophic sulfur bacteria under different growth conditions have been published by

Liebergesell et al. (1991) and are summarized in Table 2. The polymer most commonly found in purple sulfur bacteria was poly(3-hydroxybutyrate) (PHB), a homopolymer of 3-hydroxybutyric acid (R = 1). Most purple sulfur bacteria studied seem only able to synthesize PHB, regardless of the organic acid supplied. However, some organisms (*Chromatium vinosum* 1611 (DSM 182), *Chromatium purpuratum* BN5500 and *Lamprocystis roseopersicina* 3112) synthesized a copolymer of β-hydroxybutyrate (R = 1) and β-hydroxyvalerate (R = 2) when incubated in the presence of organic acids with an odd number of carbons. In some organisms the specific content of PHA was as high as 83% of the dry weight, but only when incubating the cells in nitrogen-free medium containing 20 mM acetate. When grown in regular growth medium with a lower concentration of acetate (6.8 mM) the specific content of PHA never surpassed 30% of the dry weight and, in several cases, was close to zero. Analyses carried out on samples from a population of purple sulfur bacteria in Lake Cisó, Spain, indicate a homogeneous composition of the polymer, with virtually 100% hydroxybutyrate (Mas-Castellà, 1991). Similar analyses of samples from the phototrophic layer in microbial mats from the Ebro Delta show the presence of hydroxybutyrate and of important amounts of hydroxyvalerate which in some cases can be the most abundant monomer (Mas-Castellà, 1991).

Although little information is available on the structure of the PHA inclusions of phototrophic sulfur bacteria, a considerable body of data has been gathered on the inclusions present in other organisms. Although early X-ray diffraction studies indicated that native PHB occurred in a crystalline form (Alper et al., 1963), latter studies by different authors suggest that the polymer in the granules is not crystalline, but rather exists in a less compact form, more readily accessible to the enzymes responsible for its degradation. Thus, X-ray diffraction data collected by different authors (Barham, 1990; Kawaguchi and Doi, 1990) agree in that the polymer in the granules must be in an amorphous form which only crystallizes after either heating or removal of the water and lipids present in the granule. This view is supported by C^{13}-NMR analyses of in vivo PHB in several organisms (Barnard and Sanders, 1989; Nicolay et al., 1982). The 'in vivo' density of PHB (1.155 $pg.\mu m^{-3}$) (Mas et al., 1985) is considerably lower than the density of the purified polymer (1.23-1.25 $g.cm^{-3}$) (Williamson and Wilkinson, 1958; Okamura and Marchessault, 1967). To explain these differences it was postulated that the inclusions were hydrated, with a water content of approximately 40% (Mas et al., 1985).

Electron microscopy studies show that PHA inclusions are spherical structures with a diameter of 240-720 nm in *Bacillus cereus* (Dunlop and Robards, 1973) and 200–700 nm in *Bacillus megaterium* (Ellar et al., 1968), surrounded by a single layer envelope 2–8 nm thick (Merrick, 1978). Similar studies carried out in samples from a natural population of *Chromatium minus* indicate a rather homogeneous size distribution ranging between 252 and 298 nm (Esteve et al., 1990). Chemical analysis of PHB granules (Griebel et al., 1985) shows the presence besides PHB of small amounts of protein (2% w/w) and lipid (0.5% w/w). For a discussion of the use of phototrophic bacteria for production of PHAs and the chemical properties of various PHAs, the reader is referred to Chapter 60 by Fuller.

D. Polyphosphate

Polyphosphates are linear polymers of orthophosphate with variable chain-length ($3–10^3$) which can be found as a variety of different fractions, forming intracellular granules (Kulaev, 1979), soluble in the cytoplasm (Ferguson et al., 1979), in the periplasmic space (Ostrovsky et al., 1980) or even associated to PHB and Ca^{++}, delimiting transport channels in the cytoplasmic membrane (Reusch and Sadoff, 1988). Each of these fractions probably fulfills a rather specific role (Halvorson, 1990; Kulaev, 1990). Little is known about accumulation of polyphosphates by phototrophic sulfur bacteria. Their presence has been detected in *Chlorobium limicola* (Hughes et al., 1963, Cole and Hughes, 1965), as well as in *Chromatium minutissimum* where it reached around 2% of the dry weight (Zaitseva et al., 1965). Experiments carried out with *Chromatium vinosum* (Mas and Van Gemerden, 1992) growing at a constant dilution rate in a chemostat and experiencing different degrees of phosphate limitation, revealed that the cells adjusted to the limitation with an almost threefold decrease in their phosphorus content (from 0.85 to 0.30 μmol $P.mg^{-1}$ protein). Whether the surplus P was present as poly-P or as other P containing molecules was not determined; however, the value found provides an estimate of the maximum amount of P accumulated in the cell when growing in a medium containing excess P.

Table 2. Poly(3-hydroxyalkanoate) (PHA) content of various purple sulfur bacteria expressed as a percentage of the dry weight. Data from Liebergesell et al. 1991.

Organism	Strain	Growth medium	Nitrogen-free storage medium				
		Acetate (6.8 mM)	Acetate (20 mM)	Propionate (20 mM)	Valerate (10 mM)	Heptanoate (10 mM)	Octanoate (10 mM)
Chromatium vinosum D	DSM 180	0.6	58.0	0.6	0	0.5	0
Chromatium vinosum	ABGÖ	2.3	31.3	1.2	0.7	0.8	1.1
Chromatium vinosum	DSM 182	1.5	50.4	2.0[a]	17.0[c]	0	0
Chromatium minutissimum	BN 7511	0.3	36.3	0.3	0	0	0
Chromatium purpuratum	BN 5500	0	8.9	0	14.0[d]	0	0
Chromatium okenii	1111	0.6	12.8	1.5	0.4	0.6	0.4
Chromatium warmingii	1113	11.5	22.9	3.9	5.6	5.9	5.0
Lamprocystis roseopersicina	3112	28.6	27.1	25.3[b]	21.1[e]	23.4[f]	23.1
Thiocapsa pfennigii	9111	0.9	36.2	0.3	0	0	0
Amoebobacter roseus	DSM 235	1.0	32.6	0	0.7	0.9	0.7
Amoebobacter pendens	DSM 236	24.7	30.8	20.0	22.2	17.0	20.4
Thiocystis violacea	DSM 208	28.0	83.0	27.1	24.5	25.8	37.4
Ectothiorhodospira mobilis	BN 9903	18.5	57.5	17.9	15.4	16.7	18.0
Ectothiorhodospira vacuolata	DSM 2111	17.0	35.6	16.3	18.0	16.5	16.2
Ectothiorhodospira shaposhnikovii	DSM 243	15.0	29.3	21.4	12.0	0.2	6.1

Percent of 3-hydroxyvalerate in PHA, remaining being 3-hydroxybutyrate: [a] 15.0, [b] 3.6, [c] 85.5, [d] 100.0, [e] 1.4, [f] 2.1, all other data 100% 3-hydroxybutyrate

II. Environmental Conditions Leading to the Accumulation of Storage Compounds

In aerobic heterotrophic bacteria, reserve material and energy storage often is regarded as being synonymous. However in phototrophic bacteria one has to differentiate between energy storage compounds, conservation of carbon skeletons, and conservation of reducing power. Being able to survive under anoxic conditions with the energy liberated in the degradation of glycogen makes this an energy storage compound, in contrast to poly(3-hydroxybutyrate (PHB) and other poly(3-hydroxyalkanoates) (PHA), which cannot be degraded under anoxic dark conditions. However, in the light, these compounds supply extremely valuable precursors for the synthesis of essential cell components. The so-called 'elemental sulfur' or 'zero-valent sulfur' (S°), deposited intracellularly by members of the family Chromatiaceae, provides neither carbon nor energy under anoxic conditions, but still is a powerful electron donor when sulfide becomes depleted.

In general, storage compounds are synthesized when substrates are present in excess. The nature of the storage compound very much depends on the nature of the substrate. Sulfide (and other reduced forms of sulfur) in combination with carbon dioxide, as well as organic compounds such as malate can be converted into glycogen, whereas the presence of e.g. acetate tends to result in the intracellular storage of PHB (Stanier et al., 1959).

Purple sulfur bacteria growing in batch culture on sulfide/CO_2 accumulate glycogen as long as sulfide is present. In the early stages of growth the reducing equivalents required for the conversion of carbon dioxide into structural cell material and glycogen are obtained mainly from the oxidation of sulfide to sulfur, and to a far lesser extent from the oxidation of S° to sulfate, although large differences between different strains can be observed. In *Chromatium vinosum* DSM 185, growth continues at μ_{max} after sulfide depletion, at the expense of stored S° and glycogen (Van Gemerden and Beeftink, 1978, see Fig. 1). In the stationary phase of growth the concentration of glycogen is zero. The above data clearly show that the rate-limiting step in the Calvin cycle exceeds the specific rate of growth, enabling the deposition of storage compounds. A temporary intracellular accumulation of glycogen also was observed in *Thiocapsa roseopersicina* during growth

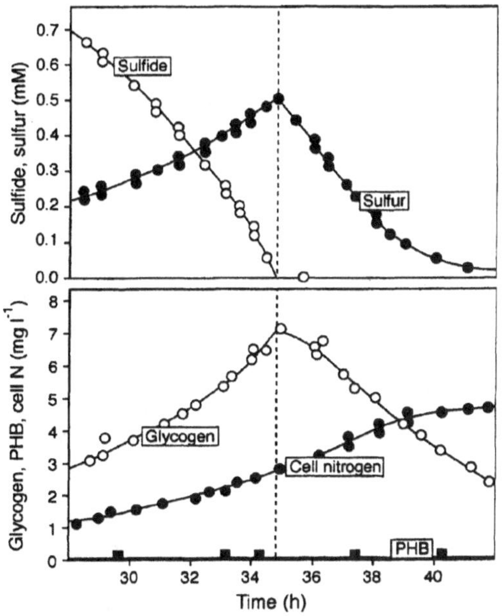

Fig. 1. Electron donor oxidation and product formation during batch growth of *Chromatium vinosum* strain DSM 185. Redrawn from Van Gemerden and Beeftink, 1978.

on sulfide (Visscher et al., 1990), thiosulfate (De Wit and Van Gemerden, 1987), polysulfide (S_3^{2-}) (Visscher et al, 1990), and dimethyl sulfide (Visscher and Van Gemerden, 1991).

When growth of *Chromatium vinosum* was inhibited by the addition of growth inhibitors, enhanced rates of glycogen formation were observed whereas substrate oxidation rates remained constant (Van Gemerden and Beeftink, 1978). Glycogen synthesis, together with the intracellular storage of S° thus enables continued growth after external substrates have been exhausted. High rates of carbon dioxide incorporation and sulfide oxidation also have been reported to occur in short-term experiments with *Chromatium okenii* (Trüper, 1964a,b; Trüper and Schlegel, 1964), *Chromatium weissei* and *Chromatium vinosum* (Van Gemerden, 1974). In view of the relatively low specific growth rates of purple sulfur bacteria (see Van Gemerden, 1984), these observations suggest that glycogen was synthesized.

Sulfide-limited cultures of *Chromatium vinosum* synthesize glycogen at all dilution rates imposed, the rate roughly being proportional to the dilution rate except for dilution rates close to μ_{max}. At dilution rates < 85% of μ_{max} about 5% of the reducing power obtained by sulfide and sulfur oxidation is channeled into the synthesis of glycogen. At higher specific growth rates this value may increase to 14% (Beeftink and Van Gemerden, 1979).

Immediately after removal of nutrient limitation (addition of sulfide directly to the culture over and above that supplied with the metering pump) increased rates of glycogen synthesis were observed. After the addition of chloramphenicol cultures showed similar high rates as observed in batch cultures, irrespective of the dilution rate imposed. Apparently, the maximum specific rate of synthesis in this organism is unaffected by the specific growth rate.

Although sulfur and glycogen are stored during sulfide limited growth of *Chromatium vinosum*, their content is highly increased when the organisms are either light or phosphate limited. A conceptual model of how these factors interact to support growth has been sketched in Fig. 2. According to this model, the metabolic rates of the different processes required for growth must be high enough to support growth at a given rate. A low supply of N or P will reduce the specific growth rate; however all the requirements for photosynthesis are present and the organism keeps synthesizing organic carbon which is then routed towards synthesis of glycogen. In the event of a reduced energy supply such as during light limitation, organic carbon will be formed at a lower rate. However, since even under these conditions glycogen is synthesized, glycogen synthesis obviously competes with growth processes for the products of photosynthesis.

Some data illustrating this point have been plotted in Fig. 3 as the amount of reducing power stored per mg protein, as either glycogen or sulfur. The results correspond to *Chromatium vinosum* growing at a constant specific growth rate (0.07 h^{-1}) but subject to three different types of limitation. The lowest contents are found when the organisms are sulfide limited. Phosphate limitation results in much higher contents of both sulfur and glycogen which share, in roughly equal proportions, the reducing power stored. Under light limitation, the content of sulfur is as high as during P limitation, however, the amount of glycogen is lower, as could be expected in a situation of insufficient energy supply.

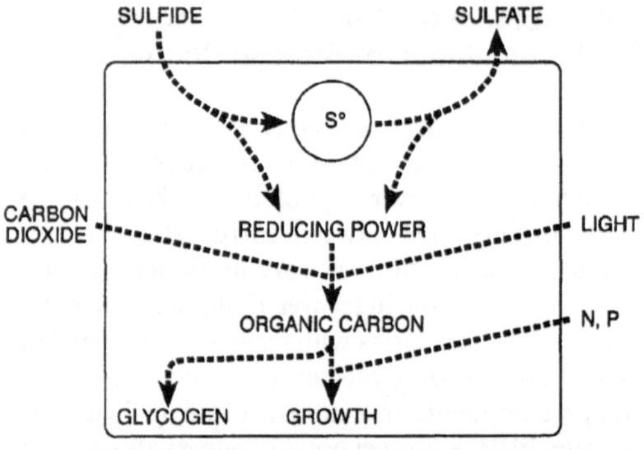

Fig. 2. Interaction of the main resources required for autotrophic growth in phototrophic sulfur bacteria. The metabolic rates of the different processes must be high enough to support growth at a given rate. A low supply of N or P will reduce the specific growth rate; however, all the requirements for photosynthesis are present and the organism accumulates glycogen. During light limitation, organic carbon will be formed at a lower rate. However, even under this condition glycogen is synthesized and thus competes with growth for the products of photosynthesis. Sulfur will be accumulated whenever sulfide is present.

III. Physiological Role of Storage Compounds

A. Experimental Studies Under Fluctuating Conditions

Oxygen exerts an inhibitory effect on the synthesis of photopigments by anoxygenic phototrophic bacteria. Incubation of *Thiocapsa roseopersicina* under oxic conditions for a prolonged period of time results in completely colorless cells which grow chemotrophically. Consequently, the yield per reducing equivalent is severely reduced to about one third of that obtained during phototrophic growth (De Wit and Van Gemerden, 1987) and is comparable to that found in genuine chemolithotrophs such as *Thiobacillus* sp. (Kelly, 1982). Once exposed to the presence of oxygen, the time course of the concentration of BChl *a* in a continuous culture of *Thiocapsa roseopersicina* followed the theoretical wash-out curve, suggesting that this photopigment was not actively degraded. When present, BChl*a* remained functional. It was observed that the protein yield during incubations in 24-h oxic/anoxic regimen, in which the oxic period was 4 to 21 h, was virtually identical to that observed during continuously anoxic conditions despite the fact that the photopigment content was decreasing in proportion to the length of the oxic period. Incubation at a 3h anoxic/21h oxic regime resulted in a BChl*a* content equal to about

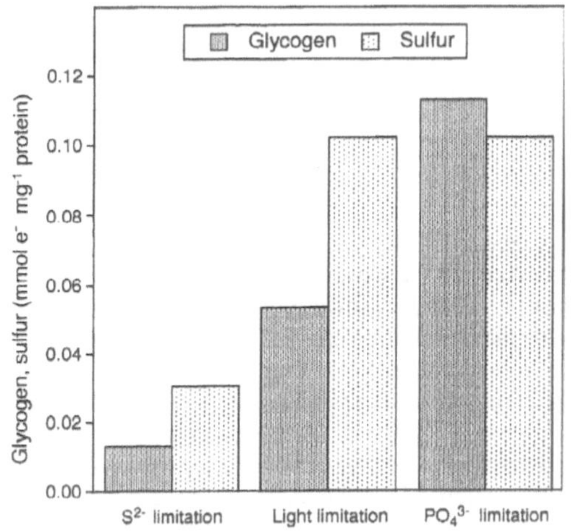

Fig. 3. Storage of reducing power as either glycogen or sulfur in sulfide, light, or phosphate limited steady state cultures of *Chromatium vinosum* strain DSM 185 growing at D = 0.07 h^{-1}. (Drawn using data from Van Gemerden et al., 1990)

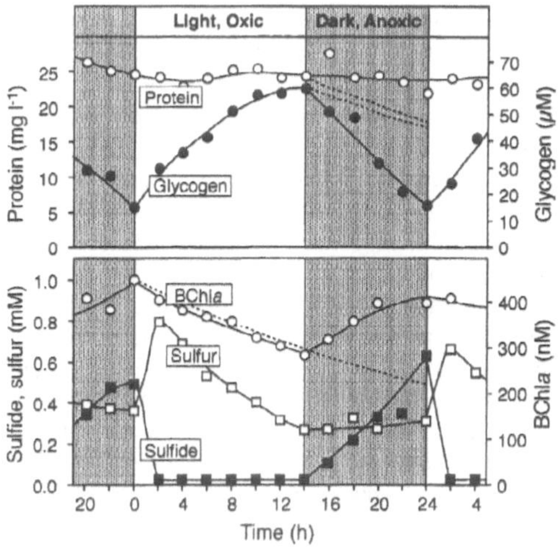

Fig. 4. Time course of the concentration of BChl*a*, protein, glycogen and S° in a continuous culture of *Thiocapsa roseopersicina* strain M1 after cultivation for five volume changes under a 14h oxic-light/10 h anoxic-dark regimen. The theoretical washout curves expected if synthesis did not occur are shown as dashed lines. (Redrawn from DeWit and Van Gemerden, 1990b.)

20% of that observed during continuously anoxic conditions. Apparently, under the latter conditions these organisms have a large surplus of photopigments (De Wit and Van Gemerden, 1990a). At even shorter anoxic periods, however, insufficient photopigments are synthesized to allow full phototrophic growth and the organism shifts to a combined photo-chemotrophic metabolism, as could be judged from the yield (Schaub and Van Gemerden, 1994).

The role of glycogen to sustain phototrophic growth becomes apparent when *Thiocapsa* is incubated in a regimen in which light conditions were combined with the presence of oxygen, and dark conditions with anoxia. After being incubated at such a regimen for a prolonged period of time, a repeating pattern of phenomena emerged. As expected, during the oxic-light periods no synthesis of BChl*a* took place, in contrast to anoxic-dark periods. Concomitantly, glycogen, accumulated during the preceding oxic-light period using the BChl still present, was degraded in the dark (Fig. 4). It was calculated on the basis of known synthetic pathways of bacteriochlorophyll and spirilloxanthin, the major carotenoid of *Thiocapsa*, that, depending on the regimen employed, degradation of 20-50% of the stored glycogen could account for the observed synthesis of photopigments in the dark (De Wit and Van Gemerden, 1990a). As was judged from the growth yield, the organisms were growing phototrophically and did not use their chemotrophic potential. These data show the functional role of glycogen during incubation under circumstances mimicking environmental conditions.

B. Dark Metabolism and Survival

Anoxygenic phototrophic bacteria, for which light is the ultimate source of energy, would be deprived of energy for maintenance purposes in the dark unless materials such as glycogen were stored in the light. It thus appears ecologically relevant that at least some glycogen is stored by cells facing sulfide limitation (see Section II).

From an energy point of view, the most efficient way to use glycogen in the dark would be to convert each glycosyl moiety to two carbon dioxide and two acetyl-CoA, followed by a conversion of the latter to acetate via acetyl phosphate. Taking into account that the initial phosphorylation does not require the participation of ATP but instead is carried out with P_i, five ATP could be produced from each glycosyl moiety. From a carbon-preserving point of view, however, this metabolic route is not very efficient, since all carbon, initially fixed at great expense through the Calvin cycle, would be excreted. The possibility of re-assimilation of acetate in the subsequent light period depends on the activities of other organisms, a fact often overlooked by scientists studying pure cultures. It appears that another way to degrade glycogen is feasible in purple sulfur bacteria. This metabolic route involves the condensation of two acetyl-CoA to form one aceto-acetyl-CoA, which, after reduction to 3-hydroxybutyryl-CoA, can be polymerized to form poly-3-hydroxybutyrate (PHB). Following this route would result in less energy to become available upon glycogen degradation (three ATP for each glycosyl moiety instead of five), however, four out of six carbon atoms present in glycogen would be preserved in PHB. In both pathways NADH is formed. In the first pathway eight [H] are produced in the breakdown of one glycosyl moiety, and in the second pathway six. With no energy input required, S° could serve as an acceptor for these electrons, resulting in the formation of sulfide (mid-point potentials: $NAD^+/NADH = -320$ mV, $S°/H_2S = -280$ mV).

Utilization of S° in the dark accompanied by the production of sulfide has been observed both in cultures of the large celled species *Chromatium okenii* (Trüper and Schlegel, 1964), and in cultures of the small-sized *Chromatium vinosum* (Van Gemerden, 1968b) in which PHB accumulated in agreement with the following stoichiometry:

$$C_6H_{12}O_6 + H_2O \rightarrow C_4H_8O_3 + 2\ CO_2 + 6\ [H]$$

$$3\ S° + 6\ [H] \rightarrow 3\ H_2S$$

$$C_6H_{12}O_6 + 3\ S° + H_2O \rightarrow C_4H_8O_3 + 2\ CO_2 + 3\ H_2S \qquad (1)$$

Hendley (1955) observed that acetate was formed in the endogenous metabolism of 'Thiorhodaceae' (Chromatiaceae). However, the organisms, grown in media containing hydrogen as the electron donor, did not contain intracellular S°, thus making the formation of sulfide inconceivable. Data on the production of hydrogen, energetically a less favorable option, were not reported. De Wit (1989) observed in *Thiocapsa roseopersicina* cultures which were repeatedly replenished with sulfide in order to reach high glycogen concentration, that dark incubation did not result in PHB accumulation. Instead, acetate was excreted in addition to sulfide. The molar ratio of glycosyl moieties to acetate and sulfide was 1 to 2.4 to 3.7. These phenomena, so far not observed in pure cultures of other anoxygenic phototrophs, suggest the following stoichiometry:

$$C_6H_{12}O_6 + 2\ H_2O \rightarrow 2\ CH_3COOH + 2\ CO_2 + 8\ [H]$$

$$4\ S° + 8\ [H] \rightarrow 4\ H_2S$$

$$C_6H_{12}O_6 + 4\ S° + 2\ H_2O \rightarrow 2\ CH_3COOH + 2\ CO_2 + 4\ H_2S \qquad (2)$$

Liebergesell et al. (1991) showed that some members of the Chromatiaceae do not (or hardly) form PHB under growing conditions although they have the metabolic capacity to do so (see Table 2). Therefore, although it is tempting to consider the two different ways of glycogen degradation described above as 'strategies', data from more organisms growing under various conditions are required before a firm statement can be made in this respect.

C. The Role of S° as Electron Storage in Purple and Green Sulfur Bacteria

As shown above, the S° in Chromatiaceae can be regarded a genuine storage compound. In fact, S° is a powerful electron donor: during its oxidation to SO_4^{2-} six electrons are released, compared to two electrons liberated during the oxidation of H_2S to S°. Due to the intracellular deposition, no other individual than the organism itself will be able to profit from the temporary accumulation of S°. This can be demonstrated using sulfide-limited chemostats by stepwise increasing the sulfide concentration in the reservoir ($S_{R\text{-sulfide}}$). According to the theory of continuous cultivation, the concentration of external substrates at any given value of the dilution rate D depends on the kinetic characteristics of the organism cultivated and is not influenced by the concentration of the limiting substrate in the reservoir solution. Accordingly, in sulfide limited continuous cultures of *Chromatium vinosum*, the steady state sulfide concentration is identical at increasing values of $S_{R\text{-sulfide}}$, however, the S° concentration is proportional to the magnitude of $S_{R\text{-sulfide}}$ as is biomass. As a consequence, the ratio of S° to protein was constant.

In Chlorobiaceae S° is deposited outside the cells. Nevertheless, the S° concentration in cultures of *Chlorobium limicola* also appeared to be affected by the magnitude of $S_{R\text{-sulfide}}$, indicating that the S° produced by one individual is not available to other individuals. This is further illustrated by the fact that in chemostat cultures of *Chlorobium*, as in cultures of *Chromatium*, the ratio of S° to protein was constant. The mechanistic explanation of this phenomenon in

Chlorobium could be that the S° is retained by the formation of extracellular capsules or pili, as described for other strains (Cohen-Bazire, 1963; K Schmidt, pers. comm.). Consequently, S° in *Chlorobium*, despite being deposited outside the cells, has to be regarded a storage product (Van Gemerden, 1986). Data on other anoxygenic phototrophs are lacking at the moment, and it would be particularly interesting to know how *Ectothiorhodospira* species would react.

IV. Storage Compounds in Natural Populations of Phototrophic Sulfur Bacteria

Storage inclusions also have been quantified in natural populations of phototrophic sulfur bacteria. In such cases, their distribution in space follows rather closely the distribution of bacterial biomass. This point is illustrated rather clearly in Fig. 5 (redrawn after Guerrero et al., 1985a), which depicts the variations with depth of bacterial abundance, sulfur, glycogen and PHB in a summer day in Lake Cisó (Spain), when the water column was stratified. The community of phototrophic sulfur bacteria consisted of green sulfur bacteria and two purple sulfur bacteria, a medium sized *Chromatium* and a smaller aggregate forming organism later identified as *Amoebobacter* (Pedrós-Alió and Guerrero, 1993). Green sulfur bacteria constituted only a small fraction of the community (Fig. 5A) and therefore, it can be assumed that the storage compounds measured resided mainly with the purple sulfur bacteria, which represented more than 90 % of the total biovolume. Although glycogen, sulfur and PHB obviously are more abundant at the peak of bacterial biomass, their distribution says little about the physiological state of the organisms. To gain some insight on this point, the data from Fig. 5 have been used as a basis to calculate the specific contents of each of the inclusions at each depth. The results have been plotted in Fig. 6 together with the vertical profile of light penetration at that moment. Light was almost extinguished at 2 m and phototrophic sulfur bacteria below that depth were facing virtual darkness. The specific contents of sulfur, glycogen and PHB changed with depth, in agreement with what could be expected from such a situation. Cells in the upper part of the bacterial plate where light irradiance is high, are loaded with glycogen and sulfur, and contain little PHB. As they enter the twilight zone and begin to sink, they start using up their intracellular stores of glycogen and sulfur, simultaneously depositing PHB as an end product of their dark metabolism. However, in most cases this is a one way trip and, unless given the chance to return to the surface layers (such as during holomictic overturn), the organisms eventually die. This loss of viability with depth has been documented by Van Gemerden et al. (1985).

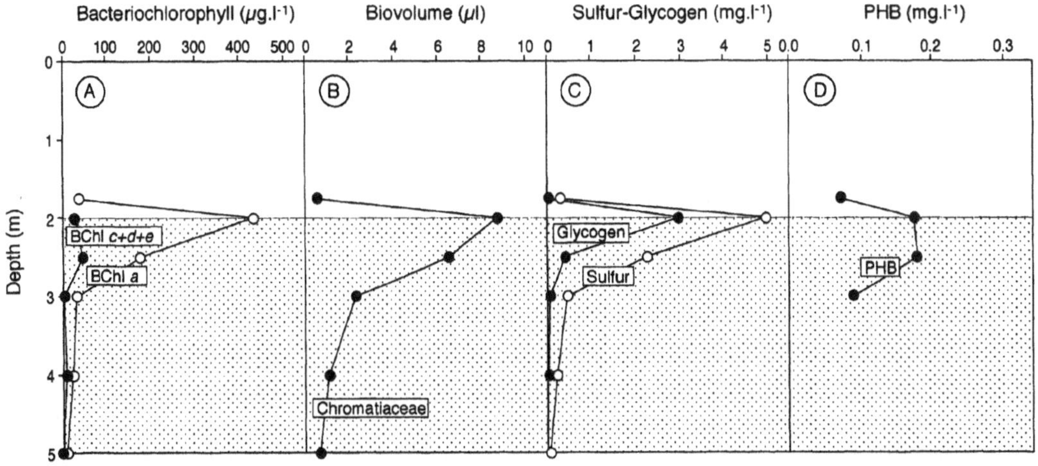

Fig. 5. Vertical distribution of bacteriochlorophylls (A), biovolume of Chromatiaceae (B), glycogen and sulfur (C) and PHB (D) in Lake Cisó (6 July, 1982, at zenith). Redrawn from Guerrero et al., 1985.

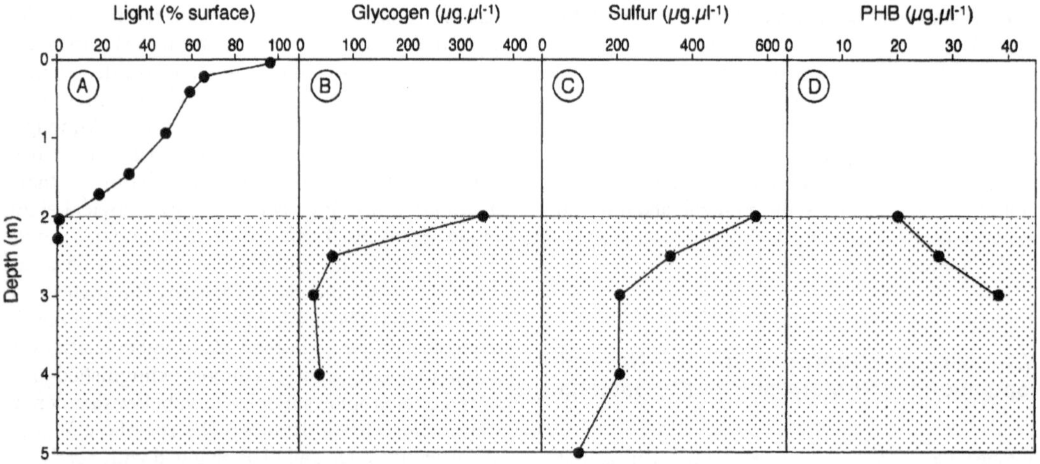

Fig. 6. Vertical distribution of irradiance (expressed as a % of the surface irradiance) (A), and specific contents of glycogen (B), sulfur (C) and PHB (D) in Lake Cisó (6 July, 1982, at zenith). Calculated and redrawn from data in Guerrero et al., 1985.

The specific content of sulfur and PHB also has been quantified in ultrathin sections corresponding to samples from Lake Cisó (Esteve et al., 1990). The results of this study indicate that the specific content of PHB increases with depth, and that this increase is not due to an increase in the size of the PHB inclusions, but rather to an increase in their number. The specific content of sulfur, however, does not change with depth, an observation which clearly disagrees with the data shown in Fig. 6. Similar differences between variations in the specific content of sulfur and variations in cell size (Mas and Van Gemerden, 1987) led the authors to postulate that utilization of intracellular sulfur might proceed without a concomitant reduction in the structure of the sulfur globule which would then remain as an empty shell until further degradation of the structure occurred. Clarification of this point requires, no doubt, further studies.

Besides their involvement in the rather hopeless task of maintaining viability during sinking to the sediment, storage inclusions carry out a remarkable role during day/night fluctuations. This has been shown in studies conducted both in Lake Cisó (Guerrero et al., 1985a, Van Gemerden et al., 1985) and in Lago di Cadagno Switzerland, Del Don et al., 1991; Bachofen et al., 1991).

Both in Lake Cisó (Fig. 7) and in Lago di Cadagno the concentration of sulfide at the depth of maximum biomass increased during the night, a phenomenon attributed to the dark metabolism of the organism. Concomitantly, glycogen concentrations were reported to decrease during the night. In Lake Cisó sulfide production and glycogen depletion were accompanied by an increase in the concentrations of PHB (Van Gemerden et al., 1985). In Lago di Cadagno, acetate, instead of PHB, was produced in the dark (Del Don et al., 1991). Thus, the phenomena observed in Lake Cisó appear to follow the stoichiometry given in equation 1, whereas those in Lago di Cadagno rather point to a stoichiometry similar to that described by equation 2 (see above). Attention is focused to the fact that the dominant organisms in these lakes belong to different species. Detailed field data, in combination with experimental studies on isolated strains are required to evaluate the possibility of different strategies.

V. Physical Consequences of the Accumulation of Storage Compounds

Storage inclusions accumulate in amounts which can constitute a large fraction of the dry weight of the organism. This accumulation has an inevitable effect on some physical properties of the organisms such as their size and specific density.

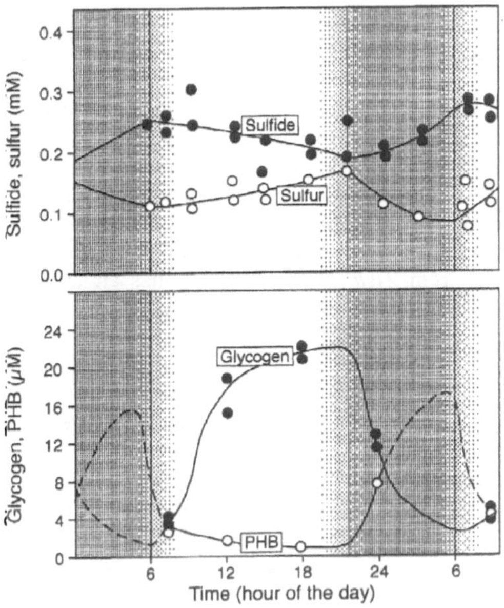

Fig. 7. Diel fluctuations in the concentrations of sulfide, sulfur, glycogen and PHB at a depth of 2 m in Lake Cisó on 6-7 July, 1982. Redrawn from Van Gemerden et al., 1985.

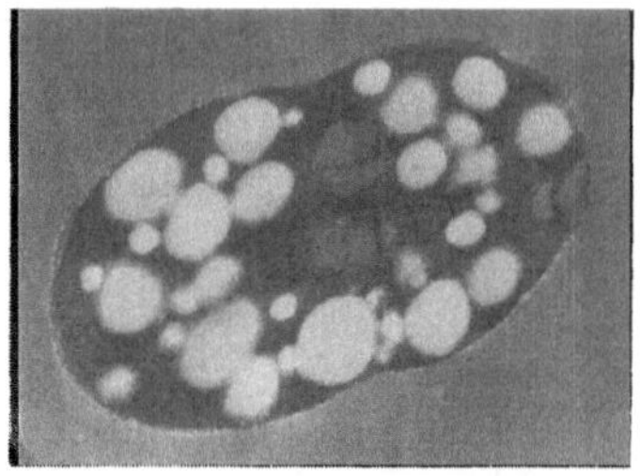

Fig. 8. Ultrathin section of a cell of *Chromatium minus* collected from the bottom of the plate at lake Cisó. The cell contains a large amount of PHB (white inclusion bodies) and a smaller amount of sulfur (dark-gray inclusions). Electron micrograph provided by I. Esteve.

A. Cell Size

Electron micrographs (Fig. 8) usually show a large part of the intracellular volume occupied by storage inclusions. Electron microscopy studies carried out by Esteve et al. (1990) in natural samples indicated that sulfur occupied as an average 7% of the total cell volume, while PHA granules filled between 0.6 and 12.2% of the intracellular space. Experiments carried out in the laboratory (Mas and Van Gemerden, 1987) under conditions which favored the accumulation of sulfur showed that cells of *Chromatium vinosum* in which the content of sulfur was 25% of the dry weight, had 25% of their total cell volume occupied by sulfur inclusions. In order to accommodate structures of this size, the volume of the cell must increase. Variations in cell size related to the accumulation or depletion of intracellular sulfur have been repeatedly reported in the literature (Van Gemerden, 1968a, Guerrero et al, 1984; Mas and Van Gemerden, 1987; Montesinos, 1987). However, the fact that cell size can also change as a function of the specific growth rate of the organism (Schaechter et al., 1958; Shehata and Marr, 1971) often has hampered the discrimination between the effect of both factors. Experiments carried out with *Chromatium vinosum* undergoing a shift from a low to a high dilution rate in a chemostat (0.029 → 0.086 h^{-1}, app. 25 → 75% of μ_{max}) showed an increase in cell size from 3.23 to 5.45 μm^3 (Fig. 9). The contribution of sulfur inclusions to this increase was 1.35 μm^3, while 0.87 μm^3 were considered a consequence of the increased specific growth rate (Mas and Van Gemerden, 1987). These observations indicate that when changes in the content of sulfur inclusions occur along with variations in the specific growth rate, the content of sulfur is still the main determinant of the variations in volume observed.

B. Cell Density

Storage inclusions are compact structures with a specific density higher than that of the surrounding cytoplasm. Therefore, their deposition inside the cell usually results in an increase in the density of the organism. This phenomenon, which has been well documented for storage carbohydrates in *Escherichia coli* (Mas et al., 1989) and for poly(3-hydroxybutyrate) in *Alcaligenes eutrophus* (Pedrós-Alió et al., 1985), is also observed in phototrophic sulfur bacteria especially in relation to the accumulation of sulfur which seems to be the main factor in density differences (Table 3). Studies carried out in *Chromatium warmingii* and *Chromatium vinosum*, show increases of 0.037 pg.μm^{-3} and 0.050 to 0.064, respectively, when undergoing sulfur accumulation. Similar studies carried out in *Chlorobium limicola* which deposits sulfur extracellularly indicate a similar effect, with an increase of 0.029 pg.μm^{-3}. This suggests that sulfur globules of green sulfur bacteria,

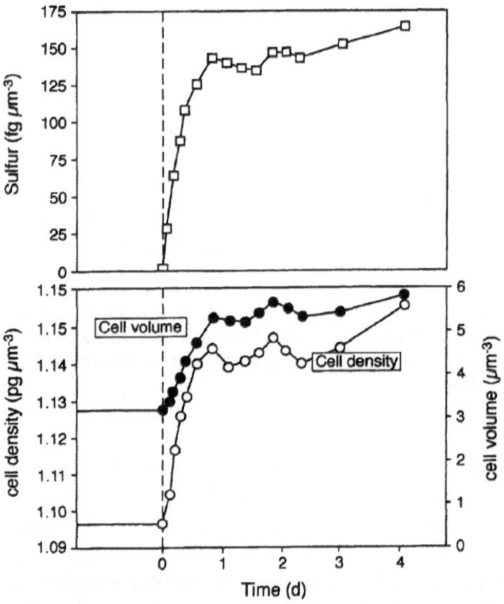

Fig. 9. Variation in the specific content of sulfur (upper panel) and cell volume and buoyant density (lower panel) in a sulfide-limited continuous culture of *Chromatium vinosum* strain DSM 185 during a shift-up from a low (0.029 h^{-1}) to a high dilution rate (0.086 h^{-1}). Redrawn from Mas and Van Gemerden, 1987.

although accumulated extracellularly, somehow might remain attached to the cell, thus supporting the idea (Van Gemerden, 1986) that sulfur in *Chlorobium* should also be considered a storage polymer. Carbohydrates, which are usually present in amounts much lower than sulfur (see Table 1), have a marginal impact on density which only becomes relevant in organisms such as *Pelodictyon phaeochlathratiforme* in which storage carbohydrates can account for as much as 35% of the dry weight (Overmann et al, 1991). In this case, the observed increase in density (0.037 pg.μm^{-3}) is comparable to the changes due to the presence of sulfur in other organisms. Variations in density as a function of the amount of inclusions are described by a hyperbolic function which asymptotically approaches a maximum value, set by the actual density of the inclusion being accumulated (Mas et al., 1985).

Experimental determinations of this 'in vivo' density for the sulfur inclusions in *Chromatium vinosum* and *Chromatium warmingii* provide different values for each organism (1.22–1.23 and 1.132, respectively) suggesting that the structure of the sulfur globule is different in each organism (Guerrero et al, 1984; Mas et al., 1985). Although no similar data on storage carbohydrates and PHA are available for phototrophic sulfur bacteria, studies performed in *Escherichia coli* (Mas et al., 1985; Mas et al., 1989) and in *Alcaligenes eutrophus* (Pedrós-Alió et al., 1985; Mas et al., 1985) indicate that the in vivo density of the storage carbohydrate in *Escherichia coli* is somewhere between 1.21–1.29 pg.μm^{-3}, while PHB from *Alcaligenes eutrophus* has an in vivo density of 1.15 pg.μm^{-3}. The values mentioned above constitute the highest densities the organisms could reach if virtually all the intracellular space were filled with inclusion bodies. However, this is hardly ever the case, and accumulation usually stops well before reaching these densities. Occasional claims of unusually high cell densities are often the consequence of the utilization of high osmolarity media such as sucrose for the construction of gradients used in density determination. While these media might prove suitable for the separation of fractions with different densities, they are completely inadequate to carry out analytical determinations because loss of water from the cell due to osmotic stress results in artificially high densities. An extensive discussion of this topic can be found in Guerrero et al. (1985b).

C. Impact on Sedimentation Rates

Many phototrophic sulfur bacteria develop in planktonic environments where the increase in size and density derived from the presence of storage inclusions potentially could increase their sinking speed and, thus, favor their removal from the water column (see Chapter 4 by Van Gemerden and Mas).

However, field studies carried out in Lake Cisó failed to find a correlation between the actual densities of the organisms in the lake and the rate at which they were sinking (Pedrós-Alió et al., 1989). Despite the fact that theoretical sinking speeds, calculated according to Stoke's law (1.6-18.4 cm.d^{-1}) were comparable to sinking speeds actually measured in the laboratory by other authors (19.2 cm.d^{-1} Matsuyama, 1991), the rates measured were well below the minimum predicted. Cell motility, water turbulence and probably, the presence of loosely attached slime layers might have had an effect on sedimentation not predicted by Stokes law, and thus, have lead to the observed discrepancies.

Although their direct effect on sinking rates seems

Table 3. Variation in the specific density of several phototrophic sulfur bacteria, related to changes in the content of different storage compounds

Organism	Main Storage	Specific content (fg.μm^{-3})	Density (pg.μm^{-3})	Max-Min[a]	Reference
Chromatium vinosum	Sulfur	0	1.097	0.050	Mas and Van Gemerden 1987
"	"	18	1.100		"
"	"	154	1.147		"
Chromatium vinosum	Sulfur	0	1.096	0.064	Guerrero et al. 1984
"	"	294	1.160		"
Chromatium warmingii	Sulfur	0	1.071	0.037	Guerrero et al. 1984
"	"	178	1.108		"
Chlorobium limicola	Sulfur	0	1.094	0.029	Guerrero et al. 1985b
"	"	>0[b]	1.123		"
Pelodictyon phaeoclathratiforme	Carbohydrate	5	1.010	0.037	Overmann et al. 1991[c]
"	"	54	1.041		"

[a]Difference between the maximum and the minimum values reported in each experiment
[b]Determination carried out in a batch culture sampled at the begining of the incubation in the light and at the moment at which sulfide was depleted and sulfur accumulation in the culture was maximum.
[c]Specific contents calculated taking into account the specific contents of carbohydrate (9.3 and 34.5% of the dry weight) as well as the specific volumes (20.1 and 6.4 *ml*.mg^{-1} DW) provided by the authors.

irrelevant, inclusions might play an important role in the modulation of buoyancy in organisms containing gas vesicles. This function has been well documented in cyanobacteria (Kromkamp and Mur, 1984; Thomas and Walsby, 1985; Utkilen et al., 1985). In some of these organisms, accumulation or degradation of carbohydrate, which actually acts as a ballast, enables the organism to switch from positive to negative buoyancy and, thus, regulate its vertical position within the water column. Studies carried out in *Pelodictyon phaeochlathratiforme* (Overmann et al., 1991) suggest that carbohydrates might play a similar role; however, density of this organism is also determined by the presence of extracellular slime layers and by a certain amount of protein ballast.

VI. Concluding Remarks

Accumulation of storage compounds constitutes one of the most remarkable adaptive features of phototrophic sulfur bacteria. The capacity to synthesize them enables the organism to take advantage of resource pulses in a limiting environment or, on the contrary, to survive short periods of starvation in a otherwise rich environment. The difference between both situations might be subtle but it certainly calls for different abilities. In the first case, the organism must posses low maintenance requirements, an efficient harvesting system which surpasses the potential for growth, together with a large storage capacity. In the second case, what is needed is the possibility to store a certain amount of substrate even when growth is limited by that substrate, in order to survive periods of complete deprivation later on. Phototrophic sulfur bacteria seem well suited to face both situations. When facing a pulse of substrates, they are able to take up and store considerable amounts of energy, carbon and reducing power. When growing in a fluctuating environment, they manage to store enough resources to be able to overcome the fluctuations.

Very often, physiological studies concentrate on how the organisms can adapt to a certain set of environmental conditions, under the assumption that these conditions will hold steady. However, in nature, this is seldom the case. Thus, physiologists should consider the artificial constraints imposed by a constant laboratory environment and try to understand the metabolic strategy of the organism living in a continuously changing environment.

Acknowledgment

JM wishes to acknowledge the support of the DGICYT (grant PB91-0075-C02-02)

References

Alper R, Lundgren DG, Marchessault RH, and Cote WA (1963) Properties of poly-β-hydroxybutyrate. I. General considerations concerning natural occurring polymer. Biopolymers 1: 545-556

Bachofen R, Israng R, Del Don C, Hanselmann K, and Tonolla M (1991) Chemo- and phototactic behavior of phototrophic bacteria under natural conditions in Lago di Cadagno, a meromictic alpine lake. Abstracts. VII Intl Symp Phototrop Prokary, p 158

Barham PJ (1990) Physical properties of poly(hydroxybutyrate) and poly(hydroxybutyrate-co-hydroxyvalerate). In: Dawes EA (ed) Novel Biodegradable Microbial Polymers, pp 81–96. Kluwer Academic Publishers, Dordrecht

Barnard GN and Sanders JKM (1989) The poly-β-hydroxybutyrate granule in vivo. A new insight based on NMR spectroscopy of whole cells. J Biol Chem 264: 3286–3291

Beeftink HH and Van Gemerden H (1979) Actual and potential rates of substrate oxidation and product formation in continuous cultures of *Chromatium vinosum.* Arch Microbiol 121: 161–167

Cohen-Bazire G (1963) Some observations on the organization of the photosynthetic apparatus in purple and green bacteria. In: Gest H, San Pietro A, and Vernon LP (eds) Bacterial Photosynthesis, pp 89–110 The Antioch Press, Yellow Springs, Ohio

Cole JA and Hughes DE (1965) The metabolism of polyphosphates in *Chlorobium thiosulfatophilum*. J Gen Microbiol 38: 65–72

Dawes EA (1992) Storage polymers in prokaryotes. In: Mohan S, Daw C, and Cole JA (eds) Prokaryotic Structure and Function: A New Perspective, pp 81–122. Cambridge University Press, Cambridge

Dawes EA and Senior PJ (1973) The role and regulation of energy reserve polymers in micro-organisms. Adv Microb Physiol 10: 135–266

De Wit R (1989) Interactions between phototrophic bacteria in marine sediments. PhD Thesis. University of Groningen. The Netherlands

De Wit R and Van Gemerden H (1987) Chemolithotrophic growth of the phototrophic sulfur bacterium *Thiocapsa roseopersicina*. FEMS Microbiol Ecol 45: 117–126

De Wit R and Van Gemerden H (1990a) Growth of the phototrophic purple sulfur bacterium *Thiocapsa roseopersicina* under oxic/anoxic regimens in the light. FEMS Microbiol Ecol 73: 69–76

De Wit R and Van Gemerden H (1990b) Growth and metabolism of the purple sulfur bacterium *Thiocapsa roseopersicina* under combined light/dark and oxic/anoxic regimens. Arch Microbiol 154: 459–464

Del Don C, Hanselmann KW and Bachofen R (1991) Metabolic responses of Chromatiaceae to diurnal cycles in a natural lake habitat. Abstracts. VII Intl. Symp Phototrop Prokary p 106

Dunlop WF and Robards AW (1973) Ultrastructural study of poly-β-hydroxybutyrate granules from *Bacillus cereus.* J Bacteriol 114: 1271–1280

Eimhjellen KE, Steensland H and Traetteberg J (1967) A *Thiococcus* sp. nov. gen., its pigments and internal membrane system. Arch Microbiol 59: 82–92

Ellar D, Lundgren DG, Okamura K and Marchessault RH (1968) Morphology of poly-β-hydroxybutyrate granules. J Mol Biol 35: 489–502

Esteve I, Montesinos E, Mitchell J G and Guerrero R (1990) A quantitative ultrastructural study of *Chromatium minus* in the bacterial layer of Lake Cisó (Spain). Arch Microbiol 153: 316–323

Ferguson SJ, Gadian DG and Kell DB (1979) Evidence from ^{31}P nuclear magnetic resonance that polyphosphate synthesis is a slip reaction in *Paracoccus denitrificans*. Biochem Soc Trans 7: 176–179

Griebel R, Smith Z and Merrick JM (1968) Metabolism of poly-β-hydroxybutyrate. I. Purification, composition, and properties of native poly-β-hydroxybutyrate granules from *Bacillus megaterium*. Biochemistry 7: 3676–3681

Guerrero R, Mas J and Pedrós-Alió C (1984) Buoyant density changes due to intracellular content of sulfur in *Chromatium warmingii* and *Chromatium vinosum*. Arch Microbiol 137: 350–356

Guerrero R, Montesinos E, Pedrós-Alió C, Esteve I, Mas J, Van Gemerden H, Hofman PAG and Bakker JF (1985a) Phototrophic sulfur bacteria in two Spanish lakes: Vertical distribution and limiting factors. Limnol Oceanogr 30: 919–931

Guerrero R, Pedrós-Alió C, Schmidt TM and Mas J (1985b) A survey of buoyant density of microorganisms in pure cultures and natural samples. Microbiologia 1: 53–65

Hageage Jr. GJ, Eanes ED and Gherna RL (1970) X-ray diffraction studies of the sulfur globules accumulated by *Chromatium* species. J Bacteriol 101: 464–469

Halvorson HO (1990) Some possible roles of polyphosphate in microorganisms. In: Dawes EA (ed) Novel Biodegradable Microbial Polymers, pp 205–211. Kluwer Academic Publishers, Dordrecht

Hara F, Akazawa T and Kojima K (1973) Glycogen biosynthesis in *Chromatium* strain D I. Characterization of glycogen. Plant and Cell Physiol 14: 737–745

Hendley DD (1955) Endogenous fermentation in Thiorhodaceae. J Bacteriol 70: 625–634

Hofman PAG, Veldhuis MJW and Van Gemerden H (1985) Ecological significance of acetate assimilation by *Chlorobium phaeobacteroides*. FEMS Microbiol Ecol 31: 271–278

Hughes DE, Conti SF and Fuller RC (1963) Inorganic polyphosphate metabolism in *Chlorobium thiosulfatophilum*. J Bacteriol 85: 577–584

Kawaguchi Y and Doi Y (1990) Structure of native poly(3-hydroxybutyrate) granules characterized by X-ray diffraction. FEMS Microbiol Lett 79: 151–156

Kelly DP (1982) Biochemistry of the chemolithotrophic oxidation of inorganic sulphur. Phil Trans R Soc London. 298: 499–528

Kromkamp JC and Mur LR (1984) Buoyant density changes in the cyanobacterium *Microcystis aeruginosa* due to changes in cellular carbohydrate content. FEMS Microbiol Lett 25: 105–109

Kulaev IS (1979) The Biochemistry of Inorganic Polyphosphates. John Wiley and Sons, New York

Kulaev IS (1990) The physiological role of inorganic polyphosphates in microorganisms: Some evolutionary aspects. In: Dawes EA (ed) Novel Biodegradable Microbial Polymers, pp 223–233. Kluwer Academic Publishers, Dordrecht

Liebergesell M, Hustede E, Timm A, Steinbüchel A, Fuller RC, Lenz RW and Schlegel HG (1991) Formation of poly(3-hydroxyalkanoates) by phototrophic and chemolithotrophic bacteria. Arch Microbiol 155: 415–421

Mas J and Van Gemerden H (1987) Influence of sulfur accumulation and composition of sulfur globule on cell volume and buoyant density of *Chromatium vinosum*. Arch Microbiol 146: 362–369

Mas J and Van Gemerden H (1992) Phosphate-limited growth of *Chromatium vinosum* in continuous culture. Arch Microbiol 157: 135–140

Mas J, Pedrós-Alió C and Guerrero R (1985) Mathematical model for determining the effects of intracytoplasmic inclusions on volume and density of microorganisms. J Bacteriol 164: 749–756

Mas J, Pedrós-Alió C and Guerrero R (1989) Variations in cell size and buoyant density of *Escherichia coli* K12 during glycogen accumulation. FEMS Microbiol Lett 57: 231–236

Mas-Castellà J (1991) Acumulación de poli-β-hidroxialcanoatos por bacterias. Distribución en la Naturaleza y Biotecnología. PhD Thesis. University of Barcelona. Spain

Matsuyama M (1991) Buoyant density of *Chromatium* sp.: Its effect on the blooming at an upper boundary of the H_2S layer in Lake Kaiike. Jap J Limnol 52: 57–63

Merrick JM (1978) Metabolism of reserve materials. In: Clayton RK and Sistrom WR (eds) The Photosynthetic Bacteria, pp 199–219. Plenum Press, New York

Montesinos E (1987) Change in size of *Chromatium minus* cells in relation to growth rate, sulfur content, and photosynthetic activity: A comparison of pure cultures and field populations. Appl Environ Microbiol 53: 864–871

Nicolay K, Hellingwerf KJ, Kaptein R and Konings WN (1982) Carbon-13 nuclear magnetic resonance studies of acetate metabolism in intact cells of *Rhodopseudomonas sphaeroides*. Biochim Biophys Acta 720: 250–258

Nicolson GL and Schmidt GL (1971) Structure of the *Chromatium* sulfur particle and its protein membrane. J Bacteriol 105: 1142–1148

Okamura K and Marchessault RH (1967) X-ray structure of poly-β-hydroxybutyrate. In: Ramachandran BM (ed) Conformation of Biopolymers, Vol 2, pp 709–720. Academic Press, New York

Ostrovsky DN, Sepetov NF, Reshetnyak VI and Sibel Dina LA (1980) Study of the localization of polyphosphates in cells of micro-organisms by high-resolution phosphorus-31 NMR at 145.78 MHz. Biokhimiya 45: 517–525

Overmann J, Lehmann S, and Pfennig N (1991) Gas vesicle formation and buoyancy regulation in *Pelodyction phaeoclathratiforme* (Green sulfur bacteria). Arch Microbiol 157: 29–37

Pedrós-Alió C and Guerrero R. (1993). Microbial ecology in Lake Cisó. Adv Microb Ecol 13: 155–209

Pedrós-Alió C, Mas J and Guerrero R (1985) The influence of poly-β-hydroxybutyrate accumulation on cell volume and buoyant density in *Alcaligenes eutrophus*. Arch Microbiol 143: 178–184

Pedrós-Alió C, Mas J, Gasol JM and Guerrero R (1989) Sinking speeds of free-living phototrophic bacteria determined with covered and uncovered traps. J Plankton Res 11: 887–905

Pfennig N and Trüper HG (1989) Anoxygenic phototrophic bacteria. In: Staley JT, Bryant MP, Pfennig N and Holt JG (eds) Bergey's Manual of Systematic Bacteriology, Vol 3, pp 1635–1709. Williams & Wilkins, Baltimore

Remsen CC (1978) Comparative cellular architecture of photosynthetic bacteria. In: Clayton RK and Sistrom WR (eds) The Photosynthetic Bacteria, pp 31–60. Plenum Press, New York

Reusch RN and Sadoff HL (1988) Putative structure and functions of a poly-β-hydroxybutyrate/calcium polyphosphate channel in bacterial plasma membranes. Proc Natl Acad Sci USA 85: 4176–4180

Schaechter M, Maaloe O and Kjeldgaard NO (1958) Dependency on medium and temperature of cell size and chemical composition during balanced growth of *Salmonella typhimurium*. J Gen Microbiol, 19: 592–606

Schaub BEM and Van Gemerden H (1994) Simultaneous phototrophic and chemotrophic growth in the purple sulfur bacterium *Thiocapsa roseopersicina* M1 . FEMS Microbiol Ecol 13: 185–196

Schmidt GL and Kamen MD (1970) Variable cellular composition of *Chromatium* in growing cultures. Arch Mikrobiol 73: 1–18

Schmidt GL, Nicolson GL and Kamen MD (1971) Composition of the sulfur particle of *Chromatium vinosum* strain D. J Bacteriol 105: 1137–1141

Shehata TE and Marr AG (1971) Effect of nutrient concentration on the growth of *Escherichia coli*. J Bacteriol 107: 210–216

Shively JM (1974) Inclusion bodies of prokaryotes. Ann Rev Microbiol 28: 167–187

Shively JM, Bryant DA, Fuller RC, Konopka AE, Stevens Jr. SE and Strohl WR (1989) Functional inclusion bodies in prokaryotic cells. Int Rev Cytol 113: 35–100

Sirevåg R and Ormerod JG (1977) Synthesis, storage and degradation of polyglucose in *Chlorobium thiosulfatophilum*. Arch Microbiol 111: 239–244

Stanier RY, Doudoroff M, Kunisawa R and Contopoulou R (1959) The role of organic substrates in bacterial photosynthesis. Proc Natl Acad Sci USA 45: 1246–1260

Steudel R, Holdt G, Visscher PT and Van Gemerden H (1990) Search for polythionates in cultures of *Chromatium vinosum* after sulfide incubation. Arch Microbiol 153: 432–437

Steudel R (1989) On the nature of the 'elemental sulfur' (S^o) produced by sulfur-oxidizing bacteria—a model for S^o globules. In: Schlegel HG and Bowien B (eds) Biology of Autotrophic Bacteria, pp 289–303. Science Tech Publ, Madison, and Springer, Berlin

Thomas RH and Walsby AE (1985) Buoyancy regulation in a strain of *Microcystis*. J Gen Microbiol 131: 799–809

Trüper HG (1964a) Sulphur metabolism in Thiorhodaceae. II. Stoichiometric relationship of CO_2 fixation to oxidation of hydrogen sulphide and intracellular sulphur in *Chromatium okenii*. Antonie van Leeuwenhoek 30: 385–394

Trüper HG (1964b) CO_2-Fixierung und Intermediärstoffwechsel bei *Chromatium okenii* Perty. Arch Mikrobiol 49: 23–50

Trüper HG and Hathaway JC (1967) Orthorombic sulphur formed by photosynthetic sulphur bacteria. Nature 215: 435–436

Trüper HG and Schlegel HG (1964) Sulphur metabolism in Thiorhodaceae. I. Quantitative measurements on growing cells of *Chromatium okenii*. Antonie van Leeuwenhoek 30: 225–238

Utkilen HC, Skulberg OM and Walsby AE (1985) Buoyancy regulation and chromatic adaptation in planktonic *Oscillatoria* species: alternative strategies for optimizing light absorption in stratified lakes. Arch Hydrobiol 104: 407–417

Van Gemerden H (1968a) Growth measurements of *Chromatium* cultures. Arch Mikrobiol 64: 103–110

Van Gemerden H (1968b) On the ATP generation by *Chromatium* in darkness. Arch Mikrobiol 64: 118–124

Van Gemerden H (1974) Coexistence of organisms competing for the same substrate: An example among the purple sulfur bacteria. Microb Ecol 1: 104–119

Van Gemerden H (1984) The sulfide affinity of phototrophic bacteria in relation to the location of elemental sulfur. Arch Microbiol 139: 289–294

Van Gemerden H (1986) Production of elemental sulfur by green and purple sulfur bacteria. Arch Microbiol 146: 52–56

Van Gemerden H and Beeftink HH (1978) Specific rates of substrate oxidation and product formation in autotrophically growing *Chromatium vinosum* cultures. Arch Microbiol 119: 135–143

Van Gemerden H, Montesinos E, Mas J and Guerrero R (1985) Diel cycle of metabolism of phototrophic purple sulfur bacteria in lake Cisó (Spain). Limnol Oceanogr 30: 932–943

Van Gemerden H, Visscher PT and Mas J (1990) Environmental control of sulfur deposition in anoxygenic purple and green sulfur bacteria. In: Dawes EA (ed) Novel Biodegradable Microbial Polymers, pp 247–262. Kluwer Academic Publishers, Dordrecht

Van Niel CB (1936) On the metabolism of Thiorhodaceae. Arch Mikrobiol 7: 323–358

Visscher PT and Van Gemerden H (1991) Photo-autotrophic growth of *Thiocapsa roseopersicina* on dimethyl sulfide. FEMS Microbiol Lett 81: 247–250

Visscher PT, Nijburg JW and Van Gemerden H (1990) Polysulfide utilization by *Thiocapsa roseopersicina*. Arch Microbiol 155: 75–81

Weast RC (1986-1987) Handbook of Chemistry and Physics. 67th edition. CRC Press Inc. Boca Raton. Florida

Williamson DH and Wilkinson JF (1958) The isolation and estimation of poly-β-hydroxybutyrate inclusions of *Bacillus* species. J Gen Microbiol, 19: 198–209

Winogradsky S (1887) Ueber Schwefelbacterien. Botanische Zeitung. 45: 489–507, 513–523, 529–539, 545–559, 569–576, 585–594, 606–610

Winogradsky S (1945) Microbiologie du sol. Masson et cie editeurs. Paris

Zaitseva GN, Gulikova OM and Kondrat'eva EN (1965) Biochemical changes in the cells of *Chromatium minutissimum* under photoautotrophic and photoheterotrophic conditions of growth. Mikrobiologiya 34: 577–583

Chapter 46

Degradation of Aromatic Compounds by Nonsulfur Purple Bacteria

Jane Gibson
Section of Biochemistry, Molecular and Cell Biology, Cornell University, Ithaca, NY 14853, USA

Caroline S. Harwood
Department of Microbiology, University of Iowa, Iowa City, IA 52242, USA

Summary

The nonsulfur purple bacteria exhibit remarkable versatility in aromatic compound transformations. In recent years it has become evident that these bacteria use a much wider range of lignin-derived monomers for photosynthetic and aerobic growth than was previously suspected. While aerobic attack on the benzene ring appears to follow patterns common to aerobic pseudomonads, the anaerobic pathway is radically different, and the reductive attack on the ring, first suggested for *Rhodopseudomonas palustris* growing on benzoate, has now become firmly established. Coenzyme A thioesters are involved in removal of sidechains from complex aromatic acids prior to attack on the ring itself, and also at all stages of degradation of the nucleus up to and including ring opening. Probable intermediates in the ring reduction reactions have been identified both in vivo and in vitro, but characterization of the enzymes and cofactors involved has yet to be achieved. Two aromatic acid CoA ligases have been purified and the corresponding genes cloned and sequenced. A regulatory gene involved in expression of the ligase needed for 4-hydroxybenzoate activation has also been identified and belongs to the cyclic AMP receptor protein family of transcriptional activators. Disruption of this gene has specific effects on aromatic acid metabolism. There is increasing evidence that the aromatic acid degradation

R. E. Blankenship, M. T. Madigan and C. E. Bauer (eds): Anoxygenic Photosynthetic Bacteria, pp. 991–1003.

process is similar in phototrophs and in denitrifying bacteria, but very little is known of the pathways followed in sulfate- or iron-reducing bacteria that utilize aromatic compounds. At present, methods for assessing which type of microorganism plays the major role in degrading natural or unnatural aromatic nucleus-containing compounds in natural environments are not adequate for determining how important the role of nonsulfur photosynthetic bacteria may be on a global scale.

I. Introduction

The photosynthetic bacteria have remarkable physiological flexibility, particularly in terms of energy source, which can originate from light, respiration using a number of electron acceptors in addition to oxygen, and even fermentation (Biebl and Pfennig, 1981). By comparison, the range of carbon sources used has appeared to be modest, with a general preference for fatty or dicarboxylic acids over the sugars favored by many conventional chemoorganotrophs. In the interests of obtaining rapid growth and large cell yields for experimental purposes, or of developing elective culture procedures, the range of organic molecules that can support growth of these bacteria has probably been greatly underestimated, however. This is well exemplified by the range of aromatic compounds that can be used as carbon source to support growth of nonsulfur purple bacteria if one is patient enough to observe cultures over long periods. It may be the case that substrates that can at best support only slow growth still serve as significant carbon sources in natural environments, particularly where there are relatively few organisms which compete for their use. This chapter will be concerned primarily with the physiological, biochemical and regulatory aspects of aromatic compound utilization by photosynthetic bacteria that have come into focus since this subject was reviewed in the Clayton and Sistrom volume on these microorganisms (Dutton and Evans, 1978).

II. Physiology

When Sher and Proctor (1960) first investigated the use of benzoate as a carbon source, most of the strains of *Rhodopseudomonas palustris*, as well as of *Rhodobacter sphaeroides*, *Rhodopseudomonas gelatinosa* (now *Rubrivivax gelatinosus*) and some of *Rhodospirillum rubrum* in the van Niel culture collection at Pacific Grove were found to grow well anaerobically in light with benzoate as substrate. Since then, a number of studies suggest that the range of photosynthetic bacteria that use aromatic compounds as carbon sources is more restricted, and the general experience has been that the ability is well-developed primarily in strains of *Rp. palustris* (e.g. Biebl and Pfennig, 1981; Harwood and Gibson, 1988; Madigan and Gest, 1988; Khanna et al. 1992). However, *Rhodospirillum fulvum* (Pfennig et al., 1965), *Rhodocyclus purpureus* (Pfennig, 1978) and strains of *Rubrivivax gelatinosus* (Rahalkar et al., 1991), *Rp. acidophila* (Yamanaka et al., 1983) as well as *Rhodomicrobium vannielii* (Wright and Madigan, 1991) have also been shown to grow at the expense of aromatic compounds.

If the range of species that can utilize aromatic compounds has decreased over the last years, the opposite is true for the range of compounds containing benzene nuclei that can be attacked by purple bacteria growing either photosynthetically or as aerobic chemoorganotrophs (Harwood and Gibson, 1988). The original work commonly explored growth only on benzoate or 4-hydroxybenzoate, but a more systematic investigation of potential substrates has since shown that hydroxylated and methoxylated aromatic acids and aldehydes are commonly used growth substrates, as are aromatic acids with extended side chains. Although most of these compounds support growth both in the presence and absence of oxygen, metabolism of some is restricted to aerobic or anaerobic phototrophic conditions. For example, benzoate, as had been recognized earlier, does not serve as an aerobic growth substrate for any strain of photosynthetic bacterium studied to date, although 4-hydroxybenzoate is commonly used under these conditions. Both benzoate and 4-hydroxybenzoate support rates of anaerobic, photosynthetic growth that are about half those obtained with more conventional carbon sources such as malate or succinate. Typical generation times for *Rp. palustris* growing on these aromatic acids are 12–14 h. ω-Phenylalkane carboxylates, with side chain lengths

Abbreviations: CoA – coenzyme A; *Rb.* – *Rhodobacter; Rp.* – *Rhodopseudomonas*

from three to eight carbons, are excellent substrates for *Rp. palustris* (Harwood and Gibson, 1988; Elder et al., 1992a; Madigan and Gest, 1988), as are several unsubstituted or 4-hydroxylated lignin-related monomers, such as *trans*-cinnamate and 4-hydroxycinnamate (4-coumarate).

It is clear from published work that there is a considerable amount of strain variation in the range of aromatic substrates utilized photosynthetically. The ability to degrade lignin-related compounds with 3-methoxyl or 3-hydroxyl substitutions is a strain-dependent trait in *Rp. palustris*, and these substrates typically support growth rates that are much lower than attained with benzoate (Harwood and Gibson, 1988). A recent report includes phenol among substrates supporting photosynthetic growth of a strain of *Rp. palustris* isolated from effluent of a petrochemical waste water treatment plant (Khanna et al., 1992). With the exception of this last report, a preexisting carboxyl group, or one readily derived from an alcohol or aldehyde substituent, appear to be essential for utilization. As a final example, a strain of *Rp. palustris* that is able to incorporate ^{14}C from isotopically labeled 3-chlorobenzoate when growing in mixtures of benzoate and the chlorinated analog, has been described (Kamal and Wyndham, 1990), although chlorobenzoates are generally toxic to this microorganism. A more versatile isolate has recently been described (van der Woude et al., 1994) which, unlike Kamal and Wyndham's (1990), is able to grow on 3-chlorobenzoate alone. When cultured with mixed chlorobenzoates, this strain also continues to metabolize 2-, 4- or 3,5- chlorobenzoates after exhaustion of the favored substrate 3-chlorobenzoate. Both of these studies suggest that chlorobenzoate-metabolizing enzymes are not normally expressed and that benzoate or 3-chlorobenzoate may be needed as cosubstrates to permit induction of these enzymes.

III. Biochemistry

A. Major Metabolic Themes of Aromatic Acid Degradation

The presence of an aryl group at the distal end of a long-chain fatty acid should offer little hindrance to a conventional oxidative attack on the molecule, although such a bulky group only a few carbons' removed from the carboxyl group probably demands some degree of enzyme specialization. The benzene ring itself is stabilized by about 125 kJ per mol of resonance energy, however, and this presents a formidable barrier to attack by any organism. Radically different biochemical strategies have evolved to meet this challenge under aerobic and anaerobic conditions. Although there are indeed differences in the details of the metabolic pathways used by different microorganisms, these represent relatively minor variations on broad general themes.

Aerobically, ring resonance is distorted by introducing hydroxyl groups, followed by dioxygenase-catalyzed ring cleavage either between or adjacent to vicinal hydroxyls. These reactions require participation of molecular oxygen and are thus restricted to aerobic conditions. Only one study of the pathway used for aerobic degradation of 4-hydroxybenzoate by *Rp. palustris* has been carried out (Hegeman, 1967), and we are not aware of detailed studies of the aerobic metabolism of aromatic compounds by other species of photosynthetic bacteria.

Anaerobically, the rings of only one, or perhaps two, aromatic acids are attacked. More complex molecules are first converted to the central intermediates benzoate or 4-hydroxybenzoate or their coenzyme A thioesters. The CoA modification serves several functions. It can facilitate uptake of substrate into the cell, since rapid conversion to a non-diffusible CoA derivative by a cytoplasmic enzyme with high affinity for its aromatic substrate will maintain a downhill concentration gradient between the cell and its environment. A high internal concentration of the substrate for the next reaction can be built up, which would assist in pushing an energetically unfavorable reaction. In fact, we have estimated the steady state intracellular benzoyl CoA concentration in cells of *Rp. palustris* growing on benzoate to be about 200μM. The increased size of the thioester also allows more interactions between these substrates and their enzymes. The ring itself is attacked reductively, and the reduced product subjected to a series of reactions leading to formation of a cyclic β-keto thioester, which is then cleaved to yield a 7-carbon dicarboxylic derivative. This molecule is readily converted to reactants in central metabolic pathways. These themes are illustrated in Fig. 1. Anaerobic transformations carried out by *Rp. palustris* are discussed in greater detail in the sections that follow.

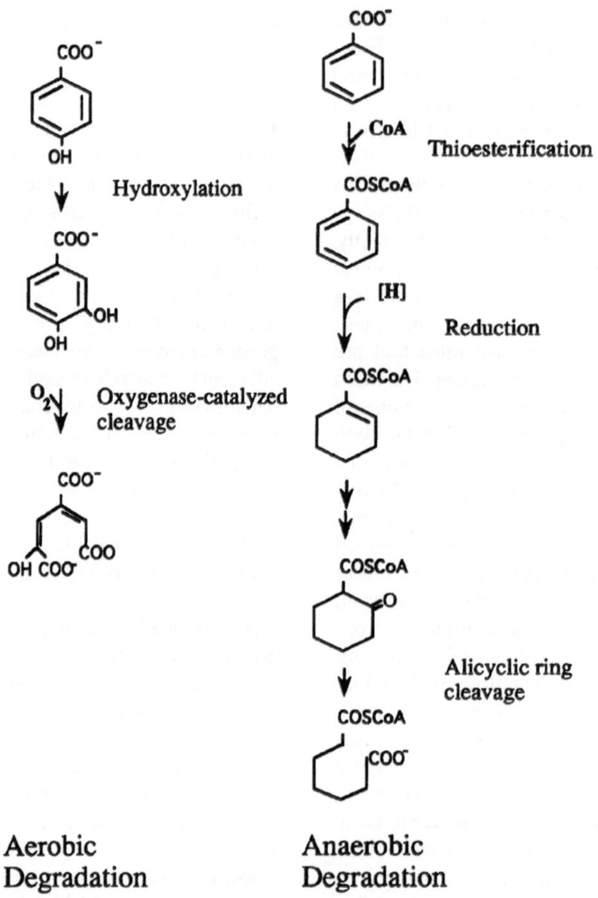

Fig. 1. Outline of major metabolic steps used by *Rp. palustris* for degrading aromatic acids aerobically and anaerobically.

B. Removal of Ring Substituents by Phototrophic Cells

Aromatic compounds with extended acyl side chains often support growth rates of *Rp. palustris* that are higher than those seen with simple aromatic acids, either aerobically or anaerobically (Harwood and Gibson, 1988; Elder et al., 1992a; Rahalkar et al., 1993). The pattern of molar growth yields in photosynthetic cultures growing on compounds with side chains of varying length strongly suggests that these are attacked by β-oxidation mechanisms (Elder et al., 1992a), since cell mass formed increases with side chain length, and ring degradation occurs only when successive removal of 2-carbon units yields benzoate or 4-hydroxybenzoate. In the strains used for these studies, phenylacetate is not attacked, and accumulates in cultures grown with aromatic substrates carrying even-numbered carbon side chains. By analogy with long-chain fatty acid β-oxidation, coenzyme A modification would be expected as the initial step, and two ligases have been detected in extracts of cinnamate-grown cells, one relatively specific for cinnamate among the substrates tested, and one also able to thioesterify compounds such as ferulate and caffeate that have additional ring substituents (Elder et al., 1992b). Expected intermediates in degradation of some of these compounds have been detected in culture supernatants. Benzoate was found transiently during growth on *trans*-cinnamate (Elder et al., 1992a) and both cinnamate and benzoate were detected when phenylvalerate was utilized (Rahalkar et al., 1993). Taken together, these findings suggest not only that the sidechain is degraded by stepwise removal of 2-carbon units, but that a complex interaction between thioesterification and hydrolysis reactions is occurring during this process (Elder and Kelly, 1994). It is also conceivable that the relatively plentiful energy supply (i.e. light) that may be enjoyed by photosynthetic bacteria permits a simple hydrolysis rather than the thiolytic cleavage typical of the β-oxidation sequence found in other aerobic and anaerobic microorganisms. Other than the initial thioesterification, these reactions have not yet been characterized in any detail. Side chain degradation evidently precedes modification of any other ring substituents, since protocatechuate was detected during growth on caffeate, and vanillate during growth on ferulate (Harwood and Gibson, 1988). The low cell yield of some strains on ferulate suggests that growth may be solely at the expense of the side chain and that, as in the case of long-chain substituents with even numbers of carbons, the aromatic nucleus itself is not always used (Harwood and Gibson, 1988; Elder et al., 1992b; Rahalkar et al., 1993). The reactions undergone by aromatic ring substituents are summarized in Fig. 2.

Although the suggested β-oxidation reactions fit well with all observations, some *Rp. palustris* strains appear also to be able to carry out an anaerobic removal of a single carbon unit, since benzoylformate and mandelate were used by our strains (Harwood and Gibson, 1988), and an unusual characteristic of the strain studied by Rahalkar et al. (1993) was photosynthetic growth on phenylacetate.

By analogy with other anaerobic systems, reductive dechlorination of 3-chlorobenzoate may be predicted to occur before reduction of the aromatic nucleus, but this has not actually been shown to be the case.

C. Anaerobic Attack on the Aromatic Ring

A novel reductive attack on benzoate during photosynthetic growth was first proposed by Dutton and Evans (1969), based on the results of classical isotope dilution experiments. Concentrated suspensions of *Rp. palustris* were incubated with radioactively labeled benzoate in the presence of several possible partially or fully reduced intermediates, which were subsequently reisolated and their specific radioactivity established. The results of these experiments suggested a pathway involving reductive ring saturation, followed by a series of reactions analogous to a β-oxidation, prior to opening the ring to yield pimelate (Figs 1 and 3). Support for this suggested pathway came from a series of mutants isolated by Guyer and Hegeman (1969) that had restricted abilities to utilize some of the proposed intermediates as growth substrates. Some years elapsed before it was fully recognized that the actual substrates involved in the benzoate degradation sequence were coenzyme A thioesters rather than the free acids (Whittle et al., 1976; Hutber and Ribbons, 1983; Harwood and Gibson, 1986; Evans and Fuchs, 1988).

Detection of excreted metabolites has played a key role in establishing routes of aerobic aromatic metabolism, as well as in supporting β-oxidation as a mechanism for degrading phenylalkanoates under anaerobic conditions in *Rp. palustris*. However, this approach has not been useful, at least to date, for detecting and analyzing intermediates of benzoate degradation. One reason for this may be that coenzyme A thioesters cannot pass through the cell membrane, and that very little intracellular hydrolysis to yield diffusible products occurs. In addition, small quantities of hydrolyzed intermediates that may have escaped from cells would be difficult to detect, even under ideal conditions, because of the low extinction coefficients of metabolites that have lost aromaticity. Several possible intermediates are also autoxidizable, and may merge with the growth substrate during preparations for analysis.

Although much remains uncertain about the pathway proposed, several recent findings have contributed some clarification to the reductive phase of the reaction scheme. Intracellularly, small quantities of two unconjugated cyclohexadiene derivatives—cyclohexa-2,5-diene-1-carboxylate and cyclohexa-1,4-diene-1-carboxylate_were detected in the putative CoA thioester pools of *Rp. palustris* growing exponentially on benzoate as carbon source (Gibson and Gibson, 1992), suggesting an initial two electron reduction analogous to the chemical reduction of the benzene nucleus in the Birch reaction (Birch and Smith, 1958). More recently, Fuchs and his colleagues have succeeded in demonstrating reduction of benzoyl CoA with cell extracts of a denitrifying pseudomonad (Koch and Fuchs, 1992), and detected the conjugated isomer cyclohexa-1,5-diene-1-carboxyl-CoA directly using NMR methods (Koch et al., 1993). They also found the same product

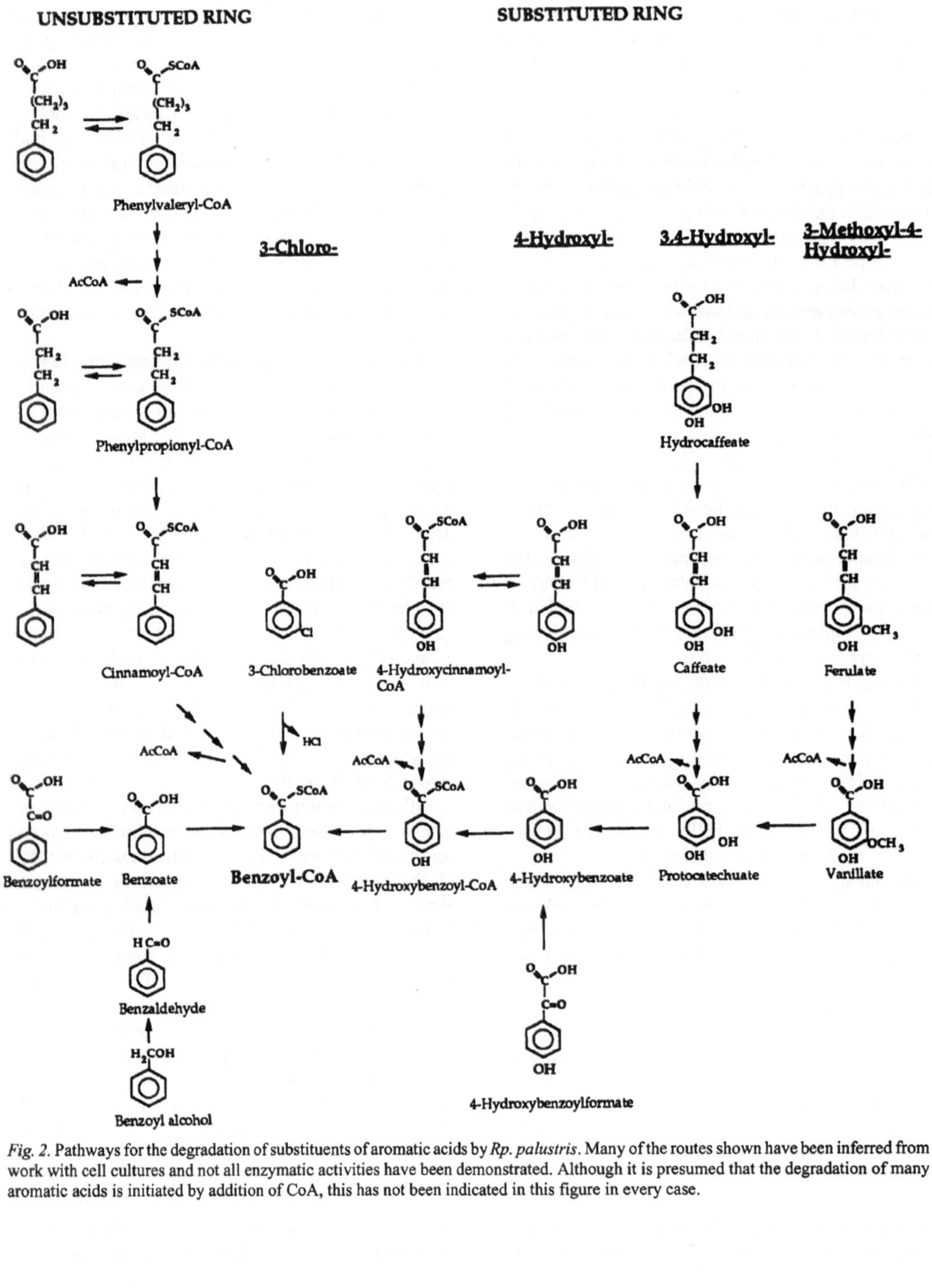

Fig. 2. Pathways for the degradation of substituents of aromatic acids by *Rp. palustris*. Many of the routes shown have been inferred from work with cell cultures and not all enzymatic activities have been demonstrated. Although it is presumed that the degradation of many aromatic acids is initiated by addition of CoA, this has not been indicated in this figure in every case.

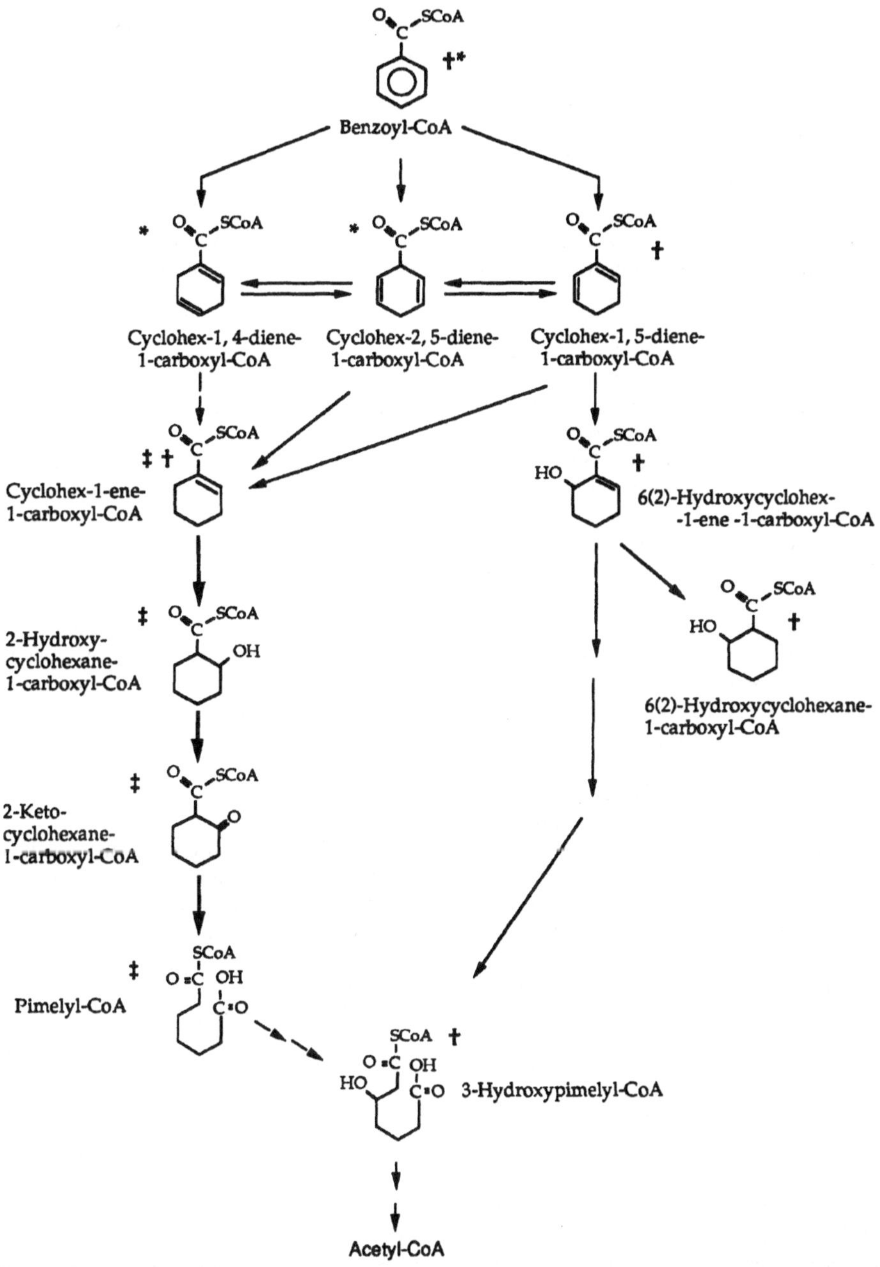

Fig. 3. A current view of the pathway proposed for anaerobic benzoate degradation in *Rp. palustris.* ‡ – metabolites from experiments of Dutton and Evans with whole cells (1969); * – detected in soluble contents of cells growing exponentially on benzoate (Gibson and Gibson, 1992); † – detected as products in cell-free reactions (Koch et al., 1993).

in experiments with *Rp. palustris* extracts. These additions to the pathway as proposed initially are shown in Fig. 3. The differences between the metabolites detected in extracts and in whole cells are not yet explained, but kinetic effects may significantly alter the distribution of intermediates in vivo and in vitro. The physiological electron donor in the initial ring reduction is not known.

It seems probable that reductive saturation of the benzene ring involves a single pathway, and that 4-hydroxybenzoate, as well as other aromatic acids with side chains in addition to a 4-hydroxyl group, such as 4-hydroxycinnamate, undergo reductive dehydroxylation before ring reduction. Free benzoate has been detected in the culture medium of some strains of *Rp. palustris* during growth on 4-hydroxybenzoate (Evans and Fuchs, 1988; unpublished observations from our laboratories), and benzoyl CoA as well as 4-hydroxybenzoyl CoA have been seen among early products in the small molecule pools of cells that had taken up ^{14}C-4-hydroxybenzoate for one to two minutes (Perrotta and Gibson, unpublished). An oxygen-sensitive dehydroxylating enzyme which carries out the reduced benzyl viologen-dependent reduction of 4-hydroxybenzoyl CoA to benzoyl CoA has been purified and characterized by Brackmann and Fuchs (1993) from extracts of a denitrifying bacerium grown anaerobically on 4-hydroxybenzoate. These workers also detected this activity in extracts of *Rp. palustris*.

The alicyclic intermediates detected in the experiments of Dutton and Evans suggested that ring reduction proceeded at least to the state of a cyclohexene carboxyl derivative. The subsequent steps that they postulated resembled those involved in β-oxidation of fatty acids, including a hydration, a dehydrogenation and a cleavage (Fig. 3), and it seems probable that the early intracellular and in vitro benzoyl CoA reduction products that have been detected (Gibson and Gibson, 1992; Koch et al., 1993) are converted to a cyclohex-1-enecarboxyl CoA. Enzyme activities catalyzing the sequential hydration and dehydrogenation of cyclohex-1-enecarboxyl CoA (Fig. 3) were demonstrated in cell extracts (Whittle et al., 1976; Hutber and Ribbons, 1983), and recent reinvestigation of these reactions has indicated that these activities are increased at least three-fold by growth on benzoate (Perrotta and Harwood, 1994). An apparent 2-ketocyclohexanecarboxyl CoA cleavage that does not require coenzyme A has also been detected, suggesting that cleavage of the ring is hydrolytic rather than thiolytic (Perrotta and Harwood, 1994).

A significantly different benzoyl CoA degradation pathway, yielding 3-hydroxypimelyl CoA on ring opening, was recently suggested for the denitrifying pseudomonads, and also for *Rp. palustris* (Koch et al., 1993; Fig. 3), although all the postulated intermediates were not detected in their experiments. Until the enzymes involved have been characterized it is not possible to distinguish clearly between products in the mainstream of carbon flow and those resulting from side reactions.

A pathway for β-oxidation of pimelyl CoA to 3-hydroxypimelyl CoA followed by cleavage to yield acetyl CoA and glutaryl CoA was suggested by Blakley (1978) based on studies of cyclohexanecarboxylate utilization by a strain of *Alcaligenes*. His further proposal that glutaryl CoA is oxidatively decarboxylated to crotonyl CoA is supported by the recent demonstration of these reactions in extracts of benzoate-grown *Rp. palustris*, *Rhodomicrobium vannielii* and *Rhodocyclus purpureus*, as well as denitrifying pseudomonads (Härtel et al., 1993). Glutaryl CoA dehydrogenases which catalyze both the oxidation and decarboxylation reactions leading to formation of crotonyl CoA have been purified from two denitrifying strains (Härtel et al., 1993).

D. Enzymology

Most of the detailed work on enzymes involved in aromatic acid metabolism has centered on the coenzyme A ligases. Two ligases, one apparently specific to 4-hydroxybenzoate degradation and the other involved in benzoate utilization have been purified and characterized from *Rp. palustris* (Geissler et al., 1988; Gibson et al., 1994). As noted above, two enzymes that activate cinnamate have also been resolved (Elder et al., 1992b). These are clearly distinct from the benzoate and 4-hydroxybenzoate enzymes studied in our laboratories, which do not react with cinnamate. Some characteristics of the aromatic CoA ligases so far known are summarized in Table 1. The benzoate and 4-hydroxybenzoate CoA ligases have quite narrow, but overlapping, substrate specificities. They differ considerably from each other in terms of molecular and kinetic characteristics, and antisera prepared against each of the purified enzymes do not cross-react. The high in vitro affinity of benzoate CoA ligase for its aromatic substrate parallels the high affinity of whole cell

Table 1. Some characteristics of *Rp. palustris* aromatic acid CoA ligases

	Benzoate CoA ligase *	4-OH-benzoate CoA ligase **	Cinnamate CoA ligases † PCS I	PCS II
Induced by growth on:	benzoate 4-OH benzoate cinnamate cyclohex-1-ene-carboxylate cyclohex-3-ene-carboxylate	4-OH benzoate 4-OH cinnamate	All growth substrates	cinnamate 3-phenylpropionate caffeate ferulate
Substrates used (K_m: μM)	benzoate (2) cyclohexadiene-carboxylates (40)	4-OH benzoate (120) benzoate (400) cyclohexadiene-carboxylates (400)	cinnamate	cinnamate 4-OH cinnamate caffeate ferulate
	ATP (3)	ATP (90)	ATP	ATP
	CoASH (100)	CoASH (400)	CoASH	CoASH
Molecular characteristics	monomer	dimer		
Mol. wt. native	61,000	117,000		
Mol. wt. in SDS	60,000	61,000		
pI	5.2	ND	4.6	5.1

* from Geissler et al., 1988
** from Gibson et al., 1994
† from Elder et al., 1992b; PCS, Phenylpropenoyl-Coenzyme A Synthetase

uptake for this substrate, and we have suggested that the enzyme plays a role in substrate acquisition from the environment as well as in its subsequent metabolism (Harwood and Gibson, 1986). A curious and as yet unexplained finding is that purified 4-hydroxybenzoate CoA ligase has a much lower affinity for 4-hydroxybenzoate than do intact cells (Merkel et al., 1989; Gibson et al., 1994). *Rp. palustris* presumably has additional CoA ligases, since the enzymes characterized so far do not use alicyclic acids as substrates. One such enzyme has recently been partially purified and characterized (Kuever et al., unpublished) and activates a range of both alicyclic and aromatic acid substrates. This enzyme is expressed during growth on alicyclic substrates such as cyclohexane carboxylate, or cyclohex-1-ene-1-carboxylate, as well as aromatic acids, but not on succinate, and is immunologically distinct from the two ligases already purified in our laboratory.

Other portions of the proposed benzoate degradation reaction sequence have so far been studied only in crude extracts. The initial reductive step(s) are clearly extremely sensitive to oxygen inactivation, and so differ both from the ligase reactions and those converting cyclohex-1-ene-carboxyl CoA to pimelyl CoA in this characteristic (Hutber and Ribbons, 1983; Perrotta and Harwood, 1994).

III. Regulation and Genetics

A. Expression of Aromatic Acid CoA Ligases

Antisera prepared against the two purified CoA ligases from *Rp. palustris* have been used to follow expression of these enzymes by Western immunoblotting. This approach avoids potential complications caused by an unexplained partial inhibition of activities in crude extracts (Geissler et al., 1988; Gibson et al., 1994). Benzoate CoA ligase is induced by growth on several metabolizable aromatic acids, including benzoate (Table 1), as well as by growth on succinate in the presence of non-metabolizable aromatic acids—for example, 3-chlorobenzoate, 2-aminobenzoate and 4-methylbenzoate— with various ring substituents. Cyclohexadiene- and cyclohexenecarboxylates also elicit expression. Antigen was not detected in extracts of cyclohexane carboxylate grown cells, nor following growth on succinate (Kim and Harwood, 1991). The diversity of aromatic compounds that can serve as inducers may reflect the role of benzoate as a critical central metabolite in aromatic compound degradation. The 4-hydroxybenzoate CoA ligase, in contrast, is fully induced only by growth on 4-hydroxybenzoate or 4-hydroxycinnamate. Of the two cinnamate CoA ligases resolved by Elder et al. (1992b), activity measurements indicated that one was constitutively expressed, even when malate was the growth substrate, but the second was found only after growth on aromatic acids bearing side chains (Table 1).

B. Genetic Approaches

As far as we are aware, genetic analyses of aromatic acid degradation systems have only been attempted in *Rp. palustris* (Guyer and Hegeman, 1969; Elder et al., 1993), and the intensive efforts to apply such approaches in our laboratories have only recently begun to yield useful information (Dispensa et al., 1992; Gibson et al., 1994). So far, only a few stable mutants have been isolated that are unable to grow on aromatic compounds. Some possible explanations for this are that *Rp. palustris* has very effective repair mechanisms, and that a measure of redundancy is built into the pathways through somewhat relaxed substrate specificities of the enzymes involved or even alternative routes of carbon flow.

1. Benzoate and 4-Hydroxybenzoate CoA Ligase Genes

We have used immunoscreening to identify and isolate *Rp. palustris* clones encoding the benzoate and 4-hydroxybenzoate CoA ligase genes (*badA* and *hbaA*) Gibson et al., 1994; Egland and Harwood, 1994). Both genes are chromosomally encoded and both have been sequenced and found to share 50% predicted amino acid identity. In addition, they are each about 20% identical at the amino acid level to a number of coenzyme A ligases that are not involved in anaerobic aromatic compound metabolism. These include acetate CoA ligases from bacteria and fungi (Connerton et al., 1990; Eggen et al., 1991), aromatic CoA ligases required for the aerobic degradation of 3-chlorobenzoate (Babbitt et al., 1992) and 2-aminobenzoate (Altenschmidt and Fuchs, 1992), and coumarate CoA ligases involved in phenylpropenoid biosynthesis in plants (Lozoya et al., 1988).

Studies with purified enzymes and with extracts of cells grown under different conditions suggest that some aromatic acids can be activated by more than one CoA ligase, and this seems to be the case for benzoate. A *badA* mutant that we constructed by gene replacement techniques (Egland et al., unpublished) had extremely low levels of ligase activity when mutant cell extracts were assayed with 10 μM benzoate, a concentration that gives maximal ligase activity in wild-type cell extracts. Despite this, the *badA* mutant had a normal growth rate on benzoate under laboratory culture conditions (3 mM benzoate). Further experiments showed that when *badA* cell extracts were assayed at high benzoate concentrations (1mM), benzoate CoA ligase activities were about the same as those seen in wild type cells. Thus *Rp. palustris* must have another enzyme with benzoate-CoA ligase activity, in addition to the *badA* product, that can function when benzoate is present at relatively high concentrations.

In contrast to the situation with benzoate, 4-hydroxybenzoate appears to be activated by a single CoA ligase. An *hbaA* mutant that was constructed lacked detectable 4-hydroxybenzoate-CoA ligase activity and was also incapable of anaerobic growth on 4-hydroxybenzoate (Gibson et al., 1994).

There are open reading frames (ORFs) adjacent to each of the ligase genes, and preliminary evidence suggests that at least one of these ORFs may be cotranscribed with the benzoate CoA ligase gene. This provides some hope that the genes for anaerobic degradation of aromatic acids may be organized into operons, a situation which would greatly simplify further genetic analysis.

2. aadR*, a Regulatory Gene Involved in Anaerobic Aromatic Degradation*

Intensive study of the few stable *Rp. palustris* mutants that we have isolated following random screening for defective growth on benzoate or 4-hydroxybenzoate has led to the identification of a gene termed *aadR*, for anaerobic aromatic degradation regulator (Dispensa et al., 1992). *AadR* mutants are unable to grow photosynthetically on 4-hydroxybenzoate and grow only very slowly on benzoate. Aerobic growth on 4-hydroxybenzoate remains unimpaired, as does anaerobic growth on succinate and other organic acids. *AadR* is required for expression of 4-hydroxybenzoate CoA ligase, and also for 4-hydroxybenzoate-induced expression of benzoate CoA ligase, as evidenced in Western blots.

Additional, as yet uncharacterized, functions needed for optimal growth on benzoate also appear to be controlled, at least in part, by *aadR*, since all these mutants grow much more slowly than the parent on this carbon source.

The deduced amino acid sequence of the *aadR* gene product indicates that it is a member of the cyclic AMP receptor protein (Crp) family of transcriptional regulators. In addition to Crp, this family includes Fnr, a regulator of genes required for anaerobic respiration in *Escherichia coli* (Fischer, 1994). Although *aadR* mutants do not have an *fnr*-like phenotype, the AadR protein contains several of the conserved cysteine residues that have been proposed to play a role in redox potential sensing by Fnr. The greatest similarity between AadR and other members of the Crp family is to the FixK proteins of nitrogen-fixing bacteria (Batut et al., 1989; Fischer, 1994), but dinitrogen-dependent growth of *Rp. palustris* is not affected by the *aadR* mutation. The very distinctive phenotype of the *Rp. palustris aadR* mutants indicates that AadR protein functions specifically as a transcriptional regulator of genes required for anaerobic aromatic acid degradation.

V. Ecology

Nonsulfur purple photosynthetic bacteria are universally distributed and it is relatively easy to isolate species and strains that will grow on aromatic compounds anaerobically in light. As already noted, the range of compounds used by such isolates includes many that may arise during anaerobic lignin degradation as well as a number of potentially toxic human products that are of environmental concern. Where this has been investigated, the affinity of whole cells for aromatic substrates is very high indeed, with apparent K_m values below micromolar levels (Harwood and Gibson, 1986; Merkel et al., 1989), and virtually complete assimilation of carbon from aromatic substrates into photosynthetic bacterial biomass has been noted repeatedly (for example, Wright and Madigan, 1991). Anaerobic utilization and mineralization of many compounds containing benzene nuclei is also carried out by some denitrifying pseudomonads, by sulfate and iron reducers and also by methanogenic consortia (see Evans and Fuchs, 1988; Fuchs et al., 1994), and one may speculate as to which physiological type plays an overall dominant role in natural environments. Significant factors that will affect which type or types will function effectively in any natural system will include their versatility in carbon source utilization and their ability to both tolerate and exploit rapidly changing environmental conditions such as oxygen levels. Substrate affinity must be another significant factor. Various metabolic characteristics of *Rp. palustris* discussed in this chapter suggest that it and its relatives may be competitive under some conditions. However, massive growth of nonsulfur purple bacteria are uncommon in natural systems, and there are as yet no good methods for determining which aspects of genetic potential are expressed in nature. Although infrared light used by photosynthetic bacteria penetrates solid substrata better than visible wavelengths, the depth to which this energy source will be available is still quite limited. Taken together, these factors suggest that anoxygenic phototrophs are likely to be only minor, although occasionally significant, participants in the anaerobic turnover of aromatic compounds in most natural systems.

VI. Comparative Aspects

Since a reductive pathway for anaerobic degradation of aromatic compounds was first proposed, evidence has accumulated for similar pathways in a number of bacteria that can use such substrates in the absence of molecular oxygen. Particularly impressive progress has come from the studies of Fuch's group with two denitrifying bacteria. It is worth pointing out, however, that very little is known about how aromatic compounds are degraded by microorganisms such as sulfate or iron reducers, in which an adequate metabolic energy supply is far less easily obtained than in phototrophs. It remains possible that as the activities of more microorganisms are explored in increasing detail, distinctly different routes for the anaerobic degradation of aromatic nuclei will be revealed, much as has occurred in the study of aerobic pathways.

Acknowledgments

The work from our laboratories was supported in part by a grant (DE-FG02-86ER13495) from the Department of Energy, Division of Energy Bio-sciences to JG, and by grants from the U. S. Army Research office (DAAL03-92-G-0104 and DAAL03-

92-G-0313) with co-funding by the Department of Energy, Division of Energy Biosciences, to CSH.

References

Altenschmidt U and Fuchs G (1992) Novel aerobic 2-aminobenzoate metabolism. Purification and characterization of 2-aminobenzoate-CoA ligase, localisation of the gene on a 8-kbp plasmid, and cloning and sequencing of the gene from a denitrifying *Pseudomonas* sp. Eur J Biochem 205: 721–727

Babbitt PC, Kenyon GL, Martin BM, Charest H, Sylvestre M, Scholten JD, Chang, K-H, Liang, P-H and Dunaway-Mariano D (1992) Ancestry of the 4-chlorobenzoate dehalogenase: Analysis of amino acid sequence identities among families of acyl:adenyl ligases, enoyl-CoA hydratases/isomerases, and acyl-CoA thioesterases. Biochemistry 31: 5594–5604

Batut J, Daveran-Mingot M-L, David M, Jacobs J, Garnerone AM and Kahn D (1989) *fixK*, a gene homologous with *fnr* and *crp* from *Escherichia coli*, regulates nitrogen fixation genes both positively and negatively in *Rhizobium meliloti*. EMBO J 8: 1279–1286

Biebl H and Pfennig N (1981) Isolation of members of the family Rhodospirillaceae. In: Starr MP, Stolp H, Trüper HG, Balows A and Schlegel HG (eds) The Prokaryotes, pp 267–273. Springer Verlag, Berlin

Blakley ER (1978) The microbial metabolism of cyclohexane-carboxylic acid by a β-oxidation pathway with simultaneous induction to the utilization of benzoate. Can J Microbiol 24: 847–855

Birch AJ and Smith H (1958) Reduction by metal-amine solutions: application to the synthesis and determination of structure. Quart Rev Chem Soc Lond 12: 17–33

Brackmann R and Fuchs G (1993) Enzymes of anaerobic metabolism of phenolic compounds: 4-hydroxybenzoyl-CoA reductase (dehydroxylating) from a denitrifying *Pseudomonas* species. Eur J Biochem 213: 563–571

Connerton IF, Fincham JRS, Sandeman RA and Hynes MJ (1990) Comparison and cross-species expression of the acetyl-CoA synthetase genes of the ascomycete fungi, *Aspergillus nidulans* and *Neurospora crassa*. Mol Microbiol 4: 451–460

Dispensa M, Thomas CT, Kim M-K, Perrotta JA, Gibson J and Harwood CS (1992) Anaerobic growth of *Rhodopseudomonas palustris* on 4-hydroxybenzoate is dependent on AadR, a member of the cyclic AMP receptor protein family of transcriptional regulators. J Bacteriol 174: 5803–5813

Dutton PL and Evans WC (1969) The metabolism of aromatic compounds by *Rhodopseudomonas palustris*. Biochem J 113: 525–535

Dutton PL and Evans WC (1978) Metabolism of aromatic compounds by Rhodospirillaceae. In: Clayton RK and Sistrom WR (eds) The Photosynthetic Bacteria, pp 719–726. Plenum Press, New York

Eggen RIL, Geerling ACM, Boshoven ABP and De Vos WM (1991) Cloning, sequence analysis, and functional expression of the acetyl coenzyme A synthetase gene from *Methanothrix soehngenii* in *Escherichia coli*. J Bacteriol 173: 6383–6389

Egland PG and Harwood CS (1994) Anaerobic aromatic compound degradation: Cloning of the benzoate-coenzyme A ligase gene from *Rhodopseudomonas palustris*. Abstr Ann Meet Am Soc Microbiol 94: Q-412

Elder DJE, Morgan P and Kelly DJ (1993) Transposon Tn*5* mutagenesis in *Rhodopseudomonas palustris*. FEMS Microbiol Lett 111: 23–30

Elder DJE and Kelly DJ (1994) The bacterial degradation of benzoic acid and benzenoid compounds under anaerobic conditions: Unifying trends and new perspectives. FEMS Microbiol Rev 13: 441–468

Elder, DJE, Morgan P and Kelly DJ (1992a) Anaerobic degradation of *trans*-cinnamate and ω-phenyl alkane carboxylic acids by the photosynthetic bacterium *Rhodopseudomonas palustris*: evidence for a β-oxidation mechanism. Arch Microbiol 157: 148–154

Elder DJE, Morgan P and Kelly DJ (1992b) Evidence for two differentially regulated phenylpropenoyl-coenzyme A synthetase activities in *Rhodopseudomonas palustris*. FEMS Microbiol Lett 98: 255–260

Evans WC and Fuchs G (1988) Anaerobic degradation of aromatic compounds. Annu Rev Microbiol 42: 289–317

Fischer H-M (1994) Genetic regulation of nitrogen fixation in Rhizobia. Microbiol Rev 58: 352–386

Fuchs G, Mohamed ME, Altenschmidt U, Koch J, Lack A, Brackmann R, Lochmeyer R and Oswald B (1994) Biochemistry of anaerobic biodegradation of aromatic compounds. In: Ratledge C (ed) Biochemistry of Microbial Degradation, pp 513–553. Kluwer Academic Publishers, Dordrecht

Geissler, JF, Harwood CS and Gibson J (1988) Purification and properties of benzoate-coenzyme A ligase, a *Rhodopseudomonas palustris* enzyme involved in the anaerobic degradation of benzoate. J Bacteriol 170: 1709–1714

Gibson J, Dispensa M, Fogg GL, Evans DT and Harwood CS (1994) 4-Hydroxybenzoate-coenzyme A ligase from *Rhodopseudomonas palustris*: Purification, gene sequence, and role in anaerobic degradation. J Bacteriol 176: 634–641

Gibson KJ and Gibson J (1992) Potential early intermediates in anaerobic benzoate degradation by *Rhodopseudomonas palustris*. Appl Environ Microbiol 58: 696–698

Guyer M and Hegeman G (1969) Evidence for a reductive pathway for the anaerobic metabolism of benzoate. J Bacteriol 99: 906–907

Härtel U, Eckel E, Koch J, Fuchs G, Linder D and Buckel W (1993) Purification of glutaryl-CoA dehydrogenase from *Pseudomonas* sp., an enzyme involved in the anaerobic degradation of benzoate. Arch Microbiol 159: 174–181

Harwood CS and Gibson J (1986) Uptake of benzoate by *Rhodopseudomonas palustris* grown anaerobically in light. J Bacteriol 165: 504–509

Harwood CS and Gibson J (1988) Anaerobic and aerobic metabolism of diverse aromatic compounds by the photosynthetic bacterium *Rhodopseudomonas palustris*. Appl Environ Microbiol 54: 712–717

Hegeman GD (1967) The metabolism of *p*-hydroxybenzoate by *Rhodopseudomonas palustris* and its regulation. Arch Mikrobiol 59: 143–148

Hutber GN and Ribbons DW (1983) Involvement of coenzyme A esters in metabolism of benzoate and cyclohexanecarboxylate by *Rhodopseudomonas palustris*. J Gen Microbiol 129: 2413–2420

Kamal VS and Wyndham RC (1990) Anaerobic phototrophic

metabolism of 3-chlorobenzoate by *Rhodopseudomonas palustris* WS17. Appl Environ Microbiol 56: 3871–3873

Khanna P, Rajkumar B and Jothikumar N (1992) Anoxygenic degradation of aromatic substances by *Rhodopseudomonas palustris*. Current Microbiol 25: 63–67

Kim M-K and Harwood CS (1991) Regulation of benzoate-CoA ligase in *Rhodopseudomonas palustris*. FEMS Microbiol Lett 83: 199–204

Koch J and Fuchs G (1992) Enzymatic reduction of benzoyl-CoA to alicyclic compounds, a key reaction in anaerobic aromatic metabolism. Eur J Biochem 205: 195–202

Koch J, Eisenreich W, Bacher A and Fuchs G (1993) Products of enzymatic reduction of benzoyl-CoA, a key reaction in anaerobic aromatic metabolism. Eur J Biochem 211: 649–661

Lozoya E, Hoffmann H, Douglas C, Schulz W, Scheel D and Hahlbrock K (1988) Primary structures and catalytic properties of isozymes encoded by the two 4-coumarate: CoA ligase genes in parsley. Eur J Biochem 176: 661–667

Madigan MT and Gest H (1988) Selective enrichment and isolation of *Rhodopseudomonas palustris* using *trans*-cinnamic acid as sole carbon source. FEMS Microbiol Ecol 53: 53–58

Merkel SM, Eberhard AE, Gibson J and Harwood CS. (1989) Involvement of coenzyme A thioesters in anaerobic metabolism of 4-hydroxybenzoate by *Rhodopseudomonas palustris*. J Bacteriol 171: 1–7

Perrotta JA and Harwood CS (1994) Anaerobic metabolism of cyclohex-1-ene-1-carboxylate, a proposed intermediate of benzoate degradation, by *Rhodopseudomonas palustris*. Appl Environ Microbiol 60: 1775–1782

Pfennig N (1978) *Rhodocyclus purpureus* gen. nov. and sp. nov., a ring-shaped, vitamin B_{12}-requiring member of the family *Rhodospirillaceae*. Intl J Syst Bacteriol 28: 283–288

Pfennig N, Eimhjellen KE and Liaaen-Jensen S (1965) A new isolate of the *Rhodospirillum fulvum* group and its photosynthetic pigments. Arch Mikrobiol 51: 258–265

Rahalkar SB, Joshi SR and Shivaraman N (1991) Biodegradation of aromatic compounds by *Rhodopseudomonas gelatinosa*. Curr Microbiol 22: 155–158

Rahalkar SB, Joshi SR and Shivaraman N (1993) Photometabolism of aromatic compounds by *Rhodopseudomonas palustris*. Current Microbiol 26: 1–9

Sher S and Proctor MH (1960) Studies with photosynthetic bacteria: anaerobic oxidation of aromatic compounds. In: Allen MB (ed) Comparative Biochemistry of Photoreactive Pigments, Vol 25, pp 387–393. Academic Press, New York

van der Woude BJ, deBoer M, van der Put NMJ, van der Geld FM, Prins RA and Gottschal JC (1994) Anaerobic degradation of halogenated benzoic acids by photoheterotrophic bacteria. FEMS Microbiol Lett 119: 199–208

Whittle PJ, Lunt DO and Evans WC (1976) Anaerobic photometabolism of aromatic compounds by *Rhodopseudomonas* sp. Biochem Soc Trans 4: 490–491

Wright GE and Madigan MT (1991) Photocatabolism of aromatic compounds by the phototrophic purple bacterium *Rhodomicrobium vannielii*. Appl Environ Microbiol 57: 2069–2073

Yamanaka K, Moriyama M, Minoshima R and Tsuyuki Y (1983) Isolation and characterization of a methanol-utilizing phototrophic bacterium, *Rhodopseudomonas acidophila* M402 and its growth on vanillin derivatives. Agric Biol Chem 47: 1257–1267

Chapter 47

Flagellate Motility, Behavioral Responses and Active Transport in Purple Non-Sulfur Bacteria

Judith P. Armitage
Microbiology Unit, Department of Biochemistry, University of Oxford , Oxford OX1 3QU, U.K.

David J. Kelly
Department of Molecular Biology and Biotechnology, University of Sheffield, Sheffield S10 2UH, U.K.

R. Elizabeth Sockett
Department of Life Sciences, University of Nottingham, Nottingham NG7 2RD, U.K.

R. E. Blankenship, M. T. Madigan and C. E. Bauer (eds): Anoxygenic Photosynthetic Bacteria, pp. 1005–1028.

Summary

Remarkably little is known about the motility, behavior and transport systems of phototrophic bacteria, perhaps because in the past there has been a general assumption that these were basic bacterial functions that would turn out to be the same as those characterized in enteric bacteria. However, we hope that the following brief review will show that in all three areas the purple non-sulfur bacteria have proved to be different from the enteric bacteria in many important ways and may well provide a better model system for species under growth limiting conditions. The work described is limited, reflecting the dearth of data in these fields and is concentrated mainly on the *Rhodobacter* species. As the latter have provided us with many unexpected surprises, what would we find if the green bacteria were studied in detail?

Many species of anoxygenic phototrophs are motile, using a variety of patterns of flagellation e.g. peritrichous, polar tufts, single flagella. However, *Rb. sphaeroides* was the first species to be found which could only rotate its flagellum in one direction, stopping rotation with a full Δp to change direction. This has implications for the mechanism of proton transport through flagella in general, and the cloning of the motor genes will be particularly important in helping characterize the energetics of motor function. In addition to a unidirectional motor, and perhaps in some way connected, chemotaxis to the dominant chemoattractants has been found to require transport and metabolism. *Rb. sphaeroides* also does have chemoreceptors suggesting two chemosensory pathways operating in parallel. It seems probable that *Rhodobacter* provides a model for one end of the sensory spectrum, linking its responses to its current growth requirement, whereas the enteric bacteria provide the other end of the spectrum, primarily using dedicated receptors. Other photosynthetic species may lie somewhere between, using dedicated and metabolism based sensing equally. The type of sensing may reflect the increased importance of the flexible electron transport mechanisms in these species compared to enteric species. Similar motility and chemosensing systems have recently been identified in a number of rhizosphere bacteria, suggesting a more widespread distribution.

The *Rhodobacter* species also show prolonged chemokinesis in response to some chemoattractants. Chemokinesis depends on transport but not metabolism; however, as little is known about transport, the mechanism involved is unclear. Investigation into different transport mechanisms in the purple non-sulfur bacteria suggests a new class of periplasmic binding protein dependent transporter, highlighting the importance of pHi in controlling transport.

It is hoped that this chapter will highlight the value of investigating what may be considered common bacterial phenomena in the phototrophic species, as it does not produce 'me too' data, but reveals the true diversity of bacterial systems and the adaptation of physiology to fit their growth environment.

I. Introduction

Why study standard physiological phenomena in species of anoxygenic photosynthetic bacteria when there is so much information from the enteric species? The limited studies undertaken so far using the purple non-sulfur bacteria have already revealed that bacteria use a far more diverse range of mechanisms than previously thought to transport solutes and to swim and sense their environment. But, in addition, there are technical advantages to working with the *Rhodobacter* species not available to those working with enterics. One advantage is related to the techniques that are available to study the bioenergetics of phototrophic bacteria, which are of use in determining the relationship between e.g. solute uptake and its driving force and the electrochemical proton gradient and flagellar rotation i.e. the mechanism of energy coupling in these systems. For example, the ability to use non-invasive methods such as carotenoid electrochromism to measure membrane potential (Jackson, 1988) has proved of great value, since the extremely fast response time of the absorbance changes can be used to measure membrane ionic currents and their dependence on membrane potential. Although the transmembrane distribution of lipophilic cations has also been used to determine membrane potential, there is continuing controversy over which method (ion distribution or electrochromism) is the more reliable (Crielaard et al., 1988). In addition, light is an almost 'instantaneous' and controllable energy source which has obvious advantages when investigating the energetics of these processes. The bioenergetic advantages are now enhanced by the advanced genetic systems available for *Rb. capsulatus* and *Rb. sphaeroides*, which, although not yet quite on a par with the enteric bacteria, has made possible the cloning and analysis of genes involved in transport, flagella synthesis and rotation, and chemotaxis.

II. Motility and Behavior

A. Historical Perspective

Motility is widespread amongst photosynthetic bacteria and, in the vast majority of cases, the motor organelle is the flagellum. Early microscopic studies of motility by Ehrenberg and Engelmann in the 19th century (Ehrenberg, 1838; Engelmann, 1881), concentrated on the photosynthetic bacteria. Ehrenberg was able to observe polar bundles of 'wave-shaped' flagella on cells of what was probably a *Chromatium* species, and he presumed that these might be required for motion. Engelmann cultured photosynthetic bacteria from the river Rhine and studied their randomly stopping and starting motility. He also noted that the bacteria could accumulate in spots of light, the first indication of phototaxis (see taxis section later in this chapter). Later work in the 1960s by Cohen-Bazire and London (1967) revealed the microscopic structure of isolated flagella from *Rhodospirillum rubrum*. They showed the flagellum, complete with the basal-body, consisting of a series of proteinaceous rings embedded in the cell wall and membranes (a typical electron micrograph of an *Rs. rubrum* flagellum is shown in Fig. 2). From this early structural work, theories for the rotational mechanism of flagellar action were developed. More recently, molecular genetic analysis in enteric and photosynthetic bacteria has been used to probe the structure and function of these bacterial 'propellers'.

B. Motility

Most species of the purple non-sulfur bacteria (on which this review will necessarily concentrate because of the dearth of data from other species) are flagellate, as are many Chromatiaceae, and fresh environmental isolates are nearly always highly motile. Most patterns of flagellation seen on non-photosynthetic species have been identified on one or more photosynthetic species (Fig. 1). *Chromatium* swims with a single tuft of flagella whereas *Rs. rubrum* has bipolar bundles (lophotrichous) (Clayton, 1957; Pfennig, 1968) *Rhodobacter* and *Rhodopseudomonas* species are uniflagellate with a medial or sub-polar flagellum (Armitage and Macnab, 1987; Tauschel, 1987). Peritrichous or lateral flagella, as found in the much studied enteric and bacillus species, have been identified on *Rhodomicrobium vannielii* and the swarming phase of *Rhodospirillum centenum*. *Rhodospirillum centenum* swims in liquid medium by means of a single, possibly sheathed, polar flagellum but when grown on solid media a large number of lateral flagella are induced. These lateral flagella appear to allow the colonies of *Rs. centenum* to swarm towards and away from the light (see later). The induction of lateral flagella appears to be the result of surface contact not viscosity as, unlike *Vibrio parahaemolyticus* or *Serratia marcescens*, lateral flagella are not induced by increasing viscosity (Ragatz et al., 1995).

1. Flagellar structure

The structure of the flagella of enteric bacteria has

Abbreviations: Che – chemotaxis proteins; CM – cytoplasmic membrane; *Cm. – Chromatium*; Dct – dicarboxylate transport system; DMSO – dimethyl sulphoxide; *Ec. – Ectothirorhodospira*; FPr – fructose protein; HPK – histidine protein kinase; HPr – histidine protein; ICM – intracytoplasmic membrane; Kdp – high-affinity potassium transport system; MCP – methyl accepting chemotaxis protein; Mod – molybdenum transport system; Mot – motor proteins; MSX – methionine sulfoxamine; ORF – open reading frame; PBP – periplasmic binding protein; PCPPTS – chemotaxis protein; PEP – phosphoenolpyruvate; pH_i – intracelluar pH; Pro – proline transport system; PTS – phosphotransferase sugar transport system; *Rb. – Rhodobacter*; *Rs. – Rhodospirillum*; *Rp. – Rhodopseudomonas*; ΔNa^+ – sodium motive force; Δp – electrochemical proton gradient; $\Delta\psi$ – membrane potential

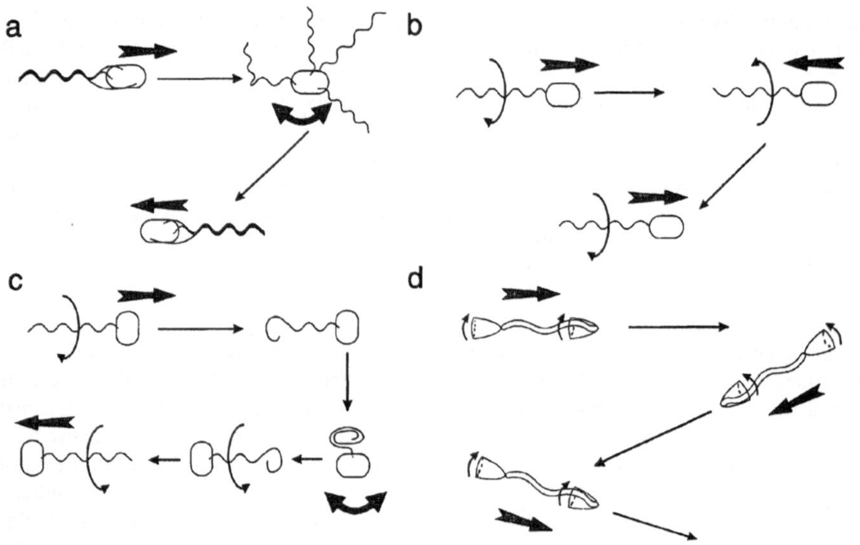

Fig. 1. Different patterns of motility (a) peritrichously flagellate species swimming with a rotating bundle, periodically the flagella switch from CCW to CW rotation, the bundle is forced apart and the cell 'tumbles', smooth swimming resumes in a new direction; (b) polarly flagellate species changing direction by briefly 'backing-up' as a result of a transient change from CCW to CW flagellar rotation; (c) direction changing as a result of stopping flagellar rotation, the loss of torque causes the flagellum to 'relax' and Brownian motion reorients the cell for future smooth swimming; (d) lophotrichously flagellate species, swimming equally well in either direction and coordinating flagellar switching.

been characterised in great detail (Iino, 1985; Macnab, 1992) and it seems likely that the overall structure and organization is similar in all bacterial species. However, the purple non-sulfur bacteria have provided a few surprises and the differences from enteric models are helping to characterise motor function in general. More recently, some other (non-photosynthetic) species have been characterized in which motility is more closely allied to that found in *Rhodobacter* than to that found in enteric bacteria.

A stylized drawing of a flagellum is shown in Fig. 2. The flagellar filament is a semi-rigid helix, 15 nm in diameter and 3–10 μm in length, made up of approximately 20,000 flagellin subunits polymerized in an 11-start-helix to form a hollow protein cylinder. Flagellins (usually about 50–60 kDa) do not show great sequence conservation between species, except at the N and C terminal regions that seem to be involved in subunit interaction; indeed an antiserum against *Rb. sphaeroides* flagellin reacts very poorly with flagellins from other photosynthetic species. The functional flagellar filament forms a helix of specific wavelength and handedness; the structure results from the two alternative interactions between the flagellin subunits, one resulting in a shorter protofilament and one a longer. The arrangement of the subunits into the 11 protofilaments that make up the flagellum means that there are 12 polymorphic forms of the flagellum. In many species the handedness and wavelength alter as the number of protofilaments forced into the short conformation changes when the direction of flagellar rotation changes. This results in a change in swimming direction as the flagellar bundle is forced apart and the cell 'tumbles' (Fig. 1). *Rb. sphaeroides* shows a dramatic change in flagellar shape during swimming. Motor rotation causes a normal functional helix, but periodically the motor stops rotating and the flagellum relaxes to a short-wavelength large-amplitude coil, during which time Brownian motion reorients the cell before rotation starts again.

The flagellar filament is connected to the rotor by a hook, so called because in most species it is hook shaped. The hook has a similar subunit arrangement to the filament and is thought to act both as a universal joint, allowing peritrichous flagella to come together as a rotating bundle and as a mechanism for transferring torque changes from the motor to the

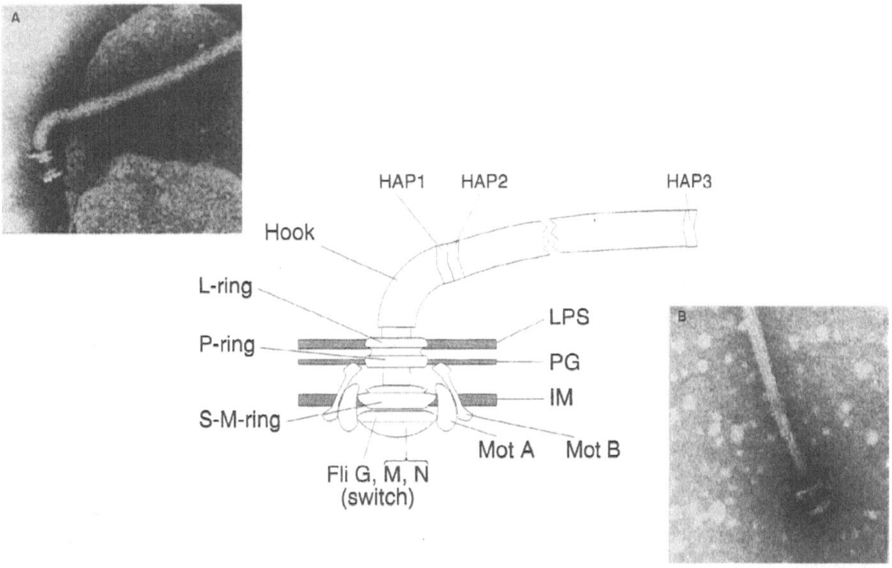

Fig. 2. Diagrammatic representation of the flagellar motor with negative stained electron micrographs of the isolated motor from (A) *Rs. rubrum* and (B) *Rb. sphaeroides*. HAP = hook associated proteins, LPS = lipopolysaccharide, PG = peptidoglycan, IM = inner membrane, MotA/B = motor proteins. (A) reproduced from Cohen-Bazire and London (1967) with permission.

filament. At the boundary between the hook and filament are hook-accessory proteins (HAPs) which may help polymerization and torque transfer.

The hook connects to the rod, and this in turn is connected to a series of membrane and wall-bound rings. The M-S ring (see Fig. 2) located in the cytoplasmic membrane and the membrane-associated proteins FliG, M, and N (switch-complex), at the cytoplasmic face of the M-S ring, are thought to be where rotation is both generated and the direction of rotation controlled (see section on taxis). The other rings probably act as 'washers' to allow passage of the rotating rod through the outer layers of the cell. Complexes of two proteins, Mot A and B are arranged around the M-S ring and these seem to be the force-generating units, acting as the proton channels and the stators of the motor.

2. Flagellar Rotation

a. Energy Source

The energy for flagellar rotation comes from the electrochemical proton gradient (Δp) across the cytoplasmic membrane; ATP is not required for flagellar rotation although it may be needed for tactic responses. Experiments on *Rp. palustris* by Tauschel (Tauschel, 1987) suggested that there was ATPase activity on the cytoplasmic side of the flagellar basal-body. The identification of ATPase activity in the flagellar assembly protein, FliI, makes it probable that the activity is involved in flagellin export rather than rotation (Dreyfus et al., 1993). FliI has now been independently identified in *Rb. sphaeroides* by two groups (G. Dreyfus, I.P. Goodfellow, C. E. Pollitt and R. E. Sockett, unpublished).

It is assumed that the protons which cause the rotation of the flagellum come from the bulk phase in the periplasm, but the presence of localized electron transport chains dedicated to the flagellum has not been tested experimentally. However, as *Rb. sphaeroides* is motile under both aerobic and photosynthetic conditions (in the latter case the predominant electron transport components are located in the ICM whereas the flagellar motor presumably remains associated with the CM), it seems probable that bulk phase protons drive rotation.

Studies with free swimming *Rb. sphaeroides* show that, as with other species, the threshold for motility is very low, below –15 mV, with a linear increase in speed as the Δp rises until the speed saturates at about –50 mV (Harrison et al., 1994). The range is somewhat lower for *Rb. sphaeroides* than *E. coli*, which may reflect either the difference between single and peritrichous flagella or a true difference in motor behavior. The maximum swimming speed varies from species to species, with peritrichously flagellated bacteria swimming at about 15 μm s^{-1} and monoflagellate species such as *Rb. sphaeroides* swimming at speeds of up to 100 μm s^{-1}; representing a flagellar rotation rate of over 100 Hz. Calculations suggest that it takes between 200 and 1000 protons per rotation, which means that motility may take up to 1% Δp for bacteria growing in minimal media (Meister et al., 1987; Brown et al., 1993). *Rb. sphaeroides* swims efficiently (using Δp not a ΔNa^+) between pH6 and pH9, reflecting the low threshold for motility; the velocity starts to fall as the pH is increased above 9 but does not stop until close to 10 (Packer et al., 1994). Interestingly, although the speed of flagellar rotation saturates at about –50 mV and remains constant as the Δp is increased to saturation, it is possible to make *Rb. sphaeroides* swim faster by stimulation with a range of chemokinetic compounds (see later), suggesting that the efficiency can be increased.

The mechanism by which the electrochemical proton gradient is transduced into rotational work is unknown, as reflected by the large number of theoretical papers on the subject (see Blair, 1990). The MotA protein in *E. coli* has been identified as the probable site of proton transport, possibly with MotB as the stator and the energy transduction occurring in either the M-S-ring or the switch complex. Recently two *Rb. sphaeroides* flagellar motor genes have been identified (Sockett et al., 1990; Sockett and Armitage, 1991); and sequenced (Shah et al., 1995; Shah and Sockett, 1995). The protein encoded by one of these genes has been found by *pho*A fusion analysis to be located in the cytoplasmic membrane with a C terminal periplasmic domain. This *Rb. sphaeroides* motor gene shows limited homology to MotB of enteric bacteria. MotB homology is strongest in the single membrane spanning domain which may form part of the proton channel, along with MotA. The periplasmic domain of *Rb. sphaeroides* MotB differs from others in that it does not contain obvious peptidoglycan binding motifs. The product of the second *Rb. sphaeroides* gene has 21% identity to MotA from enteric bacteria; it has a predicted secondary structure consisting of four membrane-spanning helices, and a central, charged, cytoplasmic domain. Although the level of identity is not high, several amino acids identified in enteric bacteria as being absolutely required for rotation (Blair and Berg, 1990), are conserved in the *Rb. sphaeroides* genes. The molecular comparison of *Rb. sphaeroides* and *E. coli* flagellar motor proteins will be of particular interest for identifying the residues important for proton transport and for helping identify the differences between a switching and a unidirectional motor.

b. Swimming Patterns

Rotation of flagella in most bacterial species, including *Rs. rubrum* and *Chromatium* spp., is bidirectional, with frequent switching from clockwise to counterclockwise. *Rs. rubrum* swims in a series of runs and reversals, the cells back-up as the rotational direction of the flagellar bundle changes (Clayton, 1957; Mitchell et al., 1991). The periodic switch in the direction of rotation from CCW to CW is what changes the swimming direction of a cell and is the response modulated during taxis to bias the usual 3 dimensional swimming pattern towards an attractant. The control of switching has been, as will be seen later, the subject of a great deal of research in enteric bacteria and in these species (and probably many others) direction changing is brought about by the binding of a small 12 kDa protein, CheY, to the switch complex when phosphorylated (phosphorylation being controlled by either environmental stimuli or modulated internally)(Barak and Eisenbach, 1992a). However a 12 kDa protein can only diffuse a limited distance (Ishihara et al., 1983) and is unlikely to be the cause of the coordinated switching of the flagellar bundles of *Rs. rubrum*, where reversal in rotation direction occurs simultaneously at each end, even in cells 60 μm long. It is possible that an electrical signal may coordinate flagellar reversal in these species, but as yet there is no definitive evidence for this (Lee and Fitzsimmons, 1976).

Although the coordination between the polar flagellar bundles may be electrical, the control of flagellar switching during chemotaxis may still involve CheY, as proteins showing homology to the enteric chemosensory proteins controlling CheY have been identified in *Rs. rubrum* (Sockett et al., 1987; Morgan et al., 1993). How the two systems interact has not been investigated, but the recent identification of the clustering of the chemosensory proteins in *E. coli* (Maddock and Shapiro, 1993) suggests the possibility that these proteins are clustered around the tufts of flagella, allowing a rapid chemosensory response. No studies have yet been carried out on switch proteins in photosynthetic bacteria, but in enteric bacteria the FliN, M and G proteins are implicated in generating flagellar reversals (Jones and Aizawa, 1991). It may well be that similar proteins are present in *Rs. rubrum,* responding to chemosensory signals, but there may be additional proteins responding to electrical signaling.

The *Rb. sphaeroides* flagellum rotates unidirectionally, intermittently stopping, then restarting rotation, the bacterium moving off in a new direction (Armitage and Macnab, 1987). The flagellar stopping frequency is increased if the bacteria swim down a gradient of a positive tactic effector (see later). Until recently it was thought that rotation was always clockwise in these bacteria, but recently a naturally occurring variant has been found in wild-type populations which shows exclusively counterclockwise rotation (H. Packer and J. P. Armitage, 1993). The CCW rotating flagellum has a different handedness and wavelength compared to the CW flagellum, and is more efficient at moving the cell through viscous environments, but whether this is of physiological relevance is not known. The rotational stopping frequency and swimming speed of cells from the same population varies quite dramatically. It is not known whether stopping and speed changes represent the action of a 'clutch' i.e. decoupling the energy supply from the flagellum; or the action of a 'brake' i.e. blocking rotation of the flagellum with the energy supply left 'plugged in'.

Studying the *Rb. sphaeroides* flagellum gives an unique opportunity to look at the activity of a single copy enzyme under different conditions and at different stages of the cell cycle. Heterogeneity in the motility of *Rb. sphaeroides* may merely reflect variation in flagellar activity which is present in all enzymes in cells, but which is averaged out by the presence of multiple copies of the enzyme.

3. Flagellar Genetics and Biosynthesis

The biosynthesis of flagella has been extensively studied in enteric bacteria by several workers (Koch et al., 1982; Iino, 1985; Macnab, 1992), and it has been established that flagellar components are synthesised and assembled in a defined order. The rings and rod are synthesised and assembled first followed by the hook, HAPs and filament. The motor and switch components are added last, after the passive structural elements of the flagellum have been assembled. The axial components of the flagellum such as the hook and filament are exported from the cell to their final position in the flagellum by a *sec-* independent pathway. Axial proteins do not have signal sequences, and it is thought that their export is mediated by a specific protein complex which includes the FliI, FlhA and FliP proteins. Recently a 5.5 kb *Eco*RI restriction fragment with homology to *fliI* (the flagellar specific, ATP-binding export protein) from *Salmonella typhimurium* was isolated from *Rb. sphaeroides* (G. Dreyfus unpublished). Transposon insertions in this clone resulted in nonmotile mutants with no flagellar filament, with one of these mutants accumulating cytoplasmic flagellin monomers (C. E. Pollitt and R.E. Sockett unpublished), suggesting that a similar flagellar export pathway exists in *Rhodobacter* to that present in enteric bacteria.

In *Rb. sphaeroides,* transposon mutations in at least 20 different genes result in a filament-minus phenotype. This indicates that flagellar assembly in this bacterium must proceed by a similar hierarchical pathway and it is probable that even more flagellar genes exist, as over 30 genes are involved in flagellar biosynthesis in enteric species. Genes encoding the flagellar filament and the proteins responsible for its export cluster to one 15 kb region not closely linked to the flagellar motor genes. All the flagellar genes cloned to date, map to one 244 kb *Ase*I fragment on the physical map of the *Rb. sphaeroides* genome (Suwanto and Kaplan, 1989) adjacent to the genes encoding the photosynthetic apparatus.

C. Behavioral Responses

Bacteria don't just swim randomly, they use environmental signals to control their direction changing frequency and bias their overall movement to either maintain themselves in, or move towards, an optimum environment for growth. There is good evidence that if a bacterial species swims, motility provides a survival advantage for that species in their particular environment. In fact, in rich or well mixed environments bacteria often lose motility as it no longer provides a survival advantage. Because it takes up to 5% of a bacterial genome to make a functional flagellum, the energy involved in its synthesis and operation, particularly under growth-limiting conditions, would favor the rapid loss of the phenotype if it did not provide an advantage.

All motile bacteria can sense and respond to a wide range of physical and chemical stimuli, but the flexible metabolism of the purple non-sulfur bacteria means that not only can they probably sense more than most other species, but the response shown to a particular stimulus will depend on the current growth conditions. Amongst stimuli sensed are light, oxygen and other terminal electron acceptors, organic acids, sugars and inorganic ions. The signals coming from the different receptor systems must be integrated to balance the different responses, ensuring, for example, that a photosynthetically growing cell does not follow a light or electron donor gradient into an oxygenated environment.

1. Chemotaxis

All motile species investigated swim towards some chemicals but the sensing mechanisms vary from species to species. Most bacteria are too small to sense a chemical (or any other) gradient across their body length; they must therefore temporally sense their environment, comparing the concentration of a present stimulus with that a period of time earlier.

Changes in the concentration of specific chemicals are sensed through a range of different mechanisms, all finally integrating to control the flagellar motor. Three chemotactic systems have been identified (i) the dedicated receptor system (MCPs), (ii) the transport dependent sugar chemotaxis system based on PTS transport and (iii) a metabolism based chemosensing system linked to the current growth limitations of the cells. It is probable that all three systems interact and the relative importance of different chemosensing pathways varies between species, possibly related to their growth environment (see later).

a. Methylation-Dependent Chemotaxis

This is probably the best characterised chemosensing system in biology, with the most data coming (of course) from enteric bacteria (Bourett et al., 1991; Armitage, 1992). Dedicated membrane-spanning chemosensory proteins (methyl-accepting chemotaxis proteins, MCPs) bind a limited number of chemoeffectors, and the conformational change caused by the effector binding transduces the information to a cytoplasmic messenger. The chemoeffector needs only to bind to the MCP for a sensory signal to occur, neither transport nor metabolism are required. There are several hundred copies (in *E. coli*) of 4 different MCPs, each responding to about 2 different effectors. The different MCPs have conserved cytoplasmic regions, responsible for signaling, and variable periplasmic regions, responsible for binding the chemoeffector. Antibodies have been raised against the Trg receptor in *E. coli* (this is the receptor responsible for sensing ribose and galactose) and these not only cross react with other MCPs from *E. coli* but also from a wide range of other unrelated species. These studies have revealed MCPs in *Rs. rubrum* and *Cm. vinosum* but not in *Rb. sphaeroides* (Sockett et al., 1987; Morgan et al., 1993). The cross-reaction with *Rs. rubrum*, however, was very weak compared to the cross-reaction shown to some more distantly related species, such as *Halobacterium halobium* (now *H. salinarium*). This implies that either the MCPs in *Rs. rubrum* are very different from those in other species, or there are fewer of them, perhaps clustered around the flagellar tufts.

Figure 3 shows the sensory pathway from the MCPs to the flagellar motor. A change in MCP conformation caused by a change in chemoeffector binding on the periplasmic side is transmitted to two proteins associated with the cytoplasmic side of the receptor, CheW and CheA. CheA is a histidine protein kinase, related to the HPKs involved in controlling transcriptional regulators in the two-component sensing systems. If there is an increase in repellent binding to a receptor, CheA autophosphorylates, but instead of phosphorylating a transcriptional regulator (as seen in the other two-component systems) it phosphorylates two intracellular non-DNA binding proteins (Bourett et al., 1991). CheY, a small diffusible protein (12 kD), is phosphorylated within milliseconds, and when phosphorylated binds to the switch on the flagellar motor causing it to rotate in the opposite direction and the bacterium to change swimming direction. The rate of dephosphorylation is controlled by CheZ which appears to be present at a constant concentration and activity; therefore the more CheY phosphorylated by CheA the more frequent the direction changing.

Being able to sense a gradient depends on the ability to adapt (Schnitzer et al., 1990). If a cell continued to respond after being stimulated it would not be able to sense a change in that or any other signal, nor would it be able to 'sense' the direction of any gradient. The second protein phosphorylated by CheA is a specific methyl esterase, CheB, involved

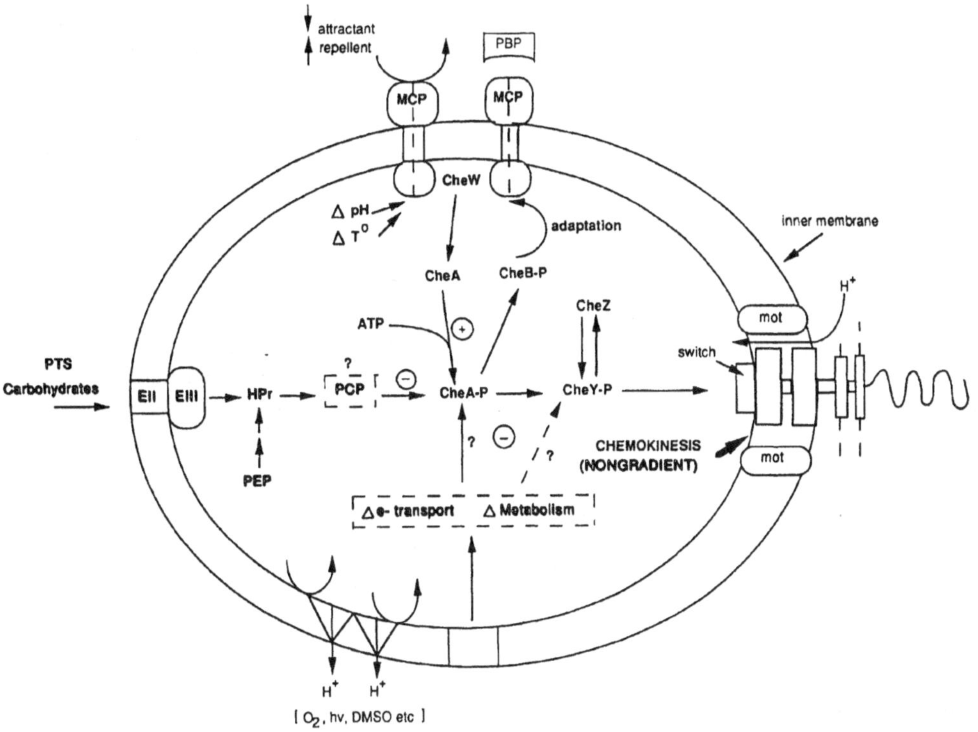

Fig. 3. Possible pathway for the integration of sensory pathways in bacteria. A reduction in chemoeffector binding to an MCP results in an increase in CheA-P and thus CheY-P and an increase in flagellar switching, whereas an increase in transport through the PTS causes a reduction in CheA-P and thus a reduction in switching. CheA-P also controls the activity of the adaptation pathway by controlling the activity of the methyl esterase, CheB. The site of interaction of the electron transport-dependent signal and the metabolism dependent pathway have not been identified. Chemokinesis appears to effect motor activity directly. MCP = methyl accepting chemotaxis protein, Che proteins = sensory pathway, PCP = PTS-chemotaxis protein, PBP = periplasmic binding protein, HPr = histidine protein.

in adaptation to new but stable environments. There are four specific glutamate residues on the cytoplasmic side of the MCP which can be methylated by a specific methyl transferase (CheR). As the methyl transferase is present at constant activity but the activity of the methyl esterase is changed by phosphorylation, an increase in CheA-P causes an increase in CheB-P, and hence its activity. If the level of CheB-P increases, methyl groups are removed from the MCP to match the change in binding and reset the signaling state of the MCP. The adaptation cycle can take several minutes and the difference in time between sensory signaling and adaptation and the extent of methylation provides the 'memory' in the system.

In enteric bacteria and many other species living in nutrient rich environments, this system appears to be the primary chemical sensing system. It has also been implicated in thermosensing and pHi sensing; possibly by directly changing the conformation of the cytoplasmic component of the MCP. However, in many other species, primarily those living in nutrient limited environments, this may be only one of several different sensing pathways, and by no means the dominant pathway.

b. Methylation-Independent Chemotaxis

Sugars. Sugars that are transported via a binding-protein-dependent system may be sensed through MCPs e.g. the galactose-charged galactose-binding protein binds not only to the transport protein but also to the MCP, Trg. However there are other sugars that do not use the MCP system but need to be transported through the PTS system for a sensory signal to occur (Postma et al., 1993). Detailed experiments on enteric bacteria have shown that it is the rate of turnover of phosphates on EI protein of the phosphotransferase pathway that causes the chemotactic signal, but how this signals to the flagellar motor is not understood. EI appears to control the activity of CheA, but the mechanism is currently unclear. Adaptation, that essential of chemotaxis, is proposed to involve the resetting of the metabolic pathway involved in controlling phosphorylation.

The role of PTS-dependent chemosensing in photosynthetic bacteria is unresolved. In enteric bacteria fructose transport through the fructose PTS does not cause chemotaxis. This appears to be because the rate of phosphorylation of FPr in the fructose-PTS, although sufficient to allow transport, does not cause a chemotactic signal unless greatly over-expressed. In *Rb. sphaeroides* the only PTS system is for fructose, and it is related to the FPr dependent fructose system; other sugars use other, as yet not well characterised, systems (see section III)(Saier et al., 1971). Fructose, glucose, mannose, and sorbitol all cause good chemoattractant responses in *Rb. sphaeroides,* suggesting that the chemotactic signal is independent of either the MCP or PTS pathways (Y. Jezore and J. P. Armitage, unpublished).

Metabolism-Dependent Chemotaxis. All the chemoattractants identified for *Rb. sphaeroides* are metabolites, including the weak organic acids (which are repellents for enteric bacteria), sugars and amino acids (Armitage et al., 1990). Most attractants are effective in the mM concentration range. Despite extensive searching by our group and by others, membrane bound MCPs have not been identified in *Rb. sphaeroides* by either methylation/demethylation reactions or by antibody hybridization (Sockett et al., 1987; Morgan et al., 1993). Methionine auxotrophs also behave normally; methionine starvation in enteric bacteria results in an inability to adapt and therefore loss of taxis. Except for fructose, the chemoattractant sugars are not transported through a PTS pathway, and even that pathway is the one not involved in chemosensing in enteric bacteria.

However, *Rb. sphaeroides* does show true chemotaxis, the removal of an attractant such as pyruvate results in a transient change in swimming behavior, showing that the cells respond and then adapt. Unlike the responses seen in enteric cells, the major response identified in individual cells of *Rb. sphaeroides* is an increase in stopping frequency when moving down a gradient; the bacterial equivalent of a pessimist! *E. coli* responds to an increase in attractant but *Rb. sphaeroides* ignores an increase, at least of the dominant attractants, and responds only to a step down, whether it is a chemoattractant, electron acceptor or light. Modeling shows that even a pessimistic system results in accumulation in regions of high attractant. In addition, the number of responsive cells of *Rb. sphaeroides* in a population depends on the growth limitation of the population– more cells in a starved population show a response compared to a well fed population. The population must also have been induced for growth on a particular chemoeffector for a chemotactic response to occur (Poole et al., 1990b); this is very different from the MCP-dependent sensing in enteric bacteria where sensing is completely separate from metabolism. Transport and, at least limited, metabolism are required for a chemotactic response in *Rb. sphaeroides* to some compounds. For example, the alanine analogue, 2-aminoisobutyrate, although transported through the alanine transport system, does not cause a tactic response (Poole et al., 1993). The inhibition of glutamine synthetase by MSX inhibits ammonia taxis without inhibiting transport (Poole and Armitage, 1989). Similarly the transportable succinate analogue, malonate, does not cause chemotaxis, although as will be seen later, malonate does cause chemokinesis. Data on the Dct transport system suggest that *Rb. capsulatus* also requires transport and metabolism for chemotactic responses to dicarboxylates, as no Dct mutants have been identified which have lost the chemotactic response to dicarboxylates without the additional loss of transport (D. J. Kelly, unpublished). Mutants in the newly identified periplasmic binding protein-dependent glutamate transport system of *Rb. sphaeroides* (Jacobs et al., 1995) have also been shown to lose their chemotactic response to glutamate. Interestingly, complementation of the transport defect by the expression of the unrelated H^+-dependent glutamate transporter from *E. coli* in *Rb. sphaeroides* also led to the restoration of chemotaxis towards glutamate (W. Konigs, personal communication). This suggests that the mechanism of transport is not important in signal transduction and supports the idea that metabolism is required for chemotaxis in *Rb. sphaeroides*. All of these data are incompatible with a transport-independent, dedicated, membrane spanning sensory protein such as the MCP being the major sensory pathway.

How does chemotactic signaling occur in *Rhodobacter* species and how does it relate to that identified in enteric bacteria? If different chemo-attractants are present at one time, unrelated compounds are found to compete. For example, either succinate or fructose can inhibit chemotaxis towards alanine (Poole et al., 1993). However, competition only occurs if the specific metabolic pathway of the successful competing compound has been induced. All these data imply that the bacteria may be sensing a change in concentration of a common metabolic intermediate, but what is this intermediate and to what does it signal? Arsenate has been shown to inhibit chemotaxis, suggesting that phosphorylation is required for a response. Calcium, acetyl adenylate, acetyl phosphate and fumarate have all been implicated in motor switching in other species, but their mechanism of interaction with the Che pathway and their role in vivo are unknown (Wolfe et al., 1988; Barak and Eisenbach, 1992b; Lukat et al., 1992; Tisa and Adler, 1992).

Homologues of the *che* genes of *E. coli* have recently been identified in *Rb. sphaeroides*. The gene arrangement in a *che/mcp* cluster is unlike that identified in enteric species, but related to that recently identified in other related species, *Rhizobium melioti* (Greck et al., 1995) and *Caulobacter crescentus* (M. R. K. Alley and L. Shapiro, personal communication). Antibody raised against *C. crescentus* MCPa has identified at least two 65 kD MCP homologues (Ward et al., 1995b). Sequencing has identified two putative *mcp* genes in the chemotaxis cluster. Why were these not identified using the *E. coli* antibody? It seems probable that the *Rb. sphaeroides* MCPs are present in low numbers, they have a sequence which in the protein has been shown to localize the *C. crescentus* MCPs to the cell pole suggesting that in *Rb. sphaeroides* they are also localized, although whether this is at the pole or close to the flagellar motor remains to be determined. This, combined with the limited similarity to the *E. coli* MCP and the dominant metabolism based chemotaxis system, probably explains why it was not identified earlier. In addition, the MCP has only one membrane spanning region, the majority of the protein being cytoplasmic, possibly related to the cytoplasmic MCP identified in *Myxococcus xanthus* (McBride et al., 1992). This is compatible with a cytoplasmic sensory signal.

Perhaps the most significant finding, however, is that of two *cheY* homologues, in addition to *cheA*, *cheW*, *cheR* and probably *cheB* (Ward et al., 1995a). No chemotaxis mutants have ever been identified by transposon mutagenesis in *Rb. sphaeroides*. It is now apparent that the loss of one *cheY* is compensated for to some extent by the second, as they are on separate operons (loss of *cheY* in *E. coli* results in smooth swimming and an obvious phenotype). Why should there be two CheY proteins? This is the protein which when phosphorylated binds to the motor switch. It could be related to the unidirectional motor, but on the other hand it may imply two different phosphorylation paths; a 'classical' pathway from MCPs to CheA and CheY2 with the phosphorylation of the second CheY controlled by a phospho-donor linked to metabolism. This remains to be resolved by detailed molecular biology and protein chemistry.

All the behavioral chemotaxis experiments on *Rb. sphaeroides* suggest that chemotaxis is directly linked to their current growth requirements. The MCP based system in enteric bacteria, with dedicated sensing proteins, means that these bacteria must respond to a change in effector whether required for growth or not, for example *E. coli* will swim towards maltose in a background of galactose. An intracellular sensing system sensing a change in an intermediary metabolite would result in a chemotactic response only if there was an increase in metabolic rate, transiently altering the concentration of that intermediate. This may be an advantage for organisms living in growth limited environments. The MCP and metabolism based systems may operate side by side, the MCPs sensing a limited range of essential chemoattractants and an intracellular system sensing a change in metabolic rate. Rather than the MCP-based system being dominant, as it appears (at least in the laboratory) for enteric bacteria, it may just be one of several systems feeding sensory information into the motor.

c. Electron-Transport Dependent Chemotaxis

For photosynthetic bacteria growing under both heterotrophic and photoheterotrophic conditions, the rate of electron transport is obviously crucial. It would therefore be expected that they would respond to changes in the concentration of electron donors and electron acceptors. The study of the mechanisms involved in these responses has been complicated by two factors: (i) any change in the rate of electron transport causes at least a transient change in the electrochemical proton gradient (Δp), which is the driving force for flagellar rotation and therefore any change in flagellar behavior may be a direct effect on the flagellum rather than a sensory response, and (ii) the complexity and flexibility of the electron transport chain in most photosynthetic bacteria.

Rb. sphaeroides shows good positive aerotaxis when grown aerobically, but negative aerotaxis when grown anaerobically in the light. This type of response was first seen by Engelmann back in the 1880s when he saw photosynthetic bacteria accumulate around air bubbles in dim light but move away from the air bubbles in bright light. Recent advances in motion analysis have made possible characterization of the aerotactic response by following the behavior of single cells in an oxygen gradient. A response equivalent to the chemotactic response has been identified, with an increase in stopping (and therefore direction) changing frequency of aerobically grown cells as the oxygen concentration falls (P. Hamblin, H. L. Packer, J. P. Armitage, unpublished).

Rb. sphaeroides grown anaerobically in the dark with DMSO as an electron acceptor shows chemotaxis towards DMSO when incubated anaerobically in the dark, but if incubated in either the light or aerobically in the dark, DMSO taxis is lost, paralleling the loss of electron transport to the DMSO reductase. DMSO reductase mutants also show no response to a gradient of DMSO. Therefore, no DMSO 'receptor' exists other than the reductase, and electron transport to the reductase is essential for a sensory response (D. Gauden, J. P. Armitage, unpublished). This supports data from *E. coli* on chemotaxis towards nitrate (Shioi and Taylor, 1984). What these data do not identify is whether the signal is a change in Δp caused by the change in electron transport or a change in the electron transport rate itself, altering either a redox sensitive protein or a component of the electron transport chain. Selective inhibition of different branches of the electron transport chain had suggested that a change in Δp was required for a sensory signal (Armitage et al., 1985). However, recent experiments exploiting the different kinetics of DMSO and TMAO binding to the DMSO reductase enzyme suggest that the rate of electron transport, presumably altering the redox state of an intermediate, is important (D. E. Gauden and J. P. Armitage, unpublished). This is supported by the unexpected finding that *Rb. sphaeroides* responded to the removal of an uncoupler by a 'step-down' stop response, equivalent to moving down a gradient; no response was seen to the addition of FCCP. The addition of FCCP would result in a decrease in Δp but an increase in the rate of electron transport, whereas its removal would produce the reverse changes. As *Rb. sphaeroides* only responds to a decrease in stimulus, the stop response on FCCP removal is probably the result of the decrease in electron transport.

These data suggest that the primary tactic signal is either the change in Δp or the change in electron transport. What is the subsequent intracellular signal and does it interact with the Che pathway? Aerotaxis in *E. coli* occurs in the absence of MCPs, but there is some evidence that at least CheY may be required in *E. coli*. Early suggestions that the Arc pathway may be involved in aerotactic signaling in *E. coli* have proved unfounded, as mutants in this pathway respond normally toward oxygen (B. L. Taylor, I. B. Zhulin, unpublished). Glagolev has suggested a 'protometer' sensing mechanism which responds to changes in Δp; this could be part of the motor itself or it may be part of the sensory pathway (Glagolev, 1984). However, as yet there are no data that can separate a Δp from a redox sensor.

2. Phototaxis

Photosynthetic bacteria accumulate in the light, the preferred wavelengths matching the photosynthetic absorption spectrum for most photosynthetic species. However, whether this can be considered a 'true' tatic response is still controversial, with some workers maintaining that, by comparison with eukaryotic microorganisms, bacteria cover too short a distance in a given time to be able to sense a gradient and simply accumulate as a result of 'trapping', changing direction as they swim over a light/dark boundary. However, new data suggest that this may be too simplistic a view. There are two crucial questions about the photoresponses in eubacteria: (i) could a simple step-down response without adaptation work for all bacterial species, and (ii) is accumulation in a light spot all that occurs or can bacteria sense a light gradient.

Important new data on the photobehavior of *Rs. centenum* support the view that bacteria can sense and respond to light gradients. Using the converging light test for phototaxis originally devised by Buder in 1917, Ragatz et al. (1995) have shown that swarming colonies of *Rs. centenum* show a true phototactic response. These authors have redefined the light responses, describing the simple response to a change in light intensity when, for example, a cell moves over a light-dark boundary, as 'scotophobia.' Using a converging beam of light they showed that *Rs. centenum* moved towards the light source, even though the intensity was decreasing, a phototactic response rather than a scotophobic response where the intensity would have been sensed.

It has been recognized for some time that the halobacteria have specific photosensory proteins that generate positive and negative phototactic responses (Oesterhelt and Marwan, 1990). Recently a similar protein, the yellow protein, has been identified in the halophilic eubacterium *Ectothiorhodospira halophila* (Sprenger et al., 1993) and swarming colonies of *Rs. centenum* move away from light at 585 nm and towards light at 800–880 nm (Ragatz et al., 1994, 1995). In addition, calculations suggest that although the random swimming pattern shown by cells may not expose them to a significant light gradient in open water, in very turbid water or mud, a significant light gradient can be formed and a 'memory' of over 30 s would allow this gradient to be sensed.

In general nonsulfur purple bacteria have not been shown to have a true 'negative' phototactic response. However, the photoactive yellow protein has recently been identified in *Rb. sphaeroides* (K. Hellingserf and W. Hoff, unpublished). In general, however, they respond to a reduction in light intensity or a shift in light quality to a less photoactive wavelength. In the 1950s Clayton did a series of seminal experiments examining the motile behavior of cells of *Rs. rubrum* to changes in light intensity (Clayton, 1953a; Clayton, 1953b; Clayton, 1953c). He carried out a range of step-down experiments, showing that *Rs. rubrum* could respond to a change in light intensity of as little as 2% over a 100 fold range of background intensities. The strength of the response depended on both the duration of the stimulus and its strength, with two sub-threshold stimuli being additive if given more than 250 ms and less than 3 s apart. These led to the idea that a perturbation in photosynthesis led to a signal (then ATP was suggested) causing the bacteria to reverse direction.

More recent work has confirmed that active photosynthetic electron transport is required (c.f. aerotaxis) for accumulation. Inhibition of electron transport in either *Rs. rubrum* or *Rb. sphaeroides* led to the loss of phototaxis (Harayama and Iino, 1977), and reaction center mutants fail to show phototaxis even though fully pigmented (Armitage and Evans, 1981). Therefore, there is not a specific sensory pigment involved, as in *Halobacterium salinarium*, but instead the direct or indirect sensing of a change in electron transport is occurring. Does the change in rate of electron transport simply make the flagella reverse rotation, and if so what is the signal and how does it relate to other electron transport dependent taxes?

The majority of photosynthetic species do have tufts of flagella and do change direction by reversal. A fall in photosynthetic electron transport when swimming over a light-dark boundary linked to flagellar reversal could result in trapping. However, *Rb. sphaeroides* does not reverse direction when given a step-down in light intensity, but increases its stopping frequency. If the response was a simple 'stop', then the cells would become trapped in the dark rather than the light. If, however, there was memory in the system and the cells stop and start for a period of time, a new direction may take them into higher light intensities. This 'memory' suggests a taxis system similar to chemotaxis. The early experiments of Clayton did show that *Rs. rubrum* reversed for up to 30 s after the light intensity was reduced, and reversals were suppressed if the light intensity was increased. *Rb. sphaeroides* shows an increase in stopping frequency for up to 2 min after a reduction in light intensity. This is consistent with the behavior of cells in gradients of oxygen or other electron acceptors (see aerotaxis).

One major physiological role of phototactic responses may be in interacting with other electron transport dependent systems. If *Rb. sphaeroides* is incubated in subsaturating light and given a light gradient, it only swims into the increased light if the medium is anaerobic; oxygenated buffer causes a repellent response. Measurement of the $\Delta\psi$ using the electrochromic response of the carotenoids showed that the addition of a pulse of oxygen to *Rb. sphaeroides* in the light resulted in a transient fall in the size of the $\Delta\psi$, presumably as electron transport was directed towards the alternative cytochrome oxidase (Armitage et al., 1985). This change in electron flow probably causes a stop signal, resulting in the bacteria not following a light gradient into an aerobic environment. Whether it is the change in electron flow through a key signaling component of the electron transport chain or the change in $\Delta\psi$ that causes the signal, has not yet been identified.

Rs. centenum shows both a positive phototactic response, with the maximum response in the region of infrared light harvesting and bacteriochlorophyll absorbance (875–800 nm) and a negative phototactic response in regions of visible light (450–590 nm), regions where carotenoids and bacteriochlorophyll also absorb (Ragatz et al., 1995). These data suggest a more complex system of light sensing and signalling, as light absorbance in both these regions would be expected to change the $\Delta\Psi$. *Rs. centenum*, and possibly other species, may have a sensing mechanism which can estimate the ratios of particular wavelengths and use this to maintain their optimum position in, for example, stratified mats.

The behavioral responses to light, oxygen and DMSO all seem to rely on electron transport. As the components of the electron transport chain are shared, it seems unlikely that it is the electron donors or acceptors that are sensed directly, rather it is the flux in electron transport that is sensed, with bacteria moving towards environments where electron transport rates are increased and away from regions where they decrease. The response will depend on the necessary metabolites to sustain or increase the rate. Therefore, unlike chemotaxis (particularly MCP-dependent chemotaxis), there would be no receptor for each effector, but a sensor responding to the changes in electron transport, be that Δp or redox change. If this is the case for most species, phototaxis is no different from any other sensory signal sensed through the electron transport chain.

3. Kinesis

The addition of certain chemoattractants cause *Rb. sphaeroides* to increase its swimming speed by up to 25%. This increase in swimming speed is caused by the flagella rotating faster, but occurs in the absence of increased metabolism and without any measurable increase in the driving force for flagellar rotation, the Δp, or any change in the rate of electron transport (Poole et al., 1990a and 1991; Brown et al., 1993). Chemokinesis (probably more accurately called orthokinesis), as a sustained increase in the rate of flagellar rotation, has only been seen in bacterial species which do not depend primarily on MCPs for chemotaxis, and it is possible that the adaptation cycle in MCP-dependent chemotaxis inhibits any but transient kinetic effects. Chemokinesis in *Rb. sphaeroides* is only seen to a limited range of chemoattractants, mainly organic acids and inorganic ions such a K^+ and Rb^+, but not to other chemoattractants e.g. sugars and most amino acids. Unlike chemotaxis, chemokinesis also occurs in response to some non-metabolizable but transportable analogues such as malonate. Evidence suggests that chemokinesis depends on specific transport pathways; benzoate did not cause chemokinesis although it did get into cells, and aspartate but not glutamate caused chemokinesis (Packer and Armitage, 1994). Understanding the role of specific transport pathways in inducing an increase in swimming speed will have to wait until there is more understanding of transport in nonsulfur purple bacteria, however it may be worth noting that (a) all chemokinetic compounds also caused some cell swelling and with the same time course, and (b) data suggest that the transport pathways involved may involve changes in pHi.

Whatever the cause, chemokinesis would cause a cell population to spread, not to accumulate; unlike chemotaxis, there was no obvious adaptation to chemokinesis. As long as a detectable concentration of effector was present, the cells swam at an increased speed. Could this have a role in vivo? Computer modelling suggests that although chemokinesis would cause a population to spread, it would not stop a tactic response from occurring when a gradient was encountered, and it may even enhance the response (Poole et al., 1991).

Nonsulfur purple bacteria therefore show tactic responses to a wide range of environmental effectors. The limited data so far suggest that some sensory pathways are similar to those identified in enteric bacteria, e.g. electron transport dependent taxis and MCP-dependent chemotaxis, but that chemotaxis is also metabolism dependent, using a *che* pathway which appears to be a modification of that found in enteric bacteria, probably modified to suit the variable growth conditions of these species.

III. Solute Transport

A. Introduction

Although many anoxygenic photosynthetic bacteria are capable of actively transporting a large variety of solutes from the environment, as with motility and taxis, detailed studies on the structure and mechanism of the transport systems responsible have mainly been performed on *Rhodobacter* species. Hopefully, this will be a temporary state of affairs, as the bioenergetic advantages involved in using photosynthetic bacteria as model systems to study transport mechanisms, in addition to the intrinsic importance of understanding transport processes in these organisms in their own right, become appreciated.

A number of the prominent and in some cases unique physiological properties which phototrophic bacteria possess are linked to the functioning of efficient solute transport systems, although this is often overlooked. For example, the biosynthesis of bacteriochlorophyll requires magnesium which has to be accumulated from the environment. Similarly, the high cytochrome content of many phototrophs requires iron transport for the biosynthesis of the heme moiety. Nitrogenase synthesis requires transport of both iron and molybdenum and nickel transport is necessary for incorporation into the hydrogenases that allow many phototrophs to grow autotrophically. Transport systems for the organic electron donors and carbon sources used during photoheterotrophic growth are essential and also seem intimately involved in eliciting chemotaxis. Here, information concerning the mechanism and structure of some of the better studied solute transport systems in phototrophs (primarily the purple bacteria) will be reviewed.

B. Inorganic Ions

1. Potassium

Potassium is now recognized to have multiple roles in bacteria, the most important of which include the regulation of pHi (Booth, 1985), osmotic regulation (Epstein, 1986) and the controlled dissipation of the proton-motive force under certain conditions (Mulder et al., 1988). Potassium transport in purple bacteria was first studied by Jasper (1978), who concluded from experiments with $^{42}K^+$ that a single K^+ transporter existed in *Rb. capsulatus*, with an apparent K_m value of 0.2 mM. This system also transported Rb^+ with lower affinity. More recent studies have revealed several important physiological roles for potassium transport in anoxygenic phototrophs. In potassium, as opposed to sodium phosphate buffer, the membrane potential in energized *Rb. sphaeroides* cells was decreased and the ΔpH increased (Abee et al., 1988), consistent with the operation of an electrogenic K^+ uptake system. Using carotenoid electrochromism to monitor membrane potential, Golby et al., (1990a) found that the addition of K^+ or Rb^+ (but not Na^+) to washed cells of *Rb. capsulatus* resulted in an increase in the ionic current across the cytoplasmic membrane and to a decrease in membrane potential upon illumination, which was attributed to the electrophoretic uptake of these ions. The response was half-maximal at about 0.4 mM K^+ or 2.0 mM Rb^+, values of the same order as the K_m values determined by Jasper (1978) and indicative of a low affinity uptake system, possibly similar to the TrkA system of enteric bacteria. This system appears to be a uniporter (Golby et al., 1990b) and is constitutive. It plays a key role in the regulation of the pHi, which is significantly increased at all external pH values in the presence of K^+ (Abee et al., 1988). Because of the excellent time resolution of the electrochromic response, it was possible to determine that the increase in the K^+ dependent ionic current was not instantaneous with the increase in membrane potential at the onset of illumination (Golby et al., 1990a), providing evidence for a $\Delta\psi$-mediated activation of the transporter ($t_{1/2}$ = 120 ms). This phenomenon is undoubtedly found in many types of bacterial membrane transporter but is particularly amenable to study in phototrophs which have electrochromically responsive carotenoids.

Recently, it has been found that *Rhodobacter* cells can induce the synthesis of a Kdp-like high-affinity K^+ uptake system when grown in media of low K^+ concentration (Abee et al., 1992 a,b). A vanadate sensitive, K^+ stimulated ATPase activity was purified from cytoplasmic membranes of *Rb. sphaeroides* and was found to be composed of three subunits of 70, 43.5 and 23.5 kDa, corresponding to the KdpB, A and C subunits of the *E. coli* complex, respectively. However, only the KdpB-like protein (the catalytic subunit) was immunologically cross-reactive. This type of high-affinity K^+ ATPase appears to be widely distributed in nonsulfur purple bacteria (Abee et al., 1992b) and is likely to be the major mechanism whereby potassium is obtained from low external concentrations.

It has long been known that a significant membrane potential is maintained in *Rhodobacter* cells incubated under strictly anaerobic conditions in the dark (Cotton et al., 1981). Several explanations have been advanced to explain the origin of this dark potential, but the key observation that KCl significantly decreased the potential in *Rb. sphaeroides* (Abee et al., 1988) led to the suggestion that it is a result of the electrogenic efflux of K^+. In experiments with *Rb. capsulatus*, where the dark anaerobic membrane potential was measured by carotenoid electrochromism (Golby et al., 1990a), it was shown that the decrease in $\Delta\psi$ was proportional to the logarithm of the external K^+ concentration, indicating a Nernstian relationship between $\Delta\psi$ and the internal and external K^+ concentration. Other electrogenic processes may also make a smaller contribution to the anaerobic dark $\Delta\psi$ in *Rb. capsulatus* (Golby et al., 1990a). However, in *Rhodospirillum rubrum*, the dark potential could be largely abolished by DCCD, suggesting its dependence on ATP hydrolysis rather than K^+ efflux (Fenoll et al., 1985).

2. Ammonium

Although widely used as a source of nitrogen, uptake of the ammonium ion in anoxygenic phototrophs has not been studied extensively. This is at least partly due to technical difficulties in measuring the flux of NH_4^+ directly. Instead, the analogue methylammonium has traditionally been used but this is known to be converted by glutamine synthetase to γ-glutamyl-methylamide in *Rb. capsulatus* (Yoch et al., 1983). Those direct measurements which have been performed indicate the presence of more than one NH_4^+ uptake system in many species; in *Rb. capsulatus*, NH_4^+ uptake was characterised by two K_m values of 1.7 μM and 11.1 μM (Sharak Genthner and Wall, 1985). In these studies, a tight coupling was observed between transport and assimilation of NH_4^+, as mutants deficient in glutamine synthetase did not take up ammonium ions. Evidence for an ammonium transporter in *Rs. rubrum* has also been obtained (Alef and Kleiner, 1982). More recently, it has become clear that potassium and ammonium ions can be transported by the same carrier in many bacteria and that this can have important physiological consequences. In *Rb. capsulatus*, several pieces of evidence indicate that the low-affinity K^+ transporter, described above, is also capable of transporting NH_4^+ (Golby et al., 1990a,b). In contrast to the high-affinity uptake systems characterised by Sharak Genthner and Wall (1985), ammonium transport via the K^+ uptake system is not inhibited by low concentrations of methylamine and the concentration dependence of the NH_4^+ dependent ionic current indicates an approximate K_m value of 3 mM (Golby et al., 1990a). The physiological significance of these observations is that, as a consequence of electrophoretic ammonium uptake, $\Delta\psi$ is decreased and ΔpH increased; this may be followed by the passive efflux of NH_3 from the cell in response to the increased ΔpH, thus leading to a partial dissipation of the Δp. This may be one example of an 'energy-spilling' reaction which could operate in photosynthetic bacteria in particular circumstances. Indeed, evidence for such reactions has been obtained with *Rb. capsulatus* cells grown anaerobically under illuminated conditions in which the energy supply exceeds the demands of anabolism (Taylor and Jackson, 1987).

3. Divalent Cations

Separate transport systems for the energy-dependent uptake of magnesium and manganese have been described in both chemoheterotrophically grown and photoheterotrophically grown cells of *Rb. capsulatus* (Jasper and Silver, 1978). The characteristics of these systems are distinct, with respective K_m values for magnesium and manganese of 55 μM and 0.5 μM. The maximum rate of transport for magnesium was up to a hundred-fold higher than that of manganese. Several divalent cations were competitive inhibitors of magnesium transport; Fe^{2+}, Mn^{2+} (K_i = 250 μM) and Co^{2+} (K_i = 270 μM). Calcium did not inhibit transport, and it was also demonstrated that *Rb. capsulatus* cells did not normally accumulate Ca^{2+}, but that an energy dependent efflux system may be present for this ion. Accumulated magnesium was not tightly bound to cellular components and under de-energized conditions was exported from the cells. The details of the efflux process have not been elucidated and there is little additional information on magnesium transport in any other anoxygenic phototroph. In contrast, much more is known about magnesium transport in enteric bacteria (Gibson et al., 1991). With the current high level of interest in bacteriochlorophyll biosynthesis and its control, it would be relevant to know more about the molecular properties of the magnesium transport systems in phototrophic bacteria.

Nickel is a component of the hydrogenase and urease of *Rb. capsulatus* and its uptake was studied by Takakuwa (1987). There appeared to be a sharp pH optimum for uptake (at pH 7.0) and the optimum temperature was 28 °C. The concentration dependence obeyed Michaelis-Menten kinetics with a measured K_m value of 5.5 μM. This indicates the operation of a high-affinity uptake system for this ion. A number of other divalent cations inhibited the uptake of nickel, particularly magnesium and cobalt, which gave K_i values of less than 100 μM. It is therefore possible that a common transport system exists for nickel, magnesium and cobalt in *Rhodobacter,* but ultimate confirmation of this will require identification and mutagenesis of the genes involved.

4. Iron

Despite the enormous interest in electron transport processes and cytochrome complexes in phototrophic bacteria, there is a surprising lack of information concerning the mechanisms by which iron is acquired and managed by these organisms. Two studies have directly addressed the question of iron transport (Moody and Dailey, 1984; Moody and Dailey, 1985), both using *Rb. sphaeroides* as the model system. It was found that growth in an iron-deficient medium under either aerobic or anaerobic conditions did not result in the production of soluble phenolate or hydroxamate siderophores. No evidence was obtained for any siderophore remaining associated with the cells. In addition, iron-limited cells were unable to remove iron from ferric transferrin unless supplemented with 2,3-dihydroxybenzoic acid (Moody and Dailey, 1984). However, radiolabelled iron was taken up when supplied as ferric chloride, ferric citrate or ferric parabactin (a siderophore from *Paracoccus denitrificans*). In the case of ferric citrate, the citrate itself was not taken up by the cells. There were no significant differences in the protein profiles of cells grown under low or high iron conditions, and no evidence for the induction of specific outer membrane proteins in response to iron starvation (Moody and Dailey, 1985). Membrane vesicles prepared from aerobically grown cells were able to transport iron in the presence of NADH as electron donor. Uptake was sensitive to triphenylmethylphosphonium, possibly suggesting a $\Delta\psi$ driven process, although a very high concentration of inhibitor (5 mM) was used. Inhibition of ferrochelatase by N-methylprotoporphyrin did not result in a decrease in iron uptake, indicating that heme synthesis and iron transport are not tightly coupled (Moody and Dailey, 1985). Radiolabelled gallium (an iron analogue which is trivalent and cannot be reduced) was not taken up by intact cells, although excess Ga^{3+} did inhibit the uptake of ferric iron. These results suggested a model for transport in which ferric iron is recognized by a transport protein, reduced to the ferrous state and then transported into the cell. No details are available concerning the molecular nature of the transport system(s) involved.

5. Molybdenum

Molybdenum is of particular importance in photosynthetic bacteria, not only because of the widespread ability of these organisms to fix nitrogen using the molybdenum containing nitrogenase, but also because some of the terminal oxidoreductases involved in anaerobic electron transport contain the molybdenum cofactor molybdopterin (Ferguson et al., 1987). Although transport is clearly essential for molybdenum incorporation into the cytoplasmic nitrogenase, the question of how this metal is incorporated into the periplasmic oxidoreductases is unresolved. Recently, the genes encoding a high-affinity transport system for molybdate from *Rb. capsulatus* have been cloned and sequenced (Wang et al., 1993). Four genes, *modABCD*, form an operon located in the *nif* region B of the *Rb. capsulatus* chromosome, downstream of the *nifB*(II) gene. These four *mod* genes encode a periplasmic binding-protein dependent transporter. ModA is a hydrophilic protein containing a typical signal sequence and was proposed to be a periplasmic molybdate-binding protein, although this has not yet been tested experimentally. ModB is a hydrophobic, presumably integral membrane, component homologous to ChlJ of *E. coli,* and ModC is a member of the ATP-binding-protein superfamily homologous with ChlD of *E. coli.* The function of the *modD* gene product is unknown. Insertion mutations in *modA*, *modB* or *modC* resulted in a 500 fold increase in the concentration of molybdenum required to repress the alternative nitrogenase, indicating that although the *mod* genes are involved in high-affinity molybdenum uptake, an additional transport system of lower affinity must also be present in *Rb. capsulatus*. Two additional open-reading frames (*mopA* and *mopB*) upstream of the *mod* operon which encode highly homologous proteins were also identified by Wang et al., (1993). They share significant sequence similarity with low molecular weight molybdenum-pterin-binding proteins from *Clostridium pasteurianum* but their role in molybdenum metabolism is at present unclear.

C. Sugars

Many photosynthetic bacteria do not grow particularly well on sugars, probably because they are sensitive to the acidification that can result from sugar metabolism (Willison, 1993), particularly under oxygen-limiting conditions. In general terms, very little information is available on the mechanisms of sugar transport, with the exception of fructose uptake. In several members of the purple non-sulfur bacteria, including *Rhodospirillum* and *Rhodobacter*, fructose uptake is catalysed by an inducible fructose specific PTS system (Saier et al., 1971). This system has turned out to be of general interest, as it is much simpler in structure than many other types of PTS. The system consists of just two proteins, a membrane bound component with molecular weight of 55,000 kDa (FruA) and a 110,000 kDa soluble protein (FruB) which is probably dimeric (Saier et al., 1971; Lolkema et al., 1985). Daniels et al. (1988) isolated Tn5 insertion mutants of *Rb. capsulatus* defective in growth on fructose. The mutants were totally lacking in fructose fermentation, fructose uptake, phosphoenolpyruvate-dependent fructose phosphorylation activity and fructose-1-phosphate dependent fructose transphosphorylation. An operon containing three genes, *fruBKA*, with the *fruK* gene encoding fructose-1-phosphate kinase was identified (Wu et al., 1990). The sequence analysis of the *fruB* gene has revealed that it encodes an unusual and complex protein which contains three distinct domains that are normally located in separate proteins in other PTS systems; a fructose specific enzyme III-like N-terminal domain (B), an FPr(HPr)-like domain (H) and a C-terminal enzyme I-like domain (I) which has striking sequence similarity to maize pyruvate : phosphate dikinase. These three functional units are joined by two flexible linker regions, rich in ala, gly and pro residues. Sequence comparisons with other known PTS proteins allowed the identification of three conserved histidine residues in each domain that are predicted to be the sites of phosphorylation. Thus the data indicate that FruB(HI) is a multi-phosphoryl transfer protein. The *fruA* gene encodes the fructose permease (enzyme II) of the system (Wu and Saier, 1990) and is very similar to the fructose enzyme II of *E. coli*. However, although it had been originally proposed that a histidyl residue was the phosphorylation site in the *E. coli* enzyme II, this residue is not conserved in the *Rb. capsulatus* FruA protein. Instead, a conserved cysteine residue present in both proteins and subsequently identified in other types of enzyme II is now thought to be the actual phosphorylation site. Thus, studies on the *Rb. capsulatus* fructose-PTS have led to important conclusions regarding the mechanism of group translocation in general.

D. Organic Acids

Organic acids have long been known to be very effective growth substrates for purple non-sulfur bacteria and can be used as both carbon and energy source under chemoheterotrophic conditions or as carbon source and electron donor under photo-heterotrophic conditions. In the latter case, if the organic acid is relatively reduced, an electron sink such as carbon dioxide or an auxiliary oxidant such as dimethylsulphoxide is required to maintain redox balance (Ferguson et al., 1987). Traditionally, the C4-dicarboxylates malate and succinate have been used routinely for the growth of purple phototrophs but pyruvate and lactate are also excellent substrates. Both D- and L-malate can be used by *Rhodobacter* strains (Stahl and Sojka, 1973).

The transport of C4-dicarboxylates was first studied by Gibson (1975) who showed that *Rb. sphaeroides* possessed an inducible transport system for malate, succinate and fumarate. Subsequently, C4-dicarboxylate transport has been studied in most detail in *Rb. capsulatus*. The major route for the transport of malate, succinate and fumarate is a high-affinity binding-protein-dependent system (Dct) which is essential for growth on these substrates under aerobic/dark conditions (Hamblin et al., 1990; Shaw and Kelly, 1991). The binding protein has been purified (Shaw et al., 1991) and its ligand binding kinetics analyzed in detail (Walmsley et al., 1992 a,b). It is able to bind L-malate, succinate and fumarate with high affinity and D-malate with somewhat lower affinity, but does not bind other organic acids significantly. The Dct system is regulated in activity by the intracellular pH value (Shaw and Kelly, 1991). Interestingly, *dct* mutants are still able to grow slowly on C4-dicarboxylates photoheterotrophically, due to the presence of additional low-affinity systems which are not present aerobically (Hamblin et al., 1990; M. C. Behrendt and D. J. Kelly, unpublished). The *dct* locus has been cloned and sequenced (Shaw et al., 1991; Hamblin et al., 1993 M. J. Hamblin and D. J. Kelly, unpublished) and this has revealed that it is not a conventional periplasmic transport system. Five *dct* genes have been identified, two of which (*dctS* and *dctR*) encode a translationally coupled two-component sensor-regulator pair which regulate expression of the structural genes. DctS and DctR share the greatest sequence similarity with the FixLJ system in *Rhizobium* (Hamblin et al., 1993). Three structural genes are present; *dctP*, which encodes a periplasmic C4-dicarboxylate binding protein, *dctQ*, encoding a small integral membrane protein, and *dctM*, which encodes a large integral membrane protein. These form an operon divergently transcribed from *dctSR*.

Insertion mutations in any of the *dct* genes leads to the abolition of chemotaxis to C4-dicarboxylates, supporting other work indicating that transport itself or an intracellular signal is necessary for chemotaxis and that the sensor-regulator pair is not directly involved. The *dct* locus has been physically mapped at high resolution on the *Rb. capsulatus* chromosome (Fonstein and Haselkorn, 1993). Significantly, there is no gene at the *dct* locus encoding the type of conserved ATP-binding protein characteristic of conventional periplasmic permeases. Moreover, operons encoding homologues of DctP, Q and M are present in *E. coli* (Sofia et al., 1994) and *Bordetella pertussis* (Willems et al., 1992) but again, there is no ATP-binding protein in these systems. There is also bioenergetic evidence to indicate that dicarboxylate transport through the Dct system is driven by the membrane potential rather than by ATP hydrolysis (Forward et al., 1993). It has been proposed that this system is a representative of a novel class of symporter which possesses a periplasmic binding-protein. It is not known how widespread this type of transporter is, but as most other purple non-sulfur bacteria are capable of efficient growth on C4-dicarboxylates (with the interesting exception of *Rs. centenum*; Stadtwald-Demchick et al., 1990), it is presumably common in this group at least. There is a report that the purple sulfur bacterium *Ectothiorhodospira shaposnikovii* contains a sodium-ion driven dicarboxylate symporter (Karzanov and Ivanovsky, 1980) but this system has not been investigated further.

The transport of other organic acids has not been studied at the molecular level in phototrophic bacteria. Gibson (1975) obtained evidence for more than one system for pyruvate transport in *Rb. sphaeroides*, and both high-affinity and low-affinity systems exist for this substrate in *Rb. capsulatus* (H. V. Modak and D. J. Kelly, unpubl).

E. Amino Acids

The uptake of certain amino acids has been studied extensively in purple non-sulfur bacteria in an effort to understand the way in which bacterial solute transport processes in general are regulated by the proton-motive force. In *Rb. sphaeroides*, the relationship between alanine transport and the magnitude and composition of the proton-motive force was determined in a number of investigations (Elferink et al., 1984, 1987). It was found that under conditions where the rate of electron transport was low, but where a significant membrane potential existed, e.g. under anaerobic dark incubation conditions, alanine transport did not occur. Conversely, high rates of electron transfer correlated with high alanine uptake rates. This formed part of the evidence which led to the proposal that there are direct interactions between components of the electron transport chain and solute transport systems (Elferink et al., 1984, 1987). However, in these studies, it was assumed that the alanine uptake system in *Rb. sphaeroides* was a typical secondary transporter. This appeared consistent with the fact that alanine uptake could be measured in membrane vesicles and could be energized by either respiratory or photosynthetic electron transport (Hellingwerf et al., 1975). However, it was later shown that in fact the alanine transporter is an osmotic-shock sensitive, binding-protein dependent system which is strongly regulated in activity by the intracellular pH value (Abee et al., 1989). This system is vanadate sensitive and is likely to be a typical ATP-driven periplasmic permease. Thus, at least some of the earlier results can be explained by the significant regulatory effect exerted by internal pH on the transporter, without the need to invoke direct interactions with the electron transport chain. The ability to detect alanine uptake in membrane vesicles is more puzzling, but may be due to residual binding-protein and ATP remaining associated with the vesicles. In addition, the presence of an additional alanine transporter cannot be ruled out.

The binding-protein dependent glutamate transport system in *Rb. sphaeroides* has recently been characterised (M. Jacobs and W. N. Konings, unpublished) and the glutamate binding-protein purified. This binding-protein has a high affinity for both glutamate and glutamine but does not bind aspartate or arginine. However, the latter amino acids do inhibit glutamate transport. This suggests that there may be two binding-proteins present which interact with the same membrane complex. Transport systems for several other amino acids (leucine, proline, glutamine and histidine) have been detected in intact cells of *Rb. sphaeroides* (Abee, 1989) and these were all found to be sensitive to inhibition by vanadate, suggesting the involvement of a phosphorylated intermediate in the uptake process. Vanadate sensitivity appears to be a characteristic of ATP-driven, binding-protein dependent systems and it is striking that, so far, only binding-protein dependent amino-acid transport systems have been found in *Rhodobacter*.

F. Nucleotides

Nucleotide transport in eukaryotic cells is well-known, but most free-living bacteria appear to be unable to transport nucleotides across the cytoplasmic membrane, although these molecules can pass through the outer membrane (Nikaido and Varra, 1985). Until recently, evidence for nucleotide transport systems in bacteria had come only from intracellular and/or parasitic species. However, it has been reported (Hochman et al., 1978, 1981) that cytoplasmic membrane vesicles isolated from *Rb. capsulatus* have the ability to transport adenine nucleotides. In this study, transport activity was not detectable in intact cells, suggesting that the outer membrane in this organism is impermeable to these molecules, i.e. the reverse of the situation normally encountered. From both direct and indirect measurements of adenine nucleotide transport, it seemed possible that an ADP-ATP exchanger was present, but it proved insensitive to inhibitors of the mitochondrial ADP-ATP exchanger (Carmeli and Lifshitz, 1989). These results are interesting but caution needs to be exercised in interpreting them, since there are difficulties in analyzing nucleotide transport; adenine nucleotide pools in cells or vesicles turn over rapidly and it is difficult to separate transport from metabolism. Nevertheless, it was suggested that this transporter could function to transfer ATP to the periplasm for various energy requiring reactions (as yet ill-defined). If confirmed, the properties and role of this system would prove of general interest.

G. Other Solutes

Some work has been performed on the transport systems involved in the responses of photosynthetic bacteria to osmotic stress (Abee et al., 1990). In *Rb. sphaeroides*, an increase in ionic strength resulting from the addition of sodium chloride to the growth medium results in a decrease in growth rate. This effect could be overcome by the addition of betaine (N,N,N-trimethylglycine), which was effective under either aerobic or anaerobic growth conditions. Choline was also identified as an efficient osmoprotectant but only under aerobic conditions; it was found that choline was converted to betaine by an oxygen-dependent system. Betaine and choline transport were examined. In cells grown and assayed either in the absence of added NaCl or in 0.3M NaCl, similar affinities for betaine were determined (K_t values of 15.1 μM and 18.2 μM respectively), but the V_{max} was increased from 3.2 to 9.2 nmol min^{-1} mg $protein^{-1}$. In each case, only one transport system could be distinguished kinetically. This system also transported proline, with lower affinity, and proline was also found to be mildly osmoprotective. Choline was transported by a separate system with similar kinetic properties in cells from high or low osmolarity growth media (K_t 2.4–3.0 μM and V_{max} 5.4–4.2 nmol min^{-1} mg $protein^{-1}$). The addition of either choline or betaine to growth media did not result in the induction of alternative transport systems. It would be interesting to know whether either of the transport systems identified are binding-protein dependent or not and how similar they are to the well studied ProU and ProP systems in enteric bacteria.

IV. Future Perspective

In recent years, investigations of transport and motility processes in purple anoxygenic phototrophs have led to important insights into bacterial solute transport, flagellar activity and tactic behavioral responses in general. In particular, the central importance of pHi as a regulator of the activity of transport systems has become evident. The relationship between potassium transport, pHi and the activity of other transport systems in phototrophs is now appreciated. The properties of periplasmic binding-protein-dependent transport systems are of particular interest in phototrophic bacteria, as the unusual nature of the *Rhodobacter* Dct system suggests that not all such transporters are members of the ABC-superfamily, but instead may represent a new class of secondary transporter. The unidirectional motor of *Rb. sphaeroides* has been identified, with its implications for flagellar rotation in general, as has a metabolism-based chemotaxis system which may well operate alongside the MCP-based system, using two independent CheY response regulators, in a wide range of species.

Photosynthetic bacteria grow under a wide range of conditions, inducing metabolic pathways in response to changing conditions. They have therefore evolved a range of systems for sensing, moving towards and transporting different metabolites, dependent on their current environment and requirements. This flexibility combined with the bioenergetic and genetic advantages of these species makes them ideal for studying the diverse transport and sensing systems that are emerging as investigation moves beyond the enteric bacteria. These systems, which are producing so many surprises, also need to be understood if we are to understand the physiology of the photosynthetic bacteria as a whole.

Acknowledgments

The authors would like to thank C. Bauer, G. Dreyfus, W. Konings, R. Schmitt, L. Shapiro and B. L. Taylor for unpublished data. The work of the authors described here was supported by the SERC (J.P.A., D.J.K. and R.E.S.) and the Wellcome Trust (J.P.A.)

References

Abee T (1989) Regulation of solute transport, internal pH and osmolarity in *Rhodobacter sphaeroides*. PhD Thesis, University of Groningen

Abee T, Hellingwerf KJ and Konings WN (1988) Effects of potassium ions on proton motive force in *Rhodobacter sphaeroides*. J Bacteriol 170: 5647–5653

Abee T, Van der Wal F-J, Hellingwerf KJ and Konings WN (1989) Binding-protein-dependent alanine transport in *Rhodobacter sphaeroides* is regulated by the internal pH. J Bacteriol 171: 5148–5154

Abee T, Palmen R, Hellingwerf KJ and Konings WN (1990) Osmoregulation in *Rhodobacter sphaeroides*. J Bacteriol 172: 149–154

Abee T, Knol J, Hellingwerf KJ, Bakker EP, Siebers A and Konings WN (1992) A Kdp-like, high-affinity, K^+-translocating ATPase is expressed during growth of *Rhodobacter sphaeroides* in low potassium media. Distribution of this K^+-ATPase among purple non-sulfur phototrophic bacteria. Arch Microbiol 158: 374–380

Abee T, Siebers A, Altendorf K and Konings WN (1992) Isolation and characterization of the high-affinity K^+- translocating ATPase from *Rhodobacter sphaeroides*. J Bacteriol 174: 6911–6917

Alef K and Kleiner D (1982) Evidence for an ammonium transport system in the N_2-fixing phototrophic bacterium *Rhodospirillum rubrum*. Arch Microbiol 132: 79–81

Armitage JP (1992) Behavioral responses in bacteria. Annu Rev Physiol 54: 683–714

Armitage JP and Evans MCW (1981) The reaction centre in the phototactic and chemotactic response of photosynthetic bacteria. FEMS Microbiol Letts: 11: 89–92

Armitage JP and Macnab RM (1987) Unidirectional intermittent rotation of the flagellum of *Rhodobacter sphaeroides*. J Bacteriol 169: 514–518

Armitage JP, Ingham C and Evans MCW (1985) Role of the proton motive force in phototactic and aerotactic responses of *Rhodopseudomonas sphaeroides*. J Bacteriol 163: 967–972

Armitage JP, Havelka WA and Sockett RE (1990) Methylation-independent taxis in bacteria. In: Armitage, JP and Lackie, JM (eds) Biology of the Chemotactic Response, pp 177–197. Soc Gen Microbiol Symp 46. Cambridge University Press, Cambridge

Barak R and Eisenbach M (1992a) Correlation between phosphorylation of the chemotaxis protein CheY and its activity at the flagellar motor. Biochemistry 31: 1821–1826

Barak R and Eisenbach M (1992b) Fumarate or a fumarate metabolite restores switching ability to rotating flagella of bacterial envelopes. J Bacteriol 174: 643–645

Blair DF (1990) The bacterial flagellar motor. Sem in Cell Biol 1: 75–85

Blair DF and Berg HC (1990) The MotA protein of *E. coli* is a proton-conducting component of the flagellar motor. Cell 60: 439–449

Booth IR (1985) Regulation of cytoplasmic pH in bacteria. Microbiol Rev 49: 359–378

Bourett RB, Borkovich KA and Simon MI (1991) Signal transduction pathways involving protein phosphorylation in prokaryotes. Ann Rev Biochem 60: 401–444

Brown S, Poole PS, Jeziorska W and Armitage JP (1993) Chemokinesis in *Rhodobacter sphaeroides* is the result of a long term increase in the rate of flagellar rotation. Biochim Biophys Acta 1141: 309–312

Carmeli C and Lifshitz Y (1989) Nucleotide transport in *Rhodobacter capsulatus*. J Bacteriol 171: 6521–6525

Clayton RK (1953a) Studies in the phototaxis of *Rhodospirillum rubrum*. I. Action spectrum, growth in green light and Weber-law adherence. Arch Mikrobiol 19: 107–124

Clayton RK (1953b) Studies in the phototaxis of *Rhodospirillum rubrum*. II. The relation between phototaxis and photosynthesis. Arch Mikrobiol 19: 125–140

Clayton RK (1953c) Studies in the phototaxis of *Rhodospirillum rubrum*. III. Quantitative relationship between stimulus and response. Arch Mikrobiol 19: 141–165

Clayton RK (1957) Patterns of accumulation resulting from taxes and changes in motility of microorganisms. Arch Mikrobiol 27: 311–319

Cohen-Bazire G and London J (1967) Basal organelles of bacterial flagella. J Bacteriol 94: 458–465

Cotton NPJ, Clark AJ and Jackson JB (1981) The effect of venturicidin on light and oxygen-dependent electron-transport, proton translocation, membrane potential development and ATP synthesis in intact cells of *Rhodopseudomonas capsulata*. Arch Microbiol 129: 94–99

Crielaard W, Cotton NPJ, Jackson JB, Hellingwerf KJ and Konings WN (1988) The transmembrane electrical potential in intact bacteria: Simultaneous measurements of carotenoid absorbance changes and lipophilic cation distribution in intact cells of *Rhodobacter sphaeroides*. Biochim Biophys Acta 932: 17–25

Daniels A, Drews G and Saier MH, Jr (1988) Properties of a Tn5 insertion mutant defective in the structural gene (*fru*A) of the fructose specific phosphotransferase system of *Rhodobacter capsulatus* and cloning of the *fru* regulon. J Bacteriol 170: 1698–1703

Dreyfus G, Williams AW, Kawagishi I and Macnab RM (1993) Genetic and biochemical analysis of *Salmonella typhimurium* FliI, a flagellar protein related to the catalytic subunit of the F_0F_1 ATPase and to virulence proteins of mammalian and plant pathogens. J Bacteriol 175: 3131–3138

Ehrenberg GS (1838) Die infusionstrierchen als vollkommene organismen. L. Voss, Leipzig

Elferink MGL, Hellingwerf KJ, van Belkum MJ, Poolman B and Konings WN (1984) Direct interaction between linear electron transfer and solute transport systems in bacteria. FEMS Microbiol Lett 21: 293–298

Elferink M L, Hellingwerf KJ and Konings WN (1987) The relation between electron transfer, proton-motive force and the energy consuming processes in cells of *Rhodopseudomonas sphaeroides*. Biochim Biophys Acta 848: 58–68

Engelmann TW (1881) Bacterium photometricum: An article on the comparative physiology of the sense for light and colour. Arch Ges Physiol Bonn 30: 95–124

Epstein W (1986) Osmoregulation by potassium transport in *Escherichia coli*. FEMS Microbiol Rev 39: 73–78

Fenoll CS, Gomes-Amores A and Ramirez JM (1985) The membrane potential of intact *Rhodospirillum rubrum* cells in the absence of light-dependent and oxygen-linked electron transfer. Biochim Biophys Acta 806: 168–174

Ferguson SJ, Jackson JB and McEwan AG (1987) Anaerobic respiration in the *Rhodospirillaceae*: characterisation of pathways and evaluation of roles in redox balancing during photosynthesis. FEMS Microbiol Rev 46: 117–143

Fonstein M and Haselkorn R (1993) Chromosomal structure of *Rhodobacter capsulatus* SB1003: Cosmid encyclopedia and high resolution physical and genetic map. Proc Natl Acad Sci USA 90: 2522–2526

Forward JA, Behrendt MC and Kelly DJ (1993) Evidence that the high-affinity C4-dicarboxylate transport system of *Rhodobacter capsulatus* is a novel type of periplasmic permease. Biochem Soc Trans 21: 3435

Gibson J (1975) Uptake of C4 dicarboxylates and pyruvate by *Rhodopseudomonas sphaeroides*. J Bacteriol 123: 471–480

Gibson MM, Bagga DA, Miller CG and Maguire ME (1991) Magnesium transport in *Salmonella typhimurium:* The influence of new mutations conferring Co^{2+} resistance on the CorA Mg^{2+} transport system. Mol Microbiol 5: 2753–2762

Glagolev AN (1984) Bacterial H^+-sensing. Trends Biochem Sci 9: 397–400

Golby P, Carver M and Jackson JB (1990a) Membrane ionic currents in *Rhodobacter capsulatus*: Evidence for electrophoretic transport of K^+, Rb^+, and NH_4^+. Eur J Biochem 187: 589–597

Golby P, Carver M and Jackson JB (1990b) Evidence that the low-affinity K^+ transport system of *Rhodobacter capsulatus* is a uniport-the effects of ammonia on the transporter. Arch Microbiol 157: 125–130

Greck M, Plazer J, Sourjik V and Schmitt R (1995) Analysis of a chemotaxis operon in *Rhizobium meliliti*. Mol Microbiol 15: 989–1000

Hamblin MJ, Shaw JG, Curson JP and Kelly DJ (1990) Mutagenesis, cloning and complementation analysis of C4-dicarboxylate transport genes from *Rhodobacter capsulatus*. Mol Microbiol 4: 1567–1574

Hamblin MJ, Shaw JG and Kelly DJ (1993) Sequence analysis and interposon mutagenesis of a sensor- kinase (DctS) and response-regulator (DctR) controlling synthesis of the high-affinity C4-dicarboxylate transport system in *Rhodobacter capsulatus*. Mol Gen Genet 237: 215–224

Harayama S and Iino T (1977) Phototaxis and membrane potential in the photosynthetic bacterium *Rhodspirillum rubrum*. J Bacteriol 131: 34–41

Harrison DH, Packer HL and Armitage JP (1994) Swimming speed and chemokinetic response of *Rhodobacter sphaeroides* investigated by natural manipulation of the membrane potential. FEBS Letts 342: 37–40

Hellingwerf KJ, Michels PAM, Dorpema J and Konings WN (1975) Transport of amino acids in membrane vesicles of *Rhodopseudomonas sphaeroides* energized by respiratory and cyclic electron flow. Eur J Biochem 55: 397–406

Hochman A, Bittan R and Carmeli C (1978) Nucleotide translocation across the cytoplasmic membrane in the photosynthetic bacterium *Rhodopseudomonas capsulata*. FEBS Lett. 89: 21–25

Hochman A, Bittan R and Carmeli C (1981) Nucleotide exchange in membrane vesicles from the photosynthetic bacterium *Rhodopseudomonas capsulata*. Arch Biochem Biophys 211: 413–448

Iino T (1985) Genetic control of flagellar morphogenesis in *Salmonella*. In: Eisenbach M and Balaban M (eds) Sensing and Response in Microorganisms, pp 83–92. Elsevier, Amsterdam

Ishihara A, Segall JE, Block SM and Berg HC (1983) Coordination of flagella on filamentous cells of *Escherichia coli*. J Bacteriol 155: 22 –237

Jackson JB (1988) Bacterial photosynthesis. In: Anthony C (ed) Bacterial Energy Transduction, pp 317–375. Academic Press Ltd, London

Jacobs MHJ, Driessen AJM and Konigs WN (1995) Characterization of a binding protein-dependent glutamate transport system of *Rhodobacter sphaeroides*. J Bacteriol 177: 1812–1816

Jasper P (1978) Potassium transport system of *Rhodopseudomonas capsulata*. J Bacteriol 133: 1314–1322

Jasper P and Silver S (1978) Divalent cation transport systems of *Rhodopseudomonas capsulatus*. J Bacteriol 133: 1323–1328

Jones CJ and Aizawa S-I (1991) The bacterial flagellum and flagellar motor: Structure, assembly and function. Adv Microb Physiol 32: 109–172

Karzanov VV and Ivanovsky RN (1980) Sodium dependent succinate uptake in purple bacterium *Ectothiorhodospira shaposhnikovii*. Biochim Biophys Acta 598: 91–99

Koch AL, Higgins ML and Doyle RJ (1982) Surface tension-like forces determine bacterial shape: *Streptococcus faecium*. J Gen Microbiol 123: 151–161

Lee AG and Fitzsimmons JTR (1976) Motility studies in normal and filamentous forms of *Rhodospirillum rubrum*. J Gen Microbiol 93: 346–354

Lolkema JS, Ten Hoeve-Duukens RH and Robillard G (1985) The phosphoenolpyruvate-dependent fructose specific phosphotransferase system in *Rhodopseudomonas sphaeroides*. Mechanism of transfer of the phosphoryl group from phosphoenolpyruvate to fructose. Eur J Biochem 149: 625–631

Lukat GS, McCleary WR, Stock AM and Stock JB (1992) Phosphorylation of bacterial response regulator proteins by low molecular weight phospho-donors. Proc Natl Acad Sci USA 89: 718–722

Macnab RM (1992) Genetics and biogenesis of bacterial flagella. Annu Rev Genet 26: 131–158

Maddock JR and Shapiro L (1993) Polar location of the chemoreceptor complex in the *Escherichia coli* cell. Science 259: 1717–1723

McBride MJ, Köhler T and Zusman DR (1992) Methylation of FrzCD, a methyl-accepting taxis protein of *Myxococcus xanthus*, is correlated with factors affecting cell behavior. J Bacteriol 174: 4246–4257

Meister M, Lowe G and Berg HC (1987) The proton flux through the bacterial flagellar motor. Cell 49: 643–650

Mitchell JG, Martinez-Alonso M, Lalucat J, Esteve I and Brown S (1991) Velocity changes, long runs and reversals in *Chromatium minus* swimming responses. J Bacteriol 173: 997–1003

Moody MD and Dailey HA (1984) Siderophore utilisation and iron uptake in *Rhodopseudomonas sphaeroides*. Arch Biochem Biophys 234: 178–186

Moody MD and Dailey HA (1985) Iron transport and its relation to heme biosynthesis in *Rhodopseudomonas sphaeroides*. J Bacteriol 161: 1074–1079

Morgan DG, Baumgartner JB and Hazelbauer GL (1993) Proteins antigenically related to methyl-accepting chemotaxis proteins of *Escherichia coli* detected in a wide range of bacterial species. J Bacteriol 175: 133–140

Mulder MM, van der Gulden HML, Postma PW and Van Dam K (1988) Energetic consequences of two mutations in *Escherichia coli* K^+ uptake systems for growth under potassium limited conditions in the chemostat. Biochim Biophys Acta 933: 65–69

Nikaido H and Varra M (1985) Molecular basis of bacterial outer membrane permeability. Microbiol Rev 49: 1–32

Oesterhelt D and Marwan W (1990) Signal transduction in *Halobacterium halobium*. In: Armitage JP and Lackie JM (eds) Biology of the Chemotactic Response, pp 219–240. Cambridge University Press, Cambridge

Packer HL and Armitage JP (1994) The chemokinetic and chemotactic behavior of *Rhodobacter sphaeroides*: Two independent responses. J Bacteriol 176: 206–217

Packer HL, Harrison DH, Dixon RM and Armitage JP (1994) The effect of pH on the growth and motility of *Rhodobacter sphaeroides* and the nature of the driving force of the flagellar motor. Biochim Biophys Acta 1188: 101–107

Pfennig N (1968) *Chromatium okenii* (Thiorhodaceae) Biokenvelztion, aero-und phototaktisches Verhalten. In: Encyclopaedia Cinematographiia, Wolf G, ed, pp 3–10. Institut für den Weissenschaftlichen film, Gottingen Film Bd. 2A, H.3

Poole PS and Armitage JP (1989) Role of metabolism in the chemotactic response of *Rhodobacter sphaeroides* to ammonia. J Bacteriol 171: 2900–2902

Poole PS, Brown S and Armitage JP (1990a) Swimming changes and chemotactic responses in *Rhodobacter sphaeroides* do not involve changes in the steady state membrane potential or respiratory electron transport. Arch Mikrobiol 153: 614–618

Poole PS, Williams RL and Armitage JP (1990b) Chemotactic responses of *Rhodobacter sphaeroides* in the absence of apparent adaptation. Arch Mikrobiol 153: 368–372

Poole PS, Brown S and Armitage JP (1991) Chemotaxis and chemokinesis in *Rhodobacter sphaeroides*: modelling of the two effects. Binary 3: 183–190

Poole PS, Smith MJ and Armitage JP (1993) Chemotactic signaling in *Rhodobacter sphaeroides* requires metabolism of attractants. J Bacteriol 175: 291–294

Postma PW, Lengeler JW and Jacobson GR (1993) The PTS as a signal transduction system in chemotaxis. Microbiol Rev 57: 543–594

Ragatz L, Jiang Z-Y, Bauer C and Gest H (1994) Phototactic purple bacteria. Nature 370: 104

Ragatz L, Jiang Z-Y, Bauer CE and Gest H (1995) Macroscopic phototactic behavior of the purple photosynthetic bacterium *Rhodospirillum centenum*. Arch Microbiol 163: 1–6

Saier MH, Jr, Fuecht BU and Roseman S (1971) Phosphoenolpyruvate-dependent fructose phosphorylation in photosynthetic bacteria. J Biol Chem 246: 7819–7821

Schnitzer MJ, Block SM, Berg HC and Purcell EM (1990) Strategies for chemotaxis. In: Armitage JP and Lackie JM (eds) Biology of the chemotactic response, pp 15–34. Cambridge University Press, Cambridge

Shah DSH, Armitage JP and Sockett RE (1995) *Rhodobacter sphaeroides* WS8 encodes a polypeptide that is similar to MotB of *Escherichia coli*. J Bacteriol 177: 2929–2932

Shah DSH and Sockett RE (1995) Cloning and expression of the *motA* flagellar gene from *Rhodobacter sphaeroides*, a bacterium with a unidirectional, stop-start, flagellum. Mol Microbiol, in press

Sharak Genther BR and Wall JD (1985) Ammonium uptake in *Rhodopseudomonas capsulata*. Arch Microbiol 141: 219–224

Shaw JG and Kelly DJ (1991) Binding-protein dependent transport of C4-dicarboxylates in *Rhodobacter capsulatus*. Arch Microbiol 155: 466–472

Shaw JG, Hamblin MJ and Kelly DJ (1991) Purification, characterisation and nucleotide sequence of the periplasmic C4-dicarboxylate binding-protein (DctP) from *Rhodobacter capsulatus*. Mol Microbiol 5: 3055–3062

Shioi J-I and Taylor BL (1984) Oxygen taxis and proton motive force in *Salmonella typhimurium*. J Biol Chem 259: 10983–10988

Sockett RE and Armitage JP (1991) Isolation, characterization, and complementation of a paralyzed flagellar mutant of *Rhodobacter sphaeroides* WS8. J Bacteriol 173: 2786–2790

Sockett RE, Armitage JP and Evans MCW (1987) Methylation-independent and methylation-dependent chemotaxis in *Rhodobacter sphaeroides* and *Rhodospirillum rubrum*. J Bacteriol 169: 5808–5814

Sockett RE, Foster JCA and Armitage JP (1990) Molecular biology of the *Rhodobacter sphaeroides* flagellum. FEMS Symp 53: 473–479

Sofia MJ, Burland V, Daniels DL, Plunkett III G and Blattner FR (1994) Analysis of the *Escherichia coli* genome. V. DNA sequence of the region from 76.0 to 81.5 minutes. Nucl Acids Res 22: 2576–2586

Sprenger WW, Hoff WD, Armitage JP and Hellingwerf KJ (1993) The eubacterium *Ectothiorhodospira halophila* is negatively phototactic, with a wavelength dependence that fits the absorption spectrum of the photoactive yellow protein. J Bacteriol 175: 3096–3104

Stadtwald-Demchick R, Turner FR and Gest H (1990) Physiological properties of the thermotolerant photosynthetic bacterium *Rhodospirillum cenentum.* FEMS Microbiol Lett 67: 139–143

Suwanto A and Kaplan S (1989) Physical and genetic mapping of the *Rhodobacter sphaeroides* 2.4.1 genome: Genome size, fragment identification, and gene localization. J Bacteriol 171: 5840–5849

Takakuwa S (1987) Nickel uptake in *Rhodopseudomonas capsulata.* Arch Microbiol 149: 57–61

Tauschel H-D (1987) ATPase activity of the polar organelle demonstrated by cytochemical reaction in whole unstained cells of *Rhodopseudomonas palustris.* Arch Microbiol 148: 159–161

Taylor MA and Jackson JB (1987) Adaptive changes in membrane conductance in response to changes in specific growth rate in continuous cultures of phototrophic bacteria under conditions of energy sufficiency. Biochim Biophys Acta 891: 242–255

Tisa L S and Adler J (1992) Calcium ions are involved in *Escherichia coli* chemotaxis. Proc Natl Acad Sci USA 89: 11804–11808

Tolner B, Poolman B and Konings W (1992) Characterisation and functional expression in *Escherichia coli* of the sodium/proton/glutamate symport proteins of *Bacillus stearothermophilus* and *Bacillus caldotenax.* Mol Microbiol 6: 2845–2856

Walmsley AR, Shaw JG and Kelly DJ (1992a) The mechanism of ligand binding to the periplasmic C4-dicarboxylate binding protein (DctP) from *Rhodobacter capsulatus.* J Biol Chem 267: 8064–8072

Walmsley AR, Shaw JG and Kelly DJ (1992b) Perturbation of the equilibrium between open and closed conformations of the periplasmic C4-dicarboxylate binding protein from *Rhodobacter capsulatus.* Biochemistry 31: 11175–11181

Willems RJL, van der Heide HGJ and Mooi FR (1992) Characterisation of a *Bordetella pertussis* fimbrial gene cluster which is located directly downstream of the filamentous haemagglutinin gene. Mol Microbiol 6: 2661–2671

Wang G, Angermuller S and Klipp W (1993) Characterisation of *Rhodobacter capsulatus* genes encoding molybdenum transport and putative molybdenum-pterin-binding proteins. J Bacteriol 175: 3031–3042

Ward MJ, Bell AW, Hamblin P, Packer HL and Armitage JP (1995a) Identification of a chemotaxis operon with two *cheY* genes in *Rhodobacter sphaeroides.* Mol Microbiol, in press

Ward MJ, Harrison DM, Ebner MJ and Armitage JP (1995b) Identification of a methyl-accepting chemotaxis protein in *Rhodobacter sphaeroides.* Mol Microbiol, in press

Willison JC (1993) Biochemical genetics revisited: the use of mutants to study carbon and nitrogen metabolism in the photosynthetic bacteria. FEMS Microbiol Rev 104: 1–38

Wolfe AJ, Conley MP and Berg HC (1988) Acetyladenylate plays a role in controlling the direction of flagellar rotation. Proc Natl Acad Sci USA 85: 6711–6715

Wu L-F and Saier MH, Jr (1990) Nucleotide sequence of the *fruA* gene, encoding the fructose permease of the *Rhodobacter capsulatus* phosphotransferase system, and analyses of the deduced protein sequence. J Bacteriol 172: 7167–7178

Wu L-F, Tomich JM and Saier MH, Jr (1990) Structure and evolution of a multidomain multiphosphoryl transfer protein. Nucleotide sequence of the *fruB(HI)* gene in *Rhodobacter capsulatus* and comparisons with homologous genes from other organisms. J Mol Biol 213: 687–703

Yoch DC, Zhang Z-M and Claybrook DL (1983) Methylamine transport and its role in nitrogenase 'switch-off' in *Rhodopseudomonas capsulata.* Arch Microbiol 134: 45–48

Chapter 48

Genetic Manipulation of Purple Photosynthetic Bacteria

JoAnn C. Williams and Aileen K. W. Taguchi
Department of Chemistry and Biochemistry, and Center for the Study of Early Events in Photosynthesis, Arizona State University, Tempe AZ 85287-1604, USA

R. E. Blankenship, M. T. Madigan and C. E. Bauer (eds): Anoxygenic Photosynthetic Bacteria, pp. 1029–1065.

Summary

Genetic manipulation has become widely used in the study of purple photosynthetic bacteria. Some species have been extensively characterized genetically and thus are readily amenable to new applications of these techniques, while in other species genetic work is being developed and may require some adaptation of existing protocols. Genetic studies in purple bacteria include both classical genetic mapping that yields information about the genomic distribution of the genetic loci under study and molecular genetic techniques that employ recombinant DNA techniques and center on analysis at a nucleotide level. As a result of these studies, a large number of genes have been characterized, most prominently those whose products are involved in photosynthesis and nitrogen fixation, but also including a variety of other processes. Techniques that have been used to identify genes range from complementation of mutant strains to hybridization probes based on homologous genes or the amino acid sequences of proteins. The functional properties of genes have been studied by the generation of random mutations by chemical or transposon mutagenesis and by site-directed insertions, deletions and alterations of nucleotide sequences. The transcription and translation of genes have been investigated through gene fusions and in vitro and heterologous expression systems. The combination of classical genetics with the advances in molecular biological techniques provides an array of tools for dissecting the underlying basis for metabolic processes in these bacteria.

I. Introduction

The goal of this chapter is to summarize genetic techniques that have been applied to purple photosynthetic bacteria, including basic manipulations such as the transfer of DNA, as well as genetic mapping, cloning, sequencing, mutagenesis and analyses of gene expression. Adaptations of microbial genetic and molecular biological methods developed in other bacteria have generally been used for purple bacteria, although the unique physiology of these bacteria has also been exploited. These applications serve as a source for the types of options available when approaching the genetic study of any particular aspect of these bacteria, and provide a foundation for further analyses of genes already defined. The majority of genetic work has been performed in the two closely related species of purple nonsulfur bacteria, *Rhodobacter capsulatus* and *Rhodobacter sphaeroides*. Other purple nonsulfur bacteria for which some genetic information is available include *Rhodobacter blasticus, Rhodobacter marinus, Rhodopseudomonas acidophila, Rhodopseudomonas palustris, Rhodopseudomonas viridis, Rhodospirillum centenum, Rhodospirillum rubrum, Roseobacter denitrificans* and *Rubrivivax gelatinosus*. In addition, the purple sulfur bacterium, *Chromatium vinosum*, has been the focus of a number of genetic studies.

Abbreviations: ALA – 5-aminolevulinic acid; *Cm. – Chromatium; E. coli – Escherichia coli;* GTA – gene transfer agent; kb – kilobase pair; PCR – polymerase chain reaction; PHB – poly(3-hydroxybutyrate); *Rb. – Rhodobacter; Ro. – Roseobacter; Rp. – Rhodopseudomonas; Rs. – Rhodospirillum; Rv. – Rubrivivax;* RubisCO – ribulose 1,5-bisphosphate carboxylase-oxygenase; TMAO – trimethylamine-N-oxide; Tris – tris(hydroxymethyl)aminomethane

II. Genetic Manipulation

A. Gene Transfer

1. Conjugation

Conjugation is the most commonly used method of transferring DNA into purple bacteria. Transfer can take place both between Escherichia coli and the purple bacteria and among various purple species. Matings are performed by mixing the donor and recipient cells and plating on a solid surface, such as an aged agar plate or a membrane filter, incubating for several hours to overnight, and then replica plating, replating or overlaying with a selection medium (reviewed in Donohue and Kaplan, 1991; Table 1). Selection for the recipient over the donor strain is usually accomplished by antibiotic resistance markers in the recipient or auxotrophic markers in the donor. Mutants with enhanced recipient activity that is thought to be due to inactivation of the restriction system have been isolated from *Rb. capsulatus* (Taylor et al., 1983) and *Rb. sphaeroides* (Sistrom et al., 1984).

Both the chromosome and plasmids can be

Table 1. Protocols for manipulation of DNA

Method	Species	References
Conjugation	*Rb. capsulatus*	Marrs, 1981; Yu et al., 1981; Youvan et al., 1982; Taylor et al., 1983; Genther and Wall, 1984; Kaufmann et al., 1984; Klug and Drews, 1984; Zsebo and Hearst, 1984; Willison et al., 1985; Youvan and Ismail, 1985; Colbeau et al., 1986; Giuliano et al., 1986; Johnson et al., 1986; Daldal et al., 1987; Kelly et al., 1988; Klipp et al., 1988; Leclerc et al., 1988; Tichy et al., 1989; Young et al., 1989
	Rb. sphaeroides	Sistrom, 1977; Miller and Kaplan, 1978; Pemberton and Bowen, 1981; Weaver and Tabita, 1983; Nano and Kaplan, 1984; Pemberton and Harding, 1986; Davis et al., 1988; Donohue et al., 1988; Hunter, 1988; Tai et al., 1988; Penfold and Pemberton, 1991; Benning and Somerville, 1992a; Suwanto and Kaplan, 1992a, b
	Rp. palustris	Dispensa et al., 1992; Elder et al., 1993
	Rp. viridis	Lang and Oesterhelt, 1989b; Laußermair and Oesterhelt, 1992
	Rs. centenum	Yildiz et al., 1991
	Rs. rubrum	Hessner et al., 1991; Saegesser et al., 1992
DNA isolation	*Rb. capsulatus*	Klug and Drews, 1984; Wright et al., 1987; Giuliano et al., 1988; Cook et al., 1989; Xu et al., 1989; Forkl et al., 1993
	Rb. sphaeroides	Fornari and Kaplan, 1982, 1983; Williams et al., 1983; Iba et al., 1987; Hunter and Turner, 1988; Brandner et al., 1989; Suwanto and Kaplan, 1989a, 1992a; Dryden and Kaplan, 1990
	Ro. denitrificans	Liebetanz et al., 1991
	Rp. palustris	Elder et al., 1993
	Rp. viridis	Michel et al., 1985, 1986; Laußermair and Oesterhelt, 1992
	Rs. rubrum	Bérard et al., 1986; Self et al., 1990; Kerby et al., 1992; Saegesser et al., 1992
	Rv. gelatinosus	Uffen et al., 1990

transferred by conjugation. Chromosomal mobilization has been induced using broad host range R plasmids in *Rb. capsulatus* and *Rb. sphaeroides* (reviewed in Scolnik and Marrs, 1987; Donohue and Kaplan, 1991; Saunders, 1992; Clewell, 1993). The conversion of these IncP plasmids (RP1, RP4, RK2, R18 and R68) to a form that can induce increased chromosome mobilization activity appears to involve the incorporation of elements capable of transposition, IS insertion sequences or transposable Tn elements. The plasmid R68.45, a derivative of R68, induces chromosome mobilization activity in *Rb. sphaeroides* (Sistrom, 1977; Sistrom et al., 1984) and *Rb. capsulatus* (Yu et al., 1981). Chromosome mobilization is thought to occur following R68.45 integration into the chromosome via transposition-mediated events involving IS*21* sequences (Willetts et al., 1981). Another conjugative plasmid, RP1::Tn*501*, has been used to map a number of biosynthetic and antibiotic resistance markers in two linkage groups of *Rb. sphaeroides* (Pemberton and Bowen, 1981). RP1::Tn*501* was derived from RP1 via the insertion of the mercury resistance transposon Tn*501*. Chromosome mobilization activity has been demonstrated in *Rb. capsulatus* with the conjugative plasmids pBLM2 (Marrs, 1981) and pTH10 (Willison et al., 1985), which appear to initiate chromosome transfer from multiple sites of origin. The plasmid pBLM2 was isolated from a derivative of RP1 for its enhanced sex factor activity. The plasmid pTH10, another mutant RP1 plasmid, contains IS*21* and Tn*1* sequences and is believed to induce chromosome mobilization activity via pTH10-chromosome co-integrate formation. R′ plasmids generated during chromosome mobilization have been useful in the identification of the photosynthetic gene clusters of *Rb. capsulatus* (Marrs, 1981; see also Chapter 50 by Alberti et al.) and *Rb. sphaeroides* (Sistrom et al., 1984; Wu et al., 1991b). Most engineered plasmid systems utilize a mobilizable but not self-transmissible plasmid vector, with the transfer functions provided in *trans* either from a helper plasmid such as pRK2013, a derivative of RK2, in a triparental mating (Ditta et al., 1980) or from a donor strain such as *E. coli* S17-1, in which the transfer genes of RP4 have been integrated into the chromosome (Simon et al., 1983).

2. Transformation

Transformation has the advantage of not requiring counterselection against the donor strain or mobilization ability of the DNA, but its success usually depends on empirical application of protocols for the preparation of competent cells or electroporation. Competence has been induced in *Rb. sphaeroides* by washing cells with $CaCl_2$, polyethylene glycol and high concentrations of Tris (Fornari and Kaplan, 1982; Donohue and Kaplan, 1991), in *Rb. marinus* by $CaCl_2$ treatment (Matsunaga et al., 1986, 1990) and in *Rs. rubrum* by $CaCl_2$ and freeze/thaw methods (Fitzmaurice and Roberts, 1991). Conditions for transformation by electroporation have been described for *Rb. sphaeroides* (Donohue and Kaplan, 1991) and *Rp. viridis* (Laußermair and Oesterhelt, 1992).

3. Transduction

Although a number of endogenous phage have been isolated from purple bacteria, none have been shown to be generally useful for transduction (reviewed in Marrs, 1978; Saunders, 1978, 1992; Donohue and Kaplan, 1991). The 'gene transfer agent' or GTA from *Rb. capsulatus* (Marrs, 1974; Solioz et al., 1975; Solioz and Marrs, 1977) consists of phage-like particles that package short (approximately 4.5 kb) linear fragments of DNA. Although DNA transfer by GTA resembles generalized transduction, GTA lacks many typical viral activities. For example, GTA does not transfer to recipients the ability to produce GTA, and the size of the DNA carried by GTA is apparently too small to package all of the genes necessary for its production. Thus, GTA particles represent either a 'pre-phage' particle that confers the advantages of genetic exchange or a defective phage population (reviewed in Marrs, 1978; Saunders, 1978). Most *Rb. capsulatus* strains are GTA donors and recipients (Wall et al., 1975b) and an overproducer mutant has been isolated (Yen et al., 1979). The ability of the GTA to transfer linear fragments of genomic DNA has been very useful in mapping genes and in making replacements of chromosomal DNA in *Rb. capsulatus*.

B. Genetic Markers

Nutritional and antibiotic resistance markers have been developed, particularly for *Rb. capsulatus* and *Rb. sphaeroides* (reviewed in Saunders, 1978; Sistrom, 1978; see also Sistrom, 1977; Miller and Kaplan, 1978; Pemberton and Bowen, 1981; Willison et al., 1985; Pemberton and Harding, 1986, 1987; Choudhary et al., 1994). Antibiotic resistance markers in *Rb. capsulatus* were isolated on complex media in the presence of antibiotics at frequencies in the range of 10^{-6} to 10^{-9} per cell plated (Willison et al., 1985). For example, rifampicin resistance has only been noted at one genetic locus that probably codes for the β-subunit of RNA polymerase. Penicillin resistance found in some strains has been shown to be due to an endogenous β-lactamase in *Rb. sphaeroides* (Baumann et al., 1989) and *Rb. capsulatus* (Campbell et al., 1989). In addition, exogenous antibiotic resistance genes are widely used to select for recipients of DNA transfer, although not all markers are expressed equally well in purple bacteria (reviewed in Donohue and Kaplan, 1991). Several laboratories have isolated nitrogen fixation (*nif*) mutants of *Rb. capsulatus* (reviewed in Willison, 1993). Many bacteriochlorophyll (*bch*) and carotenoid (*crt*) biosynthesis mutants are easily scored due to their dramatically altered colony color (Yen and Marrs, 1976).

C. Vectors

Broad host range vectors are often used to shuttle DNA fragments between *E. coli* and purple bacteria. Broad host range plasmids that have been engineered for use as DNA vectors contain unique restriction sites for cloning, antibiotic resistance markers, and usually have been reduced in size to facilitate handling. One set of widely used vectors includes derivatives of RK2 such as pRK290, pRK404 and pRK415 (Ditta et al., 1980, 1985; Keen et al., 1988). These vectors are approximately 10–20 kb in size, contain a tetracycline resistance marker, and in some cases have multiple cloning sites and insertional inactivation of the β-galactosidase (*lacZ*) gene. They have been shown to replicate in *Rb. capsulatus* (Klug and Drews, 1984), *Rb. sphaeroides* (Ditta et al., 1985), *Rp. viridis* (Lang and Oesterhelt, 1989b), *Rs. rubrum* (Saegesser et al., 1992) and *Rp. palustris* (Elder et al., 1993). The copy number of pRK404 derivatives has been estimated to be between four and nine in *Rb. sphaeroides* (Davis et al., 1988; Brandner et al., 1989) and eleven and fourteen in *Rs. rubrum* (Saegesser et al., 1992). *Rb. capsulatus* strains can be cured of R plasmids of the P incompatibility group by repeated subculturing in a rich medium lacking calcium and magnesium salts (Magnin et al., 1987). RK2 plasmids are lost from up to 50% of the cells when *Rb. sphaeroides* is grown for six to eight doublings in the absence of selection (Donohue and Kaplan, 1991). In *Rs. rubrum*, loss of RK2 derivatives was reported after ten generations of anaerobic photosynthetic growth without selection, although the plasmids were retained under aerobic growth conditions (Saegesser et al., 1992). Cosmid derivatives of pRK290 such as pLAFR1, pVK102 and pLA2917 have been used to make libraries of *Rb. capsulatus* (Avtges et al., 1985; Colbeau et al., 1986), *Rb. sphaeroides* (Weaver and Tabita, 1985; Benning and Somerville, 1992a,b; Choudhary et al., 1994) and *Rp. palustris* (Dispensa et al., 1992), and vectors that contain *lacZ* translational fusion sites and an RK2 replicon have been constructed for analysis of promoter regions in *Rb. capsulatus* (Adams et al., 1989; Hübner et al., 1991). Plasmids constructed from the fusion of derivatives of pRK290 and pBR322 that apparently result in a higher copy number in *E. coli* than pRK290 alone have been used for analysis of the *puf* operon in *Rb. capsulatus* (Bylina et al., 1989; Taguchi et al., 1993). Other vectors derived from the broad host range plasmid RSF1010 have been constructed, including a cloning vector, pNH2, modified by the addition of a tetracycline resistance gene containing restriction sites for cloning, that has been used in *Rb. sphaeroides* (Hunter and Turner, 1988), a cosmid vector, pJRD215, that has been used for making libraries of *Rb. capsulatus* (Sganga and Bauer, 1992), *Rb. sphaeroides* (Choudhary et al., 1994) and *Rs. centenum* (Yildiz et al., 1992) and a *lacZ* fusion vector that has been used in *Rb. sphaeroides* (Nano et al., 1984).

Portions of broad host range plasmids have been used to create mobilizable versions of other cloning vectors. For example, mobilizable versions of ColE1 plasmids created by addition of a *mob* site (Simon et al., 1983; Lang and Oesterhelt, 1989b) have been used in purple bacteria (reviewed in Donohue and Kaplan, 1991; Hunter and Mann, 1992). Another system for mobilizing plasmids that has been used in *Rb. capsulatus* involves the supply of the gene products needed to mobilize pBR322 in *trans* from the plasmid pDPT51, which results from the fusion of a promiscuous plasmid with a ColE1 derivative (Taylor et al., 1983). Shuttle vectors for *Rb. marinus* were constructed by combining an endogenous plasmid with *E. coli* cloning vectors (Matsunaga et al., 1986, 1990).

D. Isolation of DNA

Specific protocols for DNA isolation from purple bacteria have been described (Table 1). When isolating DNA from these bacteria, one should keep in mind that endogenous plasmids have been found in *Rs. rubrum*, *Rb. sphaeroides* and *Rb. capsulatus* strains (reviewed in Saunders, 1978; Willison, 1993). *Rs. rubrum* strains contain a single class of 55 kb plasmids. *Rb. sphaeroides* strains contain up to six plasmids and *Rb. capsulatus* strains usually contain one or two plasmids in the 42 to 150 kb size range. Additional purification steps such as selection for antibiotic resistance after transformation into *E. coli* or separation by pulsed field agarose gel electrophoresis may be necessary to isolate recombinant vectors from endogenous plasmids. While normal genomic DNA isolation procedures result in shearing of the DNA into large fragments, intact genomic DNA for use in pulsed field gel electrophoresis can be prepared after first embedding the cells in agarose (McClelland et al., 1987; Suwanto and Kaplan, 1989a; Fonstein et al., 1992). Pulsed field gels are useful for the separation of large restriction fragments generated by rare-cutting restriction enzymes. The high GC content of DNA from many purple bacteria such as *Rb. capsulatus* and *Rb. sphaeroides* tends to result in a small number (usually less than 20 per genome) of restriction sites for enzymes that have a recognition sequence that is AT rich or that contains rarely used codons, such as *Ase*I, *Dra*I, *Spe*I, *Ssp*I and *Xba*I (McClelland et al., 1987; Suwanto and Kaplan, 1989a; Fonstein et al., 1992).

III. Gene Mapping

A. Isolation of Independent Events

To begin a genetic analysis of a pool of mutants with a given phenotype, it is desirable to ascertain that the phenotype is due to a single mutation in each strain. To make this more likely, different mutant strains should be derived from different single colonies that presumably arose from different single cells. If mutagens are used, they should be at levels low enough to insure that a reasonable fraction (10% to 50%) of cells survive the treatment. A frequency of reversion to the wild type phenotype in the range of approximately 10^{-6} to 10^{-9} per cell plated is an indication that a mutant phenotype is due to a single mutation. Reversion can also be induced via recombination systems from a wild type strain (for example, via GTA transduction in *Rb. capsulatus*), and the reversion frequency should match the gene transfer frequency of the specific system. However, some types of single mutations, such as very large deletions or rearrangements, may not revert easily to wild type and thus, may act like the result of multiple mutational events. Also, closely linked mutations may act like a single mutation.

B. Linkage Groups

The next step in a genetic mapping analysis is to sort the new mutations into linkage groups that probably correspond to genes. Linkage tests may be performed using R plasmid-induced conjugation (for example, Willison et al., 1985; Colbeau et al., 1990). In *Rb. capsulatus*, GTA transduction is well suited for these kinds of tests since cotransfer is limited to closely linked markers. The ratio test of Yen and Marrs (1976) has been used in *Rb. capsulatus* to sort fourteen independently isolated *nif* mutants into six linkage groups (Wall and Braddock, 1984; Wall et al., 1984). Since this test requires an independent marker, rifampicin resistant (Rif^r) derivatives of each of their original rifampicin sensitive (Rif^s) nitrogen fixation (Nif^-) mutants were constructed. The transfer of the Rif^r marker was then used to normalize the number of recombinants in pairwise GTA crosses. The expected frequency of transfer of each Nif marker was established by determining the ratio (Nif^+:Rif^r) of the frequency of transfer of the wild type nitrogen fixation marker (Nif^+) to the frequency of transfer of the Rif^r marker in control crosses ($Nif^+ Rif^r$ x $Nif^- Rif^s$). In the experimental crosses, one rifampicin resistant, nitrogen fixation mutant ($Nif_1^- Rif^r$) was used as a donor and a second rifampicin sensitive, nitrogen fixation mutant ($Nif_2^- Rif^s$) was the recipient. If the *nif1* and *nif2* loci are very far apart and therefore are not linked within the 4.5 kb that can be packaged by GTA particles, then the Nif^+:Rif^r ratio should be similar to the ratio of the corresponding control crosses. If the two *nif* loci are very near to each other, then it should be more difficult to detect Nif^+ recombinants, and the Nif^+:Rif^r ratio in the recombinants will be much less than that of the control crosses. Since the frequencies determined by this type of linkage test are subject to variation, the results must be shown to be statistically significant.

A modernized variant of the classical linkage test was used to determine the extent of linkage of several inhibitor resistance mutations to the *pet* operon of *Rb. capsulatus*, which encodes the three structural genes of the cytochrome bc_1 complex (Daldal et al., 1989). A kanamycin resistance marker was specifically introduced at the 3′ end of the *pet* operon, and GTA transduction from the kanamycin resistant, inhibitor sensitive strain into kanamycin sensitive, inhibitor resistant mutants was performed. After selection for transfer of kanamycin resistance to the recipient strain, the frequency of co-transduction of inhibitor sensitivity was tested. Three classes of mutations were observed among mutants resistant to the inhibitor stigmatellin. One group of mutations linked to the kanamycin marker had a co-transduction frequency of approximately 8%, and a second group of linked mutations had a co-transduction frequency of approximately 20%, indicating that these mutations were located closer to the 3′ end of *pet* operon than the first group. A third group was not linked to the kanamycin resistance marker (0% co-transduction) and thus were greater than 4.5 kb from the 3′ end of the *pet* operon. The *pet* operons from representative isolates whose mutations were shown to be linked to the kanamycin marker were then cloned and sequenced to determine the molecular basis of the inhibitor resistance.

C. Linkage Mapping

The primary requirement for classical genetic mapping is the ability to transfer large stretches of the genome between the mutant strain under study and a tester strain. A second requirement is the presence of selectable markers throughout the genome. One can determine the proximity of the mutation of interest to known markers by conducting two-point crosses. The linear order of three genes and the distances separating them can be determined in three-point crosses. The chromosome of *Rb. capsulatus* can be scanned with six or so widely spaced markers. Thus, if at least two or three markers are present in any tester strain, only two to four tester strains are required as a minimal set. The *hup* (H_2-uptake hydrogenase) loci in *Rb. capsulatus* have been mapped utilizing the mutant R plasmid, pTH10 (Colbeau et al., 1990). In one mapping cross, an Ade^- His^- Hup^+ Nif^+, pTH10 donor strain was mated to an Ade^+ His^+ Hup^- Nif^- recipient strain. Since the recovery of recombinants ranges from 10^{-4} to 10^{-6}

per donor cell plated, counterselection against the parental donor and recipient strains was accomplished by requiring adenine biosynthesis (Ade$^+$) and hydrogenase activity (Hup$^+$). After selection for Ade$^+$ Hup$^+$ recombinants, the secondary marker phenotypes, His$^-$ and Nif$^+$, were examined. The cotransfer frequency of two genes from the donor into the recipient genome is a measure of the genetic map distance separating these two loci. For the His$^-$ and Hup$^+$ markers, the cotransfer frequency, defined as the percentage of His$^-$ recombinants in the Hup$^+$ recombinant population, was measured to be approximately 55%. Similarly, the Nif$^+$ Hup$^+$ and His$^-$Nif$^+$ cotransfer frequencies were measured to be approximately 25% and 53%, respectively. Thus, the linear gene order was determined to be *hup-his-nif*. The map distance may be estimated using the equation of Wu (1966) as adapted by Kondorosi et al. (1977):

$$d = 1 - c^{1/3} \quad (1)$$

where d is the map distance in arbitrary units and c is the cotransfer frequency. Map distances calculated in this way show approximate additivity for values of c larger than 0.1. In carrying out crosses of this type, one must keep in mind that the reversion frequencies of these markers to wild type is in the 10^{-7} to 10^{-9} range.

D. Chromosomal Linkage Maps

The circular genetic linkage map of *Rb. capsulatus* (see Chapter 55 in this volume by Vignais et al.) was compiled utilizing derivatives of strain B10 (Willison et al., 1985; Willison, 1993). Data largely from two and three point crosses yielded the positions of approximately 40 genetic loci. Given that the chromosome of *Rb. capsulatus* is approximately 3,700 kb and assuming an average gene size of 1 kb, one might expect to find some 4,000 genetic loci. Undoubtedly, mutations at some of these sites result in a lethal phenotype. The *Rb. capsulatus* map determined by Willison and coworkers is a total of 4.6 arbitrary map units and the chromosome has been determined to be 3,700 kb; this yields an equivalence of approximately 800 kb per map unit. The *Rb. capsulatus* genetic map is largely in agreement with the physical map of *Rb. capsulatus* strain SB1003 (Fonstein and Haselkorn, 1993; see also Chapter 49 in this volume by Fonstein and Haselkorn) except in the *nifR1-glnA-trp1* region, which appears to be inverted between the two maps. This may be due to true strain differences, a nonequivalence of marker loci, effects of repeated elements on genetic recombination in the region or a sparsity of genetic loci in the region.

Chromosomal linkage maps for regions of *Rb. sphaeroides* also exist, though these are linear and the allele designations on different maps are not necessarily equivalent (reviewed in Kiley and Kaplan, 1988). Bowen and Pemberton (1985) utilized derivatives of broad host range plasmids coupled to mercury-resistance transposons, pRP1::Tn*501* and pR751::Tn*813*, to map nutritional and antibiotic markers including *crt* and *bch* loci on the *Rb. sphaeroides* RS630 chromosome. Sistrom et al. (1984) have generated a linear linkage map of the *Rb. sphaeroides* strain WS2 chromosome utilizing the broad host range, self-mobilizable plasmid R68.45. A physical map of two chromosomes of *Rb. sphaeroides* strain 2.4.1 generated by ordering large restriction fragments has been presented by Suwanto and Kaplan (1989a,b), and a genetic map has been related to a higher resolution physical map of the small chromosome (Choudhary et al., 1994). The origin of transfer from one of the endogenous plasmids of *Rb. sphaeroides* was shown to promote chromosomal transfer, and was used to correlate several genetic markers with the physical map (Suwanto and Kaplan, 1992a,b).

IV. Gene Cloning and Sequencing

A. Genomic Libraries

A DNA library for an organism generally consists of genomic restriction fragments cloned in a vector in *E. coli*. For complete libraries, every fragment should have a reasonable probability of being represented. Based on a statistical distribution, the number of independent clones, N, necessary for a random library to have a probability, P, of having any particular fragment can be estimated by:

$$N = \ln(1-P) / \ln(1-f) \quad (2)$$

where f is the fraction of the genome represented by each cloned fragment (Clarke and Carbon, 1976). The genome sizes, including the chromosomal and endogenous plasmid DNA, of *Rb. sphaeroides* and *Rb. capsulatus* have been estimated to be approximately 4,400 kb and 3,800 kb, respectively (Suwanto and Kaplan, 1989a; Fonstein and Haselkorn, 1993). After digestion with a restriction enzyme that recognizes a six base sequence and cloning into a plasmid vector, fragments of an average size of 4 kb would be expected to be cloned. Thus, assuming that each clone contains 1/1000 of the genome, for a probability of 99% the library size would have to be approximately 4,600 clones, and for a probability of 90% the library size would have to be approximately 2,300 clones. Larger restriction fragments, usually generated by partial digestion of genomic DNA followed by size fractionation on an agarose gel or sucrose gradient, can be cloned in phage and cosmid vectors. Lambda phage vectors can accept fragments of 10 to 20 kb, and cosmids can accept fragments of 30 to 40 kb, so the number of clones required for a complete library would be correspondingly smaller. A minimal set of 192 clones covering the chromosome and an endogenous plasmid was determined for an ordered, overlapping cosmid library of *Rb. capsulatus* (Fonstein et al., 1992; Fonstein and Haselkorn, 1993). In some cases, where the size of a restriction fragment containing the target gene was determined by hybridization with digests of genomic DNA, fragments of the correct size were isolated and then cloned to create size-enriched libraries. For example, one positive plasmid containing RubisCO genes from *Cm. vinosum* was obtained after screening plasmid DNA from 60 colonies with inserts of 8 to 9.5 kb isolated after agarose gel electrophoresis of a genomic *Bgl*II digest (Kobayashi et al., 1991).

B. Identification of Genes

1. Complementation

The identity of genes contained on restriction fragments cloned in a genomic library may be determined by their ability to complement mutant strains (Table 2). Libraries of *Rb. capsulatus*, *Rb. sphaeroides* and *Rs. rubrum* DNA have been screened by individual mating of several hundred to several thousand members of the library into a target mutant strain (Klug and Drews, 1984; Ahombo et al., 1986; Pemberton and Harding, 1986, 1987; Moore and Kaplan, 1989; Benning and Somerville, 1992a,b; Hustede et al., 1992). The screening can be accomplished by replica plating members of the genomic library from microtiter plates onto lawns of the mutant recipient (Klug and Drews, 1984; Moore and Kaplan, 1989; Benning and Somerville, 1992a,b). The number of matings that must be performed can be reduced by mating pools of clones from the library, and then isolating the desired cloned fragment from the complemented transconjugant (Avtges et al., 1985; Weaver and Tabita, 1985; Biel et al., 1988; Daniels et al., 1988; Rainey and Tabita, 1989; Xu et al., 1989; Biel and Biel, 1990; Hamblin et al., 1990). Alternatively, each pool from which a complementing clone was observed can be further subdivided and mated again, until individual clones can be tested (Colbeau et al., 1986; Hunter and Coomber, 1988; Hunter and Turner, 1988). Individual crosses are also necessary for plasmids that are not stable in the mutant recipient strain (Hunter and Turner, 1988). Interspecies complementation analysis has also been used to identify genes. For example, libraries of DNA from *Rb. capsulatus* complemented an *E. coli* glutamine synthetase mutant (Scolnik et al., 1983), *Rb. sphaeroides* and *Rs. rubrum* complemented an *Alcaligenes eutrophus* PHB synthase mutant (Hustede et al., 1992), *Rs. centenum* complemented *Rb. capsulatus* photosynthesis mutants (Yildiz et al., 1992), *Cm. vinosum* complemented an *E. coli* chaperonin mutant (Ferreyra et al., 1993), *Rb. capsulatus* complemented a *Rb. sphaeroides* cytochrome mutant (Jenny and Daldal, 1993) and *Rb. sphaeroides* complemented a *Pseudomonas aeruginosa recA* mutant (Calero et al., 1994) . In addition, resistance to cleavage by *Eco*RI due to the expression in *E. coli* of the gene for the *Rsr*I methyltransferase led to the identification of a clone encoding the methyltransferase gene from *Rb. sphaeroides* (Kaszubska et al., 1989).

Once a clone has been obtained that complements a mutant, the boundaries of the gene and its promoter can be narrowed by testing smaller fragments or making progressive deletions until complementation is no longer observed. The frequency of complementation can be used to distinguish between recombination of a chromosomal point mutation with DNA from the plasmid (marker rescue) and expression of the gene from the plasmid, since recombination occurs at a low frequency. For example, in order to determine the location of nitrogen fixation mutations using fragments of cloned nitrogenase structural genes, the number of transconjugants identified by antibiotic resistance was compared to the number that exhibited the Nif^{+} phenotype (Avtges et al., 1983). For complementation from the plasmid, 100% of the antibiotic resistant colonies were Nif^{+}, whereas marker rescue usually produced Nif$^+$ colonies at a frequency of 0.5% to 1% of the antibiotic resistant colonies. Similarly, when an *E. coli* strain, harboring a plasmid with a genomic DNA insert that had been identified by hybridization with a heterologous probe as containing a *Rb. capsulatus hup* gene, was mated with a *Rb. capsulatus* Hup$^-$ mutant, less than 1% of the transconjugants were Hup$^+$, leading to the conclusion that the entire gene was probably not present on the fragment (Leclerc et al., 1988). For mutants with phenotypes that can be distinguished by inspection of colonies, such as the color of *bch* and *crt* mutations, restreaking of transconjugants without selection for plasmid maintenance will either yield unsectored, primarily wild type colonies arising from cells in which recombination has occurred, or sectored colonies arising from cells whose mutation is complemented from the plasmid (Young et al., 1989). Recombination of the complementing DNA fragment can also be tested by using recombination deficient strains such as the *recA* mutants of *Rb. sphaeroides* (Sistrom et al., 1984; Rainey and Tabita, 1989; Calero et al., 1994) or by plasmid curing of the complemented strains (Davis et al., 1988; Tai et al., 1988; Brandner et al., 1989; Xu et al., 1989). The properties of an RNA processing gene in *Rb. capsulatus* expressed in *cis* and in *trans* were compared by forcing recombination of the complementing plasmid into the chromosome by the introduction of an incompatible plasmid (Klug et al., 1994).

2. Heterologous Probes

Once a gene has been cloned from one species, it can often be used as a hybridization probe to identify homologous genes in other species (Table 2). Since the degree of similarity among homologous proteins varies considerably, the evolutionary distance between the probe and the target needs to be considered for successful hybridization. Hybridization among a large number of species has been observed for genes that are highly conserved, such as the genes encoding RubisCO (Shively et al., 1986). The conditions used will affect the amount of hybridization observed, as occasionally genes that might be expected to cross-hybridize do not (for example, Beatty and Cohen, 1983; Coomber and Hunter, 1989). Generally, the heterologous probe is hybridized against colonies or plaques from the gene library, although with an *E. coli* probe for genes encoding ATPase, too much background hybridization with genomic DNA from the *E. coli* in the library was observed, so isolated DNA from the clones was used for hybridization instead of plaques (Tybulewicz et al., 1984). Heterologous probes have been used to identify DNA fragments in physical genome maps of *Rb. sphaeroides* and *Rb. capsulatus* (Suwanto and Kaplan, 1989a,b; Fonstein et al., 1992; Fonstein and Haselkorn, 1993). Ribosomal RNA has also been used as a hybridization probe (Yu et al., 1982; Dryden and Kaplan, 1990).

3. Oligonucleotide and PCR Probes

If a complete or partial amino acid sequence has been determined for a protein, then a corresponding DNA sequence can be deduced. A synthetic oligonucleotide containing this sequence can then be used as a hybridization probe for the structural gene encoding the protein (Table 2). Due to the degeneracy in the genetic code, the deduced DNA sequence is usually ambiguous, so probes consisting of all the possible coding sequences are synthesized as a mixture. To reduce the complexity of the probe, regions of the amino acid sequence with low codon degeneracy are often selected for the probe sequence. In addition, guesses as to the correct coding sequence can be based on the codon usage for a particular bacterium (Schatt et al., 1989; Majewski and Trebst, 1990; Self et al., 1990; Kerby et al., 1992). For example, a 27 nucleotide nondegenerate probe for ferredoxin in *Rb. capsulatus* was based on the codon usage for a sequence of nine amino acid residues that contained 30 ambiguities (Schatt et al., 1989). Oligonucleotides specific for conserved amino acid sequences of homologous proteins have also been used as hybridization probes (Majewski and Trebst, 1990; Masepohl et al., 1993b). Protein sequences that are reasonably far apart (approximately 70 to 700 residues) can be used as the basis for primers in PCR reactions to amplify the portion of the gene that lies between the two sequences. PCR fragments generated from primers based on consensus sequences have been used as hybridization probes and have also been directly cloned and sequenced (Table 2).

4. Immunological Screening

Screening genomic libraries with an antibody to a particular protein can be used to identify the structural

Table 2. Identification of cloned genes

Method	Gene type	Species	References
Complementation	ALA synthase	*Rb. capsulatus*	Biel et al., 1988
	aromatic acid	*Rp. palustris*	Dispensa et al., 1992
	bc_1 complex	*Rb. capsulatus*	Daldal et al., 1987
	CO_2 fixation	*Rb. sphaeroides*	Weaver and Tabita, 1985; Rainey and Tabita, 1989; Gibson and Tabita, 1993
	chaperonin	*Cm. vinosum*	Ferreyra et al., 1993
	cytochromes	*Rb. capsulatus*	Kranz, 1989; Biel and Biel, 1990; Jenny and Daldal, 1993
	dicarboxylic acid	*Rb. capsulatus*	Hamblin et al., 1990
	ferredoxin	*Rb. capsulatus*	Saeki et al., 1993
	fructose	*Rb. capsulatus*	Daniels et al., 1988
	hydrogenase	*Rb. capsulatus*	Colbeau et al., 1986, 1990; Xu et al., 1989
	lipid synthesis	*Rb. sphaeroides*	Benning and Somerville, 1992a, b; Arondel et al., 1993
	nitrogen fixation	*Rb. capsulatus*	Scolnik et al., 1983; Avtges et al., 1985; Ahombo et al., 1986; Foster-Hartnett et al., 1993
		Rb. sphaeroides	Meijer and Tabita, 1992; Zinchenko et al., 1990
	PHB synthesis	*Rb. sphaeroides*	Hustede et al., 1992
		Rs. rubrum	Hustede et al., 1992
	photosynthesis	*Rb. capsulatus*	Taylor et al., 1983; Youvan et al., 1983; Klug and Drews, 1984; Giuliano et al., 1986; Tichy et al., 1989, 1991; Young et al., 1989; Yang and Bauer, 1990; Sganga and Bauer, 1992; Pollich et al., 1993; Mosley et al., 1994
		Rb. sphaeroides	Pemberton and Harding, 1986, 1987; Hunter and Coomber, 1988; Hunter and Turner, 1988; Coomber and Hunter, 1989; Hunter et al., 1991; Coomber et al., 1992; Gibson et al., 1992; Eraso and Kaplan, 1994
		Rs. centenum	Yildiz et al., 1992
	RNA processing	*Rb. capsulatus*	Kordes et al., 1994
	recombination	*Rb. sphaeroides*	Calero et al., 1994
Heterologous probe	ALA synthase	*Rb. capsulatus*	Hornberger et al., 1990
		Rb. sphaeroides	Tai et al., 1988
	antenna	*Ro. denitrificans*	Liebetanz et al., 1991
		Rp. acidophila	Gardiner et al., 1992
	ATPase	*Rb. blasticus*	Tybulewicz et al., 1984
		Rs. rubrum	Falk et al., 1985
	bc_1 complex	*Rb. sphaeroides*	Davidson and Daldal, 1987b; Yun et al., 1990
		Rp. viridis	Verbist et al., 1989
		Rs. rubrum	Majewski and Trebst, 1990; Shanker et al., 1992
	CO_2 fixation	*Cm. vinosum*	Viale et al., 1985, 1989; Kobayashi et al., 1991
		Rb. sphaeroides	Fornari and Kaplan, 1983; Muller et al., 1985; Gibson and Tabita, 1986, 1987, 1988
	cytochrome oxidase	*Rb. sphaeroides*	Cao et al., 1991; Shapleigh and Gennis, 1992
	DNA gyrase	*Rb. capsulatus*	Kranz et al., 1992
	hydrogenase	*Rb. capsulatus*	Leclerc et al., 1988
		Rv. gelatinosus	Uffen et al., 1990
	nitrogen fixation	*Rb. capsulatus*	Avtges et al., 1983; Scolnik and Haselkorn, 1984; Klipp et al., 1988
		Rb. sphaeroides	Fornari and Kaplan, 1983; Zinchenko et al., 1994
		Rs. rubrum	Lehman et al., 1990
	PHB synthesis	*Cm. vinosum*	Liebergesell and Steinbüchel, 1992
	reaction center	*Rb. sphaeroides*	Donohue et al., 1986a
		Ro. denitrificans	Liebetanz et al., 1991
		Rs. rubrum	Bérard and Gingras, 1991; Hessner et al., 1991

Table 2. Continued

Method	Gene type	Species	References
Oligonucleotide	ALA synthase	*Rb. capsulatus*	Hornberger et al., 1990
probe	antenna	*Rb. capsulatus*	Youvan and Ismail, 1985
		Rb. sphaeroides	Ashby et al., 1987; Kiley and Kaplan, 1987
		Rp. palustris	Tadros and Waterkamp, 1989; Tadros et al., 1993
	ATPase	*Rs. rubrum*	Falk and Walker, 1988
	bc_1 complex	*Rb. capsulatus*	Daldal et al., 1987
		Rb. sphaeroides	Iba et al., 1987
		Rs. rubrum	Majewski and Trebst, 1990
	CO dehydrogenase	*Rs. rubrum*	Kerby et al., 1992
	cytochrome c_2	*Rb. capsulatus*	Daldal et al., 1986
		Rb. sphaeroides	Donohue et al., 1986b
		Rp. viridis	Grisshammer et al., 1990
		Rs. rubrum	Self et al., 1990
	ferredoxin	*Rb. capsulatus*	Schatt et al., 1989; Saeki et al., 1991; Duport et al., 1992
		Rs. rubrum	von Sternberg and Yoch, 1993
	flavocytochrome *c*	*Cm. vinosum*	Dolata et al., 1993
	nitrogen fixation	*Rs. rubrum*	Fitzmaurice et al., 1989
	peroxidase	*Rb. capsulatus*	Forkl et al., 1993
	reaction center	*Rb. sphaeroides*	Williams et al., 1983
		Rp. viridis	Michel et al., 1986
	*Rsr*I endonuclease	*Rb. sphaeroides*	Stephenson et al., 1989
	thioredoxin	*Rb. sphaeroides*	Pille et al., 1990
PCR probe	cytochrome oxidase	*Rb. capsulatus*	Thöny-Meyer et al., 1994
		Rb. sphaeroides	Shapleigh et al., 1992
	hydrogenase	*Rb. capsulatus*	Toussaint et al., 1993
	NADH oxidoreductase	*Rb. capsulatus*	Dupuis, 1992
Antibodies	bc_1 complex	*Rb. sphaeroides*	Usui and Yu, 1991
	coenzyme A ligase	*Rp. palustris*	Gibson et al., 1994
	RubisCO	*Rs. rubrum*	Somerville and Somerville, 1984
	reaction center	*Rp. viridis*	Michel et al., 1985

gene for that protein (Table 2). Since this method requires that the target protein be synthesized in *E. coli* from the cloned gene, libraries are generally constructed in expression vectors. This method may fail if the gene is not expressed in *E. coli*, as was observed during screening for the hydrogenase gene of *Rb. capsulatus* (Colbeau et al., 1986).

C. Sequences

Over 200 genes from purple photosynthetic bacteria have been sequenced, including structural and regulatory genes for a variety of metabolic processes (Table 3). In some cases, the genes for homologous proteins have been sequenced from several species, allowing the identification of conserved residues and the construction of evolutionary trees. Sequencing DNA that has been cloned from photosynthetic bacteria is performed by standard protocols. The major problem encountered is frequent gel artifacts such as compression due to the high GC content of DNA from many of the purple bacteria. These artifacts usually occur in different places on complementary strands. They can also usually be relieved by measures that reduce the amount of secondary structure in the DNA, such as the use of dGTP analogues during polymerization, the use of thermostable polymerases that enable the reactions to be carried out at higher temperatures, and the addition of formamide to the polyacrylamide gels. The usefulness of being able to amplify by PCR any portion of DNA that has already been sequenced has been demonstrated by the amplification of specific regions of *Rb. capsulatus* and *Rb. sphaeroides* DNA for sequence and size

Table 3. Gene sequences from purple photosynthetic bacteria

Gene type	Organism	References
ALA synthase	*Rb. capsulatus*	Hornberger et al., 1990; Wright et al., 1991
	Rb. sphaeroides	Neidle and Kaplan, 1993a
Antenna	*Rb. capsulatus*	Youvan et al., 1984; Youvan and Ismail, 1985; Tichy et al., 1989
	Rb. sphaeroides	Ashby et al., 1987; Kiley and Kaplan, 1987; Kiley et al., 1987; Burgess et al., 1989; Gibson et al., 1992
	Ro. denitrificans	Liebetanz et al., 1991
	Rp. acidophila	Gardiner et al., 1992
	Rp. palustris	Tadros and Waterkamp, 1989; Tadros et al., 1993
	Rp. viridis	Wiessner et al., 1990
	Rs. rubrum	Bérard et al., 1986
	Rv. gelatinosus	Nagashima et al., 1994
Aromatic acid	*Rp. palustris*	Dispensa et al., 1992; Gibson et al., 1994
ATPase	*Rb. blasticus*	Tybulewicz et al., 1984
	Rs. rubrum	Falk et al., 1985; Falk and Walker, 1988
Bacteriochlorophyll	*Rb. capsulatus*	Bauer et al., 1988; Adams et al., 1989; Wellington and Beatty, 1989; Yang and Bauer, 1990; Bollivar and Bauer, 1992; Sganga et al., 1992; Young et al., 1989, 1992; Armstrong et al., 1993; Burke et al., 1993a, b
	Rb. sphaeroides	Hunter et al., 1991; Coomber et al., 1992; McGlynn and Hunter, 1993; Gibson and Hunter, 1994
	Rs. rubrum	Lee and Collins, 1993
bc_1 complex	*Rb. capsulatus*	Davidson and Daldal, 1987a; Tokito and Daldal, 1992
	Rb. sphaeroides	Davidson and Daldal, 1987b; Iba et al., 1987; Yun et al., 1990; Usui and Yu, 1991
	Rp. viridis	Verbist et al., 1989
	Rs. rubrum	Majewski and Trebst, 1990; Shanker et al., 1992
Carotenoid	*Rb. capsulatus*	Armstrong et al., 1989; Bartley and Scolnik, 1989; Garí et al., 1992b
	Rb. sphaeroides	Lang et al., 1994
Chaperonin	*Cm. vinosum*	Ferreyra et al., 1993
CO dehydrogenase	*Rs. rubrum*	Kerby et al., 1992
CO_2 fixation	*Cm. vinosum*	Viale et al., 1989, 1991; Kobayashi et al., 1991
	Rb. sphaeroides	Hallenbeck and Kaplan, 1987; Gibson et al., 1990, 1991; Chen et al., 1991; Gibson and Tabita, 1993; Xu and Tabita, 1994
	Rs. rubrum	Nargang et al., 1984; Leustek et al., 1988; Falcone and Tabita, 1993
Cytochrome oxidase	*Rb. capsulatus*	Thöny-Meyer et al., 1994
	Rb. sphaeroides	Cao et al., 1991, 1992; Shapleigh and Gennis, 1992; Shapleigh et al., 1992
Cytochromes	*Rb. capsulatus*	Daldal et al., 1986; Beckman et al., 1992; Beckman and Kranz, 1993; Jenny and Daldal, 1993
	Rb. sphaeroides	Donohue et al., 1986b; MacGregor and Donohue, 1991
	Rp. viridis	Grisshammer et al., 1990
	Rs. rubrum	Self et al., 1990
Dicarboxylic acid	*Rb. capsulatus*	Shaw et al., 1991; Hamblin et al., 1993
DNA gyrase	*Rb. capsulatus*	Kranz et al., 1992
DNA transfer	*Rb. sphaeroides*	Suwanto and Kaplan, 1992a
Fatty acid	*Rb. capsulatus*	Beckman and Kranz, 1991a
Ferredoxin	*Rb. capsulatus*	Schatt et al., 1989; Duport et al., 1990; Saeki et al., 1990, 1991, 1993; Grabau et al., 1991
	Rb. sphaeroides	Neidle and Kaplan, 1992
	Rs. rubrum	von Sternberg and Yoch, 1993
Flavocytochrome *c*	*Cm. vinosum*	Dolata et al., 1993
Flavodoxin	*Rb. capsulatus*	Jouanneau et al., 1990
Fructose	*Rb. capsulatus*	Wu and Saier, 1990; Wu et al., 1990, 1991a

Table 3. Continued

Gene type	Organism	References
Hydrogenase	*Rb. capsulatus*	Leclerc et al., 1988; Richaud et al., 1991; Toussaint et al., 1991; Xu and Wall, 1991; Colbeau et al., 1993; Elsen et al., 1993; Toussaint et al., 1993
	Rv. gelatinosus	Uffen et al., 1990
β-lactamase	*Rb. capsulatus*	Campbell et al., 1989
Lipid synthesis	*Rb. sphaeroides*	Benning and Somerville, 1992a, b; Arondel et al., 1993
Mannitol	*Rb. sphaeroides*	Schneider et al., 1993
NADH oxidoreductase	*Rb. capsulatus*	Dupuis, 1992
Nitrogen fixation	*Rb. capsulatus*	Schumann et al., 1986; Jones and Haselkorn, 1988, 1989; Masepohl et al., 1988, 1993a,b; Pollock et al., 1988; Alias et al., 1989; Moreno-Vivian et al., 1989a,b; Kranz et al., 1990; Preker et al., 1992; Foster-Hartnett et al., 1993; Schüddekopf et al., 1993; Schmehl et al., 1993; Wang et al., 1993; Willison et al., 1993
	Rb. sphaeroides	Meijer and Tabita, 1992; Zinchenko et al., 1994
	Rs. rubrum	Fitzmaurice et al., 1989; Lehman et al., 1990
Nucleotide	*Rb. capsulatus*	Beckman and Kranz, 1991b
Peroxidase	*Rb. capsulatus*	Forkl et al., 1993
PHB synthesis	*Cm. vinosum*	Liebergesell and Steinbüchel, 1992
Polysaccharide	*Rs. rubrum*	Ideguchi et al., 1993
Reaction center	*Rb. capsulatus*	Youvan et al., 1984
	Rb. sphaeroides	Williams et al., 1983, 1984, 1986; Donohue et al., 1986a; Arnoux et al, 1990; Lee et al., 1989a
	Ro. denitrificans	Liebetanz et al., 1991
	Rp. viridis	Michel et al., 1985, 1986; Weyer et al., 1987
	Rs. rubrum	Bélanger et al., 1988; Bérard and Gingras, 1991
	Rv. gelatinosus	Nagashima et al., 1994
Recombination	*Rb. sphaeroides*	Calero et al., 1994
Regulation	*Rb. capsulatus*	Sganga and Bauer, 1992; Pollich et al., 1993; Buggy et al., 1994a,b; Mosley et al., 1994
	Rb. sphaeroides	Eraso and Kaplan, 1994; Gong et al., 1994; Penfold and Pemberton, 1994
RNA	*Rb. capsulatus*	Höpfl et al., 1988
	Rb. sphaeroides	Dryden and Kaplan, 1990
RsrI	*Rb. sphaeroides*	Kaszubska et al., 1989; Stephenson and Greene, 1989; Stephenson et al., 1989
Thioredoxin	*Rb. sphaeroides*	Pille et al., 1990
Tryptophan	*Rb. capsulatus*	Becker-Rudzik et al., 1992

analysis from mutants (Klug and Jock, 1991; Colbeau et al., 1993; Gong et al., 1994; Kordes et al., 1994), for use in protein binding assays (Klug, 1991; Lee et al., 1993) and for plasmid constructions (Foster-Hartnett and Kranz, 1992; Arondel et al., 1993; Bollivar et al., 1994b). Generation of DNA fragments by PCR was also used to determine the position of a Tn*5* insertion in *Rp. palustris* (Dispensa et al., 1992) and to confirm a chromosomal deletion in *Rp. viridis* (Laußermair and Oesterhelt, 1992).

The identification of genes in sequenced regions enables tabulation of the codon usage for an organism (Table 4). The codon usage in turn facilitates identification of other genes by codon preference profiles of sequences, which have been used to identify ATPase genes in *Rp. blastica* (Tybulewicz et al., 1984) and *Rs. rubrum* (Falk et al., 1985; Falk and Walker, 1988) and pigment synthesis genes (Bauer et al., 1988; Young et al., 1989, 1992; Yang and Bauer, 1990) and fructose metabolism genes (Wu et al., 1991a) in *Rb. capsulatus*. Codon usage information is also useful in designing oligonucleotides for use as hybridization probes and PCR primers, and in site-directed mutagenesis. Evidence has been presented that codon usage may be different for genes expressed under phototrophic and heterotrophic growth conditions in *Rb. capsulatus* (Wu and Saier, 1991).

Table 4. Codon frequencies[1]

Amino acid	Codon	*Rb. capsulatus*	*Rb. sphaeroides*	*Rp. viridis*	*Rs. rubrum*	*Cm. vinosum*
Gly	GGG	0.25	0.21	0.09	0.14	0.06
Gly	GGA	0.03	0.05	0.08	0.04	0.03
Gly	GGT	0.08	0.06	0.21	0.17	0.23
Gly	GGC	0.64	0.68	0.62	0.65	0.68
Glu	GAG	0.54	0.76	0.57	0.58	0.72
Glu	GAA	0.46	0.24	0.43	0.42	0.28
Asp	GAT	0.40	0.27	0.32	0.41	0.27
Asp	GAC	0.60	0.73	0.68	0.59	0.73
Val	GTG	0.48	0.52	0.30	0.37	0.38
Val	GTA	0.00	0.01	0.01	0.01	0.00
Val	GTT	0.08	0.06	0.23	0.16	0.02
Val	GTC	0.44	0.41	0.47	0.46	0.60
Ala	GCG	0.49	0.42	0.40	0.21	0.25
Ala	GCA	0.03	0.04	0.05	0.00	0.04
Ala	GCT	0.03	0.04	0.26	0.08	0.05
Ala	GCC	0.45	0.50	0.29	0.70	0.66
Arg	AGG	0.02	0.04	0.00	0.00	0.01
Arg	AGA	0.01	0.02	0.01	0.01	0.02
Ser	AGT	0.02	0.02	0.01	0.05	0.04
Ser	AGC	0.22	0.25	0.22	0.27	0.34
Lys	AAG	0.74	0.88	0.91	0.88	0.93
Lys	AAA	0.26	0.12	0.09	0.12	0.07
Asn	AAT	0.24	0.16	0.17	0.32	0.09
Asn	AAC	0.76	0.84	0.82	0.68	0.91
Met	ATG	1.00	1.00	1.00	1.00	1.00
Ile	ATA	0.01	0.01	0.00	0.01	0.01
Ile	ATT	0.09	0.06	0.20	0.15	0.05
Ile	ATC	0.91	0.93	0.80	0.84	0.95
Thr	ACG	0.36	0.40	0.44	0.24	0.21
Thr	ACA	0.02	0.02	0.02	0.01	0.02
Thr	ACT	0.03	0.03	0.11	0.03	0.02
Thr	ACC	0.59	0.55	0.44	0.72	0.75
Trp	TGG	1.00	1.00	1.00	1.00	1.00
End	TGA	0.78	0.75	0.57	0.29	0.73
Cys	TGT	0.16	0.08	0.20	0.16	0.13
Cys	TGC	0.84	0.92	0.80	0.84	0.87
End	TAG	0.07	0.09	0.00	0.14	0.00
End	TAA	0.15	0.16	0.43	0.57	0.27
Tyr	TAT	0.51	0.45	0.40	0.56	0.37
Tyr	TAC	0.49	0.55	0.60	0.44	0.63
Leu	TTG	0.06	0.02	0.04	0.11	0.02
Leu	TTA	0.00	0.00	0.01	0.00	0.00
Phe	TTT	0.21	0.06	0.06	0.12	0.08
Phe	TTC	0.79	0.94	0.94	0.88	0.92
Ser	TCG	0.54	0.50	0.55	0.36	0.35
Ser	TCA	0.01	0.02	0.00	0.03	0.03
Ser	TCT	0.03	0.03	0.08	0.03	0.01
Ser	TCC	0.17	0.17	0.14	0.27	0.24

Table 4. Continued

Amino acid	Codon	*Rb. capsulatus*	*Rb. sphaeroides*	*Rp. viridis*	*Rs. rubrum*	*Cm. vinosum*
Arg	CGG	0.33	0.33	0.14	0.18	0.12
Arg	CGA	0.02	0.03	0.00	0.01	0.02
Arg	CGT	0.09	0.04	0.34	0.15	0.24
Arg	CGC	0.54	0.55	0.51	0.65	0.59
Gln	CAG	0.81	0.90	0.91	0.82	0.95
Gln	CAA	0.19	0.10	0.09	0.18	0.05
His	CAT	0.57	0.41	0.26	0.39	0.27
His	CAC	0.43	0.59	0.74	0.61	0.73
Leu	CTG	0.65	0.51	0.57	0.62	0.66
Leu	CTA	0.00	0.00	0.00	0.01	0.01
Leu	CTT	0.14	0.07	0.11	0.11	0.01
Leu	CTC	0.14	0.40	0.28	0.15	0.30
Pro	CCG	0.58	0.54	0.83	0.50	0.70
Pro	CCA	0.01	0.02	0.00	0.01	0.01
Pro	CCT	0.03	0.03	0.05	0.05	0.02
Pro	CCC	0.38	0.41	0.12	0.44	0.27

[1] Compiled from 119 genes (37,451 codons) for *Rb. capsulatus*, 55 genes (16,000 codons) for *Rb. sphaeroides*, 7 genes (1484 codons) for *Rp. viridis*, 21 genes (4920 codons) for *Rs. rubrum* and 12 genes (3448 codons) for *Cm. vinosum* in GenBank, Release 77.0, 6/93.

V. Mutagenesis

A. Enrichment by Antibiotic Selection

Antibiotics such as ampicillin and penicillin that kill actively growing cells have been used to enrich cultures for mutants (reviewed in Marrs et al., 1980). Cultures are subjected to the antibiotics under conditions in which bacteria with the wild type function can grow but those with impaired function will not grow. After this enrichment period, the cells are washed and then screened for the presence of the function. Examples of ampicillin and penicillin selection in photosynthetic bacteria include the isolation of auxotrophs in *Rb. sphaeroides* (Sistrom, 1977) and *Rb. capsulatus* (Willison et al., 1985), nitrogen fixation mutants in *Rb. capsulatus* (Wall et al., 1975a; Avtges et al., 1983; Wall and Braddock, 1984), carbon dioxide fixation mutants in *Rb. sphaeroides* (Weaver and Tabita, 1983), photosynthesis mutants in *Rb. capsulatus* (Chen et al., 1988; Sganga and Bauer, 1992; Mosley et al., 1994), *Rb. sphaeroides* (Hunter and Turner, 1988) and *Rs. centenum* (Yildiz et al., 1991) and hydrogenase mutants in *Rb. capsulatus* (Xu et al., 1989). Other antibiotics have also been used for enrichment of *Rb. sphaeroides* mutants (Schneider et al., 1993). Tetracycline is accumulated in photosynthetically growing cells, and tetracycline suicide techniques have been used to isolate photosynthetically deficient mutants in *Rb. capsulatus* (Zannoni et al., 1980; Marrs, 1981; Youvan et al., 1983) and *Rs. rubrum* (Johansson and Baltscheffsky, 1980).

B. Chemical Mutagenesis

Mutagens that have been used in photosynthetic bacteria include nitrosoguanidine, ethylmethane sulfonate and UV light (reviewed in Sistrom, 1978; Willison, 1993; Table 5). The survival of cells in a population exposed to a mutagen can be used as a measure of the extent of mutagenesis. For example, a survival curve comparing the survival of the cells to the time of exposure to UV light and to nitrosoguanidine was determined for *Rp. viridis* (Lang and Oesterhelt, 1989a). The degree of mutagenesis in such an experiment can be determined by scoring the frequency of antibiotic resistance mutants, as has been done for nitrosoguanidine mutagenesis of *Rb. sphaeroides* (Benning and Somerville, 1992a). A study of the effects of UV irradiation indicated that *Rb. capsulatus* has a DNA repair system similar to that of *E. coli* (Barbé et al., 1987) and the *recA* gene has been isolated from *Rb. sphaeroides* (Calero et al., 1994). UV irradiation-sensitive mutants that appear to be recombination-deficient have been isolated

Table 5. Examples of mutagenesis

Method	Target gene type	Species	References
Chemical	CO_2 fixation	*Rb. sphaeroides*	Rainey and Tabita, 1989
	hydrogenase	*Rb. capsulatus*	Colbeau et al., 1986; Xu et al., 1989
	lipid synthesis	*Rb. sphaeroides*	Benning and Somerville, 1992a
	nitrogen fixation	*Rb. capsulatus*	Wall et al., 1975a; Avtges et al., 1983; Wall and Braddock, 1984; Willison et al., 1985
	photosynthesis	*Rb. capsulatus*	Yen and Marrs, 1976; Mosley et al., 1994
		Rb. sphaeroides	Pemberton and Bowen, 1981; Pemberton and Harding, 1986, 1987; Ashby et al., 1987; Hunter and Turner, 1988; Wu et al., 1991b; Garí et al., 1992a
		Rp. viridis	Lang and Oesterhelt, 1989a
		Rs. centenum	Yildiz et al., 1991
Transposon	ALA synthase	*Rb. capsulatus*	Wright et al., 1987
	aromatic acid	*Rp. palustris*	Elder et al., 1993
	carotenoid	*Rp. palustris*	Elder et al., 1993
	CO_2 fixation	*Rb. sphaeroides*	Weaver and Tabita, 1983
	cytochromes	*Rb. capsulatus*	Biel and Biel, 1990
	dicarboxylic acid	*Rb. capsulatus*	Hamblin et al., 1990
	flagella	*Rb. sphaeroides*	Sockett and Armitage, 1991
	fructose	*Rb. capsulatus*	Daniels et al., 1988
	hydrogenase	*Rb. capsulatus*	Xu et al., 1989; Colbeau et al., 1990
	nitrogen fixation	*Rb. capsulatus*	Klipp et al., 1988; Schüddekopf et al., 1993
	mannitol	*Rb. sphaeroides*	Schneider et al., 1993
	nitrate reduction	*Rb. sphaeroides*	Moreno-Vivian et al., 1994
	photosynthesis	*Rb. capsulatus*	Kaufmann et al., 1984; Pollich et al., 1993
		Rb. sphaeroides	Coomber et al., 1992; Wu et al., 1991b
		Ro. dentrificans	Liebetanz et al., 1991
		Rs. centenum	Yildiz et al., 1991
	TMAO reductase	*Rb. capsulatus*	Kelly et al., 1988
Local transposon	CO_2 fixation	*Rb. sphaeroides*	Falcone et al., 1988; Rainey and Tabita, 1989; Gibson et al., 1990; Hallenbeck et al., 1990a,b
	cytochromes	*Rb. capsulatus*	Kranz, 1989
	hydrogenase	*Rb. capsulatus*	Xu et al., 1989
	nitrogen fixation	*Rb. capsulatus*	Avtges et al., 1983, 1985; Scolnik et al., 1983; Masepohl et al., 1993
		Rb. sphaeroides	Meijer and Tabita, 1992; Zinchenko et al., 1994
		Rs. rubrum	Fitzmaurice et al., 1989
	photosynthesis	*Rb. capsulatus*	Youvan et al., 1982; Zsebo and Hearst, 1984; Zsebo et al., 1984; Armstrong et al., 1990
		Rb. sphaeroides	Hunter, 1988; Coomber et al., 1990, 1992
Interposon (GTA)	bc_1 complex	*Rb. capsulatus*	Daldal et al., 1987; Atta-Asafo-Adjei and Daldal, 1991; Tokito and Daldal, 1992
	cytochromes *c*	*Rb. capsulatus*	Daldal et al., 1986; Daldal, 1988; Jenny and Daldal, 1993
	ferredoxins	*Rb. capsulatus*	Saeki et al., 1991, 1993
	hydrogenase	*Rb. capsulatus*	Cauvin et al., 1991
	nitrogen fixation	*Rb. capsulatus*	Scolnik and Haselkorn, 1984; Schumann et al., 1986; Zinchenko et al., 1990
	photosynthesis	*Rb. capsulatus*	Youvan et al., 1985; Bauer and Marrs, 1988; Bauer et al., 1988, 1991; Giuliano et al., 1988; Young et al., 1989, 1992; Yang and Bauer, 1990; Sganga and Bauer, 1992; Sganga et al., 1992; Bollivar et al., 1994a,b; Buggy et al., 1994a,b
	tryptophan	*Rb. capsulatus*	Becker-Rudzik et al., 1992

Table 5. Continued

Method	Target gene type	Species	References
Interposon	ALA synthase	*Rb. capsulatus*	Hornberger et al., 1990
		Rb. sphaeroides	Neidle and Kaplan, 1993b
	bc_1 complex	*Rb. sphaeroides*	Yun et al., 1990; Chen et al., 1994
	CO dehydrogenase	*Rs. rubrum*	Kerby et al., 1992
	CO_2 fixation	*Rb. sphaeroides*	Hallenbeck et al., 1990a, b; Falcone and Tabita, 1991; Gibson et al., 1991; Gibson and Tabita, 1993
		Rs. rubrum	Falcone and Tabita, 1993
	coenzyme A ligase	*Rp. palustris*	Gibson et al., 1994
	cytochrome oxidase	*Rb. sphaeroides*	Cao et al., 1992; Shapleigh and Gennis, 1992
	cytochromes	*Rb. capsulatus*	Beckman and Kranz, 1993; Thöny-Meyer et al., 1994
		Rb. sphaeroides	Donohue et al., 1988; Caffrey et al., 1992; Yun et al., 1994
	dicarboxylic acid	*Rb. capsulatus*	Hamblin et al., 1993
	ferredoxin	*Rb. sphaeroides*	Neidle and Kaplan, 1992
	hydrogenase	*Rb. capsulatus*	Cauvin et al., 1991; Elsen et al., 1993
	nitrogen fixation	*Rb. capsulatus*	Masepohl et al., 1988, 1993a,b; Moreno-Vivian et al., 1989a, b; Schneider et al., 1991; Gollan et al., 1993; Schüddekopf et al., 1993; Schmehl et al., 1993; Wang et al., 1993; Cullen et al., 1994
		Rb. sphaeroides	Meijer and Tabita, 1992
		Rs. rubrum	Lehman et al., 1990; Liang et al., 1991
	photosynthesis	*Rb. capsulatus*	Chen et al., 1988; Wellington and Beatty, 1989; LeBlanc and Beatty, 1993
		Rb. sphaeroides	Davis et al., 1988; Burgess et al., 1989; Farchaus and Oesterhelt, 1989; Lee et al., 1989b; Paddock et al., 1989; Sockett et al., 1989; Takahashi and Wraight, 1990; Hunter et al., 1991; Garí et al., 1992a; Gibson et al., 1992; Jones et al., 1992a,b; Barz and Oesterhelt, 1994; Penfold and Pemberton, 1994
		Rp. viridis	Laußermair and Oesterhelt, 1992
		Rs. rubrum	Hessner et al., 1991
	recombination	*Rb. sphaeroides*	Calero et al., 1994
	RNA	*Rb. sphaeroides*	Dryden and Kaplan, 1993
Oligonucleotide	antenna	*Rb. capsulatus*	Bylina et al., 1988; Dörge et al., 1990; Babst et al., 1991; Richter et al., 1991, 1992; Goldman and Youvan, 1992
		Rb. sphaeroides	Fowler et al., 1992, 1993
	bc_1 complex	*Rb. capsulatus*	Atta-Asafo-Adjei and Daldal, 1991; Davidson et al., 1992
		Rb. sphaeroides	Yun et al., 1990, 1991a; Hacker et al., 1993
	cytochrome c_2	*Rb. capsulatus*	Caffrey et al., 1991, 1992
		Rb. sphaeroides	Brandner and Donohue, 1994
	cytochrome oxidase	*Rb. sphaeroides*	Shapleigh et al., 1992b
	nitrogenase	*Rb. capsulatus*	Pierrard et al., 1993
	reaction center	*Rb. capsulatus*	Bylina et al., 1986, 1989; Bylina and Youvan, 1988; Stocker et al., 1992; Taguchi et al., 1992; Robles and Youvan, 1993
		Rb. sphaeroides	Paddock et al., 1989; Gray et al., 1990; Nagarajan et al., 1990; Takahashi and Wraight, 1990; Williams et al., 1992; Farchaus et al., 1993; Jones et al., 1994; Lin et al., 1994; Stilz et al., 1994
		Rp. viridis	Laußermair and Oesterhelt, 1992
	regulation	*Rb. capsulatus*	Chen and Belasco, 1990; Narro et al., 1990; Klug and Cohen, 1991; Ma et al., 1993
		Rb. sphaeroides	Lee et al., 1993; Gong et al, 1994
	RubisCO	*Rs. rubrum*	Niyogi et al., 1986; Soper et al., 1992

from *Rb. capsulatus* (Genther and Wall, 1984) and *Rb. sphaeroides* (Sistrom et al., 1984). In order to increase the frequency of mutation in a given region of DNA, cloned fragments can be subjected to mutagens in vitro and then transferred back into a host strain for screening, as demonstrated by Narro et al. (1990) in studying oxygen regulation of the *puf* operon in *Rb. capsulatus*.

C. Transposon Mutagenesis

Transposons can be used to make polar, random insertion mutations in chromosomal or plasmid DNA (Table 5). Transposons are generally transferred into a cell on a mobilizable suicide vector, where survival of the transposon depends on it having moved to the genomic DNA before the plasmid is lost. The most widely used system of this type is the suicide plasmid pSUP2021::Tn*5* (Simon et al., 1983). A derivative of this vector with improved transposition frequencies in *Rs. rubrum* has been constructed (Ghosh et al., 1994). Plasmids that have a temperature sensitive origin of replication and carry derivatives of Tn*5* that confer resistance to chloramphenicol (pCHR83) or gentamicin (pCHR84) were used to deliver transposons into *Rs. centenum* (Yildiz et al., 1991). Similarly, the plasmids pJB4J1 and pRK340 have been used to introduce transposons into *Rb. sphaeroides* (Weaver and Tabita, 1983). A restriction fragment containing the transposon can be cloned by selecting for the antibiotic resistance marker on the transposon, and then used as a hybridization probe for the wild type DNA (for example, Sockett and Armitage, 1991; Schüddekopf et al., 1993). Alternatively, the strain with the transposon insertion can be complemented by a genomic library to identify the gene. Additionally, the wild type *nif* genes in *Rb. capsulatus* were obtained after reciprocal exchange with cloned *nif*::Tn*5* fragments and isolation of the resulting R' plasmids (Klipp et al., 1988).

Localized transposon mutagenesis can take place in a cloned piece of DNA, which is then transferred back into the species from which the DNA originated (reviewed in Hunter and Mann, 1992; Table 5). The phenotype of the mutation on the plasmid can be revealed either after homologous recombination into the wild type chromosome or by the ability of the mutated plasmid to complement mutant strains. If the transposon is inserted into a piece of DNA for which the restriction map is known, the approximate site of the insertion can be determined by restriction mapping of the DNA containing the transposon.

D. Interposon Mutagenesis

An antibiotic resistance marker can be inserted into restriction sites in a cloned gene and transferred back into the chromosome by homologous recombination through a double crossover event to create a site-directed chromosomal insertion (reviewed in Donohue and Kaplan, 1991; Table 5). This technique of in vitro insertional mutagenesis has been termed interposon mutagenesis by Prentki and Krisch (1984), who constructed a widely used cassette called the Ω cartridge that consists of a spectinomycin/streptomycin antibiotic resistance gene flanked by transcription and translation termination signals and restriction sites. Other antibiotic resistance markers, including kanamycin and gentamicin, have been used to make such replacements. These types of insertions can be used to define the boundaries of cloned genes, to identify open reading frames, and to create background deletion strains for site-directed mutagenesis experiments.

In *Rb. capsulatus,* transfer of the marker into the chromosome can be mediated by GTA transduction (Scolnik and Haselkorn, 1984; reviewed in Scolnik and Marrs, 1987; Donohue and Kaplan, 1991). Chromosome mobilization or nonreplicating suicide plasmids can also be used to mediate insertion of the interposon into the chromosome. The mostly widely used suicide plasmids are pSUP202 and other derivatives described by Simon et al. (1983), which are mobilizable but are generally unable to replicate in strains outside the enteric bacterial group. Similar mobilizable suicide vectors have been used for interposon mutagenesis in *Rb. capsulatus* (Cauvin et al., 1991; Hübner et al., 1993). A suicide vector derived from R6K is functional in *Rb. sphaeroides* (Penfold and Pemberton, 1992, 1994). An artificial suicide vector created by deletion of the origin of replication from pRK404 was used to construct a chromosomal replacement of the *puf* operon after introduction into *Rp. viridis* by electroporation (Laußermair and Oesterhelt, 1992). Interposons have also been introduced via broad host range vectors in cases where recombination with the chromosome can be forced, such as by the introduction of an incompatible plasmid (Kranz et al., 1990; Lehman et al., 1990; Beckman and Kranz, 1993). In *Rb. capsulatus*, reciprocal exchange of the chromosomal *puf* operon with an interposon constructed in a pRK404 vector was enriched for by selecting only for the antibiotic resistance of the marker followed by penicillin selection against photosynthetically competent cells (Chen et al., 1988). Chromosomal deletions or insertions that do not involve replacement with an antibiotic resistance marker can be made if the phenotype of the interrupted gene can be selected or screened for, or by replacing a previously inserted marker. For example, mutations and a small deletion were made in the *puf* operon of *Rb. sphaeroides* by replacement of a previously inserted kanamycin resistance marker (DeHoff et al., 1988; Gong et al., 1994; Stilz et al., 1994), and a nonpolar insertion of a 22 base pair fragment in the chromosomal *nifR3* gene of *Rb. capsulatus* was screened for by replacement of a previously inserted polar kanamycin marker (Foster-Hartnett et al., 1993).

E. Oligonucleotide-Directed Mutagenesis

Specific local changes in DNA sequences, such as alteration of the coding sequence of structural genes, can be accomplished by oligonucleotide-directed mutagenesis (Table 5). In order to study structure-function questions in proteins, a mutagenesis system consisting of a chromosomal deletion of the wild type gene, a shuttle vector in which to transfer mutated copies of the gene back into the deletion strain, and a vehicle for mutagenesis of the DNA is required (Fig. 1). Mutagenesis can be accomplished on small fragments, for example by cloning into single-stranded phage vectors. After the mutagenesis, the fragment can be resequenced, inserted back into the shuttle vector, and mated into the deletion strain. To prevent selection for reversions, growth of the mutant strain should not require the function of the mutated gene.

Other types of alterations can be specifically introduced into DNA. Oligonucleotide-directed mutagenesis has been used to introduce deletions and insertions into the *puf* operon of *Rb. capsulatus* (Klug et al., 1987; Chen et al., 1988; Richter et al., 1992) and to create restriction sites for easier manipulation of DNA (Bylina et al., 1986; Pollock et al., 1988; Adams et al., 1989; Farchaus et al., 1990, 1992; Pille et al., 1990; Robles et al., 1990; Suetsugu

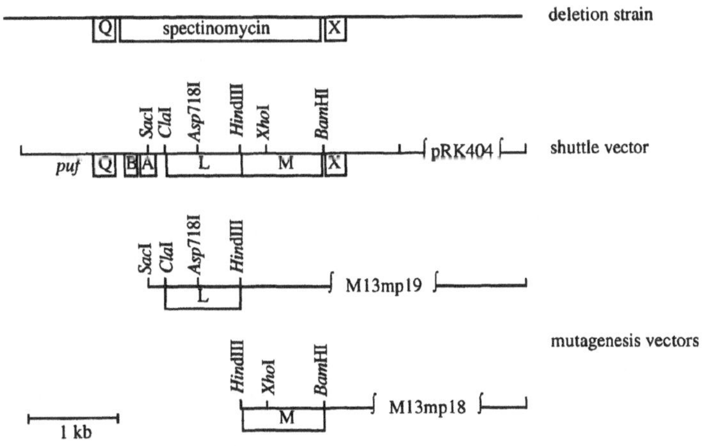

Fig. 1. Example of a mutagenesis system for reaction center genes in *Rb. sphaeroides*. A number of unique restriction sites have been added so that segments of the *pufL* and *pufM* genes can be manipulated independently. Each gene has been cloned into single-stranded M13 bacteriophage vectors to facilitate oligonucleotide-directed mutagenesis experiments. Restriction fragments containing the mutations are then moved into a broad host range plasmid in which the *puf* operon has been cloned, and the resulting shuttle vector is transferred by conjugation into a strain in which the wild type genes on the chromosome have been deleted. The *pufB, A, L* and *M* genes in the deletion strain have been replaced with an antibiotic resistance marker by interposon mutagenesis. In this deletion strain, the *pucA* and *B* genes, which encode the structural polypeptides of the B800–850 complex, have been deleted as well (Lee et al., 1989b). Since the *Sac*I site in the *pufA* gene creates a mutation that results in the loss of the B870 antenna, the only pigmented complex in the complemented deletion strain is the reaction center. (See Chapter 22 in this volume by Hunter for other examples of strains that lack antenna complexes.) The mutant strains can be grown semi-aerobically in the dark to avoid photosynthetic selection pressure while also maintaining expression of the *puf* operon.

et al., 1991; Jones et al., 1992a,b; Shapleigh and Gennis, 1992; Shapleigh et al., 1992b). Mutagenesis (Stocker et al., 1992; Ma et al., 1993; Barz and Oesterhelt, 1994) or the specific addition of restriction sites (Beckman et al., 1992; Dispensa et al., 1992; Burke et al., 1993a; Armengaud et al., 1994; Bollivar et al., 1994a; Gibson and Hunter, 1994; Gibson et al., 1994; Lang et al., 1994) has also been accomplished by PCR reactions. Replacement of DNA segments with small (50 to 60 base pairs) double-stranded fragments consisting of annealed complementary oligonucleotides has been used to create large scale mutations in the reaction center genes from *Rb. capsulatus* (Taguchi et al., 1992), to insert a transcription terminator in the *puf* operon of *Rb. sphaeroides* (Farchaus et al., 1990) and to create small insertions in the gene encoding an RNA polymerase sigma factor in *Rb. capsulatus* (Cullen et al., 1994). This type of cassette mutagenesis can also be used to make simultaneous sets of mutations at several amino acid residues. For example, the antenna and reaction center genes of *Rb. capsulatus* have been mutagenized by incorporating mutations in the chemical synthesis of one strand of short (110 to 120 base pairs) restriction fragments, which were then amplified by PCR reactions using primers from each end of the fragment and subsequently cloned into an expression vector (Goldman and Youvan, 1992; Robles and Youvan, 1993; see also Chapter 61 in this volume by Goldman and Youvan).

F. Suppressors

When selection pressure is put on a mutant strain, reversion to the wild type phenotype may occur. For single site mutations, the reversion is usually at the original mutation site. However, mutations at a second site (suppressor mutations) may be recovered that complement the original mutation. For mutants containing a deletion of the wild type gene, revertants must be due to suppressor mutations in other genes. In *Rb. capsulatus* and *Rb. sphaeroides*, reversion to photosynthetic growth has been used to select suppressors of amino acid changes in the reaction center (Robles et al., 1990; Hanson et al., 1992), *puf* regulatory mutations (Klug and Jock, 1991) and deletions of *pufX* (Lilburn and Beatty, 1992; Lilburn et al., 1992; Barz and Oesterhelt, 1994) and *pucC* (LeBlanc and Beatty, 1993). Additional copies of nitrogenase genes in *Rb. capsulatus* (Scolnik and Haselkorn, 1984; Schneider et al., 1991; Schüddekopf et al., 1993), cytochrome genes in *Rb. capsulatus* (Daldal et al., 1986; Jenny and Daldal, 1993) and *Rb. sphaeroides* (Rott and Donohue, 1990) and cytochrome *c* oxidase genes in *Rb. sphaeroides* (Shapleigh and Gennis, 1992; Shapleigh et al., 1992a) have also been found after deletion of the normally expressed genes. The ability of homologous genes to recombine under selection in vivo was used to generate a set of chimeric reaction centers in which deletions of the *puf*M gene in *Rb. capsulatus* were repaired by DNA donated from *Rb. sphaeroides* (Taguchi et al., 1993).

VI. Gene Expression

A. Gene Fusions

LacZ gene fusions have been used extensively to study expression in *Rb. capsulatus* and *Rb. sphaeroides* since the amount of the fusion protein can be measured by a quantitative colorimetric assay for β-galactosidase levels. Most commonly, specific translational fusions of the *lacZ* coding region are made near the beginning of a gene, then the construction is transferred back into the purple bacterium. Transcriptional fusions to the *lacZ* gene have also been used. After growth under the appropriate environmental conditions, the activity of β-galactosidase in cell extracts is determined. This technique is useful for identifying regions of DNA that contain promoter or regulatory regions (Belasco et al., 1985; Bauer et al., 1988, 1991; Adams et al., 1989; Tichy et al., 1989; Wellington and Beatty, 1989, 1991; Young et al., 1989; Hornberger et al., 1991; Wellington et al., 1991; Foster-Hartnett and Kranz, 1992; Preker et al., 1992; Beckman and Kranz, 1993; Duport et al., 1992; Foster-Hartnett et al., 1993; Lee and Kaplan, 1992a; LeBlanc and Beatty, 1993; Ma et al., 1993; Saeki et al., 1993; Tadros et al., 1993). *LacZ* gene fusions to open reading frames have helped to establish if the open reading frame is actually translated (Bauer et al., 1988; Foster-Hartnett et al., 1993). Gene fusions have also been used as reporters to define conditions under which genes are expressed (Kranz and Haselkorn, 1986; Bauer and Marrs, 1988; Forrest et al., 1989; Yang and Bauer, 1990; Hornberger et al., 1991; Hübner et al., 1991; Suetsugu et al., 1991; Wright et al., 1991; Colbeau and Vignais, 1992; Duport et al., 1992; Foster-Hartnett and Kranz, 1992; Armstrong et al., 1993; Forkl et al., 1993; Hübner et al., 1993; Buggy et al., 1994a,b;

Calero et al., 1994; Cullen et al., 1994; Eraso and Kaplan, 1994; Foster-Hartnett et al., 1994). As reporters of gene expression, fusions are also useful for studying the effects of mutations on the expression of genes (Moreno-Vivian et al., 1989a; Kranz et al., 1990; Hübner et al., 1991; Richaud et al., 1991; Colbeau and Vignais, 1992; Lee and Kaplan, 1992a,b; Preker et al., 1992; Schilke and Donohue, 1992; Sganga and Bauer, 1992; Colbeau et al., 1993; Gollan et al., 1993; Lee et al., 1993; Masepohl et al., 1993a; Mosley et al., 1994). In addition to assaying the effects of mutations, a *nifH::lacZ* gene fusion was used to isolate constitutive mutants in *Rb. capsulatus* by forcing the bacteria to grow on lactose under conditions where the *nif* genes are repressed (Kranz and Haselkorn, 1986). *LacZ* genes carried by Mu or miniMu bacteriophage transposons have been placed randomly into cloned fragments of DNA (Biel and Marrs, 1983; Avtges et al., 1985; Kranz and Haselkorn, 1985; Kranz, 1989) and have also been introduced into genomic DNA in *Rb. sphaeroides* (Nano and Kaplan, 1984). The expression of the *lac* operon in *Rb. sphaeroides* was studied by Nano and Kaplan (1982).

Promoterless antibiotic resistance genes provide a method for direct selection of regulatory regions. For example, transcriptional gene fusions to a kanamycin resistance marker were used to select for mutants involved in the oxygen regulation of the *puf* operon in *Rb. capsulatus* (Narro et al., 1990) and the *puc* operon in *Rb. sphaeroides* (Lee and Kaplan, 1992a,b) by selecting for kanamycin resistance under conditions where the genes are normally not expressed. A promoterless chloramphenicol transferase gene was used as a probe for promoter activity of DNA fragments in *Rb. sphaeroides* (Benning and Somerville, 1992b; Penfold and Pemberton, 1994). Expression of the *xylE* gene encoding catechol 2,3 dioxygenase has also been used as a reporter for promoter probe vectors in *Rb. sphaeroides* (Dryden and Kaplan, 1993; Xu and Tabita, 1994).

Alkaline phosphatase is a useful marker for cellular location since it is only active when secreted into the bacterial periplasm, and gene fusions with the alkaline phosphatase gene (*phoA*) have been used to study secretion and membrane topology of proteins in *Rb. capsulatus* and *Rb. sphaeroides*. The gene fusions are constructed either randomly on the chromosome or on a plasmid by a Tn*phoA* hybrid transposon (Moore and Kaplan, 1989; Brandner et al., 1991), or in specific alkaline phosphatase fusion vectors in which the *phoA* coding region is joined in frame to the amino terminal portion of a gene containing a potential signal sequence. This construction can be expressed in *E. coli* or in purple bacteria, and the activity of the alkaline phosphatase measured after cell fractionation. These types of fusions have been used to study the gene regions required for the synthesis of cytochrome c_2 (Varga and Kaplan, 1989; Brandner et al., 1991; Beckman et al., 1992; Beckman and Kranz, 1993). The topology of membrane proteins in *Rb. sphaeroides* has also been studied using *phoA* gene fusions (Yun et al., 1991b; Neidle and Kaplan, 1992). The validity of this approach was demonstrated by fusions with the gene encoding the L subunit of the reaction center, which showed that those fusions with high specific activity were in regions that are at or near the periplasmic surface as defined by the X-ray structure (Yun et al., 1991b).

An expression vector, that can be used to make transcriptional or translational fusions and allows a wide range of transcriptional levels by varying the nitrogen source in the medium, has been constructed using the *nifHDK* promoter from *Rb. capsulatus* (Pollock et al., 1988). This system was used to correlate the expression of the *pufQ* gene and the accumulation of bacteriochlorophyll (Bauer and Marrs, 1988) and to demonstrate that transcription of *pucC* is essential for the formation of the B800–850 complex (Tichy et al., 1991). A translational fusion to the beginning of the *nifH* gene in *Rb. capsulatus* was also used to confirm the function of a phytoene desaturase gene cloned from soybean (Bartley et al., 1991). A fructose dependent promoter from *Rb. capsulatus* has been used in another inducible expression vector (Duport et al., 1994).

B. In vitro Transcription-Translation

Cell-free transcription-translation systems can be used to study the expression of cloned genes. Cell extracts from *Rb. sphaeroides* capable of in vitro transcription-translation were described by Chory and Kaplan (1982) and were shown to have all the components necessary to express RubisCO genes from *Rb. sphaeroides* and *Rs. rubrum* (Chory et al., 1985). This system has been used to localize structural genes and regulatory regions by examination of the products from specific cloned DNA fragments for genes encoding reaction centers (Donohue et al., 1986a), cytochrome c_2 (Donohue et al., 1986b), phosphoribulokinase (Hallenbeck and Kaplan, 1987), antenna complexes (Kiley and Kaplan, 1987) and ALA synthase (Tai et al., 1988), and also to quantitatively characterize the products of the *puf* operon (Hoger et al., 1986; Kiley et al., 1987). Sensitivity of *Rb. sphaeroides* ribosomes to protein synthesis inhibitors was also studied in a cell-free protein synthesis system (Sánchez et al., 1994). A cell-free translation system from *Rb. capsulatus* was developed to study membrane assembly of proteins, and has been used to study membrane integration of antenna and reaction center proteins (Troschel and Müller, 1990; Troschel et al., 1992) and to define promoter activity for the ALA synthase gene (Hornberger et al., 1991). Signal peptidase, isolated from *Rb. capsulatus* membranes, has been shown to process a *Rb. sphaeroides* cytochrome c_2 precursor synthesized in an *E. coli* cell free transcription-translation system (Wieseler et al., 1992). An *E. coli* cell-free translation system was used to determine the purity of mRNA coding for antenna polypeptides from *Rs. rubrum* mRNA (Bélanger et al., 1985; Bérard et al., 1986), and RNA polymerase from *Rs. rubrum* was used in a partially defined *E. coli* system to define in vitro transcriptional start sites (Leustek et al., 1988). In addition, in vitro transcription systems have been reported using isolated RNA polymerase from *Rb. capsulatus* (Forrest and Beatty, 1987), *Cm. vinosum* (Valle et al., 1988) and *Rb. sphaeroides* (Kansy and Kaplan, 1989; Karls et al., 1994).

C. Interspecies Expression

Some genes cloned from photosynthetic bacteria have been expressed in *E. coli* or other hosts and the products identified by size, antibody reaction, or function. For example, RubisCO genes from *Rs. rubrum* (Somerville and Somerville, 1984; Larimer et al., 1986; Niyogi et al., 1986), *Rb. sphaeroides* (Muller et al., 1985; Gibson and Tabita, 1986) and *Cm. vinosum* (Viale et al., 1985, 1990; Kobayashi et al., 1991) have been expressed in *E. coli*, although expression is dependent on the use of *E. coli* promoters in front of the genes. In a double expression experiment, a regulatory protein from *Cm. vinosum* fused to the amino terminus of *lacZ* was expressed in *E. coli* and shown to affect the transcription of the gene for RubisCO from *Cm. vinosum* that was carried on a different plasmid (Viale et al., 1991). Other genes coding for enzymes involved in CO_2 fixation from *Rb. sphaeroides* have also been expressed in *E. coli* (Gibson and Tabita, 1987, 1988; Hallenbeck and Kaplan, 1987; Chen et al., 1991). Carotenoid biosynthesis genes from *Rb. sphaeroides* have been expressed in other hosts (Pemberton and Harding, 1987; Penfold and Pemberton, 1991; Hunter et al., 1994; Lang et al., 1994) and a fragment of the *crtI* gene from *Rb. capsulatus* was expressed in *E. coli* in order to generate antibodies (Bartley and Scolnik, 1989). Successful isolation of the PufQ protein from *E. coli* required fusion of the *pufQ* gene from *Rb. capsulatus* with an amino-terminal maltose binding domain and a protease recognition site, with subsequent proteolysis of the resulting fusion protein (Fidai et al., 1993). Additional genes cloned from photosynthetic bacteria that have been expressed in *E. coli* or other hosts include those coding for ALA synthase (Tai et al., 1988; Neidle and Kaplan, 1993b), the *Rsr*I restriction endonuclease (Stephenson et al., 1989), thioredoxin (Pille et al., 1990), ferredoxin (Grabau et al., 1991; Jouanneau et al., 1992; Agarwal et al., 1993; Armengaud et al., 1994), reaction centers (Söhlemann et al., 1991), bacteriochlorophyll biosynthesis enzymes (Coomber et al., 1992; Burke et al., 1993a; Bollivar et al., 1994c; Gibson and Hunter, 1994), PHB synthase (Hustede et al., 1992; Liebergesell and Steinbüchel, 1992), phosphatidylethanolamine methyltransferase (Arondel et al., 1993), chaperonin (Ferreyra et al., 1993) and nitrogenase (Durner et al., 1994). The decay of mRNA from *Rb. capsulatus* genes in *E. coli* was analyzed by Belasco et al. (1985).

The synthesis in other hosts of some proteins can be complicated, as demonstrated by several groups who have attempted to express cytochrome genes in *E. coli* and other species. Cytochrome c_2 from the *Rb. sphaeroides* gene was found in *E. coli* at low levels and only under anaerobic conditions (McEwan et al., 1989) and the cytochrome c_2 gene from *Rs. rubrum* was expressed but the cultures grew very poorly, indicating that the product may be toxic (Self et al., 1990). Expression of *c*-type cytochrome genes from *Rp. viridis* in *E. coli* resulted in aggregates of the apoprotein (Grisshammer et al., 1991), however the mature cytochrome was observed when synthesized in *Paracoccus denitrificans* (Gerhus et al., 1993). Similarly, when the cytochrome c_3 gene from the sulfate-reducing bacterium *Desulfovibrio vulgaris* was expressed in *E. coli* only the apoprotein was observed, but when this gene was expressed in *Rb. sphaeroides* the mature cytochrome was obtained (Cannac et al., 1991). Genes for *c*-type cytochromes from *Rb. capsulatus* have also been expressed in *Rb.*

sphaeroides (Caffrey et al., 1992; Jenny and Daldal, 1993).

RubisCO genes from various other species have been expressed in *Rb. sphaeroides* and *Rs. rubrum* (Falcone and Tabita, 1991, 1993). Other genes from exogenous sources expressed in photosynthetic bacteria include cellulase genes from *Cellulomonas fimi* in *Rb. capsulatus* (Johnson et al., 1986) and a carotenoid biosynthesis enzyme from soybean in *Rb. capsulatus* (Bartley et al., 1991). Genes from purple bacteria can generally be expressed in other related photosynthetic bacteria. For example genes encoding the subunits of the cytochrome bc_1 complex from *Rb. sphaeroides* and *Rs. rubrum* were expressed in *Rb. capsulatus* (Davidson et al., 1989; Shanker et al., 1992). Hybrid or chimeric reaction centers containing subunits from both *Rb. capsulatus* and *Rb. sphaeroides* have also been obtained (Zilsel et al., 1989; Taguchi et al., 1993).

VII. Conclusions

Given the diversity of metabolic processes in photosynthetic bacteria, the extent of genetic manipulation in these bacteria will almost certainly continue to increase. The variety of species being worked on is steadily expanding, as is the degree of sophistication in the techniques employed. A major change in the last fifteen years is the accessibility of genetic manipulation to a large number of laboratories. For example, the determination of the three dimensional structure of the reaction center has led to detailed mutational analyses of the role of amino acid residues in electron and proton transfer, even by researchers whose primary interests are in spectroscopy and physical chemistry. While some expertise in genetics is still required, the basic manipulations have become standard enough that they are readily obtainable. Those whose primary interest is in molecular biology are usually not limited by the techniques available, since a number of alternative methods exist for realizing most goals. In general, the emphasis has shifted from classical genetic approaches and understanding of recombination processes and genetic exchange mechanisms to more gene-specific topics, with a consequent loss of a more global organismic perspective. One of the most appealing aspects of the genetic approach lies in the possibility of detecting genic interactions that might not be discovered using a strictly molecular biological approach. It has been said that bacteria are endlessly surprising, so something new can always be expected to emerge from the study of purple bacterial genetics.

Acknowledgments

Work in our laboratories was supported by grant GM-45902 from the National Institutes of Health and grant MCB-9219378 from the National Science Foundation. This is publication No. 193 from the Arizona State University Center for the Study of Early Events in Photosynthesis.

References

Adams CW, Forrest ME, Cohen SN and Beatty JT (1989) Structural and functional analysis of transcriptional control of the *Rhodobacter capsulatus puf* operon. J Bacteriol 171: 473–482

Agarwal A, Tan J, Eren M, Tevelev A, Lui SM and Cowan JA (1993) Synthesis, cloning and expression of a synthetic gene for high potential iron protein from *Chromatium vinosum*. Biochem Biophys Res Comm 197: 1357–1362

Ahombo G, Willison JC and Vignais PM (1986) The *nifHDK* genes are contiguous with a *nifA*-like regulatory gene in *Rhodobacter capsulatus*. Mol Gen Genet 205: 442–445

Alias A, Cejudo FJ, Chabert J, Willison JC and Vignais PM (1989) Nucleotide sequence of wild-type and mutant *nifR4* (*ntrA*) genes of *Rhodobacter capsulatus*: Identification of an essential glycine residue. Nucleic Acids Res 17: 5377

Armengaud J, Meyer C and Jouanneau Y (1994) Recombinant expression of the *fdxD* gene of *Rhodobacter capsulatus* and characterization of its product, a [2Fe-2S] ferredoxin. Biochem J 300: 413–418

Armstrong GA, Alberti M, Leach F and Hearst JE (1989) Nucleotide sequence, organization, and nature of the protein products of the carotenoid biosynthesis gene cluster of *Rhodobacter capsulatus*. Mol Gen Genet 216: 254–268

Armstrong GA, Schmidt A, Sandmann G and Hearst JE (1990) Genetic and biochemical characterization of carotenoid biosynthesis mutants of *Rhodobacter capsulatus*. J Biol Chem 265: 8329–8338

Armstrong GA, Cook DN, Ma D, Alberti M, Burke DH and Hearst JE (1993) Regulation of carotenoid and bacteriochlorophyll biosynthesis genes and identification of an evolutionarily conserved gene required for bacteriochlorophyll accumulation. J Gen Microbiol 139: 897–906

Arnoux B, Ducruix A, Astier C, Picaud M, Roth M and Reiss-Husson F (1990) Towards the understanding of the function of *Rb. sphaeroides* Y wild type reaction center: Gene cloning, protein and detergent structures in the three-dimensional crystals. Biochimie 72: 525–530

Arondel V, Benning C and Somerville CR (1993) Isolation and functional expression in *Escherichia coli* of a gene encoding phosphatidylethanolamine methyltransferase (EC 2.1.1.17) from *Rhodobacter sphaeroides*. J Biol Chem 268: 16002–16008

Ashby MK, Coomber SA and Hunter CN (1987) Cloning, nucleotide sequence and transfer of genes for the B800–850 light harvesting complex of *Rhodobacter sphaeroides*. FEBS Lett 213: 245–248

Atta-Asafo-Adjei E and Daldal F (1991) Size of the amino acid side chain at position 158 of cytochrome *b* is critical for an active cytochrome bc_1 complex and for photosynthetic growth of *Rhodobacter capsulatus*. Proc Natl Acad Sci USA 88: 492–496

Avtges P, Scolnik PA and Haselkorn R (1983) Genetic and physical map of the structural genes (*nifH,D,K*) coding for the nitrogenase complex of *Rhodopseudomonas capsulata*. J Bacteriol 156: 251–256

Avtges P, Kranz RG and Haselkorn R (1985) Isolation and organization of genes for nitrogen fixation in *Rhodopseudomonas capsulata*. Mol Gen Genet 201: 363–369

Babst M, Albrecht H, Wegmann I, Brunisholz R and Zuber H (1991) Single amino acid substitutions in the B870 α and β light-harvesting polypeptides of *Rhodobacter capsulatus*. Eur J Biochem 202: 277–284

Barbé J, Gibert I, Llagostera M and Guerrero R (1987) DNA repair systems in the phototrophic bacterium *Rhodobacter capsulatus*. J Gen Microbiol 133: 961–966

Bartley GE and Scolnik PA (1989) Carotenoid biosynthesis in photosynthetic bacteria: Genetic characterization of the *Rhodobacter capsulatus* CrtI protein. J Biol Chem 264: 13109–13113

Bartley GE, Viitanen PV, Pecker I, Chamovitz D, Hirschberg J and Scolnik PA (1991) Molecular cloning and expression in photosynthetic bacteria of a soybean cDNA coding for phytoene desaturase, an enzyme of the carotenoid biosynthesis pathway. Proc Natl Acad Sci USA 88: 6532–6536

Barz WP and Oesterhelt D (1994) Photosynthetic deficiency of a *pufX* deletion mutant of *Rhodobacter sphaeroides* is suppressed by point mutations in the light-harvesting complex genes *pufB* or *pufA*. Biochemistry 33: 9741–9752

Bauer CE and Marrs BL (1988) *Rhodobacter capsulatus puf* operon encodes a regulatory protein (PufQ) for bacteriochlorophyll biosynthesis. Proc Natl Acad Sci 85: 7074–7078

Bauer CE, Young DA and Marrs BL (1988) Analysis of the *Rhodobacter capsulatus puf* operon: Location of the oxygen-regulated promoter region and the identification of an additional *puf*-encoded gene. J Biol Chem 263: 4820–4827

Bauer CE, Buggy JJ, Yang Z and Marrs BL (1991) The superoperonal organization of genes for pigment biosynthesis and reaction center proteins is a conserved feature in *Rhodobacter capsulatus*: Analysis of overlapping *bchB* and *puhA* transcripts. Mol Gen Genet 228: 433–444

Baumann M, Simon H, Schneider KH, Danneel HJ, Küster U and Giffhorn F (1989) Susceptibility of *Rhodobacter sphaeroides* to β-lactam antibiotics: Isolation and characterization of a periplasmic β-lactamase (cephalosporinase). J Bacteriol 171: 308–313

Beatty JT and Cohen SN (1983) Hybridization of cloned *Rhodopseudomonas capsulata* photosynthesis genes with DNA from other photosynthetic bacteria. J Bacteriol 154: 1440–1445

Becker-Rudzik M, Young DA and Marrs BL (1992) Sequence of the indoleglycerol phosphate synthase (*trpC*) gene from *Rhodobacter capsulatus*. J Bacteriol 174: 5482–5484

Beckman DL and Kranz RG (1991a) A bacterial homolog to the mitochondrial enoyl-CoA hydratase. Gene 107: 171–172

Beckman DL and Kranz RG (1991b) A bacterial homolog to HPRT. Biochim Biophys Acta 1129: 112–114

Beckman DL and Kranz RG (1993) Cytochromes *c* biogenesis in a photosynthetic bacterium requires a periplasmic thioredoxin-like protein. Proc Natl Acad Sci USA 90: 2179–2183

Beckman DL, Trawick DR and Kranz RG (1992) Bacterial cytochromes *c* biogenesis. Genes Dev. 6: 268–283.

Bélanger G, Bérard J and Gingras G (1985) Isolation and partial characterization of the messenger RNA encoding the B880 holochrome protein of *Rhodospirillum rubrum*. Eur J Biochem 153: 477–484

Bélanger G, Bérard J, Corriveau P and Gingras G (1988) The structural genes coding for the L and M subunits of *Rhodospirillum rubrum* photoreaction center. J Biol Chem 263: 7632–7638

Belasco JG, Beatty JT, Adams CW, von Gabain A and Cohen SN (1985) Differential expression of photosynthesis genes in *R. capsulata* results from segmental differences in stability within the polycistronic *rxcA* transcript. Cell 40: 171–181

Benning C and Somerville CR (1992a) Isolation and genetic complementation of a sulfolipid-deficient mutant of *Rhodobacter sphaeroides*. J Bacteriol 174: 2352–2360

Benning C and Somerville CR (1992b) Identification of an operon involved in sulfolipid biosynthesis in *Rhodobacter sphaeroides*. J Bacteriol 174: 6479–6487

Bérard J and Gingras G (1991) The *puh* structural gene coding for the H subunit of the *Rhodospirillum rubrum* photoreaction center. Biochem Cell Biol 69: 122–131

Bérard J, Bélanger G, Corriveau P and Gingras G (1986) Molecular cloning and sequence of the B880 holochrome gene from *Rhodospirillum rubrum*. J Biol Chem 261: 82–87

Biel AJ and Marrs BL (1983) Transcriptional regulation of several genes for bacteriochlorophyll biosynthesis in *Rhodopseudomonas capsulata* in response to oxygen. J Bacteriol 156: 686–694

Biel SW and Biel AJ (1990) Isolation of a *Rhodobacter capsulatus* mutant that lacks *c*-type cytochromes and excretes porphyrins. J Bacteriol 172: 1321–1326

Biel SW, Wright MS and Biel AJ (1988) Cloning of the *Rhodobacter capsulatus hemA* gene. J Bacteriol 170: 4382–4384

Bollivar DW and Bauer CE (1992) Nucleotide sequence of S-adenosyl-L-methionine: Magnesium protoporphyrin methyltransferase from *Rhodobacter capsulatus*. Plant Physiol 98: 408–410

Bollivar DW, Jiang ZY, Bauer CE and Beale SI (1994a) Heterologous expression of the *bchM* gene product from *Rhodobacter capsulatus* and demonstration that it encodes *S*-adenosyl-L-methionine:Mg-protoporphyrin IX methyltransferase. J Bacteriol 176: 5290–5296

Bollivar DW, Suzuki JY, Beatty JT, Dobrowolski JM and Bauer CE (1994b) Directed mutational analysis of bacteriochlorophyll *a* biosynthesis in *Rhodobacter capsulatus*. J Mol Biol 237: 622–640

Bollivar DW, Wang S, Allen JP and Bauer CE (1994c) Molecular genetic analysis of terminal steps in bacteriochlorophyll *a* biosynthesis: Characterization of a *Rhodobacter capsulatus* strain that synthesizes geranylgeraniol-esterified bacteriochlorophyll *a*. Biochemistry 33: 12763–12768

Bowen ARSG and Pemberton JM (1985) Mercury resistance transposon Tn*813* mediates chromosome transfer in *Rhodopseudomonas sphaeroides* and intergeneric transfer of pBR322. In: Helinski DR, Cohen SN, Clewell DB, Jackson DA and Hollaender A (eds) Plasmids in Bacteria, pp 105–115. Plenum, New York

Brandner JP and Donohue TJ (1994) The *Rhodobacter sphaeroides* cytochrome c_2 signal peptide is not necessary for export and heme attachment. J Bacteriol 176: 602–609

Brandner JP, McEwan AG, Kaplan S and Donohue TJ (1989) Expression of the *Rhodobacter sphaeroides* cytochrome c_2 structural gene. J Bacteriol 171: 360–368

Brandner JP, Stabb EV, Temme R and Donohue TJ (1991) Regions of *Rhodobacter sphaeroides* cytochrome c_2 required for export, heme attachment, and function. J Bacteriol 173: 3958–3965

Buggy JJ, Sganga MW and Bauer CE (1994a) Nucleotide sequence and characterization of the *Rhodobacter capsulatus hvrB* gene: HvrB is an activator of *S*-Adenosyl-L-homocysteine hydrolase expression and is a member of the LysR family. J Bacteriol 176: 61–69

Buggy JJ, Sganga MW and Bauer CE (1994b) Characterization of a light-responding *trans*-activator responsible for differentially controlling reaction center and light-harvesting-I gene expression in *Rhodobacter capsulatus*. J Bacteriol 176: 6936–6943

Burgess JG, Ashby MK and Hunter CN (1989) Characterization and complementation of a mutant of *Rhodobacter sphaeroides* with a chromosomal deletion in the light-harvesting (LH2) genes. J Gen Microbiol 135: 1809–1816

Burke DH, Alberti M and Hearst JE (1993a) The *Rhodobacter capsulatus* chlorin reductase-encoding locus, *bchA*, consists of three genes, *bchX*, *bchY*, and *bchZ*. J Bacteriol 175: 2407–2413

Burke DH, Alberti M and Hearst JE (1993b) *bchFNBH* bacteriochlorophyll synthesis genes of *Rhodobacter capsulatus* and identification of the third subunit of light-independent protochlorophyllide reductase in bacteria and plants. J Bacteriol 175: 2414–2422

Bylina EJ and Youvan DC (1988) Directed mutations affecting spectroscopic and electron transfer properties of the primary donor in the photosynthetic reaction center. Proc Natl Acad Sci USA 85: 7226–7230

Bylina EJ, Ismail S and Youvan DC (1986) Plasmid pU29, a vehicle for mutagenesis of the photosynthetic *puf* operon in *Rhodopseudomonas capsulata*. Plasmid 16: 175–181

Bylina EJ, Robles SJ and Youvan DC (1988) Directed mutations affecting the putative bacteriochlorophyll-binding sites in the light-harvesting I antenna of *Rhodobacter capsulatus*. Isr J Chem 28: 73–78

Bylina EJ, Jovine RVM and Youvan DC (1989) A genetic system for rapidly assessing herbicides that compete for the quinone binding site of photosynthetic reaction centers. Bio/Technology 7: 69–74

Caffrey MS, Daldal F, Holden HM and Cusanovich MA (1991) Importance of a conserved hydrogen-bonding network in cytochromes *c* to their redox potentials and stabilities. Biochemistry 30: 4119–4125

Caffrey M, Davidson E, Cusanovich M and Daldal F (1992) Cytochrome c_2 mutants of *Rhodobacter capsulatus*. Arch Biochem Biophys 292: 419–426

Calero S, Fernandez de Henestrosa AR and Barbé J (1994) Molecular cloning, sequence and regulation of expression of the *recA* gene of the phototrophic bacterium *Rhodobacter sphaeroides*. Mol Gen Genet 242: 116–120

Campbell MA, Scahill S, Gibson T and Ambler RP (1989) The phototrophic bacterium *Rhodopseudomonas capsulata* sp108 encodes an indigenous class A *β*-lactamase. Biochem J 260: 803–812

Cannac V, Caffrey MS, Voordouw G and Cusanovich MA (1991) Expression of the gene encoding cytochrome c_3 from the sulfate-reducing bacterium *Desulfovibrio vulgaris* in the purple photosynthetic bacterium *Rhodobacter sphaeroides*. Arch Biochem Biophys 286: 629–632

Cao J, Shapleigh J, Gennis R, Revzin A and Ferguson-Miller S (1991) The gene encoding cytochrome *c* oxidase subunit II from *Rhodobacter sphaeroides*; comparison of the deduced amino acid sequence with sequences of corresponding peptides from other species. Gene 101: 133–137

Cao J, Hosler J, Shapleigh J, Revzin A and Ferguson-Miller S (1992) Cytochrome aa_3 of *Rhodobacter sphaeroides* as a model for mitochondrial cytochrome *c* oxidase: The *coxII/coxIII* operon codes for structural and assembly proteins homologous to those in yeast. J Biol Chem 267: 24273–24278

Cauvin B, Colbeau A and Vignais PM (1991) The hydrogenase structural operon in *Rhodobacter capsulatus* contains a third gene, *hup*M, necessary for the formation of a physiologically competent hydrogenase. Mol Microbiol 5: 2519–2527

Chen CYA and Belasco JG (1990) Degradation of *pufLMX* mRNA in *Rhodobacter capsulatus* is initiated by nonrandom endonucleolytic cleavage. J Bacteriol 172: 4578–4586

Chen CYA, Beatty JT, Cohen SN and Belasco JG (1988) An intercistronic stem-loop structure functions as an mRNA decay terminator necessary but insufficient for *puf* mRNA stability. Cell 52: 609–619

Chen JH, Gibson JL, McCue LA and Tabita FR (1991) Identification, expression and deduced primary structure of transketolase and other enzymes encoded within the form II CO_2 fixation operon of *Rhodobacter sphaeroides*. J Biol Chem 266: 20447–20452

Chen YR, Usui S, Yu CA and Yu L (1994) Role of subunit IV in the cytochrome *b*-c_1 complex from *Rhodobacter sphaeroides*. Biochemistry 33: 10207–10214

Chory J and Kaplan S (1982) The in vitro transcription-translation of DNA and RNA templates by extracts of *Rhodopseudomonas sphaeroides*: Optimization and comparison of template specificity with *Escherichia coli* extracts and in vivo synthesis. J Biol Chem 257: 15110–15121

Chory J, Muller ED and Kaplan S (1985) DNA-directed in vitro synthesis and assembly of the form II D-ribulose-1,5-bisphosphate carboxylase/oxygenase from *Rhodopseudomonas sphaeroides*. J Bacteriol 161: 307–313

Choudhary M, Mackenzie C, Nereng KS, Sodergren E, Weinstock GM and Kaplan S (1994) Multiple chromosomes in bacteria: Structure and function of chromosome II of *Rhodobacter sphaeroides* 2.4.1^T. J Bacteriol 176: 7694–7702

Clarke L and Carbon J (1976) A colony bank containing synthetic Col EI hybrid plasmids representative of the entire *E. coli* genome. Cell 9: 91–99

Clewell DB (1993) Bacterial Conjugation. Plenum Press, New York.

Colbeau A and Vignais PM (1992) Use of *hupS*::*lacZ* gene fusion to study regulation of hydrogenase expression in *Rhodobacter capsulatus*: Stimulation by H_2. J Bacteriol 174: 4258–4264

Colbeau A, Godfroy A and Vignais PM (1986) Cloning of DNA fragments carrying hydrogenase genes of *Rhodopseudomonas capsulata*. Biochimie 68: 147–155

Colbeau A, Magnin JP, Cauvin B, Champion T and Vignais PM (1990) Genetic and physical mapping of an hydrogenase gene cluster from *Rhodobacter capsulatus*. Mol Gen Genet 220: 393–399

Colbeau A, Richaud P, Toussaint B, Caballero FJ, Elster C, Delphin C, Smith RL, Chabert J and Vignais PM (1993) Organization of the genes necessary for hydrogenase expression in *Rhodobacter capsulatus*. sequence analysis and identification of two *hyp* regulatory mutants. Mol Microbiol 8: 15–29

Cook DN, Armstrong GA and Hearst JE (1989) Induction of anaerobic gene expression in *Rhodobacter capsulatus* is not accompanied by a local change in chromosomal supercoiling as measured by a novel assay. J Bacteriol 171: 4836–4843

Coomber SA and Hunter CN (1989) Construction of a physical map of the 45 kb photosynthetic gene cluster of *Rhodobacter sphaeroides*. Arch Microbiol 151: 454–458

Coomber SA, Chaudhri M, Connor A, Britton G and Hunter CN (1990) Localized transposon Tn*5* mutagenesis of the photosynthetic gene cluster of *Rhodobacter sphaeroides*. Mol Microbiol 4: 977–989

Coomber SA, Jones RM, Jordan PM and Hunter CN (1992) A putative anaerobic coproporphyrinogen III oxidase in *Rhodobacter sphaeroides*. I. molecular cloning, transposon mutagenesis and sequence analysis of the gene. Mol Microbiol 6: 3159–3169

Cullen PJ, Foster-Hartnett D, Gabbert KK and Kranz RG (1994) Structure and expression of the alternative sigma factor, RpoN, in *Rhodobacter capsulatus*; physiological relevance of an autoactivated *nifU2-rpoN* superoperon. Mol Microbiol 11: 51–65

Daldal F (1988) Cytochrome c_2-independent respiratory growth of *Rhodobacter capsulatus*. J Bacteriol 170: 2388–2391

Daldal F, Cheng S, Applebaum J, Davidson E and Prince RC (1986) Cytochrome c_2 is not essential for photosynthetic growth of *Rhodopseudomonas capsulata*. Proc Natl Acad Sci USA 83: 2012–2016

Daldal F, Davidson E and Cheng S (1987) Isolation of the structural genes for the Rieske Fe-S protein, cytochrome *b* and cytochrome c_1, all components of the ubiquinol:cytochrome c_2 oxidoreductase complex of *Rhodopseudomonas capsulata*. J Mol Biol 195: 1–12

Daldal F, Tokito MK, Davidson E and Faham M (1989) Mutations conferring resistance to quinol oxidation (Q_z) inhibitors of the Cyt bc_1 complex of *Rhodobacter capsulatus*. EMBO J 8: 3951–3961

Daniels GA, Drews G and Saier MH (1988) Properties of a Tn*5* insertion mutant defective in the structural gene (*fruA*) of the fructose-specific phosphotransferase system of *Rhodobacter capsulatus* and cloning of the *fru* regulon. J Bacteriol 170: 1698–1703

Davidson E and Daldal F (1987a) Primary structure of the bc_1 complex of *Rhodopseudomonas capsulata*: Nucleotide sequence of the *pet* operon encoding the Rieske, cytochrome *b*, and cytochrome c_1 apoproteins. J Mol Biol 195: 13–24

Davidson E and Daldal F (1987b) *fbc* operon, encoding the Rieske Fe-S protein, cytochrome *b*, and cytochrome c_1 apoproteins previously described from *Rhodopseudomonas sphaeroides*, is from *Rhodopseudomonas capsulata*. J Mol Biol 195: 25–29

Davidson E, Prince RC, Haith CE and Daldal F (1989) The cytochrome bc_1 complex of *Rhodobacter sphaeroides* can restore cytochrome c_2-independent photosynthetic growth to a *Rhodobacter capsulatus* mutant lacking cytochrome bc_1. J Bacteriol 171: 6059–6068

Davidson E, Ohnishi T, Atta-Asafo-Adjei E and Daldal F (1992) Potential ligands to the [2Fe-2S] Rieske cluster of the cytochrome bc_1 complex of *Rhodobacter capsulatus* probed by site-directed mutagenesis. Biochemistry 31: 3342–3351

Davis J, Donohue TJ and Kaplan S (1988) Construction, characterization and complementation of a Puf⁻ mutant of *Rhodobacter sphaeroides*. J Bacteriol 170: 320–329

DeHoff BS, Lee JK, Donohue TJ, Gumport RI and Kaplan S (1988) In vivo analysis of *puf* operon expression in *Rhodobacter sphaeroides* after deletion of a putative intercistronic transcription terminator. J Bacteriol 170: 4681–4692

Dispensa M, Thomas CT, Kim MK, Perrotta JA, Gibson J and Harwood CS (1992) Anaerobic growth of *Rhodopseudomonas palustris* on 4-hydroxybenzoate is dependent on AadR, a member of the cyclic AMP receptor protein family of transcriptional regulators. J Bacteriol 174: 5803–5813

Ditta G, Stanfield S, Corbin D and Helinski DR (1980) Broad host range DNA cloning system for Gram-negative bacteria: Construction of a gene bank of *Rhizobium meliloti*. Proc Natl Acad Sci USA 77: 7347–7351

Ditta G, Schmidhauser T, Yakobson E, Lu P, Liang XW, Finlay DR, Guiney D and Helinski DR (1985) Plasmids related to the broad host range vector, pRK290, useful for gene cloning and for monitoring gene expression. Plasmid 13: 149–153

Dolata MM, Van Beeumen JJ, Ambler RP, Meyer TE and Cusanovich MA (1993) Nucleotide sequence of the heme subunit of flavocytochrome *c* from the purple phototrophic bacterium, *Chromatium vinosum*: A 2.6-kilobase pair DNA fragment contains two multiheme cytochromes, a flavoprotein, and a homolog of human ankyrin. J Biol Chem 268: 14426–14431

Donohue TJ and Kaplan S (1991) Genetic techniques in *Rhodospirillaceae*. Methods Enzymol 204: 459–485

Donohue TJ, Hoger JH and Kaplan S (1986a) Cloning and expression of the *Rhodobacter sphaeroides* reaction center H gene. J Bacteriol 168: 953–961

Donohue TJ, McEwan AG and Kaplan S (1986b) Cloning, DNA sequence, and expression of the *Rhodobacter sphaeroides* cytochrome c_2 gene. J Bacteriol 168: 962–972

Donohue TJ, McEwan AG, Van Doren S, Crofts AR and Kaplan S (1988) Phenotypic and genetic characterization of cytochrome c_2 deficient mutants of *Rhodobacter sphaeroides*. Biochemistry 27: 1918–1925

Dörge B, Klug G, Gad'on N, Cohen SN and Drews G (1990) Effects on the formation of antenna complex B870 of *Rhodobacter capsulatus* by exchange of charged amino acids in the N-terminal domain of the α and β pigment-binding proteins. Biochemistry 29: 7754–7758

Dryden SC and Kaplan S (1990) Localization and structural analysis of the ribosomal RNA operons of *Rhodobacter sphaeroides*. Nucleic Acids Res 18: 7267–7277

Dryden SC and Kaplan S (1993) Identification of *cis*-acting

regulatory regions upstream of the rRNA operons of *Rhodobacter sphaeroides*. J Bacteriol 175: 6392–6402
Duport C, Jouanneau Y and Vignais PM (1990) Nucleotide sequence of *fdxA* encoding a 7Fe ferredoxin of *Rhodobacter capsulatus*. Nucleic Acids Res 18: 4618
Duport C, Jouanneau Y and Vignais PM (1992) Transcriptional analysis and promoter mapping of the *fdxA* gene which encodes the 7Fe ferredoxin (FdII) of *Rhodobacter capsulatus*. Mol Gen Genet 231: 323–328
Duport C, Meyer C, Naud I and Jouanneau Y (1994) A new gene expression system based on a fructose-dependent promoter from *Rhodobacter capsulatus*. Gene 145: 103–108
Dupuis A (1992) Identification of two genes of *Rhodobacter capsulatus* coding for proteins homologous to the ND1 and 23 kDa subunits of the mitochondrial complex I. FEBS Lett 301: 215–218
Durner J, Böhm I, Hilz H and Böger P (1994) Posttranslational modification of nitrogenase: Differences between the purple bacterium *Rhodospirillum rubrum* and the cyanobacterium *Anabaena variabilis*. Eur J Biochem 220: 125–130
Elder DJE, Morgan P and Kelly DJ (1993) Transposon Tn*5* mutagenesis in *Rhodopseudomonas palustris*. FEMS Microbiol Lett 111: 23–30
Elsen S, Richaud P, Colbeau A and Vignais PM (1993) Sequence analysis and interposon mutagenesis of the *hupT* gene, which encodes a sensor protein involved in repression of hydrogenase synthesis in *Rhodobacter capsulatus*. J Bacteriol 175: 7404–7412
Eraso JM and Kaplan S (1994) *prrA*, a putative response regulator involved in oxygen regulation of photosynthesis gene expression in *Rhodobacter sphaeroides*. J Bacteriol 176: 32–43
Falcone DL and Tabita FR (1991) Expression of endogenous and foreign ribulose 1,5-bisphosphate carboxylase-oxygenase (RubisCO) genes in a RubisCO deletion mutant of *Rhodobacter sphaeroides*. J Bacteriol 173: 2099–2108
Falcone DL and Tabita FR (1993) Complementation analysis and regulation of CO_2 fixation gene expression in a ribulose 1,5-bisphosphate carboxylase-oxygenase deletion strain of *Rhodospirillum rubrum*. J Bacteriol 175: 5066–5077
Falcone DL, Quivey RG and Tabita FR (1988) Transposon mutagenesis and physiological analysis of strains containing inactivated form I and form II ribulose bisphosphate carboxylase/oxygenase genes in *Rhodobacter sphaeroides*. J Bacteriol 170: 5–11
Falk G and Walker JE (1985) Transcription of *Rhodospirillum rubrum atp* operon. Biochem J 229: 663-668
Falk G and Walker JE (1988) DNA sequence of a gene cluster coding for subunits of the F_0 membrane sector of ATP synthase in *Rhodospirillum rubrum*: Support for modular evolution of the F_1 and F_0 sectors. Biochem J 254: 109-122
Falk G, Hampe A and Walker JE (1985) Nucleotide sequence of the *Rhodospirillum rubrum atp* operon. Biochem J 228: 391–407
Farchaus JW and Oesterhelt D (1989) A *Rhodobacter sphaeroides puf L, M*, and *X* deletion mutant and its complementation in *trans* with a 5.3 kb *puf* operon shuttle fragment. EMBO J 8: 47–54
Farchaus JW, Gruenberg H and Oesterhelt D (1990) Complementation of a reaction center-deficient *Rhodobacter sphaeroides pufLMX* deletion strain in *trans* with *pufBALM* does not restore the photosynthesis-positive phenotype. J Bacteriol 172: 977–985
Farchaus JW, Barz WP, Grünberg H and Oesterhelt D (1992) Studies on the expression of the *pufX* polypeptide and its requirement for photoheterotrophic growth in *Rhodobacter sphaeroides*. EMBO J 11: 2779–2788
Farchaus JW, Wachtveitl J, Mathis P and Oesterhelt D (1993) Tyrosine 162 of the photosynthetic reaction center L-subunit plays a critical role in the cytochrome c_2 mediated rereduction of the phoooxidized bacteriochlorophyll dimer in *Rhodobacter sphaeroides*. 1. Site-directed mutagenesis and initial characterization. Biochemistry 32: 10885–10893
Ferreyra RG, Soncini FC and Viale AM (1993) Cloning, characterization, and functional expression in *Escherichia coli* of chaperonin (*groESL*) genes from the phototrophic sulfur bacterium *Chromatium vinosum*. J Bacteriol 175: 1514–1523
Fidai S, Kalmar GB, Richards WR and Borgford TJ (1993) Recombinant expression of the *pufQ* gene of *Rhodobacter capsulatus*. J Bacteriol 175: 4834–4842
Fitzmaurice WP and Roberts GP (1991) Artificial DNA-mediated genetic transformation of the photosynthetic nitrogen-fixing bacterium *Rhodospirillum rubrum*. Arch Microbiol 156: 142–144
Fitzmaurice WP, Saari LL, Lowery RG, Ludden PW and Roberts GP (1989) Genes coding for the reversible ADP-ribosylation system of dinitrogenase reductase from *Rhodospirillum rubrum*. Mol Gen Genet 218: 340–347
Fonstein M and Haselkorn R (1993) Chromosomal structure of *Rhodobacter capsulatus* strain SB1003: Cosmid encyclopedia and high-resolution physical and genetic map. Proc Natl Acad Sci USA 90: 2522–2526
Fonstein M, Zheng S and Haselkorn R (1992) Physical map of the genome of *Rhodobacter capsulatus* SB1003. J Bacteriol 174: 4070–4077
Forkl H, Vandekerckhove J, Drews G and Tadros MH (1993) Molecular cloning, sequence analysis and expression of the gene for catalase-peroxidase (*cpeA*) from the photosynthetic bacterium, *Rhodobacter capsulatus* B10. Eur J Biochem 214: 251–258
Fornari CS and Kaplan S (1982) Genetic transformation of *Rhodopseudomonas sphaeroides* by plasmid DNA. J Bacteriol 152: 89–97
Fornari CS and Kaplan S (1983) Identification of nitrogenase and carboxylase genes in the photosynthetic bacteria and cloning of a carboxylase gene from *Rhodopseudomonas sphaeroides*. Gene 25: 291–299
Forrest ME and Beatty JT (1987) Purification of *Rhodobacter capsulatus* RNA polymerase and its use for in vitro transcription. FEBS Lett 212: 28–34
Forrest ME, Zucconi AP and Beatty JT (1989) The *pufQ* gene product of *Rhodobacter capsulatus* is essential for formation of B800-850 light-harvesting complexes. Current Microbiol 19: 123–127
Foster-Hartnett D and Kranz RG (1992) Analysis of the promoters and upstream sequences of *nifA1* and *nifA2* in *Rhodobacter capsulatus*; activation requires *ntrC* but not *rpoN*. Mol Microbiol 6: 1049–1060
Foster-Hartnett D and Kranz RG (1994) The *Rhodobacter capsulatus glnB* gene is regulated by NtrC at tandem *rpoN*-independent promoters. J Bacteriol 176: 5171–5176
Foster-Hartnett D, Cullen PJ, Gabbert KK and Kranz RG (1993)

Sequence, genetic, and *lacZ* fusion analyses of a *nifR3-ntrB-ntrC* operon in *Rhodobacter capsulatus*. Mol Microbiol 8: 903–914

Fowler GJS, Visschers RW, Grief GG, van Grondelle R and Hunter CN (1992) Genetically modified photosynthetic antenna complexes with blueshifted absorbance bands. Nature 355: 848–850

Fowler GJS, Crielaard W, Visschers RW, van Grondelle R and Hunter CN (1993) Site-directed mutagenesis of the LH2 light-harvesting complex of *Rhodobacter sphaeroides*: Changing *β*Lys23 to Gln results in a shift in the 850 nm absorption peak. Photochem Photobiol 57: 2–5

Gardiner AT, MacKenzie RC, Barrett SJ, Kaiser K and Cogdell RJ (1992) The genes for the peripheral antenna complex apoproteins from *Rhodopseudomonas acidophila* 7050 form a multigene family. In: Murata N (ed) Research in Photosynthesis, Vol. I, pp 77–80. Kluwer, Dordrecht

Garí E, Gibert I and Barbé J (1992a) Spontaneous and reversible high-frequency frameshifts originating a phase transition in the carotenoid biosynthesis pathway of the phototrophic bacterium *Rhodobacter sphaeroides* 2.4.1. Mol Gen Genet 232: 74–80

Garí E, Toledo JC, Gibert I and Barbé J (1992b) Nucleotide sequence of the methoxyneurosporene dehydrogenase gene from *Rhodobacter sphaeroides*: Comparison with other bacterial carotenoid dehydrogenases. FEMS Microbiol Lett 93: 103–108

Genther FJ and Wall JD (1984) Isolation of a recombination-deficient mutant of *Rhodopseudomonas capsulata*. J Bacteriol 160: 971–975

Gerhus E, Grisshammer R, Michel H, Ludwig B and Turba A (1993) Synthesis of the *Rhodopseudomonas viridis* holo-cytochrome c_2 in *Paracoccus denitrificans*. FEMS Microbiol Lett 113: 29–34

Ghosh R, Elder DJE, Saegesser R, Kelly DJ and Bachofen R (1994) An improved procedure and new vectors for transposon Tn*5* mutagenesis of the phototrophic bacterium *Rhodospirillum rubrum*. Gene 150: 97–100

Gibson JL and Tabita FR (1986) Isolation of the *Rhodopseudomonas sphaeroides* form I ribulose 1,5-bisphosphate carboxylase/oxygenase large and small subunit genes and expression of the active hexadecameric enzyme in *Escherichia coli*. Gene 44: 271–278

Gibson JL and Tabita FR (1987) Organization of phosphoribulokinase and ribulose bisphosphate carboxylase/oxygenase genes in *Rhodopseudomonas* (*Rhodobacter*) *sphaeroides*. J Bacteriol 169: 3685–3690

Gibson JL and Tabita FR (1988) Localization and mapping of CO_2 fixation genes within two gene clusters in *Rhodobacter sphaeroides*. J Bacteriol 170: 2153–2158

Gibson JL and Tabita FR (1993) Nucleotide sequence and functional analysis of CbbR, a positive regulator of the Calvin cycle operons of *Rhodobacter sphaeroides*. J Bacteriol 175: 5778–5784

Gibson JL, Chen JH, Tower PA and Tabita FR (1990) The form II fructose 1,6-bisphosphatase and phosphoribulokinase genes from part of a large operon in *Rhodobacter sphaeroides*: Primary structure and insertional mutagenesis analysis. Biochemistry 29: 8085–8093

Gibson JL, Falcone DL and Tabita FR (1991) Nucleotide sequence, transcriptional analysis and expression of genes encoded within the form I CO_2 fixation operon of *Rhodobacter sphaeroides*. J Biol Chem 266: 14646–14653

Gibson J, Dispensa M, Fogg GC, Evans DT and Harwood CS (1994) 4-hydroxybenzoate-coenzyme A ligase from *Rhodopseudomonas palustris*: Purification, gene sequence, and role in anaerobic degradation. J Bacteriol 176: 634–641

Gibson LCD and Hunter CN (1994) The bacteriochlorophyll biosynthesis gene, *bchM*, of *Rhodobacter sphaeroides* encodes *S*-adenosyl-L-methionine:Mg protoporphyrin IX methyltransferase. FEBS Lett 352: 127–130

Gibson LCD, McGlynn P, Chaudhri M and Hunter CN (1992) A putative anaerobic coproporphyrinogen III oxidase in *Rhodobacter sphaeroides*. II. analysis of a region of the genome encoding *hemF* and the *puc* operon. Mol Microbiol 6: 3171–3186

Giuliano G, Pollock D and Scolnik PA (1986) The gene *crtI* mediates the conversion of phytoene into colored carotenoids in *Rhodopseudomonas capsulata*. J Biol Chem 261: 12925–12929

Giuliano G, Pollock D, Stapp H and Scolnik PA (1988) A genetic-physical map of the *Rhodobacter capsulatus* carotenoid biosynthesis gene cluster. Mol Gen Genet 213: 78–83

Goldman ER and Youvan DC (1992) An algorithmically optimized combinatorial library screened by digital imaging spectroscopy. Bio/Technology 10: 1557–1561

Gollan U, Schneider K, Müller A, Schüddekopf K and Klipp W (1993) Detection of the in vivo incorporation of a metal cluster into a protein: The FeMo cofactor is inserted into the FeFe protein of the alternative nitrogenase of *Rhodobacter capsulatus*. Eur J Biochem 215: 25–35

Gong L, Lee JK and Kaplan S (1994) The *Q* gene of *Rhodobacter sphaeroides*: Its role in *puf* operon expression and spectral complex assembly. J Bacteriol 176: 2946–2961

Grabau C, Schatt E, Jouanneau Y and Vignais PM (1991) A new [2Fe-2S] ferredoxin from *Rhodobacter capsulatus*: Coexpression with a 2[4Fe-4S] ferredoxin in *Escherichia coli*. J Biol Chem 266: 3294–3299

Gray KA, Farchaus JW, Wachtveitl J, Breton J and Oesterhelt D (1990) Initial characterization of site-directed mutants of tyrosine M210 in the reaction centre of *Rhodobacter sphaeroides*. EMBO J 9: 2061–2070

Grisshammer R, Wiessner C and Michel H (1990) Sequence analysis and transcriptional organization of the *Rhodopseudomonas viridis* cytochrome c_2 gene. J Bacteriol 172: 5071–5078

Grisshammer R, Oeckl C and Michel H (1991) Expression in *Escherichia coli* of *c*-type cytochrome genes from *Rhodopseudomonas viridis*. Biochim Biophys Acta 1088: 183–190

Hacker B, Barquera B, Crofts AR and Gennis RB (1993) Characterization of mutations in the cytochrome *b* subunit of the bc_1 complex of *Rhodobacter sphaeroides* that affect the quinone reductase site (Q_c). Biochemistry 32: 4403–4410

Hallenbeck PL and Kaplan S (1987) Cloning of the gene for phosphoribulokinase activity from *Rhodobacter sphaeroides* and its expression in *Escherichia coli*. J Bacteriol 169: 3669–3678

Hallenbeck PL, Lerchen R, Hessler P and Kaplan S (1990a) Roles of CfxA, CfxB, and external electron acceptors in regulation of ribulose 1,5-bisphosphate carboxylase/oxygenase expression in *Rhodobacter sphaeroides*. J Bacteriol 172: 1736–1748

Hallenbeck PL, Lerchen R, Hessler P and Kaplan S (1990b) Phosphoribulokinase activity and regulation of CO_2 fixation critical for photosynthetic growth of *Rhodobacter sphaeroides*. J Bacteriol 172: 1749–1761

Hamblin MJ, Shaw JG, Curson JP and Kelly DJ (1990) Mutagenesis, cloning and complementation analysis of C_4-dicarboxylate transport genes from *Rhodobacter capsulatus*. Mol Microbiol 4: 1567–1574

Hamblin MJ, Shaw JG and Kelly DJ (1993) Sequence analysis and interposon mutagenesis of a sensor-kinase (DctS) and response-regulator (DctR) controlling synthesis of the high-affinity C4-dicarboxylate transport system in *Rhodobacter capsulatus*. Mol Gen Genet 237: 215–224

Hanson DK, Nance SL and Schiffer M (1992) Second-site mutation at M43 (Asn → Asp) compensates for the loss of two acidic residues in the Q_B site of the reaction center. Photosynthesis Res 32: 147–153

Hessner MJ, Wejksnora PJ and Collins MLP (1991) Construction, characterization and complementation of *Rhodospirillum rubrum puf* region mutants. J Bacteriol 173: 5712–5722

Hoger JH, Chory J and Kaplan S (1986) In vitro biosynthesis and membrane association of photosynthetic reaction center subunits from *Rhodopseudomonas sphaeroides*. J Bacteriol 165: 942–950

Höpfl P, Ludwig W and Schleifer KH (1988) Complete nucleotide sequence of a 23S ribosomal RNA gene from *Rhodobacter capsulatus*. Nucleic Acids Res 16: 2343

Hornberger U, Liebetanz R, Tichy HV and Drews G (1990) Cloning and sequencing of the *hemA* gene of *Rhodobacter capsulatus* and isolation of a δ-aminolevulinic acid-dependent mutant strain. Mol Gen Genet 221: 371–378

Hornberger U, Wieseler B and Drews G (1991) Oxygen tension regulated expression of the *hemA* gene of *Rhodobacter capsulatus*. Arch Microbiol 156: 129–134

Hübner P, Willison JC, Vignais PM and Bickle TA (1991) Expression of regulatory *nif* genes in *Rhodobacter capsulatus*. J Bacteriol 173: 2993–2999

Hübner P, Masepohl B, Klipp W and Bickle TA (1993) *nif* gene expression studies in *Rhodobacter capsulatus*: *ntrC*-independent repression by high ammonium concentrations. Mol Microbiol 10: 123–132

Hunter CN (1988) Transposon Tn*5* mutagenesis of genes encoding reaction centre and light-harvesting LH1 polypeptides of *Rhodobacter sphaeroides*. J Gen Microbiol 134: 1481–1489

Hunter CN and Coomber SA (1988) Cloning and oxygen-regulated expression of the bacteriochlorophyll biosynthesis genes *bch E, B, A*, and *C* of *Rhodobacter sphaeroides*. J Gen Microbiol 134: 1491–1497

Hunter CN and Mann NH (1992) Genetic manipulation of photosynthetic prokaryotes. In: Mann NH and Carr NG (eds) Photosynthetic Prokaryotes, pp 153–179. Plenum, New York

Hunter CN and Turner G (1988) Transfer of genes coding for apoproteins of reaction center and light-harvesting LH1 complexes to *Rhodobacter sphaeroides*. J Gen Microbiol 134: 1471–1480

Hunter CN, McGlynn P, Ashby MK, Burgess JG and Olsen JD (1991) DNA sequencing and complementation/deletion analysis of the *bchA-puf* operon region of *Rhodobacter sphaeroides*: In vivo mapping of the oxygen-regulated *puf* promoter. Mol Microbiol 5: 2649–2661

Hunter CN, Hundle BS, Hearst JE, Lang HP, Gardiner AT, Takaichi S and Cogdell RJ (1994) Introduction of new carotenoids into the bacterial photosynthetic apparatus by combining the carotenoid biosynthetic pathways of *Erwinia herbicola* and *Rhodobacter sphaeroides*. J Bacteriol 176: 3692–3697

Hustede E, Steinbüchel A and Schlegel HG (1992) Cloning of poly(3-hydroxybutyric acid) synthase genes of *Rhodobacter sphaeroides* and *Rhodospirillum rubrum* and heterologous expression in *Alcaligenes eutrophus*. FEMS Microbiol Lett 93: 285–290

Iba K, Morohashi K, Miyata T, and Takamiya K (1987) Structural gene of cytochrome *b*-562 from the cytochrome bc_1 complex of *Rhodobacter sphaeroides*. J Biochem 102: 1511–1518

Ideguchi T, Hu C, Kim BH, Nishise H, Yamashita J and Kakuno T (1993) An open reading frame in the *Rhodospirillum rubrum* plasmid, pKY1, similar to *algA*, encoding the bifunctional enzyme phosphomannose isomerase-guanosine diphospho-D-mannose pyrophosphorylase (PMI-GMP). Biochim Biophys Acta 1172: 329–331

Jenny FE and Daldal F (1993) A novel membrane-associated *c*-type cytochrome, Cyt c_y, can mediate the photosynthetic growth of *Rhodobacter capsulatus* and *Rhodobacter sphaeroides*. EMBO J 12: 1283–1292

Johansson BC and Baltscheffsky M (1980) Characterization of a reaction center mutant of *Rhodospirillum rubrum*: I. isolation and growth in relation to spectral and enzymatic properties. Photobiochem Photobiophys 1: 191–198

Johnson JA, Wong WKR and Beatty JT (1986) Expression of cellulase genes in *Rhodobacter capsulatus* by use of plasmid expression vectors. J Bacteriol 167: 604–610

Jones R and Haselkorn R (1988) The DNA sequence of the *Rhodobacter capsulatus nifH* gene. Nucleic Acids Res 16: 8735

Jones R and Haselkorn R (1989) The DNA sequence of the *Rhodobacter capsulatus ntrA*, *ntrB*, and *ntrC* gene analogues required for nitrogen fixation. Mol Gen Genet 215: 507–516

Jones MR, Fowler GJS, Gibson LCD, Grief GG, Olsen JD, Crielaard W and Hunter CN (1992a) Mutants of *Rhodobacter sphaeroides* lacking one or more pigment-protein complexes and complementation with reaction-centre, LH1, and LH2 genes. Mol Microbiol 6: 1173–1184

Jones MR, Visschers RW, van Grondelle R and Hunter CN (1992b) Construction and characterization of a mutant of *Rhodobacter sphaeroides* with the reaction center as the sole pigment-protein complex. Biochemistry 31: 4458–4465

Jones MR, Heer-Dawson M, Mattioli TA, Hunter CN and Robert B (1994) Site-specific mutagenesis of the reaction centre from *Rhodobacter sphaeroides* studied by Fourier transform Raman spectroscopy: Mutations at tyrosine M210 do not affect the electronic structure of the primary donor. FEBS Lett 339: 18–24

Jouanneau Y, Richaud P and Grabau C (1990) The nucleotide sequence of a flavodoxin-like gene which precedes two ferredoxin genes in *Rhodobacter capsulatus*. Nucleic Acids Res 18: 5284

Jouanneau Y, Duport C, Meyer C and Gaillard J (1992) Expression in *Escherichia coli* and characterization of a recombinant 7Fe ferredoxin of *Rhodobacter capsulatus*. Biochem J 286: 269–273

Kansy JW and Kaplan S (1989) Purification, characterization, and transcriptional analyses of RNA polymerases from

Rhodobacter sphaeroides cells grown chemoheterotrophically and photoheterotrophically. J Biol Chem 264: 13751–13759

Karls RK, Jin DJ and Donohue TJ (1993) Transcription properties of RNA polymerase holoenzymes isolated from the purple nonsulfur bacterium *Rhodobacter sphaeroides*. J Bacteriol 175: 7629–7638

Kaszubska W, Aiken C, O'Connor CD, and Gumport RI (1989) Purification, cloning and sequence analysis of *Rsr*I DNA methyltransferase: Lack of homology between two enzymes, *Rsr*I and *Eco*RI, that methylate the same nucleotide in identical recognition sequences. Nucleic Acids Res 17: 10403–10425

Kaufmann N, Hüdig H and Drews G (1984) Transposon Tn*5* mutagenesis of genes for the photosynthetic apparatus in *Rhodopseudomonas capsulata*. Mol Gen Genet 198: 153–158

Keen NT, Tamaki S, Kobayashi D and Trollinger D (1988) Improved broad-host-range plasmids for DNA cloning in Gram-negative bacteria. Gene 70: 191–197

Kelly DJ, Richardson DJ, Ferguson SJ and Jackson JB (1988) Isolation of transposon Tn*5* insertion mutants of *Rhodobacter capsulatus* unable to reduce trimethylamine-N-oxide and dimethylsulphoxide. Arch Microbiol 150: 138–144

Kerby RL, Hong SS, Ensign SA, Coppoc LJ, Ludden PW and Roberts GP (1992) Genetic and physiological characterization of the *Rhodospirillum rubrum* carbon monoxide dehydrogenase system. J Bacteriol 174: 5284–5294

Kiley PJ and Kaplan S (1987) Cloning, DNA sequence, and expression of the *Rhodobacter sphaeroides* light-harvesting B800–850-α and B800–850-β genes. J Bacteriol 169: 3268–3275

Kiley PJ and Kaplan S (1988) Molecular genetics of photosynthetic membrane biosynthesis in *Rhodobacter sphaeroides*. Microbiol Rev 52: 50–69

Kiley PJ, Donohue TJ, Havelka WA and Kaplan S (1987) DNA sequence and in vitro expression of the B875 light-harvesting polypeptides of *Rhodobacter sphaeroides*. J Bacteriol 169: 742–750

Klipp W, Masepohl B and Pühler A (1988) Identification and mapping of nitrogen fixation genes of *Rhodobacter capsulatus*: Duplication of a *nifA-nifB* region. J Bacteriol 170: 693–699

Klug G (1991) A DNA sequence upstream of the *puf* operon of *Rhodobacter capsulatus* is involved in its oxygen-dependent regulation and functions as a protein binding site. Mol Gen Genet 226: 167–176

Klug G and Cohen SN (1991) Effects of translation on degradation of mRNA segments transcribed from the polycistronic *puf* operon of *Rhodobacter capsulatus*. J Bacteriol 173: 1478–1484

Klug G and Drews G (1984) Construction of a gene bank of *Rhodopseudomonas capsulata* using a broad host range DNA cloning system. Arch Microbiol 139: 319–325

Klug G and Jock S (1991) A base pair transition in a DNA sequence with dyad symmetry upstream of the *puf* promoter affects transcription of the *puc* operon in *Rhodobacter capsulatus*. J Bacteriol 173: 6038–6045

Klug G, Adams CW, Belasco J, Doerge B and Cohen SN (1987) Biological consequences of segmental alterations in mRNA stability: Effects of deletion of the intercistronic hairpin loop region of the *Rhodobacter capsulatus puf* operon. EMBO J 6: 3515–3520

Klug G, Jock S and Kordes E (1994) 23S rRNA processing in *Rhodobacter capsulatus* is not involved in the oxygen-regulated formation of the bacterial photosynthetic apparatus. Arch Microbiol 162: 91–97

Kobayashi H, Viale AM, Takabe T, Akazawa T, Wada K, Shinozaki K, Kobayashi K and Sugiura M (1991) Sequence and expression of genes encoding the large and small subunits of ribulose 1,5-bisphosphate carboxylase/oxygenase from *Chromatium vinosum*. Gene 97: 55–62

Kondorosi A, Kiss GB, Forrai T, Vincze E and Banfalvi Z (1977) Circular linkage map of *Rhizobium meliloti* chromosome. Nature 268: 525–527

Kordes E, Jock S, Fritsch J, Bosch F and Klug G (1994) Cloning of a gene involved in rRNA precursor processing and 23S rRNA cleavage in *Rhodobacter capsulatus*. J Bacteriol 176: 1121–1127

Kranz RG (1989) Isolation of mutants and genes involved in cytochromes *c* biosynthesis in *Rhodobacter capsulatus*. J Bacteriol 171: 456–464

Kranz RG and Haselkorn R (1985) Characterization of *nif* regulatory genes in *Rhodopseudomonas capsulata* using *lac* gene fusions. Gene 40: 203–215

Kranz RG and Haselkorn R (1986) Anaerobic regulation of nitrogen-fixation genes in *Rhodopseudomonas capsulata*. Proc Natl Acad Sci USA 83: 6805–6809

Kranz RG, Pace VM and Caldicott IM (1990) Inactivation, sequence, and *lacZ* fusion analysis of a regulatory locus required for repression of nitrogen fixation genes in *Rhodobacter capsulatus*. J Bacteriol 172: 53–62

Kranz RG, Beckman DL and Foster-Hartnett D (1992) DNA gyrase activities from *Rhodobacter capsulatus*: Analysis of target(s) of coumarins and cloning of the *gyrB* locus. FEMS Microbiol Lett 93: 25–32

Lang FS and Oesterhelt D (1989a) Microaerophilic growth and induction of the photosynthetic reaction center in *Rhodopseudomonas viridis*. J Bacteriol 171: 2827–2834

Lang FS and Oesterhelt D (1989b) Gene transfer system for *Rhodopseudomonas viridis*. J Bacteriol 171: 4425–4435

Lang HP, Cogdell RJ, Gardiner AT and Hunter CN (1994) Early steps in carotenoid biosynthesis: Sequences and transcriptional analysis of the *crtI* and *crtB* genes of *Rhodobacter sphaeroides* and overexpression and reactivation of *crtI* in *Escherichia coli* and *R. sphaeroides*. J Bacteriol 176: 3859–3869

Larimer FW, Machanoff R and Hartman FC (1986) A reconstruction of the gene for ribulose bisphosphate carboxylase from *Rhodospirillum rubrum* that expresses the authentic enzyme in *Escherichia coli*. Gene 41: 113–120

Laußermair E and Oesterhelt D (1992) A system for site-specific mutagenesis of the photosynthetic reaction center in *Rhodopseudomonas viridis*. EMBO J 11: 777–783

LeBlanc HN and Beatty JT (1993) *Rhodobacter capsulatus puc* operon: Promoter location, transcript sizes and effects of deletions on photosynthetic growth. J Gen Microbiol 139: 101–109

Leclerc M, Colbeau A, Cauvin B and Vignais PM (1988) Cloning and sequencing of the genes encoding the large and the small subunits of the H_2 uptake hydrogenase (*hup*) of *Rhodobacter capsulatus*. Mol Gen Genet 214: 97–107

Lee IY and Collins MLP (1993) Identification and partial sequence of the *bchA* gene of *Rhodospirillum rubrum*. Current Microbiol 27: 85–90

Lee JK and Kaplan S (1992a) *Cis*-acting regulatory elements involved in oxygen and light control of *puc* operon transcription in *Rhodobacter sphaeroides*. J Bacteriol 174: 1146–1157

Lee JK and Kaplan S (1992b) Isolation and characterization of *trans*-acting mutations involved in oxygen regulation of *puc* operon transcription in *Rhodobacter sphaeroides*. J Bacteriol 174: 1158–1172

Lee JK, DeHoff BS, Donohue TJ, Gumport RI and Kaplan S (1989a) Transcriptional analysis of *puf* operon expression in *Rhodobacter sphaeroides* 2.4.1 and an intercistronic transcription terminator mutant. J Biol Chem 264: 19354–19365

Lee JK, Kiley PJ and Kaplan S (1989b) Posttranscriptional control of *puc* operon expression of B800-850 light-harvesting complex formation in *Rhodobacter sphaeroides*. J Bacteriol 171: 3391–3405

Lee JK, Wang S, Eraso JM, Gardner J and Kaplan S (1993) Transcriptional regulation of *puc* operon expression in *Rhodobacter sphaeroides*: Involvement of an integration host factor-binding sequence. J Biol Chem 268: 24491–24497

Lehman LJ, Fitzmaurice WP and Roberts GP (1990) The cloning and functional characterization of the *nifH* gene of *Rhodospirillum rubrum*. Gene 95: 143–147

Leustek T, Hartwig R, Weissbach H and Brot N (1988) Regulation of ribulose bisphosphate carboxylase expression in *Rhodospirillum rubrum*: Characteristics of mRNA synthesized in vivo and in vitro. J Bacteriol 170: 4065–4071

Liang J, Nielsen GM, Lies DP, Burris RH, Roberts GP and Ludden PW (1991) Mutations in the *draT* and *draG* genes of *Rhodospirillum rubrum* result in loss of regulation of nitrogenase by reversible ADP-ribosylation. J Bacteriol 173: 6903–6909

Liebergesell M and Steinbüchel A (1992) Cloning and nucleotide sequences of genes relevant for biosynthesis of poly(3-hydroxybutyric acid) in *Chromatium vinosum* strain D. Eur J Biochem 209: 135–150

Liebetanz R, Hornberger U and Drews G (1991) Organization of the genes coding for the reaction-centre L and M subunits and B870 antenna polypeptides α and β from the aerobic photosynthetic bacterium *Erythrobacter* species OCH114. Mol Microbiol 5: 1459–1468

Lilburn TG and Beatty JT (1992) Suppressor mutants of the photosynthetically incompetent *pufX* deletion mutant *Rhodobacter capsulatus* ΔRC6(pTL2). FEMS Microbiol Lett 100: 155–160

Lilburn TG, Haith CE, Prince RC and Beatty JT (1992) Pleiotropic effects of *pufX* gene deletion on the structure and function of the photosynthetic apparatus of *Rhodobacter capsulatus*. Biochim Biophys Acta 1100: 160–170

Lin X, Williams JC, Allen JP, and Mathis P (1994) Relationship between rate and free energy difference for electron transfer from cytochrome c_2 to the reaction center in *Rhodobacter sphaeroides*. Biochemistry 33: 13517–13523

Ma D, Cook DN, O'Brien DA and Hearst JE (1993) Analysis of the promoter and regulatory sequences of an oxygen-regulated *bch* operon in *Rhodobacter capsulatus* by site-directed mutagenesis. J Bacteriol 175: 2037–2045

MacGregor BJ and Donohue TJ (1991) Evidence for two promoters for the cytochrome c_2 gene (*cycA*) of *Rhodobacter sphaeroides*. J Bacteriol 173: 3949–3957

Magnin JP, Willison JC and Vignais PM (1987) Elimination of R plasmids from the photosynthetic bacterium *Rhodobacter capsulatus*. FEMS Microbiol Lett 41: 157–161

Majewski C and Trebst A (1990) The *pet* genes of *Rhodospirillum rubrum*: Cloning and sequencing of the genes for the cytochrome bc_1-complex. Mol Gen Genet 224: 373–382

Marrs B (1974) Genetic recombination in *Rhodopseudomonas capsulata*. Proc Natl Acad Sci USA 71: 971–973

Marrs BL (1978) Genetics and bacteriophage. In: Clayton RK and Sistrom WR (eds) The Photosynthetic Bacteria, pp 873–883. Plenum, New York

Marrs B (1981) Mobilization of the genes for photosynthesis from *Rhodopseudomonas capsulata* by a promiscuous plasmid. J Bacteriol 146: 1003–1012

Marrs B, Kaplan S and Shepherd W (1980) Isolation of mutants of photosynthetic bacteria. Methods Enzymol 69: 29–37

Masepohl B, Klipp W and Pühler A (1988) Genetic characterization and sequence analysis of the duplicated *nifA/nifB* gene region of *Rhodobacter capsulatus*. Mol Gen Genet 212: 27–37

Masepohl B, Angermüller S, Hennecke S, Hübner P, Moreno-Vivian C and Klipp W (1993a) Nucleotide sequence and genetic analysis of the *Rhodobacter capsulatus* ORF6-*nifU$_I$SVW* gene region: Possible role of NifW in homocitrate processing. Mol Gen Genet 238: 369–382

Masepohl B, Krey R and Klipp W (1993b) The *draTG* gene region of *Rhodobacter capsulatus* is required for post-translational regulation of both the molybdenum and the alternative nitrogenase. J Gen Microbiol 139: 2667–2675

Matsunaga T, Matsunaga N, Tsubaki K and Tanaka T (1986) Development of a gene cloning system for the hydrogen-producing marine photosynthetic bacterium *Rhodopseudomonas* sp. J Bacteriol 168: 460–463

Matsunaga T, Tsubaki K, Miyashita H and Burgess JG (1990) Chloramphenicol acetyltransferase expression in marine *Rhodobacter* sp. NKPB 0021 by use of shuttle vectors containing the minimal replicon of an endogenous plasmid. Plasmid 24: 90–99

McClelland M, Jones R, Patel Y and Nelson M (1987) Restriction endonucleases for pulsed field mapping of bacterial genomes. Nucleic Acids Res 15: 5985–6005

McEwan AG, Kaplan S and Donohue TJ (1989) Synthesis of *Rhodobacter sphaeroides* cytochrome c_2 in *Escherichia coli*. FEMS Microbiol Lett 59: 253–258

McGlynn P and Hunter CN (1993) Genetic analysis of the *bchC* and *bchA* genes of *Rhodobacter sphaeroides*. Mol Gen Genet 236: 227–234

Meijer WG and Tabita FR (1992) Isolation and characterization of the *nifUSVW-rpoN* gene cluster from *Rhodobacter sphaeroides*. J Bacteriol 174: 3855–3866

Michel H, Weyer KA, Gruenberg H and Lottspeich F (1985) The 'heavy' subunit of the photosynthetic reaction centre from *Rhodopseudomonas viridis*: Isolation of the gene, nucleotide and amino acid sequence. EMBO J 4: 1667–1672

Michel H, Weyer KA, Gruenberg H, Dunger I, Oesterhelt D and Lottspeich L (1986) The 'light' and 'medium' subunits of the photosynthetic reaction centre from *Rhodopseudomonas viridis*: Isolation of the genes, nucleotide and amino acid sequence. EMBO J 5: 1149–1158

Miller L and Kaplan S (1978) Plasmid transfer and expression in *Rhodopseudomonas sphaeroides*. Arch Biochem Biophys 187: 229–234

Moore MD and Kaplan S (1989) Construction of Tn*phoA* gene fusions in *Rhodobacter sphaeroides*: Isolation and characterization of a respiratory mutant unable to utilize dimethyl sulfoxide as a terminal electron acceptor during anaerobic growth in the dark on glucose. J Bacteriol 171: 4385–4394

Moreno-Vivian C, Hennecke S, Pühler A and Klipp W (1989a) Open reading frame 5 (ORF5), encoding a ferredoxin-like protein, and *nifQ* are cotranscribed with *nifE*, *nifN*, *nifX*, and ORF4 in *Rhodobacter capsulatus*. J Bacteriol 171: 2591–2598

Moreno-Vivian C, Schmehl M, Masepohl B, Arnold W and Klipp W (1989b) DNA sequence and genetic analysis of the *Rhodobacter capsulatus nifENX* gene region: Homology between NifX and NifB suggests involvement of NifX in processing of the iron-molybdenum cofactor. Mol Gen Genet 216: 353–363

Moreno-Vivian C, Roldán MD, Reyes F and Castillo F (1994) Isolation and characterization of transposon Tn*5* mutants of *Rhodobacter sphaeroides* deficient in both nitrate and chlorate reduction. FEMS Microbiol Letts 115: 279–284

Mosley CS, Suzuki JY and Bauer CE (1994) Identification and molecular genetic characterization of a sensor kinase responsible for coordinately regulating light harvesting and reaction center gene expression in response to anaerobiosis. J Bacteriol 176: 7566–7573

Muller ED, Chory J and Kaplan S (1985) Cloning and characterization of the gene product of the form II ribulose-1,5-bisphosphate carboxylase gene of *Rhodopseudomonas sphaeroides*. J Bacteriol 161: 469–472

Nagarajan V, Parson WW, Gaul D and Schenck C (1990) Effect of specific mutations of tyrosine-(M)210 on the primary photosynthetic electron-transfer process in *Rhodobacter sphaeroides*. Proc Natl Acad Sci 87: 7888–7892

Nagashima KVP, Matsuura K, Ohyama S and Shimada K (1994) Primary structure and transcription of genes encoding B870 and photosynthetic reaction center apoproteins from *Rubrivivax gelatinosus*. J Biol Chem 269: 2477–2484

Nano FE and Kaplan S (1982) Expression of the transposable *lac* operon Tn*951* in *Rhodopseudomonas sphaeroides*. J Bacteriol 152: 924–927

Nano FE and Kaplan S (1984) Plasmid rearrangements in the photosynthetic bacterium *Rhodopseudomonas sphaeroides*. J Bacteriol 158: 1094–1103

Nano FE, Shepherd WD, Watkins MM, Kuhl SA and Kaplan S (1984) Broad-host-range plasmid vector for the in vitro construction of transcriptional/translational *lac* fusions. Gene 34: 219–226

Nargang F, McIntosh L and Somerville C (1984) Nucleotide sequence of the ribulose bisphosphate carboxylase gene from *Rhodospirillum rubrum*. Mol Gen Genet 193: 220–224

Narro ML, Adams CW and Cohen SN (1990) Isolation and characterization of *Rhodobacter capsulatus* mutants defective in oxygen regulation of the *puf* operon. J Bacteriol 172: 4549–4554

Neidle EL and Kaplan S (1992) *Rhodobacter sphaeroides rdxA*, a homolog of *Rhizobium meliloti fixG*, encodes a membrane protein which may bind cytoplasmic [4Fe-4S] clusters. J Bacteriol 174: 6444–6454

Neidle EL and Kaplan S (1993a) Expression of the *Rhodobacter sphaeroides hemA* and *hemT* genes, encoding two 5-aminolevulinic acid synthase isozymes. J Bacteriol 175: 2292–2303

Neidle EL and Kaplan S (1993b) 5-aminolevulinic acid availability and control of spectral complex formation in HemA and HemT mutants of *Rhodobacter sphaeroides*. J Bacteriol 175: 2304–2313

Niyogi SK, Foote RS, Mural RJ, Larimer FW, Mitra S, Soper TS, Machanoff R and Hartman FC (1986) Nonessentiality of histidine 291 of *Rhodospirillum rubrum* ribulose-bisphosphate carboxylase/oxygenase as determined by site-directed mutagenesis. J Biol Chem 261: 10087–10092

Paddock ML, Rongey SH, Feher G and Okamura MY (1989) Pathway of proton transfer in bacterial reaction centers: Replacement of glutamic acid 212 in the L subunit by glutamine inhibits quinone (secondary acceptor) turnover. Proc Natl Acad Sci USA 86: 6602–6606

Pemberton JM and Bowen ARSG (1981) High-frequency chromosome transfer in *Rhodopseudomonas sphaeroides* promoted by broad-host-range plasmid RP1 carrying mercury transposon Tn*501*. J Bacteriol 147: 110–117

Pemberton JM and Harding CM (1986) Cloning of carotenoid biosynthesis genes from *Rhodopseudomonas sphaeroides*. Current Microbiol 14: 25–29

Pemberton JM and Harding CM (1987) Expression of *Rhodopseudomonas sphaeroides* carotenoid photopigment genes in phylogenetically related nonphotosynthetic bacteria. Current Microbiol 15: 67–71

Penfold RJ and Pemberton JM (1991) A gene from the photosynthetic gene cluster of *Rhodobacter sphaeroides* induces *trans* suppression of bacteriochlorophyll and carotenoid levels in *R. sphaeroides* and *R. capsulatus*. Current Microbiol 23: 259–263

Penfold RJ and Pemberton JM (1992) An improved suicide vector for construction of chromosomal insertion mutations in bacteria. Gene 118: 145–156

Penfold RJ and Pemberton JM (1994) Sequencing, chromosomal inactivation, and functional expression in *Escherichia coli* of *ppsR*, a gene which represses carotenoid and bacterio-chlorophyll synthesis in *Rhodobacter sphaeroides*. J Bacteriol 176: 2869–2876

Pierrard J, Willison JC, Vignais PM, Gaspar JL, Ludden PW and Roberts GP (1993) Site-directed mutagenesis of the target arginine for ADP-ribosylation of nitrogenase component II in *Rhodobacter capsulatus*. Biochem Biophys Res Comm 192: 1223–1229

Pille S, Chuat JC, Breton AM, Clément-Métral JD and Galibert F (1990) Cloning, nucleotide sequence, and expression of the *Rhodobacter sphaeroides* Y thioredoxin gene. J Bacteriol 172: 1556–1561

Pollich M, Jock S and Klug G (1993) Identification of a gene required for the oxygen-regulated formation of the photo-synthetic apparatus of *Rhodobacter capsulatus*. Mol Microbiol 10: 749–757

Pollock D, Bauer CE and Scolnik PA (1988) Transcription of the *Rhodobacter capsulatus nifHDK* operon is modulated by the nitrogen source. construction of plasmid expression vectors based on the *nifHDK* promoter. Gene 65: 269–275

Preker P, Hübner P, Schmehl M, Klipp W and Bickle TA (1992) Mapping and characterization of the promoter elements of the regulatory *nif* genes *rpoN*, *nifA1* and *nifA2* in *Rhodobacter capsulatus*. Mol Microbiol 6: 1035–1047

Prentki P and Krisch HM (1984) In vitro insertional mutagenesis with a selectable DNA fragment. Gene 29: 303–313

Rainey AM and Tabita FR (1989) Isolation of plasmid DNA sequences that complement *Rhodobacter sphaeroides* mutants deficient in the capacity for CO_2-dependent growth. J Gen Microbiol 135: 1699–1713

Richaud P, Colbeau A, Toussaint B and Vignais PM (1991) Identification and sequence analysis of the *hupR*$_1$ gene, which encodes a response regulator of the NrtrC family required for hydrogenase expression in *Rhodobacter capsulatus*. J Bacteriol 173: 5928–5932

Richter P, Cortez N and Drews G (1991) Possible role of the highly conserved amino acids Trp-8 and Pro-13 in the N-terminal segment of the pigment-binding polypeptide LHIα of *Rhodobacter capsulatus*. FEBS Lett 285: 80–84

Richter P, Brand M and Drews G (1992) Characterization of LHI$^-$ and LHI$^+$ *Rhodobacter capsulatus pufA* mutants. J Bacteriol 174: 3030–3041

Robles SJ and Youvan DC (1993) Hydropathy and molar volume constraints on combinatorial mutants of the photosynthetic reaction center. J Mol Biol 232: 242–252

Robles SJ, Breton J and Youvan DC (1990) Partial symmetrization of the photosynthetic reaction center. Science 248: 1402–1405

Rott MA and Donohue TJ (1990) *Rhodobacter sphaeroides spd* mutations allow cytochrome c_2-independent photosynthetic growth. J Bacteriol 172: 1954–1961

Saegesser R, Ghosh R and Bachofen R (1992) Stability of broad host range cloning vectors in the phototrophic bacterium *Rhodospirillum rubrum*. FEMS Microbiol Lett 95: 7–12

Saeki K, Miyatake Y, Young DA, Marrs BL and Matsubara H (1990) A plant-ferredoxin-like gene is located upstream of ferredoxin I gene (*fdxN*) of *Rhodobacter capsulatus*. Nucleic Acids Res 18: 1060

Saeki K, Suetsugu Y, Tokuda K, Miyatake Y, Young DA, Marrs BL and Matsubara H (1991) Genetic analysis of functional differences among distinct ferredoxins in *Rhodobacter capsulatus*. J Biol Chem 266: 12889–12895

Saeki K, Tokuda K, Fujiwara T and Matsubara H (1993) Nucleotide sequence and genetic analysis of the region essential for functional expression of the gene for ferredoxin I, *fdxN*, in *Rhodobacter capsulatus*: sharing of one upstream activator sequence in opposite directions by two operons related to nitrogen fixation. Plant Cell Physiol 34: 185–199

Sánchez E, Teixidó J, Guerrero R and Amils R (1994) Hypersensitivity of *Rhodobacter sphaeroides* ribosomes to protein synthesis inhibitors: Structural and functional implications. Can J Microbiol 40: 699–704

Saunders VA (1978) Genetics of *Rhodospirillaceae*. Microbiol. Rev. 42: 357–384

Saunders VA (1992) Genetics of the photosynthetic prokaryotes. In: Mann NH and Carr NG (eds) Photosynthetic Prokaryotes, pp 121–152. Plenum, New York

Schatt E, Jouanneau Y and Vignais PM (1989) Molecular cloning and sequence analysis of the structural gene of ferredoxin I from the photosynthetic bacterium *Rhodobacter capsulatus*. J Bacteriol 171: 6218–6226

Schilke BA and Donohue TJ (1992) δ-aminolevulinate couples *cycA* transcription to changes in heme availability in *Rhodobacter sphaeroides*. J Mol Biol 226: 101–115

Schmehl M, Jahn A, Meyer zu Vilsendorf A, Hennecke S, Masepohl B, Schuppler M, Marxer M, Oelze J and Klipp W (1993) Identification of a new class of nitrogen fixation genes in *Rhodobacter capsulatus*: A putative membrane complex involved in electron transport to nitrogenase. Mol Gen Genet 241: 602–615

Schneider K, Müller A, Schramm U and Klipp W (1991) Demonstration of a molybdenum- and vanadium-independent nitrogenase in a *nifHDK*-deletion mutant of *Rhodobacter capsulatus*. Eur J Biochem 195: 653–661

Schneider KH, Giffhorn F and Kaplan S (1993) Cloning, nucleotide sequence and characterization of the mannitol dehydrogenase gene from *Rhodobacter sphaeroides*. J Gen Microbiol 139: 2475–2484

Schüddekopf K, Hennecke S, Liese U, Kutsche M and Klipp W (1993) Characterization of *anf* genes specific for the alternative nitrogenase and identification of *nif* genes required for both nitrogenases in *Rhodobacter capsulatus*. Mol Microbiol 8: 673–684

Schumann JP, Waitches GM and Scolnik PA (1986) A DNA fragment hybridizing to a *nif* probe in *Rhodobacter capsulatus* is homologous to a 16S rRNA gene. Gene 48: 81–92

Scolnik PA and Haselkorn R (1984) Activation of extra copies of genes coding for nitrogenase in *Rhodopseudomonas capsulata*. Nature 307: 289–292

Scolnik PA and Marrs BL (1987) Genetic research with photosynthetic bacteria. Ann Rev Microbiol 41: 703–726

Scolnik PA, Virosco J and Haselkorn R (1983) The wild-type gene for glutamine synthetase restores ammonia control of nitrogen fixation to Gln$^-$ (*glnA*) mutants of *Rhodopseudomonas capsulata*. J Bacteriol 155: 180–185

Self SJ, Hunter CN and Leatherbarrow RJ (1990) Molecular cloning, sequencing and expression of cytochrome c_2 from *Rhodospirillum rubrum*. Biochem J 265: 599–604

Sganga MW and Bauer CE (1992) Regulatory factors controlling photosynthetic reaction center and light-harvesting gene expression in *Rhodobacter capsulatus*. Cell 68: 945–954

Sganga MW, Aksamit RR, Cantoni GL and Bauer CE (1992) Mutational and nucleotide sequence analysis of S-adenosyl-L-homocysteine hydrolase from *Rhodobacter capsulatus*. Proc Natl Acad Sci USA 89: 6328–6332

Shanker S, Moomaw C, Güner S, Hsu J, Tokito MK, Daldal F, Knaff DB and Harman JG (1992) Characterization of the *pet* operon of *Rhodospirillum rubrum*. Photosynthesis Res 32: 79–94

Shapleigh JP and Gennis RB (1992) Cloning, sequencing and deletion from the chromosome of the gene encoding subunit I of the aa_3-type cytochrome *c* oxidase of *Rhodobacter sphaeroides*. Mol Microbiol 6: 635–642

Shapleigh JP, Hill JJ, Alben JO and Gennis RB (1992a) Spectroscopic and genetic evidence for two heme-Cu-containing oxidases in *Rhodobacter sphaeroides*. J Bacteriol 174: 2338–2343

Shapleigh JP, Hosler JP, Tecklenburg MMJ, Kim Y, Babcock GT, Gennis RB and Ferguson-Miller S (1992b) Definition of the catalytic site of cytochrome *c* oxidase: Specific ligands of heme *a* and the heme a_3-Cu_B center. Proc Natl Acad Sci USA 89: 4786–4790

Shaw JG, Hamblin MJ and Kelly DJ (1991) Purification, characterization and nucleotide sequence of the periplasmic C4-dicarboxylate-binding protein (DctP) from *Rhodobacter capsulatus*. Mol Microbiol 5: 3055–3062

Shively JM, Devore W, Stratford L, Porter L, Medlin L and Stevens SE (1986) Molecular evolution of the large subunit of ribulose-1,5-bisphosphate carboxylase/oxygenase (RuBisCO). FEMS Microbiol Lett 37: 251–257

Simon R, Priefer U and Pühler A (1983) A broad host range

mobilization system for in vivo genetic engineering: transposon mutagenesis in gram negative bacteria. Bio/Technology 1: 784–791

Sistrom WR (1977) Transfer of chromosomal genes mediated by plasmid R68.45 in *Rhodopseudomonas sphaeroides*. J Bacteriol 131: 526–532

Sistrom WR (1978) Lists of mutant strains. In: Clayton RK and Sistrom WR (eds) The Photosynthetic Bacteria, pp 927–934. Plenum, New York

Sistrom WR, Macaluso A and Pledger R (1984) Mutants of *Rhodopseudomonas sphaeroides* useful in genetic analysis. Arch Microbiol 138: 161–165

Sockett RE and Armitage JP (1991) Isolation, characterization and complementation of a paralyzed flagellar mutant of *Rhodobacter sphaeroides* WS8. J Bacteriol 173: 2786–2790

Sockett RE, Donohue TJ, Varga AR and Kaplan S (1989) Control of photosynthetic membrane assembly in *Rhodobacter sphaeroides* mediated by *puhA* and flanking sequences. J Bacteriol 171: 436–446

Solioz M and Marrs B (1977) The gene transfer agent of *Rhodopseudomonas capsulata*: Purification and characterization of its nucleic acid. Arch Biochem Biophys 181: 300–307

Solioz M, Yen HC and Marrs B (1975) Release and uptake of gene transfer agent by *Rhodopseudomonas capsulata*. J Bacteriol 123: 651–657

Söhlemann P, Oeckl C and Michel H (1991) Expression in *Escherichia coli* of the genes coding for reaction center subunits from *Rhodobacter sphaeroides*: Wild-type proteins and fusion proteins containing one or four truncated domains from *Staphylococcus aureus* protein A at the carboxy-terminus. Biochim Biophys Acta 1089: 103–112

Somerville CR and Somerville SC (1984) Cloning and expression of the *Rhodospirillum rubrum* ribulose bisphosphate carboxylase gene in *E. coli*. Mol Gen Genet 193: 214–219

Soper TS, Larimer FW, Mural RJ, Lee EH and Hartman FC (1992) Role of asparagine-111 at the active site of ribulose-1,5-bisphosphate carboxylase/oxygenase from *Rhodospirillum rubrum* as explored by site-directed mutagenesis. J Biol Chem 267: 8452–8457

Stephenson FH and Greene PJ (1989) Nucleotide sequence of the gene encoding the *Rsr*I methyltransferase. Nucleic Acids Res 17: 10503

Stephenson FH, Ballard BT, Boyer HW, Rosenberg JM and Greene PJ (1989) Comparison of the nucleotide and amino acid sequences of the *Rsr*I and *Eco*RI restriction endonucleases. Gene 85: 1–13

Stilz HU, Finkele U, Holzapfel W, Lauterwasser C, Zinth W and Oesterhelt D (1994) Influence of M subunit Thr222 and Trp252 on quinone binding and electron transfer in *Rhodobacter sphaeroides* reaction centres. Eur. J. Biochem. 223: 233–242

Stocker JW, Taguchi AKW, Murchison HA, Woodbury NW and Boxer SG (1992) Spectroscopic and redox properties of sym1 and (M)F195H: *Rhodobacter capsulatus* reaction center symmetry mutants which affect the initial electron donor. Biochemistry 31: 10356–10362

Suetsugu Y, Saeki K and Matsubara H (1991) Transcriptional analysis of two *Rhodobacter capsulatus* ferredoxins by translational fusion to *Escherichia coli lacZ*. FEBS Lett 292: 13–16

Suwanto A and Kaplan S (1989a) Physical and genetic mapping of the *Rhodobacter sphaeroides* 2.4.1 genome: genome size, fragment identification, and gene localization. J Bacteriol 171: 5840–5849

Suwanto A and Kaplan S (1989b) Physical and genetic mapping of the *Rhodobacter sphaeroides* 2.4.1 genome: presence of two unique circular chromosomes. J Bacteriol 171: 5850–5859

Suwanto A and Kaplan S (1992a) A self-transmissible, narrow-host-range endogenous plasmid of *Rhodobacter sphaeroides* 2.4.1: physical structure, incompatibility determinants, origin of replication, and transfer functions. J Bacteriol 174: 1124–1134

Suwanto A and Kaplan S (1992b) Chromosome transfer in *Rhodobacter sphaeroides*: Hfr formation and genetic evidence for two unique circular chromosomes. J Bacteriol 174: 1135–1145

Tadros MH and Waterkamp K (1989) Multiple copies of the coding regions for the light-harvesting B800-850 α- and β-polypeptides are present in the *Rhodopseudomonas palustris* genome. EMBO J 8: 1303–1308

Tadros MH, Katsiou E, Hoon MA, Yurkova N and Ramji DP (1993) Cloning of a new antenna gene cluster and expression analysis of the antenna gene family of *Rhodopseudomonas palustris*. Eur J Biochem 217: 867–875

Taguchi AKW, Stocker JW, Alden RG, Causgrove TP, Peloquin JM, Boxer SG and Woodbury NW (1992) Biochemical characterization and electron-transfer reactions of *sym1*, a *Rhodobacter capsulatus* reaction center symmetry mutant which affects the initial electron donor. Biochemistry 31: 10345–10355

Taguchi AKW, Stocker JW, Boxer SG and Woodbury NW (1993) Photosynthetic reaction center mutagenesis via chimeric rescue of a non-functional *Rhodobacter capsulatus puf* operon with sequences from *Rhodobacter sphaeroides*. Photosynthesis Res. 36: 43–58

Tai TN, Moore MD and Kaplan S (1988) Cloning and characterization of the 5-aminolevulinate synthase gene(s) from *Rhodobacter sphaeroides*. Gene 70: 139–151

Takahashi E and Wraight CA (1990) A crucial role for Asp^{L213} in the proton transfer pathway to the secondary quinone of reaction centers from *Rhodobacter sphaeroides*. Biochim Biophys Acta 1020: 107–111

Taylor DP, Cohen SN, Clark WG and Marrs BL (1983) Alignment of genetic and restriction maps of the photosynthesis region of the *Rhodopseudomonas capsulata* chromosome by a conjugation-mediated marker rescue technique. J Bacteriol 154: 580–590

Thöny-Meyer L, Beck C, Preisig O and Hennecke H (1994) The *ccoNOQP* gene cluster codes for a *cb*-type cytochrome oxidase that functions in aerobic respiration of *Rhodobacter capsulatus*. Mol Microbiol 14: 705–716

Tichy HV, Oberlé B, Stiehle H, Schiltz E and Drews G (1989) Genes downstream from *pucB* and *pucA* are essential for formation of the B800-850 complex of *Rhodobacter capsulatus*. J Bacteriol 171: 4914–4922

Tichy HV, Albien KU, Gad'on N and Drews G (1991) Analysis of the *Rhodobacter capsulatus puc* operon: The *pucC* gene plays a central role in the regulation of LHII (B800-850 complex) expression. EMBO J 10: 2949–2955

Tokito MK and Daldal F (1992) *petR*, located upstream of the *fbcFBC* operon encoding the cytochrome bc_1 complex, is homologous to bacterial response regulators and necessary for photosynthetic and respiratory growth of *Rhodobacter capsulatus*. Mol Microbiol 6: 1645–1654

Toussaint B, Bosc C, Richaud P, Colbeau A and Vignais PM (1991) A mutation in a *Rhodobacter capsulatus* gene encoding an integration host factor-like protein impairs in vivo hydrogenase expression. Proc Natl Acad Sci USA 88: 10749–10753

Toussaint B, Delic-Attree I, de Sury d'Aspremont RD, David L, Vinçon M and Vignais PM (1993) Purification of the integration host factor homolog of *Rhodobacter capsulatus*: Cloning and sequencing of the *hip* gene, which encodes the β-subunit. J Bacteriol 175: 6499–6504

Troschel D and Müller M (1990) Development of a cell-free system to study the membrane assembly of photosynthetic proteins of *Rhodobacter capsulatus*. J Cell Biol 111: 87–94

Troschel D, Eckhardt S, Hoffschulte HK and Müller M (1992) Cell-free synthesis and membrane-integration of the reaction center subunit H from *Rhodobacter capsulatus*. FEMS Microbiol Lett 91: 129–134

Tybulewicz VLJ, Falk G and Walker JE (1984) *Rhodopseudomonas blastica atp* operon: Nucleotide sequence and transcription. J Mol Biol 179: 185–214

Uffen RL, Colbeau A, Richaud P and Vignais PM (1990) Cloning and sequencing the genes encoding uptake-hydrogenase subunits of *Rhodocyclus gelatinosus*. Mol Gen Genet 221: 49–58

Usui S and Yu L (1991) Subunit IV (M_r = 14,384) of the cytochrome bc_1 complex from *Rhodobacter sphaeroides*: cloning, DNA sequencing, and ubiquinone binding domain. J Biol Chem 266: 15644–15649

Valle E, Kobayashi H and Akazawa T (1988) Transcriptional regulation of genes for plant-type ribulose-1,5-bisphosphate carboxylase/oxygenase in the photosynthetic bacterium, *Chromatium vinosum*. Eur J Biochem 173: 483–489

Varga AR and Kaplan S (1989) Construction, expression, and localization of a CycA::PhoA fusion protein in *Rhodobacter sphaeroides* and *Escherichia coli*. J Bacteriol 171: 5830–5839

Verbist J, Lang F, Gabellini N and Oesterhelt D (1989) Cloning and sequencing of the *fbcF*, *B* and *C* genes encoding the cytochrome bc_1 complex from *Rhodopseudomonas viridis*. Mol Gen Genet 219: 445–452

Viale AM, Kobayashi H, Takabe T and Akazawa T (1985) Expression of genes for subunits of plant-type RuBisCO in *Chromatium* and production of the enzymatically active molecule in *Escherichia coli*. FEBS Lett 192: 283–288

Viale AM, Kobayashi H and Akazawa T (1989) Expressed genes for plant-type ribulose 1,5-bisphosphate carboxylase/oxygenase in the photosynthetic bacterium *Chromatium vinosum*, which possesses two complete sets of the genes. J Bacteriol 171: 2391–2400

Viale AM, Kobayashi H and Akazawa T (1990) Distinct properties of *Escherichia coli* products of plant-type ribulose-1,5-bisphosphate carboxylase/oxygenase directed by two sets of genes from the photosynthetic bacterium *Chromatium vinosum*. J Biol Chem 265: 18386–18392

Viale AM, Kobayashi H, Akazawa T and Henikoff S (1991) *rcbR*, a gene coding for a member of the LysR family of transcriptional regulators, is located upstream of the expressed set of ribulose 1,5-bisphosphate carboxylase/oxygenase genes in the photosynthetic bacterium *Chromatium vinosum*. J Bacteriol 173: 5224–5229

Von Sternberg R and Yoch DC (1993) Molecular cloning and sequencing of the ferredoxin I *fdxN* gene of the photosynthetic bacterium *Rhodospirillum rubrum*. Biochim Biophys Acta 1144: 435–438

Wall JD and Braddock K (1984) Mapping of *Rhodopseudomonas capsulata nif* genes. J Bacteriol 158: 404–410

Wall JD, Weaver PF and Gest H (1975a) Genetic transfer of nitrogenase-hydrogenase activity in *Rhodopseudomonas capsulata*. Nature 258: 630–631

Wall JD, Weaver PF and Gest H (1975b) Gene transfer agents, bacteriophages, and bacteriocins of *Rhodopseudomonas capsulata*. Arch Microbiol 105: 217–224

Wall JD, Love J and Quinn SP (1984) Spontaneous Nif$^-$ mutants of *Rhodopseudomonas capsulata*. J Bacteriol 159: 652–657

Wang G, Angermüller S and Klipp W (1993) Characterization of *Rhodobacter capsulatus* genes encoding a molybdenum transport system and putative molybdenum-pterin-binding proteins. J Bacteriol 175: 3031–3042

Weaver KE and Tabita FR (1983) Isolation and partial characterization of *Rhodopseudomonas sphaeroides* mutants defective in the regulation of ribulose bisphosphate carboxylase/oxygenase. J Bacteriol 156: 507–515

Weaver KE and Tabita FR (1985) Complementation of a *Rhodobacter sphaeroides* ribulose bisphosphate carboxylase-oxygenase regulatory mutant from a genomic library. J Bacteriol 164: 147–154

Wellington CL and Beatty JT (1989) Promoter mapping and nucleotide sequence of the *bchC* bacteriochlorophyll biosynthesis gene from *Rhodobacter capsulatus*. Gene 83: 251–261

Wellington CL and Beatty JT (1991) Overlapping mRNA transcripts of photosynthesis gene operons in *Rhodobacter capsulatus*. J Bacteriol 173: 1432–1443

Wellington CL, Taggart AKP and Beatty JT (1991) Functional significance of overlapping transcripts of *crtEF*, *bchCA*, and *puf* photosynthesis gene operons in *Rhodobacter capsulatus*. J Bacteriol 173: 2954–2961

Weyer KA, Lottspeich F, Gruenberg H, Lang F, Oesterhelt D and Michel H (1987) Amino acid sequence of the cytochrome subunit of the photosynthetic reaction center from the purple bacterium *Rhodopseudomonas viridis*. EMBO J 6: 2197–2202

Wieseler B, Schiltz E and Müller M (1992) Identification and solubilization of a signal peptidase from the phototrophic bacterium *Rhodobacter capsulatus*. FEBS Lett 298: 273–276

Wiessner C, Dunger I and Michel H (1990) Structure and transcription of the genes encoding the B1015 light-harvesting complex β and α subunits and the photosynthetic reaction center L, M, and cytochrome *c* subunits from *Rhodopseudomonas viridis*. J Bacteriol 172: 2877–2887

Willetts NS, Crowther C and Holloway BW (1981) The insertion sequence IS*21* of R68.45 and the molecular basis for mobilization of the bacterial chromosome. Plasmid 6: 30–52

Williams JC, Steiner LA, Ogden RC, Simon MI and Feher G (1983) Primary structure of the M subunit of the reaction center from *Rhodopseudomonas sphaeroides*. Proc Natl Acad Sci USA 80: 6505–6509

Williams JC, Steiner LA, Feher G and Simon MI (1984) Primary structure of the L subunit of the reaction center from *Rhodopseudomonas sphaeroides*. Proc Natl Acad Sci USA 81: 7303–7307

Williams JC, Steiner LA and Feher G (1986) Primary structure of the reaction center from *Rhodopseudomonas sphaeroides*. Proteins: structure, function and genetics 1: 312–325

Williams JC, Alden RG, Murchison HA, Peloquin JM, Woodbury NW and Allen JP (1992) Effects of mutations near the bacteriochlorophylls in reaction centers from *Rhodobacter sphaeroides*. Biochemistry 31: 11029–11037

Willison JC (1993) Biochemical genetics revisited: the use of mutants to study carbon and nitrogen metabolism in the photosynthetic bacteria. FEMS Microbiol Rev 104: 1–38

Willison JC, Ahombo G, Chabert J, Magnin JP and Vignais PM (1985) Genetic mapping of the *Rhodopseudomonas capsulata* chromosome shows non-clustering of genes involved in nitrogen fixation. J Gen Microbiol 131: 3001–3015

Willison JC, Pierrard J and Hübner P (1993) Sequence and transcript analysis of the nitrogenase structural gene operon (*nifHDK*) of *Rhodobacter capsulatus*: Evidence for intramolecular processing of *nifHDK* mRNA. Gene 133: 39–46

Wright MS, Cardin RD and Biel AJ (1987) Isolation and characterization of an aminolevulinate-requiring *Rhodobacter capsulatus* mutant. J Bacteriol 169: 961–966

Wright MS, Eckert JJ, Biel SW and Biel AJ (1991) Use of a *lacZ* fusion to study transcriptional regulation of the *Rhodobacter capsulatus hemA* gene. FEMS Microbiol Lett 78: 339–342

Wu LF and Saier MH (1990) Nucleotide sequence of the *fruA* gene, encoding the fructose permease of the *Rhodobacter capsulatus* phosphotransferase system, and analyses of the deduced protein sequence. J Bacteriol 172: 7167–7178

Wu LF and Saier MH (1991) Differences in codon usage among genes encoding proteins of different function in *Rhodobacter capsulatus*. Res Microbiol 142: 943–949

Wu LF, Tomich JM and Saier MH (1990) Structure and evolution of a multidomain multiphosphoryl transfer protein: nucleotide sequence of the *fruB(HI)* gene in *Rhodobacter capsulatus* and comparisons with homologous genes from other organisms. J Mol Biol 213: 687–703

Wu LF, Reizer A, Reizer J, Cai B, Tomich JM and Saier MH (1991a) Nucleotide sequence of the *Rhodobacter capsulatus fruK* gene, which encodes fructose-1-phosphate kinase: evidence for a kinase superfamily including both phosphofructokinases of *Escherichia coli*. J Bacteriol 173: 3117–3127

Wu TT (1966) A model for three-point analysis of random general transduction. Genetics 54: 405–410

Wu YQ, MacGregor BJ, Donohue TJ, Kaplan S and Yen B (1991b) Genetic and physical mapping of the *Rhodobacter sphaeroides* photosynthetic gene cluster from R-prime pWS2. Plasmid 25: 163–176

Xu HH and Tabita FR (1994) Positive and negative regulation of sequences upstream of the form II *cbb* CO_2 fixation operon of *Rhodobacter sphaeroides*. J Bacteriol 176: 7299–7308

Xu HW and Wall JD (1991) Clustering of genes necessary for hydrogen oxidation in *Rhodobacter capsulatus*. J Bacteriol 173: 2401–2405

Xu HW, Love J, Borghese R and Wall JD (1989) Identification and isolation of genes essential for H_2 oxidation in *Rhodobacter capsulatus*. J Bacteriol 171: 714–721

Yang Z and Bauer CE (1990) *Rhodobacter capsulatus* genes involved in early steps of the bacteriochlorophyll biosynthetic pathway. J Bacteriol 172: 5001–5010

Yen HC and Marrs BL (1976) Map of genes for carotenoid and bacteriochlorophyll biosynthesis in *Rhodopseudomonas capsulata*. J Bacteriol 126: 619–629

Yen HC, Hu NT and Marrs BL (1979) Characterization of the gene transfer agent made by an overproducer mutant of *Rhodopseudomonas capsulata*. J Mol Biol 131: 157–168

Yildiz FH, Gest H and Bauer CE (1991) Genetic analysis of photosynthesis in *Rhodospirillum centenum*. J Bacteriol 173: 4163–4170

Yildiz FH, Gest H and Bauer CE (1992) Conservation of the photosynthetic gene cluster in *Rhodospirillum centenum*. Mol Microbiol 6: 2683–2691

Young DA, Bauer CE, Williams JC and Marrs BL (1989) Genetic evidence for superoperonal organization of genes for photosynthetic pigments and pigment-binding proteins in *Rhodobacter capsulatus*. Mol Gen Genet 218: 1–12

Young DA, Becker-Rudzik M and Marrs BL (1992) An overlap between operons involved in carotenoid and bacteriochlorophyll biosynthesis in *Rhodobacter capsulatus*. FEMS Microbiol Lett 95: 213–218

Youvan DC and Ismail S (1985) Light-harvesting II (B800-B850 complex) structural genes from *Rhodopseudomonas capsulata*. Proc Natl Acad Sci USA 82: 58–62

Youvan DC, Elder JT, Sandlin DE, Zsebo K, Alder DP, Panopoulos NJ, Marrs BL and Hearst JE (1982) R-prime site-directed transposon Tn7 mutagenesis of the photosynthetic apparatus in *Rhodopseudomonas capsulata*. J Mol Biol 162: 17–41

Youvan DC, Hearst JE and Marrs BL (1983) Isolation and characterization of enhanced fluorescence mutants of *Rhodopseudomonas capsulata*. J Bacteriol 154: 748–755

Youvan DC, Bylina EJ, Alberti M, Begusch H and Hearst JE (1984) Nucleotide and deduced polypeptide sequences of the photosynthetic reaction-center, B870 antenna, and flanking polypeptides from *R. capsulata*. Cell 37: 949–957

Youvan DC, Ismail S and Bylina EJ (1985) Chromosomal deletion and plasmid complementation of the photosynthetic reaction center and light-harvesting genes from *Rhodopseudomonas capsulata*. Gene 38: 19–30

Yu PL, Cullum J and Drews G (1981) Conjugational transfer systems of *Rhodopseudomonas capsulata* mediated by R plasmids. Arch Microbiol 128: 390–393

Yu PL, Hohn B, Falk H and Drews G (1982) Molecular cloning of the ribosomal RNA genes of the photosynthetic bacterium *Rhodopseudomonas capsulata*. Mol Gen Genet 188: 392–398

Yun CH, Beci R, Crofts AR, Kaplan S and Gennis RB (1990) Cloning and DNA sequencing of the *fbc* operon encoding the cytochrome bc_1 complex from *Rhodobacter sphaeroides*: Characterization of *fbc* deletion mutants and complementation by a site-specific mutational variant. Eur J Biochem 194: 399–411

Yun CH, Crofts AR and Gennis RB (1991a) Assignment of the histidine axial ligands to the cytochrome b_H and cytochrome b_L components of the bc_1 complex from *Rhodobacter sphaeroides* by site-directed mutagenesis. Biochemistry 30: 6747–6754

Yun CH, Van Doren SR, Crofts AR and Gennis RB (1991b) The use of gene fusions to examine the membrane topology of the L subunit of the photosynthetic reaction center and of the cytochrome *b* subunit of the bc_1 complex from *Rhodobacter sphaeroides*. J Biol Chem 266: 10967–10973

Yun CH, Barquera B, Iba K, Takamiya K, Shapleigh, J, Crofts AR and Gennis RB (1994) Deletion of the gene encoding cytochrome *b*-562 from *Rhodobacter sphaeroides* FEMS Microbiol Letts 120: 105–110

Zannoni D, Prince RC, Dutton PL and Marrs BL (1980) Isolation and characterization of a cytochrome c_2-deficient mutant of *Rhodopseudomonas capsulata*. FEBS Lett 113: 289–293

Zilsel J, Lilburn TG and Beatty JT (1989) Formation of functional inter-species hybrid photosynthetic complexes in *Rhodobacter capsulatus*. FEBS Lett 253: 247–252

Zinchenko VV, Babykin MM, Shestakov S, Allibert P, Vignais PM and Willison JC (1990) Ammonia-dependent growth (Adg^-) mutants of *Rhodobacter capsulatus* and *Rhodobacter sphaeroides*: Comparison of mutant phenotypes and cloning of the wild-type (*adgA*) genes. J Gen Microbiol 136: 2385–2393

Zinchenko V, Churin Y, Shestopalov V and Shestakov S (1994) Nucleotide sequence and characterization of the *Rhodobacter sphaeroides glnB* and *glnA* genes. Microbiology 140: 2143–2151

Zsebo KM and Hearst JE (1984) Genetic-physical mapping of a photosynthetic gene cluster from *R. capsulata*. Cell 37: 937–947

Chapter 49

Physical Mapping of *Rhodobacter capsulatus*: Cosmid Encyclopedia and High Resolution Genetic Map

Michael Fonstein and Robert Haselkorn
Department of Molecular Genetics and Cell Biology, University of Chicago, 920 East 58 Street, Chicago, IL 60637, USA

Summary

A combination of cosmid genome walking and pulse field gel electrophoresis was used to construct a high resolution physical and genetic map of the 3.8 Mb genome of *Rhodobacter capsulatus* SB1003. The mapping was done by grouping and further mapping of cosmids and bacteriophages from genomic libraries using PFGE-generated DNA fragments and SP6 and T7 specific transcripts corresponding to the ends of the cosmid inserts. Cosmid and phage clones formed two uninterrupted and ordered groups, one corresponding to the chromosome of *Rb. capsulatus*, the other to its 134 kb plasmid. *Cos* site end-labeling and partial *Eco*RV digestion of cosmids were used to construct a high resolution restriction map of the genome. Overlapping of the cosmids was confirmed by the resemblance of the cosmid restriction maps and by direct end-to-end hybridization. 34 genes or gene clusters were located in the ordered gene library and mapped with an accuracy of 1–10 kb. Three *Rb. capsulatus* strains; KB-1, St. Louis and 2.3.1., were chosen out of 14 others for a detailed comparison of their physical maps, which were partially constructed using the minimal cosmid set of *Rb. capsulatus* SB1003 as a source of ordering probes. Blots of the minimal set of 192 cosmids, covering the chromosome and the plasmid, with known map position of each cosmid, gives to *Rb. capsulatus* the same advantages that the Kohara phage panel gives to *E. coli*. Blots of this minimal cosmid set digested with *Eco*RV represent the entire genome split into gene-size pieces, and provide an opportunity for direct high resolution mapping of genes and transcripts.

R. E. Blankenship, M. T. Madigan and C. E. Bauer (eds): Anoxygenic Photosynthetic Bacteria, pp. 1067–1081.

I. Introduction

Rhodobacter capsulatus is a purple, non-sulfur photosynthetic bacterium. The biochemical versatility to choose between photo- and heterotrophic growth and nitrogen fixation are all packed in a genome that is smaller than that of *E. coli*. Ease of plating and generation of mutations, together with convenient systems for cloning and genetic analysis (Yen and Marrs, 1976; Marrs, 1981; Jonson et al., 1986), summarized in Donohue and Kaplan (1991), makes *Rb. capsulatus* a popular model system for studies of the photosynthetic apparatus and nitrogen fixation. The Gene Transfer Agent (GTA), randomly packing 4.6 kb pieces of any intracellular DNA in phage-like capsids, has been used for insertional mutagenesis, gene replacement and linkage analysis (Yen et al., 1979). Mobilization of the bacterial chromosome by an integrated R' factor produced the current genetic map of *Rb. capsulatus* (Willison, 1985; Willison, 1993; Chapter 55 by Vignais et al.). Though direct plasmid transformation can not be efficiently applied to *Rb. capsulatus*, mobilized shuttle vectors derived from broad host range plasmids RK2 and RSF1010 facilitate manipulations with cloned genes. Efficient random and site-specific mutagenesis with transposons and gene cloning by mutant complementation was done with these vectors, for example Klipp et al., (1988). Using these tools together with powerful biophysical methods, substantial progress was achieved in studying gene control, regulation and protein machinery of photosynthesis and nitrogen fixation (this volume). However, scarcity of data concerning general genetics of *Rb. capsulatus* is already slowing current research, for instance studies on regulation of *nif* and photosynthetic gene expression, where the involvement of general regulatory cascades remains to be integrated. To provide this genetic background, we are developing a broad and fast strategy aimed at producing a dense genetic map and revealing regulatory relations among its elements. The first component of this strategy is the high resolution physical map of *Rb. capsulatus* carrying most of its cloned genes, described in Fonstein and Haselkorn (1993).

Abbreviations: *E. coli* – *Escherichia coli;* GTA – Gene transfer agent; PFGE – Pulse field gel electrophoresis; *Rb.* – *Rhodobacter*

II. Low Resolution Physical Mapping of the Genome of *Rhodobacter capsulatus*

To facilitate high resolution mapping of the *Rhodobacter* chromosome via assembly of the minimal cosmid set, covering its entire genome, a low resolution restriction map of its chromosome was constructed. For optimal physical mapping, a restriction endonuclease should cut the bacterial genome into 10–30 pieces. This number of fragments can be well separated and mapped by conventional PFGE. Among the six-base cutters with high-AT recognition sequences, we found *Xba*I (15 fragments), *Ase*I (25 fragments), *Sca*I (14 fragments), *Dra*I (25 fragments) and *Spe*I (3 fragments) to be useful . The first two were chosen for mapping because they yielded a good separation of fragments with uniform distribution in the gels.

The total genomic size of the *Rb. capsulatus* chromosome was estimated by summation of the individual fragment lengths represented in each digest. Those sums for *Xba*I and *Ase*I are respectively 3679 and 3639 kb. This difference is explained in part by the presence of a large plasmid consisting of fragments Xba-11, Xba-12' and Xba-13. Its size is 134 kb. This plasmid has no *Ase*I sites and didn't enter the gel as a separate band in PFGE under the conditions used. The sum of the sizes of *Xba*I-generated fragments differ from the *Ase*I sum by the size of this plasmid.

Macrorestriction DNA fragments generated by PFGE were used as probes for blot hybridization. Groups of *Xba*I, *Ase*I and *Xba*I+*Ase*I-generated fragments were separated by PFGE and transferred to nylon membranes. Filters corresponding to each group were hybridized with a single fragment probe. If, for example, an *Xba*I fragment used as a probe identifies two *Ase*I fragments, it will link them. Using as a second probe one of those *Ase*I fragments, one can step by step map the whole genome. Troubles arise when an *Xba*I fragment overlaps more than three *Ase*I fragments or this overlapping is significantly unequal. In the latter case, the mapping sequence of steps would be interrupted. The other source of mapping troubles was repeated DNA. Using this method several stretches of restriction fragments were assembled, the longest one covering half of the *Rhodobacter* chromosome. Further mapping was done using the cosmid library as a source of clones linking these fragments.

With a specially constructed printing device, individual cosmid clones were replicated onto nylon filters as two ordered sets of 864 clones each. These sets of clones were hybridized with all individual *Xba*I and *Ase*I fragments. Forty-one restriction sites for *Ase*I and *Xba*I were mapped onto the 3.7 Mb circular genome. Complicated cases of ambiguous cosmids that linked three or more restriction fragments, due to internal repeats in the *Rhodobacter* chromosome, were resolved by hybridization with riboprobes prepared by transcription of the ends of the inserts. After such experiments, only two linked restriction fragments typically remained. These cosmids revealed eight groups of repeated sequences. Five of these were attributed to the duplicated *gyrB (parE), nifB, rrn, rpoB* and *anf* genes. The three ways of hybridization used (fragments to fragments, fragments to cosmid sets and riboprobes to sets) make it possible to vary the probe/target ratio, thus distinguishing artifacts connected with different repeated sequences. Comparison of all these data gives the final long-range map without contradictions.

At this stage the cosmid clones were grouped in about 80 subcontigs, corresponding to macrorestriction fragments and restriction sites. They formed two groups, one corresponding to the chromosome of *Rb. capsulatus,* the other to its 134 kb plasmid (Fonstein et al., 1992).

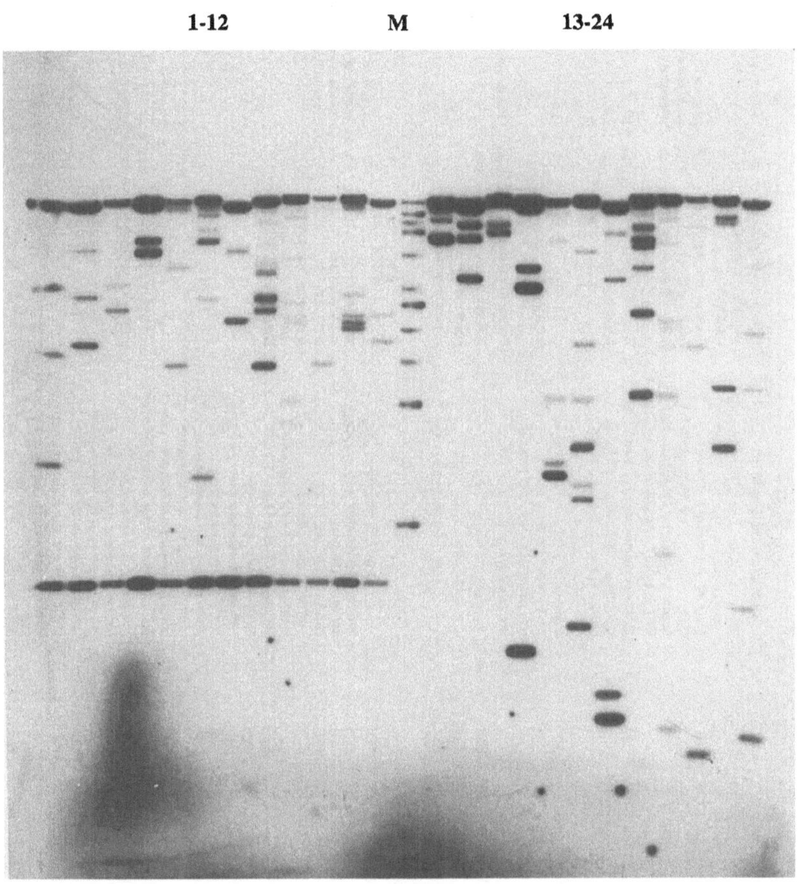

Fig. 1. Physical mapping of cosmids (previously linearized with λ terminase) by partial digestion with *Eco*RV and *cos*-site end-labeling. Lanes M: Takara DNA marker (4.6 – 48.5 kb); lanes 1–11: annealing with *cos*L; lanes 12–22: annealing with *cos*R. Measurement of the fragment sizes gives directly the distance of each *Eco*RV site from the left or right end of each cosmid, respectively.

Table 1. Rhodobacter genes mapped in this study

#	Gene/cluster	Gene mapped	Cosmid address	Sequenced DNA (nt)	# of genes/ ORFs	Function or corresponding protein	Mutant phenotype	Organ- ism	Genetic regulation	Gene bank entry	Reference
Major house keeping genes											
1	*groEL*	*groEL*	2C4-2C7	nd	1	Major chaperonin of bacterial cell	Essential for cell growth, assembly of enzymatic complexes**	sph	Induced by heat shock	Not in GenBank	R.Tabita, unpubl.
2	*dnaA, dnaN, gyrB*	*gyrB*	2G8-2G9	4 kb	3	DnaA, N - initiation of replication GyrB - β-subunit of topoisomerase II, responsible for negatively supercoiling DNA in an ATP-dependent manner.	Essential for cell growth maintaining the supercoiling balance**	cap	Regulated by changing of negative supercoiling (indirect self-regulation)	Not in GenBank	E. Sveen, unpubl Kranz et al., 1992
3	*parE*	*parE*	1A6	3.1 kb	1	Subunit of topoisomerase IV, essential for cell partition, supposedly by resolving catenated chromosomes in *E. coli* and *Salmonella*	Essential for cell growth (chromosome partition)**.	cap	nd	Not in GenBank	E. Sveen, unpubl.
4	*nusG, rpl, rpoB, rpoC*	*rpoB*	2A3	7 kb	4	RpoB, C - β, β'-subunit of RNA polymerase NusG - antitermination *rpl* - ribosomal proteins (50S subunit)	Essential for transcription**	cap	Growth rate dependent	Not in GenBank	B. Abella and L. Scappino, unpubl.
5-A9	*rrn*	S16'+S23'	**A** 1D3, 4	1470	4·2	ribosomal RNA	Essential for protein	cap	Growth rate dependent	RCRN23S X06485	Regensburger et al., 1988
6-B	4 clusters		**B** 1F6, 7	2884			synthesis (variations in	sph		RCARRDA M34129	
7-C			**C** 2A11, 2A5, 6;				number of copies are viable)**				
8-D			**D** 2G2, 3								
Genes involved in nitrogen fixation and linked with them											
9	*nif*-cluster A *(orf19, 18, 17, 16, 14, fdx P, N (C=P?), C, orf 9,10, nifENX-orf4-fdxB- orf5-nifQ-orf6-nifUSVW, nifA, nifB-orf1-nifZ)*	*nifA-2*	2F9-10 4638 638	3570 4282 2977 5540	15/26	Nitrogenase cofactor biosynthesis (major nitrogen fixation complex) and related proteins Orf18, 17 - Fe-S proteins FdxPN(C?) and B- ferrodoxins NifEN, Q and B - FeMoCo synthesis NifX- FeMoCo synthesis/regulation NifUS -iron cluster assembly NifV - homocitrate synthesis NifW - homocitrate processing NifA - nif specific transcriptional activator (final part of fixed N sensing cascade) NifZ - full activation or catalytic stability of FeMo protein of nitrogenase (nifDK)	*nifS*-Nif *nifV - leaky* Nif⁻ *nifUI, UII, W* and *orf6* - Nif⁻ *fdxN* and/or *fdxC* -Nif⁻ *nifA, B* - Nif⁻ (but in double mutant)	cap sph	*nif* regulatory cascade is induced by low oxygen pressure and low ratio glutamine/α- ketoglutorate	RCNIF X68444 RCNIFAB X07567 RCNIFS X17433115 RCAFDXC D1362523 RCAFDXN M31073 RCPTFAF1 X51316 RCANIFA M86823	Masepohl et al., 1993; Saeki et al., 1990
10	*nifHDK* cluster **B** *nifHDK,* *nifU-* *nifR4, nifA,* *nifB,* *fdxD*	*nifA-1* *nifR4*	1H3	2466 381 72 2423 1040 512 854 936 2583 2083	8	Nitrogenase and its regulation NifH - Fe protein of nitrogenase (major nitrogen fixation complex) containing [4Fe-4S] cluster NifDK - FeMo protein of nitrogenase, a tetramer containing 2 FeMo cofactors NifR4 - *nif* specific sigma factor FdxD -ferrodoxin	*nifH, D, K* and R4 -Nif⁻	cap	*nif* regulatory cascade	RCANIFHD M15270 RCANIFPRA M29400 RCANIFPRB M29401 RCFDXD X63352 RCNIFH X0786666 RCNIFK X63354 RCNIFKD X6335368. RCNIFKG X63196 RCNIFR4 X12358 RCNIFR4B X15437	Preker et al., 1992; Klipp et al., 1988 Jones and Haselkorn, 1989

Table 1. Continued

#	Gene/cluster	Gene mapped	Cosmid address	Sequenced DNA (nt)	# of genes/ ORFs	Function or corresponding protein	Mutant phenotype	Organ-ism	Genetic regulation	Gene bank entry	Reference
11	*nifR1* cluster *niR1, nifR2, nifR3*	*nifR1*	1B7-9	8030	3	*nif* regulation homologous to NtrBC NifR2 - kinase phosphorylating NifR1 in response to changing concentration of GlnB-UMP	Nif⁻	cap	*nif* regulatory cascade	RCNIFR12 X12359	Jones and Haselkorn, 1989
12	*adgA*	*adgA*	1D5,6	2432	1	ATP-binding protein involved in nitrogen metabolism	Ammonia dependent growth in *Rb. capsulatus* (homolog is essential for growth of *E. coli)*	cap	nd	RCADGA X59399	Willison, 1992
13	*glnAB*	*glnA*	1C1, 2	1360 2460	2	*nif* regulation GlnB - target of uridylylation by UTase in response to low ratio of glutamine/α-ketoglutorate GlnA - glutamine synthetase	Tn*5* in *glnB* - constitutive *nif* expression, glutamine auxotrophy (polar effect) *glnA*- glutamine auxotrophy	cap sph	Two-fold induction by nitrogen limitation. (glnBA-lacZ transtional fusion). NifR1 is responsible for 50% of the effect	RCAGLNAB M28244 . RSBLNBAA X71659	Kranz et al, 1990.
14-A	*anfGDHA*	*anfGDHA*	A-1F3, 4 B-2C1, 2B12	nd	4	Alternative nitrogenase	nd	cap	repressed by Mo, NH_4 and O_2	Not in GenBank	W. Klipp, unpubl.
15-B											
Genes of the photosynthetic apparatus											
16	*reg* cluster (*ahcY* *hvrAB, regA, regB orf5 and 7)*	*regA-hvrA*	2H3, 2G11	1642 1105	4/6	Regulation of the photosynthetic apparatus Ahc - adenosyl-homocysteine hydrolase. It controls ratio S-aden-met/ S-aden-homocys that influences methylation of protoporphyrin in BChl biosynthesis RegA - anaerobic response activator of LH and RC gene expression hvr - nd *hvrA*⁻ fails to trans-regulate LHI and RC expression in response to alterations in light intensity	*ahc*⁻ - methionine, homocysteine auxotrophy. *regA*⁻ *regB*⁻ fails to trans-activate LH and RC gene expression but not bch, produces ICM vesicles. Becomes essential in dim light	cap	*ahc* transcription is repressed by high light *regA* activity is regulated by the sensor kinase RegB via phosphorylation	RCAAHCY M80630 RCAREGA M64976 RCASENKIN L35179 RCAHURB L23836 RCHVRA X67236	Sganga et al., 1992; Sganga and Bauer, 1992 Mosley et al., 1994 Buggy et al., 1994a Buggy et al., 1994b
17	*pucABCDE*	*pucA*	2C9	600 3169 743 2846 1078 437 2846	5	LH-II *pucA, B* - α and β polypeptides of the light-harvesting complex B800-850 (LH-II) *pucC* - essential for high level transcription of the genes of the LHII complex *pucD, E* (γ subunit of LH-II) - stabilize the B800-850 complex.	LH-II⁻	cap sph	Anaerobic induction (30 fold)	RCALHII K0233734. RCAPUCOA M28510 RCAPUCAB M16777 RSPUCBAC X68796 RCALH2A M28360 RSB800AB X05200 RSPUCBAC X68796	Tichy et al., 1991; Tichy et al., 1989
18	Photosynthetic cluster	*puhA* *pufQ*	1G7-10	45959 4023 4844		Biosynthesis of chlorophylls - *bch* genes; LH and RC proteins - *puh* and *puf* genes; biosynthesis of carotenoid - *crt* genes	Detailed description below		Detailed description below	RCPHSYNG Z11165 RCARC1 K01183 RCARC2 K01184 K01185 K01186	D. Burke, unpubl. Youvan et al., 1984 Alberti et al., Chapter 50
18.1	Bacterio chlorophyll biosynthetic genes. Three subclusters *bchB-M* *bchE-I* *bchC,A*	*puhA* *pufQ*	1G7-10	1194 314 760 5606	13	*bchCA* operon *bchFNBHLM* operon *bchEJGDI* operon different stages of the biosynthesis of bacteriochlorophyll	Unable to synthesize BChl or its precursors and grow photosynthetically; accumulate corresponding precursor	cap sph	Limited (3 times) anaerobic induction through complicated mechanism: weak moderately oxygen regulated and stronger oxygen dependent promotors (anaerobic induction), cotranscribed *puh* and *puf* genes have their own regulated promotors.	RCABCHC M29966 RCABCHH M34843. RCBCHC X16164 RSBCHPUF X68795 X63320	Wellington and Beatty, 1989; Biel, 1992; Wright et al, 1991; Bauer et al., 1993; Bauer et al., 1991

Table 1. Continued

#	Gene/cluster	Gene mapped	Cosmid address	Sequenced DNA (nt)	# of genes/ ORFs	Function or corresponding protein	Mutant phenotype	Organ- ism	Genetic regulation	Gene bank entry	Reference
18.2	Carotenoid cluster *crtA-K* and *crtJ*	*puhA* *pufQ*	1G7-10	1194 1660 1794	9	Nine *crt* genes were attributed to certain stages of C_{40} biosynthesis (carotenoids protect cells from photo-oxidative damage). Totally 13 steps of C_{40} biosynthesis were proposed.	Unable to synthesize corresponding carotenoid, accumulate precursor, light sensitive	cap sph	Two-fold anaerobic and high light intensity induction. (These results are interpreted based on the protective function of carotenoids under high light intensity in the presence of O_2).	RCABCHC M2996647. J04969 J04969. RSCRTD X63204	Armstrong et al., 1989; Armstrong et al., 1990
18.3	*puhA*	*puhA*	1G7-10	783 199 711	1	H polypeptide of the RC complex	LH^-	cap sph	The expression of the genes coding for RC-L, RC-M, and RC-H is coordinately regulated by light intensity and O_2 concentration (30 times induction). An increase in light intensity causes a decrease in the expression of the genes for LH-I, LH-II, and RC proteins.	RSPUHAG X63378 RCAPUHA M14732 RCAPUFAB M15105	Zhu and Hearst, 1986
18.4	*pufQBALMX*	*pufQ*	1G7-10	932 814 59 432 199 886	5/6	LH-II and RC complex; PufBA- α and β subunits of LH-I; PufLM - L and M subunits of RC complex; PufQ - required for BChl biosynthesis	LH^-	cap sph	The same as above	RCAPUFQ M2275236. RCAPUFQBM20141 J03183 S97551 S97551 RCAIMP J05098 RCAPUHA M14732 RCARCL M10206	Zhu and Hearst, 1986
19	*ORF798*	*ORF798*	1A7	813	1	Gene involved in oxygen-regulated expression of the *puf* and *puc* operons	nd	cap	nd	RCORF798A Z21973	M. Pollich, unpubl.
Carbohydrate biosynthesis and transport											
20	*cbbL (rbcL) crfX, fbpA, prkA, fbaA, rbcS**	*cbbL*	1H2, 1H3	5628* 98* 4453*	6*	Large subunit of 1,5 bisphosphate carboxylase/oxygenase form I (RubisCO)	No phenotype due to suppression by the other form of RubisCO	sph	Light induction.	RCAFICO2F M64624 RCAPRKAA M28006. RCAPRKBA M68914	Gibson et al., 1991; Falcone et al., 1988
21	*cbbM (rbpL) fbpB, prkB, tklB, gapB, fbaB**	*cbbM*	1B5, 1B6	2099*	6*	Large subunit of 1,5 bisphosphate carboxylase/oxygenase form II (RubisCO)	No phenotype due to the suppression by the other form of RubisCO	sph	Light induction.	RCAFBPPRK J02922	Gibson et al., 1990
22 23	*cbbX*	*cbbX*	A-2A5 B-1D1		1	ORF of unknown function, linked with *cbbS*⁻ in *Rb. sphaeroides*	nd	sph	nd	no entry	R. Tabita, unpubl.
24	*ORF*	*ORF*	2F, 2		1	Gene required for photolithoautotrophic growth	nd	sph	nd	no entry	R. Tabita, unpubl.
25	*cbbE*	*cbbE*	2A7			pentose-5-phosphate-3-epimerase		sph		no entry	R. Tabita, unpubl.
26	*fruB, K, A*	*fruA*	2C10, 11	5646 5642	3	Sugar transport and uptake; FruA - integral membrane enzyme II (fructose permease); FruB - peripheral membrane enzyme I (multiphosphoryl transfer protein); FruK - soluble fructose-1-phosphate kinase	*fruA::Tn5* - fructose-negative, glucose-positive	cap	100-fold inducible (The uninduced mutant fruA exhibited measurable activities of both enzyme I and fructose-1-phosphate kinase, which were increased three-fold when grown in the presence of fructose.)	RCAFRUAK M62785 M68879 X53150 RCFRUOP X53150	Daniels et al, 1988; Wu and Saier, 1990; Wu et al, 1990

Table 1. Continued

#	Gene/cluster	Gene mapped	Cosmid address	Sequenced DNA (nt)	# of genes/ ORFs	Function or corresponding protein	Mutant phenotype	Organ-ism	Genetic regulation	Gene bank entry	Reference
27	*dctS, dctR and dctP*	*dctP*	2E8, 9	2753 1062	3	DctP - major permease for the C4-dicarboxylates (malate, succinate and fumarate) aerobic dependent transport. (There is another anaerobic-light dependent low-affinity system). Dct S predicted to be a membrane bound sensor-kinase and DctR a response-regulator (all by comparison with FixJ)	*dctS*, *R* or *P* unable to grow on malate, succinate and fumarate in the dark due to the transport deficiency	cap	nd	RCDCTSRS X64733 RCDCTP X63974	Shaw et al., 1991; Hamblin et al., 1993
Genes coding electron transfer proteins											
28	hydrogenase cluster *(hupR2-hyp-ORF20)*	*hupSL*	1G2-4	4175 2271 5456 4920 3052 4175 4674 4126 5456	18/20	HupSL - small and large subunits of hydrogenase. Its functions are H_2-uptake for authotrophic growth with H_2 as the electron source and aerobic oxidation of H_2 coupled with oxygen reduction. HypA, F, HupR1 - regulators HupD, G, C, D - hydrogenase processing HupM -e^- carrier HupJ - rubredoxin HypB - Ni incorporation HupF, H and K and HypE - unidentified but essential	Unable to consume H_2 (Hup^- phenotype)	cap	Complicated pattern responding to H_2 (induction via HypF) and various environmental switches. Process involvesHupR1 (response regulator activating expression in various conditions) and HypA (repressor of *hup*SL)	RCHUPLS X13520 RCHYPFG Z15087 RCHYPHUP X61007 RCORF1920 Z15088 RCAHYOX M55089 RCHUPR2U X57380 RCHUPXG Z15089 RCHYPHUP X61007	Colbeau et al., 1993
29	cytochrome *c* biogenesis cluster *ORF124, helABCDX, hpt, secDF*	*helAB*	1B9	760 3272 776	7/8	HelX - periplasmic disulphide oxidoreductase involved in cytochrome *c* biogenesis. (c-type cytochromes have heme covalently attached to Cys in conserved motif (c_2 is periplasmic, c_1 isa component of the bc_1 complex) HelABCD-heme transporter in cytochrome *c* biogenesis SecDF - named by homology to the *E.coli* genes essential for protein export. Hpt - homolog of hypoxanthine phosphoribosyltransferase	*hel*⁻ cannot grow anaerobically, unable to oxidize cytochrome specific electron donor TMPD It was not possible to inactivate secDF to see phenotype (presumably essential for cell growth).	cap	Induced under dark growth anaerobic conditions	RCAHELDTRX M96013 RCHEL X63462 RCHPT X60977	Beckman and Kranz, 1993; Beckman et al., 1992; Beckman and Kranz, 1991a
30	*ccl* cluster *ccl1 ccl2 ORG257 argD*	*ccl1* and *2*	1F3-5	1002 3155	4	ORF257 - enoyl-CoA hydratase homologue (related to mitochondria but not to *E. coli* and peroxisomes) Ccl - required for the biogenesis of *c*-type cytochromes, demonstrate 52% homology with ORFs from chloroplasts	*ccl*⁻ cannot grow anaerobically	cap	Induced under dark growth anaerobic conditions	RC257 X60194 RCCCL12 X63461	Beckman and Kranz, 1991b; Beckman et al., 1992
31	*cycA*	*cycA*	1E2-5	1242 1538 548	1	Soluble periplasmic cytochrome c-2 performs electron transfer between membrane proton-translocating ubiquinol:cytochrome c_2 oxidoreductase (bc_1-complex) and cytochrome oxidase during respiration; and between bc_1 and oxidized photochemical RC during photosynthesis.	Not essential in *R. capsulatus* (function is duplicated by CytY), but essential in *Rb. sphaeroides*	cap sph	Induced under dark anaerobic growth conditions	RCACYCA M12776 RCAC2AA M64777 RCACYC2 M14501	Daldal et al., 1986

Table 1. Continued

#	Gene/cluster	Gene mapped	Cosmid address	Sequenced DNA (nt)	# of genes/ ORFs	Function or corresponding protein	Mutant phenotype	Organ-ism	Genetic regulation	Gene bank entry	Reference
32	*pet* cluster *petABC* (*fbcFBC*) and *perP, R*	*petABC*	1D8, 9	315 4007 1672 316 3381 3874	5	PetA - Rieske Fe-S protein PetB - cytochrome b PetC - cytochrome c_1 - ubiquinol-cyto-chrome c_2 oxidoreductase. Electron transfer in respiration and photosynthesis by oxidation of quinols and eventual reduction of soluble electron carrier. PetR deduced amino acid sequence is homologous to various bacterial response regulators, especially OmpR subgroup.	PetABC are essential for photosynthetic growth. PetR is essential for photosynthetic and respiratory growth .	cap sph	nd	RCAPETAR M18576 RCPETG X05630 RCPETPR Z12113 RCAPETA M18577 RSFBCOPER X56157 RCFBC X03476	Davidson and Daldal, 1987a,b Tokito and Daldal, 1992
Other metabolic genes											
33	sucA	**sucA**	1G5	332 360	1	α-ketoglutarate dehydrogenase	nd	cap	nd	RCASUCAA L10207 RCASUCAB L10208	F. Dastoor, upubl.
34	trpC	**trpC**	2B9-11	801	1	Indolglycerol phosphate synthase	*trp*$^{-}$**	cap	Repression, attenuation, rel	RCATRPC M97640	Becker et al., 1992

cap –*Rb. capsulatus*
sph – *Rb. sphaeroides*
* – in *Rb. sphaeroides*
** – in *E. coli*
nd – not described

III. High Resolution Physical Mapping

To start the high resolution mapping, 40 cosmids located on different macrorestriction fragments were chosen. Individual cosmids were cleaved with λ terminase, partially digested with *Eco*RV, annealed to labeled oligonucleotides complementary to *cos*L and *cos*R independently and then separated on 0.3% agarose gels (Fig. 1). The pattern of *Eco*RV sites can be read directly from these gels for each cosmid. Riboprobes generated by SP6- and T7-specific transcription of the ends of the inserts of these cosmids were then used to reveal groups of neighboring cosmids. For the next round of cosmid walking, the overhanging ends of these secondary cosmids were used for transcription and hybridization. Finally, after about 300 riboprobe hybridizations with the first set of 864 cosmids and about the same number of *cos*-mappings, six uninterrupted contigs were mapped with high resolution.

Four of the remaining gaps had a statistical origin and were closed using the second set of 864 clones in similar hybridizations with the probes corresponding to the ends of the contigs. One of the remaining two gaps was linked by phages selected from a λDASH library, possibly corresponding to DNA unclonable in cosmids. Finally, a minimal uninterrupted set of 186 cosmids mapped with high resolution covered 3.7 Mb of the chromosome of *Rb. capsulatus* (Fig. 2). Only one gap inside fragment *Ase*7/*Xba*8 remained and its size, according to a comparison of the *Eco*RV and *Ase*7/*Xba*8 map, is less than 10 kb.

Thirty-four loci, corresponding to individual cloned genes or large gene clusters were mapped with a resolution of 1–10 kb. A description of these genes and some of their genetic characteristics are presented in Table 1. Another four genes recently cloned from *Rb. capsulatus* and 18 genes, cloned from *Rb. sphaeroides*, whose *Rb. capsulatus* homologues are not yet mapped are in the process of mapping. Using PCR-generated probes, corresponding to 16S and 23S rRNA, and digestion with *Ceu*I, a rarely-cutting endonuclease highly specific for eubacterial *rrn* operons (Ralph et al., 1993), the orientation of transcription of three of the *rrn* operons was established (Fig. 2).

IV. Comparison with the Existing Genetic Map

Aligning this physical map with the existing genetic one (Willison et al., 1985; Willison, 1993; Chapter 55), it is possible to compare eight markers. Their relative positions are the same on both maps, with one exception. On the genetic map, *glnA* is located between *trp* and *nifR1* as is the case on our earlier physical map (Fonstein et al., 1992). However, this region is now reassigned on the new map (Fig. 3). The repeated sequences responsible for our earlier mapping mistake might act the same way in the genetic experiments, altering the apparent order of the genes. Another possible explanation is that these repeated sequences induce real genome rearrangement, and the strains used in the two laboratories now differ.

V. Comparison of the Chromosomal Maps of Different *Rhodobacter* Strains

With regard to strain comparison, the defined positions of the ends of cosmids mapped in the existing ordered library can be used as ordered probes for a variety of closely related strains, drastically simplifying the construction of their high resolution maps. Points where continuity is lost can be attributed to chromosomal rearrangements that can be resolved by the same strategy used in primary mapping.

Pulse field separated restriction patterns of 14 isolates of *Rb. capsulatus* were analyzed and 3 strains, namely B1, St. Louis and 2.3.1. were chosen for detailed comparison. The evolutionary distance between strains 2.3.1 and St. Louis (40–60 % similarity in *Eco*RV fragments) revealed in the study seems to be optimal for the proposed strategy. Cosmid libraries of the chosen *Rhodobacter* strains were hybridized with probes generated by SP6 and T7 – specific transcription from the ends of the inserts of the cosmids of the *Rb. capsulatus* SB1003 minimal set.

Totally about 80 sequential probes corresponding to six areas of the genome and covering about one third of it were used. Cosmids linking the probes were mapped with high resolution, and assembled in contigs. In preliminary analysis a large number of genome-changing events were revealed. Among them are a deletion that is greater than 60 kb in strain 2.3.1, a set of 1–5 kb deletion/insertions and an inversion/translocation with the ends separated by 400 kb in strains B1 and St. Louis, relative to SB1003 and 2.3.1., and point mutations in the *Eco*RV sites. To get a full descriptive picture of strain diversity we plan to

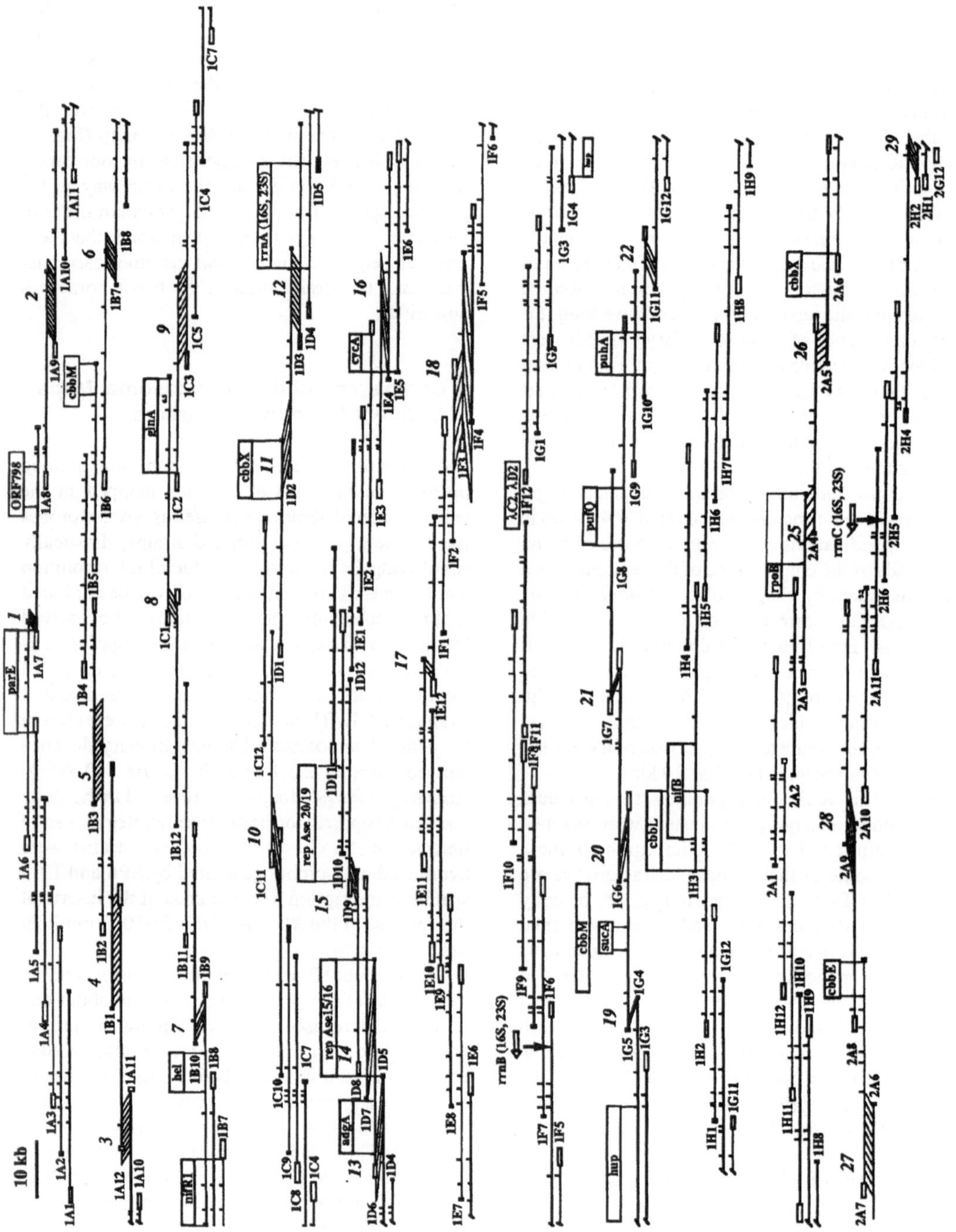

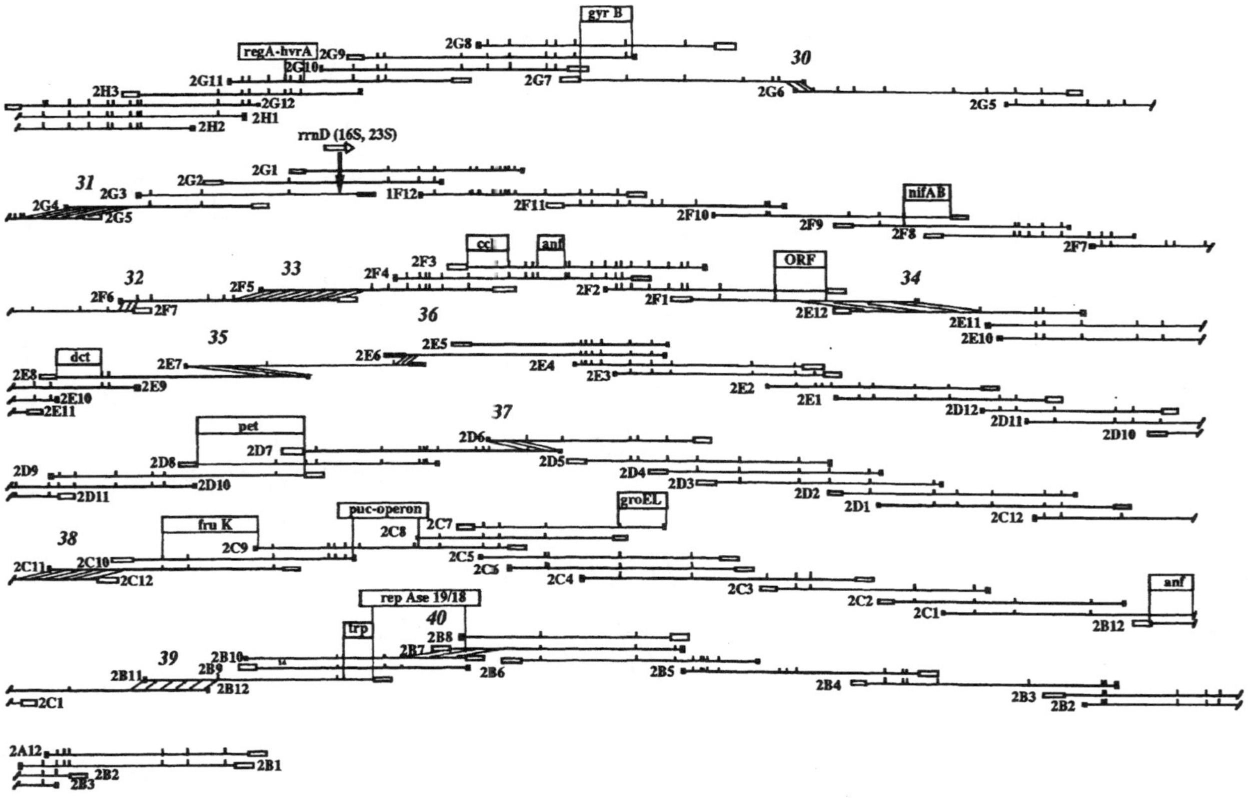

Fig. 2. Combined high resolution physical and genetic map of the chromosome of *Rb. capsulatus*. Each cosmid is represented by a horizontal line, on which the vertical ticks are the EcoRV sites. The name of each cosmid is on its left, e.g. 1A1, 1A2 etc. There are 40 places where the distribution of *Eco*RV sites does not permit the precise size of the overlaps to be determined; these are designated by diagonal connecting lines and italicized numbers from 1 to 40. On each cosmid, the L and R *cos* sites are shown by large and small boxes, respectively. Mapped genes and repeated elements are indicated by boxes above each cosmid. The physical map is read continuously from left to right. Designated gene positions correspond to the *Eco*RV fragments revealed in blot-hybridizations. In case of the *rrn* operons, their location is fixed by the *Ceu*I cleavage site, marked by an arrow. Cosmids 1A1 and 2A12, which are separated by the only remaining gap in the fine structure map, are connected by their hybridization to the restriction fragments Xba-8 and Ase-7 (Fig. 1). Note that the operons *rrnB* and *rrnC* have the same orientation, while *rrnD* has the opposite orientation.

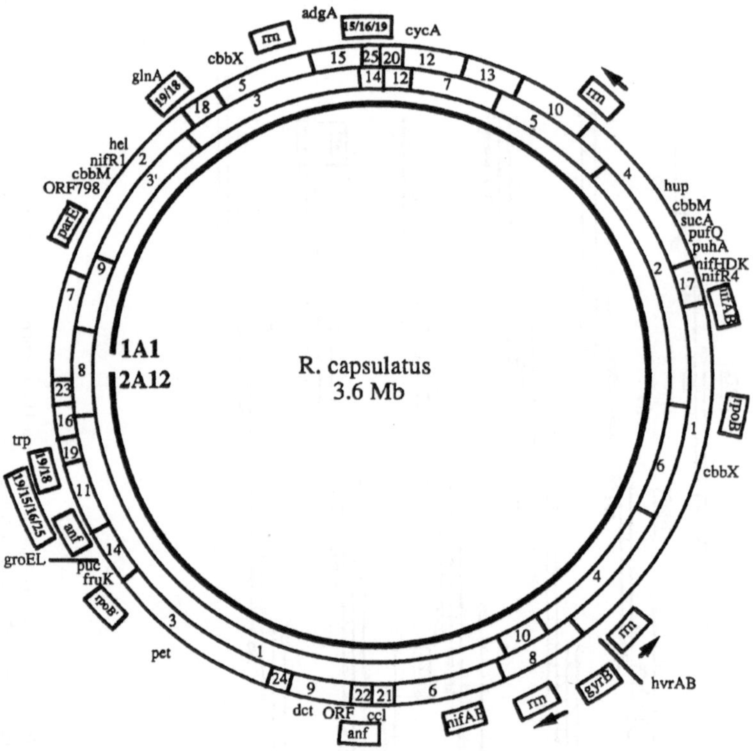

Fig. 3. Corrected long range physical and genetic map of the *Rb. capsulatus* chromosome. The inner circle shows the *Xba*I sites, the outer circle the *Ase*I sites, the numbers in these circles indicating the macro-restriction fragments. The internal arc represents an uninterrupted set of cloned fragments. Numbers at the end of the arc designate cosmids flanking the gap. Boxes represent repeated DNA, attributed to genes or to unidentified (numbered) sequences in the macro-restriction fragments.

complete the mapping of the chromosomes of these three strains to have their ends meet.

VI. Possible Applications of the Minimal Cosmid Set of *Rhodobacter capsulatus* SB1003

An *E. coli* cosmid hybridization panel was used to study the chromosomal localization of coordinately induced multigenic systems (Chuang et al., 1993) by probing with differentially expressed RNAs. Complications, solved by sophisticated computerized image processing, can be avoided by splitting the 40 kb cosmid inserts into smaller pieces, as recently shown in Chuang and Blattner (1993). DNA samples of the 192 cosmids from the *Rb. capsulatus* minimal set were digested with *Eco*RV, separated in two gels and transferred to nylon membranes (Fig. 4). The resulting blots carry about 1500 *Eco*RV fragments already mapped on the chromosomal restriction map. It divides the entire *Rhodobacter* genome into physically separated gene-size pieces, and should drastically increase specificity during probing with any specific RNAs. This approach should produce reliable data on functional genome architecture and give access to new genes involved in commonly regulated processes.

Acknowledgment

This work was supported by a grant from the Department of Energy (86 ER 13546).

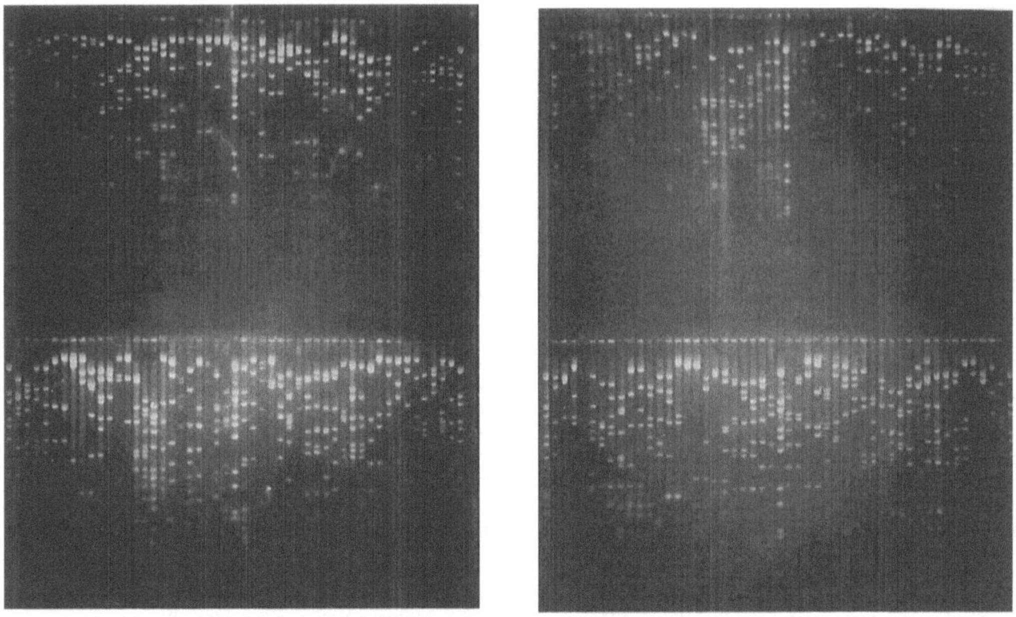

*Fig. 4. Eco*RV restriction patterns of all 192 cosmids from the minimal set, comprising the entire *Rhodobacter* genome. The center and outermost lanes on each gel contained λ DNA standards. The last six lanes on the lower right contain cosmids whose inserts derive from the 134 kb plasmid. The other lanes contain DNA from cosmid 1A1 (upper left) through 2A12 (lower right).

Note Added in Proof

The physical map of Fig. 2 was current in the summer of 1993. A revised version, with 3600 restriction sites mapped, was recently published by Fonstein et al. (1995a). A high-resolution alignment of the maps of three strains of *Rhodobacter capsulatus* was recently published by Fonstein et al. (1995b). The latter work reveals a mosaic chromosome organization, in which chromosomal regions of 15–80 kb with highly polymorphic restriction sites are interspersed with regions in which most of the sites are conserved among all three strains.

References

Armstrong GA, Alberti M, Leach F and Hearst JE (1989) Nucleotide sequence, organization, and nature of the protein products of the carotenoid biosynthesis gene cluster of *Rhodobacter capsulatus*. Mol Gen Genet 216:254–268

Armstrong GA, Schmidt A, Sandmann G and Hearst JE (1990) Genetic and biochemical characterization of carotenoid biosynthesis mutants of *Rhodobacter capsulatus*. J Biol Chem 265:8329–8338

Bauer C, Buggy J and Mosley C (1993) Control of photosystem genes in *Rhodobacter capsulatus*. Trends Genet 9:56–60

Becker RM, Young DA and Marrs BL (1992) Sequence of the indoleglycerol phosphate synthase (*trpC*) gene from *Rhodobacter capsulatus*. J Bacteriol 174:5482–5484

Beckman DL and Kranz RG (1991a) A bacterial homolog to HPRT. Biochim Biophys Acta 1129:112–114

Beckman DL and Kranz RG (1991b) A bacterial homolog to the mitochondrial enoyl-CoA hydratase. Gene 107:171–172

Beckman DL and Kranz RG (1993) Cytochromes c biogenesis in a photosynthetic bacterium requires a periplasmic thioredoxin-like protein. Proc Natl Acad Sci U S A 90:2179–2183

Beckman DL, Trawick DR and Kranz RG (1992) Bacterial cytochromes *c* biogenesis. Genes Dev 6:268–283

Biel AJ (1992) Oxygen-regulated steps in the *Rhodobacter capsulatus* tetrapyrrole biosynthetic pathway. J Bacteriol 174:5272–5274

Buggy JJ, Sganga MW and Bauer CE (1994a) Nucleotide sequence and characterization of the *Rhodobacter capsulatus hvrB* gene: HvrB is an activator of *S*-adenosyl-L-homocysteine hydrolase expression and is a member of the LysR family. J Bacteriol 176: 61-69

Buggy JJ, Sganga MW and Bauer CE (1994b) Characterization of a light responding *trans*-activator responsible for differentially controlling reaction center and light harvesting-I gene expression in *R. capsulatus*. J Bacteriol 176: 6936-6943

Chuang EC and Blattner FR (1993) Characterization of twenty six new heat shock genes of *Escherichia coli*. J Bacteriol

175:5242–5252

Chuang SE, Daniels DL and Blattner FR (1993) Global regulation of gene expression in *Escherichia coli*. J Bacteriol 175:2026–2036

Colbeau A, Richaud P, Toussaint B, Caballero J, Elster C, Delphin C, Smith RL, Chabert J and Vignais PM (1993) Organization of the genes necessary for hydrogenase expression in *Rhodobacter capsulatus*. Sequence analysis and identification of two *hyp* regulatory mutants. Mol Microbiol 8: 15–29

Daldal F, Cheng S, Applebaum J, Davidson E and Prince RC (1986) Cytochrome c2 is not essential for photosynthetic growth of *Rhodopseudomonas capsulata*. Proc Natl Acad Sci USA 83:2012–2016

Daniels GA, Drews G and Saier MJ (1988) Properties of a Tn*5* insertion mutant defective in the structural gene (*fruA*) of the fructose-specific phosphotransferase system of *Rhodobacter capsulatus* and cloning of the *fru* regulon. J Bacteriol 170:1698–1703

Davidson E and Daldal F (1987a) *fbc* operon, encoding the Rieske Fe-S protein cytochrome b, and cytochrome c1 apoproteins previously described from *Rhodopseudomonas sphaeroides*, is from *Rhodopseudomonas capsulata*. J Mol Biol 195:25–29

Davidson E and Daldal F (1987b) Primary structure of the bc1 complex of *Rhodopseudomonas capsulata*. Nucleotide sequence of the *pet* operon encoding the Rieske cytochrome b, and cytochrome c_1 apoproteins. J Mol Biol 19:513–524

Donohue TJ and Kaplan S (1991) Genetic techniques in *Rhodospirillaceae*. Methods Enzymol 204:459–485

Fonstein M and Haselkorn R (1993) Chromosomal structure of *Rhodobacter capsulatus* strain SB1003: cosmid encyclopedia and high-resolution physical and genetic map. Proc Natl Acad Sci USA 90:2522–2526

Fonstein M, Zheng S and Haselkorn R (1992) Physical map of the genome of *Rhodobacter capsulatus* SB 1003. J Bacteriol 174:4070–4077

Fonstein M, Koshy EG, Nikolskaya T, Mourachov P and Haselkorn R (1995a) Refinement of the high-resolution physical and genetic map of *Rhodobacter capsulatus* and genome surveys using blots of the cosmid encyclopedia. EMBO J 14: 1827–1841

Fonstein M, Nikolskaya T and Haselkorn R (1995b) High-resolution alignment of a 1-megabase-long genome region of three strains of *Rhodobacter capsulatus*. J Bacteriol 177: 2368–2372

Gibson JL, Falcone DL and Tabita FR (1991) Nucleotide sequence, transcriptional analysis, and expression of genes encoded within the form I CO_2 fixation operon of *Rhodobacter sphaeroides*. J Biol Chem 266:14646–4653

Hamblin MJ, Shaw JG and Kelly DJ (1993) Sequence analysis and interposon mutagenesis of a sensor-kinase (DctS) and response-regulator (DctR) controlling synthesis of the high-affinity C4-dicarboxylate transport system in *Rhodobacter capsulatus*. Mol Gen Genet 237:215–224

Johnson JA, Wong WK and Beatty JT (1986) Expression of cellulase genes in *Rhodobacter capsulatus* by use of plasmid expression vectors. J Bacteriol 167:604–610

Jones R and Haselkorn R (1989) The DNA sequence of the *Rhodobacter capsulatus ntrA, ntrB* and *ntrC* gene analogues required for nitrogen fixation. Mol Gen Genet 215:507–516

Klipp W, Masepohl B and Puhler A (1988) Identification and mapping of nitrogen fixation genes of *Rhodobacter capsulatus*: duplication of a *nifA-nifB* region. J Bacteriol 170:693–699

Kranz RG, Pace VM and Caldicott IM (1990) Inactivation, sequence, and *lacZ* fusion analysis of a regulatory locus required for repression of nitrogen fixation genes in *Rhodobacter capsulatus*. J Bacteriol 172:53–62

Kranz RG, Beckman DL and Foster HD (1992) DNA gyrase activities from *Rhodobacter capsulatus:* analysis of target(s) of coumarins and cloning of the *gyrB* locus. FEMS Microbiol Lett 72:25–32

Marrs B (1981) Mobilization of the genes for photosynthesis from *Rhodopseudomonas capsulata* by a promiscuous plasmid. J Bacteriol 146:1003–1012

Mosely CS, Suzuki JY and Bauer CE (1994) Identification and molecular genetic characterization of a sensor kinase responsible for coordinately regulating light harvesting and reaction center gene expression in response to anaerobiosis. J Bacteriol 176: 7566-7573

Preker P, Hubner P, Schmehl M, Klipp W and Bickle TA (1992) Mapping and characterization of the promoter elements of the regulatory *nif* genes *rpoN, nifA1* and *nifA2* in *Rhodobacter capsulatus*. Mol Microbiol 6:1035–1047

Regensburger A, Ludwig W, Frank R, Blocker H and Schleifer KH (1988) Complete nucleotide sequence of a 23S ribosomal RNA gene from *Micrococcus luteus*. Nucleic Acids Res 16:2344

Saeki K, Miyatake Y, Young DA, Marrs BL and Matsubara H (1990) A plant-ferredoxin-like gene is located upstream of ferredoxin I gene (*fdxN*) of *Rhodobacter capsulatus*. Nucleic Acids Res 18:1060

Sganga MW and Bauer CE (1992) Regulatory factors controlling photosynthetic reaction center and light-harvesting gene expression in *Rhodobacter capsulatus*. Cell 68:945–954

Sganga MW, Aksamit RR, Cantoni GL and Bauer CE (1992) Mutational and nucleotide sequence analysis of S-adenosyl-L-homocysteine hydrolase from *Rhodobacter capsulatus*. Proc Natl Acad Sci USA 89:6328–6332

Shaw JG, Hamblin MJ and Kelly DJ (1991) Purification, characterization and nucleotide sequence of the periplasmic C4-dicarboxylate-binding protein (DctP) from *Rhodobacter capsulatus*. Mol Microbiol 5:3055–3062

Tichy HV, Oberle B, Stiehle H, Schiltz E and Drews G (1989) Genes downstream from *pucB* and *pucA* are essential for formation of the B800–850 complex of *Rhodobacter capsulatus*. J Bacteriol 171:4914–4922

Tichy HV, Albien KU, Gad'on N and Drews G (1991) Analysis of the *Rhodobacter capsulatus puc* operon: the *pucC* gene plays a central role in the regulation of LHII (B800–850 complex) expression. EMBO J 10:2949–2955

Tokito MK and Daldal F (1992) *petR*, located upstream of the *fbcFBC* operon encoding the cytochrome bc1 complex, is homologous to bacterial response regulators and necessary for photosynthetic and respiratory growth of *Rhodobacter capsulatus*. Mol Microbiol 6:1645–1654

Wellington CL and Beatty JT (1989) Promoter mapping and nucleotide sequence of the *bchC* bacteriochlorophyll biosynthesis gene from *Rhodobacter capsulatus*. Gene 83:251–261

Willison JC (1992) An essential gene (*efg*) located at 38.1 minutes on the *Escherichia coli* chromosome. J Bacteriol 174:5765–5766

Willison JC (1993) Biochemical genetics revisited: the use of mutants to study carbon and nitrogen metabolism in the photosynthetic bacteria. FEMS Microbiol Rev 10:1–38

Willison JC, Ahombo G, Chabert J, Magnin JP and Vignais PM (1985) Genetic mapping of *Rhodopseudomonas capsulata* chromosome shows non-clustering of genes involved in nitrogen fixation. J Gen Microbiol 131:3001–3015

Wright MS, Eckert JJ, Biel SW and Biel AJ (1991) Use of a *lacZ* fusion to study transcriptional regulation of the *Rhodobacter capsulatus hemA* gene. FEMS Microbiol Lett 62:339–342

Wu LF and Saier MJ (1990) Nucleotide sequence of the *fruA* gene, encoding the fructose permease of the *Rhodobacter capsulatus* phosphotransferase system, and analyses of the deduced protein sequence. J Bacteriol 172:7167–7178

Wu LF, Tomich JM and Saier MJ (1990) Structure and evolution of a multidomain multiphosphoryl transfer protein. Nucleotide sequence of the *fruB(HI)* gene in *Rhodobacter capsulatus* and comparisons with homologous genes from other organisms. J Mol Biol 213:687–703

Yen HC and Marrs B (1976) Map of genes for carotenoid and bacteriochlorophyll biosynthesis in *Rhodopseudomonas capsulata*. J Bacteriol 126:619–629

Yen HC, Hu NT and Marrs BL (1979) Characterization of the gene transfer agent made by an overproducer mutant of *Rhodopseudomonas capsulata*. J Mol Biol 131:157–168

Youvan DC, Bylina EJ, Alberti M, Begusch H and Hearst JE (1984) Nucleotide and deduced polypetide sequences of the photosynthetic reacton-center, B870 antenna, and flanking polypeptides from *R. capsulata*. Cell 37: 949-957

Zhu YS and Hearst JE (1986) Regulation of expression of genes for light-harvesting antenna proteins LH-I and LH-II; reaction center polypeptides RC-L, RC-M, and RC-H; and enzymes of bacteriochlorophyll and carotenoid biosynthesis in *Rhodobacter capsulatus* by light and oxygen. Proc Natl Acad Sci USA 83: 7613–7617

Chapter 50

Structure and Sequence of the Photosynthesis Gene Cluster

Marie Alberti, Donald H. Burke* and John E. Hearst
Calvin Lab, Lawrence Berkeley Laboratories, Berkeley, CA 94720, USA

Summary

A natural cluster of genes containing nearly all the genetic information specifically required for photosynthesis in *Rhodobacter capsulatus* has been sequenced. Within this 46 kilobase pair cluster are 43 identified genes which include the coding sequences for the structural reaction center polypeptides (RC), the structural light harvest I complex (LHI), as well as the coding sequences for the enzymes required for the biosynthesis of bacteriochlorophyll (BChl) and for the biosynthesis of the carotenoids, spheroidene and spheroidenone.

This cluster and the accompanying DNA sequence have been extremely valuable in the identification of DNA sequences in other organisms, both bacteria and green plants, with comparable functions associated with photosynthesis. These sequence homologies have been tabulated. In addition, these homologies provide evolutionary information relating to reaction centers, light harvesting complexes, carotenoid biosynthesis and chlorophyll biosynthesis. In particular, a strong structural similarity has been identified between some reductive enzymes in the chlorophyll biosynthetic pathways and nitrogen reductases associated with nitrogen fixation. This has led to a new interpretation of Granick's hypothesis, the hypothesis that steps in biosynthetic pathways can be assumed to be linearly associated with the order of evolution of the corresponding activity in time.

Finally, the *Rhodobacter capsulatus* photosynthetic gene cluster and the corresponding sequence will clearly be the source of genes for recombinant constructs, for mutational studies on reaction centers and light harvesting complexes and for the overproduction of enzymatic activities essential for mechanistic biochemistry and for structural studies for many years to come.

*Current address: Department of Molecular, Cellular and Developmental Biology, University of Colorado, Boulder, CO 80309, USA

R. E. Blankenship, M. T. Madigan and C. E. Bauer (eds): Anoxygenic Photosynthetic Bacteria, pp. 1083–1106.

I. Introduction

Rhodobacter capsulatus (Weaver et al., 1975) has provided a unique route to the study of genes that are essential for photosynthesis, principally by being the first photosynthetic system to lend itself to genetic manipulation. This was made possible largely by the pioneering work of Barry Marrs (1978). His initial discovery of the gene transfer system and his subsequent creation of the R-prime plasmid, pRPS404 (Marrs, 1981), containing nearly all of the genes that are essential for photosynthesis in a natural cluster, provided the means to manipulate and alter these photosynthesis genes in *Escherichia coli.*

We created a large random library of transposon mutants of these genes in *E. coli*, which were conjugated back into *Rb. capsulatus*, where the phenotypes of these insertional mutations were established (Youvan et al., 1982; Zsebo et al., 1984; Zsebo and Hearst, 1984). Simultaneously, the restriction maps for many common restriction enzymes were determined, leading to a high resolution physical DNA map that was superimposed on the somewhat lower resolution genetic map (Taylor et al., 1983; Zsebo et al., 1984; Zsebo and Hearst, 1984). For example, Youvan et al., 1984a, located restriction fragments containing the light harvesting I peptides and the reaction center peptides by identifying colonies of *Rb. capsulatus* which were incapable of performing photosynthesis, but which were capable of fluorescing in the infrared, following conjugation of the randomly mutated R-prime back into *Rb. capsulatus*. Restriction fragments containing transposon inserts in these fluorescent colonies were identified and then sequenced (Youvan et al., 1984a,b). Open reading frames for both LHI peptides were identified by comparing the translated DNA sequence with protein sequences of LHI peptides in *Rb. sphaeroides,* which were kindly provided to us by Professor Herbert Zuber. Reaction center peptides L, M, and H were identified from the amino terminal sequences of these proteins (Sutton et al., 1982; Williams et al., 1983; Worland et al., 1984). The plant analogs of these peptides, the PS II reaction center peptides, were rapidly identified by sequence comparison (Doolittle, 1986; Brutlag et al., 1990) with the bacterial genes (Hearst and Sauer, 1984a,b; Youvan et al., 1984b; Hearst, 1985, 1986), leading to many fundamental studies.

Since that early achievement, we have completed the sequence of the entire photosynthetic gene cluster of *Rb. capsulatus*, a sequence of 45,959 base pairs. This sequence has provided many additional insights into the photosynthetic systems of plants and in many cases has been the basis for the initial identification of function for newly sequenced plant genes. This 45,959 base pair sequence may be accessed in the EMBL DNA sequence database using accession number Z11165 (Genbank locus RCAPHSYNG). This review consists of a broad description and a summary of the information which has been gleaned from the *Rb. capsulatus* photosynthetic gene cluster DNA sequence by our laboratory.

II. Biosynthetic Pathways

In addition to the reaction center and light harvesting peptides, the gene cluster codes for the biosynthetic enzymes for the carotenoids (Yen and Marrs, 1976; Taylor et al., 1983; Armstrong et al., 1989, 1990a,b,c; 1993b; Schmidt et al., 1989; and see Chapter 53 by Armstrong) and for the bacteriochlorophylls (Taylor et al., 1983; Burke et al., 1993a,b; Armstrong et al., 1993a). Both classes of pigment molecules are essential for normal photosynthesis in plants and photosynthetic bacteria. Sequence similarity between *Rb. capsulatus* sequences and plant genes has provided great insight. Figure 1 displays the carotenoid biosynthetic pathway for *Rb. capsulatus* as well as for the non-photosynthetic bacterium, *Erwinia herbicola* (Hundle et al., 1991,1992,1993; Hundle and Hearst, 1991; Armstrong et al., 1993b) We have studied this second system both to provide comparison with the *Rb. capsulatus* genes and because *Erwinia* more closely mimics the plant biosynthetic pathway in that its late products include lycopene (tomato red) and β-carotene (carrot orange). Unlike *Rb. capsulatus*, the *Erwinia* genes can be expressed in *E. coli* allowing mutants to be complimented and identified. Farnesyl pyrophosphate, FPP, is a precursor for many essential structural products in all life forms. The photosynthetic gene cluster contains only the genes needed for the synthesis of carotenoids and carotenoid intermediates beyond FPP. The earliest gene in the pathway codes for CrtE, geranylgeranyl pyro-

Abbreviations: E. coli – Escherichia coli; E. herbicola – Erwinia herbicola; M. polymorpha – Marcantia polymorpha; N. – Neurospora; Rb. – Rhodobacter; Rp. – Rhodopseudomonas; Rs. – Rhodospirillum; S. – Saccharomyces

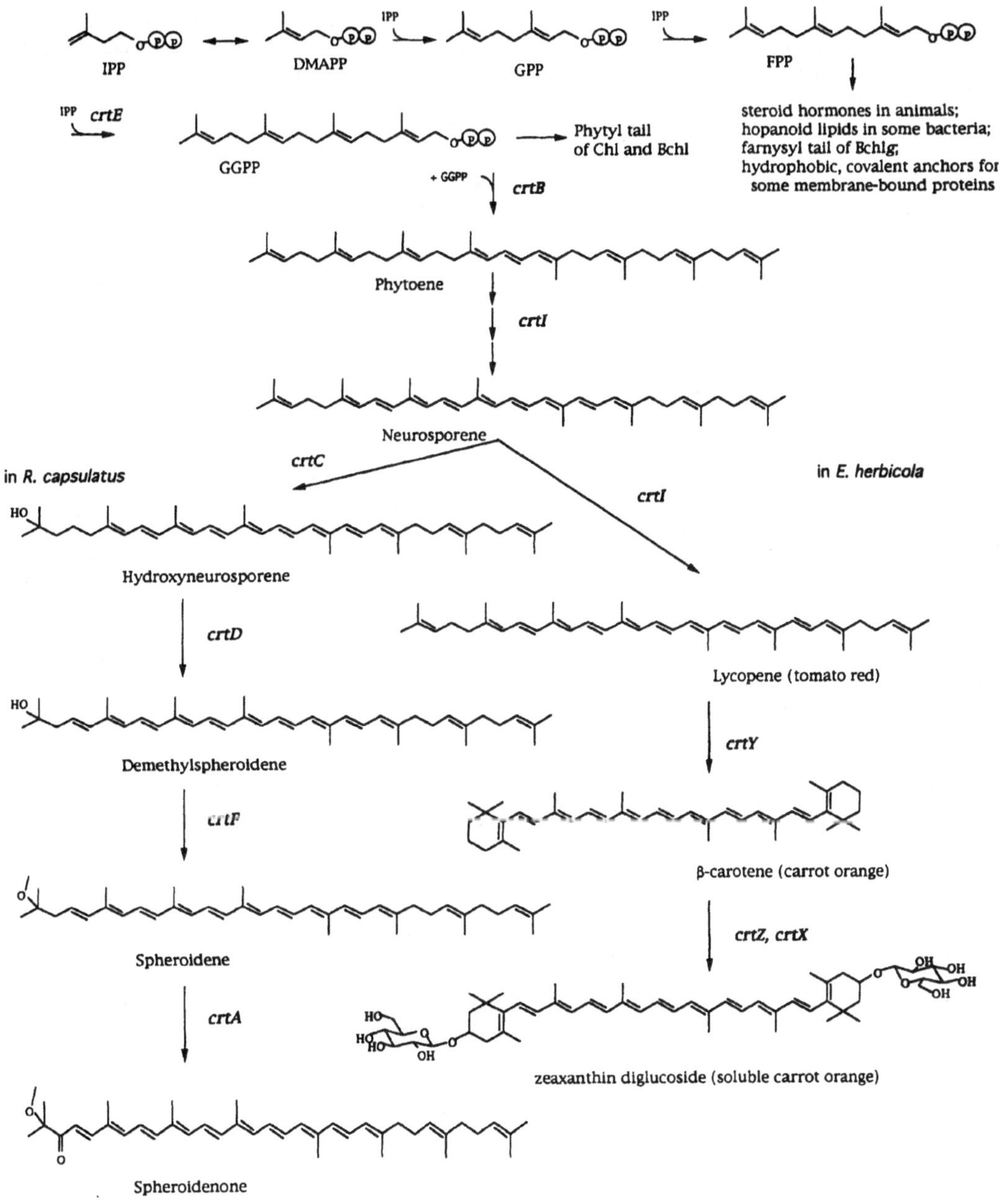

Fig. 1. The carotenoid biosynthetic pathway.

phosphate (GGPP) synthase. The next codes for CrtB, phytoene synthase (Math et al., 1992). The amino acid sequences of CrtB and CrtE have regions of strong conservation with other enzymes in the isoprene pathway. The CrtE protein also shares sequence identity with several prenyltransferases that catalyze 1'-4 condensations such as GGPP synthase from *Neurospora crassa*, hexaprenyl diphosphate synthase from *Saccharomyces cerevisiae* and FPP synthases from *E. coli*, *S. cerevisiae*, *Rattus ratus* and *Homo sapiens*, (Math et al., 1992). CrtB has regions of similarity with other enzymes which catalyze 1'-1 condensations such as several other known phytoene synthases and a yeast squalene synthase (Math et al., 1992).

The R-prime containing the photosynthesis gene cluster (pRPS404) has a point mutation (*crtD223*) that prevents activity of the mutant CrtD enzyme (Marrs, 1981). Armstrong et al., 1990c, suggested that this mutation corresponded to the first nucleotide of an AGA codon (Arg-13) that is present in the deduced amino acid sequence of crtD223 at a position which is invariably occupied by a glycine in a consensus sequence for a $\beta\alpha\beta$ fold which has been associated with FAD or NAD(P) cofactor binding. Subsequent sequencing of wild type *Rb. capsulatus* strain B100 genomic DNA has shown a guanine to be present rather than an adenine at nucleotide 37 of *crtD* (Alberti, M., Ma, D. and Burke, D. H. unpublished data). The deduced amino acid translation of the GGA codon found in wild type DNA is therefore glycine, bringing the Val-7 to Val-33 area of CrtD into good agreement with many other reported ADP-binding motifs (Armstrong et al., 1990c), including the deduced CrtD sequence from the corresponding *Rb. sphaeroides* gene (Gari et al., 1992).

Figure 2 displays the biosynthetic pathway for bacteriochlorophyll in *Rb. capsulatus* and for chlorophyll in higher plants. Here protoporphyrin IX, the only naturally occurring porphyrin without a metal, is a precursor for heme, an essential component of the electron transport pathways of non-photosynthetic bacteria. So, as is the case with the carotenoid biosynthesis pathway, the photosynthetic cluster contains only the genes for the biosynthesis of bacteriochlorophyll beyond the branch point between the synthetic pathways of compounds required specifically for photosynthesis and those shared with other functions. There remains some uncertainty relating to the gene assignments in the bacteriochlorophyll pathway, although considerable progress in gene identification has been made in recent years.

We have made greatest progress with two steps in the bacteriochlorophyll biosynthesis pathway that reduce double bonds to single bonds. The first, monovinyl protochlorophyllide *a* to chlorophyllide *a*, involves gene loci *bchL*, *bchN* and *bchB*, some of these being very recently assigned (Burke et al., 1993b). The second, chlorophyllide *a* to 2-desacetyl-2-vinyl bacteriochlorophyllide *a* (also involved in an alternate route in which BchF acts first), requires the newly assigned three enzymes coded by *bchX*, *bchY* and *bchZ* (Burke et al., 1993a). These were once thought to be a single gene, *bchA* (Yen and Marrs, 1976 and Young et al., 1989). A fourth locus located near the center of the photosynthesis gene cluster, *bchW*, seems also to be required for this step (Bauer et al., 1993b; Bollivar et al., 1994a). There is preliminary evidence for translational coupling in the *bchXYZ* gene cluster (Burke et al., 1993a). The plant and cyanobacterial loci involved with light-independent protochlorophyllide reduction are *chlL(frxC)*, *chlN(gidA) and chlB*. This has been demonstrated by disruption of these genes within the chloroplast and cyanobacterial genomes (Ogura et al., 1992; Suzuki and Bauer, 1992, for *chlL(frxC)*, Roitgrund and Mets, 1990; Choquet et al., 1992, for *chlN(gidA)* and Li et al., 1993 and Liu et al., 1993, for *chlB*). Since plant and purple bacterial enzymes show extensive amino acid identity with each other, the genes in plants and bacteria might be expected to play analogous roles. Although the *chlLNB* system is widely distributed among lower plants (Burke et al., 1993d), it appears to have been completely supplanted by a distinct light-dependent activity for the same reaction in higher plants, reviewed in Bauer et al., 1993b. [The light-dependent activity, also prokaryotic in origin, was first observed by Fujita et al., in 1992. The cyanobacterial gene is stongly homologous to the plant gene (Suzuki and Bauer, 1995b).

The subunits of the reduction complexes show a very interesting sequence conservation with genes involved in nitrogen fixation, in which NifH, a [4Fe-4S] cluster protein, donates electrons to NifD and NifK (Hearst et al., 1985; Burke et al., 1993a,b). This has led us to define a new class of chlorophyll and bacteriochlorophyll biosynthetic enzymes called the 'Chlorophyll Iron Proteins' with structural and

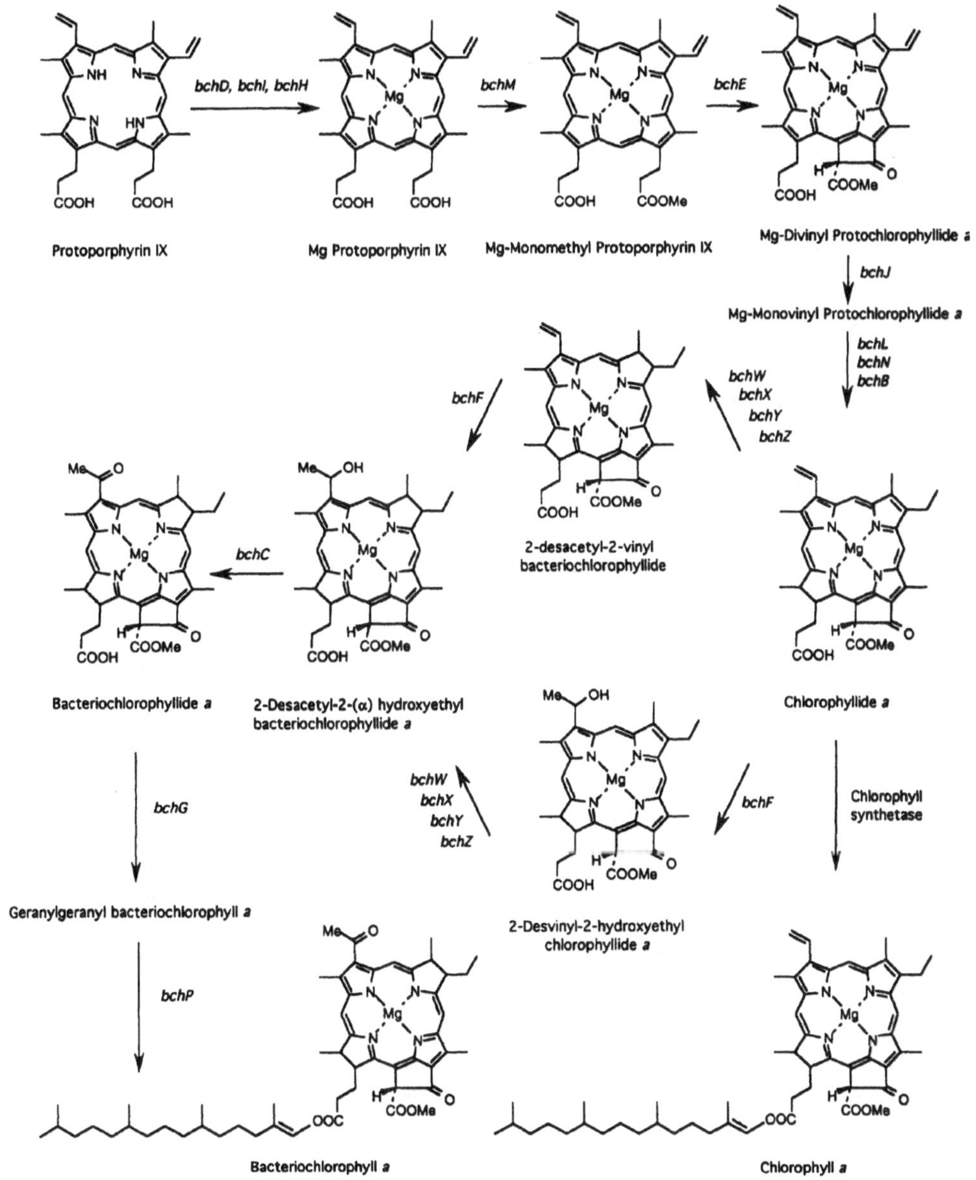

Fig. 2. The bacteriochlorophyll and light-independent chlorophyll biosynthetic pathways.

functional similarities to the nitrogenase iron proteins. Chlorophyll (Chl), according to the Granick hypothesis (Granick, 1965), has been viewed as having preceded bacteriochlorophyll (BChl) as the primary photoreceptor pigment in early photosynthetic systems because Chl synthesis requires one fewer enzymatic reduction than BChl (see Fig. 2). A statistical protein sequence analysis of the enzymes catalyzing each reduction has been conducted by Burke et al., 1993c. When nitrogenase iron proteins were used as outgroups, the chlorophyll iron protein tree is rooted on the *bchX* lineage. As a result, the gene duplication that separated *bchX* from *bchL* occurred prior to the speciation event that separated purple bacteria from cyanobacteria. Thus, this rooting strongly suggests that the last common ancestor of photosynthetic eubacteria contained both enzymes and utilized BChl in its reaction center. The Chl-containing reaction centers apparently evolved later, being unique to the cyanobacteria/chloroplast lineage. These arguments also have the enzymes which synthesize modern photosynthetic pigments evolving from nitrogenase enzymes (Burke et al., 1993c).

The conflict between this model and Granick's hypothesis can be explained by postulating that a single enzyme complex once was able to do both ring reduction steps in the early bacterial pathway, converting protochlorophyllide directly to a bacteriochlorin. These two steps have subsequently been replaced by separate and more specific enzymatic activities, only one of which (*chlLNB*) is retained in cyanobacteria and lower plants. Also, the *crtI* locus codes for a desaturase (an enzyme which converts single carbon-carbon bonds into double bonds) in the carotenoid pathway which sequentially oxidizes *three* bonds in the conversion of phytoene to neurosporene in *Rb. capsulatus*. The *crtI* homologue in *E. herbicola* sequentially oxidizes *four* bonds (converting phytoene to lycopene). Thus, alterations in the number of tandemly repeated actions of an enzyme during metabolite synthesis, followed by evolutionary partitioning of each function into separate, specialized enzymatic activities, may be a common feature of enzyme evolution. Plants are thought to require more than a single enzymatic activity for these same four oxidations, although the exact evolutionary relationship between plant and purple bacterial enzymes is not immediately apparent (Pecker et al., 1992).

III. The Genetic Map

Figure 3 contains our most recent map of open reading frames and gene assignments. The gene cluster contains approximately 43 open reading frames, 33 of which have been assigned. The very strongly expressed structural gene for the RC-H subunit is near the left-hand limit of the cluster (low map number), while the genes for the strongly expressed structural genes encoding the LHI$\alpha\beta$ and RC-LM subunits, are near the right-hand limit of the cluster (high map number). These genes are expressed in an outward direction from the remainder of the cluster, a fact that suggested that the superhelical density of the intervening DNA might be altered when reduced oxygen tension induces the transcription of the structural genes. This hypothesis, however, has been convincingly ruled out by Cook et al. (1989). There are five changes in the direction of expression over the entire gene cluster. The relationship between this fact and other elements of gene expression are not yet well established, although some aspects are discussed below. Also shown in Fig. 3 are the locations of the transposon mutations in a random mutant library created by Zsebo and Hearst, 1984. Most of these mutants are still available and are an important source of research materials. The complete description of this library is contained in the Ph. D. thesis of Krisztina M. Zsebo, 1984, University of California, Berkeley.

The map numbers in Fig. 3 correspond to the base pair number (in thousands) designated in the EMBL DNA sequence databank (accession number Z11165). The eight genes for carotenoid biosynthesis are located between 26.7 and 35.8 thousand. *crtA* is the only carotenoid synthesis gene that possibly shares an operon with bacteriochlorophyll biosynthetic genes. The evidence from polar mutants seems to exclude obligatory cotranscription of *crtA* and the downstream *bch* genes (Zsebo and Hearst, 1984; Guiliano et al., 1988). The eighteen bacteriochlorophyll biosynthetic genes are found in three large, multigene domains from 5.1 to 13.6 thousand, from 16.8 to 26.7 thousand (with some potential gaps for genes with unknown functions) and from 36.0 to 41.1 thousand. There is evidence for the carotenoid gene cluster, *crtEF;* the bacteriochlorophyll gene cluster, *bchCXYZ*; and the structural gene cluster, *pufQBALMX*, being in a 'superoperon', that might be driven by a σ^{70}-like promoter before

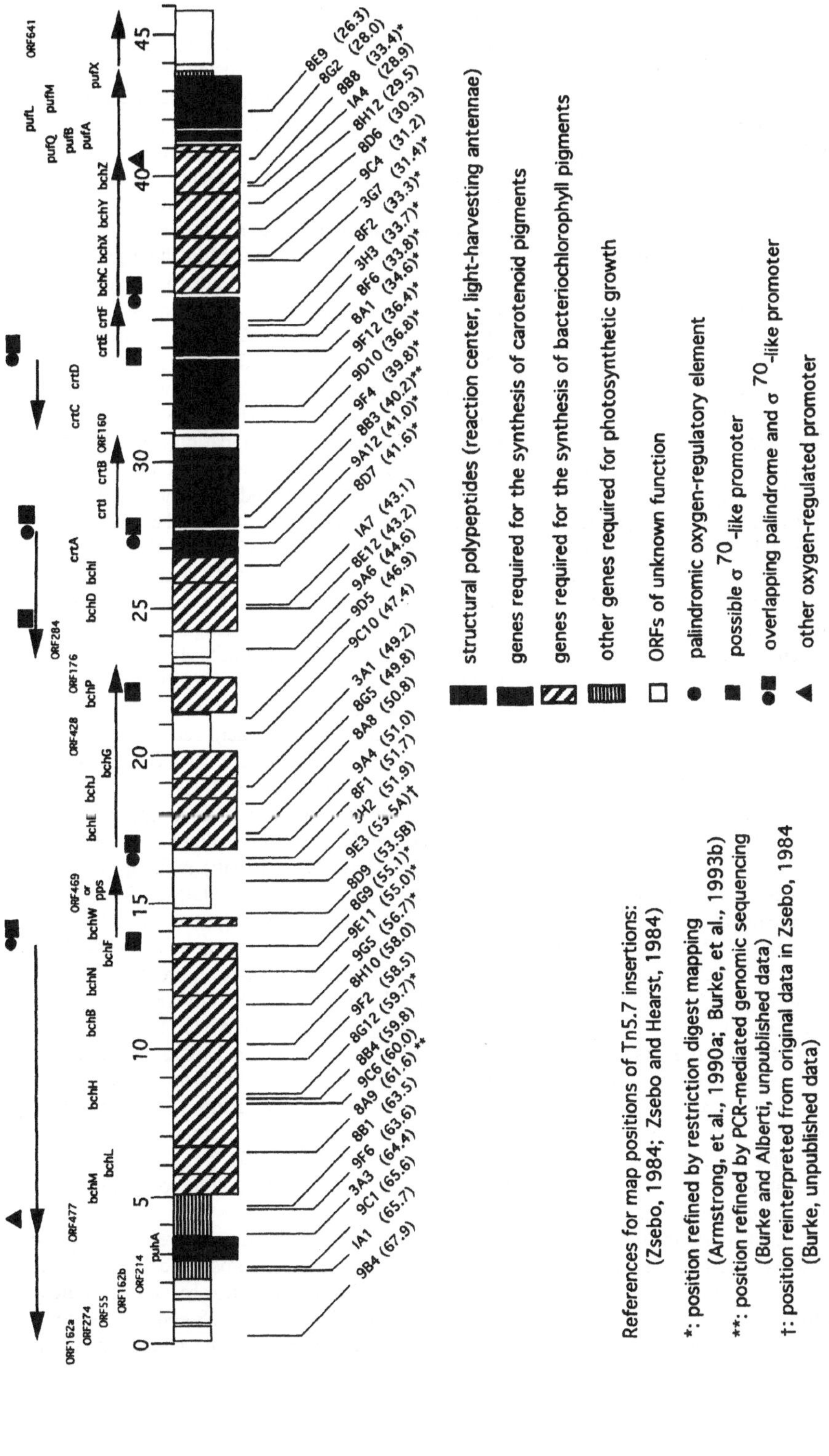

Fig. 3. Map of the *Rhodobacter capsulatus* photosynthesis gene cluster.

crtE (Table 6) and several internal promoters (Young et al., 1989; Chapter 57 by Beatty).

IV. Functional Assignments of the Open Reading Frames (ORFs) in the Sequence

Table 1 lists all the ORFs in the photosynthetic gene cluster of *Rb. capsulatus* and their sequence positions. The arrows show the direction of expression of each ORF; an arrow to the left indicates that the upper DNA strand is used as template for the messenger RNA, while an arrow to the right indicates that the lower DNA strand serves as template for the mRNA transcript. The table contains the current symbols, *eg. crtC,* given to each of these assigned genes. The number of amino acids coded by the ORF and the corresponding molecular weight of the protein product are also supplied. The enzymatic function or name, when available from the literature, is provided together with the references from which the genetic loci, proposed functions and sequence assignments were obtained. There is ambiguity about the function of some ORFs. For example, Penfold and Pemberton (1994) report strong sequence similarity between the ORF at location 14800 to 16209 in *Rb. capsulatus* and a *Rb. sphaeroides* regulatory gene, *pps*, which maps to the same location and is required for repression of pigment synthesis. Bollivar et al. (1994a), on the other hand, report that an insertion mutation in *Rb. capsulatus* causes no effect on photosynthesis. A transposon insertion within ORF214 defines the *rxcD* locus (Zsebo and Hearst, 1984). The *Rb. capsulatus* strain carrying this transposon is defective in reaction center synthesis, suggesting that the product of ORF214 may be required for the assembly or function of reaction centers; however, more work is needed in order to establish the function of this ORF.

Functional assignment of the early steps in BChl synthesis has proven to be problematic and controversal, due both to the complexity of the system and to the difficulty in properly analyzing the pigments accumulated by mutant strains. The bulk of the evidence currently available suggests that Mg-chelatase is derived from the combined activities of the products of *bchD, bchI* and *bchH,* and that *bchM* encodes the methyltransferase. Mg-chelatase activity in plants has been shown to require ATP and to be separable into at least two components (Walker and Weinstein, 1991). The amino acid sequence of BchI contains a consensus ATP-binding sequence near its N-terminus (Armstrong, et al., 1993a). Only two other proteins within the photosynthesis gene cluster contain ATP-binding sequences. These are BchL and BchX and both are involved in redox reactions of the bacteriochlorophyll biosynthesis pathway. Furthermore, BchI shares a strong sequence identity (49%) with *cs/ch42,* an *Arabidopsis* nuclear encoded chloroplast-localized protein (Armstrong et al., 1993a). *cs/ch42* is required for Mg chelation (Falbel and Staehlin, 1994). When S-adenosyl methionine (SAM) and protoporphyrin IX are added to intact cell preparations from *bchD* and *bchH* transposon mutants of *Rb. sphaeroides*, the *bchD* mutant extracts retained methyltransferase activity, while the *bchH* mutant extracts did not (Richards et al., 1991; Gorchein, personal communication). Nevertheless, mutations in a plant homolog to *bchH,* known as *olive* for the resulting altered pigmentation, resulted in a loss of Mg-chelatase activity, and both *bchH* and *olive* share sequence identities with the cobalt chelatase gene in the vitamin B_{12} synthesis pathway (Hudson et al., 1993). In vitro Mg-chelatase activity has recently been described using *E. coli* expressed BchH, BchD and BchI subunits derived from *Rb. sphaeroides* genes. Synthesis of each of these gene products was required for activity (Gibson et al., 1995). The deduced amino acid sequence of BchD has an unusual sequence feature: 10 out of 11 residues in a row are prolines (PPPPPPPPEPP).

Until recently, no mutant accumulating Mg-protoporphyrin IX had been isolated (e.g., Biel and Marrs, 1983; Burke et al., 1993b), making it difficult to separate genetically the chelatase and methyltransferase activities. This pigment has now been found in *bchM* mutants (Bollivar et al., 1994a). In addition, purified *bchM* product expressed in *E. coli* exhibits methyltransferase activity (Bollivar et al., 1994c). This gene product had been thought, based on an analysis of pigments isolated from *bchM* mutants (Yang and Bauer, 1990), to be involved in the oxidative formation of Ring V, as might have been expected from the high fraction of amino acids shared between the *bchM* gene product and mitochondrial NADH:ubiquinone oxidoreductase (Table 2).

Some sequence similarities between the amino acid translations of the photosynthesis genes and related genes from other species are listed in Table 2. For example, ORFs 477 and 428 code for proteins that have sequence similarity to each other and to the

Table 1. All the ORFs of the sequenced portion of pRPS404 (published in EMBL nucleotide sequence database)

Location	Gene name	aa	kD	Function or Enzyme Name	References detailing:		
					Genetic Locus	Proposed Function	Sequence Assignment
← 500-112	ORF162a	162 aa	18 kD	ambiguous			*
← 1434-610	ORF274	274 aa	30 kD	ambiguous			*
← 1598-1431	ORF55	55 aa	6 kD	ambiguous			*
← 2094-1606	ORF162b	162 aa	17 kD	ambiguous			*
← 2749-2105	ORF214	214 aa	24 kD	required for reaction center function or assembly	42	42	*
← 3585-2821	*puhA*	254 aa	28.5 kD	reaction center H protein	42,33,38	31,39	39,40
← 5034-3601	ORF477 also known as ORF 1696	477 aa	50 kD	required for LHI assembly	42,8	8	35,40
← 5705-5031	*bchM*	224 aa	25 kD	Mg protoporphyrin methyl transferase oxidative cyclase subunit	35	14	35, 40
← 6621-5707	*bchL*	304 aa	33 kD	protochlorophyllide reductase iron protein subunit	42	18,35,15	35,40
← 10234-6650	*bchH*	1194 aa	129 kD	Mg protoporphyrin IX chelatase subunit	10,42	12,20b	11,16
← 11801-10224	*bchB*	525 aa	57 kD	protochlorophyllide reductase subunit	42,36,10,8	10,35,18,16	16
← 13072-11798	*bchN*	424 aa	46 kD	protochlorophyllide reductase subunit	41	16	16
← 13584-13069	*bchF*	171 aa	19 kD	2-vinyl-bacteriochlorophyllide hydratase	33,42	33,10	16
14147-14725	*bchW*	192 aa	20 kD →	regulatory function		12	*
14800-16209	ORF469*(ppsR)*	469 aa	51 kD →	ambiguous (repression of pigment synthesis)		12 (27)	*(28)
16768-18495	*bchE*	575 aa	66 kD →	Mg protoporphyrin IX monomethyl ester oxidative cyclase subunit	33,42,35	33,10,35	*
18511-19152	*bchJ*	213 aa	23 kD →	4-vinyl reductase	42	9,12, 32	*
19198-20112	*bchG*	304 aa	33 kD →	geranylgeranyl bacteriochlorophyll synthase	10	12, 13	*
20109-21392	ORF428	428 aa	44.5 kD →	ambiguous	42	42	*
21397-22572	*bchP*	391 aa	43 kD →	geranylgeranyl bacteriochlorophyll reductase		12, 13	*
22575-23105	ORF176	176 aa	20 kD →	ambiguous			*
← 24171-23317	ORF284	284 aa	30 kD	ambiguous			*
← 25859-24174	*bchD*	561 aa	60 kD	Mg protoporphyrin IX chelatase subunit	33,42	10,3,20b	*
← 26913-25861	*bchI*	350 aa	38 kD	Mg protoporphyrin IX chelatase subunit	42,22	3,20b	1,3
← 27635-26910	*crtA*	241 aa	27 kD	spheroidene monooxygenase	36,42,22	36,30,22	1,3
27793-29367	*crtI*	524 aa	58 kD →	phytoene dehydrogenase	42,21,22	21,1,5	1,4
29364-30383	*crtB*	339 aa	37 kD →	phytoene synthase	36,33,22	2,26,17,29	1
30459-30941	ORF 160 formerly designated *crtK*	160 aa	17.5 kD →	ambiguous	22,1	23	1
← 32021-31176	*crtC*	281 aa	32 kD	hydroxyneurosporene synthase	36,22,2	36,30	1
← 33571-32087	*crtD*	494 aa	52 kD	methoxyneurosporene dehydrogenase	36,22	36,30,1	1

Table 1. Continued

Location	Gene name	aa	kD	Function or Enzyme Name	References detailing: Genetic Locus	Proposed Function	Sequence Assignment
33708-34577	*crtE*	289 aa	30 kD →	geranylgeranyl pyrophosphate synthase	36,22	26,17,29	1
34580-35761	*crtF*	393 aa	43 kD →	hydroxyneurosporene-O-methyltransferase	30,22, 2	30	1
35924-36868	*bchC*	314 aa	33 kD →	2-α-hydroxyethylbacteriochlorophyllide dehydrogenase	33,10,42	10	34
36868-37869	*bchX*	333 aa	35.5 kD →	chlorophyllide reductase iron protein subunit	36,33,42,15	15	15
37915-39408	*bchY*	497 aa	52.5 kD →	chlorophyllide reductase subunit	36,33,42,15	15	15
39408-40880	*bchZ*	490 aa	53 kD →	chlorophyllide reductase subunit	36,33,42,15	15,37	15
40882-41106	*pufQ*	74 aa	8.5 kD →	probable regulator of Bchl-synthesis levels	6	7	6
41263-41412	*pufB*	49 aa	5.5 kD →	light harvesting I β (B850β) polypeptide	38,33	19,39	39,40
41426-41602	*pufA*	58 aa	6.6 kD →	light harvesting I α (B850α) polypeptide	38,33	19,39	39,40
41735-42538	*pufL*	282 aa	31.5 kD →	reaction center L protein	38,33	31,39	39,40
42576-43499	*pufM*	307 aa	34.5 kD →	reaction center M protein	38,33	31,39	39,40
43512-43748	*pufX*	78 aa	8.6 kD →	required for electron transfer from RC to Cyt bc_1	24	25,20a	40
43929-45854	ORF641	641 aa	68 kD →	ambiguous			*

* Sequence assigned in the sequence database (EMBL, Accession number Z11165)

Key to references numbered in Table 1:

1 : Armstrong et al., 1989
2 : Armstrong et al., 1990a
3 : Armstrong et al., 1993a
4: Bartley and Scolnik, 1989
5: Bartley et al., 1990
6: Bauer et al., 1988
7: Bauer and Marrs, 1988
8: Bauer et al., 1991
9: Bauer et al., 1993b
10: Biel and Marrs, 1983
11: Bollivar and Bauer, 1992
12: Bollivar et al., 1994a
13: Bollivar et al., 1994b
14: Bollivar et al., 1994c
15: Burke et al., 1993a
16: Burke et al., 1993b
17: Chamovitz et al., 1992
18: Coomber et al., 1990
19: Drews, 1978
20a: Farchaus et al., 1992
20b: Gibson et al., 1995
21: Guiliano et al., 1986
22: Guiliano et al., 1988
23: Kaplan, personal communication
24: Klug and Cohen, 1988
25: Lilburn et al., 1992
26: Math et al., 1992
27: Penfold and Pemberton, 1991
28: Penfold and Pemberton, 1994
29: Sandmann and Misawa, 1992
30: Scolnik et al., 1980
31: Sutton et al., 1982
32: Suzuki and Bauer, 1995a
33: Taylor et al., 1983
34: Wellington and Beatty, 1989
35: Yang and Bauer, 1990
36: Yen and Marrs, 1976
37: Young et al., 1989
38: Youvan et al., 1983
39: Youvan et al., 1984a
40: Youvan et al., 1984b
41: Zsebo, 1984
42: Zsebo and Hearst, 1984

Table 2. Amino acid sequence similarities

Rb. capsulatus gene product	Function	Percent amino acids Identical	Gene product to be compared	Species	Function	Reference
PuhA	reaction center H protein	23%	PuhH	*Rhodospirillum rubrum*	reaction center H protein	Berard and Gingras, 1991
PuhA	"	23%	PuhH	*Rhodopseudomonas viridis*	reaction center H protein	Berard and Gingras, 1991
PuhA	"	23%	PuhH	*Rhodobacter sphaeroides*	reaction center H protein	Berard and Gingras, 1991
PuhA	"	19%	PuhL	*Rhodobacter capsulatus*	reaction center L protein	Hearst and Sauer, 1984a, 1984b
PuhA	"	19%	PuhM	*Rb. capsulatus*	reaction center M protein	Hearst and Sauer, 1984a
PufL	reaction center L protein	78%	PufL	*Rb. sphaeroides*	reaction center L protein	Belanger et al.,1988
PufL	"	70%	PufL	*Rs. rubrum*	reaction center L protein	Belanger et al.,1988
PufL	"	62%	PufL	*Rp. viridis*	reaction center L protein	Belanger et al.,1988
PufL	"		PufL	*Chloroflexus aurantiacus*	reaction center L protein	Blankenship, 1992
PufL	"	21-32%	PufM	*Rb. capsulatus*	reaction center M protein	Youvan etal., 1984b; Hearst and Sauer, 1984a,b
PufL	"	21-27%	PsbA	various plants	reaction center D1 protein	Youvan etal., 1984b; Hearst and Sauer, 1984a,b; Hearst, 1985, 1986
PufL	"	19%	PsbA	*Anabena I*	reaction center D1 protein	Hearst, 1985
PufL	"	18%	PsbA	*Euglena*	reaction center D1 protein	Hearst, 1985
PufM	reaction center M protein	76%	PufM	*Rb. sphaeroides*	reaction center M protein	Belanger et al.,1988
PufM	"	59%	PufM	*Rs. rubrum*	reaction center M protein	Belanger et al.,1988
PufM	"	51%	PufM	*Rp. viridis*	reaction center M protein	Belanger et al.,1988
PufM	"		PufM	*Cf. aurantiacus*	reaction center M protein	Blankenship, 1992
PufM	"	20-23%	PsbD	various plants	reaction center D2 protein	Youvan etal., 1984b; Hearst and Sauer, 1984a,b; Hearst, 1986
ORF160	ambiguous	35-38%		various mammals	peripheral-type benzodiazepine receptor	Baker and Fanestil, 1991
ORF160	"	45%	ORF	*Rb. sphaeroides*	ambiguous	Kaplan, personal communication
47 amino acids in ORF176	ambiguous	45%	IDI1	*Schizosaccharomyces pombe*	isopentenyl diphosphate isomerase	Poulter, personal communication
ORF428	ambiguous	32%	ORF477	*Rb. capsulatus*	required for LH-1 assembly	D. Burke (unpublished)
ORF428	"	25%	PucC	*Rb. capsulatus*	required for LH-2 assembly	D. Burke (unpublished)
ORF477	required for LH-1 assembly	60%	PucC	*Rb. capsulatus*	required for LH-2 assembly	Bauer et al., 1991
ORF477	"	49%	ORF477	*Rb. sphaeroides*	required for LH-2 assembly	Bauer et al., 1991
ORF477	"	56%	ORF-G115	*Rs. rubrum*	unknown	Berard and Gingras, 1991
ORF469	ambiguous		Pps	*Rb. sphaeroides*	regulatory	Penfold and Pemberton, 1994
ORF469	"		NifA	several bacteria	regulatory	G. Armstrong (unpublished)
C-terminus of	"		NtrC	several bacteria	regulatory	G. Armstrong (unpublished)
ORF469			PgtA	*Salmonella*	regulatory	G. Armstrong (unpublished)
				short region is similar to some DNA binding proteins		G. Armstrong (unpublished)
BchB	protochlorophyllide reductase subunit	33%	ChlB	*Chlamydomonas reinhardtii*	protochlorophyllide reductase ChlB subunit	Burke et al., 1993b
BchB	"	34%	ORF513	*Marcantia polymorpha*	by sequence similarity, probably ChlB	Burke et al., 1993b

Table 2. Continued

Rb. capsulatus gene product	Function	Percent amino acids Identical	Gene product to be compared	Species	Function	Reference
first 250 amino acids in						
BchD	Mg protoporphyrin IX chelatase subunit	23%	BchI	*Rb. capsulatus*	Mg protoporphyrin IX chelatase subunit	Falbel, personal communication
BchI	Mg protoporphyrin IX chelatase subunit	approx. 50%		various plants and algae	Mg chelatases	Armstrong et al., 1993a; Bauer et al., 1993b; Falbel, personal communication
BchI	"	49%	Ch42	*Arabidopsis thaliana*	nuclear-encoded Mg chelatase	Armstrong et al., 1993a
BchL	protochlorophyllide reductase subunit	34%	BchX	*Rb. capsulatus*	chlorophyllide reductase Fe protein	Burke et al., 1993c
BchL	"	30-37%	NifH	many eu- and archaebacteria	nitrogenase reductases	Hearst et al., 1985
BchL	"	49%	ChlL(FrxC)	*M. polymorpha*	protochlorophyllide reductase subunit	Burke et al., 1993b
BchL	"	53%	ChlL	*Chlamydomonas reinhardtii*	protochlorophyllide reductase subunit	Suzuki and Bauer, 1992
25 amino acids in						
BchM	Mg protoporphyrin methyl transferase	52%	primate mitochondrial NADH-ubiquinone oxidoreductase			D. Burke (unpublished)
48 amino acids in						
BchM	"	42%	hypothetical protein in Tn554			D.Burke (unpublished)
BchN	protochlorophyllide reductase subunit	35%	ChlN(GidA)	*Chlamydomonas reinhardtii*	protochlorophyllide reductase subunit	Burke et al., 1993b
BchN	"	36%	ORF465	*M. polymorpha*	by sequence similarity, probably ChlN	Burke et al., 1993b
BchN	"	35%	ChlN	*Pinus negra* chloroplast	protochlorophyllide reductase subunit	Burke et al., 1993b
BchN	"	19%	NifK	*Rb. capsulatus*	nitrogenase	Ogura et al., 1992
29 amino acids in						
BchP	geranylgeranyl bacteriochlorophyll reductase	53%	NifM	*Azotobacter vinelandii*	activates *nifH*	M. Alberti (unpublished)
BchX	chlorophyllide reductase Fe protein subunit	30-37%	NifH	many eubacteria and archaebacteria species	nitrogenase reductases	Burke et al., 1993a
BchX	"	34%	BchL	*Rb. capsulatus*	bacteriochlorophyll Fe protein	Burke et al., 1993c
BchX	"	32%	ChlL	*M. polymorpha*	chlorophyll Fe protein	Burke et al., 1993c
BchY	probable chlorophyllide reductase subunit	18%	BchN	*Rb. capsulatus*	protochlorophyllide reductase subunit	Burke et al., 1993c
BchY	"	19%	ChlN	*M. polymorpha*	chlorophyllide reductase subunit	Burke et al., 1993c
BchZ	chlorophyllide reductase subunit	24%	BchB	*Rb. capsulatus*	protochlorophyllide reductase subunit	Burke et al., 1993c
BchZ	"	20%	ChlB	plants and algae	protochlorophyllide reductase subunit	Burke et al., 1993c
CrtB	phytoene synthase	31%	CrtB	*Erwinia herbicola*	phytoene synthase	Armstrong et al., 1990c

Table 2. Continued

Rb. capsulatus gene product	Function	Percent amino acids Identical	Gene product to be compared	Species	Function	Reference
CrtB	phytoene synthase	28%	protein product of pTOM5	tomato	cDNA clone differentially expressed during fruit ripening	Armstrong et al., 1990c
CrtB	"	28 to 32%		many bacteria	phytoene synthases	G. Armstrong, Chapter 53
CrtD	methoxyneurosporene dehydrogenase	56%	CrtD	*Rb. sphaeroides*	methoxyneurosporene dehydrogenase	Gari et al., 1992
CrtD	"	27%	CrtI	*Rb. capsulatus*	phytoene dehydrogenase	Armstrong et al., 1989, 1990c
CrtD	"	25 to 28%	CrtI	many bacteria	phytoene dehydrogenase	G. Armstrong, Chapter 53
CrtD	"	30%	ORF4	*Myxococcus xanthus*	hydroxyneurosporene dehydrogenase	G. Armstrong, Chapter 53
CrtE	geranylgeranyl pyrophosphate synthase	30%	CrtE	*E. herbicola*	geranylgeranyl pyrophosphate synthase	Armstrong et al., 1990c
CrtE	"	36%	GGPPS	bell pepper	geranylgeranyl pyrophosphate synthase	G. Armstrong, Chapter 53
CrtF	hydroxyneurosporene O-methyl transferase	30%		bovine	hydroxyindole O-methyl transferase	G. Armstrong, Chapter 53
CrtF	"	25%		chicken	hydroxyindole O-methyl transferase	G. Armstrong, Chapter 53
CrtF	"	22 to 28%		*Streptomyces glaucescens*	various O-methyl transferases	G. Armstrong, Chapter 53
CrtI	phytoene dehydrogenase	27%	CrtD	*Rb. capsulatus*	methoxyneurosporene dehydrogenase	Armstrong et al., 1989, 1990c
CrtI	"	25%	CrtD	*Rb. sphaeroides*	methoxyneurosporene dehydrogenase	Gari et al., 1992
CrtI	"	41%	CrtI	*E. herbicola*	phytoene dehydrogenase	Armstrong et al., 1990c
CrtI	"	30%	CarC	*Myxococcus xanthus*	phytoene dehydrogenase	G. Armstrong, Chapter 53
CrtI	"	32%	Al-1	*Neurospora crassa*	phytoene desaturase	Bartley et al., 1990

product of *pucC*, a *Rb. capsulatus* gene required for the accumulation of LHII peptides and which lies outside of this gene cluster (Tichy, et al., 1989). We and others (Bauer et al., 1991) have noticed that if translation of *pucC* is initiated at a position upstream from the originally proposed translation initiation site, additional matches can be made with ORF428 and ORF477.

Rb. capsulatus has a very high GC composition at 65%. It is thus not surprising that such a bias would be reflected in the pattern of codon usage, as shown in Table 3. There are, however, dramatic biases in this pattern which might suggest a functional role, or in an extreme interpretation, the absence of a tRNA which corresponds to the associated anticodon. For example, there are 253 occurrences of CTT as a codon for leucine while there is only one instance of CTA used as a leucine codon. Similar biases exist against all codons of the form NTA. In fact all codons ending in A, i.e. NNA appear to have been selected against. Such an extreme bias may ultimately have a regulatory role, but the evidence at the present time is insufficient.

Table 4 lists the recognition sites for some restriction enzymes. These sites were central to our cloning and sequencing strategies, and have been referred to repeatedly in the literature regarding this gene cluster. This listing may therefore be useful in interpreting such references.

The potential ribosome binding sites for these 43 genes show a clear pattern of translation initiation signals in *Rb. capsulatus*. When fourteen nucleotides of the DNA template for 16S RNA are matched with the nucleotides preceding the genes' start codons (Table 5), there is a core sequence, AAAGGAGG, which occurs two to three times more frequently than predicted in a 65% C+G, 35% A+T random DNA sequence and which precedes the start codon by an interval of 4 to 12 nucleotides. Nucleotides flanking the core sequence occur with random frequency. These observations do not rule out the possibility that a particular message may make use of base complementarity outside this 8 base region to signal translation initiation.

V. Potential Regulatory Sequences

Transcription termination hairpins have been suggested in several locations (Armstrong et al., 1989). Some are presented in Fig. 4. The hairpin after *puhA*, (between 2810 and 2781), occurs at the end of the RC-H gene and may protect the corresponding mRNA from exonuclease digestion, or it may alter the probability of transcription of downstream genes. The very large hairpin after ORF469/*ppsR* (between 16310 and 16379) provides isolation for an already very isolated open reading frame.

Between map positions 23103 and 23319 there is an enigmatic pair of double hairpins (labeled 'ORF176' and 'ORF284' in Fig. 4) that are almost perfect inverted repeats of each other. These structures occur between the ends of two converging operons involved in the expression of bacteriochlorophyll genes, near the center of the 45959 base photosynthesis gene cluster. Another nearly perfect inverted repeat involves the 90 or so bases following ORF469/*ppsR*, bases 16307 to 16393, that are almost identical (again with opposite polarity) to 90 bases 15 kb distant, 31088–31001, that follow *crtC*. Some of these identical 90 bases are shown in two structures in Fig. 4: 'ORF469/*ppsR*' and the right hand side of '*crtC*'. The symmetrical appearance of these features that show DNA sequence similarity beyond stop codons and what, if any, regulatory or long range interactive role they play awaits further experimentation.

The hairpins labeled '*crtI*' and '*crtB*' have no known function. Perhaps they separate the three genes in this operon serving as a block to exonuclease digestion of messenger RNA in the same way that the structure following *pufA* allows differential expression of LHI$\alpha\beta$ and RC-LM (Belasco et al., 1985; Adams et al., 1989). The hairpin labeled 'ORF 160' is separated by 21 nucleotides from the double hairpin labeled '*crtC*'. These are possible mRNA structures between operons of opposing directions: ORF 160 terminates at 30941, *crtC* terminates at 31176. The structure labeled '*crtF*' is proposed to form right at the end of the coding region of *crtF*, (the TGA stop is within the loop,) and may serve as transcription terminator for the *crtEF* segment of the *crtEFbchCXYZpufBALMX* superoperon. The structures labeled '*bchX*' and '*bchY*' are each predicted to occlude the ribosome binding sites and translation initiation sites of their respective downstream genes (*bchY* and *bchZ*). These structures may play a role in regulating the translation of the components of the *bchXYZ* chlorin reductase through translational coupling (Burke et al., 1993a). The last gene known to be involved in the photosynthesis cluster, *pufX*, is

Table 3. Codon usage for all the genes and ORFs listed in Table 1. 14227 Codons (in 43 genes) of the *Rb. capsulatus* photosynthesis gene cluster

Amino Acid	Codon	Total Number	Percentage Use	Amino Acid	Codon	Total Number	Percentage Use
Phe	TTT	103	0.7%	Tyr	TAT	158	1.1%
	TTC	478	3.4%		TAC	144	1.0%
		581	4.1%			302	2.1%
Leu	TTA	1	<0.1%	Cys	TGT	14	0.1%
	TTG	103	0.7%		TGC	131	0.9%
	CTT	253	1.8%			145	1.0%
	CTC	234	1.6%				
	CTA	1	<0.1%	Stop	TAA	9	
	CTG	913	6.4%		TAG	2	
		1505	10.5%		TGA	32	
						43	
Ile	ATT	59	0.4%	His	CAT	168	1.2%
	ATC	581	4.1%		CAC	139	1.0%
	ATA	3	<0.1%			307	2.2%
		643	4.5%				
Met	ATG	450	3.2%	Trp	TGG	264	1.9%
	GTG	2	<0.1%				
		452					
Val	GTT	95	0.7%	Gln	CAA	81	0.6%
	GTC	468	3.3%		CAG	289	2.0%
	GTA	6	<0.1%			370	2.6%
	GTG	499	3.5%				
		1068	7.5%				
Ser	TCT	22	0.2%	Arg	CGT	105	0.7%
	TCC	104	0.7%		CGC	510	3.6%
	TCA	9	0.1%		CGA	20	0.1%
	TCG	360	2.5%		CGG	297	2.1%
	AGT	9	0.1%		AGA	6	<0.1%
	AGC	137	1.0%		AGG	12	0.1%
		641	4.6%			950	6.6%
Pro	CCT	33	0.2%	Thr	ACT	25	0.2%
	CCC	324	2.3%		ACC	426	3.0%
	CCA	11	0.1%		ACA	11	0.1%
	CCG	495	3.5%		ACG	296	2.1%
		863	6.1%			758	5.4%
Ala	GCT	76	0.5%	Gly	GGT	120	0.8%
	GCC	805	5.7%		GGC	792	5.6%
	GCA	59	0.4%		GGA	42	0.3%
	GCG	853	6.0%		GGG	287	2.0%
		1793	12.6%			1241	8.6%
Asn	AAT	70	0.5%	Lys	AAA	120	0.8%
	AAC	219	1.5%		AAG	381	2.7%
		289	2.0%			501	3.5%
Asp	GAT	264	1.9%	Glu	GAA	386	2.7%
	GAC	487	3.4%		GAG	417	2.9%

Table 4. Recognition sites for some restriction enzymes. This table can be useful for subcloning the ORFs of interest in the Photosynthesis Gene Cluster.

Restriction Enzyme		Cuts after bases number
Name	Sequence	(location in Z11165):
*Bam*H I	G/GATCC	1, 662, 2707, 6723,13826,13837,14953,15468,18791, 25244,29150,33018,33829,36280,45954
Bcl I	T/GATCA	5090, 8041,10162,13380,14526,19262,20075,24305, 24602,28262,28502,35550,35881,38300,45610
Bgl II	A/GATCT	2880, 2985,14671,17466,23053,30091
Eco RI	G/AATTC	1564, 8681, 9191, 9332,10530,14165,15433,17545, 28210,29640,30191,32552,34850,40056,41116
Hind III	A/AGCTT	234, 4301, 5172, 6751, 7496, 7937,11268,12785, 14814,18915,35321,36121,37159,37483,39325
Pst I	CTGCA/G	2316, 2403, 5119, 6365, 6419, 7935, 8556, 9966,11725, 13475,14091,14301,14322,15074,20827,22148,24532,27346, 27861,37433,38342,39742,40102,42032,44506,45163
Sal I	G/TCGAC	3082, 7230,10980,15318,15784,17452,18478,19432,21622, 21658,25812,27723,28543,29511,30825,40921,42648,43443
Xho I	C/TCGAG	3040, 3313,10782,10809,11936,13988,15280,20739, 23323,35039,40116,40314
Sph I	GCATG/C	2269, 4585, 5135, 7796, 8420, 9689,10147,14315,17074,18997, 20903,20930,24321,25196,25469,27374,30249,31278,32134, 34301,35525,36338,39842,41326,42481,44121,44213,45051
Xma I	C/CCGGG	978, 1910, 2548, 6991, 9723,13919,14910,15374,15976,16596, 18829,19479,21815,23113,23166,23999,24338,26241,26487,29345, 29881,30418,30995,31641,32169,32286,32326,32863,33059, 33343,33453,33606,34667,35789,35993,36085,37772,37887
Sst I	GAGCT/C	6687,44668
Kpn I	GGTAC/C	no *Kpn* I sites
Xba I	T/CTAGA	no *Xba* I sites

followed by what looks like a transcription terminator.

The entire photosynthetic gene cluster of *Rb. capsulatus* is regulated by oxygen tension at both transcriptional and post-transcriptional levels (Zhu and Hearst, 1986; Zhu et al., 1986; Bauer et al., 1988; Bauer et al., 1993a; Chapter 58 by Bauer). At atmospheric oxygen pressure, the gene cluster is essentially not expressed, although some ORFs are transcribed at low constitutive levels. Upon reduction of oxygen tension to 2 mm Hg, the photosynthetic membrane is formed, reaction centers are made and pigments are synthesized. A two-component sensor/effector system for oxygen regulation of the structural genes (*puf*, *puh* and *puc*) has been recently described (Sganga and Bauer, 1992; Bauer et al., 1993a; Mosley et al., 1994; Inoue et al., 1995). In bacteria, various sensor/effector systems have been identified as involved in the transcriptional regulation of genes whose expression responds to environmental signals such as nitrogen limitation (Backman et al., 1981), phosphate starvation (Wanner, 1987) and osmoregulation (Csonka, 1989). It is not yet known how many other regulatory proteins are involved in the signal transduction cascade or how reduced oxygen tension is sensed by *Rb. capsulatus*. In addition to the sensor/effector regulatory systems, bacteria some-

Table 5. Potential ribosome binding sequences

CACAAGGTGTTCCCCG	ATG	ORF162	GCTCTGGAGGCGTAACG	ATG	ORF284
TCACCGGGGGTGCCGCGC	ATG	ORF271	AGAAACGTTGCCGTAAC	ATG	bchD
TGAAGTGAGGAACCC	ATG	ORF55	GGAAAGGGAAACTGC	ATG	bchI
AATAAGGGGACCCCG	ATG	ORF162b	ACAGGGGAGGACGTAG	ATG	crtA
AATACGGAGGTCTGC	ATG	ORF214			
ACAAAGGAGGACCAAC	ATG	puhA	TGAAACTACCGAAGAAACC	ATG	crtI
TGCCTGGAGTATCGGCC	ATG	ORF477	GCCAAGGCGGCGCA	ATG	crtB
TTCTGGGATTCGACTGAG	ATG	bchM	CAACCGGAGGCCATG	ATG	ORF160
GAAAAGGAGGGCCGC	ATG	bchL			
CTCTATGAGGCCAAGGCCCATT	ATG	bchH	GAAAAGGCCTTCTCG	ATG	crtC
TGAACGGAGGCGCGGC	ATG	bchB	GTGCGGGAGCGAGCG	ATG	crtD
GGAACGGGGGACGCG	ATG	bchN			
AGAAAGGCTGGCTGCCACT	ATG	bchF	GCAGCGGAGGGCTCTGTC	ATG	crtE
			CGCCGAGAGGGGCTGATC	GTG	crtF
GGACCGGAGCTTGGCGCG	ATG	bchW	TACCGGGAGCAAGAA	ATG	bchC
GGAAAGGTTGCGGCC	GTG	ORF469	GGAAAGATGCAAAATA	ATG	bchX
GCCCTGGAGGGTCAGCC	ATG	bchE	TGCACGGAGGGCCGGGC	ATG	bchY
AAGAAGGAGTGACGGCC	ATG	bchJ	GCCGAGGAGATGATCTG	ATG	bchZ
CTTCCGGTGACCTGACCCC	ATG	bchG	GGAAGGGGGGGAACTGAA	ATG	pufQ
GATCCGGGGCTTCACGGT	ATG	ORF428	AATCCGGAGGTTGTT	ATG	pufB
GTCATGGCGGGTGAATGAA	ATG	bchP	TCTGAGGAGAAACTGAAA	ATG	pufA
GCAATGGACCTGACC	ATG	ORF176	ACAGCGGAGACACGGGC	ATG	pufL
			TGGCAGGAGGCATCA	ATG	pufM
			CGTAAGGAGAAGAGACC	ATG	pufX
			TCATCGGAGGACGCAACC	ATG	ORF641

A	G	A	A	A	G	G	A	G	G	T	G	A	T	The DNA template sequence for the 3' terminus of *Rb. capsulatus* 16S RNA (Youvan, *et al*, 1984a)
9	15	21	22	18	38	41	28	35	24	9	10	9	7	Number of matches in these 43 upstream sequences
7	14	7	7	7	14	14	7	14	14	7	14	7	7	Number of matches expected (by random statistics) for 65% G+C, 35% A+T DNA
		A	A	A	G	G	A	G	G					Consensus Ribosomal Binding Site used by these genes

times utilize alternative sigma factors to govern the initiation of transcription under environmental stresses (Stock et al., 1989). It seems likely that a novel sigma factor is used to control transcription initiation for the *puh* and *puf* operons in *Rb. capsulatus* (Bauer et al., 1993a).

Despite the fact that the gene cluster on pRPS404 shows no evidence of being expressed in *E. coli*, σ^{70} promoter-likelike sequences have been identified in *Rb. capsulatus*. These are presented in Table 6. Fusion of such a potential promoter sequence (from *bchCXYZ*) to *lacZ* in *Rb. capsulatus* has confirmed that this sequence element is essential for the initiation of transcription of *bchCXYZ* (Ma et al., 1993).

Armstrong has discovered a palindromic motif (Table 7) at six sites in the gene cluster, some of which overlap the potential σ^{70} promoters. A detailed study of the oxygen-dependent, σ^{70}-like promoter upstream of *bchCXYZ* has revealed some additional details concerning the function of the palindrome motif. It has been shown to bind an oxygen-dependent regulatory protein factor in *Rb. capsulatus* (Ma et al., 1993). These authors have also identified a possible regulatory role for an AT-rich region, the 50 base pairs immediately upstream of the suggested *bchC* σ^{70}-like promoter. Since a σ^{70}-like promoter and this palindrome sequence are found upstream of several other *bch* and *crt* operons, these sequences may be responsible for regulating oxygen-dependent pigment biosynthesis at the level of transcription in *Rb. capsulatus*. This motif is not found upstream of the *puh* and *puf* structural genes, indicating that a different regulatory mechanism is required for the production of the structural polypeptides for the photosynthetic apparatus. Interestingly, this palindrome sequence motif is found preceding the *puc* operon (Youvan and Ismail, 1985) which encodes structural polypeptides of the light harvesting complex II. The significance

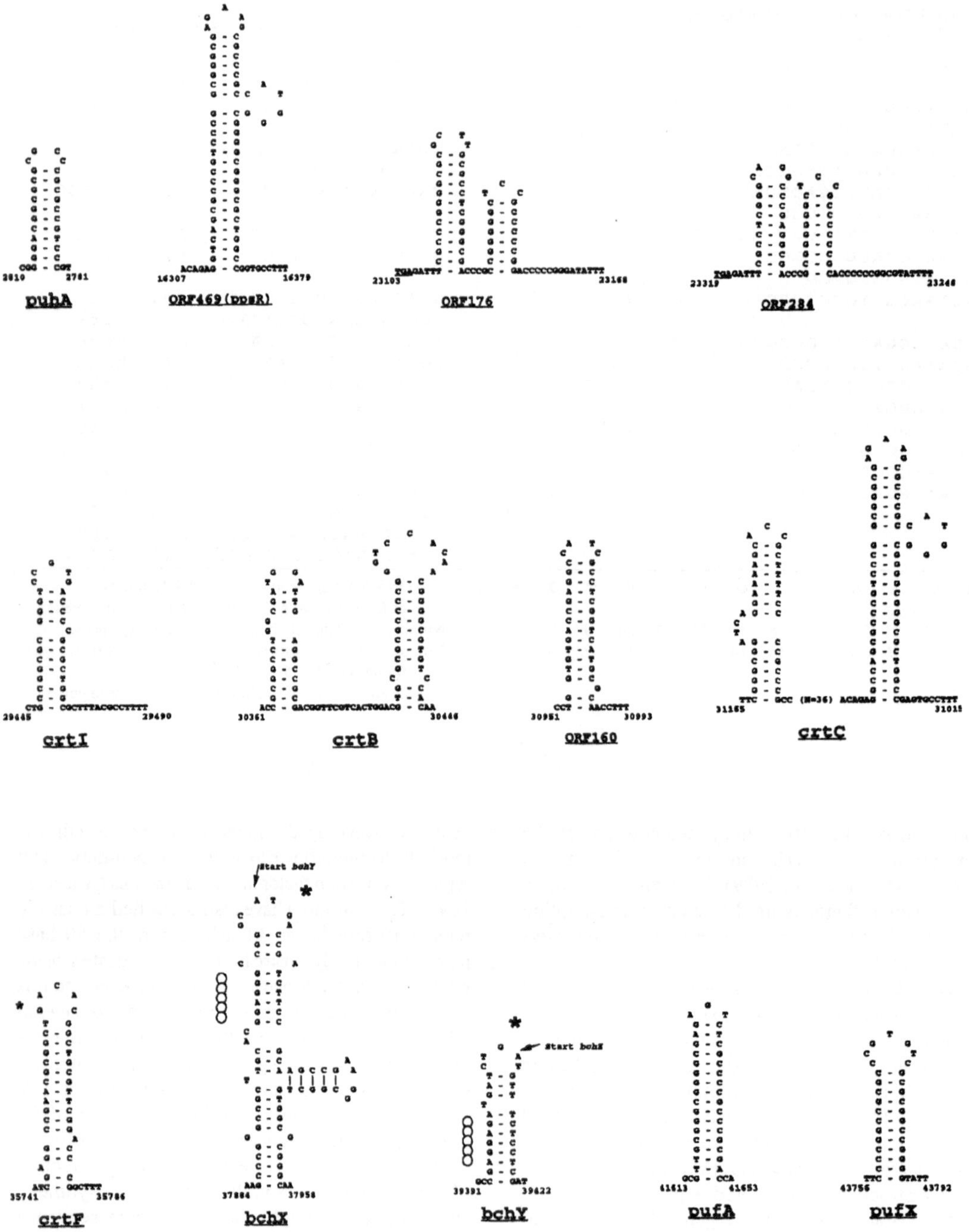

Fig. 4. Potential secondary structures near the ends of the mRNA sequences for some of the genes of the photosynthesis gene cluster. * indicates the 'TGA' stop codon; ○○○○○ is the ribosomal binding site for the next gene; start (occluded) 'ATG' is marked by arrow.

Table 6. Potential E. coli σ^{70}-like promoter sequences

TTGACA	(N=15)	ACCAGT	43 bases before *bchF*	| These three
TTGACA	(N=17)	CAGTTT	41 bases before *bchF*	| share the same
TTGACA	(N=18)	AGTTTT	40 bases before *bchF*	| –35 sequence
GTGACA	(N=15)	CAGAGT	30 bases before *bchF*	
TTGATG	(N=15)	TTTATT	251 bases before *bchW*	
TGGACA	(N=16)	CAAACT	229 bases before ORF477	
TTGAAA	(N=15)	TCAAAA	221 bases before ORF477	| These two share
TTGAAA	(N=16)	CAAAAC	220 bases before ORF477	| same –35 sequence
TCGTCA	(N=18)	TATGAC	74 bases before ORF477	
TTTTCT	(N=17)	TTTAAT	55 bases before *bchE*	
TTGACC	(N=18)	TGGGAT	44 bases before *bchE*	
ATGGCA	(N=19)	GATGAT	199 bases before ORF176	
TTGCCA	(N=18)	CATGAC	118 bases before ORF176	
TTGCCC	(N=17)	TACATT	66 bases before ORF284	
TGGACA	(N=15)	TTTCAG	138 bases before *crtA*	
ATGATA	(N=17)	TACAAC	95 bases before *crtA*	
TTTACA	(N=19)	GACAAA	72 bases before *crtA*	
GCGACA	(N=18)	TGTGAT	139 bases before *crtI*	
GGGATA	(N=17)	TGTAAT	103 bases before *crtI*	
TTTACA	(N=16)	TGGAAG	98 bases before *crtI*	
TTGTCG	(N=18)	TGTAAA	56 bases before *crtI*	
TGTAAA	(N=15)	TATCAT	35 bases before *crtI*	| These two share
TTGTAA	(N=16)	TATCAT	35 bases before *crtI*	| same –10 sequence
TTGACG	(N=17)	TGCAAC	21 bases before *crtI*	
TTGACC	(N=16)	GTTTAT	147 bases before *crtD*	| These three
TTGACC	(N=17)	TTTATC	146 bases before *crtD*	| share the same
TTGACC	(N=19)	TATCCA	144 bases before *crtD*	| –35 sequence
TTGCGA	(N=16)	TAAACT	78 bases before *crtD*	
CTTACA	(N=15)	CAAAAT	47 bases before *crtD*	
TTTACA	(N=15)	GCCAAT	31 bases before *crtE*	
CGCACA	(N=19)	TCTAAT	88 bases before *bchC*	
TTGACA	(N=15)	TCAATG	57 bases before *bchC*	| These three
TTGACA	(N=17)	AATGAT	55 bases before *bchC*	| share the same
TTGACA	(N=19)	TGATAC	53 bases before *bchC*	| –35 sequence

Consensus sequence for *E. coli* σ^{70} promoter: **TTGACA (N = 15 to 19) TATAAT**

of this palindrome sequence in regulating the *puc* operon has been confirmed in *Rb. sphaeroides* (Lee and Kaplan, 1992a) but remains to be tested in *Rb. capsulatus*. An intriguing question now being investigated is how the regulation between pigment genes and structural genes is coordinated. This coordination has been suggested to involve the *pufQ* gene product (Bauer and Marrs, 1988).

VI. Conclusion

The sequence and function of the photosynthetic gene cluster of the purple bacterium, *Rb. capsulatus*, has provided great insight into the structures of the photosynthesis reaction center and light harvesting proteins, as well as into the assignment of genes in the carotenoid and bacteriochlorophyll biosynthetic

Table 7. Palindrome motif

13584	13645	13668	14147
<—bchF	aaGTGTCAATgaaAacTTACACtc		bchW—>
16209	16725	16748	16768
end of ORF469	acATGTCAACtgaGgtTTACACct		bchE—>
27635	27653	27676	27793
<—crtA	atGTGTAACGggaTatTTACATct		crtI—>
27635	27729	27752	27793
<—crtA	agTTGTAAATcggAatTGACGAcc		crtI—>
33571	33636	33659	33708
<—crtD	ggGTGTAAGTttcAgtTTACACag		crtD—>
35761	35852	35875	35924
end of crtF	gcGTGTAAGTtcaAtgATACACac		bchC—>
	24	47	198
	caGTGTAAGCccgActTTACACtt		pucB—>
	267	290	441
	+++ + ++		
Rb. capsulatus consensus sequence:	GTGTAARTnnnAnnTTACAC		
	|||| ||||		
Consensus sequence for several transcriptional regulatory factors*:	TGTGT N6-10 ACACA		

Comparison of a palindromic motif found 5' to *Rb. capsulatus* photosynthesis genes. The genes flanking each palindrome are indicated to the left and right, respectively with arrows to show direction of transcription.
Numbers above the sequence shows position in the EMBL DNA database, accession number Z11165.
Numbers above the *pucB* palindrome motif are taken from Youvan and Ismail, 1985. Numbers below are taken from Tichy, et al., 1989who report a cytosine (position 273) replacing the adenine at position 30.
Complimentary nucleotides in the two halves of the palindromes are underlined.
Nucleotides in the positions shown in bold face in the consensus sequence are capitalized.
Nucleotides that match the consensus are given in bold type.
+ indicates the nucleotide is conserved in all cases.
*Other transcriptional regulatory factors include NifA, AraC, CAP, LacI, GalR, LexA, TrpR, LysR and lcII (Gicquel-Sanzey and Cossart, 1982).

pathways in other organisms. In addition, initial steps in oxygen regulation of this gene cluster have been elucidated. There are vast numbers of issues relating to photosynthesis remaining to be solved in which the sequence of this cluster will have increasing significance. For example, the mechanism of initiation of photosynthetic membrane formation and the sequential steps associated with photosynthetic apparatus assembly are, without a doubt, addressed by these 45,959 bases of DNA.

Acknowledgments

This work has been supported by contract #DE-ACO3-76SF00098 from the U.S. Department of Energy. We wish to thank Bob Rabson of DOE for years of sustained support for research on bacterial photosynthesis. We are grateful for the assistance of G. Armstrong, B. Hundle and D. Ma and the editors, R. E. Blankenship and C.E. Bauer for advice they offered in the writing of this review.

References

Adams CW, Forrest ME, Cohen SN and Beatty JT (1989) Structural and functional analysis of transcriptional control of the *Rhodobacter capsulatus puf* operon. J Bacteriol 171: 473–482

Armstrong GA, Alberti M, Leach F and Hearst JE (1989) Nucleotide sequence, organization and nature of the protein products of the carotenoid biosynthesis gene cluster of *Rhodobacter capsulatus*. Mol Gen Genet 216: 254–268

Armstrong GA, Schmidt A, Sandmann G and Hearst JE (1990a) Genetic and biochemical characterization of carotenoid biosynthesis mutants of *Rhodobacter capsulatus*. J Biol Chem 265: 8329–8338

Armstrong GA, Alberti M, Leach F and Hearst JE (1990b) Organization of the *Rhodobacter capsulatus* carotenoid biosynthesis gene cluster. In: Drews G and Dawes E (eds) Molecular Biology of Membrane-Bound Complexes in Phototrophic Bacteria, pp 39–46. Plenum Press, London, UK

Armstrong GA, Alberti M and Hearst JE (1990c) Conserved enzymes mediate the early reactions of carotenoid biosynthesis in nonphotosynthetic and photosynthetic prokaryotes. Proc Nat Acad Sci USA 87: 9975–9979

Armstrong GA, Cook DN, Ma D, Alberti M, Burke DH and Hearst JE (1993a) Regulation of carotenoid and bacteriochlorophyll biosynthesis genes and identification of an evolutionarily conserved gene required for bacteriochlorophyll accumulation. J Gen Microbiol 139: 897–906

Armstrong GA, Hundle BS and Hearst JE (1993b) Evolutionary conservation and structural similarities of carotenoid biosynthesis gene products from photosynthetic and nonphotosynthetic organisms. Meth Enz (Carotenoids) 214: 297–311

Backman K, Chen Y-M and Magasanik B, (1981) Physical and genetic characterization of the *glnA-glnG* region of the *Escherichia coli* chromosome. Proc Natl Acad Sci USA 78: 3743–3747

Baker ME and Fanestil DD (1991) Mammalian peripheral-type benzodiazepine receptor is homologous to CrtK protein of *Rhodobacter capsulatus*, a photosynthetic bacterium. Cell 65: 721–722

Bartley GE and Scolnik PA (1989) Carotenoid biosynthesis in photosynthetic bacteria. Genetic characterization of the *Rhodobacter capsulatus CrtI* protein. J Biol Chem 264: 13109–13113

Bartley GE, Schmidhauser TJ, Yanofsky C and Scolnik PA (1990) Carotenoid desaturases from *Rhodobacter capsulatus* and *Neurospora crassa* are structurally and functionally conserved and contain domains homologous to flavoprotein disulfide oxidoreductases. J Biol Chem 265: 16020–16024

Bauer CE and Marrs BL (1988) *Rhodobacter capsulatus puf* operon encodes a regulatory protein (*PufQ*) for bacteriochlorophyll biosynthesis. Proc Natl Acad Sci USA 85: 7074–7078

Bauer CE, Young DA, Marrs BL (1988) Analysis of the *Rhodobacter capsulatus puf* operon. J Biol Chem 263: 4820–4827

Bauer CE, Buggy JJ, Yang Z and Marrs B (1991) The superoperonal organization of genes from pigment biosynthesis and reaction center proteins is a conserved feature in *Rhodobacter capsulatus*: Analysis of overlapping *bchB* and *puhA* transcripts. Mol Gen Genet 228: 433–444

Bauer CE, Buggy JJ and Mosley C (1993a) Control of photosystem genes in *Rhodobacter capsulatus*. Trends in Genetics 9: 56–60

Bauer CE, Bollivar DW and Suzuki JY (1993b) Minireview: Genetic analysis of photopigment biosynthesis in eubacteria: a guiding light for algae and plants. J Bacteriol 175: 3919–3925

Belanger G, Berard J, Corriveau P and Gingras G (1988) The structural genes coding for the L and M subunits of *Rhodospirillum rubrum* photoreaction center. J Biol Chem 263: 7632–7638

Belasco JG, Beatty JT, Adams CW, von Gabain A and Cohen SN (1985) Differential expression of photosynthetic genes in *Rhodopseudomonas capsulata* results from segmental differences in stability within the polycistronic rxcA transcript. Cell 40: 171–181

Berard J and Gingras G (1991) The *puh* structural gene coding for the H subunit of the *Rhodospirillum rubrum* photoreaction center. Biochem Cell Biol 69: 122–131

Biel AJ and Marrs BL (1983) Transcriptional regulation of several genes for bacteriochlorophyll biosynthesis in *Rhodopseudomonas capsulata* in response to oxygen. J Bacteriol 156: 686–694

Blankenship RE (1992) Origin and early evolution of photosynthesis. Photosynth Res 33: 91–111

Bollivar DW and Bauer CE (1992) Nucleotide sequence of S-adenosyl-L-methionine: magnesium protoporphyrin methyltransferase from *Rhodobacter capsulatus*. Plant Physiol 98: 408–410

Bollivar DW, Suzuki JY, Beatty JT, Dobrowolski JM and Bauer CE (1994a) Directed mutational analysis of bacteriochlorophyll *a* biosynthesis in *Rhodobacter capsulatus*. J Mol Biol 237: 622–640

Bollivar DW, Wang S, Allen JP and Bauer CE (1994b) Molecular genetic analysis of terminal steps in bacteriochlorophyll a biosynthesis: Characterization of a *Rhodobacter capsulatus* strain that synthesizes geranylgeraniol esterified bacteriochlorophyll *a*. Biochemistry 33: 763-768

Bollivar DW, Jiang ZY, Bauer CE and Beale SI (1994c) Heterologous expression of the *bchM* gene product from *Rhodobacter capsulatus* and demonstration that it encodes S-adenosyl-L-methionine:Mg-protoporphyrin IX methyltransferase. J Bacteriol 176: 5290–5296

Brutlag DL, Dautricourt JP, Maulik S and Relph J (1990) Improved sensitivity of biological sequence database searches. Computer Applications in the Biosciences 6: 237–245

Burke DH, Alberti M and Hearst JE (1993a) The *Rhodobacter capsulatus* chlorin reductase-encoding locus, *bchA*, consists of three genes, *bchX*, *bchY* and *bchZ*. J Bacteriol 175: 2407–2413

Burke DH, Alberti M and Hearst JE (1993b) *bchFNBH* bacteriochlorophyll synthesis genes of *Rhodobacter capsulatus* and identification of the third subunit of light-independent protochlorophyllide reductase in bacteria and plants. J Bacteriol 175: 2414–2422

Burke D, Hearst JE and Sidow A (1993c) Early evolution of photosynthesis: clues from nitrogenase and from the chlorophyll iron proteins. Proc Natl Acad Sci USA 89: 7134–7138

Burke DH, Raubseson LA, Alberti M, Hearst JE, Jordan ET, Kirch SA, Valinski AEC, Conant DS and Stein DB (1993d) The *chlL (frxC)* gene: phylogenetic distribution in vascular plants and DNA sequences from *Polystichum acrostichoides* (pteridophyta) and *Synechococcus* sp. 7002 (cyanobacteria). Plant Systematics and Evolution, 187: 89–102

Chamovitz D, Misawa N, Sandmann G and Hirschberg J (1992) Molecular cloning and expression in *Escherichia coli* of a cyanobacterial gene coding for phytoene synthase, a carotenoid biosynthesis enzyme. FEBS Lett 296: 305–310

Choquet Y, Rahire M, Girard-Bascou J, Erickson J and Rochaix J-D (1992) A chloroplast gene is required for the light-independent accumulation of chlorophyll in *Chlamydomonas reinhardtii*. EMBO J 11: 1697–1704

Cook DN, Armstrong GA and Hearst JE (1989) Induction of anaerobic gene expression in *Rhodobacter capsulatus* is not accompanied by a local change in chromosomal supercoiling as measured by a novel assay. J Bacteriol 171: 4836–4843

Coomber SA, Chaudhri M, Connor A, Britton G and Hunter CN (1990) Localized transposon Tn5 mutagenesis of the photosynthetic gene cluster of *Rhodobacter sphaeroides*. Molec Microbiol 4: 977–989

Csonka LN (1989) Physiological and genetic responses of bacteria to osmotic stress. Microbiol Rev 53: 121–147

Doolittle RF (1986) Of ORFs and URFs, a primer on how to analyze derived amino acid sequences

Drews G (1978) The bacterial photosynthetic apparatus. Curr Topics Bioenerget 8: 161–207

Falbel TG and Staehelin LA (1994) Characterization of a family of chlorophyll-deficient wheat (Triticum) and barley (*Hordeum vulgare*) mutants with defects in the magnesium-insertion step of chlorophyll biosynthesis. Plant Physiology 104: 639–648

Farchaus JW, Barz WP, Grunberg H and Oesterhelt D (1992) Studies on the expression of the *pufX* polypeptide and its requirement for photoheterotrophic growth in *Rhodobacter sphaeroides*. EMBO J 11: 2779–2788

Fijita Y, Takahashi Y, Chunganji M and Matsubara H (1992) The *nifH*-like (*frxC*) gene is involved in the biosynthesis of chlorophyll in the filamentous cyanobacterium *Plectonema boryanum*. Plant Cell Physiology 33: 81–92

Gari E, Toledo JC, Gilbert I and Barbe J (1992) Nucleotide sequence of the methoxyneurosporene dehydrogenase gene from *Rhodobacter sphaeroides*: comparison with other bacterial carotenoid dehydrogenases. FEMS Microbiol Lett 93: 103–108

Gibson LCD, Willows RD, Kannangara CG, von Wettstein D and Hunter CN (1995) Magnesium-protoporphyrin chelatase of *Rhodobacter sphaeroides*: Reconstitution of activity by combining the products of the *bchH*, *-I*, and *-D* genes expressed in *Escherichia coli*. Proc Natl Acad Sci USA 92: 1941–1944

Gicquel-Sanzey B and Cossart P (1982) Homologies between different prokaryotic DNA-binding regulatory proteins and between their sites of action. EMBO J 1: 591–595

Granick S (1965) In: Bryson V and Vogel HJ (eds) Evolving Genes and Proteins, pp 67–68 Academic Press, New York

Guiliano G, Pollock D and Scolnik PA (1986) The gene *crtI* mediates the conversion of phytoene into colored carotenoids in *Rhodopseudomonas capsulata*. J Biol Chem 261: 12925–12928

Guiliano G, Pollock D, Stapp H and Scolnik PA (1988) A genetic-physical map of the *Rhodobacter capsulatus* carotenoid biosynthesis gene cluster. Mol Gen Genet 213: 78–83

Hearst JE (1985) The identification, isolation and sequence of the reaction center protein genes of the photosynthetic purple bacterium *Rhodopseudomonas capsulata*, Plenary Lecture, Ninth International Congress on Photobiology, July 3, 1984, Philadelphia. In: Longworth JW, Jagger J and Shropshire Jr W (eds) Photobiology 1984, pp 237–247. Praeger Scientific, New York

Hearst JE (1986) Primary structure and function of the reaction center polypeptides of *Rhodopseudomonas capsulata*—the structural and functional analogies with the Photosystem II polypeptides of plants. In: Van Rensen JJS (ed) Encyclopedia of Plant Physiology, Vol 19, Photosynthesis III, pp 382–389. Springer Verlag, New York

Hearst JE and Sauer K (1984a) Protein sequence homologies between portions of the L and M subunits of reaction centers of *Rhodopseudomonas capsulata* and the 32 kD herbicide-binding polypeptide of chloroplast thylakoid membranes and a proposed relation to quinone-binding sites. In: Sybesma C and Niijhoff M (eds) Advances in Photosynthetic Research, Vol III, pp 355–359. Dr W Junk Publishers, The Hague/Boston

Hearst JE and Sauer K (1984b) Protein sequence homologies between portions of the L and M subunits of reaction centers of *Rhodopseudomonas capsulata* and the QB-protein of chloroplast thylakoid membranes: a proposed relation to quinone-binding sites. Zeit Naturforsch 39C: 421–424

Hearst JE, Alberti M and Doolittle RF (1985) A putative nitrogenase reductase gene found in the nucleotide sequences from the photosynthetic gene cluster of *R. capsulata*. Cell 40: 219–220

Hudson A, Carpenter R, Doyle S and Coen ES (1993) Olive: A key gene required for chlorophyll biosynthesis in *Antirrhinum majus*. EMBO J 12: 3711–3719

Hundle BS and Hearst JE (1991) Carotenoids. Spectrum 4: 1–8

Hundle BS, Beyer P, Kleinig H, Englert G and Hearst JE (1991) Carotenoids of *Erwinia herbicola* and an *Escherichia coli* HB101 strain carrying the *Erwinia herbicola* carotenoid gene cluster. Photochem and Photobiol 54: 89–93

Hundle BS, O'Brien DA, Alberti M, Beyer P and Hearst JE (1992) Functional expression of zeaxanthin glucosyltransferase from *Erwinia herbicola* and a proposed uridine diphosphate binding site. Proc Natl Acad Sci USA 89: 9321–9325

Hundle BS, O'Brien DA, Beyer P, Kleinig H and Hearst JE (1993) In vitro expression and activity of lycopene cyclase and β-carotene hydroxylase from *Erwinia herbicola*.. FEBS Lett 315: 329–334

Inoue K, Mosley C, Kouadio J-L and Bauer C (1995) Isolation and in vitro phosphorylation of sensory transduction components controlling anaerobic induction of light harvesting and reaction center gene expression in *R. capsulatus*. Biochemistry 34: 391–396

Klug G and Cohen SN (1988) Pleiotropic effects of localized *Rhodobacter capsulatus puf* operon deletions on production of light-absorbing pigment-protein complexes. J Bacteriol 170: 5814–5821

Lee JK and Kaplan S (1992a) *cis*-acting regulatory elements involved in oxygen and light control of *puc* operon transcription in *Rhodobacter sphaeroides*. J Bacteriol 174: 1146–1157

Lee JK and Kaplan S (1992b) Isolation and characterization of trans-acting mutations involved in oxygen regulation of puc operon transcription in *Rhodobacter sphaeroides*. J Bacteriol 174: 1158–1171

Li JM, Goldschmidtclermont M and Timki MP (1993) Chloroplast-encoded ChlB is required for light-independent protochlorophyllide reductase activity in *Chlamydomonas reinhardtii*. Plant Cell 5: 1817–1829

Lilburn TG, Haith CE, Prince RC and Beatty TJ (1992) Pleiotropic effects of *pufX* gene deletion on the structure and function of the photosynthetic apparatus of *Rhodobacter capsulatus*. Biochim Biophys Acta 1100: 160–170

Liu XQ, Xu H and Huang CZ (1993) Chloroplast ChlB gene is required for light-independent chlorophyll accumulation in *Chlamydomonas reinhardtii*. Plant Molecular Biology 23: 297–308

Ma D, Cook DN, O'Brien DA and Hearst JE (1993) Analysis of the promoter and regulatory sequences of an oxygen-regulated *bch* operon in *Rhodobacter capsulatus* by site-directed mutagenesis. J Bacteriol 175: 2037–2045

Marrs BL (1978) In: Clayton RK and Sistrom WR (eds) The Photosynthetic Bacteria, pp 873–883. Plenum Press, New York and London

Marrs BL (1981) Mobilization of the genes for photosynthesis from *Rhodopseudomonas capsulata* by a promiscuous plasmid. J Bacteriol 146: 1003–1012

Math SK, Hearst JE and Poulter CD (1992) The *crtE* gene in *Erwinia herbicola* encodes geranylgeranyl diphosphate synthase. Proc Natl Acad Sci USA 89: 6761–6764

Mosley C, Suzuki J, and Bauer C (1994) Identification and molecular genetic characterization of a sensor kinase responsible for coordinately regulating light harvesting and reaction center gene expression in response to anaerobiosis. J Bacteriol 176: 7566–7573

Ogura Y, Takemura M, Yamato K, Ohta E, Fukuzawa H and Ohyama K (1992) Cloning and nucleotide sequence of a *frxC*-ORF469 gene cluster of *Synechocystis* PCC6803—conservation with liverwort chloroplast *frxC*-ORF465 and *nif* operon. Bioscience Biotechnology and Biochemistry 56: 788–793

Pecker I, Chamovitz D, Linden H, Sandmann G and Hirschberg J (1992) A single polypeptide catalyzing the conversion of phytoene to ζ-carotene is transcriptionally regulated during tomato fruit ripening. Proc Natl Acad Sci USA 89: 4962–4966

Penfold RJ and Pemberton JM (1991) A gene from the photosynthetic gene cluster of *Rhodobacter sphaeroides* induces trans suppression of bacteriochlorophyll and carotenoid levels in *R sphaeroides* and *Rb. capsulatus*. Current Microbiology 23: 259–263

Penfold RJ and Pemberton JM (1994) Sequencing, chromosomal inactivation, and functional expression in *Escherichia coli* of *ppsR*, a gene which represses carotenoid and bacteriochlorophyll synthesis in *Rhodobacter sphaeroides*. J Bacteriol 176: 2869–2876

Richards WR, Fidai S, Gibson L, Lauterbach P, Snajdarova I, Valera V, Wieler JS and Yee WC (1991) Enzymology of the magnesium branch of chlorophyll and bacteriochlorophyll biosynthesis. Photochemistry and Photobiology 53: 84S–85S

Roitgrund C and Mets LJ (1990) Localization of two novel chloroplast genome functions: trans-splicing of RNA and protochlorophyllide reduction. Curr Genet 17: 147–153

Sandmann G and Misawa N (1992) New functional assignment of the carotenogenic genes *crtB* and *crtE* with constructs of these genes from *Erwinia* species. FEMS Microbiology Lett 90: 253–258

Sanger F, Nicklen S and Coulson AR (1977) DNA sequencing with chain terminating inhibitors. Proc Natl Acad Sci USA 74: 5463–5467

Schmidt A, Sandmann G, Armstrong GA, Hearst JE and Boger P (1989) Immunological detection of phytoene desaturase in algae and higher plants using an antiserum raised against a bacterial fusion-gene construct. Eur J Biochem 184: 375–378

Scolnik PA, Walker MA and Marrs BL (1980) Biosynthesis of carotenoids derived from neurosporene in *Rhodopseudomonas capsulata*. J Biol Chem 255: 2427–2432

Sganga MW and Bauer CE (1992) Regulatory factors controlling photosynthetic reaction center and light-harvesting gene expression in *Rhodobacter capsulatus*. Cell 68: 945–954

Stock JB, Ninfa AJ and Stock AM (1989) Protein phosphorylation and regulation of adaptive responses in bacteria. Microbiol Rev 53: 450–490

Sutton MR, Rosen D, Feher G and Steiner LA (1982) Amino-terminal sequences of the L, M and H subunits of reaction centers from the photosynthetic bacterium *Rhodopseudomonas sphaeroides* R-26. Biochem 21: 3842–3849

Suzuki J and Bauer CE (1992) Light-independent chlorophyll biosynthesis: involvement of the chloroplast gene, *gidB(frxC)*. Plant Cell 4: 929–940

Suzuki JY and Bauer CE (1995a) Altered monovinyl and divinyl protochlorophyllide pools in the *bchJ* mutant of *Rhodobacter capsulatus*: Possible monovinyl substrate discrimination of light independent protochlorophyllide reductase. J Biol Chem 270: 3732–3740

Suzuki JY and Bauer CE (1995b) A prokaryotic origin for light-dependent chlorophyll biosynthesis of plants. Proc Natl Acad Sci USA 92: 3749–3753

Taylor DP, Cohen SN, Clark WG and Marrs B (1983) Alignment of the genetic and restriction maps of the photosynthesis region of the *Rhodopseudomonas capsulata* chromosome by a conjugation-mediated marker rescue technique. J Bacteriol 154: 580–590

Tichy HV, Oberle B, Stiehle H, Schiltz E and Drews G (1989) Genes downstream from *pucB* and *pucA* are essential for formation of the B800–850 complex of *Rhodobacter capsulatus*. J Bacteriol 171: 4914–4922

Walker CJ and Weinstein JD (1991) In vitro assay of the chlorophyll biosynthetic enzyme Mg-chelatase—resolution of the activity into soluble and membrane bound fractions. Proc Natl Acad Sci USA 88: 5789–5793

Wanner B (1987) Phosphate regulation of gene expression in *Escherichia coli*. In: Neidhardt FC, Ingraham JL, Low KB, Magasanik B, Schaechter M and Umbarger HE (eds) *Escherichia coli* and *Salmonella typhimurium*: Cellular and Molecular Biology, pp 1326–1333. American Society for Microbiology, Washington, DC

Weaver PF, Wall JD and Gest H (1975) Characterization of *Rhodopseudomonas capsulata*. Arch Microbiol 105: 207–216

Wellington CL and Beatty JT (1989) Promoter mapping and nucleotide sequence of the *bchC* bacteriochlorophyll biosynthesis gene from *Rhodobacter capsulatus*. Gene 83: 251–261

Williams JC, Steiner LA, Ogden RC, Simon ML and Feher G (1983) Primary structure of the M subunit of the reaction center from *Rhodopseudomonas sphaeroides*. Proc Natl Acad Sci USA 80: 6505–6509

Wood M and Kaplan S (1993) unpublished data cited in J.K. Lee and S. Kaplan, J Bacteriol. 174: 1158–1171

Worland ST, Wilson KJ, Hearst JE and Sauer K (1984) Isolation and amino-terminal sequences of subunits from the photosynthetic reaction center of *R. capsulata*. Biochim Biophys Acta 767: 651–654

Yang Z and Bauer CE (1990) *Rhodobacter capsulatus* genes involved in early steps of the bacteriochlorophyll biosynthesis pathway. J Bacteriol 172: 5001–5010

Yen H-C and Marrs BL (1976) Map of genes for carotenoid and bacteriochlorophyll biosynthesis in *Rhodopseudomonas capsulata*. J Bacteriol 126: 619–629

Young DA, Bauer CE, Williams JC and Marrs BL (1989) Genetic evidence for superoperonal organization of genes for photosynthetic pigments and pigment-binding proteins in *Rhodobacter capsulatus*. Mol Gen Genet 218: 1–12

Youvan DC, Elder JT, Sandlin DE, Zsebo K, Alder DP, Panopoulos NJ, Marrs BL and Hearst JE (1982) R-prime site-directed transposon Tn7 mutagenesis of the photo-synthetic apparatus in *Rhodopseudomonas capsulata*. J Mol Biol 162: 17–41

Youvan DC, Hearst JE and Marrs BL (1983) Isolation and characterization of enhanced fluorescence mutants of *Rhodopseudomonas capsulata*. J Bacteriol 154: 748–755

Youvan DC, Alberti M, Begusch J, Bylina E and Hearst JE (1984a) Reaction center and light-harvesting I genes from *Rhodopseudomonas capsulata*. Proc Nat Acad Sci USA 81: 189–192

Youvan DC, Bylina EJ, Alberti M, Begusch H and Hearst JE (1984b) Nucleotide and deduced polypeptide sequences of the photosynthetic reaction-center, B870 antenna and flanking polypeptides from *Rb. capsulatus*. Cell 37: 949–957

Youvan DC and Ismail S (1985) Light-harvesting II (B800–850 complex) structural genes from *Rhodopseudomonas capsulata*. Proc Natl Acad Sci USA 82: 58–62

Zhu YS and Hearst JE (1986) Regulation of expression of genes for light-harvesting antenna proteins LH-I and LH-II; reaction center polypeptides RC-L, RC-M and RC-H; and enzymes of bacteriochlorophyll and carotenoid biosynthesis in *Rhodobacter capsulatus* by light and oxygen. Proc Natl Acad Sci USA 83: 7613–7617

Zhu YS, Cook DN, Leach F, Armstrong GA, Alberti M and Hearst JE (1986) Oxygen regulated mRNAs for light-harvesting and reaction center complexes and for bacteriochlorophyll and carotenoid biosynthesis in *Rhodobacter capsulatus* during the shift from anaerobic growth to aerobic growth. J Bacteriol 168: 1180–1188

Zsebo KM, Wu F and Hearst JE (1984) Tn5.7 construction and physical mapping of pRPS404 containing photosynthetic genes from *Rhodopseudomonas capsulata*. Plasmid 11: 182–184

Zsebo KM (1984) Genetic-physical mapping of a photosynthetic gene cluster in *Rhodopseudomonas capsulata*, Ph.D. Thesis, University of California Lawrence Berkeley Laboratory, April 1984

Zsebo KM and Hearst JE (1984) Genetic-physical mapping of a photosynthetic gene cluster from *Rb. capsulatus*. Cell 37: 937–947

Chapter 51

Genetic Analysis of CO_2 Fixation Genes

Janet Lee Gibson
Department of Microbiology, The Ohio State University, 484 W. 12th Avenue, Columbus, OH 43210-1292

R. E. Blankenship, M. T. Madigan and C. E. Bauer (eds): Anoxygenic Photosynthetic Bacteria, pp. 1107–1124.

Summary

Purple photosynthetic bacteria assimilate CO_2 primarily via the Calvin reductive pentose phosphate pathway. Depending on the growth conditions, the amount of CO_2 fixation required for growth and the level of Calvin cycle enzymes can vary considerably. Although numerous biochemical studies have demonstrated induction or derepression of the enzymes, the mechanisms underlying this control have remained elusive. Cloning and expression of Calvin cycle structural and regulatory genes from different photosynthetic bacteria have led to significant advances in our understanding of CO_2 fixation at the molecular level. *cbb* genes have been shown to be clustered in several bacteria, including *Rhodobacter sphaeroides*, in which genes within two *cbb* gene clusters appear to be organized within operons thus coordinating expression at the level of transcription. Mutagenesis of *cbb* genes in *Rb. sphaeroides* has revealed independent and interdependent regulatory relationships between the two different operons that appear to be mediated through a common transcriptional regulatory protein, CbbR. Transcriptional regulation and *cbbR* genes have also been demonstrated in *Chromatium vinosum* and *Rhodospirillum rubrum*. Characterization of the CbbR proteins should provide clues to the identity of the molecular signal that serves as sensor of the environment to control expression of enzymes involved in CO_2 assimilation. Alternate routes of CO_2 fixation have unexpectedly been exposed in RubisCO deletion mutants constructed in *Rb. sphaeroides* and *Rs. rubrum* opening new avenues of research in CO_2 fixation in the photosynthetic bacteria. The RubisCO mutants have also been used as host strains for analyzing foreign RubisCO enzymes. In *Cm. vinosum*, two distinct RubisCO coding sequences were identified by hybridization and expression in *E. coli* allowed characterization of both enzymes. Finally, nucleotide sequence comparisons of *cbb* structural and regulatory genes have raised provocative evolutionary questions concerning relationships between photosynthetic bacteria and other organisms.

I. Introduction

Purple photosynthetic bacteria typically display the capacity for growth under a diverse array of environmental conditions. This metabolic versatility relies on the ability to assimilate CO_2 via the Calvin cycle, which is essential in varying capacities, for chemo- and photolithoautotrophic and photoheterotrophic growth. In the two former growth modes, cellular carbon is derived exclusively from CO_2, necessitating a relatively high level of CO_2 fixation and consequently, induced or derepressed synthesis of Calvin cycle enzymes (Tabita, 1988). On the other hand, lower levels of the same enzymes are observed during photoheterotrophic growth, in which CO_2 fixation functions primarily as an electron acceptor to balance excess reducing potential generated through metabolism of organic substrates (Tabita, 1988). Since the amount of CO_2 fixed is proportional to the reduction state of the organic compound, the level of Calvin cycle enzymes varies accordingly and is generally higher during growth on more reduced substrates (Tabita, 1988). In heterotrophic growth regimens not requiring CO_2 fixation, expression of Calvin cycle enzymes is not observed (Tabita, 1988). The regulatory network involved in the control of CO_2 fixation is expected to be extremely complex to respond to environmental signals such as light, O_2, CO_2 and organic compounds (Tabita, 1988). The situation is complicated further in some purple non-sulfur photosynthetic bacteria by the occurrence of structurally distinct forms of ribulose 1,5-bisphosphate carboxylase/oxygenase (RubisCO) that exhibit unique catalytic and inductive properties (Gibson and Tabita, 1977a,b; Shively et al., 1984; Dow, 1987). The form I RubisCO is the more ubiquitous form of RubisCO and is composed of eight large catalytic subunits and eight small subunits of undetermined function, whereas the form II RubisCO is composed of large subunits only (Tabita, 1988). Although induction of Calvin cycle enzymes has been demonstrated in photosynthetic bacteria by biochemical means, the nature and mechanisms of the molecular signals involved in regulation of CO_2 fixation enzymes has remained elusive. The application of cloning and recombinant DNA technology to this area of metabolism over the

Abbreviations: A. eutrophus – Alcalignes eutrophus; Cm. – Chromatium; E. coli – Escherichi coli; FBA – fructose 1,6-bisphosphate aldolase; FBP – fructose 1,6-bisphosphatase; GAP – glyceradehye 3-phosphate dehydrogenase; PPE – ribulose 5-phosphate-3-epimerase; PRK – phosphoribulokinase; *Rb. – Rhodobacter; Rs. – Rhodospirillum;* RubisCO – ribulose 1,5-bisphosphate carboxylase/oxygenase; TKL – transketolase past decade has had far reaching impacts. The intended focus of this chapter is to convey the current understanding of regulation and expression of enzymes involved in CO_2 fixation that has been derived from molecular biological approaches.

II. *cbb* Genes

A. Nomenclature of cbb Genes

Historically, different conventions have been used in naming genes encoding enzymes of the Calvin cycle. In photosynthetic bacteria, gene designations were originally assigned according to plant genetic nomenclature, such that three letter designations were chosen according to the enzyme, such as *rbc* for RubisCO, *prk* for phosphoribulokinase (PRK), *fbp* for fructose 1,6-bisphosphatase (FBP) and so on (Tabita et al., 1992). An entirely different system was used by groups studying chemolithoautotrophic bacteria, which employed the prefix, *cfx*, for CO_2 fixation, followed by the first letter of the enzyme encoded by the corresponding gene (Husemann et al., 1988; Meijer et al., 1991). Additional confusion stemmed from the occurrence in some organisms of more than one gene encoding a particular Calvin cycle enzyme and the nomenclature used to distinguish the structurally divergent form I and form II RubisCO large subunit genes (Tabita et al., 1992). Recently, a uniform nomenclature has been proposed by groups studying the reductive pentose phosphate cycle in bacteria that assigns to Calvin cycle genes the prefix *cbb*, for *C*alvin-*B*enson-*B*assham cycle, followed by the first letter of the corresponding enzyme, if no other enzyme begins with that letter (Tabita et al., 1992). For organisms such as *Rb. sphaeroides*, in which two Calvin cycle gene clusters have been localized to distinct genetic elements (Suwanto and Kaplan, 1989), subscripts I and II, will distinguish between *cbb* genes of each operon. Also pertinent to this chapter, is the situation found in *Cm. vinosum* in which two sets of genes encoding a form I type RubisCO have been identified, both of which are chromosomally encoded (Viale et al., 1989). It has been proposed that these genes be designated *cbbL-1cbbS-1* and *cbbL-2cbbS-2*, respectively (Tabita et al., 1992). Table 1 lists Calvin cycle enzymes and respective gene designations relevant to this chapter.

B. Identification and Isolation of RubisCO Genes

To date, *cbb* genes have been isolated from only a few photosynthetic bacteria including the purple nonsulfur bacteria *Rs. rubrum*, *Rb. sphaeroides*, and *Rb. capsulatus*, and one purple sulfur bacterium, *Cm. vinosum*. Because of the widespread interest in structure-function relationships of RubisCO, these genes were the first of the Calvin cycle genes to be cloned by recombinant DNA techniques. The *cbbM* gene encoding *Rs. rubrum* form II type RubisCO was the first procaryotic RubisCO gene to be isolated and was identified in a lambda expression library by immunological cross-reactivity (Somerville and Somerville, 1984). The first form I RubisCO genes to be isolated from a photosynthetic bacterium, were isolated from *Cm. vinosum* based on hybridization to the *A. nidulans* RubisCO genes (Viale et al., 1985). Subsequently, genes encoding a second form I type RubisCO in *Cm. vinosum* were isolated on the basis of hybridization to the first set of genes (Viale et al., 1989). Both pairs of RubisCO genes are chromosomally encoded but do not appear to be closely linked (Viale et al., 1989). Heterologous hybridization has also been used to isolate clones containing the form I and form II RubisCO genes of *Rb. sphaeroides* (Quivey and Tabita, 1984; Muller et al., 1985; Gibson and Tabita, 1986). Interestingly, the form I and form II RubisCO coding sequences map to distinct genetic elements in this organism (Suwanto and Kaplan, 1989). Form I and form II RubisCO sequences were identified by Southern hybridization analysis of *Rb. capsulatus* and *Rhodopseudomonas palustris* chromosomal DNA (Shively et al., 1986). Corresponding clones containing the form I and form II RubisCO genes from *Rb. capsulatus* have recently been isolated (J. Shively, personal communication).

C. Identification and Isolation of Other cbb Genes

One question that naturally arose upon isolation of RubisCO genes was whether or not Calvin cycle genes would be clustered. The availability of DNA sequences flanking the form I and form II *Rb. sphaeroides* RubisCO genes prompted hybridization analysis with probes encoding other Calvin cycle enzymes. In this manner FBP and PRK sequences were found linked to both form I and form II RubisCO coding sequences (Gibson and Tabita, 1987, 1988;

Table 1. Gene designations of bacterial reductive pentose phosphate cycle enzymes and regulatory proteins

Gene	Enzyme	Product
cbbL	form I RubisCO	form I RubisCO large subunit
cbbS		form I RubisCO small subunit
cbbM	form II RubisCO	form II RubisCO large subunit
cbbF	FBP	fructose-1,6-bisphosphatase/sedoheptulose-1,7-bisphosphatase
cbbP	PRK	phosphoribulokinase
cbbA	FBA	fructose-1,6-/sedopheptulose-1,7-bisphosphate aldolase
cbbG	GAP	glyceraldehyde-3-phosphate dehydrogenase
cbbK	PGK	phosphoglycerate kinase
cbbE	EPI	ribulose-5-phosphate-3-epimerase
cbbT	TKL	transketolase
cbbR	CbbR	transcriptional activator protein
cbbz	PGP	phosphoglycolate phosphatase

Hallenbeck and Kaplan, 1987) and a sequence within a form II RubisCO clone hybridized to a gene probe encoding glyceraldehyde 3-phosphate dehydrogenase (GAP)(Gibson and Tabita, 1988). Additional genes have been identified in the *Rb. sphaeroides* gene clusters primarily on the basis of nucleotide sequence comparisons. Protein database searches with deduced amino acid sequences from regions situated immediately upstream of the RubisCO genes in both clusters showed similarity to a class II fructose 1,6-bisphosphate aldolase (FBA) sequence from *E. coli* (Chen et al., 1991; Gibson et al., 1991). Similarly, the clue to the identity of the gene encoding transketolase (TKL) that is encoded within the form II cluster was its similarity to the sequence of dihydroxyacetone synthetase, an enzyme involved in methanol assimilation in yeast (Chen et al., 1991). The reaction this enzyme catalyzes is functionally similar to the TKL reaction in the Calvin cycle (Janowicz et al., 1985). Preliminary sequence determination revealed additional genes downstream of the form I RubisCO coding sequence that are similar to *cbbX*, *cbbY* and *cbbZ* in chemolithotrophic bacteria (J. Gibson, unpublished). Of these three genes, only *cbbZ* has been identified as the gene encoding phosphoglycolate phosphatase (Schäferjohann et al., 1993). Although the function of the *cbbX* and *cbbY* gene products are not known, the location of the genes within *cbb* transcriptional units in chemolithotrophic bacteria suggests a role in CO_2 fixation (Meijer at al., 1991; Kusian et al., 1992). Finally, sequence similarity to the LysR family of transcriptional regulatory proteins allowed identification of *cbbR*, a regulatory gene that is linked to the form I genes (Gibson and Tabita, 1993).

Calvin cycle genes are known to be clustered in two other photosynthetic bacteria, *Rs. rubrum* and *Rb. capsulatus*. In *Rs. rubrum*, *cbb* sequences linked to the RubisCO gene include *cbbE, cbbF, cbbP, cbbT, cbbA* and *cbbR*, all of which were identified by Southern hybridization and/or DNA sequence comparisons (Falcone and Tabita, 1993b; R. Ghosh, personal communication). Similar to *Rb. sphaeroides*, *Rb. capsulatus* has been shown to express two types of RubisCO (Gibson and Tabita, 1977b). Interestingly, hybridization analysis of clones containing the form I and form II RubisCO genes revealed *cbb* structural genes encoding FBP, PRK, and FBA linked to *cbbM* but not to *cbbLcbbS* (J. Shively, personal communication). A potential regulatory gene was also found linked to the form II cluster by hybridization to *cbbR* genes from *Rb. sphaeroides* and *Rs. rubrum* (G. Paoli and F. R. Tabita, personal communication). No other *cbb* genes have been localized to DNA sequences flanking the form I RubisCO genes. Finally, as mentioned previously, two apparently unlinked sets of genes encoding distinct L_8S_8 RubisCO enzymes have been identified in *Cm. vinosum* (Viale et al., 1989). Although no other *cbb* structural genes have been identified in this organism, a CbbR sequence has been shown to be linked to one set of RubisCO genes (Viale et al., 1991).

D. cbb *Gene Organization*

In *Rb. sphaeroides*, the nucleotide sequence of the regions flanking the RubisCO genes confirmed the gene order determined from combined restriction nuclease mapping and Southern hybridizations and established the location of genes for which no probes were available (Gibson et al., 1990, 1991; Chen et al., 1991). In addition, sequence analysis showed that genes within each cluster were oriented in the same direction and closely spaced suggesting they formed a single transcriptional unit. Potential regulatory features were also revealed such as the overlap of stop codons with ribosome binding sites of the following gene pairs: $cbbF_I$-$cbbP_I$; $cbbF_{II}$-$cbbP_{II}$; $cbbT_{II}$-$cbbG_{II}$ and $cbbL_I$-$cbbS_I$.

The arrangement of genes common to both clusters is somewhat conserved as shown in Fig. 1. The major distinctions include the 3 kbp region in the form II cluster encoding TKL and GAP that is noticeably absent from the form I cluster (Chen et al., 1991) and the regulatory gene, *cbbR*, that is divergently oriented with respect to $cbbF_I$ (Gibson and Tabita, 1993). For comparison, the gene arrangement of *cbb* sequences from a nonphotosynthetic bacterium *Alcaligenes eutrophus*, is also shown in Fig. 1. This organism contains two sets of *cbb* genes that are plasmid and chromosomally encoded and, for the most part are exact duplicates of each other (Husemann et al., 1988). It is noteworthy that the gene arrangement between organisms is similar. In spite of the different location of RubisCO genes, the gene order of other *cbb* structural genes that are present (*cbbFcbbPcbbTcbbGcbbA*) is maintained. Several genes that have been identified in *A. eutrophus* are either absent in the *Rb. sphaeroides cbb* clusters or have not yet been identified. These include *cbbE* and *cbbK* (Bowien et al., 1993).

Mapping of Calvin cycle genes in *Rs. rubrum* revealed similarities as well as differences in *cbb* gene arrangement compared to other organisms (Fig. 1). In contrast to RubisCO genes in *Rb.*

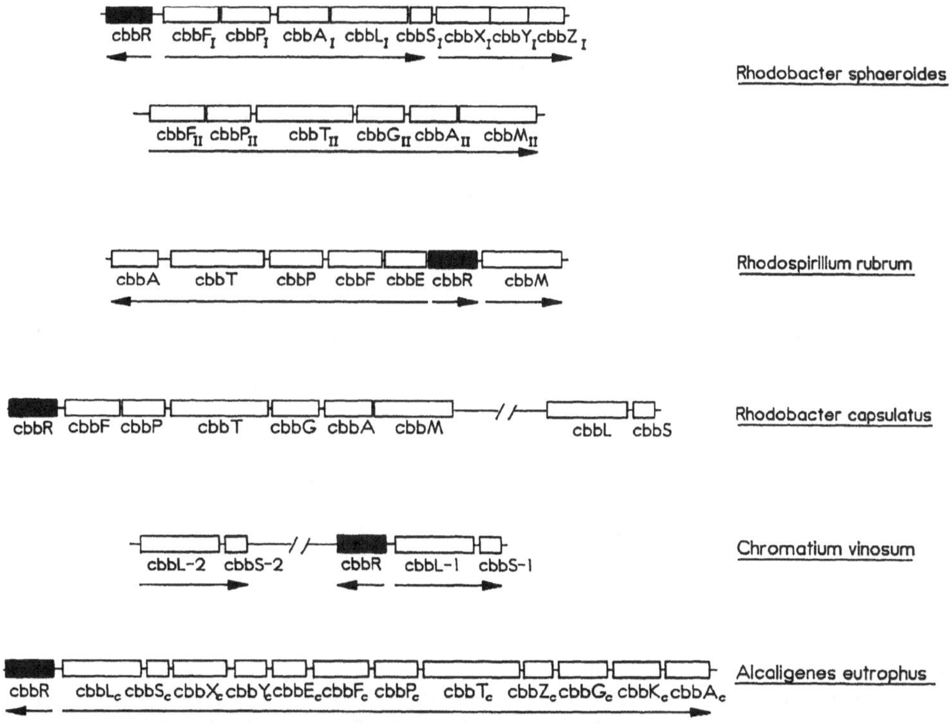

Fig. 1. cbb gene organization. For *cbb* genes of *Rs. rubrum* and $cbbX_I$, $cbbY_I$ and $cbbZ_I$, arrows indicate gene orientation only. In other cases, arrows denote transcriptional units. Gene designations for known Calvin cycle enzymes are indicated in Table 1. The functions of the proteins encoded by *cbbX* and *cbbY* are not known. The subscript C for *A. eutrophus* genes denote a chromosomal location.

sphaeroides, the *cbbM* gene is transcribed separately from other Calvin cycle genes (Falcone and Tabita, 1993b). The presumptive regulatory gene, *cbbR*, is immediately upstream from *cbbM*, but opposed to other CbbR coding sequences, is transcriptionally oriented in the same direction as the RubisCO gene (Falcone and Tabita, 1993b). Primary sequence determination of the region upstream of *cbbR* revealed additional *cbb* sequences divergently oriented with respect to *cbbR* and *cbbM* arranged in the following order: *cbbEcbbFcbbPcbbTcbbA* (Falcone and Tabita, 1993b; R. Ghosh, personal communication). The ribulose 5-phosphate-3-epimerase (PPE) coding sequence occupies the same position relative to *cbbF* in the *A. eutrophus cbb* cluster (Kusian et al., 1992).

In *Rb. capsulatus*, mapping of *cbb* genes linked to *cbbM* have thus far revealed a pattern identical to that of the form II cluster of *Rb. sphaeroides*, except that the *cbbR* homolog has been localized to a region upstream of the form II RubisCO gene (Fig. 1) (J. Shively; G. Paoli and F. R. Tabita, personal communication). The transcriptional orientation of these genes has not been determined. At present, the form I RubisCO genes appear to be isolated with respect to other Calvin cycle genes. Also illustrated in Fig. 1 are the *Cm. vinosum* RubisCO operons.

E. Amino Acid Sequence Comparisons

1. RubisCO

The deduced amino acid sequences of the form II type RubisCO proteins from *Rs. rubrum* and *Rb. sphaeroides* share 85% identity at the amino acid level (Nargang et al., 1984; Wagner et al., 1988). The relatively high degree of identity shared between the form II RubisCO enzymes is in sharp contrast to that calculated for the form II RubisCO and the large subunit of form I type RubisCO (Nargang et al., 1984; Wagner et al., 1988). Even when large subunit amino acid sequences of the form I and form II RubisCO proteins from *Rb. sphaeroides* are compared, only 23% identical residues are shared and most of these are centered around regions that have been shown to be important in either activation or catalysis (Gibson et al., 1991). Virtually the same degree of identity is observed when form II RubisCO sequences are compared to form I RubisCO large subunit sequences independent of the source of RubisCO. Numerous sequences are available for the form I type RubisCO, most of which are derived from eucaryotic algae and higher plants. Comparison of the photosynthetic bacterial form I RubisCO sequences from *Cm. vinosum* and *Rb. sphaeroides* with RubisCO sequences from other organisms has revealed interesting evolutionary relationships. First, deduced amino acid sequences of genes encoding the two RubisCO large and small subunits from *Cm. vinosum* share approximately the same degree of identity, 80% and 45%, respectively, with each other as with the amino acid sequences from cyanobacterial, green algal and terrestrial plants (Kobayashi et al., 1991). Significantly less identity, 52% between large and 24% between small subunits, is observed when either RubisCO of *Cm. vinosum* is compared to the form I RubisCO of *Rb. sphaeroides*. The *Rb. sphaeroides* form I RubisCO subunits are, in turn, more similar to those from the chemolithoautotrophic bacteria, and somewhat surprisingly, to algae belonging to the genera *Rhodophyta* and *Chromophyta* (Gibson et al., 1991; Hwang and Tabita, 1991). These comparisons revealed a previously unknown evolutionary relatedness between purple non-sulfur bacteria and non-green algae.

2. FBA, FBP, PRK

At present, sequence information for other *cbb* structural genes from photosynthetic bacteria is limited to coding sequences from *Rb. sphaeroides* and the PPE sequence of *Rs. rubrum* (Gibson et al., 1990, 1991; Chen et al., 1991; Falcone and Tabita, 1993). Comparison of the deduced amino acid sequences of gene pairs in *Rb. sphaeroides* encoding FBP, PRK and FBA is interesting from an evolutionary standpoint because, in contrast to the extreme divergence of the RubisCO genes, these sequences are highly conserved. The form I and form II FBP, PRK and FBA sequences share 66%, 86%, and 81% identical amino acid residues, respectively. Comparison of the FBP and PRK sequences to corresponding gene products from the closely related chemolithoautotrophic bacteria revealed about 20% less identity for both sequences (Gibson et al., 1990, 1991).

3. CbbR

The nucleotide sequences of several *cbbR* genes have been determined and it is interesting to note that the CbbR deduced amino acid sequences from photosynthetic and nonphotosynthetic bacteria share 40-50% identity independent of the divergence noted between RubisCO genes (Meijer et al.; 1991; Viale et al., 1991; Windhovel and Bowien, 1991; Falcone and Tabita, 1993; Gibson and Tabita, 1993; Kusano and Sugawara, 1993). Identity shared between CbbR sequences is higher than that noted between CbbR and other members of the LysR family and extends beyond the helix-turn-helix motif at the amino terminal portion of the coding sequence. It is tempting to speculate that the conserved central regions of the CbbR proteins represent sequences involved in binding a common inducer, which would implicate similar signals in the regulation of Calvin cycle gene expression in different bacteria.

III. Regulation of *cbb* Gene Expression

A. Rs. rubrum

The influence of culture conditions on expression of RubisCO in *Rs. rubrum* was first noted by Anderson and Fuller (1967). Initial investigations showed that the level of RubisCO attained under photolitho-autotrophic growth conditions with CO_2 as the sole carbon source and H_2 as the electron donor was more than 30-fold greater than that observed in cells grown photoheterotrophically with malate as carbon source. In addition, the reduction state of the organic carbon compound was shown to affect the level of RubisCO. Cells cultured on a relatively reduced substrate such as butyrate, in which exogenous bicarbonate in the media is essential for growth, expressed RubisCO at a level intermediate between that observed with either malate or CO_2 as carbon source (Tabita, 1988). During photolithoautotrophic growth, the concentration of CO_2 in the media has a profound effect on expression of RubisCO (Sarles and Tabita, 1983).

The regulation of RubisCO in *Rs. rubrum* in response to the carbon source supplied in growth media appears to be mediated, at least in part, at the level of transcription. An investigation by Leustek et al.(1988), quantitated levels of RubisCO activity and mRNA following a shift from photoheterotrophic growth on malate, to growth in an atmosphere of hydrogen containing either 1.5% CO_2 or 10% CO_2. The level of mRNA initially paralleled the increase in RubisCO specific activity in cultures grown at both concentrations of CO_2, and was four to five-fold higher in cells grown at the lower CO_2 concentration. However, later in the growth phase the specific activity of RubisCO continued to increase whereas the level of mRNA decreased until the activity of RubisCO in the cultures grown on 1.5% CO_2 was ten to thirty fold higher than the malate grown culture, although the mRNA levels had only increased 5-6 fold. Regulation at the level of translation was one possibility suggested by the authors to explain the discoordinate levels of enzyme activity and mRNA.

Northern blots of RNA extracted from cells grown under different conditions probed with a *cbbM* probe revealed, in all cases, a 1.4 kbp transcript that corresponds well to the size of the *cbbM* gene (Leustek et al., 1988). The transcriptional start site of the *cbbM* message was determined by S1 nuclease mapping of RNA isolated from *Rs. rubrum* and from *E. coli* transformed with the plasmid pRR117 that carries the *Rs. rubrum* RubisCO gene. In mRNA isolated from *Rs. rubrum,* a single transcriptional start site was detected that was situated 28 bp upstream from the *cbbM* translation initiation codon, regardless of growth condition. Interestingly, inspection of the sequence surrounding this start site showed it to be located within a region of dyad symmetry (Nargang et al., 1984). Three different start sites were mapped when the mRNA was isolated from *E. coli* harboring pRR117, none of which correspond to that mapped using *Rs. rubrum* RNA. Two of these start sites are preceded by typical *E. coli* –10, –35 consensus promoter sequences. It should be mentioned that although *cbbM* transcripts were observed in RNA isolated from *E. coli* carrying pRR117, no RubisCO protein could be detected.

B. Rb. sphaeroides

Regulation of RubisCO synthesis in *Rb. sphaeroides* in response to environmental factors was first demonstrated by Lascelles (1960) and has since been shown to be affected by such determinants as light, the concentration of O_2, CO_2, organic substrates, and alternate electron acceptors (Tabita, 1988; Hallenbeck et al., 1990a,b; Wang et al., 1993). In addition to the daunting number of environmental stimuli that affect CO_2 assimilation in this organism, the study of regulation is further complicated by the presence of at least two copies of genes encoding many of the Calvin cycle enzymes (Quivey and Tabita, 1984; Gibson and Tabita, 1986, 1987, 1988). Most of the initial work focused on the structurally and catalytically distinct form I and form II RubisCO enzymes present in this organism (Gibson and Tabita, 1977a). Induction of the form I and form II RubisCO during photolithoautotrophic growth was examined by Jouanneau and Tabita (1986). During photoheterotrophic growth on malate, although the levels of both forms of RubisCO are relatively low, the form II RubisCO is the more abundant enzyme. In an experiment in which the amount of the form I and form II RubisCO was quantitated following a shift from photoheterotrophic growth on malate to photolithoautotrophic growth in the presence of 1.5% CO_2/98.5% H_2, the level of the form I enzyme increased 17-fold replacing the form II enzyme as the predominant RubisCO. The level of the form II enzyme also increased, but only four fold. The difference in the extent of induction was one of the first indications that expression of the two enzymes was subject to independent regulation. The abundance of the form I RubisCO during growth in the presence of a low CO_2 concentration coupled with the higher affinity of this enzyme for the substrate CO_2 also led to the proposal that the form I enzyme was specialized to function in conditions of CO_2 limitation.

Zhu and Kaplan (1985) first demonstrated regulation of RubisCO at the level of transcription in *Rb. sphaeroides* by examining the effects of light, O_2, and carbon substrate on *cbbM*$_{II}$ mRNA levels. The level of the form II RubisCO transcript was highest in cells grown in high light, and when cultures were shifted from light to dark, the level of form II RubisCO mRNA decreased. Virtually no *cbbM*$_{II}$ transcript was detected in aerobically grown cells. In most cases, RubisCO activity paralleled the changes in mRNA levels. The greatest deviation was noted in malate and succinate cultures in which the level of the form II RubisCO transcript was 80-90% of that measured in cells cultured on butyrate, but the RubisCO specific activity was only 5-10% of butyrate grown cells. This discrepancy is most easily explained by the fact that the RubisCO activity measurement includes the activity of the form I RubisCO which can be quite high in butyrate grown cultures, whereas only the form II mRNA was quantitated in these experiments.

The original proposal that during photoheterotrophic growth, CO_2 fixation functions as an electron sink for excess reducing equivalents (Muller, 1933; Van Niel, 1936) was addressed in studies by Hallenbeck et al. (1990a,b) that examined the effect of CO_2 concentration on expression of PRK and RubisCO. In these investigations, photosynthetic growth of *Rb. sphaeroides* on succinate in cultures bubbled with argon was not observed, but growth did result under these conditions if cultures were supplied with either exogenous CO_2. Increasing the concentration of CO_2 in the media was shown to coordinately repress synthesis of RubisCO and PRK and the activities of both enzymes were inversely proportional with respect to growth rate and CO_2 concentration (Hallenbeck, 1989a). To distinguish between a direct effect of CO_2 on RubisCO and PRK synthesis and an indirect effect caused by perturbation of the redox balance, dimethyl sulfoxide (DMSO), an alternate electron acceptor, was added to the growth medium. Cultures supplemented with DMSO grew in the absence of exogenous CO_2 and exhibited reduced activities of PRK and RubisCO compared to cultures grown with limiting CO_2. Increasing the concentration of CO_2 in the presence of DMSO resulted in even greater repression of both PRK and RubisCO activities. Quantitation of RubisCO mRNA showed that the effects of DMSO and CO_2 concentration on PRK and RubisCO expression were exerted at the level of transcription because decreases in form I and form II RubisCO transcripts correlated well with the decreases in overall enzyme activity.

Northern hybridization of RNA isolated from *Rb. sphaeroides* detected mRNA species capable of encoding only one, or in the case of *cbbLcbbS*, two genes when probed with various *cbb* probes (Hallenbeck et al., 1990a,b; Gibson et al., 1991). Although the hybridization results revealed individual transcripts for the Calvin cycle enzymes, the results of insertional mutagenesis discussed below argue for cotranscription of genes within each cluster (Hallenbeck et al., 1990a,b; Gibson et al., 1990, 1991).

C. Cm. vinosum

In *Cm. vinosum*, RubisCO biosynthesis has been shown to be regulated by O_2, CO_2, reduced sulfur compounds, and organic compounds (Kobayashi and Akazawa, 1982). The effect of various environmental factors on transcription of RubisCO genes was determined in cultures after shift from photoheterotrophic growth, in which RubisCO synthesis is repressed, to derepressing growth conditions (Valle et al., 1988). RubisCO protein and mRNA levels were shown to increase coordinately following a shift to autotrophic media containing CO_2 as the sole carbon source. The absence of CO_2 and the presence of a reduced sulfur compound in the autotrophic media following the shift further enhanced RubisCO protein and mRNA levels. Increased RubisCO synthesis was also observed when cultures were transferred to photoheterotrophic growth conditions in the presence of a reduced sulfur compound. Transcripts were detected of 2.0 kbp, 1.5 kbp, and 1.3 kbp in Northern blots hybridized with *Cm. vinosum* RubisCO genes.

IV. Mutational Analysis

A. Autotrophic Mutants of Rb. sphaeroides

The first selection scheme for mutations affecting carbon dioxide assimilation in *Rb. sphaeroides* was designed to select strains incapable of derepressing RubisCO in growth modes that usually resulted in high levels of Calvin cycle enzymes. Hence, mutants were isolated that were incapable of growing either photoheterotrophically in butyrate-bicarbonate medium or photolithoautrophically, yet were unaffected in photoheterotrophic growth on malate (Weaver and Tabita, 1983). The mutants initially appeared to affect regulation of Calvin cycle enzymes, because in spite of the inability to derepress form I RubisCO during carbon starvation, a condition known to induce the form I enzyme in the wild type strain, the level of the form II RubisCO was normal (Weaver and Tabita, 1983). The discoordinate induction pattern in the mutant suggested independent pathways existed for regulating synthesis of the two RubisCO enzymes. Subsequent mutagenesis experiments yielded additional mutants with the same growth phenotype, but only one was incapable of derepressing form I RubisCO (Rainey and Tabita, 1989). Distinct DNA fragments from a *Rb. sphaeroides* library were found to complement three of the mutant strains and, interestingly, all three complementing fragments mapped to endogenous plasmid DNA instead of chromosomal DNA (Rainey and Tabita, 1989). The exact nature of the mutations is not understood, but based on subsequent work with RubisCO mutants described below, the growth phenotype is apparently not directly related to expression of Calvin cycle enzymes.

B. Mutations in cbb *Structural Genes*

1. Rb. sphaeroides

Gene replacement techniques have been used to obtain mutations in all of the known *cbb* structural genes in *Rb. sphaeroides* (Falcone et al., 1988; Hallenbeck et al., 1990a,b; Gibson et al., 1990, 1991). Double mutants have also been constructed in which both genes encoding RubisCO, PRK and FBA were inactivated (Hallenbeck et al., 1990a, b; Falcone and Tabita, 1991). Mutations were found to be polar on expression of downstream genes, such that all insertions eliminated expression of the corresponding RubisCO enzyme, and provided strong evidence for cotranscription of genes within each cluster (Hallenbeck et al., 1990a,b; Gibson et al., 1990, 1991). Strains containing single mutations were able to grow under photoheterotrophic as well as photolithoautotrophic growth conditions, demonstrating that neither form I nor form II enzymes are absolutely necessary for phototrophic growth (Falcone et al., 1988; Hallenbeck et al., 1990a,b; Gibson et al., 1991). Differences in growth characteristics were observed in some mutants, however. In general, these differences were more pronounced in the $cbbP_I$ and $cbbA_I$ mutants grown in the presence of low concentrations of CO_2, which assumed a wild type growth phenotype when the CO_2 concentration was increased (Hallenbeck et al., 1990a,b). Similarly, the relatively severe growth defects observed in $cbbP_{II}$ and $cbbA_{II}$ mutants in standing cultures were alleviated by addition of DMSO to culture media (Hallenbeck et al., 1990a,b). Double mutants were unable to grow photosynthetically unless cultures were supplemented with DMSO, providing direct evidence for the role CO_2 fixation plays as electron acceptor during photoheterotrophic growth (Hallenbeck et al., 1990a,b; Falcone and Tabita, 1991). It should be mentioned that the double RubisCO deletion mutant gained the capacity for photoheterotrophic growth after a prolonged incubation of two months (Wang et al., 1993). The genetic basis of the phenotype is not known, but most likely results from activation of a silent gene that in some way turns on an alternate pathway of CO_2 fixation.

Measurement of Calvin cycle enzyme activities and mRNA levels in the single mutants revealed an interesting interdependence in gene expression of the two gene clusters (Falcone et al., 1988; Hallenbeck et al., 1990a,b; Gibson et al., 1991). Altered expression of RubisCO and PRK were observed in the single PRK and FBA mutants grown under photoheterotrophic conditions in which the concentration of CO_2 was varied in the presence and absence of DMSO

(Hallenbeck et al., 1990a,b). The most dramatic case was exemplified by the $cbbP_{II}$ mutant where levels of PRK and RubisCO increased 1427-fold and 193-fold, respectively. Activities varied considerably between different mutants depending on the culture conditions (Hallenbeck et al., 1990a,b). The presence of DMSO caused severe repression of synthesis of the form II enzymes in the $cbbP_{I}$ mutant, but the level of form I enzymes was shown to increase in the $cbbP_{II}$ mutant (Hallenbeck et al., 1990b). The form I and form II RubisCO mRNA levels measured in the PRK mutants correlated well with RubisCO activity suggesting that the differences in enzyme activity reflect alterations at the level of transcription (Hallenbeck, et al, 1990a,b).

Actual RubisCO protein levels were measured in FBP and RubisCO mutants following photoheterotrophic and photolithoautotrophic growth. In contrast to activity, the immunological determination of RubisCO protein individually measures the level of form I and form II RubisCO allowing a direct comparison to levels expressed in the wild type strain (Gibson et al., 1991). In all of the mutants, increased expression of RubisCO from the unaltered gene cluster was observed (Table 2). Overexpression was most pronounced in the $cbbF_{I}$ mutant grown photolithoautotrophically in which the level of form II RubisCO increased 10-fold over that observed in the wild type strain and comprised 21% of the soluble protein (Gibson et al., 1991). Differential expression of RubisCO, PRK and FBP within each cluster was also evident from activities measured in the single mutants under different growth conditions (Table 2).

The construction and analysis of mutants in *cbb* genes substantiated previous proposals and provided invaluable insight into the nature of coordinated regulation of CO_2 fixation genes in *Rb. sphaeroides.* Polarity of the insertions on expression of downstream gene expression is evidence for cotranscription of genes within a cluster, thus providing a simple means of achieving stoichiometric amounts of *cbb* gene products. Collectively, these data illustrate two important facets of *cbb* gene regulation in *Rb. sphaeroides*: 1) the alteration of PRK and RubisCO synthesis shows that expression of one set of genes can respond to disruption of gene expression from the other operon, and 2) the different extent of induction of PRK and RubisCO demonstrates differential expression of genes within an operon. At present, posttranscriptional processing of a polycistronic message and differential stability of resultant transcripts is consistent with all of the data concerning expression of Calvin cycle enzymes in *Rb. sphaeroides*. This form of regulation has been described in other systems including that of the *puf* operon in *Rb. capsulatus* (Klug et al., 1987).

2. Rs. rubrum

Only two *cbb* mutations have been described in *Rs. rubrum*. A RubisCO deletion strain obtained by insertional mutagenesis unexpectedly retained the ability to grow photoheterotrophically, but was incapable of photolithoautotrophic growth (Falcone and Tabita, 1993b). As for the RubisCO deletion mutant of *Rb. sphaeroides*, photosynthetic growth in the absence of a functional Calvin cycle suggests the existence of an alternative pathway of CO_2 fixation. A second mutant obtained by transposon mutagenesis, was originally selected for alterations in electron transport (Ghosh et al., 1991). The mutant is characterized by impaired ability to grow phototrophically and reduced cytochrome c_2 content. However, sequence determination of regions flanking the transposon, surprisingly revealed that the mutation was in a gene corresponding to *cbbA* (R. Ghosh, personal communication). Although the effect of the mutation on electron transport is not clear, it is particularly interesting in light of the role CO_2 fixation normally plays during photoheterotrophic growth.

V. Transcriptional Regulatory Genes

A. Cm. vinosum

In *Cm. vinosum*, an open reading frame (ORF) located 226 bp upstream from and divergently oriented to *cbbL-1cbbS-1*, originally termed *rbcR*, was identified by nucleotide sequence analysis (Viale et al. 1991). No such homolog was detected upstream from *cbbL-2cbbS-2*. Since, at present, no genetic system is available for expression of genes in *Cm. vinosum*, the potential for CbbR regulation of RubisCO genes was investigated in *E. coli* (Viale et al., 1991). An expression clone was constructed in which the amino terminal portion of *lacZ* was fused to the initiation codon of CbbR and synthesis of the resultant fusion protein was directed by the *lac* promoter of the vector. RubisCO activity was lower in *E. coli* cells cotransformed with the *cbbR* construct and a plasmid containing the *cbbL-1cbbS-1* operon under control

Table 2. Levels of Calvin cycle enzymes in mutant strains

Strain	Growth Condition	Antigen[1]		FBP	PRK	RubisCO
		I	II	units/min/mg protein		
Wild type	HET[2]	1.8	0.54	0.03	0.02	0.05
	AUT[3]	5.6	1.9	0.10	0.13	0.13
cbbL$_I$cbbS$_I$	HET	–	0.73	0.05	0.05	0.04
	AUT	–	3.1	0.18	0.22	0.17
cbbM$_I$	HET	3.8	–	0.08	0.01	0.04
	AUT	5.8	–	0.08	0.09	0.09
cbbF$_I$	HET	–	2.1	0.11	0.04	0.08
	AUT	–	21.0	0.21	0.17	0.36
cbbF$_{II}$	HET	4.8	–	0.05	0.03	0.06
	AUT	14.2	–	0.12	0.13	0.17

[1] RubisCO antigen as determined by rocket immunoelectrophoresis expressed as percent total soluble protein. I, form I RubisCO; II, form II RubisCO.
[2] Photoheterotrophic growth on 0.4% malate bubbled with argon.
[3] Photolithoautotrophic growth on 1.5% CO_2/98.5% H_2.

of its own promoter than in *E. coli* cells containing the RubisCO plasmid alone. Because a reduction in RubisCO activity was observed in the presence of CbbR, the authors suggest a negative transcriptional regulatory function for the *Cm. vinosum* CbbR protein which would account for the lowered RubisCO expression observed during photoheterotrophic versus photolithoautotrophic growth. This proposal should be regarded as tentative since the promoter recognized in *E. coli* may not be the same as that in *Cm. vinosum* and gene dosage can alter properties of regulatory molecules.

B. Rb. sphaeroides

1. cbbR

As mentioned previously, nucleotide sequence determination of the region located upstream of the form I *cbb* cluster in *Rb. sphaeroides* revealed a divergently oriented gene, called *cbbR*, that showed significant identity to the LysR family of transcriptional regulatory proteins (Gibson and Tabita, 1993). Evidence that suggested CbbR regulated expression of both *cbb* operons was obtained from studies of a strain in which the *cbbR* gene was inactivated by insertional mutagenesis (Gibson and Tabita, 1993). The *cbbR* mutant was impaired in photoheterotrophic growth and was completely incapable of growing photolithoautotrophically in a CO_2/H_2 atmosphere. In cells grown photoheterotrophically on malate, no expression of form I RubisCO could be detected and the synthesis of the form II RubisCO was reduced to approximately 30% of the wild type strain (Table 3). The *cbbR* mutant could be complemented to a wild type growth phenotype with plasmids containing the form I and form II gene clusters. Normal levels of form I and form II RubisCO were observed only in the mutant carrying the plasmid encoded form I genes, which include *cbbR* (Table 3). No form I RubisCO was detected in the strain complemented with the form II genes, and although normal levels of RubisCO were measured in photoheterotrophically grown cells, no derepression or induction of RubisCO was evident in photolithoautotrophically grown cultures (Table 3). When the mutant was complemented with *cbbR*, a normal induction pattern for RubisCO was restored (Table 3). However, immunological analysis revealed that only 10% the normal level of form I RubisCO was present in photolithoautotrophic cultures and the apparent wild type activity was due to overexpression of form II RubisCO (Table 3). The inability of plasmid encoded CbbR to restore normal synthesis of the form I enzymes is not clear, but may be due to gene dosage effects. The complete absence of detectable form I RubisCO in the *cbbR* mutant suggests that CbbR is essential for transcription of the form I genes and it is evident that the CbbR gene product is required for maximum induction of the form II proteins during photolithoautotrophic growth. The low level of form II RubisCO in photoheterotrophically grown cells that is observed in the absence of CbbR may be due simply to a constitutive level of expression or may result from the involvement of additional regulatory factors. On the basis of

Table 3. RubisCO levels in wild type and *cbbR* mutant and plasmid-complemented strains

STRAIN	PLASMID encoded *cbb* genes	Growth Condition	RubisCO Protein[1]		RubisCO Specific Activity (nmol/min/mg protein)
			I	II	
Wild type		Mal[2]	1.8	1.5	26.6
		Aut[3]	9.2	2.9	363.7
cbbR		Mal	–[4]	0.5	8.0
		Aut	n.d.[5]	n.d	n.d.
cbbR	cbbR-cbb_I operon	Mal	1.7	0.7	39.5
		Aut	7.5	1.3	290.3
cbbR	cbb_{II} operon	Mal	–	2.2	32.1
		Aut	–	2.0	16.1
cbbR	cbbR	Mal	–	4.3	43.0
		Aut	0.9	28.5	271.4

[1] RubisCO protein as determined by rocket immunoelectrophoresis expressed as percent total soluble protein; the average of duplicate determinations
[2] Photoheterotrophic growth on 0.4% malate
[3] Photolithoautotrophic growth on 1.5% CO_2 in H_2
[4] –, not present or found in the strain indicated
[5] Not determined; strain does not grow under these conditions.

complementation results, the overexpression of the form II enzymes observed in response to loss of the form I gene products is mediated by CbbR as well, and can account for the apparent communication previously observed between the two operons. Interestingly, the promoters controlling expression of three open reading frames situated immediately upstream of $cbbF_{II}$ appear to be influenced by CbbR as shown by transcriptional fusions to the *xylE* gene (Xu and Tabita, 1994).

2. Aerobic Expression

In the course of examining the effect of the *cbbR* mutation on expression of the genes within the form I and form II operons, RubisCO activity was measured in aerobically grown cells of the mutant strain which carried plasmids containing the form I operon, the form II operon or *cbbR*. Significantly higher levels of RubisCO were observed in the *cbbR* mutant carrying the form II cosmid. The elevated activity was unrelated to the *cbbR* background because the same level of activity was observed in the RubisCO deletion strain harboring the form II plasmid which carries a wild type *cbbR* locus (Gibson and Tabita, 1993). This evidence, though preliminary, raises the possibility of a low copy number negative regulatory factor involved in repression of the form II genes in the presence of oxygen.

C. Rs. rubrum

A *cbbR* gene has been identified upstream of the RubisCO gene in *Rs. rubrum* (Falcone and Tabita, 1993b). Unlike other *cbbR* genes that have been described, the *cbbR* gene in *Rs. rubrum* is unique in that it is oriented in the same direction as the RubisCO gene but divergent to an operon comprised of other Calvin cycle genes (Falcone and Tabita, 1993b). Evidence for the function of the *Rs. rubrum* CbbR gene product comes primarily from plasmid directed expression of RubisCO from the *cbbM* promoter in *Rb. sphaeroides*. Phenotypic complementation of the *Rb. sphaeroides* RubisCO deletion mutant with the *Rs. rubrum* RubisCO could be accomplished only if sufficient sequence upstream of the *cbbM* gene was present that contained the entire CbbR coding sequence (Falcone and Tabita, 1993b). These data implicated *cbbR* as a requirement for expression of *Rs. rubrum* RubisCO. However, complementation of the *Rs. rubrum* RubisCO deletion strain was not achieved using either the *cbbRcbbM* containing fragment that complemented the *Rb. sphaeroides* RubisCO mutant or a series of plasmid constructs containing additional sequence upstream and downstream of the RubisCO gene. In spite of the inability of these plasmids to complement the strain to a wild type growth phenotype, expression of RubisCO was observed in *Rs. rubrum* when 2 kbp

DNA upstream from RubisCO was present, encoding the divergently oriented *cbbE* in addition to *cbbR*. In one case, RubisCO comprised up to 20% of the soluble protein, suggesting that the reason for lack of complementation was not due to insufficient RubisCO, but may instead result from an imbalance in levels of Calvin cycle enzymes.

D. Regulatory Features of Sequences 5' to cbbR

LysR regulated promoters have typically been mapped to the region separating the LysR coding sequence from the genes it regulates (Henikoff et al., 1988). The 5'-end of the *Rs. rubrum cbbM* transcript, which maps to the intergenic region between *cbbM* and *cbbR* fits this model (Leustek et al., 1988; Falcone and Tabita, 1993). Although other promoters for the photosynthetic bacterial *cbb* genes have not been located, several features of the intergenic regions are notable. All of the intergenic regions are A+T rich compared to the relatively high G+C content of the coding regions and contain T-N_{11}-A motifs common to LysR DNA binding sites (Viale et al., 1991; Falcone and Tabita, 1993b; Gibson and Tabita, 1993). In *Cm. vinosum*, the sequences, 5'-ATATTGCTT-3' and 5'-ATGTGGTTT-3', were also noted that share significant similarity with consensus sequences described for other LysR intergenic regions (Viale et al., 1991).

VI. Expression of *cbb* Genes in Heterologous Systems

A. E. coli

The *Rs. rubrum* form II type RubisCO was the first Calvin cycle enzyme to be synthesized as a fully functional protein in *E. coli* (Somerville and Somerville, 1984). High level expression has been attained by placing the *cbbM* gene under transcriptional control of the *lac* promoter (Somerville and Somerville, 1984; Pierce and Gutteridge, 1985). Characterization of the purified protein produced in *E. coli* showed it to be comparable in specific activity and migration in SDS gels to the enzyme isolated from *Rs. rubrum*, although later experiments showed the plasmid encoded enzyme was actually a fusion protein that contained an additional 25 amino acid residues of *lacZ* at its amino terminus. The plasmid was later reconstructed to eliminate the gene fusion without any apparent effect on activity (Larimer et al., 1986). The form I and form II RubisCO from *Rb. sphaeroides* and both of the form I type enzymes from *Cm. vinosum* have been successfully expressed in *E. coli* and purified recombinant proteins were subsequently shown to be identical to the authentic enzyme with respect to various catalytic and immunological properties (Quivey and Tabita, 1984; Muller et al., 1985; Viale et al., 1985, 1989, 1990; Gibson and Tabita, 1986). Almost all of the other *cbb* structural genes have been expressed and demonstrated to be fully functional enzymes in *E. coli* (Gibson and Tabita, 1987, 1988; Hallenbeck, 1987; Chen et al., 1991). These include the FBP, PRK, TKL, and GAP enzymes from *Rb. sphaeroides*. It should be mentioned that expression of TKL directed by the *lac* promoter in *E. coli* and actual measurement of TKL activity confirmed the identity of this gene (Chen et al., 1991). In addition, expression of the two PRK enzymes from *Rb. sphaeroides* in *E. coli* made possible the ability to characterize each enzyme individually before mutants were available in *Rb. sphaeroides*, because both enzymes copurify in extracts of *Rb. sphaeroides* (Gibson and Tabita, 1987). Attempts to assay FBA expressed in *E. coli* have been unsuccessful Gibson et al., 1991).

To date, the only photosynthetic bacterial Calvin cycle enzyme synthesized from its own promoter in *E. coli* is the *Cm. vinosum* RubisCO encoded by *cbbL-1cbbS-1* (Viale et al., 1990). It has not been determined whether or not *E. coli* polymerase recognizes the same signals as *Cm. vinosum* polymerase. In this context, it is interesting to note that although two RubisCO coding sequences have been identified in *Cm. vinosum*, apparently only one is expressed to a measurable extent (Viale et al., 1990). Amino terminal sequence data of RubisCO purified from *Cm. vinosum* indicated only a single RubisCO was present, corresponding to the *cbbL-1cbbS-1* gene product (Viale et al., 1990). Individual expression of the two sets of genes in *E. coli* resulted in functional RubisCO with distinct catalytic properties (Viale et al., 1990). The properties of the *cbbL-1cbbS-1* gene product were the same as the enzyme isolated from *Cm. vinosum*, further supporting the notion that only one set of RubisCO genes is expressed in the native organism.

B. Expression of cbb Genes in Heterologous Photosynthetic Bacteria

An interesting experiment was performed by Pierce et al.(1989) in which a cyanobacterial L_8S_8 RubisCO was replaced in vivo with the *Rs. rubrum* L_2 RubisCO. Initial attempts to inactivate the RubisCO genes in *Synechocystis* 6803 were unsuccessful leading to the supposition that RubisCO was essential even during growth on glucose. A variation of this experiment was carried out in which the *Synechocystis* RubisCO genes were actually replaced with the *cbbM* gene encoding the *Rs. rubrum* RubisCO fused to the cyanobacterial RubisCO promoter. The resultant cyanorubrum strain, was extremely sensitive to the ratio of oxygen to carbon dioxide and grew in the absence of glucose, provided the atmosphere was supplemented with 5% CO_2. The cyanobacterial RubisCO has a low affinity for the substrate CO_2 which is offset in part because the enzyme is housed in structures called carboxysomes, postulated to contain a specialized carbonic anhydrase to provide CO_2 at the active site (Reinhold et al., 1987). Carboxysomes were not observed in the cyanorubrum strain, a factor that may account for the observed phenotype. The construction of the cyanorubrum strain provided a unique opportunity to examine the physiological consequences of replacing the native RubisCO with a foreign enzyme.

Falcone and Tabita have utilized a RubisCO deletion strain of *Rb. sphaeroides* as a host with which to examine the effects of expression of heterologous or mutated RubisCO enzymes on the physiology of this organism as well as a means to analyze regulation of foreign *cbb* promoters (Tabita et al., 1990; Falcone and Tabita, 1991, 1993a). For the former, a broad-host range expression vector was constructed into which genes of interest could be inserted downstream from the *Rs. rubrum cbbM* promoter (Falcone and Tabita, 1991). This promoter was shown to be regulated in an expected manner in *Rb. sphaeroides* by first analyzing synthesis of the *Rs. rubrum* RubisCO itself under different growth conditions (Falcone and Tabita, 1991). Subsequently, the *Rs. rubrum* RubisCO gene was replaced with the form I and form II RubisCO genes from *Rb. sphaeroides* and the cyanobacterial *rbcLrbcS* genes from *Synechococcus* PCC6301. All of the RubisCO constructs were able to complement the RubisCO deletion mutant of *Rb. sphaeroides* to photoheterotrophic and photolithoautotrophic growth (Falcone and Tabita, 1991). Interestingly, however, the mutant harboring the cyanobacterial RubisCO genes exhibited a pronounced lag and markedly longer growth rate during photosynthetic growth on malate when cultures were bubbled with argon, although in standing cultures growth was comparable to that of the other complemented strains. The difference in growth can be attributed to the higher concentration of CO_2 in the standing culture. The growth phenotype therefore reflects the properties of the cyanobacterial RubisCO which has an unusually low affinity for CO_2 normally compensated for in its native host by carboxysomes and a CO_2 concentrating mechanism and thus illustrates the potential of using the *Rb. sphaeroides* RubisCO mutant for screening RubisCO enzymes with altered catalytic properties (Reinhold et al., 1987; Kaplan et al., 1989).

Expression of foreign *cbb* genes directed by homologous promoters was also examined in the RubisCO deletion strain (Falcone and Tabita, 1993a). In addition to the *Rs. rubrum* RubisCO, *cbb* genes from two nonphotosynthetic bacteria, *Xanthobacter flavus* and *Bradyrhizobium japonicum*, were expressed from their own promoters in the *Rb. sphaeroides* mutant. In contrast to the normal regulation that was observed for the *Rs. rubrum* promoter, the *Xanthobacter* and *Bradyrhizobium cbb* genes were regulated in a fashion exactly opposite to that for *Rb. sphaeroides* as more RubisCO was measured in cells cultured under photoheterotrophic than photolithoautotrophic growth conditions.

C. In vitro Transcription

1. Rb. sphaeroides

The inability to express RubisCO genes in *E. coli* from native promoters led to the development of in vitro transcription systems as an alternative for examining regulatory factors governing synthesis of RubisCO in photosynthetic bacteria. This approach, which has thus far met with limited success, was first used in *Rb. sphaeroides*. Plasmid encoded form II RubisCO genes from *Rs. rubrum* and *Rb. sphaeroides* were expressed in an *Rb. sphaeroides* coupled transcription-translation system using RubisCO activity as a measure of transcription (Chory et al., 1985). In vitro synthesis and assembly of the *Rb. sphaeroides* form II RubisCO was reported from a cloned DNA fragment containing only a few hundred base pairs 5' to $cbbM_{II}$. This result suggests an internal promoter for the form II RubisCO is present, a result in conflict with data derived from mutagenesis and complementation studies that are discussed above.

2. Rs. rubrum

The 5' end of *Rs. rubrum* RubisCO mRNA synthesized in an in vitro transcription system with either *E. coli* or *Rs. rubrum* RNA polymerase was determined by S1 mapping procedures (Leustek et al., 1988). Regardless of the source of polymerase, the same three 5'-ends were observed corresponding to those mapped when RNA was isolated from *E. coli* harboring plasmid encoded *cbbM,* but not when RNA was isolated from *Rs. rubrum*. The discrepancy between apparent in vitro and in vivo transcriptional start sites may reflect the absence of factors in vitro that are required for recognition of the RubisCO promoter.

3. Cm. vinosum

Expression of the first set of RubisCO genes isolated from *Cm. vinosum* in *E. coli* was dependent on an *E. coli* promoter, suggesting that the native promoter, if present, was not recognized by *E. coli* RNA polymerase. Transcription was measured by hybridizing ^{32}-P-labeled RNA synthesized in vitro to DNA fragments representing vector, *rbcL* and *rbcS* sequences blotted to nitrocellulose. With *E. coli* RNA polymerase, RubisCO RNA was only detected when synthesis was directed by the *E. coli tac* promoter. RNA synthesized in a homologous system with purified *Cm. vinosum* RNA polymerase hybridized to RubisCO genes, but the results are difficult to interpret since the plasmid used as template also contained the *tac* promoter and no control plasmid DNA lacking this promoter was used to determine whether or not it is functional in this system.

VII. Conclusions

As indicated throughout this review, expression of Calvin cycle enzymes is affected by a wide variety of environmental factors. This is not surprising given the central, yet varied role CO_2 fixation plays in cellular metabolism in photosynthetic bacteria. Over the past ten years, cloning and expression of DNA coding for Calvin cycle enzymes have provided the basis for significant advances made in our understanding of the molecular biology underlying the regulation of this important pathway. Gene replacement technology has enabled analysis of the individual roles of the two forms of RubisCO in *Rb. sphaeroides* and revealed different facets of regulation involved in coordinating expression of two different sets of Calvin cycle genes. Numerous studies have targeted transcription as a major determinant in control of Calvin cycle gene expression in phototrophic bacteria. The identification of CbbR proteins as members of the LysR family of transcriptional regulators offer the first clues into the nature of the transcriptional regulation. Deletion of RubisCO genes in *Rs. rubrum* and *Rb. sphaeroides* surprisingly unmasked alternative means of assimilating CO_2, further demonstrating the metabolic versatility of photosynthetic bacteria. Sequencing of structural and regulatory Calvin cycle genes has revealed intriguing relationships between photosynthetic bacteria and other organisms. Although numerous regulatory and evolutionary features have emerged from these studies, many questions remain unanswered. Foremost among future avenues of research will be the elucidation of promoter sequences involved in regulation of *cbb* operons and the signal(s) through which transcriptional activation or repression is mediated. Biochemical analysis of the transcriptional regulatory proteins will also be required for a complete understanding of the molecular processes involved in gene expression. Although not dealt with in this review, post-transcriptional factors affecting synthesis, assembly and activity of Calvin cycle enzymes must also be considered in examining the overall control of CO_2 assimilation. Translation is an area that remains to be explored. However, post-translational mechanisms affecting assembly and activity of RubisCO have already been identified and is discussed in Chapter 41 by F.R. Tabita. Ultimately, the question of how the regulation of CO_2 fixation is coordinated with other areas of metabolism will need to be addressed.

Acknowledgments

I gratefully acknowledge R. Ghosh, J. Shively, and G. Paoli for permission to cite unpublished results. I would also like to thank G. Paoli, T. Wahlund and F.R. Tabita for their critical review of this manuscript. Support from Support from Public Health Service Grant GM-45404 from the National Institutes of Health is acknowledged.

References

Anderson L and Fuller RC (1967) Photosynthesis in *Rhodospirillum rubrum*. III. Metabolic control of reductive pentose phosphate and tricarboxylic acid cycle enzymes. Plant Physiol 42: 497–502

Bowien B, Bednarski R, Kusian B, Windhövel U, Freter A, Schüferjohann and Yoo J-G (1993) Genetic regulation of CO_2 assimilation in chemoautotrophs. In: Murrell JC and Kelly DP (eds) Microbial Growth on C_1 Compounds, pp 481–491. Intercept Limited, Andover

Chen J-H, Gibson JL, McCue, LA and Tabita FR (1991) Identification, expression, and deduced primary structure of transketolase and other enzymes encoded with the form II CO_2 fixation operon of *Rhodobacter sphaeroides*. J Biol Chem 266: 20447–20452

Chory J, Muller ED and Kaplan S (1985) DNA-directed in vitro synthesis and assembly of the form II D-ribulose-1,5-bisphosphate carboxylase/oxygenase from *Rhodopseudomonas sphaeroides*. J Bacteriol 161: 307–313

Dow CS (1987) CO_2 fixation in *Rhodopseudomonas blastica*. Van Verseveld HW and Duine JA (eds) In: Microbial Growth on C_1 Compounds, pp 28–35. Martinus Nijhoff, Dordrecht

Falcone DL and Tabita FR (1991) Expression of endogenous and foreign ribulose 1,5-bisphosphate carboxylase-oxygenase (RubisCO) genes in a RubisCO deletion mutant of *Rhodobacter sphaeroides*. J Bacteriol 173: 2099–2108

Falcone DL and Tabita FR (1993a) Expression and regulation of *Bradyrhizobium japonicum* and *Xanthobacter flavus* CO_2 fixation genes in a photosynthetic bacterial host. J Bacteriol 175: 866–869

Falcone DL and Tabita FR (1993b) Complementation analysis and regulation of CO_2 fixation gene expression in a ribulose 1,5-bisphosphate carboxylase-oxygenase deletion strain of *Rhodospirillum rubrum*. J Bacterioleriol (in press)

Falcone DL, Quivey Jr. RG and Tabita FR (1988) Transposon mutagenesis and physiological analysis of strains containing inactivated form I and form II ribulose bisphosphate carboxylase/oxygenase genes in *Rhodobacter sphaeroides*. J Bacteriol 170: 5–11

Ghosh R, DiBerardino M, Saegesser R and Bachofen R (1991) Characterization of an unusual photosynthetically incompetent mutant of *Rhodobacter sphaeroides*. Abst. VII. Intl Symp Photo Prok, Amherst, MA, p. 195

Gibson JL and Tabita FR (1977a) Different molecular forms of D-ribulose-1,5-bisphosphate carboxylase from *Rhodopseudomonas sphaeroides*. J Biol Chem 252: 943–949

Gibson JL and Tabita FR (1977b) Isolation and preliminary characterization of two forms of ribulose 1,5-bisphosphate carboxylase from *Rhodopseudomonas capsulatus*. J Biol Chem 252: 943-949

Gibson JL and Tabita FR (1986) Isolation of the *Rhodopseudomonas sphaeroides* form I ribulose 1,5-bisphosphate carboxylase/oxygenase large and small subunit genes and expression of the active hexadecameric enzyme in *Escherichia coli*. Gene 44: 271–278

Gibson JL and Tabita FR (1987) Organization of phosphoribulokinase and ribulose bisphosphate carboxylase/oxygenase genes in *Rhodopseudomonas (Rhodobacter) sphaeroides*. J Bacteriol 169: 3685–3690

Gibson JL and Tabita FR (1988) Localization and mapping of CO_2 fixation genes within two gene clusters in *Rhodobacter sphaeroides*. J Bacteriol 170: 2153–2158

Gibson JL and Tabita FR (1993) Nucleotide sequence and functional analysis of CbbR, a positive regulator of the Calvin cycle operons of *Rhodobacter sphaeroides*. J Bacteriol 175: 5778–5784

Gibson JL, Chen J-H, Tower, PA and Tabita FR (1990) The form II fructose 1,6-bisphosphatase and phosphoribulokinase genes form part of a large operon in *Rhodobacter sphaeroides*: primary structure and insertional mutagenesis analysis. Biochemistry 29: 8085–8093

Gibson JL, Falcone DL and Tabita FR (1991) Nucleotide sequence, transcriptional analysis, and expression of genes encoded within the form I CO_2 fixation operon of *Rhodobacter sphaeroides*. J Biol Chem 266: 14646–14653

Hallenbeck PL and Kaplan S (1987) Cloning of the gene for phosphoribulokinase activity from *Rhodobacter sphaeroides* and its expression in *Escherichia coli*. J Bacteriol 169: 3669–3678

Hallenbeck PL, Lerchen R, Hessler and Kaplan S (1990a) Roles of CfxA, CfxB, and external electron acceptors in regulation of ribulose 1,5-bisphosphate carboxylase/oxygenase expression in *Rhodobacter sphaeroides*. J Bacteriol 172: 1736–1748

Hallenbeck PL, Lerchen R, Hessler P and Kaplan S (1990b) Phosphoribulokinase activity and regulation of CO_2 fixation critical for photosynthetic growth of *Rhodobacter sphaeroides*. J Bacteriol 172: 1749–1761

Husemann M, Klintworth R, Böttcher V, Salnikow J, Weissenborn C and Bowien B (1988) Chromosomally and plasmid-encoded gene clusters for CO_2 fixation (*cfx* genes) in *Alcaligenes eutrophus*. Mol Gen Genet 214: 112–120

Hwang S-R and Tabita FR (1991) Cotranscription, deduced primary structure, and expression of the chloroplast-encoded *rbcL* and *rbcS* genes of the marine diatom *Cylindrotheca* sp. Strain N1. J Biol Chem 266: 6271–6279

Janowicz ZA, Eckart MA, Drewke C, Roggenkamp RO and Hollenberg CP (1985) Cloning and characterization of the *DAS* gene encoding the major methanol assimilatory enzyme from the methylotrophic yeast *Hansenula polymorpha*. Nucleic Acid Research 13: 3043–3063

Jouanneau Y and Tabita FR (1986) Independent regulation of synthesis of form I and form II ribulose bisphosphate carboxylase-oxygenase in *Rhodopseudomonas sphaeroides*. J Bacteriol 165: 620–624

Kaplan A, Friedberg D, Schwarz R, Ariel R, Seijffers J and Reinhold L (1989) The "CO_2 concentrating mechanism" of cyanobacteria: Physiological, molecular and theoretical studies. In: Briggs WR (ed) Photosynthesis, pp 243–255. Alan R Liss, New York

Klug G, Adams CW, Belasco JG, Dörge B and Cohen SN (1987)

Biological consequence of segmental alterations in mRNA stability: effects of the intercistronic hairpin loop region of the *Rhodobacter sphaeroides puf* operon. EMBO J 6: 3515–3520

Kobayashi H and Akazawa T (1982) Biosynthetic mechanism of ribulose-1,5-bisphosphate carboxylase in the purple photosynthetic bacterium, *Chromatium vinosum*. I. Inducible formation. Arch Biochem Biophys 214: 531–539

Kobayashi H Viale AM, Takabe T, Akazawa T, Wada K, Shinozaki K, Kobayashi K and Sugiura M (1991) Sequence and expression of genes encoding the large and small subunits of ribulose 1,5-bisphosphate carboxylase/oxygenase from *Chromatium vinosum*. Gene 97: 55–62

Kusian B, Yoo J-G, Bednarski R and Bowien B (1992) The Calvin cycle enzyme pentose-5-phosphate 3-epimerase is encoded within the *cfx* operons of the chemoautotroph *Alcaligenes eutrophus*. J Bacteriol 174: 7337–7344

Kusano T and Sugawara LK (1993) Specific binding of *Thiobacillus ferrooxidans* RbcR to the intergenic sequence between the *rbc* operon and the *rbcR* gene. J Bacteriol 175: 1019–1025

Larimer FW, Machanoff R and Hartman FC (1986) A reconstruction of the gene for ribulose bisphosphate carboxylase from *Rhodospirillum rubrum* that expresses the authentic enzyme in *Escherichia coli*. Gene 41: 113–120

Lascelles J (1960) The formation of ribulose 1,5-diphosphate carboxylase by growing cultures of *Athiorhodaceae*. J Gen Microbiol 23: 499–510

Leustek T, Hartwig R, Weissbach H and Brot N (1988) Regulation of ribulose bisphosphate carboxylase expression in *Rhodospirillum rubrum*: characteristics of mRNA synthesized in vivo and in vitro. J Bacteriol 170: 4065–4071

Meijer WG, Enequist HG, Terpstra P and Dijkhuizen L (1990) Nucleotide sequences of the genes encoding fructose-bisphosphatase and photophoribulokinase from *Xanthobacter flavus* H4-14. J Gen Microbiol 136: 2225–2230

Meijer WG, Arnberg AC, Enequist HG, Terpstra P, Lidstrom ME and Dijkhuizen L (1991) Identification and organization of carbon dioxide fixation genes in *Xanthobacter flavus* H4-14. Mol Gen Genet 225: 320–330

Muller ED, Chory J and Kaplan S (1985) Cloning and characterization of the gene product of the form II ribulose-1,5-bisphosphate carboxylase gene of *Rhodopseudomonas sphaeroides*. J Bacteriol 161: 469–472

Muller FM (1933) On the metabolism of the purple sulfur bacteria in organic media. Arch Mikrobiol 4: 131–166

Nargang F, McIntosh L and Somerville C (1984) Nucleotide sequence of the ribulosebisphosphate carboxylase gene from *Rhodospirillum rubrum*. Mol Gen Genet 193: 220–224

Pierce J and Gutteridge S (1985) Large-scale preparation in ribulosebisphosphate carboxylase from a recombinant system in *Escherichia coli* characterized by extreme plasmid instability. Applied and Environ Micro 49: 1094–1100

Pierce J, Carlson TJ and Williams JGK (1989) A cyanobacterial mutant requiring the expression of ribulose bisphosphate carboxylase from a photosynthetic anaerobe. Proc Natl Acad Sci USA 86: 5753–5757

Quivey Jr RG and Tabita FR (1984) Cloning and expression in *Escherichia coli* of the form II ribulose 1,5-bisphosphate carboxylase/oxygenase gene from *Rhodopseudomonas sphaeroides*. Gene 31: 91–101

Rainey AM and Tabita FR (1989) Isolation of plasmid DNA sequences that complement *Rhodobacter sphaeroides* mutants deficient in the capacity for CO_2-dependent growth. J Gen Micro 135: 1699–1713

Reinhold L, Zwiman M and and Kaplan A (1987) Inorganic carbon fluxes and photosynthesis in cyanobacteria—a quantitative model. In: Biggins L (ed) Progress in Photosynthesis Research. Vol 4 pp 289–296 Martinus Nijhoff Publishers, New York

Sarles LS and Tabita FR (1983) Derepression of the synthesis of D-ribulose 1,5-bisphosphate carboxylase/oxygenase from *Rhodospirillum rubrum*. J Bacteriol 153: 458–464

Schäferjohann J, Yoo J-G, Kusian B and Bowien B (1993) The *cbb* operons of the facultative chemoautotroph *Alcaligenes eutrophus* encode phosphoglycolate phosphatase. J Bacteriol 175: 7329–7340

Shively JM, Davidson E and Marrs BL (1984) Derepression of the synthesis of the intermediate and large forms of ribulose-1,5-bisphosphate carboxylase/oxygenase in *Rhodopseudomonas capsulatus*. Arch Microbiol 138: 233–236

Shively JM, Devore W, Stratford L, Porter L, Medlin L and Stevens, Jr SE (1986) Molecular evolution of the large subunit of ribulose-1,5-bisphosphate carboxylase/oxygenase (RubisCO). FEMS Micro Lett 37: 251–257

Somerville CR and Somerville SC (1984) Cloning and expression of the *Rhodospirillum rubrum* ribulosebisphosphate carboxylase gene in *E. coli*. Mol Gen Genet 193: 214–219

Suwanto A and Kaplan S (1989) Physical and genetic mapping of the *Rhodobacter sphaeroides* genome: presence of two unique circular chromosomes. J Bacterioleriol 171: 5850–5859

Tabita FR, Gibson JL, Mandy WJ and Quivey Jr RG (1986) Synthesis and assembly of a novel recombinant ribulose bisphosphate carboxylase/oxygenase. Bio/Technology 4: 138–141

Tabita FR (1988) Molecular and cellular regulation of autotrophic carbon dioxide fixation in microorganisms. Micro Rev 52: 155–189

Tabita FR, Gibson JL, Falcone DL, Lee B and Chen J-H (1990) Recent studies on the molecular biochemistry of CO_2 fixation in phototrophic bacteria. FEMS Micro Rev 87: 437–444

Tabita FR, Gibson JL, Bowien B, Dijkhuizen L and Meijer WG (1992) Uniform designation for genes of the Calvin-Benson-Bassham reductive pentose phosphate pathway of bacteria. FEMS Micro Lett 99: 107–110

Valle E, Kobayashi H and Akazawa T (1988) Transcriptional regulation of genes of plant-type ribulose-1,5-bisphosphate carboxylase/oxygenase in the photosynthetic bacterium, *Chromatium vinosum*. J Biochem 173: 483–489

Van Niel CB (1936) On the metabolism of Thiorhodaceae. Arch Mikrobiol. 7: 323–358

Viale AM, Kobayashi H, Takabe T and Akazawa T (1985) Expression of genes for subunits of plant-type RubisCO from *Chromatium* and production of the enzymically active molecule in *Escherichia coli*. FEBS 192: 283–288

Viale AM, Kobayashi H and Akazawa T (1989) Expressed genes for plant-type ribulose 1,5-bisphosphate carboxylase/oxygenase in the photosynthetic bacterium *Chromatium vinosum*, which possesses two complete sets of genes. J Bacteriol 171: 2391–2400

Viale AM, Kobayashi H and Akazawa T (1990) Distinct properties of *Escherichia coli* products of plant-type ribulose-1,5-bisphosphate carboxylase/oxygenase directed by two sets of genes from the photosynthetic bacterium *Chromatium vinosum*. J Biol Chem 265: 18386–18392

Viale AM, Kobayashi H, Akazawa T and Heinkoff S (1991) *rcbR*, a gene coding for a member of the LysR family of transcriptional regulators, is located upstream of the expressed set of ribulose 1,5-bisphosphate carboxylase/oxygenase genes in the photosynthetic bacterium *Chromatium vinosum*. J Bacteriol 173: 5224–5229

Wagner SJ, Stevens Jr SE, Nixon BT, Lambert DH, Quivey Jr RG and Tabita FR (1988). Nucleotide and deduced amino acid sequence of the *Rhodobacter sphaeroides* gene encoding form II ribulose-1,5-bisphosphate carboxylase/oxygenase and comparison with other deduced form I and II sequences. FEMS Micro Ltr 55: 217–222

Wang X, Falcone DL and Tabita FR (1993) Reductive pentose phosphate-independent CO_2 fixation in *Rhodobacter sphaeroides* and evidence that ribulose bisphosphate carboxylase activity serves to maintain the redox balance of the cell. J Bacteriol 175: 3372–3379

Weaver KE and Tabita FR (1983) Isolation and partial characterization of *Rhodopseudomonas sphaeroides* mutants defective in the regulation of ribulose bisphosphate carboxylase/oxygenase. J Bacteriol 156: 507–515

Windhövel U and Bowien B (1990) On the operon structure of the *cfx* gene clusters in *Alcaligenes eutrophus*. Arch Microbiol 154: 85–91.

Windhövel U and Bowien B (1991) Identification of *cfxR*, an activator gene of autotrophic CO_2 fixation in *Alcaligenes eutrophus*. Mol Micro 5: 2695–2705

Xu HH and Tabita FR (1994) Positive and negative regulation of sequences upstream of the form II *cbb* operon of *Rhodobacter sphaeroides*. J Bacteriol 176: 7299–7308

Zhu YS and Kaplan S (1985) Effects of light, oxygen, and substrates on steady-state levels of mRNA coding for ribulose-1,5-bisphosphate carboxylase and light-harvesting and reaction center polypeptides in *Rhodopseudomonas sphaeroides*. J Bacteriol 162: 925–932

Chapter 52

Genetic Analysis and Regulation of Bacteriochlorophyll Biosynthesis

Alan J. Biel
Department of Microbiology, Louisiana State University, Baton Rouge, LA 70803, USA

Summary

This chapter focuses on recent advances in the genetic analysis of the common tetrapyrrole pathway and on the factors that regulate BChl biosynthesis. The genes for several steps in tetrapyrrole biosynthesis have been cloned, and the rest of the genes should be cloned in the near future.

This is an extremely complex pathway, with four different branches. Unlike most pathways where the endproducts are functionally very similar, the endproducts of this pathway serve very different functions, and are required in vastly different amounts. For this reason, the common tetrapyrrole pathway can be expected to have multiple layers of regulation, each one modulating carbon-flow to a different extent. The presence of oxygen, for example, has to block the magnesium branch and dramatically reduce carbon-flow over the common pathway without reducing the synthesis of the other three endproducts.

Three different environmental factors have been identified which alter the carbon-flow over the common tetrapyrrole pathway: oxygen, heme and *c*-type cytochromes. A fourth factor, light intensity, regulates BChl degradation rather than BChl synthesis. These factors appear to act independently of one another to ensure the smooth functioning of this pathway under a variety of environmental conditions.

R. E. Blankenship, M. T. Madigan and C. E. Bauer (eds): Anoxygenic Photosynthetic Bacteria, pp. 1125–1134.

I. Introduction

The purple non-sulfur photosynthetic bacteria synthesize four tetrapyrrole endproducts: heme, BChl, siroheme and vitamin B-12. These compounds are all derived from ALA via a multiply branched pathway (Fig. 1). The first branchpoint in this pathway is at uroporphyrinogen III. A small amount of uroporphyrinogen III is methylated, committing it to the synthesis of siroheme and vitamin B-12. Siroheme is the cofactor of enzymes such as sulfite reductase (Murphy and Siegel, 1973) and nitrite reductase (Murphy et al., 1974), while vitamin B-12 is the cofactor of homocysteine methyltransferase (Davis and Mingioli, 1950). Neither *Rb. capsulatus* nor *Rb. sphaeroides* appear to have the vitamin B-12-independent pathway for methionine biosynthesis (Cauthen et al., 1967; Biel, unpublished). While a small amount of the uroporphyrinogen III is diverted to these compounds, the vast majority is decarboxylated to coproporphyrinogen III, a precursor of protoporphyrin IX. At this point, either iron or magnesium is chelated into the porphyrin ring. The various hemes are formed from iron protoporphyrin, while BChl is derived from magnesium protoporphyrin.

Considerable effort has gone into understanding how various environmental factors influence tetrapyrrole biosynthesis. The factor that has the largest effect, and has received the most scrutiny, is oxygen tension. In *Rb. sphaeroides*, heme synthesis is unaffected by oxygen tension (Lascelles, 1978). However, the presence of as little as 2% oxygen dramatically reduces the BChl content of *Rb. capsulatus* and *Rb. sphaeroides* (Arnheim and Oelze, 1983; Clark et al., 1984).

In the last decade, there has been an explosion of information on BChl synthesis and how various factors regulate it. The *bch* genes, encoding the enzymes of the magnesium branch, have been cloned, physically mapped and sequenced (Chapter 50 by Alberti et al., Taylor et al., 1983; Biel and Marrs, 1983; Zsebo et al., 1984; Yang and Bauer, 1990; Bollivar and Bauer, 1992; Burke et al., 1993a; Bollivar et al., 1994a,b), and the enzymatic activities of a few of the enzymes are known (Bollivar et al., 1994c). A few of the *hem* genes, encoding the enzymes of the common tetrapyrrole pathway, have also been cloned. These advances have facilitated several studies into the regulation of BChl biosynthesis. It has become clear that the regulation of BChl synthesis is much more complex than initially realized. While it was previously assumed that the observed regulation by oxygen and heme was part of a single control mechanism, it now appears that these factors act independently of each other. In fact, this is such an important biosynthetic pathway that it now seems probable that there are multiple, independent, regulatory systems that act to modulate the intracellular levels of BChl and heme.

This review will detail the information known about the genetics of the common tetrapyrrole pathway, i.e. the pathway from succinyl CoA and glycine to protoporphyrin IX (Fig. 1). Additional information regarding the genetic analysis of the mg-tetrapyrrole branch can be obtained in Chapter 50 by Alberti et al. as well as in a review by Bauer et al., (1993).The current state of information on regulation of BChl biosynthesis by oxygen, heme, *c*-type cytochromes, and light will also be covered in this review.

Abbreviations: ALA – aminolevulinate; BChl – bacteriochlorophyll; *E. coli* – *Escherichia coli*; *Rb.* – *Rhodobacter*

II. Genetics of Protoporphyrin Biosynthesis

The *hemA* gene has been the most extensively studied of the *hem* genes in purple photosynthetic bacteria. Lascelles and Altshuler (1969) were the first to describe a *hemA* mutant in *Rb. sphaeroides*. The mutant H-5 requires exogenous ALA for both heme and BChl production, indicating that a single ALA synthase makes ALA for both pathways. However, the mutant was isolated following nitrosoguanidine mutagenesis, and may have multiple mutations. This possibility was supported by the cloning of two different genes encoding ALA synthase, *hemA* and *hemT*, from *Rb. sphaeroides* (Tai et al., 1988). Either one of the genes allows strain H-5 to grow both aerobically in the dark and photosynthetically(Neidle and Kaplan, 1993). Thus it appears that the enzyme produced by either *hemA* or *hemT* can supply all of the ALA needed for both heme and BChl production.

Wright et al. (1987) isolated a *Rb. capsulatus* strain with a Tn5 insertion in *hemA*. This strain requires exogenous ALA for aerobic growth and BChl synthesis. The *hemA* gene was cloned and subsequently sequenced (Biel et al., 1988; Hornberger, et al., 1990; Wright et al., 1991). The observation that *hemA* mutants are unable to

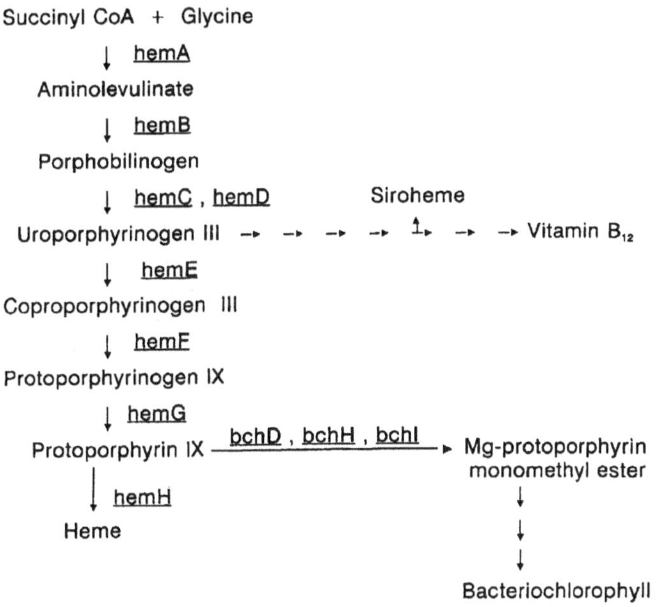

Fig. 1. The common tetrapyrrole pathway.

synthesize heme or BChl suggests that unlike the situation in *Rb. sphaeroides*, there is only one gene for ALA synthase in *Rb. capsulatus*.

The condensation of two ALA molecules to form the monopyrrole porphobilinogen is catalyzed by porphobilinogen synthase, the product of the *hemB* gene. This gene has been cloned from *Rb. sphaeroides* (Delaunay et al., 1991) and from *Rb. capsulatus* (K. Indest and A. Biel, unpublished). Both clones were isolated by complementation of an *E. coli hemB* mutant. The porphobilinogen synthase encoded by the *Rb. capsulatus hemB* gene is 43 kDa and is 54% identical to the enzyme from *Bradyrhizobium japonicum* (Chauhan and O'Brian, 1993). There are no reports of *hemB* mutants in either *Rb. capsulatus* or *Rb. sphaeroides*.

Coomber et al. (1992) isolated a *Rb. sphaeroides* mutant that accumulates coproporphyrin and does not make BChl, but does grow aerobically. They cloned a gene that complements this mutation and determined that the amino acid sequence, derived from the DNA sequence, is 44.7% similar to that of the yeast *hem13* gene, which encodes an aerobic coproporphyrinogen oxidase. The authors suggested that *Rb. sphaeroides* has two coproporphyrinogen oxidases, one for BChl synthesis and one for heme synthesis, and that the mutation inactivated the enzyme responsible for BChl synthesis. If this mutation is indeed in the *hemF* gene, the phenotype would suggest that the heme and BChl branches diverge at coproporphyrinogen instead of protoporphyrin. This would imply that there are two enzyme complexes, consisting of at least coproporphyrinogen oxidase and protoporphyrinogen oxidase, with one complex leading to heme and the other to BChl.

In *Rb. capsulatus*, the *hemE* gene has been cloned by complementation of an *E. coli hemE* mutant (G. Ineichen and A. Biel, unpublished). The predicted amino acid sequence is 34% identical to uroporphyrinogen decarboxylase from yeast (Garey et al., 1992) and from *Bacillus subtilis* (Hansson and Kederstedt, 1992). The *hemH* gene, which encodes ferrochelatase, has also been cloned from *Rb. capsulatus* (E. Kanaziereva and A. Biel, unpublished).

None of the other *hem* genes have as yet been cloned from either *Rb. capsulatus* or *Rb. sphaeroides*. However, the existence of a *hemD* mutant of *E. coli* (Sasarman et al., 1987) and strains with insertions or point mutations in each of the *Salmonella typhimurium hem* genes (Xu et al., 1992) should allow the rest of the *Rhodobacter hem* genes to be cloned by complementation of these mutations.

As the cloning and sequencing of the *Rb. capsulatus bch* genes is discussed in Chapter 50 by Alberti et al., it will not be included in this review.

III. Regulation of BChl Biosynthesis

A. Regulation by Oxygen

A thorough review of BChl biosynthesis and its regulation in *Rb. sphaeroides* was published in 1982 (Rebeiz and Lascelles, 1982). The authors proposed a simple model to explain how oxygen might regulate BChl synthesis. It was suggested that oxygen inhibits magnesium chelatase, resulting in an increase in the protoporphyrin IX pool. Ferrochelatase would convert protoporphyrin IX to heme, which would feedback inhibit aminolevulinate synthase (Fig. 1). The decrease in aminolevulinate synthase activity would prevent further synthesis of protoporphyrin IX and would explain how oxygen inhibits the production of BChl without overproducing any of the tetrapyrrole intermediates (Lascelles, 1978).

This model predicts that oxygen, either directly or indirectly, regulates two steps in the synthesis of BChl. According to this theory, the first regulated step is the formation of ALA, which can be greatly modulated depending on the amount of BChl needed, but cannot be shut off because of the constant need for heme. The second regulated step is the conversion of protoporphyrin to Mg-protoporphyrin monomethyl ester.

A recent study suggests that, as predicted, oxygen regulates the conversion of protoporphyrin to Mg-protoporphyrin monomethyl ester. A *Rb. capsulatus bchE* strain, which accumulates Mg-protoporphyrin monomethyl ester, was grown under either 23% or 3% oxygen and pigment accumulation was measured (Biel, 1992). Under high oxygen tensions no Mg-protoporphyrin monomethyl ester accumulated. Since heme synthesis must continue under these conditions, these results indicate that oxygen controls the formation of Mg-protoporphyrin monomethyl ester from protoporphyrin.

The *pufQ* gene product is essential for BChl synthesis from protoporphyrin, although its role is unclear (Bauer and Marrs, 1988). It is intriguing, however, that transcription of *pufQ* is 30-fold higher in anaerobically grown cultures than in cultures grown aerobically (Bauer et al., 1988; Chapter 58). Thereby implying that transcription of *pufQ* may have a role in this control.

Studies on the regulation of BChl biosynthesis by oxygen were facilitated by the isolation of an R'-factor, derived from RP1, which carried the entire *Rb. capsulatus* photosynthesis region (Marrs, 1981). A genetic and physical map of the photosynthesis region of the *Rb. capsulatus* chromosome, showing the positions of the genes for BChl and carotenoid biosynthesis, was then developed (Taylor et al., 1983; Zsebo and Hearst, 1984) and refined by interposon mutagenesis (Yang and Bauer, 1990; Bollivar et al., 1994a,b). Fusions between the *lacZ* gene and several of the *bch* genes were also constructed and used to determine that transcription of the *bch* genes was two- to three-fold lower in cultures grown with 20% oxygen than in those grown with 2% oxygen (Biel and Marrs, 1983; Young et al., 1989; Yang and Bauer, 1990). These results have been confirmed by measuring *bch* mRNA levels in *Rb. capsulatus* and *Rb. sphaeroides* (Clark et al., 1984; Hunter and Coomber, 1988; Armstrong et al., 1993). While transcription of the *bch* genes is regulated by oxygen tension, the magnitude of that regulation cannot account for the 50-fold lower BChl levels seen in aerobically grown cultures compared to anaerobically grown ones. In an effort to determine how oxygen regulates transcription of the *bch* promoters, Ma et al. (1993) analyzed the *bchCXYZ* promoter region. They found that it shared sequence similarity to the *E. coli* σ^{70} consensus promoter and that deletion of a 50bp AT-rich region immediately upstream of the −35 region rendered the promoter uninducible by oxygen. There is also a report of a *trans*-acting repressor in *Rb. sphaeroides* and *Rb. capsulatus* which controls expression of *bch* genes (Penfold and Pemberton, 1994; Ponnampalam et al., 1995).

When the synthesis of BChl from protoporphyrin is inhibited by oxygen, carbon-flow over the common tetrapyrrole pathway has to be reduced to prevent the overproduction of protoporphyrin. Regulation of the common tetrapyrrole pathway by oxygen was demonstrated by growing a *Rb. capsulatus bchH* mutant under high and low oxygen tensions and measuring protoporphyrin accumulation (Biel and Marrs, 1983). The low-oxygen grown culture had dramatically higher protoporphyrin levels than the high-oxygen grown culture. While this observation suggests that oxygen regulates carbon-flow over the common tetrapyrrole pathway, it does not rule out the possibility that protoporphyrin synthesis is the same under both oxygen tensions, but that under high oxygen the excess protoporphyrin is degraded. However, since a *bchH* mutant accumulates protoporphyrin when grown under high oxygen in the presence of exogenous porphobilinogen, it seems likely that oxygen does indeed regulate the synthesis of protoporphyrin (Biel, 1992).

Mutants which accumulate coproporphyrin have been isolated in both *Rb. capsulatus* and *Rb. sphaeroides* (Biel and Biel, 1990; Coomber et al., 1992). Coproporphyrin is accumulated by both mutants only when they are grown under low oxygen tension, suggesting that oxygen controls one of the steps in the formation of coproporphyrin.

Since the formation of ALA has been considered to be the first step in the common tetrapyrrole pathway, it seemed plausible that this would be the step at which oxygen acts. Evidence in support of this are the observations, made in *Rb. sphaeroides*, that ALA is rate-limiting for heme synthesis (Lascelles and Hatch, 1969), and that ALA synthase is inhibited by heme in vitro (Burnham and Lascelles, 1963).

There is no direct evidence, however, that oxygen influences ALA synthase activity. In *Rb. capsulatus*, cultures grown aerobically have the same ALA synthase activity as cultures grown photosynthetically (Biel, 1992), while in *Rb. sphaeroides*, ALA synthase specific activity varies less than five-fold with oxygen tension (Lascelles, 1959; Oelze and Arnheim, 1983). Since the *Rb. sphaeroides* enzyme is activated by trisulfides, the observed ALA synthase specific activity may not reflect the true in vivo situation (Marriott et al., 1969). Transcription of the *Rb. capsulatus hemA* gene is regulated only two-fold by oxygen (Hornberger et al., 1991; Wright et al., 1991).

If oxygen either directly or indirectly inhibits ALA formation, the addition of exogenous ALA to an aerobically growing culture should result in increased carbon-flow over the common tetrapyrrole pathway. This was tested by growing a *Rb. capsulatus bchH* mutant under high and low oxygen tensions in the presence of 1.0mM ALA and measuring the accumulation of protoporphyrin (Biel, 1992). Surprisingly, even in the presence of exogenous ALA, protoporphyrin levels were much higher in the low-oxygen grown culture than in the high-oxygen grown culture, suggesting that oxygen regulates some step after the formation of ALA. When the experiment was repeated with exogenous porphobilinogen, both the high- and low-oxygen grown cultures accumulated protoporphyrin, indicating that oxygen regulates the conversion of ALA to porphobilinogen.

An implication of this result is that either aerobically growing cells must accumulate ALA, which has not been reported, or the ALA must be metabolized by some other pathway. There are several reports in different organisms suggesting that ALA is not committed to tetrapyrrole biosynthesis. Shemin and Russell (1953), in reporting that ALA is a precursor of protoporphyrin, also indicated that ALA is incorporated into purines. Further evidence for the incorporation of ALA into purines was presented by Nemeth et al. (1957). Shigesada (1972) demonstrated that in photosynthetically grown *Rhodospirillum rubrum* the major metabolites of [4-^{14}C]ALA were not tetrapyrroles, but were instead δ-amino-γ-hydroxyvalerate, α-hydroxyglutarate and glutamate.

While the evidence presented above suggests that oxygen regulates the formation of porphobilinogen from ALA, the mechanism of this regulation is unclear. Now that the *hemB* gene has been cloned from both *Rb. capsulatus* and *Rb. sphaeroides*, it will be a simple matter to determine if oxygen regulates *hemB* transcription. It will be more difficult to determine if oxygen regulates the activity of porphobilinogen synthase. Lascelles (1959) measured porphobilinogen synthase specific activities in *Rb. sphaeroides* cultures grown under low and high oxygen tensions and found that porphobilinogen synthase activity was four-fold higher in the low-oxygen grown culture than in the high-oxygen grown culture. Porphobilinogen synthase activity does not seem to be regulated in *Rb. capsulatus* (A. Biel, unpublished). These assays may not represent the true in vivo situation, however, since no attempt was made in these studies to maintain the original oxygen tension of the culture while making the extracts. Thus if the porphobilinogen synthase in photosynthetic bacteria is oxygen-labile, as has been reported for the mammalian enzyme (Tsukamoto et al., 1979), the observed activity from anaerobically grown cultures may be considerably less than the actual in vivo activity.

B. Regulation by Heme

It has been demonstrated convincingly that hemin inhibits *Rb. sphaeroides* ALA synthase in vitro (Burnham and Lascelles, 1963; Yubisui and Yoneyama, 1972). Porphobilinogen synthase of *Rb. sphaeroides* is also very sensitive to inhibition by hemin, with 50% inhibition occurring at 33 μM (Burnham and Lascelles, 1963; Nandi et al., 1968). In contrast, the porphobilinogen synthase of *Rb. capsulatus* is insensitive to inhibition by hemin (Nandi and Shemin, 1973). One other *Rb. sphaeroides* enzyme, ferrochelatase, has been shown to be strongly inhibited by hemin, with 50% inhibition occurring at 10 μM (Jones and Jones, 1970).

In spite of these observations, the manner in which heme might regulate tetrapyrrole biosynthesis, and the physiological significance of that regulation are not clear. The role of heme in regulating tetrapyrrole biosynthesis has been investigated by starving cells for heme and monitoring the production of various tetrapyrroles. The first method used to produce heme-deficient cells was to incubate cell suspensions in iron-limited media. Cell suspensions of either *Rb. capsulatus* or *Rb. sphaeroides* incubated in the light for 24 h in iron-deficient media accumulated large amounts of coproporphyrin while synthesizing less than half the normal amount of BChl (Lascelles, 1956; Burnham and Lascelles, 1963; Cooper, 1963; A. Biel, unpublished). Jones (1963) reported that *Rb. sphaeroides* also accumulated Mg-protoporphyrin monomethyl ester when starved for iron.

A second, more specific, method for producing heme-deficient cells has been used in both *Rb. capsulatus* and *Rb. sphaeroides* (Houghton et al., 1982; A. Biel, unpublished). Cell suspensions were incubated in the presence of N-methylprotoporphyrin, an inhibitor of ferrochelatase, in the presence of excess iron. In both *Rb. capsulatus* and *Rb. sphaeroides*, N-methylprotoporphyrin had no effect on BChl production, while dramatically increasing the production of Mg-protoporphyrin monomethyl ester. Incubation of the cell suspensions with N-methylprotoporphyrin did not result in accumulation of coproporphyrin.

While both sets of experiments suggest that heme deficiency causes increased carbon-flow over the common tetrapyrrole pathway, it also appears that iron-limitation reduces the conversion of coproporphyrinogen to Mg-protoporphyrin monomethyl ester. It has been suggested that the increased carbon-flow over the common tetrapyrrole pathway is due to a loss of the normal feed-back inhibition of ALA synthase by heme (Lascelles, 1978; Houghton et al., 1982), however, there are no published reports on the level of ALA synthase in heme-deficient cells. *Rb. capsulatus* grown under 3% oxygen in the presence of N-methylprotoporphyrin had 1.4-fold more ALA synthase activity than a similar culture grown in the absence of the inhibitor (A. Biel, unpublished). It is not yet known if this increase in ALA synthase activity alone is enough to account for the increased porphyrin accumulation.

Complicating the picture is the observation that when *bch* mutants of *Rb. capsulatus* or *Rb. sphaeroides* are incubated with N-methylprotoporphyrin the pigment that accumulates is not Mg-protoporphyrin monomethyl ester, but is the pigment characteristic of the mutant (Houghton et al., 1982; A. Biel, unpublished). Houghton et al. (1982) suggested that BChl acts as an inhibitor of the conversion of Mg-protoporphyrin monomethyl ester to Mg-2,4-divinyl pheoporphyrin a_5 monomethyl ester, and that in *bch* mutants the absence of BChl allows carbon-flow to continue until the block in the pathway is reached.

It is evident that a considerable amount of work needs to be done to determine the role of heme in regulating the common tetrapyrrole pathway. The isolation of ALA synthase mutants that are insensitive to feed-back inhibition by heme might be very useful in elucidating the role of ALA synthase in heme-mediated regulation of this pathway. Presumably, these mutants would accumulate higher than normal amounts of Mg-protoporphyrin monomethyl ester, which would not be increased by heme-deficiency. These mutants will also help to determine if oxygen and heme act independently on the common tetrapyrrole pathway, as suggested by the results presented above. If, in fact, regulation by oxygen is totally independent from regulation by heme, the accumulation of Mg-protoporphyrin monomethyl ester should still be regulated by oxygen.

C. Regulation by c-type Cytochromes

Several different mutants have been isolated in *Rb. capsulatus* whose phenotypes suggest a link between *c*-type cytochrome biosynthesis and carbon-flow over the common tetrapyrrole pathway. Daldal et al. (1987) found that *Rb. capsulatus petB* and *petC* mutants, which lack the cytochrome bc_1 complex, accumulate an uncharacterized pigment. Kranz (1989) isolated mutants with alterations in three different genes (*helA*, *helB* and *helC*) required for biosynthesis of *c*-type cytochromes. These mutants also accumulated an uncharacterized pigment. Biel and Biel (1990) isolated a Tn*5* insertion mutant that completely lacks *c*-type cytochromes. Beckman et al. (1992) have sequenced the gene mutated in this strain. The gene has been designated *ccl1*, and has been shown to be involved in *c*-type cytochrome biosynthesis. This mutant accumulated nine-fold more protoporphyrin and 18-fold more coproporphyrin than the parental strain and only 19% of the normal amount of BChl. Porphyrin production in this mutant is still subject to the normal regulation by oxygen (Biel, 1991). Growth of the mutant in an iron-limited medium increases porphyrin accumulation almost three-fold (A. Biel, unpublished), suggesting that whatever regulatory mechanism has been disrupted by this mutation, it is separate from heme-mediated regulation.

The phenotypes of these mutants suggest that *Rb. capsulatus* monitors the level of one or more *c*-type cytochromes and modulates carbon-flow over the common tetrapyrrole pathway in response to changes in that level. It is unclear whether the total pool of *c*-type cytochromes is monitored, or whether it is the level of a specific *c*-type cytochrome which regulates the pathway. How carbon-flow over the common tetrapyrrole pathway is regulated is also unknown. By analyzing the cytochrome *c*-less mutant it should be possible to determine if the activities of any of the enzymes in the pathway are increased and whether transcription of the *hemA* and *hemB* genes are altered.

D. Regulation by Light

Cohen-Bazire et al. (1957) demonstrated that BChl levels were inversely proportional to light intensity in photosynthetically grown cultures. They also found that aerobically grown cultures produced very little BChl, regardless of the light intensity. It was assumed that the decreased levels of BChl found in high-light grown cultures was due to inhibition of BChl synthesis, and it was proposed that oxygen and light controlled the state of oxidation of some electron transport chain component, which in turn regulated the synthesis of BChl.

Arnheim and Oelze (1983) investigated the formation of BChl in chemostats where growth was limited by oxygen or light. They found that in *Rb. sphaeroides* the two factors acted independently of one another, and suggested that oxygen and light use separate mechanisms to control BChl formation. Oelze and Arnheim (1983) showed that while aminolevulinate synthase levels respond to changes in oxygen tension, changes in light intensity have no effect on enzyme activity. These results suggest that a single regulatory mechanism, such as that proposed by Cohen-Bazire et al. (1957), can not account for regulation by both oxygen and light. This conclusion was further strengthened when it was observed that while oxygen regulated transcription of the *bch* genes in *Rb. capsulatus*, light intensity had no effect on transcription (Biel and Marrs, 1983).

These studies indicated that oxygen and light independently regulated BChl accumulation. The assumption was made that both oxygen and light regulated the synthesis of BChl. However, several pieces of evidence suggest that in *Rb. capsulatus* light intensity controls the rate of BChl degradation rather than the rate of synthesis (Biel, 1986). A *bch*$^+$ strain, grown under 3% oxygen either in darkness or with a light intensity of 850 W m^{-2} accumulated nine times more BChl in the dark than in bright light. On the other hand, a *bchG* mutant, which synthesizes bacteriochlorophyllide *a* but cannot convert it to BChl, accumulated the same amount of bacteriochlorophyllide *a* per mg of protein under the two conditions, indicating that light intensity does not alter the synthesis of bacteriochlorophyllide *a*. This experiment does not rule out the possibility that light prevents the conversion of bacteriochlorophyllide *a* to BChl. However, when tetrapyrrole formation from [^{14}C]ALA was followed it was determined that bright light caused an increase in BChl degradation. In a *bch*$^+$ strain grown in the dark, [^{14}C]ALA was rapidly converted to BChl. When grown in bright light, the [^{14}C]ALA was converted to a colorless compound that co-migrated with BChl on silica gel TLC plates developed with either 70% ethanol or two other solvent systems. Spectroscopy and fluorimetry indicated that this compound was not a normal BChl precursor, suggesting that the compound was a degradation product of BChl. Interestingly, BChl formed by a culture grown in the dark was not degraded when the culture was shifted to bright light. Perhaps insertion of BChl into pigment-protein complexes stabilizes the BChl. Synthesis of the proteins for these complexes are known to be regulated by light intensity (Klug et al., 1985; Zhu and Hearst, 1986).

IV. Conclusions

Oxygen tension is the environmental factor with the most influence on this pathway. It prevents the conversion of protoporphyrin to Mg-protoporphyrin monomethyl ester. The *PufQ* protein appears to be involved in regulating this step, but how this regulation occurs is unknown. Oxygen also reduces carbon-flow over the common tetrapyrrole pathway. Recent evidence suggests that the site of regulation is not the formation of ALA, but rather its conversion to porphobilinogen. A considerable amount of work remains to be done to confirm this and to determine the mechanism by which oxygen regulates this step.

Heme-deprivation also causes an increase in carbon-flow over the common tetrapyrrole pathway. It is likely that regulation by heme is separate from regulation by oxygen. The isolation and characterization of a mutant with a feed-back resistant ALA synthase should help to determine if heme controls the common tetrapyrrole pathway by regulating ALA synthase activity.

This pathway is also regulated by *c*-type cytochromes. Cytochromes can be considered to be the true endproducts of the heme pathway; therefore it seems likely that the cell would monitor the level of some type of cytochrome and regulate tetrapyrrole synthesis in response to changes in the level of that cytochrome. The mechanism by which *c*-type cytochromes regulate heme synthesis are as yet totally unexplored.

In one area at least, the regulation of BChl is simpler than previously thought. Light intensity regulates the intracellular level of BChl by altering the rate of BChl degradation, rather than its rate of synthesis. Thus the brighter the light, the faster degradation occurs.

With the application of genetic approaches and molecular biological tools to this area tremendous strides have been made in the last decade. We have not yet answered the questions, but have defined them so that they can be answered. In the next decade a picture should begin to emerge of how the bacterial cell regulates such an extraordinarily complex pathway with such precision.

References

Armstrong GA, Cook DN, Ma D, Alberti M, Burke DH and Hearst JE (1993) Regulation of carotenoid and bacteriochlorophyll biosynthesis genes and identification of an evolutionarily conserved gene required for bacteriochlorophyll accumulation. J Gen Microbiol 139: 897–906

Arnheim K and Oelze J (1983) Differences in the control of bacteriochlorophyll formation by light and oxygen. Arch Microbiol 135: 299–304

Bauer CE and Marrs BL (1988) *Rhodobacter capsulatus puf* operon encodes a regulatory protein (PufQ) for bacteriochlorophyll biosynthesis. Proc Natl Acad Sci USA 85: 7074–7078

Bauer CE, Young DA and Marrs BL (1988) Analysis of the *Rhodobacter capsulatus puf* operon. J Biol Chem 263: 4820–4827

Bauer, CE, Bollivar DW and Suzuki JY (1993) Genetic analysis of photopigment biosynthesis in eubacteria: A Guiding light for algae and plants. J Bacteriol 175: 3919–3925

Beckman DL, Trawick DR and Kranz RG (1992) Bacterial cytochromes *c* biogenesis. Genes and Development 6: 268–283

Biel AJ (1986) Control of bacteriochlorophyll accumulation by light in *Rhodobacter capsulatus*. J Bacteriol 168: 655–659

Biel AJ (1991) Characterization of a coproporphyrin-protein complex from *Rhodobacter capsulatus*. FEMS Microbiol Lett 81: 43–48

Biel AJ (1992) Oxygen-regulated steps in the *Rhodobacter capsulatus* tetrapyrrole biosynthetic pathway. J Bacteriol 174: 5272–5274

Biel AJ and Marrs BL (1983) Transcriptional regulation of several genes for bacteriochlorophyll biosynthesis in *Rhodopseudomonas capsulata* in response to oxygen. J Bacteriol 156: 686–694

Biel SW and Biel AJ (1990) Isolation of a *Rhodobacter capsulatus* mutant that lacks *c*-type cytochromes and excretes porphyrins. J Bacteriol 172: 1321–1326

Biel SW, Wright MS and Biel AJ (1988) Cloning of the *Rhodobacter capsulatus hemA* gene. J Bacteriol 170: 4382–4384

Bollivar DW and Bauer CE (1992) Nucleotide sequence of *S* adenosyl methionine magnesium protoporphyrin methyltransferase from *Rhodobacter capsulatus*. Plant Physiol 98: 408–410

Bollivar DW, Suzuki JY, Beatty JT, Dobrowlski J and Bauer CE (1994a) Directed mutational analysis of bacteriochlorophyll *a* biosynthesis in *Rhodobacter capsulatus*. J Mol Biol 237: 622–640

Bollivar DW, Wang S, Allen JP and Bauer CE (1994b) Molecular genetic analysis of terminal steps in bacteriochlorophyll *a* biosynthesis: Characterization of a *Rhodobacter capsulatus* strain that synthesizes geranylgeraniol esterified bacteriochlorophyll *a*. Biochemistry 33: 12763–12768

Bollivar DW, Jiang Z-Y, Bauer CE and Beale SI (1994c) Heterologous overexpression of the *bchM* gene product from *Rhodobacter capsulatus* and demonstration that it encodes for *S*-adenosyl-L-methionine:Mg-protoporphyrin methyltransferase. J Bacteriol 176: 5290–5296

Burke D, Alberti M and Hearst JE (1993a) *The Rhodobacter capsulatus* chlorin reductase-encoding locus, *bchA*, consists of three genes, *bchX*, *bchY*, and *bchZ*. J Bacteriol 175: 2407–2413

Burke D, Alberti M and Hearst JE (1993b) *bchFNBH* bacteriochlorophyll synthesis genes of *Rhodobacter capsulatus* and identification of the third subunit of light independent protochlorophyllide reductase in bacteria and plants. J Bacteriol 175: 2414–2422

Burnham BF and Lascelles J (1963) Control of porphyrin biosynthesis through a negative-feedback mechanism. Biochem J 87: 462–472

Cauthen SE, Pattison JR and Lascelles J (1967) Vitamin B_{12} in photosynthetic bacteria and methionine synthesis by *Rhodopseudomonas spheroides*. Biochem J 102: 774–781

Chauhan S and O'Brian MR (1993) *Bradyrhizobium japonicum* δ-aminolevulinic acid dehydratase is essential for symbiosis with soybean and contains a novel metal-binding domain. J Bacteriol 157: 7222–7227

Clark WG, Davidson E and Marrs BL (1984) Variation of levels of mRNA coding for antenna and reaction center polypeptides in *Rhodopseudomonas capsulata* in response to changes in oxygen concentration. J Bacteriol 157: 945–948

Cohen-Bazire G, Sistrom WR and Stanier RY (1957) Kinetic studies of pigment synthesis by non-sulfur purple bacteria. J Cell Comp Physiol 49: 25–68

Coomber SA, Jones RM, Jordan PM and Hunter CN (1992) A putative anaerobic coproporphyrinogen III oxidase in *Rhodobacter sphaeroides*. I. Molecular cloning, transposon mutagenesis and sequence analysis of the gene. Molec Microbiol 6: 3159–3169

Cooper R (1963) The biosynthesis of coproporphyrinogen, magnesium protoporphyrin monomethyl ester and bacteriochlorophyll by *Rhodopseudomonas capsulata*. Biochem J 89: 100–108

Daldal F, Davidson E and Cheng S (1987) Isolation of the structural genes for the Rieske Fe-S protein, cytochrome *b* and cytochrome c_1, all components of the ubiquinol: cytochrome c_2 oxidoreductase complex of *Rhodopseudomonas capsulata*. J Mol Biol 195: 1–12

Davis BD and Mingioli ES (1950) Mutants of *Escherichia coli* requiring methionine or vitamin B_{12}. J Bacteriol 60: 17–28

Delaunay AM, Huault C and Balangé AP (1991) Molecular cloning of the 5-aminolevulinic acid dehydratase gene from *Rhodobacter sphaeroides*. J Bacteriol 173: 2712–2715

Garyey JR, Labbe-Bois R, Chelstowska A, Rytka J, Harrison L, Kushner J and Labbe P (1992) Uroporohyrinogen decarboxylase in *Saccharomyces cerevisiae*. FEBS Lett 205: 1011–1016

Hansson M and Hederstedt L (1992) Cloning and characterization of the *Bacillus subtilis hemEHY* gene cluster, which encodes protoheme IX biosynthetic enzymes. J Bacteriol 174: 8081–8093

Hornberger U, Liebetanz R, Tichy HV and Drews G (1990) Cloning and sequencing of the *hemA* gene of *Rhodobacter capsulatus* and isolation of a δ-aminolevulinic acid-dependent mutant strain. Mol Gen Genet 221: 371–378

Hornberger U, Wieseler B and Drews G (1991) Oxygen tension regulated expression of the *hemA* gene of *Rhodobacter capsulatus*. Arch Microbiol 156: 129–134

Houghton JD, Honeybourne CL, Smith KM, Tabba HD and Jones OTG (1982) The use of N-methylprotoporphyrin dimethyl ester to inhibit ferrochelatase in *Rhodopseudomonas sphaeroides* and its effect in promoting biosynthesis of magnesium tetrapyrroles. Biochem J 208: 479–486

Hunter CN and Coomber SA (1988) Cloning and oxygen-regulated expression of the bacteriochlorophyll biosynthesis genes *bchE*, *B*, *A* and *C* of *Rhodobacter sphaeroides*. J Gen Microbiol 134: 1471–1480

Jones MS and Jones OTG (1970) Ferrochelatase of *Rhodopseudomonas sphaeroides*. Biochem J 119: 453–462

Jones OTG (1963) The production of magnesium protoporphyrin monomethyl ester by *Rhodopseudomonas sphaeroides*. Biochem J 86: 429–432

Klug G, Kaufmann N and Drews G (1985) Gene expression of pigment-binding proteins of the bacterial photosynthetic apparatus: transcription and assembly in the membrane of *Rhodopseudomonas capsulata*. Proc Natl Acad Sci USA 82: 6485–6489

Kranz RG (1989) Isolation of mutants and genes involved in cytochromes *c* biosynthesis in *Rhodobacter capsulatus*. J Bacteriol 171: 456–464

Lascelles J (1956) The synthesis of porphyrins and bacteriochlorophyll by cell suspensions of *Rhodopseudomonas sphaeroides*. Biochem J 62: 78–93

Lascelles J (1959) Adaptation to form bacteriochlorophyll in *Rhodopseudomonas sphaeroides*: Changes in activity of enzymes concerned in pyrrole synthesis. Biochem J 72: 508–518

Lascelles J (1978) Regulation of pyrrole synthesis. In: Clayton RK and Sistrom WR (eds) The Photosynthetic Bacteria. pp 795–808 Plenum Press, New York

Lascelles J and Altshuler T (1969) Mutant strains of *Rhodopseudomonas sphaeroides* lacking δ-aminolevulinate synthase: Growth, heme and bacteriochlorophyll synthesis. J Bacteriol 98: 721–727

Lascelles J and Hatch TP (1969) Bacteriochlorophyll and heme synthesis in *Rhodopseudomonas sphaeroides*: Possible role of heme in regulation of the branched biosynthetic pathway. J Bacteriol 98: 712–720

Ma D, Cook DN, O'Brien DA and Hearst JE (1993) Analysis of the promoter and regulatory sequences of an oxygen-regulated *bch* operon in *Rhodobacter capsulatus* by site-directed mutagenesis. J Bacteriol 175: 2037–2045

Marriott J, Neuberger A and Tait GH (1969) Control of δ-aminolevulinate synthetase activity in *Rhodopseudomonas* spheroides. Biochem J 111: 385–394

Marrs B (1981) Mobilization of the genes for photosynthesis from *Rhodopseudomonas capsulata* by a promiscuous plasmid. J Bacteriol 146: 1003–1012

Murphy MJ and Siegel LM (1973) Siroheme and sirohydrochlorin: The basis for a new type of porphyrin-related prosthetic group common to both assimilatory and dissimilatory sulfite reductases. J Biol Chem 248: 6911–6919

Murphy MJ, Siegel LM, Tove SR and Kamin H (1974) Siroheme: A new prosthetic group participating in six-electron reduction reactions catalyzed by both sulfite and nitrite reductases. Proc Natl Acad Sci USA 71: 612–616

Nandi DL and Shemin D (1973) δ-Aminolevulinic acid dehydratase of *Rhodopseudomonas capsulata*. Arch Biochem Biophys 158: 305–311

Nandi DL, Baker-Cohen KF and Shemin D (1968) δ-Aminolevulinic acid dehydratase of *Rhodopseudomonas sphaeroides*. J Biol Chem 243: 1224–1230

Neidle EL and Kaplan S (1993) 5-aminolevulinic acid availability and control of spectral complex formation in hemA and hemT mutants of *Rhodobacter sphaeroides. J Bacteriol 175: 2304–2313*

Nemeth AM, Russell CS and Shemin D (1957) The succinate-glycine cycle. II. Metabolism of δ-aminolevulinic acid. J Biol Chem 111: 415–422

Oelze J and Arnheim K (1983) Control of bacteriochlorophyll formation by oxygen and light in *Rhodopseudomonas sphaeroides*. FEMS Microbiol Lett 19: 197–199

Penfold RJ and Pemberton JM (1994) Sequencing, chromosomal inactivation and functional expression in *E. coli* of *pps*, a gene which represses carotenoid and bacteriochlorophyll synthesis in *Rhodobacter sphaeroides*. J Bacteriol 176: 2869–2876

Ponnampalam S, Buggy JJ and Bauer CE (1995) Characterization of an aerobic repressor that coordinately regulates bacteriochlorophyll, carotenoid and light harvesting-II expression in *Rhodobacter capsulatus*. J Bacteriol 177: 2990–2997

Rebeiz CA and Lascelles J (1982) Biosynthesis of pigments in plants and bacteria. In: Govindjee (ed) Photosynthesis: Energy Conversion by Plants and Bacteria, Vol. I pp 699–780 Academic Press, New York

Sasarman A, Alain N, Echelard Y, Dymetryszyn J, Drolet M and Goyer C (1987) Molecular cloning and sequencing of the *hemD* gene of *Escherichia coli* K-12 and preliminary data on the uro operon. J Bacteriol 169: 4257–4262

Shemin D and Russell CS (1953) δ-Aminolevulinic acid, its role in the biosynthesis of porphyrins and purines. J Amer Chem Soc 75: 4873–4874

Shigesada K (1972) Possible occurrence of a succinate-glycine cycle in *Rhodospirillum rubrum*. J Biochem 71: 961–972

Tai TN, Moore MD and Kaplan S (1988) Cloning and characterization of the 5-aminolevulinate synthase gene(s) from *Rhodobacter sphaeroides*. Gene 70: 139–151

Taylor DP, Cohen SN, Clark WG and Marrs BL (1983) Alignment of genetic and restriction maps of the photosynthesis region of the *Rhodopseudomonas capsulata* chromosome by a conjugation-mediated marker rescue technique. J Bacteriol 154: 580–590

Tsukamoto I, Yoshinaga T and Sano S (1979) The role of zinc with special reference to the essential thiol groups in δ-aminolevulinic acid dehydrase of bovine liver. Biochim Biophys Acta 570: 167–178

Wright MS, Cardin RD and Biel AJ (1987) Isolation and characterization of an aminolevulinate-requiring *Rhodobacter capsulatus* mutant. J Bacteriol 169: 961–966

Wright MS, Eckert JJ, Biel SW and Biel AJ (1991) Use of a *lacZ* fusion to study transcriptional regulation of the *Rhodobacter capsulatus hemA* gene. FEMS Microbiol Lett 78: 339–342

Xu K, Delling D and Elliott T (1992) The genes required for heme synthesis in *Salmonella typhimurium* include those encoding alternative functions for aerobic and anaerobic coproporphyrinogen oxidation. J Bacteriol 174: 3953–3963

Yang Z and Bauer CE (1990) *Rhodobacter capsulatus* genes involved in early steps of the bacteriochlorophyll biosynthetic pathway. J Bacteriol 172: 5000–5010

Young DA, Bauer CE, Willams JC and Marrs BL (1989) Genetic evidence for superoperonal organization of genes for photosynthetic pigments and pigment binding proteins in *Rhodobacter capsulatus*. Mol Gen Genet 218: 1–12

Yubisui T and Yoneyama Y (1972) δ-Aminolevulinic acid synthetase of *Rhodopseudomonas sphaeroides*: Purification and properties of the enzyme. Arch Biochem Biophys 150: 77–85

Zsebo KM, and Hearst JE (1984) Genetic-physical mapping of a photosynthetic gene cluster from *Rb. capsulata*. Cell 37: 937-947

Zhu YS and Hearst JE (1986) Regulation of expression of genes for light-harvesting antenna proteins LH-I and LH-II; reaction center polypeptides RC-L, RC-M, and RC-H; and enzymes of bacteriochlorophyll and carotenoid biosynthesis in *Rhodobacter capsulatus* by light and oxygen. Proc Natl Acad Sci USA 83: 7613–7617

Chapter 53

Genetic Analysis and Regulation of Carotenoid Biosynthesis

Structure and Function of the *crt* Genes and Gene Products

Gregory A. Armstrong
Institute for Plant Sciences, Department of Plant Genetics, Swiss Federal Institute of Technology (ETH), CH-8092 Zürich, Switzerland

R. E. Blankenship, M. T. Madigan and C. E. Bauer (eds): Anoxygenic Photosynthetic Bacteria, pp. 1135–1157.

Summary

Carotenoids comprise a large class of natural pigments that are ubiquitous in photosynthetic organisms and are essential for photooxidative protection. A large collection of anoxygenic photosynthetic bacterial carotenoid biosynthesis mutants with novel color phenotypes have been isolated over the past 40 years. Genes involved in carotenoid biosynthesis have recently been mapped, cloned and sequenced. Seven clustered genes (*crtA, crtB, crtC, crtD, crtE, crtF, crtI*) are required for carotenoid biosynthesis in *Rhodobacter (Rb.) capsulatus* and *Rb. sphaeroides*. The *crt* gene products participate in reactions that convert FPP to spheroidene/spheroidenone. The proposed requirement for the products of *Rb. capsulatus* ORF160 and ORF 469 in carotenoid biosynthesis conflicts with experiments using their respective *Rb. sphaeroides* homologs. Nucleotide sequences have been determined for all of the *Rb. capsulatus crt* genes and for *crtD, crtI, crtB* and a portion of *crtC* in *Rb. sphaeroides*. The *Rb. capsulatus crtEF* and *crtIB* genes appear to form multigene operons, the former within the global context of a superoperon in the photosynthesis gene cluster.

Carotenoid pigment accumulation in *Rhodobacter* clearly responds to environmental stimuli, particularly light intensity and oxygen tension, and correlates with the presence and extent of the photosynthetic membrane. mRNA accumulation studies using gene-specific probes and promoter-reporter gene fusions indicate upregulation of *Rb. capsulatus crtA, crtC, crtD, crtE, crtF* and ORF 160, but not *crtB* and *crtI*, by anaerobiosis. Transcriptional regulation thus does not appear to limit the biosynthetic capacity for the conversion of GGPP to neurosporene.

Most information available on the *Rhodobacter* carotenoid biosynthesis enzymes has been deduced from the predicted amino acids sequences, rather than from biochemical or immunological studies. The carotenoid prenyltransferases (CrtB and CrtE) and phytoene dehydrogenase (CrtI) catalyze the early biosynthetic reactions in the carotenoid pathway that are conserved in other organisms. CrtI is, however, replaced in oxygenic photosynthetic organisms by a structurally distinct but functionally similar protein called CrtP/Pds. The CrtB, CrtC, CrtD, CrtE, CrtF and CrtI enzymes share conserved sequence motifs with other proteins of related function.

I. Introduction

A. Scope of This Chapter

The study of the genetics of carotenoid biosynthesis has blossomed in recent years, spurred in large part by the broad application of research results obtained with the anoxygenic photosynthetic bacteria to other organisms (reviewed in Armstrong, 1994). These advances have been achieved almost exclusively with two closely related, genetically amenable and metabolically versatile species of purple nonsulfur photosynthetic bacteria, *Rhodobacter capsulatus* and *Rhodobacter sphaeroides* (*Rhodopseudomonas capsulata* and *Rhodopseudomonas sphaeroides*, respectively, in earlier literature (Imhoff et al., 1984)). General aspects of the genetics of photosynthetic bacteria have been reviewed (Scolnik and Marrs, 1987). This chapter will focus primarily, though not exclusively, on developments since 1976 in the genetics of *Rb. capsulatus* and *Rb. sphaeroides* carotenoid biosynthesis.

Other chapters in this volume address topics such as the range of carotenoid structures and biosynthetic

Abbreviations: BChl – bacteriochlorophyll; BChlide – bacteriochlorophyllide; DMAPP – dimethylallyl pyrophosphate; *E. coli* – *Escherichia coli*; *E. herbicola* – *Erwinia herbicola*; FPP – farnesyl pyrophosphate; GGPP – geranylgeranyl pyrophosphate; GPP – geranyl pyrophosphate; IPP – isopentenyl pyrophosphate; kb – kilobase; LH I – light-harvesting I; LH II – light-harvesting II; LH – light-harvesting; *M. xanthus* – *Myxococcus xanthus*; *N. crassa* – *Neurospora crassa*; ORF – open reading frame; *P.* – *Paracoccus*; PPPP – prephytoene pyrophosphate; *Rb.* – *Rhodobacter*; RC – reaction center; *Rs.* – *Rhodospirillum*; SAH – S-adenosylhomocysteine; SAM – S-adenosylmethionine pathways encountered in the anoxygenic photosynthetic prokaryotes, the chemical and physical properties of carotenoids in their dual functions of photooxidative protection of the photosynthetic apparatus and light harvesting in the pigment-protein antenna and reaction center complexes, detailed features of the nucleotide sequence of the *Rb. capsulatus* carotenoid biosynthesis gene cluster, and those aspects of the regulation of carotenoid biosynthesis genes relating to the global organization of photosynthesis gene clusters.

B. General Background

Carotenoids comprise a diverse class of yellow, orange and red natural pigments that serve a variety of roles in photosynthetic and nonphotosynthetic prokaryotic and eukaryotic organisms. During the last 40 years the significance of carotenoid biosynthesis for photosynthesis has become evident. Biological functions of carotenoids in photooxidative protection, light absorption and the structural integrity of the photosynthetic membrane have been established in the anoxygenic photosynthetic bacteria (reviewed in Cogdell and Frank, 1987).

Carotenoid and bacteriochlorophyll (BChl) pigment molecules associate noncovalently but specifically with the photosynthetic reaction center (RC) and the light-harvesting (LH) antenna pigment-protein complexes (reviewed in Hawthornthwaite and Cogdell, 1991). In anoxygenic photosynthetic bacteria the formation of these complexes and the accompanying pigments is normally induced by anaerobic or low oxygen growth conditions. The LH antenna complexes serve as the major sites for the binding of lipophilic carotenoids in the photosynthetic membrane.

II. Early Developments in the Genetics of Carotenoid Biosynthesis

A. Rb. sphaeroides *Mutants*

The birth of the genetics of photosynthetic bacteria can be dated to the isolation of carotenoid pigment mutants of *Rb. sphaeroides* (Griffiths and Stanier, 1956). These bacteria are normally brown under anaerobic conditions and become reddish-purple upon exposure to oxygen. In this classic study, both ultraviolet-induced and spontaneous mutants were studied and classified according to the colors of the bacterial colonies. Griffiths and Stanier (1956) observed several classes of mutants, including brown, green and blue-green strains that contained BChl but lacked the normal carotenoid complement. A detailed examination of the properties of a blue-green phytoene-accumulating mutant (UV-33) provided the basis for our understanding of one of the crucial functions of carotenoids that absorb visible light, namely photooxidative protection of the organism against the deadly combination of photons, endogenous sensitizers (BChl in this case) and oxygen (Griffiths et al., 1955; Sistrom et al., 1956). The effects of light and oxygen on the formation of photosynthetic pigments in *Rb. sphaeroides* wild type and a spontaneous green mutant (Ga) were also documented in an elegant physiological study (Cohen-Bazire et al., 1957).

B. Elucidation of Biosynthetic Pathways

In the 20 years thereafter, research on carotenoid biosynthesis emphasized the identification of biosynthetic intermediates from wild type, mutant and chemically inhibited bacterial cultures (reviewed in Goodwin, 1980). Many of these early studies were performed with *Rb. sphaeroides* or *Rs. rubrum*. The classification of carotenoid pigment mutants and proposals for biosynthetic pathways in species of anoxygenic photosynthetic bacteria, based on the chemical relationships between the characterized biosynthetic intermediates, are described in detail elsewhere (Liaaen-Jensen et al., 1961; reviewed in Schmidt, 1978). In *Rb. sphaeroides* and *Rb. capsulatus*, a carotenoid biosynthetic pathway leading to the synthesis of spheroidene and spheroidenone was proposed (Fig. 1). The absence of a viable genetic system among the anoxygenic photosynthetic prokaryotes posed a major obstacle to further advances.

III. Mapping, Cloning and Sequencing of Carotenoid Biosynthesis Genes

A. Rb. capsulatus

1. Genetic Mapping

With the discovery of genetic recombination in *Rb. capsulatus*, mediated by a virus-like particle termed the gene transfer agent (Marrs 1974), the genetic mapping of carotenoid biosynthesis mutations in a

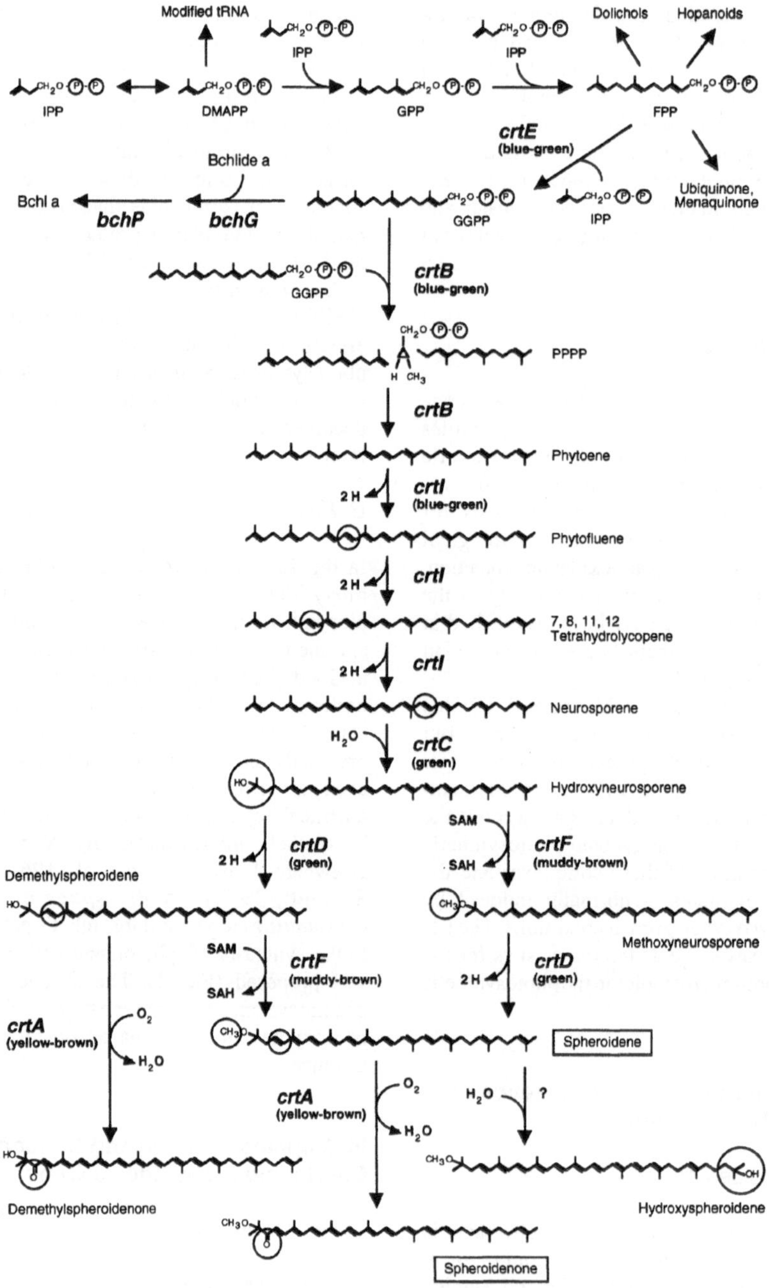

Fig. 1. Rhodobacter isoprenoid and carotenoid biosynthesis. Other biosynthetic pathways dependent on intermediates in general isoprenoid metabolism are indicated (Ourisson et al., 1987; Sherman et al., 1989; Bauer et al., 1993a). Starting with phytoene, the first stable compound unique to the C_{40} carotenoid branch of isoprenoid metabolism, the biosynthetic conversions are highlighted (circles). Intermediates up to neurosporene are common to most carotenogenic organisms. Spheroidene and spheroidenone (boxed) are the major *Rhodobacter* wild type carotenoids accumulated in the absence and presence, respectively, of oxygen in the growth medium. See Goodwin (1980) for the semisystematic nomenclature of the carotenoids and a general review of biosynthetic pathways. Specific genetic

photosynthetic prokaryote became possible (Yen and Marrs, 1976). Carotenoid (*crt*) and BChl (*bch*) biosynthesis genes were shown to be genetically linked. Five genetically distinct classes of *Rb. capsulatus crt* mutants displaying three altered color phenotypes (*crtB* or *crtE*, blue-green; *crtC* or *crtD*, green; *crtA*, yellow) were isolated (Yen and Marrs., 1976). The *Rb. capsulatus* mutants corresponded phenotypically to the blue-green, green and brown mutants previously reported in *Rb. sphaeroides* (Griffiths and Stanier, 1956). An additional class of muddy-brown mutants defined a new genetic locus, *crtF*, linked to the other *crt* genes (Scolnik et al., 1980). A carotenoid metabolic grid connecting *crtA*, *crtC*, *crtD* and *crtF* with specific biosynthetic conversions (Fig. 1) was proposed on the basis of a study of the carotenoid intermediates accumulated in *Rb. capsulatus* mutants. Another new locus associated with the green phenotype, *crtG*, was also identified and some of the previously reported *crtC* mutations (Yen and Marrs, 1976) were reassigned to it (Scolnik et al., 1980). *crtG* and *crtD* were subsequently demonstrated to be identical (Marrs, 1982).

The biosynthetic intermediates accumulated in *Rb. capsulatus crtB* and *crtE* mutants were not defined in the construction of the metabolic grid (Yen and Marrs, 1976; Scolnik et al., 1980). It was, however, recognized that these blue-green strains did not accumulate phytoene, whereas the prototypical *Rb. sphaeroides* UV-33 mutant did (Griffiths and Stanier, 1956; Sistrom et al., 1956). An *Rb. capsulatus crtH* blue-green mutant that accumulated phytoene was also mentioned in the literature but was not described in detail (Marrs 1982).

2. Physical Mapping

Shortly after the assignments of *crtA*, *crtC*, *crtD* and *crtF* to the biosynthetic grid (Scolnik et al., 1980), an R-prime plasmid, pRPS404, containing a 46 kilobase pair (kb) region of the *Rb. capsulatus* chromosome from strain SB1003 was described (Marrs, 1981). This R-prime complemented all genetic defects in photosynthesis known at that time, including blockages in carotenoid and BChl biosynthesis, upon its introduction into *Rb. capsulatus* mutants. The *puc* operon, encoding the LH II (B800–850) peripheral antenna complex, was subsequently found to lie outside of the region carried on pRPS404 (Kaufmann et al., 1984). The isolation of pRPS404 ushered in a period of intensive research to physically map, subclone and sequence genes for the photosynthetic RC and LH I (B875) pigment-protein complexes, encoded by the *puf* and *puh* operons, and the photosynthetic pigment biosynthesis enzymes, encoded by various *crt* and *bch* operons (see Burke et al., 1991; Chapter 50 by Alberti et al. and references therein).

The *crtB* locus was the first carotenoid biosynthesis function to be physically mapped on pRPS404 by a conjugation-mediated marker rescue technique (Taylor et al., 1983). Shortly thereafter, transposon mutagenesis of pRPS404 and the physical mapping of the *crtA*, *crtE*, *crtI* and *crtJ* genetic loci were also reported (Zsebo and Hearst, 1984). *crtI* and *crtJ* had not been previously described and were identified in this study by their blue-green mutant phenotypes typical of early blockages in carotenoid biosynthesis (Fig. 1). *crtJ* was surrounded by *bch* genes and was located about 12 kb away from the other *crt* genes, which formed a subcluster within the photosynthesis gene cluster (Zsebo and Hearst, 1984). The *crtC*, *crtD*, and *crtF* genes were physically mapped by interposon (i.e. antibiotic resistance cartridge) mutagenesis and in vivo complementation (Giuliano et al., 1988).

3. Assignments of Biosynthetic Blockages in Blue-Green Mutants

The phenotypes of the new *Rb. capsulatus crtI* and *crtJ* mutants coincided with those of *crtB* and *crtE* blue-green mutants both with respect to the absence of visibly absorbing carotenoids and in the pleiotropic loss of a spectroscopically detectable LH II antenna complex (Kaufmann et al., 1984; Zsebo and Hearst, 1984). Subsequently, several point mutants that

defects in carotenoid biosynthesis result in mutant color phenotypes reflecting the presence of BChl and the particular carotenoid intermediate(s) accumulated. Compounds starting with asymmetrical ζ-carotene (7, 8, 11, 12 tetrahydrolycopene) that have 7 or more conjugated double bonds absorb visible light and contribute to the overall color. *bch* genes associated with BChl biosynthesis are discussed elsewhere (Bauer et al., 1993a; Bollivar et al., 1994). *crt* genes and product functions are as follows: *crtA*, monooxygenation; *crtB*, 1′-2-3 prenylpyrophosphate condensation and 1′-1 rearrangement; *crtC*, hydration; *crtD*, 3,4 dehydrogenation; *crtE*, 1′-4 prenylpyrophosphate condensation; *crtF*, O-methylation; *crtI*, dehydrogenation. The biosynthetic functions ascribed to *crt* gene products have been deduced from precursor accumulation studies, or from in vivo or in vitro complementation of mutant phenotypes. Enzymatic activities of most of the proteins have not yet been demonstrated.

accumulated phytoene were examined and found to be complemented in vivo by DNA from the *crtI* region (Giuliano et al., 1986). A cell-free extract from a *crtI* mutant failed to synthesize colored carotenoids but could be complemented in vitro by a *crtB* mutant extract, thus defining a function for *crtI* in the biosynthetic pathway (Fig. 1).

Blue-green *crtB, crtE, crtI* and *crtJ* mutants were studied further using a cell-free *Rb. capsulatus* in vitro carotenoid biosynthesis system capable of converting ^{14}C-labeled IPP to phytoene (Armstrong et al., 1990c). Both in vitro and in vivo accumulation of phytoene were demonstrated in a *crtI* transposon mutant, in agreement with earlier results using point mutants (Giuliano et al., 1986). Although the updated biosynthetic pathway gives the all-*trans* carotenoid structures for convenience (Fig. 1), 15-*cis* and all-*trans* phytoene were present in roughly equal amounts in the *crtI* transposon mutant (Armstrong et al., 1990c).

The cell-free system also indicated an in vitro accumulation of GGPP in *crtB* and *crtJ* mutants, and GGPP plus PPPP in a *crtE* mutant. Provisional biosynthetic functions of PPPP synthesis and phytoene synthesis were therefore ascribed to the *Rb. capsulatus crtB* and *crtE* gene products, respectively (Armstrong et al., 1990c). These assignments were made under the assumption that blue-green mutants could not be defective in GGPP synthesis because this compound is essential for the synthesis of the phytyl side chain of BChl *a* in the *Rhodospirillaceae* (Fig. 1) (Biel and Marrs, 1983; Rüdiger and Schoch, 1991; Bauer et al., 1993a). Subsequent comparisons of the deduced amino acid sequences for CrtE and CrtB with those of prenyltransferases and squalene synthases led to revised proposals that CrtE might rather be the GGPP synthase and that CrtB alone might convert GGPP to phytoene via PPPP (Carattoli et al., 1991; Math et al., 1992; Kuntz et al., 1992). *In vivo* complementation and precursor accumulation studies with prokaryotic homologs of CrtB and CrtE from other organisms confirmed this hypothesis (Chamovitz et al., 1992; Math et al., 1992; Sandmann and Misawa, 1992). The complementation of an *Rb. capsulatus crtB* mutant, but not a *crtE* mutant, with a tomato phytoene synthase cDNA was also demonstrated (Bartley et al., 1992). These results together imply that a *crtE* mutant should accumulate FPP (Fig. 1) (see section VI. A).

Although a *crtJ* transposon mutant accumulated GGPP in vitro, *crtJ* was not definitively assigned to the biosynthetic pathway for several reasons (Armstrong et al., 1990c). Previous observations had suggested that this mutation could not be complemented in *trans* in *Rb. capsulatus* (Zsebo, 1984), the *crtJ* gene had not yet been characterized in detail, and an unidentified substance was found to accumulate along with GGPP in the in vitro assay (Armstrong et al., 1990c). It has also been recently demonstrated that a site-directed insertion mutation in *crtJ* does not affect carotenoid biosynthesis (Bollivar et al., 1994).

4. Nucleotide Sequences

Physical mapping of pRPS404 culminated in the nucleotide sequence determination of an 11 kb region of the *Rb. capsulatus* photosynthesis gene cluster containing a subcluster of eight genes devoted to carotenoid biosynthesis (Fig. 2) (Armstrong et al., 1989). Independent nucleotide sequencing of *crtI* and the 3′ portion of *crtF* confirmed these results (Bartley and Scolnik, 1989; Young et al., 1989). The existing genetic-physical maps of the *crt* gene cluster (Taylor et al., 1983; Zsebo and Hearst, 1984; Giuliano et al., 1986, 1988) and the high resolution mapping of previously described and newly generated transposon mutations (Armstrong et al., 1990c) permitted the alignment of open reading frames (ORFs) present in the DNA sequence with the known genetic loci (Armstrong et al, 1989). The only difficulty in this analysis came with the discovery of a previously unknown ORF located between the convergently oriented *crtB* and *crtC* genes, and predicted to be transcribed in the same direction as *crtB* (Fig. 2). An interposon mutation resulting in a *crtC* mutant phenotype (Giuliano et al., 1988) was proposed to interrupt the new ORF because of the location of a restriction site at which the interposon had been reported to be inserted (Armstrong et al., 1989). This ORF was therefore named *crtK* (Armstrong et al., 1989), to reflect the green phenotype of neurosporene accumulation reported for the interposon mutant (Giuliano et al., 1988). As with *crtJ*, the assignment of *crtK* to the biosynthetic pathway remained provisional pending a more detailed characterization of the gene function. The nucleotide sequence of the entire 46 kb *Rb. capsulatus* photosynthesis gene cluster from pRPS404, including *crtJ*, has recently become available (Burke et al., 1991; Chapter 50 by Alberti et al.).

Because pRPS404 carries the *crtD223* point mutation, the *crtD* gene sequence determined was that of the mutant allele (Armstrong et al., 1989;

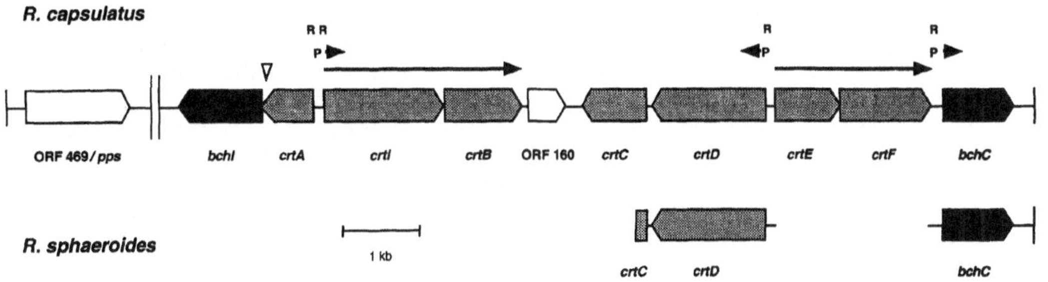

Fig. 2. Rhodobacter carotenoid biosynthesis gene clusters. Only sequenced regions are shown. *crt* genes, *bch* genes and genes of uncertain function are shaded gray, black and white, respectively. For simplicity of illustration, genes encoded in the regions flanking ORF 469/*pps*, 3′ to *bchI* and 3′ to *bchC* are not shown. Gene orientations are as indicated. The region of the *Rb. capsulatus* photosynthesis gene cluster depicted represents nucleotide positions 14800 to 16209 for ORF 469/*pps* and positions 25861 through 36868 for *bchI* through *bchC* (Armstrong et al., 1989; Burke et al., 1991). A sequencing error which revealed the existence of *bchI* is indicated (white triangle) (Armstrong et al., 1993b; Young et al., 1992). Possible σ^{70}-type promoters (P), palindromic regulatory sequences (R) (Armstrong et al., 1989) and operons are shown (arrows) (Giuliano et al., 1988; Young et al., 1989; Armstrong et al., 1990c, 1993b). Of the *Rb. capsulatus* genes depicted, most display increased (2- to 12-fold) mRNA accumulation during photosynthetic adaptation (Armstrong et al., 1993b). The exceptions are *crtI* and *crtB*, whose mRNA levels remain constant, and ORF 469/*pps and bchI*, which have not been tested. In *Rb. sphaeroides, crtD*, a portion of *crtC* and *bchC* have been sequenced (Garí et al., 1992b; McGlynn and Hunter, 1993). *crtA, crtB, crtE, crtF* and *crtI* have been physically mapped and are organized as shown for *Rb. capsulatus* (Pemberton and Harding, 1986; Coomber et al., 1990; Garí et al., 1992a).

Burke et al., 1991). Based on sequence homologies between CrtD and other proteins (see section V. B.), a specific proposal was made that the *crtD223* allele arose from a nucleotide exchange of G to A in the first position of codon 13, causing a mutation of Gly to Arg at a critical location within a putative CrtD ADP-binding fold (Armstrong et al., 1990a). Recent sequencing of a portion of the wild type *crtD* gene from genomic DNA of *Rb. capsulatus* strain B100 confirms this proposal, although the presence of additional mutations in other portions of the *crtD223* nucleotide sequence cannot be excluded (M. Alberti, personal communication).

The *Rb. capsulatus crtA* gene was recently resequenced and a frameshift error was discovered with respect to the originally reported wild type sequence (Armstrong et al., 1989, 1993b; Young et al., 1992). The results of high resolution mapping of *crtA* and *bchI* transposon mutations resulting in drastically different phenotypes motivated this effort (Armstrong et al., 1990c). The *crtA* mutant synthesized BChl while the *bchI* mutant did not, although the mutations were physically mapped to the 5′ and 3′ ends of the same ORF, respectively. In addition, higher plant (Koncz et al., 1990) and algal (Orsat et al., 1992) proteins were found to be homologous to the C-terminus but not the N-terminus of the predicted CrtA protein (Armstrong et al., 1993b). Correction of the frameshift splits the long ORF originally proposed to encode CrtA into 5′ and 3′ ORFs corresponding to *crtA* and *bchI*, respectively (Fig. 2). The ORF downstream of *bchI*, originally termed ORF H (Armstrong et al., 1989), appears to correspond to *bchD*. Evidence discussed elsewhere in detail suggests that *bchD*- *bchI* and the *Arabidopsis thaliana bchI* counterpart encode components of the Mg-chelatase, an enzyme which inserts Mg into protoporphyrin IX in the first committed step of BChl/Chl synthesis (Koncz et al., 1990, 1992; Armstrong et al., 1993b).

The translation start for *crtI* also lies farther 5′ than originally reported (Armstrong et al., 1989); this fact became clear by extending the alignment of the deduced sequences for CrtI and CrtD (Armstrong et al., 1990b), and for CrtI and *Neurospora crassa* Al-1, the fungal phytoene dehydrogenase (Bartley and Scolnik, 1989; Schmidhauser et al., 1990). This result was confirmed by complementation of an *Rb. capsulatus crtI* mutant with a construct using an upstream ATG start codon for translation of *crtI* mRNA (Bartley et al., 1990).

B. Rb. sphaeroides

1. Genetic and Physical Mapping

In parallel to the efforts with *Rb. capsulatus*, a genetic exchange system based on broad host range plasmid-mediated chromosome transfer was developed in *Rb. sphaeroides* (Sistrom, 1977; Pemberton and Bowen,

1981). This system was used to demonstrate the genetic clustering of photosynthesis genes, including *Rb. sphaeroides crtA, crtB, crtC, crtD, crtE* and *crtF*, which were identified by analogy to the corresponding *Rb. capsulatus* mutant genotypes and phenotypes. Two overlapping cosmids that between them complemented all of the *crt* mutations were isolated and used to construct a rough physical map of the *crt* genes in *Rb. sphaeroides* wild type strain RS6258 (Pemberton et al., 1986).

Transposon mutagenesis of *Rb. sphaeroides* wild type strain 8253 using a Tn5-suicide plasmid allowed the isolation of a number of mutants with photosynthesis gene defects (Coomber et al., 1990). Mutant complementation and pigment characterization led to the construction of a high-resolution genetic-physical map of the photosynthesis gene cluster, similar to that established earlier in *Rb. capsulatus* (Zsebo and Hearst, 1984). *Rb. sphaeroides crtA, crtB, crtC, crtD, crtE* and *crtI* mutants were characterized but no *crtF* mutants were isolated (Coomber et al., 1990). A physical map of the *Rb. sphaeroides crtA, crtB, crtC, crtD, crtE* and *crtF* genes from the prototypical wild type strain 2.4.1 (Griffiths and Stanier, 1956; Cohen-Bazire et al., 1957) has recently been reported (Garí et al., 1992a). Thus, the combined results of the genetic and physical mapping from three different strains of *Rb. sphaeroides* revealed the existence of a *crt* gene cluster equivalent in organization and composition to that found in *Rb. capsulatus* (Fig. 2).

2. Gene Assignments and Nucleotide Sequences

The nucleotide sequences of *crtD* and the 5′ portion of *crtC* from *Rb. sphaeroides* have recently been published (Garí et al., 1992b). *Rb. sphaeroides* and *Rb. capsulatus crtD* display 63% nucleotide sequence identity. An *Rb. sphaeroides* equivalent of *Rb. capsulatus crtK* has also been identified and sequenced (S. Kaplan, personal communication). *Rb. sphaeroides* CrtK plays, however, no role in carotenoid biosynthesis. The designation ORF 160 therefore replaces *crtK*/oxyA throughout the remainder of this chapter to emphasize this fact. Another *Rb. sphaeroides* gene, *pps*, has been recently sequenced (Penfold and Pemberton, 1994) and shown to be equivalent to *Rb. capsulatus crtJ* (Burke et al., 1991). The *pps* gene plays a role in the suppression of both BChl and carotenoid pigment levels when introduced in *trans* to a functional photosynthetic pigment gene cluster (Penfold and Pemberton, 1991, 1994). Recent experiments have demonstrated that *Rb. capsulatus crtJ* is not required for carotenoid biosynthesis and that the original *crtJ* transposon mutant described (Zsebo and Hearst, 1984) probably contains a second site *crt* mutation resulting in the observed blue-green phenotype (Bollivar et al., 1994). The nomenclature ORF 469/*pps* therefore replaces *crtJ* throughout the remainder of this chapter (see section IV. A. 2.).

C. Rs. centenum *and* Rs. rubrum

1. Genetic and Physical Mapping

Rhodospirillum centenum has recently been introduced as another suitable system in which to explore the genetics of bacterial photosynthesis (Yildiz et al., 1991, 1992). Work is in progress to obtain photosynthetic pigment mutants and to map the corresponding genes. Thus far, several *crtI*-like phytoene accumulating mutants have been isolated (Yildiz et al., 1991) and heterologous in vivo complementation of an *Rb. capsulatus crtF* mutant with a portion of the *Rs. centenum* photosynthesis gene cluster has been demonstrated (Yildiz et al., 1992). *Rs. centenum* and *Rs. rubrum,* two closely related species, use a carotenoid biosynthetic pathway thought to partially overlap that of the more distantly related *Rb. capsulatus* and *Rb. sphaeroides* (Fig. 1). Divergence at the level of neurosporene or hydroxyspheroidene gives rise to spirilloxanthin as the end product (Schmidt, 1978). Thus, homologs of *Rhodobacter crtB, crtC, crtD, crtE, crtF* and *crtI* might be expected in *Rs. centenum* and *Rs. rubrum.* Preliminary mapping of the photosynthesis gene cluster in the latter bacterium has also been described (Sägesser, 1992; Bauer et al., 1993a).

D. Genomic Organization

1. Operons and Superoperons

The nucleotide sequence of the *Rb. capsulatus crt* gene cluster (Fig. 2) revealed gene orientations and suggested the possible existence of several operons (*crtDC, crtEF, crtIB*ORF 160, *crtA*) (Armstrong et al., 1989). In *Rb. sphaeroides,* the orientations of most of the *crt* genes remain to be established although nucleotide sequencing of *crtD, crtI, crtB* and the 5′ coding region of *crtC* has revealed a gene arrangement identical to *Rb. capsulatus* (Garí et al., 1992b; Lang et al., 1994). The phenotypes resulting from interposon and transposon mutations in *Rhodobacter crtD* and *crtC* indicate that these genes do not form an obligate operon (Giuliano et al., 1988; Coomber et al., 1990; Garí et al. 1992a). Transcription of *crtF* requires the presence of the *crtE* 5′ noncoding region, suggesting the existence of a *crtEF* operon (Young et al., 1989). The situation with *crtIB*ORF 160 is somewhat more complicated; cotranscription of ORF 160 with *crtIB* does not seem to be obligatory, because the patterns of *crtIB* and ORF 160 mRNA accumulation differ substantially upon a shift from aerobic to photosynthetic growth conditions (see section IV. C. 3.) (Armstrong et al., 1993b). Differential stabilities of various portions of the same transcript cannot be excluded, however (Belasco et al., 1985). With respect to a *crtIB* operon, *Rb. capsulatus crtI* point mutants accumulate phytoene (Giuliano et al., 1986) while *crtI* interposon mutants do not (Giuliano et al., 1988). Both kinds of phenotypes are observed among *Rhodobacter crtI* transposon mutants (Armstrong et al., 1990c; Coomber et al., 1990). The simplest explanation for these results, in agreement with the physical organization of these genes (Fig. 2) is that *crtIB* form an operon and that not all *crtI* insertion mutations are polar (Armstrong et al., 1993b; Lang et al., 1994). Cotranscription of *crtA* with *bchI* does not seem to be obligatory, as *crtA* transposon and interposon mutants produce BChl and grow photosynthetically (Giuliano et al., 1988; Armstrong et al., 1990c), in contrast to a BChl-deficient *bchI* transposon mutant (Zsebo and Hearst, 1984).

The *Rb. capsulatus* photosynthesis gene cluster consists of a series of operons embedded within superoperons, which seems to provide for a low level of expression of pigment biosynthesis enzymes and pigment-binding polypeptides during aerobic growth (reviewed in Wellington et al., 1992). This mechanism apparently enables the bacteria to adapt more rapidly during a transition to anaerobic photosynthetic conditions. *crtEF* was shown to be the 5′-most operon in a superoperon encompassing *crtEFbchCXYZ-pufQBALMX* (Young et al., 1989; Burke et al., 1993; Chapter 57 by Beatty). The participation of *crtD, crtC*, ORF 160, *crtIB* in superoperons of the type mentioned above seems to be excluded because of the physical arrangement of the genes. *crtA*, on the other hand, could potentially contribute to a superoperon containing downstream *bch* genes (see section III. A. 4.).

2. Carotenoid Biosynthesis Gene Clusters

Data thus far in *Rhodobacter* indicate that the *crtA, crtB, crtC, crtD, crtE, crtF* and *crtI* genes are clustered, that this *crt* subcluster is embedded within a larger cluster of genes devoted to photosynthesis, and that the overall gene organization is very similar in two species (Fig. 2). Partial genetic-physical maps in *Rs. centenum* and *Rs. rubrum* support these conclusions (Bauer et al., 1993a). Carotenoid biosynthesis gene clusters are also found in nonphotosynthetic bacteria, including *Erwinia* species (Armstrong et al., 1990a; Misawa et al., 1990; Hundle et al., 1991; Armstrong, 1994).

All of the *crt* gene clusters described thus far contain *crtE, crtB* and *crtI* homologs that encode enzymes catalyzing the early reactions in carotenoid biosynthesis common to most carotenogenic organisms (Fig. 1) (Armstrong et al., 1990a). A comparison of the *crt* gene clusters from the nonphotosynthetic and photosynthetic bacteria reveals that the location and transcriptional orientation of the *crtE* gene, encoding GGPP synthase, varies between species. On the other hand, the arrangement of the *crtI* and *crtB* genes, encoding phytoene dehydrogenase and phytoene synthase, respectively, is identical to that observed in photosynthetic bacteria (Fig. 2). Furthermore, in the cyanobacterium *Synechococcus* PCC7942 this arrangement of the cognate genes encoding phytoene dehydrogenase (*crtP*) and phytoene synthase (*crtB*) has also been conserved (Chamovitz et al., 1992), although the phytoene dehydrogenases encoded by *crtI* and *crtP* are proposed to have evolved independently (Pecker et al., 1992). Whether the conserved physical arrangement of *crtI/crtP* and *crtB* has a functional significance has not been tested.

IV. Regulation of Carotenoid Biosynthesis

A. Carotenoid Accumulation

1. Environmental Factors

In anoxygenic photosynthetic bacteria, anaerobiosis initiates a variety of processes, including pigment, protein and lipid synthesis, accompanied by changes in gene expression (reviewed in Kiley and Kaplan, 1988). These processes culminate in the formation of a functional photosynthetic membrane. Because the coordination of membrane development ensures that the supplies of the individual components remain in balance, abnormalities in carotenoid or BChl accumulation can have pleiotropic effects on the accumulation of the RC and LH pigment-binding polypeptides or on the spectral characteristics of these complexes (Griffiths and Stanier, 1956; Marrs, 1982; Dierstein, 1983; Kaufmann et al., 1984; Zsebo and Hearst, 1984; Klug et al., 1985; Bartley and Scolnik, 1989; Yildiz et al., 1991).

The accumulation of BChl and carotenoid pigments can be influenced by environmental stimuli, particularly oxygen and light, in the anoxygenic photosynthetic bacteria. The classic study of Cohen-Bazire et al. (1957) demonstrated that both oxygen and light strongly influence carotenoid accumulation in *Rb. sphaeroides,* results which also apply to *Rb. capsulatus.* In these species a down shift in light intensity with photosynthetically growing cells triggers transient increases in carotenoid accumulation until a new steady-state with a higher specific carotenoid content has been achieved. A downshift in oxygen tension using dark aerobically grown cells produces much the same result, with or without accompanying illumination. Low oxygen tension thus initiates photosynthetic membrane synthesis, while a reduction in light intensity expands the size of the photosynthetic light-harvesting antenna (Kiley and Kaplan, 1988).

Steady-state BChl concentrations increase dramatically in *Rb. capsulatus* cells grown under decreasing oxygen tensions from 20 to 0% (Clark et al., 1984). The accompanying several-fold increase in carotenoid content has been argued to be an indirect effect of oxygen on BChl or some other component of the photosynthetic membrane (Clark et al., 1984; Biel and Marrs, 1985). Only marginal increases in carotenoid content were seen in *bch* mutants of *Rb. capsulatus* shifted to low oxygen tension (Biel and Marrs, 1985). Two problems prevent a straightforward interpretation of these data, however. First, *bch* mutations in *Rb. capsulatus* or *Rb. sphaeroides* reduce carotenoid content to about 25% of the wild type levels (Biel and Marrs, 1985; Penfold and Pemberton, 1991). In the anoxygenic photosynthetic bacteria nothing is known about carotenoid turnover or the binding sites of the carotenoids found in cells lacking a photosynthetic membrane, such as *bch* mutants or the wild type cultured aerobically under 20% oxygen tension. As *bch* mutants lack BChl, and hence stable LH I and LH II antenna complexes (Dierstein, 1983; Zsebo and Hearst, 1984; Klug et al., 1985), the major known carotenoid binding sites in the photosynthetic membrane no longer exist. Furthermore, the concentration of potential endogenous photosensitizers (i.e. BChl precursors) which might interact with the carotenoid pool differs between *bch* mutants and the wild type. Therefore, the assumption of similar carotenoid turnover rates cannot be made with confidence.

Little is known about the influence of environmental factors other than light intensity and oxygen tension on carotenoid biosynthesis and accumulation. Both iron excess and iron deficiency in the growth medium alter the carotenoid composition in *Rb. capsulatus* (Nelis and De Leenheer, 1989a). A novel mixture of *cis* and all-*trans* isomers of spheroidenone (Fig. 1) has been tentatively identified in axenically grown cultures subjected to iron excess.

2. Genetic Factors

Spontaneous green mutants of *Rb. sphaeroides,* exemplified by strain Ga, have been known from the inception of genetic studies with photosynthetic bacteria (Cohen-Bazire et al., 1957). Ga accumulates methoxyneurosporene, hydroxyneurosporene and neurosporene (Fig. 1), and therefore is classified as a *crtD* mutant (Scolnik et al., 1980). In a survey of wild type isolates of *Rb. capsulatus,* one of 33 strains examined displayed an in vivo absorption spectrum typical for a green *crtD* or *crtC* mutant (Weaver et al., 1975). Various researchers have been unable to discern any differences in growth rates or pigment-protein complexes, other than carotenoid composition, distinguishing *crtD* mutants and wild type strains accumulating spheroidene and spheroidenone (eg. Cohen-Bazire et al., 1957; Marrs, 1982; Armstrong et al., 1990c). Any selective advantage imparted by the synthesis of these wild type carotenoids may thus be subtle and/or function only under certain growth conditions. The molecular basis for the high frequency (10^{-5}) appearance and reversion of spontaneous green *crtD* mutants in the *Rb. sphaeroides* wild type strain 2.4.1 has recently been elucidated (Garí et al., 1992a). The reversible addition of a G residue to a stretch of seven G's found within the 5′ coding region of *crtD* leads to a frameshift mutation. Interestingly, this mechanism appears to be strain-specific.

In contrast to the weak or nonexistent selection against *crtD* mutations, *crtF* mutations are subject to strong negative selective pressure (Taylor et al., 1983).

Two classes of secondary mutations appear in *Rb. capsulatus crtF* mutants, namely earlier blockages in carotenoid biosynthesis, and *bch* or other mutations which completely abolish the accumulation of photosynthetic pigment-protein complexes. Although the biological significance of this phenomenon is not clear, the apparent negative selection against the accumulation of demethylspheroidene and demethylspheroidenone in *crtF* mutants may explain why this gene escaped detection in random transposon mutagenesis schemes used in *Rhodobacter* (Zsebo and Hearst, 1984; Coomber et al., 1990).

With respect to regulatory genes, no mutant with a genetic defect leading to a quantitative change in carotenoid levels independent of other components of the photosynthetic membrane has been described. The *Rb. sphaeroides pps* gene, homologous to *Rb. capsulatus* ORF 469, represses both BChl and carotenoid pigment accumulation when expressed in *trans* to a functional photosynthetic gene cluster resident in or introduced into *Rb. capsulatus, Rb. sphaeroides* or *Paracoccus denitrificans* (Penfold and Pemberton, 1991, 1994). In addition, the repression of carotenoids functions even in a *bch* mutant background and therefore independently of BChl.

An *Rb. sphaeroides* mutant, TA-R, displayed inappropriate regulation of both BChl and carotenoid accumulation by derepressing pigment synthesis under dark aerobic conditions (Lascelles and Wertlieb, 1971). More recently the *Rb. sphaeroides* $T1_a$ mutant, which is defective at a locus termed *oxyB*, was found to accumulate spectroscopically intact LH II antenna complexes containing BChl and carotenoids under aerobic growth conditions (Lee and Kaplan, 1992b). Because the stability of the photosynthetic LH and RC pigment-protein complexes depends on interactions between BChl, carotenoids and various polypeptides, it is difficult to predict what the targets of these regulatory genes might be. *Rs. centenum* provides an interesting experimental system because this bacterium possesses a functional photosynthetic membrane with all of the usual pigment and protein components when cultured under high oxygen tension (Yildiz et al., 1991). Wild type *Rs. centenum* thus resembles the *Rb. sphaeroides* TA-R and $T1_a$ mutants.

The function of ORF 160, which was originally thought to have a biosynthetic role in carotenoid synthesis, is unknown (S. Kaplan, personal communication). Interestingly, *Rb. capsulatus* ORF 160 mRNA increases 8-fold in response to the shift from aerobic growth to anaerobic photosynthesis (see section IV. C. 3.) (Armstrong et al., 1993b).

B. Carotenoid Biosynthesis Enzymes

Virtually nothing is known about the regulation of carotenoid biosynthesis enzyme levels or activities in the anoxygenic photosynthetic bacteria. The biochemical intractability of the membrane-bound enzymes which convert phytoene into later carotenoids have hindered attempts at enzyme purification (reviewed in Bramley, 1985). No carotenoid biosynthesis enzyme from these organisms has yet been purified to homogeneity.

CrtI from *Rb. capsulatus* and *Rb. sphaeroides* has been overexpressed in *Escherichia coli* (Schmidt et al., 1989; Bartley and Scolnik, 1989; Lang et al., 1994). Antibodies raised against a CrtI-LacZ fusion protein (Schmidt et al., 1989) as well as a truncated version of CrtI have been described (Bartley and Scolnik, 1989). In vitro phytoene dehydrogenation in solubilized membranes of *Aphanocapsa* PCC 6714 was inhibited by the antibody against the *Rb. capsulatus* CrtI-lacZ fusion protein (Schmidt et al., 1989). Inducible overexpression of CrtI in a *crtI* blue-green mutant of this organism produced some surprising results (Bartley and Scolnik, 1989). The uninduced bacteria carrying the expression construct contained immunologically negligible levels of CrtI, compared to the wild type, yet these cells displayed wild type pigmentation. Furthermore, the massive induced overexpression of CrtI had no effect on carotenoid levels. The roughly constant in vivo accumulation of carotenoids over a large range of CrtI concentrations strongly suggests that this enzyme does not catalyze a rate-limiting step in *Rb. capsulatus* carotenoid biosynthesis. Curiously, overproduction of CrtI resulted in higher than normal levels of BChl and LH II complex (Bartley and Scolnik, 1989). Whether this represents a specific interaction of CrtI with the LH II complex or results from unspecific interactions is unknown. The *crtI* mutant used for the in vivo complementation also carried a *crtF* mutation, so that this result was obtained in a situation leading to the synthesis of *crtF*-type carotenoids (Fig. 1). Recently, Lang et al. (1994) have also overexpressed the *Rb. sphaeroides crtI* gene product in *E. coli* and shown that a cell free extract will catalyze three desaturations of phytoene to give neurosporene. The reaction was shown to be ATP-dependent.

With respect to cofactor requirements for

carotenoid biosynthesis (Schmidt, 1978),CrtA in *Rhodobacter* mediates the conversion of (demethyl)-spheroidene to (demethyl)spheroidenone under aerobic conditions in a reaction requiring molecular oxygen (Fig. 1). The O-methylation of carotenoids, mediated by CrtF, requires SAM as a cofactor. Both oxygen and endogenous SAM levels could therefore provide regulatory points for carotenoid biosynthesis. The function of the aerobic conversion of spheroidene to spheroidenone is not known. Because *crtD* mutants of *Rhodobacter* never synthesize spheroidene and yet seem subject to little or no selective disadvantage compared to the wild type, the significance of the reactions carried out by CrtA, and for that matter CrtD, remains unclear. The *Rb. capsulatus ahcY* gene, which encodes an enzyme required for the conversion of SAH to SAM, has been shown to be essential for the presence of normal levels of BChl and LH complexes (Sganga et al., 1992). The proposal has therefore been made that the intracellular ratio of SAM:SAH affects BChl biosynthesis at the level of the methylation of Mg-protoporphyrin IX. It is conceivable that the O-methylation of carotenoids could also be controlled by a similar phenomenon.

Finally, the metabolic balance between isoprenoid requirements for carotenoid and BChl biosynthesis must be regulated in some fashion (Fig. 1). Whether the branch point in these pathways occurs at the level of GGPP, as commonly assumed, or even earlier is discussed in section VI. A.

C. crt *Gene Regulation*

1. Regulation of Other Photosynthesis Genes

The fundamental mechanisms underlying the regulation of *crt* genes in photosynthetic bacteria are unknown. In comparison, much progress has been made in studies of the regulation of the synthesis of structural polypeptides of the photosynthetic membrane in response to light and oxygen tension. One current model for the anaerobic induction of mRNAs from the highly expressed *Rb. capsulatus puf, puh* and *puc* operons incorporates a membrane-bound oxygen sensor (RegB) and a cytoplasmic regulator (RegA) (reviewed in Bauer et al., 1993b and by Bauer in Chapter 58). This model does not explain the moderate anaerobic induction of *crt* and *bch* pigment biosynthesis operons, however (Biel and Marrs, 1983; Giuliano et al., 1988; Armstrong and Hearst, 1989; Young et al., 1989; Sganga and Bauer, 1992; Armstrong et al., 1993b; Ma et al., 1993).

2. Promoter Activities

E. coli σ^{70} promoter-like sequences are present in the 5′ noncoding regions of *Rb. capsulatus crtD, crtI* and *bchC* (Armstrong et al., 1989; Young et al., 1989; Ma et al., 1993). Because σ^{70} promoter-like sequences are not found 5′ to the *regA*-regulated operons, the proposal has been made that differences in promoter structure between the *puf, puh* and *puc* genes versus the *crt* and *bch* genes may account for the *regA*-independent regulation of the latter (Sganga and Bauer, 1992; Bauer et al., 1993b; Ma et al., 1993; Chapter 58 by Bauer). Little is known experimentally about the structure of *bch* and *crt* promoters, however. Site-directed mutagenesis of a putative σ^{70} promoter for the *bchCXYZ* operon confirms that transcription initiation requires this sequence (Ma et al., 1993). On the other hand, no *crt* promoter has yet been functionally defined in any detail. Furthermore, given our current knowledge of the operon structure of the *crt* gene cluster, *crtD* and *crtI* σ^{70} promoter-like sequences would only be sufficient to direct transcription of *crtD* and *crtIB* (Fig. 2). No obvious recognition sites for alternative σ subunits have yet been observed in the 5′ noncoding regions of other *crt* genes (*crtA, crtC, crtEF*). A palindromic sequence resembling the consensus binding site for many prokaryotic DNA-binding regulatory proteins has been identified in 5′ noncoding regions of *Rb. capsulatus crtD, crtE, crtI, crtA, bchC*, and the *puc* operon (Fig. 2) (Armstrong et al., 1989; Ma et al., 1993). The palindrome seems to play a role in the regulation of the *bchC* promoter (Ma et al., 1993) and the *Rb. sphaeroides puc* promoter (Lee and Kaplan, 1992a), probably in conjunction with other *cis* elements in both cases, and could be a functional element in *crt* promoters.

The activities of several *Rb. capsulatus crt* promoters have been studied using translational or transcriptional fusions to the *E. coli lacZ* gene, although the regulatory sequences have not yet been defined in detail. Translational fusions of a *crtE* promoter fragment to *lacZ* yielded a 3-fold upregulation in response to anaerobiosis (Young et al., 1989). This promoter has been proposed to contain two transcription initiation sites, one anaerobically induced and the other constitutive (Giuliano et al., 1988). DNA fragments containing the *crtA* or *crtI* promoters were transcriptionally fused to the *lacZ* gene and demonstrated to be upregulated 2-fold or unregulated, respectively, during the adaptation from aerobic to anaerobic photosynthetic growth (Armstrong et al., 1993b). These limited data indicate modest oxygen regulation of several *crt* promoters and are qualitatively consistent with the patterns of mRNA accumulation (see below) for the corresponding genes during the adaptation from aerobic to anaerobic photosynthetic growth (Armstrong et al., 1993b). While the *crtA* promoter seems to be only 2-fold inducible by anaerobiosis, the amount of *crtA* mRNA increases about 12-fold; several possible explanations exist, including posttranscriptional regulation.

pRPS404 carries all of the known *Rb. capsulatus crt* genes but does not direct carotenoid accumulation in the noncarotenogenic bacteria *E. coli* and *Pseudomonas fluorescens* (Marrs, 1981). On the other hand, introduction of the *Rb. sphaeroides crt* gene cluster into several species of phylogenetically related noncarotenogenic bacteria (*P. denitrificans, Agrobacterium tumifaciens, Agrobacterium radiobacter, Azotomonas insolita*) results in carotenoid production (Pemberton and Harding, 1987). The *Rb. sphaeroides pps* gene exerts a negative effect on carotenoid and BChl levels when supplied in *trans* to chromosomal or plasmid-borne pigment biosynthesis genes in *Rb. capsulatus* and *P. denitrificans* (Penfold and Pemberton, 1991, 1994), and *Rs. centenum* DNA complements an *Rb. capsulatus crtF* mutation in vivo (Yildiz et al., 1992). Thus, successful heterologous expression of *crt* genes may depend on the evolutionary distance between the species involved.

3. mRNA Accumulation

The regulation of *crt* mRNA accumulation has thus far been examined only in *Rb. capsulatus.* Early mRNA accumulation studies performed before the *crt* genes had been sequenced utilized hybridization probes which overlapped several genes. These studies indicated little or no induction of *crt* and *bch* mRNA levels, in contrast to the 10 to 40-fold induction of *puf, puh* and *puc* mRNAs, during shifts from aerobic growth to low oxygen or anaerobic photosynthetic conditions (Clark et al., 1984; Klug et al., 1985; Zhu and Hearst, 1986). The most defined results, in terms of the DNA probes used, indicated qualitative upregulation of the *crtA, crtC, crtE* mRNAs and constant levels of *crtI* mRNA in response to the removal of oxygen from a bacterial culture maintained under continuous illumination (Giuliano et al., 1988). Two *crtE* mRNA species were found, one of which appeared only under anaerobic conditions.

A more quantitative study to determine the temporal patterns of *crt* mRNA accumulation with gene-specific probes has also been performed (Armstrong et al., 1993b). The shift from aerobic to anaerobic photosynthetic growth induced transient 2- to 12-fold increases in mRNA accumulation for *crtA, crtC, crtD, crtE, crtF* and ORF 160, temporally coordinated with increases in mRNAs from other photosynthesis genes (3–5 fold for *bch*; 6- to 8-fold for *puf* and *puc*; 25-fold for *puc*) (Armstrong and Hearst, 1989; Cook et al., 1989; Armstrong et al., 1993b). mRNA levels for *crtB* and *crtI* remained constant, however. Light did not qualitatively affect the accumulation of *crtA* mRNA. The *crt* mRNAs that increased under anaerobic conditions encode biosynthetic enzymes either early (*crtE*) or late (*crtC, crtD, crtF, crtA*) in the biosynthetic pathway. This suggests that transcriptional regulation of the *crt* genes could play a role in boosting overall biosynthetic capability for the synthesis of GGPP and for the conversion of neurosporene to the final products (Fig. 1). The 12-fold induction of *crtA* mRNA by anaerobiosis (Armstrong et al., 1993b) is particularly noteworthy because CrtA requires molecular oxygen as a cofactor (Schmidt, 1978). An earlier study supplied some evidence that *crtA* mRNA also increases 2-fold in response to a shift from anaerobic photosynthetic to aerobic conditions, although a gene-specific probe was not used (Zhu et al., 1986). In contrast to the lack of regulation of *crtI* mRNA accumulation in *Rb. capsulatus* (Guiliano et al., 1988; Armstrong et al., 1993b), *Rb. sphaeroides crtI* mRNA has recently been reported to increase in abundance during anaerobiosis (Lang et al., 1994).

A striking result from the above investigations is the quantitative similarity in the anaerobic accumulation of mRNAs from several *crt* operons compared to the *puf* and *puh* operons (Cook et al., 1989; Armstrong et al., 1993b). Comparisons with other studies, even those using the same approaches, are complicated by differences in the growth conditions, strains, and experimental techniques used. Nonetheless, the current conventional wisdom regarding the regulation of photosynthesis genes says that expression of the *puf, puh,* and *puc* operons occurs at much higher levels and is subject to a more stringent control by oxygen than is observed for *crt* and *bch* operons (Bauer et al., 1993b; Chapter 58 by Bauer). While the former premise is unquestionably correct (Young et al., 1989; Sganga and Bauer, 1992), the latter may not always hold true (Cook et al., 1989; Armstrong et al., 1993b).

V. Functional and Structural Conservation of the *crt* Gene Products

A. The Utility of Deduced Amino Acid Sequences

Almost nothing is known directly about the *crt* gene products because the later reactions in carotenoid biosynthesis occur in membranes and the corresponding enzymes have not proven amenable to isolation, purification or characterization (Bramley, 1985). At the time that the sequence of the *Rb. capsulatus crt* gene cluster was published, no other nucleotide sequences for carotenoid biosynthesis genes had been described in the literature (Armstrong et al., 1989). Much has since been learned by comparison of the deduced amino acid sequences of the *Rb. capsulatus* carotenoid biosynthesis enzymes with each other and with other protein sequences present in the databases (eg. Armstrong et al., 1993a; Armstrong, 1994). Probable homology between two sequences can, in general, be said to exist when the overall identity exceeds approximately 25% for sequence alignments of a few hundred amino acids. By this criterion, homologs of *Rb. capsulatus* CrtB, CrtC, CrtD, CrtE, CrtF, CrtI and ORF 160 exist in other organisms (Table 1). CrtB, CrtD, CrtE, CrtF, CrtI and ORF 469/Pps also contain specific local sequence motifs probably related to the structures and functions of these proteins (see below) (Armstrong et al., 1993a). On a cautionary note, the deduced characteristics of all of the *crt* gene products must ultimately be confirmed by studying the purified proteins. *Rb. capsulatus* CrtI has been overexpressed (see section IV. B.), though not as an active enzyme (Bartley and Scolnik, 1989; Schmidt et al., 1989). In vivo complementation of carotenoid biosynthetic mutants of *Rb. capsulatus* with heterologous eukaryotic cDNAs has been used successfully to study the conservation of phytoene dehydrogenase and phytoene synthase enzyme functions (Bartley et al., 1990, 1991, 1992). More recently, enzymatic activity has been demonstrated for *Rb. sphaeroides* CrtI that has been overexpressed in *E. coli* (Lang et al., 1994).

B. Dehydrogenases (CrtD, CrtI)

Rb. capsulatus CrtI and CrtD, the phytoene and methoxyneurosporene dehydrogenases, respectively, were the first two carotenoid biosynthesis enzymes demonstrated to share conserved features with other proteins, in this case with each other (Armstrong et al., 1989). CrtI and CrtD dehydrogenate distinct but related carotenoid intermediates and were therefore expected to exhibit some sequence similarity. Phytoene is an intermediate in carotenoid biosynthesis in most organisms (Fig. 1), whereas methoxyneurosporene and hydroxyneurosporene are restricted to a few species of bacteria (Goodwin, 1980).

Comparison of the deduced amino acid sequences for CrtI and CrtD revealed 27% overall sequence identity (Table 1), with two highly conserved regions found at the respective N- and C-termini of the proteins (Armstrong et al., 1989). Shortly thereafter, deduced phytoene dehydrogenase sequences from two nonphotosynthetic bacteria, *Erwinia herbicola* and *Erwinia uredovora* (Armstrong et al., 1990a; Misawa et al., 1990), and a fungus, *N. crassa* (Bartley et al., 1990), were found to be conserved with *Rb. capsulatus* CrtI and, to a lesser extent, with CrtD. These findings supported the proposed conserved nature of the early enzymes of carotenoid biosynthesis in photosynthetic and nonphotosynthetic prokaryotes and eukaryotes, as hypothesized in the comparison of the *Rb. capsulatus* and *Ec. herbicola crt* gene products (Armstrong et al., 1990a). Recently, the characterization of the *Rb. sphaeroides crtD* gene (Garí et al., 1992b) and genes encoding carotenoid dehydrogenases from *Myxococcus xanthus* have been reported (Table 1). The *Rb. capsulatus* and *Rb. sphaeroides* CrtD proteins are 55.6% identical. Other carotenoid dehydrogenases display between 26–30% sequence identity with *Rb. capsulatus* CrtD, whereas identities of other phytoene dehydrogenases with *Rb. capsulatus* CrtI range from 25–40%. *N. crassa* Al-1 (Schmidhauser et al., 1990), was shown to complement an *Rb. capsulatus crtI* carotenoid biosynthesis mutant in vivo, producing novel fungal carotenoids in the bacterium (Bartley et al., 1990). Despite this functional complementation, antibodies raised against Al-1 or CrtI did not crossreact with the heterologous enzyme. Antibodies raised against an *Rb. capsulatus* CrtI-LacZ fusion protein crossreacted weakly with antigens from cyanobacteria, algae and higher plants (Schmidt et al., 1989), although a bell pepper phytoene dehydrogenase antibody did not recognize antigens in a bacterial extract (Hugueney et al., 1992).

A revision of the originally proposed *Rb. capsulatus crtI* translation start (Bartley and Scolnik, 1989; Armstrong et al., 1990b) exposed the presence of putative ADP-binding folds predicted to have $\beta\alpha\beta$ structures within the conserved N-terminal sequences of the carotenoid dehydrogenases (Table 2) (Armstrong et al., 1990a; Bartley et al., 1990). ADP-binding folds interact with FAD or NAD(P) cofactors in other types of dehydrogenases that contain this structural motif (Scrutton et al., 1990). Conserved features of the *Rhodobacter*, *Erwinia*, and *N. crassa* carotenoid dehydrogenases are illustrated in sequence alignments presented elsewhere (Armstrong et al., 1993a).

Deduced protein sequences for phytoene dehydrogenases from oxygenic photosynthetic organisms, including cyanobacteria, soybean, tomato and bell pepper have recently been reported (Bartley et al., 1991,1992; Huguencey et al., 1992; Pecker et al., 1992). These proteins show little overall sequence identity to the *Rb. capsulatus, Erwinia, M. xanthus* and *N. crassa* phytoene dehydrogenases, however. Sequence conservation between the anoxygenic photosynthetic /nonphotosynthetic microbial carotenoid dehydrogenases (CrtI-type and CrtD-type) and the oxygenic photosynthetic phytoene dehydrogenases (CrtP/Pds-type) seems to be restricted to the N-terminal ADP-binding fold (Hugueney et al., 1992; Pecker et al., 1992). A major functional difference between these classes of proteins can be seen in the extent of the dehydrogenation of the carotenoid products. CrtI-type phytoene dehydrogenases introduce three or four double bonds consecutively, yielding neurosporene in *Rhodobacter* or lycopene in other organisms, respectively (Bartley et al., 1990). The CrtP/Pds-type phytoene dehydrogenases introduce two double bonds to produce symmetrical ζ-carotene, which is subsequently converted to lycopene by a dehydrogenase distinct from Pds (Bartley et al., 1991; Pecker et al., 1992). The proposal that CrtI-type and CrtP/Pds-type phytoene dehydrogenases represent two functionally similar but evolutionarily independent proteins remains to be established (see section III. D. 2.) (Pecker et al., 1992). It is not known whether the *Rhodobacter* carotenoid dehydrogenases bind FAD or NAD(P) either as a cofactor or for structural reasons, although the bell pepper Pds has recently been isolated with bound FAD (Hugueney et al., 1992).

C. Prenyltransferases (CrtB, CrtE)

Rb. capsulatus CrtE and CrtB, the putative GGPP and phytoene synthases, respectively, both utilize isoprenoid pyrophosphate substrates but carry out different types of condensation reactions (Fig. 1). Initial uncertainty as to the biosynthetic functions of CrtE and CrtB in the synthesis of phytoene has now been resolved in experiments with homologous proteins from other prokaryotes (see section III.A. 3.) (Chamovitz et al., 1992; Math et al., 1992; Sandmann and Misawa, 1992). CrtE, CrtB and CrtI were shown to be conserved within the *crt* gene clusters of *Rb. capsulatus* and *Erwinia* species (Armstrong et al., 1989, 1990a; Misawa et al., 1990), presumably because these gene products catalyze reactions in the early portion of the carotenoid biosynthetic pathway common to most organisms (Fig. 1). The *Rb. capsulatus* and *Erwinia* CrtB and CrtE proteins share 30–35% and 27–30% sequence identities, respectively, and thus are not as highly conserved as the CrtI proteins (Table 1).

The prokaryotic CrtB proteins were also found to share about 29% sequence identity with the tomato pTOM5 protein (Armstrong et a., 1990a), encoded by a fruit ripening-associated cDNA of unknown function (Ray et al., 1987). Sequence alignments of these proteins have been presented elsewhere (Armstrong et al., 1993a). Biochemical analysis of transgenic tomato plants expressing an antisense pTOM5 mRNA (Bramley et al., 1992) and in vivo complementation of an *Rb. capsulatus crtB* mutant with a pTOM5-related cDNA confirmed the role of this protein as a phytoene synthase (Bartley et al., 1992). Other phytoene synthase sequences have been identified in the *M. xanthus* carotenoid gene cluster and in two species of cyanobacteria (Table 1). Both photosynthetic and nonphotosynthetic organisms have CrtB homologs that are between 27–35% identical to *Rb. capsulatus* CrtB.

The *Rb. capsulatus* and *Erwinia* CrtE proteins were recently found to share sequence identity with several other proteins, including a possible GGPP synthase from *Cyanophora paradoxa*, FPP synthase from *E. coli*, a sporulation protein from *Bacillus subtilis*, and a hypothetical *E. coli* protein (Table 1). Features of the sequence alignments of these proteins are detailed elsewhere (Armstrong et al., 1993a). *Rb. capsulatus* and *Erwinia* CrtE also share extensive sequence identity with the GGPP synthase from bell pepper (Table 1). The sequence identities of *Rb.*

Table 1. Rb. capsulatus crt gene products and other structurally related proteins

Rb. capsulatus : related protein[1]	Function of related protein	Database entries[2, 3, 4]	AA aligned/ % identity[5,6]
Rc-CrtA:			
Rc-CrtA	spheroidene monoxygenase	S17822[2], Crta_Rhoca[3]	241 / 100
Rc-CrtB:			
Rc-CrtB	phytoene synthase	S04403[2], Crtb_Rhoca[3]	339 / 100
Eh-CrtB (Eho13)	phytoene synthase	M90698_E[4]	285 / 34.7
Eu-CrtB	phytoene synthase	E37802[2]	285 / 34.7
Eh-CrtB (Eho10)	phytoene synthase	B39273[2]	301 / 30.6
Le-Psy1	phytoene synthase	A42102[2]	271 / 29.2
Le-pTOM5	phytoene synthase	S06321[2], Pto5_lyces[3], S21981[2]	271 / 28.4
Sy-CrtB (PCC 6803)	phytoene synthase	S31489[2], F22163[4]	281 / 27.8
Mx-ORF3	phytoene synthase	S32170[2], F25974c[4]	219 / 32.0
Sc-CrtB (PCC 7942)	phytoene synthase	S20383[2]	277 / 28.5
Rc-CrtC:			
Rc-CrtC	hydroxyneurosporene synthase	S04405[2], Crtc_Rhoca[3]	281 / 100
Rs-CrtC[7]	hydroxyneurosporene synthase	S23634[2], Crtc_Rhosh[3]	44 / 79.5
Mx-ORF5	hydroxyneurosporene synthase	S32172[2], F25974e[4]	278 / 28.4
Rc-CrtD:			
Rc-CrtD[8]	methoxyneurosporene dehydrogenase	S04406[2], Crtd_Rhoca[3]	494 / 100
Rs-CrtD	methoxyneurosporene dehydrogenase	S23633[2], Crtd_Rhosh[3]	482 / 55.6
Mx-ORF2	carotenoid dehydrogenase?	S32169[2], F25974b[4]	492 / 27.6
Eh-CrtI (Eho10)	phytoene dehydrogenase	A39273[2], Crti_Erwhe[3]	494 / 27.7
Rc-CrtI	phytoene dehydrogenase	S04402[2], Crti_Rhoca[3]	493 / 27.0
Mx-CarC	phytoene dehydrogenase	S27594[2]	502 / 26.3
Eh-CrtI (Eho13)	phytoene dehydrogenase	M90698_D[4]	493 / 25.8
Eu-CrtI	phytoene dehydrogenase	D37802[2], Crti_Erwur[3]	493 / 25.6
Mx-ORF4	hydroxyneurosporene dehydrogenase	S32171[2], F25974d[4]	492 / 29.5
Rc-CrtE:			
Rc-CrtE	GGPP synthase	S04407[2], Crte_Rhoca[3]	289 / 100
Ca-GGPPS	GGPP synthase	Ggpp_Capan[3]	253 / 35.6
Ec-IspA	FPP synthase	Jq0665[2], Ispa_Ecoli[3]	267 / 33.3
Eh-CrtE (Eho10)	GGPP synthase	C33120[2], Crte_Erwhe[3]	292 / 30.1
Eu-CrtE	GGPP synthase	A37802[2], Crte_Erwur[3]	277 / 27.8
Eh-CrtE (Eho13)	GGPP synthase	M90698_A[4]	277 / 27.4
Ec-ORFX	hypothetical protein, prenyl transferase?	Yhbd_Ecoli[3]	248 / 29.4
Bs-GerC3	spore germination protein	Grc3_Bacsu[3]	229 / 29.3
Cp-CrtE	GGPP synthase?	Crte_Cyapa[3]	240 / 30.4
Rc-CrtF:			
Rc-CrtF	hydroxyneurosporene-O-methyltransferase	S04408[2], Crtf_Rhoca[3]	393 / 100
Gg-HIOMT	hydroxyindole-O-methyltransferase	S21265[2]	305 / 24.9
Bpt-HIOMT	hydroxyindole-O-methyltransferase	A42106[2]	181 / 30.4
Sg-TcmO	8-O-methyltransferase	C42276[2]	330 / 24.5
Bpt-HIOMT	hydroxyindole-O-methyltransferase	A29565[2], Hiom_Bovin[3]	306 / 28.1
Sg-TcmN	cyclase-dehydratase-3-O-methyltransferase?	B42276[2]	332 / 22.0

Table 1. Continued

Rb. capsulatus : related protein[1]	Function of related protein	Database entries[2, 3, 4]	AA aligned/ % identity[5,6]
Rc-CrtI:			
Rc-CrtI	phytoene dehydrogenase	S04402[2], Crti_Rhoca[3]	524 / 100
Eh-CrtI (Eho10)	phytoene dehydrogenase	A39273[2], Crti_Erwhe[3]	494 / 40.9
Eu-CrtI	phytoene dehydrogenase	D37802[2], Crti_Erwur[3]	492 / 41.1
Eh-CrtI (Eho13)	phytoene dehydrogenase	M90698_D[4]	492 / 40.2
Mx-ORF2	carotenoid dehydrogenase?	S32169[2], F25974b[4]	502 / 31.1
Mx-CarC	phytoene dehydrogenase	S27594[2]	521 / 30.3
Nc-Al-1	phytoene dehydrogenase	A35919[2], Crti_Neucr[3]	511 / 32.3
Rc-CrtD[8]	methoxyneurosporene dehydrogenase	S04406[2], Crtd_Rhoca[3]	493 / 27.0
Rs-CrtD	methoxyneurosporene dehydrogenase	S23633[2], Crtd_Rhosh[3]	501 / 25.0
Mx-ORF4	hydroxyneurosporene dehydrogenase	S32171[2], F25974d[4]	276 / 28.3
Rc-ORF 469/Pps:			
Rc-ORF 469	regulatory protein?	S17813[2], Crtj_Rhoca[3]	469 / 100
Rc-ORF 160:			
Rc-ORF 160	?	S04404[2], Crtk_Rhoca[3]	160 / 100
Hs-PBR	peripheral-type benzodiazepine receptor	S14257[2]	128 / 37.5
Rn-PKBS/MBR	peripheral-type benzodiazepine receptor	S32680[2], Jc1393[2], Pkbs_Rat[3]	131 / 35.1
Bpt-PBR/IBP	peripheral-type benzodiazepine receptor	A39473[2], Pkbs_Bovine[3]	125 / 35.2

[1] Abbreviations are as follows: Bpt, *Bos primigenius taurus* (cattle); Bs, *Bacillus subtilis*; Ca, *Capsicum annuum* (bell pepper); Cp, *Cyanophora paradoxa*; Ec, *Escherichia coli*; Eh, *Erwinia herbicola*; Eu, *Erwinia uredovora*; Gg, *Gallus gallus* (chicken); Hs, *Homo sapiens* (human); Le, *Lycopersicon esculentum* (tomato); Mx, *Myxococcus xanthus*; Nc, *Neurospora crassa*; Rc, *Rhodobacter capsulatus*; Rn, *Rattus norvegicus* (Norway rat); Rs, *Rhodobacter sphaeroides*; Sc, *Synechococcus*; Sg, *Streptomyces glaucescens*; Sy, *Synechocystis*. Strains are given in parentheses.
[2] NBRF-PIR Data Library, release 37.0 (June 1993)
[3] SWISSPROT Data Library, release 25.0 (April 1993)
[4] MIPSX Data Library, release 36.0 (April 1993)
[5] AA = amino acids
[6] Sequence alignments were obtained using the FASTA computer program (Genetics Computer Group (GCG) software package (version 7.0; April, 1991)). Only sequences that yielded a length-dependent *opt* score of at least 10% of the maximum value at 100% identity are listed.
[7] deduced from a partial nucleotide sequence of the corresponding gene
[8] deduced from the nucleotide sequence of the *crtD223* mutant allele carried on plasmid pRPS404

capsulatus CrtE with the related proteins range between 26–36%.

Ashby and Edwards (1990) first reported the conservation of two aspartate-rich sequence motifs, termed domains I and II, in eukaryotic prenyltransferases that carry out 1′-4 condensations between homoallylic IPP and allylic prenylpyrophosphates (Fig. 1); only domain II was found in the yeast MOD5 protein, which modifies tRNA by addition of the allylic substrate DMAPP, suggesting that domain II might be the allylic binding site. Both the aspartates and arginine residues found in domain I are thought to be important for prenyltransferase substrate binding and/or catalysis (Ashby and Edwards, 1990; Carattoli et al., 1991; Kuntz et al., 1992; Joly and Edwards, 1993).

In addition to proteins with global sequence identity to *Rb. capsulatus* CrtB and CrtE (Table 1), other proteins exhibit local sequence conservation concentrated in aspartate-rich motifs resembling domains I and II (Table 2) (Armstrong et al., 1993a). *Rb. capsulatus* CrtB, which carries out a 1′-2-3 prenylpyrophosphate condensation of two molecules of GGPP and a 1′-1 rearrangement (Fig. 1), shares a domain II-related sequence not only with phytoene synthases from other organisms but also with squalene synthases from yeast, rat and human (Math et al., 1992; Robinson et al., 1993). Squalene synthase carries out a condensation and rearrangement reminiscent of phytoene synthase but uses two molecules of FPP to produce squalene, a key intermediate in eukaryotic sterol biosynthesis

Table 2. Conserved sequence motifs related to protein structure and function

Proposed function of the motif[1]	Consensus sequence[2]	Protein[3]	Location within protein[4]
FAD/NAD(P) binding $\beta\alpha\beta$ fold	(V/I)(V/I)G(A/G)GXGG(L/I)X$_2$AX$_2$(L/A)X$_3$GX$_2$ (V/T)X(V/L)X(E/D)X$_5$GG	Rc-CrtD[5]	9 - 41
		Rs-CrtD	10 - 42
		Rc-CrtI	13 - 45
Prenyl pyrophosphate binding (domain IIa)	(L/M)GX(A/F)XQX(T/S)NIXRDX$_2$(E/D)D	Rc-CrtB	156 - 172
Prenyl pyrophosphate binding (domain I)	(E/S)X$_2$(H/Q)X$_2$(S/F)LX$_2$DDX$_{2/4}$DX$_4$RRG	Rc-CrtE	70 - 93
Prenyl pyrophosphate binding (domain II)	GX$_2$(F/Y)QX$_2$DDX$_2$(D/N)	Rc-CrtE	201 - 212
SAM-binding	D(L/V)GG(G/A)XG(A/N)X$_8$(Y/H)PX$_6$(F/Y)D(L/I) PX(V/M)	Rc-CrtF	234 - 263

[1] Domains I and II denote regions found in prenyltransferases catalyzing 1′-4 condensations. Domain IIa is used here to denote a region found in prenyltransferases catalyzing 1′-2-3 condensations and 1′-1 rearrangements.
[2] Consensus sequences reflect the alignments listed in Table 1, modifying domains I and II as previously defined (Ashby and Edwards, 1990; Armstrong et al., 1993a); X = any amino acid.
[3] species abbreviations as in Table 1
[4] inclusive amino acid residues
[5] The Rc-CrtD sequence contains a mutation which changes Gly-13 in the wild type sequence to Arg-13 (see section III. A. 4.)

(Robinson et al., 1993). *Rb. capsulatus* CrtE, which catalyzes the 1′-4 condensation between homoallylic IPP and allylic FPP, and its homologs possess both domains I and II (Tables 1 and 2). Local sequence conservation of domains I and II from *Rb. capsulatus* CrtE is seen in the *N. crassa* GGPP synthase (Al-3) (Carattoli et al., 1991) and in several other eukaryotic prenyltransferases (Armstrong et al., 1993a), which otherwise exhibit little similarity to CrtE. The functional significance of the conserved aspartate and arginine residues found in domains I and II of *Rb. capsulatus* CrtE and CrtB remains to be demonstrated.

D. Hydratases (CrtC)

Virtually nothing is known about carotenoid hydratases. *Rb. capsulatus* CrtC, hydroxyneurosporene synthase (Fig. 1), shows 80% sequence identity with a partial CrtC sequence from *Rb. sphaeroides* (Armstrong et al., 1989; Garí et al., 1992b). In addition, a 28% identical *M. xanthus* protein is thought to encode the same enzyme (Table 1). No information is yet available on sequence motifs important for the structure or function of the carotenoid hydratases.

E. O-methyltransferases (CrtF)

Rb. capsulatus CrtF, the hydroxyneurosporene-O-methyltransferase, catalyzes the SAM-dependent methylation of hydroxyneurosporene or demethylspheroidene in a few species of bacteria (Goodwin, 1980). Recently, a number of noncarotenogenic O-methyltransferases that display sequence identities of 22–30% with CrtF have been described (Table 1). Putative SAM-binding sites found in RNA, DNA and other SAM-dependent methylases have been proposed (Haydock et al., 1991). The SAM-binding sequence motif is highly reminiscent of the ADP-binding fold found in FAD and NAD(P)-containing proteins (Scrutton et al., 1990), including the carotenoid dehydrogenases. A sequence motif which is strictly conserved and may represent a SAM-binding site in *Rb. capsulatus* CrtF and in other noncarotenogenic O-methyltransferases is given in Table 2. Biochemical evidence is not yet available to confirm the binding of SAM to CrtF.

F. ORF 469 and ORF 160

Rb. capsulatus ORF 469 and ORF 160 were originally thought to be involved in carotenoid biosynthesis

(Zsebo and Hearst, 1984; Giuliano et al., 1988; Armstrong et al., 1989; 1990c; Burke et al., 1991). Ongoing studies with *Rb. sphaeroides* and *Rb. capsulatus* indicate other functions for these gene products, however (see Sections III.B.2. and IV.A.2.) (Penfold and Pemberton, 1991, 1994; Gomelsky and Kaplan, 1995; Ponnampalam et al., 1995).

ORF 469/Pps is a negative regulator of BChl and carotenoid photopigment accumulation in *Rb. sphaeroides* (Penfold and Pemberton, 1991, 1994). Comparison of *Rb. capsulatus* and *Rb. sphaeroides* ORF 469/Pps to other protein sequences present in the databases reveals no homologs or global conservation of the amino acid sequence, but does identify one interesting sequence motif. Residues at the C-terminus of CrtJ/Pps resemble conserved helix-turn-helix DNA-binding motifs found in prokaryotic transcriptional regulatory proteins (Penfold and Pemberton, 1994). The exact function of ORF 469/Pps will require further study, but this identification of a possible DNA-binding motif could explain the apparent regulatory function in *Rb. sphaeroides* and *Rb. capsulatus* (Penfold and Pemberton, 1991, 1994; Gomelsky and Kaplan, 1995; Ponnampalam et al., 1995).

Intriguingly, *Rb. capsulatus* ORF 160 (Armstrong et al., 1989) has been observed to share significant sequence identity with mammalian benzodiazepine receptor proteins (Baker and Fanestil, 1991). The responses mediated by the mammalian receptors are diverse and not well understood. Nevertheless, their 35–38% sequence identities with ORF 160 (Table 1) are impressive considering the evolutionary gap between anoxygenic photosynthetic bacteria and mammals.

VI. Perspectives for the Future

A. Unsolved Problems in Carotenoid Biosynthesis

Recent advances in our understanding of the genetics of carotenoid biosynthesis in anoxygenic photosynthetic bacteria have been dramatic, but leave many questions unanswered. Our knowledge is confined to a few species of purple nonsulfur bacteria for which carotenoid mutants and systems for genetic transfer exist. Much is now known about the structure and organization of the *crt* genes and the properties of *crt* mutants in *Rb. capsulatus* and *Rb. sphaeroides*. On the other hand, detailed studies of the regulation of *crt* genes remain in their infancy, and, outside of *Rb. capsulatus* CrtI, no biochemical or immunological data exist with respect to the *crt* gene products. The functions of several specific biosynthetic conversions are unknown. *Rhodobacter* possess CrtA, an enzyme which catalyzes a specific conversion in the presence of oxygen, yet this system is inactive with no apparent ill effects in mutants lacking CrtD activity. By the same token, a selective advantage of wild type versus *crtD*-type carotenoids is not evident. The scope for groundbreaking research into these areas remains enormous.

Another specific issue which deserves more attention is the nature of the branch point between carotenoid and BChl biosynthesis in the general isoprenoid pathway. *Rb. capsulatus* CrtE and CrtB correspond to GGPP synthase and phytoene synthase, respectively (see Section III.A.3). This situation, however, poses a problem with respect to the synthesis of the phytyl side chain of the BChl molecule (Rüdiger and Schoch, 1991). If, as currently accepted, only one pathway exists for the synthesis of GGPP (Fig. 1), then *crtE* mutants of *Rb. capsulatus* should be blocked in both carotenoid and BChl synthesis, and hence should not have a blue-green phenotype, accumulate GGPP, or be photosynthetically viable (Armstrong et al., 1990c). As this is not the case, one could postulate that only leaky *crtE* mutants can been isolated, as has been reported for *al-3* (GGPP synthase) mutants of *N. crassa* (Sandmann et al., 1993). This possibility seems highly unlikely in *Rhodobacter*, however, when one considers that a gross disruption of the *crtE* gene with a transposon or interposon should completely abolish CrtE enzymatic activity (Zsebo and Hearst, 1984; Giuliano et al., 1988; Armstrong et al., 1990c; Coomber et al, 1990). A more plausible explanation for BChl synthesis in a *crtE* mutant background would be to postulate a branch point for BChl and carotenoid biosynthesis before GGPP, at the level of FPP for example (Fig. 1). This alternative would require either (1) the ability of farnesol to directly replace geranylgeraniol in the esterification of BChlide *a* or (2) the existence of a second GGPP synthase specific for BChl biosynthesis. Esterification of BChlide with alcohols not derived from GGPP can occur in some anoxygenic photosynthetic bacteria, but has not been observed in the BChl *a*-accumulating *Rhodospirillaceae*, which use phytol or less frequently geranylgeraniol for this purpose (Rüdiger and Schoch, 1991). At present it is not possible to decide between these alternatives. It is interesting in retrospect to note that the *Rb. capsulatus* *crtE* transposon mutant studied in an in vitro cell-free phytoene synthesis system (Armstrong et al., 1990c), accumulated substantially larger amounts of FPP than the *crtB, crtI* and *crtJ* blue-green mutants examined in parallel (A. Schmidt, personal communication).

B. Biotechnological Applications

Carotenoids are highly visible natural pigments and have a wide variety of commercial uses, such as food colorants, pigmenters for fish and poultry feed, and provitamin A (reviewed in Nelis and De Leenheer, 1989b). *Rb. capsulatus* has been studied as a potential source of the red ketocarotenoid spheroidenone for eventual use as a natural coloring agent. One strategy of biotechnological interest would be to grow *Rb. capsulatus* in a continuous-flow bioreactor for biomass or single cell protein production,while simultaneously enhancing carotenoid content. Preliminary experiments to investigate the potential for reliable characterization of the accumulated pigments have demonstrated that iron concentration in the growth medium can be used to manipulate spheroidenone content (Nelis and De Leenheer, 1989a, 1989b).

In light of recent developments in the genetics of anoxygenic photosynthetic bacterial carotenoid biosynthesis, manipulation of carotenoid levels or content through genetic engineering may be an attractive option. Heterologous systems for the production of novel carotenoids in a prokaryotic host organism by introduction of a foreign prokaryotic or eukaryotic carotenoid biosynthesis gene(s) have been described (Pemberton and Harding, 1986, 1987; Bartley et al., 1990, 1991, 1992; Misawa et al., 1990, 1991; Hundle et al., 1991; Penfold and Pemberton, 1991; Chamovitz et al., 1992; Math et al., 1992; Sandmann and Misawa, 1992). One could thus envision the construction of hybrid carotenoid biosynthetic pathways with anoxygenic photosynthetic bacteria serving either as hosts or gene donors to obtain large scale microbial production of desirable natural pigments.

Acknowledgments

I would like to thank those researchers who contributed manuscripts or unpublished data, Marie Alberti and Donald Burke for their advice, and Catharina Maulbecker-Armstrong for valuable consultations during the preparation of this chapter.

References

Armstrong GA (1994) Eubacteria show their true colors: Genetics of carotenoid pigment biosynthesis from microbes to plants. J Bacteriol 176: 4795–4802

Armstrong GA and Hearst JE (1989) Regulation of the carotenoid biosynthesis genes by oxygen and light in *Rhodobacter capsulatus*. Abstract #671, VIIIth Intl Congress on Photosynthesis. Physiol Plantarum 76: A123

Armstrong GA, Alberti M, Leach F and Hearst JE (1989) Nucleotide sequence, organization, and nature of the protein products of the carotenoid biosynthesis gene cluster of *Rhodobacter capsulatus*. Mol Gen Genet 216: 254–268

Armstrong GA, Alberti M and Hearst JE (1990a) Conserved enzymes mediate the early reactions of carotenoid biosynthesis in nonphotosynthetic and photosynthetic prokaryotes. Proc Natl Acad Sci USA 87: 9975–9979

Armstrong GA, Alberti M, Leach F and Hearst JE (1990b) Organization of the *Rhodobacter capsulatus* carotenoid biosynthesis gene cluster. In: Drews G and Dawes E (eds) Molecular Biology of Membrane-Bound Complexes in Phototrophic Bacteria, pp 39–46. Plenum Press, New York, New York

Armstrong GA, Schmidt A, Sandmann G and Hearst JE (1990c) Genetic and biochemical characterization of carotenoid biosynthesis mutants of *Rhodobacter capsulatus*. J Biol Chem 265: 8329–8338

Armstrong GA, Hundle B and Hearst JE (1993a) Evolutionary conservation and structural similarities of carotenoid biosynthesis gene products from photosynthetic and nonphotosynthetic organisms. Methods Enzymol 214: 297–311

Armstrong GA, Cook DN, Ma D, Alberti M, Burke DH and Hearst JE (1993b) Regulation of carotenoid and bacteriochlorophyll biosynthesis genes and identification of an evolutionarily conserved gene required for bacteriochlorophyll accumulation. J Gen Microbiol 139: 897–906

Ashby MN and Edwards PA (1990) Elucidation of the deficiency in two yeast coenzyme Q mutants. J Biol Chem 265: 13157–13164

Baker ME and Fanestil DD (1991) Mammalian peripheral-type benzodiazepine receptor is homologous to CrtK protein of *Rhodobacter capsulatus*, a photosynthetic bacterium. Cell 65: 721–722

Bartley GE and Scolnik PA (1989) Carotenoid biosynthesis in photosynthetic bacteria. J Biol Chem 264: 13109–13113

Bartley GE, Schmidhauser TJ, Yanofsky C and Scolnik PA (1990) Carotenoid desaturases from *Rhodobacter capsulatus* and *Neurospora crassa* are structurally and functionally conserved and contain domains homologous to flavoprotein disulfide oxidoreductases. J Biol Chem 265: 16020–16024

Bartley GE, Viitanen PV, Pecker I, Chamovitz D, Hirschberg J and Scolnik PA (1991) Molecular cloning and expression in photosynthetic bacteria of a soybean cDNA coding for phytoene desaturase, an enzyme of the carotenoid biosynthesis pathway. Proc Natl Acad Sci USA 88: 6532–6536

Bartley GE, Viitanen PV, Bacot KO and Scolnik PA (1992) A tomato gene expressed during fruit ripening encodes an enzyme of the carotenoid biosynthesis pathway. J Biol Chem 267: 5036–5039

Bauer CE, Bollivar DW and Suzuki JY (1993a) Genetic analysis of photopigment biosynthesis in eubacteria: a guiding light for algae and plants. J Bacteriol 175: 3919–3925

Bauer C, Buggy J and Mosley C (1993b) Control of photosystem genes in *Rhodobacter capsulatus*. Trends in Genet 9: 56–60

Belasco JG, Beatty JT, Adams CW, von Gabain A and Cohen SN (1985) Differential expression of photosynthesis genes in *R. capsulata* results from segmental differences in stability within the polycistronic *rxcA* transcript. Cell 40: 171–181

Biel AJ and Marrs BL (1983) Transcriptional regulation of several genes for bacteriochlorophyll biosynthesis in *Rhodopseudomonas capsulata* in response to oxygen. J Bacteriol 156: 686–694

Biel AJ and Marrs BL (1985) Oxygen does not directly regulate carotenoid biosynthesis in *Rhodopseudomonas capsulata*. J Bacteriol 162: 1320–1321

Bollivar DW, Suzuki JY, Beatty JT, Dobrowlski J and Bauer CE (1994) Directed mutational analysis of bacteriochlorophyll *a* biosyntheis in *Rhodobacter capsulatus*. J Mol Biol 237: 622–640

Bramley PM (1985) The in vitro biosynthesis of carotenoids. In: Paoletti R and Kritchensky D (eds) Advances in Lipid Research, Vol. 21, pp 243–279. Academic Press, Orlando, Florida

Bramley P, Teulieres C, Blain I, Bird C and Schuch W (1992) Biochemical characterization of transgenic tomato plants in which carotenoid synthesis has been inhibited through the expression of antisense RNA to pTOM5. The Plant J 2: 343–349

Burke DH, Alberti M, Armstrong GA and Hearst JE (1991) The complete nucleotide sequence of the 46 kb photosynthesis gene cluster of *Rhodobacter capsulatus*. EMBL Data Library, accession number Z11165

Burke DH, Alberti M and Hearst JE (1993) The *Rhodobacter capsulatus* chlorin reductase encoding locus, *bchA*, consists of three genes, *bchX, bchY,* and *bchZ*. J Bacteriol 175: 2407–2413

Carattoli A, Romano N, Ballario P, Morelli G and Macino G (1991) The *Neurospora crassa* carotenoid biosynthetic gene (albino 3) reveals highly conserved regions among prenyltransferases. J Biol Chem 266: 5854–5859

Chamovitz D, Misawa N, Sandmann G and Hirschberg J (1992) Molecular cloning and expression in *Escherichia coli* of a cyanobacterial gene coding for phytoene synthase, a carotenoid biosynthesis enzyme. FEBS Lett 296: 305–310

Clark WG, Davidson E and Marrs BL (1984) Variations of mRNA levels coding for antenna and reaction center polypeptides in *Rhodopseudomonas capsulata* in response to changes in oxygen concentration. J Bacteriol 157: 945–948

Cogdell RJ and Frank. HA (1987) How carotenoids function in photosynthetic bacteria. Biochim Biophys Acta 895: 63–79

Cohen-Bazire G, Sistrom WR and Stanier RY (1957) Kinetic studies of pigment synthesis by non-sulfur purple bacteria. J Cell Comp Physiol 49: 25–68

Cook DN, Armstrong GA and Hearst JE (1989) Induction of anaerobic gene expression in *Rhodobacter capsulatus* is not accompanied by a local change in chromosomal supercoiling as measure by a novel assay. J Bacteriol 171: 4836–4843

Coomber SA, Chaudri M, Connor A, Britton G and Hunter CN (1990) Localized transposon Tn5 mutagenesis of the photosynthetic gene cluster of *Rhodobacter sphaeroides*. Mol Microbiol 4: 977–989

Dierstein R (1983) Biosynthesis of pigment-protein complex polypeptides in bacteriochlorophyll-less mutant cells of *Rhodopseudomonas capsulata* YS. FEBS Lett 160: 281–286

Garí E, Gibert I and Barbé J (1992a) Spontaneous and reversible high-frequency frameshifts originating a phase transition in the carotenoid biosynthesis pathway of the phototrophic bacterium *Rhodobacter sphaeroides* 2.4.1. Mol Gen Genet 232: 74–80

Garí E, Toledo JC, Gibert I and Barbé J (1992b) Nucleotide sequence of the methoxyneurosporene dehydrogenase gene from *Rhodobacter sphaeroides*: comparison with other bacterial carotenoid dehydrogenases. FEMS Microbiol Lett 93: 103–108

Giuliano G, Pollock D and Scolnik PA (1986) The *crtI* gene mediates the conversion of phytoene into colored carotenoids in *Rhodopseudomonas capsulata*. J Biol Chem 261: 12925–12929

Giuliano G, Pollock D, Stapp H and Scolnik PA (1988) A genetic-physical map of the *Rhodobacter capsulatus* carotenoid biosynthesis gene cluster. Mol Gen Genet 213: 78–83

Gomelsky M and Kaplan S (1995) Genetic evidence that PpsR from *Rhodobacter sphaeroides* 2.4.1 functions as a repressor of *puc* and *bchF* expression. J Bacteriol 177: 1634–1637

Goodwin TW (1980) The Biochemistry of the Carotenoids, Volume 1: Plants. Chapman and Hall, New York, New York

Griffiths M and Stanier RY (1956) Some mutational changes in the photosynthetic pigment system of *Rhodopseudomonas spheroides*. J Gen Microbiol 14: 698–715

Griffiths M, Sistrom WR, Cohen-Bazire G and Stanier RY (1955) Function of carotenoids in photosynthesis. Nature 176: 1211–1214

Hawthornthwaite AM and Cogdell RJ (1991) Bacteriochlorophyll-binding proteins. In: Scheer H (ed) Chlorophylls, pp 493–528. CRC Press, Boca Raton, Florida

Haydock SF, Dowson JA, Dhillon N, Roberts GA, Cortes J and Leadlay PF (1991) Cloning and sequence analysis of genes involved in erythromycin biosynthesis in *Saccharopolyspora erythraea*: sequence similarities between EryG and a family of S-adenosylmethionine-dependent methyltransferases. Mol Gen Genet 230: 120–128

Hugueney P, Römer S and Camara B (1992) Characterization and molecular cloning of a flavoprotein catalyzing the synthesis of phytofluene and ζ-carotene in *Capsicum* chromoplasts. Eur J Biochem 209: 399–407

Hundle BS, Beyer P, Kleinig H, Englert G and Hearst JE (1991) Carotenoids of *Erwinia herbicola* and an *Escherichia coli* HB101 strain carrying the *Erwinia herbicola* carotenoid gene cluster. Photochem Photobiol 54: 89–93

Imhoff JF, Trüper HG and Pfennig N (1984) Rearrangements of the species and genera of the phototrophic "purple nonsulphur bacteria." Int J System Bacteriol 34: 340–343

Joly A and Edwards PA (1993) Effect of site-directed mutagenesis of conserved aspartate and arginine residues upon farnesyl diphosphate synthase activity. J Biol Chem 268: 26983–26989

Kaufmann N, Hüdig H and Drews G (1984) Transposon Tn5 mutagenesis of genes for the photosynthetic apparatus in *Rhodopseudomonas capsulata*. Mol Gen Genet 198: 153–158

Kiley PJ and Kaplan S (1988) Molecular genetics of photosynthetic membrane biosynthesis in *Rhodobacter sphaeroides*. Microbiol Rev 52: 50–69

Klug G, Kaufmann N and Drews G (1985) Gene expression of pigment-binding proteins of the bacterial photosynthetic apparatus: transcription and assembly in the membrane of *Rhodopseudomonas capsulata*. Proc Natl Acad Sci USA 82: 6485–6489

Koncz C, Mayerhofer R, Koncz-Kalman Z, Nawrath C, Reiss B, Redei GP and Schell J (1990) Isolation of a gene encoding a novel chloroplast protein by T-DNA tagging in *Arabidopsis thaliana*. EMBO J 9: 1337–1346

Koncz C, Németh K, Rédei GP and Schell J (1992) T-DNA insertional mutagenesis in *Arabidopsis*. Plant Mol Biol 20: 963–976

Kuntz M, Römer S, Suire C, Hugueney P, Weil JH, Schantz R and Camara B (1992) Identification of a cDNA for the plastid-located geranylgeranyl pyrophosphate synthase from *Capsicum annuum*: correlative increase in enzyme activity and transcript level during fruit ripening. The Plant Journal 2: 25–34

Lang HP, Cogdell RJ, Gardiner AT and Hunter CN (1994) Early steps in carotenoid biosynthesis: Sequences and transcriptional analysis of the *crtI* and *crtB* genes of *Rhodobacter sphaeroides* and overexpression and reactivation of *crtI* in *Escherichia coli* and *R. sphaeroides*. J Bacteriol 176: 3859–3869

Lascelles J and Wertlieb D (1971) Mutant strains of *Rhodopseudomonas spheroides* which form photosynthetic pigments aerobically in the dark: growth characteristics and enzymic activities. Biochim Biophys Acta 226: 328–340

Lee JK and Kaplan S (1992a) *cis*-acting regulatory elements involved in oxygen and light control of *puc* operon transcription in *Rhodobacter sphaeroides*. J Bacteriol 174: 1146–1157

Lee JK and Kaplan S (1992b) Isolation and characterization of *trans*-acting mutations involved in oxygen regulation of *puc* operon transcription in *Rhodobacter sphaeroides*, J Bacteriol 174: 1158–1171

Liaaen-Jensen S, Cohen-Bazire G, and Stanier RY (1961) Biosynthesis of carotenoids in purple bacteria: a re-evaluation based on considerations of chemical structure. Nature 192: 1168–1172

Ma D, Cook DN, O'Brien DA and Hearst JE (1993) Analysis of the promoter and regulatory sequences of an oxygen-regulated *bch* operon in *Rhodobacter capsulatus* by site-directed mutagenesis. J Bacteriol 175: 2037–2045

Marrs B (1974) Genetic recombination in *Rhodopseudomonas capsulata*. Proc Natl Acad Sci USA 71: 971–973

Marrs B (1981) Mobilization of the genes for photosynthesis from *Rhodopseudomonas capsulata* by a promiscuous plasmid. J Bacteriol 146: 1003–1012

Marrs B (1982) Genetic analysis of carotenogenesis in *Rhodopseudomonas capsulata*. In: Britton G and Goodwin TW (eds) Carotenoid Chemistry and Biochemistry, pp 273–277. Pergamon Press, New York, New York

Math SK, Hearst JE and Poulter CD (1992) The *crtE* gene in *Erwinia herbicola* encodes geranylgeranyl diphosphate synthase. Proc Natl Acad Sci USA 89: 6761–6764

McGlynn P and Hunter CN (1993) Genetic analysis of the *bchC* and *bchA* genes of *Rhodobacter sphaeroides*. Mol Gen Genet 236: 227–234

Misawa N, Nakagawa M, Kobayashi K, Yamano S, Izawa Y, Nakamura K and Harashima K (1990) Elucidation of the *Erwinia uredovora* carotenoid biosynthetic pathway by functional analysis of gene products expressed in *Escherichia coli*. J Bacteriol 172: 6704–6712

Misawa N, Yamano S and Ikenaga H (1991) Production of β-carotene in *Zymomonas mobilis* and *Agrobacterium tumifaciens* by introduction of the biosynthesis genes from *Erwinia uredovora*. Appl Environ Microbiol 57: 1847–1849

Nelis HJ and De Leenheer AP (1989a) Profiling and quantitation of bacterial carotenoids by liquid chromatography and photodiode array detection. Appl Environ Microbiol 55: 3065–3071

Nelis HJ and De Leenheer AP (1989b) Microbial production of carotenoids other than β-carotene. In: Vandamme EJ (ed) Biotechnology of Vitamins, Pigments, and Growth Factors, pp 43–80. Elsevier Applied Science, London, United Kingdom

Orsat B, Monfort A, Chatellar P and Stutz E (1992) Mapping and sequencing of an actively transcribed *Euglena gracilis* chloroplast gene (*ccsA*) homologous to the *Arabidopsis thaliana* nuclear gene *cs* (*ch-42*). FEBS Lett 303: 181–184

Ourisson G, Rohmer M and Poralla K (1987) Prokaryotic hopanoids and other polyterpenoid sterol surrogates. Annu Rev Microbiol 41: 301–333

Pecker I, Chamovitz D, Linden H, Sandmann G and Hirschberg J (1992) A single polypeptide catalyzing the conversion of phytoene to ζ-carotene is transcriptionally regulated during tomato fruit ripening. Proc Natl Acad Sci USA 89: 4962–4966

Pemberton JM and Bowen ARSG (1981) High frequency chromosome transfer in *Rhodopseudomonas sphaeroides* promoted by the broad host range plasmid RP1 carrying the mercury resistance transposon Tn*501*. J Bacteriol 147: 110–117

Pemberton JM and Harding CM (1986) Cloning of carotenoid biosynthesis genes from *Rhodopseudomonas sphaeroides*. Curr Microbiol 14: 25–29

Pemberton JM and Harding CM (1987) Expression of *Rhodopseudomonas sphaeroides* carotenoid photopigment genes in phylogenetically related nonphotosynthetic bacteria. Curr Microbiol 15: 67–71

Penfold RJ and Pemberton JM (1991) A gene from the photosynthetic gene cluster of *Rhodobacter sphaeroides* induces *trans* suppression of bacteriochlorophyll and carotenoid levels in *Rb. sphaeroides* and *Rb. capsulatus*. Curr Microbiol 23: 259–263

Penfold RJ and Pemberton JM (1994) Sequencing, chromosomal inactivation and functional expression in *E. coli* of *pps*, a gene which represses carotenoid and bacteriochlorophyll synthesis in *Rhodobacter sphaeroides*. J Bacteriol 176: 2869–2876

Ponnampalam SN, Buggy JJ and Bauer CE (1995) Characterization of an aerobic repressor that coordinately regulates bacteriochlorophyll, carotenoid and light harvesting-II expression in *Rhodobacter capsulatus*. J Bacteriol 177: 2990–2997

Ray J, Bird C, Maunders M, Grierson D and Schuch W (1987) Sequence of pTOM5, a ripening related cDNA from tomato. Nucl Acids Res 15: 10587

Robinson GW, Tsay YH, Kienzle BK, Smith-Monroy CA, Bishop RW (1993) Conservation between human and fungal squalene synthetases: similarities in structure, function, and regulation. Mol Cell Biol 13: 2706–2717

Rüdiger W and Schoch S (1991) The last steps of chlorophyll biosynthesis. In: Scheer H (ed) Chlorophylls, pp 451–464. CRC Press, Boca Raton, Florida

Sägesser R (1992) Identifikation und Characterisierung des

Photosynthese-Genclusters von *Rhodospirillum rubrum*. Ph. D. Thesis, Universität Zürich, Zürich, Switzerland
Sandmann G and Misawa N (1992) New functional assignment of the carotenogenic genes *crtB* and *crtE* with constructs of these genes from *Erwinia* species. FEMS Microbiol Lett 90: 253–258
Sandmann G, Misawa N, Wiedemann M, Vittorioso P, Carattoli A, Morelli G and Macino G (1993) Functional identification of *al-3* from *Neurospora crassa* as the gene for geranylgeranyl pyrophosphate synthase by complementation with *crt* genes, in vitro characterization of the gene product and mutant analysis. J Photochem Photobiol B Biol 18: 245–251
Schmidhauser TJ, Lauter FR, Russo VEA and Yanofsky C (1990) Cloning, sequence, and photoregulation of *al-1*, a carotenoid biosynthetic gene of *Neurospora crassa*. Mol Cell Biol 10: 5064–5070
Schmidt A, Sandmann G, Armstrong GA, Hearst JE and Böger P (1989) Immunological detection of phytoene desaturase in algae and higher plants using an antiserum raised against a bacterial fusion-gene construct. Eur J Biochem 184: 375–378
Schmidt K (1978) Biosynthesis of carotenoids. In: Clayton RK and Sistrom WR (eds) The Photosynthetic Bacteria, pp 729–750. Plenum Press, New York, New York
Scolnik PA and Marrs BL (1987) Genetic research with photosynthetic bacteria. Annu Rev Microbiol 41: 703–726
Scolnik PA, Walker MA and Marrs BL (1980) Biosynthesis of carotenoids derived from neurosporene in *Rhodopseudomonas capsulata*. J Biol Chem 255: 2427–2432
Scrutton NS, Berry A and Perham R (1990) Redesign of the coenzyme specificity of a dehydrogenase by protein engineering. Nature 343: 38–43
Sganga MW and Bauer CE (1992) Regulatory factors controlling photosynthetic reaction center and light-harvesting gene expression in *Rhodobacter capsulatus*. Cell 68: 945–954
Sganga MW, Aksamit RR, Cantoni GL and Bauer CE (1992) Mutational and nucleotide sequence analysis of S-adenosyl-L-homocysteine hydrolyase from *Rhodobacter capsulatus*. Proc Natl Acad Sci USA 89: 6328–6332
Sherman MM, Petersen LA and Poulter CD (1989) Isolation and characterization of isoprene mutants of *Escherichia coli*. J Bacteriol 171: 3619–3628
Sistrom WR (1977) Transfer of chromosomal genes mediated by plasmid R68.45 in *Rhodopseudomonas sphaeroides*. J Bacteriol 131: 526–532
Sistrom WR, Griffiths M and Stanier RY (1956) The biology of a photosynthetic bacterium which lacks colored carotenoids. J Cell Comp Physiol 48: 473–515
Taylor DP, Cohen SN, Clark WG and Marrs BL (1983) Alignment of the genetic and restriction maps of the photosynthesis region of the *Rhodopseudomonas capsulata* chromosome by a conjugation-mediated marker rescue technique. J Bacteriol 154: 580–590
Weaver PF, Wall JD and Gest H (1975) Characterization of *Rhodopseudomonas capsulata*. Arch Microbiol 105: 207–216
Wellington CL, Bauer CE and Beatty JT (1992) Photosynthesis gene superoperons in purple nonsulfur bacteria: the tip of the iceberg? Can J Microbiol 38: 20–27
Yen HC and Marrs B (1976) Map of genes for carotenoid and bacteriochlorophyll biosynthesis in *Rhodopseudomonas capsulata*. J Bacteriol 126: 619–629
Yildiz FH, Gest H and Bauer CE (1991) Genetic analysis of photosynthesis in *Rhodospirillum centenum*. J Bacteriol 173: 4163–4170
Yildiz FH, Gest H and Bauer CE (1992) Conservation of the photosynthesis gene cluster in *Rhodospirillum centenum*. Mol Microbiol 6: 2683–2691
Young DA, Bauer CE, Williams JC and Marrs BL (1989) Genetic evidence for superoperonal organization of genes for photosynthetic pigments and pigment-binding proteins in *Rhodobacter capsulatus*. Mol Gen Genet 218: 1–12
Young DA, Rudzik MB and Marrs BL (1992) An overlap between operons involved in carotenoid and bacteriochlorophyll biosynthesis in *Rhodobacter capsulatus*. FEMS Microbiol Lett 95: 213–218
Zhu YS and Hearst JE (1986) Regulation of the expression of the genes for light-harvesting antenna proteins LH I and LH II; reaction center polypeptides RC-L, RC-M, and RC-H; and enzymes of bacteriochlorophyll and carotenoid biosynthesis in *Rhodobacter capsulatus* by light and oxygen. Proc Natl Acad Sci USA 83: 7613–7617
Zhu YS, Cook DN, Leach F, Armstrong GA, Alberti M and Hearst JE (1986) Oxygen-regulated mRNAs for light-harvesting and reaction center complexes and for bacteriochlorophyll and carotenoid biosynthesis in *Rhodobacter capsulatus* during the shift from anaerobic to aerobic growth. J Bacteriol 168: 1180–1188
Zsebo KM (1984) Genetic-physical mapping of a photosynthetic gene cluster in *Rhodopseudomonas capsulata*. Ph. D. Thesis, University of California, Berkeley, California, USA
Zsebo KM and Hearst JE (1984) Genetic-physical mapping of a photosynthetic gene cluster from *R. capsulata*. Cell 37: 937–947

Chapter 54

A Foundation for the Genetic Analysis of Green Sulfur, Green Filamentous and Heliobacteria

Judith A. Shiozawa
Max Planck Institut für Biochemie, Am Klopferspitz 18a, 82152 Martinsried, Germany

Summary

At the present little is known on the genetics of green sulfur, green filamentous and heliobacteria. Since, based upon the 16S RNA phylogenetic tree, none of these bacteria are closely affiliated with the cyanobacteria or purple nonsulfur bacteria it is unlikely that preexisting genetic techniques from other photosynthetic bacteria can be used directly. However, due to the improvement in the methods available and the introduction of new techniques for molecular cloning and genetic manipulation of other prokaryotes, progress in these three bacterial groups should be relatively rapid. In this chapter, I have attempted to summarize all that is currently known on the properties of the nucleic acids and their genes. Potentially relevant physiological and biochemical

R. E. Blankenship, M. T. Madigan and C. E. Bauer (eds): Anoxygenic Photosynthetic Bacteria, pp. 1159–1173.

aspects of these bacteria have also been gathered and are presented here. Genes cloned and sequenced as of November, 1994, have been tabulated and the codon frequency based upon these data has been calculated for each of these bacterial groups. One cannot yet speak of the genetic analysis of these bacteria but rather the building of the foundation to do this type of work.

I. Some Relevant Physiological and Biochemical Characteristics

A. Species Used in the Laboratory and Culture Conditions

1. Green Sulfur Bacteria

Although several species of green sulfur bacteria have been isolated and characterized, only three have been used thus far for genetic and molecular cloning work: *Chlorobium vibrioforme, Chlorobium limicola,* and

. The thiosulfate-utilizing strains of *Cb. limicola* and *Cb. vibrioforme*, designated forma specialis (f. sp.) *thiosulfatophilum* (DSM 249, 254, 257 for the former strain and DSM 263 or NCIB 8327 for the latter strain) (Pfennig, 1989) are often used.

All *Chlorobium* species are obligate anaerobic phototrophs. They are usually grown in a thiosulfate-acetate-bicarbonate medium (Sirevåg and Ormerod, 1970, Schmidt, 1980, Rieble et al., 1989, Wahlund et al., 1991) at 23–28 °C with the exception of 44–48 °C for *Cb. tepidum*.

2. Green Filamentous Bacteria

The thermophilic facultative phototroph, *Chloroflexus aurantiacus*, is the only species of filamentous and green bacteria to be biochemically genetically studied to any great extent. It is the only species which can be grown as an axenic culture at the moment. Strains J-10-fl (ATCC 29366, DSM 635), OK-70-fl (ATCC 29365, DSM 636) and B3 have been used most frequently.

Cf. aurantiacus is usually grown in a complex medium containing salts, yeast extract and glycylglycine as a buffering agent (Castenholz and Pierson, 1981; Kaulen and Klemme, 1983; and the DSM medium No. 87). Glycylglycine, which is relatively expensive especially when large volumes of cells are to be grown, has been successfully replaced by 10 mM Tris (Bingham and Darbyshire, 1982; Cox et al., 1988; Rolstad et al., 1988) or 0.1% (w/w) K-phosphate (pH 8.0) (Ivanosky et al., 1993). Madigan et al. (1974) developed defined media for aerobic and anaerobic growth. The growth temperature is usually between 48 and 55 °C. Conditions for aerobic cultivation as well as for a high to low O_2 concentration shift of the culture to induce the biosynthesis of the photosynthetic apparatus have been described by Foster et al. (1986). Cell densities achieved under dark, aerobic growth conditions are very low; this is most probably due to the low solubility of O_2 in the growth medium at 50–55 °C.

The filamentous nature and gliding movement of *Cf. aurantiacus* presents problems for the isolation of colonies. Filaments can be broken up into shorter pieces by sonication but at the expense of cell viability (Pierson et al., 1984). Under anaerobic growth conditions, cells can be partially immobilized by embedding them in soft agar (Pierson et al., 1984). Another possibility, although untested, would be to plate cells onto nitrocellulose filters laid on top of the agar surface. This would allow growth of cells under aerobic conditions. In any case, target 'colonies' must be picked and restreaked a few times to ensure that they are monoclonal.

3. Heliobacteria

In biochemical and genetic work, *Heliobacillus mobilis* (ATCC 43427) has been used preferentially over the first described heliobacteria, *Heliobacterium chlorum* (ATCC 35205, DSM 3682), perhaps because it 'grows more robustly and shows a much lower tendency to sphaeroplast and lyse' (Beer-Romero et al., 1988). Two endospore-forming species, *Hb. gestii* and *Hb. fasciculum* have been described (Ormerod et al., 1990). All species are strict anaerobic phototrophs. *Hb. chlorum* can be maintained on ATCC medium No. 112 (DSM medium No. 370) under strict anaerobic conditions between 35–37 °C (Gest and Favinger, 1983); however, it has been reported to grow more rapidly on American Type Culture Collection (ATCC) medium No. 1552 (van

Abbreviations: ATCC – American Type Culture Collection; BChl – bacteriochlorophyll; *Cb. – Chlorobium; Cf. – Chloroflexus;* DSM – Deutsche Sammlung für Mikroorganismen; *Hb. – Heliobacterium; Hc. – Heliobacillus;* NTG – N-methyl-N'-nitro-N-nitrosoguanidine; ORF – open reading frame; *Rb. – Rhodobacter;* REMI – Restriction enzyme-mediated integration de Meent et al., 1990). *Hc. mobilis* can be grown under strict anaerobic conditions between 40 and 42° C using a defined medium developed by Beer-Romero and Gest (1987), ATCC medium No. 112 plus 1.1 g sodium pyruvate per liter (Beer-Romero et al., 1988) or ATCC medium No. 1552 (van de Meent et al., 1990).

B. Inhibition of Bacteriochlorophyll Synthesis

Green sulfur and green filamentous bacteria contain very large amounts of BChl *c, d* or *e* as light-harvesting pigments. The synthesis of BChl *d* and *e* can be specifically inhibited in *Cb. vibrioforme* with anesthetic gases (Ormerod et al., 1990). It has not be reported whether these gases have any effect on BChl *c* synthesis. Gabaculin (3-amino, 2,3-dihydrobenzoic acid) has been used to inhibit the biosynthesis of BChl *a* and *c* in *Cf. aurantiacus* (Kern and Klemme, 1989); this antibiotic was first used in cyanobacteria.

C. Antibiotic Sensitivities and Resistances

Pierson et al. (1984) tested photoheterotrophically grown *Cf. aurantiacus* J-10-fl for sensitivity to carbenicillin, penicillin and ampicillin. For concentrations up to 100μg ml^{-1} carbenicillin was observed to have no effect on cell growth. On the other hand, *Cf. aurantiacus* was shown to be very sensitive to penicillin in that cell recovery was not predictable after exposure to penicillin at 100U ml^{-1} under nongrowth conditions (anaerobic dark). The authors found that the best condition for growth inhibition and subsequent cell recovery was obtained using 30–50 μg ml^{-1} ampicillin (see also III.A).

From John Ormerod's work (1988) on the natural transformability of *Chlorobium* (see III.B) one can deduce that *Cb. limicola* is normally sensitive to streptomycin and kanamycin whereas *Cb. vibrioforme* is streptomycin sensitive (Kjærulff et al., 1993).

Hc. mobilis is apparently kanamycin sensitive (Vermaas and Vu, 1993; see Section III.B).

II. The Genome

A. General

The G + C mole % content of DNA from *Chlorobium* species used for molecular cloning work ranges from 51 to 58.1% (Pfennig, 1989; Imhoff, 1992). The reported values for *Cf. aurantiacus* are 53.1 to 54.9% and for *Hc. mobilis* is 50.3% (Pierson and Castenholz, 1974; Beer-Romero and Gest, 1987; Imhoff, 1992). The genome size and presence of special structures such as the two circular chromosomes found in *Rhodobacter sphaeroides* (Suwanto and Kaplan, 1989) have not been reported for green bacteria. The presence of endogenous plasmids and bacteriophages have also not been reported. Genome sizes of other photosynthetic bacteria and cyanobacteria range between 1.6 × 10^9 to 8.6 × 10^9 daltons (Saunders, 1992).

B. DNA Restriction and Modification Systems

Thus far the activity and/or isolation of specific endogenous restriction endonucleases has been reported for only *Cf. aurantiacus*. Two type II restriction enzymes, *Cau*I and *Cau*II, were isolated by Bingham and Darbyshire (1982). The recognition sites were determined by Molemans et al. (1982) to be identical to that of *Ava*II (GG A/T CC) and *Nci*I (CC C/G GG), respectively. A third restriction endonuclease, *Cau*-B3–1, was isolated and characterized more recently by Kramarov et al. (1987). The recognition sequence and cleavage site is identical to that of *Bsp*E1, T'CCGGA. This enzyme was also found to be sensitive to methylation of adenine in the recognition sequence. A fourth restriction enzyme, *Cau*III, with the recognition sequence CTGCAG has been reported (Kessler and Manta, 1990). Methylases responsible for modifying the cognate nucleotide sequences to prevent self-restriction have not been described.

Cf. aurantiacus apparently has a *dam*-like methylation system since *Sau*3AI will digest chromosomal DNA isolated from the bacterium whereas *Mbo*I will not (Shiozawa et al., 1989). Several other restriction enzymes apparently also do not restrict *Cf. aurantiacus* DNA even though the recognition sequence for these enzymes has been found in cloned and nucleotide sequenced DNA. These enzymes include *Bam*HI, *Eco*RV, *Hind*III, *Sac*I, *Sal*I and *Xba*I (Robinson and Redlinger, 1987; Shiozawa et al., 1989, unpublished; Dracheva et al., 1991). The apparent lack of activity of these enzymes could be due to DNA modification or to a low frequency of the respective recognition sites. There are two arguments for the latter case: For mammalian DNA, *Sal*I is considered to be a rare cutting restriction enzyme with an average fragment length of 100,000 bp (see Sambrook et al., 1989). *Xba*I has been used to map the chromosomal DNA of *Rb. capsulatus* since this enzyme acts infrequently in this species (Fonstein and Haselkorn, 1993). The enzyme *Dra*III has been used to map *Rb. sphaeroides* DNA (Suwanto and Kaplan, 1989). No recognition sites for this enzyme has been found in over 15,000 bp sequenced in *Cf. aurantiacus* and in over 13,000 bp sequenced in *Chlorobium*. One complicating factor is that *Xba*I is *dam* methylation sensitive. *Dam* methylation of the *Xba*I recognition sequence (TCTAGA) occurs when the dinucleotide TC immediately follows the restriction site.

C. Genes Cloned and Sequenced

The genes cloned and sequenced as of November, 1994, are presented in Table 1 for *Chlorobium* species, in Table 2 for *Cf. aurantiacus* and in Table 3 for heliobacteria. Where available, the Genbank/EMBL data bank accession numbers are also given. In some cases, the sequence data was neither published nor available from the data bases; this is also indicated in the tables. Also indicated in the tables is whether the transcriptional organization of the genes has been analyzed.

The gene designations suggested at the EMBO Workshop on Green and Heliobacteria in Denmark August 1993 have been adopted (Bryant, 1994). An additional modification was made: The two loci of the *puf* operon in *Cf. aurantiacus* and the *psc* in *Chlorobium* species are differentiated by an Arabic number immediately following the gene locus designation as has been done for the *atp1* and *atp2* operons in cyanobacteria. This differential designation was also going to be applied to the chlorosome genes of *Chlorobium* and *Chloroflexus* but there are already three *csm* gene loci of which only one seems to occur in both bacterial groups. As more chlorosome genes are identified and sequenced, the present nomenclature system will quickly become overly complicated. As soon as functional properties can be definitively associated with these gene products, I suggest that new gene locus designations be developed.

D. Gene and Transcriptional Organization

1. Chloroflexus aurantiacus

At the moment the best characterized genes in these three bacterial groups are those of the *puf1* and *puf2* operons and the *csmA* gene of *Cf. aurantiacus*. The organization of the *Cf. aurantiacus puf* operon differs greatly from that of purple nonsulfur bacteria. Whereas the *puf* operon in purple nonsulfur bacteria forms a contiguous region of the genome and is transcribed as a single mRNA before it is processed, in *Cf. aurantiacus* the *puf* operon is divided between two loci which are separated by at least 3 kb. *Puf1* encodes the L and M subunits of the photochemical reaction center while *puf2* encodes the β and α subunits of the B806–866 light-harvesting complex and the tetraheme cytochrome *c*-554 (Shiozawa et al., 1990; Watanabe et al., 1992, 1995). How the molar relationships between the reaction center and the light-harvesting complex and between the reaction center and cytochrome *c*-554 are regulated in vivo is not known. In photoheterotrophically grown *Rb. capsulatus* differential mRNA stability is probably the major factor in maintaining the molar ratio of the *pufB* and *A* gene products to the *pufL* and *M* gene products (Klug, 1993).

The 5' end of the *puf1* mRNA begins about 130 nt upstream from the ATG start codon of *puf1L*. In the DNA region just upstream of the region corresponding to the 5' end of the *puf1* mRNA are nucleotide sequences with very high similarity to the *E. coli* RNA polymerase σ^{70} consensus sequence. In addition, an octanucleotide sequence is repeated three times and there are two regions of inverted symmetry upstream of the putative promoter region which could be regulatory protein binding sites. The electrophoretic mobility of radioactively labeled DNA fragments from this potential promoter-regulatory region after incubation with cellular extracts of *Cf. aurantiacus* is reduced, indicating DNA-protein interaction(s) (Shiozawa et al., 1990 and unpublished). The identity of the protein(s) interacting with the DNA probes is not known.

Puf2 may actually represent two semi-independent operons: one composed of *puf2B* and *puf2A*, the genes encoding the subunits of the B806–866 light-harvesting complex, and the second composed of *puf2C*, the gene encoding the tetraheme cytochrome *c*-554 (Watanabe et al., 1992, 1995). A putative σ^{70} promoter sequence precedes both *puf2B* (centered –149 bp from the start codon) and *puf2C* (centered –119 bp from the start codon). The predominant mRNA encoding *puf2B* and *puf2A* is 0.5 kb long; however, a minor transcript 2.1 kb long and encoding both genes is also observed. For *puf2C* two transcripts are observed: the 1.5 kb mRNA encodes only *puf2C*

Table 1. Compilation of genes sequenced in green sulfur bacteria

Species Gene Product	Gene Name	Accession Number	mRNA[1]	References
Cb. vibrioforme f. sp. *thiosulfatophilum* NCIB 8327				
Glutamyl-tRNA reductase	*hemA*	M59194	+	Majumdar et al. (1991)
Porphobilinogen deaminase	*hemC*	"		"
Uroporphyrinogen III cosynthase	*hemD*	M96364		Majumdar and Wyche (unpublished)
Malate dehydrogenase		X80837		Naterstad and Sirevåg (unpublished)[2]
Ribosomal protein rpl21		X79218		Naterstad and Sirevåg (unpublished)[2]
Ribosomal protein rpl27		X79219		Naterstad and Sirevåg (unpublished)[2]
Cb. vibrioforme f.*thiosulfatophilum* 8327				
Cytochrome *c*-551	*psc2C*	M95751	+	Okkels et al. (1992)
Cb. vibrioforme strain 8327D				
Chlorosome 14 and 6.3 kDa proteins	*csmC, A*	U09866		Chung and Bryant (1992, 1993)
Olson-Gerola 7.5 kDa protein	*csmB*			Chung and Bryant (1993)[2]
Soluble 2[4Fe-4S] ferredoxins	*fdxB, A*			Chung and Bryant (1993)[2]
Cb. vibrioforme				
16S rRNA		M27804 M62791		Woese et al. (1990a)
Cb. limicola f. sp. *thiosulfatophium*				
Reaction center polypeptide,	*pscIA*	M94675		Büttner et al. (1992)
Fe-S protein	*pscIB*	"		
Dolichyl phosphate-D-mannose synthase	*dpmI*	"		
Rieske Fe-S centers of bc_1 complex	*petC*			Zirngibl et al. (1992)[2]
F-ATPase β and ε-subunits	*atp2*β, ε	L08777		Xie et al. (1993)
Phosphoenol pyruvate carboxykinase	*pepck*	S56812		"
Cb. limicola				
16S rRNA		M31769		Woese, unpublished
23S rRNA		M31904 M62805		Woese et al. (1990b)
RNA component of RNase P	*rnpB*	L25703		Haas et al. (1994)
Cb. tepidum				
FMO protein	*fmoA*			Dracheva et al. (1992)
Chlorosome 14 and 6.3 kDa proteins	*csmC, A*	U09867	+	Chung and Bryant (1992, 1993)
Olson-Gerola 7.5 kDa protein	*csmB*			Chung and Bryant (1993)[2]
Nitrogenase	*nifH*			Madigan et al. (1993)[2]
Malate dehydrogenase		X80838		Naterstad and Sirevåg (unpublished)[2]
RNA component of RNase P	*rnpB*	L25704		Haas et al. (1994)
Cb. tepidum strain TLS				
16S rRNA		M58468		Wahlund et al. (1992)
Cb. phaeobacteroides DSM 266				
5S rRNA				Van den Eynde et al. (1990)
Clathrochloris sulfurica strain 1				
16S rRNA				Witt et al. (1989)

[1] Transcriptional organization examined

[2] Nucleic acid sequence not available from data bank and/or was not published as of November, 1994.

Table 2. Compilation of genes sequenced in *Chloroflexus aurantiacus* J10-fl

Gene Product	Gene Name	Accession Number	mRNA[1]	References
Reaction center L and M polypeptides	*puf1L, M*	X14979		Ovchinnikov et al. (1988a,b)
		Y00972		Shiozawa et al. (1989)
			+	Shiozawa et al. (1990)
Chlorosome 5.7 kDa protein	*csmA*	M33964	+	Theroux et al. (1990)
Tetraheme cytochrome *c*-554	*puf2C*	M77813		Dracheva et al. (1991)
			+	Watanabe et al. (1992, 1995)
B806–866 a and β polypeptides	*puf2B, A*	X73899	+	Watanabe et al. (1992, 1995)
Chlorosome 11 and 18 kDa proteins	*csmM, N*	Z34000	+	Niedermeier et al. (1994)
16S rRNA		M34116		Oyaizu et al. (1989)

Footnotes: Same as in Table 1.

Table 3. Compilation of genes sequenced in heliobacteria

Species Gene Product	Gene Name	Accession Number	mRNA[1]	References
Heliobacillus mobilis				
Reaction center polypeptide	*pshA*	L19604	+	Liebl et al. (1993)
Heliobacterium chlorum				
16S rRNA		M11212		Woese et al. (1985)

Footnotes: Same as in Table 1.

while the 2.1 kb mRNA encodes all three *puf2* genes. Primer extension experiments show that the 1.5 kb mRNA was actually composed of two species: one having a 5' terminus just downstream of the putative promoter sequence and the second, 88 nt longer, with its terminus in the potential stemloop region immediately following the stop codon of *puf2A*. There are two possible explanations for these results: All of *puf2C* is transcribed as a 2.1 kb message which later is subject to endonuclease processing. The original endonucleolytic products would be the 0.5 kb *puf2B-puf2A* mRNA and the longer 1.5 kb mRNA. Subsequent endo- or exonucleolytic activity would yield the slightly shorter 1.5 kb product. A second possible explanation is that some *puf2C* is transcribed as the 2.1 kb message, some of which is later processed to the 1.5 kb mRNA having the 5' terminus in the putative stemloop region of *puf2A*. The second shorter 1.5 kb mRNA results from the transcription of *puf2C* from the putative promoter just upstream of the mapped 5' terminus. The banding pattern from primer extension experiments support this latter explanation since no extension products between the two termini are observed (Watanabe et al., 1992, 1995). Capping experiments in conjunction with nuclease mapping would help to clarify the origin of the two 1.5 kb mRNA.

In *Cf. aurantiacus*, the 5.7 kD chlorosome protein which is encoded by the *csmA* gene is transcribed as a monocistronic message. Just upstream of the 5' end of the mRNA which was determined by primer extension are nucleotide sequences which have similarity to the alternate σ^{54} consensus sequence (Theroux et al., 1990). Two potential open reading frames (ORFs) are found upstream and downstream of *csmA* which have sequence similarity to two similarly positioned ORFs in *Cm. vibrioforme 8327D* and *Cb. tepidum* (Chung and Bryant, 1992; Chung et al., 1994; see the following section).

Two other chlorosome genes have been cloned and sequenced from *Cf. aurantiacus* recently (Niedermeier et al., 1994). The genes for the 11 and 18 kD proteins have been designated *csmM* and *csmN*, respectively, and are not in the direct vicinity of the *csmA* gene. The two genes are transcribed into one mRNA having the approximate size of their combined gene lengths. About 185 bp upstream of the ATG start codon of the *csmM* gene are nucleotide sequences resembling the σ^{70} consensus sequence.

Nucleotide sequences which could form stemloops are almost always found downstream of all stop codons of the putative final gene of a transcript in *Chloroflexus*. In *pufl* the G+C-rich region of dyad symmetry and the dA-rich region (on the DNA template strand for mRNA synthesis) is typical of rho-independent termination. A region of dyad symmetry is also found immediately following the stop codon of the cytochrome *c*-554 gene (*puf2C*) but the only dA-rich sequence proximal to this region is part of the putative stemloop or is 30 bp further downstream. Thus, it is not clear whether this is a rho-independent termination site.

The region of dyad symmetry between *puf2A* (α-subunit of the B806–866 complex) and *puf2C* lacks the poly A region on the template strand. Intercistronic stemloops have been observed, for example, in the *puf* operon of *Rb. capsulatus* and *sphaeroides* and have been shown to contribute to the stability of the message 5' to the stemloop (Belasco and Chen, 1988; Klug, 1993). The intercistronic stemloop of *puf2* may also have a similar function but this must be proven. The termination signal/mechanism of the *csmA* and *csmM-csmN* transcripts are not clear. In the former case, a region of strong dyad symmetry is present, but a dA-rich sequence is absent. In the case of the *csmM-csmN* operon only a region of weak dyad symmetry is found 3' of the *csmN* stop codon. Whether transcription in both could be terminated by a rho-like protein factor is also difficult to say. There seems to be no consensus sequence or structural features common to all rho-dependent termination sites (Beebee and Burke, 1992; Richardson, 1993). Most recently it has been suggested that a relatively G-poor and C-rich region on the transcript is required for rho binding (see Richardson, 1993). While there are nucleotide regions 50 to 100 bp upstream of the stop codon which may be described as such, whether these have a functional significance is conjecture at the moment.

2. Chlorobium *Species*

The first gene sequenced in a *Chlorobium* species was *hemA* which encodes glutamyl tRNA reductase, the first enzyme of the five carbon δ-amino levulinic acid biosynthetic pathway (Majumdar et al., 1991). In the same paper, data for the 5' end of the gene downstream, of *hemA* was also published. The deduced amino acid sequence was 42 and 46% identical to that of porphobilinogen deaminase from *E. coli* and *B. subtilis*, respectively. Subsequently, this gene, *hemC*, and *hemD* (uroporphyrinogen II cosynthase) were completely sequenced (Majumdar and Wyche; see Table 1). Based upon the size of the mRNA as seen on Northern blots, all three genes are transcribed together and form an operon (Majumdar et al., 1991). The short intercistronic space between *hemA* and *hemC*, 76 bp, and the 1 bp overlap between *hemC* and *hemD* supports this idea. A putative Pribnow box is found just 6 bp upstream of the start codon of *hemA*; two possible sequences which could correspond to the –35 consensus sequence of the RNA polymerase σ^{70} subunit were found 20 and 21 bp upstream of the Pribnow box. It is also possible that the true promoter lays further upstream and beyond the cloned DNA fragment.

The region 3' of the termination codon of each of three genes has some interesting characteristics, although the significance or function of these features is not known presently. There are stretches of DNA that are extremely dA+dT-rich, especially immediately after *hemA* and *hemD*. Twenty bp downstream of the *hemA* termination codon is a potential polyadenylation signal (Majumdar et al., 1991). Polyadenylation is typical of the 3' terminus of eukaryotic mRNA and has also been observed in other bacteria although this is the first time a signal sequence in the genomic DNA has been described. Another unusual feature of this operon is that a region of very strong dyad symmetry is centered about 90 bp into the *hemD* gene. It is not known whether a transcription termination signal occurs after the stop codon of *hemD*; the DNA fragment sequenced ends about 25 bp after the stop codon.

The genes encoding the reaction center polypeptide and the Fe-S binding polypeptide form an operon in *Cb. limicola* which has been designated *pscl* (photosystem *Chlorobium*-type; Bryant, 1994) and the genes, *A* and *B*, respectively (Büttner et al., 1992). The genes are preceded by a putative σ^{70} promoter sequence centered about 320 bp upstream of the ATG start codon. Beginning 31 bp downstream of the *psc1B* stop codon is a region of strong dyad symmetry with a interesting motif which is found in other *Chlorobium* genes: a run of dA begins the potential stemloop and a run of dT terminates and continues beyond the structure. Such termination stemloops have been observed in other prokaryotic operons and, in some cases, have been shown to be bi-directional termination sites (Platt, 1986). A third ORF which showed sequence identity to the yeast enzyme dolichyl-phosphate-D-mannose synthase (EC 2.4.1.83, Genbank/EBML data bank accession no. A32122) was found on the cloned 4.8 kb *Eag*I fragment. This ORF is located 303 bp downstream of the Fe-S binding protein compared to the 85 bp separating *psc1A* and *psc1B* and is transcribed in the same direction as but independently of the *psc1* genes (Büttner et al., 1992). A possible promoter sequence is found 68 bp upstream of the ATG start codon of *dpm1* and there is a region of dyad symmetry which encompasses the 3' terminus of the gene.

The β- and ε-subunits of F-ATPase in *Cb. limicola* which have been designated *atp2*β and *atp2*ε most likely form an operon as in the cyanobacteria (Xie et al., 1993). The authors did not specifically mention the location of possible promoter sequences; the closest σ^{70} promoter-like sequences I could find were in the noncoding region preceding *atp2*β and centered around –315 bp from the start codon. Twenty-two bp downstream of the *atp2*ε stop codon is a region of interrupted dyad symmetry. A run of dA begins and a run of dT ends this potential transcription termination structure (see preceding paragraph).

Coincidentally the gene encoding the enzyme phosphoenol pyruvate carboxykinase also lies on the cloned 4.5 kB *Bam*HI fragment. The gene, *pepck*, is almost 1 kb upstream of *atp2*β and is transcribed in the opposite direction. A potential σ^{70} promoter sequence is located between bp –38 and–67 from the start codon and the nucleotide sequences for a potential termination stemloop with a run of dA and dT forming the 3' and 5' termini are found 26 bp downstream of the stop codon.

The gene homologous to *csmA* in *Cf. aurantiacus* has been identified, cloned and sequenced in two *Chlorobium* species (Chung and Bryant, 1992, 1993). In both *Cb. tepidum* and *Cb. vibrioforme*, the *csmA* gene is preceded by a gene, now called *csmC*, which encodes the 14 kDa protein found in *Chlorobium* chlorosomes. In *Cb. tepidum csmCA* is transcribed as a single albeit not abundant transcript. A more abundant transcript which encodes only *csmA* is also found. Nucleotide sequences resembling the *E. coli* σ^{70} consensus sequence are found upstream of the 5' end of the *csmCA* mRNA. Surrounding the *csmCA* operon are conserved ORFs which also appear to be related to ORFs upstream and downstream of the *csmA* gene in *Cf. aurantiacus*.

The gene encoding the Olson-Gerola 7.5-kDa chlorosome protein in *Cb. vibrioforme* and *Cb. tepidum* has also been cloned and sequenced (Chung and Bryant, 1992, 1993). The *csmB* gene appears to be transcribed as a monocistron. An analogous protein in *Cf. aurantiacus* has not yet been identified. The *csmCA* and *csmB* represent two different loci on the genomic DNA of *Chlorobium* species. The distance separating these two loci is presently not known.

Psc2C which encodes the monoheme cytochrome *c*-551 of *Cb. vibrioforme* was identified in the genomic DNA, cloned and sequenced (Okkels et al., 1992). Centered 52 bp upstream of the start codon of the gene are nucleotide sequences resembling the consensus sequence of the σ^{70} subunit of the RNA polymerase and 45 bp downstream of the stop codon a region having the properties of a rho-independent termination site are found. Northern blot analysis also support the idea that the *psc2C* gene is transcribed as a monocistron.

The gene encoding the Fenna-Matthews-Olson protein, the polypeptide component of the water soluble BChl-protein of *Chlorobium* species, is apparently also transcribed as a monocistron in *Cb. tepidum* (Dracheva et al., 1992). This belief is based upon the presence of sequences having similarity to the RNA polymerase σ^{70} promoter sequence upstream of the start codon of *fmoA* (nt 95–123 in Dracheva et al., 1992) and the presence of a region of dyad symmetry with poly (dA) and poly (dT) beginning and ending this region just downstream of the gene stop codon.

3. Heliobacillus mobilis

Thus far, the only gene cloned and sequenced from heliobacteria is *pshA* (photosystem heliobacteria-type; Bryant, 1994), the gene encoding the reaction center polypeptide of *Hc. mobilis* (Liebl et al., 1993). The published nucleotide sequence begins –14 bp from the *pshA* start codon; the nucleotide sequence submitted to the GenBank/EMBL data bank (accession no. L19604) began over 700 bp upstream. Therefore, a visual search for promoter sequences was made. A potential Pribnow box starting –91 bp from the start codon was found; however, a reasonable –35 consensus sequence was not found 16 to 20 bases further upstream. Another possible promoter is centered −145 bp from the ATG start codon; here potential −24 and −12 sequences for a σ^{54} promoter are found. A potential G+C–rich stemloop followed by a dT-rich stretch is found starting 26 bp downstream of the stop codon. An ORF begins 147 bp downstream of *pshA*. A Pribnow box is found 38 bp upstream and, again, a −35 consensus sequence could not be identified. Beginning 28 bp downstream from the stop codon a region of dyad symmetry is found; this region is not followed by a dT-rich region. Northern blot analysis suggest that the message for *pshA* is monocistronic.

E. Translation Signals

A ribosome-binding site or Shine-Dalgarno sequence is found 8 to 12 bp upstream of the start codon of all *Cf. aurantiacus* genes thus far sequenced. The *pshA* and downstream ORF from *Hc. mobilis* were also preceded by a ribosome-binding site. In *Chlorobium* species, Shine-Dalgarno sequences just upstream or just following the start codon are seen in the *psc1* operon and in the *dpm1* and *fmoA* genes. On the other hand, ribosome-binding sites in these more typical locations have not been found in the *hemACD* and the *atp2βε* operons and in the *psc2C* and *pepck* genes. In the GenBank/EMBL submission of the complete *hemACD* sequence (accession no. M96364) Majumdar and Wyche suggest that the ribosome-binding sites lie within the genes. This has been observed in other bacteria, for instance the archaea (Zillig et al., 1988). I have found possible ribosome-binding sequences within 70 bp of the start codons of the *atp2βε* operon, but was unsuccessful in my search of the *psc2C* and *pepck* genes.

F. Codon frequency

The codon frequency for each of the three bacterial groups was calculated. The data are presented in Tables 4 through 6 and demonstrate that the codon frequencies do not parallel that of *E. coli, Rhodobacter* (Wu and Saier, Jr., 1991) or one another. The frequency for which a degenerate codon having a G or C as the third base was used over the other codons was also determined. As found previously for other prokaryotes and eukaryotes, the G/C content of the third base in a codon correlated with the G/C composition of the genome (D'Onofrio and Bernardi, 1992). Thus, *Chlorobium* species had the highest G/C frequency for the third codon position (0.70), followed by *Cf. aurantiacus* (0.66) and last by *Hc. mobilis* (0.60).

G. Summary

Almost of the genes or operons studied thus far from these three groups of bacteria have nucleotide sequences resembling the consensus motif of the σ^{70} subunit of *E. coli* RNA polymerase upstream of the start codons. Whether these putative promoter sequences are functional and the location and identity of *cis*-acting regulatory sequences, where applicable, must be demonstrated.

In most cases it appears that transcription is terminated by rho (like)-independent stemloop structures. Several of the *Chlorobium* genes have an additional characteristic—the stems of the putative termination stemloops begin with a 6–7 nt long stretch of poly (dA)·poly (dT). In *hemA* a poly-adenylation signal was found at the 3' end of the gene.

Chlorobium genes are unusual in another aspect in that about half of the genes now known have ribosome-binding sites well after the translation start codon. While this phenomenon has been observed in other bacteria, it is not always clear what the function or advantage of such binding sites are.

It is too early to say whether the genes required for photosynthesis in green sulfur, green filamentous or heliobacteria are organized into a cluster such as seen in purple nonsulfur bacteria. There does seem to be a tendency for genes that are functionally related to be grouped together into an operon. An exception here seems to be *Cf. aurantiacus* and its divided *puf* operon. The separation of the genes encoding the L and M subunits of the reaction center from the light-harvesting component to which the reaction center is most tightly coupled and the cytochrome responsible for re-reducing the oxidized special BChl pair may be an indication that *Chloroflexus* or its predecessor may have received these genes by lateral gene transfer from a purple nonsulfur bacterium or its predecessor. Blankenship (1992) has suggested that *Cf. aurantiacus* is a photosynthetic chimera. Whether this organism also received its light-harvesting apparatus and BChl synthesis genes from a green sulfur bacterium in the same manner is too early to say.

Table 4. Codon frequency in *Chlorobium* species genes calculated from available nucleic acid sequences and their translations (see Table 1)

Amino Acid	Codon	%	Pref[1]	Amino Acid	Codon	%	Pref	Amino Acid	Codon	%	Pref	Amino Acid	Codon	%	Pref
Gly	GGG	7		Arg	AGG	17		Met	ATG	100	ATG	Leu	TTG	6	
	GGA	14	GGN		AGA	9	AGR	Ile	ATA	2			TTA	1	
	GGT	31			CGG	5	CGN		ATT	26	ATY		CTG	33	CTB
	GGC	48			CGA	7			ATC	71			CTA	1	
Glu	GAG	47	GAR		CGT	23		Thr	ACG	20			CTT	25	
	GAA	53			CGC	39			ACA	12	ACN		CTC	35	
Asp	GAT	46	GAY	Ser	AGT	5			ACT	6		Phe	TTT	24	TTY
	GAC	54			AGC	23	AGY		ACC	62			TTC	76	
Val	GTG	23			TCG	30	TCN	Trp	TGG	100	TGG	Gln	CAG	87	CAR
	GTA	8	GTN		TCA	11		End	TGA	40			CAA	13	
	GTT	29			TCT	10			TAG	13	TRR	His	CAT	35	CAY
	GTC	40			TCC	21			TAA	47			CAC	65	
Ala	GCG	18		Lys	AAG	59	AAR	Cys	TGT	16	TGY	Pro	CCG	48	CCN
	GCA	21	GCN		AAA	41			TGC	84			CCA	13	
	GCT	21		Asn	AAT	24	AAY	Tyr	TAT	33	TAY		CCT	17	
	GCC	40			AAC	76			TAC	67			CCC	21	

[1] Codon preference: R = A/G; Y = C/T; M = A/C; S = G/C; W = A/T; B = C/G/T; V = A/G/C; N = A/C/G/T.
Codon useages <5% were excluded in codon preference.

Table 5. Codon frequency in *Chloroflexus aurantiacus* genes calculated from available nucleic acid sequences and their translations—see Table 2.

Amino Acid	Codon	%	Pref[1]	Amino Acid	Codon	%	Pref	Amino Acid	Codon	%	Pref	Amino Acid	Codon	%	Pref
Gly	GGG	15		Arg	AGG	0		Met	ATG	100	ATG	Leu	TTG	14	
	GGA	12	GGN		AGA	3		Ile	ATA	2			TTA	5	TTR
	GGT	36			CGG	21	CGN		ATT	46	ATY		CTG	46	CTB
	GGC	38			CGA	8			ATC	52			CTA	2	
Glu	GAG	55	GAR		CGT	28		Thr	ACG	32			CTT	8	
	GAA	45			CGC	41			ACA	17	ACV		CTC	25	
Asp	GAT	69	GAY	Ser	AGT	13			ACT	3		Phe	TTT	38	TTY
	GAC	31			AGC	28	AGY		ACC	47			TTC	62	
Val	GTG	36			TCG	38	TCR	Trp	TGG	100	TGG	Gln	CAG	70	CAR
	GTA	13	GTN		TCA	16		End	TGA	25			CAA	30	
	GTT	21			TCT	2			TAG	63	TRR	His	CAT	5	CAY
	GTC	30			TCC	2			TAA	13			CAC	95	
Ala	GCG	31		Lys	AAG	55	AAR	Cys	TGT	50	TGY	Pro	CCG	51	CCN
	GCA	18	GCN		AAA	45			TGC	50			CCA	20	
	GCT	17		Asn	AAT	57	AAY	Tyr	TAT	41	TAY		CCT	5	
	GCC	35			AAC	43			TAC	59			CCC	23	

[1] Codon preference: R = A/G; Y = C/T; B = G/C/T; V = A/G/C; N = A/C/G/T.
Codon usages <5% were excluded in codon preference.

Table 6. Codon frequency in *Heliobacillus mobilis* reaction center gene (*pshA*)

Amino Acid	Codon	%	Pref[1]	Amino Acid	Codon	%	Pref	Amino Acid	Codon	%	Pref	Amino Acid	Codon	%	Pref
Gly	GGG	0		Arg	AGG	0		Met	ATG	100	ATG	Leu	TTG	5	
	GGA	9	GGH		AGA	0		Ile	ATA	0			TTA	3	TTG
	GGT	36			CGG	18	CGB		ATT	23	ATY		CTG	55	CTS
	GGC	55			CGA	0			ATC	77			CTA	0	
Glu	GAG	20	GAR		CGT	47		Thr	ACG	9			CTT	3	
	GAA	80			CGC	35			ACA	3	ACB		CTC	33	
Asp	GAT	43	GAY	Ser	AGT	3			ACT	33		Phe	TTT	15	TTY
	GAC	57			AGC	24	AGC		ACC	55			TTC	85	
Val	GTG	13			TCG	0	TCH	Trp	TGG	100	TGG	Gln	CAG	33	CAR
	GTA	20	GTN		TCA	12		End	TGA	0			CAA	67	
	GTT	41			TCT	18			TAG	0	TRR	His	CAT	42	CAY
	GTC	26			TCC	44			TAA	100			CAC	58	
Ala	GCG	11		Lys	AAG	31	AAR	Cys	TGT	13	TGY	Pro	CCG	26	CCB
	GCA	3	GCB		AAA	69			TGC	88			CCA	4	
	GCT	64		Asn	AAT	18	AAY	Tyr	TAT	40	TAY		CCT	26	
	GCC	22			AAC	82			TAC	60			CCC	43	

[1] Codon preference: R = A/G; Y = C/T; S = G/C; B = G/C/T; H = A/C/T; N = A/C/G/T.
Codon usages <5% were excluded in codon preference.

III. Genetic Manipulation

A. Chemical and UV Light Mutagenesis

1. Chloroflexus aurantiacus

UV light radiation or treatment with N-methyl-N'-nitro-N-nitrosoguanidine (NTG) was used by Pierson et al. (1984) to generate BChl *c*- and carotenoid-deficient mutants of *Cf. aurantiacus*. Either before after the mutagenesis step, the cellular filaments were sonicated briefly to facilitate the isolation of colonies. Ampicillin sensitivity under anaerobic light (emission maximum at 740 nm where BChl *c* and not BChl *a* absorbs) growth conditions was used to reduce the number of wild type cells after NTG mutagenesis. One BChl *c* + carotenoid-deficient, one carotenoid-deficient and several BChl *c*-deficient mutants were isolated and partially characterized. Many of the mutants reverted and had to be re-isolated.

B. Transformation

1. Chlorobium *Species*

In 1988 Ormerod reported that *Cb. limicola* 8327 could be transformed by the addition of a crude DNA extract from a spontaneous streptomycin resistant mutant of the same *Chlorobium* strain. The cells were transformable in all growth phases. DNA uptake was light dependent and strain and species specific: *Cb. limicola* strain Tassajara was not transformed by the DNA from the streptomycin resistant 8327 strain. Likewise both *Cb. limicola* strains 8327 and Tassajara were not transformed to kanamycin resistance by DNA extracts from R6845 and R1822 plasmid-containing *Rb. capsulatus*. Conjugation attempts with *Rb. capsulatus* containing either of the two plasmids or *E. coli* containing RP 4 were also unsuccessful.

Kjærulff et al. (1994) have described the transformation of *Cb. vibrioforme* through electroporation. The DNA construct used contained a full copy of the *psc2C* gene encoding the monoheme cytochrome *c*-551 and the streptomycin adenyltransferase gene as a selectable marker. The transformation was dependent upon the presence of *Cb. vibrioforme* DNA indicating that homologous recombination was necessary.

2. Heliobacillus mobilis

Vermaas and Vu (1993) have been able to transform $CaCl_2$/RbCl-treated *Hc. mobilis* cells to kanamycin resistance with the broad-host range plasmid pFD666 which carries the neomycin phosphotransferase gene from Tn5.

Very recently, Wahlund and Madigan (1995) described methods for the conjugal transfer of two IncQ group plasmids into *Chlorobium tepidum* from *E. coli* and for the growth of this bacterium on agar.

3. Restriction Enzyme-Mediated Integration of Plasmid DNA into Chromosomal DNA

The method Restriction Enzyme-Mediated Integration (REMI) of plasmid DNA into chromosomal DNA was originally described for *Saccharomyces cerevisiae* (Schiestl and Petes, 1991) and was successfully used in *Dictyostelium discoideum* for identifying developmental genes (Kuspa and Loomis, 1992). Genomic regions interrupted by the presence of plasmid DNA can be easily identified and cloned as long as a restriction enzyme known to digest the host DNA but not the plasmid DNA is available (Kuspa and Loomis, 1992). REMI could probably be adapted to photosynthetic bacteria, especially those which can be cultivated under aerobic conditions, to identify structural, metabolic, and regulatory genes required for photosynthesis. If transformation methods utilizing broad-host range plasmids does not work for *Cf. aurantiacus*, then REMI would be a good alternative method to attempt.

IV. Concluding Remarks

At this time more than a dozen genes from *Chlorobium* species, eight from *Cf. aurantiacus*, and two from *Hc. mobilis* have been cloned and sequenced. In a few cases, analysis of the transcriptional organization of the genes has been initiated. For the *Cf. aurantiacus pufl* operon work on the promoter region and *cis*-acting regulatory elements has begun. Transformation systems have been established for *Cb. vibrioforme* and *Hc. mobilis* which means site-directed mutagenesis of the known genes can be carried out and the effect on the encoded proteins analyzed. Because both of these bacteria are obligate photoanaerobes and have no alternate growth possibilities, gene disruption experiments on essential genes or regulatory elements cannot be carried out. A transformation system for *Cf. aurantiacus* will probably be developed relatively soon despite the basic microbiological problems that have to be resolved concomitantly.

Acknowledgments

I would like to thank the many colleagues who sent me reprints and/or preprints of their work. This work was supported by the Deutscheforschungsgemeinschaft (SFB 143).

References

Beebee T and Burke J (1992) Gene Structure and Transcription, pp 1–34. Oxford University Press, New York

Beer-Romero R and Gest H (1987) *Heliobacillus mobilis*, a peritrichously flagellated anoxyphototroph containing bacteriochlorophyll *g*. FEMS Microbiol Lett 41: 109–114

Beer-Romero P, Favinger JL and Gest H (1988) Distinctive properties of bacilliform photosynthetic heliobacteria. FEMS Microbiol Lett 49: 451–454

Belasco JG and Chen C-Y A (1988) Mechanism of *puf* mRNA degradation: the role of an intercistronic stem-loop structure. Gene 72: 109–117

Bingham AHA and Darbyshire J (1982) Isolation of two restriction endonucleases from *Chloroflexus aurantiacus* (*Cau*I, *Cau*II). Gene 18: 87–91

Blankenship RE (1992) Origin and early evolution of photosynthesis. Photosynth Res 33: 91–111

Bryant DA (1994) Gene nomenclature recommendations for green photosynthetic bacteria and heliobacteria. Photosynth Res 41: 27–28

Büttner M, Xie D-L, Nelson H, Pinther W, Hauska G and Nelson N (1992) Photosynthetic reaction center genes in green sulfur bacteria and in Photosystem I are related. Proc Natl Acad Sci USA 89: 8135–8139

Castenholz R and Pierson BK (1981) Isolation of members of the family Chloroflexaceae. In: Starr MP, Stolp H, Trüper HG, Balows A and Schlegel HG (eds) The Prokaryotes—A Handbook of Habitats, Isolation and Identification of Bacteria, Vol 1, pp 290–298. Springer Verlag, New York

Chung S and Bryant D (1992) Genes encoding chlorosome components in the green sulfur bacteria *Chlorobium vibrioforme* 8327D and *Chlorobium tepidum*. In: Murata N (ed) Research in Photosynthesis, Vol 1, pp 69–72. Kluwer Academic Publishers, Dordrecht

Chung S and Bryant DA (1993) Genes encoding components of the photosynthetic apparatus in *Chlorobium vibrioforme* 8327D and *Chlorobium tepidum*. EMBO workshop on Green and Heliobacteria, Nyborg Denmark, abstract

Chung S, Frank G, Zuber H and Bryant DA (1994) Genes encoding two chlorosome components from the green sulfur bacteria *Chlorobium vibrioforme* strain 8327D and *Chlorobium tepidum*. Photosynth Res 41: 261–275

Cox RP, Jensen MT, Miller M and Pedersen JP (1988) Spin label studies on chlorosomes from green bacteria. In: Olson JM, Ormerod JG, Amesz J, Stackebrandt E and Trüper HG (eds) Green Photosynthetic Bacteria, pp 15–22. Plenum Press, New York

D'Onofrio G and Bernardi G (1992) A universal compositional correlation among codon positions. Gene 110: 81–88

Dracheva S, Williams JC, Van Driessche G, Van Beeumen JJ and Blankenship RE (1991) The primary structure of cytochrome *c*-554 from the green photosynthetic bacterium *Chloroflexus aurantiacus*. Biochemistry 30: 11451–11458

Dracheva S, Williams JC and Blankenship RE (1992) Cloning and sequencing of the FMO protein gene from *Chlorobium tepidum*. In: Murata N (ed) Research in Photosynthesis, Vol 1, pp 53–56. Kluwer Academic Publishers, Dordrecht

Fonstein M and Haselkorn R (1993) Chromosomal structure of *Rhodobacter capsulatus* strain SB1003: Cosmid encyclopedia and high-resolution physical and genetic map. Proc Natl Acad Sci, USA 90: 2522–2526

Foster JM, Redlinger TE, Blankenship RE and Fuller RC (1986) Oxygen regulation of development of the photosynthetic membrane system in *Chloroflexus aurantiacus*. J Bacteriol 167: 655–659

Gest H and Favinger JL (1983) *Heliobacterium chlorum*, an anoxygenic brownish-green photosynthetic bacterium containing a 'new' form of bacteriochlorophyll. Arch Microbiol 136: 11–16.

Haas ES, Brown JW, Pitulle C and Pace NR (1994) Further perspective on the catalytic core and secondary structure of ribonuclease P RNA. Proc Natl Acad Sci USA 91: 2527–2531

Imhoff JF (1992) Taxonomy, phylogeny, and general ecology of anoxygenic phototrophic bacteria. In: Mann NH and Carr NG (eds) Photosynthetic Prokaryotes, pp 53–92. Plenum Press, New York

Ivanosky RN, Krasilnikova EN and Fal YI (1993) A pathway of the autotrophic CO_2 fixation in *Chloroflexus aurantiacus*. Arch Microbiol 159: 257–264

Kaulen H and Klemme J-H (1983) No evidence of covalent modification of glutamine synthetase in the thermophilic phototrophic bacterium *Chloroflexus aurantiacus*. FEMS Microbiol Lett 20: 75–79

Kern M and Klemme J-H (1989) Inhibition of bacteriochlorophyll biosynthesis by gabaculin (3-amino, 2,3-dihydrobenzoic acid) and presence of an enzyme of the C_5-pathway of δ-aminolevulinate synthesis in *Chloroflexus aurantiacus*. Z Naturforsch 44c: 77–80

Kessler C and Manta V (1990) Specificity of restriction endonucleases and DNA modification methyl transferases—a review. Gene 92: 1–248

Kjærulff S, Diep DB, Okkels JS, Scheller HV and Ormerod JG (1994) Highly efficient integration of foreign DNA into the genome of the green sulfur bacterium, *Chlorobium vibrioforme* by homologous recombination. Photosynth Res 41: 277–283

Klug G (1993) The role of mRNA degradation in the regulated expression of bacterial photosynthesis genes. Mol Microbiol 9: 1–7

Kramarov VM, Fomenkov AI, Matvienko NI, Ubieta RKH,

Stolyaninov VV and Gorlenko VM (1987) A new sequence-specific endonuclear *Cau*-B3-1 from *Chloroflexus aurantiacus* B3. Bioorg Khim 13: 773–776 (English abstract only)
Kuspa A and Loomis WF (1992) Tagging developmental genes in *Dictyostelium* by restriction enzyme-mediated integration of plasmid DNA. Proc Natl Acad Sci USA 89: 8803–8807
Liebl U, Mockensturm-Wilson M, Trost JT, Brune DC, Blankenship RE and Vermaas W (1993) Single core polypeptide in the reaction center of the photosynthetic bacterium *Heliobacillus mobilis*: Structural implications and relations to other photosystems. Proc Natl Acad Sci USA 90: 7124–7128
Madigan MT, Kimble LK, Wahlund TM (1993) Nitrogen fixation in green sulfur bacteria and in heliobacteria. EMBO Workshop on Green and Heliobacteria, Nyborg Denmark, abstract
Madigan MT, Petersen SR and Brock TD (1974) Nutrition studies on *Chloroflexus*, a filamentous photosynthetic, gliding bacterium. Arch Microbiol 100: 97–103
Majumdar D, Avissar YJ, Wyche JH and Beale SI (1991) Structure and expression of the *Chlorobium vibrioforme hemA* gene. Arch Microbiol 156: 281–289
Molemans F, Van Emmelo J and Fiers W (1982) The sequence specificity of endonucleases *CauI* and *CauII* isolated from *Chloroflexus aurantiacus*. Gene 18: 93–96
Niedermeier G, Shiozawa JA, Lottspeich F and Feick RG (1994) Primary structure of two chlorosome proteins from *Chloroflexus aurantiacus*. FEBS Lett 342: 61–65
Okkels JS, Kjær B, Hansson O, Svendsen I, Møller BL and Scheller HV (1992) A membrane-bound monoheme cytochrome *c*-551 of a novel type is the immediate electron donor to P840 of the *Chlorobium vibrioforme* photosynthetic reaction center complex. J Biol Chem 267: 21139–21145
Ormerod JG (1988) Natural genetic transformation in *Chlorobium*. In: Olson J, Ormerod JG, Amesz J, Stackebrandt E and Trüper HG (eds) Green Photosynthetic Bacteria, pp 315–319. Plenum Press, New York
Ormerod JG, Nesbakken T and Beale SI (1990) Specific inhibition of antenna bacteriochlorophyll synthesis in *Chlorobium vibrioforme* by anesthetic gases. J Bacteriol 172: 1352–1360
Ormerod JG, Nesbakken T and Torgersen Y (1990) Phototrophic bacteria that form heat resistant endospores. In: Baltscheffsky M (ed) Current Research in Photosynthesis, Vol 4, pp 935–938. Kluwer Academic Publishers, Dordrecht
Ovchinnikov Yu A, Abdulaev NG, Zolotarev AS, Shmukler BE, Zargarov AA, Kutuzov MA, Telezhinskaya IN and Levina NB (1988a) Photosynthetic reaction centre of *Chloroflexus aurantiacus*. I. Primary structure of L-subunit. FEBS Lett 231: 237–242
Ovchinnikov Yu A, Abdulaev NG, Shmuckler BE, Zargarov AA, Kutuzov MA, Telezhinskaya IN, Levina NB and Zolotarev AS (1988b) Photosynthetic reaction centre of *Chloroflexus aurantiacus*. Primary structure of M-subunit. FEBS Lett 232: 364–368
Oyaizu H, Debrunner-Vossbrink B, Mandelco L, Studier JA and Woese CR (1987) The green non-sulfur bacteria: A deep branching in the eubacteria line of descent. System Appl Microbiol 9: 47–53
Pfennig N (1989) II. Green bacteria In: Staley JT, Bryant MP, Pfennig N and Holts JG (eds) Bergey's Manual of Systematic Bacteriology, Vol 3, pp 1682–1708. Williams and Wilkins, Baltimore
Pierson BK and Castenholz RN (1974) A phototrophic gliding filamentous bacterium of hot springs, *Chloroflexus aurantiacus* gen. and sp. nov. Arch Microbiol 100: 5–24
Pierson BK, Keith LM and Leovy JG (1984) Isolation of pigmentation mutants of the green filamentous photosynthetic bacterium *Chloroflexus aurantiacus*. J Bacteriol 159: 222–227
Platt T (1986) Transcription termination and the regulation of gene expression. Ann Rev Biochem 55: 339–372
Richardson JP (1993) Transcription termination. Critical Rev Biochem and Molec Biol. 28: 1–30
Rieble S, Ormerod JG and Beale SI (1989) Transformation of glutamate to δ-aminolevulinic acid by soluble extracts of *Chlorobium vibrioforme*. J Bacteriol 171: 3782–3787
Robinson SJ and Redlinger TE (1987) Isolation of genes encoding the photosynthetic apparatus of *Chloroflexus*. In: Biggins J (ed) Progress in Photosynthesis Research, Vol 4, pp 741–744. Martinus Nijhoff Publishers, Dordrecht
Rolstad AK, Howland E and Sirevåg R (1988) Malate dehydrogenase from thermophilic green bacterium *Chloroflexus aurantiacus*: Purification, molecular weight, amino acid composition, and amino acid sequence. J Bacteriol 170: 2947–2953
Sambrook J, Fritsch EF and Maniatis T (1989) Molecular Cloning—A Laboratory Manual, Vol 1, pp 5.1–5.95. Cold Spring Harbor Laboratory Press, Cold Spring Harbor
Saunders VA (1992) Genetics of the photosynthetic prokaryotes. In: Mann NH and Carr NG (eds) Photosynthetic Prokaryotes, pp 121–152. Plenum Press, New York
Schiestl RH and Petes TD (1991) Integration of DNA fragments by illegitimate recombination in *Saccharomyces cerevisiae*. Proc. Natl Acad Sci USA 88: 7585–7589
Schmidt K (1980) A comparative study on the composition of chlorosomes (chlorobium vesicles) and cytoplasmic membranes from *Chloroflexus aurantiacus* strain OK-70-fl and *Chlorobium limicola* f. *thiosulfatophilum* strain 6230. Arch Microbiol 124: 21–31
Shiozawa JA, Lottspeich F, Oesterhelt D and Feick R (1989) The primary structure of the *Chloroflexus aurantiacus* reaction center polypeptides. Eur J Biochem 180: 75–84
Shiozawa JA, Csiszàr K and Feick R (1990) Preliminary studies on the operon coding for the reaction center polypeptides in *Chloroflexus aurantiacus*. Drews G and Dawes EA (eds) Molecular Biology of Membrane-bound Complexes in Phototrophic Bacteria, pp 11–18. Plenum Press, New York
Sirevåg B and Ormerod JG (1970) Carbon dioxide fixation in green sulphur bacteria. Biochem J 120: 399–408
Suwanto A and Kaplan S (1989) Physical and genetic mapping of the *Rhodobacter sphaeroides* 2.4.1. genome: Presence of two unique circular chromosomes. J Bacteriol 171: 5850–5859
Theroux SJ, Redlinger TE, Fuller RC and Robinson SJ (1990) Gene encoding the 5.7 kilodalton chlorosome protein of *Chloroflexus aurantiacus*: Regulated message levels and a predicted carboxy-terminal protein extension. J Bacteriol 172: 4497–4504
Van de Meent EJ, Kleinherenbrink FAM and Amesz J (1990) Purification and properties of an antenna-reaction center complex from heliobacteria. Biochim Biophys Acta 1015: 223–230
Van den Eynde H, Van de Peer Y, Perry J and De Wachter R (1990) 5S rRNA sequences of representatives of the genera *Chlorobium*, *Prosthecochloris*, *Thermomicrobium*, *Cytophaga*, *Flavobacterium*, *Flexibacter*, and *Saprospira*, and discussion

of the evolution of eubacteria in general. J Gen Microbiol 136: 11–18

Vermaas W and Vu A (1993) *Heliobacillus mobilis:* Transformation and evolution. EMBO Workshop in Green and Heliobacteria, Nyborg Denmark, abstract

Wahlund TM and Madigan MT (1995) Genetic transfer by conjugation in the thermophilic green sulfur bacterium *Chlorobium tepidum*. J Bacteriol 177, in press

Wahlund TM, Woese CR, Castenholz RW and Madigan MT (1991) A thermophilic green sulfur bacterium from New Zealand hot springs, *Chlorobium tepidum*, sp. nov. Arch Microbiol 156: 81–90

Watanabe Y, Feick R and Shiozawa JA (1992) Cloning and sequencing of the genes encoding the polypeptides of the B806–866 light-harvesting complex of *Chloroflexus aurantiacus*. In: Murata N (ed) Research in Photosynthesis, Vol 1. pp 41–44. Kluwer Academic Publishers, Dordrecht

Watanabe Y, Feick RG and Shiozawa JA (1995) Cloning and sequencing of the genes encoding the light-harvesting B806-866 polypeptides and initial studies on the transcriptional organization of *puf2B*, *puf2A* and *puf2C* in *Chloroflexus aurantiacus*. Archiv Microbiol 163: 124–130

Witt D, Bergstein-Ben Dan T and Stackebrandt E (1989) Nucleotide sequence of 16S rRNA and phylogenetic position of the green sulfur bacterium *Clathrochloris sulfurica*. Arch Microbiol 152: 206–208

Woese CR, De Brunner-Vossbrinck BA, Oyaizu H, Stackebrandt E and Ludwig W (1985) Gram positive bacteria: Possible photosynthetic ancestry. Science 229: 762–765

Woese CR, Maloy S, Mandelco L and Raj HB (1990a) Phylogenetic placement of the Spirosomaceae. Syst Appl Microbiol 13: 19–23

Woese CR, Mandelco L, Yang D, Gherna R and Madigan MT (1990b) The case for the relationship of the flavobacteria and their relatives to the green sulfur bacteria. System Appl Microbiol 13: 258–262

Wu L-F and Saier Jr MH (1991) Differences in codon usage among genes encoding proteins of different function in *Rhodobacter capsulatus*. Res Microbiol 142: 943–949

Xie D-L, Lill H, Hauska G, Maeda M, Futai M and Nelson N (1993) The *atp2* operon of the green bacterium *Chlorobium limicola*. Biochim Biophys Acta 1172: 267–273

Zillig W, Palm P, Reiter W-D, Gropp F, Pühler G and Klenk H-P (1988) Comparative evaluation of gene expression in archaebacteria. Eur J Biochem 173: 473–482

Zirngibl S, Xie D-L, Riedel A, Nitschke W, Nelson H, Nelson N, Liebl U, LeCoutre J, Hauska G, Kellner E, Grodzitzki D, Büttner M (1992) Low potential Rieske Fe-S-centers of menaquinol oxidizing cytochrome bc complexes. In: Murata N (ed) Research in Photosynthesis, Vol 2, pp 471–478. Kluwer Academic Publishers, Dordrecht

Chapter 55

Regulation of Hydrogenase Gene Expression

Paulette M. Vignais, Bertrand Toussaint and Annette Colbeau
CEA, Laboratoire de Biochimie Microbienne (CNRS URA 1130),
Département de Biologie Moléculaire et Structurale,
Centre d'Etudes Nucléaires de Grenoble,
38054 Grenoble, cedex 9, France

Summary

In bacteria, the genes for hydrogenase (*hup* and *hyp*) are generally clustered, and the sequence of these clustered genes has been determined in several species. Among the photosynthetic bacteria, the organization of the *hup* and *hyp* genes has been the most studied in *Rhodobacter capsulatus*, and the nucleotide sequence of the entire *hup/hyp* gene cluster has been determined. This cluster includes the structural genes for the H_2-uptake [NiFe]hydrogenase (*hupSLC*) and the large number of accessory and regulatory genes. The possible roles of these genes, and the existence of other types of hydrogenase in the photosynthetic bacteria, are discussed. Hydrogenase gene expression in *Rb. capsulatus* is controlled by two main factors, namely molecular hydrogen and oxygen. Several regulatory genes have been identified in the *hup/hyp* gene cluster of *Rb. capsulatus*. The products of two of them, *hupR* and *hupT*, belong to the superfamily of two-component regulatory systems. In

R. E. Blankenship, M. T. Madigan and C. E. Bauer (eds): Anoxygenic Photosynthetic Bacteria, pp. 1175–1190.

addition, an IHF-like protein isolated from *Rb. capsulatus* has been shown to be required for in vivo hydrogenase synthesis. A molecular model of the interaction between the IHF binding site present in the promoter region of the hydrogenase structural operon and the IHF protein of *Rb. capsulatus* is presented.

I. Introduction

The potential for use of H_2 as a primary or secondary source of energy, and the possibility of transforming solar radiation into stable chemical energy, in the form of hydrogen gas, have greatly stimulated the study of H_2 metabolism in photosynthetic organisms, in particular in photosynthetic procaryotes. Many review articles on H_2 metabolism and on biotechnological aspects of hydrogen production in cyanobacteria and in photosynthetic bacteria have been published in the last 15 years. Numerous relevant references may be found in the reviews of Vignais et al. (1985) and of Sasikala et al. (1993).

A. The Discovery of Hydrogen Utilization and Production by the Photosynthetic Bacteria

1. Hydrogen Consumption for Autotrophic Growth

Early studies had shown that photosynthetic bacteria, such as *Chromatium*, can use molecular hydrogen as sole electron donor during photoautotrophic growth. The capacity to oxidize H_2 was later associated to the presence of the hydrogenase enzyme (Gest, 1952). Anaerobic photoautotrophic growth with H_2 as electron donor was demonstrated for *Rhodospirillum rubrum*, *Rhodopseudomonas palustris*, *Rhodopseudomonas gelatinosa*, and *Rhodopseudomonas capsulata* (now *Rhodobacter capsulatus*) (cf. Vignais et al., 1985). Under photoautotrophic conditions, *Rb. capsulatus* strains showed the fastest growth rates among purple non-sulfur bacteria (Klemme, 1968). *Rb. capsulatus* is also able to grow chemoautotrophically under aerobic conditions; under these conditions, H_2 serves as the source of energy and reducing power, O_2 as the terminal electron acceptor for energy transduction and CO_2 as the sole carbon source (Madigan and Gest, 1979; Seifert and Pfennig, 1979; Colbeau et al., 1980).

Abbreviations: A. eutrophus – Alcaligenes eutrophus; B. japonicum – Bradyrhizobium japonicum; E. coli – Escherichia coli; IHF – integration host factor; *Rb. – Rhodobacter; Rc. – Rhodocyclus; Rp. – Rhodopseudomonas; Rs. – Rhodospirillum; Tc. roseopersicina – Thiocapsa roseopersicina*

2. Photoproduction of Molecular Hydrogen

Photochemical production of molecular hydrogen by a photosynthetic bacterium, *Rs. rubrum*, was first observed by Gest and Kamen (1949a,b), who noted that photoevolution of H_2 proceeded readily under an atmosphere of 100% H_2 but did not occur under an atmosphere of N_2; addition of ammonium chloride or of high concentrations of yeast extract to the culture medium stimulated growth but inhibited H_2 production. Subsequent experiments with the isotope ^{15}N indicated clearly that *Rs. rubrum* contained a N_2-fixing system (Gest and Kamen, 1949b; Gest et al., 1950) as did the purple sulfur bacterium *Chromatium*, and the green bacterium *Chlorobium* (Lindstrom et al., 1949) and several other species of purple nonsulfur bacteria including *Rp. palustris*, *Rhodopseudomonas sphaeroides* (now *Rhodobacter sphaeroides*), *Rb. capsulata* and *Rp. gelatinosa* (later termed *Rhodocyclus gelatinosus*) (Lindstrom et al., 1951). Since H_2 production did not occur when N_2 fixation was repressed, there appeared to be a close relationship between these two processes. While studying nitrogen fixation by cell-free extracts of *Rs. rubrum*, Bulen et al. (1965a) demonstrated that ATP and a low potential reductant were required for the production of H_2 by these extracts as well as for N_2 fixation, and that H_2 evolution was irreversible, independent of the partial pressure of H_2 and insensitive to CO. Bulen et al. (1965b) concluded that the same enzyme, nitrogenase, catalyzes both N_2 reduction and ATP-dependent H_2 evolution.

The evidence that in purple nonsulfur bacteria photoevolution of hydrogen is mediated by nitrogenase was summarized by Meyer et al. (1978), namely: 1) H_2 evolution, like nitrogen fixation, was strongly stimulated by light and inhibited by nitrogen, the natural substrate of nitrogenase (Gest and Kamen, 1949b; Jouanneau et al., 1984, 1985); 2) unlike with classical hydrogenases, H_2 evolution in *Rb. capsulatus* was not inhibited by carbon monoxide (Jouanneau et al., 1980) similarly to ATP-dependent H_2 evolution by nitrogenase (Winter and Burris, 1968); 3) H_2 evolution was inhibited by ammonia and the biosynthesis of the H_2 evolving enzyme was repressed in the presence of ammonia (Ormerod et al., 1961;

Gogotov et al., 1974); 4) Nif⁻ strains of *Rb. capsulatus*, which lack nitrogenase activity, were incapable of H_2 evolution (Wall et al., 1975); 5) molybdenum-starved cells of *Rs. rubrum* displayed a large decrease in their ability to evolve hydrogen (Paschinger, 1974); since nitrogenases are molybdo-enzymes whereas hydrogenases are not, H_2 evolution in this bacterium could be clearly attributed to nitrogenase. Light-stimulated nitrogenase-mediated H_2 evolution was then recognized as a general property of photosynthetic bacteria (early reviews by Gest, 1972; Yoch, 1978; Meyer at al., 1978). In accordance with its high N_2 fixation capacity (Madigan et al., 1984), *Rb. capsulatus* has been found to be one of the most efficient H_2 producers (under nitrogen limited growth conditions) (Hillmer and Gest, 1977a, b). By means of its hydrogenase, *Rb. capsulatus* cells can recycle H_2 electrons to nitrogenase, which is a way for the cells to avoid energy loss in the form of H_2 (Meyer et al., 1978).

3. Production of Molecular Hydrogen in the Dark

Purple bacteria have also the capacity to ferment their endogenous substrates in the dark with production of organic acids, CO_2 and H_2 (van Niel, 1944; see Uffen, 1978 for review). Dark, fermentative H_2 production results from breakdown of formate to CO_2 and H_2 by formic hydrogenlyase (Gest, 1954). Gas production from formate by the formic hydrogenlyase reaction is not influenced by light but is completely inhibited by carbon monoxide (Gorrell and Uffen, 1977, 1978). This is a typical fermentative type of H_2 production, which is quite distinct from the CO-insensitive, ATP-dependent, production of H_2, mediated by nitrogenase.

The formic hydrogenlyase system is particularly well studied in *Escherichia coli*; it is hydrogenase 3, the product of the *hycE* gene, which transfers electrons to protons and releases H_2 (Sauter et al., 1992). A gene homolog to *hycE* has not been yet identified in a photosynthetic bacterium. *Rs. rubrum* should be a good candidate to look for it. *Rs. rubrum* can produce H_2 in the presence of ammonia, via a fermentative type of hydrogenase, and can also grow under autotrophic conditions with H_2 and CO_2 via non-fermentative pathways involving an adaptively synthesized H_2-uptake hydrogenase (Voelskow and Schön, 1980; see also Kern et al., 1994). Thus, in *Rs. rubrum*, three enzymes interact in H_2 metabolism according to the growth conditions: two hydrogenases and nitrogenase. It is not known yet whether the hydrogenase which, with carbon monoxide dehydrogenase, evolves H_2 during anaerobic oxidation of CO (Bonam et al., 1989) represents an additional type of enzyme. Our knowledge of those enzymes, at the biochemical and genetic level, is less advanced in *Rs. rubrum*, for fewer genetic tools have been developed for it than for *Rb. capsulatus* or for *Rb. sphaeroides*.

B. Genetics of Hydrogen Production and Utilization in Photosynthetic Bacteria

The development of good genetic systems to study nonsulfur purple photosynthetic bacteria (Marrs, 1974; Taylor et al., 1983; Willison et al., 1985; Donohue and Kaplan, 1991) has intensified research in this group of bacteria (reviewed by Scolnik and Marrs, 1987). Genetic studies of H_2 metabolism in the photosynthetic bacteria, undertaken only recently, are mainly restricted to members of the Rhodospirillaceae, in particular to *Rb. capsulatus*, which was the first phototrophic bacteria for which an useful genetic exchange system was discovered (Marrs, 1974). Another interesting feature of *Rb. capsulatus* is that, like *Pseudomonas* spp., it can exchange plasmids of the R type by means of conjugation, a property which has greatly contributed to the development of genetic studies in *Rb. capsulatus* (reviewed by Donohue and Kaplan, 1991; and in Chapter 48 by Williams and Taguchi).

The development of methods for the genetic analysis of the mutants affected in the enzyme systems involved in H_2 metabolism (nitrogenase and hydrogenase) has permitted the genetic mapping of *nif* (*ni*trogen *f*ixation) and *hup* (*h*ydrogen *up*take) genes on the *Rb. capsulatus* chromosome, using R plasmid-mediated conjugation (Willison et al., 1985; Willison, 1993) (Fig. 1), has led to the identification of gene clusters necessary for the biosynthesis of nitrogenase (Klipp et al., 1988) and of hydrogenase (Colbeau et al., 1993) and has provided information on the regulation of H_2 metabolism in the photosynthetic bacteria (see below).

This chapter focuses on the organization and expression of genes necessary for the expression of the H_2 uptake hydrogenase (*hup*) genes. The capacity of *Rb. capsulatus* to grow lithoautotrophically (Aut⁺ phenotype) has been used as a primary selective test to distinguish strains with an active hydrogenase (Hup⁺ phenotype) from hydrogenase-deficient strains (Hup⁻ and Aut⁻ phenotypes).

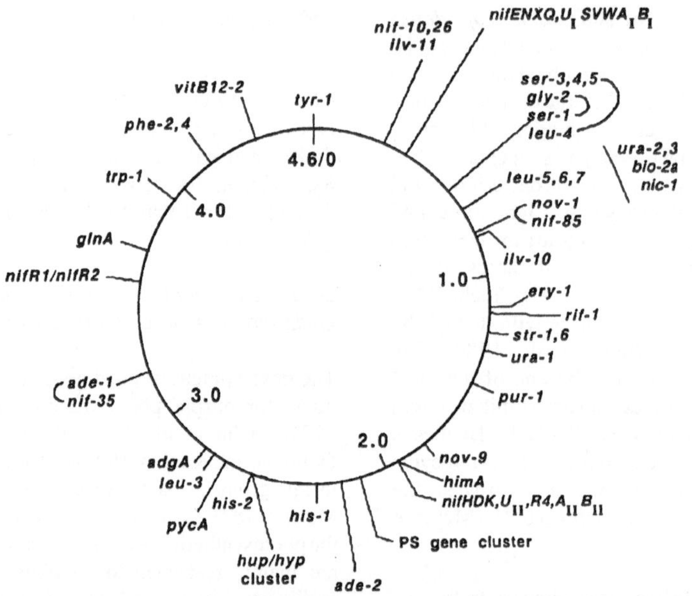

Fig. 1. Genetic linkage map of the *Rhodobacter capsulatus* chromosome. The position of genetic markers on the chromosome was determined by conjugation, using the mutant R plasmid, pTH10 (Willison et al., 1985). The positions of the *hup/hyp* gene cluster and the *himA* gene (characterized by the *aut-4* mutation) are shown. The *himA* gene is closely linked to the *nif* gene cluster containing the nitrogenase structural genes (*nifHDK*) and the $nifR_4$ regulatory gene (for further details, see Willison, 1993). The figures inside the circle show the genetic map distance in arbitrary units, with the position of the *tyr-1* marker arbitrarily chosen as zero. Physical mapping of the *Rb. capsulatus* chromosome has shown its size to be 3.7 Mbp (Fonstein and Haselkorn, 1993). The physical map shows good agreement with the genetic map for several markers, except that the positions of the *glnA* and $nifR_I$ genes are inverted with respect to *trp-1* and *adgA*. (See Chapter 49 by Haselkorn) (Adapted from Willison, 1993). The *adgA* gene has recently been shown to code for NH_3-dependent NAD synthethase (Willison and Tissot, 1994).

II. Hydrogenases of Photosynthetic Bacteria Belong to the Class of [NiFe]Hydrogenases

All hydrogenases catalyze the simple, reversible reaction of proton reduction leading to the formation of H_2 gas: $2H^+ + 2e \leftrightarrow H_2$. In addition to Fe-S clusters, the hydrogenase enzymes found in *Chromatium vinosum* (Albracht et al., 1983), in *Thiocapsa roseopersicina* (Gogotov, 1986; Bagyinka et al., 1993), as well as in *Rb. capsulatus* (Takakuwa and Wall, 1981; Colbeau and Vignais, 1983) have been shown to require Ni for activity and to contain Ni at their active site. They belong to class I of [NiFe]hydrogenases, are heterodimers (30 kDa and 65 kDa) and have a common evolutionary origin (Wu and Mandrand., 1993). The term [NiFe] is very appropriate because the crystal structure of the [NiFe] hydrogenase from *Desulfovibrio gigas* (Volbeda et al., 1995) revealed that the active site contains a nickel atom and an iron atom.

A. Mapping of *hup* Genes on the Chromosome of *Rb. capsulatus*

In contrast to *Alcaligenes eutrophus* (Eberz et al., 1986) and *Rhizobium leguminosarum* (Brewin et al., 1980), in which hydrogenase genes reside on a megaplasmid, in *Rb. capsulatus*, the hydrogenase is chromosomally encoded (Colbeau et al., 1990). Hup^- /Aut^- mutants obtained by chemical mutagenesis (Willison et al., 1984; Colbeau et al., 1986) were used to map the *hup* loci on the genetic map of the *Rb. capsulatus* chromosome. The *hup* mutations used in the mapping experiments were found to be clustered on the chromosome in the proximity of the *his-2* marker. On the other hand, the mutation *aut-4* carried by the IR4 mutant (Willison et al., 1984), in the *himA* gene (Toussaint et al., 1991) maps close to the $nifR_4$ gene, which is contiguous to the *nifHDK* structural genes of nitrogenase (Colbeau et al., 1990) (Fig. 1).

B. Organization of Hydrogenase Genes

1. Isolation of the hup Gene Clusters

By complementing Hup⁻ mutants with cosmids from gene banks of *Rb. capsulatus*, genomic DNA fragments containing *hup* determinants were isolated (Colbeau et al., 1986; Xu et al., 1989; Colbeau et al., 1990; Xu and Wall, 1991). The sequencing of those fragments revealed the presence of the cluster of *hup* genes (Leclerc et al., 1988; Richaud et al., 1990; Xu and Wall, 1991; Richaud et al., 1991; Colbeau et al., 1993) depicted in Fig. 2. The same approach, led to the isolation of the hydrogenase structural genes from *Rc. gelatinosus* (Uffen et al., 1990) and to the identification of several homologous *hup* genes in *Tc. roseopersicina* (Colbeau et al., 1994).

2. The Structural Hydrogenase Genes, hupSLC

The small and large subunits of hydrogenase are encoded by the *hupS* and the *hupL* genes, respectively. In *Rb. capsulatus*, *hupS* is capable of encoding a protein of 34,256 Da with 13 Cys residues, and the *hupL* gene can encode a protein of 65,839 Da containing 10 Cys residues. These proteins share a high degree of identity, more than 80 and nearly 70% with the small and large subunits, respectively, of the [NiFe]hydrogenases belonging to group I in the classification of Wu and Mandrand (1993) (cf. Colbeau et al., 1993).

The *hupS* gene is preceded by a sequence capable of encoding a signal peptide of 45 amino acids, which is found not only in genes encoding periplasmic hydrogenases but also is highly conserved among the membrane-bound hydrogenases (cf. Vignais and Toussaint, 1993). All the putative cleavable signal (or leader) peptides found at the N-terminal end of the small hydrogenase subunit contain a strictly conserved "RRXFXK" consensus element and are cleaved after an alanine residue which suggests a conserved mechanism for the function of these signal peptides (Voordouw, 1992). Apparently, those signal peptides may function in some cases for membrane integration, since *Rb. capsulatus* hydrogenase is an intrinsic membrane protein (Colbeau et al., 1983) and in other cases for protein export to the periplasm. A nucleotide sequence capable of encoding a putative signal peptide has been identified at the 5' end of the *hupS* gene of *Rb. capsulatus* (Leclerc et al., 1988), of *Rc. gelatinosus* (Uffen et al., 1990) and of *Tc. roseopersicina* (Colbeau et al., 1994).

Downstream from *hupL* is the *hupC* gene (formerly termed *hupM*) which, in *Rb. capsulatus*, has been shown to be expressed only from the promoter of *hupSL* and therefore to belong to the same *hupSL* transcriptional unit (Cauvin et al., 1991). Inactivation of the *hupC* (*hupM*) gene, by interposon mutagenesis, resulted in mutants which could still synthesize an active hydrogenase assayable with artificial electron acceptors (methylene blue or benzylviologen), but which could not grow autotrophically, respire with H_2, or recycle electrons to nitrogenase. In other words, in those mutants the hydrogenase was no longer functionally competent. Introduction of plasmid-borne copies of the *hupC* (*hupM*) gene into the mutant cells restored an Aut⁺ phenotype in the transconjugants and full activity to the hydrogenase enzyme (Cauvin et al., 1991). The third gene of the *hupSLC* operon is thought to encode a cytochrome *b* which links the hydrogenase to the respiratory chain, as has been demonstrated in *Wolinella succinogenes* (Dross et al., 1992).

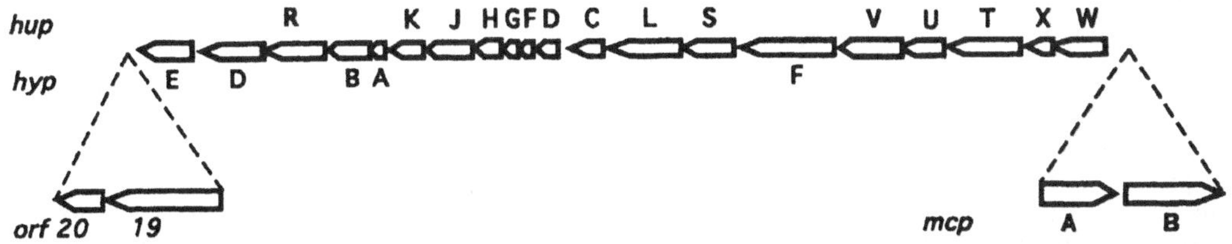

Fig. 2. Organization of genes in the *hup/hyp* gene cluster of *Rb. capsulatus* (Colbeau et al., 1993). The scheme shows also that the hydrogenase gene cluster is flanked by genes involved in chemotaxis: the product of ORF19 shares sequence similarities with *E. coli* CheB and CheR proteins, and the *mcpA* and *mcpB* genes are capable of encoding chemoreceptors of the methyl-accepting type (V. Michotey, B. Toussaint, P. Richaud and P. M. Vignais, unpublished).

3. Hydrogenase Gene Organization and Accessory Genes

The cloning and sequencing of hydrogenase gene clusters led to the surprising discovery that the number of genes necessary for hydrogenase synthesis is nearly as large as the number of *nif* genes required for the biosynthesis of nitrogenase. In the latter case, several *nif* genes have been shown to be involved in the synthesis and insertion of the metallic centers in the enzyme complex which remains in the cytoplasm as a soluble oligomeric enzyme. The membrane-bound and the periplasmic hydrogenases may require additional helper (chaperone ?) proteins and signal peptidase for membrane integration or export to the periplasm.

Comparative studies of the products of the genes which encode [NiFe]hydrogenase in various organisms suggest that there is a common core of at least 18 genes, whose products are necessary for effective hydrogenase biosynthesis. Since homologous genes have been termed differently in the various organisms, the organisation and the name of the clustered genes so far identified are indicated in Table 1. In Fig. 3, the size of the products of the best known hydrogenase gene clusters is given.

Immediately downstream from the structural hydrogenase operon *hupSLC* is *hupD*, which can encode a product homologous to *E. coli* HyaD (Table 1, Fig. 3). Inactivation of *E. coli hyaD* results in deficiency in processing of both the large and the small subunits of *E. coli* hydrogenase 1 (Menon et al., 1991). The *hyaD* product has been recently identified as a protease involved in hydrogenase maturation (Rosmann et al., 1995).

In *E. coli*, an operon necessary for the activity of the three hydrogenase isoenzymes has been cloned and sequenced. It comprises five *hyp* (for *hy*drogenase *p*leiotropic) genes, *hypABCDE* (Lutz et al., 1991). The products of the *hypB*, *hypD* and *hypE* genes are involved in the processing of the large subunits of hydrogenases 1 and 2 (Lutz et al., 1991); the *hypA* and *hypC* gene products appear to adjust the activity of hydrogenase 3, relative to those of hydrogenases 1 and 2 involved in respiration (Jacobi et al., 1992). The product of *hypB* has been recently shown by Maier et al. (1993) to catalyze some step in nickel incorporation into the 3 hydrogenases and to have GTPase activity. Homologs of the *E. coli hyp* operon genes have been found in the photosynthetic bacterium, *Rb. capsulatus* (Xu and Wall, 1991; Colbeau et al., 1993) (Table 1, Figs. 2, 3) but the exact function of their products has not yet been identified.

Genes encoding a product homologous to HypF of *Rb. capsulatus* (Colbeau et al., 1993) have been found within the *hyp* operon in *R. leguminosarum* and in *Azotobacter vinelandii* (Table 1, Fig. 3). The predicted HypF proteins contain at their N-terminal end two Cys-rich clusters which resemble zinc-finger motifs, a structure which could conceivably enable the proteins to chelate nickel atoms. The HypF protein, like HypB, is currently thought to play a role in nickel metabolism in relation with hydrogenase biosynthesis. In *Rb. capsulatus*, it has been shown also to participate in the H_2-stimulation of hydrogenase gene expression (Colbeau et al., 1993).

In summary, identification of the *hup/hyp* gene organisation has laid a firm foundation for further studies on hydrogenase biosynthesis. The gene organisation is strikingly similar in these clusters and it is anticipated that homologous genes encode proteins with analogous functions. The absence of some of the genes gives also interesting clues as to the minimum requirement for the synthesis of an effective enzyme and for the genes likely involved in specific regulation. For example, in *Tc. roseopersicina*, the *hupFG* genes appear to be lacking (Table 1), (unless they are localized elsewhere on the chromosome). The major challenge now is, on the one hand, to identify the role of those conserved genes and, on the other hand, to determine the specificity of each hydrogenase gene cluster, in terms of genes leading to the assembly of hydrogenase complexes capable of interacting with specific electron acceptors/donors, in terms of regulatory genes participating to bacterial signaling systems leading to the adaptive cellular responses to specific growth or stress conditions.

III. Regulation of Hydrogenase Genes

The remarkable metabolic versatility of the purple, nonsulfur bacteria, which permits their adaptability to environmental changes, no doubt depends on multiple regulatory circuits. Regulation of hydrogenase gene expression in response to environmental stimuli has so far been studied only in *Rb. capsulatus*.

Table 1. Gene products homologs to those encoded by the *hup* and *hyp* genes from *Rb. capsulatus*, identified in other bacteria. The Hup and Hyp proteins are ordered horizontally according to the order of their genes in the hydrogenase gene cluster in each bacterium. Empty slots mean that the corresponding gene is missing in the species. Only in *E.coli* are the hydrogenase genes dispersed on the chromosome. In *R. leguminosarum* and *A. eutrophus* the cluster of genes is plasmid-borne.

Bacterium	Ref	Name																								
Rhodobacter capsulatus	a	Hup	T	U	V		S	L	C	D		F	G	H		J	K				R					
		Hyp				F												A	B			C	D	E		
Thiocapsa roseopersicina	b	Hup					S	L	C	D				H	I						R					
Rhizobium leguminosarum	c	Hup					S	L	C	D	E	F	G	H	I	J	K									
		Hyp																A	B	F		C	D	E		
Bradyrhizobium japonicum	d	Hup		U	V		S	L	C	D		F	G	H	I	J	K									
		Hyp																A	B				D'	E		
		Hox																							X	A
Azotobacter chroococcum	e	Hup					S	L	Z	M		N	O	Q	R	T	V	A	B	Y		C	D	E		
Azotobacter vinelandii	f	Hox					K	G	Z	M		L	O	Q	R	T	V									
		Hyp																A	B	F		C	D	E		
Alcaligenes eutrophus	g	Hox					K	G	Z	M		L	O	Q	R	T	V								X	A
		Hyp																A	B			C	D	E		
Escherichia coli	h	Hya					A	B	C	D			E	F												
		Hyp																A	B			C	D	E		
		Hyd				A*																			H'	G
		Hyb					A	C	B	D						E		F				G				

* or HypF

a – Leclerc et al. (1988) (*hupSL*); Richaud et al. (1990) (*hupC*/ORFX); Colbeau et al. (1993) (*hupDFGHIJK, hypABDE, hypF*); Xu and Wall (1991) (C-ter *hupJ*/ORF1, *hupK*/ORF2, *hypA*/ORF3, *hypB*/ORF4, *N*-ter *hupR1* /ORF5'); Richaud et al. (1991) (*hupR$_1$* renamed here *hupR*); Elsen et al. (1993) (*hupT*). The third gene of the structural operon, formerly called *hupM* (Cauvin et al., 1991; Colbeau et al., 1993) has been renamed *hupC* (Vignais and Toussaint, 1993); Elsen et al., 1995 (*hupUV*).

b – Colbeau et al. (1993b) (*hupSLCDHI*).

c – Hidalgo et al. (1990) (*hupSL*); Hidalgo et al. (1992) (*hupCDEF*); Rey et al. (1992) (*hupGHIJK*); Rey et al. (1993) (*hypABFCDE*).

d – Sayavedra-Soto et al. (1988) (*hupSL*); Van Soom et al. (1993a) (*hypD'*/incomplete sequence, *hypE, hoxXA*); Van Soom et al. (1993b) (*hupCDFG*); Fu and Maier (1994a) (*hypAB*); Fu and Maier (1994b) (*hupCDF*); Fu and Maier (1994c) (*hupGHIJK*); Black et al. (1994) (*hupUV*).

e – Ford et al. (1990) (*hupSL*); Du et al. (1992) (*hupD/hypD* and *hupE/hypE*); Tibelius et al. (1993) (*hupABYC/hypABFC*); Du et al. (1994) (*hupZMNOQRTV*).

f – Menon et al. (1990a) (*hoxKGZI*) Menon et al. (1992) (*hoxMLOQ*); Chen and Mortenson (1992a) (*hoxRT*); Chen and Mortenson (1992b) (*hoxV, hupABCD*); Garg et al. (1994) (*hypE*).

g – Kortlücke et al. (1992) (*hoxKGZMLOQRTV*); Eberz and Friedrich (1991) (*hoxA*); Dernedde et al. (1993) (*hypABCDE*); Lenz et al. (1994) (*hoxX*).

h – Tomiyama et al. (1991) (*hydA/hypF*); Menon et al. (1990b) (*hyaABCDEF*); Lutz et al. (1991) and Jacobi et al. (1992) (*hypABCDE*); Stoker et al. (1989) (*hydG, hydH'* incomplete sequence); Menon et al. (1994) (*hybABCDEFG*).

A. Environmental Factors Affecting Hydrogenase Synthesis

A translational fusion of the *E. coli lacZ* gene to the *Rb. capsulatus hupSLC* promoter has been constructed and used to identify the Hup⁻ mutants mutated in regulatory genes and the environmental stimuli to which those regulatory genes responded (Colbeau and Vignais, 1992). The plasmid-borne *hupS::lacZ* fusion was transferred to the wild type strain B10 and to various Hup⁻ mutants. The expression of *β*-galactosidase and that of the chromosomally encoded hydrogenase were then measured under various physiological conditions. The factors found to affect

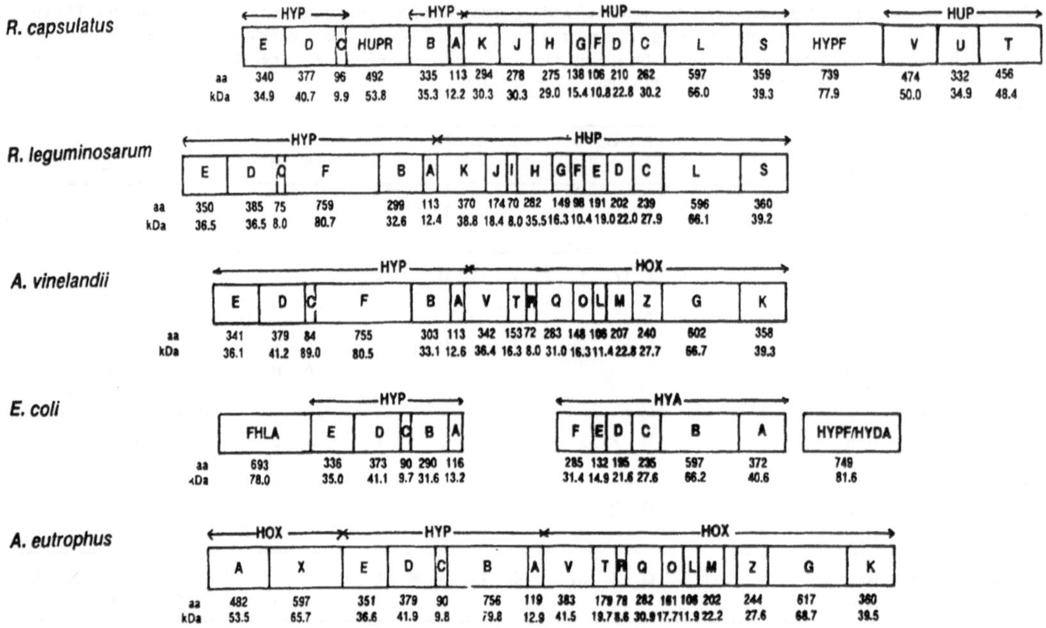

Fig. 3. Comparison of deduced hydrogenase gene products. Only the most completely sequenced clusters are shown. The size of boxes is proportional to the size (in kDa) of the predicted gene products. See legend to Table 1 for source of data.

hydrogenase expression were the following:

1. Molecular Hydrogen Stimulates Hydrogenase Gene Expression in Rb. capsulatus

Molecular hydrogen was the factor which, in the wild-type strain B10, stimulated most the expression of β-galactosidase and of hydrogenase. The highest expression was observed under conditions which favor expression of nitrogenase (anaerobiosis, light and malate-glutamate medium) or in darkness in the presence of H_2 plus O_2. Since H_2 is evolved by nitrogenase, $NifR_4^-$ mutants devoid of nitrogenase (Ahombo et al., 1986) were used to check whether the increased hydrogenase expression observed under nitrogenase-derepressing conditions was really due to H_2 and not a result of activation of the *nif* or *ntr* regulatory circuit. The *nifR*$_4$ gene of *Rb. capsulatus* encodes a protein homologous with the sigma factor σ^{54}, product of the *ntrA* (*rpoN*) gene in other bacteria (Jones and Haselkorn, 1989; Alias et al., 1989). It was found that hydrogenase and β-galactosidase were still expressed in $NifR_4^-$ mutants in a H_2-stimulatable way (Colbeau and Vignais, 1992). Therefore, firstly, the observed stimulation was indeed due to the presence of H_2, and secondly, hydrogenase expression in *Rb. capsulatus* had no absolute requirement for the factor σ^{54} encoded by *nifR*$_4$. The possibility cannot be excluded that there is another factor σ^{54} in *Rb. capsulatus*, as was found in *Bradyrhizobium japonicum* (Kullik et al., 1991) and in *Rb. sphaeroides* (Meijer and Tabita, 1992).

2. Oxygen Control

From the use of DNA gyrase inhibitors, Kranz and Haselkorn (1986) have suggested that O_2 regulation of *nif* gene expression was exerted at the level of DNA supercoiling. It is interesting to note that O_2 regulation of hydrogenase synthesis is unaffected by DNA gyrase inhibitors such as novobiocin (Willison et al., 1990). Rather unexpectedly, since hydrogenase enzymes are O_2-labile and unlike nitrogenase, hydrogenase and β-galactosidase expressed from the *hupS::lacZ* fusion were still synthesized at high levels in the form of active enzymes in the presence of oxygen provided that H_2 be present (Colbeau and Vignais, 1992). Elsen et al. (1993) have even been able to obtain mutants ($HupT^-$, see below) which overexpress active hydrogenase in the presence of oxygen, indicating that the HupT protein is involved in (O_2)-repression.

B. Transcriptional Regulation and Regulatory Genes

1. The hupR Gene Encodes a Response Regulator of the NtrC Subfamily

Measurement of β-galactosidase and of hydrogenase activity showed that both activities in the Hup^- mutant RCC8, harboring the *hupS::lacZ* fusion, were decreased to 4% of those found in the wild-type strain B10 (Colbeau and Vignais, 1992). This finding indicated that the mutation in RCC8 affects a transcriptional factor required for the expression of both enzymes. The gene, *hupR*, complementing the mutation was isolated and sequenced, together with the mutated *hupR* gene from mutant RCC8. The *hupR* gene is capable of encoding a transcriptional activator belonging to the NtrC subfamily (Richaud et al., 1991) of the superfamily of two-component regulatory systems (reviewed by Stock et al., 1989, 1991; Albright et al., 1989; Bourret et al., 1991; Parkinson and Kofoid, 1992; Parkinson, 1993). These activators act by catalyzing the isomerization of the closed RNA polymerase-DNA complex at the transcriptional start site, into an open complex, at the expense of ATP. The mutated *hupR* gene in strain RCC8 contains a point mutation in the codon of one of the amino acids belonging to the putative ATP-binding site of the HupR protein (Richaud et al., 1991). These data support the contention that HupR plays the role of a response regulator in the transcription of the *hupSLC* operon. The HupR protein shares an overall 43% identity with the HoxA protein of *A. eutrophus* shown to be a transcriptional regulator in the expression of *A. eutrophus* hydrogenase (Eberz and Friedrich, 1991).

2. The hupT Gene Product of Rb. capsulatus Represses Hydrogenase Synthesis

The *hupT* gene is located upstream from the structural *hupSLC* genes, while *hupR* is downstream in the middle of the *hyp* genes (Fig. 2). Amino acid sequence comparisons have indicated that the predicted HupT protein (456 amino acids, 48.4 kDa) belongs to the superfamily of sensor kinases which, in two-component regulatory systems, function in tandem with a response regulator. It is predicted to be a soluble sensor-kinase, since it lacks the hydrophobic region thought to maintain the protein in the membrane, as do *Rb. capsulatus* $NifR_2$ and *Klebsiella pneumoniae* NtrB with which it shares sequence similarities. HupT is expected to be a kinase molecule capable of autophosphorylation and a putative nucleotide binding site has indeed been identified in the deduced amino acid sequence (Elsen et al., 1993). HupT shares sequence similarities with components of signal transduction systems responding to O_2 such as FixL of *Rhizobium meliloti* (although no significant similarity exists with the fragment of FixL shown to bind heme and to bind oxygen (Monson et al., 1992)), or ArcB of *E. coli* which represses the *arc* modulon under anaerobiosis (Iuchi et al., 1990). However, while ArcB participates in anaerobic repression, *hupT* appears to be involved in regulation by the carbon status of the cell (S. Elsen, unpublished results). It is also of interest to note that HupT does not share sequence similarity with HoxX, the putative sensor protein thought to function in tandem with the response regulator HoxA (homologous to *Rb. capsulatus* HupR, Table 1), in *A. eutrophus* and *B. japonicum* (cf. Elsen et al., 1993).

Two other genes, *hupU* and *hupV*, contiguous with and downstream from *hupT*, belong to the same *hupTUV* operon (Fig. 2). HupU, capable of encoding a protein of 34.9 kDa, and *hupV* a protein of 50.0 kDa, share sequence similarities with the small and the large subunits of [NiFe] hydrogenases, respectively. The strongest similarity is observed with the periplasmic hydrogenase of *Desulfovibrio baculatus*, but the *hupU* gene does not comprise a sequence for a signal peptide (Elsen et al., 1993; S. Elsen, A. Colbeau and P. M. Vignais, unpublished data).

Insertional inactivation of the *hupT* gene leads to deregulation of transcriptional control so that the structural hydrogenase operon *hupSLC* becomes overexpressed in cells grown aerobically or anaerobically, either in the presence or in the absence of ammonia. The $HupT^-$ mutants are Hup^c (constitutive) mutants in which hydrogenase is maximally expressed under conditions (presence of ammonia, presence of oxygen) where hydrogenase activity is minimum in the wild-type strain B10. This indicates that the HupT protein participates in the *repression* of hydrogenase expression. This repression is restored when plasmid-borne copies of the wild-type gene are introduced in the HupT⁻ mutants (Elsen et al., 1993). It is to note that, in contrast, mutation in the *hupR* gene resulted instead in a decrease in transcription and in activity of hydrogenase indicating that the HupR protein is rather a positive transcriptional factor (Colbeau and Vignais, 1992).

Therefore, although the *hupR* gene product presents features of response regulators and the product of *hupT* those of sensor kinases indicating that they belong to the superfamily of two-component regulatory systems, it remains to be demonstrated that the two proteins form a matched pair of regulatory proteins.

3. An IHF-like Protein Activates in vivo Transcription of the hupSLC Operon

By complementation of the Hup⁻ mutant IR4, a gene called *himA* of *Rb. capsulatus* has been isolated and sequenced (Toussaint et al., 1991). The gene is capable of encoding a protein sharing 45% identical amino acids with the α-subunit of the Integration Host Factor (IHF) of *E. coli*. An IHF consensus sequence, identified in the promoter of the *hupSLC* operon, was shown to bind *E. coli* IHF by gel retardation assays, and therefore to be a true IHF binding site. Similarly, the IHF-like protein recently isolated from *Rb. capsulatus* has been shown to bind to the same IHF binding site. Like *E. coli* IHF, *Rb. capsulatus* IHF is a heterodimeric protein (10 and 11 kDa). The *hip* gene which encodes the β subunit has been isolated from *Rb. capsulatus* and sequenced; its products shares 46% identical amino acids with *E. coli* IHF β subunit. Foot-printing analyses have shown that *Rb. capsulatus* IHF and *E. coli* IHF on the one hand, and that the IHF protein isolated from the *himA* mutant IR4 on the other hand, do not interact exactly in the same manner with the IHF binding site of the *hupS* promoter region (Toussaint et al., 1993). A molecular model of the interaction between *Rb. capsulatus* IHF protein and DNA from the *hupS* promoter region is shown in Fig. 4. In this model, where DNA is strongly bent by the IHF molecule, Arg-8 is able to interact with the phosphate-ribose backbone of DNA in the flanking region of the IHF binding site (Toussaint et al., 1991, 1994). It has been shown earlier that the replacement of Arg-8 by Cys-8 in mutant IR4 affects hydrogenase transcription (Colbeau and Vignais, 1992).

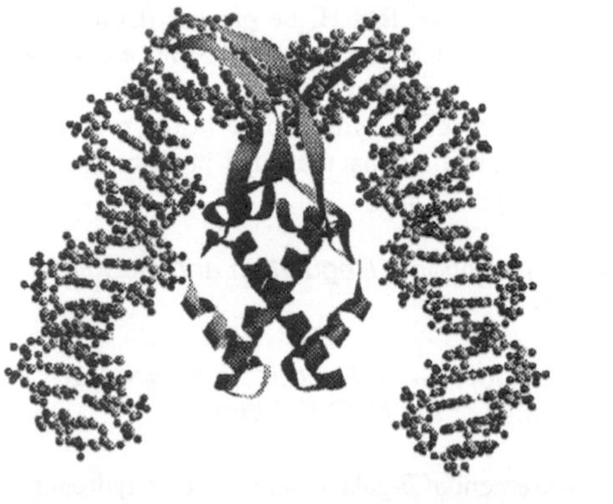

Fig. 4. A three-dimensional model of *Rb. capsulatus* IHF bound to its binding site on the *hupSL* promoter. Coordinates of the protein atoms were deduced from the crystal structure of HU of *Bacillus stearothermophilus* (Tanaka et al., 1984; White et al., 1989). The 40 bp IHF site has been modeled taking into account DNA bending. The position of IHF to its site was deduced from Dnase I foot-printing data (Toussaint et al., 1993). Arg-8 is replaced by a Cys in the IR4 mutant. The model has been assembled and refined by Laurent David using the Quanta 3 program of Molecular Simulation.

4. Sigma Factors

The *nifR₄* gene codes for a protein showing sequence homology to the *ntrA* gene product (σ^{54}) of other Gram-negative bacteria (Alias et al., 1989; Jones and Haselkorn, 1989). The *nifR₄* gene is located adjacent to, and downstream from, the nitrogenase structural genes (*nifHDK*) (Fig. 1) and is transcribed from its own promoter (Ahombo et al., 1986). Studies with a *nifR₄::lacZ* gene fusion have shown that transcription of the *nifR₄* gene requires the presence of the *nifR₁/nifR₂* (*ntrC/ntrB*) genes and of *nifA* and that the *nifR₄* gene product is also required for its own synthesis (Hübner et al., 1991). The *nifR₄* gene is therefore different from its homolog, *ntrA*, in other Gram-negative bacteria, which is generally expressed constitutively (reviewed by Thöny and Hennecke, 1989). A further difference is that in *Rb. capsulatus*, the *nifR₄* gene is exclusively involved in the regulation of nitrogen fixation, whereas in other Gram-negative bacteria, many of which are not nitrogen fixing (e.g., *E. coli*, *Pseudomonas putida*, and *A. eutrophus*) *ntrA* may be involved in the regulation of the uptake and utilization of alternative nitrogen sources, of some carbon sources, and of H_2 (Thöny and Hennecke, 1989).

As described above, nitrogen fixation and hydrogen

uptake are interrelated, since the H_2 evolved by nitrogenase may induce the synthesis of hydrogenase, and under some conditions, hydrogenase activity may allow the recycling of electrons from H_2 to nitrogenase. However, a relationship at the genetic level has not been established, and the *nifR$_4$* gene product is not required for hydrogenase gene expression (Colbeau and Vignais, 1992). Transcriptional analysis of the *hupS* promoter by primer extension or nuclease S1 protection has allowed us to determine the transcriptional start site of *hupSLC* operon. There is no direct evidence of the occurrence of the (–24)GG-N_{10}-GC(–12) consensus sequence in the promoter. On the other hand, around positions –10 and –35 we can observe the presence of hexamers, GAGAAT and TTGGCG, which contain respectively 4/6 and 5/6 of nucleotides identical to –10 –35 *E. coli* boxes recognized by sigma factor 70 (σ^{70}) (Delic-Attree I, unpublished results). Although IHF is usually implicated in transcription from σ^{54}-associated RNA polymerase, it has already been shown that expression of the *E. coli nar* and *ompF* genes (Rabin et al., 1992; Ramani et al., 1992) utilises IHF and σ^{70}-associated RNA polymerase. Nevertheless, the fact that expression of hydrogenase genes is dependent on the response regulator HupR suggests that an alternative sigma factor for RNA polymerase may also be required. (It should be noted that no additional genes homologous to *nifR$_4$* have been identified in *Rb. capsulatus* by DNA-DNA hybridization). The possible existence of a sigma factor other than $NifR_4$ has also been invoked for the regulation of the expression of the *nifA* genes in *Rb. capsulatus* (Hübner et al., 1991). In *Rb. capsulatus*, the *nifA* gene is unusually present in two copies, both of which are active (Masepohl et al., 1988). The expression of both copies of *nifA* is dependent on the *ntrBC* homologs, *nifR$_2$* and *nifR$_1$*, but is independent of *nifR$_4$*, suggesting that an additional, alternative sigma factor is required for their expression (Hübner et al., 1991).

IV. Conclusions and Perspectives

The identification of the genes necessary for hydrogenase expression in several organisms, the striking similarity of the proteins encoded by homologous genes, should accelerate our understanding of the function of those proteins and of the mechanisms of *hup* gene control and regulation. Among the photosynthetic bacteria, it is in *Rb. capsulatus* that the molecular biology of hydrogenase is the most advanced.

With respect to the environmental stimuli which have been shown to affect hydrogenase gene expression in *Rb. capsulatus*, there are clearly two main factors—namely the availability of molecular hydrogen and the levels of O_2 in the immediate environment. Synthesis of hydrogenase is stimulated by H_2 and is repressed in presence of O_2 alone, but is maximal in the presence of H_2 plus O_2.

One significant environmental factor that has received little attention until recently in terms of possible regulatory effects, is the availability of metals such as nickel. The lack of data may result from technical difficulties in addressing the problem (it is difficult to deplete *Rb. capsulatus* cells of nickel, for example), rather than from a lack of interest in the subject.

Hup gene regulation presents features typical of prokaryotic gene regulation in that it depends upon regulatory proteins belonging to the widespread family of two-component regulatory systems. It may also present other specific features since analysis of the *hup* promoter structure appears to indicate that σ^{70} rather than RpoN (σ^{54}) may be involved in *hup* gene transcription. There is still much to be learned from further studies of these systems, with respect to the detailed molecular mechanisms of activation of gene transcription, to the identification of environmental factors affecting *hup* gene expression and in the understanding of the processes of signal transduction that link gene expression to environmental changes.

Particular questions which are likely to be addressed in future studies are as follows:

(i) How is hydrogenase synthesis regulated by H_2 and by O_2? At what level does regulation occur, and are specific DNA recognition sequences involved?

(ii) Are nickel ions directly involved in regulation of expression? If they are, can they play the role of H_2 detector? In the *hup* gene cluster, which gene can encode a protein able to bind Ni and become a transcriptional activator?

(iii) Does *Rb. capsulatus* contain alternative sigma factors for RNA polymerase in addition to the constitutive sigma factor (σ^{70}) and σ^{54} ($NifR_4$/NtrA/RpoN). If it does, are these factors involved in the expression of the *hup* genes?

(iv) How are modulated the interactions between IHF and DNA? Are there additional proteins which help in the binding of IHF to its site? If yes, what are they? Is it the participation of this type of helper proteins that explains the lack of effect of the *himA* mutation, in the IR4 mutant, on *nif* gene expression?

(v) The predicted HupR and HupT proteins belong to the superfamily of two-component regulatory systems. Does the putative response regulator HupR function in tandem with the putative sensor kinase HupT? What are the signals to which each of them respond? Does HupR activate transcription in the phosphorylated form? If HupR and HupT do not form a matched pair, what are the other partners?

Elucidation of the answers to these questions will require studies both at the cellular level and in vitro, using purified components of the different regulatory gene networks.

Acknowledgments

The authors gratefully thank Drs J. Meyer and J.C. Willison for critical reading of the manuscript and J. Boyer for expert assistance in the preparation of the manuscript.

References

Ahombo G, Willison JC and Vignais PM (1986) The *nifHDK* genes are contiguous with a *nifA*-like regulatory gene in *Rhodobacter capsulatus*. Mol Gen Genet 205: 442–445

Albracht SPJ, Kalkman ML and Slater EC (1983) Magnetic interaction of nickel(III) and the iron-sulphur cluster in hydrogenase from *Chromatium vinosum*. Biochim Biophys Acta 724: 309–316

Albright LM, Huala E and Ausubel M (1989) Prokaryotic signal transduction mediated by sensor and regulator protein pairs. Annu Rev Genet 23: 311–336

Alias A, Cejudo FJ, Chabert J, Willison JC and Vignais PM (1989) Nucleotide sequence of wild-type and mutant *nifR4(ntrA)* genes of *Rhodobacter capsulatus*: Identification of an essential glycine residue. Nucleic Acids Res 17: 5377

Bagyinka C, Whitehead JP, Maroney MJ (1993) An X-ray absorption spectroscopic study of nickel redox chemistry in hydrogenase. J Am Chem Soc 115: 3576–3585

Black KL, Fu C and Maier RJ (1994) Sequences and characterization of *hupU* and *hupV* genes of *Bradyrhizobium japonicum* encoding a possible nickel-sensing complex involved in hydrogenase expression. J Bacteriol 176: 7102–7106

Bonam D, Lehman L, Roberts GP and Ludden PW (1989) Regulation of carbon monoxide dehydrogenase and hydrogenase in *Rhodospirillum rubrum*: Effects of CO and oxygen on synthesis and activity. J Bacteriol 171: 3102–3107

Bourret RB, Borkovich KA and Simon MI (1991) Signal transduction pathways involving protein phosphorylation in prokaryotes. Annu Rev Biochem 60: 401–441

Brewin NJ, DeJong TM, Phillips DA and Johnston AWB (1980) Co-transfer of determinants for hydrogenase activity and nodulation ability in *Rhizobium leguminosarum*. Nature (London) 288: 77–79

Bulen WA, Burns RC and Le Comte JR (1965a) Nitrogen fixation: Hydrosulfite as electron donor with cell free preparations of *Azotobacter vinelandii* and *Rhodospirillum rubrum*. Proc Natl Acad Sci USA 53: 532–539

Bulen WA, Burns RC and Le Comte JR (1965b) Nitrogen fixation studies with aerobic and photosynthetic bacteria. In: San Pietro A (ed) Non-heme Iron Proteins: Role in Energy Conservation, pp 261–287. Antioch Press, Yellow Springs, Ohio

Cauvin B, Colbeau A and Vignais PM (1991) The hydrogenase structural operon in *Rhodobacter capsulatus* contains a third gene, *hupM*, necessary for the formation of a physiologically competent hydrogenase. Mol Microbiol 5: 2519–2527

Chen JC and Mortenson LE (1992a) Two open reading frames (ORFs) identified near the hydrogenase structural genes in *Azotobacter vinelandii*, the first ORF may encode for a polypeptide similar to rubredoxins. Biochim Biophys Acta 1131: 122–124

Chen JC and Mortenson LE (1992b) Identification of six open reading frames from a region of the *Azotobacter vinelandii* genome likely involved in dihydrogen metabolism. Biochim Biophys Acta 1131: 199–202

Colbeau A and Vignais PM (1983) The membrane-bound hydrogenase of *Rhodopseudomonas capsulata* is inducible and contains nickel. Biochim Biophys Acta 748: 128–138

Colbeau A and Vignais PM (1992) Use of *hupS::lacZ* gene fusion to study regulation of hydrogenase expression in *Rhodobacter capsulatus*: Stimulation by H_2. J Bacteriol 174: 4258–4264

Colbeau A, Kelley BC and Vignais PM (1980) Hydrogenase activity in *Rhodopseudomonas capsulata*: Relationship with nitrogenase activity. J Bacteriol 144: 141–148

Colbeau A, Chabert J and Vignais PM (1983) Purification, molecular properties and localization in the membrane of the hydrogenase of *Rhodopseudomonas capsulata*. Biochim Biophys Acta *748*: 116–127

Colbeau A, Godfroy A and Vignais PM (1986) Cloning of DNA fragments carrying hydrogenase genes of *Rhodopseudomonas capsulata*. Biochimie 68: 147–155

Colbeau A, Magnin JP, Cauvin B, Champion T and Vignais PM (1990) Genetic and physical mapping of an hydrogenase gene cluster from *Rhodobacter capsulatus*. Mol Gen Genet 220: 393–399

Colbeau A, Richaud P, Toussaint B, Caballero FJ, Elster C, Delphin C, Smith RL, Chabert J and Vignais PM (1993). Organization of the genes necessary for hydrogenase expression in *Rhodobacter capsulatus*. Sequence analysis and identification of two *hyp* regulatory mutants. Mol Microbiol 8: 15–29

Colbeau A, Kovacs KL, Chabert J and Vignais PM (1994) Cloning and nucleotide sequences of the structural (*hupSLC*) and accessory (*hupDHI*) genes for hydrogenase biosynthesis in *Thiocapsa roseopersicina*. Gene 140: 25–31

Dernedde J, Eitinger M and Friedrich B (1993) Analysis of a pleiotropic gene region involved in formation of catalytically active hydrogenases in *Alcaligenes eutrophus* H16. Arch Microbiol 159: 545–553

Donohue TJ and Kaplan S (1991) Genetic techniques in Rhodospirillaceae. Methods Enzymol 204: 459–485

Dross F, Geisler V, Lenger R, Theis F, Krafft T, Fahrenholz F, Kojro E, Duchêne A, Tripier D, Juvenal K and Kröger A (1992) The quinone-reactive Ni/Fe-hydrogenase of *Wolinella succinogenes*. Eur J Biochem 206: 93–102

Du L, Stejskal F and Tibelius KH (1992) Characterization of two genes (*hupD* and *hupE*) required for hydrogenase activity in *Azotobacter chroococcum*. FEMS Microbiol Lett 96: 93–102

Du L, Tibelius KH, Souza EM, Garg RP and Yates MG (1994) Sequences, organization and analysis of the *hupZMNOQRTV* genes from the *Azotobacter chroococcum* hydrogenase gene cluster. J Mol Biol 243: 549-557

Eberz G and Friedrich B (1991) Three trans-acting regulatory functions control hydrogenase synthesis in *Alcaligenes eutrophus*. J Bacteriol 173: 1845–1854

Eberz G, Hogrefe C, Kortlücke C, Kamienski A and Friedrich B (1986) Molecular cloning of structural and regulatory hydrogenase (*hox*) genes of *Alcaligenes eutrophus* H16. J Bacteriol 168: 636–641

Elsen S, Richaud P, Colbeau A and Vignais PM (1993) Sequence analysis and interposon mutagenesis of the *hupT* gene, which encodes a sensor protein involved in repression of hydrogenase synthesis in *Rhodobacter capsulatus*. J Bacteriol 175: 7404–7412

Fonstein M and Haselkorn R (1993) Chromosomal structure of *Rhodobacter capsulatus* strain SB1003: Cosmid encyclopedia and high-resolution physical and genetic map. Proc Natl Acad Sci USA 90: 2522–2526

Ford CM, Garg N, Garg RP, Tibelius KH, Yates MG, Arp DJ and Seefeldt LC (1990) The identification, characterization, sequencing and mutagenesis of the genes (*hupSL*) encoding the small and large subunits of the H_2-uptake hydrogenase of *Azotobacter chroococcum*. Mol Microbiol 4: 999–1008

Fu C and Maier RJ (1994a) Nucleotide sequences of two hydrogenase-related genes (*hypA* and *hypB*) from *Bradyrhizobium japonicum*, one of which (*hypB*) encodes an extremely histidine-rich region and guanine nucleotide-binding domains. Biochim Biophys Acta 1184: 135–138

Fu C and Maier RJ (1994b) Sequence and characterization of three genes within the hydrogenase cluster of *Bradyrhizobium japonicum*. Gene 141: 47–52

Fu C and Maier RJ (1994c) Organisation of the hydrogenase gene cluster from *Bradyrhizobium japonicum*: Sequences and analysis of five more hydrogenase-related genes. Gene, in press

Garg RP, Menon AL, Jacobs K, Robson RM and Robson RL (1994) The *hypE* gene completes the gene cluster for H_2-oxidation in *Azotobacter vinelandii*. J Mol Biol 236: 390–396

Gest H (1952) Properties of cell-free hydrogenases of *Escherichia coli* and *Rhodospirillum rubrum*. J Bacteriol 63: 111–121

Gest H (1954) Oxidation and evolution of molecular hydrogen by microorganisms. Bacteriol Rev 18: 43–73

Gest H (1972) Energy conservation and generation of reducing power in bacterial photosynthesis. Adv. Microbial Physiol. 7: 243–282

Gest H and Kamen MD (1949a) Studies on the metabolism of photosynthetic bacteria. IV Photochemical production of molecular hydrogen by growing cultures of photosynthetic bacteria. J Bacteriol 58: 239–245

Gest H and Kamen MD (1949b) Photoproduction of molecular hydrogen by *Rhodospirillum rubrum*. Science 109: 558–559

Gest H, Kamen MD and Bregoff HM (1950) Studies on the metabolism of photosynthetic bacteria. V Photoproduction of hydrogen and nitrogen fixation by *Rhodospirillum rubrum*. J Biol Chem 182: 153–170

Gogotov IN (1986) Hydrogenases of phototrophic microorganisms. Biochimie 68: 181–187

Gogotov IN, Zorin NA and Bogorov LV (1974) Hydrogen metabolism and ability for nitrogen fixation in *Thiocapsa roseopersicina*. Mikrobiologiya 43: 5–11

Gorrell TE and Uffen RL (1977) Fermentative metabolism of pyruvate by *Rhodospirillum rubrum* after anaerobic growth in darkness. J Bacteriol 131: 533–543

Gorrell TE and Uffen RL (1978) Light-dependent and light-independent production of hydrogen gas by photosynthesizing *Rhodospirillum rubrum* mutant C Photochem Photobiol 27: 351–358

Hidalgo E, Leyva A and Ruiz-Argüeso T (1990) Nucleotide sequence of the hydrogenase structural genes from *Rhizobium leguminosarum*. Plant Mol Biol 15: 367–370

Hidalgo E, Palacios JM, Murillo J and Ruiz-Argüeso T (1992) Nucleotide sequence and characterization of four additional genes of the hydrogenase structural operon from *Rhizobium leguminosarum* bv. viciae. J Bacteriol 174: 4130–4139

Hillmer P and Gest H (1977a) H_2 metabolism in the photosynthetic bacterium *Rhodopseudomonas capsulata*: H_2 production by growing cultures. J Bacteriol 129: 724–731

Hillmer P and Gest H (1977b) H_2 metabolism in the photosynthetic bacterium *Rhodopseudomonas capsulata*. Production and utilization of H_2 by resting cells. J Bacteriol 129: 732–739

Hübner P, Willison JC, Vignais PM and Bickle TA (1991) Expression of regulatory *nif* genes in *Rhodobacter capsulatus*. J Bacteriol 173: 2993–2999

Iuchi S, Matsuda Z, Fujiwara T and Lin ECC (1990) The *arcB* gene of *Escherichia coli* encodes a sensor-regulator protein for anaerobic repression of the *arc* modulon. Mol Microbiol 4: 715–727

Jacobi A, Rossmann R and Böck A (1992) The *hyp* operon gene products are required for the maturation of catalytically active hydrogenase isoenzymes in *Escherichia coli*. Arch Microbiol 158: 444–451

Jones R and Haselkorn R (1989) The DNA sequence of the *Rhodobacter capsulatus ntrA*, *ntrB* and *ntrC* gene analogues required for nitrogen fixation. Mol Gen Genet 215: 507–516

Jouanneau Y, Kelley BC, Berlier Y, Lespinat PA and Vignais PM (1980) Continuous monitoring, by mass spectrometry of H_2 production and recycling in *Rhodopseudomonas capsulata*. J Bacteriol 143: 628–636

Jouanneau Y, Lebecque S and Vignais PM (1984) Ammonia and light effect on nitrogenase activity in nitrogen-limited continuous cultures of *Rhodopseudomonas capsulata*. Role of glutamine synthetase. Arch Microbiol 139: 326–331

Jouanneau Y, Wong B and Vignais PM (1985) Stimulation by

light of nitrogenase synthesis in cells of *Rhodopseudomonas capsulata* growing in N-limited continuous culture. Biochim Biophys Acta 808: 149–155.

Kern M, Klipp W and Klemme JH (1994) Increased nitrogenase-dependent H_2 photoproduction by *hup* mutants of *Rhodospirillum rubrum*. Appl Environ Microbiol 60: 1768–1774

Klemme JH (1968) Untersuchungen zur Photoautotrophie mit molekularem Wasserstoff bei neuisolierten schwefelfreien Purpurbakterien Archiv Microbiol 64: 29–42

Klipp W, Masepohl B and Pühler A (1988) Identification and mapping of nitrogen fixation genes of *Rhodobacter capsulatus*: Duplication of a *nifA-nifB* region. J Bacteriol 170: 693–699

Kortlüke CH, Horstmann K, Schwartz E, Rohde M, Binsack R, Friedrich B (1992) A gene complex coding for the membrane-bound hydrogenase of *Alcaligenes eutrophus* H16. J Bacteriol 174: 6277–6289

Kranz RG and Haselkorn R (1986) Anaerobic regulation of nitrogen-fixation genes in *Rhodopseudomonas capsulata*. Proc Natl Acad Sci USA 83: 6805–6809

Kullik I, Fritsche S, Knobel H, Sanjuan J, Hennecke H and Fischer HM (1991) *Bradyrhizobium japonicum* has two differentially regulated, functional homologs of the σ^{54} gene (*rpoN*). J Bacteriol 173: 1125–1138

Leclerc M, Colbeau A, Cauvin B and Vignais PM (1988) Cloning and sequencing of the genes encoding the large and the small subunits of the H_2 uptake hydrogenase (*hup*) of *Rhodobacter capsulatus*. Mol Gen Genet 214: 97-107. Erratum Mol Gen Genet (1989) 215: 368

Lenz O, Schwartz E, Dernedde J, Eitinger M and Friedrich B (1994) The *Alcaligenes eutrophus* H16 *hoxX* gene participates in hydrogenase regulation. J Bacteriol 176: 4385–4393

Lindstrom ES, Burris RH and Wilson PW (1949) Nitrogen fixation by photosynthetic bacteria. J Bacteriol 58: 313–316

Lindstrom ES, Lewis SM and Pinsky MJ (1951) Nitrogen fixation and hydrogenase in various bacterial species. J Bacteriol 61: 481–487

Lutz S, Jacobi A, Schlensog V, Böhm R, Sawers G and Böck A (1991) Molecular characterization of an operon (*hyp*) necessary for the activity of the three hydrogenase isoenzymes in *Escherichia coli*. Mol Microbiol 5: 123–135

Madigan MT and Gest H (1979) Growth of the photosynthetic bacterium *Rhodopseudomonas capsulata* chemoautotrophically in darkness with H_2 as the energy source. J Bacteriol 137: 524–530

Madigan M, Cox SS and Stegeman RA (1984) Nitrogen fixation and nitrogenase activities in members of the family Rhodospirillaceae. J Bacteriol 157: 73–78

Maier T, Jacobi A, Sauter M and Böck A (1993) The product of the *hypB* gene, which is required for nickel incorporation into hydrogenases, is a novel guanine nucleotide-binding protein. J Bacteriol 175: 630–635

Marrs B (1974) Genetic recombination in *Rhodopseudomonas capsulata*. Proc Natl Acad Sci USA 71: 971–973

Masepohl B, Klipp W and Pühler A (1988) Genetic characterization and sequence analysis of the duplicated *nifA/nifB* gene region of *Rhodobacter capsulatus*. Mol Gen Genet 212: 27–37

Menon AL, Mortenson LE and Robson RL (1992) Nucleotide sequences and genetic analysis of hydrogen oxidation (*hox*) genes in *Azotobacter vinelandii*. J Bacteriol 174: 4549–4557

Menon AL, Stults LW, Robson RL and Mortenson LE (1990a) Cloning, sequencing and characterization of the [NiFe] hydrogenase-encoding structural genes (*hoxK* and *hoxG*) from *Azotobacter vinelandii*. Gene 96: 67–74

Menon NK, Robbins J, Peck Jr HD, Chatelus CY, Choi ES and Przybyla AE (1990b) Cloning and sequencing of putative *Escherichia coli* [NiFe]hydrogenase-1 operon containing six open reading frames. J Bacteriol 172: 1969–1977

Menon NK, Robbins J, Wendt JC, Shanmugan KT and Przybyla AE (1991) Mutational analysis and characterization of the *Escherichia coli hya* operon, which encodes [NiFe]hydrogenase 1. J Bacteriol 173: 4851–4861.

Menon NK, Chatelus CY, Dervartanian M, Wendt JC, Shanmugan KT, Peck HD Jr and Przybyla AE (1994) Cloning, sequencing and mutational analysis of the *hyb* operon encoding *Escherichia coli* hydrogenase 2. J Bacteriol 176: 4416–4423

Meyer J, Kelley BC and Vignais PM (1978) Nitrogen fixation and hydrogen metabolism in photosynthetic bacteria. Biochimie 60: 245–260

Meijer WG and Tabita FR (1992) Isolation and characterization of the *nifUSVW-rpoN* gene cluster from *Rhodobacter sphaeroides*. J Bacteriol 174: 3855–3866

Monson EK, Weinstein M, Ditta GS and Helinski DR (1992) The FixL protein of *Rhizobium meliloti* can be separated into a hem-binding oxygen-sensing domain and a functional C-terminal kinase domain. Proc Natl Acad Sci USA 89: 4280–4284

Ormerod JG, Ormerod KS and Gest H (1961) Light-dependent utilization of organic compounds and photoproduction of molecular hydrogen by photosynthetic bacteria: Relationships with nitrogen metabolism. Arch Biochem Biophys 94: 449–463

Parkinson JS (1993) Signal transduction schemes of bacteria. Cell 73: 857–871

Parkinson JS and Kofoid EC (1992) Communication modules in bacterial signaling proteins. Annu Rev Genet 26: 71–112

Paschinger H (1974) A changed nitrogenase activity in *Rhodospirillum rubrum* after substitution of tungsten for molybdenum. Arch Microbiol 101: 379–389

Rabin RS, Collins LA and Stewart V (1992) In vivo requirement of integration host factor for *nar* (nitrate reductase) operon expression in *Escherichia coli* K-12. Proc Natl Acad Sci USA 89: 8701–8705

Ramani N, Huang L and Freundlich M (1992) In vitro interactions of integration host factor with the *ompF* promoter regulatory region of *Escherichia coli*. Mol Gen Genet 231: 248–255

Rey L, Hidalgo E, Palacios J and Ruiz-Argüeso T (1992) Nucleotide sequence and organization of an H_2-uptake gene cluster from *Rhizobium leguminosarum* bv. viciae containing a rubredoxin-like gene and four additional open reading frames. J Mol Biol 228: 998–1002

Rey L, Murillo J, Hernando Y, Hidalgo E, Cabrera E, Imperial J and Ruiz-Argüeso T (1993) Molecular analysis of a microaerobically induced operon required for hydrogenase synthesis in *Rhizobium leguminosarum* bv. viciae. Mol Microbiol 8: 471–481

Richaud P, Vignais PM, Colbeau A, Uffen RL and Cauvin B (1990) Molecular biology studies of the uptake hydrogenase of *Rhodobacter capsulatus* and *Rhodocyclus gelatinosus*. FEMS Microbiology Rev 87: 413–418

Richaud P, Colbeau A, Toussaint B and Vignais PM (1991) Identification and sequence analysis of *hupR1* gene which encodes a response regulator of the NtrC family required for hydrogenase expression in *Rhodobacter capsulatus*. J Bacteriol 173: 5928–5932

Rossman R, Maier T, Lottspeick F and Böck A (1995) Characterization of a protease from *Escherichia coli* involved in hydrogenase maturation. Eur J Biochem 227: 545–550

Sasikala K, Ramana ChV, Rao PR and Kovacs KL (1993) Anoxygenic phototrophic bacteria: Physiology and advances in hydrogen production technology. Adv Appl Microbiol 38: 211–295

Sauter M, Böhm R and Böck A (1992) Mutational analysis of the operon (*hyc*) determining hydrogenase 3 formation in *Escherichia coli*. Mol Microbiol 6: 1523–1532

Sayavedra-Soto LA, Powell GK, Evans HJ and Morris RO (1988) Nucleotide sequence of the genetic loci encoding subunits of *Bradyrhizobium japonicum* uptake hydrogenase. Proc Natl Acad Sci USA 85: 8395–8399

Scolnik PA and Marrs BL (1987) Genetic research with photosynthetic bacteria. Annu Rev Microbiol 41: 703–726

Seifert E and Pfennig N (1979) Chemoautotrophic growth of *Rhodopseudomonas* species with hydrogen and chemotrophic utilization of methanol and formate. Archiv Microbiol 122: 177–182

Stock JB, Ninfa AJ and Stock AM (1989) Protein phosphorylation and regulation of adaptive responses in bacteria. Microbiol Rev 53: 450–490

Stock JB, Lukat GS and Stock AM (1991) Bacterial chemotaxis and the molecular logic of intracellular signal transduction networks. Annu Rev Biophys Chem 20: 109–136

Stoker K, Reijnders WNM, Oltmann LF and Stouthamer AH (1989) Initial cloning and sequencing of *hydHG*, an operon homologous to *ntrBC* and regulating the labile hydrogenase activity in *Escherichia coli* K-12. J Bacteriol 171: 4448–4456

Takakuwa S and Wall JD (1981) Enhancement of hydrogenase activity in *Rhodopseudomonas capsulata* by nickel. FEMS Microbiol Lett 12: 359–363

Tanaka I, Appelt K, Dijk J, White SW and Wilson KS (1984) 3-Å resolution structure of a protein with histone-like properties in prokaryotes. Nature 310: 376–381

Taylor DP, Cohen SN, Clark WG and Marrs BL (1983) Alignment of the genetic and restriction maps of the photosynthesis region of the *Rhodopseudomonas capsulata* chromosome by a conjugation-mediated marker rescue technique. J Bacteriol 154: 580–590

Thöny B and Hennecke H (1989) The -24/-12 promoters come of age. FEMS Microbiol Rev 63: 341–358

Tibelius KH, Du L, Tito D and Stejskal F (1993) *Azotobacter chroococcum* hydrogenase gene cluster: Nucleotide sequences and genetic analysis of four accessory genes (*hupA, hupB, hupY* and *hupC*). Gene 127: 53–61

Tomiyama M, Shiotani M, Sode K, Tamiya E and Karube I (1991) Nucleotide sequence analysis and expression control of *hydA* in *Escherichia coli*. Abstracts of 3rd Intl Conference on Molecular Biology of Hydrogenases, Troia, Portugal, pp 24–25

Toussaint B, Bosc C, Richaud P, Colbeau A and Vignais PM (1991) A mutation in a *Rhodobacter capsulatus* gene encoding an integration host factor-like protein impairs in vivo hydrogenase expression. Proc Natl Acad Sci USA 88: 10749–10753

Toussaint B, Delic-Attree I, David L, de Sury d'Aspremont R, Vinçon M. and Vignais PM (1993) Purification of the integration host factor-like protein of *Rhodobacter capsulatus*. Cloning and sequencing of the *hip* gene which encodes the β-subunit. J Bacteriol 175: 6499–6504

Toussaint B, David L, de Sury d'Aspremont R and Vignais PM (1994) The IHF proteins of *Rhodobacter capsulatus* and of *Pseudomonas aeruginosa*. Biochimie 76: 951–957

Uffen RL (1978) Fermentative metabolism and growth of photosynthetic bacteria. In: Clayton RK and Sistrom WR (eds) The Photosynthetic Bacteria, pp 857–872. Plenum Press, New York

Uffen RL, Colbeau A, Richaud P and Vignais PM (1990) Cloning and sequencing the genes encoding hydrogenase subunits of *Rhodocyclus gelatinosus*. Mol Gen Genet 221: 49–58

Van Niel CB (1944) The culture, general physiology, morphology and classification of the non-sulfur purple and brown bacteria. Bacteriol Rev 8: 1–118

Van Soom C, Verreth C, Sampaio MJ and Vanderleyden J (1993a) Identification of a potential transcriptional regulator of hydrogenase activity in free-living *Bradyrhizobium japonicum* strains. Mol Gen Genet 239: 235–240

Van Soom C, Browaeys J, Verreth C and Vanderleyden J (1993b) Nucleotide sequence analysis of four genes, *hupC, hupD, hupF* and *hupG*, downstream of the hydrogenase structural genes in *Bradyrhizobium japonicum*. J Mol Biol 234: 508–512

Vignais PM and Toussaint B (1994) Molecular biology of membrane-bound H_2 uptake hydrogenases. Arch Microbiol 161: 1–10 Erratum (1994) 161: 196

Vignais PM, Colbeau A, Willison JC and Jouanneau Y (1985) Hydrogenase, nitrogenase and hydrogen metabolism in the photosynthetic bacteria. Adv Microbiol Physiol 26: 155–234

Voelskow H and Schön G (1980) H_2 production of *Rhodospirillum rubrum* during adaptation of anaerobic dark conditions. Archiv Microbiol 125: 245–249

Volbeda A, Charon MH, Piras C, Hatchikian EC, Frey M and Fontecilla-Camps JC (1995) Crystal structure of the nickel-iron hydrogenase from *Desulfovibrio gigas*. Nature 373: 580–587

Voordouw G (1992) Evolution of hydrogenase genes. Adv Inorg Chem 38: 397–422

Wall JD, Weaver PF and Gest H (1975) Genetic transfer of nitrogenase-hydrogenase activity in *Rhodopseudomonas capsulata*. Nature 258: 630–631

White SW, Appelt K, Wilson KS and Tanaka I (1989) A protein structural motif that bends DNA. Proteins: Structure, Function, and Genetics 5: 281–288

Willison JC (1993) Biochemical genetics revisited: The use of mutants to study carbon and nitrogen metabolism in the photosynthetic bacteria. FEMS Microbiology Reviews 104: 1–38

Willison JC and Tissot G (1994) The *Escherichia coli efg* gene and the *Rhodobacter capsulatus adgA* gene code for NAD synthetase. J Bacteriol 176: 3400–3402

Willison JC, Madern D and Vignais PM (1984) Increased photoproduction of hydrogen by non-autotrophic mutants of *Rhodopseudomonas capsulata*. Biochem J 219: 593–600

Willison JC, Ahombo G, Chabert J, Magnin JP and Vignais PM (1985) Genetic mapping of *Rhodopseudomonas capsulata* chromosome shows non clustering of genes involved in nitrogen fixation. J Gen Microbiol 131: 3001–3015

Willison JC, Ahombo G and Vignais PM (1990) Genetic control

of nitrogen metabolism in the photosynthetic bacterium *Rhodobacter capsulatus*. In: Ullrich WR, Rigano C, Fuggi A and Aparicio PJ (eds) Inorganic Nitrogen in Plants and Microorganisms. Uptake and Metabolism, pp 312–319. Springer-Verlag, Berlin

Winter HC and Burris RH (1968) Stoichiometry of the adenosine triphosphate requirement for N_2 fixation and H_2 evolution by a partially purified preparation of *Clostridium pasteurianum*. J Biol Chem 243: 940–944

Wu LF and Mandrand MA (1993) Microbial hydrogenases: Primary structure, classification, signatures and phylogeny. FEMS Microbiol Rev 104: 243–270

Xu HW and Wall JD (1991) Clustering of genes necessary for hydrogen oxidation in *Rhodobacter capsulatus*. J Bacteriol 173: 2401–2405

Xu HW, Love J, Borghese R and Wall JD (1989) Identification and isolation of genes essential for H_2 oxidation in *Rhodobacter capsulatus*. J Bacteriol 171: 714–721

Yoch DC (1978) Nitrogen fixation and hydrogen metabolism by photosynthetic bacteria. In: Clayton RK and Sistrom WR (eds) The Photosynthetic Bacteria, pp 657–672. Plenum Press, New York

Chapter 56

Regulation of Nitrogen Fixation Genes

Robert G. Kranz and Paul J. Cullen
Department of Biology, Box 1137, Washington University, One Brookings Drive, St. Louis, MO 63130, USA

Summary

Some prokaryotes can reduce atmospheric dinitrogen to ammonia as a source of fixed nitrogen. The biological process of nitrogen fixation (*nif*) requires significant sources of reductant and energy. Moreover, in the anoxygenic photosynthetic bacterium, *Rhodobacter capsulatus*, at least 34 proteins are involved in this process including nitrogenase, the molybdenum-containing enzyme that catalyzes this reduction. If ammonia or other source of fixed nitrogen is already present in the environment it becomes unnecessary and wasteful to express nitrogen fixation (*nif*) genes and reduce dinitrogen. Additionally, nitrogenase is irreversibly inactivated by oxygen, making nitrogen fixation possible only under anaerobic (or microaerobic) conditions in anoxygenic photosynthetic bacteria. Therefore, *Rb. capsulatus,* and other anoxygenic photosynthetic bacteria, have developed specific nitrogen and oxygen-sensing systems that activate transcription of *nif* genes only under conditions of limiting nitrogen and oxygen. This chapter reviews the current state of knowledge about the transcriptional control of *nif* genes primarily in the best studied phototroph, *Rb. capsulatus*. This remarkable

R. E. Blankenship, M. T. Madigan and C. E. Bauer (eds): Anoxygenic Photosynthetic Bacteria, pp. 1191–1208.

regulatory system involves a cascade of protein factors: UTase → GlnB → NtrB → NtrC → NifA and RpoN → transcriptional activation of other *nif* genes. Some of these factors sense nitrogen and some oxygen. The molecular basis for the activity of each of these proteins is discussed. Additional control in *Rb. capsulatus* is at the level of molybdenum; recently an alternative nitrogenase (*anf*) that appears to contain no molybdenum (but iron instead) has been discovered in *Rb. capsulatus*. Synthesis of *anf* is repressed by molybdenum but is still under the control of nitrogen and oxygen. Studies on this alternative system have contributed to our understanding of *anf* and molybdenum-based gene regulation. Anoxygenic photosynthetic bacteria rapidly fix nitrogen in the presence of light as the energy source representing one of the few ways to directly and efficiently convert the energy from sunlight into the fixation of atmospheric dinitrogen. However, there is no evidence that anoxygenic photosynthetic bacteria show any specific *nif* gene regulation in response to light, but they do exhibit amazing metabolic diversity. These capabilities include the use of a variety of sources of nitrogen and carbon under aerobic and anaerobic growth conditions. Additionally, within the aquatic habitats where photosynthetic bacteria are found, rapidly changing environmental conditions make rapid and diverse gene regulation necessary. Studies on *nif* genes in photosynthetic bacteria will continue to contribute to our own understanding of *nif* regulation in general and to our knowledge of basic mechanisms of transcriptional control.

I. Introduction: Nitrogen Fixation (*nif*) Genes; Transcriptional Control by Fixed Nitrogen and Oxygen

Nitrogen fixation is the reduction of atmospheric dinitrogen to ammonia. Although all organisms on earth require fixed nitrogen for survival and growth, only a select group of procaryotes have been shown to fix dinitrogen biologically. Anoxygenic photosynthetic bacteria comprise a large class of microorganisms that are capable of carrying out this process. Other chapters in this book describe (1) the historical perspectives and microbiology of nitrogen fixation in anoxygenic phototrophs (Chapter 42 by Madigan) and (2) the biochemical aspects and enzymology of nitrogen fixation (Chapter 43 by Ludden and Roberts). The present chapter concerns the transcriptional regulation of genes required for nitrogen fixation in photosynthetic bacteria. Biological nitrogen fixation is an energy-intense process with a major requirement for reductant and ATP. The molybdenum-containing nitrogenase enzyme composed of three polypeptides has been isolated from a number of phototrophs as have the genes that encode them (*nifHDK*). However, *Rhodobacter capsulatus* has been the model anoxygenic photosynthetic bacterium for the analysis of *nif* gene regulation and *nif* gene organization. In *Rb. capsulatus* two major clusters of *nif* genes encoding up to 34 gene products involved in nitrogen fixation have been discovered (see Fig. 1 and references in legend). Many of these *nif* genes have homologs that were described initially in *Klebsiella pneumoniae* and subsequently in other diazotrophs. An overview on the genetic organization of *nif* genes in the five best studied diazotrophs (including *Rb. capsulatus*) has recently been published (Merrick, 1993). Non-regulatory genes in *Rb. capsulatus* include the structural genes *nifHDK*; *nifE*, *nifN*, *nifX*, *nifV*, and *nifB* for synthesis of the MoFe cofactor; *nifY*, *nifU* and *NifS* for processing of the nitrogenase MoFe protein component; and *nifQ* for Mo processing. Interestingly, *Rb. capsulatus* has duplications of at least three genes, *nifU*, *nifA*, and *nifB*, although the reason for this redundancy remains obscure. Other *nif* genes or orfs within the *nif* clusters may be involved in aspects of nitrogen fixation tailored to a specific bacterial physiology. For example, photosynthetic bacteria may utilize an electron transport pathway to nitrogenase that requires specific proteins dedicated to its various reducing environments. Thus, some of the *Rb. capsulatus* orfs at the *nifENX* cluster encode ferrodoxin-like proteins, potentially involved in the electron transfer reaction to nitrogenase (Moreno-Vivian et al., 1989). Three genes within the two *nif* clusters in *Rb. capsulatus* are involved in transcriptional regulation: *nifA1*, *nifA2* and *rpoN*. Two other loci in *Rb. capsulatus* encode the regulatory genes *glnB*, *ntrB* and *ntrC* (see Fig. 1). These regulatory genes, the proteins they encode, and their functions will be described in detail.

Abbreviations: Atase – adenyltransferase; *E. coli* – *Escherichia coli;* IHF – integration host factor; *K. pneumoniae* – *Klebsiella pneumoniae; Rb. capsulatus* – *Rhodobacter capsulatus; Rb. sphaeroides* – *Rhodobacter sphaeroides;* RNAP – RNA polymerase; UAS – upstream activating sequences ; Utase – uridylyltransferase

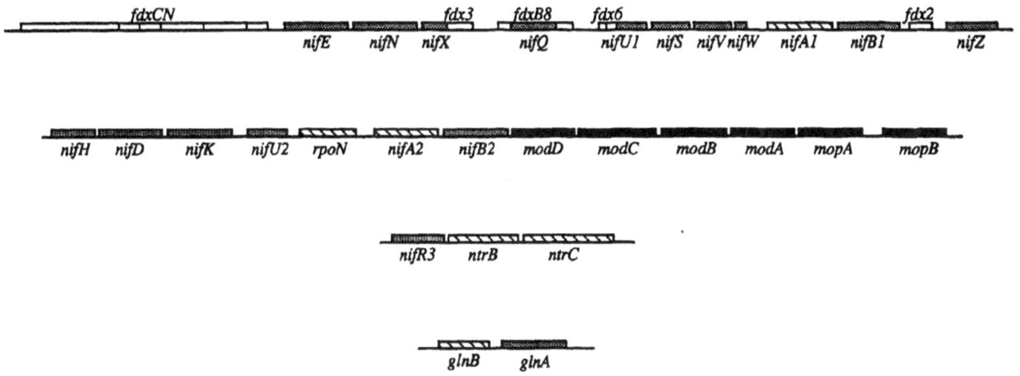

Fig. 1. Organization of *nif* genes in *Rhodobacter capsulatus*. Four separate loci contain at least 40 genes that are involved in nitrogen fixation in *Rb. capsulatus*. The *glnBA* locus (Kranz et al., 1990), *nifR3ntrBntrC* locus (Kranz and Haselkorn, 1985; and Foster-Hartnett et al., 1993), *nifHDKU2-rpoN-nifA2B2* locus (Avtges et al., 1983; Klipp et al., 1988; and Masephol et al., 1988), the molybdenum uptake locus (Wang et al., 1993) and the locus that contains the *nifENX* genes (e.g. Moreno-Vivian et al., 1989; see Merrick, 1993 for review) are shown. *nif* genes are lightly shaded, regulatory genes are hatched, *fdx* genes are not shaded, and genes involved in molybdenum uptake or repression are in black. The spacing between the genes is not to scale.

Transcription of *nif* genes is controlled by fixed nitrogen and oxygen. The biosynthesis requirement of at least 34 proteins involved in nitrogen fixation along with the energy needs to carry out the reduction reaction itself make this biological process one of the most costly enzymatic reactions in the cell. Consequently, microorganisms have evolved remarkable mechanisms to turn on *nif* genes only when they are needed and to repress transcription of *nif* genes when they are not. When free-living, nitrogen-fixing bacteria are faced with environmental conditions abundant in fixed nitrogen (e.g. ammonia) there is no need to synthesize ammonia within the cell. Thus, exogenous ammonia, or a fixed nitrogen source which is easily transported and assimilated (e.g. into amino acids, purines, pyrimidines) leads to repression of *nif* genes. Additionally, oxygen reacts with nitrogenase and irreversible inactivates this enzyme. Nitrogen-fixing bacteria have therefore developed mechanisms to (1) keep oxygen out of the cell, such as the formation of heterocysts in filamentous oxygenic cyanobacteria (see Haselkorn et al., 1986 for review), (2) protect their nitrogenase from oxygen, such as the rapid oxidase activities of *Azotobacter* species (Maier and Moshiri, 1993 for review) and/or (3) repress the expression of *nif* genes when oxygen levels are high. *Rb. capsulatus*, like all free-living unicellular diazotrophs, has evolved mechanisms to activate transcription of *nif* genes only when the concentration of these two effectors (fixed nitrogen and oxygen) are low (for review, see Kranz and Foster-Hartnett, 1990). Although *Rb. capsulatus* has been used as the model phototroph to study these mechanisms, there are clear similarities to the gene regulatory controls used by other microorganisms (non phototrophs); these will be discussed and compared.

II. General Models Describing *nif* Regulatory Mechanisms

A. Rhodobacter capsulatus

Before detailing the individual sensing and activator proteins involved in *nif* transcriptional control, an overview on the strategies used by *Rb. capsulatus* to accomplish this regulation will be presented (see *Fig. 2*). Briefly, an enzyme called uridylyltransferase (UTase) is predicted to control the uridylylation of another protein, GlnB, depending on the levels of fixed nitrogen in the cell. Under nitrogen-deficient conditions, UTase uridylylates GlnB; uridylylated GlnB would no longer interact with NtrB, NtrB consequently functioning only as a kinase. The substrate of the NtrB kinase, NtrC, is phosphorylated under this condition. (Under nitrogen-sufficient conditions a GlnB/NtrB complex is predicted to act as an NtrC~P phosphatase.) NtrC~P is predicted to be an ATPase which, upon binding to DNA upstream,

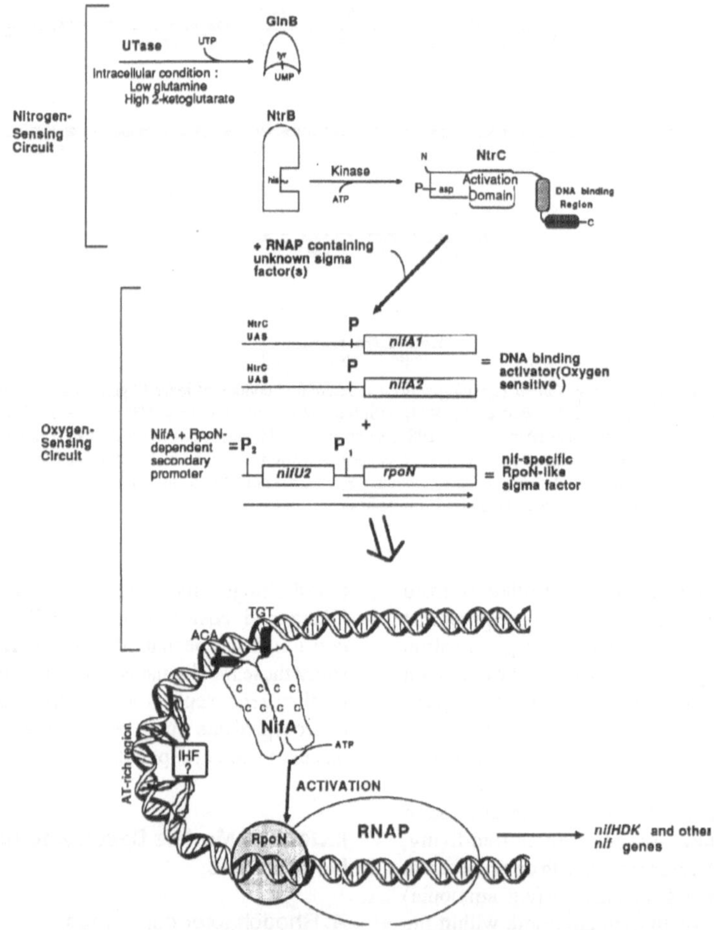

Fig. 2. Transcription regulatory cascade of nitrogen fixation genes in *Rhodobacter capsulatus;* oxygen and nitrogen control in a phototroph. Details of the cascade are described in the text. UTase, uradylytransferase; RNAP, RNA polymerase; IHF, integration host factor. UTase is proposed to sense the ratio of glutamate to 2-ketoglutarate, compounds that reflect endogenous levels of fixed nitrogen. When the ratio is low, UTase essentially inactivates GlnB by the covalent addition of uridylmonophosphate, UMP. Under these conditions, NtrB acts as a kinase that phosphorylates NtrC, a proposed transcriptional activator. The NtrC protein has three domains: An N-terminal regulatory domain, a central ATP-binding activation domain, and a C-terminal helix-turn-helix domain that is shown to bind to DNA over 100 bp upstream of the *nifA1* and *nifA2* promoters (i.e. UAS). Phosphorylated NtrC is proposed to activate transcription of the *nifA1* and *nifA2* genes via an unknown sigma factor. The *rpoN* gene is also regulated; a primary promoter is responsible for nitrogen-independent, low levels of *rpoN* expression, and an autoactivated secondary promoter increases the expression of *rpoN* under nitrogen-limiting conditions. NifA, a DNA binding transcriptional activator, and RpoN, a *nif*-specific sigma factor, together catalyze transcriptional activation of *nifHDK* and other *nif* genes, possibly in an IHF-stimulated manner. Oxygen control probably takes place at the level of NifA activity and possibly at the *rpoN* promoter.

transcriptionally activates the *nifA1* and *nifA2* genes. Transcriptional activation of *nifA1* and *nifA2* results in the nitrogen-dependent increase in NifA protein. NifA itself is predicted to possess ATPase activity (like the NtrC class of proteins) and act as a transcriptional activator of all other *nif* genes, including *nifHDK*. The transcriptional activation of *nif* genes by NifA additionally requires an RNA polymerase (RNAP) alternative sigma factor called RpoN. The *Rb. capsulatus rpoN*-encoded sigma factor is interesting because of its physical location (i.e. *nifHDK*-linkage), its expression, and its structural properties when compared to RpoN proteins from other microorganisms. Expression of the *Rb. capsulatus rpoN* gene is autoactivated (by RpoN and NifA) at a secondary promoter upstream of *nifU2*. This secondary promoter results in a more rapid and higher maximal induction of other *nif* genes which is physiologically beneficial under certain environmental conditions (see IIIf below). Oxygen regulation is not as well understood as nitrogen control, although it is thought that the NifA protein itself may be oxygen sensitive and that other oxygen-sensing control may be mediated through DNA supercoiling (see VI, Future Studies).

The activation of *nifHDK* and other *nif* genes by NifA and RNAP/RpoN may be enhanced by the DNA binding protein, integration host factor (IHF). Santero et al. (1992) have shown that the *K. pneumoniae nifHDK* promoter region is activated in vitro to 20-fold higher levels when IHF is added. This is proposed to occur by IHF-induced bending of the DNA such that the interaction frequency of RNAP/RpoN with the NifA protein, bound at upstream activating sequences (UAS), is increased. Hoover et al. (1990) have shown that *E. coli* IHF binds to the appropriate DNA within the *Rb. capsulatus nifH* upstream region, thus suggestive that a similar IHF protein and interactions may occur in *Rb. capsulatus*.

B. Other Anoxygenic Phototrophs

Concerning trans-acting factors involved in *nif* gene regulation in other phototrophs, to our knowledge, only *Rb. sphaeroides* has so far been investigated. In that organism, two *rpoN* genes and an *ntrC* gene appear to be present (see below).

III. Regulatory Components Related to Other Bacterial Proteins

A. Uridylyltransferase (= UTase = GlnD)

Studies on mutants in the enteric *glnD* gene suggested that *glnD* was involved in nitrogen control because *glnD* mutants lost their ability to deadenylate glutamine synthetase in response to the nitrogen source (e.g. Bancroft et al., 1978; Bloom et al., 1978; Foor et al., 1978). Consequently, these mutants are glutamine auxotrophs. Early studies on nitrogen control proteins in *E. coli* involved the analysis of the enzyme encoded by *glnD*, UTase (e.g. Adler et al., 1975). Biochemical experiments on the purified *E. coli* UTase indicated that (1) UTase covalently uridylylates and deuridylylates the GlnB protein (via UTP) and (2) UTase activity is allosterically effected by glutamine (which activates deuridylylation) or 2-ketoglutarate (which activates uridylylation). Since the ratio of glutamine to 2-ketoglutarate reflects the general nitrogen status of the cell, allosteric modulation of UTase by these molecules is the key sensing element of the nitrogen regulatory system in enteric bacteria. The substrate of UTase action, GlnB, is a protein of approximately 12,000 daltons (behaving as a tetramer), and the *Rb. capsulatus glnB* gene has been isolated and sequenced (Kranz et al., 1990; see below). Although the UTase gene has only been isolated and sequenced from *E. coli* and *Azotobacter vinelandii* (Conteras et al., 1991), its presence in *Rb. capsulatus* has been postulated because of the conservation of its substrate, GlnB particularly within the site of uridylylation. Cloning and inactivation of the *glnD* gene in *Rb. capsulatus* is still required to confirm this aspect of the nitrogen fixation regulatory cascade. It is important to note however that the phenotype of an *Rb. capsulatus glnD* mutant would, by the model, be predicted to be a Nif^- regulatory mutant because NtrC would always be dephosphorylated. However, if the *Rb. capsulatus glnD* gene is involved in other nitrogen-control systems the phenotype of a *glnD* strain is impossible to predict. Moreover, since some regulatory genes have already been shown to be duplicated in *Rb. capsulatus* (*nifA*; Klipp et al., 1988) and *Rb. sphaeroides* (*rpoN*; Meijer and Tabita, 1992) isolation of a double mutant in *glnD* genes, if duplicated, would be unlikely by traditional mutagenesis procedures.

B. glnB

Uridylylation of GlnB, in response to nitrogen limitation essentially inactivates the GlnB protein which allows NtrB to function in its kinase mode only (e.g. Ninfa and Magasanik, 1986; Keener and Kustu, 1988). Consequently, inactivation of the *glnB* gene would be predicted to also 'release' the NtrB protein such that it is continually a kinase, irrespective of fixed nitrogen levels. Thus, by this model, *Rb. capsulatus glnB* mutants would be Nif-constitutive with respect to nitrogen concentrations (i.e. $Nif^{c:NH_4}$). In fact, isolation of *Rb. capsulatus glnB* mutants and the *glnB* gene were isolated using this premise. *Rb. capsulatus* $Nif^{c:NH_4}$ mutants were isolated by using a lactose-selection technique and *nifH-lacZ* fusions (see Kranz and Haselkorn, 1986). Analysis of a cosmid that returned some of these $Nif^{c:NH_4}$ back to normal regulation (i.e. nitrogen repression) yielded the *glnB* gene (Kranz et al., 1990). Analysis of this locus indicated that *glnA*, encoding glutamine synthetase, was downstream from the *glnB* gene. Inactivation by Tn5 mutagenesis of the chromosomal *glnB* orf resulted in a glutamine auxotroph, presumably because of polarity on *glnA*. These *Rb. capsulatus glnB* transposon mutants are Nif^c with respect to glutamine or ammonia anaerobically but still repress synthesis of nitrogenase under aerobic conditions. Recently, transcription of the *glnBA* operon has been studied using *lacZ* fusions and primer extension (Foster-Hartnett and Kranz, 1994). These results show that (1) two promoters are upstream of *glnB*: *glnB*P1 is repressed by NtrC and *glnB*P2 is activated by NtrC and (2) a third promoter within *glnB* for *glnA* transcription may exist. Additionally, activation of *glnB*P2 is independent of RpoN, and purified *Rb. capsulatus* NtrC binds to tandem binding sites greater than 100 bps upstream of the *glnB*P2 transcription start site (but overlapping *glnB*P1). These results add a third promoter, *glnB*P2, to a short list that includes *nifA1* and *nifA2* which require NtrC for activation but not RpoN. Like *nifA1* and *nifA2*, *glnB*P2 is not an RpoN-type promoter (Foster-Hartnett and Kranz, 1994). In enteric bacteria, *glnB* is present at a different locus than *glnA*, while *glnA* is in a complex *glnA-ntrB-ntrC* operon (see below). Additionally, in enteric bacteria, *glnB* has been shown to interact with adenyltransferase (ATase), which adenylates or deadenylates glutamine synthetase depending on this interaction (i.e. nitrogen status; Stadman et al., 1980). It has not been shown whether the *Rb. capsulatus glnB* mutants are altered in their glutamine synthetase modification system, although the glutamine synthetase enzyme is apparently modified (Johansson and Gest, 1977). To summarize, in enteric bacteria the GlnB protein is a multi-faceted protein which interacts with a variety of protein substrates (e.g. NtrB, ATase and UTase) to alter their enzymatic properties. Recently it has been proposed that GlnB may interact directly with the NifA protein in *Azospirillum brasilense*, based on genetic studies (deZamaroczy et al., 1993). Thus, the versatility of GlnB as a modulator of protein activity may be greater than originally thought (see also IIIE below).

C. ntrB

Under nitrogen sufficient conditions, the GlnB protein is not uridylylated and interacts with NtrB by an unknown mechanism. Consequently, a phosphatase activity of the NtrB protein is induced resulting in dephosphorylation of the NtrC protein (e.g. Ninfa and Magasanik, 1986; Keener and Kustu, 1988). When GlnB is uridylylated by UTase (under nitrogen-limiting conditions) or the *glnB* gene is mutated (under nitrogen-limiting or sufficient conditions) NtrB only phosphorylates the NtrC protein. The *Rb. capsulatus* NtrB and NtrC proteins thus comprise a pair of the previously reviewed two-component bacterial regulatory systems (e.g. Gross *et al*, 1989; Stock *et al,* 1990). These aspects of the regulatory model are consistent with the phenotype of *Rb. capsulatus ntrB* mutants (and *ntrC* mutants); such mutants are Nif^- and unable to activate a *nifH-lacZ* fusion in *Rb. capsulatus* (Kranz and Haselkorn, 1985). The *ntrB* gene in *Rb. capsulatus* is not located in *glnA-ntrB-ntrC* operon as in the enteric bacteria. Rather, the *Rb. capsulatus ntrB* gene is located in a *nifR3-ntrB-ntrC* operon (Foster-Hartnett et al., 1993). (Note: *Rb. capsulatus ntrB* and *ntrC* were previously called *nifR2* and *nifR1* respectively.) Although the function of NifR3 is unknown (see below) the *Rb. capsulatus* NtrB sequence was predicted to be an NtrC kinase because of the phenotype of *ntrB* strains and the conservation between this protein and other kinases of the NtrB-class of protein (Jones and Haselkorn, 1988). This includes a completely conserved histidine residue which accepts the phosphate group from ATP (in an autophosphorylation reaction) and subsequently transfers this phosphate group to NtrC. Recently, the *Rb. capsulatus* NtrB, purified as a fusion protein to the maltose binding protein, has been shown to autophosphorylate and transfer the phosphate to NtrC (Cullen and Kranz, unpublished).

D. ntrC

As noted above, *Rb. capsulatus ntrC* mutants were originally isolated as Nif^- colonies (Wall and Braddock, 1984) and later shown to be *nif* regulatory mutants using a *nifH-lacZ* reporter (Kranz and Haselkorn, 1985). The *Rb. capsulatus ntrC* gene was cloned by complementation of these mutants to a Nif^+ phenotype by a wild-type DNA library (Avtges et al., 1985). Southern hybridization experiments with these DNA fragments showed that this gene was *ntrC* (Kranz and Haselkorn, 1985), and sequence analysis of this DNA showed orfs that encode NtrB- and NtrC-like proteins (Jones and Haselkorn, 1989). Recent studies have shown that the *ntrC* gene is required for the activation of both *nifA1-lacZ* and *nifA2-lacZ* fusions (Foster-Hartnett and Kranz, 1992; Preker et al., 1992). Accordingly, nitrogen control is mediated through the *nifA* regulatory genes via an NtrC-dependent process. Under nitrogen-deficient conditions, phosphorylated NtrC activates the transcription of *nifA1* and *nifA2*. Surprisingly, primer extension analyses (Foster-Hartnett and Kranz, 1992) have indicated that the *nifA1* and *nifA2* promoters do not possess the consensus RpoN-recognition sequences (i.e. GG N_{10} GC). In all other cases, members of the NtrC-class of protein activate RNAP containing the alternative RpoN sigma factor at such promoters (see Kustu et al., 1989; North et al., 1993). Moreover, $RpoN^-$ strains of *Rb. capsulatus* are still able to activate *nifA1* and *nifA2* in an NtrC-dependent manner. This is true for both deletions and point mutations in the *Rb. capsulatus rpoN* gene.

Recent surveys (Collado-Vides et al., 1991) of promoters within *E. coli* have defined two types: One class uses the housekeeping type of sigma factor (σ^{70}) and another set uses the RpoN (σ^{54}) type of sigma factor (and is always activated by an NtrC-type activator). The σ^{70} class of promoters, when activators are used, typically require DNA $\leq$ 70 bps upstream of the transcription start site for activation. The second, RpoN/NtrC class typically requires > 100 bp upstream for activation 'at a distance'. Studies performed on the *E. coli* and *Salmonella typhimurium* NtrC systems were in fact the first cases of 'enhancer' like activities discovered in procaryotic systems, describing this activation at a distance phenomenon (Reitzer and Magasanik, 1986; Reitzer et al., 1989; Su et al., 1990; Wedel et al., 1990). For *Rb. capsulatus*, deletion analysis has shown that DNA between −114 and −139 upstream of the transcriptional start site is required for NtrC-dependent *nifA2* activation (Foster-Hartnett and Kranz, 1992) and between −134 and −144 are required for *nifA1* activation (Foster-Hartnett et al., 1994). Thus, the *nifA1* and *nifA2* promoters of *Rb. capsulatus* fit the pattern of NtrC-activatable genes (i.e. > 100 bp required) but clearly do not possess the RpoN-type of promoter. Very recently, Foster-Hartnett et al. (1994) have shown by DNAse I footprint analysis that the purified *Rb. capsulatus* NtrC binds to tandem recognition sites in the same distal regions determined to be important for in vivo activation. Thus, *Rb. capsulatus* NtrC clearly activates the *nifA1* and *nifA2* promoters (and *glnB*P2) directly by binding to tandem distant sites in an RpoN-independent manner.

A comparison of selected members of the NtrC class of proteins define three regions important for function (Fig. 3; see reviews by North et al., 1993, and Mopett and Segovia, 1993, for details and extended references). First, the N-terminus of all NtrC proteins, including the *Rb. capsulatus* NtrC, possess the aspartate residues which are phosphorylated by NtrB. Second, the C-terminus of all, including the *Rb. capsulatus* NtrC, possess a typical helix-turn-helix motif that is required for binding DNA upstream of the promoters that each activates (i.e. UAS). Additionally, as proposed by North et al. (1993), the C-terminus contains extended sequence similarity to each other and to the DNA binding protein FIS. *Rb. capsulatus* NtrC also has this extended identity to FIS (Foster-Hartnett et al., 1993) and it has been proposed that this region may be involved in dimerization of NtrC to form active dimers or oligomers. In fact, recent data is suggestive that phosphorylation of the N-terminus catalyzes not only ATP-dependent hydrolysis (see next), but also oligomerization of NtrC protein (Weiss et al., 1992), both of which are required for the transcriptional activation of RNAP/RpoN. Third, the central domain of all true NtrC-type proteins contain a consensus ATP binding motif. This domain is even present in extended members of the NtrC family, such as NifA, that do not possess the N-terminal phosphorylation or FIS-like C-terminus. The central domain of the *Rb. capsulatus* NtrC protein clearly possesses the ATP binding site or 'Walker motif' (Walker et al., 1982). Studies by Weiss et al. (1991) and Austin and Dixon (1992) have shown that the hydrolysis of ATP by NtrC is required for open complex formation by the RNAP-RpoN holoenzyme; this directly distinguishes the NtrC-class of transcriptional activators from other known bacterial activator proteins. Recently, the ATP-binding site was altered by site-directed mutagenesis (of the *Rb. capsulatus* NtrC protein); this NtrC was shown to be completely inactive for *nifA1* and *nifA2* expression yet was still able to bind upstream (Foster-Hartnett et al., 1994). Thus, by these criterion, the *Rb. capsulatus* NtrC protein fits the paradigm of an RNAP/RpoN activator. However, a region between the 'Walker motif' and C-terminal domain is conserved within the other

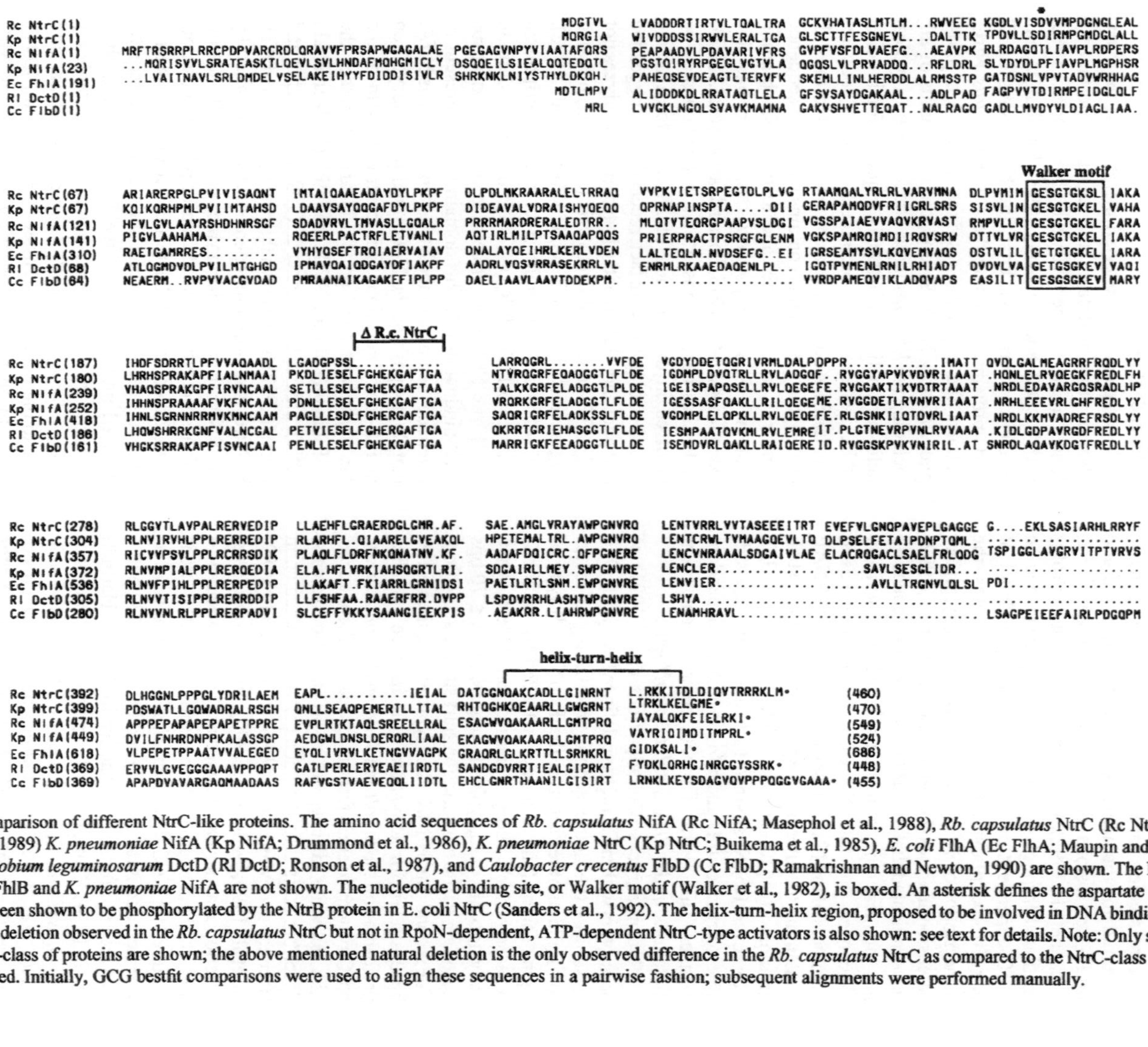

Fig. 3. Comparison of different NtrC-like proteins. The amino acid sequences of *Rb. capsulatus* NifA (Rc NifA; Masephol et al., 1988), *Rb. capsulatus* NtrC (Rc NtrC; Jones and Haselkorn, 1989) *K. pneumoniae* NifA (Kp NifA; Drummond et al., 1986), *K. pneumoniae* NtrC (Kp NtrC; Buikema et al., 1985), *E. coli* FlhA (Ec FlhA; Maupin and Shanmugam, 1990), *Rhizobium leguminosarum* DctD (Rl DctD; Ronson et al., 1987), and *Caulobacter crecentus* FlbD (Cc FlbD; Ramakrishnan and Newton, 1990) are shown. The N-terminus of the *E. coli* FhlB and *K. pneumoniae* NifA are not shown. The nucleotide binding site, or Walker motif (Walker et al., 1982), is boxed. An asterisk defines the aspartate residue (D54) which has been shown to be phosphorylated by the NtrB protein in E. coli NtrC (Sanders et al., 1992). The helix-turn-helix region, proposed to be involved in DNA binding, is marked. The natural deletion observed in the *Rb. capsulatus* NtrC but not in RpoN-dependent, ATP-dependent NtrC-type activators is also shown: see text for details. Note: Only select proteins of the NtrC-class of proteins are shown; the above mentioned natural deletion is the only observed difference in the *Rb. capsulatus* NtrC as compared to the NtrC-class of proteins so far sequenced. Initially, GCG bestfit comparisons were used to align these sequences in a pairwise fashion; subsequent alignments were performed manually.

members of the NtrC proteins but not *Rb. capsulatus* (Fig. 3). This domain may contact the RpoN/RNAPholoenzyme. Because the exact mechanisms by which NtrC~P activates transcription of RNAP/RpoN is unknown it is premature to conclude that this region interacts specifically with RpoN. Clearly, analysis of members of the NtrC-class of proteins that activate RNAP containing sigma factors other than RpoN will aid in understanding the basic molecular mechanisms behind the activation process. Do other photosynthetic bacteria contain unique NtrC members and what can be learned by analyzing exceptions to the rules? Section F below describes properties of RpoN which further distinguish the NtrC-class of protein activator systems.

E. nifA

Transcription of the *Rb. capsulatus nifA1* and *nifA2* genes depend upon an NtrC-mediated mechanism. Although this transcriptional activation is repressed by fixed nitrogen, it is oxygen insensitive (Foster-Hartnett and Kranz, 1992). Therefore oxygen repression occurs at a later stage in the regulatory cascade. This is in contrast to the regulatory cascade in *K. pneumoniae* in which a *nifLA* operon is activated by the nitrogen sensing NtrC/RpoN system only under anaerobic conditions (Dixon et al., 1980). Aerobic repression of the *K. pneumoniae nifLA* promoter is thought to occur by a DNA supercoiling mechanism whereby anaerobically grown cells show an increase in DNA negative supercoiling, a requirement for *nifLA* activation (see Whitehall et al., 1992 and references therein). The suggestion that an increase in supercoiling is required for *nifLA* activation by NtrC/RpoN/RNAP is based on results of in vivo analyses of *K. pneumoniae nif* expression and recent in vitro transcription experiments with the purified NtrC/RpoN/RNAP components.

The *Rb. capsulatus nifA1* and *nifA2* are thought to have identical orfs but differ significantly upstream of their proposed translational start sites. The *Rb. capsulatus nifA1* and *nifA2* genes were discovered by using the *K. pneumoniae nifA* gene as a probe in Southern hybridization experiments (Klipp et al., 1988). Inactivation of either *nifA1* or *nifA2* results in an *Rb. capsulatus* Nif^+ phenotype but inactivation of both *nifA1* and *nifA2* yields a Nif^- *Rb. capsulatus* strain that is unable to synthesize nitrogenase polypeptides (Masepohl et al., 1988). As indicated earlier, the physiological advantage of possessing duplicated *nifA* genes has yet to be determined.

The NifA protein from *Rb. capsulatus*, like all NifA proteins from a diverse array of diazotrophs, is a member of the NtrC class of proteins. The *Rb. capsulatus* NifA protein contains the typical NtrC-like central domain with the consensus ATP-binding site (see Fig. 3). NifA also has a C-terminal helix-turn-helix motif, presumably required for binding to the DNA with the sequence TGT-N_{10}-ACA, a typical NifA binding site (Morett et al., 1988). This consensus (or a similar binding site) has been discovered upstream of a number of NifA/RpoN activated genes in *Rb. capsulatus* (e.g. *nifH*, *nifU2*; see review by Kranz and Foster-Hartnett, 1990; Preker et al., 1992). The NifA proteins show little similarity in their C-terminus to FIS, possibly suggestive that although ATPase activity may be necessary for transcriptional activation, oligomerization may not (North et al., 1993). *Rb. capsulatus* NifA protein contains cysteine residues that are conserved among most members of the NifA family including the symbiotic organisms *Bradyrhizobium* and *Rhizobium* (Masepohl et al., 1988). It has been proposed that these cysteine residues may chelate an oxygen sensitive metal, thus explaining the oxygen sensitive NifA activity observed in these NifA proteins (Fischer and Hennecke, 1987). Interestingly, the *K. pneumoniae* and *Azotobacter* NifA proteins do not have these conserved cysteines and are not oxygen sensitive (Huala and Ausubel, 1989 and references therein). Rather, in these organisms another protein called NifL senses oxygen and by an unknown mechanism inactivates the NifA protein. Thus, one point of oxygen control within the *Rb. capsulatus nif* regulatory cascade may be the sensitivity of the NifA proteins to oxygen. Very recently, Kustu and colleagues have synthesized a Mal-NifA (from *K. pneumoniae*) fusion protein that retains in vitro activation activity, thus overcoming previous problems with the overproduction of inactive NifA protein (Lee et al., 1993). The in vitro transcription studies indicate that NifA-mediated activation requires the hydrolysis of any nucleotide triphosphate. Accordingly, although the basic mechanism of RNAP/RpoN activation by NifA may be similar to that of NtrC, specific hydrolysis of ATP is not essential, as it is with the enteric NtrC.

Another form of control in the *Rb. capsulatus* regulatory cascade may exist and Hubner et al. (1993) have placed the *Rb. capsulatus nifA1* gene behind constitutively expressed promoters to eradicate the NtrC-dependent nitrogen control. This construct was able to complement an *Rb. capsulatus nifA1/nifA2* double mutant to a Nif^{+} phenotype. However, a *nifH-lacZ* fusion in this strain is still only partially repressed by ammonia. The authors propose that a novel mechanism of nitrogen control is mediated at the level of the NifA protein. Future studies will be needed to determine if this residual control is mediated by the GlnB sensing system.

F. rpoN

As with the NifA proteins in other diazotrophs, it is thought that the *Rb. capsulatus* NifA protein binds upstream of *nif* promoters and activates RNAP containing the RpoN sigma factor. (Section IIIF-3 below describes promoters activated by *Rb. capsulatus* RpoN.) The *rpoN* gene (previously called *nifR4*) in *Rb. capsulatus* was isolated (Avtges et al., 1985) by complementation of Nif^{-} regulatory mutants (Wall et al., 1984). Surprisingly, this gene was mapped to the *nifHDK* locus, the first case of an *rpoN* gene with physical linkage to *nif* genes. In fact, previous results on other microorganisms have shown that the *rpoN*-encoded sigma factor is typically not associated only with nitrogen or *nif* control but with a wide variety of physiological responses (e.g. dicarboxycylic acid transport; Ronson et al., 1987; flagellar synthesis; Helmann, 1989, for review). The *Rb. capsulatus rpoN* gene was subsequently sequenced by Jones and Haselkorn (1988) and Alias et al. (1989).

Since the regulated expression of *rpoN* in a cell could naturally control the levels of expression of all genes that it activates, it is important to understand the controls exerted on the *rpoN* promoter(s). Moreover, the RpoN genes from many different procaryotes have been cloned and sequenced. The wide interest in this sigma factor stems from the unique aspects of NtrC-mediated activation, the plethora of environmental systems it controls, and the fact that RpoN is not homologous to the large family of sigma-70 factors that have been characterized. Such interest warrants a discussion on the expression and structure of the RpoN protein from anoxygenic phototrophs.

1. Expression of rpoN

Initial results with *Rb. capsulatus rpoN-lacZ* fusions indicated that *rpoN* expression is at least five-fold repressed by fixed nitrogen or oxygen (Kranz and Haselkorn, 1985). This repression can now be explained by the following mechanism:

(1) A primary promoter directly upstream of *rpoN* is responsible for nitrogen-independent basal levels of expression of *rpoN*. Expression from this promoter is lower than can be detected using a *rpoN-lacZ* reporter (Foster-Hartnett and Kranz, 1992). However, this promoter is responsible for sufficient *rpoN* expression to yield a Nif^{+} phenotype under normal laboratory growth conditions (see below).

(2) Upon induction of the *nifA1* and *nifA2* genes (by nitrogen-limitation) a secondary promoter, upstream of *nifU2*, that depends on NifA and RpoN, results in *nifU2-rpoN* expression in the cell (Preker et al., 1992). This raises *rpoN-lacZ* expression to measurable β galactosidase levels, at least 20- to 100-fold higher than with the primary promoter alone (i.e. above background).

An *Rb. capsulatus* strain was created (Cullen et al., 1994) that has a transcriptional terminator upstream of the *rpoN* primary promoter but downstream of the secondary promoter (and between the *rpoN* and *nifU2* genes). This strain only expresses *rpoN* via the primary promoter, and it was used to show that the secondary, autoactivated promoter is responsible for a more rapid induction of nitrogenase (with a maximal expression of approximately 3-fold higher levels when the secondary promoter is not terminated). Importantly, although this strain is Nif^{+} under normal laboratory growth conditions, it is Nif^{-} when NaCl (> 50 mM) is added or Fe is limited (< 400 nM) in the media (Cullen et al., 1994). Thus, the physiological role of the secondary promoter may be to increase levels of Nif proteins such that under stress conditions limitations on nitrogen fixation are overcome.

2. Structure of the RpoN Protein

Recent models for the enteric RpoN sigma factor have suggested that five domains are characteristic of RpoN proteins (Fig. 4):

(1) a glutamine rich N-terminus (Sasse-Dwight and Gralla, 1990),

(2) two hydrophobic heptad repeats that may form a DNA binding 'intramolecular leucine zipper' (Sasse-Dwight and Gralla, 1990),

(3) an acidic central domain located between the two leucine zippers (Sasse-Dwight and Gralla, 1990),

(4) a C-terminal helix-turn-helix motif involved in DNA binding (Coppard and Merrick, 1991; Merrick and Chambers, 1992), and

(5) a C-terminal 'RpoN box' composed of 8 amino acids that are completely conserved among all RpoN proteins (Merrick, 1993).

Some of these domains are reminiscent of eucaryotic domains that are required for transcriptional activation (Courey and Tijan, 1988; Harrison, 1991; Hahn, 1993). An alignment of the *Rb. capsulatus* RpoN with the 13 other RpoN proteins that have been sequenced is shown in Fig. 4. Recent results by Merrick and Chambers (1992) have shown that the helix-turn-helix domain in the *K. pneumoniae* RpoN is important in binding to the –13 region of the promoter (GC of the GG N_{10} GC consensus). Using linker mutagenesis, Cullen et al. (1994) have shown that the *Rb. capsulatus* RpoN helix-turn-helix is also required for function, as is the RpoN box (unknown function). A deletion of the N-terminal glutamine region (Δ2-24) also completely inactivated the *Rb. capsulatus* RpoN protein.

A recent model proposed that the 2 leucine zipper domains interact, forming a DNA binding region, that loops out the acidic domain (Sasse-Dwight and Gralla, 1990). According to the model, the acidic domain is then involved in melting the promoter to form an open complex. This model has recently been reevaluated (Merrick and Chambers, 1992; Cullen et al., 1994) for the following reasons: (1) The *Rb. capsulatus* RpoN protein (and the recently sequenced *Rb. sphaeroides* RpoN) are lacking an acidic domain yet presumably still require the NifA protein for transcriptional activation; (2) linkers in the second 'leucine zipper' of the *Rb. capsulatus* RpoN protein resulted in RpoN alleles that retained function. However, the spacing in bona-fide leucine zipper proteins is critical; insertions within the zippers would disrupt such structures. Thus, although these hydrophobic heptad repeats may form α helices, they probably do not form typical leucine zippers. Very recently, Tintut et al. (1994) have provided evidence that these hydrophobic-repeats may be involved in binding to the RNAP. In summary, the RpoN proteins from diverse microorganisms, including two photosynthetic bacteria, contain domains that are reminiscent of eucaryotic motifs important in transcription. However, only the role of the helix-turn-helix domain and heptad repeats have been critically investigated. Other domains in this sigma factor require further analyses to determine their role in transcriptional activation mediated by an ATP-dependent activator like NtrC.

3. Promoters Activated by RpoN

Promoters that have been shown (or proposed) to be activated by RpoN in *Rb. capsulatus* possess the consensus sequence GG-N_{10}-GC from –24 to –12 upstream of the transcriptional start site. S1 nuclease mapping(Pollock et al., 1988) and primer extension analyses (Foster-Hartnett and Kranz, 1992) of the *nifH* promoter and the *nifU2* promoter (Preker et al., 1992) have confirmed this suggestion. Additionally, some promoters have been shown to require RpoN for activation using *Rb. capsulatus* $RpoN^-$ strains. As mentioned above, an AT rich region, centered between the putative NifA binding site and the *nifH* promoter, may be involved in the binding of an IHF-like protein to facilitate NifA interaction with RpoN/RNAP.

IV. Regulatory Components Unique to Anoxygenic Phototrophs (Or Are They)?

A. RNA Polymerase Sigma Factors

One of the most interesting aspects of the *Rb. capsulatus nif* regulatory cascade is the possibility that NtrC activates *nifA1*, *nifA2*, *glnB*P2 and perhaps other genes (Rapp et al., 1986; Schuddlekopf et al., 1993) by a mechanism that may require ATP hydrolysis but not the RNAP sigma factor RpoN. If such is the case, then unique sigma factors are yet to be discovered. Alternatively, could the *Rb. capsulatus* NtrC protein activate a σ^{70}-type RNAP? Analyses of these alternative mechanisms would clearly contribute to our basic understanding of transcriptional activation.

An interesting difference between *Rb. capsulatus* and *Rb. sphaeroides* is that two *rpoN* genes may be present in *Rb. sphaeroides* (Meijer and Tabita, 1992).

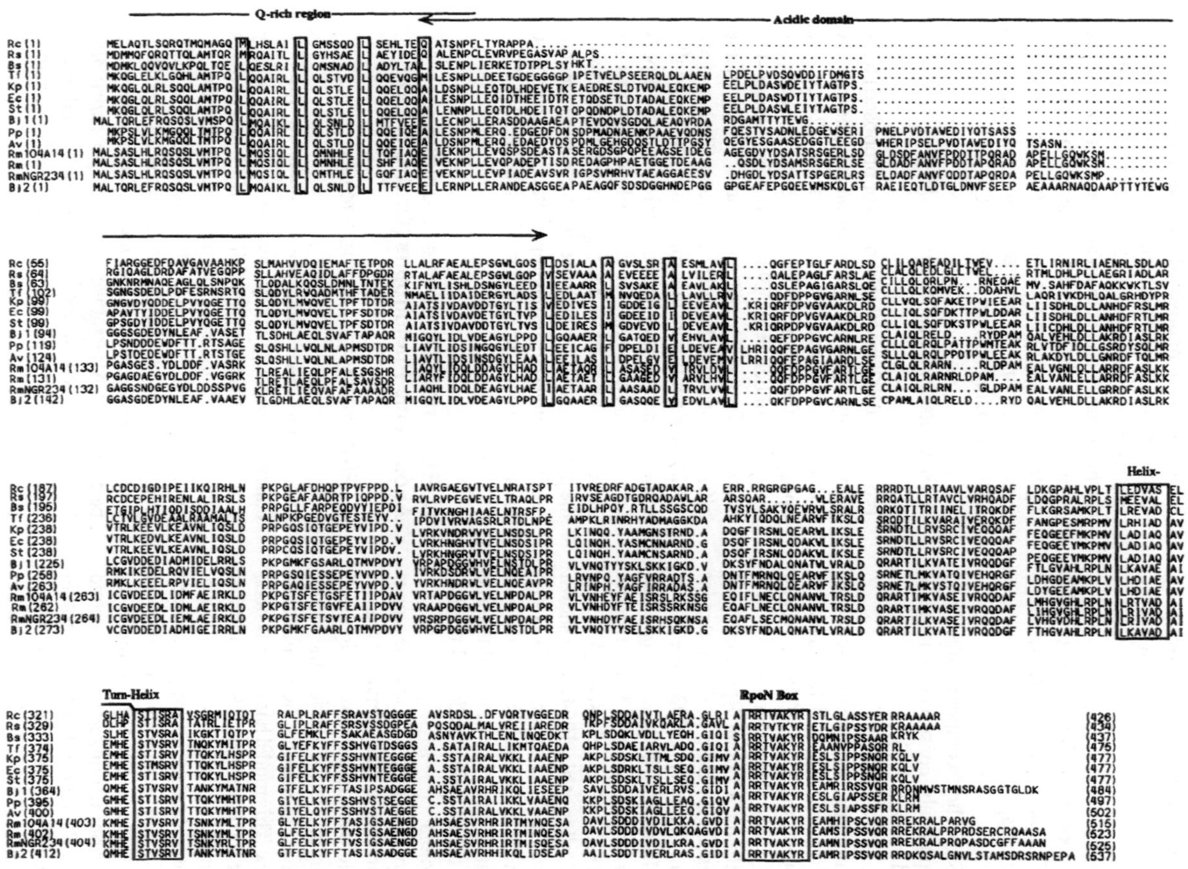

Fig. 4. Comparison of the amino acid sequences of RpoN. The RpoN amino acid sequences of *Rb. capsulatus* (Rc), *Rb. sphaeroides* (Rs), *Bacillus subtilis* (Bs), *Thiobacillus ferrooxidans* (Tf), *K. pneumoniae* (Kp), *E. coli* (Ec), *Salmonella typhimurium* (St), *Bradyrhizobium japonicum* (Bj-1 and Bj-2), *Pseudomonas putida* (Pp), *Azotobacter vinelandii* (Av), and *Rhizobium meliloti* (Rm, Rm104A14, and RmNGR234) are aligned by size. The glutamine rich region (Q-rich region), acidic domain, helix-turn-helix, and RpoN box are labeled. The hydrophobic residues that form the backbone of the two hydrophobic heptad repeats (i.e. 'leucine zippers') are boxed. See text for details. The references for these sequences are as follows: Rc, Jones and Haselkorn, 1989; Rs, Meijer and Tabita, 1992; Bs, Debarbouille et al., 1991; Tf, Berger et al., 1990; Kp, Merrick and Gibbins, 1985; Ec, Sasse-Dwight and Gralla, 1990; St, Popham et al., 1991; Bj-1 and Bj-2, Kullik et al., 1991; Pp, Kohler et al., 1989 and Inouye et al., 1989; Av, Merrick et al., 1987; Rm, Ronson et al., 1987; Rm104A14, Shatters et al., 1989 ; and Rm NGR234, vanSlooten et al., 1990.

The *Rb. sphaeroides nif*-linked *rpoN* is not essential for a Nif^+ phenotype yet the *Rb. capsulatus rpoN* is required. Presumably, the second copy of *rpoN* in *Rb. sphaeroides*, detected by Southern blot analysis, is sufficient. A second *rpoN* in *Rb. capsulatus* has not been detected by Southern blot analysis using various *rpoN* genes as probes (see review by Kranz and Foster-Hartnett, 1990). However, Colbeau and Vignais (1992) have recently suggested that because the *hupS* gene is still activated in an *Rb. capsulatus* $RpoN^-$ strain and because it contains a consensus RpoN-type promoter, that a second *rpoN* gene may be present in *Rb. capsulatus*. Nevertheless, primer extension results have not been reported for the *hupS* gene in an $RpoN^-$ background to confirm this idea. Results from the *nifA1* and *nifA2* primer extension experiments rule out the possibility of RpoN-type promoters for those genes. Possibly, photosynthetic bacteria have evolved a larger repertoire of sigma factors to respond more rapidly and with greater metabolic versatility to their changing environments. In this respect, having a single RpoN that can be activated by the large class of NtrC-like proteins places some limitations on versatility. This is particularly true if the NtrC-like proteins can activate from long distances (i.e. as enhancers), thus placing upper limits on the numbers of NtrC-like activation systems that are possible on a bacterial chromosome. From ecological or evolutionary perspectives and to contribute to our basic understanding about mechanisms of transcriptional activation, it will be important to determine what alternative sigma factors are present in these microorganisms.

B. nifR3

The *Rb. capsulatus nifR3* gene is directly upstream of *ntrB-ntrC* in an *nifR3-ntrB-ntrC* operon. *nifR3* was originally defined by Tn5 and mini-*Mu* chromosomal insertions that resulted in a Nif^- regulatory phenotype (i.e. could not activate a *nifH-lacZ* fusion; Kranz and Haselkorn, 1985 and Avtges et al., 1985). Although the original sequence of this region yielded no *nifR3* orf, additional sequence analysis has confirmed the existence of an orf upstream of *ntrB* (Foster-Hartnett et al., 1993). Insertional and deletion mutagenesis and complementation analyses of *nifR3-ntrB-ntrC* have shown that *nifR3* is not essential for nitrogen fixation and that the reason for the Nif^- phenotype was polarity into *ntrB* and *ntrC* from the Tn5 and mini-*Mu* insertions (Foster-Hartnett et al., 1993). Interestingly, NifR3 is highly homologous to an ORF (called FIS ORF1) upstream of the DNA binding protein FIS, in an ORF1-FIS operon. What is the function of NifR3? Typically, genes in operons encode proteins with functions in similar biological processes; such is the case for many *nif* genes (described here) and many of the photosynthetic and other genes described in this textbook. Does NifR3 in some way modulate the activity of NtrB or NtrC under certain environmental conditions? Very recently, the *ntrB-ntrC* locus of the plant symbiont *Rhizobium leguminosarum* has been shown to possess an upstream *nifR3* gene (Patriaca et al., 1993). Thus, *nifR3* is probably not unique to photosynthetic bacteria but understanding its function requires further study.

C. Others

A question of considerable importance during the last 50 years has been the connection between photosynthesis and nitrogen fixation in anoxygenic photosynthetic bacteria. Approximately 15 years ago it was discovered that *Rb. capsulatus* can fix nitrogen both in the dark anaerobically and the light (Madigan et al., 1979). This result suggested that photosynthesis and nitrogen fixation were not obligatorily linked. Nevertheless, anaerobiosis alone is sufficient for both the induction of *nif* (under nitrogen limitation) and photosynthetic proteins. *Rb. capsulatus* mutants that are unable to grow photosynthetically but still able to express *nifH-lacZ* fusions seem to rule out direct biosynthetic linkage (Kranz and Foster-Hartnett, 1990). A class of *Rb. sphaeroides* mutants (called *drn*) that only express their genes for nitrogen fixation in the light but not in the dark have been described (Shestakov et al., 1988). These mutants are also derepressed for nitrogenase (i.e. $Nif^{c:NH_3}$). It was proposed that another level of control, above the proteins involved in the *nif* regulatory model for *Rb. capsulatus* (Fig. 2) was operating. However, very recently, these *drn* point mutations have been shown to be located in the *Rb. sphaeroides ntrC* gene (S. Shestakov, personal communication). How could *ntrC* mutants yield the *drn* phenotype described above? One possibility is that energy charge (or nucleotide triphosphate pools) differ in dark- and light-grown cells. The NtrC mutant protein, requiring high energy phosphate for both phosphorylation and ATPase/ transcriptional activation would be sufficiently active in the light- but not dark-grown cells. Obviously, other explanations are possible, including the unlikely prospect that light is sensed in some way by the NtrC protein. Nevertheless, there is presently no data suggesting that anoxygenic photosynthetic bacteria possess specific light-sensing mechanisms involved in *nif* gene regulation.

Another class of *Rb. capsulatus* and *Rb. sphaeroides* mutant that were originally thought to be *nif* regulatory mutants are the *adg* (ammonia-dependent growth) mutants (Zinchenko et al., 1990). These strains are unable to grow on nitrogen sources other then ammonia. Although the *Rb. capsulatus* defects were originally classified as ntr (nitrogen regulatory) mutants (Allibert et al., 1987), it is clear that these strains are able to express significant levels of nitrogenase (and a *nifH-lacZ* fusion). The latter point has been shown with the mutant PA3 that has the same phenotype (and gene that complements) as *adg* mutants (Kranz and Haselkorn, 1985; Avtges et al., 1985). The gene complementing these mutants has been cloned and sequenced and has in fact been isolated from a number of different bacteria. The *adg* sequence does not offer clues to its function but it appears to be a gene essential for growth since only *adg* point mutants (but not insertions) could be isolated. Very recently, Willison and Tissot (1994) have shown that *adg* encodes an ammonia-dependent NAD synthetase. A similar class of *Rb. capsulatus* mutant was isolated by Wall and colleagues many years ago(Wall et al., 1977).

V. The Alternative Nitrogen Fixing Systems: Aspects Concerning Gene Regulation

An alternative non-molybdenum nitrogenase in *Azotobacter* was initially proposed by Bishop and colleagues (see Joerger et al., 1989, and references therein). Recently, Schneider et al. (1991) discovered that an *Rb. capsulatus* strain deleted of its *nifHDK* genes could grow under nitrogen-free conditions if molybdenum was stringently scrubbed from the media. This alternative nitrogenase was determined to be free of a heterometal (e.g. Vanadium) and probably is an Fe-nitrogenase (Schneider et al., 1991). The *Rb. capsulatus* alternative nitrogenase, encoded by *anfHDGK*, is expressed only under Mo-free, anaerobic, nitrogen-limiting conditions. A regulatory gene called *anfA* is directly upstream of *anfHDGK*, and encodes a protein that is related to NifA (Schuddlekopf et al., 1993). However, the *Rb. capsulatus* AnfA is not a member of the cysteine-containing, oxygen-sensitive NifA class. Thus, oxygen regulation of *anfHDGK* may be different than *nifHDK* in *Rb. capsulatus* (see VI). Further, it has been shown that the *Rb. capsulatus ntrB*, *ntrC* and *rpoN* but not *nifA1* and *nifA2* genes are required for *anfHDGK* expression (Schuddlekopf et al., 1993). Accordingly, one would predict that NtrC activates *anfA* in a nitrogen-dependent manner and AnfA and RpoN/RNAP subsequently activate other *anf* genes.

What represses the synthesis of *anf* genes when molybdenum is present? Wang et al. (1993) have characterized six genes (called *mol* and *mop*) linked to *nifHDKU2 -rpoN-nifA2B2*, that are involved in the repression of *anf*. Four of these genes (*molPJDX*) may encode a typical ATP-dependent, membrane-bound importer, presumably of molybdenum. When these are inactivated, transport of molybdenum does not occur and the cell derepresses *anf* expression. Two other orfs, MopA and MopB are homologous to Mo-pterin-binding proteins. A *mopA mopB* double mutant is also derepressed for *anf* expression, suggesting that MopA and/or MopB is necessary for molybdenum-dependent repression of *anf* genes, possibly via *anfA* expression. The mechanism of repression (or involvement of Mop proteins) is unknown.

VI. Future Studies and Perspectives

Although many of the *trans* and *cis* acting factors that respond to nitrogen and oxygen in *Rb. capsulatus* have been discovered, the details of specific points of control still require further experimentation. Some of these fertile areas of research have already been discussed, including a possible nitrogen-sensitive NifA system, the direct involvement of IHF in transcription and the characterization of genes not yet discovered. The later includes a *glnD* analog and the sigma factor(s) involved in the NtrC-mediated activation systems described here. Development of in vitro systems to study these novel activation systems is of clear importance. Additionally, the regulatory cascades and proteins involved in *nif* genes from other anoxygenic phototrophs warrant future studies. Why does *Rb. sphaeroides* have two different *rpoN* genes? What about other phototrophs?

An interesting difference between the *Rb.*

capsulatus ntrC mutants and enteric *ntrC* mutants is phenotype. The *Rb. capsulatus ntrC* strains can grow on many different sources of fixed nitrogen. In contrast, the enteric *ntrC* mutants are more pleiotrophic because their *ntrC* is required for activation of genes involved in many nitrogen uptake or metabolic systems (eg. histidine, proline, arginine). How does *Rb. capsulatus* or other anoxygenic phototrophs regulate their distinct nitrogen uptake systems? In fact, although some studies have been carried out (e.g. Rapp et al., 1986), nitrogen metabolism and uptake in anoxygenic phototrophs has received little attention; are there differences in photosynthetically compared to aerobically grown cells?

Finally, an area of great interest concerns the exact mechanism(s) by which oxygen represses *nif* and *anf* transcription in anoxygenic photosynthetic bacteria. It should be noted that both the primary and secondary promoters of the *Rb. capsulatus rpoN* expression system are candidate points for oxygen-mediated repression; the NifA protein itself, as described earlier is an additional candidate. As defined for the *K. pneumoniae nifLA* promoter, could DNA supercoiling be involved in oxygen sensing? In this respect, an important lesson may be learned from the analysis of the regulation of *nif* genes in different diazotrophs: in many cases that have been investigated in detail, multiple mechanisms of oxygen control seem to exist in a single cascade. For example, *K. pneumoniae* is controlled at the level of the *nifLA* promoter by DNA supercoiling and by NifL inactivation of NifA in response to oxygen. *Rhizobium meliloti* is controlled at three different levels; FixL (via a heme group) in response to limiting oxygen phosphorylates FixJ that subsequently activates transcription of *nifA* (Monson et al., 1992); *nifA* expression is also repressed by FixK in response to oxygen (Waelkens et al., 1992); NifA is oxygen-sensitive and activates other *nif* genes (Huala and Ausubel, 1989).

When complex and energy-intense processes (such as nitrogen fixation or photosynthesis) are necessarily controlled by oxygen, will multiple and overlapping regulatory mechanisms be the rule rather than the exception? Do multiple mechanisms evolve because of potential 'leakiness' or for 'fine-tuning' this regulation to respond to rapidly changing and/or distinct oxygen concentrations? Attempts to isolate Nif^c strains of *Rb. capsulatus*, that induce *nifH-lacZ* independent of oxygen, have failed (Kranz and Haselkorn, 1986). This is true even when $Nif^{c:NH_4}$ strains were used for the selections. Possibly, overlapping controls in response to oxygen, mediated by completely independent mechanisms do exist. Because of their diverse metabolic capabilities, the anoxygenic photosynthetic bacteria will no doubt continue to be excellent systems to study gene regulation both from evolutionary perspectives and from the standpoint of understanding the molecular basis of transcriptional control mechanisms.

Acknowledgments

Due to publication deadlines, only a limited number of 1993/1994 manuscripts could be cited. The fields of *nif* and gene regulatory mechanisms are extensive with many contributions from many groups. We apologize for not citing many important contributors due to page limitations. These contributions can be found within the reviews (and primary literature) that are cited. R.G.K. is supported by NIH and the USDA.

References

Alias A, Cejudo FJ, Chabert J, Willison JC and Vignais PM (1989) Nucleotide sequence of wild-type and mutant *nifR4* (*ntrA*) genes of *Rhodobacter capsulatus*: Identification of an essential glycine residue. Nucl Acids Res 17: 5377–5377

Allibert P, Willison JC and Vignais PM (1987) Complementation of the nitrogen regulatory (*ntr*-like) mutations in *Rhodobacter capsulatus* by an *Escherichia coli* gene: Cloning and sequencing of the gene and characterization of the gene product. J Bacteriol 169: 260–271

Austin S and Dixon R (1992) The prokaryotic enhancer binding protein NtrC has an ATPase activity which is phosphorylation and DNA dependent. EMBO 11: 2219–2228

Avtges P, Scolnik PA and Haselkorn R (1983) Genetic and physical map of the structural genes (*nifHDK*) coding for the nitrogenase complex of *Rhodopseudomonas capsulata*. J Bacteriol 156: 2521–256

Avtges P, Kranz RG and Haselkorn R (1985) Isolation and organization of gene for nitrogen fixation in *Rhodopseudomonas capsulata*. Mol Gen Genet 201: 353–369

Backman KC, Chen Y-M, Ueno-Nishio S and Magasanik B (1981) The product of *glnL* is not essential for regulation of bacterial nitrogen assimilation. J Bacteriol 154: 516–519

Bankroft S, Rhee SG, Neumann C and Kustu S (1978) Mutations that alter the covalent modification of glutamine synthetase in *Salmonella typhimurium*. J Bacteriol 134: 1046–1055

Berger D, Woods D and Rawlings D (1990) Complementation of *Escherichia coli* σ^{54} (NtrA)-dependent formate hydrogenase activity by a cloned *Thiobacillus ferrooxidans ntrA* gene. J Bacteriol 172: 4399–4406

Bloom FR, Levin MS, Foor F and Tyler B (1978) Regulation of glutamine synthetase formation in *Escherichia coli*: Characterization of mutants lacking uridylyltransferase. J Bacteriol 134: 569–577

Buikema WJ, Szeto WW, Lemley PV, Orme-Johnson WH and Ausubel FM (1985) Nitrogen fixation specific regulatory genes of *Klebsiella pnemoniae* and *Rhizobium meliloti* share homology with the general nitrogen regulatory gene *ntrC* of *K. pnemoniae*. Nucl Acids Res 13: 4539–4555

Colbeau A and Vignais PM (1992) Use of *hupS::lacZ* gene fusion to study regulation of hydrogenase expression in *Rhodobacter capsulatus*: Stimulation by H_2. J Bacteriol 174: 4258–4264

Collado-Vides J, Magasanik B and Gralla J (1991) Control site location and transcriptional regulation in *Escherichia coli*. Microbiol Rev 55: 371–394

Conteras A, Drummond M, Bali A, Blanco G, Garcia E, Bush G, Kennedy C and Merrick M (1991) The product of the nitrogen fixation regulatory gene *nfrX* of *Azotobacter vinelandii* is functionally and structurally homologous to the uridylyltransferase encoded by *glnD* in enteric bacteria. J Bacteriol 173: 7741–7749

Coppard J and Merrick M (1991) Cassette mutagenesis implicates a helix-turn-helix motif in promoter recognition by the novel RNA polymerase sigma factor σ^{54}. Mol Microbiol 5: 1309–1317

Courey A and Tijan R (1988) Analysis of Sp1 in vivo reveals multiple transcriptional domains, including a novel glutamine-rich activation motif. Cell 55: 887–898

Cullen PJ, Foster-Hartnett D, Gabbert K and Kranz RG (1994) Structure and expression of the alternative sigma factor, RpoN, in *Rhodobacter capsulatus*; physiological relevance of an autoactivated *nifU2-rpoN* superoperon. Mol Microbiol 11: 51–65

Debarbouille M, Martin-Verstraete I, Kunst, F and Rapoport G (1991) The *Bacillus subtilis sigL* gene encodes an equivalent of σ^{54} from Gram-negative bacteria. Proc Natl Acad Sci USA 88: 9092–9096

deZamaroczy M, Paquelin A and Elmerich C (1993) Functional organization of the *glnB glnA* cluster of *Azospirillum brasilense*. J Bacteriol 175: 2507–2515

Ditta G, Stanfield S, Corbin D and Helinski DR(1980) Broad host range DNA cloning system for Gram-negative bacteria: Construction of a gene bank of *Rhizobium meliloti*. Proc Natl Acad Sci USA 77: 7347–7351

Dixon RA, Eady RR, Espin G, Hill S, Iaccarino M, Khan D and Merrick M (1980) Analysis of regulation of Klebsiella pnemoniae nitrogen fixation (*nif*) gene cluster with gene fusions. Nature 286: 128–132

Drummond M, Clements J, Merrick M and Dixon R (1983) Positive control and autogenous regulation of the *nifLA* promoter in *Klebsiella pnemoniae*. Nature 301: 302–313

Fischer H-M, Bruderer T and Hennecke H (1988) Essential and nonessential domains in the *Bradyrhizobium japonicum* NifA protein: Identification of indispensable cysteine residues potentially involved in redox activity and/or metal binding. Nucl Acids Res 16: 2207–2224

Foor F, Cedergren RJ, Streicher SL, Rhee SG and Magasanik B (1978) Glutamine synthetase of *Klebsiella aerogenes*: Properties of *glnD* mutants lacking uridylyltransferase. J Bacteriol 134: 562–568

Foster-Hartnett D and Kranz RG (1992) Analysis of the promoters and upstream sequences of *nifA1* and *nifA2* in *Rhodobacter capsulatus*; activation requires NtrC but not RpoN. Mol Microbiol 6: 1049–1060

Foster-Hartnett D and Kranz (1994) The *Rhodobacter capsulatus glnB* gene is regulated by NtrC at tandem RpoN-independent promoters. J Bacteriol 176: 5171–5176

Foster-Hartnett D, Cullen P, Gabbert K and Kranz R (1993) Sequence, genetic, and *lacZ* fusion analysis of a *nifR3-ntrB-NtrC* operon in *Rhodobacter capsulatus*. Mol Microbiol 8: 903–914

Foster-Hartnett D, Cullen PJ, Monika E and Dranz RG (1994) A new type of NtrC transcriptional activator. J Bacteriol 176: 6175–6187

Gross R, Arico B and Rappuoli R (1989) Families of bacterial signal transducing proteins. Mol Microbiol 3: 1661–1667

Hahn S (1993) Structure and function of acidic transcription activators. Cell 72: 481–483

Harrison S (1991) A structural taxonomy of DNA-binding domains. Nature 353: 715–719

Haselkorn R, Golden J, Lammers P and Mulligan M (1986) Developmental rearrangement of cyanobacterial nitrogen-fixation genes. Trends in Genetics 2: 255–259

Helmann JD (1991) Alternative sigma factors and the regulation of flagellar gene expression. Mol Microbiol 5: 2875–2882

Hoover TR, Santero E, Porter S and Kustu S (1990) The integration host factor stimulates interaction of RNA polymerase with NifA, the transcriptional activator for nitrogen fixation operons. Cell 63: 11–22

Huala E and Ausubel FM (1989) The central domain of *Rhizobium meliloti* NifA is sufficient to activate transcription from the *Rb. meliloti nifH* promoter. J Bacteriol 171: 3354–3365

Hubner P, Masephol B, Klipp W and Bickle T (1993) *nif* gene expression studies in *Rhodobacter capsulatus*: *ntrC*-independent repression by high ammonium concentrations. Mol Microbiol 10: 123–132

Inouye S, Yamanda M, Nakazawa T and Nakazawa T (1989) Cloning and sequence analysis of *ntrA* (*rpo*N) gene of *Pseudomonas putida*. Gene 85: 145–152

Joerger RD, Jacobson MR and Bishop PE (1989) Two *nifA*-like genes required for expression of alternative nitrogenases by *Azotobacter vinelandii*. J Bacteriol 171: 3258–3267

Johansson BC and Gest H (1977) Adenylation/deadenylation of the glutamine synthetase of *Rhodopseudomonas capsulata*. Eur J Biochem 81: 365–371

Jones R and Haselkorn R (1989) The DNA sequence of the *Rhodobacter capsulatus ntrA*, *ntrB* and *NtrC* gene analogs required for nitrogen fixation. Mol Gen Genet 215: 507–516

Jouanneau Y, Lebecque S and Vignais P (1984) Ammonia and light effect on nitrogenase activity in nitrogen-limited continuous cultures of *Rhodopseudomonas capsulata*. Role of glutamine synthetase. Arch Microbiol 139: 326–331

Keener J and Kustu S (1988) Protein kinase and phosphoprotein phosphatase activities of nitrogen regulatory proteins NtrB and NtrC of enteric bacteria: roles of the conserved amino-terminal domain of NtrC. Proc Natl Acad Sci USA 85: 4976–4980

Klipp W, Masephol B and Puhler A (1988) Identification and mapping of nitrogen fixation genes of *Rhodobacter capsulatus*:

Duplication of a *nifA-nifB* region. J Bacteriol 170: 693–699

Kohler T, Cayrol J, Ramos J and Harayama S (1989) Nucleotide and deduced amino acid sequence of the RPON σ-factor of *Pseudomonas putida*. Nucl Acids Res 17: 10125

Kranz RG and Foster-Hartnett D (1990) Transcriptional regulatory cascade of nitrogen-fixation genes in anoxygenic photosynthetic bacteria: Oxygen- and nitrogen-responsive factors. Mol Microbiol 4: 1793–1800

Kranz RG and Haselkorn R (1985) Characterization of *nif* regulatory genes in *Rhodopseudomonas capsulata* using *lac* gene fusions. Gene 40: 203–215

Kranz RG and Haselkorn R (1986) Anaerobic regulation of nitrogen-fixation genes in *Rhodopseudomonas capsulata*. Proc Natl Acad Sci USA 83: 6805–6809

Kranz RG and Haselkorn R (1988) Ammonia-constitutive nitrogen fixation mutants of *Rhodobacter capsulatus*. Gene 71: 65–74

Kullik I, Fritsche S, Knobel H, Sanjuan J, Hennecke H and Fischer H (1991) *Bradyrhizobium japonicum* has two differentially regulated, functional homologs of the σ^{54} gene (*rpoN*). J Bacteriol 173: 1125–1138

Kustu S, Santero E, Keener J, Popham D and Weiss D (1989) Expression of σ^{54} (*ntrA*)-dependent genes is probably united by a common mechanism. Micro Rev 53: 367–376

Lee H-S, Berger KD and Kustu S (1993) Activity of purified NIFA, a transcriptional activator of nitrogen fixation genes. Proc Natl Acad Sci USA 90: 2266–2270

Liang YY, de Zamaroczy M, Arsene F, Paquelin A and Elmerich C (1992) Regulation of nitrogen fixation in *Azospirillum brasilense* Sp7: Involvement of *nifA*, *glnA*, and *glnB* gene products. FEMS Lett 100: 113–120

Madigan MT, Wall JD and Gest H (1979) Dark anaerobic dinitrogen fixation by a photosynthetic microorganism. Science 204: 1429–1430

Maier RJ and Moshiri F (1993) Molecular analysis of components responsible for protection of *Azotobacter* nitrogenase from oxygen damage. In: Palacios R, Mora J and Newton WE (eds) New Horizons in Nitrogen Fixation, p 383. Kluwer Academic Publishers, Boston

Masephol B, Klipp W and Puhler A (1988) Genetic characterization and sequence analysis of the duplicated *nifA/nifB* gene region of *Rhodobacter capsulatus*. Mol Gen Genet 212: 27–37

Masephol B, Angermuller S, Hennecke S, Hubner P, Moreno-Vivian C and Klipp W (1993) Nucleotide sequence and genetic analysis of the *Rhodobacter capsulatus ORF6-nifU$_I$SVW* gene region: Possible role of NifW in homocitrate processing. Mol Gen Genet 238: 369–382

Maupin JA and Shanmugam KT (1990) Genetic regulation of formate hydrogenase in *Escherichia coli*: Role of the *fhlA* gene product as a transcriptional activator for a new regulatory gene, *fhlB*. J Bacteriol 172: 4798–4806

Meijer W and Tabita F (1992) Isolation and characterization of the *nifUSVW-rpoN* gene cluster from *Rhodobacter sphaeroides*. J Bacteriol 174: 3855–3866

Merrick M (1993) Organization and regulation of nitrogen fixation genes. In: Palacios R, Mora J and Newton W (eds) New Horizons in Nitrogen Fixation, pp 48–54. Kluwer Academic Press, Boston

Merrick M and Chambers S (1992) The helix-turn-helix motif of σ^{54} is involved in recognition of the –13 promoter region. J Bacteriol 174: 7221–7226

Merrick M and Gibbins J (1985) The nucleotide sequence of the nitrogen-regulation gene *ntrA* of *Klebsiella pnemoniae* and comparison with conserved features in bacterial RNA polymerase sigma factors. Nucl Acids Res 13: 7607–7620

Merrick M, Gibbins J and Toukdarian A (1987) The nucleotide sequence of the sigma factor gene *ntrA* (*rpoN*) of *Azotobacter vinelandii*: Analysis of conserved sequences in NtrA proteins. Mol Gen Genet 210: 323–330

Monson EK, Weinstein M, Ditta GS and Helinski DR (1992) The FixL protein of *Rhizobium meliloti* can be separated into a heme-binding oxygen-sensing domain and a functional C-terminal kinase domain. Proc Natl Acad Sci USA 89: 4280–4284

Moreno-Vivian C, Hennecke S, Puhler A and Klipp W (1989) Open reading frame 5 (ORF5), encoding a ferrodoxin-like protein, and *nifQ* are cotranscribed with *nifE*, *nifN*, *nifX*, and ORF4 in *Rhodobacter capsulatus*. J Bacteriol 171: 2591–2598

Morett E, Cannon W and Buck M (1988) The DNA-binding domain of the transcriptional activator protein NifA resides in its carboxy terminus, recognizes the upstream activator sequences of *nif* promoters, and can be separated from the positive control function of NifA. Nucl Acids Res 16: 11469–11488

Morett E and Segovia L (1993) The σ^{54} bacterial enhancer-binding protein family: Mechanism of action and phylogenic relationship ot their functional domains. J Bacteriol 175: 6067–6074

Ninfa AJ and Magasanik B (1986) Covalent modification of the *glnG* product, NRI, by the *glnL* product, NRII, regulates the transcription of the *glnALG* operon in *Escherichia coli*. Proc Natl Acad Sci USA 53: 5909–5913

North AK, Klose KE, Stedman KM and Kustu S (1993) Prokaryotic enhancer-binding proteins reflect eukaryote-like modularity: The puzzle of nitrogen regulatory protein C. J Bacteriol 175: 4267–4273

Patriarca EJ, Riccio A, Tate R, Colonna-Romano S, Iccarino M and Defez R (1993) The *ntrBC* genes of *Rhizobium leguminosarum* are part of a complex operon subject to negative autoregulation. Mol Microbiol 9: 569–577

Pollock D, Bauer DE and Scolnik PA (1988) Transcription of the *Rhodobacter capsulatus nifHDK* operon is modulated by the nitrogen source. Construction of plasmid vectors based on the *nifHDK* promoter. Gene 65: 269–275

Popham D, Keener J and Kustu S (1991) Purification of the alternative σ factor, σ^{54}, from *Salmonella typhimurium* and characterization of the σ^{54}-holoenzyme. J Biol Chem 256: 19510–19518

Preker P, Hübner P, Schmehl M, Klipp W and Bickle TA (1992) Mapping and characterization of the promoter elements of the regulatory *nif* genes *rpoN*, *nifA1* and *nifA2* in *Rhodobacter capsulatus*. Mol Microbiol 6: 1035–1048

Ramakrishnan G and Newton A (1990) FlbD of *Caulobacter cresentus* is a homologue of the NtrC (NRI) protein and activates σ^{54}-dependent flagellar gene promoters. Proc Natl Acad Sci USA. 87: 2369–2373

Rapp BJ, Landrum DC and Wall JD (1986) Methylammonium uptake by *Rhodobacter capsulatus*. Arch Microbiol 146: 134–141

Reitzer LJ and Magasanik B (1986) Transcription of *glnA* in *E. coli* is stimulated by an activator bound to sites far from the promoter. Cell 45: 785–792

Reitzer LJ, Movas B and Magasanik B (1989) Activation of *glnA* transcription by nitrogen regulator I (NRI)-Phosphate in *Escherichia coli*: Evidence for a long-range physical interaction between NRI-phosphate and RNA polymerase. J Bacteriol 171: 5512–5522

Ronson C, Nixon T, Albright L and Ausubel F (1987) *Rhizobium meliloti ntrA (rpoN)* gene is required for diverse metabolic functions. J Bacteriol 169: 2424–2431

Sanders DA, Gillece-Castro BL, Burlingame AL and Koshland DE (1992) Phosphorylation site of NtrC, a protein phosphatase whose covalent intermediate activates transcription. J Bacteriol 174: 5117–5122

Santero E, Hoover T, North A, Berger D, Porter S and Kustu S (1992) Role of integration host factor in stimulating transcription from the σ^{54}-dependent *nifH* promoter. J Mol Bio 227: 602–620

Sasse-Dwight S and Gralla J (1990) Role of eukaryotic-type functional domains found in the prokaryotic enhancer receptor factor σ^{54}. Cell 62: 945–954

Schneider BL, Shiau SP and Reitzer LJ (1991) Role of multiple environmental stimuli in control of transcription from a nitrogen-regulated promoter in *Escherichia coli* with weak or no activator binding sites. J Bacteriol 173: 6355–6363

Schuddlekopf K, Hennecke S, Liese U, Kutsche M and Klipp W (1993) Characterization of *anf* genes specific for the alternative nitrogenase and identification of *nif* genes required for both nitrogenases in *Rhodobacter capsulatus*. Mol Microbiol 8: 673–684

Shestakov S, Zinchenko V, Babykin M, Kopteva A, Kameneva S, Frolova V, Shestopalov V and Bondarenko O (1988) Genetic studies on the regulation of nitrogen fixation in *Rhodobacter sphaeroides*. In: Bothe F, deBrujin FJ and Newton WE (eds) Nitrogen Fixation: 100 Years After. Proceedings of the 7th International Congress on Nitrogen Fixation, pp 163–169. Fischer, Stuttgart and New York

Shatters R, Somerville J and Kahn M (1989) Regulation of glutamine synthetase II activity in *Rhizobium meliloti* 104A14. J Bacteriol 171: 5087–5094

Stock JB, Stock AM and Mottonen JM (1990) Signal transduction in bacteria. Nature 344: 395–400

Su W, Porter S, Kustu S and Echols H (1990) DNA-looping and enhancer activity: Association between DNA-bound NtrC activator and RNA polymerase at the bacterial *glnA* promoter. Proc Natl Acad Sci USA 87: 5504–5508

Tintut Y, Wong C, Jiang Y, Hsich M and Gralla J (1994) RNA polymerase binding using a strongly acidic hydrophobic-repeat region of σ^{54}. Proc Natl Acad Sci USA 91: 2120–2124

vanSlooten J, Cervantes E, Broughton W, Wong C and Stanley J (1990) Sequence and analysis of the *rpoN* sigma factor gene of *Rhizobium sp.* strain NGR234, a primary coregulator of symbiosis. J Bacteriol 172: 5563–5574

Waelkens F, Foglia A, Morel J-B, Fourment J, Batut J and Boistard P (1992) Molecular genetic analysis of the *Rhizobium meliloti fixK* promoter: Identification of sequences involved in positive and negative regulation. Mol Microbiol 6: 1447–1456

Walker JE, Saraste M, Runswick MJ and Gay N (1982) Distantly related sequences in the α- and β-subunits of ATP synthase, myosin, kinases and other ATP-requiring enzymes and a common nucleotide binding fold. EMBO 1: 945–951

Wall J, Johansson BC and Gest H (1977) A pleiotropic mutant of *Rhodopseudomonas capsulata* defective in nitrogen metabolism. Arch Microbiol 115: 259–263

Wall JD and Braddock K (1984) Mapping of *Rhodopseudomonas capsulata nif* genes. J Bacteriol 158: 404–410

Wall JD, Love J and Quinn P (1984) Spontaneous nif$^-$ mutants of *Rhodopseudomonas capsulata*. J Bacteriol 159: 652–657

Wang G, Angermüller S and Klipp W (1993) Characterization of *Rhodobacter capsulatus* genes encoding a molybdenum transport system and putative molybdenum-pterin-binding proteins. J Bacteriol 175: 3031–3042

Wedel A, Weiss DS, Popham D, Droge P and Kustu S (1990) A bacterial enhancer functions to tether a transcriptional activator near a promoter. Science 248: 486–490

Weiss V and Magasanik B (1988) Phosphorylation of nitrogen regulator (NRI) of *Escherichia coli*. Proc Natl Acad Sci USA 85: 8919–8923

Weiss D, Batut J, Klose K, Keener K and Kustu S (1991) The phosphorylated form of the enhancer-binding protein NtrC has an ATPase activity that is essential for activation of transcription. Cell 67: 155–165

Weiss V, Claverie-Martin F and Magasanik B (1992) Phosphorylation of nitrogen regulator I of *Escherichia coli* induces strong cooperative binding to DNA essential for activation of transcription. Proc Natl Acad Sci USA. 89: 5088–5092

Whitehall S, Austin S and Dixon R DNA supercoiling response of the σ^{54}-dependent *Klebsiella pnemoniae nifL* promoter in vitro. J Mol Biol 225: 591–607

Willison J and Tissot G (1994) The *Escherichia coli efg* gene and the *Rhodobacter capsulatus adgA* gene code for NH_3-dependent NAD synthetase. J Bacteriol 176: 3400–3402

Zinchenko VV, Babykin MM, Shestakov S, Allibert P, Vignais PM and Willison JC (1990) Ammonia-dependent growth (Adg) mutants of *Rhodobacter capsulatus* and *Rhodobacter sphaeroides*: Comparison of mutant phenotypes and cloning of the wild-type (*adgA*) gene. J Gen Microbiol 136: 2385–2393

Chapter 57

Organization of Photosynthesis Gene Transcripts

J. Thomas Beatty
Department of Microbiology and Immunology, The University of British Columbia, #300-6174 University Boulevard, Vancouver, BC, V6T 1Z3, Canada

R. E. Blankenship, M. T. Madigan and C. E. Bauer (eds): Anoxygenic Photosynthetic Bacteria, pp. 1209–1219.

Summary

Clusters of genes that encode pigment biosynthesis and pigment-binding proteins have been found in several species of anoxygenic photosynthetic bacteria. Transcripts in excess of ten kilobases in length have been found to encode what were previously thought to be separate operons with independent promoters in *Rhodobacter capsulatus*. This transcriptional organization of operons is called a superoperon, and is important for efficient transition in metabolism when cells are shifted from aerobic respiratory to anaerobic photosynthetic growth conditions. Since some key features of superoperons are present in other species of photosynthetic bacteria, the clustering of photosynthesis genes may have been evolutionarily conserved because of advantages accrued to cells that have linked transcription units into superoperons. The evidence for the types of transcription units in several species of anoxygenic photosynthetic bacteria is summarized and compared, with an emphasis on unifying patterns of transcript organization. Because much of the data are not conclusive, some of the conclusions are somewhat speculative in nature. Since more is known about transcripts of *Rb. capsulatus*, it is used as a model for the other species. The reader is advised to consult recent reviews of related areas (Wellington et al., 1992; Klug, 1993).

I. Introduction

A. Overview and History

One of the first discoveries of the organization of photosynthesis genes was that pigment biosynthesis genes in *Rhodobacter capsulatus* are clustered in a relatively small region of the chromosome (Yen and Marrs, 1976). This led Yen and Marrs (1976) to suggest that these genes might be clustered to obtain their coordinate transcription. In its most extreme form, this model would consist of a single transcript of all the genes located in the photosynthesis gene cluster, now known to be about 46 kb in length (see Chapter 50 by Alberti et al.). In the years that followed these pioneering experiments it became clear that, in addition to the pigment biosynthesis genes, the reaction center, light-harvesting I (LH-I or B870) and other genes were also present in this cluster (Taylor et al., 1983; Youvan et al., 1984; Zsebo et al., 1984). The initial studies of the regulation of transcription of the clustered photosynthesis genes indicated that there were separate transcripts of bacteriochlorophyll biosynthesis genes and genes that encode LH and reaction center (RC) proteins (Clark et al., 1984; Belasco et al., 1985; Zhu et al., 1986). Other data showed that transcription of some pigment biosynthesis genes was divergent, implying an organization of many transcription units (Giuliano et al., 1988; Armstrong et al., 1989). A series of publications then appeared that showed read-through transcription from pigment biosynthesis genes, across promoter regions and into genes encoding LH and RC proteins (Wellington and Beatty, 1989; Young et al., 1989; Bauer et al., 1991; Wellington et al., 1991). More recently, the DNA sequence of a 46 kb contiguous region of the *Rb. capsulatus* photosynthesis gene cluster was generously deposited in GenBank by John Hearst (see Chapter 50 by Alberti et al.). This sequence is a valuable resource for prediction of possible transcripts, as well as for other purposes. It is now clear that, in *Rb. capsulatus*, the genuine organization of transcripts lies somewhere between the extremes of a single transcript and separate, independently regulated, transcription units of pigment biosynthesis and other photosynthesis genes.

As genetic research progressed in *Rb. capsulatus*, striking advances were made in the areas of the organization and regulation of expression of photosynthesis genes in other species, notably *Rhodobacter sphaeroides*. It now seems likely that many of the genes in photosynthetic bacteria that are homologous to those of the *Rb. capsulatus* photosynthesis gene cluster are organized and transcribed similarly to *Rb. capsulatus* (Wellington et al., 1992; Bauer et al., 1993; see below).

This chapter summarizes some of the current data and my speculations on the organization of photosynthesis genes into transcription units, and points out some unifying patterns of gene organization that seem likely to be related to transcript organization in a variety of species. The photosynthesis genes

Abbreviations: Cf. – Chloroflexus; E. coli – Escherichia coli; kb – kilobase(s); LH I – light-harvesting I complex; LH II – light-harvesting II complex; orf – open reading frame; *Rb. – Rhodobacter;* RC – reaction center; *Rp. – Rhodopseudomonas; Rs. – Rhodospirillm*

considered are those located within the *Rb. capsulatus* cluster, the LHII genes encoded by the *puc* operon, and their homologues in other species.

B. Methods Used for Analysis of Transcript Organization: Advantages and Drawbacks

DNA sequences of genes, coupled with genetic mapping, are powerful tools for obtaining the minimal number of transcription units within a cluster of genes. All of the techniques described here require cloned DNA fragments, and the ability to identify genes by complementation of mutations or by DNA sequence analysis. Unambiguous transcript analyses require the use of two or more of the methods described below.

1. Indirect Methods

The effects of polar mutations on phenotypes have been used to infer the directions of transcription and the existence of operons of two or more genes (Zsebo et al., 1984; Young et al., 1989, 1992). The omega cartridge (Prentki and Krisch, 1984) has proven to be a good polar mutagen, but its use requires the ability to return omega-disrupted fragments back into cells, preferably by homologous recombination into the chromosome. There must also be a way to distinguish between inactivation of one gene and inactivation of two or more genes.

Gene fusions have been widely used to identify the location of promoters and, therefore, the approximate location of RNA 5' ends resulting from transcription initiation. This is usually of fairly low resolution; however, see Adams et al. (1989), Lee and Kaplan (1992) and Ma et al., (1993) for the use of gene fusions to evaluate the consequences of site-directed promoter mutations. The presence of a promoter on a cloned DNA fragment does not mean that there is not another promoter further upstream, which functions in the chromosome but which may be absent from the DNA fragment.

2. Direct Methods

The crudest direct methods for analysis of transcripts are the original Southern blot method (radiolabelled RNA used to probe specific DNA fragments), or 'dot-blot' methods (Clark et al., 1984; Zhu and Hearst, 1986). Although these techniques can show that transcripts are derived from specific DNA fragments, they do not differentiate between the existence of one or more transcripts. Unless the number and sizes of hybridizing transcripts are known, it is difficult to draw quantitative conclusions that relate the relative number of transcripts to the intensities of hybridization signals.

RNA (northern) blots give the number and sizes of transcripts, and their relative steady-state amounts, when corrections are applied for specific activities of probes and the lengths of transcripts. When a single-stranded probe is used, the direction of transcription can also be obtained. This procedure is less sensitive than end-mapping methods (see below) and, by itself, yields little information about transcription initiation and termination sites. Also, unless RNA is hydrolyzed by treatment with alkali or RNA is transferred from the gel to the membrane by electroblotting, large RNA molecules do not transfer as efficiently as small ones and may not be detected.

Mung bean or S1 nuclease end-mapping methods are probably the most sensitive techniques for detection of transcripts that are present in low amounts. These procedures can be quantitative, but require care in titration of the ratio of nuclease:nucleic acids, and the optimal temperature of hybridization must be determined with each probe. In conjunction with RNA blots, end mapping experiments are usually sufficient to define specific RNA molecules to the nucleotide level. When 'bipartite' probes (Wellington and Beatty, 1991) are used, end mapping methods can reveal the existence of read-through (super-operon) transcripts. However, differentiation of transcription initiation and termination sites from ends resulting from RNA cleavage requires additional techniques such gene fusions (see above), pulse-chase analyses (Belasco et al., 1985), or 5' end-capping experiments (Zucconi and Beatty, 1988).

Primer extension mapping of a 5' end is technically easier to perform than nuclease end-mapping. However, DNA sequence information is necessary to design a oligonucleotide primer, and reverse transcriptase enzymes are not robust enough to extend very far.

II. Transcripts of *Rb. capsulatus* Photosynthesis Genes

A representation of the genes in the *Rb. capsulatus* photosynthesis gene cluster and the proposed primary transcripts is given in Fig. 1. The gene and orf

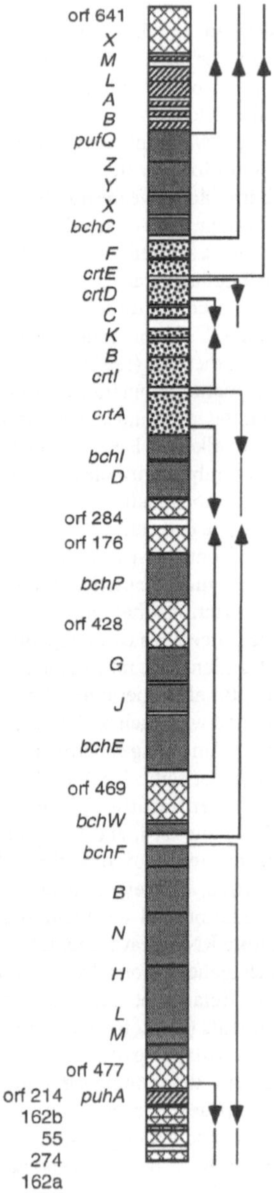

Figure 1. Representation of genes and transcripts of the *Rhodobacter capsulatus* photosynthesis gene cluster. Bacteriochlorophyll biosynthesis genes (*bch*) are designated by grey shading, carotenoid biosynthesis genes (*crt*) are shown by spots, light-harvesting and reaction center genes (*puf* and *puh*) are represented by diagonal lines, and open reading frames of uncertain function are shown as cross-hatched boxes. Proposed transcripts are designated by arrows, with possible read-through extensions shown as extensions beyond arrow heads.

designations are based on revisions (Bauer et al., 1993; Bollivar et al., 1994a,b) of those given in the Hearst laboratory's Genbank submission (accession number Z11165). Because experiments have not yet been done to directly identify some proposed transcripts, the location of many of the 5' and 3' ends in Fig. 1 is tentative. Processing products of primary transcripts are not shown.

A. The crtEF-bchCXYZ-pufQBALMX *Superoperon*

This region of the *Rb. capsulatus* photosynthesis gene cluster contains genes that give rise to the best-defined transcripts, at this time. The DNA sequence upstream of the *crtE* gene contains a motif with similarity to an *Escherichia coli*-like promoter (Armstrong et al., 1989). However, a low resolution 5' end mapping experiment indicated that there are two transcripts of *crtE* (one strongly oxygen-repressed) with 5' ends that map near the start of the *crtE* gene (Giuliano et al., 1988). Although it has been presumed that the *crtE* and *crtF* genes are co-transcribed, this supposition has not been tested. At least one of the *crtEF* primary transcripts extends into the *bchC* coding region, and transcription across the *bchC* promoter seems to stimulate transcription initiation at this promoter (Young et al., 1989; Wellington and Beatty, 1991; Wellington et al., 1991). Analogously, transcription that initiates at the *bchC* promoter continues through the *puf* operon promoter region, and seems to stimulate transcription initiation at the *puf* operon promoter (Wellington and Beatty, 1991; Wellington et al., 1991). Although it is probable that transcription which initiates at the *crtEF* promoters continues into the *puf* operon, this assumption has not been tested directly. Extensive post-transcriptional processing of the *crtEF-puf* transcripts occurs, best documented for the *puf* operon segments (described in the Chapter by Klug). Although two RNA 3' ends were mapped to two rho-independent terminator-like structures downstream of *pufX* (Chen et al., 1988), it is conceivable that there is read-through transcription from *pufX* into orf641 (and beyond).

B. The crtD *and* crtC *Genes*

These two genes are transcribed in the same direction and separated by only 70 base pairs (Armstrong et al., 1989), yet they seem to have separate promoters.

This is because polar omega cartridge insertions into the *crtD* gene did not eliminate expression of the *crtC* gene, and a low resolution S1 nuclease protection experiment showed the existence of a *crtC* RNA 5' end that mapped within the *crtD* gene (Giuliano et al., 1988). Thus, these two genes seem to form a superoperon, in the sense that a *crtC* promoter is located within the *crtD* gene. An *E. coli* promoter-like sequence upstream of *crtD* was described (Armstrong et al., 1989).

C. The crtIBK *Genes*

These genes are designated as a transcription unit because a polar mutation in *crtI* had a *crtB* phenotype (which would mask a *crtK* phenotype since *crtB* mutations cause a block earlier in the pathway), and because only about 75 base pairs separate *crtK* from *crtB* (Giuliano et al., 1988; Armstrong et al., 1989). A *crtI* RNA 5' end was mapped near the start of the *crtI* gene (Giuliano et al., 1988), and a possible promoter suggested from the DNA sequence (Armstrong et al., 1989).

D. The crtA-bchID-*orf284 Genes*

Experiments with omega interposon mutants of *crtA* indicated that a *crtA* transcript also encodes the *bchI* gene (Young et al., 1992). It is not clear if a single transcript encodes *bchD* and orf284, as well as *crtA* and *bchI*, although the DNA sequence is consistent with this likelihood (see Chapter 50 by Alberti et al.). The DNA sequence upstream of *crtA* contains an *E. coli* promoter-like sequence (Armstrong et al., 1989), and a low resolution S1 nuclease protection experiment demonstrated an RNA 5' end that mapped near the start of the *crtA* gene (Giuliano et al., 1988).

E. The bchW-*orf469*-bchEJG-*orf428*-bchP-*orf176 Genes*

These sequences could be encoded by a single transcript, although this possibility has not been tested either indirectly or directly. Two transcripts are assigned in Fig. 1 because these proposed genes all have the same direction of transcription, are flanked by genes with opposite polarities of transcription, and a sequence with similarity to an *E. coli*-like promoter has been identified between orf469 and *bchE* (see Chapter 50 by Alberti et al.).

F. The bchFBNHLM-*orf477*-puhA-*orf214-orf162b.... Superoperon*

Transcripts of this region of the *Rb. capsulatus* photosynthesis gene cluster have been fairly well characterized, although several important questions remain to be answered. A long transcript initiated at the *bchF* promoter extends into the *puhA* gene, and seems to be degraded to a more stable *puhA*-encoding segment of 1.1 kb in length (Bauer et al., 1991). Although an *E. coli* promoter-like sequence exists upstream of the *bchF* gene (see Chapter 50 by Alberti et al.), it is not clear if this sequence functions as a promoter in *Rb. capsulatus*. A promoter within orf477 (previously known as F1696) yields a 0.95 kb *puhA*-encoding transcript (Bauer et al., 1991). Unpublished data (Wong and Beatty) indicate that expression of orf214 and at least one additional gene beyond orf214 are important for photosynthetic growth, and are dependent on read-through transcription from the *puhA* gene for normal expression. It is possible that a low-level promoter for orf214 is located near the 3' end of the *puhA* gene, and that transcripts initiated at the *bchF* and *puhA* promoters extend into orf214 and beyond.

G. The pucBACDE *Operon*

This operon maps to a region of the chromosome distant from the photosynthesis gene cluster (see Chapter 49 by Fonstein and Haselkorn). Two very abundant *ca.* 0.5 kb transcripts encode the *pucBA* genes (Zucconi and Beatty, 1988). Additional molecules of 2.4, 1.0, and 0.7 kb have been detected (LeBlanc and Beatty, 1993). The results of RNA blots, end-mapping and gene fusion experiments (Zucconi and Beatty, 1988; Tichy et al., 1989, 1991; Oberlé et al., 1990; LeBlanc and Beatty, 1993), taken together, are consistent with the synthesis of a *pucBACDE* primary transcript of about 2.4 kb in length, along with degradation of this transcript to yield molecules of 1.0 and 0.7 kb (encoding *pucDE*), and 0.5 kb (encoding *pucBA*). This model remains to be tested directly, as do the possibilities of transcription attenuation between *pucBA* and *pucC*, the existence of a weak *pucDE* promoter, and of transcripts extending into or out of the *pucBACDE* region.

III. Transcripts of *Rb. sphaeroides* Photosynthesis Genes

A. The pufQBALMX *Operon*

The stable transcripts of these genes are similar in size to the homologous *puf* messages of *Rb. capsulatus*, with similar endpoints (Zhu et al., 1986). The most likely origin of *Rb. sphaeroides puf*-specific transcripts is from a single promoter region upstream of *pufQ* and within a *bchZ* gene (Hunter et al., 1991; McGlynn and Hunter, 1993), although Kaplan and coworkers have argued for the existence of two promoters immediately preceding the *pufBA* genes (Kiley and Kaplan, 1988; Gong et al., 1994). Since the *puf* operon promoter identified by Hunter *et al.* (1991) lies within the *bchZ* gene, it is likely that *bchZ* transcripts continue through the *puf* operon, as part of a superoperon.

B. The *bchCXYZ* Operon

This region of the *Rb. sphaeroides* photosynthesis gene cluster is organized very similarly to, and with the same transcriptional polarity as the homologous region from *Rb. capsulatus*. Transcripts in excess of 9.5 kb in size that hybridized to *bchC, bchCX* and *bchYZ* probes were found in RNA blot experiments, and the DNA sequences of these genes showed overlaps of stop codons with start codons at the junctures of the *bchCX, bchYZ* and *bchZ-pufQ* genes (McGlynn and Hunter, 1993). A similar overlap was observed for the *bchYZ* juncture in *Rb. capsulatus* which has been implicated in promoting translational coupling of these genes (Burke et al., 1993). As noted above, a promoter necessary and sufficient for expression of *puf* operon genes was located within the *bchZ* gene (Hunter et al., 1991).

C. The *crtE* to *puhA* Region of the Photosynthesis Gene Cluster

Although only the *crtI* and *puhA* regions have been investigated in RNA and DNA sequence analyses, the overall similarity of the order of the *Rb. sphaeroides crt* and *bch* genes to the order of the homologues in *Rb. capsulatus* (Chapter 50 by Bauer; Bauer et al., 1993; Coomber et al., 1990) invites speculation that the polarities of transcription of these genes have also been conserved.

The *crtI* and *crtB* genes were sequenced and the proposed *crtI* stop and *crtB* start codons were found to overlap (Lang et al., 1994). Earlier studies indicated that transposon insertions in *crtI* reduced *crtB* expression (Coomber et al., 1990). However, Lang et al. (1994) found that *crtI* transposon mutants were partially complemented by a plasmid containing only the *crtI* gene. Furthermore, primer extension and RNA blot experiments on *crtI* messages showed the existence of a 5' RNA end 70 nt before the start of the 1.6 kb gene, and a *ca.* 1.6 kb *crtI* message. These results exemplify the difficulties in obtaining unambiguous data about labile transcripts of genes that are closely linked, yet possibly expressed from more than one promoter or from a promoter located within a 'polar' transposon insertion. It was concluded that a weak *crtB* promoter might be located within the 3' region of the *crtI* gene, although other possibilities were entertained (Lang et al., 1994).

The *puhA* gene is preceded by a sequence that appears to be part of a homologue of the *Rb. capsulatus* orf477 (Bauer et al., 1991; Donohue et al., 1986). RNA blot analysis of the *puhA* gene revealed two RNA messages of 1.4 and 1.1 kb in length (Donohue et al., 1986), analogs to two of the three molecules that code for the *Rb. capsulatus puhA* gene. If it is assumed that the 3' ends of these messages map near the 3' end of the *puhA* gene, their 5' ends would lie within the orf477 homologue. Although a large *bch-puhA* transcript was not found in *Rb. sphaeroides* (Donohue et al., 1986), this negative result does not prove the absence of such a transcript, so it is possible that *Rb. sphaeroides* and *Rb. capsulatus* have similar superoperons that yield similar primary and processed *puhA* transcripts.

D. The pucBAC *Operon*

The DNA sequence of this region of the *Rb. sphaeroides* genome showed that it differs from the analogous region in *Rb. capsulatus* because of the absence of *pucDE* genes (Gibson et al., 1992). RNA blot and 5' end-mapping experiments demonstrated the presence of a 2.3 kb molecule with the same 5' end as a more abundant 0.6 kb *pucBA* message (Lee et al., 1989). Therefore, by analogy with the *Rb. capsulatus* model, it may be that a large (2.3 kb) transcript is processed to yield a more stable, small (0.6 kb) molecule.

IV. Transcripts of *Rhodospirillum rubrum* Photosynthesis Genes

A. The pufBALM *Operon*

RNA blot and end-mapping experiments have shown that these genes are encoded by large (2.5 kb *pufBALM*) and small (*ca.* 0.6 kb *pufBA*) messages (Bélanger and Gingras, 1988). The *puf* operon promoter is located within a *bchZ* homologue (Lee and Collins, 1993), again inviting speculation that a superoperon exists, consisting of *bch* and *puf* genes.

B. The puhA *Region*

From the results of RNA blot, 5' and 3' end-mapping experiments (Bérard et al., 1989) it was concluded that the *puhA* gene is encoded by two primary transcripts (of approximately 1.2 and 1 kb in size) that originate at two adjacent promoters located within orfG115 (homologous to orf477 of *Rb. capsulatus* [Bérard et al., 1991]). Alternatively, Bauer et al. (1991) have suggested that one of these two molecules is the product of degradation of a larger transcript that originates upstream of orfG115. Although a large transcript encoding orfG115 and upstream sequences, and extending into the *puhA* gene, was not identified by Bérard et al. (1991), rigorous experiments to exclude this possibility have not been done. Additional experiments need to be done to directly test for read-through transcription into the *puhA* gene, and from *puhA* into orfs I2372 and I3087 (which encode amino acid sequences similar to the *Rb. capsulatus* orfs 214 and 162b). Because of the great similarities in gene sequences and organization between *Rb. capsulatus* and *Rs. rubrum*, the transcripts of these sequences may also be similar.

C. The Region between puhA *and* puf *Genes*

Several carotenoid and bacteriochlorophyll biosynthesis genes have been identified by mutation or DNA sequence analyses (Sägesser, 1992; Lee and Collins, 1993), and their locations relative to *Rb. capsulatus* and *Rb. sphaeroides* homologues indicate a conservation of gene order within these species (Bauer et al., 1993). However, the transcripts of these genes have not yet been studied in *Rs. rubrum* and their transcriptional polarities relative to the *puf* operon, apart from the *bchZ* gene, are not known.

V. Transcripts of *Rhodopseudomonas viridis* Photosynthesis Genes

A. The pufBALMC *Operon*

The *Rp. viridis puf* operon genes are encoded by two large (3.6 and 3.7 kb) *pufBALMC* and two small (0.6 and 0.8 kb) *pufBA* messages, as evidenced by the results of RNA blot, 5' and 3' end-mapping experiments (Wiessner et al., 1990). The 5' ends of the 3.7 and 0.8 kb molecules mapped within a sequence that encodes the carboxy terminus of a protein that is 50% identical to the corresponding end of the *Rb. capsulatus* BchZ amino acid sequence (Wiessner et al., 1990). Although 5' ends resulting from transcription initiation were not distinguished from 5' ends arising from cleavage of a precursor, the deduced overlap between *bchZ* and *puf* transcripts, once again, invites speculation that a labile *bch-puf* primary transcript is processed to contribute to more stable *puf* messages.

B. The Reaction Center H Gene Region

The distance between the *Rp. viridis* reaction center H gene and *puf* operon has been estimated to be in excess of 100 kb, which is a much greater distance than in *Rb. capsulatus* and *Rb. sphaeroides* (Lang and Oesterhelt, 1989). The gene encoding the *Rp. viridis* reaction center H subunit (homologous to the *puhA* gene of *Rb. capsulatus*; Michel et al., 1985) is located immediately downstream of an orfS, which exhibits 'strong homology' with orf477 of *Rb. capsulatus* (Wiessner, 1990). RNA blots showed the existence of 1.2 and 1.6 kb messages, and end-mapping experiments placed the 5' end of a H subunit transcript within orfS, analogous to the results obtained with *Rb. capsulatus*, *Rb. sphaeroides* and *Rs. rubrum* (see above). Thus, all of these species may contain superoperons that include *puhA* genes.

VI. Transcripts of *Rubrivivax gelantinosus* Photosynthesis Genes

A. The pufBALMC Operon

The *Ru. gelatinosus puf* operon genes are organized similarly to the *Rp. viridis puf* genes, except that an orf1 was identified just upstream of *pufB* and an orf2 was found between *pufB* and *pufA* (Nagashima et al., 1994). This operon is encoded by 4 kp *pufBALMC* and 1 kb *pufBA* messages, as revealed by RNA blot, 5' and 3' end-mapping experiments (Nagashima et al., 1994). The 5' end of the 1.0 kb (and, presumably, the 4 kb) molecule mapped within a sequence that encodes the carboxy terminus of a protein that is similar to the corresponding end of the *Rb. capsulatus* BchZ amino acid sequence (Nagashima et al., 1994). Thus, the overlap between *bchZ* and *puf* transcripts is conserved in all species of purple photosynthetic bacteria that have been investigated.

VII. Transcripts of *Chloroflexus aurantiacus* Photosynthesis Genes

A. The pufLM Genes

A major transcript of approximately 2.1-2.3 kb, and minor species of 3.1, 1.7 and 1.5 kb in length were detected by hybridization of RNA blots with a *pufL* gene probe (Shiozawa et al., 1990; Chapter 54 by Shiozawa). Signals corresponding to molecules of 10 kb or more in length were seen, but were attributed to DNA contamination of the RNA preparations. Primer extension analysis showed a predominant 5' end that mapped about 130 bases upstream of the *pufL* gene start codon, near a sequence that was similar to an *E. coli* promoter consensus sequence (Shiozawa et al., 1990). Although the data are consistent with the existence of a single, *ca.* 2.3 kb *pufLM* primary transcript that is degraded to yield smaller quasi-stable segments, it is not clear if the *pufL* transcript 5' end that was mapped arises from transcription initiation or message processing, and if there are other promoters for *pufM* or far upstream of the *pufL* gene.

B. The Genes Encoding B806-866 LHα, LHβ and Cytochrome c-554

These three genes are located adjacent to each other, in the order LHα, LHβ and cytochrome *c*-554 (Watanabe et al., 1992). The distance from the *pufLM* genes is unknown. The results of RNA blot and primer extension experiments were interpreted as consistent with the existence of a 2.1 kb primary transcript encoding LHα/LHβ/cytochrome *c*-554 genes that is degraded to yield a 0.5 kb LHα/LHβ segment, and a 1.5 kb segment encoding cytochrome *c*-554. It was proposed that another 1.5 kb message initiates about 80 base pairs downstream of the 5' end of the 1.5 kb proposed cleavage product of the 2.1 kb transcript (Watanabe et al., 1992). Although this interpretation is plausible, additional experiments need to be done to differentiate between 5' ends that arise from initiation of transcription and 5' ends that result from cleavage of a precursor molecule.

VIII. Prospects for the Future and Concluding Remarks

Although many of the details of photosynthesis gene transcripts have been elucidated in *Rb. capsulatus*, there are a number of outstanding questions that remain. For example, the distance that the *bchF-puhA* superoperon extends beyond orf214 (see Fig. 1) needs to be better defined, higher resolution transcript analyses need to be done of the *crtD* to *bchD* region, and the possibility that the *bchW*-orf176 genes are part of a (super-) operon should be tested.

A large amount of genetic mapping, DNA sequence and transcript analyses needs to be done with *Rb. sphaeroides*, *Rs. rubrum*, *Rp. viridis* and *Cf. aurantiacus* to extend the results summarized here. It would be of great interest to fill in the gaps in the genetic maps of these species and follow up the mapping with transcript analyses, to see if some of my speculations about the existence of superoperons are correct.

Relatively low resolution genetic mapping and complementation experiments resulted in a map of *Rhodospirillum centenum* photosynthesis genes (Yildiz et al., 1992). Although the conservation in gene order within this gene cluster, as compared to the *Rb. capsulatus* map (Bauer et al., 1993), suggests a similar organization of transcripts, no promoter mapping, DNA sequence or RNA analyses have been done on this species.

A DNA fragment containing *pufBALM* genes from an *Erythrobacter* species has been cloned and sequenced (Liebetanz et al., 1991). The *pufB* gene on this DNA fragment is preceded by a sequence with similarity to the 3' terminal region of *bchZ* genes, and the *pufM* gene is followed by a sequence with similarity to the 5' end of the *Rp. viridis pufC* gene. These results further underscore the existence of conserved patterns of photosynthesis gene organization, and raise additional questions about the how these patterns relate to gene transcripts, the regulation of gene expression and the evolution of photosynthesis.

Finally, it would be extremely interesting to extend the genetic studies of anoxygenic photosynthetic bacteria to include purple sulfur bacteria. These species are likely to possess photosynthesis genes homologous to the genes identified in purple nonsulfur photosynthetic bacteria and *Cf. aurantiacus*, yet they are located within a separate, coherent group on the basis of 16S rRNA sequence analyses (Woese, 1987). Starts in this direction have been made with the cloning of reaction center genes of *Chlorobium limicola* (Buttner et al., 1992; Chapter 54 by Shiozawa), and the finding that *Chromatium* and *Thiocystis* species contain *puf* genes that are greatly homologous to the corresponding genes of purple non-sulfur bacteria. The benefits of widening the field of the genetics of photosynthetic bacteria include an improved understanding of genetic regulatory mechanisms and evolutionary processes.

Acknowledgments

I thank everyone who sent me reprints, preprints, theses and other information, and the members of my laboratory who gave helpful suggestions for modification of this chapter.

References

Adams CW, Forrest ME Cohen SN and Beatty JT (1989) Structural and functional analysis of transcriptional control of the *Rhodobacter capsulatus puf* operon. J Bacteriol 171: 473–482

Armstrong GA, Alberti M, Leach F and Hearst JE (1989) Nucleotide sequence, organization, and nature of the protein products of the carotenoid biosynthesis gene cluster of *Rhodobacter capsulatus*. Mol Gen Genet 216: 254–268

Bauer CE, Buggy J, Yang Z and Marrs BL (1991) The superoperonal organization of genes for pigment biosynthesis and reaction center proteins is a conserved feature in *Rb. capsulatus*: analysis of overlapping *bchB* and *puhA* transcripts. Mol Gen Genet 228: 438–444

Bauer CE, Bollivar DW and Suzuki JY (1993) Genetic analyses of photopigment biosynthesis in eubacteria: a guiding light for algae and plants. J Bacteriol 175: 3919–3925

Bélanger G and Gingras G (1988) Structure and expression of the *puf* operon messenger RNA in *Rhodospirillum rubrum*. J Biol Chem 263: 7639–7645

Belasco JG, Beatty JT, Adams CW, von Gabain A and Cohen SN (1985) Differential expression of photosynthesis genes in *Rb. capsulatus* results from segmental differences in stability within the polycistronic *rxcA* transcript. Cell 40: 171–181

Bérard J, Bélanger G and Gingras G (1989) Mapping of the *puh* messenger RNAs from *Rhodospirillum rubrum*. J Biol Chem 264: 10897–10903

Bollivar DW, Suzuki JY, Beatty JT, Dobrowlski J and Bauer CE (1994a) Directed mutational analysis of bacteriochlorophyll *a* biosynthesis in *Rhodobacter capsulatus*. J Mol Biol 237: 622-640

Bollivar DW, Wang S, Allen JP and Bauer CE (1994b) Molecular genetic analysis of terminal steps in bacteriochlorophyll *a* biosynthesis: Characterization of a *Rhodobacter capsulatus* strain that synthesizes geranylgeraniol esterified bacteriochlorophyll *a*. Biochemistry 33: 12763-12768

Burke DH, Alberti M and Hearst JE (1993) The *Rhodobacter capsulatus* chlorin reductase-encoding locus, *bchA* consists of three genes *bchX*, *bchY* and *bchZ*. J Bacteriol 175: 2407-2413

Buttner M, Xie DL, Nelson H, Pinther W, Hauska G and Nelson N (1992) Photosynthetic reaction center genes in green sulfur bacteria and in photosystem 1 are related. Proc Natl Acad Sci USA 89: 8135–8139

Chen C-YA, Beatty JT, Cohen SN and Belasco JG (1988) An intercistronic stem-loop structure functions as an mRNA decay terminator necessary but insufficient for *puf* mRNA stability. Cell 52: 609–619

Clark WG, Davidson E and Marrs BL (1984) Variation of levels of mRNA coding for antenna and reaction center polypeptides in *Rhodopseudomonas capsulata* in response to changes in oxygen concentration. J Bacteriol 157: 945–948

Coomber SA, Chaudhri M, Connor A, Britton G and Hunter CN (1990) Localized transposon Tn5 mutagenesis of the photosynthetic gene cluster of *Rhodobacter sphaeroides*. Mol Microbiol 4: 977–989

Donohue TJ, Hoger JH and Kaplan S (1986) Cloning and expression of the *Rhodobacter sphaeroides* reaction center H gene. J Bacteriol 168: 953–961

Gibson LCD, McGlynn P, Chaudhri M and Hunter CN (1992) A putative coproporphyrinogen III oxidase in *Rhodobacter sphaeroides*. II. Analysis of a region of the genome encoding *hemF* and the *puc* operon. Mol Microbiol 6: 3171–3186

Giuliano G, Pollock D, Stapp H and Scolnik PA (1988) A genetic-physical map of the *Rhodobacter capsulatus* carotenoid biosynthesis gene cluster. Mol Gen Genet 213: 78–83

Gong L, Lee JK and Kaplan S (1994) The *Q* gene of *Rhodobacter sphaeroides*: Its role in *puf* operon expression and spectral complex assembly. J Bacteriol 176: 2946-2961

Hunter CN, McGlynn P, Ashby MK, Burgess J G and Olsen JD (1991) DNA sequencing and complementation/deletion analysis of the *bchA-puf* operon region of *Rhodobacter sphaeroides*: *in vivo* mapping of the oxygen-regulated promoter. Mol Microbiol 5: 2649–2661

Kiley PJ and Kaplan S (1988) Molecular genetics of photosynthetic membrane biosynthesis in *Rhodobacter sphaeroides*. Microbiol Rev 52(1): 50–69

Klug G (1993) The role of mRNA degradation in the regulated expression of bacterial photosynthesis genes. Mol Microbiol 9: 1–7

Lang FS and Oesterhelt D (1989) Gene transfer system for *Rhodopseudomonas viridis*. J Bacteriol 171: 4425–4435

Lang, HP, Cogdell RJ, Gardiner AT and Hunter CN (1994) Early steps in carotenoid biosynthesis: sequences and transcriptional analysis of the *crtI* and *crtB* genes of *Rhodobacter sphaeroides* and overexpression and reactivation of *crtI* in *Escherichia coli* and *R. sphaeroides.* J Bacteriol 176: 3859-3869

LeBlanc HN and Beatty JT (1993) *Rhodobacter capsulatus puc* operon: promoter location, transcript sizes and effects of deletions on photosynthetic growth. J Gen Microbiol 139: 101–109

Lee IY and Collins MLP (1993) Identification and partial sequence of the *bchA* gene of *Rhodospirillum rubrum*. Curr Microbiol 27: 85–90

Lee JK, Kiley PJ and Kaplan S (1989) Posttranscriptional control of puc operon expression of B800-850 light-harvesting complex formation in *Rhodobacter sphaeroides*. J Bacteriol 171: 3391–3405

Lee JK and Kaplan S (1992) *cis*-acting regulatory elements invovled in oxygen and light control of *puc* operon transcription in *Rhodobacter sphaeroides.* J Bacteriol 174:1146-1157

Liebetanz R, Hornberger U and Drews G (1991) Organization of the genes coding for the reaction-centre L and M subunits and B870 antenna polypeptides a and b from the aerobic photosynthetic bacterium *Erythrobacter* species OCH114. Mol Microbiol 5: 1459–1468

Ma D, Cook DN, O'Brien DA and Hearst JE (1993) Analysis of the promoter and regulatory sequences of an oxygen regulated *bch* operon in *Rhodobacter capsulatus* by site-directed mutagenesis. J Bacteriol 175: 2037-2045

McGlynn P and Hunter CN (1993) Genetic analysis of the *bchC* and *bchA* genes of *Rhodobacter sphaeroides*. Mol Gen Genet 236: 227–234

Michel H, Weyer KA, Gruenberg H and Lottspeich F (1985) The 'heavy' subunit of the photosynthetic reaction centre from *Rhodopseudomonas viridis*: isolation of the gene, nucleotide and amino acid sequence. EMBO J 4: 1667–1672

Nagashima KVP, Shimada K and Matsuura K (1993) Primary structure and transcription of genes encoding B870 and photosynthetic reaction center apoproteins from *Rubrivivax gelatinosus*. J Biol Chem 269: 2477-2484

Oberlé B, Tichy HV and Drews G (1990) Regulation of formation of photosynthetic light-harvesting complexes in *Rhodobacter capsulatus*. In: Drews G and Dawes EA (eds) Molecular Biology of Membrane-Bound Complexes in Phototrophic Bacteria, pp 77–84. Plenum Press, New York

Prentki P and Krisch HM (1984) *In vitro* insertional mutagenesis with a selectable DNA fragment. Gene 29: 303–313

Sägesser R (1992) Identifikation und charakterisierung des photosynthese-genclusters von *Rhodospirillum rubrum*. Ph.D. thesis, Universität Zürich

Shiozawa JA, Csiszár K and Feick R (1990) Preliminary studies on the operon coding for the reaction center polypeptides in *Chloroflexus aurantiacus*. In: Drews G and Dawes EA (eds) Molecular Biology of Membrane-Bound Complexes in Phototrophic Bacteria, pp 11–18. Plenum Press, New York

Taylor DP, Cohen SN, Clark WG and Marrs BL (1983) Alignment of genetic and restriction maps of the photosynthesis region of the *Rhodopseudomonas capsulata* chromosome by a conjugation-mediated marker rescue technique. J Bacteriol 154: 580–590

Tichy HV, Oberlé B, Stiehle H, Schiltz E and Drews G (1989) Genes downstream from *pucB* and *pucA* are essential for formation of the B800–850 complex of *Rhodobacter capsulatus*. J Bacteriol 171: 4914–4922

Tichy HV, Albien KU, Gadon N and Drews G (1991) Analysis of the *Rhodobacter capsulatus puc* operon—the *pucC* gene plays a central role in the regulation of LHII (B800–850 complex) expression. EMBO J 10: 2949–2955

Watanabe Y, Feick RG and Shiozawa JA (1992) Cloning and sequencing of the genes encoding the polypeptides of the B806-866 light-harvesting complex of *Chloroflexus aurantiacus*. In: Murata M (ed) Research in Photosynthesis. Vol I, pp 41–44. Kluwer Academic Publishers, Dordrecht

Wellington CL and Beatty JT (1989) Promoter mapping and nucleotide sequence of the *bchC* bacteriochlorophyll biosynthesis gene from *Rhodobacter capsulatus*. Gene 83: 251–261

Wellington CL and Beatty JT (1991) Overlapping mRNA transcripts of photosynthesis gene operons in *Rhodobacter capsulatus*. J Bacteriol 173(4): 1432–1443

Wellington CL, Taggart AKP and Beatty JT (1991) Functional significance of overlapping transcripts of *crtEF, bchCA*, and *puf* photosynthesis gene operons in *Rhodobacter capsulatus*. J Bacteriol 173: 2954–2961

Wellington CL, Bauer CE and Beatty JT (1992) Photosynthesis gene superoperons in purple nonsulfur bacteria: the tip of the iceberg? Can J Microbiol 38: 20–27

Wiessner C (1990) Molekularbiologische analyse der gene des photosynthetischen apparates von *Rhodopseudomonas viridis*. Ph.D. thesis, Wolfgang Goethe-Universität, Fankfurt

Wiessner C, Dunger I and Michel H (1990) Structure and transcription of the genes encoding the B1015 light-harvesting complex β and α subunits and the photosynthetic reaction center L, M, and cytochrome *c* subunits from *Rhodopseudomonas viridis*. J Bacteriol 172: 2877–2887

Woese CR (1987) Bacterial evolution. Microbiol Rev 51(2): 221–271

Yen HC and Marrs B (1976) Map of genes for carotenoid and bacteriochlorophyll biosynthesis in *Rhodopseudomonas capsulata*. J Bacteriol 126: 619–629

Yildiz FH, Gest H and Bauer CE (1992) Conservation of the photosynthesis gene cluster in *Rhodospirillum centenum*. Mol Microbiol 6: 2683–2691

Young DA, Bauer CE, Williams JC and Marrs BL (1989) Genetic evidence for superoperonal organization of genes for photosynthetic pigments and pigment-binding proteins in *Rhodobacter capsulatus*. Mol Gen Genet 218: 1–12

Young DY, Rudzik MB and Marrs BL (1992) An overlap between operons involved in carotenoid and bacteriochlorophyll biosynthesis in *Rhodobacter capsulatus*. FEMS Microbiol Lett 95: 213–218

Youvan DC, Bylina EJ, Alberti M, Begusch H and Hearst JE (1984) Nucleotide and deduced polypeptide sequences of the photosynthetic reaction-center, B870 antenna and flanking polypeptides from *Rb. capsulata*. Cell 37: 949–957

Zhu YS, Cook DN, Leach F, Armstrong GA, Alberti M and Hearst JE (1986) Oxygen-regulated mRNAs for light-harvesting and reaction center complexes and for bacteriochlorophyll and carotenoid biosynthesis in *Rhodobacter capsulatus* during the shift from anaerobic to aerobic growth. J Bacteriol 168: 1180–1188

Zhu YS, Kiley PJ, Donohue TJ and Kaplan S (1986) Origin of the mRNA stoichiometry of the *puf* operon in *Rhodobacter sphaeroides*. J Biol Chem 261: 10366–10374

Zsebo KM and Hearst JE (1984) Genetic physical mapping of a photosynthetic gene cluster from *R. capsulata*. Cell 37: 937-947

Zucconi AP and Beatty JT (1988) Posttranscriptional regulation by light of the steady-state levels of mature B800-850 light-harvesting complexes in *Rhodobacter capsulatus*. J Bacteriol 170: 877–882

Chapter 58

Regulation of Photosynthesis Gene Expression

Carl E. Bauer
Department of Biology, Indiana University, Bloomington, IN 47405, USA

Summary

Regulation of photosynthesis gene expression has been a long studied area of research with anoxygenic phototrophic bacteria. This area of study predates the classical analysis by Cohen-Bazire et al., (1957) who performed a detailed measurement of the repressing effects of light and oxygen on synthesis of the purple bacterial photosystem. In their study they demonstrated that shifting photosynthetically growing cells from anaerobic to aerobic conditions resulted in the immediate cessation of photopigment biosynthesis. A similar effect was also observed when photosynthetically grown cells are shifted from low to high light intensity. We now know that the repressing effects of these environmental factors is a result, in part, to the regulation of photosynthesis gene expression. It is this topic, the repressing effects of light and oxygen on photosynthesis gene expression, that this review will be centered. This chapter covers our current understanding of *cis*- and *trans*-acting factors that controls transcription of the light harvesting and reaction center structural genes as well as genes involved in photopigment biosynthesis.

I. Introduction

Synthesis of the purple eubacterial photosystem has long been known to be influenced by such environmental factors as oxygen tension and light intensity. The first detailed analysis of the regulation of pigment biosynthesis is the oft cited study by Cohen-Bazire, Sistrom and Stanier (1957). Their work clearly established that pigment biosynthesis in photosynthetically growing *Rhodobacter sphaeroides* cells is rapidly inhibited by the addition of molecular oxygen or by an increases in light intensity. Despite numerous physiological investigations that have been undertaken since then, the underlying molecular

R. E. Blankenship, M. T. Madigan and C. E. Bauer (eds): Anoxygenic Photosynthetic Bacteria, pp. 1221–1234.

mechanism(s) responsible for controlling photopigment biosynthesis are only now being unveiled.

Part of the difficulty in analyzing regulatory events responsible for controlling synthesis of the photosystem is the complexity of the 'regulatory circuits' that control photosynthesis gene expression. For example, light harvesting I (LHI) and reaction center (RC) gene expression are known to be regulated by such events as; (i) the control of transcription initiation, (ii) the linking of genes into 'superoperons' that contain overlapping transcripts, and (iii) the control of mRNA decay rate. This chapter will be centered on the first level of regulation, namely a discussion of *cis* and *trans*-acting regulatory factors responsible for controlling the initiation of photosynthesis gene expression in response to alterations in light intensity and oxygen tension. Related chapters in this volume which also overlap with this topic include Chapter 57 by Betty which covers the superoperonal organization of the photosynthesis gene transcripts, Chapter 59 by Klug which covers the mRNA decay events, Chapter 50 by Alberti et al. which covers the photosynthesis gene cluster sequence analysis, and Chapter 53 by Armstrong which covers carotenoid genes. There have also been several recent minireviews that cover this topic (Bauer et al., 1993; Klug, 1993).

II. The Photosynthesis Gene Cluster

A. Linkage Order of Photosynthesis Genes

Analysis of genes involved in bacterial photosynthesis was primarily advanced by the work of Marrs and coworkers who in the mid 1970's discovered and exploited the use of a generalized transducing agent (GTA) from *Rhodobacter capsulatus* (Marrs, 1974; Yen and Marrs, 1976). These studies established that carotenoid (*crt*) and bacteriochlorophyll (*bch*) biosynthesis genes are present in the chromosome in a tight linkage order (Yen and Marrs, 1976). The transduction mapping studies were followed by the isolation of an R' plasmid by Marrs (1981) that was shown by marker rescue/complementation analysis (Taylor et al., 1983), as well as by transposition mapping techniques (Biel and Marrs, 1983; Zsebo and Hearst, 1984), to contain all of the essential genetic information needed by *Rb. capsulatus* to synthesize a 'core' bacterial photosystem composed of LHI and RC complexes. Structural and functional information regarding these loci has recently been refined to a high degree by the complete sequence analysis of the R' born 'photosynthesis gene cluster' by Hearst and coworkers, (discussed in detail in chapter 50 by Alberti et al.), as well as by the construction of a defined set of interposon mutations in each of the sequenced open reading frames (Giuliano, 1988; Young et al., 1989; Yang and Bauer, 1990; Bollivar et al., 1994a,b). Figure 1, summarizes the location of various open reading frames in the *Rb. capsulatus* photosynthesis gene cluster; as indicated, carotenoid genes (hatched boxes) and bacteriochlorophyll biosynthesis genes (shaded boxes) are clustered together in the central region of the photosynthesis gene cluster. The pigment biosynthesis genes are in turn flanked by the highly expressed LHI and RC structural genes that are coded by the *puf* and *puh* operons (highlighted as filled boxes in the expanded region). Unlinked to this region is the *puc* operon that codes for polypeptides involved in formation of the light harvesting II (LHII) complex (Youvan and Ismail, 1985). It is interesting that a similar clustering and linkage order of photosynthesis genes are known to occur in the closely related species *Rb. sphaeroides* (Coomber et al., 1990) as well as in more distantly related species such as *Rhodospirillum centenum* (Yildiz et al., 1992), *Rhodospirillum rubrum* (Sagesser, 1992) and possibly *Rhodopseudomonas viridis* (Wiessner et al., 1990) and *Rhodoferax gelatinosus* (Nagashima et al., 1993). The conservation of the photosynthesis gene cluster among distantly related species has raised the possibility of acquisition of the photosynthesis gene cluster by horizontal gene transfer (Yildiz et al., 1992). This possibility has more recently been supported by phylogenetic analysis of RC genes which appear to diverge from that of phylogenetic trees derived from 16 S rRNA and cytochrome c_2 (Nagashima et al., 1993).

B. Photosynthesis Gene Transcripts

The first evidence that photosynthesis gene expression was regulated at the transcriptional level was a study by Biel and Marrs (1983) who constructed *lacZ* fusions of bacteriochlorophyll biosynthesis genes

Abbreviations: bch – bacteriochlorophyll biosynthesis genes; BChl *a* – bacteriochlorophyll *a*; *crt* – carotenoid biosynthesis genes; GTA – generalized transducing agent; LHI – light harvesting I; LHII – light harvesting II; *Rb.* – *Rhodobacter*; RC – reaction center; *Rp.* – *Rhodopseudomonas*; *Rs.* – *Rhodospirillum*

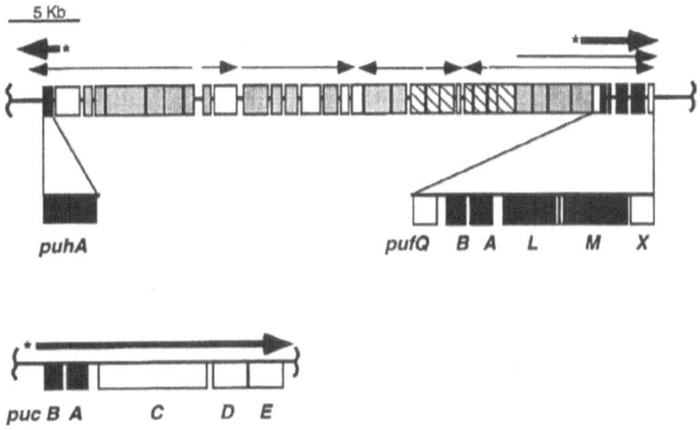

Fig. 1. Photosystem genes from *Rb. capsulatus*. Shaded boxes indicate bacteriochlorophyll biosynthesis genes (Bollivar et al., 1994a,b), hatched boxes indicate carotenoid biosynthesis genes (Armstrong et al., 1989), open boxes have indeterminate function and filled boxes code for LHI and RC structural polypeptides. The arrows denote proposed unprocessed transcripts.

with Mu d1 (Ap^r *lac*). Their study indicated that bacteriochlorophyll gene expression was repressed approximately 2-fold when cells are grown under high oxygen (23% O_2) versus low oxygen tension (3% O_2). Subsequent RNA-DNA hybridization analyses performed in several laboratories confirmed this observation and, in addition, indicated that the LH and RC structural genes coded by the *puf, puh* and *puc* operons are expressed to a much higher level, and are more highly regulated (30 to 100x), than are the photopigment biosynthesis genes (Clark et al., 1984; Klug et al., 1984; Belasco et al., 1985; Zhu and Kaplan, 1985; Klug et al., 1985; Zhu and Hearst, 1986; Zhu et al., 1986a; Zhu et al., 1986b; Kiley and Kaplan, 1987; Donohue et al., 1986; Hunter et al., 1987). In good confirmation with the physiological studies of Cohen-Bazire et al., (1957), both oxygen and light were observed to repress transcription with oxygen exhibiting a more severe repressing effect than that of light.

Promoter mapping studies, the details of which are discussed below, also revealed that transcripts within the photosynthesis gene cluster are organized into large 'superoperons' containing overlapping transcripts. The overlapping nature of these transcripts is such that transcription of the weakly expressed and poorly regulated pigment biosynthesis genes extend into the more highly regulated *puf* and *puh* operons that code for the LHI and RC structural genes (Young et al., 1989; Bauer et al., 1991; Wellington and Beatty, 1991; Young et al., 1992). These overlapping transcripts appear to ensure that there is coupled synthesis of enzymes involved in photopigment biosynthesis with that of polypeptides that ultimately bind the photopigments (Bauer et al., 1991; Wellington and Beatty, 1991). A more detailed description of the superoperonal organization of the photosynthesis gene cluster can be found in chapter 57 by Beatty and in a recent review by Wellington et al., (1992).

III. *cis*-Acting Regulatory Sites

Promoters that are responsible for expressing photosynthesis genes have been mapped using such techniques as S1 (or Mung bean) nuclease mapping, primer extension, 5' end capping and deletion analysis. Despite these efforts, there are relatively few sequenced regions for which there exists a consensus agreement among investigators that they actually contain a promoter region responsible for transcription of photosynthesis genes. Part of the historical difficulty in performing promoter mapping studies appears to reside in the difficulty in obtaining full length mRNA transcripts from photosynthetic bacteria. Many photosynthesis gene transcripts appear to undergo rapid processing events that affect the stability of the 5' end of the message (see chapter 59 by Klug; Klug, 1991a, 1993). Thus, published 5' ends of messages could either be the site of transcription initiation or produced by a processing event. Because of the problems that investigators have had in mapping the site of transcription initiation, it is critical that primer extension or nuclease mapping procedures be corroborated by a functional deletion analysis of a putative promoter region. This has only been performed for a few photosynthesis gene promoter regions as discussed below.

A. puf *and* puh *Promoters*

The first photosynthesis gene promoter to be functionally mapped was the *puf* operon promoter from *Rb. capsulatus*. Deletion mapping (Bauer et al., 1988; Adams et al., 1989) indicated that regulated *puf* operon expression was dependent on a region of DNA that was located over 300 bp upstream from the previously mapped stable 5' end for the *puf* operon (Belasco et al., 1985). Subsequent analysis of the intergenic region indicated that there existed an additional gene upstream from *pufB*, termed *pufQ*, that was shown by mutational analysis to be required for synthesis of bacteriochlorophyll (Bauer et al., 1988; Bauer and Marrs, 1988; Klug and Cohen, 1988; Adams et al., 1989). It was also shown by mutational and sequence analysis that the *puf* promoter region was located within the *bchZ* (formerly called *bchA*; Burke et al., 1993) structural gene that codes for a subunit of an enzyme involved in bacteriochlorophyll biosynthesis (Bauer et al., 1988, Young et al., 1989). High resolution Mung bean and S1 nuclease mapping, as well as 5' mRNA capping experiments, confirmed deletion analysis by indicating that there was indeed a stable *puf* transcript 5' end located within the *bchZ* gene (Bauer et al., 1988; Adams et al., 1989). A similar 5' processing story appears to occur for the *puf* operon promoter region in *Rb. sphaeroides*. In this case, initial 5' mRNA mapping studies that utilized S1-nuclease mapping technique indicated the presence of two stable 5' ends within the *pufQ* region (Zhu et al., 1986). However, like that observed for *Rb. capsulatus*, subsequent deletion analysis indicated that the *Rb. sphaeroides puf* operon is also driven by a region that is located several hundred bp upstream from the S1 mapped stable 5' end, a region that is also within the *bchZ* open reading frame (Hunter et al., 1991). There has also been 5' mRNA analysis of the *Rs. rubrum* and *Rp. viridis puf* transcripts (Belanger and Gingras, 1988; Wiessner et al., 1990). However, deletion analysis of the putative promoter regions from these additional species has not yet been undertaken.

Less is known about the *puh* operon promoter region than that of the *puf* operon promoter. Using a combination of deletion and Mung bean nuclease mapping techniques, Bauer et al., (1991) mapped the *Rb. capsulatus puh* promoter to a position 381 bp upstream from *puhA*. In an interesting analogy with the *puf* operon, the *puh* promoter is also located within an upstream open reading frame termed ORF1696. Mutational analysis indicates that ORF1696 may be involved in formation of a stable LHI complex (Bauer et al., 1991; Yang and Bauer, 1990). A stable 5' end for the *Rs. rubrum puh* message has also been mapped to be well within the upstream ORF1696 coding region (Bérard et al., 1989). The *puh* promoter region of other species has not been mapped to high resolution, however, Northern blot analysis does indicate that the *Rb. sphaeroides puhA* transcript is initiated far upstream of the start of the *puhA* structural gene (Donohue et al., 1986).

Figure 2a shows an alignment of the putative *puf* and *puh* promoter regions from *Rb. capsulatus*, *Rb. sphaeroides, Rp. viridis* and *Rs. rubrum* each of which show some level of sequence conservation. [The putative *Rs. rubrum puf* promoter region shows no identifiable sequence similarity with that of other species and was therefore left out of the alignment. It is most likely that the S1 mapped *Rs. rubrum puf* transcript is a processing site since transposition mutagenesis 'functionally ' places the *Rs. rubrum puf* promoter to within a region of the *bchZ* gene that is approximately 650 bp upstream of the mapped 5' end (Hessner et al., 1991)]. As shown in Fig. 2a, the *puh* promoter regions show a 'reasonable match' with the *puf* promoter sequences, especially when one considers that the *puh* and *puf* promoter sequences also code for polypeptides of differing function. Areas of noted similarity are the –12 and –24 regions from the start of transcription (Fig. 2a). The –12 and –24 regions share no similarity to known eubacterial promoter elements thereby indicating that they may be utilizing an alternative sigma factor for promoter recognition. Mutational and biochemical analysis for the existence for this putative sigma factor (sigma-P) is, however, still lacking. The *puf* and *puh* promoter regions also exhibit a 'pseudo' region of dyad symmetry at positions –30 to –57 relative to the start site of transcription (Bauer et al., 1988, Adams et al., 1989; Bauer et al., 1991). Deletion of this region results in loss of expression of the *puf* operon (Adams et al., 1989) indicating that this area is essential for expression. A point mutation in the region of dyad symmetry has also been shown to result in a 2–8 fold elevation of aerobic *puf* operon expression as well as

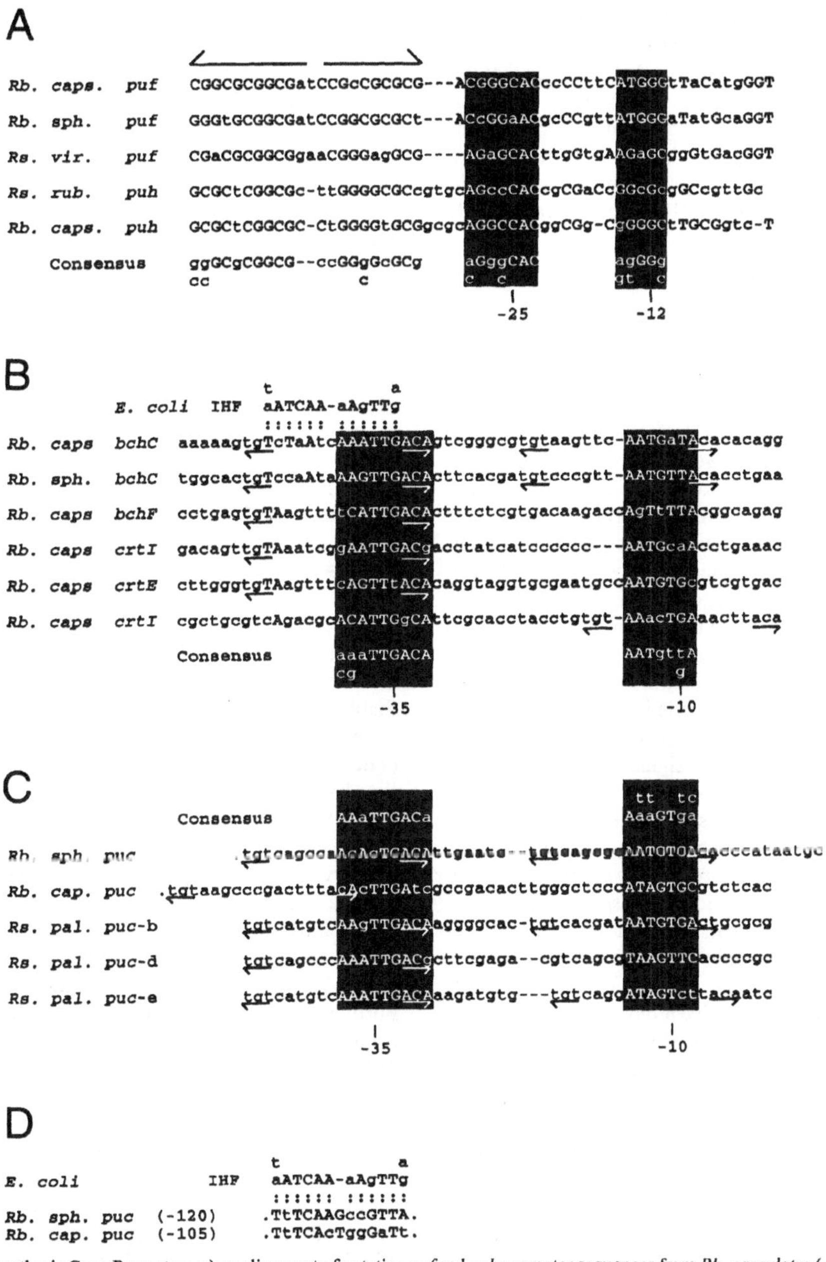

Fig. 2. Photosynthesis Gene Promoters. a) an alignment of putative *puf* and *puh* promoter sequences from *Rb. capsulatus* (Adams et al., 1989; Bauer et al., 1988; 1991), *Rb. sphaeroides* (Hunter et al., 1991) *Rs. rubrum* (Bérard et al., 1989) and *Rp. viridis* (Wiessner et al., 1990). b) an alignment of putative *bch* and *crt* photopigment promoters from *Rb. capsulatus* (Young et al., 1989; Wellington and Beatty, 1989; Ma et al., 1993; Alberti et al., chapter) and *Rb. sphaeroides* (McGlynn and Hunter, 1993). c) An alignment of putative *puc* operon promoters from *Rb. sphaeroides* (Lee et al., 1989) *Rb. capsulatus* (Zucconi and Beatty, 1988) and *Rp. palustris* (Tadros et al., 1993). d) Potential IHF binding domains (Hoover et al., 1990) in the *puc* promoter region of *Rb. sphaeroides* (Lee et al., 1993) and *Rb. capsulatus* (Nickens and Bauer, unpublished)

an effect on light mediated anaerobic expression (Narro et al., 1990; Klug, 1991b; Klug et al., 1991). Several groups have also provided evidence that the *puf* dyad region undergoes a gel mobility shift with cell free extracts in an oxygen and light dependent manner, the binding of which is affected by the phosphorylation state of protein extracts (Taremia and Marrs, 1990; Klug, 1991b; Shimada et al., 1993).

B. bch/crt *Promoters*

Perhaps one of the better characterized photosynthesis gene promoters is that of the *bchC* promoter of *Rb. capsulatus*. Deletion mapping of this promoter region was initially performed by Young et al., (1989). This was followed by high resolution primer extension 5' end mapping of the *bchC* transcript by Wellington and Beatty (1989). Unlike the *puf* and *puh* promoter sequences which exhibit a unique promoter sequence, Young et al., (1989) noted that the *bchC* promoter region contained a sequence that was similar to the canonical *E. coli* housekeeping sigma 70 promoter sequence (Fig. 2b). Recent mutational analysis of the *bchC* promoter region by Ma et al., (1993) has confirmed this observation by indicating that a sigma 70-like promoter sequence is indeed important for *bchC* promoter activity. Sequence gazing has also revealed the presence of 'sigma-70 like' promoter sequences upstream of other *Rb. capsulatus* photopigment biosynthesis genes such as *crtI, crtD crtA, bchF* and *bchE* (Armstrong et al., 1989; Chapter 50 by Alberti et al.). There is also a similar sigma-70 like promoter sequence in the *Rb. sphaeroides bchC* region Fig. 2b) that can be observed by sequence gazing of the intragenic region between *bchC* and *crtF* (McGlynn and Hunter, 1993). An more extensive discussion of these putative promoter sequences can be found in Chapter 50 by Alberti et al.

Also shown in Fig. 2b, and discussed in detail in Chapter 50 by Alberti et al., is the observation by Armstrong et al., (1989) that several of the putative *Rb. capsulatus* 'sigma-70 like' photopigment promoter regions also exhibit a nearby palindrome sequence (TGTGT -$N_{6\text{-}10}$- ACACA). Mutational analysis of the *bchC* palindromic sequence indicates that it may be involved in binding of a repressor to this region (Ma et al., 1993). Mutational analysis has also indicated that the palindromic region is not all that is needed to properly regulate *bchC* promoter activity as evidenced by the requirement for an additional upstream AT rich region for optimal activity (Ma et al., 1993). As shown in the alignment in Fig. 2b, the AT rich region contains a sequence motif that is similar to an IHF binding site that actually overlaps with the *bchC* -35 promoter region. Also shown in Fig. 2b is evidence that there is an IHF-like sequence in the same region of the putative *bchF, bchC, bchE crtI* and *crtA* promoter sequences. Regulated expression of photopigment promoters may therefore require two putative *trans*-acting factors, an IHF-like factor that binds near the –35 region as well as an additional factor that interacts with the palindromic sequence. It is also interesting to note that *puf* and *puh* promoter regions, which as discussed above exhibit a unique promoter motif, do not have an identifiable IHF or 'photopigment like' palindromic sequence. This would be consistent with mutational evidence that the *puf* and *puh* promoters are regulated independently of photopigment promoters (Sganga and Bauer, 1992; Mosley et al., 1994).

C. The puc *Promoter*

One of the more highly expressed photosystem promoters appears to be that of the *puc* operon which drives expression of the LHII structural polypeptides. Stable 5' ends have been mapped for the *puc* message from *Rb. sphaeroides* (Lee et al., 1989), *Rb. capsulatus* (Zuconni and Beatty, 1988) and more recently for the '*puc* like' antenna cluster from *Rp. palustris* (Tadros et al., 1993). As indicated in Fig. 2c, the *puc* promoter regions exhibit a strong degree of sequence conservation among these differing species and genera. Surprisingly, the *puc* promoter alignments shows little similarity with that of the *puf* and *puh* promoters, and instead, exhibits a 'good fit' with the sigma-70 type *bch* and *crt* promoters (Fig. 2). Furthermore, as is the case with the *bch* and *crt* promoters, the *puc* promoter region also exhibit the TGTGT -$N_{6\text{-}10}$- ACACA palindromic motif. There is also a putative IHF binding sequence located 120 and 105 bp upstream of the *Rb. sphaeroides* and *Rb. capsulatus puc* stable 5' ends respectively (Fig. 2d). The putative IHF binding site has been shown by footprint analysis to interact with cell free extracts derived from *Rb. sphaeroides* (McGlynn and Hunter, 1992) as well with *E. coli* IHF (Lee et al., 1993). Deletion of the IHF binding site also shows a 5-fold reduction in anaerobic *puc* expression in vivo (Lee et al., 1993).

Despite sequence similarity of the *puc* promoters with that of the *bch* and *crt* promoters, there is no strong evidence for coordinate regulation of these operons. Instead, there is evidence that the *puc* operon may utilize one or more *trans*-acting regulatory elements that interact with *puf* and *puh* promoters to which *puc* exhibits little sequence similarity. For example, gel retardation experiments indicates that the *puc* promoter region can compete for a protein which binds to the *puf* promoter region of dyad symmetry (Klug, 1991b). A mutation in the *puf* region of dyad symmetry which leads to increased expression of the *puf* operon has also been shown to cause instability of *Rb. capsulatus* strains that results in high frequency segregation of daughter cells that have reduced *puc* expression (Klug and Jock, 1991). There are also several *trans*-acting regulatory mutants that affect *puf*, *puh* as well as *puc* operon expression (Sganga and Bauer, 1992; Lee and Kaplan, 1992b; Eraso and Kaplan, 1994; Mosely et al., 1994) which provides strong evidence for the utilization of a similar set of *trans*-acting factors among these three operons.

Even though there appear to be a common set of *trans*-acting factors involved in *puf*, *puh* and *puc* operon expression, there is also several studies which indicate that *puc* expression is actually discoordinate from that of *puf* and *puh*. For example, kinetic studies involving growth shifts from aerobic to anaerobic conditions indicates that induction of *puc* operon expression lags that of *puf* and *puh* (Klug et al., 1985; Hunter et al., 1987; Kiley and Kaplan, 1987). Mutations that disrupt formation of the LHII complex, such as those which affect synthesis of carotenoids, bacteriochlorophyll or assembly of the LHII complex, also appear to reduce *puc* expression without significantly affecting *puf* expression (Klug et al., 1985; Lee et al., 1989; Tichy et al., 1991). *Puc* operon expression is also poorly regulated by the light responding *trans*-acting activator *hvrA* which coordinately regulates *puh* and *puf* operon expression in response to light intensity (Buggy et al., 1994b). Thus, a picture of discoordinate regulation of *puc* operon expression from that of other photosynthesis genes is emerging.

What then are the factors that allows discoordinate expression of the *puc* operon? One clue may be derived from promoter deletion analysis by Lee and Kaplan (1992a). In their study they concluded that expression of the *Rb. sphaeroides puc* operon involves a complex promoter region that extends up to 629 bp upstream from the point of transcription initiation. By deletion analysis they were able to separate the *puc* promoter region into an upstream regulatory region (629 to −150) and a downstream regulatory region (−150 to −1). They concluded that the downstream regulatory region contained the *puc* promoter region (as mapped by deletion and primer extension analysis) whose maximal and proper expression required the upstream elements. Presumably the 'extended' length of the *puc* promoter region may be involved in the complex expression pattern of the *puc* operon. An additional conclusion of these studies, which was not directly addressed by Lee and Kaplan (1992a), is the possibility that the upstream region contains a second distinct promoter that is differentially regulated from a promoter that is present in the downstream region. If so, it could account for several of the above noted observations on *puc* operon expression.

IV. *trans*-Acting Regulatory Circuits

The first report of a regulatory mutant that affects synthesis of the eubacterial photosystem was a report by Lascelles and Wertlieb (1971) who described the isolation of *Rb. sphaeroides* strains that synthesized elevated levels of photopigments when grown under aerobic conditions. Unfortunately these strains were lost prior to the development of recombinant DNA technology so the genetic basis of their phenotype remains unknown. It wasn't until very recently that *trans*-acting regulatory genes that control photosynthesis gene expression were reported in the literature. The section below describes the current understanding of the *trans*-acting regulatory circuits that are emerging from these studies.

A. Regulating Transcription of Light harvesting and Reaction Center Genes

1. Oxygen Control Circuit

Sganga and Bauer (1992) initially performed mutational and molecular-genetic characterization of a *trans*-acting regulatory gene, termed *regA*, that appears to be responsible for coordinating expression of the *puf*, *puh* and *puc* operons in *Rb. capsulatus*. This study was followed by the isolation of a mutation in a similar locus in *Rb. sphaeroides* (Lee and Kaplan, 1992b) which was subsequently shown to code for a gene product that exhibits 92% sequence identity with that of *regA* (Eraso and Kaplan, 1994; Phillips and Hunter, 1994). The *regA* gene product codes for a 184 amino acid protein that is homologous to the two component class of bacterial response regulators, specifically to the subclass of smaller response regulators such as CheY, SpoOF and CheB which do not interact with DNA but instead function as intermediates in a more complex multicomponent regulatory cascade (this class of prokaryotic regulatory cascades are reviewed by Parkinson and Kofoid, 1992). In a *regA* mutant background, cells fail to induce expression of *puh,* and *puf,* above basal levels, as well as for a significant (but not absent) reduction of *puc* expression (Sganga and Bauer, 1992; Eraso and Kaplan, 1994). RegA-mediated activation was initially thought to be specific only for LH and RC promoters, since RegA does not influence regulation of the nearby *bchC* or *bchH* promoters nor the expression of the oxygen regulated nitrogen fixation operon *nifHDK* (Sganga and Bauer, 1992; mosley et al., 1994). However, Eraso and Kaplan (1994) recently reported that a disruption of the *Rb. sphaeroides regA* homolog also affects transcription of the cytochrome c_2 gene the product of which shuttles electrons between the cytochrome bc_1 complex and the RC. So it appears that RegA may also affect expression of electron transfer components of photosystem.

Based on homology to other prokaryotic two-component regulatory systems, RegA activity is thought to be modulated by a phosphorylation event in which a specific cognate sensor kinase acts to phosphorylate and possibly dephosphorylate RegA in response to environmental changes in oxygen tension (Fig. 3b). A RegA-specific sensor kinase has been shown to be the product of the linked *regB* gene (Fig. 3a) (Bauer et al., 1993; Mosley et al., 1994). Mutations in the *Rb. capsulatus regB* locus have a phenotype almost identical to that of a *regA* disruption; namely a dramatic decrease in RC and LH complexes as a result of a failure to activate *puh, puf,* and *puc* transcription under anaerobic conditions. DNA sequence analysis indicates that the RegB predicted amino acid sequence has a carboxyl terminal domain that is a member of the sensor kinase family of response regulatory proteins. Like that of many other sensor kinases, the amino terminal region of RegB also has three to four transmembrane domains. In vivo biochemical analysis of a truncated version of RegB which lacks the membrane spanning domains confirms that RegB has autophosphorylation activity and can rapidly transfer the phosphate to RegA (Inoue et al., 1995). Surprisingly, the truncated RegB protein only has autophosphorylation activity under anaerobic conditions thereby indicating the redox sensing portion of RegB remains intact in the cytosolic domain of the polypeptide (Inoue, Mosley and Bauer, unpublished).

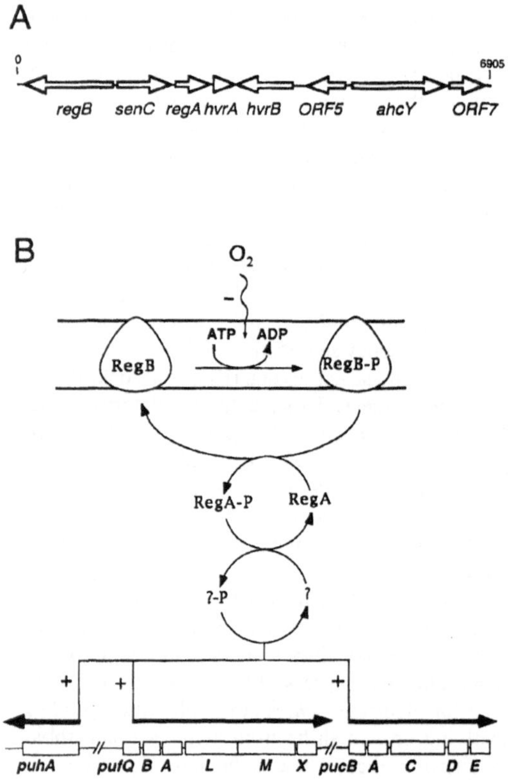

Fig. 3. The *Rb. capsulatus* 'regulatory gene cluster' that code for components of the oxygen sensory transduction cascade. a) Open reading frames of the *Rb. capsulatus* regulatory gene cluster (Sganga and Bauer, 1992; Sganga et al., 1992; Buggy et al., 1994a,b; Mosley et al., 1994). Arrows denote direction of transcription. b) Component of the sensory transduction cascade that controls expression of the *puf, puh* and *puc* operons in response to alterations in oxygen tension. RegB is proposed to be a membrane spanning kinase as based in its amino acid sequence (Mosley et al., 1994). In vitro assays indicates that RegB exhibits rapid auto kinase activity and transfer of the phosphate to RegA (Inoue et al., 1995). This activity is inhibited by molecular oxygen (Inoue, Mosley and Bauer, unpublished). Based on the absence of any noticeable DNA binding sequence motif or DNA binding activity it is proposed that RegA then transfers its phosphate to a third component which then controls activity of the *puf, puh* and *puc* promoters.

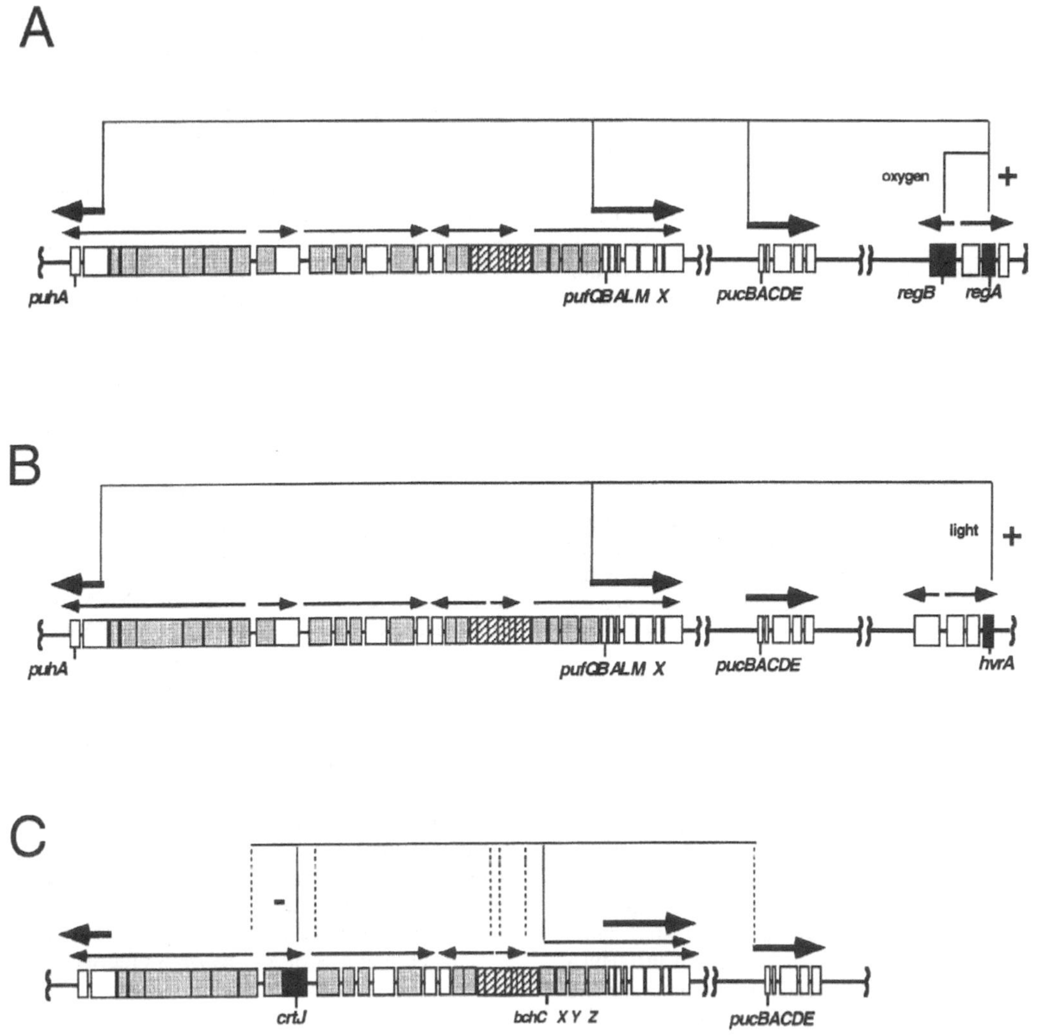

Fig. 4. Regulatory circuits that control synthesis of the *Rb. capsulatus* photosystem. a) Transcripts controlled by the RegB-RegA oxygen regulatory circuit (Sganga and Bauer, 1992; Mosley et al., 1994). b). Transcripts controlled by the hvrA mediated light regulatory circuit (Buggy et al., 1994b). c) Photopigment transcripts controlled by *crtJ* (Ponnampalam et al., 1995). The dotted lines denote proposed regulatory circuits whereas the solid lines denote those that are more firmly established.

Although RegB and RegA are the two primary *trans*-acting factors that are firmly established to be in the oxygen control circuit (as shown in Fig. 4a), there remains the possibility that the sensory transduction pathway is more complex then that of only two players. For example, inspection of the RegA primary amino acid sequence gives no indication of a DNA-binding motif. The apparent absence of a DNA-binding motif is supported by the inability to observe consistent RegA-DNA interactions by gel mobility shift (unpublished results from the C. Bauer laboratory). It is therefore possible that RegA may only be an intermediary component of a more complex phosphorylation regulatory cascade that involves a downstream DNA binding protein. This latter possibility is supported by *in vitro* observations that a protein that does not have a RegA amino terminal sequence interacts with the *puf* promoter region of dyad symmetry in a phosphorylation-dependent manner (Taremi and Marrs, 1990). There is also the possibility that there is more then one sensor kinase feeding into this signal transduction pathway. A candidate gene for additional input is *senC* which is a gene that is located immediately proceeding *regA* (Fig. 3a). The polypeptide coded by *senC* has sensor kinase features although it is yet to be established that it indeed has kinase activity (Buggy and Bauer, unpublished).

2. Light Control Circuit

In addition to molecular oxygen, light also has a repressing effect on the synthesis of the purple eubacterial photosystem. The magnitude of the repressing effects of light (2-fold) is however, greatly overshadowed by the 30- to 100-fold repressing effects of oxygen. Thus, an analysis of light regulation has proven difficult, if not intractable, to classic genetic studies. Despite these difficulties there has been some recent advances in providing a molecular genetic understanding of light regulation of LH and RC gene expression. One advance was provided by the serendipitous linkage of the oxygen regulator *regA* to an open reading frame, termed *hvrA,* (Fig. 3a) that appears to be responsible for light-mediated regulation of the *puf* and *puh* operons (Buggy et al., 1994b). A mutation of *hvrA* results in the inability to *trans*-activate *puf* and *puh* expression in response to a reduction of light (Fig. 4b). Inspection of the *hvrA* sequence indicates that it exhibits a putative helix-turn-helix motif which would indicate that HvrA could interact with the *puf* and *puh* promoter regions. This is confirmed by footprint analysis which indicates the HvrA binds to a region 90–100 bp upstream of the base of *puh* and *puf* transcription initiation (Kouadio and Bauer, unpublished results).

The mechanism whereby light affects activity of HvrA is as yet unclear, however, Shimada et al., (1992) recently performed an action spectrum of the wavelengths of light that are responsible for controlling *Rb. sphaeroides puf* operon expression by light intensity. Their results suggest that blue light in the 450 nm region exhibits the most sever repressing effects. This would indicate that HvrA, or an upstream component in a light-mediated signal transduction cascade, may be a flavin binding protein. In plant systems, there are thought to be a number of blue light photoreceptors that control plant development (Kaufman, 1993), one of which is also thought to be a flavin binding protein as based on sequence similarity to photolyase which is known to utilize flavin as a chromophore (Ahmad and Cashmore, 1993). HvrA does not exhibit any noticeable sequence similarity to other known flavin binding proteins, although as a class, they do not contain a universally conserved flavin binding sequence motif. Thus it remains unclear whether HvrA is itself a flavin binding protein or is instead a terminal component in a more complex light-mediated signal transduction cascade.

Light-mediated regulation of *puc* operon expression appears to be more complex than that of *puf* and *puh* as evidenced by some conflicting data in the literature. Several initial hybridization studies indicated that the steady state level of *puc* mRNA was reduced when photosynthetically grown cells are shifted from low to high light intensity (Zhu and Hearst, 1986; Kiley and Kaplan, 1987). However, when Zucconi and Beatty (1988) analyzed *puc* mRNA levels that were carefully normalized to rRNA levels, they observed that steady state *puc* mRNA levels were actually *higher* when the cells were grown under high light conditions than under low light conditions. This is despite the observation that the level of LHII polypeptide synthesis was inversely regulated (i.e. LHII polypeptides are present in higher amounts under low light growth conditions than under high light growth conditions). They therefore concluded that light regulation of *puc* expression involve an as yet unclear post-transcriptional control mechanism. A similar conclusion was recently obtained in a study of the 'puc-like' antenna gene cluster of *Rp. palustris* (Tadros et al., 1993).

B. Regulation of Pigment Biosynthesis Genes

Less is known about *trans*-acting factors involved in controlling bacteriochlorophyll and carotenoid gene expression in response to changes in light intensity and oxygen tension. As discussed above, gel mobility shifts by Ma et al., (1993) has indicated that there exists a protein which interacts with the palindrome sequence of putative promoters that control photopigment gene expression. The protein that interacts with the palindrome has not been purified or cloned so its identity is unclear. Penfold and Pemberton (1991, 1994) recently provided evidence that the *Rb. sphaeroides* photosynthesis gene cluster codes for a repressor of carotenoid and *bchC* expression that they have termed *pps*. Sequence analysis indicates that *pps* is analogous to the *crtJ* open reading frame described by Zsebo and Hearst (1984). Mutational analysis of this open reading frame in *Rb. capsulatus* also indicates that it has a similar role in this species (Fig. 4c) (Bollivar et al., 1994; Ponnampalam et al., 1995). The polypeptide coded by the *pps/crtJ* open reading frame has a reasonable helix-turn-helix motif (Penfold and Pemberton, 1994) which would indicate that it may interact with the photopigment promoters. Purification and footprint analysis of this polypeptide has not been undertaken so it remains to be seen if PPS/CrtJ is the factor that interacts with the palindrome sequence that is present in these promoters. It is also unclear whether the *pps/crtJ* polypeptide responds to both light and oxygen or to only one of these signals.

V. Concluding remarks

Based on the repressing effects of light and oxygen on photopigment biosynthesis, Cohen-Bazire, Sistrom and Stanier (1957) proposed a model whereby synthesis of photopigments was regulated by a central redox responding 'regulator' that tempered synthesis of photopigments relative to changes in the redox state of the cells electron transport system. As we begin to identify regulatory factors that are responsible for controlling photosystem gene expression it is becoming increasingly evident that the mechanism of regulation is much more complex than could have initially been envisioned. There appears to exist independent and overlapping light and oxygen regulatory circuits that are responsible for controlling synthesis of photopigments relative to that of the pigment binding polypeptides. These circuits provide both coordinate and discoordinate regulation of the various components which comprise the purple eubacterial photosystem. Superimposed on the *trans*-acting circuits are layers of post-transcriptional control involving the formation of superoperons (see chapter 57 by Beatty), coupled with mRNA processing events (see chapter 59 by Klug). Although much progress has recently been made in identifying the existence of these various levels of regulation, there remains much to be learned about how they coordinately regulate proper synthesis of the photosystem in response changes in the environment.

Acknowledgments

I would like to thank members of the photosynthetic bacteria community for sending reprints and preprints of their work as well as members of my own laboratory for allowing communication of unpublished results. Research from the C. Bauer laboratory is supported by the National Institutes of Health and the National Science Foundation.

References

Adams CW, Forrest ME, Cohen SN and Beatty JT (1989) Structural and functional analysis of transcriptional control of the *Rhodobacter capsulatus puf* operon. J Bacteriol 171: 473–482

Ahmad M and Cashmore AR (1993) HY4 gene of *A. thaliana* encodes a protein with characteristics of a blue-light photoreceptor. Nature 366: 162–166

Armstrong GA, Alberti M, Leach F and Hearst JE (1989) Nucleotide sequence, organization and nature of protein products of the carotenoid biosynthesis gene cluster of *Rhodobacter capsulatus*. Mol Gen Genet 216: 254–268

Bauer CE and Marrs BL (1988) *Rhodobacter capsulatus puf* operon encodes a regulatory protein (PufQ) for bacteriochlorophyll biosynthesis. Proc Natl Acad Sci USA 85: 7074–7078

Bauer CE, Young DY and Marrs BL (1988) Analysis of the *Rhodobacter capsulatus puf* operon: Location of the oxygen-regulated promoter region and the identification of an additional *puf*-encoded gene. J Biol Chem 263: 4820–4827

Bauer CE, Buggy JJ, Yang Z and Marrs BL (1991) The superoperonal organization of genes for pigment biosynthesis and reaction center proteins is a conserved feature in *Rhodobacter capsulatus*: analysis of overlapping *bchB* and *puhA* transcripts. Mol Gen Genet 228: 438–444

Bauer CE, Buggy J and Mosley C (1993) Control of photosystem genes in *Rhodobacter capsulatus*. Trends in Genet 9: 56–60

Bélanger G and Gingras G (1988) Structure and expression of the *puf* operon messenger RNA in *Rhodospirillum rubrum*. J Biol Chem 263: 7639–7645

Belasco JG, Beatty TJ, Adams CW, Von Gabain A and Cohen SN (1985) Differential expression of photosynthesis genes in *Rb. capsulata* results from segmental differences in stability within the polycistronic *rxc* transcript. Cell 40: 171–181

Bérard J, Bélanger G and Gingras G (1989) Mapping of the *puh* messenger RNA's from *Rhodospirillum rubrum*: Evidence for tandem promoters. J Biol Chem 264: 10897–10903

Biel AJ and Marrs BL (1983) Transcriptional regulation of several genes for bacteriochlorophyll synthesis in *Rhodopseudomonas capsulata* in response to oxygen. J Bacteriol 156: 686–694

Bollivar DW, Suzuki JY, Beatty JT, Dobrowlski J and Bauer CE (1994a) Directed mutational analysis of bacteriochlorophyll *a* biosynthesis in *Rhodobacter capsulatus*. J Mol Biol 237: 622–640

Bollivar DW, Wang S, Allen JP and Bauer CE (1994b) Molecular genetic analysis of terminal steps in bacteriochlorophyll *a* biosynthesis: Characterization of a *Rhodobacter capsulatus* strain that synthesizes geranylgeraniol esterified bacteriochlorophyll *a*. Biochemistry 33: 12763–12768

Buggy JJ, Sganga MW and Bauer CE (1994a) Nucleotide sequence and characterization of the *Rhodobacter capsulatus hvrB* gene: HvrB is an activator of *S*-adenosyl-Lhomocysteine hydrolase expression and is a member of the LysR family. J Bacteriol 176: 61–69

Buggy JJ, Sganga MW and Bauer CE (1994b) Characterization of a light responding *trans*-activator responsible for differentially controlling reaction center and light harvesting-I gene expression in *R. capsulatus*. J Bacteriol 176: 6936–6943

Burke D, Alberti M and Hearst JE (1993) The *Rhodobacter capsulatus* chlorin reductase-encoding locus, *bchA*, consists of three genes, *bchX*, *bchY*, and *bchZ*. J Bacteriol 175: 2407–2413

Clark WG, Davidson E and Marrs BL (1984) Variation of levels of mRNA coding for antenna and reaction center polypeptides in *Rhodopseudomonas capsulata* in response to changes in O_2 concentration. J Bacteriol 157: 945–948

Cohen-Bazire GW, Sistrom WR and Stanier RY (1957) Kinetic studies of pigment synthesis by non-sulfur purple bacteria. J Cellular Comp Physiol 49: 25–68

Coomber SA, Chaudhri M, Conner M, Britton G and Hunter CN (1990) Localized transposon Tn5 mutagenesis of the photosynthetic gene cluster of *Rhodobacter sphaeroides*. Mol Microbiol 4: 977–989

Donohue TJ, Hoger JH and Kaplan S (1986) Cloning and expression of the *Rhodobacter sphaeroides* reaction center H gene. J Bacteriol 168: 953–961

Eraso JM and Kaplan S (1994) *prrA*, a putative response regulator involved in oxygen regulation of photosynthesis gene expression in *Rhodobacter sphaeroides*. J Bacteriol 176: 32–43

Giuliano G, Pollock D, Stapp H and Scolnik PA (1988) A genetic-physical map of the *Rhodobacter capsulatus* carotenoid biosynthesis cluster. Mol Gen Genet 213: 78–83

Hessner MJ, Wejksnora PJ and Collins MLP (1991) Construction, characterization and complementation of *Rhodospirillum rubrum puf* region mutants. J Bacteriol 173: 5712–5722

Hoover TR, Santero E, Porter S and Kustu S (1990) The integration host factor stimulates interaction of RNA polymerase with NifA, the transcriptional activator for nitrogen fixation operons. Cell 63: 11–22

Hunter CN, Ashby MK and Coomber SA (1987) Effect of oxygen on levels of mRNA coding for reaction-centre and light-harvesting polypeptides of *Rhodobacter sphaeroides*. Biochem J 247: 489–492

Hunter CN, McGlynn P, Ashby MK, Burgess JG and Olsen JD (1991) DNA sequencing and complementation/deletion analysis of the *bchCA-puf* operon region of *Rhodobacter sphaeroides:* in vivo mapping of the oxygen-regulated *puf* promoter. Mol Microbiol 5: 2649–2661

Inoue K, Mosley CS, Kouadio J-LK and Bauer CE (1995) Isolation and in vitro phosphorylation of sensory transduction components controlling anaerobic induction of light harvesting and reaction center gene expression in *R. capsulatus*. Biochemistry 34: 391–396

Kaufman LS (1993) Transduction of Blue-light signals. Plant Physiol 102: 333–337

Kiley PJ and Kaplan S (1987) Cloning, DNA sequence, and expression of the *Rhodobacter sphaeroides* light harvesting B800–850-α and B800–850-β genes. J Bacteriol 169: 3268–3275

Klug G (1991a) Endonucleolytic degradation of *puf* mRNA in *Rhodobacter capsulatus* is influenced by oxygen. Proc Natl Acad Sci USA 88: 1765–1769

Klug G (1991b) A DNA sequence upstream of the *puf* operon of *Rhodobacter capsulatus* is involved in its oxygen-dependent regulation and functions as a protein binding site. Mol Gen Genet 226: 167–176

Klug G (1993) Regulation of expression of photosynthesis genes in anoxygenic photosynthetic bacteria. Arch Microbiol 159: 397–404

Klug G and Cohen SN (1988) Pleotropic effects of localized *Rhodobacter capsulatus puf* operon deletions on production of light-absorbing pigment-protein complexes. J Bacteriol 170: 5814–5821

Klug G and Jock S (1991) A base pair transition in a DNA sequence with dyad symmetry upstream of the *puf* promoter affects transcription of the *puc* operon in *Rhodobacter capsulatus*. J Bacteriol 173: 6038–6045

Klug G, Kaufman N and Drews G (1984) The expression of genes encoding proteins of B800–850 antenna pigment complex and ribosomal RNA of *Rhodopseudomonas capsulata*. FEBS Lett 177: 61–65

Klug G, Kaufmann N and Drews G (1985) Gene expression of pigment-binding proteins of the bacterial photosynthetic apparatus: transcription and assembly in the membrane of *Rhodopseudomonas capsulata*. Proc Natl Acad Sci USA 82: 6485–6489

Klug G, Gad'on N, Jock S and Narro ML (1991) Light and oxygen effects share a common regulatory DNA sequence in *Rhodobacter capsulatus*. Mol Microbiol 5: 1235–1239

Lascelles J and Wertlieb D (1971) Mutant strains of *Rhodopseudomonas spheroides* which form photosynthetic pigments aerobically in the dark. Growth characteristics and enzymatic activities. Biochem Biophys Acta 226: 328–340

Lee JK and Kaplan S (1992a) *cis*-acting regulatory elements involved in oxygen and light control of *puc* operon transcription in *Rhodobacter sphaeroides*. J Bacteriol 174:1146–1157

Lee JK and Kaplan S (1992b) Isolation and characterization of *trans*-acting mutations involved in oxygen regulation of *puc* operon transcription in *Rhodobacter sphaeroides*. J Bacteriol 174: 1158–1171

Lee JK, Kiley PJ and Kaplan S (1989) Posttranscriptional control of *puc* operon expression of B800–850 light harvesting complex formation in *Rhodobacter sphaeroides*. J Bacteriol 171: 3391–3405

Lee JK, Wang S, Eraso JM, Gardner J and Kaplan S (1993) Transcriptional regulation of *puc* operon expression in *Rhodobacter sphaeroides*: Involvement of an integration host factor-binding sequence. J Biol Chem 268: 24491–24497

Ma D, Cook DN, O'Brien DA and Hearst JE (1993) Analysis of the promoter and regulatory sequences of an oxygen regulated *bch* operon in *Rhodobacter capsulatus* by site-directed mutagenesis. J Bacteriol 175: 2037–2045

Marrs BL (1974) Genetic recombination in *Rhodopseudomonas capsulata*. Proc Natl Acad Sci USA 71: 971–973

Marrs BL (1981) Mobilization of the genes for photosynthesis

from *Rhodopseudomonas capsulata* by a promiscuous plasmid. J Bacteriol 146: 1003–1012
McGlynn P and Hunter CN (1993) Genetic analysis of the *bchC* and *bchA* genes of *Rhodobacter sphaeroides*. Mol Gen Genet 236: 227–234
Mosley CS, Suzuki JY and Bauer CD (1994) Identification and molecular genetic characterization of a sensor kinase responsible for coordinatively regulating light harvesting and reaction center gene expression in response to anaerobiosis. J Bacteriol 176: 7566–7573
Nagashima KVP, Shimada K and Matsuura K (1993) Phylogenetic analysis of photosynthetic genes of *Rhodocyclus gelatinosus*: Possibility of horizontal gene transfer in purple bacteria. Photosyn Res 36: 185–191
Narro ML, Adams CW, and Cohen SN (1990) Isolation and characterization of *Rhodobacter capsulatus* mutants defective in oxygen regulation of the *puf* operon. J Bacteriol 172: 4549–4554
Parkinson JS and Kofoid EC (1992) Communication modules in bacterial signaling proteins. Ann Rev Genet 26:71–112
Penfold RJ and Pemberton JM (1991) A gene from the photosynthetic gene cluster of *Rhodobacter sphaeroides* induces *trans*-suppression of bacteriochlorophyll and carotenoid levels in *Rb. sphaeroides* and *Rb. capsulatus*. Curr Microbiol 23: 259–263
Penfold RJ and Pemberton JM (1994) Sequencing, chromosomal inactivation and functional expression in *E. coli* of *pps*, a gene which represses carotenoid and bacteriochlorophyll synthesis in *Rhodobacter sphaeroides*. J Bacteriol In Press
Phillips-Jones MK and Hunter CN (1994) Cloning and nucleotide sequence of *regA*, a putative response regulator gene of *Rhodobacter sphaeroides*. FEMS Lett 116: 269275
Ponnampalam S, Buggy JJ and Bauer CE (1995) Characterization of an aerobic repressor that coordinately regulates bacteriochlorophyll, carotenoid and light harvesting-II expression in *Rhodobacter capsulatus*. J Bacteriol 177: 2990–2997
Sagesser R (1992) Identifikation und charakterisierung des photosynthesegenclusters von *Rhodospirillum rubrum*. Ph. D. Thesis, Universität Zürich.
Sganga MW and Bauer CE (1992) Regulatory factors controlling photosynthetic reaction center and light harvesting gene expression in *Rhodobacter capsulatus*. Cell 68: 945–954
Sganga MW, Aksamit RR, Cantoni GL and Bauer CE (1992) Mutational and nucleotide sequence analysis of *S*-adenosyl-L-homocysteine hydrolase from *Rhodobacter capsulatus*. Proc Natl Acad Sci USA 89: 6328–6332
Shimada H, Iba K and Takamiya K-I (1992) Blue-light irradiation reduces the expression of *puf* and *puc* operons of *Rhodobacter sphaeroides* under semi-aerobic conditions. Plant Cell Physiol 33: 471–475
Shimada H, Ohta H, Masuda T, Shioi Y and Takamiya K-I (1993) A putative transcription factor binding to the upstream region of the *puf* operon in *Rhodobacter sphaeroides*. FEBS Letters 328: 41–44
Taylor DP, Cohen SN, Clark WG, and Marrs BL (1983) Alignment of genetic and restriction maps of the photosynthesis region of the *Rhodopseudomonas capsulata* chromosome by a conjugation-mediated marker rescue technique. J Bacteriol 154: 580–590
Tadros MH, Katsiou E, Hoon MA, Yurkova N, and Ramji DP (1993) Cloning of a new antenna gene cluster and expression analysis of the antenna gene family of *Rhodopseudomonas palustris*. Eur J Biochem 217: 867–875
Taremi SS and Marrs BL (1990) Regulation of gene expression by oxygen: Phototrophic bacteria. In: Hauska G, Thauer R (eds) The molecular bases of bacterial metabolism, pp 146–151. Springer, Berlin Heidelberg, New York
Tichy H-V, Oberle' B, Stiehl, H, Schlitz, E and Drews, G (1991) Analysis of the *Rhodobacter capsulatus puc* operon: the *pucC* gene plays a central role in the regulation of LHII (B800–850 complex) expression. EMBO J 10: 2949–2956
Wellington CL and Beatty JT (1989) Promoter mapping and nucleotide sequence of the *bchC* bacteriochlorophyll biosynthesis gene from *Rhodobacter capsulatus*. Gene 83: 251–261
Wellington CL and Beatty T (1991) Overlapping mRNA transcripts of photosynthesis gene operons in *Rhodobacter capsulatus*. J Bacteriol 173: 1432–1443
Wellington CL, Bauer CE and Beatty JT (1992) Photosynthesis gene superoperons in purple nonsulfur bacteria: the tip of the iceberg? Can J Micro 38: 20–27
Wiessner C, Dunger I and Michel H (1990) Structure and transcription of the genes encoding the B1015 light harvesting complex β and α subunits and the photosynthetic reaction center L, M, and cytochrome *c* subunits from *Rhodopseudomonas viridis*. J Bacteriol 172: 2877–2887
Yang ZY, and Bauer CE (1990) *Rhodobacter capsulatus* genes involved in early steps of the bacteriochlorophyll biosynthetic pathway. J Bacteriol 126: 619–629
Yen H-C, Hu NT and Marrs BL (1976) Map of genes for carotenoid and bacteriochlorophyll biosynthesis in *Rhodopseudomonas capsulata*. J Bacteriol 126: 619–629
Yildiz FH, Gest H and Bauer CE (1992) Conservation of the photosynthesis gene cluster in *Rhodospirillum centenum* Mol Microbiol 6: 2683–2691
Young DA, Bauer CE, Williams JC, and Marrs BL (1989) Genetic evidence for superoperonal organization of genes for photosynthetic pigments and pigmentbinding proteins in *Rhodobacter capsulatus*. Mol Gen Genet 218: 1–12
Young DA, Rudzik MB and Marrs BL (1992) An overlap between operons involved in carotenoid and bacteriochlorophyll biosynthesis in *Rhodobacter capsulatus*. FEMS Microbiol Lett 95: 213–218
Youvan DC and Ismail S (1985) Light-harvesting II (B800–B850 complex) structural genes from *Rhodopseudomonas capsulata*. Proc Natl Acad Sci USA 82, 58–62
Zhu YS and Hearst JE (1986) Regulation of expression of genes for light-harvesting antenna proteins LHI and LHII; reaction center for polypeptides RC-L, RC-M, and RC-H; and enzymes of bacteriochlorophyll and carotenoid biosynthesis in *Rhodobacter capsulatus* by light and oxygen. Proc Natl Acad Sci USA 83: 7613–7617
Zhu YS and Kaplan S (1985) Effects of light, oxygen and substrates on steady-state levels of mRNA coding for ribulose-1, 5-bisphosphate carboxylase and light harvesting and reaction center polypeptides in *Rhodopseudomonas sphaeroides*. J Bacteriol 162: 925–932
Zhu YS, Kiley PJ, Donohue TJ and Kaplan S (1986a) Origin of the mRNA stoichiometry of the *puf* operon in *Rhodobacter sphaeroides*. J Biol Chem 261: 1036610374

Zhu YS, Cook DN, Leach F, Armstrong GA, Alberti M and Hearst JE (1986b) Oxygenregulated mRNA for light harvesting and reaction center complexes and for bacteriochlorophyll and carotenoid biosynthesis in *Rhodobacter capsulatus* during shift from anaerobic to aerobic growth. J Bacteriol 168: 1180–1188

Zsebo KM and Hearst JE (1984) Genetic physical mapping of a photosynthetic gene cluster from *Rb. capsulata*. Cell 37: 937–947

Zucconi AP and Beatty JT (1988) Posttranscriptional regulation by light of the steady-state levels of mature B880–850 light harvesting complexes in *Rhodobacter capsulatus*. J Bacteriol 170: 877–882

Chapter 59

Post-Transcriptional Control of Photosynthesis Gene Expression

Gabriele Klug
Zentrum für Molekulare Biologie Heidelberg, Im Neuenheimer Feld 282, 69120 Heidelberg, Germany

Summary

Oxygen tension and light intensity influence the formation of pigment protein complexes in facultatively photosynthetic bacteria. Multiple control mechanisms are required to regulate expression of photosynthesis genes in response to external factors and to guarantee the coordinated syntheses of components of the photosynthetic apparatus. Regulation of expression of photosynthesis genes is partly accomplished by variation of mRNA levels. Both the rate of synthesis and the rate of decay of mRNA are involved in determining these mRNA levels. Our current knowledge on the molecular mechanisms of stabilization and destabilization of mRNAs transcribed from photosynthesis genes will be summarized. Some experimental data suggest that photosynthesis genes are also regulated at later steps of gene expression. This chapter will address the 23S rRNA processing in *Rhodobacter* and its putative role in gene regulation. Other molecular mechanisms influencing the rate of translation or post-translational processes have not been investigated to date but evidence for their participation in regulation of photosynthesis genes is provided. Since regulation of bacterial photosynthesis genes has been studied most intensively in *Rhodobacter capsulatus* and *Rhodobacter sphaeroides*, this review will focus on the results obtained with these closely related bacterial species.

R. E. Blankenship, M. T. Madigan and C. E. Bauer (eds): Anoxygenic Photosynthetic Bacteria, pp. 1235–1244.

I. Introduction

Facultatively photosynthetic bacteria have the ability to use different modes of energy conversion and to adapt their cellular composition and membrane structure to environmental changes (reviewed in: Drews and Imhoff, 1991). Oxygen tension and light intensity are the two external factors known to regulate the formation of the photosynthetic apparatus. Regulation of gene expression in bacteria is often achieved by variation of transcriptional rates. Initiation of transcription is, however, not the only step in gene expression involved in the regulated formation of bacterial photosynthetic complexes. This chapter will summarize our current knowledge on the role of post-transcriptional processes that affect the expression of photosynthesis genes in *Rhodobacter*.

Under high oxygen tension facultatively photosynthetic bacteria perform aerobic respiration. When oxygen tension in the environment drops, these bacteria form photosynthetic complexes which are incorporated into intracytoplasmic membranes originating from invaginations of the cytoplasmic membrane. During phototrophic growth, light intensity affects the formation of the photosynthetic apparatus (reviewed in Drews (1992) and in Chapter 12 by Drews and Golecki). The interaction of pigments and proteins and their highly ordered structural organization is required for the formation of functional photosynthetic complexes.

All genes encoding pigment binding proteins and a large number of genes encoding the enzymes catalyzing pigment syntheses have been identified over the last years in the closely related species *Rhodobacter capsulatus* and *Rhodobacter sphaeroides* (Chapter 50 by Alberti et al.), and investigations on the regulation of the expression of these genes have been initiated (Chapter 58 by Bauer). A schematic overview on the molecular steps involved in the regulated expression of these photosynthesis genes is given in Fig. 1. The known cellular factors involved in transmitting the oxygen signal in the cell are described in Chapter 58 by Bauer, as is our knowledge on the effect of oxygen on the transcription of photosynthesis genes. Oxygen, however, also influences the rate of mRNA degradation, and some observations suggest that later steps in gene expression may also be affected by oxygen. Post-transcriptional regulation of gene expression is also involved in the coordination of the syntheses of different components of the photosynthetic apparatus and in determining the relative levels of gene products.

Abbreviations: bp – base pairs; *E.* – *Escherichia*; LH – light harvesting; *Rb.* – *Rhodobacter* ; RC – reaction center; RNase – endoribonuclease

II. Control of Photosynthesis Gene Expression on the Level of mRNA Stability

A. The Role of Segmental Differences in puf *mRNA Stability in the Formation of the Photosynthetic Apparatus of* Rb. capsulatus

The pigment-binding proteins of *Rb. capsulatus* are encoded by two polycistronic operons that are separated on the chromosome (Fonstein and Haselkorn, 1993). The *puf* operon (formerly *rxcA* operon, Belasco et al., 1985) (Fig. 2) encodes the proteins of the LHI complex (genes *pufB* and *pufA*) and of the RC complex (genes *pufL* and *pufM*) (Youvan et al., 1984). In addition it comprises the genes *pufQ* (Bauer et al., 1988) and *pufX* (Youvan et al., 1984) that are required for the formation of wild type levels of photosynthetic complexes (Bauer and Marrs, 1988; Klug and Cohen, 1988; Farchaus et al., 1992; Lilburn et al., 1992). All *puf* genes are under the control of an oxygen-regulated promoter positioned about 200 bp upstream of the first open reading frame (Bauer et al., 1988; Adams et al., 1989). This polycistronic organization allows the coordinated expression of all *puf* genes in response to oxygen.

Although the *puf* genes are part of the same transcriptional unit, the *puf* gene products are required in different amounts. 10-15 LHI complexes surround one RC complex in the membrane of *Rb. capsulatus* wild type cells (Kaufmann et al., 1982; Klug et al., 1987). The relative amounts of the *pufQ* and *pufX* gene products in the cells are not known, but there is evidence that PufQ is a regulatory protein (Bauer and Marrs, 1988) that is present in very low concentration. Synthesis of different amounts of the *puf* gene products is achieved by post-transcriptional regulation of gene expression. In *Rb. capsulatus* posttranscriptional regulation of gene expression occurs at least in part on the level of mRNA degradation.

Two *puf* specific mRNAs can be detected in *Rhodobacter* by Northern blot analysis (Belasco et

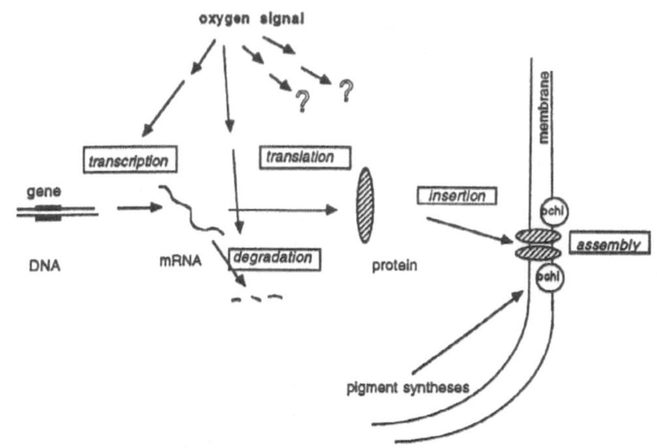

Fig. 1. Steps in gene expression involved in the formation of photosynthetic complexes. Sensing and transmission of the oxygen signal is described in Chapter 58. Oxygen affects the expression of photosynthesis gene by regulating their transcription and the stability of the corresponding mRNA. The effect of oxygen on later steps of gene expression has not been studied to date. Oxygen also affects pigment syntheses at multiple levels of gene expression (Chapter 58).

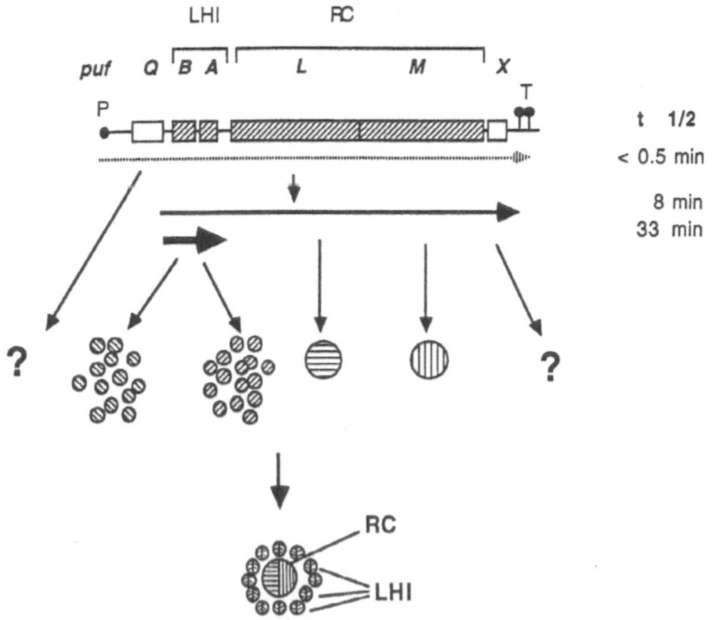

Fig. 2. Schematic genetic map of the polycistronic *puf* operon of *Rb. capsulatus*. Arrows represent mRNA species and half-lives as determined by Northern blot analysis (referring to Klug et al., 1987) are listed. The segmental differences in *puf* mRNA stability in part determine the relative amounts of the products of the individual *puf* genes in the cell. P: oxygen-regulated promoter, T: transcriptional terminators.

al., 1985; Zhu and Hearst, 1986). A 2.7 kb *pufBALMX* mRNA decaying with a half-life of about 8 min (Klug et al., 1987) and a 0.5 kb *pufBA* mRNA showing a half-life of about 30 min (Klug et al., 1987) share the same 5′ end (Fig. 2). The *puf* mRNA segment comprising the untranslated 5′ region and the *pufQ* gene is highly unstable. Due to its higher stability, the 0.5 kb *puf* mRNA encoding LHI proteins only, is present in a 10 fold molar excess over the 2.7 kb *puf* mRNA encoding LHI and RC proteins and PufX (Belasco et al., 1985). Pulse labeling experiments revealed that the 0.5 kb mRNA is a processing product of the 2.7 kb mRNA (Belasco et al., 1985). The 2.7 kb *puf* mRNA extends to either of two transcriptional terminators (Belasco et al., 1985). The 3′ end of the 0.5 kb *puf* mRNA is formed by a highly stable secondary structure located in the intercistronic region between *pufA* and *pufL* (Belasco et al., 1985). This secondary structure is required for the higher stability of the 0.5 kb *pufBA* mRNA by protecting it against 3′ to 5′ exonucleases (Klug et al., 1987; Chen et al., 1988). Removal of this secondary structure results in continuous degradation of the 2.7 kb *pufBALMX* mRNA (Klug et al., 1987; Chen et al., 1988) and in altered stoichiometry of the *puf* gene products in the membrane. The ratio of LHI to RC complexes dropped from 15:1 in a wild type control to 7:1 in the absence of the intercistronic secondary structure (Klug et al., 1987). Although the chemotrophic and phototrophic growth rates of *Rb. capsulatus* cultures were not affected by this mutation in the presence of the LHII complex, cultures lacking the stable *pufBA* mRNA segment showed a prolonged lag phase when shifted from chemotrophic to phototrophic growth (Klug et al., 1987). Thus, segmental differences in the stability of *puf* mRNA segments are involved in regulating the relative amounts of *puf* gene products (Fig. 2), but are not the only molecular basis for the excess of LHI.

B. Oxygen Affects the Stability of mRNAs Encoding Pigment Binding Proteins

When the oxygen tension in aerobically growing cultures of *Rhodobacter* is reduced, the *puf* and *puc* mRNA levels increase 15–20 fold (Clark *et al.*, 1984; Klug *et al.*, 1985; Zhu and Kaplan, 1985; Zhu and Hearst, 1986). This increase is not only due to increased *puf* operon transcription (Bauer *et al.*, 1988; Adams *et al.*, 1989; Hunter et al., 1991; Lee and Kaplan, 1992; Chapter 58), but is in part the result of altered *puf* mRNA stability (Klug, 1991). The 2.7 kb *pufBALMX* mRNA of *Rb. capsulatus* shows a half-life of about 3 min at high oxygen tension but of about 8 min at low oxygen tension, while the 0.5 kb *pufBA* mRNA decays with similar rate (half-life of about 33 min) at high and low oxygen tension. The decay of the 0.5 kb *pucBA* mRNA (encoding proteins of the LHII complex) is affected by oxygen to a lower extent (half-life 18 min under high oxygen, 29 min under low oxygen tension), whereas other mRNA species like *fbc* (cytochrome fbc complex), *puhA* (non-pigment binding subunit of the RC), and *cycA* (cytochrome c_2) decay independently of the aeration of the cultures (Klug, 1991). Hence, differences in mRNA stability not only influence the molar ratios of the *puf* gene products, but also contribute to the oxygen regulation of *puf* gene expression. The mechanisms by which oxygen can affect mRNA decay are not understood to date.

C. Stabilization and Destabilization of puf mRNA Segments

As pointed out above, and shown in Fig. 2, individual segments of the polycistronic *puf* mRNA exhibit extremely different half-lives. These differences in *puf* mRNA stability are involved in the oxygen dependent regulation of *puf* genes (A2) and in differential expression of LHI and RC genes (A1). In order to understand the regulated expression of *puf* genes, we therefore need to understand the molecular basis of mRNA degradation and stabilization.

The intercistronic secondary structure located between *pufA* and *pufL* is required for the stabilization of the 0.5 kb *pufBA* mRNA segment by protecting it against 3′ to 5′ exonucleases (Klug et al., 1987; Chen et al., 1988). In the same way, the structures that terminate *puf* transcription provide protection against exonucleolytic attack at the 3′ end of the 2.7 kb *pufBALMX* mRNA (Klug and Cohen, 1988). The 5′ end of the 2.7 kb and 0.5 kb *puf* mRNAs is also highly structured (Rothfuchs and Klug, unpublished). It was shown for the *E. coli ompA* mRNA that a secondary structure at the 5′ end of this mRNA provides protection against endonucleolytic attack (Emory et al., 1992), and it is likely that the secondary structures at the 5′ end of the stable *puf* mRNA species serve a similar function.

Decay of the 2.7 kb *pufBALMX* mRNA is initiated by endonucleolytic cleavages at multiple sites within the *pufLM* coding region (Klug and Cohen, 1990; Chen and Belasco, 1990). One of the target sites for endonucleolytic cleavage was identified (Fritsch et al., 1995) and it shows remarkable similarity to *E. coli* RNA sequences that are recognized by the endoribonucleases (RNase) E (Ehretsmann et al., 1992) or K (Lundberg et al., 1990). When *Rb. capsulatus puf* mRNA is expressed in *E. coli*, the rate of decay of the 2.7 kb *puf* mRNA indeed depends on the activity of the RNase E enzyme, whereas the 0.5 kb *puf* mRNA decays with similar rates in presence or absence of RNase E activity (Klug et al., 1992). Remarkably, the putative recognition sequence for an endonuclease in *Rb. capsulatus* is part of the coding region for a structural protein of the reaction center. This mRNA sequence therefore has to fulfill two functions: i) encoding a functional RC protein, and ii) to serve as an endonuclease recognition site. Other *puf* mRNA sequences share homology with the endonucleolytic cleavage site within the *pufL* gene (Fig. 3) but their function has not been analyzed to date. Data on *puf* mRNA decay collected over the past few years resulted in a model that attributes the segmental differences in *puf* mRNA stability to the specific distribution of RNA decay promoting elements (recognition sites for endonucleases) or RNA decay impeding elements (secondary structures at the 3′ or 5′ end of stable mRNA species) (Klug, 1993) (Fig. 3).

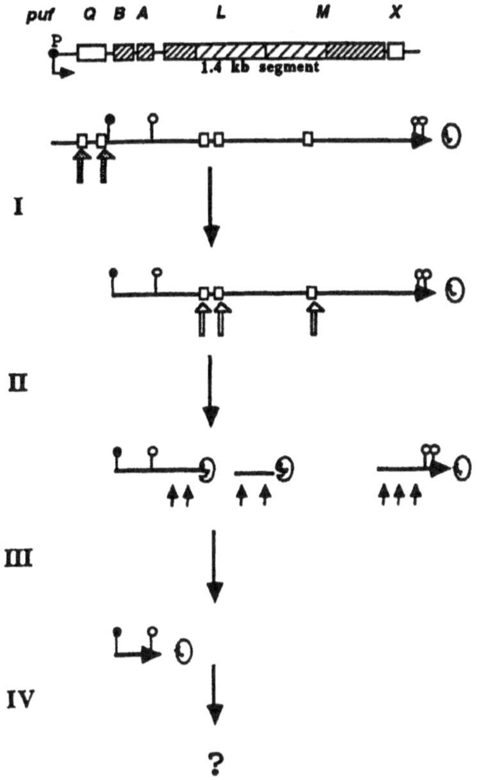

Fig. 3. Model for the degradation of the polycistronic puf mRNA. The segmental differences in mRNA stability are attributed to a specific distribution of endonuclease recognition sites (□) and mRNA secondary structures that function as stabilizing elements (↑,⚲). Degradation of the primary transcript is initiated by endonucleolytic cleavages (⇧) within the *pufQ* coding sequence (I). The 2.7 kb processing product is stabilized by several secondary structures at the 5′ end (↑) and by the two terminator structures at the 3′ end that prevent 3′ to 5′ decay by exonucleases (⚲). Degradation of the 2.7 kb mRNA (II) is initiated by multiple cleavages within the RC coding region, (⇧) and the rate of cleavage is affected by the oxygen tension. Further degradation occurs by mechanisms that have not been identified (III) and results in the occurrence of the 0.5 kb mRNA that carries highly stable secondary structures at its 3′ and 5′ end and does not carry putative recognition sites for endonucleolytic cleavage. The mechanisms initiating the final degradation of the 0.5 kb *puf* mRNA segment (IV) have not been identified to date.

D. Segmental Differences in Stability of Other mRNAs Encoding Photosynthesis Proteins

Differential decay of mRNA segments is most likely not restricted to the polycistronic *puf* operon of *Rhodobacter*. Bauer *et al.*(1991) detected 11 kb, 1.1 kb and 0.95 kb mRNA species that hybridized to the *puhA* gene (encoding the non pigment binding protein of the RC) (Fig. 4). The 11 kb mRNA also encodes the upstream *bchFNBHLM-F1696* operon, suggesting that the 1.1 kb mRNA is derived from the 11 kb precursor by processing. The 0.95 kb mRNA is initiated at the *puhA* promoter. The biological significance of the *bchFNBHLM-F1696-puhA* mRNA processing is not known.

The *puc* operon encodes the pigment-binding proteins of the LHII complex (genes *pucB* and *pucA*) and also comprises the *pucC* gene (Tichy *et al.*, 1991), which is required for high levels of *puc* expression, and the genes *pucD* and *pucE*, encoding gene products necessary for the assembly or stabilization of LHII complexes (Tichy *et al.*, 1989; Lee et al., 1989) (Fig. 4). mRNA secondary structures are predicted for the *pucA-pucC* intercistronic region (Zucconi and Beatty, 1988; Lee *et al.*, 1989; Tichy *et al.*, 1989) as for the *pufA-pufL* intercistronic region (Belasco *et al.*, 1985), and it was suggested that the

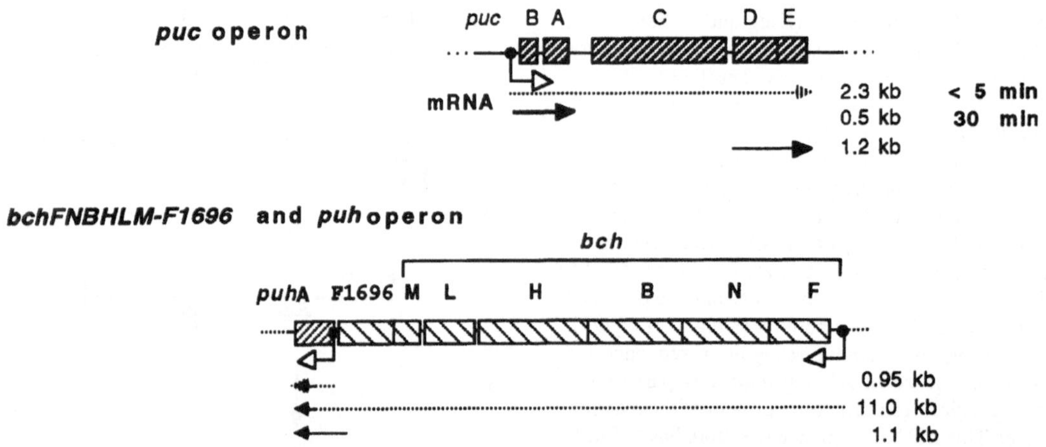

Fig. 4. Schematic maps of the *puc* operon and *puhA-bchFNBHLM* operon of *Rb. capsulatus*. mRNA species are represented by arrows and the mRNA half-lives are listed, if determined (Lee et al., 1989).

pucA-pucC intercistronic mRNA structure stabilizes the upstream mRNA rather than functioning as a transcriptional terminator (Tichy *et al.*, 1991). It was also reported that the *pucDE* mRNA segment is more stable than the *pucC* mRNA segment in *Rb. capsulatus* (Oberlé *et al.*, 1990) suggesting a similar role of mRNA processing in the regulation of *puc* gene expression as demonstrated earlier for the *puf* gene expression. No attempts have been undertaken to study the decay of *puc* mRNA, and therefore the mechanisms involved in this process remain to be solved.

III. 23S rRNA Processing in *Rhodobacter*

The bacterial 50S ribosomal subunit is generally composed of a 23S and a 5S rRNA molecule and ribosomal proteins. Some bacterial species, however, have ribosomes that do not contain an intact 23S rRNA species. In vivo fragmentation of 23S rRNA has been reported for some cyanobacteria (Doolittle, 1979), for *Agrobacterium tumefaciens* (Schuch and Loening, 1975), for *Bdellovibrio bacteriovorus* (Meier and Brownstein, 1976), for *Salmonella* (Burgin et al., 1990) and for the closely related bacteria *Rb. sphaeroides* (Marrs and Kaplan, 1970), *Rb. capsulatus* (Lessie, 1965), and *Paracoccus denitrificans* (MacKay, 1979). The biological significance of 23S rRNA fragmentation is not understood.

In *Rhodobacter* 23S rRNA is processed to 16S and 14S rRNA species (Marrs and Kaplan, 1970). The *rrn* operons of *Rb. sphaeroides* and *Rb. capsulatus* (Dryden and Kaplan, 1990; Höpfl *et al.*, 1988) show strong homology. The organization of the *Rb. sphaeroides rrn* operons is identical to that found in *E. coli* with two tRNAs in the 16S-23S spacer region. An initiator methionine tRNA was identified immediately downstream of the 5S rRNA gene (Dryden and Kaplan, 1990). Comparison of the secondary structures of the 23S rRNA molecules from *E. coli* and *Rhodobacter* and the size of the cleavage products suggests that the cleavage of 23S rRNA occurs within an rRNA region forming a stem-loop structure that is present in *Rhodobacter* but absent from *E. coli* (Dryden and Kaplan, 1990; Fig. 5). The rRNA sequences forming these extra stem-loop structures postulated for *Rb. sphaeroides* or *Rb. capsulatus* differ significantly from each other. Hybridization experiments showed that the nucleotides forming this extra stem-loop in *Rb. capsulatus* are not part of the 16S nor the 14S rRNA of the wild type rRNA suggesting their loss during processing (Kordes et al., 1994).

In *Anacystis nidulans* cleavage of 23S rRNA is induced by light indicating that it may be involved in gene regulation (Doolittle, 1973). Some early investigations reported differences in the ribosome composition of facultatively photosynthetic bacteria grown either chemotrophically or phototrophically (Mansour and Stachow, 1975; Chow, 1976). When

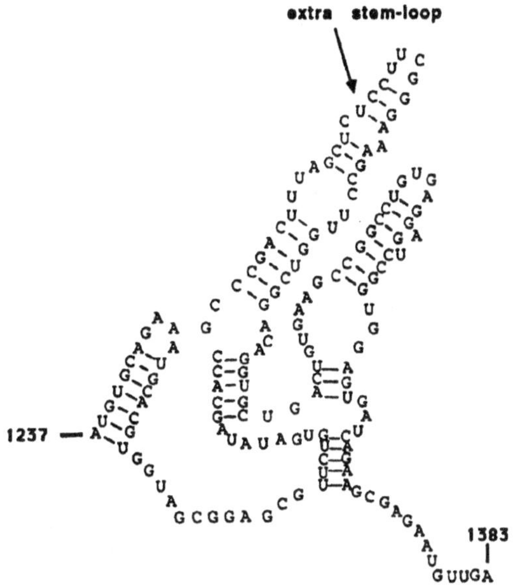

Rhodobacter capsulatus

Fig. 5. Secondary structure of the *Rb. capsulatus* 23S rRNA (Gutell et al., 1990) in which processing to 16S and 14S rRNA takes place. The nucleotides forming the extra stem-loop structure have no corresponding sequence in the *E. coli* 23S rRNA sequence and differ significantly from the *Rb. sphaeroides* rRNA sequence, despite the strong overall homology of the *rrn* operons of *Rb. capsulatus* and *Rb. sphaeroides*.

Rb. capsulatus cells are shifted from high oxygen tension to low oxygen tension or from anaerobic phototrophic growth to chemotrophic growth, an increase of their rRNA content was observed (Klug et al., 1984; Drews et al., 1986). A relative higher content of unprocessed 23S rRNA was observed in aerobic cells, when compared to cells grown under low oxygen tension (Klug et al., 1984). Based on these observations, it was under discussion that the processing of 23S rRNA in *Rb. capsulatus* may be involved in the oxygen regulation of the formation of the photosynthetic apparatus.

The chemical mutant *Rb. capsulatus* Fm65 (Klug et al., 1984) was isolated due to its lack of bacteriochlorophyll, but it was reported recently that this strain also differs from the wild type strain 37b4 by containing an intact 23S rRNA (Kordes et al., 1994). The lack of 23S rRNA processing could be attributed to reduced levels of an RNase that is also involved in processing of the primary rRNA transcript (Kordes et al., 1994). When the synthesis of bacteriochlorophyll of strain Fm65 was complemented by the transfer of wild type *bch* genes, the resulting strain formed photosynthetic complexes indistinguishable from the wild type photosynthetic apparatus but did not show processing of the 23S rRNA. When oxygen tension was reduced, photosynthetic complexes in the Fm65-derived strain and in wild type cells accumulated with the same kinetics and to the same levels (Klug et al., 1995). These data show that 23S rRNA processing is not involved in the oxygen-regulated formation of the photosynthetic apparatus in *Rhodobacter* and the biological relevance of 23S rRNA processing in this bacterial species remains obscure.

IV. Evidence for Translational or Post-Translational Regulation of Photosynthesis Gene Expression

The translation of bacterial photosynthesis genes has not been studied to date. We do not know whether individual *puf* or *puc* genes are translated with different rates and whether different rates of translation affect the rates of mRNA decay. Some data, however, provide evidence that translation or post-translational processes are involved in the regulation of photosynthesis gene expression. These regulatory processes seem to guarantee that some components of a photosynthetic complex do not accumulate in the absence of other components that are necessary to form functional photosynthetic complexes.

Many mutant strains that lack bacteriochlorophyll also lack significant amounts of pigment-binding proteins in their membrane, even under low oxygen tension. This is often due to an increased turn-over of pigment-binding proteins in the absence of bacteriochlorophyll (Dierstein, 1983; Klug et al., 1986). If bacteriochlorophyll synthesis was blocked at early steps by specific inhibitors, no RC and LHI proteins were detectable in the membrane or the soluble fraction even after pulse-labeling, although normal levels of *puf* and *puc* mRNA were detected, suggesting that an early block in bacteriochlorophyll synthesis may prevent the synthesis of pigment-binding proteins. Several strains with mutations in early steps of bacteriochlorophyll synthesis showed abnormal oxygen regulation of the *puf* and *puc* operon (Klug et al., 1986; Neidle and Kaplan, 1993).

When oxygen tension is reduced, the accumulation of photosynthetic complexes occurs with kinetics that follow the increase of *puf* and *puc* mRNA levels in *Rb. capsulatus* (Klug et al., 1985). When light intensity was varied in phototrophic growing cultures, no direct correlation of the amount of LHII complexes and *puc* mRNA levels was observed (Zucconi and Beatty, 1988). Cells grown under low light intensity contained about four-fold less *puc* mRNA but about four times as much LHII complexes as cells grown under high light intensity, suggesting a regulation by light that occurs on post-transcriptional level of gene expression (Zucconi and Beatty, 1988).

As described above (II.A) the relative levels of the *puf* mRNA species influence the stoichiometry of LHI and RC complexes in the membrane. However, in cells with a continuous degradation of the 2.7 kb *pufBALMX* mRNA (LHI specific mRNA to RC specific mRNA is about 1:1) there is still a 7 fold excess of LHI complexes over RC complexes (Klug et al., 1987) indicating that later steps in gene expression affect the relative amounts of pigment protein complexes. Further investigations on the translation of *puf* and *puc* mRNA and on post-translational processes will be necessary to identify all steps of gene expression that are affected by external factors or are involved in the coordinated formation of the components of the photosynthetic apparatus.

V. Concluding Remarks

As demonstrated in Fig. 1, multiple steps of gene expression could be involved in the regulation of the formation of photosynthetic complexes. Oxygen affects the rates of transcription of photosynthesis genes and the rate of decay of some mRNA species. The formation of functional photosynthetic complexes also depends on translation, incorporation of the pigment binding proteins into the membrane and the correct assembly of the individual components. Despite some evidence that steps in gene expression following mRNA synthesis and decay may also be affected by oxygen and contribute to the coordination of the expression of the individual components, the regulatory mechanisms acting on these later steps in gene regulation have not been investigated to date. Our current knowledge on the mechanisms involved in incorporation of pigment-binding proteins into the membrane and the assembly of photosynthetic complexes is summarized in the Chapter 12 by Drews and Golecki.

Acknowledgments

I thank Dr. Michael Nassal for reading the manuscript. Work in the author's laboratory was supported by the Deutsche Forschungsgemeinschaft (Kl 563/2-1), the Bundesministerium für Forschung und Technologie and the Fonds der Chemischen Industrie.

References

Adams CW, Forrest ME, Cohen SN and Beatty JT (1989) Structural and functional analysis of transcriptional control of the *Rhodobacter capsulatus puf* operon. J Bacteriol 171: 473-482

Bauer CE and Marrs BL (1988) *Rhodobacter capsulatus puf* operon encodes a regulatory protein (PufQ) for bacteriochlorophyll synthesis. Proc Natl Acad Sci USA 85: 7074-7078

Bauer CE, Young DA and Marrs BL (1988) Analysis of the *Rhodobacter capsulatus puf* operon. Location of the oxygen-regulated promoter region and the identification of an additional *puf*-encoded gene. J Biol Chem 263: 4820-4827.

Bauer CE, Buggy JJ, Yang Z and Marrs BL (1991) The superoperonal organization of genes for pigment biosynthesis and reaction center proteins is a conserved feature in *Rhodobacter capsulatus*: analysis of overlapping *bchB* and *puhA* transcripts. Mol Gen Genet 228: 433–444

Belasco JG, Beatty JT, Adams CW, von Gabain A and Cohen SN (1985) Differential expression of photosynthetic genes in *Rhodopseudomonas capsulata* results from segmental differences in stability within a polycistronic transcript. Cell 40: 171–181

Burgin AB, Parados K, Lane DJ, Pace NR (1990) The excision of intervening sequences from *Salmonella* 23S ribosomal RNA. Cell 60: 405–414

Chen C-YA and Belasco JG (1990) Degradation of *pufLMX* mRNA in *Rhodobacter capsulatus* is initiated by nonrandom endonucleolytic cleavage. J Bacteriol 172: 4578–4586

Chen C-YA, Beatty JT, Cohen SN and Belasco JG (1988) An intercistronic stem-loop structure functions as an mRNA decay terminator necessary but insufficient for *puf* mRNA stability. Cell 52: 609–619

Chow CT (1976) Functional and structural differences between photosynthetic and heterotrophic *Rhodospirillum rubrum* ribosomes and S-100 fractions. Can J Microbiol 22: 1522–1539

Clark WG, Davidson E and Marrs BL (1984) Variation of levels of mRNA coding for antenna and reaction center polypeptides in *Rhodopseudomonas capsulata* in response to changes in oxygen concentration. J Bacteriol 157: 945–948

Dierstein R (1983) Biosynthesis of pigment-protein complex polypeptides in bacteriochlorophyll-less mutant cells of

Rhodopseudomonas capsulata YS. FEBS Lett 160: 281–286
Doolittle WF (1973) Postmaturational cleavage of 23S ribosomal ribonucleic acid and its metabolic control in the blue-green alga *Anacystis nidulans*. J Bacteriol 113: 1256–1263
Drews G (1992) Intracytoplasmic membranes in bacterial cells: Organization, function and biosynthesis. In: Mohan S, Dow C and Cole A (eds) Prokaryotic Structure and Function: A New Perspective, pp 249–274. Cambridge University Press
Drews G and Imhoff JF (1991) Phototrophic purple bacteria. In: Shively JM and Barton LL (eds) Variations in Autotrophic Life, pp 51–97. Academic Press, New York
Drews G, Klug G, Liebetanz, R and Dierstein R (1986) Regulation of gene expression and assembly of photosynthetic pigment-protein complexes. In: Biggins J (ed) Progress in Photosynthetic Research, Vol IV, pp 691–697. Martin Nijhoff Publishers, Dordrecht
Dryden SC and Kaplan S (1990) Localization and structural analysis of the ribosomal RNA operons of *Rhodobacter sphaeroides*. Nucleic Acid Res 18: 7267–7277
Ehretsmann CP, Agamemnon JC and Krisch HM (1992) Specificity of *Escherichia coli* endoribonuclease E: In vivo and in vitro analysis of mutants in a bacteriophage T4 mRNA processing site. Genes Develop 6: 149–159
Emory SA, Bouvet P and Belasco JG (1992) A 5′ terminal stem-loop structure can stabilize mRNA in *Escherichia coli*. Genes Develop 6: 135–148
Farchaus JW, Barz WP, Grünberg H and Oesterhelt D (1992) Studies on the expression of the *pufX* polypeptide and its requirement for photoheterotrophic growth in *Rhodobacter capsulatus*. EMBO J 11: 2779–2788
Fonstein M and Haselkorn WR (1993) Chromosomal structure of R. *capsulatus* strain SB1003: Cosmid encyclopedia and high-resolution physical and genetic map. Proc Natl Acad Sci USA 90: 2522–2526
Fritsch J, Rothfuchs R, Rauhut R and Klug G (1995) Identification of an mRNA element promoting rate-limiting cleavage of the polycistonic *puf* mRNA in *Rhodobacter capsulatus* by an enzyme similar to RNase E. Mol Microbiol, in press
Gutell RR, Schnare MN and Gray MW (1990) A compilation of large subunit (23S-like) ribosomal RNA sequences presented in a secondary structure format. Nucleic Acid Res 18: 2319–2329
Höpfl P, Ludwig W and Schleifer KH (1988) Complete nucleotide sequence of a 23S ribosomal RNA gene from *Rhodobacter capsulatus*. Nucleic Acid Res 16: 2343
Hunter CN, McGlynn P, Ashby MK, Burgess JG and Olson JD (1991) DNA sequencing and complementation/deletion analysis of the *bchA-puf* operon region of *Rb. sphaeroides*: in vivo mapping of the oxygen-regulated *puf* promoter. Mol Microbiol 5: 2649–2661
Kaufmann N, Reidl HH, Golecki J, Garcia AF, Drews G (1982) Differentiation of the membrane system in cells of *Rhodobacter capsulatus* after transition from chemotrophic to phototrophic growth conditions. Arch Microbiol 131: 313–322
Klug G (1991) Endonucleolytic degradation of *puf* mRNA in *Rb. capsulatus* is influenced by oxygen. Proc Natl Acad Sci USA 88: 1765–1769
Klug G (1993) The role of mRNA degradation in the regulated expression of bacterial photosynthesis genes. Mol Microbiol 9: 1–7
Klug G and Cohen SN (1988) Pleiotropic effects of localized *Rhodobacter capsulatus puf* operon deletions on the production of light-absorbing pigment-protein complexes. J Bacteriol 170: 5814–5821
Klug G and Cohen SN (1990) Combined action of multiple hairpin loop structures and sites of rate-limiting endonucleolytic cleavage determine differential degradation rates of individual segments within polycistronic *puf* operon mRNA. J Bacteriol 172: 5140–5146
Klug G and Drews G (1984) Construction of a gene bank of *Rhodopseudomonas capsulata* using a broad host range cloning system. Arch Microbiol 139: 319–325
Klug G, Kaufmann N and Drews G (1984) The expression of genes encoding proteins of the B800-850 antenna pigment complex and ribosomal RNA of *R. capsulata*. FEBS Lett 177: 61–65
Klug G, Kaufmann N and Drews G (1985) Gene expression of pigment-binding proteins of the bacterial photosynthetic apparatus: Transcription and assembly in the membrane of *Rhodopseudomonas capsulata*. Proc Natl Acad Sci USA 82: 6485–6489
Klug G, Liebetanz R and Drews G (1986) The influence of bacteriochlorophyll biosynthesis on formation of pigment-binding proteins and assembly pf pigment-protein complexes in *Rhodopseudomonas capsulata*. Arch Microbiol 146: 284–291
Klug G, Adams CW, Belasco JG, Dörge B and Cohen SN (1987) Biological consequences of segmental alterations in mRNA stability: Effects of the intercistronic hairpin loop region of the *Rhodobacter capsulatus puf* operon. EMBO J 6: 3515–3520
Klug G, Jock S and Rothfuchs R (1992) The rate of decay of *Rhodobacter capsulatus*-specific *puf* mRNA segments is differentially affected by RNase E activity in *Escherichia coli*. Gene 121: 95–102
Klug G, Jock S and Kordes E (1994) 23S rRNA processing in *Rb. capsulatus* is not involved in the oxygen-regulated formation of the bacterial photosynthetic apparatus. Arch Microbiol 162: 91–97
Kordes E, Jock S, Fritsch J, Bosch F and Klug G (1994) Cloning of a gene that is involved in rRNA precursor processing and in 23S rRNA cleavage in *Rb. capsulatus*. J Bacteriol 176: 1121–1127
Lee KJ and Kaplan S (1992) *cis*-acting regulatory elements involved in oxygen and light control of *puc* operon transcription in *Rhodobacter sphaeroides*. J Bacteriol 174: 1146–1157
Lee KJ, Kiley PJ and Kaplan S (1989) Posttranscriptional control of *puc* operon expression of B800-850 light-harvesting complex formation in *Rhodobacter sphaeroides*. J Bacteriol 171: 3391–3405
Lessie, TG (1965) The atypical ribosomal RNA complement of *Rhodopseudomonas sphaeroides*. J Gen Microbiol 39: 311–320
Lilburn TG, Copper EH, Prince RC, Beatty JT (1992) Pleiotropic effects of *pufX* gene deletion on the structure and function of the photosynthetic apparatus of *Rhodobacter capsulatus*. Biochim Biophys Acta 1100: 160–170
Lundberg U, von Gabain A and Meleförs Ö (1990) Cleavages in the 5′ region of the *ompA* and *bla* mRNA control stability: Studies with an *E. coli* mutant altering mRNA stability and a novel ribonuclease. EMBO 9: 2731–2741
MacKay RM, Zablen LB, Woese CR and Doolittle WF (1979) Homologies in processing and sequence between the 23S

ribosomal ribonucleic acids of *Paracoccus denitrificans* and *Rhodopseudomonas sphaeroides*. Arch Microbiol 123: 165–172

Mansour JD and Stachow LS (1975) Structural changes in the ribosomes and ribosomal proteins of *Rhodopseudomonas palustris*. Biophys Res Commun 62: 276-281

Marrs BL and Kaplan S (1970) 23S precursor ribosomal RNA of *Rhodopseudomonas sphaeroides*. J Mol Biol 49: 297–317

Meier JR and Brownstein BH (1976) Structure, synthesis and post-transcriptional modification of ribosomal ribonucleic acid in *Bdellovibrio bacteriovorus*. Biochem Biophys Acta 454: 86–96

Neidle EL and Kaplan S (1993) 5-aminolevulinic acid availability and control of spectral complex formation in *hemA* and *hemT* mutants of *Rhodobacter sphaeroides*. J Bacteriol 175: 2304–2313

Oberlé B, Tichy H-V, Hornberger U and Drews G (1990) Regulation of formation of photosynthetic light-harvesting complexes in *Rhodobacter capsulatus*. In: Drews G and Dawes E (eds) Molecular Biology of Membrane-Bound Complexes in Phototrophic Bacteria, pp 77–84. Plenum Press, New York.

Schuch, W and Loening, UE (1975) The ribosomal acid of *Agrobacterium tumefaciens*. Biochem J 149: 17–22

Tichy H-V, Oberle B, Stiehle H, Schiltz E and Drews G (1989) Genes downstream from *pucA* are essential for the formation of the B800-850 complex of *Rhodobacter capsulatus*. J Bacteriol 171: 4914–4922

Tichy H-V, Albien KU, Gad′on N and Drews G (1991) Analysis of the *Rhodobacter capsulatus puc* operon: the *pucC* gene plays a central role in the regulation of LHII (B800-850 complex) expression. EMBO J 10: 2949–2956

Youvan DC, Bylina EJ, Alberti M, Begusch H and Hearst JE (1984) Nucleotide and deduced polypeptide sequence of the photosynthetic reaction center, B870 antenna, and flanking polypeptides from *Rhodopseudomonas capsulata*. Cell 37: 949–957

Zhu YS and Hearst JE (1986) Regulation of expression of genes for light-harvesting antenna proteins LH-I and LH-II; reaction center polypeptides RC-L, RC-M, and RC-H; and enzymes of bacteriochlorophyll and carotenoid biosynthesis in *Rhodobacter capsulatus* by light and oxygen. Proc Natl Acad Sci USA 83: 7613–7617

Zhu, YS and Kaplan S (1985) Effects of light, oxygen and substrates on steady-state levels of mRNA coding for ribulose-1,5-bisphosphate carboxylase and light-harvesting and reaction center polypeptides in *Rhodopseudomonas sphaeroides*. J Bacteriol 162: 925–932

Zucconi AP and Beatty JT (1988) Posttranscriptional regulation by light of the steady state levels of mature B800-850 light-harvesting complex in *Rhodobacter capsulatus*. J Bacteriol 170: 877–882

Chapter 60

Polyesters and Photosynthetic Bacteria

From Lipid Cellular Inclusions to Microbial Thermoplastics

R. Clinton Fuller
Department of Biochemistry and Molecular Biology, University of Massachusetts, Amherst, MA 01003, USA

Summary

Poly-β-hydroxyalkanoates, PHAs, are synthesized in bacteria and form unique intracellular inclusions. These inclusions can be found in a wide variety of anaerobic, aerobic and facultative chemotrophic and phototrophic bacteria, including cyanobacteria. Inclusions do not occur in the enteric bacteria of any known animal species. PHAs represent a family of aliphatic polyesters which are isotactic, biodegradable and biocompatible, thermoplastic elastomers. The organisms will utilize a wide variety of both hydrocarbons and fatty acid substrates of varying pendant chain lengths of three to twelve carbons to synthesize random copolymers of mixed-length, pendant carbon side chains. They will utilize functionalized substrates (phenyl, cyano, bromo, unsaturated and others) to synthesize chemically and biologically functional polyesters. Using the process of cometabolism, they will incorporate unnatural and even toxic substrates into polymer when co-fed a related natural substrate. Using the wide diversity of bacteria in nature, literally dozens of new and unusual biosynthetic PHAs have been produced. All PHA polyesters occur in bacterial cells as unique fluid inclusions surrounded by a single-layered, protein phospholipid coat. This coat is a monolayer and not a unit membrane. Both the metabolic pathways and the molecular genetics of the enzymes concerned with PHA synthesis, polymerization

R. E. Blankenship, M. T. Madigan and C. E. Bauer (eds): Anoxygenic Photosynthetic Bacteria, pp. 1245–1256.

and depolymerization (biodegradation) have been worked out in several microbial systems, including the phototrophic bacterium *Rhodospirillum rubrum*, an organism in which some of the first research on PHAs was done.

In order to cope with the overwhelming onslaught on the environment of plastic waste disposal into the next century, we must turn to biodegradable plastics. In order to compete with and supplant the petroleum-based polyester industry's production of non-biodegradable products, we must turn to large scale production from inexpensive resources. Using the results of the microbial studies on the synthesis, regulation, material science and molecular genetics of polyesters, we shall turn again to photosynthesis. Using genetic engineering technology from microorganisms, genes will be transferred into plant species such as sugar beet, potato and rape seed. New, unusual, practical and inexpensive polyesters can be made from CO_2 and the sun, completing three and a half billion years of the evolution of the production of biosynthetic and biodegradable thermoplastics.

I. Introduction

All organisms contain products of the polymerization of simple water-soluble or lipid-soluble monomeric molecules which, in one form or another, appear as repeating units in polynucleotides, polysaccharides, polypeptides, polyisoprenoids and a second class of hydrophobic biopolymers, the poly-β-hydroxyalkanoates, PHAs (Findlay and White, 1983, 1984). Most of these biopolymers have a relatively high molecular weight in the range of 100,000 to 1,500,000. PHAs exist in prokaryotic cells as inclusion bodies representing a family of aliphatic polyesters with the chemical structure shown below:

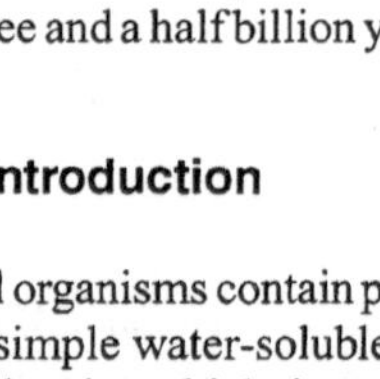

$R = (CH_2)xCH_3$

$x = 0 - 8$ or higher

Polyesters from microbial inclusions show a wide variety of repeating units, have high molecular weights and, like all biological polymers, are isotactic. These properties cannot be reproduced from petroleum-based synthetic products. Because of the wide variety of repeating units and crystalline properties, these polyesters, unlike any other biopolymers, have the properties of thermoplastics and elastomers and, of course, are inherently (i.e. at least within the cell) biodegradable. Fig. 1A shows a thin section of cells of *Rhodospirillum rubrum* (× 200,000) and Fig. 1B and 1C show an optical image of Nile blue stained cells of *Bacillus cereus*.

Abbreviations: A. – Alcaligenes; B. – Bacillus; P. – Pseudomonas; PHA – Poly-β-hydroxyalkanoate; PHB – Poly-β-hydroxybutyrate; *Rb. – Rhodobacter; Rs. – Rhodospirillum*

II. PHAs as Prokaryotic Inclusions

PHAs, originally thought to be primarily poly-β-hydroxybutyrate, PHB, were first observed and characterized by Lemoigne at the Institute Pasteur in 1925 from *Bacillus megaterium* (Lemoigne, 1925). His characterization of the polymer and its potential functions was truly innovative and, from a chemical and biological point of view, way ahead of his time (Lemoigne, 1926). Chemically, polyester compounds had not yet been characterized in biological systems and the concept of these polymeric materials as energy storage products in cells had not been suggested.

Since that time, PHAs, as carbon and energy storage inclusions, have been described in a wide variety of prokaryotes. The most recent inventory has been compiled in a thorough review by Steinbüchel (1991), listing over 200 bacterial species known to contain PHA inclusions. Interestingly, in spite of the wide variety of microbial families of prokaryotes that contain PHA as storage inclusions, none of the Enterobacteriaceae or their relatives appear to contain them. However, PHB does occur in another polymeric form in the Enterobacteriaceae, in other bacterial families, and in many eukaryotic plant and animal cells (Reusch et al.; 1983, 1986, 1987, 1988, 1989). The role of this polymer, containing phosphorus and calcium and located in the cytoplasmic membrane, is not clear but may involve genetic competence. It is not a storage product in any event, and will not be discussed in this review. However, it should be noted for its ubiquitous occurrence and another important role that PHAs may play in nature.

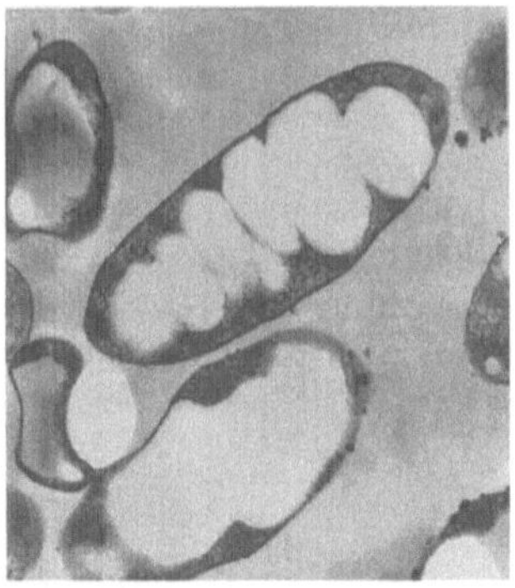

A

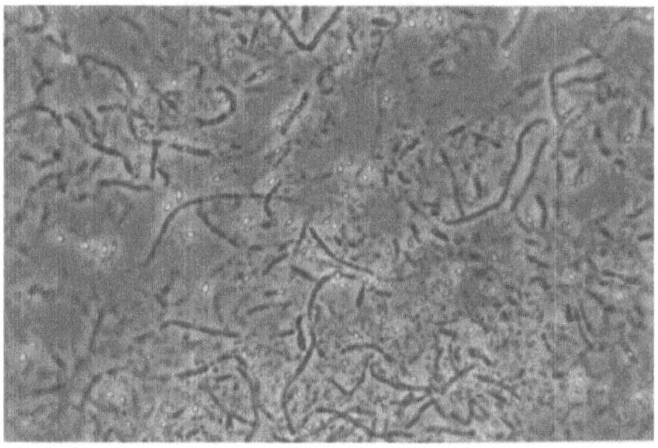

B

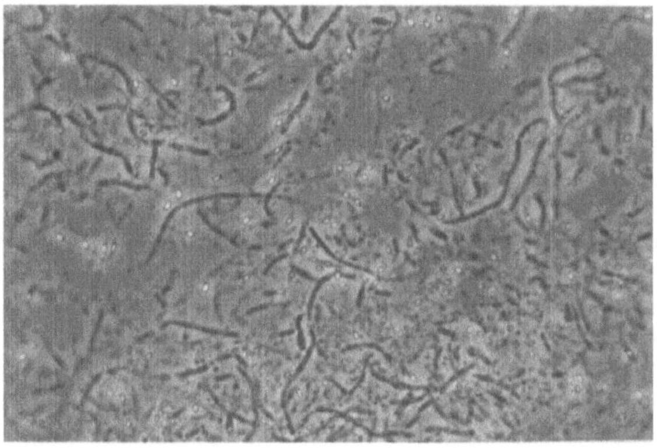

C

Fig. 1. (See also Color Plate 3.) A: A thin section of ***Rs. rubrum*** showing PHA occupying 70–90% of the cell, its 'fluid nature' and independence of intracytoplasmic membranes (chromatophores) and cytoplasmic membranes. B: Phase contrast optical microscope image of Nile Blue stained cells of *Bacillus cereus*. C: Same as B, but activated by 450 nm light showing specific fluorescence of PHA. Spores do not fluoresce and note that not all cells contain PHA. This is a very specific reaction and can give a quantitative measure of PHA from fluorescence intensity.

III. Occurrence of PHAs in Anoxygenic Phototrophic Bacteria

PHAs were first described in the phototrophic bacterium, *Rs. rubrum* in a series of papers by Doudoroff and Stanier (1959) and Merrick and Doudoroff (1961). Since that time, PHAs have been shown to be present in at least 14 facultative nonsulfur purple bacteria and 14 purple sulfur bacteria (see Tables 1 and 2).

A. Purple Bacteria

Tables 1 and 2 show the published and some unpublished references of the occurrence of PHAs in a variety of facultative nonsulfur purple bacteria and the obligately anaerobic purple sulfur bacteria. These data are reported by Liebergesell et al. (1991) and have been slightly modified and updated for this chapter. This list shows the wide distribution of PHAs in purple bacteria. However, it should be noted that of the organisms listed, 22 show a polymer content that is less than 5% of the cell mass. In addition, those with 5% polymer or less contain 100% of that polymer as PHB. This observation could mean that under the growth conditions used (acetate and succinate), the polymer measured could well be present only in the cytoplasmic membrane structures as described by Reusch in which *only* PHB, not a variety of PHAs, is observed.

In the case of two organisms, *Rs. rubrum* and *Rhodobacter sphaeroides*, large amounts of storage PHAs can be synthesized under controlled growth conditions on specific substrates. Brandl et al. (1991) have shown that *Rb. sphaeroides* when grown on acetate alone or acetate in combination with malate, pyruvate, glucose, crotonate, or propionate, can produce from 50–65% of its cell mass as PHA. In all cases, 98% of the polymer was PHB and the remainder consisted of a small amount of 3-hydroxyvalerate.

Table 2 (Liebergesell et al., 1991) summarizes the presence of PHAs in purple sulfur bacteria. In these organisms, the polymers consist of 100% butyrate repeating units when the cells are grown on acetate alone. Again, many contain less than 5% polymer and this may not be stored in inclusions as described above.

B. Green and Heliobacteria

At this point little is known about the presence of storage PHAs in the green sulfur and nonsulfur bacteria. Several authors have reported PHA in *Chloroflexus aurantiacus* (Pierson and Castenholz, 1974; Sirevåg and Castenholz, 1979; Krasil'nikova et al., 1986). At the present time, it has not been demonstrated that the Chlorobiaceae accumulate PHA inclusions. Cells of *Heliobacillus mobilis* grown on acetate contained no detectable PHA as measured by gas chromatography (Madigan and Fuller, unpublished).

IV. Biosynthesis of PHAs in *Rhodospirillum rubrum*

The biosynthetic pathways of PHA synthesis in the various phototrophic bacteria are relatively unexplored. However, Moskowitz and Merrick (1969) undertook the study of the enzymes involved in the synthetic pathway of PHB synthesis in *Rs. rubrum* (Merrick and Yu, 1966 and Merrick et al., 1965). From this and comparative research on *Alcaligenes eutrophus* and *Pseudomonas oleovorans*, the overall biosynthetic pathway in *Rs. rubrum* can be represented as shown in Fig. 2.

The above pathway is the only well established synthetic route in any of the phototrophic bacteria. It is clear that in order to make PHAs other than PHB there must be a great deal of flexibility in at least two steps in the synthesis of PHA. The first is in the ability of the ketothiolase to form and condense the CoA derivatives of substrates from C-2 to C-7 monomers. The second, of course, is the polymerase reaction which, in the case of *Rs. rubrum*, can make a random co-polymer containing repeating units of 4, 5, or 6-chain monomers from both hexanoate and heptanoate (Brandl et al., 1989).

V. Regulation of PHA Metabolism

The enzymology of the biosynthetic polymerization and depolymerization reactions and the control mechanisms invoked in PHA synthesis have been thoroughly researched, primarily in the laboratories of Schlegel and Dawes using the system as it exists in *A. eutrophus*. This work has recently been reviewed and updated in an excellent review by Anderson and Dawes (1990) and will not be rereviewed here, but the same or similar control mechanisms are almost certainly applicable to the phototrophic bacteria in

Table 1. PHA content of nonsulfur purple bacteria and composition of the polyester

Organism	Source (investigator) or culture #	Cultivation in growth medium			Further incubation in nitrogen-free media		
		Succinate (6.2 mM) + Acetate (6.5 mM)			No added carbon source		
		PHA content % of cellular DW	Composition (mol%) 3 HB	3 HV	PHA content % of cellular DW	Composition(mol%) 3 HB	3 HV
Rhodobacter capsulatus Kb1	DSM 155	17.3	100	0	5.0	100	0
Rhodobacter sphaeroides 17023	ATCC 17023	31.6	96.5	3.5	17.5	86.3	13.7
Rhodobacter sphaeroides Y	DSM 160	31.0	97.7	2.3	18.4	96.7	4.3
Rhodocyclus gelatinosus 2150	DSM 149	6.1	100	0	4.5	75.6	24.4
Rhodocyclus tenuis 2761	DSM 109	1.9	100	0	2.3	100	0
Rhodomicrobium vannielii GP	K. Schmidt	3.9	100	0	2.6	100	0
Rhodopseudomonas acidophila 10050	DSM 145	5.7	80.7	19.3	2.2	100	0
Rhodopseudomonas blastica NCIB 11576	DSM 2131	8.1	88.2	11.8	2.6	84.6	15.4
Rhodopseudomonas palustris 1850	DSM 126	1.7	100	0	0.7	100	0
Rhodopseudomonas palustris 17001	N. Pfennig	2.8	100	0	1.3	100	0
Rhodospirillum fulvum Forst	Own isolate	2.7	100	0	1.2	100	0
Rhodospirillum molischianum S	DSM 120	2.4	100	0	2.7	100	0
Rhodospirillum rubrum Ha	DSM 107	9.6	94.8	5.2	8.1	93.8	6.2
Rhodospirillum rubrum 25903	ATCC 25903	1.3	100	0	0	0	0

The numbered cultures in Tables 1 and 2 are from the Deutsche Sammlung von Mikroorganismen (DSM), Braunschweig, Germany, and the American Type Culture Collection (ATCC), Rockville, MD (USA). DW – dry weight; HB – hydroxybutyrate; HV – hydroxyvalerate. Note that *Rhodocyclus gelatinosus* has been renamed *Rubrivivax gelatinosus.*

Table 2. PHA content of sulfur purple bacteria and composition of the polyester

Organism	Source (investigator) or culture #	Cultivation in growth medium			Further incubation in nitrogen-free media		
		Acetate (6.8 mM)			No added carbon source		
		PHA content % of cellular DW	Composition (mol%) 3 HB	3 HV	PHA content % of cellular DW	Composition(mol%) 3 HB	3 HV
Chromatium vinosum D	DSM 180	0.6	100	0	0	0	0
Chromatium vinosum ABGÖ	Own isolate	2.3	100	0	0.8	100	0
Chromatium vinosum 1611	DSM 182	1.5	100	0	0	0	0
Chromatium minutissimum BN 7511	J.F. Imhoff	0.3	100	0	0	0	0
Chromatium okenii 1111	N. Pfennig	0.6	100	0	0.5	100	0
Chromatium warmingii 1113	N. Pfennig	11.5	100	0	5.2	100	0
Lamprocystis roseopersicina 3112	N. Pfennig	28.6	100	0	8.3	100	0
Thiocapsa pfennigii 9111	N. Pfennig	0.9	100	0	0	0	0
Amoebobacter roseus 6611	DSM 235	1.0	100	0	0	0	0
Amoebobacter pendens 1314	DSM 236	24.7	100	0	5.4	100	0
Thiocystis violacea 2311	DSM 208	28.0	100	0	3.0	100	0
Ectothiorhodospira mobilis BN 9903	J.F. Imhoff	18.5	100	0	16.6	100	0
Ectothiorhodospira vacuolata BN 9512	DSM 2111	17.0	100	0	9.2	100	0
Ectothiorhodospira shaposhnikovii N1	DSM 243	15.0	100	0	4.3	100	0

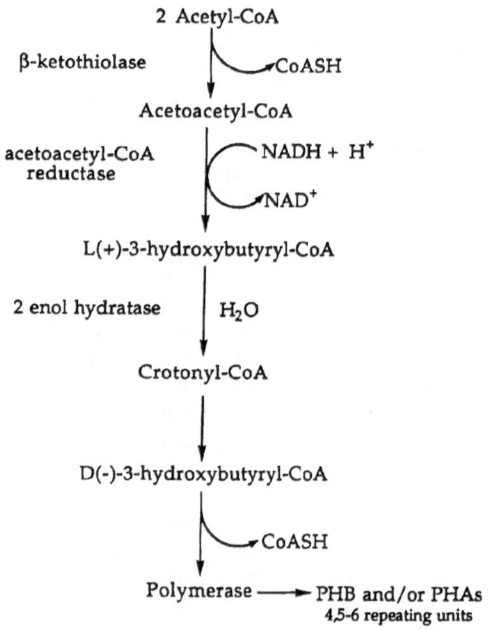

Fig. 2. Rhodospirillum rubrum PHA-biosynthetic pathway.

general, and to *Rs. rubrum* specifically (see Fig. 3). The same enzymes and cofactors seem to function in most organisms studied (Steinbüchel, 1991).

The *Rs. rubrum* (Merrick and Yu, 1966; and Moskowitz and Merrick, 1969), as well as *A. eutrophus* (Anderson and Dawes, 1990) and *P. oleovorans* (Huisman et al., 1991) PHA biosynthetic pathways include an NADH-dependent reductase that forms L(+)-3-hydroxybutyryl-CoA which is converted to the D(-)-3-hydroxybutyryl-CoA by two enoyl-CoA hydrolases. Parts of the pathway have been expressed in *Escherichia coli* (Fidler and Dennis, 1992). Balanced cell growth is dependent on NADH and the presence of acetyl-CoA. During balanced growth, cell synthesis and energy production is in competition with the PHA synthetic pathway for both NADH and acetyl-CoA. In addition, the accumulation of NAD^+ represses enzymes of the Krebs Cycle and would favor the PHA synthetic pathway. Thus, there is probably an NADH $\leftrightarrow$ NAD^+ redox control system involved in the regulation of PHA synthesis during balanced or unbalanced growth conditions in these organisms.

PHAs are produced and degraded continuously in most microorganisms at low levels during cell growth. However, during periods of metabolic stress (e.g., low O_2 in an aerobe, or a shortage of nitrogen or sulfur in the environment, thus restricting protein synthesis) the cells convert any excess carbon present to PHA. This shift from the energy and growth-promoting Krebs Cycle to PHA synthesis is tightly regulated as shown in Fig. 4.

As diagrammed above, under conditions of balanced cell growth and ATP production (energy excess), NADH levels inhibit both the depolymerization of PHAs and the Krebs Cycle. The resulting reduced level of free CoA means that the acetyl-CoA thiokinase reaction can catalyze the condensation of acetyl-CoA to acetoacetyl CoA and free CoA and then polymer synthesis proceeds. When the control is reversed, depolymerization of polymers occurs.

VI. Polymer Production and Material Properties

At the time of the initial discovery and early characterization of PHAs in bacteria it appeared that the polymer consisted exclusively of four carbon *β*-hydroxy repeating units (PHB). It was first demonstrated by Findlay and White (1983) that *B. megaterium* had the ability to form random copolymers. In addition, both *P. oleovorans* and *A. eutrophus* (Brandl et al., 1988) form a wide variety of such copolymers (Anderson and Dawes, 1990). Bacteria can produce polyesters with pendant side chains of 8 carbons or higher. The physical and material properties of these copolymers such as stiffness, strength, melting point and brittleness, as well as degradability by extracellular lipases, vary widely depending on the monomer composition and the resulting degree of crystallinity. Whereas *P. oleovorans* will only make PHAs when grown on substrates containing six or more carbon atoms in the chain, both *A. eutrophus* and *Rs. rubrum* make shorter chain PHA copolymers. *A. eutrophus* makes a copolymer of PHB-PHV which is currently being commercially produced as 'Biopol.' *Rs. rubrum,* which can synthesize PHAs from substrates ranging from acetate to heptanoate, can make a number of copolymers including an interesting terpolymer consisting of C_4, C_5, and C_6 repeating units (Brandl et al., 1989). This variability and/or flexibility of the polymerase enzyme(s) to polymerize a wide variety of substrates yielding PHAs with differing material properties may prove very useful for the production

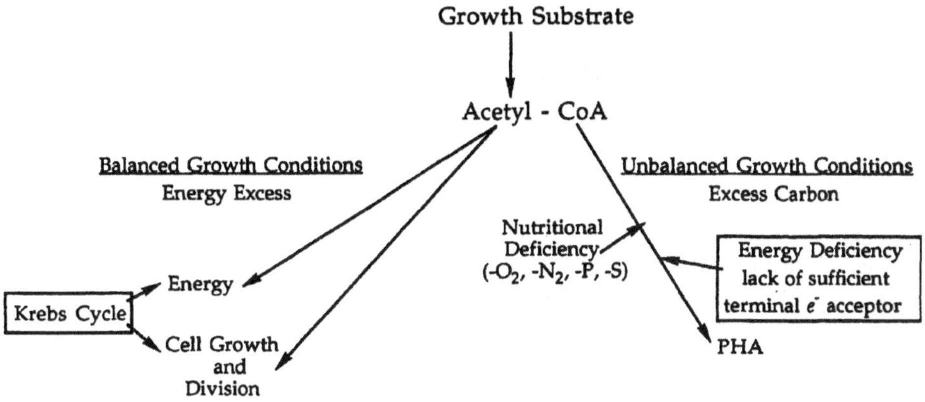

Fig. 3. The regulation and control of energy production, cell growth and PHA production in bacteria.

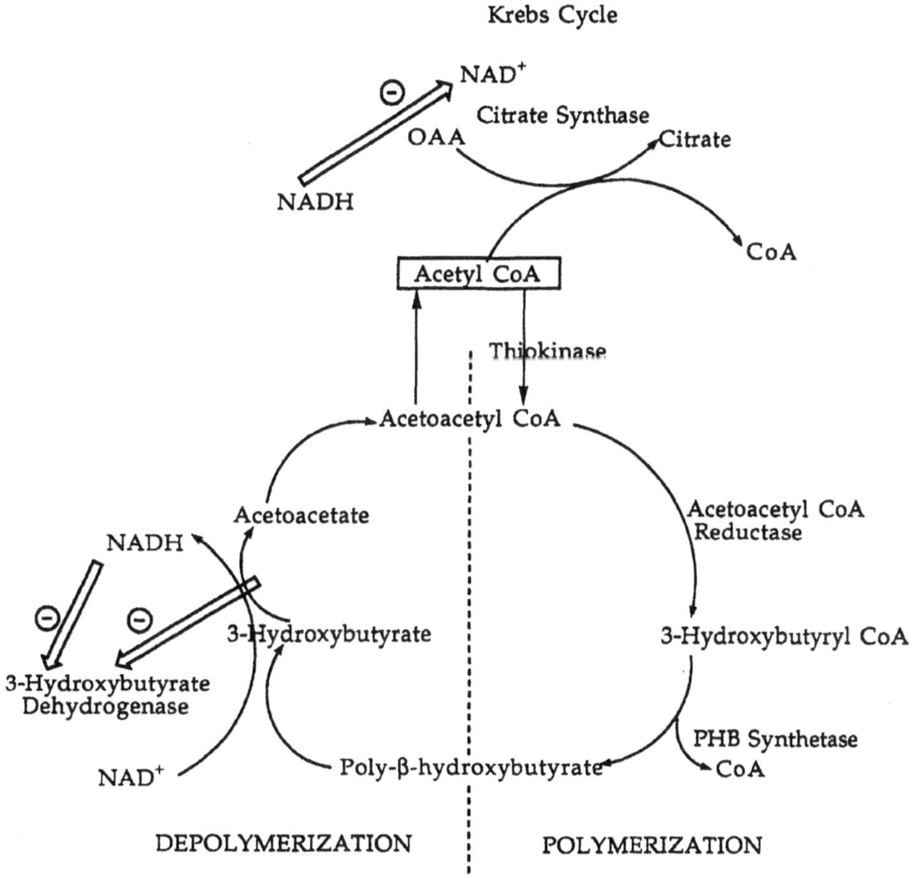

Fig. 4. Proposed regulatory mechanisms of PHA polymerization-depolymerization reactions (alkyl pendant group can be 0-8 or higher)

of thermoplastic polyesters of both commercial and medical value.

In addition to the polymerization of a wide variety of alkanoic acids, it has been shown that both *P. oleovorans* and *Rs. rubrum* are capable of producing unusual PHAs in which the substituent at the *β*-position is a structure other than an n-alkyl group (Lenz et al., 1992) and that contains functional or reactive groups. Since the pendant side chains are relatively inactive hydrocarbons, the addition of functional groups allows the production of polymers that can be further chemically modified. *Rs. rubrum* has been shown to be capable of polymerizing unsaturated compounds, thus positioning a reactive double bond in the resulting PHA. *P. oleovorans* shows a high degree of flexibility in incorporating functional groups into polymer. Methyl, cyano and phenyl groups, as well as substrates with varying degrees of unsaturation, can all be incorporated into the PHAs produced by this organism.

It has also been shown that when the compounds 5-methyloctanoic acid, 6-bromohexanoic acid and 11-cyanoundecanoic acid (none of which can be formed in polymers by themselves) are present in growing cultures with either octanoic or nonanoic acid as cometabolites, the previously unusable monomer is incorporated into polymer. This phenomenon of cometabolism (Leadbetter and Foster, 1960) was first described for the methanotrophic bacteria and no clear-cut mechanism for the process has been described. However, since cofeeding with a natural substrate can allow the incorporation of functional and even toxic groups into PHAs, this offers further flexibility in bacterial polyester synthesis. It is clear that at least *Rs. rubrum* and *P. oleovorans* have the ability to produce a wide variety of functional and unusual polyesters and, of course, contain the enzyme(s) for degradation of these polymers.

VII. Structure and Function of PHA Inclusions

A. State of PHA in Inclusion Bodies

Since the time of the initial identification of *β*-hydroxybutyrate in *B. megaterium* (Lemoigne, 1925), considerable interest has been focused on the physical and chemical state of the PHA inclusions and associated macromolecules in a wide variety of prokaryotes. That PHAs could occur in an amorphous as well as a crystalline state was also first proposed in the initial research of Lemoigne, who suggested that during isolation of the inclusions, a lipid material was lost. Lundgren et al. (1965) also suggested that PHAs could occur in the crystallized state. However, recent work of Bernard and Sanders (1988, 1989) using ^{13}C NMR spectroscopy has indicated that PHAs are in a predominantly mobile state in the living cell and that this amorphous condition has water associated as an integral component of the inclusion. Bernard and Sanders stated that the 'enzyme(s) responsible for PHB biosynthesis and consumption operate only on mobile hydrated material, and that solid granules characteristic of dried cells are partially artifactual.' The studies of Alper et al. (1963), Merrick and Doudoroff (1964) and Lundgren et al. (1965) demonstrated the presence of proteins and enzyme activity associated with isolated inclusions. Ellar et al. (1968) has also described a role for proteins in, or at the surface of, the PHA inclusion. It is clear that PHA in the native state is not a solid nor a mobile liquid, nor is it in solution; rather it is present in cells as an amorphous elastomer (Horowitz et al., 1993). Whether complete enzymatic biodegradation can occur *only* in a hydrated state still is not clear and indeed a point of controversy that needs considerable clarification. How the amorphous state is maintained in native granules is not clear, nor are the association and distribution of enzymatic proteins on or within the granule fully understood or clearly defined.

B. Other Inclusion Components

Boatman (1964) described a limiting 'membrane' surrounding the inclusions in *Rs. rubrum* when observed in an electron microscopic thin section. Its composition was not determined. However, Griebel et al. (1968) and Griebel and Merrick (1971) purified PHB inclusions in density gradient systems and determined that PHB constituted 97–98% of the dry weight, the remainder being made up of protein (~2%) (Merrick et al., 1965) and lipid (~0.5%). It was assumed that these constituents were components of the outer shell of the inclusion.

1. Lipids

Although the evidence for the presence of lipids and/or phospholipids associated with the PHA was reported in 1968 (Griebel et al., 1968), further analyses were fragmentary. Kanaguchi and Doi (1990) have used X-ray diffraction studies to evaluate the state of PHAs in whole cells and isolated inclusions from *A. eutrophus* and, like others, concluded that in both whole cells (wet) and in native isolated granules the polymer was still in a non-crystalline state. Treatment with hypochlorite, acetone or a lipase enzyme caused crystallization, leading to the conclusion that the removal of a lipid or lipids from the inclusion caused this change of state. Thin layer chromatographic analysis of acetone extracts of inclusions demonstrated the presence of at least two lipid components. Recently, de Koning and Lemstra (1992) have clearly shown phospholipid as a constitutive of PHA inclusions and have suggested their role in nucleation polymerization of PHAs during inclusion formation. These observations have been recently confirmed in our laboratory (E. S. Stuart, A. Tehrani and R. C. Fuller, unpublished).

2. Proteins

That enzymatic activity and proteins were associated with PHA inclusions was first demonstrated by Merrick and Doudoroff (1961, 1964). Both PHA synthetase and depolymerase activities were detected in cell-free systems prepared from *Rs. rubrum*. However, the clear separation and definitive cellular partition of the enzymes concerned in these preparations with the synthesis of the CoA substrate monomers and the actual polymerase reaction itself was not clearly established.

Using primarily a molecular genetic approach, Peoples and Sinskey (1989a and 1989b), Messner and Sleytr (1992) and Huisman et al. (1991) have established the enzymatic pathway of synthesis of PHAs in a number of chemotrophic organisms. They demonstrated that the polymerase enzyme (one or two depending on the organisms) and the depolymerase occurs along with a promoter as a single genetic operon. In all probability the polymerase-depolymerase system is associated with the PHA inclusions and the ketothiolase and NADH oxidation-reduction and hydratase enzymes are associated with the cytoplasmic membrane cell components.

Recent work using *P. oleovorans* grown on octanoic acid as a substrate has established protein association with the PHA inclusion. In this system, the PHA formed is a random co-polymer of C_6, C_8 and C_{10} repeating units. When isolated, the inclusions maintained their amorphous hydrated state as a fluid elastomer. Both whole cells and highly purified granules were examined in this native state using Nomarski computer-enhanced imaging light microscopy. This technique involves differential interference contrast using polarized light. Fig. 5a shows an image of unstained, unfixed living cells which show the internal PHA granules. Fig. 5b is an image of highly purified native granules from these cells (E. S. Stuart, R. W. Lenz and R. C. Fuller, unpublished).

The size of a typical PHA granule is around 0.5 μm. There is no sign in this, or many other such preparations, of significant contaminating membrane fragments. This is confirmed by sodium dodecyl sulfate-polyacrylamide gel electrophoresis (SDS-PAGE) peptide analysis after granule purification. This, of course, is an important consideration when assaying for inherent enzymatic activity of the proteins associated with granule preparations.

In addition to Nomarski computer-enhanced imaging and negative staining, freeze-fracture electron microscope examination of the surface structure of the granule was carried out. Again, the granules were maintained in their native state until they were rapidly frozen, fractured and either etched or stained with uranyl acetate. Examination of uranyl acetate-stained granules demonstrated a highly organized paracrystalline network located on the surface of the granule (not shown). Because of its staining properties, this network was clearly proteinaceous. A freeze etch of these same granule preparations is shown in Fig. 6 and dramatically emphasizes this surface network.

The image in Fig. 6 displays a network pattern with repeating units measuring 135 Å in width which can be seen on the surface of the highly purified granules. The pattern resembles the regularly arranged outer membrane proteins (rOmp) of *P. acidovorans*. This kind of structure has never been observed in association with internal structures of prokaryotes. Purified inclusions were analyzed by SDS-PAGE and a characteristic pattern is shown in Fig. 7 lane 1. Fig. 7, lane 2 shows the granule fraction before final purification.

After granule purification, SDS-PAGE revealed four major polypeptides with molecular masses of 59 kDa, 55 kDa, 43 kDa, and 18 kDa (Fig. 7, lane 1). A minor protein band at 32 kDa was also consistently observed. Huisman et al. (1991) have used a recombinant strain of *Pseudomonas* to express the genes responsible for the two polymerases and found protein bands at 58 kDa and 56 kDa in the recombinant cells. In addition, the DNA sequence of esterase

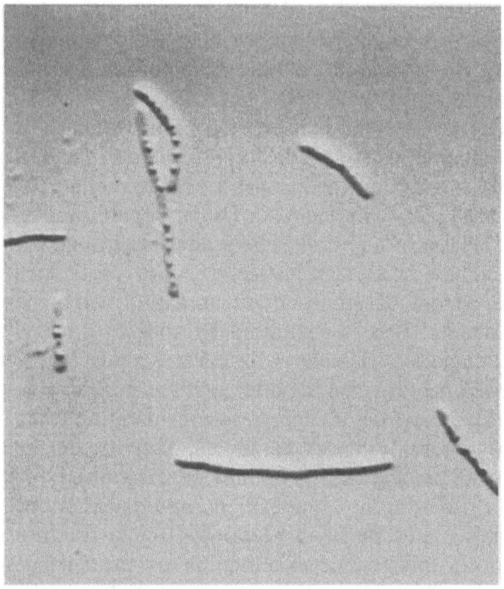

A

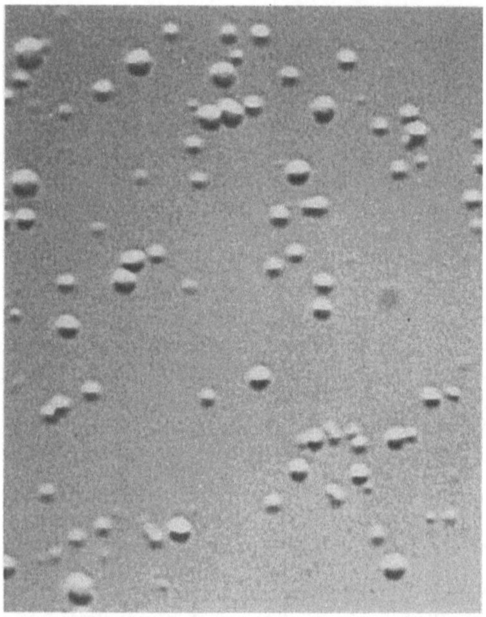

B

Fig. 5. (A) Living *P. oleovorans* cells under computer-enhanced Nomarski light microscopy. (B) Isolated inclusions as in 5a.

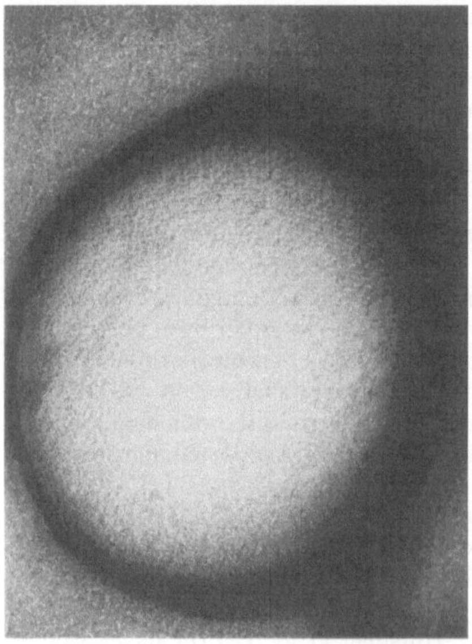

Fig. 6. An electron micrograph of a single PHA granule (~0.5μm diameter) showing an organized paracrystalline array of surface proteins that control structural, functional and probably material properties of the in vivo amorphous PHA.

suggests that the 32-kDa polypeptide could be the depolymerase system. These DNA sequence and derived molecular weights agree well with the above observations of native structural proteins on the purified granules of *P. oleovorans* (E. S. Stuart, R. W. Lenz and R. C. Fuller, unpublished).

These proteins could be removed from the granule using the NaOH extraction procedure developed by Merrick (1978). Thus it seems that in *P. oleovorans* the two polymerases from this organism, as well as the intracellular depolymerase, are associated with the PHA inclusion as in *Rs. rubrum* as was first suggested by Merrick.

However, the most abundant polypeptides on the inclusion are the 43-kDa and the 18-kDa bands. These molecular weights do not correspond to any of the biosynthetic or polymerase-depolymerase systems of either *P. oleovorans* or *A. eutrophus* (Peoples and Sinskey, 1989a and b and Huisman et al., 1991).

Since the protein surface of the inclusion shows a paracrystalline array much in the form of the S protein on the outer surface layer of most bacteria (Messner and Sleytr, 1992), it has been suggested that the 43-kDa polypeptide might play a similar structural as well as a developmental role for the

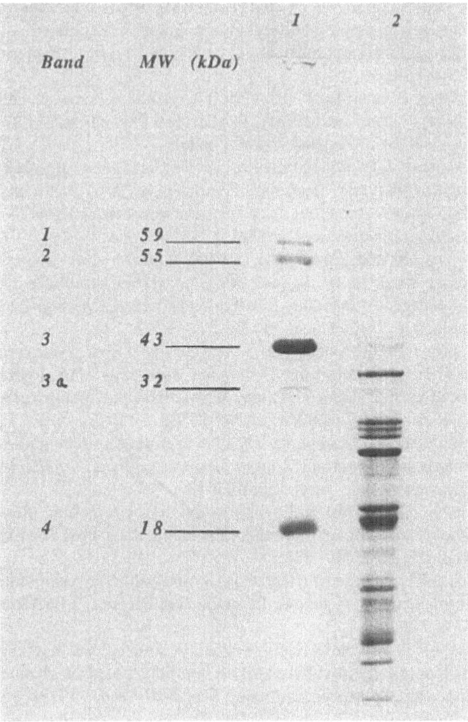

Fig. 7. S.D.S. Gel of purified inclusions from cells of *P. oleovorans*. See text for explanation.

PHA inclusion in *P. oleovorans*. However, the exact role of the 43-kDa and 18-kDa polypeptides remains to be determined. Liebergesell et al. (1992) have shown by gel electrophoresis that proteins isolated from PHA inclusions of *Chromatium* are also distinct from the polymerase systems from various chemotrophic organisms. An update of information on polymer inclusion proteins is pesented in Stuart et al. (1995).

Acknowledgments

I wish to thank my colleagues, Dr. R. W. Lenz, Dr. Helmut Brandl, and Dr. Elizabeth Stuart, for their active collaborative work on the research of the phototrophic and chemotrophic polyester systems and the use of unpublished data. I also wish to thank Ms. Ann Brainerd for her thoughtful and thorough editorial contribution. The support of the National Science Foundation MRL Grant # DMR90–23848, Professor Christopher Woodcock and the NSF-supported Microscopy and Imaging Center # NSF BBS 871–423335, NSF MCB–92022419, and the Johnson and Johnson Co. is gratefully acknowledged. I would also like to thank Dr. Alexander Steinbüchel and *FEMS Microbiol. Letters* and *FEMS Microbiol. Reviews* for use of previously published material.

References

Alper R, Lundgren DG, Marchessault RH and Cote WA (1963) Properties of poly-β-hydroxybutyrate. I. General considerations concerning the naturally occurring polymer. Biopolymers 1: 545–556

Anderson AJ and Dawes EA (1990) Occurrence, metabolism, metabolic role, and industrial uses of bacterial polyhydroxyalkanoates. Microbiol Rev 54: 450–472

Bernard GN and Sanders JKM (1988) Observation of mobile poly-β(hydroxybutyrate) in the storage granules of *Methylobacterium* AM1 by in vivo ^{13}C-NMR spectroscopy. FEBS Lett 231: 16–18

Bernard, GN and Sanders KM (1989) The poly-β-hydroxybutyrate granule in vivo. J Biol Chem 24: 3286–3291

Boatman ES (1964) Observations on the fine structure of spheroplasts of *Rhodospirillum rubrum*. J Cell Biol 20: 297

Brandl H, Gross RA, Lenz RW and Fuller RC (1988) *Pseudomonas oleovorans* as a source of poly (β-hydroxyalkanoates) for potential applications as biodegradable polyesters. Appl Environ Microbiol 54: 1977–1982

Brandl H, Knee EJ, Fuller RC, Gross RA and Lenz RW (1989) The ability of the phototrophic bacterium *Rhodospirillum rubrum* to produce various poly(β-hydroxyalkanoates): potential sources for biodegradable polyesters. Int J Biol Macromol 11: 49–56

Brandl H, Gross RA, Lenz RW, Lloyd R and Fuller RC (1991) The accumulation of poly(3-hydroxyalkanoates) in *Rhodobacter sphaeroides*. Arch Microbiol 155: 337–340

de Koning, GJM and Lemstra, PJ (1992) The amorphous state of bacterial poly[(*R*)-3-hydroxyalkanoate] in vivo. Polymer 33: 3292–3294

Doudoroff M and Stanier RY (1959) Role of poly-β-hydroxybutyric acid in assimilation of organic carbon by bacteria. Nature 183: 1440

Ellar D, Lundgren TG, Okanawa K and Marchessault RH (1968) Morphology of PHB granules. J Mol Biol 35: 489–502

Fidler S and Dennis D (1992) Polyhydroxyalkanoate production in recombinant *Escherichia coli*. FEMS Microbiol Rev 103: 231–236

Findlay RH and White DC (1983) Polymeric beta-hydroxyalkanoates from environmental samples and *Bacillus megaterium*. Appl Environ Microbiol 45: 71–78

Findlay RH and White DC (1984) *In situ* determination of metabolic activity in aquatic environments. Microbiol Sci 1: 90–95

Griebel R and Merrick JM (1971) Metabolism of poly-β-hydroxybutyrate: effect of mild alkaline extraction on native poly-β-hydroxybutyrate granules. J Bacteriol 108: 782–789

Griebel R, Smith Z and Merrick JM (1968) Metabolism of poly-β-hydroxybutyrate. I. Purification, composition and properties of native poly-β-hydroxybutyrate granules from *Bacillus megaterium*. Biochemistry 7: 3676–3681

Horowitz DM, Clauss J, Hunter B and Sanders J (1993) Amorphous polymer granules. Nature 363: 781

Huisman GW, Wonink E, Meima R, Kazemier B, Terpstra P and Witholt B (1991) Metabolism of poly(3-hydroxyalkanoates) (PHAs) by *Pseudomonas oleovorans*. J Biol Chem 266: 2191–2198

Kanaguchi Y and Doi Y (1990) Structure of native (3-hydroxybutyrate) granules characterized by X-ray diffraction. FEMS Microbiol. Lett 70:151–156

Krasil'nikova EN, Keppen OI and Kondrat'yeva EN (1986) *Chloroflexus aurantiacus* growth in media with different organic compounds and the pathways of their metabolism. Mikrobiologiya 55: 425–430

Leadbetter ER and Foster JW (1960) Bacterial oxidation of gaseous alkanes. Arch Mikrobiol 35: 92–104

Lemoigne M (1925) Etudes sur l'autolyse microbienne acidification par formation d'acide β-oxybutyrique. Ann Inst Pasteur Paris 39: 144

Lemoigne M (1926) Products of dehydration and of polymerization of β-hydroxybutyric acid. Bull Soc Chem Biol 8: 770–782

Lenz RW, Kim YB and Fuller RC (1992) Production of unusual bacterial polyesters by *Pseudomonas oleovorans* through cometabolism. FEMS Microbiol Rev 103: 207–214

Liebergesell M, Hustede E, Timm A, Steinbüchel A, Fuller RC, Lenz RW and Schlegel HG (1991) Formation of poly(3-hydroxyalkanoates) by phototrophic and chemolithotrophic bacteria. Arch Microbiol 155: 415–421

Liebergesell M, Schmidt B and Steinbüchel A (1992) Isolation and identification of granule-associated proteins relevant for poly(3-hydroxyalkanoic acid) biosynthesis in *Chromatium vinosum* D. FEMS Microbiol Lett 99: 227–232

Lundgren DG, Alper R, Schnaitman C and Marchessault RH (1965) Characterization of poly-β-hydroxybutyrate depolymerase extracted from different bacteria. J Bacteriol 89: 245–251

Merrick JM (1978) Metabolism of reserve materials. In: Clayton RK and Sistrom WR (ed) The Photosynthetic Bacteria, pp 199–219. Plenum Press, New York

Merrick JM and Doudoroff M (1961) Enzymatic synthesis of poly-β-hydroxybutyric acid in bacteria. Nature (London) 189: 890

Merrick JM and Doudoroff M (1964) Depolymerization of poly-β-hydroxybutyrate by an intracellular enzyme system. J Bacteriol 88: 60

Merrick JM and Yu CI (1966) Purification and properties of a D(-)-β-hydroxybutyric dimer hydrolase from *Rhodospirillum rubrum*. Biochemistry 5: 3563

Merrick JM, Lundgren DG and Pfister RM (1965) Morphological changes in poly-β-hydroxybutyrate granules associated with decreased susceptibility to enzymatic hydrolysis. J Bacteriol 89: 234–239

Messner P and Sleytr UB (1992) Crystalline bacterial cell-surface layers. In: Advances in Microbial Physiology, Vol. 33, pp 213–275. Academic Press, London

Moskowitz GJ and Merrick JM (1969) Metabolism of poly-β-hydroxybutyrate. Enzymatic synthesis of D-(-)-β-hydroxybutyryl coenzyme A by an enoyl hydrase from *Rhodospirillum rubrum*. Biochemistry 8: 2748–2755

Peoples OP and Sinskey AJ (1989a) Poly-β-hydroxybutyrate biosynthesis in *Alcaligenes eutrophus* H16. Characterization of the genes encoding β-ketothiolase and acetoacetyl-CoA reductase. J Biol Chem 264: 15293–15297

Peoples OP and Sinskey AJ (1989b) Poly-β-hydroxybutyrate (PHB) biosynthesis in *Alcaligenes eutrophus* H16. Identification and characterization of the PHB polymerase gene (*phbc*). J Biol Chem 264: 15298–15303

Pierson BK and Castenholz RW (1974) A phototrophic gliding filamentous bacterium of hot springs, *Chloroflexus aurantiacus, gen.* and *sp.* nov. Arch Microbiol 100: 5–24

Reusch RN (1989) Poly-β-hydroxybutyrate/calcium polyphosphate complexes in eukaryotic membranes. Proc Soc Exp Biol Med 191: 377–381

Reusch RN and Sadoff HL (1983) D(-)-Poly-β-hydroxybutyrate in membranes of genetically competent bacteria. J Bacteriol 156: 778–788

Reusch RN and Sadoff HL (1988) Putative structure and functions of a poly-β-hydroxybutyrate/calcium polyphosphate channel in bacterial plasma membranes. Proc Natl Acad Sci USA 85: 4176–4180

Reusch RN, Hiske TW and Sadoff HL (1986) Poly-β-hydroxybutyrate membrane structure and its relationship to genetic transformability in *Escherichia coli*. J Bacteriol 168: 553–562

Reusch RN, Hiske TW, Sadoff HL, Harris R and Beveridge T (1987) Cellular incorporation of poly-β-hydroxybutyrate into plasma membranes of *Escherichia coli* and *Azotobacter vinelandii* alters native membrane structure. Can J Microbiol 33: 435–444

Sirevåg R and Castenholz R (1979) Aspects of carbon metabolism in *Chloroflexus*. Arch Microbiol 120: 151–153

Steinbüchel A (1991) Polyhydroxyalkanoic acid. In: Byrom D (ed) Biomaterials, pp 125–213. Stockton Press, New York

Stuart ES, Lenz RW and Fuller RC (1995) The ordered macromolecular surface of polyester inclusion bodies in *Pseudomonas oleovorans*. Can J Microbiol, in press

Chapter 61

Imaging Spectroscopy and Combinatorial Mutagenesis of the Reaction Center and Light Harvesting II Antenna

Ellen R. Goldman
CuraGen Corporation, 322 East Main St., Branford, CT 06405, USA

Douglas C. Youvan*
Palo Alto Institute of Molecular Medicine, 2462 Wyandotte St., Mountain View, CA 94043,USA

Summary

The increasing sophistication of mutagenesis and screening technologies has led to the isolation of many biophysically important mutants of the photosynthetic reaction center and light harvesting antennae. Site-directed mutagenesis and structural motif rearrangements of these proteins have accelerated our understanding of the fundamental mechanisms of the light reactions of photosynthesis. Knowledge of structural requirements for pigment binding proteins has also been acquired through mutants. Combinatorial cassette mutagenesis promises to produce additional biophysically interesting mutants. Since protein structure and function can not yet be accurately predicted from protein sequence, this method has the ability to generate phenotypes that would not be found through point mutations or global rearrangements. Libraries generated through combinatorial cassette mutagenesis can be rapidly screened by digital imaging spectroscopy, and mutants with interesting absorption spectra identified for further characterization. Currently, combinatorial cassette mutagenesis

* Author for correspondence.

R. E. Blankenship, M. T. Madigan and C. E. Bauer (eds): Anoxygenic Photosynthetic Bacteria, pp. 1257–1268.

experiments have focused on the reaction center and light harvesting II antennae. Specific amino acid residues in the reaction center were randomly substituted, and the library subjected to a photosynthetic selection prior to screening. Using such methodology, mutants can be isolated which occur with a frequency of approximately one-in-a-million. In the light harvesting system, strategies for more efficiently searching sequence space are being explored. Methods have been implemented which limit the amino acids coded for at each mutated position. These methods enhance for the desired phenotype over random mutagenesis. Goals for mutants generated via combinatorial methods include searching for reaction centers that perform wrong way electron transfer, and redesigning light harvesting antennae to perform charge separation.

I. Introduction

Light reactions of photosynthesis are mediated by membrane-bound pigment-protein complexes. Light is absorbed by the bacteriochlorophylls and carotenoids of light harvesting proteins which direct the energy to the photosynthetic reaction center (RC) where charge separation takes place. Extensive structural, spectroscopic, and genetic techniques exist for characterizing and manipulating the photosynthetic proteins of the purple nonsulfur bacteria, Rhodospirillaceae. The photosynthetic reaction center contains a bacteriochlorophyll (BChl) dimer, two BChl monomers, two bacteriopheophytin (Bphe) monomers, two quinones, and a ferrous non-heme iron atom attached to two quasi-symmetrically arranged protein subunits (L and M). The structure of the *Rhodopseudomonas viridis* RC has been determined through X-ray crystallography (Deisenhofer et al., 1985). Light harvesting I (LHI), the core antennae complex, is modeled to contain two transmembrane alpha helical polypeptides (α and β) which bind a BChl dimer. The peripheral antenna complex, light harvesting II (LHII) is also modeled to include two transmembrane alpha helical polypeptides (α and β) which bind a BChl dimer and a BChl monomer (Zuber, 1986). The bound pigments can be exploited as reporter groups; near infrared (NIR) and visible spectral information can be used to assay for protein assembly, structure, and function.

Molecular biological systems have been developed for the manipulation of both RC and LH proteins. Screening technologies and mutagenesis techniques have undergone a dramatic increase in sophistication with the development of digital imaging spectroscopy and combinatorial cassette mutagenesis (Youvan, 1994; Youvan et al., 1994). Historically many biophysically important mutants have been isolated through advances in screening and mutagenesis, and the newest techniques promise to yield more interesting and informative phenotypes.

Abbreviations: BChl – Bacteriochlorophyll; Bphe – Bacteriopheophytin; CCD – charge coupled device; CCM – combinatorial cassette mutagenesis; DIS – digital imaging spectroscopy; EEM – exponential ensemble mutagenesis; LH I – light harvesting I; LH II – light harvesting II; NIR – near infrared; NIR – near infrared; PG – group probability; *Rb.* – *Rhodobacter;* RC – reaction center; REM – recursive ensemble mutagenesis; *Rp.* – *Rhodopseudomonas* ; SDM – site-directed mutagenesis; SSD – sum-of-the-squares-of-the-differences; TSM – target set mutagenesis; WT – wild type

II. The Role of Mutagenesis in Photosynthesis Research

A. Spontaneous Mutations

Spontaneous mutants were used to isolate and/or confirm the identity of the genes for the photosynthetic apparatus. An R-factor with enhanced chromosomal mobilizing ability in *Rhodobacter capsulatus* was isolated and used to generate R-prime derivatives that carry a photosynthetic gene cluster (Marrs, 1981). R-prime plasmids carrying parts of the chromosome specifying the photosynthetic apparatus were confirmed by their ability to restore photosynthetic growth when mated to photosynthetically defective mutants. Mutants incapable of photosynthetic growth were isolated by subjecting them to a tetracycline suicide procedure. Mutants isolated from the tetracycline suicide were screened for enhanced fluorescence in the NIR; enhanced fluorescence mutants are typically defective in LHI or RC but have functional LHII (Youvan et al., 1983). An R-prime plasmid, pRPS404, complemented all the enhanced fluorescent mutants. The location of the genes for LHI and the RC were narrowed down further by complementation by pBR322 derivatives with smaller fragments of *Rb. capsulatus* DNA.

The LHII structural genes are not within the photosynthetic gene cluster, and were isolated using synthetic probes based on the amino-terminal sequences of α and β LHII which were used in Southern hybridizations (Youvan and Ismail, 1985). A 6 kb *Eco* RI fragment that hybridized to the probes was subcloned into pBR322 (to form pRPSLH2). This construction was able to complement the strain MW422, which has a chromosomal mutation that inactivates LHII.

B. Site-Directed Mutagenesis

Introducing specific alterations in the protein sequence has allowed biophysicists to study some of the fundamental mechanisms of the light reactions of photosynthesis. In the mid 1980s, a series of genomic deletion backgrounds and plasmids (Youvan et al., 1985; Bylina et al., 1986) facilitated the use of site-directed mutagenesis (SDM) to modify the photosynthetic proteins of *Rb. capsulatus*. Sequence modifications have been made in the vicinity of all the prosthetic groups in the RC, and at a variety of non-symmetric amino acid residues in the RC (see Coleman and Youvan, 1990 and references therein). Many of these mutations lead to changes in spectral properties or differences in the rate of electron transfer. Both LHI and LHII antennae have been altered by SDM in attempts to correlate amino acid sequence with protein structure and function. A few examples of the SDM experiments that have been performed on both RCs and LH antennae are summarized below.

The heterodimer mutants are examples of SDM experiments which resulted in significant advances in our understanding of the molecular mechanisms of photosynthesis. Replacement of either His ligand to the BChls in the special pair (His^{L173}, His^{M200}) with an aliphatic amino acid residue (Leu) leads to a BChl/Bphe heterodimer (Bylina and Youvan, 1988, 1990). The M-side heterodimer mutant has been extensively characterized and resulted in a greater understanding of the very early steps involved in charge separation. The overall quantum yield is 50% (as opposed to 100% for wild type (WT)); the decrease in quantum yield is attributed to unproductive decay of the excited dimer to the ground state (Bylina et al., 1989). Picosecond transient absorption spectroscopy (Kirmaier et al., 1988, 1989), low temperature ground state and linear dichroism absorption spectra (Breton et al., 1989), electron paramagnetic resonance (Bylina et al., 1990), and Stark spectroscopy (DiMango et al., 1990) suggest the existence of an intradimer charge transfer state that mixes with the excited singlet state. A charge transfer state has been postulated to exist in WT, but at a higher energy, making it harder to observe. The observation of the charge transfer state in the heterodimer mutant suggests that these states probably facilitate efficient charge separation in the RC.

In a similar type of pigment switching experiment, the Bphe acceptor was replaced with a BChl by substituting a Leu with a His which can act as a metal coordinating ligand for the Mg in BChl (Kirmaier et al., 1991). The resulting RCs from this Leu$^{M214}\rightarrow$His mutant, undergoes charge separation from the special pair through the BChl intermediate acceptor to the quinone, but with a reduced quantum yield of 60%. The electron transfer from the acceptor to the quinone is responsible for the decrease in quantum yield, as charge recombination is much faster in the mutant and competes with electron transfer to the quinone. The heterodimer and Leu$^{M214}\rightarrow$His mutants suggests that the WT RC achieves high quantum yield by avoiding situations where there is significant electronic coupling to the ground state, by limiting participation of states where charge is separated between strongly interacting chromophores (Kirmaier et al., 1991).

A strategy to determine which residues are responsible for the unidirectionality of electron transfer has been to symmetrize the RC by changing amino acid residues that differ between L and M. The first example of this type of experiment focused on Glu^{L104}, since this protonated amino acid is thought to hydrogen bond to the ring V C-9 keto group of the primary acceptor, which would be expected to lower the energy of the anion, thus facilitating unidirectionality. There is no analogous residue on the M side. The Glu$^{L104}\rightarrow$Leu mutant resulted in only minor changes in the kinetics of electron transfer. The initial electron transfer step is less than two fold slower in the mutant and the second step is only marginally slower. However Glu^{L104} was found to be responsible for the spectroscopic red shift of the active branch Bphe relative to the inactive branch Bphe (Bylina et al., 1988b).

To study BChl-peptide interactions in LH antennae, mutations were made in the putative BChl binding sequence, Ala-X-X-X-His (Theiler and Zuber, 1984; Theiler et al., 1984), of the α subunit in LHI from *Rb. capsulatus*. No LHI assembly, as judged by absorption spectroscopy, was detected when the His is changed to any other amino acid residue, suggesting that in LH, no other residue can function in place of His as a ligand to the Mg of the BChl. Mutagenesis of the Ala residue showed protein assembly occurred only when it is changed to an amino acid with a small molar volume (Gly, Ser, Cys) indicating that there are molar volume constraints for BChl binding at this position (Bylina et al., 1988).

In an investigation of residues at the N-terminus of the α and β subunits of LHI, charged amino acids were exchanged with those of opposite charge. It was believed that the positively charged α subunit N-terminal segment helped to direct the protein into the membrane and influenced its orientation, while the negative charges on the β subunit N-terminal were thought to help stabilize the complex by interacting with the α subunit. When four positively charged amino acid residues in the α subunit were exchanged with negatively charged amino acids no formation of the LH complex was observed. However, when four negative amino acids in the β subunit were changed to positive amino acids, LHI assembly was impaired but not completely blocked (Stiehle et al., 1990). This experiment shows that charged amino acids in α and β influence LHI formation in different ways, with β being much more tolerant of the mutations than the α subunit.

Extensive point mutations in LHI at amino acid positions which are highly conserved across species were performed to obtain information about their structural and functional role. Eighteen mutants with single amino acid substitutions were constructed. All the substitutions resulted in structural effects judged from characterization based on quantification of core complexes and the LHI spectral characteristics (Babst et al., 1991). For example, exchanges at $\alpha 8$ ($Trp^{\alpha 8} \rightarrow$ Leu, Ala, Tyr) resulted in the absence of core complex, the absence of core antennae and a reduction in the carotenoid content of the complex respectively. While substitutions at $\alpha 43$ ($Trp^{\alpha 43} \rightarrow$ Ala, Leu, Tyr) resulted in 8–11 nm blue shifts of the absorption peak.

SDM on charged and aromatic residues of LHII from *Rhodobacter sphaeroides* resulted in blue-shifts of the dimer band. A correlation was found between two Tyr residues in the α subunit of LHII and the dimer absorption band. By constructing the single ($Tyr^{\alpha 44} \rightarrow$ Phe) and double ($Tyr^{\alpha 44} \rightarrow$ Phe and $Tyr^{\alpha 45} \rightarrow$ Leu) mutant, the band shifted 11 and 24 nm respectively (Fowler et al., 1992). This effect is interpreted as being due to interactions of the BChls with the aromatic amino acids. Changing $Lys^{\beta 23} \rightarrow$ Gln resulted in an 18 nm blue shift of the dimer band (Fowler et al., 1993). This amino acid is near the putative BChl monomer binding site and is conserved in a number of LHII complexes. The monomer band was not affected in this mutant, and the interpretation given to the spectral shift of the dimer peak is that the relative orientation of the polypeptide backbone in the membrane might have been altered.

C. Structural Motif Rearrangements

Large scale rearrangement of RC sequences has been necessary to engineer phenotypes that could not be found by spontaneous or site-directed mutagenesis. Global changes have been made in both the RC and LH antennae.

Structurally important segments of the RC have been duplicated and exchanged between the L and M subunits. The D helices are good candidates for this type of manipulation as they have interactions with all the prosthetic groups. Mutants were constructed with two M subunit D helices (D_{MM}), two L subunit D helices (D_{LL}), and with the helices exchanged (D_{LM}) in the hope of identifying the asymmetric region responsible for the unidirectionality of electron transfer. The D_{MM} mutant resulted in a photosynthetically functional, but severely impaired RC. The D_{LL} mutant is photosynthetically inactive, and was found to be missing the primary electron acceptor. The D_{LL} RC has a total side chain molar volume greater than WT, which probably interferes with the Bphe binding pocket. Photosynthetically competent revertants of both D helix duplications have been isolated. All the D_{LL} revertants decrease the molar volume near the Bphe binding site, and all bind the Bphe. The positions at which reversions or secondary site compensatory mutations take place under photosynthetic selection point to what may be critical amino acid asymmetries (Robles et al., 1990a, 1990b). The extended lifetime of the excited dimer in the D_{LL} mutant allowed the observation of oscillations in the decay of this state which suggests that vibrational coherence may modulate electron transfer (Vos et al., 1991).

A second series of helix exchange and duplication experiments involved the amphipathic cd helixes which provide the axial ligands to the accessory BChls and form part of the special pair binding pocket. None of the constructions resulted in photosynthetically competent mutants; however, functional revertants were obtained from both helix duplications. Compensatory mutations were found in both the mutagenized and nonmutagenized subunits, and several pairs of symmetrical suppressors (compensatory mutations that occur at homologous positions in either subunit) were isolated (Robles et al., 1992). A speculative possibility is that the symmetrization of the RC caused the two chromophore branches to have the same potential for electron transfer, and that either branch could be activated by a secondary mutation. Further experiments will be needed to search for aberrant electron transfer in these mutants.

Structural motif mutagenesis has been used to investigate pigment binding and the criteria for proper helix formation in LHI antennae. Five amino acid residues in the α subunit of LHI (flanking the His thought to be the ligand to a BChl of the dimer) were changed to Leu. This resulted in a stretch of fourteen residues all of which are Leu except the Ala and the His in the Ala-X-X-X-His BChl binding motif. This 'poly Leu' mutant rapidly mutates to loose either carotenoid or LH expression (Arkin and Youvan, 1993). The α subunit of LHI seems to have only a limited tolerance to change in the vicinity of the BChl binding site.

III. Massively Parallel Screening of Mutants

A. Searching Protein Sequence Space

Combinatorial mutagenesis has a greater potential to generate novel phenotypes that can not be found using either SDM or structural motif mutagenesis, because of the inability to accurately predict the structure and function of proteins from their primary amino acid sequence. Combinatorial cassette mutagenesis (CCM) (Oliphant et al., 1986; Reidhar-Olson et al., 1991) allows the exploration of a large number of mutations in a protein. Multiple amino acid residues are simultaneously mutagenized in a random fashion resulting in a library of proteins, some of which may exhibit desired phenotypes. The complexity of these libraries grows exponentially with the number of sites mutagenized.

It is advantageous to reduce the total number of proteins in a combinatorial mutagenesis library to increase the efficiency of searching 'sequence space' for mutants with the desired phenotypes. Instead of using the codon NNN (i.e., N = 25% each A,T,G,C) at each mutagenized position, the nucleotide composition can be restricted. A simple example is NN(G/C), which is an alternate way to randomize a sequence position using all 20 amino acids but only half as many codons. In more sophisticated doping schemes, criteria such as physicochemical parameters, expert rules, structural, and phylogenetic data can be used to construct 'target sets' of amino acids to be used at each mutagenized sequence position (Arkin and Youvan, 1992b). One of several algorithms can be used to convert these target sets into nucleotide mixtures amenable to DNA synthesis (Youvan et al., 1992).

Two equations for adjusting nucleotide dopes have been used experimentally: 1) Group Probability P_G:

$$P_G = \prod_i P_D[i] \tag{1}$$

where $P_D[i]$ is the probability of the ith amino acid in a target set occurring based on a specific doping scheme; 2) Sum-of-the-Squares-of-the-Differences (SSD):

$$SSD = \sum_i (P_D[i] - P_T[i])^2 \tag{2}$$

where $P_T[i]$ is the fractional representation of the ith amino acid in the target set. The best dope is found by either maximizing P_G or minimizing SSD. The P_G function forces every amino acid in the target set, regardless of the frequency at which they occur, to be encoded by the cassette. In contrast, the SSD algorithm takes into consideration how many times each amino acid is found in the target set and may drop infrequently used amino acids from the dope. Computer simulations show SSD generates a higher throughput of positive mutants than P_G (Youvan et al., 1992), but this is potentially at the expense of phenotypic diversity.

B. Digital Imaging Spectroscopy

Too many mutants are generated by combinatorial mutagenesis to screen individual isolates by conventional spectroscopy. Digital imaging spectroscopy (DIS) provides a method for the parallel screening of the ground state (visible and NIR) spectra from up to five hundred colonies directly on a Petri dish (Yang and Youvan, 1988; Arkin et al., 1990; Arkin and Youvan, 1994; Youvan et al., 1994; Youvan 1994). Spectra acquired from DIS are often superior to spectra recorded from purified chromatophore membranes taken with a conventional spectrophotometer because DIS is less sensitive to light scattering in these turbid samples.

1. Instrumentation

All current configurations of DIS use a charged coupled device (CCD) camera as a detector to image Petri dishes mounted on the exit port of an integrating sphere. The sphere provides uniform light across the Petri dish(es). In a second generation imager, up to 25 Petri dishes can be illuminated simultaneously. The light source uses Fabry-Perot interference filters or an 1/8 meter monochromator to illuminate the target at different wavelengths. Typically, spectra can be obtained at 5–10 nm resolution which can detect 2 nm band shifts. Images are transferred to a Silicon Graphics Personal Iris computer where they are stored, manipulated, and compiled into absorption spectra. Commercial imaging spectrophotometers have been recently reviewed (Youvan, 1994).

Fluorescence images and images taken at a single wavelength are useful for rapid pre-screening of colonies. Fluorescence images can be obtained by illuminating with broad band blue-green light, and placing an 830 nm long pass filter in front of the CCD camera. Screening at a single wavelength is a fast method to select mutants with strong absorption characteristics at an important wavelength. For example, a library of light harvesting antennae can be scanned by absorbance at 860 nm before acquiring the full DIS spectra.

2. Spectral Display

Spectra acquired by DIS can be displayed in either 'tile mode' or as 'color contour maps'. In tile mode, each spectrum is displayed as a conventional two dimensional absorption spectrum. However, after a few hundred colonies have been imaged, this type of display is not optimum for rapid inspection of the data. In the color contour mode, all the spectral data from a single Petri dish are presented as a two dimensional display. The vertical axis corresponds to different colonies and the spectra of each colony is represented by a horizontal row. Absorption is color-coded at each wavelength along the row (white = high absorption, black = zero absorption). Spectra can be sorted according to similarity or absorption at various wavelengths. This type of display makes it easier to identify and compare spectral classes. In either mode, the spectra can be scaled relative to the lowest and highest absorption anywhere in the image, or each spectra can be scaled between its own minimum and maximum absorption (Arkin and Youvan, 1993; Youvan et al., 1994).

Images of Petri dishes screened by absorption at a single wavelength or by fluorescence can be displayed as either monochrome or pseudocolored images. Single wavelength images are divided by a blank image at the same wavelength to correct for uneven light intensity from the monochrome illumination. Gray scale values from the ratioed single wavelength absorption (or fluorescence) image are rescaled to enhance contrast. After establishing the low and high gray scale values, such monochrome images can be linearly mapped by pseudocoloring schemes to enhance differences between mutants.

IV. Combinatorial Cassette Mutants

A. Random Mutagenesis of the RC

Nine amino acid residues in the vicinity of the monomeric BChl on the active branch of the RC were randomly mutagenized using CCM. By manipulating the environment around the monomer BChl (and therefore its electronic states) it should be possible to affect the rate of electron transfer. Three libraries were constructed: one having five sites in the L subunit cd helix mutagenized (L146, L150, L153, L154, L157), another with four sites in the M subunit D helix mutagenized (M195, M201, M204, M205), and a third with all nine sites randomized simultaneously (Robles and Youvan, 1993). According to the X-ray structure of the *Rp. viridis* RC (Deisenhofer et al., 1985), all of these positions are in van der Waals contact with the active branch monomer BChl. Photosynthetically competent mutants were selected by growing the libraries under light: 1/3500 of the five site cd library, 1/15 of the four site D library and 1/50,000 of the nine site (combined) library were found to be functional. These frequencies show that criteria for functional residues mutagenized in the cd helix library are more stringent than in the D helix library. Plates having between 50 and 500 photosynthetically grown colonies were imaged for each library and representative colonies were chosen for sequencing. There is spectral diversity in the mutants constructed in this experiment; Fig. 1 shows the spectra from representative mutants from each library. In the most

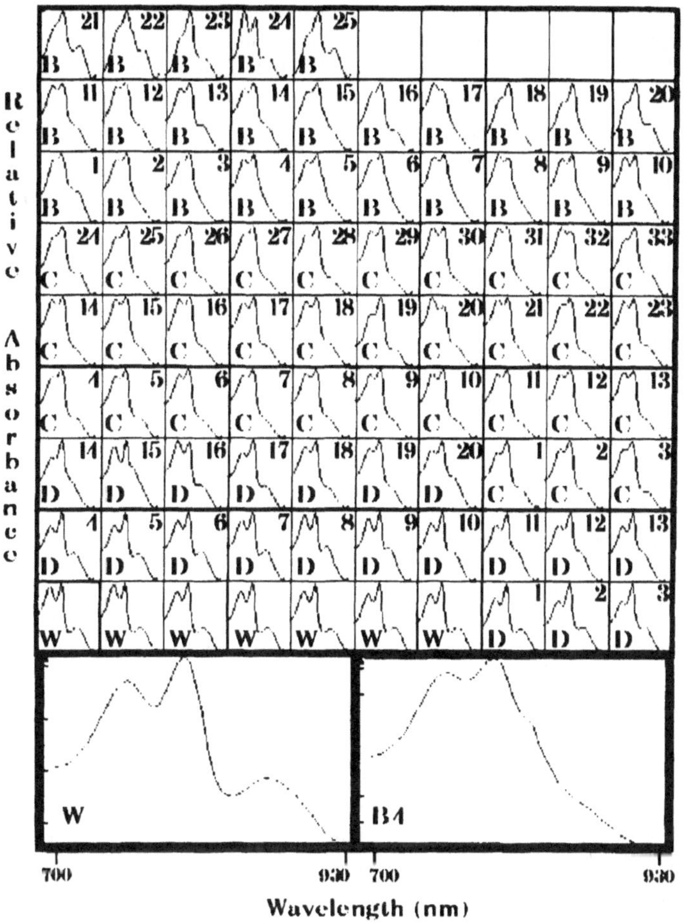

Fig. 1. Digital imaging spectrophotometer data (tile mode) showing the spectral diversity of combinatorial mutants of the photosynthetic reaction center from *Rb. capsulatus*. Boxes labeled D and C correspond to mutants from the four site D helix and five site cd helix libraries, respectively. Boxes labeled B were obtained from the nine site library in which both helixes were mutagenized simultaneously. The spectra of seven WT controls are labeled W. Spectra are shown from 700 to 930 nm from left to right. The spectra of a WT control and mutant B4 (with a blue shift of the dimer peak) are enlarged in the bottom panels.

striking spectral change, four unique mutants (out of four million selected and screened from the nine site library) have a well defined band at 825 nm that can be reversibly bleached with actinic light (Robles and Youvan, 1993).

B. Target Set Mutagenesis of LHII

1. Using the P_G Algorithm

Phylogenetic data from 29 homologous LH antennae (Zuber, 1990) were used as a database to construct target sets for seven amino acid positions on one face of a transmembrane alpha helix comprising a BChl binding site of the LHII antennae. Amino acid residues in the β subunit of LHII, one and two turns of the alpha helix away from the His modeled to be the ligand to a BChl of the dimer (positions −7, −4, −3, 0, 3, 4, 7 relative to the His) were mutagenized. Specific nucleotide mixtures were calculated by a computer program to maximize the probability of occurrence of amino acids in the target set according to the equation for P_G. To compare throughput of BChl binding mutants from target set mutagenesis (TSM) and random libraries, CCM libraries were constructed with the seven amino acid positions randomized, and with six of the positions (–7, –4, –3, 3, 4, 7) randomized (Goldman and Youvan, 1992).

DIS was used to screen 10,000 TSM mutants directly on Petri dishes. Three phenotypes were observed: 1) pseudo LHII with absorbance at 800 nm and 860 nm, 2) pseudo LHI with reduced 800 nm absorbance and maximal absorption near 875 nm, and 3) null mutants with absorption characteristics of membranes with free pigments. Approximately 6% of the library assembled pigment-protein complexes as judged from absorption spectra (i.e., 4% pseudo LHII; 2% pseudo LHI). Representatives from each phenotype were sequenced. It appears that the molar volume of the –7 position residue is important in determining which of the two BChl binding phenotypes is expressed. When an amino acid with a small molar volume is in this position, the pseudo LHII phenotype is favored, while the pseudo LHI phenotype tends to be associated with amino acids of larger molar volume.

The TSM library showed a 100–600 fold improvement in throughput versus the random (NN(G/C)) library. In a comparison to conventional CCM, only 3 positives were observed out of 10,000 colonies screened in the seven site random library (see Color Plate 10A,B). A significant amount of the difference in gain is probably due to the fact that the phylogenetic data restricts the 0 position of the TSM library to His. The probability of finding positive mutants in a fully random CCM library is greatly diminished when a critical sequence position accepts only a few amino acid substitutions. Because of this, TSM throughput will be highest (versus conventional CCM) when sequence positions are stringent. A six site random library was constructed in which the His was not randomized. This yielded the expected result of a 1% throughput because His is a critical sequence position. As a first approximation, the stringency of an amino acid position can be estimated from phylogenetic data.

2. Using the SSD Algorithm

The SSD equation was used to construct a phylogenetically based cassette to mutagenize seventeen amino acid residues (β27–β43) in the region of the binding site of the dimer BChl associated with the β subunit of LHII. The amino acid complexity of the resulting library is 7×10^7 as opposed to 3×10^{13} if the P_G algorithm had been used to construct the cassette. Extrapolating from the results of the seven site phylogenetically based P_G experiment in this region, a dope which used the P_G algorithm on seventeen sites would generate a 0.03% throughput of positives. DIS was used to screen 5000 colonies; the library showed an extremely high throughput (5–10%) of mutants expressing LH, with good phenotypic diversity (see Color Plate 10C). Thus, while SSD reduced the total number of possible protein sequences in the library by orders of magnitudes, different classes of BChl binding mutants were still observed.

3. Comparing SSD and P_G

Seventeen amino acid positions in the β subunit in the vicinity of the BChl monomer binding site were mutagenized simultaneously. SSD was used with phylogenetically based target sets except at position β21 where a dope was used that omitted the WT amino acid (Tyr) but still encoded the other two aromatic amino acids and other amino acids found in the core antenna phylogeny (Table 1). Ten thousand colonies were screened by DIS, and even with the inclusion of a non-WT position, 2% of the library was found to assemble LH antennae. Most of the 'positives' showed a WT like absorption spectrum in the NIR, albeit with lower optical density (see Color Plate 11). Sixteen of these pigment binding mutants were sequenced, and each one was found to have a unique sequence.

A second cassette was constructed (using the same seventeen sites) according to the equation for P_G. This library used a phylogenetically based target set at all of the mutagenized positions (Table 2). Out of 10,000 colonies screened no positive recombinants were found as judged by imaging spectroscopy. This result (in conjunction with the 2% throughput from the modified SSD library) indicates that SSD gives at least a 200 fold enhancement in the number of BChl binding proteins. When using TSM on a large number of sites (> seven), it is advantageous to use the SSD algorithm to construct cassettes because it increases the throughput of the library over P_G.

C. Recursive Ensemble Mutagenesis of LHII

Recursive ensemble mutagenesis (REM) (Arkin and

Table 1. SSD based nucleotide dopes for β monomer region

Site	Dope(SSD)	Amino Acid Residues
β8	(AT)C(GC)	ST
β9	GG(GC)	G
β10	(TC)T G	L
β11	A(GC)(GC)	ST(R)
β12	(GC)(ATC)(GC)	L*DEPAV(QH)
β13	(GAC)(AC)G	K*AEQ(PT)
β14	(GC)AG	EQ
β15	(GT)C(GC)	AS
β16	(GA)AG	EK
β17	GAG	E
β18	(ATC)TC	I*LF
β19	CAC	H
β20	(GAT)(AC)G	S*AEK(TX)
β21	**(GATC)(GT)(GC)**	**LIMVFW(RSCG)**
β22	(GAC)TC	L*FV
β23	(GA)(AT)(GC)	IKMV(DEN)
β24	(GA)(GATC)C	DSTV(AGIN)

Nucleotide dopes used for the SSD based mutagenesis of the BChl monomer binding site of LHII. Amino acids listed in parenthesis are not in the target set, but are unavoidably coded for in the dope, because of the structure of the genetic code. The asterisks indicates that the WT amino acid had to be entered into the algorithm more often than it occurred in the phylogeny to ensure that it was encoded in the dope. Target sets were based on phylogeny except at position β21 (in bold).

Youvan, 1992a; Youvan et al., 1992) uses sequence information acquired from random CCM to construct target sets for subsequent cassettes (Fig. 2). Six amino acid positions on the carboxyl-terminus of the β subunit of LHII were mutagenized using REM (Delagrave et al., 1993). Sites were randomized using NN(G/C) codons; mutants were pre-screened by fluorescence before evaluation by DIS. Only one out of 10,000 mutants screened had the desired fluorescence characteristics. Five positives recombinants were isolated and sequenced. This sequence data showed that the mutants were not merely trivial variations of the WT sequence. The sequences did not recapitulate the known phylogeny; one mutant showed an inversion in a completely conserved sequence motif that resulted in a 10 nm blue shift of the dimer peak. The next iteration of REM cassettes were designed using the P_G algorithm. REM returned a thirty-fold increase in the number of positive mutants over random CCM. Twelve mutants were sequenced and all were found to have unique peptide sequences that differed from both WT and the set of mutants used to generate the first target sets (Delagrave et al., 1993). This is a useful method when there is not an extensive phylogenetic data base. When mutagenizing larger stretches of amino

Table 2. P_G based nucleotide dopes for β monomer region

Site	Dope(PG)	Amino Acid Residues
β8	(GA)(GAC)(GC)	STEAKNR(DG)
β9	(GAC)(GACT)C	GNPTV(ADHILRS)
β10	(GAC)TC	LVI
β11	(TA)(CT)C	STF(I)
β12	(TACT)(ACT)C	LAVIEDPNS(FHKMQSYX)
β13	(GAC)(GAC)(GC)	KAEQGNS(HPRST)
β14	(GC)AG	EQ
β15	(GT)(GAC)C	ASCD(GY)
β16	(GATC)(GA)G	EKWQR(GX)
β17	(GA)A(GC)	EN(DK)
β18	(GATC)(TC)C	ILFVA(LPST)
β19	CAC	H
β20	(GA)(GAC)(GC)	SKEAGP(NRT)
β21	(GATC)(AT)(GC)	YQHIMVL(DEFKNX)
β22	(GATC)(AT)C	LVFNY(DHIL)
β23	(GA)(ATC)(GC)	IKVTM(ADEN)
β24	(GATC)(ATC)(GC)	DVLKQST(AEFHIMNPYX)

Nucleotide dopes used for the phylogenetic P_G based experiment in the vicinity of the BChl monomer binding site of LHII. This algorithm forces each amino acid in the target set to be coded for in the dope regardless of its frequency. Amino acid residues in parenthesis are not in the target set, but were unavoidably encoded because of the genetic code.

acids, short segments can be randomized independently and these sequences can then be combined. This defines exponential ensemble mutagenesis (EEM) (Delagrave and Youvan, 1993).

D. Exponential Ensemble Mutagenesis of LHII

An efficient method for generating combinatorial libraries with a high percentage of unique and functional mutants has been developed in recent studies. Exponential ensemble mutagenesis (EEM) should be advantageous in cases where many residues must be changed simultaneously to achieve a specific engineering (Delagrave and Youvan, 1993). With the enhanced functional mutant frequencies obtained by this method, entire proteins could be mutagenized combinatorially.

Combinatorial libraries have been successfully used in the past to express ensembles of mutant proteins in which all possible amino acids are encoded at a few positions in the sequence. However, as more positions are mutagenized the proportion of functional mutants is expected to decrease exponentially. Small groups of residues were randomized in parallel to identify, at each altered position, amino acids which lead to functional proteins. By using optimized nucleotide mixtures deduced from the sequences selected from the random libraries, we have

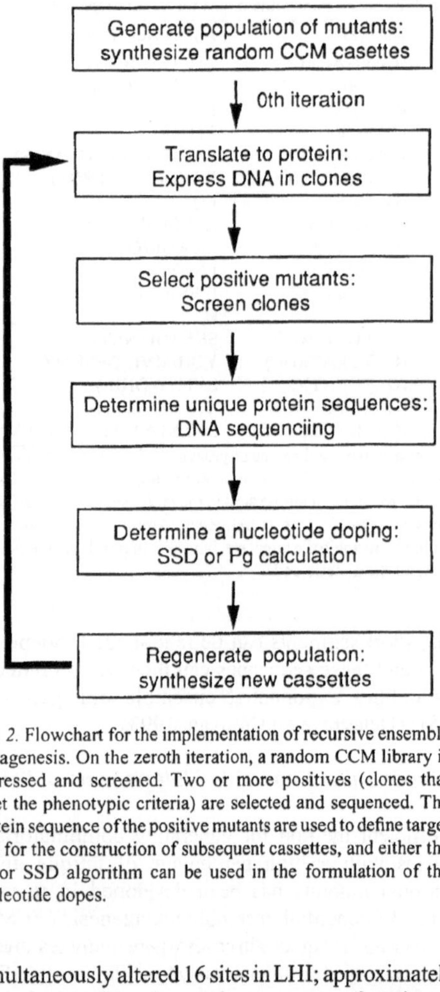

Fig. 2. Flowchart for the implementation of recursive ensemble mutagenesis. On the zeroth iteration, a random CCM library is expressed and screened. Two or more positives (clones that meet the phenotypic criteria) are selected and sequenced. The protein sequence of the positive mutants are used to define target sets for the construction of subsequent cassettes, and either the P_G or SSD algorithm can be used in the formulation of the nucleotide dopes.

simultaneously altered 16 sites in LHI; approximately one percent of the observed mutants were functional. Mathematical formalization and extrapolation of our experimental data suggests that a 10^7-fold increase in the throughput of functional mutants has been obtained relative to the expected frequency from a random combinatorial library.

Recently, the comparison of two different combinatorial mutagenesis experiments on the LHII protein of *Rhodobacter capsulatus* indicates that heuristic rules relating sequence directly to phenotype are dependent on which sets, or groups, of residues are mutagenized simultaneously (Delagrave et al, 1995). Previously reported combinatorial mutagenesis of this chromogenic protein (based on both phylogenetic and structural models) showed that substituting amino acids with large molar volumes at $Gly^{\beta31}$ caused the mutagenized protein to have a spectrum characteristic of LHI (Goldman et al, 1994). The six residues that underwent combinatorial mutagenesis were modeled to lie on one side of a transmembrane α-helix that binds bacteriochlorophyll. In a second CCM experiment, a set of six contiguous residues was selected for combinatorial mutagenesis. In this latter experiment, the residue substituted at $Gly^{\beta31}$ was not a determining factor in whether LHII or LHI spectra were obtained; therefore, we conclude that the heuristic rules for phenotype prediction are context dependent. While phenotype prediction is context dependent, the ability to identify elements of primary structure causing phenotype diversity appears not to be. This strengthens the argument for performing combinatorial mutagenesis with an arbitrary grouping of residues if structural models are unavailable.

V. Conclusions and Prospectus

We reviewed RC and LH mutants in *Rb. capsulatus* and *Rb. sphaeroides* that have been produced through increasingly sophisticated mutagenesis techniques. At present, combinatorial cassette mutagenesis of RC and LH antennae has only been accomplished in *Rb. capsulatus*. Although there is currently no crystal structure for the *Rb. capsulatus* RC, a chemoheterotrophic *Rp. viridis* genetic system is being developed in the Bylina laboratory (E. J. Bylina, personal communication). This could lead to a potentially useful combination of experiments in these two species: Sequence space can be screened by combinatorial mutagenesis in *Rb. capsulatus*, and when single interesting mutants are found they can be reconstructed in the *Rp. viridis* strain for crystallography.

Future goals include using cassette mutagenesis to engineer a RC mutant that performs wrong-way electron transfer. Instead of randomly mutagenizing sections of the RC, regions of the protein in contact with the pigments could be mutagenized using TSM. Phylogenetically based target sets are being constructed using the amino acid residues at homologous sites in both the L and M subunits, which are thought to be the product of a gene duplication (Youvan and Marrs, 1984).

A second goal is to alter LH antennae to perform charge separation. Combinatorial cassette mutagenesis could be used to change the protein environment (i.e., the electronic environment) around the pigments. An antenna that was modified to perform charge separation would allow biophysicists to study electron transfer in a much simpler protein system than the RC.

LHII is an excellent model system in which to attempt to establish a correlation between sequence and phenotype. Computer programs can be used to correlate sequence data with phenotypic information to generate phenotypic estimators. This obliterates the need for a solution to the protein folding problem, and facilitates various engineering projects on pigment-protein complexes.

Mutant RC and LH complexes have helped biophysicists better understand the events in photosynthesis. In the future, increasingly complex mutagenesis techniques should contribute still more to our understanding of the molecular mechanisms of photosynthesis and more general problems in protein folding and design. Data already tabulated into this system are being used to train neural nets for 'phenotypic estimation'.

Acknowledgments

This work was supported in part by NIH GM42645, DOE DE-FG02-90ER20019, DOE DE-FG05-91ER79031, and by a grant from the Human Frontiers Science Program. ERG received an NIH Biotechnology Training Grant Award. We thank Dr. Mary Yang at KAIROS Scientifiic Inc. (Mountain View, CA) for her help with the imaging systems and for providing the computer programs to calculate nucleotide dopes, and Matthias Scholl for his assistance in the laboratory.

References

Arkin AP, and Youvan DC (1992a). A combinatorial optimization procedure for protein engineering: Simulation of recursive ensemble mutagenesis. Proc Natl Acad Sci USA 89: 7811–7815

Arkin AP, and Youvan DC (1992b) Optimizing nucleotide mixtures to encode specific subsets of amino acids for semi-random mutagenesis. Bio/Technology 10: 297–300

Arkin AP, and Youvan DC (1993) Digital imaging spectroscopy. In Deisenhofer H and Norris JR. (eds) The Photosynthetic Reaction Center, Vol. 1 pp 133–155 Academic Press, New York

Arkin A, Goldman E, Robles S, Coleman W, Goddard C, Yang M, and Youvan DC (1990). Applications of imaging spectroscopy in molecular biology: Colony screening based on absorption spectra. Bio/Technology 8: 746–749

Babst M, Albrecht H, Wegmann I, Brunisholz R, and Zuber H (1991). Single amino acid substitutions in the B870 α and β light-harvesting polypeptides of *Rhodobacter capsulatus*. Eur J Biochem 202: 277–284

Breton J, Bylina EJ, and Youvan DC (1989). Pigment orientation in genetically modified reaction centers of *Rhodobacter capsulatus*. Biochemistry 28: 6423–6430

Bylina EJ, and Youvan DC (1988). Directed mutations affecting spectroscopic and electron transfer properties of the primary donor in the photosynthetic reaction center. Proc Natl Acad Sci USA 85: 7226–7230

Bylina E J, and Youvan DC (1989). Mutagenesis of reaction center histidine L173 yields an L-side heterodimer. In Baltscheffsky M (ed) Current Research in Photosynthesis, pp 53–59. Kluwer Academic Publishers, Dordrecht

Bylina EJ, Ismail S, and Youvan DC (1986). Plasmid pU29, a vehicle for mutagenesis of the photosynthetic *puf* operon in *Rhodopseudomonas capsulatus*. Plasmid 16: 175–181

Bylina EJ, Robles S, and Youvan DC (1988a). Directed mutations affecting the putative bacteriochlorophyll-binding sites in the light-harvesting antenna of *Rhodobacter capsulatus*. Israel J Chem 28: 73–78

Bylina EJ, Kirmaier C, McDowell L, Holten D, and Youvan DC (1988b). Influence of an amino acid residue on the optical properties and electron transfer dynamics of a photosynthetic reaction center complex. Nature 336: 182–184

Bylina EJ, Kolaczkowski SV, Norris JR, and Youvan DC (1990). EPR characterization of genetically modified reaction centers of *Rhodobacter capsulatus*. Biochemistry 29: 6203–6210

Coleman WJ, and Youvan DC (1990). Spectroscopic analysis of genetically modified photosynthetic reaction centers. Ann Rev Biophys Chem 19: 333–367

Coleman W and Youvan DC (1993) Atavistic reaction centre. Nature 366: 517–518

Deisenhofer J, Epp O, Miki K, Huber R, and Michel H (1985). Structure of the protein subunits in the photosynthetic reaction center of *Rhodopseudomonas viridis* at 3Å resolution. Nature 318: 618–624

Delagrave S, and Youvan DC (1993). Searching sequence space to engineer proteins: Exponential ensemble mutagenesis. Bio/Technology 11: 1548–1552

Delagrave S, Goldman ER, and Youvan DC (1992). Recursive Ensemble Mutagenesis. Protein Eng 6: 327–331

Delagrave S, Goldman E and Youvan DC (1995) Context dependence of phenotype prediction and diversity in combinatorial mutagenesis. Protein Engineering, in press

DiMango TJ, Bylina EJ Angerhofer A, Youvan DC, and Noris JR (1990) The stark effect in wild-type and heterodimer reaction centers from *Rhodobacter capsulatus*. Biochemistry 29: 6201–6210

Fowler GJS, Visschers RW, Grief GG, van Grondelle R, and Hunter CN (1992). Genetically modified photosynthetic antenna complex with blue shifted absorption bands. Nature 355: 848–850

Fowler GJS, Crielaard W, Visschers RW, van Grondelle R, and Hunter CN (1993). Site-directed mutagenesis of the LH2 light-harvesting complexes of *Rhodobacter sphaeroides*: Changing βlys23 to gln results in a shift in the 850 nm absorption peak.

Photochem and Photobiol 57: 2–5

Goldman ER, and Youvan DC (1992). An algorithmically optimized combinatorial library screened by digital imaging spectroscopy. Bio/Technology 10: 1557–1561

Goldman E, Fuellen G and Youvan DC (1994) Estimation of protein function from combinatorial mutagenesis by decision algorithms and neural networks. Drug Dev Res 33: 125–132

Kirmaier C, Holton D, Bylina EJ, and Youvan DC (1988). Electron transfer in a genetically modified reaction center containing a heterodimer. Proc Natl Acad Sci USA 85: 7562–7566

Kirmaier C, Bylina EJ, Youvan DC, and Holten D (1989). Subpicosecond formation of the intradimer charge transfer state $[BChl_{LP}^{+}BPh_{MP}^{-}]$ in reaction centers from the $His^{M200} \rightarrow$Leu mutant of Rhodobacter capsulatus. Chem Phys Lett 30: 609–613

Kirmaier C, Gaul D, Debey R, Holten D, and Schenck C (1991). Charge separation in a reaction center incorporating bacteriochlorophyll for photoactive bacteriopheophytin. Science 251: 922–927

Marrs B (1981). Mobilization of the genes for photosynthesis from *Rhodopseudomonas capsulata* by a promiscuous plasmid. J Bacteriol 146: 1003–1012

Oliphant AR, Nussbaum AL, and Struhl K (1986). Cloning of random-sequence oligodeoxynucleotides. Gene 44: 177–183

Reidhaar-Olson JF, Bowie JU, Breyer RM, Hu JC, Knight KL, Lim WA, Mossing MC, Parsell DA, Shoemaker KR, and Sauer RT (1991). Random mutagenesis of protein sequences using oligonucleotide cassettes. Meth Enzym 208: 564–587

Robles SJ, and Youvan DC (1993). Hydropathy and molar volume constraints on combinatorial mutants of the photosynthetic reaction center. J Mol Biol 232: 242–252

Robles SJ, Breton J, and Youvan DC (1990a). Transmembrane helix exchange between quasi-symmetric subunits of the photosynthetic reaction center. In Michel-Beyerle M-E (ed)Reaction Centers of Photosynthetic Bacteria pp 283-291 Springer Verlag, Berlin Heidelberg

Robles SJ, Breton J, and Youvan DC (1990b). Partial symmetrization of the photosynthetic reaction center. Science 248: 1402–1405

Robles SJ, Ranck T, and Youvan DC (1992). Symmetrical intragenic suppressors of the bacterial reaction center cd-helix exchange mutants. In Breton J and Vermeglio A (eds) Structure of the bacterial photosynthetic reaction center (II). Plenum press, New York

Stiehle H, Cortez N, Klug G, and Drews G (1990). A negatively charged N terminus in the α polypeptide inhibits formation of light-harvesting complex in *Rhodobacter capsulatus*. Eur J Biochem 202: 277–284

Theiler R, and Zuber H (1984)The light-harvesting polypeptides of *Rp. sphaeroides* R-26.1: II. Conformational analysis by attenuated total reflection infrared spectroscopy and the possible molecular structure of the hydrophobic domain of the B850 complex. Hoppe-Seyler's Z Physiol Chem 365: 721–729

Theiler R, Suter F, Wiemken V, and Zuber H (1984). The light-harvesting polypeptides of *Rp. sphaeroides* R-26.1: I. Isolation, purification and sequence analyses. Hoppe-Seyler's Z Physiol Chem 365: 703–719

Vos MH, Lambry J-C, Robles SJ, Youvan DC, Breton J, and Martin J-L (1991). Direct observation of vibrational coherence in bacterial reaction centers using femtosecond absorption spectroscopy. Proc Natl Acad Sci USA 88: 8885–8889

Yang MM, and Youvan DC (1988). Applications of imaging spectroscopy in molecular biology. I. Screening photosynthetic bacteria. Bio/Technology 8: 746–749

Youvan DC (1994) Imaging sequence space. Nature 369: 79–80

Youvan DC, and Ismail S (1985). Light-harvesting II (B800–B850 complex) structural genes from Rhodopseudomonas capsulata. Proc Natl Acad Sci USA 82: 58–62

Youvan DC, and Mars BL (1984). Molecular genetics and the light reactions of photosynthesis Cell 39: 1–3

Youvan DC, Hearst JE, and Marrs BL (1983). Isolation and characterization of enhanced fluorescence mutants of *Rhodopseudomonas capsulata*. J Bacteriol 154: 748–755

Youvan DC, Ismail S, and Bylina EJ (1985). Chromosomal deletion and plasmid complementation of the photosynthetic reaction center and light harvesting genes from *Rhodopseudomonas capsulatus*. Gene 38: 19–30

Youvan DC, Arkin AP, and Yang MM (1992). Recursive ensemble mutagenesis: A combinatorial optimization technique for protein engineering. In: Manverik B (ed) Parallel problem solving from Nature, 2 pp 401–410 Elsevier publishing Co. Amsterdam

Youvan DC, Goldman E, Delagrave S, and Yang MM (1994). Digital imaging spectroscopy for massively parallel screening of mutants. Meth Enzym 246: 732–748

Zuber H (1986). Structure of light harvesting antenna complexes of photosynthetic bacteria, cyanobacteria, and red algae. TIBS 11: 414–419

Zuber H (1990). Consideration on the structural principles of the antenna complexes of phototrophic bacteria. In: Drews G and Dawes EA (eds) Molecular biology of membrane-bound complexes in phototrophic bacteria. pp 161–180 Plenum press, New York

Chapter 62

Waste Remediation and Treatment Using Anoxygenic Phototrophic Bacteria

Michiharu Kobayashi
The International Research Institute for Applied Biology, Kyoto 606, Japan

Michihiko Kobayashi
Department of Agricultural Chemistry, Kyoto University, Kyoto 606, Japan

Summary

Based on ecological studies of anoxygenic phototrophic bacteria as to their presence and their role in the natural world, we have developed this knowledge to the level of an applied study connected with their practical utilization. After explaining these processes, Methods of biological treatment and applications related to bioremediation of various organic wastes are described.

I. Distribution and Interrelationship with Other Organisms

Anoxygenic phototrophic bacteria are widely distributed in aquatic and some terrestrial habitats in nature (Table 1)(Kobayashi et al., 1967). Phototrophic bacteria grow favorably in polluted environments created from human activities (Fig. 1)(Kobayashi et al., 1978). Cultures of these bacteria are also used to feed animal plankton. In general, phototrophic bacteria are, as shown in Fig. 1, capable of assimilating both carbon dioxide and molecular nitrogen (nitrogen fixation), using light as an energy source.

Cultures of phototrophic bacterial cells are used as feed for small organisms in water and soil, and the excretions of phototrophic bacteria are used by other organisms including heterotrophic bacteria and algae. For example, in paddy soils containing sulfur, as shown in Fig. 2, sulfate reducing bacteria grow

R. E. Blankenship, M. T. Madigan and C. E. Bauer (eds): Anoxygenic Photosynthetic Bacteria, pp. 1269–1282.

Table 1. Cell counts of anoxygenic phototrophic bacteria in various samples

		(number per g)
Ditch	(B.O.D. 250 ppm)[a]	$10^6 - 10^7$
Lake	(B.O.D. 10 ppm)	$10^2 - 10^3$
River	(B.O.D. 1.0 ppm)	$+ - 10$
Aeration tank (activated sludge method)	(B.O.D. 150 ppm)	$10^6 - 10^7$
Soil of paddy field		$10^5 - 10^6$
Soil of seaside		$10^3 - 10^4$

[a] B.O.D – biological oxygen demand

vigorously yielding hydrogen sulfide in the reproductive growth period of the rice plant (the young ear formation period: early in August); this leads to lowering the oxidation-reduction potential at the rhizosphere of the rice plant which can cause enormous damage to their roots. At the appropriate time, however, phototrophic bacteria grow and oxidize the toxic hydrogen sulfide, yielding favorable effects on the rice plant (see Fig. 3). Figure 4 shows the cycling of sulfur in the paddy field. It has been known that some useful substances excreted from phototrophic bacteria give favorable effects to the plants, including an increase in the number of grains per rice plant. Such useful substances were studied in detail, which resulted in our discovery of a substance accelerating formation of flower buds as well as the setting and thickening of fruits; these have been developed to the practical level, contributing much to improved agricultural productivity (Kobayashi et al., 1966) (high productivity of rice will be covered later, see Table 18).

II. Growth Acceleration by Symbiosis

A phenomenon was found that when *Rhodobacter capsulatus*, a nonsulfur purple bacterium, coexists with aerobic heterotrophs, the growth of both organisms is markedly accelerated (Okuda and Kobayashi 1961, 1963).

Figure 5 shows growth in co-culture of *Rb. capsulatus* and *Bacillus megaterium*. In an environment limited in combined nitrogen, coexistence of these bacteria accelerated their growth more than that in obtained in axenic culture. *Rb. capsulatus cannot* use glycerol although it is able to fix nitrogen, but nitrogen fixation is feasible by *Azotobacter* making use of glycerol. *Bacillus megaterium* and *B. subtilis* are able to use glycerol, but are unable to fix nitrogen. However, in various mixed cultures of these organisms, nitrogen fixation and carbon

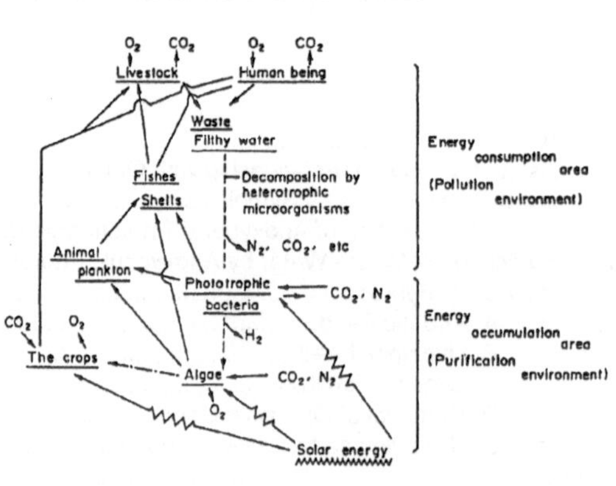

Fig. 1. Diagram showing roles of phototrophic bacteria in natural environments.

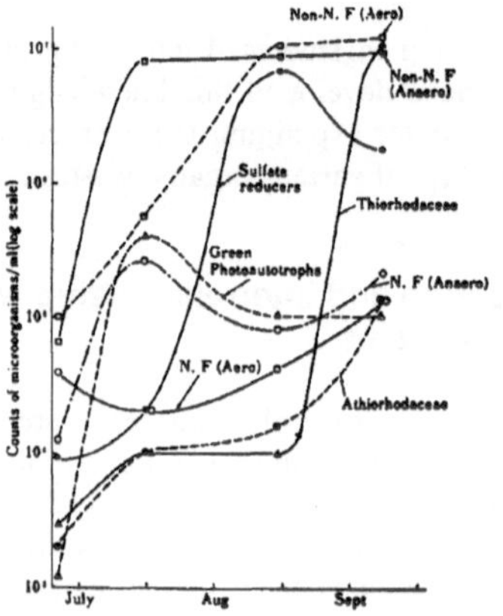

Fig. 2. Seasonal changes of microbial counts in the rhizosphere of rice plants. Non-N.F. – Non nitrogen-fixing microorganisms; N.F. – Nitrogen-fixing microorganisms; Aero – Aerobes; Anaero – Anaerobes; Thiorhodaceae – Chromatiaceae; Athiorhodaceae – Nonsulfur purple bacteria

Abbreviations: B. – Bacillus; PTB – phototrophic bacteria; *Rb. – Rhodobacter*

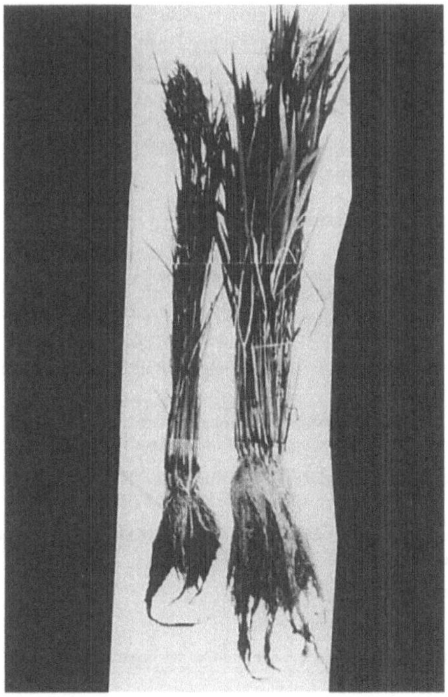

Fig. 3. Rice plant growth in early August. Left: Affected by sulfate-reducing bacteria with a custom cultivation. Right: Affected by phototrophic bacteria with an early cultivation.

Table 2. Nitrogen fixation in pure and mixed culture of *Rhodobacter capsulatus* with *Azotobacter agilis*, *Bacillus subtilis* or *Bacillus megaterium* (in shaking culture under aerobic conditions with illumination)

	Fixed nitrogen:N mg/100 ml*
Rhodobacter capsulatus	0
Bacillus megaterium	0
Azotobacter agilis	2.69
Bacillus subtilis	0
Rb. cap.-B. mega.-Mix	4.41
Rb. cap.-Azotobacter-Mix	4.95
Rb. cap.-B. sub.-Mix	0.54

*0.3% glycerol medium; nitrogen free.

dioxide assimilation are markedly accelerated (Fig. 5 and Tables 2, 3).

III. Ecological Variation of Anoxygenic Phototrophic Bacteria in Organic Sewage

When rice straw and a small amount of rice field soil are placed in a glass bottle under waterlogged

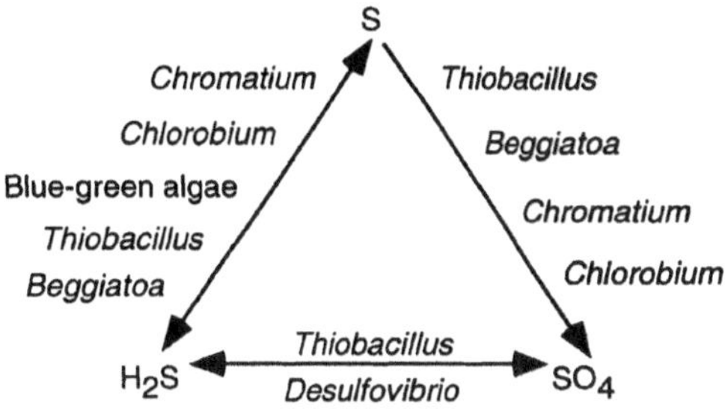

Fig. 4. Sulfur cycle by microbial action in paddy field.

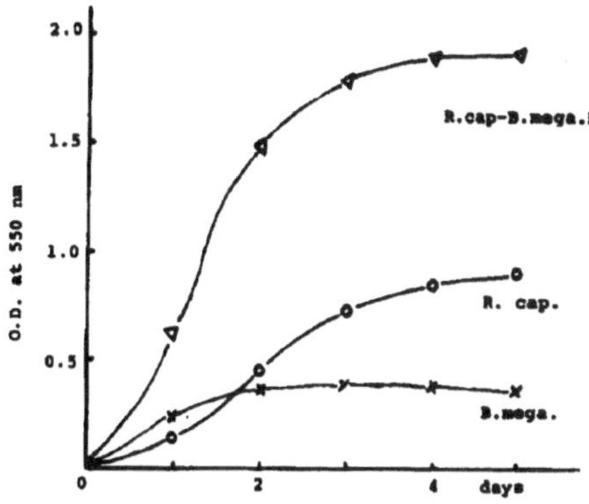

Fig. 5. Growth in pure culture and in mixed culture of *Rhodobacter capsulatus* with *Bacillus megaterium* (nitrogen deficient, 0.3% glucose medium).

Table 3. Assimilation of carbon dioxide by *Rhodobacter capsulatus* and *Azotobacter agilis* or *Rb. capsulatus* and *Bacillus megaterium* in pure and mixed culture

	Specific activity c.p.m./mg protein	Activity ratio
Azotobacter or *B. megaterium*	trace	trace
Rb. capsulatus	0.65×10^2	1.0
Rb. cap.-Azotobacter-Mix	11.88×10^2	18.4
Rb. cap.-B. megaterium-Mix	8.34×10^2	12.9

Exposed for 1 hour at 27 °C under light.
0.1 millicurie $^{14}CO_2$ was added. (Carrier CO_2 free.)

conditions for a month (25–30 °C), gas-bubbles are produced due to the abundant growth of heterotrophs accompanied by decomposition and foul odor. With the decrease of the substrate and accumulation of the decomposed products, the growth of the heterotrophs becomes poorer, and in such conditions the growth of phototrophic bacteria becomes rapid as shown in

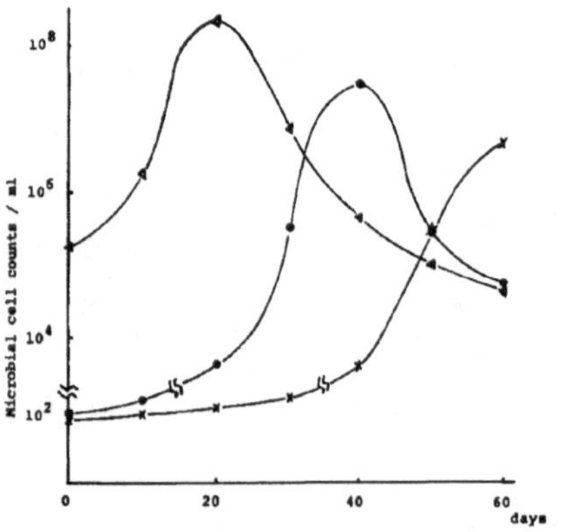

Fig. 6. Microbial changes during the decomposing process of organic materials under submerged natural conditions.
-△-△- Heterotrophic microorganisms
-●-●- Phototrophic bacteria
-x-x- Algae

Fig. 6. In two or three weeks, phototrophic bacteria gradually stop growing and green algae appear. Examination of the change in the biochemical oxygen demand (BOD) during this process showed that the degree of putrefaction was high (greater than ten thousand ppm BOD) when the heterotrophs were growing but the BOD decreased to only several hundred ppm when phototrophic bacteria were growing. This indicates that purification of the medium progressed by the removal of soluble organic compounds, and with the appearance of green algae (and subsequent O_2 production), the purification could be accelerated even further (see Table 4) (Kobayashi and Haque 1971).

IV. Purification of Waste Water by Anoxygenic Phototrophic Bacteria

It was found that phototrophic bacteria play a major role in the natural purifying process of various kinds of waste water containing high organic concentrations (see Tables 5 and 6). Based on the theory of purification, various experiments and pilot plant tests were carried out. The commercialization was made about ten years ago (in Japan) and many purification treatment plants are being operated at present. A flow sheet (describing this process) is illustrated in Fig. 7, and some data on purification in Tables 7 and 8 (Kobayashi 1970, Kobayashi and Tchan 1973). The advantages of this system over the activated sludge

Table 4. Purification of polluted water, including organic materials (animal excrement), under natural biological processes[a]

	Ammonia ppm	B.O.D. ppm
Heterotrophs growing stage	over 5,000	over 5,000
After growth of anoxygenic phototrophic bacteria	200–500	200–600
Photoautotrophs (green algae) growing stage	10–50	10–60

[a] Values represent concentrations present following growth of the microorganisms listed.

Table 5. Source of waste materials purified by phototrophic bacteria

Various kinds of microbial industry wastes (beer, antibiotics, amino acids, nucleic acids, fermentations, etc.)
Various kinds of chemical synthesizing industry wastes (synthetic fibers, synthetic resins, chemical fertilizers, chemicals, etc.)
Various kinds of food industry wastes (canned food, bottled food, cakes, miso, tofu: bean cake, etc.)
Petroleum industry
Starch and woolen industry
Others: activated sludge, excrement, other organic materials

process are: a) dilution is not required so that a photosynthetic bacterial system can be used in areas where water is a scarce resource, b) unlike the activated sludge method which creates a secondary sludge disposal problem, this system produces phototrophic bacteria and green algae which are excellent and useful byproducts, c) because very little dilution is required, the space required for treatment facilities is much less in comparison with the activated sludge method.

V. The Use of Byproducts

A. Aquatic Feed

In outdoor experimental tanks, a rapid development of animal plankton was noted after increased growth of anoxygenic phototrophic bacteria; therefore, a series of detailed experiments using ^{14}C-labelled phototrophic bacteria was carried out to study the feeding process. As a result, it was confirmed that animal plankton are direct predators of phototrophic bacteria and, upon addition of phototrophic bacteria, growth of zooplankton is stimulated much more than when green algae serve as food source (Fig. 8).

Table 6. Treatment of organic waste solution; BOD values at different stages of purification and yield of microbial cells

	BOD values in ppm				
Type of waste solution	Original waste solution	After treatment in Tank 2	After treatment in Tank 3	Yield of phototrophic bacteria cells, etc[a] (g l^{-1})	Yield of algae cells, etc[a] (g l^{-1})
Starch industry	>10, 000	600–1000	50–100	1–5.0	0.2–1.0
Wool washing factory	5000–10, 000	200–500	50–100	1–3.0	0.2–1.0
Canned food factory (fish meat)	2000–5000	100–400	20–50	0.5–1.0	0.1–0.5
Pharmaceutical fermentation industry (penicillin, erythromycin)	2000–5000	500–800	20–80	0.8–3.0	0.3–1.0

[a] Including other heterotrophs

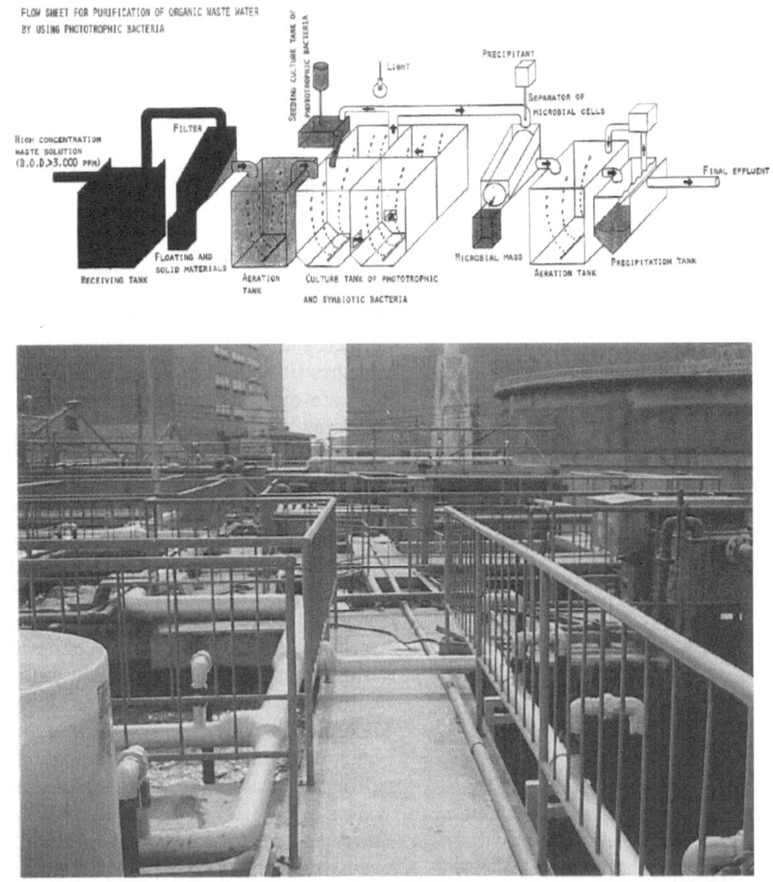

Fig. 7. (a) Flow sheet for purification of organic waste water by using anoxygenic phototrophic bacteria. (b) Example of practical plant.

Table 7. Example of purification of waste solution from a swinery

Property	Units	Original	Supernatant in precipitation tank after PTB[a] treatment	Purified effluent
BOD	(ppm)	6600	380	15
COD[b]	(ppm)	3364	354	64
SS[c]	(ppm)	6540	450	17
Kjeldahl nitrogen (as NH_4^+)	(ppm)	915	32.8	7.8
pH		6.8	7.3	7.1

[a] PTB – Phototrophic bacteria
[b] COD – Chemical oxygen demand
[c] SS – Suspended solids

Table 8. Example of purification of waste solution from bean cake factory

Property	Units	Original	Supernatant in precipitation tank after PTB[a] treatment	Discharged water
BOD	(ppm)	11300	340	15
COD	(ppm)	9800	270	17
SS	(ppm)	3930	23	5
Kjeldahl nitrogen (as NH_4^+)	(ppm)	3850	280	11
pH	6.4	7.8	7.2	

[a] PTB: Phototrophic bacteria

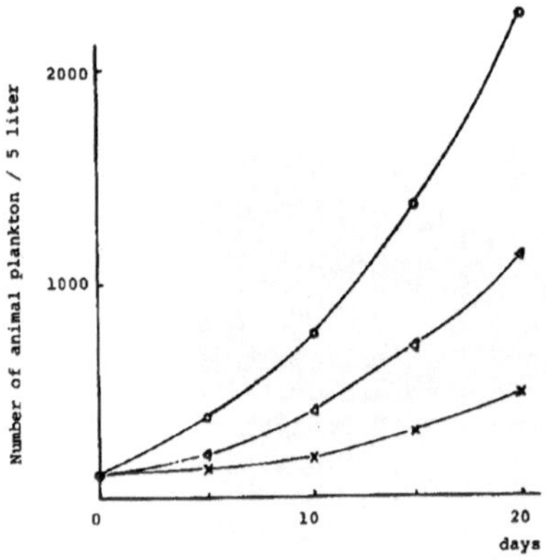

Fig. 8. Comparison of animal plankton growth by feeding with various microbial cells.
-O-O- Phototrophic bacteria
-Δ-Δ- Yeast
-×-×- *Chlorella*

In particular it was found that anoxygenic phototrophic bacteria are very useful for growth of brine shrimp, one of the most important salt water plankton, which is used as feed for fish and shellfish, and for collection of their eggs. Thus, a favorable outlook for completely artificial fish culture, which thus far was considered infeasible, has now been formed. Further, the fact was also identified that fry of fishes (loach, goldfish, carp, ark shell, sweetfish, etc.) are direct predators of phototrophic bacteria soon after hatching, resulting in an increase in weight and survival rate of more than two fold within 2 to 4 weeks after hatching (nearly no death was noted in some experiments). Table 9 gives data for one of these experiments. It is shown that when live cells of phototrophic bacteria are added at 0.1% to the formula feed given to the fry of crucian carp soon after hatching, the rate of survival highly increased and nearly no mortality was observed. Further, purple sulfur- and green sulfur-bacteria are phototrophic bacteria that use hydrogen sulfide, which is highly toxic, and the activities of these phototrophs convert H_2S into nonpoisonous sulfur compounds. Therefore, sulfur phototrophs are utilized for cleansing of waste waters high in sulfide or to clean the bottom environment containing high levels of sludge. For example, in the culture pond for eels, abnormal death was noted from the deteriorating water quality

Table 9. Effect of phototrophic bacteria on survival of young fry of crucian carp[a]

	Survival numbers after one month	Survival ratio (%)
Control	2772	69.3
Treatment (with 0.1% PTB cells)[b]	3860	96.5

[a] The experiment was carried out in a tank of 2 tons capacity with an initial number of 4000 fry.
[b] The PTB cells were obtained from the waste treatment plant of a fish meat industry. The bacterial powder contained only 50 per cent of PTB cells; the other half contained heterotrophic bacteria (contaminants).

in winter, whereas treatment with phototrophic sulfur bacteria made the environment suitable for continued growth.

Special attention has been played to the use of phototrophic bacteria in the culture of prawn. As shown in Fig. 9, 50% of the prawn consumed in Japan is cultured in such large water tanks (prawns amounting to 20,000 million ¥en per month are consumed domestically), while another 50% depends on importation. Thus, prawns worth 100 million ¥en per annum are cultured in one of such water tanks (more than 100 tanks are operating in Japan). In the past prawns in such culture tanks were frequently affected by gill disease causing great economic damage, however this has now been completely prevented by supplementing the tanks with anoxygenic phototrophic bacteria. In addition, such effects have also been displayed in suppression of virus diseases noted on others including swellfish (Okamoto et al. 1988, Hirotani et al. 1991). Perhaps the presence of phototrophs creates unfavorable competitive effects for the pathogens, preventing establishment of the latter.

B. Livestock Feed

Since cells of *Rhodobacter capsulatus* contain much protein (showing a good balance of amino acids) vitamins and other effective substances, addition of cells at a rate of only a ten-thousandth increases the egg-laying rate of chickens by 15–20%. Tables 10–12 give data in this connection. Such effect to improve egg-laying is especially apparent in the sickly seasons (such as midwinter or midsummer). While testing the effect of phototrophic bacteria mixed in the feed, a favorable result was obtained in experiments with more than 2,000 hens that the egg-laying ratio was increased roughly ten percent by mixing fresh phototrophic bacteria into the formula feed at a rate of 1/10,000. The nutrient effect tests were carried out as follows. Phototrophic bacteria were given to chicks

Fig. 9. A general culture pond of prawn using PTB cells in Japan.

Table 10. Composition of general component of phototrophic bacteria, green algae, and yeast cells (g/100 g dry weight)

	Phototrophic bacteria	Algae	Yeast
Crude protein	60.95	55.52	50.5
Crude fat	9.91	8.07	1.1
Soluble carbohydrates	20.83	21.04	39.3
Crude fiber	2.92	12.09	2.1
Ash	5.39	3.28	7.0

Phototrophic bacteria – *Rhodobacter capsulatus*
Green algae – *Chlorella vulgaris*
Yeast – *Saccharomyces anomalus*

Table 11. Amino acid composition of cell hydrolysates of phototrophic bacteria, green algae and yeast (g/100 g dry weight)

	Phototrophic bacteria[a]	Algae[b]	Yeast[c]
Lysine	2.86	2.71	3.76
Histidine	1.25	1.06	0.90
Arginine	3.34	3.24	2.50
Aspartic acid	4.56	4.74	3.11
Threonine	2.70	2.28	2.65
Serine	1.68	2.12	2.75
Glutamic acid	5.34	4.62	6.21
Proline	2.80	2.12	1.77
Glycine	2.41	2.28	2.18
Alanine	4.65	2.98	2.86
Valine	3.51	3.02	3.20
Methionine	1.58	0.27	0.51
Isoleucine	2.64	2.44	2.63
Leucine	4.50	4.46	3.54
Tyrosine	1.71	0.96	1.30
Phenlyalanine	2.60	2.65	2.20
Tryptophan	1.09	0.64	0.66
NH_3	4.01	2.58	5.30

[a] Phototrophic bacteria – *Rhodobacter capsulatus*
[b] Green algae – *Chlorella vulgaris*
[c] Yeast – *Saccharomyces anomalus*

on a continuous basis immediately after hatching, and data on growth until they laid their first eggs and for 6 months of the laying period were collected. 144 chicks (leghorn pullets) were divided into 4 groups, and given assorted food shown in Table 13 (a, b): the control was given only assorted food shown in Table 13 (see Table 14); Group I was given feed containing 0.01% fresh phototrophic bacterial cells (Treatment I in Table 14); Group II was given feed containing 0.02% cells (Treatment II in Table 14); and Group III was given feed containing 0.04% cells (Treatment III in Table 14). The recording of the results was started

Table 12. Vitamin, pigment, and true metal analysis of phototrophic bacterial cells[a]

Vitamin	μg/100g
B_2	3600
B_6	3000
Folic acid	2000
B_{12}	200–2000
C	20,000
D	10,000 I.U.
E	31,200
RNA	4.9%
DNA	1.0%
Bacteriochlorophyll	5.61%
Carotenoid pigments	4.17%
Element	%
N	9.75
P	2.49
K	0.21
SiO_2	0.82
Ca	0.87
Na	0.31
Fe	0.13
Mg	5.0
Mn	0.001
Cu	0.0021
Zn	0.11

[a] Phototrophic bacteria – *Rhodobacter capsulatus*

one month after the beginning of egg-laying. The period of data collection was from October to March, including the winter season when the egg-laying rate decreases. The Haugh-Unit Coefficient (Table 15) is a coefficient representing the quantity of thick white and freshness of an egg.

As shown in Table 14, it was observed that the egg-laying rate tended to increase by increasing the amount of phototrophic bacterial cells added. The hens in the groups fed on phototrophic bacterial cells appeared to begin laying eggs a little earlier than the control, and the term until the egg-laying rate decreased was slightly prolonged. As shown in Table 15, in the groups with the added cells, both the yolk index and the Haugh-Unit Coefficient increased. Larger coefficients mean that the eggs maintain their freshness better in storage. Also, in the groups with the added photosynthetic bacterial cells, it was confirmed that the color of the yolk was improved and the carotene content was increased depending on the quantity of added bacterial cells. This suggests

Table 13a. Composition of hen feed

		%
Crushed corn		50.0
Milo[a]		10.0
Oil cake of soybean		10.0
Fish meal		7.5
Rice bran		12.5
Alfalfa meal		2.0
$CaCO_3$		3.8
$Ca_3(PO_4)_2$		1.5
NaCl		0.3
Vitamins and mineral mixture		0.35
Fat mixture (plant origin)		2.0
Ethoxyquin-25[b]		0.05
Component	Crude protein	17.0
	Total digestible nutrients	67.4

[a] A kind of corn; crushed Sorghum
[b] Anti-oxidizing agent containing 1,2 dehydro-6-ethoxy-2,2,4-trimethylquinoline; 25%

Table 13b. Composition of vitamins and mineral mixture·kg^{-1} feed

Vitamin A	2650 I.U. (International Unit)
Vitamin D	200 I.C.U. (International Chick Unit)
Vitamin B_1	2.5 mg
Vitamin B_2	5.5 mg
Nicotinic acid	29 mg
Pantothenic acid	9.3 mg
Vitamin B_6	6.7 mg
Biotin	0.09 mg
Choline	1800 mg
Folic acid	0.55 mg
Vitamin B_{12}	0.009 mg
Mn	50 mg
I	1 mg
Mg	440 mg
Fe	18 mg
Cu	1.8 mg
Zn	40 mg

Table 14. Average ratio of egg laying[a]

Month after egg laying started	Control[b]	Treatment I[c]	Treatment II[d]	Treatment III[e]
	%	%	%	%
1	27.2 ± 2.8	37.8 ± 8.2	28.8 ± 4.2	39.8 ± 6.0
2	59.0 ± 4.0	65.6 ± 3.6	63.8 ± 7.0	71.6 ± 3.4
3	63.9 ± 7.6	72.6 ± 7.7	74.5 ± 4.4	77.0 ± 5.0
4	64.2 ± 5.5	73.1 ± 5.6	74.7 ± 3.5	66.7 ± 3.0
5	67.8 ± 1.3	67.2 ± 2.9	74.6 ± 0.9	70.1 ± 2.7
6	63.1 ± 7.0	65.5 ± 3.8	69.1 ± 1.2	68.2 ± 0.0
Average	56.2 ± 3.4	62.5 ± 3.7	62.8 ± 4.3	64.3 ± 0.9
Index[f]	100	111.2	111.7	114.4

[a] No. of egg laying hens/total no. of hens/day
[b] Fed hen feed described in Table 13
[c] As in (b) but supplemented with 0.01% fresh phototrophic bacterial cells
[d] As in (b) but supplemented with 0.02% fresh phototrophic bacterial cells
[e] As in (b) but supplemented with 0.04% fresh phototrophic bacterial cells
[f] Compared with control of 100

that components from phototrophic bacterial cells can be well absorbed and transfered into yolk. In addition, the feed efficiency, the body weight of hens, and the weight of total eggs produced per hen during a 6 month period was found to increase much more than 10% by the addition of the phototrophic bacterial cells (Kobayashi and Kurata, 1978).

In addition, as phototrophic bacteria contain antiviral substances, they may effectively suppress Marek's virus diseases, which may stimulate the weight gains and increase production rates of chickens observed in these experiments. If true, photosynthetic bacteria could be considered biocontrol agents of diseases in chickens.

C. Organic Fertilizer

As shown in Table 5, the kinds of liquid waste to which treatment with phototrophic bacteria is actually applicable, are noted in a fairly wide range. It is,

Table 15. Comparison of egg quality

	Month after egg laying started	Control	Treatment I[a]	Treatment II[a]	Treatment III[a]
Color tone (Roche yolk color fan)	1	7.6 ± 0.6	9.6 ± 0.2	10.5 ± 0.4	11.3 ± 0.4
	2	8.0 ± 0.3	10.9 ± 1.0	11.1 ± 1.2	12.2 ± 0.7
	3	9.4 ± 1.1	11.5 ± 0.4	12.0 ± 0.4	11.8 ± 0.2
	4	9.5 ± 1.6	12.7 ± 0.3	12.6 ± 0.2	13.2 ± 0.1
	5	10.2 ± 0.5	12.5 ± 0.3	12.9 ± 0.3	13.2 ± 0.7
	6	9.5 ± 0.9	12.5 ± 0.2	12.7 ± 0.1	13.0 ± 0.6
Yolk index (usual index 0.45)[b]	1	0.45 ± 0.02	0.46 ± 0.02	0.45 ± 0.01	0.46 ± 0.01
	2	0.47 ± 0.01	0.47 ± 0.02	0.47 ± 0.01	0.47 ± 0.01
	3	0.46 ± 0.01	0.46 ± 0.02	0.48 ± 0.01	0.46 ± 0.01
	4	0.45 ± 0.00	0.46 ± 0.01	0.46 ± 0.00	0.47 ± 0.01
	5	0.46 ± 0.01	0.46 ± 0.01	0.47 ± 0.01	0.48 ± 0.02
	6	0.47 ± 0.01	0.48 ± 0.01	0.49 ± 0.01	0.47 ± 0.02
Haugh unit coefficient[b]	1	90.3 ± 8.9	91.9 ± 1.5	92.6 ± 0.3	93.8 ± 3.7
	2	93.3 ± 0.4	91.1 ± 2.7	93.0 ± 1.8	93.3 ± 1.6
	3	86.1 ± 6.4	90.2 ± 2.4	90.2 ± 1.0	87.9 ± 2.2
	4	80.0 ± 2.0	84.5 ± 4.6	84.4 ± 2.2	85.9 ± 2.6
	5	82.6 ± 4.3	80.0 ± 1.2	82.1 ± 1.0	83.8 ± 1.9
	6	71.3 ± 2.5	80.6 ± 2.9	81.8 ± 2.1	84.8 ± 2.1

[a] See footnotes c–e to Table 14
[b] See text

however, likely that markedly contaminated water would certainly not be processed to obtain such byproducts to be used for livestock feed, but instead could be better utilized by returning it to the farming land to be used as an organic fertilizer, since this would serve as a natural fertilizer. Organic matter treated by phototrophic bacteria while passing through layers of the soil in the farmland could be used as the source of nutrition for the roots of higher plants, being useful for growth directly and indirectly. Further, being degraded and cleansed by other soil microorganisms, they stream again into the rivers, lakes and seas after passing through subsurface waters. Such technology of processing contaminated water is obviously an ideal resources- and energy-saving method, especially when it is noted that the energy to drive these conversions comes from sunlight.

As noted from the flow sheet in Fig. 7a, livestock liquid waste can be partially degraded in a first aeration tank, followed by secondary treatment by phototrophic bacteria grown in a second tank. This is an efficient way of processing wastes to return the bacterial suspension to the farmland. It is of special interest that when this material is added to soil previously damaged by continuous croppings of tomato, cucumber, eggplant and others, it is feasible to correct nutritional problems in the soil. This may occur by stimulation of the growth of actinomycetes, which as antagonistic microorganisms, spontaneously suppresses growth of pathogenic fungi (Fig. 10).

Besides, as shown in Table 16, when the suspension of phototrophic bacteria is applied as an organic fertilizer to the persimmon tree, it not only improves the yield of fruit markedly, but also elevates the sugar level of the fruit and improves taste as well as color and gloss (Table 17).

At present in Okinawa (Japan), a plant is under operation wherein the electric power is developed from the energy of methane gas obtained from livestock barn liquid waste. After that, the waste water is processed making use of the phototrophic bacteria method, and is then returned to the farmland which helps to improve the yield of crops.

In addition, a technology has recently been established wherein dung and urine are not separated in the pighouse but instead are mixed with running water into a strong aeration tank to be foamed under excessive aeration, whereto highly activated phototrophic bacteria are added. Deodorizing and making it adsorbed into foams, the mixture is especially useful for treating fallen out hair from animal hides, undegradable soil and sand, etc., while denitrification and decarboxylation proceed nearly to completion. Such technology is suitable for cleansing small to moderate-sized pighouses with

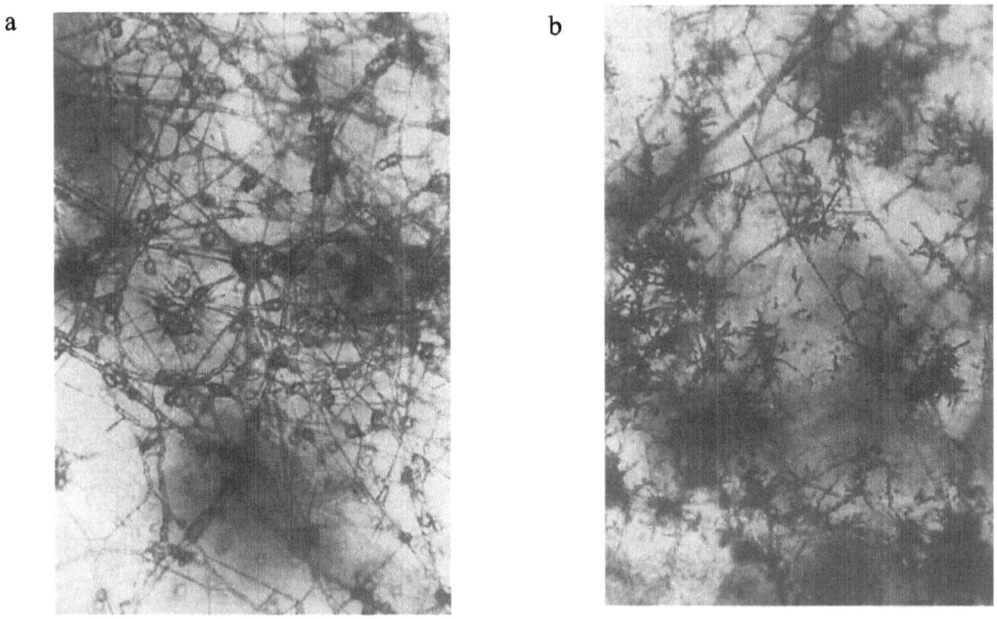

Fig. 10. Microscopic photographs of the pathogenic fungus, *Fusarium* and the antagonistic microorganism, *Streptomyces*. (a) Hyphae with durable spore of *Fusarium oxysporum*. (b) *Streptomyces fradiae* attack the hyphae of *Fusarium* and kill them. Growth of *S. fradiae* in soil may be stimulated by supplementation of the soil with phototrophic bacterial cells. (× 600)

Table 16. Weight and chemical components of persimmon fruit

	Control (Inorganic fertilizer)	Treatment (Organic fertilizer)[a]
Total No. of fruit	32	43
Total Wt of fruit, Kg	7.1	8.2
Average Wt/fruit	222	191
Fruit color[b] H.C.C.	12	13
Water contents %	84.7	83.3
Degree of sugar in Brix	14.7	16.4
Acid contents %	0	0
Reducing sugar contents %	10.82	12.56
Non-reducing sugar contents %	2.34	2.57
Total sugar contents %	13.16	15.13

[a] Phototrophic bacterial cells.
[b] Fruit color: 12 orange; 13 Saturn red.

prompt and economical advantages, and its utilization has been further enlarged.

As stated at the beginning of this chapter, photosynthetic bacteria technology has been applied to high-yielding culture of rice; as compared with the conventional method, referred to as the control, significant increases in yield can be expected (Table 18)(Yoshida et al. 1991). It can be seen in Fig. 3 that roots of control plants show root rot due to the presence of hydrogen sulfide, while those treated with phototrophic sulfur bacteria show good root development and growth, which is closely connected with increase of the yield.

D. Removal of Offensive Odors

Offensive odors develop from livestock barns including poultry houses, as well as from garbage. Phototrophic bacteria are able to absorb, degrade and eliminate such substances responsible for these offensive odors (Lee and Kobayashi, 1992). For example, acetic acid, propionic acid, butyric acid, iso-butyric acid, valeric acid and iso-valeric acid present in pig wastes were eliminated rapidly in the presence of phototrophic bacteria as shown in Fig. 11.

Putrescine and cadaverine, secondary amines with serious unpleasantness and high toxicity, and hydrogen sulfide, can also be eliminated, as shown in Table 19. Especially, those with the utmost unpleasantness (like mercaptans) are eliminated, too, as shown in Table 20. A higher removal can be obtained in the light, but even in the dark a certain

Table 17. Contents of carotenoid pigment in persimon peels (mg/100 g fresh weight)

	β-carotene	Lycopene	Cryptoxanathin	Zeaxanthin	Total contents of carotenoid pigment
Control (Inorganic fertilizer)	3.108	2.773	13.018	7.682	26.581
Treatment (Organic fertilizer)[a]	2.929	4.237	15.667	8.970	31.803

[a] Phototrophic bacterial cells

Table 18. Yield of rice plants in the presence and absence of phototrophic bacteria[a]

	Ear number (Stub No.)	Weight of unpolished rice Kg/10 acres	Index[b]
Control (customary cultivation with inorganic fertilizer)	12.5	425.1	(100)
Treatment A (compost with aerobic fermentation)	27.5	877.4	(206.3)
Treatment B (compost with aerobic fermentation + PTB cells[a])	28.1	1056.7	(248.4)

[a] *Rhodobacter capsulatus*
[b] Compared with control of 100

Table 19 a. Removal of putrecsine and cadaverine by cells of *Rhodobacter capsulatus*

	Condition (after 7 days) Putrescine mg/100 ml	
	Dark	Light
Control	5.0	5.0
Treatment (added *Rb. capsulatus*)	1.8	0.1
	(after 7 days) Cadaverine mg/100 ml	
Control	5.0	5.0
Treatment (added *Rb. capsulatus*)	2.1	0.1

Table 19 b. Removal of hydrogen sulfide by *Chromatium*

	Hydrogen sulfide (mg/100 ml) present after 7 days	
	Dark	Light
Control	5.0	5.0
Treatment (added *Chromatium*)	3.8	0.8

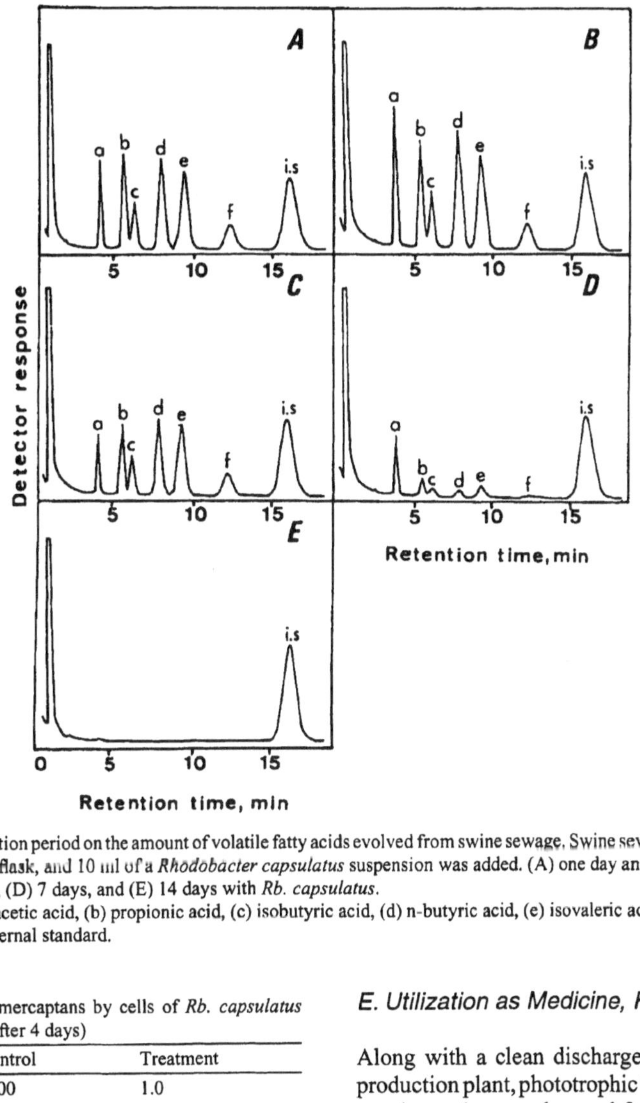

Fig. 11. Effect of incubation period on the amount of volatile fatty acids evolved from swine sewage. Swine sewage (50 g) was transferred to a 300 ml Erlenmeyer flask, and 10 ml of a *Rhodobacter capsulatus* suspension was added. (A) one day and, (B) 14 days without *Rb. capsulatus*. (C) one day, (D) 7 days, and (E) 14 days with *Rb. capsulatus*.
Peak identification: (a) acetic acid, (b) propionic acid, (c) isobutyric acid, (d) n-butyric acid, (e) isovaleric acid, (f) n-valeric acid. (i.s.) isocaproic acid as an internal standard.

Table 20. Removal of mercaptans by cells of *Rb. capsulatus* (light condition) (ppm after 4 days)

	Control	Treatment
CH_3SH	1000	1.0
CH_3SCH_3	800	0.0
$(CH_3)_2S_2$	3400	3.0

level of elimination can be expected; therefore, photosynthetic bacteria are frequently used in livestock barns, being promoted as one of the solutions for offensive odor pollution, and thereby displaying its effect on sanitation and protection of the environment.

E. Utilization as Medicine, Food and Energy

Along with a clean discharge out of the foodstuff production plant, phototrophic bacterial cells yielded as a byproduct can be used for medicinal purposes and as a source of single cell protein. Since abundant ubiquinone, UQ_{10}, is contained in phototrophic bacteria, UQ_{10} extracted and purified from bacterial cells obtained from growth on liquid wastes has been placed on the market as a remedy for myocardial infarction, being very useful for saving human life.

Many other useful components have been extracted from phototrophic bacteria. They are nonpoisonous to normal cells and contain some antiviral substances; therefore, further development of their application is very much anticipated. Especially in Brazil, where automobiles run with alcohol nowadays, the liquid waste of such alcohol fermentation can be considered not waste but an important resource for culture of phototrophic bacteria. From the standpoint of food protein and biomass production other than medicines, alcohol wastes can be a very hopeful processing technology. Further, as phototrophic bacteria can evolve hydrogen gas (via nitrogenase, see Chapter 43 by Ludden and Roberts), a technology of large-scale hydrogen gas development from such concentrated organic waste has successfully been established together with the hydrogen fueled engine technology. This technology is expected to be practically applied in the future as existing energy supplies from petroleum and other sources diminish.

References

Hirotani H, Ohigashi H, Kobayashi M and Koshimizu K (1991) Inactivation of T_5 phage by an antivirus substance from *R. capsulata*. FEMS Microbiol Lett 77: 13–18

Kobayashi M, Katayama T and Okuda A (1966) Seasonal changes of microbial counts in paddy field. J Sci Manure, Japan, 37: 441–446

Kobayashi M, Takahashi E and Kawaguchi K (1967) Distribution of nitrogen-fixing microorganisms in paddy soils of southeast Asia. Soil Science 104: 113–118

Kobayashi M (1970) Treatment and remediation of waste solution by photosynthetic bacteria. Chemistry and Biology 8: 604–613 (Japanese Soc of Agricultural Biological Chemistry)

Kobayashi M and Haque MZ (1971) Contribution to nitrogen fixation and soil fertility by photosynthetic bacteria. Plant and Soil, Special Vol: 443–456

Kobayashi M and Tchan YT (1973) Treatment of industrial waste solutions and production of useful byproducts using a photosynthetic bacterial method. Water Research 7: 1219–1224

Kobayashi M, Fujii K, Shimamoto I and Maki T (1978) Treatment and re-use of industrial waste water by phototrophic bacteria. Prog Water Tech 11: 279–284.

Kobayashi M and Kurata S (1978) The mass culture and cell utilization of photosynthetic bacteria. Process Biochem 13: 27–30

Lee MG and Kobayashi M (1992) Deodorization of swine sewage by addition of phototrophic bacteria. Soil Sci Plant Nutr 38: 767–770

Okamoto N, Hirotani H, Sano T and Kobayashi M (1988) Antiviral activity of extracts of phototrophic bacteria to fish viruses. Nippon Suisan Gakkaishi 54: 2225

Okuda A and Kobayashi M (1961) Production of slime substance in mixed culture of *R. capsulata* and *Azotobacter*. Nature 192: 1207–1208

Okuda A and Kobayashi M (1963) Symbiotic relation between *R. capsulata* and *Azotobacter*. Mikrobiologiya 32: 936–945

Yoshida T, Tabata T, Saraswati R and Kobayashi M (1991) Study on resourceful disposal of organic waste and high-yielding culture of rice plant. J Resource Soc of Environmental Technology 20: 607–610

Index

Symbols

A

B

C

D

F

G

H

J

K

L

N

O

P

Q

R

S

T

U

V

W

X

Y

Z